ORCHIDS
from the Botanical Register 1815–1847

The Texts

Edited by SAMUEL SPRUNGER

Birkhäuser Verlag
Basel · Boston · Berlin

All orchid descriptions from the *Botanical Register 1815–1847* have been arranged chronologically by volumes 1–33. Reproductions of page facsimiles have been reduced in scale. Each volume is preceded by its original title page. Intervals between the lines in the text indicate pagination breaks in the original version.

A special word of thanks is due to the Basel University Library in kindly making available all archival copies of the *Botanical Register* for the production of the photolithographic plates and the microfilm for the text.

Library of Congress Cataloging-in-Publication Data

Orchids from the Botanical register, 1815–1847 / edited by Samuel Sprunger in collaboration with Phillip J. W. Cribb and William T. Stearn.
p. cm.
"The illustrations and accompanying descriptive texts reprinted here originally appeared in Edwards' botanical register" – Verso t. p.
Includes bibliographical references and indexes.
Contents: The illustrations – The texts.
ISBN 0-8176-2479-1 (U.S. : set)
1. Orchids. 2. Orchids – Pictorial works. 3. Orchid culture. 4. Botanical illustration. I. Sprunger, Samuel. II. Cribb, Phillip. III. Stearn, William T. (William Thomas), 1911– IV. Edwards' botanical register.
SB409.0693. 1991
635.9'3415–dc20

Deutsche Bibliothek Cataloging-in-Publication Data

Orchids from the Botanical Register 1815 [eighteen hundred and fifteen] – 1847 / ed. by Samuel Sprunger in collab. with Phillip J. W. Cribb and William T. Stearn. - Basel ; Boston ; Berlin : Birkhäuser.
ISBN 3-7643-2479-1 (Basel...) Gb. in Schuber
ISBN 0-8176-2479-1 (Boston) Gb. in Schuber
NE: Sprunger, Samuel [Hrsg.]
The texts. - 1991

Printed in Germany
ISBN 3-7643-2479-1
ISBN 0-8176-2479-1

CONTENTS

(in chronological order)

THE

BOTANICAL REGISTER:

CONSISTING OF

Coloured Figures

OF

EXOTIC PLANTS,

CULTIVATED IN

BRITISH GARDENS;

WITH THEIR

HISTORY AND MODE OF TREATMENT.

THE DESIGNS BY

Sydenham Edwards,

FELLOW OF THE LINNEAN SOCIETY.

VOL. I.

——viret semper——nec fronde caducâ
Carpitur.

LONDON:
PRINTED FOR JAMES RIDGWAY, PICCADILLY.
1815.

17

EPIDENDRUM nutans.
Nodding Epidendrum.

GYNANDRIA *MONANDRIA.*

EPIDENDRUM. *Cal.* 0. *Cor. Pet.* 5, oblonga, patentia. *Labellum* basi ecorne, tubulatum, columnam amplectens, *lamina* erecta, patula. *Columna germini* teretiusculo infero imposita, gibba, apice supra foveam adfigens *Antheram* 2-4-locularem, singulo loculo includentem *Corpuscula pollinis* 1-4, pedicello communi anticè inserto, affixa. *Stig.* ante foveam Antheræ, concavum. *Caps.* oblonga, 3-gona, 1-locularis, 3-valvis. *Sem.* numerosa, minutissima, subrotunda: *Arillo* scobiformi induta. *Swartz. Fl. ind. occid.* 3. 1485. Herbæ *perennes, parasiticæ.*

E. *nutans,* caule simplici, foliis ovato-lanceolatis amplexicaulibus, floribus subspicatis nutantibus, lamina labelli triloba, lobo intermedio tridentato. *Swartz Nov. Act. ups.* 6. 68. *prod.* 121. *flor. ind. occid.* 3. 1499. *Willd. sp. pl.* 4. 117. *Hort. Kew. ed.* 2. 5. 219.

Radices *crassæ, filiformes, longæ, albidæ.* Caulis *(nunc plures aggregati)* 1-2 *pedalis, crassus, simplex, foliosus, subteres, suberectus, s. arboribus horizontali situ adnascens.* Folia *sessilia, vaginantia, amplexicaulia, alterna, obliquè patula, ovato-lanceolata, acuta, plana, subtùs levissimè carinata, integra, lineata, glabra, subcoriacea.* Vaginæ *foliorum caulem arcte cingentes, teretes, lineatæ, glabræ, rubicundæ.* Pedunc. *terminalis, solitarius, longitudine ferè foliorum (spithamæus), compressiusculus, vaginis aliquot remotis lanceolatis albido-purpurascentibus munitus, glaber.* Flores *terminales, subspicati, approximati,* 20-30, *sessiles, majusculi, lutei, spicam nutantem referentes.* Bracteæ *lanceolato-lineares, carinatæ, longit. sæpè germinis, pallidæ.* Pet. 5, *isometra, patula:* 3 exter. *ovato-lanceolata, acuta, lateralia, subfalcata, planiuscula, integra, elevato-striata, extùs luteo-rufescentia:* 2 inter. *lineari-acuminata.* Labellum *columnæ basi adnatum, amplectens.* Lamina *horizontalitèr deflexa,* 3-*loba*; lobis *lateralibus ovatis, integris, convexis, intermedio* 3-*dentato: dentibus lateralibus cornuum instar erectis, medio deflexo.* Calli 2 *propè faucem, undè ad apicem lobi intermedii sulci* 3 *elevati.* Columna *labello brevior, erecta, teres, apice dilatata, marginibus biloba.* Anth. 4 *loc. in foveâ apicis columnæ,* corpuscula *solitaria oblonga, pedicellata tegens.* Caps. *oblonga,* 6-*carinata, magna.* Swartz.

Native of the mountains of the western side of the Island of Jamaica, where it grows (as the generic name implies) on trees, striking root in the fissures of their bark. First brought to England by Admiral Bligh, in 1793.

The cultivation of tropical parasitic vegetables was long regarded as hopeless with us; it appeared a vain attempt to find substitutes for the various trees each species might affect, within the limits of the hot-house. But experience has shown, not only that they succeed as easily as any other plants from the same climate; but that, from the little space they occupy, they are peculiarly suited to the stove, for which their curious structure and beauty render them most desirable acquisitions. We are told they should be planted superficially in proportionate pots of hazel-loam, and placed on the shelf or the flue of the hot-house, but never plunged into the bark-bed.

The genus is of Jussieu's natural order of Orchideæ, rearranged a few years since with great perspicuity and judgment by the learned Dr. Olof Swartz; and still more recently revised and enlarged by Mr. Brown, in his Prodromus of the Flora of New Holland, with the accuracy for which he is justly praised. The bloom of this species is of a lemon-colour, darkening as it decays to a reddish brown; the spike is constantly bent on one side, and offers a characteristic name for the species. The whole plant measures from one to two feet in height. It has been elaborately described by Swartz, among many others of the same tribe, in his Flora Indiæ occidentalis.

The drawing was made at the nursery of Messrs. Lee and Kennedy, at Hammersmith, in February last.

a The upper part of the shaft of the fructification, showing the 4 oblong parallel pollen-masses, as they present themselves, on the removal of the lidshaped moveable anther from the aperture of the cavity in which they have been formed: magnified. *b* The pollen-masses withdrawn from the cavity: magnified. *c* The same in a different position, showing their granulate filiform pedicles: very much magnified. The moveable lidshaped anther, frontwise: magnified. This forms the brown spot seen at the apex of the shaft in the flower. *e* Its converse, with the base of the partitioning of its four cells: magnified. *f* Stigma: magnified.

67

EPIDENDRUM fuscatum.
Brown Epidendrum.

GYNANDRIA *MONANDRIA.*

Sect. V. *Anthera* terminalis, mobilis, operculiformis, decidua. *Pollinis massæ* cereaceæ, læves. *Brown prod.* 330.

EPIDENDRUM. (*Cal.* 0. *Cor.* 5-petala, patens. *Labellum* ecalcaratum, laminâ patulâ). *Columna* cum ungue labelli longitudinalitèr connata in tubum (quandòque decurrentem ovarium). *Massæ pollinis* 4 parallelæ, septis completis persistentibus distinctæ, basi filo granulato elastico auctæ. *Brown in Hort. Kew. ed.* 2. 5. 217. *Herbæ perennes, parasiticæ. Caps.* 3-*valvis: sem. plurima, minutissima.*

E. *fuscatum,* caule simplici, foliis oblongis acuminatisve, pedunculo terminali elongato, spica globosa, columna petalis breviore. *Swartz in Nov. act. ups.* 6. 69.

Epidendrum fuscatum. *Smith spicil.* 21. *t.* 23. *Andrews's reposit.* 441. *Willd. sp. pl.* 4. 120. *Hort. Kew. ed.* 2. 5. 218.

Epidendrum anceps. *Jacq. amer.* 224. *t.* 138.

Epidendrum secundum. *Swartz obs.* 325; *(nec aliorum).*

Radix *crassè fibrosa.* Caudex *tereti-compressus tectus vaginis foliorum, basi subbulboso-tumidus.* Folia *coriacea, alterna, disticho-patentia, vaginantia, carinata, laminâ oblongâ cum apiculo brevi obliquè reflexo, modòve sublanceolatâ, glabrâ, subtilitèr striatâ, saturatè viridi, subtùs pallidiore v. purpurascente.* Caulis *continuus, strictus,* 1-2 *pedalis, compressus, totus vaginis alternis aridis carinatis acutis striatis internodia æquantibus vestitus.* Spica *brevis, secunda, in corymbum digesta, refracta, nutans, terminalis v. ex vaginâ lateralis (in vegetiore plantâ fortè plures?).* Bracteæ *solitariæ, lanceolatæ, germine triplo breviores.* Pet. *explanata, isometra,* 3 *latiora ovali-lanceolata,* 2 *linearia,* 3-*plo angustiora, modò deflexa.* Labellum *petala superans, rotundatum, trilobum, lobo medio emarginato-retusum apiculo in sinu.* Columna *brevis, inclinata apice alis binis lateralibus truncatis suprà antheram inflexim conniventibus.* Germ. *subteres, striatum.*

The above generic character allots a far narrower extent to the genus, than is admitted by that we had prefixed to EPIDENDRUM *nutans* in the seventeenth article of this work.

Fuscatum, like its congeners, is found growing on the trunks and branches of the trees, somewhat in the manner of the Misletoe with us. It is a perennial herbaceous plant, seldom exceeding a foot and half in height, with the fibres of the *root* thick and fleshy; *stem* sheathed by the lower portions of the foliage; *leaves* coriaceous, distich, patent, alternate, varying from half an inch to an inch and half in breadth, sometimes rounded with a short point at the top, sometimes lanceolate, sometimes of a silvery hue beneath and thickly covered with minute green dots, at others purplish. *Flower-stalk* long, terminal, upright, entirely clothed by alternate sphacelate deciduous sheaths, in strong plants bearing several spikes? *Spike* refracted, many-flowered, pointing one way, with the flowers disposed in the form of a corymb. *Corolla* sessile, about three parts of an inch long; *petals* of a dusky reddish brown; *lip* of a yellowish herbaceous colour, with a firmness and gloss that gives it the appearance of being formed of wax; scent very faint.

Native of the West Indies. Found by Swartz in Jamaica, on trees growing on the mountains. Introduced by Lord Gardner in 1790. Produces several flowering stems from the same root at various times of the year. Requires the treatment which we have already recommended for another species (see fol. 17) from the same regions.

The drawing was made this summer from a specimen in the hothouse at the nursery of Messrs. Lee and Kennedy, Hammersmith.

a A front view of the summit of the shaft of fructification with a portion of the lip, showing the 4 pollen-masses as they present themselves upon the removal of the anther from which they have been excreted. *b* The 4 pollen-masses extracted from the recess in the summit of the shaft. *c* The anther frontwise. *d* The converse of the same, showing its 4 partitions. All somewhat magnified.

80

EPIDENDRUM umbellatum.
Umbel'd Epidendrum.

GYNANDRIA *MONANDRIA.*

EPIDENDRUM. Suprà fols. 17, 67.

E. *umbellatum*, caule simplici, foliis oblongis subemarginatis, floribus in sinu folii terminalis confertis, lamina labelli triloba, lobo intermedio emarginato. *Swartz in Nov. act. ups.* 6. 68.
Epidendrum umbellatum. *Swartz prod.* 121. *Fl. ind. occid.* 3. 1501. *Willd. sp. pl.* 4. 117. *Hort. Kew. ed.* 2. 5. 218.
E. difforme. *Jacq. amer.* 223. *t.* 136.

Radices *simplices, repentes, rigidæ, albidæ.* Caules *(nunc unicus) aggregati, semipedales, simplices, compressiusculi subflexuosi, foliosi, glabri.* Folia *alterna, ovato-oblonga, basi vaginantia, obliqua, patula, subindè recurvata, integra, apice parùm emarginata, avenia, glaberrima, crassiuscula:* vaginæ *arctæ, apice apertæ, compressiusculæ, margine ancipites, (hinc folia subtùs basi carinata,) striatæ.* Flores *(sesquiunciam excedentes) è sinû folii terminales, aggregati, brevissimè pedunculati, (plures) numerosi, majusculi, (viridi-lutescentes;)* vaginæ (bracteæ) *spathæformes, plures ad basin pedunculorum, oblongæ, acutæ, pallidæ (involucelli instar.)* Swartz. Petala *isometra, lanceolata, patula, striata, 2 interiora plurimùm angustiora linearia.* Labellum *petala subexsuperans;* lamina *deflexa (flos enìm torsione germinis resupinatus), transversìm latior, oblato-subcordata, triloba, venosa, costulâ rectâ transeunte medium de fauce, cujus orificio astant glandulæ 2 elevatæ virentes;* lobi laterales *rotundati, posticè secundùm utrumque latus columnæ profundiùs deflexi,* medius *multò minor bifidus lobulis obtusatis planis.* Columna *fermè duplo brevior corollâ; alæ laterales marginis paulò ultra antheram productæ v. obsolescentes.* Germen *petalis plus duplo longius.* Anthera *fusca, loculis 4 parallelo-convergentibus:* Massæ pollinis *globosæ, pallidè luteæ, appendicibus filiformibus brevibus alterâ fine connexæ.*

A species which is very rarely met with in our collections, and remarkable in the genus for an aggregated sub-sessile inflorescence, issuing from the bosom of the uppermost leaf on the stem, separated at the base by small enclosed spathaceous bractes, and standing upon very short nearly obsolete pedicles.

Flowers several, an inch and half long, upright, of a green-yellow colour. *Stem* simple, little more than half a foot high, sheathed the whole length by the lower attenuated portion of the *leaves*; these alternate, distich, patent, oblong, substantial, firm, smooth. A native of the West Indies, where it grows on the trunks of trees, sometimes horizontally, sometimes perpendicularly downwards. Introduced in 1793 by Rear-admiral William Bligh, in the Providence.

Professor Jacquin having described in the specific phrase to *difforme* the *column* of the flower as equal to the corolla, and the *labellum* as obcordate; Dr. Swartz has had the precaution to rename our plant, subjoining Jacquin's merely as a probable synonym. In his general description, however, the former will be found to speak of the *column*, only as *almost* equal to the corolla. We have reviewed attentively what each of them has said of his subject, and inspected both specimens of their plants in the Banksian Herbarium (where Jacquin's indeed is flowerless), and are persuaded of the specific identity of the two.

The *corolla*, when full blown, by a contortion of the lower part of the germen which supports it, is always resupinate, that is, with the directions of the upper and lower halves of its plane reversed, as in the european Violets. *Germen* together with the short continuous pedicle about an inch high, cylindrical, tapering downwards, prominently three-ribbed, unilocular, with three parietal placentiform receptacles, attaching numerous *ovula*. *Corolla* superior, five-parted, segments lanceolate, equal in length, varicosely nerved, *two inner* very narrow and linear. *Labellum (Nectary* Linn.) placed between the two outer lateral segments of the corolla, which it rather exceeds and differs from in form and consistence, conjoined for the length of its narrow upright turbinately tubular spurless *unguis* with the column in front; *lamina* broad, patent, cordate, rounded, transversely broader, conspicuously veined, 3-lobed, two *lateral lobes* large, rounded, descending far down the sides of the column; *middle one* small, cleft into two blunt flat lobules; a straight prominent nerve passes to the apex from the mouth at the base, on each side of which is placed a raised green-coloured glandule. *Column* (or *gynandrous style)* upon the summit of the germen, upright, semicylindrical, one third or more shorter than the corolla, edge of the summit scarcely raised beyond the anther. *Stamen* an *anther* inserted nearest the nether side of the summit of the column, lid-shaped, moveable, deciduous, brown, hemisphærical; *cellules* 4, parallel, convergent, standing on the inner front of the lid-shaped receptacle, and immersed in the cavity at the summit of the column. *Pollen-masses* 4, globular, compact, smooth, parallel, each tailed by a short granulated thread, by which it becomes fixed to the stigma, when the case of the anther from a cell of which it has been excreted, falls off. *Stigma* concave, fronting the labellum at the top of the column, close to the anther. *Capsule* 3-valved, fenestrate, that is, with lateral openings, its frame remaining connected at both ends. *Seeds* numerous, minute, resembling filings; coated by an *aril*, pointed at each end.

Requires to be kept in the hothouse; and will do with the treatment we have recommended for the other species in this work.

The drawing was made from a plant which flowered this autumn in the collection of Mr. Griffin, at South Lambeth.

a The labellum and summit of the column, showing the pollen-masses in the position they retain after the case of the anther has fallen: magnified. *b* The four pollen-masses removed from their position. *c* The inner front of the quadrilocular anther. *d* Its converse.

THE

BOTANICAL REGISTER:

CONSISTING OF

Coloured Figures

OF

EXOTIC PLANTS,

CULTIVATED IN

BRITISH GARDENS;

WITH THEIR

HISTORY AND MODE OF TREATMENT.

THE DESIGNS BY

Sydenham Edwards,

FELLOW OF THE LINNEAN SOCIETY.

VOL. II.

——viret semper——nec fronde caducâ
Carpitur.

LONDON:
PRINTED FOR RIDGWAYS, PICCADILLY.
1816.

148

POGONIA ophioglossoides.

Adder's-tongue-leaved Pogonia.

GYNANDRIA *MONANDRIA.*

Nat. ord. ORCHIDEÆ. *Jussieu gen.* 64. *Brown prod.* 309.

Div. III. *Anthera* terminalis, inserta, persistens. *Massæ pollinis* v. pulvereæ v. è corpusculis angulatis: basi v. infra apicem affixæ. *Brown in Hort. Kew. ed.* 2. 5. 201.

POGONIA. *Labellum* sessile, cucullatum, intùs cristatum. *Petala* 5 distincta, eglandulosa. *Pollen* farinaceum. *Brown in Hort. Kew. ed.* 2. 5. 203.

P. *ophioglossoides,* radice fibrosa, scapo dissitè bifoliato 1-2-floro, foliis ovali-lanceolatis, labello fimbriato. *Pursh amer. sept.* 2. 590; *(sub* ARETHUSA).

Arethusa ophioglossoides. *Linn. sp. pl.* 2. 1346. *Michaux bor. amer.* 2. 159. *Lamarck illustr. t.* 729. *f.* 2; *(de sicco). Willd. sp. pl.* 4. 80.

Arethusa radice fibrosa, scapi folio ovali, foliolo spathaceo lanceolato. *Gron. virg. ed.* 2. 139.

Cypripedium folio caulino ovato-oblongo, terminali lanceolato plano. *Gron. virg. ed.* 1. 110.

Helleborine virginiana ophioglossi folio D. Banister. *Pluk. alm.* 182. *t.* 93. *f.* 2.

H. aquatica flore in summo caule unico carneo, barba purpurea fimbriata, foliis ophioglossi, radice fibrosa. *Clayt. MSS. n.* 77.

Radix *crassè fibrosa, simplicior, rigida, horizontalis.* Caulis *erectus, simplex, bifoliatus, uncias* 5 *altus.* Folia 2 *semivaginanti-decurrentia, laminâ ovali-lanceolatâ* 3-5-*nervi,* inferioris *subsesquiunciali,* superioris *vix semunciali.* Flos 1 *(vix 2?) semuncialis, nutans* 1 *oblongo-semicampanulatus, semiringens:* pet. 5, *dilutiùs lilacina, fornicato-ascendentia, isometra,* 3 exteriora *divergentia lineari-lanceolata duplo angustiora magìsve, concava,* 2 interiora *obovato-oblonga sub summo medio imbricato-conniventia.* Labellum *vix longius, pandurato-obovatum, fimbriâ brevi patente pectinato-incisâ transversè purpureo-striatâ barbatum, intùs pilis densis suprà æquatis infernè versùs decrescentibus cristatum, basi subventricosum.* Columna *transversè striata lineis purpureis, duplo brevior et appressa labello, semiclavata, dorso rotundata, facie plana, suprà utrinque dentato-fissa lobulis contiguis, infernè complexa marginibus involutis unguis labelli. Operculum* antheræ *antrorsùm columnæ continuo-connexum.*

This singular little plant was introduced by Mr. Nuttall, to whom our collections are indebted for several of their rarest specimens. Native of North America, from Canada to Carolina, and grows in mossy bogs and swamps. Although well known in our books, and represented by two figures from dried spontaneous specimens; it has, we believe, never till now been introduced into any european garden, nor been represented anywhere from the living plant. *Root* perennial, fibrous, fibres thickish, simple, wiry, horizontal. *Stem* under six inches high, upright, simple, two-leaved, one-flowered. *Leaves* halfstem-embracing and decurrent, blade oval-lanceolate, three-nerved, that of the lower leaf an inch and half long, of the upper scarcely half an inch. *Flower* nutant, oblong, semicampanulate, semiringent, about half an inch long: *petals 5,* of a palish pink or lilac colour, ascendent, vaulted, of one length, *three outer ones* divergent, linear-lanceolate, concave, twice the narrowest or more; *inner two* obovately oblong, converging imbricately under the arch of the middle of the three outermost. *Lip* scarcely longer than these, obovate, somewhat contracted at the sides below the middle, bearded by a narrow pectinated purple-streaked horizontal fringe round the front edge, beset within by a level-topped crest or comb of compact hairs, growing shorter as they descend inwards, unguis or narrow part somewhat tumid or enlarged at the base. *Column* twice shorter than the lip to which it is pressed down, semicylindrically clavate, streaked transversely by purple lines, rounded at the back, flat in front, clasped at the base by the inflected sides of the unguis or narrow portion of the lip, at the top dentately cleft on each side, lobules contiguous. *Anther* persistent. *Pollen* mealy. We had not an opportunity of examining the anther and stigma so completely as we wished.

The drawing was made at the nursery of Messrs. Colville, in the King's Road, Chelsea. The plant is most probably quite hardy; but in order to secure its flowering it was planted in bog-earth, and the pot placed in a pan of water in the hothouse, as soon as the roots arrived from America. Without this treatment we should most likely not have been enabled to have procured the present figure of it.

a The outline of an artificially extended flower. *b* The column with the anther turned up out of its place of insertion.

153

CRYPTARRHENA lunata.

Crescent-lipped Cryptarrhena.

GYNANDRIA *MONANDRIA.*

Nat. ord. ORCHIDEÆ. *Jussieu gen.* 64. *Brown prod.* 309.

Div. IV. *Anthera* terminalis mobilis decidua. *Massæ* pollinis demùm cereaceæ. *Brown in Hort. Kew. ed.* 2. 5. 205.

CRYPTARRHENA. *Petala* 5, distincta patentia. *Labellum* ecalcaratum: *lamina* dilatata explanata. *Columna* libera, aptera. *Anthera* apice cucullato columnæ inclusa, porrecto-adscendens. *Massæ* pollinis quatuor. *Brown MSS.*

Cryptarrhena *lunata.* Brown MSS.

Herba humilis. Folia *radicalia numerosa, multifariàm ambientia, patentia, crassiuscula, carnosa, remotiùs nervosa,* exteriora *recumbentia, subovato-lanceolata, acuminata, subtriuncialia, lætè virentia,* interiora *erectiora, lineari-lanceolata.* Caulis *longior istis, striato-teres, crassitudine pennæ corvinæ, infernè foliolis paucis lineari-subulatis erectis vagè stipatus.* Spica *subbiuncialis, inclinata, laxiùs digesta, nuda.* Flores *viridi-lutei, vix tertiam partem unciæ transversi.*

We had not the opportunity of inspecting the flower of this plant after the drawing was made; but to the greater advantage of Botany, Mr. Brown has had that opportunity. He has found it an unrecorded vegetable, and to possess features so distinct from those already known of the tribe, as to make it expedient that it should form the foundation of a new generic group.

All we know of its history is, that it was lately received by Messrs. Lee and Kennedy, of the Hammersmith nursery, from Jamaica, and that it flowered in one of their hothouses in the course of last summer.

A humble herbaceous species; and, judging by analogy, probably parasitic. *Leaves* radical, numerous, ambient, multifariously patent, fleshy, thickish, loosely nerved, of a bright grass-green colour: outer recumbent broadest, lanceolate, and somewhat ovate, acuminate, three parts of an inch across, almost three inches in length; inner more upright, linear-lanceolate. *Stem* higher than these, simple, round, angularly striate, bearing a few linearly subulate leaflets, placed irregularly towards its base. *Spike* many-flowered, loose, bent. *Corolla* hardly the third of an inch over, green and yellow; and, as we are told, having scarcely any scent.

---

a The outline of the front view of a magnified flower, with the lid of the anther forced back, showing the loculaments. *b* The outline of a magnified flower viewed sideways, showing the cowled part of the column which covers the anther.

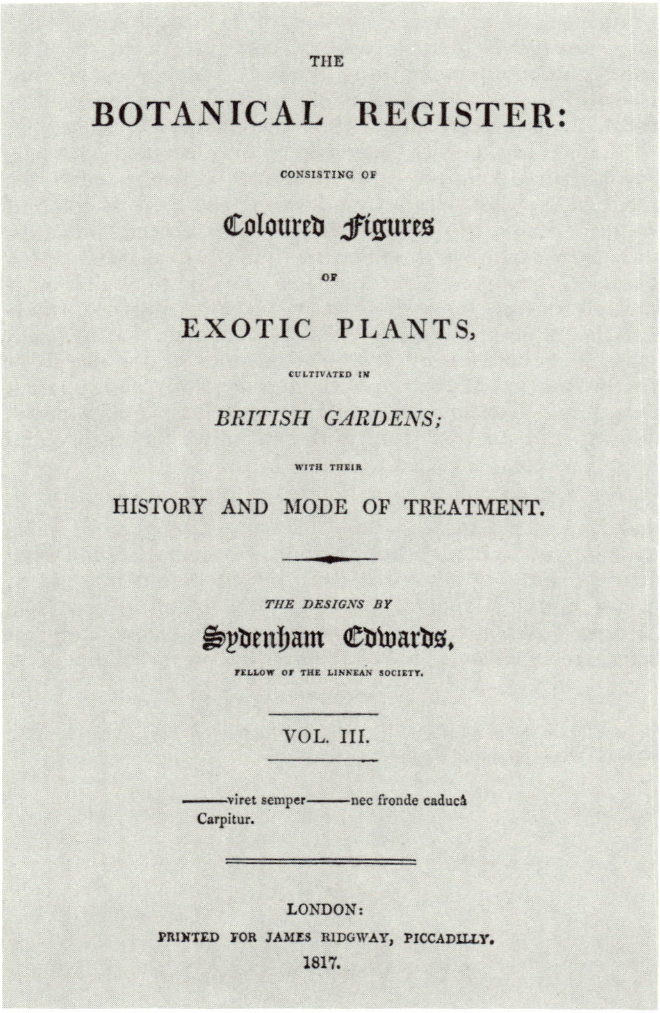
THE

BOTANICAL REGISTER:

CONSISTING OF

Coloured Figures

OF

EXOTIC PLANTS,

CULTIVATED IN

BRITISH GARDENS;

WITH THEIR

HISTORY AND MODE OF TREATMENT.

THE DESIGNS BY

Sydenham Edwards,

FELLOW OF THE LINNEAN SOCIETY.

VOL. III.

——viret semper——nec fronde caducâ
Carpitur.

LONDON:
PRINTED FOR JAMES RIDGWAY, PICCADILLY.
1817.

202

ORCHIS longicornu.
Long-spurred Orchis.

GYNANDRIA *MONANDRIA.*

Nat. ord. ORCHIDEÆ. *Jussieu gen.* 64. *Brown prod.* 309.
Div. I. *Anthera* adnata subterminalis persistens. *Pollinis massæ* è lobulis angulatis elasticè cohærentibus; basi affixæ. *Brown in Hort. Kew. ed.* 2. 5. 188.
ORCHIS. *Cor.* ringens. *Labellum* basi subtùs calcaratum. *Glandulæ* (1-2) pedicellorum pollinis inclusæ cucullo unico. *Br. loc. cit.*

O. *longicornu,* labello trilobo crenulato, lobis obtusis, medio breviore, petalis obtusis conniventibus, cornu ascendente compresso spathulato germen subæquante. *Willd. sp. pl.* 4. 19.
Orchis longicornu. *Poiret voy. en barb.* 2. 247. *Lamarck encyc.* 4. 591. *Desfont. atl.* 2. 317. *t.* 246.

Rad. *didymo-tuberosa.* Fol. *immaculata, radicalia, plurifaria, basi imbricato-compacta, indè patentia,* 5-6, 3-*uncialia,* ¾ *partes unciæ lata, lanceolata, micaceo-pruinosa, medio canaliculata, distantiùs lineata; caulina* 4-5, *vaginantia.* Caulis *scapiformis bis altior foliis, supernè purpurascens.* Flores *laxi, patentes,* 10 *v. plures, violaceo-purpurei, cum germine simul subunciales, odore grato amaricante fragrantes, resupinato-explicantes.* Bracteæ *vaginantes, coloratæ, lanceolatæ, breviores germine.* Petala *ordine duplici;* 5 *(ex resupinatione) superiora galeato-conniventia, rosea, intùs pallentia, obtusa, bìs tèrve breviora calcare; trium* exteriorum lateralia *obliquata, inæquilateri-elliptica, nervis* 5 *viridantibus, carinâ eccentricâ;* medium *oblongum antrorsùm fornicatum trinerve;* interiora *extimis-accumbentia, angustiora, invicèm approximata.* Labellum *dimidio longius istis, oblatum, porrectum, disco pallescens punctis atropurpureis,* 3-*lobum lobis lateralibus obscuro-coloratis externo margine denticulatis, intermedio admodùm minore sæpiùs retuso:* calcar ⅔ *unciæ longum, subæquale germini, assurgens, roseo-punicans, transversè compressum, cuneato-elongatum, truncatum, erugatum, quandòque retusum.* Germen *fusco-purpureum,* ¾ *partes unciæ longum, pro flore resupinando tortum, suprà curvatum.* Massæ pollinis *stipitato-clavatæ, indivisæ, thecis* 2 *marginalibus cuculli erecti inclusæ, stipites septo incompleto perpendiculari intergerino albo duplicato distincti:* pollen *virens, graveolens.*

A species very recently introduced by Mr. William Swainson from the Botanic Garden at Palermo. Native of the coast of Barbary, and probably of Sicily. Found by Messrs. Poiret and Desfontaines in swamps near the sea. The latter has published an engraving of it in his celebrated Flora Atlantica, but from a dried specimen.

It comes very close to our own common Meadow Orchis *(Morio),* but the leaves are more numerous, the flowers larger and brighter and very fragrant; the spur is longer in proportion, flat, and truncate; the two outside petals are not so green, the bractes are much shorter than the germen, and the side lobes of the labellum reflectent. There seem to be other differences; but not having a fresh plant of *Morio* at hand, we cannot rely with confidence on our view of them.

The drawing was made at the Chelsea Garden, where the species of this tribe, generally reputed as scarcely to be reclaimed to culture, are treated by Mr. Anderson with as much success as those of any other vegetable family. Among many plants of this species which flowered at the same time, we perceived several varieties. The two we have represented in the plate seemed to be, however, the extremes of the variation; the detached flower being that which receded the farthest from *Morio.*

Root of twin tubers. *Leaves* unspotted, *radical ones* 5-6, multifariously patent, scarcely 3 inches long, lanceolate, chagreened with minute crystalline papillæ, flat, channelled down the middle, with distant lines; *cauline ones* 4-5 sheathing. *Stem* twice as high as the leaves, purple upwards. *Flowers* 10 or more, in a loose spreading spike, violet-purple, very fragrant, about an inch long including the germen, when open resupinate. *Bractes* coloured, sheathing the much shorter germens. *Upper lip* of the open corolla pentapetalous, convergent, vaulted; *petals* in two ranks, all pale purple, obtuse, 2 or 3 times shorter than the spur, the lateral ones of the three outermost marked with five greenish nerves, elliptic with unequal sides, the middle one oblong pointing forward arched 3-nerved; *two innermost* narrower, convergent, overlaid by the others. *Label* half as long again as these, broader than long, pointing forwards, paler along the middle, where it is spotted with dark purple; 3-lobed, side-lobes dark-purple rounded outwards and denticulate, middlemost a great deal smaller, generally retuse: *spur* about $\frac{2}{3}$ of an inch long, about equal to the germen, assurgent, rose-purple, horizontally compressed, cuneately linear, truncate, unwrinkled. *Germen* brownish red, $\frac{3}{4}$ of an inch long bent towards the top. *Pollen-masses* 2, clavate, pedicled, undivided, enclosed within the two marginal cists of the headpiece of the *column* formed by the filaments and style, separated above their glandular base by an incomplete upright white bilaminar partition.

205

OPHRYS tenthredinifera.

Saw-fly Ophrys.

GYNANDRIA *MONANDRIA.*

Nat. ord. ORCHIDEÆ. *Jussieu gen.* 64. *Brown prod.* 309.
Div. I. *Anthera* adnata subterminalis persistens. *Pollinis massæ* è lobulis angulatis elasticè cohærentibus; basi affixæ. *Brown in Hort. Kew. ed.* 2. 5. 188.
OPHRYS. *Cor.* subpatens. *Labellum* ecalcaratum. *Glandulæ pollinis* cucullis distinctis inclusæ. *Id. l. c.* 195.

O. *tenthredinifera*, caule folioso, labello villoso obovato bilobo appendiculato, petalis patentibus, 3 exterioribus oblongis obtusis, binis interioribus brevissimis. *Willd. sp. pl.* 4. 64.
Ophrys insectifera. *var.* A. *rosea.* *Desfont. atl.* 2. 320.
Ophrys villosa. *Desfont. in ann. du Muséum* 10. 225. *t.* 14.
Orchis orientalis, fucum referens, flore parvo, villosissimo, scuto azureo. *Tournef. cor. inst.* 30. *Vélins du Muséum.*

Tuber *didymum, indivisum.* Folia *subtùs lineata, radicalia-4-5, elliptico-lanceolata, subtriuncialia, latiora unciam ferè transversa; caulina angustè lanceolata vaginantia.* Caulis *ferè duplo altior.* Spica *laxa 3-8-flora.* Bracteæ *pallido-carneæ lineis viridibus, convoluto-lanceolatæ, erectæ,* germen *semunciale viride tortum curvatum superantes.* Cor. *nutans, cruciato-ringens, subpatens rictû subunciali:* pet. 5 *lilacino-rosea,* 3 exteriora *distantia, isometra, glabra, obtusa, nervo medio carinata, medium erectum ellipticum fornicatum, lateralia ovato-elliptica patentissima horizontalia: 2 interiora istis alterna minima angulari-ovata intùs tomentosa, summæ columnæ acclinia:* labellum *dependens, amplum, subpandurato-obcordatum, bilobo-retusum acuminulo terminali crasso rigido uncatìm infracto, intùs sericeum at gibbo medio hirsutiùs ac si barbatum, viridi-luteum, convexum, margine depressum, pictum maculâ magnâ pectoraliformi castaneâ suprà includente alteram scutiformem polyedram glabratam cyaneam albo iridiatam ipsam continentem cum glandulâ elevatâ thoraciformi rubro-fuscâ splendenteque.* Columna *anticè fornicata, margine flavo villoso ad utrumque latus antheræ ocello nigro notato.* Cucullus antheræ *dithecus rostrato-terminalis erectus virens compressus aduncus thecis 2 collateralibus anticè fenestratis.* Pollinis massæ *flavæ erectæ clavatæ, stipite longiore capitello.*

Many species in this genus recall forcibly the images of various insects, as flies, bees, spiders, &c.; of which we have instances in the few indigenous ones of our own country. The present is a native of the Coast of Barbary, the Levant, and probably of Sicily, having been lately introduced from the botanic garden of Palermo, by Mr. William Swainson. In the Flora Atlantica it is included among the varieties of the *insectifera* of Linnæus, but now stands distinct, as well as the others of that promiscuously combined species. What has suggested to Willdenow the resemblance of its flower to any of the Saw-flies, we cannot guess, as he does not appear to have seen either figure or specimen. The synonym from a later publication by Prof. Desfontaines we have added upon our own judgment.

Tubers twin, undivided. *Lower* leaves 4-5, elliptically lanceolate, about 3 inches long, upper 2-3 narrowly lanceolate, convolutely folded upright. *Stem* about twice their height. *Spike* loose, 3-8-flowered. *Bractes* of a faded pink-colour with green lines, upright, lanceolate, folding round the germens which they overtop. *Corolla* nodding, cruciformly ringent, rather spreading, nearly an inch from top to bottom; *petals* 5, pink; *exterior* 3 of one length, standing apart, smooth, obtuse, green-keeled, middle one of these upright elliptic vaulted, side ones horizontally extended; *interior* 2 alternate, diminutive, tomentose at the inner surface, inclining over the column. *Label* broadly obcordate, contracted above the base, bilobedly retuse with a small intermediate fleshy rigid flat-pressed turned-up point; on the inner surface greenish yellow, covered with a velvet-like pubescence, marked with a large chestnut-coloured stain, comprehending at the base a smaller smooth polyedrously scutiform one, edged with white and surmounted by a reddish brown fulgid glandular prominence, resembling the thorax of some species of insect. *Anther* in two distinct cells, terminal, upright, compressed, narrow, rostrate, crooked; *cells* adjacent, open in front; *pollen-masses* with a separate glandular base to each; not as in ORCHIS with a glandular base common to both. Mr. Brown observes, that both genera, as now defined, are wholly extratropical, and confined chiefly to Europe.

210

DISA prasinata.

Mr. Griffin's Disa.

GYNANDRIA *MONANDRIA.*

Nat. ord. ORCHIDEÆ. *Jussieu gen.* 64. *Brown prod.* 309.
Div. I. *Anthera* adnata subterminalis persistens. *Pollinis massæ* è lobulis angulatis elasticè cohærentibus; basi affixæ. *Brown in Hort. Kew. ed.* 2. 5. 188.
DISA. *Cor.* ringens: galea basi unicalcarata v. saccata, petala interiora columnæ accreta. *Labellum* ecalcaratum. *Idem loc. cit.*

Div. Calcare longiore.

D. *prasinata*, galea obtusa; calcare oblongo dorso subcarinato-convexo, ventre concaviusculo appresso germini sublongiori, labello lineari acutiusculo, spica laxa, bracteis floribus subbrevioribus.

Caulis *sesquipedalis, crassitudine pennæ mediocris, totus foliosus, erectus, virens.* Fol. *amplexicaulia, lanceolatìm ligulata, erectiuscula, sparsa, distantia, intervallis longiora, lætè virentia, decrescentia, carinâ crassiore, infernè rubentia præsertìm inferiora, longiora sub9uncialia, latitudine $\frac{3}{4}$ partium unciæ.* Spica *laxè subcylindracea, 6-uncialis v. ultrà.* Bracteæ *convoluto-lanceolatæ, erectæ, cuspidatæ, margine baseos membranaceo, apiculo rubro.* Flores *inodori, cum germine simul unciales.* Cor. *viridans, resupinata, ringens, patentissima:* galea *lata, cuneato-obovata, erecto-fornicata, parùm brevior calcare, intùs saturatè rosea, rubedine virorem externè perlucente;* calcar *virens compressiusculum:* pet. lateralia 2 inferiora *(utì galea, exteriora) lineari-oblonga, obtusula, depresso margine convexa, labellum versùs obliquata, retroflexa, modò ut ex apicibus ponè germen contingant, chloroleuca basi intùs maculis et punctis roseis:* labellum *istis æquilongum, duplo angustius, viride, deflexum, lævigatum, sensim versùs apicem incrascescens, convexum:* pet. lateralia 2 superiora *(utì labellum, interiora) crassiuscula, viridia, galeâ inclusa, erecta, à plano opposita, basi lata, indè abruptiùs angustata, lanceolata, obliquato-obtusa, emarginata, ultra basin punticulis irrorata.* Anthera *per summam columnam supinata; cuculli paries erecta, à parte galeæ proximâ incompleta, latere utroque replicato-auriculata atque cum petalis 2 intimis commissa:* pollinis massæ *clavatæ, luteæ, ex vesiculis variè pressis loricatìm dispositis.* Stigma *obliquè prominens trans pedem columnæ supra basin labelli.*

A genus confined to the Cape of Good Hope and the adjoining regions. Twenty-eight species have been enumerated by Willdenow. Two only appear to have found their way into the collections of this country, before the present unrecorded one, lately introduced by Mr. Griffin, with whom it flowered this spring. It comes near to both *rufescens* and *bracteata*; but is at once distinguished from the first, by an oblong spur which is shorter than the germen; from the latter by a spike, in which the bractes are shorter than the flowers; from both by a narrow linear subacuminate label.

We did not inspect the root; but learned from Mr. Griffin, that it was a round undivided tuber. *Stem* about a foot and half high, with the thickness of a common pen, upright, green, leafy quite to the spike. *Leaves* stemclasping, lanceolately ligulate, scattered, almost upright, standing pretty far asunder, but much longer than the intervals, grass-green, decrescent upwards, keeled by a thickish midrib, red at the lower part, especially those nearest the root, longer ones about 9 inches in length, and about $\frac{3}{4}$ of an inch broad. *Spike* many-flowered, subcylindrically but loosely elongated. *Bractes* convolute, lanceolate, pointed, upright, somewhat shorter than the flowers, having near the base a colourless membranous margin, at their end a red point. *Flowers* without smell, including the germen about an inch long. *Corolla* greenish, resupinate, ringent, spreading: *galea* broad, obovate, upright, arched, but little shorter than its spur, a deep rose-colour on the inside, which colour shines through the green of the outside, *spur* slightly compressed green: *two lateral petals* of the lower lip (like the galea, exterior) linearly oblong, greenish white, somewhat blunt, depressed at the margin, convex, slanting towards the labellum, sometimes bent so far back as for their points to meet behind the germen, stained and dotted at their bases with purple: *labellum* of the same length as these but twice as narrow, green, dependent in front of the germen, linear, polished, thickening towards the end and convex: *two upper lateral petals* interior (as is the labellum, with which they agree in colour and substance) placed within the arch of the galea, and but little shorter than that, upright, standing broadways in relation to each other, with their inner edge towards the galea, very broad at the base, and from thence abruptly tapered, lanceolate, with a slantedly blunt point, emarginate, dotted above the base. *Anther* reversed over the summit of the column; its outer wall upright, incomplete at the end towards the galea, eared on each side, ears reversely folded, and let into a hollow at the base of the two inner petals.

a An outline of the corolla, with the galea forced back. *b* Anther.

220

AERIDES paniculatum.

Sir Joseph Banks's Aerides.

GYNANDRIA *MONANDRIA.*

Nat. ord. ORCHIDEÆ. *Jussieu gen.* 64. *Brown prod.* 309. *Div. II. Anthera* terminalis mobilis decidua. *Massæ pollinis* demum cereaceæ. *Brown Hort. Kew. ed.* 2. 5. 205.

AERIDES. Labellum calcaratum v. saccatum, insertum apice processûs unguiformis, cujus lateribus petala antica exteriorum adnata. *Massæ pollinis* 2 posticè bilobæ, processû communi stigmatis medio affixæ. *Brown loc. cit.* 212.

A. *paniculatum*, caule paniculato; calcare recto præpendente germini vix longiori, lobo terminali labelli difformi-trilobulato: foliis inæquali-retusis.

Planta *parasitica herbacea perennis:* caudex *rhizomatosus ramoso-radicatus, compressus, erectus, totus foliosus;* fibræ *intervallis foliorum laterales, vermiculari-elongatæ, teretes, crassæ, tortuosæ, nodoso-rugatæ, læves, cinereæ, fibrillis nudæ, apice virentes.* Folia *plurima, alterna, disticha, patentissima, firma, crassiuscula, lineari-lorata, canaliculata, subsesuncialia, lætè viridia, inæqualitèr retusa lobo altero productiore.* Caulis *aphyllus, inter folia lateralis, ascendens, purpureo-fusco varius, teres, lævis, rigidus, sesquipedalis, subgeniculato-flexuosus, articulis* bracteâ *brevi sterili cylindricâ vaginatus, paniculatus* spiculis *paucis multifloris alternè distantibus patentissimis, terminali arrecto-continua.* Flores *flavi, rubro-fusco picti, inodori, erecti, nec resupinati, sessiles,* bracteis *singularibus ovato-acuminatis luteis appressis germini eoque tèr quatèrve brevioribus.* Germ. *curvo-ascendens,* 6-*striato-teres, luteo-viridans,* $\frac{1}{3}$ *unciæ longum.* Cor. *bilabiata, semiringens, rigidiuscula:* pet. 5 *semicirculari-patula, æquilonga,* $\frac{1}{3}$ *breviora germine;* summum medium *(ex exterioribus unum) distantius, obovato-oblongum, subfornicatum;* lateralia superiora 2 *(ex interioribus) angustiora, lineari-oblonga, obtusa;* inferiora 2 *(ex exterioribus) duplo latiora, ab imis lateribus connata cum ungue labelli, ovali-oblonga, acutula suprà maculâ atro-sanguineâ picta, sursùm obliquantia;* labellum *porrectum, rigidum, carnosum, calcaratum, subtùs intrusum, ab ungue connexum cum columnâ, 3-lobum palato medio canaliculato albo pubescente, lobis 2 lateralibus inflexis erectis dentiformibus puniceis, terminali calcarigero ochroleuco trilobulato lobulo medio rostellato-capitato lateralibus subulatis alato-divergentibus (simul subsimulantibus volucrem);* calcar *pallidum oblongo-conicum subdidymo-inflatiusculum, obtusum, germini vix longiori præpendens, in longum æqualitèr bicameratum. Ante anthesin labellum totum super columnam reversum est.* Columna *oblonga, erecta, subcylindracea, virens, infrà punicea, parùm brevior petalis, foramine medio antico orificio oblongo marginato excavata, apice bidentata (rudimentis filamentorum?).* Anthera *terminalis, mobilis conceptaculo operculiformi subdidymo convexo intùs biloculari:* pollinis massæ *sessiles, collaterali-geminæ, hemisphæricæ, vitellinæ, durissimæ, homogeneæ, transversè bilobo-dissectæ, ex parte plana didymo-contingentes, posticè glandulæ sagittatæ stigmatis affixæ.* Stigma *ex lamellis 2 verticalibus dentiformibus parallelis diaphano-pallentibus juxtà infra antheram et supra cavitatem columnæ prominentibus.*

A parasitic genus, instituted by Loureiro in his account of the plants of Cochinchina, and named in reference to the faculty its species possess of growing when suspended, so as to be cut off from all sustenance but that derived immediately from the atmosphere. Plants of other genera of this tribe, and even some of a different tribe *(see fol.* 105 *of this work)*, are endowed with a like faculty; in none however can such insulation be considered as the state of existence which suits them best, but merely as one that they are enabled to endure, as a Carp is known to do that of being suspended out of water in a damp cellar. The most successful mode of treating plants of this nature in these climates, has been devised by Sir Joseph Banks, to whom we are obliged for the opportunity of representing the present specimen, which flowered in the hothouse of his garden at Smallberry-Green, and had been introduced by himself. The method he pursues, is, to place the plants separately in light cylindrical wicker baskets or cages of suitable widths, of which the frame-work is of long slender twigs wattled together at the bottom and shallowly round the side, the upper portion being left open that the plant may extend its growth in any direction through the intervals, and yet be kept steady in its station, the ends of the twigs having been tied together by the twine that suspends the whole to the woodwork of the stove, a thin layer of vegetable mould is strewed on the floor of the basket, on which the rootstock is placed, and then covered lightly over with a sufficiency of moss to shade it, and preserve a due degree of moisture, water being occasionally supplied. The diminished outline in the corner of the annexed plate is intended to exemplify he appearance of our plant under that treatment.

The present is an unpublished species, and native of China. We do not believe that any of the genus has been figured from a plant which has flowered in Europe till now. It is closely allied to the Chinese portion of CYMBIDIUM, in which the label, however, has not a spurlike pouch as here.

AERIDES *paniculatum* is an herbaceous perennial; the caudexlike *rootstock* leafy from top to bottom, compressed, upright, branchingly radicant, *roots* lateral in the intervals of the foliage, round, thick, tortuous, knottedly wrinkled, smooth, of a colour resembling that of the silkworm, green at the points. *Leaves* many, alternate, in two ranks, spreading, firm, rather thick, linearly lorate, channelled, about six inches long, bright green, unequally retuse, the lobe of one side being higher than the other. *Stem* leafless, lateral between the leaves, ascendent, mottled with dark-purple, round, smooth, rigid, about a foot and half high, flexuose and slightly kneed, sheathed at the joints by a short cylindrical sterile *bracte*, panicled; *spikelets* few, many-flowered, alternate, distant, wide spread, the end one upright and continuous. *Flowers* deep yellow, marked with brownish red, without scent, upright, not reversed or resupinate, sessile; *bractes* one to each flower, ovately acuminate yellow, placed against the germen, which is 3 or 4 times higher. *Germen* upright with a curve, round and sixstreaked, yellowish green, $\frac{1}{3}$ of an inch long, *Corolla* bilabiate, semiringent, stiffish: *petals* 5, spreading in a semicircle, of one length, a third shorter than the germen; *the upper middle one* somewhat aloof from the rest, obovately oblong, and slightly vaulted; *two upper lateral ones* narrower, linearly oblong, obtuse: *two lower ones* twice broader, grown to the unguis of the label by the lower part of each side, rather pointed, marked with a red stain above, slanting upwards: *label* outstretched, stiff, fleshy, spurred, dinted underneath, connected by its unguis with the column, 3-lobed with a white channelled furred palate or boss in the middle, the *two side lobes* turned up, toothshaped, crimson, *end one* spurbearing yellowish white trifid, the central lobule capitate and beaked, the two side ones subulate and outspread, the three composing an image that may be fancied to resemble a bird in miniature; *spur* pale-coloured oblong, conical, slightly inflated and didymously divided by a shallow furrow, hanging before the germen which it nearly equals, vertically two-chambered within. The whole label is reversed over the column before the flower opens. *Column* oblong, upright, nearly cylindrical, green, crimson below, but little shorter than the petals, having a central cavity in front, with an oblong bordered orifice, bidentate at the apex (the teeth being probably the rudiments of two filaments?). *Anther* terminal, moveable, with a convex lid-shaped bilocular receptacle: *pollen-masses* two sessile, very hard, homogeneous, hemisphærical, transversely bisected, smooth, deep yellow, placed sideways to each other by the plane part, inserted at the forked gland of the stigma. *Stigma* of two vertical parallel teethlike, pale, diaphanous laminæ, jutting forwards just below the anther and above the cavity of the column.

a The back of the entire anther. *b* The pollen-masses removed from their receptacle.

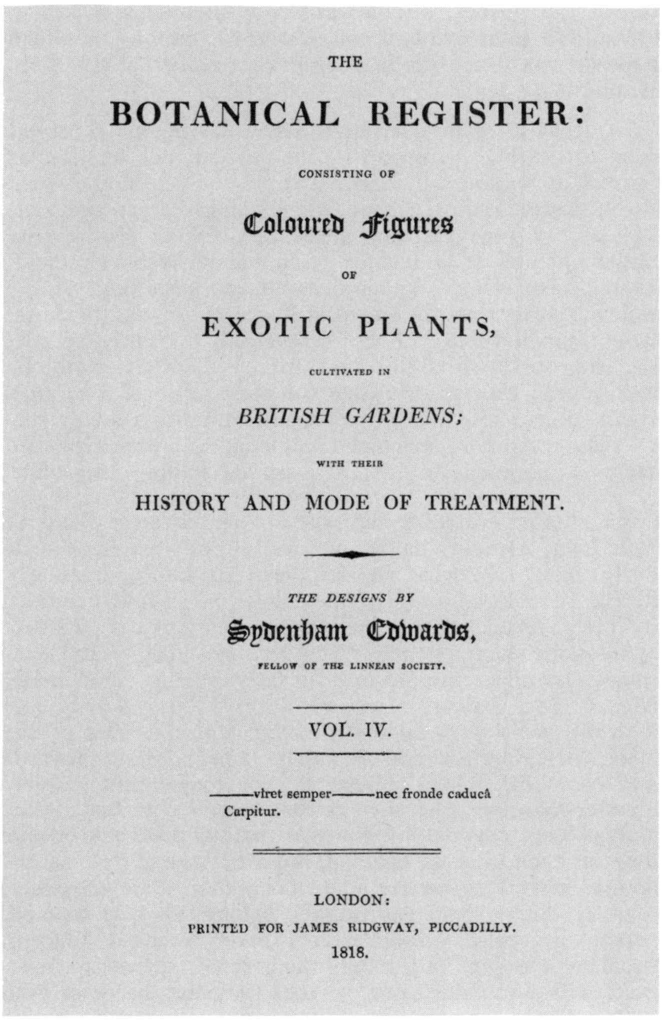

THE

BOTANICAL REGISTER:

CONSISTING OF

Coloured Figures

OF

EXOTIC PLANTS,

CULTIVATED IN

BRITISH GARDENS;

WITH THEIR

HISTORY AND MODE OF TREATMENT.

THE DESIGNS BY

Sydenham Edwards,

FELLOW OF THE LINNEAN SOCIETY.

VOL. IV.

——viret semper——nec fronde caducâ
Carpitur.

LONDON:
PRINTED FOR JAMES RIDGWAY, PICCADILLY.
1818.

271

GOODYERA discolor
Brasil Goodyera.

GYNANDRIA *MONANDRIA.*

Nat. ord. ORCHIDEÆ. *Jussieu gen.* 64. *Brown prod.* 309.

Div. II. Anthera stigmati parallela persistens. *Massæ pollinis* v. farinaceæ v. è corpusculis angulatis; apicibus stigmati affixæ. *Brown in Hort. Kew. ed.* 2. 5. 197.

GOODYERA. Cor. ringens: petalis exterioribus anticis labello infernè gibboso supernè indiviso suppositis. *Columna* libera. *Pollen* angulatum. *Brown loc. cit.*

Div. II. Petalis exterioribus anticis horizontali-patentibus.

G. *discolor*, rhizomate horizontali, caudescente, foliis discoloribus ovalibus crassiusculis, spica multiflora, ungue labelli ad basin bullato-didymo adnatoque laminâ transversè oblongâ.

Rhizoma *carnosum subrepens, radicato-caudescens, cylindricum crassitudine digiti, foliorum vestigiis annulatum, fusco-sanguineum, supernè foliosum, indivisum.* Folia *ambientia, sparsa, discolora, vaginoso-petiolata vaginâ sanguineo-rubente;* lamina *crassiuscula, ovalis, utrinque attenuata, acuminata, subtriuncialis, suprà papillis atomoideis densis lucido-pruinata, atro-virens aliquâ irrepente rubedine, subtùs glabrata punicea longitudinalitèr nervosa nervorum intervallis flexuoso-venosa.* Scapus *simplex erectus cylindraceus, subpedalis, teres crassitudine tubi pennæ scriptoriæ vel majore, albo-hirsutus, bracteis aliquibus vacuis florum majoribus appressis hìc indè munitus.* Spica *simplex, erecta, cylindracea, subtriuncialis, laxiùs multiflora.* Bracteæ *membranosæ, rubicundæ, pruinoso-micantes, lanceolato-acuminatæ, convolutæ, longiores germine, 3-nerves, erectæ, cuspidatæ, subciliatæ.* Flores *supinati, nutantes, sessiles, nivei, cruciato-ringentes, ¾ partes unciæ transversi:* germ. *semunciale, album, hirsutum, tortum, apice flexum, sublongius corollâ:* cor. *tetrapetaloidea;* pet. *æquilonga, exteriora pruinoso-opaca, interiora pruinoso-lucida;* galea *ovato-elliptica, erecta, obtusa, disco extrorsùm gibbosa, formata petalis* 2 lateralibus interioribus *dimidiato-lanceolatis angustis cum* mediâ *ovali duplo latiori ad suum marginem utrinque imbricato-subcohærentibus;* pet. 2 exteriora lateralia *oblongo-ovata acumine obtuso, horizontali-divaricata, rictum floris exactè decussantia, angustiora galeâ, deorsùm obliquata lateribus modo tali depressis ut promineat discus, extùs apice maculâ parvâ cinnamomeo-rubente notata;* labellum *difforme, indivisum, concolor, unguiculatum;* unguis *3-plo magìsve longior laminâ, lineari-oblongus, subtubuloso-convolutus, à sacculo brevissimo didymo-hemisphærico anticè bibullato-protuberante adnatus basi columnæ;* lamina *brevis transversè oblonga integerrima apiculo obsoleto, basi levissimè subcordata, glabra.* Columna *alba, duplo ferè brevior corollâ, libera, erectiuscula, infernè cylindrica, suprà posticè subdidymo-gibbosa, flavescens, lucida, vertice depresso:* anth. *opercularis, persistens, bilocularis, receptaculo oblongo acuminato antrorsùm incumbente processû glanduloso rostellatìm producto atque areolam papulosam secernentem* stigmatis *paralleli obumbrante:* massæ pollinis *binæ sulphureæ pedicellato-clavatæ, è lobulis obversis pressione angulosis axim elasticam glutinosam imbricatìm contegentibus conflatæ. Differt* GOODYERIS *veris habitû valdè robustiore atque carnosiore, charactereque tenùs ex petalis 2 extimis alarum instar dispansis, neque labello suppositis atque collaterali-conniventibus; aliàs verò plurimùm cum genere consentit.*

An unrecorded orchideous species, recently introduced from the Brasils by Messrs. Lee and Kennedy, of the Hammersmith Nursery, where the drawing was taken from a plant that flowered in the hothouse in January last. We have enrolled it under GOODYERA, notwithstanding the disagreement in regard to a characteristic that has been proposed as essential to that genus; viz. the convergence of the two exterior front segments of the corolla under the label, which two segments are here widely divergent, and extended like wings. In general resemblance, however, the plant approaches too near to GOODYERA to suggest the expediency of originating a new generic group from it, especially as that genus is still sparingly stocked with species. A very slight modification of the present generic character will make room for our plant.

Rootstock fleshy horizontal, cylindrical, about the thickness of a man's finger, scored in circles by the scars of the fallen leaves, blackish red or liver-coloured, caudescent and radicant upwards, unbranched, leafy at the top. *Leaves* scattered all round, two-coloured, sheathingly petioled, sheaths red, *blade* thickish, oval, tapered both ways, acuminate, about 3 inches long, finely and shiningly chagreened, blackish green on the upper side, on the under smooth, and crimson-coloured, longitudinally nerved, flexuosely veined in the intervals of the nerves. *Scape* terminal, upright, stiff, about a foot long, of the thickness of a pen or thicker, round, covered by a white fur, furnished here and there with barren close-pressed bractes larger than those of the inflorescence. *Spike* simple, upright, cylindrical loosishly manyflowered, about 3 inches long. *Bractes* membranous, red, glitteringly frosted, lanceolately acuminate, convolute, longer than the germen, 3-nerved, upright, sharp-pointed, slightly fringed. *Flowers* reversed, nodding, sessile, white, crucially ringent, about ¾ of an inch across: *germen* twisted half an inch long, bent at the top, rather longer than the corolla, white-furred. *Corolla* appearing as of 4 segments only; *segments* of one length, outer opaquely frosted, inner shiningly frosted; *casque* ovately elliptic, upright, obtuse gibbous at the back, formed of 2 lateral narrow parabolically lanceolate interior segments that lap over and adhere by their inmost edge to each side of a middle oval one of twice their own breadth; 2 *exterior lateral segments* oblongly ovate with a rounded point, horizontally spread, intersecting the ricture of the flower transversely, slanting downwards, narrower than the casque, disk convex, edge depressed, with a small cinnamon spot at the top on the outside: *label* all white, biformed, undivided, unguiculated, *unguis* 3 times as long as the blade or more, oblong and straight-sided, subtubularly convolute, affixed to the foot of the front of the column by a very shallow didymous pouch, resembling the heads of two very small studs; *blade*, shallow transversely oblong, entire, with a scarcely perceptible point, shallowly cordate at the base, smooth. *Column* white, almost twice shorter than the corolla, quite detached, nearly upright, cylindrical at the lower part, upwards at the back enlarged by a double protuberance, yellowish, polished, depressed at the summit: *anther* lidshaped, permanent, of two cells, with an oblong acuminated receptacle, the point of which rests upon the beakshaped process that overhangs the parallel slanted *stigma: pollen-masses* 2, pale-yellow, pedicled, clavate, composed of obversely oblong angularly pressed lobules collected round an elastic glutinous axis.

233

LIMODORUM falcatum.

Sickle-leaved Limodorum

GYNANDRIA *MONOGYNIA.*

Nat. ord. ORCHIDEÆ. *Jussieu gen.* 64.
ORCHIDEÆ. *Brown prod.* 309. *Div. IV. Anth.* terminalis, mobilis, decidua. *Mass. Poll.* demùm cereaceæ.

LIMODORUM. *Cor.* hexapetala, plerumque patens, rariùs resupinata: *pet.* 3 exteriora, 2 interiora sæpè minora: *labellum* (ex interioribus unum) basi concavum, posticè in calcar variæ figuræ et longitudinis prominens: *lamina* patula, indivisa s. lobata. *Genitalia* et *Capsula* ut in CYMBIDIO et EPIDENDRO, à quibus differt labio cornuto. Ab ORCHIDE corollâ non ringente et structurâ columnæ genitalium. *Swartz in Schrader's journ.* 1799. 2. 228. *Id. in ejusdem neues journ.* 1. 84.

L. *falcatum*, foliis subradicalibus ensiformi-canaliculatis falcatis, scapis paucifloris, cornu filiformi longissimo. *Swartz in nov. act. ups.* 6. 79.
Limodorum falcatum. *Thunb. ic. pl. jap. t.* 6. *Swartz in Schrader's journ.* 1799. 2. 231. *Id. in ejusd. neues journ.* 1. 85. *Willd. sp. pl.* 4. 125.
Orchis falcata. *Thunb. jap.* 26.

Herba perennis, rigida, stolonifera, subtriuncialis, terrestris. Radix *fibris horizontalibus crassis subsimplicibusque.* Fol. *radicalia, subquina, falcato-disticha, equitantia, crassiuscula, angustiùs lorata, acuminata, arctiùs complicato-canaliculata, opaco-viridia:* stipulæ *plures radicales arido-membranaceæ lanceolato-vaginosæ.* Scapus *radicalis, brevior foliis, rigidus, viridis, angulosus, flexuosus, subquadriflorus.* Flores *resupinati, erecto-spicati alternè approximati, corymbum simulantes, albi, sesquiunciales v. circitèr, fulvido-emarcesentes; singuli basi* bracteâ *arido-membranaceâ fuscâ triplo breviore lanceolato-conduplicatâ trinervi stipati.* Germ. *unciam cum* $\frac{1}{4}$ *alterius longum, chloroleucum, gracile, subclavatum, subsigmoideo-flexuosum, anguloso-striatum, tortum.* Cor. *unicolor, subnutans, radiato-patens,* $\frac{3}{4}$ *v. circitèr unciæ transversa, subæqualis:* pet. 5 *similia, distantia, summum medium versûs subascendentia, lineari-oblonga, acuminata, angusta, plana;* 3 exteriora *revoluta;* 2 lateralia inferiora *infernè labelli lateribus à facie interiore applicita, indèque revoluto-divergentia:* labellum *recurvo-porrectum, oblongatum, difforme, non longius reliquis petalis vixve latius,* ungue *cuneato-oblongo, convoluto-concavo, libero, lobo dentiformi acuto suprà ad utrumque latus, cum plicâ transversâ ubi abit in laminam;* lamina *vix longior ungue, angustior, oblonga, subattenuata, truncato-retusa, subdeflexa:* calcar *liberum, filiforme, æquale vel longius germini, gyro antrorsum ascendens modo ut finis ejus æquet vel supereemineat reliquam corollam, apice lutescens, obtusulum.* Columna *alba, oblonga, subtriplo brevior corollâ, libera, tota anticè in longum cavata cavitate incumbente sulco parallelo unguis labelli.* Anthera *hemisphærica, bilocularis, decidua;* massæ pollinis 2 *luteæ, pruinoso-opacatæ, solidæ, cartilagineo-duræ depressiùs hemisphæricæ v. subpatelliformes, horizontales, utrinque pedicello membranaceo subsagittatìmque dilatato apici columnæ anticè agglutinato à marginibus suis interioribus in sitû horizontali collaterali-annexæ, subtùs planiusculæ ibìque intra peripheræriam laterum foramine cæco obliquo compresso triangulari ad centrum ducente pro dimidio suo circitèr pertusæ.* Stigma *anticum, album, subbicolle, sursum spectans, basi columnæ intra cavitatem ad pedem unguis labelli positum.*

If we regard the present plant as of the type that is to determine the genus LIMODORUM, it is clear that in conformity with the standard now used in restricting the degrees of generic affinity, many species included by Dr. Swartz as congeners of the group, must fall to be considered as too widely anomalous to remain together within the same generic pale. Accordingly we find that several of the species have been already removed, and formed into other genera. Among these, the fine one by which the genus had been till of late the most familiarly exemplified in our gardens, viz. the LIMODORUM (now BLETIA) *Tankervilliæ*. We are not aware that any species, which would be now deemed a suitable generic associate for the subject of this article, has been ever introduced into our European collections before the present specimen, which was received about four years ago through Dr. Roxburgh, from the East Indies, by Sir Abraham Hume, and cultivated in the hothouse at Wormleybury, a source from which many rare and curious vegetables have been liberally distributed among the gardens of this country. The specimen which afforded our drawing, was sent to Mr. Edwards from thence, by the direction of the owner, in March last, and had bloomed at the same period the three preceding years. The species was first observed by Thunberg in Japan, near the port of Nagasaki, growing on hills among the bushes; and is probably likewise a native of China. At some points it comes near to AERIDES, of which a species has been given in a foregoing number of this publication.

A stiff-leaved herbaceous stoloniferous perennial, about three inches high, growing on the ground, not on trees as many of its kindred do. *Root* with thick horizontal undivided fibres. *Leaves* radical, about five, falcately distich, equitant, thickish, narrowly lorate, acuminate, folded into a deep closish channel, of an opaque green colour: *stipules* radical, several, membranous, withered, lanceolate, sheathing. *Scape* radical, shorter than the leaves, stiff, green, angular, flexuose, about four-flowered. *Flowers* reversed, uprightly spiked, but being set at small distances from each other, owing to their length, they afford rather the appearance of a corymb than a spike, white, but as they fade turning to a tawny yellow colour, about an inch and a half long, each with a sphacelately membranous brown three-nerved folding lanceolate thrice shorter *bracte* at the base. *Germen* more than an inch long, greenish white, slender, slightly clavate, shallowly bent somewhat in the form of an S, angularly fluted, twisted. *Corolla* of one colour, very slightly nutant, radiately spread, about $\frac{3}{4}$ inch across, nearly equal: 5 *petals* of the same shape, standing asunder, ascending slightly towards the upper middle one, linearly oblong, acuminate, narrow, flat; 3 *exterior ones* revolute; 2 *lower lateral ones* with the lower part applied broadways to the sides of the label, diverging upwards revolutely: *label* pointing forwards and recurved, oblong, biformed, not longer and scarcely broader than the other five petals; *unguis* detached, cuneately oblong, convolutely concave, with a pointed toothlike lobe on each side at the top, marked where it joins the lamina with a transverse plait; *lamina* or *blade* narrower, scarcely longer than the unguis, oblong, slightly tapered, truncately retuse, subdeflectent; *spur* detached, filiform, equal to or rather longer than the germen, turned up frontways with a circular bend, so that the point becomes even with or higher than the rest of the corolla, yellow and bluntish at the end. *Column* white, oblong, about thrice shorter than the corolla, detached, hollowed its whole length in front, the cavity incumbent upon the parallel one of the unguis of the label. *Anther* hemispherical, terminal, moveable, deciduous, bilocular; *pollen-masses* two, yellow, opaquely frosted, solid, of a cartilaginous consistence, depressedly hemispherical, something of the shape of the knee-pan in miniature, horizontal, fixed side by side at their inner margins in a horizontal direction to a membranous subsagittately dilated pedicle, which adheres to the front edge of the apex of the column, flattish underneath, where within the edge of the periphery of their outer sides they are pierced for about half their diameter obliquely with a compressed triangular puncture which points towards their centre. *Stigma* placed in front at the base and within the hollow of the column, looking upwards, white, resembling two small studs or bosses, opposite the foot of the label.

a The two pollen-masses with their pedicle, removed from the receptacle of the anther, in the state they appear when they have left the summit of the column to apply themselves to the stigma at the foot.

324

DISA bracteata.
Small-flowered Disa.

GYNANDRIA *MONANDRIA.*

Nat. ord. ORCHIDEÆ. *Jussieu gen.* 64. *Brown prod.* 309. *Div. I. Anthera* adnata subterminalis persistens. *Pollinis massæ* è lobulis angulatis elasticè cohærentibus; basi affixæ. *Brown in Hort. Kew. ed.* 2. 5. 188.
DISA. Suprà vol. 3. *fol.* 210.

Div. Calcare longiore.

D. *bracteata*, galeâ obtusâ, calcare oblongo, labello lineari apice latiore, spicâ cylindricâ, bracteis erectis floribus longioribus. *Swartz act. holm.* 1800. 211.

Disa bracteata. *Willd. sp. pl.* 4. 48. *Swartz in Schrader's neues journ. fur die bot.* 1. 28. *Exemp. spont. in Herb. Banks. repostum.*

Planta *spithamæa.* Tuber *subrotundum.* Folia *caulina plura* (5-6) *lanceolata erecta concaviuscula.* Spica *multiflora bipollicaris.* Bracteæ *altitudine circitèr florum, foliaceæ, lanceolatæ acumine brevi.* Germ. *cylindraceum tortum in flore expanso, subsessile.* Flores *parvi, luteo-virides galeâ pallidè fuscâ, erecti.* Galea *erecta ovalis obtusa modicè concava,* calcari *ipsâ galeâ ter breviore descendenti obtuso fusco.* Petala 2 inferiora exteriora *arctè reflexa lanceolato-subovata acutiuscula plana:* interiorum 2 superiora *crassiuscula, conniventia lanceolato-linearia viridia basin versus sensim latiora ibique columnæ latera amplexantia apice lineari integro citiùs marcescente.* Labellum *lingulato-lineare obtusum integerrimum planum dependens primùm viride deindè superiore parte citiùs marcescente.* Brown MSS.

Of the four or five species, of this curious South African genus, introduced into the gardens of Europe, two are due to Mr. Griffin; viz. the present and the *prasinata* of the 210th article of this publication.

We have to thank Mr. Brown for his very liberal communication of a description of the plant made by himself at South Lambeth, where it was in flower in June last in the conservatory, the root having been very recently received by Mr. Griffin from the Cape of Good Hope.

Our only guide to the species intended by Swartz by his DISA *bracteata* is the above short specific phrase, unaccompanied by any note: with this our plant, however, in Mr. Brown's opinion, agrees too closely to justify a separation; and being moreover clearly of the same species with the specimen in the Banksian Herbarium, deemed to be the *bracteata* of Swartz, we have no hesitation in giving it under the present title.

Plant a span in height. *Tuber* roundish. *Leaves* cauline, several (5-6) lanceolate upright slightly concave. *Spike* manyflowered two inches long. *Bractes* about the height (a little higher) than the flower; foliaceous, lanceolate with a short point. *Germen* cylindrical, twisted when the flower is expanded, subsessile. *Flowers* small, yellowish-green with a pale brown casque, upright. *Casque* upright oval obtuse moderately concave, with a *spur* three times shorter than the casque itself inclining downwards obtuse brown. *Two lower outside petals* closely reflected lanceolately-subovate rather pointed flat: *two inner upper ones* thickish connivent lanceolately linear green gradually broader towards the base, where they embrace the column of the fructification, the top part linear entire withering before the lower portion. *Label* lingulately linear obtuse quite entire flat dependent, at first green, afterwards decaying earlier at the upper part than the other.

THE

BOTANICAL REGISTER:

CONSISTING OF

Coloured Figures

OF

EXOTIC PLANTS,

CULTIVATED IN

BRITISH GARDENS;

WITH THEIR

HISTORY AND MODE OF TREATMENT.

THE DESIGNS BY

Sydenham Edwards,

AND OTHERS.

VOL. V.

——viret semper——nec fronde caducâ
Carpitur.

LONDON:
PRINTED FOR JAMES RIDGWAY, PICCADILLY.
1819.

357

ORCHIS longibracteata.
Winter Orchis of Sicily.

GYNANDRIA *MONANDRIA.*

Nat. ord. ORCHIDEÆ. *Jussieu gen.* 64. *Brown prod.* 309. *Div. I. Anthera* adnata subterminalis persistens. *Pollinis massæ* è lobulis angulatis elasticè cohærentibus; basi affixæ. *Brown in Hort. Kew. ed.* 2. 5. 188.
ORCHIS. Suprà vol. 3. *fol.* 202.

O. *longibracteata*: bulbis indivisis; petalis conniventibus; nectarii labio (labello) trifido: laciniâ mediâ exporrectâ bilobâ; bracteis flore longioribus. *Biv. Bernardi sic. pl. cent.* 1. 57. *n.* 66. *tab.* 4.

Orchis myodes liliacea hircina flore rubro porphyrographi fimbriata. *Bonan. t.* 33.

Monorchis myodes liliacea hircina, flore magno rubro porphyrographi. *Cup. Panph.* 1. *t.* 200.

Orchis myodes hyemalis liliacea, hircina, fimbriato flore magno rubro porphyrographi. *Cup. hort. cath.* 157. *et suppl. alt.* 67.

Bulbi *duo subrotundi.* Scapus *teres, subflexuosus, vaginatus foliis ovatis vel oblongo-lanceolatis, striatis, paginâ superiore præcipuè lucidis.* Flores *magni graveolentes in spicâ pyramidali compactâ.* Petala *in galeam conniventia, extùs purpurea nervis viridibus, intùs albo-viridia punctis rubris pulchrè notata: opercula (interiora 2) petalis (exterioribus) angustiora, herbida.* Nectarii labium (labellum) *maculis albis purpureisque variegatum, trifidum; segmento medio productiori, bilobo cum appendiculâ internâ laterali brevissimâ, aliquotiès nullâ; segmentis lateralibus margine externo undulatis.* Cornu (Calcar) *germine trigono intorto brevius.* Columna antherifera *brevis, apice obtusa.* Stigma *obcordatum.* Bracteæ *virides, lanceolatæ, cucullum sive galeam superantes.* Biv. Bernardi loc. cit.

The English botanist is indebted to Mr. Swainson, of Elm Grove, near Liverpool, for the introduction of this fine species. During an official residence in Sicily, the native place of our plant, the leisure time of this gentleman was dedicated to Natural History, and the exploring of the productions of that Island. His pursuits have led to the introduction and cultivation of many new and rare vegetables.

ORCHIS *longibracteata* has been described in a masterly way by M. Bivona Bernardi, a Sicilian naturalist of eminence, and author of two botanical publications of great merit; copies of which have been lately received by Mr. Bohn, Bookseller in Frith Street.

We shall avail ourselves of Mr. Swainson's communication to us by quoting the following passage from his letter which accompanied the present specimen on the 4th of January last: "ORCHIS *longibracteata* is one of the rarest species " found in Sicily, nor had I ever met with it in six years " residence on that island, until it was pointed out to me " by my friend Bivona Bernardi, growing on a rich soil be- " hind the convent of St. Francesco di Paola, near Pa- " lermo, the only spot he had ever found it on. In its na- " tive place it is the first that flowers among the many little " known species found in Sicily, generally opening its blos- " soms in the beginning of February, the depth of a Medi- " terranean winter. The present plant was not taken into " the greenhouse, from the open air, until after one or two " of its flowers had expanded. This circumstance, as well " as others convince me that we should find the (southern) " ORCHIDEÆ much less difficult to cultivate, were they " treated for the most part in a more hardy manner. Nearly " thirty roots of species belonging to this tribe, brought " three years ago from the Mediterranean, are now growing " in my garden, under a common frame, as vigorously as " in their native climate."

The following description is the version of that in the first " *Centuria sicularum plantarum*," above quoted.

Bulbs two, roundish. *Scape* cylindrical, subflexuose, sheathed by the lower parts of the ovate or oblongly lanceolate *leaves*, which are streaked and shining, especially on the upper side. *Flowers* large, with a rank smell; in a pyramidal closeset spike. *Petals* converging so as to form a casque, rose-purple with green nerves on the outside, whitish green prettily marked with red dots on the inside: the two inner ones narrower than the exterior ones, herbaceous. *Label* variously marked with white and purple, trifid, *middle segment* longer than the others, two-lobed, with an interior very short lateral appendage, sometimes with none; *lateral segments* undulated on the outside at the edge: *spur* shorter than the three-cornered twisted germen. *Column* short, obtuse at the summit. *Bractes* green, lanceolate, overtopping the casque of the corolla.

We had no opportunity of inspecting the flower for description after the drawing was made.

There is an ORCHIS deposited in the Banksian Herbarium, by the title of *foliosa,* collected by Mr. Masson in the Island of Madeira, and which, we have little doubt, is of the same species with the present.

367

ORCHIS variegata.

Pointed-petalled Orchis.

GYNANDRIA *MONANDRIA.*

Nat. ord. ORCHIDEÆ. *Jussieu gen.* 64. *Brown prod.* 1. 309. *Div. I. Anthera* adnata subterminalis persistens. *Pollinis massæ* è lobulis angulatis elasticè cohærentibus; basi affixæ. *Brown in Hort. Kew. ed.* 2. 5. 188.

ORCHIS. Suprà vol. 3. *fol.* 202.

O. *variegata*, bulbis subrotundis, caule folioso; labello tripartito lævi, lobis apice dentatis, medio majori emarginato cum mucronulo interjecto; cornû bracteisque germine brevioribus. *Biv. Bern. sic. pl. cent.* 2. 44. *n.* 60.

Orchis variegata. *Allion. pedem. n.* 1828. *D'Asso flor. arag.* 129. *n.* 888. *Jacq. ic. rar.* 3. *t.* 599. *Coll.* 2. 267. *Swartz in Schrader's neues journ. fur die bot.* 1. 15. *Willd. sp. pl.* 4. 21. *Bertol. pl. genuen.* 119.

Orchis acuminata. *Desfont. atl.* 2. 318. *t.* 247. *Swartz in Schrader's neues journ. fur die bot.* 1. 14. *Willd. sp. pl.* 4. 17. *Curtis's magaz.* 1932.

Orchis militaris. γ. *Lin. sp. pl. ed.* 2. 2. 1334.

Orchis radicibus subrotundis, spicâ brevissimâ, labello brevitèr 4-fido circumserrato punctato. *Hall. helv. n.* 1275. *t.* 30.

Orchis militaris pratensis elatior, floribus variegatis. *Segu. veron.* 2. 123. *t.* 15. *fig.* 3.

Orchis militaris. *Riv. hex. t.* 15.

Orchis militaris minor et minima Rivini. *Rupp. jen.* 279. *cum tab.*

Orchis latifolia hiante cucullo minor. *Vaill. paris. t.* 31. *figg.* 22, 23, 24.

Orchis seu Cynosorchis galeata, purpurea leucostincta, sponsam ornatam effigiens. *Bonan. t.* 35. *Cup. hort. cath. suppl. alt.* 68. *Panphyt.* 2. *t.* 165.

Rad. *didymo-tuberosa, fuscescens, tuberibus ovali-oblongis.* Fol. *subseptena, plurifaria, erecto-patentia, subglauco-v. potiùs cinereo-virentia, opaca, 3-5-uncialia, rariùs unciam latitudine exsuperantia modò submaculata, laminâ lanceolato-oblongâ acutâ striatâ;* interiora *longiora, prolixiùs vaginantia.* Scapus *spithamæus, teres, sulcis striatus, pallidè virens.* Spica florum *curta cylindrica conferta patens,* seminum *elongata laxa appressa.* Flores *mediocres, remissiùs odori, dilutiùs incarnati, nutantes,* $\frac{1}{3}$ *unciæ vel circà transversi.* Bracteæ *albidæ v. nunc coloratæ, membranosæ, lanceolato-acuminatæ, germini sublongiori appressæ.* Germ. *sessile, tortum, pallido-virens, tereti-angulare, vix longius petalis.* Cor. *nutans, semiringens:* pet. 5 *galeato-conniventia, infernè fulvo-viridia, supernè carneo-albida, sanguineo-striata, lanceolata, cuspidatìm acuminata:* 3 exteriora *æquilonga; lateralia inæquilatera obliquo-assurgentia, striis 4 incompletè percursa; medium rectum unistriatum:* 2 interiora *opposita, inclusa, plurimùm minora, linearia, acuta, unistriata.* Labellum *propendens, vix longius petalis, oblato-subrotundum, carneo-albicans, micans, laxiùs sanguineo-punctatum, glabrum, denticulato-erosum, profundè trilobum, brevitèr unguiculatum;* lobi *laterales breviores cuneato-oblongi margine summo obliqui, medius latior obcordatus corniculo v. mucrone minuto in fundo emarginaturæ:* calcar *pallidum, descendens, germine cui parallelum brevius, paululùm suprà basin dilatatum, acumine obtusulo.* Columna *duplo brevior petalis v. ultrà, infernè dilatata, alba:* cucullus *sanguineus, terminalis, erectus anticè retrorsùm obliquatus,* thecis *sulco profundo glandulam albam intergerinam incompletam infernè comprehendente distinctis:* glandula antherifera *obliquato-protuberans obtusa:* pollinis massæ *clavâ indivisâ brevi æruginea, stipite longo flavo lucido.* Stigma *areola didyma antica madida purpurascens parieti interiori cavitatis filamenti v. columnæ partis inferioris sub glandulâ antheriferâ adnata.*

Another of the species of this interesting genus which have been introduced into our collections by Mr. Swainson, of Elm Grove, Liverpool. It was found by that gentleman growing plentifully on the mountains near Palermo, particularly on those of Capreto and Monreale, where it flowers in the spring. That it is the ORCHIS *variegata* of his friend Bivona Bernardi, Mr. Swainson tells us he can have no doubt, that he has scarcely any of its being the O. *acuminata* of Desfontaines; but not having seen the figure in Jacquin's works, he was less able to satisfy himself of its being also the plant of that author. We have carefully reviewed the synonymy adduced above, and are convinced that the whole belongs to this species; which extends itself widely over the South of Europe, and is known to reach the Coast of Barbary. It comes the nearest to the ORCHIS *militaris,* of this country, of any other species we are aware of; and was deemed a variety of that by Linnæus; but afterwards properly distinguished from it by Allioni.

Root tuberous, twin, tubers oval, oblong, brownish. *Leaves* about 7, multifariously disposed, uprightly spreading, of a dullish glaucous or rather cinereous green, sometimes marked with a few irregularly disposed spots, opaque, 3-5 inches long, seldom exceeding an inch in breadth, sometimes with a few spots, blade lanceolately oblong, sharp-pointed, streaked: *inner ones* longest, and sheathing the scape to a greater height. *Scape* 6 or 7 inches high, round, fluted, pale green. *Spike in flower* short close cylindrical spreading, *in seed* elongated distant adpressed. *Flowers* of a middling size, slightly scented, very palely flesh-coloured, nutant, about one third of an inch in diameter. *Bractes* white, sometimes coloured, membranous, lanceolately subulate, pressed close to the germen, which is rather longer. *Germen* sessile, twisted, pale green, round, angular, scarcely longer than the petals. *Corolla* nutant, semiringent: *petals* 5, converging into a casque, at the lower part of a tawny green, at the upper of a rosy-white, striped with deep purple, lanceolate, cuspidately long-pointed: 3 *outer ones* of the same length; side ones uneven-sided, erecting themselves obliquely, marked with 4 incomplete stripes, middle one straight, one-striped: 2 *inner ones* facing each other, enclosed, much smaller, linear, pointed, one-striped. *Label* hanging forwards, scarcely longer than the petals, oblately roundish, rosy white, glittering, loosely spotted with purple, smooth, denticulately eroded round the margin, deeply three-lobed, shortly unguiculate; side lobes cuneately oblong slanted at the top, middle one broader, obcordate, with a minute cornicle or mucro at the bottom of the notch: *spur* pale coloured, pointing downwards, shorter than the germen with which it is parallel, bluntish at the end, a little dilated below the tip. *Column* twice shorter than the petals or more, dilated at the lower part and white; *headpiece* or *cucullus* deep purple-red upright slanting backwards in front, *cists* parted by a wide furrow which contains the white glandular incomplete partition in its lower portion; *antheri-*

ferous gland slanting and protuberant obtuse: *pollen masses* with a short copperas-green head and a longer tawny yellow shining stalk. *Stigma* two-lobed, shining, humid, purplish, affixed to the front of the interior wall of the cavity in the lower part of the column or filament underneath the antheriferous gland.

370

OPHRYS Speculum.
Mirror-lipped Ophrys.

GYNANDRIA *MONANDRIA.*

Nat. ord. ORCHIDEÆ. *Jussieu gen.* 64. *Brown prod.* 1. 309. *Div. I. Anthera* adnata subterminalis persistens. *Pollinis massæ* è lobulis angulatis elasticè cohærentibus; basi affixæ. *Brown in Hort. Kew. ed.* 2. 5. 188. *OPHRYS. Suprà vol.* 3. *fol.* 205.

O. *Speculum*; folia caulina lanceolata. Petala 3 exteriora expansa, acutiuscula, viridia, interiora 2 acuta, purpurea. Labelli laciniæ laterales erectæ, media rotundata emarginata non appendiculata, medio glaberrima, cærulescens, limbo villoso fusco. *Link in Schrader's journ. fur die bot.* 1799. 2. 324; *(malè à Willdenovio ad* OPHRYDEM Scolopacem *citata; cui tamen verè pertinet* OPHRYS Speculum *Biv. Bern. sic. pl. cent.* 1. 61. *n.* 70. *t.* 4.)
Ophrys ciliata. *Biv. Bernardi sic. pl. cent.* 1. 60. *n.* 69.
Ophrys insectifera; *myodes.* ♄. *Lin. sp. pl. ed.* 2. 2. 1343.
Orchis muscam cæruleam majorem representans. *Breyn. cent.* 100. *t.* 44. *Moris. hist.* 3. 494. *sect.* 12. *t.* 13. *fig.* 11.
Orchis Ricinum villosum referens. *Cup. panph.* 1. *t.* 175. *Hort. cath.* 158; *et. suppl. alt.* 68.—*Bonan. t.* 28.
Nil pulchrius. Radix *bulbi 2 subrotundi.* Caulis 3-6-*uncialis, foliis lanceolatis vaginatus, pauciflorus.* Flores *in spicâ laxâ muscas circumvolantes mirè referunt.* Petalum *superius fornicatum, columnam tegens, marginibus revolutis; 2 lateralia patentia, ovata, viridia, fasciâ roseâ per medium.* Opercula (petala 2 interiora) *petalis (exterioribus) duplo breviora, angustissima, acuminata, intensè purpurea, recurvata.* Nectarii labium (Labellum) *trilobum, cæruleum, splendens, limbo barbato barbâ purpureâ; lobo medio exporrecto, obovato, convexo, emarginato, mutico; lateralibus angustioribus, suspensis alarum instar.* Columna *obtusa.* Biv. Bern. l. c.

One of the prettiest of the genus, and now first introduced by Mr. Swainson; by whom the excellent design, from which our engraving has been made, was taken from a sample that flowered in his garden at Elm Grove, near Liverpool, in February last. The species is native of Portugal and Sicily, in the latter of which countries Mr. Swainson tells us that it is very rare, and that he never found it in any other place except in the hilly meadows behind the Convent of Santa Maria di Gesù, near Palermo. Link speaks of it as growing in Portugal, and abundantly in the neighbourhood of Setuval.

Our plant has been erroneously adduced by Willdenow for a synonym of OPHRYS *Scolopax;* to which however the OPHRYS *Speculum* of Bivona Bernardi really belongs.

The specific name has been suggested by the lustre of the ceruleous convex disk in the lip of the corolla.

Root of two roundish tubers. *Stem* 3-6 inches high, few-flowered. *Leaves* several, ambient, spreading, lanceolate, sheathing. *Flowers* in a loose spike, with the appearance of so many flying insects. *Uppermost petal* vaulted, covering the column, revolute at the edges; *two outer side-ones* spreading, ovate, green, with a red stripe along the middle; *two inner side ones* twice shorter, very narrow, long-pointed, deep purple, recurved. *Labellum* three-lobed, ceruleous, bright, with a purple bearded border; *middle lobe* pointing forwards, obovate, convex, notched, pointless; *side-ones* narrower, hung like wings. *Column* obtuse.

It is not improbable but that the OPHRYS *vernixia* of the Flora Lusitanica of Professor Brotero may be the same with the present species. But the description is far too scanty and pointless for us to decide by.

375

ORCHIS tephrosanthos; *β. undulatifolia.*
Curled-leaved Orchis.

GYNANDRIA *MONANDRIA.*

Nat. ord. ORCHIDEÆ. *Jussieu gen.* 64. *Brown prod.* 1. 309. *Div. I. ORCHIS. Suprà vol.* 3. *fol.* 202.

Div. Radice bituberosâ.

O. *tephrosanthos,* labello 3-partito, laciniis linearibus, mediâ 2-furcâ cum mucrone inter crura, petalis acutis liberè conniventibus, bracteis lato-subulatis cornûque duplo brevioribus germine.
Orchis tephrosanthos. *Villars delph.* 2. 32. *Desfont. atl.* 2. 319. *Willd. sp. pl.* 4. 21. *Swartz in schrad. neues journ.* 1. 15.
Orchis longicruris. *Link in schrad. journ.* 1799. 2. 323. *Willd. sp. pl.* 4. 22.
Orchis militaris; ε. *Lin. sp. pl. ed.* 2. 2. 1334.
Orchis zoophora cercopithecum exprimens oreades. *Column. ecphr.* 319. *t.* 320.
Orchis flore simiam referens. *Vaill. paris. t.* 31. *figg.* 25, 26.
(β) flore majore; foliis undulatis.
Orchis undulatifolia. *Biv. Bernardi sic. pl. cent.* 2. 44. *n.* 61. *t.* 6.
Tubera *mole nucis avellanæ majoris.* Fol. 8-9, *plurifaria, erecto-patentia,* 3-5-*uncialia, latiora* ¼ *unciæ vix excedentia, oblongo-et ligulato-lanceolata, remotiùs lineata, acuta, interiora interdùm undulata.* Caulis *spithamæus ad dodrantalem, striato-teres.* Spica *densa, brevis, conica, albo-incarnata,* floribus *odoris.* Bracteæ *albo-membranosæ, lato-subulatæ, germini appressæ duploque breviores.* Germ. *galeæ semunciali æquale, tortum, curvatum, rubido-virens.* Cor. *semiringens:* pet. 3 exter. *galeato-conniventia, libera, ovato-lanceolata, cuspidata, lateralia surrecto-obliquata* 4-*lineata, medium* 3-*lineatum:* 2 inter. *duplo minora, concolora, lineari-oblonga acumine abrupto brevi, columnam intercludentia:* labellum ⅕ *parte v. circà longius petalis, propendens; lamina oblongiuscula, angustiùs 5-fido-3-loba, intùs disco convexiusculo subtilitèr scabrata, lobi laterales lineari-attenuati, medius longior 3-fidus laciniis lateralibus lobos laterales exactè referentibus mediâ minimâ mucroniformi:* calcar *germini rectè præpendens duploque brevius, oblongum, supernè compresso-subdilatatum, obsoletè emarginatum.* Columna *brevis, alba,* thecâ *puniceâ:* pollinis massæ *indivisæ, æruginosæ.* Stig. *bulla subcordata, subpunicans in cavitate anticâ columnæ propè infra glandulam antheriferam.*

We do not think our plant will bear a separation from this species, solely on account of the crispature in the foliage, and we are not aware of any other distinctive mark. *Tephrosanthos* itself does not appear to us to afford any very certain difference from *militaris*; the coloured pencilled spots upon the label have been relied on, though they are now known to be the objects of mere variety. The principal difference we can detect between the two, lies in the longer narrower and more tapered segments of the label of the former. The shortness of the bractes in the latter, we suspect from a review of samples in the Banksian Herbarium, is occasional and will not hold good as a character. If the outer petals are really confluent in *militaris*, this might afford a valid distinction, but we believe these will be found merely to converge, as in our plant, but not to cohere.

The drawing was taken from a sample remitted by Mr. Swainson, with whom it had flowered at Elm Grove, in February last. The root was imported by that gentleman from Sicily, where the plant grows with a longer and more luxuriant spike than in the present specimen, and is common on the mountains of Messina, but more rare on those of Palermo.

Tubers twin, ovately oblong, about the size of largish acorns. *Rootsheath* white, membranous, nerved. *Leaves* 8-9, alternate, compassing, uprightly spreading, 3-5 inches long, where widest about ¼ of an inch across, smooth, lineated, ovately and ligulately lanceolate, sharp, midrib keeled, inner ones gradually longer sheathing farther more upright curled, one or two of the lower ones and the upper stem-one plain. *Stem* 5-9 inches high, pale green, round, fluted. *Spike* short, dense, cylindric, conical; *flowers* white and flesh-coloured, scented. *Bractes* white, membranous, broadly subulate long-pointed, as if awned, close-pressed to the germen than which they are twice shorter. *Germen* half an inch long, roundedly hexagonal, equal to the casque of the corolla, twisted. *Corolla* semiringent: 3 *outer petals* converging into a casque, ovately lanceolate, long-pointed, concave, 2 side ones obliquely erected, with 4 deep red streaks, middle one 3-streaked: 2 *inner petals* twice smaller, of the same colour, linearly oblong with a short abrupt point, lapping together over the summit of the column: *label* pale-pink, about ⅓ longer than the petals, propendent, lamina or broad part nearly oblong, narrowly fivecleft and threelobed, disk oblong slightly raised minutely roughened, two side lobes linear and tapered narrow divergent, middle lobe about ⅓ longer trifid, side segments of the same shape as the two lateral lobes, middle segment

minute mucroniform: *spur* pale, twice shorter than the germen, descendent, oblong, compressed, slightly dilated upwards, obscurely notched at the end. *Column* short, oblate, white; *anthercase* dark crimson: *pollen-masses* verdigrise-coloured. *Stigma* raised, heart-shaped, purplish, in front of the inner wall of the cavity of the column, close below the gland on which the anther stands.

405

HABENARIA fimbriata.

Purple fringed Habenaria.

GYNANDRIA *MONANDRIA.*

Nat. ord. Orchideæ. *Jussieu gen.* 64.
Orchideæ. *Brown prod.* 1. 309. *Sect. I. Anthera* adnata subterminalis persistens. *Pollinis massæ* è lobulis angulatis elasticè cohærentibus; basi affixæ. *Id. in Hort. Kew. ed.* 2. 5. 188.

HABENARIA. Cor. ringens, petalis 3 *v.* 5 in galeam conniventibus. *Labellum* basi subtùs calcaratum v. subsaccatum. *Anthera* terminalis, loculis adnatis, basibus quandoque solutis elongatis. *Massæ pollinis* pedicellatæ, pedicellis singulis basi glandulæ respondenti nudæ insertis. *Brown prod.* 1. 312.

Div. Loculis antherarum columnæ juxta apicem longitudinalitèr adnatis.

H. *fimbriata*, cornu filiformi germine longiore, labello tripartito laciniis cuneiformibus fimbriatis. *Brown in Hort. Kew. ed.* 2. 5. 193.

Orchis fimbriata. *Hort. Kew.* 3. 297. *Willd. sp. pl.* 4. 39. *Pursh amer. sept.* 2. 588.

Radix *fasciculata.* Caulis *erectus glaber ex ancipiti acutè tetragonus.* Folia *caulina nonnulla* (3-5) *alterna sessilia oblonga acuta glabra integerrima nervosa carinata, basi vaginantia biuncialia.* Spica *ovato-oblonga, multiflora.* Flores *è cæruleo purpurascentes.* Bracteæ *lanceolatæ nervosæ germinibus paulò longiores.* Petala *quinque plana longitudine æqualia trilinearia, supremum seu dorsale ovatum obtusum erectum, lateralia exteriora ovata acuta patentissima, lateralia interiora oblonga obtusa juxta petalum dorsale erecta, infra medium dilatata ibique denticulata, basi attenuata.* Labellum *petalis paulo longius tripartitum, laciniæ latæ cuneiformes æquales planæ ad medium subdivisæ in cilias subulatas, laterales divaricatæ, intermedia patens.* Germen *semiunciale.* Solander in Hort. Kew. 3. 297.

Native of North America; where it is found in low meadows and high mountain bogs from Newfoundland to Pensylvania. Introduced by Dr. W. Pitcairn in 1777.

The drawing was taken at Mr. Knight's nursery, in the King's Road, Little Chelsea; and affords, we believe, the only representation of the species yet published.

The feature, mainly relied upon by Mr. Brown for the technical distinction of the present genus from Orchis, is, the circumstance of the glandular supports of the pollen-masses being naked or uncovered in this, not enclosed within the case or hood of the anther as in that. Some of the species are remarkable for the elongation of the detached bases of the cells of the anther-case, and have suggested the generic

appellation. Orchis is scarcely met with beyond the limits of Europe; no where within the tropics or the southern hemisphere. Habenaria is represented by various species in the four quarters of the world, as well as in our own country.

From one to two feet high. *Root* fascicled. *Stem* upright, smooth, proceeding from two-edged to four-edged. *Leaves* cauline, several (3-5) alternate, sessile, oblong, sharp-pointed, quite entire nerved, keeled, at the base sheathing two inches long. *Spike* ovately oblong, many-flowered. *Flowers* blue-purple. *Bractes* lanceolate, nerved, but little longer than the germens. *Petals* 5, flat, of the same length, 3 lines long, the uppermost or dorsal one

ovate obtuse upright, outer lateral ones ovate sharp-pointed outspread, inner lateral ones oblong obtuse near to the dorsal petal upright, widened below the middle, where they are denticulate, tapered at the base. *Labellum* but little longer than the other petals, tripartite, segments broad cuneiform or wedge-shaped equal flat divided to the middle into subulate strips, lateral ones divaricate, middle one spreading. *Germen* half an inch long.

416

SATYRIUM cucullatum.

Cucullate Satyrium.

GYNANDRIA *MONANDRIA.*

Nat. ord. Orchideæ. *Jussieu gen.* 64. *Brown prod.* 1. 309. *Div. I. Anth.* adnata subterminalis persistens. *Pollinis massæ* è lobulis angulatis elasticè cohærentibus; basi affixæ. *Brown in Hort. Kew. ed.* 2. 5. 188.

SATYRIUM. Cor. ringens; *pet.* 5 antica basi connata. *Labellum* posticum, fornicatum, basi bicalcaratum v. bisaccatum. *Anth.* resupinata. *Stig.* 2 labiatum. *Brown in loc. cit.* 196.

S. *cucullatum*, foliis radicalibus binis cordato-subrotundis concavis, caulinis remotis cucullatis subretusis, floribus cernuis. *Swartz in act. holm.* 1800. 216.

Satyrium cucullatum. *Willd. sp. pl.* 4. 54. *Swartz in Schrader's neues journ.* 1. 33. *Hort. Kew. ed.* 2. 5. 196. *Thunb. flor. cap.* 1. 88.

Satyrium bicorne. *Thunb. prod.* 6.

Orchis bicornis. *Lin. sp. pl. ed.* 2. 2. 1330. *Houtt. nat. hist.* 2. *D.* 12. *St.* 455. *t.* 86. *fig.* 1. *Hort. Kew.* 3. 294. *Andrews's reposit.* 315.

Orchis lutea, caule geniculato. *Buxb. cent.* 3. 6. *t.* 8.

Fol. *radicalia bina, ovato-subrotunda, basi vaginata, amplexicaulia, concava, arcuato-nervosa, margine membranacea, subtùs pallidiora.* Scapus *pedalis, erectus, teres, purpureo-maculatus, inanis, vaginatus:* vaginæ 2-3, *remotæ, cucullatæ apice apertæ, ovatæ, subretusæ, rigidiusculæ, nervosæ.* Spica 2 *v.* 3-*pollicaris, erecta, floribus alternis, distinctis, cernuis.* Bracteæ *sub singulo flore sessiles, lanceolatæ, integræ, glabræ, reflexæ, longitudine floris.* Cor. *submonopetala, subringens, s.* petala 5, *basi coalita;* 3 *exteriora quorum unum superius magnum s. gulea ovato-subrotunda fornicata apice producto obtuso, acutè carinata, posticè basi bicalcarata:* calcaria *cylindrico-subulata, deflexa, parùm curva, germine ferè longiora:* pet. 2 lateralia *linearia integra obtusa,* 2 interiora *paulò minora linearia obtusa stylo petalisque exterioribus ad basin accreta:* labellum *petalis interioribus consimile sed longius illisque basi connatum.* Anth. *globosa didyma, scrotiformis versùs apicem styli adnata 2-loc. loculis prope stigma dehiscentibus.* Stylus *elongatus, supernè dorso gibbus, sub galeâ reconditus, apice dilatatus concavus excisus:* stig. *infra apicem dilatatum, supra insertionem antheræ concavum.* Thunb. fl. cap. l. c.

Introduced from the Cape of Good Hope by Mr. Masson. Flowered at Kew Gardens in 1786, as we learn by a manuscript note of Mr. Dryander's, in Sir Joseph Banks's library, which also notices the fragrance of the blossom. The drawing was taken from a specimen that flowered this summer in Mr. Griffin's collection at South Lambeth, where it had been lately imported.

Mr. Brown has remarked the inaccuracy of preceding authors, in calling the flower of the *Orchideæ* "reversed," it

being in fact originally "upright," and the change of posture merely the consequence of a subsequent twisting either in the pedicle or in the germen which supports it. Hence the *labellum* from a back petal becomes a front one.

Satyrium is however an exception, for the middle front petal, designated by Swartz, as the labellum, has been found by Mr. Brown not to be one; the casque at the back of the flower being the true labellum, a part which is always determined by its position in relation to the stamens and placenta of the germen, by being the middlemost of the 3 inner petals, and by facing the middlemost of the 3 outer.

The genus (consisting of about 12 or 13 known species from the Cape of Good Hope, of which the present and another are all that have been seen in our gardens), according to Mr. Brown, has a ringent *corolla* with 5 front *petals* connate at the base; a vaulted two-spurred or two-pouched *labellum* situated at the back of the flower; an adnate subterminal permanent reversed *anther, pollen-masses* (of angular lobules held together by an elastic substance) affixed at the base, and a two-lipped *stigma.*

The flowers in all the spontaneous specimens we have seen, have changed to a blackish purple hue in drying, and if not remembered in the fresh state would pass for the representatives of a purple inflorescence in the live subjects. To this we impute Thunberg's describing the flower of the species as purple; he always describing from a dried sample whenever he treats of Cape plants. We missed the opportunity of examining the blossom while alive.

The plant published by Messrs. Loddiges for SATYRIUM *cucullatum* in the Botanical Cabinet, is a very distinct species.

THE

BOTANICAL REGISTER:

CONSISTING OF

Coloured Figures

OF

EXOTIC PLANTS,

CULTIVATED IN

BRITISH GARDENS;

WITH THEIR

HISTORY AND MODE OF TREATMENT.

THE DESIGNS BY

Sydenham Edwards,

AND OTHERS.

VOL. VI.

——viret semper——nec fronde caducâ
Carpitur.

LONDON:
PRINTED FOR JAMES RIDGWAY, PICCADILLY.
1820.

506

VANDA Roxburghii.

Chequer-flowered Vanda.

GYNANDRIA *MONANDRIA.*

Nat. ord. **ORCHIDEÆ.** ***Jussieu gen.*** **64.** ***Brown prod.*** **1. 309.** ***Div. IV. Anthera*** terminalis mobilis decidua. ***Massæ pollinis*** demùm cereaceæ. ***Brown in Hort. Kew. ed.*** **2. 5. 205.**

VANDA. ***Labellum*** calcaratum, cum basi simplici (breviusve productâ) *columnæ* apteræ continuum, trifidum, lobo medio carnoso. *Petala* patentia, distincta. *Massæ pollinis* 2, obliquè bilobæ. ***Brown MSS.***

OBS. AERIDES paniculatum, ***(suprà v. 3. f. 220.) generis pristini species impar nimìs huic (monente D. Brown) aptè satis associari potest. Parasiticæ ambæ.***

Vanda *Roxburghii,* ovariis contortis, petalis oblongo-obovatis undatis, foliis obliquè tridentatis. ***Brown MSS.***
Cymbidium tesselloides. ***Roxburgh MSS. cum icone ined.***
Vandâ. ***(Sir William Jones in) asiat. res.*** **4. 302.**

The generic character now given has been formed by Mr. Brown, so as to include AERIDES *paniculatum,* published provisionally by that generic name in our third Volume (fol. 220); but now it will be seen that this species is not a very close congener of the one before us. The CYMBIDIUM *tessellatum* of Roxburgh is not included in the present genus.

Mr. Brown had some scruples in not ranking the group under ANGRÆCUM of M. Du Petit-Thouars (of which a species is given in voy. de Bory de St. Vincent. 1. 359. tab. 19.), not being able from want of sufficient detail in the description of the structure of the parts of that genus, to decide satisfactorily, whether he ought to do so or not; and has at last been determined principally by the *label* in ANGRÆCUM being undivided and probably membranous, while in VANDA it is three-lobed and fleshy. He expects that several species of Swartz's genus LIMODORUM will be found to rank under the present.

This beautiful parasitic plant was imported by Sir Joseph Banks, and flowered last autumn in the hothouse at Springrove, now belonging to Lady Banks. It is common in most parts of Bengal, and grows on various trees, but principally upon one which bears the Mango. We have already spoken of the culture best suited to tropical parasites in the 220th article of this Register.

The following description is from the manuscript of Dr. Roxburgh.

"*Stem* creeping, sending forth long thick round branching fleshy ash-coloured *roots,* which fasten firmly to the trunk or branches of the tree the plant grows on. The plants are seldom more than two or three feet in length, for they decay at the base as fast as they shoot from the top. *Leaves* sheathing, bifarious, near together, recurved, linear, keeled, forebitten, about five or six inches long. *Scapes* generally axillary, solitary, naked, supporting from six to twelve large beautiful flowers. *Petals* 5, nearly equal, expanded, oblong, waved at the edge and here and there a little inflected, upper surface chequered with yellow and dusty ferruginous purple, under surface white: *label* shorter than the petals; *spur* conical, protruding towards the germen between the two lower petals; *lamina* (broad part) oblong, turgid, apex 3-lobed, sides inflected, so as to be convex above and deeply concave underneath, colour blueish purple towards the apex: *upper lip* or portion which forms the attachment of the base and horn to the base of the column of fructification, with two lateral obliquely broad lanceolate lobes having acute points incurved towards the summit of the column. *Column of fructification* thick, short, obtuse, open at the inner margin next the summit; this mouth or opening tapers down through the column into a point, which ends in the cavity of the germen; *anther* lidded, suborbicular, with two lidded cells for the two globular *pollen masses* which rise, when the lid is removed gently, with a jerk on their broad cordate filament, which is inserted on the anterior part of the top of the column, by a large funnelform base. If removed with less care, and before the pollen masses are ripe, these remain in their cells, and the funnelshaped base of the filament rises erect. *Stigma* or channel for conveying the subtle essence of the pollen to the germen, a clammy opening in the fore part of the column near its top. *Pericarp* clavate, with six sharp ridges running the whole length." *Roxb.*

NOTES.

Aerides *paniculatum.* *Vol.* 3. *fol.* 220.
This species has been lately included by Mr. Brown in his genus VANDA.

(See VANDA *Roxburghii,* fol. 506 of the present Volume). So that the above article should be now inscribed
VANDA paniculata.
The germen is not twisted, nor the corolla reversed, as in *Roxburghii.*

THE

BOTANICAL REGISTER:

CONSISTING OF

Coloured Figures

OF

EXOTIC PLANTS,

CULTIVATED IN

BRITISH GARDENS;

WITH THEIR

HISTORY AND MODE OF TREATMENT.

THE DESIGNS BY

Sydenham Edwards,

AND OTHERS.

VOL. VII.

——viret semper——nec fronde caducâ
Carpitur.

LONDON:
PRINTED FOR JAMES RIDGWAY, PICCADILLY.
1821.

529

CYMBIDIUM xiphiifolium.

Xiphium-leaved Cymbidium.

GYNANDRIA *MONOGYNIA.*

Nat. ord. ORCHIDEÆ. *Brown prod.* 1. 309. *Sect. V.* Anthera terminalis, mobilis, operculiformis, decidua. Pollinis massæ cereaceæ, læves. *Parasiticæ, paucæ Terrestres.* Brown prod. 1. 330.

CYMBIDIUM. Labellum ecalcaratum, concavum, cum basi (simplici nec productâ) columnæ articulatum. *Petala* patentia distincta. *Massæ pollinis* duæ, posticè bilobæ. *Brown in Hort. Kew. ed.* 2. 5. 213.

OBS. *Genus nimis artificiale, posthâc certè dividendum.* Brown prod. 1. 331.

Cymbidia vera.

C. *xiphiifolium*, foliis crassiusculis lineari-subulatis canaliculatis nervoso-striatis crenulatis æquantibus scapum; spicâ pauci(4)florâ, labello immaculato.

Scapus *foliorum longitudine*, 4-*florus.* Flores *virides, majusculi.* Cor. 5-*petala:* pet[s]. 3 *exterioribus ovato-lanceolatis, patentibus,* 2 *interioribus conformibus columnâ parallelis, exterioribus latioribus.* Labellum *cum columnâ articulatum, eâque brevius, cymbiformi, columnæ applicito, medio intùs* 2-*calcarato;* laminâ *ovato-undulatâ deflexâ ungue breviore.* Columna *libera* 1-*pteris, laciniis paulò brevior, anticè concava.* Anth. *terminalis persistens opercularis,* 1-*locularis utrinque intùs juxta cardinem membranâ auriculata.* Massæ pollinis 4, *per paria cohærentes, cereaceæ, obtusè triangulares, glandulæ transversæ angustæ utrinque attenuatæ adhærentes.* Stigma *latum transversum infra antheram margine superiore emarginatum pro glandulæ receptione.* Lindley MSS.

Our plant flowered late last autumn in the hothouse at Wormleybury, having been newly introduced by Sir Abraham Hume, as believed, from China. The sample obligingly sent us for the use of this work, was requested by Mr. Lindley, who offering us his notes concerning it, we took none ourselves. Nor have we had any opportunity of comparing it with *ensifolium*, to which it appears the nearest; but as far as recollection serves us in regard to our plant, it differs from that in having a fleshy rigid regularly tapered prominently nerved subcartilaginously edged crenulated pale green foliage. The flowers are of a yellowish green throughout, and the label unspotted. Further differences may offer when samples of the two shall have been compared. We have not much hesitation however in separating them, although not confident. The leaves remind us of those of IRIS *Xiphium*.

According to Baron Humboldt, the shady vallies of the Andes of Mexico, New Granada, Quito, and Peru, situated between the elevations of 800 and 1100 fathoms, where refreshing breezes and vernal warmth prevail the year through, are rivalled by no part of the globe in the number of the species, the beauty and fragrance of the blossom, as well as the fullness and brilliancy of the foliage of their *Orchideæ*. He calculates that not a twentieth part of the species belonging to those regions have been observed by the botanist; and remarks, that while the whole of Europe affords about 70 or 80, the tropical portion of America alone, although yet so cursorily explored in its mountainous districts, has presented us with 244, of which 61 were discovered by himself and M. Bonpland.

Scarcely any of the species with a spurred label are found in tropical regions, that section of their type belonging principally to the temperate and frigid zones. ORCHIS, HABENARIA, CYPRIPEDIUM, OPHRYS, SERAPIAS, EPIPACTIS, &c. &c. are the prevailing groups in our northern hemisphere*; SATYRIUM, PTERYGODIUM, DISPERIS, CORYCIUM, DISA, PTEROSTYLIS, CALADENIA, &c. &c. of the southern.

The equatorial *Orchideæ* are principally of the Epidendrous section, and differ from their coordinates of the temperate zones in being gregarious and parasitical.

* Excellent and accurate figures of several of these genera, so rare in our collections, may be seen in "THE JOURNAL OF SCIENCE AND THE ARTS."

548

DENDROBIUM cucullatum.

Lady Banks's Dendrobium.

GYNANDRIA *MONANDRIA.*

Nat. ord. ORCHIDEÆ. *Jussieu gen.* 64. *Brown prod.* 1. 309. *Sect. V.* Monandræ. *Anthera* terminalis, mobilis, operculiformis, decidua. *Pollinis massæ* cereaceæ, læves. *Parasiticæ, paucæ Terrestres.* Brown loc. cit. 330.

DENDROBIUM. Labellum ecalcaratum, articulatum cum apice processûs unguiformis, cujus lateribus petala antica adnata, calcar emulantia. *Massæ pollinis* 4, parallelæ. *Brown in Hort. Kew. ed.* 2. 5. 212.

OBS. *Genus forsan dividendum. Species enim labello postico (floribus resupinatis secund. Auct.) aliquantulum differunt structurâ antheræ. Ad hanc phalangem ferè omnes species Novæ Hollandiæ pertinent, in quibus racemi oppositifolii, nec axillares vel verè terminales sed instar ramorum ex opposito latere articuli caulini orti.* Brown prod. 1. 332.

D. *cucullatum*, caulibus pendulis, foliis bifariis lanceolatis acuminatis, pedunculis oppositifoliis subbifloris, labello indiviso circumscriptione ovato basi cucullato. *Brown MSS.*

We were favoured with this curious and handsome plant from the collection of Lady Banks at Springrove; where it is cultivated, like the other tropical parasites of that establishment, by being suspended in a wicker basket from the rafters of the hothouse; and, we believe, has never been known to have flowered any where else in this country.

It is said to be in several collections, and to have been generally conceived to be DENDROBIUM *Pierardi* of the unedited part of Roxburgh's Flora Indica; and which, according to Mr. Brown, to whom, as in so many instances, we are indebted for all we have to say of the subject, it is possible it may really be; but finding by both figure and description, the flowering stems of that species represented as leafless, the lamina of the labellum of an obovate form, and leaves as emarginate, while in the present plant the flowering stems are in full leaf, its labellum of an ovate outline, and its leaves taper-pointed, he has deemed it safer to regard the present sample as belonging to a distinct species.

DENDROBIUM ranks in Mr. Brown's fifth section of *Orchideæ*, distinguished by " a terminal moveable lidshaped deciduous anther; *pollen-masses* of a waxen consistence and smooth surface; and by its species being nearly all parasitical." In the sectional group, the genus is marked by the same author, as having " a spurless label connected by a joint with the top of the unguislike process, to the sides of which the front-petals are attached in such way as to afford the appearance of a true orchideous spur; and 4 parallel *pollen-masses*." *Cucullatum* is specifically distinguished by Mr. Brown, " as having loosely hanging stems, lanceolate taper-pointed leaves disposed in two ranks, peduncles placed opposite to the leaves, and generally 2-flowered, and an undivided label with an ovate outline and cowled base."

The stem of our plant had been tied up to a garden stick for support on its removal to London, and in this position it is shown by our draughtsman; but when growing naturally in its station on the trunks or branches of the vegetables it affects in its native India, the stems are said to hang loosely down.

578

ist richtig 573

LISSOCHILUS speciosus.

Mr. Griffin's Lissochilus.

GYNANDRIA *MONANDRIA.*

Nat. ord. ORCHIDEÆ. *Jussieu gen.* 64. *Brown prod.* 1. 309. *Div. IV. Anthera* terminalis mobilis decidua. *Massæ pollinis* demùm cereaceæ. *Brown in Hort. Kew. ed.* 2. 5. 205.

LISSOCHILUS. Massæ pollinis 2, obliquè bilobæ. *Labellum* basi saccatum: laminâ sessili indivisâ convexâ, posticè excisâ et cum basi columnæ apteræ edentulæ connatâ. *Perianthii foliola exteriora* calycina (reflexa:) *interiora* petaloidea (divaricata). *Brown MSS.*

L. *speciosus,* petalis interioribus divaricatis duplo majoribus exterioribus reflexis, labello altè cordato. *Brown MSS.*

Herba 4-*pedalis et ultra.* Folia *radicalia ensiformia acutissima crassiuscula rigidula viridia nec glauca, marginibus carinâque lævibus: basibus dilatatis incrassatis in bulbum emersum connatis, scapo aliquotiès breviora.* Scapus *vaginatus teres viridis solidus, fasciculis vasorum sparsis.* Vaginæ *distantes strictæ foliaceæ, in foliola appressa lanceolata acuta longitudine vix ipsius vaginæ productæ.* Spica *racemosa simplicissima erecta laxa sesquipedalis.* Bracteæ *foliaceæ lanceolatæ, patulæ sæpiùsve divaricatæ, immersè nervosæ, marginibus lævibus, infimæ in scapum paulò decurrentes ovarium cum pedicello suo superantes, reliquæ sessiles, superiores ovario breviores.* Flores *speciosi è torsione pedicelli baseosque ovarii verè resupinati labello antico.* Perianthii foliola 3 exteriora *utrinque viridia calycina, distincta arctè reflexa ovata acutiuscula immersè nervosa, postico plano, lateralia paulò latiora levissimè concava:* interiorum lateralia *petaloidea, flava, immaculata, divaricatissima, oblongo-ovata, obtusiuscula limbo parùm reflexo, disco hinc convexiusculo; nervosa nervis lateralibus alternatìm simplicibus bifidisque, ipsa basi alba.* Labellum *lateralibus interiorum vix brevius, basi subtùs calcarato-saccatum cornu brevissimo compresso albo:* lamina *sessilis, circumscriptione ovato-oblonga, flava basi albicante venis purpureis, apice subretuso, basi altè cordata lobis posticis rotundatis dimidio inferiore liberis superiore cum basi columnæ cohærentibus suturis manifestis, lateribus subparallelo-reflexis plicatis plicis paulò tantùm elevatis obtusis sulcis intersticiliabus angustis: disco convexo axi gibbosiusculo.* Columna *proportionata semiteres è basi alba pallidè virens, aptera apice edentulo posticè rotundato.* Anthera *terminalis mobilis decidua, in expanso flore apicem anticè declivem columnæ operiens, didyma dorso incrassato dilutè viridi apice retuso, bilocularis loculis approximatis septulo partiali incompleto contrario in latere exteriore instructis.* Massæ pollinis *in singulo loculo singulæ, in gemmâ florali juniori minutissimè granulosæ, in expanso flore cereaceæ, ovatæ, post migrationem in apice obliquo columnæ (clinandro) inversæ, posticè (respectû antheræ anticè) instructæ lobulo minore operiente majoris cavitatem illitam materiâ viscidâ elastica flavescenti processû cujus massæ insertæ lingulæ communi brevi lato-lineari membranaceæ albæ è glandulâ subrotundâ apicis stigmatis ortæ.* Stigma *anticum cavum subrotundum viridè secernens.* Brown MSS.

The plant here described is the only certain species of Lissochilus; *Cymbidium giganteum,* however, according to Thunberg's description, may also belong to the same genus.

This genus in affinity comes nearest to those plants of South Africa at present referred to Limodorum and Cymbidium, namely, *Limodorum barbatum, triste, longicorne,* and perhaps also *hians* of Thunberg (not however Satyrium hians *Linn.* which seems to be a Disa), *Cymbidium pedicellatum* and *aculeatum.* These along with several Indian species also referred by Swartz and Willdenow to Limodorum, especially *L. virens, carinatum,* and perhaps *epidendroides,* form a genus (EULOPHUS), related on the one hand to Dipodium (vide *prodr. flor. nov. holl. p.* 330), and essentially agreeing with Lissochilus in the structure of anthera, but sufficiently different from it in its trilobed crested labellum, which is neither cordate nor connected at base with the column. In both these characters Lissochilus equally differs from *Angræcum* of M. du Petit Thouars, which has also an elongated spur and a considerably different habit.

Among the plants at present referred to, Limodorum, another very distinct genus may be noticed, consisting of *Limodorum veratrifolium,* and judging from Kæmpfer's figure, *L. striatum* also. This genus (CALANTHE) agrees with Bletia in having eight pollen masses, but differs from it in the claw of the labellum being connected with the column. It is not unlikely that this may be the *Cyanorkis* of M. du Petit Thouars, who will probably hereafter publish it under a different name. *Brown MSS.*

Our plant is one of the handsomest of its natural family, and until introduced from the Cape of Good Hope by Mr. Griffin, unnoticed by any writer, and a stranger to both our gardens and herbariums. It flowered this summer in the hothouse at South Lambeth, throwing up a stem more than four feet high, which reminded us of that of the Tuberose, and producing a spike of fragrant blossom, that by successive expansion became nearly two feet long, and lasted almost two months in beauty.

We feel pride that this work has been the means of communicating the above learned and elaborate document.

602

NEOTTIA australis; β. *chinensis.*

Chinese Neottia or Ladies' Traces.

GYNANDRIA *MONANDRIA.*

Nat. ord. ORCHIDEÆ. *Jussieu gen.* 64. *Brown prod.* 1. 309. *Div. II. Anthera* stigmati parallela persistens. *Massæ pollinis* v. farinaceæ v. è corpusculis angulatis; apicibus stigmati affixæ. *Brown in Hort. Kew. ed.* 2. 5. 197.

NEOTTIA. Cor. ringens: petalis exterioribus anticis labello imberbi suppositis; interioribus conniventibus. *Columna* aptera. *Pollen* farinaceum. *Brown in Hort. Kew. ed.* 2. 5. 198.

N. *australis,* labello indiviso oblongo crispato: basi bilingulatâ, columnæ lobulis lateralibus stigmate longioribus, spicâ tortili, foliis radicalibus lanceolatis, bulbis oblongis. *Brown prod.* 1. 319.

(β) SPIRANTHES *pudica:* foliis lineari-lanceolatis, labello subsessili apicem versùs crenulato, callis subrotundis; perianthii laciniis ovario rachique glaberrimis. *Lindley collect.* 30.

Folia *erecta, lineari-lanceolata, acuta, plana, enervia.* Scapus *glaberrimus, teres, subflexuosus, bracteis vaginantibus acuminatis vestitus.* Flores *parvi, albi, erubescentes, in spicâ tortili glaberrimâ horizontalitèr dispositi.* Laciniæ *glaberrimæ, ovato-lanceolatæ, obtusæ, cum labello parallelæ.* Labellum *brevitèr pedicellatum, oblongum, apice dilatatum, pubescens, columnæ marginibus oppositis amplectens; basi callis 2 villosis globosis munitum.* Columna *porrecta, clavata, anticè* gynizum (stigma, *v. si mavis* areola secernens stigmatis) *hemisphæricum viridissimum glabrum ferens, apice in linguam lanceolatam obtusiusculam dorso uniglandulosam productum.* Anthera *postica erecta bilocularis,* massas pollinis 4 *farinaceas per paria cohærentes glandulæ gynizi adhærentes promens.* Id. loc. cit.

Drawn in Mr. Griffin's conservatory at Lambeth. Introduced by the Horticultural Society from China. We had no opportunity of inspecting the plant; but comparing our figure and the description we have quoted above with the character given by Mr. Brown of NEOTTIA *australis*, a native of India as well as New Holland, we can as yet see no sufficient ground for determining a difference in their species. At all events no distinction has been pointed out, nor the two compared by any author; and until that has been done they are evidently too near akin to be idly disposed of by different names.

The Chinese plant is very like the "Ladies' Traces" (NEOTTIA *spiralis)* of our own country. Whether the ARISTOTELEA *spiralis* of Loureiro belongs here or not, can scarcely be satisfactorily decided from the description given by that author. We should think it cannot be far off from the present species.

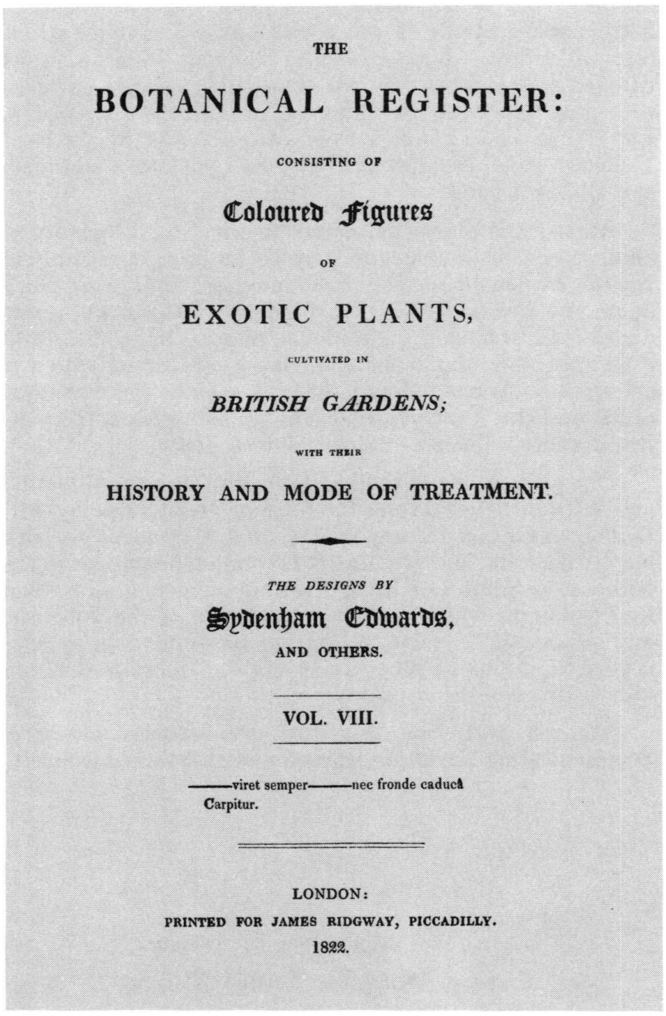
THE

BOTANICAL REGISTER:

CONSISTING OF

Coloured Figures

OF

EXOTIC PLANTS,

CULTIVATED IN

BRITISH GARDENS;

WITH THEIR

HISTORY AND MODE OF TREATMENT.

THE DESIGNS BY

Sydenham Edwards,

AND OTHERS.

VOL. VIII.

——viret semper——nec fronde caducâ
Carpitur.

LONDON:

PRINTED FOR JAMES RIDGWAY, PICCADILLY.

1822.

An unobserved species, now first introduced by Mr. Griffin, who received the plant, represented in the annexed plate, from Trinidad. It flowered last December in the hothouse at South Lambeth, and was communicated to Mr. Brown, who has made it the foundation of the present genus, to be placed next to ONCIDIUM, from which it differs by a convolutely concave undivided taper-pointed *label*, a corolla with the *petals* all distinct, a *column* with the two top-lobes converging convolutely (not extended), and by two furrowless or unindented *pollen-masses* (furrowed or indented in ONCIDIUM).

A humble plant (about four or five inches high), with two or three upright lanceolate leaves rather longer than the 5-flowered stem. *Flowers* scentless, about an inch in diameter, subbilabiately stellate. *Bulb* green, oblong, ovate, above ground. So far we have taken from the drawing, having had no opportunity of seeing the plant; what follows is from the Latin of Mr. Brown. *Corolla* dingy yellow, marbled on the inside with brownish purple; *petals* distinct, spreading: *label* sessile, shorter than the petals, obovate, convolutely concave, entire, taper-pointed with a sharp linear reflex appendage, marked on the inside with broken purple lines. *Column* detached, nearly of the length of the label, semicylindrical, gradually widened upwards with the side lobes of the top converging into a hood that infolds the anther. *Anther* extending from a base which is rather more protuberant and 2-celled within, into a broadly linear concave point several times longer than the base itself and infolding the subulate process of the stigma, the gland, and its appendage. *Pollen-masses* two, furrowless, wedge-shaped, moderately compressed, after they leave the anther fixing themselves by their pointed tips to the disk of the widened summit of the elongated appendage of the *proscollum* (gland of the stigma).

A and B are two views of the flower, slightly magnified. The two other outlines are two views of the column and its parts, very much magnified.

612

MACRADENIA lutescens.

Trinidad Macradenia.

GYNANDRIA *MONANDRIA.*

Nat. ord. ORCHIDEÆ. *Jussieu gen.* 61. *Brown prod.* 1. 309. *Div. IV. Anthera* terminalis mobilis decidua. *Massæ pollinis* demùm cereaceæ. *Brown in Hort. Kew. ed.* 2. 5. 205.

MACRADENIA. *Labellum* sessile cucullato-concavum indivisum acuminatum. *Corollæ* petala distincta patula. *Columna* libera, lobis apicis conniventibus. *Massæ pollinis* duæ, exsulcæ. *Brown MSS.*

Macradenia lutescens. *Brown MSS.*

Cor. *sordidè lutea, intùs purpureo-fusco-marmorata.* Labellum *corollâ brevius, ecalcaratum, album, intùs purpureo interruptè lineatum, obovatum, appendice lineari acuto reflexo.* Columna *longitudine ferè labelli, semiteres, suprà sensìm latior lobis lateralibus apicis in cucullum antheram amplexantem conniventibus.* Anthera *è basi paulò gibbosiore intùsque biloculari producta in acumen lato-lineare concavum ipsâ basi aliquotiès longius processum subulatum stigmatis, glandulam, ejusque appendicem involvens.* Massæ pollinis *cuneiformes modicè compressæ, post migrationem apicibus suis acutis affixæ disco apicis dilatati appendicis elongati proscolli.* Brown MSS.

618

ANGRÆCUM maculatum.

Spotted-leaved Angræcum.

GYNANDRIA *MONOGYNIA.*

Nat. ord. ORCHIDEÆ. *Jussieu gen.* 61. *Brown prod.* 1. 300. *Div. IV. Anthera* terminalis mobilis decidua. *Massæ pollinis* demùm cereaceæ. *Brown in Hort. Kew. ed.* 2. 5. 205.

ANGRÆCUM. *(Aubert du Petit Thouars in nouv. bullet. de scien. par la soc. philom.* 1. 518.) *Flores* supinati. *Perianthium* (*cor.* nob.) irregulare connivens. *Laciniæ (petala* nob.) subæquales, 3 superiores galeatæ. *Labellum* dependens, calcaratum, trilobum, lobis lateralibus minoribus adscendentibus. *Columna* libera, aptera, elongata, glabra. *Anthera* terminalis, opercularis, semibilocularis, decidua. *Massæ pollinis* 2, integræ. *Stig.* concavum, transversum. *Herbæ in truncis arborum parasiticæ. Folia plana.* Lindley collect. 15.

A. *maculatum* foliis lanceolatis maculatis planis integerrimis. *Lind. loc. cit.*
Bulbi *obscuro-virides, glauciusculi, ad collum pallidiores, squamis mar-*

cescentibus primùm vestiti. Fol. *carnosa, plana, enervia, integerrima, maculis confluentibus inquinata.* Scapus *foliorum longitudine, teres, glaber, pallens, viridis,* bracteis *paucis scariosis.* Racemus *pauciflorus quaquaversus.* Flores *parvi, resupinati, galeati.* Laciniæ (petala) *subæquales;* exteriores *lineari-oblongæ, apice rotundatæ, inferioribus margine inferiore in lobo pallidiore dilatato;* interiores *paulò latiores, pallidè rosaceæ, margine imbricantes, super columnam fornicatæ.* Labellum *planum, oblongum, calcaratum, 3-partitum, lobis lateralibus intùs sanguineo-venosis, erectis; intermedio emarginato; disco bilamellato, sanguineo, axe et apice flavidis.* Columna *labello dimidio brevior, anticè plana, sursùm dilatatâ.* Anth. *terminalis, decidua, bilocularis loculis discretis, in apice aliquantisper excavato columnæ insidens.* Massæ pollinis 2, *glandulâ nullâ.* Stig. *viscidum transversum excavatum.* Ovar. (germ.) *pedicellatum, contortum.* Lind. l. c.

Having missed the opportunity of inspecting the inflorescence of the present plant, we trust to the work above cited for its description. The species is there attributed to a genus formed by M. Aubert du Petit Thouars for the reception of a considerable orchideous group belonging to the islands that lie off the shores of Southern Africa. The genus however is one with which we suspect the species before us, a native of the Brazils, will not be found to combine so naturally as is requisite in a well-assorted group.

The drawing was taken from a plant in the hothouse of Messrs. Colvill, in the King's Road, Chelsea.

Bulbs dark green, slightly glaucous, paler at the neck, shrouded at first by scales which wither away. *Leaves* fleshy, flat, without nerves, entire, marked with deeper spots that run into one another. *Scape* the length of the leaves, round, smooth, pale, green, furnished with a few scariose *bractes. Raceme* few-flowered, scattered. *Flowers* small, reversed, casqued, irregular. *Petals* nearly equal, converging; *outer* linearly oblong, rounded at the end, the lower ones of which have their lower edge widened into a paler lobe; *inner* but little wider, pale pink, lapping at the edge, forming a vault over the column. *Label* dependent, flat, oblong, spurred, 3-parted, with smaller side-lobes veined with crimson and upright; the middle one notched at the end; disk bilamellated, crimson, reddish-yellow at the axis and top. *Column* detached, half the length of the label, smooth, flat in front, widened upwards. *Anther* terminal, lid-fashioned, half two-celled, with the cells disjoined from each other, placed upon the slightly hollowed apex of the column, deciduous. *Pollen-masses* 2, entire. *Stigma* concave, transverse, viscid. *Germen* pedicled, twisted.

NOTE.

The following corrections should be made in the article MACRADENIA *lutescens* of the last fasciculus.

Fol. 612. l. 9. pro " *Corollæ* petala" lege " *Petala.*"
l. 18. pro " *glandulam*" lege " *proscollam.*"
l. 20. pro " proscolli" lege " proscollæ."
l. 27. dele the words " a corolla with the petals all distinct." And last line overleaf for " *proscollum*" read " *proscolla.*"

639

NEOTTIA procera.

Nepal Neottia.

GYNANDRIA *MONANDRIA.*

Nat. ord. ORCHIDEÆ. *Jussieu gen.* 64. *Brown prod.* 1. 309. *Div. II.* Anthera stigmati parallela persistens. *Massæ pollinis* vel farinaceæ vel è corpusculis angulatis, apicibus stigmati affixæ. *Brown in hort. kew. ed.* 2. 5. 198.

NEOTTIA. Suprà vol. 7. *fol.* 602.

N. *procera*, foliis lanceolatis acuminatis; spicæ axi villosâ, labello ovato-acuminato erecto basi ventricoso apice recurvo, intùs infernè barbato, supernè in palatum pruinosum secùs medium sulcatum intumescente.

Neottia procera. *Titulus quo hort. bot. calcutt. venit.*

Rhizoma *pluriceps, crassiùs fibrosum.* Scapi *erecti, teretes, foliorum petiolis vaginati, tripedales ultràve, basi digitum vel circitèr crassi.* Fol. *plurima, decrescentia, caulina, sparsa;* lamina *lanceolata, acuminata, subundulata, patentissima, 8-uncialis v. ultrà latitudine quater ferè minore, suprà pruinoso-micans nervis paucis conspicuioribus:* petiolorum *vaginæ membranosæ, pallidæ, subtriunciales, nervis longitudinalibus lineatæ.* Spica *parviflora, simplex, elongato-cylindrica, numerosa, laxiùs sparsa, erecta, patula, axi tereti villosâ.* Bracteæ *virentes, germen arctè semi-involventes subsuperantesque, acuminatæ, extùs villosæ.* Flores *chloroleuci, glabri, torsione germinis supinati, subdiaphani.* Germ. *sessile, virens, adscendens, cylindricum,* ¼ *unciæ longum.* Cor. *nutans, germine* ¼ *brevior, campanulata, semiringens:* petala 3 exteriora *æquilonga, oblongo-ovata, acuminata, extùs infernè subvirentia,* unum medium *summum porrecto-fornicatum,* duo infernè lateralia *labello supposita, patentia, divergentia, plana, suprà recurva, intùs albo-micantia;* interiora 2 *supernè lateralia, tenerrima, albo-diaphana, latiora, apice rotundata, cum summo medio exteriore secundùm latus interius utrique incumbente in galeam posticam fornicato-porrectam cohærentia, à margine exteriore supernè recurva.* Labellum *erectum, carnosum, petalis æquilongum latitudine duplâ, sessile, ovatum cum acumine brevi, integerrimum, columnam breviorem à marginibus involutis comprehendens, infernè extùs ventricoso-convexum, à cavo internè respondente pilis robustis glandulosis obtusis ab utrinque versùs medium conniventibus barbatum, supernè brevè recurvatum cum disco interiore albido pruinoso levitèr pulvinato sulcoque longitudinali intersecto.* Columna *aptera, erecta, libera, ovato-acuminata, brevis, labello longiori occultata, diaphano-albicans, posticè convexa, anticè concava:* anthera *reversa, opercularis, castaneo-fuscescens, apice posticè declivi columnæ inserta, stigmate parallela, persistens, convexa, altiùs à parte posticâ didymo-fissâ immersa, anticè dehiscens, bilocularis loculis contiguis et à septulo incompleto septi duplicati principalis parallelo secùs medium interceptis.* Massæ pollinis 2, *grossè granulosæ, ochroleucæ, clavato-didymæ, bilobo-partitæ, anticè* proscollæ *immediatè annexæ, peragente anthesi sitù reversæ.* Stigma *anticum, ovato-acuminatum, concavum frontem fermè totum columnæ tenens, albicans;* areola secernens *inferior glandulosa glabra profundiùs cavata, in* rostellum *superius membranaceum acuminatum (post peractam anthesin ob elapsam proscollam bifidum) continuata.* (N. anticè vel posticè *semper respectû floris expansi supinati intelligimus.*)

Samples of this species collected in Nepal are deposited in Mr. Lambert's Herbarium; and seed of it has been recently sent from the botanic garden at Calcutta, by Dr. Wallich, under the present name.

The drawing was made at the Nursery of Messrs. Colvill, in the King's Road, Chelsea, where the plant is cultivated in the hothouse, and flowers during the early part of the summer.

Whether it is a genuine NEOTTIA or not, we do not pretend to decide, never having critically inspected any one of the genus. According to the latest definition, the *labellum* should be beardless in NEOTTIA, which is not the case here; but that may be a circumstance not essential to the group.

Scapes several simple 2-3 feet high, sheathed by the petioles of the numerous spreading scattered lanceolate foliage. *Spikes* terminal, simple, cylindrically elongated, from 6 inches to a foot in length, with numerous small open-ranked greenish white inconspicuous flowers: axis round, villous. *Germen* about ¼ of an inch long, about ¼ longer than the corolla, half inveloped by a rather longer greenish *bracte*, which is pointed and villous on the outside. *Label* fleshy, nearly of the colour of the rest of the corolla but fading to a rusty white, of the same length as the other petals but much broader, sessile, upright, ovate, shortly pointed, entire, ventricose at the base where it is bearded within, top shortly recurved and bluntly pointed, palate frosted, slightly pulvinate, and intersected by a single furrow.

675

GEODORUM dilatatum.

Shovel-lip'd Geodorum.

GYNANDRIA *MONANDRIA.*

Nat. ord. ORCHIDEÆ. *Brown prod.* 1. 309. *Sect. V.* Anthera terminalis, mobilis, operculiformis, decidua. Pollinis massæ cereaceæ, læves. *Parasiticæ, paucæ Terrestres.* Brown l. c. 330.

GEODORUM. Jackson in Andrews's reposit. 626. *Labellum* cucullato-ventricosum (nunc basi calcaratum), sessile, cum columnâ non articulatum. *Petala* conformia subsecunda. *Massæ pollinis* 2, posticè lobulo auctæ. *Brown in Hort. Kew. ed.* 2. 5. 207. Plantæ (Indiæ orientalis, Novæ Hollandiæ) terrestres.

OBS. Genus MALAXIDI proximum habitû, nec multùm absimile structurâ columnæ florisque expansione. *Br. prod.* 1. 332; *sub fine observationis* CYMBIDIO picto *(nunc in* GEODORUM *transponendum) appensæ.*

G. *dilatatum*, scapo foliis breviore, spicâ pendulâ: floribus congestis, labello subcalcarato; apice dilatato crenulato. *Brown in Hort. Kew. ed.* 2. 5. 207.

Limodorum recurvum. *Roxburgh corom.* 1. 33. *t.* 39. *Swartz nov. act. ups.* 6. 79. *Willd. sp. pl.* 4. 124.

Terrestris, bulbo *subrotundo, fibris crassis radicato.* Fol. *subtrina,* 6-*uncialia ad pedalia, erecta, lanceolata, latitudine* 3-4-*unciali,* 5-*nervia, basi vaginoso-complexa, laminâ submembranaceâ;* scapus *radicalis, extra foliaceus, erectus, virens, teres, cavus, glaber,* 5-9-*uncialis, foliis brevior, crassitie ferè pennæ scriptoriæ, foliis paucis* (3?) *vaginosis suprà in bracteas descrescentibus acuminatis instructus; supernè reflexus.* Racemus *cernuus, multiflorus, imbricato-sparsus, subcapitato-abbreviatus:* bracteæ *lineari-lanceolatæ, germen æquantes.* Germ. *sessile, rectum, viride, corollâ brevius, rotundatè costatum.* Cor. *nutans, carneo-albicans, sublabiato-campanulata, semunciâ vix longior:* petala 5, *concolora, in labium summum patulum imbricato-convergentia,* exteriora *trina, sibi similia, cuneato-oblonga, dorso subcarinata, apiculo acuto:* interiora *bina duplo fermè latiora, sublongiora, pallidiora, teneriora, lineari-oblonga, plana, obtusata, dorso baseos carinato-gibbosa:* labellum *petalis exterioribus æquilongum, roseo-pictum, rectum, latum, retusum, supernè areolâ internâ luteâ calloso-incrassatâ in disco prominulâ notatum, infernè in ventrem obliquè cavum convolutum, extùs supra ventrem scrobiculo utrinque impressum indèque carinatum, suprà in laminam summam brevem patulam rotundato-retusam utrinque reflexam crenulatamque extensum; quoad axin spicæ interius.* Columna *libera, petalis* 3-*plo ferè brevior, ochroleuco-pallens, rectum, obovatum, cum brevi acumine, posticè concaviusculum, anticè planiusculum curvulum:* anth. *operculiformis, oblato-rotunda, convexiuscula, lutea, acumini erecto columnæ posticè inserta, decidua, extùs pruinosa* (clinandrium *cicatricibus binis puniceis sub situ loculorum pollinis conspicuè notatum!).* Stigma *area cava secernens oblato-subrotunda columnæ concolora, antheræ immediatæ subjectum, suprà margine prominulâ cinctum.*

The plant from which the drawing has been made, flowered this summer in the hothouse of Dr. Hood, at South Lambeth. It is a very rare species in our collections; and has been no where represented except in the work on the Coromandel Plants, by Dr. Roxburgh; where it is spoken of as native of the moist vallies among the hills in India, and as growing on the ground, not on trees, like the great portion of this natural order in the tropical climates.

Introduced into our gardens by Sir Joseph Banks in 1800, along with its congeners *purpureum* and *citrinum,* all natives of the same country.

We observed two very singular crimson scars upon the bed of the pollen-masses, one under each of them, and evidently connected with the early stage of their production. Mr. Brown, to whom we showed them, thought them a remarkable appearance.

GEODORUM was originally instituted by the late Mr. Jackson, then librarian to Mr. Lambert; but has been subsequently remodelled by Mr. Brown. It comes near to MALAXIS; but the pollen-masses are there four; here two augmented by a lobe behind.

676

VANDA teretifolia.

Cylindric-leaved Vanda.

GYNANDRIA *MONANDRIA.*

Nat. ord. ORCHIDEÆ. *Jussieu gen.* 64. *Brown prod.* 1. 309. *Div. IV. Anthera* terminalis mobilis decidua. *Massæ pollinis* demùm cereaceæ. *Brown in Hort. Kew. ed.* 2. 5. 205.

VANDA. Suprà vol. 6. *fol.* 506.

V. *teretifolia,* perianthii (corollæ) laciniis interioribus brevioribus. *Lindley collect.* 6. *tab.* 6.

Epidendrum teretifolium. *Link enum. hort. berol.* 2. 372?

Rami *stricti, teretes, foliosi, radices pallidas tortuosas promentes.* Folia *teretia, biuncialia, internodiis* 2-4-*plo longiora, cum caule et vaginâ articulata.* Spica *simplex, horizontalis, foliorum longitudine (vel plurimùm longior)* 5- *(v. laxiùs multi-) flora.* Flores *majusculi, ante anthesin cuneati, resupinati: laciniæ oblongæ, reflexæ, lurido-luteæ, obtusæ, dorso et tribus nervis intùs sanguineis;* 2 interioribus *brevioribus.* Labellum *album, ovatum, dependens, laciniis longius, cum processû villoso columnæ elasticè articulatum, basi in calcare obtuso, sanguineo-lineato productum, versùs apicem utrinque unidentatum, dentibus erectis apice coloratis.* Calcar *dimidiâ inferiore* 2-*loculare, membranâ transversâ foraminibus duobus pertusâ, pro receptione baseos corporis carnosi compressi triangularis, margine et apice bicorni pilosis, dorso interioris calcaris affixi.* Columna *laciniis multò brevior, lurido-lutea, parte anteriore villosâ, planâ, apice utrinque unidenticulatâ, lineâ pilosâ ab utrâque dente decurrente; anticè circà faucem stigmatis membranâ quadratâ erectâ cartilagineâ.* Anthera *terminalis, opercularis, persistens, anticè elongata,* 2-*locularis, loculis discretis.* Massæ pollinis 2, *cereæ, duræ, integræ (posticè lobato-circumscissæ,* Nob.), *posticè obscurè foveatæ, pedicello affixæ apice falcato compresso; basi in membranâ cartilagineâ trilaterali dilatato, quæ, ejus lateribus membranis faucis stigmatis applicatis, fit quoad stigma horizontalis.* Stigma *infundibulare, purpureum, viscidum: glandulâ massarum pollinis pedicello medio proximâ.* Ovarium (germen) *perianthio (corollâ) longius, contortum (rectum* Nob.). Lindley loc. cit.

Figured and described by Mr. Lindley in the publication we have quoted.

Native of China, and but lately introduced into our gardens, where it grows readily and blossoms in perfection. The annexed figure is from a sample we were favoured with by Sir Abraham Hume, from his hothouse at Wormleybury.

The flower is curious in the structure, as well as ornamental. The foliage is very different from that of the two species already published in our Register. The roots are produced from all parts of the stem or branches from between the leaves, and serve as aboveground stays or props, adhering where requisite to the smoothest surface by the same process that holds the misletoe to the oak and the sea-weed to the rock.

According to the character of this division of the order, the anther should be deciduous; but Mr. Lindley has characterized it as permanent; and it is true, that in our hothouses we have never seen it spontaneously removed at any stage of the existence of the flower from the bud state to its decay. By a slight pressure, however, from the point of a needle it falls off entire, while the pollen-masses, by means of the remarkable process which supports them, start with considerable elasticity from their places; and we suspect that in the native country of the plant the anther is spontaneously deciduous, and not of the nature of that intended by Mr. Brown by a permanent anther. We do not mean to imply, by this remark, that such removal is in any way necessary for the due access of the pollen-masses to any portion of the pistil.

In terming the pollen-masses entire *(integræ),* if Mr. Lindley means, that they are without division or incision, he is certainly mistaken, for each is scored or cut round into two unequal lobes or segments of its orb, and in that respect comes directly within the definition of the character attributed to the genus by Mr. Brown.

Mr. Lindley speaks of the germen being twisted; to us it appeared straight, as in the VANDA *paniculata* of this publication, to which it comes near in many respects.

686

EULOPHIA guineensis.

Sierra Leone Eulophia.

GYNANDRIA *MONANDRIA.*

Nat. ord. ORCHIDEÆ. *Jussieu gen.* 61. *Brown prod.* 1. 300. *Div. IV.* Anthera terminalis mobilis decidua. Massæ pollinis demùm cereaceæ. *Brown in Hort. Kew. ed.* 2. 5. 205.

EULOPHIA. Brown suprà vol. 7. *fol.* 578; *in textû anglicè. Petala* 5 distincta, conformia, adscendentia, patentia. *Labellum* basi calcaratum; laminâ sessili cristatâ trilobâ, posticè indivisâ. *Massæ pollinis* 2, bilobæ lobulo postico.

E. *guineensis,* foliis lanceolatis: labelli calcare subulato adscendente laminam subæquante; laminæ lobis lateralibus cum columnâ in faucem lineato-cristatam convolutis; terminali grandiore ovato-rotundato convexè explanato, disco erugato.

Bulbus *externus, viridis, oblongo-ovatus, è dilatatis foliorum basibus compactus.* Fol. *plura, radicalia, sparsa, lætè virentia (non glauca), petiolata; laminâ lentâ (non rigente), lanceolatâ, nervosâ, acuminatâ, spithamæâ v. magis latitudine biunciali v. circitèr:* petiolus *multotiès angustior, aliquotiès brevior, pallidus.* Scapus *imo bulbo ortus, solidus, teres, erectus, virens, infernè imbricatè supernè remotè vaginatus, foliis altior (sesquipedalis).* Vaginæ *membranaceæ, purpureo-lineatæ, laminâ lanceolatâ erectâ.* Spica *stricta, multiflora, sparsa, laxa, simplex.* Bracteæ *herbaceo-pallentes, erectiusculæ, lanceolatæ, nervosæ, germine subbreviores.* Flores *pulchelli, subsesquiunciales, resupinati, nutantes, semiringentes, inodori.* Germ. *unciâ sublongius, gracile, virens, adscendens, tortum, teretiusculum, obitèr costatum, supernè incrascescens rubro-fuscescens atque recurvum.* Pet^a^. 5, *distincta, viridi-purpurea, lurida, lineari-oblonga, acuta, recta, conformia, intùs convexiuscula, extùs obsoletè nervosa, ⅔ unciæ longa, in labium radiato-semiforme æquabile erectiusculum adscendentia, basi vividè purpurea, margine pallentia undulata,* 2 interiora *latiora (latitudine subtrilineari):* Labellum *deflexum, difforme, basi calcaratum;* calcar *rectè continuum, adscendens, laminâ subæquilongum, subulatum, pallidè virens, obtusum, suprà dilatatum* dorso *carinato-compresso, fine sursùm flexo;* lamina *sessilis, petalis vix brevioribus pluriès latior, oblonga, carneo-albicans, triloba, convexula, infrà in faucem cristis pluribus lineæformibus calloso-verruculatis viridibus parallelis parùm elevatis symmetricè inscriptam convoluta, suprà explanata;* lobi laterales *abruptè truncati erosuli cum lateribus extenuatis incumbentis columnæ involuto-conniventes et infernè connati;* medius *magnus recto-continuus ovato-rotundatus explanatus convexulus undulatus erugatus, intùs radiis pluribus purpureis de basi citra medium usque pergentibus pictus.* Columna *deflexa, subalato-dilatata, lobis lateralibus labelli æqualis, posticè convexa subvirescens et apice in marginem rubidam brevem bicrenatam extensa, anticè acclivi-obliquata plana nitida alba.* Anthera *chloroleuca, decidua, oblata, convexa, summam columnam operculans, secundùm latus posterius impage carnosâ pruinosâ brevi transversè dilatatâ trifidâ (lobulis 2 extimis intra crenas marginis posticæ columnæ immissis) munita, bilocularis septo (ob retractas induplicatas valvas) deficiente, antrorsùm dehiscens:* massæ pollinis 2, *flavæ, cereaceæ, transversè collaterales, globosæ, compressæ, bilobæ (lobus minor posticus cavo centrali glutinoso incumbens), ligulâ communi membranaceâ albâ transversè dilatatâ processûs stigmatosi continuâ deciduâ è puncto viscoso elastico baseos affixæ.* Stigma *anticum cavum transversum chloroleucum, summæ columnæ facie juxta antheram insculptum.*

For the outline of the genus and enumeration of the species we refer to Mr. Brown's observations on Lissochilus *speciosus*, vol. 7. fol. 578. of this work; where the name is written " Eulophus," and has been now altered in the termination on the suggestion of its author.

The recorded species are from the Cape of Good Hope and the East Indies; the unrecorded one before us from Sierra Leone, whence it was introduced last summer into the garden of the Horticultural Society by Mr. George Don, collector for that public-spirited, well-directed association.

The crested character of the label is, perhaps, less conspicuously marked in the present species, than in any mentioned by Mr. Brown; and is moreover strictly confined to the funnelform portion of that petal, where it appears in the form of several green slightly raised even parallel callously tuberculated symmetrically disposed vertical lines, easily mistaken for as many varicose nerves.

The lid of the anther is remarkable for a thickened fleshy frosted three-lobed shallow border extended along the hinder edge; the two lateral divergent lobes of which are let into corresponding indentations in a raised margin at the back of the apex of the column. Upon the flower being fully expanded this lid falls off, and when viewed in the microscope is not unlike the empty upper shell of the diminutive Crabs so frequent on our shores.

The two uncoloured outlines in the plate are front and side views of the flower when deprived of the five upper petals, and are intended to show the relative positions of the column and label.

THE

BOTANICAL REGISTER:

CONSISTING OF

Coloured Figures

OF

EXOTIC PLANTS,

CULTIVATED IN

BRITISH GARDENS;

WITH THEIR

HISTORY AND MODE OF TREATMENT.

THE DESIGNS BY

Sydenham Edwards,

AND OTHERS.

VOL. IX.

——viret semper——nec fronde caducâ
Carpitur.

LONDON:
PRINTED FOR JAMES RIDGWAY, PICCADILLY.
1823.

Sub t. 700 (1823)

From fol. 703.

List of Cape *Orchideæ* figured in " The Journal of Science and the Arts;" with references to the Volumes and Plates of that work. Added in this place for want of room in article 703.

Bartholina pectinata *(malè* Burmanniana). *vol.* 4. *Pl.* 6. *fig.* 2.
Disa grandiflora. *vol.* 4. *Pl.* 6. *fig.* 1.
Disa spathulata. *vol.* 4. *Pl.* 6. *fig.* 3.
Disa porrecta. *vol.* 5. *Pl.* 1. *fig.* 1.
Disa graminifolia. *vol.* 6. *Pl.* 1. *fig.* 2.
Disperis capensis. *vol.* 5. *Pl.* 1. *fig.* 2.
Disperis secunda. *vol.* 5. *Pl.* 1. *fig.* 3.
Disperis villosa. *vol.* 6. *Pl.* 1. *fig.* 5.
Disperis cucullata. *vol.* 6. *Pl.* 1. *fig.* 4.
Corycium bicolor. *vol.* 6. *Pl.* 1. *fig.* 1.
Corycium orobanchioides. *vol.* 8. *Pl.* 3. *fig.* 3.
Pterygodium catholicum. *vol.* 6. *Pl.* 1. *fig.* 3.
Pterygodium alatum. *vol.* 8. *Pl.* 3. *fig.* 2.
Pterygodium Volucris. *vol.* 9. *Pl.* 4. *fig.* 2.
Pterygodium inversum. *vol.* 9. *Pl.* 4. *fig.* 1.
Satyrium bracteatum. *vol.* 8. *Pl.* 3. *fig.* 1.

Not more than three of the above species have been introduced into our gardens. It is worthy of notice, considering the frequent intercourse with the colony, that hundreds of species of fine and curiously flowered *Liliaceæ*, *Ensatæ*, and *Orchideæ*, growing in the neighbourhood of Cape Town, should never have been imported into our botanical establishments; while from far more distant and sequestered quarters there has been introduced a much larger relative proportion of the species of those natural orders.

701

NEOTTIA orchioides.

Jamaica Neottia.

GYNANDRIA *MONANDRIA.*

Nat. ord. Orchideæ. *Jussieu gen.* 64. *Brown prod.* 1. 309. *Div. II. Anthera* stigmati parallela persistens. *Massæ pollinis* vel farinaceæ vel è

corpusculis angulatis; apicibus stigmati affixæ. *Brown in Hort. Kew. ed.* 2. 5. 197.
NEOTTIA. Suprà vol. 7. *fol.* 602.

N. *orchioides,* scapo stricto foliis glauciusculis præcociore floribusque pubescentibus; corollâ sacco recto oblongo obeso antico ad basin auctâ; columnâ glaberrimâ; stigmate infrà unilabiato-lobato.
Neottia orchioides. *Swartz prod.* 118. *flor. ind. occid.* 3. 1411. *Willd. sp. pl.* 4. 75. *Curtis's magaz.* 1036? *Hort. Kew. ed.* 2. 5. 198.
Satyrium. 7; aphyllum, scapo erecto simplici subsquamoso spicato. *Browne jam.* 325.
Fol. *plura, elongato-lanceolata, glabra, glauca, post defloratam spicam procrescentia, basi vaginantia.* Scapus *bipedalis, crassitudine calami, sublanato-villosus, melino-pallescens, strictus, erectus, teres, infernè vaginatus, indè spicâ tenùs aphyllus.* Vaginæ *glabræ, foliaceæ, distantes, cucullato-convolutæ, acuminatæ acumine erecto.* Spica *numerosa, erecta,* 4-5-*uncialis, sparsa, patens, floribus resupinatis, melino-erubescentibus, viscosiusculis, laxiùs approximatis, oblongis, suprà bilabiato-patulis, cum germine simul uncialibus v. circitèr, extùs sublanato-villosis.* Bracteæ *lanceolatæ longitudine florum, pallentes, pubescentes.* Petala 5, *subæqualia, lanceolata;* summum exterius *paulò latius, oblongo-lanceolatum, erectum, acutum, lateralibus* 2 *interioribus à latere utroque coadunato-incumbens et cum iis labium superius erectum concavum efformans, basi in gibbum brevem abruptum germen posticè decurrentem terminans;* lateralia 2 interiora *pallidiora, subhyalina, subdimidiato-lanceolata acuminata margine interiore rectiore;* lateralia 2 exteriora *lanceolata, subulato-acuminata, subbreviora, erecta, supernè patentia, labellum inter se recipientia, germini secùs frontem profundè adnata inque saccum rectum obesum obtusum cum convolutè productâ labelli basi conniventia nec non ad ultimum finem inter se connata:* labellum *inclusum, carneo-candicans, crassius firmiusque sed æquale petalis, ovato-lanceolatum, rostrato-acuminatum, convoluto-concavum, porrectum, apice recurvum, columnam complectens, basi demissè productum convolutum lanato-barbatum et intra petalorum exteriorum saccum conferruminatè inclusum.* Columna *inclinata, carneo-albicans, labelli æquilongi involutis lateribus comprehensa, oblonga, semicylindrica, tota glabra, anticè convexa et faciei germinis intra basin cornutam labelli inclusæ obliquè adnata, suprà in rostellum rectum gracile aciculari-elongatum abruptè producta, posticè antherifera:* anth. *melino-fulvescens lineari-lanceolata convexa, dorsum cavum columnæ omninò occupans, biloc., septo duplicato longitudinali intercepta;* massæ pollinis 2, *reversæ, farinaceæ, ochroleucæ, clavato-elongatæ, parallelæ, angustæ, sursùm pedicellato-attenuatæ pediculis supernè in unum coalescentibus,* retinaculo *fusco cartilagineo lineari-elongato* proscollam *terminalem* rostelli *parallelè affigente affixæ*

Stig. *in summâ columnâ anticum, cavum, secernens, album, orbiculato-dilatatum, obliquum, lobulo decurvo infrà prominulo labiatum, suprà rostellatum.* Germ. *oblongum, subsemiunciale, pubescens, melino-subrubescens, obsoletè costatum, basi semigyrato-tortum.*

Introduced into our hothouses from Jamaica about 1806, by the late Mr. E. I. A. Woodford, but continues a very scarce plant; nor had we met with it for many years before this winter, when it flowered in the garden of the Horticultural Society, where the drawing was taken.

The species comes the nearest of any to *speciosa,* in which however the leaves are undulate and not glaucous, the stem and flowers entirely smooth, the corolla of a much brighter red without the elongated pouch in front, the column bearded in front, the stigma without a prominent nether lip, and the stem at least three times shorter.

In *orchioides* the stem is two feet high or more, and as well as the inflorescence covered with a frizzly pubescence and appears before the leaves, which are glaucous and not undulated. The corolla terminates downwards in a short thick oblong pouch parallel with the front of the germen, which pouch is formed by the elongated bases of the two outer side-petals enclosing the elongated frizzly bearded base of the label. The column is entirely smooth in front; and the stigma has a prominently recurved nether lip. In *speciosa* the anther does not occupy the whole space in the back of the column, but in *orchioides* it does.

The drawing in Curtis's Magazine has been taken from a plant in a very different state from the one represented in the present plate, if it is really of the same species. Ours is clearly the *orchioides* of Swartz.

703

SATYRIUM coriifolium.

Golden-flowered Satyrium.

GYNANDRIA *MONOGYNIA.*

Nat. ord. ORCHIDEÆ. *Jussieu gen.* 64. *Brown prod.* 1. 309. *Div. I. Anthera* adnata subterminalis persistens. *Pollinis massæ* è lobulis angulatis elasticè cohærentibus; basi affixæ. *Brown in Hort. Kew. ed.* 2. 5. 188.
SATYRIUM. Suprà vol. 5. *fol.* 416.

S. *coriifolium,* foliis ovatis acuminatis subreflexis vaginantibus coriaceis, margine membranaceo-crenatis, floribus galeâque cernuis. *Swartz in act. holm.* 1800. 216.
Satyrium coriifolium. *Willd. sp. pl.* 4. 54. *Curtis's magaz.* 2172.
Satyrium erectum. *Thunb. flor. cap.* 1. 91; *(vix tamen Swartzii.)*
Satyrium cucullatum. *Loddiges's bot. cab.* 104; *(non aliorum.)*
Diplectrum coriifolium. *Persoon syn.* 2. 509.
Orchis lutea, caule purpureo-maculato. *Roxb. cent.* 3. 7. *t.* 10.
Bulbus *ovatus, indivisus.* Fol. *plura bifaria, ima erecto-patentia, coriacea, ovato-acuminata, basi convoluto-vaginantia, decrescentia.* Scapus *foliis plurimùm altior, infrà spicam foliaceo-vaginatus, foliorumque bases sæpiùs purpureo-maculati.* Spica *erecta, plurima, sparsa,* 3-4-*uncialis,* floribus *proximis flavissimis uncialibus,* bracteis *foliaceis ovato-lanceolatis florem æquantibus v. superantibus, retrofrangendis.* Cor. *erecto-nutans, neque supinata, semiringens, nunc passim rubro ardente tincta:* labellum *posticum, erectiusculum, petalis altiùs plurimùmque amplius, ovato-galeatum obtusum fornice profundâ intùs subbarbatâ lateribus antrorsum productis, fronte summâ brevi reflexâ; posticè ad basin bicalcaratum,* calcaribus *rectis subulatis parallelis cavis secundùm dorsum germinis subæqualis descendentibus:* petala *distincta, æquilonga, concolora, germine sublongiora,* labium *inferius reflexum undè efformantia;* exteriora 3 *arctiùs inter se approximata,* lateralia *lineari-oblonga, acuminata, medium lineari-ligulatum planum obtusissimum cæteris decurvius duploque angustius;* interiora 2 *divaricato-opposita deorsùm obliquata, exteriorum medio subangustiora, cæterum cum eo exactè conformia.* Columna *adscendens, galeâ inclusa paulòque brevior, chloroleuca, glabra, libera, lineari-elongata, tereti-compressa:* stigma *terminale, posticum, bilabiatum, obcordato-dilatatum, antrorsùm flexum,* labio superiore *subtruncatè retuso eroso-denticulato, facie internâ secernente;* inferiore *æquali recurvo pro excipiendis retinaculis bicrenato cum rostello medio, utrinque margine membranaceâ undulatâ patulâ decurrente alato (de imo rictûs interlabialis angulo prominet lamella parva concolor subrotunda brevis, nedùm aliis observata!):* anth. *reversa, anticè columnæ infrà stigma adnata, oblongato-didyma, bilocularis, septo medio longitudinali;* pollinis massæ *luteæ clavatæ stipite robusto longo, è particulis lineari-oblongis angulosè pressis confectæ,* retinaculis 2 *albis glutinosis triangulari-cordatis à crenis labii inferioris stigmatis exceptis è basi affixæ.* Germ. *virescens, sessile, labello brevius, oblongum, appressum, rectum, posticè planiusculum ecostatum, anticè convexum costatum.*

In the great majority of orchideous genera the back of the corolla comes round to the front, its position being reversed by a half turn of either the germen or pedicle, a secondary process consequent upon expansion. In the present genus no such process takes place, and the corolla retains its connatural position till it decays.

It was in this genus that the middle petal of the lower lip had been mistaken by Swartz for the label of the corolla, and the mistake rectified by Mr. Brown, who has shown the helmet-shaped petal at the back of the flower to be the true label; a part always determinable by its position in relation to the germen, being the middle petal of the three inner ones, all opposite to the three valves of the germen, while the three outer ones are opposite to the three intervening prominent riblike sutures. The label is usually different from the other petals, either in consistence, form, or colour; sometimes in all three ways.

We are convinced that the species before us is the same with SATYRIUM *erectum* of Thunberg's Flora Capensis, though possibly different from the plant intended by Swartz under that name.

The drawing of this rare and handsome-flowered vegetable was taken in the greenhouse of Mr. Lee at Hammersmith; where it had been imported from the Cape of Good Hope, the native place of the species.

We doubt much whether SATYRIUM *parviflorum* of Swartz (ORCHIS *bicornis. Jacq. hort. Schœnb.* 2. 26. *t.* 179) is distinct from SATYRIUM *cucullatum* (fol. 416) of this work, which is certainly ORCHIS *bicornis* of the first edition of Hortus Kewensis, and SATYRIUM *cucullatum* of the second. We believe several of the recorded species of this genus are mere iterations.

Dispersed in some of the earlier volumes of "The Journal of Science and the Arts" will be found sixteen coloured representations taken from the living plants of as many rare and curious species of *Orchideæ* belonging to the Cape of Good Hope, all of them executed with great accuracy. The original drawings are in Mr. Brown's library; but the name of the meritorious Dutch artist, who drew them for Mr. Masson, the then King's collector, is unknown. *(See last page of fol. 700.).*

720

CALANTHE veratrifolia.

Hellebore-leaved Calanthe.

GYNANDRIA *MONOGYNIA.*

Nat. ord. ORCHIDEÆ. *Brown prod.* 1. 309.

Div. IV. Anth. terminalis mobilis decidua. *Massæ pollinis* demùm cereaceæ. *Brown in Hort. Kew. ed.* 2. 5. 205.

CALANTHE. Brown infrà vol. 7. fol. 573; *(in textû anglico.) Labellum* porrecto-explanatum latum, latus inferius columnæ ab apice continuans (appendicem simulans), porrectum, explanatum, 3-lobum, basi dentato-barbatum. *Columna* corollâ explanatâ omninò libera, duploque brevior. *Massæ pollinis octo.*

C. *veratrifolia,* foliis lanceolatis subplicato-nervosis: scapo radicali foliis breviore aphyllo, spica numerosa conferta oblonga cylindrica, bracteis parvis lanceolatis.

Limodorum veratrifolium. *Willd. sp. pl.* 4. 122. *Persoon syn.* 2. 520.

Folia *subbipedalia, latitudine 5-unciali, subplicato-nervosa, lanceolata, basin versùs petiolatìm attenuata, canaliculata, radicalia, plura, erecto-patentia, luteo-virentia, non glauca, subundulata.* Scapi *plures, radicales, aphylli, villosi, tripedales v. ultra, stricti, teretes, simplicissimi, vaginis remotis lanceolatis appressis sparsè vagantibus.* Spica *semi-ad-pedalem, numerosa, confertiuscula, horizontali-patens, recta, terminalis, cylindrica;* axi *viridi-albicante, sulcato-striata.* Flores *candidissimi, inodori, vulnerati plagæ loco ærugineo-virentes, biunciales vel longiores, porrecto-nutantes, extùs villosi opaci.* Germ. *sesquiunciale, album, clavato-elongatum sessile, porrecto-patens, infernè teres esulcatum, supernè recurvum, crassiùs striatum, seminiferum;* bracteæ *germine aliquotiès breviores lanceolatæ, gradatìm decrescentes, herbaceæ.* Cor. *radiato-explanata, ad basin usque distanter-distincta diametro unciali, nutans, subregularis:* petala 5, *æquidistantia obtusa, mucronata;* exteriora 3 *majora semuncialia,* medium summum *lateralibus parùm grandius, erectum, obovato-oblongum,* lateralia 2 *inferiora, labellum declinato-sequentia, trapezoidi-oblongata v. inæquilateri-oblonga;* interiora 2 *superiora summum medium approximantia, cæteris æquilonga inflexioneque simili, duplo tamen angustiora.* Calcar *tereti-elongatum, germine gracilius triploque brevius, descendens, corollâ parum longius, obtusum, rectum, columnæ basin directione pari continuans.* Labellum *basi et in fauce columnam continuante fulvo-barbatum, barbâ triseriatâ seriebus è dentibus confertis prominulis ordine distinctissimo atque patente in longum positis (linguæ felinæ spinulis non absimilibus), columnæ latus inferius porrecto-continuans, triploque longius, explanatum, ochroleuco-emarcescens, tripartitum, exungue, margini deflexum, corollâ duplo longius;* lobis 2 *inferioribus divaricatis, medio plurimum brevioribus oblongis obtusis integris,* terminali *recto furcato medium citra bifido (literam Y quodammodo referens) segmentis lineari-oblongis recurvis, ungue longioribus.* Columna *corollâ omninò libera duplo brevior concolorque obovato-turbinata obesa, summo dorso antherifera, porrecta, apicis lateribus alato-productis brevibus erectis,* rostello *medio longiori truncato-obtuso suprà stigma prominente, à supino scrobiculo longitudinali pro retinaculo antheræ impresso:* anthera *dorso summo columnæ incumbens paulò immersa, turbinata corollæ ferè concolora, bilocularis loculis scariosis subtus convoluto-conniventibus; loculus quisque* massulas pollineas 4 *includens;* massæ pollinis 8, *parallelæ, lineari-spathulatæ albidæ durissimæ.* Stigma *sub lobo medio elongato columnæ latens, bilobum, recurvato-prominens, vitreo-splendens, corrugatum sed lævissimum, atque nitidissimum lobis apice rotundatis, areolam secernentem subtus abscondens.*

Leaves radical, two feet high, yellowish green, plaitedly nerved, lanceolate, five inches over, channeled and petiolately tapered toward the base. *Scapes* several, 2-4 feet high, simple, upright, round, villous, green, opaque, with several close-pressed lanceolate short scattered sheaths. *Spike* terminal, numerous, close, upright, six inches to a foot long, cylindrical, horizontally spreading: *axis* whitish green, sulcate; *bractes* several times shorter than the germen, green, lanceolate, gradually smaller. *Flowers* villous outside, uniformly white, turning green where wounded, 1½-2 inches long. *Germen* 1½ inch long, more than an inch longer than the corolla, and one third longer than the spur. *Corolla* outspread, radiate, nearly regular, two inner segments narrowest. *Label* twice the length of the corolla, continuous with the lower side of the apex of the column, and resembling an appendage to that part, outspread, 3-parted, lower side-lobes divaricate, entire, shorter, middle furcately bifid like the letter Y, segments blunt, the whole turning to a cream-colour before it fades away. *Column* short, thick, turbinate: *anther* incumbent along the upper side of the summit of the column; bilocular: *pollen masses* 8, parallel, linear-spatulate (the shape of Harlequin's sword in miniature), compressed, white; *stigma* 2-lobed, shining, recurvedly prominent, hid beneath the prominent pistil.

The name of CALANTHE was proposed for this genus, and its separation from LIMODORUM and BLETIA, in a former article of this work, by Mr. Brown.

Drawn from a fine plant in Mr. Colvill's hothouse at Chelsea, where several stems, three feet and more in length, were thrown up from the same root.

727

ONCIDIUM luridum.

Mr. Griffin's Oncidium.

GYNANDRIA *MONANDRIA.*

Nat. ord. ORCHIDEÆ. *Brown prod.* 1. 309. *Div. IV.* Anthera terminalis mobilis decidua. Massæ pollinis demùm cereaceæ. *Brown in Hort. Kew. ed.* 2. 5. 205.

ONCIDIUM. Labellum explanatum lobatum basi tuberculatum. *Petala* patentia (2 antica nunc connata). *Columna* alata. *Massæ pollinis* 2, posticè bilobæ, medio affixæ processû communi stigmatis. *Brown in loc. cit.* 215.

O. *luridum,* foliis ellipticis acutis, scapo stricto ramoso, perianthii laciniis patentibus undulatis retusis subæqualibus, labello reniformi, columnæ alis rotundatis. *J. L.*

An unrecorded species; drawn from the collection of Mr. Griffin, at South Lambeth. Native of South America.

It is very nearly related to *Oncidium carthaginense,* from which it is nevertheless essentially distinguished by the different shape of its flowers, and by having its upper petal shaped like those next it, and not galeate and incumbent upon the column as in *O. carthaginense.* It also seems to differ from the latter species in not having the two lateral outer petals smaller than the corresponding inner petals, and in having a colourless, not red, columna. The *Oncidium olivaceum* of M. Kunth, which probably resembles this in some particulars, is obviously distinguished by its reflexed petals and trifid labellum. *J. L.*

ONCIDIUM is distinguished in its division of the order by " a flat-lobed label, knobby at the base; spreading *petals* (the two front ones of which are sometimes grown together); a winged *column;* a pair of *pollen-masses,* double-lobed behind, and fixed at the middle by a common process belonging to the *stigma.*"

The coloured leaf and uncoloured scape, in our figure, are much diminished: the coloured spike of flowers and uncoloured leaf are of the natural size.

732

DENDROBIUM squalens.

Dingy-coloured Dendrobium.

GYNANDRIA *MONANDRIA.*

Nat. ord. ORCHIDEÆ. *Jussieu gen.* 61. *Brown prod.* 1. 300. *Div. IV.* Anthera terminalis mobilis decidua. Massæ pollinis demùm cereaceæ. *Brown in Hort. Kew. ed.* 2. 5. 205.

DENDROBIUM. Labellum ecalcaratum, articulatum cum apice processûs unguiformis, cujus lateribus petala antica adnata, calcar emulantia. *Massæ pollinis* quatuor, parallelæ. *Brown in Hort. Kew. ed.* 2. 5. 212.

D. *squalens,* terrestre bulbis conicis truncatis, floribus resupinatis confertis, foliis lanceolatis plicatis sub-3-nervibus scapo duplo longioribus. *Lindley MSS.*

Herba *terrestris bulbis conicis nudis truncatis.* Folia *cuique bulbo* 2 *v.* 3 *suberecta rigida lanceolata* 1-½ *pedalia læto-viridia tri-plurinervia, plicata.* Scapus *rigidus erectus foliis duplo brevior basi vaginatus.* Spica *conferta* 8-10-*flora.* Bracteæ *ovarii longitudine ovatæ acuminatæ membranaceæ.* Flores *majusculi, pallidè rufescentes, recti (sec. autores resupinati), perianthii laciniis oblongis obtusis recurvis,* labello *carnoso canaliculato viridi lobo medio obtuso purpureo crassiore.* Columna *et ceteræ partes generis.* Lindley MSS.

This species of Dendrobium appears to be nearly related to the D. *longifolium* of M. Kunth, from which, however, it seems to be perfectly distinct. In the latter plant the scape is described as being twice as long as the leaves; but in the species before us the scape is almost hidden by the leaves, than which it is at least twice as short. The D. *longifolium* is moreover a native of Popayan; while the present plant was found growing in woods near Rio Janeiro, whence it was sent to England by Mr. John Forbes, a collector in the service of the Horticultural Society, in 1822.

The period of its flowering is May and June.

Whether the Dendrobia *undulatum et variegatum* of the authors of the Flora Peruviana are related to this it is scarcely possible to judge, with the very imperfect materials afforded by those botanists; but it is probable that the above-mentioned plants belong, on account of the other species among which they are placed, to some of the parasitical sections of the genus. *Lindley MSS.*

Grows on the ground, not on trees; *bulb* coated; conical. Two-three *leaves* on each bulb, which are nearly upright, stiff, lanceolate, a foot and a half high, lively green, plicate; with three or several nerves. *Scape* twice shorter than the leaves, stiff, upright, sheathed at the base. *Spike* thickly 8-10-flowered. *Bractes* the length of the germen, ovate, taper-pointed, membranous. *Flowers* longish, pale, brownish, straight, according to authors, reversed. *Segments* of the corolla oblong, obtuse, recurved. *Label* fleshy, channelled, green, middle lobe obtuse, purple, still thicker. The *column* and other parts as usual in the genus.

742

EULOPHIA gracilis.

Slender Eulophia.

GYNANDRIA *MONANDRIA.*

Nat. ord. Orchideæ. *Jussieu gen.* 61. *Brown prod.* 1. 300. *Div. IV.* Anthera terminalis mobilis decidua. Massæ pollinis demùm cereaceæ. *Brown in Hort. Kew.* 2. 5. 205.

EULOPHIA. Brown suprà vol. 8. *fol.* 686.

E. *gracilis,* scapo gracillimo, foliis lanceolatis trinerviis 3-plo longiore, calcare clavato, labelli lobo medio obsoleto. *Lindley MSS.*

Herba *terrestris.* Bulbi *conici vestigiis asperis foliorum vestiti.* Folia 3-5, *disticha, rigida, angustè lanceolata suberecta pedalia basi inflatâ bulbos vaginantia, nervis tribus primariis,* 4 *secundariis, illis subtùs prominentibus.* Scapus *teres, gracillimus multiflorus foliis triplo longior post anthesin bractearum vestigiis tuberculatus.* Flores *in spicâ sparsim dispositi, luteo-virides, intùs pallidiores, rubro paululùm suffusi* bracteis *lanceolatis acuminatis ovario duplo breviori: deciduis, suppositis.* Corollæ *petala patentia, ovata, acuta, interiora paulò minora.* Labellum *pendulum infundibuliforme,* calcare *apice clavato,* lobo *medio obsoleto fimbriato.* Columna *libera, anticè plana, margine subciliato, ovata, à fronte in rostellum productum desinens.* Gynizus *transversus excavatus.* Anthera *terminalis, opercularis, decidua, unilocularis, infrà apicularis, anticè rostro suo glandulam polliniorum tegens, posticè juxta cardinem elongata purpurea.* Pollinia 2, *dorso semibiloba, glandulæ filiformi affixa.* Lindley MSS.

Obviously distinguished from Eulophia *guineensis (vol.* 8. *fol.* 686) by its narrow rigid leaves and very long scapes, which remain in flower for many months. The middle lobe of the labellum is obsolete, and its place supplied by a few minute processes resembling a fringe.

The plant was sent from Sierra Leone, in 1822, by Mr. George Don, collector in the service of the Horticultural Society; and has been in flower in the garden of this Society at Chiswick for nearly all the summer through. *Lindley MSS.*

A ground species (not parasitic: not growing upon trees or other bodies). *Bulbs* conical, enveloped in the hardened rugged permanent bases of the leaves. *Leaves* 3-5, two opposite stiff, narrowly lanceolate, nearly upright, a foot high, covering the bulbs by the enlargement of their permanent remnants; primary nerves 3, secondary 4, the former varicosely permanent at the under surface of the foliage. *Scape* round, very slender, many-flowered, three times the length of the leaves: as the flowers drop knobbed with the remains of the bractes. *Flowers* disposed in a scattered spike, subtended by lanceolate taper-pointed deciduous *bractes,* twice shorter than the germen, of a yellowish green colour, paler on the inside and slightly suffused with red. *Petals* of the *corolla* spreading, ovate, pointed, inner a little smaller. *Label* pendulous, funnelform; *spur* clubbed at the top, the middle lobe nearly obliterated and its place supplied by a sort of fringe. *Column* detached, flat in front, slightly ciliate at the edge, ovate, ending at the front in an elongated rostel or beaklike appendage. Secreting surface of the *stigma (gynizus)* a transverse hollow. *Anther* terminal, lid-shaped, deciduous, of one cell, with a projecting apex beneath, covering by its beaklike point the gland of the *pollen-masses,* elongated at the back near the fringe or joint by which it is connected with the column, where it is also purplish. *Pollen-masses* 2, divided into two lobes for half their length at the back, held together by a filiform glandular corpuscle.

745

ISOCHILUS linearis.

Linear Isochilus.

GYNANDRIA *MONANDRIA.*

Nat. ord. Orchideæ. *Brown prod.* 1. 300. *Div. IV.* Anthera *terminalis mobilis decidua.* Massæ pollinis *demùm cereaceæ.* Brown in Hort. Kew. ed. 2. 5. 205.

ISOCHILUS. Labellum petalis distinctis conniventibus subconforme. *Massæ pollinis* quatuor, parallelæ. *Brown in Hort. Kew. ed.* 2. 5. 209.

I. *linearis,* spicâ terminali, foliis distichis linearibus obtusis emarginatis, caule simplici. *Brown loc. cit.* 209.

Cymbidium lineare. *Willd. sp. pl.* 4. 97.

Epidendrum lineare. *Jacq. amer.* 221. *t.* 131. *f.* 1.

A West Indian species, introduced by Mr. Elcock in 1791.—Drawn at the Bayswater collection belonging to Comtesse des Vandes.

A genus detached and defined by Mr. Brown, who distinguishes it from other genera of the same division, by a *label* nearly of similar shape and dimensions as the rest of the petals of the corolla (distinct from each other), and four parallel *pollen-masses.*

Two species are known in our gardens.

We had no opportunity of examining the flowers; and refer to the annexed figure.

759

PLEUROTHALLIS punctata.

Trowel-lipped Pleurothallis.

GYNANDRIA *MONANDRIA.*

Nat. ord. Orchideæ. *Jussieu gen.* 64. *Brown prod.* 1. 309. *Div. IV.* Anthera terminalis mobilis decidua. Massæ pollinis demum cereaceæ.

PLEUROTHALLIS. Labellum articulatìm connexum cum basi simplici vel brevissimè producta columnæ. *Petala* 2 antica exteriorum infernè connata. *Massæ pollinis* 2, exsulcæ. *Brown in Hort. Kew. ed.* 2. 5. 211.

P. *punctata*, foliis (folio?) oblongis utrinque obtusis scapo deflexo ramoso brevioribus, floribus sparsis distantibus.

Parasitica? radice *fibrosa*. Folia *(folium?)* 1-3, *oblonga, biuncialia v. circà, latitudine subsesquiunciali, glauco-opacata, utrinque obtusata:* petiolus *radicalis, brevis, robustus.* Scapus *folio longior, radicalis, declinatus, crassiusculus, virens, teres, subflexuosus, rigidus, nudus, ab ortû circà divisus, remanentibus bractearum basibus tandem tuberculosus.* Flores *spicati, numerosi, laxiùs sparsi, parvuli, viridi-pallentes, inodori.* Bracteæ *minutæ, ovato-acutatæ, germine aliquotiès breviores, basi carnosâ, atro-viridi, persistente.* Germ. *clavato-gracile, flexum, viride, petalis æquale.* Cor. *cruciato-4-petaloidea, bilabiata:* petala *distincta, æquilonga:* exteriorum *medium galeato-adscendens, concavum, ima* 2 *(lateralium inferiora) labello supposita, infernè connata, supernè soluta atque recurva:* interiora 2 *(lateralium superiora) punctis trinis? flavis, parallelis, notata, duplo angustiora, pallidiora, lineari-subfalcata, obtusula, incurvo-patentia, labello transversè opposita (v. labelli margines laterales ex facie interiori respicientia).* Labellum *albidum, trullæforme, porrectum, è basi carnosâ tumidiusculâ cum columnæ pede articulatum?, petalis distinctum et æquale:* unguis *strictus, glaber, angustus, à facie supinâ carinatus vel ab axi elevatâ utrorsum declivis:* lamina *concolor, latior, membranaceè extenuata, cuneato-ovata, recta, opaca, apice rotundata (rariùs èrosa v. truncata).* Columna *conica, erecta, libera, pallida, obtusa, petalis ferè æqualis.* Anthera *reversa, decidua, bilocularis, summæ secùs columnæ dorsum recumbens, spathulatè oblonga, convexa, albicans,* anticè *caudato-attenuata, cassa,* rostellum *cristato-compressum pruinosum virescens stigmatis obtegens*; posticè *subrotunda, pollinisque massis* 2 *fœta;* ligula pollinifera *alba, membranacea, horizontalis, circumscriptione thecæ antherinæ, massas pollinis à parte posticâ dilatatâ è nexû glutinoso elastico affigens,* glandulâ (proscollâ?) *fuscâ, parvâ, mobili, in apice:* massæ pollinis *binæ, parallelæ, ab invicèm spatio angusto semotæ, aureæ, cereaceæ, semiovato-orbiculares, compressæ, perpendiculares, integerrimæ.* Stigma *viridissimum, columnam totam anticè occupans, hujusque lateribus involuto alatis, solâ relictâ* areolâ nudâ secernente (gynizo) *rimæformi perpendiculari, velatum.*

It was only the other day a very general opinion, that the migration of pollen-masses in Orchideous plants, was accessary to the fertilization of the germen, and the mode by which those bodies attained a consummatory contact with the stigma. This opinion is now, we believe, consigned to the catalogue of vulgar errors. A nearer view and a more philosophical research have demonstrated the probability of a clandestine fertilization, previous to the opening of the flower and removal of the pollen-masses; at a period when these bodies are endued with a very different nature and substance from those they are found with at the migratory stage. A theory, which, as the result of more accurate investigation and sounder reasoning, may safely replace that which has been found groundless, at least until itself shall have been refuted in turn, or modified by future observation. The original notion, we suspect, had no better foundation than the fact of the inert and exhausted residua of the pollen (after projection by an elastic dehiscence of the anther) having been frequently observed to cling in masses of a determinate form to the neighbouring stigma, where they are retained (probably for the use of bees or of some other insects) by a viscid moisture secreted at this period from the surface of that organ.

The waxen or horny state in which the pollen-masses are found in a great proportion of this family, is never that of their pristine consistence, but a consistence induced after parting with their fertilizing principle, and indicatory of exhaustion. For this reason, when these concrements enter into the definitions of the secretions of *Orchideæ* by Mr. Brown, we find them designated "*demùm cereaceæ*" *(finally waxen)*. And in this state only, from their permanence, greater evidence, and easier accessibility, could they have been adopted for characteristic marks. Previous to that period, besides the continual change both in form and consistence which is more or less in progress, the precise but fugitive moment at which these bodies might be deemed perfect, that is, mature and still pregnant with the fertilizing principle, could not have been easily seized for practical discrimination, even if such point of their existence was held a truer and safer ground of distinction.

It is by the singular habit of the pollen that *Orchideæ* are distinguished among Monocotyledons, as the *Asclepiadeæ* are by one nearly similar among Dicotyledons.

In the Asclepiadeous family the anthers are five, all with a pollen concreting into a determinate number of masses, which, on issuing from the mould or case of the anther, attach themselves to certain adjoining processes of the stigma, where they are held, probably with the same intent as that which has been surmised in *Orchideæ;* and the recomposition they undergo is not improbably analogous.

The summary of the natural character of *Orchideæ* stands in the *Prodomus* of Mr. Brown as follows: "*Corolla* superior. *Filaments* and *style* united either at the base or along their whole length. *(Anther* one, seldom two): *pollen* resolving into concretions of determinate shape. *Capsule* one-celled, with three many-seeded receptacles attached to the axis of the valves. *Seeds* minute, with an *albumen,* but no *vitellus;*" (from the Latin.) DENDROBIUM *ruscifolium* and *racemiflorum* of Swartz were both included in this genus by Mr. Brown.

The drawing was taken at Mr. Colvill's Nursery, in the King's Road, from a plant sent from the Governor's garden at Trinidad by Mr. David Lockhart; and is the first of the species that has appeared in our hothouses.

NOTE.

Of the two plates belonging to No. 720, one represents the whole plant in miniature; the other, a part of the spike and leaf, of the natural size.

In the plate attached to No. 742, the figure with the coloured leaves is a diminished representation of the whole plant; that by the side of the spike, of flowers of the natural size.

760

PONTHIEVA petiolata.

Long-stalked Ponthieva.

GYNANDRIA *MONANDRIA.*

Nat. ord. ORCHIDEÆ. *Sect. II.* Anthera stigmati parallela persistens. Massæ pollinis v. farinaceæ v. e corpusculis angulatis; apicibus stigmati affixæ. *Brown in Hort. Kew. ed.* 2. 5. 197.

PONTHIEVA. Cor. irregularis. *Labellum* posticum cum petalis interioribus columnæ insertum. *Pollen* farinaceum. *Brown l.c.* 5. 199.

P. *petiolata*, spica laxa erecta, foliis petiolatis erectis crispis glabris, floribus discoloribus.

Folia *radicalia, erecta, longè petiolata, elliptica, acuta, margine crispa,* 5-7 *nervia, glaberrima: caulina acuta, vaginantia, pilosa.* Scapus *erectus, teres, pilosus, foliis duplò longior.* Bracteæ *ovario breviores, obtusæ.* Flores *in spica laxa dispositi, patentes, recti, (non resupinati,) discolores.* Petala *glabra,* 3 inferiora *ovata, acuta, galeata, cohærentia, interioribus Platantheræ modo obliquis, unguiculatis, basi hinc cordatis, columnæ dorso acretis,* 2 superiora *oblonga, fortiter* 5*-nervia, alæformia, reflexa, marginibus revolutis.* Labellum *rufo-brunneum, ovatum, cymbiforme, glabrum, supra columnam incumbens, petalis duplò brevior,* ungue *luteo, carnoso, utrinque marginato, e basi columnæ continuo.* Columna *brevissima, obesa:* gynizo *cavo, lunato;* rostello *subulato, apice uniglanduloso.* Anthera *postica, linearis, acuta, in clinandrio quoad axin columnæ ferè verticali supra gynizum incumbente apposita; obsoletè* 4*-locularis, valvarum exteriorum marginibus scariosis.* Pollinia 4, *pulverea, geminata, in clinandrio ope glandulæ rostelli retenta.* Ovarium *rectum, glandulosum.*

The genus PONTHIEVA was established by Mr. Brown, in the second edition of the Hortus Kewensis, upon the *Neottia glandulosa* of the Botanical Magazine; a singular plant, which we fear is now lost to our gardens. With NEOTTIA it has little in common, besides the peculiar characters of the section of the order in which it is included.

For the new species which we are now enabled to add to the genus, we are indebted to the Horticultural Society, in whose garden at Chiswick our drawing was made in August last. The plant had been recently brought to that establishment, with many other interesting plants, from the island of St. Vincent, by Mr. James McRae; and we understand it is expected that among them is at least one other species of the same genus.

In foliage this plant is so very similar to the NEOTTIA *adnata* of Swartz, that, before it produced its inflorescence, we supposed it would prove at least of the same genus. That, however, is a NEOTTIA with a spur, and therefore wholly distinct from PONTHIEVA. It constitutes an undescribed genus (COLLĒA *Nob.)*, to which we believe the N. *calcarata* of Swartz may also be referred.

Leaves radical, erect, with long stalks, elliptical, acute, curled at the edges, with 5 or 7 nerves, quite smooth; the leaves of the stem acute, sheathing, hairy. *Scape* erect, round, hairy, twice as long as the leaves. *Bracteæ* shorter than the ovaria, obtuse. *Flowers* disposed in a loose spike, spreading, not twisted, of a brown colour. *Petals* smooth, the 3 *lower* ones ovate, acute, cohering into a sort of hood, the two interior being oblique like those of *Platanthera*, stalked, cordate on one side at the base, and fastened to the back of the column; the 2 *upper* ones oblong, strongly 5-nerved, wing-shaped, reflexed, with the edges rolled back. *Labellum* brownish red, ovate, boat-shaped, smooth, lying upon the column, twice as short as the petals: its *stalk* yellow, fleshy, edged on each side, and continuous with the base of the column. *Column* very short, thick; *gynizus* hollow, lunate; *rostellum* subulate, with a gland at the tip. *Anther* posterior, narrow, pointed, lying upon the *clinandrium*, which with respect to the column is vertical, and lies immediately upon the gynizus; obscurely 4-celled, with the outer valves of the cells scariose at the edge. *Pollen-masses* 4, powdery, in pairs, retained in the clinandrium by means of the gland on the point of the rostellum. *Ovarium* straight, glandular.

J. L.

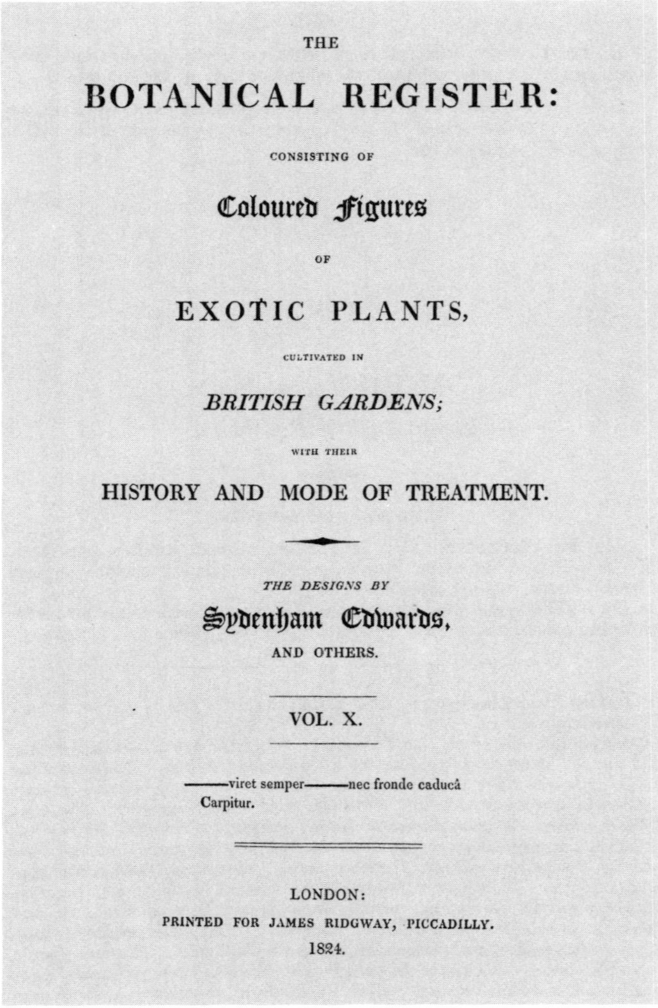
THE

BOTANICAL REGISTER:

CONSISTING OF

Coloured Figures

OF

EXOTIC PLANTS,

CULTIVATED IN

BRITISH GARDENS;

WITH THEIR

HISTORY AND MODE OF TREATMENT.

THE DESIGNS BY

Sydenham Edwards,

AND OTHERS.

VOL. X.

——viret semper——nec fronde caducâ
Carpitur.

LONDON:
PRINTED FOR JAMES RIDGWAY, PICCADILLY.
1824.

783

EPIDENDRUM cuspidatum.

Yellow fringed-flowered Epidendrum.

GYNANDRIA *MONANDRIA.*

Nat. ord. ORCHIDEÆ. *Jussieu gen.* 61. *Brown prod.* 1. 300. *Div. IV.* Anthera terminalis mobilis decidua. Massæ pollinis demùm cereaceæ. *Brown in Hort. Kew. ed.* 2. 5. 205.
EPIDENDRUM. Suprà vol. 1. *fol.* 67.

E. *cuspidatum*, foliis sæpè trinis, spica remota pauciflora; limbi laciniis duabus interioribus latioribus subrhombeo-lanceolatis cuspidatis; labelli lamina triloba, lobis lateralibus fimbriato-incisis, medio lineari integerrimo.

Epidendrum cuspidatum. *Loddiges's bot. cab.* 10.
Epidendrum ciliare. *Curtis's magaz.* 463; *(nec Linnæi.)*

Scapus *erectiusculus, simplex tereti-compressus, virens, cicatricibus remotis annulatus infernè versus attenuatus, vaginis compressis alternis sphacelatis densè parallelèque nervosis ad flores usque vestitus.* Vaginæ *ex imis longioribus(biuncialibus) supernè versus descrescentes, lineari-lanceolatæ, rectissimæ, bifariam alternæ.* Folium (2-3?) *coriaceum, firmum, erectiusculum, scapo subæquale, supra virens lucidum, subtus opacum pallidum, 4-unciale latitudine* $\frac{2}{3}$ *vel* $\frac{3}{4}$ *partes unciæ, costâ media prominente, apice obtuso mucronulato.* Spica *remotæ pauci-(3-) flora, axi geniculato-flexuosâ:* bracteæ *geminatæ, arido-membranaceæ tenues, germine breviores, lanceolato-lineares, cuspidatæ.* Germen *viride biunciale vel magis, obsoletè trigonum, crassitudine pennæ minoris, esulcum, glabrum, erectiusculum, corollæ lacinias exsuperans.* Corollæ *laciniæ exteriores ligulato-lanceolatæ elongatæ cuspidatæ striatæ, interioribus duabus tenuioribus duplo fermè angustiores.* Columna *alba erecta clavato-cylindracea, diametro ferè pennæ columbinæ, subtrigono-teres, apice anticè protuberans cum cavo infundibuliformi labellifero, posticè antherifera membranula alba laciniata marginata.* Labellum *columnam anticè cuniculato-decurrens, limbi laciniis æquale:* unguis *brevis, latus truncato-lobatus, supernè patens utrinque infernè canaliculato-connivens:* lamina *tripartita,* lobis lateralibus *deflexis, semiovato-rotundis, orâ externâ laciniato-fimbriatâ lacinulis angustissimis profundissimis;* medio *istis æquali integerrimo subulato-lineari, apice non dilatato, lateribus deflexis, filo longo præfixo.* Anthera *convexa.* Massæ pollinis *flavæ complanatæ inæquilateri-conoideæ, filis binis supra furcatis longioribus per paria connexa.* Stigma *anticum, intrà frontem capitelli columnæ reconditum.*

Received by Messrs. Loddiges, of the Hackney Nursery, in 1808, from the island of Dominica.

Drawn at the Chelsea Nursery, where it is cultivated by Mr. Colvill in the hothouse.

In Curtis's Botanical Magazine the plant has been, perhaps too hastily, confounded with EPIDENDRUM *ciliare* of Linnæus; from which it is distinguished in the following article of the Botanical Cabinet, and, as it seems, with some reason.

" Differs from *ciliare* in often having three leaves; " whereas there are never more than two in that. The pe- " tals also are very differently formed in *ciliare;* these be- " ing all equal, slender, and much curved. In the present " species the three exterior petals are equal, and remarkably " stiff and pointed; the two interior very broad, spreading " in the form of wings: this disposition of the petals pro- " duces a very remarkable appearance. The middle seg- " ment in the lip of the nectary in this sort is quite linear; " in *ciliare* it is subulate and much longer. The whole " plant is of much larger growth." *Loddiges's bot. cab.* 1. *n.* 10.

The flower in *cuspidatum* is yellow; in *ciliare*, greenish white.

We had no opportunity of comparing the two plants while in blossom, having only received a sample of the one before us. As far as we can judge, they seem to belong to distinct types; and Mr. Brown appeared to incline to the same opinion.

784

EPIDENDRUM ciliare.

Fringed Epidendrum.

GYNANDRIA *MONANDRIA.*

Nat. ord. ORCHIDEÆ. *Jussieu gen.* 61. *Brown prod.* 1. 300. *Div. IV.* Anthera terminalis mobilis decidua. Massæ pollinis demum cereaceæ. *Brown in Hort. Kew. ed.* 2. 5. 205.

EPIDENDRUM. Suprà vol. 1. *fol.* 67.

E. *ciliare*, foliis constanter binis, spica remota pauciflorâ, labelli lamina triloba lobis duobus exterioribus fimbriato-incisis, medio subtili aristiformi setaceo integerrimo: limbi laciniis duabus interioribus lineari-lanceolatis.

Epidendrum ciliare. *Linn. sp. pl. ed.* 2. 2. 1349. *Jacq. amer.* 224. 179. *fig.* 89. *Swartz nov. act. ups.* 6. 69. *Willd. sp. pl.* 4. 119. *Hort. Kew. ed.* 2. 5. 219; *(exclusâ citatione bot. magaz.* 463. *rectiùs ad E.* cuspidatum *pertinente). Redouté liliac.* 82.

Helleborine graminea, foliis rigidis carinatis. *Plum. sp.* 9. *ic.* 179. *fig.* 2.

Native of the West Indies. Introduced in 1790 by Mr. Edward Elcock. Drawn at the Chelsea Nursery, where it is cultivated by Mr. Sweet, the superintendant in the hothouse.

Stems several, cylindrical, knobby, a little thicker upwards than downwards, owing to the remains of the by-gone foliage; smooth like the rest of the plant. *Leaves* cauline, alternate, sheathing at the base, spreading oblong, obtuse, firm, deep green. *Flowers* from two to five, large, white, in a thin loose spike: *axis* or *rachis* cylindrical, green, smooth, branching: *bractes* oblong, half embracing, becoming sphacelate, placed at the base of the spike and the branches. *Corolla* sixparted: 3 *outer segments* exactly alike, oblong, narrow, pointed, slightly revolute: 2 *inner ones* resembling the three outer, but whiter, a little sharper pointed, slightly waved and slanted: *Labellum* adhering by its base to the column, spurless; *unguis* involute, semicylindrically channelled; *limb* spreading, deeply three-parted, middle division long, linear, fine, simple, pointed, two lateral divisions a little shorter, much wider, entire at the inner edge, deeply fringed on the outer by the fine linear pointed slips into which it is cut. *Germen* running into the pedicle: *column* upright, semicylindrical, connected in front with the labellum, and terminated at the summit by a hollow containing the *stamen:* in front, beneath this hollow, the *stigma* is borne: the *anther* is contained within an orbicular lidshaped case or shell; is of an hemispherical form, with two roundish cells.

The plant has been ascertained, by an authentic sample, to belong to the EPIDENDRUM *ciliare* of Linnæus, with which *cuspidatum* of the preceding article has been pretty generally confounded.

788

CYPRIPEDIUM venustum.

Spotted-leaved Ladies' Slipper.

GYNANDRIA *DIANDRIA.*

Nat. ord. ORCHIDEÆ. *Jussieu gen.* 64. *Brown prod.* 1. 300. *Div.* Diandria.

CYPRIPEDIUM. Labellum ventricosum, inflatum (nunc saccatum). *Columna* posticè terminata lobo petaloideo (stamine sterili), antheras distinguente. *Petala dua antica* sæpiùs connata. *Brown in Hort. Kew. ed.* 2. 5. 220.

C. *venustum*, foliis distachyo-patentibus, equitantibus, lanceolato-oblongis, acute canaliculatis enerviis, supra triste glaucescentibus, maculis obscuris informibus laxè notatis, subtus purpureo-punctatis, scapo villoso.

Cypripedium venustum. *Wallich MSS. Curtis's magaz.* 2129.

Acaulis. Folia firmiuscula, radicalia, plura ad plurima, alternè et imbricatè bifaria, basi equitantia supernè recurvo-patentia, oblonga, acutata, altè et acutissimè carinata, 4-5-uncialia latitudine subunciali, ex glauco tristè virentia, opaca, papillulis atomoideis veluti in Araneæ epidermidi densè pruinata, cartilagine subtili marginata, supra maculis saturatioribus subliteratis laxè sparsa, avenia, enervia, subtùs glabriora pallidiora et punctis purpureis rotundis varie approximatis vel passìm confluentibus et modò in lineas catenatis sparsa. Scapus *rigidus foliis parum altior, erectus, teres, marmorato-maculosus, villosus, pennâ corvinâ crassior.* Germen *sessile oblongum villosum spatha brevius, costis rotundatis,* 3 *alternis minoribus.* Corollam *recentem completam nequimus habere; sed ex tabula* CYPRIPEDII insignis *pluribus videtur diversam.*

Our plant is a recent acquisition from the East Indies, and the first that has accrued to the genus from that country. Independent of novelty and beauty, the paramount importance of conformity in the structure of those organs which have been adopted for generic standard, receives from this species an unprecedented illustration, even when contending with extreme anomaly of habit. Who of us, if he could have been apprized of this difference alone, that would not have prognosticated such consequent distinction in those parts, termed from their influence essential, that would at least have justified the removal of plants of so heterogeneous an appearance into separate groups? But he will find, upon inspection, that the present species can, with no more propriety, be kept from a place in CYPRIPEDIUM than those which rank in that genus by almost indefinable gradation.

The drawing was taken in the hothouse of Mr. Colvill's Nursery at Chelsea; where the plant grows in company with many others of the tribe from tropical countries, in a column formed of the rinds of cocoa-nuts, intended by Mr. Sweet, who superintends that extensive establishment, as a substitute for the decaying trunks of the forest, on which they are usually found in their native places.

Leaves radical, of a rather firm consistence, but not thick; several or many, alternately and imbricately two-ranked, equidistant (or alternately and mutually interclasping), recurvedly spreading, oblong, pointed, sharply and deeply keeled, 4-5 inches long by about one broad, apparently without veins or nerves, with a fine cartilaginous margin of a dull glaucous and opaque green, with the epidermis finely shagreened like the coat of a spider, loosely spotted on the upper side with darker shapeless marks that remind us of the letters of some oriental alphabet, freckled at the paler under side with deep purple dots, sometimes confluent and sometimes linearly catenated. *Stem* a little taller than the leaves, marbled with purple stains, rigid, upright, round, and somewhat roughly furred. *Germen* oblong, rather more than half an inch in length, furred, six-furrowed, the ridges alternately larger. We could not obtain the fresh perfect corolla for inspection; but, judging from the figures, we think it must differ, in several material points, from the *insigne* subsequently introduced from the same country.

The genus is the only one of its natural order that comes into the Order DIANDRIA, of the Class GYNANDRIA, which comprizes all the rest in the Order MONANDRIA. It is also remarkable as the only group of the natural order where it belongs, that has a middle sterile filament with two fertile side ones. In all others the middle filament bears the only anther present; while those at the sides are obliterated in the view of certain Botanists, and their existence questionable.

794

NEOTTIA bicolor.

Two-coloured Neottia.

GYNANDRIA *MONANDRIA.*

Nat. ord. ORCHIDEÆ. *Jussieu gen.* 64. *Brown prod.* 1. 309. *Div. II.* Anthera stigmati parallela persistens. *Massæ pollinis* vel farinaceæ vel è corpusculis angulatis, apicibus stigmati affixæ. *Brown in Hort. Kew. ed.* 2. 5. 198.

NEOTTIA. Suprà vol. 7. *fol.* 602.

N. *bicolor*, foliis plurimis, lanceolatis nervosis petiolatis glabris, scapo villoso infra foliolis spathaceis obsito brevioribus; racemo numeroso floribus cernuis, labello oblongo, lamina summâ brevi oblatâ obsoletè trifida undulatâ crenatâ, intùs minutè papillosâ.

Introduced, by the Horticultural Society, from the Island of Trinidad.

The plant was brought home by Mr. George Don, one of the Society's collectors, on his return from his mission in 1823.

817

AERANTHES grandiflora.

Large-flowered Aeranthes.

GYNANDRIA *MONANDRIA.*

Nat. ord. ORCHIDEÆ.—*Sect. IV.* Epidendreæ.

AERANTHES. Labellum calcaratum, membranaceum, integrum, cum processu unguiformi *columnæ* cui laciniæ anticæ adnascuntur articulatum. *Perianthium* ringens. *Pollinia* 2, cava, hinc perforata, filo nullo! glandulâ duplice. Herbæ *parasiticæ, perennes, radicibus filiformibus.* Folia *disticha.* Flores *maximi, solitarii, pallidi.* Sepala *cornuta, æqualia.*

A. *grandiflora*, foliis apice bilobis valde inæqualibus scapo radicali vaginato debili brevioribus, calcare emarginato.

Herba *parasitica, non bulbosa.* Radices *fibrosæ, glabræ.* Folia *disticha, ligulata, undulata, apice valde inæqualiter biloba, glaucescentia.* Scapus *radicalis, debilis, foliis longior, vaginis aridis, duris, lineatis vestitus, serie continuatâ florifer.* Flos *solitarius, magnus, inodorus.* Perianthium *ringens, pallide flavum.* Sepala *æqualia, carnosa, lucida, glaberrima, semidiaphana, ad basin pallide viridia, versus apicem flava, patentia, canaliculata, acuminata, quasi cornuta;* exteriora *ovato-lanceolata, lateralibus ad basin connatis antice usque ad articulum labelli productis, rotundatis, cum ungue columnæ subtus connatis;* interiora *breviora, cordata, abruptè acuminata, marginibus posticis conniventibus, anticis cum columnâ connatis.* Labellum *album, liberum, lanceolatum, integerrimum, ad basin subrepandum, cum calcare et pede columnæ articulatum;* calcar *e processu unguiformi columnæ productum, breve, incurvum, ore abruptè constricto, basi inflato, depresso, emarginato, viridi, intus piloso.* Columna *brevis, teres, in pedem membranaceum, alatum, cum laciniis interioribus ad marginem, anticis subtùs connatum, demum calcaratum producta, antice utrinque alata, alis semiovatis, conniventibus, supra gynizum horizontaliter porrectis.* Gynizus *subrotundus, excavatus, infundibularis.* Anthera *terminalis, opercularis, bilocularis, decidua.* Pollinia 2, *rotunda, cava, hinc perforata, filo nullo, glandulis duabus.*

Roots of this singular plant were sent, in a living state, from St. Mary's, Madagascar, by the late Mr. Forbes, in 1823, to the Horticultural Society; in whose garden our drawing was taken in June last. It is there successfully cultivated in decomposed woody soil, in the same manner as other plants of the same family.

So little has hitherto been done in describing the Orchideous plants of tropical countries, that new forms are continually presenting themselves, and requiring the establishment of new genera for their reception. The distinctions upon which these are to be founded, however minute they may occasionally appear, are singularly permanent and decisive. Thus the genus now proposed, which, in many respects, is allied to *Aerides*, of which A. *odoratum* must be considered the type, has a decidedly different habit, and, to a certain extent, geographical distribution, which are accompanied by peculiar modifications of the parts of fructification. In *Aerides* the spur is an integral portion of the Labellum; in *Aeranthes*, of the Columna! *Aerides* has the solid pollen-masses attached to a long elastic process, as in *Vanda;* while *Aeranthes* has its pollen-masses of the same number, but hollow, with a small perforation on one side, and no process whatever, at least not when the flower is fully expanded.

Among the curious Orchideous plants, for a knowledge of which the world is indebted to the perseverance of M. du Petit Thouars, are two plants, native of the isles of France, Madagascar, and Bourbon, which are referable to this genus. The first resembles A. *grandiflora* strongly in habit, but is essentially distinct. It may be called,

A. *arachnitis*, foliis apice emarginatis scapo radicali suberecto longioribus.
Dendrobium Arachnitis. Pet. Thouars hist. orch. afr. t. 88.

The second is a plant with large white flowers, and native of Madagascar. It may be called,

A. *sesquipedalis*, cornu longissimo filiformi, spicis vaginatis axillaribus.
Angræcum sesquipedale. Pet. Th. l. c. tt. 66 et 67.

The following is a list of the other genera of that section of *Epidendreæ* to which *Aerides* and *Aeranthes* belong:

EPIDENDREÆ. Sect. IV. *Calcaratæ, vel saccatæ; polliniis duobus.*

VANDA. *Brown in bot. reg. f.* 506.

SARCANTHUS. *Nob.* Labellum carnosum, integrum, saccatum, sepalis patentibus, regularibus, subconforme. (*Cymbidium præmorsum.* Willd.)

AERIDES. *Swartz in Schrad. diar.* 1799. *p.* 233. *t.* 2. *f.* 4.

AERANTHES. *Nob. in pag. opp.*

IONOPSIS. *Humb. et Bonpl. nov. gen. et sp.* 1. *t.* 83.
We were formerly, judging from the figure in the work just cited, disposed to think with Meyer, that this genus was referable to *Oncidium*. We now believe, however, that the original species must be placed with *Dendrobium utriculari-oides* and *testiculatum* of Swartz, which certainly belong to this section.

EULOPHIA. *Brown in bot. reg. t.* 686.

ANGRÆCUM. *Pet. Thouars hist. orch. afr. passim.*

ÆONIA. *Nob.* Labellum membranaceum, cucullatum, 3-lobum, horizontale. Perianthium explanatum, liberum, secundum. Pollinia duo postice 2-loba, glandulis propriis, filo nullo.—Sp. Æonia Auberti *Nob. Epidendrum volucre* Pet. Th. l. c. t. 81.

CRYTOPUS. *Nob.* Labellum membranaceum horizontale, liberum. Sepala heteromorpha, lobata, unguiculata, explanata, basi distantia. Pollinia 2, integra, filis et glandulis propriis in bursis totidem clinandrii occultis!—Sp. Cryptopus elata *Nob. Angræcum elatum* Pet. Th. l. c. tt. 79, 80.

J. L.

823

SPIRANTHES cernua.

Nodding Lady's Traces.

GYNANDRIA *MONANDRIA.*

Nat. ord. ORCHIDEÆ. *Sect. II. Br.*

SPIRANTHES Rich. Sepala in cylindro conniventia. *Labellum* porrectum, planum, basi bicallosum, marginibus columnam cum eo parallelam arctè amplectentibus.—*Herbæ graciles. Folia angusta, radicalia. Spicæ densæ, tortiles.*

S. *cernua*, foliis lanceolatis trinervibus, caule vaginato pubescente, floribus recurvato-cernuis, labello oblongo integerrimo acuto: callis depressis.
Ophrys cernua. *Linn. sp. pl.* 1340.
Neottia cernua. *Willd. sp. pl.* 4. 75. *Bot. mag.* 1568. *Sweet Flower Gard.* 42. *Pers. syn.* 2. 511.

Radix *fibris plurimis, crassis.* Folia radicalia *linearia, longa,* caulina *vaginantia, brevissima.* Spica *densa, oblonga.* Flores *arcte recurvati, cernui* (Willd. sp. l. c.). *Sepala superiora parallela, pubescentia, non undulata, labello lanceolato, obtuso, concavo, integerrimo, venoso, sessili, callis baseos depressis.* Columna *brevissima.* Gynizus *magnus, altè bifidus; laciniis longitudine rostri.*

This kind of Spiranthes has been long known in our gardens, although it is, like other North American Orchideæ, impatient of cultivation. The specimen from which our drawing was taken was communicated by Robert Barclay, Esq. from his fine collection at Bury Hill, near Dorking.

As it has been repeatedly decided, that the generic appellation Neottia belongs to the *Ophrys Nidus Avis* of older writers, there can be no reason why the name *Spiranthes*, applied by Richard to this set of plants, should not be retained; the *Gyrostachys* of Persoon never having, as far as we know, been employed by any author. *Spiranthes* can only be confounded with the *Stenorhynchus* of Richard, which, although not distinguished from it by any very decisive character, has a peculiarity of habit which renders it desirable that it should, if possible, be retained distinct. *Pelexia* of Poiteau, which we formerly (fol. 760), without being aware of that name having been published, called *Collea*, is more distinct in character, but less in habit.

The following are the species of Spiranthes at present known; of which M. Richard mentions only six:

1. S. autumnalis. *Rich.* (v. v. sp.)
2. S. æstivalis. *Rich.* (v. v. sp.)
3. S. australis. *Nob.* (v. s. sp.) *Neottia australis.* Brown.
4. S. pudica. *Lindl.* (v. v. c. et s. sp.) This is quite distinct from S. *australis*, with which it has been confounded in fol. 602 of this work.

We are well acquainted with both species. Of S. *pudica* we have specimens from Davuria, collected by Adams.

5. S. flexuosa. *Nob.* (v. s. sp.) *Neottia flexuosa.* Smith in Rees.
6. S. parviflora. *Nob.* (v. s. sp.) *Neottia parviflora.* Smith in Rees.
7. S. africana. *Nob.* foliis ligulatis oblongis radicalibus humifusis uninervibus, scapo vaginato glabro, spica secunda labelloque 3-lobo acuto plano saccato glaberrimis.—Hab. in insulis Africæ australis. Flores virides. (v. fig.) *Satyrium spirale.* Pet. Th. hist. orch. afr. t. 9.
8. S. congesta. *Nob.* foliis lanceolatis suberectis, scapo glabro vaginis obtusis, spica oblonga congesta subspirali puberula, sepalis obtusis, labello integerrimo.—Hab. in Sibiria. Flores rosei? (v. s. sp. comm. cel. Prescot.)
9. S. tortilis. *Rich.* (v. s. sp.) *Neottia tortilis.* Willd.
10. S. cernua. *Rich.* (v. v. c.)
11. S. elata. *Rich.* (v. v. c.) *Neottia elata.* Bot. mag. 2026.
12. S. picta. *Nob.* (v. v. c.) *Neottia picta.* Bot. mag. 1562.
13. S. bicolor. *Nob.* (v. v. c.) *Neottia bicolor.* Suprà fol. 794; where the leaves are inaccurately represented as whole-coloured. They are variegated with lighter and darker green, like those of S. *picta.*
14. S. diuretica. *Nob.* *Neottia diuretica.* Willd. *Spiranthes Nuil.* Rich.
15. S. quadridentata. *Nob.* *Neottia 4-dentata.* Willd.
16. S. strateumatica. *Nob.* *Orchis strateumatica.* Willd.

Many of the above are nearly related to each other in characters; but they are all, in our judgment, truly distinct.

J. L.

825

ISOCHILUS prolifer.

Proliferous Isochilus.

GYNANDRIA *MONANDRIA.*

Nat. ord. ORCHIDEÆ. Div. IV. *Epidendreæ.* Sect. 2. Ecalcaratæ v. ad maximum saccatæ. Pollinia 4.
ISOCHILUS. Br. V. suprà vol. 9. fol. 745.

I. *prolifer*, floribus axillaribus, foliis distichis lanceolato-oblongis, caule prolifero; bulbis axillaribus diphyllis. *Br. in Hort. Kew. ed.* 2. 5. 209.
Epidendrum proliferum. *Swartz prodr.* 124.
Cymbidium proliferum. *Swartz nov. act. ups.* 6. *p.* 71. *Willd. sp. pl.* 4. 95. *Pers. syn.* 2. 514.

Herba *humilis, parasitica.* Caules *proliferi, teretes, striati, glabri, basi vaginati, radicantes.* Folia *solitaria, lineuria, emarginata, enervia, patentia, in caulibus junioribus disticha, sæpius* 4. Flores 2-3, *axillares, albi, sessiles, basi bracteis* 6, *imbricatis, sphacelatis, acutis involucrati.* Perianthium *resupinatum, regulare, connivens, basi obliquum.* Sepala *libera, parallela,* exteriora *oblonga, obtusa, lateralibus labello suppositis, basi antice connatis,* interiora *angustiora.* Labellum *sepalis concolor, lineare, apice cordatum, planum, emarginatum, disci apice discolore, ungue columnæ parallelo carnosiore, apice transversim calloso.* Columna *teres, glabra.* Gynizus *infundibularis, apice obliquus.* Anthera *minima, terminalis, opercularis, bilocularis?, margine exteriore membranaceo.* Pollinia 4, *compressa, parallela, (filo tenuissimo in polliniis replicato?)*

This is a very rare plant. It appears to have been introduced in 1793, but was probably soon lost again. Within a few months it has been imported from the West Indies, and exists for the present in two or three collections about London. Our drawing was made at Mr. Colvill's.

The plant is preserved, with great difficulty, by being planted in decayed wood and peaty soil. It is found in various islands of the West Indies.

The genus Isochilus is one of those published by Mr. Brown, in his celebrated arrangement of Orchideæ, in the 2d edition of the Hortus Kewensis: one species has been already figured at p. 745 of this work, and we shall soon take an opportunity of giving a figure of a third species. It belongs to a section of Epidendreæ, which we characterize by having four pollen-masses, and a perianthium either entirely destitute of a spur, or, at most, with nothing more than a pouch. To the same section belong ***Dendrobium, Stenoglossum, Restrephia, Ornithidium, Telipogon, Broughtonia, Cattleya, Epidendrum, Polystachya,*** and ***Cryptarrhena.***

The other Orchideous genera with four cereaceous pollen masses, of which *Malaxis* is the representative, must be referred to a section of the Order distinct from Epidendreæ, which we shall take another opportunity of defining. This section, which may be called MALAXIDEÆ, is formed of ***Prescotia, Microstylis*** badly defined by Nuttall, ***Malaxis, Liparis, Dienia, Empusa,*** and, as a separate tribe, of ***Pedilea*** and ***Stelis.***

Empusa is a most remarkable plant, which can scarcely be compared with any known genus. We are acquainted with only a single species, for specimens of which from Nepal we are indebted to Mr. Lambert. Its character will be,

EMPUSA. *Perianthium* sepalis exterioribus lateralibus erectis obliquis basi labelli adnatis, cæteris linearibus patentibus. *Labellum* posticum, bituberculatum, sagittatum, facie cum basi columnæ elongatæ apice alatæ connatâ. *Pollinia* 4 per paria cohærentia.

Sp. 1. Empusa paradoxa.
Herba terrestris, pedalis, radice bulboso. *Folia* lineari-lanceolata, vaginantia, plicata. *Scapus* foliis longior. *Racemus* laxus, floribus patentibus, bracteis ovario bis torto! paulò brevioribus.
Hab. in Nepalia *(v. s. comm. cel. Lambert).*

The genus ***Dienia*** is founded upon plants from Nepal with the aspect of ***Malaxis,*** to which genus it is probable that some of them have already been referred by Sir Jas. Smith, in Rees's Cyclopædia. It may be characterized thus:

DIENIA. *Perianthium* explanatum, sepalis linearibus liberis. *Labellum* cucullatum, excavatum, cum columna parallelum. *Columna* clavata, apice alata. *Pollinia* 4, semiteretia, per paria cohærentia.

Sp. 1. Dienia congesta.
Folia caulescentia, ovato-lanceolata, *petiolis* vaginantibus. *Scapus* foliis longior. *Spica* terminalis, cylindracea, densissima. *Flores* minutissimi, sepalis interioribus filiformibus.
Hab. in Nepalia *(v. s. sp. comm. cel. Lambert).*

Prescotia we have referred to Epidendreæ, chiefly on account of its habit. In the nature of its pollen masses it approaches Mr. Brown's third section of Orchideæ, as it also does in the structure of its column; but it is in both these respects of ambiguous character.

J. L.

832

BRASSIA caudata.

Long-tailed Brassia.

GYNANDRIA *MONANDRIA.*

Nat. ord. ORCHIDEÆ.—*Sect. IV.* Epidendreæ.
BRASSIA. R. Br. Labellum explanatum indivisum. *Petala* patula distincta. *Columna* aptera. *Massa pollinis* 2, posticè bilobæ; medio affixæ processui communi stigmatis. *R. Br. in Ait. Kew. ed.* 2. 5. 215.

B. *caudata*, sepalis ovato-lanceolatis acuminatis: inferioribus caudatis, labello acuminato, bulbo ancipite.
Helleborine ramosissima, caulibus et floribus maculosis. *Plum. spec.* 9. *ic.* 277.
Epidendrum caudatum. *Linn. sp. pl.* 1349.
Malaxis caudata. *Willd. sp. pl.* 4. 93. *Pers. syn.* 2. 514.

Folia *bina, oblonga, erecta, 5-nervia, in bulbo ancipite insidentia.* Scapus *radicalis, vaginatus, foliis longior.* Spica *rigida, erecta, glabra, 9-flora.* Flores *resupinati, magni, subsecundi; bracteis ovatis, acutis, brevibus.* Perianthium *explanatum, patens, regulare, luteo-viride, fusco purpureo intus maculatum;* sepala *libera, carnosa, basi vix imbricata,* exteriora *ovato-lanceolata supremo erecto acuminato, inferioribus longissimis dependentibus caudatis* (4½ *uncialibus);* interiora *ovato-lanceolata, acuminata, supremo duplo breviora.* Labellum *dependens, liberum, longitudine, sepalorum interiorum, ovato-lanceolatum, acuminatum, margine membranaceum, ad apicem viridi-luteum, infra medium fusco maculatum, basi paulo excavatum, tuberculis* 2, *albis, linearibus, compressis, parallelis, pubescentibus: vertice luteo, in facie excavatâ columnæ recumbentibus. Cis tuberculos adsunt processus duo, parvi, cornuti, albi.* Columna *erecta, libera semi-teres, anticè excavata ad alterum tuberculorum labelli finem recipiendem.* Gynizus *excavatus, subquadratus, marginatus.* Anthera *terminalis, opercularis, cito decidua, 1-locularis, pileiformis.* Pollinia 2, *oblonga, lutea, postice obsoletè biloba, filo brevi communi insidentia, glandula crassa, brunnea gynizi apici affixa.*

For an opportunity of publishing this curious and beautiful plant we are obliged to Mr. Lee, of the Hammersmith Nursery, by whom it was imported from the West Indies. It requires the heat of a bark-stove, and is cultivated in the same way as other plants of the family.

The genus Brassia, as established by Mr. Brown, in the Hortus Kewensis, consisted of a single species, a native of

Jamaica. It is principally distinguished from Cymbidium by the labellum, which is entire, and not elastically jointed with the columna. It constitutes part of a large section of

Epidendreæ, which may be combined under the common character of having no spur, or at most a little bag occasioned by the extension of the columna, and only two pollen-masses. Nearly half the known genera of Epidendreæ are referable to this section, and may be subdivided thus:

EPIDENDREÆ. Sect. I. *Ecalcaratæ; polliniis duobus.*

§ 1. Pollinia filo unico v. tantum glandulâ distincta.

1. Trichoceros, *Kunth.*—2. Rodiguezia, *Fl. per.* from which Gomeza of Mr. *Brown* is scarcely distinct.—3. Cymbidium, *Swartz.*—4. Brassia, *R. Br.*—5. Anguloa, *Fl. per.*—6. Catasetum, *Rich.*—7. Lissochilus, *Br.*—8. Geodorum, *Jackson.*—9. Fernandezia, *Fl. per.*—10. Sobralia, *Fl. per.*—11. Megaclinium, *Nob. Bulbophyllum clavatum, conicum, &c. &c. Pet. Th. orch. afr.*—12. Trizeuxis, *Lindley.*—13. Pleurothallis, *R. Br.*—14. Cyrtochilus, *Kunth.*—15 Odontoglossum, *Kunth.*—16. Oncidium, *Swartz.*—17. Cyrtopodium, *Br.*—18. Pachyphyllum, *Kunth.*—19. Cœlogyne, *Lindley.*—20. Macradenia, *R. Br.*

§ 2. Pollinia filis duobus v. (sub anthesi) nullo.

21. Masdevallia, *Fl. per.*—22. Humboldtia, *Fl. per.*—23. Gongora, *Fl. per.*—24. Dipodium, *R. Br.*—25. Sunipia, *Buchanan, in Rees, article Stelis.*—26. Tribrachia, *Nob.* Labellum unguiculatum cum basi sepalorum exteriorum lateralium connatum. Columna bicirrhosa. *Nepalensis.*—27. Bulbophyllum, *Pet. Th.*—28. Cestichis, *Pet. Th.*—29. Cirrhopetalum, *Nob. Bulbophyllum longiflorum,* Pet. Th. which is C. Thouarsii, *Nob.*

The genus Malaxis, as it stands in Willdenow, is a strange medley.

J. L.

840

CATASETUM Claveringi.

Captain Clavering's Catasetum.

GYNANDRIA *MONANDRIA.*

Nat. ord. ORCHIDEÆ.—Div. IV. *Epidendreæ.* Sect. I. Ecalcaratæ v. ad maximum saccatæ. Pollinia 2. *Suprà fol.* 832.

CATASETUM, Rich.—*Perianthium* rectum, sæpius globosum. *Labellum* saccatum, concavum, sepalis difforme. *Columna* aptera bicirrhosa. *Anthera* cornuta, bilocularis. *Pollinia* postice biloba, in filo maximo, denudato, transverso, demum elastice contractili, inserta.

Div. I. *Perianthium globosum. Labellum saccatum, galeatum.*

C. *Claveringi*, spicâ foliis breviore, labello carnoso apice tridentato, sepalis oblongis obtusis: interioribus maculatis.

Bulbi *epigæi, aggregati, ovales, subfusiformes, vaginis foliorum rigidis pallidis undique vestiti.* Folia 4-5, *lanceolata, glaberrima, ad basin attenuata, pallide viridia, 5-nervia, subplicata, nervis lateralibus approximatis à medio distantibus.* Bulbi *juniores et bases foliorum purpureo hic illic maculati.* Scapus *radicalis, teres, glaucus, foliis multo brevior, 6-florus, vaginis brevibus, pallidis, substriatis vestitus.* Flores *maximi, spicati, globosi, Myrrhidis odore, cito decidui.* Sepala *secunda, ovato-lanceolata, glaberrima, extus viridia apices versus rubro suffusa, exteriora intus purpuro obscuro maculata, interiora maculis purpuri intensi variegata.* Labellum *maximum, posticum, carnosum, inter sepala exteriora erectum, saccatum, cucullatum, apice viride, limbo et intus intense luteum, medio margine tridentatum.* Columna *cum labello continua, recta, clavata, cornuta, semiteres, dorso carinata, purpuro punctata et maculata, antice cirrhis duobus filiformibus ex angulis inferioribus gynizi dependentibus, in basi labelli incurvatis.* Gynizus *quadratus, excavatus, labio inferiore tenui, porrecto, transverso.* Clinandrium *posticè in cornu columnæ ferè dimidiâ longitudine productum.* Anthera *terminalis, opercularis, 2-locularis, elastice dissiliens, cristâ cornutâ cornu clinandrii respondente.* Pollinia 2, *pallidè-lutea, cava, postice fissa,* filo *annexa corneo, rhomboideo, partem clinandrii extra antheram tegente, demum elastice contractili et convoluto;* glandulâ *maxima, ovata, parti superiori gynizi adhærente.*

This fine new species of Catasetum was brought, in 1822, from Bahia de St. Salvador, in BRAZIL, to the Horticultural Society, by Mr. George Don. It is a far more remarkable plant than the Catasetum tridentatum of Hooker, and altogether the most singular Orchideous plant which has yet been seen in a cultivated state. Its flowers are soon past; while in perfection, they expand slightly in the middle of the day, and emit a faint smell, resembling Myrrh. Our drawing was made last September.

We have named the species in compliment to Captain Clavering, the commander of His Majesty's Sloop of war, the Pheasant, in which the Horticultural Society's collector was carried from the coast of Africa to South America. To the protection and liberal assistance of this officer, and of Captain Sabine, who accompanied him, the happy result of the Society's mission is principally to be ascribed.

The genus Catasetum is an instance of the rapid progress our knowledge of Orchideous plants has made within a few months. In 1823, one species only had been published, the *Catasetum maculatum* of M. Kunth, upon which the genus may be said to have been founded, the C. *macrocarpon* of Richard being scarcely known except by name. Before the publication, however, of the work of M. Kunth, in which the genus was established, two other plants had been seen in this country, both of which still remain to be recorded; the one was a plant which, some time before 1817, flowered in the garden at Kew, where it was called Lockhartia, but which has not been since seen in any other collection; the second was sent from Brazil, by Mr. Swainson, to Dr. Hooker, in whose garden, at Halesworth, it produced its blossoms soon after its arrival. Of the first, our recollection is not sufficient to enable us to speak more particularly. The second is a plant with globose greenish flowers in a long raceme, with a bright green labellum mottled with purple in the inside: we long ago proposed to call this species C. *Hookeri.* In the autumn of 1823, a plant from Bahia de St. Salvador, in every respect agreeing with Catasetum in habit, flowered in the garden of the Horticultural Society. In the floral envelope, however, it offered some striking variations from the characters which had been observed to be common to the species previously noticed. The perianthium was not closed, but in some measure spread open, and by no means globose; the labellum was neither concave, nor fleshy, nor even in its natural position, but generally anterior with respect to the other parts of the flower, nearly flat, and strongly crested with fleshy processes; in other particulars, however, especially in the two horns of the columna, the cornute form of the anthera, and the peculiar structure of the pollen-masses and appendages, it agreed with the character of the genus. This species we have called C. *cristatum.*—In the 8th part of Dr. Hooker's Exotic Flora, a

remarkable species, native of Trinidad, was figured in March last: it resembles, in a considerable degree, the subject of this article, but is less beautiful; and is distinguished by the long points to the sepals, and a degree of flaccidity in the flower. Dr. Hooker called it C. *tridentatum.*—Finally, in September last, the blossoms of C. *Claveringi* expanded in all their grandeur.

We have thus traced the history of the genus during its first year, in which period alone it has increased from one certain and one uncertain species to five certain and one uncertain species, which may be thus arranged:

Div. I. Perianthium globosum. Labellum saccatum galeatum.

1. C. maculatum. *Kunth.*
2. C. tridentatum. *Hooker.*
3. C. Claveringi. *Nob.*
4. C. Hookeri. *Nob.*
5? C. macrocarpum. *Rich.* (which is possibly either No. 2 or 3.)

Div. II. Perianthium patens. Labellum, explanatum, saccatum, cristatum.

6. C. cristatum. *Nob.*

J. L.

844

CAMARIDIUM ochroleucum.

Pale Yellow Camaridium.

GYNANDRIA *MONANDRIA.*

Nat. ord. ORCHIDEÆ.—Div. IV. *Epidendreæ.* Sect. 2. Ecalcaratæ v. ad maximum saccatæ. Pollinia 4. *Suprà fol.* 825.

CAMARIDIUM. *Perianthium* resupinatum, explanatum; sepalis liberis. *Labellum* liberum, sessile, cucullatum, trilobum. *Columna* teres. *Gynizus* fornicatus. *Pollinia* parallela, compressa; filo sub anthesi nullo.

C. *ochroleucum.*

Herba *parasitica, caulescens.* Caules 1½ *pedales, ascendentes, ancipites, vaginis foliorum rigidis emarcidis vestiti, ad axillas bulbiferi: bulbis ovalibus compressis ancipitibus unifoliis, ad apices foliosi.* Folia *ligulata, dorso acutè carinata, luteo-viridia, contorta, apice obliquè emarginata, bulborum solitaria, caulis disticha abbreviata.* Flores *solitarii, brevè persistentes, axillares, bracteis 2-3, imbricatis, lanceolatis, vaginantibus, acutis.* Perianthium *explanatum, patens, ochroleucum, glabrum, odore debili;* sepalis *obovato-lanceolatis, carnosis, subæqualibus.* Labellum *sepalis brevius, cum columnæ basi producto articulatum, luteum, marginibus albidis, cucullatum, trilobum: lobis subæqualibus; medio subcordato, rotundato, acuto; disco barbato.* Columna *libera, erecta, semiteres; gynizo fornicato, rotundato.* Anthera *terminalis, opercularis, decidua, semibilocularis, septis obsoletis.* Pollinia 4, *compressa, parallela, libera.* Ovarium *glabrum, ecostatum.*

This is a genus principally distinguished from Ornithidium, by not having the labellum united with the base of the columna, nor a connivent perianthium, nor a fleshy discus to the labellum, and especially characterized by its small labellum and expanded perianthium. It is easily distinguished from Isochilus and Telipogon by the obvious difference in form between the sepals and labellum.

The plant from which our figure was taken, was sent, this year, by his Excellency Sir Ralph Woodford, from Trinidad, to the Horticultural Society; in whose garden at Chiswick our drawing was made in last July. It appears to have been in a wild state parasitical upon the trunks of trees, and requires the cultivation applicable to other plants of a similar description.

We find it nowhere described. *Cymbidium vestitum,* of Swartz, is possibly of the same genus, but differs, according to the description of that writer, in some important particulars.

A parasitical, caulescent plant. *Stems* 1½ foot high, ascending, two-edged, at the ends leafy, clothed with the rigid withered remains of the leafsheaths, bearing at the axillæ oval, compressed, two-edged bulbs, each bearing one leaf. *Leaves* ligulate, acute, keeled at back, yellowish green, a little twisted, at the end obliquely emarginate; those on the bulbs solitary, on the stem two-ranked, shorter than those of the bulb. *Flowers* solitary, remaining in perfection for a short time only, axillary, with two or three imbricated, lanceolate, sheathing, acute bracteæ. *Flower* spread-open, pale yellow, smooth, with a faint smell; *sepals* obovate-lanceolate, fleshy, nearly equal. *Lip* shorter than the sepals, jointed with the lengthened base of the columna, yellow with white edges, cucullate, three-lobed; the lobes nearly equal, the middle one being somewhat cordate, rounded, acute; its disk bearded. *Columna* free, erect, half rounded, with a vaulted rounded gynizus. *Anther* terminal, like a lid, deciduous, ½-2-celled, with obsolete divisions. *Pollen-masses* 4, compressed, parallel, loose. *Ovary* smooth, without ribs.

851

POLYSTACHYA puberula.

Pubescent Polystachya.

GYNANDRIA *MONANDRIA.*

Nat. ord. ORCHIDEÆ. Div. IV. Epidendreæ. Sect. 2. Ecalcaratæ v: ad maximum saccatæ. Poll. 4. *Suprà fol.* 825.

POLYSTACHYA, Hook. —— Perianthium rectum cuneatum clausum. Filum simplex nudum glandulâ impositum.

P. *puberula,* spica paniculata thyrsiformi, foliis lanceolatis 7-nervibus scapo longioribus, floribus ovariisque pubescentibus, bulbis ovatis.

Bulbi *epigæi, ovati.* Folia *6-12-uncialia, circa tria, lanceolata, patentia, 7-nervia, glaberrima, scapo multo longiora.* Scapus *terminalis, teres, pubescens, superne ramosus, bracteis parvis, membranaceis, acuminatis sub axillis.* Spica *paniculata, thyrsiformis, undique cum floribus pubescens.* Perianthium *rectum, cuneatum, clausum, luteum, pubescens, ovario subæquale.*

This undescribed species of Polystachya was received by the Horticultural Society, from Sierra Leone, in 1822. In generic character it perfectly agrees with the *Dendrobium polystachyon* of Swartz, upon which the genus Polystachya was founded by Dr. Hooker; but differs specifically in its leaves, which are very distinctly veined, and in its flowers, which are covered all over with a minute yellow pubescence, and disposed in a sort of thyrsoid branched spike.

The following are the species at present known of this genus, which was divided by Dr. Hooker from *Dendrobium* with great propriety.

1. Polystachya *luteola.*
   α. occidentalis. Flores viridi-lutei.
   P. luteola. *Hook.*
   Dendrobium polystachyon. *Lindl.*
   β. mauritiana. Flores luteo-purpurei.
   Dend. polystachyon. *Pet. Th. Orch. afr. t.* 85.
   Hab. α in India occidentali, *v. v.;* β in Mauritio, *v. s. sp.*

2. Polystachya *puberula.* Tab. nostr.
   Hab. in Sierra Leone, *v. v.*

3. Polystachya *fusiformis.*
   Dendrobium fusiforme. *Pet. Th. l. c. t.* 86
   Hab. Ins. Borboniæ. *Pet. Th.*

4. Polystachya *cultrata.*
   Dendrobium cultriforme. *Pet. Th. l. c. t.* 87.
   Hab. in Mauritio. *Pet. Th.*

Of this species the *bulbs* stand on the surface of the ground and are ovate. The *leaves* are about three to each bulb, 6-12 inches long, spreading, 7-nerved, very smooth, much longer than the scape. The *Scape* is terminal, rounded, pubescent, branched upwards, with little acuminate, membranous bracteæ under the axillæ. The *Spike* is panicled and thyrse-shaped, all over covered with pubescence like the flowers. The *Perianth* upright, wedge-shaped, closed, pubescent, about as long as the ovary.

J. L.

862

GLOSSULA tentaculata.

Long-horned Glossula.

GYNANDRIA *MONANDRIA.*

Nat. ord. ORCHIDEÆ. Sect. 1. *Orchideæ veræ.* Pollen granulosum. Anthera terminalis erecta persistens.

GLOSSULA.—Sepala in galeam conniventia: supremo ecalcarato. *Labellum* anticum calcaratum tripartitum: calcare inflato. *Pollinia* 2, tripartita, glandulis 2, cucullis duobus discretis inclusis. Herba *chinensis,* radice *simplice.* Habenariæ cucullatæ *aspectu,* floribus *herbaceis spicatis vix spiralibus.*

G. *tentaculata.*

Radix *tuber fusiformis, simplex.* Folia 3, *radicalia, cæsia, plana, elliptica, basi amplexicaulia, subquinquenervia, horizontalia, glaberrima.* Scapus *erectus, spithamæus, glaber, teres, squamis, quarum infima foliacea, vaginatus.* Flores *spicati, parvi, herbacei, vix spirales, sed undique versi.* Bracteæ *ovatæ, acuminatæ, ovarii longitudine.* Ovarium *tortum, angulis, sepalis exterioribus oppositis, costatis.* Sepala *in galeam conniventia, ovato-elliptica, æqualia, interiora exterioribus pallidiora.* Labellum *cum basi columnæ connatum, 3-partitum, lacinia intermedia lingulata, calcare longiore, dependente, lateralibus multo longioribus, porrectis, variè flexis, filiformibus; disco inter sinus laciniarum gibboso; calcar e basi labelli ortum, sepalorum longitudine, ovale, inflatum, compressum, ad collum constrictum.* Anthera *brevis, terminalis, ovata, erecta, persistens, bilocularis, loculis discretis.* Gynizus *area madida infra antheram, superne in processu inter loculos erecto, convoluto, antheræ fere longitudine, deorsum in cuculla duo desinente productus.* Pollinia 2, *bipartita, granulosa, quoque filo brevi glandulæ affixo, in sinu cucullorum jacentis.* Tuberculum *granulosum antheram ad basin utrinque adstat.*

This very curious little Orchideous plant was brought from China, in 1824, by Mr. John Damper Parks, an assiduous collector in the service of the Horticultural Society. Our drawing was made at the Chiswick Garden, last December.

The character of this plant depending entirely upon the structure of its minute green flowers, we have caused a more detailed representation of them to be prepared than can usually be afforded in publications so cheap as this is. Fig. 1, shows a side view of the flower, much magnified. Fig. 5, the same, seen in front. Fig. 3, is a representation of the anther and spur, with the front part of the labellum cut off. At fig. 4, the anther, with its curious pouches, and the process between the cells of the anther, are shown. Fig. 2, represents a pollen mass and its gland.

The general aspect of the plant is that of the Habenaria cucullata; but from all the calcarate genera it is as obviously as curiously distinguished, as Ophrys is from all those without spurs, by having two distinct pouches for the reception of the glands of the pollen masses instead of one.

Root a simple, fleshy, fusiform tuber. *Leaves* 3, radical, blueish, flat, elliptical at the base, stem-clasping, about five-nerved, horizontal, quite smooth. *Scape* erect, a span high, smooth, rounded, sheathed with scales, of which the lowest is leafy. *Flowers* spiked, small, herbaceous, scarcely spiral, but directed all ways. *Bracts* ovate, acuminate, the length of ovary. *Ovary* twisted, with the angles opposite the outer sepals, ribbed. *Sepals* conniving into an helmet, ovate-elliptical, equal, the inner paler than the outer. *Lip* united with the base of the columna, 3-parted, the intermediate segment lingulate, longer than the spur, hanging down; the side ones much longer, porrect, variously bent, filiform; the disk of the lip gibbous between the recesses of the segments; the spur proceeding from the base of the lip, the length of the sepals, oval, inflated, compressed, contracted at the neck. *Anther* short, terminal, ovate, erect, persistent, 2-celled, with separate cells. *Gynizus* a moist area below the anther, with its upper edge produced between the cells of the anther into an erect convolute process, nearly as long as anther, and extended downwards on each side into a little bag. *Pollen masses* 2, 2-parted, granular, each with a short thread attached to the gland which lies in the little bags of the gynizus. A granular tubercle is fixed on each side of the anther at its base.

J. L.

NOTES.

Angræcum maculatum. *Suprà Vol. 8. fol.* 618.
The variety of this species, which is represented in the above article, is a native of Brazil; another variety with pale flowers has been found in Sierra Leone. The latter differs from the former in no essential particular; but proves the genus, of which this species forms a part, to be common to the continents of Africa and America. Whether it is a genuine species of Angræcum it is difficult to decide. There are certainly no retinacula to the pollen-masses, in which respect it materially differs from Angræcum; but, on the other hand, it appears to be of that genus in the other parts of its structure, and the objection which might have been taken to its being an Angræcum, on account of its supposed geographical difference, is now, as above stated, removed.

Neottia procera. *Suprà Vol. 8. fol.* 639.
This plant proving, upon a more careful examination, referable to the genus Goodyera, the head of our article should be changed to Goodyera procera, and the following synonym added:

Goodyera procera. *Hooker's Exotic Flora, t.* 39.

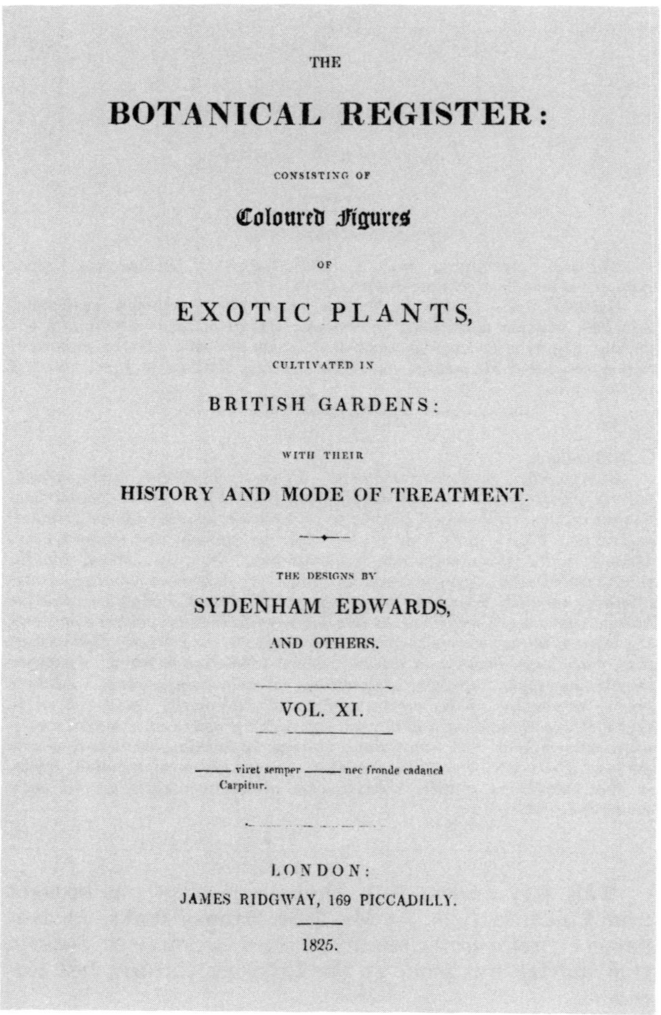
THE
BOTANICAL REGISTER:
CONSISTING OF
Coloured Figures
OF
EXOTIC PLANTS,
CULTIVATED IN
BRITISH GARDENS:
WITH THEIR
HISTORY AND MODE OF TREATMENT.
THE DESIGNS BY
SYDENHAM EDWARDS,
AND OTHERS.
VOL. XI.
—— viret semper —— nec fronde cadanci
Carpitur.
LONDON:
JAMES RIDGWAY, 169 PICCADILLY.
1825.

868

CŒLOGYNE fimbriata.

Fringed Cœlogyne.

GYNANDRIA *MONANDRIA.*

Nat. ord. ORCHIDEÆ. Div. IV. Epidendreæ. Sect. II. suprà fol. 851.
CŒLOGYNE, Lindl. *Perianthium* resupinatum, patens. *Labellum* trilobum, cucullatum, cum columna articulatum. Columna alata ultra antheram producta. *Anthera* lateralis, opercularis, bilocularis. *Massæ pollinis* duæ bipartitæ (filo nullo). *Gynizus* infundibularis, bilabiatus. Herbæ *parasiticæ* (Indiæ Orientalis) *bulbosæ; foliis coriaceis, glabris, nervosis.* Flores *pauci, racemosi, e spathâ squamosâ, radicali. Lindl. Coll. Bot.* 33.

C. *fimbriata,* foliis binis oblongo-lanceolatis patentibus, floribus terminalibus solitariis: sepalis interioribus filiformibus; labello fimbriato bicristato.
Caudex *repens, squamosus, ramosus.* Bulbi *ovales, glabri, solitarii, bifolii, basi squamis duabus vaginati.* Folia *oblongo-lanceolata, basi attenuata,* 5-*nervia, glabra, patentia.* Flores *solitarii, terminales, in pedunculo brevi undique squamoso insidentes.* Perianthii *sepala exteriora ovato-lanceolata, basi libera, glaberrima, lateralibus labello suppositis; interiora linearia, tenuia, uninervia.* Labellum *liberum, cucullatum, cum columnâ articulatum, bicristatum,* 3-*lobum, lobis lateralibus erectis, sub-ciliatis, intermedio retuso, fimbriato, a cristis ad latera fulvo.* Columna *libera, erecta, obovata, alata, gynizo excavato, bilabiato: labio inferiore lunato recurvo, superiore magno, porrecto, transverso, demum incumbente.* Anthera *infra apicem columnæ inserta bilocularis; loculis bivalvibus completis; antice denticulata, marginata.* Pollinia 4, *non collateralia, sed antica et postica, quorum postica minora; basi farinosa nec glutinosa nec glandulifera.*

This genus is very different from Cymbidium, from which it is clearly distinguishable by habit, winged column below the summit of which the anthera is inserted; and also by the remarkable two-lipped gynizus, which is almost exactly the same as that of *Arethuseæ.* We established it sometime ago in our Collectanea Botanica, defining two certain species from Nepal, and indicating another, the Cymbidium nitidum of Wallich, as probably belonging to the same genus. Our definition of the genus having been drawn up from dried specimens only, it is peculiarly gratifying to us to find, from the inspection of a new living species, that our account of the genus, and of its essential characters, was in all respects correct, with the exception of the anthera being one-celled. That organ being, in the present species, distinctly two-celled, we rather believe that the contraction, in the same year, of its interior thin valves in drying, may have deceived us.

Our drawing was made in July 1824, at the garden of the Horticultural Society, where the plant had been received, by means of Mr. John Damper Parks, from China. It is a delicate stove plant, requiring the cultivation applicable to similar subjects.

Stem creeping, scaly, branched. *Bulbs* oval, smooth, solitary, two-leaved, at the base sheathed by two scales. *Leaves* oblong-lanceolate, attenuate at the base, 5-nerved, smooth, spreading. *Flowers* solitary, terminal, with a short stalk, clothed with many scales. *Divisions of the flower;* exterior ovate-lanceolate, distinct, smooth; the lateral ones being placed under the lip; interior slender, linear, one-nerved. *Lip* distinct, hooded, jointed with the column, 2-crested, 3-lobed, the lateral lobes erect, minutely ciliated, the middle one blunt, fringed, from the crests to the sides of a brown colour. *Column* distinct, erect, obovate, winged, the gynizus hollowed out, 2-lipped: the lower lip lunate, the upper large, porrect, broad, at length incumbent upon the lower. *Anther* inserted below the end of the column, 2-celled: *cells* 2-valved, complete; in its front edge toothletted. *Pollen masses* 4, not lying side by side, but anterior and posterior; those behind being smaller than those in front; at their base powdery, and not glutinous or glandular.

J. L.

882

LIPARIS foliosa.

Many-leaved Liparis.

GYNANDRIA *MONANDRIA.*

Nat. ord. ORCHIDEÆ. Sect. V. Malaxideæ. *Lindl. supra, fol.* 825, *in textu. Anthera* terminalis, opercularis, decidua. *Pollinia* 2, v. 4 per paria cohærentia, cereacea, filo et glandulis nullis.—Herbæ *sæpissimè terrestres, foliis membranaceis.* Scapi *terminales.* Flores *herbacei, inconspicui.* Labellum *sæpius posticum.*

LIPARIS, Rich. Labellum planum, explanatum, integrum, variè versum. Columna alata. *Pollinia* 4.

L. *foliosa;* foliis radicalibus inæqualibus lanceolatis integris acutis carnosis racemo subæqualibus, labello oblongo retuso, clinandrio integerrimo.

Herba *terrestris, acaulis, bulbis nullis.* Folia 3, *v.* 4, *suberecta, crassa, oblongo-lanceolata, subcomplicata, glaberrima, acuta; ultimo spicæ longitudine; extus ad basin foliis duobus abortivis vaginata.* Racemus *terminalis, erectus,* 8-12-*florus.* Flores *parvi, virides, distantes, undique versi.* Ovarium *bractea multoties longius, cum pedicello articulatum.* Perianthium *explanatum, sepalis liberis, subæqualibus, lineari-lanceolatis, reflexis;* labello *viridi-luteo, libero, oblongo, erecto-patente, integro, retuso.* Columna *tenuis, erecta, libera, marginata,* clinandrio *marginato, integro.* Anthera *bilocularis, terminalis, opercularis, decidua.* Pollinia 4, *cereacea, filo nullo, v. saltem citissimè obsoleto.*

For this curious new species of Liparis we are obliged to Robert Barclay, Esq. by whom it was received from the Isle of France in 1823. Our drawing was made last August, from a plant cultivated in the stove of that gentleman's collection; which is unquestionably the most extensive and useful ever formed by a private individual in this country.

The genus Liparis, which was established by the late M. Richard upon the Ophrys Löselii of Europe, has now assumed an extent which certainly could not have been contemplated by that most excellent and learned botanist, who appears to have been acquainted only with the European Ophrys and the *Malaxis liliifolia* of N. America.

The following is the arrangement which we propose for the ten species of Liparis, at present known to us:

LIPARIS, *Richard.* (Genus Malaxide *certissime* distinctum, *R.*)

Sect. I. Folia membranacea. Terrestres.

1. L. *Löselii,* Rich.
Ophrys Löselii, *L.*
Malaxis Löselii, *Smith.*
Hab. in Europæ borealis palustribus. (v. v. sp.)
2. L. *liliifolia,* Rich.
Ophrys liliifolia, *Bot. rep.*
Malaxis liliifolia. *Br. in Ait. Kew. ed.* 2.
Hab. in Americæ septentrionalis palustribus. (v. v. sp.)
3. L. *nepalensis;* foliis oblongis acuminatis membranaceis, caudice repente bulboso, sepalis interioribus tenuissimis, labello acuminatissimo.
Hab. in Nepaliâ, *Wallich.* (v. s. sp.) *Facies* Liparidis liliifoliæ. *Bulbi* foliorum vaginis marcescentibus tecti. *Folia* 2-na, plana, oblonga, acuminata, nervosa, membranacea, spicâ paulo breviora. *Sepala* filiformia, tenuissima; *labello* maximo, obovato, acuminatissimo.
4. L. *flavescens.*
Malaxis flavescens. *Pet. Th. orch. afr. t.* 25.
Hab. in Mauritio, et Insulâ Borboniæ.
5. L. *purpurascens.*
Malaxis purpurascens. *Pet. Th. l. c. tt.* 26 *et* 27.
Hab. in Mauritio, et Insulâ Borboniæ.
6. L. *bituberculata.*
Cymbidium ? bituberculatum. *Hooker, ex. fl. t.* 116.
Hab. in Nepalia (v. s. c. in herb. Hooker).

Sect. II. Folia carnosa. Parasiticæ.

Cestichis Pet. Th. et Lindl. supr. 837, *in textu.*

7. L. *foliosa.* Supra.
Hab. in Mauritio.
8. L. *disticha.*
Malaxis disticha. *Pet. Th. l. c. t.* 89.
Hab. in Mauritio et Insulâ Borboniæ.
9. L. *cæspitosa.*
Malaxis cæspitosa. *Pet. Th. l. c. t.* 90.
Hab. in Mauritio et Insulâ Borboniæ.
10. L.? *reflexa.*
Cymbidium reflexum. *Brown, prodr.* 331.
Hab. in Novâ Hollandiâ ad portum Jackson (v. s. sp. sine polliniis).

At our folio 825 we have noticed the other genera of *Malaxideæ,* to which we have at present none to add.

J. L.

897

MAXILLARIA Harrisoniæ.

Mrs. Harrison's Maxillaria.

GYNANDRIA *MONANDRIA.*

Nat. ord. ORCHIDEÆ: *tribus* Vandeæ. *Lindl. coll. bot. ined.*

MAXILLARIA Fl. Per.—Perianthium patens, resupinatum. *Labellum* cum processu unguiformi columnæ articulatum, trilobum. *Sepala* lateralia exteriora basibus cum processu columnæ connata. *Pollinia* 4, basibus connata, glandulosa.—— Herbæ *parasiticæ, bulbosæ, Americæ meridionalis.* Folia *coriacea plicata.* Racemi *radicales.*

Maxillaria *Harrisoniæ;* foliis solitariis lanceolatis plicatis, racemo bifloro, perianthio maximo cerino patente, labelli venosi disco glanduloso piloso, lobis recurvis crispis.

Dendrobium Harrisoniæ. *Hooker's Exotic Flora, tab.* 120.

From an inspection of the specimen from which Dr. Hooker's figure of his Dendrobium Harrisoniæ was taken, we have ascertained that it is the same as the plant now published, notwithstanding the difference in the colour of their flowers. The plant which Dr. Hooker examined appears to have been in a less vigorous state of health than that before us, and to have produced only one flower, instead of two, or probably a greater number.

A native of South America. The plant from which our drawing was taken was kindly communicated by William Cattley, Esq., from his Conservatory at Barnet. A robust stove parasite, flowering in September.

No group of plants has undergone greater changes, in consequence of accurate investigation, and the application of modern analytical principles of botany, than the genus Dendrobium. As it stands in Willdenow's Species Plantarum, published twenty years ago, it consists of 25 species. Of these, D. *sanguineum* has been detached, by Mr. Brown, under the name of Broughtonia; D. *graminifolium*, by the same distinguished Botanist, as Octomeria; and D. *ruscifolium*, as Pleurothallis. To the latter genus, D. *racemiflorum* has been referred by the writer of this article. D. *polystachyon* is the type of Dr. Hooker's genus Polystachya; D. *utricularioides*, and probably *testiculatum*, are species of Ionopsis; D. *palmifolium* and *Barringtoniæ* belong to the genus now reformed. Of the few remaining after these deductions, the East Indian kinds are probably genuine species of Dendrobium, and most of the others require further examination.

The genus Dendrobium, as limited by Mr. Brown in the 2d edition of the Hortus Kewensis, should include no species except those with 4, parallel, distinct pollen-masses; and will consequently contain only the East Indian species, which form a numerous group, distinctly characterised by the character above mentioned, and naturally connected by habit. They are trailing or erect parasites, without bulbs, but with long, leafy, or naked round stems, producing the flowers from their axillæ, either singly, or in few-flowered fascicles. From these, the American species, agreeing with them in general structure of perianthium, are now divided. They are distinguished by having only two pollen-masses, each divided down the mdidle, and connected at the base by a common glandular process; and by being all bulbous. Besides the kind now published, D. Barringtoniæ (*Maxillaria Barringtoniæ* Nob.) and D. palmifolium (*M. palmifolia* Nob.), belong to the genus Maxillaria, which will be increased by the examination of the numerous Dendrobium-like plants of the continent of South America. It is distinguished from Polystachya by the structure of its perianthium, and by its radical inflorescence.

Nearly related to Maxillaria, in general characters, and in geographical position, is a genus of which the plant published at tab. 732 of this work, under the name of Dendrobium squalens, must be considered the type. This plant has only two pollen masses, furrowed on each side, and its flowers are in their natural position. It may be called *Xylobium*, and thus characterised:

XYLOBIUM.

Perianthium rectum, patens. *Labellum* cum processu unguiformi columnæ articulatum, trilobum, in columna incumbens. *Sepala* lateralia exteriora basibus cum processu columnæ connata. *Pollinia* 2, hinc sulcata. ——Herbæ *habitu Maxillariæ*.

Sp. 1. Xylobium squalens.

Dendrobium squalens. *Suprà t.* 732.

J. L.

904

ERIA stellata.

Starry Eria.

GYNANDRIA *MONANDRIA*.

Nat. ord. ORCHIDEÆ. *Div.* Epidendreæ.

ERIA Lindley.—Perianthium lanatum, connivens. *Labellum* trilobum, articulatum cum processu unguiformi columnæ cujus lateribus sepala antica adnascuntur. Pollinia octo. *Collectanea Botanica* 42. *B. inedita.*

E. *stellata;* foliis lanceolatis carnosis 5-nervibus, sepalis ovato-lanceolatis acuminatis, labelli lobo medio acuminato, scapo angulato, ovario sepalisque exterioribus pube ferrugineâ lanatis. *Lindl. l. c.*

Dendrobium javanicum. *Swartz Act. Holm.* 1800 *p.* 247? *Willd. sp.* 4.137?

Herba *parasitica, repens, robusta.* Folia 1½-2 *pedalia, erecta, carnosa, glaberrima, lanceolata, tri-quinquenervia, basi squamis ovato-lanceolatis, vaginantibus, nervosis circumdata.* Scapus 2-*pedalis, teres, angulatus, undique pube lanatâ, rarâ, ferrugineâ obductus.* Flores *spicati, apice stellatim patentes, luteo-ferruginei, extùs eâdem lanâ ac scapus obducti; basi bracteâ ovatâ, concavâ, scariosâ, siccâ, deciduâ, ovario longiore, extùs villosâ suffulti.* Perianthium *connivens, sepalis apice patentibus, lineari-lanceolatis, acuminatis, æqualibus, 5-nervibus,* exterioribus *basi connatis, extùs ferrugineo lanatis,* interioribus *glaberrimis:* anticis *basibus productis cum basi elongata columnæ connatis.* Labellum *erectum, cum columna parallelum, ejusque basi articulatum, ovato-lanceolatum, basi cucullatum, 3-lobum:* disco *carnoso, tricostato:* lobis lateralibus *abbreviatis, rotundatis, pallidis, venosis:* intermedio *elongato, acuminato, sepalis breviore.* Columna *brevis, semiteres.* Gynizus *anticus, concaviusculus, apice ovatus.* Anthera *carnosa, cristata, apice dilatata truncata, omnino bilocularis, loculis discretis, bivalvibus.* Pollinia *cuique loculo 4, in paribus duobus, collateralia, basibus ad faciem mediam loculi adversis, et materie viscidâ connexis.*

This genus has been founded, in the work above cited, upon a remarkable plant from Nipal, the Dendrobium pubescens of Professor Hooker, and the subject of this article. It differs from all the other genera, having the flower of Dendrobium, in the number of its pollen masses, with the exception of Octomeria, which we have had no opportunity of examining; but which, as an American genus, and with a peculiar habit, we cannot doubt is sufficiently distinct.

The native country of this species is said to be Java; but we are not acquainted with the time or circumstances of its introduction to this country. For our drawing we have to thank Mr. Cattley, by whom fine specimens were communicated in February last.

A stout, creeping, parasitical plant. *Leaves* 1½ or 2 feet long, erect, fleshy, smooth, lanceolate, 3-5 nerved, surrounded at base with ovate-lanceolate, sheathing, nerved scales. *Scape* 2 feet high, round, angular, covered all over with thin rusty wool. *Flowers* spiked, at the end spreading in a stellate manner, of a yellow ferruginous colour, and covered on the outside with the same sort of wool as the scape; at the base supported by an ovate, concave, scarious, dry, deciduous bract, longer than the ovary. *Perianthium* connivent, spreading at end; sepals linear-lanceolate, acuminate, equal, 5-nerved; the *outer* connate at base, and covered with ferruginous wool outside, the *inner* quite smooth; those in front united by their dilated bases with the elongated foot of the column. *Labellum* erect, parallel with the column, and jointed with its base, ovate-lanceolate, cucullate at base, 3-lobed; *disk* fleshy, 3-ribbed; *lateral lobes* short, rounded, pale, veiny; *intermediate* long, acuminate, shorter than sepals. *Column* short, half-round. *Gynizus* in front, a little concave, ovate at end. *Anther* fleshy, crested, with a dilated, truncate end, completely 2-celled, with distinct, 2-valved cells. *Pollen masses* 4 in each cell, placed side by side, in two pairs, their bases pointing towards the front of the cell, and united by a viscid substance.

J. L.

908

POGONIA pendula.

Pendulous Pogonia.

GYNANDRIA *MONANDRIA.*

Nat. ord. ORCHIDEÆ. *Div.* Neottieæ.
POGONIA. *Suprà, vol.* 2. *fol.* 148.

P. *pendula;* foliis ovatis squamiformibus amplexicaulibus, floribus subsolitariis cernuis, labelli lobo medio oblongo crispo, caule angulato.
Helleborine mariana, flore pallido purpureo trianthophoros. *Pluk. mant.* 100. *t.* 348. *f.* 6.
Arethusa parviflora. *Mich. am.* 2. *p.* 160.
Arethusa trianthophoros. *Swartz act. holm.* 1800. *p.* 230.
Arethusa pendula. *Willd. sp. pl.* 4.82. *Pursh. am. sept.* 2. *p.* 590.
Triphora pendula. *Nuttall am. gen.* 2. 193.

Herba *vix spithamæa, nuda, simplex.* Caulis *erectus, rubescens, angulatus, basi bulbosus;* foliis 3 *v.* 4, *parvis, ovatis, trinervibus, vaginantibus, rubro marginatis.* Flores *in axillis foliorum superiorum,* 2-3, *pedunculati, primùm cernui, demùm erecti.* Ovarium *trigonum, v. inæqualiter hexagonum, angulis marginatis.* Perianthium *ringens, sepalis albis, rubescentibus, liberis, exterioribus linearibus patentibus, interioribus oblongo-lanceolatis erectis, ad dorsum columnæ adstantibus.* Labellum *album, trilobum, cucullatum, unguiculatum, cum columnâ articulatum eâque parallelum; lobis lateralibus parvis, erectis, intermedio ovato, rotundato, crispo, patente, disco viridi granuloso.* Columna *libera, elongata, lobis lateralibus labelli brevior, semiteres, spatulata;* gynizo *elliptico, marginibus introflexis sursùm fornicato.* Anthera *terminalis, suberecta, cristata, carnosa, bilocularis, cum gynizo parallela, valvis purpureis membranaceis.* Pollinia 2, *sulco longitudinali exarata, linearia, reniformia, è particulis angulatisconstantia.*

Roots of this curious little plant were collected in Canada, by Mr. David Douglas, for the Horticultural Society, and flowered in an open border in the Garden at Chiswick in August 1824, at which time our drawing was made. It is a very distinct species, much smaller than any other published kind, and easily recognised in any state by its elegant little pendulous white and pink flowers.

We cannot agree with Mr. Nuttall in separating this plant from the genus Pogonia, with which it, in our judgment, agrees in all essential particulars. From Arethusa it is well distinguished by the segments of the flower being all distinct, and having a nearly equal insertion.

This species seems to be distributed over nearly all North America, from Canada to Georgia. Mr. Nuttall says it is parasitic round the roots of beech trees from New York to Kentucky, and that he has also found it in Canada. Our own specimens were collected near the Alleghany River, by Mr. Goldie.

A small plant, scarcely a span high, naked, simple. *Stem* erect, tinged with red, angular, bulbous at base. *Leaves* 3 or 4, small, ovate, 3-nerved, sheathing, edged with red. *Flowers* in the axillæ of the upper leaves 2-3, stalked, at first drooping, afterwards erect. *Ovary* 3-cornered, or unequally 6-angled; the angles edged. *Perianthium* ringent, with white, pinkish, distinct divisions; outer linear spreading, inner obl. lanceolate, erect, standing on each side of the back of the columna. *Lip* white, 3-lobed, cucullate, clawed, jointed with the column, and parallel with it; lateral lobes small, erect; intermediate ovate, rounded, crisp, spreading, with a green granular disk. *Column* distinct, elongated, a little shorter than the lateral lobes of lip, ½-round, spatulate; *gynizus* elliptical, vaulted upwards by the inflexed edges. *Anther* terminal, sub-erect, crested, fleshy, 2-celled, parallel with the gynizus; valves purple, membranous. *Pollen-masses* 2, furrowed by a longitudinal line, linear, uniform, powdery, formed of angular particles.

J. L.

910

ONCIDIUM papilio.

Trinidad Butterfly-Plant.

GYNANDRIA *MONANDRIA.*

Nat. ord. ORCHIDEÆ. *Div.* Epidendreæ.
ONCIDIUM. *Suprà fol.* 727.

O. *papilio;* foliis solitariis ovalibus pictis patentibus, scapo articulato ancipite paucifloro, sepalis superioribus linearibus longissimis: inferioribus distinctis ovato-lanceolatis undulatis, columnâ bicorni: alis fimbriatis.

Folia *ovalia, patentia, carnosa, in bulbo compresso subangulato solitaria, suprà fasciis interruptis maculata, subtùs purpureo creberrimè punctata: maculis rotundis viridibus.* Scapus *radicalis,* 3-*pedalis, articulatus, nudus, articulis inferioribus teretibus, superioribus ancipitibus,* 3-4-*uncialibus, margine et disco rubro maculatis, basi squamâ scariosâ instructis.* Perianthium *patentissimum, diametro (in exempl. spont.)* 5-*unciarum ab apice sepali postici ad apicem labelli;* sepalis 3 *superioribus lineari-spatulatis, longè unguiculatis, erectis, atro-aurantiacis, nunc luteo interruptè fasciatis; inferioribus ovato-lanceolatis, falcatis, undulatis, crispis, labelli longitudine, et cum eo luteis aurantiaco nubilis.* Labellum *panduriforme, basi cordatum, apice dilatatum, emarginatum, undulatum, disco baseos glanduloso, glandulis pallidis purpureo fasciatis, et formam ranæ cubantis referentibus.* Columna *brevis, erecta, utrinque alata; alis patentibus, versùs basin quadratis, carnosis, margine laceris, versùs apicem angustioribus fimbriatis, ad ipsam apicem cornubus duobus, subulatis, porrectis, apice glandulâ capitatis, instructis.* Anthera *terminalis, opercularis, unilocularis, ovata, ad cardinem foveata.* Pollinia 2, *cereacea, oblonga, collateralia, postice biloba, basi glandulâ latâ rostello affixa.*

For this highly curious parasite, the public is indebted to his Excellency Sir Ralph Woodford, Governor of Trinidad, by whom living plants have been sent to several collections in this country. The plant from which our drawing was taken flowered in a stove at Mr. Colvill's Nursery, last March; but not in perfection, the blossoms in our only native specimen having a diameter nearly twice as great as is represented in the accompanying figure, and being much more completely expanded.

We have not succeeded in tracing this plant in any work within our reach. The Butterfly-plant of Santa Cruz, described by West, for which the "Helleborine flore papilionaceo" of Plumier is quoted by authors, is a totally different plant from this. The name has doubtless been suggested by the brilliant colours of the flower, its singular form, which may easily be likened to the wings, body, antennæ, and tongue of a butterfly, and its fluttering motion when hanging from its stalk at the extremity of the weak, elastic, jointed scape.

Leaves oval, spreading, fleshy, seated upon a compressed, somewhat angular, solitary bulb, above spotted with interrupted fasciæ, beneath dotted all over with very numerous purple points, with round, green spots among them. *Scape* from the root, 3 feet long, jointed, naked, the lower joints round, the upper 2-edged, 3 or 4 inches long, spotted with red at the edge and disk, and at the base having a scariose scale. *Flowers* spreading, 5 inches across (in wild specimens) from the tip of the upper sepal to the point of the labellum; the upper sepals linear, spatulate, with long stalks, erect, dark orange, sometimes interrupted with yellow; the lower ovate-lanceolate, falcate, wavy, curled, the length of lip, and like it yellow clouded with orange. *Lip* fiddle-shaped, cordate at base, dilated and emarginate at end, wavy, glandular at base, the glands pale, variegated with purple, and representing the figure of a couchant frog. *Column* short, erect, with spreading, fringed wings.

J. L.

ONCIDIUM pumilum.

Mr. Herbert's Dwarf Oncidium.

GYNANDRIA *MONANDRIA.*

Nat. ord. ORCHIDEÆ. *Div.* Epidendreæ.
ONCIDIUM. Suprà fol. 727.

Div. 1. *Perianthii foliola omnia discreta.* Lindl. Coll. bot. xxvii.

O. *pumilum;* foliis rigidis ovalibus obliquis, paniculâ erectâ thyrsoideâ foliorum longitudine, laciniis perianthii obovatis: superioribus maculatis, labello trilobo cristato: lobis æqualibus ovatis, alis columnæ integerrimis.

Herba *parasitica, pumila, sempervirens, bulbis nullis.* Folia *disticha, patentia, ovalia, rigida, avenia, sæpiùs apice et basi ob torsionem obliqua.* Scapus *terminalis, erectus, paniculatus, thyrsoideus, brachiatus.* Flores *luteo-brunnei, parvi.* Perianthium *patens, 5-partitum, laciniis obovatis, subundulatis, apice incurvis: superioribus fusco maculatis, inferioribus immaculatis.* Labellum *luteum, planum, subrotundum, trilobum: lobis ovatis, obtusis, subæqualibus: intermedio paulò minore, disco callis duobus longitudinalibus sinubus loborum oppositis cristato.* Columna *alis parvis, integerrimis, rotundatis.* Anthera *rostrata.*

This pretty new species of Oncidium was communicated to us by the Honourable and Rev. Wm. Herbert, from his collection at Spofforth. It was found by one of his collectors upon the trunk of a Bombax Ceiba, between Rio Janeiro and Boto Fogo. It flowers in June, and grows readily in a stove, in a mixture of dead wood and moss, as indeed does the whole genus, which is, perhaps, the most easily cultivated of all the tribes of Orchideæ.

It has been contended by some recent Botanists, that the modern principles upon which Orchideous plants are divided into genera are inadmissible; and we believe that their judgment is influenced, in a great degree, by perceiving, that a large number of recently formed genera consist of one or two species only. But to such objections, it appears to us that the genus Oncidium itself is a sufficient reply. It consists of a considerable number of species, we believe 23, to which there can be no doubt that many more will be added; the species have decidedly-marked combining characters in their habit, as well as in the structure of their floral organs, and their union in one genus is in no instance affected by modern principles of analysis, the application of which to some other genera, even of Swartz himself, has sometimes shaken his combinations into almost as many genera as species. But if the mode of limiting genera to which we allude were unsatisfactory, and tended only to disunion, it would scarcely admit so large an assemblage of individuals as Oncidium presents, in a state of combination, nor would the newly-discovered species of Orchideæ be continually reinforcing the identical genera, which are pronounced unnecessary, upon the ground of their thinness of species.

A dwarf, evergreen, parasitical *herb*, without bulbs. *Leaves* distichous, spreading, oval, rigid, veinless, generally having an oblique direction, on account of a twist at their two ends. *Scape* terminal, erect, panicled, thyrsoid, brachiate. *Flowers* yellowish-brown, small. *Perianthium* spreading, 5-parted, with obovate segments, which are a little wavy and incurved at the end: the 3 upper mottled with brown, the lower not mottled. *Lip* yellow, flat, roundish, 3-lobed; with ovate, blunt, nearly equal lobes; the middle one being rather smaller than the others, and having on its disk two longitudinal protuberances opposite the recesses of the lobes. *Column* with small entire rounded wings. *Anther* beaked.

J. L.

DISA grandiflora.

Large-flowered Disa.

GYNANDRIA *MONANDRIA.*

Nat. ord. ORCHIDEÆ. Sect. 1. *Orchideæ veræ.*

DISA L. — *Perianthium* explanatum; laciniis exterioribus liberis: posticâ galeatâ 1-calcaratâ; interioribus nanis basi columnæ accretis. *Labellum* basi columnæ insertum liberum integrum. *Columna* utrinque alata: alis petaloideis. *Clinandrium* dilatatum anticè suprà stigma prominens trilobum: lobo intermedio inflexo. *Anthera* verticalis, loculis basi divaricatis solutis in clinandrio incumbentibus. *Glandulæ* nudæ. *Stigma* ad basin columnæ.——OBS. Disa cornuta *quæ clinandrio non dilatato et præsertim anthera horizontali differt, fortè genus proprium* Pterygodio *affine* (REPANDRAM) *indicat, cui species multæ aliæ spicatæ referendæ sunt.*

D. *grandiflora;* laciniis perianthii exterioribus maximis ovali-lanceolatis acuminatis, calcare conico nutante, caule subbifloro.
Disa uniflora. *Berg. cap.* 348. *t.* 4. *f.* 7.
Disa grandiflora. *Linn. suppl.* 406. *Swartz. act. holm.* 1800. *p.* 210. *Willd. sp. pl.* 4.45. *Pers. syn.* 2. 507. *Ker in Journ. sc. vol.* 4. *p.* 205. *tab.* 5. *f.* 1.
Satyrium grandiflorum. *Thunb. prodr.* 4.

Caulis *glaber, pedalis, foliosus, erectus.* Folia *lanceolata, acuminata, patentia, basi vaginantia, caulis longitudine, superiora sensim minora.* Flores *terminales,* 1-3, *maximi, speciosi.* Bracteæ *ovato-lanceolatæ, acuminatæ, nervosæ, ovarii longitudine.* Perianthium *explanatum, patens,* laciniis exterioribus *maximis, ovali-lanceolatis, acuminatis, lateralibus coccineis planis, superiore (galeâ) rubescente, venosâ, cucullatâ, suprà basin calcare brevi obtuso vix ipsâ galeâ longiore;* interioribus *minimis ovatis ad basin columnæ.* Labellum *anticum, lineare, acuminatum, laciniis exterioribus brevius.* Columna *erecta utrinque appendiculata:* appendicibus *ovalibus dimidiatis erectis cum antherâ parallelis, margine laterali replicato, dentato, et cum dorso columnæ basi adnato;* clinandrium *magnum dilatatum, anticè declive, carnosum, suprà stigma prominens, trilobum: lobo medio inflexo.* Anthera *terminalis, erecta, apiculata, in parte posteriore clinandrii inserta,* connectivo *carnoso rigido,* loculis *parallelis, basi divaricatis, solutis, in lobis lateralibus clinandrii incumbentibus.* Pollinia *gracilia, granulosa, sulcata: glandulis nudis.* Stigma *carnosum, concavum, ad basin columnæ.* Obs. *In æstivatione columna arcuatur et clinandrii* complicati *lobi laterales tali modo suprà stigma* invertuntur, *ut glandulæ polliniorum faciem stigmatis viscosam tangunt.*

For this truly splendid plant, which has now flowered, probably, for the first time in Europe, we are indebted to Wm. Griffin, Esq., of South Lambeth, by whom it was received from the Cape of Good Hope. Our drawing was made in his hot-house, in June last.

It must be considered the type of the genus Disa, from which we believe it will be found necessary to separate D. *cornuta*, and several other species with spiked flowers, on account of their simple clinandrium and horizontal anthers, in the nature of which they approach the curious form of Pterygodium.

If the columna and stigma of this genus be examined when the flower is expanded, it seems impossible that any communication can take place between the pollen and the stigma, on account of the fleshy dilatation of the clinandrium which is interposed. But if the flower is dissected in an early state, it will then be found that the two lobes of this dilated process are folded together, and inverted over the stigma in such a manner, by the contraction of the columna, that the glands are applied immediately to the stigmatic surface. At this stage also of the flower, the grains of pollen, which eventually cohere in the form of an indefinite number of waxy, wedge-shaped bodies, are all separable by the aid of a little nitric acid, and appear to be in their greatest state of perfection.

A bulbous-rooted orchideous plant, of extremely difficult cultivation. It will probably succeed best in very fine sandy peat, never allowed to become saturated with moisture, and, during the period when the plant is at rest, kept quite dry. The *stem* is a foot high, clothed with spreading, lanceolate, acuminate leaves, sheathing at the base. The *flowers* grow two or three together, and are very large and showy; of the outer segments the two lower are oval-lanceolate, and bright crimson, the upper of a delicate pink colour, and furnished with a short conical spur behind. The two inner segments of the flower are very small, and placed at the base of the column; while their place in the flower seems to be supplied by two petal-like appendages adhering to the column on each side. The *labellum* is small, and linear-lanceolate.

J. L.

930

RODRIGUEZIA secunda.

Side-flowered Rodriguezia.

GYNANDRIA *MONANDRIA.*

Nat. ord. ORCHIDEÆ. Sect. *Epidendreæ.*
RODRIGUEZIA. Fl. Peruv. Perianthium tetraphyllum ringens: foliolis superioribus erectis; inferiore labello supposito. *Labellum* integrum anticum unguiculatum basi cornutum medio callosum: ungue columnâ parallelo. *Columna* teres apice barbata. *Anthera* unilocularis. *Pollinia* 2 posticè excavata filo elastico adhærentia. *Gynizus* lobatus.——Herbæ *epiphytæ.* (Americæ æquinoctialis.) Folia *carnosa avenia disticha vaginantia.* Spicæ *axillares secundæ.* Flores *speciosi.*

R. *secunda;* foliis oblongo-lanceolatis apice obliquis emarginatis basi laxis vaginantibus.
R. secunda. *Humboldt, Bonpl. et Kunth. nov. gen. et sp. pl.* 1 *p.* 367. *t.* 92. *synops.* 1.348.
R. lanceolata. *Bot. Cab.* 676.
Pleurothallis? coccinea. *Hooker, exot. flora,* 129.
Planta epiphyta, acaulis. Folia *disticha, basi vaginantia, decidua, patentia, ovali-lanceolata, complicata, enervia, apice obtusa, mucronulata, obliqua; superiùs basi incrassato, compresso, ancipite, corrugato, bulboideo.* Spicæ *axillares, multiflori, cernui, pedunculati, foliorum longitudine ante expansionem tritici spicam referentes.* Flores *pulchrè rosei, secundi.* Perianthium *4-partitum, ringens, laciniis superioribus ovalibus, conniventibus, inferiore cymbiformi, emarginato, labello supposito.* Labellum *liberum, unguiculatum; ungue faciei columnæ applicito, marginato, basi extùs carnoso, cornuto;* laminâ *cuneata, emarginata, obsoletè trinervi, medio carnosâ, basi callosâ, callo elevato marginato.* Columna *teres, clavata, pubescens, apice bidentata.* Clinandrium *nudum, retrorsùm declive, anticè pulvinatum.* Pollinia *duo, albida, cereacea, posticè excavata, filo communi elastico glandulæ affixa.* Anthera *unilocularis, carnosa, anticè truncata.* Gynizus *subquadratus, apertus, utrinque cornutus, apice demùm emarginatus.*

A beautiful parasitical orchideous plant, which was introduced about the year 1818, from Trinidad. It is more easily cultivated than the greater part of its tribe, growing freely, and flowering in abundance in a damp stove among rotten tan and decayed vegetable mould.

Whether this is the original species upon which the genus was founded by the authors of the Flora Peruviana, we dare not decide. There can be no doubt, however, that it is the same as the *R. secunda* of M. Kunth, and as that excellent Botanist considers his plant distinct from that of the Flora Peruviana, we have no hesitation in following him. It was found by Messrs. Humboldt and Bonpland in cool temperate places, in the neighbourhood of Carthagena, growing upon trunks of Crescentia Cujete, and flowering in October. It produces its beautiful spikes at the same time of the year in our stoves in great profusion. The labellum seems to be subject to some diversity of form and colouring in different individuals, but the differences are too inconstant to characterise even varieties. The spikes are usually nodding, as is elegantly represented in Dr. Hooker's Exotic Flora. Occasionally they are erect, as in the plant now figured, and as in that examined by M. Kunth.

The genus Gomeza of Mr. Brown is nearly allied to Rodriguezia, from which it is chiefly distinguished by its prominent stigma, by its 3-lobed labellum, and by the absence of any elongation from the base of the same part. The value of the latter character is, however, diminished by the consideration, that the slight elongation of the base of the labellum in Rodriguezia cannot properly be considered a spur, because it is solid, without any corresponding excavation between the labellum and columna.

The only other published genus with which it is necessary to compare Rodriguezia is Pleurothallis, which is essentially distinguished by the absence of an elastic filum bearing the pollen masses, by its axillary inflorescence, and by its solitary, or nearly solitary leaves, which neither sheath at the base, nor are seated on a bulbous base. It is necessary, however, to explain, that we consider the singular plant represented at fol. 759 of this work under the name of Pleurothallis punctata, and about the same time by Messrs. Loddiges, in their Botanical Cabinet, as a species of Gomeza (G. tenuiflora), to be essentially distinct from both those genera, from which it differs in several important particulars. Not to mention its radical inflorescence, which is a character of some value in distinguishing many genera of Orchideæ, the labellum is not parallel with the columna in any portion of its length; the anther, instead of being terminal, as in nearly all the genera of true Epidendreæ, is placed at the back of the columna! exactly as in Neottieæ; there is a large projecting callus at the back of the point of the columna, to which we have seen nothing similar in this tribe of plants; and the columna is tapered off to a point, in which this plant again resembles Neottieæ; while the secreting stigmatic surface is not transverse with respect to the columna as in Epidendreæ, but perpendicular, and like a slit in the face of the columna, occupying half its length. This curious plant may be named

NOTYLIA.

Perianthium tetraphyllum explanatum: foliolis superioribus æqualibus distantibus; inferiore oblongo bilobo labello supposito. *Labellum* divaricatum integrum medio callosum. *Columna* libera teres acuminata. *Anthera* postica unilocularis cum stigmate parallela. *Pollinia* duo integra filo elastico glandulæ affixa. *Clinandrium* planum apice callosum, posticè marginatum. *Stigma:* rima antica verticalis.——Herbæ *epiphytæ* (Americæ æquinoctialis). Folia *solitaria nervosa stipitata.* Spicæ *radicales.* OBS. Huic affinis est *Cymbidium dependens* Lodd.; planta nimirùm structurâ in Orchideis inauditâ, et genus proprium (CIRRHÆAM Nob.) efformans.
Sp. 1. Notylia punctata.
Pleurothallis punctata. *Suprà fol.* 759.
Sp. 2. Notylia multiflora.
Pleurothallis sagittifera. *Humb. Bonpl. et Kunth. nov. gen. et sp. pl.* 1.365. *t.* 91.

A stemless, parasitical plant. *Leaves* distichous, sheathing at the base, deciduous, spreading, oval-lanceolate, folded together, nerveless, blunt at the end, mucronulate, and oblique; the upper one thickened at the base, compressed, two-edged, wrinkled, bulb-like. *Spikes* axillary, many-flowered, cernuous, stalked, the length of the leaves before expansion resembling a wheat-ear. *Flowers* bright rose-coloured, one-sided. *Perianthium* 4-parted, ringent; the upper segments oval, conniving, the lower boat-shaped, emarginate, and placed under the labellum. *Labellum* unguiculate, separate; the *claw* applied to the face of the columna, edged, at the base outside fleshy and cornute; the *lamina* cuneate, emarginate, obsoletely 3-nerved, fleshy in the middle, and bearing an elevated margined callus at the base. *Columna* round, clavate, downy, 2-toothed at end. *Clinandrium* sloping backwards, naked, elevated like a cushion in front. *Pollen-masses* 2, whitish, waxen, hollowed out behind, fixed to the common elastic filum of the gland. *Anther* 1-celled, fleshy, truncate in front. *Gynizus* nearly square, open, with a little inflexed horn on each side, finally becoming emarginate at tip.

J. L.

935

STELIS ophioglossoides.

Narrow-leaved Stelis.

GYNANDRIA *MONANDRIA.*

Nat. ord. ORCHIDEÆ: *tribus* Malaxideæ. *Lindl. Coll. Bot. ined.*
STELIS Swartz. Pollinia 2 cereacea integerrima; v. omninò libera, v. apicibus viscidis cohærentia, caudiculâ glandulâque nullis. *Anthera*

terminalis opercularis 1-locularis decidua. *Stigma* subrotundum excavatum: rostello apiculato. *Columna* semiteres mutica. *Labellum* parvum sepalis interioribus nanis liberis subconforme. *Sepala* conniventia: exteriora basibus connata.——Herbæ *teneres, epiphytæ,* (Americæ æquinoctialis) *non bulbosæ.* Folia *solitaria.* Racemi *subsimplices.* Flores *minuti, herbacei, inodori et quandoque nocturni.*

S. *ophioglossoides;* foliis tenuibus lanceolatis apice sæpiùs recurvis, spicis laxis, floribus cernuis quaquaversis obsoletè hexagonis, sepalis interioribus truncatis, labello basi subcordato.
Epidendrum ophioglossoides. *Jacquin Amer.* 225. *t.* 133. *f.* 2.
Epidendrum trigoniflorum. *Swartz prodr.* 135.
Dendrobium ophioglossoides. *Swartz nov. act. ups.* 6.83.
Stelis ophioglossoides. *Swartz in Schrad. diar. bot.* 1799. *p.* 239. *Willd. sp.* 4.138. *Lodd. bot. cab. t.* 442.

Folia *solitaria, oblonga, glabra, plana, enervia, apice ferè integra, in petiolo carnoso, canaliculato attenuata.* Spica *terminalis, filiformis, multiflora, folio longior, vaginis parvis, truncatis, acuminatis interruptè vestita.* Flores *vespertini, minutissimi, resupinati, subsecundi, subrotundi, angulis* 6 *obsoletis.* Perianthium *subglobosum;* sepala exteriora, *ovata, obtusa, coriacea, basi connata, sub-fornicata, æqualia,* interiora *nana, carnosa, truncata.* Labellum *carnosum, cum columnâ parallelum, sepalis interioribus subconforme, laminâ depressâ, cordatâ.* Columna *brevissima semi-teres, sepalorum interiorum longitudine.* Anthera *terminalis, opercularis, unilocularis, decidua.* Pollinia 2, *libera, caudiculâ nullâ.*

A native of Trinidad, whence plants were sent by Mr. D. Lockhart to Mr. James Colvill, in whose nursery our drawing was made. A tender stove plant, requiring the treatment of other epiphytal Orchideæ.

We find much difficulty in ascertaining the true characters by which *S. ophioglossoides* and *micrantha* are to be distinguished from each other; the points noticed by Swartz appearing to be altogether inapplicable. To this species, which we have compared with authentic specimens from himself in the Banksian Herbarium, he attributes triangular flowers, while, in fact, they possess that character in a much less degree than S. micrantha, to which he assigns hexangular flowers. The latter species, elegantly figured by Dr. Hooker in his Exotic Flora, differs from the subject of this article in having a rigid obtuse leaf, and a compact raceme of nodding flowers, which, when closed, present a distinctly triangular appearance. In all which characters it is opposed to S. ophioglossoides. The leaves of the latter are more membranous, and are almost always recurved at the end, where they are sometimes toothed; the flowers are larger and less compactly arranged in the spike; there is also a difference in the form of the inner floral segments and labellum. S. micrantha would therefore be more satisfactorily characterised thus:

S. *micrantha* (Swz.), foliis oblongo-lanceolatis obtusis coriaceis, floribus nutantibus secundis trigonis, sepalis interioribus apiculatis labelli conformibus.

Of S. ophioglossoides we possess native specimens from St. Vincents, communicated to us by our friend Dr. Hooker; they in no essential point differ from the plants in our gardens.

Leaves solitary, oblong, smooth, flat, nerveless, nearly entire at the end, narrowed into a channelled fleshy stalk. *Raceme* terminal, filiform, many-flowered, longer than the leaf, dotted at intervals with small truncate acuminate vaginæ. *Flowers* very minute, resupinate, somewhat on one side, roundish, with 6 obsolete angles. *Perianth,* with the outer sepals, ovate, obtuse, coriaceous, connate, and somewhat fornicate at the base, equal; inner, dwarf, fleshy, and truncated. *Lip* fleshy, parallel with the column, nearly of the same form as the inner sepals, with a depressed cordate lamina. *Column* very short, half-round, the length of the inner sepals. *Anther* terminal, opercular, 1-celled, deciduous. *Pollen-masses* 2, loose, with no caudicula.

J. L.

CATTLEYA Forbesii.

Forbes's Yellow Cattleya.

GYNANDRIA *MONANDRIA.*

Nat. ord. ORCHIDEÆ. Tribus VI. Epidendreæ *Lindl. Coll. Botanica in Append.*

CATTLEYA Lindl.— *Perianthium* resupinatum patens: laciniis subæqualibus. *Columna* libera, semiteres, labello eroso cucullato amplexa. *Anthera* infrà-apicilaris, oprecularis, persistens, columnæ apice subulato supertecta, 4-locularis; septis completis membranaceis marginatis. *Massæ pollinis* 4, lenticulares, per pares filo elastico granulato in ipsis reflexo connexæ.——Herbæ *parasiticæ* (Americæ æquinoctialis); bulbis *fasciculatis;* foliis *solitariis, carnosis, enervibus;* floribus *terminalibus, geminis, grandibus, subodoris.* Lindl. l. c. tab. 33.

C. *Forbesii;* foliis oblongis obtusis planis, sepalis lanceolatis subæqualibus; interioribus angustioribus undulatis obtusis, labelli trilobi lobo medio cordato lunato argutè dentato apice saccato. *Lindl. l. c.*

Planta epiphyta, è caudice tortuoso, cicatrizato, repente, promens caules plures, unipedales, teretes, sursùm clavatos, vestigiis membranaceis arctè appressis, striatis, albidis, internodiis longioribus vestitos. Folia *in fastigio, bina, plana, coriacea, sessilia, ad marginem scabrella, enervia, ovali-lanceolata, subopposita, obtusa, cum apiculo.* Flores *bini, inodori cum pedunculo terete foliis longiores.* Bractea *parva, non spathacea, sphacelata, rigida, acuminata, vaginans, pedunculo multò brevior.* Perianthii *laciniæ lineari-lanceolatæ, acutæ, extùs luteo-fuscæ, intùs ochraceo-virides; interiores angustiores, ad marginem crispæ.* Labellum *pallidissimè flavidum, intùs luteum, rubro venosum et punctatum, ad apicem trilobum; lobis lateralibus rotundatis, porrectis, basi interiore sub-cordatis; intermedio cordato, rotundato, explanato, crispo, argutè dentato, apice saccato.* Labellum *per axin lineis duabus elevatis.* Columna *inclusa, pallidè ochroleuca, purpureo lineata.* Stigma *purpureo marginatum.*

Native of Rio Janeiro, where it was found growing upon decayed trees, by the late Mr. Forbes, by whom it was transmitted to the Horticultural Society in 1823. Our drawing was made in the Chiswick Garden, in June 1824.

The genus Cattleya now contains four well-ascertained species, of which this is the least beautiful. They are all elegant epiphytes, growing among the decayed vegetable matter found upon trees. If planted in a similar substance, and kept in a hot damp stove, they may be easily subjected to cultivation. They are all, however, at present, exceedingly rare plants. Cattleya Loddigesii, which is now the commonest, only exists in a few collections; of C. labiata, which is the finest, we are acquainted with but two plants, both in the possession of Mr. Cattley. The present species is in no other collection than that of the Horticultural Society, where there is but one certain plant. A single individual of C. citrina is also supposed to exist in the latter establishment, but it has not yet blossomed.

It will have been remarked by our readers, that we have for some time been endeavouring to establish divisions in the extensive tribe of Orchideæ, comprising the last section in Mr. Brown's system, and commonly called Epidendrums. Indeed, so long since as the year 1820, we remarked (Coll. Bot. t. 15.) that the parasitical Orchideæ, with waxy pollen-masses, were capable of being separated into at least two divisions, distinguished by certain peculiar and important modifications of the pollen-masses and their appendages. Having at length, in another publication, applied these principles to all the Orchideous plants of which we have any satisfactory information, we think a brief exposition of their final arrangement may not be unacceptable to the readers of the Botanical Register, especially as the work of which we speak is in the hands of but few persons.

The Orchideous plants with waxy pollen are susceptible of the three following modifications of that substance. In the first state the pollen-masses are seated upon a transparent, generally elastic, body, adhering to the upper edge of the stigma by a gland. The elastic body has been named by Richard *caudicula*, and answers to the extensile pedicel of the common Orchis, and to the filamentous axis of Neottieæ. The nature of the gland is more doubtful; but it exists in several tribes of Orchideæ, the pollen of which is not waxy. Plants so characterised we term *Vandeæ*. The second form of pollen-mass consists in the absence of the gland, and in the caudicula being divided into two or four filiform segments, covered with pollen not in a state of cohesion, and folded back over the margins of the pollen-masses in such a way as to be interposed between them and the Clinandrium, or Bed of the Anther. These constitute true *Epidendreæ*. The third state of the pollen-masses depends upon the total absence of both caudicula and gland. Pollen of this nature may be considered to vary the most from the general structure which obtains in Orchideæ, and to differ from ordinary pollen only in the intimate cohesion of all the granules. Plants with this character are called *Malaxideæ*.

The following is the general arrangement of Orchideæ which has been proposed in the work above mentioned:—

1. §. *Neottieæ.* Pollen simplex v. è granulis laxè cohærentibus.
   Tribus I. Neottieæ. Anthera stigmate parallela erecta.
   Tribus II. Arethuseæ. Anthera terminalis opercularis.
2. §. *Orchideæ.* Pollen in granulis demùm cereaceis numero indefinitis cohærens.
   Tribus III. Gastrodieæ. Anthera terminalis opercularis.
   Tribus IV. Ophrydeæ. Anthera terminalis erecta v. inversa. Pollinia caudiculata.
3. §. *Epidendreæ.* Pollen in granis demùm cereaceis numero definitis cohærens.
   Tribus V. Vandeæ. Pollinia caudiculâ diaphanâ v. glandulâ stigmati affixa.
   Tribus VI. Epidendreæ. Pollinia caudiculis filiformibus pulvereis replicatis stigmati affixa.
   Tribus VII. Malaxideæ. Pollinia libera; nunc ad apicem materie viscidâ v. pulvereâ aut granulosâ cohærentia.
4. §. *Cypripedieæ.* Antheræ laterales fertiles: intermediâ sterili petaloideâ.
   Tribus VIII. Cypripedieæ.

Obs. *Affinitas* Ordinis summa cum Scitamineis per Neottieas; inferior magisque obscura cum Asphodeleis per Thelymitram et Junceis per Malaxideas.

J. L.

THE

BOTANICAL REGISTER:

CONSISTING OF

Coloured Figures

OF

EXOTIC PLANTS,

CULTIVATED IN

BRITISH GARDENS:

WITH THEIR

HISTORY AND MODE OF TREATMENT.

THE DESIGNS BY

SYDENHAM EDWARDS,

AND OTHERS.

VOL. XII.

——viret semper——nec fronde caducâ
Carpitur.

LONDON:

JAMES RIDGWAY, 169 PICCADILLY.

1826.

963

TRIBRACHIA pendula.

Drooping Tribrachia.

GYNANDRIA *MONANDRIA.*

Nat. ord. Orchideæ. *Tribus* Malaxideæ *Lindl. coll. bot. in app.*

TRIBRACHIA. Suprà, t. 832.—*Pollinia* duo cereacea posticè sulcata: caudiculâ et glandulâ nullis. *Anthera* terminalis, opercularis, decidua, semibilocularis membranacea. *Columna* apice bicirrhosa. *Labellum* posticum, integrum, cum basi productâ columnæ unguiculatum. *Sepala* patentia: lateralibus exterioribus cum basi columnæ connatis; interioribus nanis.—— Herbæ *epiphytæ, acaules, bulbosæ* (Asiæ temperatæ et Africæ æquinoctialis); caudicibus *reptantibus*; scapis *radicalibus*; floribus *parvis*. Lindley *collect. bot.* tab. 41, A.

T. *pendula;* foliis ovalibus subtùs discoloribus apice inæqualibus emarginatis in bulbis ovatis compressis solitariis, scapis pendulis, spicis imbricatis, sepalis ovatis acuminatis: interioribus obovatis. *Lindl. l. c. in textu.*

Bulbi *ovati, carnosi, subteretes.* Folia *solitaria, 2-3-uncialia, ovalia, patentia, apice obliquè emarginata, carnosa, plana, subtùs purpureo maculata.* Scapi *radicales, penduli, bracteis distantibus angustis acutis squamosi.* Flores *in spicâ quadrifariàm imbricati, parvi, recti.* Bracteæ *florum longitudine, ovatæ, acuminatæ, diaphanæ, rubicundæ.* Ovarium *breve, rectum, turbinatum, ecostatum.* Perianthium *regulare, patens, ovario longius.* Sepala exteriora *æqualia, triangularia, acuta, valvata, enervia, viridia, ciliata: lateralia basibus cum columnæ pede connatis;* interiora *abbreviata, obovata, pallidè viridia, obtusa.* Labellum *parvum, rubro-purpureum, ovatum, integerrimum, per axin carnosum, basi excavatum, cum sinu sepalorum lateralium exteriorum articulatum, in columnâ incumbens.* Columna *brevis, recta, semiteres, ad basin producta usque ad sinum sepalorum exteriorum lateralium.* Stigma *excavatum, marginibus inflexis semiclausum, apice utrinque aristatum.* Anthera *terminalis, opercularis, persistens, ut credo 1-locularis.* Pollinia *duo, subrotunda; quodque intùs sulco exaratum obscuro, lobum minorem simulante.* Caudicula et glandula *nullæ.*

Sent to the Horticultural Society in 1822, from Sierra Leone, by Mr. George Don. Our drawing was made at Mr. Colvill's nursery in October 1824. Like nearly all the epiphytal orchideous plants of Sierra Leone, this plant is very impatient of culture. We believe it will succeed better in a hot, damp, shady frame, than in any other situation.

All the other species of this genus which have been published are natives of Nepal; but we have reason to believe, that there are several other Sierra Leone species in the gardens, besides that which is now for the first time figured.

A very tender stove plant, with small ovate, fleshy, roundish bulbs. *Leaves* solitary, two or three inches long, oval, spreading, obliquely emarginate at the end, fleshy, flat, beneath spotted with purple. *Scapes* radical, pendulous, scaly, with distant narrow acute bracteæ. *Flowers* imbricated in a 4-cornered spike, small, not inverted. *Bractes* the length of the flowers, ovate-acuminate, transparent, reddish. *Ovarium* short, straight, turbinate, without ribs. *Perianthium* regular, spreading, longer than the ovary. *Outer sepals* equal, triangular, acute, valvate, nerveless, green, ciliated; the lateral ones united at their bases with the foot of the columna: *inner* shorter, obovate, pale green, obtuse. *Labellum* small, red-purple, ovate, entire, fleshy in the axis, excavated at the base, jointed with the sinus formed by the cohesion of the lateral outer sepals, incumbent upon the columna. *Columna* short, straight, half-round, elongated at the base as far as the sinus of the lateral exterior sepals. *Stigma* hollowed out, half-closed up by the inflexed margins, awned at the upper end on each side. *Anther* terminal, opercular, persistent, as it appears to me, one-celled. *Pollen-masses* two, roundish, each furrowed on the inside so as to appear as if furnished with a smaller lobe. Caudicula and gland none.

J. L.

966

CATASETUM cristatum.

Crested Catasetum.

GYNANDRIA *MONANDRIA.*

Nat. ord. ORCHIDEÆ. *Tribus* Vandeæ *Lindley.*
CATASETUM.—Suprà, vol. 10. *tab.* 840.

C. *cristatum;* sepalis patentibus; superioribus conniventibus, labello cristato saccato expanso. *Lindley in hort. trans. vol.* 6, *p.* 83.

Habitus *omninò C. tridentati, sed folia paulò latiora et magis nervosa.* Spica *radicalis, suberecta, multiflora, foliis brevior.* Bracteæ *parvæ, ovatæ.* Ovarium *unciale longiusve teres.* Perianthium *explanatum, viride;* sepalis *duobus* anterioribus *patentibus, oblongo-lanceolatis, canaliculatis, tribus* posticis *erectis subparallelis: interioribus brevioribus tenuioribus.* Labellum *cum columnâ articulatum, patens, basi saccatum, in margine et disco processubus plurimis carnosis albidis cristatum.* Columna *omninò Cataseti.*

The first account which appeared of this singular plant was in the sixth volume of the Transactions of the Horticultural Society, in a Report upon the New or Rare Plants which had flowered in their Garden at Chiswick, from its first formation to March 1824. To this establishment the plant was introduced from Bahia de St. Salvador, in 1823, by Mr. George Don. Our drawing was made in the Society's garden, in August 1824.

The present plant differs from all the other species of Catasetum in the form and nature of its labellum, and in the manner in which the sepals are expanded. But in every other respect, in its habit, foliage, inflorescence, and especially in the structure of its columna, it entirely agrees with Catasetum.

The unimportance of the peculiarity which exists in the labellum is manifested in a singular manner by a curious monster of this plant which we observed on an individual in the Horticultural Society's Garden. Among flowers of the ordinary structure, two or three others were observed in which the labellum was precisely of the same nature as that of Catasetum tridentatum, that is to say, destitute of the crested appendages, and perfectly galeate and naked.

An extremely rare stove plant, easily cultivated in decayed earth, and propagated very slowly by the fleshy bulbs on which the leaves are seated.

The plants have altogether the appearance of C. *tridentatum;* but the leaves are rather broader, and more nerved. *Spike* radical, nearly erect, many-flowered, shorter than the leaves. *Bracteæ* small, ovate. *Ovary* an inch or more long, rounded. *Perianth* expanded, green; the *two anterior sepals* spreading, oblong-lanceolate, channelled; the *three posterior* erect, and nearly parallel; the innermost being shorter and thinner. *Labellum* jointed with the columna, spreading, saccate at base, crested on the margin and disk with numerous fleshy, whitish processes. *Columna* in all respects that of Catasetum.

J. L.

978

ERIA rosea.

Pink Chinese Eria.

GYNANDRIA *MONANDRIA.*

Nat. ord. ORCHIDEÆ. *Tribus* Malaxideæ *Lindl.*
ERIA.—Suprà, vol. 11. *fol.* 904.

E. *rosea;* bulbis costalis rugosis, foliis solitariis coriaceis lanceolatis, spicâ axillari pauciflora, sepalis glabris.

Herba *cæspitosa, bulbosa, bulbis epigæis, ovatis, irregulariter costatis, rugosis, vestigiis squamarum parcè vestitis, sæpiùs omninò denudatis.* Folia *terminalia in apice bulborum, solitaria, coriacea, atro-viridia, lanceolata, petiolata; petiolo crasso terete hinc canaliculato.* Spica 2-3 *flora, foliis multò brevior, ex axillâ folii terminalis, ante squamarum inferiorum decessum.* Flores *rosei.* Sepala exteriora *ovato-oblonga: anteriora latâ basi dorso carinata, basi subsaccata; interiora membranacea oblonga subunguiculata.* Labellum *posticum, in columnâ pronum, trilobum: lobis lateralibus erectis, rubidis, venosis, intermedio obtuso patente;* ungue *cristis duabus elevatis pallidis;* laminâ *cristis tribus luteo-aurantiacis.* Anthera *terminalis, opercularis, decidua, subrotunda, sublutea, posticè purpurascens, bilocularis, loculo utroque semibipartito.* Pollinia 8, *apice materie viscida cohærentia, glandula nulla.* Stigma *oblongum, transversum.*

This pretty addition to the genus Eria was brought from China for the Horticultural Society by Mr. J. D. Parks, in 1824. Our drawing was made from a plant which flowered in a stove in the Chiswick Garden, in October last.

The species is easily cultivated in moss and decayed vegetable mould, in which it flourishes more than almost any other plant of the family.

This is the first species of Eria in which the flowers are free from a greater or less degree of downiness; all the others having their flowers protected by hairs in a remarkable manner. The carinate mid-rib of the exterior sepals is also peculiar to this.

A dwarf plant, forming little patches by its fleshy, ovate, ribbed, rugose bulbs, which are generally quite naked. *Leaves* terminal, on the end of the bulbs, solitary, coriaceous, dark green, lanceolate, stalked; the stalk thick, round, channelled on the upper side. Spike 2 or 3-flowered, much shorter than leaves, proceeding from the axilla of the terminal leaf before the scales of the bulb fall off. *Flowers* pink. *Outer sepals* ovate-oblong: the lateral with a broad base, carinate at back, somewhat saccate at base; the *inner* membranous, oblong, with a short claw. *Labellum* at the back, prostrate on the columna, three-lobed; the lateral lobes erect, reddish, veiny, the intermediate obtuse and spreading; the *claw* with two raised pale crests; the *lamina* with 3 yellowish orange-coloured crests. *Anther* terminal, opercular, deciduous, roundish, yellowish, a little purple at back, 2-celled; each cell being half divided in two. *Pollen-masses* 8, cohering at the end by a viscid matter, but with no gland. *Stigma* oblong, transverse.

J. L.

981

SARCANTHUS rostratus.

Long-beaked Sarcanthus.

GYNANDRIA *MONANDRIA.*

Nat. ord. ORCHIDEÆ. *Tribus* Vandeæ *Lindl.*

SARCANTHUS.—Pollinia 2, cereacea, posticè sulcata v. lobata: caudiculâ variâ. *Anthera* bilocularis. *Stigma* excavatum v. quadratum: rostello vario. *Columna* semiteres inappendiculata. *Labellum* subintegrum difforme cum columnâ articulatum, calcaratum: calcare intùs appendiculato. *Sepala* patentia subæqualia. *Herbæ epiphytæ* (*Indiæ orientalis et Chinæ*) *caulescentes, radicibus caulinis tortuosis, foliis distichis planis v. teretibus, floribus vittatis v. fasciatis, racemis foliis oppositis.* Lindl. coll. bot. t. 39. B.

S. *rostratus;* foliis lanceolatis planis subrecurvis, spicâ simplice horizontali foliis subæquali, sepalis ovalibus patentibus subæqualibus, labello antheràque rostratis. *Lindl. l. c.*
Vanda rostrata. *Lodd. bot. cab. tab.* 1008.
Vanda recurva. *Hook. ex. fl. tab.* 187.

Epiphyta ramosa. Caules *crassi, purpurei, teretes.* Folia *lanceolata, carnosa, subrecurva.* Spica *pedunculata, oppositifolia, horizontalis, foliis subæqualis.* Ovarium *teres, album.* Sepala *patentia, æqualia, ovalia, lutescentia, rubro-vittata.* Labellum *carnosum, amœne purpureum, calcaratum, apice rostratum, incurvum;* clacar *obtusum, ovario brevius, intùs 1-loculare, anticè valdè carnosum* appendice *operculari glabro simplice.* Columna *erecta, clavata, semiteres.* Stigma *subrotundum, excavatum, rostello in rostro longo producto.* Anthera *conformis, bilocularis.* Pollinia 2, *biloba;* caudiculâ *subulatâ elongatâ;* glandula *parva.* Lindl. l. c.

This plant belongs to the same genus as the Aerides paniculata of fol. 220 of this work, and has been separated from Vanda, to which it was subsequently referred, by the writer of this article.

The species of Sarcanthus " differ from *Vanda* in the form and structure of their labellum, which is never saccate, but has always a spur with one or more appendages in its inside, in the texture of their perianthium, and in habit. The plants which agree with *Sarcanthus* as above defined are *Vanda teretifolia* and *paniculata*, an unpublished species (*Sarcanthus succisus*) from China, and probably some plants at present referred to *Aerides. Vanda trichorhiza* of Hooker, which is nearly allied to, if not the very same as, *Epidendrum triste* of Forster; and the *Aerides Arachnitis* of Swartz, appear to be species of *Cymbidium*, a genus which differs from *Vanda* in scarcely any thing beyond the absence of a spur from the lip, and in the articulation of the latter with the columna."

A native of China, whence it was imported by the Horticultural Society in 1821. Our drawing was made at Mr. Colvill's Nursery in June 1824.

A branching epiphyte. *Stems* thick, purple, rounded. *Leaves* lanceolate, fleshy, somewhat recurved. *Spike* stalked, opposite a leaf, horizontal, as long as leaves. *Ovary* round, white. *Sepals* spreading, equal, oval, yellowish, banded with red. *Labellum* fleshy, bright purple, spurred, rostrate at point, incurved; *Spur* obtuse, shorter than ovarium, one-celled inside, very fleshy in front, with an opercular, smooth, simple appendage. *Columna* erect, clavate, half-round. *Stigma* roundish, hollowed out, with a rostellum lengthened into a long beak. *Anther* of the same figure, 2-celled. *Pollen-masses* two, 2-lobed; *caudicula* subulate, elongated; gland small.

J. L.

985

PELEXIA spiranthoides.

Spiranthes-like Pelexia.

GYNANDRIA *MONANDRIA.*

Nat. ord. ORCHIDEÆ. *Sect.* Neottieæ *Lindl.*

PELEXIA.—Poiteau in Richard. Orch. annot. p. 37. *Sepala* in cylindro conniventia: lateralibus exterioribus dependentibus basi labello connatis. *Labellum* integrum, porrectum, marginibus columnam cum eo parallelam amplectentibus, basi in calcare cum ovario connato productum.—— Herba *habitu omninò* Spiranthis elatæ.

P. *spiranthoides.*
Satyrium adnatum. *Swartz prodr.* 118.
Neottia adnata. *Swartz fl. ind. occ.* 3. *p.* 1409. *Willd. sp. pl.* 4. 75.

Herba *habitu omninò Spiranthis elatæ.* Folia *radicalia, ovato-lanceolata, subundulata,* 3-5 *nervia, glaberrima, longè petiolata, more* Ponthievæ petiolatæ. Scapus *centralis, pedalis et ultrà, teres, basi squamis vaginatus, apicem versùs pilosus.* Bracteæ *ovato-lanceolatæ, acuminatæ, apice glabræ.* Flores *sessiles, bractearum circiter longitudine.* Ovarium *leviter pubescens.* Sepala exteriora *pallidè viridia: lateralibus linearibus intùs albis, dependentibus, et cum basi labelli connatis;* interiora *alba superiori arctè appressa.* Labellum *basi calcaratum: calcare cum ovario connato; marginibus cum columnâ parallelis; lamina ovatâ acutâ integrâ.* Columna *Spiranthis.*

This exceedingly rare plant was brought from the island of St. Vincent's, by Mr. James M'Rae, in 1823, and by him presented to the Horticultural Society, in whose garden, at Chiswick, our drawing was made in March last.

It is an herbaceous plant, with the habit of *Spiranthes elata,* and requires the same mode of treatment. It is propagated slowly by dividing the roots.

The leaves grow from the root, and are ovate-lanceolate, somewhat wavy, 3 or 5-nerved, very smooth, on long stalks, are very like those of Ponthieva petiolata. *Scape* a foot and more in height, round, sheathed with scales at the base, hairy towards the upper part. *Bracteæ* ovate-lanceolate, acuminate, smooth at tip. *Flowers* sessile, about as long as bracteæ. *Ovarium* slightly downy. *Outer sepals* pale green: the lateral white inside, linear, hanging down, and connate with the base of the labellum; the *inner* white, and closely pressed against the upper. *Labellum* calcarate at base; the spur being connate with the face of the ovarium; its margins parallel with the columna; its lamina ovate, acute, and entire. The *Columna* as in Spiranthes.

J. L.

989

MEGACLINIUM falcatum.

Falcate Megaclinium.

GYNANDRIA *MONANDRIA.*

Nat. ord. ORCHIDEÆ. *Tribus* Malaxideæ, *Lindl.*

MEGACLINIUM.—Lindl. suprà, vol. 10. *fol.* 832, *in notis. Coll. bot. append. Orch. scel. p.* x.—Bulbophylli pars. *Pet. Th.*—Ephippium; sect. 1. *Blume. Flor. Nederl.* 308—*et fortè* Diphyes *ejusdem, sed habitus obstat.*

Pollinia 4 cereacea æqualia geminatim cohærentia, caudiculâ glandulâque nullis. *Anthera* terminalis, opercularis persistens minuta unilocularis. *Stigma* parvum intrusum, rostello emarginato. *Columna* plana, abbreviata, apice bicuspidata. *Labellum* integerrimum cum pede columnæ elasticè articulatum. *Sepala* exteriora basi connata: superiore difformi, interioribus nanis.——Herbæ *epiphytæ, oligophyllæ, cæspitosæ, bulbosæ* (Africæ et Asiæ inter tropicos). Scapi *radicales simplices;* rachidibus *dilatatis.*

Sect. I. PLATYCLINIUM. *Lindl. ined.*

M. *falcatum;* foliis binis ovalibus emarginatis biplicatis, rachide compressa falcata crenata, perianthii lacinia superiore obtusa apice utrinque callosa, lateralibus exterioribus reflexis bidentatis, interioribus minimis subulatis obtusis.

Bulbi *aggregati, cæspitosi, oblongi vel ovati, demùm subtetragoni, squamis rigidis striatis sphacelatis ad basin imbricati.* Folia *bina, erecta, ovalia, emarginata, utrinque plicata, margine purpureo, suprà viridia, subtùs pallidiora: junioribus subtùs purpureo puncticulatis.* Scapus *radicalis, teres, squamis paucis vaginantibus, clavatus, in rachide desinens compressâ, atropurpureâ, falcatâ, crenatâ, utrinque in facie floriferâ.* Flores *sessiles, parvi, in dentibus faciei rachidis inserti, facie suâ rachidem respiciente.* Perianthium *ringens, laciniis* exterioribus *basi connatis: superiore erecto, oblongo, obtuso, apiculato, apice utrinque callis duobus luteis; inferioribus concavis, basi connatis, latè ovatis, reflexis, apice bidentatis obliquis; interioribus parvis, subulatis, obtusis, viridibus, apice luteis.* Labellum *ovatum, obtusum, basi excavatum, cum columnâ elasticè articulatum.* Columna *magna, marginata, in ovario incumbens, cum basi sepalorum lat. ext. connata, plana, nitida basi producta, apice tridentata, dentibus lateralibus rotundatis, intermediâ antheriferâ, in medio stigma minutum intrusum fovente.* Anthera *minima, planiuscula, persistens, unilocularis, cristata.* Pollinia 4, *cereacea, binatim cohærentia, lutea, compressa,* glandula *et* caudicula *nullis.*

Sent to the Horticultural Society from Sierra Leone, in 1822, by Mr. George Don. It flowered, for the first time, in last April, in the stove, where it grows more readily than most of its compatriots, which are usually extremely impatient of cultivation. It should be planted in rich vegetable mould.

This genus differs from Bulbophyllum of Du Petit Thouars, in the absence of both gland and caudicula from the pollen masses, in the more manifest articulation of the labellum, which is never fringed, nor much unguiculate, with the columna, and in the curiously dilated state of the rachis. In the latter character, the 1st section of Dr. Blume's genus Ephippium appears to offer a transition to Bulbophyllum, to which his second section of the same genus may be actually referable. Whether Diphyes of the same author is really distinct from Megaclinium, we should almost doubt, were it not that the natural habit of the plants referred to it appears to be very different.

J. L.

1002

EULOPHIA streptopetala.

Twisted-petaled Eulophia.

GYNANDRIA *MONANDRIA.*

Nat. ord. ORCHIDEÆ. *Tribus* V. Vandeæ. *Lindl. coll. et Orch. scel.*

EULOPHIA.—R. Br. Pollinia 2, posticè sulcata; *caudiculâ* lineari planâ, *glandulâ* reflexâ. *Anthera* semibilocularis, terminalis, opercularis, decidua. *Stigma* excavatum, *rostello* acuto. *Columna* libera, semiteres, aptera. *Labellum* cucullatum trilobum calcaratum: lobo medio cristato, cum columna articulatum. *Sepala* subæqualia, libera.——Herbæ (*Africæ tropicæ* et *Indiæ orientalis*) foliis *rigidis, lanceolatis, nervosis, radicalibus,* scapis *radicalibus,* floribus *sparsis spicatis.*

E. *streptopetala;* foliis lineari-lanceolatis nervosis, scapis simplicibus, sepalis exterioribus oblongis obtusis: interioribus duplo majoribus coloratis basi tortis, labelli lobo medio rotundato emarginato: calcare conico abbreviato.

Folia *è basi bulbosâ squamosâ assurgentia, lineari-lanceolata, acuta, 3-nervia, basi vaginantia,* venis primariis *parallelis, contiguis,* venulis propriis *paucissimis interjectis:* communibus *nullis.* Scapus *radicalis, erectus, teres, vaginatus.* Spica *elongata, multiflora.* Flores *solitarii, speciosi.* Bracteæ *ovato-lanceolatæ, ovario breviores.* Perianthium *patens.* Sepala *exteriora oblonga, obtusa, patentia, olivaceo-viridi maculata, basi, nisi ad imum labellum, omninò distincta; duo interiora majora, lutea, subrotunda, venosa, unguis torsione reversa, basi omninò libera.* Labellum *luteum, liberum, cum basi columnæ articulatum, calcaratum, trilobum; lobis lateralibus erectis, intermedio oblongo, retuso, suprà 3-costato;* calcare *recto, ovario duplo breviore.* Columna *libera, erecta, semiteres, subclavata;* clinandrium *planiusculum, anticè declive;* rostellum *ovatum;* stigma *excavatum.* Anthera *ovata, ecristata, anticè in rostro pallido cartilagineo acuminata, semibilocularis, valvulis inferioribus subcompletis, membranaceis.* Pollinia *duo, oblonga, posticè sulcata;* caudiculâ *planâ, lineari, cartilagineâ,* glandulâ *parvâ, brunneâ.*

A fine new addition to the genus Eulophia, for a drawing of which we have to thank Mr. Colvill, at whose Nursery it was made in March last. A tender stove plant, requiring to be cultivated in well-drained pots of light vegetable mould, in which a small portion of loam and silver sand has been mixed.

Said to be a native of Brazil; but this we conceive must be an error; it is more probably an inhabitant of either tropical Africa or India.

Leaves rising from a scaly, bulbous base, linear-lanceolate, acute, 3-nerved, sheathing at base; *venæ primariæ* parallel, contiguous, *venulæ propriæ* very few, *communes* none at all. *Scape* radical, erect, round, sheathed. *Spike* long, many-flowered. *Flowers* solitary, shewy. *Bracteæ* ovate-lanceolate, shorter than the ovarium. *Perianthium* spreading. *Exterior sepals* oblong, obtuse, spreading, spotted with olive-green, at the base, excepting at the bottom of the labellum, entirely distinct; the two inner larger, yellow, roundish, veiny, twisted round by a turn in their unguis, quite separate at base. *Labellum* yellow, distinct, jointed with the base of the columna, calcarate, 3-lobed; lateral lobes erect, middle one oblong, retuse, with 3 ribs above; *spur* straight, twice as short as ovarium. *Columna* distinct, erect, half-round, somewhat clavate; *clinandrium* flattish, sloping forwards; *rostellum* ovate; *stigma* excavated. *Anther* ovate, without a crest, lengthened in front into a pale cartilaginous beak, half two-celled, with the lower valves nearly complete, and membranous. *Pollen masses* 2, oblong, furrowed behind; *caudicula* flat, linear, cartilaginous; *gland* small, brown.

J. L.

1007

ONCIDIUM pubes.

Olive-green Oncidium.

GYNANDRIA *MONANDRIA.*

Nat. ord. ORCHIDEÆ. Tribus V. Vandeæ *Lindl.*

ONCIDIUM Swz.—*Pollinia* duo cereacea posticè sulcata: *caudiculâ* planâ subulatâ; *glandulâ* minutâ. *Anthera* infrà-apicilaris terminalis opercularis decidua ½-bilocularis clinandrio marginato. *Stigma* excavatum apertum, *rostello* vario. *Columna* libera semiteres apice alata. *Labellum* cum columna continuum indivisum disco tuberculato. *Sepala* libera patentia subæqualia: 2 inferioribus labello suppositis nunc connatis.——Herbæ, *rarò bulbosæ,* (Americæ æquinoctialis), foliis *planis carnosis,* scapis *radicalibus paniculatis,* floribus *sæpius luteis maculatis inodoris.*

---

§. ** *Perianthii foliola* 2 *anteriora connata.* Lindl. coll. bot. 27.

O. *pubes;* bulbis subcylindricis monophyllis, foliis lanceolatis nervosis, paniculâ simplice multiflorâ subsecundâ, sepalis 4 fasciatis: inferiore minore bidentato, labello pandurato, columnæ alis linearibus obtusis, stigmate rostelloque pubentibus.

Bulbi *cæspitosi,* 6 *uncias longi, teretes, compressi, corrugati.* Folia *solitaria, bulbis circiter duplo longiora, elliptico-lanceolata, utrinque acuta, coriacea, plicata, obsoletè nervosa.* Scapus *radicalis, debilis, teres, foliis longior, squamis paucis distantibus membranaceis vestitus.* Panicula *simplex, subsecunda, patens, multiflora.* Bracteæ *ovatæ, membranaceæ.* Flores *vix resupinati, extùs olivacei, intùs spadiceo-fasciati et maculati.* Sepala 4, *conniventia; 3 superiora obovata obtusa: lateralibus subundulatis; inferius dependens, labello suppositum, oblongo-lanceolatum, canaliculatum, apice bilobum, sepalis superioribus multò minus.* Labellum *porrectum, unguiculatum, fusco-rubrum, panduratum, cum columnæ basi continuum;* ungue *carnoso, margine recurvo, suprà bicristato:* cristâ posteriore *depressâ, transversìm rugosâ, anticè bicorni,* anteriore *cordatâ, anticè obsoletè 3-dentatâ, dente medio submembranaceo fornicato,* lobo *lineari canaliculato utrinque è regione cristæ anterioris dependente:* ungue *suo subrotundo undulato subrhomboideo, rugis plurimis in disco uniserialibus elevatis carnosis.* Columna *compressa, pubescens, basi convexa, apice utrinque alata, alis linearibus integris porrectis, apice carnosis, truncatis.* Clinandrium *marginatum, denticulatum.* Stigma *excavatum, ovale, basi villis inflexis clausum!* Anthera *infra-apicilaris, subrotunda, apice recurva, bicornis: cornubus ascendentibus obtusis fimbriatis.* Caudicula *subulata.* Glandula *minuta.* Rostellum *nullum.*

Native of Rio Janeiro, whence roots were sent to the Horticultural Society by Mr. David Douglas, in 1824. The plant from which this drawing was taken flowered in the stove in March of the present year.

Easily cultivated in decayed vegetable soil, in a good stove, not exposed to the direct rays of the sun. Being one of the bulbous tribe of Epidendrums, it is not advisable to cultivate it in baskets suspended in the air, such treatment being applicable to those kinds only which are not bulbous.

This species offers the only instance we at present recollect, in the tribe, of hairs being produced around the margin of the stigma; they exist here in the form of a soft, conspicuous, thin, villous border.

Bulbs forming a tuft, about 6 inches long, round, compressed, wrinkled. *Leaves* solitary, about twice as long as bulbs, elliptical-lanceolate, acute at each end, coriaceous, plaited, obsoletely nerved. *Scape* radical, weak, round, longer than leaves, clothed with a few distant membranous scales. *Panicle* simple, rather one-sided, spreading, many-flowered. *Bracteæ* ovate, membranous. *Flowers* scarcely resupinate, olive-green outside, banded and spotted with light brown in the inside. Two lower *sepals* connate, smaller than the rest. *Labellum* panduriform, much tuberculated on the disk. *Columna* with two little linear blunt fleshy wings. *Clinandrium* with a toothletted margin. *Stigma* and *rostellum* downy.

J. L.

1014

SARCANTHUS succisus.

Bitten-leaved Sarcanthus.

GYNANDRIA *MONANDRIA.*

Nat. ord. ORCHIDEÆ. *Tribus* Vandeæ *Lindl.*
SARCANTHUS. Suprà, fol. 981.

S. *succisus*: foliis oblongis apice præmorsis dentatis, spicâ simplice horizontali foliis longiore: rachi compressâ, sepalis obtusis, calcare apice didymo inflato, stigmate fornicato.
S. succisus. *Lindley coll. bot. sub fol.* 39. *B. sine charactere.*
Caulis *compressus, erectus, foliosus, flexuosus, purpureo-maculatus, internodiis nunc planis, nunc sulcatis, radices pallidas tortuosas promentibus.* Folia *disticha, vaginantia, cum vagina articulata, oblonga, subundulata, apice emarginata, succisa et erosa, avenia, purpureo paululum ad costam et margines superiorum maculata.* Spica *oppositifolia, foliis longior, horizontalis, v. dependens, pedunculatus, pedunculo purpureo terete, distanter vaginato, rachi compressâ, clavatâ.* Flores *magnitudine S.* teretifoliæ, *extùs lividè purpurei, intùs pallidè lutei, paulò purpureo maculati.* Sepala 2 *exteriora lateralia patentia, subrotunda, subunguiculata, supremo erecto fornicato.* Sepala *interiora superiori appressa et conformia paululum minora.* Labellum *cucullatum, calcaratum, dependens, cum columna non articulatum, pallidè luteum, trilobum, lobis lateralibus columnæ longitudine truncatis anticè vittâ unicâ sanguineâ, intermedio ovato acuto carnoso convexo, caput aviculi simulante.* Calcar *ad faucem fornicibus duabus semiclausum, apice didymum ventricosum.* Columna *parva, erecta, cum basi labelli connata, posticè sursùm arcuata, marginibus inflexis anticè carnosis luteis approximatis fornicem super stigma formantibus, ad latera bicallosa; posticè multò evectior.* Clinandrium *excavatum, arcuatum, modo currûs antiquæ.* Anthera *anticè elongata, subquadrata, unilocularis, membranacea.* Pollinia *duo, oblonga, dura, cereacea, intùs altè sulcata apicibus caudiculæ adhærentibus filiformi cartilagineæ diaphanæ pallidè luteæ in glandulam parvam denudatam desinenti.*

Native of China, whence it was brought for the Horticultural Society, in 1824, by Mr. John Damper Parks. Flowers in the stove in June and July. It thrives in a mixture of rotten wood and decayed vegetable mould, and may be increased by cuttings, which will root in the tan-bed, like the suckers of a Pine plant.

This species is distinguished from *Sarcanthus rostratus*, fol. 981, by its leaves, which appear as if their end was bitten off, and by the spur of the labellum, which has a double inflated apex.

Stem compressed, erect, leafy, spotted with purple and protruding tortuous pale roots. *Leaves* oblong, somewhat wavy, emarginate at the end, which is bitten off, and eroded. *Spike* opposite to a leaf, horizontal, or hanging down, with a round purple stalk. *Flowers* the size of S. teretifolius and rostratus, dull purple externally, pale yellow inside, with a few purple spots. *Lip* pale yellow, 3-lobed, the lateral lobes as long as the columna, truncate, with a single crimson stripe in front; spur half closed up with two scales. *Stigma* overshadowed by a projecting arch.

J. L.

1028

HETEROTAXIS crassifolia.

Thick-leaved Heterotaxis.

GYNANDRIA *MONANDRIA.*

Nat. ord. ORCHIDEÆ. Tribus II. Arethuseæ. *Lindl. coll. bot. append. Orchid. Scelet. p.* 10.
HETEROTAXIS.—Pollinia 2, (v. 4? binatim connata) linearia, pulverea; caudiculâ glandulâque nullis. *Anthera* terminalis opercularis decidua unilocularis, septis duobus incompletis. *Stigma* oblongum excavatum; rostello obsoleto. *Columna* semiteres apice alata. *Labellum* ovatum liberum integrum carnosum disco callosum. *Sepala* subæqualia conniventia carnosa; 2 inferioribus labello suppositis.——Herba *terrestris acaulis* (Americæ æquinoctialis), foliis *carnosis aveniis*, scapo *radicali vaginato.*

Heterotaxis crassifolia.
Herba *acaulis, terrestris, an epiphyta? foliis radicalibus, erectis, carnosis, oblongo-lanceolatis, obliquè tortis, planis, aveniis, apice obliquè uncinato-apiculatis, basi vaginantibus.* Flos *luteus, solitarius, radicalis, bracteâ unicâ oblongâ vaginatus, subsessilis ad basin foliorum, vix suprà humum elevatus.* Sepala *carnosa, conniventia, oblonga, concava, subæqualia; exteriora apice viridia, apiculata; interiora breviora, tenuiora.* Labellum *anticum, liberum, integrum, ovatum, acutum, sepalis subconforme, disco medio calloso, sepalis anticis exterioribus suppositum.* Columna *semiteres, clavata, truncata, apice utrinque alata;* stigma *excavatum, oblongum, quadratum;* clinandrium *parvum, rostello obsoleto.* Anthera *terminalis, opercularis, decidua, carnosa, verruculosa, hemispherica, unilocularis: septis duobus incompletis.* Pollinia *duo, v. quatuor, ut potiùs opinor, binatim cohærentia, linearia, hinc crassiora, è granulis plurimis minutis, angulatis, filo axili elastico cohærentibus.*

This very remarkable undescribed plant was introduced from Jamaica by the late Mr. Lee, of the Hammersmith Nursery, in 1823. Our drawing was made in the month of June of the following year. Not having however had, at that time, an opportunity of examining the pollen-masses, we were obliged to defer the publication of the plant till the present time, when we are enabled to complete its history by means of a fresh specimen with which we have recently been obligingly supplied by Mr. John Lee, in whose possession the original plant still remains.

The immediate affinity of this genus is undoubtedly with Arethusa and its allies, with which it agrees essentially in the structure of its columna and pollen-masses; but from which it differs remarkably in habit, which is quite unlike that of any species previously known as belonging to Arethuseæ. It may be considered a connecting link between *Arethuseæ* and *Epidendreæ*.

A stemless, terrestrial, or perhaps epiphytous plant, with radical, erect, fleshy leaves, which are oblong-lanceolate, obliquely twisted, flat, veinless, with an oblique hooked end, sheathing at the base. *Flower* yellow, solitary, radical, sheathed with one oblong bractea, nearly sessile at the base of the leaves, scarcely elevated above the soil. *Sepals* fleshy, conniving, oblong, concave, nearly equal; the outer green, with a little point at the end; the inner shorter and thinner. *Labellum* placed in front, separate, entire, ovate, acute, of nearly the same form as the sepals, with the disk callous in the middle, supported by the anterior outer sepals. *Columna* half-round, clavate, truncate, winged on each side at the end; *stigma* hollowed out, oblong, nearly square; *clinandrium* small, with an obsolete rostellum. *Anther* terminal, opercular, deciduous, fleshy, covered over with little warts, hemispherical, one-celled; with two incomplete dissepiments. *Pollen-masses* two, or rather perhaps four, cohering in pairs, linear, thicker at one end than the other, formed of many minute angular granules, adhering to an elastic thread passing through their axis.

J. L.

1043

SPIRANTHES grandiflora.

Large-flowered Spiranthes.

GYNANDRIA *MONANDRIA.*

Nat. ord. ORCHIDEÆ. *Tribus* Neotteæ *Lindl.*
SPIRANTHES. Suprà, vol. 10. *fol.* 823.

S. *grandiflora;* foliis obovato-lanceolatis obtusis glabris maculatis, scapo piloso, bracteis ovato-lanceolatis ovarii longitudine, perianthii laciniis inferioribus longissimè cum ovario connatis, labello cuneato-cucullato viridi-venoso apice deltoideo basi sagittato.

Folia *radicalia sublucida, hic illic colore pallidiore nebulosa, obovato-lanceolata, obtusa, patentia.* Scapus *erectus, 2-pedalis, basi glaber vaginatus, sursùm cum floribus pilosus.* Bracteæ *ovato-lanceolatæ, rigidæ, pilosæ, basi glabræ, viridi venosæ, ovarii longitudine.* Ovarium *unciam ferè longum villosum.* Perianthium *connivens, laciniis carnosis, parallelis; superiore basi corrugata, oblonga, apice recurva, intùs nitida; lateralibus interioribus paulò brevioribus, obtusis, appressis, lucidis, atro-viridibus; lateralibus inferioribus basi in ovario per totam longitudinem decurrentibus, margine antico discretis, lamina oblonga, obtusa, labello longiore.* Labellum *cucullatum, cuneatum, pedicellatum, basi sagittatum, marginibus columnæ arctissimè agglutinatis, apice deltoideo, viridi-venoso.* Columna *brevis posticè ad marginem dentatus, anticè pubescens.* Stigma *verticale, reniforme, semiclausum, rostratum, emarginatum.* Anthera *pedunculata, rostrata, bilocularis.* Pollinia *duo, biloba, caudiculata, glandulæ emarginatæ adhærentia.*

This new species was found near Rio Janeiro, by Mr. David Douglas, by whom roots were sent to the Horticultural Society, in 1824. It grows freely in rotten mould in the stove, and flowers in October. It is the largest and most remarkable of the genus, except Spiranthes elata, from which it is obviously distinguished.

The cohesion of the labellum with the sides of the columna, which is very firm in all the species, is in this so powerful, that it may easily be mistaken for actual union of the parts.

Leaves radical, somewhat shining, clouded here and there with a paler colour, obovate-lanceolate, obtuse, spreading. *Scape* erect, two feet high, at the base smooth and sheathing, upwards hairy, as well as the flowers. *Bracteæ* ovate-lanceolate, rigid, hairy, smooth at base, with green veins the length of the ovarium. *Ovarium* about an inch long, villous. *Perianthium* conniving, with fleshy parallel segments; the upper shrivelled at the base, oblong, recurved at the apex, shining inside; the inner lateral ones a little shorter, obtuse, appressed, shining, dark green; the lower lateral ones at the base running down the ovarium nearly the whole of its length, with their front edge separate, with an obtuse, blunt lamina, longer than the labellum. *Labellum* cucullate, cuneate, pedicellate, sagittate at base, with the edges firmly adhering to the sides of the columna, at the end deltoid, with green veins. *Columna* short, toothed at the back edge, pubescent in front. *Stigma* vertical, like a slit half-closed, rostrate, emarginate. *Anther* stalked, rostrate, 2-celled. *Pollen-masses* two, 2-lobed, with a caudicula adhering to an emarginate gland.

J. L.

1050

ONCIDIUM divaricatum

Cushion-lipped Oncidium.

GYNANDRIA *MONANDRIA.*

Nat. ord. ORCHIDEÆ. *Tribus* Vandeæ *Lindl.*
ONCIDIUM. Suprà, vol. 9. *fol.* 727.

Div. I. *Perianthii foliola omnia discreta.* Lindl. Coll. Bot. xxvii.

O. *divaricatum;* bulbis subrotundis ancipitibus compressis, foliis ovalibus apiculatis carnosis, labello crenulato: lobo medio transverso emarginato angustiore; disco pubescente pulvinato, stigmate mutico, columnæ alis semilunatis integerrimis, paniculâ divaricatissimâ.

β. *cupreum;* perianthii foliolis angustioribus cupreis vix maculatis.

Bulbi *cæspitosi, compressi, ancipites, squamis magnis acutis vaginantibus venosis sphacelatis vestiti, monophylli.* Folia *crassa, carnosa, pallidè lutea, suberecta, ovalia, obtusa cum acumine, ad basin ferè semper concava, ad marginem sæpiùs hic illic fissa.* Scapus *radicalis,* 1½ *pedalis, teres, purpureus.* Panicula *multiflora, divaricatissima, flexuosa, internodiis subæqualibus.* Flores *magni, lutei, paululum olivacei, labello excepto.* Perianthii *foliola* 5, *discreta, patentia, spatulato-oblonga, subæqualia, basin versùs sanguineo maculata.* Labellum *planum, trilobum, crenulatum, guttatum, lobis lateralibus rotundatis, basi altè cordatis, intermedio minore emarginato sæpiùs complicato; disco pubescente, pulvinato, cruciato, fulvo.* Columna *glabra, alis semilunatis integerrimis.* Stigma *muticum.* Anthera *ecristata.* Pollinia 2, *caudiculata, glandulâ parvâ.*

The empire of Brazil seems to be inexhaustible in new forms of Orchideous plants. Every year is bringing us acquainted with species and genera previously unknown; so little indeed has the rich vegetation of that fertile land been investigated, that almost every epiphyte proves, upon its first appearance in our gardens, to be new to Science. To this fact, the various volumes of this work bear ample testimony. The subject of the present plate was sent to the Horticultural Society by A. J. Heatherly, Esq., his Britannic Majesty's Vice-Consul at Rio Janeiro. It blossomed, in October 1826, in the stove, where it grows readily in decayed vegetable matter. The variety β. *cupreum,* above indicated, was received from Mr. Heatherly at the same time. It differs nothing in foliage; but the divisions of the flower are much narrower, of a bright copper-colour towards the base, and scarcely at all varied with spots.

Oncidium is the most distinctly characterised of all the genera of epiphytous Orchideæ: the spreading segments of the perianthium; the 3-lobed labellum, of various figures, but always having a fleshy crested discus; the winged columna, and the two pollen masses hollowed out at their back, and attached to a slender pedicel, together with the radical inflorescence,—afford characters which have never yet been materially affected by the discovery of Transition Species. We may, however, take this opportunity of remarking, that the genus *Fernandesia* of the Flora Peruviana, of which the *Lockhartia elegans* of the Botanical Magazine, fol. 2715, is a species, and from which *Pachyphyllum* of M. Kunth is probably not distinguishable, differs from Oncidium in little besides its axillary inflorescence, imbricated stems, dilated bracteæ, and solid pollen masses. By the latter character it is neatly distinguished from Oncidium, which has hollow pollen masses cut open at the back.

Of Fernandesia there are three, if not a greater number of species, in our gardens. Mr. Lambert is the fortunate possessor of some from Pavon; and we were a few months ago obliged by a sight of fresh specimens, in flower, of a species perhaps not different from F. subbiflora, with which we were favoured by Messrs. Loddiges.

Bulbs in clusters, compressed, two-edged, one-leaved. *Leaves* thick, fleshy, pale yellow, nearly erect, oval, obtuse with a point, generally concave at base, and split here and there at the margin. *Scape* about a foot and half high, round, purple. *Panicle* many-flowered, very much divaricating, flexuose. *Flowers* large, yellow, rather olive-

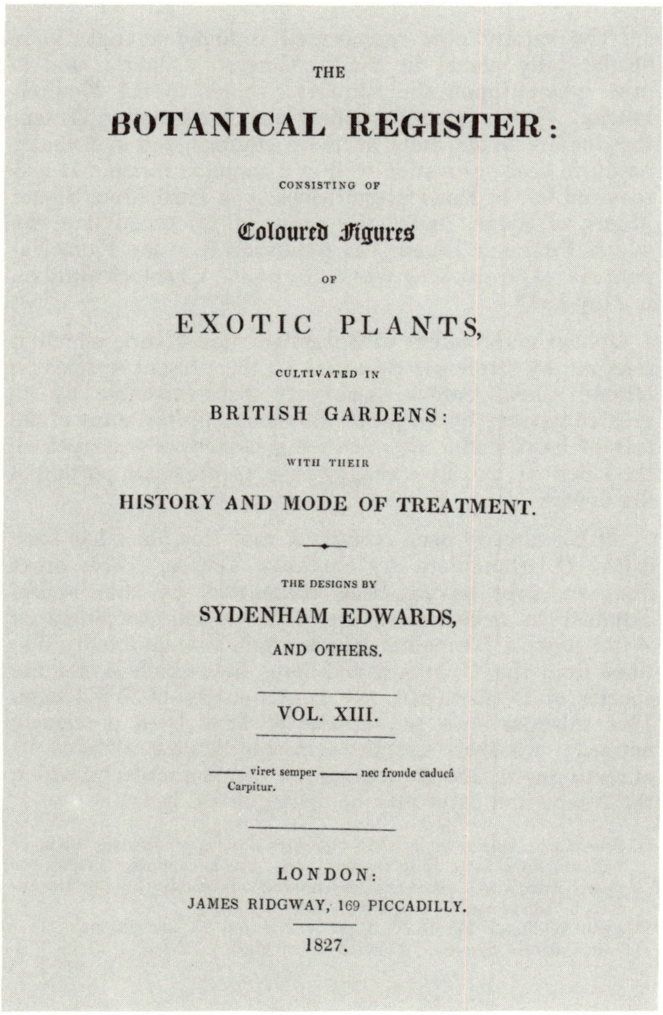
THE

BOTANICAL REGISTER:

CONSISTING OF

Coloured Figures

OF

EXOTIC PLANTS,

CULTIVATED IN

BRITISH GARDENS:

WITH THEIR

HISTORY AND MODE OF TREATMENT.

THE DESIGNS BY

SYDENHAM EDWARDS,

AND OTHERS.

VOL. XIII.

—— viret semper —— nec fronde caducâ
Carpitur.

LONDON:

JAMES RIDGWAY, 169 PICCADILLY.

1827.

1071

OPHRYS fusca.

Dull-purple Ophrys.

GYNANDRIA *MONANDRIA.*

Nat. ord. Orchideæ. *Tribus* Ophrydeæ.
OPHRYS. Suprà, vol. 3. *fol.* 205.

§. *Sepala interiora glabra.*

O. *fusca;* foliis oblongo-lanceolatis suberectis obtusis: caulinis acutioribus, labello villoso trilobo; lobis lateralibus oblongis obtusis deflexis; intermedio retuso emarginato mutico; disci macula transversa flexuosa lævigata, sepalis interioribus obtusis glabris superiore duplò breviore.
" Ophrys insectifera ξ. *Sp. pl.* 1343."
" Orchis myodes fusca lusitanica. *Breyn. cent.* 41."
O. fusca. *Link. in Schrad. diar.* 1799. 2. 324.
O. *fusca;* caule folioso, labello villoso trilobo, lobis lateralibus oblongis obtusis, medio oblongo emarginato, petalis patentibus lanceolatis obtusiusculis, binis interioribus parùm brevioribus. *Willd. sp. pl.* 4. 69.

coloured, except the lip. *Segments of the perianthium* 5, separate, spreading, spatulate-oblong, nearly equal, spotted with crimson towards the base. *Lip* yellow, flat, 3-lobed, crenulate, spotted; the lateral lobes rounded, deeply cordate at base, the intermediate one smaller, emarginate, often folded together.

J. L.

Folia *glauca, subtrinervia, obtusa, caulinis angustioribus, acutioribus.* Bracteæ *oblongæ, concavæ, floribus breviores.* Sepala *herbacea, exterioribus lateralibus ovato-oblongis, obtusis, labelli longitudine, supremo oblongo, fornicato, breviore, lateralibus interioribus hoc duplò brevioribus, obtusis, glabris.* Labellum *oblongum, atro-purpureum, anticè hirsutum, posticè glabrum, lobis lateralibus obtusis deflexis, intermedio longiore, emarginato; disci macula transversa, cærulea, flexuosa, lævigata.*

Roots of this plant were collected upon the rock of Gibraltar, by Colonel Chapman, and sent to William Atkinson, Esq., of Grove End, St. John's Wood, in whose green-house our drawing was made in February last. It was cultivated in a pot, in common light garden mould, in which it grew luxuriantly. Most of the European Orchideæ, however cold the countries may be in which they grow naturally, seem in the garden to prefer as much protection from weather as is afforded by a good frame or cool green-house.

It is extremely difficult to distinguish the species of this genus, which are probably more numerous than is even now supposed: we hope to be able, in time, to make this work a vehicle for conveying to the public accurate ideas respecting the greater part of them.

Leaves glaucous, somewhat three-nerved, obtuse, those of the stem narrower, and more acute. *Bracteæ* oblong, concave, shorter than the flowers. *Sepals* green, the outer lateral ones ovate-oblong, obtuse, as long as the labellum; the uppermost oblong, arched, and shorter; the inner ones twice as short as the latter, obtuse, and smooth. *Labellum* oblong, dark-purple, hairy in front, smooth behind; the lateral lobes obtuse deflexed, the middle one longer and emarginate; the spot of the disc transverse, light-blue, wavy, and polished.

J. L.

1087

OPHRYS atrata.

Dark-lipped Ophrys.

Nat. ord. Orchideæ. *Tribus* Ophrydeæ *Lindl.*
OPHRYS. Suprà, vol. 3. *fol.* 205.

O. *atrata;* labello emarginato integerrimo convexo villoso inappendiculato bivittato versùs basin bicorni, sepalis herbaceis: interioribus ovatis pubescentibus discoloribus exterioribus obtusis duplò brevioribus, foliis glaucis.
Caulis *pedalis v. ultrà, foliosus.* Folia *glauca, ovali-lanceolata, plana.* Sepala *exteriora ovata, obtusa, subherbacea, inferioribus latioribus et longioribus; interiora duplò breviora, acutiuscula, pubescentia, purpureo discolora.* Labellum *atro-purpureum, subrotundum, convexum, integerrimum, emarginatum, appendiculâ nullâ, in disco glabriusculum, vittis duabus pallidè fuscis sublucidis notatum, versùs basin utrinque cornutum, ad ipsam basin lucidum, glaberrimum.* Columna *pubescens, sepalis interioribus subæqualis v. brevior.*

This plant was sent from Rome to the Horticultural Society, in 1826, by Signor Mauri, under the name of

Ophrys araneifera, to which it is undoubtedly closely allied. But there are some differences between the plants, in the structure of their flowers, which, as the species of Ophrys are at present constructed, induce us to consider them as two kinds, hitherto confounded with each other.

The true *O. araneifera,* which is well represented in English Botany, tab. 65, is distinguished from *O. arachnites,* firstly, by the absence of an appendage, or mucro, from the apex of its labellum; secondly, by the roundish outline of the same part, which in *O. arachnites* is always more or less cuneate; and, thirdly, by its smoother and more elongated interior sepals. In these points, the plant now represented agrees with *O. araneifera,* but it differs in having a very entire shaggy labellum, which in *O. araneifera* is downy, and usually 3-lobed; in its inner sepals being more decidedly pubescent; and in the presence of two large horn-like elevations towards the base of the labellum. The marking of the latter is nearly the same in both species.

In the cornute processes of its labellum, *O. atrata* agrees with O. *arachnites.*

We have fine specimens of this species, collected in the neighbourhood of Trieste, by Dr. Hornschuch, from whom we received it as a new species.

Our drawing was made in the Garden of the Horticultural Society, at Chiswick, in May 1827: the plant was in a pot, and had been preserved during the winter in a common wooden frame.

We may here remark, that this was received by the Horticultural Society, from Signor Mauri, at the same time with several other Orchideous plants of the South of Europe: their roots had been dried, and then packed in paper, like seeds, and they have all succeeded perfectly, although when the roots arrived in England they were so shrivelled in appearance that it was not expected they would have survived. It is to be hoped, that if this statement should meet the eyes of Botanists or Amateurs who have the means of procuring supplies of these plants from the South of Europe, or from Barbary, the very simple mode just mentioned, of transmitting them to this country, will be adopted.

J. L.

1093

OPHRYS tenthredinifera; β. minor.

Dwarf large-flowered Ophrys.

GYNANDRIA *MONANDRIA.*

Nat. ord. ORCHIDEÆ. *Tribus* Ophrydeæ *Lindl.*
OPHRYS. Suprà, vol. 3. *fol.* 205.

O. *tenthredinifera;* labello integerrimo subquadrato cuneato villoso appendiculato basi auriculato cornuto sub apice barbato, disco glabro cinnamomeo areolâ quadratâ cyaneo-marginatâ, sepalis coloratis: interioribus parvis ovatis acutis tomentosis; exterioribus oblongis retusis patentissimis, bracteis coloratis spatulatis florum longitudine, caule squamoso.
Ophrys *insectifera A. rosea* labello villoso oblongo-obovata, apice bilobo appendiculato. *Desf. atl.* 2. 320.
Ophrys tenthredinifera. *Willd. sp. pl.* 4. 67. *Suprà, vol.* 3. *fol.* 205. *Bot. mag. fol.* 1930. *Spreng. syst.* 3. 702.
β. *minor;* caule semper humiliore 3-floro.
? Orchis orientalis fucum referens flore parvo villosissimo scuto azureo. *Tourn. cor.* 30.
? Ophrys villosa. *Desf. in. ann. mus.* 10. 225. *t.* 14.
Orchis ornifuciflora genata rubiginea ambitu viridi. *Cupani, Panphyt.* 1. *t.* 175. *Bonanni t.* 28. fide *Bivonæ.*
Orchis orniflora, amplo labello, genato, rubigineo, ambitu viridi, larvulam fictitante. *Cupani Hort. Cathol. p.* 158. fide *Bivonæ.*
O. tenthredinifera. *Bivon. Bernard. sic. plant. centuria,* 2. *p.* 39. *t.* 4. *Presl. fl. sic.* 1. *p. xli. Viviani Fl. Cors. Prodr.*
O. grandiflora. *Tenore app. alt. p.* 83. *Fl. Napolit.* 308. *t.* 94.

There is already an accurate description of one variety of the present species at fol. 205 of this work; it is therefore needless to repeat it on this occasion.

The Ophrys tenthredinifera divides into two well-marked varieties; the first, which we judge to be that originally described by Desfontaines, and upon which Willdenow founded the species, is well represented at fol. 205 of our third volume. It is a native of Barbary and Turkey, and of all the African side of the basin of the Mediterranean; and is chiefly distinguished by its greater stature, more numerous and larger flowers, and spreading bracteæ. It may be considered the most genuine form of the species.

The variety now represented is found abundantly in shady, hilly places in Sicily, Corsica, Calabria, and in other places upon the European shores of the Mediterranean. It is a beautiful little plant, representing O. tenthredinifera in its more northern stations, and sufficiently hardy to bear cultivation well in a common frame. It was received by the Horticultural Society in 1826, from Signor Mauri, of Rome, under the name of O. grandiflora, by which Professor Tenore has published it in the Flora Napolitana. Our drawing was made at the Chiswick Garden, in May 1827.

Ophrys crabronifera of Sebastiani and Mauri, which is referred by Professor Sprengel to the present species, is closely allied, but is apparently distinguishable by its greater stature, by its brown labellum, by the want of the tuft of hairs under the little appendage of the apex of the labellum, and by a shaggy base to the same portion of the flower.

It has already been remarked, that this plant has been called O. grandiflora by Professor Tenore. This arose from an error having been committed by that skilful Botanist in referring to the Ophrys tenthredinifera of Willdenow, a Neapolitan plant which is undoubtedly distinct from the Ophrys grandiflora, but which is not the species of Desfontaines, nor consequently of Willdenow. This mistake does not appear to have been previously noticed; we shall use the opportunity now afforded us of rectifying it, and of pointing out the characters by which the Neapolitan plant may be distinguished in future.

O. *Tenoreana;* labello integerrimo cuneato villoso appendiculato sub apice barbato, disco fusco linea flexuosa lutea, sepalis coloratis: interioribus parvis ovatis acutis tomentosis, exterioribus subrotundis, bracteis lanceolatis herbaceis floribus longioribus, caule squamoso.
O. tenthredinifera. *Tenore Fl. Napol. p.* 308. *tab.* 93. nec aliorum.
Hab. in Calabriæ dumetis. Floret Aprili et Maio. *Tenore.*

J. L.

THE

BOTANICAL REGISTER;

OR,

ORNAMENTAL FLOWER-GARDEN

AND SHRUBBERY.

CONSISTING OF

COLOURED FIGURES OF PLANTS AND SHRUBS,

CULTIVATED IN BRITISH GARDENS;

ACCOMPANIED BY THEIR

History, Best Method of Treatment in Cultivation, Propagation, &c.

THE DESIGNS BY

SYDENHAM EDWARDS,

AND OTHERS.

VOL. XIV.

——viret semper——nec fronde caducâ
Carpitur.

LONDON:
JAMES RIDGWAY, 169, PICCADILLY.

M.DCCC.XXVIII.

1129

SOPHRONIA cernua.

Drooping Sophronia.

GYNANDRIA *MONANDRIA.*

Nat. ord. ORCHIDEÆ. *Tribus* Vandeæ *Lindl.*

SOPHRONIA. *Pollinia* 8, anticè et posticè parallela, caudiculâ duplici pulvereâ, glandulâ 0? *Anthera* terminalis opercularis 8-locularis, cardine crasso. *Stigma* concavum rostello obtuso. *Columna* libera apice utrinque alata alis integris conniventibus super cristam labelli. *Labellum* integrum cucullatum linguiforme, basi cum columnâ connatum, medio cristâ simplici transversâ. *Sepala* subæqualia, imbricata, basi libera. —— Herba *epiphyta* (Braziliensis) *monophylla, non bulbosa,* racemis *axillaribus effusis paucifloris,* floribus *rubris.*

S. cernua.

Epiphyta, inter muscos vegetans. Caulis *brevissimus, vaginatus, anceps.* Folium *solitarium, ovale, apiculatum, avenium, carnosum, purpureo discolor.* Racemus *axillaris, abbreviatus, effusus,* Pleurothallidis *cujusdam modo.* Sepala *rubra, basi lutea, subæqualia, ovata, acuta, patentia, interioribus imbricatis.* Labellum *basi luteum, apice rubrum.* Columnæ *albæ* alæ *atro-purpureæ.* Pollinia *olivacea.*

Found upon a tree at Botofogo, three miles from Rio Janeiro, by William Harrison, Esq., by whom it was transmitted to Mrs. Arnold Harrison, of Aigburgh, near Liverpool, whence it was obligingly sent with a sketch, in December 1826. It is a very remarkable little epiphyte, growing readily in decayed vegetable soil among moss, in a hot humid shady part of the stove.

In generic affinity it is most nearly akin to Octomeria, which differs in having sessile pollen masses, a column without wings, the two lower sepals united at the base, and a 3-lobed labellum. Eria, which is also nearly allied, belongs to Malaxideæ.

An epiphyte, about 3 inches high, growing among moss. *Stem* very short, sheathing, two-edged. *Leaf* solitary, oval, apiculate, veinless, fleshy, discoloured with purple.

Raceme axillary, short, effuse, in the way of some Pleurothallis. *Sepals* red, yellow at base, nearly equal, ovate, acute, spreading, the inner ones imbricated. *Labellum* yellow at base, red at apex. *Wings* of the white columna dark purple. *Pollen masses* olive green.

J. L.

1131

RENANTHERA coccinea.

Scarlet Air-Plant.

GYNANDRIA *MONANDRIA.*

Nat. ord. ORCHIDEÆ. § *Vandeæ* Lindl.

RENANTHERA Lour.—*Pollinia* duo posticè biloba; *caudiculâ* medio constrictâ, non elasticâ, *glandulâ* triangulari. *Anthera* terminalis, opercularis, decidua, unilocularis, posticè bivalvis. *Stigma* subrotundum cavum; rostello obtuso. *Columna* aptera, libera, margine anteriore super stigmate inflexo. *Labellum* nanum, sessile, liberum, cum columnâ articulatum, 3-lobum basi saccatum, lobis lateralibus erectis. *Sepala* patentia, subæqualia, basi distincta, unguiculata, spatulata, undulata.——Epiphyta (Chinensis) *caulescens, radicans, ramosa.* Folia *carnosa, avenia.* Flores *magni, paniculati.*

R. coccinea. *Loureiro Flora cochinchinensis* 2. *p.* 521.

Caulis *longus, sæpè* 8-10-*pedalis, radicibus paucis longis, tortuosis, super arbores v. saxa decumbens, teres, foliosus, basi foliorum vestigiis vaginatus, paululum ramosus.* Folia *carnosa, disticha, avenia, plana, apice obliqua, emarginata, atroviridia, sublucida,* 4-5 *uncias longa.* Paniculæ *laterales, laxæ, multifloræ,* $2\frac{1}{2}$ *pedes longæ, ramis teretibus, duris.* Bracteæ *breves, ovatæ, subcorrugatæ.* Ovarium *cum pedunculo brevi continuum, pallidè coccineum,* 6-*sulcatum.* Sepala *patentia, basi libera, non imbricata, carnosa;* 3 superiora *linearia, erecta, intermedio majore spatulato, coccinea, maculis luteis nubilis fasciata;* 2 inferiora $1\frac{1}{2}$ *uncialia superioribus longiora, dependentia, collateralia unguiculata, lanceolata, obtusa, medio abruptè undulata, coccinea, fasciis quibusdam obscuris pallidioribus.* Labellum *nanum,* 3 *lineas tantum longum, saccatum, cum columnâ articulatum, trilobum,* lobis lateralibus *erectis, truncatis, luteis, coccineo marginatis et vittatis,* intermedio *linguæformi, reflexo, coccineo, basi luteo;* sacco *conico, obtuso, intùs luteo, glabro, extùs coccineo punctato, fauce utrinque calloso.* Columna *coccinea, semiteres, labelli longitudine, aptera, anticè lutea vittata, margine anteriore super stigmate inflexo.* Stigma *excavatum, subrotundum.* Anthera *terminalis, opercularis, atrococcinea, obtusa, unilocularis, posticè semibivalvis.* Pollinia *duo, reniformia, posticè biloba,* glandulâ *triangulari,* caudiculâ *diaphanâ vix elasticâ, medio constrictâ.*

The work of the missionary Loureiro, published in 1790, a drawing in the possession of the Horticultural Society of London, and the reports of some travellers who have visited China, have been, up to the present time, the only evidence to Europeans of the existence of this truly magnificent plant, the beauty of whose blossoms surpasses every thing known in the vegetable world. That the Chinese suspend in baskets from the ceilings of their rooms several of the tribe of plants to which this belongs, some for the sake of their fine flowers, and some on account of their delightful fragrance, is familar to every one. Many, if not all, of this description, have been introduced from time to time; and a few have flowered with us. Among those which bid defiance to the skill of the cultivator, has long been recognised a species, introduced sometime previous to 1817, with long leafy stems, sometimes attaining the height of 8 or 10 feet, and fleshy, veinless leaves; it is frequently imported from China, and is now to be found in almost every collection of which parasitical Orchideæ form a part, attaching itself by means of its long tortuous roots to a damp wall, columns, or other bodies placed as its support. This is Renanthera coccinea.

The cause of previous want of success in inducing it to flower, has resided in its having been cultivated in too dry an atmosphere. Mr. Fairbairn, gardener to His Royal Highness Prince Leopold, at Claremont, impressed with this opinion, tried the effect of tying moss around the stems, and keeping it constantly damp, exposed as much as possible to the influence of the sun. With what success his experiment has been rewarded, appears from the accompanying representation of a portion of a panicle, 2½ feet long, which was finally produced in the Hothouse at Claremont, in October 1827.

To Botanists it has been as little known as to the rest of the world,—almost every systematist having omitted it. And yet the language of Loureiro is far from unsatisfactory, allowance being made for certain peculiarities of diction.

A native of woods in Cochinchina, where it climbs over trees. Propagated without difficulty by cuttings. When in flower, the plant may be safely removed to a dwelling apartment, where the blossoms, which are very durable, will remain in perfection many weeks.

Stem often 8 or 10 feet in length, round, leafy, slightly branched, the lower part sheathed with the persistent bases of the leaves, pushing forth a few long, tortuous roots, by which it clings to trees or stones. *Leaves* fleshy, distichous, veinless, flat, obliquely emarginate at the apex, dark-green, sublucid, 4-5 inches long. *Panicles* lateral, loose, many-flowered, 2 feet and a half long, with hard, round branches. *Bracteæ* short, ovate, somewhat shrivelled. *Ovarium* continuous with the short peduncle, pale red, with 6 furrows. *Sepals* spreading, distinct at the base, not imbricated, fleshy, the 3 *upper* linear, erect, the middle one being larger and spatulate, scarlet, banded with yellow cloudy spots; the 2 *lower* an inch and half in length, longer than the upper, hanging down, collateral, unguiculate, lanceolate, obtuse, abruptly undulated in the middle, scarlet, with a few obscure paler bands. *Labellum* dwarf, only 3 lines long, bagged, articulated with the column, 3-lobed, the lateral lobes erect, truncate, yellow, bordered and striped with scarlet, the *middle one* tongue-shaped, reflexed, scarlet, yellow at the base; *saccus* conical, obtuse, yellow and smooth inside, scarlet and dotted outside, with the throat callous all round. *Columna* scarlet, half round, the length of the labellum, apterous, striped with yellow in front, its anterior margin incurved over the stigma. *Stigma* hollowed out, roundish. *Anther* terminal, opercular, dark scarlet, obtuse, one-celled, half two-valved at back. *Pollen masses* 2, reniform, two-lobed at back, with a triangular *gland*, and a diaphanous scarcely elastical *caudicule*, contracted in the middle.

J. L.

1147

EULOPHIA ensata.

Sword-leaved Eulophia.

GYNANDRIA *MONANDRIA.*

Nat. ord. ORCHIDEÆ. § Vandeæ *Lindl.*
EULOPHIA. Suprà, vol. 12. fol. 1002.

E. *ensata;* foliis strictis ensiformibus plicatis, scapo radicali vaginato, racemo capitato, labelli lobo medio linguiformi barbato, calcare recto conico abbreviato, bracteis membranaceis venosis.

Radix *tuberosus, semisubterraneus, subrotundus, depressus, cicatrizatus.* Caules *simplices, stricti, pedales, foliis ensiformibus, strictis, plicatis, Bletiæ more provenientibus.* Scapus *radicalis, pedalis, vaginatus, supernè nudus, vaginis bracteisque membranaceis venosis: his ovario longioribus.* Flores *lutei, ferè unciam longi, in racemo arcto subcapitato glomerati.* Perianthium *patens.* Sepala 3 exteriora *basi imbricata, ovali-lanceolata, æqualia: lateralibus anticè cordatis, margine anteriore labello supposito;* 2 interiora *minora, conformia.* Labellum *anticum, oblongum, calcaratum, cucullatum, trilobum, lobis lateralibus abbreviatis, obtusis, glabris, intermedio linguiformi disco barbato, calcare brevi, recto, conico, retuso.* Columna *labello multò brevior, calcare paulò longior, semiteres, marginata.* Stigma *transversum, excavatum.* Anthera *terminalis, opercularis, decidua, emarginata, bilocularis, septuli rudimento nullo.* Pollinia 2, *altè sulcata, cereacea, caudiculâ brevi, glandulâ transversâ.*

We received this plant, in July 1827, from Mr. Tate, of the Sloane Street Nursery, as a native of the Cape of Good Hope. We believe, however, that it is peculiar to the colony of Sierra Leone, where roots were originally found by Mr. George Don, and by him transmitted to the Horticultural Society in 1822. It is a tender stove plant, extremely difficult to cultivate, being liable to rot from various causes. In growth it resembles the West Indian Bletias, like them dying down to the ground periodically.

Root tuberous, semi-subterraneous, roundish, depressed, covered with scars. *Stems* simple, straight, a foot high, with ensiform, straight, plaited leaves. *Scape* from the root, about a foot high, sheathed, at the upper end naked, the sheaths and bracteæ membranous and veiny; the latter longer than the ovarium. *Flowers* yellow, about an inch long, crowded in a close capitate raceme. *Perianthium* spreading. The three *external sepals* imbricating at the base, oval-lanceolate, equal; the lateral cordate in front, with their anterior margin placed underneath the labellum; the *two inner* smaller, but of the same figure. *Labellum* anterior, oblong, calcarate, cucullate, 3-lobed, the lateral lobes short, obtuse, smooth, the middle one tongue-shaped, with a bearded disk; spur short, straight, conical, retuse. *Column* much shorter than the labellum, rather longer than the spur, half round, margined. *Stigma* transverse, excavated. *Anther* terminal, opercular, deciduous, emarginate, 2-celled, with no rudiment of a dissepiment. *Pollen masses* 2, deeply furrowed, waxy, with a short caudicula and a transverse gland.

J. L.

NOTE upon *Sophronia modesta,* fol. 1129.

At the time this article was written, we had not received the Botanical Division of Captain Freycinet's Voyage, in which there is a genus of Fungi named Sophronia. This being the case, it is necessary, in order to avoid confusion, that we should alter our name; our readers are therefore requested to substitute *Sophronitis* for *Sophronia.*

1155

ORCHIS* papilionácea.

Purple Butterfly Orchis.

GYNANDRIA *MONANDRIA.*

Nat. ord. ORCHIDEÆ. § Ophrydeæ *Lindley.*

---

* Labello radicibusque indivisis.

O. *papilionacea;* labello obovato subemarginato crenulato venoso, sepalis obtusiusculis erectis nervosis, germine cornu inflexum acutum superante, bracteis coloratis breviore. *Sprengel syst.* 3. 684.

O. papilionacea. *Linn. sp. pl.* 1331. *Scop. carn. n.* 1103. *Willd. sp. pl.* 4. 24. aliorumque.

O. rubra. *Jacquin ic. rar.* 1. *t.* 183. *Collect.* 1. *p.* 60.

Radices *carnosi, indivisi.* Folia *brevia, oblongo-ensiformia, patentia, obtusa.* Caulis *erectus, pedalis v. minor, vaginis submembranaceis venosis, apice virescentibus.* Spica 5-7-*flora.* Bracteæ *membranaceæ, dilatatæ, roseæ, venosæ, ovario longiores.* Sepala *rubro-purpurea, venosa, ascendentia, inferioribus retrorsùm subfalcatis.* Labellum *pallidè purpureum, oblongum, obtusum, indivisum, subundulatum, crenulatum.* Calcar *crassum, conicum, rectum, subacutum, ovario brevius.*

A half-hardy species, introduced into this country by M. Mauri, by whom it was sent from Rome to the Horticultural Society, in 1826: our drawing was made in the Chiswick Garden, in May 1827. It thrives very well in light soil, if planted in a pot, and kept in a frame during the winter; but it would not bear our climate with less attention.

A native of most parts of the South of Europe, in Corsica, Liguria, near Monte Nuovo, Rome, and Naples, and also in Barbary.

By most authors, the O. rubra of Jacquin is considered

---

* Orchis is a primitive Greek word, which, being interpreted, signifies a small oval fleshy body; and was applied to the plants that bear its name, in consequence of the figure of the roots according with that interpretation.

identical with O. papilionacea; Sprengel asserts, that the figure in the Icones is a mere error, arising from its having been copied from a wretched drawing. But M. de Candolle, in his *Flore Française*, entertains a different opinion, and distinguishes O. rubra, that is to say, the Piedmontese plant described by Allioni under that name, by its smaller labellum, which is very slightly crenulated, and neither truncate nor emarginate at its apex. We have specimens called O. rubra from Granddidier's Herbarium, which answer to this statement; but they are otherwise so similar to our O. papilionacea from Naples, that we do not think it advisable to distinguish them even as varieties.

Roots fleshy, undivided. *Leaves* short, oblong-ensiform, spreading, obtuse. *Stem* erect, a foot high, or something less, with membranous venous sheaths, which are greenish at the apex. *Spike* 5-7-flowered. *Bracteæ* membranous, dilated, pink, venous, longer than the ovarium. *Sepals* reddish purple, venous, ascending; the lower ones somewhat falcate backwards. *Labellum* pale purple, oblong, obtuse, undivided, somewhat undulated, crenulate; *spur* thick, conical, straight, somewhat acute, shorter than the ovarium.

J. L.

1172

CÁTTLEYA* críspa.

Curled-petaled Cattleya.

GYNANDRIA *MONANDRIA.*

Nat. ord. Orchideæ. § *Epidendreæ* Lindl.
CATTLEYA. Suprà, vol. 11. *fol.* 953.

C. *crispa;* perianthii laciniis exterioribus lanceolatis unguiculatis: interioribus latioribus undulatis crispis, labello crispo acuminato.

Folium *solitarium, bulbo subcylindraceo, angulato insidens, oblongo-lanceolatum, emarginatum.* Spica, *ex axillâ folii, 4-5-flora.* Perianthii *laciniæ albæ, exteriores lineari-lanceolatæ, unguiculatæ, interioribus undulatis crispis angustiores.* Labellum *acuminatum, intùs atropurpureum, margine quàm maximè crispa.*

This splendid epiphyte was sent from Rio Janeiro to the Horticultural Society by Sir Henry Chamberlayne, Bart. in 1826. It flowered in the stove in the Chiswick Garden in August 1827. It is very distinct from all the other species of Cattleya, both in colour and in the form of the labellum and other segments of the flower. There are now five species of Cattleya upon record.

Cultivated in decayed vegetable mould, in which it grows freely.

Leaf solitary, seated on a subcylindrical, angular bulb, oblong-lanceolate, emarginate. *Spike* of 4 or 5 flowers, from the axilla of the leaf. *Segments* of the perianthium white, the outer linear-lanceolate, unguiculate, narrower than the inner, which are wavy and curled at the edges. *Labellum* acuminate, deep purple inside, with an exceedingly curled margin.

J. L.

---

* So called in compliment to William Cattley, Esq., of Barnet, in Hertfordshire, a great patron of Botany, and the most ardent collector of rare plants of his day.

1175

LÍPARIS* eláta.

Tall Liparis.

GYNANDRIA *MONANDRIA.*

Nat. ord. Orchideæ. § *Malaxideæ.*
LIPARIS.—Suprà, vol. 11. *fol.* 882.

L. *elata;* foliis oblongo-lanceolatis acuminatis undulatis plicatis, bracteis foliaceis reflexis, labello obcordato bituberculato, caulibus ovatis.

Herba *terrestris, caules ovatos, breves, cicatrizatos efformans.* Folia *oblongo-lanceolata, acuminata, undulata, plicata, glabra, basi vaginantia, pedales v. circiter.* Scapus *erectus, simplex, bipedalis, angulatus, angulis alatis.* Spica *cylindracea, sensim elongata, apice nutans.* Bracteæ *ovatæ, virides, reflexæ, ovario breviores.* Perianthium *patens;* sepala exteriora *fusca, supremo lineari apice recurvo, inferioribus brevioribus, oblongis, obtusis, labello suppositis, collateralibus, basi connatis, apice recurvis;* interiora *linearia, convexa, divaricata, supremi longitudine.* Labellum *carnosum, ungue viridi, suberecto, canaliculato, bicorni, disco reflexo, obcordato, fusco.* Columna *erecta, arcuata, semiteres, antice antè stigma parvum concavum alata.* Pollinia 4, *cereacea, glandulâ retinaculoque nullis.*

For this species of Liparis we are indebted to the researches of Sir Henry Chamberlain, who discovered it near Rio Janeiro, whence he sent it to the Horticultural Society in 1826. It is a tender stove herbaceous plant, growing freely in decayed wood, and flowering in July and August.

In the genus Liparis there is a section, consisting of the Cymbidium bituberculatum of Dr. Hooker, the Malaxis odorata of Willdenow, the Malaxis Rheedii of the same author, and the present plant, the species of which are extremely difficult to distinguish from each other. The present is particularly characterised by its short ovate

---

* Perhaps derived from λιπαρὸς, unctuous, in allusion to the soft surface of the leaves of some species: M. Richard does not explain the name.

stems, or bulbs, as they are often called, its obcordate labellum, and long green bracteæ.

These species of Liparis have already been in part indicated at fol. 882 of this work, in which all the Isle of France and Madagascar species of M. du Petit Thouars are also taken up. We have, therefore, been surprised at finding the same species lately described with the same names, by our good friend M. Achille Richard, who was of course ignorant of the prior claim on our part to the names of the species he has distinguished.

J. L.

1189

SERÁPIAS* cordígera; var. *longipétala.*

Pale heart-lipped Serapias.

GYNANDRIA *MONANDRIA.*

Nat. ord. ORCHIDEÆ. § *Ophrydeæ.*
SERAPIAS Linn. — *Pollinia* 2 granulata, *caudiculis* duabus *glandulæ* unicæ affixis, cucullo communi inclusæ. *Anthera* antica, erecta, persistens. *Stigma* subrotundum, concavum. *Columna* semiteres, ultrà antheram in cuspide producta. *Labellum* trilobum, lobis lateralibus erectis, medio dependente acuminato. *Sepala* acuminata, conniventia, subæqualia.——Herbæ *terrestres* (Europæ); radicibus *tuberosis;* caulibus *foliosis;* spicâ *simplici* paucìflorâ.

S. *cordigera;* bracteis coloratis floribus longioribus, labello basi bicalloso.
α. *labello atro-fusco.*
S. cordigera. *Linn. sp. pl.* 1345. *Bot. rep.* 475, &c.
Helleborine cordigera. *Sebast. rom. pl. fasc.* 1. *p.* 13. *t.* 4. *f.* 2. *Tenore fl. nap.* 2. 315.
β. longipetala; *labello pallido.*
Helleborine longipetala. *Tenore fl. nap. prodr. p.* 53. *Sebast. fl. rom. prodr.* 312.
Helleborine pseudo-cordigera. *Sebast. rom. pl. fasc.* 1. *p.* 14. *t.* 4. *f.* 1.
Serapias lancifera. *St. Amans fl. agen. p.* 378. *D'Urville in act. Linn. par.* 1. 377.

We cannot agree with the Italian and other Botanists quoted above, in distinguishing this plant from S. cordigera, to which it seems to us to belong in every essential respect. The paleness of the flowers is, indeed, the only character that can be readily seized to distinguish them.

It is a beautiful spring flower, common about Rome in very dry soil. It is also a native of the neighbourhood of Agen, where it was found by M. de St. Amans; and Tenore speaks of it as abounding at the foot of Vesuvius, near Ottojano, Boscotrecase, Mauro, Mortelle, Portici, and elsewhere, always in very dry meadows.

Our drawing was made in April 1827, in the Garden of the Horticultural Society, where it had been received from Signor Mauri, in the same year

J. L.

* So called after the Egyptian divinity Serapis.

1197

ÓPHRYS* araneífera; *var.* limbáta.

Bordered Spider Orchis.

GYNANDRIA *MONANDRIA.*

Nat. ord. ORCHIDEÆ. § Ophrydeæ. *Lindley.*
OPHRYS. — *Suprà, vol.* 3. *fol.* 205.

O. *araneifera;* labello emarginato integro convexo pubescente inappendiculato subrotundo mutico: disco maculato, perianthio herbaceo: petalis ovatis pubescentibus sepalis duplò brevioribus, foliis glaucis.
O. araneifera *Auctorum.*
Var. *limbata;* maculâ disci sinuatâ limbatâ anticè bilobâ medio ocellatâ.

This pretty variety of O. araneifera was sent to the Horticultural Society, in 1826, by Signor Mauri, of Rome. It is figured chiefly for the purpose of shewing the differences that exist between it and Ophrys atrata, represented at Plate 1087. It is perhaps the same as Sir James Smith's O. fucifera; but of that species we have not seen any authentic specimens which will justify us in absolutely referring it hither. The Kentish O. araneifera varies so much in the figuring, and outline of its labellum, that we can scarcely doubt O. fucifera being a mere variety of it.

A hardy, perennial spring plant, when wild loving to grow in chalky soil among short grass; or, if cultivated, in a pot in a well-aired frame.

J. L.

* The form of the flowers of some of the plants formerly referred to Ophrys is said to have a resemblance to the eyebrow, ὀφρὺς in Greek. In what the resemblance consists, we confess our inability to discover; especially if the original Ophrys was, as is supposed, the plant now called Listera ovata.

1206

MAXILLÁRIA* ciliáta.

Fringed-lipped Maxillaria.

GYNANDRIA *MONANDRIA.*

Nat. ord. ORCHIDEÆ. § Vandeæ *Lindley.*
MAXILLARIA. — *Suprà, vol.* 11. *fol.* 897.

M. *ciliata;* pseudo-bulbis ovatis ancipitibus, foliis lanceolatis 3-pluricostatis, scapo unifloro, labio abbreviato fimbriato.
Dendrobium ciliatum. *Swartz. Pers. syn.* 2. 523.
Maxillaria ciliata. *Flor. peruv. syst. p.* 221.
Caules *bulbiformes, ovati, sub-ancipites, diphylli.* Folia *lanceolata, 3-pluricostata, acuminata.* Scapi *radicales, uniflori, flexuosi, vaginati, pseudo-bulbis paulò longiores.* Flores *magni, herbacei.* Sepala *ovato-lanceolata, subæqualia; inferiora cum processu elongato columnæ connata;* petala *conformia, minora.* Labellum *cum columnâ articulatum, disco lobi medii fimbriati cristâ latâ sulcatâ abruptè terminato.* Columna *marginata, clavata.* Pollinia 4, *per retinaculum glandulæ affixa.*

A native of woods in South America, flowering in the autumn. It is among the most curious of the Maxillaria tribe, and is readily distinguished by its short, fringed lip. With us it is a tender epiphyte, requiring the heat of a good stove.

Our drawing was made in Mr. Lee's Nursery, in May 1828.

The *stems* are short, ovate, compressed, and bulb-like, each bearing two lanceolate, plaited, sharp-pointed *leaves,* which have from three to a greater number of ribs. The *scapes* arise from the root, each bearing one flower, flexuose, sheathed with convolute, obtuse scales, a little longer than the bulb-like stems. *Flowers* large, green. *Sepals* ovate-lanceolate, nearly equal, the lower connate with the elongated process of the column; *petals* of the same shape, but smaller. *Lip* articulated with the column, the disc of the middle lobe, which is fringed, terminated abruptly by a broad, furrowed crest. *Column* bordered, clavate. *Pollen masses* 4, attached to the gland by a common stalk.

The bulbs, as they are improperly but commonly called, of this tribe of Orchideæ, are in fact stems in a particular

* From the *maxillæ* or jaws of an insect, to which some resemblance has been found in the column and lip of the genus.

state: Botanists have as yet given no definite name to them, wherefore we propose hereafter to distinguish them by the denomination of *pseudo-bulbi*.

J. L.

1213

PHOLIDÓTA* imbricáta.

Imbricated Pholidota.

GYNANDRIA *MONANDRIA.*

Nat. ord. ORCHIDEÆ. § *Malaxideæ.* Lindley.

PHOLIDOTA Lindley. — *Pollinia* 4, cereacea, libera, glandulâ nullâ. *Anthera* terminalis, decidua, bilocularis, 4-valvis! *Stigma* cuneatum, in processu saliente columnæ insertum. *Columna* brevis, utrinque alata, carnosa. *Labellum* saccatum, cucullatum, tridentatum. *Petala* linearia, erecta. *Sepala* conniventia, concava, duobus inferioribus carinatis.—— Spicæ *imbricatæ terminales.* Folia *coriacea in pseudo-bulbis solitaria.*

P. *imbricata;* pseudo-bulbis sulcatis ovatis obtusis fasciculatis, foliis oblongis coriaceis plicatis, racemis distichis pendulis imbricatis.

P. imbricata. *Lindley in Hooker's exotic Flora, fol.* 138.

Cymbidium imbricatum. *Roxb. MSS. Hort. Beng. p.* 63.

Ornithidium imbricatum. *Wallich MSS.*

Ptilocnema bracteatum. *Don prodr. fl. Nep.* 33.

Wellia-Theka-Maravara, vel Mau-Maravara. *Rheede Malab.* XII. 47. *t.* 24.

Planta *perennis, omnibus partibus lævis, latè suprà arbores serpens, frequentissimè dependens, nunc nutans.* Caules *consistunt bulbis fasciculatis, numerosis, carnosis, bipollicaribus, oblongo-cylindricis, v. subovatis, obtusis, obsoletè angulatis, unifoliis, indutis bracteis aliquot magnis, inæqualibus, lanceolatis, emarcescentibus.* Folium *è vertice bulbi petiolatum, coriaceum, crassum, valdè rigidum, lanceolatum, utrinque attenuatum, acutum, sulcato-plicatum, pedale ad sesquipedale, quinquenervium, suprà atro-viride, nitidum, subtùs opacum, costâ robustâ valdè elevatâ.* Petiolus *crassus, firmus, suprà canaliculatus, bi-tri-pollicaris.* Racemus *solitarius, laxus, pedunculatus, è sinu folii nutans v. pendulus, lineari-oblongus, complanatus, distichus, folii apicem plerumque attingens.* Pedunculus *gracillimus, lignosus, filiformis, nudus, exceptis bracteis racemi infimis vacuis, vaginantibus, imbricatis.* Rachis *compressiuscula, leviter flexuosa.* Flores *numerosi, utrinque circiter 40, inter minores, exactissimè bifarii, alterni, approximatè patentes, pallidè flavicantes, inodori, pedicellis brevibus, clavatis.* Bracteæ *infrà singulum florem solitariæ, magnæ, ferè unguiculares, patentissimæ, latè ovatæ, denticulatæ, concavæ, scariosæ, antè anthesin in spicam subcylindricam imbricatæ, persistentes, infimæ paucæ abortivæ.* Perianthium *ringens, bilabiatum, pentaphyllum, membranaceum;* foliola *distincta;* exteriorum *alterum ovatum, acutum, fornicatum, cum interioribus falcatis occultis in galeam posticam connivens, altera parùm majora, dissimilia, prismatico-scaphoidea, marginibus interioribus parallelis, dorso gibboso-carinata, labello supposita, cumque illo labium inferius efficientia.* Labellum *sessile, cucullato-ventricosum, subinclusum, basi semicirculari columnæ haud productæ accretum, ore truncatum, fundo intùs cristulis duabus elevatis;* lamina *brevissimè unguiculata, recurvata, biloba, lobis semicordatis, obtusis, crenulatis.* Columna *medio carnosa (elevata), lateribus apiceque marginata membranulâ latiusculâ, scutiformis.* Anthera *terminalis, rotundata, subdidyma, rufescens, apici columnæ adnata, bilocularis, loculis prominulis bivalvibus, valvulis inferioribus minoribus.* Massæ pollinis *ovatæ, cereaceæ, posticè sulculo notatæ, in singulo loculo geminatæ, pedicellis propriis apice convexis.* Stigma *ad apicem columnæ, horizontale, vertice valvatum, demùm clausum.* Wallich MSS.

---

* Derived from φολὶς, a scale, and οὖς ὠτὸς, an ear; in allusion to the scaly ear-like bracteæ of the spike.

The preceding elaborate and most excellent description, made from plants observed in India, has been kindly communicated to us by our valued friend Dr. Wallich, by whom an account of this and a second species was long ago prepared for publication. At that time he proposed to refer them to Ornithidium, remarking, however, at the same time, that certain peculiarities of structure, which had not escaped his accurate observation, " would perhaps require the plant to be formed into a new genus."

This measure we had ourselves adopted, so long since as 1824, and having communicated the name now retained to Dr. Hooker, the species was published in the *Exotic Flora*, in January 1825; but at that time no knowledge was possessed in this country of Dr. Wallich's MS. name. The genus Ornithidium, to which Pholidota is undoubtedly very nearly allied, is distinguished by its labellum being partially connate with the column, by the column being taper, not round, by the anther having only two valves, and by its pollen masses being fixed upon a gland. In this last point the figure in the *Exotic Flora* of Ornithidium coccineum, otherwise excellent, is defective, the pollen masses being there represented as lying loose in the cells of the anther; and, therefore, as having the structure of Malaxideæ, instead of that of Vandeæ.

A native of several parts of India. Dr. Wallich informs us, that he has found it on the mountains west of Ava, also high up the rivers in Martaban, and on the mountains upon the Tenasserim coast; and that he does not find it differ in any respect from the species as found in Nipal, where it is exceedingly common, as well as in Sylhet and Chittagong, growing parasitically upon trees in mountain forests.

Our drawing was made last November, in the Garden of the Horticultural Society, to which establishment it had been presented by the Honourable Court of Directors of the East India Company, along with a great quantity of other most rare and valuable plants personally brought to England in the summer of 1828 by Dr. Wallich.

It is a stove plant, requiring the same treatment as other similar epiphytes, but by no means difficult to cultivate. At the same time with this, the other species of Pholidota above alluded to (*undulata* Wall.) was brought home; and is thus, it is to be hoped, secured to our Gardens: it is a much more delicate plant.

Upon the subject of its cultivation, Dr. Wallich remarked in the paper from which the foregoing extracts have been made, that, " Like most members of this lovely tribe, it is easily made to grow on the trunks of old trees, taking care to place some vegetable mould under its roots, and tying it so as to retain its situation. It requires constant humectation, which is easily effected by means of a small vessel suspended over it, with a perforated bottom, through which the water is led down upon the plant by means of a string, the upper end of which fills, though it does not quite shut up, the aperture of the vessel. It is propagated by separating its bulbous stems, which generally form dense tufts, ornamented with evergreen, dark-coloured, shining leaves, of a peculiarly firm and leathery texture."

J. L.

EDWARDS'S

BOTANICAL REGISTER:

OR,

ORNAMENTAL FLOWER-GARDEN

AND SHRUBBERY:

CONSISTING OF

COLOURED FIGURES OF PLANTS AND SHRUBS,

CULTIVATED IN BRITISH GARDENS;

ACCOMPANIED BY THEIR

History, Best Method of Treatment in Cultivation, Propagation, &c.

CONTINUED

BY JOHN LINDLEY, F.R.S. L.S. AND G.S.

PROFESSOR OF BOTANY IN THE UNIVERSITY OF LONDON,
&c. &c. &c.

VOL. XV.

——viret semper——nec fronde caducâ
Carpitur.

LONDON:

JAMES RIDGWAY, 169, PICCADILLY.

M.DCCC.XXIX.

1239

DENDRÓBIUM* ánceps.

Two-edged Dendrobium.

GYNANDRIA *MONOGYNIA.*

Nat. ord. ORCHIDEÆ. § *Malaxideæ.* Lindley.
DENDROBIUM.—*Suprà, vol.* 7. *fol.* 548.

D. *anceps;* caule ancipiti simplici, foliis distichis scapelliformibus planis, pedunculis binis è basi vaginarum brevissimis. *Swartz act. holm.* 1800. *p.* 246. *Willd. sp. pl.* 4. 136. *Spreng. syst.* 3. 738.

Caules *numerosi, compressi, carnosi, penduli.* Folia *disticha, carnosa, compressa, ovato-oblonga, acuta, pallidè viridia.* Flores *solitarii, herbacei.* Sepala *ovata, erecta, acuta, interiora minora, inferiora cum basi productâ columnæ longè connata.* Labellum *unguiculatum, ecallosum, inappendiculatum, articulatum, cuneatum, emarginatum, crenulatum, paululùm coloratum.* Stigma *bicallosum intrà cavitatem.* Pollinia 4, *didyma, libera, collateralia.* Anthera *pedicellata.*

An inhabitant of the trunks of trees in swampy, low situations, in the æstuaries of the rivers of Bengal and Pegu, according to Dr. Wallich, to whom the Gardens of England are indebted for the introduction of this curious species. In its natural position it is pendulous; but in our drawing it is represented erect,—the plant in the Garden of the Horticultural Society from which the figure was taken, having at that time been tied to a stake. It flowers at uncertain seasons, and grows more freely than other plants with a similar habit.

In appearance it is very like the Herba supplex quinta of Rumphius, vol. 6. p. 111. t. 51. f. 2; but that plant has spiked flowers, and, Dr. Wallich informs us, is quoted by Roxburgh in his MSS. to his Dendrobium acinaciforme.

* From δένδρον, a tree, and βίος, life. All the genuine species are found upon trees, in the hot parts of the East Indies.

Stems numerous, compressed, fleshy, pendulous. *Leaves* distichous, fleshy, compressed, ovate-oblong, acute, pale green. *Flowers* solitary, herbaceous. *Sepals* ovate, erect, acute; the inner ones smallest, the lower ones connate with the long base of the column. *Lip* unguiculate, with neither callosities nor appendages, articulated, cuneate, emarginate, crenulate, a little coloured. *Stigma* with two callosities within its cavity. *Pollen masses* 4, twin, loose, collateral. *Anther* with a little stalk.

J. L.

1290

MICROSTÝLIS* ophioglossoídes; β. *mexicana.*

Mexican Snake's-tongue-leaved Microstylis.

GYNANDRIA *MONANDRIA.*

Nat. ord. ORCHIDEÆ. § Malaxideæ *Lindley:* * Pleurothalleæ.
MICROSTYLIS Nutt.—*Sepala* patentia, libera; lateralia basi æqualia sæpiùs breviora. *Petala* patentia, linearia v. filiformia. *Labellum* patentissimum, cum columnâ angulum rectum formans, basi excavatum, sagittatum v. auriculatum, integerrimum v. dentatum, tuberculis nullis. *Columna* minima, raró paululùm elongata, apice dentibus seu auribus duâbus instructa. *Pollinia* 4, collateralia.——Herbæ *terrestres v. epiphytæ, habitu* Liparis, foliis *plicatis v. membranaceis, basi raró incrassatis.* Flores *herbacei, nunc flavescentes v. discolores.*—Lindley, Genera and Species of Orchideous Plants, part 1. ined.

M. *ophioglossoides;* caule unifolio, racemo obtuso capitato multifloro, labello tridentato. *l. c.*
Malaxis unifolia. *Michaux.*
Malaxis ophioglossoides. *Willd. no.* 3. *Lodd. bot. cab. t.* 1146. excellent.
Microstylis ophioglossoides. *Nutt. gen.* 2. 196.
β. *mexicana;* caule pedali, racemo densiore, labelli dente mediâ lateralibus minore. *l. c.*

This genus was instituted by Mr. Nuttall, but with a very erroneous description. It is, however, most distinctly characterised, and has been since proposed by Dr. Blume under the name of Crepidium; and by Dr. Wallich, in his manuscript papers, under that of Thyreochilus. Fourteen species are known to us, of which five are unpublished. Malaxis monophyllos *Willd.*, diphyllos of *Chamisso*, umbellulata and spicata of *Swartz*, acuminata of *Don*, Rheedii *Willd.*, Crepidium flavescens *Blume*, and Ophrys macrostachya *Llave*, all belong to the genus. We shall give the characters of these and other species in a work specially devoted to the subject, which will appear in the spring of the following year, with illustrations drawn from the admirable sketches of Mr. Bauer.

* So named from μικρὸς, small, and στύλος, a column; in allusion to the very small size of what is called the column in this plant.

The subject of the accompanying plate was raised in the Garden of the Horticultural Society, from roots transmitted from Mexico by Mr. John Brown. It grows in the greenhouse, in a pot, coming up every year in the spring, flowering for five or six weeks, and then dying down. The inflorescence is remarkable, when it first appears, for its very depressed state. The axis afterwards elongates; and what was at first an umbel, becomes a corymbose raceme. In the figure this is pretty well shewn; but the leaf is represented too cordate, and the form of the labellum is, as may be seen from the magnified figure, inaccurate.

It is impossible to distinguish this from the M. ophioglossoides of North America, which is well figured in Mr. Loddiges' *Botanical Cabinet*. It differs in its greater stature, it is true, and also in the proportion borne by the middle lobe of the labellum to the side ones; but we doubt whether the species of Orchideous plants are to be characterised by marks so slight as these.

J. L.

1291

DENDRÓBIUM* secúndum.

One-sided Dendrobium.

GYNANDRIA *MONANDRIA.*

Nat. ord. ORCHIDEÆ. § Malaxideæ *Lindl.* * Dendrobieæ.
DENDROBIUM.—Suprà, vol. 7. *fol.* 548.

§ *Caules undique foliosi, cylindracei, sæpè penduli.*

D. *secundum;* caulibus pendulis, foliis oblongis apice obliquè retusis, racemis lateralibus terminalibusque secundis, sepalis ovatis: lateralibus basi in cornu longo obtuso incurvo connatis, petalis supremo paulò minoribus, labello integerrimo acuto apice tumido.—*Lindley, Genera and Species of Orchideous Plants, part* 1. ined.
D. secundum. *Wall. in cat. herb. ind.*
Pedilonum secundum. *Blume Bijdrag. p.* 322.

For an opportunity of presenting the public with a figure of this fine species, we have to thank Mr. Tate, of the Sloane Street Nursery, by whom a plant in flower was communicated in July last. It had been collected at Rajah Bassa, in Sumatra, by William M'Killigin, Esq.

It appears, from Dr. Wallich's drawings, to be a most lovely species in its native places of growth, forming long pendent stems, which throw out an abundance of one-sided racemes of purple flowers 5 or 6 inches long. The specimen now represented blossomed imperfectly, as its stems had been allowed to remain in an erect, instead of pendent position. The habit of this, and many other Dendrobriums, being to hang down from the trees on which they grow, it is impossible to cultivate them with any success unless they are suspended in the air in pots, or otherwise so managed that they can shoot freely in the way that is natural to them.

If we are right in Dr. Blume's synonym, which we can scarcely doubt, this species is a native of mountainous places in Java. Dr. Wallich found it in one of the Islands in the Straits of Malacca, and also in the province of Martaban, as appears from that part of the *Catalogue of Dried Plants* distributed under the orders of the Honourable Court of Directors of the East India Company, which has just appeared. This catalogue already comprehends upwards of 2000 species, of the greater part of which specimens have been, or will be, presented to scientific institutions and persons in every part of the civilised world. Large as is the number already enumerated, it is to be considered as a mere fragment of what the catalogue will eventually contain, so stupendous are the stores from which the collection is derived, and so unreserved is the liberality of the generous and enlightened donors. We trust that this splendid example will be followed by other bodies in whose possession are similar extensive collections, whether Botanical or belonging to any department of Natural History whatsoever. To keep the duplicates, triplicates, and multiplicates of collections which have often cost the public large sums of money, and which have always been formed at great personal risk and trouble,—we say, to keep such collections locked up in chests, deposited in cellars, or tied up in bundles, in public museums, where they can only become the food of insects, or the victims of dust and time, is cruel towards those by whom they were procured, unjust towards the community, unworthy of men of liberal minds, and most injurious to the best interests of science. It is highly to the honour of Great Britain that this system of distributing the duplicates of public collections should have originated with her; and we are sure that, whether this example is followed by the British and other Governments, as we trust that it speedily will be, or not, the name of the English East India Company will stand in the records of science as far above that of all other associations of individuals, as it already does in the annals of commerce, and in the history of political affairs.

Dr. Blume's genus Pedilonum is characterised by the cohesion of its lateral sepals into a spur; but as this character is not in any degree connected with habit, and can frequently not be determined with accuracy, in consequence of the numerous and insensible gradations of union between the sepals, and is, moreover, unaccompanied by any kind of secondary character, we are obliged to reject it.

J. L.

* So called from δένδρον, a tree, and βιόω, to live; the species all growing upon trees.

NOTE.

In the last Number of the *Botanical Magazine* a fine Orchideous plant was published, under the name of *Stanhopea insignis*. As this is likely to attract attention, and will probably be soon a common plant in collections, we take the earliest opportunity of giving notice, that the name assigned to it in the *Botanical Magazine* cannot be retained, as it is a species of Ceratochilus, a genus long since published in Mr. Loddiges' *Botanical Cabinet.*

1298

PLEUROTHÁLLIS* prolífera.

Proliferous Pleurothallis.

GYNANDRIA *MONANDRIA.*

Nat. ord. ORCHIDEÆ. § Malaxideæ *Lindley.* * Pleurothalleæ.
PLEUROTHALLIS.—Suprà, vol. 9. *fol.* 759.

P. *prolifera;* caule ancipiti, folio oblongo cochleato carnoso prolifero racemo duplò longiore, bracteis cuneatis cuspidatis cucullatis pedicello brevioribus, sepalis conniventibus lateralibus semiconnatis petalis labelloque conformibus duplò longioribus, clinandrio dentato.—*Lindley Gen. and Sp. of Orchideous Plants, part* 1. *ined.*
P. prolifera. *Herbert in litteris.*

Epiphyta, caulibus *monophyllis, flexuosis, palmaribus, basi teretibus, apice compressis obtusè marginatis.* Folia *crassa, carnosa, avenia, cucullata, ovato-oblonga, acuta, semper prolifera.* Racemus *simplex, brevis, ex axillâ folii in cujus sinu recumbit, pauciflorus, folio brevior, basi squamis paucis, scariosis, acutis;* rachis *flexuosa;* bracteæ *solitariæ, membranaceæ, herbaceæ, cucullatæ.* Flores *lurido-purpurei, intùs punctati, brevè pedunculati.* Ovarium *brevissimum, turbinatum.* Sepala *carnosa, conniventia, acuminata, basi connata, lateralibus semicohærentibus.* Petala *linearia, acuminata, columnæ longitudine.* Labellum *lanceolatum, integerrimum, sepalis paulò brevius.* Columna *clavata, semiteres.* Clinandrium *alatum, denticulatum.* Anthera *bilocularis.* Pollinia 2, *teretia, basi materie pulvereâ cohærentia.*

For this curious plant we are indebted to the Honourable and Rev. William Herbert, by whom we were favoured with specimens in January last. Mr. Herbert informs us, that it was found at Boto Fogo, near Rio Janeiro, growing on a steep rock, which the sun could rarely shine upon.

At Spofforth it flowers six or seven months in the year, producing successive racemes year after year. It likes a shady end of the stove, and to throw its roots about in the air. It will probably grow better in moss half decayed than in any other material.

It is remarkable for the proliferous character of its leaves. These constantly produce young rooting plants from their axillæ; and, what is singular, the first leaf of each new individual is produced from the same side of the axis

* So named from πλευρά, a side or rib, and θάλλω, to flower; in allusion, we presume, either to the one-sided disposition of the flowers of some species, or to the developement of the inflorescence from what appears to be the rib of the leaves.

of the mother plant as the old leaf from which it sprung. This apparent exception to the universal laws under which leaves are developed, is due to the abortion of the first leaf that is developed, which appears in the form of a withered scale, while the second leaf is that which is finally and fully developed.

J. L.

1299

DENDRÓBIUM* chrysánthum.

Golden-flowered Dendrobium.

GYNANDRIA *MONANDRIA.*

Nat. ord. ORCHIDEÆ. Tribus *Malaxideæ.* * Dendrobieæ *Lindl.*
DENDROBIUM.— Suprà, vol. 7. fol. 548.

§ *Caules undique foliosi, cylindracei, sæpè penduli. Flores fasciculati v. solitarii.*

D. *chrysanthum;* caulibus teretibus pendulis, foliis contortis ovato-lanceolatis acuminatis, floribus ternatis inter folia nascentibus, sepalis carnosis oblongis obtusis: venis extùs tuberculatis, petalis obovatis retusis carnosis sepalo supremo latioribus, labello cucullato denticulato retuso obsoletè trilobo. *Lindl. Gen. et Sp. Orch. ined.*

D. chrysanthum. *Wallich cat. no.* 2012.

Flores *intensè lutei, carnosi, extùs tuberculis crebris minuti;* petala *tamen minùs tuberculata.* Labellum *intùs maculâ atrosanguineâ.* Ibid.

This beautiful species is a native of Nipal, whence it was introduced into the Botanic Garden, Calcutta, by Dr. Wallich, and by that zealous Botanist brought to England in the year 1828. The plant from which the accompanying figure was taken blossomed in a stove in the Horticultural Society's Garden, in February 1829: it had been presented to that establishment by the Honourable Court of Directors of the East India Company.

It is difficult to conceive a plant at once more graceful and beautiful than this; its pendulous stems, which hang from the rugged, deep brown, moss-clad trunks of trees, are clothed with lucid leaves of the most lively green, and its flowers are of the richest and deepest yellow. At first sight the species might be mistaken for D. fimbriatum, but it will be found very different upon comparison.

* See fol. 1239.

Most cultivators of stove Orchideous plants find a difficulty in managing the particular tribe to which this belongs; that difficulty is, however, completely overcome in the Garden of the Horticultural Society, in which these flourish more than almost any others. In that establishment they are treated thus: they are planted in perfectly rotten wood in small pots, which are covered with moss tied securely about them; these pots are suspended *obliquely* from the rafters of the front part of a small stove, in such a way that the plants are not compelled to grow upright, but are allowed to assume the pendulous or horizontal position, which is natural to them. Thus treated, species of the true Dendrobium habit, such as D. chrysanthum, flourish in a degree which is at least equal to that of their native woods. The temperature of such a stove should never fall below 75°, and the dew point should be always near saturation.

J. L.

EDWARDS'S

BOTANICAL REGISTER:

OR,

ORNAMENTAL FLOWER-GARDEN

AND SHRUBBERY:

CONSISTING OF

COLOURED FIGURES OF PLANTS AND SHRUBS,

CULTIVATED IN BRITISH GARDENS;

ACCOMPANIED BY THEIR

History, Best Method of Treatment in Cultivation, Propagation, &c.

CONTINUED

BY JOHN LINDLEY, F.R.S. L.S. AND G.S.

PROFESSOR OF BOTANY IN THE UNIVERSITY OF LONDON,
&c. &c. &c.

VOL. XVI.

——viret semper——nec fronde caducâ
Carpitur.

LONDON:

JAMES RIDGWAY, 169, PICCADILLY.

M.DCCC.XXX.

1314

DENDRÓBIUM* monilifórme.

Necklace-stemmed Dendrobium.

GYNANDRIA *MONANDRIA.*

Nat. ord. ORCHIDEÆ. § *Malaxideæ.* * Dendrobieæ.
DENDROBIUM.— Suprà, vol. 7. fol. 548.

Sect. *Caules undique foliosi, versùs apicem sensim incrassati, unde clavati fiunt.* Lindl. Gen. and Sp. of Orchideous Plants, ined.

D. *moniliforme;* caulibus erectis clavatis ramosis: internodiis tumidis, foliis oblongis obliquè emarginatis obtusis, floribus geminatis foliis longioribus, sepalis petalisque oblongis acutis venosis, labello cucullato acuto conformi.

Fu Ran. *Kæmpf. amœn. t.* 865.

Epidendrum moniliforme. *Linn. sp. pl.* 1352.

Dendrobium moniliforme. *Swartz act. Holm.* 1800. *p.* 246. *Willd. no.* 19.

Caulis *erectus, bipedalis, lævigatus, ramosus, internodiis tumidis, pallidè viridibus.* Folia *oblonga, subdisticha, apice obliquè biloba, basibus amplexicaulibus, brevibus, membranaceis.* Flores *gemini, pedunculo communi insidentes, è caule versùs apicem erumpentes, dilutè rosei, venis rubris picti.* Bracteæ *oblongæ, obtusæ, membranaceæ, læviter pilosæ.* Labellum *maculis duabus luteis in fauce.*

A native of China and Japan, from the former of which countries it was introduced several years since by the Horticultural Society.

The figure now given was obtained from a specimen that flowered in the collection of William Cattley, Esq., at Barnet, in November last. By this gentleman it has been cultivated with great success, his plants having attained the height of two feet, and being in a state of the most vigorous health. In general it is unhealthy, grows slowly, and never flowers. It is particularly dis-

* See fol. 1239.

tinguished by the tumid joints of the erect stem, of which the contractions become when old so considerable, that the stem acquires something the appearance of a necklace.

Thunberg describes, in his *Flora Japonica* (p. 30), an Epidendrum monile, to which he refers the Fu Ran of Kæmpfer; but he adds, that the leaves are acute, and the flowers white, which renders it probable that he intended some other species. Kæmpfer tells us, that it is suspended by the Japanese in baskets before the doors of their houses, in consequence of some vulgar superstition, the nature of which, however, he did not ascertain.

Stem erect, 2 feet high, polished, branched, with pale-green, tumid joints. *Leaves* oblong, somewhat distichous, obliquely 2-lobed at the apex, with short, membranous, stem-clasping bases (petioles). *Flowers* in pairs, seated on a common peduncle, proceeding from the stem towards its apex, pale rose-coloured, marked with red veins. *Bracteæ* oblong, obtuse, membranous, slightly hairy. *Labellum* with two yellow spots in the throat.

J. L.

1315

DENDRÓBIUM* longicórnu.

Long-horned Dendrobium.

GYNANDRIA *MONANDRIA*.

Nat. ord. ORCHIDEÆ. § *Malaxideæ.* * Dendrobieæ.
DENDROBIUM. — *Suprà, vol.* 7. *fol.* 548.

Sect. *Caules undique foliosi, cylindracei, sæpè penduli.* Lind. Gen. et Sp. of Orchideous Plants, ined.

D. *longicornu;* caulibus erectis hispidis flexuosis, foliis ovato-lanceolatis apice valdè obliquis, floribus fasciculatis v. solitariis terminalibus, bracteis ovatis acuminatis hispidis pedicello multò brevioribus, sepalis lateralibus in calcare longo acuminato connatis, petalis sepalo supremo conformibus, labello infundibulari cucullato integro dentato basi cum pede columnæ accreto.

Dendrobium longicornu. *Lindl. in Wallich cat. no.* 1997.

Arborum parasiticum, sæpiùs dependens. Radix *constans fibris cylindricis copiosis, carnosis, fasciculatis.* Caules *plures, graciles, sulcati, flexuosi, semipedales seu pedales, quin bipedales, pilis paleiformibus copiosis, atris, subdecumbentibus, solubilibus conspersus, ætate glabriores.* Folia *alterna, bifariè patentia, interstitiis triplò longiora, lineari-lanceolata, extrorsùm attenuata, apice valdè obliquè bidentata, dentibus angustis, obtusis, altero abbreviato; basi acutiuscula, sessilia, brevissimè vaginantia ;* 3-4-*pollicaria, glabra, obsoletè* 3-*nervia, lineata, planiuscula.* Flores *magni, albi, bipollicares, inodori, glabri, pauci, terminales, nunc laterales in caulibus aphyllis, suffulti pedunculo carnoso, cylindrico, paleaceo-pilosulo, cum ovario oblongo-clavato pollices duos emetientibus.* Bracteæ *duæ, lanceolatæ, acutæ, carinatæ, pilosulæ,* ½-*unguiculares ad basin pedunculi.* Sepala *erecto-patentia, lanceolata, acuta, subcarinata ;* superius *levissimè fornicatum ;* lateralia *columnæ adnata, deorsùm valdè dilatata, cumque basi columnæ producta in calcar longum infundibuliforme, attenuatum, apice teres, ovarium æquans.* Petala *sepalis subconformia, infra superius subconniventia.* Labellum *magnum, infundibuliforme, rectum, laminâ brevi, ovatâ, obtusâ, subretusâ, ciliato-fimbriatâ ; marginibus obtusis, conniventibus; disco papilloso-glanduloso, lineisque parallelis flavis ornato ; basi valdè attenuatâ, desinens in calcar floris.* Columna *columnaris crassa, anticè plana ; infernè*

* See fol. 1239.

sepala lateralia adfigens ; parte superiore brevi libera, apice cavâ lateraliter obtusè bilobatâ. Anthera *opercularis, conica, obtusa, ope fili brevissimi postice adfixa, decidua, bilocularis.* Pollen *pulvereum, flavum, inclusum membranulâ subtilissimâ, divisâ in massas duas oblongas, sulco longitudinali notatas, facilè bipartibiles.*—Wallich MSS.

For the foregoing valuable description of this rare plant we are obliged to our friend Dr. Wallich, by whom it was brought from India in 1828, and deposited in the Garden of the Horticultural Society, by permission of the Honourable Court of Directors of the East India Company.

Dr. Wallich adds: "This fine species is a native of most of the mountains in Nipal, where it blossoms during the rainy season. I have also received it from the late Mr. Smith, whose people found it on the mountains bordering on the district of Sylhet. It thrives well at the Honourable Company's Botanic Garden at Calcutta, into which it has been introduced from these countries."

Our drawing was made in the Garden of the Horticultural Society in May last. It grows pretty well in decayed vegetable mould, among moss, in the stove.

An *epiphyte*, usually hanging down. *Root* formed of many cylindrical, fleshy, fasciculated fibres. *Stems* numerous, slender, furrowed, flexuose, 6 inches, a foot, or even two feet long, covered with copious, chaff-like, dark, decumbent, separable hairs, but becoming nearly smooth when old. *Leaves* alternate, spreading bifariously, thrice as long as the intervals, linear-lanceolate, tapering outwards, very obliquely 2-toothed at the apex; the teeth narrow, obtuse, unequal; at the base rather acute, sessile, with very short sheaths, 3 or 4 inches long, smooth, obsoletely 3-nerved, striated, nearly flat. *Flowers* large, white, 2 inches long, scentless, smooth, terminal, or sometimes lateral, on leafless stems, placed on a fleshy, cylindrical, hairy peduncle, measuring, with the oblong clavate ovarium, about 2 inches. *Bracteæ* 2, lanceolate, acute, keeled, rather hairy, half a nail long, at the base of the peduncle. *Sepals* erect, spreading, lanceolate, acute, somewhat keeled; the lateral ones adnate to the column, very much dilated downwards, and, together with the elongated base of the column, produced into a long, funnel-shaped spur. *Petals* shaped like the sepals, somewhat conniving under the upper one. *Labellum* large, funnel-shaped, straight, with a short, ovate, blunt lamina, fringed, with a papillose glandular disk, and a few yellow lines on it.

This belongs to Dr. Blume's genus Pedilonum.

J. L.

EDWARDS'S

BOTANICAL REGISTER:

OR,

ORNAMENTAL FLOWER-GARDEN

AND SHRUBBERY:

CONSISTING OF

COLOURED FIGURES OF PLANTS AND SHRUBS,

CULTIVATED IN BRITISH GARDENS;

ACCOMPANIED BY THEIR

History, Best Method of Treatment in Cultivation, Propagation, &c.

CONTINUED

BY JOHN LINDLEY, F.R.S. L.S. AND G.S.

PROFESSOR OF BOTANY IN THE UNIVERSITY OF LONDON,
&c. &c. &c.

VOL. XVII.

——viret semper——nec fronde caducâ
Carpitur.

LONDON:
JAMES RIDGWAY, 169, PICCADILLY.

M.DCCC.XXXI.

1401

BLÉTIA* flórida.

Pallid Bletia.

GYNANDRIA *MONANDRIA.*

Nat. ord. ORCHIDEÆ *Juss.—(Introduction to the natural system of Botany, p.* 262.)

BLETIA Fl. Peruv.—*Sepala* patentia, æqualia. *Petala* nunc patentia, nunc conniventia, sepalis subæqualia. *Labellum* cucullatum, cum columna articulatum, nunc basi saccatum, trilobum, disco sæpiùs lamellato v. tuberculato. *Columna* elongata, semiteres. *Anthera* carnosa, 8-locularis. *Pollinia* 8, æqualia, ligulis 4 pulvereis cohærentia.——Herbæ *subterrestres,* foliis *ensiformibus plicatis,* scapis *racemosis multifloris,* floribus *sæpiùs speciosis.*

B. *florida;* sepalis ovali-lanceolatis obtusiusculis subpatentibus, petalis latioribus obtusis supra columnam fornicatis, labelli disco costis quinque rectis indivisis: lobis lateralibus rotundatis abbreviatis, intermedio cuneato crispo lamellis tribus undulatis apice abruptis, scapo simplici radicali, foliis ensiformibus.

Cymbidium floridum. *Salisb. prodr.* 9.
Gyas florida. *Salisb. in act. soc. hort.* 1. 261.
B. florida. *R. Brown in hort. Kew. ed.* 2. 5. 206.
B. pallida. *Loddiges bot. cab. t.* 629.

Herba *et* pseudobulbi *ferè omninò B. verecundæ.* Scapus *radicalis, aphyllus, squamis vaginatus, bitripedalis, gracilis, nunquam (quantum vidi) ramosus.* Racemus *sub-12-florus.* Flores *pallidè rosei.* Sepala *oblongo-lanceolata, obtusa, septemvenia, patula.* Petala *conformia, sed multò latiora, magis membranacea, venis tantum quinque, quarum laterales venulas plures sub-ramosas versùs marginem promunt, supra columnam in fornicem duplicem s. galeam conniventia.* Labellum *pallidius, trilobum, oblongum, lobis lateralibus, rotundatis, abbreviatis, intermedio cuneato, truncato, crispo, lateralibus multò longiore;* costis *medii quinque luteis, rectis, quarum* laterales *majores, nunc divisæ, vix basin lobi medii attingunt,* intermediæ *ad apicem labelli tendunt dimidiâ inferiore parùm elevatæ, rectæ, planæ, superiore petaloideæ, crispæ, subtruncatæ.*

* Named after Don Luis Blet, a Spanish medical man, much attached to Botany.

A native of Trinidad, whence it was introduced by the late Mr. Salisbury in the year 1786. It requires the heat of a very good stove, and is cultivated in a mixture of peat, sand, and loam.

It has been generally supposed that this is a species distinct from B. florida, a plant which no cultivator has been able, for many years, to satisfy himself that he possessed. For ourselves, we at one time were disposed to believe that B. florida was either a slight variety of Bletia verecunda, or absolutely the same plant; a supposition to which we were led by the account given of it in the last edition of the *Hortus Kewensis.* The diagnosis of it in that work is made to depend upon the cuneate form of the middle lobe of the labellum, its ridges being simple (not branched), and its lateral segments broadest (not narrowest) at the point; to which is added, as a character which it possesses in common with B. verecunda, a somewhat branched scape; Limodorum purpureum of Redouté's *Liliacées* is quoted as a synonym; and finally, the English name is Purple Bletia. But this latter name is perfectly applicable to B. verecunda; the plant figured by Redouté is unquestionably that species; and the ridges of the labellum are often perfectly simple in the same; so that of the characters and synonyms of B. florida, nothing is left to distinguish it except the form of the middle lobe of the labellum, and a not very intelligible difference in that of its lateral lobes. The character, however, in the *Hortus Kewensis* being very precisely worded, and the plant itself not belonging to a tribe that is difficult of cultivation, and therefore not likely to be lost, it appeared to us more probable that some inaccuracy should exist in the *Hortus Kewensis,* than that the species should no longer be found in any of the great collections near London. In renewing the search for it, we have found nothing to which it is probable that the name ought to be attached, except the plant now represented, the B. pallida of Mr. Loddiges; and, after much consideration, we have come to the conclusion, that, notwithstanding certain differences that exist between this and the account in the *Hortus Kewensis,* they are really the same species. It is true, that the scape of B. pallida is, as far as we have ever seen, unbranched, that the name *purple* is inapplicable to it, and that the ridges of its labellum are sometimes divided; but then the middle segment of its labellum is much as described in B. florida; and if the expression "lobis lateralibus apice latioribus," is merely intended to indicate the rounded form of the lateral lobes of that species, as contrasted with their somewhat acute termination in B. verecunda, it becomes difficult to resist the conclusion, that B. pallida is the same as B. florida, the apparent differences between them depending upon an incautious citation of a synonym, and an imperfect diagnosis, in the *Hortus Kewensis.*

Our drawing was made some years since at Mr. Colvill's; and we think that at that time the possibility of the plant being Bletia florida was suggested by Mr. Sweet.

The generic character here given has been constructed so as to exclude Spathoglottis of Blume, and also Phaius of Loureiro, or Pachyne of Salisbury, of which Bletia Tankervilliæ is the type.

J. L.

1406

CÁTTLEYA* guttáta.

Spotted Cattleya.

GYNANDRIA *MONANDRIA.*

Nat. ord. ORCHIDEÆ *Juss.*—(*Introduction to the natural system of Botany*, *p.* 262.)
CATTLEYA.—*Suprà*, *vol.* 11. *fol.* 953.

C. *guttata;* floribus carnosis, sepalis lineari-oblongis obtusis, petalis conformibus paulò latioribus undulatis, labelli trilobi lobis lateralibus ovatis obtusis: intermedio cuneato obcordato disco tuberculato, foliis concavis, spathâ obsoletâ.

Caulis *teres, squamis membranaceis fuscis, arctè appressis vestitus, nullo modo tuberosus, palmaris v. ultra, diphyllus.* Folia *carnosa, oblonga, concava, patentia, apice obliquè emarginata.* Racemus *spathâ nullâ,* 4-5-*florus, erectus, rachi terete maculatâ.* Bracteæ *brevissimæ, acutæ, scariosæ.* Flores *pallidè virides, sanguineo guttati.* Labelli *lobi laterales albi, intermedius purpureus, lineis disci purpurei tuberculatis.*

This beautiful species was sent to the Horticultural Society from Brazil by the Right Honourable Robert Gordon, along with other interesting plants. The spotting of the flower is remarkably different from any thing that has yet been seen in the same genus.

It may be interesting to cultivators to know that what are called the stems of this genus, and indeed of many other Orchideous plants, that is to say, the erect stalks that bear the leaves, are analogous to tubers, and really the branches of a rhizoma, or prostrate stem, which creeps upon the surface of the ground, resembling a root, and that consequently each of the leaf-bearing branches may be safely cut off with a portion of the rhizoma attached, for the purpose of propagation. The gardener of Mr. Harrison, of Liverpool, has practised this method with great success.

J. L.

* See fol. 1172.

1415

EPIDÉNDRUM* odoratissimum.

Sweet-scented Epidendrum.

GYNANDRIA *MONANDRIA.*

Nat. ord. ORCHIDEÆ *Jussieu.*—(§ Epidendreæ; *Introduction to the natural system of Botany*, *p.* 262.)
EPIDENDRUM.—*Suprà*, *vol.* 1. *fol.* 17.

§ *Foliis in pseudobolbos insidentibus.*

E. foliis in pseudobolbos ovatos corrugatos binis ensiformibus, racemo subsimplici longiore, sepalis oblongis petalisque cuneatis patentibus, labelli ferè liberi trilobi lobis oblongis obtusis integris subæqualibus disco calloso depresso.

Encyclia patens. *Hooker in Bot. mag. t.* 3013.
Macradenia lutescens. *Bot. cab. non Bot. reg.*

Pseudobolbi *ovati, corrugati, sulcati, atrovirides, nunc induviati.* Folia *lineari-ligulata, coriacea, atroviridia, obtusa, racemo erecto, terminali, multifloro breviora.* Flores *odoratissimi, viridi-lutei, purpureo submaculati.* Sepala *ovata, acuta, patentia.* Petala *patentia, cuneato-obovata.* Labellum *sæpiùs posticum, cum columna parallelum, basi adnatum, trilobum, lobis oblongis, conformibus, lateralibus, erectis, intermedio marginibus reflexis; disco albo, elevato, carnoso.* Columna *semiteres, apice antrorsùm subauriculata;* anthera *bilocularis, carnosa; loculis semibipartitis;* pollinia 4, *geminata, filis* 4 *replicatis.*

This delightfully fragrant epiphyte is a native of woods near Rio Janeiro, whence it was introduced to our Gardens a few years since. The plant from which our drawing was taken had been received by the Horticultural Society from the Right Hon. Robert Gordon, and flowered in the Garden at Chiswick last summer. It is easily cultivated in decayed moss and wood well drained, and placed in a hot damp part of the stove.

This is certainly a genuine Epidendrum, from which it differs in nothing but the imperfect union of the labellum with the columna; a circumstance not common indeed, but far from unknown, in that genus. We therefore cannot assent to its being referred to Encyclia, which, although very nearly related to Epidendrum, appears to have been judiciously distinguished by Dr. Hooker on account of its connivent sepals and petals, and its cucullate labellum, which has no adhesion with the columna.

It is scarcely to be doubted, that the Macradenia lutescens of our friend Mr. Loddiges is also this plant, and not the species figured under that name in the Botanical Register. The latter belongs to the tribe called Vandeæ.

J. L.

* From ἐπί, upon, and δένδρον, a tree; the branches of trees being the usual habitations of such plants.

1428

MAXILLÁRIA* tetragóna.

Four-cornered Maxillaria.

GYNANDRIA *MONANDRIA.*

Nat. ord. ORCHIDEÆ *Juss.* § Vandeæ *Lindl.*—(*Introduction to the natural system of Botany*, *p.* 262.)
MAXILLARIA.—*Suprà*, *vol.* 11. *fol.* 897.

M. *tetragona;* pseudobulbis ovatis tetragonis, foliis oblongo-lanceolatis plicatis solitariis, floribus radicalibus solitariis pseudobulbis brevioribus, sepalis oblongis obtusis patulis, petalis conformibus paulò minoribus, labello carnoso ventricoso trilobo erecto: lobis lateralibus parvis acutis intermedio ovato extùs convexo, disci appendice carnoso tabulari incumbente.

Pseudobulbi *ovati, tetragoni, rugosi, apice nigri.* Folia *oblongo-lanceolata, plicata, acuta, petiolata.* Flores *odorem violæ spirantes, virides, solitarii, radicales, inflati, pseudobulbis breviores.* Sepala *subglobosa, concava, apice recurva, acuta, carnosa, purpureo leviter tincta; lateralia basi in cornu producta, cum basi producta columnæ connata, fundo atropurpureo.* Petala *acuta, sepalis paulò minora.* Labellum *sepalis minus, ventricosum, carnosum, utrinque basi cordatum, apice trilobum: lobis lateralibus brevibus, acutis, intermedio ovato, mucronato; extùs subviride marginibus maculatis, intùs atropurpureum, fasciatum, subpubescens; appendix disci magna, carnosa, truncata, incumbens.* Columna *carnosa, cuneata.* Anthera *unilocularis, loculo marginato, anticè membranacea.* Pollinia 4, *obliquè collateralia, inæqualia, caudiculâ lineari, glandulâ obovatâ, cuspidatâ.*

A native of Rio Janeiro, whence it was imported by the Horticultural Society. Our drawing was made in the Chiswick Garden in June 1830.

This species is rather a shy grower, and is by no means remarkable for beauty; the latter quality is compensated for by its fragrance, which resembles fresh violets. The structure of the labellum is highly curious: in the inside grows a large, fleshy, deep-purple body, which gradually passes into the labellum at the lower margin; but anteriorly it projects into a distinct lobe, resembling a shovel, glued to the face of the labellum.

J. L.

* So named in consequence of the resemblance between the column and labellum, and the jaws or *maxillæ* of some animal.

1433

EULÓPHIA* Mackaiána.

Mr. Mackay's Eulophia.

GYNANDRIA *MONANDRIA.*

Nat. ord. Orchideæ *Jussieu.* § Vandeæ *Lindley.* (*Introduction to the natural system of Botany, p.* 262.)
EULOPHIA. — *Suprà, vol.* 7. *fol.* 686.

E. *Mackaiana;* foliis lorato-lanceolatis striatis apice recurvis racemo brevioribus, bracteis ovatis concavis ovario brevioribus, sepalis petalisque oblongo-lanceolatis acutis maculatis undulatis, labelli lobo medio maximo obcordato versùs basin callo transverso lunato, calcare brevissimo.
Zygopetalum Mackaii. *Hooker in Bot. mag. t.* 2748.

This most lovely species is, of all the Orchideous tribe, one of the most deserving cultivation; not only because of the size and rich colours of its flowers, but also because it succeeds well with ordinary care, and blossoms frequently.

We have not seen it thriving better than in the Garden of the Horticultural Society, where it is grown in pots filled with moss or rotten wood, and hung by wires from the rafters of the stove.

It is readily increased by separating the pseudobulbs, with a few roots attached to them.

A native of Brazil, whence it appears to have been originally introduced by Mr. Mackay, of the Dublin College Botanic Garden. Flowers in winter and spring.

The leaves are a foot and a half long, deep green, and rather fleshy, arising from a very short hidden stem, which gradually swells, and becomes a green pseudobulb, of an ovate or oblong figure. The scape is radical, and a foot and a half or two feet high, sometimes bearing as many as 6 or 8 flowers expanded at once. They are less perishable than those of many of the Orchis tribe; their sepals and petals are greenish, spotted with irregular blotches of brownish purple; the labellum is a bright blue, deeply stained with darker lines, and has at its base a remarkable fleshy, elevated, lunate, transverse ridge.

Eulophia is readily known among the genera most nearly allied to it, by its ascending sepals and crested calcarate labellum: the crest is seen under various forms, sometimes as a few elevated lines, sometimes as a number of thick ridges, or, as in the present case, in the form of a thick transverse hump. The length of the spur is very variable.

Our drawing was made in the Garden of the Horticultural Society in November 1828.

J. L.

---

* So named by Mr. Brown, from εὔλοφος, having a handsome crest; in allusion to the usual character of the labellum, bearing elevated lines or ridges.

1443

SARCÁNTHUS* guttátus.

Spotted-flowered Sarcanthus.

GYNANDRIA *MONANDRIA.*

Nat. ord. Orchideæ *Juss.* § Vandeæ.—(*Introduction to the natural system, p.* 262.)
SARCANTHUS Lindl.—*Pollinia* 2, cereacea, posticè sulcata v. lobata: *caudiculâ* variâ. *Anthera* bilocularis. *Stigma* excavatum v. quadratum: rostello vario. *Columna* semiteres, inappendiculata. *Labellum* subintegrum difforme cum columna articulatum, calcaratum: calcare intùs appendiculato. *Sepala* patentia subæqualia.——Herbæ *epiphytæ* (Indiæ orientalis *et* Chinæ) *caulescentes;* radicibus *caulinis tortuosis;* foliis *distichis planis v. teretibus;* floribus *vittatis v. fasciatis;* racemis *foliis oppositis.* — Collect. botanic. t. 39. B.

S. *guttatus;* foliis loratis carinatis apice truncatis subæqualibus, spicis cylindraceis pendulis densifloris foliis subæqualibus, sepalis lateralibus ovatis obtusis petalis sepaloque supremo duplò majoribus, labelli limbo ovato-oblongo obtuso: calcare obtuso ovario breviore.
Aerides guttatum. *Roxb. MSS.*
Caulis *pedalis v. ultra, à ramis arborum dependens, densè foliis v. foliorum basibus vestitus.* Folia *pedalia v. longiora, carnosa, carinata, lucida, apice truncata, sed subæqualia, submucronata.* Spicæ *cylindraceæ, axillares, multifloræ, foliis subæquales, scapo brevi squamuloso.* Bracteæ *minimæ.* Flores *albi, violaceo et roseo guttati, carnosi, expansione* 6-7-*lineari.* Labellum *roseum, calcare albo.* Columna *semiteres,* rostello *ovato acuto, stigmate lineari longitudinali.* Pollinia 2, *posticè sulcata, caudiculâ lineari, elasticâ, glandulâ minutâ.*

This species was long since transmitted to England from India: in 1820 we saw, in Sir Joseph Banks's library, a specimen in full flower that had been sent from the Royal Garden at Kew; but we have not heard of any European Botanist, except Mr. Brown, having examined it, nor of any of the noble collections in this country having been permitted to participate in the possession of so beautiful a plant. It is to be hoped that it is now before the public under more favourable auspices: the Horticultural Society, in whose Garden is the specimen from which the accompanying drawing was taken, will be eager to distribute it as quickly as its slow growth and difficult propagation will permit.

The plant in the Chiswick Garden was brought to England by Dr. Wallich, and presented by the Hon. Court of Directors of the East India Company. It is cultivated in the stove, in a very hot damp atmosphere, in a pot full of moss, suspended from the roof by a wire, and a little overshadowed by climbing and other plants: it flowers in April.

By the kind permission of Dr. Wallich, we are enabled to give the following extract relating to the species, from Dr. Roxburgh's MS. *Flora Indica:*—

" This lovely parasitical species was found growing on trees in the vicinity of Dacca by the Hon. Charles Andrew, who sent plants to the Botanic Garden at Calcutta: flowering time the hot and rainy season. It has great resemblance to the figures in Rheede's *Hortus Malabaricus,* quoted by Willdenow for Aerides retusum and præmorsum, but the capsule differs from both, and, I believe, the horn and labellum also; however, they are certainly very nearly allied, and not inferior in beauty, while in flower, to any in the whole tribe of Orchideæ."

" *Root* of several thick, fleshy, obtuse fibres, which issue from the stem through the sheaths of the lower leaves,

---

* So named on account of the fleshiness of the flowers of all the species; from σὰρξ, flesh, and ἄνθος, a flower.

and adhere to the tree, which gives support to the plant. *Stems* perennial, creeping, invested in the sheaths of the leaves. *Leaves* bifarious, imbricate, sheathing, recurvate, linear, channelled, thick, firm, and smooth; apex obliquely præmorse, from 6 to 12 inches long by one broad when laid flat. *Racemes* axillary or lateral, solitary, drooping, longer than the leaves. *Flowers* numerous, approximate, diverging, pretty large; colour a beautiful mixture of red and white spotted. *Bracts* ovate-cordate, one-flowered, embracing the insertion of the pedicel. *Corol* (*calyx* Swartz) five-petalled, upper three oblong, the middle one broader; lower two semi-cordate. *Labellum* longer than the petals, and more deeply coloured, ascending, concave, cuneiform, with the apex slightly 3-lobed, behind protruded into a large open obtuse bag or horn. *Columnæ* of fructifications short; pollen masses 2, covered with a deciduous lid. *Capsule* pedicelled, oblong, 3-sided; angles sharp; sides with a broad flat keel."

J. L.

1465

BRASAVÓLA* nodósa.

Knotty Brasavola.

GYNANDRIA *MONANDRIA.*

Nat. ord. Orchideæ *Juss.* Tribus *Epidendreæ* Lindl. (*Introduction to the natural system of Botany, p.* 262.)

BRASAVOLA R. Br.—*Sepala* et *petala* subæqualia, libera, acuminata. *Labellum* cucullatum, integrum, columnam involvens. *Columna* marginata, clavata, stigmate infundibulari, clinandrio posticè tridentato. *Pollinia* 8, subæqualia, quibusdam aliis parvis interjectis. *Anthera* 4-locularis, septis marginatis, loculis semibipartitis.——Herbæ *caulescentes, epiphytæ, apice folium unicum alterumve semicylindraceum, carnosum, suprà sulcatum, apice subulatum gerentes.* Flores *terminales, magni, speciosi.*—Gen. et sp. Orchid. part 2. p. 114.

B. *nodosa*; labello ovato acuminato integerrimo, petalis sepalisque linearibus acuminatis, dentibus lateralibus clinandrii integris posticâ emarginatâ.
Epidendrum curassavicum Orchidi affine. *Herm. parad.* 187. *t.* 187.
Viscum delphinii flore minus, etc. *Sloane jam.* 1. 251. *t.* 125. *f.* 1.
Epidendrum nodosum. *Linn. sp. pl.* 1350. *Jacq. Amer.* 226. *t.* 140. *Swartz obs.* 328.
Cymbidium nodosum. *Swartz nov. act. ups.* 6. 73. *Willd. no.* 24.
Brassavola nodosa. *Lindl. gen. et sp. Orchid. part* 2. *p.* 114.

Rhizoma *repens inter muscos supra arbores, caules basi clavatos, sursùm teretes monophyllos emittens, squamâ membranaceâ fuscâ involutos.* Folia *semipedalia, solitaria, carnosa, subtùs convexa, suprà canaliculata.* Spicæ 3-4-*floræ, foliis breviores, bracteis minutis, squamiformibus.* Flores *die inodori, noctu odoratissimi, pallidi, luteo-virides.* Sepala *linearia, acuminata, extùs lævia, sanguineo punctata, patentissima.* Petala *conformia, patentissima, angustiora, impunctata.* Labellum *album, limbo ovato-subrotundo, acuto, integerrimo, plano, ungue dimidio breviore, arctè convoluto, margine eroso, intùs sanguineo guttato.* Columna *ungue labelli duplò brevior, clinandrii dente posticâ emarginatâ, anticis integerrimis.* Anthera 4-*locularis, dissepimentis marginatis, loculis bipartitis.* Pollinia 8, *quaternatim paululùm minoribus, massulis* 2-3 *interjectis.*

* Named after Antonio Musa Brasavolo, a noble Venetian, who was one of the most enlightened Botanists of his day. A worthy pupil of Leonicenus and Manardus, he corrected the errors of Dioscorides and Pliny, in his *Examen omnium simplicium Medicamentorum*, a work of much reputation in its day, published at Leyden in 1537.

This very rare epiphyte has long been known as a native of trees in different parts of the West Indies. The elder Jacquin discovered it in the small island of Baru, off the coast of Carthagena, filling the woods at night with its fragrance; Sloane observed it in Jamaica; and lastly, the plants from which this figure was taken were received from Mexico by Sir John Lubbock, Bart., and by him presented to the Horticultural Society in 1828.

It requires a hot damp stove, where it grows freely among moss and decayed vegetable matter. Its habits are the same as those of Cattleya.

Drawn in the Garden of the Horticultural Society last October.

At the time of publication of the second part of the Genera and Species of Orchideous Plants, we had not seen any specimen of this. We now improve its character from the living plant.

J. L.

EDWARDS'S

BOTANICAL REGISTER:

OR,

ORNAMENTAL FLOWER-GARDEN

AND SHRUBBERY:

CONSISTING OF

COLOURED FIGURES OF PLANTS AND SHRUBS,

CULTIVATED IN BRITISH GARDENS;

ACCOMPANIED BY THEIR

History, Best Method of Treatment in Cultivation, Propagation, &c.

CONTINUED

By JOHN LINDLEY, F.R.S. L.S. and G.S.

PROFESSOR OF BOTANY IN THE UNIVERSITY OF LONDON, &c. &c. &c.

VOL. XVIII.

——viret semper——nec fronde caducâ
Carpitur.

LONDON:

JAMES RIDGWAY, 169, PICCADILLY.

M.DCCC.XXXII.

1485

AERIDES* cornútum.

Horn-flowered Air-plant.

GYNANDRIA *MONANDRIA.*

Nat. ord. Orchideæ *Juss.* (*Introduction to the natural system of Botany, p.* 262.)

§ *Vandeæ* Lindley.

AERIDES.—Suprà, vol. 3. *fol.* 220.

A. *cornutum*; foliis coriaceis apice obtusis obliquis, racemis pendulis multifloris foliis longioribus, labelli cucullati infundibularis laciniis lateralibus erectis cuneatis rotundatis: intermediâ ovatâ acutâ inflexâ: calcare incurvo.
A. cornutum. *Roxb. hort. beng. p.* 63.

Caulis *ramosus, crassus, lætè viridis, radices longas aereas tortuosas emittens.* Folia *coriacea,* 6-7 *uncias longa, apice obliqua, obtusa, lætè viridia, subdisticha.* Bracteæ *ovatæ, rigidæ.* Flores *carnei, fragrantissimi.* Sepala *patentia; inferiora ovata, latâ basi columnæ pedi adnata, supremum ovatum, basi paululùm angustatum, obtusum.* Petala *sepalo superiori subconformia, patentia.* Labellum *infundibuliforme, cum pede columnæ articulatum, 3-lobum, laciniis conniventibus: lateralibus erectis, cuneatis, rotundatis, intermediâ lineari-lanceolatâ; calcaratum, calcare conico, incurvo, apice viridi.* Columna *brevis, basi valdè producta, canaliculata, intra lacinias labelli abscondita.* Clinandrium *rostratum, acutum, ad caudiculam subulatam pollinis recipiendam.* Anthera *ovata, rostrata, basi bilocularis.* Massæ pollinis *duæ, globosæ, intùs sulcatæ;* glandula *obovata.*

This most lovely plant was originally obtained by Dr. Roxburgh from Dacca, and must have been sent many years since to the Garden at Kew; for a blossom of it was brought to us on the 8th of July, 1822, at which time it was flowering in great beauty.

Subsequently Dr. Wallich obtained it from Noakote, where it grows upon trees, blossoming in June. In the Botanic Garden, Calcutta, it is cultivated successfully; and has there gained the name of the Jamaica pomatum plant, from the rich fragrance of its flowers resembling the unguent so called. For ourselves, we compared it rather to the odour of the Tuberose.

* So called from the power that this, like many other Orchideous genera, possesses of living almost entirely upon the matter it absorbs from the atmosphere, into which it emits its long tortuous absorbent roots.

It is certainly the most interesting of its tribe that has yet been introduced, whether we consider the great mass of its blossoms, their curious form, or delicate colour, or long duration, or delicious perfume. It flourishes in a very damp Hothouse, planted in moss in a pot suspended from the rafters; but, as it branches rather unwillingly, it is slow of propagation.

Our drawing was made last August in the Garden of the Horticultural Society, from a plant that had been brought to England by Dr. Wallich, and presented by the Hon. Court of Directors of the East India Company.

J. L.

1499

HERMÍNIUM* cordátum.

Heart-leaved Herminium.

GYNANDRIA *MONANDRIA.*

Nat. ord. ORCHIDEÆ *Juss.* Sect. Ophrydeæ *Lindl.* (*Introduction to the natural system of Botany, p.* 262.)

HERMINIUM R. Br.—*Sepala* et *petala* patentia v. conniventia subæqualia. *Labellum* vix sepalis longius, tridentatum trifidum integrumve, basi nunc planiusculum, sæpiùs saccatum v. gibbosum; saccus v. apertus v. ore constricto vesiciformis. *Anthera* erecta, minuta, loculis basi divergentibus, polliniorum glandulis nudis, sæpiùs utrinque squamâ (antherâ sterili) instructa. —— Herbæ; radicibus *carnosis fasciculatis, lobo uno alterove incrassato.* Flores *sæpiùs herbacei, nunc albi.*—OBS. *Gymnadeniæ* et *Platantheræ* quàm maximè affinis, illâ loculis antheræ basi divergentibus, hâc labello ecalcarato tantùm saccato, difficillimè distinguendum.

H. *cordatum;* caule diphyllo, foliis cordatis acutis, spicâ secundâ, sepalis petalisque ovato lanceolatis acutis subæqualibus, labello trilobo basi saccato, squamis antheræ lateralibus elongatis clavatis.
Satyrium diphyllum. *Link in Schrad. diar.* 1799. *p.* 323.
Orchis cordata. *Willd. sp. pl.* 4. 28.
Habenaria cordata. *Br. prodr. p.* 312. *Spreng. syst. v.* 3. 691. *Hooker in bot. misc. v.* 1. *p.* 270. *t.* 55. *Bot. mag.* 3164.

A native of the north-west of Africa and south-west of Europe; Link and Brotero have found it in Portugal; we have specimens from shady hills near Tangier, collected by Salzmann; and the Rev. Mr. Lowe found it on walls and rocks in Madeira. For the specimen from which our drawing was taken, we are obliged to the Rev. Mr. Berkeley, at whose request it was forwarded to us from the rich collection of Lord Milton, by Mr. Joseph Henderson. To the latter excellent cultivator we are indebted for the following note upon its habits:—

* An unexplained name of Linnæus, mentioned in his *Philosophia Botanica*, under the head of words derived from the titles of divinities; from which it is to be supposed that it has some reference to the god Mercury; but, as far as we know, not applied by that Botanist.

"The plant grows very well in the Greenhouse; but it requires a little more heat at this season, which seems to be its flowering season (November), than the Greenhouse affords; and I find that it flowers best in the coolest part of the stove."

The genus Habenaria was first separated from Orchis by Willdenow, who limited it to those remarkable West Indian plants, *Orchis Habenaria* and *monorhiza* of Swartz, in which the cells of the anther are distinct from the elongated lateral processes of the stigma. In this limitation he was wrong, as there are several of the species still retained by him in Orchis, to which the characters of those species equally apply.

Afterwards Dr. R. Brown separated from Orchis all the species in which the glands of the pollen-masses are naked, referring the greater part of them to Habenaria, in two distinct sections, and indicating the existence of the genus now called Gymnadenia. The same learned Botanist also separated from the genus Ophrys, as it stands in Willdenow, the O. *monorchis* and *alpina;* to which he afterwards applied the unappropriated Linnean name Herminium.

At a subsequent period the late M. Richard restored the genus Habenaria as it was originally understood, referred the greater part of Dr. Brown's Habenarias to a genus he called Platanthera, and distinguished *Ophrys alpina* under the name of Chamorchis.

If particular species be taken as the types of these genera, as, for example, *Orchis monorhiza* of Habenaria, O. *bifolia* of Platanthera, O. *conopsea* of Gymnadenia, and *Ophrys monorchis* of Herminium, there can be no difficulty in understanding the differences upon the existence of which these genera are established. But if a more extended view is given to the subject, the differences insensibly vanish, till at last it becomes doubtful whether the whole of the genera do not rather belong to a single one.

Thus Gymnadenia is said to be the same as Orchis, except that its pollen-glands are naked; that is, it has the cells of the anther parallel, and approximated at their bases; but then Dr. Brown excludes from it the old *Satyria viride* and *albidum,* while Richard refers those species to it; and in the numerous Indian plants of a similar structure, it is often extremely difficult to say whether the cells are parallel, or diverging at the base.

Habenaria has its cells with their bases more or less separated from the stigmatic processes; but so many degrees of adhesion and separation occur, that this is by no means a precise character.

Platanthera is said to differ from Herminium in nothing except the presence of a spur; but *Herminium monorchis* has a bag to its lip, almost as deep as that of the subject of our plate; and yet the latter, with all the habit of Herminium, and with no tangible differences in its structure, except that its lip is a very little more bagged, is referred to Habenaria or Platanthera.

Taking these and many more circumstances into consideration, it has appeared to us that it is absolutely necessary to recast the genera of Ophrydeæ with naked pollen-glands, in the following manner:—

HABENARIA. *Loculi* antheræ basi producti, à processubus stigmatis elongatis soluti. *Labellum* calcaratum.
PLATANTHERA. *Loculi* antheræ basi divaricati, stigmati adnati. *Labellum* calcaratum, sepalis longius.
HERMINIUM. *Loculi* antheræ basi divaricati. *Labellum* basi planum, gibbosum v. saccatum.
GYMNADENIA. *Loculi* antheræ basi approximati. *Labellum* calcaratum (v. saccatum?).

To which are to be added the very distinct Bonatea, and the following remarkable genus in Dr. Wallich's Indian Herbarium, with a horizontal anther, the bases of which are exceedingly divaricated, and connected by a broad erect membrane, opposite which is an erect fleshy body proceeding from the mouth of the spur.

DIPLOCHILOS. *Loculi* antheræ basi divaricati, stigmate dilatato bilabiato soluti. *Labellum* calcaratum.

Thus limited, the genus Herminium will consist of at least the following species, to which will possibly have to be added some Gymnadenias, when they shall have been more completely examined. The Benthamia of M. Achille Richard is clearly an Herminium, the pollen and anther of which have been altogether misunderstood by that Botanist, although they have been correctly figured by Du Petit Thouars. There also seems to be no difference of any moment between Chamorchis and Herminium.

HERMINIUM.

1. Herminium monorchis (*R. Br. in Hort. Kew.* 5. 191).
   Ophrys monorchis. L.
   Hab. in pratis Europæ. (v. v. sp.)
2. Herminium congestum (*Lindl. in Wall. cat. no.* 7068); foliis erectis oblongis apice latioribus spicæ densæ subæqualibus, bracteis squamæformibus ovario brevioribus, sepalis ovatis obtusis, petalis ovalibus subæqualibus, labello ovato integro basi gibboso et marginato.
   Hab. in Napaliâ, *Wall.* (hab. s. sp. è museo Anglo-indico).
3. Herminium latifolium.
   Satyrium latifolium. Thouars orch. t. 10.
   Benthamia latifolia. Ach. Rich. orch. maur. t. 7. no. 2.
   Habenaria chlorantha. Spreng.
   Hab. in Borboniâ, *Thouars;* Mauritio, *Bouton.* (hab. s. sp.)
4. Herminium cordatum. (*Suprà.*)
5. Herminium plantagineum; foliis (3) erectis oblongo-lanceolatis acutis spicæ ferè æqualibus, sepalis ovatis, petalis subrotundis obtusis, labello rotundato obtusè tridentato, sacco sphæroideo.
   Hab. in Zeylonâ, *Macrae.* (hab. s. sp.)
6. Herminium Hamiltonianum (*Lindl. in Wall. cat. no.* 7069); foliis (pluribus) lanceolatis acuminatis spicâ longioribus, bracteis acuminatis floribus longioribus, sepalis petalisque obtusis conformibus æqualibus, labello rotundato tridentato, sacco subrotundo.
   Hab. in collibus Morang Indiæ orientalis, *Hamilton.* (hab. s. sp. è museo Anglo-indico.)
7. Herminium constrictum; foliis pluribus oblongo-lanceolatis scapo brevioribus, sepalis lineari-lanceolatis petalis duplò angustioribus, labello trifido: laciniis lateralibus angustioribus et longioribus.
   Hab. in Napalia, *Wallich.* (hab. s. sp. è museo Anglo-indico et v. ic. pict.)
8. Herminium goodyeroides (*Lindl. in Wall. cat. no.* 7066).
   Habenaria goodyeroides. Don prodr. fl. nap. 25.
   β floribus minoribus, spicis foliisque brevioribus.
   Habenaria affinis. Don l. c.
   Hab. in Napalia, *Wallich.* (hab. s. sp. α, è museo Anglo-indico; β comm. cel. Lambert).
9. Herminium alpinum.
   Ophrys alpina. L.
   Chamorchis alpina. Rich. annot. p. 35.
   Hab. in alpibus Europæis. (hab. s. sp.)
10. Herminium reniforme (*Lindl. in Wall. cat. no.* 7067); folio solitario suborbiculato cordato, sepalo supremo petalisque conniventibus: inferioribus subparallelis porrectis, labello lineari esaccato.
   Hab. in Napaliâ, *Wallich* (hab. s. sp. et ic. pict.)

The only known species of DIPLOCHILUS are the following:—

1. Diplochilos hirsutum (*Lindl. in Wall. cat. no.* 7065), caule monophyllo unifloro, folio oblongo hirsuto, calcare incurvato.
2. Diplochilos longifolium (*Lindl. in Wall. cat. no.* 7064), caule 2-3-phyllo subunifloro, foliis ensiformibus glabris, calcare recto.

J. L.

1508

CYRTOPÓDIUM* Woodfórdii.

Woodford's Cyrtopodium.

GYNANDRIA *MONANDRIA.*

Nat. ord. ORCHIDEÆ *Juss.* Sect. Vandeæ *Lindl.* (*Introduction to the natural system of Botany, p.* 262.)

CYRTOPODIUM.—*Labellum* ungue geniculato; lamina 3-partita. *Petala* 5, distincta. *Massæ pollinis* 2 posticè bilobæ.—*R. Brown in Hort. Kew.* 5. 216.

C. *Woodfordii;* bracteis subulato-lanceolatis, sepalis lanceolatis patentibus secundis, petalis oblongis conniventibus brevioribus, labello basi saccato: lobis lateralibus rotundatis intermedio emarginato basi bicorni.

C. Woodfordii. *Bot. mag.* 1814.

Caulis *2-3-pedalis, fusiformis, carnosus, succo scatens viscido ad calceos agglutinandos utili.* Folia *basi vaginantia, plurima, lanceolata,* $1\frac{1}{2}$-2-*pedalia, plicata, glabra.* Scapus *radicalis,* 2-3-*pedalis, teres, squamis elongatis plicatis vaginatus.* Racemus *elongatus, multiflorus.* Bracteæ *subulato-lanceolatæ, ovario longiores.* Sepala *angustè lanceolata, pallidè viridia, secunda.* Petala *breviora, viridi-fusca, oblonga, obtusa, erecta, apice conniventia.* Labellum *basi lividum, apice roseo-purpureum, ventricosum, basi saccatum,* 3-*lobum, lobis lateralibus rotundatis abbreviatis, intermedio obtuso, undulato, emarginato, disco glanduloso basi bicorni.*

This beautiful plant is a native of the island of Trinidad, of Guiana, and of Brazil, where it grows in shady damp places, rooting among the mass of decayed vegetable soil with which the earth is thickly covered in such places. It is easily cultivated in a good stove, but very seldom flowers. The specimen from which the accompanying figure was taken, was produced in the Garden of the Horticultural Society, in September last, upon plants sent from Rio Janeiro by the Right Hon. Robert Gordon. It is increased by the separation of the thick spindle-shaped stems with a portion of the rooting Rhizoma attached, or even by cuttings of the stems themselves; the latter must, however, be well dried before they are planted, or they will absorb moisture by their wounded extremity with much greater rapidity than they can part with it by perspiration, and will perish from a sort of vegetable dropsy.

This is no Cyrtopodium. Nevertheless we have left it with the name already given to it; because all the genera allied to it, viz. Eulophia, Cymbidium, Lissochilus, Maxillaria, and the like, required to be very carefully studied before their real limits can be precisely determined. In all such cases, an old name, although incorrect, is preferable to a change, which may be premature, and may only lead to further changes.

J. L.

* So called from κυρτὸς, curved, and ποὺς, a foot; in allusion to the curved stalk of the labellum of Cyrtopodium Andersonii.

1510

MAXILLÁRIA* víridis.

Green Maxillaria.

GYNANDRIA *MONANDRIA.*

Nat. ord. ORCHIDEÆ *Juss.* Sect. Vandeæ *Lindl.* (*Introduction to the natural system of Botany, p.* 262.)

MAXILLARIA.—*Suprà, vol.* 11. *fol.* 897.

M. *viridis;* pseudo-bulbis nullis, foliis lanceolatis undulatis plicatis, floribus solitariis radicalibus globosis, sepalis petalisque subrotundis obtusis, labello brevi: lobo medio transversè rhomboideo unguiculato medio refracto.

Flos *radicalis, solitarius, globosus.* Sepala *subrotunda, ovata, obtusa, viridia, immaculata.* Petala *minora, conformia, purpureo intùs confertissimè punctata.* Labellum *cum basi parùm productâ columnæ elasticè articulatum, unguiculatum, trilobum, medio refractum, lilacinum, lobis lateralibus minoribus erectis, intermedio transversè rhomboideo plano.* Columna *libera, semiteres, basi producta.* Stigma *subrotundum, excavatum.* Anthera *carnosa, bilocularis; loculorum valvulis transversè dehiscentibus.* Pollinia 4, *albida, geminata, caudiculâ sulcatâ, fuscâ!, utrinque alatâ, dorso purpurascente, viscido, prominulo.*

A native of Rio Janeiro, whence it was sent by the late Sir Henry Chamberlain. Our drawing was made in the stove of the Horticultural Society's Garden in May 1831.

It is rather a weak-growing plant, requiring shade, much moisture to its leaves, and little to its roots, together with a high temperature and decayed vegetable mould. It seems to have no tendency to form those pseudo-bulbs which are generally so characteristic of the genus.

J. L.

* Some of the species have flowers the segments of which gape like the open jaws, *maxillæ*, of some grinning animal.

1522

ANGRÆCUM* ebúrneum.

Ivory Angræcum.

GYNANDRIA *MONANDRIA.*

Nat. ord. ORCHIDEÆ. § Vandeæ *Lindl.* (*Introduction to the natural system of Botany, p.* 262.)

ANGRÆCUM Thouars.—*Perianthium* patens. *Sepala* et *petala* subæqualia, libera. *Labellum* sessile, cum basi columnæ continuum, carnosum, indivisum, petalis multò latius; calcare recto cornuto, sæpiùs subcylindraceo, perianthio multò longiore, raRò obconico. *Columna* nana, subteres, raRò elongata, semiteres. *Anthera* 2-locularis, truncata. *Pollinia* 2, bipartibilia, caudiculâ brevi angustâ, glandulâ triangulari.—— Epiphytæ, caulescentes. Folia coriacea, ligulata, apice obliqua. Flores solitarii v. racemosi, albi, nunc citrini, v. herbacei.

A. *eburneum;* caule simplici, foliis coriaceis lucidis apice obliquis 7-striatis, spicis multifloris secundis, labello cordato subrotundo cuspidato basi jugo elevato cristato, calcare sepalo supremo parallelo et dimidio longiore, ovario scabro.

Limodorum eburneum. "*Bory Voyage*, 1. 359. *t.* 19." *Willd. sp. pl.* 4. 125.

Angræcum eburneum. *Thouars orch. afr. t.* 65. *Ach. Richard orch. maurit. p.* 71.

Aerobion eburneum. *Spreng. syst. veg. v.* 3.

Caulis *erectus, simplex, crassus, sesquipedalis, hic illic radices tortuosas promens.* Folia *subdisticha, ligulata, lucida, valdè coriacea, apice obliqua, quandoque emarginata.* Spica *è basi caulis, ascendens, caulis longitudine, teres, secunda, multiflora.* Flores *inodori, explanati, diametro* $3\frac{1}{2}$*-unciali.* Sepala *et* petala *subæqualia, lineari-lanceolata, coriacea, basi convexa, omninò libera, pallidè herbacea.* Labellum *eburneum, cordatum, subrotundum, abruptè acuminatum, concavum, basi cristâ obcuneatâ elevatâ longitudinali instructum, calcaratum.* Calcar *ferè tres uncias longum, cornutum, ad apicem sensim attenuatum, intùs pubescens, cum sepalo dorsali parallelum.* Columna *nana, carnosa, subteres, truncato-conica.* Clinandrium *planum, immarginatum, apice angustatum; margine antico altè excavato.* Stigma *sub fornice clinandrii occultum.* Anthera *subrotunda, truncata, bilocularis; loculis paulò disjunctis, longitudinaliter dehiscentibus.* Pollinia 2, *bipartita, incumbentia, basi approximata, caudiculâ lineari emarginatâ glandulæ triangulari affixâ.* Ovarium *breve, sexsulcatum, verruculosum, costis rotundatis.*

* An alteration of Angurek, the Malayan name of such plants.

This remarkable plant is not uncommon in the Island of Bourbon, growing upon trees, where it was found both by Colonel Bory de St. Vincent and by the late Mons. Du Petit Thouars. It was also met with at St. Mary's, Madagascar, by the unfortunate Forbes, by whom the only plant that is known to exist in Europe was sent to the Horticultural Society. From that plant the accompanying drawing was made in November last, when it flowered for the first time, continuing in beauty for nearly two months. It grows slowly, but is in a perfectly healthy state, without, however, manifesting the smallest disposition to branch, or provide means for propagating.

It is very much to be regretted that some more efficient means are not taken to procure the plants of this description, which abound in all the tropical parts of the East. They are very tenacious of life, and require no care in collecting, it being only necessary to strip them off the trees on which they grow, and to suspend them in the cabin, never watering them, but moistening them occasionally with a wet sponge. Captains of ships touching at the Isle of France might succeed in importing them without difficulty; and yet of the numerous species figured by Du Petit Thouars, not more than three or four have ever been seen in our gardens. The only precaution that requires to be taken in importing them, is that they should not be over-watered; if this is done, they are sure to die; it would be much better to give them no water whatever. They should also, if possible, be collected *in the dry season,* at which period they are naturally in a state of torpor.

This plant offers a striking instance of the precaution Nature seems to have taken to prevent direct contact between the pollen and the stigmatic surface, and at the same time of the skilful means by which an indirect communication is substituted; so that no impediment may really exist to the process of fertilisation, notwithstanding the apparent obstacles to its taking effect. The pollen is locked up in an anther so firmly united to the top of the column, that nothing but actual force can remove it; while the stigma is situated upon the face of the column far beyond the reach of contact with the pollen. To establish the requisite communication, a narrow slip of the top of the column, proceeding from the edge of the stigma to beneath the anther, gradually separates from the surrounding tissue; the face of the anther opens along the middle of the cells, contracts, and allows the pollen masses to glue themselves to the narrow slip, which is the caudicula; at the same time, while this operation has been in progress, a triangular portion of the front of the edge of the stigma has been adhering to the caudicula; and thus establishes the direct communication which is required, afterwards separating with the caudicula under the form of the gland. From this statement it will be obvious to Botanists that the caudiculæ of Ophrydeæ, and that of Vandeæ, are not analogous; the former belonging to the pollen, as was demonstrated by Mr. Bauer so long since as the year 1800, and the latter to the stigma. These phenomena, the progress of which we remarked last winter with great care, will form the subject of two plates that have been prepared by Mr. Bauer for the next Part of the *Illustrations of Orchideous Plants.*

The genus Angræcum was first established by Du Petit Thouars in 1822. It consists, as far as is at present known, exclusively of plants native of trees in the islands of Bourbon, Mauritius, and Madagascar, and of the south-eastern part of the continent of Africa. Strictly limited, that is to say excluding certain species admitted by its founder, it is known by its undivided lip, which is neither cucullate, nor articulated with the column; by its spreading perianthium, which never has the segments turned upwards as in Eulophia; by its long taper-pointed spur, which is rarely enlarged at the base; and finally, by all these characters being connected with coriaceous leaves, that are never ribbed or plaited. Such being the definition of Angræcum, it will be apparent that, independently of the species already separated under the names of Aeranthes, Œonia, &c., certain others, such as our Angræcum maculatum, and a few of those of Du Petit Thouars, must be excluded; these form a genus nearly related to Eulophia, from which they are to be distinguished by their coriaceous leaves, by the perianthium never being secund, and by the want of a crest upon the lip. They may be thus defined:

ŒCEOCLADES.

Perianthium patens, v. ringens, v. connivens, subæquale, liberum, non secundum. Labellum liberum, sessile, cum columna articulatum, calcaratum; calcare curvo sæpè obconico, laminâ lobatâ cucullatâ, v. complicatâ, v. concavâ. Columna sæpè elongata, semiteres. Anthera 2-locularis. Pollinia 2, posticè sulcata, caudiculâ angustâ, glandulâ minutâ.—— Epiphytæ, acaules v. caulescentes, nunc pseudobolbosæ. Folia coriacea, aut rarò membranacea, nunquam plicata. Flores sæpiùs minuti.

Sp. Angræcum maculatum *Lindl.* —parviflorum, caulescens, carpophorum, gracile, *Thouars*—Limodorum pusillum, filiforme? funale, falcatum, W. et plures Indicæ ineditæ.

Among the collections of Dr. Wallich are the following new genera, belonging to the same tribe as Angræcum and Œceoclades, with which our Botanical readers will be glad to be acquainted.

DIPLOCENTRUM.

Perianthium subclausum, liberum. Sepalum supremum lateralibus paullò obliquis multò minus. Petala sepalo supremo paululùm majora. Labellum cum columna articulatum, indivisum, bicalcaratum. Columna minuta, semiteres, rostello obtuso. Pollinia 2? caudiculâ subulatâ, glandulâ maximâ concavâ.——Epiphyta, caulescens. Folia complicata, carnosa, recurva. Racemi paniculati, recurvi. Flores minuti. Calcaria brevia, obconica, incurva.

Sp. 1. Diplocentrum recurvum.

MICROPERA.

Perianthium æquale, patens, sepalis lateralibus basi labelli adnatis. Labellum cum basi columnæ continuum, calceiforme, leviter trilobum, lobo intermedio minimo, lateralibus majoribus erectis. Columna brevis, rostello maximo inflexo. Pollinia 2, caudiculâ subulatâ.——Epiphyta, caulescens. Folia disticha, linearia, arcuata, apice æqualia, emarginata. Spicæ laterales, horizontales, foliis breviores. Flores citrini.

Sp. 1. Micropera pallida=Aerides pallidum *Roxb.*

CHILOSCHISTA.

Perianthium subpatens. Petala sepalis majora, cum sepalis lateralibus basi longè productæ columnæ adnata. Labellum cum ungue columnæ articulatum, bipartitum, medio cristatum. Columna minima, erecta, semiteres. Pollinia 2, caudiculâ brevi, subulatâ, glandulâ minutâ.——Herba humilis, epiphyta, hirsuta, aphylla; radicibus complanatis viridibus (quasi foliaceis!). Spicæ erectæ. Flores albi, odorati.

Sp. 1. Chiloschista usneoides=Epidendrum usneoides *Don prodr.*

TETRAPELTIS Wall. MSS.

Perianthium subpatens. Sepala libera, æqualia. Petala angustissima, æquilonga. Labellum liberum, sessile, saccatum, geniculatum, trilobum, lobis lateralibus abbreviatis, intermedio indiviso. Columna erecta, petalorum longitudine, semiteres, clavata; clinandrio cucullato; rostello rostrato recto. Pollinia 4, globosa, posticè foveata; geminatim caudiculis duabus angustis glandulâ communi connexis adhærentia.——Herba epiphyta, caulescens. Caules squamis vaginati, curvi. Folia geminata, lanceolata, plicata. Spica erecta, multiflora, foliis brevior. Flores albi.

Sp. 1. Tetrapeltis fragrans *Wall. MSS.*

J. L.

1525

ÁCERAS* secundiflóra.

One-sided Aceras.

GYNANDRIA *MONANDRIA.*

Nat. ord. Orchideæ. § Ophrydeæ *Lindl.* (*Introduction to the natural system of Botany, p.* 262.)

ACERAS R.Brown.—*Perianthium* clausum, æstivatione valvatâ. *Petala* sepalis multò angustiora. *Labellum* brevissimè saccato-calcaratum, v. gibbosum, pendulum v. patens, tripartitum. *Columna* erecta, bilocularis, loculis parallelis basi approximatis, caudiculis glandulæ solitariæ cucullo unico inclusæ adhærentibus.——Herbæ, radicibus carnosis fasciculatis, lobo uno alterove incrassato. Flores luteo-herbacei, v. obscurè violacei, labello sæpè maculato discolore.

A. *secundiflora;* foliis maculatis, spicâ densâ secundâ parviflorâ, petalis subulatis, labello patente oblongo sepalis æquali basi gibboso trilobo: lobo intermedio obovato mucronulato.
Satyrium maculatum. *Desf. fl. atl.* 2. 319, *fide Cambessédes.*
Ophrys densiflora. *Desf. coroll. p.* 11. *t.* 16.
Orchis secundiflora. "*Bertolon. amæn. Ital.* 82." *Cambess. enum. pl. balear.* 140. *Spreng. syst. veg.* 3. 687.

Radix carnosus *oblongus, indivisus.* Folia radicalia *oblonga, erecta, plicata, carnosa, maculata.* Scapus *in cultâ spithamæus, in spontaneâ humilior, erectus, apice spicam densam secundam florum parvorum obscurè violaceorum gerens.* Sepala *ovata, acuta, æstivatione valvari.* Petala *subulata, sepalorum longitudine.* Labellum *patens, sepala non excedens, oblongum, basi concavum, sacculatum, trilobum; lobis lateralibus linearibus acutis, intermedio obovato mucronato longiore.* Columna *minima, generis.*

This plant seems to have an unusually extensive geographical range for an Orchis. Bertoloni found it in Provence and Liguria; Cambessedes gathered it on the Puig-dî-Torrella, in Majorca; it has been collected in Corsica by Viviani, and also in Calabria; Desfontaines brought it from near Belis, in the Atlantis; and finally, it was discovered in Madeira by the Rev. Mr. Lowe, to whom we owe its introduction to our gardens.

It is a neat little plant, requiring the same kind of treatment as Ixias and other Cape bulbs: that is to say, to be kept quite dry and quiescent during summer. Under such management, Mr. Henderson, at Lord Milton's, succeeds in making it flower freely every spring. For the specimens from which the drawing was taken we are obliged to Mr. Henderson, and also to the Rev. Mr. Berkeley, by whom they were communicated in April last. The parts of fructification are very minute, and difficult to make out. Mr. Berkeley first directed our attention to the agreement between their structure and that of Aceras anthropophora, an opinion which our own observation has subsequently confirmed.

We have wild specimens from the Balearic islands, for which we have to thank Mr. Bentham; but they are not one-third the size of the cultivated plant.

Orchis parviflora of Willdenow, referred to this genus by Sprengel, in pursuance of a hint of the elder Richard, is, according to Tenore, nothing but a starved specimen of O. ustulata, and must be expunged from the list. In its place may be inserted the following very distinct Indian species, viz.: —

A. *angustifolia* (Lindl. in Wall. cat. ind. no. 7061), foliis lineari-lanceolatis acuminatis, spicâ rarâ elongatâ secundâ parviflorâ, petalis subulatis, labello pendulo lineari sepalis duplò longiore, apice trifido: laciniâ intermediâ breviore.

Hab. in *Gossam Than, Wallich.* (*hab. s. sp. e museo Anglo-Indico.*)

J. L.

---

* So called by Dr. Brown, from α, priv., and κέρας, a horn; in allusion to the *primâ facie* difference between the original species and Orchis. Afterwards Richard altered it to Loroglossum, because one of the species was found to have a horn. Then Professor Sprengel changed the name to Himantoglossum, because he did not like Loroglossum. Finally, we retain Dr. Brown's name, because it was the first, and is as good as the others, and because we do not approve of these idle changes.

1529

STANHÓPEA* ebúrnea.

Ivory-lipped Stanhópea.

GYNANDRIA *MONANDRIA.*

Nat. ord. Orchideæ *Juss.* § Vandeæ *Lindl.* (*Introduction to the natural system of Botany, p.* 262.)

STANHOPEA Hooker.—*Perianthium* membranaceum, patentissimum v. reflexum. *Sepala* libera, subundulata, mole suâ ruentia. *Petala* conformia angustiora. *Labellum* liberum, anticum, ecalcaratum, carnosum, utrinque cornutum: dimidio superiore (*epichilio*) convexo, inferiore (*hypochilio*) excavato. *Columna* longissima, petaloideo-marginata. *Anthera* 2-locularis. *Pollinia* 2, elongata, fissa, caudiculâ quam glandula biloba stipitata breviore.——Epiphytæ *pseudo-bulbosæ.* Folia *plicata.* Scapi *radicales, vaginati, pauciflori.* Flores *maximi, magis minùsve maculati.*

S. *eburnea;* labello oblongo non medio constricto, hypochilio pone basin bicorni, epichilio ovato-oblongo obtuso, metachilio duplò longiore solido plano-convexo anticè truncato bidentato, scapo pendulo bifloro petalis duplò longiore.

Scapi *penduli, biflori, cum ovario* $4\frac{1}{2}$ *poll. longi.* Vaginæ *laxæ, ovatæ, obtusæ, internodiis vix æquales.* Sepala *oblongo-lanceolata,* $2\frac{1}{2}$ *poll. longa,* $1\frac{1}{3}$ *lata.* Petala *conformia, angustiora.* Labellum *cornubus è medio margine hypochilii, metachilioque lineis quibusdam interruptis purpureis notatis.*

We are indebted for our knowledge of this noble epiphyte to James Bateman, Esq., who most obligingly sent us specimens from his rich collection of Orchideous plants at Knypersley Hall, near Congleton, in July last. They were accompanied by a faithful drawing from the amateur pencil of Miss Jane Edwards, and by several important notes, which have together enabled us to prepare an account of the species.

It is a native of Rio Janeiro, whence it was originally imported by Messrs. Loddiges. Mr. Bateman informs us that "the plant from which the drawing was taken produced at the same time from the base of the same bulb another scape, which, coming in contact with the side of its pot, *suâ vi* was broken off: the other was only preserved from a similar fate by breaking an opening through the sides of the pot. The flowers were slightly fragrant, and of short duration; the lip when fresh appeared to be formed of the most pure and highly-polished ivory. The plant in question differs decidedly from S. oculata and insignis, which have both flowered here; and in foliage from a plant I possess of S. grandiflora, being much smaller, more pointed and pliant in its leaves. Not, however, having seen a flower of the latter, I know not exactly in what respect it differs from it."

To this last species it is indeed very closely allied; but it appears to differ from it in some points of importance. In the first place, its flowers are not more than two-thirds of the size; secondly, the horns of the base of the lip proceed from the middle of the margin of the hypochilium, and not from the front of the margin; and thirdly, the scape in S. grandiflora is shorter than the sepals, so that the flowers are erect, while in S. eburnea the scape is twice as long, and pendulous. Such at least are differences that are to be discovered upon comparing this with Messrs. Loddiges' figure of Ceratochilus grandiflorus, and with a few notes upon that species which we formerly made when a blossom was communicated to us by those gentlemen in August 1828. We, however, should not perhaps have attached so much importance to these peculiarities, if S. eburnea had not been a native of Rio Janeiro, and S. grandiflora of Trinidad; but we know that it rarely happens that the same species of Orchideous plant inhabits such very distant stations.

When we originally examined this plant, we suggested to Messrs. Loddiges as a name for the genus that of Ceratochilus, which was accordingly adopted; but we unfortunately did not at the time advert to the existence of the same name in Dr. Blume's Observations upon Java Plants. The genus of the last-mentioned Botanist being, however, apparently distinct, it has become necessary to adopt the name of Stanhopea, which was subsequently applied to another species of this genus, the Epidendrum grandiflorum of Humboldt and Bonpland, by Dr. Hooker. We the more readily do this now, because on a former occasion, in objecting to the reception of the name Stanhopea, we suffered ourselves to be betrayed into unkind expressions, which should not have been applied to any one, and least of all to so amiable and excellent a man as our long-tried friend the Professor of Botany at Glasgow.

J. L.

---

* So called in compliment to the present Earl Stanhope, president of the Medico-Botanical Society.

1530

CYMBÍDIUM* marginátum.

Red-edged Cymbidium.

GYNANDRIA *MONANDRIA.*

Nat. ord. ORCHIDEÆ *Juss.* § Vandeæ *Lindl.* (*Introduction to the natural system of Botany*, p. 262.)
CYMBIDIUM.—Suprà, vol. 7. *fol.* 529.

Sect. 4. *Rhizoma repens, pseudo-bolbos* 1-3-*phyllos gerens.* (Bolbidium.) —*Lindl. gen. et sp. Orch. part* 3. *ined.*

C. pseudo-bulbis ovato-oblongis teretibus 1-3-phyllis, foliis lineari-lanceolatis coriaceis erecto-recurvis, scapis unifloris gracilibus vaginatis foliis brevioribus, sepalis oblongo-lanceolatis, petalis minoribus conformibus, labello oblongo-lanceolato obtuso cucullato subrepando indiviso.

? Maxillaria gracilis. *Lodd. bot. cab.* 1837.

Folia 4-8 *poll. longa*, 6-7 *lineas lata, atroviridia.* Scapi *palmares, vaginis longis fuscis distantibus.* Flores *lutei.* Sepala *pollicem longa, rubro-marginata; supremum cum petalis parallelum labium superius, lateralibus divergentibus inferius constituentibus.* Labellum *interruptè umbrino limbatum.*

An air-plant, native of Rio Janeiro, whence it has been received by the Horticultural Society. It grows freely under the usual management required by this tribe of plants, but does not often flower. Our drawing was made in a stove in the Chiswick Garden in November 1830. Young plants may be obtained by dividing the creeping stem, when the pseudo-bulbs will establish for themselves an independent life by means of their little white and green roots.

The genus Cymbidium in its most definite state comprehends none but plants the lower sepals of whose flowers are spreading, the upper with the petals forming a sort of superior lip; the labellum being articulated with the column, and either entire or 3-lobed, but always cucullate. With these characters are combined a pair of 2-lobed pollen masses, sessile upon a transverse gland. The genus thus restricted will be found to comprehend two natural groups, one of which consists of plants that, like C. aloifolium, form neither stem nor pseudo-bulbs, but throw up several leaves from the surface of the soil; the other of species with a creeping rhizoma, from which at intervals spring pseudo-bulbs, bearing from 1 to 3 half-coriaceous leaves. It is to the latter that the subject of this Plate is referable. Besides these, it will probably not be expedient, with our present imperfect knowledge of the structure of those plants, to exclude such species as C. diurnum and utriculatum, both of which are types of additional groups; and perhaps our Camaridium ochroleucum will be better reduced to Cymbidium than retained distinct.

A very large portion of the Willdenovian species of Cymbidium are referable to other genera, especially C. echinocarpum and its allies, which form a genus we call Dichæa. The Maxillaria gracilis of the Botanical Cabinet is either this in a sickly state, or a nearly allied species.

J. L.

* From κύμβος, a hollow recess; in allusion to the lip.

1534

CYPRIPÉDIUM* macránthos.

Large-flowered Lady's Slipper.

GYNANDRIA *DIANDRIA.*

Nat. ord. ORCHIDEÆ *Juss.* § Cypripedieæ *Lindl.* (*Introduction to the natural system of Botany*, p. 262.)
CYPRIPEDIUM.—Suprà, vol. 10. *fol.* 788.

C. *macranthos;* lobo styli cordato-acuminato subsessili, labello petalis longiore, ore contracto crenato.—*Swartz act. holm.* 1800, *p.* 251.

Calceolus petalis nectario æqualibus aut minoribus. *Gmel. sib.* 1. *p.* 2. *t.* 1. *f.* γ.

Calceolus purpureus speciosus. *Amman. ruth. p.* 132. *n.* 176. *t.* 21.

C. macranthos. *Willd. sp. pl.* 4. 145. *Hooker in bot. mag. t.* 2938.

Roots of this fine species have been frequently sent to England by Dr. Fischer and other Russian Botanists, but they rarely have produced their blossoms, or even maintained a feeble existence. Those from which our drawing was made grew under a north wall, in a peat border, in the Garden of the Horticultural Society, where they flowered in May 1829: they had been presented by John Prescott, Esq., but have now disappeared.

In its native country it inhabits the northern parts of Asia as far as 58° north latitude, especially plains over which birch-trees are thinly scattered: it also occurs in rather dense birch-woods. Gmelin found it in all Siberia; Ammann met with it blossoming beautifully in June in a birch-wood on the eastern bank of the Irtisch, above the Tartar village of Tebendrinsk; we have been favoured by Dr. Fischer with specimens gathered in the Altai, and as far eastward as Dahuria, and with others from as much to the west as the village of Krasnoi-rog, in the government of Chernigow, in the Ucrain.

From this fine species, Swartz, and, following him, Willdenow, have distinguished a plant which they call C. ventricosum, but which Gmelin does not consider more than a variety. Of C. ventricosum we possess a specimen from the Herbarium of Stephan, with no locality assigned to it, but with a ticket marked "*C. grandiflorum β, var. Lin. Sp. Pl. ed. Willd. Gen.* 1613, *spec.* 7. 8;" whence it appears that this Botanist considered the two supposed species as varieties of one. In this he possibly was correct; for it must be confessed, that the differences which exist between them are not of a very decided nature. The principal separating characters appear to us to consist in C. ventricosum having a rather larger flower, with very acuminate sepals and petals, being generally 2-flowered, and having the petals nearly twice as long as the lower double sepal. The first of these distinctions, namely, in the size of the flowers, is probably of no moment; for our Ucrain C. macranthos is nearly as large as C. ventricosum; the others may perhaps be of greater value.

J. L.

* Literally Venus's slipper, from Cypris, one of her names, and πόδιον. It is thought to have been the Cosmosandalum of some Greek writers.

1538

CIRRHÆA* Loddigésii.

Loddiges' Cirrhæa.

GYNANDRIA *MONANDRIA.*

Nat. ord. ORCHIDEÆ *Juss.* § Vandeæ *Lindl.* (*Introduction to the natural system of Botany, p.* 265.)

CIRRHÆA Lindl. suprà sub folio 930.—*Perianthium* patens. *Sepala* libera, æqualia. *Petala* multò angustiora, linearia, flexuosa. *Labellum* longè unguiculatum, columnâ continuum, tripartitum; laciniis angustis intermediâ minore. *Columna* erecta, clavata, teres, stigmate subquadrato, horizontali! rostello cirrhato. *Anthera* dorsalis! membranacea, sub-unilocularis. *Pollinia* 2, parallela, oblonga, compressa, elasticè prosilientia, caudiculâ brevi corneâ, glandulâ incurvâ.——Herbæ *epiphytæ, pseudo-bulbosæ.* Folia *nervata.* Racemi *penduli, multiflori, radicales.* Flores *maculati.*

C. *Loddigesii;* foliis oblongo-lanceolatis utrinque acuminatis, petalis linearibus flexuosis, labelli lobo medio lineari-spatulato lateralibus acutis.
Cymbidium dependens. *Lodd. bot. cab.*
Pseudo-bulbi *ovati, costati,* 1-2½ *poll. longi, sæpiùs nudi.* Folia *solitaria, pedalia, oblongo-lanceolata, utrinque acuminata, costis* 7 *subæqualibus.* Racemi *radicales, penduli, multiflori, foliis minùs longi.* Sepala *oblonga, obtusa, æqualia, et tali modo patentia ut triangulum æquilaterum efficiunt; apice maculata.* Petala *linearia, flexuosa, basi cum sepalo supremo parallela.* Labellum *ungue longo tereti, lobis lateralibus acuminatis, erectis, intermedio lineari spatulato obtuso duplò breviore.* Columna *libera, arcuata, teres, clavata, apice truncata et stigmatifera.* Stigma *subhorizontale, obliquum; rostello subulato, elongato, cirrhoso.* Anthera *postica, opercularis, unilocularis, decidua, in clinandrio dorsali incumbens, valvis membranaceis involutis.* Pollinia 2, *cereacea, parallela, oblonga, compressa, elasticè prosilientia, glandulæ unicæ affixa, quæ processui tenui corneo elastico annectitur.*

This plant has been hitherto only known by the figure in the Botanical Cabinet, and a short note at fol. 930 of this work. Its structure is, however, so remarkable, that we trust the present account of it will not be wholly uninteresting.

* The flowers are remarkable for what is called the rostellum being prolonged in the form of a small tendril or *cirrhus.*

According to Messrs. Loddiges, its native country is *supposed* to be China; but it is more probably Brazil; for not only are its affinities altogether American, but two other Brazilian species are now known.

From all the genera of the tribe to which it belongs it is distinguished by its stigma occupying the apex, and not the face of the column, while its anther is situated at the back. In this last respect it agrees with Notylia, which is, however, otherwise very different.

It grows tolerably well in vegetable mould, if placed where the air is humid, and the drainage complete. In such situations it flowers in August.

The two other species above alluded to are the Gongora viridi-purpurea of the Botanical Magazine, and an undescribed plant from the Corcovado, of which we find a drawing, and some dried flowers, among a multitude of other interesting Orchideous plants, which, with the liberality that marks the man of true science, have been placed in our hands by Dr. Hooker. These two species may be thus named and defined:

C. *viridi-purpurea;* foliis oblongo-lanceolatis, petalis linearibus arcuatis, labelli lobo medio acuminato lateralibus æquali.
Gongora viridi-purpurea. *Hooker in bot. mag. t.* 2978.
Hab. supra arbores in sylvis ad sinum *Botofogo* in *Brazilia.*

The beautiful green colour of the outside of the sepals and petals, together with their rich spotted orange-yellow inside, distinguishes this at once from both the other species.

C. *fusco-lutea;* foliis lanceolatis acuminatis, petalis lineari-lanceolatis, labelli lobo medio ovato, lateralibus obtusis.
Hab. supra arbores *Braziliæ* ad Corcovado (*exam. s. sp. et v. ic. pict. in herb. Hooker.*)

This very much resembles C. Loddigesii; but the flowers are twice as large, the sepals are dusky brown, with a few reddish lines, and the petals greenish yellow; while all the parts are destitute of spots except the middle lobe of the labellum.

In the same rich collection we find two other genera nearly allied to Cirrhæa, of which the following will be the names and characters:—

MYANTHUS.

Perianthium explanatum. Sepala libera, æqualia, lateralibus paululùm ascendentibus. Petala conformia, angustiora, sepalo supremo supposita. Labellum planum, obovatum, 3-dentatum, sepalis brevius. Columna erecta, teres, basi bicirrhosa, posticè ad cardinem antheræ longè producta. Anthera et pollinia Cataseti.——Epiphyta, Cataseti omninò vegetatione.

1. Myanthus cernuus. Found on trees in the neighbourhood of Rio Janeiro. The racemes are cernuous, and about a foot long, or more. The flowers have a greenish yellow ground, closely covered with rich reddish brown spots. —— Perhaps Catasetum cristatum would be better referred to this genus.

MONACHANTHUS.

Perianthium explanatum. Sepala et petala æqualia, deorsùm versa. Labellum posticum, carnosum, indivisum, ventricosum, sepalis multò majus. Columna brevis, crassa, mutica. Anthera .... Pollinia .....——Epiphyta, Cataseti habitu.

1. Monachanthus viridis. Found on trees in the Corcovado.

The fleshy stems are exactly those of Catasetum, as are the leaves, except that they are shorter. In the drawing we have examined, the flowers are represented in pairs from the apex of an erect scape about 9 inches high: they are fleshy, and of a deep uniform green, with a labellum that looks something like a hood drawn over a little face represented by the column.

This genus differs from Catasetum in the want of cirrhi upon its short column, and in the expanded perianthium, the five segments of which are turned all one way, like those of Eulophia.

J. L.

1541

NANÓDES* discolor.

Purple Nanodes.

GYNANDRIA *MONANDRIA.*

Nat. ord. ORCHIDEÆ *Juss.* § Vandeæ *Lindl.* (*Introduction to the natural system of Botany, p.* 265.)

NANODES.—*Perianthium* ringens. *Sepalum* supremum fornicatum; lateralia labello supposita eoque basi connata. *Petala* sepalis lateralibus conformia iisque subparallela, libera. *Labellum* carnosum, indivisum, cum columna connatum (more Epidendri). *Columna* ad utrumque marginem clinandrii alata. *Anthera* 2-locularis, rostrata. *Pollinia* 4, compressa, collateralia, in glandulam ovatam sessilia.——Epiphyta *humilis, cæspitosa, caulescens.* Folia *coriacea, disticha, vaginantia, approximata.* Flores *discolores, in apice caulium sessiles, solitarii.*

Nanodes discolor.
Caules *aggregati,* 2-3 *pollices alti, simplices, densè foliosi, nihil bulbosi.* Folia *ovato-oblonga, emarginata, basi amplexicaulia et vaginantia, internodiis duplò longiora, viridi-purpurea.* Flores *solitarii, sessiles, terminales, obscurè purpurei, inter folia immersi.* Sepalum *supremum ascendens, lineari-oblongum, convexum; petala declinata, cum sepalis lateralibus labello suppositis parallela.* Labellum *carnosum, ovatum, minutissimè crenulatum.*

A native of the woods near Rio Janeiro, whence it was received, some years since, by the Horticultural Society. Our drawing was made in the month of August 1829.

Curious as are very many of the species of epiphytal Orchideæ, we do not remember one that is much more remarkable than this, which possesses a habit quite its own. Its flowers are so like the leaves from among which they spring, and by which they are embosomed, that you would scarcely discover the plant to be in flower even if every branch was blossoming.

We presume its cultivation should be that applied to the more delicate of the same tribe, namely, plenty of heat and atmospheric moisture, with no watering upon the mould or stem. It is, however, no doubt what gardeners call "a very shy plant," for it has long been lost in the Horticultural Society's Garden.

J. L.

* From νανώδης, pigmy.

1542

ONCÍDIUM* cornígerum.

Horned Oncidium.

GYNANDRIA *MONANDRIA*.

Nat. ord. Orchideæ *Juss.* § Vandeæ *Lindl.* (*Introduction to the natural system of Botany*, *p.* 265.)
ONCIDIUM.—Suprà, *vol.* 13. *fol.* 1050.

§ *Labellum trilobum. Folia plana, v. complicata. Sepala lateralia connata. Labelli laciniæ laterales nanæ v. obliteratæ.*—Gen. et sp. Orch. ined.

O. *cornigerum;* pseudo-bulbis oblongis sulcatis monophyllis, foliis ovalibus acutis sessilibus striatis scapo simplici decumbente paucifloro brevioribus, sepalo supremo petalisque obovatis concavis undulatis obtusis inferioribus minoribus angustioribus basi connatis, labelli lobis lateralibus linearibus connatis intermedio obovato subrepando undulato, cristâ anticè verrucosâ truncatâ posticè lamellâ crenatâ transversâ cornutâ, columnæ alis linearibus obtusis porrectis.

Folia 2-3 *poll. longa, patentissima.* Scapus *debilis,* 5 *poll. longus.* Anthera *membranacea, truncata; clinandrio valdè proclivi.*

A rare little species of Oncidium, native of Brazil, for which we are indebted to the Hon. and Rev. W. Herbert. The drawing has been in our portfolio ever since August 1830. We have never met with it in the collections round London, neither has it been received from any of our numerous country correspondents. It may, therefore, be supposed to be extremely uncommon.

Perhaps the smallest flowered of all the genus of Oncidium properly so called. The truncated callosity of the crest, with a sort of 2-horned screen at its back, and the two lateral erect lobes of the lip at its side, is, when viewed in profile, not very unlike a bull's head in miniature.

J. L.

* So named from ὄγκος, a tumour; the genus being composed wholly of species the labellum of which bears at its base warts, tumours, or other excrescences.

1549

MAXILLÁRIA* décolor.

Pale Yellow Maxillaria.

GYNANDRIA *MONANDRIA*.

Nat. ord. Orchideæ. § Vandeæ. (*Introduction to the natural system of Botany*, *p.* 262.)
MAXILLARIA.—Supra, *vol.* 11. *fol.* 897.

§ Scapigeræ. *Pedunculi radicales; labellum posticum.* Xylobium.—*Gen. et sp. of Orch. ined.*

M. *decolor;* pseudo-bulbis oblongis compressis, foliis solitariis oblongo-lanceolatis plicatis utrinque acuminatis, scapo radicali multifloro vaginato foliis breviore, sepalis ovato-oblongis obtusis patentibus, petalis duplò minoribus conniventibus, labello postico obsoletè trilobo obtuso cucullato: callis 5 clavatis parallelis æquilongis ultra medium procurrentibus.

Folia *diversæ formæ et magnitudinis, nunc* 8 *poll. longa,* 3 *p. lata, nunc* 18 *p. longa,* 2 *p. lata.* Scapus 4-*pollicaris.* Flores *ochroleuci, immaculati,* M. squalentis *magnitudine.*

A new and very distinct species, native of Jamaica, whence it was imported by Mr. Lee, in whose Nursery at Hammersmith it was drawn in January 1831. It is a tender stove-plant, requiring the same treatment as other species of the same family. In our Plate, *a* is a magnified representation of the inside of the labellum.

The genus Maxillaria, as originally constituted by the authors of the *Flora Peruviana,* consisted of epiphytal caulescent Orchideæ, having the flowers proceeding singly, or in small numbers, from the axillæ of the leaves, as in the genus Ornithidium, of which they have in many respects the habit. Subsequently other species were added, having a radical inflorescence, and pseudo-bulbs bearing one or two coriaceous or plaited leaves. Of these, some have the labellum anterior, and were formerly called by us Colax, while a few have the labellum posterior, and were considered a distinct genus, to which we gave the name Xylobium. An examination of many new species, and a reconsideration of the whole subject, has convinced us that the two last-named genera, together with Maxillaria, form one natural group, which is defined by its having two entire or split pollen masses attached by a broad caudicula and distinct gland to the rostellum, and a labellum that is neither inflated nor spurred, articulated with the elongated base of the column. It is by the latter character chiefly that it is distinguished from Cymbidium.

Of the genus thus limited, thirty-nine species are known, of which the following is a brief enumeration: —

§ 1. Axillifloræ. *Pedunculi axillares* (Eumaxillaria).

1. M. undulata. *Fl. Peruv. syst.* 221.
2. M. variegata. *Ib.*
3. M. ligulata. *Ib.*
4. M. hastata. *Ib.*
5. M. paniculata. *Ib.*
6. M. cuneiformis. *Ib.*
7. M. bicolor. *Ib.*
8. M. tricolor. *Ib.*
9. M. triphylla. *Ib.*
10. M. undatiflora. *Ib.*
11. M. prolifera. *Ib.*
12. M. ramosa. *Ib.*
13. M. longipetala. *Ib.*

*13.†M. ochroleuca. *Lodd. bot. cab. ined.*

14. M. platypetala. *Ib.*

*15. M. alba = Dendrobium album. *Hooker exot. fl. t.* 142.

§ 2. Scapigeræ. *Pedunculi radicales.*
*Labellum anticum (Colax).

* See fol. 1428.

*16. M. stapelioides. *Link et Otto abbild. p.* 111. *t.* 52.
*17. M. subulata. *Formerly cultivated in the Kew Garden.*
*18. M. tetragona. *Lindl. in bot. reg. t.* 1428.
*19. M. viridis. *Ib. t.* 1510. = M. placanthera. *Hooker in bot. mag. t.* 3173.
*20. M. Deppii. *Lodd. bot. cab. t.* 1612.
*21. M. Parkeri. *Hooker in bot. mag. t.* 2729.
*22. M. aromatica. *Graham in Hook. exot. fl. t.* 219. = Colax aromaticus. *Spreng.*
*23. M. Barringtoniæ. *Lodd. bot. cab. t.* 1824. = Epidendrum Barringtoniæ *Smith.* = Dendrobium Barringtoniæ *Swartz.* = Colax Barringtoniæ *Lindl.* = Dendrobium ciliatum *Swartz.* = Maxillaria ciliata *Fl. Peruv. syst. Bot. reg. t.* 1206

*24. M. picta. *Hooker bot. mag. t.* 3154.
*25. M. parvula. *Hooker exot. fl. t.* 217. = Colax parvulus. *Spreng.*
26. M. grandiflora = Dendrobium grandiflorum. *Humb. et Kunth. n. g. et sp. pl.* 1. 359. *t.* 88.
*27. M. Harrisoniæ. *Lindl. in bot. reg. t.* 897. = Dendr. Harrisoniæ. *Hooker.* = Colax Harrisoniæ. *Lindl.*
*28. M. racemosa. *Hooker in bot. mag. t.* 2789.
29. M. maculata = Dendr. maculatum. *Humb. et Kunth. n. g. et sp.* 1. 359.
*30. M. pallidiflora. *Hooker in bot. mag. t.* 2806.
31. M. latifolia = Dendr. latifolium. *Humb. et Kunth. l. c. p.* 360.
32. M. longifolia = Dendr. longifolium. *Id.* = Xylobium ? longifolium. *Lindley.*
*33 M. palmifolia = Epidendr. palmifolium. *Swartz.* = Dendr. palmifolium. *Swartz.* = Colax palmifolius. *Lindl.*
*34. M. Warreana. *Lodd. bot. cab. ined.*

** Labellum posticum (Xylobium).

*35. M. decolor. *Suprà.*
*36. M. squalens. *Hooker bot. mag. t.* 2955. = Dendrob. squalens *et* Xylobium squalens. *Lindl.*

§ 3. Spathaceæ. *Pedunculi terminales è spathâ vaginante erumpentes. Caules cylindracei, non bulbosi. Species habitu omninò Cateleyæ.* Nothium.

37. M. ? aggregata = Dendr. aggregatum. *Humb. et Kunth. n. g. et sp.* 1. 358.
38. M. spathacea. *Lindl. in herb. Hooker.*

Those marked * are cultivated in the collections of this country.

J. L.

1552

SACCOLÁBIUM* papillósum.

Pimpled Saccolabium.

GYNANDRIA *MONANDRIA.*

Nat. ord. Orchideæ. § Vandeæ *Lindl.* (*Introduction to the natural system of Botany, p.* 262.)
SACCOLABIUM Blume.—*Perianthium* explanatum, patens. *Sepala* petalis æqualia, lateralia sæpiùs majora. *Labellum* indivisum, calcaratum, basi columnæ accretum. *Columna* erecta, semiteres, rostello subulato. *Anthera* semibilocularis. *Pollinia* 2, subglobosa, caudiculâ elongatâ, glandulâ minutâ.——Herbæ *epiphytæ, caulescentes.* Folia *disticha, coriacea, apice sæpiùs obliqua.* Flores *axillares, racemosi v. solitarii.*

S. *papillosum;* foliis ligulatis apice obliquis cuspidatis, racemis brevissimis capitatis, sepalis carnosis lineari-ovatis obtusis, labelli calcare obconico obtuso intùs villoso laminâ ovatâ carnosâ papillosâ recurvâ.
Thalia Maravara. *Rheede hort. malab.* 12. *t.* 4.
Cymbidium præmorsum. *Swartz nov. act. ups.* 6. 75. *Willd. sp. pl.* 4. 103.
Epidendrum præmorsum. *Roxb. corom.* 1. 34. *t.* 43.
Aërides undulatum. *Smith in Rees suppl.*
Saccolabium papillosum. *Lindl. in Wall. cat. no.* 7305.
Caulis *arcuatus, subpedalis, distichè foliosus, radices tortuosas promens.* Folia *atroviridia,* 4-6 *poll. longa.* Racemi *parvi, foliis triplò breviores.* Flores *albi, luteo purpureoque picti.*

The genus Saccolabium consists of a large number of caulescent epiphytal species, inhabiting the continent and archipelago of India. It comprehends some of the most interesting and beautiful of the tribe, all of which are cultivated without difficulty by being suspended, in pots filled with potsherds intermixed with a little earth, from the back wall of a damp and hot stove.

We believe that at least four other genera will be reduced to it, namely *Rhyncostylis* of Blume; *Gastrochilus* of Don, which, however, forms a curious section, with a ventricose spur; *Robiquetia* of Gaudichaud; and *Gussonea* of Achille Richard.

Found upon trees in Malabar, and in various other parts of the continent of India. We have it from Dr. Wallich from Prome. The plant in the Garden of the Horticultural Society, from which our drawing was taken, was introduced by that indefatigable Botanist. Flowers in August and September.

J. L.

* From *saccus,* a bag, and *labium,* a lip; in allusion to the bagged labellum of all the species.

1561

BRASAVÓLA* Perrínii.

Perrin's Brasavola.

GYNANDRIA *MONANDRIA.*

Nat. ord. Orchideæ *Juss.* Tribus *Epidendreæ* Lindl. (*Introduction to the natural system of Botany, p.* 262.)
BRASAVOLA.—*Suprà, vol.* 17. *fol.* 1465.

B. *Perrinii;* labello ovato acuminato integerrimo: ungue convoluto columnæ longitudine, petalis sepalisque linearibus acuminatis glabris, dentibus lateralibus clinandrii integris postico serrato.
Habitus omninò *B. nodosæ.* Præcipuè differt characteribus suprà datis.

When we received a specimen of this plant in the month of September last from Mrs. Arnold Harrison, by whom it had been obtained from Rio Janeiro, we were disposed to consider it the same as Br. nodosa. A more careful examination has, however, satisfied us that it is essentially distinguished by its smaller flowers, the short stalk of the labellum, and the toothing of the back lobe of the clinandrium. It is by no means so handsome as that species; but as it grows very freely, it will be esteemed an acquisition by the lovers of Epiphytes.

At the request of Mr. Harrison it is named after Mr. Perrin, his Gardener, under whose skilful management the collection of Epiphytes at Aighburgh has arrived at a state of great perfection.

J. L.

* See fol. 1465.

EDWARDS'S

BOTANICAL REGISTER:

OR,

ORNAMENTAL FLOWER-GARDEN

AND SHRUBBERY:

CONSISTING OF

COLOURED FIGURES OF PLANTS AND SHRUBS,

CULTIVATED IN BRITISH GARDENS;

ACCOMPANIED BY THEIR

History, Best Method of Treatment in Cultivation, Propagation, &c.

CONTINUED

BY JOHN LINDLEY, Ph.D. F.R.S. L.S. AND G.S.

PROFESSOR OF BOTANY IN THE UNIVERSITY OF LONDON,
&c. &c. &c.

New Series.

VOL. VI.

OR VOL. XIX. OF THE ENTIRE WORK.

——viret semper——nec fronde caducâ
Carpitur.

LONDON:

JAMES RIDGWAY, 169, PICCADILLY.

M.DCCC.XXXIII.

1566

MAXILLÁRIA* racemósa.

Raceme-flowered Maxillaria.

GYNANDRIA *MONANDRIA.*

Nat. ord. ORCHIDEÆ. § Vandeæ. (*Introduction to the natural system of Botany*, *p.* 262.)
MAXILLARIA.—Suprà, vol. 11. *fol.* 897.

§ 2. Scapigeræ. *Pedunculi radicales.* Labellum anticum (COLAX). Scapi 2-multiflori.

M. *racemosa;* pseudobulbis ovatis compressis tetragonis monophyllis, foliis oblongo-lanceolatis tricostatis scapis gracilibus strictis multifloris brevioribus, sepalis oblongis acutis: lateralibus basi valdè elongatis in calcar spurium connatis, petalis linearibus spatulatis triplò minoribus, labello oblongo cucullato indiviso undulato emarginato in axi calloso, columnâ pubescente.—*Lindl. gen. et sp. orch. p.* 149. *ined.*

M. racemosa. *Hooker in bot. mag. t.* 2789. *Lodd. bot. cab. t.* 1318.

Pseudobulbi 2 *poll. longi.* Folia 4 *poll. longa, basi in petiolum angustata.* Scapus *pedalis.* Flores *sordidè flavi, membranacei.* Sepala *patentia: superius cum petalis parallelum Cymbidiorum more.* Labellum *cum sinu sepalorum lateralium articulatum, album v. sordidè flavum purpureis maculis pallidis notatum.* Anth. 1-*locularis.* Pollinia *posticè sulcata, caudiculá brevi, glandulá parvá.*

This rare little Maxillaria was originally sent from woods near Rio Janeiro by Mr. William Harrison, and afterwards flowered in several collections. It has now become very scarce, being apparently too delicate to bear the imperfect cultivation that such plants receive even at the most skilful hands. We have seen it thrive for a short time planted in moss, and suspended in a pot from the rafter of a stove; in such a situation, in the Garden of the Horticultural Society, the plant flowered from which our drawing was taken in October 1829.

Its flowers are pale green, or greenish yellow, and have not much beauty; their lip is prettily marked with light crimson dots in the inside. J. L.

* See fol. 1428.

1569

ONCÍDIUM* Harrisoniánum.

Mrs. Arnold Harrison's Oncidium.

GYNANDRIA *MONANDRIA.*

Nat. ord. ORCHIDEÆ. § *Vandeæ* Lindl. (*Introduction to the natural system of Botany*, *p.* 262.)
ONCIDIUM.—Suprà, vol. 16. *fol.* 1349.

§. *Folia plana; v. complicata. Sepala lateralia libera. Labelli laciniæ laterales nanæ, v. obliteratæ.*

O. *Harrisonianum;* pseudobulbis subglobosis monophyllis, foliis carnosis lineari-oblongis acutis recurvis, paniculâ ramosâ multiflorâ, sepalis petalisque linearibus obtusissimis, labelli laciniis lateralibus minutis auriculiformibus intermediâ unguiculatâ transversâ emarginatâ, cristâ 5-lobâ centro pubescente, columnæ alis deltoideis rectiusculis.

Pseudobulbi *ovi passerini magnitudine sed globosi, lætè virides, læves.* Folia 4-6 *poll. longa, omninò avenia.* Scapus *è squamâ tubulosâ proveniens, ascendens, pedalis v. ultrà.* Panicula *composita, apice cernua, ramulis divaricatis, v. decurvis.* Flores *vitellini.* Sepala *petalis paulò majora, atrofusco maculata.* Labellum *unicolor, sepalis lateralibus longius.* Crista *è tuberculis 5 constat, quorum 4 lateralia, compressa, rotundata, leviter incurvantur; quintum anticum multò est angustius et cornu parvum refert; inter omnia adest area pulvinata pubescens.*

For what purpose can the world have been adorned with these Orchideous plants? To man or animals they are scarcely ever of any known use. No honey is secreted by their flowers; neither poison, medicine, nor food, are collected in the recesses of their stems; and their very seeds seem unfit for feeding even the smallest bird. We can scarcely suppose them provided for the purification of the unwholesome atmosphere of the forest recesses in which they delight, for their organization is that of plants whose leaves perform their vital actions too slowly to effect such a purpose. For what then can they have been formed, unless to delight the sense of man, to gratify his eye by their gay colours and fantastic forms, and to shew the inexhaustible fertility of that creative power which we recognise every where in Nature.

* See fol. 1349.

If this be not the object of those countless changes of form and colour which the Orchis tribe exhibits, we shall scarcely comprehend why in this very genus Oncidium the lip bears at its base a collection of tubercles which are not only different in every species, but so strangely varied, that

" Eye of newt, and toe of frog,"

are the least singular of the forms that lie cowering in the bosom of their petals; the heads of unknown animals, reptiles of unheard-of figures, coils of snakes rising as if to dart upon the curious observer, may all be seen in the blossoms of the various species, whose very flowers may be likened to unearthly insects on the wing.

This very distinct species was discovered on the Organ Mountains of Brazil by Mr. William Harrison of Rio Janeiro.

We have taken the liberty of naming it after a family more distinguished than any other for the number of species they have introduced, and for the success with which they have cultivated them. Whenever Horticulture shall again find an historian, he will have to record the period when the difficulty of cultivating tropical Orchideæ, which was once considered insuperable, was successfully overcome; in such a history the names of Mr. William Harrison, of Mrs. Arnold Harrison, and of Mr. Richard Harrison, will stand among the foremost.

O. *Harrisonianum* is easily recognised by its fleshy, slightly channelled, recurved leaves, each of which is placed upon a little pseudobulb, not much bigger than a sparrow's egg, but round and shining. The panicles of flowers are about a foot high, and arranged in a graceful manner, something in the way of Oncidium flexuosum. Our figure has been made from a drawing and specimen communicated by Mrs. Arnold Harrison in October 1832.

Cultivated among moss and rotten wood, in which it grows freely.

Fig. 1 is a magnified representation of the face of the column and the base of the lip, shewing the way in which the tubercles are arranged. One might fancy they were a bat's head, of which the downy centre forms the forehead, the back tubercles the ears, the horn in front the snout, and the two lateral tubercles a pair of flaccid cheeks.

J. L.

1610

DENDRÓBIUM* speciósum.

Shewy Dendrobium.

GYNANDRIA *MONANDRIA.*

Nat. ord. ORCHIDEÆ *Juss.* § Malaxideæ *Lindl.* (*Introduction to the natural system of Botany, p.* 262.)
DENDROBIUM.—Suprà, vol. 7. *fol.* 548.

D. *speciosum;* caulibus erectis apice 2-3-phyllis, foliis ovali-oblongis integerrimis racemo terminali multifloro brevioribus, perianthii foliolis angusto-oblongis, labello infra divisuram carinâ unicâ, lobo intermedio latiore quàm longo ecarinato. *R. Br. prodr.* 332. *Lindl. gen. et sp. Orch.* 87. *Hooker Bot. mag.* 3074. *Bauer's illustrations of Orchid. tab. VI. genera.*

Scarcely any plant is more common than this in collections near London; so readily is it cultivated, that it rarely happens that a Gardener does not succeed in keeping it healthy, and multiplying it abundantly: tied to a stick or planted in earth, preserved in a greenhouse or stimulated in a stove, in all cases it preserves the deep green of its leaves and its aspect of robustness. But it seldom flowers. This is owing to the weakness of the specimens, compared with what is natural to them in their own country: it is not enough to keep their leaves green and plump, it is also necessary that the plants should grow rapidly, and to at least twice the size we are accustomed to see them; then and then only will they yield their stately and beautiful racemes of pale-yellow wax-like blossoms.

The specimen from which the accompanying drawing was taken, was one of five produced upon the same cluster of stems, in the garden of the Rev. Mr. Huntly, of Kimbolton: each stem was nine inches long, stout in proportion, and emitted from its base many scores of long white healthy roots, which extended to the distance of near a foot all round it. It flowered in March of the present year.

* See fol. 1239.

The method pursued in its cultivation was merely that repeatedly recommended in this work, namely, to keep such plants growing in a warm damp atmosphere, in pots extremely well drained; and not exposed to the direct light of the sun. Under such treatment the stems soon begin to throw out a number of little green-tipped delicate roots, which gradually turn white and lengthen, adhering to the surface of garden-pots, or any thing they may be near. If no accident befalls such roots, nothing is likely to interfere with the future progress of the plant; but slugs and woodlice are great enemies of them, feeding upon the youngest and tenderest tips; and it is indispensable that these animals should be completely kept away; for a root of this sort once injured rarely recovers.

J. L.

1616

GONGÓRA* maculáta.

Spotted Gongóra.

GYNANDRIA *MONANDRIA.*

Nat. ord. ORCHIDEÆ *Juss.* (*Introduction to the natural system of Botany, p.* 262.)
GONGORA Fl. *Peruv.—Perianthium* explanatum. *Sepala* lateralia libera divaricata, supremum cum dorso columnæ connatum. *Petala* minora, columnæ medio adnata. *Labellum* cum basi columnæ continuum, liberum, unguiculatum, hypochilio explanato utrinque cornuto, epichilio verticali ancipiti (faciebus oppositis complicatis connatis) acuminato. *Columna* longissima, arcuata, clavata, marginata. *Anthera* subbilocularis. *Pollinia* 2, linearia, in caudiculam cuneatam sessilia.——Herbæ *epiphytæ, pseudobulbosæ.* Folia *plicata.* Racemi *longissimi, flexuosi, multiflori.* Lindl. gen. et sp. Orchid. *pars III. p.* 158.

G. *maculata;* foliis 5-plicatis obovato-oblongis basi valdè angustatis, sepalis lateralibus è latâ basi angustatis, hypochilio oblongo subtùs convexo basi obtusè bicorni apice truncato angulis acutis in cirrhis duobus producto, epichilio acuminato.

Epiphyta, *acaulis,* pseudobulbis *ovatis angulatis.* Folia *pedem et ultra longa,* 4 *pollices lata, læte viridia.* Racemi *radicales, penduli, laxissimi,* $2\frac{1}{2}$ *pedes longi.* Scapus *purpureus, glaber, squamis quibusdam parvis longè distantibus vaginatus.* Pedicelli *divaricati, cum ovario gracili* 2 *poll. longi, purpurei.* Sepala *vitellina, purpureo interruptè fasciata;* supremum *lineari-lanceolatum, erectum, dorso columnæ accretum;* lateralia *multò majora, reflexa, marginibus revolutis, à latâ basi angustata, et à basi supremi longo intervallo sejuncta.* Petala *pallidè purpurea, obscurè maculata, lineari-lanceolata, acuminata, antrorsùm arcuata, sepalo supremo minora.* Columna *viridis, obscurè maculata, semiteres, clavata, ferè sigmoidea, tali modo arcuata ut in ovarium obliquissimè recumbit, et in pedem productissimum sepala gerit utrinque labellumque in medio.* Clinandrium *valdè declive, ut in Cirrhæa.* Labellum *compressum, colore sepalorum, hypochilio quàm epichilium subsagittatum duplò minore.*

* Named by the authors of the *Flora Peruviana* in honour of D. Antonio Caballero y Gongora, Bishop of Cordova, who, while Archbishop of Santa Fé and Viceroy of New Granada, was a zealous patron of the celebrated Mutis, and an ardent promoter of all branches of Natural History.

Many a strange figure has been met with among Orchideous plants, and numerous are the animal forms which Botanists have fancied they could recognise among their singular flowers. Some are said to bear little men and women swinging below their canopy of petals; others have appeared to carry the likeness of lizards, frogs, and other reptiles, crouching among their leaves; while some have been compared to Oberons and Titanias hanging by their tiny arms from the bells, where they have concealed themselves. To what the flowers of the plant now figured can be likened, we profess not to know, unless to some of the fantastic animals of heraldry; a griffin *segreant,* as they term it, would do as well as any other for a comparison.

This most curious species was sent us by Richard Harrison, Esq., from his Garden at Liverpool: it was originally introduced from Demerara, in 1832, by Mr. Thomas Moss, of Otterspool. It flowered in the hothouse in May; its bunches of flowers were two feet and a half long, and hung down most gracefully from the pot in which the plant was suspended; of this our diminished figure in the background is intended to be a representation.

A few years ago the genus Gongora was so little known, that some doubts were even entertained of its existence. Our Gardens now possess two species, neither of which is the kind originally figured in the *Flora Peruviana*.

J. L.

1618

SAUROGLÓSSUM* elátum.

Tall Lizard's Tongue.

GYNANDRIA *MONANDRIA.*

Nat. ord. ORCHIDEÆ. § Neottieæ *Lindl.* (*Introduction to the natural system of Botany, p.* 262.)

SAUROGLOSSUM.—*Perianthium* connivens, sepalis lateralibus linearibus, arcuatim patentibus, basi in ovarium decurrentibus. *Labellum* anticum, lineare, canaliculatum, ecallosum, basi dilatatum. *Columna* elongata, semiteres, basi producta, apice subspatulata; clinandrio declivi immarginato, rostello ovato. *Pollinia* 2, biloba, caudiculâ brevissimâ.—— Herbæ *terrestres* (Americæ æquinoctialis) foliis *radicalibus,* floribus *racemosis herbaceis scapo insidentibus,* radicibus *carnosis fasciculatis.*

S. *elatum.*

Radices *terrestres, fasciculati, carnosi.* Folia radicalia *oblongo-lanceolata, suberecta, carnosa, plicis nullis, scapo* 3-*plò breviora.* Scapus *strictus, ferè* 2-*pedalis, pubescens, squamis foliaceis viridibus distantibus vaginatus.* Racemus *pedalis et ultra, densus, cylindraceus.* Bracteæ *canaliculatæ, subulatæ, florum ferè longitudine.* Ovarium *pubescens.* Perianthium *cylindraceo-convolutum,* sepalis *viridibus lineari-lanceolatis, versus apicem latioribus,* superiori *petalis agglutinato,* lateralibus *arcuatis, basi valdè obliquis, anticè paululùm sub labello connatis.* Labellum *album, cum columna parallelum, lineare, canaliculatum, sessile, basi ecallosum, paulò dilatatum, medio pubescens, apice ovatum, dilatatum.* Columna *libera, elongata, semiteres, anticè producta, apice ovata,* stigmate *cordiformi,* rostello *acuto, vix bifido,* clinandrio *retrorsùm declivi, immarginato.* Pollinia 2, *bisulca, pulverea,* caudiculâ *brevi,* glandulâ *ovali.*

For the specimen from which our drawing of this new Orchideous plant has been taken, we are indebted to Mrs. Arnold Harrison, who obligingly communicated it in March last. It is a native of the woods of Brazil, whence it was sent by Mr. Henry Harrison.

A stove plant, requiring to be cultivated in earth, like other terrestrial Orchideæ of the Neottia tribe, to some of which, such as Spiranthes grandiflora and Pelexia spiranthoides, it bears a good deal of resemblance. Its structure is, however, distinctly different from that of any published genus.

As there are several other genera of the Neottia division of Orchideous plants, which are at present only known by name, we avail ourselves of the present opportunity of defining some of those which are the most remarkable, in order that no confusion may be introduced into our nomenclature, in consequence of Botanists being unacquainted with the plants upon which such genera are founded.

SYNASSA *Lindl. Sceletos.*

Perianthium connivens? *Sepala superiora* et petala agglutinata, *lateralia* labello supposita, basi producta in calcare cum ovario connato. *Labellum* cum columnâ omnino connatum, apice crispum, dilatatum, callis duobus infra apicem, basi cuniculatum. *Pollinia* 2, bisulca.—Sp. 1. Synassa corymbosa; in Peru.

CNEMIDIA.

Perianthium connivens. *Sepala* lateralia semiconnata, basi in calcare spurio producta. *Labellum* posticum, liberum, calcaratum, canaliculatum, acuminatum. *Columna* teres, apice acuminata, clinandrio declivi immarginato. *Pollinia* duo, caudiculâ subulatâ. Anthera subulata. (DECAISNIA *Lindl. in Wall. cat.*; nomen libenter mutatum ob Decaisniam Ad. Brongniartii nuperius editam, sed melius cognitam et multò magis divulgatam.)—Sp. 2. Cnemidia angulosa et C. semilibera.

TRIPLEURA *Lindl. in Wall. cat.*

Perianthium explanatum, cruciatum, quasi tetraphyllum, sepalo supremo lato petalis agglutinato, lateralibus linearibus patentissimis. *Labellum* posticum, liberum, concavum. *Columna* teres, brevissima, antice processubus 2 membranaceis (stam. steril.) corniformibus instructa. *Anthera* subrotunda, apiculata, terminalis.

MYODA *Lindl. in Wall. cat.*

Perianthium reflexum, petalis sepalo superiori agglutinatis. *Labellum* anticum, cum columnâ connatum, apice unguiculatum, hastato-tripartitum. *Columna* teres, clavata, clinandrio carnoso cucullato, rostello acuminato obliquo. *Stigma* punctum parvum, bilabiatum. *Anthera* ovata. *Pollinia* 2.

ZEUXINA *Lindl. Sceletos.*

Perianthium connivens. *Sepalum* supremum basi saccatum, lateralia connata. *Labellum* anticum, basi membranaceum, bicallosum, cum columnâ connatum, apice unguiculatum, carnosum, dilatatum, concavum. *Columna* brevissima, teres, clinandrio membranaceo, antheram involvente. *Pollinia* 4 (v. sec. iconem Hamiltoni 8), in duas massas connata, caudiculâ brevi ligulatâ, glandulâ uncinatâ.

GEORCHIS *Lindl. in Wall. cat.*

Perianthium cylindrico-connivens, basi ventricosum. *Labellum* anticum, sessile, basi intus pilosum, cucullatum, apice convolutum. *Columna* brevissima. *Clinandrium* longissimum, acuminatissimum, immarginatum. *Anthera* acuminatissima. *Pollinia* 4, minuta, caudiculis totidem longissimis setaceis separabilibus.

HYLOPHILA *Lindl. in Wall. cat.*

Perianthium connivens. *Sepalum* supremum arcuatum, lateralia obliquè recurva, ferè uncinata. *Labellum* nanum, anticum, liberum, saccatum, dependens, apiculatum. *Columna* brevissima, clinandrio proclivi, rostello curvo subulato. *Anthera* subulata. *Pollinia* 2, caudiculâ subulatâ.

HÆMARIA *Lindl. Sceletos.* Gonogona *Link.*

Perianthium explanatum, sepalis lateralibus reflexis. *Labellum* liberum, anticum, basi bicallosum, testiculato-saccatum, apice obliquum, unguiculatum, subsagittatum, planum. *Columna* teres, clavata, clinandrio carnoso, cucullato, obliquè torto, rostello acuminato. *Anthera* bilocularis, acuminata, clinandrii cucullo semioccultato. *Pollinia* 2, sulcata.

TROPIDIA *Lindl. in Wall. cat.*

Perianthium connivens. *Sepala* lateralia majora, basi obliquè producta, calcar mentientia. *Labellum* liberum, sessile, ventricosum, compressum, lateribus reflexis lamellatis, apice constrictum, cucullatum. *Columna* teres, clinandrio longè infra stigma excavato. *Anthera* dorsalis, mutica. *Pollinia* 4, in massas 2 bilobas cohærentia, caudiculâ brevi.

HERPYSMA *Lindl. in Wall. cat.*

Perianthium connivens, sepalo supremo paulò sejuncto, lateralibus et petalis antrorsùm arcuatis. *Labellum* anticum, calcaratum, cum columnâ connatum, basi dilatatum, apice abruptè subrotundum, sub apice bicallosum. *Columna* semiteres, clinandrio marginato. *Pollinia* 2, caudiculis distinctis, polline sectili.

J. L.

* So named from σαῦρα, a lizard; because many of the parts of this curious plant may be likened to the tongue of some reptile: the leaves may be compared to the tongues of antediluvian Saurians, and the sepals to those of modern species.

1623

EPIDÉNDRUM* oncidioídes.

Oncidium-flowered Epidendrum.

GYNANDRIA *MONANDRIA.*

Nat. ord. ORCHIDEÆ. § Epidendreæ *Lindl.* (*Introduction to the natural system of Botany, p.* 262.)

EPIDENDRUM.—*Suprà, vol.* 17. *fol.* 1415.

§. Foliis in pseudobulbos insidentibus. *Lindl. gen. et sp. of Orchideous plants, p.* 97.

E. *oncidioides,* foliis in pseudobulbos ovales ancipites 2-3 ensiformibus paniculâ terminali ramosâ brevioribus, sepalis petalisque obovatis unguiculatis patentibus, labelli liberi trilobi lobis lateralibus nanis oblongis columnæ longitudine intermedio subrotundo cuspidato basi 5-costato.

Pseudobulbi *oblongi, ancipites, pallidè virides.* Folia 2-3, *latè ensiformia, subobtusa, ferè bipedalia,* $1\frac{1}{2}$ *poll. longa, pallidè viridia.* Scapus *ferè à basi paniculatus, terminalis, 3 pedes longus, divaricatim ramosus.* Flores *odoratissimi.* Sepala *et* petala *sordidè lutea, versus apicem fusco-purpurea.* Labellum *sordidè luteum, lineis quibusdum brevibus in discum.*

For this noble plant we are indebted to our indefatigable correspondent Richard Harrison, Esq. of Liverpool, who favoured us with a splendid specimen, along with a sketch by Mrs. Arnold Harrison, in June last. It has so much the appearance of Oncidium luridum, that we at first sight took it for a new species of that genus: it is, however, a genuine Epidendrum, nearly allied to E. *odoratissimum*, with which it agrees in its delicious and powerful fragrance: when the specimen was unpacked, the whole air of the room became perfumed as if a basket full of violets and roses had been scattered over the floor.

No doubt it is a native of some part of South America, but of what country in particular is uncertain. Mr. Harrison informs us, that he received it from Mr. Cattley, who knew nothing of its history. It has been many years in the country, and is in almost every collection; but it never flowered before. Generally the plants one sees are so weak as to resemble the very different East Indian species called Liparis longipes; and the history of Mr. Harrison's having at last succeeded in flowering it, is no doubt his superior mode of cultivation. The leaf we received with the specimen was near two feet long; and the panicle itself fully three feet from one extremity to the other.

We think it is one of the most interesting species we have yet figured; for it is not only handsome, singular, and fragrant, but its flowers are very durable, remaining in perfection for at least a fortnight.

J. L.

---

* See fol. 1415.

1625

LEPTÓTES* bícolor.

Two-coloured Leptotes.

GYNANDRIA *MONANDRIA.*

Nat. ord. Orchideæ. § Epidendreæ. (*Introduction to the natural system of Botany, p.* 262.)
LEPTOTES.—Sepala et *petala* linearia, subæqualia, patentia. *Labellum* posticum, trilobum, cum columna parallelum, laciniis lateralibus nanis circa columnam convolutis, intermediâ oblongâ marginibus reflexis. *Columna* brevis, crassa, semiteres. *Anthera* .... *Pollinia* sex, incumbentia, quorum 2 superiora pyriformia obliqua compressa, 4 inferiora inæqualia tenuiora, anterioribus duplò minoribus.——Caules *breves, teretes, squamis vaginati.* Folia *teretia. Scapus* terminalis, simplex.

Leptotes bicolor.
Caules *unciales, inter muscos provenientes, teretes, squamis membranaceis acuminatis vaginati, quorum superiores caulem sæpiùs superant.* Folia *solitaria,* 3-4 *poll. longa, atroviridia, rigida, cylindracea, supra sulcata.* Racemus *triflorus, folio brevior, bracteis minimis*; pedicelli *erecti, pedunculo longiores.* Sepala *et* petala *candida, æqualia, linearia, patentia.* Labelli *laciniæ laterales, candidæ, lineari-ovatæ, obtusæ; intermedia sanguinea, apice et margine candida, oblongo-lanceolata, acuta, pone basin quasi fracta, et ibi fascia angusta, pallida, notata.* Columna *viridis.* Pollinia *lutea, filis nullis replicatis, nec glandulâ ullâ separabili adhærente.*

It is to Mrs. Arnold Harrison that we are obliged for this pretty little plant, which flowered in her stove in April last. It is a native of the Organ mountains of Brazil.

Its habit seems to be to grow among broken potsherds, decayed vegetable matter, and moss; treated in this way, a small portion, sent to us with the sketch from which our drawing was made, struck root, and has now become a little healthy plant.

---

* From λεπτὸς, slender; in allusion to the leaves.

Leptotes differs from Tetramicra,* next to which it should be placed in any system, firstly, in the singular structure of its pollen-masses; secondly, in the dwarf character of the lateral lobes of the labellum; and thirdly, in the want of a spur adhering to the ovarium. It is also nearly related to Brasavola, from which it is known by its pollen-masses, and by its 3-lobed labellum.

J. L.

* This genus, founded on the *Cymbidium rigidum* of Willdenow, is the *Brasavola elegans* of the Botanical Magazine, t. 3098; a plant which we believe has not yet been introduced.

1627

CYRTOCHÍLUM* flavéscens.

Straw-coloured Cyrtochilum.

GYNANDRIA *MONANDRIA.*

Nat. ord. Orchideæ *Juss.* § Vandeæ *Lindl.* (*Introduction to the natural system of Botany, p.* 262.)
CYRTOCHILUM Humb. et Kunth.—*Perianthium* explanatum. *Sepala* libera, lateralia unguiculata. *Petala* paulò minora. *Labellum* ecalcaratum, indivisum, ungue tuberculato cum basi columnæ continuo. *Columna* brevis, alata. *Anthera* bilocularis. *Pollinia* 2, caudiculâ filiformi, glandulâ minutâ. —— Herbæ *epiphytæ, v. terrestres, pseudobulbosæ.* Folia *coriacea.* Scapi *radicales, paniculati.* Flores *speciosi.* Genera et sp. of Orchideous plants, p. 210.

C. *flavescens*; foliis lineari-ensiformibus geminatis scapo æqualibus, scapo compresso arctissimè vaginato, racemo multifloro, bracteis glumaceis (flavescentibus) floribus subæqualibus canaliculatis, labello sessili ovato-lanceolato repando crispo basi pubescente inappendiculato.

Rhizoma *epigæa, teres, tortuosa, repens.* Folia primordialia *brevia, lineari-oblonga, è medio pseudobulbum, oblongum, angustum, diphyllum promentia;* perfecta *lineari-ensiformia, obtusa, apiculata, pallidè viridia, patula, racemi longitudine.* Vaginæ *scapi arctissimè imbricatæ, acuminatæ, carinatæ, flavo-virides.* Bracteæ *ultra* 2 *poll. longæ, glumaceæ, stramineæ, canaliculatæ, acuminatæ, floribus parùm breviores.* Sepala *et* petala *flava, lineari-lanceolata, acuminata.* Labellum *sessile, flavum, sanguineo-maculatum, basi pilis crebris loco tuberculorum munitum.*

A native of Mexico, whence it was imported by Mr. Tate about three years ago.

It flowered for the first time in June last in the collection of Richard Harrison, Esq. of Aighburgh, to whom we are indebted for a specimen, along with a sketch from the pencil of Mrs. Arnold Harrison.

It is interesting not only for its beauty, but also as being the first species of the genus which has yet blossomed in Europe. Like the other kinds, its flowers turn orange-colour in drying.

A tender stove plant, growing upon a branch of a tree, or in a pot among decayed vegetable matter and potsherds.

J. L.

---

* From κυρτὸς, convex, and χεῖλος, a lip; in allusion to the form of the labellum.

1646

CALÁNTHE* densiflóra.

Clustered Calanthe.

GYNANDRIA *MONANDRIA.*

Nat. ord. ORCHIDEÆ *Juss.* (*Introduction to the natural system of Botany, p.* 262.)
CALANTHE.—Suprà, vol. 9. *fol.* 720.

C. *densiflora;* scapo foliis multò breviore, squamis laxis ventricosis, racemis densis multifloris corymbosis, labelli cum columnâ parùm connati lobo medio cuneato bilobo, disco bilamellato, calcare longo recto pendulo apice clavato.
C. densiflora. *Lindl. gen. et sp. Orch. p.* 250.
Rhizoma *subterraneum, brevissimum, pullulationibus junioribus duris rigidis cornutis sanguineo pallido punctatis.* Folia *oblongo-lanceolata, plicata.* Scapus *semipedalis et ultrà, squamis* 2-3 *ventricosis sanguineo-punctatis laxe vaginatus.* Flores *pallidè lutei, in corymbum densum multiflorum congesti.* Sepala *et* petala *conniventia, æqualia.* Labellum *cum columnâ parùm connatum, trilobum; lobis lateralibus ascendentibus rotundatis, intermedio cuneato rotundato bilobo basi lamellis* 2 *cornutis brevibus ornato;* calcar *curvum, pendulum, apice clavatum.*

A native of the mountains of Sylhet, whence it was obtained by Dr. Wallich. The specimen from which our drawing was taken was communicated by W. W. Salmon, Esq., in October last, and was said to have been procured from Ava. When we first saw it, the flowers were drooping and injured, and we supposed it might prove a new species; but a fine flower-spike subsequently given us by Messrs. Loddiges, has satisfied us that it is *Calanthe densiflora,* of which we had previously examined nothing better than a few badly-dried flowers.

A terrestrial species, growing very freely in loam and decayed vegetable matter in a damp stove; propagated by division of the crown of the root. J. L.

* Formed from two Greek words, signifying literally " Pretty-blossom."

1651

ONCÍDIUM* altíssimum.

Lofty Oncidium.

GYNANDRIA *MONANDRIA.*

Nat. ord. ORCHIDEÆ *Juss.* (*Introduction to the natural system of Botany, p.* 262.)
ONCIDIUM.—Suprà, vol. 13. *fol.* 1050.

O. *altissimum;* pseudobulbis oblongis compressis, foliis ensiformibus rigidis carinatis carnosis scapo (6-pedali) erecto ramoso multò brevioribus, sepalis petalisque labelli longitudine lineari-lanceolatis undulatis, labello cordato utrinque introrsùm arcuato apice dilatato subreniformi emarginato, cristâ enneadactylâ, columnæ alis abbreviatis. *Lindl. gen. et sp. Orch. p.* 200.
α. alis columnæ rotundatis.
Epidendrum altissimum. *Jacq. amer.* 229. *t.* 141.
O. altissimum. *Swartz. fl. ind. occ.* 1481. *Willd. sp. pl.* 4. 112. *Hooker in bot. mag.* 2990.
β. alis columnæ truncatis.
O. Baueri. *Illustrations of the gen. and sp. Orch. t.* 7. Genera.
Epidendrum gigas. *Richard in herb. Vahl.*

One of the most gigantic of Epiphytes. The specimen from which our drawing was made received a medal from the Horticultural Society, and had a flowering stem from six to seven feet high: it was grown in the nursery of the late Mr. Colvill. It is described by Richard as frequently attaining the height of nine feet, *with leaves as long,* in its native woods in Guiana; and we have actually had a specimen from Mr. Harrison of Liverpool, the flower-stem of which exceeded ten feet in length.

The blossoms are beautifully marked with green, yellow, and scarlet, but are not of large size: the appearance they present may be judged of from our diminished figure at the back of the single branch drawn of its natural size.

* See fol. 1542.

Many of the Orchideous epiphytes are found to succeed best if tied to short pieces of the branches of trees with rugged bark; none succeed better upon that plan than the different species of Oncidium; and provided so large a plant as this, with its long heavy leaves, can be made fast to a branch, we have no doubt it will also be found to like that kind of treatment. The specimen, however, from which our drawing was taken had been grown in a pot in decayed vegetable mould.

It is a native of most of the tropical parts of America, from the eastern to the western shores.

J. L.

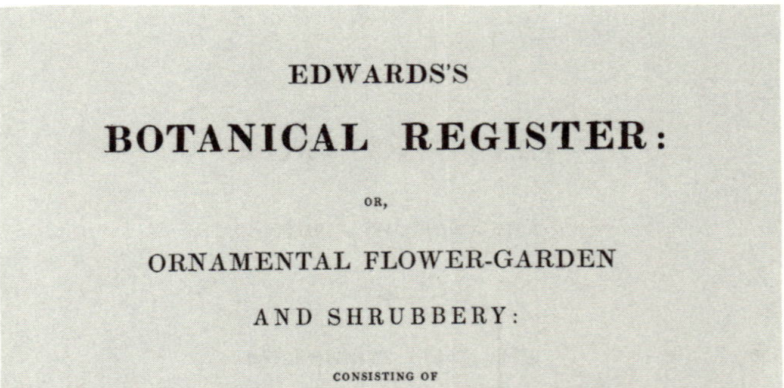
EDWARDS'S

BOTANICAL REGISTER:

OR,

ORNAMENTAL FLOWER-GARDEN

AND SHRUBBERY:

CONSISTING OF

COLOURED FIGURES OF PLANTS AND SHRUBS,

CULTIVATED IN BRITISH GARDENS;

ACCOMPANIED BY THEIR

History, Best Method of Treatment in Cultivation, Propagation, &c.

CONTINUED

BY JOHN LINDLEY, Ph. D. F.R.S. L.S. AND G.S.

PROFESSOR OF BOTANY IN THE UNIVERSITY OF LONDON,

&c. &c. &c.

New Series.

VOL. VII.

OR VOL. XX. OF THE ENTIRE WORK.

——viret semper——nec fronde caducâ
Carpitur.

LONDON:

JAMES RIDGWAY AND SONS, PICCADILLY.

M.DCCC.XXXV.

1653

BARTHOLÍNA* pectináta.

Pectinated Bartholina.

GYNANDRIA *MONANDRIA.*

Nat. ord. ORCHIDEÆ. § Ophrydeæ *Lindl.* (*Introduction to the natural system of Botany*, *p.* 262.)

BARTHOLINA R. Br.—*Perianthium* ringens. *Calyx* basi tubulosus, secundus, laciniis æqualibus. *Petala* sepalis parallela, falcata, infernè labello subconnata. *Labellum* calcaratum, explanatum, patens, 3-lobum: lobo medio multifido lateralibus trilobis: laciniis omnibus filiformibus. *Anthera* erecta, elongata, lobis parallelis cucullatis; *cuculli* distincti, basibus valvularum inflexis absconditi. *Pollinia* parva, caudiculis longissimis canaliculatis, glandulis intra cucullos inclusis.——Herba *pusilla, uniflôra, unifolia, hirsuta,* radicibus *tuberculiformibus.*

Bartholina pectinata. *R. Br. in Hort. Kew.* 5. 194. *Smith in Rees append.*
Bartholina Burmanniana. *Ker in Brande's journal*, 4. 204. *t.* 5. *fig.* 2.
Orchis pectinata. *Willd. sp. pl.* 4. 11.
Orchis Burmanniana. *Linn. sp. pl.* 1334. *Swartz in Web. et Mohr Archiv.* 1. 55. *t.* 3.
Arethusa ciliaris. *Linn. suppl.* 405.

Folium *reniforme, lobulis baseos imbricatis, pilosum ut omnes partes virides.* Caulis *vix palmaris, erectus, fuscus, teres, uniflorus.* Bractea *unica, ovata, cucullata,* ovarium *tegens.* Calyx *basi tubulosus, viridis, laciniis secundis, erectis, parallelis, lanceolato-linearibus.* Petala *falcata, acuminata, alba, violaceo colore tincta, supra antheram curva, basi cum labello levissimè connata, glabra.* Labellum *unciam latum, medio viride, basi calcare brevi conico pubescente cavo prædituum, laminâ explanatâ, circumscriptione orbiculari, trilobum; laciniis lateralibus trilobis, intermedio multifido, omnibus violaceis filiformibus.* Anthera *erecta, loculis elongatis, basi contiguis: suturis ob torsionem valvularum basi approximatis;* cucullis *distinctis, sub basibus valvularum exteriorum antheræ absconditis.* Pollinia *parva, aurantiaca; caudiculis longissimis fulvis linearibus canaliculatis; glandulis parvis in cucullis suis latentibus.*—Obs. *foramen, v. areola transparens muco repleta in cucullo utroque adest sub glandulis.* Stigma *areola madida, oblonga, ad basin columnæ, ubi calcar ab ovario separatur.*

* Dedicated by Dr. Brown to the memory of the great Danish anatomist and physiologist Thomas Bartholin, whose various writings relating to plants, in the old Copenhagen Transactions, entitle us to adorn the history of the science with his truly illustrious name.—*Smith.*

We believe this to be, with the exception of a very imperfect representation in Brande's Journal, the first figure from the live plant that has been published of this rare and curious species, which appears from the *Hortus Kewensis* to have been introduced so long ago as 1787, but which we never met with till we were favoured with it in August 1832 by the Messrs. Rollisson, of Tooting. It is a native of the Cape of Good Hope, where it was found by Thunberg on the sides of hills in Roode Sand, and near Cape Town, flowering in October, November, and December.

We presume that this, like all the Cape Orchideous plants, is incapable of being cultivated permanently by any means hitherto discovered; for the roots, although when first imported they flower, afterwards disappear. They should be planted in sandy loam, and kept in as light a greenhouse as possible; for it is probable that the reason of their disappearing is the want of light during their growing season in this country.

In the accompanying figures *a* is the plant of its natural size; *b*, a flower seen in front; *c*, the same viewed from behind; *d*, the same shewn in half profile, the lip being cut away, 1. is the spur; *e*, a front view of the petals and anther, 1. the spur, and 2. the anther: the two last figures are magnified.

1660

ONCÍDIUM* ciliátum.

Ciliated Oncidium.

GYNANDRIA *MONANDRIA.*

Nat. ord. ORCHIDEÆ *Juss.* § Vandeæ. (*Introduction to the natural system of Botany*, *p.* 262.)

ONCIDIUM.—*Suprà, vol.* 13. *fol.* 1050.

O. *ciliatum;* pseudobulbis ovatis compressis monophyllis, foliis complicatis lineari-oblongis obtusis scapo erecto flexuoso apice paucifloro pluries brevioribus, sepalis undulatis lineari-oblongis obtusis: anteriore bilobo, petalis obovatis crispis, labello æqualiter tripartito sepalis breviore: laciniis obovatis sinubus suis latissimis fimbriatis, crista 5-corni cornubus posticis divergentibus anticis collateralibus tuberculis quibusdam interjectis, columnæ alis ovatis acutis. *Lindl. Gen. et Sp. Orchid. Plants, p.* 200.

This plant is probably not uncommon in some parts of Brazil. It originally reached the Kew Garden, where it flowered in February 1819, at which time a sketch of it was made by Mr. Bauer; it was afterwards obtained by Mr. Cattley, who had it for several years. Mr. Knight's rich collection in the King's Road supplied the specimen from which our figure was made, in October last: about the same time Mr. Booth sent us a drawing and specimen from Sir Charles Lemon's Garden; and two or three patches of it were lately received by the Horticultural Society from Brazil.

It is very nearly related to Oncidium barbatum, from which it may, however, be certainly distinguished by all the divisions of the lip being equal, and by its very dwarf habit. In colour its flowers vary, sometimes being yellow spotted with red, and sometimes the brownish orange of the accompanying figure.

* See fol. 1542.

It will only grow in a hot damp stove; but succeeds best if tied to a piece of the branch of a tree. The following is Mr. Booth's account of the species:—

" This pretty little species of Oncidium has been cultivated for the last three years in the collection of Sir Charles Lemon, Bart., M.P., at Carclew, Cornwall, where it has annually, in November, produced its interesting and curiously-formed flowers. I do not know who has the merit of bringing it to this country, but believe it to have been received through some of the officers of his Majesty's Packet Establishment at Falmouth.

" *Pseudobulbs* ovate, compressed, one-leaved. *Leaves* slightly complicate at the base, oblong-obtuse, about 2 inches long and half an inch broad, with a little hook at the point. Their colour is a rich shining green, excepting where they are connected with the bulbs, which is brownish red. *Scape* nearly erect, small and round, from 4 to 6 inches high, and thickly covered with brownish spots; usually two-flowered. *Sepals* much undulated, brownish green, the two lower ones pointed, the three upper emarginate and rather shorter than the rest. *Labellum* bright yellow, and about the length of the upper sepals. The lower division is reniform and unguiculate; the other two are oblique, broad at their extremity, and somewhat crescent-shaped. The fleshy part in the centre of the flower is tuberculated, and beautifully marked with blood-red spots. The lower edge is finely ciliated. The column in front is of the same colour as the labellum; behind and at its base it is marked with deep red."

1666

* CYPRIPÉDIUM spectábile.

Large white Lady's Slipper.

GYNANDRIA *DIANDRIA.*

Nat. ord. CYPRIPEDIEÆ. *Lindl. Nixus Plantarum, p.* 22, *No.* 234.
CYPRIPEDIUM. Suprà, vol. 10, *fol.* 788.

C. *spectabile ;* sepalis subæqualibus oblongis obtusis, petalis lanceolatis planis longioribus, labello petalis longiore ventricoso subsulcato, stamine sterili cordato-ovato obtuso.
C. spectabile. *Swartz in act. holm.* 1800, *p.* 251. *Salisb. in act. Linn.* 1. 78. *t.* 3. *f.* 3. *Willd. sp. pl.* 4. 144.
C. album. *Ait. Kew. ed.* 1. 3. 303. *Bot. Mag. t.* 216.
C. canadense. *Mich. Amer* 2. 161.
C. reginæ. *Walt. Carol.* 222. *fide Pursh.*

A native of the low meadows and bogs of North America, particularly in the mountainous tracts from Canada to Carolina, flowering in May and June, according to Pursh. Our wild specimens were gathered in Canada by Mr. Gouldie.

This remarkable plant is the finest of the North American Cypripediums, and is not by any means uncommon in the gardens of this country, its roots being periodically imported. It has, however, resisted all attempts at propagating it, and seldom lives above a year or two after its arrival. Those who manage it the best treat it as a greenhouse plant, keeping it constantly under glass, in a moderately warm atmosphere, and very near the light until its leaves have withered, when it is removed to a dry shelf till its growing season returns. Except in the colour of its flower, it is very nearly the same as the Siberian *C. macranthos,* already figured at fol. 1534 of this work.

* See fol. 1534.

1667

* CATASÉTUM lúridum.

Lurid Catasetum.

GYNANDRIA *MONANDRIA.*

Nat. ord. ORCHIDEÆ. *Juss.* § Vandeæ *Lindl.* (*Introduction to the Natural System of Botany, p.* 262).
CATASETUM. Supra, vol. 10, *fol.* 840.

C. *luridum ;* caulibus defoliatis angustis sulcatis, perianthio subgloboso parum maculato, sepalis petalisque oblongis apice rotundatis, labello cucullato carnoso mutico apice paulò producto truncato, racemo brevi nutante.
C. luridum. *Lindl. Gen. & Sp. Orch. p.* 156.
Anguloa lurida. *Link in Verhandl. des Vereins. z. bef. des Gartenb. in dem Königl. Preuss. Staat.* 1. *p.* 289. *t.* 6.

Caules *deflorati angusti, cinerei, sulcati,* 3-5 *uncias longi.* Folia *plicata, subundulata, acuminata.* Racemus *vix pedalis, apice nutans, pauciflorus ;* bracteis *brevibus canaliculatis obtusis.* Perianthium *globosum ;* sepalis petalisque *herbaceis oblongis apice rotundatis maculis quibusdam pallidis notatis.* Labellum *carnosum, cucullatum, lutescens, anticè truncatum subemarginatum, marginibus atro-purpureis maculatis, nec ciliatis, nec dentatis, nec inflexis.* Columna *bicirrhosa generis.*

After having been imported into Prussia some years since, and been apparently lost, this rare plant has suddenly made its appearance almost at the very same time in several collections. The first notice we had of it was from Messrs. Loddiges, with whom it flowered in September, 1833. Almost immediately afterwards, it was communicated to us by our amiable and much lamented correspondent the late Mrs. Arnold Harrison, whose truly splendid collection of Orchideous epiphytes has lately been added to the stock of Mr. Knight, of the King's Road ; it was obtained by that lady from Bahia. We had scarcely secured a drawing of Mrs. Harrison's specimen, before a sketch and specimens reached us from Mr. Bateman ; and before many more days had elapsed it burst into blossom in the stove of the Horticultural Society.

* A name, the meaning of which is unexplained.

It is a native of the woods not only of Bahia, but probably of the greater part of Brazil, the Prussian specimens and those of the Horticultural Society having been imported from Rio Janeiro. Like all the species with similar habits, it grows freely in decayed vegetable matter, mixed with a little pure loam, among quantities of potsherds, and it probably will soon become common. Although it cannot be compared for beauty with Catasetum tridentatum, it is nevertheless a very interesting species ; the spots on the margin of the lip are of the deepest and richest ruddy-brown ; while the horns of the column may be compared to the forelegs of some spider lurking in the bosom of the flower to seize upon the victims that may enter it.

1671

* LÍPARIS guineénsis.

S. Leone Liparis.

GYNANDRIA *MONANDRIA.*

Nat. ord. ORCHIDEÆ. *Juss.* § Malaxideæ *Lindl.* (*Introduction to the Natural System of Botany, p.* 262.)
LIPARIS. Supra, vol. 11, *fol.* 882.

L. *guineensis ;* foliis pluribus oblongis acutis plicatis racemo paucifloro brevioribus, scapo angulato, labello postico cuneato bilobo plano basi bituberculato, petalis sepaloque supremo linearibus patentibus, sepalis lateralibus subrotundo-ovatis labello brevioribus.

Pseudobulbi *ovati, nucis avellanæ magnitudine, exuviis membranaceis foliorum vestiti.* Scapus *semipedalis.* Flores *parvi virides sessiles, maculis* 2, *lutescentibus in apice labelli.* Ovarium *alatum, bracteis lineari-subulatis longius.*

A native of Sierra Leone, whence plants were brought in 1832 by Mr. Whitfield. Our drawing was made in September last in the garden of the Horticultural Society.

It requires to be kept in a damp stove while growing, but to be removed into a cooler and dry place as soon as its leaves decay.

The *Liparis Löselii* of this country is nearly related to it, but is abundantly distinct in the leaves and the form of the labellum.

* See fol. 1175.

1681

* BLÉTIA grácilis.

Slender Bletia.

GYNANDRIA *MONANDRIA.*

Nat. ord. ORCHIDEÆ, § EPIDENDREÆ *Lindl.* (*Introduction to the Natural System of Botany, p.* 262.)
BLETIA. Supra, vol. 17, *fol.* 1401.

B. *gracilis;* sepalis petalisque subæqualibus lanceolatis acuminatis, labelli trilobi lobis lateralibus nanis rotundatis intermedio transverso emarginato undulato: lamella solitaria in medio, foliis plicatis oblongo-lanceolatis acutis subtus discoloribus.

B. gracilis; *Lodd. Bot. Cab.* 1681.

Pseudobulbi *ovati, aggregati, purpurei.* Folia *plicata, membranacea, latitudine variabilia, sæpius unico tantum perfecto cuique pseudobulbo.* Scapus *terminalis, gracilis, purpuratus,* 1½ *pedem longus, erectus, calami corvini crassitudine, parcissime squamatus, apice racemum brevem* 3-4-*florum gerens.* Flores *subringentes, sepalis petalisque sordidè luteis, roseo versus apices suffusis; labello basi roseo venis picto apice luteo.*

A native of Mexico, whence it was introduced by the Messrs. Loddiges, of Hackney. Our drawing was communicated by James Bateman, Esq. from his collection at Knypersley, in July, 1833; we also received it in flower at nearly the same time from the Hon. and Rev. William Herbert.

It probably requires the treatment of *Bletia verecunda, acutipetala,* and *Shepherdi,* to all which it is nearly related; and it ought, when at rest, to be kept where it is in no degree exposed to circumstances that are favourable to its growth. Dryness, and a cool place at the back of a

* See fol. 1401.

green-house, or even a common pit, protected from cold and wet, would probably suit it until the season for its growth returns, at which time it should be removed to a hot damp-stove, among tropical epiphytes, to remain there till its leaves decay; when that happens, it should once more be restored to a resting-house.

The species is very distinct from any previously described.

1687

* GEODÓRUM fucátum.

Painted Geodorum.

GYNANDRIA *MONANDRIA.*

Nat. ord. ORCHIDEÆ. *Juss.* § VANDEÆ, *Lindl.* (*Introduction to the Natural System of Botany, p.* 262.)

GEODORUM. Supra, vol. 8, *fol.* 675.

G. *fucatum;* scapo florifero foliis duplò breviore, spicâ pendulâ congestâ, labello gibboso ovato emarginato integerrimo: lineis duabus callosis elevatis.

Folia *e tubere subterraneo annulatim cicatrizato erumpentia, oblongo-lanceolata, acuta, plicata, pedem longa, scapo florifero duplo longiora, frugifero subæqualia.* Scapus *radicalis, erectus, vaginatus, apice recurvus, et spicam ideo pendulam, partibus omnibus inversis, compactam, brevem gerens.* Bracteæ *lineares, acutæ, ovarii longitudine.* Flores *subcampanulati, magnitudine et facie* G. dilatati. Sepala *lineari-oblonga, acuta, rosea, apice paululum recurva petalis paulò latioribus omninò conformia et parallela.* Labellum *ob spicam inversam, anticum, revera posticum, ovatum, concavum, subtus gibbosum, cum columna parallelum et continuum nec articulatum, integerrimum, emarginatum; roseum, venis lateralibus intensioribus pictum; lineis duabus latis, elevatis, parallelis, contiguis, ochraceo-sanguineis, in medio.*

A single plant of this new species of Geodorum sent to the Horticultural Society from Ceylon by Mr. Watson, in 1832, flowered in the Chiswick garden last July. It resembles *G. dilatatum* figured in tab. 675 of this work; but has rather smaller flowers, and a very different labellum.

It thrives in a hot, damp stove, but requires to be rested after its leaves have withered.

Fig. 1. represents the labellum seen from above; fig. 2. the column and labellum after the other parts of the flower have been removed.

* From γῆ the earth, and δῶρον a gift.

1695

* DENDRÓBIUM aggregátum.

Clustered Dendrobium.

GYNANDRIA *MONANDRIA.*

Nat. ord. ORCHIDEÆ. §. Malaxideæ *Lindl.* (*Introduction to the Natural System of Botany, p.* 262.)

DENDROBIUM. Supra, vol. 7, *fol.* 548.

D. *aggregatum;* pseudobulbis monophyllis fusiformibus cinereis corrugatis, folio oblongo coriaceo obtuso racemo laterali cernuo multifloro duplò breviore, petalis ovatis sepalis duplò latioribus, labello integerrimo latiore quam longo medio transversè plicato basi pubescente.

D. aggregatum *Roxb. Fl. Ind.* 3. 477. *Lindl. in Wallich. cat.* 7411.

Pseudobulbi *fusiformes, cæspitosi,* 2-*pollicares, squamis cinereis pellucidis vestiti.* Folia 3-*pollicaria, obscurè* 5-*striata.* Racemi *e latere pseudobulbi provenientes,* 9 *pollices et ultra longi, laxi, cernui, v. nutantes,* 10-15-*flori.* Bracteæ *minimæ, ovatæ, purpurascentes.* Flores *pallidi, subcrocei, pedicellis gracilibus flexuosis inserti.* Sepala *explanata, angustè ovata, anterioribus valde obliquis in pseudocalcar obtusum conniventibus, more generis.* Petala *ovata, sepalis duplò latiora.* Labellum *versus basin intensius croceum, tali modo dilatatum et plicatum ut marginibus suis posterioribus petalorum dimidiam inferiorem imbricat; cum columna articulatum.* Columna *parva, albida, basi processu carnoso squamiformi deflexo aucta; clinandrio obtusè tridentato.*

Received, according to Dr. Roxburgh, into the Botanic Garden, Calcutta, from Mr. Pierard, who found it growing on the trunk of *Lagerstræmia Reginæ,* on the northern border of Arracan, and observed it in the woods exclusively on that tree; it was, however, found to thrive on the Mango tree in the Botanic Garden. It is also a native of the banks of the Chappadong river in the Gulf of Martaban, whence it was brought some years since by Dr. Wallich. It belongs to a curious section of the genus, all the species of which have their stems shortened into the pseudobulbous form of Bolbophyllum, and thus form a natural transition to that genus.

---

* See fol. 1239.

It appears to require as much heat and moisture as any of the Indian species, a circumstance which is explained by its inhabiting, when wild, the damp and sultry woods of Martaban. We believe it was originally distributed by the Horticultural Society; but it has hitherto flowered only in the collections of Mr. Harrison and Mr. Bateman, from both of whom we have received specimens. It blossoms in March and April.

The lip of this species is remarkable for being much broader than long, and plaited in the middle in such a manner as to form a projection round the hollowed part which lies against the column; it is moreover more distinctly articulated with the column than is usual in the genus. Its column has a fleshy reflexed appendix near its base.

In the accompanying figure 1 represents the lip, viewed from above: and 2 a section of the column and part of the lip, for the purpose of shewing at *a* the bed of the anther, at *b* the scale near the base of the column, at *c* the hairiness of the cavity of the lip, and at *d* the point of articulation between the lip and the column.

1699

* ONCÍDIUM ampliátum.

Broad-lipped Oncidium.

GYNANDRIA *MONANDRIA.*

Nat. ord. ORCHIDEÆ. *Juss.* (*Introduction to the Natural System of Botany, p.* 262.)

ONCIDIUM. Supra, vol. 13. *fol.* 1050.

O. *ampliatum;* sepalis omnibus liberis, labello bilobo subrotundo transverso: laciniis lateralibus brevissimis, callo baseos 5-lobo: lobis lateralibus patentissimis planis truncatis intermediis teretibus centrali compresso, alis columnæ cuneatis dentatis reflexis, pseudo-bulbis subrotundis compressis, foliis planis oblongo-lanceolatis, scapo erecto apice ramoso.

O. ampliatum. *Lindl. in Hook. Bot. misc. v. 3. p. Gen. et sp. orch. part 3. p.* 202.

Folia *et* pseudobulbi *facie omnino* O. papilionis. Scapus *ascendens, radicalis,* $1\frac{1}{2}$*-2-pedalis, apice ramosus.* Flores *lutei, labelli dorso albo.*

First found in central America by Mr. Cuming, and afterwards procured in a living state by Richard Harrison, Esq. from whom the beautiful specimen now figured was received in March last.

Peculiar as are its flowers, and distinct as the species is in most respects, it is curious that its leaves and pseudo-bulbs should be so like those of *O. Papilio,* that we have known the latter to be mistaken for it.

Like all the rest of its genus, it requires the hot damp atmosphere of a stove, in which, if we may judge by Mr. Harrison's specimens, it finds itself perfectly at home. We have not yet heard of it in any other collection.

It is well known that the most considerable part of the Epiphytal Orchideæ is found in the greatest vigour in damp sultry woods of tropical countries; and accordingly we endeavour in our artificial cultivation, to form an atmosphere for them as nearly as possible that which they would naturally breathe in such stations. That this is attended with very great success is obvious from such plants as the one now figured, and from the numerous splendid specimens which are from time to time appearing in the collections of Earl Fitzwilliam, Lord Grey of Groby, the Messrs. Harrison, Bateman, Huntley, Loddiges, and Knight, and the Horticultural Society.

But it is sufficiently evident that although this kind of treatment is admirably suited to a considerable number, there are others which grow most unwillingly, or scarcely survive, under such circumstances. For instance, *Dendrobium speciosum* languishes in situations where the *Stanhopeas* are in their greatest splendour; and the Chinese *Bletias* almost perish by the side of *Eulophia* and *Zygopetalum.* This arises from the great difference in their respective constitutions, which are each adapted to distinct conditions of life, and our failure arises from our mistaking a general principle for an universal law. If a great majority of Epiphytal Orchideæ swarms in damp tropical forests, there is a considerable minority which lives in an entirely different climate, of which a few examples will not be without instruction. Thus in the genus Oncidium itself, where almost all the species are of tropical habits, *O. nubigenum* is only found on the cool mountains of Peru, at the height of 14,000 feet; it will therefore require a treatment altogether distinct from that of the mass of the genus. *Dendrobium moniliforme* and *catenatum,* again, occur only in Japan, as far north as 37° or 38°, or the parallel of Lisbon, and are periodically subject to a very low temperature.

* See fol. 1542.

But the most remarkable instances of a disposition on the part of some Orchideous Epiphytes to depart from the ordinary habits of the tribe are found in Australia and its dependency New Zealand. In some extremely valuable observations upon the geographical distribution of the Orchideous plants of New Holland, which have been placed in our hands by Mr. Allan Cunningham, we find a passage which bears so directly upon this subject, that we cannot do better than quote it entire.

"There are two, if not three plants of this family," says this enterprising and scientific traveller, "that grow on trees or rocks in New South Wales, whose natural constitution should, in cultivating them, form exceptions to the uniformly adopted mode of treatment of Epiphytes generally in our English stoves; namely, that in which high temperature and considerable humidity are employed. These are *Dend. æmulum, Br.,* an Epiphyte uniformly found upon the rugged trunk of Eucalyptus resinifera or Ironbark, in the open very dry forest grounds of the older colony at Port Jackson;—*Cymbidium canaliculatum, Br.,* which of late years has been observed beyond the Tropic, both at Moreton Bay and still farther to the southward at Hunter's River, growing upon the principal limbs of several of the Eucalypti in the dry open shadeless forest. These two Epiphytes flourish most luxuriantly in an extremely dry atmosphere, and flower usually in the summer season in their native wilds, the high temperature of which is oftentimes greatly increased by the blighting hot winds, which not unfrequently prevail at that period from the north-west. The third is *Dendrobium undulatum* of Mr. Brown, a handsome species, originally discovered by Sir Joseph Banks at Bustard Bay, and which has been lately found on barren hills, naturally clear of timber, upon the banks of the Brisbane River at Moreton Bay, where the plant forms tufts on bare rocks exposed to the full heat of the sun, which during nine months of the year is very considerable on that part of the coast. These species were some years since received alive at Kew, from New South Wales; and with them was communicated, as a guide to their culture, a note of the particular situations, with regard to exposure to drought, &c. which they naturally occupy and delight in, in their native wilds. These particulars were, however, in all probability wholly unheeded in the King's Gardens—the plants were associated with other Epiphytes of this vast and variable family, from Equinoctial America and the West India Islands, desiring a humid air with warmth to luxuriate in, amongst whom the Australians soon shewed sickness, in consequence of the excess of moisture to which they were constantly subjected; and eventually dying, were not only lost to Kew, but I may add to Europe! Had they been placed in the dry stove among Cacti, Stapeliæ, &c. with but an occasional light sprinkle of water afforded them, they would have fared better! D. æmulum was, I find, notwithstanding, induced to flower, and thus shewed by its delicate blossoms that it was well worthy of better treatment: and might afterwards have been retained, had the notes communicated with the plant from the Colony, and its *look* and constitution, so to speak, been at all consulted. I would just observe, in this place, that it is to be greatly regretted, that collectors of these beautiful vegetables in foreign countries, are not more careful to note and communicate home with the collections they form, the particular localities of the species, which would be of great use to the experienced cultivator; inasmuch as it would enable him to treat them in a way, as nearly accordant with their habits in their respective native countries, as would secure their lives in the Garden, and probably induce them to flower when fully established in their new situations."

To these instances of Orchideous Epiphytes may be added two others, which are worthy of still more attention than those just cited. One is the beautiful little *Gunnia australis,* which has much the aspect of *Chiloschista usneoides* found in the jungle of Nipal; it grows on the branches of shrubs in Emu Bay, in Van Diemen's Land, in about 41° S. Lat. and 146° E. Long. *Earina mucronata* is the other example. This plant, although occurring as far to the Northward as 35° S. Lat. in humid forests at the Bay of Islands, in New Zealand, exists also in abundance in the "very (permanently) damp woods which clothe the shores of Dusky Bay, (Lat. 45° 45′ S.) on the western side of the Larger or Middle Island of New Zealand," where it was originally observed by Forster, in Cook's Second Voyage, and where it has been since met with by Mr. Cunningham, whose words we have quoted.

Considering the lower rate of temperature which prevails in the Southern hemisphere, as compared with that of the Northern in corresponding latitudes, the station of *Earina* in New Zealand is not naturally different from the damper parts of the south-west coast of Ireland.

These remarks will we trust suffice to cause a greater degree of attention to be paid to the differences of constitution of particular species of Orchideous Epiphytes; for although we have only cited extreme cases, we may be assured that minor peculiarities, which it is not less important to study, exist in abundance.

One of the plants just mentioned being imperfectly known to Botanists, and the other not at all, we subjoin the following brief account of them for the use of our systematic friends.

Gunnia.—Perianthium ringens. Sepala herbacea, lateralia postica, subfalcata, erecta, ungui labelli lineari longe producto adnata. Petala herbacea, sublanceolata, obtusa, ab ungue columnæ omninò libera, cum sepalo altero distincto pendulo parallela. Labellum carnosum, ungue longè producto lineari erecto, cum basi columnæ continuum, bilobum, anticè cornutum, mucrone inflexo, disco tuberculatum. Columna nana, semiteres, aptera. Pollinia 4, in paribus globosis coadunata; retinaculo lineari. Rostellum bifidum.——Herba epiphyta; radicibus longis tortuosis supra fruticum ramos repentibus. Folia lanceolata, falcata, disticha, basi, articulata. Racemus simplex, strictus, foliorum longitudine.

Sp. 1. Gunnia australis.

Hab. in *Insula Van Diemen,* in sinu Emu, *Backhouse. (hab. s. sp. comm. cel. Gunn.)*

Tota planta vix 2 pollices excedens. Sepala et petala luteo-viridia. Labellum verosimiliter album, v. leviter rubescens, lobis lateralibus oblongis obtusis; tuberculis 4, luteis, quorum 2 exteriores majores.

We have named this most curious plant after our liberal correspondent, Ronald L. Gunn, Esq. who is now examining the vegetation of Van Diemen's Land, with equal skill and assiduity. The genus is nearly related to *Chiloschista.*

Earina. (*εαρινος.*)—Sepala erecta, æqualia, acuta, membranacea, carinata. Petala carnosa, obtusata. Labellum carnosum, posticum, cucullatum, trilobum, disco nudo, cum columna continuum et subparallelum. Columna teres, nana, stigmatis obliqui labio inferiore prominulo. Clinandrium proclive. Anthera bilocularis. Pollinia 4, per paria cohærentia, collateralia.——Herba caulescens, rhizomate articulato, repente. Folia linearia, disticha, vaginantia. Flores parvi, paniculati, bracteis cartilagineis, striatis, cucullatis.

Sp. 1. Earina mucronata.

Epidendrum autumnale. *Forst. prodr. n.* 319.

Cymbidium autumnale. *Swartz. nov. act. ups.* 6. 72. *Willd. sp. pl.* 3. 98.

Hab. in *Nova Zelandia. (hab. s. sp. comm. cel. Cunningham.)*

Rhizomata inter muscos mortuos repentia, articulata. Caules ascendentes, palmares, pedalesque, pennæ corvinæ crassitudine, maculati, basi vestigiis laceris vaginarum vestiti. Folia lineari-ensiformia, apice leviter obliqua, mucronulata. Flores e bracteis rigidis cucullatis mucronatis erumpentes, parvi, ovario recto costato bracteolarum longitudine. Sepala omnia libera, angusta, mucronata, carinulata. Petala latiora, carnosa, obtusiora, æquilonga. Labellum posticum, carnosum, cucullatum, inappendiculatum, basi liberum, cum columna continuum et parallelum, trilobum, nudum; lobo intermedio bipartito laciniis oblongis crenulatis mucrone intermedio. Columna suberecta, nana, antica, teres, clinandrio subcucullato membranaceo-marginato; stigmate excavato, rostello obtuso. Anthera ovata, 2-locularis. Pollinia 4, per paria cohærentia, cereacea, collateralia, materie viscidâ rostello adhæreutia.

For fine specimens of this we are indebted to Mr. Cunningham, who observed it "growing commonly in moist woods upon the shores of the Bay of Islands, New Zealand, on mossy rocky banks, and on the limbs of trees, flowering in September and October, which in New

Zealand is the season of spring." From the latter circumstance we have contrived the generic name.

This genus belongs to Malaxideæ, and is related to *Cœlogyne* and *Dilochia*, of the latter of which in particular it has much the habit, only on a smaller scale. From the former its wingless column, and from the latter the number of its pollen masses, sufficiently distinguish it, independently of other points of difference.

1701

* ÓRCHIS foliósa.

Leafy-spiked Orchis.

GYNANDRIA *MONANDRIA.*

Nat. ord. ORCHIDEÆ. *Juss.* (*Introduction to the Natural System of Botany, p.* 262.)
ORCHIS. Supra, vol. 3, *fol.* 202.

§ 1. MASCULÆ. *Sepalis lateralibus reflexis v. patentibus.* Lindl. gen. et sp. orch. pars. 4. ined.

O. *foliosa* ; foliis oblongo-lanceolatis acuminatis laxè vaginantibus, spicâ oblongâ multiflorâ, sepalis ovatis acutis, labello latiore quam longo obsolete trilobo plano : laciniis lateralibus emarginatis intermediâ acutâ multò majoribus, calcare pendulo cornuto labello duplo breviore, bracteis herbaceis acuminatis flore sæpè longioribus, tuberculis palmatis.
O. foliosa. *Soland. MSS. Lowe Primit. Fl. Mader. p.* 13.
Omnino *O. latifoliam* refert ; sed omnibus partibus major est, labello plano manifeste trilobo nec rhomboideo, calcare breviore graciliore, caule elatiore.

A fine species of Orchis, native of woods and copses in Madeira; very much like the European *O. latifolia*, from which it differs in being larger in all its parts, having a distinctly three-lobed flat lip, instead of a lozenge-shaped convex one, a shorter and more slender spur, and a taller stem.

We were favoured with our specimens by Messrs. Young and Penny of Milford, near Godalming, in whose collection, so rich in Canary plants, it has for some time been cultivated. It succeeds, we are informed, extremely well, either in well-drained pots, or a turf pit, in a soil composed of the turfy portions of heath mould, with a mixture of moss and sand.

Like many others, this species varies with spotted and spotless leaves. In this country it flowers in May.

* See fol. 1155.

The genus Orchis is usually divided into sections, characterized by the nature of the tubercles of the roots, a distinction which obviously is inconvenient, and which is also incompatible with a natural distribution of the species. A better character may be obtained from the calyx, which in some converges into a sort of casque or helmet (§. MILITARES), and in others is spread open or reflexed (§. MASCULÆ.) The sections of Orchis thus obtained are so extremely distinct that it may be doubted whether, upon the principles on which other genera are constituted, they should not be considered each a genus. But so much inconvenience would attend an alteration in the names of a great many common species, that less disadvantage will perhaps arise from retaining the *militares* as a mere form of Orchis ; although the inconvenience might be diminished by the creation of such a name as *Herorchis.*

As some time must elapse before the appearance of the fourth part of the genera and species of Orchideæ, it may be interesting to Botanists to know in what way we propose to distribute the genera of Ophrydeæ. We therefore avail ourselves of the present opportunity to explain our present views upon the subject. The following list will shew, firstly, what the general arrangement of the tribe will be, and secondly, what new genera we propose to establish.

* *Antheræ loculi contigui paralleli, basi nullo modo elongati v. divaricati.*

1. Orchis.—2. Anacamptis.—3. Nigritella.—4. Glossaspis.—5. Gymnadenia.—6. Scopularia *nob.*—7. Aceras, as limited by Dr. Brown.—8. Serapias.—9. Ophrys.

** *Antheræ loculi sejuncti, basi divergentes, sæpè elongati.*

10. Monotris *nob.*—11. Holothrix.—12. Bartholina.—13. Disa.—14. Repandra *nob.* (=Disa cornuta, &c.)—15. Corycium.—16. Dryopeia.—17. Disperis.—18. Pterygodium.—19. Satyrium.—20. Bonatea.—21. Habenaria.—22. Diplomeris Don (=Diplochilus *Lindl.*)—23. Bilabrella *nob.*—24. Cynorchis.—25. Tryphia *nob.* (=Orchis secunda Thunb.)—26. Bicornella *nob.*—27. Cœloglossum *nob.*—28. Peristylus *Blume* (the calcarate Herminia.)—29. Herminium (as limited by Dr. Brown.)—30. Aopla *nob.*—31. Platanthera, (=Mecosa *Blume*, and including most of the North American Habenarias.)—32. Perularia *nob.* (=O. fuscescens *L.*)

The following are the essential characters of the new genera in the foregoing list :

6. SCOPULARIA.—Calyx membranaceus, connivens, sepalis lateralibus minoribus. Petala membranacea, lacero-multifida, circa labellum convoluta. Labellum erectum, convolutum, membranaceum calcaratum apice lacero-multifidum. Anthera libera, erecta ; loculis approximatis, parallelis, basi dilatatis ascendentibus cucullum magnum inæquilateralem glandulas tegentem formantibus. Glandulæ polliniorum sub basi dilatatâ loculorum antheræ absconditæ, cucullo nullo (quantum videre potui) inclusæ.—Radices testiculatæ. Folia radicalia membranacea. Caulis aphyllus. Spica elongata subsecunda.—Sp. 1. *S. Burchellii.* C. B. S.

10. MONOTRIS.—Sepala membranacea, postice connata, lateralibus anticè distinctis, labello petalisque duplo minora. Petala carnosa, acuminata. Labellum liberum, apice carnosum, trifidum, cucullatum, basi cornutum. Anthera libera, erecta, loculis basi divergentibus : valvulis exterioribus dilatatis, incurvis, glandulam tegentibus.—Radices testiculatæ. Folium minimum radicale. Caulis retrorsum hispidus. Flores parvi secundi. Bracteæ et sepalorum margines hispida.—Sp. 1. *Monotris secunda.* C. B. S.

23. BILABRELLA.—Sepala patentia, carnosa, petalis multò minora; supremum 3 partitum. Petala membranacea, reflexa, maxima. Labellum carnosum, 3-partitum, calcaratum. Anthera loculis basi elongatis, solutis, divergentibus, distantibus, ascendentibus, glandulis nudis. Rostellum lineari-lanceolatum, recurvum, antheræ longitudine, Processus carnosi magni ascendentes.——Radices . . . . Caulis foliosus. Folia angusta membranacea. Racemus elongatus Habenariæ facie.—Sp. 1. *B. falcicornis.* C. B. S.

26. BICORNELLA.—Sepala inæqualia ; lateralia majora, labello extus obliquè adnata. Petala sepalo supremo in galeam convexam agglutinata. Labellum calcaratum, indivisum, angustum, canaliculatum, columnæ nanæ adnatum. Anthera ferè horizontalis, lobis ascendentibus, basi elongatis, lobis lateralibus rostelli adnatis, extus appendice linguæformi (stamine sterili) auctis. Rostelli lobus medius ovatus, planus, abbreviatus. Pollinia minima ; retinaculo longo lineari basi antheræ elongatâ tecto. — Herbæ (Mascarenenses) caule folioso v. subfolioso. Flores parvi, spicati.

27. —CŒLOGLOSSUM.—Sepala conniventia, æqualia, libera. Petala conformia, ungui labelli adnata. Labellum carnosum, unguiculatum, calcaratum, tripartitum, disco sæpius tuberculatum ; ungue valde carnoso, concavo, ascendente, margine nunc eroso et glanduloso, sæpiùs (an semper) processus 2, carnosos, clavatos, ex ore calcaris ortos, parallelos, adnatos gerente. Anthera parva, cavitate unguis brevior, lobis basi ascendentibus, rostello tridentato brevi adnatis. Glandulæ nudæ —Herbæ, (Indicæ,) radicibus testiculatis, caule folioso aut vaginato. Flores omnium minuti. Sp. 5. adhuc notæ quarum duæ sunt Gymnadenia ? tenuis *Wall. Cat. No.* 7057, et G. secunda. *ib. No.* 7054.

30. AOPLA.—Calyx bilabiatus. Sepala lateralia deflexa, supremum erectum cum petalis agglutinatis galeam formans. Labellum lineare, ecalcaratum. Anthera brevis, erecta, lobis brevibus ascendentibus, rostello decurvo elongato. Glandulæ nudæ.—Herba (Indica) radicibus testiculatis. Folium solitarium, radicale. Spica laxa secunda. Flores herbacei.—Sp. 1. *A. reniformis.* (=Herminium reniforme Wall. Cat. no. 7067.)

1708

* CATASÉTUM semiapértum.

Half-open Catasetum.

GYNANDRIA *MONANDRIA.*

Nat. ord. ORCHIDEÆ. *Juss.* §. Vandeæ. *Lindl.* (*Introduction to the Natural System of Botany, p.* 262.)
CATASETUM. Supra, vol. 10. *fol.* 840.

C. *semiapertum* ; foliis oblongo-lanceolatis undulatis racemo compacto longioribus, perianthio subpatente secundo : laciniis lanceolatis, labello apice incurvo cucullato marginibus ciliato-denticulatis.
C. semiapertum. *Hooker Exot. fl. t.* 213. *Lindl. gen. et sp. Orch. p.* 156.

Habitus, pseudobulbi, *et* folia *omnino* C. tridentati. Flores *diversissimi, odorati, herbacei, immaculati: sepalis petalisque angustè lanceolatis sæpius dependentibus ob labellum posticum.* Labellum *subcompresum, carnosum, lobo apicis rotundato incurvo, lateralibus ciliâ tenui, denticulatâ tamen, marginatis.*

First introduced by Mr. Bell Edward Lloyd, who sent it from Brazil to Miss Falkner of Fairfield, about 8 or 9 years ago. More recently it has been transmitted to Mr. Harrison of Liverpool by Dr. Dundas, an eminent medical gentleman residing at Bahia.

It is not so shewy as *C. tridentatum*, but it is peculiarly fragrant, a quality which all the other known species are destitute of. It requires the same treatment; and when in rapid growth will thrive the better if its roots are actually allowed to immerse themselves in water.

Our specimen was communicated to us by Mr. Harrison in February last.

---

* See folio 1667.

1714

* BATEMÁNNIA Cólleyi.

Colley's Batemannia.

GYNANDRIA *MONANDRIA.*

Nat. ord. ORCHIDEÆ § VANDEÆ. (*Introduction to the Natural System of Botany, p.* 262.)

BATEMANNIA. Flores ringentes. *Sepala* patentia, lateralia unguiculata basi æqualia. *Petala* sepalis latiora, basi obliqua, pedi producto columnæ adnata. *Labellum* cum columna articulatum, trilobum, cucullatum. *Columna* semiteres, basi elongata, clinandrio marginato. *Anthera* parva, bilocularis, membranacea. *Pollinia* 2, postice biloba, glandulâ triangulari, caudiculâ nulla.

B. *Colleyi.*

Epiphyta. Pseudobulbi *ovati, subtetragoni, lucidi, olivacei, subcorrugati, ovi gallinacei mugnitudine.* Folia 2-3, *obovato-oblonga, plicata, basi angustata.* Racemus *radicalis, pendulus, semipedalis, laxè* 5-10-*florus.* Bracteæ *rhombeæ, striatæ, cucullatæ, inflatæ.* Flores *pedicellati,* ovario *pedicello breviore.* Sepala *et* petala *poll. longa, intus fusco-purpurea, extus viridi vitta notata, apice quoque viridia;* sepalum *posterius oblongum obtusiusculum:* lateralia *spatulata divergentia petalis paulo longiora;* petala *a basi triangulari oblonga, sepalo supremo latiora.* Labellum *oblongum, ultra medium disci bidentatum, album, intus lutescens, extus levissime rubescens; lobis rotundatis, serrulatis, lateralibus intermedio subcuneato brevioribus.* Anthera *membranacea, depressa, quadrata,* 2-*locularis.* Rostellum *subulatum.* Clinandrium *marginatum, dentatum.* Stigma; *rima parva transversa.*

Sent from Demerara to James Bateman, Esq. by Mr. Colley, his collector in that country, who has just returned from a successful mission with a considerable number of epiphytes, which are new to our gardens.

Mr. Colley states that this plant, when in perfection, bears double the number of flowers in the accompanying plate. Even in its present state, it is a very handsome species, and extremely different from all genera previously described. It belongs to the Maxillaria set of Vandeæ, which is characterised by the projecting base of the column. But in this genus the petals are not only much larger than the sepals, but seem as if they had pushed the lateral ones out of their place. It is generally the sepals which grow to the lengthened base of the column, hiding the lip until the flower opens; but in Batemannia the sepals have a narrow contracted base, and the office of protecting the lip is performed by the broad-based petals.

Our specimens were communicated, along with a drawing by Mr. Holland, in August last.

---

* We name this genus in compliment to James Bateman, Esq. of Knypersley, an ardent collector and successful cultivator of Orchideous epiphytes. Mr. Colley, after whom the species is called, was Mr. Bateman's collector in Demerara.

Fig. 1 is a view of the column and back sepal, the other parts being cut away; 2 is the inside of the lip; 3 is a view of the pollen masses from above; 4 one of the pollen masses with its posterior lobe.

1721

* MYÁNTHUS cérnuus.

Drooping Fly-wort.

GYNANDRIA *MONANDRIA.*

Nat. ord. ORCHIDEÆ § VANDEÆ. (*Introduction to the Natural System of Botany, p.* 262)

MYANTHUS. Lindl. Supra, fol. 1538.

M. *cernuus. Gener. and Sp. of Orch. pl. part* 3. *p.* 155.
Catasetum trifidum. *Hooker. Bot. Mag. t.* 3262.

This very curious plant is found growing upon the trunks of trees near Rio Janeiro. It seems first to have found its way to this country through Trinidad, whence it was sent to the Glasgow Botanic Garden, and figured in the Botanical Magazine.

For the beautiful drawing which we now publish, we are indebted to Miss M. A. Huntley, the daughter of the Rev. J. T. Huntley, in whose rich collection it flowered in great beauty in May last.

Like all the species with the habit of the now common Catasetums, this is very easily cultivated, provided it is rested for some months by being kept *cool* and dry, when not growing, and is vigorously forced when in full vegetation. We know not whether it is as avid of water as some of the Catasetums, but we find those plants thrive best when their roots are actually immersed in water while growing.

[In the dissections, fig. 1. represents a side view of the flower, to shew that when it is expanded the two petals are connivent with the upper sepal; fig. 2. shews the labellum jointed with the column; fig. 3. a back view of the pollen-masses with their caudicula and gland. *M. A. H.*]

---

* From μυῖα a fly; the flowers look when dried very much like a fly pressed flat.

1730

* ACANTHOPHÍPPIUM bícolor.

Two-coloured Barrel-Orchis.

GYNANDRIA *MONANDRIA.*

Nat. ord. ORCHIDEÆ § VANDEÆ. *Lindl.* (*Introduction to the Natural System of Botany, p.* 262.)

ACANTHOPHIPPIUM. Blume. Perianthium ventricosum. *Sepala* agglutinata, lateralia ungui columnæ adnata, dorsale cum petalis spatulatis fornicato. *Labellum* unguiculatum, cum basi longè productâ columnæ articulatum, limbo trilobo indiviso complicato, disco lamellato. *Anthera* carnosa, bilocularis. *Pollinia* 8, inæqualia, sessilia.——Herba *terrestris, subcaulescens.* Caulis *infernè bulbosus, vaginatus.* Folia *oblongo-lanceolata, plicata.* Pedunculus *vaginatus pauciflorus.* Flores *speciosi.* Lindl. Gen. et Sp. Orch. 177.

A. *bicolor;* petalis oblongo-lanceolatis acutiusculis, labelli lobis lateralibus rotundatis, perianthio ovato.

Planta *terrestris;* pseudobulbis *oblongo-ovatis, corrugatis, atroviridibus, reliquiis foliorum vestitis, paulo in collum angustatis.* Folia 2-3, *oblongo-lanceolata, utrinque acuta, plicata, erecta, basi angustata, sed petiolo nullo.* Pedunculus *radicalis;* squamis *ovatis, concavis, brunneis, magnis, vaginatus,* 2-4-*florus.* Perianthium *carnosum, unciam et dimidiam longum, ovatum, vel subconicum, flavum, apice patulum sanguineum.* Sepala *omnia conglutinata, oblonga, obtusa, lateralia basi obliqua, basi longè productæ columnæ inserta.* Petala *æquilonga, apice minùs maculata, lineari-oblonga, acuta.* Labellum *cum pedè longâ columnæ unguem efficiente articulatum, inflexum, complicatum, trilobum ; lobis lateralibus rotundatis intermediâ magis luteâ et carnosâ, asperiusculâ ; lamellis disci 2 tantum, concavis, lineâ elevatâ sejunctis.* Columna *semiteres, acuta ;* stigmate *marginato ;* clinandrio *immarginato proclivi.* Pollinia 8, *geminata, in glandulam anticè emarginatam sessilia.*

A highly curious and extremely rare epiphyte, found in Ceylon by Mr. Watson, the Superintendant of the Government Garden at Peradenia, and transmitted by him to the Horticultural Society. It has very much the habit of a Geodorum, only it has pseudo-bulbs, instead of tubers.

* A name, the meaning of which is not explained by its author, Dr. Blume.

It flowers in June, and succeeds extremely well in a mixture of peat and sand, mixed with broken pots, provided it has a great deal of heat and moisture during the growing season, and a few months rest annually in a cool and dry atmosphere.

Of the only two other known species, one is a native of Java, the other of Sylhet. Neither has yet been imported.

1735

* MONACHÁNTHUS díscolor.

Dingy Monk-flower.

GYNANDRIA *MONANDRIA.*

Nat. ord. ORCHIDEÆ, § VANDEÆ, *Lindl.* (*Introduction to the Natural System of Botany, p.* 262.)

MONACHANTHUS. Perianthium explanatum. *Sepala* et *petala* deorsùm versa. *Labellum* posticum, carnosum, indivisum, ventricosum, sepalis multo majus. *Columna* brevis, crassa, mutica. *Anthera* et *Pollinia* Cataseti. —Epiphyta *Cataseti* habitu. *Genera et Sp. Orch.* 157.

M. *discolor;* racemo laxo multifloro, labello hemispherico marginibus planis medio fimbriatis.

Epiphyta, pseudobulbis *3-poll. longis, oblongis, teretibus, leviter corrugatis, cicatricibus foliorum annulatis.* Folia *oblongo-lanceolata, plicata.* Racemus *cylindraceus, laxus, multiflorus, spithamam longus.* Sepala *linearia, obtusa, reflexa, viridi-fusca.* Petala *recta, conformia, fusco-purpurea.* Labellum *posticum, cucullatum, acutum, carnosum, viridi-purpureum, intus viridi-luteum, margine reflexo medio fimbriato.* Columna *brevis, viridi-lutea, mutica, stigmatis excavati margine inferiore convexo prominente, lateralibus deorsum productis.*

A very rare plant in Demerara, whence Mr. Bateman received a single bulb by his collector, Mr. Colley. We are enabled to figure it by means of a specimen from the collection at Knypersley, which we received in November last.

Although not handsome, this is an interesting plant, as confirming the genus *Monachanthus,* which before consisted of but a single species found in the Brazils, of which we shall give a figure in some future number of this work. It differs from *M. viridis* in the number of the flowers, in their colour, and especially in the form of the lip, which looks more like an old rusty iron scull-cap, than any thing else we can compare it to.

* *Monks flower,* which is the meaning of this generic name, will not be considered unapplicable, when you compare the labellum of the original species to a cowl.

1740

* GRÓBYA Amhérstiæ.

Lady Amherst's Grobya.

GYNANDRIA *MONANDRIA.*

Nat. ord. ORCHIDEÆ, § VANDEÆ, *Lindl.* (*Introduction to the Natural System of Botany, p.* 262.)

GROBYA. Perianthium explanatum, bilabiatum. *Sepala* lateralia basi connata, sigmoidea, labello supposita, supremo erecto breviora. *Petala* dilatata, sepalo multo majora, erecta, conniventia. *Labellum* liberum, lobatum, nudum, cum basi columnæ articulatum, ascendens, sepalis minus, (nanum). *Columna* erecta, semiteres, arcuata, basi incrassata ; *antherâ* proclivi ; *stigmat* fornicato. *Pollinia* 2, posticè lobata, caudiculis duabus brevibus glandulæ ovali adnatis. Herba *Braziliensis, pseudobulbosa,* foliis *gramineis,* racemo *pendulo radicali.*

G. *Amherstiæ.*

Pseudobulbi *ovati, virides teretes, cicatricibus* 1-2 *foliorum annulati.* Folia 3-4 *e vagina squamata, linearia, striata, acuta, debilia.* Racemus *pendulus, densus,* 3 *pollices longus, pseudobulbis duplò longior.* Sepala *pallide ochracea, unicolora.* Petala *maculis fusco-purpureis seriatis notata.* Labellum *cuneatum, apice 5-lobum, lobo quinto cæteris exteriore, disco nudum, atropurpureum.* Columna *pallida, facie purpureo fasciata.*—1. Labellum. 2. Columna. 3. Pollinia *cum* glandulâ *suâ.*

For this curious species we are obliged to the Countess Amherst, in whose collection at Montreal it flowered for the first time in September last. It had been sent home by Mr. Hayne, a Commissioner in Brazil, from the interior of that country in 1829. To the pencil of Lady Sarah Amherst we are indebted for a sketch of the manner in which the plant grows.

* The genus is named in compliment to the Right Honourable Lord Grey of Groby, a munificent patron of Horticulture, and a most zealous cultivator of Orchideous Epiphytes; the species records the sense we entertain of the claims of the noble family of Amherst upon the gratitude of Botanists, for the countenance afforded by them to natural history so long as their power continued in the Eastern world.

Grobya is most nearly allied to *Cymbidium,* from which it is distinguished, firstly, by the lateral sepals being united at the base; secondly, by the large size of the petals; thirdly, by the lip having no parallel elevated lines ; and fourthly, by the pollen masses being united to the gland by two distinct caudiculæ.

EDWARDS'S

BOTANICAL REGISTER:

OR,

ORNAMENTAL FLOWER-GARDEN

AND SHRUBBERY:

CONSISTING OF

COLOURED FIGURES OF PLANTS AND SHRUBS,

CULTIVATED IN BRITISH GARDENS;

ACCOMPANIED BY THEIR

History, Best Method of Treatment in Cultivation, Propagation, &c.

CONTINUED

BY JOHN LINDLEY, Ph. D. F.R.S. L.S. AND G.S.

PROFESSOR OF BOTANY IN THE UNIVERSITY OF LONDON,
AND THE ROYAL INSTITUTION OF GREAT BRITAIN.
&c. &c. &c.

New Series.

VOL. VIII.

OR VOL. XXI. OF THE ENTIRE WORK.

——viret semper——nec fronde caducâ
Carpitur.

LONDON:
JAMES RIDGWAY AND SONS, PICCADILLY.

M.DCCC.XXXVI.

1742

*CYCNÓCHES Loddigésii.

Loddiges' Swanwort.

GYNANDRIA *MONANDRIA.*

Nat. ord. ORCHIDEÆ § VANDEÆ, (*Introduction to the Natural System of Botany*, p. 262.)

CYCNOCHES, Lindl. Perianthium explanatum. *Sepala* lateralia lanceolata, basi paululùm sub labello connata; supremo angustiore. *Petala* latiora, falcata, decurva. *Labellum* liberum, ecalcaratum, columnâ continuum, lanceolatum, integerrimum, ungue abrupto calloso. *Columna* elongata, arcuata, teres, apice clavata, auriculis 2 falcatis ad latera clinandrii. *Anthera* bilocularis. *Pollinia* 2, postice sulcata, subpedicellata; *caudiculâ* lineari; *glandulâ* grossâ.—Habitus *Cataseti* (sed racemus lateralis). *Lindl. Gen. & Sp. Orch.* p. 154.

Cycnoches Loddigesii, *Lindl. l. c. Bot. Cab. t.* 2000.

Caulis *erectus, carnosus, cylindraceus, subpedalis, pallidè viridis, corrugatus, basi vaginis membranaceis, apice foliis vestitus.* Folia *omninò* Cataseti *cujusdam.* Racemus *multiflorus, pendulus, e latere caulis erumpens, flexuosus.* Flores *maximi ordinis, initio inodori, mox Vanillam gratissimè spirantes, præsertim manè.* Bracteæ *ovatæ, cucullatæ, obtusæ, pedicellis breviores.* Sepala *viridia apice fusca, et obscure maculata, mox rufescentes; supremum lineari-oblongum, arcuatum 3-pollicare; lateralia 2-poll. longa imâ basi connata, ovato oblonga, subundulata, pendula.* Petala *ejusdem coloris cum sepalis, sed minùs conspicuè maculata, oblongo-lanceolata, inæquilatera, decurva, 2¼-poll. longa.* Labellum *patentissimum, cum columnâ continuum, angustè oblongum, convexum, carnosum, medio album, apicem versus tenuius, sordidè lutescens, undique maculis latis inæqualibus sanguineis notatum;* ungue *brevi alato maculoso, alis lutescentibus, disco convexo albo.* Columna *gracillima, arcuata, utrinque clavata; dimidio inferiore atropurpureo, superiore complanato virescente purpureo maculoso.* Anthera *citò decidua, membranacea, bilocularis, ferè pellucida, omninò mutica.* Clinandrium *posticè bicorne, cornubus falcatis compressis supra antheram incurvis.*

For the discovery of this most extraordinary Epiphyte the public is indebted to John Henry Lance, Esq. who discovered it, and a great number of other new plants in the woods of Surinam. It was originally sent by that gentleman to Messrs. Loddiges, with whom it flowered imperfectly three or four years ago. Other and more healthy plants were afterwards presented by Mr. Lance to the Horticultural Society, the produce of one of which was the singular raceme now represented.

* So called from κύκνος a swan, and αὐχὴν the neck; in allusion to the column of this plant, which is gracefully curved, like the neck of the swan.

In size the flowers of this plant are the largest that have yet been met with among Orchideous plants, measuring not less than five inches from the tip of the back sepal, to the point of the lip. They are more singular than beautiful, but they compensate in fragrance for their want of brilliant colours: the purest odour of Vanilla is exhaled by them when they have been open a short time, especially in the morning.

This species flowers in the months of June and July. It resembles a Catasetum in habit, but its flowers originate from the side, and not the base, of the fleshy stem. It grows very freely under the same management as Catasetums, but seems to require the greatest heat that is ever obtained in a damp stove. It is disposed to increase itself pretty freely, by multiplication of its fleshy stems, and will probably cease, after a few years, to be so rare a plant as it now is.

1751

*LÆLIA ánceps.

Two-edged Lælia.

GYNANDRIA *MONANDRIA.*

Nat. ord. ORCHIDEÆ § EPIDENDREÆ, *Lindl.* (*Introduction to the Natural System of Botany*, p. 262.)

LÆLIA, Lindl. Sepala explanata, lanceolata, æqualia. *Petala* majora paulò difformia. *Labellum* (posticum), 3-partitum, lamellatum, circa columnam convolutum. *Columna* aptera carnosa, anticè canaliculata. *Anthera* 8-locularis. *Pollinia* 8, caudiculis 4 elasticis.——Herbæ *epiphytæ*, rhizomate *pseudobolbophoro.* Scapi *terminales, pauci v. multiflori.* Flores *speciosi, odorati.* Gen. & Sp. Orch. p. 115.

L. *anceps;* foliis binis aut solitariis lanceolatis, scapo ancipiti bifloro squamis carinatis vestito, ovario viscoso, labelli disco lineari elevato apice trilobo, pseudobolbis ovatis distantibus tetraquetris.

Rhizoma *repens, squamosum.* Pseudobulbi *ovati, distantes, tetraquetri,* 2 *poll. longi, juventute squamis acuminatis vestiti.* Folia *solitaria, rarò bina, lanceolata, acuminata, coriacea, atroviridia.* Scapus *ex apice pseudobulbi ortum ducens, sesquipedalis, gracilis, anceps, squamis carinatis arctè vaginantibus vestitus, apice biflorus.* Ovarium *viscosum, bracteâ lato-lanceolatâ, acuminatissimâ, carinatâ, membranaceâ, fuscâ, involvente brevius, apice teres, cuniculo a basi labelli orto brevi instructum.* Sepala *violacea, membranacea, dorso subherbacea, lanceolata, bipollicaria, patentissima.* Petala *oblongo-lanceolata, ejusdem coloris texturæ, et longitudinis, sed lineâ mediâ dorsi tantum herbacea, et duplò latiora.* Labellum *aut anticum, aut posticum, cucullatum, trilobum, basi columnam involvens et pallidè violaceum, intus luteum sanguineo-venosum; trilobum, lobis lateralibus rotundatis, intermedio atropurpureo basi albo, oblongo, acuto, subundulato, plano, lineâ latâ elevatâ crassâ luteâ antrorsum trilobâ in medio; lobis lateralibus abruptis, intermedio productiore in laminam abruptam verticalem attenuato.* Columna *semiteres, marginata, clavata, cum labello continua.* Anthera 8-*locularis.* Pollinia 8, *cuneata, utrâque extremitate caudicularum quatuor inserta et inflexa.*

A very beautiful Orchideous plant, imported from Mexico by Messrs. Loddiges, in whose collection our drawing was made last December. We at first took it for *Lælia grandiflora*, but Mr. George Loddiges has satisfied us that it is

not that species, which it is, however, highly probable that he possesses in a plant from Xalapa, with pear-shaped pseudobulbs, each of which has two or three lanceolate purplish leaves. The pseudo-bulbs of this are remarkable for their ovate outline, and for their four-angled figure, produced by a sharp ridge being planted upon each face of a compressed body.

When we say that this plant is equal in beauty to any of the Cattleyas, that it has a far more graceful mode of growth, in consequence of the length of its slender scaly stems, from the point of which the flowers swing, and that it diffuses an agreeable fragrance, we shall have said that it is one of the most interesting of the tribe that has yet made its appearance in our stoves.

It will probably succeed, without difficulty, in any hothouse which is adapted for the cultivation of Maxillarias and plants of that description.

1752

* MONACHÁNTHUS víridis.

Green-flowered Cowlwort.

GYNANDRIA *MONANDRIA.*

Nat. ord. ORCHIDEÆ § VANDEÆ, *Juss.* (*Introduction to the Natural System of Botany, p.* 262.)

MONACHANTHUS.—Supra, vol. 20. *fol.* 1735.

M. *viridis;* racemo multifloro, labello oblongo cuspidato margine lævi, sepalis petalisque rigidis ovatis.

M. viridis, *Lindl. Gen. & Sp. Orch. p.* 157.

Planta *facie omnino* Cataseti tridentati *quo foliis et caule tantum convenit ut floribus ademptis eandem diceres.* Flores *carnosi, herbacei, labelli margine vitellino.* Sepala *ovata, acuta, rigida, petalis parallela.* Petala *conformia, sed latiora et maculis quibusdam purpureis obscuris notata.* Labellum *posticum ventricosum, carnosum,* $1\frac{1}{2}$*-poll. longum, ore obliquo lævi cuspidato.* Columna *brevis, ovata, carnosa, mutica.*

This is the original species on which the genus *Monachanthus* was founded. In habit it is so like *Catasetum tridentatum* that we long doubted whether it ought to be generically separated. The absence of the long tendril-like processes, which are so conspicuous upon the column of *Catasetum*, was, however, so remarkable a circumstance, that we came to the conclusion that it must be an essentially distinct form; and the discovery of the curious *Monachanthus discolor*, figured at tab. 1735, has shewn that we judged correctly.

A native of Brazil, growing upon trees in the Corcovado, whence Dr. Hooker received the drawing and specimen which first made the genus known. Our figure was made from a specimen communicated last November from Wentworth, by permission of Lord Fitzwilliam. We are not aware at what time, or by whom it was imported, but it has probably been taken for a green form of the common *Catasetum tridentatum*, and consequently no record has been kept of it.

It requires precisely the same treatment as *Catasetums*, and may be cultivated with facility by any one who has a damp stove.

* See fol. 1735.

1754

* BRÁSSIA Lanceána.

Mr. Lance's Brassia.

GYNANDRIA *MONANDRIA.*

Nat. ord. ORCHIDEÆ § VANDEÆ, *Lindl.* (*Introduction to the Natural System of Botany, p.* 262.)

BRASSIA.—Supra, vol. 10. *fol.* 832.

B. *Lanceana;* sepalis ovato-lanceolatis acuminatis, petalis minoribus, labello oblongo acuminato undulato sepalis lateralibus duplò breviore.

Color *totius herbæ amœnè et diffuso viridis.* Pseudobulbi *densi, oblongi, macri, compressissimi, ancipites, longitudinaliter sulcati, haud rarò arcuati.* Folia *bina aut solitaria, oblongo-lanceolata, macra, valdè striata.* Racemi *radicales, foliis longiores, floribus magnis luteis secundis odorem Primulæ suavissimum spirantibus onusti.* Bracteæ *ovatæ, concavæ, breves, membranaceæ.* Sepala *patentissima, lineari-lanceolata, purpureo paululum maculata, basi virescentia; supremum* $1\frac{1}{2}$, *lateralia* 2 *poll. longa.* Petala *ejusdem formæ et coloris sed sepalo supremo duplo minora.* Labellum *luteum, immaculatum, oblongum, undulatum, ferè repandum, acuminatum, sepalis lateralibus duplò brevius; basi tuberculis* 2 *albis oblongis contiguis parallelis inanibus pubescentibus, dentibusque totidem membranaceis in fronte.*

A native of Surinam, in woods, where it was found growing upon trees by John Henry Lance, Esq.; and by him presented to the Horticultural Society in 1833. It is also wild in Brazil, where it was found by Dr. von Martius upon the trees on the banks of the river Jui, one of the tributaries of the Japurà in the Province of Rio Negro, flowering in February. It blossomed at nearly the same time in the Chiswick Garden, and with Messrs. Loddiges, in whose collection our drawing was made last August.

It is of course, like other Surinam plants, a very tender species, and requires the hottest and dampest part of a stove; but there is no Orchideous plant which is more easy to cultivate, or more ready to multiply. It thrives in decayed leaf-mould, better than in any other compost, and may be readily known when out of flower, by the bright light green colour of its leaves, and by its peculiarly thin pseudobulbs, which often curve down upon themselves, as is represented in our figure.

If this is not equal to *Lælia* in the size of its flowers, and the brilliancy of its colours, it far surpasses that plant in its exquisite fragrance, which is like nothing so much as newly gathered cowslips and primroses.

* Thus named by Mr. Brown in due commemoration of the late Mr. Brass, a skilful botanical traveller and draughtsman, who collected seeds, plants, and dried specimens on the Guinea Coast for Sir Joseph Banks, Dr. Fothergill and Dr. Pitcairn, and whose sketches being most liberally lent by Sir Joseph Banks to Dr. Afzelius in his visit to Sierra Leone, were maliciously damaged, and partly destroyed out of characteristic and wanton brutality by some piratical slave-mongers under the French flag, during the late war.—*Smith in Rees' Cyclopædia.*

1756

* DENDRÓBIUM Pierárdi.

Mr. Pierard's Dendrobium.

GYNANDRIA *MONANDRIA.*

Nat. ord. ORCHIDEÆ § MALAXIDEÆ, *Lindl.* (*Introduction to the Natural System of Botany,* p. 262.)

DENDROBIUM.—Supra, vol. 7. *fol.* 548.

D. *Pierardi;* caulibus pendulis glabris, foliis ovato-lanceolatis acutis, floribus geminatis racemum spurium formantibus, sepalis acuminatis membranaceis, petalis sepalo supremo majoribus acuminatis, labello cucullato dilatato subtruncato pubescente ciliato. *Gen. et Sp. Orch. p.* 79.

D. Pierardi. *Roxb. Fl. Ind.* 3. 482. *Hooker Exot. Fl. t.* 9.

Many years since an Orchideous Epiphyte was sent from Chittagong with some other species to the Botanical Garden, Calcutta, when Dr. Roxburgh named one of them *Dendrobium Pierardi*, in compliment to the gentleman who discovered it. In his Flora Indica Dr. Roxburgh adds that it is also a native of various parts of the Delta of the Ganges, where it is generally found on Mangoe Trees.

No one has however been able to discover any plant which answers exactly to Dr. Roxburgh's description of his *D. Pierardi*; and what is now cultivated under that name in the Calcutta Garden appears from Dr. Wallich's specimens to be the plant defined by Dr. Brown in the seventh volume of the first series of this work under the name of *D. cucullatum*.

When the first part of the *Genera and Species of Orchideous Plants* was published, we had satisfied ourselves that in all probability Dr. Roxburgh had confounded two or three different species under the common name of *D. Pierardi*; and accordingly we took the species represented by Dr. Hooker in his Exotic Flora, as the authority for the name, and we considered Dr. Brown's *Dendrobium cucullatum*, to be a mere form of it.

We have since had an opportunity of comparing *D. Pierardi* and *cucullatum* side by side, in the utmost perfection in the stove of the Messrs. Loddiges, where these lovely species flowered in great splendour in January last, forming festoons two or three feet long, quite covered with the most delicate pink and yellow blossoms. At first sight they are so entirely the same that one does not immediately perceive in what their differences consist; but upon a more attentive inspection it is found that *D. cucullatum* has larger flowers, with a broad roundish-ovate lip, the base of which is rolled up into a sort of short stalk, while in *D. Pierardi* the lip is rounded and very blunt, and its base is rolled into a much longer stalk, which quite conceals the column; it is moreover remarkably incurved. These

* See folio 1239.

differences are expressed at the bottom of the accompanying plate, where fig. 1. represents *D. Pierardi*, and fig. 2. *D. cucullatum*. Whether or not the latter is really a distinct species may possibly be doubted; but it is certainly so remarkable a form, as to have amply justified Dr. Brown in separating it from *Pierardi*.

Allied to these plants are some new species, which we avail ourselves of the present opportunity of making known to Botanists.

1. D. *ochreatum* (Lindl. in Wall. Cat. no. 7410); caulibus brevibus e squamis membranaceis ventricosis erumpentibus, foliis ovato-lanceolatis acutis, vaginis ventricosis, floribus geminatis sepalis petalisque majoribus lanceolatis acuminatis, labello rhomboideo-ovato unguiculato cucullato intùs tomentoso.

Hab. in Chittagong, *Wallich* (*hab. s. sp. e Museo Anglo-Indico*).

Species vaginis laxis, alioquin facie D. Pierardi, primo intuitu distinguenda. Flores expansi ferè 3 uncias lati, pallidi. Labellum purpureo-maculatum; ungue intùs supra medium calloso.

A beautiful plant.

2. D. *Cunninghamii*; caulibus gracilibus pendulis ramosis vaginis foliorum corneis transversim corrugatis squamatis, foliis ovato-linearibus obtusiusculis, pedunculis oppositifoliis bifloris foliis multò brevioribus, sepalis ovatis acutis, petalis oblongis acutis latioribus, labelli trilobi lobo intermedio subrotundo undulato basi 5-lamellato lateralibus nanis acutis.

Hab. supra truncos Callistemonis elliptici (*A. C.*) in insula septentrionali *Novæ Zelandiæ*, juxta mare, incolis *Raumangha*; R. Cunningham (*hab. s. sp. comm. cel. A. Cunningham*).

This species is nearly allied to *D. biflorum*, of which I had examined no specimens at the time the *Gen. et Sp. Orch.* was published. Having since received that plant in a good state from Mr. Mathews, who found it in Otaheite hanging from the branches of trees, I am enabled to offer a correct definition of that very rare plant.

D. *biflorum* Swartz; caule pendulo gracili tereti, foliis lineari-lanceolatis acuminatis planis, pedunculis bifloris lateralibus e paleis corneis erumpentibus; sepalis petalisque acuminatissimis, labello rhomboideo medio bilamellato trilobo lobis lateralibus acutis nanis intermedio deltoideo acuminato margine fimbriato.

3. D. *Griffithianum*; caulibus erectis elongatis clavatis sursum tetragonis apice diphyllis, foliis lanceolato-oblongis obtusis, racemis laxis flexuosis multifloris pendulis e latere caulium natis, bracteis oblongis membranaceis pedicellis 4-plò brevioribus, petalis oblongis ciliolatis sepalis duplò latioribus, labello ovato pubescente denticulato subrepando, capsulis pyriformibus angulatis.

Hab. in Regno *Burmano* supra arbores, *W. Griffith*. (*hab. s. sp. comm. cel. Griffith*.)

A beautiful species with much the appearance of *D. aggregatum*.

4. D. *extinctorium*; pseudobulbis depressis orbiculatis aggregatis, foliis . . . ., pedunculis terminalibus unifloris, sepalis lateralibus posticis basi longè in extinctorii forma productis, labelli ungue longissimo tenui, limbo trilobo: lobis lateralibus erectis acutis intermedio oblongo rotundato, capsula pyriformi angulata.

Hab. in Regno *Burmano* locis humidis supra truncos Careyæ arboreæ, *W. Griffith* (*hab. s. sp. comm. cel. Griffith*.)

This is next *D. pusillum* Blume, and has very much the appearance of a Bolbophyllum. Its leaves have not been seen.

1758

* ONCÍDIUM citrínum.

Lemon-coloured Oncidium.

GYNANDRIA *MONANDRIA*.

Nat. ord. ORCHIDEÆ § VANDEÆ, *Lindl.* (*Introduction to the Natural System of Botany*, *p.* 262.)

ONCIDIUM.—Supra, vol. 13. *fol.* 1050.

O. *citrinum*; pseudobulbis oblongis compressis, foliis ensiformibus rigidis scapo simplici brevioribus, sepalis petalisque labelli longitudine lineari-oblongis undulatis, labello cordato utrinque introrsùm arcuato apice dilatato subreniformi, cristâ 8-tuberculatâ pubescente, alis minimis, stigmate orbiculari.

Planta O. altissimo (*fol.* 1651.) *valdè affinis et fortè mera varietas. Diversa tamen videtur scapo non ramoso, floribus parum maculatis, sepalis petalisque minùs acuminatis, cristâ tuberculatâ potius quam digitatâ, demum alis minimis, et stigmate orbiculari nec angusto compressoque.*

A native of Trinidad, whence it was introduced by Messrs. Loddiges, in whose collection our drawing was made last November. Unfortunately the plant soon after sickened and died; so that it is for the present lost to the country.

It approaches very nearly to *O. altissimum*, figured at fol. 1651 of this work; and is principally distinguished by the following characters. Its flowering stem is simple and not branched; its flowers are of a pale lemon colour, very distant from each other, and by no means so much spotted; the crest of the lip consists of about eight warts, which are slightly downy, and not of nine smooth finger-like processes; its stigma is nearly orbicular, and not long and narrow, and the wings of the column are exceedingly small; and finally, both the pseudo-bulb and the leaves have a singularly yellow tint, as we are informed by Mr. George Loddiges.

Like the rest of the genus it requires a hot and damp stove.

* See folio 1542.

1760

* BLÉTIA refléxa.

Reflexed Bletia.

GYNANDRIA *MONANDRIA*.

Nat. ord. ORCHIDEÆ § EPIDENDREÆ, *Lindl.* (*Introduction to the Natural System of Botany*, *p.* 262.)

BLETIA.—Supra, vol. 17. *fol.* 1401.

B. *reflexa*; sepalis lineari-lanceolatis lateralibus reflexis, petalis cuneato-lanceolatis supra columnam conniventibus, labelli trilobi lobis lateralibus rotundatis planis intermedio angusto undulato lamellis 5 altis parallelis indivisis ad pedem columnæ decurrentibus, foliis angustis ensiformibus plicatis.

Folia *angusta, ensiformia, plicata.* Scapus $1\frac{1}{2}$-2-*pedalis, teres, distanter vaginatus*, 2-4-*florus.* Bracteæ *oblongo-lineares, membranaceæ, ovario duplò breviores.* Sepala *angusto-lanceolata, ex viridi rubescentia basi alba; lateralia reflexa, supremum paulò latius magis erectum, apice tantùm reflexum.* Petala *ex viridi rubescentia, cuneato-lanceolata, erecta, acuta, collateralia columnam a tergo tegentia.* Labellum *cucullatum, parte inferiore album, cum basi columnæ articulatum, oblongum, trilobum; lobis lateralibus erectis rotundatis, albis, margine roseis, intermedio elongato pariter rotundato, crispo, atropurpureo, patente, lamellis 5 altis integris subundulatis totum axim labelli percurrentibus.* Columna *purpurea, clavata, apice alata, dente unico incurvo post cardinem antheræ.* Pollinia 8, *quorum* 4 *cæteris minora.*

Drawn in the hothouse of the Messrs. Loddiges, in November 1834. It is a terrestrial species, and a native of Mexico. The greenish colour of its flowers, and the remarkably narrow reflexed sepals, distinguish it at once from the remainder of the genus.

The specimen figured was less brilliant in its colours, and altogether smaller than is to be expected hereafter; for wild specimens now before us, collected by Baron de Karwinsky, and forming part of the Royal Bavarian Museum, have a rich purple cast upon every part of the flowers, and the scapes each bear from two to four blossoms.

Requires the same treatment as other Bletias.

* See folio 1401.

1762

* LEPÁNTHES tridentáta.

Three-toothed Lepanthes.

GYNANDRIA *MONANDRIA.*

Nat. ord. ORCHIDEÆ § MALAXIDEÆ, *Lindl. (Introduction to the Natural System of Botany, p.* 262.)

LEPANTHES Swz. Sepala patula, basi, præsertim lateralia, connata. *Petala* 2, libera, nana. *Labellum* apice sub-bilobum, cum columna connatum. *Columna* teres, nana, aut elongata. *Pollinia* 2.——Herbæ *epiphytæ*, caulibus *filiformibus, monophyllis, vaginatis*, spicis *terminalibus*, Pleurothallis *omninò habitu.* Flores *parvi, lutescentes vel rubescentes.*

L. *tridentata;* folio ovato acuto apice tridentato, pedunculis subbifloris folio longioribus, sepalis ovatis lateralibus semiconnatis, petalis semisagittatis tomentosis, labelli ciliati lobis lateralibus rotundatis, intermedio brevissimo acutiusculo incurvo, columnâ nanâ.

L. tridentata. *Swartz Prodr. p.* 125. *fl. Ind. occ. p.* 1561.

Caules *cæspitosi, vaginis lævibus laxis cincti, staturâ a lineis* 2 *usque ad uncias totidem variantes.* Folia *ovato-oblonga,* $1\frac{1}{2}$ *unciam longa, crassa, apice acutè tridentata, subtus pallidiora.* Pedunculi *terminales, bracteati, subbiflori, foliis longiores.* Flores *parvi, bilabiati, oculo nudo purpureo-lutei.* Sepala *ovata, acuta, æqualia;* supremum *purpureum luteo-marginatum;* lateralia *semiconnata lutescentia, parum purpurata.* Petala *atropurpurea, tomentosa, semisagittata: lobo postico rotundato, sepalis dimidio breviore.* Labellum *minus, lutescens, basi concavum, purpurascens, cum basi columnæ connatum eâque parallelum, imò longius; trilobum, lobis lateralibus ascendentibus ciliatis columnam amplectantibus, intermedio incurvo minimo, minùs manifestè ciliato.* Columna *teres, minima, bicornis, basi tantum labello adnata; stigmate depresso subcuneiformi.* Pollinia 2, *parallela, conica, caudiculis apice incrassato aurantiaco cohærentibus.*

This which is one of the most pigmy of Orchideæ, not much exceeding in size the moss amongst which it grows, will nevertheless amply repay the observer for a careful examination of it. In order to explain its singular structure we have given magnified figures of the more important parts of its fructification.

The first thing that strikes us is the singular termination of its leaves, which do not end in a sharp point as it seems to the naked eye; but have three little delicate teeth (fig. 2.). The flowers are covered externally by three sepals, the uppermost of which is purple with a yellowish border, and the lateral ones yellow with a little stain of purple; these are frosted over with brilliant little tubercles, and glitter in the sunshine as if encrusted with emeralds. The petals are two deep purple downy parts (fig. 2.), standing upright in the centre of the blossom, and forming a sort of back for the column to rest against; the column itself (fig. 5.) is a minute two-horned body, which buries its head between the lobes of the labellum (fig. 3.); and the latter is a yellowish three-lobed downy organ, whose two lateral divisions stand up on each side of the column, in the form of tiny bucklers ready to defend it from aggression, while the centre lobe projects in the fashion of a little horn (fig. 3. and 4.).

This is the first species of the genus which has been seen alive in Europe. It is a native of the highest parts of the mountains of Jamaica, where it is found growing on the bark of trees, among mosses. It can only be preserved alive with great care by being kept under a bell glass among damp moss in a cool part of a stove. We received it from Messrs. Loddiges in January last.

* So called from λεπος bark, or λεπις small, and ανθος a flower; because the plants of this genus have very small flowers and grow upon the bark of trees.

1765

* EPIDÉNDRUM grácile.

Graceful Epidendrum.

GYNANDRIA *MONANDRIA.*

Nat. ord. ORCHIDEÆ, *Juss. (Introduction to the Natural System of Botany, p.* 262.)

EPIDENDRUM.—Supra, vol. 1. *fol.* 17.

E. *gracile;* foliis in pseudobulbos ovatos corrugatos pluribus levato-ensiformibus, racemo simplici longissimo, sepalis oblongis petalisque cuneatis patentibus, labelli ferè liberi trilobi lobis lateralibus semiovatis intermedio oblongo crispo obtusissimo duplò minoribus disco bicostato.

Labellum *tripartitum, cum basi columnæ leviter connatum; laciniis lateralibus erectis, semiovatis, obtusis, subundulatis, flavis sanguineo venosis, intermediâ subrotundâ crispâ ejusdem coloris brevioribus; in medio crassum, album venis purpureis, bicostatum; costis utrinque infra medium obsolete unidentatis.* Ovarium *brevè cuniculatum.* Columna *semiteres, marginata apice utrinque auriculata.* Gynizus *deorsum trilobus;* rostillum *crassum, breve, rotundatum.* Anthera *complete* 4-*loculares; dissepimentis membranâ marginatis.* Pollinia 4, *compressa basi unidentata; caudiculis totidem pulvereis replicatis.*

A native of the Bahamas, whence it was sent to the Horticultural Society by John Campbell Lees, Esq. in 1833. From a plant which flowered last September the accompanying drawing has been taken.

The plant is very nearly allied to *E. odoratissimum*, but its leaves are three times as long, and the flowering stem shoots up to the length of three feet, without a branch or a leaf beyond the little scales that appear upon it here and there; the flowers moreover are larger, and the form of the segments of the lip materially different, while the callosities upon the disk of the latter distinguish it positively. In *E. odoratissimum* the disk is simply thickened and flattened and gradually thins away towards the base; in *E. gracile* it is distinctly two-ribbed, and has two obsolete teeth a little below the middle.

The want of beauty in this plant is amply compensated for by its delicious fragrance, early in the morning and at night. It seems as if it were more terrestrial than epiphytal, and grows freely in any light well drained soil. The patch sent home by Mr. Lees consisted of a mass of pseudobulbs nearly two feet in diameter.

* See folio 1415.

1772

* ANGRÆCUM micranthum.

Small-flowered Angræcum.

GYNANDRIA *MONANDRIA.*

Nat. ord. ORCHIDEÆ § VANDEÆ, *Lindl. (Introduction to the Natural System of Botany, p.* 262.)

ANGRÆCUM.—Supra, vol. 18. *fol.* 1522.

A. *micranthum;* caule brevi, foliis oblongis trinerviis apice obliquis, spicis secundis multifloris horizontalibus congestis foliis brevioribus, sepalis petalisque subæqualibus angustè ovatis apice patulis, labello conformi basi bilobo bidentato medio pubescente, calcare obtuso incurvo postico.

Caulis 2-*pollicaris.* Folia 1½-*p. longa,* 1½-*p. lata, plana, crassa.* Flores *minuti, albi.*

A very small and unattractive species, lately introduced by Messrs. Loddiges from Sierra Leone. It is quite distinct from any previously described, but retains the peculiar characters of the genus without deviation. It flowered in February.

It is curious that Angræcum should be so exclusively an African genus, that no certain species has yet been found beyond its continent or immediately dependent islands, although it evidently stretches across the whole of the tropical part of Africa itself. In addition to this from Sierra Leone, three or four other species are now in our gardens from the same place, but all of them equally inconspicuous with the present. Of these *A. clandestinum, teretifolium* and *distichum* have already flowered with Messrs. Loddiges. Fig. 1. represents a lip with the spur attached; 2. an entire flower; and 3, a pair of pollen masses with their gland.

---

* See folio 1522.

1777

* PHOLIDÓTA imbricáta.

Imbricated Pholidota.

GYNANDRIA *MONANDRIA.*

Nat. ord. ORCHIDEÆ § MALAXIDEÆ. (*Introduction to the Natural System of Botany, p.* 262.)

PHOLIDOTA.—Supra, vol. 14. *fol.* 1213.

P. *imbricata;* pseudobulbis oblongis corrugato-sulcatis obtusis, foliis solitariis oblongo-lanceolatis plicatis acutis, spicis pendulis foliorum longitudine, bracteis membranaceis concavis acutis, sepalis lateralibus ovatis carinatis, labello subgloboso cucullato: lobis lateralibus parvis erectis intermedio bilobo: laciniis rotundatis.

P. imbricata. *Lindl. in Hook. Exot. Fl. t.* 138. *Loddiges Bot. Cab. t.* 1934.

It appears that there are two different species confounded under the name of *Ph. imbricata;* namely, this one and that which is figured at fol. 1213 of this work.

So long a time has elapsed since we saw the latter in flower that we are scarcely able to point out any very satisfactory characters to distinguish them, although we have no difficulty in acceding to the opinion of cultivators that they are in fact entirely distinct. We therefore propose to call *Ph. pallida* the plant before figured by the name of *Ph. imbricata,* and to retain the latter designation for this, which was originally figured by Dr. Hooker in the Exotic Flora, and afterwards by Messrs. Loddiges in their Botanical Cabinet.

They are both natives of India, and possibly occupy similar regions; they principally differ in the following particulars. *P. pallida* has very round blunt bracts, white flowers, smaller leaves, and grows very weakly and unwillingly under ordinary circumstances. *P. imbricata* has pointed bracts, yellowish flowers with a dash of violet, very long strong leaves, and grows and flowers most freely.

* See fol. 1213.

Our specimens were communicated by Richard Harrison, Esq., in May, 1834.

1778

* MYÁNTHUS barbátus.

Bearded Flywort.

GYNANDRIA *MONANDRIA.*

Nat. ord. ORCHIDEÆ §. VANDEÆ. (*Introduction to the Natural System of Botany, p.* 262.)

MYANTHUS.—Supra, vol. 1. *fol.* 1538.

M. *barbatus;* labello in pilis succulentis barbæformibus dissoluto basi supra unicorni.

Caules *vetusti fusiformi-cylindracei,* 4-5-*poll. longi, parum induviati, luteo-olivacei.* Folia *atroviridia, undulata, Cataseti, oblongo-lanceolata, versus basin angustata.* Racemus *radicalis, prostratus vel pendulus, flores* 9-10 *inversos, i. e. labello quoad axin superiore gerens.* Bracteæ *lineari-lanceolatæ membranaceæ, pedicellis purpurascentibus breviores.* Perianthium *bilabiatum; sepalis petalisque lineari-lanceolatis, herbaceis, fusco-sanguineo maculatis.* Sepalum *supremum, cum petalis in galeam agglutinatum;* lateralia *patentissima.* Labellum *cum columna subarticulatum, lineare, medio infractum et saccatum; margine in fila tenuia alba numerosissima dissoluto, cristam aviculæ cujusdam simulantia; cornu deorsum falcatum, candidum, basi superiore tridentatum gerens.* Columna *herbacea, fusco-sanguineo maculata, cornu apicis galeæ longitudine, cirrhis duabus rectis purpurascentibus apice pellucidis cornu labelli amplectentibus.* Glandula *maxima, cartilaginea, oblonga, stigmate cum vi elasticá dissiliens, et convoluta.*

This curious new species is a native of Demerara, near the falls of Wapopekai on the Massarony river, some hundreds of miles from its mouth, where it was found by Mr. John Henchman growing in the clefts of trees in shady situations. It was first sent us in February last by Mr. Lowe, with whom it first flowered, but almost immediately after we received it from Chatsworth, and from the garden of Mr. Willmore, of Oldford, near Birmingham. The first specimens were pallid, as if they had suffered from want of light, but Mr. Willmore's were much darker. In the latter the sepals were of a deep green spotted with purple, the petals were a deep rich brownish red; and the lip itself, with its innumerable thread-like feelers, was tinged with red in all parts, except a strong white horn, which stands upon it near its base, curving downwards towards the little pit which occupies the centre of the lip.

* See folio 1721.

This plant has altogether the habit of Catasetum, and no doubt requires the same sort of management.

1779

* DENDRÓBIUM cúpreum.

Copper-coloured Dendrobium.

GYNANDRIA *MONANDRIA.*

Nat. ord. ORCHIDEÆ §. MALAXIDEÆ, *Lindl.* (*Introduction to the Natural System of Botany, p.* 262.)

DENDROBIUM.—Supra, vol. 7. *fol.* 548.

D. *cupreum;* caulibus teretibus pendulis, foliis oblongis obtusis emarginatis, racemis lateralibus laxis multifloris, bracteis ovato-linearibus canaliculatis obtusis, sepalis ovatis obtusiusculis petalis oblongis explanatis obtusis minoribus, labello unguiculato cochleariformi obtusissimo extus villoso intus nudo.

D. cupreum. *Herb. MSS.*

A native of the East Indies, whence it was sent by Dr. Wallich to the Hon. and Rev. W. Herbert, about the year

1825. It flowered at Spofforth for the first time at Midsummer 1834, when the drawing was made by Mr. Herbert, which we have now the opportunity of publishing.

It has been suspected that there are two Dendrobia in the East Indies, which agree with each other in a sort of tessellated appearance of the flowers, and a remarkably inflated slipper-like hairy labellum; one of these, called *D. moschatum*, has pale orange coloured blossoms, and a lip with 5 crested veins near its base; the other, to which the name of *D. Calceolaria* has been applied by Dr. Hooker, has rose coloured flowers with a little yellow in the centre, and is said to be destitute of the crested veins of the lip. There is, moreover, a *D. clavatum* of Dr. Wallich's Indian Catalogue, the specimens that were distributed of which were too imperfect to be described in the Genera and Species of Orchideous plants; but whether this plant is one of the preceding, or something altogether distinct is unknown. Furthermore, the plant now represented belongs to the same little cluster, and appears as if it might form a fourth species, unless indeed it should prove to be this *D. clavatum*. We have had no opportunity of examining it, but Mr. Herbert describes it to us in such a manner as to shew that it must be different from *D. Calceolaria* at least. "Its flowers are of a pale copper colour, veined with a redder tinge, and have two brown red blotches inside the lip. The spike of ten large flowers all expanded simultaneously, and the progress was so rapid that only about a week or ten days elapsed between its shewing bud and bursting into bloom. Independently of the colour of the flowers this species differs from *D. Calceolaria* in not having such long shoots, and in having the leaves less attenuated and shorter. *D. Calceolaria* under the same treatment makes shoots above four feet long; this plant under three feet. It is curious that these Dendrobiums if they miss flowering put forth a young plant instead of a spike of flowers at the point of inflorescence, as *D. Calceolaria* is doing here now."

* See folio 1239.

1781

*ANGRÆCUM dístichum.

Two-rowed Angræcum.

GYNANDRIA *MONANDRIA.*

Nat. ord. ORCHIDEÆ § VANDEÆ. (*Introduction to the Natural System of Botany, p.* 262.)

ANGRÆCUM.—Supra, vol. 18. *fol.* 1522.

A. *distichum*; caule imbricato, foliis distichis compressis recurvis obtusis suprà canaliculatis, floribus solitariis axillaribus pedunculis foliis subæqualibus, sepalis ovatis petalisque angustioribus secundis obtusis, labello postico oblongo concavo apice tridentato calcare tereti horizontali pedunculo breviore.

Caules 3-4 *pollices longi, atrovirides, foliis densissimè imbricati.* Flores *parvi, lactei, carnosi, inodori.* Labelli *dentes laterales rotundati intermedio recto acuto.*

A remarkably neat and pretty looking species, on account of its small closely packed deep bright green leaves and tufted stems, but having no beauty of a striking nature in its flowers.

It is a native of Sierra Leone, where it is found on the bark of trees. Imported by Messrs. Loddiges, in whose Nursery our drawing was made in October last. It appears to grow freely in the damp hot atmosphere which so many similar plants delight in.

* See folio 1522.

1787

*ONCÍDIUM pulchellum.

Pretty Oncidium.

GYNANDRIA *MONANDRIA.*

Nat. ord. ORCHIDEÆ, *Juss.* § VANDEÆ, *Lindl.* (*Introduction to the Natural System of Botany, p.* 262.)

ONCIDIUM.—Suprà, vol. 13. *fol.* 1050.

§. 2. Folia triquetra v. teretia.

O. *pulchellum*; foliis acutè triquetris carinatis subfalcatis integris, scapo confertissime multifloro subcernuo, sepalis cymbiformibus acuminatis lateralibus connatis, petalis ovatis undulatis acutis, labelli lobis lateralibus rotundatis intermedio subrotundo sessili retuso v. emarginato subæqualibus, cristâ trilobâ anticè 1-callosâ, alis columnæ acinaciformibus denticulatis.

O. pulchellum. *Hooker in Bot. Mag. t.* 2773. *Lindl. Gen. & Sp. Orch. p.* 206. *Lodd. Bot. Cab. t.* 1984.

A beautiful species found in several parts of the West Indies, and probably mistaken for *O. variegatum* till Dr. Hooker distinguished it. When in flower its panicle is so loaded with white blossoms tinged with yellow and pink as to be weighed down with their profusion.

It requires the treatment usually given to its tribe.

* See folio 1542.

1789

*ONCÍDIUM Lemoniánum.

Sir C. Lemon's Oncidium.

GYNANDRIA *MONANDRIA.*

Nat. Ord. ORCHIDEÆ, *Juss.* § VANDEÆ, *Lindl.* (*Introduction to the Natural System of Botany, p.* 262.)

ONCIDIUM.—Suprà, vol. 13. *fol.* 1050.

§ 2. Folia triquetra v. teretia.

O. *Lemonianum*; foliis compressis acuminatis suprà sulcatis, scapo stricto paucifloro, sepalis parvis spatulatis apiculatis omnibus liberis, petalis oblongis undulatis, labelli laciniis lateralibus linearibus abbreviatis intermediâ maximâ reniformi bilobâ: lobis dentatis ungue brevi margine denticulato, columnæ alis subquadratis truncatis oblique unidentatis.

"This curious little epiphyte was among a collection imported from the Havannah in March last, by Capt. Sutton of Flushing near Falmouth, and by him presented to Sir Charles Lemon, Bart., M. P., in whose garden at Carclew it flowered in May. It appears to us to be totally different from any species hitherto noticed, and is readily distinguished by its peculiar foliage."

"*Pseudobulbs* very small, scarcely discernible, roundish, one-leaved. *Leaves* from one to three inches long, and about a quarter of an inch broad, dagger-shaped, two-sided or edged, and sharp at the point; pale green, slightly striated, and covered with a profusion of minute, roughish, pale dots or warts. *Scape* about 9 inches high, round and slender, nearly erect, and in our specimen 5-flowered. Its colour is a pale brownish green, unless at the joints where it is a

* See folio 1542.

dark brown. *Sepals* yellowish, marked along the back with reddish spots. The upper one which rises erect is spathulate, and appears to have generally 2 spots on the inside behind the column. The two lower ones are small and acuminate. *Labellum* bright yellow, beautifully spotted. The lower division is crescent-shaped, two-lobed, slightly unguiculate and fringed at the outer edge. The upper divisions are panduriform and of the same colour as the lower, but without the spots. The centre is tuberculated and ciliated on each side. *Column* deep yellow, with comparatively large spreading wings."

For the previous note we are indebted to Mr. Booth, under whose successful management this and several other curious new Orchideous plants have been raised at Carclew. The species is nearly allied to O. Cebolleta, from which, however, its characters abundantly distinguish it.

1793

* CORYANTHES maculàta.

Spotted Coryanthes.

GYNANDRIA *MONANDRIA.*

Nat. ord. Orchideæ, § Vandeæ, *Lindl. (Introduction to the Natural System of Botany, p.* 262.)

CORYANTHES, Hooker. Perianthium patentissimum. *Sepala* dilatata, flexuosa, conduplicata; lateralibus maximis, basi distinctis. *Petala* multò minora, erecta. *Labellum* unguiculatum, maximum, galeatum, cum basi columnæ continuum, nullo modo articulatum, tridentatum, in medio unguis appendice poculiformi circumdatum. *Columna* teres, basi bicornis, elongata, apice recurva, bialata. *Stigma* rima transversa. *Anthera* bilocularis. *Pollinia* 2, compressa, posticè sulcata, caudiculâ lineari arcuatâ, glandulâ lunatâ apicibus approximato-recurvis.——Herbæ *epiphytæ, pseudo-bulbosæ.* Folia *striata.* Racemi *penduli.* Flores *maximi.* Lindl. Gen. et Sp. Orch. 159.

C. *maculata;* foliis lato-lanceolatis, sepalis lateralibus semicordatis rectis, labello intùs purpureo maculato cornubus elongatis.

C. maculata. *Hooker Bot. Mag. t.* 3102. *Lindl. l. c.*

Sepala *membranacea, tenuia, citò flaccida, supremum lanceolatum acuminatum undulatum, lateralia maxima* (2½*-pollicaria*) *libera, basi angusta, semicordata, primùm explanata vespertilionis alarum more, dein flaccida.* Petala *lanceolata undulata flexuosa in labelli galeam prona.* Labellum *carnosum,* 2½ *poll. longum, unguiculatum, luteo purpureoque variegatum; hypochilio dilatato lateribus inflexis galeato extus rotundato, in mesochilio angusto canaliculato transeunta; epichilio maximo minus carnoso galeato apice ovato inflexo.* Columna *libera, basi cornubus duobus elongatis succum stillantibus instructa, apice clavata et recurva marginibus in alasproductis.* Stigma *rima transversa.*

A native of the woods in Demerara, where it is not uncommon hanging from the branches of trees, and suspending in the air the singular lips of its flowers like fairy buckets, as if for the use of the birds and insects that inhabit the surrounding foliage.

---

* Literally, Helmet-flower; in allusion to the peculiar form of both the hypochilium and epichilium.

There certainly is not a more singular genus than this in the whole vegetable kingdom, nor one whose flowers are less like flowers to the eye of the ordinary observer. The sepals are of the most delicate texture; when young they spread equally round the centre, but after a few hours they collapse, and assume the appearance of a bat's wing half closed. The lip is furnished near its base with a yellow cup, over which hang two horns constantly distilling water into it, and in such abundance as to fill it several times; this cup communicates by a narrow channel formed of the inflated margin of the lip, with the upper end of the latter, and this also is a capacious vessel very much like an old helmet, into which the honey that the cup cannot contain may run over.

C. maculata differs from *C. speciosa*, not only in the colour of the flowers and the greater breadth of its leaves, but also in its lateral sepals being almost exactly half-cordate, without any bend in the middle, and in the horns of the base of its lip being much longer; the middle lobe of its lip too is less distinctly trifid.

C. macrantha, which has lately flowered at Mr. Knight's, surpasses both the others in the richness and magnitude of its flower, and is decidedly distinguished by the mesochilium being irregularly plaited and ridged.

Our drawing of this species was made from a specimen communicated by the Horticultural Society in July 1833.

1795

* GOVÉNIA supérba.

Superb Govenia.

GYNANDRIA *MONANDRIA.*

Nat. ord. Orchideæ, § Vandeæ, *Lindl. (Introduction to the Natural System of Botany, p.* 262.)

GOVENIA, Lindl. Perianthium bilabiatum. *Sepala* lateralia falcata, labello supposita, basi paululum connata, supremo paulò majora. *Petala* sub sepalo supremo conniventia, breviora, obliqua. *Labellum* integerrimum, ecalcaratum, concavum, cum basi parum productâ columnæ articulatum, sessile. *Columna* basi paulò producta, teres, subfusiformis, apice utrinque marginata. *Anthera* calyptriformis, 1-locularis. *Pollinia* 4, solida, incumbentia, caudiculâ brevi, glandulâ minore triangulari.——Terrestres. Folia *plicata.* Spicæ *radicales, multifloræ.* Flores *speciosi.* Lindl. gen. et sp. Orch. 153.

G. *superba;* labello ovato-cordato, spicâ cylindraceâ, bracteis acuminatis, foliis oblongis acuminatis basi angustatis scapo æqualibus.

G. superba, *Lindl. in Lodd. Bot. Cab. t.* 1709. *Gen. & Sp. Orch. l. c.*

Maxillaria superba. *Lexarsa et Llave Orch. Mex.* 2. 13.

Caulis *cum foliis 5-pedalis.* Folia 3*-pedalia, et ultra, basi vaginis purpurascentibus vestita.* Racemus 1-1½*-pedalis, cylindraceus, floribus onustus.* Sepala 7 *lineas longa, lateralia semi-oblonga.* Petala *obliquè rhomboidea acuminata.* Labellum 3 *lineas tantum longum, lividum, unguiculatum, cordatum, ovatum, obtusum, suprà canaliculatum.*

This noble species was originally discovered by Messrs. Lexarza and La Llave on the mountains near Valladolid, a town of Mexico, flowering in August. The native name is *Azuzena amarilla.* It is one of the handsomest plants of the whole order of Orchideæ, and is especially remarkable for its stately appearance, the rich orange of its flowers, and the long time they continue to open. The specimen in Messrs. Loddiges' collection grows to the height of a man; that from which our drawing was made was furnished by Mr. Bateman, in February 1834.

---

* So called in compliment to James Robert Gowen, Esq. under whose care were conducted many of the curious experiments upon cross fertilization at Highclere, the seat of the Earl of Carnarvon.

It is found to bear the hottest and dampest stove, but like all the terrestrial species requires a season of repose.

Beautiful as is this plant it must be equalled in the showiness of its blossoms by the *Maxillaria liliacea,* or *Azuzena del Monte* of the people of Valladolid, in the neighbourhood of which town it grows beneath the shade of trees. This plant is described as having a scape about a foot high, terminated by a thryse of snow-white flowers, whose segments are variegated with purple lines, and whose lip is spotted at the base with yellow. It is also in all probability a *Govenia,* and may be named *G. liliacea.*

Of a similar noble aspect is a species collected in Mexico by Count Karwinski, near Atotovilco el Chico, in a cool climate, flowering in June and July. Specimens of this plant, belonging to the Royal Herbarium of Munich, are now before us. The leaves are oblong-lanceolate, tapering to each end, especially to the point, a foot long and about 5 inches wide, and very like those of *Govenia superba*, except that they are smaller. The stem is stated to be five feet high. It is terminated by a short and very dense headed raceme of large flowers, to each of which is an oblong obtuse bract, not quite so long as the ovary. The raceme is 4 inches long, and as much broad at the base. In the form of the flowers it resembles *Govenia superba* very much, but they are three times as large; the sepals are an inch and half long and the other parts in proportion. The lip has an ovate form, but is not cordate at its base, it is not quite half so long as the sepals, and is marked towards its end with largish round very dark dots. The colour of the flowers cannot be ascertained from the dried specimen, but seems to have been pale and uniform; probably pale yellow. This plant may be thus distinguished —

G. *capitata*; labello ovato obtuso, racemo capitato, bracteis oblongis obtusiusculis, foliis oblongis acuminatis basi angustatis scapo multò brevioribus.

1797

* PLEUROTHÁLLIS Gróbyi.

The Groby Pleurothallis.

GYNANDRIA *MONANDRIA.*

Nat. ord. ORCHIDEÆ, §. MALAXIDEÆ, *Lindl.* (*Introduction to the Natural System of Botany, p.* 262.

PLEUROTHALLIS.—Suprà, vol. 9. *fol.* 759.

P. *Grobyi*; folio obovato emarginato caule triplo longiore racemo laxo erecto multoties breviore, bracteis minimis membranaceis, sepalis costatis oblongis acutis lateralibus apice tantùm sejunctis, petalis lanceolatis acutis, labello lineari obtuso carnoso superne unisulcato.

P. Grobyi. *Bateman in litt.*

Folia *cum caulibus brevibus secundariis cæspitem efficientibus, atroviridia.* Racemi *laxi flexuosi.* Sepala *intus vitellina extus sanguineo vittata.* Petala *flava medio sanguinea.* Labellum *petalis paulò longius, sanguineum apice luteum.* Columna *semiteres, marginata, clinandrio anticè bidentato.* Pollinia 2, *parva, in unum ferè confluentia, basi* (*apice*) *pellucida sed glandulæ nullæ se affigentia.*

A native of Demerara, whence it was imported by Mr. Bateman, who named it in compliment to Lord Grey of Groby, of whose skill and ardour in the cultivation of these curious plants we have already offered our own feeble acknowledgment. (See Grobya Amherstiæ, fol. 1740.)

It is nearly related to *P. picta* from the same country, and with it belongs to a very pretty set of species, all of which are characterized by having their leaves so much longer than the secondary stems as to conceal them and to form dense tufts.

It is readily cultivated in the Orchideous house; but where the atmosphere is not very damp, will require to be kept under a bell-glass.

Our drawing was made at Messrs. Loddiges in April last.

* See folio 1298.

WE take this opportunity of filling a spare leaf with the characters of some unpublished species of Pleurothallis, together with a new arrangement of the genus, and a few remarks upon *Specklinia* and *Physosiphon.*

PLEUROTHALLIS.

§. 1. *Acaules. Flores aggregati v. solitarii.*

Sp. 1. ruscifolia; 2. laurifolia; 3. succosa; 4. tribuloides; 5. cordata.

6. P. *grandiflora*; folio perfoliato oblongo-lanceolato caudato apice tridentato, pedunculis unifloris fasciculatis, sepalo supremo ventricoso acuminato lateralibus in unum connatis recurvis triplo latiore, petalis lineari-lanceolatis hispido-ciliatis, labello breviore unguiculato subrotundo carnoso ciliato supra unguem excavato.——In *Peruvia*; supra arbores in convalle Lloæ, alt 8000-ped, *Hall.* (*hab. s. sp. comm. cel. Hooker.*)

7. P. *angustifolia*; caulibus ascendentibus folio lineari-lanceolato longioribus, floribus pedicellatis subsolitariis, sepalis carinatis, petalis lanceolatis denticulatis, labello lineari acuto marginibus pone basin incrassatis.——In *Mexico*, prope Xalapam, *Deppe.* (*h. s. sp.*)

8. P. *bidentata*; caulibus ascendentibus folio oblongo-lanceolato apice bi-tridentato longioribus, floribus pedicellatis subsolitariis, sepalis ecarinatis, labello lineari obtuso apice ciliato.——Prope *Rio Janeiro, Forbes.* (*h. s. sp.*)

9. P. *papillosa*; caulibus strictis folio angustè oblongo apice tridentato paulò longioribus, pedunculis subaggregatis unifloris flore vix longioribus, sepalis campanulatis lateralibus semiconnatis supremo oblongo brevioribus, labello subhastato obtusissimo apice vittisq. duabus elevatis carnosis papillosis——In *Braziliæ* prov. Minarum in arboribus ad S. Joâo Baptista, *Martius* (*ex. s. sp. in hb. M.*) Flores majusculi.

10. P. *discoidea*; caulibus strictis folio ovato-lanceolato brevioribus, pedunculis solitariis unifloris, sepalis oblongis acutis lateralibus omnino connatis, petalis linearibus acuminatis decurvis, labello cordato-ovato undulato obtusiusculo sepalis subæquali.——In insula *Trinitatis* (*v. ic. pict. in hb. Hooker.*) Facies *P. ruscifoliæ.* Flores flavi maculâ sanguineâ oblongâ anticè bidentatâ in medio labelli.

11. P. *punctata*; caulibus strictis folio crasso ovali plano-convexo marginato longioribus, floribus solitariis breviter pedicellatis, petalis lanceolatis apice pubescentibus, labello unguiculato cordato-sagittato obtuso, columna elongata.——In *Brazilia* (*exam. v. c.*). Folia subtus purpureo punctata. Flores atropurpurei. Petala herbacea. Tota planta vix $1\frac{1}{4}$ poll. alta.

§. 2. *Acaules. Flores racemosi.*

a. *Folia caulibus secundariis longiora* (Cæspitosæ).

12. alpestris; 13? laxa; 14. Grobyi.

15. P. *picta*; folio spathulato marginato retuso racemis laxis duplò breviore, bracteis minimis, sepalis acuminatis lateralibus apice tantùm sejunctis, petalis lineari-lanceolatis acutis, labello lineari obtuso carnoso supra 1-sulcato.——In *Demerara, Loddiges* (*exam. v. c.*).

16. P. *hymenantha*; folio lineari-lanceolato apice obsoletè 3-dentato racemis flexuosis capillaribus subæquali, bracteis minimis, sepalis acutis lateralibus semi-sejunctis, petalis acuminatis parum brevioribus, labello oblongo obtuso subundulato complicato.——In *Brazilia*; in arboribus vetustis prope Retiro provinciæ Sebastianop. *Martius.* (*exam. s. sp. in hb. M.*)

17. P. *ochreata*; folio crasso lanceolato complicato mucrone acuto racemo denso flexuoso nutante duplò longiore, bracteis cucullatis imbricatis acutis, sepalis carinatis acuminatis lateralibus semiconnatis, petalis lineari-oblongis nanis mucronatis, labello . . . . .——In *Brazilia*; in rupibus Serra do Monte Santo provinciæ Bahiensis, *Martius.* (*exam. s. sp. in hb. M.*) Calyx ruber v. miniatus. Petala duplo minora flava, omnia subæquali. *Mart. mss.*

18. P. *rupestris*; folio tereti mucronato anticè sulcato racemo stricto rigido paulò breviore, bracteis parvis rigidis concavis, sepalis acutis lateralibus semiconnatis, petalis lanceolatis ciliatis, labello subtrilobo membranaceo obtuso appendice magno carnoso bifurco: cruribus parallelis porrectis acutis.——In *Brazilia*, inter rupes M. Itacolumi et Moro de V^a. R^a. provinciæ Minarum, *Martius.* (*exam. s. sp. in hb. M.*) Scapus rubenti-virens. Sepala purpurascentia. Petala purpurea. *Mart. mss.*

19. P. *teres*; folio tereti racemo gracili breviore, bracteis membranaceis ochreatis, floribus pendulis, sepalis acuminatis lateralibus apice tantum liberis, petalis minimis oblongis acutis, labello lineari canaliculato acuto marginibus incrassatis, columnâ basi pubescente.——In *Brazilia, Loddiges* (*exam. v. c.*). Flores cinnamonei. Folia omnino *Leptotis bicoloris.*

20. P. *hians*; caule brevi, folio oblongo basi angustato obtuso racemo flexuoso 6-9-floro duplò breviore, bracteis minutis ochreatis, floribus bilabiatis: labio postico 2-dentato antico oblongo obtuso, petalis cuneatis, labello ovato acuto complicato.——In *Brazilia*, in montibus Organ (*v. ic. pict. et exam. fl. s. in herb. Hooker.*) Caulis pollicaris. Folium 3-pollicare. Scapus fusco purpureus. Flores ejusdem coloris, aciculatìm maculati, extùs pallidè cinnamonei. Sepalum supremum anticum, lateralibus connatis posticis æquilatum. Petala labellumq. nana atropurpurea. Columna ejusdem coloris clinandrii margine membranaceo latissimo.

21. spiralis. 22. racemiflora.

b. *Folia caulibus secundariis breviora v. æqualia.* (Macropodæ.)

23. prolifera; 24. saurocephala; 25. quadrifida (=Dendrobium quadrifidum. *La Llave*); 26. pulchella; 27. Lanceana.

28. P. *Jamiesoni*; folio lineari-oblongo carnoso submucronato racemo secundo stricto duplò breviore, bracteis densis brevibus ochreatis cucullatis, sepalis oblongis subæqualibus leviter carinatis lateralibus basi tantum connatis, petalis duplo brevioribus apice rotundatis, labello breviore membranaceo rhombeo-trilobo plicâ unguis transversâ.——In *Peruvia*, in præcipitiis Pichinchæ haud procul a Quito, *Jamieson.* (*exam. s. sp. in hb. Hooker.*) P. pulchellæ valdè similis, sed minor.

29. P. *capillaris*; folio lineari-lanceolato racemis capillaribus fasciculatis subæquali, sepalis acuminatis lateralibus basi connatis, petalis ferè æquilongis serratis acuminatis, labello lanceolato basi angustato acuto membranaceo 3-venio. ——In *Brazilia*, in arborum truncis provinciæ Sebastianop. *Martius.* (*exam. s. sp. in hb. M.*) Facies P. *ruscifoliæ.*

30. lanceolata.

31. P. *sclerophylla;* folio oblongo petiolato obtuso carnoso racemis multò breviore, bracteis ovatis cucullatis, sepalis acuminatissimis omnibus ferè liberis, petalis nanis obtusis, labello oblongo conduplicato apice piloso-glanduloso.——In *Brazilia;* in Serra de Piedade, prov. Minarum, etiam in rupibus campestribus prov. S. Pauli, *Martius. (exam. s. sp. in hb. M.)*

32. P. *rubens;* folio oblongo-lanceolato obtuso, racemis longissimis, bracteis oblongis ochreatis, sepalis lanceolatis acuminatis lateralibus basi tantùm connatis petalis obtusis, labello obovato repando obtuso lineis 2 elevatis flexuosis carnosis. ——In *Brazilia;* in rupibus ad Opp. S. Joao d' El Rey, *Martius. (exam. s. sp. in hb. M.)*

§. 3. *Caulescentes;* i. e. rhizomate nullo sed caule folioso.

33. macrorhiza. 34. caulescens.

Under the name of *Stelis tubata,* Messrs. Loddiges have published a very curious plant with the habit of Pleurothallis, but with its sepals all united into a tube, which is inflated at the base and contracted at the mouth. Otherwise its fructification has the structure of *Stelis.* I propose to distinguish this as a peculiar genus, with the following name and character.

PHYSOSIPHON.

Calyx tubulosus basi ventricosus, apice 3-fidus. Petala in fundo calycis carnosa, nana. Labellum et columna Stelidis. Pollinia 2, sphærica.——Herbæ epiphytæ habitu Pleurothallidis.

1. P. Loddigesii; (=Stelis tubata, *Lodd. Bot. Cab. t.* 1601.)
2. P. emarginata; (=Pleurothallis emarginata, *Lindl. Gen. et. Sp. Orch. Pl. p.* 6.)
3. P. *spiralis;* folio angustè oblongo subemarginato carnoso spicâ spirali confertâ duplò breviore, bracteis minimis ovatis acutis cucullatis, calyce semitripartito laciniis erectis acutis, petalis cuneatis carnosis, labello ovato-subrotundo membranaceo apiculato trivenio.——In *Brazilia,* ad Portum S^ae^. Catharinæ, *Tweedie. (hab. s. sp. comm. Cel. Hooker.)*

SPECKLINIA.

To this genus is to be added the following:—

S. *atropurpurea;* vaginis caulis 2-3 ventricosis, folio oblongo plano 7-9-venio, flore solitario subsessili, petalis 3-cuspidatis, labelli laminâ sagittatâ.——In *Jamaica. (Exam. s. cult. in herb. Hooker. ex horto Liverp.)* Folium 2½ poll. longum, basi angustatum. Vaginæ infundibulares, pollicem longæ, in caulem 2-pollicarem. Flores atropurpurei. Ovarium lineam longum, hexapterum; alis apice truncatis petalinis minoribus. Alabastrus quem tantum vidi semunciam longus, angustus, falcatus. Sepala anteriora basi gibbosa, oblongo-linearia. Petala nana, vix 2 lineas longa, ovata, 3-cuspidata, obliqua, sæpiùs venâ alterâ excurrente, alterâ in dimidio majore petali abbreviatâ. Labellum unguiculatum, in columnam pronum, cum ejus basi articulatum; ungue lineari, 3-venio axi cristato, laminâ sagittatâ apicibus posticis obtusis. Columna petaloideo-alata, apice tridentata, basi paulò producta; pedis marginibus carnosis rotundatis medio unidentatis. Anthera parva 1-locularis. Pollinia 2, pyriformia, atropurpurea, apice materie diaphanâ granulosâ stigmati adhærentia.

And probably several spurious Dendrobia of authors, especially—

S. retusa	=	Dendr. retusum, *La Llave.*
S. scariosa	=	Dendr. scariosum, *Do.*
S. pusilla	=	Dendr. pusillum, *Humb. et Kunth.*
S. acuminata	=	Dendr. acuminatum, *Do.*
S. elegans	=	Dendr. elegans, *Do.*

1799

* MAXILLÁRIA crócea.

Saffron-coloured Maxillaria.

GYNANDRIA *MONANDRIA.*

Nat. ord. ORCHIDEÆ, § VANDEÆ, *Lindl. (Introduction to the Natural System of Botany, p.* 262.)

MAXILLARIA.—Suprà, vol. 11. *fol.* 897.

---

§. 2. *Scapigeræ.* Pedunculi radicales. Labellum anticum. Scapi uniflori.

M. *crocea;* pseudobulbis oblongis compressis foliatis, foliis oblongis undulatis obtusis emarginatis latè vaginantibus, scapis erectis unifloris laxè vaginatis foliis duplo brevioribus, sepalis petalisque angustè triangularibus apice elongatis abruptè acutis, labello obovato obtuso levissimè trilobo apice carnoso crispo supra medium unituberculato.

We have received the drawing and following account of this plant from Mr. Booth.

"We owe the introduction of this curious plant to Capt. Sutton, of his Majesty's Packet establishment at Falmouth, who imported it from Rio Janeiro in the spring of 1833, and presented it, with others, to Sir C. Lemon, Bart., M.P., in whose collection at Carclew, Cornwall, the subject of the present notice produced its flowers in August 1834.

"*Root* fibrous. *Pseudo-bulbs* small, oblong, compressed, nearly covered by the three sheathing, strap-shaped leaves, which are each thick and rigid, from four to six inches long, and an inch in width, a little twisted, with an oblique sharp point. The leaf on the bulb is similar to the others but much longer. All of them are of a deep green above, and a pale silvery green beneath. *Scapes* four inches high, one-flowered, round and slender, rising from the base of the leaves, and covered with three or four, thin, pale green, acuminate bracts, forming a joint at their junction with the stem. *Flowers* before expansion having some resemblance to the beak of a bird, pale green with a brownish tinge at the spur. *Perianth* of a greenish yellow, becoming of an orange colour when expanded. *Sepals* narrow and pointed, the three exterior ones being each about the same size and length, with their edges turned back. The two interior ones are rather shorter than the others, somewhat curved, and crossing so as almost to conceal the column. *Labellum* about half the length of the sepals, thick and fleshy, connected to the lengthened base of the column, which forms the spur; cucullate and crisp at the margin, which is rather paler than the sepals, otherwise it is of a brownish red colour, together with the column.

"It is obviously allied to *M. picta* and *punctata,* from both which the character assigned to it, abundantly distinguish it.

"Like most of the terrestrial Orchideous plants from Tropical countries, it requires the constant heat of the stove, and to be kept rather moist. It seems to flourish in the soil we find in the hollow decayed trunks of trees, mixed with some finely chopped moss; and promises to increase freely by off-sets."

* See folio 1428.

1800

* STANHÓPEA oculáta.

Eyed Stanhopea.

GYNANDRIA *MONANDRIA.*

Nat. ord. ORCHIDEÆ, § VANDEÆ, *Lindl. (Introduction to the Natural System of Botany, p.* 262.)

STANHOPEA.—Suprà, vol. 18. *fol.* 1529.

S. *oculata;* labello medio constricto, hypochilii elongati lateribus cornutis erectis basi bimaculatis, epichilio ovato acuminato cornubus duobus baseos incurvis. *Lindl. gen. et sp. Orch. Pl. p.* 158.
Ceratochilus oculatus. *Lodd. Bot. Cab. t.* 1764.

Imported by Messrs. Loddiges. It is the most interesting of this splendid genus, on account of the extremely delicate waxy appearance of its surface, the softness of its ground colour, and the richness of the deep purple spots, which lying upon a bright yellow field so very conspicuously ornament the base of the lip and the petals.

Our drawing was taken from a most noble specimen produced in July 1834, in the stove of Mr. Bateman. The leaves, including the petioles, were more than two feet long, and being of the darkest green had a most imposing appearance. Mr. Bateman informs us that he finds it requires eight such leaves to form a flowering plant; and that those who wish to succeed in blossoming it must suffer it to remain a long time undisturbed, since a good many pseudo-bulbs are required to furnish one really fine specimen.

The singular form of the labellum of this and some other Orchideous plants has rendered it necessary to distinguish it into three parts for the convenience of description.

---

* See folio 1529.

The lowest part of the lip is the *hypochilium*, the upper end the *epichilium*, and the intermediate portion the *mesochilium*. All these parts are, however, nothing more than divisions in a very fleshy lip, produced by contractions of its sides, and by the lobes, so common in the order, variously arranged and combined.

The species is at once known by the narrowness and length of the hypochilium, independently of all other circumstances.

It is said to be a native of Brazil; but we suspect some mistake in this statement. It is certainly wild in Mexico, for it exists among Count Karwinski's plants in the Royal Munich Herbarium.

1802

* MAXILLÁRIA picta.

Painted Maxillaria.

GYNANDRIA *MONANDRIA.*

Nat. ord. ORCHIDEÆ, § VANDEÆ, *Lindl.* (*Introduction to the Natural System of Botany, p.* 262.)

MAXILLARIA.—Suprà, vol. 11. *fol.* 897.

---

§. 2. *Scapigeræ.* Pedunculi radicales. Labellum anticum. Scapi uniflori.

M. *picta;* pseudobulbis ovatis subsulcatis 1-2-phyllis, foliis oblongo-lanceolatis planis, scapis erectis unifloris distanter vaginatis foliis duplò brevioribus, sepalis petalisque lanceolatis (abruptè acutis) sub-conniventibus, labello trilobo medio unituberculato lobis oblongis obtusis lateralibus brevioribus.

M. picta. *Hooker in Bot. Mag. t.* 3154. *Lindl. gen. & sp. Orch. p.* 146.

A truly beautiful plant found wild in the Organ mountains of Brazil, and originally introduced by the late lamented Mrs. Arnold Harrison. When well managed it throws up a profusion of its gay speckled flowers, which remain perfect for some time.

Our drawing was made in the stove of the Messrs. Loddiges in November 1834.

There are two other unpublished species of Maxillaria, belonging to this section, which we take this opportunity of naming.

1. M. *rufescens;* pseudobulbis ovatis subtetragonis monophyllis, foliis lanceolatis utrinque acuminatis, scapis unifloris prostratis vaginis minimis distantibus, sepalis petalisque oblongis conformibus obtusis, labello oblongo trilobo etuberculato laciniis lateralibus minimis acutis intermediâ elongatâ emarginatâ.

A native of Trinidad, whence it was imported by Mr. Lowe of Clapton. Our description is taken from a drawing and flower transmitted from Chatsworth in December 1834.

---

* See folio 1428.

It is a species of no beauty. The sepals are a dull greenish purple, the petals and lip yellow, the latter speckled with dull purplish red. Possibly the specimen was bleached, and the colours would become brighter if the plant had flowered in brighter weather. Stands next M. *picta*.

2. M. *graminea;* pseudobulbis nullis, foliis gramineis recurvis, perianthio campanulato, sepalis oblongis obtusis lateralibus paululum connatis, petalis angustioribus, labello cum pede longè producto columnæ articulato trilobo: lobis lateralibus semi-ovatis acutis ascendentibus intermedio majore reniformi, tuberculo disci magno carnoso truncato posticè bilobo.

A native of Demerara; imported by Mr. Lowe; flowered in Messrs. Loddiges' collection, January 1835. A small and inconspicuous species. The blossoms are pale yellow, with the segments a little banded with purple near the base. Not very nearly related to any known species.

1804

* MAXILLARIA densa.

Dense-flowered Maxillaria.

GYNANDRIA *MONANDRIA.*

Nat. ord. ORCHIDEÆ, § VANDEÆ, *Lindl.* (*Introduction to the Natural System of Botany, p.* 262.)

MAXILLARIA.—Suprà, vol. 11. *fol.* 897.

---

§. 1. *Axilliflorœ.* Pedunculi axillares.

M. *densa;* pseudobulbis oblongis compressis axillaribus monophyllis, foliis oblongo-lanceolatis obtusis emarginatis, racemis axillaribus densissimè aggregatis, bracteis cucullatis, perianthiis bilabiatis, sepalis lineari-lanceolatis acuminatis carinatis, petalis paulo minoribus, labello oblongo indiviso apice recurvo et canaliculato, medio lineâ transversâ elevatâ.

A native of Mexico, whence it was imported by the Messrs. Loddiges, in whose stove it flowered in January last.

It belongs to that set of Maxillarias on which the genus was founded by the authors of the Flora Peruviana, and which in some respects appear to differ from those with which we are but acquainted in Gardens. The stems rise some height above the ground, and are closely invested with brown withered scales, from the axils of which spring the pseudo-bulbs and flowers; in most Maxillarias on the contrary, there seems to be no stem, but the pseudo-bulbs sit close upon the ground, as in *M. picta* and *ochroleuca*. In truth, however, there is a stem in both kinds; only it is erect and above ground in the one, and prostrate or underground in the other. And so it is with all the pseudo-bulbous Orchideæ; their true stem is in most cases a rhizoma, like that of Iris, rooting on the side that touches the ground, and pushing up leaves and pseudo-bulbs on the other.

---

* See folio 1428.

1806

FERNANDÉZIA acúta.

** Sharp-leaved Fernandezia.*

GYNANDRIA *MONANDRIA.*

Nat. ord. ORCHIDEÆ, §. VANDEÆ, *Lindl.* (*Introduction to the Natural System of Botany. p.* 262.)

FERNANDEZIA, Fl. Peruv. Perianthium patens. *Sepala* libera. *Petala* conformia, sub sepalo supremo conniventia. *Labellum* anticum v. posticum, ecalcaratum, liberum, trilobum, disco tuberculatum. *Columna* brevis, utrinque aurita. *Anthera* bilocularis. *Pollinia* 2, solida, obovata, *glandulâ . . . . . caudiculâ . . . . . .*—Epiphytæ *caulescentes.* Folia *disticha, equitantia, imbricata.* Racemi *pauciflori, terminales v. laterales.* Flores *parvi lutei.* Lindl. Gen. et Sp. Orch. p. 207.

F. *acuta;* foliis acuminatis carinatis, corymbo laxo multifloro, bracteis obtusis membranaceis, labello lineari 3-lobo laciniis lateralibus rectis abbreviatis intermediâ emarginatâ obtusâ, disco pulvinato tuberculo minimo pone basin.

Omnia fere F. elegantis *nisi folia et labellum. Hoc pallidè luteum disco prominulo pulvinato anticè bilobo sanguineo-marginato; laciniis lateralibus nullo modo hastæ effigiem referentibus sed laciniâ intermediâ parallelis.*

A native of Trinidad, whence it was imported by Mr. Knight, in whose collection it flowered in June last.

In habit is very like *F. elegans*, from which it is principally distinguished by the tapering form of its leaves, and the shape of its lip, which, instead of having a sort of hastate form, in consequence of the divergence of its lateral lobes, and a number of irregular unequal tubercles upon its disk, has the lateral lobes parallel with the middle one, only a single small tubercle at its base, and a slightly elevated disk which is 2-lobed, and bordered with crimson in front.

A hot damp stove is necessary to keep this rare species in health.

---

* So called after George Garcias Fernandez, a Spanish Botanist, who is unknown, except through the medium of this genus.

1809

*VÁNDA téres.

Taper-leaved Vanda.

GYNANDRIA *MONANDRIA.*

Nat. ord. ORCHIDACEÆ, § VANDEÆ. (*Introduction to the Natural System of Botany, p.* 262.)

VANDA.—Suprà, vol. 6. *fol.* 506.

V. *teres;* foliis teretibus, racemis ascendentibus subbifloris foliis æqualibus, sepalis oblongis obtusis: supremo erecto lateralibus semitortis labello suppositis, petalis majoribus suborbiculatis undulatis, labello basi conico: laciniis lateralibus ascendentibus subtruncatis intermediâ pubescente apice dilatatâ truncatâ emarginatâ. *Lindl. gen. et sp. Orch. p.* 217.

This superb epiphyte was originally discovered upon trees in Sylhet by Dr. Wallich, and it has more recently been met with by Mr. William Griffith abundantly near Medown in the Burmese Empire, also growing on trees in the woods.

When Dr. Wallich came to England, this species formed one of the numerous living plants brought home under his care, and blossomed during the voyage; the very flowers which were then produced we are so fortunate as to possess in spirits, through the kindness of their liberal discoverer; they measure nearly four inches and a half from the tip of one petal to that of the other, but only two were formed upon the spike. The specimens from Mr. Griffith are not quite so large in the flower, but there are three blossoms to each spike.

---

* Vandá is the Sanscrit name of the original species of this genus.

That from which our drawing was taken was produced at Syon, in the collection of his Grace the Duke of Northumberland, by whose permission we are enabled to publish it. It was not quite so fine as what we have represented, but considering what the native specimens are we have felt quite justified in forming our figure between the two. The flowers in the accompanying plate, though larger than those on the plant at Syon, are considerably smaller than those we possess in spirits.

Nothing can exceed the flowers of this plant in delicacy of texture or softness of colour, the deep purple of the petals softens away to the margin, and seems to melt as it were into the purer white of the sepals, while the rich crimson and yellow of the lip renders the brilliancy of the other parts still more conspicuous.

We found the plant had been managed by having its stem covered with moss, which if kept damp would maintain the surface in a uniform state. We do not, however, perceive any trace of the roots in our wild specimens having been growing among moss, and we consider it probable that it will succeed as well if the roots are exposed freely to the air.

1811

*MAXILLÁRIA cristáta.

Crested Maxillaria.

GYNANDRIA *MONANDRIA.*

Nat. ord. ORCHIDACEÆ, § VANDEÆ. (*Introduction to the Natural System of Botany, p.* 262.)

MAXILLARIA.—Suprà, vol. 11. *fol.* 897.

M. *cristata;* pseudobulbis ovatis sulcatis monophyllis, foliis oblongo-lanceolatis plicatis, scapo pendulo bifloro squamis laxiusculis vaginato, floribus explanatis, sepalis petalisque lanceolatis acutis æqualibus, labello multo minore tripartito carnoso: laciniis lateralibus falcatis intermediâ rotundatâ cristato-fimbriatâ utrinque unidentatâ, ungue subcristato disco bidentato.

Sepala $1\frac{3}{4}$ *unciam longa, alba, sanguineo interruptè striata et fasciata.* Petala *æqualia et conformia, apice sanguinea basi maculata dorso alba.* Labellum *purpureum, ungue viridi cristâ dentibusque albis;* crista laciniæ intermediæ *fimbriata pilis submoniliformibus:* unguis *e cirrhulis* 4-6 *rectis submoniliformibus constans; dentes disci compressi divaricati margine crenati posteriore duplò majore.* Columna *basi viridis, apice lutea, et utrinque subalata, rostello longissimo subulato.* Caudicula *polliniorum longissima, glandulâ parvâ subtriangulari.*

For the opportunity of publishing this remarkably beautiful epiphyte, we are obliged to Mr. Knight, of the King's Road Nursery, with whom it flowered in the stove in July last.

It is a native of Trinidad, growing on old decayed branches of trees near the Mud Lake. We have seen a drawing of it taken on the spot, and it was in no degree superior in beauty to the present specimen.

The lip is a most curious organ. It is of a firm fleshy texture, and is deeply divided into three parts; of these the two lateral ones are falcate and smooth, while the middle one is rounded, has a strong tooth on each side, and is bordered by a deep white fringe of necklace-shaped hairs; then the short stalk of the lip has four or more such hairs growing straight from its border, and the disk is furnished with a large white two-lobed flat tooth, the divisions of which diverge from each other.

---

* See folio 1206.

The striping, banding, and painting of the delicate white flowers with rich crimson, produces a very rich and striking effect.

1815

* MACRADÉNIA triándra.

Triandrous Long-gland.

GYNANDRIA *MONANDRIA.*

Nat. ord. ORCHIDACEÆ, § VANDEÆ, *Lindl.* (*Introduction to the Natural System of Botany, p.* 262.)

MACRADENIA.—Suprà, vol. 8. *fol.* 612.

M. *triandra;* foliis coriaceis lineari-oblongis acuminatis, racemo prostrato, labello in medio trilamellato, clinandrio serrato antheris duabus sterilibus.

Pseudobulbi *cylindracei paulo attenuati.* Folia *plurivenia. Racemi foliis multò breviores.* Sepala *et* petala *intus sanguinea herbaceo limbata.* Labelli *lamellæ rotundatæ, costæformes, antrorsùm crassiores.* Antheræ steriles *ovales, integræ, sanguineo marginatæ.*

The *Macradenia lutescens* of this work (folio 612) we have never seen; but as it was described by Dr. Brown we have no difficulty in deciding, upon the evidence that the description affords, that it is specifically different from this.

No notice is taken by Dr. Brown of the existence of any lamellæ in the middle of the lip, nor of the abortive anthers which are always present; besides which the clinandrium here is regularly and strongly serrated, the leaves much narrower, and the raceme pendulous or prostrate instead of erect.

A native of Surinam, whence it was brought by Mr. Lance. It flowered in the possession of the Horticultural Society in May last. A strong damp stove heat is required to keep it in health.

* So named in allusion to the long caudicula of the pollen masses, fig. 5. from μακρος long, and αδην a gland.

1821

* EULÓPHIA lúrida.

Lurid Eulophia.

GYNANDRIA *MONANDRIA.*

Nat. ord. ORCHIDACEÆ, § VANDEÆ, *Lindl.* (*Introduction to the Natural System of Botany, p.* 262.)

EULOPHIA.—Suprà, vol. 17. *fol.* 1433.

E. *lurida;* foliis (lineari-) lanceolatis scapo ramoso multò brevioribus, bracteis minimis subulatis, sepalis lineari-spatulatis obtusis, petalis paulò latioribus, labelli tripartiti basi callosi lobis lateralibus obtusis recurvis: intermedio obcordato, calcare cylindraceo inflexo obtuso. *Lindl. gen. et sp. Orch. p.* 182.

One of the easiest of all stove Orchideous plants to cultivate, growing upon a damp wall, or a rough stone, or under almost any kind of condition, provided the air be hot and damp, and uniform. It then flowers profusely at intervals throughout the whole year. Our drawing was made in the noble collection of the Messrs. Loddiges in January last.

A native of Sierra Leone, whence great quantities are occasionally brought. It appears to be extremely common upon the trunks of trees in some parts of that colony.

* See folio 1433.

1825

* PLEUROTHÁLLIS pícta.

Painted Pleurothallis.

GYNANDRIA *MONANDRIA.*

Nat. ord. ORCHIDACEÆ, § MALAXIDEÆ.

PLEUROTHALLIS.—Suprà, vol. 9. *fol.* 759.

P. *picta;* folio spathulato marginato retuso racemis laxis duplò breviore, bracteis minimis, sepalis acuminatis lateralibus apice tantùm sejunctis, petalis lineari-lanceolatis acutis, labello lineari obtuso carnoso supra 1-sulcato. *Supra fol.* 1797.

Very near the *P. Grobyi* already figured in this work (fol. 1797), from which it principally differs in its spathula-shaped retuse leaves, and in the narrower form of all the parts of the flower; as will be seen by comparing the dissections in the two plates.

This also is a native of Demerara, and was sent us by Mr. Loddiges first in October 1834, and next in March 1835, when our drawing was made. It is a graceful pretty species, and well deserves an attentive examination.

Its tufts are rapidly formed under good management, but it seems to require the close atmosphere of a bell-glass.

Fig. 1. is a view of the column, petals, and lip. 2. The column by itself. 3. The upper side of the lip. 4. The pollen mass in situ. 5. The same separated from each other.

* See folio 1298.

1828

* DENDRÓBIUM densiflórum.

Dense-flowered Dendrobium.

GYNANDRIA *MONANDRIA.*

Nat. ord. ORCHIDACEÆ, § MALAXIDEÆ.

DENDROBIUM.—Suprà, vol. 7. *fol.* 548.

D. *densiflorum;* caulibus articulatis clavatis pendulis apice foliosis, foliis oblongis acutis nervosis, racemis lateralibus multifloris foliis longioribus; junioribus strobiliformibus, bracteis oblongis plicatis recurvis pedicellis longioribus, sepalis patentissimis ovatis obtusis, petalis conformibus majoribus, labello majore rhomboideo unguiculato serrulato retuso. *Lindl. Gen. & Sp. Orch. p.* 90. *Wall. Pl. as. rar. no.* 40.

Planta pendula, lævis, cæspitosa, arborum parasitica, pedalis-2-pedalis. Radix constans fibris numerosis, fasciculatis, teretibus, indivisis. Caules *plures, rigidi, cylindraceo-clavati, lucidi, ætate flavicantes, maximâ parte nudi, sulcis aliquot profundis totidemque angulis obtusissimis notati, vestigiis vaginarum remotis annulati, extrorsùm digitum crassi, apice foliosi.* Folia *valdè approximata, patentia, alterna, ovato-oblonga, acuta,* 5-*pollicaria, firma, subcarnosa, lucida, plana, nervosa, basi contractâ obsoletissimè petiolatâ abientia in* vaginas *breves, adpressas, deciduas. Racemus magnus, solitarius, laxè pendulus, versus apicem caulis infra folia lateralis, pedunculatus,* 6-8-*pollicaris, ovato-oblongus, obtusus, multiflorus; novellus, longè ante expansionem, sessilis, ovatus, strobiliformis,* 2-*pollicaris, valdè regulariter imbricatus bracteis adpressis, planis, alternantibus. Pedunculus crassus, cylindricus,* 2-3-*pollicaris,* bracteis *sparsis, oblongis, recurvis, nervosis, membranaceis, ferè pollicaribus, persistentibus. Flores valdè numerosi, majusculi, flavi, approximati, undique sparsi, patentissimi, insidentes* pedicellis *teretibus, pollicaribus, suffultis* bracteâ *magnâ, obovatâ, nervosâ, persistente. Sepala oblonga, obtusa, semipollicaria;* lateralia *parùm majora, basique dilatatâ infra basin columnæ in* calcar *brevissimum, subcylindricum, obtusum connata, uti* petala *ovata, subæquantia, patentissima. Labellum planum, patens, retusum, ciliato-denticulatum, suprà pubescens, basi desinens in unguem brevem, canaliculatum. Columna brevis, conica, crassa, basi leviter producta. Anthera ovata.* Wallich.

---

* See folio 1249.

" This lovely *Orchidea* comes so near to Roxburgh's *Dendrobium clavatum* (Hort. Beng. p. 63), that I should consider them as identical, if he had not ascribed bulbs to his plant, of which ours is entirely destitute. In the Calcutta garden the two plants maintain that character, although in other respects they are alike. It is possible that under particular circumstances, the bulbs or pseudo-bulbs may become elongated into clavate stems, and if so the Silhet plant would be the same species with ours from Nipal."

The above we borrow from Dr. Wallich's splendid Plantæ Asiaticæ.

Beautiful as is the specimen of this plant, represented in the accompanying figure, it is still inferior to what is produced in India, so that cultivators have still a point to gain in respect to this charming species. Our drawing was made in the stove of Messrs. Loddiges in May last.

Since the last addition to the published species of Dendrobium, we have received from Mr. Allan Cunningham specimens of a most remarkable new kind, with the following memorandum.

"*Dendrobium cassythoides;* aphyllum, caule glabro stolonifero, floribus racemosis, racemis (stolonibus oppositis) subtribrachiatis, bracteis lanceolato-ovatis appressis, labello convoluto indiviso apice incurvo crispato, [callo lineari in medio et altero ovato acuminato sub apice."]

"A leafless stoloniferous plant, with tribrachiate racemes of flowers, of which the perianths are of a brownish or golden yellow colour, and the labellum white, and elegantly pencilled within, as in D. *Pierardi.*"

EDWARDS'S

BOTANICAL REGISTER:

OR,

ORNAMENTAL FLOWER-GARDEN

AND SHRUBBERY:

CONSISTING OF

COLOURED FIGURES OF PLANTS AND SHRUBS,

CULTIVATED IN BRITISH GARDENS;

ACCOMPANIED BY THEIR

History, Best Method of Treatment in Cultivation, Propagation, &c.

CONTINUED

BY JOHN LINDLEY, Ph. D. F.R.S. L.S. AND G.S.

PROFESSOR OF BOTANY IN THE UNIVERSITY OF LONDON,
AND THE ROYAL INSTITUTION OF GREAT BRITAIN,
&c. &c. &c.

New Series.

VOL. IX.

OR VOL. XXII. OF THE ENTIRE WORK.

——viret semper——nec fronde caducâ
Carpitur.

LONDON:
JAMES RIDGWAY AND SONS, PICCADILLY.

M.DCCC.XXXVI.

1830

ONCIDIUM Russelliánum.

The Duke of Bedford's Oncidium.

Nat. ord. ORCHIDACEÆ, § VANDEÆ.
ONCIDIUM.—Suprà, vol. 13. *fol.* 1050.

B. *Labellum integerrimum.*

O. *Russellianum;* pseudobulbis ovatis costatis diphyllis, foliis ligulato-lanceolatis patentibus, racemo paucifloro radicali, sepalis petalisque conformibus ovato-oblongis subundulatis, labello postico oblongo-cuneato retuso apiculato subsinuato, lamellis disci truncatis.

Folia *lætè viridia.* Caulis *fusco-purpureus.* Sepala *et* petala *fusco-purpurea viridi marginata et costata.* Labellum *lilacinum lamellis disci purpureis albo-marginatis, quarum series tres geminatæ truncatæ sensim antepositæ pares duas includunt angustissimas pariter truncatas.* Columnæ *alæ semiovatæ erosæ acutæ luteæ ; gynizus margine purpureo basi utrinque bidentato (fig.* 1.) *; anthera et pollinia Oncidii.*

That this is a genuine species of Oncidium I by no means aver; on the contrary it differs from the general form of that genus in its undivided posterior lip, and in the two teeth which are placed below the gynizus on either side: its colour too, notwithstanding the example of O. Lanceanum, is different from what prevails in the principal part of the species of this very natural and extensive genus. Nevertheless I do not feel justified in forming a new genus out of the materials I at present possess. The plant may possibly be a transition species; and it ranges well enough with the verbal character of Oncidium.

I have named this species in compliment to his Grace the Duke of Bedford, one of the many liberal patrons of Botanical science of whom England now can boast, and in whose stove at Woburn it first appeared in Europe. It was obtained from the garden of Mrs. Moke at Tejuca near Rio Janeiro, by the Hon. Capt. J. Roos, R.N. who sent it to Woburn along with many other valuable plants in 1835.

" Found growing from the crevices of sandstone rocks, on the eastern side of Sydney Cove, Port Jackson, where it was observed in flower in October 1834."—*R. C.*

" The above is extracted from a letter from my brother, Mr. Richard Cunningham, dated August 2, 1834. He adds,

"I send you in a bottle a Dendrobium, which, not finding noticed in either printed or manuscript enumerations in my possession I have called Cassythoides, from the *primâ facie* resemblance it has to the genus Cassytha, not only in its leafless character and short racemes of flowers, but in its peculiar chocolate bronze or japanned papulose stems—it may be found, that it also resembles it in its climbing propensities."

It is not a little remarkable that so highly curious a plant as this should so long have been overlooked, although a native of a locality which, as Mr. Allan Cunningham remarks, has doubtless been traversed by Botanists of many countries in Europe, who have visited Port Jackson in ships of discovery since the Colony was founded in January 1788, viz., French, Spanish, German, Swedish, Russian, &c., besides many of our own countrymen.

1832

SARCOCHILUS falcatus.

Falcate-leaved Sarcochilus.

Nat. ord. Orchidaceæ § Vandeæ.

SARCOCHILUS, R. Br. *Perianthium* patens. *Sepala* lateralia cum ungue labelli subtùs connata. *Petala* conformia. *Labellum* ecalcaratum, cum ungue columnæ continuum, calceiforme; lobo intermedio càrnoso solido, laterali-bus ascendentibus petaloideis. *Columna* brevissima, marginibus tenuibus inflexis. *Anthera* bilocularis, valvis antheræ inferioribus deflexis erosis. *Pollinia* caudi-cula lineari affixa glandulâ deltoideâ.——Epiphyta *acaulis, vel brevissimè cau-lescens.* Folia *disticha, lineari-lanceolata, subcoriacea.* Racemi *axillares, erecti,* 3-6-*flori, secundi.* Bracteæ *breves, latæ, ovatæ.* Flores *mediocres.*

S. *falcatus.* R. Br. Prodr. 332. Lindl. g. & sp. orch. 142.

This rare plant has been sent me both by Mr. Bateman and Messrs. Loddiges. The drawing was made in the collection of the latter gentlemen in April last.

It is a native of New Holland and near Hunter's River; but so rare that I have never yet received a single native specimen except an imperfect one for which I was some years since indebted to Dr. Brown, who however saw it only in an imperfect state.

The flowering specimens, with which my garden friends have supplied me, while they have enabled me to complete my account of it, have also shown that this hitherto little known species is really an extremely pretty little plant, which, without any of the gaudy colours of many of its tribe, is so neat and simple in its appearance as to be sure to captivate the feelings of every lover of nature.

It must be treated just like other Orchideous Epiphytes.

1837

STANHÓPEA insígnis.

Noble Stanhopea.

Nat. ord. Orchidaceæ, § Vandeæ.

STANHOPEA.—Suprà, vol. 18. *t.* 1529.

S. *insignis*; labello medio quam maximè constricto, hypochilii subrotundi late-ribuscornutis incurvis, epichilio ovato medio depresso cornubus duobus baseos incurvis. *Gen. et Sp. Orch.* 157.

Epidendrum grandiflorum. *Humb. & Bonpl. Pl. æquinoct. p.* 94, *t.* 27.
Anguloa grandiflora. *H. B. K. nov. gen. & sp. pl.* 1. 343. *Kunth Synops.* 1. 332.
Stanhopea insignis. *Hooker in Bot. Mag.* 2948-9. *Lodd. Bot. Cat. t.* 1985.

Bracteæ *membranaceæ, spathaceæ, subscariosæ, striatæ, ovarii longitu-dine, ut et ovarium ipsum leviter furfuraceæ.* Ovarium *obtuse triquetrum, obclavatum, angulis intermediis costatibus minoribus.* Sepala 3 *patentia, circiter* 4 *uncias inter suos apices, ovata, carnosa, concava, obtusa, flava, lucida, intus sanguineo-guttata;* 2 *lateralibus basibus connatis.* Petala *lanceolata, undulata, reflexa, magis lutea, sanguineo interruptè fasciata.* Labellum *carnosum, cum basi columnæ non articulatum, a sepalis distinctum, basi ventricosum, albidum, atro-sanguineo confertè guttatum, limbo* 3-*lobo, lobo intermedio subcordato-ovato carnoso sanguineo-guttato, lateralibus falcatis incurvis acuminatis cornuformibus.* Columna *maxima,* 2 *uncias longa, cum ovario continua, basi semiteres sursum alata, apice obtusè triloba, sanguineo densissimè guttata.* Stigma *clausum, rostello subulato acumina-tissimo incumbente.* Anthera *apice membranacea, bilocularis.* Pollinia 2, *obovata, basi acuminatissima, extrorsum fissa.* Glandula *subrotunda, anticè subulata, posticè magis membranacea, subbiloba;* caudicula *brevis membra-nacea.*

The drawing of this plant was made from a specimen sent me in 1830 by the late Mr. Cattley; at that time it was a rarity, and the specimen was thought a fine one, but the cultivation of this genus has since become so much improved that still handsomer specimens are not uncommon. Some apology may perhaps be required for again figuring what has been well represented in the Botanical Magazine and the Botanical Cabinet; but upon comparing the accompanying plate with those of Dr. Hooker and Mr. Loddiges, I find enough to induce me not to suppress the drawing that was made for this work.

Stanhopea insignis was originally found by Messrs. Humboldt and Bonpland on the trunks of old trees in shady woods near Cuença in Quito; it has since been procured abundantly from various districts of the north-eastern part of South America. There it fixes its pseudo-bulbs upon branches, clinging to them with its numerous creeping roots, and suspending in the air its stout zig-zag spikes of fleshy wax-like flowers. The species grows freely in a mixture of coarse peat, sand, and broken tiles or potsherds, provided it is kept constantly growing; but it does not like to be periodically dried up as many others do. In order to see its curious blossoms in perfection the young spikes should be watched for, and as soon as they appear they should be artificially led over the edge of the pot or basket; otherwise their tendency to turn downwards is so great that they are apt to force themselves into the earth and so to become smothered.

In the annexed plate, the dissections are taken from a specimen, for which I am obliged to Lord Fitzwilliam, with much richer colours than usual; fig. 1. represents a lip seen from the upper side, and 2. the same part viewed from beneath; these figures show the cup-shaped base of this organ, the spotted tongue that terminates it, and the two long fleshy horns that project on each side of it.—Who shall imagine the use that such curious parts are intended for? 3. and 4. are the hardly less curious pollen-apparatus.

1841

CORYÁNTHES macrántha.

Large-flowered Coryanthes.

Nat. ord. Orchidaceæ, § Vandeæ.

CORYANTHES.—Suprà, vol. 21. *t.* 1793.

C. *macrantha;* foliis angusto-lanceolatis, pseudo-bulbis ovato-conicis altè sulcatis, labello utrinque quater plicato: plicis deflexis.
Gongora macrantha. *Hooker Bot. Misc.* 2. *p.* 151. *t.* 80.
Coryanthes macrantha. *Hooker in Bot. Mag. fol.* 3102 *in textu. Gen. & Sp. of Orchideous Pl.* 159.

Accustomed as we are now become to strange forms among Orchideous plants, I doubt whether any species has yet been seen more remarkable for its unusual characters than that now represented.

It was first figured by Dr. Hooker in the Botanical Miscellany, from a specimen in spirits sent him from the Caraccas by Mr. Lockhart. When the plant blossomed in Trinidad, the flowers appeared so extraordinary to those who saw them that the visitors to the Botanic Garden supposed them to be artificial. It has, however, lately flowered in the collection of Mr. Knight in the King's Road; and it proves to be in all respects the same as the specimen sent to Dr. Hooker.

The plant has the habit of a Stanhopea, or a Gongora; and pushes forth from the base of its pseudo-bulbs a pendulous scape, on which two or three flowers are developed. Each flower is placed at the end of a long stiff cylindrical furrowed ovary, and when expanded, measures something more than 6 inches from the tip of one sepal to that of the opposite one. In colour the sepals are an ochrey yellow, spotted irregularly with dull purple; they have a most delicate texture; the upper sepal falls back from the tip of the ovary, is narrow and not above one half the length of the two lateral ones, which, instead of applying themselves to the lip as is usually the case, turn directly away from it, placing themselves at an acute angle with the upper sepal, and after a while collapsing at their sides till they look something like bats' wings half at rest. The petals, which are narrowly lanceolate, very weak and much curved at the edge, have the same colour and texture as the sepals nearly, and are intermediate in length between the upper one and those at the side; they hang nearly parallel with the column, but are so placed as to conceal in no degree the lip; nature taking most especial care to exhibit this strange part in the most conspicuous manner. The lip is as fleshy and solid in its texture as the sepals and petals are delicate; it is seated on a deep purple stalk, nearly an inch long, and forming an obtuse angle with the column, and consequently an acute one with the ovary; this stalk terminates in a hemispherical greenish-purple cup, or rather cap considering its position, and the latter, contracting at its front edge, extends forward into a sort of second stalk of a very vivid blood colour, the sides of which are thinner than the centre, turned back, and marked with 4 or 5 very deep solid sharp-edged plaits. These plaited edges again expand and form a second cup, less lobed than the first, thinning away very much to the edges, of a broadly conical figure, with a diameter of at least two inches at the orifice; this second cup is of an ochrey yellow, streaked and spotted with pale crimson, and seems intended to catch a watery secretion which drips into it from two succulent horns which take their origin in the base of the column, and hang over the centre of the cup.

Of course this species will require the heat of a damp stove.

1844

* ANGRǼCUM caudátum.

Long-tailed Angræcum.

Nat. ord. Orchidaceæ, § Vandeæ.
ANGRÆCUM.—Suprà, vol. 18. *fol.* 1522.

A. *caudatum;* foliis loratis canaliculatis emarginatis, spicâ radicali pendulâ flexuosâ 4-florâ, labello obovato rostrato serrulato, calcare longissimo apice bilobo.

Epiphyta radices *crassas simplices nebulosas promens.* Caulis *simplex, brevis,* foliis *loratis canaliculatis apice fissis* 6 *poll. longis, distichè imbricatis.* Spica *pedalis, ex axilla imorum foliorum pendula; pedunculo gracili, atroviridi, ad nodos tumido; rachi flexuosa, internodiis circiter* 2 *poll. longis.* Ovarium *sessile, fusco-maculatum, basi tortum.* Sepala *et* petala *explanata, lineari-lanceolata, acuminata, herbacea, æqualia, basi fulva, ferè* 2 *poll. longa.* Labellum *ejusdem longitudinis, album, petaloideum, serrulatum, cuneatum angulis rotundatis v. potius obovatum, apice mucrone longo viridi rostratum, imâ basi involutum et denticulatum, in calcar productum* 9 *poll. longum, fulvum, subulatum apice (fig.* 3.) *obtusum et bilobum.* Columna *fulvo-viridis, brevis, erecta, basi (fig.* 1.) *crassior et angulata, sursum attenuata et marginata, gynizo plano fulvo, rostello subulato gynizo longiore.* Pollinia 2, *posticè paulò sulcata, in caudiculam longam cuneatam sessilia.*

A most remarkable new species of Angræcum, imported from Sierra Leone by the Messrs. Loddiges, in whose collection the accompanying drawing was made in August last. At present it is exceedingly rare, and is likely to remain so; for it seems to be one of the most difficult of the tribe to manage successfully. In the Nursery at Hackney it is attached to a piece of wood suspended from the roof of the stove for epiphytes.

---

* See folio 1522.

The most curious point of structure in this species is the unusual length of its spur, which measures nine inches from its base to its two-lobed apex. The only parallels to this among all the Orchideous plants I am acquainted with are those of *Habenaria longicauda* figured in the Botanical Magazine, t. 2957, and of *Angræcum sesquipedale* of Du Petit Thouars's Mascaren Orchideæ, t. 66 and 67. For what wise purpose these extraordinary appendages may have been destined by nature, we may well be unable to imagine. It would seem that they must be added to the vast list of objects which, to our confined apprehension, appear merely intended to exhibit the endless diversity of power of the Creator.

1848

* MAXILLÁRIA rufèscens.

Brownish Maxillaria.

Nat. ord. Orchidaceæ, § Vandeæ.
MAXILLARIA.—Suprà, vol. 11. *fol.* 897.

M. *rufescens;* pseudobulbis ovatis subtetragonis monophyllis, foliis lanceolatis utrinque acuminatis, scapis unifloris (prostratis) vaginis distantibus, sepalis petalisque oblongis conformibus obtusis, labello oblongo trilobo etuberculato laciniis lateralibus minimis acutis intermediâ elongatâ emarginatâ. *Suprà, fol.* 1802. *in textu.*

A native of Trinidad, whence it was imported by Mr. Lowe of Clapton. It first flowered at His Grace the Duke of Devonshire's at Chatsworth, in December 1834, whence I was favoured with a sketch; since that time it has appeared in many other collections.

It requires the usual management of a hot damp stove, in which it grows freely.

By no means one of the prettiest of the genus, nevertheless its yellow labellum richly spotted with crimson, is a beautiful object when closely examined.

This species also occurs in gardens under the name of M. fucata.

---

* See folio 1428.

1851

* ONCÍDIUM altíssimum.

Tallest Oncidium.

GYNANDRIA *MONANDRIA.*

Nat. ord. ORCHIDACEÆ, § VANDEÆ.

ONCIDIUM.—Suprà, vol. 13. *fol.* 1050.

O. *altissimum ;* pseudobulbis subrotundis compressis ancipitibus, foliis distichis ensiformibus carinatis acutis scapo decurvo multò brevioribus, racemo simplici, sepalis petalisque labelli longitudine lineari-lanceolatis undulatis, labello apice dilatato bilobo medio constricto basi auriculato, cristâ enneadactylâ, columnæ alis rotundatis crenulatis.

Epidendrum altissimum. *Jacq. stirp. amer.* 229. *t.* 141.

Planta elegans, parasitica arborum. Radices *teretes, fibrosæ, cinereæ, numerosæ.* Folia *acuta, ensiformi-oblonga, avenia, nitida, integerrima, crassiuscula, pollicem lata, sesqui-pollicem longa, basi carinata, cæterum plana : orta singula e nodo vel tubere ovato, compresso, glabro, ovi interdum anserini magnitudine ; qui ipse insidet basi folii alius radicalis, unius alteriusve, et similis. Inter hoc nodumque exsurgit* scapus *solitarius, teres, glaber, colore ferrugineo, tenuis, inclinatus, quadripedalis, superne racemosus ; qui ad pedunculos et nodos vestitur spathis membranaceis, lanceolatis, cinereisque.* Pedunculi *biflori vel triflori plerumque, distichè alterni.* Flores *inodori, flavi cum maculis fuscis, numerosi. His sunt* petala *oblonga, distincta, utrinque acuta, undulata, subæqualia.* Nectarii *labii inferioris lacinia media est subquadrata, flavaque tota sine maculis. Reliqua in charactere obtinent, sicuti in Epidendro* (Oncidio) *carthaginensi.* Jacq. l. c. 229.

I long ago suspected that two distinct species were confounded under the name of O. *altissimum,* and I even distinguished, by the name of O. *Baueri,* what appeared to be a different species from the Epidendrum altissimum of Jacquin, which is the original authority for the former name. I was afterwards led to suppose that O. *Baueri* was a mere form of O. *altissimum,* and I accordingly reduced it to a simple variety, in the genera and species of Orchideous plants. In this however I am satisfied that I was wrong, as will I think be evident from the following comparison of the two species as they were seen last year in the stove of the Messrs. Loddiges.

* See folio 1542.

O. *altissimum.* Pseudobulbs nearly round, very much compressed, and two-edged. Leaves acute. Raceme decumbent nearly simple. Colours of the flower very bright. Wings of the column rounded and a little crenelled.

O. *Baueri.* Pseudobulbs oblong, a little compressed, only slightly two-edged. Leaves broader, and rather acuminate. Raceme erect, very compound. Colour of the flowers rather dingy. Wings of the column very remarkably truncated.

It is impossible, now that these differences are made out, to combine the two plants as I once proposed. I therefore avail myself gladly of an opportunity afforded me by Messrs. Loddiges of figuring the true O. *altissimum ;* and for the satisfaction of those who may not have access to the work in which it was first described, I have added the very words of Jacquin in speaking of that species.

The O. altissimum figured at fol. 1651 is the true O. *Baueri.*

1857

* ZYGOPÉTALUM cochleáre.

Spoon-lipped Zygopetalum.

GYNANDRIA *MONANDRIA.*

Nat. ord. ORCHIDACEÆ, § VANDEÆ.

ZYGOPETALUM, Hooker. *Perianthium* explanatum, sepalis petalisque ascendentibus subæqualibus, cum ungue producto columnæ connatis. *Labellum* muticum, indivisum, patens, ungue ascendente : cristâ magnâ transversâ carnosâ. *Columna* brevis, arcuata, semiteres. *Anthera* subbilocularis. *Pollinia* 2, bipartibilia, in glandulam transversam subsessilia.——Herbæ *terrestres, subacaules,* foliis *plicatis patentibus.* Flores *speciosi,* labello *cœruleo.* Gen. et Sp. Orch. 187.

Z. *cochleare ;* foliis plicatis pedunculis unifloris radicalibus solitariis duplò longioribus, sepalis petalisque ovato-lanceolatis conniventibus inferioribus majoribus, labello cochleato bilobo cristâ transversâ crenatâ.

Pseudobulbi *nulli.* Folia *pallidè viridia obovata vel oblonga, plicata, pedem longa, infima abbreviata petiolaria.* Pedunculi *ex axillis foliorum infimorum, basi vaginati, uniflori, erecti, apice sub ovario bracteâ cucullatâ oblongâ.* Ovarium *albidum, incurvum.* Perianthium *semi-explanatum.* Sepala *ovata, subundulata, acuta, albo-viridia, lateralia majora.* Petala *sepalo supremo conformia et æqualia.* Labellum *cum columna basi paululum elongata articulatum, concavum, unciam latum,* $1\frac{1}{4}$ *unc. longum, indivisum, emarginatum, extùs albidum, intus venis creberrimis atrocœruleis velutinis pictum ; brevè unguiculatum, supra unguem callo lato rotundato convexo plurièsplicato et cristato instructum.* Columna *teres clavata, dorso viridi-alba, fronte purpureo-striata, basi paululum producta.* Anthera *bilocularis valvis bilabiatis.* Pollinia 4, *per paria incumbentia.*

Beautiful as all the species of Zygopetalum are, without exception, this is perhaps upon the whole the most attractive, not only on account of the delicate waxy surface of the petals and sepals, and the peculiarly rich veining of the Lapis lazuli blue of its lip, but because of its delicious fragrance. If Lilies of the Valley were growing intermingled with the plants, the air could not be more perfumed with their pure and delightful odour than it is after the curious flowers have unfolded.

* So named by Sir William Hooker from ζευγνυω, to join ; in allusion to the adhesion of the segments of the perianth by their bases, in the original species.

Like all the other species of the genus, this is easily cultivated in earth in a damp stove. It is a native of Trinidad. The drawing was made from a specimen supplied by Mr. Knight, in August last.

1. represents a front view of the column, with the bases of the petals and lips adhering to it ; 2. is a view of the fleshy ridge of the lip ; and 3. shews the pollen masses and their glands.

1858

* HABENÁRIA procéra.

Tall Habenaria.

GYNANDRIA *MONANDRIA.*

Nat. ord. ORCHIDACEÆ, § OPHRYDEÆ.

HABENARIA, Willd. Perianthium Orchidis. Calcar elongatum. *Columna* libera reclinata. *Anthera* basibus loculorum solutis divergentibus canalibus stigmaticis adhærentibus. *Glandulæ* nudæ. *Rostellum* planum, antheræ adnatum. *Processus carnosi* 2 stigmatici, ultra antheram projicientes, formâ varii.—Habitus *Orchidis.* Gen. & Sp. Orch. p. 306.

A. Erostres ; *ovariis pedunculatis subcylindraceis aut fusiformibus, nunquam rostratis.*

§ 2. *Petala indivisa.*

a. Labellum trifidum, laciniis filiformibus indivisis.

H. *procera ;* caule folioso, foliis oblongis basi cucullatis patentibus sensim in bracteis decrescentibus, racemo multifloro, bracteis herbaceis inferioribus foliaceis ovarii longitudine superioribus parvis ovatis, labelli tripartiti laciniis lateralibus linearibus intermediâ latiore paulò brevioribus, calcare pendulo clavato ovario duplo longiore.

Orchis procera. *Swartz in Pers. syn.* 2. 506.

Habenaria procera. *Lindl. Gen. et Sp. Orch.* 318.

Caulis *bipedalis,* foliis *circiter* 5 *æquidistantibus dimidiam inferiorem vestientibus.* Racemus *laxus, multiflorus,* 8 *poll. longus.* Ovaria $2\frac{1}{2}$*-poll.* Sepala *ovata, alba apice viridia, lateralia latiora patentissima, supremum concavum cum petalis ovatis erectis albis dorso viridulis galeam referentibus.* Labellum *ferè pollicem longum, album, laciniis apice luteo-viridibus.* Calcar $3\frac{1}{2}$*-poll. longum, basi album, cæterùm viride.* Columna *parva, alba, carnosa, reclinata, auriculis* (*staminibus ster.*) *rotundatis.* Anthera *ochracea, loculis distantibus basibus elongatis, paululum incurvis, a canalibus stigmaticis facile separabilibus, caudiculam longam filiformem in gremio suo foventibus* *cui glandula parva pallida adnascitur.* Canalia stigmatica *linearia alba, truncata, crassitiei æquabilis.* Rostellum *planum antheræ adnatum.* Processus carnosi *herbacei, ultra os calcaris arcuatim projicientes, et canalibus stigmaticis paulò longiores.*

---

* From *habena* a rein or thong, in allusion to the long strap-shaped spur.

This rare species is a native of Sierra Leone, where it was found by Afzelius many years since. It was afterwards introduced with a brief character into Persoon's Synopsis, and from that time remained unknown, until it was imported last year by Messrs. Loddiges, in whose stove it flowered in August.

It offers an excellent illustration of the characters of the curious genus *Habenaria,* as limited in the *genera and species of* Orchideous plants, and will shew the student in a distinct manner what the points are in which it differs from the genus *Platanthera,* whither I refer our British Butterfly Orchis, to which this bears a striking resemblance. In order to make this clear, attention should be paid to the magnified figure of a column extracted from the flower, and placed at the right hand corner of the accompanying plate. In this the lower white portion to the left is the column, with an auricle or sterile stamen at its upper corner to the right. Immediately proceeding from this in a curved direction upwards are the white stigmatic canals, in whose hollow the lengthened bases of the anther are placed when in their natural position. The upper yellow body which divides downwards into two legs is the anther; the legs are its lobes, which lengthen at their lower end and fit into the stigmatic canals, enclosing the pollen masses in their upper portion, and in their lower keeping the caudicle of the pollen in such a position that it must inevitably come in contact with the gland which once formed the tip of the stigmatic canal, but which eventually separates from the latter, adhering to the caudicle, as is seen in the thread-shaped processes, which in the figure rise up from the anther-bases. All these parts equally exist in the genus Platanthera. But in *Habenaria* we find an addition of two greenish horns, which spring from the lower edge of the stigma, skirting the orifice of the spur, and finally project beyond the latter, as is seen in the figure. These horns, which are considered to be processes of the stigma, do not occur in *Platanthera,* unless in a very rudimentary state, while in *Habenaria* they are always so fully and obviously developed as to form conspicuous objects, even when the flowers are dried.

It may appear to some Botanists that this is but a slight distinction upon which to found a genus. But it is to be remembered, firstly, that it is a decided organic difference, inasmuch as it is the developement of a new organ in the apparatus for reproduction; secondly, that it is a constant and obvious character which in many cases is far more remarkable than even in the species before us; and moreover, that after being limited within the comparatively narrow bounds that I have assigned it, and after striking off the genera Bonatea, Peristylus, Platanthera, Aopla, and another or two, the genus Habenaria still comprehends no fewer than eighty-five well ascertained species, to which many more will doubtless have to be added.

This plant must have the heat of a damp stove, when in a growing state, but will doubtless partake of the habits of its kindred species in requiring a long period of coolness and dryness, while its roots are at rest, after the leaves have perished.

1859

* CATTLEYA labiáta.

Crimson-lipped Cattleya.

GYNANDRIA *MONANDRIA.*

Nat. ord. ORCHIDACEÆ, § EPIDENDREÆ.
CATTLEYA.—Suprà, vol. 14. *fol.* 1172.

C. *labiata;* sepalis lineari-lanceolatis, petalis membranaceis lato-lanceolatis acutis subundulatis, labello obovato undulato obtuso indiviso, pseudobulbis oblongis angulatis, spathâ maximâ foliaceâ. *Gen. & Sp. Orch. pl. p.* 116.
C. labiata. *Lindl. Collect. Bot. t.* 33. *Hooker Exot. fl.* 157. *Lodd. Bot. Cab. t.* 1856.

A native of Brazil, whence it was introduced about 18 years ago by Mr. William Swainson. It has since that time been represented in most of our Botanical periodicals, so that I fear I shall hardly stand excused for reproducing it here. Nevertheless, all the plates above quoted are deficient in the richness of colour that is so peculiarly characteristic of the species, and that constitutes its chiefest ornament; and the knowledge of the existence of so truly beautiful species cannot be too widely diffused, the more especially as this, like the rest of its genus, requires the excessive heat and dampness of an Orchideous house in a less degree than many other kinds.

It is one of those plants which flourish so remarkably in the hothouse at Wentworth, with no greater dampness and heat than can be endured by human beings without inconvenience.

---

* See folio 1406.

The specimen selected for representation in this place is a small one, with its colours remarkably rich and well developed; it is figured in the Botanical Cabinet with four flowers in a cluster, and I have seen it with six. In such a state, and with several stems, each laden with flowers in a similar manner, there is certainly no plant of which I have any knowledge that can be said to stand forth with an equal radiance of splendour and beauty. For it is not merely the large size of the flowers, and the deep rich crimson of one petal contrasted with the delicate lilac of the others that constitute the loveliness of this plant, it owes its beauty in almost an equal degree to the transparency of its texture, and the exquisite clearness of its colours, and the graceful manner in which its broad flag-like petals wave and intermingle when they are stirred by the air, or hang half drooping half erect when at rest and motionless.

The drawing was made in the garden of the Horticultural Society in October last.

1861

* MORMÓDES atropurpúrea.

Dark-purple Mormodes.

GYNANDRIA *MONANDRIA.*

Nat. ord. ORCHIDACEÆ, § VANDEÆ.

MORMODES. Lindl. Sepalum superius subfornicatum, angustum; *lateralia* conformia reflexa. *Petala* latiora, conformia, erecta. *Labellum* sellæforme, ascendens, trilobatum, subcuneatum, apiculatum, cum columna articulatum. *Columna* semiteres, mutica; *gynizus* longus angustus; *clinandrium* postice acuminatum. *Pollinia* 4, per paria connata, caudiculæ crassæ affixa, glandulæ carnosæ crassæ adhærenti.—Habitus Cataseti. *Lindl. Nat. Syst. of Botany, ed. 2. p.* 446.

Mormodes *atropurpurea.*

Pseudobulbi *caulescentes,* Cataseti *facie, polyphylli, oblongi, basibus foliorum distantibus vaginantibus imbricati.* Folia *plicata* (Cataseti), 3-5-*costati, erecti, apice recurvi.* Racemus *lateralis, densus, oblongus, pedunculatus, pseudobulbo altior.* Flores *atropurpurei.* Sepala *lineari-oblonga æqualia, reflexa, basibus.lateralium paulò obliquis et ungui labelli adnatis.* Petala *ovata, erecta, supra columnam conniventia.* Labellum *replicatum, retrorsum arcuatum, circumscriptione cuneatum, leviter unguiculatum, trilobum, lateribus deflexis venosis, lobo intermedio magis carnoso, cuspidato, subtrilobo.* Columna *compressa, rostrato-acuminata, cum labello continua, erecta, mutica,* antherâ *posticè rostratâ, gynizo oblongo.* Pollinia 4, *per paria connata, caudiculæ subcucullatæ adnata, glandulâ concavâ crassâ.*

Sent to me in December last from the garden of John Willmore, Esq. of Oldford near Birmingham, with whom it then flowered for the first time. It had been imported from the Spanish Main in 1834. It is a new genus, differing from Catasetum and Myanthus in the want of cirrhi upon the column, and from Monachanthus in its lip (fig. 1.) being membranous and curved upwards, with the sides turned downwards, like the sides of a saddle, instead of being fleshy and helmet-shaped.

* From μορμὼ a frightful-looking object, a goblin, in allusion to the strange appearance of the flowers.

The leaves are pale green; the flowers one uniform rich purple.

A tender stove plant, requiring the same treatment as Catasetum, Cycnoches, &c. With reference to Orchideous plants, with this habit, it may in general be observed, that they require to be kept cool and dry when not in a growing state, to be forced gently into growth, and when in the full vigour of their vegetation to have a copious supply of moisture. They will at that season even introduce their roots into water, if they are allowed, and flourish the more under such treatment.

1863

TRICHOPÍLIA tórtilis.

Twisted-petalled Trichopilia.

GYNANDRIA *MONANDRIA.*

Nat. ord. Orchidaceæ, § Vandeæ.

TRICHOPILIA. Lindl. *Sepala* et *petala* æqualia, patentia, angusta. *Labellum* magnum, petaloideum, convolutum, c. columna parallelum, trilobum, lobo intermedio sub-bilobo planiusculo; intus nudum. *Columna* teres, clavata. *Clinandrium* cucullatum, 3-lobum, villoso-fimbriatum. *Anthera* 1-locularis, compressa, antice convexa. *Pollinia* 2, postice sulcata, caudiculæ tenui cuneatæ adhærentia; glandulâ minimâ.—Pseudobulbi *carnosi, vaginis maculatis supertecti, monophylli, coriacei.* Flores *solitarii axillares.* Lindl. Natural System of Botany, ed. 2. p. 446.

Trichopilia *tortilis.*

Pseudobulbi *oblongi, sulcati, compressi, vaginis fusco-maculatis arctè vestiti, aliquandò folii fere longitudine.* Folia *solitaria, oblonga, coriacea, acuta, plana, v. leviter complicata.* Flores *solitarii, axillares, horizontales, sessiles.* Sepala *et* petala *æqualia, lineari-lanceolata, patentissima, spiraliter torta, margine crispatula, fusco-lutea, disco latentia.* Labellum 2-*poll. longum, circa columnam convolutum, album, maculis pluribus magnis inæqualibus ad interius; limbo* 3-*lobo intermedio subbilobo.* Columna *cum ovario continua, teres, clavata, alba; clinandrio cucullato trilobo; lobis ascendentibus, falcatis, ciliato-laceris.* Anthera *compressa, apiculata.* Pollinia 2, *parva, pyriformia, posticè sulcata, caudiculâ cuneatâ inserta, glandulâ minimâ ovali.* Gynizus *excavatus, paululum obliquè retrorsum versus.*

A beautiful and highly curious plant, introduced from Mexico in 1835, and communicated in January last by George Barker, Esq. of Springfield near Birmingham. In many respects the genus approaches Maxillaria, but differs in the column not being reclinate upon the ovary and subtended by the partially united lateral sepals, in the regular expansion of both sepals and petals, and especially in the singular column, (fig. 1.) terminated by three little plume-like lobes which unite at their bases into a sort of hood, that covers over a remarkably compressed anther (fig. 2.).

* From θρὶξ, τριχὸς hair, and πιλίον a cap; the anther of this genus is concealed below a cap surmounted with three tufts of hair.

The white of the lip, which is very clear and pure, forms a brilliant contrast with the rich blotches of deep crimson that ornament the interior of the little funnel formed by the rolling of the lip round the column.

From the habit of this plant it may be conjectured that it will thrive in the stove, under the same treatment as Maxillarias.

1865

* DENDRÓBIUM macrostáchyum.

Long-spiked Dendrobium.

GYNANDRIA *MONANDRIA.*

Nat. ord. Orchidaceæ, § Malaxideæ.

DENDROBIUM.—Supra, vol. 7. *fol.* 548.

D. *macrostachyum;* caulibus teretibus pendulis flagelliformibus, foliis ovato-lanceolatis submembranaceis, floribus ternatis racemum spurium formantibus, sepalis ovatis acutis, petalis lanceolatis sepalo supremo subæqualibus, labello cucullato venoso: limbo ovato obtuso ciliato intus pubescente. *Gen. & Sp. Orch.* 78.

A native of Ceylon, where it was discovered by the late Mr. James Macrae, who some years ago sent me dried specimens and a drawing of it. Upon the former I found a minute blanched portion that seemed still alive; this was fastened by a nail and shred to a damp shady wall in a stove in the garden of the Horticultural Society, where it gradually recovered its green colour and began to grow. By tending it carefully, and not feeding it until it had recovered the effects of its long fast while buried between two sheets of brown paper in a dry chest, it gradually recovered and grew into a plant, the offspring of which has been distributed. From one of them, which flowered in the garden of Mr. Bateman, the accompanying drawing was prepared in June last.

The species approaches to *D. Pierardi, cucullatum,* and *pulchellum,* than all of which it is less beautiful, and it requires precisely the same treatment as those species.

* See folio 1249.

1867

* EPIDÉNDRUM armeniácum.

Apricot-coloured Epidendrum.

GYNANDRIA *MONANDRIA.*

Nat. ord Orchidaceæ, § Epidendreæ.

EPIDENDRUM.—Suprà, vol. 1. *tab.* 17.

E. *armeniacum;* caulibus teretibus, foliis lanceolatis coriaceis acutis subplicatis, racemis pedunculatis cylindraceis nutantibus, sepalis patulis ovatis acutis, petalis setaceis, labelli subcucullati laciniis lateralibus rotundatis intermediâ ovatâ acuminatâ: callo magno oblongo in disco.

Caules *erecti, compressi, semipedales, foliis* 3-4*ve distantibus in spatham abeuntibus vestiti.* Racemus 3-4 *poll.* Flores *minuti armeniaci coloris.* Bracteæ *setaceæ ovario triplò breviores.*

A native of Brazil, where it was found in company with Grobya Amherstiæ, figured at fol. 1740 of this work. It was first seen in England in flower in the year 1835, at one of those splendid exhibitions in the Garden of the Horticultural Society, which attest more strongly than even the country residences of our nobility and gentry, the skill and perseverance of English gardeners. There, in the midst of the dazzling scarlet or pink of various kinds of Cacti, and surrounded by the brilliant plumes of Chinese Azalea flowers, that weighed down their graceful branches, which really seemed as if they were proud of their lovely burthen, from a basket of humble moss, a little tuft of stems of this species was seen to rear its modest head, as if in hopelessness of attracting notice in so gay a company. The neatness however of its tiny flowers, the pleasing tint of its apricot-coloured petals, the elegant form of their slightly nodding or even drooping clusters, and the novelty of their form in so well known a genus as

* See folio 1415.

Epidendrum, arrested the curious observer, who soon found the symmetry and simple elegance of the little blossoms of Epidendrum armeniacum compensate for the absence of those more obvious beauties that adorned its gaudier rivals.

It is a stove plant, increasing readily by division of its tufted stems, like E. elongatum, and the kindred of that common species. It was imported by Messrs. Rollissons of Tooting, to whom I was indebted for a specimen in the month of June.

Fig. 1, is a profile view of the lip, with its column; Fig. 2, is the lip cut from the column, and viewed from above, with the great callus that occupies its middle; Fig. 3, represents the pollen masses, with their powdery reflexed caudicle.

1870

EPIDÉNDRUM ~~Skinnéri~~. clavatum

Mr. Skinner's Epidendrum.

GYNANDRIA *MONANDRIA.*

Nat. ord. Orchidaceæ, § Epidendreæ.
EPIDENDRUM.—Suprà, vol. 1. *fol.* 17.

E. *Skinneri;* caule clavato in pseudobulbum ovale desinente diphyllo, foliis lanceolatis patulis, racemo simplici subæquali, bracteis ovatis canaliculatis acutis ovariis inferioribus duplò brevioribus, sepalis petalisque lanceolato-linearibus æqualiter patentibus, columnâ clavatâ, labelli tripartiti basi bicallosi laciniis lateralibus ovatis subfalcatis margine posteriore denticulato: intermediâ unguiculatâ laminâ ovatâ acuminatâ.
E. Skinneri. *Bateman's MSS.*

Caules *vetusti duri, clavati, subarticulati, erecti, in pseudobulbum ovale desinentes vestigiis foliorum vaginantium vestiti, diphylli, foliis lanceolatis, patulis, coriaceis.* Pedicelli *florum inferiorum elongati sed non corymbosi.* Ovaria *filiformia.* Sepala *et* petala *viridia ferè unciam longa, angusta.* Columna *virens.* Labelli *lamina nivea.*

Found in August, 1834, near Cumana, by Mr. Skinner. Communicated to this work in July, 1835, by the late Lord Grey of Groby. It is not a pretty species, but it is very distinct from any previously described, and is remarkable for its stems being dilated at the upper end, like some of the species of Dendrobium.

The station of the plant will be near Epidendrum concolor.

* See folio 1415

1871

* MAXILLÁRIA aromática.

Aromatic Maxillaria.

GYNANDRIA *MONANDRIA.*

Nat. ord. Orchidaceæ, § Vandeæ.
MAXILLARIA.—Suprà, vol. 11. *fol.* 897.

M. *aromatica;* pseudobulbis ovatis compressis, foliis pluribus oblongo-lanceolatis plicatis scapis unifloris erectis longioribus, sepalis ovato-oblongis petalisque conformibus acutis, labelli semicylindracei laciniis lateralibus porrectis subulatis; intermediâ bilabiatâ! labio superiore truncato nano inferiore spathulato apice recurvo serrulato. *Gen. & Sp. Orch. pl. p.* 146.
M. aromatica. *Graham in Hooker's Exot. fl.* 219.
Colax aromaticus. *Spreng. cur. post.* 307.

A fragrant stove plant, breathing cinnamon and sweet spices, found in Mexico, whence it was sent by Lord Napier to the Botanic Garden, Edinburgh, previous to the year 1826. The species is now not uncommon in good collections, flowering abundantly in the month of May.

The callosity of the disk of the lip of Orchideous plants, which sometimes appears in the form of lamellæ, following the course of the veins, and sometimes is a thick tubercle, in the present species is a truncated plate, occupying the base of the middle segment of the lip, and looking like the half of a petal laid over the true lip. What is the real nature of this truncated plate?

* See folio 1428.

1872

* CRÝBE rósea.

Pink-flowered Crybe.

GYNANDRIA *MONANDRIA.*

Nat. ord. Orchidaceæ, § Arethuseæ.
CRYBE, Lindl. Sepala et *petala* similia, lanceolata, conniventia; lateralibus basi obliquis. *Labellum* multo majus, membranaceum, cucullatum, nunquam expansum, cum columna clavata marginata semi-connatum. *Lindl. Nat. Syst. of Botany, ed.* 2. *p.* 446.

Crybe *rosea.*
Pseudobulbi *subrotundi, virides, leviter angulati.* Folia *ex apice erumpentia, lanceolata, plicata, subterna, lætè viridia, pedalia.* Spica 3-4-*flora, scapo laterali, basi purpureo, squamis circiter* 3 *distantibus vaginati. Flores penduli, ovario brevi, arcuato, bracteâ subulato-ovatâ* 3-*plò longiore.* Flores *fere* 2 *pollices longi, clavati, nunquam expandentes.* Sepala *æqualia, obovato-lanceolata, basi adhærentia, pallidè viridia, apice purpureo-maculata.* Petala *subæqualia, subalba, purpureo tincta, membranacea.* Labellum *longius, obovatum, acutum, atropurpureum, marginibus plicato-crispis inflexis, basi cum columna semiadnatum.* Columna *clavata, basi teres, ultra insertionem labelli marginata, paulo supra labellum bidentata, apice paululum cucullata.* Anthera *terminalis, opercularis,* polline *granulari.*

A native of Mexico, whence it was imported by Messrs. Loddiges, in whose stove it blossomed in June last.

This plant is remarkable for never expanding its singular club-shaped flowers, which always remain as much closed as is represented in the accompanying figure, the edges of the lip turning inwards, and forming a sort of dish at the end of the flower.

The genus is nearly allied to the North American Arethusas, from which the adhesion of its labellum to the base of

* From κρύπτω to conceal, whence κρυβεὶς concealed, in allusion to the manner in which the column is hidden by the floral envelopes.

the column sufficiently distinguishes it, independently of the remarkably property of always keeping its flowers closed.

A stove plant, requiring the same management as the common Bletia verecunda, and the like.

1875

* BIFRENÁRIA aurantiáca.

Orange-coloured Bifrenaria.

GYNANDRIA *MONANDRIA.*

Nat. ord. ORCHIDACEÆ, § VANDEÆ.

BIFRENARIA, Lindl. Sepala patula, libera, subæqualia; lateralia cum basi productâ columnæ connata vix basi obliqua. *Petala* sepalis duplò minora. *Labellum* cum pede mucronato columnæ articulatum, cucullatum, trilobum, medio callosum. *Columna* brevis, semiteres, mutica. *Anthera* mutica, subcristata. *Pollinia* 4, per paria incumbentia, caudiculis duabus distinctis materiei viscidæ rostelli adhærentibus, glandulâ (oblongâ).——Epiphyta *pseudobulbosa, Maxillariæ (Colacis)* habitu. *Gen. et Sp. Orch.* 152.

B. *aurantiaca;* pseudobulbis subrotundis compressis diphyllis, foliis oblongis plicatis racemi erecti longitudine, petalis erectis, labelli lobis lateralibus semicordatis intermedio transverso ovali subundulato basi bicalloso, columnâ pubescente.

Perianthium *bilabiatum.* Petala *obovata, rotundata, crenata, erecta, cum sepalo supremo oblongo acuto labium supremum formantia.* Sepala *lateralia ovata, obtusa, supra pubescentia, patentia, basi obliqua, supremo duplò latiora.* Labellum *cum columnæ pede producto articulatum, unguiculatum; unguis cuneatus, carnosissimus, limbi longitudine, callo transverso truncato ad apicem ubi in limbum abit;* limbus *tripartitus laciniâ intermediâ transversâ apiculatâ indivisâ subundulatâ, lateralibus erectis semicordatis, obtusis utrâque flexurâ elevatâ baseos callum mentiente.* Columna *semiteres, pubescens, clinandrii dorso acuminato.* Anthera *triangularis: angulis pilosis, lateralibus productioribus.* Pollinia 2, *biloba,* caudiculis *totidem* glandulæ *communi adhærentibus.*

A pretty epiphyte, native of Demerara. For the opportunity of publishing it I am indebted to his Grace the Duke of Devonshire, in whose hot-house at Chiswick it flowered in October, 1835.

* So named in allusion to the double strap or frænum that connects the pollen masses with their gland.

The colours of the flowers being deep orange-yellow, mottled with deep brown spots, the aspect of this species is rather handsome. Fig. 1. represents the appearance of the lip separated from the column, and especially of the double callus at the base of its middle lobe. Fig. 2. shews the face of the column with the bases of the sepals and petals. Fig. 3. is a view of the double strap, connecting the pollen masses to the gland. Fig. 4. shews the anther removed from the column, and viewed in front.

1879

* EPIDÉNDRUM bifidum.

Hare-lipped Epidendrum.

GYNANDRIA *MONANDRIA.*

Nat. ord. ORCHIDACEÆ, § EPIDENDREÆ.
EPIDENDRUM.—Suprà, vol. 1. *fol.* 17.

E. *bifidum;* foliis in pseudobulbos subternis lanceolatis, scapo ramoso foliis multò longiore, sepalis oblongis acutis petalisque lanceolatis patentibus, labelli cuneati lobis lateralibus ovatis: intermedio maximo apice dilatato subreniformi sulcato basi in disco biappendiculato (a columna ferè libero). *Gen. et Sp. Orch. pl.* 100.
Helleborine flore papilionaceo. *Plum. sp.* 9. *ic.* 186. *f.* 1.
Epidendrum bifidum. *Aubl. guian. p.* 824. *Swartz Fl. Ind. Occ.* 3. 1489. *Willd. Sp. pl. no.* 3. *Redout. Liliac.* 84.
E. papilionaceum. *West. St. Cruc. p.* 230. sec. Willd.

In this state E. bifidum appeared when, in July, 1835, a specimen was communicated to me by Messrs. Loddiges. It was then very pretty, on account of the beautiful veining of its lip; but it cannot have been any thing like so handsome as Swartz describes it to be, with all the divisions of the flower bright purple, and a flowering stem three feet high and branching.

It is a very remarkable and distinct species, with a peculiar slit lip, by which it is readily known from all, except E. auropurpureum, a kind that appears to be very nearly related to it.

It is described as found upon branches of trees in the West India islands, especially St. Christophers, St. Bartholomews, and Santa Cruz. The Messrs. Loddiges had it from Tortola; Aublet found it in Cayenne.

* See folio 1415.

1881

* EPIDÉNDRUM Skinnéri.

Mr. Skinner's Epidendrum.

GYNANDRIA *MONANDRIA.*

Nat. ord. ORCHIDACEÆ, § EPIDENDREÆ.
EPIDENDRUM.—Suprà, vol. 1. *fol.* 17.

E. *Skinneri;* foliis distichis lanceolatis acuminatis, caule apice longè aphyllo squamoso, racemo cylindraceo multifloro, floribus cernuis, sepalis lineari-lanceolatis, petalis ovalibus acutis, labello ovato acuminato integerrimo basi callo sulcato cristato.
E. Skinneri. *Bateman's MSS.*

Caulis *erectus, teres, distichè foliosus, apicè aphyllus vaginatus.* Folia 5 *poll. longa.* Bracteæ *lineari-lanceolatæ, pedicellis capillaribus paulò brevioribus.* Flores *pallidè purpurei,* $1\frac{1}{2}$ *poll. lati.* Labellum *cum columnâ semiconnatum, concavum, basi luteum, cuniculatum.*

For this beautiful species of Epidendrum I am obliged to James Bateman, Esq. jun. in whose hot-house at Knypersley it flowered last January. He states, "that it was sent to him in the summer of 1835, from the neighbourhood of Guatemala, by his most excellent friend G. U. Skinner, Esq., to whose enthusiasm in the cause of science he is already indebted for many new and interesting plants. Ep.

Skinneri is among the most free-flowering of its tribe; every one of its shoots, both great and small, having been invariably succeeded by a spike of flowers."

The perfect state of the specimens, so short a time after the importation of the species, attests the skill of Mr. P. N.

* See fol. 1415.

Don, Mr. Bateman's gardener, in the management of these curious plants.

The species is in the way of *E. elongatum*, and I presume requires the same management as that plant. I have also received it from Messrs. Loddiges.

1887

* ONCÍDIUM Lanceánum.

Mr. Lance's Oncidium.

GYNANDRIA *MONANDRIA.*

Nat. ord. ORCHIDACEÆ. § VANDEÆ.
ONCIDIUM.—Suprà, vol. 13. *fol.* 1050.

A. Labellum trilobum.
§ 1. Folia plana, v. complicata.
α. *Sepala lateralia libera.* * *Labelli laciniæ laterales nanæ v. obliteratæ.*
O. *Lanceanum;* ebulbe, foliis oblongis acutis planis substriatis carnosis, scapo racemoso composito erecto rigido racemulis confertifloris, sepalis petalisque conformibus oblongis obtusis carnosis concavis margine undulatis, labelli lobo medio dilatato subcuneato integerrimo basi hastato: lobis lateralibus semi-ovatis, cristâ trilobâ carnosâ jugoque elevato proclivi, columnæ alis carnosis rotundatis, antherâ cristatâ.
O. Lanceanum. *Lindley in Hort. Trans. n. ser. vol.* 2. *p.* 100. *tab.* 7.

This remarkable plant has lately been published in the Transactions of the Horticultural Society of London, whence the following account of it is extracted:—

"In the year 1834, John Henry Lance, Esq. upon his return to England from Surinam, where he had been residing several years, brought with him a considerable collection of Orchideous Epiphytes, which he presented to the Society. Among other interesting species was the subject of the following memorandum; a plant, than which a more acceptable addition to the hot-houses of this country has rarely been made.

"The genus Oncidium already includes several very beautiful species, in particular *O. flexuosum, bifolium, ampliatum* and *crispum*, but all these have flowers in which yellow or brown are the only colours, they owe their beauty to the graceful arrangement of their branches, and to the singular form of their petals, rather than to their colour, and moreover their blossoms are destitute of fragrance. When, therefore, it was ascertained that among Mr. Lance's plants there existed an Oncidium, with violet-coloured sweet-scented flowers, great interest was excited, and no pains were spared to ensure its successful cultivation. Many plants were soon distributed by Mr. Lance's orders, one of which blossomed in the hot-house of the Messrs. Loddiges, and another shortly after in that of the Society. The plant was found by no means difficult to manage; and in point of beauty and fragrance it more than answered all the expectations that had been entertained of it.

"The roots are flexuose, slender, simple elongations of the base of the stem, evidently intended to grow upon places where the quantity of mould is insufficient to cover them; they lengthen independently of their growth at the point, like the aerial roots of other Epiphytal Orchideæ, and differ from those of other Oncidiums only in being of a greenish yellow colour. The leaves spread from a very short, woody, annulated root-stock, and are about a foot in length on the average; they are of a broadly oblong figure, of a leathery consistence, are nearly flat, a little curved back at the point, and have a light green colour faintly mottled

* See folio 1542.

with purple. The flowers are disposed in a short-branched rigid panicle, elevated on a stalk not quite so long as the longest leaves; it is about six or nine inches long, and densely covered with flowers, which sometimes assume a corymbose, sometimes a racemose arrangement. The flowers when expanded measure an inch and three-quarters from the tip of their back sepal to the point of their lip; they emit a delicious fragrance resembling that of the garden pink. The sepals are oblong, concave, obtuse, a little waved and greenish yellow at the edge, bright yellow in the middle, and regularly marked with broad blotches of crimson which run together near the base. The two petals are similar to the sepals. The lip is bright violet, darkest at the lower half; at the base it is prolonged on each side into a triangular tooth, and in the middle of the base there are three nearly equal tubercles which towards the column terminate a ridge that gradually lowers and then disappears at the expanded portion of the lip; above the base it is narrow, it then expands again into a broad, thin, light purple, somewhat truncated and toothed extremity. The column has an oblique, rounded, ear-like appendage on each side, and is capped by a rich crimson anther.

"Such was the specimen from which the annexed drawing was taken; but it was far inferior to one which I have just seen (June 29, 1836), in the rich collection of epiphytes belonging to the Messrs. Rollinsons of Tooting. This plant, which I regard as the most perfect instance of successful cultivation I have yet witnessed among epiphytes, had leaves eighteen inches long, and upwards of thirty flowers two inches and a quarter in expansion, with all the markings of the sepals and petals of the richest chocolate brown, and of the lip of the deepest violet. In fragrance there was a resemblance to the spicy odour of that sweetest of all flowers Aerides cornutum.

"In the Society's garden this plant is cultivated along with other epiphytes in a damp hothouse facing the north; it is planted in a mixture of sandy peat, potsherds, and decayed wood; and under these circumstances it thrives very well.

"Mr. Lance has favoured me with the following account of the discovery and subsequent management of this remarkable plant in its native country.

"'The first specimen of this splendid Epiphyte I discovered, was growing on the trunk of a large tamarind tree, in a noble avenue of those trees close to the Government House in Surinam. I took it home with me and planted it in a pot filled with rotten pieces of wood and a little light earth; but though it remained alive and flowered once or twice, it did not thrive, but wasted away and became less. I afterwards found a great number of the plants in different parts of the colony; they were generally attached to the stems or branches of the Tamarind, the Sapodilla, or the Calabash trees, appearing to prefer those to any other; however, on being tied to the branches of the Orange, the Soursop, the Mammee, and even the Brugmansia arborea, it grew well upon them all and produced vigorous stems with upwards of twenty blossoms on each stem. The scent is extremely fragrant, and is retained after the flower is dried, only becoming fainter and more of a spicy flavour than when fresh. The plant remains in full beauty ten or twelve days, a long period in that climate, and I found that it always required a shady situation and a living stem to grow upon, without which it would not produce its flowers in the highest perfection.'

"Although the Society's sense of the importance of Mr. Lance's endeavours to introduce new plants to this country has already been recognized by the Council having awarded him the Society's Large Silver Medal, yet I trust it will not be considered improper in me to indicate this in a more specific manner, by naming after that gentleman one of the most beautiful of the plants we owe to his exertions."

1889

* CIRRHǼA tristis.

Sad-coloured Cirrhæa.

GYNANDRIA *MONANDRIA.*

Nat. ord. ORCHIDACEÆ, § VANDEÆ.
CIRRHÆA.—Suprà, vol. 18. *fol.* 1538.

C. *tristis;* foliis oblongo-lanceolatis basi parum angustatis, petalis lineari-spathulatis, labelli sagittati lobo intermedio lineari abruptè acuto lateralibus acuminatis parallelis breviore.

Pseudobulbi *ovati, sulcati, angulis rotundatis.* Folia *solitaria, subplicata, coriacea, oblongo-lanceolata vel oblonga, basi in petiolum brevem convolutum angustata.* Racemi *penduli, foliis paulò longiores, radicales, densè multiflori.* Flores *odoratissimi tristè purpurei, sanguineo suffusi, apicibus herbaceis.* Sepala *linearia, patentissima, obtusa, postremo sigmoideo, reflexo.* Petala *lineari-spatulata sepalis paulò breviora et duplò angustiora.* Labellum *atropurpureum;* ungue *arcum continuum cum columnâ resupinatâ efficiente; limbo sagittato, laciniis lateralibus acuminatis subparallelis carnosis intermediâ ferè duplò longioribus.* Columna *arcuata, clavata, truncata,* antherâ *posticâ,* stigmate *obliquo, verticali, cirrhifero, ut in genere.*

A deliciously scented species, native of Mexico, whence it was obtained by Messrs. Loddiges, in whose hot-house the drawing was made in June, 1835. The form of its lip is quite different from that of any of the other three species now known.

Like its other botanical allies it grows upon the branches of trees, and apparently in a pendulous manner.

* See folio 1538.

1896

* MYÁNTHUS deltoídeus.

Triangular lipped Flywort.

GYNANDRIA *MONANDRIA.*

Nat. ord. ORCHIDACEÆ, § VANDEÆ.
MYANTHUS.—*Suprà, vol.* 18. *fol.* 1538.

M. *deltoideus;* labello imberbi sagittato triangulari, angulis posticis rotundatis dentatis apice dilatato calloso margine recurvo basi tuberculato.

Planta *omninò facie* M. cernui; *foliis quodammodò latioribus.* Rachis *purpurascens, angulata.* Bracteæ *parvæ, membranaceæ, acuminulatæ.* Petala *et* Sepala *subæqualia, lineari-lanceolata, virescentia, purpureo-maculata et interruptè fasciata.* Labellum *sepalis duplò brevius, versus basin saccatum, planum, carnosum, atro-purpureum, disco virescente, tuberculo unico oblongo pone basin, et callo virescente ad apicem. Cætera* M. cernui.

This makes the fourth species of the present curious genus, and in all probability many more will be added. It is a native of trees in the neighbourhood of the great waterfall of the Demerara river, where it was found by Mr. Joseph Hubbard, who sent it to his friend Mr. Booker of Liverpool, by whom it was presented to Richard Harrison, Esq. of Aighburgh. To the latter gentleman I am indebted for the fine specimen from which the accompanying drawing was made in October, 1835.

Its distinction from the other species resides in its lip, which has none of the fringe-like hairs of M. barbatus and cristatus, but is flat, of a thick fleshy consistence, and of a rich purple colour with a dash of green on its centre; its figure is arrow-headed, with the barbs rounded off and serrated, and the point a little dilated with the edges reflexed.

* See folio 1721.

When the third part of the *Genera and Species of Orchideous Plants* was published in 1833, I was only acquainted with Myanthus cernuus, and cristatus, the latter of which I considered a Catasetum, and the former as the only genuine species of the genus; to a certain degree mistaking the real generic character of Myanthus, in consequence of the imperfect materials of which only I was then in possession. Now however that four species are known in a living state, it has become necessary to alter the original character of the genus so as to include Catasetum cristatum. This I think is more advisable than to combine Myanthus with Catasetum, as my learned friend Sir William Hooker has recommended (*Bot. Mag. fol.* 3514). If the latter measure were to be adopted it would be equally necessary to suppress the genera *Monachanthus, Mormodes, Cycnoches,* &c. the effect of which would be to form a heterogeneous collection of species, the principal combining character of which would reside in the peculiar succulent stems. As the genera now stand they have each a clear distinction, and each already possesses as many species as are usually assembled under newly discovered types of structure; *Myanthus* has already 4, *Cycnoches* 2, *Monachanthus* 2, *Catasetum* 5; and *Mormodes* is the only one that comprehends only a solitary species.

Like the rest of its allies a tender stove plant.

1898

* EPIDÉNDRUM æmulum.

Emulous Epidendrum.

GYNANDRIA *MONANDRIA.*

Nat. ord. ORCHIDACEÆ, § EPIDENDREÆ.
EPIDENDRUM.—*Suprà, vol.* 1. *fol.* 17.

§ 1. *Foliis in pseudobulbos insidentibus.*

E. *æmulum;* rhizomate repente, pseudobulbis ovalibus compressis, foliis solitariis oblongo-lanceolatis coriaceis, sepalis petalisque lineari-lanceolatis æqualibus patentibus, labello subrotundo mucronato cochleato lineato integerrimo, racemo paucifloro.

Rhizoma *pennam anserinam crassum, repens, apice squamis membranaceis deciduis vestitum.* Pseudobulbi 2 *poll. longi, glabri, rugosi, oblongi, compressi.* Racemus 3-4-*florus, folii longitudine.* Flores *E. fragranti simillimi, sed 3-plò minores; petala non sepalis latiora, et labellum mucronatum nec cuspidatum.*

A very rare little plant, closely allied to the variable *E. fragrans,* from which it is distinguished by its pseudobulbs being very exactly oval, and not tapered to each end, by its more leathery less acuminate leaves, and very much smaller flowers, the petals of which are the same width as the sepals.

A native of Para, whence it was sent by Mr. Hesketh, the English Consul, to Richard Harrison, Esq. who favoured me with a specimen in February, 1834.

A tender stove species.

* See folio 1415.

1901

* SCAPHYGLÓTTIS violácea.

Violet Boatlip.

GYNANDRIA *MONANDRIA.*

Nat. ord. ORCHIDACEÆ, § VANDEÆ.
SCAPHYGLOTTIS. Pöppig et Endlicher Nova Genera et Species Plantarum, 1. 58. *Sepala* conniventia, lateralia basi paulo producta cum pede columnæ connata labello supposita, supremum lineare convexum. *Petala* conformia sed paulo breviora. *Labellum* oblongum canaliculatum cum pede paululum producto columnæ continuum eique in parallelum, margine leviter repandum. *Columna* marginata. *Pollinia* 4. teretia, glandulam cuneatam sessilia.—— Herbæ *epiphytæ, caulescentes, nunc pseudobulbosæ in axillis.* Folia *coriacea, angusta.* Flores *parvi, axillares, pedunculis densè vaginatis.*

S. *violacea;* pseudobulbis nullis, foliis linearibus apice emarginatis, floribus subgeminis, labello lineari apiculato leviter repando.

Cladobium violaceum. *Lindl. Nat. syst. of Botany, p.* 446.

Caules *teretes, striati, articulati, verosimiliter penduli, ad nodos proliferi.* Folia 2-3 *poll. longa, linearia v. lineari-lanceolata, emarginata, fortè in pseudobulbos insidentia dum planta vegetior pullulat.* Flores *minuti, violacei, pedunculis dense vaginatis brevissimis, geminati.* Sepala *lateralia basi producta et obliqua, supremo lineari duplò latiora.* Petala *supremo conformia, sed paulo breviora et minùs colorata.* Labellum *album, cum basi leviter producto columnæ continuum, carnosum, canaliculatum, lineare, apiculatum, apice purpurascens, margine leviter repandum ideoque subtrilobum.* Columna *semiteres, alba, apice utrinque unidentata.* Pollinia 4, *distincta, teretia, in glandulam triangularem sessilia.*

A native of Demerara, whence it was imported by Messrs. Loddiges. It is not a species of any beauty, but it is a great

Botanical rarity, flowering in the Orchideous stove in February.

* Σκάφος a boat, and γλῶττα a tongue, in allusion to the usual form of the labellum.

It constitutes, along with some other South American plants with a similar habit, a genus representing among Vandeæ *Isochilus* in the Epidendreous section; and distinctly characterized by its flowers having the calyx and corolla of *Specklinia*, combined with pollen masses of a very unusual structure. When four pollen masses are combined in pairs, they are usually pressed so close together that the one is flattened against the other, and they seem as if in reality each pair were produced by the slitting of one; in this genus, on the contrary, each mass is rounded and separate from its neighbours.

When I published this plant in the work above quoted, under the name of Cladobium violaceum, I had not seen the late parts of Pöppig and Endlicher's Nova genera, which appeared in the end of 1835; and consequently I was unaware of the genus having been already named Scaphyglottis in that work. Perhaps as the name Cladobium was in circulation, from me, so long since as February, 1835, I might on that ground set up a claim to priority, and preserve it; but as Mr. Pöppig has described several species, I think the interests of science will be best consulted by cancelling Cladobium and adopting Scaphyglottis.

Of the dissections, 1. is an entire flower seen in profile and magnified; 2. is a side view of the column and lip; 3. a view of the lip from above; 4. a set of pollen masses with their gland.

1904

* IONÓPSIS ténera.

Delicate Ionopsis.

GYNANDRIA *MONANDRIA.*

Nat. ord. ORCHIDACEÆ, § VANDEÆ.

IONOPSIS, Kunth. — *Perianthium* clausum. *Sepala* lateralia basi approximata, (v. connata) cum basi labelli connata. *Petala* sepalis conformia. *Labellum* sepalis multò majus, basi saccatum, ungue bicalloso columnâ parallelo, laminâ explanatâ patente bilobâ. *Columna* erecta, aptera, semiteres, rostello rostrato. *Anthera* 1-locularis rostrata. *Pollinia* 2, posticè sulcata, caudiculâ lineari inclusâ, glandulâ obovatâ.——Herbæ *epiphytæ, acaules.* Folia *coriacea.* Scapi *paniculati, vaginati.* Flores *parvi, pallidi.* Lindl. Gen. & Sp. Orch. 193.

I. *tenera;* foliis carinatis acuminatis, scapo subsimplici, sepalis acutis lateralibus liberis labello cuneato obliquè truncato bilobo crenulato duplò brevioribus.

The species of this genus are but little known. It is seldom that they appear in gardens, in consequence of the difficulty attending their preservation on shipboard, and when imported they are difficult to manage and soon are lost. In their native woods they grow upon the smaller branches of trees, or upon dead sticks, which their white slender delicate roots quickly overspread. When cultivated they require all the aid of a hot and damp stove.

Materials to illustrate this genus are so uncommon, that there is much difficulty in ascertaining what species it contains. I think, however, four may be safely characterized, exclusively of I. pulchella and testiculata. The first of these is figured in the Collectanea Botanica, under the name of I. utricularioides; it is a smaller plant in all respects than the other three, has a panicled scape, and its labellum is not much longer than the sepals, which are acute; a second is Sir William Hooker's Iantha pallidiflora, which has a large lip with a little irregular toothing at its margin, and extremely obtuse sepals and petals. The third is that now represented, which differs from I. pallidiflora in its sepals being acute, and its leaves sharp-pointed and carinate, without any intermediate tooth between the lobes of the lip. The fourth is a very remarkable species, with which I am acquainted through an extensive unpublished collection of excellent water-colour drawings, by Mons. Descourtilz, of Brazilian Orchideous plants, obligingly lent me by Baron Delessert. This species has purple panicled scapes a foot and half long, bending gracefully beneath the weight of the delicate snow-white flowers. It was found in the ancient forests of Brazil, in the province of St. Paul. M. Descourtilz mentions a variety of it with flowers of a delicate and pure rose colour. "Son inflorescence," he says, "a lieu dans les mois de Septembre et d' Octobre, et se prolonge jusqu'en Mars *sans que les fleurs se fanent;* elles sont inodores."

* Literally Violet-faced; ἴον a violet and ὄψις look; why so named I do not know.

The systematic characters of these species are given below. For the following note, and the drawing of the accompanying plate, I have to thank Mr. Booth, the very intelligent gardener of Sir Charles Lemon, Bart.

"This elegant little plant flowered in May last, in Sir Charles Lemon's collection at Carclew, to which it was presented by Captain Sutton, of Flushing, near Falmouth, who brought it from Havannah in March, 1835."

"*Pseudo-bulbs*, very small and nearly hidden by the leaves, oval, deep green. *Leaves* oblong-lanceolate, of a brownish green colour, striated, sheathing at the base, thick and fleshy, from 2 to 4 inches long, and about half an inch broad, slightly carinate, and tapering to a narrow point. *Scape* produced from the bottom of the small pseudo-bulb, and rising from 8 to 10 inches high; jointed, round and slender, nearly erect, of a pale brownish green, bearing a number of delicately marked flowers arranged alternately in a loose sort of panicle. *Sepals* very small and thin, closely laid over the petals which they resemble in colour, but are rather shorter and more acuminate. *Petals* oblong, rounded at the edge, concealing the column. *Labellum* large, and spreading, two-lobed, almost obcordate, with a spur at the base, of a pale pinkish colour, beautifully marked with bright violet-coloured veins, darkest towards the claw, which has two, raised, fleshy, yellow appendages, parallel to, and about twice the length of, the column: the latter is pale green and roundish oblong."

I. *utricularioides* (Collect. Bot. t. 39. A.) foliis carinatis acuminatis, scapo ramoso, sepalis acutis lateralibus liberis, labello cuneato oblique truncato bilobo integerrimo.——Labellum sepalis dimidio tantum longius. Flores minimi in genere.——Ins. Trinitatis.

I. *pallidiflora* (Iantha pallidiflora, Hooker Exot. Fl. t. 113.) foliis linearibus obtusis planiusculis, scapo subsimplici, sepalis obtusissimis lateralibus liberis labello cuneato obliquè truncato bilobo cuspidulato crenulato duplò brevioribus.——Ins. Trinitatis.

I. *paniculata* (Epidendre paniculé, Descourtilz Epidendr. Bras. ined. in bibl. Delessert t. 54.) foliis lineari-lanceolatis carinatis, scapo ramosissimo, sepalis acutis lateralibus ultra medium connatis labello obcordato apiculato plano integerrimo duplò brevioribus.——Flores majores quam in I. tenera, albi. Labellum medio purpureo lineatum, basi flavum. Variat floribus roseis.——In sylvis primævis Brasiliæ, prov. S^t. Pauli.

1907

* ASPÁSIA variegáta.

Variegated Aspasia.

GYNANDRIA *MONANDRIA.*

Nat. ord. ORCHIDACEÆ, § VANDEÆ.

ASPASIA. Lindl. — *Perianthium* patens, æquale. *Sepala* lateralia libera; supremum cum petalis basi et dorso columnæ connatum. *Labellum* oblongum, concavum, ecalcaratum, obsoletè 4-lobum, cum columna semi-connatum. *Columna* labello parallela, semiteres, marginata. *Anthera* bilocularis. *Pollinia* 2, pyriformia, posticè sulcata, caudiculâ planâ cuneatâ, glandulâ parvâ.——Herbæ *epiphytæ, caulescentes, pseudobulbosæ.* Folia *subcoriacea.* Spicæ *radicales, breves.*

A. *variegata;* pseudobulbis oblongis ancipitibus, sepalis lineari-oblongis petalisque subrhomboideis acutis, labelli lobis lateralibus recurvis intermedioque carnosis serratis.

Bracteæ *ovatæ, cucullatæ, herbaceæ, coriaceæ.* Sepala *coriacea, herbacea, sanguineo fasciata.* Petala *herbacea luteo-marginata sanguineo interruptè striata, cum basi sepali supremi connata ideoque obliquè inserta.* Labellum *carnosum, serratum, album, violaceo-maculatum, basi limbi bicallosum.* Pollinia *pyriformia, posticè sulcata caudiculâ simplici inserta.*

A native of the tropical part of South America. I received specimens of it for the first time from Mr. Joseph Knight of the King's Road, in February, 1836, and subsequently from Mr. Bateman. The flowers are deliciously sweet in the morning. It will probably prove a plant of easy cultivation, and if so it will deserve to be in every collection.

In most respects it is very like A. epidendroides, for which, before I carefully examined it for publication, I had mistaken

* From ασπαζομαι I embrace, in allusion to the manner in which the column is embraced by the labellum.

it. There are however some essential differences between them, especially in the form of the labellum, and in the far greater degree of obliquity in the insertion of the back sepal in A. epidendroides; it is moreover probable from the dried specimens of the latter that its flowers are whole coloured; I have subjoined a character by which it may be distinguished.

In the same collection of unpublished drawings, belonging to Baron Delessert, to which I have already referred (fol. 1904), there is a figure of a third species of this genus. It has oblong smooth not two-edged pseudo-bulbs, solitary much larger scentless flowers, of which the sepals and petals are yellowish green blotched with crimson, and the lip and column pure white, with a faint purplish crescent-shaped stain in the middle of the lip. M. Descourtilz found it on the Cedrela in Brazil, in the district of Bananal. A variety of it is mentioned by him with a pale lilac lip, the stain on which is deep violet. He also represents the pollen masses as each furnished with a separate caudicula; if this is correct it will be an additional reason for doubting how far that character is of importance in distinguishing Genera.

Of the dissections in the accompanying plate that in the centre represents a magnified view of the column and the base of the labellum; the other the pollen masses with their caudicle and gland seen from behind.

A. *epidendroides* (Gen. & Sp. Orch. p. 139); pseudobulbis oblongis ancipitibus, sepalis lineari-oblongis acutis, petalis obtusis concavis, labelli lobis lateralibus rotundatis integerrimis, intermedio crenulato emarginato.——Panama and Columbia occidentalis.

A. *lunata;* pseudobulbis oblongis compressis lævibus, sepalis petalisque linearibus obtusis patentissimis, labelli lobis lateralibus nanis intermedio plano subquadrato undulato, floribus solitariis.——Brazilia.——Sepala et petala luteoviridia, sanguineo-maculata. Labellum album maculâ lunatâ in medio. Flores inodori.

1911

* ONCÍDIUM iridifólium.

Pigmy Oncidium.

GYNANDRIA *MONANDRIA.*

Nat. ord. ORCHIDACEÆ, § VANDEÆ.
ONCIDIUM.—Suprà, vol. 13. *fol.* 1050.

C. *iridifolium;* foliis ensiformibus brevibus equitantibus, scapo simplici subunifloro, sepalo supremo obtuso: lateralibus acutis collateralibus (herbaceis), petalis obtusis undulatis majoribus, labelli lobis lateralibus parvis subrotundis unguiculatis: intermedio multò majore subrotundo bilobo utrinque versus apicem emarginato, cristâ (depressâ 5-lobâ apice truncatâ), columnæ alâ crenulatâ circumdante. *Gen. & Sp. Orch.* 203.

Folia nunc equitantia, nunc explanata, semper acuta. Crista labelli apice truncata crenulata, basi disciformis, lobulis duobus in vertice disci.

This curious little species of Oncidium seems to be common in many of the hotter parts of America; for it has been found from Mexico, New Grenada, and Surinam, as far south as Brazil. M. Descourtilz, in his manuscript account of Brazilian Orchideæ, observed it in the neighbourhood of the town of Bom Jesus de Bananal in the province of St. Paul's, growing exclusively upon the branches of Orange and Lemon trees; it was very abundant there, and *constantly preferred dry places exposed to the sun.*

The specimen from which the accompanying drawing was taken was sent me by Lord Fitzwilliam's desire, from the rich collection at Wentworth, in August, 1835.

The outline figure at the bottom of the plate represents the column, with its two serrated wings, and the lower part of the lip, with the tuberculated disk, by which the species of Oncidium are remarkably well distinguished.

* See folio 1542.

1914

* BRASAVÓLA cordáta.

Heart-lipped Brasavola.

GYNANDRIA *MONANDRIA.*

Nat. ord. ORCHIDACEÆ, § EPIDENDREÆ.
BRASAVOLA.—Suprà, vol. 17. *fol.* 1465.

B. *cordata;* labello cordato acuminato integerrimo ungue parum longiore, sepalis petalisque linearibus acuminatis, clinandrio integerrimo posticè in dentem subulatum producto.

A species of Orchidaceous plant, closely allied to B. nodosa, already figured at fol. 1465 of this work, from which it differs in its flowers being only half the size, with a cordate labellum, and a very different clinandrium.

It was imported from Brazil by Messrs. Loddiges, who obligingly furnished me with the specimen from which the drawing was made in January 1836.

There will be no certainty in the cultivation of epiphytal Orchidaceæ till we become more precisely acquainted with the habits of the different species than we now are. At present it is usual to consider them all natives of trees in damp shady woods. It is however quite certain that such is the habit of only some of them. The whole genus Brasavola, for example, grows upon stones and rocks, *never* upon trees, in open forest glades, fully exposed to the sun.

It is not a little remarkable that no species of this genus should occur in the extensive series of drawings of Brazilian Orchidaceæ by M. Descourtilz, with which Baron Delessert has favoured me. Neither indeed do more than two species exist in Dr. Von Martius's extensive Brazilian Herbarium, one of which is Br. tuberculata, and the other the following new species.

* See fol. 1465.

B. *Martiana;* labello ovali acuminato ciliato-dentato sessili, petalis sepalisque lineari-lanceolatis acuminatis longioribus, clinandrio cucullato inciso.——Brazil, on the banks of the Rio Negro, *Martius.*—Leaves subulate, channelled, 1½ foot long. Flowers white.—Next B. cucullata.

1916

* PRESCÓTTIA colórans.

Purplish Prescottia.

GYNANDRIA *MONANDRIA.*

Nat. ord. ORCHIDACEÆ, § NEOTTIEÆ.

PRESCOTTIA. Lindl. Sepala reflexa, basi paululum connata. *Petala* minora, reflexa aut erecta. *Labellum* erectum, posticum, cucullatum, carnosum, integerrimum. *Columna* nana, teres, aptera, libera, clavata. *Anthera* opercularis, rotundata, loculis completis divaricatis, connectivo carnoso. *Stigma* obtusum. *Pollinia* 4, geminata.

P. *colorans ;* folio solitario ovato-oblongo acuminato basi cucullato petioli longitudine, spicâ densâ cylindraceâ, petalis subulatis ascendentibus.
Scapus *bipedalis, glauco-purpurascens.*

A native of Brazil, whence it was imported by Messrs. Loddiges. The drawing was made in January 1834.

It is a stove herbaceous plant, growing readily in sandy peat, flowering in its season, and then dying down for the remainder of the year.

Little is as yet known about the species of this genus; the following certainly belong to it, besides *P. plantaginea.*

1. P. *stachyodes.*=Cranichis stachyodes. *Swartz Prodr.* 3. 1437.

2. P. *petiolaris ;* foliis latis ovato-lanceolatis ovalibusque petiolo brevioribus, spicâ densissimâ elongatâ multiflorâ, petalis sepalisque revolutis.——Peru *Mathews* (*No.* 1875).——Principally distinguished from P. plantaginea by its widely different leaves.

* Named in compliment to John Prescott, Esq. of St. Petersburgh, a learned and indefatigable botanist, with whose attainments it is to be regretted that the public should not be as well acquainted as his personal friends.

3. P. *micrantha ;* foliis ovatis subsessilibus rosulatis, scapo erecto filiformi subaphyllo, spicâ nutante pauciflorâ, bracteis subulatis ovario multò brevioribus, sepalis petalisque patulis.——Brazil, Sierra d'Estrella, *Dr. Griesbach.*——Scapus palmaris. Flores minuti.

4. P. *leptostachya ;* folio oblongo acuto petiolo longiore, scapo subaphyllo, spicâ gracili rariflorâ, bracteis brevissimis subulatis, petalis sepalisque revolutis. ——Bahia, in fruticetis sabulosis, *Salzmann.*

1919

* CÁTTLEYA intermédia; *var.* pallida.

Pale-flowered intermediate Cattleya.

GYNANDRIA *MONANDRIA.*

Nat. ord. ORCHIDACEÆ, § EPIDENDREÆ.
CATTLEYA.—Suprà, vol. 11. *fol.* 953.

C. *intermedia ;* foliis oblongis emarginatis cauli æqualibus, sepalis petalisque lineari-oblongis subæqualibus subundulatis obtusiusculis, labello trilobo lamellis pluribus in medio carnosis cristato : laciniis lateralibus ovatis obtusis planiusculis intermedio crispo rotundato denticulato.
α. *floribus roseis.*
C. intermedia. *Graham in Bot. Mag. t.* 2851. *Gen. et Sp. Orch. p.* 117.
β. *floribus subalbis rubescentibus, disco labelli sanguineo.*

This very pretty Cattleya was imported from the Brazils some years since by the Horticultural Society, in whose Garden the drawing was made so long since as June 1834. The mixture of white and crimson in its flowers gives it a beautiful appearance, and renders it much more interesting than the original variety, which is too like C. Loddigesii.

The only wild specimens I am acquainted with are in Sir Wm. Hooker's Herbarium, collected near Buenos Ayres by Mr. Tweedie, and sent home with the following note:—"This is by far the handsomest of the tribe in this country, and grows equally well on the sea-beaten rock, and the moss-covered tree in the heart of the forest. It is to be found in bloom at all seasons. There are many varieties of it; their colour pink and crimson." I suspect, however, that Mr. Tweedie confounds different things, and that several species of Cattleya are to be found in the neighbourhood of Buenos Ayres. One at least I am able to describe below, and I avail myself of the same opportunity to characterise two or three others with which I am acquainted. Beautiful as are C. labiata, crispa, and Loddigesii, C. coccinea and bicolor described below, are not at all inferior to them, and moreover there is in the possession of Messrs. Loddiges a very distinct crimson-flowered *fragrant* species, found in British Guiana by Mr. Schomburgh, with which I have not sufficient acquaintance to define it.

* See folio 1172.

C. *ovata ;* foliis binis ovatis obtusis caule cylindraceo longioribus (?) sepalis petalisque lanceolatis acuminatis, labello nudo trilobo laciniis lateralibus apice intermediâque oblongâ obtusâ denticulatis undulato-crispis.——St. Catharine's, Brazil *(Mr. Hindes).*——A very fine species in the way of C. labiata.

C. *bicolor ;* foliis oblongo-loratis caule tereti elato triplò brevioribus, sepalis lanceolatis falcatis acutis, petalis parum latioribus subundulatis obtusis, labello indiviso plano apice dilatato rotundato crenato convexo.——Brazil, *Descourtilz in Bibl. Deless. t.* 49.——Stems 2-3 feet long. Sepals and petals tawny; labellum bright purple, with a lanceolate streak in the centre, white slightly spotted with purple. The flowers are slightly fragrant.

C. *maritima ;* foliis binis ovalibus obtusis spatha parum longioribus, caule subclavato, sepalis oblongis acutis, petalis lanceolatis falcatis obtusis, labello trilobo (nudo?); laciniis lateralibus erectis rotundatis intermediâ dilatatâ denticulatâ emarginatâ.——Sea-beaten rocks, Buenos Ayres, (*Tweedie, in herb. Hooker.*) Flowers fine, rose-coloured, usually in threes; many varieties. Its small leaves sufficiently mark this, of which I have only seen one specimen.

C. *coccinea ;* foliis solitariis oblongis acutis caulibus ovalibus teretibus longioribus, floribus solitariis, spathâ nullâ, sepalis lineari-oblongis obtusis rectis, petalis triplò latioribus, labello ovato basi cucullato indiviso apice plano sepalis breviore.——Brazil, *Descourtilz in bibl. Deless. t.* 10.——Stems 2 or 3 inches high. Flowers bright scarlet, 3 inches across. A most remarkable and beautiful species.

C. *Harrisoniana* (*Bateman in litt.*); foliis angustè lanceolatis, racemo 1-4-floro, sepalis petalisque patentibus his subrotundo-ovalibus, labello subtrilobo verrucoso. *Bateman.*——Brazil, *Mr. Harrison.*——Very near C. Loddigesii.

EDWARDS'S

BOTANICAL REGISTER:

OR,

ORNAMENTAL FLOWER-GARDEN

AND SHRUBBERY:

CONSISTING OF

COLOURED FIGURES OF PLANTS AND SHRUBS,

CULTIVATED IN BRITISH GARDENS;

ACCOMPANIED BY THEIR

History, Best Method of Treatment in Cultivation, Propagation, &c.

CONTINUED

BY JOHN LINDLEY, Ph. D. F.R.S. AND L.S.

PROFESSOR OF BOTANY IN UNIVERSITY COLLEGE, LONDON,

AND THE ROYAL INSTITUTION OF GREAT BRITAIN,

&c. &c. &c.

New Series.

VOL. X.

OR VOL. XXIII. OF THE ENTIRE WORK.

——viret semper——nec fronde caducâ
Carpitur.

LONDON:

JAMES RIDGWAY AND SONS, PICCADILLY.

M.DCCC.XXXVII.

1920

* ONCÍDIUM críspum.

Curled-flowered Oncidium.

GYNANDRIA *MONANDRIA.*

Nat. ord. ORCHIDACEÆ, § VANDEÆ.

ONCIDIUM. *Suprà, vol.* 13. *fol.* 1050.

A § 1. *a.* * Gen. et Sp. Orch. 197.

O. *crispum;* pseudobulbis oblongis sulcatis rugosis diphyllis, foliis lanceolatis coriaceis acutis, scapo simplici multifloro, sepalis (obovatis) recurvis undulatis obtusis lateralibus semiconnatis, petalis duplò majoribus subrotundis undulatis unguiculatis, labelli lobis lateralibus cornuformibus recurvis nanis intermedio maximo unguiculato subrotundo cordato undulato, cristâ subhastatâ acuminatâ tuberculis subuniserialibus circumdatâ, columnæ alis rotundatis denticulatis carnosis.

O. crispum. *Lodd. Bot. cab. t.* 1854. *Gen. et Sp. Orch.* 197. *Hooker in Bot. Mag. t.* 3499.

The first notice I had of the existence of this species was the finding, in the Herbarium of Sir William Hooker, a drawing and one single dried flower of it, which had been sent from the Organ Mountains in Brazil, with a memorandum that from fifty to sixty flowers frequently grow on a stalk. A small specimen of it shortly after flowered with Messrs. Loddiges, and it has subsequently blossomed in many collections, but never with the vigour that it possesses in its wild state.

Even as we know it, it is a stately, noble looking object, with its very large deep chesnut flowers; but if it can be brought to its full degree of vigour, we shall have little in its tribe that can vie with it in appearance.

The plant that furnished the accompanying drawing grew in the hothouse of Richard Harrison, Esq., Liverpool, in May 1836.

The genus Oncidium is one of the most distinct in the whole of the extensive natural order to which it belongs, and in general it is easily enough determined; but even here, as in all such cases, there are species that do not exactly belong to it, or to any other genus. Some instances of these I shall give in the next number of the Register; for the present I confine myself to the genuine and certain species. These, which abound in the tropical parts of America, extend also so far into colder regions, that one species, O. nubigenum, occurs on the mountains of Peru, at the height of 14,000 feet above the sea, where the air must nearly freeze. A large number of them occur in our gardens; of these many are among the older inhabitants of stoves, others are of such recent introduction as hardly yet to be known even by name; among the last are more particularly to be named the lovely rose-coloured *O. ornithorhynchum* from Mexico, and a beautiful little species, *O. lunatum*, a native of Demerara. The genus Oncidium will not however be seen in all its glory till we possess O. tigrinum from Mexico, O. pictum from Peru, and O. macranthum from Guayaquil. The first and last of these equal O. crispum in the size of their flowers; the other is probably the finest of the yellow species.

I am acquainted with the following in addition to those already published.

A. § 1. *a.* *

1. O. *maculosum;* pseudobulbis ovalibus compressis, foliis lanceolatis acutis, scapo stricto maculoso apice paniculato, sepalis ovalibus acutis reflexis lateralibus basi connatis, petalis oblongis obtusis, labello maximo basi pubescente obsolete quadrilobo reniformi: laciniis lateralibus minimis auriculæformibus, cristâ oblongâ e tuberculis pluribus digitiformibus constante, columnæ alis inferioribus truncatis superioribus falcatis integerrimis majoribus.——Brazil, von Martius.——A fine species in the way of Oncidium bifolium.

2. O. *varicosum;* pseudobulbis oblongis subtetragonis diphyllis, foliis rigidis spathulato-lanceolatis scapo gracili pyramidali subsimpliciter racemoso ter brevioribus, floribus distantibus, petalis sepalisque quorum lateralia semiconnata acutis reflexis, labello maximo subrotundo obsoletè quadrilobo: lobis lateralibus rotundatis, cristâ posticè tridentatâ fornicatâ anticè cuneatâ varicibus venarum in posticis confluentibus circumdatâ, alis superioribus columnæ rotundatis denticulatis inferioribus integerrimis. -——Brazil, Prince Maximilian of Wied Neuwied.——A very fine species in the way of O. bifolium, but with large spreading compound racemes.

3. O. *isopterum;* pseudobulbis ovalibus compressis diphyllis, foliis angustis canaliculatis recurvis scapo apice paniculato flexuoso quadruplò brevioribus, sepalis acutis supremo fornicato lateralibus semiconnatis divaricatis, petalis obovatis reflexis, labelli laciniis lateralibus brevibus angustis obtusis intermediâ transversâ angulatâ bilobâ, cristâ tuberculosâ, columnæ alis 4 subæqualibus rotundatis.——Brazil, Von Martius.——A small species very near O. flexuosum.

A. § 1. *b.* *

4. O. *Martianum;* pseudobulbis, ... foliis ...., scapo erecto paniculato, sepalo supremo petalisque obovatis obtusis concavis lateralibus linearibus acutis brevioribus, labello bilobo reniformi: laciniis lateralibus rotundatis nanis, cristâ falcatâ compressâ utrinque tuberculis 3 linearibus $\frac{2}{1}$ suffultâ, alis columnæ denticulatis subquadratis angulis rotundatis.——In Brazil, von Martius.——A beautiful yellow species, with a scape two feet high; it resembles in general appearance O. ampliatum.

5. O. *lunatum;* pseudobulbis oblongis compressis 1-2-phyllis, foliis angustè oblongis planis obtusis scapo brevioribus, scapo racemoso, sepalis petalisque spathulatis retusis, labello pubescente lunato basi piloso: laciniis lateralibus minimis inflexis, cristâ lineari apice depressâ utrinque bidentatâ: dentibus glanduligeris, alis columnæ cuneatis integris, clinandrio posticè dentato.——A pretty little species, with a white crescent-shaped lip, blotched with dull orange, and white sepals blotched with deep orange.——Messrs. Loddiges imported this plant from Demerara. It will be figured in the Botanical Register, folio 1929.

6. O. *retusum;* pseudobulbis ...., foliis lineari-lanceolatis, scapo paniculato divaricato, sepalis petalisque subæqualibus spathulatis retusis subcarnosis, labelli lobo medio reniformi bilobo lateralibus paulò majore, cristâ e lamellis 5 brevibus tuberculos 6 elongatos circumdantibus, columnæ alis maximis acinaciformibus dentatis.——Peru, Mathews in herb. Hooker.——A beautiful species with deep chesnut and yellow flowers, and a yellow lip. Its mode of growth is that of O. Baueri.

7. O. *gracile;* pseudobulbis ovatis compressis diphyllis, foliis lanceolatis acutis brevibus, scapo gracili apice racemoso, 3-6-floro, sepalis lineari-lanceolatis, petalis paulò latioribus, labello cuneato emarginato basi auriculato lamellis duabus linearibus cristato, columnæ alis oblongis integris.——Brazil, von Martius.——The scape of this species is about a foot high, with from 3 to 6 rather small flowers at the end. The leaves are not more than two inches long, and the pseudobulbs about $\frac{1}{3}$rd that length. The lip is yellow, the remainder of the flower reddish brown and green.

8. O. *ramosum;* pseudobulbis ...., foliis ...., paniculâ strictâ densâ ramosâ, sepalis angustis acutis supremo erecto fornicato lateralibus unguiculatis angustioribus, petalis oblongis subundulatis, labello subrotundo bilobo basi auriculato, cristâ digitato-tuberculatâ squamis $\frac{3}{2}$ carnosis interjectis, columnæ alis rotundis integerrimis.——Brazil, von Martius.——Scape a foot and half long, much branched, very compact, not at all flexuose. Flowers extremely numerous, yellow and olive, rather smaller than those of O. altissimum. A beautiful species.

9. O. *leucochilum* (Bateman in litt.); foliis .... pseudobulbis ...., scapo elato paniculato, sepalis petalisque oblongis obtusis subæqualibus patentissimis, labello reniformi altè bilobo utrinque emarginato: laciniis lateralibus retusis nanis, cristâ 3-corni basi utrinque dentatâ, columnæ alis acinaciformibus dentatis.——Mexico, Mr. Bateman.——A noble species with the habit and stature of Oncidium Baueri. Flowers greenish, banded with crimson; lip pure white.

10. O. *reflexum;* pseudobulbis ovatis monophyllis, foliis angustè lanceolatis acutis, racemis subcompositis longissimis pendulis multifloris, sepalis petalisque lineari-lanceolatis undulatis acutis reflexis, labello reniformi emarginato: laciniis lateralibus rotundatis dilatatis planis, cristâ depressâ tridentatâ apice bicorni, columnæ alis falcatis dentatis.——Mexico, Count Karwinski.——A beautiful species related to O. altissimum; there are specimens in the Royal Bavarian Museum, now in my care, and the species is cultivated by Messrs. Loddiges. I mistook a specimen of this plant, preserved in Mr. Lambert's Herbarium, for O. funereum of La Llave, but that species, having lenticular *depressed* pseudobulbs, must be different.

A. § 1. *b.* **

11. O. *raniferum;* pseudobulbis ovatis sulcatis diphyllis, foliis lato-linearibus acutis scapo paniculato brevioribus, sepalis petalisque oblongis acutis patulis, labelli laciniis lateralibus triangularibus intermediâ cuneatâ emarginata, cristâ depressâ medio constrictâ: dimidiâ posteriore quadratâ basi utrinque callosâ medio tuberculatâ anteriore triangulari apice emarginata utrinque foveatâ, alis columnæ integris.——Brazil.——A small yellow-flowered species, with the lower half of the lip chesnut coloured; it exists in Mr. Knight's collection.

A. § 1. *b.* ***

12. O. *deltoideum;* pseudobulbis angustis angulatis, foliis lanceolatis canaliculatis, scapo paniculato ramis multifloris flexuosis divaricato, sepalo supremo unguiculato obovato lateralibus longioribus spathulato-lanceolatis, petalis latioribus obovatis crispis, labello deltoideo angulis rotundatis, cristâ basi tuberculatâ apice trilamellatâ, columnæ alis maximis acinaciformibus dentatis.——Peru, Mathews.——A remarkable species, with somewhat the habit of a dense-flowered O. altissimum.

* See folio 1542.

1923

* TRIGONÍDIUM obtúsum.

Blunt-petalled Trigonidium.

GYNANDRIA *MONANDRIA.*

Nat. ord. ORCHIDACEÆ, § VANDEÆ.

TRIGONIDIUM. *Sepala* æqualia, semiherbacea, in cyathum trigonum isoscelem cohærentia, apice patula. *Petala* corollina, duplo minora, venosa. *Labellum* nanum, trilobum, cum columnâ articulatum, medio carnosum. *Columna* nana, libera, semiteres. *Anthera* unilocularis. *Pollinia* 4, cohærentia, minoribus dorsalibus, in glandulam triangularem stigma parvum triangulare tegentem insidentia.——Herbæ *epiphytæ*, *Americanæ*, rhizomate *repente florifero ! et pseudobulbifero.* Folia *coriacea.* Pedunculi *erecti, uniflori, vaginati.*

T. *obtusum ;* foliis lineari-lanceolatis, vaginis acutis, sepalis obovatis, petalis obtusis, labello dorso tuberculato, anthera glanduloso-pilosâ.

Pseudobulbi *compressi, oblongi, obsoletè striati, in plantâ spontaneâ per duas uncias distantes, in rhizoma scapos plurimos promens insidentes.* Folia *cuique pseudobulbo duo, angustè lanceolata, pedalia, sesquiunciam lata, ferè enervia.* Bateman in litt. Pedunculi *palmares, erecti, foliis breviores ;* vaginis 5, *convolutis, acutis, pedunculo appressis, superiore ovario breviore.* Sepala *obovata, obtusa, basi quasi agglutinata in cyathum trigonum petalorum longitudine, luteo viridia, versus apicem fuscescentia.* Petala *alba, apice aurantiaco-fusca et callosa, venis purpureis ornata.* Labellum *petalis plus quam duplò brevius, trilobum, axi et lobo medio carnosis, infra apicem extùs tuberculatum.* Anthera *glandulis piliformibus cristata, 1-locularis.*

For the opportunity of publishing this curious plant I am indebted to Mr. Bateman, who sent it me in August 1836, with the following note:—

" This singular plant I owe to the exertions of Mr. Colley in Demerara. I received it from thence in the summer of 1834; the entire plant, at the time of its importation, consisting of only two pseudo-bulbs, connected by a very stout rhizoma. To this rhizoma a prodigious number of withered flower-stems were attached; a circumstance so remarkable, that I confidently expected the new-comer would prove a distinct genus, a suspicion which the curious flowers, now for the first time produced, have confirmed. The species grows slowly, but is of easy cultivation, and of a free flowering habit. It has not yet increased."

As a genus it belongs to the set of Maxillaria-like plants, from all which, however, its curious flowers sufficiently distinguish it. Among M. De Lessert's Brazilian drawings is a second and very fine species, with larger flowers and much broader leaves, found by Descourtilz on the trees that surround the foot of the mountains of *La Bucanha*, and on the borders of the *Parahyba*, below the town of *Rezende*, flowering in January. This species may be distinguished as follows:—

T. *latifolium ;* foliis obovato-oblongis, vaginis acuminatis supremâ ovario longiore, sepalis lanceolatis acutiusculis, petalis acutis, (labello dorso nudo ?) antherâ glabrâ.——Epidendre à chainons *Descourtilz Brasilian drawings, No.* 33.

In the plate fig. 1. shews the petals and lip, magnified; 2. is the lip by itself more magnified; 3. is the column with its gland-crested anther; 4. the pollen masses and gland; and 5. a transverse section of the latter.

---

* Named in allusion to the triangular form of several parts; the sepals form a 3-cornered cup, the gland on which the pollen masses rest is an obtuse-angled triangle; and the stigma is a triangular excavation.

1927

* BURLINGTÓNIA cándida.

Snow-white Burlingtonia.

GYNANDRIA *MONANDRIA.*

Nat. ord. ORCHIDACEÆ, § VANDEÆ.

BURLINGTONIA. *Perianthium* membranaceum, convolutum, obliquum. *Sepala* unguiculata, labello breviora, basi a petalis distincta; *lateralia* basi concava, connata, labello supposita. *Petala* unguiculata, labello parallela, sepalis longitudine æqualia, sed latiora. *Labellum* unguiculatum, bilobum, basi cornutum aut muticum, cum columnâ parallelum, apice dilatatum; ungue canaliculato, lamellato. *Columna* teres, longè clavata, nunc apice appendicibus duabus coloratis aucta; clinandrio dorsali, stigmate utrinque cornuto. *Anthera* unilocularis. *Pollinia* 2, posticè excavata, caudiculæ subulatæ elasticæ adnata.——Herbæ *epiphytæ*, pseudobulbis 1-2-*phyllis, basi foliatis.*

B. *candida ;* racemis pendulis, sepalo anteriore apice bilobo supremo petalisque obtusis, labello seriebus duabus lamellarum carnosarum in disco.
Rodriguezia candida. *Bateman in litt.*

Folia *oblonga, apicem versus paulò latiora, subcanaliculata, solitaria.* Racemus *pendulus, 5-florus.* Flores *candidi, membranacei, semidiaphani ; venis quibusdam luteis in labello, duos pollices longi.* Sepalum *supremum obovatum, planum, emarginatum ; inferius canaliculatum, bifidum, basi concavum, ad calcar incurvum labelli recipiendum, a petalis distinctum.* Petala *obovata, apice patula, circa columnam voluta.* Labellum *cum columnâ parallelum, ungue canaliculato, apice utrinque unidentato, apice dilatatum, subcuneatum, bilobum, venis lutescentibus ornatum ; lamellis pluribus distichis carnosis in disco, quarum anteriores multò longiores.* Columna *gracilis, teres, clavata, apice dentibus duabus carnosis stigmati adstantibus ; clinandrio declivi immarginato.*

The vegetable kingdom comprehends nothing more perfectly lovely than the delicate flowers of this plant, in which not a tinge of colour sullies the snow-white transparency of the petals, unless it be a faint dash of straw-colour on the lip. For its introduction to this country we are indebted to Mr. Bateman, who imported it from Demerara, and flowered it in his rich collection at Knypersley, in April 1835.

In many respects it agrees with the genus Rodriguezia, especially in its lower sepals being united in one, in its lip having a short horn at the base, in the column being taper, and in the structure of the pollen masses; and I was at first disposed to refer it thither; but its membranous and convolute (not herbaceous, or coloured and spreading) flowers, its unguiculate (not sessile) petals and sepals, its long slender (not dwarf) column, and its membranous 2-lobed lip, so much longer than the other parts, have satisfied me that it is a truly distinct genus.

The species now represented does not stand alone in solitary beauty, but it belongs to a little family, at present consisting of five, each of which vies with the other in loveliness. One of them, *Burlingtonia venusta*, a native of Brazil, may be literally said to sink beneath its long heavy clusters of snow-white blossoms just tinged with pink; a second, seated upon the highest branches of the Cedrela tree, perfumes the forests of the same country with the odour of Jonquils: it is *Burlingtonia fragrans ;* while a third, *Burlingtonia rubescens*, delicately spotted with rose colour, inhabits the trunks of the Calabash tree in the mission of Yurimaguas, in the province of Maynas, the hottest part of Peru, and is loaded with blossoms all the year long; and the fourth, *Burlingtonia rigida*, with many peculiarities, to which I shall immediately advert, breathes the delicious fragrance of the violet.

At the end of the column of Burlingtonia fragrans there are two little brownish purple teeth, which are probably rudiments of anthers; even in Burlingtonia venusta there is a trace of them, but a less distinct one. These bodies, in Burlingtonia rigida become long purple ears, fringed with gland-like hairs, and standing erect like a pair of curved pen-knife blades. Nothing indeed is more common throughout the whole of the order than such rudimentary stamens in one state or other. In Zygostates they attain their greatest degree of developement, separating from the column to the very base, and spreading away from it at a right angle. One species of the latter genus exists in Dr. von Martius's Brazilian Herbarium; the other I find among M. Descourtilz's drawings.

Characters of the four additional species of Burlingtonia, and of the last mentioned genus, are given below.

Burlingtonia venusta ; racemis pendulis, sepalo anteriore integerrimo supremo acuto, labello infra medium tranversè plicato.——Brazil; Descourtilz's drawings, No. 83.

Burlingtonia fragrans ; racemis erectis, sepalo anteriore integerrimo supremo acuto, labello postico cucullato supra unguem bilamellato disco pubescente elevato apice tridentato.——Brazil, in the districts of Morro-Quemado and Macahé, and in the neighbourhood of the town of Bom Jesus de Bananal; Descourtilz's drawings, No. 12.

Burlingtonia rubescens ; racemis erectis, sepalo anteriore integerrimo lineari-lanceolato patente supremoque acuminatis, labello basi auriculato bilamellato.—— Peru; Pöppig (*nov. gen. et sp. vol.* 1. *p.* 41. *t.* 70 Rodriguezia Batemanni).

Burlingtonia rigida ; caulibus teretibus rigidis durissimis erectis per longa intervalla proliferis, pseudobulbis obovatis compressis monophyllis, foliis oblongis, scapo longo rigido apice racemoso, labello basi mutico disco sparsè lamellato, columnâ pubescente apice appendicibus 2 falcatis erectis pilosis auritâ.——Found in Brazil, near Villa Nova de Almeida, by Prince Maximilian of Wied Neuwied, and preserved in the Brazilian Herbarium of Dr. Von Martins.——This singular plant has stiff, apparently erect stems, the thickness of a crow's quill, which throw out pseudo-bulbs at intervals of from 1½ to 2 feet, each pseudo-bulb pushes forth its own roots, which are extremely long and as fine as hairs; and gives birth to a scape of the same thickness and texture as the main stem, a foot and half long, at the tip of which are

---

* Those who have the honour to be acquainted with the Countess of Burlington will feel that no compliment was ever more justly conveyed, than in the dedication of these most lovely flowers to the amiable and accomplished Lady whose name they are henceforth to bear.

clustered 5 or 6 flowers, the size of those of Burlingtonia nivea, with the perfume of violets. I believe Messrs. Loddiges possess living plants of this.

ZYGOSTATES. *Sepala* membranacea, libera, reflexa. *Petala* dilatata fimbriata, sepalis multò majora, patula. *Labellum* cymbiforme, membranaceum, patens, basi intus cornutum aut dentatum, cum columna articulatum. *Columna* teres, anticè excavata, basi utrinque brachio aucta lineari, apice globoso, sepalis æquali, patentissimo (stamine sterili). *Rostellum* longissimum, subulatum, arcuatum. *Anthera* rostrata, unilocularis. *Pollinia* 2, caudiculâ subulatâ, glandulâ minimâ.——Herbæ *epiphytæ. acaules*, Brazilienses, racemis *pendulis*, floribus *parvis viridi-luteis*.

1. *Zygostates cornuta;* cornu unico incurvo ad basin labelli, petalis rhomboideis.—— Brazil, Prince Maximilian of Wied Neuwied.
2. *Zygostates lunata;* dente duplici ad basin labelli, petalis lunatis.——On trees on the highest wooded mountains of the district of Bananal in Brazil, in the midst of dense cold fogs, which are almost perpetual, flowering in March; Descourtilz's drawings, No. 4.——Sepals whitish. Petals greenish yellow.

1929

* ONCÍDIUM lunâtum.

Crescent-lipped Oncidium.

GYNANDRIA *MONANDRIA.*

Nat. ord. ORCHIDACEÆ, § VANDEÆ.
ONCIDIUM.—Suprà, vol. 13. *fol.* 1050.

O. *lunatum;* pseudobulbis oblongis compressis 1-2-phyllis, foliis angustè oblongis planis obtusis scapo brevioribus, scapo racemoso, sepalis petalisque spathulatis retusis, labello pubescente lunato basi piloso: laciniis lateralibus minimis inflexis, cristâ lineari apice depressâ utrinque bidentatâ: dentibus glanduligeris, alis columnæ cuneatis integris, clinandrio posticè dentato. *Suprà, fol.* 1920. *in textu.*

Perianthium *flavum, laciniis croceo maculatis.* Labellum *album, cristâ medio sanguineo-lineatâ, limbo obscurè croceo maculato.* Alæ *columnæ sanguineo fasciatæ.*

This very pretty species of Oncidium was imported from Demerara by Messrs. Loddiges, with whom it flowered in their stove for epiphytes in June last.

It is in some measure related to O. Harrisonianum, from which it is however abundantly distinct. The crescent-shaped lip is, as far as I know, quite peculiar to this species; and the very blunt flat sepals conspicuously point it out.

Fig. 1. is a magnified representation of the lip, with the crest fringed with glands at the angles. Fig. 2. is the upper end of the column, with the wedge-shaped banded appendages, and the toothed clinandrium.

* See folio 1920.

Notes upon Burlingtonia candida, fol. 1927.

Since the appearance of the last number of this work, I have been favoured by the following memorandum from Mr. Bateman:—

"The plant to which you formerly gave the name of Rodriguesia Batemanni, and which you now call 'Burlingtonia candida,' is a native of Demerara, from whence it was sent to me in 1834 by Mr. Colley, and it is, I think, without exception, the most beautiful new species that has hitherto made its appearance amongst his collections.

"It produces its flower-spikes freely, twice every year, and their large blossoms, suspended over the margin of the pot in which the plant grows, have a singularly delicate effect, resembling nothing so much as white satin trimmed with gold.

"While at Vienna last year, I was shewn, by the kindness of M. Endlicher, some of the drawings belonging to his and Dr. Pœppig's work on the plants of Peru; among the number was a species bearing so strong a resemblance to our present subject, that I did not hesitate to pronounce it to be Rodriguesia Batemanni, and it was accordingly published by those gentlemen under that name in their admirable 'Genera et Species Plantarum.' I have now before me, through the liberality of Dr. Pœppig, wild specimens gathered in Maynas by that intrepid traveller, and on comparing these with specimens of the Demerara species produced in my stove, I am satisfied that I erred in supposing them to be the same. The following are among the more prominent differences. Dr. Pöppig's Rodriguesia Batemanni (now Burlingtonia rubescens) has stiff, upright many-flowered spikes; B. candida few-flowered, entirely pendulous ones. The flowers of the latter are nearly twice the size of those of the former, and streaked with gold instead of being spotted with red. The column of B. candida is nearly half the length of the lip; while in B. rubescens it is not one-third. In B. rubescens, too, there is a sort of tooth placed on the under side, at the extremity of the lip, which is wholly wanting in B. candida."

To this I would add, that it must have been difficult for the most experienced Botanist to have distinguished Burlingtonia rubescens from B. candida, under the circumstances by which alone Mr. Bateman could compare them. They are very similar in appearance: no one could have anticipated that so many beautiful species, distinct as they nevertheless are from each other, would have been suddenly discovered to form a new genus; and only an actual comparison of specimens could have shewn that the Rodriguesia Batemanni of Pöppig and Endlicher was distinct from Burlingtonia candida, to which I once applied the former name.

1934

* SPIRANTHES bracteósa.

Long-bracted Lady's Traces.

GYNANDRIA *MONANDRIA.*

Nat. Ord. ORCHIDACEÆ, § NEOTTIEÆ.
SPIRANTHES. Suprà, vol. 10. *fol.* 1043.

S. *bracteosa;* foliis rosulatis oblongis acutis, scapo vaginato pubescente, bracteis lineari-lanceolatis foliaceis floribus longioribus, sepalis ovatis pubescentibus acutis lateralibus basi connatis, petalis glabris sepalo supremo agglutinatis et conformibus, labello pubescente: ungue obovato canaliculato mutico limbo subcuneato rotundato apiculato.

An addition to the very difficult genus of which our own sweet Lady's Traces forms a part. It was imported from Saint Catharine's by Messrs. Loddiges, in whose collection the drawing was made in May 1835.

It is readily known by its long leafy bracts, downy stems and flowers, spreading oval root leaves, and peculiar lip.

Of course a stove herbaceous plant.

Fig. 1. is a magnified view of an entire flower; fig. 2. is a still more highly magnified view of the back of the column and the lip, the sepals and petals being cut away.

* In allusion to the spiral manner in which the flowers of many species are arranged.

1937

* CHÝSIS aúrea.

Golden-flowered Chysis.

GYNANDRIA *MONANDRIA.*

Nat. Ord. ORCHIDACEÆ, § EPIDENDREÆ.

CHYSIS. Sepala paulò connata, patula; lateralia pedi producto columnæ adnata et calcar simulantia. *Petala* sepalis conformia. *Labellum* trilobum, patulum, venis basi callosis. *Columna* marginata, canaliculata, mutica. *Anthera* subrotunda, opercularis, glabra. *Pollinia* 8, in laminam luteam semifusa; quatuor exterioribus tenuibus quatuor interiora crassiora abscondentibus. *Rostellum* laminatum, convexum.——Herbæ *epiphytæ, occidentales ab arboribus pendulæ;* caulibus *Cyrtopodii depauperatis,* foliis *nervosis basi vaginantibus,* racemis *lateralibus multifloris.*

Chysis *aurea.*

Caules *penduli, subclavati, articulati, squamis fuscis, membranaceis vestiti.* Folia *ovato-lanceolata, basi vaginantia, nervosa, undulata.* Racemi *pedunculati, ex axillâ folii inferioris erumpentes, penduli, multiflori.* Sepala *ovata, obtusa, subundulata, basi alba, cæterùm crocea ; lateralium margine anteriore infracto.* Petala *ovato-oblonga, obtusa, magnitudine et colore sepalis similia.* Labellum *album, concavum, venis sanguineis, trilobum ; lobo laterali oblongo, obtuso, crispo, apice subsaccato ; venis 5 mediis tumidis callosis, tribus lateralibus utrinque pubescentibus elevatis.* Pollinia *sunt hujusmodi ; lamina adest tenuis, lutea, pulverulenta, bipartita, utrinque biloba, in clinandrium prona, marginibus suis incrassatis, et incurvis (= pollinia 4 basi in laminis duabus geminatìm cohærentia.) Præterea ad sunt 4 alia minora, et prioribus supertecta, quorum duo ab angulis posterioribus, et duo ab anterioribus lamellæ supradictæ fissæ oriuntur (= pollinia 4 alia minora). Horum corporum vera indoles latet ; sed verosimiler pollinia 8 adsunt quorum 4 exteriora in laminam fissam margine bilobam deliquescunt.*

The first knowledge I had of this beautiful epiphyte was from Messrs. Lowe and Co. of Clapton, who sent it to me in June 1836, with a note, from which the following is an extract: —

---

* Χύσις a melting. The pollen masses of this plant are as it were fused together.

"It was collected in 1834 by Mr. Henchman, in the valley of Cumancoa, in Venezuela. Mr. H. describes it as growing suspended by long fibrous roots from the lateral branches of trees, so that its pseudo-bulbs, which in their growing state are uncommonly brittle, hang downwards and wave in the wind, which would otherwise be sufficient to break them. It has a very delicate perfume in the morning, but appears to lose it in the heat of the day. The remains of spikes have been seen, which had produced ten flowers."

Shortly afterwards Mr. Bateman forwarded it from Knypersley, with this memorandum :—

"The plant which produced the specimen from which your drawing is made, I derived in September (1835) from Messrs. Lowe and Co. of the Clapton Nursery. It had then been only recently imported, and was just beginning to push a shoot, which has ever since continued to increase in length and strength. It is now, although apparently far from completed, upwards of two feet long, and equal in girth to the stem of an ordinary Cyrtopodium. It is suspended from a rafter in a pot (filled with broken potsherds and turfy peat), and its stems are completely pendulous, like those of some eastern Dendrobia. Common as is this pendent habit in the Old World, our present subject presents, as far as I am aware, the only example of its occurrence in the New. It may also be worthy of remark that the fleshy stems of this species, although apparently similar in structure to those of Cyrtopodium, Catasetum, Myanthus, Mormodes, Cycnoches, and the like, seem utterly incapable of pushing a shoot after the *radical eye* is gone, several healthy pieces having remained without *breaking* for more than a year ; whereas from any of the above-named genera, under similar circumstances, plants would have been long since obtained."

With regard to the affinity of this genus, although it has in so many respects the habit of certain genuine Dendrobia, and in others that of Cyrtopodium and its allies, it is in reality essentially different from those genera, and much more nearly akin to Epidendrum and its section. Its pollen masses consist of two yellow plates, placed side by side in the bed of the anther, united at the back and slightly notched on the outer edge, so that it is in reality four-lobed, the lobes being extremely unequal ; each lobe has a thickened margin, and rising up, overlies and conceals four other lobes of a thicker texture and smaller size, two of which arise from the back and two from the front of the inner edge of the principal lobes of the plate above described. This remarkable structure may be theoretically described as being equivalent to eight pollen masses, of which the straps of connection, such as exist in all Epidendreæ, are run together into two plates, from an expansion of the edges of which the pollen masses appear to spring.

This is a very showy plant ; the colours of the flower being very bright, its texture firm, and its surface even and waxy. The lip, with its crimson veins, and narrow elevations radiating from the base, is especially worthy of attention.

Fig. 1. is a representation of the upper portion of the column ; 2. is the lip laid open ; 3. are the pollen masses spread open, to shew their structure.

1942

* BOLBOPHÝLLUM barbígerum.

Bearded Bolbophyllum.

GYNANDRIA *MONANDRIA.*

Nat. Ord. ORCHIDACEÆ.

BOLBOPHYLLUM, Thouars. Sepala erecta, acuminata, subæqualia, lateralibus cum pede columnæ connatis et basi obliquis. *Petala* nana (rarissimè sepalis subæqualia). *Labellum* cum pede columnæ articulatum, unguiculatum, sæpiùs integrum et posticum. *Columna* nana, anticè bidentata et bicornis. *Anthera* 1-2-locularis. *Pollinia* 4, libera, valdè inæqualia, nunc in uno connata, nunc per paria cohærentia, altero cujusvis paris minuto lobuliformi.—— Herbæ *epiphytæ*, rhizomate *repente pseudobulbifero.* Folia *coriacea, avenia.* Racemi *radicales.* Lindl. Gen. & Sp. Orch. p. 47.

B. *barbigerum ;* pseudobulbis lenticularibus, foliis solitariis racemo erecto brevioribus, bracteis ovatis amplexicaulibus ovario subæqualibus, sepalis lineari-lanceolatis acuminatis, petalis subulatis columnâ brevioribus, labello lineari-lanceolato acuminato villoso apice stuposo-barbato.

A most curious plant introduced from Sierra Leone, by Messrs. Loddiges, with whom it flowered in June 1836. It grows pretty freely under the hot damp system of cultivating epiphytes.

That a drawing is altogether incapable of representing so strange a conformation as exists in this species, will be evident in the course of the following description. The pseudo-bulbs are small, lenticular, very pale green bodies, from one edge of which springs an oblong, firm, smooth, veinless leaf. From the opposite edge of the pseudo-bulb there rises a raceme of flowers, about six inches long, the base of which is protected by brown, narrow, imbricated scales. The number of flowers in each raceme is from 15 to 20. The bracts are, for the size of the flowers, rather large, broad, ovate, a little stem-clasping, very pale green, and stained with crimson at the points. The three sepals are narrow and taper to a point, pale green externally, dull chocolate brown in the inside *(fig. 2. a. a. a.).* The petals are minute, slender-pointed scales, shorter than the column, and not discoverable without disturbing the sepals. The column is dwarf, and terminated in part by two long curved horns *(fig. 1. a. a.,* and *fig. 2. c.).* The anther is a little round lid, beautifully studded with crystalline points *(fig. 1.).* The lip *(fig. 2. b.)* is one of the most extraordinary organs known even among Orchidaceous plants ; it is a long, narrow, flexuose, sharp-pointed body, closely covered with a yellow felt; just within its point there is a deep purple beard of exceedingly fine compact hairs ; on the under side, at a little distance from the point of the lip, is another such beard ; and besides these there is, at the end of the lip, a brush consisting of very long, purple threads, so excessively delicate, that the slightest disturbance of the air sets them in motion, when they wave gently to and fro, like a tuft of threads cut from a spider's web ; of the last mentioned hairs some are of the same thickness throughout, others terminate in an oblong club, so that when the hairs are waving in the air, and I do not know that they ever are at rest, a part float along gracefully and slowly, while the others are impelled by the weight of their glandular extremities to a more rapid oscillation.

* From βολβος a bulb, and φυλλον a leaf, in allusion to the leaves universally arising from a bulb-like stem or pseudo-bulb.

Nor is this all; the lip itself, with its yellow felt, its two beards, and its long purple brushes, is articulated with the column by such a very slight joint, that to breathe upon it is sufficient to produce a rocking movement, so conspicuous and protracted, that one is really tempted to believe that there must be something of an animal nature infused into this most unplant-like production.

Messrs. Loddiges possess another species, with similar habits.

1947

* LÆLIA ánceps; *var.* Barkeriana.

Mr. Barker's variety of Lælia anceps.

GYNANDRIA *MONANDRIA.*

Nat. ord. ORCHIDACEÆ, § EPIDENDREÆ.
LÆLIA. Suprà, vol. 21. *fol.* 1751.

L. anceps. *Suprà l. c.*
Var. *Barkeriana;* sepalis petalisque subæqualibus, labelli lobo medio angusto acutissimo.

A native of Mexico, whence it was procured by Messrs. Lowe and Co., as I am informed by George Barker, Esq. of Birmingham, who favoured me with a specimen and drawing from his rich collection in December last. I believe it to be common in some parts of that country, and that considerable quantities have been recently imported.

It is, in general appearance, like *Lælia anceps;* from which it differs in two of the opposite ribs of its four-angled pseudo-bulbs being smaller than the other two; in its petals being so much narrower as not to differ materially in size from the sepals; and in the middle lobe of the lip being both narrower and sharper. Mr. Barker tells me, that when the flower first opened, the lilac of the sepals and petals was lighter, and possessed the transparent character of Cattleya labiata, but in a more striking degree.

I do not, however, think these differences will justify its being considered more than a striking variety of Lælia anceps.

It will, doubtless, be easily cultivated upon the same plan as the Cattleyas.

1951

* TRICHOCÉNTRUM fúscum.

Brown-flowered Trichocentrum.

GYNANDRIA *MONANDRIA.*

Nat. ord. ORCHIDACEÆ, § VANDEÆ.
TRICHOCENTRUM. Pöppig Nova Gen. & Sp. Pl. 2. 11. *Perianthium* patens, liberum, æquale. *Labellum* sessile, cum basi columnæ connatum, calcaratum, planum, bilobum, basi lamellosum. *Columna* nana, semiteres, crassa, utrinque alata. *Anthera* bilocularis, mutica. *Pollinia* 2, complicata; *caudiculâ* cuneatâ; *glandulâ* minutâ.——Herbæ *epiphytæ Americanæ acaules;* foliis *planis (aut nullis?);* floribus *radicalibus.* Genus *Æceocladi* proximum, labello cum columnæ continuo et subconnato, necnon columnâ nanâ utrinque alatâ, diversum.

T. *fuscum;* foliis oblongis acutis obliquè tortis racemo longioribus, labello glabro bilobo basi bilamellato venis quibusdam callosis, alis columnæ serratis.
Acoidium fuscum. *Mihi in hortis.*

Imported from Mexico by Mr. Knight, of the King's Road, from whom I received it in flower in July of last year. Shortly after it was sent me by Mr. Bateman. It is a stove epiphyte, and apparently by no means of difficult cultivation.

When I first received it the genus had not been distinguished, and I called it *Acoidium fuscum,* from ακοη, an ear, in allusion to the two ears of the column; but Mr. Pöppig having subsequently figured and described a second species under the name of *Trichocentrum pulchrum,* the former must give way.

* From θριξ τριχος a hair, and κεντρον either a spur or a centre; but the applicability of the word is not apparent, nor is it explained by its author, Mr. Pöppig.

As a genus, *Trichocentrum* must rank next to *Æceoclades,* from which it principally differs in having the labellum united at the base with the column, and in the latter having two ears or wings. It is also close upon *Angræcum* and *Mystacidium,* with which it agrees in habit, but from which it is abundantly distinct.

To the two species now admitted into the genus *Trichocentrum,* must, I think, be added, as a third, the *Limodorum* funale of Swartz, a very curious *leafless* Jamaica epiphyte, which no modern botanist seems to have met with alive, and which I only know from Swartz's description. He states it to grow upon old trunks of trees on the mountains of Jamaica, emitting numerous, simple, stiff, thick, pale green roots, from two to three feet long, and having, in the spring, large whitish flowers in pairs. This plant ought to be sought for again; if it were not found, there would be other species to reward the trouble of the traveller, for I see there are not fewer than twenty species of Jamaica epiphytes described by Swartz, of which none have as yet reached England alive; not to mention the Cymbidium utriculatum, a terrestrial species inhabiting moist woods on the island, having a tuber *as large as the largest potatoe,* a scape about two feet high, Calanthe-like leaves, and pretty large (majusculi) white succulent flowers.

1951*

MONACHANTHI ET MYANTHI cristati

proles biformis.

In November, 1836, His Grace the Duke of Devonshire was so kind as to put into my hands the extraordinary flower represented in the accompanying plate, which may be regarded as one of the greatest curiosities that our gardens ever produced. Accustomed, as Botanists now are, to the freaks and masqueradings of Nature, and to the strangest departures from all rules, at every step among Orchidaceous plants, there is certainly nothing upon record to be for a moment compared with the case before us. It is that of a plant of *Myanthus cristatus* changing into a *Monachanthus,* related to *Monachanthus viridis,* and combining in its own proper person no fewer than three supposed genera, *Myanthus, Monachanthus,* and *Catasetum.*

I doubt very much whether any one would have believed in the possibility of such transmutations upon weaker evidence than that I am about to produce. At least, for my own part, I am much in the position of the person who, upon being assured of the truth of an improbable story, exclaimed, "Why, Sir, I would not believe it if I saw it myself." I am the first Botanist who ever witnessed any of these changes; my observation was put upon record several years ago, and yet, when I read it again in 1833, I really believed I must have been mistaken, and doubted my own positive testimony. In this very Botanical Register, vol. xii. fol. 966, in April, 1826, is the following note under Catasetum cristatum:—

"The unimportance of the peculiarity which exists in the labellum (namely, its flattened or fringed and crested state) is manifested in a singular manner by a curious monster of this plant, which we have observed on an individual in the Horticultural Society's garden. Among flowers of the ordinary structure two or three others were observed, in which the labellum was precisely of the same nature as that of Catasetum tridentatum; that is to say, destitute of the crested appendage, and perfectly galeate and naked."

This, I repeat, appeared to me so extraordinary a statement, especially as after seven years it had never been corroborated by any other case of the same kind, that I concluded I must have made some mistake, and I accordingly formed the genus *Myanthus* out of a species nearly allied to the very Catasetum cristatum, which, in 1826, I had seen sporting back to C. tridentatum.

Not content with this, I added the genus *Monachanthus*, distinguishing it from Catasetum by the want of cirrhi on its column, and by its perianth being turned back; and when the original species, *M. viridis*, was sent me from Wentworth, previously to publication in this work, fol. 1752, I felt no doubt of its being an entirely distinct plant. Even when Lord Fitzwilliam assured me that it was beyond all doubt an accidental sport of *Catasetum tridentatum*, I still adhered to my idea that an imported plant of *Monachanthus viridis* had been accidentally taken for the latter common species. Nor do I think that as a Botanist I was to be blamed for these errors: the genera being founded upon characters which were apparently important, and which most assuredly no one could, *à priori*, have suspected could pass into each other in the manner that has now been seen. If, however, it should be thought that I ought to have been aware of such metamorphoses, I at least have lost no time in acknowledging the mistakes, and putting others on their guard against them for the future.

With regard to the particular case now before us, Mr. Paxton has furnished me with the following note:—

"The monster sent by his Grace the Duke of Devonshire, was produced on a plant of Myanthus cristatus, which also produced a perfect spike of Myanthus cristatus at the same time; the same plant has flowered twice before, but did not sport: a high state of cultivation appears to favour the production of monsters of this kind. There are more than twenty plants of Myanthus barbatus at Chatsworth, which are almost all different in colour, but none of them have yet sported; there are also, at least, twenty plants of Monachanthus discolor, which keep true. Mr. Hillhouse (a gentleman residing in Demerara), has informed the Duke of Devonshire that a flower spike of Catasetum often has on it several distinct kinds of flowers."

To make the reader more distinctly aware of the changes that the flowers of this specimen exhibit, I will describe one of those of the species in its ordinary state.

It has narrow, ovate, lanceolate, green sepals, of which the two lateral ones are placed obliquely on each side of the lip; the petals are narrow, spotted with red, and placed almost in front of the upper sepal; the lip is a narrow, flattish, fleshy body, having a sac or conical depression (*fig*. 4. *c*.) in its middle, a large, curved, white, fleshy horn (*fig*. 4. *b*.) projecting forwards from its base, and a row of stiff, short, fleshy teeth standing up all round the edge. The column has a long horn rising up from its back, and forming a narrow bed for the anther, while two other horns or cirrhi (*fig*. 4. *a*.) hanging downwards over the lip.

In other flowers, which may be said to exhibit the extreme of difference, the sepals and petals are all equally spotless, of the same short, ovate, obtuse figure, and spread equally round the column; the lip is a fleshy hood-like, or helmet-like, greenish yellow body, a little turned back at the point, entirely without sac, horn, or marginal teeth, and standing erect so as to overshadow the column. The column itself has an exceedingly short horn at the back, and no cirrhi in front. I think I may safely say, that until these two sorts of flowers had been seen on the raceme, no Œdipus could have guessed them to belong to either the same species, or even genus.

The way in which they pass into each other is sufficiently curious, as far as regards the lip. All the fringe-like teeth of this organ disappear, except a few at the base on one side (*fig* 3. *a*.), the curved horn shortens (*fig*. 3. *b*.), and the remainder contracts about the sac, which widens and deepens till the commencement of a helmet is shadowed out. In another case all the fringes disappear except a few minute bristles (*fig*. 1. *a*.), the lateral cirrhi are not formed, the curved horn is wanting, and the helmet-like form in the lip is completed. All these transitions were perfectly shewn in the Duke of Devonshire's specimen, and have been faithfully copied by Miss Drake. On the same specimen was also a case of a flower with two lips, as is represented at *figure* 2.

The necessary consequence of this is, that the supposed genera *Myanthus* and *Monachanthus* must be restored to *Catasetum*, and I have no doubt now, although no proof has been seen of it, that *Mormodes* must share the same fate. But which of the species have their masks on, and which shew their real faces, I certainly will not at present presume to guess.

Mr. Schomburgh has lately sent to the Linnæan Society, from Demerara, a specimen of another Monachanthus, sporting to a crested Myanthus; of which, I presume, some account will in due time be published. And I am acquainted with the following example of this tendency in a very different genus.

In August 1836, Mr. Willmer, of Oldfield, near Birmingham, sent me a specimen of a Cycnoches, which had broad petals, a short column, hooded and dilated at the apex, and a broad roundish lip, gibbous at the base, and with its stalk much shorter than the column. It was, however, destitute of scent, while Cycnoches Loddigesii has, as is well known, a delicious odour of Vanilla. I had no doubt of its being distinct species, and called it *C. cucullata*. But in the autumn of 1836, in the garden of the Horticultural Society, a plant of Cycnoches produced from the opposite sides of the same stem two racemes; those of one raceme were the well-known fragrant flowers of Cycnoches Loddigesii, and of the other the scentless flowers of the new C. cucullata.

1953

* PERISTÉRIA cerína.

Waxen Dove-flower.

GYNANDRIA *MONANDRIA*.

Nat. ord. ORCHIDACEÆ, § VANDEÆ.

PERISTERIA, *Hooker*. *Perianthium* globosum. *Sepala* basi subconnata, concava, basi labello connata. *Petala* conformia, paulò minora. *Labellum* erectum, medio articulatum; dimidio superiore obovato, truncato, medio pulvinato, inferiore bilobo columnâ continuo. *Columna* erecta, semiteres, basi magnâ dilatatâ. *Anthera* ecristata, bilocularis. *Pollinia* 2, posticè fissa, glandulâ sessili nudâ rostellum involvente.——Herba *subterrestris*, *pseudo-bulbosa*. Folia *plura*, *plicata*. Scapi *vaginati*, *radicales*, *multiflori*. Flores *speciosi*. Lindl. gen. et sp. Orch. p. 160.

P. *cerina;* scapo brevi pendulo, racemo denso, labelli lobo medio margine crispo, columnâ apterâ.

Pseudobulbus *oblongus, vel ovatus, atro-viridis*, 3-*pollicaris, apice* folia 3-4 *gerens, oblongo-lanceolata, plicata, basi vaginantia, pedalia et ultrà.* Scapus *brevis, vaginatus, pendulus, rigidus,* racemo *denso*, 9-10-*floro, secundo onustus.* Sepala *et* petala *colore et texturâ ceræ lutescentis, immaculata, obtusa.* Labellum *trilobum, cum basi productâ columnæ articulatum, ejusque faciei appressum; lobis lateralibus ovatis, acutis, planis, intermedio abruptè inflexo, ovato, emarginato, margine crispo et croceo, cæterum colore sepalorum.* Columna *crassa, ovata, semiteres, aptera.* Pollinia 2, *biloba, in glandulam triangularem sessilia.* Flores *juniperum spirant.*

A new species of the curious genus Peristeria, imported from the Spanish Main by Mr. Knight of the King's Road, in whose Nursery the accompanying figure was made in June last.

It is allied to the *P. pendula* of the Botanical Magazine, from which it differs in its spotless, smaller flowers, in the crisped border of the middle lobe of the labellum, and especially in the absence of wings from the column. This latter

* From περιστερα a dove: the column in the original species, *P. elata*, resembles a dove hovering with expanded wings.

circumstance brings the genus *Peristeria* extremely near *Maxillaria Warreana* on the one hand, and *Maxillaria cristata* on the other.

The flowers have a strong smell of Juniper.

Fig. 1. is a side view of a column and lip; shewing how the middle lobe of the latter is bent inwards, so as to press against the face of the column. *Fig.* 2. a front view of the column. *Fig.* 3. the pollen masses with their glands.

1959

* MEGACLÍNIUM máximum.

Largest Megaclinium.

GYNANDRIA *MONANDRIA.*

Nat. ord. ORCHIDACEÆ, § MALAXIDEÆ.
MEGACLINIUM. Suprà, vol. 12. *fol.* 989.

M. *maximum;* sepalo supremo acuto lateralibus margine involutis acuminatissimis reflexis, labello lineari revoluto.
M. maximum. *Lindl. gen. & sp. Orch. p.* 47; *sine char.*

Facies *herbæ ferè* M. falcati, *sed paulo robustior, et folia minùs emarginata et margine subrevoluta.* Rachis *ensiformi-falcata, dentata, multiflora,* 9 *poll. longa, pallidè viridis.* Flores *herbacei.* Pedicelli *et* calyx *extus sanguineo-maculati.* Sepalum *supremum erectum, oblongum, acutum, denticulatum, utrinque callosum, lateralibus duplò longius;* lateralia *triangularia, e latâ basi acuminatissima, reflexa, marginibus involutis.* Petala *parva, plana, linearia, acuta, pallidè viridia.* Columna *lata, discoidea, tenuis, marginata, basi cum labello lineari fusco revoluto mobili elasticè articulata.* Anthera *unilocularis.* Pollinia 2 *solida, geminata, nec lobata nec lineâ exaratâ.*

Nearly a thousand plates ago, the first species of this most singular genus was published in the present work. At a later period I found the imperfect remains of a second in the Banksian Herbarium, collected by Smeathman in Sierra Leone; and this has at last been added to our gardens by the correspondents of the Messrs. Loddiges, in whose collection the drawing was made in August last.

In general appearance it is a good deal like *M. falcatum*, from which species it differs, in having larger leaves, which are less deeply notched at the point, and more revolute at the edges; in its larger sabre-shaped rachis, which is as much as nine inches long; and in the form of all the parts

* From μεγας large, and κλίνη a bed, in allusion to the broad, sword-shaped bed or rachis of the flowers.

of the flower. Its lip is hardly so moveable as in *M. falcatum*, in which that part swings up and down with considerable rapidity, like the heads of some Chinese images, when set in motion by a sudden jar.

I believe this will prove one of the easier kinds to cultivate.

Fig. 1. is a highly magnified figure of an entire flower; 2. is a pair of pollen masses.

1961

* EPIDENDRUM noctúrnum; β. latifólium.

Broad-leaved Night-smelling Epidendrum.

GYNANDRIA *MONANDRIA.*

Nat. ord. ORCHIDACEÆ, § EPIDENDREÆ.
EPIDENDRUM, Suprà vol. 1. *fol.* 17.

E. *nocturnum;* foliis distichis oblongis acutis, floribus subgeminis terminalibus, sepalis petalisque linearibus acuminatis patentibus, labelli lobis lateralibus ovatis integerrimis intermedio setaceo brevioribus. *Lindl. Gen. & Sp. Orch.* 105. cum synonymis.
β. *latifolium;* foliis oblongis obtusis subemarginatis, labelli lobo medio lineari-lanceolato, floribus duplò majoribus.

The original Epidendrum nocturnum was found by Jacquin, filling the mountainous woods of Martinique with its fragrance at night, and is now common in our hot-houses. The plant now figured, for which I am indebted to His Grace the Duke of Devonshire, differs from it in several respects, especially in its much greater size, broader leaves which are of an oblong figure, and larger flowers. Mr. Paxton is, however, certainly right in considering it a mere variety of the species, under which it is here arranged.

I received it from Chatsworth in Sept. 1836.

* See folio 1415.

1964

* BOLBOPHÝLLUM cocoínum.

The Cocoa-nut Bolbophyllum.

GYNANDRIA *MONANDRIA.*

Nat. ord. ORCHIDACEÆ, § MALAXIDEÆ.
BOLBOPHYLLUM. Suprà, vol. 23. *fol.* 1942.

B. *cocoinum;* pseudobulbis ovatis tetragonis, foliis solitariis lanceolatis spicâ strictâ multiflorâ brevioribus, bracteis membranaceis linearibus acutis ovario longioribus, sepalis aristatis, petalis linearibus denticulatis columnâ bicorni longioribus, labello ovato-lanceolato obtuso basi ciliato sub apice foveato.
B. *cocoinum. Bateman in litt.*

A pretty little species of this extensive genus, imported by Messrs. Loddiges from Sierra Leone.

It is related to Bolb. recurvum, tetragonum, and the others in their neighbourhood, but is readily known by the pale flesh-coloured flowers, the serrated petals, and the concave short lip, delicately ciliated towards its base.

It flowered at Hackney, in January 1835, filling the house with the scent of a cocoa-nut, and I received it about the same time from Mr. Bateman.

In the dissections fig. 1. is a front view of a column with its two horns arising from either margin, and the serrated petals; fig. 2. is the upper side of the lip; fig. 3. represents the pollen-masses.

* See folio 1942.

1968

* PLEUROTHALLIS saurocéphala.

Lizard-headed Pleurothallis.

GYNANDRIA *MONANDRIA.*

Nat. ord. ORCHIDACEÆ, § MALAXIDEÆ.
PLEUROTHALLIS, Suprà vol. 9. *fol.* 759.

P. *saurocephala;* folio coriaceo oblongo cauli vaginato subæquali racemis breviore, bracteis ovatis ovario brevioribus, sepalis coriaceis oblongis pubescentibus lateralibus semi-connatis, petalis nanis acutis, labello oblongo medio excavato utrinque bilamellato: lamellis marginantibus incurvis.
P. saurocephala, *Lodd. Bot. Cab.*

The accompanying drawing was sent me, with the following description, by Mr. W. B. Booth from the Garden of Sir Charles Lemon, Bart. M.P. of Carclew. The plant has been some years in the gardens near London, but has always remained a scarce species.

" The species now before us was, with some other tropical plants, added to the collection of Sir Charles Lemon, by Mrs. Lake of Falmouth, and flowered for the first time in September last.

" *Stem* jointed, round, smooth and upright, from six to eight inches high, and nearly covered by the sheathing permanent bracteæ which proceed from the joints. *Leaves* ovate-oblong, obtuse, very thick and fleshy, about five inches in length and two in breadth, of a deep green on both sides, nearly flat, excepting at the base, which is slightly compressed or cupped. The *flower spikes* spring from a kind of spathe in the axil, formed by the leaf and stem; whether they always come in pairs we are unable to say with certainty, our specimen had only two, but on other parts of the same plant we found the remains of five, and even as many as eight of these spikes. They are each from four to six inches long, with numerous cucullate, roundish acuminate bracteæ, so closely imbricated at first as almost to conceal the flowers, which are small and produced on round, pale green, short foot-stalks. *Sepals* fleshy, conniving, the upper one arched, and rising nearly erect, three veined on the outside, which is of a dull green. The inside is a yellowish green, beautifully marked with small dark brown spots. The two lower sepals are partly connected together, and of a much darker colour and more glossy than the upper. *Petals* small, embracing the column, or closely joined to it; somewhat cordate and pointed, of a dull green with their edges tinged with brown. *Labellum* entire, rounded at the point and slightly undulated at the margin, which is also tinged like the petals. *Column* round, somewhat bent and club-shaped, largest towards the point."

* See folio 1298.

1970

* BOLBOPHÝLLUM saltatórium.

Dancing Bolbophyllum.

GYNANDRIA *MONANDRIA.*

Nat. ord. ORCHIDACEÆ, § MALAXIDEÆ.
BOLBOPHYLLUM. Suprà, vol. 23. *fol.* 1942.

B. *saltatorium;* pseudobulbis ovatis compressis ancipitibus, foliis solitariis acutis racemis brevioribus, bracteis ovatis membranaceis appressis emarcidis ovario subæqualibus, sepalis ovato-oblongis pubescentibus subæqualibus, sepalis subulatis columnâ longioribus, labello ovato acuminato sepalis longiore villoso apice stuposo.

This is the plant alluded to at folio 1942 as resembling Bolbophyllum barbigerum. Like that species it was imported from Sierra Leone by Messrs. Loddiges, in whose collection the drawing was made in December last.

Fig. 1. is a magnified view of the whole flower; 2. the column and petals; 3. the base of the lip; and 4. the pollen masses.

It is worthy of remark, that in all this genus Bolbophyllum the spiral vessels are particularly tough and numerous. I have noticed this circumstance in the calyx of the plant now figured, and Mr. Griffith has observed it in the sheaths of the leaves of Bolbophyllum auricomum in Burma.

* See folio 1942.

I take this opportunity of defining a new genus of Mexican Orchidaceæ received by the Horticultural Society from the neighbourhood of Vera Cruz, where it was found by Mr. Theodore Hartweg, after whom, as the original Hartwegia proves to be nothing but Chlorophytum, I have named it.

HARTWEGIA *(Orchidaceæ § Epidendreæ).*

Perianthium patulum, coloratum; sepalis lateralibus basi productis labello adnatis. *Labellum* cum columna connatum, basi gibbosum, limbo ovato basi calloso. *Anthera* 4-locularis. *Pollinia* 4, olivaceo-purpurea, filis replicatis. ——Caulis *monophyllus.* Pedunculus *longissimus, capillaris, arctè vaginatus, apice floridus.*

H. *purpurea.* Folium solitarium, coriaceum, ovato-lanceolatum lentiginosum, cauli tereti æquale, pedunculo capillari multotiès brevius. Flores parvi, purpurei. Sepala acuta petalis paulò majora. Labelli limbus basi albus callosus.

1972

* EULÓPHIA macrostáchya.

Long-spiked Eulophia.

GYNANDRIA *MONANDRIA.*

Nat. ord. ORCHIDACEÆ, § VANDEÆ.
EULOPHIA, Suprà vol. 7. *fol.* 686.

E. *macrostachya;* foliis oblongis utrinque acuminatis plicatis subtricostatis, scapo simplici radicali (laterali) foliis longiore, sepalis lineari-lanceolatis acuminatis, petalis conformibus latioribus undulatis, labelli suborbicularis trilobi lobis lateralibus intermedio altè bilobo abbreviato subæqualibus: lamellis duabus nanis ad basin, calcare subrotundo inflato obtuso. *Gen. et sp. Orch. p.* 183.

A handsome species of this extensive genus, inhabiting shady woods in Ceylon, whence it was sent to the Horticultural Society some years ago, by Mr. Watson, the Superintendent of the Botanic Garden at Peradenia.

It is one of the easiest of Orchidaceous plants to cultivate, and produces its graceful racemes of green and yellow flowers abundantly towards the latter part of the year. They go on growing and producing fresh flowers till Christmas.

The stems are in the form of long irregular erect cones, and when old are covered by the withered or ragged remains of the leaves; they are analogous to the pseudo-bulbs of other Orchidaceæ, and to those horizontal tuberous rhizomata which in some species of this genus yield a kind of Salep.

Fig. 1. is a section of the lip showing one of the scales ,that stand at the base of its limb.

* See folio 1433.

1976

* CYMBIDIUM ensifólium; *var.* estriátum.

Sword-leaved Cymbidium; streakless variety.

GYNANDRIA *MONANDRIA.*

Nat. ord. ORCHIDACEÆ, § VANDEÆ.

CYMBIDIUM. Suprà, vol. 7. fol. 529.

C. *ensifolium;* foliis lineari-ensiformibus striatis acutis scapis erectis multifloris, bracteis ovario multò brevioribus, sepalis petalisque lineari-oblongis acutis, labello indiviso planiusculo oblongo undulato revoluto obtuso medio sulcato, sulco pone apicem bidentato.

Epidendrum ensifolium. *Linn. sp. plant.* 1352. "*Smith. spicileg. botanic.* 22. *t.* 24."

Limodorum ensatum. *Thunb. fl. jap.* 29. *Banks. ic. Kæmpf. t.* 3.

C. ensifolium. *Swartz. nov. act. ups.* 6. 77. *Gen. et sp. Orch.* 162.

a. striatum; sepalis petalisque concoloribus purpureo-lineatis. *Bot. mag. t.* 1751.

β. estriatum; sepalis virescentibus petalisque albidis obsoletè striatìm sanguineo punctatis, lineis nullis.

The variety of this plant originally introduced from China, we have not seen for some years; it had flowers distinctly lined with purple, as in the figure in the Botanical Magazine. That which is now more generally cultivated is what is here represented with greener sepals, and pale whitish petals, both without lines; and seems intermediate between the first and the spotless *Cymbidium xiphiifolium.*

It is a very easy plant to cultivate, requiring nothing more than good greenhouse management, when it produces quantities of its slightly fragrant flowers in the spring. The drawing was made in the garden of the Horticultural Society in May last.

* See folio 1530.

C. *xiphiifolium* with spotless pallid flowers is very near this species, and is perhaps a variety; but it appears to be distinguished by its lower bracts being nearly as long as the ovary, and by the ovate flat labellum, in which no undulation exists.

Fig. 1. is the labellum of C. ensifolium *β*; fig. 2. of C. xiphiifolium.

Messrs. Loddiges and Mr. Barker of Springfield have lately flowered a most beautiful new Brazilian Orchidaceous plant, which will soon be published in this work. It will be called *Miltonia spectabilis,* the character of the genus being very briefly the following:—

Perianthium explanatum, conforme. *Labellum* indivisum dilatatum cum ovario continuum basi cuniculatum. *Columna* et *pollinia* Oncidii.

Sp. 1. *M. spectabilis.* Pseudobulbosa. Vaginæ scapi uniflori equitantes. Sepala et petala pallide viridicitrina. Labellum grande violaceum.

1980

* DIPÓDIUM punctátum.

Dotted Dipodium.

GYNANDRIA *MONANDRIA.*

Nat. ord. ORCHIDACEÆ, § VANDEÆ.

DIPODIUM, R. Br. Perianthium patens, petalis sepalisque æqualibus. *Labellum* subconforme, utrinque infra medium auriculatum, disco barbatum, basi saccatum et cum columna connatum. *Columna* erecta, marginata, semiteres. *Anthera* membranacea, bilocularis? *Pollinia* 2, obliquè biloba, caudiculis duabus glandulæ communi affixis.——Herbæ *terrestres, aphyllæ.* Radix *crassa, ramosa.* Caules *squamosi.* Racemi *speciosi.* Gen. et sp. Orch. 186.

D. *punctatum;* squamis radicalibus distantibus, labello recto disco pubescente. *Ibid.* 186.

D. punctatum. *R. Brown prodr.* 331.

Dendrobium punctatum. *Smith exot. bot.* 1. 21. *t.* 12.

Radix *fibrosa; fibris quibusdam horizontalibus, crassis, succulentis, cæteris filiformibus perpendicularibus.* Caulis *fusco-purpureus, glaber, teres,* $1\frac{1}{2}$-2-*pedalis, distanter vaginatus, squamis brevibus obtusis.* Racemus *cylindraceus, laxus; pedunculis lætè purpureis subhorizontalibus.* Perianthium *fusco-purpureum, maculis sanguineis conspersum;* sepalis petalisque *lineari-oblongis subæqualibus.* Labellum *oblongum, trilobum, basi pubescens sulco exaratum, saccatum, cum basi columnæ breviter connatum, trilobum;* lobis lateralibus *angustis falcatis recurvis disco pilosis,* intermedio *oblongo, plano, apiculato, basi angustato, disco villoso.* Columna *erecta, semiteres, truncata, margine glandulosa; clinandrio denticulato;* stigmate *brevi, transverso.* Pollinia *duo, postice exarata, caudiculis binis glandulæ sublunatæ affixis.*

A most curious, leafless, terrestrial Orchidaceous plant, with thick fleshy fibrous roots, and purple spotted flowers; for which I have to thank the inexhaustible collection of the Messrs. Loddiges.

* Δις *two,* and πους ποδος, *a foot;* in allusion to the two stalks of the pollen masses.

Dr. Robert Brown found it both in New Holland and Van Diemen's Land; but it would appear to be rare in the latter island, as Mr. Gunn has sent very little of it home in his rich collections, of which it forms No. 127. Mr. Allan Cunningham met with it in sandy forest ground in the colony of Port Jackson, and beyond the Blue Mountains, flowering in December.

Fig. 1. is the column; 2. the labellum seen from above; 3. the pollen masses, half of one of them being cut away.

1986

* MAXILLÁRIA Steélii.

Mr. Steel's Maxillaria.

GYNANDRIA *MONANDRIA.*

Nat. ord. ORCHIDACEÆ, § VANDEÆ.

MAXILLARIA. Suprà, vol. 11. *fol.* 897.

M. *Steelii;* caule pendulo nano ramoso squamato: ramulis monophyllis, foliis longissimis flagelliformibus canaliculatis, floribus solitariis axillaribus, sepalis petalisque oblongis obtusiusculis, labello trilobo, lobis lateralibus erectis rotundatis: intermedii bilobi laciniis rotundatis subcuneatis divergentibus callo baseos elevato apice obtuse tridentato.

M. Steelii. *Hooker in bot. mag. t.* 3573.

Folia 3-4 *pedes longa, apice subulata.* Flores *flavi, maculis atro-purpureis irregularibus tigrini; labellum potiùs interruptè venosum quam maculatum.* Columna *semiteres.* Pollinia 4, *in glandulam transversam utrinque aristatam sessilia, per paria distincta glandulæ agglutinata.*

A native of Demerara, where it would seem to be common if we are to judge from the large quantity of it that has been introduced within these two last years. When it first came over I called it *Maxillaria flagellifera*, in allusion to the thong-like leaves, and under that name it exists in many collections; but the name not having been published must, of course, give way to that of *M. Steelii*, under which it was first described, in May last, by Sir W. Hooker.

It is not only a very remarkable plant on account of the strange appearance of the leaves, but really a beautiful species, as must be obvious from the accompanying figure taken in September, 1836, from a plant in the possession of Messrs. Loddiges.

It has no very obvious affinity with any known species; but the truth is that we know as yet so little of the genus, which no doubt abounds in species, that the determination of affinity is of no great importance. The most useful thing to do at present is to describe the species, and when the means of doing so are more exhausted it will be time enough to think of their final arrangement. As a contribution to this labour I take the present opportunity of noticing the following.

M. *Rollissonii;* acaulis, pseudobulbis subrotundis compressis subbifoliis, foliis oblongo-lanceolatis acutis apice recurvis, scapis unifloris diffusis laxè vaginatis unifloris, sepalis carinatis lateralibus basi subæqualibus petalisque acutissimis, labelli lobis lateralibus ovatis angustis acutis intermedio oblongo membranaceo apiculato margine deflexo: callo disci elevato carnoso anticè transverso lobos laterales labelli conjungente medio producto truncato tridentato et denticulis utrinque reflexis posticè fornicato truncato bilobo.—— Brazil; in Messrs. Rollisson's collection. Flowers pale lemon colour; labellum dotted with purple in the middle. *Next M. stapelioides.*

M. *acicularis*; (Herbert mss.) subcaulescens, pseudobulbis subfusiformibus sulcatis diphyllis basi squamosis, foliis linearibus acuminatis canaliculatis dorso convexis, pedunculis arctissimè squamosis axillaribus unifloris, perianthio connivente, sepalis petalisque ovatis acutis, petalis obtusioribus, labello indiviso oblongo obtuso disco illinito: callo lineari apice rotundato integro.——Brazil; in the collection of the Hon. and Rev. W. Herbert. ——Flowers purplish chocolate. I only know this from a drawing by Mr. Herbert. The structure of the flower requires to be re-examined. The pollen I have not seen at all. *Affinity uncertain, except with M. uncata.*

M. *uncata;* caulescens, pseudobulbis in axillis squamarum membranacearum iisque brevioribus teretibus sulcatis monophyllis, foliis linearibus canaliculatis obtusis, lateralibus basi maximè productis petalis sepaloque supremo quadruplò majoribus, labello oblongo obtuso carnoso margine crispo membranaceo basi angustato disco unicalloso, columnâ apice utrinque uncatâ.—— Demerara; in the collection of Messrs. Loddiges.

M. *chlorantha;* acaulis, pseudobulbis ovalibus compressis monophyllis, foliis ligulato-oblongis obtusis lucidis basi in petiolum canaliculatum angustatis, sepalis patulis subæqualibus linearibus acutis lateralibus basi parùm obliquis, petalis linearibus conniventibus, labelli trilobi oblongi lobis lateralibus nanis obtusis planis intermedio producto ovato carnoso granulato crispo: callo solitario ovato.——Demerara; in the collection of Messrs. Loddiges.——Flowers small, yellowish green, sweet-scented.

M. *variabilis;* (Bateman mss.) caulescens, pseudobulbis ovalibus compressis squamis membranaceis longioribus monophyllis, foliis lineari-ligulatis obtusis emarginatis planiusculis, floribus axillaribus solitariis, sepalis lineari-oblongis acutis petalis subæqualibus lateralibus basi parùm productis, labello oblongo retuso carnoso basi membranaceo venoso medio utrinque contracto, callo disci parvo ovato obtuso integerrimo, columnâ clavatâ, antherâ pubescente.——Mexico; in various collections.——Flowers small, deep purple. Known in the gardens under the names of M. atropurpurea and M. concinna. *Near M. platypetala.*

M. *tenuifolia;* caulescens, pseudobulbis ovato-oblongis compressis monophyllis squamis longioribus, foliis lineari-lanceolatis acutis recurvis, pedunculis axillaribus solitariis basi squamatis, ovario denudato arcuato, floribus cernuis, sepalis ovato-lanceolatis margine revolutis subæqualibus reflexis lateralibus basi subæqualibus, petalis ovatis obtusis conniventibus, labello oblongo indiviso apice ovato reflexo infra apicem utrinque contracto, callo disci oblongo integerrimo.——Mexico; found near Vera Cruz by Mr. Hartweg. ——Flowers rich purple, spotted and broken into small yellow patches. *Near M. platypetala.*

With regard to the limits of the genus Maxillaria, the reader will find some observations at the end of folio 1991.

* See folio 1428.

1991

* CYPRIPÉDIUM purpurátum.

Purple-stained Lady's Slipper.

GYNANDRIA *DIANDRIA.*

Nat. ord. Orchidaceæ, § Cypripedieæ.

CYPRIPEDIUM. Suprà, vol. 10. *fol.* 788.

C. *purpuratum;* foliis oblongis acutis striatis maculatis basi equitantibus, scapo aphyllo pubescente, sepalo dorsali acuminato ciliato margine revoluto, petalis oblongis subundulatis pubescentibus, stamine sterili lunato.

A further addition to the tropical section of the genus Cypripedium, of which C. insigne and venustum constitute the other species. It is a native of the Malayan Archipelago, and was imported by Mr. Knight, of the King's Road Nursery. The accompanying drawing was made in Nov. 1836, in Messrs. Loddiges' collection.

Not having had an opportunity of examining personally the only flower that was produced, I am unable to say any thing more of the plant than is shewn in the figure. It is manifestly distinct from the other two species in its crescent-shaped sterile stamen. Its leaves are most like those of C. venustum, but are more oblong, and far less rigid, besides having a pallid colour, quite unlike that of the species alluded to; its petals moreover want the long bearded fringe, and the dorsal sepal is convex, not concave.

(*Continued from folio* 1986.)

Whether or not the genus Maxillaria is well limited as it is at present understood is uncertain; nor, till a larger number of species shall have been accurately examined, will it be safe to arrive at any conclusion upon the subject. The value of the numerous modifications of the pollen masses and glandular apparatus is, in particular, still to be ascertained. But, in the meanwhile, there is one character at least, without which it may be safely asserted that no plant can be a Maxillaria; namely, a labellum attached to the column by a moveable joint, or, as we technically say, articulated with the column. This is found in every true Maxillaria, and therefore I have no hesitation in separating from it a Demerara plant which flowered lately with Mr. Barker at Birmingham, in which this character is wanting. It may be called *Stenia* in allusion to the singular narrowness of the pollen masses.

* See folio 1534.

Stenia.

Perianthium explanatum, subæquale, sepalis lateralibus basi obliquis. Labellum cum basi columnæ continuum, (indivisum), concavum, disco appendiculatum. Columna semiteres, basi producta, apice rotundata, mutica. Anthera bilocularis. Pollinia 4, linearia, geminata, basi sejuncta, dorsalia minima. Caudicula subquadrata acuta (pubescens).——Herba acaulis, ebulbis. Folia oblonga, basi angustata, apice recurva. Flores solitarii, radicales, basi ovarii bracteolati.

1. Stenia pallida. *Bot. reg. ic. ined.*

There is moreover an unpublished genus, called Huntleya by Mr. Bateman, in compliment to the Rev. J. T. Huntley, an ardent cultivator of Orchidaceous plants (see fol. 1721 of this work), of which the original species was found, according to a memorandum I have received from Mr. Bateman, in Demerara by Mr. Schomburgk, and of which another and very fine species occurs among Mons. Descourtilz's Brazilian drawings. These plants manifestly differ from Maxillaria in combining with the distichous habit of a Vanda, a flat, spreading labellum, which is unguiculate with the free base of the column, and furnished with a fine fringed appendage at its base, and a column broadly winged towards the apex. This genus may be established upon the following character, which I take from the species of M. Descourtilz.

Huntleya.

Perianthium explanatum, subæquale, sepalis lateralibus basi anticè involutis vix obliquis. Labellum planum, unguiculatum, rhomboideum, patens, basi longè fimbriatum, cum basi libera producta columnæ articulatum. Columna clavata, apice cucullata, margine alata. Anthera bilocularis mutica. Pollinia 4. ——Herbæ caulescentes. Folia ligulata, disticha. Flores axillares (solitarii).

1. H. *sessiliflora* (Bateman ined.); "floribus sessilibus." In Guiana Schomburgk.
2. H. *Meleagris;* floribus longè pedunculatis. In Brasilia. Epidendre fritillé *Descourtilz ic. Orch. Bras. ined. t.* 8.——Planta Vandæ omnino facie. Folia pedalia, flabellatim disticha. Pedunculi cum ovario 5 pollices longi. Flores diametro 3-pollicari, medio candidi, ultra medium Fritillariæ Meleagridis aspectu et colore vinoso. Labellum ovato-rhomboideum, acutum. Columna apice virens.

1992

* MILTÓNIA spectabilis.

Showy Miltonia.

GYNANDRIA *MONANDRIA.*

Nat. ord. ORCHIDACEÆ, § VANDEÆ.

MILTONIA. Perianthium explanatum, petalis revolutis sepalisque lateralibus basi connatis sessilibus conformibus. *Labellum* maximum, dilatatum, indivisum, sessile, cum columnâ leviter connatum, basi lamellatum. *Columna* nana, semiteres, apice aurita. *Pollinia* 2, caudiculæ oblongæ adnata.—— Herbæ *epiphytæ, pseudobulbosæ.* Scapi *uniflori, vaginati, squamis equitantibus.* Flores *speciosissimi.*

M. spectabilis. *Suprà, fol.* 1976 *(August* 1, 1837.)
Macrochilus Fryanus. *Floral Cabinet, t.* 45. (*September* 1, 1837.)

Scapus *teres, vaginatus; squamis carinatis, cartilagineis, equitantibus, obtusis, virescenti-fuscis, superioribus majoribus ovario subæqualibus.* Perianthium *patentissimum,* 4-*pollices longum, tres circiter latum, petalis explanatis.* Sepala *pallida, viridi-alba, oblonga, apiculata, subundulata, margine reflexa.* Petala *conformia et ejusdem coloris, sed latiora, margine crispata, revoluta.* Labellum *maximum, subrotundo-cuneatum, undulatum, cum columna continuum, violaceum, basi intensiùs margine pallidiùs,* 7-*nervatum: venis arcuatis, convergentibus, pictis, basi atropurpureis; tribus centralibus basi cristatis: cristis integerrimis anticè truncatis, intermediâ breviore, crassiore, magis elevatâ, lutescente.* Columna *brevis, erecta, compressa, petalorum colore, anticè gibbosa, lutescens, et cum basi labelli connatum; alis duabus purpureis, carnosis, acinaciformibus.* Anthera. . . . . . Pollinia.

A most beautiful Brazilian Orchidaceous plant, which I originally received from Messrs. Loddiges, and subsequently, at the end of July last, from George Barker, Esq. of Birmingham. I had promulgated the name of Miltonia spectabilis previously to hearing that Messrs. Knowles and Westcott had called the plant Macrochilus Fryanus, and as the former name was published a month earlier the latter will have to give way.

The genera Brassia, Miltonia, Cyrtochilum, Odontoglossium, and Oncidium are closely related, and no doubt form the nucleus of a group of Vandeæ, the limits of which remain to be ascertained. Of these, Oncidium has a column with two ears, and a distinct lobed lip; Miltonia a column with two ears, and an entire lip partially united to it at the base; Odontoglossum a winged column, and an entire lip partially united to it at the base; Cyrtochilum a winged column and a distinct entire lip; and finally, Brassia has a column that is neither winged nor eared, and a distinct entire lip. I say nothing of the tubercular processes upon the lip of all these genera, for I do not see how they will serve with any certainty to distinguish them. Moreover, Oncidium and Cyrtochilum should have unguiculate sepals and petals, while all the other genera have them sessile. While, however, such are the real distinctions between these genera, I am by no means sure that all the species now stationed under them are rightly placed. But that is a question upon which I cannot enter at this time.

Miltonia differs from Oncidium in its lip being slightly connected with the column, much dilated and undivided; and from Odontoglossum, Brassia, and Cyrtochilum, in its column being auricled, and its lip not only much larger than the sepals, but also altogether different in form.

The following are additional species of these genera.

Cyrtochilum Karwinskii; scapo paniculato, sepalis petalisque oblongo-lanceolatis apiculatis, labello obovato rigido ecristato venis centralibus incurvantibus lamellatis, margine columnæ serrato.——Teoxomulco, prope Oaxaca, Karwinski. *(Hb. reg. Monac.)*——A noble species. Flowers $2\frac{1}{2}$ inches in diameter. Sepals and petals blotched with brown, upon a yellow ground.

Odontoglossum angustatum; scapo paniculato, sepalis lineari-lanceolatis acuminatis, petalis ovato-lanceolatis unguiculatis undulatis, labello ovato-lanceolato acuminato: cristâ baseos lamellatâ multifidâ, columnâ muticâ.——Super arbores in valle Lloa, regni Quitensis, alt. 8000. ped. Hall. *(Hb. Hooker.)*

Odontoglossum Hallii; scapo paniculato, sepalis petalisque ovato-lanceolatis acuminatis unguiculatis, labello oblongo acuminato pectinato; cristâ baseos lamellatâ multifidâ, columnæ alis bidentatis dente superiore aristato.—— Supra arbores in valle Lloæ regni Quitensis, alt. 8000 ped. Hall. *(Hb. Hooker.)*—— A most beautiful species. Flowers blotched with brown upon a yellow ground, $3\frac{1}{2}$ inches in diameter.

* In some countries there is a difficulty in finding individuals whose love for Natural History would justify such little compliments, as the naming after them rare or beautiful plants. In this country, on the contrary, the crowd of claimants is so great as to render a selection of names an embarrassing office: and it sometimes happens that those who ought to be the earliest recognised are accidentally lost sight of. Such is the case in the present instance, where the name of a Nobleman, one of the oldest and steadiest friends of Natural Science in this country is only now, for the first time, fixed among the imperishable records of Botany. Let me hope, however, that the present beautiful genus, selected from among his favourite flowers, will be some atonement to Lord Fitzwilliam for the obliviousness of which I at least have to confess myself guilty.

1994

* ONCIDIUM Cebolléta.

Chive-leaved Oncidium.

GYNANDRIA *MONANDRIA.*

Nat. ord. ORCHIDACEÆ, § VANDEÆ.
ONCIDIUM. Suprà, vol. 13. *fol.* 1050.

O. *Cebolleta;* foliis radicalibus teretibus subulatis, scapo paniculato rigido glutinoso, sepalis petalisque obovatis acutis concavis unguiculatis, labelli laciniis lateralibus obovatis retusis intermediâ reniformi undulatâ emarginatâ, cristâ unidentatâ truncatâ basi tuberculosâ, alis columnæ parvis carnosis.
Epidendrum Cebolleta. *Jacq. stirp. Amer. p.* 230. *t.* 131. *fig.* 2.
Oncidium Cebolleta. *Swartz. act. Holm.* 1800. *p.* 240. *Willd. sp. pl.* 4. 113. *Gen. & Sp. Orch. p.* 206.

It is, I presume, impossible to determine with certainty to which, out of two or three taper-leaved Oncidiums, the name of Cebolleta was applied by Jacquin; for he neither describes nor figures the flowers of his plant. In this uncertainty let the present, which agrees with the description of that Botanist as far as it goes, bear the name. It is a native of the Spanish Main, whence it was obtained by Messrs. Lowe and Co. of Clapton, who flowered it in February, 1836.

It is clearly distinguished from O. nudum by the lateral lobes of its lip being obovate, and by the nature of the tuberculated crest at its base. In all probability O. juncifolium combined by me with O. Cebolleta will form another species, but I do not possess the materials for defining it.

Jacquin found O. Cebolleta in the woods of Carthagena, especially near the coast; and he says that the young pseudo-bulbs, pounded with salt, are of great repute as an external application in extensive bruises, broken bones, and other surgical cases.

O. *nudum* (Bateman); facie Cebolletæ sed sepalis petalisque subsessilibus rotundatis acutis concavis, labelli lobis lateralibus linearibus recurvis intermediâ obovatâ emarginatâ, cristâ callis 3 linearibus in fronte lunulâque crenatâ convexâ ad basin.——Hab. in Caraccis, unde in horto regio Berolinensi receptum, a Dom. Otto Batemanno meo benevolè communicatum est.

* See folio 1830.

1998

* CYNÓRCHIS fastigiáta.

Corymbose Dog-Orchis.

GYNANDRIA *MONANDRIA.*

Nat. ord. ORCHIDACEÆ, § OPHRYDEÆ.

CYNORCHIS. Thouars. Sepala æqualia, conniventia. *Petala* sub sepalo superiore agglutinata. *Labellum* columna connatum, calcaratum, sæpiùs 4-partitum, sepalis multò majus et texturæ diversissimæ. *Columna* nana. *Anthera* horizontalis v. resupinata, loculis distantibus elongatis. *Rostellum* planum (v. plicatum), dilatatum, tripartitum. *Pollinia* glandulis nudis. *Stigma* lobis duobus carnosis rostelli lobis lateralibus suppositis.——Herbæ, *Africæ australis, et insularum Mascarensium ;* radicibus *testiculatis.* Flores *nunquam herbacei.* Caules *sæpius squamâ vaginante in medio.* Gen. & Sp. Orch. 331.

C. *fastigiata ;* foliis binis radicalibus oblongo-lanceolatis acuminatis horizontalibus, caule unisquamato, racemo corymboso, labello quadrifido: laciniis æqualibus truncatis, calcare longissimo filiformi. *Gen. & Sp. Orch. l. c.* cum synonymis.

Sepala *extus rosea intùs virescentia.* Petala *rosea, sepalo supremo agglutinata.* Labellum *pallidè herbaceum basi extùs glandulosum, sepalis 4-plò majus, roseo levissimè tinctum, basi intensiùs; calcar roseum.* Anthera *atropurpurea, brevis.* Rostellum *virescens.* Stigmatis *lobi laterales pulvinati, carnei, papillosi.* Rostellum *compressum plicatum.*

This is one of the representatives in tropical countries of the terrestrial Orchises of Europe, growing in similar situations, and having knobby roots of the same kind. It occurs in damp places in the Isle of France, and has a smell very like that of Orchis mascula.

The drawing was made in the stove of Messrs. Loddiges, in April of the present year.

The analytical figures of the plate will help the student to understand the complicated structure of such plants. Fig. 1. is a view of the column and lower part of the lip seen in profile; fig. 2. is the same part seen in front; *a. a.* are the lateral stigmatic lobes; *b. b.* the glands of the pollen-masses; *c.* the middle lobe of the rostellum ; *d. d.* the base of the lip.

* Literally Dog-Orchis.

2005

* CIRRHǼA obtusáta.

Blunt-petalled Cirrhæa.

GYNANDRIA *MONANDRIA.*

Nat. ord. ORCHIDACEÆ, § VANDEÆ.

CIRRHÆA. Suprà, vol. 18. *fol.* 1538.

C. *obtusata ;* foliis oblongo-lanceolatis utrinque acuminatis, petalis sepalisque lineari-oblongis obtusiusculis planis, labelli lobo medio obovato acuto inflexo latè unguiculato lateralibus acuminatis.

This species of Cirrhæa appears distinct from any of the others yet described. It is a good deal like *C. fusco-lutea,* for which I at first mistook it; but it seems sufficiently distinguished by its much more obtuse petals and smaller flowers. I am, however, far from being satisfied with the goodness of such distinctions among these plants, and am rather inclined to fear that we sometimes trust them too much. I believe, however, the form of the middle lobe of the lip to be essentially different in this species from all, except one which flowered in Messrs. Loddiges' collection in March last, and which I then named *C. pallida.*

That plant had most of the characters of *C. obtusata,* except that the sepals and petals were acuminated, and the middle lobe of the labellum placed on a slender unguis. The real value of these distinctions must remain for future enquiry.

This *C. obtusata* was sent me from Liverpool, in September 1836, by Richard Harrison, Esq. who had received it from Mr. Moke of Rio de Janeiro ; it had been collected by that gentleman at Tejuca.

* See folio 1538.

2006

* ONCÍDIUM deltodíeum.

Triangular-lipped Oncidium.

GYNANDRIA *MONANDRIA.*

Nat. ord. ORCHIDACEÆ, § VANDEÆ.

ONCIDIUM. Suprà, vol. 13. *fol.* 1050.

§ 1. Sepala lateralia libera.

*** Labelli laciniæ laterales intermediâ majores.

O. *deltoideum ;* pseudobulbis oblongis compressis sulcatis 2-3-phyllis, foliis lineari-lanceolatis canaliculatis, scapo paniculato ramis multifloris flexuosis divaricato, sepalo supremo unguiculato obovato lateralibus longioribus spathulato-lanceolatis apice revolutis, petalis latioribus obovatis crispis, labello deltoideo angulis rotundatis, cristâ tuberculatâ serie subduplici tuberculorum juxta basin et quatuor versus apicem majoribus distinctis biseriatis minore interjecto, columnæ alis maximis acinaciformibus.

O. deltoideum. *Suprà, vol. . fol.*

Omnes partes virides leviter glaucescentes. Panicula *secunda, nutans, in apice scapi sesquipedalis. Sepala et petala lutea. Labellum apice acutiusculum reflexum. Crista e tuberculis brevibus oblongis composita, ferè totum discum labelli occupans ; tubercula baseos interiora in circulum confluentia unico centrali magis prominulo, exteriora in lineas duas interruptas divergentes ordinata ; ultra hæc versus apicem adsunt unum oblongum in ipso apice circuli et utrinque duo alia majora patentia et magis versus apicem tendentia : Circulus et tubercula exteriora pallide flavescunt maculis quibusdam sanguineis interjectis. Alæ columnæ, in planta culta integerrimæ, in spontanea denticulis quibusdam marginantur.*

I was very unexpectedly gratified at finding this rare plant a few weeks since flowering in the greenhouse of Wm. Joseph Myers, Esq. of Aighburgh, near Liverpool. It had been sent him from Lima by John Maclean, Esq. a great friend to science, now resident in that distant station. Mr. Maclean probably received it from Mr. Mathews, who first discovered it in the province of Chacapoyas in Peru, and from one of whose specimens, without a number, I had previously described it. The arrival of this species in a living state should be a fresh stimulus to those who have mercantile relations with Peru, to procure Orchidaceous plants from the west side of the Cordilleras, for many of them, especially those from the valley of Lloa, are of singular beauty. Hitherto scarcely any of them have found their way into our gardens.

I believe that Mr. Myers and Mr. Richard Harrison are at present the only possessors of this species, which is very graceful and pretty.

O. deltoideum is nearly related to O. divaricatum, from which its narrow leaves and differently formed labellum sufficiently separate it. To the little known O. macranthum,

* See folio 1050.

from Guayaquil, with flowers three inches across, it is also closely allied; but the hastate labellum and peculiar crest of that plant are, independently of the many other circumstances, satisfactory distinctive characters.

It may be expected that this species will grow well with the heat of a greenhouse. I found it among a general collection of greenhouse plants in Mr. Myers's garden, and it was in such good health that I can hardly doubt its preferring the cool temperature in which Cape and New Holland plants thrive, to the excessive heat and damp to which Orchidaceæ are commonly subjected.

2010

* ANŒCTOCHÍLUS setâceus.

Fringed Anæctochilus.

GYNANDRIA *MONANDRIA.*

Nat. ord. ORCHIDACEÆ, § NEOTTICEÆ.

ANŒCTOCHILUS, Blume. Perianthium ringens. *Sepalum* supremum cum petalis in galeam agglutinatum; lateralia patula, basi inæqualia labello supposita. *Labellum* cum columna basi connatum, ventricoso-calcaratum, ungue rigido canaliculato inflexo fimbriato, limbo bilobo patente. *Columna* brevis, marginibus membranaceis dilatatis involutis appendiculam bilamellatam simulantibus. *Stigma* basi bicallosum. *Pollinia* 2, pulverea, subbiloba, caudiculis brevibus, glandulâ oblongâ.——Herbæ *tropicæ, Asiaticæ, radice fibrosâ,* rhizomate *filiformi subterraneo.* Caules *foliosi.* Folia *nervosa, discolora, venis sæpè auromicantibus.* Flores *spicati, laxiusculi, glanduloso-pilosi.* (Anæctochilus, *Blume Prefat. fl. jav.* vi. *sec. Endlicher. gen. pl. no.* 1569. Anecochilus, *Blume Bijdr.* 411. *f.* 15. Chrysobaphus, *Wallich fl. nep. tent. t.* 17. "Orchipedum, *Kuhl & Hasselt Orchid. ed. Breda, t.* 10." *sec. cl. Endl.*

A. *setaceus;* foliis ovatis vel ovalibus acutis discoloribus, spicâ sub-4-florâ sepalisque exterioribus pubescentibus, labello infernè ad margines setaceo-multifido, sacco labelli subdidymo. *Blume Bijdr.* 1. 412.

This extremely curious plant is a native of damp, shady places, among stones, on the mountains of Java, and of similar situations in the woods of Ceylon, whence it was sent to the Duke of Northumberland, in whose collection at Syon it flowered in June last, soon after its arrival.

It, or a nearly related species, is included by Rumf among his most alpine Amboyna plants, "ubi nebulosis nubibus adeo cincti sumus ut ipsa meridie vix sol dignosci possit, qui nor perpetuò fluens cum frigore perpetuo quam maxime delassat et horridum reddit prætereuntem." He says, it always grows in the highest and most inaccessible mountains, where the scattered trees just afford it a little shade, and where the soil is but scantily supplied with moisture. He adds that it is called by the natives *Daun petola*, or Petola leaf, on account of its resemblance to the Petola, a precious silken vestment of many gaudy colours.

The beauty of the plant consists in this character, for its flowers are white and green, and very inconspicuous. But the leaves are singularly painted with golden veins upon a brownish purple ground, as if, in the words of Rumf, "some cunning painter had traced them over with unknown characters."

The plant will require the same treatment as Goodyera, Hæmaria, Spiranthes, Pelexia, and other terrestrial Orchidaceous plants of the same division of the order.

I have it from Mr. Macrae, and Sir W. Hooker from Mrs. Walker, from Ceylon.

Fig. 1. represents a flower after the sepals and petals have been cut off; 2. the column with the curious involute membranous margin; 3. the same unrolled and seen in a front view.

* Apparently from ἀνοικτὸς open, and χεῖλος a lip, I presume in allusion to the spreading apex of that organ.

EDWARDS'S

BOTANICAL REGISTER:

OR,

ORNAMENTAL FLOWER-GARDEN AND SHRUBBERY:

CONSISTING OF

COLOURED FIGURES OF PLANTS AND SHRUBS,

CULTIVATED IN BRITISH GARDENS;

ACCOMPANIED BY THEIR

History, Best Method of Treatment in Cultivation, Propagation, &c.

AND

MONTHLY CHRONICLE

OF

BOTANICAL AND HORTICULTURAL NEWS.

CONTINUED

BY JOHN LINDLEY, Ph. D. F.R.S. AND L.S.

PROFESSOR OF BOTANY IN UNIVERSITY COLLEGE, LONDON,
AND THE ROYAL INSTITUTION OF GREAT BRITAIN,
VICE-SECRETARY OF THE HORTICULTURAL SOCIETY,
&c. &c. &c.

1838.

OR VOL. XXIV. OF THE ENTIRE WORK.
OR VOL. XI. OF THE NEW SERIES.

——viret semper——nec fronde caducâ
Carpitur.

LONDON:

JAMES RIDGWAY AND SONS, PICCADILLY.

M.DCCC.XXXVIII.

2

* CĀTTLĔYA Perrīnii.

Mr. Perrin's Cattleya.

GYNANDRIA *MONANDRIA.*

Nat. ord. ORCHIDACEÆ, § EPIDENDREÆ.

CATTLEYA. Bot. Reg. vol. 11. *fol.* 953.

C. *Perrinii;* foliis oblongis coriaceis cauli æqualibus, sepalis oblongis linearibus obtusis lateralibus falcatis petalis obtusis angustioribus, labello oblongo-lanceolato trilobo unguiculato nudo basi in cuniculum ventricosum producto; laciniis lateralibus erectis acutis intermediâ oblongâ undulatâ obtusâ elongatâ sinubus intermediis rectangulis.

I have for some years been acquainted with this Cattleya, which was originally sent me from the garden of Richard Harrison, Esq. with a drawing by Mr. Arnold Harrison. Since that time I have received it from Mr. Bateman, and more recently from Mr. Knight of the King's Road; on which last occasion the accompanying drawing was made.

This species is a native of Brazil, and is not unlike C. labiata, although inferior to it in beauty. It is distinguished readily by there being an inflated fistular cavity proceeding from the base of the lip to that of the ovary, and by the peculiar form of the lip, which is perfectly destitute of plaits, ridges, or other projections in its middle. It is probable that other species of this genus are furnished with a similar fistular passage, but if so it is not inflated, and consequently is not observable upon external inspection. Although this indicates the presence of a spur to the lip, yet it does not appear to be of more than specific importance, for we find in the genus Epidendrum some species with and others without

* See Bot. Reg. vol. 14. fol. 1172.

the same structure. Fig. 1, which represents a section of the lower part of the flower, explains the peculiarity now adverted to.

The species has been named after Mr. Perrin, Mr. Harrison's intelligent gardener, under whose care so many fine South American epiphytes have been for the first time brought into flower in this country.

Like the rest of the genus this requires to be grown in a moist stove, the temperature of which may be kept from 60 to 70 degrees of Fahr. in winter, and from 70° to 90°, or even 100° with sun heat, in summer. It is propagated, like the other plants of this order, by divisions of the rhizoma, or root-stock, with a stem adhering to them. The soil should consist of good peat, broken, or cut into pieces one inch, or one inch and a half square. The pots should be about half filled with broken bricks, or something of that description, to carry off superfluous water, and if they are plunged in a tan-bed, this will allow the heat to rise more freely than if the pots were wholly filled with soil. It is of the greatest importance to preserve and encourage the roots, and as they are generally protruded near the surface of the soil, it should be raised several inches above the level of the pots, in a pyramidal form, in order that they may have full room to push out.

5

* STANHŌPĔA quadrĭcōrnis.

Four-horned Stanhopea.

GYNANDRIA *MONANDRIA.*

Nat. ord. ORCHIDACEÆ, § VANDEÆ.
STANHOPEA. Bot. Reg. vol. 18. *t.* 1529.

S. *quadricornis;* foliis oblongis utrinque acutis brevè petiolatis, labello medio constricto, hypochilii subrotundi basi bicornis lateribus rotundatis, epichilio ovato cornubus 2 baseos incurvis.

An exceedingly pretty species of Stanhopea, allied to S. oculata, from which it differs very obviously in the general want of spotting upon the flower, and especially in the two short horns which project from the crimson base of the lip. In some circumstances it also approaches S. insignis, especially in the colour and form of the sepals and petals, but the lip is altogether different.

I am indebted for my knowledge of the species to Sigismund Rucker, Esq. jun. of Wandsworth, who received it from the Spanish Main, and who informs me that the pseudo-bulbs and leaves are extremely like those of S. grandiflora.

It is with great pleasure that I lay before the readers of the Botanical Register the following excellent directions by Mr. Paxton for the cultivation of such plants, extracted from the first number of the Sertum Orchidaceum, just published.

"The success with which epiphytes are cultivated by Mr. Paxton is wonderful, and the climate in which this is effected, instead of being so hot and damp that the plants can only be seen with as much peril as if one had to visit them in an Indian jungle, is as mild and delightful as that of Madeira. As to luxuriance of growth, never have they been seen in their native woods in such perfect beauty. It, therefore, affords me no little satisfaction to be enabled, by the permission of the Duke of Devonshire, to publish the following account of the management of Orchidaceæ at Chatsworth, drawn up by Mr. Paxton himself.

* See Bot. Reg. vol. 18. fol. 1529.

"The following treatment is not only applicable to the growth of Stanhopeas, and others of like habit, but an advantage in the growth of any species of Orchidaceæ (the terrestrial, and those that grow in moss excepted).

"Over the drainage hole of the pot to be used, is inverted one of a smaller size, generally covering about half the bottom of the pot; over this is carefully thrown a quantity of broken pots, sufficient to fill the former to within one-third of the top. A sufficient quantity of fibrous, moderately sandy peat, is next selected and placed on the top of the drainage, being first broken into various forms and sizes, but none of them less than a walnut; in placing these, care is taken to dispose of each, so as to leave a passage for the escape of water; this is more effectually secured by putting in, as the process of potting goes on, a few pieces of broken pots, say between every layer, more or less, according to the size of the plant; indeed, I find it an excellent plan to continue a connection of broken pots all the way up the centre, to the bottom of the pseudo-bulbs. After the peat becomes level with the pot, the successive external layers are made fast by means of small pegs, varying from four to six inches long, these pegs run through the layers of peat, and thus secure the whole firmly together. At eight inches above the line the plant is placed on the top; the roots are carefully laid out and covered up to the place of the bulbs very carefully with smaller pieces of peat and potsherds, continuing to fasten the peat as before described, until the whole is finished, when it will be a foot or fifteen inches from the top of the pot;—small plants are not potted so high. At each shifting the plant is raised a little higher. When I commence potting a small plant it is not raised more than three or four inches at first, but as it grows larger it is progressively raised in building up as here described with peat: it does not terminate in the shape of a cone, but is carried up nearly square, being merely rounded a little at the top. Unless the plants are very healthy, but very little water is given at the roots, and in winter very little or none, the great desideratum in the cultivation of Orchidaceæ being *to preserve the roots,* which, by over-watering, especially in winter, are almost sure to be destroyed.

"The general temperature of the house ranges from 60 to 85 degrees; in the afternoon, during the growing season, it is shut up early and the paths well watered, and once or twice a week a little water is sprinkled on their heads. I find great advantage in having a tan-bed in the house to plunge the plants in, the heat from the tan circulates through the peat and potsherds, and causes the plant to grow with great luxuriance. It might be objected to this method of growing Orchidaceæ, that wood-lice would damage the plants; I have followed this plan for two years, and do not know an instance of damage by them; indeed, the tan is too moist to harbour any kind of insects. In order to make this account as intelligible as I can, I will detail the manner in which a young plant was treated:—On the 20th of May last year I received a very small damaged plant of a new Stanhopea; I allowed it to get perfectly dry, it was then potted, and placed in a strong bottom heat, with a strong heat above; the plant began to grow in about a fortnight, and at the end of July had perfected a small bulb; the plant was then kept dry about a fortnight, and was again placed in a strong bottom heat, and in a temperature never lower than 70 degrees, but often amounting in the day to from 90 to 100. By the end of September it had perfected a second bulb, considerably larger than the first. The plant was again dried on a hot flue for a fortnight, and then removed into a larger pot, and elevated a little above the surface; it was again replunged into a strong bottom heat, and, by the end of December, had perfected two more bulbs, making four since its commencement. I should here observe that the plant had but one bulb when I received it;—the plant was now dried for a month, then re-potted, and placed, as before, in a strong heat; about the first week in April the plant had made two more perfect bulbs, the process of drying was again gone through, and the plant replaced in strong heat;—it has on it now, August 31, nine bulbs, made in a short space of fifteen months. I expect to have the plant in a state for flowering next season. This plant was cultivated with a number of other small ones, in a small house that could be kept very hot.

"I cannot conclude this statement better than by recommending those who wish to grow Terrestrial Orchidaceæ well, to attend to the following brief rules, in applying the four great elements of vegetable life, viz. air, light, heat, and water.

"*Air.*—Terrestrial Orchidaceæ should never have a great volume of external air admitted at once, however fine the weather may be; to prevent the house becoming too hot, a thick canvas shading should be covered over it during sunshine.

"*Light.*—The best aspect for an Orchidaceous-house is due south, and the house should be made to admit as much light as possible. In summer a thick canvas is always put on the house to prevent the bright sun damaging the plants. In winter every ray of light is advantageous to the plants.

"*Heat.*—During the growing season, Orchidaceæ require a moderately moist heat, varying from 65 to 85 degrees; in the dormant season from 60 to 75 is quite sufficient; in the season of rest the house should be kept dry.

"*Water.*—With this element more damage is done than by all the others put together. Orchidaceæ in pots should be sparingly watered in the growing season; in the dormant state little or no water should be given. The secret of growing these plants is to take care *never* to kill the old roots; when too much water is given while the plants are not in a growing state, almost all the old roots invariably perish.

"N. B. The brief account here given refers entirely to plants potted in peat soil; those grown in moss and on bits of wood require quite a different treatment."

11

* CIRRHŎPĒTĂLUM Thouārsĭĭ.

Insular Cirrhopetalum.

GYNANDRIA *MONANDRIA.*

Nat. ord. ORCHIDACEÆ, § MALAXIDEÆ.

CIRRHOPETALUM, Lindl. *Sepala* ringentia; lateralia acuminata, longissima, valdè obliqua, basi productæ columnæ adnata, supremo multò longiora. *Petala* nana apiculata. *Labellum* integrum, cum basi columnæ articulatum. *Columna* minima, basi longè producta, apice bicornis. *Anthera* bilocularis. Pollinia 4, quorum duo interiora multo minora, per paria connata.—— Herbæ epiphytæ, rhizomate repente, pseudobulbos monophyllos gerente. Folia coriacea, avenia. Flores densè racemosi, nunc radiati, in apice scapi radicalis. *Gen. & sp. orch. p.* 58, quibusdam mutatis.

C. *Thouarsii;* petalis ciliatis sepaloque supremo aristatis, foliis oblongis obtusis emarginatis scapo brevioribus. *Lindl. l. c. cum. synonymis.*

This very curious plant is one of the most extensively diffused of all epiphytal Orchidaceæ. I have specimens gathered in the Society Islands by Mr. Mathews, Reinwardt found it in Java, Thouars in the isles of France and Madagascar, and Mr. Cumming has lately sent it from Manilla to Messrs. Loddiges, with whom it flowered last July.

Nothing can be more singular than the long strap-shaped sepals growing from one side of the flowers, and almost bearing them down with their weight; they offer a singular instance of the unequal force with which the phenomena of developement are exerted in one and the same flower. The petals, represented at fig. 1. of the magnified dissections, are very small, yellow, spotted finely with red, bordered by bristle pointed teeth, and terminated in a long awl-shaped point.

* From κιῤῥος tawny, and πεταλον, in allusion to the prevailing colour of the flowers.

It is a very pretty little species and well worth cultivation, for if in good health it will bear a good many such umbels as are here represented upon the same plant. Like most other Orchidaceæ it requires to be grown in a moist stove. The soil best suited to its growth is rough pieces of decayed peat, amongst which broken pieces of pots and bricks may be mixed with advantage. A quantity of drainage should always be put in the bottom of the pots, and very little water given in winter when the sun's rays are weak, and the weather generally cloudy. The rhizoma of this plant elongates very much, and is in a growing state always at its extremity, leaving the parts formed previously in a state of inaction; on this account it is necessary at every potting to keep the active part nearest the side of the pot which it grows away from. It may easily be propagated by taking off the pseudobulbs, with a portion of the rhizoma adhering to them on each side; although inactive when upon the old plant, they will push when made to depend upon their own resources.

13

* GOVĒNĬĂ lilĭăcĕă.

Lily-flowered Govenia.

GYNANDRIA *MONANDRIA.*

Nat. ord. ORCHIDACEÆ, § VANDEÆ.
GOVENIA, Bot. Reg. vol. 21. *fol.* 1795.

G. *liliacea;* labello ovato basi imâ cucullato, spicâ oblongâ, scapo univaginato, bracteis oblongis cucullatis, foliis obovato-lanceolatis plicatis radicalibus scapo brevioribus, radice tuberosâ.
G. liliacea. *Bot. Reg.* 21. *fol.* 1795 *in textu.*
Maxillaria liliacea. *La Llave et Lexarza nov. veg. descr.* 12.

A small tuberous plant, with a stem about a foot high, imported from Mexico by George Barker, Esq. of Birmingham, along with a great quantity of still more interesting plants and seeds of all descriptions. It is an Orchidaceous plant, with the habits of a Bletia, flowering in the month of July. Its flower-stem is hardly above a foot high, and rises directly from the tuber, to about the length of the plaited leaves. The colour of the flowers is not a good or pure white, but they are delicately streaked upon the petals with lines of rather pale purple.

From G. superba it differs not only in colour, but in the form of the labellum, which is not at all cordate, or channelled along the middle.

The first notice we have of the existence of this species is from Hernandez who, in the Madrid edition of his work, calls it by the many-syllabled name of "Iztactepetzacuxochitl Icohueyo." It was many years afterwards described and named more scientifically by Paul de la Llave and John Lexarza, in their *Novorum Vegetabilium Descriptiones*, under the title of Maxillaria liliacea. These authors state that it is met with near Valladolid, in Mexico, under the shade of

* See Bot. Register, vol. 21. fol. 1795.

trees or sometimes growing over their roots, and flowering in the summer months; its vernacular name is Azuzena del Monte.

It is at present extremely rare; I have not heard of it in any collection except that of Mr. Barker. It will require the same treatment as Bletias.

20

* STÉNĬĂ pállĭdă.

Pale Stenia.

GYNANDRIA *MONANDRIA.*

Nat. ord. ORCHIDACEÆ, § VANDEÆ.

STENIA. Lindl. Perianthium explanatum, subæquale, sepalis lateralibus basi obliquis. *Labellum* cum basi columnæ continuum (indivisum), concavum, disco appendiculatum. *Columna* semiteres, basi producta, apice rotundata, mutica. *Anthera* bilocularis. *Pollinia* 4, linearia, geminata, basi sejuncta, dorsalia minima. *Caudicula* subquadrata acuta (pubescens).——Herba *acaulis, ebulbis.* Folia *oblonga, basi angustata, apice recurva.* Flores *solitarii, radicales, basi ovarii bracteolati.* Bot. reg. fol. 1991. in textu.

S. pallida. *Bot. reg. l. c.*

Caulis *nullus, nec rhizoma conspicuum, nec pseudobulbus.* Folia 2-3, *oblonga, acuta, avenia, basi paulò angustata et carinata, è squamis spathaceis brunneis vaginantibus erumpentia; texturâ rigido-chartacea.* Scapi *radicales, prostrati, uniflori, apice bracteis duabus oppositis ovatis instructi, ovario breviores, saltèm haud longiores.* Perianthium *explanatum, pallidè citrinum, diametro sesquiunciali et ultrà; sepalis petalisque æqualibus, ovatis, lineatis, acutis.* Labellum *saccatum, integrum, carnosum, ovatum, cum pede producto columnæ continuum, pallidè citrinum, rubro-punctatum, cristâ transversâ dentatâ.* Columna *semiteres, clinandrio proclivi, facie pubescens.* Pollinia 4, *linearia, obtusa, dorsalia minima, caudiculæ brevi villosæ adnata, glandulæ rhomboideæ cartilagineæ medio insertæ.*

A rather pretty epiphyte, imported from Demerara by Mr. Barker of Birmingham, in whose very extensive collection of Orchidaceous plants, this drawing was made last August. It differs essentially from Maxillaria, which it resembles in habit, in the labellum not being articulated with the foot of the column, a distinction, the importance of which there is at present no cause to suspect. In general appearance it is more like Maxillaria Rollissoni than any plant I am acquainted with, but it wants the pseudo-bulbs of that species.

* From στενος narrow, in allusion to the form of the pollen-masses.

No one has deserved better of Botanists than Mr. Barker, and I therefore see with pleasure a genus named after him by Messrs. Knowles and Westcott, with the following character.

BARKERIA.

Sepala patentia vel reflexa, basi subconnata. *Petala* subæqualia vel paulò latiora. *Labellum* liberum, cum columnâ parallelum, limbo indiviso, disco costato. *Columna* elongata, plana vel compressa, sulcata, alata, super labellum incumbens. *Anthera* carnosa, 4-locularis, septorum marginibus membranaceis. *Pollinia* 4, caudiculis totidem ligulatis reflexis, per paria subconnatis.——Herba *epiphyta, Mexicana, pseudobulbosa* (?), foliis *alternis, vaginantibus, subcarnosis,* scapo *terminali racemoso.* Floral Cabinet, 7.

The only species mentioned is *B. elegans*, a beautiful Mexican plant with delicate lilac flowers. I believe the Cattleya domingensis, *Gen. & Sp. Orchid. p.* 118. also belongs to the genus.

23

* CRYPTOCHĪLŬS sanguĭnĕă.

Blood-coloured Cryptochilus.

GYNANDRIA *MONANDRIA.*

Nat. ord. ORCHIDACEÆ, § EPIDENDREÆ.

CRYPTOCHILUS. Perianthium tubulosum, fauce contractum, basi anticè globosum. *Sepala* ferè omninò connata, apice tantùm libera. *Petala* libera, paulò minora. *Labellum* indivisum, liberum, cum basi productâ columnæ continuum. *Columna* semiteres, clinandrio dentato. *Anthera* bilocularis. *Pollinia* 8, paribus 4 materiei pulvereæ adhærentia.——Terrestris; pseudobulbis *aggregatis.* Folia *coriacea, plana, in sicco striata.* Spica *secunda, multiflora.*

C. sanguinea. *Wallich tentam, p.* 36. *t.* 26. *Gen. & Sp. Orch. p.* 193.

It is needless to redescribe a plant which has already been so carefully illustrated by Dr. Wallich in the work above quoted.

A very pretty species with roundish pseudo-bulbs forming tufts, and producing each a single broad coriaceous recurved leaf. When the pseudo-bulbs are forming they are enveloped in broad, green, carinate sheaths, tinged with crimson at the edge; but the latter soon wither away. The flowers were neither so numerous nor was the spike so long in the cultivated specimen as in the wild ones in my possession from Sylhet; they are of a brilliant scarlet, and form a compressed or one-sided spike, which, after the flowers have fallen off, is pectinated as it were with the long narrow acuminate rigid bracts.

It is a native of rocks in the northern provinces of India. Dr. Wallich found it only once, and then in a single spot on stones, on the summit of Chandaghiry, a mountain of Nepal; he also received it from his collectors, who sent it from the Cachar mountains on the frontier of Sylhet.

* The name refers to the concealed lip, or labellum, which cannot be easily seen, in consequence of the contraction of the mouth of the calyx.

The drawing was made in June 1837, in the collection of Messrs. Loddiges.

When this genus was placed in the Genera and Species of Orchidaceous plants, I had only examined a young dried bud, and I had been led to suppose it a member of the section called Vandeæ, but the recent plant shows that it in fact belongs to Epidendreæ, and that it must be placed, along with Acanthophippium in the neighbourhood of Phaius and Bletia.

Fig. 1. represents the petals and labellum after the calyx has been cut away; 2. is the column seen in half-front, with the lacerated erect margin of the anther-bed (clinandrium); 3. shews the pollen-masses adhering to a shapeless granular disk.

26

* EPIDÉNDRŪM ochráceŭm.

Yellow-Ochre-coloured Epidendrum.

GYNANDRIA *MONANDRIA.*

Nat. ord. Orchidaceæ, § Epidendreæ.

EPIDENDRUM. Bot. Reg. vol. 1. *fol.* 17.

E. *ochraceum;* pseudobulbis obversè pyriformibus cæspitosis sursum attenuatis 1-3-phyllis, foliis linearibus acutis recurvis scapo æqualibus, spicâ terminali laxiflorâ, bracteis rigidis acutis squamiformibus, sepalis petalisque subæqualibus lineari-oblongis obtusis, labello postico sublibero trilobo medio calloso: laciniis lateralibus inflexis truncato-rotundatis denticulatis intermediâ brevi emarginatâ, callo plano apice tridentato, columnâ apice tricorni, clinandrio denticulato, ovario triptero. *Bot. Reg.* 1838. *misc. no.* 15.

Calycina foliola 5, *patentia, subæqualia, fusca, duobus interioribus lateralibus paulo angustioribus et basi angustatis, omnibus liberis.* Labellum *basi cum gynostemio connatum, (respectu germinis) erectum, flavum, laminâ trilobâ, lobis lateralibus gynostemii partem superiorem amplectentibus, intermedio basi in prono denticulis tribus luteis instructo.* Gynostemium *apice tripartitum, laciniis laceris.* Anthera *terminalis operculata* 4-*locularis, corpusculis polliniferis* 3 *cereaceris fœta, pedicellatis parallelis.* Schiede mss.

This pretty little plant inhabits several parts of tropical America. Messrs. Loddiges received it from Oaxaca; Mr. Skinner found it in Guatemala; and I have specimens from Schiede and Deppe, for which I am indebted to Professor Schlechtendahl, collected near the Hacienda de la Laguna, in Mexico, flowering in August.

In July 1836, I first received specimens from Sir Charles Lemon's garden, along with an account of the plant by Mr. Booth, which I take this opportunity of subjoining.

"This singular little plant was gathered in Guatemala, by George Ure Skinner, Esq. and brought to this country in June 1835, by Captain Sutton, who presented it with others from the same place to Sir Charles Lemon, Bart. in whose collection it flowered in July 1836.

"*Pseudo-bulbs*, when young, round and slender, one inch and a half long, gradually swelling at the base as they get old, and diminishing from it upwards; sometimes surmounted by *three* narrow, thinnish, rigid *leaves*, but more frequently, to judge from the original specimens, by only *one* flat, oblong-lanceolate leaf, from three to five inches in length, and a quarter of an inch in breadth, nearly upright, or but slightly curved towards the point. *Scape* about half the length of the leaves, slender, somewhat angular, and curved, three-flowered, with a small acuminate bractea to each. *Pedicels* three-angled, seldom exceeding one-fourth of an inch in length. *Sepals* spreading, concave, roundish oblong, of a dingy brown outside, becoming paler towards the point, which is acute and of a deep green colour; they are each about one-fourth of an inch long, and one-eighth broad, except the two inner divisions which are rather smaller than the others; the inside is of a pale brownish green. *Labellum* three-lobed, nearly white, or a very pale yellow; the middle lobe the smallest; the other two rise nearly upright, so as to embrace the column, which is triangular, and about the length of the sepals, green at the base, becoming of a pale yellow as it thickens outwardly, and having a small projecting acute point.

"In cultivating this species, and others of a like nature which grow on the trunks of trees in their native country, I have found nothing equal to tying them, by means of a piece of small copper wire, to the branches of an old *apple* or *pear tree*, sufficiently large to protect them from the scorching heat of the sun, covering them with moss, and keeping them in a close, moist stove."

Fig. 1. is a magnified view of the front of the column, with the labellum, &c. cut away; it shews the three teeth of the column, and the curious fringed anther-bed within them; fig. 2. is a view of the upper part of the labellum; fig. 3. is a transverse section of the ovary, shewing the position of the three flattened wings that cover its sides; fig. 4. represents the pollen-masses.

* See Bot. Reg. vol. 17. fol. 1415.

33

* PHĀIŬS ālbŭs.

White Phaius.

GYNANDRIA *MONANDRIA.*

Nat. ord. Orchidaceæ, § Epidendreæ.

PHAIUS. Lour. Sepala et *petala* subæqualia, patentia, libera. *Labellum* sæpiùs cucullatum, cum basi columnæ adnatum, calcaratum, integrum v. trilobum, sæpiùs supra carinatum lamellosum v. cristatum. *Columna* erecta, cum ovario continua, semiteres, marginata, elongata. *Anthera* 8-locularis. *Pollinia* 8, subæqualia.——Herbæ *terrestres* (Asiaticæ), *caulescentes v. acaules,* foliis *latis plicatis.* Scapi *radicales.* Flores *speciosi.* Gen. & sp. orch. 126.

P. *albus;* caulescens, foliis oblongo-lanceolatis acutis subtùs glaucis, sepalis petalisque oblongo-lanceolatis acutis subæqualibus, labello oblongo cucullato denticulato apice rotundato: disco 5-cristato, calcare rectiusculo emarginato (bracteis cucullatis herbaceis persistentibus imbricatis floribus æqualibus). *Lindl. in Wallich plant. as. rar.* 2. *t.* 198. *gen. et sp. orch.* 128.

This lovely plant was originally found by Dr. Wallich upon Mount Chandaghiry in Nepal, growing on trees; and subsequently it was sent down from the frontiers of Silhet by Francis de Silva, a collector in the pay of the Botanic Garden, Calcutta. From drawings made in that establishment, the figure in the Plantæ asiaticæ rariores was taken; but it must be confessed that it does not give a stranger a correct idea of the beauty of the original, which ranks among the most showy of the order. So indeed do all the species of Phaius, if well grown; the finest of which, P. bicolor, a native of Ceylon, with yellow and pink flowers, is still to introduce.

In some respects the plant now figured, and that of Dr. Wallich, are apparently so different, that some doubt might be entertained about their identity, if it were not for dried specimens in my collection, which place the question beyond all doubt. There was no sign of the yellow colour, which is so conspicuous a mark of the labellum in the Indian drawing, the pink veins were much deeper, and the flowers were arranged in a dense, imbricated, nodding, many-flowered raceme.

In a young state the plant is very remarkable for the light blue appearance of all its parts, and for the broad, round, amplexicaul, scale-like leaves which appear upon the stem when it first begins to lengthen.

The accompanying drawing was made at Messrs. Loddiges in July 1837.

It requires to be grown in a moist stove, or orchidaceous house, which is always shaded during bright sunshine in summer, otherwise the leaves will become yellow and have a sickly appearance.

The pots into which it is put must be well drained, in order to carry off superfluous moisture. The best soil is a rich brown peat, well chopped up and mixed with some sub-

* From φαιὸς brown, in allusion to the colour of the original species.

stance which will keep it open, and allow the water to pass freely through; clean cinders or broken pots answer the purpose very well. It will bear a free supply of water during the growing season, and may then be syringed once or twice every day; but late in autumn, or during the dull weather in winter, very little water is required, the object at that season being to keep the plant as torpid as possible.

The best season for dividing it for purposes of propagation is in spring, just before it begins to grow.

34

* PHALÆNOPSIS amabilis.

The Indian Butterfly-plant.

GYNANDRIA *MONANDRIA.*

Nat. ord. ORCHIDACEÆ, § VANDEÆ.

PHALÆNOPSIS. Blume. Perianthium explanatum, patens, sepalis liberis, petalis majoribus dilatatis. *Labellum* cum basi paulò productâ columnæ connatum, liberum, basi callosum, trilobum; lobis lateralibus ascendentibus petaloideis, intermedio angustiore bicirrhoso. *Columna* in ovarium recumbens, semiteres, rostello gladiato. *Anthera* bilocularis. *Pollinia* 2, subglobosa, caudiculâ planâ spathulatâ, glandulâ maximâ cordatâ.——Herba *epiphyta.* Caules *radicantes, simplices.* Folia *rigida, lato-lanceolata, apice obliquè retusa.* Flores *paniculati.* Gen. et Sp. Orch. 213.

P. amabilis. *Blume bijdragen* 294. *tabellen* 44. *Lindl. gen. et sp. orch.* 213.
Epidendrum amabile. *Linn. sp. pl.* 1351.
Angræcum album majus. *Rumph. herb. amboin. vol.* 6. *p.* 99. *t.* 43.

This very rare and beautiful epiphyte flowered a few weeks since in the Epiphyte-house of Messrs. Rollissons of Tooting, I believe for the first time in Europe. It had been sent to them from Manilla by Mr. Cuming.

The curious form of the flowers, the graceful way in which they hang down from below the leaves, their large size, and the brilliant whiteness of their broad leathery petals give this species a most striking and uncommon appearance.

Dr. Blume met with it on the wooded coast of Nusa Kambanga. Rumph, who first noticed it, speaks of its growing in Amboyna, on short thick trees, covered with moss, up which, he says, it turns like a rope, and from which it hangs down in entangled tufts. He mentions a variety whose petals are a deep rich purple on the outside.

* From φάλαινα a moth, and ὄψις resemblance.

The Dutch Colonists, in the Malayan Archipelago, call this plant the *Vliegende Duive;* the Malays themselves *Angrec poeti besaar, Bombo terbang, Angrec colan,* and *Wanlecu.*

It belongs to that class of Orchidaceous plants which succeed best when fixed to pieces of wood, along with a little turf or moss, and suspended from the roof or pillars of the stove. It grows very well in this manner in the Orchidaceous-house at Tooting, which is kept saturated with moisture, and scarcely ever receives any air. The wood on which it is fixed should either be covered with rough bark, or be in a decaying state, to enable the roots to fix themselves to it; it may be a foot or eighteen inches in height, and four or five inches in diameter. The plant does not appear to throw out side shoots freely, and consequently will be difficult to propagate.

40

* MAXILLARIA Rollissonii.

Messrs. Rollisson's Maxillaria.

GYNANDRIA *MONANDRIA.*

Nat. ord. ORCHIDACEÆ, § VANDEÆ.
MAXILLARIA. Bot. Reg. vol. 11. *fol.* 897.

M. *Rollissonii;* acaulis, pseudobulbis subrotundis compressis subbifoliis, foliis oblongo-lanceolatis acutis apice recurvis, scapis subunifloris diffusis laxè vaginatis, sepalis carinatis lateralibus basi subæqualibus petalisque acutissimis, labelli lobis lateralibus ovatis angustis acutis intermedio oblongo membranaceo apiculato margine deflexo: callo disci elevato carnoso anticè transverso lobos laterales labelli conjungente medio producto truncato tridentato et denticulis utrinque reflexis posticè fornicato truncato bilobo. *Bot. Reg. fol.* 1986.

Planta *omninò facie* M. stapelioidis, *colore minùs pallido cæsio.* Folia *nunc oblonga apice acuta recurva, nunc magis lanceolata; semper sessilia.* Bracteæ *cucullatæ, acuminatæ, ovariis longiores.* Sepala *et* petala *pallidè flava, immaculata, conformia et subæqualia.* Labellum *flavescens, sanguineo guttulatum; lobis lateralibus columnam utrinque aures ad instar adstantibus, margine posteriore crenulato.* Crista *labelli elevata, transversa, anticè cornubus duobus clavatis armata altero minimo interjecto, posticè biloba fornicata utrinque dentata, denticulis quibusdam columnam respicientibus in latera jugi.* Pollinia 4 (⁘) *per paria adhærentia glandulæ tenui rhomboideæ.*

A curious little species inhabiting the woods of Brazil, whence it was imported by Messrs. Rollissons, with whom it flowered in August, 1837. In many respects it is very like *M. stapelioides,* and when out of flower the two species may be mistaken for each other; but the blossoms are strikingly different in colour and in the form of the curious crest which stretches across the lip from one side to the other.

* See Bot. Reg. fol. 1428.

Fig. 1. is a magnified view of the lip, with its crested appendage. Fig. 2. represents the pollen masses, with their gland; half one of the pairs being cut away.

44

* CYRTOCHILUM maculatum.

Spotted Cyrtochilum.

GYNANDRIA *MONANDRIA.*

Nat. ord. ORCHIDACEÆ, § VANDEÆ.
CYRTOCHILUM. Bot. Reg. vol. 19. *fol.* 1627.

C. *maculatum;* pseudobulbis ovatis compressis subangulatis diphyllis basi foliosis, foliis latè ligulatis acuminatis striatis apice obliquè emarginatis, scapo simplici (?), bracteis brevissimis squamæformibus, sepalis petalisque carnosis obovato-lanceolatis acutissimis, labello membranaceo oblongo apiculato utrinque dentato lamellis duabus ad basin et corniculo utrinque, alis columnæ falcatis integerrimis. *Suprà in miscell. no.* 39. *April.*
C. maculatum. *Knowles and Westcott, Floral Cabinet, t.*

Planta *habitu* Oncidii *cujusdam pseudobulbosi,* foliis *planis basi tantum complicatis, spithamæis et ultra.* Flores *ferè* $1\frac{1}{4}$ *unciam lati, luteo-virides, fusco purpureo maculati;* sepalis petalisque *lanceolatis acutis subcarnosis æqualibus patentibus.* Labellum *oblongum, columnæ perpendiculare eâque continuum, apiculatum, convexum, utrinque dente obtuso auctum, medio album, apice lutescens, margine sanguineo maculatum et reflexum; lamellis duabus elevatis collateralibus apice liberis pone basin et denticulo uno libero utrinque.* Columna *brevis, crassa, semiteres, candida; alis brevibus falcatis integerrimis.* Anthera *flava, lineâ elevatâ cristata.* Pollinia 2, *pyriformia, posticè excavata, caudiculâ oblongâ, glandulâque parvâ, fuscâ, ovali.*

This plant first flowered in the garden of the Horticultural Society, where it had been received from Mr. Hartweg, who found it near Vera Cruz. It was subsequently communicated by Messrs. Rollissons, and I have also received it from Chatsworth, and from Mr. Barker of Birmingham. When it first flowered there were those who took it for the

* See folio 1627.

Oncidium tigrinum of La Llave and Lexarza, a very different plant, with a reniform lip placed upon a long stalk.

Fig. 1. represents a magnified view of the column and labellum; fig. 9. shews the pollen masses, with their caudicula and gland.

45

* CORYCIUM orobanchoides.

Broomrape Corycium.

GYNANDRIA *MONANDRIA.*

Nat. ord. ORCHIDACEÆ, § OPHRYDEÆ.

CORYCIUM, Swartz. *Perianthium* ringens. *Sepala* basi æqualia, erecta, lateralia connata. *Petala* libera, concava. *Labellum* unguiculatum, cum columnæ facie connatum, anticè appendiculatum. *Anthera* dorsalis, resupinata, loculis ungue accreto labelli interposito sejunctis. *Glandulæ* nudæ.

C. *orobanchoides.* Swartz in act. holm. 1800. p. 220. Thunb. Fl. Capens. ed. Schultes, 1. 20. Ker in Brande's Journal, vol. 8. t. 3. f. 3.??
Satyrium orobanchoides. *Linn. suppl.* 402.

Sepalum supremum *lineare, angustum, acutum, inter petala saccatu, flavescentia, apice purpurea rotundata interpositum;* inferius *paulò brevius, oblongum, concavum, ascendens, bidentatum, dentibus sanguineis.* Labelli *appendix (lobus intermedius) alba, deflexa, cuneata, biloba, laciniis patentibus rotundatis; lobi laterales erecti, virides, angustè lanceolati, furcam referentes, petalis breviores.* Antheræ *lobi purpurei, oblongi, breves, per totum unguis adnati labelli interpositi latitudinem sejuncti, inversi; connectivo hippocrepico carnoso subincurvo.* Pollinia *granulosa, sectilia;* glandulis *nudis in apices loborum lateralium rostelli incurvos jacentes.* Rostellum *resupinatum, trilobum; lobo medio connectivi formâ eique appresso, lateralibus linearibus, obtusis divergentibus.*

This most curious little plant inhabits sandy places in the colony of the Cape of Good Hope, flowering in the months of September and October. It agrees pretty well with the species described by Thunberg under this name, but not entirely: for that author speaks of the upper sepal being somewhat notched at the end, and of a plant a foot high. It is therefore possible that two species nearly allied to each other may exist at the Cape of Good Hope; and that while this figure represents one of them, the barbarous

* From κόρυκος a little bag, in allusion to the form of the flower.

drawing in Brande's Journal, as above quoted, may have been intended for the other.

The drawing was made in October, 1837, from a plant that flowered in the collection of John Rogers, Esq. Jun., at Streatham. It is, as far as I know, the first instance of a Corycium having blossomed in Europe.

The dissections in the accompanying plate will serve to shew the nature of the very unusual organization found in this genus. Fig. 1. is a flower seen in profile, with none of the parts removed. Fig. 2. is a front view of the same, the petals and upper sepal having been cut off; the principal part of the figure represents the lip, with the point of the lower double sepal just visible at the base. Fig. 3. shews the column in profile with the lip adhering to it, and one lobe of the purple anther. Fig. 4. is a front view of the column, the lip having been cut off, as is shewn by the scar represented on the upper edge; the glands of the pollen masses are seen lying in the cavities of the rostellum. Fig. 5. is a sketch of a pollen mass sticking to the rostellum, and pulled out of the anther. Fig. 6. are the same parts seen in a different direction, with the pollen in its cell.

48

* ONCIDIUM raniferum.

Frog Oncidium.

GYNANDRIA *MONANDRIA.*

Nat. ord. ORCHIDACEÆ, § VANDEÆ.

ONCIDIUM. Bot. Reg. vol. 13. *fol.* 1050.

O. raniferum. *Suprà fol.* 1920.

Rather a pretty species of this extensive genus, with a very peculiar habit. It evidently hangs down from the branches of trees, instead of growing erect as is most usual.

Its name is derived from the curious form of the tubercles which grow at the base of the lip, and of which a magnified figure is given in the accompanying plate. Looked at from above it resembles the figure of a frog *couchant*, the double lower tubercle representing the creature's haunches, and the anterior emarginate one his head.

I originally received it from Mr. Knight of the King's Road; and since that time it has been sent me by various persons. Among the specimens is one, of which a separate figure is given, with all the scale-like bracts of the panicle converted into long narrow leaves.

The species is a native of Brazil. It occurs among Mr. Gardner's Organ Mountain plants (No. 637), and M. Descourtilz met with it in damp forests near Bananal. This traveller observed that it fixes itself to branches not exceeding twelve feet in distance from the earth. Such pieces of

* See Bot. Reg. fol. 1050.

information are invaluable to the cultivator, who rarely has any idea of the precise circumstances under which his plants grow naturally, and who, consequently, can only learn after long and dear experience how to treat them. For this reason I hope the following additional particulars concerning the habits of other species of Oncidium, extracted from M. Descourtilz' MSS. will prove acceptable.

O. *divaricatum.* Trunks of the most lofty trees, on the high mountains of the Serra das Agoâs, in the district of Ilha Grande.

O. *ciliatum.* Common in the woods which cover the low plains among the hills near Bananal.

O. *iridifolium.* Branches of Orange and Lemon trees only. Very common in the neighbourhood of the town of Bom Jesus de Bananal. It prefers dry places, exposed to the sun.

O. *pubes.* Thin forests clothing the table-land near Bom Jesus de Bananal.

This species must be cultivated in a very damp stove, where it may either be tied to a piece of wood and suspended from the roof, or grown in a pot. In the latter case the soil should consist of rough pieces of peat mixed with broken bricks, or something which will form a communication with the drainage below, and carry off superfluous water. The only other thing to be attended to in its cultivation, is the giving of water. This must be done cautiously, especially when the young shoots begin to grow, for they are very apt to suffer at that period. Afterwards it may be given more freely, until the end of the growing season, when the plant must be allowed a period of rest.

It is propagated in the usual manner, namely by division.

53

* EPIDĒNDRŪM Schombŭrgkiī.

Mr. Schomburgk's Epidendrum.

GYNANDRIA *MONANDRIA.*

Nat. ord. ORCHIDACEÆ, § EPIDENDREÆ.
EPIDENDRUM. Bot. Reg. vol. 1. *fol.* 17.

E. *Schomburgkii;* foliis distichis oblongis obtusis margine sanguineo-punctatis, caule simplici apice aphyllo, sepalis petalisque lineari-lanceolatis acutis æqualibus patentissimis, columnâ clavatâ elongatâ, labelli trilobi basi bicallosi lineâ medianâ elevatâ, lobis lateralibus latis rotundatis laceris; intermedio cuneato apice triangulari crispo utrinque acuminato. *Suprà no.* 16. *misc.*

The first knowledge I obtained of this beautiful species was through a drawing and a few dried flowers sent home by Mr. Schomburgk, who met with it during his second expedition into the interior of British Guayana. Since that time it has blossomed with Messrs. Loddiges, and forms one of the most striking of the "elongated" division of the large genus Epidendrum.

In a work like this the brilliant vermilion red of the flowers cannot be imitated; they are of the deep rich tint that is found in Silene laciniata, Lychnis Bungeana, and plants of that kind.

In his drawing Mr. Schomburgk makes the leaves deeply and distinctly marked with blood-red dots all round the margin. In the cultivated plant this did not make its appearance; nevertheless I leave that character unaltered, because it is possible that the spots may hereafter be developed.

* See Bot. Reg. fol. 1415.

57

* BOLBOPHȲLLŪM bracteolātum.

Bracteolate Bolbophyllum.

GYNANDRIA *MONANDRIA.*

Nat. ord. ORCHIDACEÆ, MALAXIDEÆ.
BOLBOPHYLLUM. Bot. Reg. vol. 23. *fol.* 1942.

B. *bracteolatum;* pseudobulbis ovatis obtusis tetrapteris diphyllis, foliis angustè oblongis carnosis patentibus, scapo radicali pendulo rachi fusiformi sparsè multiflorâ pluries breviore, floribus carnosis extùs bibracteolatis, sepalis ovatis acutis lateralibus basi connatis, petalis ovatis multò brevioribus, labello solido triquetro acuto canaliculato dorso serrato.

Flores *parvi, rachi duplò angustiores.* Bracteolæ *ovatæ, minutæ, petalorum basi sitæ, ejusdem colore et textura sepali supremi carnei apice atropurpurei.* Sepala *lateralia lutescentia dorso sanguinea.* Petala *lætè purpurea.* Labellum *luteum, sanguineo-punctatum, apice violaceum.*

A most curious little epiphyte, native of Demerara, whence it was obtained by Messrs. Loddiges, with whom it flowered in July 1837. It and B. setigerum, and probably some others offer the singular instance of the existence in America of a genus hitherto discovered only in the Old World.

In habit this species is much like *B. pusillum* and *clavatum*, two species not yet introduced from the Mauritius.

When magnified the flowers are beautifully variegated with flesh-colour, yellow, red, and violet, but what is more remarkable there exists on each side of the flower, at the base of the petals, a small ovate bract, the rudiment of which is also met with in B. setigerum, in the form of a minute tubercle. This additional part has never before been noticed in the order, and is possibly the explanation of the true nature of the exterior series of floral organs found in Epistephium. It would, therefore, seem as if the ordinary condition of the flowers of Orchidaceæ were in a sort of middle state between two extremes, of which Epistephium is the most complete, and Monomeria, in which there are no petals, the most imperfect.

* See folio 1942.

Fig. 1. represents a flower viewed from the side; 2. is the same, with the sepals cut away so as to expose the petals, lip, and column; 3. is a labellum, seen a little from within; 4. are the four unequal pollen-masses.

60

* PAXTŌNĬĂ rŏsĕă.

Pink Paxtonia.

GYNANDRIA *MONANDRIA.*

Nat. ord. ORCHIDACEÆ, § MALAXIDEÆ.

PAXTONIA. Perianthium patens, hexaphyllum, æquale; labello petalis conformi. *Columna* erecta, teres, subclavata, petalis paulò brevior. *Anthera* terminalis, opercularis, decidua. *Pollinia* 8, angusta, clavata, apice cohærentia. *Stigma* rima transversa, verticalis, sub rostello.——*Herba* cæspitosa, pseudobulbosa, *foliis* plicatis, *racemo* multifloro radicali foliorum longitudine.

P. rosea. *Suprà misc. p.* 61. *no.* 113.

Pseudobulbi *densè aggregati, oblongi, cicatricibus foliorum annulati, 3-phylli.* Folia *lanceolata, plicata, basi angustata, subpedalia, basi squamis ovatis concavis acutis fuscis vaginata.* Scapus *erectus, radicalis, foliorum longitudine, calami scriptorii crassitudine, squamis 2-3 ovatis distantibus vaginati.* Bracteæ *ovatæ, acutæ, patulæ, pedicellorum longitudine.* Flores *rosei, ferè sesquiunciam lati; pedicellis gracilibus ovarii longitudine.* Sepala et petala *conformia, æqualia, oblonga, acuta, 3-4 lineas lata;* labello *antico.* Columna *erecta, teres, anticè paulò complanata, apice subclavata, rosea, petalis brevior;* clinandrium *margine crasso sublobato.*

This curious plant was sent to Messrs. Loddiges, from Manilla, by Mr. Hugh Cuming, who has been for some time engaged in exploring the Philippine Islands. It is so entirely different from all Orchidaceous genera hitherto discovered, that I know not with what to compare it.

In this natural order there are two organs essentially at variance with the usual structure of plants, namely, the column and the lip. The former, consisting of stamens and style consolidated, has never yet been found in a state of disintegration; but the latter, to whose numerous forms, and irregular varieties, the grotesque appearance of Orchidaceous

* Suprà misc. p. 61. no. 113.

flowers is usually owing, assumes a regular structure in a few cases, of which this is one of the most striking. In this genus the lip is so much like the other petals, that it is only to be recognized by its position. Thelymitra, Hexisea, Isochilus, and such others as have been previously discovered with this character, are so extremely unlike Paxtonia in other respects that it is not worth instituting any comparison between them.

63

* CATASĒTŬM atrātŭm.

Dark-flowered Catasetum.

GYNANDRIA *MONANDRIA.*

Nat. ord. ORCHIDACEÆ, § VANDEÆ.
CATASETUM. Suprà vol. 10. *fol.* 840.

C. *atratum;* racemo decurvo, sepalis petalisque patentibus ovatis acutis, labello carnoso cucullato margine tenui pectinato apice rotundato reflexo crasso denticulato. *Bot. Reg.* 1838. *misc. no.* 114.

A very distinct species of this curious genus, imported by Messrs. Loddiges from Brazil. The flowers are gracefully drooping, and are among the handsomest of the genus.

Had this been known when the now-abolished genus *Myanthus* was proposed, that error would not have been committed, for the species is neither exactly a *Myanthus* nor exactly a *Catasetum.*

It is unnecessary to do more than refer to folio 1951 of this work for an explanation of the reasons which led to the union of these genera, and to the further suppression of *Monachanthus* and *Mormodes.* But there is a circumstance observed by Mr. Schomburgk in connection with them which is very curious, and deserves to be recorded. In a letter I received from him some time since, he says, alluding to *Monachanthus,* " are you aware that *Catasetum* and *Myanthus* are not seed-bearing, but that Monachanthus bears seed abundantly?" I do not know what conclusion to draw from

* See folio 1667.

this statement; but it would be a most curious fact if, as Mr. Schomburgk's observation would appear to imply, the species of Catasetum and Myanthus should prove to be sterile states of Monachanthus.

65

DENDRŌBĬŬM sulcātŭm.

Furrowed Dendrobium.

GYNANDRIA *MONANDRIA.*

Nat. ord. ORCHIDACEÆ, § MALAXIDEÆ.
DENDROBIUM. Bot. Register, vol. 7. *fol.* 548.

§ IV. *Caules undique foliosi versus apicem sensim incrassati undè clavati fiunt.*
D. *sulcatum;* caule clavato sulcato subflexuoso compresso, foliis oblongis acutis 3-nerviis, pedunculis lateralibus trifloris, bracteis minimis acutis appressis, petalis oblongis obtusis glabris sepalis subæqualibus, labello obcordato hirsuto ungue canaliculato sulcato.

A native of India, whence it was obtained by Mr. Gibson for His Grace the Duke of Devonshire. The drawing was made from a specimen which flowered at Chatsworth in April last.

It is a fine species, nearly related to D. Griffithianum, from which it differs in its three-flowered peduncles, and in the form of the lip,

Its cultivation differs from that of many species belonging to this order. At certain seasons the plants will manifest an inclination for growth; they must then have plenty of water and be freely syringed over-head. When the growing season is over, the leaves will become yellow, and finally drop off; water must then be discontinued, the temperature lowered, and the plants allowed a season of rest. After remaining in this state for a few weeks, the temperature may again be raised, and the plants—which by this treatment will have been rendered exciteable—will come freely into flower. Water need not be given until the plant begins to shoot from below, when it will require the same treatment as before. Of course it is to be cultivated in a stove or orchidaceous house. The soil is the same as is used for other plants of this kind, namely, turfy peat.

It is propagated by division; but where only a small piece of a pseudo-bulb can be obtained, it will frequently send out young plants from its joints, which can be taken off, potted, and treated in the same manner as is recommended above.

68

* COMPARĔTTĬĂ coccĭnĕă.

Scarlet Comparettia.

GYNANDRIA *MONANDRIA.*

Nat. ord. ORCHIDACEÆ, § VANDEÆ.
COMPARETTIA. Pöppig et Endlicher. Perianthium ringens. *Sepalum* intermedium et *petala* nana, libera, subgaleata; lateralia in unum connata calcaratum labello suppositum. *Labellum* liberum, explanatum, obcordatum, unguiculatum, basi calcaribus duobus intra sepalorum calcar absconditis. *Columna* libera, erecta, mutica. *Pollinia* 2, posticè foveata, caudiculæ cuneatæ rostratæ adnata.——Herbæ *epiphytæ, pseudobulbosæ,* racemis *terminalibus.*

C. *coccinea;* foliis angustis coriaceis apice obliquis acutis, sepalo antico acuto calcare falcato duplò breviore, labello obcordato subrepando basi bilamellato calcaribus subulatis ciliatis, columnâ medio sub stigmate elevatâ.

This beautiful and very rare epiphyte flowered with Messrs. Loddiges in August last, and is said to be a native of Brazil; it however agrees so entirely with dried specimens collected near Xalapa, and now before me, that some mistake is to be suspected in its reputed country.

The scarlet of its lip is very brilliant, its flowers are arranged gracefully, and they have a most curious structure. Nothing is more common for Orchidaceæ than to produce spurs from their sepals, and labellum, but it is very unusual for the same organ to produce two spurs. Satyrium and Diplocentron were almost the only cases of this structure. Here, however, the labellum not only has two spurs, but they are hidden within the spur of the united pair of lateral sepals, so that they are not discovered till the latter is cut open. In this respect we find almost the same structure for Orchidaceæ as we have in Aconitum among Ranunculaceæ.

* Named after Andreas Comparetti, Professor at Padua, who first understood the real nature of spiral vessels, and who wrote with great skill upon Vegetable Physiology. *Pöppig and Endlicher.*

Comparettia consists at present of this and two other species inhabiting trees in Peru. Of these *C. falcata* approaches the present subject in many respects. It, however, appears to differ in having broader leaves, larger flowers, no elevated plate at the base of the stigma, and the spurs of the lip smooth.

Fig 1. represents the pollen-masses, with their caudicula. Fig 2. is a view of the column and spurs, all the sepals and petals being cut off; *a.* is the elevated plate below the stigma; *b.* shews the lamellæ at the base of the lip; *c.* represents the labial spurs, a little pulled out of the sepaline spur *d.*

MISCELLANEOUS NOTICES.

⁂ It is intended to publish monthly, under this head, some account of the most rare or curious plants that may from time to time be discovered or introduced, and of which there is no immediate possibility of giving plates. To this will be added such occasional notes upon plants already figured here or elsewhere, as will serve to make them better known, or to correct their history. It is expected that by this plan the present work will be rendered more useful and efficient than ever as a Register of information concerning Horticultural Botany. It must be evident to any one at all acquainted with the extent to which the introduction of new plants to this country is now taking place, that no periodical, the price of which will procure it a remunerating sale, can at all keep pace with discovery, if its contents are limited to anything like a hundred figures a year; although that number of plates may be amply sufficient to comprehend what new plants are strikingly beautiful, or extremely curious in their structure. Upon the arrangement now commenced, while the latter only will find a place among the plants actually figured, others will not be neglected, but on the contrary will be gradually brought before the reader by means of the descriptive letter-press to be given monthly, under the title which stands at the head of this page.

3. EPIDENDRUM Pastōris.

La Llave Gen. & Sp. Orch. p. 98.

A Mexican epiphyte, with narrow pale green leaves, growing in pairs or threes upon pale, yellowish green, fusiform, erect pseudo-bulbs. The flowers are two or three in number, upon a smooth yellowish-green slender scape, hardly so long as the leaves. They are about the size of those of Epid. odoratissimum. The sepals and petals are very narrow, sharp-pointed, and striped with dull chocolate-purple, upon a dingy olive-coloured ground, and slightly bent back. The lip is white, united to the column for about half its length, folded up, so that its edges almost meet, with two narrow sharp-pointed purple-striped lobes, one on each side at the base, and three round wrinkled elevated ribs along its middle. It is fully described in La Llave and Lexarza's work on Mexican plants; but I do not find the scape ancipitous as those authors represent; it is a little compressed and that is all. They speak of the flowers having a powerful Vanilla fragrance, but of this I could not judge on account of the coldness of the weather when the specimen reached me. A correct figure will be found in Link and Otto's Abbildungen Neuer und Seltener Gewächse, plate 12.

Originally sent from Mexico, by Dr. Deppe, to the Royal Garden of Berlin, about the year 1828. The plant before me came from the garden of Richard Harrison, Esq. having been sent from Mexico by Mr. Bates to Mr. Tayleure, of Parkfield, near Liverpool. *April* 8, 1837.

4. STANHŌPĔA Lindlēyi. *Zuccar.*

A Mexican plant of this supposed species flowered in August last, in the collection of the Earl of Derby. It does not appear to be distinct from S. oculata, differing principally in the colour of its flowers, which are a dull pale brownish red on the sepals and petals, while the spots on those parts are comparatively inconspicuous. The bright yellow, so great an ornament of the lip of the original S. oculata, is wanting, and is replaced by the dull vinous (port-wine) stain of the other parts. The flowers are perhaps a little smaller than in S. oculata.

5. SACCOLĀBĬUM bĭfĭdum.

S. *bifidum;* racemo pendulo gracili multifloro, petalis basi obscurè hastulatis, labello recurvo hastato apice setaceo fisso: laciniis basi linearibus obtusis erectis.

A pretty species of this interesting genus of epiphytes, with the habit of a small Vanda, small oblong, flat, distichous leaves, and a slender pendulous raceme, about four inches long, of pale pink and yellow small flowers, in the form of their lip not unlike our wild Man-Orchis. It was received by Messrs. Loddiges from Manilla, where it had been collected by Mr. Cuming, and flowered in December.

The following is a technical description of the plant.

Flores parvi, Ribesios rubri magnitudine. *Perianthium* explanatum, ochraceum, *sepalis* linearibus obtusis; *petalis* minoribus, basi obsoletè hastulatis. *Labellum* albidum, subrecurvum, subtriangulare, apice setaceo-bifidum, basi hastatum: lobis linearibus, obtusis, apice purpureis, erectis, columnâ longioribus; calcare intus nudo, cylindraceo, ochraceo, apice compresso, pendulo, ovario duplò breviore. *Columna* nana, alis duabus linearibus obtusis porrectis stigma marginantibus. *Anthera* hemisphærica, mutica, pellucida, anticè truncata. *Pollinia* 2, posticè excavata, ab apice brevi glandulæ tenuis cuneatæ acuminatæ pendula.

6. PESOMĒRĬA tetrăgōna.

In a work consisting of indifferent plates, without explanatory letter-press, published upon the Orchidaceæ of the Isle of France, Madagascar, and Bourbon, by the late M. Aubert du Petit Thouars, there occurs a curious terrestrial plant, native of two of these islands, and called Epidendrum tetragonum. It is represented as growing about two feet high, with a four-cornered stem, and many oblong plaited acuminate leaves like those of a Phaius, to which moreover the flowers bear some resemblance both in size and form. In the very rare coloured copy of this work, in the possession of the Horticultural Society, the flowers are represented purple, like those of a Bletia; and upon the whole little doubt could exist of its being related to those two genera. Nevertheless it has not been again noticed by any original writer, but has been left among the obscure plants of the work in which it first appeared.

I was therefore not a little gratified at having an Isle of France plant brought me a few weeks since by Mr. Loddiges, which upon comparison appeared to be identical with the species in question, differing from it in nothing except the leaves being narrower, and the flowers of a pale yellowish brown, instead of a bright rose-colour; to the latter I attach no importance, as it is by no means improbable that the colour in Du Petit Thouars's book was given from an examination of dried specimens, or from memory. The plant of course was no Epidendrum, a genus unknown in the old world, as the Asiatic Dendrobium is in the new world; on the contrary it proved a very curious and distinct genus, differing from Bletia, Phaius, and their immediate allies, in having only four, instead of eight pollen masses, and moreover in the labellum being united to the column. The most curious circumstance about it, however, is a phenomenon which I have never seen in any other Orchidaceous plant. The sepals are spontaneously thrown off from the flower shortly after they have expanded, just as leaves are thrown off the stems of many of these plants when they receive a sudden check, and then the petals and labellum only remain to constitute the flower. It even appears from Du Petit Thouars's figure that the same thing happens to the large convolute bracts. This unusual singularity has suggested the name Pesomeria (πιπτω, I fall off, and μεροs a part of any thing); of which the following is a character and description.

PESOMERIA.

Nat. ord. ORCHIDACEÆ, § EPIDENDREÆ.

Sepala subæqualia, libera, spontè decidua. *Petala* conformia, basi columnæ adnata, persistentia. *Labellum* posticum, cum columna connatum, basi gibbosum, limbo indiviso convoluto. *Columna* clavata, semiteres, clinandrio dentato. *Pollinia* 4, cuneata.——Herba *terrestris; caule tetragono; foliis membranaceis plicatis; racemo laterali simplici*, bracteis *spontè secedentibus; radicibus crassis, simplicibus, fibrosis.*

1. P. tetragona.

Epidendrum tetragonum. *Thouars Orch. Afr. tt.* 33. 34.

Sepala carnosa, patula, oblonga, carinato-acuminata, luteo-viridia, intùs fusca, lineis intensioribus, omninò libera, æqualia, spontè decidua! ferè pollicem longa, 3 lineas lata. *Petala* aliquotiès breviora et angustiora, obovata, acutissima, intùs luteo-fusca, cum columna parallela ejusque basi utrinque adnata, haud decidua. *Labellum* posticum, convolutum, carnosum, cum columna semiconnatum, imâ basi gibbosum, cuniculo intus pubescente; laminâ obscurè luteâ, venis divergentibus fusco-purpureis, latiore quam longa, apice cuspidata et crispa, intùs sparsè villosâ, lineâ mediâ elevatâ latâ carnosâ sursùm sulcatâ. *Columna* clavata, semiteres, clinandrio dentato. *Stigma* lunatum, rostello plicato? *Anthera* obtuse conica, pilis albis erectis villosa, apice membranacea, bidentata, suberosa, obsoletè 8-locularis. *Pollinia* 4, cuneata, cereacea, materie pulverulentâ elasticâ cohærentia (ulteriùs indaganda).

7. EPIDĒNDRŪM Boothĭānŭm.

E. *Boothianum;* pseudo-bulbis ovatis subdiphyllis, foliis ligulato-oblongis undulatis apice subobliquis, racemo terminali laxo sub-7-floro basi spathaceo foliis longiore, sepalis petalisque subæqualibus patulis ovali-linearibus acutis fusco-variegatis, labello semilibero rhomboideo acuto utrinque deflexo.

I have had an account of this pretty species of Epidendrum in my possession for nearly two years, without being able to satisfy myself about its being certainly new, so many are the species of this extensive genus. Upon a recent examination of the evidence it does appear distinct, and therefore it is named in compliment to Mr. W. B. Booth, one of the most intelligent gardeners in this country, to whom I am obliged for a drawing and the following description of it; Sir Charles Lemon has supplied me with some excellent specimens.

" This curious plant is a native of the Havannah, from whence it was brought by Capt. Sutton of Flushing, near Falmouth, in the spring of 1835, and added to Sir Charles Lemon's collection at Carclew, where it flowered in September.

" *Pseudo-bulbs* flat and compressed, growing in large masses, and laid over one another with great regularity. Each bulb is nearly round, and about an inch in diameter, tapering a little towards the crown, which is terminated sometimes by *one*, but more generally by *two*, thinnish, rigid, smooth, slightly recurved, yellowish green leaves, from four to six inches long, and an inch broad, reflexed and undulated at the margin with an obtuse, oblique, and emarginate point. *Scape* about six inches high, in our specimen seven-flowered, round, erect and pale green, issuing from a thin, oblong, two-edged, pointed kind of sheath, which rises from between the two leaves, and which in old decayed specimens is seen to proceed with the flower-stem from the crown of the pseudo-bulbs. *Flowers* produced on pedicels of about an inch in length, small and round at their junction with the scape, but otherwise three-angled and twisted. *Sepals* spreading, oblong acuminate, nearly uniform in size and colour, excepting the two inner ones which are somewhat narrower than the rest. All of them are of a dingy yellowish green, irregularly marked on the inside with dull reddish brown spots, and a little reflexed at the edges. *Labellum* pale yellow, nearly white, cordate acute, with the sides bent downwards and much recurved, about as long as the sepals and projecting horizontally from them with the column to which it is partly attached. *Column* about half the length of the labellum, rather paler in colour, becoming greenish and faintly spotted at the base, from which it thickens outwardly, and has on each side a deep longitudinal groove that gives it a triangular appearance. *Anthers* deep yellow, seated in a depression at the extremity of the column.

" On first comparing the plant with the figure of *Epidendrum variegatum* of the Botanical Magazine, I was almost inclined to consider them identical, but a closer examination has led me to alter my opinion, although it must be acknowledged that the two are very closely allied. The principal points of difference appear to be in the size and form of the pseudo-bulbs, which in *E. variegatum* are oblong, whereas in the present species they are round and flat. The leaves of the former are striated, sometimes in *threes*, and have a blotched or variegated appearance—of the latter they are quite plain, and do not appear to be ever more than *two*. The scape of the plant before me issues from a long thin sheath, whereas the one referred to has no sheath. The column and lip are also different from Dr. Hooker's figure, but in general character and appearance they nearly agree; except that E. variegatum is a much larger plant."

The following new Epidendra form a part of the many undescribed species now in our gardens; the account of them is from the pen of James Bateman, Esq. the author of one of the most splendid works on Natural History that has ever issued from the press; it is needless to say that I speak of the " Orchidaceæ of Mexico and Guatemala," of which one part has appeared, and a second is in an advanced state of readiness for publication.

8. EPIDENDRUM papillosum. *Bateman Mss.*

" E. *papillosum;* foliis strictis linearibus carinatis acutis in pseudobulbos pyriformes glaucescentes ternatim insidentibus, scapo erecto simplici multifloro foliis subæquali, floribus distantibus, sepalis ovato-lanceolatis acutis, petalis paulo angustioribus obtusis, labello semi-libero 3-partito, laciniis lateralibus obovatis, intermediâ 3-plo majore obscure quadrilobâ versus basin callosâ, columnâ bidentatâ, pedunculis papillosis."

Hab. in Guatemalâ. *Skinner, v. v. c.*

" Sepals and petals of a greenish brown, lip white, with three minute pink veins. Column orange-coloured at its apex. Next E. squalidum, from which its habit alone would at once distinguish it."

9. EPIDENDRUM tessellatum. *Bateman Mss.*

" E. *tessellatum;* pseudobulbis ovalibus compressis 2-raro 3-phyllis, foliis lineari-lanceolatis scapo simplici flexuoso multifloro paulo brevioribus, floribus subpatentibus, sepalis lanceolatis acutis mucronatis, petalisque minoribus subspathulatis; labelli liberi trilobi laciniis lateralibus rotundatis, intermediâ majore oblongâ cucullatâ callis tribus cristigeris longitudinalibus munita."

Hab. in Guatemalâ. *Skinner.*

" Sepals and petals greenish on the outside, but brown on the inside, and marked with regular streaks of a darker shade which produces a tessellated appearance. Lip beautifully streaked with purple. Column dark coloured at base, tipped with orange. Near E. pastoris."

10. EPIDENDRUM rhizophorum. *Bateman Mss.*

"E. *rhizophorum;* caulibus elongatis foliosis radices crassos utrinque promens (Renantheræ more), foliis alternis lineari-oblongis racemo terminali paucifloro paulo brevioribus; sepalis petalisque subæqualibus lanceolatis, labello tripartito laciniis lateralibus laceratis acinaciformibus, intermediâ unguiculatâ apice dilatatâ alte emarginatâ."

Habitat in Guatemalâ. *Skinner v. s. s.* Flores expansione 1½ uncias lati, exsiccatione aurantiaci.

"A most remarkable species, near E. cinnabarinum."

11. EPIDENDRUM aurantiacum. *Bateman Mss.*

"E. *aurantiacum;* foliis oblongis planiusulis coriaceis caule clavato diphyllo duplo brevioribus, racemo brevi subcernuo densifloro, sepalis petalisque lanceolatis acutis subæqualibus, labello libero oblongo cucullato basi columnam involvente."

Habitat in Oaxacâ, *Karwinski;* in Guatemalâ, *Skinner.*

"Sepals and petals of a uniform rich orange colour; lip the same, but with a few delicate streaks of the richest crimson. This is a most remarkable species, and I was for some time doubtful whether it were an Encyclia, Cattleya, or Epidendrum. On the whole, however, it seems to approach nearest to the latter genus. Flowering specimens and a beautiful drawing were obligingly communicated to me by Sir Chas. Lemon, and it has also flowered at Knypersley. Some of the wild specimens have borne as many as thirteen flowers in a spike. In habit it approaches E. clavatum."

12. EPIDENDRUM tibicinis. *Bateman Mss.*

"E. *tibicinis;* caulibus cylindraceis 3-4-phyllis foliis ovatis crassissimis triplo longioribus, scapo altissimo giganteo in racemum multiflorum desinente, floribus ——."

Habitat in Honduras. *Skinner.*

"By far the most magnificent species of the genus. Flowers of the size and colour of Cattleya labiata, but I have not the means of giving their specific character. Scape *three yards* long. The hollow cylindrical stems are used as rumpets by the native children: hence the name."

14. LISSŎCHĪLŪS parvĭflōrus.

Gen. & Sp. Orch. p. 191.

This rare plant flowered in December, in the collection of Messrs. Loddiges, who obtained it from Algoa Bay. The leaves were plaited like those of a Bletia, and varied in breadth from half an inch to 1½ inch. The sepals are a dull greenish purple; the petals were much broader, and of a rich brownish red, rather strongly veined. The labellum was yellow, with the lateral lobes light purple.

Not to be compared in point of beauty with L. speciosus and streptopetalus.

15. EPIDĒNDRŪM ochrăcĕum.

E. *ochraceum;* pseudobulbis obverse pyriformibus cæspitosis sursum attenuatis 1-3-phyllis, foliis linearibus acutis recurvis scapo æqualibus, spicâ terminali laxiflorâ, bracteis rigidis acutis squamiformibus, sepalis petalisque subæqualibus lineari-oblongis obtusis, labello postico sublibero trilobo medio calloso: laciniis lateralibus inflexis truncato-rotundatis denticulatis intermediâ brevi emarginatâ, callo plano apice tridentato, columnâ apice tricorni denticulatâ, ovario triptero.

I have received this little plant from several collections, and have long since called it Encyclia ochracea; but I am now convinced that the genus Encyclia must be reduced to Epidendrum. Mr. Booth first sent it from Sir Charles Lemon's garden at Carclew, it was afterwards drawn for this work in Messrs. Loddiges' stove in July 1837, and the figure will hereafter appear; it is, therefore, only necessary to state that it is a small species, hardly exceeding six inches in height, forming tufts of slender inverted pear-shaped pseudo-bulbs, and with yellowish brown flowers. It was first found in Guatemala by Mr. Skinner, and Messrs. Loddiges received it from Oaxaca. It is near E. tessellatum.

16. EPIDĒNDRŪM Schombūrgkĭi.

E. *Schomburgkii;* foliis distichis oblongis obtusis margine sanguineo punctatis, caule simplici apice aphyllo, sepalis petalisque lineari-lanceolatis acutis æqualibus patentissimis, columnâ clavatâ elongatâ, labelli trilobi basi bicallosi lineâ medianâ elevatâ, lobis lateralibus latis rotundatis laceris: intermedio cuneato apice triangulari crispo medio et utrinque acuminato.

A noble species in the way of E. elongatum, with bright scarlet flowers, according to Mr. Schomburgk, its discoverer, from one of whose letters the following is an extract:—

"In both cases that I had opportunity of meeting with this beautiful plant, it grew in company of Coryanthes, on a tree on the banks of the river, exposed to full light; the bright colour of its flowers, the three-lobed finely fringed lip, the long period it remains in flower, the form of its leaves, spotted on the margin, will make it, should I have been correct in my supposing it to be new, a great favourite with the admirers of Orchidaceæ."

The characters of this species have been drawn up from a coloured figure and some dried flowers, sent home by Mr. Schomburgk. It seems to grow about two feet high; and will be readily known by its deep green leaves, richly dotted with crimson at the edge. The diameter of the head of flowers appears from the drawing to be about five inches.

17. EPIDĒNDRŪM fucātūm.

E. *fucatum;* pseudobulbis subrotundo-ovatis cæspitosis monophyllis, foliis ligulatis coriaceis obtusis scapo brevioribus, paniculâ nutante multiflorâ, bracteis ovatis acutis squamiformibus, sepalis petalisque lineari-oblongis tessellatis æqualibus obtusis conniventibus, labelli liberi tripartiti lobis lateralibus erectis linearibus apice rotundatis intermedio acuto ovali multò brevioribus, callo sulcato plano elevato basi lobi intermedii.

"This curious plant was imported from Havannah in the spring of 1835, by Captain Sutton, and added to Sir Charles Lemon's collection at Carclew, where it flowered for the first time in July 1837, and continued in perfection for several weeks.

"*Pseudo-bulbs,* one-leaved, nearly round, about the size of a large marble, enveloped at first in a thin brown covering, which afterwards dies off. They are then of a deep green, rather wrinkled, with a tinge of reddish brown on the exposed side. *Leaves,* upright, thick and rigid, from four to eight inches long, and about three-fourths of an inch broad; oblong lanceolate; a little keeled at the base and twisted, but otherwise flat with a roundish obtuse point. *Scape,* solitary, rising from the crown of the bulb to the height of from twelve to fifteen inches, round, wiry, and slender, with several joints, each of which is surrounded by a small, thin, brown, acuminate, persistent bractea. The *flowers* are from twenty to thirty or more on each scape, arranged in a curved, loose, branched panicle, each branch supporting from three to seven flowers. *Pedicels,* nearly an inch long, small and round, thickening outwardly, where they become slightly angular, and have a brownish spot in the middle. *Sepals,* spreading, oblong, lanceolate acute; about half an inch long, and one-eighth of an inch broad, thick and fleshy, a little recurved at the margin, and incurved at the point. *Petals,* rather smaller and thinner than the sepals, but very similar in form and colour: the latter is a dingy yellowish green on both sides, with four or five brownish, longitudinal veins, so much branched at the extremity, as to give both sepals and petals a reticulated appearance. *Labellum,* deeply three-lobed; the middle one, which is nearly the length of the sepals, is flat, ovate-acuminate, and incurved at the point; of a pale yellowish colour, almost white, streaked with pink in the centre, and having two oblong fleshy processes at its base. The two lateral lobes are only half the length of the middle one, roundish oblong, converging at the point, so as to embrace and conceal the column, and marked inwardly with four or five deep purple lines or veins. *Column,* slightly curved, three-sided, rounded above, hollowed beneath, and thin at the edges. *Anthers,* reniform, pinkish at the margin, two-celled, with two pollen masses in each.

"The plant requires the same treatment as other Epidendra, and flourishes either in vegetable earth, or attached to a piece of wood."

A very pretty species of this extensive genus, for the preceding account of which I am indebted to Mr. Booth. It is nearly related to E. odoratissimum, from which it is known by its nodding panicle of tessellated flowers having a pinkish lip, and by the solitary leaves. Moreover the bract of E. odoratissimum are comparatively large and sheathing; here they are minute and do not half surround the rachis by their base.

18. EPIDENDRUM lacerum.

E. *lacerum;* foliis distichis lineari-oblongis obtusis, caule simplici apice aphyllo squamato, sepalis petalisque patentibus lineari-lanceolatis æqualibus acutis, labelli trilobi lobis æqualibus laceris pectinatis intermedio apice mucronato integro: lamellis duabus membranaceis sub apice conniventibus duabusque callosis parallelis e basi ortis.

Communicated by Mr. Booth, with the following note, in January, 1837 :—

" This is the third species of Epidendrum introduced from Havannah, in the spring of 1835, by Captain Sutton, and added to Sir Charles Lemon's collection at Carclew, where it flowered during November and December 1836. Although at first sight it may be mistaken for *E. elongatum*, to which it is unquestionably very nearly allied, yet, when compared with that species, it will be found to be of a more lax and slender habit. The leaves are much narrower, and thicker. The stem is erect, not zigzag, as in E. elongatum, and the labellum is larger and deeper fringed, and the two converging membranous lamellæ are quite peculiar.

" *Stem* from two to three feet high, round and nearly erect; destitute of leaves at the top, which is covered with long, thin, brownish scales. *Leaves* distichous, sheathing, oblong-lanceolate, obtuse, thick and fleshy, about 3½ inches long, and half an inch broad, of a rich shining green. *Flowers* in a terminal panicle, pale pink. *Pedicels* an inch long, somewhat angular and twisted, brownish green. *Sepals* spreading, linear-lanceolate acute, rather larger than the *petals*, but all nearly of equal length. *Labellum* three-lobed, the two lateral lobes roundish and deeply fringed at the edge, as well as the intermediate one, which is emarginate, with a small point. *Column* about two-thirds the length of the sepals, small, roundish-oblong, and concave, widening towards the labellum, to which it is attached.

20. HUNTLEYA meleagris. *Bot. Reg. fol.* 1991.

This rare epiphyte, described very briefly in the Botanical Register, under fol. 1991, has blossomed with Messrs. Rollissons. The flower was three inches across; the colour pale yellowish white, with a brownish purple tint towards the upper part of the sepals and petals. The lip was of the same pale colour as the base of the petals. Across the bend of the lip is stretched a broad yellow-fringed crest, which is very remarkable. The colours were all very much less brilliant than if the plant had flowered at a brighter season of the year.

24. BOLBOPHYLLUM setigerum.

B. *setigerum;* pseudobulbis ovatis monophyllis, foliis ovalibus acutis scapo erecto radicali spicâque pendulâ multò brevioribus, rachi filiformi, sepalis ovatis acutis erectis, petalis setaceis patentissimis basi extus tuberculatis, labello oblongo recurvo medio incrassato et elevato apice tridentato, columnâ bicorni, antherâ pedicellatâ.

A very curious little epiphyte, obtained by Messrs. Loddiges from Demerara; the flowers are small and dull purple, upon a light green ground. One very remarkable circumstance in its structure is the presence of a minute tubercle at the base of the petals on the outside, which in another species, B. *bracteolatum*, from the same country, is in the form of a distinct scale.

26. MAXILLARIA variabiles; *var.* unipunctata.

M. *var. unipunctata;* caule ascendente squamato, pseudobulbis ovalibus compressis monophyllis, foliis ligulatis obtusis apice obliquè emarginatis, floribus solitariis pedicellis pseudobulborum longitudine, sepalis lineari-oblongis acutis subæqualibus lateralibus basi parùm productis, petalis conformibus paulò brevioribus, labello oblongo retuso obsolete trilobo apice carnoso: glandulâ succulentâ sanguineâ obtusâ in medio.

" A plant of this singular little epiphyte was, with others, received from the Horticultural Society in March, 1837, by Sir Charles Lemon, Bart. in whose collection it flowered in January, 1838.

" *Stem* creeping, covered with numerous sheathing, acute, deep brown scales. *Pseudo-bulbs* ovate-oblong, thick and fleshy, from one to two inches in length, and about half an inch in breadth, one-leaved. *Leaves* nearly erect, thin and flat, ligulate-lanceolate, dark shining green, from four to five inches long, and three-eighths or half an inch broad, a little compressed and narrowed at the base, with an oblique, obtuse, emarginate point. *Scape* one-flowered, very short, almost hid by the sheathing scales. *Peduncle* round, slightly striated, nearly an inch long, pale green. *Flowers* deep yellow, shaded with a faint greenish tinge. *Sepals* oblong-lanceolate, acute, all about one size, and not spreading very wide. *Petals* scarcely so long as the sepals, and rather narrower, reflexed at the point. *Labellum* 3-lobed, the middle one about the length of the petals, but a good deal broader, with a roundish spatulate recurved point, slightly indented in the centre; lateral lobes very small, their edges curved outwards, and having in the hollow between them, which is of a brownish yellow, a roundish oblong reddish tubercle, extending towards the base of the column, which is marked with several deep red spots. *Column* about two-thirds the length of the petals, a little curved, and rounded above, hollowed beneath and somewhat angular, with a brownish point."

The foregoing note is from Mr. Booth. The plant was sent to the Horticultural Society of London, from Mexico, by Mr. Hartweg, their collector in that country, and was found in the neighbourhood of Vera Cruz. I have also had it from the very rich collection of Thomas Harris, Esq. of Kingsbury. It is scarcely more than a yellow-flowered variety of M. variabilis.

27. PLEUROTHALLIS circumplexa.

P. *circumplexa;* folio coriaceo ovato emarginato in petiolum carinatum canaliculatum ipsi æqualem angustato, caule subnullo, pedunculo basi folii immerso et circumplexo, floribus capitatis ovariisque pubescentibus, capitulo disticho sexfloro basi vaginato, sepalis hiantibus erectis inferioribus in unum oblongum rotundatum supremo brevius connatis: supremo angustiore subpanduriformi, petalis lineari-oblongis ciliatis, labello crasso calceiformi serrulato margine calloso.

A curious new species of this genus, obtained from Mexico by Messrs. Loddiges, with whom it flowered in February, 1838. It approaches P. *saurocephala* and *prolifera*, differing from both in the peduncle being strictly embraced by the base of the leaf, so as to give the flowers an epiphyllous appearance. The latter are small and of a dull dirty brownish yellow colour.

28. *EPIDENDRUM chloranthum.

E. *chloranthum;* pseudobulbosum, foliis coriaceis ligulatis apice rotundatis obscurè bilobis inæqualibus, racemo erecto paniculato, sepalis petalisque subæqualibus lineari-lanceolatis obovatis, labelli trilobi liberi lobis lateralibus linearibus obtusis inflexis intermedio ovato acuminato crispulo multò brevioribus: disco venis elevatis calloso.

A green flowered species, allied to the Encyclia viridiflora of the Botanical Magazine, t. 2831, from which it differs in having the leaves obtusely and obliquely two-lobed, in the side divisions of the labellum being much shorter than the middle lobe, and in the flowers being destitute of the dull purple tinge found in that species. It is a native of Demerara, whence it was sent by Mr. Schomburgk to Messrs. Loddiges. The flowers are a pale green without spots, and appear in March.

29. MILTONIA candida.

M. *candida;* sepalis petalisque lanceolatis acuminatis, labello subrotundo obsoletè quadrilobo undulato, columnâ pubescente basi bicorni.

A most charming new species of this beautiful genus, of which a single specimen flowered imperfectly with Messrs. Loddiges a few weeks since. The sepals and petals are a rich yellowish brown. The labellum is pure white, with some darker markings, of a bright pink, in the middle. The flower is nearly three inches in diameter. A figure of it will be given at some future time.

* The reader is requested to make the following corrections of typographical errors in the last sheet.

Page 23, *for* 26. Maxillaria variabiles *read* Maxillaria variabilis.

31. BRÁSSĬĂ macrostáchȳă.

B. *macrostachya;* pseudobulbis compressis margine obtusis 2-3-phyllis, foliis ligulato-oblongis striatis abruptè acutis, scapo nutante multifloro, sepalis linearibus acuminatis lateralibus longissimis, labello oblongo-lanceolato acuminato petalis longiore. *Sertum Orchidaceum, t.* 6.

A most beautiful species obtained by Messrs. Loddiges from Demerara, and resembling B. caudata, *Bot. Reg. t.* 832, which differs in the following particulars. Its pseudo-bulbs are acute at the margin not obtuse: its flowers are smaller, greener, and much more mottled with deep brown; and its labellum is ovate, acuminate, and the same length as the petals, instead of being oblong-lanceolate, and longer than the petals.

The *Sepals* are linear, acuminate, spreading, pale yellow, with a very few spots of crimson; the upper about two inches long, the lower hanging down, and six inches long. The *Petals* are the same form and colour as the sepals, curving inwards till their points cross each other, rather more than an inch long. The *Labellum* is rather more than two inches long, pale cream-colour, oblong-lanceolate, acute, crisp at the edge, with a few crimson spots at the base, where it is furnished with two elevated downy lamellæ, in front of which stand three horns, the lateral of which are erect and rather recurved, the middle one much smaller and pointing forward.

This plant is figured in the *Sertum Orchidaceum, t.* 6.

32. CĂLÁNTHĒ díscŏlŏr.

C. *discolor;* racemo laxo pubescente, sepalis petalisque acutis, labelli trilobi columnæ omninò accreti basi pubescentis bilamellati lobo intermedio bilobo 3-carinato, calcare pubescente acuto limbo breviore. *Sertum Orchidaceum, sub t.* 6.

This and the following species have been introduced lately from Belgium. They are handsome greenhouse species, and therefore well adapted to the means of those who wish to cultivate Orchidaceæ, but are not possessors of a stove. I do not know from what country they come, but they no doubt belong either to Java or Japan. Both are rather small species, at present not much exceeding a foot in height. In this the lip is of a delicate white, with a few dots of pink near the base. The sepals and petals are on the contrary of a deep reddish brown, a little disposed to be striped.

33. CĂLÁNTHĒ bícŏlŏr.

C. *bicolor;* racemo laxo pubescente, sepalis petalisque acutis, labelli trilobi columnæ omninò accreti lobis subæqualibus: intermedio cuneato apiculato trilamellato basi convexo pubescente bicorni, calcare acuto limbo duplò breviore glabro. *Sertum Orchidaceum, sub t.* 6.

Flowers larger than in the preceding, bright yellow inside, and rich orange-red externally; when spread open they are nearly two inches in diameter. Mr. Auguste Mechelynck sent it to me under the name of Amblyglottis flava of Blume, which is a very different plant.

34. CĂLÁNTHĒ fūrcátă. *Bateman Mss.*

C. *furcata;* foliis ovato-lanceolatis profundè plicatis scapo densè capitato subæqualibus, labello columnæ omninò adnato tripartito basi glanduloso-cristato laciniis lateralibus semiovatis apice rotundatis intermediâ cuneatâ divaricatim bilobâ lobis apice serratis, calcare incurvo pubescente apice furcato labello breviore.

A white-flowered species, with a scape about a foot high, received by Mr. Bateman from the Luzon islands, where it was collected by Mr. Cuming.

35. BLÉTĬĂ hăvānénsĭs.

B. *havanensis;* scapo simplici foliorum longitudine, sepalis patulis ovatis acutis elamellatis, petalis supra columnam conniventibus ovatis obtusiusculis undulatis, labello subrotundo-ovato cucullato trilobo: laciniis lateralibus obtusis intermediam retusam crispam longiorem basi imbricantibus, cristis tribus lamellatis versus apicem interruptis basi evanescentibus venis duabus utrinque arcuatis varicosis.

This species differs from B. verecunda in the colour of its flowers, which are much paler, in its unbranched scape, and in the form of its lip, whose crested plates are interrupted about the base of the middle lobe, and bounded on each side by a pair of curved varicose veins. I have the following account of it from Mr. Booth.

"Introduced from the Havannah by Capt. Sutton, in the spring of 1835, and added to Sir Charles Lemon's collection at Carclew, where it flowered in March 1837.

"*Bulb* terrestrial, nearly round, about the size of a large marble. *Leaves* several, 3-5, sheathing, plaited, ensiform, from one and a half to two feet long, and an inch broad; so thin and slender as to be almost unable to support themselves upright, of a light green colour, with five or more longitudinal ribs, which are rather prominent on the under side. *Scape* upwards of two feet and a half high, nearly erect, round and woody, of a glossy, reddish brown colour and spotted; bearing ten or eleven peach-coloured flowers, in a loose upright panicle. *Pedicels* an inch long, of a deep brown colour, slightly channelled, with a small cordate, acuminate bractea at the base of each. *Sepals* spreading, oblong-lanceolate, a little recurved at the point and edges; the upper one is of a paler colour, and somewhat longer, and narrower than the other two, which are crescent-shaped, and have the lower half of each deeper tinged than the upper. *Petals* oblong-obtuse and convex, closely covering the column as if it were with a hood. They are of a similar colour to the sepals, but rather paler at the edges, and of a thinner texture, with a pink coloured line up the centre of each. *Labellum* three-lobed; the two lateral ones are rounded at the edges, which are erect and of a redder colour than either petals or sepals, and finely veined. The middle lobe is reflexed and pendent, of an irregular figure, nearly as long as it is broad, with a white undulated margin, and rather deeply indented in the centre. The colour is a deep reddish pink. Along the middle are seven, longitudinal, elevated ribs, of a yellow colour, except the outer one on each side which is nearly white. *Column* very much curved, rounded on the upper side, a little hollowed beneath, two-edged, widest at the point. Anther-case eight-celled. *Pollen-masses* 8, united in pairs.

"The plant requires the same treatment as the other Bletias, and seems to thrive in a strong loamy soil, with plenty of water when in a growing state."

36. EPIDÉNDRŪM áspĕrūm. *Lindley,*

In Hooker's Journal of Botany, p. 6.

E. *asperum;* pseudobulbis ovatis basi subangulatis diphyllis, foliis coriaceis ligulatis carinatis acutis, scapo ramoso pedunculisque subasperis, sepalis patentibus obovato-lanceolatis obtusis, petalis magis cuneatis, labello semilibero cucullato flabellatim cuneato crispo basi callo elevato pubescente obovato medio depresso apice tridentato.

This species was first known from dried specimens, collected in the western parts of the republic of Columbia by Mr. Cuming, and received its name in consequence of the scape and peduncles being covered all over with hard elevated points. It has now flowered in the valuable collection of Thomas Harris, Esq. of Kingsbury, and I am enabled to amend the character. The roughness of the branches is still visible, but chiefly in the form of little specks, and it is only after being dried that this peculiarity becomes distinct. It has dull, dingy, yellowish brown flowers, with a dull yellow lip, neatly streaked with red veins, which are a little elevated, but not enough so to be of any importance. Mr. Harris received his plant from Mexico.

37. EPIDÉNDRŪM vărĭcósūm. *Bateman Mss.*

E. *varicosum;* pseudobulbis diphyllis, foliis coriaceis ligulatis acutis, scapo simplici gracili brevioribus, sepalis petalisque subæqualibus cuneato-lanceolatis carnosis vernicosis planis, labello libero unguiculato non cucullato trilobo basi pubescente: laciniis lateralibus nanis triangularibus obtusis, intermediâ reniformi emarginatâ venis tuberculatis et varicosis pictâ, callo pulvinato ovato pubescente, columnæ apice trilobæ lobo dorsali crasso cylindraceo obtuso.

One of the many small-flowered dull-coloured Epidendra with which the tropical parts of America abound. It is a native of Guatemala, whence it was imported by Mr. Bateman through his friend Mr. Skinner. The form of the lip and the varicose veins are very remarkable.

38. CYRTOCHĪLŪM mystacīnūm.

C. *mystacinum;* pseudobulbis ovalibus compressis corrugatis monophyllis basi polyphyllis, foliis ligulatis acutis planiusculis carinatis scapo ramoso multò brevioribus, bracteis lanceolatis pedunculis duplò brevioribus, sepalis petalisque ovatis acuminatis, labello unguiculato cordato obovato-lanceolato subrepando plano apice reflexo medio pubescente basi obsoletè lamellato, columnæ alis multifidis.

A Peruvian Orchidaceous plant which flowered in the stove of Richard Harrison, Esq. of Aighburgh, in October, 1837. It has a branched scape like that of an Oncidium, bright yellow white-coloured flowers, with a most curious fringed or whiskered column.

39. CYRTOCHĪLŪM maculātūm.

C. *maculatum;* pseudobulbis ovatis compressis subangulatis diphyllis basi foliosis, foliis latè ligulatis acuminatis striatis apice obliquè emarginatis, scapo simplici (?) bracteis brevissimis squamæformibus, sepalis petalisque carnosis obovato-lanceolatis acutissimis, labello membranaceo oblongo apiculato utrinque dentato lamellis duabus ad basin et corniculo utrinque, alis columnæ falcatis integerrimis.

An Orchidaceous plant found near Vera Cruz by Mr. Hartweg, and by him sent to the Horticultural Society. It has rather large flowers, with a greenish yellow ground colour, blotched with deep rich purple, and a whitish lip, with a few stains of dull red. I have also received it from Messrs. Rollissons.

40. SPECKLĪNĬĂ cĭlĭārĭs.

S. *ciliaris;* folio angustè lanceolato obtuso cauli subæquali, spicis secundis fasciculatis folio duplò brevioribus, sepalis carinatis ciliatis inferiore bifido, petalis ellipticis obtusis apice ciliatis, labello obovato obtuso ciliato trinervi, columnâ apice fasciculatim ciliatâ.

A small plant, resembling a Lepanthes, with purplish green leaves, and dull green purple-spotted flowers; imported from Mexico by Messrs. Loddiges.

41. SPECKLĪNĬĂ orbĭcŭlārĭs.

S. *orbicularis;* folio subrotundo-ovato emarginato caule breviore, floribus fasciculatis, sepalis lineari-oblongis glabris inferiore bipartito, petalis ellipticis acutis serratis, labello ovato rostrato basi concavo apice ciliato obtuso, columnâ apice serratâ.

A small plant, resembling the preceding in habit, but with both the leaves and flowers more purple. It was imported by Messrs. Loddiges from Demerara. Both these have the loose sheaths on the stem, such as are characteristic of Lepanthes, but neither of them offers the slightest transition to that curious genus.

42. EPIDĔNDRŪM pachyānthūm.

E. *pachyanthum;* pseudobulbosum, foliis lato-ligulatis subundulatis apice obliquè obtusis dorso rotundatis, perianthio carnoso herbaceo, sepalis lanceolatis, petalis obovato-lanceolatis apice complicatis, labelli liberi trilobi laciniis lateralibus ascendentibus truncatis intermediâ spathulatâ acutâ basi callosâ convexâ inappendiculatâ multò brevioribus.

A large green-flowered species, sent to Messrs. Loddiges from Guayana, by Mr. Schomburgk. Its leaves are thinner and broader than is usual among the pseudobulbous Epidendra, and a little wavy at the margin. The flowers are fully two inches in diameter, thick and fleshy, dull green, stained with a dirty reddish brown towards the ends of the sepals and petals. The labellum is a pale straw-colour, streaked along the middle with violet.

43. EPIDĔNDRŪM pīctūm.

E. *pictum;* pseudobulbosum, foliis ligulatis coriaceis obtusis dorso rotundatis, racemo erecto paniculato, sepalis petalisque obovato-linearibus subæqualibus, labelli trilobi liberi lobis lateralibus linearibus acutiusculis subfalcatis columnam amplexantibus margine anteriore plicato intermedio ovali acuto crispo multo brevioribus, disco venis elevatis calloso.

One of the numerous unpublished species resembling *E. odoratissimum;* with dull yellow flowers, neatly striped with crimson. It is very nearly related to *E. chloranthum,* from which its leaves readily distinguish it. Messrs. Loddiges received it from Demerara.

44. EPIDĔNDRŪM smarāgdĭnūm.

E. *smaragdinum;* caule gracili ad fastigium ipsum folioso, foliis lineari-lanceolatis acuminatis subdistichis, spicâ terminali paucíflorâ rigidâ nutante foliis breviore, bracteis acuminatis squarrosis ovario ventricoso cuniculato brevioribus, sepalis lineari-oblongis acutissimis, petalis angustissimis acuminatis, labello cuneato cordato carnoso apice tridentato basi bicalloso: laciniis æquilongis lateralibus rotundato-truncatis intermediâ acutâ decurvâ.

A species closely allied to *E. orchidiflorum,* with small bright green flowers of no beauty whatever, and hardly distinguishable from the leaves. It is a native of Demerara, whence it was obtained by Messrs. Loddiges. It flowers in March.

45. PHYSĪNGĂ prostrātă.

PHYSINGA G. nov. (Epidendreæ). Sepala membranacea, æqualia, basi connata. Petala minora, basi sepalis obliquè adnata. Labellum carnosum, tuberculatum, indivisum, cum basi columnæ connatum, sacco vesiciformi basi auctum. Columna carnosa, nana, biloba, basi imâ antherifera. Pollinia 4, filis duobus geminatis pulvereis adnata. Stigma, area minuta, madida, bidentata, supra faucem vesicæ.

P. *prostrata.* Caulis prostratus, brevis, anceps, densè et distichè foliosus. Folia crassissima, ovato-lanceolata, acuta, amplexicaulia, dilutè purpurascentia, striata. Scapus terminalis, filiformis, prostratus, ramosus, 8-9 pollices longus, vaginis angustis, fusco-purpureis, acutis, carinatis, striatis, arctè vestitus. Flores pauci, parvi, apice ramorum, herbacei, dilutè purpureo suffusi. Calycis laciniæ acuminatæ, patentes. Petala angustè lanceolata, acutissima. Labellum albidum, obovatum, acutum, apice recurvum, basi auriculis carnosis inflexis auctum; lineâ mediâ callosâ tuberculisque duobus anticè procurrentibus; vesica sphæroidea, viridescens. Columna alba, auriculis imis labelli brevior, basi ipsâ tantum labello adnata.

This is an Orchidaceous plant of no beauty, but one of the most curious that I am acquainted with. It has much of the habit of an Epidendrum, and is not at all unlike E. orchidiflorum in its manner of growth; but it belongs to no genus hitherto described. The sepals are united at the base into a cup as in Masdevallia; the labellum has a little bladder-like pouch at its base, in room of a spur: and the column, which is very small, is more like that of a Cypripedium than anything else. Messrs. Loddiges obtained it from Demerara.

46. CLEISŌSTŎMĂ tridentātă.

C. *tridenta;* foliis lineari-lanceolatis subcarinatis apice oblique tridentatis, racemis horizontalibus paucifloris foliis brevioribus, sepalis lineari-oblongis rectiusculis, petalis linearibus falcatis, labello trilobo intus pubescente laciniâ intermediâ obsoletâ obtusâ lateralibus ovatis acuminatis marginibus involutis: dente calcaris obtuso inflexo hirsuto, calcare pendulo apice obtuso subventricoso.

A small flowered epiphyte, of no beauty, with the habit of a Saccolabium. It is a native of New Holland, from whence it was obtained by Messrs. Loddiges. The flowers are very small, of a dull, dirty, reddish white, mixed with a little yellow. The pollen-masses are remarkable for having each a concave membrane, in the form of a watch glass, adhering to them at the back; these are undoubtedly the back valves of the anther case, which separate all round and adhere slightly to the apex of the caudicula, below the pollen-masses.

47. EPIDĔNDRŪM cucullātūm.

E. *cucullatum;* caule compresso folioso, foliis lineari-lanceolatis acutis, pedunculo terminali squamato paucifloro foliis multò breviore, rachi brevi flexuosâ, sepalis petalisque erectis ovato-lanceolatis acutis, labello obovato acuto adnato: lineâ media tuberculisque duobus basilaribus callosis, antherâ longè infra apicem clinandrii cucullati dentati truncati insertâ.

One of the most unattractive of this large genus. The flowers are small and white, at the top of a leafy stem about nine inches high; the anther is remarkable for being inserted far below the apex of the column, in which respect the species approaches the genus *Physinga.* Obtained from Para by Richard Harrison, Esq. of Liverpool.

48. PLEUROTHÁLLĬS ophiocéphălă.

P. *ophiocephala*; folio oblongo emarginato concavo carnoso caule breviore, flore solitario e spathâ duplici membranaceâ carinatâ erumpente, sepalis oblongis obtusis hiantibus intus papillosis extùs pubescentibus inferiore latiore emarginato, petalis lineari-oblongis intus villosis labello duplò minore lineari-ovato obtuso carnoso glabro canaliculato marginibus elevatis truncatis.

A most curious Mexican species, for which I have to thank Messrs. Loddiges and Mr. Barker of Birmingham. It has no beauty, but is remarkable for having one solitary flower protruding from the base of the leaf, and so similar to the head of a snake with the jaws open, that it is difficult at first sight to believe it really a flower that one looks upon. To add to the deception, the small labellum is shaped like a tongue, and moves up and down when you peep into the mouth of the flower. The stem of the leaf is strong, terete, six inches long, with one long striated brown sheath in the middle; the leaf is remarkably thick and fleshy, and above four inches long; the flower is dull yellowish brown, spotted with dull purple, and nearly three-quarters of an inch long.

49. EPIDÉNDRŪM longicóllĕ.

E. *longicolle*; caule erecto compresso folioso, foliis linearibus apice angustatis, floribus aggregatis terminalibus nutantibus foliis brevioribus, sepalis lineari-lanceolatis acuminatis patentibus, petalis linearibus acutis supra columnam convergentibus, labelli trilobi lobis lateralibus semiovatis acutis integerrimis intermedio lineari-acuminato paulo brevioribus: lamellis 2 callosis ad basin, collo ovarii elongato.

A species nearly allied to *E. nocturnum*, but much less handsome. The sepals and petals are pale yellow; the lip white, with two yellow plates at the base. It was obtained by Messrs. Loddiges from Demerara.

50. SARCOCHÍLŬS parviflórŭs.

S. *parviflorus*; foliis oblongis undulatis chartaceis apice acute et oblique bidentatis, racemis nutantibus quadrifloris, sepalis lineari-obovatis acutis lateralibus basi dilatatis, petalis falcatis obtusis, labelli cum columnâ articulati lobo medio minimo transverso rotundato: disco callis variis crassis rotundatis tuberculato.

A small flowered epiphyte introduced from New Holland by Messrs. Loddiges. It has no claim to beauty, but is interesting as a second species of the genus Sarcochilus. The flowers are green, with a few spots of dull purple in the sepals. The labellum is less green, beneath indeed almost white, banded with dull purple, *articulated* with the column, and covered over in the inside with various crowded tubercles, of which the three largest form a crescent next the apex of the lip.

53. CIRRHOPÉTĂLŪM cæspĭtósūm. *Wall. Mss.*

C. *cæspitosum*; pseudobulbis ovatis monophyllis, foliis angustè ovalibus obtusis basi angustatis scapo longioribus, scapo filiformi erecto vaginis paucis membranaceis carinatis acutis distantibus laxè vestito apice umbellato, sepalo supremo obovato fornicato acuto margine scabro lateralibus acuminatis convolutis triplò longioribus, petalis ovatis apiculatis pectinato-serratis, labello minimo solido triquetro apice recurvo angulis superioribus marginatis.

A little epiphyte, forming part of the noble collection of plants imported sometime since from the East Indies, by His Grace the Duke of Devonshire, and now preserved at Chatsworth. It has small, pale, yellow-ochre coloured flowers, without any thing striking about them. It flowered in the beginning of April.

54. DENDRÓBĬŪM cándĭdūm. *Wall. Mss.*

D. *candidum*; caulibus erectis flexuosis teretibus, foliis ovato-lanceolatis apice obliquis obtusis, pedunculis axillaribus erectis bifloris, sepalis patulis lateralibus basi parùm productis, petalis conformibus, labello ovato-lanceolato obtuso medio barbato supra basin calloso lateribus erectis cucullatis.

For flowers of this I am indebted to His Grace the Duke of Devonshire, who sent them to me in the beginning of April; I have since been favoured by Mr. Paxton with a sketch of the plant. It was found by Mr. Gibson, his Grace's collector in India, at Nungclow, on the north side of the Khoosea Hills, growing on rocks, sand, and decayed trees. The flowers, which are pure white, are most exquisitely scented. The habit of the plant is that of *Dendrobium nobile;* that is to say, erect, not pendulous as in *D. Pierardi* and that race. The stems are about a foot high, and lose the principal part of their leaves before flowering.

55. OCTOMÉRĬĂ grácĭlĭs. *Loddiges Mss.*

O. *gracilis*; folio lineari-carinato apiculato recurvo, floribus paucis fasciculatis, labelli emarginati lobis lateralibus erectis rotundatis.

An inconspicuous species of this genus, having flowers very like those of *O. Baueri*, but smaller and with a differently shaped lip, and very narrow, channelled, recurved leaves. It was received by Messrs. Loddiges from Rio Janeiro.

56. ONCĬDĬŪM tetrapétălūm. *Gen. & Sp. Orch.* 198.

A small plant of this beautiful little species, imported from Jamaica by John Henry Lance, Esq. has lately flowered in the garden of the Horticultural Society. It proves to have triquetrous leaves, and therefore has been arranged in its wrong place in the work above quoted. Although the flowers are among the smallest in the genus, they are extremely pretty. The labellum is a pure dead alabaster white, except at its base, where it is covered with yellow and brown tubercles. The sepals and petals are gaily barred and spotted with brown, while the column has two large pink-spotted spreading wings. Jacquin describes the species as having a flowering scape two feet high; that which appeared in the Horticultural Garden was scarcely more than six inches high, for the plant, though healthy, was weak, and had suffered from its voyage. When in perfection this species will be found among the handsomest of the small-flowered epiphytes.

58. VANÍLLĂ bícŏlŏr.

V. *bicolor*; foliis ovato-oblongis subsessilibus acutis striatis margine rubescentibus, sepalis lineari-lanceolatis acutis patentibus dorso rotundatis, petalis conformibus dorso carinatis, labello membranaceo semilibero convoluto venis ramentaceis medio densè ramentaceo-stuposo, columnâ barbatâ: auriculis crenulatis.

A deliciously fragrant climbing epiphyte, sent by Mr. Schomburgk to Messrs. Loddiges from Guayana. Its flowers are full three inches long, with dull red sepals and petals, and a cream-coloured lip. It is very different from any species yet described in this most curious and interesting, but little studied genus. It flowered freely in the month of April.

59. CYMBÍDĬŪM virescēns.

C. *virescens*; (Eucymbidium) foliis gramineis recurvis serrulatis subtrinerviis, scapo laxè vaginato unifloro, pedunculo tereti bracteâ duplò longiore, sepalis linearibus obtusis concavis patentibus, petalis brevioribus conformibus supra columnam conniventibus, labello oblongo indiviso apice revoluto imberbi plicis duabus carnosis lamellatis apice liberis foveâ truncatâ irregulari utrinque infra apicem plicarum.

A native of Japan, whence it was brought to Europe by Dr. Siebold. It flowered in April in the nursery of Messrs. Rollissons of Tooting. It has greenish sepals and petals, about an inch and half long, and a pale dull yellow lip slightly blotched with dull red. I did not remark any smell. No doubt this will prove a greenhouse species.

60. CHÆNÁNTHĒ Barkérī.

CHÆNANTHE, *G. n.* (*Nat. ord.* Orchidaceæ, § Vandeæ). Perianthium ringens. Sepalum superius erectum; lateralia connata, basi productæ columnæ adnata, saccata, porrecta. Petala basi valdè obliqua, et columnæ productæ adnata, revoluta, sepalis majora et magis membranacea. Labellum cum columna connatum, basi in calcar cum basi columnæ productum, trilobum. Anthera anticè truncata. Pollinia 2, pyriformia, in caudiculam linearem deflexa.
Chænanthe Barkeri.

This most singular plant is a native of Para, whence it was imported by Geo. Barker, Esq. of Birmingham, who sent me specimens in April last. In no plant that I am acquainted with is the obliquity of the flower more striking than in this. In the first place the lateral sepals are lengthened into a bag, which hangs down in front of the ovary; then the petals are stretched out at their base to adapt themselves to this; further, the labellum is connate with the column for nearly all its length, the latter being lengthened so as to fill up the bag of the sepals; and, finally, the anther is abruptly truncated in front, and the two pollen-masses are bent down upon the caudicula so as to lie almost parallel with it. The genus is nearest allied to Notylia, in natural affinity.

61. EPIDÉNDRŪM altíssimūm. *Bateman in litt.*

E. *altissimum;* pseudobulbis elongatis teretibus 2-3-phyllis, scapis ramosis longissimis, sepalis lineari-oblongis acutis, petalis conformibus basi angustatis, labelli liberi lobis lateralibus dimidiatis erectis tortis obtusis intermedio dilatato undulato recurvo apiculato basi bicostato.

Communicated by Mr. Bateman with the following note: "Found in rocky parts of the Bahamas by the indefatigable Mr. Skinner, from whom I received it in the summer of 1837. This and other pseudo-bulbous Epidendra grow in the greatest profusion in the situation described; and though at the period of Mr. Skinner's visit they were in a parched and torpid state, he was informed by a resident that during the season of their flowering they scented the air to an extraordinary and almost insupportable degree. To this 'excess of sweets' E. altissimum contributes a powerful odour resembling bees-wax; but in potency it is far surpassed by another unpublished species from the same quarter, for which I am also indebted to the zeal and liberality of Mr. Skinner, and which yields a perfume at once delicate and powerful, and so closely resembling that of our wild English primroses, that I have in consequence named it 'E. primulinum.'"

I have also received it from Messrs. Rollissons. It is very like Epid. oncidioides, *Bot. Reg. t.* 1623, from which its long pseudo-bulbs, gigantic stature, and labellum, alone distinguish it.

63. ONCÍDĬŪM stramíněūm. *Bateman in litt.*

O. *stramineum;* ebulbe, foliis crassis carnosis ovato-lanceolatis acutis dorso rotundatis scapo paniculato rigido erecto brevioribus, sepalis subrotundis unguiculatis concavis liberis integerrimis, petalis duplò majoribus oblongis obtusis emarginatis margine crispis, labelli lobis lateralibus oblongis carnosis margine revolutis basi columnæ proximâ nectariferis intermedio reniformi plano emarginato majoribus, tuberculis disci 4 geminatis, columnæ alis carnosis linearibus obtusis elongatis genuflexis decurvis.

A beautiful stove epiphyte, sent from the neighbourhood of Vera Cruz to the Horticultural Society by their collector, Mr. Hartweg. It has pale straw-coloured flowers, about as large as those of *O. flexuosum,* with a faint smell of primroses. The base of the lower sepals, the lower part of the lip, the column, and a line along the origin of the petals are neatly dotted with brown. Honey is secreted by the lip at the base of the lateral lobes next the column. Young plants are very like *O. pumilum,* for which they may be mistaken: but they have much thicker leaves.

64. MASDEVÁLLĬĂ infrácta.

Gen. & Sp. Orch. p. 193.

This curious plant has just flowered with Messrs. Loddiges, who imported it from the Brazils. It has pale, whitish-yellow flowers, slightly tinged with pink, and yellow cirrhi; and it proves that my conjecture was right, that the genus would belong to Malaxideæ and not to Vandeæ. I find the pollen-masses to be two, half-pyriform, without any gland. It seems to be an abundant flowerer. M. Descourtilz found it on the cold wooded mountains which separate Canta Gallo from the province of Rio Janeiro, with bright purple flowers. Sir W. Hooker received it from the Organ Mountains; and I find it in a collection of dried plants gathered near Rio Janeiro by Mr. Gardner, (No. 647.)

66. EPIDÉNDRŪM sellígěrūm. *Bateman in litt.*

E. *selligerum;* pseudobulbis 2-3-pollicaribus, scapo subramoso, sepalis patentissimis obovatis concavis obtusis, petalis unguiculatis spathulatis acutis, labello semilibero laciniis lateralibus obtusis recurvis planis intermediâ ovatâ acutâ crispatulâ: disco elevato carnoso, antherâ bilobâ terminali.

A native of Guatemala. Mr. Bateman's note about the species is as follows:—"Another of the many interesting discoveries of Mr. Skinner, who kindly sent it to me in 1836. It is a very pretty and distinct species, a free grower, and yields a perfume as delightful as that of the Tuberose, and very similar to it. Two distinct species have already flowered with me, one of which has a white lip, the lateral lobes of which are nearly upright, while in the other the lip is rose-coloured, and its lateral lobes bent back. The flowers of both varieties are very durable." The sepals and petals are a dull, dingy, brownish purple.

67. BRASÁVŎLĂ angustáta.

B. *angustata;* sepalis petalisque linearibus acuminatissimis collo ovarii longioribus, labello ovato acuminato serrulato basi haud cucullato, clinandrii trilobi lobo medio incurvo integro lateralibus cuneatis flabellatim incisis.

A fine new species, imported from Demerara by — Wilmore, Esq., of Oldford, near Birmingham. The flowers are large and long-stalked, of a pale yellowish green, with a narrow fringed white lip. It differs from B. cucullata in the lip not being at all cucullate at the base. The leaves are very long and slender.

68. ACANTHOPHÍPPĬŪM striátūm.

A. *striatum;* petalis oblongis acutis sepalisque apice revolutis, labelli lobis lateralibus rotundatis intermedio acuto carnoso utrinque plicato: disci callo solitario elevato longitudinali.

For this I am obliged to Mr. Bateman, who received it as a Nepal plant from Her Majesty's Botanic Garden at Kew. It is very inferior in appearance to A. bicolor, having pale "French white" flowers, with dull longitudinal stripes, and not a single stain of brilliant colour. The flowers are smaller than in the species just mentioned, and their lower sepals are so much contracted into a pouch at the bend of the column-foot, as almost to form a spur; the labellum has long rounded side-lobes, a small, contracted, tawny-spotted, acute middle lobe, and a single elevated callosity running down its middle. Mr. Bateman adds, that it has larger, narrower, and more attenuated pseudo-bulbs, a three-flowered scape, and leaves like those of A. *bicolor.*

I take this opportunity of stating, that the genus Acanthophippium does not belong to *Vandeæ,* as I formerly supposed, from the examination of an imperfect specimen, but in reality should be stationed in *Epidendreæ,* where it represents the Maxillaria of Vandeæ, and the Dendrobium of Malaxideæ; an interesting fact with reference to systematical purposes. It should be placed near Bletia and Phaius.

69. STÉLĬS trístўlă.

S. *tristyla;* folio oblongo concavo subundulato spicâ breviore, spicâ erectâ laxâ subtortili, bracteis membranaceis cucullatis cuspidatis, calyce triangulari, petalis truncatis carnosis, labello truncato medio sulcato, columnâ depressâ angulis tortis incurvis facie superiore viscidâ stigmaticâ.

A species imported from Brazil by Messrs. Loddiges, and, like the rest of the genus, not at all remarkable for beauty, although one of the largest that we yet have seen. It flowered in April.

70. PLEUROTHÁLLĬS margináta.

P. *marginata;* folio obovato biconvexo marginulato caule pluries longiore racemo laxo erecto multò breviore apice obsoletissimè tridentato, sepalis subfalcatis carinatis lateralibus semiconnatis, petalis linearibus obtusis apice callosis, labello lineari obtuso canaliculato.

A very small species with the habit of P. *Grobyi*, and of no beauty. It was sent to Mr. Bateman from Guatemala, by Mr. Skinner, and grows in dense tufts on the branches of trees.

71. PLEUROTHÁLLĬS aphthósă.

P. *aphthosa;* folio ovali coriaceo subtùs discolore cauli carinato triquetro æquali, spathâ bivalvi pauciflorâ, floribus subsessilibus tomentosis carnosis fragilibus, sepalis oblongis convexis subæqualibus intus papillosis lateralibus basi connatis, petalis ovatis acutis glabris subdiaphanis, labello ovato carnoso obtuso antice papilloso basi excavato glabro, columnâ petalorum longitudine, clinandrio cucullato crenulato.

A Mexican species of this extensive genus, for which I am obliged to Mr. Bateman, who received it from the Birmingham Botanic Garden, without a name. Its leaves are between three and four inches long, dull purple at the edge, and brighter underneath; the flowers are dull yellow, tomentose externally, and scarcely longer than the two-valved spatha. The sepals are very thick and brittle, smooth inside, and covered at the end and towards the edges with little elevated purple warts; the petals are almost transparent and a very light dull yellow; the labellum is fleshy, and dull deep purple.

73. BLÉTĬĂ Shephérdĭi. *Bot. mag. t.* 3319.

This plant has lately flowered in the garden of the Horticultural Society, and I find it is only a dark-flowered variety of Bletia verecunda; it does not appear to be distinguishable by any character of importance.

74. MAXILLÁRĬĂ mádĭdă.

M. *madida;* caulescens, pseudobulbis ovatis elongatis subteretibus sulcatis diphyllis secus caulem imbricatis, foliis lanceolatis linearibusque acutis subplicatis canaliculatis, floribus solitariis intra spatham siccam squamosam subsessilibus, sepalis subparallelis acutis lateralibus basi parùm productis, petalis conformibus; labelli trilobi medio callosi lobis lateralibus inconspicuis intermedio rotundato plano obtuso recurvo disco madido et discolore, columnâ clavatâ arcuatâ labello longiore.

The species now described is one of the least showy of the large genus to which it belongs; it is nearly related to *M. acicularis*, which, with some others in the possession of Messrs. Loddiges, forms a well marked group. The flowers are a dull dirty yellow, spotted obscurely with yellowish brown. The labellum has a broad chocolate-brown blotch at the end, where it is so shining as to look as if wetted; this circumstance has suggested the specific name. It is a native of Brazil.

75. CŒLŎGȲNĒ prolífĕră. *Gen. & Sp. Orch.* 40.

This plant has flowered at Chatsworth, among the numerous species brought to the Duke of Devonshire by Mr. Gibson, from the East Indies. Its flowers are small, pale, brownish yellow green, with brown veins. It corresponds very well with the character assigned it in the above work, except that the pseudobulbs are not angular, but quite terete, or only slightly compressed when old.

76. EPIDÉNDRŪM ĕquĭtans.

E. *equitans;* caule folioso ancipiti, foliis equitantibus ancipitibus lanceolatis acuminatis, flore solitario terminali pendulo, pedunculo ancipiti, spatha diphylla foliolo inferiore erecto foliaceo flore longiore, sepalis linearibus acuminatis patulis, petalis conformibus paulò brevioribus, labello ovato-lanceolato complicato cum columnâ basi connato medio trilamellato; lobo medio carnoso semitereti recurvo lateralibus erectis membranaceis integris.

A very curious species in its habit, which resembles Fernandezia so much that no one could have doubted its belonging to that genus, until it flowered. It, however, proves a genuine Epidendrum. The single flower is of a dull chocolate brown. Mr. Hartweg sent it to the Horticultural Society from near Vera Cruz in 1836.

80. GRAMMATOPHȲLLŪM multiflōrūm.

G. *multiflorum;* racemo erecto longissimo multifloro, bracteis ovato-oblongis obtusis squamiformibus dorso convexis, sepalis oblongis obtusiusculis planis, sepalis acutis subconformibus angustioribus, labelli trilobi pubescentis medio hirsuti lobo intermedio plano oblongo rotundato lateralibus erectis subfalcatis, jugo carnoso elevato in medio basi lobi intermedii interrupto cis apicem evanescente, columnæ margine supra basin elevato flexuoso incurvo foveam altam obconicam circumdante.

This fine plant was sent home in 1837, from Manilla by Mr. Cuming; and was believed to be the famous *Letter-plant* of the Malayan Archipelago. It, however, appears from a noble specimen of the flowers sent me from Knypersley, by Mr. Bateman, that, although belonging to the genus it is a very different species, far inferior in beauty to the Letter-plant. Such remarks as I may have to make upon it are reserved until the publication of a figure which has been prepared for that purpose.

81. EPIDÉNDRŪM tridáctȳlūm.

E. *tridactylum;* caule fusiformi gracili apice distichè folioso, foliis angustè oblongis apice paulo angustatis obtusis cum mucronulo, spicâ terminali tripartitâ multiflorâ foliis longiore, sepalis subrotundo-ovatis patentibus incurvis lateralibus majoribus, petalis spathulatis linearibus apice incurvis, labello ascendente tripartito basi bicalloso: laciniis linearibus carnosulis lateralibus margine involutis intermediâ breviore planâ, columnâ brevi crassâ cuneatâ labello omninò adnatâ.

A curious Brazilian Orchidaceous plant, with smaller flowers than any other species of the genus yet in the gardens; I am obliged for it to Stephen Cannon, Esq. of Stratford Green. The flowers are a pale brownish yellow, except the column which is green, short, thick, and wedge-shaped. Flowered in the stove in May, for the first time.

82. EPIDÉNDRŪM cauliflōrūm.

E. *cauliflorum;* caule tereti, foliis ovato-oblongis planis acuminatissimis, corymbis brevibus axillaribus e latere caulis erumpentibus, sepalis angustis oblongis patentissimis subæqualibus concavis, petalis linearibus apice cuneatis acutis reflexis, labello columnæ arcuatæ clavatæ omninò accreto cuneato-subrotundo trilobo; lobo intermedio truncato tridentato lateralibus rotundatis repandis, callis tribus linearibus in medio labelli lateralibus majoribus.

A native of Rio Janeiro, whence it was obtained by Messrs. Loddiges. The flowers are about the size of those of E. *nutans*, of a pale straw-colour, and are remarkable for appearing from the side of the stout cylindrical stem, bursting forth from among the dry sheaths with which it is closely invested.

86. DENDRŌBĬŪM formōsūm.

Wallich pl. as. rar. 1. *t.* 39.

This noble plant, whose large ivory-white flowers are unrivalled for beauty, in even the rich Flora of India, was sent from Chatsworth on the 12th of May: having been imported for the Duke of Devonshire, by Mr. Gibson, His Grace's collector in the East Indies. The flowers grow at the end of a leafy stem, and are between three and four inches in expansion, with every part of the purest and most transparent white, except one delicate lozenge-shaped, buff-coloured blotch in the centre of the lip. Neither the form nor the colour of this is correctly given in Dr. Wallich's work above quoted. It has been sometimes said that the Flora of South America is richer in beautiful Orchidaceous plants than that of India, and an appeal has been made to the Cattleyas, Epidendra, Oncidiums, and Maxillarias of the former country. But I know of no South American species so admirably formed and coloured as India can produce in the case of Dendrobia of various kinds, Saccolabiums, and species of the genera Phaius, Vanda, Cœlogyne, and Grammatophyllum. Of these Dendrobium formosum must stand among the foremost in point of beauty.

87. EPIDÉNDRŪM ionōsmūm.

E. *ionosmum ;* pseudobulbis diphyllis scapo apice simplici racemoso brevioribus, sepalis petalisque coriaceis obovatis concavis obtusis subæqualibus, labelli trilobi liberi lobis lateralibus erectis oblongis columnam amplexantibus apice rotundatis undulatis coloratis striatis intermedio subrotundo emarginato crispo: lineis duabus elevatis prope basin unicâ angustiore apici propiore, columnâ anticè bicorni.

The Western world wants no violets where this charming plant is found, for it fills the air with a fragrance as delicate and delicious as that of our favourite wild-flower. The blossoms are rather large for an Epidendrum of this class, a dull reddish green, with the lip delicately streaked with deep lilac. Messrs. Loddiges imported it from Essequibo.

88. SACCOLĀBĬŪM gemmātūm.

S. *gemmatum ;* foliis distichis semiteretibus canaliculatis subtortis apice inæqualiter tridentatis, spicâ ramosâ, floribus minutis carnosis, sepalis conniventibus ovatis obtusis lateralibus multò majoribus, petalis brevioribus subrotundis, labello carnoso ovato crystallino dorso tuberculato marginibus calcaris conici penduli dilatatis carnosis rotundatis, columnâ brevissimâ, apice antheræ hemisphæricæ reflexo.

This new species was discovered in 1837, on the Khoseea hills of India by Mr. Gibson, and was brought by him to the Duke of Devonshire, in whose collection at Chatsworth it flowered in May. The blossoms are the smallest of the genus, not being larger than a grain of mustard-seed, but the finest amethysts are not of a more brilliant purple, and the tips of the labellum and sepals are quite white.

89. EPIDÉNDRŪM vesicātūm.

E. *vesicatum ;* caule elongato, foliis inflatis equitantibus carinatis acutis glaucis, floribus terminalibus fasciculatis carnosis foliis vix longioribus, sepalis lineari-oblongis acutis petalis conformibus angustioribus, labello subrotundo cordato lineâ mediâ tuberculisque duobus basilaribus elevatis, collo ovarii medio vesicato.

A curious Brazilian species, for which I am indebted to Messrs. Loddiges. In habit it approaches E. equitans, but in the structure of both flowers and leaves it is widely different. The latter are covered with a glaucous bloom, imbricated, and more like inflated carinate bracts than true leaves. The flowers are greenish white, and offer some analogy with those of Physinga, but are in reality different from those of a genuine Epidendrum in nothing, except the neck of the ovary having an elevated semi-transparent blister near its middle. This blister is in fact the lower extremity of the cuniculus of the flower.

90. ODONTOGLŌSSŪM cordātūm.

O. *cordatum ;* scapo ascendente radicali, sepalis lineari-lanceolatis acuminatis petalis conformibus longioribus, labello cordato acuminato unguis lamellâ carnosâ apice bilobâ basi bicristatâ, columnâ pubescente clavatâ.

As yet only a small specimen of this has flowered with George Barker, Esq. of Birmingham, who imported it from Mexico; but I have no doubt that it will become much larger, and more abundantly furnished with flowers. The latter are very handsome; having the sepals and petals richly blotched with brown, upon a yellowish green ground, while the lip is white, with the crest at the base purplish, and the apex spotted and blotched with rich brown. It is very different from any of the species previously described.

91. EPIDÉNDRŪM lĭvĭdūm.

E. *lividum ;* pseudobulbis angustè ovalibus compressis diphyllis, foliis lineari-lanceolatis subundulatis obtusiusculis, scapo terminali paucifloro foliis breviore, ovario triquetro, sepalis erectis herbaceis linearibus mucronatis, petalis spathulatis obtusis æquilongis, labello libero lineari-oblongo obtuso crispo medio tomentoso serie triplici dentium elevatorum intermediâ majore, columnâ triquetrâ apice tricallosâ.

An obscure unattractive species, imported from Columbia by Messrs. Loddiges. The flowers are small, dull dingy purple, a little tessellated; their lip pale dirty yellow, with a few dull purple veins.

92. ONCĪDĬŪM confrăgōsūm.

O. *confragosum ;* pseudobulbosum; scapo simplici glaucescente apice nutante, sepalis ovatis acuminatis undulatis lateralibus semiconnatis, petalis majoribus oblongis undulatis acutis, labelli lobis lateralibus nanis recurvis intermedio unguiculato altè bilobo subangulato reniformi: disco maximo digitato confragoso, columnæ alis rotundatis denticulatis.

A native of Mexico, whence it was imported by George Barker, Esq. The flowers are of just the same colour as O. stramineum, for which the species might be mistaken upon a casual examination; the colour is a very delicate straw, faintly spotted with pale purple.

93. MICRŌSTЎLĪS excavātă.

M. *excavata ;* floribus corymbosis, scapo hexagono, petalis filiformibus reflexis, labello ovato carnoso basi acutissimè sagittato medio utrinque altè excavato apice tridentato dente intermediâ majore.

A green-flowered Orchidaceous plant, resembling M. ophioglossoides, but with a corymbose inflorescence. It was imported by Mr. Barker from Mexico.

94. DENDRŌBĬŪM stūpōsūm.

D. *stuposum ;* caule erecto tereti, foliis lineari-oblongis obliquè emarginatis, pedunculis bifloris, bracteis obtusis cucullatis, sepalis petalisque erectis lateralibus carinatis, labello oblongo cucullato obtuso sub apice calloso stuposo.

A native of India, whence it was obtained by his Grace the Duke of Devonshire; I have also received it from Messrs. Loddiges. It is an erect species, with the habit and general appearance of Dendrobium candidum; the flowers are of the same white colour, and but little smaller; the labellum, however, has a deep orange callus below its point, where it is thickly covered with a coarse tow-like hairiness.

95. MAXILLĀRĬĂ Boōthĭī.

M. *Boothii ;* rhizomate repente, pseudobulbis ovalibus compressis diphyllis, foliis angustis obtusis, flore terminali intra bracteas carinatas submembranaceas distantes basi incluso, sepalis petalisque conformibus erectis angustis acutis, labello oblongo obtuso margine membranaceo.

" This plant was collected in Guatemala, by George Ure Skinner, Esq., and introduced, in 1835, by Capt. Sutton, who added it to Sir Charles Lemon's collection at Carclew, where it flowered in May 1838.

" *Stem* creeping, and imbricated with small, thin and dry, ovate, acuminate, brown scales, of which the two outer ones are the largest and embrace the bulb. *Pseudo-bulbs* one or two-leaved, ovate, oblong, a little compressed and tapering towards both ends. *Leaves* bright green, smooth and shining, from four to six inches long, and from three-eighths to half an inch broad, thin and rigid, oblong-lanceolate, with a bluntish point. *Scape* one-flowered, about two inches high, thin and compressed, and having two large, thin, sheathing, acuminate bractes; it issues from the centre of an upright, keel-shaped, narrow leaf, which is imbricated at the base, and forms a sort of sheath that protects the flower until it is ready to expand. *Pedicels*, an inch long, dull green, marked with brownish spots. *Sepals* spreading, lanceolate, acute, pale greenish yellow. *Petals* of the same colour as the sepals, but conniving and about a third shorter, with a blunter point. *Labellum*, the length of the petals, slightly recurved, rounded at the point, and keeled below, with the outer edge very thin and more delicate than the rest. Towards the base are two fleshy processes of a deep yellow. *Column* about half as long as the petals, rounded above, and having the edge, immediately below the anthers, deep brown."

The species is new, and I have named it after Mr. Booth, to whom I am indebted for my knowledge of it.

101. LUĬSĬĂ alpīnă.

L. *alpina ;* foliis distichis coriaceis canaliculatis obliquè bilobis mucrone interjecto, sepalis oblongis obtusis lateralibus carinatis labello suppositis, petalis paulò angustioribus, labello oblongo concavo inappendiculato utrinque versus basin emarginato obsolete bilobo sub apice gibboso.

A very distinct species, with coriaceous distichous leaves, resembling those of an Aerides. Its sepals and petals are light green; the lip is strongly streaked internally with deep purple. It was collected for his Grace the Duke of Devonshire, by Mr. Gibson, at Nungklow, on the Khoseea hills, at an elevation of 4,000 feet above the sea. Snow frequently falls at this place in the cold season.

102. BOLBOPHȲLLŪM umbellātūm.

Lindl. Gen. et Sp. Orch. 56.

This curious plant has lately flowered at Chatsworth, having been sent to the Duke of Devonshire from the Botanical Garden at Calcutta. Its flowers are a dull dirty yellow, spotted with brown.

103. SACCOLĂBĬŪM densiflōrūm.

Lindl. Gen. et Sp. Orch. 220.

Among some Orchidaceæ, collected in Manilla by Mr. Cuming, this plant has made its appearance, having flowered with Messrs. Loddiges. It has small, pale, dull brownish yellow flowers, of a very fleshy consistence. The specimen that flowered was a small one; the Penang specimens brought home by Dr. Wallich were nearly two feet high.

113. PAXTŌNĬĂ rŏsĕă.

One of the most curious plants sent from Manilla by Mr. Cuming is this, which flowered in June in the collection of Messrs. Loddiges, and which will be shortly figured in this work. In the mean while the following short character will enable Botanists to identify it, and will secure the name to Mr. Paxton, whose claim to be permanently associated with Orchidaceæ will be readily admitted by all who know anything of the admirable cultivation of such plants at Chatsworth.

PAXTONIA (Nat. ord. *Orchidaceæ* §. *Malaxideæ.*) *Perianthium* 6-phyllum, petaloideum, patens, æquale. *Columna* libera, clavata, semiteres. *Pollinia* 8, angusta, clavata. *Stigma* verticale.—The leaves are long, narrow, and slightly plaited, and proceed from an oblong pseudo-bulb, which is marked with circular scars indicating whence they fell. The flowers are a purplish-lilac, rather larger than a shilling, and grow upon a stem about a foot high, in a somewhat corymbose raceme. They look like those of a *Thelymitra*, but the structure of their column is entirely different.

114. CATASĒTŪM atrātūm.

C. *atratum;* racemo decurvo, sepalis petalisque patentibus ovatis acutis, labello carnoso cucullato margine tenui pectinato apice rotundato reflexo crasso denticulato.

A curious dark-flowered species, obtained from Brazil by Messrs. Loddiges. It will soon be figured in this work.

115. ONCĪDĬŪM pulvinātūm.

O. *pulvinatum;* paniculâ ramosissimâ divaricatâ, sepalis obovatis lateralibus liberis, petalis conformibus acutis, labelli lobis subæqualibus intermedio bilobo undulato lateralibus crenatis rotundatis crispis, disco pulvinato villosissimo, columnæ alis rotundatis.

A very fine plant, resembling *O. divaricatum*, with a panicle eight or nine feet long, imported from Brazil by Richard Harrison, Esq., of Aighburgh, to whom a medal was awarded for his exhibition of it at a recent meeting of the Horticultural Society of London. A figure will appear in this work.

116. MAXILLĀRĬĂ vitellīnă.

M. *vitellina:* pseudobulbis ovatis obtusè angulatis monophyllis, foliis lanceolatis in petiolum canaliculatum angustatis, racemo cernuo radicali foliorum longitudine, labelli cuneati trilobi lobis lateralibus acutis anticè crenulatis intermedio bilobo rotundato cordato crenulato, tuberculo disci trilobo obtusissimo, ungue pubescente.

I am unwilling to delay noticing this pretty yellow Maxillaria till a figure can be published. It is a native of Brazil, whence it was imported by Messrs. Loddiges, and is remarkable for having a rich deep brown spot in the middle of its yellow lip.

119. SPIRĀNTHĒS diurĕtĭcă.

This little plant, inhabiting the mountainous parts of Chile, where it is common, has lately been received from Valparaiso by Capt. James Mangles, R.N. Its flowers are green and white, unattractive to the careless observer, but studded in the most beautiful manner with crystalline points, particularly at the apex of the lip. It is a greenhouse Orchidaceous plant, and flowers in August.

124. ONCĪDĬŪM hĭāns.

O. *hians;* sepalis petalisque æqualibus ovalibus obtusis leviter concavis, labello angusto auriculato medio contracto apice bilobo, callo disci bilobo utrinque dentato carnoso papilloso erecto columnâ parallelo eique longitudine æquali, alis columnæ carnosis acutis vix falcatis.

I have only seen flowers of this little species, which comes near to O. carinatum. Messrs. Rollissons received it from the vicinity of the gold mines in Brazil. It has small yellow and brown flowers, with an extraordinary appendage to the lip, erect, white, fleshy, as long as the column, parallel with that organ, and resembling the four fingers of the hand a little hollowed out and closed together. This is quite a new modification of structure.

125. VĀNDĂ lamellātă.

V. *lamellata;* foliis distichis coriaceis obliquè et acutè bidentatis, spicâ multiflorâ, sepalis petalisque obovatis obtusis undulatis inferioribus subincurvis majoribus, labello basi mammoso, limbo obcuneato retuso auriculato bilamellato pone apicem bituberculato.

A fine species of epiphytal Orchidaceæ, received by Messrs. Loddiges from Manilla. It has pale yellow flowers stained with red, and as large as those of Vanda Roxburghi. The lip with its two red elevated plates, and a pair of red tubercles just below the apex, distinguishes this with certainty. It has something the habit of *Vanda spathulata*, a common and handsome East Indian species which no one seems yet to have imported.

132. PHYSŎSĬPHŌN carinātŭs.

P. *carinatus;* folio oblongo angusto obtuso aut emarginato racemis binis breviore, calycis tubo triquetro angulis carinatis, labelli lobo intermedio serrato scabro.

A plant resembling Physosiphon Loddigesii in the colour of its flowers, and the form of the leaves, but differing in having a serrated labellum, whose surface at the point is broken up into little sharp teeth. It has been recently obtained in a live state from Mexico by George Barker, Esq.; but was originally met with by Schiede, growing upon the trunks of trees near Sosocola in Mexico, as I learn from a dried specimen (No. 8), for which I am obliged to Professor Schlechtendahl.

133. PLEUROTHĂLLĬS vittātă.

P. *vittata;* folio ovali crassissimo, caule tereti, spicâ distichâ folio multò breviore, floribus tomentosis, sepalis oblongis planiusculis intus lævibus superiore longiore et angustiore sub apice calloso lateralibus acutissimis falcatis semiconnatis, petalis obovatis acutis serratis, labello ovato carnoso obtuso lævi basi excavato biauriculato margine scabro, columnâ petalorum longitudine, clinandrio cucullato crenulato.

A Mexican species imported by Messrs. Loddiges, and one of the handsomest of this not handsome genus. The flowers are slightly stained with dull purple; the lower sepals are broad and spotted with deep purple; and the upper sepal is striped with the same colour. It is very like *P.*

aphthosa, from which it differs in the perfect smoothness of the sepals on their inside, in their form and proportion, and in the form of the petals.

134. DIĒNĬĂ cordātă.

D. *cordata;* folio solitario subrotundo cordato membranaceo-marginato, racemo tenui elongato, sepalis lateralibus ovatis obtusis intermedio lineari-oblongo, petalis linearibus, labello carnoso trilobato; laciniis lateralibus auriculæformibus intermedio ovato.

A native of Mexico, whence it was procured by Mr. Barker. It has a single cordate leaf, a slender spike of green small flowers, and narrow ovate pseudobulbs. It is a plant of no beauty, and differs from *D. Myurus* in the form of its lip, the number and form of its leaves, and in its long slender spike.

135. TRIGONĪDĬŪM Egertoniānūm.

Bateman in litt.

T. *Egertonianum;* pseudobulbis ovalibus compressis sulcatis diphyllis aggregatis, foliis ensiformibus pedunculis subæqualibus vel brevioribus, sepalis ovali-lanceolatis acutis (lateralibus reflexis), petalis lineari-lanceolatis acutiusculis apice callosis, labello trilobo medio calloso petalis triplo breviore.—*J. B.*

"A native of the Bay of Dulce in Honduras, where it was discovered by G. U. Skinner, Esq. to whom I am indebted for its possession. It is a very distinct species of a most singular genus, and I have, therefore, not hesitated to name it after Sir P. de M. Grey Egerton, Bart. It approaches nearest to T. obtusum, from which its acute petals, and narrow leaves (frequently fully a foot and a half long), and clustered pseudo-bulbs abundantly distinguish it. Its flowers are of a pale liver-colour, dashed and veined with brown, and from a resemblance which they are supposed to bear to a "Dragon's mouth," the plant has received that appellation from the inhabitants of Honduras. It is of the easiest cultivation."—*J. B.*

136. TRIGONĪDĬŪM acumināt̄um.

Bateman in litt.

T. *acuminatum;* pseudobulbis ovatis sulcatis monophyllis, foliis linearibus pedunculis longioribus, sepalis acuminatis apicibus recurvis, petalis ovali-lanceolatis mucronatis, labello trilobo petalis duplo breviore.—*J. B.*

"Found in Demerara by Mr. Colley and other collectors. It is a small though interesting species. The flowers are of a dull straw-colour externally, but are most elegantly pencilled on the inside with a rich brown. The petals in this, as in all the other species, have each a dark-coloured callosity at their apex, which is placed in such a manner near the entrance of the triangular cup which the sepals form, that the appearance of a pair of eyes peeping out is produced."—*J. B.*

138. CIRRHOPĔTĂLŪM cornūtūm.

C. *cornutum;* pseudobulbis ovatis angulatis scapo paulò brevioribus, sepalis lateralibus in cornu connatis superiore petalisque ovatis ciliatis, labello angusto triangulari suprà sulcato subtùs carinato.

A remarkable species of this genus, with the lateral sepals united above their base into a kind of horn. The leaves are six or eight inches long, and the flowers dull purple. It was found by Mr. Gibson at Nungclow, on the Khoseea hills, growing upon rocks, and flowered at Chatsworth in August last.

139. SACCOLĂBĬŪM calceolārĕ.

Gen. et Sp. Orch. p. 223.

This plant has flowered at Chatsworth, having been found by Mr. Gibson at Chirra, on the Khoseea hills, at an elevation of 400 feet, growing on trees. It has small yellow flowers, blotched with reddish brown.

142. POLYSTĂCHYĂ ramŭlōsă.

P. *ramulosa;* ebulbis, foliis . . . . . , scapo paniculato, floribus glaberrimis, sepalorum basi conicâ elongatâ, labello cuneato trifido apice carnoso margine involuto: lobis acutis intermedio minore, disco plano imberbi.

A native of Sierra Leone, whence it was imported by Messrs. Loddiges, who flowered it in September, 1838. It is a small green-flowered plant with the habit of Polystachya luteola, but more branched, and with branchlets at the base of its ramifications. The smooth flowers prolonged at the apex (that is to say, at the base of the sepals) into a long cone, and the wedge-shaped naked lip, distinguish this from the other species previously known. The apparatus connected with the pollen-masses is very remarkable in this plant; the masses are two waxy bodies obliquely divided half-way into two very unequal lobes, and adhering to a long white wedge-shaped separable process, which has all the appearance of the caudicula of a Vandeous genus, and which is moreover attached to a minute separable gland; a trace of this structure also exists in *P. luteola*, and is probably what Sir W. Hooker saw when making the drawing afterwards published in his Exotic Flora. At first sight a fact like this would appear either to weaken the value of the distinction by which Orchidaceous plants of the Vandeous are separated from those of the Malaxideous division, or to render it necessary to transfer Polystachya from the latter to the former. But upon an attentive examination of the structure of the process in question, it is seen that it is not a cartilaginous elastic strap, adhering to a hard gland with a well defined outline, but a collection of large cells, loosely cohering, very convex, and filled with air; while the representation of the gland is a small hemispherical succulent mass, to which the cellules adhere. This process may therefore be regarded as a mere modification of the shapeless viscid matter to which the pollen-masses of many Malaxideous genera are attached.

Mr. Loddiges, in sending the above plant, accompanied it with specimens of Polystachya luteola from the West Indies and Ceylon, with the remark that those plants were improperly considered as the same species. Upon a careful reconsideration of their structure I have come to the conclusion that this opinion is well founded, and that they are to be distinguished upon sufficient grounds. In fact the probability of a difference existing between the Western and Eastern plants had occurred to me when writing the first part of the Genera and Species of Orchideous Plants, but I had no materials which would enable me to point out in what the difference consisted, except in colour. Upon a comparison of the live plants with each other, it now appears that the Eastern plant is much taller, with long lateral branches, that its flowers are smaller and yellower, with a tinge of purple, while the Western plant has pale watery green flowers; that its lip is wider, with the middle lobe rounded, and the disk furnished with a smooth oblong callosity projecting from among the down, which otherwise covers it, while the other has the middle lobe more wedge-shaped and the disk completely buried in down; finally, that the spaces between the ribs of the ripe fruit are plainly reticulated in the species from Ceylon, but free from reticulations in that from the West Indies. The two plants may therefore be separated for the future by the following diagnoses.

143. POLYSTĂCHYĂ lutĕŏlă.

Hooker Exotic Flora, t. 105.

P. *luteola;* caulescens, foliis oblongo-lanceolatis, panicula racemosâ ramis abbreviatis, floribus glabris, sepalorum basi triangulari, labelli lobo medio subcuneato apiculato disco furfuraceo, capsulâ inter costas subaveniâ.= *Dendrobium polystachyon* Swartz.

The plant figured by Sir W. Hooker is certainly that from the West Indies; and there must have been some mistake in the Liverpool Garden, where it was said to have been received from Dr. Wallich from the East Indies. It does not appear among any of Dr. Wallich's collections, and probably was never in his possession.

144. POLYSTĂCHȲĂ zeylănĭcă.

P. *zeylanica;* caulescens, foliis oblongo-lanceolatis, paniculâ racemosâ ramis elongatis, floribus glabris, sepalorum basi triangulari, labelli lobo medio rotundato apiculato disco furfuraceo in medio nudo calloso, capsulâ inter costas reticulatâ.=*Dendrobium polystachyum* Thouars Orch. Afr. t. 85.

In both these plants the disk of the labellum is covered with a fine frost-like mealiness, which is removed by the least touch. This mealiness is a curious modification of the hairs found in other plants. When undisturbed it consists of threads with egg-shaped joints, which are filled with air; the surface of each joint is marked with wavy oblique striæ, and the interior uniformly contains a nucleus, to which there appears to belong a circulating apparatus of the same nature as that in the hairs of Tradescantia and other plants; I have not however succeeded in actually observing any circulation. But the slightest touch suffices to destroy the cohesion between the joints of these singular necklace-shaped bodies, so that when they are placed on the field of the microscope the latter appears as if covered with the eggs of some insect; it is only when they are removed from the labellum with great care that their real articulated structure, and their analogy with such hairs as those of Tradescantia, is made out.

145. BRYŎBĬŪM pubēscēns.

Under this name I long since distinguished a little green-flowered East Indian Orchidaceous plant which, though in many gardens, has never yet been published, except with a very short character in the Natural System of Botany, ed. 2. It has oval fleshy stems an inch long, closely covered by membranous scales, and terminated by about two narrow-oblong, fleshy, veinless, emarginate leaves. The flowers are a line and a half or two lines in diameter, green, membranous, and collected in small stalked heads very much longer than the leaves. The plant is naturally related to Phreatia and Eria, from both which it differs in having the lateral sepals equal at the base and not so prolonged, in conjunction with the foot of the column, as to resemble a spur. Artificially it will stand next Octomeria.

BRYOBIUM. Flores subvillosi. Sepala conniventia, conformia, lateralia basi æqualia. Petala angustiora et breviora, inter sepala reflexa. Labellum indivisum, inappendiculatum, basi constrictum. Columna nana. Pollinia Eriæ. ——Sp. 1. *B. pubescens.* Herba pusilla; caule carnoso vaginato unciali. Folia 2, carnosa, avenia, angustè oblonga, emarginata. Flores minuti, herbacei, in capitulum brevipedunculatum congesti, pilis stellatis rigidis obtusis aëre repletis laxè vestiti. Sepala ovata. Petala tenuiora et dimidio breviora, lineari-lanceolata, truncata. Labellum ovatum, basi lævissimè cordatum. ——Eriæ Phreatiæque affine, sepalis lateralibus basi æqualibus diversum.

146. EPIDĒNDRŪM dichŏtŏmūm.

E. *dichotomum;* fruticosum, caule fruticoso decumbente filiformi dichotomo, foliis angustè lanceolatis acutissimis corymbo terminali longioribus, sepalis lineari-lanceolatis, petalis conformibus angustioribus, labello cuniculato cordato obtusiusculo basi bicalloso.

A green-flowered species obtained from Demerara by Messrs. Loddiges, and found in the Organ Mountains of Brazil by Mr. Gardner, (No. 631.) It has no beauty, but is remarkable for its hard, wiry, forking stem.

147. ĔRĬĂ pūmĭlă. *Gen. et Sp. Orch. p.* 68.

This little plant has flowered with Messrs. Loddiges, who received it without name from the Botanic Garden, Calcutta. It has small capitate flowers, membranous, and slightly tinged with pink: the horns of the labellum, the column and anther being also pink. It was originally described in the above work from bad and broken dried specimens, and is stated to have the middle lobe of the labellum ovate; but in reality that part is two-lobed with an intermediate point, each lobe being ovate. The specific character should therefore be amended thus:

E. *pumila;* foliis angustè lanceolatis acuminatis caulibus teretibus erectis subæqualibus, spicis capitatis lateralibus subsessilibus, ovario villoso, sepalis petalisque ovatis acutis basi tantùm pilosis, labelli trilobi basi bilamellati pubescentis lobis lateralibus subulatis patentibus; intermedio quadrato bilobo apiculo interjecto.

148. CĂTTLĒYĂ bĭcŏlŏr.

Bot. Reg. fol. 1919 *in textu.* *Sertum Orchidaceum, t.* 5. *f.* 1.

This beautiful species has just flowered with Messrs. Loddiges. It agrees perfectly with the figure above quoted, and is very remarkable for its labellum wanting the side lobes, so that it is only curved downwards, and cannot wrap up the column as in the other species of the genus. The sepals and petals are a dull tawny olive green, the labellum is of the deep violet which we find in the centre of that of Cattleya labiata.

149. CATASĒTŪM Mīllērī. *Loddiges.*

Under this name, given in compliment to Dr. Miller, of H. M. ship Victory, a most zealous collector of rare plants, Mr. Loddiges has distinguished a Catasetum from Brazil, with a stem two feet high, and dull purple spotted flowers with a half green lip, but otherwise very like *C. semiapertum.*

150. CLEISŎSTŎMĂ rŏsĕă.

C. *rosea;* caule folioso, foliis angustè lanceolatis coriaceis acutis, corymbis paucifloris subsessilibus, sepalis lateralibus obliquè ovatis acutis supremo petalisque linearibus obtusissimis, labello carnoso lobo medio rotundato lateralibus crassissimis transversè truncatis, dente operiente lineari acuto.

This little plant has small pale straw-coloured flowers with a pink lip, and in habit resembles Sarcanthus rostratus. It was sent by Mr. Cuming to Messrs. Loddiges from Manilla.

153. SATȲRĬŪM cāndĭdūm.

S. *candidum;* foliis binis subrotundo-ovatis glabris, vaginis caulis utriculatis inflatis distantibus margine pellucidis lævibus, bracteis oblongo-lanceolatis acutis reflexis, sepalis linearibus obtusis patentibus, petalis conformibus minoribus ascendentibus apice recurvis, labello inflato obtuso apice inflexo dorso carinato, calcaribus pendulis ovario longioribus.

One of the terrestrial Orchidaceæ of the Cape of Good Hope, concerning which so little is as yet known in Europe. It was brought home by Sir John Herschel, with whom it flowered in Hanover Terrace, Regent's Park, in October; its flowers are pure white, and emit a most delightful aromatic fragrance. In many respects it is like *S. cucullatum,* which Sir John Herschel has also succeeded in flowering, especially in having the sheaths of the stem inflated, and so grown together at the edges as to form a kind of cup, capable of holding water. But it differs in the colour of the flowers, which are pure white, not green, in their scent which is aromatic and not fetid, and in their size, for they are larger; the edge of the cups of the stem is quite smooth and pellucid, not fringed with a coarse short hairiness; the labellum has a sharp-keeled back, is inflated, and the apex is reflexed, all which circumstances are at variance with *S. cucullatum;* finally the dorsal lip of the stigma of the latter species is short, linear, and emarginate, while in *S. candidum* it is short, round, and entire. These are not all the distinctions between these species, but they are abundantly sufficient to enable Botanists to recognize them; in a dried state they are easily confounded. The first knowledge I had of this species was from the collection of M. Drége, who found it near Gnadenthal and Riebokscasteel; I afterwards met with it in Sir W. Hooker's Herbarium, from Mr. Harvey. It appears from the specimens sent home by the latter Botanist, that its cups are occasionally enlarged into leaves, a circumstance which also happens in *S. cucullatum* itself.

154. SATȲRĬŪM papīllōsūm.

Illustrations of the Genera and Species of Orchidaceous Plants, t. 14. fructification.

Of this lovely plant I have also received fresh specimens, through the kindness of Sir John Herschel. The flowers are of a deep clear rose colour, melting into white, and richly spotted with purple in the throat; they smell like sweet-vernal grass. I know no Orchidaceous plant prettier than this.

It is particularly deserving of notice that the first of these species was planted *in the open ground* in May last; and that it was under such circumstances that the fine specimens I examined were produced. Upon this subject Sir John Herschel has favoured me with the following note.

"The specimen of S. candidum was flowered *in a box* of bog earth, openly exposed, except in heavy rains when it has been taken in; and in cold nights has been protected by mats. But there are also several specimens of the S. candidum now flowering in an open bed under a south wall, with no protection whatever. Of these, however, the spikes are not yet so fully developed. Some specimens of that very fine Satyrium, the "carneum," have also flowered (though very pale in colour) in garden pots of bog earth, exposed as in the case of the S. candidum; some have come to very handsome heads, though far inferior to what they do with a very little care in South Africa, where specimens have been obtained with 120 or more flowers in the spike.

"The Satyrium papillosum was brought over in a box of Cape soil in a growing state, and though generally exposed, in London, has been occasionally brought in doors in cold nights; and, since the flower-stalks have become fully developed, has been entirely kept in the house, and sometimes placed near a fire.

"Satyrium chrysostachyum (a fine orange-flowered kind) similarly imported, has grown well and gone through its course, but without throwing up a single flower-stalk. It is now dormant, but the new tubers are satisfactorily formed, and promise well for next year."

155. SATȲRĬŪM cărnĕūm.

Brown in hort. Kew. ed. 2. 5. 196.

This is the very fine species alluded to by Sir John Herschel in the previous note. It has been well figured in the *Botanical Magazine, t.* 1512. In herbaria it is very rare. I have only seen one wild specimen, collected on sandy hills about Groenevalei, at the Cape of Good Hope, by Drége. It bears large, oblong, dense spikes of whitish flowers tinged with pink, and is in all respects a beautiful plant. It is not a little singular that so conspicuous a species should have been overlooked by all the writers upon the South African Flora.

156. DENDRŌBĬŪM denŭdāns.

Don Prodr. fl. nep. 34. *Gen. et Sp. Orch. p.* 84.

This species was received by His Grace the Duke of Devonshire upon Mr. Gibson's return from his Indian mission; and having been subsequently imparted to others, has flowered in several collections. The finest specimens I have seen were sent by Mr. Bateman to Messrs. Loddiges. The stems are erect, about six inches high, not very unlike a small state of *Dendrobium nobile*, and are covered by a profusion of nodding racemes of rather small green and white flowers.

157. CŒLŎGȲNĒ Wallichiānă.

Lindl. Gen. et Sp. Orch. 43.

At last a plant of the beautiful division of Cœlogyne, called Pleione by Professor Don, has appeared in the collection at Chatsworth, whither it had been brought by Mr. John Gibson. It has large, handsome, scentless, deep rose-coloured flowers, growing close to the ground, from within some hard tuberculated sheaths, proceeding from the base of depressed flask-shaped green and purple speckled pseudobulbs. In its native country this and allied species cover the ground with a pavement of their curious stems, which wither up in the dry season, but change into a brilliant carpet of rosy flowers upon the approach of rain. A figure is prepared for this work.

161. MAXILLĀRĬĂ Cōllĕyī.

Bateman in litt.

M. *Colleyi;* pseudobulbis sphæricis, pedunculis radicalibus multifloris, labello postico oblongo apice triangulari obtuso medio 1-dentato.

This species, although no doubt new, is very near *M. squalens*, from which it differs in the form of the pseudobulbs and labellum. Its dingy flowers have a disagreeable smell, like that of an over-ripe melon.

163. EPIDĒNDRŪM calamārĭŭm.

E. *calamarium;* pseudobulbis teretibus calamiformibus diphyllis, foliis linearibus planis abruptè acutis, racemo simplici erecto, bracteis squamiformibus acuminatis, sepalis patentissimis linearibus acutis apice recurvis, petalis angustioribus subspathulatis, labello postico subrotundo acuto convexo.

A Brazilian species, allied to *E. fragrans*, imported by Messrs. Loddiges, with whom it flowered in October. It has pale, whole-coloured, yellowish green flowers, with five small violet-coloured spots at the base of the lip. It is a plant of little beauty, and no fragrance.

164. CATASĒTŪM porĭfĕrūm.

C. *poriferum;* labello plano cordato-ovato truncato crenato, lobo apicis elevato carnoso ovato medio nectarifero, dente baseos ovato incurvo, columnâ cirrhosâ.

This is a remarkable species, sent to Messrs. Loddiges from Demerara by Mr. Schomburgk. In its manner of growth and general appearance it agrees with Catasetum deltoideum, like which its flowers are richly spotted with deep purple broken fasciæ, closely arranged upon a clear green ground. The difference between these two species resides in the lip, which, in the present plant, is dull green, nearly flat, deeply cordate, truncate at the apex, with an obscurely crenated margin; at its base is an elevated yellow-tipped broad tooth, and at its apex an ovate, yellow, fleshy, elevated lobe, in the middle of which is a pore secreting honey. The sepals are oblong, the petals narrow-lanceolate, the column just as in *C. deltoideum.*

165. PLEUROTHĀLLĬS muscoīdĕă.

P. *muscoidea;* cæspitosa, acaulis, folio ovali biconvexo, pedunculo setaceo bifloro, perianthio explanato ringente, sepalis lineari-lanceolatis lateralibus basi imâ connatis, petalis multò minoribus conformibus, labello oblongo-lineari obtuso canaliculato, polliniis 2 pyriformibus connatis.

This is the most tiny Orchidaceous plant yet discovered. It has no stem; the leaves are two lines and a half long; the peduncle as fine as a hair, and about four lines long. The flower is dull purple, with a pale orange-coloured margin to the sepals and petals; the lip is linear, obtuse, richly stained with purple, dull orange along the middle and at the edge, very slightly cordate, and contracted for a short space in the middle. For a specimen I am indebted to Messrs. Loddiges.

166. NOTȲLĬĂ punctātă. *Gen. et Sp. Orch.* 192.

N. *punctata;* labello unguiculato cordato-ovato obtuso basi calloso margine reflexo, sepalis petalisque obtusis; rachi flexuosâ.

I believe the original species to which this name belongs is lost to our gardens, and its place has been taken by other species, resembling it enough to be mistaken for it. At least I must confess that I had regarded specimens sent me from time to time by Mr. Bateman, Mr. Barker, Mr. Loddiges, and others, as mere varieties of Notylia punctata. I am however now satisfied, from the examination of specimens placed in my hands by Mr. Loddiges, that the supposed varieties constitute at least five well marked species; all agreeing in habit, but differing in the form of their labellum, &c. in the size of their flowers, in smell, and in the length of their racemes. I hope the specific characters now given will prevent confusion in future.

The true *Notylia punctata* has whiter flowers than any of the others, with a short flexuose raceme, and obtuse sepals, petals, and labellum, at the base of the latter of which is an elevated callus.

167. NOTŸLĪĂ incurvă.

N. *incurva;* labello unguiculato cordato-ovato acuminato sub apice carinato basi ecalloso margine reflexo, sepalis lateralibus apice rectis petalisque lanceolatis acutis columnâ recurvâ.

Of this the flowers are larger than any of the others, a pale straw colour, with two or three yellow spots near the base of each petal; the divisions are lanceolate and acute, not acuminate; the ends of the lateral sepals are straight, not reflexed; and the column is rather abruptly curved inwards towards the dorsal sepal. Messrs. Loddiges obtained it from Trinidad.

168. NOTŸLĪĂ Barkerī.

N. *Barkeri;* labello unguiculato cordato-ovato basi calloso margine subreflexo, sepalis lateralibus apice reflexis petalisque lineari-lanceolatis acutis, columnâ recurvâ.

Very like the last, but the flowers are smaller and yellower, the labellum is not acute, and it has a distinct elevated callosity at the base. Mexico, Mr. Barker, 1837.

169. NOTŸLĪĂ tĕnŭĭs.

N. *tenuis;* labello unguiculato cordato-ovato setaceo-acuminato basi ecalloso margine reflexo, sepalo supremo falcato inflexo lateralibus apice revolutis petalisque linearibus acuminatis, columnâ rectâ.

This species is readily known by its very narrow acuminate sepals and petals, erect column, and acuminate lip, which is not carinate at the point. Its flowers are smaller than those of N. incurva. Native of Demerara, Messrs. Loddiges.

170. NOTŸLĪĂ micrānthă.

N. *micrantha;* labello subsessili plano ovato acuto basi calloso, sepalis petalisque linearibus acutiusculis, columnâ rectâ.

The flowers of this are not more than half the size of the smallest of the others; they are pale green with a yellowish lip, and the petals have no spots. Obtained from Demerara by Messrs. Loddiges.

171. CŒLŎGȲNĒ ovālĭs.

C. *ovalis;* pseudobulbis fusiformi-ovalibus striolatis, foliis geminis ovali-lanceolatis acutis spicâ sublongioribus, pedunculo basi vaginato sub-4-floro, bracteis concavis cartilagineis floribus longioribus, petalis linearibus reflexis, labelli trilobi lobis lateralibus in fronte pectinatis intermedio ovato emarginato pilis fuscis villoso et fimbriato, lamellis 2 elevatis crispis, columnâ apice subintegrâ.

In arranging the dried specimens in Dr. Wallich's Indian herbarium some years since, I met with a plant, from Nepal and Kamaon, without flowers, which I took for *C. fimbriata*, and under that name it was eventually catalogued. Lately however a plant of this species, sent by Dr. Wallich to Messrs. Loddiges, has flowered, and proves, although very near *C. fimbriata*, to be a different species. Its general appearance is the same; but the pseudo-bulbs are narrow and oval, not roundish-oblong; delicately striated, not covered with a thick dull veinless cuticle. The flower is twice the size, of nearly the same colour, but the margin of the middle lobe is more decidedly shaggy with brown hairs, and the two elevated lamellæ which pass along its middle from end to end are crisp, not straight. An opportunity having occurred of examining fresh flowers of *C. fimbriata* at the same time, it enabled me to amend the specific character of that species, as follows:

172. CŒLŎGȲNĒ fimbriătă.

Gen. et Sp. Orch. 41.

C. *fimbriata;* pseudobulbis subrotundo-oblongis estriatis, foliis geminis ovali-lanceolatis undulatis acutis spicâ longioribus, pedunculo basi vaginato paucifloro bracteis concavis cartilagineis deciduis floribus longioribus, petalis linearibus reflexis, labelli trilobi lobis lateralibus in fronte denticulatis intermedio ovato pilis fuscis fimbriato, lamellis 2 elevatis rectis, columnâ apice denticulatâ.

173. MAXILLĀRĬĂ porrēctă.

M. *porrecta;* foliis oblongo-lanceolatis trinerviis, pedunculo erecto unifloro vaginato, sepalis linearibus apice concavis obtusis lateralibus incurvis, petalis conformibus, labello angusto apice trilobo medio calloso pubescente: lobo medio convexo rotundato margine crispo revoluto.

An uninteresting species, obtained by Messrs. Loddiges from Rio Janeiro. The flowers are about the size of those of *Maxillaria picta*, of a pale buff, with the sepals and petals tipped with dull red. The labellum is the same colour on the outside below the point.

174. MAXILLĀRĬĂ macrophȳllă.

Pöppig et Endl. nova genera et species pl. vol. 1. *t.* 64.

M. *macrophylla;* foliis oblongo-lanceolatis plicatis scapo unifloro ascendente laxè vaginato duplò longioribus, bracteâ herbaceâ cucullatâ acutâ ovarii longitudine, sepalis oblongis undulatis patentibus apice recurvis basi intùs pilosis, petalis erectis columnâ longioribus oblongis carnosis apice recurvis margine postico versus apicem sublobato, labello breviore oblongo concavo apice trilobo: laciniâ intermediâ subrotundâ serratâ, appendice linguæformi concavo adnato inter lacinias laterales rotundatas incurvas, antherâ villosâ.

Very like *M. Deppii*, but much larger. The sepals are green outside, and dull purplish brown inside; the petals are very pale straw-colour; the labellum is slightly sprinkled with crimson dots. The flowers are about three inches across, and have a disagreeable smell, resembling apples beginning to ferment. Messrs. Loddiges imported it from Columbia; it was originally found by Pöppig in dry thickets in the transandine parts of Peru, in the district of Chihuamccala, near Cuchero, always growing in the ground, and never upon trees.

Very nearly related to this is a species, inhabiting the same country, but not yet introduced, of which I have specimens gathered by Mr. Mathews (No. 1026), the character of which is the following:

175. MAXILLARIA costata.

M. *costata;* foliis lanceolatis plicatis scapo unifloro ascendente laxè vaginato longioribus, bracteâ herbaceâ cucullatâ acutâ ovario multò longiore, sepalis petalisque M. macrophyllæ similibus, labello trilobo concavo laciniâ intermediâ rotundatâ serratâ, appendice carnoso emarginato adnato 5-costato inter lacinias laterales, antherâ glabrâ.

176. MORMŌDĒS pardīnă. *Bateman in litt.*

M. *pardina;* pseudobulbis turbinatis foliis strictis 4-plo brevioribus; racemo nutante multifloro foliis breviore, sepalis petalisque subæqualibus ovato-lanceolatis acutis conniventibus, labelli trilobi lobis lateralibus acutis decurvis intermedio elongato acuminato. *Orchid. Mex. et Guatemala, tab.* 14.

"A fine new species of this remarkable genus, discovered by Baron Karwinski in Oaxaca, and by him communicated some years since to my collection at Knypersley, where it flowered in July last. Its flowers are of a primrose colour, spotted with reddish purple in every part, and they exhale a most delightful perfume. Its habit is very robust, being at least three times as large as that of the old M. atropurpurea.

"Shortly after M. pardina flowered with me, I received from Mr. Barker specimens of a remarkable variety of it, the flowers of which were entirely self-coloured." *Note from Mr. Bateman.*

177. BIFRENĀRĬĂ? longicŏrnĭs.

B. *longicornis;* pseudobulbis elongatis tetragonis, foliis oblongo-lanceolatis subplicatis nitidis, racemo laxo multifloro, sepalis lateralibus ovatis acutis basi connatis in calcar longum clavatum productis, petalis ovatis acutis, labello longè unguiculato spathulato apice trilobo laciniis rotundatis medio pubescente disco elevato calloso, polliniis sphæricis subsessilibus glandulis duabus discretis.

If this plant is really to be referred to Bifrenaria, it will be necessary to modify the character of that genus very considerably; for the lateral sepals are extended into a long slender clavate spur, and there are two glands as well as two caudiculæ to the pollen-masses. Its relationship to *B. aurantiaca* is however such as to make me unwilling to separate it at present. The flowers are orange spotted with brown, and in a raceme very like that of the species just mentioned. Messrs. Loddiges imported it from Demerara.

178. TRICHOCĒNTRŌN iridifŏlĭum.

G. Loddiges in litt.

T. *iridifolium*; foliis distichis lineari-lanceolatis carnosis, labello ovali obtuso indiviso basi bilamellato, columnæ alis parvis obtusis integris.

A small species, with pale yellow flowers, having the lip delicately streaked with darker yellow. It was imported by Messrs. Loddiges from Demerara.

179. ÆTHĒRĬĂ occŭltă.

Goodyera occulta, *Thouars orch. afr. t.* 28. Platylepis goodyeroides, *Ach. Rich.*

A specimen of this plant, obtained from the Mauritius, has flowered with Messrs. Loddiges. It proves to belong to the genus Ætheria of Blume: that is to the Goodyeras, with the lip and the column united to each other by their edges. The plant is about a foot high, with a dense oval spike of membranous downy bracts, from among which the small white and green flowers are just protruded.

180. LĬPĂRĬS pendŭlă.

L. *pendula*; pseudobulbis elongatis compressis diphyllis, racemo terminali longissimo pendulo, sepalis reflexis obtusis lateralibus oblongis intermedio antico angustiore et longiore, petalis linearibus reflexis, labello postico ovato concavo basi 2-calloso.

A native of the continent of India, whence it was obtained by Messrs. Loddiges. The flowers are small, green, and arranged in a pendulous raceme full a foot long. All the parts of the flower abound in compound 5-threaded spiral vessels, and raphides lodged in cubical parcels in the inside of cells larger and more transparent than those surrounding them. The latter give the parts the appearance of having transparent dots. The species ranges near *L. longipes.*

181. IONŌPSĬS tĕrĕs.

I. *teres*; foliis teretibus, scapo simplici apice paucifloro, sepalis acutis lateralibus semiconnatis, labello obovato-lanceolato apice undulato quasi trilobo medio bilamellato.

A curious little plant, with delicate lilac-striped flowers. Imported from Demerara by Messrs. Loddiges.

182. PLEUROTHĂLLĬS stenopĕtălă.

G. Lodd. in litt.

P. *stenopetala*; folio oblongo apice rotundato basi in petiolum planum angustato racemo multò breviore caule longiore, bracteis membranaceis cucullatis, sepalis acuminatissimis linearibus intus pubescentibus omnibus ferè liberis, petalis nanis obtusis dorso carinatis labelloque oblongo conduplicato glabris.

A native of Brazil, whence it was obtained by Messrs. Loddiges. The flowers are pretty, although of a very pale green, and they have a powerful but agreeable smell of balsam. The midrib of the petals and the middle of the back of the labellum are crimson. It is very near *P. sclerophylla.*

183. BOLBOPHȲLLŪM cŭprēŭm.

B. *cupreum*; folio angusto solitario scapo longiore, racemo brevi oblongo nutante, petalis setaceo-acuminatis serrulatis, labello ovato dente utrinque setaceo, columnæ angulis aristatis.

A native of Manilla, whence Messrs. Loddiges received it from Mr. Cuming. The flowers are copper-coloured, and have a smell extremely like that of Valerian root.

EDWARDS'S

BOTANICAL REGISTER:

OR,

ORNAMENTAL FLOWER-GARDEN AND SHRUBBERY:

CONSISTING OF

COLOURED FIGURES OF PLANTS AND SHRUBS,

CULTIVATED IN BRITISH GARDENS;

ACCOMPANIED BY THEIR

History, Best Method of Treatment in Cultivation, Propagation, &c.

AND

MONTHLY CHRONICLE

OF

BOTANICAL AND HORTICULTURAL NEWS.

CONTINUED

BY JOHN LINDLEY, Ph. D. F.R.S. AND L.S.

PROFESSOR OF BOTANY IN UNIVERSITY COLLEGE, LONDON,
AND THE ROYAL INSTITUTION OF GREAT BRITAIN,
VICE-SECRETARY OF THE HORTICULTURAL SOCIETY,
&c. &c. &c.

1839.

OR VOL. XXV. OF THE ENTIRE WORK.
OR VOL. XII. OF THE NEW SERIES.

——viret semper——nec fronde caducâ
Carpitur.

LONDON:
JAMES RIDGWAY, PICCADILLY.

M.DCCC.XXXIX.

1

STANHŌPĒĂ tigrīnă.

Tiger-flowered Stanhopea.

GYNANDRIA *MONANDRIA.*

Nat. ord. ORCHIDACEÆ, § VANDEÆ.
STANHOPEA. Bot. Reg. fol. 1529.

S. *tigrina*; hypochilio subrotundo intus lamellis glandulosis radiato, metachilii cornubus falcatis porrectis epichilii tridentati longitudine, sepalis lateralibus maximis subrotundo-oblongis petalis multò latioribus.
S. tigrina. *Bateman Orchid. Mex. et Guatem. t.* 7.

The species of Stanhopea are so much alike, except in their flowers, that it is rarely necessary to introduce their organs of vegetation into the description of them. It is in the flowers that their differences are apparent, and especially in the labellum, if colour is disregarded.

The present beautiful species is characterised by having the epichilium shallowly 3-lobed, in which respect it corresponds with no other hitherto discovered, except S. *saccata*, which is extremely different. The inner surface of the hypochilium will also be found very remarkable, being broken up into glandular lamellæ, which radiate from the base of a kind of ovate tooth which is itself directed towards the cavity they occupy. Fig. 1. represents this structure.

The flowers of S. *tigrina* are larger and handsomer than those of any other known species, even exceeding those of the magnificent S. *Devoniensis*; this is sufficiently apparent from the annexed figure, in which nevertheless the colours are by no means so brilliant as in the plate of this plant in Mr. Bateman's magnificent work on the Orchidaceæ of Mexico and Guatemala.

For the opportunity of publishing it I have to thank Messrs. Rollissons of Tooting, with whom it flowered in August last. Mr. Bateman states that it was originally imported from the neighbourhood of Xalapa by Messrs. Lowe and Co.; and that it is among the easiest of the genus to cultivate.

The fragrance of the flowers is very peculiar, resembling a mixture of Melon and Vanilla.

If this plant is cultivated in a pot, it must be raised considerably above the level of the rim, not only to prevent its suffering from too much water, but also that its flower, which is pendulous, may be seen with advantage. The best method is to hang it up in a basket. The soil should be the same for this as for other orchidaceous plants, but when put into a basket it is found useful to place a little moss (sphagnum) round the soil; this keeps it moist, and the roots seem to thrive in it. The only other circumstance necessary to be noticed is, that it must be kept perfectly dry when not in a growing state; it will be found to push much more vigorously when this is attended to.

5

BRASAVŌLĂ Martiānă.

Dr. Von Martius' Brasavola.

GYNANDRIA *MONANDRIA.*

Nat. ord. ORCHIDACEÆ, § EPIDENDREÆ.

BRASAVOLA. *Bot. Register, fol.* 1465.

B. *Martiana*; labello ovali (aut ovato) acuminato ciliato-dentato sessili; petalis sepalisque lineari-lanceolatis acuminatis longioribus, clinandrio cucullato inciso. *Bot. Reg. fol.* 1914. *in textu.*

Folia *longa, teretia, suprà sulcis tribus exarata, racemo subcorymboso duplò longiora.* Pedunculi *læves, teretes, sepalis longiores, pone basin fusco-purpureo tincta.* Sepala *et* petala *lineari-lanceolata, subæqualia, patula.* Labellum *ovatum aut ovale, acuminatum, fimbriatum, basi luteum; ungue brevi columnæ appresso.* Columna *apice cucullata, inflexa, fimbriata.*

This very distinct species of a small but interesting genus, imported from Berbice by Messrs. Loddiges, was originally discovered by Dr. Von Martius on the banks of the Rio Negro in Brazil, and it was from dried specimens in his herbarium that I first described it.

B. *cucullata* and B. *amazonica* are the only other species yet known with a fringed labellum; the former has much larger flowers, and a lip of an entirely different form; the latter has a one-sided raceme, and a labellum contracted in the middle so as to be distinctly divided into a hypochilium and epichilium.

All the Brasavolas yet described by Botanists now exist in this country, with the exception of B. *subulifolia*, a fine species inhabiting Nevis, with very slender subulate leaves, and the B. *amazonica* above named. The former might easily be procured; the latter is beyond the reach of ordinary travellers, occupying the branches of trees surrounding lake "Egen," one of the offsprings of the Amazons.

Fig. 1. represents the column, seen in front, with the fringed clinandrium or *anther-bed.*

This genus seems to delight in a rough and stony soil, not too retentive of moisture. This circumstance should therefore be kept in view when the species are potted or shifted. The soil should consist of rough peat, well mixed with broken bricks or small stones, and the pots must be well drained at the bottom. They do not seem to require so much water as is commonly given to most plants belonging to this order, but in other circumstances the treatment should be the same. Some cultivators prefer tying their plants to pieces of wood, and suspending them from the roof of the stove, but upon the whole they can hardly be said to succeed so well under that treatment as in pots, particularly if they have plenty of pot-room. All the species are propagated by division of the rhizoma.

8

MAXILLĀRĬĂ tenuifŏlĭă.

Slender-leaved Maxillaria.

GYNANDRIA *MONANDRIA.*

Nat. ord. ORCHIDACEÆ, § VANDEÆ.

MAXILLARIA. *Bot. Register, fol.* 897.

M. *tenuifolia*; caulescens, pseudobulbis ovato-oblongis compressis monophyllis squamis longioribus, foliis lineari-lanceolatis acutis recurvis, pedunculis axillaribus solitariis basi squamatis, ovario denudato arcuato, floribus cernuis, sepalis ovato-lanceolatis margine revolutis subæqualibus reflexis lateralibus basi subæqualibus, petalis ovatis obtusis conniventibus, labello oblongo indiviso apice ovato reflexo infra apicem utrinque contracto, callo disci oblongo integerrimo. *Bot. Reg. sub folio,* 1986.

A native of Mexico, in the vicinity of Vera Cruz, where it was found by Mr. Theodore Hartweg, an excellent naturalist, employed by the Horticultural Society of London, in Mexico, upon a mission which would have already produced most important additions to our gardens, had not his collections been unfortunately detained at Tampico in consequence of the French blockade of that port.

The species inhabits trees in Mexico, and probably is local, as it does not appear in any of the collections brought from the interior of the country. It belongs to the first or axilliforal section of the genus, the distinctive character of which is to have caulescent stems, covered with pseudo-bulbs, and having the flowers appearing from the axils of scales covering the stem. These, which are the most genuine form of Maxillaria, are the least beautiful part of the genus, and constitute a perfectly natural group, at first sight very different from the kinds which, like *M. aromatica*, have naked pseudo-bulbs rising immediately from the surface of the earth. But, upon comparing the two sections with each other, it will be found that the only essential difference between them consists in the one having erect and the other prostrate rhizomata; the scales, or imperfect leaves, of the axilliflorous section being more developed than in the scapigerous species, in consequence of their being more exposed to light. Perhaps the genus should be limited to the species which form the two sections now mentioned, the spathaceous species being excluded, and such as *M. Warreana* and *costata* being either formed into a new genus or referred to *Peristeria*, whose character would then require to be modified; perhaps also such genera as *Bifrenaria* and *Dicrypta* would be better reduced to Maxillaria, with which, if newly limited, they would agree in habit. But these are points upon which it will be more easy to decide when a larger number of species shall have been correctly studied.

The present species is very pretty when in flower, and, from its freshness and greenness when in leaf only, is well worth cultivation, especially as it is one of the easiest to manage. It succeeds in a warm damp stove, in a pot with a block of wood thrust into the soil, and the long branching rhizoma tied to it. It grows almost equally well when tied to a wooden block, and suspended from the rafters of the stove. It bears without injury a quantity of water at its roots, and must also be freely syringed overhead. Amongst orchidaceous plants, none are more easily multiplied, as it throws out numerous pseudo-bulbs and roots, which, if taken carefully off, and subjected to the above treatment, will soon become vigorous growing plants.

11

EPIDĒNDRŪM variegātum.

Variegated Epidendrum.

GYNANDRIA *MONANDRIA.*

Nat. ord. ORCHIDACEÆ, § EPIDENDREÆ.
EPIDENDRUM. Bot. Reg. fol. 17.

E. *variegatum ;* pseudobulbis oblongis compressis 2-3-phyllis, foliis oblongo-lanceolatis obtusis, racemo simplici terminali, sepalis petalisque obovatis obtusis coriaceis, labello postico subrotundo acuto : callo baseos concavo emarginato dente columnæ posticâ obtusâ.
E. variegatum. *Bot. Mag. t.* 3151.
E. coriaceum. *Id. t.* 3595. a variety.

A native of Brazil, whence it has long since been imported, so that it now common in collections of Orchidaceæ, where it is valued for the delicious fragrance of its flowers, resembling nothing so much as Lily of the Valley. The figure now published was made many years ago in the collection of Messrs. Loddiges.

In many respects it agrees with *E. lancifolium* a Mexican, and *E. crassilabium* a Peruvian species ; but it differs from both in having the fleshy tooth, placed at the back of the anther, quite entire, blunt and rounded, not to mention other marks of distinction. Like *E. fragrans* it is very subject to variation, as must be obvious to any one who has remarked the peculiarities of the imported plants, which from time to time flower in collections, scarcely any two of which are exactly alike. Of these one is figured in the Botanical Magazine under the name of *E. coriaceum.*

In general the sepals and petals are marked with distinct purple blotches, on a pale green ground ; but sometimes they are richly dotted, and occasionally are cream-coloured, with only a few specks of purple. I am also disposed to regard as another variety a plant found by M. Descourtilz near Ilha Grande in Brazil, on fallen trees, in the midst of sandy plains, fully exposed to the sun. In this the leaves and pseudo-bulbs are very narrow, and the flowers are a pale dull dirty yellow, with narrow linear-lanceolate segments. It forms t. 67 of Baron Delessert's unpublished figures of Brazilian Orchidaceæ.

12

MAXILLĀRĬĂ vitellīnă.

Yellow racemose Maxillaria.

GYNANDRIA *MONANDRIA.*

Nat. ord. ORCHIDACEÆ, § VANDEÆ.
MAXILLARIA. Fl. Peruv.

M. *vitellina ;* pseudobulbis ovatis obtusè angulatis monophyllis, foliis lanceolatis in petiolum canaliculatum angustatis, racemo cernuo radicali foliorum longitudine, labelli cuneati trilobi lobis lateralibus acutis anticè crenulatis intermedio bilobo rotundato cordato crenulato, tuberculo disci trilobo obtusissimo, ungue pubescente. *Bot. Reg.* 1838. *misc. no.* 116.

A Brazilian epiphyte, allied to *M. racemosa,* which differs in having an undivided labellum, a hairy column, and in many other circumstances.

M. aureo-fulva, another beautiful plant related to this, and well figured in the Botanical Magazine, t. 3269, has an acuminate even lip, little differing in form from the sepals and petals. It is the " Epidendre Limodore " of Descourtilz, and was found by that traveller in great abundance upon fallen trees encumbering the sandy plain through which the great public road passes from Bananal to Ilha Grande.

The exact locality of this is unknown. It was imported by Messrs. Loddiges, with whom it flowered in June, 1838.

It requires the temperature and humidity of the moist stove. After it has perfected its pseudo-bulbs for the season, it should be kept perfectly dry for a considerable length of time ; and, if convenient, removed to a cooler house. This will make it grow and flower freely when it is brought back to the moist stove. In all other respects, its treatment should be the same as is practised with other tropical plants of this very extensive order.

Any of the back pseudo-bulbs, such as are represented in the plate, by being taken off will form plants, but the front ones are by far the best.

This genus is now so extensive, that it is difficult even for those most familiar with the species to avoid errors in publishing what are supposed to be new ones. I therefore may be permitted to observe, not for the purpose of criticizing, but for the sake of preventing, if possible, the inconvenience of multiplying names, that the *M. Henchmanni* of the Bot. Magazine, t. 3614, is one of the forms of *M. variabilis. M. pumila* of the same work, t. 3613, is so very near *M. uncata,* that although from the figure it seems different, it is desirable to re-examine it, and I shall be glad to receive it from some of my correspondents.

14

* HŪNTLĔYĂ Meleāgrĭs.

Speckled Huntleya.

GYNANDRIA *MONANDRIA.*

Nat. ord. ORCHIDACEÆ, § VANDEÆ.
HUNTLEYA. Bot. Reg. fol. 1991 *in textu.*

H. *Meleagris ;* sepalis petalisque ovatis acuminatis tessellatis, labello subconformi unguiculato concavo cristâ baseos fimbriatâ, columnæ cucullo crenato.
Huntleya Meleagris. *Bot. Reg. l. c.* 1838. *misc. no.* 20.

This is at present one of the rarest of the epiphytes in cultivation, the only specimen I have seen being that now figured, which flowered with Messrs. Rollissons in July, 1838. Its blossoms are much yellower and less tessellated with purple than in the Brazilian drawing from which the species was first described, and it is not improbable that it will be found to vary in this respect. The whole surface of the flowers had quite the appearance of being glazed.

The following is M. Descourtilz's account of the plant, a little reduced from his manuscript in M. Delessert's copy.

Rootstock as thick as the little finger, green, cylindrical, with white rootlets on the under-side. *Leaves* alternate, in two opposite rows, forming a very much compressed fan; above they are bright green and smooth, beneath they are bluish green, with paler and projecting longitudinal veins. These leaves are a foot or more long, and about an inch wide; from the axil of the lowest of them rises a cylindrical pale green peduncle, with two opposite bracts near the middle.

The *flower* is large, terminal, solitary, having five petals, broad at the base, with a white claw, and a claret-coloured ground on the inside, which is sometimes speckled with greenish pink, and always marked by longitudinal lines connected by other transverse ones, which thus form numerous elevations, and make the flower look like a draught-board. The two lower sepals have their inner edge at the base rolled inwards like a horn. The *labellum* is triangular, tongue-shaped, of a pure ivory white, bordered with deep purple, and nerved with a deeper tint. Its edges are turned downwards, and it is attached to the base of the column by a narrow white claw. At the origin of the claw is a crescent-shaped plate, hollowed out at its upper edge, and fringed with long stiff white hairs, which furrow downwards all the part that supports them.

This charming plant is found in gloomy damp woods on the banks of the Rio de Pirapitinga, in the district of Bananal. It is scentless, and flowers in June.

Mr. Rollisson assures me that he received his specimen from the same country as that which produced the *Zygopetalon cochleare*, figured in this work, plate 1857; if so, I must have been misinformed as to Trinidad being the native country of the latter. I am the more disposed to believe that such was really the fact, because I find a drawing, of what is apparently a luxuriant specimen of Z. cochleare, among M. Descourtilz's collection, gathered in Brazil, on the high mountains separating the province of St. Paul's from that of Minas Geraes, and known by the name of the Mantiqueiras, where it flowers in the hottest season.

It thrives very well in the orchideous-house at Tooting, where the atmosphere is kept saturated with moisture, and the temperature is in winter from 60° to 70° Fahr., and in summer from 70° to 90°. The house is of course well shaded from the bright rays of the summer's sun. Like other kinds of orchidaceous plants with thick fleshy roots, this requires a considerable quantity of water, and should be freely syringed. It is grown in a pot, but would probably succeed quite as well if hung up, as the greater part of those with fleshy roots do much better in that way.

It is propagated by taking off the young shoots, which it sends out rather slowly.

* So called by Mr. Bateman, in compliment to the Rev. Mr. Huntley, a zealous collector of rare plants.

16

ONCĬDĬUM lūrĭdūm; *var.* guttātūm.

Mr. Boyd's Oncidium.

GYNANDRIA *MONANDRIA.*

Nat. ord. Orchidaceæ, § Vandeæ.
ONCIDIUM. Swartz.

O. *luridum.* *Gen. et Sp. Orch. p.* 201.
β guttatum; sepalis petalisque luteis sanguineo-maculatis.
Epidendrum guttatum. *Linn. sp. pl.* 1351.
Cymbidium guttatum. *Willd. sp. pl.* 4. 102.
Oncidium Boydii. *hort.*

That this is the long lost *Epidendrum guttatum* of Linnæus, a Jamaica plant, unknown to Swartz, who so carefully investigated the Botany and especially the Orchidaceous plants of the island, I do not doubt. The name *luridum* ought therefore in strictness to be abolished; but that of *guttatum* applies so generally to the whole genus, that there would be more inconvenience than advantage in the measure.

It was imported from Jamaica by Messrs. Rollissons, to whom I am obliged for the specimen from which the figure has been taken. Certainly, *O. Lanceanum* and *Forbesii* alone excepted, this is the most beautiful plant of the genus yet in our gardens; the richness of its colours, the profusion of its flowers, and its stately growth, for it is from three to four feet high, would even make it doubtful whether it does not rival the former species, only it wants the aromatic odour.

In structure it differs nothing from *O. luridum,* so far as I can ascertain.

It requires to be cultivated in the orchidaceous-house or moist stove, either suspended from the roof, or elevated above the surface of the pot. It should be placed in the warmest part of the stove, and its roots in particular freely syringed. The soil should be turfy peat, kept open with broken bricks or pots, to allow a free passage for the water. Water must be given more freely at certain seasons than at others; but it does not require such a long period of rest as Catasetums and plants of that kind, and therefore may be kept almost continually growing.

It is multiplied by division, the front shoots making the best plants.

So little is correctly known concerning the native habits of the great order of Orchidaceous epiphytes, that I gladly avail myself of a vacant space to give the following localities of some other Oncidiums, as stated by M. Descourtilz in his MSS.

Oncidium ciliatum. Flowers in September, in *low woods,* surrounding the table-land (plateaux) of the neighbourhood of Bananal.

Oncidium iridifolium. Found in Brazil, exclusively on branches of the Orange and Lemon, flowering in April. The fruit is large, and has six large transparent wings. Very common in the province of St. Paul's, about the town of Bom Jesus de Bananal, choosing in preference *dry places exposed to the sun.*

Oncidium pubes. *Thin forests,* surrounding the table-land near Bom Jesus de Bananal, flowering in the month of May.

Oncidium divaricatum. Trunks of the highest trees, in elevated mountains of the Serra das Argoas in the district of Ilha Grande, flowering in February.

17

MAXILLĀRĬĂ stapelioīdēs.

Stapelia-like Maxillaria.

GYNANDRIA *MONANDRIA.*

Nat. ord. ORCHIDACEÆ, § VANDEÆ.
MAXILLARIA. Fl. Peruv.

M. *stapelioides;* pseudobulbis ovatis tetragonis 1-2-phyllis, foliis tenuibus lanceolatis patentibus pallidè glaucis reticulatis, pedunculo diffuso bifloro, sepalis petalisque subrotundo-ovatis acutis patulis subæqualibus, labello oblongo trilobo: laciniis lateralibus erectis linearibus obliquis obtusis intermediâ ovato-oblongâ basi cucullatâ, cristâ transversâ flexuosâ carnosâ intùs dente carnoso ovato auctâ.
M. stapelioides. *Lind. et Otto abbild.* 111. *t.* 52. *Gen. et Sp. orch.* 146.

This is one of the Brazilian Maxillarias, whose pseudobulbous habit separates them in appearance so widely from the caulescent species upon which the genus was originally founded. For remarks upon this point, the reader is referred to plate 8 of the present volume.

The species whose singular speckled flowers have suggested the comparison with a Stapelia, inhabits the Organ Mountains, where it was found by Mr. Gardner, (No. 651); but it was long before obtained from Brazil by the director of the Berlin Garden, by whom it was named. It is now common in collections, where it is at once recognized by its pallid glaucous thin leaves, which look as if suffering under the attack of red spider. It is one of the most easy species to cultivate.

Brazil will doubtless be found to contain many such plants. *M. Rollissoni* (*Bot. Reg.* 1838, *t.* 40) is one, and a plant called by Descourtilz " Epidendre Jonquille," because of its colour not its smell, for it is scentless, is another. This was also found in the Organ Mountains by Mr. Gardner, from whom I have a specimen (No. 652), and as it is probably already in this country, it may as well be named and characterized. The unpublished name of Jonquil is so likely to lead to misconception, that I make no excuse for changing it to

M. *xanthina;* pseudobulbis ovalibus tetragonis 1-2-phyllis, foliis angustè lanceolatis, pedunculis ascendentibus unifloris pedicello sterili terminatis, bracteâ ovatâ mucronatâ cucullatâ, sepalis petalisque oblongis acutis patulis subæqualibus, labello oblongo trilobo laciniis lateralibus erectis linearibus obtusissimis integris intermediâ bilabiatâ: labio superiore carnoso abbreviato truncato 5-dentato inferiore oblongo acuto.

This is very different from both *M. aromatica* and *Rollissonii.* M. Descourtilz found it on the high mountains of Ilha Grande, on the side towards the sea, and exposed to the rising sun. He does not say whether it occurs on trees or on the ground.

20

DENDRŌBĬŪM aurēum; *var.* pallidum.

Golden-flowered Dendrobium; pale variety.

GYNANDRIA *MONANDRIA.*

Nat. ord. ORCHIDACEÆ, § MALAXIDEÆ.
DENDROBIUM. Bot. Reg. vol. 7. *fol.* 548.

D. *aureum;* caulibus teretibus clavatis internodiis brevibus, foliis lineari-oblongis apice obliquis emarginatis, pedunculis subbifloris aggregatis lateralibus, sepalis lineari-ovatis acuminatis obtusis, petalis latioribus ovatis acutis undulatis, labello ovali undulato obtusiusculo indiviso medio serrulato per axim pubescente.
D. aureum. *Gen. & Sp. Orch. p.* 78.

A native of Ceylon, where it was first found by Mr. Macrae growing upon trees, near Nuera Ellia, flowering in January. It has since been frequently imported, and is occasionally seen in the collections of this country. The accompanying drawing was made in the Nursery of the Messrs. Loddiges, in March 1838.

The species varies with pale yellow and white flowers, the latter being what is here represented. In both varieties the fragrance is remarkable, forming a something intermediate between violets and primroses.

The materials from which it was first described were very imperfect, and consequently various alterations and emendations in the specific character have become necessary, and are now made.

The magnified figure represents the labellum seen from the inside.

It is propagated in the same way, and requires precisely the same treatment as D. crumenatum, described at t. 22 of this volume. The side shoot with the roots represented in the figure, if taken carefully off, would make an excellent plant.

22

DENDRŌBĬŪM crumēnātūm.

Sweet Club-stemmed Dendrobium.

GYNANDRIA *MONANDRIA.*

Nat. ord. ORCHIDACEÆ, § MALAXIDEÆ.
DENDROBIUM. Bot. Register, vol. 7. *fol.* 548.

D. *crumenatum;* caulibus cæspitosis erectis basi incrassatis teretibus, foliis ovato-oblongis obtusis emarginatis, racemo terminali (3-5) multifloro, sepalis petalisque ovatis acuminatis subundulatis conformibus, labello cucullato trilobo: lobis lateralibus truncatis intermedio ovato acuto, disco lamellato.
Angræcum crumenatum. *Rumph. herb. amb.* VI. 105. *t.* 47. *f.* 2.
Onychium crumenatum. *Blum. Bijdr. p.* 326.
Dendrobium crumenatum. *Swartz. Willd. sp. pl. no.* 20. *Hort. Trans.* VII. *p.* 70. *Gen. et Sp. orch. p.* 88.

A native of various parts of the Indian Archipelago, where it inhabits the branches of trees: this species has long been known to Botanists from the figure given by Rumphius, and by specimens which travellers, attracted by its delicious perfume, have from time to time sent to Europe.

Dr. Blume found it in Java, near Batavia, and on the coast of the little island of Nusa Kambanga; and the late Sir Stamford Raffles met with it in Sumatra. Rumphius gives no locality for it, referring from the body of his work to a description in the Appendix or Auctuarium, where, however, nothing is to be found except a back reference to the body of the work; it is doubtless however an inhabitant of Amboyna. The specimen now figured was sent by Mr. Nightingale from Ceylon to his Grace the Duke of Northumberland, in whose collection at Syon it flowered in August 1837.

According to Blume it varies with white and pink flowers, and with leaves more or less oblong and coriaceous. It is one of the easiest of the genus to manage, and well repays the cultivator for the trouble he bestows upon it.

Why it is called *crumenatum,* (literally purse-shaped) I am unable to state. Rumphius, with whom the name originated, gives no explanation; but in one place he calls it A. *crumenatum*, and in another *Angræcum angustis crumenis*.

Fig. 1. represents a side view of the labellum; 2. the column, and 3. the pollen-masses.

The species is easily multiplied by taking side shoots from the old plant; sometimes young shoots will grow from the stem instead of flowers; these, if taken off carefully and laid upon warm damp moss, will, in a short time, make excellent plants. Its cultivation is simple. It belongs to that class of plants which have a period of growth and a period of rest. All that is requisite, is the temperature of the stove, and a plentiful supply of water during the growing season. When this season is past, the plant should be removed to a cooler house, or at least to the coolest part of the house, and kept perfectly dry, when its wood will harden, it will lose its leaves, and form its flower-buds. It may afterwards be brought into the warmest part of the stove, when the result will be a profusion of flowers.

26

LÆLĬĂ furfurăcĕă.

Scurfy-stalked Lælia.

GYNANDRIA *MONANDRIA.*

Nat. ord. ORCHIDACEÆ, § EPIDENDREÆ.
LÆLIA. Botanical Register, vol. 21. *fol.* 1751.

L. *furfuracea;* pseudobulbis ovatis striatis submonophyllis, foliis angustè oblongis erectis acutis scapo multò brevioribus, scapo unifloro (?) tereti, bracteis oblongis membranaceis acutis, sepalis lanceolatis acuminatis patentissimis, petalis subrhombeis lanceolatis undulatis sublobatis, labelli trilobi bilamellati lobis lateralibus erectis rotundatis truncatis intermedio oblongo revoluto, ovario glandulis nigris furfuraceo.

This plant was imported by Mr. Barker of Birmingham, from whom I received it in November 1838, as a new species. It is very like L. autumnalis, represented in the next plate, but its pseudo-bulbs are merely ovate and slightly furrowed, instead of having a long neck and being deeply furrowed; the leaves are solitary or in pairs, and not in twos or threes; they are erect and straight, not spreading and curved; the flowers have little or no smell; the petals are so much more undulated as to appear lobed, and they are distinctly rhomboidal, and finally the ovary is closely covered with black mealy glands. It would seem moreover that the scape does not bear more than one flower instead of several, but of this I cannot so well judge.

It was found near Oaxaca, by Count Karwinski, and is probably not uncommon in collections, large quantities having been received by various persons from Mexico, especially by the Horticultural Society, who have distributed it among their fellows.

A. and B. are varieties differing in colour, but apparently in nothing more. The species is figured in the next plate to *L. autumnalis*, in order that the differences between the two species may be the more readily perceived.

The cultivation of this species, in so far as our knowledge extends, is rather difficult. It probably proceeds from the nearly uniform temperature of our stoves in this country, differing so much from the temperature to which plants are subjected, at considerable elevations in tropical regions.

There are many plants which belong to this order, the peculiar treatment of which is now perfectly understood, and which consists in removing them from the stove, and placing them in a cooler house for several months every season; some of the species of Bletia are examples of this.

The present species should be cultivated in a cooler house than is generally used for Orchidaceous plants, and subjected to considerable variations of temperature by being at certain seasons kept in the greenhouse.

27

LÆLĬĂ autumnalis.

Autumnal Lælia.

GYNANDRIA *MONANDRIA.*

Nat. ord. ORCHIDACEÆ, § EPIDENDREÆ.
LÆLIA. Botanical Register, vol. 21. *fol.* 1751.

L. *autumnalis;* pseudobulbis ovatis teretibus costatis apice attenuatis 2-3-phyllis, foliis oblongo-linearibus patentissimis scapo multò brevioribus, scapo tereti apice subsexfloro, bracteis oblongis membranaceis acutis, sepalis lanceolatis acuminatis patentissimis, petalis oblongo-lanceolatis undulatis, labelli trilobi bilamellati lobis lateralibus erectis rotundatis truncatis intermedio oblongo-lanceolato apice reflexo, ovario glabro.
Bletia autumnalis. *La Llave et Lexarz. nov. veg. descr.* 2. 19.
L. autumnalis. *Gen. et Sp. Orch. p.* 115. *Bateman Orch. Mexic. et Guatemal. t.* 9.

A very fragrant and beautiful plant, imported from Mexico of late years, and now not uncommon in gardens. A considerable number of it has been given away among the Fellows of the Society by order of the Council of the Horticultural Society of London.

The specimen now figured was taken from a plant which flowered at Woburn, and which was sent me by the Duke of Bedford.

I have nothing to add to the following account extracted from Mr. Bateman's magnificent work on the Orchidaceous plants of Mexico and Guatemala.

"The genus Lælia may be regarded as one of the most ornamental of its tribe, since pleasing colours, graceful habit, long duration, and delicious perfume, in short, all the essentials of floral beauty, seem to be combined in its various species. Of these, five or six are already known, of which the one now represented, however charming it may be, is, perhaps, the least interesting; for it is far surpassed by *L. grandiflora* (the Flor de Corpus of Mechoacan) in the magnitude of its flowers, and by *L. anceps* and some unpublished species, in the brilliancy of its colours. Being found at a considerable elevation they all thrive best in a moderate temperature, and require to be high-potted, as by that means the roots are more likely to be retained in a healthy state, and are better able to withstand the extremes of heat and moisture which, even in the most carefully conducted establishments, will sometimes occur, and which we have found excessively injurious to Lælias, Cattleyas, and species of some allied genera. In winter they should be very sparingly watered, and kept in almost a dormant state. *L. autumnalis* flowers both in this country and its own, at the season which its name implies."

In the Garden of the Horticultural Society its cultivation is found extremely simple. When plants are received they are tied to a block of wood, and kept perfectly dry until they begin to send out roots, and manifest other signs of growth. They are then freely syringed two or three times a day, and this practice is continued until the growing season is past. They are then removed to a cooler house with a drier atmosphere, in which they are allowed to winter, and when this season is over, the above treatment is again renewed.

It is multiplied in the usual way; the front shoots make the best plants.

35

ĔRĬĂ ferrugĭnĕă.

Rusty Eria.

GYNANDRIA *MONANDRIA.*

Nat. ord. Orchidaceæ, § Malaxideæ.
ERIA. Botanical Register, vol. 11. *fol.* 904.

E. *ferruginea*; caulibus teretibus articulatis vaginatis, foliis oblongis obtusis, racemo erecto laterali multifloro basi squamato, bracteis ovatis acutis ovario villosissimo brevioribus, labelli lobis lateralibus erectis truncatis intermedio ovato serrato subplicato: disci cristis 4 dentatis intermediis carnosis incurvis.

This very distinct species was imported from Calcutta by Messrs. Loddiges, and flowered at Hackney in March, 1838. I have no information as to the part of India which it inhabits.

It is not handsome, but it is very distinct, has a particularly deep green foliage, and the smooth delicate pink petals form a striking contrast with the coarse green shaggy sepals. The lip (*fig.* 1.) is most singularly crested, and looks more like the edge of some cowrie shell than the petal of a flower.

It is propagated like all other orchidaceous plants, namely, by divisions of the rhizoma. The soil used in its cultivation is turfy peat, well mixed with broken bricks; and the pot should have plenty of drainage. Its treatment generally is similar to that which has been frequently recommended for plants of this description. The house in which it is grown must be well shaded from bright sunshine in summer.

37

DENDRŎBĬŪM Jenkīnsĭĭ.

Captain Jenkins's Dendrobium.

GYNANDRIA *MONANDRIA.*

Nat. ord. Orchidaceæ, § Malaxideæ.
DENDROBIUM. Botanical Register, vol. 7. *fol.* 548.

D. *Jenkinsii*; pseudobulbis aggregatis oblongis tetragonis monophyllis, foliis oblongis coriaceis marginatis retusis, pedunculis subradicalibus unifloris vel racemosis, sepalis ovatis obtusis petalis multò minoribus, labello cucullato rotundato basi parùm producto limbo dilatato serrato villoso emarginato.
D. Jenkinsii. *Wallich. in litt.*

This pretty plant has been sent by Dr. Wallich to many persons in this country, and is now by no means uncommon. The accompanying drawing was made in September, 1838, from a plant in the possession of Messrs. Loddiges, and I have since received it from other places. To Sir Charles Lemon I am indebted for the following account of the species from the pen of Dr. Wallich.

"This elegant little Orchidaceous plant grows in large tufts on trees. The pseudo-bulbs are densely aggregated on a creeping rhizoma, oblong, marked with one or two rings and vestiges of sheaths, about an inch long, each bearing an oblong, shining, coriaceous, obtuse, sometimes slightly retuse, sessile leaf, about 1½ inch long, sometimes a little longer. Flowers spreading, large, yellow, inodorous, on long and slender peduncles, geminate from the side of the pseudo-bulb on a very short cylindric common peduncle, having a few scales at its base. Partial peduncles filiform, two or three inches long. Sepals and petals obtuse; the latter broadest, oval. Lip very large, reniform, retuse, slightly pubescent and ciliate, measuring nearly an inch across, almost sessile, a little channelled at the base, otherwise spreading flat.

"I received the plant from Capt. Jenkins in November, 1836. He had obtained it at Gualpara. It flowered finely at this garden in the middle of March following. I have since had abundant supplies from the same liberal and indefatigable source.

"I am very happy in dedicating this very distinct species to my valued friend Capt. Jenkins, to whom this garden, the cause of botany, and science generally, are deeply indebted.

"The flowers are larger than those of D. aggregatum of Roxburgh, to which it bears a slight resemblance."

The species shews in a striking manner the impropriety of generically separating the pseudo-bulbous Dendrobia from those with slender stems, notwithstanding the dissimilarity in their appearance. Here we have a plant with the pseudo-bulbs consisting, in the majority of instances, of a single internode, hardened, become four-cornered, and as dissimilar as possible from the same part of the stem of Dendrobium Pierardi; but in D. aggregatum, figured in this work, t. 1695, several internodes (3) together constitute a body altogether intermediate in nature between a pseudo-bulb and ordinary stem, and similar transitions from one to the other occur in D. densiflorum, fol. 1828, and D. Griffithianum, all which are so nearly allied to each other that a person unacquainted with all of them may possibly mistake one for the other. To prevent this the following distinctions in the labellum will be found certain.

D. *Jenkinsii.* Labellum broader than long, repand, slightly two-lobed, shaggy, serrated.

D. *aggregatum.* Labellum broader than long, scarcely wavy, undivided, downy only near the base.

D. *densiflorum.* Labellum cordate, repand, two-lobed, reflexed at the point, serrated.

D. *Griffithianum.* Labellum ovate, slightly hastate, serrated, downy, except near the edges.

This species is more difficult to cultivate than those kinds with long free-growing stems. It is frequently seen in an unhealthy state, owing to its being grown in a pot, and subjected to a uniform high degree of temperature. The best way to insure its success, is, to tie it to a block of wood with a piece of turfy peat attached to it, and suspend it from the rafter of the house. There it must be well syringed at least twice a day, so long as it continues to grow, and afterwards it may be removed to a cooler house. In fact it never requires so much heat as those species with long trailing stems.

It is propagated in the usual way.

42

ONCIDIUM pulvinatum.

The Cushion Oncidium.

GYNANDRIA *MONANDRIA.*

Nat. ord. Orchidaceæ, § Vandeæ.
ONCIDIUM. Swartz.

O. *pulvinatum*; paniculâ ramosissimâ divaricatâ, sepalis obovatis lateralibus liberis, petalis conformibus acutis, labelli lobis subæqualibus intermedio bilobo undulato lateralibus crenatis rotundatis crispis, disco pulvinato villosissimo, columnæ alis rotundatis. *Bot. Reg.* 1838. *misc. no.* 115.

Panicula *ramosissima, intricata, glabra,* 8-9-*pedes longa; axi percurrente rectiusculo pennæ corvinæ crassitudine, ramis simplicibus ramosisque divaricatis geniculatim flexuosis, internodiis pollicaribus bracteis refractis membranaceis acutis pluries longioribus.* Flores *O. divaricati simillimi, lutei in medio sanguineo guttati.*

One of the largest of the Oncidia, vieing with *O. altissimum* in stature, its panicle being 8 or 9 feet long, of which one of the smaller branches only is represented in the accompanying plate, and resembling *O. divaricatum* in the structure, colour, and size of its flower. It however differs from that species in its lip having the middle lobe largest, not smallest; and in the cushion at its base being much more villous and equally convex, not divided into equal quarters. The lateral lobes of the lip are also crisp in this species, not plain, as in O. divaricatum.

In the specimen before me there is one central axis, in the middle, about as thick as a crow's quill, gently wavy, and so weak as to be unable to support its flowers without the assistance of neighbouring objects. At regular distances from this proceed the branches, which are either simple or themselves branched, and zigzag in a remarkable manner; the whole forming an entangled mass of inflorescence.

It is a native of Rio Janeiro, whence it was sent to Mr. Richard Harrison, of Aighburgh, in the year 1834, by his brother, Mr. William Harrison.

The treatment frequently recommended for orchidaceous plants of this kind, will be found to suit the present. Like the greater number of species in the extensive genus to which it belongs, it succeeds best if hung up in a basket, or tied to a block of wood, and suspended from the rafters of the stove. The only things to be noticed in this system of cultivation, is, first to tie some good turfy peat to the block along with the plant; and secondly, to syringe freely, as the plants are apt to get too dry.

44

BURLINGTŌNĬĂ maculātă.

Spotted Burlingtonia.

GYNANDRIA *MONANDRIA.*

Nat. ord. ORCHIDACEÆ, § VANDEÆ.

BURLINGTONIA. Botanical Register, vol. 23. *fol.* 1927.

B. *maculata;* racemis pendulis, sepalo anteriore emarginato: supremo sepalisque ovato-oblongis undulatis acutis, lamellis labelli cristatis dentatis, labello bilobo undulato denticulato versus unguem lacero, foliis lineari-lanceolatis.

Pseudobulbi *ovales compressi monophylli basi aphylli; foliis lineari-lanceolatis striatis rectis.* Pedunculi *penduli, racemosi, foliorum longitudine; bracteis ovario dimidio brevioribus.* Flores *odorati, lutei, cinnamomeo colore maculati, in genere parvi;* labellum *cæteris longius, basi album, lamellis* 3 *inæqualiter dentatis carnosis parallelis anticè abruptè truncatis.* Columna *ungue labelli brevior eique appressa, clavata, semiteres, apice utrinque bicornis sanguinea.*

A sweet-scented epiphyte, obtained from Brazil by Messrs. Loddiges, with whom it flowered in May, 1838. It is very distinct from all the species of this genus previously known, and is, I hope, an omen of there being many more yet to discover. It confirms the generic character originally given, and removes all doubt about the distinctness of Burlingtonia from Rodriguezia.

In cultivating this plant, it should be tied to a block of wood and suspended from the rafters of the orchidaceous house. If a little piece of turfy peat is tied to the block along with it, it is of considerable service in keeping the roots moist. In the growing season it will require to be syringed freely two or three times every day, and shaded from bright sunshine. When it is not growing, water must be given very sparingly.

If this treatment is practised it will flower freely.

Fig. 1. is a view of the two connate anterior sepals; 2. is the column and lip, shewing the lamellæ and lacerated margin of the latter, together with the crimson teeth at the apex of the column; 3. are the caudicula, gland, and pollen masses; one of the latter being divided across to shew that it is really a plate rolled up.

48

* ODONTOGLŌSSŪM Rōssĭī

Ross's Odontoglossum.

GYNANDRIA *MONANDRIA.*

Nat. ord. ORCHIDACEÆ, § VANDEÆ.

ODONTOGLOSSUM, Kunth. Sepala lateralia patula libera, *Labellum* planum, unguiculatum, ascendens, limbo reflexo diviso dentato apice angustato; basi concavum cristâ bilamellatâ rarò fimbriatâ sæpiùs anticè bidentatâ auctum. *Columna* elongata, apice auriculata aut aptera. *Sertum Orchidaceum, t.* 25.

Odontoglossum *Rossii;* pseudobulbis ovatis cæspitosis ancipitibus monophyllis, foliis oblongo-lanceolatis scapo radicali subbifloro longioribus, bracteis membranaceis carinatis acuminatis, sepalis lineari-lanceolatis carinatis acuminatis patentibus, petalis oblongis obtusis revolutis, labello subrotundo-ovato emarginato undulato lamellis unguis confluentibus rotundatis denticulis 2 anterioribus obtusis, columnâ apterâ pubescente. *Sertum Orchid. t.* 25.

Pseudobulbi *compressi, vaginis marcescentibus breviores.* Folia *erecta, bipollicaria, lanceolata, striata, vix coriacea, scapo longiora.* Scapus *erectus, radicalis, medio nudus aut squamâ solitariâ auctus, subbiflorus; pedunculis medio vaginatis.* Flores 2 *uncias lati, erecti; sepalis patentissimis, lanceolatis, luteo-viridibus fusco maculatis; petalis albis basi purpureo guttatis; labello candido, purissimo, sub lente minutissimè pubescente, ungue 3 lineas longo columnâ apterâ velutinâ paulò breviore.* Lamellæ labelli *carnosæ, anticè connatæ, fronte bidentatæ, luteæ, intus sanguineo striatæ.* Stigma *purpureo-marginatum.*

A charming plant, sent to Mr. Barker from Mexico by his collector Mr. Ross, after whom it is named. The bright white lip, lying as it were in the centre of a rich green yellow and blue star of three points, produces a peculiarly beautiful and unusual appearance.

Many of the species of this genus are however handsomer, and few less beautiful, than this; let us therefore hope that Mr. Hartweg will succeed in procuring them for the Horticultural Society, now that he is engaged in exploring the rich neighbourhood of Oaxaca, the head-quarters of Mexican Orchidaceæ. Odontoglossum nebulosum has flowers nine inches in circumference, those of O. Cervantesii are much like O. Rossii, but larger and richer coloured, and both these are from the south-west of Mexico.

It requires to be cultivated in a warm damp stove, where it may either be potted in the usual way, or, which is preferable, suspended from the roof upon a block of wood.

Like other Orchidaceæ it may be multiplied by dividing its pseudo-bulbs.

* From οδους a tooth, and γλωσσα a tongue, in allusion to the toothings of the labellum at the base.

51

GONGŌRĂ fūlvă.

Tawny-flowered Gongora.

GYNANDRIA *MONANDRIA.*

Nat. ord. ORCHIDACEÆ, § VANDEÆ.
GONGORA, Fl. Peruv. Botanical Register, vol. 19. *fol.* 1616.

G. *fulva;* hypochilii convexi cornubus lateralibus elongatis capitatis aristis setaceis, epichilio acuminato æquilongo, pedicellis columnâ triplò longioribus.

The drawing of this plant has been lying in my portfolio since the month of July, 1836, when I received a specimen from Mr. Barker with the following memorandum.

"The leaves of the plant are very similar in form to the Gongora maculata, but are somewhat larger, and perhaps a little finer or thinner. The bulbs are deeply ribbed, and a little longer than maculata. It is highly fragrant, scenting the whole house; the scent approaches nearer that of the violet than any thing I know."

Mr. Barker considered the plant a variety of Gongora maculata, already figured in this work at fol. 1616; and perhaps rightly. But at the same time, till we have a more certain knowledge of the value of the differences found in the flowers of Orchidaceous plants, it seems necessary to distinguish it specifically. In addition to the peculiarities adverted to in the above memorandum, and the darker colour, the flowers of this are not more than half the size of *G. maculata*, and the raceme is far more contracted, in consequence of the shortness of the pedicels. I also find that the lateral horns of the hypochilium are terminated by a round dilated head; as is represented in the magnified flower of the accompanying plate.

54

LÆLĬĂ ālbĭdă.

White-flowered Lælia.

GYNANDRIA *MONANDRIA.*

Nat. ord. ORCHIDACEÆ, § EPIDENDREÆ.
LÆLIA. Botanical Register, vol. 21. *fol.* 1751.

L. *albida;* pseudo-bulbis ovalibus diphyllis, foliis linearibus acutis spicâ multiflorâ 4-plò brevioribus, sepalis oblongo-lanceolatis acutis vel subacuminatis petalisque latioribus acutis, omnibus apicibus reflexis fortiter mucronulatis; labelli altè trilobi tricostati lobis lateralibus erectis rotundatis intermedio duplo majore subrotundo obscure apiculato reflexo. *Bateman in litt. Botanical Register,* 1839, *misc. no.* 4.

Pseudobulbi *ovati, sulcati, diphylli.* Folia *angustè lanceolata, erecto-patula, scapo breviora.* Racemus *erectus,* 3-5-*florus.* Bracteæ *ovatæ, duræ, obtusæ, sessiles, nanæ.* Sepala *candida, apice rosea, oblongo-lanceolata, patula.* Petala *breviora, oblonga, obtusiora, revoluta, concolora.* Labellum *obovatum, trilobum, 3-lamellatum: laciniis lateralibus obtusis intermedio obtuso apiculato undulato rubescente multò brevioribus; lamellæ labelli luteæ, purpureo-punctatæ.* Columna *elongata, glabra.*

Oaxaca is the head-quarters of Orchidaceæ, in the kingdom of Mexico. It was there that Count Karwinski found the greater part of the species, of which specimens were brought home by him to Munich; it is there that the majority of European collectors have obtained the best part of what they have sent home; and where, lately, Mr. Hartweg has got together not fewer than 140 species for the Horticultural Society.

Oaxaca is the native country of this lovely plant, a notice of which has already been inserted in the present work, and than which none is more worthy of a figure; for its flowers are as sweet as a bed of primroses, which in fact they much resemble in odour. Count Karwinski found it near St. Pedro, in cool places, according to the herbarium of Dr. Von Martius, in which wild specimens exist.

It was originally sent me by Mr. Bateman; and I have since had it from Thomas Harris, Esq. of Kingsbury, who bought it, along with numerous Cacti, from a French collector who visited London two years ago. The pseudo-bulbs were so much like those of *Lælia autumnalis,* that, when the latter reached London from Mr. Hartweg, it was thought to be the same species.

It is the only Lælia as yet discovered with white flowers.

57

ONCĬDĬŬM trullĭfĕrūm.

Trowel-lipped Oncidium.

GYNANDRIA *MONANDRIA.*

Nat. ord. ORCHIDACEÆ, § VANDEÆ.
ONCIDIUM. Botanical Register, vol. 13. *t.* 1758.

O. *trulliferum;* pseudobulbis elongatis ovalibus compressis 2-3-phyllis foliis ovalibus ferè æqualibus, scapo radicali rigido semel ramoso, sepalis lateralibus distinctis superioribus petalisque obtusis concavis, labelli lobis lateralibus nanis rotundatis intermedio trulliformi serrulato, cristâ verrucosâ medio depressâ lævi anticè dente ascendente serrato apice appendiculato, columnæ alis integerrimis ovatis obtusis.

This addition to the extensive genus *Oncidium* is a native of Brazil, whence it was imported by Messrs. Loddiges, with whom it flowered in September, 1838.

The pseudo-bulbs are represented much diminished in the accompanying plate; they are in fact six inches long, and the leaves themselves are scarcely longer; they are pressed almost flat, so that although fourteen or fifteen lines broad, they are not more than three lines thick (fig. 2.).

The scape is stiff, erect, and branched in a simply pinnated manner, each branch being covered with flowers from the base to the apex. The figure of the lip (fig. 1.) is very uncommon; its middle lobe is produced into a segment much like a bricklayer's trowel, at the base of which there stands a strong tooth ascending, notched at the side, and terminated by a little notched appendage; this tooth rests by its back upon a tumour covered with regular warts, and forming the centre of three such tumours, the lateral ones standing nearer the base of the labellum, with a smooth wartless hollow between them.

In cultivation it requires the damp stove. It is easily managed; growing freely either in a pot with the usual soil, or suspended from the roof upon a block of wood. In either way, but particularly in the latter, it must be freely syringed during the growing season.

It is propagated in the same way as other plants of this kind.

62

CYRTOCHĪLŪM mystacīnūm.

Whiskered Curvelip.

GYNANDRIA *MONANDRIA.*

Nat. ord. ORCHIDACEÆ, § VANDEÆ.
CYRTOCHILUM. Botanical Register, vol. 19. *fol.* 1627.

Cyrtochilum *mystacinum;* pseudobulbis ovalibus compressis corrugatis monophyllis basi polyphyllis, foliis ligulatis acutis planiusculis carinatis scapo ramoso multò brevioribus, bracteis lanceolatis pedunculis duplò brevioribus, sepalis petalisque ovatis acuminatis, labello unguiculato cordato obovato-lanceolato subrepando plano apice reflexo medio pubescente basi obsoletè lamellato, columnæ alis multifidis. *Bot. Register,* 1838. *misc. no.* 38.

Pseudobulbi *ovales, corrugati, compressi, intra bases foliorum latentes vaginantium; ipsi monophylli.* Folia *ensiformia, carinata, acuta, avenia, erecta, scapo composito-racemoso multò breviora.* Bracteæ *acuminatæ, sphacelatæ, ferè naviculares, majores 6-7 lin. longæ.* Pedicelli *bracteis majoribus æquales, minoribus longiores.* Ovarium *pedicello suo multò brevius.* Flores *lutei, unicolores, distantes.* Sepala *linearia, acuminata; lateralibus labello suppositis.* Petala *conformia, sed latiora.* Labellum *unguiculatum, cordatum, obovatum, acumine recurvo; columnâ subparallelum eique basi adnatum; ungue calloso obsoletè lamellato apice dentato.* Columna *clavata; alis laceris, appendice ante stigma emarginato.*

A notice of this rare plant has already been given in the volume of the present work for 1838, among the miscellaneous matter, no. 38; where by a misprint the flowers are described as being "bright yellow *white*-coloured," instead of "bright yellow *whole*-coloured."

It is a native of Peru, whence it was obtained by Richard Harrison, Esq. of Aighburgh, with whom it produced its flowers in the latter end of 1837. It is a genuine species of the genus, having the base of the lip united to the face of the lower part of the column, a circumstance overlooked in the separate figure of the column in the accompanying figure, which is otherwise correct.

This species will require to be cultivated in the moist stove, but will probably succeed best when the temperature is rather lower than it commonly is in this kind of house.

It should be potted in brown turfy peat well drained, and treated as other plants of this kind.

It is multiplied by division.

64

DENDRŌBĬŪM formōsūm.

Beautiful Tree-bloom.

GYNANDRIA *MONANDRIA.*

Nat. Ord. ORCHIDACEÆ, § MALAXIDEÆ.
DENDROBIUM. Botanical Register, vol. 7. *fol.* 548.

D. formosum; caulibus teretibus pendulis pilosis, foliis distichis ovatis apice obliquè emarginatis obtusis, racemo brevi terminali 4-5-floro, bracteis brevibus ovatis, (floribus maximis), sepalis oblongis acutis: lateralibus basi longè productis, petalis duplò latioribus acutis, labello obovato dilatato retuso cum basi columnæ in calcar obtusum connato. *Lindley in Wall. Pl. As. Rar. p.* 34. *t.* 39.
D. formosum. *Roxb. Fl. Ind.* iii. 485. *Wall. Cat. no.* 1998.

This magnificent species flowered at Chatsworth, in May 1838, and the drawing on the opposite page was then made from a specimen given me by his Grace the Duke of Devonshire. As a white flowered epiphyte, it is almost unrivalled among Asiatic Orchidaceæ, the *Phalænopsis amabilis* being the only one that can enter into comparison with it.

It has been already amply described by Drs. Roxburgh and Wallich; the former tells us he found it on trees in the forests of Sylhet, and in the Garrow mountains, flowering in April and May.

It was gathered by Dr. Wallich on the mountains of Nepal and Sylhet; also in the province of Martaban near Moulmein, and in Tavoy on the Tenasserim coast, flowering and fruiting in both the dry and rainy seasons. Mr. Griffith also met with it on trees in damp places in the neighbourhood of Moulmein; so that its range is unusually extensive for plants of this kind; it does not however appear to belong to the Flora of the Indian Archipelago. Dr. Wallich says it grows generally in large tufts upon trees, or sometimes on rocks; the flowers possessing a delightful though faint perfume.

The best way to cultivate the plant, is to suspend it from the rafters of the stove. It should have plenty of good turfy peat to nourish its roots, and must be freely watered and syringed during its growing season. When that period is over, it should be kept dry and cool. This treatment will retain it in health and vigour, and make it flower freely.

65

* GRAMMATOPHȲLLŪM multiflōrūm.

Many-flowered Letter-leaf.

GYNANDRIA *MONANDRIA.*

Nat. ord. ORCHIDACEÆ, § VANDEÆ.
GRAMMATOPHYLLUM *Blume; Perianthium* explanatum, patens, sepalis petalisque subæqualibus. *Labellum* cum columnâ articulatum, nanum, trilobum, cucullatum. *Columna* arcuata, erecta, semiteres, basi callosa. *Anthera* subbilocularis. *Pollinia* 2, globosa, basi sulcata, in extremitatibus glandulæ arcuatæ sessilia.——Herba *epiphyta, caulescens.* Caules *simplices, incrassati.* Folia *linearia, disticha, striata.* Pedunculi *radicales, longissimi, (v. terminales?) multiflori.* Flores *speciosissimi.* Gen. et Sp. Orch. p. 173.

G. *multiflorum;* racemo erecto longissimo multifloro, bracteis ovato-oblongis obtusis squamiformibus dorso convexis, sepalis oblongis obtusiusculis planis, petalis acutis subconformibus angustioribus, labelli trilobi pubescentis medio hirsuti lobo intermedio plano oblongo rotundato lateralibus erectis subfalcatis, jugo in medio carnoso elevato ad basin lobi intermedii interrupto in lamellas 4 simplices cis apicem evanescentes producto, columnæ margine supra basin elevato flexuoso incurvo foveam altam obconicam circumdante. *Botanical Register,* 1835, *misc. no.* 80.

For this noble Orchidaceous plant now figured we are indebted to the exertions of Mr. Hugh Cuming, who discovered it in Manilla, and sent it to his customers in England. I am not aware of its having flowered anywhere except with Mr. Bateman, who sent me in May, 1838, the specimen, of which the annexed plate represents the upper part; the whole raceme was upwards of two feet long, and bore forty-eight flowers, each about an inch and half in diameter.

The plant has very much the aspect of a gigantic *Cymbidium,* with long coriaceous leaves, distichous at the base, and in fact there is not much to separate Grammatophyllum from that genus; the principal mark of distinction yet remarked consists in the gland of the pollen-masses, which in Cymbidium is triangular, and in the present genus is crescent-shaped, with one pollen-mass on each extremity of

* From γραμμα a letter, and φυλλον a leaf, in allusion to the marking of the leaves of the flower.

the crescent. I observe however that the base of the column is rolled up so as to form a fistular cavity, or cuniculus, near the base of the labellum; but I am uncertain whether to regard this as a generic character or not.

It was hoped, when this plant was imported, that it would prove the famous Letter-plant of Amboyna, Java, and the neighbouring coast, so called because its flowers are marked with deep brown stains arranged upon a pale ground so as to resemble grotesque characters. In this however we have been disappointed, as, notwithstanding the noble appearance of this, it is very inferior to the Letter-plant. Of that I have before me a single flower from Dr. Wallich's Herbarium, gathered at Pulo Dinding, in Cochin China, by Mr. Finlayson, which must have been four inches from the tip of one sepal to that of the opposite petal, or a foot in circumference!

As the species is in this natural order among the easiest to cultivate, it is well worth possession, even in a small collection, notwithstanding that the flowers want richness of colour: it is probable however that they will improve in this respect.

MISCELLANEOUS NOTICES.

*** The Botanical memoranda published in the last volume of this work under the above title, have enabled me to bring before the public no fewer than 183 plants, the greater part of which are new, in addition to the 68 of which figures were published; so that no fewer than 251 plants have been the subject of illustration during a single year. I have reason to believe that this arrangement has proved advantageous to the purchasers of the Botanical Register, since it has been the means of informing them what the real character is of the new plants whose names are found in sale catalogues, and of enabling them to judge how far they may be deserving of being purchased.

The experiment having thus far succeeded, it is now proposed to commence what I hope will be found an improvement upon the plan, by adding to the notices of plants a short account of such *new books* or *new discoveries*, &c. in Horticulture and Botany, as are of sufficient importance or interest to deserve to be recorded. In order to gain space for this addition, a little alteration in the typographical arrangements has been found necessary.

---

1. PLEUROTHĀLLĬS *pectinā̆tă;* folio oblongo acuto cochleato glauco caule ancipiti breviore, spicâ simplici distichâ in folium pronâ eoque breviore, bracteis membranaceis cucullatis ovarii longitudine, sepalis pubescentibus elongatis intermedio lineari lateralibus latioribus basi ventricosis approximatis omninò liberis, petalis lineari-lanceolatis, labello unguiculato oblongo basi pectinato apice truncato denticulato.

A curious species, resembling P. prolifera in habit. It was obtained from Rio Janeiro by Messrs. Loddiges. The flowers are sea green, with a few deep purple spots at the base of the labellum. The leaf is so firm, and so much hollowed out, that it is capable of holding water, as if it were made of metal.

2. MAXILLĀRĬĂ *foveā̆tă;* foliis lanceolatis undulatis plicatis racemo triplò longioribus, bracteis linearibus sphacelatis ovario longioribus, sepalis petalisque lineari-oblongis obtusiusculis, labello postico oblongo apice trilobo: laciniâ mediâ rotundatâ carnosâ excavatâ, disco ter aut quinquies sulcato.

A new species from Demerara, for which I am obliged to Messrs. Loddiges. It is very near M. squalens, but has a different lip, and its flowers are of a pale uniform dull straw colour. They have a faint, and not unpleasant, smell.

3. PLEUROTHĀLLĬS *strupifŏlĭă;* folio longissimo coriaceo obtuso lorato plano, racemis brevibus fasciculatis pendulis, floribus hiantibus pubescentibus, sepalo intermedio oblongo apice carnoso revoluto lateralibus semiconnatis majoribus porrectis intùs maculis pilosis (!) notatis, petalis rhombeolinearibus acutis, labello oblongo obtuso concavo nudo.

This is the most remarkable species of the genus I have yet seen. Its leaves are like leather thongs, and full eighteen inches long. The flowers are dull purple and white, with patches of short deep purple hairs on the inside of the lower sepals. It was imported from Rio by Messrs. Loddiges.

4. LǼLĬĂ *ălbĭdă;* pseudo-bulbis ovalibus diphyllis, foliis linearibus acutis spicâ multiflorâ 4-plò brevioribus, sepalis oblongo-lanceolatis acutis vel subacuminatis petalisque latioribus acutis, omnibus apicibus reflexis fortiter mucronulatis; labelli alte trilobi, tricostati lobis lateralibus erectis rotundatis intermedio duplo majore subrotundo obscure apiculato reflexo. *Bateman in litt.*

"A native of the environs of Oaxaca, whence it was sent to me this spring by the Messrs. Sadler of that place. Its colour is quite a novelty in the genus, all the other species bearing rose-coloured or lilac flowers. It has the graceful appearance of L. autumnalis, from which it can scarcely be distinguished in habit. Its flowers are very different in their form (as well as colour) from all the other species; they are about two inches across, sweet-scented, and excepting a bright yellow streak down the centre of the lip, and a few crimson dots at its base, are of a uniform semi-transparent white. It appears to be of easy cultivation, and the most free flowering individual of the genus." The foregoing memorandum has been communicated to me by Mr. Bateman. I have also received the plant from Mr. Harris. It will be figured hereafter in this work.

5. BOLBOPHȲLLŪM *fŭscūm;* pseudobulbis oblongis obtusè tetragonis diphyllis, foliis lineari-oblongis patentibus emarginatis, spicâ pedunculatâ distichâ ancipiti glabrâ pendulâ, bracteis ovatis cucullatis acutis coloratis distantibus florum longitudine sepalis triangularibus acutis, labello trilobo carnoso laciniâ intermediâ convexâ rotundatâ lateralibus brevioribus magis membranaceis acutis serrulatis.

Nearly related to B. tetragonum, like which species it is a native of Sierra Leone. The flowers are a deep dull chocolate colour, and are chiefly remarkable for the beauty of their anatomical structure. Like Liparis pendula they are studded with large transparent cells, containing raphides in cubical parcels, and they are moreover filled with short spiral vessels, so closely filled with air that it is difficult to drive it out even with the aid of the air-pump. Imported by Messrs. Loddiges.

6. QUEKĒTTĬĂ *microscŏpĭcă.* A very singular little plant, with the habit of a Pleurothallis, and the pollen-masses of a Vandeous Orchidacea; more nearly allied to Rodriguezia than to any thing else, but quite different in habit, and essentially distinguished by its cylindrical perianth, and labellum not only parallel with the auriculate column throughout its whole length, but excavated at the base, and furnished there with two callosities like those of Spiranthes. Its leaves are terete, subulate, about three inches long, and beautifully mottled with light green, deep green, and purple.

Although this little plant is only a few inches high, and has no attractions for the vulgar eye, it is in some respects one of the most interesting I know, if examined microscopically. Nothing can be more beautiful than the fabric of the leaves below the epidermis, and it undoubtedly deserves more examination in this respect than I am at present able to give it. The flowers abound in raphides, clustered in cells larger than those which surround them, and of a different colour, so that the flower, when examined with low powers of the microscope, looks as if it were dotted. I have observed this already in Liparis pendula (*Bot. Reg.* 1838, *misc. no.* 128.), and in Bolbophyllum fuscum, and it will probably be found a common structure in the sepals and petals of Orchidaceæ, as we already know it is in their leaves and stems. The caudicula is excellently adapted to shew the cellular nature of that part, and to explain upon what plan it is formed in other cases.

QUEKETTIA (§ Vandeæ). Perianthium cylindraceum, sepalis linearibus æqualibus basi gibbosis lateralibus connatis, petalis linearibus æquilongis. Labellum oblongum, integrum, muticum, cum labello parallelum, basi excavatum bicallosum. Columna semiteres, erecta, apice utrinque auriculata. Anthera unilocularis. Pollinia 2, posticè excavata; caudiculâ lineari, glandulâ minutâ.——Folia teretia maculosa. Panicula capillaris, 3-pollicaris. Flores parvi, flavi. Sepala linearia obtusa et labellum oblongum acutum cellularum majorum lutearum in medio raphidophorarum copiâ repleta. Columna linearis, petalorum ferè longitudine, auriculis acutis inflexis.

It gives me great pleasure to name this curious plant after Edwin J. Quekett, Esq. F.L.S. an excellent Botanical observer, and one of our most skilful vegetable anatomists. I am already indebted to that gentleman for some valuable facts concerning raphides; a plant, therefore, in which these crystals form a conspicuous part of the structure, may not be inappropriately selected to bear his name.

7. CYCLŌSĬĂ *maculātă*, (Klotzsch in Allgem. Gartenzeitung, no. 39. 1838.) a supposed new Mexican Orchidaceous plant, appears to be Mormodes pardina *Bateman in Bot. Reg.* 1838. *misc. no.* 176.

8. OBERŌNĬĂ *recŭrvă;* acaulis, foliis brevibus acutis, racemo recurvo multifloro, bracteis ovatis integris, petalis obovatis subdentatis, labello subrotundo quadrilobo denticulato mucrone interjecto.

Bombay has produced this curious little plant, which was sent overland to Messrs. Loddiges. It has a pendulous raceme, scarcely more than an inch long, consisting of minute densely imbricated green flowers. Its nearest affinity seems to be with *O. Wightiana, m.* an unpublished species from Madras, of which the following is the character.

9. OBERŌNĬĂ *Wightiană;* acaulis, foliis ensiformibus, racemo erecto multifloro, bracteis ovatis integris, petalis linearibus, labelli quadrilobi lobis lateralibus rotundatis integris intermediis elongatis truncatis apice denticulatis.

10. MEGACLĪNĬŪM *oxypterum;* pseudobulbis monophyllis acutissimè 4-5-gonis oblongis, folio oblongo coriaceo caulis sine rachi longitudine, rachi ensiformi arcuatâ crispatâ apice tetragonâ pyramidali, bracteis reflexis, sepalis lateralibus ovatis intermedio lineari acuminatissimo, petalis nanis linearibus falcatis, labello crasso linguiformi apice recurvo margine pone basin fimbriato.

A fine species of this very curious genus, obtained from Sierra Leone by Messrs. Loddiges. It is most nearly related to *M. maximum,* from which its sharp-angled pseudo-bulbs at once distinguish it.

11. PLEUROTHĀLLĬS *bicarinātă;* folio oblongo coriaceo planiusculo leviter carinato basi subcordato petiolo acutè canaliculato univaginato breviore, sepalis linearibus æqualibus lateralibus carinatis ad apicem ferè connatis, sepalis lineari-obovatis minutè serratis glabris, labello obovato carnoso medio exarato denticulo inflexo utrinque prope basin.

A native of Brazil, and imported by Messrs. Loddiges. The leaf is five inches long, the petiole six, with a large withered sheath in the middle. The flowers are dull greenish yellow, not unlike those of *P. saurocephala.*

12. SCHOMBŪRGKĬĂ *marginātă.*

This most beautiful Orchidaceous plant, of which there is a figure in the Sertum Orchidaceum, t. 13, and of which so many plants were brought to England in 1834 by Mr. Lance, has at length flowered in this country, with Thomas Brocklehurst, Esq. of the Fence near Macclesfield, who recently imported it from Surinam. The flowers were much paler in their colours than those of the plant in its native country, but this was doubtless owing to the dark season of the year.

13. EPIDĔNDRŪM (HORMĪDĬŪM) *uniflōrŭm;* rhizomate moniliformi, floribus solitariis, sepalis petalisque linearibus acuminatis incurvis, labello trilobo columnæ adnato laciniis lateralibus rotundatis erectis intermediâ triangulari acuminatâ, sepalis lateralibus labello suppositis eique adnatis.

A Mexican plant of no beauty, with yellowish green flowers, imported by George Barker, Esq. of Birmingham.

The genus Epidendrum I once thought very natural; but it is now becoming so very extensive, that it comprehends plants with extremely different habits, and it is daily becoming more desirable for it to be divided. But great as is the diversity of appearance among the species, there is a singular uniformity in the structure of their flowers, and it is not a little remarkable that such differences as exist can scarcely be said to be connected with corresponding differences in the organs of vegetation; so that, as far as I am at present able to discover, if Epidendrum is broken up, by means of characters taken from such modifications as are employed for the definition of other genera, the new groups are scarcely more natural than the old one. For this reason I have suppressed the genera Auliza and Amphiglottis of Salisbury, and the Encyclia of Sir William Hooker, as genera depending upon mere differences of habit and not of fructification. For instance, Encyclia has the labellum separate from the column; and if this were always connected with the pseudo-bulbous stem and panicled inflorescence of many of the species, it would be an excellent character; but *E. equitans, smaragdinum,* and others, have the disunited lip with an entirely different mode of growth; this is especially the case with the beautiful *E. bicornutum.* So with Auliza of Salisbury, the type of which is *E. ciliare;* this supposed genus was distinguished by there being a long fistular cavity proceeding from the base of the lip down the ovary, or, in other words, by its labellum being calcarate, and the calcar consolidated with the ovary, as happens in Pelargonium; but the same character exists in *E. vesicatum,* which has no resemblance to *E. ciliare.* With regard to the Mexican plant recently published by Messrs. Knowles and Westcott under the name of *Prosthecia* or *Epithecia glauca,* I fear it must be united with Epidendrum, for it does not differ from the greater part of the Encyclia division in any essential circumstance; the columna is by no means "nana" in the sense in which that word is employed by Botanists, and the process at the back of the apex of the column is common in numerous species of *Epidendrum,* (*rigidum* for instance); and is what, when thinned away, forms a hard or petaloid scale at the back of the column in such plants as *E. ciliare, clavatum, nocturnum, cucullatum,* and many others.

It is for reasons of this kind that I hesitate actually to separate the curious little plant which has given rise to these remarks. Its necklace-shaped pseudo-bulbs placed upon a creeping rhizoma give it a peculiar habit, and have suggested its name, (ὅρμος, a necklace); and it is capable of being defined with apparent precision by the union of the anterior sepals with the base of the labellum, to which I know of no parallel in the genus Epidendrum, except in the case of *Epidendrum pygmæum* and *E. cæspitosum,* plants with a similar habit, and no doubt belonging to the same section, subgenus, or genus, whichever Hormidium may eventually become.

14. BRASAVŌLĂ *grandiflōră;* folio (plano ?) angustè lanceolato rigido labello maximo subrotundo-quadrato acuminato ungue longiore, sepalis petalisque linearibus acuminatissimis.

Flowers of this, which is much the finest species of Brasavola yet known, have been lately received from Honduras by the Hon. W. F. Strangways. The limb of the labellum, which is white, is considerably larger than a half-crown, and the sepals and petals are nearly three inches long. Those who have commercial relations with Belize should make a point of obtaining this beautiful plant from their correspondents. It resembles a gigantic specimen of *Br. nodosa.*

15. PONĒRĂ *graminifōlĭă,* (Nemaconia graminifolia, *Knowles and Westcott, Floral Cabinet, p.* 127); foliis lineari-lanceolatis planis, labello oblongo acuto apice recurvo.

In the *Genera and Species of Orchidaceous plants,* p. 113, a genus was established under the name of Ponera, (from πονηρος, miserable, vile, alluding to the wretched appearance of the species) from a Mexican plant in the herbarium of Mr. Lambert. As I had originally no opportunity of critically examining the structure of the plant, further than was afforded by the dissection of a single dried flower, its character was not very complete; and I presume the trifling differences which exist between this species and the description in the above work, are owing to imperfect observations in the latter case. They have however led Messrs. Knowles and Westcott to suppose that this plant forms a new genus, an error which I hasten to correct. Having received flowers of it from Mr. Barker, I find that it agrees in every essential particular with the structure of *P. juncifolia;* so nearly indeed that, if the leaves of the two were not very different, some doubt might be entertained of their being specifically different. *P. graminifolia* is a plant of no beauty, and is chiefly remarkable for having the scabrous stems, which are so conspicuous in the elegant *Arpophyllum spicatum* from the same country.

16. ARPOPHYLLUM *spicatum*, La Llave.

One of the most graceful and beautiful of the Mexican Orchidaceæ; it has recently reached England in a living state. The stem is slender, and a foot or foot and half high, with the sheaths as rough as shagreen leather; it is terminated by one long curved leaf, from the axil of which there curves, in an opposite direction, a dense spike of pink or pale purple flowers. It is a species of great rarity even in Mexico, where it has hitherto been only seen at Sultepec and near Arimbaro, growing upon trees. The genus belongs to Epidendreæ, near Ponera, and not to Vandeæ.

17. HUNTLEYA *violacea*. *Sertum Orchid. t.* 26. *ined.*

Of this superb plant a specimen has flowered with Messrs. Loddiges, which will be figured in the Sertum in the course of a few months. The flowers are a deep rich violet, darker than the darkest part of *Zyzopetalum Mackaii;* they are between two and three inches in diameter, and are tipped with greenish yellow, melting downwards into white, which soon confounds itself with the general tint of rich violet.

20. CHEIROSTYLIS *parvifolia;* foliis ultra vaginam petiolatis ovatis acutis undulatis, labelli apice subrotundo 4-dentato callis baseos bipartitis, columnâ minimâ processibus rectis liberis ipsi æqualibus.

Herba tenera, zeylanica, 3-pollicaris; *caule* stricto, tereti, retrorsum piloso apice nudo bracteato. *Folia* 4, distantia, supernè sensim minora; vaginata, glabra, ultra vaginam petiolata; limbo ovato, undulato, acuto, rubro-viridi, 4 lineas longo. *Flores* 3, terminales, minuti, albi, pedicellati, corymbosi; bracteis linearibus acuminatis rufis, pedicellis longioribus. *Ovarium* obpyramidale, piloso-glandulosum. *Sepala* clausa ultra medium connata, glabra, basi paulo ventricosa, obtusiuscula, apice virentia. *Petala* retrorsum falcata, retusa, sepalo supremo agglutinata. *Labellum* sepalis parum longius, liberum, canaliculatum, versus apicem constrictum, apice subrotundum, concavum, trilobum: lobis lateralibus emarginatis; callis baseos incurvis, lucidis, subvirentibus, bipartitis. Ante columnam *processus* duo, liberi, erecti, carnosi, stigmatis longitudine, eique omninò paralleli. *Columna* minima, libera, basi labelli fere abdita, *stigmate* bipartito. *Pollinia* 4, pulverea, glandulæ lineari acutæ inter brachia stigmatis prominenti, agglutinata.

This very interesting, but inconspicuous plant was obtained by Messrs. Loddiges from Ceylon. It is a new species of the genus Cheirostylis, of which one only had been before described by Dr. Blume, and of that I have never been so fortunate as to see specimens. In general aspect it is like a minute *Goodyera, G. querceticola* for instance; but it differs from that genus, firstly, in having the sepals united into a tube enclosing the petals and labellum; secondly, in the absence of a pouch from the lip, and, as far as has yet been observed, from all other genera of Neottieæ in having a pair of fleshy processes, analogous to what we find in Habenaria, standing freely in front of the column. This fact is the more interesting to those who occupy themselves with organography, because it proves that the calli found so constantly at the base of the lip of Spiranthes are not analogous to those processes as might be suspected: for in this Cheirostylis the calli and processes are both present.

21. SCAPHYGLOTTIS *reflexa;* foliis semiteretibus supra planis sulcatis apice integerrimis, sepalis lateralibus ovato-triangularibus reflexis, petalis linearibus obtusis, labello oblongo-emarginato obtusissimo undulato medio refracto.

A branching, straggling plant, with slender leaves, and solitary, pale, dull-yellow flowers, with a crimson stain along the middle of the lip. Like the rest of the genus it is a species quite destitute of beauty. It was obtained by Messrs. Loddiges from Demerara.

22. MACRADENIA *mutica;* foliis coriaceis lineari-lanceolatis dorso convexis, racemo prostrato trifloro bracteis oblongis obtusis cucullatis pedicellis æqualibus, labello cordato-ovato acuto basi cucullato medio transversè calloso, clinandrio serrato, rostello mutico.

Folia solitaria, lineari-lanceolata, coriacea, avenia, dorso convexa, caulibus brevibus compressis teretibus insidentia. *Racemus* radicalis, 3-florus, debilis; *bracteis* brevibus, cucullatis, striatis, sphacelatis. *Flores* sordidè albi, rubro levissimè tinctæ. *Sepala* et *petala* æqualia, linearia, acuta, explanato-patentia. *Labellum* cum columnâ parallelum, nec articulatum, cordato-ovatum, acutum, basi cucullatum; callo elevato canaliculato transverso. *Columna* semiteres, labello duplò brevior; *clinandrio* cucullato denticulato. *Anthera* parva, in fundo cuculli, 2-locularis, cristata. *Pollinia* 2, *caudicula* brevi lineari. *Rostellum* nullo modo elongatum.

A small plant, with dingy white flowers, which flowered with Mr. Knight, of the King's Road, in August 1835. It is said to be a native of Trinidad. It is chiefly remarkable for being destitute of the attenuated rostellum and consequent prolongation of the anther point, from which the name of the genus was formed by Dr. Brown.

ORCHIDACEÆ OF BRAZIL.

The following is a free translation of M. Descourtilz's MSS. account of the Orchidaceæ of Brazil, and I am sure it will be read with interest by every one occupied, in however slight a degree, in the cultivation of this singular and beautiful race of plants.

It is in the bosom of the vast solitudes of America that these, the most diversified of plants, spring up, flower, and perish. The entire life of a man, though devoted to their special study, would never finish their examination, so prodigious is the variety of their species, many of which are only seen after the fall of the protector upon which they lived.

There is no part of Brazil, no latitude, no elevation above the sea, where are not to be found Orchidaceæ as different from each other as the conditions under which they grow. Some bask in the heat of the plains, others luxuriate in the agreeable freshness of a stream of water, attaching themselves to the branches of the trees which cover the waves with a verdant grotto; and others, real children of the mist, delight in a drizzling atmosphere, and support with ease the violence of stormy winds, and the often icy coldness of the Serras, whether stationed within a few feet of the earth, or swinging in the air from the boughs of the ancient patriarchs of the forest. Some grow in deep recesses and gloomy arcades, where there is a perpetual circulation of a damp and heated atmosphere; others, on the contrary, prefer the open glades, or Roças, where some fallen trees, whose own foliage has perished, supply them with a scanty but sufficient nourishment.

It is impossible to form an idea of a tropical forest by the woods of Europe, where the ivy is the only parasite which finds a permanent support. The *Sertoes*, or virgin woods, which cover a part of America, present the traveller with scenery incomparable for its majestic character, and rich variety. Who is there that would not be astonished at finding himself amidst a vegetation, of which each individual struggles with its neighbour for existence, darting up, eagerly searching for the light of a cloudless sun and a purer air, only to be found at a prodigious elevation, and leaving darkness and water at their feet. It is here that trees of patriarchal age perish in the embrace of enormous climbers which overwhelm and bear them down, and which are sometimes carried overhead like cables, in other cases interlaced like the meshes of a net, and not unfrequently stand like lofty leaf-capped columns of spiral open-work, after the trees about which they have writhed themselves have fallen to decay within their grasp.

Amidst this forest of ropes of sylvan rigging, grow innumerable Ferns, which hang down in plumes, or festoons, or the gayest lacework, vast quantities of Araceous plants, and especially Tillandsias, forming broad patches of verdure upon a sombre ground. In the midst of airy garlands of Aristolochias, Bignonias, Convolvuluses, and Passion-flowers live the Orchidaceæ, each particular species of which seems to haunt its own peculiar plant. Thus the Epidendrum of the Cinchona refuses to live in the branches of the Lecythis and Couratari, notwithstanding that the seeds of these epiphytes are scattered indiscriminately by the wind. Other tribes again are always from free Orchidaceæ, as the huge trunks of Malvaceous trees, Isoras, Carolineas, Plantains, and Palms.

It is chiefly at the time of flowering that Orchidaceæ become remarkable in their native haunts, and then less for the diversity of their forms, or the gaudiness of their colours, than for the exquisite perfume which most of them exhale. It is thus that I have often been led to the discovery of charming species, lurking amidst the foliage, and which my eye would never have detected.

Changes of seasons are announced with the greatest regularity by the Orchidaceous epiphytes, many of which expand their blossoms amidst hurricanes and torrents of rain which deluge the earth at certain seasons; but seem struck with torpor when the sky recovers its serenity.

25. CŒLŎGȲNĒ *ocellātă.* (Gen. et Sp. Orch. 40.)

This beautiful plant has just flowered imperfectly with Messrs. Loddidges, who imported it from India. The sepals and petals are pure white; the lip is also white, but it has two very bright orange yellow spots on each lateral lobe, and two others smaller, and of the same colour at the base of the middle lobe, besides which there are some lateral streaks of brown. The column is bordered with brilliant orange yellow. The crests of the lip are three, which converge towards the base of the middle lobe, and there the lateral ones diverge again over a pair of convexities, beyond which they disappear; on each side of them, at the base of the said convexities, and on the outside, is an additional short curled crest. The flowers grow in erect racemes about six inches long, and are themselves an inch or more long.

26. DENDRŌBĬŪM *linguæfōrmĕ.* (Swartz.)

I have formerly received this curious plant in flower from various collections, but never in such perfection as from the garden of Richard Harrison, Esq. of Aighburgh, who cultivates it upon the decayed branch of a tree. It inhabits the country near Sydney, in New South Wales, where it was found by the earliest Botanists who visited that colony. It has hard, thick tongue-shaped leaves, lying flat upon the rhizoma which creeps upon the rock or tree it grows on; they have the texture of an Aloe, and probably indicate that the species does not inhabit damp shaded places, but rather such hot dry situations as Mr. Cunningham assigns to *Dendrobium æmulum* and *undulatum;* see Bot. Register, fol. 1699, for observations upon the culture of these plants. The flowers are greenish white, with long slender sepals and petals, and appear from the young ends of the rhizomata. They have no smell, nor any feature of beauty.

27. SARCOCHĪLŪS *olivăcĕŭs;* foliis oblongis membranaceis falcatis apice acutis integris, racemo flexuoso angulato 4-5-floro nutante, sepalis lineari-obovatis obtusis: intermedio latiore petalisque conformibus minoribus columnæ dorso adnatis, labelli cum columnâ productâ articulati lobo medio minimo transverso rotundato lateralibus acuminatis; disco callis variis crassis rotundatis tuberculato.

A New Holland epiphyte of no beauty, imported by Messrs. Loddiges. It is very nearly the same as S. *parviflorus,* from which it principally differs in its leaves being broader and not emarginate at the point. The flowers are small and yellowish orange green.

29. DENDRŌBĬŪM *teretifŏlĭŭm.* (R. Brown prodr. nov. holl. 189.)

This curious plant, which is well defined by Dr. Brown, has lately flowered with Messrs. Loddiges. It is a small creeping species, with deep green fleshy taper leaves, between two and three inches long, and solitary flowers of a dull yellow, streaked and spotted with dull purple; the labellum is white. It is a plant of no beauty.

As the importation of New Holland Orchidaceous epiphytes has taken place to some extent, it is desirable that there should now be made known an account of several species first discovered by Mr. Allan Cunningham, and of which notes were given me by my indefatigable friend in June 1834. The credit of his discoveries will thus be secured to their author, those who may import the plants will be able to determine their names, and to ascertain whether they are new or not, and there will be the additional advantage of proving that no species of striking beauty are to be expected from this part of the world, as none of those found by Mr. Cunningham can be considered more than Botanical curiosities.

NOTES OF ORCHIDACEOUS EPIPHYTES OF NEW SOUTH WALES.

By Allan Cunningham, Esq.

30. DENDROBIUM *tetragonum;* caulibus gracilibus erectis elongatis quadrangularibus apice clavatis 2-3-phyllis, foliis ovato-lanceolatis acutis undulatis striatis s. nervosis glabris racemo terminali 2-4-floro longioribus, foliolis perianthii elongato-linearibus patentibus, labello vittato: disco tricarinato, lobo medio dilatato subcordato acuto.

An epiphyte, hanging loosely from the stems of small trees, in *dry* shaded woods, Moreton Bay. June 1828.

31. DENDROBIUM *tortile.* (Perhaps a Polystachya.)?

On the upper branches of the loftiest trees of Flindersia australis, 100 feet high; in shaded woods, Brisbane River, Moreton Bay. Oct. 1824. June 1828.

32. DENDROBIUM? *pygmæum.* (*D. Caleyi,* Cunn. in Bot. Mag. Compan. II. 377.)

On rocks in the Illawarra district. Flowers not known.

33. DENDROBIUM *elongatum;* caulibus articulatis teretibus sulcatis elongatis erectis 4-5-phyllis, foliis ovato-lanceolatis acutiusculis apice obliquis emarginatis, racemo terminali 3-10-floro, foliolis perianthii ovatis obtusiusculis, labelli disco 3-carinato lobo intermedio lato cordiformi.

On trees in shaded dark woods, on the banks of the Brisbane River. Sept. 1828.

34. DENDROBIUM *pugioniforme;* caulibus repentibus nodoso-articulatis radicalibus setoso-stipulatis, foliis elliptico-lanceolatis apice attenuatis acutis carnosis infra fere convexis carinatis lævibus nitentibus, floribus (resupinatis) subsolitariis axillaribus, perianthii foliolis oblongo-lanceolatis acutis patentibus, labello longitudinaliter tricarinato, lobo intermedio dilatato undulato-crispato apice acuto recurvato.

A beautiful epiphyte, hanging loosely from the stems of trees, so as to swing freely to the wind that sweeps through the forests on the summit of the belt of mountain bounding the coast district of Illawarra, on the west (lat. 34½ S.) and flowering in August. 1818.

Obs. This species approaches so near D. *rigidum* Br. prodr. (not of Gen. et Sp. Orch. p. 86. n. 51) that I have often viewed it as the same plant which was described by Mr. Brown, from a figure in the Banksian collection. It now appears to me distinct.

35. DENDROBIUM? *crassulæfolium.*

Of this I have no specimens or description. It is a plant growing on trees in ravines of the Blue Mountains. Never seen in flower. The pseudobulbs are oval and one-leafed; the leaf is small and oval.

36. DENDROBIUM? *complanatum.*

An epiphyte growing in tufts. It has a flattened base, and cultriform distichous leaves: the whole plant, although healthy, is of a yellow green colour. On dead trees, in shaded woods, at Moreton Bay. 1828. Flowers not seen.

37. CYMBIDIUM *iridifolium;* foliis elongato-linearibus, racemis multifloris bracteatis, perianthii foliolis patentibus (exterioribus tribus ellipticis obtusis) labello 3-lobato: lobo intermedio linguiformi disco ecarinato glanduloso-punctato nitido.

On decayed trunks of trees in large masses, damp shaded woods on the Brisbane River. July 1828. This plant was alive at Kew.

38. CYMBIDIUM ——?; caule radicante scandente, foliis alternis lanceolatis acutis subcarnosis, racemis axillaribus laxis, fol. perianthii conniventibus labello carnoso 3-lobato, basi cavo.

A Vanda looking plant, in habit very much like V. teretifolia. *Lindl. Coll. bot. t.* 6. Can it be Epidendrum triste, *Forst.* which that Botanist discovered in New Caledonia? My plant, of which I only detected a single specimen, was hanging to the thin laminated bark of the trunk of *Callistemon rigidum,* in small open savannahs, subject to inundation; near the banks of the Brisbane River. Sept. 1829.

39. CALANTHE *veratrifolia. R. Br.*

In August 1822, whilst on an excursion to the Illawarra, a coast district on the south of Port Jackson, I met with a plant in dark shaded woods, which I introduced to Kew, in the following year, considering it a Bletia. It soon afterwards flowered in that collection, and was then ascertained to differ in no one respect from the Java plant. They have repeatedly been in flower together since that period, and on a close critical examination of the two plants, no difference could be discovered, excepting that the Australian plant is not so purely white in the flower as the Java one; their identity is therefore clearly determined. Illawarra district, near Port Jackson, (lat. 34½. S.) is therefore another locality.

40. PHAJUS *grandifolius.* Loureiro.

This plant I discovered in Sept. 1824, growing in extensive swamps at the back of the beach, on the shores of Moreton Bay, in lat. 27½, from whence I sent a large tub of the plant to Kew, where, on flowering, it proved to be identical with the old Limodorum Tankervilliæ; and of its flower M. Bauer made a drawing. Moreton Bay therefore is another locality.

41. DENDRŌBĬŪM *Heynĕānŭm.* Gen. & Sp. Orch. p. 90.

This very pretty species has flowered imperfectly with Messrs. Loddiges, who received it from Bombay by the overland conveyance. Its exact locality was not before known, the specimens in Heyne's herbarium, where I found the species, affording no information upon that point. In a healthy state the plant forms a tuft of slender clavate stems, from four to six inches high, loosely covered with the withered sheaths of leaves that have fallen off. From all sides there appear in the flowering season slender spreading racemes about three inches long, having each from five to nine smallish white flowers, with a greenish-yellow lip beautifully streaked with violet; they have no smell.

42. LÆLIĂ *majālis.* Lindl. mss. (*Cattleya Grahami.* Gen. & Sp. Orch. 116.)

This plant, the *Flor de Mayo* of the natives of San Bartolo, and the adjoining parts of Mexico, has lately been received by the Horticultural Society of London from Mr. Hartweg, who found it on the mountains near Leon, growing upon oak trees, at the height of 8000 feet, where it sometimes freezes. I possess specimens (No. 3.) given me by Prof. Schlechtendahl, gathered by Dr. Schiede in the same situations. It is one of the most beautiful of the whole order; a dried flower now before me, of a bright violet colour, measures nearly five inches from the tip of one petal to that of another, and when fresh I have no doubt the expansion of the flower was as much as six inches. The labellum is still nearly two inches and a half long. Many plants of this magnificent species have been given away by the Horticultural Society; but it proves exceedingly difficult of cultivation.

43. OCTOMĔRĬĂ *tridentāta;* folio ovali-lanceolato crassissimo acuto, floribus fasciculatis, labello oblongo basi angustato apice tridentato auriculâ utrinque rotundatâ inflexâ: lamellis obsoletis.

A Demerara plant of no beauty. The flowers are bright yellow; the leaves are remarkably thick and hard.

44. POLYSTĂCHȲĂ *affinis.* Gen. et Sp. Orch. p. 73.

This has lately flowered with Messrs. Loddiges, who imported it from Sierra Leone. It proves extremely different from *P. puberula,* of which I once thought it might be a variety, and has larger flowers than any of the racemose species of the genus. Sir W. Hooker has observed a gland and caudicula in my *Polystachya macrantha* (Bot. Mag. t. 3707.), which is probably the *Dendrobium galeatum* of Swartz; I had previously noted the same thing in *P. ramulosa* (Bot. Reg. 1838. misc. no. 144.), and I find a similar structure in the present species; it is therefore probable that the genus Polystachya should be placed in Vandeæ, among the genera belonging to which subdivision a better station may be found for it than next to Dendrobium.

45. ISOCHĪLŬS *līvĭdŭm:* pseudobulbis fusiformibus attenuatis squamis pallidis membranaceis arctè vestitis, foliis solitariis linearibus apice obscurè 2-3-dentatis, racemo capillari secundo nutante subtrifloro folio multò breviore, sepalis petalisque ovatis concavis patulis, labello cordato revoluto apice foveato basi mellifero.

A small dingy-flowered plant, imported from Mexico by George Barker, Esq. of Birmingham. Its livid semitransparent flowers, and slender pedicels, are accompanied by spindle-shaped pseudobulbs, tightly covered with sheaths, resembling in colour the external skin of the onion which gardeners call "the silver-skinned."

46. DENDRŌBĬŪM *macrophȳllūm;* foliis ovato-oblongis obtusis nervosis basi subcordatis, sepalis lanceolatis lateralibus parùm productis, petalis oblongis acutis, labello pubescente convoluto denticulato subunguiculato ovato callo baseos elevato transverso obsoletè trilobo.

This, the handsomest of the Dendrobia, has been received by Messrs. Loddiges from Manilla, where it was found by Mr. Cuming. The flowers are nine inches in circumference, aud will probably be still larger when the plant becomes more healthy. The sepals and petals are a clear and bright rose-colour, the lip is downy and deeply stained with two large broad blood-red blotches at its base.

47. CĂTTLĒYĂ *supĕrbă;* foliis ovato-oblongis obtusis coriaceis marginatis caule clavato brevioribus, sepalis oblongis acutiusculis, petalis lanceolatis acutis membranaceis duplò latioribus, labelli trilobi cucullati lobis lateralibus acutis: intermedio transverso plano denticulato emarginato subunguiculato basi venis elevatis rugoso; callis duobus pone basin. *Sertum Orchidaceum, t.* 22.

This magnificent sweet-scented Cattleya has been found in British Guayana by Mr. Schomburgk, who sent a live plant of it to Messrs. Loddiges, and a drawing to the Linnean Society, by permission of which a figure has been published in the Sertum Orchidaceum. The flowers, if not so large as those of Cattleya Mossiæ, are, from the richness of their colours, inferior to none in beauty. According to Mr. Schomburgk, the plant appears peculiar to the 3rd or 4th degree of N. Lat.; it is not to be met with in the Essequibo north of the mouth of the Rupununy; from thence it is found southwards on trees which skirt the banks of the brooks and rivers which meander through the savannahs. He discovered only a few solitary specimens in the Essequibo south of the Cayuwini, and none at the equator. The Caribees call it *Oponopodoli,* or *Ducksmouth,* the Macoosees *Masame.* It is very fragrant; the odour in the morning is said to become too powerful in a confined place; the splendid flowers last from three to four weeks.

Although only now brought into notice, the species was many years since discovered by Dr. Von Martius, who found it near Taruma on the banks of the Rio Negro, in woods at the Barra de Rio Negro, and in forests near Parà. It is readily distinguished from all previously described species by its three-lobed lip with acute lateral segments, the middle lobe being flat, toothletted and emarginate, and by the cluster of elevated veins at the junction of the epichilium and hypochilium.

50. EPIDĔNDRŪM *glumăcĕŭm;* pseudobulbis ovatis apice angustatis diphyllis, foliis angustè oblongis patentibus, racemo terminali cylindraceo e squamis glumaceis acuminatissimis pedunculo longioribus erumpente, sepalis linearibus petalisque lineari-lanceolatis acuminatissimis, labello obovato acuminato convexo integerrimo basi unicalloso.

A Brazilian plant very near *E. fragrans,* from which it differs in the form of the lip, and in the colour of the flowers, which are white delicately striped with pink. It is a pretty species, with a raceme about five inches long.

51. GOVĒNĬĂ *Gardnĕrī* (Hooker in Bot. Mag. t. 3660); scapo obtusè tetragono medio vaginato, racemo elongato floribus post anthesin refractis, bracteis oblongis herbaceis ovarii dimidio longitudine, sepalis petalisque ovatis obtusiusculis, labello ovato acuto nudo apice maculis 5 marginalibus notato lineis duabus convexis convergentibus in medio, anthera cornu brevi inflexo.

The figure of this plant in the Botanical Magazine seems to have been taken from a specimen out of health, for the colour of the flowers and their markings are by no means such as I find them in a specimen now before me, for which I have to thank my friend John Miers, Esq. F. L. S. of 14, Hans Place, Chelsea. The flowers are pure white; the petals are delicately spotted with violet dots inside; the lip is downy at the base, and yellowish-green, except at the point, where it is white, with five small equidistant spots, of which the three in the middle are bright orange colour, and the two side ones, which are much more indistinct, dull purple. Mr. Miers tells me that he found it flowering in the month of February on the Organ Mountains, at two distant intervals from 3000 to 3500 feet above the level of the sea; at the lower one in an exposed situation on a bank, at the higher, in a wood upon a quantity of rich mould in the hollow of a decaying tree.

52. SACCOLĀBĬŪM *micrănthŭm* (Gen. et Sp. Orch. p. 220.)

This curious little plant has flowered with Messrs. Loddiges. Its flowers are pale violet, with the limb of the labellum much darker. It should rather be referred to the genus Cleisostoma.

54. CYRTOCHĪLŪM *stellatum;* pseudobulbis diphyllis ovalibus compressis striatis inter squamas lanceolatas carinatas axillaribus, foliis ligulatis obtusis aveniis scapo multò brevioribus, scapo tereti erecto basi vaginato, racemo disticho multifloro, bracteis carinatis convolutis acuminatis glumaceis ovario longioribus, sepalis petalisque lineari-obovatis acutis stellatis, labello oblongo undulato acuto basi canaliculato striato, alis columnæ acinaciformibus integerrimis. *Sertum Orchidaceum, plate* 7.

This noble species, nearly related to C. flavescens of the *Botanical Register,* t. 1627, but differing in its much larger flowers, the sepals of which are by no means acuminated, but only drawn to a sharp point; in the greater breadth of the pseudo-bulbs; in its stature being four times as great; and in the labellum being white instead of yellow, has lately flowered in the Nursery of Mr. J. Youell, Nurseryman, of Great Yarmouth.

M. Descourtilz found it dispersed through the districts of Macahé and Bananal. It flowers in September and remains in that state till the end of January. It diffuses but a weak perfume, but the beautiful spikes, which seen at a distance make it resemble a mass of verdure strewed with large stars, render it a most remarkable object.

56. DENDRŌBĬŪM *Paxtōni*; caulibus teretibus sulcatis, foliis ovato-lanceolatis acuminatis apice hinc obsoletè emarginatis, pedunculis bifloris, sepalis oblongis acutis lateralibus basi parùm productis, petalis latioribus obovatis acutis serrulatis, labello unguiculato ovato concavo indiviso villoso margine multifido fimbriato.

This beautiful new Dendrobium has orange-yellow flowers, with a deep brown spot in the middle of the lip. It is related to *D. chrysanthum*, from which it differs in having the petals serrated, and in the surface and margin of the labellum. It was found at Pundua, at the foot of the Khoseea hills of India, by Mr. John Gibson, at that time employed as collector to His Grace the Duke of Devonshire, and at whose request it was named after Mr. Paxton. It was sent me from Chatsworth in April last.

57. PHOLIDŌTĂ *articulātă*; Gen. et Sp. Orch. p. 38.

This plant has been introduced to Chatsworth by Mr. Gibson. It is of no beauty; its flowers are small and dirty white, with a little yellow. It is different in habit from the common *Pholidota imbricata*, the stem being jointed like an *Otochilus*, and not pseudo-bulbous.

58. PHĀĬŬS *Wallĭchĭi*. Lindl. in Wall. pl. as. rar. t. 158.

Another addition to the Chatsworth collection, made by Mr. Gibson during his stay in India. The specimen sent me was a good deal damaged, but it appeared to be as stately and beautiful a plant as the common *P. grandifolius*.

59. TRIGONĪDĬŪM *tĕnŭĕ* (*Lodd. cat. no.* 582.); pseudobulbis ovalibus compressis monophyllis, folio ensiformi acutissimo scapo erecto tenui longiore, sepalis reflexis acutissimis, labello oblongo obtuso trilobo glabro apice reflexo medio appendice plano obovato emarginato adnato acuto.

A brownish purple species with a slender habit, introduced from Demerara by Messrs. Loddiges, with whom it flowered in May.

60. SCAPHYGLŎTTĬS *stellātă*; (Loddiges in litt.) pseudobulbis fusiformibus, foliis linearibus canaliculatis obtusis emarginatis, fasciculis sessilibus terminalibus, sepalis linearibus secundis patentibus petalis conformibus angustioribus, labello cuneato trilobo lobis lateralibus rotundatis intermedio acutiusculo.

This species nearly resembles Scaphyglottis violacea, (Bot. Reg. t. 1901) from which it differs in having larger flowers with spreading segments, and the lateral lobes of the lip as large as the middle lobe. It is a native of Demerara, whence it was obtained by Messrs. Loddiges.

65. AGANĬSĬĂ *pulchellă*.

A very pretty new genus of Vandeous Orchidaceæ, sent by Mr. Brotherton from Demerara to Messrs. Loddiges. It has a creeping rhizoma, with distant slender pseudo-bulbs, each tipped by a single leaf, and cream-coloured delicate flowers, resembling those of some Maxillarias in form. The genus is distinguished from Maxillaria by the brown sepals not being oblique at the base, and by the nature of the pollen-masses; from Encnemis it differs in the form of the flowers, and in their regularity. It may be thus defined.

AGANISIA. Perianthium patens, æquale; sepalis lateralibus haud basi productis. Labellum liberum, mobile, indivisum, concavum, hypochilio parvo concavo, ab epichilio cristâ transversâ glandulosâ diviso. Columna erecta, semiteres, marginata, apice utrinque brachio acuto patulo aucta. Anthera ecristata. Rostellum elongatum. Pollinia 4, collateralia, per paria connata (oo oo) caudiculâ lineari, glandulâ parvâ ovali.——Rhizoma repens, pseudobulbosa. Pseudobulbi monophylli. Racemus erectus, radicalis, 3-4-florus, foliis multò brevior.——1. *Aganisia pulchella*. Pseudobulbi attenuati. Folia oblongo-lanceolata, acuta, plicata. Sepala et petala oblonga, acuta, ochroleuca. Labellum oblongum, obtusum, concavum, cristâ luteâ. Brachia columnæ obliquè bidentata, acuta.

66. GOVĒNĬĂ *lagenŏphŏră*; pseudobulbo ovato vaginâ lagenæformi utriculatâ incluso, petiolis tetragonis, racemo longissimo multifloro, labello ovato obtuso omninò glabro, sepalis petalisque obtusis.

For this very distinct and curious species of Govenia I am indebted to John Rogers, Esq. Jun. who imported it from Mexico, and has obligingly furnished me with the following note concerning it.

"Root a solitary tuber; the old not perishing until the new one is nearly full grown, but then dying away completely; about the size of a duck's egg, forming aboveground, and of a bright green, marked with the scars of three or four sheaths. The innermost sheath which surmounts the tuber is entire, and resembles a Florence flask in shape; it is about eight inches high, two to three in diameter at the base, and three-quarters at its throat; translucent, or semi-transparent, containing about one-third of a pint of water.

"Leaves two, opposite, lanceolate ovate, eighteen inches long by four to five broad; articulated with their petioles, just at the top of the pitcher. The petioles are acutely four-angled, sheathing, so that their transverse section is an equilateral rhomboid, with concave sides. The flower-stem rises from the bulb, within the pitcher, and *opposite* to the midrib of the outer leaf; about three feet high, bearing from forty to fifty flowers, which expand rapidly, and continue long in perfection, exhaling in the forenoon the odour of Habenaria bifolia.

"The pitcher is generally full of water, all the rain and dew which falls on the leaves being conducted into it; and it is apparently absorbed by the plant, as, if not replenished, it disappears more rapidly than evaporation would account for.

"The fibres are simple, proceeding from the base of the tuber, and have a tendency to rise and run on the surface of the ground. The plant evidently delights in water when growing, and is apparently a native of bogs or swamps."

Upon comparing this with Swartz's account of his *Cymbidium utriculatum*, no doubt can remain of that plant being another species of Govenia, differing from the present chiefly in having a pubescent scape, and succulent white flowers; it will therefore have to be added to this genus under the name of *Govenia utriculata*.

67. BRASAVŌLĂ *glaucă*; foliis coriaceis oblongis obtusis planiusculis glaucis, spathâ uniflorâ, sepalis petalisque lineari-lanceolatis obtusis herbaceis, labello subsessili subrotundo acuto margine lobato, clinandrio dentato denti dorsali apice glanduloso.

A most curious Orchidaceous plant, obtained near Vera Cruz for the Horticultural Society by Mr. Hartweg. Its habit is so much that of a Cattleya that till it flowered it was expected to belong to that genus. It however proves to be a Brasavola, with very large flowers. A figure of it is in preparation for Mr. Bateman's noble work on the plants of this order from Mexico and Guatemala.

68. ONCĬDĬŪM *sanguĭneŭm*; ebulbosum, foliis oblongis coriaceis dorso carinatis, scapo longissimo paniculato, sepalis subrotundis unguiculatis lateralibus basi connatis petalisque crispis sublobatis, labelli trilobi subcrispi vernicati lobis subæqualibus intermedio retuso cuneato, cristâ ovatâ convexâ corrugatâ, columnæ alis rotundatis sublobatis antherâ puberulâ.

A noble species of this showy genus, with the habit of Oncidium carthaginense, but with straw-coloured flowers stained with crimson blotches. It was imported from La Guayra by Messrs. Loddiges.

69. CYMBĪDĬŪM *bicolor*. Gen. et Sp. Orch. p. 164.

This has flowered with Messrs. Loddiges, who imported it from Ceylon. It proves to be a handsome species, with flowers like those of C. alorifolium, only streaked and stained with very deep crimson. The species is readily distinguished by having a sac at the base of the lip.

71. DICHÆĂ *ochracea*; foliis linearibus acutis carinatis, pedunculis foliorum ferè longitudine, bracteâ cucullatâ ovarii longitudine, sepalis petalisque oblongis acutis, labello subrotundo rhomboideo sessili, columnâ pilosâ, clinandrio membranaceo-marginato, antherâ bigibbosâ.

A small Demerara plant, with narrow leaves, and pale yellow-ochre-coloured flowers. It is next *D. graminoides*, which differs in having smaller flowers with very short peduncles, and both shorter and flatter leaves. Messrs. Loddiges obtained it from Demerara.

76. ONCĬDĬUM *unicorne:* pseudobulbis ovalibus compressis diphyllis, foliis oblongo-linearibus recurvis, racemo composito ramis divaricatis rectiusculis, sepalis lateralibus in unum concavum emarginatum connatis, petalis obovatis undulatis, labelli lobis lateralibus nanis intermedioque emarginato rotundatis, disco basi transversè elevato anticè cornu ascendente compresso subfalcato acuto, alis columnæ angustis obsoletis.

This is a pretty little species, with a compound straggling raceme of pale yellow flowers. The singular horn on the lip, to which it owes its name of the "Unicorn," at once distinguishes it from all species previously described. Messrs. Rollissons imported it from Rio, and flowered it three weeks since.

77. EPIDĒNDRŪM *Candollei;* pseudobulbis sphæricis, scapo paniculato, sepalis petalisque obovato-oblongis, labelli liberi trilobi cucullati lobo medio crispo acuminato, disco elevato calloso sulcato pubescente.

The first knowledge I had of this plant was from M. Alphonse DeCandolle, who shewed me a drawing of a very imperfect specimen that had flowered at Geneva, and which I took for a variety of Ep. *asperum.* Subsequently the same species has flowered with Messrs. Loddiges, and I find that it is distinct. The flowers are dull brown, with a dull yellow lip, striped with the same colour. It is a Mexican plant.

85. DENDRŌBĬUM *bicameratum;* caulibus fusiformibus, foliis lineari-lanceolatis apice obliquè bidentatis, pedunculis lateralibus squamatis 4-floris, floribus concavis subcarnosis, sepalis petalisque subrotundo-ovatis acutis, labelli trilobi laciniâ intermediâ rotundatâ apiculatâ carnosâ lateralibus triangularibus acutis breviore, columnæ facie excavatâ semibiloculari.

A native of the north of India, whence it was brought by Mr. Gibson for the Duke of Devonshire. I am indebted to George Barker, Esq. of Birmingham, for a specimen. The stems appear to grow nearly upright, and are rather short, fusiform and furrowed when old. The flowers are a little smaller than those of Maxillaria stapelioides, and like them in form; their colour is dull yellow, spotted and streaked with purple. In the specimen I examined the flowers grew in fours, on a very short peduncle, closely covered with ovate concave bracts, speckled with purple. The excavation of the face of the column, and its division into two cells are remarkable.

86. GONGŌRĂ *nigrita;* hypochilii convexi cornubus lateralibus ascendentibus abbreviatis aristis setaceis, epichilio acuminato apice uncinato breviore.

This is much the darkest of the Gongoras, the appearance of the flowers being like that of the deepest puce-coloured velvet. It is very nearly the same as G. atropurpurea, but the lower half of the lip is convex not concave, longer than the upper half not shorter, and the horns at its sides are very short. It was imported from Demerara by S. Rucker, Esq. Jun. who informs me that the smell is quite different from that of G. atropurpurea.

91. PHAĬŬS bicolor. *Lindl. g. et sp. orch. p.* 128. *Sertum Orchidaceum, t.* 25.

This, which is one of the handsomest species in the fine genus to which it belongs, has flowered with Messrs. Loddiges, who imported it from Ceylon. Its sepals and petals are, in that island, a very bright deep red, and the lip is yellow; but owing to unhealthiness or some other cause the colours of Messrs. Loddiges' plant were by no means of their native brilliancy: for it is certain that they are fully as bright and rich as in the figure in the Sertum Orchidaceum.

92. GOODYERĂ *rubicunda* (Neottia rubicunda. *Blume Bijdr. p.* 408).

This plant has flowered with Messrs. Loddiges, who received it from Manilla, from Mr. Cuming. It has the habit of Goodyera procera, but is smaller; the scape and spike are rather more than a foot high, downy, and of a dull cinnamon brown. The flowers are also downy and of the same colour, with the exception of the lip, which is white, and densely fringed inside with glandular hairs. The whole of Blume's 3rd section of Neottia appears to belong to Goodyera.

93. MAXILLĀRĬĂ *lentiginosa;* bracteis latè ovatis acuminatis, labelli lobo medio ovato-oblongo obtuso, cristâ transversâ medio processu quadrato tridentato auctâ, antheræ apice incurvo; alioquin *M. stapelioidi* similis.

A plant very like Maxillaria stapelioides; but the sepals are more acute, the purple spots redder, more distinct, and less run into bars; the lip is of the same colour as the petals, and its transverse crest has a square three-toothed process in the middle. Imported from Brazil by Messrs. Loddiges.

94. VANDA *congesta;* foliis oblongis coriaceis apice obliquis mucronulatis, spicis capitatis subsessilibus, perianthio carnoso fragili, sepalis oblongo-linearibus petalisque lineari-spathulatis obtusissimis, labello oblongo limbo subdeltoideo crenulato papilloso basi excavato pubescente, columnæ angulis mucronatis.

A small yellow and brown-flowered species from Ceylon, communicated by Messrs. Loddiges. It is near V. multiflora in the structure of the flowers, but the dense inflorescence and bearded sac of the labellum are obvious marks of distinction.

109. ANGRÆCŪM *armeniacum;* caulescens, foliis distichis canaliculatis apice obliquè et acutè bidentatis, spicis lateralibus horizontalibus secundis, sepalis ovatis, petalis linearibus, labelli 3-lobi laciniis lateralibus acuminatis intermediâ triangulari longioribus, calcare pendulo clavato basi compresso calyce duplo longiore.

This very singular plant is a native of Sierra Leone, whence it has been obtained by Messrs. Loddiges. Its flowers are of a uniform apricot-colour, small, secund, and closely arranged in a horizontal lateral spike. The habit is that of a Saccolabium. It is very distinct from all that have before been seen, but has some affinity with Angræcum mystacinum. The spur is twice as long as the calyx, compressed at the base, and then inflated a little so as to appear clavate.

110. MALACHADENIA (Vandeæ).

Flos resupinatus. Sepala lateralia connata, apice reflexa libera, galeam formantia; dorsale cordatum acuminatum. Petala minima, squamæformia, rotundata. Labellum carnosum, margine revolutum, basi mucronatum, cum pede elongato columnæ sepalis lateralibus galeatis adnatæ articulatum. Columna antice bicirrhosa, basi longè productâ, stigmate lineari oblongo. Anthera 2-locularis, decidua. Pollinia 2, ceracea, sessilia, glandulâ molli, cubicâ, nudâ.——Herba repens, pseudobulbis monophyllis, scapo radicali, floribus carnosis galeatis.

MALACHENIA *clavata.*

Rhizoma repens. Pseudobulbi subglobosi, monophylli. Folia angustè ovalia, carnosa. Scapus radicalis, foliis ter longior, erectus, teres, apice clavatus, subbiflorus; axi spicæ articulatâ: internodiis clavatis. Spica terminalis, pauciflora. Bracteæ membranaceæ, ferrugineæ, reflexæ. Ovarium brevissimum. Flores obscurè virides purpureo maculati. Sepala lateralia acuminata, apice revoluta, basi obscurè semisagittata: supremum subcordatum: sinu lato inter sepala. Petala squamiformia, rotundata, obsoleta. Glandula polliniorum mollis, subcubica, cinnamomea, cui pollinia leviter annectantur.

This very singular plant was given to Mr. Bateman by Mr. Wm. Hooper of Lambeth, who received it from Rio in 1836. It is a remarkable genus, resembling Megaclinium in some respects, but belonging in reality to Vandeæ, among which it is marked by the nearly total absence of petals, the cirrhate column, and the soft fleshy cinnamon-coloured gland, to which a pair of reniform pollen-masses are slightly attached. In this division of Orchideæ it is uncertain where it must stand; probably other genera still undiscovered will connect it with the system better than can at present be done.

The scape is about nine inches long; the flowers are fleshy, dull green, slightly spotted with purple. The labellum is an ovate fleshy body with revolute edges, concave only on the upper side, and with two little mucronate processes at the base, one on each side.

113. STANHŌPĔĂ *oculata.* Botanical Register, t. 1800.

Var. *Barkeriana;* sepalis petalis et columnæ dorso purpureo maculatis, hypochilii sacco discolore.

This is a remarkable variety of *S. oculata,* obtained from Mexico by Mr. Barker. It looks like *S. insignis* with the lip of S. oculata, and is very handsome. The sepals, petals, and column are covered with numerous purple freckles rather than spots, which, as the flower fades, run together, as if their colouring matter were dissolved; so that at last the flower becomes of a dull wine-red tint.

118. CIRRHOPĒTĂLŪM *nutans;* pseudobulbis ovato-subrotundis rugosis, foliis ovato-subrotundis emarginatis coriaceis humi pronis, scapo erecto elongato, umbellâ multiflorâ nutante, bracteis linearibus acuminatis, sepalis glaberrimis: supremo acuminato lateralibus linearibus ligulatis vix acutis, petalis ovatis acutis serrulatis, labello obtuso convexo bicristato, columnæ angulis obsoletè bidentatis.

A pretty little epiphyte, sent to Messrs. Loddiges from Manilla by Mr. Cumming. It has a nodding umbel of pale straw-coloured flowers, at the end of a weak scape about six inches high. The leaves are from an inch and half to two inches long, very thick, emarginate, and lying almost flat upon the ground. The species is very near C. Wallichii, a Nepalese plant, the specific character of which in the Genera and Species of Orchidaceous plants is erroneous, in consequence of bad specimens and an Indian drawing having been misunderstood. It is necessary to correct the definition of that species as follows.

119. CIRRHOPĔTĂLŬM *Wallichii* (Gen. et sp. orch. p. 59.); foliis lanceolatis apice fissis scapi erecti longitudine, racemo multifloro pendulo, bracteis linearibus acuminatis, sepalis glaberrimis supremo acuminato lateralibus linearibus ligulatis vix acutis, petalis acuminatis subciliatis, labello . . . .

120. CIRRHOPĔTĂLŬM *fimbriatum*; pseudobulbis ovatis subtetragonis, foliis ovalibus . . . . . , scapo erecto, bracteis linearibus acuminatis, umbellâ multiflorâ, sepalis lateralibus ligulatis vix acutis cohærentibus supremo petalisque ovatis acuminatis fimbriatis, labello crasso linguiformi nudo obtuso, columnæ angulis cornutis edentulis, antherâ papillosâ.

A very pretty new species of this curious genus, with the long lower green sepals united into a channelled rather stiff strap, while the upper sepal and the petals are broken up at the margin into beautiful purple fringes. It is extremely different from all the species previously known, and was imported from Bombay by Messrs. Loddiges. In both this and C. nutans the pollen-masses are four in number, and collateral, all adhering together, with the interior pair much smaller than the two outer.

121. CIRRHÆĂ *saccata*; pseudobulbis ovatis obtusè tetrapteris inter angulos costatis, foliis oblongis plicatis sub-septem-costatis in pseudobulbos omninò sessilibus, scapo pendulo trigono, racemo multifloro, sepalis oblongis obtusis lateralibus margine revolutis, petalis lineari-lanceolatis obtusis basi teretibus compressis, labelli lobo medio galeato saccato.

This is a very distinct species of the curious genus Cirrhæa, for which I have to thank His Grace the Duke of Bedford, by whose directions it was sent me from Woburn, where it flowered in August. From all the previously known species it differs in the middle lobe of the lip being concave, and having very much the form of that of many Satyria. It has the general appearance of the other species, but its raceme is nearly a foot long, and the dull yellow green flowers are twice as large as those of any before discovered. It has been figured in the Botanical Magazine under the name of C. *fusco-lutea*, which is a different plant.

135. EPIDĔNDRŬM *inversum*; pseudobulbis elongatis compressis, foliis loratis canaliculatis obtusis, spicâ terminali sub-6-floro, bracteis brevibus ovatis acuminatis, ovario triquetro, sepalis petalisque patentibus lineari-lanceolatis convexis subæqualibus, labello adnato oblongo acuminato basi convexo ecalloso, columnâ obtusè 3-dentatâ, antheræ cardinis appendice oblongâ denticulatâ.

A Brazilian epiphyte, nearly related to *Epidendrum fragrans*, for which I am obliged to Messrs. Loddiges. The flowers are straw-coloured, with a few purple streaks on the column and at the base of the lip, and have a heavy not very pleasant smell, something like that of Ground ivy (*Glechoma*). Of this form of the genus Epidendrum, of which *E. fragrans* may be selected as the type, there are now several species on record, and it is probable that many more remain to be discovered; I am already acquainted imperfectly with more than one undescribed species. It will therefore be necessary to provide a distinct section for such species, to which the name of OSMOPHYTUM may be assigned, in allusion to their being usually scented plants.

137. SPECKLĬNĬĂ *obovata*; folio coriaceo obovato emarginato basi angustato canaliculato caule longiore, spicis brevibus fasciculatis, floribus glaberrimis, sepalis petalisque linearibus acuminatis, labello lineari abruptè acuto medio paululùm constricto, columnâ cucullatâ alatâ subdentatâ.

A small Brazilian plant, with the appearance of a Pleurothallis. The leaves are obovate and very thick; the flowers are small, pale yellow, scentless, and in numerous fascicled short spikes.

138. RODRIGUĔZĬĂ *laxiflora*; pseudobulbis ancipitibus ovalibus, foliis lineari-lanceolatis acutissimis, racemo laxo cernuo, bracteis ovario subæqualibus, labelli recurvi obsoletè bicristati appendice ungui æquali, sepalo antico angusto cuneato bifido postico petalisque planis.

I received this plant as long since as the year 1834, from Mr. Bateman, who first distinguished it from R. planifolia and recurva; since that time it has been found in the Organ mountains of Brazil by Mr. Gardner, of whose herbarium it is no. 654; and I have recently observed it in the collection of Messrs. Loddiges. It is a pale green-flowered species, much smaller than those just mentioned, with a very lax nodding spike, on which the flowers are generally arranged at considerable intervals; not always however, for I have specimens with the inflorescence as compact as that of *R. secunda.*

139. RODRIGUĔZĬĂ *crispa*; sepalis omnibus liberis petalisque undulato-crispis, labelli bicristati appendice ungue multò breviore.

This is the finest of the green-flowered species, and is remarkable for the crisped appearance of its flowers, which are sea-green bordered with yellow. Their fragrance is delicious, resembling that of Primroses. It is a native of Brazil.

140. CATASĒTŬM *proboscideum*; labello deflexo multifido: laciniis filiformibus laceris, columnâ apice proboscidiformi: cirrhis deflexis columnâ longioribus.

For this novelty I am obliged to George Wailes, Esq. of Newcastle, who received it from Mr. Gardner, marked "No. 2, new, found growing on a small species of Palm, near Sertâo." The plant is nearly related to *Catasetum cernuum* and *barbatum*, from which its deeply lacerated lip, and the longer proboscis of the column, seem to distinguish it; it may however be a mere variety of the latter species.

Mr. Wailes observes, that "it is somewhat like *Catasetum trifidum* in growth; the flower-stem is about ten inches high, rather drooping, with the upper part of a pinkish hue. The flowers in two specimens are nine in number, though in one the rudiments of two or three others are apparent, but probably owing to the weakness of the plant they have not come forward. When well established we may expect finer specimens."

The same plant has been communicated to me by Mr. Herbert.

143. LÆLĬĂ *flava*; pseudobulbis ovalibus 1-2-phyllis foliis carnosis coriaceis planis subconvexis brevioribus, scapo erecto foliis longiore squamulis quibusdam distantibus vaginato apice paucifloro bracteis minimis acutis, sepalis petalisque oblongo-linearibus obtusis, labelli lobo medio crispo recurvo lateralibus obtusis undulatis multò longiore.

"This plant, I believe, is a native of Mexico, from whence it was brought several years ago, and added to Sir Charles Lemon's collection at Carclew, where it flowered for the first time in the autumn of 1839.

"*Pseudo-bulbs* erect, roundish oblong, from two to three inches high, and about two and half or three inches in circumference at the base, from which they taper upwards and become one or two-leaved. They are smooth and of a deep shining green, nearly all concealed by several large, imbricated, thin, brown-coloured leafy scales. *Leaves* from three to five inches long, and about an inch broad, oblong lanceolate acute, very thick and stiff, recurved both at the point and edges, and of a dark green colour. *Scape* about a foot high, rising from the crown of the pseudo-bulb between the two leaves, round and erect, pale green, bearing three or four flowers near the top, with a small, persistent, acute bractea at the base of each, and a single brown-coloured sheath an inch long at its junction with the bulb. *Pedicels* nearly erect, an inch in length, somewhat clavate and channelled, and of a pale green. *Flowers* bright yellow, opening one at a time, and remaining expanded for a fortnight or three weeks. *Sepals* spreading, oblong-lanceolate, bluntish at the point, about an inch or more in length, and a quarter of an inch in breadth. *Petals* similar in form and colour to the sepals, but exceeding them a little in size. *Labellum* rather shorter than the petals, three-lobed, the middle one is recurved and undulated at the margin in the way of Cattleya crispa, with two raised longitudinal processes forming a sort of groove down the centre. The two lateral lobes are erect, with their edges plain, and meet above the column so as almost to conceal it. *Column* triangular, or nearly so, the upper edge only being a little rounder than the other. *Pollen-masses* eight, contained in four distinct cells, which on being opened exhibit on either side of the middle partition two pairs of roundish compressed waxy bodies attached to one another by a small elastic membrane.

"The plant requires the same treatment as other Orchidaceæ, thriving pretty well in moss and decayed vegetable mould."

For the above memorandum and accurate description of this quite new species I am indebted to Mr. Booth. The plant is nearly allied to *L. cinnabarina*, but it appears to be of a smaller size, with much more coriaceous leaves; the flowers are a clear pure yellow, and not a brilliant cinnabar colour, and the lateral lobes of the lip are obtuse not acute.

145. DICRŸPTĂ *discŏlŏr* (G. Loddiges *mss.*); foliis latis papyraceis subtus discoloribus, labello linguæformi lineâ mediâ elevatâ villosâ, antheræ margine angulisque papillosis.

For this addition to the genus Dicrypta I am indebted to Messrs. Loddiges, who imported it from Demerara. It is remarkable for the deep purple colour of the underside of the leaves, and its lip has a thick villous line running from the base almost to the apex. The flowers are orange-coloured, about the size of those of *D. Baueri*, and the species is altogether handsomer.

146. OCTOMĔRĬĂ *diăphănă*; folio ovato convexo acuto, squamis caulis equitantibus superioribus majoribus, floribus solitariis, sepalis petalisque ovatis acutissimis diaphanis, labello tridentato margine subcrispo lineis duabus elevatis.

A small but pretty species of this genus, with white transparent scentless flowers. It is readily known by the form of the leaves, and the large equitant sheaths that invest the upper part of the stem. Imported from Brazil by Messrs. Loddiges.

147. FERNANDĔZĬĂ *lŭnĭfĕră*; foliis obtusis petalis incurvis falcatis reduplicatis obtusis emarginatis, labelli 5-lobi disco tuberculato laciniis infimis linearibus falcatis petalorum longitudine lateralibus nanis rotundatis intermediâ dilatatâ bilobâ, columnâ a dorso compressâ latiore quam longâ dentatâ, stigmate verticali rimæformi.

A pretty and quite new species, resembling F. elegans in its foliage, but the flowers are thrice as large, and quite differently formed, especially as regards the great extension of a pair of supernumerary lobes situated at the base of the labellum, where they stand erect like two curved horns. It was imported by Messrs. Loddiges from Brazil, where it was first discovered by M. Descourtilz, who in his MSS. states that it grows upon trees, forming tufts of verdure, which from a distance remind one of the box of Europe. It is exceedingly common near Bananal. From the singular form of the lip he called it *Epidendre anthropomorphe*, or the Man Epidendrum, in allusion to the Man Orchis of Europe. Dr. von Martius also found it in woods near Ilheos.

148. MAXILLĀRĬĂ *acūtĭfŏlĭă*; pseudobulbis angustis ovalibus compressis monophyllis, folio oblongo-lanceolato acuminato, pedunculis unifloris radicalibus vaginatis, bracteâ carinatâ ovario longiore, sepalis petalisque oblongis, labello lineari-oblongo retuso emarginato medio pubescente apice glabro utrinque versus basin laciniâ brevi acutâ incurvâ aucto.

A species resembling *M. rufescens*, and having brownish orange flowers with little beauty. In the collection of Messrs. Loddiges from Demerara.

149. ONCĪDĬŬM *Forbĕsĭī*; Hooker in Bot. Mag. t. 3705.

For a specimen of this rare and handsome plant I am indebted to James W. Buller, Esq. of Downes near Exeter, who observes that it resembles Oncidium crispum more than any of the species he has seen, but is much more beautiful, having the petals gaily edged with bright yellow, and some of the sepals barred with the same colour. The plant has a large panicle and very gay appearance. I must however add, that I find the ground colour of the flowers very much browner than in the figure in the Botanical Magazine, in my copy of which it is brick red! The bidentation of the wings of the column is a variable circumstance, but the appendages of the base of the lip are constantly as represented and described by Sir Wm. Hooker.

150. ONCĪDĬŬM *excavātŭm*; Sertum Orchidaceum, sub t. 25.

This fine Peruvian plant has flowered with Messrs. Loddiges; it has yellow flowers, spotted with brown, and is easily known by the base of the labellum being very convex, a little hollowed out in front, and excavated with a deep pit on the under side. The petals and lower sepals are much more acute than in my wild specimens from Chachapoyas, and it is probable that the specific character will require modification.

151. CŒLŎGȲNĒ *ēlūtă*; Genera and Species of Orchidaceous plants, p. 40.

This fine species has lately flowered in the garden of the Horticultural Society, where it had been received from Dr. Wallich. The leaves are more than a foot long; the scape is terminal upon a large oblong pseudo-bulb, and is terminated by a sheath formed of numerous imbricated bracts, out of which appear eight or nine white flowers, stained with yellow near the point of the lip, and having an unpleasant smell, very like that of the Barberry blossom.

153. ODONTOGLŌSSŬM *Clŏwesĭī*; pseudobulbis ovalibus diphyllis, foliis ensiformibus angustis erectis scapo longioribus, racemo paucifloro corymboso, bracteis minimis setaceis, sepalis petalisque lanceolatis æqualibus, labelli cordati medio constricti apice subrotundo acuto basi lamellis 5 inæqualibus abruptis quincuncialibus aucto.

A very handsome Orchidaceous plant, for which I am indebted to the Rev. John Clowes of Broughton, a most zealous and successful Horticulturist, who received it from the Organ mountains of Brazil a few years since. Its flower-stem is about a foot high, and is terminated by four or five large starry flowers, yellow mottled with brown, while the lip is white with a rich violet base. A drawing of it, by Miss Mearns, will appear shortly in this work.

154. CATASĒTŬM *longĭfŏlĭŭm*; foliis longissimis gramineis, racemo cylindraceo pendulo multifloro, sepalis ovatis subrotundis petalorum conformium dorso applicitis, labello urceolari a tergo incurvo limbo truncato apiculato intus cereaceo glabro margine fimbriato.

This plant is in several collections where Demerara Orchidaceæ are grown, and is known as the long-leaved Catesetum which never flowers. It has however at last yielded to the good management of Valentine Morris, Esq. of the Retreat, Battersea, where it has produced its blossoms abundantly. It is too large a plant to suit the pages of the Botanical Register, and will therefore appear in the 7th fasciculus of the Sertum Orchidaceum, now in preparation. The flowers are bright orange, a little bordered with violet, and appear in a drooping raceme, over which they are closely packed for the length of a foot or more; they are extremely beautiful, and the species is beyond all comparison the handsomest of its genus.

155. PLEUROTHĀLLĬS *scăbrĭpes*; folio lineari-lanceolato apice tridentato caule longiore, vaginâ pilosâ caulis medium subæquante, flore solitario pubescente, ovario tomentoso, sepalis quam petala triplo-longioribus acutis lateralibus connatis, labello spathulato rotundato.

A curious little plant, transmitted to me by Mr. Booth with the following note.

"For specimens of this singular plant I am indebted to Michael Williams, Esq. of Trevince, who informed me that he received it in 1837, with some other Brazilian plants, from Lieut. Downey of H. M. Packet establishment at Falmouth, and that it has been successfully cultivated in a shaded part of the stove, attached to a small bit of wood, and kept very moist.

"The whole plant does not exceed five inches in height. The *stem*, which is one-leaved, is about two inches long, hard and round, with a groove on one side, and having for half its length a thin, pubescent, brownish covering, thickly marked with darker spots. *Leaves* upright, from two to three inches long, lanceolate acute, thick and leathery, nearly flat, of a deep green above, a little paler beneath, and somewhat rusty. *Flowers* one or two, very small, of a dull dingy yellow, striated with reddish purple lines, and issuing from a thin, keel-shaped, acute, brownish, spotted spathe at the base of the leaf. *Pedicels* short and round, nearly sessile, with a small acuminate, spotted bractea. *Sepals* two, the upper one slightly arched, lanceolate acute; the lower one similar to the other, but rather larger and more hollowed at the base. *Petals* ovate, acuminate, thin, and shining, striated like the sepals, and about half their length. *Labellum* dull reddish purple, a little longer than the petals, spathulate, and rounded at the point. *Column* very short, nearly concealed by the upper sepal and the two petals which support it on either side."

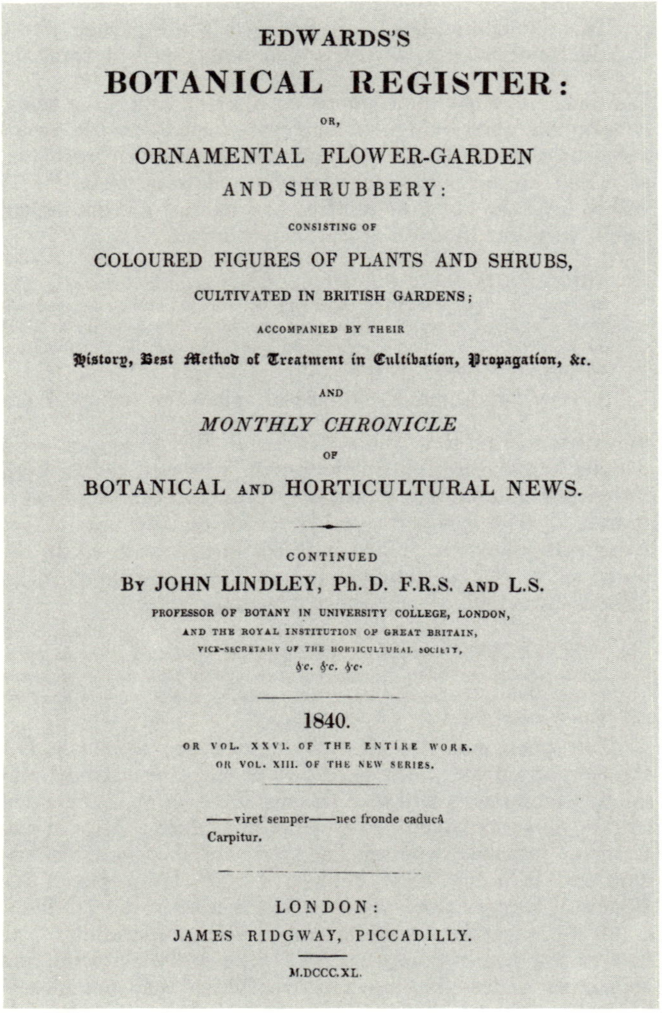
EDWARDS'S

BOTANICAL REGISTER:

OR,

ORNAMENTAL FLOWER-GARDEN

AND SHRUBBERY:

CONSISTING OF

COLOURED FIGURES OF PLANTS AND SHRUBS,

CULTIVATED IN BRITISH GARDENS;

ACCOMPANIED BY THEIR

History, Best Method of Treatment in Cultivation, Propagation, &c.

AND

MONTHLY CHRONICLE

OF

BOTANICAL AND HORTICULTURAL NEWS.

CONTINUED

BY JOHN LINDLEY, Ph. D. F.R.S. AND L.S.

PROFESSOR OF BOTANY IN UNIVERSITY COLLEGE, LONDON,
AND THE ROYAL INSTITUTION OF GREAT BRITAIN,
VICE-SECRETARY OF THE HORTICULTURAL SOCIETY,
&c. &c. &c.

1840.

OR VOL. XXVI. OF THE ENTIRE WORK.
OR VOL. XIII. OF THE NEW SERIES.

——viret semper——nec fronde caducâ
Carpitur.

LONDON:

JAMES RIDGWAY, PICCADILLY.

M.DCCC.XL.

6

EPIDĔNDRŬM glumācēŭm.

Glumaceous Epidendrum.

Nat. ord. ORCHIDACEÆ, § EPIDENDREÆ.
EPIDENDRUM. Botanical Register, vol. 1. *fol.* 17.

§ OSMOPHYTUM, *Bot. Reg.* 1839. *misc. no.* 135.

E. *glumaceum;* pseudobulbis ovatis apice angustatis diphyllis, foliis angustè oblongis patentibus, racemo terminali cylindraceo e squamis glumaceis acuminatissimis pedunculo longioribus erumpente, sepalis linearibus petalisque lineari-lanceolatis acuminatissimis, labello obovato acuminato convexo integerrimo basi unicalloso. *Bot. Reg. misc. no.* 50.

This very pretty plant was imported from Brazil by Messrs. Rollissons. It has the habit and sweet odour of E. *fragrans,* from which it differs in the colour of its flowers, in the long tapering figure of the flower-buds, which is caused by the peculiar form of the sepals and petals, and in the shape of the labellum, which is convex not concave, obovate, and delicately striped with pink at its base.

It derives its specific name from the long, withered, sharp-pointed, ribbed scales, resembling the *glumes* of grasses, out of which the raceme of flowers grows.

It submits readily to cultivation in the moist stove. The soil in which it grows best is rich brown turfy peat, well mixed with the materials of drainage. It requires a good supply of water, both at its roots and also over its leaves, and is not so easily injured from this element as many other plants belonging to the order. This treatment however is of course only applicable during its growing season, for, like other plants of the kind, it should be kept nearly dry at certain periods.

It is propagated by division of the rhizoma or prostrate ringed stem, out of which the pseudo-bulbs grow. Every ring is produced by the fall of a scale, or rudimentary leaf; in the axil of every one of those scales there is a bud, capable of becoming a young plant; under ordinary circumstances however, a few only of such buds develope and become pseudo-bulbs, the rest remaining dormant. But although dormant they are not dead; and if the rhizoma is cut to pieces, and thus the power of lengthening at the end arrested, the plant will seek to repair its losses by the developement of the dormant buds.

10

ONCĪDĬŬM ornithorhy̆nchŭm.

Bird-billed Oncidium.

GYNANDRIA *MONANDRIA.*

Nat. ord. ORCHIDACEÆ, § VANDEÆ.
ONCIDIUM. Botanical Register, vol. 13. *fol.* 1050.

O. *ornithorhynchum;* pseudobulbis ovatis diphyllis, foliis ensiformibus recurvis scapo paniculato brevioribus, sepalis lineari-oblongis undulatis reflexis omninò liberis, labelli panduriformis lobis lateralibus acutis intermedio bilobo, cristâ e lamellis 5 crenatis apice rostratis constante, columnæ alis cuneatis dentatis, stigmate longi-rostrato.
O. ornithorhynchum. *Humboldt, Bonpland, & Kunth. Nov. gen. et sp. plant. I.* 345. *t.* 80. *Genera & Species Orchid.* 204. *Bateman Orchid. Mex. et Guatemal. t.* 4.
O. roseum. *Hort.*

This beautiful epiphyte was originally found by Humboldt and Bonpland in the temperate parts of Mexico, near El Puerto de Andaracuas, between the towns of Guanaxato and Valladolid de Mechoacan, on mountains, at the elevation of 6,000 feet above the sea, flowering in September. It has also been met with by Mr. Hartweg, at a place called Llano Verde, loaded, in the month of July, with larger and finer flowers than I have yet seen in our garden specimens. Mr. Skinner again found it in Guatemala, and sent it to Mr. Bateman, who published the first figure from living plants in his magnificent work on the *Orchidaceæ of Mexico and Guatemala,* t. 4.

In their natural state the panicles of rosy flowers are pendulous, and extremely ornamental; their beauty is much impaired when they are tied up to a stake, and so forced into a position alien to their habits. They have an agreeable odour, which Mr. Bateman compares with justice to new hay.

In this species we have another instance to add to those of *O. Russellianum, Lanceanum,* and *pulchellum* of violet flowers in a genus so very generally yellow. In the coloured copies of Humboldt's genera and species they are really made yellow; which should be a warning to all Botanists how they judge of the colours of flowers from dried specimens.

At the upper left-hand corner of the accompanying plate is a magnified view of the anther, from which the specific name of bird-billed has been taken, and with it the toothed wings of the column, and the peculiar crest of the lip, which forms one of the most prominent characters of the species.

It must have the temperature of the stove, and succeeds best if suspended from the roof upon a block of wood. In this situation its flowers hang down and look very graceful. When the roots are thrown out for nourishment they must be freely syringed, and preserved from wood-lice or other insects. It must have a period of rest, and at that time it should be kept cool and dry.

12

MAXILLĀRĬĂ cucullāta.

Hooded Maxillaria.

GYNANDRIA *MONANDRIA*.

Nat. Ord. ORCHIDACEÆ, § VANDEÆ.

MAXILLARIA. Botanical Register, vol. 11. *fol.* 897.

§ 2. SCAPIGERÆ; *scapis radicalibus imbricatìm squamatis unifloris.*

M. *cucullata;* pseudobulbis ovalibus compressis monophyllis, folio lato scapis longiore, vaginis imbricatis inflatis, bracteâ cucullatâ ovario longiore, sepalis ovatis acuminatis erectis, petalis conformibus paulò minoribus, labello oblongo carnoso trilobo petalorum longitudine: lobis lateralibus nanis intermedio elongato obtuso apice incrassato apiculato, callo disci spathulato loborum lateralium longitudine.

A native of equinoctial America, whence it is said to have been brought by Mr. Henchman. For my specimens I am indebted to His Grace the Duke of Devonshire, who ordered them to be sent from Chatsworth in September, 1838. The species is one of the less interesting of the genus, and in its habit and general appearance approaches the Trigonidia. It belongs to a rather large section of true Maxillarias, characterized by their peduncles being constantly one-flowered, densely clothed with scale-like sheaths, and proceeding immediately from the base of the pseudo-bulbs which lie upon the earth.

Of this section several have been defined in the *Genera and Species of Orchidaceous Plants,** and others have appeared in Pöppig and Endlicher's *Nova Genera et Species Plantarum,* so that among my collections of unpublished plants belonging to the Vandeous section of Orchidaceæ I find only one to add. That one is a native of Mexico, where it was first found by Count Karwinski, at Teoxomulco, near Oaxaca. As I believe it exists in this country in a living state I subjoin* its character, taken from specimens given me by Mr. Bateman, and from others in the herbarium of Dr. von Martius.

Fig. 1. shews the petals and labellum, seen from one side; 2. is a view of the labellum from the inside; 3. is the column, with the pollen-masses adhering to the apex; 4. are the pollen-masses themselves, placed considerably below the upper end of their caudicula.

In cultivating this species great care must be taken not to let too much water into the scales of the young shoots, for it is very apt to rot them, and so spoil the pseudo-bulbs. The temperature of the stove in which it is grown should be kept low in the cold and cloudy weather of winter, and gradually raised as spring advances. In fact it is very unnatural, and highly injurious to grow plants of this kind in a temperature of 70° or 80° in the dull weather of winter, for it is applying the stimulus of heat without the corresponding action of light.

It should be potted in brown turfy peat well mixed with drainage, and treated generally as other plants of this kind.

* I am happy to state that a sixth part of this work, comprehending *Arethuseæ, Gastrodieæ,* and *Vanilleæ,* is in the printer's hands, and may be expected in about a month.

* MAXILLARIA *rhombea;* acaulis, pseudobulbis ovalibus ancipitibus, foliis . . . . . , vaginis acutis carinatis, bracteâ ovario æquali, sepalis petalisque acutis, labello subrhombeo-trilobo: laciniâ intermediâ oblongâ apiculatâ medio exaratâ lateralibus nanis rotundatis appendice planâ retusâ longioribus.——Mexico, Oaxaçæ, *Karwinski.*

14

ONCĪDĬŪM stramĭnĕūm.

Straw-coloured Oncidium.

GYNANDRIA *MONANDRIA*.

Nat. ord. ORCHIDACEÆ, § VANDEÆ.

ONCIDIUM. Botanical Register, vol. 9. *fol.* 727.

O. *stramineum;* ebulbe, foliis crassis carnosis ovato-lanceolatis acutis dorso rotundatis scapo paniculato rigido erecto brevioribus, sepalis subrotundis unguiculatis concavis liberis integerrimis, petalis duplò majoribus oblongis obtusis emarginatis margine crispis, labelli lobis lateralibus oblongis carnosis acutis margine revolutis basi columnæ proximâ nectariferis intermedio reniformi plano emarginato longioribus, tuberculis disci 4 geminatis, columnæ alis carnosis linearibus obtusis elongatis genuflexis decurvis. *Bot. Reg.* 1838. *misc. no.* 63.

Of this pretty species a short account has already been given in this work among the miscellaneous matter for the year 1838. It was one of the first plants sent from Vera Cruz to the Horticultural Society by Mr. Hartweg, who found it at a place called Zacuapan, where some other fine things, especially the rare and beautiful Berberis tenuifolia, were obtained.

This species is readily known when out of flower by its rigid fleshy unspotted leaves, rounded, not keeled, at the back. When in flower the stiff panicles, and pale straw coloured blossoms, smelling slightly of primroses, together with the peculiar form of the wings of the column, and the almost orbicular sepals, offer sufficient means of recognition.

Fig. 1. represents the lip and column, the latter with the narrow falcate deflexed wings peculiar to this species; fig. 2. is a transverse section of a leaf to show its thickness and peculiar form.

O. stramineum will not submit readily to the same treatment as is given to the West Indian species of this genus. It is injured by that degree of heat in which they flourish; the leaves which it forms under it are small, and the flowers in many cases are not fully developed. It is a curious fact that two plants—one in the garden of the Horticultural Society, and the other in the collection of George Barker, Esq., of Birmingham—threw up flower stems, which instead of producing flowers formed leaves; and this was probably owing to over excitement, for, another plant which was generally kept at the coolest end of the stove was always healthy, and perfected flowers in abundance. It may either be potted or suspended upon wood from the rafters of the stove, where it does not require so much water as the free-growing species, but it may be frequently syringed in the growing season with good effect.

18

SATȲRĬŪM pustulātūm.

Pustular Satyrium.

GYNANDRIA *MONANDRIA.*

Nat. Ord. ORCHIDACEÆ, § OPHRYDEÆ.

SATYRIUM. Botanical Register, vol. 5. *fol.* 416.

S. *pustulatum*;* foliis binis radicalibus cordato-orbicularibus planis papillosis, vaginis caulis appressis apice acutis foliaceis patentibus complicatis, spicâ oblongâ obtusâ multiflorâ, sepalis linearibus convexis recurvis obtusis petalisque conformibus minoribus patulis, labello oblongo acuto margine revoluto dorso cristato apice subcrispo recurvo: omnibus altè connatis, calcaribus ovario paulò brevioribus. *Lindl. genera et sp. orch. p.* 341.

S. pustulatum. *Lindl. in Bauer's Illustrations of Orchidaceous Plants, t.* 12. 13. 14. Fructification.

A native of the Cape of Good Hope, whence it was received at the Royal Botanic Garden at Kew so long since as the year 1800, when Mr. Bauer prepared the beautiful drawings which form the subject of the 12th, 13th, and 14th plates of Fructification in the *Illustrations of Orchidaceous plants.* It does not however appear to have been noticed by any systematical writer until it was introduced into the *Genera and Species of Orchidaceous plants.* Its name has been given in consequence of the leaves appearing as if covered with watery pustules. The flowers smell like the delicious *Sweet Vernal Grass* of our meadows.

Sir John Herschell found it growing in the Cape Colony in clayey places, baked hard by the sun, and never in sandy soil; according to Drège however it occurs on the sand-hills near Saldanha Bay, as well as on the Camiesbergen, and near the villages of Donkerhoek and Liefde.

Nothing is more vexatious to lovers of flowers than the difficulty of preserving the terrestrial species of Orchidaceæ found at the Cape of Good Hope, where there are great numbers of such plants of the gayest colours and most singular forms. They live indeed for a year or two after their arrival, but then perish, from debility brought on by some mismanagement in their cultivation, the nature of which is not understood. Some very valuable hints relating to this matter have been given by Sir John Herschell, in a communication published in the *Proceedings of the Horticultural Society, vol.* 1, *p.* 56; among which the following relates to the subject of this notice, which was flowered by that distinguished individual, by whose permission the accompanying figure was made.

* This is printed papillosum, by an accidental oversight, in the work here referred to.

"I have never found this in the sandy flats about the Cape. My roots were all taken from *clay*, baked by the sun nearly to the consistence of a brick, at 'De Koch's,' a place about 40 miles east of Cape Town, in the district of 'Hottentot's Holland.' They were then in flower, rather past their maximum. Nevertheless they grew well enough in the peaty-sand of which my garden consisted, and to admiration in a fine black sand enriched with vegetable matter, from the shrubby hills in the neighbourhood. Plenty of water and moderate temperature while leafing, diminished supply of wet and increased heat as the flower rises, and total dryness *with heat*, when all is withered, seem to be the conditions."

Practically the cultivation of Cape Orchidaceæ has always been found extremely difficult in this country. The best method of cultivating the present species is to plant it out in a pit, with a south aspect fully exposed to the sun's rays at all seasons of the year. The pit must be well drained, and the soil should consist of peat well mixed with sand and leaf mould. A great deal of harm may be done by injudicious watering. When it is not growing it should be kept perfectly dry; as soon as it begins to push through the soil a little water may be given, not amongst the leaves, for this will probably injure them, but round the tubers; this may be gradually increased as the plant grows, until the flower-stem makes its appearance, when water must be discontinued. Heat and light are now the agents which will perfect the flower, and ripen the tubers for the following year.

When propagation is attempted it must be done when the tubers are ripe, and the plant in a state of inaction.

Those who have not pits to grow this plant in, may be equally successful by cultivating it in pots, if the above suggestions are attended to.

24

*CŒLŎGȲNĒ Wallichiānā.

Wallich's Cœlogyne.

GYNANDRIA *MONANDRIA.*

Nat. Ord. ORCHIDACEÆ, § MALAXIDEÆ.

CŒLOGYNE. Botanical Register, vol. 11. *fol.* 868.

C. *Wallichiana*; pseudobulbis ampullaceis vaginisque duris tuberculatis, pedunculis radicalibus unifloris basi vaginatis, labelli trilobi basi saccati lobis lateralibus integerrimis cum intermedio denticulato crispo apice truncato integerrimo plicato apiculato confluentibus, disci cristis 4-5 incompletis denticulatis, columnâ apice dentatâ.

C. Wallichiana. *Gen. & Sp. Orch. p.* 43. *Wall. Plant. As. rar.* 1. 46. *t.* 54. *Bot. Reg.* 1838. *misc. no.* 157.

This beautiful little plant, is, according to Dr. Wallich, a native of the lofty range of mountains which confine Bengal towards the district of Sylhet, in an easterly and northerly direction, inhabiting rocks and the trunks of trees among moss. It is not mentioned by Dr. Royle, but that distinguished traveller found a nearly allied species, C. præcox, ornamenting with its large richly coloured flowers the branches of oaks on Lundour, at 7,500 feet of elevation, in 30° N. lat.; but only during the moisture of the rainy season. (*Illustrations of the Botany of the Himalayas*, p. 364.)

The figure now given was taken from specimens communicated by order of His Grace the Duke of Devonshire, from Chatsworth, to which place it had been brought by Mr. John Gibson, as has been already stated in the miscellaneous matter of this work for 1838, No. 157. Other flowers have been sent me by Mr. Bateman, with the following note.

"This beautiful plant I owe to the liberality of the Duke of Devonshire. It is now (Oct. 1838) flowering with me in great perfection, and in the brilliancy and magnitude of its blossoms far surpasses the figure of it in Wallich's Plantæ Asiaticæ rariores; in other respects the representation there given is singularly characteristic. The pseudo-bulb has much the form and hue of a truffle, and loses its leaf before the appearance of the flowers; which come up, one on either side. They are almost as large as those of Cattleya labiata, and are of a uniformly rich transparent rose-colour, except in the interior of the lip which is decorated with a broad streak of yellow on its disk, and traversed by five parallel ridges of white tubercles. There are also a few deep crimson stains scattered on its surface. The plant appears to be of the easiest cultivation, and its flowers very durable."

* So named from κοιλος hollow, and γυνη, used in the sense of stigma; in allusion to the form of that organ.

The examination of fresh specimens of this and other species of Cœlogyne has satisfied me that the genus belongs to *Epidendreæ* rather than to *Malaxideæ*, and that it should be stationed near Phaius; for the pollen-masses do not lie loose and free in the inside of the anther, on the contrary they adhere to two thin plates of pulverulent matter, bent back upon themselves, as is represented at fig. 3. of the accompanying plate; where fig. 1. is a view of the interior of the labellum, and fig. 2. of the column with the projecting hollow stigma, and the anther lifted up, and turned back. In the

Indian figure published by Dr. Wallich the lower lip of the stigma is represented 3-toothed, and this supposed peculiarity forms part of the specific character originally framed for the species; I now find, however, that no such structure exists.

25

CYMBĬDĬŬM pĕndŭlŭm.

Thick-leaved Cymbidium.

GYNANDRIA *MONANDRIA.*

Nat. Ord. ORCHIDACEÆ.

CYMBIDIUM. Botanical Register, vol. 7. fol. 529.

C. *pendulum;* foliis ensiformibus distichis coriaceis obliquè obtusis, racemis pendulis multifloris, bracteis minutis, petalis sepalisque lineari-oblongis obtusis, labelli trilobi lobis lateralibus acutis intermedio oblongo apiculato: lamellis continuis approximatis apice confluentibus.

Epidendrum pendulum. *Roxb. corom. plants.* 1. 35. *t.* 44.

C. pendulum. *Swartz. nov. act. ups.* 6. 73. *Willd. Sp. Pl.* 4. 101. *no.* 30. *Roxb. fl. ind.* 3. 458. *Lindl. gen. et sp. orch. p.* 165.

C. crassifolium. *Wall. cat. no.* 7357.

Originally observed by Dr. Roxburgh on trees in the province of Sylhet, and in the forests which cover the Circar mountains as well as those of Bengal; afterwards met with at Noakote in Nepal by Dr. Wallich.

The latter Botanist considered his plant different from that of Roxburgh, and it has smaller flowers; but I do not perceive in what the distinction otherwise consists. The specimen now figured was sent by Dr. Wallich to the Hon. and Rev. W. Herbert, who forwarded a fine specimen, nearly three feet long, with a dozen flowers upon it, to the Horticultural Society in August, 1838.

It is very distinct from all others. Its leaves are from two to three feet long, stiff, leathery, obliquely obtuse, and strongly furrowed; at the base they form a distichous tuft, like that produced by the equitant leaves of an Iris. The flowers exceed in size those of any of the species nearly allied to this. They are however of a dirty yellowish brown colour, which diminishes their beauty, notwithstanding the clear red and white of the labellum. As they hang downwards, the plant should be suspended from the roof of the stove like Cymbidium aloifolium, to which it is most nearly allied; but, independently of the large size of the flowers, this is readily known by the lamellæ of the lip being continuous, straight, and touching at the point, instead of being broken in two and curved. I fear however that some at least of the specimens distributed by Dr. Wallich upon my authority under the number 7352, and the name of C. aloifolium, are this C. pendulum.

30

ODONTOGLŎSSŬM maculātŭm.

Yellow and Brown Odontoglossum.

GYNANDRIA *MONANDRIA.*

Nat. ord. ORCHIDACEÆ, § VANDEÆ.

ODONTOGLOSSUM. Botanical Register for 1839, *t.* 48.

O. *maculatum;* pseudobulbis oblongis compressis monophyllis, foliis oblongis nervosis acutiusculis racemis pendulis multifloris brevioribus, bracteis navicularibus herbaceis ovario brevioribus, sepalis lineari-lanceolatis acuminatis discoloribus, petalis oblongis undulatis acuminatis, labello cordato acuminato subcrenato: appendice unguis bivalvi concavâ cochleari apice productâ emarginatâ per medium argutè serrulatâ, columnâ pubescente.

Folia *præter illud pseudobulbo innatum plura, cum petiolo equitante articulata,* 5-7-*nervia, pergamenea, lætè viridia.* Sepala *extus viridia, intus fusca,* petalis *teneris luteis versus basin sanguineo-nebulosis crassiora.* Labellum *ejusdem coloris texturæque ac petala;* appendice *bivalvi, subpubescente, intùs purpureo-venosâ.* Columna *versus apicem utrinque obsoletè auriculata.*

Another rarity from the rich storehouse of Mr. Barker, who imported it from Mexico. It is one of the prettiest of the family, because of its large two-coloured spotted flowers and drooping habit, and seems to have much the manner of growth and constitution of an Oncidium.

When it was first sent to town I mistook it for the *Odontoglossum Cervantesii* of La Llave, for it bears a great resemblance to that plant as described by its author; but upon a more attentive consideration I see it must be different, for the latter is said to have snow-white flowers, and this is not a circumstance in which plants of Orchidaceæ are likely to vary. It is no doubt the *O. maculatum* of this author.

It was also found in the west of Mexico by Count Karwinski, as appears from a wild specimen for which I am indebted to Mr. Bateman.

Odontoglossum cordatum, now figured in the Floral Cabinet, t. 100, approaches this very nearly; but, independently of the difference in the colour and size of its flowers, it has exceedingly acuminate sepals and petals, and quite another kind of appendage to the stalk of the labellum. Before I had seen the plant now figured I was disposed to believe that species to be the *O. maculatum* of La Llave, as is stated in the Sertum Orchidaceum, so very difficult is it to judge correctly of the insufficient descriptions of the Mexican Botanist. I now find that they are truly distinct.

32

AGANĪSĬĂ pulchēllă.

Pretty Aganisia.

GYNANDRIA *MONANDRIA.*

Nat. Ord. ORCHIDACEÆ § VANDEÆ.

AGANISIA. Perianthium patens, æquale; sepalis lateralibus haud basi productis. *Labellum* liberum, mobile, indivisum, hypochilio parvo concavo, ab epichilio cristâ transversâ glandulosâ diviso. *Columna* erecta, semiteres, marginata, apice utrinque brachio acuto patulo aucta. *Anthera* ecristata. *Rostellum* elongatum. *Pollinia* 4, per paria connata, caudiculâ lineari, glandulâ parvâ ovali.—— *Rhizoma* repens, pseudobulbosa. *Pseudobulbi* monophylli. *Racemus* erectus, radicalis, foliis brevior.

Aganisia pulchella. Bot. Register, 1839, misc. no. 65.

Rhizoma *repens, pennæ anserinæ crassitie, squamis vaginantibus arctè vestitum;* pseudobulbis *parvis acuminatis intra squamas latentibus.* Folia *solitaria, oblonga, membranacea,* 5-*costata, in petiolum angustata.* Racemi *radicales. i. e. e squamis pseudobulbos fulcientibus, erecti,* 3-6-*flori, foliis breviores.* Flores *candidi,* 1¼ *unciam lati;* sepalis petalisque *subæqualibus, ovato-oblongis, acutis, patulis.* Labellum *cum columnâ articulatum; hypochilio concavo sanguineo maculato, epichilio latiore, ovato, integro, medio luteo, basi cristâ lutescente glandulosâ ab hypochilio diviso.* Columna *erecta, basi ne minimè quidèm producta, semiteres, apice brachiis duobus patentibus incurvis aucta;* rostello *lineari producto.* Pollinia 4, *incumbentia* (8 8) *nec collateralia ut olim dixi; caudiculâ lineari membranaceâ, glandulâ parvâ, subrotundâ tenerâ.*

This pretty Demerara plant has hitherto flowered no where that I am aware of except with Messrs. Loddiges, who imported it. A short account of it was given in this work in the place above quoted; and the present figure, with a correction or two, completes its history.

If its column were produced into a foot, and the lower sepals unequal at the base, it would be a Maxillaria; but as

* From ἀγανὸς, quiet or desirable, in allusion to the pretty, neat appearance of the plant.

there is no trace of that character, which is essential to Maxillaria, the genus seems sufficiently distinct.

Fig. 1. represents the column, with the labellum pulled downwards to shew its true form, and the crest that separates the upper lip from the lower. Fig. 2. shews the pollen-masses in their true position, one half of the right hand pair being cut away. By some accident the pollen-masses of the first flower I examined had been pressed out of their natural position, which formerly led me to describe them as collateral instead of incumbent.

In order to cultivate this plant successfully it should be suspended upon a block of wood from the rafters of the stove, and its thick fleshy roots allowed to hang in the air and imbibe its moisture. A damp atmosphere, syringing its roots and leaves freely when in a growing state, and shade during bright sunshine, are the principal requisites in its cultivation. In other respects it may receive the same treatment as the rest of this tribe.

35

EPIDĒNDRŪM vitellīnum.

Yolk-of-egg Epidendrum.

GYNANDRIA *MONANDRIA.*

Nat. ord. ORCHIDACEÆ, § EPIDENDREÆ.

EPIDENDRUM. L:

Sect. ENCYCLIA. *Pseudobulbosa, labello sublibero.*

E. *vitellinum;* pseudobulbis ovatis acuminatis diphyllis, foliis oblongo-ligulatis acutis basi vaginantibus racemo terminali erecto multifloro brevioribus, sepalis petalisque ovato-lanceolatis acutis subæqualibus patulis, labello lineari apice angustato abruptè acuto basi calloso et bifoveato semilibero.

E. vitellinum. *Genera & Species Orch. page* 97.

A dried Mexican specimen in Mr. Lambert's herbarium conveyed the first knowledge of the existence of this beautiful plant, and for many years no additional information concerning it was received. It was afterwards found without a locality among Karwinski's dried plants in the Munich museum, and in September, 1839, the specimen now figured produced its flowers in Mr. Barker's collection at Birmingham. Neither this, however, nor any of the dried materials, conveyed a true idea of the natural appearance of this species when in health and flowering under favourable circumstances.

It was not till specimens, collected on the Cumbre of Tetontepeque in Mexico, at the elevation of 9000 feet above the sea, were received from Mr. Hartweg, that I had any conception of the superb appearance of the plant. One of them now before me has a spike covered with fifteen large orange-coloured flowers, all expanded at once, over a space of more than six inches in length, and forming a most conspicuous object. The flowers are fully thrice as large as those now represented.

To the roots of this specimen are sticking fragments of a Jungermannia and a foliaceous Lichen, so that no doubt can be entertained of its inhabiting a damp, cool, shady situation. This certainly reveals the secret of its cultivation, and shews that a very different climate from that of East and West Indian Orchidaceæ is requisite for it. Probably it will like the management recommended for Lælias at pages 26 and 27 of this work for 1839, and most undoubtedly it must be guarded against exposure to a high temperature at any period of the year.

With reference to this subject cultivators will probably be glad to see the following extracts from Humboldt's account of the Mexican climate.

"Mexico in lat. 19° 25′, at the height of 7008 feet, has a mean temperature of 63°; that of the hottest months is from 60° to 70°; of the coldest months 50° to 59°.

"Toluca in lat. 19° 16′, at the height of 8280 feet, has a mean temperature of 59°.

"On the Puerta del Volcan, a volcanic mountain, near Toluca, at the height of 10,494 feet, the temperature of a rivulet flowing down from it is about 49°."

Humboldt does not mention Orchidaceæ in his list of plants growing within the boundary of this tierra fria; but he speaks of Peperomia umbilicata, Cheirostemon platanoides, Rosa Montezumæ which is our R. canina, Hoitzia coccinea, a Strawberry, and common Horehound, among many other plants.

39

BRASAVŌLĂ venōsă.

Vein-lipped Brasavola.

GYNANDRIA *MONANDRIA.*

Nat. Ord. ORCHIDACEÆ § EPIDENDREÆ.

BRASAVOLA. *Botanical Register, fol.* 1465.

Brasavola *venosa;* folio lanceolato semicylindraceo supra canaliculato, sepalis petalisque linearibus, labelli ungue longo complicato laminâ subrotundo-ovatâ subtrilobâ acuminatâ basi serratâ: venis elevatis. *Bot. Reg.* 1840. *misc. no.* 24.

Folium *semipedale, carnosum, lineari-oblongum, acutum, atroviride, canaliculatum marginibus rotundatis, nec ut in speciebus pluribus in cylindrum convolutum.* Scapus *folio brevior, spathâ nullâ conspicuâ, sed squamis paucis vaginatus, 3-florus; bracteis parvis rigidis acutissimis.* Sepala *et* petala *linearia, acuminata, virentia, 2-pollicaria, inter se æqualia.* Labelli *lamina candida, venis elevatis corrugata, subrotundo-ovata, acuminata, subcarnosa, integra v. obsoletè triloba, basi serrata; unguis laminâ parùm brevior, convoluta, margine serrata, basi integerrima et maculis quibusdam sanguineis notata.*

A fine species resembling *B. nodosa* in habit, but with much larger flowers, the sepals being more than two inches long. The lip is white, the other parts greenish. It is at once distinguished from *B. nodosa* and all the allied species by the firmness of the lip, which is more or less evidently lobed at the side, and has the veins distinctly elevated. The flowers are deliciously sweet at night. Messrs. Loddiges imported it from Honduras.

The leaf is intermediate between that of *Br. glauca* and the common terete species; and shews that in the latter the leaf owes its peculiar appearance in part to a general thickening of its parenchyma, and in part to the edges turning inwards, meeting and growing together. And this is no doubt also the origin of the terete leaves found in Vanda, Luisia, and Dendrobium.

It has been recommended to grow the plants of this genus in well-drained pots, and they grow very freely under such circumstances; but they certainly do better when hung from the roof or pillars of the stove. The only thing to be attended to in the latter mode of culture is, to tie upon the block of wood along with the plant some pieces of turfy peat to keep the roots moist; the roots will soon fix themselves into the peat, and throw their tender points into the air and feed upon it.

41

LÆLIA rubescens.

Blushing Lælia.

GYNANDRIA *MONANDRIA.*

Nat. ord. ORCHIDACEÆ, § EPIDENDREÆ.

LÆLIA. Botanical Register, vol. xxi. *fol.* 1751.

Lælia *rubescens;* pseudo-bulbis subrotundis compressis utrinque angulo elevato, foliis oblongis obtusis scapo tereti vaginato brevioribus, racemo multifloro, bracteis ovario plus duplò brevioribus, sepalis linearibus, petalis lanceolatis subundulatis, labello conformi auriculato medio pubescente: lineis duabus elevatis. *Bot. Reg.* 1840. *misc. no.* 25.

Pseudobulbi *oblongi, compressè tetragoni, vix sesquiunciam longi.* Folium *solitarium, coriaceum, oblongum, planiusculum, quasi emarginatum, scapo gracili erecto squamato multò brevius.* Racemus *brevis, multiflorus.* Bracteæ *lineares, murcescentes, ovario plus duplò breviores, carinatæ, pilis brevissimis minutissimis fasciculatis pubescentes.* Flores *L. primulinâ paulò minores et minus patentes.* Sepala *linearia, obtusa, pollicaria, basi rubescentia, apice virescentia, recta, petalis paululùm longiora.* Petala *lanceolata, acuta, subundulata, secus margines rubentia.* Labellum *ovato-lanceolatum, versus basin auriculatum, apicem versus undulatum, acutum, medio luteum et subpubescens, lineis duabus elevatis tertiâque intermediâ minus conspicuâ, basi atropurpureum convolutum.*

This pretty species of Lælia has the smallest flowers of any yet discovered, and they are wholly scentless. It was bought at the nursery of Mr. Joseph Knight, in the King's Road, by Mr. Barker of Birmingham, to whom I am indebted for a specimen. No wild specimen has yet been seen, and I am unacquainted with the source from which Mr. Knight obtained it.

It is most nearly related to the fragrant *L. primulina*, from which it differs in the form of its pseudo-bulbs, leaves and labellum, in the size of the flowers, and in the petals not being at all revolute.

In the specific character originally given it the bracts were described as pubescent. This appearance arises from the presence of minute patches of extremely delicate hairs, which soon wither up, and leave nothing but a stain to indicate their having been present.

The best and most natural way to cultivate it, is to hang it from the roof of the stove, and allow its roots to imbibe the moisture of the atmosphere. This is one of those orchidaceous plants which will succeed in a lower temperature than that which most of them are grown in, but like the others it requires a moist atmosphere and shade. It must be kept perfectly dry when in a state of rest, but should be freely watered and syringed when growing.

44

BRASAVOLA glauca.

Glaucous Brasavola.

GYNANDRIA *MONANDRIA.*

Nat. ord. ORCHIDACEÆ, § EPIDENDREÆ.

BRASAVOLA. Botanical Register, fol. 1465.

B. *glauca;* foliis coriaceis oblongis obtusis planiusculis glaucis, spathâ uniflorâ, sepalis petalisque lineari-lanceolatis obtusis herbaceis, labello subsessili subrotundo acuto margine lobato, clinandrio dentato dente dorsali apice glanduloso. *Bot. Reg.* 1839, *misc. no.* 67.

B. glauca. *Bateman's Orchidaceæ of Mexico and Guatemala, t.* 16.

It is said that this charming plant was originally found near Xalapa in Mexico, by Henchman, and that it was afterwards imported more abundantly by a Frenchman of the name of Deschamps, who brought over a large quantity of Orchidaceæ and Cacti for sale some years ago.* To me it was first made known by its having been gathered near Vera Cruz by Mr. Hartweg, who sent it among his earliest consignments to the Horticultural Society. It has since been received from Guatemala, where Mr. Skinner found it growing on the oaks of the country in company with *Cyrtochilum maculatum* var. *Russellianum,* like which he directs it to be treated. He calls it "a splendid white flower, with a most extraordinary strong aromatic fragrance." It is certainly very handsome and sweet; but not so much with us as it appears to be in Guatemala.

Its aspect is not that of a *Brasavola,* but rather of a *Cattleya,* to which genus it was thought that it would belong: but as soon as the flower appeared it was found to be a true *Brasavola,* with all the characters that peculiarly mark that genus. Like others it has the anther-bed lacerated, but the dorsal tooth is tipped with a glandular swelling. Fig. 1. represents this; and fig. 2. the pollen-masses, artificially separated so as to show the large size of the powdery elastic straps that hold them together.

This is found as easy to cultivate as any of the other Mexican Orchidaceæ, requiring a damp warm atmosphere when growing, and to be kept cool and dry when in a state of inaction. It will succeed either suspended from the roof of the house upon a block of wood, or in a well drained pot, the only difficulty in its cultivation being in inducing it to blossom. Upon this subject I have received the following note from Mr. Fortune, who has succeeded in making it flower freely in the garden of the Horticultural Society.

"At the base of every leaf there is a bud, and from the leaf itself the flower springs, which in many instances proves abortive, apparently owing to the luxuriance of the bud at its base. As a proof of this—after many fruitless attempts to make this plant flower—one of these buds was removed, which allowed the sap intended for the nourishment of that bud to go to the formation of the flower, and the result was the production of the subject of the present plate. In the following season the plant was covered with flowers upon the same principle, though not at the expense of its buds. This was done by keeping it dry and not allowing the buds at the base to grow much until the flower stems were so far advanced as to be out of danger."

It is easily propagated by division.

* Upon this subject Mr. Bateman has given the following note in his splendid work above quoted. "This importation was quite unique in its way, and formed a sort of epoch in the history of the *Orchido-mania.* A vessel came into port freighted, almost exclusively, with Epiphytes and *Cacti,* and such was their abundance, that it was found necessary to engage an extensive suite of apartments, for their accommodation, in Hungerford Market! The plan pursued by M. Deschamps was, to parcel out his plants in small collections of about twenty species, for which, in the first instance, he asked and obtained very high prices; but the London market being at length exhausted, similar collections were distributed through the provinces, and offered at greatly reduced rates. The author himself purchased, in a country town, a set of at least twenty kinds for a sum which, in the metropolis, he had in vain tendered for only two! In case of any future inundation of Orchidaceæ, this little fact should be borne in mind."

48

CATTLEYA Aclandiæ.

Lady Acland's Cattleya.

GYNANDRIA *MONANDRIA.*

Nat. Ord. ORCHIDACEÆ § EPIDENDREÆ.

CATTLEYA. Botanical Register, vol. 12. *fol.* 927.

§. 2. *Labelli lobis lateralibus abbreviatis, columnam haud supertegentibus.*

C. *Aclandiæ;* foliis oblongis, floribus subsolitariis, sepalis petalisque herbaceis lanceolatis æqualibus incurvis maculatis, labelli plani calvi hypochilio dilatato patulo subrepando epichilio orbiculari reniformi emarginato.

Of this very distinct and pretty species of the handsomest of all the genera of Orchidaceæ I have only seen a single flower, which I owe to the kindness of Lady Acland of Killerton, by whom the drawing, from which the annexed figure was prepared, was also supplied.

It was received from Brazil in October, 1839, having been discovered by Lieut. James of H. M. ship Spey, and flowered in the stove at Killerton in the month of July, 1840, under the able management of Mr. Craggs, Sir Thomas Acland's gardener.

It is probable that the plant, when in a more vigorous state, will produce several flowers of a larger size, in a cluster; even in the present state the large purple lip and column render it very handsome.

Together with *Cattleya bicolor* it forms a distinct section of the genus, distinguished by the base of the lip being too narrow, and too spreading, to cover over the column.

49

* GALEĀNDRA Bāuĕrī.

Bauer's Casquewort.

GYNANDRIA *MONANDRIA.*

Nat. ord. ORCHIDACEÆ, § VANDEÆ.

GALEANDRA. Illustr. of Orch. Pl. gen. t. 8. *Gen. & Sp. Orch. p.* 186.

Perianthium patens v. connivens, *sepalis petalisque* æqualibus ascendentibus, liberis. *Labellum* infundibuliforme, calcaratum, indivisum, sessile, intùs læve, margine nunc fimbriato. *Columna* erecta, membranaceo-alata, clinandrio declivi. *Anthera* galeæformis, cristâ recurvâ cum dorso clinandrii cardinatâ. *Pollinia* 2, posticè excavata, caudiculâ brevi, cum glandulâ elongatâ basi divergenti-bilobâ articulatâ.—— Herbæ *terrestres,* foliis *plicatis,* scapis *radicalibus* v. *racemis terminalibus.*

G. *Baueri.* (G. & Sp. Orch. 186.) caule simplici, foliis lanceolatis 3-nerviis, corymbo terminali pedunculato nutante foliis breviore, pedunculo squamis lineari-lanceolatis membranaceis vaginato, sepalis petalisque lineari-oblongis acutiusculis, labello maximo antico emarginato apiculato crenulato, calcare ovario æquali.
G. Baueri. *Bateman Orch. Mex. & Guat. t.* 19.

This plant was originally discovered in French Guiana by Martin, among whose dried plants it was found when they were purchased by Mr. Lambert and others. A drawing was subsequently made from it by Mr. Francis Bauer, the details of which were published in the *Illustrations of Orchidaceous plants;* and from those materials exclusively the species was adopted in the *Genera and Species of Orchidaceæ.* Subsequently the identical specimen from which Mr. Bauer's drawing was made was given by Mr. Lambert to Mr. N. B. Ward, who most liberally added it to my herbarium. At last a Mexican plant was received by Mr. Barker, from his collector Mr. Ross, who met with it at a place called "Kisatipa, ten leagues from Melacatapec, and growing at the upper end of a dry ravine, terminating half way up the mountains, in a south-west aspect. The temperature varying from 69° to 77° Fahr. by day, and at night as low as 59°. The mountain surrounding this ravine, that is on the north-east side, is covered with a great variety of Orchideæ, whilst on the south side of the ravine there are none to be found but a few Tillandsias. On the top of the mountain there is a continual mist all the year, from December to February excepted. The atmosphere is particularly moist and warm." From this plant the accompanying drawing was made, and it is a faithful likeness.

* From *galea,* a casque, and ανηρ an anther; in allusion to the crested anther. At first sight this would seem to be a hybrid name, being constructed of two words belonging to different languages; but as galea itself is from the Greek γαλῆ, I hope critics will consider the name allowable.

The same plant has subsequently reached the Horticultural Society through Mr. Hartweg; but the flowers it produced were of a much more dull colour than those of Mr. Barker. That all the specimens now adverted to are identical I am certain; so that the plant ranges over an unusual extent of country, the South of Mexico on the one hand, and French Guiana on the other, being its northern and southern limits. It is at present excessively rare in gardens.

In the garden of the Horticultural Society it is treated in the same manner as Catasetums and plants of that kind. It is grown in a well-drained pot, has plenty of water when forming its leaves and flowers, and when that period has passed by, is allowed gradually to dry up and winter itself. It is found by this treatment to push with more vigour the next year than if it is kept in a continual state of excitement. Another plant now to be found in several collections, and called Galeandra Baueri, but which is much more branched, and has not yet flowered, certainly grows better when fixed to a block of wood than when grown in a pot.

54

RODRIGUĔZĬĂ crīspă.

Crisped Sweet-scented Rodriguezia.

GYNANDRIA *MONANDRIA.*

Nat. Ord. ORCHIDACEÆ § VANDEÆ.

RODRIGUEZIA. Botanical Register, vol. 11. *fol.* 930.

R. *crispa;* pseudobulbis ovatis elongatis compressis diphyllis, foliis oblongo-lanceolatis patentibus undulatis, racemo elongato denso pyramidato nutante, sepalis omnibus liberis petalisque undulato-crispis, labello sigmoideo lanceolato bicristato appendice ungue multò breviore.
R. crispa. *Bot. Reg.* 1839, *misc. no.* 139.

One of the sweetest plants I know; its fragrance resembling that of Primroses. It is an Orchidaceous plant from the Organ mountains of Brazil, and flowered with Messrs. Loddiges in October, 1839. Its singularly crisped flowers, of a dull sea-green, bordered with yellow, have an uncommon appearance.

It is evidently very near *R. suaveolens,* the *Pleurothallis foliosa* of the Botanical Magazine, t. 2746, a plant I have never seen. But according to the figure and description the two lateral sepals of that species are united at the base, and the flowers are very little crisped.

It is one of those plants which succeed either in a well drained pot, or suspended in the air of the stove upon a block of wood, as it suits the fancy or convenience of the cultivator. It should never be kept so dry as Catasetums, or plants with thick and fleshy pseudo-bulbs. It is easily propagated by division, the front bulbs making the best plants.

55

CALĀNTHĒ dĭscŏlŏr.

Discoloured Fairbloom.

GYNANDRIA *MONANDRIA.*

Nat. ord. ORCHIDACEÆ, § VANDEÆ.

CALANTHE. Botanical Register, vol. vii. *fol.* 571.

C. *discolor;* racemo laxo pubescente, sepalis petalisque acutis, labelli trilobi columnæ omninò accreti basi pubescentis bilamellati lobo intermedio bilobo 3-carinato, calcare pubescente acuto limbo breviore. *Sertum Orchidaceum, sub. t.* 9. *Bot. Reg. misc.* 1838. *no.* 32.

Folia *oblonga, venis tribus majoribus, cæterisque minoribus interjectis, scapo elongato racemoso multifloro breviora.* Scapus *palmaris ad sesquipedalem floribus* 8-20. Sepala *et* petala *vinoso-purpurea;* labellum *album subroseum.*

This very pretty species was first communicated to me by M. Auguste Mechelinck of Ghent, who had flowered it and *C. bicolor.* It was subsequently sent to this country, and has now blossomed in several places; no where however with more beauty than with Mr. Bateman, whose specimen was a foot and half high. Its rich wine-red sepals and petals form so good a back-ground for the white lip, which they so much relieve, that this species is to be regarded as one of the handsomer species of the genus, and is certainly much more worth cultivating than *C. veratrifolia, furcata,* or *densiflora,* which have whole-coloured blossoms.

I am unacquainted with the native country of this species; but as it comes to us through the Dutch, it is probably either Japan or Java, and most likely the former country.

In general the species of the genus Calanthe require to be cultivated in the stove or orchidaceous house. It consists of plants which appear to derive a considerable quantity of nourishment from the soil in which they grow; and the species should therefore be planted in pots in preference to being suspended in the atmosphere of the stove. Their soil should consist of good brown peat, and the pots must be well drained. Whether or not there is any thing specially requisite in the management of this species I am not informed.

58

CĂTTLĔYĂ labiātă; *var.* Mŏssiæ.

Mrs. Moss's Cattleya.

GYNANDRIA *MONANDRIA.*

Nat. Ord. ORCHIDACEÆ § EPIDENDREÆ.

CATTLEYA. Botanical Register, vol. xi. *fol.* 953.

C. *labiata:* sepalis lineari-lanceolatis, petalis membranaceis undulatis multò latioribus, labello obovato crispo-undulato obtuso v. emarginato disco lævi, pseudobulbis oblongis sulcatis.

α. petalis lanceolatis undulatis, labelli disco sanguineo.

C. labiata. *Lindl. coll. bot. t.* 33. *Gen. et Sp. orch.* 116. *Hooker exot. fl. t.* 157. *Lodd. Bot. Cab. t.* 1856.

β. petalis oblongis subcrispis, labelli disco luteo sanguineo punctato aut picto.

C. Mossiæ. *Hooker in Bot. Mag. t.* 3669.

A native of La Guayra, whence it was imported by the London Nurserymen in considerable quantity some time since; and it is consequently dispersed through the country more plentifully than was imagined when it first flowered in Mr. Moss's garden at Liverpool. Since that period it has been sent me by several of my correspondents, and has appeared in the London exhibitions. The specimen from which the annexed figure was taken was communicated by Sir Philip Egerton, Bart., who bought it of Mr. Joseph Knight, in the King's Road.

Beautiful a plant as this is, and different as it sometimes appears to be from *C. labiata,* I am satisfied that it is a mere variety of that species, from which it differs principally in the lip being yellow, mottled with crimson, instead of deep blood red. The characters relied upon to separate it are —1. Its branching stem; 2. its furrowed pseudobulbs; 3. its broader sepals and petals and larger flowers; 4. its unguiculate petals; and 5. its colour: but none of these distinctions can be relied upon.

1. The branching stem is only a trifling modification of the ordinary habit of Cattleyas, whose rhizoma is the stem, of which the pseudo-bulbs are the branches. 2. The furrowed pseudo-bulbs exist equally in C. labiata. 3. Nothing can well be more variable than the size of the sepals and petals, and consequently of the flowers themselves in different individuals, and under different circumstances. Sir W. Hooker describes the original C. Mossiæ as being seven inches and a half in diameter, from the tip of the upper sepal to the end of the labellum; from tip to tip of the two opposite petals eight inches and a half, each petal being a little more than four inches long, and two inches and a half broad; twenty-four inches in the circumference of the entire blossom! One such specimen I have seen, in the collection of Mrs. Lawrence, and it is probable that similar cases will not be uncommon when the imported plants shall have recovered their full health; but up to the present time the greater part of the specimens I have seen are not larger than the flowers in the annexed plate, and are therefore less rather than larger than C. labiata. 4. I do not find the petals more unguiculate in the one than in the other; but they are broader and more crisp in *C. Mossiæ* than in its prototype. 5. To colour alone of course no importance can be assigned.

It is not merely in the case of C. Mossiæ that the supposed species of the genus require reconsideration. It is probable that C. Harrisonii is a var. of C. Loddigesii; and C. intermedia, is certainly nothing more than a state of *C. Forbesii,* with the yellow exchanged for white and purple.

62

CATASĔTŪM maculātūm.

Spotted Feelerbloom.

GYNANDRIA *MONANDRIA.*

Nat. ord. ORCHIDACEÆ, § VANDEÆ.

CATASETUM. Botanical Register, vol. 20. *fol.* 1667.

C. *maculatum;* sepalis petalisque conniventibus, labello carnoso galeato basi inflexo lateribus suis invicem imbricantibus.

α. *labello basi ciliato.*

C. maculatum. *Bateman Orchid. of Mexico & Guatemala, t.* 11. *an Kunthii?*

C. integerrimum. *Hooker in Bot. Mag. t.* 3823.

β. *labello basi lævi.*

This Catasetum is very nearly the same as the old *C. tridentatum,* from which it differs in its helmet-shaped lip having the lower edges brought together so as to press upon the column, instead of being wide apart. Whether or not it is the *C. maculatum* of Kunth I am unable to say. It probably is, it possibly is not. That plant was never seen by the Botanist who described it, but it was taken up from M. v. Humboldt's notes. Those notes were made upon the living plant as found at Turbaco, and are therefore less liable to error than if they had been made up from a dried specimen. Now Baron Humboldt says that in his plant the petals are somewhat serrated, and the leaves only three inches long, arising from the *apex* of the bulb; but here the petals are not serrated, the leaves are nearer a foot in length, and they arise from the sides, not apex, of the pseudo-bulb. Are these differences unimportant? Serrated petals are not uncommon in the order, but they have not yet been seen in Catasetum; in this plant the lip is sometimes serrated, sometimes perfectly even at the edge; perhaps therefore the petals may be also both serrated and entire in different individuals. It is probable that there is some mistake about the length of the leaves, because no such size either has been or seems likely to be seen in a genus like Catasetum. Whether the leaves arise from near the apex or sides of the stem or pseudo-bulb is certainly unimportant in this genus.

At all events the plant is certainly the *C. maculatum* of Mr. Bateman, and the *C. integerrimum* (why integerrimum?) of Sir William Hooker.

According to Mr. Skinner this species grows in a mean temperature of 75° to 80° Fahr.

If the cultivator of this species chooses to suspend it from the roof of the stove instead of growing it in a pot, he should be careful to surround its roots with some substance capable of retaining moisture, as it requires to be kept in a very moist state during the time in which the flowers are forming, and if this is not attended to they will frequently wither without opening. But the better way is to keep it in a pot, almost dry until it has begun to grow, and even then water must be given sparingly until the pseudo-bulbs begin to form, because at this period too much water will certainly damage the young shoots. In other respects as regards soil, temperature, propagation and general treatment, it must be managed in the same way as C. tridentatum and other plants of this kind.

66

ODONTOGLŌSSŪM Bictoniensĕ.

The Bicton Tooth-tongue.

GYNANDRIA *MONANDRIA.*

Nat. ord. ORCHIDACEÆ, §. VANDEÆ.

ODONTOGLOSSUM. *Botanical Register, vol.* 1839. *fol.* 48.

O. *Bictoniense;* pseudobulbis oblongis compressis 2-3-phyllis, foliis ensiformibus undulatis patentibus scapo racemoso duplò brevioribus, bracteis herbaceis lanceolatis acuminatis ovario duplò brevioribus, sepalis petalisque subæqualibus lineari-lanceolatis maculatis, labelli ungue bilamellato laminâ cordatâ acuminatâ undulatâ, columnæ alis integris.

α. labello roseo.

O. Bictoniense. *Sertum Orchidaceum sub. t.* 25.

Cyrtochilum Bictoniense. *Bateman, Orch. Mexic. & Guatemala, t.* 6.

β. labello subalbido.

Zygopetalum africanum. *Hooker in Bot. Mag. t.* 3812.

A native of Guatemala, whence it was sent to Mr. Bateman by his correspondent Mr. Skinner. Having however been also communicated at the same time to Lord Rolle and Sir Charles Lemon, it first flowered at Bicton, whence the beautiful specimen figured in Mr. Bateman's work, t. 6, was sent. That now represented was given me by Mr. Bateman in April, 1837.

It has already been observed, in the Miscellaneous notices of this work, No. 139, for the present year, that the *Zygopetalum africanum*, sent to Sir W. Hooker from Mr. Forbes of Woburn, as a Sierra Leone plant, collected by Mr. Whitfield, is no other than a pale variety of this species. Upon turning to Mr. Bateman's work I see that he speaks of a variety "with a pure white lip, and pale green sepals and petals faintly blotched with a darker colour:" and it is probable that the Woburn plant belongs to that very variety. Most assuredly it never came from Sierra Leone. Those who have charge of imported plants ought to be very careful that they make no mistakes regarding so very important a subject.

With regard to the genus Zygopetalum itself, it is distinctly separated from Odontoglossum by the crest of its lip being solid and transverse, by its column having no ears, and by its sepals and petals being directed upwards above the level of the centre of the flower.

Those who cultivate the thin leaved species of Oncidium, such as O. altissimum, will find the same treatment applicable to this plant. It will grow very well in a pot, but better if hung up; if treated in the former way it should be planted upon the top of the soil, and its roots allowed to scramble over the surface of the pot. In either case its roots and leaves must be well watered and syringed during the growing season, and must never at any time be kept so dry as Catasetums and plants of that description.

68

ANGRÆCŬM gladiifŏlĭŭm.

Sword-leaved Angrec.

GYNANDRIA *MONANDRIA.*

Nat. ord. ORCHIDACEÆ, § VANDEÆ.

ANGRÆCUM. *Botanical Register, vol.* 18. *fol.* 1522.

A. *gladiifolium;* caulescens, foliis distantibus lanceolatis acutis, pedunculis axillaribus unifloris, floris patuli phyllis omnibus ovato-lanceolatis acutis, calcare gracili pendulo pedunculo subæquali.

Orchis Mauritiana. *Poir. dict.* 4. 601. sec. Ach. Rich.

Aerobion gladiifolium. *Spreng. syst.*

Angræcum gladiifolium. *Thouars orch. afr. t.* 53. *Ach. Rich. orch. maurit. p.* 69. *Lindl. Genera and Species of Orchidaceous plants, no.* 5.

A native of the islands of Bourbon, Madagascar, and the Mauritius, where it seems to have been first met with by Commerson. It was afterwards collected by Du Petit Thouars and others, flowering in the month of February. Like the greater part of the genus it is a plant of little beauty. For the opportunity of making a drawing of it I am indebted to Messrs. Loddiges.

The figures 1 and 2 represent the pollen-masses, which adhere each to a separate oblong gland, in which respect this species differs essentially from *A. eburneum*. It will, however, be found that the greater number of Angrecs possess two glands, either quite separate, or touching each other so as to form a slight adhesion. To determine the real value of this character requires a complete revision both of this and of the neighbouring genera, particularly of Œceoclades, some of the plants referred to which are certainly true Angrecs, as for example *Œ. funalis* and *Retzii*.

Among the minute species of Angrec are many still unknown to systematic Botanists. The following specific distinctions of a few will serve to fill a vacant page.

ANGRÆCUM *tenue;* aphyllum, acaule, radicibus longissimis intricatis, spicis tenuibus filiformibus erectis simplicibus, sepalis petalisque acuminatis apice reflexis, labello ovato acuto calcare erecto oblongo carinato.—— A multitude of very long entangled roots, without leaves, form the principal part of this plant, which seems wholly destitute of leaves. The spikes are clustered, and are from two to four inches high. The flowers are the smallest in the genus.——*On trees near Pará, Martius.* (herb. reg. monac.)

ANGRÆCUM *Fasciola;* aphyllum, acaule, radicibus longissimis flexuosis, spicis erectis multifloris distichis distantibus, sepalis petalisque linearibus acutis, labello membranaceo petalis adnato! apice brevi triangulari, calcare pendulo obtuso inflato lævi.——A small leafless plant, with the habit of *A. tenue*, only the flowers are much larger, although very small, and distinctly distichous. It answers to the short character we possess of *Limodorum Fasciola*, a plant I have never seen; but as the latter is a native of the Society Islands it is probably distinct.—— *Demerara, Schomburgk* (herb. propr.)

ANGRÆCUM *ornithorhynchum*; foliis acicularibus, spicis oppositifoliis brevibus paucifloris (5-6), sepalis petalisque lineari-lanceolatis acutis, labello ovato apice triangulari subcarnoso, antheræ apice elongato obtuso membranaceo, calcare tereti ascendente ovario longiore.——Resembling *A. subulatum* in habit, but quite different in its flowers.——*Saint Catharines in South Brazil, Tweedie* (herb. Hooker).

ANGRÆCUM *polystachyum;* caulibus ramosis, foliis distichis ovalibus coriaceis obliquè bidentatis, spicis oppositifoliis flexuosis multifloris foliis paulò longioribus, bracteis orbiculatis ciliatis, sepalis basi et apice ciliatis, labello subrotundo cuspidato calcare vesiciformi.——A cæspitose plant, throwing out long roots from the lower part of its stem, whose branches are only two or three inches long. A spike of very minute flowers proceeds from opposite every leaf.——*Peru, Mathews,* 1905. (herb. Hooker).

ANGRÆCUM *brevifolium;* foliis coriaceis lineari-oblongis obtusis obliquè emarginatis, spicis oppositifoliis brevissimis, sepalis petalisque obtusis, labello obsoletè trilobo obtuso, calcare clavato ovario longiore.—— Stems long, rooting, with a few leaves at their extremity. Flowers exceedingly small.——*Brazil, where it was found by the Prince Maximilian of Wied Neuwied.* (herb. reg. monac.)

MISCELLANEOUS NOTICES.

4. ERIA planicaulis (*Wallich*); caule compresso folioso erecto, foliis coriaceis aveniis obtusis emarginatisque, floribus glabris intra bracteas siccas striatas subsessilibus, labello reniformi supra unguem bicalloso, petalis linearibus sepalis ovatis acutis multò angustioribus.

Upon this plant Mr. Booth has favoured me with the following communication :—

"This singular species was forwarded by Dr. Wallich from the Honourable East India Company's Botanic Garden, Calcutta, in 1838, and added to Sir Charles Lemon's collection at Carclew, where it flowered during the autumn of 1839. It has nothing to recommend it to the notice of cultivators, but to the Botanist it is a highly curious and interesting subject. It requires the constant heat of the stove, and seems to thrive pretty well in a pot of finely chopped moss and decomposed vegetable earth.

"*Stem* erect, fleshy and compressed, of a deep yellowish green, widening from the root upwards, where it forms a kind of flat pseudo-bulb, partly covered by the sheathing appendage to the leaves, and, at the base, by the imbricated, sheathing, ovate-acuminate, deep brown, persistent scales. *Leaves* three or four near the top, distichous, alternate, nearly erect, of a deep shining green, ovate oblong with a round emarginate point, comparatively thin and rigid, varying from three to five inches long, and about an inch broad, twisted near the base. *Flowers* very small and numerous, cream-coloured, produced in a recurved massy cluster at the extremity of the stem, nearly concentrated by the brown scales and bracteas in which they are enveloped. *Pedicells* short and round, pale green, slightly channelled, bearing two flowers on each. *Sepals* very thin, ovate-acuminate, nearly all of the same form and size, but the two lateral ones are more oblique and surround the spur-looking end of the labellum, whilst the upper one extends over and embraces the column. *Petals* very small, linear, spreading and recurved. *Labellum* concave, reniform, slightly undulated at the margin with two deep yellow processes or glands near the centre, from which it diminishes to a narrow claw that connects it with the end of the column, and gives it the appearance of being spurred. *Column* very short and nearly round, with a fleshy projection at its base. *Anthers* roundish cordate, apparently two-celled. *Pollen masses* 8, ovate, pointed, deep yellow."

5. SACCOLABIUM *compressum*; caule juniore ancipiti, foliis distichis amplexicaulibus undulatis obtusis obliquè 3-dentatis, racemis cylindraceis pendulis, labelli calcare falcato obtuso laminâ carnosâ minimâ dentiformi sepalisque triplò longiore.

A fine Manilla plant imported by Messrs. Loddiges. The flowers are in drooping cylindrical racemes, with small sepals blotched with crimson, and a long ivory-white spur. A figure of it is preparing for the 7th No. of the Sertum Orchidaceum.

6. CYMBIDIUM *madidum*; foliis ensiformibus basi canaliculatis, racemis pendulis, sepalis ovalibus obtusis apiculatis patulis, petalis conformibus erectis, labello nudo per medium madescente oblongo : lobis lateralibus minimis intermedio oblongo obtuso basi paululùm angustato.

A new species of East Indian Cymbidium imported by Messrs. Rollissons. The leaves are like those of C. sinense; the flowers are a dull yellowish olive green, with sepals about half an inch long; the labellum is yellower, with a deep purple stain inside, and is altogether destitute of the lamellæ so usual in this genus; in room of which there is a shining exudation all along the axis. I did not remark any smell in this species.

7. DENDROBIUM *plicatile* (D. *fimbriatum*, Lindl. gen. & sp. pl. p. 76.)

This plant has flowered with Messrs. Rollissons, imported from Manilla. It has solitary, rather large flowers, of a dull yellow a little tinged with red, and a curious plaited labellum. The original name of this plant being the same as that applied to a very different species, it has become necessary to substitute another, which I have taken in allusion to the very conspicuous plicatures of the labellum.

8. CYCNOCHES *maculatum*; racemo longissimo multifloro, labello lineari-lanceolato, hypochilio lineari, metachilio basi cornuto glandulis utrinque teretibus elongatis genuflexis pinnatifidè marginato, epichilio lanceolato membranaceo acuto margine incurvo.

A very fine species, with long, pendulous, many-flowered racemes of dull yellowish-brown flowers spotted with brown. It has lately flowered in Mr. Barker's invaluable collection of epiphytes, and will be immediately published in the Sertum Orchidaceum; the page of the Botanical Register is too small to do justice to so noble a flower.

9. MORMODES *buccinator*; sepalis lineari-oblongis lateralibus reflexis dorsali petalisque ovali-lanceolatis erectis, labello unguiculato carnoso nudo subrotundo-cuneato apiculato utrinque emarginato lateribus in buccinæ formam revolutis, columnæ dorso acuto.

Flowers pale green, with an ivory-white lip, whose sides are so rolled back as to give it the appearance of a trumpet. The column is twisted sometimes to the right, sometimes to the left. The habit and general appearance of the flowers, except in colour, is that of *M. atropurpurea*, figured in the Botanical Register, t. 1861. Specimens were sent me by Mr. Thomas Williams, gardener to J. Willmore, Esq. of Oldford, near Birmingham.

18. SOPHRONITIS *violacea*; pseudobulbo ovali, folio solitario lineari scapo terminali basi vaginato 1-floro longiore, labello obovato acuto nudo basi gibboso, columnæ alis maximis carnosis obtusis falcatis.

A pretty little epiphyte with solitary violet flowers, for which I am indebted to Mrs. Cannon, of Stratford Green, in whose hothouse it flowered in the beginning of February. It is destitute of the brick red colour found in the other species of this genus.

¶ 19. CATASETUM *Russellianum*; pseudobulbo elliptico magno, foliis latolanceolatis, racemo amplo, labello submembranacco anticè inflato ore contracto margine anteriore producto undulato fimbriato disco membranâ cristato, columnâ nudâ (ecirrhosâ). *Hooker in Bot. Mag. t.* 3777.

A large species with green flowers, a little relieved by the greater whiteness of the labellum; sent from Guatemala by Mr. Skinner to the late Duke of Bedford. The labellum is described by Sir W. Hooker as being almost membranous. It is very near *Catasetum laminatum*, a Mexican species.

20. EPIDENDRUM *falcatum*. Lindley in Taylor's Annals of Natural History, Feb. 1840. (*E. Parkinsonianum*, Hooker in Bot. Mag. Feb. 1840. t. 3778.)

A very fine species of the caulescent tribe of this extensive genus, with fleshy branching short stems, long channelled falcate leaves, and large yellow flowers, springing from within a long pale yellowish-green membranous spathe. In Sir Wm. Hooker's figure only two flowers are represented, but in my wild specimens there are four or more to each spathe. The latter were found by Mr. Hartweg among rocks and loose stones, at the Hacienda de Santa Ana, in the neighbourhood of Oaxaca, flowering in May. The plant is not very uncommon in collections of Mexican epiphytes.

21. ONCIDIUM *Insleayi* (Barker in litt.); pseudo-bulbis ovatis compressis diphyllis, foliis erectis coriaceis oblongo-ensiformibus subundulatis apice recurvis racemo simplici erecto rigido brevioribus, sepalis petalisque oblongis subæqualibus undulatis infimis basi connatis, labello obovato retuso basi sagittato disci tuberculo apice depresso dilatato bilobo utrinque in medio unidentato lamellâque unicâ retrofractâ aucto, columnæ alis cirrhatis.

This beautiful Oncidium is a native of Mexico, whence it was imported by George Barker, Esq., of Birmingham, who has named it after Mr. Insleay, his gardener. It has flowers similar in size, colour, and spotting, to those of O. Papilio, but their form is quite different, and they are produced upon a stiff erect spike. It is among the finest of the genus, and will form one of the plates in Mr. Bateman's magnificent work on the Orchidaceæ of Mexico and Guatemala.

22. BROUGHTONIA *aurea*; sepalis linearibus acutis, petalis conformibus, labello ovato convoluto acuminato.

A native of Mexico, in the neighbourhood of Valladolid, whence it was obtained by Mr. Barker.— The flowers are a bright yellowish red colour, like those of *Epidendrum vitellinum*, rather larger than in *Broughtonia sanguinea*, and scentless.

23. OBERŌNĬĂ *cylindrica*; spicâ densâ cylindraceâ, bracteis canaliculatis subciliatis, sepalis petalisque reflexis intùs labelloque cordato truncato fimbriato pubescentibus.

A most curious little species of this strange genus of epiphytes; imported from Manilla by Messrs. Loddiges. Its flowers are small, green, very densely arranged in a perfectly cylindrical spike about three inches long; each flower appears to consist of a labellum only, the sepals and petals being so closely turned back as to be unobserved until carefully looked for. The minute flowers are beautiful microscopical objects.

24. BRASAVŌLĂ *venosa*; folio lanceolato semicylindraceo supra canaliculato, sepalis petalisque linearibus, labelli ungue longo complicato laminâ subrotundo-ovatâ subtrilobâ acuminatâ basi serratâ : venis elevatis.

A fine species resembling *B. nodosa* in habit, but with much larger flowers, the sepals being more than two inches long. The lip is white, the other parts greenish. It is at once distinguished from *B. nodosa* and all the allied species by the firmness of the lip, which is more or less evidently lobed at the side, and has the veins distinctly elevated. The flowers are deliciously sweet at night. Messrs. Loddiges imported it from Honduras.

25. LÆLĬĂ *rubescens*; pseudo-bulbis subrotundis compressis utrinque angulo elevato, foliis oblongis obtusis scapo tereti vaginato brevioribus, racemo multifloro, bracteis pubescentibus ovario plus duplò brevioribus, sepalis linearibus, petalis lanceolatis subundulatis, labello conformi auriculato medio pubescente: lineis duabus elevatis.

For a specimen of this pretty new species I am indebted to Mr. Barker, who purchased it some time ago from Mr. Joseph Knight, Nurseryman, in the King's Road. Its native country is unknown, but is probably Mexico. It forms a patch of roundish compressed pseudo-bulbs, looking like those of a Maxillaria, and bearing each a single broad leaf, much shorter than the slender terminal scape. The latter is about a foot long, and bears at its extremity a loose raceme of delicate whitish flowers tinged with pink, and about the size of those of *L. albida*. The two first flowers which opened with Mr. Barker exhibited a singularly monstrous structure, with only two sepals and two petals, the lip and one of the sepals being missing; all the other parts were in the usual state.

¶ 28. STANHŌPEĂ *maculosa*; hypochilio rotundato saccato intùs verrucis glandulosis anticè obsito, metachilio brevi constricto cornua 2 falcata porrecta gerente, epichilio oblongo obtusè 3-dentato apice subreflexo. *Knowles & Westcott, Floral Cabinet, t.* 121.

A fine plant, imported (from Mexico ?) by Mr. Barker. It is said to resemble *S. tigrina* in several points, but to differ in the much smaller size of the flowers, in its general markings, in the warty, not lamellated, inner surface of the hypochilium, &c. "The inner surface of the hypochilium, although not broken up into lamellæ, has the warts arranged in a radiated manner."

¶ 33. PERISTĔRĬĂ *guttata*; scapo brevi pendulo multifloro, racemo denso secundo, labelli dimidio inferiore disco crasso concavo lobis lateralibus vix conspicuis : superiore ovato margine denticulato apice integro abruptè incurvo intùs tuberculis minutissimis obsito, columnâ apterâ. *Knowles & Westcott, Floral Cabinet, t.* 70.

A beautiful plant allied to P. pendula, "from which it differs in the colour, the shape, and the smaller size of the flowers, in the absence of wings from the column, and in the scarcely developed lateral lobes of the labellum. It is much more nearly allied to the *P. cerina* of the Botanical Register, with which it agrees in the general shape of the flowers, and the denticulated margin of the middle lobe of the labellum, but differs from it in the colour and copious markings of the flower, in the entire not emarginate point of the middle lobe, and in the absence of distinct lateral lobes. Still however the resemblance between them, except in colour, is so great that we question whether our plant may not eventually be considered merely a spotted variety of P. cerina." Said to have been imported from Rio Janeiro—but this is doubtful. It flowered with Mr. Barker.

¶ 35. EPIDĒNDRŬM *crispatum*; pseudo-bulbis ovatis diphyllis, foliis lineari-lanceolatis carnosis obtusis submucronatis, scapo multifloro, petalis et sepalis linearibus striatis, petalis sepalis minoribus, labello tripartito laciniis lateralibus columnam involventibus intermediâ elongatâ crispatâ (duplò longiore), ovario varicoso. *Knowles & Westcott, Floral Cabinet, v.* 2. *p.* 79.

Imported from Mexico by Mr. Barker. "The contrast between the long crisped white labellum, and sepals and petals, is very pleasing and striking." It is one of the many species of the Encyclia division of the genus, all of which require a careful comparison with each other. It must be very near E. pictum, *Bot. Reg. misc.* 1838. *no.* 42.

¶ 37. BRĂSSĬĂ *cochleata*; sepalis petalisque subæqualibus linearibus acuminatis, labello elongato cochleato acuminato. *Knowles & Westcott, Floral Cabinet, t.* 53.

A native of Demerara, where it was found by Henchman. It has green flowers, with long narrow spotted segments, and a lip narrow, and hollowed out below the point, where it terminates in a slender bristle. It is not handsome, but is very distinct.

¶ 40. PLEUROTHĀLLĬS *villosa*; foliis ovatis coriaceis, petiolis sulcatis, scapo flexuoso, bracteis spathosis, floribus villosis purpureo-maculatis, supremo sepalo subspathulato margine reflexo carinato, sepalis lateralibus connatis basi gibbosis, petalis spathulatis minutis, labello linguiformi recurvato sulcato. *Knowles & Westcott, Floral Cabinet, v.* 2. *p.* 78.

Imported from Mexico by Mr. Barker. "It grows about four inches high; the scape is flexuose, and at each bend there is a flower, very woolly, and beautifully spotted with purple; the lateral sepals are gibbous at the base."——It therefore seems to be a Specklinia.

¶ 41. PLEUROTHĀLLĬS *ciliata*; caule adscendente subtereti sulcato, folio coriaceo elliptico-lanceolato emarginato, racemo recurvo folio breviore, floribus alternis secundis, sepalis minutissimè ciliatis lateralibus nisi apice connatis superiore erecto, petalis lineari-lanceolatis ciliatis columnâ paulò longioribus, labello sub-ciliato linguæ-formi obtuso concavo, clinandrio tridentato. *Knowles & Westcott, Floral Cabinet, t.* 19.

A cinnamon-flowered species, collected by Henchman on the Masseroni river in Demerara, and first flowered by Mr. Willmore.

¶ 43. MAXILLĀRĬĂ *aureo-fulva*; pseudo-bulbis rotundato-ovatis angulatis rugosis monophyllis, foliis oblongo-lanceolatis costatis acutis scapo radicali multifloro brevioribus, floribus longè pedicellatis apice reflexis, sepalis lanceolatis acuminatis lateralibus basi in calcar spurium productis, petalis striatis lineari-lanceolatis acuminatis, labello unguiculato trilobo in medio striato: lobis lateralibus acutis intermedio lanceolato acuminato, gynizo subrotundo margine superiore crasso. *Knowles & Westcott, Floral Cabinet, t.* 83. *Hooker in Bot. Mag. t.* 3629. (M. stenopetala, *Flor. Cab. vol.* 2. *p.* 112.)

A very pretty species from the Organ mountains of Brazil, with racemes of orange-coloured flowers. From the observations of the authors of the Floral Cabinet it has the technical character of Bifrenaria.

¶ 45. ONCĪDĬŬM *carinatum*; pseudo-bulbis ovatis, foliis lanceolatis, scapo erecto, floribus secundis racemosis, petalis conniventibus maculatis, pedicellis flexuosis, labello cristato integerrimo, bracteis scariosis. *Knowles & Westcott, Floral Cabinet, vol.* 2. *p.* 30.

A pretty little species obtained from Xalapa by Mr. Barker. "It differs from Oncidium in having the wings of the column in the centre instead of at the apex.——The labellum is also entire and keeled from the apex."

¶ 46. ONCĪDĬŬM *intermedium*; foliis oblongis acutis rigidis carnosis, scapo flexuoso paniculato ramoso multifloro, sepalis undulatis unguiculatis subintegris supremo rotundato subcochleato lateralibus spathulatis subreflexis, petalis latè unguiculatis subrotundatis crenulato-crispis, labelli lobis lateralibus nanis revolutis intermedio reniformi undulato emarginato, crista posticè lobis 2 crassis tuberculatis anticè callis tribus intermedio elevato elongato lateralibus divergentibus, columnæ alis carnosis obliquis suprà margine repando-sinuato infrà auriculâ elongatâ. *Knowles & Westcott, Floral Cabinet, t.* 60.

Imported from Cuba by George Barker, Esq. A very beautiful plant, allied both to *O. carthaginense* and *luridum*, from both which it is said to be distinguished by many characters. The flowers are bright yellow spotted with crimson, as in *O. guttatum*.

¶ 47. ONCĪDĬŬM *unicornutum*; pseudo-bulbis oblongis compressis costatis diphyllis, foliis lineari-lanceolatis, scapo simplici gracili glauco multifloro, sepalis petalisque latè lineari-spathulatis inferioribus in unum connatis, labelli lobis lateralibus rotundatis denticulatis intermedio subcuneato basi unicornuto, cristâ elevatâ transversâ, columnæ alis abbreviatis. *Knowles & Westcott, Floral Cabinet, vol.* 2. *p.* 143.

A native of the Organ mountains, from Mr. Willmore's collection. Nothing further is said about it.

¶ 49. EPIDĔNDRŬM *stenopĕtălŭm;* pseudo-bulbis cylindraceo-oblongis; foliis lanceolatis; floribus solitariis; sepalis lanceolatis; petalis angustè linearibus; labello trilobo, lobo medio cuspidato-purpureo, lobis lateralibus subrotundis albidis columnam involventibus. *Knowles & Westcott, Floral Cabinet, vol. 2. p.* 175.

"A delicate and pretty species, native of Mexico, and imported by Mr. Barker in 1837. Its sepals and petals are brownish; the labellum white, with the upper portion tinged with purple. It appears to produce only a solitary flower at a time."

56. EPIDĔNDRŪM (Encyclia) *glaūcūm;* undique glaucedine obductum, pseudobulbis ovalibus compressis monophyllis, foliis ensiformibus acutis pergameneis basi canaliculatis scapo pendulo paniculato breviori-bus, sepalis petalisque ovalibus obtusis, labelli linearis apice carnosi trilobi lobis lateralibus rotundatis intermedio ovato medio calloso, columnæ cardine appendice ovatâ obtusâ recurvâ aucto. (Epithecia glauca, *Knowles & Westcott, Floral Cabinet, t.* 87.

Mr. Barker, who imported this curious plant from Mexico, has favoured me with specimens, which enable me to state that it cannot be generically separated from the genus Epidendrum. It is not a showy plant, the flowers being small, scentless, and green stained with dull purple. The drooping panicle is somewhat remarkable, and the glaucous covering of every part is much more so; but the species has no claims upon the attention of those cultivators who seek after ornamental plants.

57. BARKĔRĬA *ēlĕgāns.* Knowles & Westcott, Floral Cabinet, t. 49.

An elegant little Mexican Orchidaceous plant, with tapering fleshy stems like those of a meagre Cycnoches, narrow-lanceolate acuminate membranous leaves, and a terminal inflorescence, in the form of a raceme of a few large bright pink flowers. It is well figured in the Floral Cabinet, and is one of the few plants which rival the Cattleyas in gaiety of appearance. It is however said to be difficult of cultivation, and is at present of great rarity. The genus is well distinguished from Cattleya, to which it approaches most nearly, by its lip being flat and undivided, instead of cucullate, and more or less completely three-lobed, as well as by its peculiar habit. As the genus has not yet found its way into books generally used by Botanists, a short technical character of it is added.

BARKERIA. Sepala & petala æqualia, libera, membranacea, patentissima. Labellum planum, integerrimum, nudum, (cuneatum cum apiculo) columnæ adpressum. Columna petaloidea. Anthera 4-locularis, carnosa, septorum marginibus membranaceis. Pollinia 4, caudiculis totidum ligulatis reflexis, per paria connatis.——Herba epiphyta, caulibus fusiformibus carnosis. Folia angusta, membranacea, nervosa. Pedunculi longi, graciles, squamati, terminales, apice racemosi. Flores magni, conspicui, nutantes.

66. BRĀSSĬĂ *verrūcōsă.*

At the last meeting of the Horticultural Society in Regent Street this beautiful plant was exhibited by Messrs. Rollissons of Tooting. As it is intended for one of the plates in Mr. Bateman's "Orchidaceæ of Mexico and Guatemala," it is only necessary here to state that it is a plant of a particularly graceful habit, and that the flowers are of a clear greenish yellow, with a labellum curiously covered over all the lower part with green warts.

67. CLEISŌSTŌMĂ *maculōsă.* Lindl. gen. et sp. orch. 227.

Messrs. Loddiges have flowered this Orchidaceous plant, from a specimen imported from Ceylon. It has the habit of a small Vanda, with long-stalked spikes of small yellowish-brown and pink roundish fleshy blossoms.

75. CORYĀNTHĒS *speciōsă;* var. alba.

A pale, indeed almost white, variety of this extraordinary plant has lately flowered in the collection of Thomas Brocklehurst, Esq. of the Fence, near Macclesfield, who obtained it from Demerara. Mr. Appleby, the gardener at the Fence, informs me that the pseudo-bulbs are four in number, five inches long, tapering from the root to the insertion of the leaf, and deeply furrowed; the leaves are sixteen inches long, of a very long oval shape in the widest part, which is near the middle, one inch and three-quarters wide, deeply ribbed, and of a light green colour. The flower-stalk is one foot to the first flower and four inches beyond it, bearing four flowers. He thinks it probable that the flowers will be more numerous and larger as the plant becomes stronger. The appearance of the plant is described as being intermediate between C. macrantha and maculata.

GUATEMALA ORCHIDACEÆ.

Mr. Skinner, a gentleman resident in Guatemala, and well known as a very liberal and enthusiastic naturalist, has lately sent several valuable collections of Orchidaceous plants to his correspondents in this country, which have generally arrived in the most healthy state. Amongst others Mr. Harris of Kingsbury has received a portion, which Mr. Beaton informs me were in such excellent condition as to be little the worse for their journey; a circumstance probably owing to the plants having been nailed to the sides of the boxes, with the interstices packed with Tillandsia. Mr. Beaton has found dry sawdust answer the same purpose. This at least is certain that moss, from its hygrometrical quality, is the worst of all materials. Mr. Harris having placed in my hands the dried specimens transmitted with the plants, together with Mr. Skinner's valuable observations upon the climate which they affect, I have made the following selection from the correspondence, which cannot fail to possess the greatest interest for all growers of Orchidaceæ.

79. ONCĪDĬŪM *leucochīlum;* }
80. STANHŌPĔĂ *oculātă;* } both growing on the same branch.

"This plant inhabits the higher temperatures, and I should recommend for its cultivation in Europe that it should not be kept in a temperature colder than 55°, or warmer than 70°; well watered from June to September every afternoon; and from October to May inclusive only slightly watered every evening at sundown, to resemble our dews, not, it must be recollected, so heavy as people represent them in Europe—the region being high, and very different to a coast climate. The seasons here are the same as in England, the coldest weather December, January, and February; on the 25th, 26th, and 27th of December, 1839, the thermometer at six o'clock in the morning in the open air for three days averaged 36° Fahr. and yet *Oncidium leucochilum* continued to shoot its young stems. Flowers in February and March."

81. EPIDĔNDRŪM *Skinnĕrī.*

"This plant inhabits also a middling temperature, and will thrive best in a climate graduated from 56° to 70°. Do not put this plant in earth, but permit it to have free scope for young roots and shoots, which it will throw out in August."

82. EPIDĔNDRŪM *aurantĭăcūm.*

"Same habitat as Oncidium leucochilum, always found together; only that this plant seeks exposure, and therefore is subject to greater extremes of heat and cold; the finest masses however are always found on the steep brows of rocky barrancas—thus however rarely '*come-at-able.*'" My specimen bore last month fifteen flowers on one stem.

83. CĀTTLĔYĂ *Skinnĕrī.*

"This plant inhabits the *hot damp* COASTS, and will require a very different treatment to any of the foregoing; it is always found on very high trees, and most difficult to get at, except after a storm that may have chanced to throw down some of the large forest trees. It should be well watered daily, to represent the heavy dews and the rains, which latter are from May to November. I should recommend on whatever you may grow this plant, it may not imbibe too much of the extra moisture, as its habitat being on branches of large trees seldom having any lichen, the heavy rains do not lay.

This flower does not seek too much shade, but rather, like 'Epid. aurantiacum,' exposed places. Climate 80° to 85°, and sometimes 95°, during the day. Flowers in January and February; vulgarly called 'Flor de San Sebastian,' from its being in season, and adorning the altars on that saint's day (20th of January)."

84. EPIDĒNDRŪM (Encyclia) *incumbēns;* floribus densè paniculatis, sepalis linearibus patentissimis basi angustatis, petalis conformibus sed paulò latioribus, labelli postici lobis lateralibus triangularibus acuminatis intermedio subrotundo-ovato apiculato venis elevatis cristato, callis duobus oblongis secus unguem.

A plant "with a peculiarly graceful flower," having the same locality as Oncidium leucochilum. It is very near O. alatum, oncidioides, and the rest of the panicled Encyclias.

85. EPIDĒNDRŪM *macrochīlūm.*

"This plant inhabits the *coast*, where only it thrives in perfection. It flowers in February, March, and April. A constant attendant on Cattleya Skinneri, and will require much the same treatment. Vulgarly called '*Boca del dragón*,' '*Dragon's mouth.*'"

86. CYRTOCHĪLŪM *maculātūm;* var. Russelianum.

This is certainly one of the many fine varieties of Cyrtochilum maculatum, with very large richly spotted flowers. Mr. Skinner calls it *C. Russelii*, and gives the following account of it. "Its habitat is cold climate, and its treatment will be the same as that of Oncidium leucochilum; it luxuriates amongst pines, but only attached to oaks. No plants of the Orchideæ tribe are ever found on pines, except Catasetum towards the north coast, but then these plants form such a mass of hairy roots as to secure them completely from the influence of the pitch that, in all hot climates, constantly oozes from the pines. Climate 65° to 70°, generally."

87. LÆLĬA *superbiēns;* scapo longissimo multifloro, sepalis petalisque lineari-oblongis obtusis membranaceis labello longioribus, labelli lobo medio obtusissimo: disco lamellis quinque maximis subserratis anticè truncatis aucto, antheræ cristâ biauritâ, clinandrio denticulato.

Of this fine new species Mr. Skinner speaks in the following manner. "*Saw you ever any thing like this!!!* This most magnificent of all plants I have sent several times, it is decidedly Epidendrea, but I expect it will get a new title in Europe. It flowers in November, and in some instances bears from eighteen to twenty flowers on stems from nine to twelve feet long!! Mr. Hartweg says he found this plant in great abundance in the State of Quesaltenango, near Chantla, and that he went forty leagues out of his route to find it, and now he has sent splendid masses of it to England, and at least thirty dried specimens. I have found it most rarely and evidently planted by the Indians before their doors, in 'Acatenango,' from whence I brought the specimen now sent, and in 'Sumpango,' where I found it twelve months ago; also from Costa Rica. Chantla, its native habitat, is very cold, in fact this plant I expect will require only a greenhouse warmth. Its habitat in Costa Rica is 68° to 70°, where I have found it here 70°, but Chantla is 55° to 65° generally."

88. EPIDĒNDRŪM *Stamfordiānum.*

"This plant inhabits only the coast, shady, very moist lands. Flowers in November, and in different temperatures until March, I having found it in flower at all times in different journeys. The north coast is very hot and wet, and the dry weather commences in March; I found it in flower in April there. On this side (south coast) dry weather commences in October; I found it in flower in November. You may therefore treat this plant as you would the Cattleya, its habitat on this side being from the same temperature, only it requires more wet."

89. BRASSAVŌLĂ *glaūca.*

"A splendid large white flower, and a most extraordinary strong aromatic; is the companion on oaks of the Cyrtochilum Russelii, and must be treated in the same manner."

90. HEXŌPĬĂ *crucĭgĕră.* Bateman mss.

"Climate 68° to 75°, and 65° to 70°.

91. EPIDĒNDRŪM *rhizŏphŏrūm.*

"Climate 70°."

92. ASPĂSĬĂ *epidendroīdes.*

"Very sweet; habitat 'Escuintla,' and 'Medio Monte.' Flowers in January and February. Climate 75° and 80°."

This same plant had previously been sent by Mr. Skinner from Costa Rica.

93. EPIDĒNDRŪM *arŏmătĭcūm.*

"Flowers in February and March. Habitat 'Acatenango,' 'Duenas,' 'Mixco,' 'Naranjo,' &c. Climate 60° and 75°."

94. ODONTOGLŌSSŪM *grandĕ;* sepalis lanceolatis lateralibus convexis falcatis petalisque oblongis obtusiusculis latioribus subundulatis, labello subrotundo basi auriculato sepalis plus duplò breviore: tuberculis basi tribus corrugatis aliisque lateralibus dentiformibus minoribus, columnæ tomentosæ marginibus rotundatis convexis incurvis.

"This is the finest plant we have in Orchidaceæ in this country. The present specimen is a fine mass, and I hope may arrive safe. Climate 60° and 70°. I recommend not too much heat; it seldom gets a climate reaching 70°, and thrives best in damp shades, with perhaps 60°, rather than beyond that."

Certainly this is a most extraordinary plant. Its habit is altogether that of Odontoglossum, with which its unguiculate lip and peculiar column also correspond; but the tubercles at the base of the lip are those of Oncidium. The dried flowers measure *six inches and a half* from the tip of the petals, and are doubtless shrunk in the process of drying. They look something like an enormous Maxillaria, and appear to have been of a uniform yellow or white colour; but of this I cannot very well judge. Two flowers grew together on a scape, which in the specimens before me is not more than five inches long. Mr. Skinner thinks that the reason why the plants he has formerly sent home have not flourished, is because they have been kept too warm.

95. ONCĪDĬŪM *ornithorhynchūm.*

"Habitat shady damp places, climate 68° to 70° and 75°; flowers in December and November."

96. HARTWĔGĬĂ *purpŭrĕă.*

"Climate varied from 56° to 75°."

97. ONCĪDĬŪM *ampliātūm.*

"From Costa Rica; found there on the sea shore in the Gulf of Nicaya; since found throughout the coasts of Nicaragua, and also in the Escuintla, 15 leagues from this city; climate 80° to 85°; flowers in February."

98. CYCNŌCHĒS *ventricōsūm.*

"Habitat 80° to 85°, but found in a climate of 70° to 75° much superior.

99. CATASĒTŪM *maculātūm.*

"Habitat 75° and 80. Flowers in July."

100. TRIGONĪDĬŪM *Egertoniānūm.*

"Habitat 75° and 80°. Flowers in April."

101. MAXILLĀRĬĂ *Skīnnĕrī* (Bateman mss.); pedunculis radicalibus? bracteâ flori proximâ ovatâ cucullatâ acuminatâ sepalis duplò breviore, sepalis oblongis acutis glabris lateralibus majoribus basi parùm productis intùs villosis, petalis subrhombeis velutinis sepalo supremo brevioribus, labelli lobis lateralibus truncatis glabris intermedio ovato obtuso apiculato pubescente: laminâ retusâ transversâ infra lobum medium venisque sparsè villosis, columnâ elongatâ tomentosâ.

"This is the finest Maxillaria I ever saw. It has a gentle odour; but the lip is magnificent purply white, with centre tigered in deep rose colour. Flowers in December and January, and even in April have I seen it. Habitat 68° and 75°. Terrestrial."

Nearly allied to M. Harrisoniæ. Diameter of the flower from tip to tip of the lateral sepals four inches and a half.

---

102. POLYSTĂCHYĂ *bractĕōsa;* foliis obovato-oblongis acutis in petiolum carinatum angustatis, racemo tomentoso elongato cernuo basi ramoso,

bracteis foliaceis acuminatissimis squarrosis, sepalis villosis, labelli ciliati trilobi per medium villosi lobo medio ovato acutiusculo.

A drooping flowered Polystachya, with orange coloured flowers, imported from Sierra Leone by Messrs. Loddiges. Its leafy tapering squarrose bracts and downy flowers readily distinguish it from the rest of the genus.

103. EPĬPHŎRĂ *pubēscēns.* Lindl. in Comp. to Bot. Mag. v. 2. p. 201.

This very rare little Orchidaceous plant, originally found by Dr. Burchell in the Caffre country, and afterwards collected by Drège, has lately flowered with Messrs. Loddiges, who imported it from Delagoa Bay. It grows about 6 inches high, and has an ancipitous flexuose scape, terminated by a few fragrant bright yellow flowers streaked with red. It approaches very near to the genus Polystachya, next to which, among Vandeæ, it must be placed in any future arrangement. It is an epiphyte. The inside of the little trident-shaped lip is bearded with long hairs.

106. CIRRHOPĔTĂLŪM *picturātum* (G. Loddiges); pseudobulbis oblongis angulatis, folio oblongo convexo emarginato, floribus umbellatis, sepalo supremo ciliato setaceo-acuminato lateralibus oblongis obtusis, petalis acuminatissimis ciliatis villosis, labello lineari recurvo secus medium elevato, columnæ auriculis obtusis integris.

A pretty little Indian plant with purple flowers, growing in flat umbels, and deeply stained with dark red. Its habit is that of a Bolbophyllum, its scape from 5 to 6 inches high. Communicated by Messrs. Loddiges.

107. CIRRHOPĔTĂLŪM *aurātum;* pseudobulbis oblongis angulatis, folio oblongo convexo, floribus umbellatis, sepalo supremo petalisque setaceo-acuminatis fulvo-ciliatis lateralibus acutis, labello lineari recurvo, columnæ auriculis rotundatis integris.

Sent to Messrs. Loddiges from Manilla, by Mr. H. Cuming. It is very like the last, but much larger in the flowers, which are fringed with golden yellow hairs, which gives them an appearance distinct from that of the last, where they are deep purple.

108. ONCĬDĬŪM *pallĭdum;* pseudobulbis obcordatis compressis marginatis, foliis acutè carinatis convexis oblongis cœsiis, scapo erecto paniculato, sepalis petalisque oblongis obtusis liberis, labelli auriculis rotundatis lobo intermedio cuneato-obovato emarginato angustioribus, disci basi glandulosi appendice 5-partitâ: laciniis cornutis æqualibus pubescentibus, columnæ alis brevissimis obliquè truncatis.

A very pretty Orchidaceous plant, obtained from the Brazils by Messrs. Lucombe, Pince, and Co. of Exeter. The leaves are sea-green like those of Maxillaria Rollissonii; the panicle is almost a foot long, and the flowers are in colour similar to O. divaricatum. It is very near O. Harrisonianum.

109. STANHŌPĔĂ *Martiăna* (J. B.); foliis anguste lanceolatis acutis racemo paucifloro longioribus, sepalis petalisque paulo angustioribus ovatis obtusis, hypochilio brevi sessili saccato utrinque cornubus maximis porrectis apice cirrhosis sub-incurvis instructo, epichilio oblongo lævigato obscurè 3-dentato apice subreflexo cornubus hypochilii breviore vel subæquali, columnæ subclavatæ marginibus parùm dilatatis. *Bateman in litt.*

"A native of Mexico, discovered by Baron Karwinski in 1827, and sent by him to Knypersley, where it flowered for the first time in May of the present year. It is one of the most distinct and magnificent species of the extraordinary genus to which it belongs, and in the magnitude of its blossoms is second only to *S. tigrina.* The sepals are straw-coloured, faintly and sparingly marked with clusters of little vinous dots, the petals appear transparent white, with large spots of intense crimson, the lip is also a clear ivory white, except a slight discoloration at the base. It is however in the horns of the latter member that the most striking peculiarity of the species consist. These are of great size and strength, and might with propriety be likened to elephant tusks, their extremities moreover are twisted into small *cirrhi,* a circumstance wholly without precedent in the genus Stanhopea. The nearest affinity of the plant is perhaps *S. saccata,* but the points of distinction between the two species are so numerous and obvious, that it is not necessary to contrast them. The name of *Martiana,* which I have given to the species, is designed, I need scarcely say, to compliment the illustrious Professor of Botany at Munich. Mrs. Withers has prepared a figure which will appear in an early number of 'The Orchidaceæ of Mexico and Guatemala.'"

For the foregoing note I am indebted to Mr. Bateman.

110. DENDRŎBĬŪM *revolūtum;* caulibus obtusè ancipitibus, foliis ovato-oblongis obtusis apice obliquis emarginatis basi subcarinatis, floribus solitariis oppositifoliis, sepalis petalisque acutissimis revolutis, labello carnoso convexo obtuso subtrilobo aut rhombeo per medium exarato (s. lamellis duabus inflexis parallelis instructo) lineis tribus discoloribus.

A new species imported from Sincapore, by Mr. Cuming. I have received it both from Mr. Barker of Birmingham, and Messrs. Loddiges. The flowers are straw-coloured, and about the size of those of D. Pierardi, but their lip is fleshy, convex, almost lozenge-shaped, and marked with 3 brown lines. It is not very pretty, but it is very distinct from any previously described.

111. DENDRŎBĬŪM *tĕrĕs;* caulibus gracilibus glabris, foliis carnosis teretibus obtusis, racemo terminali nudo, bracteis coriaceis spathaceis, floribus semiclausis longè cornutis, sepalis petalisque erectis acuminatis, labello cuneato apice truncato serrulato longiùs cuspidato: venis tribus pone apicem subcristatis, columnâ versus basin dente deflexo auctâ.

Another importation from Sincapore, for which I am indebted to Messrs. Loddiges. It is a slender plant, with the leaves of *Vanda teres,* and whitish fragrant flowers in terminal naked racemes. The lip is deeply stained with orange inside near the apex.

112. DINĒMĂ *pălĕācĕum;* pseudobulbis ovalibus compressis monophyllis, foliis ensiformibus obtusiusculis planis spicâ paucìflorâ longioribus, bracteis lineari-lanceolatis acuminatis paleaceis ovario lentiginoso brevioribus, floribus secundis, sepalis petalisque lineari-lanceolatis patulis, labello petalis æquali et subconformi carnoso apice dilatato per medium sulcato.

A Guatemala plant, imported by Mr. Bateman, to whom I am indebted for a specimen. It has pale straw-coloured flowers, of little beauty, and is only interesting as confirming the goodness of the genus *Dinema,* whose character however depends upon the number of pollen straps being 2, not 4 as in Epidendrum; the horn-like processes of the column, on which I formerly placed some reliance, are apparently only of specific value.

113. DENDROCHĪLUM *filifŏrmĕ;* pseudobulbis conicis (ex Dom. Bateman), scapo filiformi terminali elongato, racemi multiflori rachi angulatâ subflexuosâ, bracteis paleaceis convolutis ovario æqualibus, petalis obovatis, labello cuneato rotundato basi auriculato integro intùs bilineato, columnæ appendicibus subulatis glabris liberis columnæ longitudine.

This, the first living specimen seen in Europe, of Blume's genus Dendrochilum, has flowered with Mr. Bateman, from a plant discovered in Manilla, by Mr. Cuming. It has the habit of a Bolbophyllum, with very minute greenish brown flowers, arranged in a spike, 6 inches long, upon a terminal thread-like scape about one-half that length. In affinity the genus is nearest Liparis; but the column is furnished on each side with a lateral process, apparently representing the two stamens usually suppressed in Orchidaceæ; and this is a sound mark of distinction, independent of the flowers being nearly regular, and not ringent. The species is very near *D. simile,* from which the form of the labellum separates it, according to Blume's figure; and it must also be closely allied to *D. cornutum* of the same author; but it has neither fusiform pseudobulbs, nor elongated lateral column-processes.

120. BLĔTĬĂ *secŭndă;* foliis oblongo-lanceolatis acuminatis 7-nerviis, floribus racemosis, sepalis petalisque herbaceis subæqualibus cuneato-oblongis secundis collateralibus, labelli trilobi laciniis lateralibus nanis erectis intermediâ membranaceâ obovatâ emarginatâ margine inflexâ secus medium bilamellatâ.

Flowers green slightly dotted with crimson, with a straw-coloured labellum. A species of no beauty, imported from Mexico by Messrs. Loddiges. It is remarkable for the sepals and petals being all directed into a plane parallel with the labellum and overlying the column.

121. TRIGONĬDĬŪM *ringens;* pseudo-bulbis compressis subrotundis, foliis oblongis recurvis coriaceis nitidis, scapo filiformi stricto distanter vaginato foliorum longitudine, perianthio bilabiato: sepalo postico arcuato petalisque oblongis margine revolutis labium superius lateralibus oblongis apice obtuse carinatis inferius formantibus, labelli trilobi nani ciliati pubescentis lobis lateralibus dentiformibus: intermedio revoluto in medio lucido basi callo reniformi anticè obsoletè tridentato aucto, columnâ pubescente.

A singular species of Orchidaceous plant, forming dense patches of fine vigorous deep green pseudobulbs and leaves, promising something much better than the little insignificant scentless yellowish-green flower which they produce. The latter stands singly at the end of a very slender erect scape, and is distinctly divided into two lips. The labellum is very finely downy, ciliated at the edge, and slightly bearded along four of the veins near its base on the upper side. The column is moreover downy, and has a small collection of minute shining hairs near its base. The species was introduced from Mexico by Mr. Hartweg; but had been previously discovered in shady places in the Barranca de Tioselo, by Schiede, whose MS. description, together with a specimen, has been obligingly communicated to me by Prof. Schlechtendahl, and is here subjoined.

Fol. calycina omnia libera, 5, patentia, obscure flavescentia, lineis atropurpureis percursa, oblonga; 3 galeam formantia conniventia, 2 lateralia externa obliquiuscula deflexa. *Labellum* liberum erectiusculum, basi parum angustatum, reliquis calycinis foliolis multo brevius, trifidum, lobo intermedio multo longiore, apice reflexo, atropurpureum, in medio laminæ trituberculatum. *Gynostemium* labello paulo brevius, erectiusculum, paulo anterius curvatum, non alatum. *Anthera* terminalis anticam partem apicis truncati occupans. *Operculum* univalve transverse sese solvens, valde deciduum. *Corpuscula pollinifera* 4, 2 anteriora labello propriora majora, posteriora minora, omnia membrânulæ lunatæ insidentia, sine pedicello; membranulæ cornua in anteriore parte gynostemini affixa sunt.—*Schiede mss.*

122. EPIDĔNDRŬM (Encyclia) *bractĕscēns;* pseudobulbis ovatis cæspitosis 3-4-phyllis, foliis linearibus, scapo debili 3-4-floro, bracteis infimis foliaceis floribus longioribus supremis obsoletis, floribus nutantibus longè pedunculatis, sepalis petalisque lineari-lanceolatis acuminatis discoloribus labello longioribus, labelli liberi lobis lateralibus apice recurvis obtusis subdentatis intermedio unguiculato subrotundo-ovato multò longiore secus unguem elevato sulcato pubescente.

This is one of the prettiest of the small species of Epidendrum, and is quite distinct from all hitherto described. Mr. Hartweg found it in Mexico in the vicinity of Oaxaca. The pseudobulbs are exactly ovate, closely clustered, and about as large as a pigeon's egg. The flowers have a beautifully but delicately painted white lip, the gay effect of which is heightened by the contrast with the dingy purple of the long narrow sepals and petals. Its nearest affinity is with *E. papillosum* and *E. pastoris* of Link and Otto, the *E. Linkianum* of Klotzsch. The flowers have no smell.

123. SARCĂNTHŬS *oxyphy̆llŭs*. Wallich mss.

This plant, which has been lately received from Calcutta by several persons, has flowered in the garden of the Horticultural Society, and proves to be nothing more than a narrow-leaved variety of *Sarcanthus rostratus*, a species of no beauty, long since introduced by the Horticultural Society from China.

125. STANHŌPĔĂ *grăvĕŏlēns;* petalis ovato-lanceolatis undulatis basi carnosis, labelli hypochilio sub-compresso saccato intus glaberrimo antice bidentato et inter dentes profundè sulcato, metachilii cornubus acuminatissimis incurvis, epichilio subrotundo-ovato integerrimo, columnæ apice truncato-bilobæ alis latissimis subquadratis.

This is a noble species with the habit and general appearance of *S. saccata*, but far handsomer. The sepals and petals are of the most delicate straw colour; the lip at the base, and the central parts of the flower generally, are of a deep rich apricot yellow, while the horns and upper end of the lip are like ivory turning yellow. It differs from *S. saccata* in its hypochilium being much deeper from front to back, the petals smaller, the epichilium entire, and the column much more broadly winged. It was purchased from Mr. Tate who imported it from Peru, by the Hon. and Rev. W. Herbert, who informs me that its leaves are 4 inches wide, and 17 inches long besides the petiole, which is $2\frac{1}{2}$ inches long, acute with seven strong ribs. Its odour is so powerful that it communicates itself to the fingers after touching the flowers, and like many other smells, though agreeable in itself, is offensive from its intensity.

126. APŎRŪM *Lĕōnĭs;* (A indivisum *Gen. & Sp. Orch. p.* 70. nec Blumii) foliis coriaceis ovatis v. brevissimè cultratis obtusis, floribus solitariis terminalibus, labello lineari-oblongo emarginato ecristato apice ciliato-dentato et minutissimè pubescente.

A native of Sincapore, where it was originally found by Mr. Prince, who communicated it to Dr. Wallich with a rude drawing, upon which it was admitted by me into the *Genera and Species of Orchidaceous plants* as synonymous with the *A. indivisum* of Dr. Blume. It having lately been brought home from the same place in a living state by Mr. Cuming, under the name of "Lion's mouth," and Messrs. Loddiges having sent it to me in flower, I am now able to state that it is in reality a species distinct from that of Blume, in its labellum not having a transverse crest, and probably in other circumstances. It has therefore received a name in allusion to that just mentioned, which has arisen from a fancied resemblance between the reddish brown ringent flower and the narrow blunt undivided lip, which are compared to the jaws and tongue of a lion. However, notwithstanding this high sounding comparison, the plant is of no beauty.

127. CLEISŌSTŎMĂ *lătĭfŏlĭŭm;* foliis oblongis obtusis apice subæqualibus bilobis, floribus paniculatis: ramis simplicibus rigidis, petalis sepalisque linearibus obtusis, labelli calcare ventricoso laminâ reniformi dente membranaceo bilobo.

A Vanda-like Orchidaceous plant from Sincapore, for which I am indebted to Messrs. Loddiges. It is very like *C. maculosum*, but has broader leaves, and there are differences in the form of the labellum, &c. Its flowers are yellow bordered with red, small and densely arranged. It is pretty, but not strikingly so.

128. EPIDĒNDRŬM (Amphiglottis) *Trĭnĭtātĭs;* caulibus foliosis ancipitibus, foliis lineari-lanceolatis glaucescentibus apice obtusis obliquè emarginatis, racemo terminali nutante pedunculato basi vaginis herbaceis equitantibus imbricato, sepalis petalisque lineari-lanceolatis, petalis piliformibus, labelli lobis lateralibus concavis denticulatis intermedio lineari-lanceolato acuminato basi 3-calloso.

Flowers small, in a long raceme, pale greenish yellow, with a deep apricot-yellow lip. Not very ornamental, but pretty. I owe my knowledge of it to Messrs. Loddiges, who imported it from Trinidad.

130. CHY̆SĬS *lævis;* bracteis brevibus ovatis pedicelli longitudine, sepalo dorsali lineari-oblongo lateralibus acuminatis, petalis falcatis, labelli lobis lateralibus falcatis apice rotundatis supra columnam convergentibus intermedio membranaceo crispo subrotundo emarginato lamellis 5 carnosis glaberrimis parallelis lateralibus minoribus, columnâ basi altè excavatâ.

Of this singular Orchidaceous genus two new species have appeared in the collection of Mr. Barker. The present has very large yellow flowers, whose lip is spotted with dull brown, and more membranous than in the others; and it is remarkable for its column becoming hard and very tumid after fertilization has occurred, and eventually acquiring a green colour. The other species has white flowers, with a deep yellow fleshy lip, which is divided into two lobes, each folded twice, so as to give the whole the appearance of being four-lobed. It may be named and defined as follows:

131. CHY̆SĬS *bractĕscēns;* bracteis cucullatis venosis foliaceis ovario longioribus, sepalis petalisque ovatis obtusis, labelli lobis lateralibus obtusis intermedio carnoso bilobo plicato lamellis 5 carnosis subæqualibus parallelis basi pubescentibus, columnâ latissimâ carnosâ cymbiformi anticè pubescente.

134. ¶ EPIDĒNDRŬM *densĭflōrŭm*. Hooker in Botanical Mag. t. 3791.

This plant, with densely spiked, panicled flowers of a greenish yellow colour, relieved by a white 4-lobed lip, appears to be the same as the E. *floribundum* of Humboldt and Kunth.

135. ¶ CATASĒTŬM (monachanthus) *roseo-album;* labello triangulari acutissimo concavo-hemisphærico (non ventricoso) intùs pubescente margine basi longissimè ciliato. *Hooker in Botanical Magazine, t.* 3796. sub Monachantho.

A pretty species from Para, which has flowered in the Botanical Garden, Glasgow. The flowers are described as delicate, of a white colour, with a lip tipped and banded with red; but they are represented as greenish yellow. As the genus *Monachanthus* is suppressed I am obliged to alter the designation under which the plant has been published in the Botanical Magazine. It is very near *Catasetum* (Monachanthus) *discolor*.

136. ¶ CATASĒTŬM (Myanthus) *spīnōsŭm;* petalis versus apicem serratis, labelli infra medium saccati margine pilis tenuibus succulentis albis longè fimbriato apice angustè attenuato recurvo supra ad basin spinâ tripartitâ infra apicem spinâ magnâ dentatâ porrectâ. *Hooker in Botanical Magazine, t.* 3802. sub Myantho.

Found by Mr. Gardner in the province of Ceará in Brazil, and very nearly the same as *Catasetum* (Myanthus) *barbatum*, from which its brighter coloured and larger flowers, serrated petals and horned labellum seem to distinguish it sufficiently. From the Glasgow Botanic Garden.

137. ONCĪDĬUM *Huntiānum*. Hooker in Botanical Magazine, t. 3806, is the same as *O. sanguineum*, Sertum Orchidaceum, t. 27.

138. ONCĪDĬUM *pachyphyllum*. Hooker in Botanical Magazine, t. 3807, is the same as the *O. Cavendishianum* of Mr. Bateman's Orchidaceæ of Mexico and Guatemala, t. 3.

139. ¶ ZYGOPĔTĂLŬM *africānum*. Hooker in Botanical Magazine, t. 3812.

This, which was sent to Sir Wm. Hooker from Woburn as a Sierra Leone Orchidaceous plant, is certainly American, and in all appearance is a pale variety of *Odontoglossum Bictoniense*, the *Cirtochylum Bictoniense* of Mr. Bateman's work.

143. SARCOCHĪLŬS *unguiculātus;* racemo 3-4-floro, labelli unguiculati trilobi ungue convexo lineari lineâ mediâ exaratâ, laminâ cavâ, lobis lateralibus semi-ovatis obtusis intermedio supra trinervi subtus globoso apice papillæformi.

A native of Manilla, where it was discovered by Mr. Cuming. The flowers are light straw-colour, the side lobes of the labellum white streaked with crimson, and the middle. lobe rounded, fleshy, and dotted with crimson. It is rather a pretty species, for which I am indebted to Mr. Bateman.

145. MAXILLĀRĬA *Skinneri*. Supra no. 101.

This species, already noticed at the place above referred to in the present volume, has flowered with Sir Charles Lemon at Penrhyn. It has very much the appearance of a large form of *M. aromatica*, but the structure of the flowers is different. They are a deep dull yellow. The lip is a dull olive-brown except the middle lobe, and the column is a little variegated with crimson. I find nothing to correct in the specific character already given from the dried specimen.

146. PLEUROTHĂLLĬS *pachyglossa;* folio ovato-lanceolato apice tridentato caule suo longiore scapo breviore, scapo filiformi erecto distanter 4-floro, sepalis lineari-lanceolatis acuminatis membranaceis inferiore bicarinato bidentato, petalis carnosis obovatis convexis trinerviis apice rotundatis dorso sub apice verrucosis, labello petalis duplò longiore unguiculato crassissimo obtuso linguiformi basi auriculis duabus membranaceis aucto medio villoso dorso verrucoso.

This is the largest flowered species I have yet seen in the genus. It has purple semitransparent sepals six lines long, and about four flowers on a slender scape, divided from each other by intervals of nearly an inch. It stands next *P. fusca* and *ephemera*. Mr. Barker imported it from Mexico.

147. STANHŌPĔA Wardii. *Sertum Orchidaceum, t.* 14.

Of this fine species several varieties are now in the gardens, among which an exceedingly handsome one is that named *S. Barkeri*, without the eye-like spots, and with the anterior part of the lip of a delicate ivory white. Of this.the fragrance is very agreeable, which is more than can be said of *S. Wardii* itself and some of the other varieties, of which I now fear even *S. graveolens*, no. 125, is one, so many forms have I examined within the last few weeks.

151. ANGRÆCŪM *bilobum;* caule brevissimo, foliis cuneato-obovatis oblique bilobis reticulatis racemo verrucoso pendulo multifloro multò brevioribus, sepalis petalisque lanceolatis patentibus, labello conformi paulò majore calcare filiformi emarginato breviore.

A lovely Orchidaceous plant, with long drooping racemes of snow-white flowers, just tipped with pink, and slightly sweet-scented. It has singular, netted obovate leaves, cloven at the point, and slightly sprinkled with small tubercles. It was imported by Messrs. Loddiges from Cape Coast Castle.

152. EPIDĒNDRŬM *lancifolium*. Lindl. Gen. & Sp. Orch. p. 98.

This plant has been imported from Mexico by Messrs. Loddiges, with whom it has flowered. It is very like *E. cochleatum*, but its lip is striated with deep purple radiating lines upon a pale yellow ground. It has a slight perfume.

153. DENDRŌBĬŬM *herbaceum;* caulibus ramosis teretibus glabris, foliis lineari-lanceolatis acuminatis apice obliquè bilobis, racemo paucifloro terminali, sepalis lateralibus basi paulo productis dorsali petalisque linearibus, labello lanceolato integerrimo nudo, clinandrio biauri.

A small branching Orchidaceous plant, with leaves from two to three inches long, and inconspicuous green flowers. It was imported from the East Indies by Messrs. Loddiges.

154. ONCĪDĬUM *ramosum*. Lindl. in Bot. Reg. sub fol. 1920.

A very fine species, which has flowered with Messrs. Loddiges, imported from Gongo Soco in Brazil. It has gay pale flowers in a branched panicle, already a foot square, upon a scape five feet high, and in its native state the inflorescence is much larger. Several finger-like streaks of brown radiate from the base of the tuberculated crest, whose elevations are disposed in two tiers, the upper consisting of five short wavy ridges, and the lower of two, with a truncated four-lobed tubercle between them.

155. MAXILLĀRĬĂ *Macleei;* sub-caulescens, pseudobulbis confertis oblongis monophyllis subcompressis, foliis aciculatis, pedunculis axillaribus solitariis basi squamatis, floribus rectis, sepalis ovato-lanceolatis acutis conniventibus, petalis multò brevioribus acutis paululùm recurvis, labello oblongo indiviso margine plano apice haud reflexo, callo disci oblongo integro. *Bateman in litt.*

"This plant is nearly related to, but undoubtedly distinct from, *M. tenuifolia;* the only species with which it is necessary to contrast it. The following are the more prominent marks of distinction. The pseudo-bulbs of *M. Macleei* in form are a true oblong, instead of tapering as do those of *M. tenuifolia;* the leaves too of the former are not more than one-third the length of those of the latter. In *M. tenuifolia* the sepals and petals have their edges recurved, in *M. Macleei* they are perfectly plane; again, in *M. tenuifolia* the sepals are turned backwards, and the petals are connivent, whereas in *M. Macleei* it is exactly the reverse; the lip moreover of the latter is not recurved. The colouring of the two plants is also different, the sepals and petals of *M. Macleei* being internally of a uniform rich reddish brown, and its lip whitish, with intense maroon spots, while in *M. tenuifolia* there is a considerable admixture of yellow, both in the sepals and petals, and in the lip.

"The species is a native of the inland parts of Guatemala, whence it was sent to me in the spring of the present year, by my indefatigable correspondent Mr. Skinner, after whose worthy partner, Mr. McKlee, I have much pleasure in naming it; the more particularly, as it is to his exertions that I have frequently been indebted for the safety of his friend's collections."—*Copy of a note from Mr. Bateman.*

157. CATASETUM *deltoideum*. Bot. Reg. fol. 1896. *(Myanthus.)*

In a former volume of this work it has been related how the species of Orchidaceous plant which was called *Myanthus barbatus*, with a revolute bearded lip, and a column with two feelers, changes by a marvellous metamorphosis, into what was also called *Monachanthus viridis*, thus proving not only that the supposed genera *Myanthus* and *Monachanthus* are the same, but that they are mere forms of Catasetum itself. A new case of this kind of variation, previously unheard of and unsuspected in the Vegetable Kingdom, has been observed by Mr. Dunsford, who has brought me a scape of *Catasetum deltoideum*, figured at fol. 1896 of this work, in a similar state of alteration. The scape itself has become three times as stout as usual, the length of the raceme is much reduced, the sepals and petals retain their form and colour; but the labellum, instead of being arrow-headed, flat, deep purple, toothed at the base, and placed in front of the flower, has become of exactly the same form as that of *Monachanthus viridis*, hooded, undivided, and of a dull greenish colour tinged with dull purple. The column too, has in like manner lost its cirrhi, has shortened, and its lengthened beak has also disappeared.

171. BOLBOPHYLLŬM *limbātum;* pseudobulbis ampullaceis depressis, foliis . . . . , racemo cylindraceo multifloro longè pedunculato, squamis pedunculi sursum nudi duabus distantibus parvis vaginantibus, bracteis minimis, sepalis ovatis obtusis ciliatis, petalis multò brevioribus obovatis margine superiore villoso, labello ovato crasso obtuso petalis longiore basi bidentato: disco lævi nitido limbo scabritie minutâ opacato.

The flowers of this plant are deep dull purple, in a rather loose raceme. The sepals and petals are both fringed with whitish hairs. The labellum is remarkable for being polished in the middle, and surrounded by a broad belt of minute points, which give it an opaque shagreen appearance. Like most of the genus it is a plant of little beauty. Messrs. Loddiges received it from Sincapore.

172. DENDRŌBĬŪM (Desmotrichum) *longicollĕ;* pseudobulbo compresso in collum longissimum extenso, folio oblongo concavo, flore solitario, ovario filiformi c. pedicello articulato, sepalis petalisque patentibus a latâ basi in acumen lineare attenuatis, cornu brevi conico, labello trilobo: lobis lateralibus nanis margine anteriore crispis intermedio apice ovato plicato crispatoque sublobato, lamellis 2 crispis conniventibus per axin.

A singular *Dendrobium,* belonging to the same section as *D. amplum,* which is remarkable for combining the habit of *Bolbophyllum* with the entire structure of the former genus. The neck of the pseudo-bulb is nearly eight inches long; the leaf five inches; the ovary two inches and a half. The flower is pale straw colour, with the ends of the sepals and petals tapering into a fine purple point. Sent to Messrs. Loddiges from Sincapore by Mr. Cuming.

173. CIRRHOPĔTĂLŪM vaginātūm. *Lindl. Gen. & Sp. orch. p.* 59.

Messrs. Loddiges have received this plant from Mr. Cuming, who collected it at Sincapore. It has the pale straw-coloured flowers in heads, with their anterior sepals prolonged into a fine thread nearly two inches long. The pseudo-bulbs are in the form of a truncated pyramid; the leaves coriaceous and emarginate.

174. ONCĪDĬŪM *incūrvūm* (Barker in litt.); pseudobulbis ovatis ancipitibus utrinque tricostatis diphyllis, foliis ensiformibus acutis, scapo elongato racemoso-paniculato, sepalis lineari-lanceolatis undulatis liberis, petalis conformibus incurvis, labelli laciniis lateralibus rotundatis nanis intermediâ subrotundâ concavâ acutâ, cristâ ovatâ depressâ dimidiâ inferiore lineatâ superiore tricostatâ, columnâ apterâ.

A pretty species, with pale pink flowers mottled with white. It has at first sight the appearance of *O. ornithorhynchum,* but it wants the bird's-beaked anther, and the column-wings of that species. Its panicled scape is long, narrow, and about three feet high. It was originally sent me by Mr. Barker, and I have subsequently seen it in the collection of Messrs. Lee and Co. of Hammersmith.

175. PLEUROTHĂLLĬS *seriātă;* folio carnoso plano ovali emarginato caule multò longiore, pedunculo filiformi paucifloro folii longitudine, sepalo dorsali carinato margine reflexo antico latiore acuto emarginato, petalis brevibus spathulatis carnosis lævibus, labello oblongo obtusissimo pubescente medio exarato lobulis lateralibus minimis rotundatis.

An inconspicuous species of this large genus, with pale yellowish-green flowers marked with rows of purple dots. Like all the genus it is a beautiful object when examined microscopically, but to the naked eye it offers no attraction. The Horticultural Society received it from Rio Janeiro, whence it was procured by John Hearne, Esq.

176. CATASĒTŪM *Trullă;* sepalis petalisque patentibus ovalibus planis, labello latè ovato acuminato obtuso subcordato concavo fimbriato apice lævi, columnâ brevi cirrhatâ.

Flowers about thirty in a spike, green with a brown stain upon the lip. This species is distinct from all previously published, but has little beauty. It was communicated to me by Mr. J. Rigby of the Stanhope Nursery, Old Brompton. A figure will shortly be published in this work.

177. CYMBĪDIŪM *pubēscens;* foliis ensiformibus striatis apice obliquè bidentatis, racemo brevi pendulo, bracteis minimis squamæformibus, sepalis petalisque linearibus acutiusculis, labelli trilobi basi saccati intùs pubescentis laciniis lateralibus acutis intermediâ oblongâ subundulatâ obtusâ, lamellis rectiusculis medio interruptis.

A beautiful species, collected at Sincapore by Mr. Cuming, and flowered by Messrs. Loddiges. It has a short raceme of rich purple flowers, bordered and spotted with brilliant yellow. It is nearly allied to *C. Finlaysonianum* and *bicolor,* but its flowers are smaller, its raceme shorter, and its downy lip is quite different. A figure will appear in this work.

178. CŒLŎGȲNĒ *Cumingĭi;* pseudobulbis ovatis, foliis geminis lanceolatis 5-nerviis utrinque acuminatis racemo paucifloro longioribus, scapo basi nudo, bracteis convolutis floribus longioribus, petalis lineari-lanceolatis patentibus, labelli trilobi lobis lateralibus rotundatis intermedio ovato acuto revoluto basi crispo, lamellis 3 crispis continuis.

A beautiful Orchidaceous plant found at Sincapore, and brought home by Mr. Cuming. It has white flowers, with a lip yellow in the middle. It is allied to *C. trinervis.* Messrs. Loddiges have communicated specimens, of which a figure is in preparation.

179. CATASĒTŪM *saccātum;* sepalis lanceolatis patentibus dorsali petalisque fornicatis, labello subrotundo abruptè acuminato fimbriato medio saccato: ostio contracto reniformi postice dentato, columnâ cirrhatâ.

Of all the strange forms presented by the various species of Catasetum this is one of the most extraordinary. It has very large flowers, with rich purple-spotted sepals and petals, and a bright yellow lip covered closely with crimson dots. The latter is pierced in the middle by a narrow aperture, which leads into a conical chamber or bag, which is not observed till the back of the lip is turned up. Messrs. Loddiges obtained it from Guayana. A figure of it is destined for the Sertum Orchidaceum.

182. CATASĒTŪM (Myanthus) *cornūtūm;* petalis maculatis lineari-lanceolatis sepalo dorsali conformi suppositis, labello subcordato-ovato basin versus saccato supra saccum cornu valido inflexo instructo processibus teretibus rigidis fimbriato basi dentato, columnæ acuminatæ cirrhis cornu labelli attingentibus.

A native of Demerara, with the habit of Catasetum barbatum, formerly Myanthus. There are sixteen or more flowers in a raceme, of a dull green, richly spotted with deep blackish purple. The lip is light green, spotted with the same dark colour; above the base it is hollowed out; above the hollow it is furnished with a strong inflexed white horn, which rises from a somewhat rugged base; and the margin is broken up into slender stiff processes, which are evidently an incomplete state of the fringes found on the lip of Catasetum cristatum, barbatum, &c. Communicated by Messrs. Loddiges, (no. 544).

183. CATASĒTŪM *callōsūm;* petalis concoloribus lineari-lanceolatis sepalo dorsali conformi suppositis, labello ovato-oblongo obtuso basin versus saccato supra saccum callo magno (aurantiaco) instructo margine obsoletè crenato, columnæ acuminatæ cirrhis vix ultra callum extensis.

In habit this plant is exactly like Catasetum tridentatum, *var.* floribundum, but its flowers are quite different. The sepals and petals are of a dull reddish brown, without spots; the column is of the same colour, which may perhaps be best compared to that of old spoiled port wine. The lip is green, flat, with a yellow tubercle near the base above the hollow, and a stain of the same colour near the apex. It was imported by Messrs. Loddiges from La Guayra, (no. 553).

184. MYCARĀNTHĒS *oblīquă;* foliis lato-linearibus carnosis canaliculatis apice obliquè emarginatis scapo brevioribus, sepalis lateralibus carinatis, labelli cuneati 5-lobi callo apicali majore axi pube densâ deciduâ farinam referente vestitâ.

A fleshy-leaved plant from Sincapore, with very small white flowers, covered slightly with rusty down. It was imported by Mr. Cuming, and flowered with Messrs. Loddiges, (no. 121).

185. SARCĀNTHŬS *pāllĭdus;* foliis distichis coriaceis ligulatis apice obliquè retusis paniculâ multiflorâ ramosâ multò brevioribus, labelli lobo medio solido tereti incurvo.

Brought to His Grace the Duke of Devonshire from India by Mr. Gibson. It is a plant with the foliage of Aerides odoratum, or some such plant; but its flowers are small, pale, and by no means beautiful, although they are extremely numerous, and arranged in a panicle a foot and a half long. They are about the size of *S. teretifolius,* of a dirty greenish white, with a faint purple streak through the middle of each sepal and petal, and with the intermediate lobe of the lip dull yellow.

186. COMPARĔTTĬĂ *rŏsĕă;* foliis subsessilibus, racemo pendulo laxo paucifloro, labelli laminâ subrotundo-oblongâ calcare subulato breviore.

A delicate little Orchidaceous plant from the Spanish Main, with a slender drooping stem, bearing four or five flowers of a bright rich rose colour near its apex. I am indebted for it to Messrs. Loddiges, (no. 752).

187. PERĬSTȲLŬS *goodyeroīdĕs.* Lindl. gen. & sp. orch. p. 299.

A herbaceous species of the Orchidaceous order, inhabiting the north of India. It has a long spike of pure white flowers, about the size of the Lily of the Valley, the fragrance of which it rivals. A specimen before me from Ceylon, which has flowered with Messrs. Loddiges (no. 965), does not appear to differ from the plant of the Himalayas.

188. DENDRŌBĬŪM (Onychium) *acicularĕ*; caule basi conico angulato sursum tereti, foliis acicularibus, flore solitario ? terminali galeato, labelli postici trilobi per medium tricarinati lobis lateralibus rotundatis anticè et intermedio ovato serrulatis.

A curious little species, with the base of the stems angular and conical, while the whole of the upper part suddenly tapers and becomes very slender. It bears about three acicular leaves, from the axil of the uppermost of which springs the short peduncle, bearing at its end a cluster of small dry scales, from which proceeds a single yellowish flower slightly tinged with pink. Messrs. Loddiges received it from Mr. Cuming, who gathered it at Sincapore (no. 174).

189. LĪPĂRĬS *spathulātă*; pseudobulbis ovato-oblongis diphyllis, foliis ensiformibus basi angustatis scapo ancipiti longioribus v. subæqualibus, racemo denso erecto cylindraceo, petalis filiformibus, sepalis lineari-oblongis, labello ovato-lanceolato canaliculato recurvo basi ecalloso.

A plant of no beauty, imported by Messrs. Loddiges from India, with a long raceme of minute green flowers. It was originally found in the Burmese empire by Mr. Griffith, and is the no. 772 of that indefatigable Botanist's Burmese herbarium.

190. EPIDĒNDRŪM (Aulizeum) *vīscĭdūm*; caule tereti monophyllo vaginâ alterâ longiore infra folium, folio canaliculato mucronulato, racemo paucifloro, bracteis herbaceis glutinosis carinatis ovarii dimidium æquantibus, petalis sepalisque linearibus acuminatis herbaceis, labelli tripartiti basi bicallosi laciniis lateralibus semirhombeis acuminatis setaceo-fimbriatis intermediâ subulatâ brevioribus, clinandrio cucullato dentato.

A native of Mexico, and nearly allied to E. ciliare, from which its single narrow channelled leaf, smaller flowers, and glutinous bracts distinguish it. The flowers have a weak smell like Cucumbers. It was imported by Messrs. Loddiges (373).

191. MAXILLĀRĬĂ *macrophȳllă*. Poppig. Bot. Reg. 1838. misc. 175.

Of this fine plant Messrs. Loddiges have flowered a new variety from Bolivia (487), with the inside of the sepals and the tips of the petals stained with purple. It does not otherwise differ from the original form of the species.

192. DENDRŌBĬUM *gemēllŭm*. Lindl. gen. & sp. orch. no. 28.

This plant has been imported from Sincapore by Messrs. Loddiges, and forms their no. 158 of Mr. Cuming's collections. It is a long-stemmed grassy-leaved plant, with small pale yellowish green flowers, growing in pairs from short rigid two-valved truncated compressed spathes, placed opposite the leaves. In this respect it is like *D. biflorum*, of which it has all the habit; but the lip is perfectly entire and not three-lobed.

193. ONCĪDĬŪM *microchīlŭm* (Bateman in litt.); pseudobulbis lenticularibus brevibus monophyllis, folio ensiformi carinato carnosissimo acuto quam scapus erectus versus apicem paniculatus quadruplò breviore, sepalis liberis lateralibus longiùs unguiculatis petalisque oblongis subundulatis retusis, labello subrotundo cordato tridentato sepalis duplò breviore, cristâ subrotundâ 5-crenatâ dente intermedio labelli parùm breviore, columnæ nanæ alis oblique truncatis.

Of this curious Oncidium, sent by Mr. Skinner from Guatemala, Mr. Bateman possesses live specimens. He truly calls it a most distinct and remarkable species; the smallness of its lip, which is not half so long as the lateral sepals, being a singular feature. The leaves are about eight inches long; the flowers the colour of *O. crispum.*

194. ONCĪDĬŪM *Wentworthiānŭm* (Bateman in litt.); pseudobulbis nebulosis oblongis compressis ancipitibus diphyllis, paniculâ angustâ elongatâ ramulis 3-floris, sepalis liberis petalisque oblongo-lanceolatis acutis, labelli cordati laciniis lateralibus rotundatis grossè crenatis intermediâ multò brevioribus; intermediæ ungue basi lato apice angustiore lateribus rectis laminâ reniformi denticulatâ basi ipsâ labelli duplo angustiore, cristâ 5-dentatâ denticulis 2 anticis auctâ, columnæ alis erosis brevibus.

Mr. Bateman has favoured me with a fine specimen of this beautiful species, which he has received from Mr. Skinner and named after Lord Fitzwilliam, and which approaches *O. Baueri* and *altissimum* in its general appearance. The flowers are however very differently shaped, and richly stained with crimson upon a yellow ground. The specimen was accompanied by the following note.

"This is quite a distinct and most beautiful species; it forms festoons sometimes twice the length of the specimen sent, and is much used in adorning altars. The beautifully mottled pseudo-bulbs at once distinguish it from every other species; it has flowered here, and I have had it drawn for the Orchidaceæ of Mexico and Guatemala. In habit it is never half the size of *O. Baueri* and *O. altissimum,* though it rivals them in the length of its stems; neither does it ever form compound *lateral* branches from the spike."

195. BOLBOPHȲLLŪM *flāvĭdŭm*; pseudobulbis oblongis compressis vaginis 2 membranaceis vestitis, folio basi angustato spicis laxis longiore, sepalis acuminatis, petalis oblongis obcordatis acutis, labello bicarinato recurvo: lobis lateralibus undulatis obsoletis, columnâ bicorni.

A pale yellow-flowered Orchidaceous plant, imported from Sierra Leone by Messrs. Loddiges (107). It has flowers in size and form like those of B. cocoinum, but they are arranged in a lax spike.

196. ERĬĂ *nūtans*; caule tereti folioso, foliis oblongis undulatis substriatis basi canaliculatis, flore campanulato solitario terminali nutante glabriusculo, bracteâ planâ patente carnosâ subrotundo-ovatâ, sepalis lateralibus duplo latioribus basi rotundatis apice petalisque obtusis, labelli trilobi lobis lateralibus erectis intermedio transverso carnoso apiculato: lineâ unâ per axin productâ infra apicem interruptâ et magis elevatâ duabus lateralibus abbreviatis prope sinus.

A pretty species of Orchidaceous Epiphyte, with a single large nodding terminal white flower, beneath which is a roundish-ovate flesh-coloured thick bract. The tip of the petals and labellum is yellow. Collected at Sincapore by Mr. Cuming, and flowered by Messrs. Loddiges (129).

197. GRŌBYĂ *galeātă*; petalis oblongis obliquè rhombeis apice rotundatis cum sepalo dorsali breviore in galeam dispositis, sepalis lateralibus deflexis basi connatis, labelli tripartiti laciniis lateralibus linearibus intermediâ cuneatâ truncatâ: disco basi 4-dentato tuberculis nitentibus verrucoso.

A Brazilian Orchidaceous plant, of which I have received specimens from both Messrs. Lucombe, Pince, and Co. of Exeter, and Mr. Barker. Its habit is quite that of Grobya Amherstiæ, but the flowers are different. They are dull green, stained a little with purple; and the lip is indistinctly banded with the same colour. The petals are much larger than the dorsal sepal, beneath which they are so placed as to resemble a helmet overshadowing the lip. This species indicates the affinity of the curious genus Malachadenia, which is apparently allied to Grobya.

198. PHOLIDŌTĂ *conchŏĭdĕă*; spicâ pendulâ distichâ imbricatâ, bracteis oblongis concavis apiculatis, sepalorum lateralium carinâ limbo æquali, epichilio 3-lobo laciniis lateralibus rotundatis intermedio obsoleto, hypochilio utrinque in lobum ascendentem producto basi intus lineis tribus elevatis distantibus.

This curious plant was found in Manilla by Mr. Cuming, and raised at Knypersley, where it blossomed a month ago, after having been occupied in forming its flower spike during the space of six months. It is very like *P. imbricata*, but its flowers are nearly twice as large, and the keels of the lateral sepals are so deep and concave as to give the lower side of the flower the appearance of the inside of a bivalve shell.

208. POLYSTĂCHYĂ *cērĕă*; pseudobulbis ovatis, foliis oblongo-linearibus undulatis racemo simplici obtuso brevioribus, floribus cernuis, labelli trilobi pone basin callosi per axin pubescentis lobis lateralibus nanis intermedio obtuso undulato.

A small species, imported from Oaxaca by Messrs. Loddiges (265). The flowers, when full blown, have the colour and texture of old wax; they are distinctly drooping and arranged in a very dense obtuse raceme about an inch long. In habit, size, and general appearance the species bears much resemblance to the Encyclia nana of Pöppig, which, like the E. polystachya of that author, is apparently a species of Polystachya, notwithstanding the manner in which the pollen-masses are represented, about which I think there must be some error. In this plant there is a pair of pollen-masses, excavated at the back, and attached to a short caudicula and small gland.

209. ĔRĬĂ *velutīnă* (G. Loddiges in litt.); caule tereti velutino, foliis distichis carnosis ovato-lanceolatis margine revolutis junioribus utrinque velutinis adultis suprà glabris, flore solitario oppositifolio inter squamas

herbaceas obtusas villosas sessili, sepalis petalisque linearibus apice recurvis extus hirsutis, labello cuncato elongato canaliculato apice rotundato subtrilobo supra pubescente inappendiculato.

A singular plant of no beauty, brought to Messrs. Loddiges from Sincapore by Mr. Cuming. It has pale dirty yellow flowers, growing singly among hairy bracts opposite the leaves. It is probable that Wallich's *Dendrobium vestitum*, of which I have seen only fragments of flowers, is a species of Eria allied to this, and it is certain that among the distributed specimens of that plant are branches of the present species without flowers.

214. MICRŌSTȲLĬS *histionănthă;* (Link, Klotzsch, & Otto, ic. t. 5.) pseudobulbis elongato-ovatis diphyllis, foliis membranaceis latè ovatis acutis undulatis nitidis basi attenuato-vaginantibus apice recurvatis, scapo 6-8-angulato, floribus æreo-olivaceis depresso-corymbosis, sepalis reflexis anticis connatis, petalis filiformibus circinatis, labello oblongo-orbiculari integerrimo obtuso.

A plant found in La Guayra by Moritz, and sent to the Berlin Garden in 1836. It has small green flowers, at the end of a scape a foot and half high, and arranged in a depressed corymb. The leaves are large, thin, and green. It is a stove plant, and is apparently the same as Malaxis Parthoni, *Morren in Bullet. acad. Bruxell.* 5. 486. *with a figure*, a native of Brazil, from which however Dr. Klotzsch thinks it may be distinguished.

215. ONCĪDĬŬM *carthaginĕnsĕ.* Link, Klotzsch, & Otto, ic. t. 6.

This is one of the many varieties that now abound in our gardens, concerning whose limits as species it is difficult to arrive at any positive conclusion. It is certainly not distinct from *O. sanguineum*, which however may itself be a variety of *O. carthaginense.*

216. ONCĪDĬŬM *pelicānŭm.* Hort. Monac.

A specimen from Mr. Bateman, who received his plant from the botanic garden of Munich, shows this plant to be very closely akin to *O. reflexum*, from which it differs principally in the sepals and petals being less blotched, in the lateral lobes of the lip being smaller in proportion to the intermediate segment, and in the tubercles of the crest, which is smooth, not downy, being rather differently arranged. I am unable to judge how far it may be entitled to rank as a distinct species. The name has doubtless been given in allusion to the column, which is not unlike a pelican pecking her breast.

217. BOLBOPHȲLLŪM *sŏrdĭdŭm;* pseudobulbis tetragonis . . . . , spicâ carnosâ acuminatâ conicâ, floribus sessilibus carnosissimis, bracteis ovatis obtusis adpressis ovario longioribus, sepalis ovatis acutissimis, petalis linearibus apice rotundatis, labello unguiculato ovato plano convexo basi auriculato, columnâ utrinque bisetâ.

A native of Guatemala, very near the Bolbophyllum clavatum of Thouars. It grows with a spike 6 inches long, upon a scape about half as long again. The flowers are very fleshy, a dull olive brown externally, but brightly mottled with purple in the inside. I owe my knowledge of it to Mr. Bateman.

218. RODRIGUĒZĬĂ *macŭlātă;* racemis nutantibus basi laxè et distanter vaginatis, sepalo supremo cum petalis acuto lateralibus basi tantùm connatis paulò breviore, labello oblongo convexo retuso secus axin depresso infra medium cristâ pubescente utrinque elongatâ aucto basi 4-dentato, columnâ brevi utrinque brachium protrudente.

A native of Guatemala, whence it was sent to Mr. Bateman by Mr. Skinner. It has small flowers, faintly spotted with red, and has no beauty. Communicated by Messrs. Loddiges.

219. DENDRŌBĬŬM (§ Eudendrobium, foliis planis, floribus subgemellis, labello trilobo) *calcarātŭm;* caulibus nigro-pubescentibus, foliis linearibus obliquè bilobis, floribus resupinatis, sepalis ovatis lateralibus revolutis basi in cornu conicum pedicelli longitudine productis, petalis lineari-lanceolatis, labelli linearis carnosi lævis trilobi laciniâ intermediâ subrotundâ marginibus incurvis ungue cum columnæ pede in calcar verum connato.

A slender inconspicuous species, allied to *D. revolutum* and *biflorum;* with green flowers growing in pairs opposite the leaves. It was found at Sincapore by Mr. Cuming, and flowered with Messrs. Loddiges, (no. 158).

220 ERIĂ (Tonsæ) *clavicaulĭs* (Wallich in litt.); caulibus clavatis, foliis lanceolatis acutissimis, bracteis ovato-oblongis concavis recurvis, floribus glabris, sepalis lateralibus triangularibus acutis dorsali petalisque subæqualibus ovalibus subundulatis, labelli trilobi lobis lateralibus ascendentibus rotundatis intermediâ oblongâ suprà carinatâ tomentosâ infra sinus 2-tuberculatâ.

A very pretty white Indian epiphyte belonging to a division of the genus Eria, in which the sepals are perfectly smooth, or nearly so. Its lip is downy along the middle, and all over the centre lobe, and is bordered with pink round the lateral segments. It was sent to Messrs. Loddiges by Dr. Wallich, with the name above cited.

EDWARDS'S

BOTANICAL REGISTER:

OR,

ORNAMENTAL FLOWER-GARDEN

AND SHRUBBERY:

CONSISTING OF

COLOURED FIGURES OF PLANTS AND SHRUBS,

CULTIVATED IN BRITISH GARDENS;

ACCOMPANIED BY THEIR

History, Best Method of Treatment in Cultivation, Propagation, &c.

AND

MONTHLY CHRONICLE

OF

BOTANICAL AND HORTICULTURAL NEWS.

CONTINUED

BY JOHN LINDLEY, Ph. D. F.R.S. AND L.S.

PROFESSOR OF BOTANY IN UNIVERSITY COLLEGE, LONDON,
AND THE ROYAL INSTITUTION OF GREAT BRITAIN,
VICE-SECRETARY OF THE HORTICULTURAL SOCIETY,
&c. &c. &c.

1841.

OR VOL. XXVII. OF THE ENTIRE WORK.
OR VOL. XIV. OF THE NEW SERIES.

——viret semper——nec fronde caducâ
Carpitur.

LONDON:
JAMES RIDGWAY, PICCADILLY.

M.DCCC.XLI.

The following characters of some Orchidaceæ will serve to occupy an empty space in our pages.

MICROSTYLIS *caulescens;* caule elongato folioso, foliis lanceolatis basi angustatis, racemo laxo multifloro cernuo, pedicellis filiformibus bracteis longioribus, labello acuminato intra basin biaurito.——The only caulescent species yet described. The stem is about four inches long, and is covered with ten or twelve distichous leaves. The flowers are very small, green, in a thin raceme, about three inches long.——*Found by the late Colonel Hall in Peru, in the valley of Lloa, at the elevation of* 8000 *feet above the sea.* (Herb. Hooker.)

ISOCHILUS *grandiflorum;* vaginis imbricatis, foliis distichis lineari-lanceolatis acutissimis, floribus solitariis axillaribus, bracteâ acuminatâ sepalorum dimidio æquali, sepalis acuminatissimis, petalis duplo brevioribus conformibus, labello lineari-oblongo utrinque emarginato basi nudo, columnâ petalis parum breviore.——This is very like I. graminifolium, but the flowers are four times as large, independently of the distinctions included in the foregoing character.——Good specimens exist in the Royal Herbarium of Munich, collected in *Peru by Hœnke;* but I find nothing like it in the Reliquiæ Hœnkeanæ.

ISOCHILUS *graminifolium* (Humb. Bonpl. & Kunth, nov. g. et sp. pl. 1. 340. t. 78.); vaginis imbricatis foliis distichis lineari-lanceolatis acutissimis, floribus solitariis axillaribus, bracteâ acuminatâ pedunculo breviore, sepalis aristatis, petalis conformibus duplò brevioribus, labello lineari-oblongo utrinque emarginato basi callo magno duro oblongo, columnâ petalis duplò breviore.——When not in flower this is undistinguishable from I. *grandiflorum.* It is well figured in Humboldt and Bonpland's work, excepting the analysis, in which the form of the labellum is inaccurate, and the presence of a hard oblong callus, called a purple stain in the description by M. Kunth, is overlooked.——*Peru, Mathews,* 1064; *Trunks of trees near Lloa, Jameson.* (Herb. propr. & Hooker.)

LÆLIA *caulescens;* folio coriaceo lineari-oblongo caule tereti longiore, scapo elongato tereti e spathâ membranaceâ cylindraceâ erumpente, racemo subdecemfloro, bracteis squamæformibus erectis rigidis striatis, sepalis petalisque subæqualibus lineari-lanceolatis acutis, labelli postici nudi lobo intermedio obtuso crispo lateralibus parùm longiore.——A species very near L. cinnabarina. The flowers are apparently purple, and about the size of L. rubescens. The lip is perfectly destitute of all elevations or inequalities. In the herbarium of von Martius is a smaller plant from the same locality, with a three-flowered raceme, and much shorter leaves; apparently it is a mere variety.——*From the Serra de Piedade in the province of Minas Geraes in Brazil* (herb. Martius).

2

GONGŎRĂ bufŏnĭă.

Toad-skinned Gongora.

GYNANDRIA *MONANDRIA.*

Nat. ord. ORCHIDACEÆ, § VANDEÆ.

GONGORA. Botanical Register, vol. 19. *fol.* 1616.

G. *bufonia;* hypochilii convexi cornubus lateralibus obsoletis, aristis setaceis, epichilio acuminato paulò breviore, pedicellis columnâ triplò longioribus.

In May, 1839, I received this plant from Mr. Rucker, with the following note:—

"The accompanying flower of a *Gongora,* which was sent me last year from Hamburgh by a friend who had received it from the Brazils, appears to me different from any of the species I have seen. The pseudo-bulbs are as large as in *G. maculata,* and are almost white; the leaves are also of a pale whitish green colour, and the whole appearance of the plant is different from any I know."

Upon comparing it with the other species of the genus already in cultivation it is found to differ, not only in its singular colour, which is a dull purple, like wine-lees, in the form of very irregular stains and spots, upon a dirty yellow ground, but also in the minuter parts of its structure.

It may be distinguished from *G. fulva, maculata,* and *atropurpurea,* in wanting the lateral horns which project from the sides of the lower part of the lip in those species. From *G. nigrita,* which equally wants those horns, it differs in having much longer bristles, and a lip altogether longer, especially at the lower half.

It should be cultivated with other Orchidaceous plants in the stove. The soil should consist of brown turfy peat well mixed with drainage. If the plant is kept a little above the surface of the pot, the roots will soon spread themselves in all directions in the soil, and also in the atmosphere: sure signs of health and luxuriance; the flowers also will hang down over the soil and pot and have a curious appearance. Although it must not be kept too wet in winter, yet extreme dryness is hurtful, and should be guarded against; as the spring and summer advance it should be kept in the warmest and most moist part of the house.

5

VARIOUS SPECIES OF CATASETUM.

GYNANDRIA *MONANDRIA.*

Nat. ord. ORCHIDACEÆ, § VANDEÆ.

CATASETUM. Botanical Register, vol. 10, *fol.* 840.

It happens in some genera of plants that the different species, or supposed species, present so great a similarity in their foliage and general appearance as, in those respects, to seem identical when figured. It is therefore my intention occasionally to occupy a plate with flowers only of such plants, and thus to present a better comparative view of species whose only distinctions reside in their blossoms.

On the present occasion the genus Catasetum affords five illustrations.

Fig. 1. CATASĒTŬM callŏsŭm.

The Tumour-lipped Catasetum.

C. *callosum;* petalis concoloribus lineari-lanceolatis sepalo dorsali conformi suppositis, labello ovato-oblongo obtuso basin versus saccato supra saccum callo magno (aurantiaco) instructo margine obsoletè crenato, columnæ acuminatæ cirrhis vix ultra callum extensis. *Bot. Reg.* 1840, *misc.* 183.

Exactly like Catasetum tridentatum, *var.* floribundum in habit, but its flowers are different. The sepals and petals are of a dull reddish brown, without spots; the column is of the same colour, which may perhaps be best compared to that of old spoilt port wine. The lip is green, flat, with a yellow tubercle near the base above the hollow, and a stain of the same colour near the apex. It was imported by Messrs. Loddiges from La Guayra.

Fig. 2. CATASĒTŬM cornūtūm.

Horned Catasetum.

C. *cornutum;* petalis maculatis lineari-lanceolatis sepalo dorsali conformi suppositis, labello subcordato-ovato basin versus saccato sub sacco cornu valido inflexo instructo processibus teretibus rigidis fimbriato basi dentato, columnæ acuminatæ cirrhis cornu labelli attingentibus. *Bot. Reg.* 1840. *misc.* 182.

A native of Demerara, with the habit of Catasetum barbatum. There are sixteen or more flowers in a raceme, of a dull green, richly spotted with deep blackish purple. The lip is light green, spotted with the same dark colour; above the base it is hollowed out; above the hollow it is furnished with a strong inflexed white horn, which rises from a somewhat rugged base; and the margin is broken up into slender stiff processes, which are evidently an incomplete state of the fringes found on the lip of Catasetum cristatum, &c. It was imported by Messrs. Loddiges.

Fig. 3. CATASĒTŬM barbātŭm, *var.* proboscideum.

Long-beaked Catasetum.

C. *proboscideum;* labello deflexo multifido: laciniis filiformibus laceris, columnâ apice proboscidiformi: cirrhis deflexis columnâ longioribus. *Bot. Reg.* 1839. *misc. no.* 140.

I had unadvisedly supposed that this plant was a new species; and it is perhaps as much entitled to such a character as some others about which no doubt has hitherto been raised. But I confess my inability, upon a formal examination of it, to distinguish it specifically from C. barbatum. It is however a different looking plant, owing to the beard of the lip being green and not pink. Mr. Wailes of Newcastle received it from Dr. Gardner, who found it growing on a small species of Palm near Sertao. I have also received it from the Hon. and Very Reverend the Dean of Manchester.

Fig. 4. CATASETUM laminatum, *var.* eburneum.

White-lipped Knife-blade Catasetum.

C. laminatum. *Sertum Orchidaceum, t.* 30.
β. *eburneum;* labello eburneo columnâ petalisque immaculatis. *Ibid.*

The original *C. laminatum,* figured in the Sertum Orchidaceum, had its lip and petals spotted with purple. This, which is a mere variety, is remarkable for a total absence of spots from those parts, and for its lip being a pure ivory white. Both varieties were sent from Mexico to the Horticultural Society by Mr. Hartweg.

Fig. 5. CATASETUM lanciferum.

Lance-bearing Catasetum.

C. *lanciferum;* petalis maculatis lineari-lanceolatis sepalo dorsali conformi suppositis, labello subcordato-ovato fimbriato basin versus saccato sub sacco cornu tripartito instructo, sub apice laminâ lineari-lanceolatâ aucto, columnæ cirrhis vix columnâ longioribus.

A Brazilian species, for which I am indebted to the Hon. and Very Reverend Wm. Herbert. It was collected by Dr. Gardner in Brazil, and is perhaps a variety of *C. barbatum*; but it is very different in its lip.

8

CYRTOPŎDĬŪM Andersōnii.

Anderson's Curve-foot.

GYNANDRIA *MONANDRIA*.

Nat. ord. ORCHIDACEÆ, § VANDEÆ.

CYRTOPODIUM. R. Brown Hort. Kew. 5. 216. *Lindl. Gen. & Sp. Orch. no.* CXIV.

C. *Andersonii*; bracteis oblongis patulis concavis, sepalis petalisque obovato-oblongis planis, labelli lobis rotundatis lateralibus erectis intermedio concavo basi sulcato apice crenulato: disco nudo.

C. Andersonii. *R. Br. l. c. Lindl. l. c. Bot. Mag. t.* 1800.

Cymbidium Andersonii. *Bot. Rep. t.* 651.

Tylochilus flavus. *Nees in verhandl. des vereins zu bef. des Gartenb.* viii. 191. *t.* 3.

Caules steriles (*seu bulbi*) *fusiformes, compressi, alii pedales, alii bi-tripedales, pollicem sesquialterum in medio crassi, striati, articulati, articulis brevibus, geniculis linea arcuata fusca notatis, neque contractis nec tumidis: iuniores dense vaginati, vetustiores denudati, virescentes.* Vaginæ *arctissime imbricatæ, distichæ, equitantes, scariosæ, nervoso-striatæ, griseæ; inferiores mucronatæ, superiores foliiferæ.* Folia *pedis unius vel duorem pedum longitudine, erecta, lanceolata, utrinque attenuata, acuta, glabra, nervosa, plicata, saturate viridia.* Scapus *ad basin caulis iunioris eiusdemque altitudine radicalis, tri-quadripedalis, crassitie digiti minimi, teres, lævis, glaber, articulatus, vaginis tribus aut quatuor aphyllis obtusis internodio brevioribus subventricosis lineatis pallide viridibus præditus.* Racemus *compositus terminalis, sesqui-aut bipedalis, multiflorus, speciosus. Rami patentes, subflexuosi, superiores fastigiati; singuli suffulti vagina singula, caulinis simili, sed iam in bracteam explicata, concava vel plana, ovata obtuse mucronata.* Flores *alterni, solitarii pedicellati, flavi cum virore.* Bractea *sub pedicello proprio patens, ovata, acute mucronata, concava, subundulata, lutescenti-viridula, pedicello longior neque vero florem attingens.* Pedicellus *pollicaris, patens, teretiusculus, glaber, transiens in germen sexangulare viride, nullis limitibus conspicuis a pedicello discretum.* Flos *ob pedicelli torsionem obliquus, labello infero, diametro pollicis.* Perianthium *patentissimum, sepalis subæqualibus, basi liberis, apice inflexis, nervosis; horum exteriora* (*tria*) *ovata, obtusa cum mucrone subtilissimo acuto, dorso apiceque virescentia, basi intusque flava, margine undulata; duo interiora paulo latiora et longiora, obovata, obtusa, infra apicem complicatum concava, reliquo ambitu æquali, flava, in doro medio virescentia.* Labellum *magnitudine sepalorum interiorum, cum processu brevi obliquo gynostemii articulo coniunctum et ab eo ad angulum rectum stigma versus inflexum, crassiusculum, staturate luteum, basi planum trisulcum, ambitu trilobum, lobis subæqualibus; lateralibus erectis obovatis rotundatis, medio latiori reflexo subquadrato ex utroque latere emarginato, apice arcu rectave fere linea circumscripto subretuso lævissimo, basi depresso sulcatoque.* Callus *quadricrenatus, utrinque litura aurantiaca amplexus, inter lobos laterales, ubi medius ab iisdem deflectitur, situs est.* Gynostemium *germini incumbens, oblongum, antice planiusculum pallidum, prope labelli articulum, qui ipse fulvi coloris, brunneo irroratum; pars superior gynostemii lobis lateralibus labelli brevior, dorso convexo obtuse carinato; apice trilobo viridi et pro antheræ insertione mucronato.* Clinanthium *late trigonum, convexum.* Rostellum *tridentatum dentibus subæqualibus, medio acuto.* Stigma *transversale, ovale, profunde excavatum, glutinosum, viride.* Anthera *opercularis, triangularis, mucrone brevi truncato terminata, unilocularis, utrinque obtuse auriculata margine infero rotundato membranaceo, area media pellucida pallida, lateralibus opacis rufescentibus.* Glandula (*Retinaculum seu potius, Proscolla*) *rostello incumbens, prominula, triangularis, plana, albida.* Pollinia *quatuor, contigua, ceracea, duriuscula, fulva, bina ex utroque latere tam arcte, sibi incumbentia, ut unum corpus ovale, postice sulculo insculptum, constituere videantur; quorum anterius maius ovale antice convexum postice excavatum in cavum suum recipiens alterum seu posticum, minus angulosumque.* Caudicula *filiformis, bipartibilis, elastica, lutea, basi cruribus duobus glandulæ stigmatis imposita, constans e filis binis, ex angulo, quo pollinia utriusque lateris inter se cohærent, proficiscentibus basique in crura illa, de quibus iam sermo fuit, rursus divergentibus. Gynizus, magna copia exsudans, peracta anthesi profluens, lutescens, odoris ingrati aciduli.* Nees ab Esenb.

None of the published figures of this common plant convey a good representation of it. Both that in Andrews' Repository and the Botanical Magazine represent the sepals and petals as being wavy, which is the case only after the plant begins to wither. The figure in the Prussian Horticultural Transactions, under the name of *Tylochilus flavus*, is dingy, and the flowers are too small.

It is found wild in the tropical parts of America, where from the fleshy stems the shoemakers obtain a kind of paste or glue, which they use for the purposes of their art.

The cultivator of Orchidaceous plants finds no difficulty in keeping this in a healthy condition by potting it in well drained turf, and treating it like any of the common Catasetums; but the art of making it flower regularly and freely is not at present understood about London. It is probable that our houses for the cultivation of these plants are too uniform in temperature and moisture.

There are two varieties in our collections; one which is that now figured, has a branched inflorescence; the other has a perfectly simple one, with much larger flowers; the latter may possibly be the *C. glutiniferum*, a plant I am unacquainted with.

C. Wilmorei of the Flora Cabinet is nothing but *C. punctatum* ill flowered.

From Demerara I have an unpublished species collected by Mr. Schomburgk, with a simple raceme, small flowers, and a large fleshy oblong crest between the lateral lobes of the lip; this may be called Cyrtopodium cristatum and thus defined.

C. *cristatum*; racemo simplici, bracteis concavis patulis setaceo-acuminatis, labelli lobis subæqualibus lateralibus erectis intermedio concavo integro disco nudo, cristâ oblongâ carnosâ inter lobos laterales.

17

* SOBRĀLĬĂ sessĭlis.

Sessile-flowered Sobralia.

GYNANDRIA *MONANDRIA*.

Nat. ord. ORCHIDACEÆ, § ARETHUSEÆ.

SOBRALIA. Fl. Peruv. Perianthium maximum, petaloideum, subæquale; sepalis patentibus vel reflexis, petalis erectis. *Labellum* cucullatum, columnam amplexans, basi angustatum, disco plicato-barbatum, apice bilobum. *Columna* elongata, marginata, clavata, apicis trifidi lobo medio cucullato antherifero. *Stigma* marginatum, basi gibbere gemino nectarifero. *Anthera* terminalis, stipitata, semiquadrilocularis. *Pollinia* farinacea, 4, compressa, posticè cohærentia et contortuplicata, ecaudiculata.——*Herbæ* Americæ *æquinoctialis, terrestres, simplices, sæpe triorgyales, foliosissimæ*; foliis *plicatis*; floribus *racemosis terminalibus, vel axillaribus geminis, niveis, roseis, sanguineis, violaceisve. Omnes ferè species loca rupestria, sicca, aprica, calidissima eligunt, dumeta sæpius ampla formantes.* Gen. & Sp. Orch. p. 430.

S. *sessilis*; caule foliisque subtùs nigro-pubescentibus, foliis sessilibus oblongo-lanceolatis acuminatis 2 terminalibus squamæformibus acuminatis herbaceis, floribus sessilibus, labello rhombeo-oblongo glabro lamellis 2 intramarginalibus pone basin. *Bot. Reg. misc.* 1841. *no.* 11.

Some of the finest Orchidaceæ known are species of this genus, which inhabits Peru, Brazil, Mexico, Demerara, and no doubt the intermediate districts. They are like Evelynas in their manner of growth, or to use a more familiar comparison, they resemble reeds loaded with large red, or white, and often fragrant flowers which always grow from the extremity of the reed among the large plaited grassy leaves. One species, S. Liliastrum, is figured in the Sertum Orchidaceum, from drawings made by Mr. Schomburgk, and has most lovely white or rosy blossoms, but it does not exist in our gardens. Another species with stems from twelve to twenty feet high is "Flower of Paradise" of the Peruvians, and bears large flowers, white without and violet within, smelling of Wallflowers. Another has the inflorescence of a Heliconia. They are all said to love dry sunny rocky places, where the heat is excessive, and where they often form large thickets.

That now figured is the least pretty of any we know. It was flowered by Messrs. Loddiges last December, having been received from Mr. Schomburgk.

I previously possessed wild specimens of it, and they show that the garden plant is quite as perfect as in its native

* So called by the authors of the Flora Peruviana, after Don Francisco Martin Sobral, a botanist of their acquaintance.

meadows. Its stem is covered with small black hairs and stiff-ribbed taper-pointed leaves. From the summit of the stem there appears a single rose-coloured flower, which is very fugacious. The lip is many degrees darker than the other parts.

Fig. 1. represents the columns and anther; fig. 2. exhibits the inside of the lip. I had no opportunity of examining the former.

18

BRĀSSĬĂ Lawrenceānă.

Mrs. Lawrence's Brassia.

GYNANDRIA *MONANDRIA*.

Nat. ord. ORCHIDACEÆ, § VANDEÆ.

BRASSIA. Botanical Register, vol. 21. *fol.* 1754.

B. *Lawrenceana;* sepalis lateralibus elongatis, labello oblongo apice lanceolato subundulato: callo baseos simplici canaliculato truncato pubescente.

Perhaps no genus of the Orchidaceous order is more worth cultivation than Brassias, for, without a single exception, they are handsome, and very easy to manage; most of them too are sweet-scented. But they are difficult to distinguish from each other, especially those which, like the species now figured, have the lateral sepals much lengthened out.

At first sight, in examining this, one would refer it to *B. macrostachya;* but the sepals are not half so long, and the tumour at foot of the lip is simple and truncated, not sloping forward, and supported by three advanced tubercles.

In like manner *B. Lanceana*, which agrees with it in the truncated tumour, has two tubercles in front of it, and its lateral sepals are much shorter. When dried the sepals of B. Lawrenceana become bright brown, while those of B. Lanceana remain pale yellow.

Finally, *B. caudata* has the lip of quite a different form.

We are informed by Mrs. Lawrence that the species is a native of Brazil. It has very sweet-scented flowers.

Fig. 1. represents the hairy tumour at the base of the lip.

Like the other species this requires the heat of the stove, and should be treated in the same manner as Brassia maculata.

It requires a plentiful supply of water at its roots and over its leaves when they are fully formed, but great care must be taken in applying this when the leaves and pseudo-bulbs are young, otherwise they are sure to be damaged, and to decay.

23

* CHȲSĬS bractēscēns.

Bracteated Chysis.

GYNANDRIA *MONANDRIA*.

Nat. ord. ORCHIDACEÆ, § EPIDENDRIÆ.

CHYSIS. Botanical Register, vol. 23. *fol.* 1937.

C. *bractescens;* bracteis cucullatis venosis foliaceis ovario longioribus, sepalis petalisque ovatis obtusis, labelli lobis lateralibus obtusis intermedio minore carnoso bilobo hypochilio plicato lamellis 5 carnosis subæqualibus parallelis basi pubescentibus, columnâ latissimâ carnosâ cymbiformi anticè pubescente. *Bot. Reg.* 1840. *misc. no.* 131.

Caules *fusiformes, carnosi, palmares.* Folia *ovato-lanceolata, basi cucullata, imbricantia, undulata, acuminata, subplicata.* Racemus *lateralis, horizontalis, foliis brevior,* 4-5-*florus.* Bracteæ *magnæ, foliaceæ, reticulatæ, obtusæ, concavæ, cucullatæ, laxæ, ovario longiores.* Flores *magni, cerei, candidi, coriacei.* Sepala *lateralia dorsali latiora, obtusa, margine incurva.* Petala *oblonga, erecta, obtusa, basi angustata.* Labellum *intùs luteum, extùs candidum; hypochilii lobis lateralibus erectis rotundatis, cristis* 5 *lævibus parallelis basi pubescentibus, epichilio bilobo nudo complicato.* Columna *latissima, cymbiformis, carnosa, antice pubescens, in pedem producta.*

A fine epiphyte from Mexico, whence it was imported by George Barker, Esq. who flowered it in April 1840. It forms the third of the genus now known, and is readily distinguished by its large white, not yellow, flowers, and great inflated leafy bracts. Its pollen-masses (fig. 1) show yet more strongly than those of the original species the singular structure upon which the genus is founded.

In this, as in most other Orchidaceæ, the most exact attention should be paid to the markings and elevations of the labellum, for they furnish the most constant and exact characters for distinguishing species. Figures or descriptions in which these are neglected, or represented inexactly, are worthless in the present state of our knowledge. In the original *Chysis aurea* there are five principal ridges on the hypochilium, and three minor ones on each side, all downy, and diverging; in the whole nine. In *C. bractescens* there are five equal ridges only, all smooth and parallel; and in *C. lævis*, there are three large somewhat confluent ridges, and one smaller on each side, the whole being smooth. The whole of those species of Orchidaceæ which stand in books without a *precise* description of the tubercles of the lip, require to be re-examined critically.

* See Botanical Register, vol. 23. fol. 1937.

24

LÆLĬĂ acumĭnātă.

Tapering Lælia.

GYNANDRIA *MONANDRIA*.

Nat. ord. ORCHIDACEÆ, § EPIDENDREÆ.

LÆLIA. Botanical Register, vol. 21. *fol.* 1751.

L. *acuminata;* pseudo-bulbis ovatis compressis rugosis, foliis solitariis emarginatis scapo erecto brevioribus, floribus corymbosis, bracteis linearibus acuminatis ovario duplò brevioribus sepalis linearibus petalisque lanceolatis undulatis acuminatis, labelli lobis lateralibus rotundatis intermedio lanceolato undulato acuminato.

When a short notice of this plant was published at p. 17 of the miscellaneous notices of this volume, nothing more was known of it than that it had produced a scape with two flowers upon it, in the collections of both Sir Charles Lemon and the Horticultural Society; and that a supposed variety with violet flowers had blossomed in the same way with Messrs. Loddiges.

It now appears, from a native specimen sent home by Mr. Hartweg, that it forms a fine corymb of eight flowers, and that it is in fact so very beautiful that the Guatemalese call it "Flor de Jesus." He found it on the trunk of Crescentia Cujete, the Calabash tree, a favourite haunt of Orchidaceæ, at a place called Retatulen, in the month of December.

From *L. rubescens*, its nearest ally, it differs in its larger wrinkled pseudo-bulbs, larger and more corymbose flowers, and in the different form of the labellum.

The following is Mr. Booth's account of the plant as it flowered at Carclew.

"This plant was received by Sir Charles Lemon, Bart. from the Horticultural Society in May, 1840, and flowered at Carclew in January, 1841. Although a weak specimen it produced two scapes with a couple of flowers on each, sufficient to shew that, like the others of this delightful genus, it is not only deserving attention but will amply repay with its truly delicate and fragrant flowers any extra trouble that may be taken in its cultivation. It has been hitherto grown in a pot of decayed vegetable lumps, but I suspect it would have thriven better if it had been tied to a branch of such as the Cork tree, and suspended from the rafter of the stove.

" *Pseudo-bulbs* ovate oblong, of a pale yellowish green, covered with large brown scales when young, tapering, compressed, wrinkled and furrowed when old, about two inches and a half long, and rather more than an inch and a half broad, one-leaved. *Leaves* rigid, erect, slightly twisted and keeled at the base, oblong lanceolate, oblique and emarginate at the point, varying from three to five inches long, and nearly an inch broad; of a deep shining green. *Scape* two-flowered, about eight or nine inches high, issuing from the crown of the pseudo-bulb and embraced by the keeled part of the leaf at its base, round and slender, nearly erect, of a pale green, with five or six joints, having to each a persistent brown coloured, sheathing, acuminate bract, an inch long. *Flowers* of a very delicate semitransparent white, with a faint lilac tinge, excepting the centre of the labellum which is yellowish, and the inner part of the lateral lobes which is a deep purplish red. *Pedicels* round and slightly channelled, an inch and a half long, and of a pale green. *Sepals* spreading, oblong lanceolate acute, an inch and a half long, and a quarter of an inch broad, slightly recurved at the edges. *Petals* the same length as the sepals, but twice as broad, with the margin undulated and recurved. *Labellum* three-lobed, the middle one similar in size and form to the petals, but more undulated and contracted towards the base, where it is slightly tinged with yellow in the centre. The two lateral lobes are scarcely half the length of the middle one, and from being erect and arched with their edges meeting in a ridge above the column, they form a kind of wide-mouthed tube, the margin of which is recurved and undulated, and its inner surface of a deep purplish red beautifully veined. *Column* about two-thirds the length of the lateral lobes, rounded on the upper side, hollowed below, and somewhat triangular, nearly white, unless at the point which is a very pale pink. *Anther case* almost spherical, two-celled, the hollow part of the division between them deep pink on the outside. It seems to be held in its position, or cup, by means of a small hooked elongation of the upper ridge and the angular parts of the column, from which it is easily displaced by touching, and then displays the double series of *pollen-masses* arranged in the two cells, four in each, ovate, compressed and pointed, deep yellow."

29

CŒLŎGȲNĒ Cumingii.

Mr. Cuming's Cœlogyne.

GYNANDRIA *MONANDRIA*,

Nat. ord. ORCHIDACEÆ, § EPIDENDREÆ.
CŒLOGYNE. *Botanical Register, vol.* 11. *fol.* 868.

C. *Cumingii;* pseudobulbis ovatis, foliis geminis lanceolatis 5-nerviis utrinque acuminatis racemo paucifloro longioribus, scapo basi nudo, bracteis convolutis floribus longioribus, petalis lineari-lanceolatis patentibus, labelli trilobi lobis lateralibus rotundatis intermedio ovato acuto revoluto basi crispo, lamellis 3 crispis continuis. *Bot. Reg.* 1840. *misc.* 178.

Often as Sincapore has been visited, it still seems to abound in handsome unknown species of Orchidaceous plants. Among those collected by Mr. Cuming, and by him sold to Messrs. Loddiges, was the present, which, when in good flower, is a very pretty species, with fine white flowers, and a bright yellow blotch on the labellum. The elevated ridges which decorate it are three in number, the intermediate one being the shortest; they are terminated by a deep orange stain, and have a small outlying toothed appendage on each side of the outer ridges towards the termination. It is to these little peculiarities in the surface of the labellum that many tribes of Orchidaceous plants owe their varied aspect and much of their beauty; they also, as has been before observed, furnish the best characters for distinguishing the species.

C. *Cumingii* is closely allied to C. *trinervis,* which is however readily known by its very long narrow leaves, its shorter bracts, smaller flowers, and much shorter middle lobe to the lip, which also appears to want the outlying processes on either side of the ridges.

It should be cultivated in a very warm and moist stove, such as suits Dendrobiums and plants of that kind. The pot must be well drained, and filled up with turfy peat or sphagnum, to which the creeping stem should be fastened with wooden pegs, and the pseudo-bulbs left uncovered. If water is allowed to lodge about the young shoots, they are very apt to be injured by it.

31

CŒLŎGȲNĒ flaccida.

Drooping Cœlogyne.

GYNANDRIA *MONANDRIA.*

Nat. ord. ORCHIDACEÆ, § EPIDENDREÆ (*nec* MALAXIDEÆ).
CŒLOGYNE. *Botanical Register, vol.* 11. *fol.* 868.

C. *flaccida;* pseudo-bulbis oblongis angulatis squamis coriaceis acuminatis ustulatis vestitis, foliis lanceolatis v. oblongo-lanceolatis longè petiolatis, racemo flexuoso nutante, bracteis deciduis, petalis lineari-lanceolatis, labello ovato trilobo: lineis tribus elevatis flexuosis: lobo medio ovato duplò angustiore basi denticulato, columnâ apice dentatâ. *Lindl. gen. & sp. Orch. p.* .
C. flaccida. *Bot. Mag. t.* 3318.

Not an uncommon Epiphyte, having been in the country for several years. It is a native of Noakote, in Nepal, where Dr. Wallich found it growing on trees. Its long stalked narrow leaves, and the stiff scales that surround the base of the scapes and pseudo-bulbs, mottled with pitch brown, as if they were scorched, readily point out this species, which is one of the least attractive of its genus.

The flowers grow in long drooping racemes, and are originally enclosed in bracts, which however are soon thrown off and leave the flowers naked. They have a peculiar disagreeable odour. The sepals are white, linear-oblong, scarcely acute. The petals are the same colour, but not more than one-third the breadth, and are curved backwards. The lip is channelled, ovate, three-lobed, with the middle lobe ovate, acute, wavy, toothed at the base, and turned back at the point. Near its base are several crimson veins; towards the apex it is yellow in the centre; along the middle run three raised wavy lines which terminate abruptly a little within the base of the middle lobe. The column is bordered with a toothed membrane, originally described by me as entire, in consequence of an error in Dr. Wallich's Indian drawings.

Fig. 1. shews the column, with the lip bent downwards and flattened; fig. 2. is the pollen-masses, with their granular base.

The species requires the same management as C. Cumingii t. 29 of this volume; namely, a very moist stove, and the pots well drained. Turfy peat or sphagnum is the best material for potting. Water must be liberally given during the growing season, but must not be allowed to lodge in the leaves of the young shoots when they are tender, as it is sure to rot them. There is nothing worse for plants of this kind than allowing their leading shoot to damp off; it is some time before they form another, and then it is generally much weaker than the former one.

34

CATASĒTŪM Trūllă.

The Trowel-shaped Feeler-wort.

GYNANDRIA *MONANDRIA*.

Nat. ord. ORCHIDACEÆ, § VANDEÆ.
CATASETUM. Richard.

C. *Trulla;* sepalis petalisque patentibus ovalibus planis, labello latè ovato acuminato obtuso subcordato concavo fimbriato apice lævi, columnâ brevi cirrhatâ. *Lindl. in Bot. Reg.* 1840. *misc. no.* 176.

No doubt a native of some of the tropical parts of America, but I have been unable to learn from what country it was received. I am only acquainted with it indeed from a single specimen brought to me in September 1840, by Mr. James Rigby, of the Stanhope Nursery, Old Brompton. It had the general appearance of Catasetum maculatum, and bore thirty such flowers as are here represented. They had a most singular appearance, but it must be confessed not a beautiful one; none of the colours being at all lively. The lip has much the form of a trowel, and is not at all hollowed out into a bag. It is merely concave like the bowl of a spoon.

All the species of Catasetum require nearly the same cultivation. They must have a moist stove to grow in, should be kept nearly dry during their season of rest, and should have plenty of water when growing. When the young shoots begin to grow, they are very apt to be injured by water lodging about them, and care must be taken to prevent this.

As there is some confusion among the names of the various species of this genus, the following catalogue of all that I at present know may be acceptable to the readers of the Botanical Register.

§ 1. CATASETUM. Lip hooded, convex, placed at the back of the flower.

1. C. tridentatum. *Hooker.*
 Varieties.—C. macrocarpum, *Rich.* —Claveringi, *Lindl.* — floribundum, *Hooker.*
 Monster. Monachanthus viridis, *Lindl.*
2. C. maculatum, *Kunth.*
 Variety.—C. integerrimum, *Hooker.*
3. C. Hookeri, *Lindl.*
4. C. semiapertum, *Hooker.*
5. C. purum, *Nees.*
6. C. luridum, *Lindl.* (Anguloa lurida, *Link.*)
7. C. longifolium. *Lindl.*
8. C. discolor. (Monachanthus discolor, *Lindl.*)
 Variety.—Monachanthus Bushnani, *Hooker.*
9. C. roseo-album, *Lindl.* (Monachanthus roseo-albus, *Hooker.*) Probably a variety of the last.
10. C. atratum, *Lindl.*

§ 2. MYANTHUS. Lip flat, placed in front of the flower.

11. C. cristatum, *Lindl.*
 Varieties.—C. spinosum, *Lindl.* (Myanthus spinosus, *Hooker.*) C. proboscideum, *Lindl.*—C. barbatum, *Lindl.*
12. C. cornutum, *Lindl.*
13. C. lanciferum, *Lindl.*
14. C. deltoideum, *Lindl.* (Myanthus deltoideus, *Lindl.*)
15. C. trifidum, *Hooker.* (Myanthus cernuus, *Lindl.*)
16. C. saccatum, *Lindl.*
17. C. laminatum, *Lindl.*
18. C. Russellianum, *Hooker.*
19. C. Trulla, *Lindl.*
20. C. poriferum, *Lindl.*
21. C. callosum, *Lindl.*

35

ANGRÆCŬM bĭlōbūm.

Two-lobed Angurek.

GYNANDRIA *MONANDRIA*.

Nat. ord. ORCHIDACEÆ, § VANDEÆ.
ANGRÆCUM. Botanical Register, vol. 18. *fol.* 1522.

A. *bilobum;* caule brevissimo, foliis cuneato-obovatis obliquè bilobis reticulatis racemo verrucoso pendulo multifloro multò brevioribus, sepalis petalisque lanceolatis patentibus, labello conformi paulo majore calcare filiformi emarginato breviore. *Bot. Reg.* 1840. *misc. no.* 151.

Anthera *cristata, unilocularis.* Pollinia *dorso excavata, alba.* Glandula *alba.* Rostellum *trans stigma, quod anticè in sinum recedit, arctè deflexum, marginique sinus attingens superficiem stigmaticam in duabus partibus dividere videtur.*

A strikingly pretty epiphyte, with a very short stem, and a few distichous obovate leaves, deeply two-lobed at the apex, and streaked with firm veins from the apex to the base, which veins are connected by strong transverse bars. The flowers grow in pendulous simple racemes, and are slightly but sweetly perfumed; their colour is white, with a slight tinge of blush.

The column is remarkable enough; the anther (fig. 1.) is decorated by a glandular crest, like that of an ancient helmet. Below this is a broad shallow stigma. Over the stigma projects the beak of the column, which protrudes so far as to rest on the anterior edge of the stigma, which itself recedes in order to meet the gland; and thus the stigma seems to be divided into two parts.

Messrs. Loddiges received this species from Mrs. Lee, the widow of the late Mr. Bowdich, who found it at Cape Coast Castle.

To be cultivated successfully, it should be suspended from the rafters of the Orchidaceous house upon a block of wood. The block should be surrounded with a little turfy peat or sphagnum, to retain the moisture upon which the roots feed. If sphagnum is used, care must be taken that it does not come in contact with the more tender part of the plant, as it is apt to rot them. In this condition it should be freely syringed during the growing season, and never at any time be kept too dry.

38

CYMBĪDĬŪM pubēscēns.

Downy-lipped Cymbidium.

GYNANDRIA *MONANDRIA*.

Nat. ord. ORCHIDACEÆ, § VANDEÆ.
CYMBIDIUM. Botanical Register, vol. 7. *fol.* 529.

C. *pubescens;* foliis ensiformibus striatis apice obliquè bidentatis, racemo brevi pendulo, bracteis minimis squamæformibus, sepalis petalisque linearibus acutiusculis, labelli trilobi basi saccati intùs pubescentis laciniis lateralibus acutis intermediâ oblongâ subundulatâ obtusâ, lamellis rectiusculis medio interruptis. *Bot. Reg.* 1840. *misc.* 177.

Although the woods of Sincapore have been so often examined by Botanists it seems as if the forms of vegetation there were inexhaustible. The species now figured was met with by Mr. Cuming who sent it to Messrs. Loddiges. It is quite distinct from all the species previously published, although no doubt allied to C. *Finlaysonianum* and *bicolor*. From the former it differs in having short racemes, smaller flowers, and a hairy lip, with the lamellæ nearest the end destitute of any appendage. From *C. bicolor* also the short racemes and hairy lip divide it; but it corresponds with that species in the remarkable character of a shallow bag being present at the base of the lip.

Although not very conspicuous, the rich crimson green and yellow markings give the flowers a gay appearance. Fig. 1. is a view of the lip magnified.

It must be cultivated along with other orchidaceous plânts in the moist stove. In potting it the roots need not be raised above the surface of the pot, as its habits appear to be rather of a terrestrial nature. The pots should be well drained, and water should be freely given during the growing season, and at no season must it be kept entirely dry.

42

ERĬĂ armenĭăcă.

Apricot-coloured Eria.

GYNANDRIA *MONANDRIA*.

Nat. ord. ORCHIDACEÆ, § MALAXIDEÆ.
ERIA. Botanical Register, vol. 9. *fol.* 904.

E. *armeniaca* (Lanatæ); pseudobulbis ovalibus compressis triphyllis, foliis oblongo-lanceolatis coriaceis lævibus, racemo tomentoso radicali basi vaginato stricto foliis æquali v. longiore, foliis scapi bracteisque lanceolatis acuminatis subcoriaceis (armeniacis), labelli lobis lateralibus dentiformibus intermedio rhombeo crispo; lamellis 5 rectis haud crispatis lateralibus abbreviatis divergentibus intermediâ versus apicem labelli productâ. *Bot. Reg.* 1841. *misc.* 70.

Of the many species of this large Orchidaceous genus there is no great number handsome enough to claim the attention of the cultivator; for their flowers are either pale, or small, or dingy. That however now represented is a decided exception to the general character of the species, its gay apricot-coloured bracts amply compensating for the dullness of the flowers themselves.

It belongs to a set in which the bracts are always richly coloured; but no others of which have yet been seen in Europe. Dr. Blume enumerates several; one with yellow, and another with red bracts changing to brilliant orange. Even here the interior of the flower will repay a careful examination, in consequence of the beautiful form and rich colouring of the lip, as shewn at fig. 1.

Mr. Cuming found it in the Philippine islands, in the I. dos Negros and elsewhere, and sent it to Messrs. Loddiges, with whom it has flowered. Its flowering stem is about a foot high.

It should be potted in turfy peat or sphagnum, and grown in the warmest end of a damp stove. It does not require so much heat and moisture during the resting season; but the cultivator must not go to extremes with this, because it is very often injurious to the health and vigour of the plant.

48

ODONTOGLŌSSŪM pulchĕllŭm.

Pretty Tooth-tongue.

GYNANDRIA *MONANDRIA*.

Nat. ord. ORCHIDACEÆ, § VANDEÆ.
ODONTOGLOSSUM. Botanical Register, vol. 48. *fol.* 1839.

O. *pulchellum* (Bateman in litt.); pseudobulbis oblongis compressis ancipitibus diphyllis, foliis linearibus (vix semuncialibus pedalibus) apice obliquè emarginatis, scapo foliis æquali ancipiti debili apice racemifero 6-7-floro, sepalis ovatis acutis, petalis obovatis acutis subundulatis, labelli trilobi laciniis lateralibus triangularibus intermediâ oblongâ subquadratâ apice recurvâ: callo baseos carnoso antrorsum hippocrepico retrorsùm trilobo, columnæ alis et clinandrio laceris.

The first notice I received of the existence of this pretty epiphyte was from Mr. Bateman, who favoured me with a specimen in February last. It was afterwards sent from Chatsworth. It is a native of Guatemala.

The more striking features in it are the long thin pseudobulbs, the very narrow grassy leaves, and the singular protuberance at the base of the white lip. This protuberance is deep yellow spotted with crimson, very fleshy, firm and shining, and is almost horseshoe-shaped in front, while it is distinctly three-lobed behind, in consequence of two deep depressions. Fig. 1. shows this structure; 2. is the pollen-masses with a part of one of the masses cut off.

It is now well known that many Orchidaceous plants from Mexico and Guatemala require less heat than is generally kept in our moist stoves, and to many of them it is particularly injurious if applied at those seasons when the plants are in a resting state. This is a plant belonging to this class, and therefore there is nothing worse for it than over-excitement caused by the application of a high temperature and moisture at those seasons when it ought to be kept cool and dry. It grows either potted in turfy peat or suspended from the rafters of the house, and is propagated like all other plants of this kind by division.

52

DENDRŌBĬŪM dĭscŏlŏr.

Dull-coloured Dendrobium.

GYNANDRIA *MONANDRIA*.

Nat. ord. ORCHIDACEÆ, § MALAXEÆ.
DENDROBIUM. Botanical Register, vol. 15. *fol.* 1291.

§ ONYCHIUM: *caulibus erectis basi incrassatis.*

D. *discolor;* caulibus erectis fusiformibus, foliis oblongis obtusè emarginatis distichis, racemo terminali multifloro, sepalis petalisque lineari-oblongis patulis crispis, labelli crenulati lobis lateralibus acutis intermedio lanceolato acuto: lamellis 5 undulatis. *Lindley in Bot. Reg.* 1841. *misc.* 50. *Lodd. cat. ed. ult. no.* 323.

Caulis *erectus, 4-pedalis et ultra, in racemum longum strictum productus, basi squamatus, paulo altiùs fusiformis, dein annulatus, apice foliosus.* Folia *oblonga, emarginata, obtusa, patentia, coriacea, disticha.* Racemus *strictus, multiflorus, purpureus, flexuosus; bracteis brevibus acutis.* Flores 10-16 *in omni racemo, inter majores, horizontales.* Sepala *et* Petala *conformia, lineari-lanceolata, patentia, undulato-crispa, fusco-aurantiaca, margine pallidiora.* Labellum *pallidius, in cornu brevi emarginato productum, ovato-oblongum, concavum, trilobum; laciniis lateralibus erectis acutis, intermediâ lanceolatâ multò angustiore, acuminatâ, recurvâ, omnibus subcrenulatis seu crispatis;* lamellæ 5, *erectæ, undulatæ, parallelæ, contiguæ, pallidè violaceæ medium occupant, quarum laterales breviores, intermedia ad laciniæ intermediæ dimidiam flexuosè producitur.* Columna *brevis emarginata, antherâ infra columnæ humeros immersâ.*

A native of Java, whence Messrs. Loddiges obtained plants in 1838; it does not however appear to be included among those known to Dr. Blume, to whose genus Onychium it belongs. It has a most singular appearance, with stout erect stems four feet high, swollen in the middle, and terminal racemes of about 16 dingy yellowish brown flowers, as much curled and wavy as those of a Gloriosa. The lip has the same dull dirty colour, except along the middle, where it is decorated with five deep wavy plates of a light violet.

No plants are easier managed than Dendrobiums, and this one is as tractable as any of them. A high moist temperature is required during their period of growth, and when this is past, and the shoots formed, the house may be kept more cool and dry. In resting the plants, however, it is a bad plan to go to extremes, and hence we frequently find persons injuring them by putting in practice certain notions which they have formed upon this subject. This species should be grown in a pot, and the soil should consist of open turfy peat. The stems and leaves should be frequently well syringed.

57

CŒLŎGȲNĒ cristătă.

Crested Cœlogyne.

GYNANDRIA *MONANDRIA.*

Nat. ord. ORCHIDACEÆ, § EPIDENDREÆ.
CŒLOGYNE. Botanical Register, vol. 11. *fol.* 868.

C. *cristata;* pseudo-bulbis oblongis demùm angulatis rhizomate crasso squamoso insidentibus, foliis lineari-lanceolatis, racemo radicali erecto foliis breviore basi squamis corneis vaginato, bracteis oblongis divaricatis persistentibus, petalis oblongo-lanceolatis undulatis, labelli trilobi cristâ triplici interruptâ: infimâ brevi 5-lamellatâ intermediâ multifidâ pectinatâ supremâ bilamellatâ denticulatâ, lobo medio rotundato parvo apice crenato.

? Cymbidium strictum. *Don prodr.*
Cœlogyne cristata. *Lindl. Coll. Bot. p.* 33. *Wall Cat. no* 1958. *Gen. & Sp. orch. no.* 2.

Labellum *candidum, basi sub-auriculatum, oblongum, trilobum; lobis lateralibus rotundatis ascendentibus intermedio minore rotundato apice crenato. Crista lutea, e venis quinque orta, quæ basi totidem lamellas breves dentatas truncatas gerunt, tùm ferè ad sinus labelli usque in pilos breves flexuosos simplices bifidosve pectinatim dispositos solvuntur; demum vendè intermedià lateralibusque inappendiculatis lamellæ duæ dentatæ truncatæ tantum restant.*

Among the showy Orchidaceæ of India some of the Cœlogynes are the handsomest, and of them the present C. cristata stands pre-eminent. It is indeed a most beautiful species, fragrant, free-flowering, and having large blossoms of the purest white, except the lip, which in its centre is decorated with bright yellow fringes and plates. It has recently flowered with George Barker, Esq. of Birmingham, and being exhibited at a meeting of the Horticultural Society in Regent Street, obtained a Knightian medal.

The specimen, however, now represented must not be taken as a fair sample of its appearance; on the contrary, the wild specimens from Nepal, which are now before me, bear five flowers instead of three in a raceme, and are more than four inches in diameter.

Dr. Wallich found it on rocks and trees in Nepal. Fig. 1. shews the lip spread open, and two pollen-masses. It should be grown in the warmest and dampest end of the stove, and potted in turfy peat, having the pot well drained. The supply of water must be regulated by the state of the plant and the season of the year, giving less in winter and when in a resting state, than in summer, when growing vigorously. I would again hint at the impropriety of carrying the drying and resting practice too far, more particularly with such plants as the present one, as I believe many persons have injured their collections by doing so.

59

CYRTOCHĪLŪM fīlĭpēs.

Thread-stalked Curvelip.

GYNANDRIA *MONANDRIA.*

Nat. ord. ORCHIDACEÆ, § VANDEÆ.
CYRTOCHILUM. Botanical Register, vol. 19. *fol.* 1627.

C. *filipes;* scapo longissimo simplici filiformi apice ipso paucifloro, sepalis petalisque lanceolatis conformibus acutis planis, labello cuneato bifido apice dilatato basi dente porrecto subsimplici subtuberculato aucto, columnæ alis minimis cuneatis truncatis. *Bot Reg.* 1841, *misc.* 72.

Pseudobulbi *ovati,* 2-3-*phylli, basi etiam foliosi.* Folia *graminea, basi invicem vaginantia, vaginis in collum aggregatis, pedalia,* racemo *filiformi, gracili, erecto, apice florido breviora.* Scapus *distanter vaginatus, internodiis sesquipollicaribus.* Flores *C. maculati facie, sed lutei.* Sepala *et* petala *æqualia, lanceolata, planiuscula, tigrina.* Labellum *luteum, basi cuneatum integerrimum, apice dilatatum, subrotundum, bilobum, et subcrenatum.* Crista *elevata, anticè in cornu brevi producta, utrinque dente parvo aucta, adjectis tuberculis uno alterove inconspicuis.* Columna *cum labello angulum acutum formans, brevis, utrinque antice in aciem rotundatam producta, alis prætereà minimis truncatis.*

A native of Guatemala, whence it was originally sent by Mr. Skinner to Mr. Bateman and others, and whence it has also been received by the Horticultural Society from Mr. Hartweg. In consequence of its colour being yellow it has the aspect of an Oncidium, from which genus it is distinguished by its labellum being narrowed to the base, where it forms an acute, not an obtuse, angle with the column. It has a very slender stem, above two feet long, perfectly simple, on the extreme point of which are four or five flowers about the size of those of C. maculatum.

Fig. 1. represents the column seen from the side, and 2. the base of the column with the peculiar crest.

It is not necessary to keep this species in such a hot place as some of the West Indian Orchidaceæ, but a temperature considerably higher than a greenhouse is nevertheless indispensable. Like others it delights in a brown turfy peat, and may either be grown in a well-drained pot or suspended from the rafters of the stove. Very little water should be given when not growing.

The following are the species of Cyrtochilum at present known, of which those enclosed in brackets are not in our gardens yet.

(1. C. undulatum. *New Grenada.*)
(2. C. flexuosum. New Grenada.)
(3. C. pardinum. Peru.)
4. C. mystacinum. Peru.
5. C. maculatum. Mexico. Of this there are several varieties, of which C. m. *Russellianum* with very deep spots, figured in the Botanical Magazine, t. 3880, and C. m. *parviflorum* noticed at no. 87 of the miscellaneous matter of the present volume, are the most remarkable.
6 C. graminifolium. Mexico.
7. C. filipes. Mexico.
8. C. flavescens. Brazil.
9. C. stellatum. Brazil.
(10. C. ixioides. New Grenada.)
(11. C. volubile. Peru.)

62

ĔRĪĂ convallarioīdēs.

The Close-headed Woolwort.

GYNANDRIA *MONANDRIA.*

Nat. ord. ORCHIDACEÆ, § MALAXEÆ.
ERIA. Botanical Register, vol. 11. *fol.* 904.

§ Tonsæ ; *inflorescentiâ glabrâ v. leviter pubescente.*

E. *convallarioides;* caulibus compressis junioribus densè et laxè vaginatis, foliis oblongo-lanceolatis multinerviis, racemis densissimis oblongis cernuis breviter pedunculatis, floribus subglobosis leviter pubescentibus, labello subcordato ovato acuto inappendiculato.
Eria convallarioides. *Lindl. in Wall. Cat. no.* 1975. *Gen. & Sp. Orch. no.* 25.
Pinalia. *Lindl. orch. scel. n.* 71. *p.* 23. *c. ic.*
Pinalia alba. *Hamilton mss. apud Don Prodr.*
Octomeria spicata. *Don Prodr.* 31.
Octomeria convallarioides. *Wallich in ic. Bibl. Anglo-Indicæ, n.* 1141.

For fresh specimens of this plant I am indebted to Mr. Rogers and Messrs. Loddiges, both of whom have succeeded in flowering it. From the examination of dried specimens I was at one time inclined to regard it as the type of a genus distinct from Eria; but that was evidently an erroneous impression, for there are no marks in the plant at all at variance with those proper to Eria in its common state. It has small whitish flowers, collected in close heads in the axils of broad striated leaves; they have no smell, and the species proves much less pretty than was expected. The specific character in the work above quoted, made up chiefly from bad Indian drawings, requires considerable correction, especially as regards the flowers, which are nearly destitute of hairs, and the lip, which is entire, and not 3-toothed. I also find that the anther is not dorsal, but merely contracts towards the back of the column, so as to expose the pollen-masses.

The specific name *Convallarioides* is by no means well applied; alluding to the form and not the smell of the flowers; in the first respect there is very little resemblance to the Lily of the Valley, and in the second none at all.

Fig. 1. represents a profile view of the lip and column. Fig. 2. shows the pollen-masses holding together by a little granular viscid matter.

In cultivation this requires a hot damp stove. It should have abundance of water when growing, and should never be kept so dry at any season as plants with large pseudo-bulbs. In other respects it may have the same treatment as other orchidaceous plants.

69

* HOULLĔTĬĂ vittātă.

Striped Houlletia.

GYNANDRIA *MONANDRIA.*

Nat. ord. ORCHIDACEÆ, § VANDEÆ.

HOULLETIA. Ad. Brongniart in Annales des Sciences Naturelles, vol. 15. *new series, p.* 37. *Lindley in Bot. Reg.* 1841, *misc p.* 47. *Perianthium* patens, sepalis sub-liberis: petalis paulo minoribus, unguiculatis. *Labellum* cum basi columnæ continuum, patens; hypochilio angusto, basi excavato quasi bilabiato, apice utrinque in laciniam producto, lobulo nauo interjecto; metachilio nullo; epichilio angulari dilatato cum hypochilio articulato. *Columna* erecta, arcuata, clavata, semiteres, labello paulò brevior. *Anthera* bilocularis depressa. *Pollinia* 2, posticè fissa, caudiculâ lineari-lanceolatâ in glandulum acutam elongatâ, nec infixâ.——Herbæ *epiphytæ, pseudobulbosæ, Americæ æquinoctialis, foliis solitariis plicatis.* Scapi *radicales, erecti, apice racemosi.* Flores *speciosi, luteo-fusci, bracteis parvis nec spathaceis.* Sertum Orchidaceum, t. 43.

H. *vittata;* sepalis linearibus acutis, petalis lineari-lanceolatis basi tortis, hypochilii lobis ovatis obtusis rectis, epichilio rhombeo apice rotundato angulis lateralibus acutis. *Bot. Reg. misc.* 1841. *no.* 100.

This very curious plant was received by Messrs. Loddiges from Mr. Schomburgk. It has the erect raceme of Houlletia Brocklehurstiana, but the flowers are much smaller, form a pyramidal mass, and are yellow, streaked with so much deep chocolate-colour that the former is hardly seen except on the lip. From the similarity in habit between these plants and Maxillaria Warreana one would have thought they must have belonged to the same genus, but an examination of the fructification does not confirm that opinion.

The opinion of Mr. Adolphe Brongniart concerning the affinity of this genus has already been given at no. 47 of the miscellaneous matter of the present volume. It is not advisable to state any thing further upon the subject until the whole classification of Vandeous Orchidaceæ can be brought under consideration.

Figure 1, in the accompanying plate, represents the upper side of the labellum. At the base of the hypochilium is a downy tubercle, which projects towards a small three-toothed plate, which is directed backwards from the base of the metachilium. The latter has two falcate lobes which curve forwards, and have, in the recess formed by their junction, a triangular fleshy tooth. The epichilium is membranous, lozenge-shaped, and rounded at the point.

* So named after Mr. Houllet, a gardener, who accompanied Mr. Guillemin upon his mission to Brazil in search of evidence concerning the cultivation of Tea.

MISCELLANEOUS NOTICES.

1. PLEUROTHĂLLĬS *recūrvă;* folio obovato coriaceo subobliquo petiolo pluries longiore, spicâ brevi recurvâ, floribus pubescentibus, sepalis lateralibus connatis, petalis membranaceis rhombeis acutis, labello membranaceo ovato utrinque sinuato obtuso, columnæ margine petaloideo incurvo.

A small creeping plant, with short recurved spikes of dull pale purple flowers. Communicated by Mr. Geo. Graves from the garden of Mrs. Cannon of Stratford.

2. PLEUROTHĂLLĬS *lutĕŏlă;* folio ovato canaliculato acuto tridentato petiolo suo breviore, pedunculo trifloro folio adpresso eoque breviore, floribus glabris, sepalis lateralibus connatis omnibus carinatis, petalis spathulatis emarginatis, labello lineari obtuso canaliculato glabro.

Another new plant from the same collection as the last, with small yellow flowers. Mr. Graves informs me that both exhale the smell of a sliced cucumber in the middle of the day.

3. APŎRŬM *sinuātŭm;* foliis lanceolatis æquilateris approximatis acutis, floribus solitariis axillaribus, labello cuneato elongato intra apicem lineâ hippocrepicâ crassâ sinuatâ circumdato.

From Sincapore, where it was collected by Mr. Cuming. It has the habit of A. anceps, but its leaves are much narrower and longer, and the flowers are pale yellowish green. Communicated by Messrs. Loddiges, (132).

4. GONGŌRĂ *fulvă;* var. *vitellina.*

There is a very pretty Gongora in several collections, with bright yellow flowers less spotted than usual, and altogether of a more lively aspect. It is from Mexico, and differs from *G. fulva* in having smaller leaves, and the upper half of the labellum distinctly shorter than the lower. It however agrees in so many other respects with that species, that I hesitate to regard it as more than a variety. Fine specimens have lately been received by Messrs. Loddiges from Oaxaca.

5. ARUNDĪNĂ *bambŭsæfŏlĭă.* Genera & Species of Orchid. Plants, p. 125.

This is an epiphyte, with the foliage and habit of a small Bamboo and the flowers of a Cattleya. It is a native of the hotter parts of India, especially of Assam and the Burmese territories. Roxburgh speaks of the blossoms as numerous, large, rosy, with the lip of a lively red purple and very beautiful; he also says that the stems are, when in flower, three, four, or even five feet high. It is certainly a charming plant. It has lately flowered with Messrs. Loddiges. Its natural station is said to be well-shaded places, on the face of moist rocks.

6. BRASSIA *Lawrenceana*; sepalis lateralibus elongatis, labello oblongo apice lanceolato subundulato: callo baseos simplici canaliculato pubescente.

A pretty species from Brazil, introduced by Mrs. Lawrence. It has pale yellow flowers of the size of *Brassia Lanceana*, and is very sweet-scented. A figure of it is in preparation.

7. APŎRŪM *cuspidātŭm* (Wallich in litt.); foliis lanceolatis, floribus solitariis axillaribus, labello emarginato apice crispo per medium obsoletè bilineato.

A plant of no beauty, with the habit of *Aporum anceps*, but with much smaller flowers. It was sent by Dr. Wallich to Messrs. Loddiges with the name it now bears.

8. DENDRŌBĬŪM *tetragonŭm*. Cunningham in Botanical Register, 1839, misc. no. 30.

This plant has flowered with Messrs. Loddiges. It is a pretty species with spider-like flowers, whose sepals and petals are produced into long tapering points. They are yellowish green, bordered with brownish red. The lip is pale yellow, streaked with narrow bands of crimson. Messrs. Loddiges cultivate it on a piece of dry wood in their Cactus house.

11. SOBRĀLĬĂ *sessĭlĭs*; caule foliisque subtùs nigro-pubescentibus, foliis sessilibus oblongo-lanceolatis acuminatis 2 terminalibus squamæformibus acuminatis herbaceis, floribus sessilibus, labello rhombeo-oblongo glabro lamellis 2 intramarginalibus pone basin.

At length Messrs. Loddiges have succeeded in flowering a species of Sobralia, which was found in Demerara by Mr. Schomburgk; unfortunately, however, not one of the finest species. I previously possessed wild specimens of it from Mr. Schomburgk, and they show that the garden plant is quite as perfect as in its native meadows. It has a reed-like stem covered with small black hairs; and stiff-ribbed taper-pointed leaves. From the summit of the stem there appears a single rose-coloured flower, which is very fugacious. The lip is many degrees darker than the other parts. The plant will shortly be figured with a further account of it.

12. MONACHĀNTHŪS *Bushnānī*. Hooker in Botanical Magazine, t. 3832.

This plant is clearly a slight variety of *Catasetum* (Monachanthus) *discolor*, with the colour more yellow-green than usual, and the apex of the lip produced into a sharp point. If care is not taken we shall have the genus Catasetum a most prolific source of spurious species, as indeed it has already become. In some instances it is difficult to judge whether a new form is specifically distinct or not; but in the present instance there can be no doubt upon the subject. This is a rather pretty variety.

14. CYRTOCHĪLŪM *maculātŭm*, var. *ecornutum*. Bot. Mag. t. 3836.

This does not appear to be in any respect different from the common state of this pretty species.

15. DENDRŌBĬŪM *moschātŭm*. Wallich in Lindl. Gen. & Sp. Orch. p. 82.

Of this noble plant there is a good figure in the Botanical Magazine, t. 3837, where Sir William Hooker states his opinion that the *D. Calceolaria* of his Exotic Flora is a mere variety of it.

19. PHOLIDŌTĂ *undulātă*. Lindl. Gen. & Sp. Orch. p. 37.

A native of the East Indies, whence it has been sent to Messrs. Loddiges by Dr. Wallich. It has ovate pseudobulbs tapering to the base; a pair of narrow leaves; and a drooping raceme of small pale dull reddish flowers. It possesses little beauty.

20. EPIDĔNDRŪM *gladiātŭm*. Lindl. Gen. & Sp. Orch. p. 106.

A Mexican species with green flowers like those of E. nutans, but not panicled, nor proceeding from a spathe, and smelling very like a sliced cucumber. Messrs. Loddiges have received it from Bolanos. Some of the old stems were three feet long. This is a much greater stature than has been before observed; for neither the specimens gathered by Mr. Hartweg at the Rancho del Ojo de Agua, nor those of Pavon in Mr. Lambert's herbarium are more than a foot high. The species belongs to the section AMPHIGLOTTIUM.

21. LÆLĬĂ *ălbĭdă*.

Mr. Barker has communicated a curious monster of this pretty plant, with the dorsal sepal united to the back of the column by means of a thin plate resembling a small knife-blade. What is very singular every flower of the specimen was in the same state.

25. ODONTOGLŌSSŪM *stellātŭm*; pseudobulbo ovali compresso, folio solitario lanceolato recurvo, scapo subbifloro, bracteis ovarii triquetri medium haud attingentibus, sepalis petalisque æqualibus linearibus acuminatis herbaceis, labello rhomboideo grossè dentato appendice baseos truncato quadridentato libero.

A new species of epiphyte, imported by the Horticultural Society from Mexico, where it was found by Mr. Hartweg. It is a plant of nearly the same size and habit as *O. Rossii*, from which it differs in the sepals and petals being equally herbaceous and narrow, of a dull olive green, faintly spotted or rather clouded with purple. It has a pure white lip, and may be regarded as one of the handsomer species of epiphytes.

26. DENDRŌBĬŪM *calamiformĕ* (G. Loddiges); caule repente annulato, foliis teretibus pungentibus, racemo laxo terminali foliis breviore, bracteis minimis, sepalis petalisque erectis acuminatissimis, labello trilobo lineari-lanceolato: lobo medio setaceo acuminato crispato: lamellis axeos 3 membranaceis planis versus apicem crispis.

A pretty and graceful epiphyte from New Holland, imported by Messrs. Loddiges. Its cylindrical leaves are eight inches long. The flowers are seated on footstalks nearly an inch in length, including the ovary, and are a light yellow or cream colour, with the base stained with purple. The labellum is purple, beautifully crisped and spotted towards the upper end.

27. EPIDĔNDRŪM (Aulizeum) *vĭvĭpărŭm*; caule fusiformi diphyllo pedunculo squamato pluries breviore, foliis oblongis horizontalibus subundulatis obtusis, racemo denso paucifloro, bracteis membranaceis canaliculatis ovario cuniculato æqualibus, sepalis petalisque linearibus acutis patentissimis, labelli trilobi basi bicallosi laciniis lateralibus subrotundo-rhomboideis intermedio lineari spathulato acuminato convexo triplò longiore, columnæ cucullo fimbriato.

A native of Guayana, where it was found by Mr. Schomburgk, who sent it to Messrs. Loddiges. This species has white flowers closely arranged at the end of a flower-stem two feet long; and the latter when old produces young plants at almost every joint; a curious habit which proves that the leaves of plants in whatever form they may be, whether true leaves, scales of the stem, bracts or otherwise, have a bud, either latent or manifest, in their axil, which buds are capable of being stimulated into growth under circumstances which vary in different species. The flowers have no smell.

28. MAXILLĀRĬĂ? *Brocklehurstiānă*; foliis solitariis oblongo-lanceolatis petiolatis plicatis scapo multifloro brevioribus, sepalis patentibus oblongis inferioribus paulo connatis, petalis angustioribus basi angustatis, labello membranaceo tripartito: laciniis lateralibus linearibus acuminatis reflexis intermediâ ovato-triangulari subhastato, columnâ basi infundibulari depressâ.

This is a charming plant, imported from Rio Janeiro by T. Brocklehurst, Esq. of the Fence, near Macclesfield. It has a scape a foot and a half high, terminated by several rich cinnamon-brown spotted flowers, more than two inches in diameter, and deliciously sweet. It has the habit of *M. Warreana*, and is one of the finest species of the genus. It may indeed be doubted whether it really belongs to Maxillaria, and characters may doubtless be found to separate it. I am however averse to making genera in the Maxillaridous group of Vandeæ, until a larger number shall have been examined in a living state.

31. STANHŌPĔĂ *aŭrĕă*. G. Loddiges.

Imagine a plant whose flowers have the size of *S. insignis*, the form of *S. venusta*, the smell of *S. oculata*, the colour of *Maxillaria aromatica*, and are arranged in a spike two feet long, and the reader will have a tolerably distinct conception of this beautiful thing, which Messrs. Loddiges received from Mr. Bateman, who obtained it from Guatemala. It has the two dark spots of *Stanhopea oculata*, but they are in some manner lost in the flood of yellow that surrounds them.

32. VĀNDĂ *vĭolācĕă*; foliis canaliculatis obliquè abscissis rotundatis, racemis multifloris densis pedunculatis, sepalis obovato-oblongis obtusis planis incurvis, petalis multò angustioribus, labello oblongo apiculato plano: venis 5 crassis elevatis, sacco intus pubescente.

A very pretty epiphyte, sent from Manilla to Messrs. Loddiges by Mr. Cuming. The flowers are in short racemes, rather large, with an ivory white ground spotted with light violet; of the latter colour there is one large blotch within the extremity of each sepal and petal, and a few small dots are scattered over their surface; the lip is wholly violet. The flowers have a faint and rather disagreeable smell.

33. ONCĪDĬŪM *macranthērum.* Hooker in Bot. Mag. t. 3845.

This little plant possesses no beauty, even when represented upon paper. It is either a bad specimen of *Leochilus oncidioides* of Knowles and Westcott, or a species so nearly allied that I am not at present able to point out the difference between them. With *Oncidium carinatum* and another species or two, it forms a group readily known from Oncidium proper, by the large anther, the two-armed column, and some peculiarities in the lip. If it should be eventually separated from Oncidium, which seems probable, the name of Leochilus, although not very appropriate, must be retained for it.

38. SPIRĀNTHĒS *Lindleyana.* Link, Klotzsch & Otto, ic. t. 11.

This is a variety of *S. elata*, with the leaves spotted with light glaucous green.

41. BURLINGTŌNĬĂ *rĭgĭdă.* Sertum Orchidaceum, t. 34.

One of the many fine plants inhabiting the woods of Brazil, our knowledge of which was confined to dried specimens until the enterprise of British cultivators succeeded in transferring it to our gardens. It has been at length procured in a living state by the Messrs. Loddiges, in whose stove it flowered some months ago. It is a beautiful species, with a habit unlike that of any other genus hitherto discovered. It first forms a tuft of two or three leaves, of an ovate lanceolate form and rigid texture, whose petiole is thin, folded together in an equitant manner, and articulated with the lamina. Subsequently in the middle of these leaves appears a short branch, in the form of a pseudo-bulb, oval, thin, and furrowed, on whose apex arise one, or occasionally two leaves, like the first in form, but without the equitant petiole. The plant having advanced to this point, and succeeded in establishing itself on the branch of a tree by means of numerous fine rather stiff roots, it next produces, from the axil of one of the lower leaves, a rigid stem, slender and as thick as a crow's quill, which rises erect into the air, forming two or three membranous sheaths upon its surface, and ceasing to grow as soon as it has acquired the length of eight or ten inches. At its apex it develops just such a tuft of leaves as that from which it sprang; and thus the plant continues to live till the period of flowering has arrived. At that time it emits from the axils of one of its lower leaves, a flowering stem or scape, six or eight inches long, having a few distant membranous scales ensheathing it, and bearing at the apex a very short umbel-like raceme of several large drooping white flowers, delicately tinged with pink. When the column is deprived of all the parts that surrounded it, and so placed as to be seen in front, it bears far more resemblance to a bat's head and neck, than to any part of a flower. Travellers in Brazil report this species to have a delicious scent of violets, but it was not perceived in Messrs. Loddiges' specimen.

42. LÆLĬĂ *acuminātă*; pseudo-bulbis ovatis compressis rugosis, foliis solitariis emarginatis, scapo bifloro, sepalis linearibus petalisque lanceolatis undulatis acuminatis, labelli lobis lateralibus rotundatis intermedio lanceolato undulato acuminato.

Found in Guatemala by Mr. Hartweg, and distributed by the Horticultural Society. It flowered at the same time in the garden at Chiswick and at Carclew, in the collection of Sir C. Lemon. It is a curious Orchidaceous plant, very near L. rubescens, from which it differs in the scape being two-flowered not racemose, in the sepals being acuminate, and the flowers at least twice as large. It is a pretty species, with pale blush flowers. What seems to be a variety, from Costa-rica, no. 620, has produced a pair of pale violet blossoms with Messrs. Loddiges.

43. POLYSTĂCHYĂ *reflexa*; pseudo-bulbis conicis subcompressis annulatis, foliis . . . . , scapo subclavato apice racemoso, sepalis expansis sub apice emarginato mucronatis lateralibus maximis, labello pandurato medio crenulato genuflexo pulvinato apice dilatato membranaceo.

This singular Orchidaceous plant is a native of Sierra Leone, where it was found by Mr. Whitfield. It flowered in February in the garden of Chiswick House, under the care of Mr. Edmunds. Its leaves are at present unknown. The flowers are white with a faint tinge of pink, and spread flat, so as to look at first sight very unlike those of a Polystachya. The lip is slightly tinged with green, abruptly bent back in the middle, and below the bend furnished with a bright yellow cushion of the most curious jointed hairs, which are arranged so compactly that the whole seem to form one solid mass.

46. ĔRĬĂ *bractēscens* (Tonsæ); pseudobulbis brevibus ovatis apice subdiphyllis, foliis oblongis undulatis racemis subæqualibus, bracteis margine revolutis superioribus linearibus reflexis, racemis erectis, labelli trilobi lamellis duabus abbreviatis intermediâ productâ lobo medio truncato rugoso obtusè apiculato.

A Sincapore Orchidaceous plant, which has flowered in the collection of Messrs. Loddiges, (no. 214). It has erect racemes of delicate straw-coloured flowers, and is readily distinguished by its large bracts. It is nearly allied to *E. flavescens.* The same plant has been found in Burma by Mr. Griffith, from whence I have specimens, no. 1055. It is rather pretty, but not strikingly handsome.

49. CYRTOPŎDĬŪM *Andersonī.* Botanical Register, ante t. 8.

Upon the subject of this plant I have received the following note from Mr. Appleby, gardener to Thomas Brocklehurst, Esq. M.P., of the Fence, near Macclesfield.

"In the Botanical Register for this month is an observation that the genus Cyrtopodium does not flower freely in the Orchidaceous houses near London. As I have been successful in blooming several of that genus last season, and they are now showing flowers again, I send you the particulars of the method by which I have succeeded. As soon as I perceive the buds springing at the bottom of the pseudo-bulbs, I take the plants and carefully shake off all the old soil, and cut off all the decayed roots; I then pot them in large pots well drained, in a compost of turfy loam chopped into pieces about the size of pigeon eggs, and peaty turf broken in the same manner, and leaf-mould about half rotten, in equal parts; to which I add about an eighth of bones, also broken into small pieces; I mix these all well together, and place the plants as near as possible level with the rims of the pots, and finish by giving a good watering to settle the compost. The plants are put in the warmest part of the house, and watered very moderately at first, increasing the quantity as the plants advance in growth until the leaves are fully developed, when I give them manure water once a week to encourage the production of strong pseudo-bulbs, without which it is in vain to look for flowers. In this I succeeded to my satisfaction; and last year had the pleasure to perceive the flower-stems appearing at the same time as the bulb shoots. I had flower-stems five feet high, with numerous side branches, making a bundle of flower-stems on one shoot more than eighteen inches diameter. They are coming up this year equally strong. As soon as the pseudo-bulbs are perfected I gradually reduce the water, and when they are at rest I give them no more. To induce more perfect quiescence I have them removed to a cool dry house, the average temperature of which is about 55°. The essentials of this method are, to use a rather rich but open compost, to give plenty of water during growth, and a season of complete rest. Those who attend to all this need not fear flowering Cyrtopodiums. We have now in flower here Dendrobium nobile, a fine specimen with spikes of ten and twelve blossoms each, D. cœrulescens, Cyrtochilum maculatum, C. Bictoniense, Epidendrum aurantiacum, and a fine var. Epidendrum ciliare, var. latifolium, E. capitatum, E. nutans, E. nocturnum, Leptotes bicolor, minor and major, Cymbidium sinense, Brassavola nodosa, B. angustata, B. tuberculata, Oncidium Cebolleta, and a var. O. ampliatum, Brassia caudata, Gongoras several species, Lissochilus parviflorus, Bletias, Cypripediums, &c. &c. altogether making our Orchidaceous houses very gay, forming a strange yet pleasant contrast to the savage winter now howling around us."

50. DENDRŌBĬŪM *discolor;* caulibus erectis fusiformibus, foliis oblongis obtusè emarginatis distichis, racemo terminali multifloro, sepalis petalisque lineari-oblongis patulis crispis, labelli crenulati lobis lateralibus acutis intermedio lanceolato acuto: lamellis 5 undulatis.

Although obtained by Messrs. Loddiges from the Botanical Garden at Buitenzorg, I cannot recognize this species among those mentioned by Dr. Blume. It has a most singular appearance, with stout erect stems four feet high, swollen in the middle, and terminal racemes of about 16 dingy yellowish brown flowers, as much curled and wavy as those of a Gloriosa. The lip has the same dull dirty colour, except along the middle, where it is decorated with five deep wavy plates of a bright violet. It is of the same section as the sweet-scented Dendrobium crumenatum, now so rare in our gardens, if it exist at all.

53. DENDRŌBĬŪM *elongatum.* (Eudendrobium, foliis planis, floribus racemosis, labello trilobo.) *A. Cunningh. in Bot. Reg.* 1839. *misc.* 33.

This plant has flowered with Messrs. Loddiges. It proves to have erect stems about a foot and a half high, at the end of which grow about four lanceolate leaves, from the midst of which springs a raceme of yellowish flowers spotted with red. If they would but open, the species would be rather pretty, but they seem unwilling or unable to expand.

54. CŒLŌGYNĒ *cristata.* Gen. & Species of Orchidaceous Plants, p. 39.

A most beautiful species, fragrant, free-flowering, and having large blossoms of the purest white, except the lip, which in its centre is decorated with bright yellow fringes and plates. It has recently flowered with George Barker, Esq. of Birmingham, and was exhibited at the last meeting of the Horticultural Society in Regent Street, where it gained a Knightian medal.

56. ONCĪDĬŪM *longifolium;* (Cebolletæ) foliis teretibus longissimis diffusis, scapo erecto densè paniculato, sepalis petalisque apiculatis obtusis concavis, labelli lobis lateralibus obliquè ovatis intermedio obovato bifido minoribus, tuberculo basi depresso anticè tricorni glabro, columnæ alis brevibus rotundatis.

Although this fine species has the foliage of O. Cebolleta, it is really very different, forming dense panicles, three feet long, of very large and showy yellow and brown flowers. Its leaves are often three feet long, and hang down or spread upon the ground instead of standing stiff and erect, as in the allied species, several of which are as yet but little known.

It has been imported abundantly by the Horticultural Society from Mexico, and has flowered both in their garden and with Messrs. Loddiges. It is very handsome, and well worth growing.

57. ONCĪDĬŪM *Wrayæ;* pseudobulbis ovatis compressis sulcatis, foliis 2 terminalibus lineari-lanceolatis, scapo radicali elongato ramoso, floribus paniculatis, petalis sepalisque conformibus ovato-acuminatis patentibus ferrugineo maculatis, labello unicolori latè cuneato-obovato 3-lobo cristato, cristâ elevatâ acutè trilobâ. *Botanical Magazine, t.* 3854.

A Mexican species introduced by Mrs. Wray of Oakfield, and very handsome. This would seem to be new, but is evidently very near O. reflexum, Wentworthianum and Tayleurii. The figure in the Botanical Magazine, by which alone I know it, is so evidently inaccurate in those minute but important details, a correct knowledge of which is indispensable in this large and difficult genus, that it is not easy to place any confidence in it.

58. DENDROCHĪLŪM *glumaceum;* spicâ conicâ elongatâ apice nutante, bracteis imbricatis distichis glumaceis, sepalis petalisque acuminatis, columnâ apice pectinatâ dentibus lateralibus æquali, labelli trilobi basi bilamellati lobis lateralibus abbreviatis inflexis acutis intermedio orbiculari.

A singular Orchidaceous plant looking very much, both in foliage and flowers, like a grass. Indeed, before the latter expand, and while they are still covered over by the bracts, one would not suspect the plant to be any thing else. Mr. Cumming found it in the Philippines, and Messrs. Loddiges have flowered it. The blossoms are a pale watery green.

59. MAXILLĀRĬĂ *candida* (G. Loddiges in litt.); pedunculis radicalibus unifloris, sepalis petalisque linearibus acuminatis, labello oblongo obsoletè trilobo medio furfuraceo ad incisuram unicalloso: lobo medio ovato obtuso denticulato.

A little Brazilian plant of small beauty. The flowers are white and about as large as those of M. ochroleuca. It is distinguished readily by the whole inside of the lip being covered with a fine deciduous mealiness. It flowered at Hackney with Messrs. Loddiges.

62. EPIDĒNDRŪM (Hormidium) *miserum;* caulibus ovalibus exuviatis compressis aggregatis, foliis . . . . ., scapo brevissimo subbifloro, spathâ bivalvi pedunculis subæquali membranaceâ, sepalis ovatis acutis, petalis linearibus, labelli trilobi lobis lateralibus rotundatis inconspicuis intermedio truncato emarginato, callis nullis.

A miserable little epiphyte, whose stems, resembling pseudo-bulbs, are about half-an-inch high from the ground on which they grow. The scape, peduncles, and ovaries, taken together, are about the same length. The flowers are a dull, dingy, greenish-brown, not unlike those of E. musciferum. Messrs. Loddiges imported it from Oaxaca.

63. EPIDĒNDRŪM *leiobulbon.* Hooker's Journal of Botany, vol. iii. t. 10, is a small specimen of *Epidendrum varicosum.*

64. PLEUROTHĀLLĬS *peduncularis.* Hooker's Journal of Bot. vol. iii. t. 9, is *Pl. aphthosa.*

65. BOLBOPHȲLLŪM *imbricatum;* scapo erecto distantèr obtusè vaginato, spicâ fusiformi subtetragonâ vix compressâ, bracteis ovatis obtusis imbricatis demum marcescentibus, alabastro cornuto pubescente, sepalo supremo lineari lateralibus ovalibus brevioribus, petalis filiformibus hirsutis, labello ovato basi pectinato.

This singular plant has been obtained by Messrs. Loddiges from Sierra Leone (171). It approaches closely to the Mauritian B. occultum, and with it and some others forms an analogy with Megaclinium. The flower-stem is a foot high, stout, and nearly erect; the flowers are deep dull purple. Like many others of the genus, the spiral vessels are extremely strong and tough in the sepals and petals, and, which is very remarkable, *they are of a rich purple colour.* I am not at present acquainted with any similar case, for in general the spiral thread of the tracheæ is peculiarly colourless and silvery.

68. MAXILLĀRĬĂ *Harrisoniæ.*

A delicate white variety of this beautiful epiphyte has been imported by Messrs. Loddiges. The sepals and petals are a clear cream colour, just tinged with violet at the tip of the former. The lip is the same as in the original variety, but somewhat paler.

69. ĒRĬĂ *longilabris* (Tonsæ); Eriæ bractescentis facie sed sepalis petalisque magis acuminatis, labelli trilobi lamellis tribus ad apicem ferè productis æqualibus lateralibus abbreviatis lobo medio ovato acuminato.

A native of Panay in the Philippines, whence it was sent to Messrs. Loddiges by Mr. Cuming. It is very like Eria bractescens, but is a finer species, and bears more flowers; and is distinguished at once by its lip, which is not truncate, and has three equal wavy ridges prolonged almost as far as the tip of the middle lobe, which is long and acuminated.

70. ĔRĬĂ *armeniaca* (Lanatæ); pseudobulbis ovalibus compressis triphyllis, foliis oblongo-lanceolatis coriaceis lævibus, racemo tomentoso radicali basi vaginato stricto foliis æquali v. longiore, foliis scapi bracteisque lanceolatis acuminatis subcoriaceis (armeniacis), labelli lobis lateralibus dentiformibus intermedio rhombeo crispo; lamellis 3 rectis haud crispatis lateralibus abbreviatis divergentibus intermediâ versus apicem labelli productâ.

A beautiful epiphyte, sent to Messrs. Loddiges by Mr. Cuming, who found it abundantly in the Philippines, in the island of Negros, and elsewhere. Its flowering stem is a foot high, and bears a great number of large tough apricot-coloured scales and bracts. The flowers, however, are a dull dirty brown, not at all pretty. This is the first of the Erias with large coloured bracts that we have seen; several exist in Java according to Dr. Blume. The species is near the Dendrolirium ornatum of that author, but the bracts are not red, nor revolute at the margin, the leaves are not strongly ribbed, nor are the lamellæ of the lip at all the same.

71. EPIDĒNDRŪM (Lanium) *microphyllum*. Lindley in Hooker's Journal of Botany, 3. 85.

A curious little plant, with small woolly dull purple flowers. It and another constitute a well marked section of this great genus, to which the name of Lanium has been applied, in allusion to the singular circumstance of the flowers being woolly. Messrs. Loddiges flowered it the other day from Schomburgk's Demerara collections, and I have since received it from Mr. Bateman, who obtained it from Berbice.

72. CȲRTŎCHĪLŪM *filipes*; scapo longissimo simplici filiformi apice ipso paucifloro, sepalis petalisque lanceolatis conformibus acutis planis, labello cuneato bifido apice dilatato basi dente porrecto subsimplici subtuberculato aucto, columnæ alis minimis cuneatis truncatis.

A Cyrtochilum, with the yellow flowers of an Oncidium, imported from Guatemala by Mr. Bateman. It has a very slender stem, above two feet long, perfectly simple, on the extreme point of which are four or five flowers about the size of those of C. maculatum.

77. NOTȲLĬĂ *aromatica*; (Barker in litt.) labello unguiculato ovato-rhomboideo acuminato basi utrinque deflexo ecalloso, sepalis apice revolutis inferiore bilobo, petalis linearibus rectis acutis.

A small inconspicuous species, with pale watery-green very sweet-scented flowers, a native of Para, whence it was obtained by Mr. Barker. There are two dull yellow spots near the base of each petal; otherwise the flower is unspotted. It is very near N. Barkeri, from which it differs in the form of the labellum, which has no callus at the base; in this species the labellum is falcate, convex, taper-pointed, with the base bent considerably downwards on each side.

82. MEGACLĪNĬŪM *Bufo*; rachi maximâ lanceolatâ, floribus nigro-pilosis, sepalis acutissimis supremo lævi lateralibus intùs pubescentibus, petalis minimis acutis glabris, labello ovato carnoso rugoso.

Let the reader imagine a green snake to be pressed flat like a dried flower, and then to have a row of toads, or some such speckled reptiles, drawn up along the middle in single file, their backs set up, their forelegs sprawling right and left, and their mouths wide open, with a large purple tongue wagging about convulsively; and a pretty considerable approach will be gained to an idea of this strange plant, which, if Pythagoras had but known of it, would have rendered all arguments about the transmigration of souls superfluous. It is a native of Sierra Leone, whence Messrs. Loddiges received it. A figure of it will be published in some future number of the Gardener's Chronicle.

85. ODONTOGLŌSSŪM *Ehrenbergii* (Link, Klotzsch, & Otto, p. 38. t. 16).

This Mexican epiphyte is surely the same as O. *Rossii* figured in this work, but not so fine a specimen.

86. DENDRŌBĬŪM *acerosum*; caulibus brevibus teretibus glabris, foliis carnosis teretibus obtusiusculis, racemo terminali brevi paucifloro, bracteis membranaceis convolutis, floribus semiclausis longè cornutis, sepalis petalisque erectis acutis, labello cuneato emarginato apice crispato calvo.

A small uninteresting plant from Sincapore, imported by Messrs. Loddiges, who bought it of Cuming, (357). It is near O. teretifolium, but has smaller flowers, shorter leaves, and quite a different lip. The flowers are dull yellowish pink, with pink stripes.

87. CYRTOCHĪLŪM *maculatum*; var. *parviflorum*; labello subhastato laciniis lateralibus majoribus.

Messrs. Loddiges have flowered a Mexican plant, (no. 794) which in so many respects corresponds with the now common C. maculatum, that I cannot regard it as more than a variety, but which nevertheless differs in having smaller flowers, with the lateral lobes of the lip so much larger than usual as to give it a hastate appearance, and with a frequent union between the lower sepals more or less complete. This circumstance breaks down yet more the limits between Cyrtochilum and Oncidium.

90. EPIDĒNDRŪM (Encyclium) *hastatum*, Lindley in Hooker's Journal, vol. 3, p. 82. (pseudobulbis ovatis, foliis linearibus acutis.)

A very fragrant species from Bolanos, imported by Messrs. Loddiges. There appear to be two varieties, one with purple and the other with green sepals. In both the lip is white, and the column dull deep purple. It produces a simple raceme of flowers about a foot long.

98. EPIDĒNDRŪM (Encyclium) *aciculare*; (Bateman in litt.) pseudobulbis oblongis diphyllis, foliis linearibus canaliculatis acutis racemo simplici æqualibus, sepalis petalisque lineari-lanceolatis æqualibus acutis, labelli laciniis lateralibus ascendentibus linearibus obtusis apice recurvis intermediâ ovato-oblongâ subundulatâ (pictâ) acutâ.

A pretty little species, obtained from the Bahamas by Mr. Skinner, who gave it to Mr. Bateman. It is in the way of Epidendrum Pastoris, has long narrow leaves, a slender erect raceme of six or seven flowers, whose sepals and petals are dull purple, and lip white, enlivened with rosy veins.

99.

In the last number of the Annales des Sciences, M. Adolphe Brongniart has proposed a new genus of Orchidaceous plants, founded on a specimen that has flowered in the Garden of Plants, and which he calls Houlletia, assigning it the following character:—

Perianthium membranaceum patens (non reflexum). Sepala concava libera, inferioribus vix basi columnæ connatis. Petala basi angustata, unguiculata, apice rotundata. Labellum cum basi incurvatâ columnæ articulatum, liberum, carnosum, medio constrictum articulatum, parte basilari (*hypochilio*) oblongâ planâ, lobis seu processubus lateralibus membranaceis angustis retrorsùm arcuatis corniformibus, parte apicilari (*epichilio*) infra basilarem articulatâ integrâ.—Columna perianthio brevior arcuata semi-teres, antice plana, vix superne alato-marginata. Stigma transversè excavatum. Anthera opercularis depressa; pollinia duo oblonga angusta, externè sulcata, caudiculâ membranaceâ oblongâ basi connata, glandulâ angustâ lanceolatâ.

Epiphyta, pseudo-bulbis ovato-conicis, monophyllis, foliis longissime petiolatis, plicatis, membranaceis, scapo radicali ascendente, floribus racemosis nutantibus, ovario plus minusve contorto, perianthio membranaceo pallidè maculato, fragrante, labello carnoso maculis atro-violaceis insperso, lævem odorem stercoreum olente.

He considers it near Stanhopea, from which it differs in its sepals not being reflexed, its lip articulated in the middle, the lower half bearing two membranous horns directed backwards and upwards towards the column, by the column being short and not membranous at the edge, by the petals having a stalk, by the lanceolate form of the gland of the pollen-masses, and by the flowering stem being erect, not pendulous. Upon considering this character it is evident that Houlletia is either the same as Maxillaria? Brocklehurstiana, already mentioned in this volume, p. 10. no. 28, or a plant very nearly related to it. As the genus is no doubt a good one, it will be necessary to change the name of my species to Houlletia Brocklehurstiana. At the same time I have a new species to add.

100. HOULLĒTĬĂ *vittata*; sepalis linearibus acutis, petalis lineari-lanceolatis basi tortis, hypochilii lobis ovatis obtusis rectis, epichilio rhombeo apice rotundato angulis lateralibus acutis.

This very curious plant was received by Messrs. Loddiges from Mr. Schomburgk. It has the erect raceme of Houlletia Brocklehurstiana, but the flowers are much smaller, form a pyramidal mass, and are yellow, streaked with so much deep chocolate colour that the former is hardly seen except on the lip. From the similarity in habit between these plants and Maxillaria Warreana one would have thought they must belong to the same genus. I cannot, however, find any combining character, and their differences are very numerous.

101. GONGŎRĂ *macŭlātă*; var. *alba*.

Among the Orchidaceous plants received by the Horticultural Society from Oaxaca, a singular variety of Gongora has made its appearance, with all the habit of the original, but with flowers almost white, except some pale dull wine-coloured spots on the sepals. It is a very graceful thing, and forms a good contrast with the old yellow and purple kinds.

103. MAXILLĀRĬĂ *placanthĕră*; (Hooker in Bot. Mag. t. 3173.) sepalis lineari-oblongis obtusis intus secus medium seriatim maculatis, petalis angustioribus omninò conformibus maculatis, labelli angusti trilobi minutè pubescentis lævis laciniis lateralibus acutis nanis intermediâ dilatatâ rotundatâ cuneatâ, columnâ apice lobatâ versus basin bisulcâ pubescente, antherâ planâ acutâ.

The inspection of a specimen of this from Messrs. Loddiges convinces me that it is a species distinct from M. viridis to which I formerly referred it, although it is certainly very near it. Its flowers have however much narrower sepals and petals, are far less green, and the form of the middle lobe of the lip is distinct.

104. MAXILLĀRĬĂ *jugŏsa*; flore subgloboso, sepalis oblongis incurvis obtusis, petalis conformibus concavis basi angustatis, labelli sigmoidei oblongi unguiculati pilosi jugosi lobis lateralibus nanis intermedio semicirculari, columnâ apice lobatâ bisulcâ villosissimâ, antherâ depressâ acutâ.

A Brazilian species imported by Messrs. Loddiges (no. 828.) very near M. placanthera, from which it differs in having the flower of a globose figure, with much less linear sepals and concave oblong petals narrowed to the base. These parts are of a rich cream colour speckled with crimson. The lip has quite a different form, being semicircular at the tip, deeply furrowed and closely covered with short hair. The column has two deep furrows in front almost buried in hairs. In both these plants the caudicula is obcordate with a broad membranous border, and the gland is acuminate and continuous with it, reposing upon a slit rostellum.

105. CIRRHOPĔTĂLŪM *Macræi*. Lindl. Gen. & Sp. Orch. no. 6.

This plant has been obtained from Ceylon by Messrs. Loddiges (175). It has dull brownish yellow sepals, and purple petals, which are falcate with the point turned forwards. Many of the other species are prettier.

106. ĔRĬĂ *pulchĕllă*. (Lindl. in Wall. Cat. no. 7407. *Lanatæ*); pseudobulbis ovatis stipitatis diphyllis, foliis oblongo-lanceolatis obtusis, racemo lanato erecto terminali et laterali, bracteis ovato-subrotundis patentibus, perianthio explanato subæquali incano, labello subrotundo-ovato apiculato plano medio convexo discolore, basi trilineato.—Rhizoma repens, pennæ anserinæ crassitudine. Racemi 2 uncias longi. Perianthium luteum, intùs et extùs lanatum. Petala sepalis parùm minora. Labellum medio elevatum, fuscum, limbo luteo, subcordatum, apiculatum, planum. Pes columnæ luteæ rugosus, fuscus.

This little species was originally described from a specimen brought from India by Dr. Wallich, who had received it from Mr. Finlayson without a locality. It was then gathered at Sincapore by Mr. Cuming, from whom we have some fragments; and it has lately flowered with Messrs. Loddiges (204). It has oval pseudobulbs tapering at the base into a stalk and terminated by a pair of leaves about four inches long. The flowers are in spikes, which in Messrs. Loddiges specimen proceeded from the side of the pseudobulbs, but which also appear to be terminal; they are dull clayey-yellow of no great beauty.

107. MORMŌDĒS *lineātūm* (Bateman in litt.); sepalis petalisque oblongo-linearibus abruptè acutis margine reflexis, labello lineari incurvo carnoso sparsim piloso versus basin utrinque dente nunc brevi nunc elongatâ aucto, columnæ dorso et margine pubescente.

This curious species is a native of Guatemala, whence it has been sent by both Mr. Skinner and Mr. Hartweg. It has dull olive-green flowers, striped and spotted with dull brown. It is very distinct from the rest of the genus, but by no means pretty. The lip has two lateral lobes which vary in length from half a line to two lines.

109. EPIDĒNDRŪM (Euepidendrum ?) *lacertīnūm*; floribus racemosis, ovariis subsecundis longissimis pendulis bracteis setaceis multoties longioribus, sepalis lanceolatis acuminatis, petalis lineari-lanceolatis, labelli adnati trilobi basi bilamellâti laciniis lateralibus triangularibus intermediâ lineari acuminatâ elongatâ, antherâ immersâ, ovario cuniculato.

A singular plant from Guatemala, for which I am obliged to Mr. Bateman. I have only seen the flowers; but as it is stated to have the habit of Galeandra Baueri it probably belongs to the section of this genus which I have called Euepidendrum. The flowers hang down on long stalk-like ovaries from one side of a short raceme. They are bright green with the exception of the column, which is yellow, and the lip which is a little stained with purple, and has very much the appearance of a lizard's tail, the hind legs being outside the flower, and the head and shoulders buried within the cup.

110. CYPRĪPĔDĬŪM (5. Acaulia) *barbātūm*; acaule, foliis coriaceis acutis canaliculatis maculatis, scapo pubescente, sepalo dorsali cuspidato obtuso ciliato margine revoluto, petalis lineari-oblongis subundulatis fimbriatis margine superiore verrucoso, stamine sterili lunato pubescente.

A very fine herbaceous plant, allied to C. venustum, insigne, and purpuratum, gathered at Sincapore, on Mount Ophir by Mr. Cuming. It has flowers richly stained with purple, and streaked with green veins upon a white ground. It has flowered with Messrs. Loddiges.

114. ĔRĬĂ *polyūră* (Tonsæ); caule elongato folioso, foliis lanceolatis acutis patentibus striatis, spicis oppositifoliis multifloris nutantibus foliis æqualibus v. brevioribus, bracteis ovario duplò brevioribus adpressis, sepalis petalisque ovatis acutis erectis lævibus, labello cordato ovato acuto basi bicarinato.

A Manilla Orchidaceous plant, found by Mr. Cuming, and sent to Messrs. Loddiges; it is not a species with brilliant colours, for the flowers are small and white, except a deep crimson lip; it is however a graceful plant, in consequence of every leaf producing a long drooping spike, from the opposite side of the stem. It is very near Eria multiflora and micrantha.

115. SACCOLĂBĬŪM *Blūmeĭ*; foliis longis canaliculatis arcuatis acutis mucronatis racemis pendulis densifloris obtusis æqualibus, sepalis ovatis, petalis oblongis duplò angustioribus, labelli calcare compresso obtusiusculo intùs pubescente laminâ oblongâ emarginatâ utrinque costatâ.

A most beautiful species obtained from Java by Messrs. Loddiges, and in all probability the Rhynchostyles retusa of Blume. It is very near Saccolabium guttatum, but the racemes are shorter and thicker, the leaves are acute not unequally truncate, and the flowers are very differently painted; each sepal and petal has a bright streak of violet below the apex, and the lip is deeply stained with the same colour except the tip which is white. A figure has been prepared for the Sertum Orchidaceum.

116. AĔRĪDĒS *Brookeĭ*; foliis distichis obliquè obtusis racemis nutantibus multifloris duplo brevioribus, sepalis ovatis sub-acutis petalis ovalibus acutis duplo latioribus, labelli maximi laciniis lateralibus erectis acutiusculis, intermediâ 6-plo majore obsolete trilobâ rotundatâ deflexâ, calcare cornuto incurvo. *Bateman in litt.*

"This most beautiful species has lately flowered in the collection of Sir Richard Brooke, Bart. of Norton Priory, after whom I have ventured to name it. In the magnitude of its blossoms it infinitely surpasses any other species with which I am acquainted, the lip alone being upwards of an inch long, beautifully tipped with rose-colour. The other parts of the flower are of a brilliant white. In the form of the labellum the species slightly resembles

Aërides affine, but it is nevertheless so entirely distinct in other respects that it is unnecessary to contrast the species. In its habit it is stiff and erect, the leaves being only five inches long, while the racemes are more than double the length. Mr. Jones, the skilful gardener at Norton, is unfortunately unable to give me any account of the native country of the plant, but it doubtless comes from the 'gorgeous east.' I am not aware of its existing in any other collection."—J. B.

119. SCHOMBŪRGKĬĂ *tibīcĭnis* (Bateman in litt.); pseudobulbis conicis corniformibus annulatis sulcatis 3-phyllis, foliis oblongis coriaceis patentibus, scapo longissimo tereti distanter squamato apice paniculato, paniculâ pyramidali laxiflorâ, sepalis petalisque undulatis crispis, labello oblongo venis per medium 5 elevatis approximatis: laciniis lateralibus apice rotundatis intermediâ subrhombeâ emarginatâ, antherâ emarginatâ.

This noble plant, the cows-horn orchis of Honduras, has been for some years in our gardens, whither it was originally sent by Mr. Skinner. It nearly produced its flowers some time since with Mr. Bateman, but the gigantic flowering stem was accidentally broken before the blossoms expanded. Lady Acland has at last been so fortunate as to obtain it at Killerton in perfection, by the good management of Mr. Craggs, her Ladyship's gardener, and it was exhibited at the last great meeting in the Horticultural Society's Garden. The flowers when fully expanded are about two and half inches wide, a deep pink speckled with white on the outside, rich chocolate red within. The lip is white in the middle, but deep rose-colour at the sides, with a short chocolate red middle lobe. Upon the whole perhaps it disappoints the expectation formed by its huge pseudobulbs and flowering stem eight or nine feet long; but it is a very fine thing.

120. EPIDĒNDRŪM (Encyclium) *phœnĭcēūm*; pseudobulbis subrotundo-ovatis diphyllis, foliis oblongo-linearibus erectis, scapo paniculâque scabris, sepalis petalisque subæqualibus coriaceis obovato-lanceolatis, labelli laciniis lateralibus oblongis erectis obtusis apice recurvis intermediâ maximâ membranaceâ subrotundâ undulatâ emarginatâ basi bilamellatâ, columnâ obovatâ marginatâ utrinque unidentatâ.

A noble plant from Cuba, introduced by Messrs. Loddiges. Large pseudobulbs roundish ovate and two-leaved, a panicle from two to three feet high, and large scentless flowers are its characteristics. The sepals and petals are of a leathery texture and deep purple, a little mottled with green specks both inside and out. The lip is nearly an inch and a half long, of the clear bright violet of Cattleya labiata, and with the same deep crimson veins and stains in the middle; it is much undulated, and of a thin, not leathery texture. This plant is still finer than Epidendrum macrochilum.

121. ĔRĬĂ *convallarioīdĕs*. Lindl. gen. & sp. Orch. 70.

This plant has at last been obtained by Messrs. Loddiges. It has small whitish flowers, collected in close heads in the axils of broad striated leaves; they have no smell, and the species proves much less pretty than was expected. The specific character in the work above quoted, made up chiefly from bad Indian drawings, requires considerable correction, especially as regards the flowers, which are nearly destitute of hairs, and the lip, which is entire, and not 3-toothed. I also find that the anther is not dorsal, but merely contracts towards the back of the column, so as to expose the pollen-masses. The name *convallarioides* is unfortunate, for it can only bring to mind in India the sweet Lily of the Valley of Europe, and it is a libel on the latter to compare it with this species.

122. EPIDĒNDRŪM *rănĭfĕrūm*. Lindl. gen. & sp. Orch. no. 64.

A fine plant in the way of Ep. nutans, with large greenish flowers spotted with deep brown. It is a native of Mexico, and was imported by Mr. Barker. In some respects it disagrees with the character assigned the species in the work above quoted; but the differences arise only from the plant having been originally described from bad dried specimens.

123. EPIDĒNDRŪM (Osmophytum) *radiātum*; pseudobulbis ovatis compressis utrinque tricostatis triphyllis, foliis angustis elongatis acutis, racemo denso multifloro, ovariis trialatis, sepalis linearibus petalisque lanceolatis patentissimis, labello cochleato crenato crispo, clinandrio trilobo laciniis lateralibus erectis carnosis intermediâ truncatâ apice serratâ.

A Mexican species, allied to E. fragrans, imported by Messrs. Loddiges. It has the same habit, but the pseudobulbs are strongly ribbed at the side, and the lip is formed like that of E. cochleatum, only the margin is crisp and crenated. The flowers are pale pea-green, with the exception of the lip, which is striated with bright deeper purple rays. It is a pretty novelty, and will no doubt become a favourite, because its cultivation may be expected to be so easy.

124. POGŌNĬĂ *plicāta*. Lindl. gen. & sp. orch. p. 415.

This little terrestrial Orchidaceous plant was found in soil from the Mauritius, and has flowered in the collection of His Grace the Duke of Northumberland at Sion. Its flowers are dull olive green, with a pale lilac labellum, emarginate at the point, rolled over the column, and strongly painted with green veins, five of which are parallel with each other and occupy the centre of the lip. The leaves, which are independent of the flowers, are plaited, dull purplish brown, and covered with soft bristles.

125. PLEUROTHĀLLĪS *brevĭflōră*; folio carnoso obovato-oblongo apice rotundato subtus purpurascente, floribus sessilibus paucis fasciculatis pilosis, sepalis ovatis lateralibus connatis intùs lævibus, petalis ovatis apiculatis, labello carnoso obovato lævi basi excavato pone apicem angulato et subdenticulato.

A Mexican Orchidaceous plant imported by Messrs. Loddiges, and very near P. aphthosa, with which it agrees in the colour of its flowers. The principal differences are the following; P. breviflora has, as its name implies, much shorter and broader flowers; they are more hairy; and they are destitute of the tubercles inside the sepals; the labellum is obovate, and angular, as if about to produce a tooth, on each side near the apex; finally, the leaves are obovate and stained with purple, instead of being oval and sharp pointed.

128. EPIDĒNDRŪM (Encyclium) *pterocārpūm*. Lindley in Hooker's Journal of Botany, vol. 3. p. 82.

A singular species from Mexico which has lately flowered with Mr. Barker. It has a long thin raceme of pinkish yellow flowers, and a heart-shaped lip streaked with crimson. It is remarkable for the seed-vessel, when ripe, being extended into three broad sharp wings.

129. EPIDĒNDRŪM (Encyclium) *articulātum*. Klotzsch in Allgemeine Gartenzeitung, Sept. 22, 1838.

This species has also flowered with Mr. Barker. It is in the way of Epidendrum pastoris, but it is readily known by its lip having the middle lobe furnished with a triple row of tubercles, among which are intermixed many small projections.

141. MAXILLĀRĬĂ *barbata*; pseudo-bulbis inæqualitèr quadrangularibus, corrugatis, vaginatis: foliis ovatis, costatis, acutis: floribus racemosis; sepalis inæqualibus, petalis majoribus, lateralibus explanatis, apice subinvolutis; labello cucullato, trilobato, barbato, margine crispo, apice gibboso, disco carnoso. *Phytologist, p.* 7.

"Pseudobulbs quadrangular, wrinkled; at the apex of each is a cup-like cavity, which surrounds the base of the leaves when present; sheaths twice as long as the pseudo-bulb. Leaves solitary, ovate, ribbed, acute. Scape arising from the base of the pseudo-bulb, slender, smooth, and about six inches high. Flowers about eight, yellow, disposed in a raceme. Bracts about two lines long, brown and scarious at the edges, and closely embracing the peduncle. Peduncle about one inch long. Sepals unequal, yellow, involute at the apex, the lateral ones spread out, the upper one more or less connivent. Petals obtuse, of the same colour as the sepals, but much smaller. Labellum bearded, cucullate, three-lobed, and having a fleshy excrescence on the disk; darker in colour than the sepals and petals, and marked in the centre and on the lateral lobes with a dark purple spot. Column short and minutely pubescent on the back, in the upper part. Stigmatic cavity angular, membranous and three-lobed, the central lobe subulate. Clinandrium entire. Caudiculæ two, distinct. Gland transverse. Pollen-masses compressed, sulcate. Anther-case one-celled, and minutely pubescent like the column."

"This species in habit, in the form of the pseudo-bulbs, and in the foliage, is very similar to *Max. tetragona*, but the flowers are altogether different, and bear resemblance to those of *Max. aureo-fulva*, Hook., figured in the 'Floral Cabinet,' ii. plate 83." *l. c.*

This plant appears to be identical with M. vitellina. With respect to its supposed double caudiculæ I must, however, observe that they are no more than the elastic centre which holds the pollen-masses of all Orchidaceous plants of the Vandeous division to their common caudicula, and that they are very different from those truly double caudiculæ, properly so called, which exist in Bifrenaria.

142. **MAXILLĀRĬĂ** *purpurascēns*; pseudo-bulbis ovatis vel conicis, sulcatis: foliis duobus, lineari-lanceolatis, apiculatis: floribus solitariis, glabris; sepalis æqualibus, lanceolatis, apiculatis, supremo connivente, lateralibus labello parallelis: petalis conniventibus, linearibus, sepalis brevioribus; labello trilobato, margine purpureo maculato, glabro, disco carnoso; antherâ subpubescente. *Phytologist, p.* 8.

"Rhizoma growing along the top of the pot and sending off pseudo-bulbs, which are ovate or conical and furrowed. Leaves two, about four inches long, linear-lanceolate, and pointed at the apex. Scape about six inches high, smooth, tinged with purple and bearing a solitary flower at the apex. Flower smooth, internally yellow, externally purple. Bracts about half an inch long. Sepals about an inch long, apiculate at the apex; the lateral ones arranged parallel with the labellum, the upper one connivent over the petals and column. Petals much smaller and shorter than the sepals, and as long again as the column. Labellum smooth, three-lobed, the middle lobe recurved at the margin, the apex of the lateral lobes beautifully spotted with purple. Column smooth. Clinandrium smooth or slightly pubescent. Stigmatic cavity round. Caudiculæ two. Gland transverse, crescent-shaped. Pollen-masses two, each of which is divided into two distinct parts. Anther one-celled, somewhat compressed and slightly pubescent."

"This species is very different in habit from the preceding. It possesses a creeping rhizoma, which throws out ovate or subconical pseudo-bulbs from its upper surface; the leaves are also much narrower, and the scape is of a pleasing purple colour. In size and form the flowers nearly resemble those of *barbata*, but are solitary and of a different colour. *Maxillaria aureo-fulva*, *barbata* and *purpurascens* possess double caudiculæ (as is the case also in the genus Bifrenaria) and their inflorescence in other respects differs so much from that of any other species of *Maxillaria* with which I am acquainted, that they may be considered as forming a very distinct group of the genus.—Fred. Westcott."

Introduced from Brazil for the Birmingham Horticultural Society, by W. Fry, Esq. in 1840. This plant I have not seen and do not recognize.

143. **EPIDĒNDRŪM** (Encyclium) *tripunctātūm*; pseudo-bulbis ovalibus compressis diphyllis, foliis ligulatis tenuibus apice recurvis scapo bifloro pluriès longioribus, sepalis patentissimis lineari-oblongis, petalis duplò angustioribus, labello semiadnato obovato rugoso utrinque ultra columnam emarginato, columna atropurpurea apice triloba et luteo tripunctata.

For this very distinct new species I am indebted to Richard Harrison, Esq. of Aighburgh. It has long grassy leaves, and a very short two-flowered scape. The sepals are twice as broad and much larger than the petals; both of them are widely spreading and a dull yellow green. The column is deep purple, with three lobes, each of which is tipped with one pale orange-coloured spot. The lip is rather more than half an inch long, obovate, rugose, without being warted, and distinctly emarginate near the apex of the column; its colour is pale citron. Ep. hastatum is the species to which this seems to approach most nearly.

145. **EPIDĒNDRŪM** *Grahămī*; pseudo-bulbis ovatis diphyllis, scapo terminali, racemo plurifloro, perianthio patente, sepalis latò-linearibus, petalis spathulatis, labello trilobo basi longè bilamellato, laciniis lateralibus ovatis acutis, intermediâ rotundatâ crispatâ, columnâ apice utrinque dente obtuso, antherâ profundè sulcatâ. *Botanical Magazine, t.* 3885.

This is stated to be a native of Mexico, and to have flowered in the Botanic Garden, Edinburgh. It has a white lip streaked with crimson veins, and a yellowish-green calyx, and corolla stained with dull purple at the ends. I do not see how it differs from Epidendrum altissimum, except in being a small imperfectly formed specimen. All the panicled Epidendrums are simply racemose when weak, and *vice versa*.

147. **STANHŌPĔĂ** *Martiana*. Bot. Register, 1840, misc. 109.

Among some Orchidaceous plants collected in Mexico by M. Galeotti, this rare plant has flowered in the collection of Richard Harrison, Esq. of Aighburgh. It agrees extremely well with the character and description given by Mr. Bateman in the place above quoted; to which I may add that the epichilium is strictly linear, the two edges being as nearly as possible parallel with each other, and not a great deal broader than the column; a mark by which the species is immediately recognized.

148. **EPIDĒNDRŪM** (Spathium) *bisētūm*; racemo nutante, sepalis obovato-oblongis obtusis, petalis setaceis, labelli trilobi laciniis lateralibus subcuneatis retusis intermediâ bilobâ basi tuberculis tribus munitâ, ovario scabro.

A native of Guatemala, imported by the Horticultural Society, and flowered by Messrs. Loddiges. It is a plant allied to Ep. nutans, with much smaller flowers, more compactly arranged, and with a strong smell of cowslips. The ovaries are rough with elevated points, the sepals are brownish orange, the lip dull fawn colour. The petals are in the form of two fine bristles.

160. **ONCĬDĬUM** *mŏnŏcĕrăs*. Botanical Magazine, t. 3890.

This does not appear to be different from O. unicorne.

162. **MORMŌDĒS** *arŏmătĭcŭm*; sepalis petalisque ovatis acuminatis secundis concavis, labello angustè cuneato convexo laciniâ intermediâ triangulari acuminatâ cucullatâ, columnæ dorso subulato.

I can hardly say whether this is really distinct from Mormodes pardinum, not having a specimen at hand for comparison. It is, however, very different in colour and general appearance, having a pale pinkish ground sown thickly with dull wine-red specks. It is a Mexican plant, specimens of which have appeared in several collections. That before me is from Messrs. Loddiges. The flowers have a powerful odour like that of aromatic vinegar.

163. **EPIDĒNDRŪM** (Euepidendrum) *latilābrūm*; foliis ovatis obtusis brevivaginantibus internodiis sæpè brevioribus, sepalis petalisque angustioribus lineari-oblongis obtusis patentibus, labello repando sublobato emarginato ferè 4-plò latiore quam longo, pedunculis bifloris sessilibus.

A Brazilian plant, allied to Epidendrum umbellatum, of which it has the habit. But its lip, which is four times broader than long, and curved downwards on each side, so as to bear no little resemblance to a stiffly starched lady's apron, gives it a most singular appearance. I have long been acquainted with it through specimens in Sir W. Hooker's Herbarium; that now before me is from the collection of Messrs. Loddiges.

164. **EULŎPHĬĂ** *squālĭdă*; foliis lato-ensiformibus 5-nerviis, spicâ multiflorâ floribus inferioribus remotioribus, bracteis ovatis acuminatis ovario acutangulo brevioribus, labello oblongo trilobo laciniis rotundatis intermediâ crispâ emarginatâ calvâ venis pluribus parallelis elevatis, calcare brevi, antherâ obsoletè emarginatâ,

A terrestrial Orchidaceous plant of no beauty, brought home by Cuming. It has leaves like those of a Bletia, and a stout erect scape 1½ foot high, with dingy pallid flowers, the lowermost of which are very remote from the uppermost The lip has no appendages, but is strongly marked by elevated veins, and it is entirely confluent with the column at the base. Lately flowered by Messrs. Loddiges.

165. DENDRŌBĬŪM (Eudendrobium) *excīsum;* foliis lineari-oblongis obliquè retusis, floribus (solitariis?) oppositifoliis basi squamatis, sepalis petalisque acuminatis, labello oblongo acuto carnoso basi utrinque exciso: disco latè glanduloso, cornu pedicello ferè duplò breviore.

A small species with the habit of D. pulchellum, but with white inconspicuous flowers. It is a native of Sincapore, whence it was brought by Mr. Cuming. It has flowered with Messrs. Loddiges (No. 331). The nearest affinity is with D. revolutum and calcaratum.

166. BOLBOPHȲLLŪM *clandestīnum;* rhizomate repente squamis membranaceis cucullatis vestito distanter pseudo-bulbifero, pseudo-bulbis minimis 1-2-phyllis, foliis mucronulatis oblongis carnosis dorso convexis, floribus minimis geminis è squamis rhizomatis, sepalis clausis ovatis æqualibus in setam mollem productis, petalis ovalibus obtusis, labello angustè ovato obtuso lævi, antherâ 1-loculari, polliniis 2.

A very curious, but inconspicuous little Orchidaceous plant from Sincapore, which has flowered with Messrs. Loddiges. It is a form of the genus Bolbophyllum, in which the pseudo-bulbs almost disappear, and the flowers are only just protruded from among thin withered scales, with which the rhizoma is covered. It appears to be allied to B. tortuosum, a plant however which I have never seen. The minute flowers are very pale straw-colour, with the ends of the sepals lengthened into a soft yellow bristle. The leaves are hard, thick, and rounded at the back. The whole habit of the plant is that of some creeping Fern, such as Marginaria (Polypodium) piloselloides.

168. CATASĒTŪM *fuligīnōsum;* C. tridentati vultu, sepalis petalisque oblongis acutis reflexis, labello cucullato carnoso integerrimo v. minutè serrulato patente, columnâ brevi ecirrhosâ apice in setam productâ, antherâ parvâ (effœta?).

This singular plant has lately flowered at Sion, in the possession of His Grace the Duke of Northumberland. Its habit is quite that of C. tridentatum, but its flowers are in a dense erect raceme, and of a deep green colour, spotted with a dull blackish purple, so as to look as if they were soiled with soot. The sepals and petals are spotted, oblong, acute, and reflexed, so as to hang downwards. The lip, on the other hand, is fleshy, hooded, stained with pale purple, and either entire at the margin, or very slightly serrated; but it does not cover over the column as is usual in the hooded Catasetums: on the contrary it spreads away at almost right angles. The column itself is short, deep green, and produced at the point into a straight bristle, in front of which is placed a small and imperfect anther. In this masquing genus it is impossible to say what is a species and what is not. Judging by rules found good in other genera this is quite distinct from all species previously described; judging from the evidence we possess concerning C. tridentatum, cristatum, and viride, we should suspect it to be a male form of C. atratum, or some such species. That, however, must remain for future enquiry.

169. DENDRŌBĬŪM *secūndum.* Botanical Register, t. 1291.

A specimen of this pretty plant, which has recently flowered with Mr. Knight of the King's Road, has given me an opportunity of examining its structure carefully, and it proves so remarkable as to deserve especial notice. In the first place the lateral sepals and the base of the lip are so united into a spur, that their separate nature is entirely concealed towards the point of the spur. In the next place the rostellum is a deep two-lobed lip curved down over the stigmatic surface. The pollen-masses are in two pairs, and deep purple; and, which is extremely curious, they lie upon a loose hard transverse crustaceous gland, which seems to replace the gland and caudicula of Vandeæ, and which projects beyond the anterior edge of the anther, resting upon the two-lobed rostellum. The anther-bed itself is deeply excavated, and three-lobed, the anther adhering to the back lobe which is the narrowest.

171. DENDRŌBĬŪM *Cambridgeānum;* (Paxton's Mag. Bot. vi. 265.) caulibus pendulis medio incrassatis, foliis ovalibus acutissimis, pedunculis bifloris, petalis lanceolatis sepalis latioribus, labello cucullato rotundato indiviso supra piloso.

This plant is only known to me from the figure in Mr. Paxton's Magazine of Botany. It is said to be a native of the Khoseea hills of India, whence it was brought to Chatsworth by Mr. J. Gibson in 1837; it was found at an elevation of 4000 feet, attaching itself to rocks and trees. It is a beautiful species, with large rich yellow flowers, having a deep purple stain in the middle of the lip; but seems very near D. Paxtoni and chrysanthum.

173. CIRRHOPĒTĂLŪM *maculōsum;* floribus geminis solitariisque, sepalo supremo oblongo acuto lateralibus lineari-oblongis obtusis margine anteriore pone basin uniplicato, petalis sepalo supremo brevioribus oblongis obtusis, labello ovato obtuso convexo apice recurvo, columnâ bidentatâ? (polliniis 4 subæqualibus).

This is an Indian species of Orchidaceous plant, sent to Messrs. Loddiges by Dr. Wallich (158), with the habit of Bollophyllum affine, or leopardinum, but with the long lateral sepals of a Cirrhopetalum. It has the flower-stalks and flowers of a very pale green, finely speckled with dull purple. The lateral sepals are nearly an inch long, and have a single fold on their inner margin near the base. The species has little beauty.

174. ONCĪDĬŪM *Barkĕri;* pseudobulbis ovalibus compressis parum angulatis diphyllis, foliis angustis erectis membranaceis manifestè petiolo vaginante articulatis racemi cernui simplicis pedunculo brevioribus, sepalis liberis petalisque lanceolatis undulatis æqualibus patentibus, labelli lobis lateralibus parvis subquadratis intermedio transverso apice inflexo vix emarginato, tuberculo oblongo basi 2-dentato apice obsoletè trilobo anticè excavato, columnæ alis brevibus rotundatis.

A fine species, imported from Mexico by Mr. Barker. It has very large flowers of a clear but pale yellow on the lip, and rich brown spotted sepals and petals. The lip is of unusual size, being more than one inch and a half across. It is readily known by its short leaves having a distinct articulation, with a sheathing petiole nearly an inch long. The raceme is simple, drooping, of five or six flowers, and altogether a foot long.

175. ONCĪDĬŪM *nebulōsum;* pseudobulbis ovalibus ancipitibus nebulosi diphyllis utrinque 3-costatis, foliis angustis chartaceis paniculâ angustâ brevioribus, sepalis liberis petalisque lanceolatis undulatis patentibus, labelli lobis lateralibus nanis rotundatis intermedio subrotundo bilobo, tuberculo sulcato tridentato utrinque unidentato, columnæ elongatæ alis acinaciformibus denticulatis.

A fine species, with something the appearance of O. reflexum, but readily known by its clouded pseudo-bulbs. The flowers are large, rather pale yellow, with faint spots of brown at the base of the lip and on the sepals and petals. It is a native of Guatemala, whence it has been sent to the Horticultural Society by Mr. Hartweg.

178. CŒLŌGȲNĒ *coronāria;* pseudobulbis teretibus angustissimis diphyllis, foliis lanceolatis acuminatis undulatis coriaceis obsoletè trinerviis racemi pedunculo basi tantum squamato longioribus, bracteis setaceis apice sphacelatis, labelli lobis rotundatis lateralibus planis intermedio undulato, cristæ lamellis crispis in lobum intermedium 5 in hypochilium 4 tantum geminatis, clinandrio trilobo, antherâ antice rotundatâ fornicatâ.

A pretty species of this interesting genus, from the Chirree district of the Khosea hills of India, where it was found by Mr. Gibson, when in that country on his botanical mission from his Grace the Duke of Devonshire. It has firm, bright-green, not shining, leaves of a leathery texture, and very pale greenish yellow flowers, whose lip has a yellow centre, and a border beautifully streaked, and spotted with crimson. We shall take an early opportunity of publishing a figure of it.

179. ĔRĬĂ *bipunctătă;* (Tonsæ) caulibus ovalibus compressis, foliis lanceolatis acuminatis 11-nerviis, racemis axillaribus cylindraceis multifloris, bracteis ovatis acuminatis reflexis, alabastris globosis, sepalis petalisque ovatis obtusiusculis, labelli trilobi lævis laciniis lateralibus intermedio carnoso convexo rotundato angustioribus.

This plant was found by Mr. Gibson in the same locality as the last, and flowered with it at Chatsworth. It has the flowers of a Liparis, and the foliage of an Eria, so far as general aspect goes. The stems are flat, oval, partially covered by the sheaths of fallen leaves, when full blown between two and three inches long; at first they are green, but they become when old of a light purplish brown colour. The leaves, in size and form, are very like those of Eria floribunda, to which this species is nearly allied. The flowers are arranged in cylindrical drooping racemes, rather shorter than the leaves; they are perfectly free from downiness, except just at the base of the raceme, and are about the size of a grain of pepper, very pale, yellowish white, with the tip of the column purple, and the point of the labellum deep yellow, by which means the flowers seem as if they had two spots upon them when they are looked at from the front.

180. CYRTOCHĪLŬM *graminifŏlĭŭm;* Lindl. in Annals of Natural History, vol. 4. p. 384.

This plant, which resembles C. maculatum in habit, but has a yellow lip, the form of which is that of a wedge, with rounded angles, has flowered with Messrs. Loddiges, who imported it from Oaxaca (No. 1265). It is most nearly allied to C. filipes, figured in the present number, but differs in the form of the lip.

181. EPIDĔNDRŬM (Encyclium) *calocheīlum;* pseudo-bulbis ovato-oblongis apice diphyllis, foliis ensiformibus obtusis coriaceis obsoletè striatis paniculâ elatâ multiflorâ brevioribus, sepalis petalisque lineari-oblongis spathulatis uniformibus patentissimis, labello libero suborbiculari profundè trilobo basi carinato, lobis lateralibus latè ovatis intermedio latissimo lineato margine undulato crispato, columnâ supernè hinc alâ obtusâ. *Hooker in Bot. Reg. t.* 3898.

A Guatemala plant, from the Woburn collection, very near Epidendrum altissimum, but apparently distinct. The flowers, in a large panicle, are light greenish yellow with the sepals and petals tinged with purple at the points, and the lip crimson-veined with a yellow border.

182. PLEUROTHĂLLĬS *pictă;* folio oblongo coriaceo caule vaginato racemoque longiore, spathâ diphyllâ, racemo plurifloro, sepalis reflexis inferiore trilineato extus basi villoso duobus superioribus ultra medium unitis maculatis, petalis columnâ longioribus ovato-lanceolatis, labello ovato carnoso velutino maculato obtuso intus basi profundè canaliculato, clinandrio dentato, anthera ovarioque pubescenti-tomentosis. *Hooker in Bot. Reg. t.* 3897.

A Mexican Orchidaceous plant, with purple flowers arranged in a spike about one-third the length of the leaf. It is very near *P. strupifolia* and *aphthosa* (which latter is the P. peduncularis of Hooker's Journal of Botany, vol. 3. t. 9.) but apparently distinct from both.

184. BROMHĔADĬĂ *palustris.*

When Mr. Finlayson was in the Malay archipelago, he found a curious Orchidaceous plant at Sincapore with the habit of Epidendrum elongatum; and from specimens of it, almost destroyed by insects, which I examined some years since in Dr. Wallich's herbarium, I referred it with great doubt to Grammatophyllum, under the name of G. Finlaysonianum, (*Gen. & Sp. Orch. t.* 173.) This plant has lately flowered at Penllergan in South Wales with J. D. Llewelyn, Esq. who received it from Cuming, with the memorandum that it had been "dug out of a bog in Sumatra." Having now had the advantage of examining a perfect specimen in flower, I find that although nearly allied to Grammatophyllum, it is in fact very distinct. I therefore avail myself of the opportunity of adding to the list of genera the name of Sir Edward Ffrench Bromhead, Bart. F.R.S. whose investigations of the natural affinities of plants are well known to systematical Botanists. It may be distinguished by the following character.

Perianthium (candidum) sesquipollicare, cylindraceo-connivens, laciniis omnibus lineari-oblongis curvis canaliculatis obtusiusculis. *Labellum* cucullatum, trilobum, c. columnâ omninò parallelum, basi inarticulatum; (laciniâ intermediâ retusâ flavescente disco luteo-glandulosâ, lateralibus ovatis brevioribus violaceis; axi elevatâ pubescente). *Columna* latè alata, obtusa, carnosa. *Anthera* 2-locularis, longitudinaliter dehiscens, dorso conica et cum columnâ articulata. *Pollinia* duo, reniformia, posticè excavata, in glandulam latam triangularem membranaceam sessilia.——Herba parasitica, caulescens, ebulbis. Folia disticha, oblongo-linearia, emarginata. Spica terminalis, disticha, flexuosa, multiflora, longè pedunculata, bracteis brevissimis, rigidis, dentiformibus.

In appearance the plant has the aspect of Epidendrum elongatum, as has been already stated; and like it has the whole of the upper part of the stem provided with closely pressed distant sheaths instead of leaves, on which the spike of flowers is arranged. The latter is very rigid, between two and three inches long, regularly zigzag, with a short hard tooth-like bract at each bend, so that the spike without the flowers resembles a coarsely-toothed narrow doubled-edged saw. The flowers are about an inch long, white, and rather drooping, not spreading open, but with the divisions converging in a cylindrical manner. The labellum, in which alone any colour resides, is yellow in the middle, straw-coloured on the middle lobe, and violet at the tips of the lateral lobes.

As a genus Bromheadia is distinctly characterized by its broadly winged column, which is parallel with the labellum, and its contiguous pollen-masses. Its habit too is very peculiar.

186. PLEUROTHĂLLĬS *gĕlĭdă;* folio oblongo carnoso concavo basi attenuato caule tereti laxè 1-ochreato æquali v. longiore, spicis geminis erectis secundis folio subæqualibus e spathâ brevi univalvi exsertis, floribus cylindraceis, sepalis carnosis intus pilosis lateralibus semiconnatis, petalis minimis glabris subrotundo-oblongis apice denticulatis, labello etiam breviore lineari cuneato bilamellato apice rotundato plano, columnâ petalis breviore.

A plant with all the habit of Pl. racemiflora, but the spikes appear in pairs, and are hardly so long as the leaf, and the structure of the flowers is different. The whole plant is fully a foot high. Messrs. Loddiges obtained it from Jamaica.

187. PLEUROTHĂLLĬS *sĭcārĭă;* folio oblongo angusto plano in caulem triplò longiorem triquetrum canaliculatum deorsùm attenuatum angustato, pedunculis brevibus 2-3-floris cernuis aggregatis, sepalis lineari-lanceolatis patulis chartaceis lateralibus ad apicem usque cohærentibus sed facilè separabilibus, petalis ovalibus acutissimis serratis, labello lanceolato acutissimo bilamellato utrinque unidentato, clinandrio serrato.

A native of Trinidad. In habit this species is most remarkable, its stem being exactly like a bayonet blade (sica) with the point downwards. It grows nearly a foot high. The flowers are greenish yellow with pale purple stripes along the sepals, much deeper stripes on the petals, and none at all on the labellum.

188. PLEUROTHĂLLĬS *fragilis;* folio ovato-oblongo plano carnoso caule compresso canaliculato duplò breviore, pedunculis aggregatis brevibus (2-floris) folio adpressis, floribus carnosis fragilibus glaberrimis, sepalis lateralibus ad apicem usque cohærentibus supremo lineari recto, petalis cuneatis apiculatis columnæ apice denticulatæ subæqualibus, labello lineari apice rotundato basi utrinque paululum dilatato et denticulato superficie lævigatâ.

This little plant has the habit of P. prolifera; but its leaves are flat, not folded together, and it is not proliferous. The flowers are bright orange-yellow, and very brittle, like wax; with a double row of minute purple dots along the middle of the labellum. Messrs. Loddiges have obtained it from Rio (11).

189. EPIDĔNDRŬM (Encyclium) *virgātum* (Lindl. in Hooker's Journ. of Botany, v. 3. p. 83.); pseudobulbis ovatis oblongisve sub-compressis rugosis, foliis binis ternisque convexis subundulatis acutis glaucis unciam latis, paniculâ virgatâ ramis longis gracilibus, sepalis lanceolatis petalisque duplo angustioribus patentibus discoloribus, labelli hastati lobis lateralibus acutis patentibus intermedio subrotundo-obovato acuto; callo maximo rotundato pone basin.

A free-growing Epiphyte with the habit of E. vitellinum, but with more glaucous leaves. Its flowers are small, dirty green stained with brown, arranged in a very long lax graceful panicle, the branches of which are simple, and sometimes as much as a foot long, with nearly twenty flowers on each.

The lip is whitish yellow. It has flowered in the garden of the Horticultural Society, from plants sent from Mexico by Mr. Hartweg. It was first found near Teoxomulco, in the province of Oaxaca, by Karwinski.

191. MORMŌDĒS *buccinătŏr*. Botanical Register, 1840. misc. no. 9.

Of this singular plant Messrs. Loddiges have obtained a variety from La Guayra, which, with all the form of the original, has the flowers dull orange spotted with brown. It is handsomer than the first sort, and looks at first sight something like M. aromaticum.

EDWARDS'S

BOTANICAL REGISTER:

OR,

ORNAMENTAL FLOWER-GARDEN AND SHRUBBERY:

CONSISTING OF

COLOURED FIGURES OF PLANTS AND SHRUBS,

CULTIVATED IN BRITISH GARDENS;

ACCOMPANIED BY THEIR

History, Best Method of Treatment in Cultivation, Propagation, &c.

AND

MONTHLY CHRONICLE

OF

BOTANICAL AND HORTICULTURAL NEWS.

CONTINUED

BY JOHN LINDLEY, Ph. D. F.R.S. AND L.S.

PROFESSOR OF BOTANY IN UNIVERSITY COLLEGE, LONDON,
AND THE ROYAL INSTITUTION OF GREAT BRITAIN,
VICE-SECRETARY OF THE HORTICULTURAL SOCIETY,
&c. &c. &c.

1842.

OR VOL. XXVIII. OF THE ENTIRE WORK.
OR VOL. XV. OF THE NEW SERIES.

——viret semper——nec fronde caducâ
Carpitur.

LONDON:
JAMES RIDGWAY, PICCADILLY.

M.DCCC.XLII.

1

CĀTTLĔYĂ granulōsă.

Rough-lipped Cattleya.

GYNANDRIA *MONANDRIA*.

Nat. ord. ORCHIDACEÆ, § EPIDENDREÆ.
CATTLEYA. *Botanical Register, vol.* 11, *fol.* 953.

C. *granulosa;* caulibus teretibus gracilibus diphyllis, foliis oblongo-lanceolatis obtusis, sepalis oblongis obtusis, petalis obovato-spathulatis undulatis obtusissimis, labello cucullato tripartito: laciniis lateralibus semiovatis intermediâ sinu lato divulsâ ungue æquilateri lævi laminâ dilatatâ rotundatâ plicatâ granulosâ denticulatâ.

Of this, at present the rarest of the genus, a single small specimen exists in the garden of the Horticultural Society, where it was sent by Mr. Hartweg from Guatemala, without any intimation as to its native habits.

It has a slender stem, terminated by two narrow leaves, something like those of C. bicolor, but shorter. When it flowered only one blossom was formed, such as here represented; but it will probably bear a larger number when in health. The sepals were two inches and a half long, oblong, olive green, mottled with rich brown spots. The petals were obovate, very much rounded at the point, and narrowed to the base; wavy at the margin, and of the same colour as the sepals. Strongly contrasted with the dingy colours of those parts is the lip, of a pure white at the sides and point, and of a glowing orange spotted with crimson in the middle; its lateral lobes curve over the column, but are divided very deeply from the middle lobe by a wide slit; the middle lobe itself has its surface broken up into numerous granulations, something in the way of C. guttata.

The high temperature and excessive moisture which suits so well the Indian Dendrobiums is most injurious to this

Cattleya. A night temperature of 55° in winter and 60° in summer is quite high enough for it; and by planting it in well drained turfy peat, and keeping it rather dry when not growing, it will be found one of the easiest to manage.

4

ONCĪDĬŬM longifŏlĭŭm.

Long-leaved Oncidium.

GYNANDRIA *MONANDRIA.*

Nat. ord. ORCHIDACEÆ, § VANDEÆ.
ONCIDIUM. *Botanical Register, vol.* 9, *fol.* 727.

O. *longifolium;* (Cebolletæ) foliis teretibus longissimis diffusis, scapo erecto densè paniculato, sepalis petalisque apiculatis obtusis concavis, labelli lobis lateralibus patentibus obliquè oblongis subquadratis intermedio obovato bifido basi angustato minoribus, tuberculo basi depresso anticè tricorni glabro, columnæ alis sublunatis brevibus rotundatis. *Botanical Register*, 1841. *misc.* 56.

Folia *tripedalia, flagelliformia, prostrata.* Scapus 3-4-*pedalis paniculatus.* Sepala *et* Petala *lutea sanguineo maculata; labellum luteum sanguineo prope basin marginatum.*

Under the name of Oncidium Cebolleta many very different species exist in our gardens, among which the finest is that now figured, which, although it has the foliage of that species, is really very different, forming dense panicles, three feet long, of very large and showy yellow and brown flowers. Its leaves are often three feet long, and hang down or spread upon the ground instead of standing stiff and erect. It has been imported abundantly by the Horticultural Society from Mexico, and has flowered both in their garden and with Messrs. Loddiges. It is very handsome, and well worth growing.

From all the allied species it differs in the great length of its whip-like leaves, and in the form of its labellum, whose lateral lobes are something like bird's wings in shape, while the intermediate lobe has an unusually long unguis gradually dilating into a two-lobed extremity.

In cultivation it requires a warm damp atmosphere, where such plants as Cattleyas, Lælias, and Stanhopeas flourish. A temperature ranging from 55° to 65° with artificial heat,

will be quite sufficient. It will grow very well amongst turfy peat in a pot; but we prefer suspending it in a forked block of wood, and allowing the long leaves to hang gracefully down, as they do in its native woods. If grown in this way, the cultivator must fix it firmly to the block, and pack the roots all round with turfy peat or sphagnum, and take care that it is freely watered, particularly when it is growing. Plants which are hung up in this manner are not likely to suffer from too much water.

It is propagated in the usual way, by division.

The other species of this section are the following; viz.

O. *brachyphyllum;* foliis crassissimis brevibus erectis acutis, paniculâ laxâ pauciflorâ, sepalis petalisque obovatis acuminatis longè unguiculatis, labelli lobis lateralibus patentibus ovalibus acutis intermedio reniformi emarginato longioribus, tuberculo tricorni subverrucoso, columnæ alis sublunatis brevibus rotundatis.

This is a Mexican species, with very short stiff leaves, and smaller panicles, not above eighteen inches high.

O. *ascendens;* foliis erectis junceis, scapo stricto paniculato æqualibus, sepalis petalisque obovatis obtusis concavis, labelli lobis lateralibus nanis erectis intermedio reniformi emarginato, tuberculo ovato 7-dentato, columnæ alis linearibus integerrimis incurvis.

This is from Guatemala, and has much the habit of O. Cebolleta; but the linear wings of the column, the small erect lateral lobes of the lip, and the many-toothed tubercle distinguish it at once.

O. *Cebolleta.* Suprà fol. 1994.

Of this there are several varieties, one of which has the back of the lip covered with crimson spots, and another has no spots at all.

12

CIRRHOPĔTĂLŬM Medūsæ.

The Medusa's-head Orchis.

GYNANDRIA *MONANDRIA.*

Nat. ord. ORCHIDACEÆ, § MALAXEÆ.

CIRRHOPETALUM. Botanical Register, 1838, *t.* 11.

C. *Medusæ;* pseudobulbis ovatis tetragonis, folio oblongo convexo emarginato, vaginis scapi oblongis ventricosis supremo pedunculo longiore, floribus densè capitatis, sepalo supremo acuminato inferioribus longissimis pendulis, petalis triangularibus acuminatis, labello ovato acuminato bicarinato, columnæ auriculis acutis.

This most singular plant is a native of Sincapore, whence it was obtained by Messrs. Loddiges. In some respects it resembles C. vaginatum, but is a much larger plant, and very different in the form of its flowers, as will be apparent upon comparing the preceding character with our amended one of C. vaginatum given below. Certainly if ever there was a Medusa this must be the prototype, before her Gorgonship's beautiful tresses were changed into serpents; nor are wanting the scales with which her form was safely guarded. We believe that this young lady was carried out of harms way into the ocean of India by Neptune, and that all they tell us about Perseus having cruelly killed her is a fable; for is not here the proof?

Fig. 1. represents the column, petals, and labellum; 2. is the labellum separate.

To grow it well the temperature and moisture of the Orchidaceous stove is required. It may be fixed to a wooden block, and suspended from the rafters, but if treated in this manner a quantity of sphagnum, or turfy peat, ought to be packed round the stem to retain moisture. It ought never to be kept too dry, but of course it requires more water when growing than when resting.

The *Cirrhopetalum vaginatum* was taken up, in the Genera and Species of Orchidaceous plants, from specimens in Dr. Wallich's Indian herbarium; they were, however, not very good, and the inspection of living plants, in flower with Messrs. Loddiges, enables me to amend, as follows, the character formerly given of that species.

C. *vaginatum;* pseudobulbis pyramidatis truncatis, folio oblongo convexo emarginato, vaginis scapi oblongis ventricosis distantibus sepalo supremo ovato inferioribusque longissimis pendulis ciliatis, petalis oblongis ciliatis, labello oblongo bicarinato, columnæ auriculis setaceis.

From this it appears that C. vaginatum differs from C. Medusæ in its flowers being ciliated, the petals oblong, not triangular and acuminate, and in the distant sheaths of the scape; to which may be added that its flowers are not speckled with pink as in C. Medusæ.

13

MAXILLĀRĬĂ cruēntă.

Blood-stained Maxillaria.

GYNANDRIA *MONANDRIA.*

Nat. ord. ORCHIDACEÆ, § VANDEÆ.

MAXILLARIA. Botanical Register, vol. 11. *fol.* 897.

M. *cruenta;* pseudobulbis compressis rugosis, foliis pluribus oblongis membranaceis, pedunculis radicalibus unifloris (rarò bifloris), vaginis distantibus obtusis cucullatis, sepalis ovatis obtusis lateralibus basi parùm productis, petalis minoribus conformibus, labello sepalis duplò breviore concavo trilobo laciniis rotundatis intermediâ crispâ pubescente: tuberculo parvo plano, columnâ pubescente.

M. Skinneri, *hortorum nec Batemanni.*

It has been already stated (p. 11) that the plant current in our gardens under the title of M. Skinneri is not the species to which Mr. Bateman intended to apply that name, but one more nearly allied to M. aromatica. The accompanying figure represents the latter.

It was found by Mr. Skinner in Guatemala, and sent to his friends as a large variety of M. aromatica, to which it certainly bears much resemblance. The following circumstances however will readily distinguish it. The leaves are broader; the flowers are four times as large when in health; the lip has quite a different form, with a deep crimson blotch at its base, and is not half the length of the sepals; the middle lobe is rounded not unguiculate, and only a small tubercle in the middle instead of the large concave appendage that occurs in M. aromatica. No doubt then can be entertained of the specific difference between these plants, which may be stated as follows for M. aromatica.

M. *aromatica* (Hooker Ex. Flo. t. 219. Gen. & Sp. Orch. no. 20. Bot. Reg. t. 1871. *Colax aromaticus,* Spreng.); pseudobulbis compressis rugosis, foliis pluribus oblongo-lanceolatis membranaceis, pedunculis radicalibus unifloris erectis, vaginis distantibus obtusis cucullatis; sepalis ovato-oblongis petalisque conformibus acutis, labelli semicylindracei sepalis æqualis laciniis lateralibus acuminatis obtusis intermediâ cuneatâ apice serrulatâ; appendice magnâ concavâ carnosâ truncatâ, columnâ facie villosâ——*Mexico.*——The flowers are yellowish-orange colour, scarcely spotted even inside the lip, which has two rows of hairs along its inner face.

In the annexed plate, Fig. 1. represents the lip spread open; 2. is the column, which is hairy, although the engraver has neglected to shew the hairs.

As may be expected, the habits of this very fine species are the same as those of *M. aromatica,* and it requires the same treatment. It does very well in the cool orchidaceous stove, where plants from Mexico and Guatemala are now generally grown, and wants a season of dryness after it has made its pseudo-bulbs, but plenty of water during its growth. The flowering time is winter and spring.

The finest specimen I have yet seen flowered with Sir C. Lemon at Carclew, where one of the flower-stalks bore two blossoms, a very unusual circumstance among the species with this habit.

17

CYPRIPĔDĬŬM barbātum.

Bearded Lady's Slipper.

GYNANDRIA *DIANDRIA.*

Nat. ord. ORCHIDACEÆ, § CYPRIPEDEÆ.

CYPRIPEDIUM. *Botanical Register, Vol.* 10. *fol.* 788.

SECT. V. Acaulia; *foliis omnibus radicalibus.* Gen. & Sp. Orch. 531.

C. *barbatum;* acaule, foliis coriaceis acutis canaliculatis maculatis, scapo pubescente, sepalo dorsali cuspidato obtuso ciliato margine revoluto, petalis lineari-oblongis subundulatis fimbriatis margine superiore verrucoso, stamine sterili lunato pubescente.—*Bot. Reg.* 1841, *misc. no.* 110.

Folia *coriacea, oblonga, acuta, canaliculata, rigida, viridi intensiore interruptè vittata.* Scapus *pubescens, uniflorus, purpurascens, foliis longior.* Bractea *ovata, cucullata, acuta, ovario multò brevior.* Flores *erecti.* Sepalum dorsale *album, purpureo viridique lineatum, subrotundum, ciliatum, marginibus reflexis;* lateralia *in unum parvum obtusum connata.* Petala *oblonga, marginibus subparallelis, apice atro-purpurea, basi viridia, ciliata, margine superiore verrucoso, verrucis atropurpureis vernicatis piliferis.* Labellum *fusco-purpureum, obtusum, glabrum, petalis longitudine æquale.* Anthera *sterilis lunata, vertice verrucosa, angulis acuminatis, margine posteriore emarginata.*

There is something in the habit of the Indian Lady's Slippers so peculiar, that it was for a long time expected that they would be found to possess characters sufficient to separate them altogether from their associates. The fullest examination however shews this expectation to be fallacious, and that no peculiarity of organization exists among them. In fact, the genus itself has been found to vary so much in the aspect of its species, that the common European and North American kinds convey no idea of its nature. In addition to the Indian forms, of which this may be regarded as a type, the hotter parts of America have furnished, on the one hand, the singular C. Lindleyanum, a hairy rufous plant with panicled flowers, and the scarcely less curious C. palmifolium, which bears the aspect of a Sobralia, and has its flowers in long racemes.

The species now represented is a native of the Straits of Malacca, where Cuming found it on Mount Ophir. Messrs. Loddiges are, as far as I am aware, the only persons who have flowered it. The purple hairy shining warts which border the upper edge of its petals distinguish it immediately from C. venustum and purpuratum, which are most like it.

Fig. 1. in the plate represents the head of the sterile stamen.

Persons who are unskilful in the cultivation of Orchidaceous plants, manage the more common kinds of Cypripedium extremely well, and therefore it is only necessary to say that this species succeeds perfectly with the same treatment as C. venustum or insigne. The cultivator should not go to extremes with resting these plants, and should water with caution when they are just beginning to grow, as the young shoots are apt to damp off. They are terrestrial in their habits, and do best when grown in pots.

21

TRICHŌSMĂ suāvĭs.

Sweet-scented Hair Orchis.

GYNANDRIA *MONANDRIA.*

Nat. ord. ORCHIDACEÆ, § EPIDENDREÆ.

TRICHOSMA. *Sepala* patula, æqualia, lateralia basi producta columnæ pedi adnata. *Petala* conformia, erecta. *Labellum* trilobum, cristatum, cucullatum, cum pede columnæ articulatum. *Columna* carnosa, marginata. *Anthera* carnosa, fornicata, bilocularis, valvis verticalibus margine sphacelatis. *Pollinia* 8, quaternatim cohærentia, duobus cuique phalangi minoribus difformibus.——Epiphyta, ebulbis, caule *squamis verrucosis vestito diphyllo.* Folia *subcarnosa, petiolata.* Spica *terminalis, basi spathacea.*

Trichosma suavis.
Cœlogyne coronaria. *Bot. Reg.* 1841. *misc.* 178.

Caules *teretes, angustissimi, diphylli.* Folia *lanceolata, acuminata, undulata, coriacea, obsoletè trinervia, racemi pedunculo basi tantum squamato longiora.* Bracteæ *setaceæ apice sphacelatæ.* Labelli *lobi rotundati lateralibus planis intermedio undulato;* cristæ *lamellis crispis in lobum intermedium* 5, *in hypochilium* 4 *tantum geminatis.* Columna *semiteres, carnosa, marginata.* Anthera *carnosa, anticè projiciens, rotundata, fornicata, bilocularis; valvis verticalibus, margine nigro-sphacelatis.* Pollinia 8, *cereacea, libera, quaternatim cohærentia; duobus cuique phalangi ovalibus duobus cuneatis emarginatis.*

When this curious plant was sent me from Chatsworth I too hastily referred it to the genus Cœlogyne, misled by a certain similarity of appearance and structure. It is however so very different in several circumstances, that I feel obliged to recognize in it a genus distinct from any hitherto described.

From Cœlogyne it differs in the projecting foot of the wingless column, in the singular fleshy anther, and in the eight pollen-masses. It is more nearly allied to Ceratium, but that genus, according to Blume, has a two-celled anther, four of the pollen-masses very small, a siliquose fruit, and the inflorescence opposite the leaves. Indeed, as it is described as caulescent with sessile lanceolate coriaceous leaves, it seems to be altogether a different kind of plant.

* From θρίξ, hair, and κόσμος, ornament.

This species was brought from the East Indies to Chatsworth by Mr. Gibson, collector to his Grace the Duke of Devonshire, and he has favoured me with the following particulars concerning it.

"It inhabits the Chirra district of the Khoseea and grows upon trees in densely shaded woods near the summit of the hills. It has a strong perfume, somewhat resembling Melic-grass, on which account it is highly esteemed by the wood-cutting natives, who are fond of adorning their hair with its blossoms, and which circumstance gave me a clue to its discovery. It is growing here luxuriantly upon a block of wood, the interior of which from decay has become hollow, and is now filled with dead and living vegetable matter, the latter consisting of the roots and stems of live native Ferns, which are luxuriating with this and other Orchideous plants on the same block of wood. The pseudobulbs are invariably two-leaved, spindle-shaped, gradually tapering towards the top, and covered with dark brown sheathing scales; their length is about six inches. On the young shoots the scales are thick and covered with green warts. The roots are thick and fleshy, and covered with a woolliness like those of old Eria stellata."

Fig. 1. represents the anther in its natural position; 2. one set of pollen-masses; 3. the lip spread open.

25

EPIDĔNDRŬM cinnabarīnŭm.

Cinnabar Epidendrum.

GYNANDRIA *MONANDRIA.*

Nat. ord. ORCHIDACEÆ, § EPIDENDREÆ.

EPIDENDRUM. *Botanical Register, vol.* 1. *fol.* 80.

Sect. AMPHIGLOTTIUM, *Lindl. in Hooker's Journ. of Botany,* 3. 81.

E. *cinnabarinum;* (L. p. 106.) foliis distichis oblongis apice sub-recurvis, sepalis petalisque lanceolatis subæqualibus, labello trilobo medio unilamellato basi bituberculato: laciniis lateralibus inciso-laceris intermediâ basi obcuneatâ sub apice constrictâ apice cuneatâ truncatâ angulis acuminatis simplicibus fissisque.

Among those Western Orchidaceous plants which Salisbury called Amphiglottis, but which I regard as a mere section of Epidendrum, there are some species of great beauty, so nearly related to each other that their limits are with difficulty determined. I particularly allude to *E. Schomburgkii* figured at t. 53 of the volume of this work for 1838, this *E. cinnabarinum*, and *E. rhizophorum* of Mr. Bateman, which is no doubt the *E. radicans* of Pavon, badly described from a mutilated specimen. They all have bright scarlet or crimson flowers, a pair of tubercles at the base of the labellum, and a ridged line running from between the tubercles to nearly the apex of the lip. They however differ thus: *E. radicans* has the lateral lobes of the labellum rounded and toothletted only, not lacerated, and it produces coarse pale green roots from its stems; *E. cinnabarinum* has the lateral lobes of the labellum deeply lacerated, while the central lobe is contracted in the middle, and then *suddenly* wedge-shaped, with its angles prolonged into one or two fine teeth; *E. Schomburgkii* has the lateral lobes only toothed, with the centre lobe gradually widened to the point, and there toothletted, without being at all truncate; in the latter, indeed, the lobes of the lip are sometimes confluent!

All these plants are in cultivation; but *E. Schomburgkii* is the only one that has been figured up to the present time. The beautiful specimen now represented was obtained from Pernambuco by Messrs. Loddiges, with whom it flowered in May, 1840. Salzmann, a German collector, originally met with it in sandy thickets near Bahia; and I find it among Martius' plants from the same province, marked "high ground of the Serra de Sincorá, and on rocks near Villa Rica, in the province of Mines, growing four to five feet high."

It ought to be grown in the Orchidaceous house or moist stove. The pots should be well drained; this is very simply done by inverting a small pot in a larger one, and also allows the heat to rise readily to the roots. It grows well in brown turfy peat, and requires a liberal supply of water during the growing season. Its general treatment may be much the same as that of E. elongatum.

Although unwilling to recognize a genus in these Amphiglottia, they certainly form a most natural group. The following is a list of such as I am acquainted with.

[*L.* in all cases signifies Lindley's Genera and Species of Orchidaceous Plants.]

Sect. 1. *Flowers in racemes.*

E. Skinneri, *Bateman in Bot. Reg. t.* 1881.
setiferum, *Lindley in Ann. nat. hist. v.* 4.
fruticosum, *L. no.* 22.
orchidiflorum, *L. no.* 29.
gladiatum, *L. no.* 47.
fuscatum, *L. no.* 38. (Of this E. virescens, *Loddiges*, and E. musciferum, *Lindl.* are varieties.)
pallidiflorum, *L. no.* 41.
biforatum, *Lindley.*
ovalifolium, *L. no.* 39.
elongatum, *L. no.* 54. (E. secundum, *Linnæus*, is no doubt the same plant.)
crassifolium, *L. no.* 55. } Are these distinct?
ellipticum, *L. no.* 56. }
lacerum, *Lindl. in Bot. Reg.* 1838. *misc.* 18.
imatophyllum, *L. no.* 52.
dichotomum, *L. no.* 58.
cochlidium, *Lindley in Ann. nat. hist. v.* 4.
cinnabarinum, *L. no.* 52.
Schomburgkii, *Lindl. in Bot. Reg. t.* 53. 1838.
radicans, *L. no.* 35. (E. rhizophorum, *Bateman.*)
Ibaguense, *L. no.* 53.
flexuosum, *L. no.* 34.

Sect. 2. *Flowers in panicles.*

E. paniculatum, *L. no.* 61.
laxum, *Pöppig & Endl.* 2. *p.* 2.
Hænkeanum, *L. no.* 40.
anisatum, *L. no.* 62.
Martianum, *Lindley in Ann. nat. hist. v.* 4.
thyrsiferum, *Lindley.*
porphyreum, *Lindley in Hooker's Journal*, 3. 86.
polyanthum, *L. no.* 50. (E. bisetum, *Lindl. in Bot. Reg.* 1841. *misc.* 148.)

30

ONCĪDĬŬM sphacelātŭm.

Scorched Oncidium.

GYNANDRIA *MONANDRIA.*

Nat. ord. ORCHIDACEÆ § VANDEÆ.
ONCIDIUM. Supra vol. 9. *fol.* 727.

Sect. Euoncidium; †† Heteranthium; * Folia plana; ‡‡ Pentapetala; ¶¶ Micropetala; ☞ ☞ Labellum panduratum. *Lindl. in Sertum Orchidaceum sub t.* 48.

O. *sphacelatum;* pseudobulbis elongato-ovatis ancipitibus a latere planiusculis, foliis longis ensiformibus apice recurvis, scapo stricto, racemo composito multifloro, sepalis petalisque lineari-lanceolatis undulatis labello æqualibus, labello medio constricto basi auriculato apice dilatato rotundato bilobo, cristâ depressâ pubescente a latere dentatâ apice trilobâ, columnæ alis elongatis truncatis erosis sphacelatis. *Lindl. l. c.*

All who are acquainted with the genus Oncidium are aware of the difficulty of distinguishing the larger of the yellow-flowered species, and of determining even any of them by books. In the number of the Sertum Orchidaceum just published I have attempted to diminish those difficulties by new classification of the species, and by a very careful correction of their characters. One hundred and one species are there admitted, and many repetitions are referred to their proper places as synonyms.

Among the new species, described in that place for the first time, is the plant now figured, one of the most beautiful yet known, which has been sent from Guatemala to his friends by Mr. Skinner, by Mr. Hartweg from the same country and Mexico to the Horticultural Society, and from Honduras to Messrs. Loddiges, with whom the species first flowered in February, 1841, when the accompanying figure was taken.

In habit it resembles O. reflexum, Pelicanum, and Baueri; but it is immediately known from them by the wings of the column, which are long, notched, and bordered with brown, as if scorched, as is shewn at fig. 1.

There are two varieties in cultivation, one much handsomer than the other, with larger flowers. This is the inferior sort.

32

ĔRĬĂ polyūră.

Many-tailed Eria.

GYNANDRIA *MONANDRIA.*

Nat. ord. ORCHIDACEÆ § MALAXEÆ.
ERIA. Botanical Register, vol. 11. *fol.* 904.

E. polyura (Lindl. in Bot. Reg. 1841. misc. 144.) (Tonsæ.); caule elongato folioso, foliis lanceolatis acutis patentibus striatis, spicis oppositifoliis multifloris nutantibus foliis æqualibus v. brevioribus, bracteis ovario duplò brevioribus adpressis, sepalis petalisque ovatis acutis erectis lævibus, labello cordato ovato acuto basi bicarinato.

Of the now very numerous species of Eria this is one of the more graceful, producing from the sides of long leafy fleshy stems a profusion of delicate tails of flowers, each nearly six inches long. The flowers resemble little semitransparent bells just tinged with pink, and having a rich purple and yellow centre.

The latter appearance arises from the labellum, which is ovate, heart-shaped, and a rich port-wine purple, with a bright yellow tip.

Mr. Cuming found it in Manilla, and sent it to Messrs. Loddiges, with whom it flowered in October 1841.

Eria floribunda, to which it nearly approaches, differs in having smaller flowers, erect racemes, and a narrow wedge-shaped lip, with a solitary tooth on each side.

It is a plant of the easiest cultivation, but requires a hot damp atmosphere, such as that of an Orchidaceous house where Dendrobiums and such plants flourish. The general treatment ought to be the same as is given generally to the extensive race to which it belongs.

36

CŒLĬĂ Bauerānă.

Mr. Bauer's Cœlia.

GYNANDRIA *MONANDRIA.*

Nat. ord. ORCHIDACEÆ, § EPIDENDREÆ.

CŒLIA. Lindl. Sepala distincta, conniventia, subæqualia: lateralibus basi obliquis. *Petala* subæqualia. *Labellum* cum columnâ subparallelum, apice patulum, basi excavatum, indivisum. *Columna* nana, erecta, semiteres, basi paulò producta; *stigmate* rotundato excavato prominente, rostello inflexo. *Anthera* ovata, bilocularis. *Pollinia* 8, quaternatìm materiê pulvereâ in 2 paribus cohærentia. *Ovarium* 9-alatum.—— Herbæ *epiphytæ pseudobulbosæ, basi squamatæ; foliis gramineis subplicatis,* scapo *radicali squamato,* bracteis *longissimis.*

C. *Bauerana* (L. 36. Bauer's Illustrations, Genera, t. 3. *Epidendrum tripterum,* Smith. *Cymbidium tripterum,* Swartz.); foliis gramineis racemo multò longioribus, bracteis floribus subæqualibus, labello unguiculato ovato obtuso margine calloso.

The West Indies and Mexico produce this little epiphyte, which, although white and inconspicuous, is, like our own Lily of the Valley, so sweet that it must take precedence of most of its race. No Hawthorn hedge is more fragrant than a bed of this Cœlia.

The genus was founded upon a drawing made in February 1810 by the late Mr. Francis Bauer, and published in his illustrations of Orchidaceous plants. In that place the pollen-masses were represented as being four in number, and concavo-convex, so that, lying in pairs side by side, each pair formed a hollow body narrower at back than in front. It was by this circumstance that the name of the genus was suggested (from κοιλος hollow). To my very great surprise, however, I find, upon examining the plant myself, which I have now for the first time been enabled to do by the communication of a fresh specimen from Mr. Rogers, that no such structure as that represented by Bauer exists. On the contrary the pollen-masses are eight in number, placed in fours in double pairs, and of the supposed hollowing out no trace is discoverable; but they are as usual plano-convex, and are bound together by a powdery strap. The whole character of the genus has in consequence to be changed, as well as its position in the arrangement, which is not very far from Eria.

In consequence of the erroneous representation given in Bauer's illustrations, the analyses in the accompanying plate are more than usually copious. Fig. 1. represents the column and labellum seen in profile; 2. is the upper side of the labellum; 3. is the back of the column; 4. is the anther seen from above; 5. is the eight pollen-masses seen in front; 6. is two pairs in their true position; 7. is a transverse section of the ovary.

I never saw this plant growing so well as when cultivated in a cool stove, and suspended from the rafters on a block of wood. Almost all who grow many of this tribe are now dividing those which require a hot temperature from others which grow best in a cooler place. Amongst the latter, therefore, this should be placed, and it will grow and flourish with little care. It grows fast and is easily multiplied.

Only one more species of this genus is known, and has been largely imported from Guatemala. It has red flowers, and is a far more conspicuous plant than this. The following is its character:—

C. *macrostachya* (Lindl. in Bentham pl. Hartweg.); pseudobulbis ovatis, foliis ensiformibus plicatis, racemo longissimo multifloro, bracteis lineari-lanceolatis acuminatissimis squarrosis, labello lanceolato basi bisaccato. ——*Guatemala.* —— Folia pedalia et ultra, racemo 6-8-poll. longo. Scapus basi squamis 3-4 ventricosis adpressis vaginatus.

38

ARUNDĪNĂ dēnsă.

Close-flowered Reedorchis.

GYNANDRIA *MONANDRIA.*

Nat. ord. ORCHIDACEÆ § EPIDENDREÆ.

ARUNDINA. Blume. Sepala lineari-lanceolata, æqualia, patula, basibus paululùm connatis. *Petala* conformia, latiora. *Labellum* circa columnam convolutum, trilobum v. integrum, medio cristatum aut lamellatum. *Columna* recta, cum labello parallela, semiteres, clavata. *Anthera* 4-locularis, truncata, rotundata. *Pollinia* 8, æqualia, materie granulosâ cohærentia. *Stigma* fornicatum, prominens, sublobatum.——Herbæ *caulescentes, arundinis facie; foliis distichis ensiformibus acuminatis plicatis.* Flores *terminales, membranacei, purpurei, speciosi.*

A. *densa;* labello obovato-subrotundo quadrilobo apiculato laciniis rotundatis, lamellis 3 subæqualibus crispatis rectis, petalis oblongis, foliis subæqualibus. *Supra misc. no.* 26.

Of the fine Oriental genus Arundina we now possess two species in gardens, this and *A. bambusifolia.* Two only, *A. chinensis* and *minor* remain to be introduced; the former is found on the islands near Macao, the latter is from Ceylon.

The species now figured is a native of Sincapore, whence it was sent to Messrs. Loddiges by Cuming; among whose dried plants however I do not find it. It differs from A. bambusifolia in its leaves being nearly equal, and not gradually diminishing into small sheaths as they approach the flowers: and in the flowers being very closely, but loosely, arranged; and in the small size of the middle lobe of the lip, which is almost blended with the large lateral lobes. There is a difference too in the crested veins of the lip, which are all three crisped in *A. densa,* while the middle one is straight and even in *A. bambusifolia.*

The peculiar habit of this genus renders it easy to recognize. In structure it is near Phaius, from which it differs in the want of a spur to the lip, and in its column not being at all extended at the base into a foot.

Fig. 1. represents the lip spread open; 2. is the column; 3. the two sets of pollen-masses.

Not only are the flowers of this very handsome, but they emit a most agreeable perfume.

It requires to be grown in the orchidaceous stove. It should be potted in brown turfy peat, and should have the pot well drained. Water ought to be liberally given, particularly when it is growing freely, and at all times it requires more than plants with large pseudo-bulbs, and will suffer much if it is withheld for any length of time.

42

EPIDĔNDRŬM ranĭfĕrūm.

Frog Epidendrum.

GYNANDRIA *MONANDRIA.*

Nat. ord. ORCHIDACEÆ, § EPIDENDREÆ.

EPIDENDRUM, § SPATHIUM. *Caulis* foliosus erectus. *Pedunculus* elongatus e spathâ erumpens. *Labellum* adnatum. *Lindl. in Hooker's Journal,* 3. 85.

E. *raniferum;* (L. no. 64.) foliis distichis lanceolatis, racemis terminalibus lateralibusque divaricatis multifloris, sepalis ligulatis convexis obtusis petalisque linearibus cuneatis mucronulatis patentissimis, labelli sexlobi basi 3-callosi laciniis integerrimis dentatisque: posticis rotundatis—intermediis linearibus—anticis linearibus obtusis retrorsum falcatis

Although a good deal like E. nutans, this species is really very distinct, and indeed far handsomer: in consequence of the rich purplish brown spots with which the sepals and petals are profusely decorated. It was originally described from bad dried specimens from Mexico, in Mr. Lambert's Herbarium, labelled Ophrys tigrina. Within a few years it has appeared in several collections of Mexican Orchidaceæ; the specimen now figured was supplied by Geo. Barker, Esq. of Birmingham.

Like so many of the order this species varies a good deal in the amount of toothing observable in its lip. In the original specimens the divisions were perfectly entire; in the accompanying drawing they are represented as if they were all toothed. In a cultivated plant, now before me, they are both the one and the other.

The tubercles at the base of the lip are really three in number, not two as was at first supposed by me; but the middle one is much smaller, and is hidden by the other two.

The species is not exclusively from Mexico. I have it also from Mr. Schomburgk (424), who found it in British Guayana, with a branched raceme, just as in the original specimens of Pavon.

The section SPATHIUM, to which this species is referred, is very near AMPHIGLOTTIUM, from which it is however distinguished by the branches not being excessively lengthened and covered with close pressed scales below the flowers. In their stead there is usually a long sheathing spathe, like that of Cattleya, out of which the inflorescence proceeds: occasionally there are several of such spathes, as in the present instance. The species with which I am acquainted, belonging to the section SPATHIUM are: 1. *E. raniferum;* 2. *nutans;* 3. *grandiflorum;* 4. *spathaceum;* 5. *adenoglossum;* 6. *ventricosum;* 7. *cornutum;* 8. *Trinitatis;* and 9. *armeniacum,* which is the Encyclia macrostachya of *Pöppig.* Of these the first, second, eighth and ninth are in cultivation.

This succeeds well in a pot amongst turfy peat, and should be liberally supplied with water during its period of growth. It will be found of easy management if cultivated in the cool orchidaceous stove. When the growing season is past, it does not require so much water, but should never at any time be kept too dry.

43

MORMŌDĔS lĭnĕātŭm.

Streaked Mormodes.

GYNANDRIA *MONANDRIA.*

Nat. ord. ORCHIDACEÆ, § VANDEÆ.

MORMODES. Supra, vol. 22. *fol.* 1861.

M. *lineatum* (Bateman in litt.); sepalis petalisque oblongo-linearibus abruptè acutis margine reflexis, labello lineari incurvo carnoso sparsim piloso versus basin utrinque dente nunc brevi nunc elongato aucto, columnæ dorso et margine pubescente. *Bot. Reg.* 1841, *misc. no.* 107.

This curious species is a native of Guatemala, whence it has been sent by both Mr. Skinner and Mr. Hartweg, who were attracted to its seat among the almost inaccessible branches of trees overhanging a deep stream, by its delicious fragrance. The flowers, when they first appeared, were dull olive green, and by no means handsome; they have since acquired a bright warm tint, and the markings upon them have increased in intensity till they have become quite ornamental.

The lip is variable in form; always indeed covered with long loose straggling hairs, but having its lateral lobes sometimes not more than half a line long, and occasionally as much as two lines. Fig. 1. represents the more usual appearance of this part.

One of the most striking features in the flowers of this genus consists in an appearance of distortion, which however is natural to them. In all it exists in the labellum and column, which look as if they had had their joints broken, and then unskilfully set again; in some it appears in the sepals as well; in Mormodes luxatum it is found in every part; the whole flower appearing as if all its limbs had been dislocated.

This is one of the species which ought not to be cultivated in a high temperature, but which succeed admirably in a house which is kept between a greenhouse and stove. In such a house it grows and flowers freely with very little care, if planted amongst turfy peat in a well-drained pot, or suspended from the rafters in a basket or forked block of wood. Great care must be taken of the young shoots, as they are very easily rotted with too much water, when they commence their growth. It is easily multiplied, as it forms numerous pseudo-bulbs on the sides and top of the old ones, as well as at their base.

The species of Mormodes now in our gardens are:

1. M. atropurpureum.
2. — aromaticum.
3. — pardinum (which seems to be the same as Cyclosia maculata of *Klotzsch*).
4. — buccinator.
5. — citrinum (which is Catasetum citrinum).
6. — lineatum.
7. — luxatum.

48

VĀNDĂ cristātă.

Crested Vanda.

GYNANDRIA *MONANDRIA.*

Nat. ord. ORCHIDACEÆ § VANDEÆ.

VANDA. Supra vol. 6. *fol.* 506.

V. *cristata;* foliis canaliculatis recurvis apice truncatis obliquè excisis tridentatis, racemo erecto trifloro foliis breviore, sepalis oblongis obtusis fornicatis, petalis angustioribus incurvis, labelli lobis lateralibus brevibus acutis intermedio vittato oblongo convexo apice saccato inæqualiter tricorni, cornu brevi conico.

V. cristata. *Lindl. in Wall. cat. no.* 7328. *Gen. & Sp. Orch. no.* 9. *Sertum Orchiduceum, f.* 3. *in fronte.*

Surculus *crassitie digiti minimi, fere subsimplex, inferiori parte emittens ad latera fibras teretes crassas carnosas cortici arborum more* Vandæ Roxburghii *adhærentibus; huicque cæterùm planta habitu foliis et surculo simillima.* Folia *rigida, disticha, patentia, basi alternatìm imbricantia, nitida, canaliculata subtùs carinata marginibus acutissimis glaberrimis, apice truncata et obliquè 3-dentata, 5-6 pollicaria ½-pollicem lata.* Racemi *axillares, solitarii, (plures 3-4 ex eâdem plantâ, axillis alternis) vix folia superantes pauciflori.* Pedunculi *carnosi 2-3-pollicares, teretes, indivisi; basi bracteis vaginantibus truncatis aliquot.* Flos *magnus, carnosus, viridiflorus, labello maximo purpureo.* Peranthii *foliola crassa, lanceolata, patentia, obtusiuscula, ½ pollicaria, sub distincta; lateralia exteriora, infra insertionem labelli basi columnæ brevissimè productæ adnata, omnia patentia, demùm antica erecta cætera depressa; interiora sublinearia.* Labellum *crassissimum, basi saccatum et productum in cornu acutiusculum basi columnæ incrassatæ adnatum.* Faux *utrinque lobulo ovato obtuso recto.* Lamina *oblonga, porrecta, apice subtus cornu conico solido brevi suprà cristâ s. margine transversali elongatâ, in processus* 3. *v.* 4 *cylindricas abeunte irregulares; suprà verrucoso-lineata.* Columna *brevissima, crassa, conica.* Anthera *terminalis, subtùs bilocularis, loculis distinctis remotis.* Massæ *duæ, globosæ, posticè bilobæ, umbilico apicis columnæ adfixæ ope membranulæ solubilis, filo valde elastico connexæ.* Ovarium *longum,* 6 *angulare, angulis carinatis prominentibus.*——Obs. *Flos exquisitæ pulchritudinis.* Labelli *consistentia crassissima; color atropurpureus, præcipuè intus, ut etiam holosericeus.* Wallich MSS.

An epiphyte, with the habit of Vanda Roxburghii, found in 1818 on trees in Nepal, by Dr. Wallich, flowering in the month of April. He describes it in a manuscript memorandum now before me as a flower of exquisite beauty. It however hardly deserves so much praise, because, notwithstanding its rich marking and velvety surface, its flowers are too small and dull coloured.

The specimen now figured flowered with Messrs. Rollissons in March last.

It succeeds well when grown at the warmest and dampest end of the Orchidaceous stove, and may be fastened on a block of wood, or placed in a basket and suspended from the rafters. In either case it ought to be well supplied with water. A quantity of bog moss put round the block or in the basket will be found very useful in keeping the plant in a vigorous and healthy state. Its general treatment should be the same as is given to such things as Aërides and Saccolabiums.

50

EPIDĔNDRŬM lancifŏlĭum.

Lance-leaved Epidendrum.

GYNANDRIA *MONANDRIA.*

Nat. ord. Orchidaceæ § Epidendreæ.
EPIDENDRUM. Supra 1838. *fol.* 53.

E. lancifolium. *Lindl. Gen. & Sp. Orch. p.* 98.

This plant, which has so much the habit of E. cochleatum, is readily known by its regularly ovate sharp-pointed lip, which is streaked with deep purple radiating lines upon a pale yellow ground.

It is one of that section of Epidendrum to which I have elsewhere proposed the name of Osmophytum, in allusion to the fragrance of many of the species, a quality in which this also slightly participates. It consists of those species which have an oval or fusiform pseudo-bulb, bearing a few leaves at the point, racemose flowers, and a lip united firmly to the column, usually shell-shaped, and rarely lobed. Below is an enumeration of such species of the section as I am at present acquainted with.

Any one who can grow the old Epidendrum cochleatum, or fragrans, can easily manage this with similar treatment. It should be kept in a cool stove along with other plants of like habits, many of which have been lately imported from the high land of Mexico and Guatemala. It grows well amongst brown turfy peat in a pot which is well drained, and may be watered and syringed freely during its period of growth, and at no time kept too dry.

Osmophytum. Caulis pseudobulbosus v. fusiformis apice foliosus. Flores racemosi. Labellum adnatum sæpiùs indivisum et cochleatum. *Lindl. in Hook. Journ.* 3. 81.

1. E. *cochleatum* (Linn. L. p. 97.); foliis in pseudobulbos ovales geminatim insidentibus oblongis acuminatis subundulatis, sepalis petalisque paulò angustioribus lineari-lanceolatis reflexis, labello cordato cochleato subrotundo, columnâ brevissimâ.—— *West Indies.*——There are two varieties of this common plant; one with much paler flowers than the other.

2. E. *bracteolatum* (Presl. L. p. 98.); pseudobulbis sulcatis, foliis oblongo-lanceolatis obtusis, scapo tereti simplici, sepalis petalisque lanceolatis patentibus, labello obcordato obtuso crenulato basi cristato.—— *Guayaquil.*

3. E. *lancifolium* (Pavon. L. p. 98. Bot. Reg 1842. t. 50.); pseudobulbis elongatis teretibus, foliis oblongo-lanceolatis acutiusculis, racemo stricto terminali, sepalis petalisque lanceolatis acuminatis reflexis, labello cordato oblato-ovato cochleato.——*Mexico.*

4. E. *radiatum* (Lindl. in Bot. Reg. misc. 1841. no. 123.); pseudobulbis ovatis compressis utrinque tricostatis triphyllis, foliis angustis elongatis acutis, racemo denso multifloro, ovariis trialatis, sepalis linearibus petalisque lanceolatis patentissimis, labello cochleato crenato crispo, clinandrio trilobo laciniis lateralibus erectis carnosis intermediâ truncatâ apice serratâ.——*Mexico.*——Allied to E. cochleatum; it has the same habit, but the pseudobulbs are strongly ribbed at the side, and the lip has the margin crisp and crenated. The flowers are pale pea-green, with the exception of the lip, which is striated with bright deeper purple rays.

5. E. *venosum* (L. p. 99.); pseudobulbis angustis fusiformibus, foliis ensiformibus scapi longitudine, racemo 4-5-floro, sepalis lineari-lanceolatis petalisque angustioribus patentissimis, labello semilibero tripartito: laciniis lateralibus ovatis acutis intermediâ subrotundâ apiculatâ multò majore, callo baseos lineisque tribus disci subramosis elevatis.——*Mexico.*—— Lip white, with raised violet veins.

6. E. *fragrans* (Swartz. L. p. 97.); foliis in pseudobulbum angustum fusiforme (sæpiùs) solitariis lanceolatis apice recurvis, scapo folio breviore, sepalis lineari-lanceolatis petalisque ovato-lanceolatis acuminatis patentibus, labello subrotundo cuspidato lineato integerrimo.——*Tropical America* everywhere.——This plant is extremely variable; and occurs in almost all collections from the east coast of the American Tropic. Possibly the Brasilian form, of which Gardner's 624 is an example, may be a distinct species.

7. E. *æmulum* (Lindl. in Bot. Reg. t. 1898.); rhizomate repente, pseudobulbis ovalibus compressis, foliis solitariis oblongo-lanceolatis coriaceis, sepalis petalisque lineari-lanceolatis æqualibus patentibus, labello subrotundo mucronato cochleato lineato integerrimo.——*Para.*——A dwarfer plant than E. fragrans, from which it is also distinguished by the form of its pseudobulbs, its more leathery leaves, and its petals the same width as the sepals.

8. E. *glumaceum* (Lindl. in Bot. Reg. 1839. misc no. 50. 1840. t. 6.); pseudobulbis ovatis apice angustatis diphyllis, foliis angustè oblongis patentibus, racemo terminali cylindraceo e squamis glumaceis acuminatissimis pedunculo longioribus erumpente, sepalis linearibus petalisque lineari-lanceolatis acuminatissimis, labello obovato acuminato convexo integerrimo basi unicalloso.——*Brazil.*——This has the habit and sweet odour of E. fragrans, from which it differs in the colour of its flowers, in the long tapering figure of the flower-buds, which is caused by the peculiar form of the sepals and petals, and in the shape of the labellum, which is convex not concave, obovate, and delicately striped with pink at its base.

9. E. *inversum* (Lindl. in Bot. Reg. 1839. misc. 135.); pseudobulbis elongatis compressis, foliis loratis canaliculatis obtusis, spicâ terminali sub-6-florâ bracteis brevibus ovatis acuminatis, ovario triquetro, sepalis petalisque patentibus lineari-lanceolatis convexis subæqualibus, labello adnato oblongo acuminato basi convexo ecalloso, columnâ obtusè 3-dentatâ, antheræ cardinis appendice oblongâ denticulatâ.——*Brazil.*——Nearly related to E. fragrans. The flowers are straw-coloured, with a few purple streaks on the column and at the base of the lip, and have a heavy not very pleasant smell, something like that of Ground-ivy (*Glechoma*).

10. E. *variegatum* (Hooker Bot. Mag. t. 3151. Bot. Reg. 1839. t. 11. E. *coriaceum*, Bot. Mag. t. 3595. E. *crassilabium*, Pöppig and Endlich. nov. gen. &c. 2. 1. t. 102.); pseudobulbis fusiformibus compressis subdiphyllis, foliis oblongo-lanceolatis obtusis, racemo multifloro, sepalis petalisque subrotundo-obovatis coriaceis obtusis, labello subrotundo brevissimo acuto intùs sub columnâ pubescente: callo baseos concavo carnoso, dente columnæ postico obtuso.——*Brazil, Peru.*——A variable plant, readily known by its very short lip, on the base of which is placed a large concave callus, which presses against the column and gives the lip the appearance of being divided horizontally into two layers. On very strong specimens there are occasionally three or four leaves on a pseudobulb.

11. E. *lividum* (Lindl. in Bot. Reg. 1838. misc. 91.); pseudobulbis angustè ovalibus compressis diphyllis, foliis lineari-lanceolatis subundulatis obtusiusculis, scapo terminali paucifloro foliis breviore, ovario triquetro, sepa-

lis erectis herbaceis linearibus mucronatis, petalis spathulatis obtusis æquilongis, labello libero lineari-oblongo obtuso crispo medio tomentoso serie triplici dentium elevatorum intermediâ majore, columnâ triquetrâ apice tricallosâ. ——*Columbia.*——Flowers small, dull dingy-purple, a little tessellated; their lip, pale dirty-yellow, with a few dull purple veins.

12. E. *calamarium* (Lindl. in Bot. Reg. 1838. misc. 163.) pseudobulbis teretibus calamiformibus diphyllis, foliis linearibus planis abruptè acutis, racemo simplici erecto, bracteis squamiformibus acuminatis, sepalis patentissimis linearibus acutis apice recurvis, petalis angustioribus subspathulatis, labello postico subrotundo acuto convexo.——*Brazil.*——Flowers pale, whole coloured, yellowish-green, with five small violet spots at the base of the lip. Not fragrant.

54

ONCĪDĬŪM urophy̆llūm.

Tail-leaved Oncidium.

GYNANDRIA *MONANDRIA.*

Nat. ord. ORCHIDACEÆ.

ONCIDIUM. Bot. Reg. vol. 10. *fol.* 804.

SECTION ** *Folia equitantia.*

O. *urophyllum* (Lodd. cat. ed. 2. no. 1381. Sertum Orchid. sub tab. 48.); foliis equitantibus margine plano apice acutissimis, scapo longo pendulo paniculato, sepalis linearibus acuminatis lateralibus ad apicem ferè connatis, petalis obovato-subrotundis apiculatis, labelli lobis lateralibus brevibus convexis obtusis intermedio unguiculato reniformi emarginato, cristæ tuberculis duobus uno ante alterum trilobis, columnæ alis maximis acinaciformibus rotundatis.

This is really a charming species of Oncidium, with quite a peculiar habit. Its leaves are shaped like a pen-knife curved backwards, so as to have the edge on the convex side. This is caused by the two sides of the leaf being brought into contact, and then growing together, the back of the knife-shaped leaf consisting of their edges. From this structure to the rush-leaved species, such as *O. Cebolleta*, is but a step.

Such species constitute a distinct section of the genus, consisting of the following seven; viz. *O. iridifolium, urophyllum, triquetrum, pulchellum, Lemonianum, variegatum,* and *tetrapetalum,* all of which are, or lately have been, in cultivation.

That now figured has a graceful drooping panicle of clear yellow flowers, four feet long, almost white at the back, and with a few crimson stains near the centre on the upper side. It was imported from Brazil by Messrs. Loddiges, and flowered with them in March, 1841.

Fig. 1. represents a flower, enlarged and seen in full face; 2. is the same in profile.

It requires a damp stove in cultivation. It may be suspended from the rafters in a basket, or on a block of wood, and planted in brown turfy peat; or it may be grown successfully in a pot if it is well drained. In either case the greatest care must be exercised in watering it, particularly when it commences its growth, for the young shoots at this stage are easily rotted.

55

AĔRĬDĔS crīspūm.

Sir Richard Brooke's Air plant.

GYNANDRIA *MONANDRIA.*

Nat. ord. ORCHIDACEÆ § VANDEÆ.

AERIDES. Bot. Reg. vol. 18. *fol.* 1485.

A. crispum; foliis planis apice obtusis obliquis bilobis racemis multifloris nutantibus duplò brevioribus, sepalis petalisque subæqualibus obtusis, labelli maximi laciniâ intermediâ multò majore ovatâ retusâ serrulatâ retusâ basi bidentatâ lateralibus erectis acutis nanis, calcare cornuto incurvo.

A. crispum. *Lindl in Wall. cat. no.* 7319. *Gen. & Sp. Orch. no.* 6.

A. Brookei. *Bateman in Bot. Reg.* 1841, *misc.* 116.

"This most beautiful species has lately flowered in the collection of Sir Richard Brooke, Bart., of Norton Priory, after whom I have ventured to name it. In the magnitude of its blossoms it infinitely surpasses any other species with which I am acquainted, the lip alone being upwards of an inch long, beautifully tipped with rose-colour. The other parts of the flower are of brilliant white. In the form of the labellum the species slightly resembles Aërides affine, but it is nevertheless so entirely distinct in other respects that it is unnecessary to contrast the species. In its habit it is stiff and erect, the leaves being only five inches long, while the racemes are more than double the length. Mr. Jones, the skilful gardener at Norton, is unfortunately unable to give me any account of the native country of the plant, but it doubtless comes from the 'gorgeous east.' I am not aware of its existing in any other collection."

Such was the account given by Mr. Bateman of this lovely plant when it was first discovered in this country. At that time he supposed it to be new, and dedicated it to Sir Richard Brooke. I regret however to state that a beautiful specimen, for which I am indebted to Mr. Bateman, proved to be nothing more than A. crispum, a native of Courtallum, a small valley near the southern point of the Peninsula of India.

It is certainly one of the finest of its race, and worthy of a place in all collections of merit. Very few discrepancies are found between it and the published character of A. crispum, and they are only such as are usual when plants are described from dried specimens.

Fig. 1. represents the column and labellum; 2. shews the pollen-masses.

In cultivation it requires a hot and damp stove. When suspended upon a block of wood, or in a copper wire basket, from the rafters, it will grow luxuriantly with very little care, provided it is well supplied with water. If there is a cistern in the house it ought to be hung above it, as it evidently likes the moisture which is continually evaporating from such a place. When not in a growing state less water should be given than at other times, but it should never be kept too dry.

62

LÆLĬĂ flāvă.

Yellow Lælia.

GYNANDRIA *MONANDRIA.*

Nat. ord. ORCHIDACEÆ § EPIDENDREÆ.

LÆLIA. Botanical Register, vol. 21. *fol.* 1751.

L. *flava;* pseudobulbis ovatis 1-2-phyllis foliis oblongis carnosis coriaceis planis subconvexis brevioribus, scapo erecto foliis longiore squamulis quibusdam distantibus vaginato, racemo cylindraceo, bracteis minimis acutis, sepalis petalisque oblongo-linearibus obtusis, labelli oblongi lobo medio sessili oblongo crispo recurvo lateralibus obtusis undulatis longiore venis 4 elevatis apice divergentibus. *Lindl. in Bot. Reg.* 1839. *misc.* 143.

L. caulescens. *Lindl. in Bot. Reg.* 1841. *sub t.* 1.

A pretty epiphyte, the accompanying drawing of which was made from a specimen in the possession of Sigismund Rucker, Esq. It originally appeared in the garden of Sir Charles Lemon, and an account of it was published in this work, at the place above referred to, from materials supplied by Mr. Booth, the gardener at Carclew. At that time it was said to be a native of Mexico, and it flowered in a very imperfect manner. I therefore trust I may be excused for not having recognized it when, in the year 1841, I found it among the dried Brazilian specimens belonging to the Herbarium of Dr. Von Martius, and republished it under the name of *L. caulescens;* upon the supposition, moreover, that the flowers were purple.

It is in reality near *L. cinnabarina*, a noble species, with orange-coloured flowers, also from Brazil. From that plant it is distinguished by the shortness of the pseudo-bulbs, and of the sepals and petals; and also by the bluntness of the middle lobe of the lip, whose veins are more elevated and diverge considerably at the point, instead of being nearly straight as in L cinnabarina.

The native situation of this species appears from Dr. Von Martius' herbarium to be the Serra de Piedade, in the province of Minas Geraes; I also possess a fine specimen from Prof. Gardner, collected in elevated rocky places in the Serra do Frio, in the Diamond District of Brazil.

Fig. 1. gives a view of the upper side of the lip.

This is one of those plants which ought to be grown in a cool stove, or at least at the cool end of the common Orchidaceous house. A temperature which suits Dendrobiums, and plants of that kind is very injurious to this species, and hence most persons who grow many of it have a cool-house and a warm one. It should be tied to a block of wood, or placed in a basket, and hung from the rafters. Like the rest of the family it is propagated by dividing the pseudobulbs.

As the genus Lælia is one of the finest in cultivation, and now consists of many species which have not lately been brought into one view, the following enumeration of them will probably be found useful.

LÆLIA. *Lindl.*

Sect. 1. GRANDIFLORÆ. Petals distinctly larger than the sepals; all but one are *natives of Mexico and Guatemala.*

1. L. *grandiflora* (L. no. 1. *Bletia grandiflora*, Llave Nov. Veg. Descr. 2. 17.); foliis binis aut ternis lanceolatis, bracteis membranaceis amplexicaulibus, sepalis lanceolatis, petalis ovatis majoribus carnosis venosis lateraliter emarginatis, labello purpureo-violaceo ad basin et medium albicante profundè 3-partito; laciniâ intermediâ maximâ emarginatâ subcrenatâ, limbo undulato purpurascente, prominentiis variis petaloideis lutescentibus sulcatâ, punctis coloratis diversimodè maculatâ.—— *Mexico.* ——An obscure plant, inhabiting the province of Mechoacan, on trunks of trees, and called by the inhabitants *Lirio, Flor de Corpus*, and *Itzumaqua.*

2. L. *majalis* (Lindl. Bot. Reg. 1839. misc. 42. Bateman Orch. Mex. t. 18. *Cattleya Grahami*, L. 116.); folio angusto scapo bifloro æquali, bracteis membranaceis ovatis, sepalis lanceolatis, petalis oblongo-lanceolatis duplò latioribus, labelli lobo medio rotundato emarginato plano laterali-bus parvis obtusis.——*Mexico*, where it is called *Flor de Mayo* by the natives of San Bartolo, and the adjoining parts. —— Mr. Hartweg found it on the mountains of Leon, growing upon oak trees, at the height of 8,000 feet, where it sometimes freezes. It is one of the most beautiful of the whole order; the flowers are of a bright violet colour, and measure six inches from the tip of one petal to that of another. The labellum is dotted and veined with rich purple.

3. L. *autumnalis* (L. 115. Bateman Orch. Mex. t. 9. Lindl. in Bot. Reg. 1839. t. 27. *Bletia autumnalis*, La Llave); pseudobulbis ovatis teretibus costatis apice attenuatis 2-3-phyllis, foliis oblongo-linearibus patentissimis scapo multò brevioribus, scapo tereti apice subsexfloro, bracteis oblongis membranaceis acutis, sepalis lanceolatis acuminatis patentissimis, petalis oblongo-lanceolatis undulatis, labelli trilobi bilamellati lobis lateralibus erectis rotundatis truncatis intermedio oblongo-lanceolato apice reflexo, ovario glabro.——*Mexico*, near Oaxaca.——Scape sometimes 18 inches high. Flowers very deep rose-colour. Lip white, and spreading open at the base, where it has two elevated ridges.

4. L. *anceps* (Lindl. in Bot. Reg. t. 1751); pseudobulbis ovatis distantibus tetraquetris, foliis binis aut solitariis lanceolatis, scapo ancipiti bifloro squamis carinatis vestito, ovario viscoso, labelli disco lineari elevato apice trilobo.——*Mexico.*——Flowers bright rose-colour. Lip rolled up, with a very rich crimson point. The variety, called Barkeriana, figured at fol. 1947 of the Botanical Register, has narrow petals, and a sharper pointed lip.

5. L. *superbiens* (Lindl. in Bot. Reg. 1840. misc. 87.); scapo longissimo multifloro, bracteis lanceolatis maculatis, sepalis petalisque lineari-oblongis obtusis membranaceis labello longioribus, labelli lobo medio obtusissimo: disco lamellis quinque maximis subserratis anticè truncatis aucto, antheræ cristâ biauritâ, clinandrio denticulato —— *Guatemala* and *Costa Rica.*——Of this species Mr. Skinner says, "this most magnificent of all plants flowers in November, and in some instances bears from eighteen to twenty flowers on stems from nine to twelve feet long!!" Mr. Hartweg found it in abundance in the state of Quesaltenango, near Chantla. It is planted by the Indians before their doors. Chantla, its native habitat, is very cold. The temperature of Costa Rica is 68° to 70°. Chantla is 55° to 65° generally.

6. L. *furfuracea* (Lindl. in Bot. Reg. 1839. t. 26.); pseudobulbis ovatis striatis submonophyllis, foliis angustè oblongis erectis acutis scapo multò brevioribus, scapo paucifloro tereti, bracteis oblongis membranaceis acutis, sepalis lanceolatis acuminatis patentissimis, petalis subrhombeis lanceolatis undulatis sublobatis, labelli trilobi bilamellati lobis lateralibus erectis rotundatis truncatis intermedio oblongo revoluto, ovario glandulis nigris furfuraceo.——*Mexico.*——In the colour and size of the flowers this is most like L. autumnalis: it is readily known by the black scurfiness that clothes the ovary.

7. L. *Perrinii* (*Cattleya Perrinii*, Lindl. Bot. Reg. 1838. t. 2.); foliis oblongis coriaceis cauli æqualibus, sepalis oblongis linearibus obtusis angustioribus, labello oblongo-lanceolato trilobo unguiculato nudo basi in cuniculum ventricosum producto; laciniis lateralibus erectis acutis intermediâ oblongâ undulatâ obtusâ elongatâ sinubus intermediis rectangulis.——*Brazil.*——A plant with club-shaped pseudobulbs, and a large green spathe, from within which the scape makes its appearance; so that it has quite the habit of a Cattleya. The flowers are pale lilac, with a deep purple tip to the labellum. It would be a Cattleya if it had not eight pollen-masses.

8. L. *peduncularis* (Lindl. in Bot. Reg. 1842. misc. 10.); pseudobulbis subrotundis compressis lineâ utrinque elevatâ, foliis oblongis obtusis scapo tereti vaginato brevioribus, corymbo 4-floro, bracteis pedunculo pluries brevioribus, sepalis linearibus obtusis pedunculo brevioribus, petalis oblongis obtusis planis, labelli lobis lateralibus rotundatis intermedio oblongo rotundato plano glabro multo brevioribus; lineis duabus elevatis.—— *Mexico.*——A species of some beauty, with the habit of L. rubescens, but with rich violet whole-coloured flowers. The peduncles are $1\frac{3}{4}$ inch long, and the flowers themselves are nearly of the same length when closed. The species is near L. rubescens and acuminata, but the flowers are much larger, the flower-stalks much longer in proportion, and the labellum quite round at the point and flat. The pseudobulbs are not at all tetragonous, but are compressed, with only a slight line in the room of the intermediate angle.

9. L. *rubescens* (Lindl. in Bot. Reg. 1840. misc. 25. 1841. t. 41.); pseudobulbis subrotundis compressis utrinque angulo elevato, foliis oblongis obtusis scapo tereti vaginato brevioribus, racemo multifloro, bracteis ovario plus duplò brevioribus, sepalis linearibus, petalis lanceolatis subundulatis, labello conformi auriculato medio pubescente: lineis duabus elevatis.——*Mexico.*——Flowers white, tinged with pink, the petals and lip wavy, the latter with a yellow disk, and a deep purple spot near the base.

10. L. *acuminata* (Lindl. in Bot. Reg. 1841. t. 24.); pseudobulbis ovatis compressis rugosis, foliis solitariis emarginatis scapo erecto brevioribus, floribus corymbosis, bracteis linearibus acuminatis ovario duplò brevioribus, sepalis linearibus petalisque lanceolatis undulatis acuminatis, labelli lobis lateralibus rotundatis intermedio lanceolato undulato acuminato. —— *Guatemala*, where it is called *Flor de Jesus.*——Resembles the L. rubescens; but the flowers are larger, the lip less yellow, and the purple spot at its base less intense.

11. L. *albida* (Lindl. in Bot. Reg. 1839. misc. 4. t. 54.) pseudobulbis ovalibus diphyllis, foliis linearibus acutis spicâ multiflorâ 4-plò brevioribus, sepalis oblongo-lanceolatis acutis vel subacuminatis petalisque latioribus acutis, omnibus apicibus reflexis fortiter mucronulatis; labelli altè trilobi tricostati lobis lateralibus erectis rotundatis intermedio duplo majore subrotundo obscure apiculato reflexo.——*Mexico.*——Flowers white, tinged with pink, very sweet-scented, like Primroses. Petals and lip reflexed. Mr. Barker has communicated a curious monster of this plant, with the dorsal sepal united to the back of the column by means of a thin plate resembling a small knife-blade. What is very singular every flower of the specimen was in the same state.

Sect. II. PARVIFLORÆ. Petals the same size as the sepals. *Brazilian.*

12. L. *flava* (Lindl. in Bot. Reg. 1839. misc 143. 1842. t. 62. *L. caulescens*, Lindl. in Bot. Reg. 1841. sub. t. 1.)——*Brazil.*

13. L. *cinnabarina* (Bateman in Sertum Orchidaceum, t. 28.); pseudobulbis cylindraceo-ampullaceis elongatis, foliis binis basi discretis oblongis subrecurvis et undulatis, scapo tenui ascendente foliis multò longiore 4-5-floro, sepalis petalisque oblongo-linearibus obtusis æqualibus, labelli oblongi convoluti recurvi lobis lateralibus acutis intermedio ovali crispato, lineis 3 elevatis in axin.——*Brazil.*——Flowers rich orange-colour.

14. L. *rupestris;* pseudobulbo elongato monophyllo, folio oblongo erecto canaliculato scapo racemoso multifloro breviore, sepalis petalisque oblongis acutis, labelli circularis lobo intermedio obtuso margine crispo disco plano lateralibus obtusis, lineis 2 elevatis rectis.——*Brazil.*——Found by Professor Gardner in elevated rocky tracts in the Diamond District. "Flowers violet." Very near L. flava.

GRAMMATOPHYLLUM multiflorum; *var.* tigrinum.

The Tiger-spotted Letter-leaf.

GYNANDRIA *MONANDRIA.*

Nat. ord. Orchidaceæ § Vandeæ.

GRAMMATOPHYLLUM. Supra t. 65. 1839.

G. *multiflorum.* Botanical Register, 1839. p. 65.
Var. tigrinum; floribus lutescentibus maculis tigrinis.

We have already given the history of this fine plant at plate 65 of the Botanical Register for the year 1839. At that time, however, the variety that had flowered was but dingy and insignificant in comparison with what is now represented, in which we have almost as rich a colouring as in Oncidium papilio. We are aware that the species varies a good deal in the markings of its flowers, and, if we may judge from the dried specimens brought home by Mr. Cuming, there are kinds yet handsomer than this.

Our plate gives a reduced view of the habit and general appearance of this stately plant: fig. 1. shews a flower of the natural size.

It is a stove plant, and succeeds easily in the hot-damp Orchidaceous house. It may either be grown in turfy peat, in a pot, or suspended from the rafters; if cultivated as an air plant, the roots should be surrounded with some substance, such as sphagnum, or peat, and liberally supplied with water. When well managed it flowers regularly every year, and continues in flower longer than any plant of its race with which I am acquainted. It is propagated in the usual manner.

MISCELLANEOUS MATTER

OF THE

BOTANICAL REGISTER,

1842.

2. EPIDENDRUM polyanthum.

Gen. & Sp. Orch. no. 50.

A specimen of this noble species has lately flowered in the garden of the Horticultural Society, from plants sent from Guatemala by Mr. Hartweg. The flowers are of a rich salmon colour, and about the size of those of E. Schomburgkii. The species in its natural state is most beautiful; a specimen before me from Mr. Hartweg, bears a many-flowered panicle, loaded with close-set nodding racemes. In consequence of having examined an imperfect specimen from Messrs. Loddiges, I lately committed the error of regarding this as a new species, which I named Ep. bisetum. The latter name has therefore to be cancelled.

3. ERIA profusa.

E. *profusa* (Lanatæ); pseudobulbis ovalibus apice diphyllis, foliis oblongis striatis racemis pubescentibus multifloris brevioribus, bracteis squamæformibus pedicellis brevioribus, sepalis tomentosis, petalis lineari-oblongis obtusis, labello trilobo laciniis lateralibus erectis lævibus intermediâ subrotundâ rugosâ apice tuberculo solitario oblongo cristatâ.

A species from Ceylon, where it was found by Mr. Nightingale, who sent it to his Grace the Duke of Northumberland, in whose garden at Syon it flowered in Nov. 1841. The flowers appear in numerous racemes, nearly a foot long, and are soft with brown down. The labellum is bright light green. I perceived no smell in the specimens I examined. It is a species of little beauty.

4. SACCOLABIUM ochraceum.

S. *ochraceum;* foliis distichis coriaceis apice obliquè bilobis, racemis subramosis, sepalis petalisque linearibus obtusis carnosis (luteis rubro-fasciatis), labelli laminâ subsagittatâ recurvâ papillosâ margine erosâ, calcare obtuso.

A native of Ceylon, whence it was sent by the late Sir Wilmot Horton to Messrs. Loddiges. It has much the appearance of Cleisostoma maculosum, but it is a true Saccolabium. The flowers are small, of an ochrey yellow colour, marked with transverse broken bands of dull red.

5. BARKERIA Lindleyana.

B. *Lindleyana* (Bateman mss.); foliis ovalibus acutis, bracteis linearibus pedunculo multò brevioribus, labello exactè oblongo apiculato bicarinato carinâ alterâ sub apice interjectâ, unguiculato plano basi columnæ adnato, columnâ clavatâ alatâ apice tridentatâ immaculatâ.

This beautiful plant has exactly the appearance of Epidendrum Skinneri, with which, at first sight, it would perhaps be confounded. It appears however to be a Barkeria as far as that genus has any character to separate it from Epidendrum; that is to say, it has the winged column, and the labellum with a thin membranous border down far below the head of the column. It is however to be observed, that in the original Barkeria the labellum was absolutely free and sessile; while in the present species it has a distinct unguis connate with the base of the column.

The flowers of this plant are of a rich deep purple; the sepals and petals are of equal size, linear-lanceolate, and disposed in the same plane, directed forwards in the same manner as in the original species; the petals are however rather broader than the sepals. The labellum is rather more than three quarters of an inch long, exactly oblong, and rounded at each end, terminating in front in a small mucro, and abruptly narrowed at the base into a short unguis united to the base of the column; it is of the same rich purple as the other parts, but somewhat deeper, except in the centre, where it is nearly white; two elevated keels pass upwards along the middle, and lose themselves near the point of the lip; and a third keel much shorter is placed between them towards the point; so that the lip is bicarinate in the middle and tricarinate below the apex. The column is slightly winged, pretty exactly wedge-shaped, and 3-toothed at the point. The anther is sunk within those three teeth, and is depressed in the middle, while it is elevated at each side; it is very fleshy, and contains four cells, in which are located as many pollen-masses, each having a powdery strap bent back upon it. The plant is a native of Costa Rica, whence it was sent to Mr. Bateman by Mr. Skinner. In a letter received from the former of these gentlemen it is stated that this most charming species "has been fully seven weeks in flower, and yet the blossoms first opened are as fresh as ever, a case quite without parallel among orchidaceous flowers of its colour."

6. DENDROBIUM discolor.

Bot. Reg. 1841. *t.* 38.

Upon re-examining the species of this genus I have found that the plant published lately under the above name, although from Java, is the same as D. undulatum found by Dr. Brown on the north coast of New Holland, and before him by Sir Joseph Banks in the same country. The reader will therefore be good enough to expunge the name of *D. discolor,* and to substitute that of *D. undulatum.*

7. ONCĪDĬŪM pergamĕnĕūm.

Lindl. in Benth. Plant. Hartweg. ined.

This is a pretty species from a situation in Guatemala called Medio Monte, where Mr. Hartweg found it in August, 1839. It has very thin pseudo-bulbs, sharp-edged, and somewhat furrowed on each side. The leaves are in pairs, from five to six inches long by one and a half to two inches wide, of a thin parchment-like texture, oblong, narrowing down to their base and sharp-pointed. The flowers are in a weak loose panicle, about two or three feet long, and have much the appearance of O. Baueri. The sepals and petals are narrow, sharp-pointed, wavy and brown tipped with yellow; the lip is all yellow, except a dull brown stain on each side of the base of the middle lobe. The lip is in form like that of O. Wentworthianum; that is to say, it is broader at the base than the apex, which is transversely oblong, with a small point in front. The wings of the column are small, short, and curved like a bird's bill, with a few obscure toothings on the upper edge.

8. ONCĪDĬŪM Suttōnī.

O. *Suttoni* (Bateman mss.); pseudobulbis ovatis compressis sulcatis diphyllis, foliis linearibus gramineis sæpe scapo æqualibus, scapo simplici v. ab ipsâ basi paniculato angusto ramulis 3-5-floris, sepalis petalisque lineari-lanceolatis planis, labello oblongo basi auriculato angustiore, tuberculo cristæ 7-9-dentato, columnæ alis triangularibus.

A very distinct species from the same locality as the last, where it was found by Mr. Skinner, to whose kindness I am indebted for excellent dried specimens. The grassy leaves, often as long as the erect panicle, although sometimes much shorter, give it a peculiar appearance, added to which the sepals and petals are a dull olive brown without any spots, except at the point, where they are yellow, while the lip has also but one dull olive brown spot occupying the centre of the base. The triangular form of the wings of the column will readily distinguish this from O. pergameneum, which has also the lip broadest, not narrowest, at the base, and entirely different leaves. Mr. Bateman has named it in compliment to Captain Sutton, R.N. who has brought so many valuable plants from America to England. In his note Mr. Bateman writes of the plant thus:

"The habit is more distinct than the flowers, though they have an aspect which cannot be mistaken. Pseudo-bulbs from an inch to an inch and a half high, ovate, compressed, deeply furrowed, bearing two rather long narrow linear leaves; flower-stems few flowered, and usually both in native and cultivated specimens shorter than the leaves, and drooping; occasionally the spike is longer, and a little branched, but this is very rare. The *buff* lip gives it a characteristic *look*. The species is very pretty."

10. LÆLĬĂ pedunculāris.

L. *peduncularis;* pseudobulbis subrotundis compressis lineâ utrinque elevatâ, foliis oblongis obtusis scapo tereti vaginato brevioribus, corymbo 4-floro, bracteis pedunculo pluries brevioribus, sepalis linearibus obtusis pedunculo brevioribus, petalis oblongis obtusis planis, labelli lobis lateralibus rotundatis intermedio oblongo rotundato plano glabro multo brevioribus; lineis duabus elevatis.

A Mexican species of great beauty, with the habit of L. rubescens (Bot. Reg. 1840, t. 41), but with rich violet whole coloured flowers. The peduncles are $1\frac{3}{4}$ inch long, and the flowers themselves are nearly of the same length when closed. The species is very near both L. rubescens and L. acuminata (Bot. Reg. 1841. t. 24), but the flowers are much larger, the flower-stalks much longer in proportion, and the labellum quite round at the point and flat. The pseudobulbs are not at all tetragonous, but are compressed, with only a slight line in the room of the intermediate angle. For specimens I am indebted to Geo. Barker, Esq. of Birmingham. Probably it is the same as a supposed variety of L. acuminata, mentioned at t. 24 for 1841.

11. DENDRŌBĬŪM jūncĕūm.

D. *junceum;* caulibus fusiformibus erectis apice ramosis, foliis teretibus acutis pungentibus, pedunculis terminalibus bifloris, sepalis obtusis, labelli trilobi lobo medio rotundato emarginato subundulato disco scabro.

Another addition to the set of Dendrobia with terete leaves, imported from Sincapore by Messrs. Loddiges, of whose catalogue it is 356*. It has rather large green flowers, with faint purple veins on the side lobes of the labellum. The stems are fusiform, erect, and branched above the swelling; many of the ramifications being terminated by a pair of flowers.

12. STĔLĬS crassifŏlĭă.

S. *crassifolia;* folio angusto carnoso semitereti spicâ breviore, bracteis membranaceis cucullatis truncatis, floribus cernuis, sepalis subrotundo-ovatis omninò liberis trinerviis, petalis oblatis membranaceis, labello subtrilobo carnoso concavo laciniâ intermediâ elongatâ acuminatâ.

A singular little plant, imported from the West Indies in August 1841, by Sir C. Lemon, Bart., in whose collection at Carclew in flowered in the December following. At first sight it resembles S. *ophioglossoides*, from which its half terete fleshy leaves distinguish it at once. The flowers are extremely small and just project beyond the rim of membranous truncate bracts; the sepals are roundish ovate, three-veined, and quite distinct from each other. The petals, which is unusual in this genus, are as thin as the sepals or thinner, and broader than long; the labellum is fleshy, ovate, acuminate, with two small rounded side lobes. The plant has no beauty.

13. MAXILLĀRĬĂ Skīnnĕrī.

M. *Skinneri;* pseudo-bulbis subrotundo-ovatis triphyllis foliis lanceolatis acutis plicatis scapo laxè vaginato ascendente duplo longioribus, bracteâ herbaceâ acutâ cucullatâ ovario multò longiore sepalis patentibus oblongo-lanceolatis acutis, petalis 2-plo brevioribus ovalibus erectis supra columnam convolutis apicibus reflexis; labelli trilobi lobis lateralibus erectis truncatis, intermedio longiore ovato rotundato deflexo, appendice carnosâ linguæformi inter lacinias laterales locatâ; columnâ subtus pubescente. *Bateman MSS.*

"This, the *facile princeps* of all known Maxillarias, has at length flowered in the collection of the Rev. John Clowes, with a vigour and beauty that could not be exceeded in its native haunts. The flowers actually measure upwards of six inches across, from the tips of the lateral sepals, while the latter are nearly an inch and half wide in the broadest part. The colours of this flower are peculiarly delicate, the sepals being pure white, faintly tinged with crimson at the base;—the petals of a more rosy hue, while the lip is almost covered with spots and streaks of the most brilliant carmine. The column again is pure white at the apex, and mottled with crimson spots at the base; while a number of woolly hairs are scattered on its under side.

"The habit of the plant is stately, and its growth free and vigorous, more nearly resembling *M. Deppii* than any other species. It will shortly be figured in the Orchidaceæ of Mexico and Guatemala.

"It is a native of Guatemala, and is another of the brilliant discoveries of the gentleman to whom I have ventured to dedicate it, and who, after an absence of four years in the most glorious countries of the New World, has lately returned once more in safety to the shores of his native land; in which, I must be allowed to remark, there is scarcely a collection of any note that is not more or less indebted to his enterprise and generosity."—*James Bateman.*

I gladly avail myself of Mr. Bateman's permission to publish the foregoing memoranda. The readers of the Botanical Register will remember that another Maxillaria Skinneri has already been described in it (1840, miscellaneous matter, no. 101); but that plant, nearly allied to M. aromatica, is not what Mr. Bateman had called after Mr. Skinner, but a species far inferior in beauty, though very handsome. A figure of it will appear in the Register of next month.

14. ONCĬDĬŪM bicallōsūm.

Lindl. in Benth. Plant. Hartweg. ined.

Among the numerous fine things found in Guatemala by Mr. Skinner, there is not many which are more deserving cultivation than this, which has so much the habit of O. Cavendishianum as to seem a mere variety of it, but which is in reality a quite distinct species. It was also met with by Mr. Hartweg. The flowers are fully two inches in diameter, which is four times the size of O. Cavendishianum; they appear in a dwarf erect raceme, not panicle; they are of a rich yellow, with the sepals and petals bordered with cinnamon colour; the labellum has two very small lateral lobes, and for its crest it has two distinct tubercles, the posterior double, the anterior 3-lobed, and the two separated by a considerable space. The flowers are moreover slightly scented, which is not the case with O. Cavendishianum. Many plants are doubtless in the country; but no one has succeeded in flowering them except Mr. Bateman, who sent a fine specimen to the last meeting of the Horticultural Society.

15. ONCĬDĬŪM ensātūm.

O. *ensatum;* pseudobulbis subrotundis compressis, foliis ensiformibus strictis carinatis, paniculâ longissimâ multiflorâ, bracteis membranaceis acutis canaliculatis, sepalis petalisque lanceolatis acuminatis subundulatis (olivaceis luteo-marginatis), labello (luteo) reniformi emarginato basi auriculato (disco olivaceo), cristâ pubescente subseptemtuberculato tuberculis elongatis, columnæ alis emarginatis utrinque acutis.

A Guatemala plant with singular long sword-shaped erect leaves, and a panicle like that of O. altissimum. The flowers are yellow in the lip, and on the border of the segments, but olive brown in the middle of the lip and the face of the segments. It is a very fine looking plant, the panicle being as much as eight feet long. The downy crest of the lip, with its long tubercles, forms a more obvious mark of distinction than anything else in the flower. Messrs. Loddiges have blossomed it.

17. PONĒRĂ striātă.

A very singular Orchidaceous plant has lately flowered with Mrs. Wray of Cheltenham, among a collection sent to that lady from Guatemala by Mr. Skinner. It has the roots of a Neottia, a slender stem three or four feet high, unbranched, and covered with narrow grassy leaves, from four to six inches long, the points of which are obliquely emarginate, and the sheaths rough, like those of Arpophyllum, with little green or purple tubercles. When old the stems become leafless, are closely covered with rugged sheaths, and produce here and there from their axils clusters of two or three sessile flowers, of a pale buff colour, and less than half an inch long. In form they are exactly like a Maxillaria or a Dendrobium; both sepals and petals being striped with bright reddish-brown. The column is short, taper, with a pair of small auricles on each side of the anther-bed, and a dorsal tooth curved over the anther, which is membranous, 4-celled, and contains four pollen-masses adhering in pairs by means of single straps of powdery matter. The labellum is wedge-shaped, slightly downy, curved downwards in the middle, and two-lobed at the apex. It evidently belongs to the rare and little known genus Ponera, (so called from πόνηρος, unhappy, in allusion to the starveling appearance of the species), and represents among Epidendreæ the tribes of Malaxeæ and Vandeæ, of which Dendrobium and Maxillaria are respectively the types. The present opportunity induces me to offer an improved character of the genus, and an enumeration of the species hitherto discovered.

PONERA. Sepala erecta, carnosa, lateralia majora, basi producta cum pede elongato columnæ connata; dorsale planum. Petala ovalia, basi angustata, libera. Labellum cuneatum, bilobum, ascendens, arcuatum, cum basi columnæ articulatum, omninò nudum. Columna brevis, teres. Anthera membranacea, depressa, 4-locularis. Pollinia 4, per paria adhærentia; caudiculis binis pulvereis incumbentibus. Stigma subrotundum, excavatum, subbilabiatum.——Herbæ *terrestres, gramineæ, simplices.* Folia *linearia disticha, vaginis verrucosis.* Flores *fasciculati, terminales* (?) *in caule folioso, v. axillares in caule defoliato.* Ovarium *breve, conicum, rectum, sessile.*

Sp. 1. P. *juncifolia* (Gen. & Sp. Orch. p. 113.); foliis subulatis canaliculatis secundis, spicâ subbiflorâ terminali in caule folioso, labello retuso, clinandrio mutico.——*Mexico.*——A single specimen of this rare plant exists in Mr. Lambert's herbarium. It has a stem about a span high, and extremely narrow leaves which, in the dried state at least, are all directed towards one side. The flowers are separated from each other by an interval of nearly half an inch.

Sp. 2. P. *graminifolia* (Bot. Reg. 1839. misc. 15. *Nemaconia graminifolia,* Floral Cabinet, p. 127.); foliis lineari-lanceolatis obliquè emarginatis, spicâ subbiflorâ terminali in caule folioso, labello acuto apice recurvo crenulato, clinandrio mutico.——*Mexico.*——A grassy plant of no beauty, resembling P. striata in the form of its leaves, but different in the structure of the flowers.

Sp. 3. P. *striata;* foliis lineari-lanceolatis obliquè emarginatis, spicâ subbiflorâ axillari in caule defoliato, labello bilobo integro, clinandrii dente dorsali in antheram incurvo.——*Guatemala.*——This differs from the preceding in the lip being two-lobed, not acute and recurved, in its margins being entire and not crenulated, in the anther-bed having a dorsal tooth which curves over the anther, and two much smaller blunt teeth on each side of the stigma; it has also the property of flowering on the old leafless stems instead of the young leafy ones. The flowers are pale buff, striped with brownish orange.

18. ASPĂSĬĂ epidendroīdĕs.

Gen. & Sp. Orch. p. 139. *Bot. Reg.* 1839. *sub t.* 1907.

A great many plants of this species are now in this country, having been sent from Guatemala by Mr. Skinner to his friends, and by Mr. Hartweg to the Horticultural Society; but they do not at present seem disposed to flower with the beauty that is natural to them. I have not seen more than three or four blossoms open at a time, nor do I find that a much larger number have hitherto made their appearance; nevertheless, in the wild specimens from Mr. Skinner the spike is nine inches long, clear of the leaves, and has borne about a dozen flowers. The following accurate account of the species has been sent by Mr. Booth, and renders further remark unnecessary.

"This plant was forwarded to Sir Charles Lemon, Bart., in June, 1840, by George Ure Skinner, Esq., who collected it in Guatemala, near the coast, in a damp climate, varying in temperature from 75° to 80°, where it flowers in January and February. It was also collected by Mr. Hartweg for the Horticultural Society, by whom plants of it and others were distributed to various collections, including Carclew, where it flowered in 1842, about the same time that it usually does in its native country. *Pseudo-bulbs* roundish ovate, deep green, three inches long and rather more than two inches broad, roundish on one side and almost flat on the other; two-edged and two-leaved, with three smaller leaves at the base of each bulb. *Leaves* from six to eight inches long, and about two inches broad; oblong-lanceolate acute. *Scape* nearly as long as the leaves, deep green, two-edged, issuing from the base of the pseudo-bulb, and bearing from three to six flowers at its extremity. *Pedicels* curved at the base so as to become erect by the time the flowers are fit to open. *Flowers* not very showy, but remarkably fragrant. *Sepals* an inch long, the two lower ones reflexed so as to cross one another at the points; the upper one concave and pointed, spreading so as to form nearly a right angle with the column; outwardly they are each of a pale green, inwardly of a dull brown with two or three greenish stripes across, and a little red towards the base. *Petals* erect, similar in size and form to the upper sepal, but slightly notched at the point, and of a pale reddish brown. *Labellum* large, somewhat cordate, bent in the middle and a good deal undulated at the margin, of a white ground with a large irregular blotch of pale pink down the centre, and two white elevated lines extending nearly half its length. At the throat are two small, yellow, fleshy projections. *Column* erect, two-edged, round at the back; the upper part of a pale pink, the lower white; the inner surface appears grooved, with two pink-coloured stripes on either side resembling wings. *Anther* case same colour as the column, nearly round, with a sharp edge on the top."

19. SPIRĂNTHĒS cerīnă.

S. (Sarcoglottis) *cerina;* fusco-olivacea, hirsuta, foliis . . . . hysteranthiis squamis caulis distantibus lanceolatis circa scapum subspiralibus, spicâ laxâ 7-8-florâ, bracteis lanceolatis acuminatis incurvis ovario hirsuto longioribus, sepalis glaberrimis supremo retuso lateralibus obtusis apice concavis, labello basi imâ glabro sagittato supra basin intùs villoso apice obovato obtuso concavo nitido subcrenato.

A singular terrestrial Orchidaceous plant, sent from Guatemala to the Horticultural Society by Mr. Hartweg. It belongs to that section of the genus Spiranthes, of which the old Neottia speciosa is the type, and which has been called by Presl Sarcoglottis, among all which it is at once known by its flowering without leaves, and its dull olive-brown aspect. The stem is altogether about $1\frac{1}{2}$ foot high, covered with white hairs and clothed by rather distant scales which are long, narrow, lanceolate, taper-pointed, and turned round the stem somewhat in a spiral direction. Although the other parts are hairy, the sepals and petals, which are more than an inch long, are perfectly smooth, and shining as if varnished, both inside and outside. Externally they are of the same dull olive-brown as the stem; but inside they are of a clear bright greenish-yellow, the tip being beautifully streaked with bright green veins. This organ is hairy inside below the middle, and at the base is extended downwards into two narrow barbs, which are the calli of Spiranthes proper extending beyond the edge. The lip adheres to the column with so much force that it is difficult to detach it. The column is taper, at the very base smooth, higher up papillose, then hollowed into a narrow smooth cavity, next furnished with a prominent smooth vesicle looking like a tubercle, and finally terminating in an oblong arched stigma, the lower part of which is viscid and the upper dry and membranous. There is nothing horny in this part, or in any of the appendages of the pollen-masses.

22. LEOCHĪLŬS oncidioīdĕs.

In the miscellaneous matter of 1840 (no. 218) a plant was described under the name of Rodriguezia maculata, which is now become common in collections of Orchidaceous plants, a considerable quantity having been distributed by Mr. Skinner and the Horticultural Society. It had also been called Oncidium macrantherum by Sir Wm. Hooker in the Botanical Magazine, t. 3845, to which allusion was made in this work at no. 33 of the miscellaneous matter for 1841. A further examination of the plant has led to the conviction that its name still requires correction. In many respects it is extremely like a Rodriguezia, and its habit is not at variance with that genus; but there are some circumstances that distinguish it. In the first place, the lip is not parallel with the column, and secondly, the two arms of the column proceed from below the stigma, and not from the upper edge of that organ. In addition to this, the anther is disproportionally large as compared with the column, and there is a honey pore near the base of the lip. All these peculiarities also exist in Leochilus oncidioides, a little known Mexican plant described by Messrs. Knowles and Westcott, for specimens of which I am indebted to the kindness of Mr. Barker. They also occur in the Oncidium carinatum of the same authors, my knowledge of which is again owing to the liberality of Mr. Barker. I also find every thing, except the large anther, in a Dominica plant, formerly described by me in the Annals of Natural History, under the name of Rodriguezia cochlearis; and my specimens will not shew what proportion the anther bears to the column in that species. These circumstances induce me to adopt the genus Leochilus, although the name (smooth-lip) is not applicable, none of the species having it smooth, and one at least being remarkably tuberculated. From Oncidium this genus is well distinguished by the position of the arms of the short column, and the honey-pore. The following characters will enable Botanists to arrange the genus and its species in their systematical works.

LEOCHILUS, *Knowles & Westcott.* Sepala et petala patula, basi æqualia, lateralia nunc connata. Labellum indivisum, basi tuberculatum aut laminatum et foveâ melliferâ auctum. Columna brevis, teres, libera, brachio porrecto sub stigmate utrinque aucta. Rostellum ovatum. Anthera columnâ latior ob rostellum prominens, unilocularis, membranacea. Pollinia 2, caudiculâ elongatâ angustâ, glandulâ minutâ ovali.——Herbæ *epiphytæ pseudo-bulbosæ*, Americam tropicam *habitantes.* Flores *parvi racemosi.*

* Sepala lateralia connata.

Sp. 1. L. *oncidioides* (Knowles & Westc. *Rodriguezia maculata,* Lindl. in Bot. Reg. 1840. misc. 218. *Oncidium macrantherum,* Hooker in Bot. Mag. t. 3845.); racemis nutantibus foliis subæqualibus basi laxè et distanter vaginatis, sepalis lateralibus basi tantùm connatis, labello oblongo convexo retuso v. apice angustato rotundato secus axin depresso infra medium cristâ pubescente utrinque elongatâ aucto basi mellifero.——*Mexico and Guatemala.*——Flowers spotted with pale purple, on a dirty semi-transparent ground.

Sp. 2. L. *cochlearis* (Rodriguezia cochlearis, *Lindl. in Ann. nat. hist.* 4. 116.); foliis oblongis sessilibus horizontalibus racemo erecto radicali subcorymboso paucifloro duplò brevioribus, sepalo supremo cum petalis obtuso lateralibus acutis semiconnatis breviore, labello obovato cochleari obtuso basi cristâ hippocrepicâ utrinque elongatâ bicarinato.——*Dominica.*——Flowers apparently yellow.

** Sepala omninò libera.

Sp. 3. L. *carinatus* (Oncidium carinatum, *Knowles & Westcott, Floral Cab.* 2. p. 30.); pseudo-bulbis ovatis, foliis lanceolatis, scapo erecto, floribus secundis racemosis, sepalis petalisque linearibus disjunctis, labello lineari-oblongo emarginato basi mellifero cristato: lamellis 2 brevibus truncatis utrinque glandulis 2 filiformibus auctis.—— *Xalapa in Mexico.*——Flowers orange, the petals spotted, the sepals whole coloured; the lip yellow with a few orange spots.

23. SCELOCHILUS Ottonis.

Klotzsch in Allgem. Gartenz. 1841. *Aug.* 14.

Under this name has been published the following account of a new plant of the Orchidaceous order, which has flowered in the Berlin Garden. It appears to be very near Pleurothallis.

"SCELOCHILUS, *Klotzsch.* Perigonii foliola exteriora conniventia, carinata, basi leviter coalita, lateralia labello supposita, usque ad apicem connata, ad basin in calcar obtusum producta; interiora exteriori postico æqualia. Labellum integrum, inferne attenuatum, basi bifidum, infra basin columna obsolete continuum, ad marginem infra medium bicorne, disco calloso, puberulo, antice bidentato, dentibus obtusis conniventibus. Columna semiteres, apice subemarginata. Anthera unilocularis. Pollinia 2, solida, sphærica.——Herba caracasana, epiphyta; rhizomate cæspitoso; pseudobulbis subnullis; foliis solitariis, coriaceis, carinatis, basi vaginis squamæformibus, conduplicatis involucratis; racemo radicali; floribus compressis flavidis.

"S. *Ottonis,* n. sp. Foliis oblongis, coriaceis, læte-viridibus, margine acutis, subundulatis, apice conduplicato-acutissimis, recurvis; racemo radicali, subramoso, folio parum longiore; foliolis perigonii interioribus oblongis, obtusis, intus purpureo-striatis, sparsimve pubescentibus."

"This small Epiphyte has, with the exception of the flowers, entirely the structure of Oncidium carthaginense, but the leaves are only five inches long and one inch and a half broad. The flower spike is a little branched, slightly longer than the leaf, round, thread-like, smooth, and covered with sessile, dry, membranaceous, lanceolate and acuminate bracts. The flowers are short-stalked, yellow, compressed, 7 lin. long. The column is without colour, twice as short as the floral envelopes. It was introduced in the year 1840 into the Botanic Garden of Berlin by Mr. Edward Otto. He discovered it on his journey upon the Silla of Caracas, where he found it 6600 feet above the level of the sea, in thick woods, on the trunks of trees. It does not require a very hot temperature, as the thermometer, in the elevation where it grows, seldom rises above $72\frac{1}{2}$° Fahr.; it seems, likewise, to prefer the shade to the sun. It developed its small yellow flowers for the first time in the month of July."

24. CATASETUM abruptum.

Hooker in Bot. Mag. t. 3929.

This is certainly identical with Catasetum luridum.

25. VANDA cristātă.

Lindl. Gen. & Sp. Orch. p. 216.

This has flowered with Messrs. Rollissons of Tooting; not however very perfectly—for the flowers are greenish instead of being nearly white, and they are solitary instead of being on 3-flowered peduncles. It is a pretty species, from the warm vallies of Nepal. The sepals and petals were greenish, the lip yellow, striped with rich purple asperities and furrows; at the point it is divided into two acuminate lobes.

26. ARUNDĪNĂ dēnsă.

A. *densa;* labello obovato-subrotundo quadrilobo apiculato laciniis rotundatis, lamellis 3 subæqualibus crispatis rectis, petalis oblongis, foliis subæqualibus.

A beautiful Orchidaceous plant from Sincapore, sent to Messrs. Loddiges by Cuming. It has flowers as large as those of A. bambusifolia, rosy violet, with a crimson-bordered lip, sweet-scented and very handsome, forming a close head, and not a long loose raceme.

27. ERIA mucronata.

E. *mucronata* (Tonsæ); caulibus elongatis flexuosis, foliis distichis lineari-lanceolatis acuminatis, floribus solitariis glabris oppositifoliis, bracteis 3-5 ovato-lanceolatis patentibus coloratis, sepalis petalisque ovatis erectis, labelli subrotundi 3-lobi laciniâ intermediâ emarginatâ ungue scopulifero, lamellis 3 membranaceis integerrimis quarum intermedia in mucrone libero abit lateralibus abbreviatis.

An Orchidaceous plant, with the habit of Dendrobium chrysanthum, found at Sincapore by Cuming, has just flowered with Messrs. Loddiges. Its leaves are long, fleshy, finely acuminate, deep green. The flowers are white, with a faint tinge of pink, deliciously sweet like violets, and produced singly opposite the leaves, on stalks about half an inch long, covered with dull red spreading bracts. The lip is nearly round, 3-lobed, with the middle segment emarginate; in its middle it has three membranous plates, the lateral of which are very short, the intermediate extended into a short horn, which runs up from the surface of the lip a little below its apex. On the foot of the column, just beyond the joint of

the lip, is placed a small brush consisting of numerous fine yellow glandular hairs. In addition to the usual number of pollen-masses are a few extra of a very small size.

EPIDENDRUM.

The genus Epidendrum is one of the most extensive of the Orchidaceous order, and already exists in our gardens to the amount of at least two-thirds of the described species. They are however so much increased since the year 1831, when I last reduced them to order, that the genus requires to be entirely remodelled. My views on that subject have already

been printed in Hooker's Journal of Botany, volume 3, page 81, to which the reader is referred. Among other sections in that place proposed, is one characterized by the presence of pseudobulbs and the more or less complete separation of the lip from the column, to which, as it includes Sir William Hooker's Encyclia, the name of *Encyclium* was assigned. As the species of that division are common in our gardens, and may possibly be regarded by some as a groupe of generic value, and as they are a good deal confused, and demand a more precise sub-division than that temporarily assigned them in the work above quoted, the present opportunity is taken of laying a revision of the whole of the Encyclia before the reader.

La Llave and Lexarza divided these plants into two sections, distinguished by the racemose and panicled inflorescence, a division I formerly adopted; but I find that although the distinction would probably be valid, if we could always have the species before us in a perfect state, yet that it leads to confusion in consequence of the panicled species often being merely racemose when weak, and ill grown. It is therefore now proposed to found the divisions upon the form of the lip, as a more certain and uniform mark of distinction. This again affords three principal, and three subordinate characters. It is—1. perfectly undivided, or it is, 2. slightly 3-lobed, or it is, 3. split into three deep divisions; these are the differences regarded of the first degree of importance. The third of these sections may be itself subdivided by the form of the middle lobe of the lip, which is—1. acute, 2. rounded, or 3. two-lobed. By this method the following will be the arrangement of the Encyclium division of Epidendrum, as far as I am at present acquainted with the species.

ENCYCLIUM.

Lindl. in Hooker's Journal of Botany, 3. 81.
(*Encyclia*, Hooker in Bot. Mag. t. 2831. Lindl. Gen. et Sp. Orch. p. 111.)

§ I. *Labello integerrimo.*

1. E. *vitellinum* (L. p. 97.* Bot. Reg. 1840, t. 35. Sertum Orch. t. 45.); pseudobulbis ovatis acuminatis diphyllis, foliis oblongo-ligulatis acutis

* In all future pages of the Register, the reader will be so good as understand that L. succeeded by a page only, means Lindley's Genera and Species of Orchidaceous plants.

basi vaginantibus racemo erecto multifloro brevioribus, sepalis petalisque ovato-lanceolatis acutis subæqualibus patulis, labello lineari apice angustato abruptè acuto basi calloso et bifoveato semilibero.——*Mexico*, in the region of Lichens and Jungermanniæ.——A most superb plant with flowers of a brilliant orange.

2. E. *ligulatum* (Llave. L. p. 98.); pseudobulbis ovatis depressis, foliis lanceolatis subcoriaceis striatis, scapo filiformi, sepalis petalisque ligulatis reflexis, labello obcordato limbo subcrenato, capsulâ glabrâ trigonâ.——*Mexico.*——Flowers purplish brown with a yellowish lip.

3. E. *lividum* (Lindl. in Bot. Reg. 1838. misc. 91.); pseudobulbis angustè ovalibus compressis diphyllis, foliis lineari-lanceolatis subundulatis obtusiusculis, scapo paucifloro foliis breviore, ovario triquetro, sepalis erectis herbaceis linearibus mucronatis, petalis spathulatis obtusis æquilongis, labello libero lineari-oblongo obtuso crispo medio tomentoso serie triplici dentium elevatorum intermediâ majore, columnâ triquetrâ apice tricallosâ.——*Colombia.*——Flowers small, dull dingy purple, a little tessellated; their lip pale dirty yellow, with a few dull purple veins.

*3. E. *articulatum* (Klotzsch in Allg. Gartenz. Sept. 22, 1838.); caule repente radicante, pseudobulbis elongato-oblongis gracilibus compressis 2-3 phyllis, foliis linearibus acutis subcarinatis tenuibus, floribus subarticulatim-racemosis, sepalis linearibus acutis erecto-patentibus, petalis spatulatis obtusis extus sordide pallide-ochraceis intus purpureo-variegatis, labello indiviso libero linguiformi basi angustato albido longitudinaliter triseriatim cristato-gibboso ante columnam villoso-barbato margine erecto subinflexo undulatim crispo transversaliter rubro-venoso, columnâ obtuse-trigonâ conicâ apice glandulis tribus flavis instructâ labello duplo breviore, germine trialato. ——*La Guayra.*——I am unacquainted with this plant, which, if its lip were not stated to be undivided, I should have supposed to be E. tessellatum.

4. E. *tripunctatum* (Lindl. in Bot. Reg. 1841. misc. 113.); pseudobulbis ovalibus compressis diphyllis, foliis ligulatis tenuibus apice recurvis scapo bifloro pluriès longioribus, sepalis patentissimis lineari-oblongis, petalis duplò angustioribus, labello semiadnato obovato rugoso utrinque ultra columnam emarginato, columnâ atropurpureâ apice trilobâ et luteo tripunctatâ.——*Mexico.*——This has long grassy leaves, and a very short scape. The sepals are twice as broad, and much larger than the petals are, and a dull yellow green. The column is deep purple, with three lobes, each of which is tipped with one pale orange-coloured spot. The lip is rather more than half an inch long, not warted, and distinctly emarginate near the apex of the column; its colour is pale citron.

5. E. *asperum* (Lindl. in Hooker's Journal, p. 6. Bot. Reg. 1838. misc. 36.); pseudobulbis ovatis basi subangulatis diphyllis, foliis coriaceis ligulatis carinatis acutis, scapo ramoso pedunculisque subasperis, sepalis patentibus obovato-lanceolatis obtusis, petalis magis cuneatis, labello semilibero cucullato flabellatim cuneato crispo basi callo elevato pubescente obovato medio depresso apice tridentato.——*Mexico and Columbia.*——Flowers dull dirty yellow, with crimson veins pencilled on the lip.

§ II. *Labello trilobo;* nempè *laciniis lateralibus nanis aut brevibus rotundatis nec intermediæ æqualibus.*

6. E. *Candollei* (Lindl. in Bot. Reg. 1839. misc. 76. *Ep. cepiforme*, Hooker in Bot. Mag. t. 3765.); pseudobulbis sphæricis, scapo paniculato,

sepalis petalisque obovato-oblongis, labelli liberi trilobi cucullati lobo medio crispo acuminato, disco elevato calloso sulcato pubescente ——*Mexico.*——The flowers are dull brown, with a dull yellow lip, striped with the same colour.

7. E. *primulinum* (Bateman mss.); pseudobulbis . . . . . , foliis . . . . . , scapo paniculato, sepalis petalisque patulis oblongis acutis, labelli laciniis lateralibus nanis erectis acutis intermediâ obovatâ apiculatâ; callo duplici ad basin elevato plano carnoso.——*Mexico.*——Flowers rather large, in a close erect panicle, smelling of primroses.

8. E. *Linkianum* (Klotzsch in Allgem. gartenzeit. Sept. 22, 1829. *E. pastoris*, Link et Otto abbild. t. 12.); pseudobulbis fusiformibus 2-3-phyllis, foliis ensiformibus recurvis racemo paucifloro longioribus, sepalis patentissimis lineari-lanceolatis, petalis conformibus angustioribus, labelli lobis lateralibus minutis erectis intermedio ovato-oblongo crispo venis elevatis sub columnâ pubescente.——*Mexico.*——Flowers small dull yellow, streaked with purple. Lip nearly white.

9. E. *hastatum* (Lindl. in Hooker's Journal, 3. 82); pseudobulbis . . . , foliis . . . , racemo striato 6-7-floro, sepalis petalisque discoloribus lanceolatis acutissimis patentibus, labello subrotundo emarginato subangulato basi utrinque supra unguem lobulo aucto venis baseos elevatis. ——*Mexico*, in the temperate region.——A species with deep purple striated sepals and petals, and a broad ivory white lip. The latter has generally a short lateral lobe on each side of its base so as to obtain a hastate form, but occasionally the lobes are almost wanting.

10. E. *Boothianum* (Lindl. in Bot. Reg. 1838 misc. 7) ; pseudobulbis compressis imbricatis ovatis subdiphyllis, foliis ligulato-oblongis undulatis apice subobliquis, racemo laxo sub-7-floro basi spathaceo foliis longiore, sepalis petalisque subæqualibus patulis ovali-linearibus acutis fuscovariegatis, labello semilibero rhomboideo acuto utrinque deflexo.—— *Cuba*.——Flowers small, greenish yellow, nodding, marked with rich brown spots. Lip nearly white.

11. E. *pterocarpum* (Lindl. in Hooker's Journal, 3. 82. Bot. Reg. 1841. misc. 128) ; pseudobulbis ovalibus compressis diphyllis, foliis . . . racemo angusto, sepalis petalisque subæqualibus linearibus acuminatis patentibus, labelli subrotundi trilobi cordati laciniis lateralibus rotundatis intermediâ multò longiore acutâ basi callo pubescente obscurè tridentatâ auctâ, capsulâ ovatâ trialatâ.——*Mexico*.——This has a long thin raceme of pinkish yellow flowers, and a heart-shaped lip streaked with crimson. It is remarkable for the seed-vessel, when ripe, being extended into three broad sharp wings.

12. E. *tessellatum* (Bateman in Bot. Reg. 1838. misc. 9. Hooker Bot. Mag. t. 3638.) ; pseudobulbis ovalibus compressis 2-raro 3-phyllis, foliis lineari-lanceolatis scapo simplici flexuoso multifloro paulo brevioribus, floribus subpatentibus, sepalis lanceolatis acutis mucronatis petalisque minoribus subspathulatis, labelli liberi trilobi laciniis lateralibus brevibus rotundatis intermediâ majore oblongâ cucullatâ callis tribus cristigeris longitudinalibus munitâ sub columnâ pubescente.—— *Guatemala and Mexico*.——Flowers dingy olive brown : with a pale lip tessellated with crimson.

13. E. *viridiflorum* (*Encyclia viridiflora*, Hooker in Bot. Mag. 15. t. 2831. L. p. 111.) ; pseudobulbis ovatis diphyllis, foliis ensiformibus recurvis acutis paniculâ brevioribus, sepalis lateralibus falcatis petalisque linearibus acutis erectis, labello postico apice 3-lobo laciniis lateralibus planis

31

intermediæ ovatæ crispæ æqualibus, callo basi duplici oblongo carnoso. ——*Brazil*. ——Flowers dull green, marked with dull purple.

14. E. *glaucum*. (Lindl. in Bot. Reg. 1840. misc. 56. *Epithecium glaucum*. Knowles and Westcott, Floral Cabinet, t. 87. ;) undique glaucedine obductum, pseudobulbis ovalibus compressis monophyllis, foliis ensiformibus acutis pergameneis basi canaliculatis scapo pendulo paniculato brevioribus, sepalis petalisque ovalibus obtusis, labelli linearis apice carnosi trilobi lobis lateralibus rotundatis intermedio ovato medio calloso, columnæ cardine appendice ovatâ obtusâ recurvâ aucto.——*Mexico*.—— Flowers small, scentless, green stained with dull purple. The drooping panicle is remarkable, and the glaucous covering of every part much more so.

15. E. *squalidum* (Llave. L. p. 99.) ; floribus racemosis, sepalis petalisque oblongis acutis patulis, labello rotundato trilobo : laciniis lateralibus obtusis intermedià subreniformi undulatâ medio crassissimâ 3-costatâ. ——*Mexico*.——Flowers dull yellowish brown, purplish outside. Lip dorsal, whitish.

§ III. *Labello tripartito ;* nempè *laciniis lateralibus elongatis intermediæ æqualibus.*

* Labelli lobo medio acuto.

16. E. *Pastoris* (L. no. 7. Klotzsch in Allgem. gartenz. Sept. 22, 1838.); caule repente radicante, pseudobulbis oblongis compressis 2-3-phyllis, foliis linearibus acutis carinatis laxiusculo-subtortuosis, floribus racemosis, perianthii foliolis patenti-subincurvis margine recurvis extus sordide flavis intus lineis longitudinalibus purpureo-fuscis striatis, sepalis lineari-subspatulatis acuminatis, petalis spatulatis acutis, labello trilobo albido dein luteo lobis lateralibus majoribus basi semilunatis integerrimis glabris lævibus basin columnæ orbiculatim amplectentibus lituris transversalibus purpureis medio cordato deflexo minore glabro acuto margine basique recurvo punctis minutis purpureis ornato, columna semitereti fusca ad apicem lutea tridentata dentibus obtusis, pericarpiis elongatis acuto-triquetris.——*Mexico* ——Flowers fragrant, like Vanilla.

17. E. *concolor* (L. no. 12.) ; foliis in pseudobulbos confertos lenticulares solitariis lato-lanceolatis acutis, scapo filiformi 5-floro, sepalis ligulatis, petalis linearibus, labello tripartito laciniis integris intermediâ majore.—— *Mexico*.——A slender plant. Flowers pale yellow, whole coloured, with a striated labellum.

18. E. *fucatum* (Lindl. in Bot. Reg. 1828. misc. 17.) ; pseudobulbis subrotundo-ovatis cæspitosis monophyllis, foliis ligulatis coriaceis obtusis scapo brevioribus, paniculâ nutante multiflorâ, bracteis ovatis acutis squamiformibus, sepalis petalisque lineari-oblongis tessellatis æqualibus obtusis conniventibus, labelli liberi tripartiti lobis lateralibus erectis linearibus apice rotundatis intermedio acuto ovali multò brevioribus, callo sulcato plano elevato basi lobi intermedii.—— *Cuba*. ——Flowers small, dull yellow, tessellated, with a pink spot in the centre of a white lip.

19. E. *pictum* (Lindl in Bot. Reg. 1838. misc. 43.) ; pseudobulbosum, foliis ligulatis coriaceis obtusis dorso rotundatis, racemo erecto paniculato, sepalis petalisque obovato-linearibus subæqualibus, labelli trilobi liberi lobis lateralibus linearibus acutiusculis subfalcatis columnam amplexantibus margine anteriore plicato intermedio ovali acuto crispo multo brevioribus, disco venis elevatis calloso. ——*Demerara*.——Resembles *E.*

32

odoratissimum ; with dull yellow flowers, neatly striped with crimson. It is nearly related to *E. chloranthum,* from which its leaves readily distinguish it.

20. E. *crispatum* (Knowles & Westcott, Floral Cabinet, v. 2. p. 79.) ; pseudobulbis ovatis diphyllis, foliis lineari-lanceolatis carnosis obtusis submucronatis, scapo multifloro, petalis et sepalis linearibus striatis, petalis sepalis minoribus, labello tripartito laciniis lateralibus columnam involventibus intermediâ elongatâ crispatâ (duplò longiore), ovario varicoso. ——*Mexico*.——" The contrast between the long crisped white labellum, and sepals and petals, is very pleasing and striking."

21. E. *chloroleucum* (Hooker in Bot. Mag. t. 3557. E. *chloranthum*, Lindl. in Bot. Reg. 1838. misc. 28.) ; pseudobulbosum, foliis coriaceis ligulatis apice rotundatis obscurè bilobis inæqualibus, racemo erecto paniculato, sepalis petalisque subæqualibus lineari-lanceolatis obovatis, labelli trilobi liberi lobis lateralibus linearibus obtusis inflexis intermedio ovato acuminato crispulo multò brevioribus : disco venis elevatis calloso.——*Demerara*.——Flowers pale green without spots, and a white lip.

22. E. *odoratissimum* (L. no. 10.) ; foliis in pseudobulbos ovatos corrugatos binis ensiformibus, racemo subsimplici, sepalis oblongis petalisque cuneatis patentibus, labelli ferè liberi trilobi lobis lateralibus linearibus obtusis intermedio ovali multò latiore.——*Brazil*.——Flowers dull olive brown, bordered with yellow, very sweet.

23. E. *selligerum* (Bateman in Bot. Reg. 1838, misc. 66.) ; pseudobulbis 2-3-pollicaribus, scapo subramoso, sepalis patentissimis obovatis concavis obtusis, petalis unguiculatis spathulatis acutis, labello semilibero laciniis lateralibus obtusis recurvis planis intermediâ ovatâ acutâ crispatulâ : disco elevato carnoso medio depresso, antherâ bilobâ terminali.——*Mexico*. ——Flowers smell of violets.

24. E. *graniticum* (Lindl. in Hook. Journ. 3. 83.) ; pseudobulbis ovatis attenuatis 2-phyllis, foliis ensiformibus paniculâ multiflorâ brevioribus, sepalis petalisque patentibus lanceolatis subæqualibus acutis, labelli trilobi laciniis lateralibus lineari-oblongis obtusis intermediâ unguiculatâ obovatâ apice inflexo acuto : callo elevato acuminato secus medium canaliculato columnâ sub apice auriculatâ.—— *Guayana*.—— A fine species closely allied to E. flavum. It has a panicle regularly branched up to the apex, nearly a foot and a half long, with each side-branch having from 2-4 flowers. According to M. Schomburgk, the sepals and petals are green dotted with purple, the labellum white with a purple stain at its base, the flowers aromatic, the stem six feet high.

25. E. *pachyanthum* (Lindl. in Bot. Reg. 1838, misc. 42.) ; pseudobulbosum, foliis lato-ligulatis subundulatis apicè obliquè obtusis dorso rotundatis, perianthio carnoso herbaceo, sepalis lanceolatis, petalis obovato-lanceolatis apice complicatis, labelli liberi trilobi laciniis lateralibus ascendentibus truncatis intermediâ spathulatâ acutâ basi callosâ trilineatâ convexâ inappendiculatâ multò brevioribus. —— *Guayana*. —— A large green-flowered species. Its leaves are thinner and broader than is usual among these Epidendra, and a little wavy at the margin. The flowers are fully two inches in diameter, thick and fleshy, dull green, stained with a dirty reddish brown towards the ends of the sepals and petals. The labellum is a pale straw-colour, streaked along the middle with violet.

26. E. *nemorale* (Lindl. in Hooker's Journal, 3. 32.) ; pseudobulbis . . . . foliis . . . . . scapo apice densè racemoso sub-10-floro pedunculisque scabris, sepalis petalisque lineari-lanceolatis acuminatis æqualibus paten-

33

tibus, labelli trilobi laciniis lateralibus semiovatis acutiusculis cucullatis intermediâ maximâ ovatâ ; callo ad laminæ basin sito obsoleto antrorsùm evanescente.——*Mexico*.——A beautiful plant, with very large flowers, apparently pink or purple. The sepals and petals are two inches long, and the middle lobe of the lip is an inch long and three-quarters of an inch broad.

** Labelli lobo medio rotundato.

27. E. *bractescens* (Lindl. in Bot. Reg. 1840. misc. 122.) ; pseudobulbis ovatis cæspitosis 3-4-phyllis, foliis linearibus, scapo debili 3-4-floro, bracteis infimis foliaceis floribus longioribus supremis obsoletis, floribus nutantibus longè pedunculatis, sepalis petalisque lineari-lanceolatis acuminatis discoloribus labello longioribus, labelli liberi lobis lateralibus apice recurvis obtusis subdentatis intermedio unguiculato subrotundo-ovato multò longiore secus unguem elevato sulcato pubescente.——*Mexico*.—— This is one of the prettiest of the small species of Epidendrum. The pseudobulbs are exactly ovate, closely clustered, and about as large as a pigeon's egg. The flowers have a beautifully but delicately painted white lip, the gay effect of which is heightened by the contrast with the dingy purple of the long narrow sepals and petals.

28. E. *aciculare* (Bateman in Bot. Reg. 1841. misc. 98) ; pseudobulbis oblongis diphyllis, foliis linearibus canaliculatis acutis racemo simplici æqualibus, sepalis petalisque lineari-lanceolatis æqualibus acutis, labelli laciniis lateralibus ascendentibus linearibus obtusis apice recurvis intermediâ ovato-oblongâ subundulatâ (pictâ) acutâ.——*Bahamas*.——A pretty little species, with long narrow leaves, a slender erect raceme of six or seven flowers, whose sepals and petals are dull purple, and lip white, enlivened with rosy veins.

29. E. *gracile* (Lindl. in Bot. Reg. t. 1765) ; foliis in pseudobulbos ovatos corrugatos pluribus levato-ensiformibus, racemo simplici longissimo, sepalis oblongis petalisque cuneatis patentibus, labelli ferè liberi trilobi lobis lateralibus semiovatis intermedio oblongo crispo obtusissimo duplò minoribus disco bicostato.——*Bahamas*.——Flowers green, lip yellow, lined with purple.

30. E. *venosum* (L. no. 13.) ; foliis ensiformibus obtusis supra et sub pseudobulbos fusiformes nascentibus, racemo striato simplici, sepalis lineari-lanceolatis petalisque angustioribus patentissimis, labello semilibero tripartito : laciniis lateralibus ovatis acutis intermediâ subrotundâ apiculatâ multo majore callo baseos et lineis tribus disci subramosis elevatis.—— *Mexico*.——Scape a foot long. Lip half united to the column, white, with elevated violet veins.

31. E. *ionosmum* (Lindl. in Bot. Reg. 1838. misc. 87.) ; pseudobulbis diphyllis scapo apice simplici racemoso brevioribus, sepalis petalisque coriaceis obovatis concavis obtusis subæqualibus, labelli trilobi liberi lobis lateralibus erectis oblongis columnam amplexantibus apice rotundatis undulatis coloratis striatis intermedio subrotundo emarginato crispo : lineis duabus elevatis prope basin unicâ angustiore apici propiore, columnâ anticè bicorni.——*Essequibo*.——The world wants no violets where this plant is found. The blossoms are rather large, a dull reddish green, with the lip delicately streaked with deep lilac.

32. E. *flavum* (Lindl. in Hooker's Journal, 3. 83.) ; pseudobulbis ovatis attenuatis 3-phyllis, foliis ensiformibus paniculæ paucifloræ subæqualibus,

sepalis petalisque patentibus subæqualibus lineari-oblongis obtusis, labelli trilobi laciniis lateralibus linearibus truncatis intermediâ unguiculatâ obovatâ nudâ, columnâ sub apice auriculatâ. *Brazil.*——Leaves of this rather more than a foot long. Flowers pale yellow, about an inch and a half in diameter. The inflorescence is only panicled at the base, and is probably very often simple.

33. E. *tripterum* (Lindl. in Hooker's Journal, 3. 83.); pseudobulbis ovalibus compressis diphyllis, foliis lineari-oblongis obtusis racemo paucifloro (4—6) subæqualibus, floribus erectis, sepalis petalisque lineari-lanceolatis patulis, labelli trilobi lobis lateralibus linearibus obtusis planis intermedio subrotundo basi angustato undulato venis rugosis elevatis, capsulâ angustâ clavatâ tripterâ.——*Mexico.*——The whole plant when in bloom little more than six inches high. Flowers apparently dull purple, with a pale lip, on long peduncles, and erect not drooping.

34. E. *virgatum* (Lindl. in Hooker's Journ. of Botany, v. 3. p. 83.); pseudobulbis ovatis oblongisve sub-compressis rugosis, foliis binis ternisque convexis subundulatis acutis glaucis unciam latis, paniculâ virgatâ ramis longis gracilibus, sepalis lanceolatis petalisquè duplo angustioribus patentibus discoloribus, labelli hastati lobis lateralibus acutis patentibus intermedio subrotundo-obovato acuto; callo maximo rotundato pone basin. ——*Mexico.*——The habit of E. vitellinum, but with more glaucous leaves. Flowers small, dirty green stained with brown, arranged in a very long lax graceful panicle, the branches of which are simple, and sometimes as much as a foot long, with nearly twenty flowers on each. The lip is whitish yellow. Scape sometimes seven feet high.

35. E. *aromaticum* (Bateman Orch. Mex. t. . E. *incumbens*, Lindl. in Bot. Reg. 1840. misc. 84.); floribus densè paniculatis, sepalis linearibus patentissimis basi angustatis, petalis conformibus sed paulò latioribus, labelli postici lobis lateralibus triangularibus acuminatis intermedio subrotundo-ovato apiculato venis elevatis cristato, callis duobus oblongis secus unguem.——*Guatemala.*——Flowers very sweet; in large pale dull yellow panicles. It inhabits a climate whose temperature varies from 60° to 75°.

36. E. *oncidioides* (Lindl. in Bot. Reg. t. 1623.); pseudobulbis ancipitibus, foliis 2-3 ensiformibus paniculâ laxâ brevioribus, sepalis petalisque obovatis unguiculatis, labelli trilobi lobis lateralibus nanis intermedio subrotundo cuspidato basi 5-costato.—— .....?——A fine showy plant, with panicles of yellow and brown flowers.

37. E. *alatum* (Bateman Orch. Mex. t. 18. E. *calocheilum*, Hooker in Bot. Mag. t. 3898.); pseudobulbis ovato-oblongis diphyllis, foliis ensiformibus obtusis coriaceis obsoletè striatis paniculâ elatâ multiflorâ brevioribus, sepalis petalisque lineari-oblongis spathulatis uniformibus patentissimis, labello suborbiculari profundè trilobo basi carinato, lobis lateralibus latè ovatis intermedio latissimo lineato margine undulato crispato, columnâ supernè hinc alâ obtusâ.——*Guatemala.*——The flowers, in a large panicle, are light greenish yellow with the sepals and petals tinged with purple at the points, and the lip crimson-veined with a yellow border.

38. E. *altissimum* (Bateman Bot. Reg. 1838. misc. 61.); pseudobulbis elongatis teretibus 2-3-phyllis, scapis ramosis longissimis, sepalis lineari-oblongis acutis, petalis conformibus basi angustatis, labelli liberi lobis lateralibus dimidiatis erectis tortis obtusis intermedio dilatato undulato recurvo apiculato basi bicostato.——*Bahamas.*——Flowers scented with beeswax. Very like E. oncidioides.

*** Labelli lobo medio bilobo.

39. E. *polystachyum* (L. no. 20); foliis oblongis obtusis striato-nervosis geminatim in pseudobulbum insidentibus, scapo polystachyo, sepalis reflexis suboblongis acutis, petalis lanceolato-spatulatis, labelli trilobi lobis lateralibus irregulariter obovatis crenulatis: intermedio minore profundè emarginato.——*Peru.*——Flowers yellow.

40. E. *phœniceum* (Lindl. in Bot. Reg. 1841. misc. no. Sertum, t. 46.); pseudobulbis subrotundo-ovatis diphyllis, foliis oblongo-linearibus erectis, scapo paniculâque scabris, sepalis petalisque subæqualibus coriaceis obovato-lanceolatis, labelli laciniis lateralibus oblongis erectis obtusis apice recurvis intermediâ maximâ membranaceâ subrotundâ undulatâ emarginatâ basi bilamellatâ, columnâ obovatâ marginatâ utrinque unidentatâ.——*Cuba.*——Pseudobulbs roundish-ovate and two-leaved, panicle from two to three feet high, large scentless flowers. The sepals and petals are of a leathery texture and deep purple, a little mottled with green specks both inside and out. The lip is nearly an inch and a half long, of the clear bright violet of Cattleya labiata, and with the same deep crimson veins and stains in the middle; it is much undulated, and of a thin, not leathery texture.

41. E. *adenocarpum* (L. no. 16. E. *papillosum*, Bateman Bot. Reg. 1838, misc. 8. Hooker in Bot. Mag. t. 3631.); foliis strictis linearibus carinatis acutis in pseudobulbos pyriformes glaucescentes ternatim insidentibus, scapo erecto simplici multifloro foliis subæquali, floribus distantibus, sepalis ovato-lanceolatis acutis, petalis paulo angustioribus obtusis, labello semi-libero 3-partito, laciniis lateralibus obovatis, intermediâ 3-plo majore obscure quadrilobâ versus basin callosâ venis 5 tenuibus elevatis radiantibus, columnâ bidentatâ, pedunculis papillosis.——*Guatemala and Mexico.*——Flowers greenish-yellow. Lip white, with three short purple streaks.

42. E. *adenocaulon* (L. no. 15); foliis in pseudobulbis binis v. ternis ensiformibus coriaceis, scapo simplici glanduloso, sepalis petalisque lineari-lanceolatis subæqualibus patentissimis, labello tripartito laciniis lateralibus acutis involutis: intermediâ cordatâ emarginatâ obtusâ.——*Mexico.* ——Scape 2-3 feet high. Flowers pale violet, with a veiny lip.

43. E. *macrochilum* (Hooker Bot. Mag. t. 3534. Bateman Orch. Mex. t. 17.); pseudobulbis ovatis rugosis diphyllis, foliis lineari-oblongis coriaceis obtusiusculis, sepalis petalisque obovato-lanceolatis patentibus apicibus incurvis, labelli trilobi lobis lateralibus ovatis acutis in columnam incumbentibus intermedio maximo obcordato, disco calloso ecristato sellæformi, columnâ apterâ.——*Guatemala and Panama.*——A very fine plant; one variety having green sepals and a yellow lip, the other having purple sepals with a deep rose-coloured lip. It is very near *E. auropurpureum.* It inhabits the coast, where only it thrives in perfection, flowering in February, March, and April. It is a constant attendant on Cattleya Skinneri. Vulgarly '*Boca del Dragón*,' 'Dragon's mouth.'

44. E. *auropurpureum* (L. no. 14.); foliis in pseudobulbos ternis lanceolatis margine revolutis, scapo erecto multifloro, sepalis petalisque patentibus oblongis cuneatis subæqualibus bicoloribus, labelli tripartiti laciniis lateralibus parvis ovatis integris, intermediâ maximâ cuneatâ dilatatâ altè bilobâ.——*Hispaniola.*——Flowers yellow, with the tips purple. Lip all purple, with a deep red callus in the middle.

45. E. *bifidum* (L. no. 17. Bot. Reg. t. 1879.); foliis in pseudobulbos subternis lanceolatis, scapo ramoso foliis multò longiore, sepalis oblongis acutis petalisque lineari lanceolatis patentibus, labelli cuneati lobis lateralibus ovatis: intermedio maximo apice dilatato subreniformi sulcato basi in disco biappendiculato.——*West Indies.*——Sepals and petals light green, with a dull purple spot near the end. Lip rose-colour, with the lateral lobes yellow. This is very near E. auropurpureum, and is perhaps the same.

46. E. *varicosum* (Bateman Bot. Reg. 1838. misc. 37. E. *leiobulbon*, Hooker Journ. Bot. iii. t. 10.); pseudobulbis diphyllis, foliis coriaceis ligulatis acutis, scapo simplici gracili brevioribus, sepalis petalisque subæqualibus cuneato-lanceolatis carnosis vernicosis planis, labello libero unguiculato non cucullato trilobo basi pubescente: laciniis lateralibus nanis triangularibus obtusis, intermediâ reniformi emarginatâ venis tuberculatis et varicosis pictâ, callo pulvinato ovato pubescente, columnæ apice trilobæ lobo dorsali crasso cylindraceo obtuso.——*Guatemala and Mexico.*—— A small-flowered dull-coloured species. The form of the lip and the varicose veins are very remarkable.

47. E. *Michuacanum* (L. no. 21.); foliis in pseudobulbos pyriformes ternis quaternisve ensiformi-lanceolatis, scapo ramosissimo, sepalis petalisque angustioribus cuneato-lanceolatis, labello tripartito laciniâ intermediâ reniformi bilobâ.——*Mexico.*——Pseudobulbs large. Flowers small, scentless, pale yellow, with a white lip.

48. E. *ochraceum* (Lindl. in Bot. Reg. 1838. misc. 15. t. 26.); pseudobulbis obverse pyriformibus cæspitosis sursum attenuatis 1-3-phyllis, foliis linearibus acutis recurvis scapo æqualibus, spicâ terminali laxiflorâ, bracteis rigidis acutis squamiformibus, sepalis petalisque subæqualibus lineari-oblongis obtusis, labello postico sublibero trilobo medio calloso: laciniis lateralibus inflexis truncato-rotundatis denticulatis intermediâ brevi emarginatâ, callo plano apice tridentato, columnâ apice tricorni denticulatâ, ovario triptero.——*Mexico and Guatemala.*——Leaves grassy. Flowers small, orange yellow, inconspicuous.

*** *False Encyclia.*

Encyclia nana } Pöppig, 2. t. 113. are species of Polystachya.
——- *polystachya* }
Encyclia macrostachya. Id. 2. t. 114. is Epidendrum armeniacum.

29. CIRRHOPĔTĂLŪM chinēnsĕ.

C. *chinense;* foliis lanceolatis, umbellâ multiflorâ, sepalo supremo petalisque oblongis brevioribus obtusis serrulatis apiculatis: lateralibus lanceolatis, labello linguiformi obtuso carnoso convexo lævi tremulo.

There is no longer any occasion for speculative minds to occupy themselves with the important investigation of the cause that may have led the Chinese to invent strange figures of men and women with their chins in perpetual motion, for here is the explanation of it. We have here a plant from China, one of whose lobes is exactly a tongue and chin, which are so unstable as to be in a state of continual oscillation. The flowers are very large for the genus, in general appearance like those of C. maculosum. The petals and upper sepal are purple, the lateral sepals are yellowish green. The flowers are arranged in a circle, and all look outwards; so that on whatever side the umbel is regarded, it still presents to the eye the same row of grinning faces and wagging chins. Messrs. Loddiges imported it from China.

30. ONCĪDĬŪM nānūm.

O. *nanum;* pseudobulbis nullis, foliis rigidis ovalibus concavis, racemo stricto foliorum longitudine, sepalis petalisque obovatis obtusis incurvis, labello lucido auriculato apice cuneatìm dilatato emarginato, callis disci duobus basilari transverso altero longitudinali trisulcato, alis columnæ angustissimis decurvis apice glaudulosis.

This singular plant has the habit of O. pumilum, to which it is nearly allied. It differs in having very much larger flowers, and in the form of its lip, which is varnished as it were, and bears two large warts, one transverse and one longitudinal, placed at right angles to each other. The form of the wings of the column is almost that of a straight knife-blade directed downwards, and tipped with a lucid gland. Messrs. Loddiges obtained it from Guayana, (1327).

31. ĂNĬĂ bicŏrnĭs.

A. *bicornis;* folio oblongo lanceolato carnoso scapo breviore, labelli lobo medio emarginato apiculato ecalcarato basi bilamellato lamellâ alterâ versus apicem interjectâ, antherâ bicorni.

A singular plant, of the Orchidaceous order, imported from Ceylon by the Rev. J. Clowes, to whom I am obliged for a specimen. It has something the appearance of Eulophia macrostachya; the flowers are green except the lip which is bright yellow.

32. ĔRĬĂ acutifŏlĭă.

E. *acutifolia* (Tonsæ); caulibus oblongis abbreviatis, foliis lanceolatis coriaceis acutissimis scapi longitudine, racemo erecto paucifloro pubescentes bracteis oblongis reflexis floribusque glabris, labelli lobis lateralibus, acutissimis erectis intermedio oblongo obtuso concavo glabro: venis tribus elevatis carnosis, clinandrio profundè excavato.

A small unattractive species, imported from India by Messrs. Loddiges, (no. 209). Its leaves are fleshy, deep green, very acute, and between two and three inches long; the raceme is erect, composed of five or six distant flowers, of a dirty white, with violet veins. The lip has a purple base, and a dull yellow tip. It must be classed near E. clavicaulis.

33. ĔRĬĂ panicŭlātă.

Lindl. in Wall. Plant. Asiat. rar. vol. 1. *p.* 32. *t.* 36.

Specimens of this plant have flowered with Messrs. Loddiges. They have exactly the appearance of Dr. Wallich's figure, with long narrow taper-pointed leaves, and panicles of small greenish flowers slightly freckled with purple. They have a line of fine white powder formed all along the middle of the lip, and burying the tubercle which occurs at the base. This powder consists of a vast multitude of pyriform bodies visible only under a microscope, and of the same nature as those which occur on Polystachya. They are in fact extremely short hairs of one joint tapering to the base; in appearance they resemble fine powdery dust.

34. CŒLŎGȲNĒ testācĕă.

C. *testacea;* pseudobulbis ovalibus angustis angulatis, foliis lanceolatis, 3-costatis, racemo pendulo, bracteis ovatis cucullatis ovarii longitudine, sepalis petalisque subæqualibus, labelli oblongi lobis lateralibus rotundatis obsoletis intermedio obtuso: venis 4 papilloso-cristatis.

An Orchidaceous plant from Sincapore, introduced by Messrs. Loddiges. The flowers hang down in racemes about six inches long, and have a dull clay colour, which renders them anything rather than handsome.

45. BARKĔRĬĂ spectābĭlĭs.

Bateman in litteris.

B. *spectabilis;* racemo multifloro laxo, labello ovato-lanceolato tricarinato, sepalis lineari-lanceolatis, petalis ovatis acuminatis.

Under the name of Flor de Isabal this beautiful plant forms one of the votive offerings of the Guatemalese. Mr. Skinner long since sent living plants to England, and more recently Mr. Hartweg, one of whose specimens has just flowered in the garden of the Horticultural Society. It forms a tuft of cylindrical stems about four or five inches high, each of which bears two fleshy lanceolate acute leaves, separated from each other by intervals of about an inch. The raceme rises out of some brown dry sheaths, and in the plants that have flowered bears about six most lovely nodding blossoms: but according to a memorandum from Mr. Skinner now before me it varies in length from three inches to a foot: and indeed I possess a wild specimen from that gentlemen with as many as twelve flowers in a raceme. The expanded flowers are nearly three inches and a half wide, their colour is a bright lilac; the sepals are linear-lanceolate, the petals ovate-lanceolate in form and unspotted: but the labellum is white at the base and in the middle, lilac at the edge and point, and richly marked with small blood-red spots. Along its middle, below the column, are five purple lines, which pass into three elevated colourless ridges, beyond the place where the anther touches the lip. It is with Cattleyas, and such beautiful plants that this charming species is worthy to be arranged.

46. HEXADĒSMĬĂ fascicŭlātă.

Adolphe Brongniart in Ann. Sc. xvii. *p.* 44.

Under this name has been just described by M. A. Brongniart, a Mexican Orchidaceous plant, found in that country by Linden, and which has flowered in the Garden of Plants. There is no doubt about its being the same as Mr. Bateman's unpublished genus Hexopia, whose name must therefore be cancelled. The following is the description of the plant in the Annales des Sciences.

HEXADESMIA. Sepala subconniventia, supremo oblongo-lanceolato, lateralibus triangularibus infra cum basi columnæ in pseudo-calcare unitis. Petala sepalis subæqualia oblonga. Labellum cum basi productâ columnæ articulatum, subintegrum, plicatum. Columna petalis brevior, semiteres, apice clavata, clinandrio auriculis lateralibus et dente posticâ uncinatâ marginato, basi foveolâ nectariferâ excavata. Anthera operculæ formis, denti marginis clinandrii affixa, carnosa, sexlocularis. Pollinia 6 æqualia, 4-posticis caudiculis longioribus replicatis, 2 anterioribus caudiculis brevissimis, tria ad utrumque lobum antheræ pertinentia et interse caudiculis unita, facillimè glutine apicis caudiculorum in fasciculo unico connexis.

Herba mexicana ab indefesso viatori Linden Musæo Parisiensi allata, habitu Dendrobia quædam asiatica et formâ floris Apora præsertim referens, sed polliniorum structurâ cum Epidendreis quadrans eorumque numero ab aliis hujusce familiæ generibus, excepto Hexameriâ R. Br. è tribu Malaxidearum, distincta.

H. *fasciculata.* Caules e rhizomate nascentes, erecti, basi graciles squamis vestiti, superius incrassato-subspathulati bulbosi; pseudobulbis stipitatis compressis elliptico-lanceolatis apice, diphyllis. Foliis oblongis 5-6 pollicaribus, pollice lata, coriacea, plana, apice emarginata. Flores terminales 2-4 fasciculati subsessiles, e spatha communi brevi exeuntes, pedicellis ovariisque gracilibus incurvis, squamis obtectis.

48. CATASĒTŪM globiflōrŭm.

Hooker in Bot. Mag. t. 3942.

C. *globiflorum;* spicâ elongatâ multiflorâ, perianthio globoso, sepalis petalisque subconformibus ovatis acutis concavis arctissimè imbricatis, labello hemisphærico-globoso, ore contraeto oblongo inferne dilatato basi columnæ longitudine denticulato, columnæ brevis setis rectiusculis.—*Hooker, l. c.*

A Brazilian plant, very much like *C. Hookeri,* and perhaps a variety of it. The sepals and petals are olive-brown, and closely applied to a glaucous globular labellum, the greater part of which is uncovered. It is stated in the Botanical Magazine to have flowered in the Glasnevin Botanic Garden.

52. CALĀNTHĒ Masūcă.

Lindl. Gen. & Sp. Orch. p. 249.

C. *Masuca;* scapo erecto foliis latis oblongis petiolatis acuminatis longiore, racemis multifloris, labello tripartito basi tuberculis seriatis 5-cristato, seriebus intermediis elevatioribus; laciniis lateralibus linearibus subfalcatis intermediâ cuneatâ emarginatâ in unguem linearem laciniis lateralibus æqualem angustatâ, calcare longissimo falcato clavato, columnâ brevi obliquâ anticè bifoveatâ, ovario pubescente.

This very fine terrestrial Orchidaceous plant has flowered with Messrs. Rollinsons. It has deep violet flowers, with a rich purple lip, and may be regarded as the handsomest of the violet flowered species. It is found in many parts of India, viz. Nepal, Bengal, Ceylon, and probably Java.

55. DENDRŌBĬŪM Scōpă.

D. *Scopa;* caule erecto clavato compresso ramoso, ramis monophyllis, folio oblongo concavo undulato obtuso, floribus in fasciculis axillaribus squamatis dispositis petiolatis, sepalis petalisque lineari-lanceolatis, labelli trilobi trilamellati laciniis lateralibus hinc serratis intermediâ elongatâ triangulari basi serratâ apice in filis intricatis dissolutâ.

A very singular but not handsome species from Manilla. It is no. 273 of Messrs. Loddiges' collection, in which it flowered a few days since. The stems are erect, about a foot high, stiff, and strong. The leaves are from five to six inches long, and two and a half broad. The flowers are small, whitish, in axillary fasciculi. Their lip is very curious, being broken up at the point into a kind of broom, consisting of long entangled curling threads.

62. DENDRŌBĬŪM adūncum. *Wallich.*

D. *aduncum;* caulibus pendulis, foliis lineari-lanceolatis acutis integris, floribus ternis patentibus, sepalis petalisque ovatis obtusis lateralibus duplò latioribus, cornu rotundato, labello unguiculato ovato concavo apiculato columnæ arctè appresso intùs villoso disco glabro, columnâ apice bialatâ sub stigmate villosâ, antherâ glandulosâ.

A beautiful species allied to D. Pierardi, received by Messrs. Loddiges from Dr. Wallich. The flowers are almost transparent, and of the most delicate pink; they are nearly as large as in D. moschatum.

63. DENDRŌBĬŪM cucumerīnūm.

W. MacLeay in litt.

D. *cucumerinum;* nanum, intricatum, cæspitosum, ramis brevissimis articulatis cylindraceis monophyllis, foliis oblongis teretibus seriatim tuberculatis, pedunculis brevissimis trifloris, sepalis petalisque linearibus acuminatis obtusis, labelli trilobi lobis lateralibus triangularibus intermedio ovato crispato lamellis 5 undulatis in medio, clinandrio denticulato.

Of all the queer things which this strange order produces this is one of the very oddest. Only fancy a handful of little stunted cucumbers lying in a heap, and producing a few pale dirty-yellow striped flowers in the midst, and you have this plant before the eye. But it will be better to reserve all explanation of its appearance until we figure it, which will be shortly. It was sent from New South Wales to Messrs. Loddiges by Mr. William MacLeay.

65. SOBRALIA macrantha.

Lindl. in Sertum Orch. sub. t. 29. Gen. & Sp. Orch. p. 431.

A flower of the most delicate texture, of the richest crimson, and between seven and eight inches in diameter, growing on the summit of a reedy stem, will give the reader some notion of the magnificence of this rare plant, as it is now flowering in the garden of the Horticultural Society, from specimens sent from Guatemala by Mr. Hartweg. Or it may be compared, for the flowers, to a huge Cattleya. It is one of the terrestrial species, and apparently easy enough to cultivate in a cool stove; but it is at present in few collections, no one liking to divide the roots until something more shall have been heard of its habits. It is stated by Mr. Skinner that each flower-head will go on blossoming for many weeks successively, provided the faded blooms are pulled off before they decay. We have nothing like *this* in our gardens, yet; nor is there any thing finer of the race to be introduced. It is the leader of the *crême* of its order.

66. MORMŌDĒS luxātūm.

M. *luxatum;* foliis longissimis angustis subtùs glaucis racemo multifloro pluriès longioribus, sepalis ovato-lanceolatis petalisque oblongis margine subscariosis carnosis incurvis, labello hemisphærico concavo obsoletè trilobo apiculato supra columnam cucullato, polliniorum glandulâ anticè cornutâ.

When Mr. Ross, the collector to George Barker, Esq. of Birmingham, was in the neighbourhood of Valladolid, in Mexico, there was brought to him some masses of a plant reported to be of great beauty. They arrived safely in England, and one of them has at last flowered, proving to be this species; a most extraordinary plant, of large size, of a stately habit, with a very delicious fragrance, although powerful, but with no brilliant colours to render it what *we* call handsome. Its flowers are three inches in diameter, pale lemon-colour, fleshy, rather globular, but so distorted by the complete dislocation of all the parts, that it would be difficult to ascertain their real nature, if it were not for the token given by the labellum. The latter has a deep brown streak drawn down its middle, and covers over the column like a hood. The leaves of this plant are about three feet long, narrow, deep green, with a very fine glaucous bloom upon their underside. It is worthy to be associated with even Sobralia macrantha in the choicest of all collections of these plants.

67. CYMBĪDĬŪM pēndŭlūm; *var.* brevilābrĕ.

In the Register for 1840, t. 25, was published a figure of this plant, obtained from Bengal by the Dean of Manchester. Another variety of it has lately appeared in the collection of Messrs. Loddiges, derived from Cuming's Sincapore collections; it differs materially in having a much wider and shorter lip, the middle lobe of which is almost round; otherwise it appears to be the same.

68. ODONTOGLŌSSŪM citrōsmūm.

O. *citrosmum;* pseudobulbis subrotundis compressis lævibus monophyllis, folio oblongo-ligulato obtuso racemo paulò breviore, sepalis oblongis obtusis petalisque conformibus subæqualibus, labello unguiculato reniformi basi bituberculato, columnæ alis lateralibus subtruncatis dorsaliq. rotundato-denticulatis.

This was the most remarkable novelty exhibited at the garden of the Horticultural Society in July, 1842. It is a pseudo-bulbous plant, with a close raceme of very large pale rose-coloured lemon-scented flowers, in form resembling Oncidium Lanceanum. It had been given by Mr. Barker to T. Brocklehurst, Esq. of the Fence, from whose garden it was contributed. At first sight this species would seem to be an Oncidium; but its lip has two parallel fleshy plates at the base, which forms an acute angle with the column, and the latter again has three wings, of which two are lateral and one dorsal; all which circumstances refer it to Odontoglossum, in which, in the neighbourhood of O. pulchellum, it must be placed.

69. SARCĀNTHŪS filifōrmĭs.

S. *filiformis;* foliis filiformibus, racemis 7-8-floris, sepalis petalisque reflexis, labello apice ovato fauce utrinque bidentatâ, apice calcaris didymo, columnæ facie glabrâ.

A species of no beauty, sent to Col. Feilding from India by Dr. Wallich. It has very small flowers, with chocolate-coloured sepals and petals, and a pale yellow lip tipped with pink. It is near S. teretifolius.

70. EPIDĒNDRŪM latilābrĕ.

E. (Euepidendrum) *latilabre;* caule ancipiti folioso, foliis oblongis supremis rotundatis latioribus, floribus umbellatis, labello duplò latiore quam longo obsoletè trilobo basi bicalloso, clinandrio laciniato.

A plant of little beauty, resembling Epidendrum umbellatum; from which it differs in its lacerated anther-bed, and very broad lip, which is at least twice as wide as long. It is chiefly remarkable for its extensive geographical range. The plants in our gardens are all I believe from Brazil, where it has been found by many travellers, (it is Prof. Gardner's 628). But it is also a native of Dominica, whence I have specimens from Prof. Henslow; and Hartweg found it in the woods near Samborondan in Guatemala.

71. POLYSTĂCHŸĂ clavātā.

P. *clavata;* foliis lineari-oblongis obtusis recurvis canaliculatis, paniculâ brevi strictâ spicatâ ramis clavatis, labelli carnosi lobo medio emarginato crenato lateralibus brevibus acutis, disco farinoso basi convexo.

An inconspicuous species from Belize, introduced by Messrs. Loddiges. The flowers are pale yellow, and small. The branches of a short panicle are fleshy, club-shaped, and studded with sessile flowers.

72. NOTŸLĬĂ pubēscēns.

N. *pubescens;* foliis oblongis subundulatis subconcavis, racemo folio longiore pendulo, labello rhomboideo basi carinato et columnâ pubescentibus, sepalis lateralibus basi tantum connatis.

A Brazilian species with dull orange-coloured, rather sweet-scented, flowers. It is readily known by its downy lip and column. Imported by Messrs. Loddiges.

73. DENDRŌBĬŪM sanguinolēntūm.

D. (Eudendrobium) *sanguinolentum*; caulibus teretibus pendulis, foliis ovato-lanceolatis, floribus gemellis, sepalis petalisque ovatis obtusiusculis patulis, labello trilobo glabro: laciniâ intermediâ retusâ nunc utrinque plicatâ; ungue concavo brevi appendice cornuformi pubescente retrorsâ aucto.

Another novelty sent from Ceylon to His Grace the Duke of Northumberland by Mr. Nightingale. It is a plant of great beauty; with pendulous stems like those of D. Pierardi, but of a delicate purple when young; the leaves too are stained underneath and at the edges with the same colour. The flowers are as large as those of D. aggregatum, of a clear fawn colour, with the tips of the segments and lip stained with a deep rich violet. There is moreover a *scarlet* spot in the middle of the lip. Of this fine plant there is a variety without the violet spots, and with somewhat larger flowers. Both kinds have recently flowered at Syon, and will soon find a place among the figures of this work.

74. ONCĪDĬŪM barbātūm.

Lindl. Gen. and Sp. Orch. p. 200.

This very pretty and long-lost species has again appeared in the Glasgow garden, where it had been sent from Pernambuco. It has a graceful panicled habit, and a curious bearded bright yellow lip, the side lobes of which are large and yellow, while the middle lobe is obsolete, fringed and spotted with crimson. It must be a very rare plant, for Professor Gardner never met with it during his long botanical journeys in Brazil; nor have I ever found it in any dried collection.

75. VANĪLLĂ Palmārūm.

Lindl. Gen. and Sp. Orch. p. 436.

V. *Palmarum*; foliis ovatis subcordatis brevi-petiolatis succulentis, floribus geminis, sepalis petalisque angustis oblongis erectis canaliculatis, labello membranaceo obovato emarginato plano basi ter elevatâ pubescente, columnâ clavatâ facie barbatâ, ovario calyculato.

This plant was first made known by Salzmann's dried specimens, gathered near Bahia. Since that time it has been discovered in Surinam by Hostman. It has now flowered, for the first time in Europe, in the collection of Messrs. Loddiges. The blossoms are pale green, of a delicate texture, and about two inches long. The lip is hairy at the base, has an elevated broad line in the middle, and one on each side where it touches the column. At the summit of the ovary is a narrow calycle, obscurely three-lobed, and of a fleshy texture. It is one of the few Orchidaceous plants which inhabit Palm Trees.

76. DENDROBIUM compressum.

D. (Dendrocoryne) *compressum*; caule obovato compresso 2-6-phyllo, foliis ovalibus acutis striatis basi dilatatis membranaceis amplexicaulibus, racemis subquadrifloris cernuis, sepalis petalisque ovatis erectis, cornu elongato obtuso, labello cuneato lævi per axin sulcato.

A curious species with yellow flowers of middle size, and singular compressed stems not more than three or four inches long. The lip is singularly truncated. Mr. Nightingale sent it from Ceylon to His Grace the Duke of Northumberland at Syon. We shall publish a figure of it as soon as we can find room.

78. STĔLĬS argentātă.

S. *argentata*; folio lineari-oblongo coriaceo emarginato in petiolum canaliculatum angustato caule brevi longiorem, racemo elongato multifloro, sepalis æqualibus ovatis subdiaphanis pubescentibus, petalis labelloque subæqualibus truncatis apice particulis argenteis micantibus.

A native of Guayana, where it was found by Mr. Schomburgk, who distributed it under the number 427. It has lately flowered with Messrs. Loddiges, (no. 75). The flowers are in long racemes, dull greenish purple, slightly hairy, rather large for this genus; the petals are purple, and, as well as the labellum, have their truncated extremities covered with an appearance like minute particles of silver. The plant forms large tufts, and flowers very abundantly.

79. ĔRĬĂ pānnĕă.

E. *pannea*; (Lanatæ) ebulbis, foliis distichis pugioniformibus scapo unifloro(?) densè lanato quasi panneo longioribus, sepalis lateralibus triangularibus supremo petalisque pluriès majoribus, labello lanceolato acuminato carnoso pubescente disco farinaceo.

An Orchidaceous plant from Sincapore, imported by Messrs. Loddiges, (252). It has narrow dagger-shaped leaves of a pale yellow green, and a single greenish-yellow flower, woolly on the outside, and seated on a short scape covered with a whitish cloth-like substance. The labellum is purplish, slightly hairy, and has a broad patch of a mealy hairiness a little below its point.

80. OCTOMĔRĬĂ grandiflōră.

O. *grandiflora*; folio longo lanceolato arcuato, vaginis caulis arctè appressis, labelli trilobi unguiculati laciniis lateralibus brevibus subfalcatis acutis intermediâ obovato-cuneatâ denticulatâ fissâ; lamellis duabus carnosis.

This is the largest Octomeria yet seen. Its leaf is as much as eight inches long, and the stem is about the same length. The flowers are about four times as large as those of O. Baueri, and of the same pale yellow colour. Messrs. Loddiges imported it from Brazil, (60).

67

PLEUROTHALLIS.

This genus is one of those large assemblages of species, which has been so much increased by modern discoveries as no longer to resemble what it was only ten years ago. Of the numerous kinds defined since the publication of the Genera and Species of Orchidaceous Plants in 1830, the greater part have been made known in a desultory manner, so that a Botanist finds it difficult to compare them with each other. They have moreover been insufficiently classified, in consequence of which the difficulty of their determination has been much increased. It therefore seems to me that a complete recapitulation of all that are known up to the present time cannot be otherwise than useful; the more especially as their smallness, and the minuteness of the characters by which they are known, are sufficiently embarrassing even when the exposition of the species is as clear and well arranged as art can make it. Let me be permitted, however, before going into an enumeration of the species of this genus, to offer a few general observations on the genera allied to it.

Among Malaxeous Orchidaceæ, Pleurothallis undoubtedly forms the type of a very natural group; which, however, in the Genera and Species of Orchidaceous Plants is not sufficiently

68

circumscribed. The genera properly belonging to Pleurothallidæ are, *Pleurothallis*, *Stelis*, *Lepanthes*, *Octomeria*, *Stenoglossum*, *Restrepia*, *Physosiphon*, and *Masdevallia*; to which must be added *Myoxanthus* and *Specklinia*, if those two genera can be retained. But the former I have formerly reduced to Pleurothallis, and a more extensive knowledge of the species has led me to adopt the same conclusion respecting Specklinia, which is only characterized by an extension of the base of the lateral sepals, in which particular there are so many insensible gradations that it cannot be practically employed; all the species of Pleurothallis are more or less gibbous at the base of the lateral sepals.

Among these genera Pleurothallis is by far the most extensive, comprising species with considerable difference in habit, and in the subordinate details of their fructification. As the latter do not appear to be in any way connected with the former, it seems desirable to employ for natural purposes the habit alone, which furnishes some well marked distinctions. Some of the species have long spikes of flowers, considerably extending beyond the foliage; these form one section. Others have the spikes visibly shorter than the foliage, and form a second section. A third consists of those whose flowers are arranged in a fascicle at the base of the leaf, and not in spikes. Finally, a fourth section consists of small species with solitary or nearly solitary flowers growing on long capillary peduncles.

Taking this mode of distribution, the species may be conveniently arranged as follows.

Section I. ELONGATÆ. Spicæ v. racemi foliis evidenter longiores, v. omninò æquales.

* *Racemosæ*; nempè pedicellis quam bracteæ evidenter longioribus.

† Disepalæ (nempè sepalis lateralibus ferè ad apicem connatis); labello indiviso.

1. P. *Matthewsii* (Lindl. in Comp. Bot. Mag. v. 2. *Acronia phalangifera*, Presl. Symb. t. 57.); folio oblongo acuminato basi rotundato racemis breviore caule longiore, sepalis petalisque æquilongis setaceo-acuminatis, labello transverso obtusissimo medio carnoso cordato margine dilatato membranaceo.——*Peru.*

2. P. *Restrepioides*; folio oblongo coriaceo subsessili spicis subgeminis strictis duplò breviore, floribus ringentibus disepalis glabris longè pedunculatis, sepalis ovatis acuminatis inferiore cymbiformi bidentato, petalis confor-

69

mibus duplò minoribus, labello apice lineari carnoso obtuso basi oblongo canaliculato sessili.——*Peru.*——A very fine species, found in the province of Chachapoyas by Matthews. The flowers are three-quarters of an inch long, and, apparently, purple and green. The length of the leaves from three to five inches, their breadth one to one and a half. This requires to be compared with Dendrobium acuminatum of H. B. K., the No. 13. of this enumeration.

3. P. *hians* (Lindl. in Bot. Reg. sub. t. 1797.); caule brevi, folio oblongo basi angustato obtuso racemo flexuoso 6-9-floro duplò breviore, bracteis minutis ochreatis, pedicellis arcuatis elongatis, floribus bilabiatis: labio postico 2-dentato antico oblongo obtuso, petalis cuneatis, labello ovato-lanceolato acuminato pubescente basi concavo, columnâ clavatâ clinandrio ciliato.——*Brazil.*——A fine species from the Organ Mountains. The leaf is three inches long, the scape nine to twelve. The ovary and pedicel together are about half an inch, and the sepals nearly the same length. The flowers are purple and green.

4. P. *oblongifolia* (Lindl. in Comp. Bot. Mag. v. 2.); folio oblongo obtuso carnoso racemo erecto laxo duplò breviore, sepalis acuminatis lateralibus connatis, petalis truncatis triveniis, labello angusto glabro apice carnoso contractiore.——*Jamaica.*——Resembles P. racemiflora. Flowers red. Fruit narrow, obovate, six-ribbed, very smooth. Bracts cucullate, acute.

5. P. *macrorhiza* (Lindl. in Hooker's Bot. Miscell.); foliis lineari-oblongis acutis carnosis basi valdè angustatis cum petiolo articulatis, pedunculo 2-3-floro erecto capillari foliis subæquali v. longiore basi cum petiolo laxè vaginato, sepalis ovato-oblongis: lateralibus ferè ad apicem connatis, petalis cuneatis, labello ovato obtuso basi cucullato petalis duplò longiore.——*Peru.*——Flores verosimiliter flavi, pro genere magni.

6. P. *fusca* (Lindl. in Comp. to Bot. Mag. v. 2.); folio obovato-lanceolato caule longiore, racemo solitario laxo paucifloro folio longiore, sepalis 2 oblongis acutis inæqualibus, labello obovato lanceolato acuto plano, clinandrio subcucullato denticulato.——*Brazil.*——Flores fusco-purpurei intus lutei maculati.

7. P. *flexuosa* (*Specklinia flexuosa*, Pöppig, t. 90); caule brevissimo monophyllo, folio latè lanceolato acuto, racemo capillari erecto tandem flexuoso, sepalis intùs villosis infimis ultra medium connatis, petalis semilunatis, labello oblongo supernè angulato, columnâ apice dentatâ. ——*Peru.*——On trees.

8. P. *picta* (Lindl. in Bot. Reg. sub f. 1797.); folio spathulato marginato retuso racemis laxis duplò breviore, bracteis minimis, sepalis acuminatis lateralibus apice tantùm sejunctis, petalis lineari-lanceolatis acutis, labello lineari obtuso carnoso supra 1-sulcato.——*Demerara.*

9. P. *Grobyi* (Lindl. in Bot. Reg. t. 1797.); folio obovato emarginato caule triplo longiore recemo laxo erecto multoties breviore, bracteis minimis membranaceis, sepalis costatis oblongis acutis lateralibus apice tantùm sejunctis, petalis lanceolatis acutis, labello lineari obtuso carnoso supernè unisulcato.——*Brazil.*——In some specimens the raceme is less zigzag, and the flowers smaller than in others.

†† Disepalæ; labello trilobo v. repando.

10. P. *pachyglossa* (Lindl. in Bot. Reg. 1840. misc. 146.); folio ovato-lanceolato apice tridentato caule suo longiore scapo breviore, scapo filiformi erecto distanter 4-floro, sepalis lineari-lanceolatis acuminatis membrana-

70

ceis inferiore bicarinato bidentato, petalis carnosis obovatis convexis trinerviis apice rotundatis dorso sub apice verrucosis, labello petalis duplò longiore unguiculato crassissimo obtuso linguiformi basi auriculis duabus membranaceis aucto medio villoso dorso verrucoso.——*Mexico.* ——This is a large-flowered species. It has purple semitransparent sepals six lines long, and about four flowers on a slender scape, divided from each other by intervals of nearly an inch.

11. P. *pandurifera*; cæspitosa, folio oblongo marginato basi angustato apice tridentato caulis longitudine, pedunculo setaceo folio triplò longiore apice flexuoso 5-6-floro, sepalis diaphanis ovatis infimo bidentato, petalis nanis obovatis obtusissimis, labello pandurato lobo intermedio angustiore rotundato, clinandrio latè marginato membranaceo.——*Brazil.*——An inhabitant of the Organ Mountains, where it was found by Mr. Miers. It looks like a small starved specimen of P. Grobyi.

12. P. *quadrifida* (*Dendrobium quadrifidum*, La Llav. 2. 40); caule elongato, folio cuneato-oblongo carnoso lucido vaginante, racemis fasciculatis longissimis erectis, bracteis acutis vaginantibus, perianthio 4-partito: sepalis apice reflexis, petalis sublinearibus erectis acutis, labello 3-lobo carinato cum columnâ elasticè articulato.——*Mexico.*——Flores lutescentes molles odorem citrinum vespere spirantes.

††† Trisepalæ (nempè sepalis lateralibus ultra medium liberis.)

13. P. *acuminata* (*Dendrobium acuminatum*, H. B. K. 1. 357.); "caule monophyllo; folio oblongo acuto; spicis subgeminis; foliolis calycinis exterioribus lanceolato-linearibus acuminatis." ——*Peru.*—— Caulis, adjectâ spicâ, sesquipedalis, vaginatus. Vaginæ acutæ, apice laxiusculæ, subpollicares. Folium basi angustatum, coriaceum, bipollicare, 8-9-lineas latum. Spicæ terminales, 4-6-pollicares, strictæ, basi spathis brevibus involutæ. Flores pedicellati, bracteolis subtruncatis. Flores vix patuli. Sepalum supremum concavum; lateralia plana, basi cum ungue labelli connata, semipollicaria. Petala triplò quadruplòve breviora. Labellum lanceolato-oblongum, obtusum, basi unguiculatum, subcanaliculatum, plicis tribus longitudinalibus instructum. Ex *Kunth.*

14. P. *elegans* (*Dendrobium elegans*, H. B K. 1. 358.); "caule monophyllo, folio oblongo obtuso, spicis quaternis vel quinis, foliolis calycinis exterioribus oblongis obtusiusculis."——*New Granada.*——Caulis 8-9-poll. Vaginæ submarcidæ, subbipollicares. Folium 5-6-poll. sesquipoll. latum. Spicæ erectæ, 6-7-poll. basi bracteâ ovato-oblongâ membranaceâ glabrâ sesquipollicari fuscescenti cinctæ. Flores subsecundi cernui. Sepala diaphana, alba, nervis punctisq. violaceis picta: lateralia trinervia, basi ungui labelli adnata, 3 lineas longa; supremum latius 5-nervium. Petala duplò breviora, rotundata. Labellum unguiculatum, sepalis duplò brevius, basi sepalis lateralibus connatum; lamina libera, cum ungue articulata, ovato-elliptica, obtusa, carinata, trinervia, violaceo-punctata. Pollinia 2, pedicellata. Ex *Kunth.*

15. P. *marginata* (Lindl. in Bot. Reg. 1838. misc. 70.); folio obovato biconvexo marginulato caule pluriès longiore racemo laxo erecto multò breviore apice obsoletissimè tridentato, sepalis subfalcatis carinatis lateralibus semiconnatis, petalis linearibus obtusis apice callosis, labello lineari obtuso canaliculato.——*Guatemala.*——A very small species with the

71

habit of P. Grobyi, and of no beauty. It grows in dense tufts on the branches of trees.

16. P. *scariosa* (*Dendrobium scariosum*, Llave 2. 39.); caule brevissimo folio elliptico carnoso suprà hyalino margine scarioso, pedunculis 3-4 filiformibus flexuosis secundifloris, sepalis concavis, petalis planis, labello tripartito.——*Mexico.*——Planta exigua. Folia enervia. Spatha membranacea brevissima in sinu folii. Sepala luteo-virescentia. Petala nivea, lineâ longitudinali limboque purpurascente variegata, ex *La Llave.*

17. P. *aristata* (Lindl. in ann. nat. hist. 4. 116. 2. 329. t. 15.); cæspitosa, folio ovali acuto apiculato basi angustato caule longiore, racemo capillari flexuoso erecto 4-6-floro folio ter longiore, bracteis vaginantibus cuspidatis, pedicellis elongatis, sepalis aristatis subciliatis lateralibus semiconnatis, petalis lanceolatis acuminatis semipinnatifidis duplò brevioribus, labello lineari glanduloso-hispido apice rotundato basi glabro utrinque auriculato.——*Demerara.*——Flowers purple, apparently streaked with yellow.

18. P. *laxa* (L. 13.); "caule monophyllo, folio oblongo acuminato, racemis laxis, capsulis nudis."——*Jamaica.*——P. alpestri valdè affinis; utraque bracteis cucullatis distincta. Racemi foliis subæquales. Flores secundi, atropurpurei, lineati. Bracteæ cucullatæ. Sepala lateralia semiconnata. Labellum ovatum, canaliculatum, apice reflexum, cœrulescens.

19. P. *alpestris* (L. 14.); "caule monophyllo, folio ovato-lanceolato, racemis laxis, carinis capsularum muricatis."——*Jamaica.*——Racemi folio sæpè longiores. Flores parvi, secundi, luteo-virides. Sepala lateralia semiconnata. Labellum ovatum, obtusum, concavum, minutum, atropurpureum.

** *Spicatæ*; nempè pedicellis quam bracteæ brevioribus v. vix longioribus.

† Disepalæ.

20. P. *racemiflora* (L. no. 8. Ex. Fl. t. 123.); folio oblongo emarginato racemo secundo multifloro duplo breviore, bracteis vaginatis appressis, sepalis ovato-lanceolatis: lateralibus connatis petalis ovatis subæqualibus, labello subunguiculato linguæformi obtuso plano, clinandrio dentato. ——*Jamaica.*

21 P. *glumacea* (Lindl. in Comp. to Bot. Mag. 2.); folio lanceolato petiolo longiore, spicâ nutante multiflorâ foliis 3-plò longiore, sepalis linearibus acuminatis subæqualibus inferiore bicarinato, petalis oblongis, labello ovato concavo obtuso.——*Brazil.*——Flowers yellowish. The narrow sharp pointed sepals have something the appearance of the glume of a grass.

22. P. *lanceolata* (L. no. 5.); caule ascendente tereti folio ovato-lanceolato erecto basi angustato ferè triplo longiore, racemo solitario folio longiore apice nutante e spathâ maximâ coriaceâ dimidiam folii ferè æquante. ——*Peru.*——Taking the stem and spike together this plant is a foot and half high, while the leaf is about seven inches long. The synonyms of Humboldtia and Stelis lanceolata, quoted in the Gen. and Sp. belong to Stelis disticha of Pöppig, according to a specimen in Willdenow's herbarium.

23. P. *cauliflora* (Hook. ic. pl. t. 50.); folio subrotundo-ovato petiolato racemis secundis fasciculatis duplò breviore, spathâ oblongâ obtusâ vagi-

72

natâ, sepalo inferiore oblongo emarginato superiore ovato obtuso multò minore, petalis linearibus, labello subrotundo membranaceo concavo retuso cum ungue producto columnæ articulato.——*Peru.*

24. P. *truncata* (Hook. ic. t. 55.); folio ovato-lanceolato acuminato racemis geminis æquali, caule pluries longiore, sepalis lateralibus in unum ovatum connatis superiore cucullato duplò minoribus, petalis lineari-oblongis, labello truncato involuto. —— *Peru.*——Flowers orange-colour.

25. P. *diffusa* (Pöppig & Endl. Gen. & Sp. t. 86.); caule anguloso folium ovato-oblongum acutum erectum excedente, racemis compluribus longissimis diffusis, spathâ compressâ acuminatâ brevi, floribus secundis, sepalis latè ovatis inferioribus ad apicem connatis, labello unguiculato emarginato cucullato.——*Peru.*

26. P. *longicaulis*; folio coriaceo angustè oblongo obtuso caule apice alato basi tereti quadruplò breviore, spicis numerosis multifloris cernuis folio longioribus, sepalis acutis lateralibus apice liberis basi parùm productis, petalis lanceolatis sepalis duplò brevioribus, labello bilamellato oblongo acuto basi angustato acutè trilobo.——*Brazil.*——A remarkable species with very long stems, found by Mr Miers. The flowers appear to be yellowish. It is much like Pl. rubens, but differs in the form of the petals and lip.

27. P. *plumosa*; folio coriaceo ovali basi haud angustato spicâ paulò breviore, floribus pubescentibus, sepalo dorsali lineari sub apice intùs convexo lateralibus connatis obtusis, petalis nanis argutè inciso-serratis, labello concavo rhombeo-ovato serrulato levitèr papilloso, clinandrio serrato. ——*Trinidad.* (Loddiges no. 13.)——Flowers green, with a little purple dotting on the inside. The petals have such long delicate fringes as to look like small feathers.

28. P. *spiralis* (L. no. 10. *Stelis connata*, Presl. Reliq. Hænk. 1. 103.); caule tereti laxè vaginato folio oblongo æquali, racemo erecto spirali folio longiore, bracteis acuminatis, sepalis ovatis obtusiusculis: lateralibus in unum cymbiforme connatis, petalis columna labelloque truncatis nanis. ——*Peru.*——This very curious species, like P. hians, has the habit and petals of Stelis.

29. P. *tricarinata* (Pöppig and Endl. Gen. and Sp. t. 87.); caule compresso folium lineare oblongum obtusatum adæquante, spicâ solitariâ pedunculatâ folium superante, floribus distichis resupinatis, sepalis ultra basin connatis lineari-lanceolatis lateralibus ad apicem concretis margine revolutis tricarinatis, labello unguiculato oblongo.——*Peru.*

†† Trisepalæ.

30. P. *stenopetala* (Lindl. in Bot. Reg. 1838.); folio oblongo apice rotundato basi in petiolum planum angustato racemo multò breviore caule longiore, bracteis membranaceis cucullatis, sepalis acuminatissimis linearibus intus pubescentibus omnibus ferè liberis, petalis nanis obtusis dorso carinatis labelloque oblongo conduplicato glabris.——*Brazil.*——A pretty species with long slender cream-coloured flowers.

31. P. *sclerophylla* (Lindl. in Bot. Reg. sub t. 1797.); folio oblongo petiolato obtuso carnoso racemis multò breviore, bracteis ovatis cucullatis, sepalis acuminatissimis omnibus ferè liberis, petalis nanis obtusis, labello oblongo conduplicato apice piloso-glanduloso. —— *Brazil.*

32. P. *arbuscula*; caule erecto ramoso folioso vaginis striatis vestito, foliis ovalibus mucronatis in petiolum brevem angustatis, spicis solitariis se-

73

cundis foliis duplò longioribus, sepalis linearibus elongatis obtusis, petalis nanis rotundatis, labello lineari obtuso intùs bicarinato.——*Peru.* ——Found near Loxa by Mr. Hartweg. The stems are erect, slightly branched, and covered with firm striated sheaths, from among which appear the leaves, which are about 1½ inch long. The sepals are about half an inch long.

33. P. *gelida* (Lindl. in Bot. Reg. 1841. misc. 186.); folio oblongo carnoso concavo basi attenuato caule tereti laxè 1-ochreato æquali v. longiore, spicis geminis erectis secundis folio subæqualibus e spathâ brevi univalvi exsertis, floribus cylindraceis, sepalis carnosis intus pilosis lateralibus semiconnatis, petalis minimis glabris subrotundo-oblongis apice denticulatis, labello etiam breviore lineari cuneato bilamellato apice rotundato plano, columnâ petalis breviore.——*Jamaica.*——A plant with all the habit of Pl. racemiflora, but the spikes appear in pairs, and are hardly so long as the leaf, and the structure of the flowers is different. The whole plant is fully a foot high.

34. P. *rubens* (Lindl. in Bot. Reg. sub t. 1797.); folio oblongo-lanceolato obtuso, racemis longissimis, bracteis oblongis ochreatis, sepalis lanceolatis acuminatis lateralibus basi tantùm connatis petalis obtusis, labello obovato repando obtuso lineis 2 elevatis flexuoso carnosis.——*Brazil.*—— This plant seems to vary a good deal in size, some specimens not being more than six inches high, others more than a foot. The flowers are apparently pale green.

35. P. *Jamiesoni* (Lindl. in Bot. Reg. sub t. 1797.); folio lineari-oblongo carnoso submucronato racemo secundo stricto duplò breviore, bracteis densis brevibus ochreatis cucullatis, sepalis oblongis subæqualibus leviter carinatis lateralibus basi tantum connatis, petalis duplo brevioribus apice rotundatis, labello breviore membranaceo rhombeo-trilobo plicâ unguis transversâ.—— *Peru.*

36. P. *pulchella* (L. no. 12. H. B. K. t. 90.); spicis geminis folio oblongo-lanceolato obtuso coriaceo duplo longioribus, floribus nutantibus, sepalis oblongis apice angustatis triveniis: superiore duplo latiore.——*Peru.*

37. P. *saurocephala* (Lodd. Bot. Cab. t. 1571. Bot. Mag. t. 3030. Bot. Reg. t. 1968.); folio coriaceo oblongo cauli vaginato subæquali racemis breviore, bracteis ovatis ovario brevioribus, sepalis coriaceis oblongis pubescentibus lateralibus semi-connatis, petalis nanis acutis, labello oblongo medio excavato utrinque bilamellato: lamellis marginantibus incurvis. ——*Brazil.*

38. P. *Lanceana* (Lodd. Bot. Cab. t. 1767.); foliis carnosis ovalibus acutis, spicis solitariis pendulis strictis foliis longioribus, sepalis conniventibus acutis lateralibus ad apicem usque connatis, petalis subulatis ciliatis, labello ovato emarginato unguiculato.——*Surinam.*——Flowers yellow, crimson at the base inside.

39. P. *capillaris* (Lindl. in Bot. Reg. sub t. 1797.); folio lineari-lanceolato racemis capillaribus fasciculatis subæquali, sepalis acuminatis lateralibus basi connatis, petalis ferè æquilongis serratis acuminatis, labello lanceolato basi angustato acuto membranaceo 3-venio.——*Brazil.*

40. P. *floribunda* (*Specklinia floribunda*, L. no. 4.); caule ancipiti folio oblongo-lanceolato duplò longiore, spicis flexuosis multifloris folio æqualibus et longioribus, sepalis ovatis lateralibus basi tantum connatis, petalis duplò brevioribus serrulatis, labello trilobo bilamellato: laciniis acutis lateralibus nanis, clinandrio dentato.——*Martinique.*——A small plant

74

with drooping zigzag spikes. The P. floribunda of Pöppig is the same as P. macrophylla, no. 44.

41. P. *caulescens* (Lindl. in Hooker's Bot. Misc. 2.); caulibus ascendentibus foliosis gracillimis, foliis membranaceis patentibus vaginantibus lineari-lanceolatis vix costatis, racemis 2-3 terminalibus erectis subsecundis caulium longitudine, bracteis ochreatis ovatis acutis, sepalis lanceolatis: lateralibus subconnatis, petalis ovatis acutis duplò brevioribus, labello postico unguiculato ovato obtuso obsoletè trilobo complicato.——*Peru.* ——This is another of the species which bear many leaves on a stem instead of one; they do not however appear to possess any other character in common.

42. P. *teres* (Lindl. in Bot. Reg. sub t. 1797.); folio tereti racemo gracili breviore, bracteis membranaceis ochreatis, floribus pendulis, sepalis acuminatis lateralibus apice tantum liberis, petalis minimis oblongis acutis, labello lineari canaliculato acuto marginibus incrassatis, columnâ basi pubescente.——*Brazil.*——Flowers cinnamon-coloured. The leaves are very like those of Leptotes bicolor.

43. P. *rupestris* (Lindl. in Bot. Reg. sub t. 1797.); folio tereti mucronato anticè sulcato racemo stricto rigido paulò breviore, bracteis parvis rigidis concavis, sepalis acutis lateralibus semiconnatis, petalis lanceolatis ciliatis, labello subtrilobo membranaceo obtuso appendice magno carnoso bifurco: cruribus parallelis porrectis acutis.——*Brazil.*——Scapus rubenti-virens. Sepala purpurascentia. Petala purpurea.

Section 2. Effusæ. Spicæ v. racemi folio multò breviores.

* *Læves*; nempè, floribus omnino pube orbatis.

44. P. *macrophylla* (L. no. 11. P. *floribunda*, Pöpp. & Endl. t. 84.); caule semitereti folium oblongo-lanceolatum petiolatum coriaceum erectum duplò excedente, racemis fasciculatis numerosissimis patulis folio triplò brevioribus spathâ saccatâ basi inclusis, sepalis ovato-oblongis inferioribus ad basin liberis, labello subrotundo basi apiceque acuminato.—— *Peru.*——A specimen in the herbarium of Willdenow identifies Pöppig's P. floribunda and the macrophylla of Humboldt and Kunth.

45. P. *decurrens* (Pöppig & Endl. t. 83.); caule anguloso folium subrotundum utrinque acutum ad medium cordatum lobis adnatis decurrens triplò excedente, spicis in sinu medii folii paucis abbreviatis oliganthis, sepalis æquilongis inferiore ex duobus ad apicem connatis consistente latiore bimucronato.——*Peru.*——A large species looking like P. Mathewsii.

46. P. *secunda* (Pöppig & Endl. t. 85.); caule tereti folium oblongum acuminatum basi acutum triplò excedente, racemis folio brevioribus capillaribus paucifloris, floribus erectis secundis, sepalis latè ovatis inferioribus ad apicem connatis, labello cordato subrotundo longe unguiculato.—— *Peru.*

47. P. *prolifera* (L. no. 7.); caulibus ancipitibus, folio oblongo cochleato carnoso prolifero racemo duplo longiore, bracteis cuneatis cuspidatis cucullatis pedicello brevioribus, sepalis conniventibus: lateralibus semiconnatis, petalis labelloque conformibus duplo longioribus, clinandrio dentato.——*Brazil.*

75

48. P. *luteola* (Lindl. in Bot. Reg. 1841. misc. no. 2.); folio ovato canaliculato acuto tridentato petiolo suo breviore, pedunculo trifloro folio adpresso eoque breviore, floribus glabris, sepalis lateralibus connatis omnibus carinatis, petalis spathulatis emarginatis, labello lineari obtuso canaliculato glabro.——*Brazil?*——The small yellow flowers exhale the smell of a sliced cucumber in the middle of the day.

49. P. *bicarinata* (Lindl. in Bot. Reg. 1839. misc. 11.); folio oblongo coriaceo planiusculo leviter carinato basi subcordato petiolo acutè canaliculato univaginato breviore, sepalis linearibus æqualibus lateralibus carinatis ad apicem ferè connatis, sepalis lineari-obovatis minutè serratis glabris, labello obovato carnoso medio exarato denticulo inflexo utrinque prope basin.——*Brazil.*——Flowers dull greenish-yellow. The leaf is five inches long, the stem six, with a large withered sheath in the middle.

50. P. *ochreata* (Lindl. in Bot. Reg. sub t. 1797.); folio crasso lanceolato complicato mucrone acuto racemo denso flexuoso nutante duplò longiore, bracteis cucullatis imbricatis acutis, sepalis carinatis acuminatis lateralibus semiconnatis, petalis lineari-oblongis nanis mucronatis, labello . . . ——*Brazil.*——Calyx red or vermilion colour. Petals only half the size, yellow.

51. P. *obovata* (*Specklinia obovata*, Lindl. in Bot. Reg. 1839, misc. 137.); folio coriaceo obovato emarginato basi angustato canaliculato caule longiore, spicis brevibus fasciculatis, floribus glaberrimis, sepalis petalisque linearibus acuminatis, labello lineari abruptè acuto medio paululùm constricto, columnâ cucullatâ alatâ subdentatâ.——*Brazil.*——The leaves are obovate and very thick; the flowers are small, pale yellow, scentless, and in numerous fascicled short spikes.

52. P. *hypnicola*; folio obovato lanceolato basi angustato apice 3-dentato caule longiore, racemo capillari folio breviore paucifloro, pedicellis elongatis, sepalis obtusis lateralibus liberis, petalis obovatis rhombeis nanis, labello ovato acuminato unguiculato. ——*Brazil.*——Resembles P. fusca. Mr. Miers found it in Brazil among Hypna. The height of the largest plants is four inches.

53. P. *auriculata;* folio oblongo lanceolato in caulem paulo longiorem decurrente, spicâ cernuâ 4-5-florâ folio multò breviore, sepalis lævibus ovatis acutis concavis lateralibus semiconnatis, petalis brevioribus oblongis obtusis, labelli tripartiti lobo medio ovato lateralibus angustissimis falcatis acuminatis apice glandulosis.——*Brazil.*——Flowers dull brown. Petals white. Lip brown, with a very dark globular gland at the point of each lateral lobe. I only know this from the drawings of M. Descourtilz.

54. P. *barbata* (Westcott in Phytologist, 1. 54.); " foliis lanceolatis ovatis apiculatis, sepalis subæqualibus patentibus lateralibus reflexis basi connatis supremo subreflexo, petalis sepalis multò minoribus subciliatis acutis, labello integerrimo linguæformi medio incurvo apice acuto carnoso recurvo suprà glabro subtùs barbato."——*Para.*——Plant about three inches high. Leaves fleshy, keeled. Flowers minute, varying in number from two to many. Sepals ovate-lanceolate, rather obtuse, the lateral ones combined half-way up; colour purple, except at the base, and, if examined with a glass of moderate magnifying power, papillose. Petals much smaller than the sepals, rather ciliate, acute, or acuminate. Labellum entire and strongly bearded beneath, incurved in the middle; apex fleshy, acuminate, reflexed, and in colour similar to the sepals and petals. Clinandrium slightly laciniate.

55. P. *pruinosa;* folio lineari-lanceolato obtuso canaliculato arcuato caule

76

breviore, racemis laxis capillaribus paucifloris folio subæqualibus et brevioribus, sepalis ovatis concavis extùs pruinosis inferiore obtuso emarginato, petalis linearibus æquilongis, labello nano rhombeo-ovato acuto lævi. ——*Surinam.*——Found by Dr. Splitgerber growing on the branches of the Crescentia, a favourite haunt of Orchidaceæ in that colony. The flowers are very small and yellowish.

** *Pubescentes;* nempè floribus, aut saltèm ovario, pubescentibus.

56. P. *strupifolia* (Lindl. in Bot. Reg. 1839, misc. no. 3.); folio longissimo coriaceo obtuso lorato plano, racemis brevibus fasciculatis pendulis, floribus hiantibus pubescentibus, sepalo intermedio oblongo apice carnoso revoluto lateralibus semiconnatis majoribus porrectis intùs maculis pilosis (!) notatis, petalis rhombeo-linearibus acutis, labello oblongo obtuso concavo nudo,——*Brazil.*——Leaves like leather thongs, full 18 inches long. Flowers dull purple and white.

57. P. *bicolor* (P. *picta*, Hooker Bot. Mag. t. 3897. nec Lindl.); folio oblongo coriaceo caule vaginato racemoque longiore, spathâ diphyllâ, racemo plurifloro, sepalis reflexis inferiore trilineato extus basi villoso duobus superioribus ultra medium unitis maculatis, petalis columnâ longioribus ovato-lanceolatis, labello ovato carnoso velutino maculato obtuso intus basi profundè canaliculato, clinandrio dentato, anthera ovarioque pubescenti-tomentosis.——*Mexico.*——Flowers purple, arranged in a spike about one-third the length of the leaf.

58. P. *pectinata* (Lindl. in Bot. Reg. 1839, misc. 1.); folio oblongo acuto cochleato glauco caule ancipiti breviore, spicâ simplici distichâ in folium pronâ eoque breviore, bracteis membranaceis cucullatis ovarii longitudine, sepalis pubescentibus elongatis intermedio lineari lateralibus latioribus basi ventricosis approximatis omninò liberis, petalis lineari-lanceolatis, labello unguiculato oblongo basi pectinato apice truncato denticulato.——*Brazil.*——Flowers sea-green, with a few deep purple spots at the base of the lip. The leaf is so firm, and so much hollowed out that it is capable of holding water, as if it were made of metal.

59. P. *pubescens* (Lindl. in Comp. to Bot. Mag. 2.); folio lanceolato coriaceo canaliculato racemis brevibus nutantibus fasciculatis duplò longiore, calycis pubescentis sepalo supremo lanceolato lateralibus apice tantùm liberis æquali, petalis lanceolatis serratis, labello oblongo membranaceo glabro acutiusculo lobis lateralibus brevibus rotundatis inflexis, clinandrio serrulato.——*Mexico.*——Flowers dull purple, and downy externally; striped with bright purple, and smooth in the inside.

60. P. *inæqualis;* folio oblongo obtuso cauli tereti pluriès breviore, spicâ densâ folio breviore, ovario pubescente, sepalis liberis lateralibus ovatis dorsali oblongo duplò majore, petalis nanis semilanceolatis, labello oblongo trilineato apice rotundato fimbriato.——*Brazil.*——The fringed obtuse lip, half lanceolate petals, and very unequal sepals are the characteristic marks of this species which was discovered by Mr. Miers.

61. P. *casapensis;* folio cordato-oblongo in caulem vix duplò longiorem decurrente, spicâ folio duplò breviore, bracteis cucullatis cuspidatis ovario pubescente longioribus, sepalis carnosis acutis lateralibus ad apicem ferè connatis glabris, petalis lanceolatis serratis diaphanis, labello carnoso obovato acuto concavo margine elevato leviter serrulato.——*Peru*, near Casapi, Mathews (1830).——Leaves $2\frac{1}{4}$ inches, stem $3\frac{1}{2}$ inches long.

77

Flowers yellow, quite smooth, except the ovary, which is downy. Near P. decurrens.

62. P. *ciliata* (Knowles and Westcott, Floral Cabinet, t. 19.); folio oblongo-lanceolato emarginato coriaceo caule longiore, racemis nutantibus multifloris folio brevioribus, sepalis lineari-lanceolatis acutissimis ciliolatis lateralibus connatis, petalis conformibus multò minoribus serratis, labello linguiformi concavo serrulato intra marginem incrassato et verrucoso.——*Demerara.*——Flowers very slightly pubescent, orange-coloured.

63. P. *recurva* (Lindl. in Bot. Reg. 1841, misc. 1.); folio obovato coriaceo subobliquo petiolo pluries longiore, spicâ brevi recurvâ, floribus pubescentibus, sepalis lateralibus connatis, petalis membranaceis rhombeis acutis, labello membranaceo ovato utrinque sinuato obtuso, columnæ margine petaloideo incurvo.——*Brazil.* ?——A small creeping plant, with short recurved spikes of dull pale purple flowers, which smell like sliced cucumber.

64. P. *villosa* (Knowles & Westcott, Fl. Cab. v. 2. p. 78. *Specklinia ciliaris*, Lindl. in Bot. Reg. 1838. misc. no. 40.); folio coriaceo ovali trinervi cauli æquali laxè vaginato, spicis fasciculatis densifloris folio duplò brevioribus, sepalis carinatis longè ciliatis lateralibus ferè ad apicem usque connatis basi saccatis, petalis oblongis nanis apice ciliatis, labello obovato-oblongo obtuso ciliato.——*Mexico.*——The whole plant is about four inches high. The flowers are very small and purple. The sheaths of the stem are altogether like those which are so characteristic of Lepanthes.

Section 3. Aggregatæ. Pedunculi in sinu foliorum aggregati uniflori breves.

* *Pubescentes.*

65. P. *scabripes* (Lindl. in Bot. Reg. 1839. misc. 155.); folio lineari-lanceolato apice tridentato caule longiore, vaginâ pilosâ caulis medium subæquante, flore solitario pubescente, ovario tomentoso, sepalis quam petala triplo-longioribus acutis lateralibus connatis, labello spathulato rotundato. ——*Brazil.*——Remarkable for the rough hairy sheath of the stems.

66. P. *vittata* (Lindl Bot. Reg. 1838. misc. 133.); folio ovali crassissimo, caule tereti, spicâ distichâ folio multò breviore, floribus tomentosis, sepalis oblongis planiusculis intus lævibus superiore longiore et angustiore sub apice calloso lateralibus acutissimis falcatis semiconnatis, petalis obovatis acutis serratis, labello ovato carnoso obtuso lævi basi excavato biauriculato margine scabro, columnâ petalorum longitudine, clinandrio cucullato crenulato.——*Mexico.*

67. P. *breviflora* (Lindl. Bot. Reg. 1841. misc. 125.); folio carnoso obovato-oblongo apice rotundato subtus purpurascente, floribus sessilibus paucis fasciculatis pilosis, sepalis ovatis lateralibus connatis intùs lævibus, petalis ovatis apiculatis, labello carnoso obovato lævi basi excavato pone apicem angulato et subdenticulato.——*Mexico.*——Very near P. aphthosa, with which it agrees in the colour of its flowers. The principal differences are the following: P. breviflora has much shorter and broader flowers; they are more hairy; and they are destitute of tubercles inside the sepals; the labellum is obovate, and angular, as if about to produce a tooth, on each side near the apex; finally, the leaves are obovate and stained with purple, instead of being oval and sharp pointed.

78

68. P. *aphthosa* (Lindl. in Bot. Reg. 1838. misc. 71. *P. peduncularis*, Hooker, Journ. Bot. III. t. 9.); folio ovali coriaceo subtùs discolore cauli carinato triquetro æquali, spathâ bivalvi pauciflorâ, floribus subsessilibus tomentosis carnosis fragilibus, sepalis oblongis convexis subæqualibus intus papillosis lateralibus basi connatis, petalis ovatis acutis glabris subdiaphanis, labello ovato carnoso obtuso antice papilloso basi excavato glabro, columnâ petalorum longitudine, clinandrio cucullato crenulato. ——*Mexico.*——Its leaves are between three and four inches long, dull purple at the edge, and brighter underneath; the flowers are dull yellow.

69. P. *ophiocephala* (Lindl. in Bot. Reg. 1838. misc. 48.); folio oblongo emarginato concavo carnoso caule breviore, flore solitario e spathâ duplici membranaceâ carinatâ erumpente, sepalis oblongis obtusis hiantibus intus papillosis extùs pubescentibus inferiore latiore emarginato, petalis lineari-oblongis intus villosis labello duplò minore lineari-ovato obtuso carnoso glabro canaliculato marginibus elevatis truncatis.——*Mexico.* ——It has one solitary flower protruding from the base of the leaf, and so similar to the head of a snake with the jaws open, that it is difficult at first sight to believe it really a flower that one looks upon. To add to the deception, the small labellum is shaped like a tongue, and moves up and down. The stem is strong, terete, six inches long, with one long striated brown sheath in the middle; the leaf is remarkably thick and fleshy, and above four inches long; the flower is dull yellowish brown, spotted with dull purple, and nearly three-quarters of an inch long.

** *Læves;* a, *labello lineari v. oblongo.*

70. P. *grandiflora* (Lindl. in Bot. Reg. sub fol. 1797.); folio perfoliato oblongo-lanceolato caudato apice tridentato, pedunculis unifloris fasciculatis, sepalo supremo ventricoso acuminato lateralibus in unum connatis recurvis triplò latiore, petalis lineari-lanceolatis hispido-ciliatis, labello breviore unguiculato subrotundo carnoso ciliato supra unguem excavato. ——*Peru.*——Leaf from four to five inches long, and from one and a half to three-quarters of an inch in breadth.

71. P. *octomerioides* (Lindl. in Comp. to Bot. Mag. 2.); folio oblongo-lanceolato subemarginato caule (pedali) duplo breviore, floribus fasciculatis, sepalis liberis carnosis obtusis, petalis æquilongis lineari-lanceolatis carnosis obtusis, labello minimo ovato canaliculato integerrimo glabro.——*Mexico.* ——A tall species, very much like an Octomeria.

72. P. *fragilis* (Lindl. in Bot. Reg. 1841. misc. 188.); folio ovato-oblongo plano carnoso caule compresso canaliculato duplò breviore, pedunculis aggregatis brevibus (2-floris) folio adpressis, floribus carnosis fragilibus glaberrimis, sepalis lateralibus ad apicem usque cohærentibus supremo lineari recto, petalis cuneatis apiculatis columnæ apice denticulatæ subæqualibus, labello lineari apice rotundato basi utrinque paululum dilatato et denticulato superficie lævigatâ.——*Brazil.*——This little plant has the habit of P. prolifera; but its leaves are flat, not folded together, and it is not proliferous. The flowers are bright orange-yellow, and very brittle, like wax; with a double row of minute purple dots along the middle of the labellum.

73. P. *circumplexa* (Lindl. in Bot. Reg. 1838. misc. 27.); folio coriaceo ovato emarginato in petiolum carinatum canaliculatum ipsi æqualem angustato, caule subnullo, pedunculo basi folii immerso et circumplexo,

79

floribus capitatis ovariisque pubescentibus, capitulo disticho sexfloro basi vaginato, sepalis hiantibus erectis inferioribus in unum oblongum rotundatum supremo brevius connatis: supremo angustiore subpanduriformi, petalis lineari-oblongis ciliatis, labello crasso calceiformi serrulato margine calloso.——*Mexico.*——Flowers small, dull dirty brownish yellow.

74. P. *Pöppigii* (Lindl. in Comp. to Bot. Mag. 2. *Myoxanthus monophyllus*, Pöppig, t. 88.); folio lanceolato emarginato, sepalis setaceo-acuminatis lateralibus basi connatis, petalis conformibus, labello lineari-lanceolato obtuso pedi producto columnæ adnato, ovario pubescente.——*Peru.*——Myoxanthus differs from Pleurothallis in nothing except the lengthened base of the column; but this is variable in degree throughout the most genuine species of Pleurothallis, and in P. succosa, Mathewsii, petiolaris, &c. exists as strongly as in Myoxanthus.

75. P. *affinis* (Lindl. in Comp. to Bot. Mag. 2.); folio lanceolato tridentato, sepalis setaceo-acuminatis liberis, petalis conformibus, labello ovato-oblongo obtuso pedi producto columnæ adnato, ovario pubescente.——*Peru.*

76. P. *ruscifolia* (L. no. 1. Hook. Ex. fl. t. 197.); folio ovali-lanceolato caulis longitudine, floribus fasciculatis, spathâ bivalvi, pedicellis 2-3-bracteatis filiformibus floribus longioribus, sepalis ovato-lanceolatis: lateralibus connatis, petalis lineari-lanceolatis minoribus, labello oblongo undulato obtuso.——*Jamaica.*

77. P. *muscifera;* caulis vaginis maculatis, folio oblongo-lanceolato obtuso, sepalo supremo acuminatissimo obtuso lateralibus brevioribus ultra dimidium connatis lanceolatis, petalis linearibus clavatis patentissimis, labello lineari repando basi concavo apice carnoso.——*Guatemala.*——A very curious species, with the flowers more patent than in any other.

78. P. *bidentata* (Lindl. in Bot. Reg. sub t. 1797.); rhizomate repenti filiformi, caulibus ascendentibus folio oblongo-lanceolato apice bi-tridentato longioribus, floribus pedicellatis subsolitariis, sepalis ecarinatis, labello lineari obtuso apice ciliato.——*Rio Janeiro.*——A creeping species, with very small flowers.

79. P. *angustifolia* (Lindl. in Bot. Reg. sub t. 1797.); rhizomate repenti filiformi, caulibus ascendentibus folio lineari-lanceolato longioribus, floribus pedicellatis subsolitariis, sepalis carinatis, petalis lanceolatis denticulatis, labello lineari acuto marginibus pone basin incrassatis.——*Mexico.*——Very like the last, but much smaller.

80. P. *glanduligera* (Lindl. in Comp. to Bot. Mag. 2.); folio ovali carnoso tridentato racemis brevibus 2-3-floris fasciculatis triplo longiore, calycis bilabiati sepalis carnosis basi connatis lateralibus agglutinatis supremo oblongo brevioribus, petalis lanceolatis serratis, labello ovato unguiculato obtuso apice piloso basi utrinque auriculato: auriculis minimis triglandulosis.——*Rio Janeiro.*——Flowers pale yellowish green, faintly streaked with dull purple.

81. P. *orbicularis* (*Specklinia orbicularis*, Lindl. in Bot. Reg. 1838. misc. 41.); folio subrotundo-ovato emarginato caule breviore, floribus fasciculatis, sepalis lineari-oblongis glabris inferiore bipartito, petalis ellipticis acutis serratis, labello ovato rostrato basi concavo apice ciliato obtuso, columnâ apice serratâ.——*Demerara.*

82. P. *discoidea* (Lindl. in Bot. Reg. sub t. 1797.); caulibus strictis folio ovato-lanceolato brevioribus, pedunculis solitariis unifloris, sepalis oblongis acutis lateralibus omnino connatis, petalis linearibus acuminatis decurvis, labello cordato-ovato undulato obtusiusculo sepalis subæquali.——*Trinidad.*——Flowers yellow with a sanguine spot, which is oblong and two-toothed, in the middle of the lip.

80

83. P. *Lingua;* cæspitosa, folio spathulato acuto emarginato carnoso caule multò longiore, floribus solitariis aut aggregatis, sepalis linearibus acuminatis lateralibus basi valdè productis apice liberis, petalis multo brevioribus obovatis obtusis apiculatis, labello lineari obtuso apice convexo.——*Brazil.*——Found on the Organ Mountains by Mr. Miers. It has altogether the appearance of Pl. Lanceola. The leaves are about an inch high, including the short stem.

84. P. *tribuloides* (L. no. 6.); "caule brevissimo monophyllo, folio lanceolato obtuso emarginatove, pedunculo brevissimo unifloro, capsula globosa echinata."——*Jamaica.*——Flores minuti rubri, bracteis geminis ovatis parvis sub germine. Sepala ovata: lateralibus semiconnatis intus punctatis. Petala colore saturatiore. Labellum oblongum, obtusum, concavum, recurvatum, striatum, margine ciliato.

** *Læves;* b. *labello cordato.*

85. P. *tentaculata* (Lindl. in Comp. to Bot. Mag. 2. *Restrepia tentaculata*, Pöppig & Endl. gen. & sp. t. 59.); folio oblongo obtuso plano coriaceo caule subbreviore, floribus 2-4 longè pedunculatis disepalis, sepalis lineari-lanceolatis acuminatis subæqualibus infimo emarginato, petalis patentibus filiformibus æquilongis, labello lineari basi utrinque lobulo rotundato acuto, clinandrio alto dentato.——*Peru.*——Stem and leaf each about eight inches long. Flower-stalks, including the ovary, two inches; and the sepals the same length. Sepals white and transparent. Petals and lip dark purple. This is not a Restrepia, because it has only two pollen-masses.

86. P. *Hartwegii;* folio lanceolato coriaceo cauli valido breviore, floribus densè aggregatis subsessilibus disepalis, sepalo inferiore cymbiformi obtuso superiore multò latiore, petalis lanceolatis æquilongis, labello nano hastato-cordato unguiculato basi altè canaliculato tum deflexo utrinque supra flexuram lobato-crispato apice ovato concavo serrato, columnâ tereti.——*Loxa.*——This is a very remarkable plant. Its thick leathery leaves are nine inches and more in length, and the stem that bears them is much longer, and as stout as a reed. The flowers are large, and apparently of some pale colour, densely aggregated. The column seems quite terete, and has an anther dorsal with respect to the stigma. The pollen-masses are two.

87. P. *laurifolia* (L. no. 2.); "caule monophyllo, folio oblongo acuminato basi angustato, floribus fasciculato-congestis, foliolis calycinis exterioribus lateralibus apice liberis."——*Popayan.*——Perhaps the same as P. succosa.

88. P. *succosa* (L. no. 3. *P. multicaulis*, Pöppig & Endl. n. g. & sp. t. 82.); caule tereti folio ovato-lanceolato succoso triplò longiore, floribus fasciculatis subsessilibus, sepalis lanceolatis: lateralibus omninò connatis, petalis linearibus subulatis, labello minimo cordato ovato carnoso recurvo columnæ longitudine.——*Peru.*——Caulis semipedalis. Folium 2 uncias longum. Petala integra, sepalis paulò breviora. Columna nana, anthera horizontali subglobosa.

89. P. *tridentata* (Klotzsch in Allg. Gartenz. no. 37. 1840); caule secundario tereti monophyllo, folio oblongo-lanceolato superne attenuato apice tridentato basi rotundato, spathâ univalvi conduplicato-carinatâ, floribus minutis aggregatis flavis brevi-pedicellatis, pedicellis 1-bracteatis, bracteis amplexicaulibus vaginatis apice obsoletè 2-3-dentatis pedicello brevioribus, perigonii foliolis nudis exterioribus carinatis lateralibus con-

81

natis, interioribus linearibus acutis, labello parvo subcordato-orbiculari. ——*Caraccas.*

90. P. *scabrilinguis* (Lindl. in Comp. to Bot. Mag. 2.); folio cordato-lanceolato apice tridentato caule duplò longiore, floribus subsolitariis, sepalis ovatis convexis subæqualibus lateralibus omninò connatis, petalis brevioribus lineari-subulatis serrulatis, labello unguiculato oblongo sagittato piloso venâ intermediâ apice varicosâ.——*Peru.*——The habit of this plant is the same as of P. succosa, but the leaves are more horizontal.

91. P. *cordata* (L. no. 4.); caule tereti sulcato folio subrotundo coriaceo horizontali altè cordato acuminato multo longiore, floribus fasciculatis e spatha brevi coriacea subexsertis.——*Peru.*——One of the largest species, with stems a foot high, and almost destitute of sheaths.

92. P. *undulata* (Pöppig & Endlich. 1. p. 48.); caule semitereti folium ovato-lanceolatum acuminatum cordatum undulatum membranaceum horizontale triplò excedente, floribus in sinu folii fasciculatis paucis pedunculos spathâ truncatâ membranaceâ inclusos æquantibus, sepalis oblongo-lanceolatis acuminatis inferioribus ad apicem usque connatis, labello oblongo apice obtuso bidentato erecto plano.——*Peru.*——This is very near P. cordata, if not the same. Pöppig and Endlicher state that it differs in the leaves being coriaceous and undulated.

93. P. *atropurpurea* (*Specklinia atropurpurea*, Lindl. in Bot. Reg. sub t. 1797.); vaginis caulis 2-3 ventricosis, folio oblongo plano 7-9-venio, flore solitario subsessili, petalis 3-cuspidatis, labelli laminâ sagittatâ.——*Jamaica.*——Folium $2\frac{1}{2}$ poll. longum, basi angustatum. Vaginæ infundibulares, pollicem longæ, in caulem 2-pollicarem. Flores atropurpurei. Ovarium hexapterum; alis apice truncatis petalinis minoribus. Alabastrus semunciam longus, angustus, falcatus. Sepala anteriora basi gibbosa, oblongo-linearia. Petala nana, vix 2 lineas longa, ovata, 3-cuspidata, obliqua, sæpiùs venâ alterâ excurrente, alterâ in dimidio majore petali abbreviatâ. Labellum unguiculatum, in columnam pronum, cum ejus basi articulatum; ungue lineari, 3-venio axi cristato, laminâ sagittatâ apicibus posticis obtusis. Columna petaloideo-alata, apice tridentata.

94. P. *papillosa* (Lindl. in Bot. Reg. sub fol. 1797.); caulibus strictis folio angustè oblongo apice tridentato paulò longioribus, pedunculis subaggregatis unifloris flore vix longioribus, sepalis campanulatis lateralibus semiconnatis supremo oblongo brevioribus, labello subhastato obtusissimo apice vittisq. duabus elevatis carnosis papillosis.——*Brazil.*——Flowers of rather large size.

** *Læves;* c. *labello trilobo.*

95. P. *sicaria* (Lindl. in Bot. Reg. 1841. misc. 187.); folio oblongo angusto plano in caulem triplò longiorem triquetrum canaliculatum deorsùm attenuatum angustato, pedunculis brevibus 2-3-floris cernuis aggregatis, sepalis lineari-lanceolatis patulis chartaceis lateralibus ad apicem usque cohærentibus sed facilè separabilibus, petalis ovalibus acutissimis serratis, labello lanceolato acutissimo bilamellato utrinque unidentato, clinandrio serrato.——*Trinidad.*——This species is remarkable, in its stem being exactly like a bayonet blade with the point downwards. It grows nearly a foot high. The flowers are greenish yellow with pale purple stripes along the petals, and none at all on the labellum.

96. P. *retusa* (*Dendrobium retusum*, La Llave, 2. 40.); caule elongato, folio lanceolato carnoso retuso longitudinaliter sulcato margine integro revoluto, pedunculis plurimis filiformibus in sinu folii fasciculatis.——

82

Mexico.——"Flowers pale purplish brown, somewhat spurred." A very doubtful plant, of which nothing more is known.

Section 4. Muscosæ. Pedunculi filiformes, simplices aut bifidi, rariùs apice multifidi (folio plerumque longiores). Species maximâ pro parte minores, nonnunquam minimæ.

97. P. *ephemera;* folio oblongo emarginato basi paulò angustiore pedunculo capillari paucifloro nunc gemino subæquali, ovario trialato, sepalis linearibus acuminatis carinatis lateralibus basi connatis, petalis nanis lanceolatis acuminatis, labello lineari recurvo.——*Brazil*, Descourtilz.——Flowers orange-yellow, nearly an inch long. Petals pink. *Obs.* The lateral sepals are described by M. Descourtilz as distinct, except at the base; they are represented in his drawing as being connate

98. P. *linearis* (*Specklinia? linearis*, L. no. 5.); caule ancipiti folio lineari-oblongo subtriplò longiore, spicis capillaribus bifloris: altero folio subæquali, altero duplò breviore.——*Mexico* and *Peru.*——Caules cæspitosi, 2-3-unciales. Folia uncialia, apice æqualia, obtusa.

99. P. *parvifolia* (Lindl. in Comp. to Bot. Mag. 2.); folio minimo ovali acuto emarginato subsessili, pedunculo solitario folio longiore, sepalis oblongis acutis inferiore emarginato majore, petalis nanis obovatis, labello oblongo obtuso apice maculato.——*Brazil.*

100. P. *muscoidea* (Lindl. in Bot. Reg. 1838, misc. 165.); cæspitosa, acaulis, folio ovali biconvexo, pedunculo setaceo bifloro, perianthio explanato ringente, sepalis lineari-lanceolatis lateralibus basi imâ connatis, petalis multò minoribus conformibus, labello oblongo-lineari obtuso canalicu-

lato, polliniis 2 pyriformibus connatis.——*Brazil.*——The smallest Orchidaceous plant known. Leaves 2½ lines long. The peduncle is as fine as a hair and about four lines long. Flowers purple.

101. P. *Lanceola* (Spreng. Syst. 3. 731. *Specklinia Lanceola,* L. no. 2.); caulibus brevissimis cæspitosis, folio lanceolato subpetiolato, pedunculo capillari bifloro longitudine folii, sepalis obtusis patentibus, petalis apice subciliatis, labello oblongo obtuso concavo recurvo (purpureo margine luteo).——*Jamaica.*

102. P. *trifida;* cæspitosa, foliis lineari-lanceolatis marginatis apice tridentatis cauli æqualibus, pedunculis foliorum longitudine apice plurifloris, sepalis obtusis antico bidentato concavo, petalis spatulatis apice rhombeis, labello oblongo retuso scabro a latere repando, columnâ petaloideâ 3-fidâ ——*Brazil.*——The Organ Mountains furnished Mr. Miers with this small species, whose leaves and stems together are not more than an inch and a half high.

103. P. *plantaginea (Specklinia plantaginea,* Pöppig & Endl. 1 t. 89 A.); cæspitosa, caule brevissimo, foliis petiolatis spathulato-linearibus, pedunculis capillaribus unifloris, sepalis ovato-oblongis acutis ad medium connatis, labello lanceolato obtuso.——*Peru.*——Very near Lanceola, perhaps not distinct. The Specklinia graminea of Pöppig and Endlicher seems to be rather a Maxillaria. It cannot be even of the Pleurothallidous division.

104. P. *pusilla (Dendrobium pusillum,* H. B. K. 1. 357.); caule monophyllo, folio elliptico acutiusculo, pedunculis solitariis aut ternis uni vel bifloris, sepalis oblongo-lanceolatis acuminato-filiformibus.——*Loxa.*——Herba minuta. Caules simplices, apice monophylli, subsemipollicares. Folium

coriaceum, nervoso-striatum, 3 lineas longum. Pedunculi folio duplò longiores. Flores 2 lineas longi, flavescentes. Sepalum supremum latius, concavum, liberum; lateralia plana. Petala oblonga, sepalis triplò breviora. Labellum oblongum, subunguiculatum.

105. P. *hymenantha* (Lindl. in Bot. Reg. sub fol. 1797.); folio lineari-lanceolato apice obsoletè 3-dentato pedunculis flexuosis capillaribus subæquali, bracteis minimis, sepalis acutis lateralibus semisejunctis, petalis acuminatis parum brevioribus, labello oblongo obtuso subundulato complicato.——*Brazil.*——Resembles P. Lanceola in habit.

106. P. *uniflora* (Lindl. in Comp. to Bot. Mag. 2. 355.); folio lineari-lanceolato subsessili, pedunculo unifloro folii longitudine, sepalis oblongo-lanceolatis acutis subæqualibus inferiore emarginato, labello lineari-lanceolato obtuso basi concavo.——*Brazil.*——Near P. Lanceola, which it quite resembles in habit.

107. P. *sertularioides* (Spreng. syst. 3. 731. *Specklinia sertularioides,* L. no. 1.); rhizomate repenti articulato, caulibus brevissimis, folio lanceolato subpetiolato, pedunculis unifloris folio brevioribus, sepalis liberis lineari-lanceolatis patentibus, petalis conformibus brevioribus, labello lanceolato, clinandrio integerrimo.——*Jamaica.*——Rhizoma Sertulariæ modo articulatum, unde nomen. Pedunculi capillares, longitudine ferè foliorum, squamis spathaceis duabus in medio. Flores albidi, apicibus flavis. Labellum recurvum, basi concavum.

108. P. *Gardneri;* rhizomate repenti intricato, foliis ovalibus marginatis basi angustatis, pedunculis capillaribus foliis multò longioribus, sepalis lineari lanceolatis lateralibus ultra medium connatis, petalis conformibus æquilongis apice serrulatis, labello unguiculato rhombeo acuto lamellis duabus elevatis supra basin.——*Brazil.*——A discovery of Professor Gardner's, with much the appearance of P. sertularioides.

109. P. *punctata* (Lindl. in Bot. Reg. sub fol. 1797.); caulibus strictis folio crasso ovali plano-convexo marginato longioribus, pedunculis capillaribus paucifloris, petalis lanceolatis apice pubescentibus, labello unguiculato cordato-sagittato obtuso, columna elongata.——*Brazil.*——A very small plant, not much more than an inch high. Flowers dark purple. Leaves dotted with purple on the under side.

110. P. *corniculata (Specklinia emarginata,* L. no. 3. *Epidendrum? monophyllum,* Hooker Exot. Fl. t. 109.); caulibus folio oblongo emarginato subæqualibus, pedunculis bifloris folio brevioribus, sepalis lanceolatis, petalis duplò brevioribus, labello lineari-lanceolato basi gibboso.——*Jamaica.*

111. P. *seriata* (Lindl. in Bot. Reg. 1840, misc. 175.); folio carnoso plano ovali emarginato caule multò longiore, pedunculo filiformi paucifloro folii longitudine, sepalo dorsali carinato margine reflexo antico latiore acuto emarginato, petalis brevibus spathulatis carnosis lævibus, labello oblongo obtusissimo pubescente medio exarato lobulis lateralibus minimis rotundatis.——*Brazil.*——An inconspicuous species, with pale yellowish-green flowers marked with rows of purple dots. It is a beautiful object when examined microscopically.

112. P. *articulata;* rhizomate repenti intricato articulato polyphyllo, foliis ovalibus mucronatis marginatis basi angustatis caulibus brevissimis multò longioribus, pedunculis longissimis filiformibus bifloris, sepalis lateralibus dorsalibus semiconnatis, petalis linearibus acuminatis sepalis æqualibus, labello parvo pubescente obovato trilobo; laciniâ intermediâ carnosâ lateralibus membranaceis nanis obtusis.——*Brazil.*——A very remarkable species, found by Prince Maximilian of Wied Neuwied, creep-

ing over the trunks of trees, and covering them with a dense entangled mass of small foliage. In habit it accords entirely with P. sertularioides.

113. P. *Miersii;* rhizomate filiformi repenti intricato, foliis ovalibus mucronatis marginatis 3-nerviis, pedunculis setaceis unifloris folio longioribus, bracteâ rotundatâ, sepalo dorsali ovato acuminato lateralibus ovatis semiconnatis, petalis lanceolatis serrulatis, labello concavo oblongo acuto. ——*Brazil.*——This very minute species was found upon the Organ Mountains by John Miers, Esq. The flowers are the smallest of the genus, and nearly transparent. The leaves are not more than two lines long.

---

92. MAXILLĀRĬĂ bractēscēns.

M. *bractescens;* pseudobulbis conicis elongatis sulcatis, foliis latis plicatis, scapo trivaginato, racemo elongato paucifloro, bracteis linearibus acuminatis scariosis floribus longioribus, sepalis oblongo-lanceolatis margine revolutis, petalis conformibus minoribus, cornu brevi rotundato, labello trilobo apice recurvo obtuso venis omnibus elevatis sinuosis rugosis callo elevato rugoso apice subtrilobo.

This is a plant with all the habit of M. squalens. The flowers are however twice as large, from an inch to an inch and half apart, of a dull yellow, with a reddish brown lip. The scape bears five or six flowers, and is fully a foot and half high. The very long and curious narrow bracts readily indicate it. Sent to the Horticultural Society from Loxa by Mr. Hartweg. Flowered in October last.

94. DENDROBIUM macrophyllum.

Lindl. in Bot. Reg. 1839. *p.* 36. *Sertum Orchidaceum, t.* 35.

The *Dendrobium macranthum* of the Botanical Magazine, t. 3970, is the same plant.

95. BOLBOPHY̆LLŬM adenopĕtălŭm.

B. *adenopetalum;* foliis oblongis obtusis petiolatis basi vix bulbosis, spicâ secundâ basi laxè vaginatâ multiflorâ folio sublongiore, sepalis acuminatis, petalis spathulatis acutis intùs glandulosis, labello ovato angusto obtusè acuminato ciliato basi canaliculato.

A native of Sincapore, whence it was received by Messrs. Loddiges. The flowers are yellowish and slightly sweet-scented. It resembles B cocoinum, from which its want of a pseudo-bulb, or at most its extremely small one, and its petals not serrated but glandular on the inside, readily distinguish it. Possibly it may be one of the plants included by Dr. Blume under his Diphyes flavescens; at least it must be very near that species.

96. LYCĀSTĒ plănă.

L. *plana;* bracteâ supremâ ventricosâ cucullatâ ovario longiore, sepalis oblongis planis basi in cornu brevi conico connatis, petalis conformibus apice tantùm recurvis, labelli trilobi lobis lateralibus apice crenulatis intermedio subrotundo serrato callo elevato obtuso obsoletè trilobo, columnâ pubescente, antherâ villosâ.

This plant is nearly allied to the Maxillaria, or Lycaste, macrophylla, of which it has quite the habit. It differs in the petals being quite even not undulated, and in the lateral sepals being much more exactly oblong; the tubercle on the lip is also much more obtuse. In colour too they are different. *L. macrophylla* has olive green sepals, and petals almost colourless. *L. plana* has the sepals a deep rich madder red inside, and the petals are richly tipped with crimson.

100. STELIS atropurpŭrĕă.

Hooker in Bot. Mag. t. 3975.

This is a good figure of *St. ciliaris,* Lindl. in the Companion to the Botanical Magazine, vol. ii. p. 353., without the long ciliæ, which, although very conspicuous when fresh, almost disappear in a dry state.

EDWARDS'S

BOTANICAL REGISTER:

OR,

ORNAMENTAL FLOWER-GARDEN

AND SHRUBBERY:

CONSISTING OF

COLOURED FIGURES OF PLANTS AND SHRUBS,

CULTIVATED IN BRITISH GARDENS;

ACCOMPANIED BY THEIR

History, Best Method of Treatment in Cultivation, Propagation, &c.

AND

MONTHLY CHRONICLE

OF

BOTANICAL AND HORTICULTURAL NEWS.

CONTINUED

BY JOHN LINDLEY, Ph. D. F.R.S. AND L.S.

PROFESSOR OF BOTANY IN UNIVERSITY COLLEGE, LONDON,
AND THE ROYAL INSTITUTION OF GREAT BRITAIN,
VICE-SECRETARY OF THE HORTICULTURAL SOCIETY,
&c. &c. &c.

1843.

OR VOL. XXIX. OF THE ENTIRE WORK.
OR VOL. XVI. OF THE NEW SERIES.

——viret semper——nec fronde caducâ
Carpitur.

LONDON:
JAMES RIDGWAY, PICCADILLY.

M.DCCC.XLIII.

3

ODONTOGLOSSUM citrosmum.

Lemon-scented Odontoglossum.

GYNANDRIA *MONANDRIA*.

Nat. ord. ORCHIDACEÆ § VANDEÆ.

ODONTOGLOSSUM. Kunth.

O. *citrosmum* (Lindl. Bot. Reg. 1842. misc. 68.); pseudobulbis subrotundis compressis lævibus monophyllis, folio oblongo-ligulato obtuso racemo paulò breviore, sepalis oblongis obtusis petalisque conformibus subæqualibus, labello unguiculato reniformi basi bituberculato, columnæ alis lateralibus subtruncatis dorsaliq. rotundato denticulatis.

Although such plants as this seem as if they weakened the genus Odontoglossum, on account of their similarity in habit to Oncidium, yet in truth they rather confirm that genus by showing that although the appearance of Oncidium may be assumed, yet the main points of structure remain unaffected. One of the most important of these points is the partial parallelism of the base of the labellum with the column, and the presence of a pair of parallel raised plates at that part. This occurs in the original Odontoglossum, and runs through all the numerous species published and unpublished with which I am acquainted; and it does not occur in any Oncidium.

While however, notwithstanding its habit, this *Odontoglossum citrosmum* confirms, rather than diminishes, the propriety of separating Odontoglossum from Oncidium; it has a peculiarity of its own, which may possibly suggest to some minds the propriety of forming it and such plants as O. brevifolium and pulchellum into a separate genus, characterized by the presence of a toothed wing or membrane at the back of the anther; and the similarity of those plants in habit might seem to justify the measure. In that case the generic name of TRYMENIUM might be taken for them. I do not however at present see the necessity of regarding them as more than a section of Odontoglossum.

It was imported by George Barker, Esq., of Birmingham, from Mexico, and given to Thomas Brocklehurst, Esq., of the Fence near Macclesfield, by whose gardener it was exhibited at one of the great meetings of the Horticultural Society at Chiswick in 1842. It has large snow-white and rose-coloured flowers, of great beauty, exhaling a delicate smell of lemons.

Fig. 1. represents the column with its three wings and the base of the lip. Fig. 2. shews the masses of pollen, their strap and gland.

6

DENDROBIUM sanguinolentum.

Blood-stained Dendrobium.

GYNANDRIA *MONANDRIA*.

Nat. ord. ORCHIDACEÆ, § MALAXEÆ.

DENDROBIUM. Swartz.

Sect. EUDENDROBIUM. Caulis teres. Folia plana. Flores membranacei patuli.

D. *sanguinolentum* (Lindl. in Bot. Reg. 1842. misc. 73.); caulibus teretibus pendulis, foliis ovato-lanceolatis, floribus gemellis, sepalis petalisque ovatis obtusiusculis patulis, labello trilobo glabro: laciniâ intermediâ retusâ nunc utrinque plicatâ; ungue concavo brevi appendice cornuformi pubescente retrorsâ aucto.

Probably there is no plant among all the species of Dendrobium now known to our gardens which is more delicately beautiful than this. Its colours too are so singular as on that account alone to render it an object of much interest; for here we have the cyanic and xanthic tints in one and the same flower. We are always prepared to find red spots on a yellow ground, or vice versâ, but it is a most unusual thing to find clear pure violet on petals the whole remainder of whose tint is yellow. In this plant, however, the combination occurs, producing a very gay and unexpected effect.

It was sent from Ceylon, to His Grace the Duke of Northumberland, by Mr. Nightingale, and flowered at Sion in August last. It has pendulous stems like those of D. Pierardi, but of a delicate purple when young; the leaves too are stained underneath and at the edges with the same colour. The flowers are as large as those of D. aggregatum, of a clear fawn colour, with the tips of the segments and lip stained with a deep rich violet. There is moreover a *scarlet* spot in the middle of the lip.

Fig. 1. represents the outline of the lip of this plant. Fig. 2. shows the form of what I take to be a variety, larger, and having the middle scarlet spot, but destitute of the violet tips. This is not distinguishable from the other in foliage, nor in any other way than in the manner now indicated.

Both species have a strong downy tooth near the base of the lip, parallel with the surface and directed towards the foot of the column.

9

CĂTASĔTUM planĭceps.

Flat-headed Catasetum.

GYNANDRIA *MONANDRIA.*

Nat. ord. ORCHIDACEÆ § VANDEÆ.

CATASETUM. Botanical Register, vol. 20. *fol.* 1667.

C. *planiceps;* sepalis petalisque ovatis conniventibus, labello carnoso galeato rotundato compresso apiculato indiviso margine serrato.

Notwithstanding our unwillingness to add to the doubtful species of this strange and changeable genus we can hardly hesitate about regarding this as new. It is true that its habit is that of *C. maculatum, tridentatum,* and *semiapertum,* from the second of which it differs in its serrated lip, from the first and last in its lip not having the edges incurved, and from all in the singular truncate form of this helmet-shaped organ, which is flattened from front to back and not laterally.

It is a native of the Spanish Main, whence it was imported by Messrs. Loddiges, with whom it flowered in June 1841. It is the No. 934 of their catalogue.

It may be cultivated in a similar way to the other species; by being potted in mossy turf and potsherds mixed with some small pieces of wood. Like other fleshy-stemmed Orchidaceæ, this requires three seasons, namely, that of growth, of flowering, and of rest. In spring, while in a vigorous state, plenty of water and a humid atmosphere is requisite, at a temperature not less than 70° at night, nor yet above 80° in day, and shaded during sunny weather. As the flowering season advances, the water should be diminished gradually; and in winter, when the plant is in a dormant state, withheld altogether for a few weeks, and the temperature not allowed to rise above 60° or to fall below 50°.

12

ONCIDĬŬM bicallōsum.

Two-warted Oncidium.

GYNANDRIA *MONANDRIA.*

Nat. ord. ORCHIDACEÆ § VANDEÆ.

ONCIDIUM. Swartz.

EUONCIDIUM; Heteranthium; Folia plana; Pentapetala; Micropetala; labello pandurato; i. e. medio constricto basi angustiore. *Sertum Orchidaceum sub t.* 48.

O. *bicallosum;* bracteis ovatis membranaceis obtusis, sepalis liberis obovatis concavis, petalis oblongis obtusis, labelli lobis lateralibus abbreviatis intermedio maximo transverso emarginato subcordato, cristâ bicallosâ, tuberculis distantibus uno ante alterum posito rugosis subtrilobis, columnæ auriculis linearibus falcatim recurvis. *Lindl. in Bentham pl. Hartw. p.* 94.

Among the numerous fine things found in Guatemala by Mr. Skinner, there are not many which are more deserving of cultivation than this plant, which has so much the habit of *O. Cavendishianum* as to seem a mere variety of it, although in reality quite a distinct species. It was also met with by Mr. Hartweg. The flowers are fully two inches in diameter, which is four times the size of *O. Cavendishianum;* they appear in a dwarf erect raceme, not panicle; they are of a rich yellow, with the sepals and petals bordered with cinnamon colour; the labellum has two very small lateral lobes, and for its crest it has a pair of distinct tubercles, the posterior double, the anterior 3-lobed, and the two separated by a considerable space. The flowers are moreover slightly scented, which is not the case with *O. Cavendishianum.* Many plants are doubtless in the country; but no one has succeeded in flowering them except Mr. Bateman, who sent a fine specimen to a late meeting of the Horticultural Society.

The figure at the corner of the plate shews the peculiar arrangement of the warts on the crest of the lip. Besides which it indicates the unusual form of the narrow incurved wings of the column, and a thickening of the margin of the lip near its base.

It is a stove plant, and requires to be grown in a mixture of turfy peat and potsherds; so that the whole may be efficiently drained. A smaller pot, inverted in the one in which it is potted, allows the warmth to rise through the soil, and makes the whole lighter than when it is drained at the bottom with potsherds. It may also be grown suspended from the rafters of the house like other air plants, provided it is kept moist enough during the growing season. Like all these plants it requires a humid atmosphere, a liberal supply of water when growing, and partial shade during bright sunshine in summer. A night temperature of 60° in winter and 70° in summer will be amply sufficient for its growth.

17

DENDROBĬŬM rhombĕum.

Rhomb-lipped Dendrobium.

GYNANDRIA *MONANDRIA.*

Nat. ord. ORCHIDACEÆ, § MALAXEÆ.

DENDROBIUM. Botanical Register, vol. 7. *fol.* 548.

D. *rhombeum;* caulibus teretibus foliosis, foliis lanceolatis acutis, racemis brevibus 4-floris, sepalis ovalibus obtusis, petalis duplò latioribus ovatis, labello rhombeo undulato acuto medio pubescente, columnâ utrinque tuberculatâ pone basin cornu deflexo auctâ, antherâ pubescente.

This pretty species has much general resemblance to D. aureum, from which however it differs in having smaller blossoms, a labellum without serratures, and the flowers in short racemes instead of pairs.

It is a native of Manilla, whence it was sent by Mr. Cuming to Messrs. Loddiges, in whose catalogue it stands as no. 319, and with whom it flowered in August last.

The imported specimen is less vigorous than it may be expected to become hereafter, in which case the number of flowers will probably be augmented.

Here, as in some other cases, the column is furnished, near the base, with a deflexed horn lodged in a niche just above the insertion of the lip. This is a remarkable process, and is represented at fig. 1. of the accompanying plate; its nature requires to be investigated.

Like Maxillaria, the whole genus Dendrobium, with those which are near it, requires a careful revision; but in the absence of more complete materials from the Indian islands, and especially of Dr. Blume's Java species, it seems premature to undertake the operation.

18

PERISTĔRIA Humboldti.

Humboldt's Peristeria.

GYNANDRIA *MONANDRIA*.

Nat. ord. ORCHIDACEÆ, § VANDEÆ. Div. MAXILLARIDÆ.
PERISTERIA. Botanical Register, vol. 23. *fol.* 1953.

P. *Humboldti;* racemo elongato pendulo multifloro, calyce patulo, sepalis oblongis concavis obtusis, petalis rhombeis obtusis planis minoribus, labelli inflexi hypochilio carnoso oblongo concavo, metachilio bilobo laciniis erectis cuneatis carnosis margine inferiore valdè incrassato medio appendice bilobâ carnosâ divaricatâ utrinque bidentatâ aucto, epichilio unguiculato canaliculato apice dilatato truncato, columnâ utrinque in marginem rotundatum alatum productâ, stigmate et columnæ dorso villosis, caudiculâ lineari, glandulâ lunatâ.
Anguloa superba. *Humboldt, Bonpland & Kunth, nov. gen. & sp. pl.* 1. 343. *t.* 93. *Lindl. Gen. & Sp. Orch. p.* 160.

For this noble plant, which has a pendulous raceme a couple of feet long, we are indebted to John Wilmore, Esq., of Oldford, near Birmingham, who imported it from Porto Cabullo, in the province of Venezuela, about three years since. It flowered for the first time in March, 1842, and in April was exhibited to the Horticultural Society where it was distinguished by a silver medal. In many respects it has the habit of Peristeria Barkeri, but the leaves have longer footstalks.

When the genus Peristeria was founded nobody suspected that the long sought Anguloa superba of Humboldt could belong to it; still less did we imagine, when this noble species was sent from Birmingham, that it could be the very same. Nevertheless, upon attentively considering Humboldt's figure, no doubt can be entertained of the fact; or that Anguloa superba is nothing more than this Peristeria with the raceme made to grow erect instead of pendulous. The plant of that great traveller is described as an inhabitant of trees in the temperate parts of the province of Tumbez, near Zaruma, in Peru, and a village called Catacocha; it was also found cultivated in the gardens of Loxa, at an elevation of between 6 and 7000 feet above the sea. Its Spanish name is said to be *Periquito.*

Upon comparing our plant with the original figure of Anguloa in the Flora Peruviana it is obvious that it cannot be referred to that little known genus, one of the most distinctive characters of which is having what Ruiz and Pavon call a chrysalis-shaped lip (that is, we presume, a lip rolled up in the form of a chrysalis) *seated on a long stalk;* by which circumstance in particular it is distinguished from Maxillaria.

Thus it appears that neither of Humboldt's Anguloas belong to the genus; *A. superba* being this Peristeria Humboldti, and *A. grandiflora* bring Stanhopea insignis. With respect to Pöppig's *Anguloa squalida*, the figures of this author are so bad that it is difficult to say what it is; it may really be an Anguloa.

Fig. 1. represents a side view of a lip and column; 2. the lip seen from above: 3. the column in half-face, the lip being cut off.

22

CYCNŌCHES pentadactȳlon.

Five-fingered Swan-neck.

GYNANDRIA *MONANDRIA*.

Nat. ord. ORCHIDACEÆ § VANDEÆ.
CYCNOCHES. Botanical Register, vol. 21. *fol.* 1742.

C. *pentadactylon;* racemo brevi stricto, sepalis petalisque lanceolatis reflexis supremo incurvo, labelli unguiculati margine revoluto dorso adnato, hypochilio cornu recto incurvo aucto, metachilio 4-lobo medio foveato laciniis 2 posticis obtusis carnosis incurvis anticis subulatis, epichilio linguiformi acuto lævi. *Supra misc. no.* 26.

The fleshy-stemmed Orchidaceæ, consisting of Catasetum, Cycnoches, Mormodes, and Cyrtopodium, probably form a group (Catasetidæ), among which we find the most astonishing deviations from ordinary structure, and the most startling variations from what appears to be the rule in other parts of the organic world. If we were informed that the Camelopard in the Zoological Gardens had shortened the vertebræ of its neck till it was no longer than a cow's, or that a Kangaroo had exchanged its tail for the switch of a Shetland pony, a more surprising thing would not be announced than those changes with which we are now familiar in this group of Orchidaceæ.

How Myanthus, Monachanthus, and Catasetum have been seen to change into each other has been already told in this work, (Vol. 23, t. 1951.) An instance is also mentioned at the same place, of Cycnoches Loddigesii having been caught in the very act of playing similar tricks; and with the forthcoming number of Mr. Bateman's Orchidaceæ of Mexico and Guatemala, a yet more startling case will be recorded.

All these things are here mentioned for the sake of shewing how difficult it is in such plants to judge correctly as to what may be a species and what a sport. Influenced by these considerations I have till lately refrained from saying any thing about the plant now figured. It has in some respects so much the appearance of C. maculatum, that when I first received it from Mr. Veitch of Exeter, in March 1842, I hesitated whether to regard it as a variety or a distinct species. A plant, however, obtained from Brazil direct by Messrs. Loddiges (Cat. no. 890) has given me courage to regard it as the latter. It has a short raceme of much larger greenish-yellow flowers, with broad chocolate-brown blotches, and its lip is quite remarkable, having 5 finger-like lobes, and no more, instead of the lateral comb-like fringes of C. maculatum.

23

ONCIDĬŪM microchilum.

Small-lipped Oncidium.

GYNANDRIA *MONANDRIA*.

Nat. ord. ORCHIDACEÆ § VANDEÆ.
ONCIDIUM. Swartz.

EUONCIDIUM; Heteranthium; Folia plana; Pentapetala; Micropetala; labello bilobo. *Sertum Orchidaceum, sub t.* 48.

O. *microchilum* (Bateman in Bot. Reg. 1840, misc. 193.); pseudobulbis lenticularibus brevibus monophyllis, folio oblongo carinato carnosissimo acuto quam scapus erectus versus apicem paniculatus quadruplò breviore, sepalis liberis lateralibus longiùs unguiculatis, petalis oblongis subundulatis retusis, labello duplò latiore quam longo lobo intermedio nano triangulari lateralibus rotundatis planis, cristâ reniformi crenatâ, columnæ nanæ alis subulatis apice glandulosis.

This most curious plant was introduced some years ago from Guatemala by Geo. Ure Skinner, Esq. and was afterwards found in the same country by Mr. Hartweg. When I first published it I knew it only from report, and a single damaged dried flower, and consequently several inaccuracies crept into its description, which led to its being misplaced in the enumeration of the genus published in the Sertum Orchidaceum. Its real situation is near O. deltoideum.

Although not a very splendid plant, it is far from being undeserving notice. Its very glaucous flower-stem, the snow-white lip and crimson petals, lying as it were in the middle of dusky brown sepals, produce a singular and pleasing effect. For the opportunity of figuring it, we are indebted to J. C. Harter, Esq. of Broughton, near Manchester, who was so obliging as to forward specimens, along with a drawing by Mrs. Powell, in September last. Mr. William Ashton, the gardener to Mr. Harter, also deserves to be mentioned, as having succeeded in flowering the plant before any one else. We understand that the Orchidaceæ of Mr. Harter's collection are hardly surpassed in health by anybody's. Arpophyllum is reported to exist there, with many new species from Oaxaca.

Mr. Skinner, who discovered the species, has been so kind as to give us the following information upon that subject.

"The Oncidium microchilum I first found on the top of the 'Cuesta' of Puentezuelas, some thirteen leagues from Guatemala, and sent it to Mr. Bateman in 1838. It was growing on a bare rock, with a quantity of dead leaves and grasses about its bulbs, and its roots woven into the interstices of the rock and mould about it; very much exposed to the sun, except during the middle of the day, when a ledge of rock seemed to afford it a little shade. I afterwards found it in great abundance on the rocky banks of the river Michatayal, from whence the present one now flowered came; it was sent to Mr. Harter and Mr. Clowes in 1840, and arrived in June of that year. I never saw it except in such situations, generally exposed, and always among rocks. It is very fragrant; and in its native habitat I have always observed the sepals and petals darker and more marked than this now flowered here. The temperature generally of the above habitats is 68° to 70°; and, from being exposed, cold at nights."

Fig. 1. represents a side view of the lip; 2. a front view of the same part.

It may be potted in turfy heath-mould, mixed with a few pieces of small potsherds, which will cause superfluous water to pass off freely. The pot should be half filled with potsherds, broken smaller towards the top, and the soil considerably elevated above its brim. During the summer months, while the plant is in a growing state, plenty of water should be given, and the temperature allowed to rise to 80° by day, and fall to 68° at night—giving a slight shade in bright sunny weather. In the winter season very little water is required, and the temperature may be allowed to fall as low as 50° or 55°.

28

DENDROBĬUM taurĭnum.

Bull-headed Dendrobium.

GYNANDRIA *MONANDRIA.*

Nat. ord. ORCHIDACEÆ § VANDEÆ.

DENDROBIUM. Swz.

§ *Spatulata* (Lindl. in London Journ. of Botany, 2. 235.). Cornu breve. Petala elongata. Racemus pedunculatus oppositifolius. Caulis erectus teres.

D. taurinum.

Whether or not it is possible to divide the great, and unnatural genus Dendrobium by any really stable characters is uncertain. It certainly appears to contain several distinct types of structure, but up to the present time I have sought in vain for any thing sufficiently precise on which to found generic characters.

Among the sections of the genus no one is better marked than this, which consists of species with a stiff erect habit, racemes of flowers placed on a long peduncle opposite the leaves, large showy flowers, and the petals remarkably longer than the sepals. But beyond this I can find nothing that differs from Dendrobium; and as the three first characters are merely of habit, the last can hardly be regarded of enough importance to authorise the establishment of a genus. For this reason I proposed in the London Journal of Botany to combine them, under the name of *Dendrobia spatulata*, into a new section of the genus.

Of plants with this structure I know six completely; and a seventh is probably to be added. Of these the following is an enumeration.

1. D. *taurinum* (the subject of the accompanying plate); foliis oblongis obliquè emarginatis, racemo oblongo, sepalis ovatis obtusiusculis, petalis linearibus contortis duplò longioribus, labello oblongo apice crispo per axin lineis 3 elevatis intermediâ apice flexuosâ aucto.

A native of Manilla, whence Mr. Cuming sent it to Messrs. Loddiges. It has large flowers, with yellowish green sepals, rolled back at the points, very long deep purple twisted petals, and a paler purple lip, with three elevated lines along the middle, and a few small tubercles near the apex (fig. 2.). The column (fig. 1.) is short and somewhat pouched at the base, in consequence of the lip being adherent to it at that place. The plant is five feet high. It flowered at Hackney in October last. I have wild specimens from Mr. Cuming; but they are not nearly so handsome as that now figured from Messrs. Loddiges.

2. D. *Mirbelianum* (Gaudich. Voyage, t. 38.); foliis ovato-oblongis obtusis, racemis ovatis longè pedunculatis, sepalis ovato-lanceolatis acutis erectis, petalis spathulatis unguiculatis obtusis longioribus, labelli trilobi ovato-lanceolati subsessilis lobis lateralibus rotundatis intermedio lanceolato undulato acuto, lineis tribus elevatis lateralibus versus basin duplicibus.

A native of New Guinea, whence I have specimens from Mr. Hinds. Its racemes, including their peduncle, are more than a foot long. The petals are an inch and half long. Gaudichaud's figure is not a good one.

3. D. *veratrifolium* (Lindl. l. c. L.); foliis oblongis obtusis amplexicaulibus 9-11-nerviis, racemo terminali elongato multifloro, sepalis undulatis acutis, petalis spathulatis obtusis planis vix duplò longioribus, labello oblongo obtuso membranaceo venis tribus elevatis per axin duabusque minoribus lateralibus: lobis lateralibus nanis obtusis intermedio oblongo undulato.

Another species from New Guinea, where it was found by Mr. Hinds. It is a most beautiful plant, with racemes a foot and a half long, loaded with flowers, whose spatula-shaped petals are an inch and more in length. The inflorescence is very loose in this; in D. Mirbelianum it is far more compact.

4. D. *macranthum* (A. Rich. Sert. Astrolab. p. 15. t. 6.); foliis ellipticis apice obliquo subbilobis, racemo longipedunculato versus apicem monophyllo, sepalis ovato-lanceolatis, petalis paulò longioribus unguiculatis lanceolatis acutis, labello lanceolato trilobo per axin tricarinato laciniâ intermediâ acuminatâ lateralibus nanis rotundatis.

A native of the island of Vanikoso. The stem is described as two feet and more high; the leaves from three to four inches long, the raceme about a foot long, the flowers three inches in dameter. It is very near my D. veratrifolium, but differs in having much shorter petals and a very acute lip.

5. D. *antennatum* (Lindl. l. c.); foliis lanceolatis carnosis obliquè emarginatis racemo oppositifolio brevioribus, sepalis acuminatis, petalis linearibus duplò longioribus reflexis, labello trilobo venis 5 elevatis rectis per axin; lobo medio ovato acuto plano 3-costato.

Also among Mr. Hind's New Guinea collections. It is a most curious thing, with flowers having their long horn-like petals directed backwards, and their sepals forwards. The petals are two inches long, and scarcely half a line wide. The leaves are succulent, brittle, and veinless when fresh.

6. D. *undulatum* (*R. Brown Prodr.* 332. *L. no.* 62. D. discolor, *Lindl. in Bot. Reg.* 1841, *t.* 38.); foliis ovato-oblongis emarginatis, racemis longissimis, sepalis petalisque undulatis, labelli trilobi lobis acutis intermedio lanceolato, lineis quinque elevatis intermediâ apice flexuosâ.

A tropical New Holland species, with singular dingy yellow-brown flowers, of little beauty. It has also been found in Java.

7. ? D. *affine* (Onychium affine, *Decaisne herb. timor.* 37.); "foliis lineari-oblongis acutiusculis coriaceis, pedunculo foliis triplo longiore oppositifolio? laxifloro, floribus spicatis; perianthio erecto, segmentis exterioribus lineari-lanceolatis acutis, interioribus subobovatis rotundatis mucronulatis; labelli unguiculati lobis 3, medio lineari-lanceolato, lateralibus subrotundis."

As M. Decaisne compares this with D. Mirbelianum, it probably belongs to the present section, notwithstanding that he conjectured the stem to be bulbous. The petals are said to be obovate and mucronate. It is a native of Timor, with acute linear-lanceolate leaves.

33

MORMŌDES luxātum.

Dislocated Mormodes.

GYNANDRIA *MONANDRIA*.

Nat. ord. ORCHIDACEÆ § VANDEÆ.

MORMODES. Supra vol. 22. *fol.* 1861.

M. *luxatum*. (See the enumeration further on.)

The genus Mormodes has so entirely the habit of Catasetum that we have no means of distinguishing those two genera except when in flower. At that time Mormodes presents some peculiarities of so strange a nature that, if they were not found constant in several distinct species, we should be tempted to regard them as monstrosities. In particular the column, instead of being straight and standing erect in the centre of the flower, is bent over to one side, just as if it had been subjected to violence. There is also a great irregularity of direction and proportion in the parts that surround the column.

No where are these singularities so strongly marked as in the plant now before us, whose sepals and petals, instead of standing in an obviously alternate position with respect to each other, are so completely broken up and twisted out of their places that they can hardly be recognised, and the whole floral apparatus is as it were dislocated. For example, of the three sepals, the back one is placed almost opposite one of the petals; the other petal is shifted to one side, so as to stand half behind the first; and the lip, instead of being stationed exactly between the two petals and two lateral sepals, turns its back to the left-hand sepal, and its face to the right-hand petal. And then the column is bent to the left as well as the lip, but not in the same degree, so that even these two organs are not, as they usually are, opposite each other.

In mentioning the existence of this singular plant we some time since spoke of it thus:—" When Mr. Ross, the collector to George Barker, Esq. of Birmingham, was in the neighbourhood of Valladolid, in Mexico, there was brought to him some masses of a plant reported to be of great beauty. They arrived safely in England, and one of them has at last flowered, proving to be this species; a most extraordinary plant, of large size, of a stately habit, with a very delicious fragrance, although powerful, but with no brilliant colours to render it what *we* call handsome. Its flowers are three inches in diameter, pale lemon-colour, fleshy, rather globular, but so distorted by the complete dislocation of all the parts, that it would be difficult to ascertain their real nature, if it were not for the token given by the labellum. The latter has a deep brown streak drawn down its middle, and covers over the column like a hood. The leaves of this plant are about three feet long, narrow, deep green, with a very fine glaucous bloom upon their underside. It is worthy to be associated with even Sobralia macrantha in the choicest of all collections of these plants."

Fig. 1. represents the lip; 2. the column; 3. gland, caudicula, and pollen-masses; 4. a pollen-mass seen from behind.

With regard to its cultivation, it is a stove-plant, which should be potted in a compost of turfy heath-mould, mixed with a portion of small potsherds Water should be liberally given in fine weather, during the summer, and the temperature kept as high as 80° by day, and 70° by night. As the season of rest advances, watering should gradually be dispensed with, so that in winter it may be treated like a Catasetum, that is to say, have little or no water at all for a few weeks, while the temperature is allowed to fall to 58° by day, and 50° at night. As the operation of watering is gradually diminished in autumn, so it should be increased in spring, according to the state the plant may be in; and the house should be slightly shaded in bright sunny days.

The following are the specific characters of all the species at present known to us.

MORMODES.

1. M. *atropurpureum* (Lindl. in Bot. Reg. t. 1861.); racemo oblongo denso, sepalis lineari-oblongis æqualibus reflexis basibus lateralium paulo obliquis, petalis ovatis erectis supra columnam conniventibus, labello replicato retrorsum arcuato cuneato trilobo: lobis lateralibus deflexis venosis intermedio carnosiose cuspidato subtrilobo.—— *Central America.*——Flowers deep purple, on a close erect raceme.

2. M. *lineatum* (Bateman in Bot. Reg. 1841, misc. no. 107. 1842, t. 43.); racemo elongato multifloro, sepalis petalisque oblongo-linearibus abruptè acutis margine reflexis, labello lineari incurvo carnoso sparsim piloso versus basin utrinque dente nunc brevi nunc elongato aucto, columnæ dorso et margine pubescente.——*Guatemala.*——The flowers are deliciously fragrant; when they first appear they are dull olive green; they afterwards acquire a bright warm yellow tint, and the markings upon them increase in intensity till they have become orange-red.

3. M. *aromaticum* (Lindl. in Bot. Reg. 1841, misc. 162.); racemo brevi erecto, sepalis petalisque subrotundo-ovatis acutis secundis concavis, labello angustè cuneato convexo laciniâ intermediâ triangulari acuminatâ cucullatâ.—— *Mexico.*——This has flowers with a pale pinkish ground sown thickly with dull wine-red specks, and a powerful odour like that of aromatic vinegar. It differs from M. pardinum in the small size of its flowers, and the very dissimilar form of the lip and floral envelopes.

4. M. *pardinum* (Bateman Orch. Mexic. t. 14. Hooker Bot. Mag. t. 3900. *Cyclosia maculata*, Klotzsch in Gartenzeit. no. 39. 1838.—Var. *Unicolor*, Hooker l. c. t. 3879. *Catasetum citrinum*, Hort.); foliis elongatis lanceolatis, racemo ascendente elongato multifloro, sepalis petalisque lanceolatis acuminatis subsecundis, labello plano conformi acutè tridentato unguiculato.——*Mexico.* ——A beautiful species, with bright yellow flowers, spotted with rich brown in one variety; whole coloured with no spots whatever in the other.

5. M. *luxatum* (Lindl. in Bot. Reg. 1842, misc. 66. 1843, t. 33.); foliis longissimis angustis subtùs glaucis racemo oblongo pluriès longioribus, sepalis ovato-lanceolatis petalisque oblongis concavis margine subscariosis carnosis incurvis, labello hemispherico concavo obsoletè trilobo apiculato supra columnam cucullato, polliniorum caudiculà apice tuberculato.——*Mexico.*——Flowers large, as much as $3\frac{1}{2}$ inches in diameter, pale lemon-colour, powerfully aromatic.

6. M. *buccinator* (Lindl. in Bot. Reg. 1840, misc. 9.); racemo stricto oblongo, sepalis lineari-oblongis lateralibus reflexis dorsali petalisque ovali-lanceolatis erectis, labello unguiculato carnoso nudo subrotundo-cuneato apiculato utrinque

emarginato lateribus in buccinæ formam revolutis.—— *Mexico.*——Flowers pale green, with an ivory-white lip, whose sides are so rolled back as to give it the appearance of a trumpet. The column is twisted sometimes to the right, sometimes to the left. The habit and general appearance of the flowers, except in colour, is that of *M. atropurpureum.*

35

* LYCÂSTE plāna.

Even-flowered Lycaste.

GYNANDRIA *MONANDRIA.*

Nat. ord. ORCHIDACEÆ, § VANDEÆ. MAXILLARIDÆ.
LYCASTE. *Lindl. in hujus voluminis miscell. p.* 30.

L. *plana* (Lindl. in Bot. Reg. 1842, misc. 96.) bracteâ supremâ cucullatâ ovario longiore, sepalis oblongis planis basi in cornu brevi conico connatis, petalis conformibus apice tantùm recurvis, labelli trilobi lobis lateralibus apice crenulatis intermedio subrotundo serrato callo elevato obtuso obsoletè trilobo, columnâ pubescente, antherâ villosâ.

A Bolivian plant imported by Messrs. Loddiges, with whom it flowered in October last. It is conspicuous for the large size of its leaves, and is in fact very near L. macrophylla, from which it differs in the petals being quite even, not undulated, and in the lateral sepals being much more exactly oblong. Added to which is a greater degree of bluntness on the tubercle of the lip. The beauty of the flowers of L. plana is far greater than in L. macrophylla, which wants the rich red-wine colour of the plant before us.

Fig. 1. represents the lip spread open; 2. the column, and 3. the pollen-masses with their slender caudicula.

Like other species of Lycaste this requires to be grown in turfy peat; the pot to be half filled with potsherds, and the soil considerably elevated above its brim. Care must be taken not to have its pseudobulbs imbedded in the soil, or they will damp off. In summer, after the plant has commenced growing, plenty of water should be given to its roots, and a slight syringing over head once or twice a day as the weather permits. The house should be shaded in sunny days, and the temperature kept about 80° by day, and 70° by night. In winter, for a few weeks, very little water is required, especially in cloudy weather; if the house is kept moist it will be quite sufficient, and the temperature may be allowed to fall as low as 58° by day, and 50° by night.

* A fanciful name. Lycaste was a beautiful woman.

37

DENDROBIUM cucumerīnum.

Cucumber Dendrobium.

GYNANDRIA *MONANDRIA.*

Nat. ord. ORCHIDACEÆ, § MALAXIDEÆ.
DENDROBIUM. Swartz.

D. *cucumerinum* (W. MacLeay in litt. Lindl. in Bot. Reg. 1842, misc. 63.); nanum, intricatum, cæspitosum, ramis brevissimis articulatis cylindraceis monophyllis, foliis oblongis teretibus seriatim tuberculatis, pedunculis brevissimis trifloris, sepalis petalisque linearibus acuminatis obtusis, labelli trilobi lobis lateralibus triangularibus intermedio ovato crispato lamellis 5 undulatis in medio, clinandrio denticulato.

A native of New Holland, whence it was sent to Messrs. Loddiges from Mr. Wm. MacLeay. Of the locality for it in that country we are uninformed. It very much resembles a heap of little cucumbers, whence the name has been derived. Those bodies are *apparently* leaves, terminating the short articulated stems: but they require further examination, for they *may* be of the nature of pseudobulbs. This, however, is a point our opportunities do not permit us to settle.

The flowers appear from the base of the cucumbers in threes. They are dirty white, with long narrow sepals and petals striped with pink, and a three-lobed lip, whose middle division is crisped very much, and five wavy elevated ridges along its middle.

Fig. 1. represents the inside of the lip; 2. the column; and 3. the pollen-masses.

It is very much to be wished that the genus Dendrobium could be subdivided upon some such principles as have been lately applied to Maxillaria; for it consists of plants having an extraordinary diversity of appearance; and the New Hollanders in particular are quite a set apart. But we have sought in vain, up to the present time, to find good generic characters, unless the surface of the labellum is taken advantage of; and for such a measure Botanists are hardly yet prepared. Instead, therefore, of speculating upon what may possibly be done hereafter with the genus, we take advantage of the present opportunity to print a sketch of the geographical distribution of New Holland Orchideceæ by the late Allan Cunningham, the well known and lamented New Holland Botanist.

Notes, for the most part geographical, on the Orchidaceous plants of Australia. By the late Allan Cunningham.

In viewing the limited number of Orchidaceous plants hitherto observed by Botanists in Australia, and their geographic range on that extraordinary continent, as far as its coasts and internal country are known, I have, in order to account in some measure for the fact of a country, possessing so considerable an extent of intertropical coasts, which as far as temperature goes, may be said to be favourable to the production of the order, furnishing nevertheless so few of its epiphytic division, been led to consider the general configuration of the surface of the country; the open character generally of its forests; the attenuated ramification of its prevalent timbers; and the geological structure of the several coasts and regions in their vicinity—considerations equally referable to Filices and other portions of its cryptogamic vegetation—the same physical causes, that limit the existence of the one, especially of its Epiphytes, maintaining doubtless a like influence in regard to the extent and diffusion of species of the other.

It has been ascertained by Navigators and inland-travellers that the *highlands* of that continent are situated upon or near its shores, and that the most elevated country is upon the eastern coast. There they present a range of mountains which extends in the direction of the meridian, and stretching from Wilson's Promontory on the south (lat. 39°) to about Cape Weymouth of Cook on the north (lat. 12½°), exhibit a continuous chain, which has been happily termed, the spine or backbone of the continent. This main range, in some latitudes, closely invests the coast-line, whilst in other parallels it retires to some distance from the shore, and there gives space for the windings of streams which originate in its eastern flanks, ere they are discharged into the ocean. On other coasts, as upon one part of the north—upon the western (as inland from the Swan River)—and on parts of the southern shores of that continent, have been observed, ridges of hills; but these are isolated, or rising from the surface of a comparatively level country, present no trace of connexion, either with the great eastern chain, or with each other: and they are all of inferior elevation, no one exceeding 3000 feet of perpendicular height above the level of the sea, and most of them of not half that elevation. It may be here remarked, moreover, that with reference to the face of the interior, every observation of the traveller goes to support the theory, that although detached hills and even some ridges have been noticed on its ample surface, neither a chain of mountains, nor any elevated points to form the nucleus of a second main-range exist in the central regions of the continent, which will one day be rather found a vast level, through which its rivers, if they exist far from its eastern side, have, from the prevalent disposition of the country generally to drought, much to combat in their efforts to gain any sea coast.

From these brief remarks of the structure of the Australian continent, it will be seen that the eastern coast, or that of New South Wales, within and beyond the tropic, is the only shore to which we can look for Epiphytic Orchidaceæ and for Filices and other of the Cryptogamic class—in fact, the only one (if we except one or two points of the north coast, strictly so called) on which these orders of the vegetation of that great country have hitherto been found—its main chain, which in some parallels has been found to measure 6000 feet above the sea shore, furnishing in its ravines and rocky flanks, ample shade and humidity to the sustenance of those families; but in several islands in the Gulf of Carpentaria, says Mr. Brown, " having a flora of phœnogamous plants exceeding 200 species, I did not observe a single species of moss"—and this, evidently, because of the ordinary elevation of those isolated spots; the consequent little shade they afford; and the extreme dryness of the circumambient atmosphere.

With the requisite conditions of high temperature and considerable humidity, which the country on its eastern coast, especially within the tropic, provides, it is very remarkable that so few of the Orchidaceæ growing on trees or rocks, should have come within the observation of botanists attached to voyages of discovery along that lengthened shore. In the whole line of eastern coast, only eleven Epiphytes were remarked by Sir Joseph Banks and Mr. Brown; and these have been referred by the latter able botanist to three genera; viz. Cymbidium, Dendrobium, and to a genus intermediate to them, named Sarcochilus. It may, however, be observed, that in the voyages of both Cook and Flinders, and indeed in that more recently, of King, upon that extensive coast, although landings were effected on many of its head-lands, and opportunities afforded for an examination of the botany of the immediate shores or adjacent country, yet perhaps no instance occurred of the remoter mountainous regions having been visited, where for the most part those remarkable vegetables are more probably to be found; and that many, yet to be discovered do exist, may be perhaps fairly inferred from the fact of there having been found at the Illawarra—a coast district on the south of Port Jackson,—and at Moreton-bay (in 27½ S°.), *nine* Epiphytes not known to botanists, prior to the exploration and examination of those most interesting points on the coast.

Of the terrestrial portion of the order, 104 species (now including Phajus grandifolius, *Lour.*, and Calanthe veratrifolia, *R. Br.*, lately found in the districts just mentioned) are known to be indigenous to the Australian continent; and of these three-fifths are natives of Port Jackson and the neighbouring country.

But ten species, of the genera Habenaria, Thelymitra, Microtis, Caladenia, Pterostylis, Cymbidium (Geodorum) and Phajus, have been detected in the warmer or intertropical parts of the shores of New South Wales; and as these plants are only to be met with during the rainy season, or in the period immediately subsequent, when every vegetable becomes reanimated and exerts fresh life, not a single terrestrial individual of the order was found on the north-western coast during the several visits of Captain P. P. King, whose surveys could alone be carried on during the easterly or dry Monsoon, when the wind blows steadily *off* shore, and when all minor vegetation is wholly burnt up. Such however is the extreme aridity of that particular line of intertropical coast, that not a single Epiphyte was observed, and but two ferns; a Blechnum and a Gleichenia; and one Palm, a Livistona of Mr. Brown. Of the West coast, properly so denominated, little or nothing is known of its Orchidaceous vegetation. We however learn from the correspondence of Mr. Drummond, who has charge of the public garden at the Swan River colony, that many fine terrestrial species have been observed, in their respective seasons, in the vicinity of Perth and other parts of that settlement; but of what genera, we have yet to be informed. [This was written long before the publication of our Sketch of the Vegetation of Swan River.]

During the voyage of the Investigator in 1802, Mr. Brown discovered twenty species on the south coast belonging wholly to genera frequent at Port Jackson, with the exception of the beautiful Epiblema—a genera closely allied to Forster's Thelymitra. No Epiphyte has been seen upon that, generally speaking, very arid coast.

From what I have advanced, it will readily appear that the existence of Orchidaceous Epiphytes is confined to the eastern and perhaps to the northern shores of New South Wales; and as not the semblance of one (not even Dendrobium æmulum, *R. Br.*, so frequent on the trunks of Eucalyptus resinifera, and in the dry forests around Port Jackson) has been seen, either by myself or other travellers in the interior of that country, at least between the parallels of 28°. and 35°., their range westerly from the actual sea coast may be truly said to be limited to the main chain, or dividing range, beyond which, moreover, in a westerly direction, neither Alsophila australis, nor the Corypha of Port Jackson have ever been observed.*

Considering the extent of interior known, at the present day, in or about the parallel of Port Jackson, as also on the north and south of that degree of latitude, the terrestrial part of the family may be said to be comparatively rare in the inland country. In their journeys through it at various seasons, botanic travellers have observed but thirteen species; and the whole of them are of genera, and indeed for the greater part, of species, frequent on the sea coast at or near Sidney.

These genera are Diuris, Orthoceras, Calochilus, Caladenia, Lyperanthus, Pterostylis, Gastrodia and Dipodium. In fine, as a general remark on the geographical range of the family in Australia, it may be observed, that as the terrestrial species are greatly influenced by rains that *may* fall in the *season* when they would under such favourable circumstances appear above the soil, and as those whose localities are inland, beyond the range of those genial coast-showers which occasionally fall in the midst of a long period of dry weather, are wholly prevented from appearing above ground for two or three years, to which extent the droughts in that country have continued; these facts are sufficient to explain why it is that this group also of the order in Australia, is so much more abundant on its immediate shores, than it is either in the up-country, as those parts of the colony a little distant from the coast are termed, or the more remote interior.

* The absence of arborescent Ferns and Palms in the interior on the western side of the Blue Mountain-ranges, is to be attributed to the want of shade from high lands, (those hills that are scattered on its surface being of inconsiderable elevation, and generally of a sandstone rock,) and to the nonexistence of dense humid forests. The whole internal country traversed, beyond the meridian of 149°., declining westerly to a low level region, being exceedingly thinly and generally lightly wooded, and thus its ample surface being greatly exposed to the rays of the sun, an extreme dryness of atmosphere is engendered, by no means favourable to the existence of a shade-loving vegetation, affecting a *lower* temperature, but *cooled* simply by the surrounding air being, to a certain degree, permanently charged with humidity.

39

CLOWESIA rosea.

Pink-flowered Clowesia.

GYNANDRIA *MONANDRIA.*

Nat. ord. ORCHIDACEÆ, § VANDEÆ.

CLOWESIA. Supra misc. 39. *Flos* subglobosus, patulus. *Sepala* subæqualia; lateralia paulò obliqua, in mentum breve producta, basibus connatis. *Petala* conformia, latiora, fimbriata. *Labellum* concavum, carnosum, cum labello continuum nec articulatum, obsoletè trilobum, margine in fimbriam glandulosam laceram solutum, disco læve. *Columna* semiteres, clavata, utrinque apice obtusa cornuta; clinandrio alto carnoso serrato. *Stigma*: sinus transversalis. *Pollinia* 2, linearia, dorso sulcata, glandulâ subrotundâ, caudiculâ membranaceâ clepsydræformi!——Caulis *carnosus, foliosus.* Scapus *radicalis, multiflorus, erectus.*

Clowesia rosea.

A short time since we gave some account of this beautiful plant, which we then knew only from some flowers and a description furnished by a correspondent. Since then we have had better means of examining it.

It is a native of Brazil, and first flowered at Broughton Hall, near Manchester, with the Rev. Mr. Clowes, a zealous cultivator of Orchidaceæ, after whom it is named. At a later period (March last) we received it from Sion Gardens, by permission of His Grace the Duke of Northumberland, and from that plant the accompanying figure was taken. Up to the present time we have heard of it nowhere else.

It is very like a Catasetum in habit. The stems are from $2\frac{1}{2}$ to 4 inches long, ovate, clothed with the remains of the bases of leaves. The leaves, which I have not seen, are said to be three, lanceolate, ovate, acuminate, and at the point twisting a little to one side. The inflorescence proceeds from the base of the stems (not apex as was formerly stated on the authority of Mr. Clowes's gardener, who thought apex signified the bottom), and consists of five or six, probably more, erect delicate white flowers tinged with pink. They are remarkable for having their petals and the end of the lip broken up at the margin into numerous delicate glandular fringes, which give them a very rich and beautiful appearance.

As a genus Clowesia is perfectly distinct from every thing previously described. Its flowers being extended a little into a chin in front, suggest its belonging to the Maxillaridous division; but its whole habit and the singular apparatus of its pollen-masses oppose such an arrangement. The latter organs rest on a broad viscid gland like that of a Catasetum, but the caudicula, or part that connects the gland and pollen-masses, is broad, thin, and contracted in the middle so as to resemble an hour-glass; but whether that is the usual structure, or as we suspect merely consequent upon the separation of the caudicula from the anther-bed, we have not had an opportunity of ascertaining. Upon the whole it is probable that Clowesia must stand in the same division as Catasetum.

Fig. 1. represents the lip seen from above; 2. the column; 3. the pollen apparatus seen in front, and 4. the same behind.

41

RENANTHÈRA matutīna.

Morning Renanthera.

GYNANDRIA *MONANDRIA.*

Nat. ord. ORCHIDACEÆ. § VANDEÆ—SARCANTHIDÆ.

RENANTHERA. Bot. Reg. vol. 14. *t.* 1131.

R. *matutina* (Lindl. Gen. & Sp. Orch. p. 218. *Aerides matutina?* Blume Bijdr. 1. 366. tab. 24.); racemis densis cylindraceis longipedunculatis paniculatis simplicibusque, sepalis petalisque obovato-linearibus obtusis, labelli limbo ovato acuto calcari obtuso subæquali.

Although this plant bears no comparison with the magnificent Chinese species on which the genus was founded, yet it is very far from being unworthy of cultivation. Indeed, by the number of its flowers and the richness, though not brilliancy, of their spots, it makes up in some measure for their want of size.

The first mention that is made of it is in Blume's *Bijdragen*, where it is referred with doubt to the genus Aerides, and said to be found in flower in the month of September, on trees at the foot of Mount Salak, in Java. Mr. Cuming afterwards gathered it in the Philippines, but by no means in so luxuriant a state as the specimen now represented, which flowered at Chatsworth in December last, and was kindly placed at our disposal by His Grace the Duke of Devonshire. We have also received it from Messrs. Rollissons.

Fig. 1. represents the lip; and 2. the pollen apparatus of this plant.

43

ONCIDĬUM uniflōrum.

One-flowered Oncidium.

GYNANDRIA *MONANDRIA.*

Nat. ord. ORCHIDACEÆ § VANDEÆ.

ONCIDIUM. Swartz.

Sect. II. Euoncidium; Heteranthium. Folia plana; tetrapetala; micropetala. Sertum Orchidaceum, t. 48.

O. *uniflorum* (Booth in litt.); cæspitosum, pseudobulbis angustis linearibus squamatis sulcatis monophyllis, foliis lineari-lanceolatis obtusis undulatis apiculatis, scapo unifloro folio multò breviore, sepalis lateralibus basi connatis petalisque obovato-oblongis undulatis herbaceis, labello pandurato lobo intermedio maximo emarginato lateralibus nanis rotundatis sinu fimbriato, cristâ pubescente polydactylâ laciniâ utrinque liberâ solitariâ, columnæ alis truncatis erosis.

This curious little plant, allied to the rare Oncidium barbatum, inhabits trees in the forests of the Organ Mountains of Brazil, where it was found in April 1841, by Mr. Gardner, to whom I am obliged for specimens.

For its introduction to gardens we are indebted to Sir Charles Lemon, who received it from Brazil in August 1841, through Lieutenant Turner, of H. M. Packet "Ranger," and in whose collection at Carclew it flowered in November 1842. Mr. Booth has favoured us with a drawing, and the following account of it.

"*Pseudobulbs* when young roundish oblong, covered with several ovate, acuminate, sheathing scales; but when full grown they are ovate-oblong, compressed, rather more than an inch in length, and about 3-8ths of an inch in width, becoming slightly furrowed, one-leaved. *Leaves* oblong, lanceolate, acute, from three to four inches long, and scarcely half an inch broad, thin, and a good deal recurved, of a deep green. *Scape* one-flowered, issuing from the base of the leaf, and so short as to be almost hidden by the upper scale of the pseudobulb which partly embraces it. *Pedicel* round and slender, an inch long, somewhat curved and slightly channelled, of a pale green, with a small, thin, pale brown-coloured bractea at its base. *Flowers* large in proportion to the size of the plant and very handsome, continuing a long time in perfection. *Sepals* of a dingy brown, slightly spotted, oblong lanceolate acute, much undulated at the margin, and spreading so as to become recurved. They are all nearly of the same size and form. The two lower ones are undivided and narrow at the base. *Petals* similar in colour to the sepals, but rather shorter, and almost twice as wide, recurved at the point, and undulated at the margin. *Labellum* 3-lobed, of a bright yellow, excepting around the crest, which is blotched and spotted with blood-red spots. The middle lobe is large and spreading, somewhat cordate and apiculated, having a part of the outer edge on one side of the point folded a little way over that of the other. The margin is finely marked and undulated back to the crest, where it becomes of a brown colour, and is delicately fringed on either side. The lateral lobes are also much undulated and very short. They are nearly as broad at the extremity as they are long, and diminish gradually towards the crest at their base. *Column* erect, rounded at the back, and hollowed out in front with the spreading undulated wing on each side of the opening. In front of this is a yellowish waxy process 2-lobed, and a little elevated. *Anther case* 1-celled, hooded and conical, so as to resemble a small cup, with a minute red spot in front, where it terminates in a point. *Pollen-masses* 2, roundish-ovate, deep yellow, attached by a pale coloured elastic membrane to the small red coloured body just mentioned.

"The plant appears to grow in large masses on the branches of trees, to which it is attached by its numerous round, filiform, ash-coloured roots. It is grown at Carclew in a warm moist stove, suspended from one of the rafters."

44

STANHOPEA Martiana; *var.* bicŏlor.

Two-coloured Von Martius' Stanhopea.

GYNANDRIA *MONANDRIA.*

Nat. ord. ORCHIDACEÆ, § VANDEÆ—MAXILLARIDÆ.
STANHOPEA. Supra fol. 1800.

S. *Martiana;* labello medio parùm constricto, hypochilio brevi sessili scrotiformi, epichilio oblongo-lineari obscurè 3-dentato, cornubus rectiusculis apice cirrhosis, columnæ pubescentis subclavatæ marginibus parum dilatatis.
S. Martiana. *Bateman in Bot. Reg.* 1840, *misc.* 109. *Orch. Mex. and Guat. t.* 27.

A native of Mexico, discovered by Baron Karwinski in 1827, and afterwards by M. Galeotti. It is one of the most distinct and magnificent species of the genus, and in the magnitude of its blossoms is second only to *S. tigrina.* The sepals are straw-coloured, or almost white, faintly and sparingly marked with clusters of little vinous dots; the petals appear transparent white, with large spots of intense crimson; the lip is also a clear ivory white, except a slight discoloration at the base. The horns are of great size and strength, and taper into a kind of tendril, besides which they are exactly parallel with the epichilium, the form of which is almost linear, the two edges being as nearly as possible parallel with each other, and not a great deal broader than the column; a mark by which the species is immediately recognized.

The present variety is a lovely plant, with large pure white flowers, richly but sparingly spotted with crimson. In the original S. Martiana the sepals are straw-coloured, and much more dotted with purple. Messrs. Rollisson have lately flowered it, and believe they obtained it from Mexico. It is as fine a thing as S. tigrina would be if its flowers were white, and is very sweet-scented.

Now that the number of species of Stanhopea has been greatly increased, and that opportunities of examining them in a living state have occurred, it has become desirable that the genus should be reconsidered, partly for the sake of defining the species with more exactness, partly for the sake of determining which of the garden plants are mere varieties, and also with a view to the correction of a few errors. For this reason the following enumeration of the species has been prepared, in drawing up which I am most particularly indebted to Mrs. Lawrence, Mr. Rucker, and Messrs. Loddiges.

* *Labello apice integro.*

1. S. *insignis* (Frost in Bot. Mag. tt. 2948 & 9. Lodd. Bot. Cab. t. 1985. Lindl. in Bot. Reg. t. 1837. Gen. & Sp. Orch. no. 1.); hypochilio globoso anticè intruso basi et apice fisso mutico intus radiatim verruculoso extus ecarinato epichilio æquali, epichilio subrotundo-ovato integro, cornubus subtriangularibus falcatis incurvis brevioribus, columnâ latissimè alatâ.——*Brazil.*——This species is remarkable for the globose form of the base of its lip, by which all its varieties are readily distinguished. It is also in almost all cases sprinkled with small violet spots, which reach even to the tip of the labellum. In colour, however, it varies very considerably. The winged margin of the column is wider than in any other known species.

2. S. *Devoniensis* (Lindl. Sert. Orch. t. 1. *S. maculosa*, Floral Cabinet, t. 121? *Maxillaria lyncea*, Lindl. Gen. & Sp. Orch. p. 151. *Coatzonte Coxoahitl seu Lyncea*, Hernand. Thesaur. Rer. Med. Nov. Hisp. p. 266. *Anguloa Hernandezii*, Kunth. Synops. 1. 332.); hypochilio subgloboso anticè gibboso basi mutico epichilio æquali, epichilio ovato integro v. obsoletè tridentato, cornubus falcatis incurvis æqualibus, columnâ marginatâ.——*Mexico.*——Very much like *S. tigrina*, from which it is distinguished by the undivided middle lobe of the lip, and the almost wingless column. To *S. insignis* it approaches in form, although so different in colour; but the lower half of the lip is prominent in front instead of being pressed backwards. The flowers are yellow, with deep crimson-brown blotches; the lip is white with a few spots here and there, and a deep purple stain over half the lower part. Little doubt can be entertained that *S. maculosa* is a bad specimen of this, for although its epichilium is described as 3-lobed, it is represented as perfectly entire.

3. S. *oculata* (Lindl. Gen. & Sp. no. 5. Bot. Reg. t. 1800. *Ceratochilus oculatus*, Lodd. Bot. Cab. t. 1764. *S. Lindleyi*, Zuccarini mss.); bracteis ovariis longè acuminatis duplò brevioribus, hypochilio unguiculato elongato cymbiformi anticè intruso apice carnoso apertè sulcato mutico basi ecorni intus lævi extus bicarinato, epichilio ovato integro, cornubus semiteretibus ascendentibus acutis, columnâ latè alatâ.——*Mexico* (not Brazil).——The flowers are usually lemon-coloured, with a large number of lilac spots on the sepals, a smaller number on the petals, a deep yellow eye, and two, or occasionally four, large dark brown spots on the side of the lower part of the lip, which is very much lengthened out as if unguiculate; by the latter circumstance it is chiefly known from *S. Wardii.* Many supposed varieties occur in the gardens, varying in colour, and in the spotting of the lip. There is one, called *Barkeriana*, which is more remarkable than the others; it looks like *S. insignis* with the lip of *S. oculata*, and is very handsome; the sepals, petals, and column are covered with numerous purple freckles rather than spots, which, as the flower fades, run together, as if their colouring matter were dissolved; so that at last the flower becomes of a dull wine-red tint. This is apparently the *S. Lindleyi* of Zuccarini. Most of the varieties are very sweet-scented.

4. S. *Bucephalus*, (Lindl. Orch. no. 2. *Epidendrum grandiflorum*, Humb. & Bonpl. Pl. Æq. p. 94. t. 27. *Anguloa grandiflora*, H. B. K. nov. gen. et sp. 1. 345.); bracteis ovario subæqualibus, hypochilio unguiculato cymbiformi anticè intruso apice carnoso apertè sulcato mutico basi longè angustato ecorni intus lævi extùs bicarinato, epichilio subrotundo-ovato cuspidato integro breviore, cornubus gracilibus teretibus brevioribus, columnâ basi angustissimâ sursùm alatâ.——*Quito.*——This has deliciously sweet-scented flowers, which are pale yellow, with a pair of dark eyes at the base of each petal, and a few crimson dots elsewhere on them and the sepals. The lip is deep yellow, and spotless on the hypochilium, while the epichilium and column are thickly sown with crimson dots. It is certainly the Epidendrum grandiflorum of Humboldt, hitherto erroneously referred to S. insignis. Mr. Hartweg, who found it near Quito, has introduced it into the garden of the Horticultural Society. It is very near *S. oculata*, from which it differs in the form of the lip and especially in the very short ovaries, a character pointed out to me by Mr. Loddiges.

5. S. *Wardii* (Loddiges in Lindl. Sertum Orchidaceum, t. 20.); bracteis ovariis longè acuminatis duplò brevioribus, hypochilio exactè oblongo depresso sessili antice intruso apice carnoso fisso (nec apertè sulcato) vix dentato, basi utrinque angulato, epichilio subrotundo-ovato integro, cornubus semiteretibus falcatis incurvis subcirrhosis, columnâ latè alatâ.——*Central America.*——This very fine plant is quite distinct from any of the supposed varieties of *S. oculata;* differing in the furrow of the anterior part of the hypochilium being closed up and not open, and in the exactly oblong form of the hypochilium; also from *S. Ruckeri* in the absence of a strong tooth turned inwards from the apex of the cavity of the hypochilium; besides which the angular condition of the base of the hypochilium indicates an approach to *S. quadricornis.* The lip is pale yellow, with a few fine specks, a deep yellow hypochilium, the whole lower part of which is a deep chocolate brown. The flowers are deliciously sweet.

6. S. *Ruckeri;* bracteis ovariis longè acuminatis duplò brevioribus, hypochilio obovato sessili anticè intruso basi mutico apice carnoso apertè sulcato et dente valido inflexo aucto, epichilio subrotundo-ovato integro, cornubus semiteretibus incurvis, columnâ latè alatâ.——*Mexico.*——This is a noble species, with the habit of *S. Wardii,* and its general colour, except that it is paler; but the epichilium is beautifully stained with pink, and the eyes of the hypochilium are very faint. It is distinctly separated by the peculiar form of the hypochilium, which instead of being oblong is so much narrowed to the base as to acquire an obovate form; by the entire want of lateral teeth on the margin; and by the presence of a very strong inflexed tooth, in which the wide, not closed up, fissure of the apex of the hypochilium terminates. We have received it from Messrs. Loddiges, and have named it in compliment to Sigismond Rucker, Esq. Junr., of Wandsworth, whose skill in the management of this noble genus is well known to the cultivators of Orchidaceæ.

7. S. *graveolens* (Lindl. in Bot. Reg. 1840, misc. 125); hypochilio subcompresso saccato intus glaberrimo anticè bidentato et inter dentes profundè sulcato, epichilio subrotundo-ovato integerrimo, cornubus acuminatissimis incurvis, columnæ alis latissimis subquadratis.——*Peru?*——A species with the habit of *S. saccata.* The sepals and petals are of a delicate straw colour; the lip at the base, and the central parts of the flower generally are of a deep rich apricot yellow, while the horns and upper end of the lip are like ivory turning yellow. Its odour is so powerful that it communicates itself to the fingers after touching the flowers, and like many other smells, though agreeable in itself, is offensive from its intensity. These remarks are borrowed from the notice formerly given of the plant in the Botanical Register, for by some accident neither drawing nor specimen were preserved of the species.

8. S. *quadricornis* (Lindl. in Bot. Reg. 1838, t. 5.); hypochilio oblongo basi bicorni antice saccato apice carnoso excavato mutico, epichilio ovato integro æquali, cornubus incurvis teretibus brevioribus, columnâ latè alatâ.——*Central America.*——This has something the appearance of *S. oculata;* but its lip has a rich crimson stain at the base, instead of a pair of deep brown spots, and it has two prominent horns standing erect on the lower edge of the cavity of the hypochilium. This brings the species near *S. eburnea,* which is however very different in other respects. It is one of the finest of the genus.

9. S. *eburnea* (Lindl. in Bot. Reg. t. 1529. Gen. & Sp. no. 4. Bot. Mag. t. 3359. *S. grandiflora,* Lindl. l. c. no. 3. *Ceratochilus grandiflorus,* Lodd. Bot. Cab. t. 1414.); labello oblongo vix medio constricto, hypochilio pone basin bicorni, epichilio ovato-oblongo obtuso, metachilio duplò longiore solido plano-convexo anticè truncato bidentato, columnâ longissimâ apice tantum latè alatâ.—*Brazil* and *Trinidad.*——This very distinct and beautiful plant always has the sepals and petals pure ivory white; but in the colour of its lip it varies, sometimes being nearly colourless, except in the cavity of the hypochilium, which is crimson, and occasionally having deep rich purple stains all over the lip, except the middle lobe. Flowers extremely sweet.

* * *Labello apice tridentato.*

10. S. *tigrina* (Bateman Orch. Mex. & Guat. t. 7.); hypochilio subrotundo intus lamellis glandulosis radiato basi mutico apice sulco brevi aperto dente valido inflexo, epichilio ovali æqualiter tridentato cornubus falcatis æquali, columnâ lanceolatâ latissimè alatâ.——*Mexico,* near Xalapa.——Much the finest of the genus. Flowers deep orange yellow, as much as eight inches in diameter, richly blotched with purplish brown. The excessively broad column, and the radiating toothed lamellæ within the cavity of the lip, are peculiar characters.

11. S. *Martiana* (Bateman in Bot. Reg. 1840. misc. 109. 1843. t. 44. Orch. Mex. & Guat. t. 27.); labello medio parùm constricto, hypochilio brevi sessili scrotiformi, cornubus apice cirrhosis rectiusculis, epichilio oblongolineari obscurè 3-dentato, columnæ pubescentis subclavatæ marginibus parùm dilatatis.——*Mexico.*——A most lovely plant, and quite different from all others. In the form of the base of its lip it approaches *S. saccata,* in its markings *S. tigrina,* but it is readily known by its downy, wingless column, the linear middle lobe of its lip, and its bristle-pointed horns. The white flowers with rich crimson spots are quite peculiar.

12. S. *saccata* (Bateman Orch. Mex. & Guat. t. 15.); labello medio paulò constricto, hypochilio scrotiformi anticè omninò aperto subtus gibboso ecarinato, apice et basi mutico intus radiatim glanduloso lamellato, epichilii ovalis trilobi laciniâ intermediâ nanâ, cornubus semi-lanceolatis latis planis incurvis æqualibus, columnâ glabrâ angustè marginatâ.——*Guatemala.*——This has the smallest flowers in the genus, and almost always has the sepals and petals turned completely back on the ovary. They are greenish-yellow, regularly speckled, but not blotched, with brown, and are deep yellow at their base. The same colour extends to the lip, whose hypochilium is singularly saccate.

There are probably other species even in cultivation; more especially there remain for examination *S. venusta,* which is apparently a variety of S. Wardii, *Harrisoniæ* from Brazil, *aurea* from Guatemala, and *aurantia* from La Guayra, all mentioned in Messrs. Loddiges rich catalogue, but undescribed and unknown to me.

49

CIRRHOPETĂLUM chinense.

Chinese Cirrhopetalum.

GYNANDRIA *MONANDRIA.*

Nat. ord. **Orchidaceæ, §. Malaxeæ.**
CIRRHOPETALUM. ***Supra*** **1838. *t.* 11.**

C. *chinense.* Supra 1842. misc. 29.

To what was formerly stated, concerning this singular Chinese plant, in the place above quoted, we have nothing to add, and therefore, instead of repeating those observations, we fill our space with some account of the genus itself and its numerous species.

Cirrhopetalum is a name proposed in the year 1824 for the Bulbophyllum longiflorum of Du Petit Thouars, a Mauritius plant, remarkable for the great length of its lower sepals. In 1825, Professor Reinwardt gave to the same or some nearly allied species the name of Zygoglossum; but of course Cirrhopetalum took precedence on account of its priority of date. In 1830 the genus had increased in number of species to six, of which two had not been examined by me. Now, by the discoveries of Mr. Cuming and others, I am able to bring together twenty-one; of which fourteen have been seen alive, two in a dried state, and five are taken up upon the credit of others. They form a most singular group, and a complete collection of them would be a very interesting feature in a stove. The following is an enumeration of them.

** Flowers in umbels or racemes.*

1. C. *vaginatum* (L. no. 3. Bot. Reg. 1842. sub t. 12.); pseudobulbis pyramidatis truncatis, folio oblongo convexo emarginato, vaginis scapi oblongis ventricosis distantibus sepalo supremo ovato inferioribusque longissimis pendulis ciliatis, petalis oblongis ciliatis, labello oblongo bicarinato, columnæ auriculis setaceis.——*Sincapore.*——C. vaginatum differs from C. Medusæ in its flowers being ciliated, the petals oblong, not triangular and acuminate, and in the distant sheaths of the scape; to which may be added that its flowers are not speckled with pink as in C. Medusæ.

2. C. *Medusæ* (Lindl. in Bot. Reg. 1842. t. 12); pseudobulbis ovatis tetragonis, folio oblongo convexo emarginato, vaginis scapi oblongis ventricosis supremo pedunculo longiore, floribus densè capitatis, sepalo supremo acuminato inferioribus longissimis pendulis, petalis triangularibus acuminatis, labello ovato acuminato bicarinato, columnæ auriculis acutis.——*Sincapore.*——This in some respects resembles C. vaginatum, but is a much larger plant, and very different in the form of its flowers.

3. C. *cæspitosum* (Wallich in Bot. Reg. 1838. misc. 53.); pseudobulbis ovatis monophyllis, foliis angustè ovalibus obtusis basi angustatis scapo longioribus, scapo filiformi erecto vaginis paucis membranaceis carinatis acutis distantibus laxè vestito apice umbellato, sepalo supremo obovato fornicato acuto margine scabro lateralibus acuminatis convolutis triplò longioribus, petalis ovatis apiculatis pectinato-serratis, labello minimo solido triquetro apice recurvo angulis superioribus marginatis.——*East Indies.*——Flowers small, pale, yellow ochre-coloured.

4. C. *cornutum* (Lindl. in Bot. Reg. 1838. misc. 138.); pseudobulbis ovatis angulatis scapo paulò brevioribus, sepalis lateralibus in cornu connatis superiore petalisque ovatis ciliatis, labello angusto triangulari suprà sulcato subtùs carinato.——*E. Indies.*——Leaves six or eight inches long. Flowers dull purple.

5. C. *maculosum* (Lindl. in Bot. Reg. 1841. misc. 173.); floribus geminis solitariisque, sepalo supremo oblongo acuto lateralibus lineari-oblongis obtusis margine anteriore pone basin uniplicato, petalis sepalo supremo brevioribus oblongis obtusis, labello ovato obtuso convexo apice recurvo, columnâ bidentatâ? (polliniis 4 subæqualibus).——*East Indies.*——This is a species with the habit of Bolbophyllum affine, or leopardinum, but with the long lateral sepals of a Cirrhopetalum. It has the flower-stalks and flowers of a very pale green, finely speckled with dull purple. The lateral sepals are nearly an inch long, and have a single fold on their inner margin near the base.

6. C. *chinense* (Lindl. in Bot. Reg. 1842. misc. 29. 1843. t. 49.); foliis lanceolatis, umbellâ multiflorâ, sepalo supremo galeato petalisque oblongis brevioribus obtusis serrulatis apiculatis: lateralibus lanceolatis, labello linguiformi obtuso carnoso convexo lævi tremulo.——*China.*——Flowers rather large, pale fawn colour, with dorsal helmet-shaped sepals and petals spotted with crimson.

7. C. *Thouarsii* (L. no. 1. Bot. Reg. 1838. t. 11. *Epidendrum umbellatum*, Forst. *Bulbophyllum longiflorum*, Thouars. *Zygoglossum umbellatum*, Reinw. *Cymbidium umbellatum*, Spreng.); pseudobulbis tetragonis, folio oblongo petiolato, umbella subdimidiatâ, sepalo dorsali setaceo-acuminato lateralibus oblongo-linearibus, petalis ovatis serratis setaceo-acuminatis, labello ovato retuso, columnæ auribus bidentatis.——*Otaheite, Java, Mauritius, Madagascar.*——Flowers pale cinnamon colour, with a little purple dotting about the sepals and petals.

8. C. *Wallichii* (L. no. 5. Bot. Reg. 1839. misc. 119.); foliis lanceolatis apice fissis scapi erecti longitudine, racemo multifloro pendulo, bracteis linearibus acuminatis, sepalis glaberrimis supremo acuminato lateralibus linearibus ligulatis vix acutis, petalis acuminatis subciliatis, labello . . . ——*Nepal.*

9. C. *Roxburghii* (L. no. 2.); foliis lanceolatis, umbellâ radiatâ multiflorâ, petalis ciliatis acutis, sepalis omnibus acuminatis, lateralibus dorsali sexiès longioribus.——*East Indies.*——Leaves three or four inches long. Flowers yellow; the upper sepal and the petals veined with purple.

10. C. *Macræi* (L. no. 6. Bot. Reg. 1841. misc. 105.); folio oblongo petiolato, umbellâ pauciflorâ, petalis oblongis apiculatis falcatis nudis, sepalis acuminatissimis, labello e lata basi angustato carnoso canaliculato obtuso recurvo.——*Ceylon.*——This has the habit of C. *Thouarsii.* It has dull brownish yellow sepals, and purple petals, which are falcate and the point turned forwards.

11. C. *picturatum* (G. Loddiges in Bot. Reg. 1840. misc. 106.); pseudobulbis oblongis angulatis, folio oblongo convexo emarginato, floribus umbellatis, sepalo supremo ciliato setaceo-acuminato lateralibus oblongis obtusis, petalis acuminatissimis ciliatis villosis, labello lineari recurvo secus medium elevato, columnæ auriculis obtusis integris.——*E. Indies.*——A little plant with purple flowers, growing in flat umbels, and deeply stained with dark red. Its habit is that of a Bolbophyllum, its scape from five to six inches high.

12. C. *auratum* (Lindl. in Bot. Reg. 1840. misc. 107.); pseudobulbis oblongis angulatis, folio oblongo convexo, floribus umbellatis, sepalo supremo petalisque setaceo-acuminatis fulvo-ciliatis lateralibus acutis, labello lineari recurvo, columnæ auriculis rotundatis integris.——*Manilla.*——Very like the last, but much larger in the flowers, which are fringed with golden yellow hairs, giving them quite a distinct appearance.

13. C. *Cumingii;* pseudobulbis tetragonis, folio oblongo obtuso scapo breviore, umbellâ dimidiatâ multiflorâ, sepalo dorsali petalisque ovatis acuminatis fimbriatis, lateralibus lineari-lanceolatis, labello ovato 3-sulcato pone basin bituberculato.——*Philippines.*——Flowers deep purple, whole-coloured, with long yellow hairs on the petals and back sepal. Flowered in March, 1841, with Messrs. Loddiges.

14. C. *nutans* (Lindl. in Bot. Reg. 1839. misc. 118.); pseudobulbis ovato-subrotundis rugosis, foliis ovato-subrotundis emarginatis coriaceis humi pronis, scapo erecto elongato, umbellâ multiflorâ nutante, bracteis linearibus acuminatis, sepalis glaberrimis: supremo acuminato lateralibus linearibus ligulatis vix acutis, petalis ovatis acutis serrulatis, labello obtuso convexo bicristato, columnæ angulis obsoletè bidentatis.——*Manilla.*——This has a nodding umbel of pale straw-coloured flowers, at the end of a weak scape about six inches high. The leaves are from an inch and half to two inches long, very thick, emarginate, and lying almost flat upon the ground. The species is near C. Wallichii.

15. C. *fimbriatum* (Lindl. in Bot. Reg. 1839. misc 120.); pseudobulbis ovatis subtetragonis, foliis ovalibus . . . . scapo erecto, bracteis linearibus acuminatis, umbellâ multiflorâ, sepalis lateralibus ligulatis vix acutis cohærentibus supremo petalisque ovatis acuminatis fimbriatis, labello crasso linguiformi nudo obtuso, columnæ angulis cornutis edentulis, antherâ papillosâ.——*Bombay.*——A very pretty species, with the long lower green sepals united into a channelled rather stiff strap, while the upper sepal and the petals are broken up at the margin into beautiful purple fringes.

16. ? C. *elongatum* (*Ephippium elongatum*, Blume Bijdr. p. 309.); "foliis ebulbis oblongo-lanceolatis basi carinatis, floribus in pedunculo elongato superne dense spicatis, sepalis acuminatis."——*Java.*

17. ? C. *capitatum* (*Ephippium capitatum*, Blume Bijdr. p. 309.); "bulbis oblongis compressiusculis, foliis oblongo-lanceolatis, floribus capitatis, sepalis elongatis acuminatis, margine revolutis."——*Java.*

*** Flowers solitary.*

18. C. *antenniferum* (Lindl. in Bot. Reg. 1843. sub t. 49.); folio . . . ., scapo unifloro, sepalo dorsali lineari-lanceolato abruptè acuto lateralibus cohærentibus canaliculatis falcatis haud multùm longioribus, petalis in setam tenuissimam apice clavatam productis.——*Philippines*, (Cuming).——The flower is the largest in the genus, the back sepal measuring an inch and half in length. I have only seen a single flower.

19. C. *maxillare* (Lindl. in Bot. Reg. 1843. sub t. 49.); folio oblongo scapi uniflori longitudine, sepalo dorsali setaceo-acuminato ciliato lateralibus obtusè acuminatis disjunctis, petalis acutissimis.——*Philippines*, (Cuming).——Upper sepal 9 lines, lateral 17 lines long. Petals about $1\frac{1}{2}$ line.

20. C. *Blumii* (L. no. 4. *Ephippium ciliatum*, Blume Bijdr. 209. tab. 65.); folio lineari-lanceolato, scapo unifloro, petalis pubescentibus acutis, sepalis omnibus acuminatis, dorsali canaliculato infernè ciliolato, labello orbiculato longissimè cuspidato, columnâ bicorni.——*Java.*

21. ? C. *compressum* (*Ephippium uniflorum*, Blume Bijdr. p. 309.); "bulbis oblongis compressis, foliis lato-lanceolatis, pedunculo unifloro, sepalis ovatis acuminatis."——*Java.*——Probably of this genus, since it is associated by Blume with his Ephippium ciliatum.

54

DENDROBIUM aqueum.

Watery Dendrobium.

GYNANDRIA *MONANDRIA.*

Nat. ord. ORCHIDACEÆ. § MALAXEÆ.

DENDROBIUM. Swz.

Sect. EUDENDROBIUM; *foliis planis, floribus gemellis, labello trilobo.*

D. *aqueum;* foliis ovato-oblongis undulatis acuminatis, floribus patulis cornu brevi obtuso, sepalis petalisque ovatis, labelli trilobi pubescentis laciniâ mediâ ovatâ denticulatâ basi altè excavatâ lateralibus latioribus rotundatis margine anteriore serrulato. *Lindl. in Bot. Reg.* 1843. *misc.* 6.

Among the crowd of Indian species belonging to this large genus, or group of genera, the plant now figured seems to be hitherto unknown to the Botanists of India. With the manner of growth of *D. Pierardi*, its pale watery green flowers are quite destitute of the attractive colours of that gay species, and are entirely different in the structure of the lip, which is furnished with a large cavity, almost a pouch, at the base of the middle lobe, as is shewn at fig. 1. in the plate. It is also a much stouter plant, with wavy leaves.

The figure was made in November, 1842, in the nursery of Messrs. Loddiges, who imported it from Bombay.

This species may be grown in the same way as many other Dendrobiums, potted in rough turfy peat, well mixed with pieces of broken pots. For drainage, the pot should be nearly half filled with potsherds, and the soil elevated one fourth the height of the pot above its brim. Being thus potted, too much water can scarcely be given during the growing season. The house should be slightly shaded in sunny weather, taking care to keep the temperature as near 80° by day as possible, and about 68° by night. In autumn, as the young shoots become matured, water should gradually be withheld, so that in winter the plant may only receive it in fine weather. The temperature may then be allowed to fall as low as 50° or 55°.

56

MORMŌDES aromaticum.

Aromatic Mormodes.

GYNANDRIA *MONANDRIA.*

Nat. ord. ORCHIDACEÆ, § VANDEÆ—CATASETIDÆ.

MORMODES. Botanical Register, vol. 22. *fol.* 1861.

M. *aromaticum;* racemo brevi erecto, sepalis petalisque subrotundo-ovatis acutis secundis concavis, labello angustè cuneato convexo laciniâ intermediâ triangulari acuminatâ cucullatâ. *Lindl. in Bot. Reg.* 1841. *misc.* 162.

We have so recently (t. 33.) made some general observations upon the genus Mormodes, that we have nothing further to do on the present occasion than to give a figure of this plant, and to point out in what respect it differs from M. pardinum, the only species to which it approaches closely.

In habit the two are similar, but *M. aromaticum* is the smaller, and has shorter leaves. The spike of *M. pardinum* is much longer and bears three times as many flowers; the sepals and petals are narrower, and more taper-pointed: their difference in colour is obvious; the labellum of *M. pardinum* has the same form as the sepals, except that it has three sharp-pointed lobes, and a kind of stalk, which *M. aromaticum* wants.

This species, although only introduced a few years since from Mexico, is now common, and although of little beauty, is valued for the peculiar fragrance, which is like that of aromatic vinegar.

Fig. 1. represents the labellum.

60

DENDROBIUM Ruckĕri.

Mr. Rucker's Dendrobium.

GYNANDRIA *MONANDRIA.*

Nat. ord. ORCHIDACEÆ. § MALAXEÆ—DENDROBIDÆ.

DENDROBIUM. Supra, vol. 15. *fol.* 1291.

Sect. EUDENDROBIUM. *Caule tereti, foliis planis, floribus gemellis, labello trilobo.*

D. *Ruckeri* (Lindl. in Bot. Reg. 1843. misc. 38.); caulibus teretibus, foliis ovato-lanceolatis acuminatis planis, floribus gemellis, sepalis patulis obtusiusculis convexis margine reflexis lateralibus subtriangularibus, labelli trilobi lobis rotundatis intermedio undulato axi elevatâ villosâ.

We presume this plant to be one of Mr. Cuming's discoveries in the Philippines, although no trace of it is to be found among his dried specimens.

It belongs to the same set as *D. Pierardi*, among which it is one of those with yellow flowers, the others being *D. sanguinolentum, chrysanthum, Cambridgeanum, rugosum, salaccense, Paxtoni,* and *aureum.* Although it may not be quite so handsome as some of these, yet it is a very fine species, and perfectly distinct, both in colour, habit, and the structure of the flowers.

Its leaves are exactly lanceolate, very sharp-pointed, and a little disposed to curve backward at the end. The flowers, although of a rich yellow nankin colour when expanded, are almost white externally; their lower sepals, which are really ovate in form, are rolled back at the edge, near the middle, so as to look as if contracted there. The petals and upper sepal are nearly of the same size and form, linearly obovate, acute and spreading. The labellum is deep orange, with a white edge and a pale pink outside; when spread flat it is roundish obovate, with three rounded lobes, of which the middle one is crisp, and has a hairy ridge running along its middle; the side lobes are hairy too near the base. The flowers are deliciously sweet.

We owe the opportunity of figuring it to Sigismond Rucker, Esq. Jun., with whom it flowered in February last.

Fig. 1. represents a front view of the column and base of the labellum; 2. shews the latter spread open.

It should be potted in rough turfy peat, mixed with potsherds; the pot to be half filled with broken pots, and the soil considerably elevated above its brim, in order that the roots may sustain no injury from stagnant water. In summer, while the plant is growing, plenty of water should be given to its roots, and it may be slightly syringed once or twice a day. The house at this time should be shaded in sunny weather, and kept as moist as possible, taking care not to let the temperature rise above 80° by day, nor above 70° at night. In autumn water should be very much withheld, so that the plant may gradually receive its state of rest. In winter the temperature should never be raised above 58° with fire heat.

61

CIRRHOPETĂLUM aurātum.

Gold-edged Cirrhopetalum.

GYNANDRIA *MONANDRIA.*

Nat. ord. ORCHIDACEÆ, § MALAXEÆ.

CIRRHOPETALUM. Supra 1838. *t.* 11.

C. *auratum;* pseudobulbis ovatis sulcatis, folio oblongo convexo, floribus umbellatis, sepalo supremo petalisque setaceo-acuminatis fulvo-ciliatis lateralibus acutis, labello lineari recurvo, columnæ auriculis rotundatis integris. *Lindl. in Bot. Reg.* 1840. *misc.* 107.—1843. *sub t.* 49.

Among the singular species of this genus the present is one of the most interesting. It hangs down from the branch of a tree, or a piece of charred wood, which it soon overruns with its delicate green roots and egg-shaped furrowed pseudobulbs.

The leaves are very thick, deep green above, and convex; stained with purple beneath. The flower-stem is as slender as a small thread, and too weak to bear the umbels of flowers, which therefore hang down gracefully, and are balanced in the air.

The umbels, as in many others of this genus, are so arranged that the flowers are all on one plane, and diverging equally from the centre form a circle, whose interior is occupied by the lower part of the flowers, and whose circumference is formed by the long flat strap-shaped lateral sepals, which look like so many party-coloured ribbons collected into a balloon.

The flowers themselves have a yellowish ground, striped and mottled with crimson. The upper sepal and two petals, badly drawn in the figure, are fringed with golden hairs, and tapered into a fine point. The lateral sepals are quite destitute of hairiness, and only faintly stained with purple.

It differs from C. picturatum in its party-coloured, not purple, flowers; in its petals being far less taper-pointed, and not villous; and in the lip not having a central ridge, which is conspicuous in C. picturatum.

A native of Manilla, whence it was received by Messrs. Loddiges, with whom our drawing was made in March 1841.

Fig. 1. represents a flower much magnified, with the long lateral sepals cut off; the petals are by no means well figured; they, as well as the upper sepal, taper into a fine bristle-like point. Fig. 2. is the labellum.

It may be potted in turfy heath-mould, mixed with pieces of small potsherds; or it may be suspended on a block of wood, with a little sphagnum about its roots to retain moisture. Like many species of Pleurothallis, Stelis, &c. this requires a humid atmosphere in winter as well as in summer; but it should always be remembered, that the more light and heat, the more water is required. During the warm summer months, when the temperature of the house (although shaded) can scarcely be kept below 80°, the water required will be twice as much as is necessary in winter, when the temperature should never be raised above 50° or 55° by artificial means.

66

ONCIDIUM bicŏlor.

Two-coloured Oncidium.

GYNANDRIA *MONANDRIA.*

Nat. ord. ORCHIDACEÆ. § VANDEÆ.

ONCIDIUM. Supra, vol. 14. *fol.* 1073.

§ EUONCIDIUM, HETERANTHIUM; *planifolia, tetrapetala, micropetala.*

O. *bicolor;* pseudobulbis ovalibus compressimis utrinque 3-costatis, folio solitario oblongo sessili striato, paniculâ divaricatâ, sepalis lateralibus basi connatis petalisque ovatis acutis, labello maximo bilobo, disci tuberculo hastato 3-partito, columnæ alis rotundatis dentatis.

This charming species of Oncidium, one of the rarest of its genus, was obtained by Messrs. Loddiges from the Spanish Main, and is 1243 of their catalogue. It flowered with them in September, 1842, when the accompanying figure was made.

Among other peculiarities of the species may be mentioned its very thin pseudo-bulbs, and three-lobed crest of the labellum, the front lobe of which projects forward, while the laterals are at right angles to it, and slightly toothed. It has moreover a very large lip, which, although deep clear yellow on the upper side, is almost white underneath, as also happens in some of the Chive-leaved species.

Its nearest affinity is with O. bifolium.

MISCELLANEOUS MATTER

OF THE

BOTANICAL REGISTER.

1843.

1. CATASĒTŬM Wailēsiĭ.

Hooker in Botanical Magazine, t. 3937.

C. *Wailesii;* "foliis oblongo-lanceolatis, perianthiis ovatis compressis conniventibus, sepalis petalisque acuminatis, labello subconico cucullato ore contracto integerrimo, antheris hemisphærico-compressis umbonatis umbone subtùs squamifero."

This is a green-flowered species from Honduras, with all the appearance of C. tridentatum; Sir W. Hooker, however, assigns the following reasons for regarding it as a distinct species.

"The strange forms exhibited in the different parts of the flower of the several kinds of Catasetum have often attracted notice. In the present instance, the most striking peculiarity is to be found in the anther-case, which, instead of being carried out into a very long point or beak, is singularly short and flattened, in the centre of which is an umbo or elevated tubercle, and beneath this, a pretty large, membranaceous, spreading scale. Whether this be simply a sport of nature, or a permanent character, indicating a truly distinct species, I will not take upon me to say."

2. BRĂSSĬĂ brachīată.

Lindley in Bentham Plant. Hartweg. p. 94.

This fine plant has flowered in the collection of Messrs. Rollissons, where it exists under the garden name of *Brassia Wrayæ.* It has larger flowers than any other species, with very dark brown spots on a pale ground. The lower sepals are three inches long, and the lip is half the length, a good deal waved at the edge, and rather lobed. It is a very distinct species.

3. RENANTHĒRĂ matutīnă.

Lindley in Gen. and Sp. Orch. p. 218.

A very pretty plant with small cinnamon and scarlet flowers, collected in a stiff panicle. It has the habit of the old R. coccinea, but is not to be compared with it for beauty, the flowers not being above half an inch across. It was exhibited at the last meeting of the Horticultural Society, from His Grace the Duke of Devonshire, at Chatsworth, and we have also received it from Messrs. Rollissons. It is understood to have been brought from Manilla by Mr. Cuming, from whom we have a wild specimen.

4. EPIDĒNDRŪM aurītum.

E. (Encyclium) *auritum;* pseudobulbis elongatis compressis squamis vaginatis, folio angusto ligulato obtuso, racemo debili 3-floro folio breviore, bracteis marcidis carinatis internodiis longioribus, ovario nigro-furfuraceo, sepalis lineari-lanceolatis acutis patentibus, petalis conformibus duplò brevioribus erectis, labello lineari convexo arcuato medio carnoso subrugoso apice membranaceo, columnâ apice tridentatâ, antherâ utrinque auriculâ simplici bifidâ subulatâ auctâ.

A little Epidendrum from Guatemala, of which we have specimens from Mr. Skinner. The flowers are very pale green, almost white, with no spots or stains upon them. It has not much beauty. It is near E. tripunctatum. A very singular feature in it is the presence of one or two fine pointed horns on the anther on each side, which give it, when seen in front, the appearance of some insect's head.

5. CLEISŌSTŎMĂ dealbātum.

C. *dealbatum;* foliis linearibus canaliculatis carnosis dealbatis, spicis simplicibus glaucis, sepalis petalisque rotundatis, labelli calcare ovato emarginato lobis lateralibus acutis intermedio ovato carnoso multò majore incurvo, columnæ dente bilobo.

An Orchidaceous plant of little beauty, obtained from Manilla by Messrs. Loddiges, through Mr. Cuming. The flowers before they expand are almost white, when open they are bright yellow, with a white lip; but they are too small to produce much appearance.

6. DENDRŌBĬŪM ăquĕum.

D. *aqueum*, (§ foliis planis, floribus gemellis, labello trilobo) foliis ovato-oblongis undulatis acuminatis, floribus patulis cornu brevi obtuso, sepalis petalisque ovatis, labelli trilobi pubescentis laciniâ mediâ ovatâ denticulatâ basi altè excavatâ lateralibus latioribus rotundatis margine anteriore serrulato.

A native of Bombay, whence it was imported by Messrs. Loddiges, with whom it flowered in November, 1842. The habit is that of D. Pierardi: the flowers are rather larger, and of a pale watery green colour.

7. PLEUROTHĀLLĬS fœtēns.

P. *fœtens;* folio ovali coriaceo subtus concolori cauli carinato triquetro æquali, spathâ univalvi acutâ coriaceâ striatâ pauciflorâ, floribus subsessilibus tomentosis coriaceis, sepalis oblongis convexis subæqualibus intùs papillosis lateralibus connatis, petalis oblongis acutis glabris denticulatis subdiaphanis, labello lineari-oblongo obtuso lævi lineis duabus elevatis carnosis basi excavato.

A species very near the Mexican *P. aphthosa*, with an unpleasant odour. It has no beauty, and is a native of Brazil, whence it was obtained by Messrs. Loddiges. The leaves are rather narrower than in P. aphthosa, and green not purple underneath.

8. OBERŌNĬĂ miniātă.

O. *miniata;* caulescens, foliis equitantibus brevibus subfalcatis acutis, spicâ longissimâ pendulâ laxiflorâ, bracteis linearibus convolutis laceris, pedicellis filiformibus glanduloso-pilosis, sepalis ovatis erectis acutis, petalis conformibus minoribus, labello oblongo concavo versus apicem angustato retuso.

A singular epiphyte, belonging to a section of this strange genus, of which *O. anceps* is the type, and of which there are many species. That now figured was imported from Sincapore by Messrs. Loddiges. It has the habit of Aporum anceps; its flowers are extremely small, very brittle, vermilion red, and loosely arranged in a nodding spike sometimes as much as eight inches long.

9. ANGRÆCŪM vesicātŭm.

A. *vesicatum;* acaule, foliis canaliculatis recurvis spicæ pendulæ æqualibus, floribus distantibus divaricatis, sepalis ovatis, petalis linearibus acutis, labello ovato acuto, calcare arcuato ovario longiore apice inflato diaphano.

From the Ashantee country, whence it was obtained by Messrs. Loddiges. It bears white, inconspicuous flowers, whose spurs look like a horn with a semi-transparent bladder at the end. It seems to range next to A. caulescens from the Isle of France.

10. ACĬANTHĒRĂ punctātă.

Scheidweiler in Gartenzeitung, 1842, *p.* 292.

"Acianthera. Sepala conniventia, lateralibus connatis, supremo galeato. Petala minora cuneiformia, apice dilatata; labellum cum columna articulatum, 3-lobum; lobis lateralibus acutis, intermedio plano obtuso; Columna alata membranacea apice lacerata; anthera apiculata, unilocularis; pollinia 2, materie pulverea ad basin cohærentia.

"A. *punctata* (Scheidw.); foliis solitariis, ovatis, obtusis, glaucis, utrinque purpureo-punctatis; sepalis extus villosis, intus purpureo-punctatis vel ocellatis; labello purpureo, petalis punctatis. Flores racemosi, racemi pauciflori cernui, basi spathacei."

An orchidaceous plant, stated in the work above quoted to be a native of Brazil, and to belong to the Malaxeous division. It is described as resembling, at first sight, a Pleurothallis; very small; with oval leaves, and five or six greenish grey flowers on a common flower-stalk.

11. CENTRANTHĒRĂ punctātă.

Scheidweiler l. c. p. 293.

"Centranthera. Sepala lateralia connata, supremum galeatum; petala minora cuneiformia. Labellum cum columna articulatum, 3-lobum, lobis lateralibus minimis acutis, intermedio ovato obtuso. Columna alata, sub-membranacea apice lacerata. Anthera apiculata, unilocularis, pollinia 2 materie pulverea cohærentia.

"C. *punctata* (Scheidw.); folio ovato, obtuso, glauco, utrinque fusco punctato. Sepalis extus villosissimis, intus fusco-ocellatis; labellum purpureum. Flores racemosi, racemi radicales cernui. —Planta epiphyta brasiliensis, caule semiunciali, folio carnoso unciale exacte ovato."

Another new genus of Orchidaceæ, according to Mr. Scheidweiler, found in Brazil, and cultivated in the German gardens. It, like the last, is represented to be very like a Pleurothallis; its leaves are glaucous with brown pits.

12. LIPĂRĬS alātă.

Scheidweiler l. c. p. 293.

"L. *alata* (Scheidw.); foliis 6, oblongo-ovatis, acutis, plicatis, petiolatis, spica erecta multiflora brevioribus. Scapo alato, squamoso, purpureo; alis 6, membranaceis. Sepalis petalis linearibus latioribus; labello mucronato revoluto atropurpureo. Gynostemium viride; flores ante anthesin purpurei."

A Mexican plant, which has flowered at Lacken, in the garden of His Majesty the King of the Belgians. It is stated to be rather prettier than the majority of its race, with a spike of flowers six inches long, and purple flowers half an inch long, with a crimson lip.

13. MAXILLĀRĬĂ galeātă.

Scheidweiler l. c. p. 309.

"M. *galeata* (Scheidw.); pseudobulbis ovatis compressis diphyllis; foliis ovato-oblongis, acutis, plicatis; scapis radicalibus unifloris; sepalo supremo petalisque lateralibus conniventibus subgaleatis, sepalis lateralibus patentibus, omnibus lanceolatis acutis, labelli trilobi lobo intermedio lingulato, lateralibus parvis acutis; anthera galeata, loculorum valvulis membranaceis transversim dehiscentibus; pollinia 4, per paria in glandulam hyalinam transversam sessilia.

"Scapus duo pollices longus, purpureus; bracteæ quatuor, ventricosæ, acutæ, margine purpureæ. Flores purpurei, striato-punctati; labellum cum basi producta columnæ articulatum, atropurpureum, basi dilute roseum."

A Brazilian species, described from the stove of Count Aremberg. It is said to be scentless, and to have dirty purple flowers.

14. ONCĪDĬŪM Forkēlĭī.

Scheidweiler l. c. p. 309.

"O. *Forkelii* (Scheidw.); pseudobulbis monophyllis; foliis coriaceis, oblongis, acutis; vaginis ciliatis; scapo erecto punctato, trifloro, foliis breviore. Sepalis lateralibus liberis, oblongo-ovatis, acuminatis, superiore ovato undulato; petalis oblongis margine undulatis; labello trilobo, lobis lateralibus minoribus, intermedio maximo rotundato baseos callis duobus instructo, anteriore compresso bicorni, posterioris lamina plicata; gynostemio alato, alis lilacinis: Anthera carnosa galeata, pollinia duo in glandulam bicornutam sessilia.

"Folia 10-12 pollices longa, 3-4 poll. lata. Sepala extus virentia intus fusco-maculata, petala apice virentia basi fuscata, labellum lilacinum."

A Mexican species that has flowered in the collection of His Majesty the King of Belgium, at Lacken. It has been named after Mr. Forkel, the chief gardener there. It is said to be one of the handsomest of the genus, with flowers an inch in diameter, greenish yellow, spotted with crimson; the petals clear violet, the lip large and round, with small side lobes.

15. ONCĪDĬŪM cuneātŭm.

Scheidweiler l. c. p. 309.

"O. *cuneatum* (Scheidw.); pseudobulbis subcylindricis, arcuatis, rugosis, apice truncatis, monophyllis; foliis lanceolatis acutis; paniculis radicalibus cernuis; sepalis lateralibus cuspidatis in unum cucullatum connatis, superiore truncato cucullato; petalis cuneatis apice minutissime crenulatis; labello trilobo, lobis lateralibus acutis, intermedio cuneato emarginato; crista triplici; posteriore quadridentata, intermedio lamellis quatuor parallelis composita, anteriore bicornuta; columnæ alis linearibus, denticulatis, recurvis, acutis."

A Brazilian plant from Count Aremberg's collection. It is said to be a small species with white flowers spotted with crimson, and purple wings to the column. The leaves are three inches long and six inches broad, the flower-stalk four inches long.

16. LÆLĬĂ ălbĭdă.

A beautiful variety of this, with a violet lip, painted at the base with purple veins, has flowered with Mrs. Lawrence at Ealing Park.

17. TRICHOCĒNTRŪM recurvūm.

T. *recurvum*; bracteis distichis cucullatis ovatis acutis, sepalis petalisque ovatis acutis, labello oblongo basi purpureo bidentato, calcare recurvo ovarii longitudine, columnæ alis cuneatis denticulatis, antherâ pubescente.

A Guayana plant, imported by Messrs. Loddiges. It resembles Tr. fuscum in habit, but is smaller. The lip is white, with a deep purple spot at its base; afterwards it changes to pure yellow. The spur is long, and curved backwards and upwards, so as to arch over the back of the flower.

18. TRICHOCĔNTRUM candĭdum.

T. *candidum*; bracteis ovatis acutis, sepalis petalisque ovatis acutis, labello oblongo emarginato basi obsoletè bidentato ecalcarato gibboso, columnæ alis ovatis acutis, antherâ villosâ.

A little Guatemala plant, with white flowers, slightly tinged with yellow. It is remarkable for the want of a spur, whose place is supplied by a short projection.

19. ODONTOGLOSSUM Rossii.

Lindley in Bot. Reg. 1839. *t.* 48.

A pretty variety of this charming plant has flowered with Mr. Barker. It has narrower and darker coloured sepals, smaller flowers, and the callus at the base of the lip white, not yellow. It seems almost intermediate between O. Rossii and O. stellatum; especially as the lip is rather more ovate than in the former species.

20. EPIDĔNDRUM rubrocīnctum.

E (SPATHIUM) *rubrocinctum*; . . . . . paniculâ amplâ cernuâ, sepalis oblongis concavis acutis coriaceis, petalis angustè linearibus, labello transverso cordato trilobo suprà tricarinato; lobi medii trilobi dente intermedio minuto, columnâ inappendiculatâ.

Of this I know no more than that it is a plant with the

habit of E. nutans, from the garden of Mr. Brocklehurst, where nothing is recorded of its origin. Mr. Bateman, who sent it to me, states that it has a large branching panicle. The sweet-scented flowers are a dull yellowish green, bordered with dull purple; the lip is more yellow than the sepals. It is a plant of more beauty than E. nutans.

MAXILLARIA.

The original genus of this name was so loosely defined that it was understood to comprehend all Orchidaceous plants whose floral envelopes are so arranged that they have a ringent appearance, and a decided extension in front in the form of a chin. I have no intention of discussing in this place the history of the genus, the additions made to it, or the limits of the new genera already formed in its vicinity; but I shall confine myself to the condition in which it was left upon the publication of the "Genera and Species of Orchidaceous Plants" in the year 1833. At that time the number of species, known or guessed at, was inconsiderable, and consisted of plants extremely different in appearance, as well as structure, but connected by the common character of a pair of simple or two-lobed pollen-masses, having a distinct gland with a single caudicula, and belonging to a flower whose lateral sepals, oblique at their base, were adherent to the lengthened foot of the column. In fact, Maxillaria, among Vandeæ, answered in many respects to Dendrobium among Malaxeæ.

But experience has shewn that such a character as this brings into association plants of very different appearance, that it includes many other characters, of as much importance as itself, and brings together a group of species inconveniently large for systematical purposes. Such being the case it has now become desirable to reconsider the distinctive marks of Maxillaria, to see how far they are capable of sub-division, and to endeavour to circumscribe Maxillaria proper within more definite limits than were assigned to it in 1833.

The true Maxillarias, that is to say those intended by the authors of the Flora Peruviana, who founded the genus, are evidently the species with radical inflorescence, and a pseudobulbous growth, such as *M. picta*, *punctata*, *squalens*, and the like; they form a tolerably natural group, and have, in all

cases, a pair of double pollen-masses resting on a crescent-shaped gland, without any distinct caudicula; to them it is desirable strictly to limit the name. But a great variety of other plants have been gradually associated with them in consequence of their having the chin which so strongly marks, among Vandeæ, Maxillaria proper; and, in fact, this chin must now be considered more an indication of a division of Vandeæ than of a genus.

For example, *Maxillaria Warreana*, has a globular flower, expanded indeed, but only a little oblique, and by no means ringent: this I would call WARREA.

Then those species which are near *M. lentiginosa*, having also a flower with nothing ringent about it, have an apparatus of a singularly rugged nature, or at least much tuberculated, on the lip, and a gland of an ovate form bearing two double pollen-masses sessile; to these the name PROMENÆA may be assigned. Allied to them, but widely different in the small roundish gland and long setaceous caudicula on which the two double pollen-masses are seated, is *Maxillaria cristata*, which may be called PAPHINIA. Another set, with a similar condition of gland, caudicula, and pollen-masses, but well distinguished by the surface of the lip, is formed by such species as *M. aromatica*, *macrophylla*, &c. and these I would call LYCASTE. As for *Maxillaria Steelii*, with its long thonged leaves and deficient pseudo-bulbs, it has nothing of the aspect of a Maxillaria, and having a pair of double pollen-masses sitting on a gland tapering to each end with the form of a gliding serpent, it may be advantageously struck off under the name of SCUTICARIA.

These changes having been effected, the genus Maxillaria will remain associated with Dicrypta, Xylobium, Camaridium, and Siagonanthus, the true value of which I shall shortly endeavour to settle.

But it is not merely as an old genus, into which far too much alien blood has been infused, that Maxillaria has to be considered. We must certainly regard it as the type of a *Division* of Vandeæ, to which a good number of other genera will have to be associated. These genera are all characterized by having the lateral sepals more or less oblique at the base, the consequence of which is that the flower-bud has always, more or less visibly, a chin, and by having a labellum which is destitute of a spur, or of any direct approach to one.

It is also strictly an occidental division, saving the genus Polystachya, unless we admit *Appendicula* and *Cryptoglottis*, which are insular Indian. Those genera however require further examination.

I do not profess to be able of myself to settle at present the true limits of what may be called the Maxillaridous division of Vandeæ; but, in order to assist others who may be working in the Orchidaceous mine, I will just put down what I at present think will be the genera to be referred to it. The tabular form being the most convenient for study, I have adopted it, printing in italics those genera which seem certain, and in Romans such as require much further examination.

MAXILLARIDÆ. Sepala lateralia plus minus obliqua, sæpius majora, unde alabastrus hinc gibbus. Labellum columnæ appressum, ecalcaratum. Columna semiteres, basi producta. Pollinia 2-4 (-8 ?).

[The genus Zygopetalum, whose lateral sepals are very little oblique, joins this Division of Vandeæ to that of Sarcanthidæ, in consequence of its near affinity to Eulophia in the last mentioned division.]

* Labellum medio interruptum. *Stanhopea, Houlletia, Peristeria, Anguloa* (Cuitlauzina ?).

* Labellum continuum.

a. Sepala lateralia divaricata incurva basi angustata; *Govenia, Batemannia.*

b. Sepala lateralia ascendentia ; *Zygopetalum.*

c. Sepala lateralia recta, sæpiùs triangularia.

‖ Pollinia 2 ; *Acropera, Chænanthe, Malachadenia.*

‖ ‖ Pollinia 8 ; v. 4 plura ; Appendicula, Cryptoglottis.

‖ ‖ ‖ Pollinia 4 ; v. 2 biloba.

‡ Flores regulares clausi ; *Ornithidium.*

‡ ‡ Flores regulares expansi; *Trigonidium, Psittacoglossum, Stenia, Promenæa, Warrea.*

‡ ‡ ‡ Flores ringentes.

1. Caudiculæ 2 ; *Bifrenaria.*

2. Caudicula 1.

¶ Columna maxima navicularis, clinandrio cucullato. *Huntleya.*

¶ ¶ Columna clavata, angusta, clinandrio nudo.

☞ Glandula lunata, caudiculis brevissimis v. nullis ; *Maxillaria,* Dicrypta, Xylobium, Camaridium, Siagonanthus, *Scuticaria.*

☞☞ Glandula ovata, caudiculâ brevissimâ, v. 0. *Scaphyglottis.*

☞☞☞ Glandula minuta ; caudiculâ setaceâ. *Polystachya, Paphinia, Lycast*

13

It is not improbable that, besides these, *Maxillaria tetragona,* and a few others, may require to be further separated.

Having thus explained my notions of the manner in which Maxillaria and the genera allied to it may be best systematised, I proceed to add very brief characters of the new genera and their species.

PROMENÆA. Sepala patula. Labellum trilobum, medio cristatum v. multo-tuberculatum. Columna brevis, semiteres. Glandula ovata. Pollinia 4, geminata, sessilia.

1. P. *stapelioides* (Maxill. stapelioides, *L.* 146. *B. R.* 1839. *t.* 17.); pseudobulbis ovatis tetragonis 1-2-phyllis, foliis tenuibus lanceolatis patentibus pallidè glaucis reticulatis, pedunculo diffuso bifloro, sepalis petalisque subrotundo-ovatis acutis patulis subæqualibus, labello oblongo trilobo: laciniis lateralibus erectis linearibus obliquis obtusis intermediâ ovato-oblongâ basi cucullatâ, cristâ transversâ flexuosâ carnosâ intùs dente carnoso ovato auctâ.——*Brazil.*——Flowers green externally, yellowish internally, speckled and banded with purple.
2. P. *xanthina* (Maxill. xanthina, *Lindl. Bot. Reg.* 1839. *sub t.* 17.); pseudobulbis ovalibus tetragonis 1-2-phyllis, foliis angustè lanceolatis, pedunculis ascendentibus unifloris pedicello sterili terminatis, bracteâ ovatâ mucronatâ cucullatâ, sepalis petalisque oblongis acutis patulis subæqualibus, labello oblongo trilobo laciniis lateralibus erectis linearibus obtusissimis integris intermediâ bilabiatâ: labio superiore carnoso abbreviato truncato 5-dentato inferiore oblongo acuto.—*Brazil.*—Flowers yellow.
3. P. *lentiginosa* (Maxill. lentiginosa, *Lindl. in Bot. Reg.* 1839, *misc.* 93.); bracteis latè ovatis acuminatis, labelli lobo medio ovato-oblongo obtuso, cristâ transversâ medio processu quadrato tridentato auctâ, antheræ apice incurvo ; alioquin *M. stapelioidi* similis.——*Brazil.*——Very like P. stapelioides; but the sepals are more acute, the purple spots redder, more distinct, and less run into bars; the lip is of the same colour as the petals, and its transverse crest has a square three-toothed process in the middle.
4. P. *Rollissonii* (Maxill. Rollissonii, *Bot.Reg. sub fol.* 1986.et 1838, *t.* 40.); pseudobulbis subrotundis compressis bifoliis, foliis lanceolatis apice recurvis, scapis subunifloris diffusis laxè vaginatis, sepalis carinatis lateralibus basi subæqualibus petalisque acutissimis, labelli lobis lateralibus ovatis angustis acutis intermedio oblongo membranaceo apiculato margine deflexo ; callo disci elevato carnoso anticè transverso lobos laterales labelli conjungente medio producto truncato tridentato et denticulis reflexis posticè fornicato truncato bilobo.——*Brazil.*——Flowers pale yellow, with purple spots on the lip.
5. P. *graminea* (Maxill. graminea, *Bot. Reg. sub fol.* 1802.) ; pseudobulbis nullis, foliis gramineis recurvis, perianthio campanulato, sepalis oblongis obtusis lateralibus paululum connatis, petalis angustioribus, labello cum pede longè producto columnæ articulato trilobo : lobis lateralibus semiovatis acutis ascendentibus intermedio majore reniformi, tuberculo disci magno carnoso truncato posticè bilobo.——*Guayana; Brazil, in the prov. of Bahia.*—— Flowers pale yellow, banded with dull purple.

14

SCUTICARIA. Flores ringentes anticè in mentum rotundatum producti, expansi. Labellum continuum, membranaceum, trilobum, medio tuberculatum. Columna semiteres. Pollinia 4, per paria in glandulam utrinque acuminatam medio dilatatam sessilia, extus angulata.——Rhizoma articulatum, ramosum, ebulbe. Folia flagelliformia.

1. S. *Steelii* (Maxill. Steelii, *Bot. Mag. t.* 3573. *Bot. Reg. t.* 1986.).——*Guayana.*——Flowers large, yellow, spotted with purple.

WARREA. Flores subglobosi, subregulares, mento brevi rotundato. Labellum continuum, indivisum, lineis elevatis carnosis in medium. Columna semiteres, clavata. Pollinia 4, per paria in caudiculam brevem linearem inserta, glandulâ triangulari.——Herba terrestris, pseudobulbosa. Folia arundinacea. Scapus radicalis elatus apice racemosus. Flores magni speciosi.

1. Warrea *tricolor* (Maxill. Warreana, *Lodd. Bot. Cab. t.* 1884. *L no.* 30.) ; Folia oblongo-lanceolata acuminata plicata, scapo radicali erecto vaginato multifloro foliis longiore. Sepala ovata concava acuta basi subæqualia, petalis minoribus conformia ; labello brevissimè unguiculato obovato-oblongo indiviso cucullato obtuso jugis tribus in medio elevatis carnosis, superficie laminæ seriatim corrugatâ marginibus planis. Sepala fulvo-lutea : lateralibus basi parùm inæqualibus. Labellum basi luteum, apice pallidum, medio purpureum, cum columna sub-articulatum.——*Brazil.*

PAPHINIA. Flores subregulares, expansi, petaloidei, parum in mentum producti. Labellum parvum, unguiculatum, tripartitum, glandulis filiformibus apice alibique obsitum. Columna clavata, elongata, semiteres, apice auriculata. Pollinia 4, per paria caudiculæ elongatæ apice setaceæ affixa, glandulâ minutâ subtriangulari ; rostello subulato. ——Herba pseudobulbosa, scapo pendulo paucifloro.

1. Paphinia *cristata* (Maxill. cristata, *Lindl. in Bot. Reg. t.* 1811.); flores pulcherrimi extus albi, intus purpureo interrupte fasciati ; petalis apice omnino purpureis. ——*Trinidad and Guayana.*

LYCASTE. Flores ringentes, petalis sæpius dissimilibus, in mentum breve producti. Labellum medio appendice transverso carnoso integro v. emarginato auctum. Columna elongata, semiteres, sæpius pilosa. Pollinia 4, per paria caudiculæ angustæ elongatæ adnata ; glandulâ parvâ subrotundâ ; rostello subulato.——Herbæ pseudobulbosæ, foliis plicatis. Scapi radicales, erecti, uniflori. Flores semper speciosi bracteâ magnâ spathaceâ suffulti.

1. L. *macrophylla* (Maxill. macrophylla, *Pöppig. gen. pl.* 1. *t.* 64. *Bot. Reg.* 1838, *misc.* 175.); bracteâ herbaceâ cucullatâ acutâ ovarii longitudine,

15

sepalis oblongis undulatis patentibus apice recurvis basi intùs pilosis, petalis erectis columnâ longioribus oblongis carnosis apice recurvis margine postico versus apicem sublobato, labello breviore oblongo concavo apice trilobo : laciniâ intermediâ subrotundâ crenatâ pilosâ, appendice linguæformi concavo adnato inter lacinias laterales rotundatas incurvas, antherâ villosâ.——*Peru.*——Flowers greenish, with a little purple on the lip. According to Pöppig it inhabits dry thickets.
2. L. *plana* (Lindl. in Bot. Reg. 1842. misc. 96.) ; bracteâ supremâ ventricosâ cucullatâ ovario longiore, sepalis oblongis planis basi in cornu brevi conico connatis, petalis conformibus apice tantùm recurvis, labelli trilobi lobis lateralibus apice crenulatis intermedio subrotundo serrato callo elevato obtuso obsoletè trilobo, columnâ pubescente, antherâ villosâ. ——*Peru.*——Nearly allied to *Lycaste macrophylla,* of which it has quite the habit. It differs in the petals being quite even not undulated, and in the lateral sepals being much more exactly oblong; the tubercle on the lip is also much more obtuse. In colour too they are different. *L. macrophylla* has olive-green sepals, and petals almost colourless. *L. plana* has the sepals a deep rich madder-red inside, and the petals are richly tipped with crimson.
3. L. *costata* (Maxill. costata, *Bot. Reg.* 1838. *misc.* 175.); bracteâ herbaceâ cucullatâ acutâ ovario multò longiore, sepalis petalisque M. macrophyllæ similibus, labello trilobo concavo laciniâ intermediâ rotundatâ serratâ, appendice carnoso emarginato adnato 5-costato inter lacinias laterales, antherâ glabrâ.——*Peru.*——Quite with the habit of L. macrophylla, but certainly distinct in the foregoing characters. Its colours are unknown.
4. L. *lanipes;* bracteâ obtusâ cucullatâ ovarii longitudine, sepalis petalisque oblongis elongatis, labelli lævis lobis lateralibus ovatis obtusis brevibus intermedio oblongo obtuso basi serrato, tuberculo obtuso concavo costato glabro, columnæ pede villoso.——*Guayaquil.*——Found by Mr. Hartweg at Paccha, a village on the Andes, in the vicinity of Loxa, and sent to the Horticultural Society. It has pale green flowers, two inches and a half long before they expand, without a trace of any other colour. It differs from L. costata in the flowers being much larger, while the bract is not longer than the ovary ; and from L. gigantea in the very short bract, ribbed tubercle, and rounded not acuminate middle segment of the lip.
5. L. *gigantea ;* bracteâ herbaceâ sepalis subæquali, sepalis oblongo-lanceolatis lateralibus falcatis, petalis conformibus paulò minoribus, labello lanceolato acuminato laciniis lateralibus acutis intermediâ ovatâ acuminatâ serratâ, appendice carnoso emarginato.——*Guayaquil.*——Flowering stem more than two feet high. Sepals full three inches long ; green according to Hartweg.
6. L. *Deppii* (Maxill. Deppii, *Lodd. Bot. Cab. t.* 1612. *L. no.* 24.); scapo foliis breviore: vaginis ventricosis acuminatis, sepalis oblongo-lanceolatis patentissimis, petalis minoribus oblongis undulatis conniventibus, labello cucullato 3-lobo apice recurvo in axi calloso margine piloso subcrenato : laciniis lateralibus rotundatis intermediâ oblongâ obtusâ, callo elevato ovato.——*Mexico.* ——Sepals green and chocolate. Petals white. Lip yellow.
7. L. *Skinneri* (Maxill. Skinneri, *Bateman in Bot. Reg.* 1842, *misc.* 13.); bracteâ herbaceâ acutâ cucullatâ ovario multò longiore, sepalis patentibus oblongo-lanceolatis acutis, petalis 2-plo brevioribus ovalibus erectis supra

16

columnam convolutis apicibus reflexis; labelli trilobi lobis lateralibus erectis truncatis, intermedio longiore ovato rotundato deflexo, appendice carnosâ linguæformi inter lacinias laterales locatâ; columnâ subtus pubescente ——*Guatemala*.——The flowers measure upwards of six inches across, from the tips of the lateral sepals, while the latter are nearly an inch and half wide in the broadest part. The sepals are pure white, faintly tinged with crimson at the base;—the petals of a more rosy hue; the lip is almost covered with spots and streaks of the most brilliant carmine. The column is pure white at the apex, and mottled with crimson spots at the base; while a number of woolly hairs are scattered on its under side.

8. L. *aromatica* (Maxill. aromatica, *Hooker Ex. Flor. t.* 219. *L. no.* 20. *Bot. Reg. t.* 1871. Colax aromaticus, *Spreng.*); vaginis distantibus obtusis cucullatis; sepalis ovato-oblongis petalisque conformibus acutis, labelli semicylindracei sepalis æqualis laciniis lateralibus acuminatis obtusis intermediâ cuneatâ apice serrulatâ; appendice magnâ concavâ carnosâ truncatâ, columnæ facie villosâ.——*Mexico*.——The flowers yellowish-orange, scarcely spotted even inside the lip, which has two rows of hairs along its inner face.

9. L. *cruenta* (Maxill. cruenta, *Lindl. in Bot. Reg.* 1842. *t.* 13.); pedunculis rarò bifloris, vaginis distantibus obtusis cucullatis, sepalis ovatis obtusis lateralibus basi parùm productis, petalis minoribus conformibus, labello sepalis duplò breviore concavo trilobo laciniis rotundatis intermediâ crispâ pubescente: tuberculo parvo plano, columnâ pubescente.——*Guatemala*.——Resembles L. *aromatica*, but its leaves are broader; the flowers are four times as large when in health; the lip has quite a different form, with a deep crimson blotch at its base, and is not half the length of the sepals; the middle lobe is rounded not unguiculate, and has only a small tubercle in the middle instead of the large concave appendage that occurs in L. *aromatica*.

With regard to the species that belong to true Maxillaria, now that it has been weeded of these species, I must take another opportunity of examining them.

---

21. HEXADESMIA fasciculata.

Ad. Brongn. in Ann. des Sciences, xvii. *p.* 44.

This plant has flowered in the garden of the Horticultural Society, where it has been saved out of a collection formed by Mr. Hartweg at Quezaltenango in Guatemala. It has small green flowers of no beauty, and the habit of Epidendrum clavatum.

22. ONCĪDĬŪM suave.

O. *suave*; pseudobulbis ovatis compressis utrinque bicostatis, foliis membranaceis lineari-oblongis canaliculatis acutissimis, paniculâ elongatâ ramosâ, sepalis petalisque lineari-lanceolatis acutis undulatis revolutis, labelli lobis lateralibus rotundatis intermedio transverso apiculato minoribus, tuberculo pubescente anticè 5-lobo apice bidentato, columnæ alis rotundatis denticulatis.

A Mexican species, from the nursery of Messrs. Loddiges, very near O. reflexum. The flowers are much smaller; the sepals and petals chocolate colour, tipped with yellow; the lip yellow, with a cinnamon-brown middle. The wings of the column are yellow. It has a faint and agreeable odour. It was sent to Messrs. Loddiges by Deppe in the year 1835.

23. ONCĪDĬŪM sphegĭferum.

O. *sphegiferum*; paniculâ ramosâ divaricatâ, sepalis distinctis acutis, petalis retusis, labelli lobo intermedio bilobo transversè oblongo unguiculato lateralibus serratis subæqualibus, tuberculo villoso ovato pulvinato, columnæ alis firmis rotundatis.

A Brazilian species, intermediate as it were between Oncidium *divaricatum* and *pulvinatum*, of which it has quite the appearance and colour: it is however clearly distinguished by the lateral lobes of the lip being serrated, and by the rounded firm purple wings of the column. Flowered with Messrs. Loddiges.

24. EPIDENDRUM cubēnsĕ.

E. *cubense* (Encyclium); pseudobulbis teretibus elongatis, foliis rigidis horizontalibus subæquilongis angustè ovalibus scapo debili paucifloro duplò brevioribus, bracteis minimis, sepalis petalisque conniventibus lineari-lanceolatis, labello obovato apiculato lineis tribus ramentaceis in medio, clinandrio glabro.

A Cuba plant, rare, delicate, and beautiful. Imported by Messrs. Loddiges. The pseudobulbs are two inches long; the scape six inches high; the flowers white, with a lip folded up, white at the end, yellow in the middle, and purple, as is the column, near the base.

25. ODONTOGLOSSUM constrictum.

O. *constrictum*; paniculâ laxâ basi foliosâ, sepalis petalisque expansis lineari lanceolotis acuminatis, labello medio constricto: hypochilio oblongo, epichilio subpandurato cuspidato serrato basi lamellis 2 serratis aucto, columnâ bicirrhosâ.

A native of La Guayra, with the habit of Oncidium sphacelatum or some such plant. The flowers are yellow spotted with brown, except the lip which is white stained with violet. We have received it from Mr. Rucker, who first flowered it, and from Broughton Hall.

26. CYCNŌCHĒS pentadactўlŏn.

C. *pentadactylon*; racemo brevi stricto, sepalis petalisque lanceolatis reflexis supremo incurvo, labelli unguiculati margine revoluto dorso adnato hypochilio cornu recto incurvo aucto, metachilio 4-lobo medio foveato laciniis 2 posticis obtusis carnosis incurvis anticis subulatis, epichilio linguiformi acuto lævi.

This singular species has in some respects so much the appearance of C. maculatum, that when I first received it from Mr. Veitch of Exeter, in March last, I hesitated whether to regard it as a variety or a distinct species. A plant, obtained from Brazil direct by Messrs. Loddiges (Cat. no. 890) has decided me in regarding it as the latter. It has a short raceme of much larger greenish yellow flowers with broad chocolate-brown blotches, and its lip is quite remarkable, having 5 finger-like lobes, and no more, instead of the lateral comb-like fringes of C. maculatum. It is a very curious thing.

36. MAXILLĂRĬA acutipetălă.

Hooker in Bot. Reg. t. 3966.

M. *acutipetala*; pseudo-bulbis oblongo-ovatis angulatis diphyllis, foliis lato-linearibus acutis, scapis radicalibus uni-bifloris, sepalis petalisque oblongis acutis patentibus subconformibus, labello oblongo trilobo centro striato basique lineis elevatis subquinque. lobis lateralibus brevibus columnam involventibus intermedio acuto reflexo. *Hooker, l. c.*

According to Sir William Hooker this species, although allied on the one hand to M. tenuifolia and on the other to M. picta, is abundantly distinct from both. It was collected in Central America by Mr. Barclay, a gardener attached to the Sulphur surveying ship, and has flowered in the Botanic Garden, Kew. It is very near M. picta. The flowers are pale orange, spotted and blotched with blood colour. The petals are smaller, but of nearly the same shape and colour as the sepals. The lip is of a paler colour below, but coloured and spotted like the other parts. The column is deep red purple.

37. LISSOCHĪLUS rosĕus.

(Dendrobium roseum, *Swartz. in Persoon synops. p.* 523.)

L. *roseus*; foliis lato-lanceolatis erectis plicatis, scapo squamis lanceolatis acutis membranaceis distantibus vaginato, racemo denso oblongo, bracteis ovatis acuminatis ovario brevioribus, sepalis spathulatis acutis concavis reflexis, petalis oblongis apiculatis, labelli trilobi lobis rotundatis intermedio emarginato cum mucrone, disco lamellis tribus undulatis serrulatis tuberculo parvo utrinque.

A very remarkable terrestrial Orchidaceous plant, imported by Sigismund Rucker, Esq. from Sierra Leone. It has large plaited leaves, and rich rose-coloured flowers in a close raceme at the end of a scape between three and four feet high. Although the other species of the genus have flowers in which yellow is the predominant colour, this has scarcely any trace of it except in the middle of the lip. We shall take an early opportunity of figuring it.

38. DENDROBĬUM Ruckĕrī.

D. *Ruckeri*; caulibus teretibus, foliis ovato-lanceolatis acuminatis planis, floribus gemellis, sepalis patulis obtusiusculis convexis margine reflexis lateralibus subtriangularibus, labelli trilobi lobis rotundatis intermedio undulato axi elevatâ villosâ.

A Manilla (?) plant, which has lately flowered with Mr. Rucker. It is sweet-scented, handsome, and distinguished by its clear yellow nankin-coloured flowers, the lip of which is a little stained with rose-colour.

39. CLOWESIA rosea.

We have lately received from the Rev. John Clowes, a most zealous cultivator of Orchidaceous plants, specimens of a Brazilian species with the habit of a Catasetum, but with a structure quite different from that of any genus yet described. Its flowers have the chin of the Maxillaridous division, but they differ entirely from all the genera belonging to it in the peculiar structure of the pollen-masses and of the petals. To these points, however, we do not now advert more particularly, because we hope soon to have the opportunity of publishing a figure of the plant. The technical characters by which it is to be recognized by Botanists are the following.

CLOWESIA. Flos subglobosus, patulus. Sepala sub-æqualia; lateralia paulò obliqua, in mentum breve producta, basibus connatis. Petala conformia, latiora, fimbriata. Labellum concavum, carnosum, cum labello continuum nec articulatum, obsoletè trilobum, margine in fimbriam glandulosam laceram solutum, disco læve. Columna semiteres, clavata, utrinque apice obtusa cornuta; clinandrio alto carnoso serrato. Stigma: sinus transversalis. Pollinia 2, linearia, dorso sulcata, glandulâ subrotundâ, caudiculâ membranaceâ clepsydræformi!

Sp. 1. Clowesia rosea.

The account which Mr. Clowes's gardener gives of the appearance of the plant, which I have not yet seen, is this. Pseudo-bulb 2½ inches long, ovate, and tapering to a point, covered with a thin brownish white sheathing; each pseudo-bulb has two or three small dark brown bands, which entirely surround it, and are apart from each other about half an inch. Leaves three, resting on the summit, seven inches long and one inch across, lanceolate, ovate, and acuminate, at the point twisting a little on one side. Both leaves and pseudo-bulbs are in the way of Catasetum; it was imported a few years ago from the interior of Brazil. The scape rises from the apex of the pseudo-bulb.

41. THE GENUS SARCOGLOTTIS.

In Presl's reliquiæ Hænkeanæ was proposed a genus of Orchidaceous plants called Sarcoglottis, which was reduced in the Genera and Species of Orchidaceous plants to a section of Spiranthes. Dr. Klotzsch having however expressed his opinion that it really is distinct, chiefly upon the ground that the pollen-masses of Spiranthes are sessile while those of Sarcoglottis are stalked, and that in Spiranthes both sepals and petals spring directly from the top of the ovary, while in Sarcoglottis the two lateral sepals are decurrent in the form of a bag as far as the base of the ovary, we think it desirable that his views should be placed before our readers. We therefore extract the following characters of the genus and its species:—

SARCOGLOTTIS, *Presl. l. c.* Perigonii foliola exteriora lateralia majora, reflexa, labello supposita, in saccum ampullaceum antice rima solutum, ovario per totam longitudinem decurrentia, supremum cum interioribus arcte-adhærens. Labellum apice dilatatum, infra rimam anticam cum paginâ interiore sacci connatum, elongato-concaviusculum, columnam amplectens, basi liberum, sagittatum. Columna brevis. Anthera dorsalis, stipitata, bilocularis. Pollinia 2, pedicellata, lineari-clavata, longitudinaliter profunde sulcata, glandulæ communi affixa.

Herbæ epigææ Americæ tropicæ; radicibus fasciculatis, crassis, cylindricis, tomentosis; foliis radicalibus, magnis, lanceolatis; scapo vaginato; spica multiflora; floribus magnis erectis basi tortuosis. *Klotzsch in Allg. Gartenz. April* 2, 1842.

1. S. *picta;* foliis radicalibus oblongis, acuminatis, basi attenuatis; scapo versus apicem floribusque villosis; floribus odoratis, perigonii foliolis albis, acutis; labello revoluto, apice obtuso.

α variegata; foliis maculatis, roseo-costatis.

Spiranthes (Sarcoglottis) picta α Lindley the Genera and Species of Orchidaceous plants, p. 475. Neottia picta R. Brown Hort. Kew. 5. p. 199. Sims Botanical Magazine, t. 1562.

β immaculata; foliis saturate-viridibus unicoloribusque albido-costatis. —— Both these varieties come from the West Indies.

2. S. *grandiflora;* foliis radicalibus spathulatis, acutis, læte-viridibus; scapo versus apicem floribusque pubescentibus; perigonii foliolis flavido-viridibus, exterioribus inferioribusque falcatis, obtusis; labello deflexo, tridentato.

Spiranthes picta β Lindley, the Genera and Species of Orchidaceous plants, p. 475. Spiranthes grandiflora, Hooker Bot. Mag. t. 2730.——Found in Brazil and in the Caraccas.

3. S. *rufescens;* foliis radicalibus spathulatis, acutis; scapo glabro; floribus villosis; perigonii foliolis angustis, sordide-albidis, dein rufescentibus, versus apicem glabris; labello patentissimo, acuminato.

Spiranthes rufescens Fischer Hortus Petropolitanus.——Native country unknown.

The last species is S. speciosa of Presl, which Dr. Klotzsch considers distinguished by its lip being emarginate; unless Spiranthes cerina should also belong to Sarcoglottis; (about which we entertain no doubt if the genus Sarcoglottis is eventually adopted.)

48. AERĬDES vĭrens.

A. *virens;* foliis latis obliquè retusis, racemis pendulis multifloris, sepalis petalisque obovatis obtusis, labelli cornu acuminato ascendente lobis lateralibus apice denticulatis intermedio lanceolato medio canaliculato versus apicem denticulato.

A very fine plant from Java, with sweet-scented flowers as large as those of A. odoratum, white, stained and spotted with deep lilac. Its leaves are remarkable for their bright green colour. Messrs. Loddiges have recently flowered it. Aerides odoratum, quinquevulnera, crispum, affine and virens, would form such a cluster of beautiful flowers, as it would be very difficult to match if all the vegetable kingdom were ransacked.

52. GONGŌRĂ truncātă.

G. *truncata;* sepalis lateralibus rotundato-oblongis supremo obovato apiculato carinato, petalis minimis ovatis acutis decurrentibus quinquenerviis, labelli vernicati hypochilio medio compresso (unde bicamerato) margine lævi apice bicorni, epichilio ovato canaliculato.

A Mexican species, quite distinct from any previously described; introduced from Mexico by Mr. Rucker, who received it from Linden in 1840. The flowers are pale straw colour, with some brownish purple speckles and a yellower lip. Before expansion they are almost of the form of a bean; which is owing to their sepals being so blunt that when flattened they are nearly half oblong. The lip has no speckles at all, and looks as if varnished. It has a very peculiar scent, and is a really fine thing.

54. EPIDENDRUM arbuscŭla.

Lindley in Benth. Plant. Hartweg. p. 93.

For some years there has been in our gardens, to which it has been introduced by the Horticultural Society from Mexico, a singular Orchidaceous plant, with a large branching stem, leathery leaves three or four inches long, and many pale stout roots, and no one could tell what it was. It has now flowered with the Society, and proves to be the Epidendrum arbuscula, a species discovered by Mr. Hartweg, flowering in April, near a place called San Juan Sacatepequez. It is more curious than beautiful, resembling E. nutans in some respects, and quite destitute of brilliant colours. The flowers are a dull chocolate, only relieved by a patch of yellow in the middle of the labellum. The latter organ is so folded back at the sides and front, that although it is really almost circular, it has altogether the form of an old fashioned three-cocked hat.

56. ERIA floribunda.

Lindl. in Wall. Cat. no. 7408.

E. *floribunda* (Tonsæ); caulibus carnosis subflexuosis teretibus, foliis lanceolatis acuminatis, racemis oppositifoliis patulis multifloris pubescentibus foliis brevioribus, bracteis defloratis ovatis concavis retrorsis, sepalis petalis 3-plò latioribus, labello angusto nudo basi saccato: laciniis lateralibus ascendentibus abbreviatis intermediâ cuneatâ tridentatâ.

A caulescent species, with lanceolate leaves seven or eight inches long. The flowers are small, pink, smooth, in close spreading racemes much shorter than the leaves. The column is dark purple at the top. Mr. Prince originally found it at Sincapore, and communicated it to Dr. Wallich. Messrs. Loddiges, have since received it from the same place through Mr. Cuming. It is a pretty plant, worth cultivation.

58. HARTWEGĬĂ purpureă; *var.* angustifolia.

"This plant was gathered in Guatemala by George Ure Skinner, Esq. and added to Sir Charles Lemon's collection at Carclew in August, 1841. Although very distinct from the original form of this species in its leaves and flowers, yet the general habit and appearance of the two plants are so very similar, that the present one can only be regarded as a well-defined variety of the other. Its *leaves* are from three to four inches long, and from half an inch to an inch in breadth, oblong lanceolate acute, very thick and fleshy, hollowed along the middle, and tapering to both ends; of a brownish green colour, singularly marked with numerous deep brown irregularly formed spots. The *scape* is erect, about a foot long, round and wiry, with many joints, each of which is surrounded by a pale brown-coloured bract, which closely embraces the stem, and covers it half way between the joints. The *flowers* are terminal, of the same bright pinkish colour as those of the old variety; from which, however, they differ in the form and size of the sepals, which are not obliquely cordate, but ovate oblong, and rather smaller than in the other. The *labellum* is also very different, being much larger in every respect, and having a notch at the point; whereas in the original species it is small, roundish, cordate, and forms a kind of spur where it joins the column. In this variety there is no appearance of a spur, but it is narrowed round the column, and has a singular indentation just opposite its point. The *column* is curved, and about the same length as the sepals and petals. It is also of a deeper pink, and is rounded on the upper side; beneath it is two-edged, from being slightly hollowed out along the middle. The *anther case* is somewhat reniform, brownish pink, much darker than any other part of the flower."—*Wm. B. Booth.*

60. EPIDENDRUM lamellātūm.

Westcott mss.

E. *lamellatum*; (§ Euepidendrum) caule erecto articulato; foliis lanceolatis emarginatis glabris subcarnosis, corymbo sessili paucifloro, sepalis lanceolatis acutis, petalis obovato-lanceolatis, columnâ apice alatâ, labello obovato integerrimo vel obscurè emarginato basi lamellato.

"Stem about a foot high, articulated, and much resembling that of Dendrobium. The sepals and the petals are of a delicate pink colour. The column is very short, not more than two lines long, and having appendages at the apex as long as the column. The labellum is united to the column the whole length, and is entirely undivided or very slightly notched at the end. The appendages at the apex of the column are of a darker pink colour, and the labellum is still darker and more brilliant, having the disc covered with a row of scaly yellowish plates."

For the above memorandum we are indebted to Mr. Westcott, who first named the plant in Mr. Barker's collection, where its origin was unknown. It has been subsequently communicated by Sir Charles Lemon, to whom it had been sent from Honduras. It is a pretty species, with flowers about the size of E. Schomburgkii.

62. PLEUROTHALLIS peduncularis.

P. *peduncularis* (§ aggregatæ pubescentes); folio elongato lanceolato plano basi et apice carinato, floribus aggregatis, pedicellis calycibus duplo longioribus, bractea et sepalorum basi pubescentibus, sepalis acuminatis liberis, petalis ovatis setaceis sepalis brevioribus, labello ovato bicarinato juxta apicem lineari-cuneato.

Imported from Rio Janeiro by Messrs Loddiges. It has the habit of Pleurothallis ruscifolia, but is a much larger plant. The flowers are pale straw-colour.

63. BIFRENĀRĬA inodōra.

B. *inodora*; pseudobulbis tetragonis, folio oblongo acuto plicato margine crispo basi in petiolum brevem canaliculatum angustato, pedunculis unifloris pseudobulbis brevioribus, sepalis oblongis obtusis cornu elongato clavato, petalis subconformibus paulò undulatis, labelli trilobi cucullati lobis rotundatis undulatis subdentatis intermedio piloso, callo disci cuneato carnoso emarginato.

"This plant was imported from Rio in 1839, and added to Sir Charles Lemon's collection at Carclew, where it flowered in April, 1843. In habit and general appearance it approaches so near to Maxillaria tetragona, as to have been mistaken for it, but on flowering it proved to be widely different from that species, and much handsomer, although the flowers are destitute of that peculiar fragrance which *M. tetragona* is said to possess. It requires similar treatment, and flourishes in a warm, moist stove, potted in a mixture of rotten wood and other decayed vegetable substances.

"*Pseudo-bulbs* of a dull brownish green, ovate-oblong, four-angled, a good deal hollowed between them, upwards of three inches long, and nearly the same in circumference at the base, from which they taper to the point, where there is a black coloured band, indicating the joint at which the leaf is united to the pseudo-bulb. *Leaves* solitary, oblong-lanceolate acute, on short foot-stalks, spreading and recurved, measuring when full grown a foot in length, and four inches across. They are thin and rigid, of a pale green below, rich deep green above, slightly undulated, or crimped, at the margin, with a prominent vein on either side of the midrib. *Flowers* large and spreading, of a brownish green, with a hairy, red labellum. *Scape* radical, very short, scarcely half the length of the pseudo-bulb, round and of a pale green, bearing two and sometimes three flowers. *Pedicels* about two inches long, slightly curved, and similar in size and colour to the scape, with a brown-coloured acuminate bract at the base of each. *Sepals* of a dull green, slightly tinged with red. The two lateral ones are roundish oblong, a little recurved at the point, upwards of two inches long, and an inch wide, and ending at the base in a kind of spur, half the length of the pedicel.

The upper sepal is about the same size as the other, but is more concave and pointed. *Petals* oblique, ovate-acuminate, of the same colour, but rather shorter than the sepals, and about three-fourths of an inch wide, having their edges slightly undulated and the point recurved. *Labellum* three-lobed, outwardly of purplish-red colour, but inwardly more pale and delicate towards the base, where it tapers and forms a sort of pouch. The two lateral lobes are erect and of a pale brownish-red, marked with dark coloured veins. The middle lobe is roundish oblong, recurved and undulated at the margin. Outside it is smooth, but the inner surface is thickly covered with pale soft hairs. In the centre of the lip is an elevated process, similar to those in the flowers of *M. aromatica, cruenta, tetragona,* and other allied species. *Column* about half the length of the labellum, slightly curved, of a pale yellowish colour above where it was rounded, and slightly tinged with red below, where it is hollowed out so as to appear two-edged. *Anther-case* large and fleshy, one-celled, containing two collateral pollen-masses, each of which easily splits into two roundish ovate, slightly pointed waxy bodies of a dull yellow."—*W. B. Booth.*

For this memorandum, a drawing, and specimens of the highly curious plant, I am obliged to Mr. Booth, Sir Charles Lemon's intelligent gardener. The species is much the finest yet known, its flowers being as large as those of Maxillaria Harrisoniæ.

The Maxillaria tetragona above referred to, as resembling this in habit, belongs to the genus Lycaste, in my recent enumeration of whose species, it was unsettled. I therefore avail myself of the present opportunity of adding its amended character.

64. LYCASTE tetragōna.

L. *tetragona* (Maxillaria tetragona, *Lindl. in Bot. Reg. t.* 1428. *Hooker Bot. Mag. t.* 3146); pseudobulbis ovatis tetragonis, foliis oblongo-lanceolatis plicatis basi in petiolum angustatis solitariis, scapis 1-4-floris pseudobulbis brevioribus, sepalis petalisque oblongo-ovatis obtusiusculis patulis subæqualibus, labello carnoso ventricoso trilobo erecto: lobis lateralibus parvis acutis intermedio ovato extùs convexo, disci appendice carnoso tabulari incumbente.—— *Brazil.* ——Very sweet-scented. Flowers greenish, streaked with crimson. Lip white and purple in one variety, and green and purple in another. The appendage of the disk is very conspicuous, and shovel-shaped.

65. COLAX.

Several years ago I proposed the establishment, under this name, of a genus which eventually proved the same as Maxillaria; and which was therefore abandoned. I would now propose to apply it to a small set of plants of the Maxillaridous division of Vandeæ, at present included in the genus Maxillaria itself. These species, consisting of my M. viridis and jugosa, and the M. placanthera of Sir W. Hooker, have the closed, scarcely ringent, flower of Promenæa and Warrea, but they have a caudicula quite unlike any thing at present known among their race. It has no distinct gland, but consists entirely of a thin wavy membrane, strengthened by an elevated line in the middle, and gradually narrowing to the point, where the gland is usually found. It has also a thick fleshy crested anther, whose cells are planted on its lower side. I cannot but regard these circumstances as sufficient to justify the separation of the plants in which they are found as a distinct genus. The following characters will indicate the limits of the genus and its species.

COLAX. Flores subglobosi, vix ringentes, in mentum breve producti. Sepala et petala subæqualia. Labellum unguiculatum, trilobum, inappendiculatum, planiusculum (vix cucullatum). Columna paulò elongata semiteres, clinandrio marginato. Anthera carnosa, cristata. Pollinia 4, in paribus globosis colligata, caudiculæ obovatæ membranaceæ adnata; glandulâ nulla; rostello fisso.——Herbæ pseudobulbosæ; folia terminalia et radicalia, plicata. Pedunculi radicales, erecti, uniflori, vaginati. Flores virescentes.

1. C. *viridis* (Maxillaria viridis, *Lindl. in Bot. Reg. t.* 1510.); sepalis petalisque conniventibus oblongo-subrotundis obtusis subæqualibus, labelli brevis trilobi lobo medio transversè rhomboideo unguiculato plano.——*Brazil.*——Flowers green, with a dingy violet lip. That it differs from *C. placanthera*, in the manner stated under that species, I know from having preserved the specimen from which the figure in the Botanical Register was taken; which figure well represents the peculiar rounded form of the sepals. I believe it is no longer in our gardens; for *C. placanthera* is usually cultivated under its name.

2. C. *placanthera* (Maxillaria placanthera, *Hooker in Bot. Mag. t.* 3173. *Lindl. in Bot. Reg.* 1841. *misc.* 103.); sepalis lineari-oblongis obtusis intus secus medium seriatim maculatis, petalis angustioribus omninò conformibus maculatis, labelli angusti trilobi minutè pubescentis lævis laciniis lateralibus acutis nanis intermediâ dilatatâ rotundatâ cuneatâ, columnâ apice lobatâ versus basin bisulcâ pubescente.——*Brazil.*——

I was certainly wrong in referring this plant to *C. viridis*, from which it differs in its sepals and petals being much narrower, the flowers less green, the lip larger, and of a different form in its upper lobe.

3. C. *jugosus* (Maxillaria jugosa, *Lindl. in Bot. Reg.* 1841. *misc.* 104.); sepalis oblongis incurvis obtusis, petalis conformibus concavis basi angustatis, labelli sigmoidei oblongi unguiculati pilosi jugosi lobis lateralibus nanis obtusis, intermedio semi-circulari, columnâ apice lobatâ bisulcâ villosissimâ.——*Brazil.*——Near C. placanthera, from which it differs in having the flower of a globose figure, with much less linear sepals and concave oblong petals narrowed to the base. These parts are of a rich cream colour, speckled with crimson. The lip has quite a different form, being semicircular at the tip, deeply furrowed and closely covered with short hair. The column has two deep furrows in front almost buried in hairs.

67. BIFRENARIA.

This genus differs from Maxillaria proper in having its pollen-masses attached to a short gland, by a pair of distinct straps, or caudiculæ, instead of one; a character, minute indeed, but constant and readily detected, although often overlooked. On that account several species have been referred to Maxillaria, as I have now ascertained, and it is not improbable that others may still lie unknown among that large and troublesome genus. The following are characters of all of which I possess any certain information.

BIFRENARIA, Lindl. Sepala patula, libera, subæqualia; lateralia cum basi productâ columnæ connata. Petala sepalis duplò minora. Labellum cum pede columnæ articulatum, cucullatum, trilobum, medio callosum. Columna brevis, semiteres. Anthera subcristata. Pollinia 4, per paria incumbentia, caudiculis duabus distinctis, glandulâ oblongâ. ——Herbæ epiphytæ, pseudobulbosæ. Folia terminalia subsolitaria, plicata, cartilaginea. Pedunculi radicales, uniflori v. racemosi, sæpe elongati.

1. B. *atropurpurea* (*L. no.* 1. Maxillaria atropurpurea, *Lodd. Bot. Cab. t.* 1877.); pseudobulbis ovatis obtusis tetragonis foliis oblongo-lanceolatis plicatis solitariis, racemis radicalibus 3-floris pseudobulbis paulò longioribus, labelli lobis lateralibus brevibus divergentibus, intermedio transverso sub-3-lobo revoluto suberoso; callo in medio anticè tridentato. ——*Brazil.*——Flowers dark purple, very sweet-scented.

2. B. *inodora* (*Lindl. in Bot. Reg.* 1843. *misc.* 63.); pseudobulbis tetragonis, folio oblongo acuto plicato margine crispo basi in petiolum brevem canaliculatum angustato, pedunculis unifloris pseudobulbis brevioribus, sepalis oblongo-obtusis, cornu elongato clavato, petalis subconformibus paulò undulatis, labelli trilobi cucullati lobis rotundatis undulatis subdentatis intermedio piloso, callo disci cuneato carnoso emarginato.——*Brazil.*——Flowers as large as those of Maxillaria Harrisoniæ, green, with a bright violet lip.

3. B. *racemosa* (Maxillaria racemosa, *Hooker in Bot. Mag. t.* 2789. *Lodd. Bot. Cab. t.* 1318.); pseudobulbis ovatis compressis tetragonis monophyllis, foliis oblongo-lanceolatis tricostatis scapis gracilibus strictis multifloris brevioribus, sepalis oblongis acutis; lateralibus basi valdè elongatis, petalis linearibus spatulatis triplò minoribus, labello oblongo cucullato indiviso undulato emarginato in axi calloso, columnâ pubescente.——*Brazil.*——Flowers pale, dirty straw-colour, with a white lip, slightly speckled with pale crimson. It is well figured in the Botanical Cabinet; the representation in the Botanical Magazine appears to have been taken from a withered specimen.

4. B. *aureo-fulva* (Maxillaria aureo-fulva, *Knowles and Westcott Floral Cab. t.* 83. *Hooker Bot. Mag. t.* 3629. M. stenopetala, *Knowles & Westcott*, 2. *p.* 112.?); pseudobulbis rotundato-ovatis angulatis rugosis monophyllis, foliis oblongo-lanceolatis costatis acutis scapo radicali multifloro brevioribus, floribus longè pedicellatis, sepalis lanceolatis acuminatis lateralibus apice reflexis, petalis striatis lineari-lanceolatis acuminatis, labello unguiculato trilobo in medio striato: lobis lateralibus acutis intermedio lanceolato acuminato.——*Brazil.*——Very pretty. Flowers orange-coloured. It is the "Epidendre Limodore" of Descourtilz, and was found by that traveller in great abundance upon fallen trees encumbering the sandy plain through which the great public road passes from Bananal to Ilha Grande.

5. B. *vitellina* (Maxillaria vitellina, *Lindl. in Bot. Reg.* 1838. *misc.* 116. 1839, *t.* 12. M. barbata, *Westcott in Phytologist, p.* 7.); pseudobulbis ovatis obtusè angulatis monophyllis, foliis lanceolatis in petiolum canaliculatum angustatis, racemo cernuo foliorum longitudine, labelli cuneati trilobi lobis lateralibus acutis anticè crenulatis intermedio bilobo rotundato cordato crenulato, tuberculo disci trilobo obtusissimo, ungue pubescente.——*Brazil.*——Flowers bright yellow, with a purple spot in the middle of the labellum.

6. B. *aurantiaca* (*Lindl. in Bot. Reg.* 1836, *t.* 1875.); pseudobulbis subrotundis compressis diphyllis, foliis oblongis plicatis racemi erecti longitudine, sepalo dorsali fornicato lateralibus oblongis acutis, petalis linearibus, labelli unguiculati lobis lateralibus semicordatis intermedio majore transverso ovali subcrenulato basi bicalloso, columnâ pubescente.——*Demerara.*——Pseudobulbs and leaves beneath spotted with crimson. Flowers orange-yellow, spotted with crimson.

68. STENOCORŸNE.

With respect to the Bifrenaria (?) longicornis, formerly described in this work, it must be regarded as still another genus, agreeing with Bifrenaria in having two straps (caudiculæ) to its pollen-masses, but differing in having two distinct glands also, as is the case with a part of the plants now referred to Angræcum. For this I would propose the name of Stenocoryne.

STENOCORYNE. Flores ringentes, clausi, longè cornuti. Sepala basi longissimè producta, et connata. Petala conformia. Labellum longissimè unguiculatum, trilobum, disco callosum. Pollinia 4, per paria connata. inæqualia, semiglobosa; caudiculis 2, glandulisque totidem ovalibus. ——Herbæ epiphytæ pseudobulbosæ. Folia cartilaginea, solitaria. Flores racemosi, radicales.

1. S. *longicornis* (Bifrenaria longicornis, *Lindl. in Bot. Reg.* 1838, *misc.* 177.); pseudobulbis elongatis tetragonis, foliis oblongo-lanceolatis subplicatis nitidis, racemo laxo multifloro, sepalis lateralibus ovatis acutis, petalis ovatis acutis, labello longè unguiculato spathulato apice trilobo laciniis rotundatis medio pubescente disco elevato calloso.——*Demerara.* ——Flowers racemose, orange, spotted with brown.

70. DENDROBĬUM planibulbe.

D. (Onychium) *planibulbe*; caule compresso distichè folioso erecto basi ovali complanato, foliis oblongis obtusis emarginatis, pedunculo vaginato aphyllo, floribus terminalibus e paleis glumaceis erumpentibus, sepalis petalisque acuminatis, labelli trilobi lobis lateralibus dilatato-rotundatis intermedio lineari pectinato basi appendice emarginatâ aucto.

A singular plant, found in Manilla by Mr. Cuming, and by him sent to Messrs. Loddiges, who have just succeeded in flowering it. The stems are about three inches high, flattened at the base into an oblong thin-edged pseudobulb. The leaves are a dull greenish-red. The flowers small white, veined with purple, placed at the very extremity of a leafless peduncle. It is very near D. Blumei.

71. EPIDENDRUM Ovŭlum.

E. (Encyclium) *Ovulum*; pseudobulbis oviformibus diphyllis, foliis linearibus canaliculatis acutis, scapo filiformi foliis paulò longiore 3-floro, sepalis linearibus 3-veniis, petalis angustioribus spathulatis, labelli trilobi lobis lateralibus acutis intermedio dilatato rotundato venis radiantibus glandulosis variegato, columnæ tridentatæ dentibus lateralibus rotundatis denticulatis.

A curious little plant, in the way of E. Pastoris, or bractescens, or aciculare, from Bolanos, for which we are indebted to Messrs. Loddiges The sepals and petals are olive-green; the lip white, with crimson glandular radiating veins.

72. ERIA multiflŏra.

Lindl. Gen. and Sp. Orch. p. 68.

E. *multiflora* (Tonsæ); caulibus teretibus clavatis 2-3-phyllis, foliis lineari-lanceolatis canaliculatis, spicis multifloris, bracteis minimis sphacelatis, floribus pubescentibus, labello cuneato-obovato indiviso basi excavato nectarifluo supra foveam dentibus 2 membranaceis erectis aucto.

A rare little plant, for which we are indebted to Mr. Van Houtte, of Ghent, who received it from Java. It has small white flowers, covered externally with fine down. The only coloured part is the column, which is deep violet. In habit it resembles a small specimen of Eria floribunda.

74. DENDROCHĪLUM latifolium.

D *latifolium;* foliis oblongo-lanceolatis coriaceis trinerviis, pedunculis elongatis apice longè spicatis, labelli lævis lobis basalibus lineari lanceolatis acuminatis ciliatis, laciniis columnâ dentatâ brevioribus setaceis e basi ortis.

A plant with the habit of Dendrochilum glumaceum, but with much broader leaves and quite a different lip. It has long graceful spikes of green flowers, and was imported from Manilla by Messrs. Loddiges. It is not among Cuming's dried specimens.

75. ANGRÆCUM Ashantēnse.

A. *Ashantense;* caulibus repentibus, foliis distichis coriaceis ovatis apice oblique erosis, spicis foliis subæqualibus, floribus resupinatis, sepalis petalisque galeatis denticulatis ovato-lanceolatis obtusiusculis, labello pandurato obtuso denticulato basi unidentato, calcare arcuato subcompresso basi constricto labelli longitudine, polliniorum caudiculâ acuminatâ bipartibili.

A most curious plant with creeping stems, ovate leaves obliquely toothed at the end, as if they had been eaten away, and light cinnamon coloured flowers in spikes about four inches long. It was imported from Ashantee by Messrs. Loddiges, and flowered on the 15th June, 1843.

76. ONCIDIUM candĭdum.

O. *candidum;* pseudobulbis ovalibus compressis monophyllis, folio lineari-lanceolato basi canaliculato, racemo paucifloro flexuoso, sepalis oblongis lateralibus connatis, petalis majoribus patentibus planis obtusis, labello ovato-oblongo plano basi lineâ convexâ elevatâ transversâ aucto, columnæ alis rotundatis serratis tuberculo baseos elevato labello adnato.

A very curious plant sent by Mr. Hartweg from Guatemala. It belongs to the tetrapetalous division of the genus; and has ivory white flowers, with a couple of small violet dots at the base of each petal, and the usual prominence at the base of the column, bright yellow and adnate to the lip. The flowers are an inch in diameter.

79. PLEUROTHALLIS Smithiana.

P. *Smithiana* (§ effusæ pubescentes) folio oblongo coriaceo racemum dimidio excedente, sepalis oblongis lateralibus ad apicem cohærentibus, petalis obovatis et clinandrio denticulatis, labello obovato utrinque dentato medio excavato utrinque callis 2 elevatis.

"This plant was imported from Rio in December, 1841, by Lieut. Christopher Smith, of H. M. packet "Star," and added to Sir Charles Lemon's collection at Carclew, where it flowered in May, 1843. Like others of this genus it proves to be only interesting as a subject for the Botanist, who, on dissecting the various parts of the flower and examining them through the microscope, cannot fail to be highly delighted with their singular construction, and the beautiful frosted appearance they present.

"*Stem* round, erect, from two to three inches high, brownish green, with a single joint about an inch from the root, surrounded and embraced by a brown sheathing bract, which extends nearly to the base of the leaf. *Leaves* ovate-oblong, slightly curved and hollowed in the centre, about two inches and a half long, and an inch broad, very thick and leathery, of a brownish green colour. *Flowers* six, sometimes seven, arranged alternately in a close raceme, issuing from a two-valved brown spathe at the base of the leaf, and extending about an inch and a half along the midrib. *Sepals* fleshy, striated and covered on the outside with dense brownish pubescence. The upper one is oblong lanceolate, with three brownish purple lines along the centre, and another round the margin. The lower sepals are united, and appear as one roundish ovate acuminate body, of a brownish green, marked with numerous elevated puce-coloured spots. *Petals* very small, thin and delicate, about one-fourth the length of the upper sepal, oblong obtuse, having a purplish line along the centre and the margin, fringed and slightly marked with the same colour. *Labellum* roundish obtuse, somewhat spathulate, and spotted in the same manner as the lower sepals; attached by a small elastic membrane to the end of the column, which is curved and elongated so as to have the appearance of a little spur. *Column* about the same length as the petals, round and striated above, hollowed on the under side, with a thin pellicle at the extremity on either side of the anther case, so as to make it have the appearance of being winged. *Anther-case* deep brown, containing two very small waxy pollen-masses.

"The plant requires the constant heat of a moist stove, and thrives either on a piece of wood or in a pot of decayed vegetable matter."—W. B. Booth.

81. STANHOPEA Martiana; *var.* bicŏlor.

Bateman in Bot. Reg. 1840. *misc.* 19.

Although it is probable that some of the so-called species of Stanhopea are mere varieties of oculata, and that others may be wild mules, there seems no reason to doubt that this is really a distinct plant, for it has a lip unlike any other that we have seen, remarkably short and saccate at the base. The present variety is a lovely plant with large pure white flowers richly but sparingly spotted with crimson. In the original S. Martiana the sepals are straw-coloured, and much more dotted with purple. Messrs. Rollisson have lately flowered it, and believe they obtained it from Mexico. It is as fine a thing as S. tigrina would be if its flowers were white, and is very sweet-scented.

84. SPIRANTHES rosulata.

S. (Sarcoglottis) *rosulata;* foliis oblongis coriaceis rosulatis glabris, scapo aphyllo hirsuto 2-3-vaginato, spicâ ovatâ hirsutâ, bracteis lineari-lanceolatis dorso glabris, labelli oblongi rotundati venosi ungue apice excavato basi bicruri.

A Guatemala plant, not unlike Spiranthes picta. It has a scape about nine inches high, and a close spike of green flowers. The leaves are most beautifully coated on the underside with vesicular cells, which give them a peculiar frosted appearance. Mr. Hartweg sent it to the Horticultural Society from Guatemala.

85. EPIDENDRUM collāre.

E. (Aulizeum) *collare;* caulibus fusiformibus elongatis apice triphyllis, foliis coriaceis canaliculatis patentibus, racemo laxo paucifloro, ovario cuniculato, sepalis petalisque conformibus petaloideis lanceolatis, labello ovato obtuso undulato apice recurvo basi trilamellato, clinandrio margine elevato denticulato cincto.

From Guatemala, where Mr. Hartweg found it. Its stems are 18 inches long, strong and deeply furrowed. The flowers are white, changing to yellow and brown as they go off. The anther-bed is surrounded by an elevated edge, which stands up something like the collar of a coat. It is not pretty.

88. EARĪNA suavĕŏlens.

E. *suaveolens;* spicâ oblongâ densâ basi subcompositâ, labello transverso rhombeo versus basin bicalloso obsoletè trilobo laciniâ intermediâ rotundatâ undulatâ emarginatâ.

This extremely rare Orchidaceous plant was sold the other day among a collection of New Zealand varieties, brought to this country by Mr. Bidwill, and is now in the possession of Messrs. Loddiges. It is a tufted species, with stems four to twelve inches high, clothed with narrow distichous somewhat rigid leaves, about three inches long. The stems are terminated by dense oblong spikes of white flowers, having a double yellow spot on the lip. In a memorandum now before me from Mr. Bidwill, I learn that it grows near Roturoa in New Zealand, on trees not very densely covered with leaves; that it is very rare even there, exceedingly beautiful, and most deliciously perfumed. From Earina mucronata it differs altogether in the form of the lip and the size of the flowers, which are collected in dense spikes, and not in slender branched ones. As the temperature of New Zealand is rarely above 75°, that circumstance will have to be attended to in its cultivation.

97. EPIDENDRUM diŏtum.

E. (Encyclium) *diotum;* pseudobulbis ovatis corrugatis monophyllis, foliis coriaceis ensatis patulis racemo duplo brevioribus, floribus distantibus, sepalis petalisque obovatis unguiculatis undulatis acutis, labelli obtusè rhombei tripartiti axi elevato carnoso in medio foveato, laciniis lateralibus planis erectis rotundatis, intermediâ rotundatâ undulatâ cuspidatâ convexâ, clinandrio integro.

From Guatemala, where it was found by Mr. Hartweg. Its leaves are about a foot long, spreading, and very stiff. The raceme is about two feet high. The flowers are an inch and a half in diameter, of a dull cinnamon colour, with a little yellow and some chocolate veins in the lip. It is one of the many species in the way of E. asperum. The flowers are either scentless, or have a slight odour of Elder leaves.

99. PERISTERIA.

Upon reconsidering the generic character of this genus, and carefully studying its species, we find that it has not yet been defined with sufficient exactness, and that we ourselves lately, deceived by habit and a loose definition, have admitted into it a species which ought to have been excluded. The type of the genus is the Dove Plant, or Spirito Santo, of Panama, in which we find an erect scape; globose fleshy flowers; equal and regular sepals united at the base, but projecting forwards with the chin usual in the Maxillaridous section; a lip continuous with the column, fleshy, arrow-headed at the base, distinctly articulated in the middle, and having its epichilium undivided and bent down over the face of the column; a column short, fleshy, and wingless; and finally a pair of furrowed pollen-masses, sessile on a narrow gland. With these peculiarities, most of the Peristerias afterwards published agree, except that their scapes are pendulous, instead of being erect.

But I find that Peristeria Humboldti, figured at t. 18 of this volume, is in some important respects different; its upper sepal stands a little apart, so as to give the flower somewhat a two-lipped appearance; its labellum has no articulation in the middle; and its pollen-masses are placed at the end of a narrow caudicle, terminating in a crescent-shaped gland; added to which the column is furnished with a pair of very broad wings. All these circumstances induce us now to separate Peristeria Humboldti under the name of ACINETA, in allusion to the immoveable jointless condition of the lip. Peristeria Barkeri is a second species of this genus.

In addition to these we have a plant in cultivation, discovered by Mr Hartweg in Guatemala, the habit of which is exactly that of a Peristeria, but which has the lip not only articulated in the middle but also with the column, and has the caudicle and gland of Lycaste, from which it differs, independently of habit, in having two pollen-masses and not four. To this plant I propose to give the old classical name of Lacæna, which the incision between the upper and lower half of the lip renders applicable.

The following will be the technical characters of the plants to which these observations apply.

PERISTERIA. *Hooker Bot. Mag.* 3116. Perianthium globosum. Sepala basi connata concava. Petala conformia, paulò minora. Labellum cum columnâ continuum, epichilio carnoso incumbente indiviso cum hypochilio sagittato levitèr articulato. Columna erecta, semiteres, aptera, c. hypochilio continua. Anthera ecristata, bilocularis. Pollinia 2, posticè fissa, in glandulam linearem sessilia.——Herbæ subterrestres, pseudobulbosæ. Folia plicata. Scapi vaginati, radicales, multiflori. Flores speciosi.

1. P. *elata* (Hooker in Bot. Mag. t. 3116); scapo erecto orgyali, racemo laxo elongato, epichilio rotundato serrulato medio elevato calloso, hypochilio maximo lobis obtusis disco unidentatis columnâ longioribus.——*Panama.*——Flowers white, with lilac specks on the base of the lip, very sweet-scented. Stem sometimes six feet high.
2. P. *pendula* (Hooker in Bot. Mag. t. 3479. P. *maculata*, Hort.); scapo brevi pendulo, racemo denso, epichilio rotundato integerrimo supra bilamellato, hypochilio elongato lobis rotundatis callo maximo lunato in medio, columnâ bicorni, rostello truncato. ——*Demerara.*——Flowers pale purple, spotted with lilac, with a scarcely pleasant aromatic odour.
3. P. *guttata* (Knowles & Westcott, Floral Cabinet, t. 70.); racemo brevi denso pendulo, epichilio rhombeo-oblongo serrulato leviter verrucoso, hypochilio brevi lobis acutis, columnâ muticâ, rostello brevi.——*Demerara.* ——Flowers deeply speckled with brownish purple, on a pale dingy ground, rather agreeably sweet-scented with an aromatic resinous odour.
4. P. *cerina* (Lindl. in Bot. Reg. t. 1953.); scapo brevi pendulo, racemo denso, epichilio ovato emarginato crispo serrulato, hypochilii lobis brevibus acutiusculis, columnâ muticâ, rostello truncato.——*Central America.* ——Flowers yellow, unspotted, with the smell of bruised juniper leaves, only more aromatic.

100. ACINETA.

Perianthium carnosum, patulum, sepalis basi connatis supremo subascendente. Petala conformia, paulo minora. Labellum cum columnâ continuum, carnosissimum, hypochilio oblongo concavo, epichilio haud articulato tripartito ascendente, basi appendice carnosâ aucto. Columna erecta utrinque in marginem latum rotundatum producta. Anthera ecristata, bilocularis. Pollinia 2, postice sulcata, caudiculâ lineari, glandulâ lunatâ.——Herbæ Peristeriæ habitu.

1. A. *Humboldtii* (*Peristeria Humboldtii*, Lindl. in Bot Reg. 1843, t. 18. *Anguloa superba*, H. B. K. nov. & sp. 1. 343. t. 93. Lindl. g. & sp. orch. 160.); petalis quam sepala duplò minoribus, callo labelli lineari simplici laciniâ intermediâ bilobâ lateralibus cuneatis, columnâ brevi alis latissimis.——*Central America.*——Racemes very long, pendulous. Flowers of a rich purplish brown, enlivened by deep spots and blotches of the same.
2. A. *Barkeri* (*Peristeria Barkeria*, Bateman Orch. Mex. et Guat. t. 8.); petalis sepalis æqualibus, callo labelli duplici inferiore lineari villoso superiore subhastato apice 3-dentato basi multidentato, laciniâ intermediâ emarginatâ lateralibus acinaciformibus, columnâ elongatâ alis angustioribus.——*Mexico.*——Scape long, pendulous, lax, many-flowered. Flowers yellow, whole coloured. The lip has certainly no articulation in the middle, although from Mrs. Withers's figure in Mr. Bateman's splendid plate such would seem to be the case.

101. LACÆNA.

Perianthium carnosum patens, sepalis subæqualibus basi subconnatis. Petala conformia minora. Labellum cum columnâ et medio articulatum, hypochilio unguiculato cuneato bilobo apice pulvinato, epichilio integro. Columna erecta semiteres Pollinia 2, posticè fissa, caudiculâ setaceâ, glandulâ minutâ.——Herbæ, Peristeriæ habitu.

1. Lacæna *bicolor.*——*Guatemala*——Racemes long, drooping. Flowers pale yellow, with the lip spotted with dark purple.

102. CYMBIDIUM chloranthum.

C. *chloranthum;* foliis ensiformibus supra pseudobulbum breve oblongum equitantibus obtusis supra medium planis recurvis, racemo stricto foliis breviore, bracteis minutissimis, sepalis petalisque obtusis, labello basi pubescente apice retuso emarginato laciniis lateralibus nanis triangularibus, lamellis distantibus arcuatis verrucosis.

A Nepalese species, which has lately flowered with Messrs. Loddiges. Its strictly erect racemes, such as those of C. ensifolium, separate it widely from the majority of Indian species. The sepals and petals are green; the lip yellowish spotted with crimson at the base. After flowering the blossoms soon become of a dusky wine colour.

104. EPIDENDRUM limbātum.

E. *limbatum* (Encyclium); omninò facie *E. glauci*, sed sepalis petalisque subrotundo-ovatis, columnæ appendice truncatâ dentatâ tenui, floribus duplò majoribus.

A Guatemala plant, purchased by Messrs. Loddiges at one of Mr. Skinner's sales. It has quite the habit of E. glaucum, but is much larger, and its flowers have roundish sepals and petals, like those of E. squalidum. The appendage at the back of the anther is thin, truncate and toothed, not rounded, fleshy and entire. It flowered in August last. The name alludes to a narrow yellowish border which surrounds the dull purplish brown sepals and petals. It has a very slight smell, like that of cobler's wax.

105. ACRIOPSIS picta.

A. *picta;* pseudobulbis ovatis, foliis solitariis linearibus canaliculatis patulis emarginatis, scapo paniculato, labelli laciniis lateralibus triangularibus intermediâ elongatâ lineari disco bilamellato, columnæ brachiis 2 glandulosis.

A rather pretty small flowered Orchidaceous plant, with the habit of Sarcanthus pallidus. Its flowers are white and green, with a dull purple stain at the point of the four segments of the perianth, and the same colour on the disk of the lip, leaving a white border all round. We have received it from M. Van Houtte of Ghent, who informs us that it was obtained at Bantam, in the virgin forests of Datarbjaay and Bodjomanik, on the river Tjileman, by Mr. Adolph Papeleu, one of the partners in M. Van Houtte's Nursery.

107. GOVENĬA fasciāta.

G. *fasciata;* spicâ elongatâ cylindraceâ, bracteis ovario æqualibus, sepalis angustis acutis, petalis obovato-lanceolatis duplò latioribus, labello oblongo apiculato intus lævi, antheræ mucrone inflexo.

A discovery by Mr. Linden, who sent it from Mexico to Mr. Rucker. It is one of the prettiest of this genus of Orchidaceæ, having clear yellow flowers, whose sepals and petals are beautifully marked by fine broken bands of crimson. The long bracts, thin narrow spike of flowers, and oblong, not ovate, lip, are the marks by which it is best recognized.

109. BOLBOPHYLLUM calamarium.

B. *calamarium;* pseudobulbis tetragonis, folio oblongo plano coriaceo obtuso basi canaliculato, scapo stricto longissimo apice florido, bracteis membranaceis emarcidis ovarii longitudine, sepalis angustè triangularibus margine pubescentibus, petalis cornuformibus subteretibus glabris, labello sessili mobili lineari basi sulcato margine omni tomentoso supra medium villis longissimis crinitum, columnæ cornubus falcatis.

A very singular plant, allied to B. saltatorium, and like that species having a moveable lip fringed with long purple hairs, which rises and falls spontaneously as the stem sways about in the wind. Its scape is full two feet long, stiff and erect. The flowers are of a dirty yellow ochre colour with a little purple upon them; the lip is dull purple. For our specimens we are indebted to Col. Fielding, who procured the species from Sierra Leone.

110. MILTONIA candida; *var.* grandiflora.

Lindl. in Bot. Reg. 1838. *misc.* 29. *Sertum Orch. t.* 21.

Of this fine plant a most beautiful variety has flowered with Mr. Rucker, who purchased it in Hamburgh. Its flowers are twice as large as those of the original variety, the lip is of a most brilliant white, and the sepals and petals are of a deep rich brown spotted towards the extremities only with yellow.

111. CATASĒTUM Naso.

C. *Naso;* spicâ brevi erectâ, sepalis oblongo-lanceolatis complicatis petalisque lanceolatis ascendentibus æqualibus, labello hemisphærico apice abruptè in appendicem carnosum ovatum obtusum producto: marginibus basi tenuibus laceris amplexicolumnibus: lineâ intramarginali carnosâ inflexâ connivente aperturam cordiformem efficiente, columnâ bicirrhosâ.

This singular plant was received last year from Linden by Mr. Rucker, with whom it has lately flowered. It is a plant whose flowers before opening might be mistaken for C. tridentatum, but when expanded they are totally different. The sepals and petals, which are very pale dull green outside, are slightly pink inside, and richly spotted, in irregular bars, with a deep crimson-purple. The lip is a most singular organ, and very difficult to describe. Viewed from the side it has a hemispherical form, and is green except at the base, where it is extended into a black-purple lacerated margin embracing the column, and at the point where it is extended into a long flat horn or *nose.* Seen in front it is almost wholly of the same rich black-purple, and looks as if it were a solid hemisphere pierced in the middle with a large heart-shaped hole; but this appearance is owing to a thick fleshy rim which rises from within the true edge of the lip, and directing itself inwards horizontally with an uneven outline, at last touches in front of the column, and produces the appearance of a junction. The flowers have no smell.

112. MASDEVALLIA floribunda.

M. *floribunda;* folio carnoso spathulato obtuso apice tridentato, scapo longiore unifloro, flore bilabiato, sepalis lateralibus oblongis dorsali duplò breviore ovato omnibus in acumen linearem obtusum productis, petalis truncatis antice infra medium unidentatis lineis duabus brevibus carnosis versus apicem incrassatis, labello ovato-oblongo lævi lineis duabus elevatis carnosis rotundatis obsoletis.

A little Mexican Orchidaceous plant, from the collection of John Rogers, Esq. of Sevenoaks, flowering profusely during a large part of the year. The leaves are olive green, about three inches and a half long, including the stalk, of a thick texture, and a rounded, not acute, margin. The flowers are a dull brownish yellow, and grow singly on slender stalks five or six inches long. The length of the sepals is about three quarters of an inch. It is the only species of this singular genus not found in the northern hemisphere.

113. RODRIGUEZIA carnea.

R. *carnea;* pseudobulbis compressis ovalibus, foliis lanceolatis canaliculatis acutis, racemo secundo foliis breviore ascendente, sepalo supremo fornicato inferiore bidentato, petalis ovatis obtusis, labello juxtâ basin aurito disco interruptè bilamellato apice cuneato emarginato, columnâ elongatâ basi lanatâ.

A discovery of Mr. Linden in Columbia, in the year 1842, very like a small pale specimen of R. secunda. It differs however not merely in size and colour, but in having only two instead of three raised plates on the bend of the labellum, and in its column being unusually long and woolly. We are indebted for our specimen to the Rev. J. Clowes. The foliage is much the same as that of R. secunda, but narrower.

116. STANHOPEA guttulata.

S. *guttulata;* sepalo supremo oblongo concavo apice abruptè recurvo, labelli hypochilio angusto a latere compresso apice carnoso dente valido inflexo, epichilii cornubus lateralibus teretibus apiculatis lobo intermedio ovato indiviso brevioribus, columnâ apice tantùm abruptè alatâ.

Of this very singular and most distinct species I have only seen a single flower, from the collection of J. H. Wanklyn, Esq. of Crumpsell House, and I am unacquainted with its history. The flower was rather small for the genus, of a clear very pale nankin colour, closely covered all over with small crimson and brown spots and dots, even up to the tip of the labellum. It is to be hoped that further information concerning it will soon be gained. It is a plant of the greatest interest.

117. CYCNOCHES ventricosum, and CYCNOCHES Egertonianum.

"Strange things," says Mr. Bateman, in his magnificent work on Orchidaceæ, now alas concluded—"and no less strange than true—have already been recorded of Orchidaceous plants, but the case which is represented in the accompanying plate casts into the shade all former frolics of this Protean tribe. The facts are briefly as follow.

"Among Mr. Skinner's earliest Guatemala collections, attention was particularly directed to the specimens of a plant which to the habit of a *Cycnoches* joined the long pendulous stems of a *Gongora,* and for the possession of which, in a living state, no small anxiety was entertained. Some plants were speedily transmitted by Mr. Skinner, but these, on flowering, proved to be merely the old *C. ventricosum.* A mistake was of course suspected, and Mr. Skinner being again applied to, sent over a fresh supply of plants, for the authenticity of which he vouched; but these were scarcely settled in the stove, when flowers of *C. ventricosum* were again produced. Mr. Skinner being importuned for the third time, and being then on the point of returning to this country, determined to take one of the plants under his special protection during the voyage, which, flowering on the passage, seemed to preclude the possibility of further confusion or disappointment. The specimens produced at sea were exhibited, and the plant itself placed in the stove at Knypersley, where it commenced growing with the utmost vigour. The season of flowering soon arrived, but brought with it a recurrence of the former scene of astonishment and vexation, for the blossoms, instead of those of the coveted novelty, were not distinguishable from the old *C. ventricosum.* These were still hanging to the stem when the inexplicable plant sent forth a spike of a totally different character, and which was, in fact, precisely similar to the specimens gathered in Guatemala, and to those produced on the voyage. It is, at present, impossible to attempt any explanation of so strange a phenomenon, especially on the supposition that the two forms of flower are analogous to the male and female blossoms of other tribes, for *C. ventricosum* alone not unfrequently perfects seeds."

The plant here alluded to is a Cycnoches, which at one time bears large green flowers, in a short spike, with broad flat sepals and petals, and a convex white ovate undivided labellum; and at another produces small blackish simple flowers in a very long drooping spike, with narrow sepals and petals rolled back upon the stalk, and a circular purple labellum, with the edge broken up into many fingers, a lancet-shaped appendage at the point, and a kind of horn springing from the middle. These things are so totally different that, notwithstanding the strange sportings represented at fol. 1951 of this work, and the unquestionable authority of Mr. Bateman, there were many persons, well skilled in the habits of Orchidaceæ, who felt convinced that some mistake had been made, and that in reality it was impossible that such totally different flowers could have been borne by one and the same plant.

But what is impossible in nature?

On the 15th of September last, I received from Robert Steyner Holford, Esq. of Westonbirt, near Tetbury in Gloucestershire, a flower-spike bearing flowers of Cycnoches ventricosum and C. Egertonianum *intermixed;* it was exhibited at a meeting of the Horticultural Society, and I now produce a figure of it. The plant which bore this specimen had been purchased by Mr. Holford of Messrs. Rollissons of Tooting as *Cycnoches ventricosum.*

Here it will be seen that fig. 2. is nearly *Cycnoches ventricosum*, but its lip is here and there raised into warts, which are the beginning of the lobes of *C. Egertonianum*, and moreover some of the dark purple of the latter is appearing at the base of the column and the tips of the sepals. At fig. 3. the purple of *Egertonianum* is displacing the green of *ventricosum*, the sepals are rolling back, and the labellum is almost wholly changed, but the sepals are still those of *C. ventricosum.* At fig. 4. and 5. the transformation is complete.

Another curious point in this instance is that the transformations occur in no certain order. The lowest flower on the spike, No. 1, is more *Egertonianum* than *ventricosum;* the next above it, No. 2, is almost wholly *ventricosum;* that which succeeds, No. 3, is more *ventricosum* than *Egertonianum;* and 4 and 5, the last on the spike, are wholly *Egertonianum.*

What with such cases as this, the Dean of Manchester's Narcissi, and the singular hybrids with which botanists are becoming familiar, all ideas of species and stability of structure in the vegetable kingdom, are shaken to their foundation.

119. EPIDĔNDRUM dichrōmum.

E. (Encyclium) *dichromum;* pseudobulbis ovato-fusiformibus diphyllis, foliis ligulatis planis apice rotundatis racemo laxo paucifloro brevioribus, sepalis lineari-lanceolatis, petalis obovato-lanceolatis latioribus, labelli trilobi lobo medio obcordato cum mucrone pluries costato lateralibus duplò brevioribus rotundatis apice patulis, columnâ apice biauritâ.

A specimen of this plant was exhibited to the Horticultural Society on the 3rd Oct. by Mr. Quesnel of Havre, who stated that it came to him from Pernambuco. Unfortunately its colours were much changed, and it was in such a bad state that no positive opinion could be formed of it except that it is new. If, as we are informed, it had, when fresh, pure white sepals and petals, and a rich rose-coloured labellum, it must be handsome; but when we saw it, the flowers were of a dirty yellow. They are a little larger than those of *Epidendrum bifidum*, to which indeed the species nearly approaches; but its flowers do not seem to be panicled, the labellum wants the two appendages found in that species, the column has two large ears, and the latter organ is perfectly free from the labellum.

120. EPIDENDRUM leucochīlum.

Klotzsch in Gartenzeit. 1843. *p.* 145.

E. (Spathium) *leucochilum;* foliis distichis, coriaceis, patenti-recurvis, obtusis, emarginatis, subtus costato-carinatis, basi articulato-vaginatis; racemo simplici, terminali, 6 floro, e spatha ancipiti pedunculo breviore orto; floribus magnis, arcuato-pedicellatis, odoratis; perigonii foliolis linearibus, acutis, flavo-viridibus, versus basin attenuatis, margine recurvis, interioribus patenti-arcuatis, exterioribus deflexis; labello trilobo albido, lobis lateralibus brevioribus, integerrimis, oblique orbicularibus, intermedio elongato, acuminato, venis baseos 3, elevatis instructo; columna elongata, candida, inferne ad apicem biloba.

This is a plant found in the Caraccas by Mr. Edward Otto, in March, 1840, growing on the stems of trees at the height of 4500 feet above the sea. It is said to prefer a cool house, and to have been shewn at the Botanical exhibition at Prague by Professor Tausch. It is fully described by Messrs. Klotzch and Edward Otto in the above mentioned place.

121. MAXILLARIA rugōsa.

Scheidweiler in Gartenzeit. 1843. *p.* 101.

M. *rugosa;* pseudobulbis oblongis compressis sulcatis monophyllis vaginatis; foliis lanceolotis coriaceis rugosis apice obliquis emarginatis mucronatis, basi plicatis; scapo unifloro vaginato erecto pseudobulbo longior, scapi vaginis 6-8 ventricosis, marginatis, carinatis acutis, sepalis lateralibus lanceolatis explanatis, supremo fornicato, petalis angustioribus conniventibus; labello trilobo carnoso, lobis lateralibus integris, intermedio oblongo apice sulcato, in axi callo obcordato ad basin callo oblongo crassiore munito.

A Brazilian plant, according to its author allied to his Maxillaria galeata, the flowers of which have appeared in Count Aremberg's collection. Its pseudo-bulbs are described as being two inches high and nine lines broad; the leaf nine inches long and fifteen lines broad; the scape $3\frac{1}{2}$ inches high. The sepals are lanceolate, acute, ochraceous, purple towards the base; the petals are purple; the lip dark purple; the pollen-masses four, with a transverse transparent gland, and a membranous white caudicula.

122. CRYPTOSANUS scriptus.

Scheidweiler in Gartenzeit. 1843. *p.* 101.

Under this name is described a Brazilian Orchidaceous plant, said to be botanically intermediate between Cymbidium and Maxillaria. The pseudo-bulbs are one-leaved, compressed and two-edged. The leaf is lanceolate, coriaceous, opaque, and mucronate. The racemes are axillary; the bracts subulate and very small; the flowers two or three on each stalk, scentless, small, green, with sanguine spots and lines. Mr. Scheidweiler gives the following definition of his new genus.

Char. gen. Perianthium explanatum, sepalis petalisque subæqualibus, sepalo supremo fornicato, labello indiviso oblongo apiculato, disco calloso, cum basi columnæ accreto, saccato; sacculi ore ciliis densis cincto; columna brevis clavata apice auriculata. Anthera unilocularis, pollinia duo, pyriformia, postice sulcata; caudicula subulata, glandula oblonga.

123. CATTLEYA Arembergii.

Scheidweiler in Gartenzeit. 1843. *p.* 109.

C. *Arembergii;* sepalis lateralibus falcatis obtusiusculis, supremo lanceolato, petalis latioribus undulatis, labelli trilobi lobis undulato-crispis, intermedio margine lilacino lamina lutea, lateralibus margine lutescentibus, spatha herbacea compressa obtusa, pseudobulbis cylindraceis nitidis, foliis ovatis carnosis, obsure viridibus. Caulis 8 poll. altus, fol. 4 poll. longa, 2 poll. lata, flores inodori.

This species, says Mr. Scheidweiler, although having much resemblance to some others, is nevertheless distinguished by its beautiful dull green ovate leaves, and its great lilac flowers. It is a Brazilian species, and we presume described from Count Aremberg's garden, though that is not stated.

124. EPIDENDRUM glutinōsum.

Scheidweiler in Gartenzeit. 1843. *p.* 110.

E. *glutinosum;* foliis in pseudobulbos pyriformes tunicatos glabros, binis linearibus coriaceis oblique truncatis, racemo subsimplici pedicellisque glutinosis, sepalis oblongis acuminatis petalisque spathulatis patentibus, labelli fere liberi trilobi lobis lateralibus oblongis obtusis integris erectis, intermedio ovato crispato, disco calloso depresso, columna bidentata.—Scapus terminalis pedalis, petala et sepala viridi-purpurea, extus lineis purpureis notata, labellum albo-lutescens, lobo intermedio lineis purpureis ornato.

This plant, which comes from the neighbourhood of Rio Janeiro, and is remarkable for the glutinous condition of its scape and peduncles is, according to Mr. Scheidweiler, very near Epidendrum odoratissimum, which he considers identical with the Encyclia patens of Hooker and Macradenia lutescens of Loddiges. Its scape is a foot high. The petals and sepals are greenish purple, marked outside with purple lines. The lip is whitish yellow, its middle lobe being marked with purple lines.

125. MASDEVALLIA cuprĕa.

M. *cuprea;* folio obovato-lanceolato basi angustato sub apice levissimè tridentato carinato, scapo paulò longiore triquetro bifloro, bracteis carinatis pedicellorum longitudine, sepalis lateralibus ovato-lanceolatis tricarinatis semiconnatis supremo lineari-lanceolato paulò longiore marginibus reflexis, petalis linearibus truncatis apiculatis, labello basi concavo oblongo versus apicem convexo lineari apice obtuso reflexo.

An unknown correspondent has sent this from Hamburgh, as a native of Cayenne. Its leaf was four inches long; the scape a little longer; the upper sepal copper-coloured, the lower dark purple, copper-coloured at the point. It manifestly differs from M. guayanensis in its sepals not being caudate.

EDWARDS'S

BOTANICAL REGISTER:

OR,

ORNAMENTAL FLOWER-GARDEN
AND SHRUBBERY:

CONSISTING OF

COLOURED FIGURES OF PLANTS AND SHRUBS,

CULTIVATED IN BRITISH GARDENS;

ACCOMPANIED BY THEIR

History, Best Method of Treatment in Cultivation, Propagation, &c.

AND

MONTHLY CHRONICLE

OF

BOTANICAL AND HORTICULTURAL NEWS.

CONTINUED

BY JOHN LINDLEY, Ph. D. F.R.S. AND L.S.

PROFESSOR OF BOTANY IN UNIVERSITY COLLEGE, LONDON,
AND THE ROYAL INSTITUTION OF GREAT BRITAIN,
VICE-SECRETARY OF THE HORTICULTURAL SOCIETY,
&c. &c. &c.

1844.

OR VOL. XXX. OF THE ENTIRE WORK.
OR VOL. XVII. OF THE NEW SERIES.

——viret semper——nec fronde caducâ
Carpitur.

LONDON:
JAMES RIDGWAY, PICCADILLY.

M.DCCC.XLIV.

2

ANGRÆCUM pellucĭdum.

Transparent Angrec.

GYNANDRIA *MONANDRIA.*

Nat. ord. ORCHIDACEÆ. § VANDEÆ.
ANGRÆCUM. Thouars.

A. *pellucidum;* acaule, foliis distichis oblongis subundulatis recurvis apice obliquis basi equitantibus, racemis densis nutantibus axillaribus foliis brevioribus, sepalis petalisque lineari-lanceolatis, labello fimbriato cordato-ovato truncato, calcare brevi lanceolato porrecto, caudiculis 2 subulatis, glandulâ communi didymâ.

We lament to see how little justice our artists have been able to do to this beautiful plant, whose flowers are as delicate and transparent as if they were flakes of snow fixed by frost in the very act of melting. Each part of the lip is studded and bordered with little crystalline elevations, and the whole fabric of the blossom is as fragile as thin plates of glass.

It was imported from Sierra Leone by Messrs. Loddiges, with whom it flowered in November 1842. It is number 1572 of their catalogue.

Fig. 1. represents the column and spur of the labellum, the limb of which, spread flat to shew its true form is seen at 2. Fig. 3. represents the pollen-masses with their straps; and 4. is a single pollen-mass seen from behind where the strap fitted on.

We still regret the want of leisure to effect that reformation in the genus, the necessity of which was alluded to at fol. 68 of our volume for 1840. A comparison of the details of this plate and of those of *A. gladiifolium* at the place just quoted, will shew that the genus Angrec, as it now stands, cannot be maintained entire.

5

CATTLEYA pumila.

Bordered Dwarf Cattleya.

GYNANDRIA *MONANDRIA.*

Nat. ord. ORCHIDACEÆ. § EPIDENDREÆ.
CATTLEYA. Lindl.

C. pumila, *Hooker in Bot. Mag. t.* 3656.

This pretty little species is stated in the Botanical Magazine to be a native of the Essequibo; but we cannot confirm that statement, which has probably originated in some mistake. It is most assuredly Brazilian, being the No. 657 of Mr. Gardner's herbarium, and having been imported by a French dealer, from that country, under the name of C. marginata, a very good name, expressing the appearance of a beautiful pale border to the blood-red blotch of the lip.

Our drawing was made in Mr. Loddiges' Nursery.

It should be potted in turfy heath-mould, mixed with a few pieces of potsherds, to keep it as porous as possible. The pot should also be well drained, and the soil considerably elevated above its brim. In cultivation the genus Cattleya differs from many Orchidaceous plants; the species require very little water; even in spring, when the plants are growing, water once or twice a week will be sufficient, and if syringed it must be so slightly that no water may lodge in the axils of the leaves; but where steam can be admitted once a day, it will be found much better. During summer the house should be slightly shaded in sunny weather, and the temperature allowed to rise to 80° or 85° by day, but never above 70° at night. In winter, when little or no water will be required, except to keep the pseudo-bulbs from shriveling, the temperature may be as low as 65° by day and 58° by night.

There is in cultivation a plant called C. Pinellii, of which a specimen has been received from Messrs. Rollissons, and of which the top of the column is figured at fig. 1. of the accompanying plate, which is nearly allied to this, but appears distinct. It has white sepals, and rose-coloured straight not curved petals. It requires further examination before its true value can be determined; it is, however, very pretty.

Considering the number of species, or supposed species, of this glorious genus, now in our gardens, and the length of time that has elapsed since any arrangement of them has taken place, it seems desirable that they should be brought into one view; especially as nearly all of them exist in our gardens. And I am the more induced to rearrange the genus, because I find some of the remarks made at t. 22 of the Sertum erroneous,—especially the statement that C. maxima is a Lælia, a mistake the origin of which I cannot trace. The division there suggested among the species is too artificial; that now proposed will be found more natural.

There is no character by which Cattleya and Lælia can be distinguished except the number of pollen-masses, which are four in the former, and eight in the latter. The flowers of Lælia are usually indeed in a raceme elevated on a long graceful stalk; but such is not the case in *Lælia Perrinii*, originally described by me as a Cattleya, and subsequently regarded by Sir W. Hooker (Bot. Mag. t. 3711) as a variety of C. intermedia; nor in *L. virens.*

CATTLEYA.

Section I. Lip rolled round the column.

* Sepals of the texture of petals, the lateral quite straight.

1. C. *superba* (Lindl. Sert. Orch. t. 22. C. *Schomburgkii*, Lodd. cat. no. 434.); caulibus clavatis sulcatis, foliis oblongis marginatis caule brevioribus, sepalis oblongis acutiusculis, petalis lanceolatis acutis membranaceis ferè duplò latioribus, labelli trilobi lobis lateralibus acutis intermedio transverso plano emarginato denticulato basi venis elevatis rugoso, callis duobus pone basin.——*Demerara.*——Flowers deep rose-coloured, sweet scented, with a deep crimson lip.

2. C. *Skinneri* (Bateman Orch. Mex. Guat. t. 13.); caulibus clavatis foliis binis ovalibus obtusis, spathâ abbreviatâ, sepalis angustis oblongo-lanceolatis rectis, petalis oblongis 3-plò latioribus, labello integerrimo convoluto emarginato plano, "columnâ nanâ 3-plò longiore."——*Guatemala.*——Flowers deep rich rose colour, with a crimson lip, resembling those of C. Skinneri. This species inhabits the hot damp coast, on very high trees, and is most difficult to get at, except after a storm that may have chanced to throw down some of the large forest trees. It should be well watered daily, to represent the heavy dews and the rains, which latter are from May to November. Mr. Skinner recommends care to be taken that it may not imbibe too much moisture, as its habitat is on branches of large trees seldom having any lichen, where the heavy rains do not lie. It does not seek too much shade, but rather, like 'Epid. aurantiacum,' exposed places. Climate 80° to 85°, and sometimes 95°, during the day. Flowers in January and February; vulgarly called 'Flor de San Sebastian,' from its being in season, and adorning the altars on that saint's day (20th of January).

3. C. *Walkeriana* (Gardner in London Journ. Bot. 2. 662); foliis oblongo-ellipticis coriaceis marginatis caule cylindrico longioribus, sepalis oblongo-lanceolatis acutiusculis calloso-apiculatis, petalis ovato-lanceolatis acutis duplò latioribus, labelli trilobi cucullati lobis lateralibus apice valdè et obliquè truncatis basi rotundatis, intermedio lato rotundo emarginato edenticulato plano basi venis elevatis rugosis.——*Brazil;* on the stem of a tree overhanging a small stream which falls into the Rio S. Francisco, beyond the Diamond district.——Flowers about 4 inches in diameter. Near C. superba; distinguished by its shorter pseudobulbs, smaller leaves, larger and rounder middle lobe of the labellum, but particularly by the obliquely truncated lateral lobes which envelope only the lower half of the broadly winged column, and not the whole of it.—*Gardner.*

4. C. *maxima* (Lindl. Gen. et Sp. orch. no. 4.); caulibus obovato-clavatis angulatis, foliis 1-2 ovato-oblongis, spathâ pedunculo multò breviore, sepalis lineari-oblongis obtusis, petalis subrotundo-ovalibus undulatis membranaceis, labelli maximi crispi oblongi obsoletè trilobi lobo medio unduiato altè emarginato, disco lævi.——*Guayaquil* and *Colombia.*——Fine specimens, and live plants brought home by Mr. Hartweg shew this to be a rival of C. labiata. Its flowers are as large, and are described as being of a beautiful dark pink.

5. C. *labiata* (Lindl. coll. t. 33.); caulibus clavato-fusiformibus sulcatis, foliis solitariis oblongis, spatha pedunculi longitudine, sepalis lineari-lanceolatis acutis, petalis membranaceis oblongo-lanceolatis undulatis multò latioribus, labello obovato crispo-undulato emarginato disco lævi. *Varietates sunt.* α. petalis oblongo-lanceolatis undulatis, labelli disco sanguineo. (C. *labiata*, Lindl. l. c. Gen. et Sp. orch. 116. Hooker exot. fl. t. 157. Lodd. Bot. Cab. t. 1856. Bot. Mag. t. 3998.) β. petalis latioribus subcrispis, labelli disco luteo sanguineo punctato aut picto. (C. *Mossiæ*, Hooker in Bot. Mag. t. 3669. Bot. Reg. 1840, t. 58.)——α *Brazil.* β *La Guayra.*——One of the most noble species of this fine genus. The varieties of what has been called C. Mossiæ are numerous, and seem to prove that no reliance can be placed on the supposed distinctions between it and C. labiata. It must however be confessed that the question is open to further consideration.

6. C. *crispa* (Lindl. Bot. Reg. t. 1172. orch. no. 1. Hooker in Bot. Mag. t. 3910.) caulibus crassis oblongis clavatis, foliis solitariis, spathâ magnâ herbaceâ, sepalis lineari-obovato-lanceolatis acutis, petalis latioribus oblongo-lanceolatis undulatis crispis, labelli indivisi limbo ovato acuminato quam maximè undulato-crispato.——*Brazil.* ——Flowers very large, pure white, with a large crimson blotch in the middle of the lip.

7. C. *citrina* (Lindl. Gen. et Sp. orch. no. 8. Bot. Mag. t. 3742. " *Corticoatzontecoxochitl.* Hernand. Mex. 1. p. 240." *Sobralia citrina.* Llave Nov. Veg. Descr. 2. 21.); caulibus ovatis squamis laxis albis membranaceis vestitis, foliis lanceolatis glaucis, pedunculis longissimis solitariis, floribus carnosis pendulis, sepalis oblongo-ellipticis, petalis conformibus paulò latioribus, labelli trilobi lobo intermedio ovato undulato emarginato lineâ mediâ latâ elevatâ.—— *Mexic.*——A mountain plant remarkable for its rich clear yellow flowers, which are very sweet scented.

8 C. *pumila* (Hooker in Bot. Mag. t. 3656. Lindl. Bot. Reg. 1844. t. 5. C. *marginata*, Hort.); caulibus brevibus ovalibus sulcatis, foliis solitariis ovato-oblongis acutis, pedunculo unifloro, sepalis lineari-lanceolatis rectis, petalis ovalibus 3-plò latioribus, labello obovato indiviso apice plicato, lineis 3 elevatis in medio.——*Brazil,* (Gardner, 657).——This is said to be from the Essequibo, but that appears to be a mistake. It is a very distinct species, with peculiarly thick acute solitary leaves, and deep rose-coloured flowers. The lip is bordered with a pale colour, which is sometimes almost white.

9. C. *Pinellii* (Hort.); facie C. pumilæ sed sepalis *acutis albidis,* petalis *ovatis* roseis, labello magis crispo, columnâ apice dentibus 2 crenatis auctâ.——*Brazil.*——Very near C. pumila, but the sepals are white, the petals almost exactly ovate, the lip more crisp and of a brighter colour. Requires further examination.

** Sepals somewhat herbaceous, or more coriaceous than the petals, the latter manifestly falcate.

10. C. *Loddigesii* (Lindl. Coll. Bot. t. 37. Gen. and Sp. no. 5. Hooker Bot. Misc. t. 186. *Epidendrum violaceum,* Lodd. Bot. Cab. t. 337. *Cattleya ovata,* Lindl. in Bot. Reg. t. 1919.)—*Var.* floribus pallidioribus; (*C. intermedia,* Graham in Bot. Mag. t. 2851. L. no. 6.)—*Var.* floribus subalbis; (*C. intermedia pallida,* Lindl. in Bot. Reg. t. 1919. *C. vestalis* Hoffsg. verzeich?); caulibus elongatis teretibus, foliis 2 ovato-oblongis, spathâ brevi membranaceâ, sepalis oblongis lateralibus falcatis, petalis subconformibus, labelli lobis lateralibus rotundatis intermedio dilatato crispo 2½ longioribus, lineis pluribus elevatis lamellatis pone basin versus apicem evanescentibus. ——*Brazil* and *Buenos Ayres.*——Flowers sometimes clear lilac with a whitish lip, sometimes nearly white with a crimson lip, with many intermediate gradations. It always appeared doubtful whether C. *intermedia* could be distinguished, and more experience in estimating the value of characters among these plants leads to the conclusion that C. *ovata* must also be reduced to the same species. It seems to be very common in Brazil, and to occur as far south as Buenos Ayres. Mr. Gardner found it on trees in *marshes* at the foot of the Organ mountains, (no. 5635 of his herbarium). The size of the leaves is much affected by situation; in all cases, however, they seem to be broader at the base than at the point. So far as the character of Count Hoffmannsegg's *C. vestalis* can be judged of, that plant is the pale C. intermedia.

11. C. *Harrisoniana* (Bateman in Bot. Reg. sub t. 1919.); " foliis angustè lanceolatis," sepalis oblongis apiculatis, petalis ovalibus, labelli trilobi lobis lateralibus rotundatis intermedio angusto crispo 1½ longioribus: lineis pluribus elevatis per totam axin lamellis nullis.——*Brazil.*——I only know this plant from two flowers given me by Mr. Bateman. They have much shorter and broader sepals than C. Loddigesii, the petals are much broader, and the proportion between the middle lobe of the lip and its lateral lobes is quite different. Besides which, certain elevated lines run all the way from the base to near the apex of the lip. The flowers are lilac; the tip with a deep blotch within the margin.

12. C. *Forbesii* (Lindl. Coll. Bot. sub t. 37. Bot. Reg. t. 953. Lindl. no. 7. Bot. Cab. t. 1152.); caulibus elongatis teretibus, foliis 2 oblongis, sepalis petalisque lineari-oblongis obtusis subæqualibus, labelli trilobi lobo medio cordato subrotundo-ovato argutè dentato undulato apiculato: lateralibus minoribus rotundatis planis, lineis 4 elevatis pone basin lamellâque membranaceâ utrinque, columna basi lineis 5 elevatis quarum duæ majores cristata, clinandrii cardine tuberculo corrugato acuto.——*Brazil.*——Flowers greenish yellow. Lip with bright orange brown veins. Very near C. Loddigesii.

13. C. *maritima* (Lindl. in Bot. Reg. sub t. 1919.); pedunculo unifloro, caule subclavato, foliis binis ovalibus obtusis spatha parum longioribus, sepalis oblongis acutis, petalis lanceolatis falcatis obtusis, labello trilobo (nudo ?); laciniis lateralibus erectis rotundatis intermediâ dilatatâ denticulatâ emarginatâ.——Sea-beaten rocks, *Buenos Ayres.*—— Flowers fine, rose-coloured, usually in threes; many varieties. Its small leaves, dwarf habit, and one-flowered peduncle, mark this, of which I have only seen one specimen with a single flower. Possibly it is one of the many varieties of C. Loddigesii.

14. C. *Arembergii* (Scheidweiler in Gartenzeit. 1843. p. 109.); "sepalis lateralibus falcatis obtusiusculis, supremo lanceolato, petalis latioribus undulatis, labelli trilobi lobis undulato-crispis, intermedio margine lilacino lamina lutea, lateralibus margine lutescentibus, spatha herbacea compressa obtusa, pseudobulbis cylindraceis nitidis, foliis ovatis carnosis, obscure viridibus."——*Brazil.*——This species is only known from the above description. It is said to bear much resemblance to some others, but to be distinguished by its beautiful dull green ovate leaves, and its great lilac flowers. The stem is said to be eight inches high, the leaves four inches long and two inches broad; the flowers sweet-scented. It may be C. *Harrisoniana,* or *maritima*; but it seems to be distinct.

15. C. *guttata* (Lindl. Bot. Reg. t. 1406. Orch. no. 10. Hort. Trans. 2 ser. 2. t. 8. C. *elatior,* Lindl. Orch. no. 9.); caulibus elongatis teretibus, foliis 2 oblongis concavis basi paulò angustatis, spathâ brevi, floribus carnosis, sepalis lineari-oblongis acuminatis, petalis conformibus paulò latioribus undulatis, labelli trilobi lobis lateralibus ovatis: intermedio cuneato bilobo disco tuberculato.——*Brazil.*——Flowers greenish yellow, beautifully spotted with crimson; lip white with a lilac tip. Raceme sometimes as large as a man's head.

16. C. *granulosa* (Lindl. in Bot. Reg. 1842, t. 1. C. *guttata* β *Russelliana,* Hooker in Bot. Mag. t. 3693 ?); caulibus teretibus gracilibus diphyllis, foliis oblongo-lanceolatis obtusis, sepalis oblongis obtusis, petalis obovato-spathulatis undulatis obtusissimis, labello cucullato tripartito: laciniis lateralibus semiovatis intermediâ sinu lato divulsâ ungue æquilateri lævi laminâ dilatatâ rotundatâ plicatâ granulosâ denticulatâ.——*Guatemala.*——This has a slender stem, terminated by two narrow leaves, something like those of C. bicolor, but shorter. Sepals two inches and a half long, oblong, olive green, mottled with rich brown spots. Petals obovate, very much rounded at the point, and narrowed to the base; wavy at the margin, and of the same colour as the sepals. Strongly contrasted with the dingy colours of those parts is the lip, of a pure white at the sides and point, and of a glowing orange spotted with crimson in the middle; its lateral lobes curve over the column, but are divided very deeply from the middle lobe by a wide slit; the middle lobe itself has its surface broken up into numerous granulations, something in the way of C. guttata. It is scarcely possible to doubt that the plant figured in the Botanical Magazine as a variety of C. guttata, is this

species. The Brazilian origin attributed to it is no doubt erroneous, like that of many plants from the Woburn collection.

Section 2. Lip without the lateral lobes, and flat below the column.

17. C. *Aclandiæ* (Lindl. in Bot. Reg. 1840, t. 48.); caulibus cylindraceis striatis decumbentibus, foliis 2 oblongis, floribus subsolitariis, sepalis petalisque herbaceis lanceolatis æqualibus incurvis maculatis, labelli plani calvi hypochilio dilatato patulo subrepando epichilio orbiculari-reniformi emarginato.——*Brazil.*——Flowers dull olive green, nearly the colour of C. granulosa. Lip dull violet. The smallest species yet known.

18. C. *bicolor* (Lindl. in Bot. Reg. sub t. 1919.); foliis ovato-oblongis angustis caule tereti elato triplò brevioribus, sepalis lanceolatis falcatis acutis, petalis parum latioribus subundulatis obtusis, labello indiviso plano apice dilatato rotundato crenato convexo.——*Brazil.*——Stems two to three feet long. Sepals and petals tawny; labellum bright purple, with a lanceolate streak in the centre, white slightly spotted with purple. The flowers are slightly fragrant. This was introduced by Mr. Pontey of Plymouth, and flowered in his nursery in 1838.

---

19. C. *Domingensis* (Lindl. Orch. no. 11.); caule brevi oblongo annulato squamoso, folio ovali-oblongo coriaceo, scapo terminali longissimo stricto apice racemoso 7-8 floro, sepalis lineari-lanceolatis acutis petalis oblongis obtusiusculis triplò angustioribus, labello indiviso obovato obtuso plicato-crispo emarginato cucullato.——*St. Domingo.* ——In the absence of sufficient evidence as to this species, which is no. 231 of Jaeger's collections, it may be conjectured to be a Lælia rather than a Cattleya. It is found on logwood trees in the wood near Miragoane, in St. Domingo, where it flowers in April.

8

ANIA bicornis.

Two-horned Ania.

GYNANDRIA *MONANDRIA*.

Nat. ord. ORCHIDACEÆ. § EPIDENDREÆ.
ANIA. Lindl. gen. & sp. Orch. p. 129.

A. *bicornis* (Lindl. in Bot. Reg. 1842. misc. 31.); folio oblongo-lanceolato carnoso scapo breviore, labelli lobo medio emarginato apiculato ecalcarato basi bilamellato lamellâ alterâ versus apicem interjectâ, antherâ bicorni.
Labellum *cum basi producta columnæ articulatum.* Anthera 2-*locularis.*

To the Rev. J. Clowes, of Broughton Hall, we are indebted for a specimen of this terrestrial Orchidacea, which he received from Ceylon, and flowered in March 1842.

It belongs to a little group of the Epidendreous section, of which Bletia is the type, and is nearly allied to A. *latifolia*, a Sylhet plant, at present known only from dried specimens; from that it differs in having much smaller flowers and leaves, an entirely different labellum, and a 2-celled anther; that of A. latifolia being 8-celled. In the latter circumstance, indeed, it corresponds with the neighbouring genus Cytheris, whose distinctive character is therefore not to be taken from the cells of the anthers, but from its resupinate flowers, and truly calcarate labellum. In Ania the labellum, if it appears to have a spur, as in A. angustifolia, owes that appearance to the extension of the foot of the column.

Fig. 1 is a view of the lip of this plant; 2 shews the column, with its lengthened foot and two-horned anther; 3 gives the appearance of the pollen-masses.

At first sight this plant, when in flower, resembles a starved specimen of Eulophia macrostachya.

12

LISSOCHĪLUS roseŭs.

Rose-coloured Lissochilus.

GYNANDRIA *MONANDRIA*.

Nat. ord. ORCHIDACEÆ. § VANDEÆ.
LISSOCHILUS. Supra, vol. 12. *vol.* 1002.

L. *roseus* (Lindl. in Bot. Reg. 1843. misc. 37. *Dendrobium roseum,* Swartz. in Persoon synops. p. 523.); foliis lato-lanceolatis erectis plicatis, scapo squamis lanceolatis acutis membranaceis distantibus vaginato, racemo denso oblongo, bracteis ovatis acuminatis ovario brevioribus, sepalis spathulatis acutis concavis reflexis, petalis oblongis apiculatis, labelli trilobi lobis rotundatis intermedio emarginato cum mucrone, disco lamellis tribus undulatis serrulatis tuberculo parvo utrinque.

Although the terrestrial Orchidaceæ of hot countries are too often very inferior to the epiphytes, yet there are many exceptions to that rule, among which few are more striking than the present, which will not suffer by comparison with the Vandas, Saccolabiums, and Dendrobiums of India.

It is a native of Sierra Leone, whence it was received by Mr. Rucker, in whose magnificent collection of Orchidaceæ it flowered in February 1843. The leaves are broad, stiff, and plaited like a reed; the flower-stem is between 3 and 4 feet high. The petals are of the brightest rose colour, set off with velvety-brown sepals, and a yellowish stain on the lip.

There is another plant in the same part of Africa, which would perhaps excel even this in beauty, and it is much to be regretted that no one should have yet been able to send it home. Its petals, even dried, are an inch and a half long. Sir William Hooker has a specimen in his herbarium, collected near Boney, which we named some years ago Lissochilus macranthus.

Fig. 1. shews the inside and form of the lip; 2. the column; and 3. the pollen-masses.

While we are writing on this subject, we would also mention another most noble plant, specimens of which might certainly be procured by any of our African merchants. When Mr. Ansell was ill from the effects of the Niger expedition, at Fernando Po, he found in Clarence Cove, growing on the stems of the Oil Palm (Elais guineensis), an epiphyte with a slender jointed stem about two feet long, having at the upper end many stiff, plaited, lanceolate, 5-ribbed leaves, and a terminal panicle of flowers as large as those of Vanda Roxburghii, with dark purple spots on a pale ground. Of that plant we possess a dried specimen, with one of the lower branches of the panicle in good preservation, and as it proves to be a new genus we take this opportunity of naming it after its discoverer

ANSELLIA.

Sepala oblonga, carnosa, æqui-patentiâ, libera. *Petala* conformia, recta, patula, duplò latiora. *Labellum* sessile, patulum, trilobum, bilamellatum, lobo medio minore verrucoso. *Columna* elongata, marginata, basi utrinque auriculata. *Anthera* bilocularis. *Pollinia* 4, sessilia, basi contigua, duobus dorsalibus multo minoribus; glandula angusta utrinque acuminata.——Caulis *elongatus, teres, apice tantum foliosus.* Folia *plicata, coriacea.* Panicula *terminalis.*

Sp. 1. Ansellia africana.

It appears that this genus must be referred to that set of Vandeæ of which Cymbidium must be taken as the type. In fact, it is very near that genus in technical characters, though extremely different in its manner of growth. The auricles at the base of the column, the four pollen-masses, and the very narrow gland fining away to each side will serve to distinguish it.

13

BOLBOPHYLLUM macranthum.

Large-flowered Bolbophyllum.

GYNANDRIA *MONANDRIA.*

Nat. ord. ORCHIDACEÆ. § MALAXEÆ.
BOLBOPHYLLUM. Supra, vol. 23. *t.* 1942.

B. *macranthum*; foliis petiolatis oblongis planis coriaceis, floribus solitariis, pedunculo petiolo longiore, flore plano resupinato, sepalo dorsali plano ovato acuminato lateralibus petalisque subconformibus hinc tortis, labello minimo unguiculato subtrilobo acuminato.

This singular plant was imported by Messrs. Loddiges from Sincapore, and is closely allied to both B. leopardinum and affine. From each it differs in its much more fleshy and larger flowers, whose stalk is considerably longer than the petiole.

The flowers appear in March, and expand so flat that they seem as if they had been pressed between paper. In the centre they are a pale lemon colour, but towards the tips they are much mottled with dark chocolate-coloured spots.

Like the rest of the genus it succeeds best when tied to a block of wood, and suspended to a rafter in a moist stove. If the wood is charred enough to burn off the bark the block will be found to answer the purpose much better, and all insects that harbour about it will be destroyed. In summer the plant should receive water twice a day at least, and the temperature should be kept about 80° by day, and 68° at night. In winter for a few weeks very little water will be required, only as much as to keep the pseudo-bulbs from shriveling. The temperature then should never be raised above 50° or 55° with fire heat.

18

BROMHEADIA palustris.

Marsh Bromheadia.

GYNANDRIA *MONANDRIA.*

Nat. ord. ORCHIDACEÆ. § VANDEÆ.

BROMHEADIA, Lindl. in Bot. Reg. 1841. *misc.* 184. *Sepala* et *petala* subæqualia patentissima. *Labellum* cucullatum, trilobum, cum columnâ omninò parallelum, basi inarticulatum, secus axin elevatum pubescens. *Columna* latè alata, carnosa, elongata. *Anthera* 2-locularis, dorso conica, cum columnâ articulata. *Pollinia* 2, reniformia, posticè excavata, in glandulam latam, triangularem, membranaceam sessilia.——Herba *caulescens, ebulbis.* Folia *disticha, coriacea, emarginata.* Spica *terminalis, disticha, flexuosa, multiflora, longè pedunculata,* bracteis *brevissimis rigidis, dentiformibus.* Flores *speciosi, recti.*

Bromheadia palustris, *Lindl. l. c. Hooker. Bot. Mag. t.* 4001.
Grammatophyllum ? Finlaysonianum, *Lindl. in Wall. Cat. no.* 7561. *Gen. & Sp. Orch. p.* 173.

We formerly introduced this curious plant to our readers with a few observations which we beg permission to repeat, with some corrections.

"When Mr. Finlayson was in the Malay archipelago, he found a curious Orchidaceous plant at Sincapore with the habit of Epidendrum elongatum; and from specimens of it, almost destroyed by insects, which I examined some years since in Dr. Wallich's herbarium, I referred it with great doubt to Grammatophyllum, under the name of G. ? Finlaysonianum, (*Gen. & Sp. Orch. t.* 173.) This plant has flowered at Penllergare in South Wales with J. D. Llewelyn, Esq. who received it from Cuming, with the memorandum that it had been "dug out of a bog in Sumatra." Having now had the advantage of examining a perfect specimen in flower, I find that although nearly allied to Grammatophyllum, it is in fact very distinct. I therefore avail myself of the opportunity of adding to the list of genera the name of Sir Edward Ffrench Bromhead, Bart. F.R.S. whose investigations of the natural affinities of plants are well known to systematical Botanists."

In appearance the plant has the aspect of Epidendrum elongatum, as has been already stated; and like it has the whole of the upper part of the stem provided with closely pressed distant sheaths, instead of leaves, on which the spike of flowers is arranged. The latter is very rigid, between two and three inches long, regularly zigzag, with a short hard tooth-like bract at each bend, so that the spike without the flowers resembles a coarsely-toothed narrow double-edged saw. The flowers are about an inch long, white, and rather drooping, spreading quite open. The labellum, in which alone any colour resides, is straw-coloured on the middle lobe, and violet at the tips of the lateral lobes; along the middle, as far as the separation of the lobes, it is convex and covered with purple down; while the disk of the middle lobe is broken up into yellow granulations.

We shall take another opportunity of pointing out in what way Bromheadia most differs from its allied genera, Cymbidium, Grammatophyllum, Renanthera, &c.

Our drawing was made from a specimen communicated by His Grace the Duke of Northumberland, in November 1841.

Fig. 1. represents a front view of the column; 2. the inside of the lip; and 3. gland and pollen-masses.

20

ERIA floribunda.

Many-flowered Eria.

GYNANDRIA *MONANDRIA.*

Nat. ord. ORCHIDACEÆ. § MALAXEÆ.
ERIA. Lindl. supra, vol. 15. *fol.* 1654.

§ Tonsæ; *perianthio glabro v. parùm pubescente.*

E. *floribunda* (Tonsæ) (Lindl. in Wall. Cat. no. 7408. Bot. Reg. 1843, misc. 56. 1844, t. 20.); caulibus carnosis subflexuosis teretibus, foliis lanceolatis acuminatis, racemis oppositifoliis patulis multifloris pubescentibus foliis brevioribus, bracteis defloratis ovatis concavis retrorsis, sepalis petalis 3-plò latioribus, labello angusto nudo basi saccato: laciniis lateralibus ascendentibus abbreviatis intermediâ cuneatâ tridentatâ.

Labellum *basi cyathiforme, melliferum, bidentatum,* columnæ *adpressum, sursum unguiculatum, cuneatum, apiculatum, inflexum, columnâ paulò brevius.*

Although the flowers of this plant cannot boast of large size and rich colours, they are by no means destitute of beauty. Arranged as they are in long drooping racemes, glassy in texture, and delicately touched with crimson, they are among the prettiest of the smaller kinds.

It is a native of Sincapore, whence it has been received by Messrs. Loddiges.

Along with *E. polyura, bipunctata, profusa*, and some others, it forms a small group of species, which are very different in habit from such plants as *E. pannea, armeniaca*, &c.

Fig. 1. represents the column and lip seen in profile; 2. shows the pollen-masses adhering at their points by a granular mucilaginous substance.

It is a stove plant, and succeeds best when potted in turfy peat, mixed with a few potsherds. During the summer a liberal supply of water should be given, and the house be kept as moist as possible. In sunny weather the plant will be much benefited if slightly shaded, for although it enjoys a high temperature, yet its leaves will become pale and contracted if too much exposed to the rays of the sun. In winter very little water is required, especially if the house can be filled with steam once a day, which is always preferable to syringing.

23

SCHOMBURGKIA crispa.

Crisp-flowered Schomburgkia.

GYNANDRIA *MONANDRIA*.

Nat. ord. ORCHIDACEÆ. § EPIDENDREÆ.

SCHOMBURGKIA, Lindl. Sertum Orchidaceum, t. x. *Sepala* et *petala* conformia, patentia, omninò libera, basi æqualia. *Labellum* difforme, membranaceum, trilobum, semicucullatum, basi cum margine columnæ connatum, supra basin tumidum (intrusum): venis lamellatis. *Columna* alata. *Pollinia* octo.——Rhizoma *repens, nudum, annulatum, pseudobulbigerum.* Pseudobulbi *magni, elongati*, 2-3-*phylli*. Folia *coriacea*. Scapi *terminales vaginati*. Bracteæ *magnæ, siccæ, spathaceæ*. Flores *speciosi, racemosi, congesti*.

S. crispa, *Lindl. in Sert. Orchid. t.* 10.

This fine genus is remarkable for the large size of its pseudo-bulbs, which are occasionally as much as two feet long. It differs from Epidendrum and its allies in having eight pollen-masses, and evidently constitutes an extremely well marked group. But its species have been little examined except in a dried state, for which their fleshiness renders them ill-suited, and hence a difficulty in determining their true limits. It is hoped that the enumeration, at the end of this article, will render their distinctions clearer.

For the opportunity of figuring this we are indebted to Mrs. Marryat, who received it from Mr. Schomburgk himself. Its yellow-brown flowers distinguish it from all the others.

In cultivation it demands the same treatment as many of the genus Cattleya. It may either be potted in turfy peat in the usual way, or it may be tied to a block of wood (with a little sphagnum to retain moisture) and suspended to a rafter in a stove. In any case, care should be taken not to water over head when the plant is in a growing state, otherwise the glutinous substance on the scales which surround the young shoots will retain the water and cause them to damp off. In summer the plant should be slightly shaded, and the house kept as moist as possible at a temperature of 80° by day, and about 68° at night. In winter, when little water is required, the temperature need not be raised above 56° by artificial means.

Fig. 1. represents the lip; 2. the column seen in profile.

The species of Schomburgkia are,—

1. S *crispa* (Lindl. in Sert. Orchid. t. 10. Bot. Reg. 1844. t. 23. S. *marginata, var.* Hooker in Bot. Mag. t. 3729.); pseudobulbis fusiformibus, floribus racemosis, bracteis ovario subæqualibus, labello ovato-oblongo obtuso vix cucullato obsoletè trilobo sepalis petalisque oblongis crispis subæquali: lamellis 5 undulatis duabusque lateralibus tenuibus rectis.——*Demerara.*——This plant has yellowish brown flowers, with scarcely a trace of purple; and by that circumstance alone it may be distinguished. Its lip is nearly flat, very little three-lobed, and broadest at the base: other characters peculiar to itself. Although the author of the species, I must plead guilty to having on one occasion, (Bot. Reg. misc. Feb. 1839) confounded it with S. marginata, and thus misled Sir Wm. Hooker, whose figure in the Bot. Magazine, t. 3729, undoubtedly belongs to this species. Probably the colours in the Sertum, taken from a drawing by Mr. Schomburgk, are much too yellow.
2. S. *marginata* (Lindl. in Sertum Orchid. t. 13.); pseudobulbis clavato-fusiformibus, floribus racemosis, bracteis ovario subæqualibus, labello ovali acuto basi angustato trilobo parùm cucullato sepalis petalisque oblongis undulatis subæquali, lamellis crispis 5 continuis lateralibus interruptis.——*Surinam, Demerara.*——Formerly in cultivation, but now lost; unless the "Spread Eagle" plant of Jamaica should prove to be it. It differs from S. crispa in its flowers being distinctly purple, with only a little yellow at the edge of the sepals and petals, and in the form and appendages of the lip, which tapers to the base, is very distinctly three-lobed, has an acute oval middle lobe, and is furnished with wavy plates, which are broken up at the sides.
3. S. *undulata* (Lindl. in Bot. Reg. 1844. misc. 21.); pseudobulbis fusiformibus, floribus dense racemosis, bracteis longissimis spathaceis, sepalis petalisque æqualibus linearibus undulato-crispis labello longioribus, labelli cucullati lobis lateralibus rotundatis intermedio ovali acuto v. obtuso, lamellis 5 undulatis duabusque lateralibus rectis tenuibus.——*La Guayra.*——This has fine rich brownish purple sepals and petals, and a clear violet-purple lip. From S. crispa and marginata, it is distinguished by its very long narrow crisp sepals and petals. The flowers are as large as those of S. tibicinis, but the inflorescence is quite different. There is no yellow in them, and the lip is distinctly curved up towards the column. Some fine specimens have been lately flowered by Mrs. Lawrence, one of which has browner flowers, and the middle lobe of the lip wedge-shaped and obtuse, not oval and acute; but they do not otherwise seem to differ. The very long narrow crisp sepals and petals render this very obviously different from the two last, which it quite resembles in habit.
4. S. *tibicinis* (Bateman. Orch. Mex. & Guat. t. 30. Lindl. in Bot. Reg. 1841. misc. 119. *Epidendrum tibicinis*, Bateman in Bot. Reg. 1838. misc. 12.); pseudobulbis conicis corniformibus annulatis sulcatis 3-phyllis, foliis oblongis coriaceis patentibus, scapo longissimo tereti distanter squamato apice paniculato, paniculâ pyramidali laxiflorâ, sepalis petalisque undulatis crispis, labello oblongo cucullato venis per medium 5 elevatis approximatis: laciniis lateralibus apice rotundatis intermediâ subrhombeâ emarginatâ, antherâ emarginatâ.——*Honduras.*——This noble plant is the cow's-horn orchis of Honduras. The flowers when fully expanded are about two and a half inches wide, deep pink speckled with white on the outside, rich chocolate red within. The lip is white in the middle, but deep rose-colour at the sides, with a short chocolate-red middle lobe. The flowering stem is eight or nine feet long. Its pseudo-bulbs, between one and two feet long, are quite hollow, and as smooth inside as the chamber of a Bamboo; at their base there is always a small hole, which leads to the interior, and furnishes access to colonies of ants, which are constantly found inhabiting the plant.

24

CYMBIDIUM pendŭlum, *var.* brevilābre.

Short-lipped thick-leaved Cymbidium.

GYNANDRIA *MONANDRIA*.

Nat. ord. ORCHIDACEÆ. § VANDEÆ.

CYMBIDIUM. Botanical Register, vol. 7. *fol.* 529.

C. *pendulum* (Swartz. nov. act. ups. 6. 73. Willd. Sp. pl. 4. 101. Roxb. Fl. Ind. 3. 458. L. p. 165. Bot. Reg. 1840. t. 25. *C. crassifolium*, Wall. Cat. no. 7357. *Epidendrum pendulum*, Roxb. corom. plants, 1. 35. t. 44.); foliis ensiformibus distichis coriaceis obliquè obtusis, racemis pendulis multifloris, bracteis minutis, petalis sepalisque lineari-oblongis obtusis, labelli trilobi lobis lateralibus acutis intermedio oblongo apiculato: lamellis continuis approximatis apice confluentibus.

Var. *brevilabre;* labello latiore lobis lateralibus acutioribus intermedio subrotundo-oblongo. *Bot. Reg.* 1844. *misc.* 67.

Having already figured this plant, it may perhaps be thought that the variety now given is too trifling to deserve a separate representation. We are however rather anxious to shew that when these Orchidaceous plants do run into ordinary varieties, it is only within recognizable limits, as happens in other plants, and that the masquerading dress under which such plants as Catasetum and Cycnoches occasionally appear, are not, as some suppose, to be taken as indications of a disposition to vary, which throws suspicion upon all specific distinctions in the order.

As far as our experience goes, the ordinary variations to which Orchidaceæ are subject, are in all respects analogous to what is met with in other plants, and as is exemplified by the plant before us from Sincapore, in which, while the lip becomes shorter, broader, and with a much blunter middle lobe, every thing else remains so exactly the same, that nobody can entertain a doubt about the specific identity of the plant with Cymbidium pendulum. The vertical plates of the lip, in particular, are quite unchanged, shewing, as we find it always shewn, that the elevations and processes of the surface of the lip are of the utmost importance in considering the limits of species. Eria bractescens and longilabris, published in the present number, t. 29, furnish the converse of the rule. They are much alike, and their labella vary in form in a manner not unlike that of C. pendulum and C. p. brevilabre; but with a change in the form of the lip occurs in these plants a most material alteration in the form of the labellar processes, and both are connected with other peculiarities in the appearance of the two species.

The variations that experience tells us occur in the structure of the same species of tropical Orchidaceæ are principally in colour and size, just in fact as happens in those of Europe. Our common wild Orchises have purple or white flowers indifferently, and in some specimens they are much larger than in others, as is more particularly shewn by Orchis latifolia. Just so with the epiphytes. The flowers of Catasetum tridentatum, for instance, are spotted or quite green, and much larger in some varieties than in others. The well known Oncidia ornithorhynchum and ampliatum exhibit great differences in the size and depth of colour of their flowers; so do Lycaste Skinneri and Cattleya Forbesii; indeed, if one can judge from Mr. Hartweg's collection, it is very common for species found on the west of the Cordilleras to have much smaller flowers than when they occur on the east side. As to differences in form again, the greater or less breadth of the petals and the lobes of the lip is very uncertain in Orchis militaris and its allies; and in the same way Catasetum tridentatum, Cyrtochilum maculatum, and other epiphytes differ among themselves. But so far as is yet known, there is nothing peculiar in the tendency to variation among tropical Orchidaceæ, beyond what we find in all other plants, with the exception, as we have already said, of the masquerading species of Catasetum and Cycnoches.

The plant now figured was drawn in July 1842, from a specimen, received by Messrs. Loddiges from Mr. Cuming, who found it at Sincapore. Fig. 1. represents the lip and its peculiarities.

It should be grown in turfy heath-mould, of rather closer texture than that commonly used for Orchidaceous plants. The pot should be well drained, in order that all superfluous water may pass off freely, otherwise the roots will perish. Like some other species of the genus, this requires an ample supply of water at all times; and the atmosphere to be kept as moist as possible, especially during the growing season. To prevent the leaves from being scorched, the house should be slightly shaded in sunny weather. In summer the temperature should never be allowed to rise much above 80° by day, nor to fall below 68° at night; but in winter it should never be raised higher than 64° by artificial means.

29

ERIA bractescens.

Long-bracted Eria.

GYNANDRIA *MONANDRIA.*

Nat. ord. ORCHIDACEÆ. § MALAXEÆ.

ERIA. Botanical Register, vol. 11. *fol.* 904.

§ Tonsæ; floribus calvis v. parum pubescentibus.

E. *bractescens* (Lindl. in Bot. Reg. 1841. misc. 46.); pseudobulbis brevibus oblongis apice subdiphyllis, foliis oblongis undulatis basi angustatis racemis subæqualibus, bracteis membranaceis coloratis margine revolutis superioribus linearibus reflexis, racemis erectis, labelli trilobi lamellis duabus abbreviatis intermediâ productâ lobo medio truncato rugoso obtusè apiculato.

Among the extensive genus Eria we find a few species particularly distinguished by their short fleshy stems, and the membranous coloured bracts which accompany their hairless flowers. Of these the best known are the present species, *longilabris, obesa,* and a Philippine plant that may be called *ovata.* They are natives of the hotter parts of India, and are so much alike that an incautious observer might almost regard them as varieties. They are, however, most truly distinct, as the following definitions of them will shew.

1. E. *bractescens.*

Mr. Cuming found this at Sincapore, and Mr. Griffith in Burma, near Moulmain. It has a fleshy oblong stem, which bears at the summit two or three leaves, from one and a half to two inches broad, and gradually tapering to the base. Its flowers are in the Sincapore plant greenish white, with a lip crimson except at the end; in the Burma plant they are more straw colour than green. The lip is three-lobed, has an abruptly truncated extremity, and is marked with three elevated ridges, of which the two side ones are very short, while the middle one reaches to the end of the lip. Fig. 1. shews this structure, and fig. 2. the pollen-masses.

2. E. *longilabris* (Lindl. in Bot. Reg. 1841. misc. 69.); Eriæ bractescentis facie sed sepalis petalisque magis acuminatis, labelli trilobi lamellis tribus ad apicem ferè productis æqualibus lateralibus abbreviatis lobo medio ovato acuminato.

This is a native of Panay in the Philippines, whence it was sent to Messrs. Loddiges by Mr. Cuming. It is very like Eria bractescens, but is a much finer species, and bears more flowers. It is distinguished at once by its lip, which is not truncate, and has three equal wavy ridges prolonged almost as far as the tip of the middle lobe, which is long and acuminated. The form of the lip is shewn at fig. 3.

3. E. *obesa* (Lindl. in Wall. Cat. no. 1976. Gen. & Sp. no. 15.); foliis . . . ., caulibus crassis ovalibus, sepalis petalisque lanceolatis acutis subglabris, labelli trilobi ecallosi lobis lateralibus obsoletis rotundatis: intermedio ovali retuso, racemis paucifloris pubescentibus, bracteis ovato-lanceolatis pedicello capsularum longissimarum vix æqualibus.

This was originally taken up from imperfect Martaban specimens in Dr. Wallich's herbarium. It was afterwards met with at Moulmain and Merquy by Mr. Griffith; always however without leaves. The pseudo-bulbous stems are about twenty-seven inches long; the bracts ovate, reflexed, greenish dull purple; the flowers white with a tinge of pink, and a yellow lip; they are arranged in short spreading racemes.

4. E. *ovata;* foliis oblongis obtusis basi angustatis, racemis pubescentibus multifloris, bracteis parvis ovalibus reflexis, sepalis petalisque acuminatis, labello ovato-oblongo indiviso basi bicalloso.

This plant, found in the Philippines by Mr. Cuming, evidently differs from the three others in the shape of the lip, which has no lobes, but an ovate-oblong form and a couple of little diverging plates near the base. It is nearest to E. longilabris in general appearance.

All these should be potted in turfy heath-mould, mixed with a few pieces of potsherds. Water should be liberally given during the growing season, and the atmosphere kept as humid as possible. In sunny weather the house should be slightly shaded, for although this plant succeeds well in a high temperature, it is soon injured by the rays of the sun. In winter very little water is required, and where steam cannot be admitted a slight syringe over head will be sufficient for two or three months.

30

LÆLIA majālis.

The May-flower Lælia.

GYNANDRIA *MONANDRIA.*

Nat. ord. ORCHIDACEÆ. § EPIDENDREÆ.
LÆLIA. Lindl. supra, 1839. *misc.* 42.

§ 1. Grandifloræ; *Petals distinctly larger than the sepals.*

L. *majalis;* folio angusto scapo bifloro æquali, bracteis membranaceis ovatis, sepalis lanceolatis, petalis oblongo-lanceolatis duplò latioribus, labelli lobo medio rotundato emarginato plano lateralibus parvis obtusis.

L. majalis, *Lindl. in Bot. Reg.* 1839. *misc.* 42. *Bateman, Orch. Mex. & Guat. t.* 23.

Cattleya Grahami, *Lindl. Gen. & Sp. Orch. p.* 116.

The history of this charming plant is so well given by Mr. Bateman, that we beg to extract it from his magnificent work on the Orchidaceæ of Mexico and Guatemala.

"This lovely plant abounds in the most temperate parts of Mexico, where its exquisite beauty has rendered it a prime favourite with the natives, from whom it has received the familiar appellation of 'Flor de Mayo.' It does not however appear to have been long known to botanists, Dr. Schiede having been the first to send specimens to Europe, which, through the liberality of Professor Schlechtendahl have been extensively distributed. Living plants were first obtained by Mr. Barker from Oaxaca through the instrumentality of Mr. Ross: and more recently, a large supply has been received by the Horticultural Society of London, collected by Mr. Hartweg in San Bartolo, in situations so elevated, that the temperature sometimes falls below the freezing point. This habitat, so unusual for an Orchidaceous plant, will go far to explain the ill success that has hitherto attended its cultivation, for while it is comparatively easy to imitate the close and humid atmosphere in which most of the tribe are found, it is infinitely more difficult to provide a substitute for the pure air and frequent changes of temperature in which these mountain epiphytes would seem to delight. Indeed, so signal, in the case of Lælia majalis, has been failure of even the most experienced cultivators, that although there was scarce a collection that did not contain one or more specimens of the plant, still did it obstinately refuse to flower, except in the solitary instance about to be described, when it yielded to the skilful treatment of Mr. Dillwyn Llewelyn, of Penllergare, who has favoured us with the following note:—

"'This plant was purchased,' says Mr. Llewelyn, 'about three years since from M. Deschamps, and soon afterwards potted in very rough fibrous peat, being kept nearly dry in a cool plant house, until its new buds began to swell, when it was removed to a hot and damp stove, kept exclusively for Orchidaceæ. Water was regularly given until its buds had acquired their full size, when it was discontinued by degrees, and the plant carried back to a lower temperature. This treatment has been repeated, and under it the entire genus seems to thrive, although, perhaps, from the vigour of some specimens of L. anceps and L. autumnalis, which I have attached to pieces of rough barked wood (with a little peat tied round the root) I should prefer that course of treatment to the more usual one of potting them.'

"The mode pursued by Mr. Llewelyn is certainly the only one under which success can be expected; but there is yet a point to be gained, for while the specimens at Penllergare, however beautiful, consisted of only a solitary flower, in a wild state three or four are borne upon a spike."

The method of managing the plant in the Gardens of the Horticultural Society is as follows.

It is tied to a block of wood and suspended to a rafter in a moist stove. When removed from one block to another, which is sometimes necessary, it will be found beneficial to place a slice of turfy peat or a small quantity of sphagnum between the wood and the roots of the plant, this retains moisture in summer so that less syringing is required. The quantity of water necessary for it on the bare block of wood would prove injurious to plants in pots below it. In the growing season the temperature does not exceed 80° by day nor does it fall below 70° at night. In winter 56° with fire heat is quite sufficient.

34

EPIDENDRUM pterocārpum.

Wing-fruited Epidendrum.

GYNANDRIA *MONANDRIA.*

Nat. ord. ORCHIDACEÆ. § EPIDENDREÆ.
EPIDENDRUM. Bot. Reg. 1838. *fol.* 53.

Subg. Encyclium. § III. *Labello trilobo.*

E. *pterocarpum* (Lindl. in Hooker's Journal, 3. 82. Bot. Reg. 1841. misc. 128.); pseudobulbis ovalibus compressis subdiphyllis, foliis ensiformibus acutis, racemo angusto, sepalis petalisque subæqualibus linearibus acuminatis patentibus, labelli subrotundi trilobi cordati laciniis lateralibus rotundatis intermediâ multò longiore acutâ basi bilineatâ callo pubescente obscurè tridentato auctâ, capsulâ ovatâ tri-alatâ.

We cannot say much as to the beauty of this plant, which flowered with Messrs. Loddiges in December 1842. It however presents a remarkable instance of the formation of broad wings by the fruit, which, when ripe, has three of them of considerable size.

The flowers are a brownish green, with a broad, rolled up yellow variegated lip whose middle lobe is whitish. It is a native of Mexico, and closely allied to E. tessellatum.

Fig. 1. represents the column and lip.

37

CALANTHE Masūca.

Lilac Calanthe.

GYNANDRIA *MONANDRIA.*

Nat. ord. ORCHIDACEÆ. § VANDEÆ.
CALANTHE. Botanical Register, t. 720.

C. *Masuca* (Lindl. Gen. & Sp. Orch. p. 249. *Bletia Masuca*, Don. Prodr. fl. nep. 30. *Amblyglottis veratrifolia?* Blume Bijdr.); scapo erecto foliis latis oblongis petiolatis acuminatis subtus pubescentibus longiore, racemis multifloris, labello tripartito basi tuberculis seriatis 5-cristato, seriebus intermediis elevatioribus; laciniis lateralibus linearibus subfalcatis intermediâ multò majore cuneatâ emarginatâ in unguem linearem laciniis lateralibus æqualem angustatâ, calcare longissimo falcato clavato, columnâ brevi obliquâ anticè bifoveatâ, ovario pubescente.

For the figure of this charming plant we have to thank Messrs. Rollisson of Tooting, with whom it flowered in June, 1843.

It is a native of Nepal, and may be regarded as the gayest of the terrestrial species of that rich country, being conspicuous both for the large size of their flowers and their fine colour.

From the other purple species allied to it, this is readily known by the leaves as well as by the flowers. *C. versicolor* has leaves smooth on both sides; *C. purpurea* downy on both sides, especially beneath; while this has down only on the under side. *C. versicolor* has white sepals and petals; *C. purpurea*, and this, purple ones. While however *C. purpurea* agrees in the colour of its flowers, its lip is altogether different, being very narrow, with the lateral lobes quite round. In order that there may be no confusion among these nearly allied plants, the characters of C. purpurea and versicolor are given below.

C. Masuca should be potted in turfy heath-mould, mixed with a few pieces of potsherds. In summer it should receive an ample supply of water at its roots; and where it can be avoided, little should be allowed to fall on its leaves, otherwise the young shoots will damp off. It enjoys a humid atmosphere and a high temperature; but as the leaves are very delicate, they will soon become scorched if shading is not carefully attended to. In winter little water will be required; still it is necessary to keep the soil damp enough to preserve the bulbs from shrivelling. This is one of the most difficult of Orchidaceous plants to grow well.

C. *purpurea* (Lindl. Gen. & Sp. Orch. no. 2); scapo erecto foliis latis oblongis petiolatis acuminatis supra pilosis subtus tomentosis longiore, racemis laxis multifloris, labelli trilobi lobis lateralibus nanis rotundatis patentibus intermedio angusto cuneato divergenti-bilobo, calcare ovario vix breviore.

C. *versicolor* (Lindl. Sert. Orch. t. 42); foliis oblongo-lanceolatis erectis concavis 7-9-nerviis utrinque glaberrimis scapo apice pubescente brevioribus, racemo denso pyramidato, labelli columnæ brevi accreti trilobi lobis lateralibus ovatis nanis intermedio cuneato bilobo multò majore basi trituberculato secus lineam mediam verrucoso, calcare glabro ovarii pubescentis longitudine.

39

ODONTOGLOSSUM læve.

Smooth-lipped Odontoglossum.

GYNANDRIA *MONANDRIA*.

Nat. ord. ORCHIDACEÆ. § VANDEÆ.
ODONTOGLOSSUM. *Botanical Register*, 1840. *fol.* 30.

O. *læve*; pseudobulbis compressis sulcatis, foliis oblongo-ensiformibus obtusis apice obliquis, floribus paniculatis, bracteis laxis membranaceis, sepalis petalisque oblongo-linearibus acutis planis, labelli laminâ panduriformi apiculatâ ungue lævi obsoletissimè bidentato, columnæ alis apice rotundatis crispis basi planis.

Both Mr. Skinner and Mr. Hartweg found this plant abundantly in Guatemala, and sent it home to their friends, so that it is by no means rare in collections. It is not, however, much esteemed, because of the flatness of colour in its flowers. Nevertheless, they have cinnamon-brown blotches on the yellow ground of the petals and sepals; and the lip, which is white, is banded with violet across the middle; besides which they are fragrant.

The name Odontoglossum, or Tooth-tongue, has been given to these plants, because they are usually furnished with strong teeth on the lower part of the lip, and these teeth are not unfrequently very conspicuous. Here, however, they are reduced to a couple of trifling notches, and might be easily overlooked.

Our drawing was made in the garden of the Horticultural Society in May, 1842. Fig. 1. represents the column, seen in front, with its wings.

This species should be potted in turfy peat, and placed in a moist stove. A liberal supply of water should be given during the growing season, and the house be slightly shaded in sunny weather, in order to keep the temperature as near 80° by day as possible. In winter, like many other bulbous Orchidaceous plants, it requires very little water for a few weeks, but would be much benefited (especially in clear weather) if the house in which it is grown could be filled with steam once a day. The temperature should at that time never be raised above 60° by fire heat.

41

AERĬDES virens.

Green-leaved Air plant.

GYNANDRIA *MONANDRIA*.

Nat. ord. ORCHIDACEÆ. § VANDEÆ.
AERIDES. *Supra*, 1842. *fol.* 55.

A. *virens* (Lindl. in Bot. Reg. 1843. misc. 48); foliis latis obliquè retusis, racemis pendulis multifloris, sepalis petalisque obovatis obtusis, labelli cornu acuminato ascendente lobis lateralibus apice denticulatis intermedio lanceolato medio canaliculato versus apicem denticulato.

This is a beautiful addition to that set of Aerides of which A. odoratum was the first discovered. Like the flowers of that species, these are deliciously and very peculiarly sweet-scented, and not at all inferior in size.

Each sepal and petal has a deep purple blotch at the end, while the remainder is a delicate soft French white. The lip is speckled with crimson, and bears in the middle an inflated, sanguine, serrated tongue; as is represented at fig. 1.

The leaves are much alike in all these plants, but here they are of a peculiarly bright green, which circumstance has suggested the name.

Our drawing was made in April, 1843, in the stove of Messrs. Loddiges, who imported it from Java.

45

EPIDENDRUM radiātum.

Ray-flowered Epidendrum.

GYNANDRIA *MONANDRIA*.

Nat. ord. ORCHIDACEÆ. § EPIDENDREÆ.
EPIDENDRUM. *Supra*, 1838. *fol.* 53.

§ OSMOPHYTUM. Caulis pseudobulbosus v. fusiformis apice foliosus. Flores racemosi. Labellum adnatum, sæpius indivisum et cochleatum. *Lindl. in Hook. Journ* 3. 81. *Bot. Reg.* 1842. *sub folio* 50.

E. *radiatum* (Lindl. in Bot. Reg. misc. 1841. no. 123); pseudobulbis ovatis compressis utrinque tricostatis triphyllis, foliis angustis elongatis acutis, racemo denso multifloro, ovariis trialatis, sepalis linearibus petalisque lanceolatis patentissimis, labello cochleato crenato crispo, clinandrio trilobo laciniis lateralibus erectis carnosis intermediâ truncatâ apice serratâ.

Allied to both *E. cochleatum* and *lancifolium*, this pretty species is distinctly known by the round form and crisp margin of its lip, which moreover is almost notched out in the middle, and very delicately streaked with purple. Its petals too are both shorter and broader than in either of those two species.

Its habit is precisely theirs; and the three, considering their constant flowering and very neat appearance, deserve a place in all collections. Besides, this species smells deliciously of cinnamon.

The accompanying figure is from a plant in the garden of the Horticultural Society, received from Mexico from Mr. Hartweg. We also possess specimens gathered in that country on rocks, at the Hacienda de la Laguna, by Dr. Schiede.

It should be potted in turfy heath-mould, mixed with small pieces of potsherds. During the growing season an ample supply of water should be given, and the atmosphere kept as moist as possible. In summer when the sun is strong, the house should be slightly shaded, in order to keep the temperature as near 80° by day as possible, without admitting much air. For a few weeks in winter little or no water should be given, providing the house can be filled with steam once a day.

50

* LACÆNA bicŏlor.

Two-coloured Lacæna.

GYNANDRIA *MONANDRIA.*

Nat. ord. ORCHIDACEÆ. § VANDEÆ; MAXILLARIDÆ.

LACÆNA, Lindl. Perianthium carnosum patens, *sepalis* subæqualibus basi subconnatis. *Petala* conformia minora. *Labellum* cum columnâ et in medio articulatum, hypochilio unguiculato cuneato bilobo apice pulvinato, epichilio integro. *Columna* erecta semiteres. *Pollinia* 2, posticè fissa, *caudiculâ* setaceâ, *glandulâ* minutâ.—*Herba,* Peristeriæ *habitu.*

L. bicolor, *Lindl. in Bot. Register,* 1843. *misc.* 101.

This fine plant was sent to the Horticultural Society from Guatemala by Mr. Hartweg, who found it in the mountains of Salama, in the province of Vera Paz, also near the village of Sunil, near Quezaltenango, in Guatemala, growing on rocks, at an elevation of about 7000 feet above the sea. In May, 1843, it produced its flowers, and ever since has been in such bad health that it is doubtful if it will survive.

In habit the species is so like some of the Peristerias with pendulous racemes, that it would certainly be mistaken for them; and in structure it approaches them no doubt very closely. It is, however, essentially distinguished by the labellum and pollen apparatus. In Peristerias the labellum, although having a moveable joint in the middle, is continuous with the column by a thick fleshy base; here the articulation at that part is very conspicuous. The Peristerias have four pollen-masses; here we have but two. And, finally, the strap by which the pollen-masses cohere is long and slender, like that of a Lycaste, not deficient, so as to render the pollen-masses sessile upon a crescent-shaped gland as in Peristerias.

* This was one of the names of Helen, and may be applied to this plant because of its beauty: but it may also be derived from λακις, a cleft, in allusion to the divisions of its lip.

The flowers of this plant are formed in a pendulous raceme, as much as eighteen inches long; the lower part being clothed with distant short blunt scales, which extend into bracts about half the length of the pedicels. Each raceme consists of nine or ten flowers of a dull greenish yellow, covered externally with short hairs; their petals have three streaks of violet; their lip is downy all over the upper surface, dark purple in the middle, with two or three spots of the same colour near the point. In the centre of the lower half is a large shaggy hump.

Fig. 1, represents the column and petals; 2, the labellum spread open; 3, pollen-apparatus.

It may be grown in a wire basket like a Stanhopea, or tied to a block of wood and suspended to a rafter, in a moist stove. If potted in the usual way, the flower-spike will be apt to go down into the soil and perish. Like many other Orchidaceous plants it requires an ample supply of water during the growing season, and shade in sunny weather, at a temperature between 80° and 90° by day, but not above 70° at night. Towards the end of October water should almost be withheld for a few weeks, no more being given than will prevent the pseudo-bulbs from shriveling.

51

EPIDENDRUM verrucŏsum.

Warted Epidendrum.

GYNANDRIA *MONANDRIA.*

Nat. ord. ORCHIDACEÆ. § EPIDENDREÆ.

EPIDENDRUM. Supra, 1838. *fol.* 53.

E. *verrucosum* (Encyclium, § labello trilobo); pseudobulbis ovatis, foliis ensiformibus obtusis, scapo pedicellis ovariisque verrucosis, racemo nutante, sepalis petalisque lineari-lanceolatis acuminatis, labelli trilobi lobis lateralibus subfalcatis acutis nanis intermedio ovali crenulato basi serrato bilamellato, columnæ alis 2 nanis truncatis.

This beautiful and fragrant plant has been obtained from Mexico by Messrs. Loddiges with whom it flowered in July last. Among all the Encycliums not more than two or three are superior to it in attractiveness.

It is not very nearly related to any kinds hitherto discovered: belonging to the same set as E. tessellatum and Candollii, from all which its stems and flower-stalks, closely covered with minute asperities, and its rich crimson flowers, which are as much as three inches in diameter, readily distinguish it. Its closest affinity is perhaps with E. phœniceum, a native of Cuba, and figured in the Sertum Orchidaceum, t. 46, and *E. Hanburii,* mentioned in this work at no. 60 of the miscellaneous matter of the present year; but both those plants have the middle division of the lip 2-lobed; and they are not, that we are aware of, fragrant.

Fig. 1. represents a front view of the column; 2. the labellum spread open.

This plant should be potted in turfy heath-mould, mixed with a few pieces of potsherds. When this operation is performed, care should be taken to keep the pseudo-bulbs always above the soil. In spring and summer, while the plant is in a growing state, an ample supply of water should be given, and the temperature allowed to rise as high as 80° or 85° by day, but not above 70° at night. In winter very little water will be required, providing a humid atmosphere can be maintained. The temperature should never be raised above 66° by fire heat.

53

DENDROBIUM compressum.

Flat-stemmed Dendrobium.

GYNANDRIA *MONANDRIA.*

Nat. ord. ORCHIDACEÆ. § MALAXEÆ—DENDROBIDÆ.
DENDROBIUM. Supra, vol. 15. *fol.* 1291.

Sect. DENDROCORYNE (*Lindl. in Bot. Reg.* 1842. *misc.* 76); caulibus clavatis, sæpius apice tantum foliatis.

D. *compressum* (Lindl. in Bot. Reg. 1842. misc. 76); caule obovato compresso 2-6-phyllo, foliis ovalibus acutis striatis basi dilatatis membranaceis amplexicaulibus, racemis subquadrifloris cernuis, sepalis petalisque ovatis erectis, cornu elongato obtuso, labello cuneato lævi per axin sulcato.

This singular species was discovered in the island of Ceylon by Mr. Nightingale, and by that gentleman sent to his Grace the Duke of Northumberland, from whose collection at Syon we were permitted to obtain a figure in August 1842. The year of its importation was 1840.

The curious flattened stems are not more than three or four inches long, and resemble those of no species hitherto discovered. They appear however to bring the plant into a section of the genus to which the name Dendrocoryne (κορυνη a club) may be applied, and which will also contain D. *densiflorum, tetragonum, Griffithianum,* and *Macræi.*

The singular form of the labellum, fig. 1, which is that of a wedge drawn out at the point, and furrowed along the middle, distinguishes this plant, independently of its habit.

It should be potted in turfy heath-mould. The pot in which it is grown, should be half filled with potsherds, in order that all superfluous water may pass off; the soil must be considerably elevated above its brim. During the growing season an ample supply of water should be given, and the atmosphere kept as moist as possible. This, like many other Orchidaceous plants, requires to be shaded in sunny weather, to prevent the leaves from being scorched, as well as to keep the temperature about 85° by day, without admitting much air. In winter a humid atmosphere is necessary, but for a few weeks water should only be given to prevent the plant from shriveling.

60

*ANGULŌA uniflōra.

One-flowered Anguloa.

GYNANDRIA *MONANDRIA.*

Nat. ord. ORCHIDACEÆ. § VANDEÆ. || MAXILLARIDÆ.

ANGULOA, Fl. Peruv. prodr. 118. *t.* 26. *Flores* subglobosi, nunquam patentes. *Sepala* lateralia invicem imbricantia, basi valde convexa, nec in cornu producta; alterum nunc anticum nunc posticum, conforme, basi planum. *Petala* sepalo dorsali æqualia, et similia. *Labellum* coriaceum, unguiculatum, subconvolutum, trilobum, laminâ carnosâ latâ planâ supra medium auctum, hinc quasi bilabiatum. *Columna* teres, clavata, libera; *clinandrio* nunc mutico, nunc laciniâ acutâ porrectâ utrinque aucto. *Anthera* galeata, valvis membranaceis nunc in lacinulas acutas productis. *Pollinia* 4, plana, inæqualia, *caudiculâ* longâ lineari, et *glandulâ* acutâ.——Herbæ *epiphytæ Granatenses* & *Peruvianæ,* Lycastes *facie.*

A. *uniflora;* pedunculo unifloro radicali squamis 2 inflatis imbricatis vaginato, flore bracteâ spathaceâ herbaceâ parum longiore, sepalis lateralibus anticis petalisque acuminatis, labelli glabri lobo medio angustissimo reflexo lateralibus rotundatis, laminâ appendiculari retusâ duplò latiore, clinandrio lacinulis 2 acutis aucto.
A. uniflora, *Fl. Peruv. syst. p.* 228 *Lindl. gen. & sp. Orch. p.* 160.

We some time since announced the appearance in this country of a new species of the long lost genus Anguloa, concerning which so many mistakes have been made. We are now able, by the kindness of Mr. Barker of Birmingham, to publish the very Anguloa uniflora itself, which he was so fortunate as to flower in April last. We believe he received it among Linden's collection in Columbia; according to Ruiz and Pavon it is found in precipitous places about Muña and Chincao in Peru, and profusely in the woods of Tarma, where it is called *Carpales.*

It is a sweet-scented plant, with large white flowers slightly tinged with yellow, and the whole habit of a Lycaste, to which genus this approaches very nearly. In what respect it differs we shall endeavour to show next number, upon figuring Anguloa Clowesii; a species remarkably different in some respects.

Fig. 1. represents a section of the labellum, with the plate that lies over the middle cut in two. Fig. 2. shows the column with the two thin sharp-pointed processes which project forward; it will be remarked, upon comparing this figure with that in the centre of the flower, that there are in the latter two additional plates, looking like lappets. Those bodies were missing in the first flower that we saw, and were only present in the second; they belong to the sides of the anther, and are, it is to be supposed, accidental developments. Fig. 3. shows the pollen-masses, with their caudicle and glands.

* Dedicated to Don Francisco de Angulo, director-general of the mines (of Peru), and greatly attached to Botanical pursuits, (*Ruiz & Pavon.*)

63

ANGULŌA Clowēsii.

Mr. Clowes's Anguloa.

GYNANDRIA *MONANDRIA.*

Nat. ord. ORCHIDACEÆ. § VANDEÆ—MAXILLARIDÆ.
ANGULOA, Supra t. 63. *hujus voluminis.*

A. *Clowesii* (Lindl. in Bot. Reg. 1844. misc. 29.); pedunculo unifloro radicali laxè squamato, flore carnoso resupinato, sepalis petalisque ovatis convexis conniventibus, labelli trilobi lobo medio piloso infundibulari bilabiato: labio altero emarginato altero tridentato, columnâ integrâ.

Among a lot of plants collected in Columbia by Linden in 1842, this fine thing was received by the Rev. J. Clowes of Broughton Hall near Manchester, with whom it flowered for the first time in Europe in March, 1844. The old plants had as many as five flower-scapes to a pseudo-bulb. Thus, by the enterprise of a few English gentlemen, who had the spirit to defray the cost of a collector in the rich country once known as the kingdom of New Granada, has one of the greatest botanical puzzles been at last obtained. Mr. Barker's species was figured last number, that before us is from Mr. Clowes, and a third species, with spotted flowers, still undescribed, has blossomed with Mr. Rucker, all of whom were the supporters of Mr. Linden's expedition.

The two characters upon which the authors of the Flora Peruviana must be considered as having most relied for the distinction of their genus Anguloa, were the "chrysalis-shaped" lip, and the two-horned column.

But that rolled up form of the lip, to which the name chrysaloid was applied, is in reality common to the greater part of the Maxillaridous division; and the two horn-like processes which distinguish the column of the original Anguloa are altogether of too doubtful a nature to be relied on for generic distinction; for they have no relation to such organs as the wings of Oncidium, and it is by no means certain that they are constantly present in the same species. If, as seems probable, they are analogous to the cirrhi of a Catasetum, experience warns us to distrust their importance.

This being so, it becomes a question whether the genus is distinct from Lycaste, to which it undoubtedly approaches very nearly. The pollen-masses and gland of the two, although dissimilar, if A. Clowesii is compared with Lycaste Deppii, are nevertheless not so different when A. uniflora is the subject of comparison. The funnel-shaped condition of the middle lobe of the lip is at first sight peculiar to Anguloa, but it is in reality only an exaggerated condition of that kind of lip which we have in L. aromatica and its allies, in which there is a large flat appendage resting on the surface of the lip; the main difference consists in that appendage being attached to the lip at the base only, while in Anguloa it is united by the sides also. This, however, is a difference which may be regarded as available for generic distinction. The main difference, however, between *Anguloa* and *Lycaste* consists in this; that in *Lycaste* the lateral sepals are placed edge to edge in the manner of a true *Maxillaria*, but in *Anguloa* they overlap each other very considerably; this peculiarity causes a striking difference in the appearance of the flowers of the two genera, and, in fact, gives that of Anguloa somewhat the look of a Mormodes.

What Anguloa squalida of Pöppig may be, I cannot say. I have never seen the plant, and the barbarous analyses given by that author preclude all hope of coming to any conclusion about it. It may, however, be safely asserted that no such plant exists as is represented by Mr. Pöppig.

Fig. 1. represents the lip of this plant cut through the axis, to shew its funnel-shaped structure; 2. is the pollen apparatus, two of the pollen-masses being half cut away.

MISCELLANEOUS MATTER

OF THE

BOTANICAL REGISTER.

1844.

1. CŒLOGY̆NE fuscescens.

Lindl. Gen. & Sp. Orch. p. 41.

A plant of this rare species flowered at Sion in the end of last November. Its blossoms are perhaps the largest in the genus, but want the brilliancy necessary to render them very striking. They are of a pale greenish yellow, without any markings except on the lip, which has a few brown spots towards the base, a broad brown band along the middle of each side lobe, and three vermilion coloured streaks in the centre. We fear that brighter suns than those of November will hardly destroy the green sufficiently to render this comparable with some species, notwithstanding its large flowers.

2. LÆLIA virens.

L. *virens;* sepalis suberectis ovatis petalisque lanceolatis subæqualibus, labello oblongo obsoletè trilobo cucullato apice ovato crispo lineâ obsoletâ elevatâ versus basin, columnæ cardine unidentato.

A Brazilian plant, of which I have only seen a single flower. Mr. Loddiges, whose number 647 it is, informs me that it has quite the habit of Cattleya crispa. The flowers are very pale yellowish green, of no beauty, and about the size of Maxillaria alba, which they are something like. The number of pollen-masses is certainly eight, which makes the plant a Lælia and not a Cattleya.

3. WARRĔA cyanĕa.

W. *cyanea;* spicâ brevi, bracteis ovarii longitudine, sepalis ovatis acutis, petalis subconformibus, labello subrotundo-cuneato apiculato undulato lineis quinque elevatis.

The purest blue known in the vegetable kingdom colours the lip of this beautiful flower, which is otherwise white. It has quite the habit of Warrea tricolor, but is very much smaller in all its parts. Its most distinctive character is found in the form of its lip, which has a distinct point, and five ribs, not three, near the base. Messrs. Loddiges imported it from Columbia, and it is no. 860 of their last catalogue.

4. HEXADESMĬA crurigĕra.

(Hexopia crurigera, *Bateman mss.*)

H. *crurigera;* caule fusiformi, foliis exacte linearibus, racemis flexuosis paucifloris, bracteis ovatis membranaceis, mento valdè producto, labello obovato altè bilobo.

A small-flowered inconspicuous Epiphyte, of no beauty, introduced from Guatemala to Mr. Bateman, with whom it flowered some years since, when it received the name of Hexopia. Its flowers are not half the size of H. fasciculata.

5. HEXADESMIA micrantha.

H. *micrantha;* racemo multifloro, bracteis linearibus acuminatis membranaceis, sepalis acuminatis, labelli trilobi laciniis lateralibus rotundatis intermediâ apiculatâ.

The flowers of this are very much smaller than even those of H. crurigera, and white with a very little green. It is from Guatemala, and is the no. 389 of Messrs. Loddiges' catalogue.

9. MAXILLARĬA Meleāgris.

M. *Meleagris;* caulescens; pseudobulbis ovalibus ancipitibus, foliis solitariis angustis obtusis emarginatis undulatis, sepalis petalisque acuminatis, labelli oblongi lobo medio subemarginato sub apice carinato lateralibus nanis duplò longiore, tuberculo transverso subcrenato basi 3-costato.

A caulescent species of little beauty. Its flowers are spotted with purple upon a yellowish white ground; the lip is dark purple. Our specimen is from Messrs. Loddiges, who imported it from Oaxaca. We have also received it from Mr. Brocklehurst's garden at the Fence.

11. ONCIDIUM oblongatum.

O. *oblongatum;* facie et structurâ O. reflexi; sed petalis oblongis obtusis apiculatis patulis planis, labello altè bilobo, columnæ alis rotundatis crenulatis dimidiatis.

It is almost impossible to find words that will well distinguish the Oncidiums allied to O. reflexum; and yet they appear to merit distinction. This, which is from Messrs. Loddiges, has very much the appearance of O. Wentworthianum, with the same clouded pseudo-bulbs and speckled flower-stem; but its petals are perfectly flat, not at all reflexed or wavy, obtuse with a little point and clear yellow, with a few reddish brown bars near the base. It is a very handsome species, on account of its flowers having as much yellow in them, and being as large, as the best variety of O. sphacelatum.

12. MAXILLARIA concāva.

M. *concava;* acaulis, racemosa, pseudobulbis oblongis altè sulcatis, foliis geminis 3-costatis nitidis in petiolum angustatis, bracteis setaceis, sepalis lateralibus falcatis acuminatis sub apice carinatis, petalis duplo minoribus obtusis apice plano-convexis, labello oblongo obsoletè trilobo apice carnoso concavo rotundato extùs scabriusculo, tuberculo lineari apice obtusè 3-dentato.

At first sight this species might be mistaken for M. *bractescens;* but its flowers are smaller; the bracts very small and setaceous, and the lip of quite another form. It has little beauty, and was found in Guatemala by Mr. Hartweg, by whom it was sent to the Horticultural Society in 1841.

13. BRASSIA.

This genus consists of American Epiphytes, with pseudo-bulbs and a radical spiked inflorescence, of which many species are in our gardens. Its great features are a very dwarf wingless column, a sessile flat undivided labellum, with two (or occasionally more) short plates at its base, and a spreading perianth. Recent discoveries have considerably enlarged it, so that from consisting of a single species, as it did when first defined in 1813, or of two, as was the case when the Genera and Species of Orchidaceæ were published in 1833, we have now fourteen well defined and clearly known. Of these the following is an enumeration:

1. B. *Lanceana* (Lindl. in Bot. Reg. t. 1754.); sepalis lateralibus elongatis labello oblongo undulato acuminato subrepando vix duplò longioribus: callo baseos simplici canaliculato truncato pubescente dentibus 2 liberis membranaceis in fronte.——*Surinam.*——A charming species, with bright yellow flowers spotted with lively brown, and very sweet-scented.

2. B. *Lawrenceana* (Lindl. in Bot. Reg. 1841. misc. 6. t. 18.); sepalis lateralibus elongatis labello oblongo apice lanceolato subundulato plus duplò longioribus: callo baseos simplici canaliculato truncato pubescente.——*Brazil?*——This plant is I fear only a variety of the last, and is perhaps from the same country, for there is no proof of its being Brazilian, as is stated. The principal differences that distinguish it from B. Lanceana are, the lateral sepals being more than twice as long as the lip, and the want of free extra tubercles in front of the callus at the base of the lip. The flowers are pale yellow, faintly spotted with brown, and very sweet-scented.

3. B. *angusta;* pseudobulbis oblongis ancipitibus, foliis binis oblongis sessilibus subundulatis racemi longitudine, sepalis linearibus acuminatis lateralibus caudatis, petalis linearibus acuminatis, labello lineari-lanceolato acuminato lamellis baseos connatis pubescentibus anticè denticulo auctis.——*Brazil.*——The first knowledge I had of this plant was derived from a specimen sent by Mr. W. Masters of Canterbury, who had received it from Brazil; at that time I took it for a starved specimen of B. Lawrenceana. Since that time it has flowered in the garden of the Horticultural Society, with the same peculiarity of structure; and hence I conclude it to be a distinct species. It is readily known by its having extremely narrow sepals and lip, which are of a pale dull yellow, with little marking, except a few brown spots at the base of the sepals, petals, and lip. In this respect it varies, for in some cases the spots at the base of the petals are few and distinct, in others they are run together into one brown stain.

4. B. *macrostachya* (Lindl. Sertum. Orch. t. 6.); pseudobulbis compressis margine obtusis 2-3-phyllis, scapo nutante multifloro, sepalis linearibus acuminatis lateralibus longissimis, labello oblongo-lanceolato acuminato petalis longiore: lamellis baseos liberis villosis dentibus tribus liberis in fronte.——*Demerara.*——No species can be compared with this most graceful and brilliant plant, whose long nodding racemes of flowers bend gently over the rich and verdant foliage, while the slender sepals are so long, so light, and so delicate as to be agitated by every impulse given them by the air. They are of a clear golden yellow, slightly spotted with brown, and many degrees darker than the lip.

5. B. *caudata* (Lindl. in Bot. Reg. t. 832. *Epidendrum caudatum,* L. *Malaxis caudata,* W.); pseudobulbis compressis margine acutis, scapo erecto, sepalis ovato-linearibus acuminatis lateralibus longissimis, labello ovato acuminato petalorum longitudine, lamellis baseos liberis villosis denticulis 2 liberis in fronte.——*West Indies.*——Very nearly allied to *Br. macrostachya,* from which it is distinguished by its pseudo-bulbs being acute at the margin, not obtuse, by its flowers being smaller, greener, and much more mottled with deep brown, and by its labellum being ovate, acuminate, and the same length as the petals, not oblong-lanceolate, and longer than them. The little free teeth, too, in front of the calli at the base of the lip, are much smaller.

6. B. *bidens;* pseudobulbis angustis ancipitibus, foliis oblongo-lanceolatis scapo stricto multifloro æqualibus, sepalis acuminatis lateralibus labello duplo longioribus, labello rhombeo undulato petalis breviore: lamellis baseos lævibus rectis appendice membranaceâ bilobâ in fronte.——?——The native country of this species is unknown. I had it from the collection of Mr. R. Harrison in 1837. It is allied most nearly to B. caudata, from which it differs in the form of its lip, and in having two large teeth connected at their base in front of the calli, which are quite smooth. The lip is spotted with brown on a yellow ground. The sepals and petals appear from the dried specimen to be unspotted.

7. B. *cochleata* (Knowles & Westcott, Floral Cabinet, t. 53.); sepalis petalisque subæqualibus linearibus acuminatis, labello elongato cochleato acuminato.——*Demerara.*——Now known only from the above figure. We have seen no specimen. Possibly it is the *B. Henchmanni* of Mr. Loddiges' catalogue, or his *B. cuspidata,* (no. 1467), both of which are unknown to me.

8. B. *verrucosa* (Lindl. in Bot. Reg. 1840. misc. 66. Bateman Orch. Mex. & Guat. t. 22.); foliis scapis gracilibus duplo brevioribus, sepalis lateralibus acuminatis labello unguiculato obovato apiculato verrucoso plus duplo longioribus; lamellis baseos liberis villosis apice recurvis.——*Guatemala.*——Flowers on long slender scapes, pale green except the lip, which is white with green warts. The pseudo-bulbs are rounder at the edge, and more furrowed than is usual in this genus.

9. B. *guttata* (Lindl. in Plant. Hartweg. p. 94. *B. Wrayæ,* Bot. Mag. t. 4003.); pseudobulbis oblongis ancipitibus diphyllis, foliis oblongis obtusis racemo multifloro brevioribus, bracteis membranaceis patentibus squmæformibus, sepalis linearibus acuminatis, petalis conformibus brevioribus, labello supra basin cordato ovato crispo, lamellis baseos connatis pubescentibus edentulis apice divergentibus.——*Guatemala.*——A fine species with broad leaves, and large greenish yellow flowers spotted with green; they vary greatly in size; in Mr. Hartweg's wild specimens from the Chono they are very small, in a specimen found by Mr. Skinner at San Salvador twice as large, in the plant figured in the Botanical Magazine larger still, and in a specimen before me from the garden of the Horticultural Society they are considerably larger than even the last. In the gardens the name of Br. Wrayæ is also applied to B. brachiata, a far handsomer species.

10. B. *brachiata* (Lindl. in Plant. Hartweg. p. 94.); pseudobulbis oblongis angustis compressis diphyllis, foliis obtusis racemo multifloro brevioribus, bracteis patentibus squamæformibus, sepalis petalisque brevioribus linearibus acuminatis longissimis, labello supra basin cordato subrhomboideo acuminato crispo, lamellis baseos bidentatis obtusis.——*Guatemala.*——A most noble species, with very large flowers, having brown spots on a pale ground. In a specimen which flowered with Mr. Bateman the lower sepals were fully five inches long. The lip is much waved at the edge, and rather lobed. It is sometimes called *B. Wrayæ.*

11. B. *aristata;* racemo secundo nutante multifloro, sepalis linearibus strictis acuminatis lateralibus labello quadruplò longioribus, petalis ex ovatâ basi acuminatis aristatis, labello obovato basi angustato sub apice aristato secus medium verrucoso: lamellis baseos glabris apice liberis divergentibus rotundatis.——*Guatemala.*——Flowers the smallest in the genus, apparently whole-coloured. Very distinct in the petals, ending in bristle points, and in the presence of a distinct awn beneath the point of the labellum. Only known to me by a dried specimen from Mr. Skinner, without leaves.

12. B. *peruviana* (Pöppig & Endl. gen. & sp. 2. p. 12. t. 117.); foliis elongatis acutissimis, "sepalis anguste linearibus acuminatissimis æqualibus; floribus resupinatis, labello oblongo lanceolato acuto undulato crenulato" sepalis subæquali.——*Peru,* in dry thickets on the eastern face of the Andes, in the district of Chihuamccala, towards Cuchero.——The flowers are represented as growing in one-sided spikes, not exceeding ten in number, about an inch apart, yellowish-green spotted with purple.

13. B. *maculata* (Brown. in Hort. Kew. 5. 215. Bot. Mag. t. 1691.); sepalis petalisque linearibus acuminatis æqualibus, labello postico subrotundo.——*Jamaica.*——Sepals and petals dull olive brown, with purple blotches. Lip cream colour, very large, spotted with purple.

* * *

14. B. *Clowesii* (*Miltonia Clowesii,* Lindl. in Sert. Orchid. t. 34. *Odontoglossum Clowesii,* Lindl. in Bot. Reg. 1839. misc. 153.); pseudobulbis ovalibus diphyllis, foliis ensiformibus angustis erectis scapo longioribus, racemo paucifloro laxo, bracteis minimis setaceis, sepalis petalisque lanceolatis æqualibus, labelli cordati in medio constricti apice subrotundo acuto basi lamellis 5 inæqualibus abruptis quincuncialibus auctâ.——*Brazil.*——A fine species, with the sepals blotched with chocolate brown upon a yellow ground. Lip white at the tip, violet-coloured at the base. Now that the true limits of the genera allied to Brassia are better understood, this plant it is hoped will rest without further change. It is evidently a transition from Brassia to Miltonia; but upon the whole is better placed in the former than the latter genus, on account of its column having no trace of wings.

14. MAXILLARIA corrugata.

M. *corrugata;* pseudobulbis ovatis subimbricatis, foliis lanceolatis solitariis breviter petiolatis, racemis paucifloris petiolo multò longioribus, bracteis minimis, sepalis petalisque obtusiusculis, labello oblongo utrinque emarginato (ideoque obsoletè trilobo) venis elevatis flexuosis corrugato, tuberculo mediano obtusè tricarinato.

This plant, belonging to the same division of Maxillaria as M. squalens, was found by Linden between Maracaibo and Bogota, and has just flowered in Mr. Barker's collection. It has pale brownish purple flowers of no beauty; and a lip with numerous purple elevated zigzag veins on a pale yellow ground. The long tubercle in the middle, which in these plants is usually undivided, or merely 3-lobed at the extremity, is here broken up into 3 distinct contiguous elevated ribs.

15. ZYGOPETALUM.

This genus, founded by Sir W. Hooker on the Z. *Mackaii,* is extremely near Eulophia, from which it principally differs in having a deep ridge or bridge lying across the labellum near the base. The union of the sepals and petals, from which the name is derived, is too inconsiderable to merit attention, and is not constant among the species now known. The presence of blue, or some marked shade of that colour, upon the labellum, is another characteristic feature, by which the genus is to be distinguished from Eulophia, in which that tint is, as far as I am aware, unknown.

In consequence of the additions that have been of late years made to the genus, all of which, with one exception, are in our gardens, it seems desirable to amend the generic character, and to bring together all that is now known concerning them.

ZYGOPETALUM.

(*Hooker. in Bot. Mag. t.* 2748. *Lindl. Orch. no.* 113.)

Perianthium explanatum, sepalis petalisque ascendentibus, subæqualibus, cum ungue producto columnæ connatis. Labellum muticum, planum, indivisum, patens, ungue ascendente: cristâ magnâ transversâ carnosâ. Columna brevis, arcuata, marginata, sæpius utrinque subdilatata, nunc in cucullum expansa. Anthera subbilocularis carnosa, nunc vertice rostrato. Pollinia 2, bipartibilia, in glandulam transversam subsessilia. ——Herbæ terrestres, subacaules, foliis plicatis patentibus. Flores speciosi, suaveolentes, labello cœrulescente.

1. Z. *Mackaii* (Hooker. Bot. Mag. t. 2748. Lod. Bot. Cab. t. 1664. *Eulophia Mackaiana*, Lindl. in Bot. Reg. t. 1433.); foliis lorato-lanceolatis striatis apice recurvis racemo brevioribus, sepalis petalisque oblongo-lanceolatis acutis, labello obcordato glaberrimo, callo crassissimo bilobo. ——*Brazil.*——Known from all the genus by its blue-veined perfectly smooth lip, and two-lobed crest. The blotches on the sepals and petals are somewhat smaller than in the other species.

2. Z. *intermedium* (Lodd. Cat. no. 1136. Z. *velutinum*, Hoffmannsegg in Bot. Zeit. 1. 835. ?); foliis ensiformibus racemo brevioribus, sepalis petalisque oblongis acutissimis, labello subrotundo undulato basi angustato altè bilobo pubescente, callo crenulato indiviso.——*Brazil.*—— This is generally confounded in collections with Z. *Mackaii*, from which it is readily known by its downy labellum. It is perhaps the finest of the genus.

3. Z. *brachypetalum*; foliis ensiformi-lanceolatis scapo elato multifloro brevioribus, sepalis petalisque oblongis obtusis, labello transverso subrotundo emarginato basi vix angustato pubescente, callo integerrimo. ——*Brazil.*——Flowers most like those of Z. *intermedium*, but less than half the size. The sepals and petals are short, stiff, convex, and hardly acute, very much more brown than green, in consequence of the blotches running together. The lip is entirely covered with blue veins and spots, firm, hardly at all narrowed to the base, and very little emarginate. I received it from Mr. Waterhouse, of Halifax, in December, 1840.

4. Z. *crinitum* (Lodd. B. Cab. t. 1687. Bot Mag. t. 3402. Z. *pubescens*, Hffsgg. in Bot. Zeit. 1. 835. ?); foliis lato-lanceolatis, bracteis cucullatis, sepalis petalisque lineari-lanceolatis acutissimis, labello obovato emarginato basi longè angustato venis villosissimis, callo angusto incurvo emarginato. ——*Brazil.*——There are varieties of this with pink, blue, and almost colourless veins to the lip. They all agree in having those veins quite shaggy, and a lip which, instead of being broader than long, narrows very sensibly towards the base. From the manner in which Count Hoffmannsegg describes his Z. pubescens, "with the tracery of the lip scarcely at all verging upon blue, and having a broad white margin," I presume he must have had in view some variety, perhaps the pink one, of this.

5. Z. *stenochilum* (Lodd. Bot. Cab. t. 1923.); foliis latioribus racemo æqualibus v. longioribus, sepalis oblongis acuminatis, petalis brevioribus duplò angustioribus, labello angusto oblongo obtuso villoso, callo emarginato.——*Brazil.*——Of this very distinct species the sepals are blotched, and somewhat striped with brown. The lip is white, with blue or violet veins in the middle only. Its small petals and very narrow lip readily indicate it.

6. Z. *maxillare* (Lodd. Bot. Cab. t. 1776. L. no. 2. Bot. Mag. t. 3686.); foliis lineari-lanceolatis undulatis acuminatis racemo flexuoso longioribus, sepalis petalisque ovato-oblongis acutis, labello glabro obovato, cristâ unguliformi maximâ crenatâ.——*Brazil.*——Sepals and petals green, with blotches and broken bands of chocolate brown. Lip violet, with a deep blue bridge. According to Mr. Gardner this species always occurs on Tree-ferns.

7. Z. *Murrayanum* (Hooker in Bot. Mag. t. 3674.); pseudobulbis ovatis profundè sulcatis, racemis foliis brevioribus, sepalis petalisque ovato-lanceolatis, labelli 3-lobi lobis lateralibus ovatis erectis intermedio ovato-lanceolato reflexo glabro, callo integro sulcato.——*Brazil.*——A species of little beauty. The sepals and petals are pale green and unspotted. The lip is white, with some dark claret-coloured spots at the base. It was found on the Organ Mountains, "at the height of about 4000 feet above the sea."

8. Z. *gramineum*; foliis lineari-lanceolatis, scapo subunifloro capillari brevi ascendente, vaginulis laxis unâ in medio duabus sub pedicello, labello subrotundo fimbriato tenui supra basin appendicibus 2 ovato-lanceatis acutis aucto. Flores 3 Z. Murrayani magnitudine, magis membranacei. Scapus 2-pollicaris. Folia spithamæa.——*Popayan*; in woods rare. ——This is much smaller than the other species, and is the only one yet found west of the Andes. It is not in cultivation.

9. Z. *cochleare* (Lindl. in Bot. Reg. t. 1857.); foliis obovato-oblongis pedunculis unifloris multò longioribus, sepalis petalisque ovato-lanceolatis conniventibus inferioribus multò majoribus, labello cochleato subrotundo bilobo velutino, cristâ arcuatâ crenatâ, antherâ ecristatâ.—— *Trinidad.* ——Leaves unusually broad in this genus. Flowers pure white, except the lip, which is of a rich lapis lazuli blue, and very fragrant. Remarkable in this genus for its solitary flowers.

10. Z. *rostratum* (Hooker in Bot. Mag. t. 2819.); foliis lato-lanceolatis patentibus scapo paucifloro longioribus, sepalis petalisque lineari-lanceolatis undulatis acuminatis, labello subrotundo-ovato, callo baseos lunato crenato, columnâ apice cucullatâ dentatâ, antherâ rostratâ.——*Demerara.*——The broad short leaves of this plant readily distinguish it. The sepals and petals are greenish, dashed with purple, but not at all blotched. The lip is white, with a violet crest, and a few pink radiating lines. It has no scent.

16. CLEISOSTŎMA decipĭens.

C. *decipiens*; foliis distichis loratis obliquè emarginatis leviter undulatis, spicis recurvis compositis, sepalis petalisque oblongis rotundatis, labelli laminâ rotundatâ transversâ sacco subventricoso pubescente.

This little epiphyte, which was received from Ceylon by Col. Fielding, is so like Saccolabium micranthum, that it might easily be mistaken for it. The one is, however, a genuine Saccolabium, the other a true Cleisostoma. S. micranthum has pink flowers with a tooth proceeding from the base of the blade of the labellum; C. decipiens has dirty ochre-coloured flowers, with a tooth springing from the back of the spur just below the column. It is an insignificant plant. Our specimen was communicated by Messrs. Loddiges.

17. SPIRANTHES lobāta.

S. *lobata* (Sarcoglottis); foliis oblongis acutis immaculatis, scapo rufescente ovarioque pubescente, sepalis lateralibus deflexis, labello trilobo cucullato lobo medio reniformi lateralibus ascendentibus rotundatis; ungue utrinque villoso: callis elongatis liberis gyratis.

A fine species of that section of the terrestrial genus Spiranthes to which the name of Sarcoglottis has been applied. It is near S. picta, but has yellow flowers, spotless leaves, a brown hairy scape, and a three-lobed lip, the lateral segments of which are erect, in the manner of a Bletia. For a specimen we are indebted to Sir C. Lemon, who received it from Mr. Rule, of the Real del Monte mines.

18. DENDROBIUM Kingiānum.

Bidwill MSS.

D. (Desmotrichum) *Kingianum*; pseudobulbis ovatis in collum longum extensis apice bifoliis, foliis ovalibus emarginatis, pedunculo terminali (2-floro foliis æquali ?), sepalis ovatis mento emarginato, petalis obovatis apiculatis duplò brevioribus, labelli trilobi pubescentis laciniis lateralibus acutis intermediâ paulo longiore transversè rhombeâ angulis lateralibus rotundatis apiculi acuto axi elevatâ trilineatâ apice tridentatâ.

This curious epiphyte was bought by the Messrs. Loddiges at the sale of Mr. Bidwill's New Holland Plants. It has pseudo-bulbs between four and five inches long, tapered from an ovate base into a very long and narrow neck, on the top of which stand two oblong emarginate dark-green rather wavy leaves. Between these is a flower-stalk having two pink flowers gaily spotted with crimson in the inside. It will probably flower more profusely when in better health, and will then be a plant of considerable interest.

20. SPIRANTHES diaphăna.

S. *diaphana*; foliis hysteranthiis (?), vaginis inflatis acuminatis glabris diaphanis, floribus capitatis extùs tomentosis, labello apice plano oblongo integerrimo.

For a fresh specimen and drawing of this new species of terrestrial Orchidaceæ I am obliged to Mr. Van Houtte of Ghent. It is a curious plant, flowering without leaves, about 8 inches high, smooth, with 3 or 4 large semi-transparent loose sheaths, streaked with rose-colour. The flowers are enclosed in a head of such sheaths, are downy outside, white, with green streaks outside, and a yellow blotch below the point of the white lip. Probably a native of Mexico.

21. SCHOMBURGKIA undulāta.

S. *undulata*; floribus dense racemosis, bracteis longissimis spathaceis, sepalis petalisque æqualibus undulato-crispis labello longioribus, labelli lobis lateralibus rotundatis intermedio ovali acuto trilamellato lineis minoribus elevatis interjectis.

This is a beautiful species received by Mr. Rucker from Linden, who gathered it in La Guayra. It has fine rich brownish purple sepals and petals, and a clear violet-purple lip. From S. crispa it differs in having the middle lobe of the lip oval acute, and only 3-ribbed, while the side lobes are much smaller and flatter; from S. marginata, with which it corresponds in the form of the lip, it is distinguished by its very long narrow crisp sepals and petals. The flowers are as large as those of S. tibicinis, but the inflorescence is quite different. Unless my memory fails me, there is a representation of this plant among Plumier's figures, which I have not just now an opportunity of consulting.

23. ERIA cochleata.

E. *cochleata* (Tonsæ); foliis lanceolatis coriaceis 5-costatis, sepalis glabriusculis petalisque lineari-lanceolatis, labello basi costis 5 apice 3 undulatis lineato: lobis lateralibus obtusis intermedio oblongo-spathulato, scapo elongato radicali multifloro ovarioque altè costato cochleari glabris.

A plant from Manilla, with very much the appearance of E. stellata, from which it differs in having smaller flowers, a smooth scape, and a shining ovary with high ribs, which are twisted so as to have the appearance of a very deep-threaded screw. The sepals are very pale green, the petals and lip white, the latter painted with crimson veins and margin. From Messrs. Loddiges.

24. The Section of EPIDENDRUM named AMPHIGLOTTIUM.

Many years ago, in the Transactions of the Horticultural Society, Richard Anthony Salisbury proposed the name of Amphiglottis for the Epidendrum elongatum of Jacquin. More recent writers have, however, preferred to retain that species in the genus where it was first stationed. In the year 1841, in attempting to form natural subdivisions of the great genus Epidendrum, I proposed, in Hooker's Journal of Botany (vol. 3. p. 81.) several sections, among which was one of which the aforesaid Epidendrum elongatum was taken as the type, and to which the name of Amphiglottium was applied: Salisbury's name being a little altered, so as to be of the same gender as Epidendrum, in order to enable those who prefer to regard the plants collected under it as a distinct genus, to do so without at all disturbing the nomenclature.

The characters by which it was proposed to distinguish the section Amphiglottium were the long leafy stem, with distichous leaves, the want of any tendency to form pseudobulbs, a terminal peduncle covered with close sheaths, and a labellum entirely united to the column. Three years additional experience has brought me acquainted with many more species than I then possessed, and, not having induced me to modify my views, it will now be as well to give an enumeration of what is at present known about this section, more especially as the greater part of the species is to be found in cultivation.

§. AMPHIGLOTTIUM.

AMPHIGLOTTIS, *Salisbury in Hort. Trans.* 1. 261. AMPHIGLOTTIUM, *Lindl. in Hooker. Journ. Bot.* 3. 81. (1841.)

A. *Flowers in spikes or racemes.*

* Lip in no degree lobed.

1. E. *Skinneri* (Bateman in Bot. Reg. fol. 1881.); foliis distichis lanceolatis acuminatis, caule apice longè aphyllo squamoso, racemo cylindraceo multifloro, floribus cernuis, sepalis lineari-lanceolatis, petalis ovalibus acutis, labello ovato acuminato integerrimo basi callo sulcato cristato.——*Guatemala.*——One of the most beautiful of its race. Flowers large, deep rose colour, in long racemes. It is difficult to cultivate. Mr. Skinner says that it inhabits a middling temperature, and will thrive best in a climate graduated from 56° to 70°. He directs us not to put it in earth, but to permit it to have free scope for the young roots and shoots, which it throws out in August.
2. E. *biforatum* (Lindl. in Lodd. cat. ed. 2. no. 566. Bot. Reg. 1842. sub t. 25.); foliis elongato-lanceolatis pedunculo squamato longioribus, racemo paucifloro cernuo, bracteis carinatis rigidis floribus brevioribus, labello subrotundo-ovato per medium elevato basi biforato.——*Brazil.*——The habit of this is something like that of E. fuscatum, but its entire lip, with a central ridge and two pores at the base distinguish it. The only specimen I have examined was out of health, and possibly the flower-stem, which in it was very short, may lengthen out under more favourable circumstances.
3. E. *setiferum* (Lindl. in Ann. Nat. Hist. vol. 4.); foliis distichis lanceolatis acutis, caule simplici squamis lineari-lanceolatis acuminatis sub floribus foliaceis vaginato, racemo cernuo, bracteis longissimis setaceis, petalis linearibus obovatis obtusis sepalis angustioribus, labello cordato integerrimo reticulato acuto basi tritubcrculato. ——*Brazil.*——With the habit of E. fuscatum and biforatum, but with much shorter leaves, and longe bracts than the latter. The flower-stem is about ten inches long in the Brazilian specimens. (Not in cultivation.)

** Lip more or less lobed, but in no degree fringed.

4. E. *cornutum* (Lindl. in Hooker's Journal, 3. 86.); foliis gramineis lineari-lanceolatis acutissimis, racemo elongato cylindraceo cernuo, sepalis lineari-lanceolatis acuminatis striatis, petalis filiformibus, labelli trilobi laciniis lateralibus nanis rotundatis intermediâ cornutâ basi 3-callosa. ——*Peru, Columbia.*——Spathaceous bracts acuminate, imbricated, as long as the flower-stem. Flowers white, very fragrant. (Not in cultivation.)
5. E. *ovalifolium* (L. no. 39); foliis distichis ovali-lanceolatis acutis, sepalis oblongo-lanceolatis, petalis linearibus, labelli trilobi cordati lobis lateralibus rotundatis: intermedio lineari bilobo disco 3-carinato basi bituberculato, racemo elongato.——*Mexico.*——A Mexican species with green flowers like those of E. nutans, but not panicled, nor proceeding from a spathe, and smelling very like a sliced cucumber. Some of the old stems are three feet long.
6. E. *Clowesii* (Bateman in litt.); foliis lanceolatis racemo longioribus (an semper?), bracteis minimis mucronatis, sepalis oblongo-linearibus, petalis filiformibus reflexis, labelli trilobi 3-carinati ecallosi lobis subundulatis rotundatis intermedio bilobo.——*Guatemala.*——Nearly allied to E. ovalifolium, from which it differs in having three distinct raised lines on the lip, but no basal calli, and a very short raceme overtopped by the leaves. Flowers yellowish white.
7. E. *fuscatum* (Swartz, L. no. 38. B. R. t. 67. B. M. t. 2844. Bot. Cab. t. 887. *Ep. anceps*, Jacq. Amer. 224. t. 138. *Amphiglottis lurida*, Salisb.—VAR. β, floribus minoribus magis viridibus. E. *virescens*, Lodd. Bot. Cab. t. 1867. E. *musciferum*, Lindl. in Hooker's Journal, 1. 6.—VAR. γ, floribus multo majoribus, labello violaceo luteo-marginato. E. *viridipurpureum*, Hooker in Bot. Mag. t. 3666.); foliis oblongis undulatis apice recurvis, sepalis oblongis, petalis linearibus, labello carnoso cordato trilobo laciniis rotundatis intermediâ emarginatâ apice calloso basi 2-tuberculato, spicâ globosâ recurvâ.——*Jamaica, Panama, Martinique, Guayana, Grenada.*——One of the most common of all plants in some of the tropical parts of America, and varying greatly in the size of its flowers and length of the flower-stem. Hence have arisen several species, proposed by myself and others, which a better acquaintance with these plants induces me to believe mere varieties. In

colour the flowers vary from dull greenish purple to green and almost yellow. Sir W. Hooker's E. *viridipurpureum* has unusually large flowers, and a flower-stem shorter than common; but in real structure I perceive no difference. There is a little difference among these plants in the degree of lobing observable in the lip; but I find no limit to that kind of variation.

8. E. *orchidiflorum* (L. no. 29.); foliis distichis lineari-oblongis obtusis, caule apice nudo obtusè vaginato, racemo brevi terminali, floribus carnosis, sepalo supremo obovato lateralibus oblongis dimidiatis, petalis linearibus cuneatis, labello subrotundo basi bicalloso apice tridentato: dente intermediâ minore.——*Bahia.*——Leaves very small, fleshy and shining. Flower-stem a foot long, stout, clothed with blunt somewhat rough sheaths. Flowers as large as in the Ep. viridipurpureum of Hooker. (Not in cultivation.)

9. E. *fruticosum* (L. no. 22.); foliis distichis ovato-oblongis acutis, caule ramoso fruticoso, racemo terminali nutante, sepalis cartilagineis petalisque membranaceis ovato-lanceolatis acutis erectis, labello subrotundo cordato retuso apiculato undulato callis 5 magnis confluentibus in disco. ——*Mexico.*——Stem branching, without a trace of sheaths, as thick as a crow-quill. Bracts linear-lanceolate, about as long as the ovary. (Not in cultivation.)

10. E. *campestre*; foliis distichis lineari-lanceolatis, pedunculi elongati squamis distantibus appressis, racemo cylindraceo, sepalis lateralibus falcatis labello duplò brevioribus, petalis conformibus angustioribus, labello subrotundo obsoletè 4-lobo margine integerrimo basi lineis duabus elevatis confluentibus alterâ intermedia axin percurrente, clinandrio 4-lobo: laciniis quadratis.——*Brazil*; in dry upland Campos, Serro do Frio, Diamond District, Gardner 5207.——Nearly related to Epidendrum elongatum. Remarkable for its short falcate lateral sepals and round lip, without any sort of fringing. (Not in cultivation.)

*** Lip distinctly fringed more or less, and usually deeply lobed.

|| *Lip with a concave lobed tubercle occupying the disk.*

11. E. *xanthinum* (E. *ellipticum* β *flavum*, Lindl. in Ann. Nat. Hist. vol. 4.); foliis distichis oblongis carnosis, caule longè aphyllo, sepalis petalisque patentibus lineari-lanceolatis æqualibus, labelli trilobi lobis lateralibus truncatis grossè serratis intermedio cuneato denticulato, callo concavo acuto serrato.——*Brazil.*——This is very like E. ellipticum, of which I formerly considered it a variety. Its yellow flowers are however peculiar; and upon further examination I find the central tubercle of the lip destitute of the shoulders that run off into the side lobes in E. ellipticum. According to Martius its stems are three feet high. Mr. Gardner's (no. 5205), from the province of Mines, are not half that size. (Not in cultivation.)

12. E. *elongatum* (Jacq. B. M. t. 611. L. no. 54. *Amphiglottis secunda*, Salisb. *E. secundum*, Linn. L. no. 57.); foliis distichis oblongo-lanceolatis acutiusculis, caule apice longè aphyllo nunc ramoso squamoso, sepalis petalisque patentibus lineari-lanceolatis æqualibus, labelli trilobi lobis lateralibus truncatis denticulatis: intermedio duplò majore dilatato cuneato emarginato biloboque denticulato basi calloso: callo carnoso concavo acuminato crenato utrinque verrucâ crenatâ aucto.——*West Indies.*——Flowers bright rose colour. Leaves longer and thinner than in E. ellipticum. It can hardly be doubted that the Ep. secundum of Linnæus belongs here; its character having been taken from a crushed and damaged specimen.

13. E. *ellipticum* (Graham in Hook. Exot. Bot. t. 207. L. no. 56. Bot. Cab. t. 1216. *E. crassifolium*, L. no. 55.); foliis distichis ellipticis obtusis concavis succulentis, caule apice longè aphyllo nunc ramoso squamoso, sepalis petalisque patentibus lineari-lanceolatis æqualibus, labelli trilobi lobis lateralibus rotundatis laciniatis: intermedio minore truncato apice denticulato basi calloso, callo concavo plicato acuminato utrinque in lobos laterales decurrente.——*Brazil.*——Very near E. elongatum, from which it differs in its elliptical blunt succulent leaves, small middle segment of the labellum, paler flowers, and especially in the form of the tubercle, which is deeply furrowed, or plaited, and shoulders off to the lateral lobes of the lip, and is not furnished there with a free tubercle as in E. elongatum. Flowers light rose colour.

14. E. *ibaguense* (Humb. et Kunth. Nov. Gen. et Sp. 1. 352. L. no. 53.); foliis distichis oblongis obtusis carnosis, caule simplici apice aphyllo nunc ramoso, sepalis suboblongis acutis patentibus, petalis conformibus paulò minoribus, labelli trilobi lobo intermedio obcordato crenulato-fimbriato lateralibus fimbriatis ovatis dimidiatis apice rotundatis, callo concavo carnoso trilobo apiculato. ——*Peru.*——A very fine species, with orange-coloured flowers. The leaves are an inch and a quarter broad, and the stems as thick as a swan's quill, and apparently often branched. When old they become as smooth as bamboo. (Not in cultivation.)

15. E. *cochlidium* (Lindl. in Ann. Nat. Hist. vol. 4.); foliis distichis ovato-oblongis obtusis emarginatisque coriaceis, sepalis petalisque lineari-lanceolatis patentibus æqualibus, labelli laciniis laceris subæqualibus callo carnoso maximo excavato trilobo parùm majoribus.——*Peru.* ——Only known from a specimen found by Mathews in Peru. The flowers *seem* to be yellow. The very large tubercle which covers the whole centre of the lip is very remarkable. (Not in cultivation.)

16. E. *dichotomum* (Presl. Rel. Hænk. 101. L. no. 58.); foliis distichis lanceolato-oblongis obtusis, caule apice aphyllo squamoso ramoso, sepalis petalisque lanceolatis acutis reflexis, labelli tripartiti laciniis fimbriatis: lateralibus rotundatis patentissimis intermediâ majore obcordatâ, callo concavo crenulato indiviso tuberculis 2 prope basin.——*Peru.*——A very fine species with yellowish? flowers, which are perhaps the largest of any in this section. Mathews found it in Peru, and it is no. 1024 of his collections. The unlobed tubercle, with a pair of calli near its base, affords a good mark of distinction. (Not in cultivation.)

17. E. *lacerum* (Lindl. in Bot. Reg. 1838. misc. 18.); foliis distichis lineari-oblongis obtusis, caule simplici apice aphyllo laxè squamato, sepalis petalisque patentibus lineari-lanceolatis æqualibus acutis, labelli trilobi lobis æqualibus laceris pectinatis intermedio apice mucronato integro: lamellis duabus membranaceis sub apice conniventibus duabusque callosis parallelis e basi ortis.—— *Cuba.*——A very slender species with narrow grassy leaves, pale rosy flowers, and a distinct entire point to the middle lobe of the lip. The tubercle is very peculiar, not fleshy as in the allied species, but consisting of a pair of plates converging just below the apex, and another pair of more fleshy calli near the base.

|| || *Lip with a distinct middle elevated ridge, and two calli at the base.*

18. E. *Blepharistes* (Barker in litt.); sepalis ovalibus, petalis acuminatis, labelli quadrilobi laciniis æqualibus linearibus apice multifidis lineâ mediâ elevatâ callisque duobus semiliberis arcuatis ad basin.——*La Guayra.*——This plant was receivèd by Mr. Barker from Linden. It is closely allied to E. elongatum; but differs in the structure of the labellum, which approaches nearer to that of E. Schomburgkii. The sepals and lip are a rich rose colour; the column has the deepest tint of the garnet, which gives the flowers a very gay appearance.

19. E. *imatophyllum* (L. no. 52.); foliis distichis ligulatis obtusis subemarginatis, caule simplici apice aphyllo vaginato, sepalis lineari-lanceolatis patentibus, petalis duplo latioribus serratis, labelli trilobi carinati basi bicornis lobis lateralibus nanis laciniatis: intermedio cuneato-rotundato subundulato basi serrato.——*Demerara.*——Flowers pale rose colour, larger than in E. ellipticum. Leaves very long, strap-shaped. The column is sometimes green, and the same colour occurs to a small extent in the sepals.

20. E. *radicans* (Pavon. Mss. L. no. 35. E. *rhizophorum*, Bateman in Bot. Reg. 1838. misc. 10.); foliis distichis subcordatis ovatis obtusis, caule simplici apice aphyllo vaginato, sepalis petalisque lanceolatis acutis patentibus, labelli carinati basi bicallosi lobis lateralibus rotundatis denticulatis intermedio cuneato apice fimbriato emarginato utrinque integro.——*Mexico, Guatemala.*——One of the finest of its race; with flowers as much as $1\frac{1}{2}$ inch across, according to Mr. Skinner. It is in gardens, and has long white roots proceeding from the sides of the stem; but it has not flowered.

21. E. *cinnabarinum* (L. p. 106. Bot. Reg. 1842, t. 25.); foliis distichis oblongis apice sub-recurvis, sepalis petalisque lanceolatis subæqualibus, labello trilobo carinato basi bituberculato: laciniis lateralibus inciso-laceris intermediâ basi obcuneatâ sub apice constrictâ apice cuneatâ truncatâ angulis acuminatis simplicibus fissisque.——*Brazil.* ——In this section of Amphiglottium are three species of great beauty, so nearly related to each other that their limits are with difficulty determined. They all have bright scarlet or crimson flowers, a pair of tubercles at the base of the labellum, and a ridged line running from between the tubercles to nearly the apex of the lip. They however differ thus: *E. radicans* has the lateral lobes of the labellum rounded and toothletted only, not lacerated, and it produces coarse pale green roots from its stems; *E. cinnabarinum* has the lateral lobes of the labellum deeply lacerated, while the central lobe is contracted in the middle, and then *suddenly* wedge-shaped, with its angles prolonged into one or two fine teeth; *E. Schomburgkii* has the lateral lobes only toothed, with the centre lobe gradually widened to the point, and there toothletted, without being at all truncate; the lobes of the lip are confluent in what I take to be a variety of that species.

22. E. *Schomburgkii* (Lindl. in Bot. Reg. 1838, misc. no. 16. t. 53.); foliis distichis oblongis obtusis margine sanguineo-punctatis, caule simplici apice aphyllo, sepalis petalisque lineari-lanceolatis acutis æqualibus patentissimis, columnâ clavatâ elongatâ, labelli trilobi carinati basi bicallosi lobis lateralibus latis rotundatis laceris: intermedio cuneato apice triangulari crispo medio et utrinque acuminato.——*Demerara.*——Flowers rich scarlet. Leaves sometimes bordered with crimson. Very handsome. I have a dried specimen of what I suppose to be a variety of this plant, with all the lobes of the lip run together into one, a little toothed at the edge, and more so at the point. It may however be distinct.

---

23. E.? *flexuosum*, (Meyer Fl. Essequeb. p. 260. L. no. 34.); foliis distichis carnosis lanceolatis emarginatis, caule flexuoso, sepalis lanceolatis, petalis rhomboideis, labello acuminato subquadrato margine lacero.——*Essequebo.* ——Stem two or three feet high, compressed, zigzag from sheath to sheath. Leaves six inches long. Flowers purple. A doubtful plant, of which little is known. (Not in cultivation.)

B. *Flowers panicled.*

24. E. *Martianum* (Lindl. in Ann. Nat. Hist. v. 4.); foliis distichis angustis lanceolatis, caule apice ramoso squamis concavis obtusiusculis vaginato, racemis corymbosis, petalis linearibus obovatis obtusis sepalis multo angustioribus, labello cordato subrepando basi bituberculato axi elevatâ. ——*Brazil.*——Allied to E. fuscatum. Stem from $1\frac{1}{2}$ to 2 feet high. Flowering stem stiff, erect, branched, much longer than the leaves, which are narrow and sharp pointed. (Not in cultivation.)

25. E. *laxum*, (Pöppig et Endlich. n. g. et sp. 2. p. 2.); "caulibus basi repentibus, adscendentibus, ancipitibus; foliis lanceolato-oblongis, acutis; panicula terminali elongata, laxa, multiflora; sepalis erectis, subæqualibus, lanceolatis, acutis; labelli columna triplo longioris, ovalis, subcordati, apice breviter trifidi, nudi laciniis integerrimis, intermedia majore, acutiore."——*Peru.*——Stems above a foot high on a long creeping rhizoma. Leaves seldom more than three on each, with a spiny point. Panicle a foot long or more. Flowers small, pale green. (Not in cultivation.)

26. E. *rubrocinctum* (Lindl. in Bot. Reg. 1843, misc. 20. E. *densiflorum*, Hooker in Bot. Mag. t. 3791??); foliis oblongo-ligulatis acuminatis apice carinatis paniculâ amplâ cernuâ subsessili parum brevioribus, sepalis oblongis concavis acutis coriaceis, petalis angustè linearibus, labello transverso cordato trilobo suprà tricarinato; lobi medii trilobi dente intermedio minuto, columnâ inappendiculatâ.——*Brazil.*——If E. densiflorum is the same as this, which I suspect, then the name of rubrocinctum must be changed, but I am not certain. The sweet-scented flowers are a dull yellowish green, bordered with dull purple; the lip is more yellow than the sepals. It has much the habit of E. nutans, but is a plant of more beauty. Not having seen sufficient specimens, I formerly regarded it as a species of the section Spathium. It is however an Amphiglottium with a nearly sessile panicle, and only two or three herbaceous sheaths at the base. It connects Spathium and Amphiglottium.

27. E. *pallidiflorum* (Hooker Bot. Mag. t. 2980. L. no. 41.); foliis distichis oblongis linearibus obtusis pedunculo distanter vaginato brevioribus, floribus paniculatis, sepalis petalisque lanceolatis obtusis patentibus, labelli trilobi lobis subæqualibus rotundatis: intermedii trisulci 2-callosi marginibus involutis.——*St. Vincents.*——Flowers pale yellowish white, with a purple stain on the end of the column. The specimen figured in the Botanical Magazine was very weak. Its inflorescence is really a loose panicle.

28. E. *gladiatum* (L. no. 47.); foliis paucis distichis gladiiformibus, racemis subsessilibus terminalibus fasciculatis corymbosis recurvis, sepalis oblongis obtusis margine reflexis petalisque spatulatis patulis, labelli trilobi carinati basi bituberculati lobis lateralibus subquadratis rotundatis intermedio multò minore emarginato.——*Mexico.*——A small species, not more than eight or twelve inches high. The leaves are three or four inches long, and as many lines broad. Mr. Hartweg found it in Mexico, and therefore the habitat of Peru given by me formerly, on authority of Pavon's herbarium, was probably a mistake.

29. E. *anisatum* (La Llave. L. no. 52.); foliis distichis elliptico-lanceolatis, sepalis petalisque patentibus linearibus obtusis, labelli tripartiti laciniis lateralibus obtusis intermediâ profundè bilobâ: lobis acutis.——*Mexico.*——This is very near E. gladiatum, and is perhaps the same species. The flowers are described as being of a dingy colour, and emitting a smell of Anise at night. The lip is said to have three purple lines at the base. (Not in cultivation.)

30. E. *polyanthum* (L. no. 50. Bateman Orch. Mex. et Guat. t. 34. *E. bisetum*, Lindl. in Bot. Reg. 1841. misc. 148.); foliis distichis ovali-lanceolatis acuminatis, caule ramoso, racemis plurimis multifloris cernuis basi spathaceis, sepalis ovato-lanceolatis acutis striatis, petalis linearibus reflexis, labelli trilobi tricostati lobis lateralibus subcuneatis retusis: intermedio lineari retuso, ovario scabro.——*Mexico, Guayaquil.*——Flowers bright orange, or salmon-colour, with a strong smell of cowslips. The ovaries are rough with elevated points, the sepals are brownish orange, the lip dull fawn colour. The petals are in the form of two fine bristles.

31. E. *Hænkeanum* (Presl. Rel. Hænk. 100. L. no. 40.); foliis distichis ovato-lanceolatis nervosis acutis, spicâ paniculatâ, sepalis oblongo-lanceolatis acutiusculis in basin attenuatis, petalis spatulatis, labelli trilobi lobis lateralibus rotundatis, medio minore obcordato.——*Peru.* ——Scape a foot high. Sepals ten lines long. Leaves ten lines broad. (Not in cultivation).

32. E. *porphyreum* (Lindl. in Hook. Journal, 3. 86.); foliis distichis oblongis acutissimis, squamis spathaceis dense imbricatis acuminatis pedunculo longioribus, paniculâ simplici multiflorâ, sepalis oblongis acutis lateralibus falcatis, petalis lineari-spathulatis, labelli trilobi laciniis lateralibus rotundatis intermediâ quadratâ bidentatâ: disci axi elevatâ apicem versus bicallosâ basi bicornutâ.

——*Peru.*——A fine species, with an oval panicle and large flowers like those of E. nutans, but purple or dark orange. (Not in cultivation).

33. E. *læve;* foliis distichis lanceolatis apiculatis, paniculâ simplici multiflorâ, sepalis oblongis lævibus, petalis filiformibus, labelli quadrilobi laciniis posticis cuneato-rotundatis anticis linearibus divaricatis, disco callis tribus et lamellis duabus.——*Popayan, Pasto.*——A very fine species in the way of E. paniculatum; but its leaves are lanceolate, the panicle simple, the flowers larger, and the sepals smooth. It is one of Mr. Hartweg's discoveries. (Not in cultivation).

34. E. *paniculatum* (Fl. Peruv. L. no. 61.); foliis distichis oblongis acuminatis, caule apice vaginis acuminatissimis membranaceis vestito, paniculâ confertâ compositâ, sepalis oblongis venis extùs varicosis, petalis filiformibus, labelli quadrilobi laciniis posticis brevibus oblongis obtusis: anticis linearibus divaricatis, disco callis tribus et tuberculis duabus.——*Guayaquil, Peru.*——A most noble looking species when dried; its panicle, which is eight or nine inches long and nearly a foot broad, being crowded with flowers. Mr. Hartweg found it in woods on the western declivity of Pichincha. (Not in cultivation).

28. MILTONIA cuneāta.

M. *cuneata;* pseudobulbis ovato-oblongis, foliis oblongis striatis subundulatis, racemo plurifloro, sepalis petalisque lanceolatis undulatis, labello cuneato rotundato basi bilamellato utrinque subrepando, columnâ medio bidentatâ, clinandrio integerrimo.

A beautiful Epiphyte, allied to M. candida, with flowers nearly four inches in diameter; the sepals and petals are a rich brown tipped with green. The lip is pure white, with a tinge of pink near the base. It has lately flowered with Messrs. Rollissons.

29. ANGULOA Clowesii.

A. *Clowesii;* pedunculo unifloro radicali laxè squamato, flore carnoso resupinato, sepalis petalisque ovatis convexis conniventibus, labelli trilobi lobo medio piloso infundibulari bilabiato: labio altero emarginato altero tridentato, columnâ integrâ.

At last a genuine species of the genus Anguloa, which has hitherto puzzled every one, has made its appearance in the collection of the Rev. J. Clowes, of Broughton Hall near Manchester, who obtained it from Linden's Columbian collections. It is indeed a noble plant. The flowers are four inches in diameter, of a clear lemon colour, with a pure white lip. We refrain from doing more than announcing it, because it will soon be published with a plate in an early number of this work.

30. GONGORA maculata, *var.* tricolor.

This also is from Mr. Clowes's collection, where it was received from Liverpool, with a statement that it came from Peru. It is a most beautiful variety of G. maculata. The ground colour of every part of the flower, except the lip, is clear yellow; the column and petals are delicately banded with rich sienna brown, and a few large clear distinct blotches of the same colour occur on the sepals. The lip itself is white, with a cinnamon stain on the ends of the lateral tubercles and the sides of its upper half.

34. DENDROCHILUM? abbreviātum.?

Blume Bijdragen, 398. *Lindl. Gen. & Sp.* 35.

D. *abbreviatum;* pseudobulbis ovatis diphyllis, foliis oblongis basi angustatis racemo æqualibus, ovario hexaptero bracteæ striatæ oblongæ acutæ citò deciduæ longitudine, labello saccato cordato apice dilatato retuso denticulo interjecto internè bicalloso, columnæ truncatæ denticulatæ angulis anticis dentiformibus.

Messrs. Loddiges having imported this little Orchidaceous plant from Java, it is to be presumed that it must be one of those enumerated by Dr. Blume; and as none of his definitions suit it well, except that of D. abbreviatum, we must conclude that this must be what he intended by that name. Nevertheless, it is difficult to reconcile his statement that the lip of that plant has a reflexed tooth on each side at the base, unless we suppose that he intended to describe the heart-shaped base of the lip. The plant has little beauty. The flowers are small, green, with a white lip, having a yellow stain in its middle. The column is a deep cinnamon brown, truncated and notched, with the front angles a little lengthened into teeth. The lateral lobes of the column, found in other species, are deficient in this. The pollen is white, separates into 4 soft masses, having as many thread-like tails by which they adhere to the point of the stigma, which will even come off with them, like the gland of Vandeæ. This structure requires to be investigated with more care than we have yet been able to give it.

38. ONCIDIUM lacĕrum.

O. (TERETIFOLIA) *lacerum;* foliis longis teretibus carinatis, paniculâ contractâ multiflorâ, sepalis petalisque conformibus obovatis concavis, labelli elongati laciniis lateralibus linearibus refractis intermediâ longè et angustè unguiculatâ bilobâ margine lacerâ, cristæ dente altero transverso altero majore compresso a fronte, columnæ brevis pubescentis alis semiovatis.

Panama has furnished this very pretty species to Messrs. Loddiges, who flowered it in April 1844. It has very much the habit of O. longifolium, but has rather denser flowers, with a lacerated and rather crisp middle lobe to the lip, and a crest consisting of one transverse tubercle, like that of O. nudum, and another larger, compressed, rounded, and at right angles to it in front.

40. The Species of CATASETUM.

This genus was founded by Louis Claude Richard upon two plants from tropical America, one of which he called C. *macrocarpon,* the other C. *maculatum.* Among the peculiarities of the genus were two long feelers or cirrhi directed downwards from the column into the cavity of the lip, and a large fleshy helmet-shaped lip.

As soon as the attraction of European gardeners began to be turned to the cultivation of tropical Orchidaceæ, other species of the genus were made known, and now it has become a very considerable group, the greater part of which is scattered through various writings. In the course of gathering together these materials certain plants were obtained, which appeared to form distinct genera; one was called *Myanthus,* the other *Monachanthus. Myanthus* was thought to be distinguished by having a flat lip, and *Monachanthus* by want of feelers on the column. But, strange to say, both these flowers were afterwards ascertained to be monsters of genuine Cataseta, as was fully explained at tab. 1951 of the Botanical Register.

The following are, we believe, all the Cataseta now on record, divided into two sets, one of which has a plane lip, and answers to *Myanthus;* the other a helmetted lip, answering to the original *Catasetum.*

§ 1. *Labellum galeatum;* Catasetum.

1. C. *tridentatum* (Hooker Exot. Fl. t. 90. 91. L. no. 2. C. *macrocarpon* H. B. K. 7. 158. t. 631. C. *Claveringii* Lodd. Bot. Cab. t. 1344.——var. *floribus majoribus sepalis petalisque acutis.* C. *Claveringii* Lindl. Bot. Reg. t. 840. C. *floribundum* Hooker Exot. Fl. t. 151.——var. *floribus viridibus concoloribus.* C. *Wailesii,* Hooker in Bot. Mag. t. 3937.); foliis oblongo-lanceolatis acuminatis, perianthiis compressis conniventibus, sepalis petalisque acuminatis, labello galeato aperto apice tridentato margine lævi.——*All parts of tropical South America, east of the Andes.*——There is no doubt that C. macrocarpon is this plant, whose name should therefore, according to the strict laws of priority, be changed. But the name macrocarpon is unmeaning, and much inconvenience would arise from the restitution of it. It is a

very common variable plant, sometimes even losing its spots, as in C. *Wailesii* from Honduras.——MONSTROUS FORMS. *Monachanthus viridis*, Lindl. no. 1. Bot. Reg. t. 1752, from *Brazil* is a monster with no cirrhi to the column, and the lip perfectly undivided. C. *cristatum* L. no. 7. is another monster, as is proved, at plate 1951 of the Botanical Register.

2. C. *maculatum* (H. B. K. 7. 157. t. 630.); "foliolis calycinis duobus interioribus maculatis, labello ciliato."——*Turbaco in New Grenada.*——All that can be said about this species is that it certainly is not the C. maculatum of our gardens. At the time I referred the latter to it I was not aware that it had been figured. Nothing like it in a cultivated state has yet been seen.

3. C. *integerrimum* (Hooker Bot. Mag. t. 3823. C. *maculatum* Lindl. Bot. Reg. 1841. t. 62. Bateman Orch. Mex. et Guat. t. 11. nec H. B. K.); sepalis petalisque conniventibus, labello carnoso galeato basi inflexo lateribus suis invicem imbricantibus margine lævi vel serrato. ——*Guatemala.*——This is very nearly the same as C. *tridentatum*, from which it differs in its helmet-shaped lip, having the lower edges brought together so as to press upon the column, instead of being wide apart, and being generally serrated, although sometimes smooth. It is certainly not the species so named by Humboldt, which was found near Turbaco in New Grenada. That plant is represented with the labellum quite open, not contracted and curved inwards near the column, and surrounded by a broad plaited border, called in the description "ciliatum."

4. C. *planiceps* (Lindl. in Bot. Reg. 1843. t. 9.); sepalis petalisque ovatis conniventibus, labello carnoso galeato rotundato compresso apiculato indiviso margine serrato. ——*Spanish Main.*——Flowers green and yellow. The habit is that of C. *integerrimum, tridentatum,* and *semiapertum*, from the second of which it differs from its serrated lip, from the first and the last in its lip not having the edges incurved, and from all in the singular truncate form of this helmet-shaped organ, which is flattened from front to back and not laterally.

5. C. *fuliginosum* (Lindl. in Bot. Reg. 1841. misc. no. 168); C. tridentati vultu, sepalis petalisque oblongis acutis reflexis, labello cucullato carnoso integerrimo v. minutè serrulato patente, columnâ brevi ecirrhosâ apice in setam productâ, antherâ parvâ (effœta ?).—— ? ——This singular plant has the habit of C. tridentatum, but its flowers are in a dense erect raceme, and of a deep green colour, spotted with a dull blackish purple, so as to look as if they were soiled with soot. The sepals and petals are spotted, oblong, acute, and reflexed, so as to hang downwards. The lip, on the other hand, is fleshy, hooded, stained with pale purple, and either entire at the margin, or very slightly serrated; but it does not cover over the column as is usual in the hooded Catasetums: on the contrary it spreads away at almost right angles. The column itself is short, deep green, and produced at the point into a straight bristle, in front of which is placed a small and imperfect anther. Judging from the evidence we possess concerning C. tridentatum and cristatum, we should suspect it to be a male form of C. atratum, or some such species. That, however, must remain for future inquiry.

6. C. *luridum* (Lindl. Gen. & Sp. Orch. p. 156. Bot. Reg. t. 1667. *C. abruptum,* Hooker in Bot. Mag. t. 3939. *Anguloa lurida,* Link in Verhandl. des Vereins. z. bef. des Gartenb. in dem Königl. Preuss. Staat. 1. p. 289. t. 6.); caulibus defoliatis angustis sulcatis, perianthio subgloboso parum maculato, sepalis petalisque oblongis apice rotundatis, labello cucullato carnoso mutico apice paulò producto truncato, racemo brevi nutante.—— *Brazil.*——Flowers green, globose, with a blunt shovel-shaped lip, having some brown spots on a yellowish ground.

7. C. *semiapertum* (Hooker. Exot. Fl. t. 213. Bot. Reg. t. 1708.); foliis oblongo-lanceolatis multiplicatis racemo compacto longioribus, perianthiis subpatentibus, labello galeato apice incurvo: lateribus denticulatis.——*Brazil.* ——Flowers small, green, whole coloured.

8. C. *purum* (Nees ab Esenb. Plantarum in h. med. Bonnens. Icon. p. 1. t. 1. L. no. 5.); foliis lineari-lanceolatis, perianthio patente obsoletè punctato, labello ventricoso apice incurvo concolore ciliato integerrimo.——*Brazil.* ——Much like the two last, from which its very narrow leaves seem to distinguish it. Flowers green, and scarcely spotted, about as large as those of C. Hookeri.

9. C. *Hookeri* (Lindl. Coll. Bot. t. 40. Gen. & Sp. no. 3.); foliis lanceolatis triplicatis, perianthiis globosis, labello apice incurvo: lateribus denticulatis.——*Brazil.*—— Flowers small, yellowish brown. Petals green. Lip green, yellow at the point, spotted with purple inside.

10. C. *globiflorum* (Hooker in Bot. Mag. t. 3942.) spicâ elongatâ multiflorâ, perianthio globoso, sepalis petalisque subconformibus ovatis acutis concavis arctissimè imbricatis, labello hemisphærico-globoso, ore contracto oblongo inferne dilatato basi columnæ longitudine denticulato, columnæ brevis setis rectiusculis.——*Brazil.*——Much like C. Hookeri, and perhaps a variety of it. The sepals and petals are olive-brown, and closely applied to a glaucous globular labellum, the greater part of which is uncovered.

11. C. *longifolium* (Lindl. in Bot. Reg. 1839. misc. 154. Sertum Orchid. t. 31.); foliis longissimis gramineis, racemo cylindraceo pendulo multifloro, sepalis ovatis subrotundis petalorum conformium dorso applicitis, labello urceolari a tergo incurvo limbo truncato apiculato intus cereaceo glabro margine fimbriato.——*Demerara.* ——Leaves very long and narrow. Flowers bright orange, a little bordered with violet, in a drooping raceme, over which they are closely packed for the length of a foot or more; they are extremely beautiful, and the species is beyond comparison the handsomest of its genus.

12. C. *discolor* (*Monachanthus discolor* Lindl. in Bot. Reg. t. 1735. *M. Bushnani* Hooker, Bot. Mag. t. 3832. *M. roseo-albus* Hooker, l. c. t. 3736.); racemo laxo multifloro, labello hemispherico marginibus planis medio fimbriatis.——*Brazil.*——Leaves rather narrow. Flowers in a loose erect raceme, dull pale purple, with the inside of the lip yellow. They vary in colour to yellowish green, or French-white; and sometimes the lip, which is usually whole coloured, is tipped and banded with red. In M. roseo-albus the fringes of the lip are unusually long.

13. C. *atratum* (Lindl. in Bot. Reg. 1838. misc. 114. t. 63.); racemo decurvo, sepalis petalisque patentibus ovatis acutis, labello carnoso cucullato margine tenui pectinato apice rotundato reflexo crasso denticulato.——*Brazil.* ——Flowers deep green outside; sepals rich purple inside; petals spotted with the same colour; lip dull green, fringed, with a yellow recurved flap at the end.

14. C. *Russellianum* (Hooker in Bot. Mag. t. 3777.); pseudobulbo elliptico magno, foliis lato-lanceolatis, racemo amplo, labello membranaceo anticè inflato ore contracto margine anteriore producto undulato fimbriato disco membranâ cristato, columnâ nudâ (ecirrhosâ).——*Guatemala.*——A large species with green flowers, a little relieved by the greater whiteness of the labellum, which is almost membranous. It is very near C. *laminatum.* The absence of cirrhi from the column, if constant, is remarkable.

§ 2. *Labellum planum, sæpius foveatum v. saccatum;* Myanthus.

15. C. *laminatum* (Lindl. in Ann. nat. hist. vol. 4. p. 384. Bentham, Plantæ Hartwegianæ, p. 72. Sertum Orchidaceum, t. 38.—*Var.* 1. *maculatum;* labello, columnâ petalisque purpureo-fusco maculatis, *Sert. Orch. t.* 38.— *Var.* 2. *eburneum;* labello eburneo columnâ petalisque immaculatis, *Bot. Reg.* 1841. *t.* 5. *f.* 4.); labello lanceolato basi saccato apice marginibusque incurvo basin versus fimbriato per axin lamellâ unicâ carnosâ altâ integrâ v. denticulatâ basi bilobâ instructo, columnâ cir-

rhatâ.——*Mexico.*——Flowers spread open, large, purple, or greenish spotted with that colour. Lip sometimes pure white, sometimes speckled with purple, with a deep plate running along its middle from end to end.

16. C. *tabulare;* vegetatione sepalis petalisque omninò C. laminati, labello oblongo antico concavo basi cornuto fimbriato extùs sub apice lineis serrulatis elevatis striato secus medium cristâ tabulari latâ transversè corrugatâ anticè dentatâ aucto.——*Grenada.*——This is very like the Guatemala C. laminatum; but the structure of the lip is different. In place of the thin knife-like crest of that species, there is a broad raised oblong fleshy table of a yellowish brown colour, closely wrinkled across and broken up into asperities, which, at the front of the table where it forms a kind of cliff above the lip, are extended into strong teeth. On the outside of the lip near the point the veins are raised and somewhat toothed. The colour and size of the flowers are much the same as in C. laminatum.

17. C. *saccatum* (Lindley in Botanical Register, 1840. misc. 179. Sertum Orch. t. 41.); sepalis lanceolatis patentibus dorsali petalisque fornicatis, labello subrotundo abruptè acuminato fimbriato medio saccato: ostio contracto reniformi posticè dentato, columnâ cirrhatâ.——*Demerara.*——This has very large flowers, with rich purple-spotted sepals and petals, and a bright yellow lip covered closely with crimson dots. The latter is pierced in the middle by a narrow aperture, that leads into a conical chamber or bag, which is not observed till the back of the lip is turned up.

18. C. *Naso* (Lindl. in Bot. Reg. 1843. misc. 111.); spicâ brevi erectâ, sepalis oblongo-lanceolatis complicatis petalisque lanceolatis ascendentibus æqualibus, labello hemisphærico apice abruptè in appendicem carnosum ovatum obtusum producto: marginibus basi tenuibus laceris amplexicolumnibus: lineâ intramarginali carnosâ inflexâ connivente ostium cordiforme efficiente.——*Caraccas.*——This is a singular plant whose flowers before opening might be mistaken for C. tridentatum, but when expanded they are totally different. The sepals and petals, which are very pale dull green outside, are slightly pink inside, and richly spotted, in irregular bars, with a deep crimson-purple. The lip is a most singular organ, and very difficult to describe. Viewed from the side it has a hemispherical form, and is green except at the base, where it is extended into a black-purple lacerated margin embracing the column, and at the point where it is extended into a long flat horn or *nose.* Seen in front it is almost wholly of the same rich black-purple, and looks as if it were a solid hemisphere pierced in the middle with a large heart-shaped hole; but this appearance is owing to a thick fleshy rim which rises from within the true edge of the lip, and directing itself inwards horizontally with an uneven outline, at last touches in front of the column and presents the appearance of a junction.

19. C. *Trulla* (Lindl. in Bot. Reg. 1840. misc. 176. 1841. t. 34.); sepalis petalisque patentibus ovalibus planis, labello latè ovato acuminato obtuso subcordato concavo fimbriato apice lævi, columnâ brevi.——*Tropical America.*——Flowers green, unspotted; lip of the same colour, but whitish near the base, with a deep brown spot in the middle.

20. C. *callosum* (Lindl. in Bot. Reg. 1840. misc. 183. 1841. t. 5. f. 1.); petalis concoloribus lineari-lanceolatis sepalo dorsali conformi suppositis, labello ovato-oblongo obtuso basin versus saccato supra saccum callo magno (aurantiaco) instructo margine obsoletè crenato, columnæ acuminatæ cirrhis vix ultra callum extensis.——*La Guayra.*——Exactly like Catasetum tridentatum, *var.* floribundum in habit, but its flowers are different. The sepals and petals are of a dull reddish brown, without spots; the column is of the same colour, which may perhaps be best compared to that of old spoilt port wine. The lip is green, flat, with a yellow tubercle near the base above the hollow, and a stain of the same colour near the apex.

21. C. *poriferum* (Lindl. in Bot. Reg. 1838. misc. 164.); labello plano cordato-ovato truncato crenato, lobo apicis elevato carnoso ovato medio nectarifero, dente baseos ovato incurvo, columnâ cirrhosâ.——*Demerara.*——Similar in appearance to C. deltoideum, like which its flowers are richly spotted with deep purple broken fasciæ, closely arranged upon a clear green ground. The lip is dull green, nearly flat, deeply cordate, truncate at the apex, with an obscurely crenated margin. At its base is an elevated yellow-tipped broad tooth, and at its apex an ovate yellow fleshy raised lobe, with a pore in the middle secreting honey.

22. C. *deltoideum* (Lindl. Bot. Reg. 184 . misc. 157. *Myanthus deltoideus* Bot. Reg. t. 1896.); labello imberbi sagittato triangulari, angulis posticis rotundatis dentatis apice dilatato calloso margine recurvo basi tuberculato.——*Demerara.*——Flowers deep green, spotted with purple. Lip purple, flat, fleshy, with a dash of green in the centre.——Monstrous State. This has been observed to sport in the same way as C. tridentatum. A case is mentioned in the Botanical Register for 1840, misc. 157. where the scape became three times as stout as usual, the length of the raceme much reduced, the sepals and petals retaining their form and colour; but the labellum, instead of being arrow-headed, flat, deep purple, toothed at the base, and placed in front of the flower, was of exactly the same form as that of *Monachanthus viridis,* hooded, undivided, and of a dull greenish colour tinged with dull purple. The column too, in like manner lost its cirrhi, shortened, and its lengthened beak disappeared.

23. C. *trifidum* (Hooker in Bot. Mag. t. 3262. *Myanthus cernuus,* Lindl. p. 155. Bot. Reg. t. 1721.); racemis elongatis cernuis multifloris, sepalis petalisque lanceolatis convergentibus, labello transverso plano altè trifido laciniâ intermediâ minore.——*Brazil, Trinidad.*——Flowers green, with a few purple spots.

24. C. *cornutum* (Lindl. Bot. Reg. 1840, misc. 182. 1841. t. 5. f. 2,); petalis maculatis lineari-lanceolatis sepalo dorsali conformi suppositis, labello subcordato-ovato basin versus saccato sub sacco cornu valido inflexo instructo processibus teretibus rigidis fimbriato basi dentato, columnæ acuminatæ cirrhis cornu labelli attingentibus.——*Demerara.*——Flowers dull green, richly spotted with deep blackish purple.

25. C. *lanciferum* (Lindl. in Bot. Reg. 1841. t. 5. f. 5.); petalis maculatis lineari-lanceolatis sepalo dorsali conformi suppositis, labello subcordato-ovato fimbriato basin versus saccato sub sacco cornu tripartito instructo, sub apice laminâ lineari-lanceolatâ aucto, columnæ cirrhis vix columnâ longioribus.——*Brazil.*——Very like C. cristatum; but the margin only of the lip is broken up into fringes; and at the point is a broad lancet-shaped spine.

26. C. *barbatum* (*Myanthus barbatus,* Lindl. Bot. Reg. t. 1778 *C. spinosum,* Lindl. Bot. Reg. 1841. misc. 136. *Myanthus spinosus,* Hooker Bot. Mag. t. 3802. *C. proboscideum,* Lindl. in Bot. Reg. 1839. misc. 140. 1841. t. 5. f. 3.); petalis versus apicem serratis, labelli infra medium saccati margine pilis tenuibus succulentis albis longè fimbriato apice angustè attenuato recurvo supra ad basin spinâ tripartitâ infra apicem spinâ magnâ dentatâ porrectâ, cristæ pilis opacis subulatis apice sæpe bifidis per paria tria quatuorve basi connatis.——*Brazil, Demerara.*——Sepals and petals narrow, green, blotched with purple. Lip green or pink, with narrow delicate fibres arising from its margin. No doubt the supposed species quoted above are mere varieties of the same natural form.

27. C. *cristatum* (Lindl. in Bot. Reg. t. 966.); foliis oblongo-lanceolatis perianthiis explanatis, sepalo supremo petalisque conniventibus, labello fimbriato cristatoque expanso cristæ pilis crassis lucidis compressis obtusissimis emarginatisque simplicibus.——*Brazil.*——Both this and the last have a strong smell of Tarragon; they are probably varieties of each other.——Monstrous State.

—Has been found to sport into *C. tridentatum;* see Botanical Reg. t. 1951.

41. LYCASTE crinīta.

L. crinita; sepalis petalisque intus crinitis, labello intus villoso altè trilobo laciniis angustis obtusis intermediâ ovali, appendice lineari adnato; facie L. cruentæ.

This plant is much like L. cruenta, and has flowers of nearly the same colour; but it differs in being smaller, with the interior clothed with very long silky hairs, and in the very different form of the lobes of the lip; which is moreover destitute of the crimson stain that gives its name to L. cruenta. The appendage of the lip is moreover long and elevated, not very short and inconspicuous. We have received a specimen from Messrs. Loddiges.

42. VANDA furva.

V. furva (Lindl. Gen. & Sp. Orch. p. 215. *V. Roxburghii unicolor,* Hooker in Bot. Mag. t. 3416. *Angræcum furvum,* Rumph. Amb. 6. t. 46. f. 1. *Epidendrum furvum,* Linn. sp. pl. 1348. *Cymbidium furvum,* Willd. Sp. Pl. 4. 103.); caule alto, foliis laxis membranaceis apice obliquè tridentatis, racemis lateralibus plurifloris, floribus distantibus, sepalis petalisque oblongo-obovatis undulatis unicoloribus obtusis, labelli trilobi lobis lateralibus obtusis intermedio cuneato bilobo.

I am indebted to Mr. Loddiges for suggesting that this is in all probability the Vanda furva about which there has been so much uncertainty. It certainly corresponds very closely with Rumphius's figure; but it comes from China, and not from Amboyna. It has the habit of V. Roxburghii, to which Sir W. Hooker has referred it; but it differs not only in the whole colour of its sepals and petals, and the other characters above indicated, but also in being a large lax growing plant, five or six feet high, with much thinner and longer leaves. The former is a short growing plant, with the leaves very closely set together. The middle lobe of the lip of V. Roxburghii is oblong and contracted at the point; here it is wedge-shaped and two-lobed.

44. HEXADESMIA bicornis.

H. bicornis; caulibus fusiformibus elongatis, foliis linearibus obliquè bidentatis, floribus subsolitariis cernuis, labello ovato retuso subserrato, columnâ utrinque bicorni.

An addition to the insignificant genus Hexadesmia, sent to Mr. Loddiges from Columbia by Linden. It is very like H. fasciculata, but differs in the leaves and column.

45. DENDROBIUM crinifĕrum.

D. criniferum; caulibus teretibus erectis ramosis, foliis oblongis subundulatis emarginatis, floribus axillaribus lateralibusque solitariis cernuis, sepalis ovatis, petalis linearibus acuminatis, labelli trilobi crenulati lobis lateralibus nanis intermedio apice in filis intricatis multifido lamellis binis undulatis.

An orchidaceous plant with small yellowish flowers of little beauty, introduced by His Grace the Duke of Northumberland, who received it from Mr. Power in Ceylon. It is closely allied to D. Scopa, but is readily distinguished by its slender stems, wavy leaves, and the shortness of the lateral lobes of the labellum. The breaking up the point of the lip into numerous threads is remarkable, and resembles what occurs in the bearded Bolbophylla, which are no doubt the Desmotricha of Blume.

47. LYCASTE aromatica; *var.* retŭsa.

L. aromatica (Supra 1843. misc. p. 16.; var. *retusa;* floribus vitellinis, labelli lobo medio dilatato retuso.)

We have received a flower of this from Geo. Barker, Esq. of Birmingham, who obtained it from Lima. It differs from L. aromatica in having very clear yolk-of-egg-yellow flowers, without a tinge of green, and the middle lobe of the lip dilated and almost 2-lobed.

48. LYCASTE gigantēa.

Lindl. in Bot. Reg. 1843. *misc. p.* 16.

This very fine plant, formerly described from Mr. Hartweg's dried specimens, has just flowered with Mr. Barker. It proves to have a very different aspect from any of the species associated with it; its flowers, whose sepals are 3½ inches long, being of a warm yellowish olive, and its lip a deep rich morone, bordered with orange, and looking like the finest velvet. As we shall soon figure it, a more particular account is deferred.

49. ASPASIA lunāta.

Lindl. in Bot. Reg. sub t. 1907.

We have received this curious Brazilian epiphyte from J. C. Lyons, Esq. of Ladiston near Mullingar, by whom it was obtained from the country about Rio. It has exactly the structure of the other Aspasias, and not a double caudicula, as appeared from the drawing of M. Descourlitz, on which the species was founded. The flowers have long narrow sepals and petals, yellowish blotched with brown; the lip is large and white, and is stained in the middle with pale purple, which had not, in the specimens sent us, that distinct crescent form from which the name has been derived. The anther has a large tumour in front, and a serrated sharp-ridged crest, which gives the column much the appearance of a parrot's head.

51. LYCASTE Barringtoniæ.

L. Barringtoniæ (*Epidendrum Barringtoniæ,* Smith. ic. pict. t. 25. *Dendrobium Barringtoniæ,* Swartz. Nov. Act. Ups. 6. 82. Willd. Sp. Pl. 4. 132. Hooker Exot. Fl. t. 119. *Colax Barringtoniæ,* Lindl. Bot. Reg. 897. *Dendrobium ciliatum,* Swartz. Pers. Syn. 2. 523. *Maxillaria ciliata,* Fl. Peruv. Syst. p. 221.? Lindl. in Bot. Reg. 1206. *Maxillaria Barringtoniæ,* Lodd. Bot. Cab. t. 1824. Lindl. Gen. & Sp. Orch. no. 23.); pseudobulbis oblongis compressis costatis, foliis subternis oblongo-lanceolatis acuminatis plicatis, scapo radicali vaginato unifloro erecto pseudobulbo vix longiore, bracteâ ovarii longitudine, sepalis petalisque ovato-lanceolatis obtusis, labelli trilobi lobo medio ovato rotundato fimbriato apice concavo, callo lato omninò adnato altè sulcato apice dentato.

An examination of fresh specimens of this old plant, enables me to state that it is a true Lycaste, readily distinguished by its stout flower-stems, drooping flowers, and very deeply furrowed callus. The blossoms appear to vary in colour from green to a tawny yellow.

54. DENDRŌBIUM chlorops.

D. chlorops; caulibus teretibus, floribus laxè corymbosis, pedicellis filiformibus glaucescentibus, sepalis lineari-oblongis, petalis duplo latioribus obovatis, labelli trilobi plani lobis lateralibus nanis acutis intermedio lineari-oblongo apice paulò latiore basi villoso, cornu brevi conico.

A species from Bombay, with the habit of D. Heyneanum. It has small flowers of a pale nankin colour, while the base of the lip is a light pea green; this gives the flower the appearance of having a green eye, whence the specific name. It has lately flowered with Messrs. Loddiges.

55. CATASĒTUM ochrāceum.

C. ochraceum; sepalis petalisque ovatis secundis, labello cucullato integerrimo lævi apice in rostrum brevem latum obtusum carnosum contracto, cirrhis brevibus.

A new species in the way of C. luridum, with deep yellow ochre-coloured flowers. The lip is hooded, but contracted at the point into a kind of beak, thick at the edges and short. Mr. Hartweg sent it to the Horticultural Society from the Hacienda del Hospicio, in the province of Bogota.

56. BOLBOPHYLLUM Cheiri.

B. Cheiri; foliis oblongis coriaceis basi canaliculatis apice emarginatis, floribus solitariis (magnis), sepalis lanceolatis lateralibus basi ventricosis petalisque e latâ basi linearibus acuminatis in manûs speciem velut digiti conniventibus, labello mobili basi ovato concavo apice longè rostrato, columnâ edentulâ, polliniis 4 geminatis.

This very singular species was sent from Manilla to Mr. Loddiges, with whom it blossomed a few days since. The flowers are very large for the genus, the sepals being as much as an inch and a half long, a clear olive green, neatly marked with brownish stripes. The lip is jointed so loosely with its support that it falls forward every time the flower is waved by the wind; a phenomenon common, if not universal, among the Bolbophylla, but in none of them more remarkable than in this case. Both sepals and petals are so arranged that they converge very much like the human fingers when they are brought together without being bent; this circumstance has suggested the specific name.

57. VANDA parviflōra.

V. *parviflora;* racemo simplici, sepalis oblongis, petalis linearibus spathulatis, labelli trilobi lobis lateralibus ascendentibus acutis intermedio oblongo canaliculato spongioso bilamellato apice circulari denticulato, calcare angusto obtuso.

This little plant is a native of Bombay, whence Messrs. Loddiges received it. The flowers are small, pale ochre-coloured, with a lip sprinkled all over with extremely fine purple points; the middle lobe of the lip is rather spongy, has two broad ridges, between which runs a channel, and at the point it is almost exactly circular, with a a few small toothings. The four pollen-masses adhere to a narrow strap connected with a broad gland; but the rostellum is not in any degree extended into a beak. The habit of the plant is, I am informed, that of Vanda lamellata.

58. SPATHOGLOTTIS plicāta.

Blume Bijdragen, p. 401. *t.* 76.

This very pretty plant has flowered with Messrs. Loddiges, who received it from Penang. It has light purple flowers, and has quite the appearance of Paxtonia.

60. EPIDĒNDRUM Hanburĭi.

E. (Encyclium) *Hanburii;* pseudobulbis ovatis, foliis ensiformibus coriaceis recurvis, floribus racemosis distantibus, sepalis petalisque spathulatis patentibus, labelli tripartiti lobis lateralibus falcatis obtusis nanis erectis intermedio oblongo bilobo venis elevatis dichotomis sulcato ungue obtusè bicarinato glabriusculo.

A Mexican plant, with something the appearance of E. porphyreum, but far less handsome. The leaves are very coriaceous, shorter than the raceme, which is about two feet long. The sepals and petals are deep dull purple, the lip is pale rose, with crimson radiating veins. We owe it to R. Hanbury, Esq. of Stamford Hill.

61. PHYSŪRUS pictus.

Lindl. Gen. & Sp. Orch. p. 504.

Messrs. Loddiges have flowered this beautiful little thing, which rivals the Wana Rajah of Ceylon (Anæctochilus setaceus) in the singular beauty of the foliage. The leaves are curiously veined, and appear as if covered with a film of silver, especially when they are turned so as to allow the light to strike them obliquely. The flowers are small, white, in a short spike, and have a bar of blackish brown along the middle of each petal and sepal. It is a native of Brazil, where it has been found by sevĕral travellers. Martius met with it on the hills of Botofogo near Rio Janeiro, growing among decayed leaves.

62. DENDROBIUM.

When we some time since succeeded in breaking up the old genus Maxillaria, we had hoped to find some means of performing the same good office for Dendrobium (the Maxillaria of the East), which, as at present limited, is composed of species exceedingly different from each other in general appearance. We must confess, however, that up to the present time, we have met with little or no success; the peculiarity of habit being accompanied with almost no differences in the structure of the flowers. Indeed, it is not an easy matter to say how Bolbophyllum and Cirrhopetalum are to be strictly defined.

Under these circumstances, we think the most useful course to be taken will be to collect the species into groups, which shall be so limited as to bring together no species that do not quite correspond in habit; and then it is to be hoped that the arrangement will be but little disturbed, in case distinctions good enough for genera should be hereafter made out. In doing this I shall endeavour to observe as nearly as possible the course already taken by Blume, whose plants, however, I have never had the advantage of examining.

Among the crowd of species which constitute this genus, there is a large set, of which D. Pierardi may be taken as the example, all which have thin flat leaves, slender stems, and flowers growing uniformly in pairs from the sides of the stem opposite the leaves. They may be regarded as constituting Dendrobium proper, and therefore the first step will be to place them in one group, of which the following is the enumeration.

Sect. 1. EUDENDROBIUM. *Stems slender; Leaves, thin and flat; Flowers in pairs, or threes, opposite the leaves.* (*Grastidium* Blume.)

A. Lip undivided.

1. D. *macrophyllum* (Lindl. in Bot. Reg. 1839. p. 36. Sertum Orchidaceum, t. 35. *Dendrobium macranthum,* Bot. Mag. t. 3970); caulibus pendulis, foliis ovato-oblongis obtusis nervosis basi subcordatis, sepalis lanceolatis lateralibus parùm productis, petalis oblongis acutis, labello pubescente convoluto denticulato subunguiculato ovato callo baseos elevato transverso obsoletè trilobo.——*Manilla.* ——This is one of the handsomest species. The flowers are nine inches in circumference. The sepals and petals are a clear and bright rose-colour, the lip is downy and deeply stained with two large broad blood-red blotches at its base.
2. D. *moniliforme* (*Fu Ran,* Kæmpf. amœn. t. 865. *Epidendrum moniliforme,* Linn. sp. pl. 1352. *Dendrobium moniliforme,* Swartz act. Holm. 1800. p. 246. Willd. no. 19. Bot. Reg. t. 1314. L. no. 40); caulibus erectis ramosis, internodiis tumidis, foliis oblongis obliquè emarginatis obtusis, floribus geminatis foliis longioribus, sepalis petalisque oblongis acutis, labello cucullato acuto serrulato intus pubescente.——*Japan.*——This is very near D. cœrulescens, from which it is mainly distinguished by its stem contracted at every joint, and its serrated lip.
3. D. *cœrulescens* (Lindl. Sertum, t. 17); facie omnino D. nobilis, floribus nunc ternatis, petalis angustioribus margine reflexis valdè undulatis, labello ovali acuminato intus villoso——*E. Indies.*——The form of the petals, and especially of the lip, principally distinguish this from D. nobile. Its flowers are also smaller.
4. D. *nobile* (Lindl. Gen. no. 24. Sertum, t. 3); caulibus teretibus erectis, foliis oblongis obliquè emarginatis obtusis, floribus geminatis, sepalis ovalibus, petalis duplo majoribus undulatis, labello ovato subrotundo cucullato villoso.——*China.*——Flowers large, showy, violet, with a very dark purple stain in the interior of the lip.

* * *

5. D. *pulchellum* (Roxb. L. 35. Lodd. Bot. Cab. t. 1935); caulibus teretibus pendulis, foliis oblongo-lanceolatis subplicatis, racemis lateralibus strictis multifloris, bracteis brevibus ovatis obtusis, sepalis ovatis acuminatis subcostatis, petalis oblongis obtusis tenuioribus et latioribus, labello unguiculato cochleariformi obtusissimo ciliato. ——*Sylhet.*——Sepals whitish. Petals pink. Lip pink at the edge with a yellow centre. A most charming species.
6. D. *Devonianum* (Paxton Mag. 7. 169); caulibus pendulis, foliis ovato-lanceolatis acutis, pedunculis 2-3-floris, sepalis oblongis acutis integris, petalis duplò latioribus fimbriatis, labello maximo cochleariformi margine

plumoso.——*Khoseea Hills of India.*——The King of Dendrobiums. Sepals rose-colour. Petals and lip lemon-colour, tipped with purple, the latter with a broad yellow blotch in the middle.

7. D. *Pierardi* (Roxb.! Fl. ind. 3. 482. Hooker Exotic Flora, t. 9. Bot. Reg. t. 1756); caulibus pendulis glabris, foliis ovato-lanceolatis acutis, floribus geminatis, sepalis acuminatis membranaceis, petalis sepalo supremo majoribus acuminatis, labello cucullato dilatato subtruncato pubescente ciliato.——*E. Indies.*——Flowers delicate pink, with a pale yellow lip, not sweet-scented.

8. D. *cucullatum* (R. Brown in Bot. Reg. t. 548. ibid. t. 1756, fig. 2); facie omnino D. Pierardi sed labello subrotundo-ovato basi in unguem brevem convoluto——*E. Indies.*——Extremely like D. Pierardi, but it has larger flowers, with a broad roundish ovate lip, the base of which is rolled up into a sort of short stalk; while in D. Pierardi the lip is rounded and very blunt, with its base rolled into a much longer stalk which quite conceals the column. It is moreover remarkably incurved.

9. D. *transparens* (Wallich. Cat. no. 2008); caulibus pendulis teretibus, foliis ovato-lanceolatis acuminatis apice obliquis, floribus geminatis, sepalis acuminatis, petalis obtusis sepalo supremo majoribus, labello oblongo undulato ciliolato obtuso intùs levissimè pubescente.——*Nepal.*——Flowers pale rose-colour, with a lip of the same tint. Very like D. Pierardi, from which it is distinguished by its leaves being longer and emarginate at the apex, by the petals being obtuse and the labellum much less hairy, and pink not yellow. The stems are also far shorter.

10. D. *amœnum* (Wall. L. no. 21. *Limodorum aphyllum*, Roxb. corom. 1. t. 41. *Cymbidium aphyllum*, Willd. sp. f. 4); caulibus pendulis gracilibus nebulosis floriferis aphyllis, foliis lanceolatis acutis, floribus geminatis racemum spurium formantibus longè pedunculatis divaricatis, sepalis oblongis obtusis, petalis obtusis sepalo supremo paulò majoribus, labelli cucullati limbo ovato obtuso crenulato leviter ciliato: disco villoso.——*Nepal.*——This lovely species is abundantly distinguished by its slender stems, leafless, and clouded with dusky brown when in blossom; and also by its delicate white flowers, the tips of whose segments are remarkably blunt and tinged with delicate violet. Dr. Wallich remarks in his MSS. that its fragrance is exquisite, not unlike that of Olea fragrans.

11. D. *heterocarpum* (Wall. L. no. 20); caulibus teretibus pendulis, foliis oblongis acutis planis, floribus geminatis odoratis racemum spurium formantibus, sepalis lineari-oblongis acutis, petalis ovatis acutis sepalo supremo majoribus, labello unguiculato, limbo subpanduriformi holosericeo medio elongato acuminato plano.——*Nepal.*——Very near D. cucullatum from which it is distinguished by its pale yellow fragrant flowers, and the form of the labellum.

12. D. *macrostachyum* (L. no. 19. Bot. Reg. t. 1865); caulibus teretibus pendulis flagelliformibus, foliis ovato-lanceolatis submembranaceis, floribus ternatis racemum spurium formantibus, sepalis ovatis acutis, petalis lanceolatis sepalo supremo subæqualibus, labello cucullato venoso: limbo ovato obtuso ciliato intùs pubescente.——*Ceylon.*——Flowers pale yellow, smaller than in D. Pierardi. Lip streaked with purple veins.

* * *

13. D. *foliosum* (Ad. Brongn. in Duperr. Voy. t. 41).

14. D. *gemellum* (L. no. 28. *Pedilonum biflorum*, Blume, p. 322); foliis membranaceis lanceolatis acuminatis, pedunculis oppositifoliis subbifloris, labello ovali acuto apice subcrispo lamellis tribus serrulatis inconspicuis.——*Sincapore, Java.*——A long-stemmed grassy-leaved plant, with small pale yellowish green flowers, growing in pairs from short rigid two-valved truncated compressed spathes, placed opposite the leaves. In this respect it is like *D. biflorum*, of which it has all the habit; but the lip is perfectly entire and not three-lobed.

15. D. *rugosum* (L. no. 57. *Grastidium rugosum*, Blume, p. 333); foliis subcoriaceis lanceolatis apice emarginatis, sepalis et labelli limbo acuminatis, labello intùs membranaceo-rugoso.——*Java.*——Flowers pale yellow.

16. D. *salaccense* (L. no. 56. *Grastidium salaccense*, Blume, p. 333); foliis membranaceis lineari-lanceolatis obliquè retusis, sepalis acutis, labello obtuso intùs 1-lineato.——*Java.*——Flowers deep yellow.

17. D. *acuminatissimum* (L. no. 55. *Grastidium acuminatissimum*, Blume, p. 335); foliis membranaceis linearibus sepalisque acuminatissimis.——*Java.*——Flowers greenish.

* * *

18. D. *chrysanthum* (Wall. Cat. no. 2012. Bot. Reg. 1299 & 25); caulibus teretibus pendulis, foliis contortis ovato-lanceolatis acuminatis, floribus inter folia nascentibus, sepalis carnosis oblongis obtusis: venis extùs tuberculatis, petalis obovatis retusis carnosis sepalo supremo latioribus, labello cucullato denticulato retuso obsoletè trilobo.——*Nepal.*——Flowers dark yellow, occasionally in threes, marked externally with distinct warts. Lip with a deep purple spot on each side.

19. D. *Paxtoni* (Lindl. in Bot. Reg. Misc. 1839, no. 56); caulibus teretibus sulcatis, foliis ovato-lanceolatis acuminatis apice hinc obsoletè emarginatis, pedunculis bifloris, sepalis oblongis acutis lateralibus basi parùm productis, petalis latioribus obovatis acutis serrulatis, labello unguiculato ovato concavo indiviso villoso margine multifido fimbriato.——*Khoseea Hills of India.*——This has orange-yellow flowers, with a deep brown spot in the middle of the lip. It differs from D. chrysanthum in having the petals serrated, and the surface and edge of the lip hairy.

20. D. *ochreatum* (Lindl. in Wall. Cat. no. 7410. Bot. Reg. sub fol. 1756. *D. Cambridgeanum*, Paxton, Mag. Bot. vi. 265); caulibus pendulis medio incrassatis, foliis ovalibus acutissimis, pedunculis bifloris, petalis lanceolatis sepalis latioribus, labello cucullato rotundato indiviso supra piloso.——*Khoseea Hills of India.*——A beautiful species, with large rich yellow flowers, having a deep purple stain in the middle of the lip, which is entire and hairy.

21. D. *aureum* (Lindl. Gen. & Sp. Orch. p. 78. Bot. Reg. 1839. t. 20); caulibus erectis teretibus clavatis internodiis brevibus, foliis lineari-oblongis apice obliquis emarginatis, pedunculis subbifloris aggregatis lateralibus, sepalis lineari-ovatis acuminatis obtusis, petalis latioribus ovatis acutis undulatis, labello ovali undulato obtusiusculo indiviso medio serrulato per axim pubescente.——*Ceylon.*——The species varies with pale yellow and white flowers. In both varieties the fragrance is remarkable, forming a something intermediate between violets and primroses.

22. D. *sulcatum* (Lindl. in Bot. Reg. 1838. t. 65); caule erecto clavato sulcato subflexuoso compresso, foliis oblongis acutis 3-nerviis, pedunculis lateralibus trifloris, bracteis minimis acutis appressis, petalis oblongis obtusis glabris sepalis subæqualibus, labello obcordato hirsuto ungue canaliculato sulcato.

* * *

23. D. *nutans* (L. no. 73.); caulibus erectis subclavatis hispidis apice foliosis, foliis ovato-lanceolatis apice valdè obliquè unciuato-bilobis, pedunculis subterminalibus bifloris foliis brevioribus, bracteis minimis, petalis sepalisque lineari-lanceolatis acuminatis æqualibus membranaceis, labello conformi carnoso crenulato indiviso, capsulis oblongis nutantibus.——*Ceylon.*——Leaves small. Flowers small, white, slightly tinged with green; lip yellow.

24. D. *candidum* (Wallich in Bot. Reg. 1838, misc. 54); caulibus erectis flexuosis teretibus, foliis ovato-lanceolatis apice obliquis obtusis, pedunculis axillaribus erectis bifloris, sepalis patulis lateralibus basi parùm productis, petalis conformibus, labello ovato-lanceolato obtuso medio barbato supra basin calloso lateribus erectis cucullatis.

---*Khooseea Hills of India.*——Flowers small, pure white, sweet-scented. Stems about a foot high.

25. D. *stuposum* (Lindl. in Bot. Reg. 1838. misc. 94); caule erecto tereti, foliis lineari-oblongis obliquè emarginatis, pedunculis bifloris, bracteis obtusis cucullatis, sepalis petalisque erectis lateralibus carinatis, labello oblongo cucullato obtuso sub apice calloso stuposo.——*E. Indies.* ——Flowers white, small. Lip with a deep orange callus below its point, where it is thickly covered with a coarse tow-like hairiness.

B. Lip 3-lobed.

26. D. *Ruckeri* (Lindl. in Bot. Reg. 1843. misc. 38. t. 60); caulibus teretibus, foliis ovato-lanceolatis acuminatis planis, floribus gemellis, sepalis patulis obtusiusculis convexis margine reflexis lateralibus subtriangularibus, labelli trilobi lobis rotundatis intermedio undulato axi elevatâ villosâ. —— *Philippines?* —— A fine species. Flowers large, yellow. Lip white outside, yellow and brown inside.

27. D. *sanguinolentum* (Lindl. in Bot. Reg. 1842. misc. 73. 1843. t. 6); caulibus teretibus pendulis, foliis ovato-lanceolatis, floribus gemellis, sepalis petalisque ovatis obtusiusculis patulis, labello trilobo glabro: laciniâ intermediâ retusâ nunc utrinque plicatâ; ungue concavo brevi appendice cornuformi pubescente retrorsâ aucto.——*Ceylon.*——This has pendulous stems like those of D. Pierardi, but of a delicate purple when young; the leaves too are stained underneath and at the edges with the same colour. The flowers are as large as those of D. aggregatum, of a clear fawn colour, with the tips of the segments and lip stained with a deep rich violet. There is moreover a *scarlet* spot in the middle of the lip.

* * *

28. D. *revolutum* (Lindl. in Bot. Reg. 1840, misc. 110); caulibus obtusè ancipitibus, foliis ovato-oblongis obtusis apice obliquis emarginatis basi subcarinatis, floribus solitariis oppositifoliis, sepalis petalisque acutissimis revolutis, labello carnoso convexo obtuso subtrilobo aut rhombeo per medium exarato (s. lamellis duabus inflexis parallelis instructo) lineis tribus discoloribus.——*Sincapore.*——Flowers straw-coloured, and about the size of those of D. Pierardi, but their lip is fleshy, convex, almost lozenge-shaped, and marked with 3 brown lines.

29. D. *excisum* (Lindl. in Bot. Reg. 1841. misc. 165); foliis lineari-oblongis obliquè retusis, floribus (solitariis?) oppositifoliis basi squamatis, sepalis petalisque acuminatis, labello oblongo acuto carnoso basi utrinque exciso: disco latè glanduloso, cornu pedicello ferè duplò breviore. ——*Sincapore.*——A small species with the habit of D. pulchellum, but with white inconspicuous flowers.

30. D. *aqueum* (Lindl. in Bot. Reg. 1843. misc. 6. t. 54); foliis ovato-oblongis undulatis acuminatis, floribus patulis cornu brevi obtuso, sepalis petalisque ovatis, labelli trilobi pubescentis laciniâ mediâ ovatâ denticulatâ basi altè excavatâ lateralibus latioribus rotundatis margine anteriore serrulato.——*Bombay.*——With the manner of growth of D. Pierardi, the pale watery green flowers of this are quite destitute of the attractive colours of that gay species, and are entirely different in the structure of the lip, which is furnished with a large cavity, almost a pouch at the base of the middle lobe. It is also a much stouter plant with wavy leaves.

31. D. *longicornu* (Lindl. in Wall. Cat. no. 1997. Bot. Reg. t. 1315. Gen. et Sp. Orch. no. 26); caulibus erectis hispidis flexuosis, foliis ovato-lanceolatis apice valdè obliquis, floribus fasciculatis v. solitariis terminalibus, bracteis ovatis acuminatis hispidis pedicello multò brevioribus, sepalis lateralibus in calcare longo acuminato connatis, petalis sepalo supremo conformibus, labello infundibulari cucullato integro dentato basi cum pede columnæ accreto.——*Nepal.*——Flowers large, white. Lip yellow, with orange-coloured veins. A very fine species.

* * *

32. D. *tridentiferum* (Lindl. in Hooker's Journal of Botany, vol. 3); foliis oblongo-lanceolatis obliquè emarginatis, gemmis paleaceis, floribus geminis, sepalis lateralibus ovatis carnosis obliquis, petalis lanceolato-oblongis acutis membranaceis, labello carnoso trifido basi tuberculo carnoso lineâque utrinque elevatâ acuto: lobis acutis lateralibus antrorsum curvis intermedio ovato, cornu brevi obtuso.——*New Guinea.*—— A broad-leaved species looking like D. biflorum to which it is nearly allied. It has fleshy flowers as large as those of D. Pierardi.

33. D. *Luzonense*; foliis distichis angustè lanceolatis apice oblique emarginatis, pedunculis brevibus 2-floris, labello spathulato obtusissimo laciniis lateralibus obsoletis denticulatis.——*Luzon.*——A small-flowered plant, with very closely set distichous leaves.

34. D. *Cunninghamii* (Lindl. in Bot. Reg. sub t. 1756. D. *biflorum*, Ach. Rich. sub Astrolab. t. 26); caulibus gracilibus pendulis ramosis vaginis foliorum corneis transversim corrugatis squamatis, foliis ovato-linearibus obtusiusculis, pedunculis oppositifoliis bifloris foliis multò brevioribus, sepalis ovatis acutis, petalis oblongis acutis latioribus, labelli trilobi lobo intermedio subrotundo undulato basi 5-lamellato lateralibus nanis acutis.——*New Zealand.* ——Stems very slender, much branched. Flowers white.

35. D. *biflorum* (*Epidendrum biflorum*, Forst.! Prodr. n. 318. *Dendrobium biflorum*, Swartz. act. Holm. 1800. p. 246. Willd. no. 17. and no. 18); caule pendulo gracili tereti, foliis lineari-lanceolatis acuminatis planis, pedunculis bifloris lateralibus e paleis corneis erumpentibus, sepalis petalisque acuminatissimis, labello rhomboideo medio bilamellato trilobo lobis lateralibus acutis nanis intermedio deltoideo acuminato margine fimbriato.——*Society Islands.*——Leaves very narrow and grassy. Flowers very small, with setaceous points.

36. D. *calcaratum* (Lindl. in Bot. Reg. 1840. misc. 219); caulibus nigro-pubescentibus, foliis linearibus obliquè bilobis, floribus resupinatis, sepalis ovatis lateralibus revolutis basi in cornu conicum pedicelli longitudine productis, petalis lineari-lanceolatis, labelli linearis carnosi lævis trilobi laciniâ intermediâ subrotundâ marginibus incurvis ungue cum columnæ pede in calcar verum connato.——*Sincapore.*——A slender inconspicuous species, allied to D. *revolutum* and *biflorum;* with green flowers growing in pairs opposite the leaves.

37. D. *bilobum* (Lindl. in Hook. Journ. Bot. vol. 3); foliis linearibus obtusis apice subæqualibus bilobis, floribus membranaceis minutis (solitariis?), sepalis obtusis, petalis duplò minoribus apiculatis, labello libero elongato obovato apice carnoso trilobo: laciniâ mediâ verrucosâ, cornu elongato obtuso.——*New Guinea.*——A small inconspicuous species, with the appearance of Isochilus linearis.

* * *

38. D. *pugioniforme* (A. Cunn. in Bot. Reg. 1839. misc. 34); caulibus repentibus nodoso-articulatis radicalibus setoso-stipulatis, foliis elliptico-lanceolatis apice attenuatis acutis carnosis infra fere convexis carinatis lævibus nitentibus, floribus (resupinatis) subsolitariis axillaribus, perianthii foliolis oblongo-lanceolatis acutis patentibus, labello longitudinaliter tricarinato, lobo intermedio dilatato undulato-crispato apice acuto recurvato.——*N. Holland.*——A beautiful epiphyte, hanging loosely from the stems of trees, so as to swing freely to the wind that sweeps through the forests on the summit of the belt of mountain bounding the coast district of Illawarra, on the west (lat. 34½ S.)

VERY nearly allied to this section of the genus are those Dendrobia, which, with slender stems and thin flat leaves, have flowers growing in racemes. They form a group which may be subdivided upon the same principle as the last.

Sect. II. STACHYOBIUM. *Stems slender; Leaves thin and flat; Flowers in racemes.*

A. Lip undivided.

39. D. *aduncum* (Lindl. in Bot. Reg. 1842. misc. 62); caulibus pendulis, foliis lineari-lanceolatis acutis integris, floribus ternis patentibus, sepalis petalisque ovatis obtusis lateralibus duplò latioribus, cornu rotundato, labello unguiculato ovato concavo apiculato columnæ arctè appresso intùs villoso disco glabro, columnâ apice bialatâ sub stigmate villosâ, antherâ glanbulosâ.——*East Indies.* ——This is a beautiful species allied to D. Pierardi, near which it should perhaps be placed, rather than in this division. The flowers are almost transparent, of the most delicate pink, and nearly as large as in D. moschatum.

40. D. *formosum* (Roxb. fl. ind. 3. 485. L. p. 81. Wall. pl. as. rar. t. 39. Bot. Reg. 1839. t. 64); caulibus teretibus pendulis pilosis, foliis distichis ovatis apice obliquè emarginatis obtusis, racemo brevi terminali 4-5-floro, bracteis brevibus ovatis, (floribus maximis), sepalis oblongis acutis: lateralibus basi longè productis, petalis duplò latioribus acutis, labello obovato dilatato retuso cum basi columnæ in calcar obtusum connato.——*East Indies.*——One of the finest of the genus, with very large white flowers in terminal racemes.

41. D. *rhombeum* (Lindl. in Bot. Reg. 1843. t. 17); caulibus teretibus foliosis, foliis lanceolatis acutis, racemis brevibus 4-floris, sepalis ovalibus obtusis, petalis duplò latioribus ovatis, labello rhombeo undulato acuto medio pubescente, columnâ utrinque tuberculatâ pone basin cornu deflexo auctâ, antherâ pubescente.——*Manilla.*——This species has much general resemblance to D. aureum, from which however it differs in having smaller blossoms, a labellum without serratures, and the flowers in short racemes instead of pairs. Here, as in some other cases, the column is furnished, near the base, with a deflexed horn lodged in a niche just above the insertion of the lip. This is a remarkable process, whose nature requires to be investigated. Is it the unguiculus of Blume?

42. D. *fimbriatum* (Hooker. Exot. Fl. 71. Wall. Cat. no. 2011. L. no. 38); caulibus teretibus pendulis, foliis ovato-lanceolatis, racemis lateralibus multifloris, sepalis oblongis undulatis patentissimis, petalis majoribus undulatis ciliatis, labello indiviso rotundato convexo fimbriato, fimbriis laceris.——*Nepal, Burma.*——Flowers very deep yellow, sometimes stained with crimson at the base of the lip. It has much the appearance of Dendrobium chrysanthum, from which it is distinguishable both by its inflorescence and the different shape of its petals.

43. D. *polyanthum* (Wall. Cat. no. 2009. L. no. 31); caulibus teretibus pendulis, foliis oblongo-lanceolatis apice integris, racemis 3-4-floris lateralibus foliis brevioribus, sepalis lanceolatis: lateralibus supremo subæqualibus, petalis majoribus oblongis undulatis, obtusis, labello obovato cucullato ciliato, capsulis fusiformibus apteris. ——*Moolmayne.*——This plant is evidently very near D. moschatum. It however appears to have smaller flowers, with far narrower segments, and, as far as can be ascertained from dried specimens, they are nearly colourless, except the lip, which is yellow.

44. D. *moschatum* (Wallich in L. no. 36. Hooker in Bot. Mag. t. 3837. *Cymbidium moschatum*, Willd. *Epidendrum moschatum*, Hamilt. in Symes' Ava, 3. 315. c. ic. *D. Calceolus*, Hooker ex. fl. t. 184. L. no. 37. *D. cupreum*, Herbert in Bot. Reg. t. 1779. *D. clavatum*, Wall. cat. no. 2004); caulibus teretibus pendulis, foliis lineari-lanceolatis oblongisque, racemis lateralibus laxis 5-6-floris, bracteis pedicellis 3-plò brevioribus, sepalis patentissimis petalisque oblongis obtusis latioribus venosis reticulatis, labello unguiculato calceolari obtusissimo margine abruptè inflexo pubescente.——*Burma, Pegu, Ava, Sylhet.*——This beautiful plant, with large nankin coloured flowers, stained and veined with crimson, is readily known by its labellum, whose edge is turned inwards and delicately fringed with hairs. Its odour is compared by some to musk, by others to rhubarb, and by others to wood-roof.

45. D. *calcaratum* (A. Richard Sert. Astrolab. t. 7); caule ramoso quasi articulato florifero aphyllo, foliis lineari-lanceolatis acutis, floribus spicato-corymbosis, bracteis linearibus acutis glabris, sepalis lateralibus in cornu clavatum fissum productis, labello lineari subspathulato plano acuto.——*Vanikoro.*——Known to me only by the figure and description in the Voyage of the Astrolabe. It has the habit of D. ramosum. Possibly it should be referred to the § Pedilonum.

46. D. *flavescens* (L. no. 48. *Onychium flavescens*, Blume, p. 325); ebulbe, foliis paucis lineari-lanceolatis obtusis basi canaliculatis, floribus alternis subspicatis terminalibus pubentibus, labello intùs pubescente, limbo emarginato plicato.——*Java.*——Flowers small, yellow.

47. D. *nudum* (L. no. 49. *Onychium nudum*, Blume, p. 325); foliis lineari-lanceolatis inæquali-acuminatis membranaceis, racemis paucifloris, sepalis interioribus lanceolatis dorsali angustioribus.——*Java.*——Stems jointed, zigzag at the upper part. Flowers pale purple, changing to yellow. Labellum with a little claw at the base; according to Blume.

48. D. *ramosum* (L. no. 34); caulibus pendulis gracilibus ramosis glabris, foliis lanceolatis acuminatis apice integerrimis, floribus subracemosis, bracteis membranaceis reflexis obtusis, sepalis lateralibus magnis in cornu elongatum rectum connatis, petalis minoribus, labello cucullato acuto crispo.——*India.*——This plant is only known by some imperfect specimens distributed by Dr. Wallich. It is much more slender than D. Pierardi, has an evident tendency to branch, and appears to be a distinct species. It however requires further examination. It is near D. herbaceum; but the flowers have a long and conspicuous spur.

49. D. *herbaceum* (Lindl. Bot. Reg. 1840, misc. 153); caulibus ramosis teretibus glabris, foliis lineari-lanceolatis acuminatis apice obliquè bilobis, racemo paucifloro terminali, sepalis lateralibus basi paulo productis dorsali petalisque linearibus, labello lanceolato integerrimo nudo, clinandrio biauri.——*East Indies.*——A small branching plant, with leaves from two to three inches long, and inconspicuous greenish flowers.

50. D. *mutabile* (L. no. 50. *Onychium mutabile*, Blume, p. 324); foliis lineari-lanceolatis obtusis obliquè retusis membranaceis.——*Java.*——Flowers pale rose-colour, in very close racemes. Column downy in front. Petals broader than the upper sepal. Labellum with three sulphur yellow glands, furnished at the base with a free claw, according to Dr. Blume.

51. D. *sclerophyllum* (*D. rigidum*, L. no. 51. nec R. Br. *Onychium rigidum*, Blume, p. 324); foliis lato-lanceolatis obliquè retusis rigidis.——*Java.*——Flowers whitish, with three yellow glands. Dr. Blume doubts whether it may not be a variety of his D. mutabile, but he says that its parts are more firm.

52. D. *auriferum* (L. no. 39); caulibus teretibus pendulis, foliis oblongis obtusis integris, racemis lateralibus horizontalibus clavatis foliorum longitudine, bracteis cucullatis densè imbricatis, sepalis petalis labelloque lineari-lanceolatis conniventibus acuminatissimis.——*China.* ——Stem slender, throwing out very long roots from its sides. Leaves oblong-lanceolate. Flowers about an inch long, bright yellow, arising from short broad imbricated bracts, which are much longer than the spaces between them. Known only from a drawing in the possession of the Horticultural Society.

B. Lip 3-lobed.

53. D. *barbatulum* (L. no. 44. *D. chlorops*, Lindl. in Bot. Reg. 1844. misc. 54); caulibus teretibus, floribus laxè corymbosis, pedicellis filiformibus glaucescentibus, sepalis lineari-oblongis, petalis duplo latioribus obovatis, labelli trilobi plani lobis lateralibus nanis acutis intermedio lineari-oblongo apice paulò latiore basi villoso, cornu brevi conico.——*Bombay.*——A species with the habit of D. Heyneanum. It has small flowers of a pale nankin

colour, while the base of the lip is a light pea green; this gives the flower the appearance of having a green eye. When dried the parts of the flower become extremely thin and difficult to examine; and I was thus led to suppose that the lip was undivided. I have however ascertained that it is really three-lobed, and that my D. chlorops is the same species, more correctly defined.

54. D. *lancifolium* (Ach. Richard Sert. Astrolab. t. 8); caule erecto ramoso folioso tereti glabro, foliis angustè lanceolato-linearibus acutis, floribus purpurascentibus sæpiùs quaternis pedicellatis subspicatis, sepalis lanceolatis acutis, labelli unguiculati trilobi lobo medio latiore ovato margine subcrispo lateralibus obtusis subcrispis planis.——*Moluccas*——Only known to us by M. Richard's description. It is a branching species, having little resemblance to the plants here placed near it.

55. D. *bicameratum* (Lindl. in Bot. Reg. 1839. misc. 85); caulibus fusiformibus, foliis lineari-lanceolatis apice obliquè bidentatis, pedunculis lateralibus squamatis 4-floris, floribus concavis subcarnosis, sepalis petalisque subrotundo-ovatis acutis, labelli trilobi laciniâ intermediâ rotundatâ apiculatâ carnosâ lateralibus triangularibus acutis breviore, columnæ facie excavatâ semibiloculari.——*North of India.*——The stems grow nearly upright, and are rather short, fusiform and furrowed when old. The flowers are a little smaller than those of Maxillaria stapelioides, and like them in form; their colour is dull yellow, spotted and streaked with purple. In the only specimen I have ever seen the flowers grew in fours, on a very short peduncle, closely covered with ovate concave bracts, speckled with purple. The excavation of the face of the column, and its division into two cells are remarkable.

56. D. *elongatum* (A. Cunn. Bot. Reg. 1839. misc. 33); caulibus teretibus sulcatis elongatis erectis 4-5-phyllis, foliis ovato- lanceolatis acutiusculis apice obliquis emarginatis, racemo terminali multifloro, foliolis perianthii ovatis obtusiusculis, labelli disco 3-carinato lobo intermedio lato cordiformi.——*Shaded dark woods, on the banks of the Brisbane River, New Holland.*——This plant has erect stems about a foot and a half high, at the end of which grow about four lanceolate leaves, from the midst of which springs a raceme of yellowish flowers spotted with red. They seem unwilling or unable to expand.

57. D. *bicolor* (L. no. 74); caulibus erectis elongatis subclavatis apice foliosis, foliis lanceolatis acutis, racemo brevi erecto terminali v. subterminali 3-4-floro foliis multò breviore, bracteis oblongis obtusis petaloideis venosis, petalis sepalisque oblongis obtusis æqualibus, labello trilobo obtuso: lobis lateralibus erectis rotundatis.——*Ceylon.*——Bracts large, oblong, petaloid, veiny, two-coloured, green at the point, brownish red at the base. Flowers nearly white. This is scarcely known except from a drawing, and I am by no means sure that it is not an Eria.

58. D. *catenatum* (L. no. 41. *Epidendrum monile*, Thunb. Fl. Jap.); caulibus teretibus erectis: internodiis tumidis, foliis lineari-lanceolatis acutis, racemis lateralibus 4-5-floris, sepalis petalisque lanceolatis acuminatis, labello oblongo apice trilobo.——*Japan, China.*——This has regularly jointed stems, 5 or 6 inches high, small narrow leaves and flowers with green sepals and white petals. I only know it from a Chinese drawing in the possession of the Horticultural Society.

59. D. *denudans* (Don. Prodr. 34. Wall. Cat. no. 2014. L. no. 42); caulibus brevibus cæspitosis erectis vaginis laxis foliorum vestitis, foliis oblongo-lanceolatis submembranaceis apice obliquis emarginatis, racemis lateralibus terminalibusque nutantibus multifloris, sepalis acuminatis: lateralibus basi incurvis, petalis minoribus, labelli trilobi oblongi lobo medio acuminato crispo lateralibus serratis, disco bilamellato.——*Nepal.*——Flowers greenish white. Lip green, whole-coloured. This species is the most variable that I know in point of size. Some of Dr. Wallich's specimens are not more than an inch and half high, with two or three leaves and as many flowers on each individual, while others are six or seven inches high, with numerous distichous leaves, and three or four long nodding racemes, each consisting of eight or nine flowers. The species is always distinguishable by the loose membranous sheaths which clothe the stem when the leaves have fallen off.

60. D. *alpestre* (Royle Illustr. t. 88. fig. 2); caulibus brevissimis ovatis 3-4-phyllis, foliis lineari-oblongis, racemis 3-5-floris erectis, sepalis acuminatis cornu brevi obtuso, petalis minoribus, labelli trilobi lanceolati lobo medio acuminato crispo basi angustato lateralibus laceris, disco bilamellato.——*Himalayas.*——Stems and leaves only a few inches high. Flowers white, with a purplish lip.

61. D. *cuspidatum* (Lindl. no. 43); caulibus brevibus erectis 3-5-phyllis, foliis lineari-oblongis apice obliquis emarginatis, racemis terminalibus 3-4-floris foliis brevioribus, bracteis lineari-ovatis acutis, petalis sepalisque lineari-lanceolatis acuminatis æqualibus: horum lateralibus basi incurvis cornu referentibus, labelli trilobi petalis duplò brevioris lobis lateralibus maximis erectis rotundatis integerrimis: intermedio ovato acuminato.——*Tavoy.*——This resembles in many respects D. denudans, from which it is abundantly distinguished by the loose sheaths of the fallen leaves not covering the stem, and by the peculiar form of the lip. My specimens are from three to four inches high. The flowers appear to be white.

62. D. *microbolbon* (Ach. Richard. Ann. Sc. n. s. 15. t. 8); caulibus parvis ovatis aggregatis diphyllis, foliis oblongo-linearibus submembranaceis acutis, scapo gracili foliis longiori apice trifloro, racemo trifloro, petalis linearibus obtusis, labello crasso erecto arcuato subcanaliculato quasi trilobo, lobis lateralibus minimis medio rotundato obsoletè crenulato.——*Nilgherries.*——About three or four inches high. Lip pale green. Sepals whitish flesh-colour.

63. D. *pygmæum* (Lindl. no. 45. *Dendrobium pusillum*, Don. Prodr. 35); caulibus erectis cæspitosis vaginis laxis foliorum vestitis, foliis lineari-lanceolatis, racemis suberectis terminalibus multifloris foliorum longitudine, rachi angulato, bracteis membranaceis acuminatis, sepalis inferioribus ovatis acuminatis ad basin cornu incurvo connatis: supremo petalisque lineari-spatulatis acutis, labello unguiculato cuneato integro: callis duobus longitudinalibus in disco; margine anteriore carnoso crispo plicato lobo intermedio ovato.——*Martaban.*——A very small plant, forming tufts. Very near D. microbulbon, but different in the form of the lip; and the flowers are much smaller.

Nearly allied to the racemose Dendrobia just described is that curious race, already enumerated at t. 28 of our volume for last year, under the name of Spatulata. They are plants with hard stiff slender stems, clothed with coriaceous distichous leaves, and bear their flowers in long lateral many-flowered racemes. They are all remarkable for having the petals considerably longer than the sepals. In order to preserve some uniformity in nomenclature I would propose the name Ceratobium for them.

Sect. III. Ceratobium. *Stems erect, slender. Leaves coriaceous, distichous. Racemes lateral, many-flowered. Petals much longer than the sepals.*

64. D. *taurinum* (Lindl. in Bot. Reg. 1843. t. 28.); foliis oblongis obliquè emarginatis, racemo oblongo, sepalis ovatis obtusiusculis, petalis linearibus contortis duplò longioribus, labello oblongo apice crispo per axin lineis 3 elevatis intermediâ apice flexuosâ aucto.——*Manilla.*——This has large flowers, with yellowish green sepals, rolled back at the points, very long deep purple twisted petals, and a paler purple lip, with three elevated lines along the middle, and a few small tubercles near the apex. The stem is five feet high.

65. D. *Mirbelianum* (Gaudich. Voyage, t. 38.); foliis ovato-oblongis obtusis, racemis ovatis longè pedunculatis, sepalis ovato-lanceolatis acutis erectis, petalis spathulatis

unguiculatis obtusis longioribus, labelli trilobi ovato-lanceolati subsessilis lobis lateralibus rotundatis intermedio lanceolato undulato acuto, lineis tribus elevatis lateralibus versus basin duplicibus.——*New Guinea.*——The racemes, including their peduncle, are more than a foot long. The petals are an inch and a half long.

66. D. *veratrifolium* (Lindl. in London Jour. Bot. 2. 235.); foliis oblongis obtusis amplexicaulibus 9-11-nerviis, racemo terminali elongato multifloro, sepalis undulatis acutis, petalis spathulatis obtusis planis vix duplò longioribus, labello oblongo obtuso membranaceo venis tribus elevatis per axin duabusque minoribus lateralibus: lobis lateralibus nanis obtusis intermedio oblongo undulato. ——*New Guinea.*——A most beautiful plant, with racemes a foot and a half long, loaded with flowers, whose spatula-shaped petals are an inch and more in length. The inflorescence is very loose in this; in D. Mirbelianum it is far more compact.

67. D. *macranthum* (A. Rich. Sert. Astrolab. p. 15. t. 6.); foliis ellipticis apice obliquo subbilobis, racemo longipedunculato versus apicem monophyllo, sepalis ovato-lanceolatis, petalis paulò longioribus unguiculatis lanceolatis acutis, labello lanceolato trilobo per axin tricarinato laciniâ intermediâ acuminatâ lateralibus nanis rotundatis.——*Vanikoro.*——The stem is described as two feet and more high; the leaves from three to four inches long, the raceme about a foot long, the flowers three inches in diameter. It is very near D. veratrifolium, but differs in having much shorter petals, and a very acute lip.

68. D. *antennatum* (Lindl. in London Journ. Bot. 2. 235); foliis lanceolatis carnosis obliquè emarginatis racemo oppositifolio brevioribus, sepalis acuminatis, petalis linearibus duplò longioribus reflexis, labello trilobo venis 5 elevatis rectis per axin; lobo medio ovato acuto plano 3-costato.——*New Guinea.*——This is a most curious thing, with flowers having their long horn-like petals directed backwards, and their sepals forwards. The petals are two inches long, and scarcely half a line wide.

The leaves are succulent, brittle, and veinless when fresh.

69. D. *undulatum* (R. Brown, Prodr. 332. L. no. 62. *D. discolor*, Lindl. in Bot. Reg. 1841. t. 52); foliis ovato-oblongis emarginatis, racemis longissimis, sepalis petalisque undulatis, labelli trilobi lobis acutis intermedio lanceolato, lineis quinque elevatis intermediâ apice fluxuosâ.——*Java, Tropical New Holland.*——A species with singular dingy yellow-brown flowers, of little beauty.

70? D. *affine* (*Onychium affine*, Decaisne herb. timor. 37); foliis lineari-oblongis acutiusculis coriaceis, pedunculo foliis triplo longiore oppositifolio? laxifloro, floribus spicatis; perianthio erecto, segmentis exterioribus lineari-lanceolatis acutis, interioribus subobovatis rotundatis mucronulatis; labelli unguiculati lobis 3, medio lineari-lanceolato, lateralibus subrotundis.——*Timor.*——Since M. Decaisne compares this with D. Mirbelianum, it probably belongs to the present section. The petals are said to be obovate and mucronate.

NEXT after these species must be placed those plants which Dr. Blume calls Pedilonum, of which one species only is known to me. In that I long since (Bot. Reg. 1839, misc. 169) stated that the structure is of the following singular nature:—In the first place the lateral sepals and the base of the lip are so united into a spur, that their separate nature is entirely concealed towards the point of the spur. In the next place the rostellum is a deep two-lobed lip curved down over the stigmatic surface. The pollen-masses are in two pairs, and deep purple; and, which is extremely curious, they lie upon a loose hard transverse crustaceous gland, which seems to replace the gland and caudicula of Vandeæ, and which projects beyond the anterior edge of the anther, resting upon the two-lobed rostellum. The anther-bed itself is deeply excavated, and three-lobed, the anther adhering to the back lobe which is the narrowest.

Should these characters, all or any of them, prove common to the plants now collected into this section, there will be no doubt about their forming a perfectly distinct genus. In the mean while I would not extend its definition beyond that of Dr. Blume.

Sect. IV. PEDILONUM (Blume, Sect. 2.) *Stems erect. Racemes lateral. Lateral sepals united into a pouch, with which the base of the labellum is entirely coherent, without any articulation.* The flowers are said to be in all cases purplish.

71. D. *secundum* (Wallich. Cat. no. 1996. L. no. 29. Bot. Reg. t. 1291. *Pedilonum secundum*, Blume, p. 322); caulibus erectis, foliis oblongis apice obliquè retusis, racemis lateralibus terminalibus quesecundis, sepalis ovatis: lateralibus basi in cornu longum obtusum incurvum connatis, petalis supremo paulò minoribus, labello integerrimo acuto apice tumido.—— *Malacca, Java, Burma.*——Flowers very closely arranged in a one-sided spike, of various shades of purple. Pollen purple.

72. D. *erosum* (L. no. 52. *Pedilonum erosum*, Blume, p. 323); foliis subcoriaceis lanceolatis obtusis, racemis brevibus oppositifoliis subterminalibusve, sepalis obtusis, dorsali interioribus latiore, labelli limbo rotundato concavo eroso.——*Java.*——

73. D. *hymenophyllum* (L. no. 53. *Pedilonum undulatum*, Blume, p. 322. t. 36); foliis membranaceis lanceolatis inæquali-acuminatis, pedicellis brevibus, sepalo dorsali interioribus obtusis æquali, sacco calcariformi longissimo, limbo labelli undulato.——*Java.*——Sepals purplish. Labellum yellow.

74. D. *Kuhlii* (L. no. 58. *Pedilonum Kuhlii*, Blume, p. 321); foliis submembranaceis lanceolatis apice inæquali-obtusis, sepalo dorsali interioribus lato-lanceolatis latiori, labelli limbo spatulato acuto.——*Java.*——Flowers pale purplish.

75. D. *Hasseltii* (L. no. 59. *Pedilonum Hasseltii*, Blume, p. 321); foliis rigidis lanceolatis profundè obliquè emarginatis, sepalo dorsali interioribus lato-lanceolatis æquali, labelli limbo lineari-acuto.——*Java.*——Flowers deep purple.

We shall take an early opportunity of going on with the enumeration of this genus.

66. MAXILLARIA scabrilinguis.

M. *scabrilinguis* (*Cyrtopera scabrilinguis*, Lindl. Gen. & Sp. no. 3.); pseudobulbis ovatis elongatis sulcatis, foliis basi angustatis scapo erecto multifloro longioribus, bracteis acutissimis ovario brevioribus, sepalis petalisque lineari-oblongis acutis lateralibus falcatis, labelli trilobi lobis lateralibus rotundatis intermedio ovato carnosissimo subtereti obtuso longiore: utrâque facie papillis densissimis versus callum oblongum tricostatum decrescentibus obsitâ, antherâ pubescente.

By an error, resulting from the examination of insufficient specimens, this plant, which is nearly allied to M. squalens, was referred to Cyrtopera in the "Genera and Species of Orchidaceous Plants." It is a native of the neighbourhood of Loxa, where Mr. Hartweg found it, and it has flowered recently in the Garden of the Horticultural Society. The flowers are dull purplish yellow. The labellum is remarkable for having all its end, inside and out, covered over with stiff short pale tubercules, terminated by a purple stain.

67. DICHÆA glauca.

Lindl. Gen. & Sp. Orch. no. 7.

This, the finest of its curious genus, and delightfully sweet-scented, has flowered with Messrs. Loddiges, who have received it from Oaxaca, where, as in other parts of Mexico, it is common. The leaves, which are placed in two rows, in an imbricated manner, are quite blue with bloom on the underside. The flowers are pure white, except a spot of yellow at the very base of each division.

69. GALEANDRA cristāta.

G. *cristata;* sepalis petalisque lineari-lanceolatis reflexis, labello convoluto margine crispo intus pubescente et basi bicarinato calcare acuminato horizontali, antheræ cristâ rhombeâ unguiculatâ.

A native of Cayenne, whence it was received by Messrs. Loddiges (Cat. 1308) in 1840. It has the habit of G. Devoniana, but the flowers are much smaller and paler, and the anther has a very singular crest, resembling a purple lozenge mounted on a white shaft.

70. BOLBOPHYLLUM recurvum.

Lindl. Gen. & Sp. Orch. p. 53. no. 30.

It is a very curious fact that this little plant, originally brought from Sierra Leone, should prove to be also an inhabitant of the New World. That it is so has been ascertained by Messrs. Loddiges, who received it from Guiana in 1839; the specimens that have just flowered having proved to be identical in all respects with the African plant. *Angræcum* or *Œceoclades maculatum* is another instance of the same kind.

71. CAMAROTIS obtūsa.

C. *obtusa;* labello calceiformi lobulo terminali truncato obsoletè tridentato inflexo intus appendice bilobâ canaliculatâ aucto, columnâ tortâ, rostello ungue labelli vix longiore.

I have only seen a few flowers of this, which was given by Mr. Bateman to Messrs. Loddiges. It is of Indian origin, and has dull dirty rose-coloured flowers, with a yellow lip. They are much larger than in C. purpurea, but by no means so handsome. The little lobe at the end of the lip is fleshy, slightly 3-lobed, and furnished with a singular 2-lobed channelled appendage. The column is twisted half round, as in Mormodes.

73. BOLBOPHY̆LLUM pileātum.

B. *pileatum;* . . . . . floribus solitariis glabris, sepalis oblongis obtusis lateralibus convexis longioribus, petalis conformibus duplò angustioribus, labello linguiformi obtuso lævi pone basin bijugato inter juga subaspera nectare lucido obducto, columnâ tereti muticâ, antherâ conicâ cum dente postico columnæ subulato articulatâ, polliniis 4 inæqualibus.

A native of Sincapore, and No. 178 of Messrs. Loddiges' last catalogue, by whom it was received from Sincapore in the year 1840. It has a large yellow-ochre-coloured flower, with a nearly flat moveable labellum stained with two purple ridges near the base, between which is a little bright yellow valley, secreting a mucilaginous matter along its whole length. The anther is conical, like an old-fashioned high-crowned hat.

74. PILUMNA laxa.

Among the curious plants that have flowered this year with Mr. Barker is the present, which is said to have been purchased at Mr. Skinner's sale, and which, if there is no mistake in the matter, must have come from Guatemala. It is, however, so exactly like specimens found in Popayan by Mr. Hartweg, that we cannot help suspecting some error. Its general appearance may be understood by its having been mistaken for a Trichopilia. Its flowers are produced in loose erect racemes, out of broad obtuse short membranous spotted bracts. The stalks and ovary are an inch and half long; the latter with three very stout and strong ribs. The sepals and petals are a pale watery green, erect, linear-lanceolate, equal, faintly tinged with purple. The lip is cream-colour, rolled round the column at the base, to which it also is united at the lower end. The column has a singular fringed hood, overlying the anther, and a nearly vertical stigma, closed in by fleshy inflected cheeks. In these circumstances it differs from Aspasia, to which the genus is nearly akin. Mr. Hartweg found in the same country another species, of which and the present the following are the technical characters.

Pilumna. *Ovarium* tricostatum. *Sepala* et *petala* æqualia patula obliquè inserta. *Labellum* basi columnæ adnatum, subintegrum, unguiculatum, convolutum, inappendiculatum. *Columna* clavata, teres, clinandrio cucullo dentato membranaceo circumdato; buccis duabus carnosis semiclausum. *Stigma* verticale. *Pollinia* 2, posticè fissa, caudiculæ brevi et glandulæ ovatæ adnata.—— Herbæ *epiphytæ*, pseudobulbis *vaginatis*, foliis *coriaceis*, pedunculis *radicalibus*.——Genus Aspasiæ proximum, clinandrio cucullato, columnâ tereti, nec non stigmate verticali nec faciali diversum.

Sp. 1. Pilumna *laxa;* folio lineari-oblongo, racemo laxo multifloro breviore, bracteis laxis cucullatis obtusis, sepalis petalisque lineari-lanceolatis, labello oblongo indiviso rotundato medio constricto per axin unilamellato. ——*In the woods of Timbio near Popayan.*

Sp. 2. Pilumna *fragrans;* folio lato oblongo, racemo 2-3-floro breviore, bracteis lanceolatis erectis obtusis, sepalis petalisque oblongo-lanceolatis acuminatis, labello oblongo apiculato subtrilobo lævi.——*Near the city of Popayan.* ——Flowers large, white, sweet-scented, with an orange-coloured spot on the lip, according to Mr. Hartweg.

75. EPIDENDRUM purum.

E. *purum* (§ Eucpidendrum); caule elongato tereti, foliis ensiformibus obtusis, floribus paniculatis ramis (3) racemosis gracilibus foliis paulo erectioribus, sepalis angustè lanceolatis, petalis linearibus, labelli tripartiti basi 5-costati laciniis ovatis acutis.

A Caraccas plant, sent by Linden to Mr. Rucker, with whom it flowered in the beginning of September. The leaves are not half an inch wide, and six or eight inches long. The flowers are in a thin panicled raceme, of a light pale green colour, and about the size of those of E. *nutans.*

76. ONCIDIUM spilopt̆erum.

O. *spilopterum* (Euoncidia, Pentapetala, micropetala, pandurata); pseudobulbis compressis oblongis diphyllis, foliis erectis oblongis scapo erecto racemoso brevioribus, sepalis ovatis acutis liberis, petalis conformibus latioribus, labello maximo trilobo subrotundo emarginato apiculato lobis lateralibus nanis obtusis, cristâ anticè trilobâ jugis quibusdam verrucisque utrinque pone basin quasi 5-lobâ, columnæ alis semicordatis crenulatis maculatis.

A handsome Oncidium, with quite a peculiar habit, imported from Brazil by Messrs. Loddiges. Its flowers are large and yellow, with small brownish purple sepals and petals; the base of the labellum is the same colour; while the wings of the column are clear yellow, spotted with crimson. The flowers grow in an erect raceme, longer than the leaves. It may be stationed between O. Lanceanum and Carthaginense.

77. HABENARIA candĭda.

H. *candida;* caule diphyllo, foliis undulatis acuminatis, spicâ pauciflorâ (4-5), bracteis acuminatis herbaceis ovario brevioribus, sepalis ovatis acutis subæqualibus dorsali horizontali, petalis indivisis galeatis obtusis, labello integerrimo ensiformi, calcare pendulo ovario duplò longiore apice bilobo.

The flowers of this rare Orchidaceous plant are scentless and snow-white, without the least trace or stain, except on the spur, which is pale green. They are of the size of those in Cynorchis fastigiata, of which the plant has the habit. Messrs. Loddiges received it from Sierra Leone, and it is No. 1901 of their last catalogue.

78. WARREA bidentāta.

W. *bidentata;* bracteis pedicello 4-plo brevioribus, labelli apice bidentati venis valdè convexis flabellatis lamellis altis intermediâ duplò majore.

At first sight we took this for Warrea tricolor. Its lip is however regular, slit at the end, the veins are much more convex, and the central plates thinner and deeper than in that species. The bracts too are not half the length. Sent from the Caraccas to Sigismund Rucker, Esq. with whom it flowered in the beginning of September.

79. ERIA vestita.

(Dendrobium vestitum, *Wall. Cat. no.* 2005. *Lindl. gen. & sp. no.* 33.)

E. caulibus pendulis villis deciduis densè vestitis, foliis coriaceis lanceolatis apice obliquis obtusis integris suprà sparsè subtùs densissimè villosis, racemis elongatis multifloris flexuosis bracteis ovatis coriaceis persistentibus floribus capsulisque villosis, sepalis lanceolatis lateralibus in cornu obtuso porrecto connatis mucronulatis, petalis conformibus obtusis brevioribus glabris, labelli trilobi laciniis lateralibus obtusis intermediâ subrotundâ crispâ emarginatâ pilosâ per axin lineâ elevatâ interruptâ velutinâ dentibusque paucis auctâ, lamellis 7 quarum duæ lanceolatæ sinus fere attingentes et quinque parvæ dentiformes.

This singular plant, with the structure of a woolly Eria, and the habit of a pendulous Dendrobium, is a native of the Indian Archipelago. Dr. Wallich's collectors had it from Sincapore, and Mr. Cuming sent it from Manilla (?) to Messrs. Loddiges. It is singularly clothed with a thick reddish brown hairiness. The flowers are reddish brown externally, white inside, and hang down in pendulous spikes, which are longer than the leaves.

82. The Section of EPIDENDRUM named SPATHIUM.

This is one of the divisions of Epidendrum proposed in *Hooker's Journal of Botany, vol.* 3. *p.* 81, and is distinguished by its stem being slender leafy and erect, its flowers placed on a long peduncle, which proceeds from a spathe consisting of one or more equitant bracts, and by its labellum being wholly attached to the corolla. It differs from § AULIZEUM in its stem being covered with leaves and not pseudobulbous, with two or three leaves at the end only. It is also very near § AMPHIGLOTTIUM, from which it is distinguished by the branches not being excessively lengthened and covered with close pressed scales below the flowers. In their stead there is usually a long sheathing spathe, like that of Cattleya, out of which the inflorescence proceeds: occasionally there are several of such spathes. E. (Amphiglottium) cornutum is one of the connecting links between the two sections.

§ SPATHIUM.

SPATHIUM, *Lindl. in Hooker's Journal of Botany*, 3. 81. (1841).

* *Lip undivided.*

1. E. *grandiflorum* (Lindl. in Hook. Journ. 3. 86); foliis distichis ensiformibus obtusis, racemo denso terminali basi flexuoso e spathâ duplici orto, sepalo dorsali ovali lateralibus duplo latioribus dimidiatis, petalis linearibus, labello subrotundo cordato emarginato margine postico crispo venis baseos 2 elevatis.——*Peru.*——A plant with the inflorescence of E. variegatum. Flowers coriaceous, about twice as large as in that species.—(Not in cultivation.)

2. E. *Scutella;* foliis coriaceis ovato-lanceolatis distichis, spathâ brevi latâ acinaciformi pedunculo æquali, racemo umbellato paucifloro, sepalis petalisque lineari-lanceolatis, labello oblongo cordato plano ecalloso.——*In the woods of Guayan, on the western slope of Pichincha,* rare. (Hartweg.)——Flowers greenish yellow, very large. Lip full an inch long. Spathe very broad, about half an inch long.

3. E. *mancum;* foliis 3-4 oblongis obtusis carnosis obliquè emarginatis supremo spathæ longitudine, racemo stricto multifloro foliis multò longiore, floribus carnosis, sepalis subrotundo-oblongis obtusissimis, petalis spathulatis apice concavis, labello minimo postico carnoso concavo indiviso callo magno margine membranaceo in medio.——*On the Cordillera, near Loxa, flowering in July.* (Hartweg.)——Flowers rather small, orange and yellow.

4. E. *adenoglossum* (Lindl. in Hook. Journ. 3. 86); foliis carnosis lineari-oblongis obtusis, racemo elongato simplici terminali e spathâ ancipiti pedunculo breviore orto, sepalis ovatis reticulatis, petalis linearibus 3-veniis acutis, labello lineari basi callis 3 instructo.——*Peru.*——Flowers small, in a thin spike.—(Not in cultivation.)

5. E. *ventricosum* (Lindl. in Hook. Journ. 3. 86); foliis lineari-lanceolatis acutis, racemis angustis multifloris e spathâ lineari ortis, floribus membranaceis, sepalis subæqualibus oblongis acutis, petalis filiformibus, columnâ ventricosâ, labello ovato cordato acutissimo basi bicalloso.——*Peru.*——A slender plant, with the stem about six inches up to the commencement of the spathe. Racemes from 4—5 inches long, including the spathe which covers the whole peduncle. Flowers purple, small, membranous.—(Not in cultivation.)

** *Lip more or less three-lobed.*

6. E. *brachyglossum;* foliis lineari-lanceolatis apice rotundatis obliquis, racemo elongato simplici terminali nutante e spathâ ancipiti pedunculo breviore orto, floribus carnosis, sepalis obovatis concavis, petalis linearibus 3-veniis acutis, labello brevi trilobo: laciniis lateralibus nanis intermediâ ovatâ: callis duobus lamellæformibus tuberculoque interjecto instructo.——*On the western face of the Andes, near Nanegal.* (Hartweg.)——A slender creeping rooted plant, with small flowers and the appearance of E. adenoglossum, but the leaves are narrower and the lip has a different form.

7. E. *leucochilum* (Klotzsch in Gartenzeit. 1843. p. 145); foliis distichis coriaceis patenti-recurvis obtusis emarginatis subtus costato-carinatis, racemo simplici terminali 6-floro e spathâ ancipiti pedunculo breviore orto floribus magnis arcuato-pedicellatis odoratis perigonii foliolis linearibus acutis versus basin attenuatis margine recurvis interioribus patenti-arcuatis exterioribus deflexis, labelli trilobi lobis lateralibus brevioribus integerrimis oblique orbicularibus intermedio elongato acuminato venis baseos 3 elevatis instructo, columna elongata candida inferne ad apicem biloba.——*Caraccas.*——Found by Mr. Edward Otto, growing on the stems of trees at the height of 4500 feet above the sea. It is said to prefer a cool house, and to have been shewn at the Botanical exhibition at Prague by Professor Tausch. It is fully described by Messrs. Klotzsch and Edward Otto in the above-mentioned place. The flowers are yellowish green with a white lip.

8. E. *armeniacum* (Lindl. in Bot. Reg. t. 1867. *Encyclia macrostachya*, Pöppig. & Endl. n. g. et sp. 2. t. 114); caulibus teretibus erectis indivisis, foliis lanceolatis coriaceis acutis subplicatis, racemis pedunculatis cylindraceis nutantibus, sepalis patulis ovatis acutis, petalis setaceis, labelli subcucullati laciniis lateralibus rotundatis intermediâ ovatâ acuminatâ: callo magno oblongo in discum.——*Brazil and Peru.*——Flowers small, apricot coloured, in a slender drooping spike.

9. E. *nutans* (Swartz Prodr. 121. Fl. Ind. Occ. 1. 1499. Willd. no. 13. Hooker Exot. Bot. 1. t. 50 optimè. Lindl. gen. et sp. orch. no. 44); foliis oblongis distichis undulatis obtusis, paniculâ nutante multiflorâ, sepalis oblongo-lanceolatis petalisque lineari-lanceolatis obtusis patentibus, labelli trilobi lobis lateralibus cordatis ovatis: intermedio transverso truncato apiculato basi bicalloso venis tribus elevatis.——*West Indies, and Brazil.*——This well known plant has been found by Dr. Von Martius on the Corcovado in Brazil, in rocky places. Its green flowers are sweet-scented in an evening. Many varieties are found in our gardens.

10. E. *refractum* (Lindl. in Ann. Nat. Hist. xii.); folio carnoso ovato-oblongo coriaceo obtuso, pedunculo elongato spathis plurimis falcatis obtusis distinctis vaginato, racemo brevi 4—5-floro cernuo, floribus carnosis, sepalis oblongis acutis dorsali refracto, petalis ovalibus subconformibus, labello subrotundo cordato trilobo basi biverrucoso per medium calloso, lacinia intermedia abbreviata truncata.——*Caraccas.*——A very singular plant, with six or seven herbaceous, equitant, distinct falcate spathes on a peduncle about nine inches long. The sepals are about three-quarters of an inch long.

11. E. *geminiflorum* (Humb. et Kunth. Nov. Gen. et Sp. Pl. 1. 354. Lindl. Gen. et Sp. no. 26); caule repente ramoso, foliis oblongis obtusis coriaceis, pedunculis subbifloris, sepalis lanceolatis acuminatis patulis margine

revolutis, petalis conformibus planis, labello ovato cordato subcucullato trilobo: lobis lateralibus rotundatis ab intermedio sinu brevi sejunctis.——*Popayan.*——Stem with dwarf, stiff, erect side branches. Flowers middle-sized, very fleshy, with extremely sharp pointed divisions. (Not in cultivation.)

*** *Lip three-parted.*

12. E. *longiflorum* (Humb. et Kunth. Nov. Gen. et Sp. Pl. 1. 354. Lindl. Gen. et Sp. no. 45); foliis oblongis obtusis striatis coriaceis, racemi nutantis pedunculo spathâ maximâ breviore, sepalis petalisque lanceolato-linearibus acuminatis margine revolutis, labelli tripartiti laciniis lateralibus ovatis obliquè truncatis revolutis: intermediâ lanceolatâ acuminatâ margine revolutâ duplò longiore. ——*New Granada and Popayan.*——A most noble plant, with its yellow flowers more than four inches across. Some of the leaves are about a foot long. (Not in cultivation.)

13. E. *spathaceum* (Lindl. in Hook. Journ. 3. 85); foliis . . . . , racemis alternis densissimis pendulis spathis foliaceis falcatis conduplicatis vix longioribus, sepalis rigidis striatis acutis, petalis filiformibus, labelli trilobi laciniis lateralibus subintegris intermediâ ovali obtusâ basi bilamellatâ brevioribus.——*Peru. Obtained by Mr. Mathews out of the herbarium of Ruiz and Pavon, preserved at Lima.*—— The masses of inflorescence of this plant are upwards of one and a half foot long, and consist of dense racemes proceeding from the axil of falcate spathes, so as to have a great resemblance to that of some Palm. Perhaps this ought to be rather referred to Amphiglottium; but its large sheathing spathes, full six inches long, under each branch of the inflorescence, seem to justify its being placed in this section. (Not in cultivation.)

14. E. *excisum;* foliis latis oblongis apice rotundatis excisis, spathâ latâ coriaceâ pedunculo longiore, floribus laxè paniculatis, petalis filiformibus, labello hastato basi bicalloso laciniis lateralibus cordatis crenulatis intermediâ lineari divergenti bilobâ.——*On the ascent to Sotara, near the village of Totoro, in Popayan.* (Hartweg.)—— This is very much like E. floribundum, of which it has quite the habit, but it has a great sheathing spathe.—(In cultivation?)

15. E. *cylindraceum;* caule ramoso, ramulis basi vaginatis glabris, foliis cuique ramulo 2 ovato-oblongis, pedunculo spathis 2-3 equitantibus elongatis herbaceis vestito, racemo elongato cylindraceo, sepalis obovatis acutis dorso scabris, petalis linearibus lævibus, labelli 3-partiti basi tricallosi laciniis lateralibus angustis antrorsum falcatis postice serrulatis intermediâ lineari basi latiore apice in lobos 2 subrotundos serrulatos divisâ.——*In the woods of Pitayo, in the province of Popayan, at the height of 10,500 feet.* (Hartweg.)——Flowers small, in a very dense cylindrical raceme, about six inches long. (Not in cultivation?)

16. E. *Trinitatis;* caulibus foliosis ancipitibus, foliis lineari-lanceolatis glaucescentibus apice obtusis obliquè emarginatis, racemo terminali nutante pedunculato basi vaginis herbaceis equitantibus imbricato, sepalis petalisque lineari-lanceolatis, petalis piliformibus, labelli lobis lateralibus concavis denticulatis intermedio lineari-lanceolato acuminato basi 3-calloso.——*Trinidad.*—— Flowers small, in a long raceme, pale greenish yellow, with a deep apricot-yellow lip. Not very ornamental, but pretty.

**** *Lip six-parted.*

17. E. *raniferum* (L. no. 64. Bot. Reg. 1842. t. 42.); foliis distichis lanceolatis, racemis terminalibus lateralibusque divaricatis multifloris, sepalis ligulatis convexis obtusis petalisque linearibus cuneatis mucronulatis patentissimis, labelli sexlobi basi 3-callosi laciniis integerrimis dentatisque: posticis rotundatis—intermediis linearibus —anticis linearibus obtusis retrosum falcatis.——*Mexico and British Guayana* (Schomburgk, no. 424.)—— Although a good deal like E. nutans, this species is really very distinct, and indeed far handsomer; in consequence of the rich purplish brown spots with which the sepals and petals are profusely decorated. Like so many of the order this species varies a good deal in the amount of toothing observable in its lip. In the original specimens the divisions were perfectly entire; in other plants they are all toothed, or both the one and the other.

90. LEOCHĪLUS herbacĕus.

L. *herbaceus;* labello obovato emarginato apice denticulato medio convexo lævi basi excavato villoso.

An inconspicuous plant from La Guayra, imported by George Wailes, Esq., of Newcastle. Its sepals and petals are green, with a single red stripe along the middle. The lip is white slightly tinged with green, and irregularly banded with pale purple. It differs from L. oncidioides in the form and surface of the lip. I believe, too, that the flowers are in pairs; but I have only seen fragments.

91. LEOCHĪLUS sanguinolentus.

L. *sanguinolentus;* labello pandurato sepalis multò majore apice bilobo: laciniis subrotundis imbricatis superficie planâ lævi supra basin 3-tuberculatâ basi ipsâ villosâ, columnæ alis abbreviatis, clinandrii dorso elevato, antherâ parvâ.

This also comes from La Guayra, whence Mr. Barker received it. Although the flowers are small they are very beautiful, having a deep crimson lip richly studded with clear purple spots. In the smallness of its anthers, the extension of the anther-bed behind into an elevated rim, and in the shortness of the column wings, it is somewhat different from the rest of the genus. I have only seen a couple of flowers.

92. EPIDENDRUM ceratistes.

E. *ceratistes* (Encyclium III.*); pseudobulbis conicis 2-3-phyllis, foliis ensiformibus strictis scapo duplo brevioribus, paniculâ longâ cernuâ racemiformi ramulis lævibus divaricatis, sepalis petalisque obovatis acutissimis subconformibus (herbaceis), labelli trilobi lobis lateralibus erectis obtusis ovatis apice recurvis intermedio subrotundo unguiculato acuminato crispo striato, ungue medio depresso versus basin 4-costato, columnâ bicornutâ, antherâ tricostatâ.

A native of the Spanish Main, whence it was brought by Mr. Hartweg to the Horticultural Society, with whom it flowered in October last. Its panicle is very narrow, and about three feet long; the leaves resemble those of E. virgatum, but are not so glaucous; the flowers are extremely similar to those of E. selligerum, but they are a clear green with a whitish lip, and the column has two horn-like arms. It is rather sweet-scented.

EDWARDS'S

BOTANICAL REGISTER:

OR,

ORNAMENTAL FLOWER-GARDEN

AND SHRUBBERY:

CONSISTING OF

COLOURED FIGURES OF PLANTS AND SHRUBS,

CULTIVATED IN BRITISH GARDENS;

ACCOMPANIED BY THEIR

History, Best Method of Treatment in Cultivation, Propagation, &c.

AND

MONTHLY CHRONICLE

OF

BOTANICAL AND HORTICULTURAL NEWS.

CONTINUED

BY JOHN LINDLEY, Ph. D. F.R.S. AND L.S.

PROFESSOR OF BOTANY IN UNIVERSITY COLLEGE, LONDON,
AND THE ROYAL INSTITUTION OF GREAT BRITAIN,
VICE-SECRETARY OF THE HORTICULTURAL SOCIETY,
&c. &c. &c.

1845.

OR VOL. XXXI. OF THE ENTIRE WORK.
OR VOL. XVIII. OF THE NEW SERIES.

——viret semper——nec fronde caducâ
Carpitur.

LONDON:
JAMES RIDGWAY, PICCADILLY.

M.DCCC.XLV.

2

ERIA vestita.

Furred Eria.

GYNANDRIA *MONANDRIA.*

Nat. ord. ORCHIDACEÆ. § MALAXEÆ=DENDROBIDÆ.
ERIA. Lindl.

E. *vestita* (Lindl. in Bot. Reg. 1844. misc. 79. *Dendrobium vestitum*, Wall. Cat. no. 2005. Lindl. gen. & sp. no. 33); caulibus pendulis villis deciduis densè vestitis, foliis coriaceis lanceolatis apice obliquis obtusis integris suprà sparsè subtùs densissimè villosis, racemis elongatis multifloris flexuosis bracteis ovatis coriaceis persistentibus floribus capsulisque villosis, sepalis lanceolatis lateralibus in cornu obtuso porrecto connatis mucronulatis, petalis conformibus obtusis brevioribus glabris, labelli trilobi laciniis lateralibus obtusis intermediâ subrotundâ crispâ emarginatâ pilsoâ per axin lineâ elevatâ interruptâ velutinâ dentibusque paucis auctâ, lamellis 7 quarum duæ lanceolatæ sinus fere attingentes et quinque parvæ dentiformes.

Of the shaggy Erias this is one of the more remarkable, having so much the habit of a Dendrobium that it was so considered by Dr. Wallich; for at the time of the publication of that indefatigable Botanist's laborious Catalogue, it was not known that all the species of Dendrobium are hairless or nearly so.

What is not a little singular is that these plants have very frequently tawny or reddish-brown hairs, if they have any. The history of the production of such a colour would be a good subject for examination by some phyto-chemist. The flowers too are reddish-brown externally, white inside, and hang down in pendulous spikes, which are longer than the leaves.

The species is a native of the Indian Archipelago. Dr. Wallich's collectors had it from Sincapore, and Mr. Cuming sent it from Manilla(?) to Messrs. Loddiges. It also occurs, in a small flowered state, among Mr. Cuming's dried plants from the Philippines, marked "Laguna." The specimen

4

EPIDENDRUM dipus.

Two-footed Epidendrum.

GYNANDRIA *MONANDRIA.*

Nat. ord. ORCHIDACEÆ. § EPIDENDREÆ—LÆLIADÆ. *Natural System of Botany, ed.* 3. (ined.)
EPIDENDRUM. Linn.

§. Spathium. *Bot. Reg.* 1844. misc. 82.

E. *dipus;* foliis longis flaccidis distichis subacuminatis, paniculâ nutante densâ multiflorâ e spathis 2 herbaceis erumpente, sepalis oblongolanceolatis petalisque linearibus obtusis patentibus, labelli trilobi lobis lateralibus semicircularibus planis intermedio bilobo laciniis linearibus recurvis dente minuto nunc interjecto; basi bicallosi venis 2 elevatis.

One of those innumerable species inhabiting South American forests, to the enumeration of which there seems no end. It was imported by Messrs. Loddiges, from Brazil, and produced its densely clustered panicles of sweet-scented green, brown and white flowers in January.

In many respects it approaches E. nutans, but its panicle is very much more compact, its colour is more that of E. paniculatum, and the form of its lip is different, the two terminal lobes being very narrow, and bowed back like the fore legs of the splay-footed truffle dogs.

When we lately published an enumeration of the species of the section Spathium, to which this belongs, it was overlooked, having been mistaken for the very different E. rubrocinctum (or densiflorum ?).

before me has a Jungermannia clinging to its flower-spike, shewing that it must have come from some damp and shady station.

It should be potted in turfy heath-mould mixed with potsherds. In summer it should receive an ample supply of water, while the atmosphere should be kept as moist as possible. In sunny weather it should be shaded to keep the temperature about 80° by day, without giving much air. The temperature at night, during the growing season, may be from 68° or 70°. In winter, little water will be required, provided the house can be filled with steam once a day. At this period the temperature requires to be kept up to about 65° with fire heat.

8

MILTONIA cuneata.

Wedge-lipped Miltonia.

GYNANDRIA *MONANDRIA.*

Nat. ord. ORCHIDACEÆ. § VANDEÆ — BRASSIDÆ, *Lindl. Natural System of Botany, Ed.* 3. *ined.*
MILTONIA. Botanical Register, fol. 1992.

M. *cuneata* (Lindl. in Bot. Reg. 1844. misc. 28.); pseudobulbis ovatooblongis, foliis oblongis striatis subundulatis, racemo plurifloro, sepalis petalisque lanceolatis undulatis, labello cuneato rotundato basi bilamellato utrinque subrepando, columnâ medio bidentatâ, clinandrio integerrimo.

This very handsome addition to the beautiful genus Miltonia, has already been noticed at No. 28 of the miscellaneous matter of last year; its figure completes its history.

It is a most beautiful Epiphyte, allied to M. *candida*, with flowers nearly four inches in diameter; the sepals and petals a rich brown tipped with green. The lip is pure white, with a tinge of pink near the base, in form quite different from M. *candida*, for it is scarcely at all curled at the edge, is very much narrowed to the base, and has only one pair of plates instead of two and a half. The wings of the column too are scarcely divided or at all events not at all notched. Fig. 2 represents the pollen-masses with their appendages.

It should be potted in turfy heath mould mixed with potsherds. During the summer an ample supply of water should be given to its roots, and it should be syringed over head once or twice a day. To prevent the leaves from being scorched as well as to keep the temperature between 80° and 90° by day without admitting much air, shading will be necessary in sunny weather. In winter for a few weeks little or no water should be given, but the house should be filled with steam once every fine day. The temperature should never be raised above 66° by fire heat.

The following is a classed list of the species of Miltonia at present known.

1. M. *spectabilis* (Lindl. in Bot. Reg. sub t. 1976. Aug. 1. 1837. t. 1992. *Macrochilus Fryanus*, Floral Cabinet, t. 45. Sept. 1, 1837); pseudobulbis ovalibus ancipitibus lævibus, foliis ligulatis patentissimis, pedunculis unifloris squamis magnis fuscis striatis imbricatis carinatis densè imbricatis, sepalis ovalibus planis, petalis conformibus revolutis, labello maximo cuneato rotundato basi trilamellato, columnæ alis angustis acutissimis——*Brazil.* ——Flowers very large, solitary, with pale straw-coloured sepals and petals, and a broad violet purple lip ribbed with crimson.

2. M. *flavescens* (Lindl. Sertum, sub t. 48. *Cyrtochilum flavescens*, Lindl. in Bot. Reg. t. 1627); pseudobulbis angustissimis ancipitibus lævibus, foliis lineari-ensiformibus binis scapo æqualibus, scapo compresso arctissimè vaginato, racemo multifloro, bracteis glumaceis flavescentibus carinatis pedunculi longitudine, sepalis petalisque linearibus acuminatis, labello ovato-lanceolato repando crispo basi pubescente inappendiculato.——Said to inhabit Mexico, but most probably *Brazil.*—— Flowers bright yellow with no spots, but a few crimson specks and streaks on the lip. Wings of the column crimson.

3. M. *stellata* (Lindl. Sertum, sub t. 48. *Cyrtochilum stellatum*, Lindl. Sertum, t. 7); pseudobulbis diphyllis ovalibus ancipitibus, foliis ligulatis obtusis scapo multò brevioribus, scapo tereti erecto vaginato, racemo disticho multifloro, bracteis carinatis convolutis acuminatis glumaceis ovario longioribus, sepalis petalisque lineari-obovatis acutis stellatis, labello oblongo undulato acuto basi canaliculato striato, alis columnæ acinaciformibus integerrimis——*Brazil.*——Flowers pale yellow with no spots, lip white with yellow streaks. Wings of the column yellow streaked with crimson. Bracts rose-coloured. This has much larger flowers than the last species, its sepals and petals are not taper-pointed, its pseudo-bulbs are much broader, its stature four times as great, and the colours of the parts different, as will be seen upon comparing the above descriptions.

4. M. *candida* (Lindl. in Bot. Reg. 1838. misc. 29. Sertum Orchidaceum, t. 21.); pseudobulbis ovatis apice angustatis diphyllis, foliis angustis racemo brevioribus, bracteis brevibus ovatis membranaceis concavis squamæformibus, sepalis petalisque oblongis æqualibus, labello subrotundo crispo circa columnam convoluto basi 5-lamellato, columnâ pubescente basi biauri, clinandrio crispo membranaceo-marginato utrinque in alam decurrente.——*Brazil.*——Flowers in long erect racemes, very large; sepals and petals spotted with brown; lip white. There is a large flowered variety in cultivation.

5. M. *cuneata*.

6. M. *Russelliana* (Lindl. in Sertum, sub t. 48. *Oncidium Russellianum*, Lindl. in Bot. Reg. t. 1830); pseudobulbis ovatis costatis diphyllis, foliis ligulato-lanceolatis patentibus, racemo paucifloro, bracteis acuminatis ovario 3-plo brevioribus, sepalis petalisque conformibus ovato-oblongis subundulatis, labello postico oblongo-cuneato retuso apiculato subsinuato, lamellis disci pluribus truncatis, columnæ bidentatæ alis magnis falcatis, stigmatis margine incrassato pubescente colorato.——*Brazil.*—— Flowers dingy greenish purple. Lips dull violet; wings of the column yellow; border of the stigma crimson.

19

SPATHOGLOTTIS Fortūni.

Mr. Fortune's Spathoglottis.

GYNANDRIA *MONANDRIA.*

Nat. ord. ORCHIDACEÆ. § EPIDENDREÆ—BLETIDÆ.

SPATHOGLOTTIS, Blume. Perianthium subregulare, nec galeatum. *Sepala* patentia, libera, æqualia. *Petala* paulò latiora magisque membranacea, patentia v. conniventia. *Labellum* cum basi columnæ articulatum, nunc saccatum, tripartitum, laciniâ intermediâ unguiculatâ basi tuberculatâ v. cristatâ sæpiùs utrinque unidentatâ. *Columna* alata v. petaloidea. *Anthera* bilocularis. *Pollinia* 8.——Herbæ *Asiaticæ, terrestres,* cormis *subterraneis,* foliis *ensiformibus plicatis.* Flores *nunc lutei nunc violacei.*

Sp. Fortuni; foliis binis lanceolato-linearibus scapo pubescente brevioribus, racemo secundo pubescente, bracteis acuminatis ovario duplo brevioribus, sepalis ovatis obtusis, petalis oblongis subsessilibus latioribus, labelli laciniis lateralibus oblongis erectis intermediâ cuneatâ emarginatâ, ungue utrinque dente aucto, lamellis tribus basi villosis carnosis subdentatis lateralibus truncatis intermediâ ad apicem fere procurrente, columnâ oblongâ apice denticulatâ, polliniis inæqualibus omnibus acutis.

One of the first plants which Mr. Fortune met with on the granitic mountains of Hong Kong, was the pretty little Bletia-like plant figured in this plate. From some corms of it, which he sent home in his first despatch, the specimen sprang up of which the annexed drawing was made in the Garden of the Horticultural Society in January last.

Like the Bletias it has thin plaited leaves, and fleshy tubers, or corms, which lie dormant for some months after the foliage has disappeared. The genus, indeed, differs from Bletia principally in having the middle lobe of the lip stalked, with some deep plates at its base, and in its anther having but two cells instead of eight.

It appears that there are three species of this genus very much alike in their general appearance and yellow flowers; the only one of which hitherto published has been *S. pubescens*, found by Dr. Wallich on the Sylhet mountains, at Prome, and on the Avan mountain called Tong Dong. This plant differs from the present in having much smaller bracts, and perhaps larger leaves, and especially in the form of its lip, which wants the pair of teeth at the bottom of its middle segment; its pollen-masses too appear, from my memoranda, to be bluntly cuneate and not acute. Another species I owe to Mr. Griffith's Khasiyan collections. This has very small narrow grassy leaves, acute sepals, flowers on much longer stalks, and a lip without either hairs, teeth, or tubercles at the base of its middle lobe, while of the three plates those at the sides are truncate and conspicuous, but the middle plate is merely a raised line; the column of this species is moreover remarkably long and narrow. The technical distinctions of them will be found below.

I also find among Mr. Cuming's Manilla plants, a fine species gathered on Mindanao, in the province of Miscamis, with as many as twenty flowers in a raceme, and the stem and very obtuse bracts coated with a close fur. The specific character of that is added under the name of Sp. tomentosa. If any are possessors of a Manilla Spathoglottis, resembling S. plicata, or Paxtonia rosea in herbage, they would do well to take care of it, for it may be this S. tomentosa, which seems to be really a fine thing.

S. *pubescens* (Lindl. gen. & sp. orch. p. 120); foliis binis lineari-lanceolatis basi angustatis scapo pubescente longioribus, racemo secundo pubescente, bracteis minutis acutis, sepalis ovatis acutis, petalis oblongis obtusis, labelli laciniis lateralibus oblongis erectis intermediâ cuneatâ emarginatâ, lamellis tribus basi villosis carnosis lateralibus truncatis intermediâ parum elevatâ cis apicem evanescente, columnâ oblongâ apice denticulatâ (?), polliniis obtusis inæqualibus 4 cuneatis.——*Sylhet mountains, Prome, and Tong Dong mountain in Ava.*

S. *parvifolia;* foliis solitariis gramineis scapo pubescente multò brevioribus, racemo trifloro pubescente, bracteis acuminatis glabriusculis pedicellis brevioribus, sepalis ovatis acutis, petalis oblongis manifeste unguiculatis, labelli laciniis lateralibus oblongis erectis intermediâ lineari subcuneatâ indivisâ, lamellis tribus basi glabris lateralibus truncatis intermediâ filiformi, columnâ angustâ elongatâ.——*Khasiya hills.*

S. *tomentosa;* foliis binis lato-lanceolatis scapo tomentoso longioribus, racemo tomentoso multifloro, bracteis obtusissimis concavis rigidis velutinis pedicellis multo brevioribus, sepalis petalisque obtusissimis his latioribus, labelli laciniis lateralibus erectis lineari-oblongis truncatis intermedia apice reniformi ungue elongatâ lineari angustissimâ basi hastatâ, lamellis 2 ad basin lanceolatis semiconnatis erectis, columnâ angustâ elongatâ.——*Mindanao.*

The analyses in the accompanying plate represent, 1. a lip spread open so as to shew the lamellæ, and the lateral teeth of the middle segment; 2. the column; 3. the pollen-masses.

24

STANHOPEA Bucephălus.

Bull-horned Stanhopea.

GYNANDRIA *MONANDRIA.*

Nat. ord. ORCHIDACEÆ. § VANDEÆ—MAXILLARIDÆ.

STANHOPEA, Supra. fol. 1800.

S. *Bucephalus;* bracteis ovario subæqualibus, hypochilio unguiculato cymbiformi anticè intruso apice carnoso apertè sulcato mutico basi longè angustato ecorni intus lævi extùs bicarinato, epichilio subrotundo-ovato cuspidato integro breviore, cornubus gracilibus teretibus brevioribus, columnâ basi angustissimâ sursùm alatâ. *Supra, fol.* 1843. *sub t.* 44.
S. Bucephalus, *Lindl. Gen. & Sp. Orch. no.* 2.
Epidendrum grandiflorum, *Humb. & Bonpl. Pl. Æquinoct. p.* 94. *t.* 27.
Anguloa grandiflora, *Humb. Bonpl. & Kunth, Nov. gen. & sp.* 1. 345.

This is one of the rarest and finest of the Stanhopeas, in some measure resembling S. oculata, especially in the long narrow hypochil. Its flowers are deliciously scented, and their bright golden colour produces a very rich effect.

At first sight it might be mistaken for a mere variety of S. oculata, but Mr. Loddiges long since pointed out the shortness of its ovary as a decisive mark of distinction. The effect of this shortness is to make the inflorescence of S. Bucephalus very narrow, while in S. oculata it is broad and straggling.

The species is a native of the woods of Paccha, a small village in the Andes on the ascent from Guayaquil to Loxa, at an elevation of 6,000 feet above the level of the sea, where it was found by Mr. Hartweg. It first flowered in the garden of the Horticultural Society in August, 1843. Its stem is spotted as well as its blossoms.

It may either be potted in turfy heath-mould mixed with potsherds, in the usual way, or it may be suspended to a rafter, in a wire basket filled with sphagnum. It is a very free growing species, which will prosper under ordinary treatment. During the growing season an ample supply of water should be given, and the house shaded in sunny weather. In summer the temperature should be kept about 80° or 85° by day, and about 70° at night. For a few weeks in winter, no more water should be given than keeps the pseudo-bulbs from shrivelling, when the temperature should never be raised above 65° with fire heat.

28

WARRĔA cyanĕa.

Blue-lipped Warrea.

GYNANDRIA *MONANDRIA.*

Nat. ord. ORCHIDACEÆ. § VANDEÆ—MAXILLARIDÆ.

WARREA, Lindl. Flores subglobosi, subregulares, mento brevi rotundato. *Labellum* continuum, indivisum, lineis elevatis carnosis in medium. *Columna* semiteres, clavata. *Pollinia* 4, per paria in caudiculam brevem linearem inserta, *glandulâ* triangulari.——Herba *terrestris, pseudobulbosa.* Folia *arundinacea.* Scapus *radicalis apice racemosus.* Flores *speciosi.*

W. cyanea, *Supra,* 1844. *misc. no.* 3.

When, in the year 1843, I proposed to establish a new genus upon the Maxillaria Warreana, it certainly did not occur to me that two new and quite distinct species would be added to it in the course of as many years. Such, however, has been the fact; in the beginning of 1844 this beautiful species blossomed with Messrs. Loddiges, and a few months since another appeared in the collection of Mr. Rucker. Both the latter are from the Spanish main, and it is not improbable that others may lurk in the unexamined forests of that vast region.

Warrea cyanea is remarkable for the intense porcelain-blue colour of its lip, to which it is not easy to find a parallel in the order; for pure blue is scarcely known among Orchids. The plant has quite the habit of Warrea tricolor, but is very much smaller in all its parts. Its most distinctive character is found in the form of its lip, which has a distinct point, and five ribs, not three, near the base. Messrs. Loddiges imported it from Colombia, and it is no. 860 of their last catalogue.

Being a terrestrial species this requires treatment very similar to Phaius maculatus. The soil best suited is turfy heath mould with a mixture of silver sand; nor is it necessary to have the soil elevated above the brim of the pot, for it will not succeed well if the roots are too much exposed. In summer it requires an ample supply of water and a moist atmosphere, at a temperature of 80° or 85° by day, and about 70° at night. Like all Orchidaceous plants this should be kept rather dry in winter, and repotted in spring as soon as it shows any symptoms of growth. While in a dormant state the temperature should never be raised above 65° with fire heat.

It may be useful to bring together into one view the three species now known to exist. They are as follows :—

1. Warrea *tricolor* (Lindl. in Bot. Reg. 1843. misc. p. 14. *Maxill. Warreana,* Lodd. Bot. Cab. t. 1884. L. no. 30.); scapo multifloro foliis longiore, sepalis ovatis basi subæqualibus petalis minoribus conformibus, labello brevissimè unguiculato obovato-oblongo indiviso cucullato obtuso jugis tribus in medio elevatis carnosis, superficie laminæ seriatim corrugatâ marginibus planis.——*Brazil.* ——Sepals yellowish brown, lip yellow at base, pallid at the end, rich purple in the middle.

2. W. *cyanea* (Lindl. in Bot. Reg. 1844. no. 3.—1845. t. 28.); spicâ brevi, bracteis ovarii longitudine, sepalis ovatis acutis, petalis subconformibus, labello subrotundo-cuneato apiculato undulato lineis quinque elevatis.——*Colombia.*

——Very much smaller than the last, and conspicuous with its porcelain-blue lip.

3. W. *bidentata* (Lindl. in Bot. Reg. 1844. misc. 78.); bracteis pedicello 4-plo brevioribus, labelli apice bidentati venis valdè convexis flabellatis lamellis altis intermediâ duplò majore.——*Caraccas.*——Much like Warrea tricolor. Its lip is however regular, and slit at the end, the veins are much more convex, and the central plates thinner and deeper than in that species. The bracts too are not half the length.

30

SCHOMBURGKIA tibicinis, *var.* grandiflora.

Large-flowered variety of the Trumpet Schomburgkia.

GYNANDRIA *MONANDRIA.*

Nat. ord. ORCHIDACEÆ=EPIDENDREÆ. § LÆLIADÆ, *Lindl. veg. kingd. p.* 181. (ined.)

SCHOMBURGKIA. Supra 1844. *t.* 23.

S. *tibicinis;* pseudobulbis conicis corniformibus annulatis sulcatis 3-phyllis, foliis oblongis coriaceis patentibus, scapo longissimo tereti distanter squamato apice paniculato, paniculâ pyramidali laxiflorâ, sepalis petalisque undulatis crispis, labello oblongo cucullato venis per medium 5 elevatis approximatis: laciniis lateralibus apice rotundatis intermediâ subrhombeâ emarginatâ, antherâ emarginatâ.—*Lindl. l. c.*
S. tibicinis, *Bateman Orch. Mex. & Guat. t.* 30.
Var.; grandiflora; floribus duplò majoribus labello extus pallido intus lobo medio luteo, albo v. violaceo-limbato.

That was a noble specimen from which the accompanying figure was taken,—the pseudobulbs being fifteen inches long, and the flowering stem five feet high. It was produced in the collection of Robert Hanbury, Esq. in May, 1844.

It is certainly the same species as that named by Mr. Bateman "tibicinis," because the hollow pseudobulbs are used as trumpets by the Indian children of Honduras; a scene in which they are so occupied forms the subject of one of the exquisite wood-cuts in Mr. Bateman's magnificent work.

But although the same species it is very different in the size and colour of the flowers, which are very much larger, far paler on the outside, and have a broader lip, whose middle lobe is not rich violet but yellow, with a white or purple border. In this instance the plant realises the expectations that had been formed of it: in other cases it has disappointed them.

It may either be tied to a block of wood and suspended to a rafter, or it may be potted in turfy heath-mould, mixed with potsherds, and treated in the same manner as Cattleyas. During the growing season an ample supply of water should be given to its roots, but as little over head as possible, except in the form of vapour; for water is apt to lodge in the axils of the leaves and cause the young ones to damp off. To prevent the leaves from being scorched in summer, as well as to keep the temperature about 80° by day, it will be necessary to use shading. In winter, for a few weeks, if the atmosphere is kept moist, very little water will be required, and then the temperature should not be raised above 60° by artificial means.

34

LYCASTE gigantea.

Large-flowered Lycaste.

GYNANDRIA *MONANDRIA.*

Nat. ord. ORCHIDACEÆ. § VANDEÆ—MAXILLARIDÆ.

LYCASTE. Supra, vol. for 1843. *fol.* 35.

L. *gigantea;* bracteâ herbaceâ sepalis subæquali, sepalis oblongo-lanceolatis lateralibus falcatis, petalis conformibus paulò minoribus, labello lanceolato acuminato laciniis lateralibus acutis intermediâ ovatâ acuminatâ (obtusâ) serratâ, appendice carnoso emarginato.—*Lindl. in Bot. Reg.* 1843. *misc.* 16.
Maxillaria Heynderycxii, *Morren in Annales de la Société Royale d'Agriculture et de Botanique de Gand. vol.* 1. *t.* 9.

Although the colours of this species are not gay, yet its large size and the great height to which its flowering stem rises (full two feet), are remarkable features; besides which its flowers are among the largest in the race of Orchids.

Mr. Hartweg found it in Guayaquil, flowering in the month of August, at a place called the Quebrada de las Juntas; and from his dried specimens it was described in this work in the year 1843. The solitary flower, however, which we examined had a labellum very much drawn to a point; but the specimens which have been produced in our stoves had that part abruptly rounded off, as is shewn in the annexed plate (fig. 1.); and the former therefore was perhaps some accidental form. This is the more probable, since the species has also flowered in Belgium, whither it had been introduced from La Guayra.

M. Morren, who called it Maxillaria Heynderycxii "after Senator Heynderycx, vice-president of the Royal Agricultural and Botanical Society of Ghent," was apparently unacquainted with the enumeration of Lycastes published in this work in the year 1843.

It should be potted in turfy heath-mould mixed with potsherds. To prevent the water from becoming stagnant about its roots the pot should be well drained, and the soil considerably elevated above its brim. In summer, while the plant is growing, plenty of water should be given to its roots, and it must be syringed over head once or twice a day. Like many other Orchids this requires a humid atmosphere, and a shade in sunny weather. During the season of growth the temperature should be kept about 80° by day, and about 70° at night. In winter, the season of rest, very little water will be required for a few weeks, nor is it necessary to raise the temperature above 66°, with fire heat.

36

ODONTOGLOSSUM Cervantesii.

Cervantes' Tooth-tongue.

GYNANDRIA *MONANDRIA.*

Nat. ord. ORCHIDACEÆ. § VANDEÆ—BRASSIDÆ. *Lindley's Vegetable Kingdom, p.* 181. *ined.*

ODONTOGLOSSUM, Bot. Reg. vol. 1839. *t.* 48.

O. *Cervantesii;* pseudobulbis ovatis angulatis, foliis solitariis oblongis in petiolum canaliculatum angustatis, scapo paucifloro, bracteis vaginisque membranaceis acutissimis equitantibus elongatis, sepalis membranaceis oblongo-lanceolatis acutis, petalis latioribus subunguiculatis acutis, labello subcordato-ovato acuto unguiculato ungue carnoso cyathiformi pubescente antice bidentato medio tuberculato processubus 2 pilosis ante cyathum, columnæ pubescentis auriculis rotundatis.

Of the beautiful things belonging to the white-lipped section of this genus not one is more perfectly lovely than that which is now figured from a specimen communicated last February from Messrs. Loddiges, who imported it from Oaxaca. It is the No. 1613 of their catalogue, and in the way of the well-known O. Rossii; but its large thin delicately pink flowers, banded with crimson near the centre, are in all respects much handsomer.

Among other valuable attributes the plant has that of being very sweet-scented, emitting a delicious odour something like bitter almonds.

Although it has just been compared with O. Rossii, yet in truth it is much more like O. membranaceum, (the No. 1614* of Messrs. Loddiges' catalogue,) a plant found at Oaxaca by Count Karwinski, and also in the great Hackney collection. That plant, however, has perfectly white flowers, with the exception of concentric broken crimson bands which occupy the lower part of all the segments of the flower; its petals and lip are very blunt, and the latter is very deeply heartshaped; while in O. Cervantesii on the other hand the lip is scarcely heartshaped, and has no band-like markings; it and the petals are particularly acute, and the ground tint of the flower is a delicate flesh colour. The teeth, too, which stand in front of the saucer-shaped stalk of the lip, are longer in the latter species.

In no respect does it appear to differ from the plant described by La Llave under the name of O. Cervantesii, except in its lip not being somewhat lacerated. It is however probable that the slight undulations of the lip are what the Mexican Botanist called lacerations.

It should be potted in turfy heath-mould in the same way that Orchidaceous plants generally are. In summer an ample supply of water should be given, and the temperature kept up to about 80° by day and about 70° at night. In autumn as the pseudo-bulbs become matured, water should gradually be withheld, and for a few weeks in winter little or none will be required, providing the house can be filled with steam for about half an hour every fine day. Although many plants in this order succeed well on blocks of wood, without any soil whatever, still those under pot culture are much benefited by having it renewed every season; this should be done early in spring when the plants begin to grow.

40

ONCIDIUM spilopterum.

Spot-winged Oncidium.

GYNANDRIA *MONANDRIA.*

Nat. ord. ORCHIDACEÆ. § VANDEÆ—BRASSIDÆ. *Vegetable Kingdom, p.* 181. *ined.*

ONCIDIUM, Swartz.

Sect. II. EUONCIDIUM.

†† Sepala v. petala basi angustata.

* Folia plana.

‡‡ *Pentapetala;* sepalis lateralibus omnino disjunctis.

¶¶ *Micropetala;* petalis sepalis lateralibus subæqualibus et similibus.

☞☞ *Labellum panduratum;* i. e. medio constrictum basi angustius.

O. *spilopterum* (Lindl in Bot. Reg. 1844. misc. 76); pseudobulbis compressis oblongis diphyllis, foliis erectis oblongis scapo erecto racemoso brevioribus, sepalis ovatis acutis liberis, petalis conformibus latioribus, labello maximo trilobo subrotundo emarginato apiculato lobis lateralibus nanis obtusis, cristâ anticè trilobâ jugis quibusdam verrucisque utrinque pone basin quasi 5-lobâ, columnæ alis semicordatis crenulatis maculatis.

O. gallopavinum, *Morren in Ann. Soc. Hort. Gand.* 1. 13.

We have to thank Mr. Loddiges for specimens of this plant, and also for pointing out its identity with Prof. Morren's O. gallopavinum. It had been previously described in the miscellaneous matter of this work as a native of Brazil, and it probably was this circumstance which prevented the learned professor at Liège from identifying his plant with ours. Prof. Morren says that it comes from Mexico; and Messrs. Loddiges informs us that their statement of Brazil being its habitat was made on the authority of a correspondent, the plant not having been imported by themselves.

It is really a beautiful species, in an artificial arrangement seeming to be intermediate between O. Lanceanum and carthaginense, but naturally standing near O. reflexum and the Mexican species allied to it. Its flowers are large and yellow, with small brownish purple sepals and petals; the base of the labellum is the same colour; while the wings of the column are clear yellow, spotted with crimson. The flowers grow in an erect raceme, longer than the leaves.

Fig. 1. is a magnified view of the column and base of the lip, shewing the wings of the former, and the peculiar nature of the tubercles that form the crest of the latter.

It should be grown in turfy heath-mould, mixed with potsherds. The pot in which it is grown should be well drained, and the soil considerably elevated above its brim. During the summer months when a humid atmosphere is necessary an ample supply of water should be given, and the house shaded in sunny weather, which will regulate the temperature without admitting much air. In winter, while the plant is in a state of rest, and the temperature kept about 60° or 65° by fire heat, little water will be required for a few weeks; but as soon as it begins to grow it should be repotted and the heat and moisture gradually increased.

44

BOLBOPHYLLUM umbellatum.

Umbelled Bolbophyl.

GYNANDRIA *MONANDRIA.*

Nat. ord. ORCHIDACEÆ. § MALAXEÆ—DENDROBIDÆ. *Lindley's Vegetable Kingdom, p.* 181. *ined.*

BOLBOPHYLLUM, Thouars.

B. *umbellatum* (Lindl. in Wall. Cat. no. 1984. Gen. & Sp. Orch. no. 43); rhizomate repente, pseudo-bulbis oblongis angulatis, foliis solitariis oblongis loratis obtusis subemarginatis, scapis foliis æqualibus, floribus umbellatis, sepalis lateralibus obliquis falcatis obtusis majoribus supremo rotundato nano, petalis ovatis obtusis, labello cordato ovato complicato integerrimo emarginato, columnâ marginatâ setis duabus hinc unidentatis auctâ, polliniis 4 reniformibus posticis minimis muco apice cohærentibus.

The mountains of India yield this little Bolbophyl. Dr. Wallich's collectors found it in Nepal in 1821; it was one of the plants collected in the Khasiya hills, by Mr. Gibson, for his Grace the Duke of Devonshire, and flowered at Chatsworth in 1838; and now has been reintroduced by Messrs. Loddiges, with whom our drawing was made in September, 1844.

The pale straw-coloured flowers spotted with purple, are rather large, and for a Bolbophyl, pretty. The column has two long horns, which have had a single tooth on the upper side in all the specimens we have examined. The pollen-masses are, as is customary in the genus, of very unequal size, the front pair almost concealing that behind; and in this case are held together by some soft mucus, which resembles the gland of the Vandeous Orchids, but is destitute of its firmness and permanency.

Like many others of the same genus this species may either be potted in turfy heath-mould in the usual way, or it may be tied to a block of wood and suspended to the rafter. It is a stove plant which requires a humid atmosphere, and an ample supply of water during the growing season. Instead of giving air in summer to regulate the temperature, shading should be used, which will also prevent the leaves from being scorched by the sun. In winter, the season of rest, little or no water will be required for a few weeks, provided the house can be filled with steam for half an hour every fine day.

Fig. 1, represents the column and lip seen from the side; 2, the lip seen from above; 3, the pollen-masses with the connecting mucus.

49

CHLORÆA virescens.

Green-veined Chloræa.

GYNANDRIA *MONANDRIA.*

Nat. ord. ORCHIDACEÆ. § ARETHUSEÆ.

CHLORÆA, Lindl. *Perianthium* ringens, basi obliquum, *petalis* sepalo supremo suppositis et galeam referentibus. *Sepala* lateralia basi producta sed vix gibbosa, apice sæpius callosa. *Labellum* unguiculatum, cordatum, cucullatum, membranaceum, integrum aut trilobum, lobo medio lateralibus parum angustiore, venis sæpius cristatis aut glandulosis. *Columna* erecta, membranaceo-marginata, clavata, petalis fere æqualis iisdem dorso adnata, antice utrinque producta. *Stigma* oblongum, prominens. *Anthera* terminalis, opercularis, semiquadrilocularis. *Pollinia* 4, vel 2 bipartita.—Herbæ *terrestres Americæ australis extra tropicam,* radicibus *fasciculatis carnosis (gummoso-farinaceis,* Pöppig). Folia *radicalia.* Scapus *vaginatus.* Racemi *densi, multiflori.* Flores "*sæpe permagni, speciosi, albidi, virescentes, flavidi, vel crocei, nunquam purpurei*" *colore nigro nonnunquam adjecto,* "*odore nonnullis aromaticus.*"—Lindl. gen. & sp. orch. p. 399.

§ 2. Labellum manifestè trilobum.

c. *Labelli venæ centrales appendicibus biformibus.*

C. *virescens;* labelli trilobi lobis ovatis obtusis intermedio duplò majore: venis 9 basi lamellis totidem parallelis æqualibus auctis deinde appendicibus seriatis elongatis v. falcatis, sepalis apice obtusis concavis incrassatis extus verrucosis, petalorum venis varicosis.—*Lindl. l. c. p.* 204. char. emendato.

C. chrysantha, *Pöppig & Endlicher, nov. gen. & sp.* 1. 31. ?

At length we have the satisfaction of producing a figure of one of those beautiful terrestrial Orchids which, under the names of Pichiguen, Gavilu, Azuzena, Pica de Loro, &c., are most charming ornaments of the subalpine pastures of the Cordilleras of Chili. The present, which is one of the finest, has been reared by Mr. Cameron, the intelligent and skilful Curator of the Botanic Garden, Birmingham, by whom it was exhibited at the last May exhibition, in the garden of the Horticultural Society.

Mr. Cameron regarded it as the Chloræa chrysantha of Pöppig, in which he is probably correct; but it is also the Chl. virescens, so called on account of the green veins which are drawn over the orange ground colour of the gaudy flowers, and the latter name being the oldest must be retained.

Fig. 1. represents the inside of the lip of this plant; 2. its column; and 3. its pollen-masses. At the foot of the column are visible two honey pores, which had been previously overlooked in the descriptions of the genus.

Now that the possibility of cultivating these plants has been shewn, and that they are found not to be more difficult of management than a Spiranthes or Sarcoglottis, it is to be hoped that some pains will be taken to procure their roots. They are as common over all the subalpine country between Conception and Valparaiso, as the Meadow Orchises here. It will be necessary, however, to mark their stations when in flower, and to take them up only when the stem and leaves are withered.

This species requires to be grown in a warm greenhouse. The soil best adapted is rough sandy peat. Like most of the terrestrial Orchids, this requires an ample supply of water during the growing season. After flowering the stem and leaves will naturally die off; this being its season of rest little or no water will be required, until it begin to put forth new leaves, when it should be re-potted and encouraged to grow.

50

CYMBIDIUM Mastersii.

Masters's Cymbid.

GYNANDRIA *MONANDRIA.*

Nat. ord. ORCHIDACEÆ.

CYMBIDIUM, Botanical Register, vol. 3. *fol.* 217.

C. *Mastersii;* foliis distichis angustè ensiformibus obtusis, pedunculo erecto squamis herbaceis equitantibus acutissimis imbricato, spicâ brevi pauciflorâ squamis immersâ, sepalis petalisque lineari-oblongis acutis, labello obovato trilobo intùs pubescente, lamellis continuis apice confluentibus in tuberculum subtridentatum expansis, laciniâ intermediâ oblongâ undulatâ lateralibus rotundatis planis.

C. Mastersii, *Griffiths in hort. Bot. Calc. Loddiges' Catalogue, no.* 1233.

Pollinia *apice pilosa;* glandulâ *quadratâ.*

We have no information concerning this plant, further than that it was received from the East Indies by Messrs. Loddiges in the year 1841, and blossomed in December, 1844.

It is a very distinct species, with snow-white flowers, sweet-scented, with the fragrance of almonds. Its erect flower-stalk, closely covered with long green sharp-pointed equitant imbricated sheaths, is quite unlike that of any other species.

It was, we understand, named by the late Mr. Griffith after Mr. Masters, one of the principal assistants in the Botanical Garden, Calcutta. Fig. 1. represents the inside of the lip, and fig. 2. the pollen-masses.

It should be potted in turfy heath-mould, well mixed with potsherds, and liberally supplied with water during its period of growth. To maintain a humid atmosphere, so very essential to the growth of all Orchids, it will be necessary to use shading in sunny weather; by this means the temperature, which should be kept about 80° by day, may be regulated without admitting much air. In winter it does not require much water, but should not at any time be kept too dry.

53

SCHOMBURGKIA undulāta.

Wavy Schomburgkia.

GYNANDRIA *MONANDRIA.*

Nat. ord. ORCHIDACEÆ. § EPIDENDREÆ.

SCHOMBURGKIA, Botanical Register, 1844. *t.* 23.

S. *undulata* (Lindl. in Bot. Reg. 1844. misc. 21. necnon sub t. 23.) pseudobulbis fusiformibus, floribus densè racemosis, bracteis longissimis spathaceis, sepalis petalisque æqualibus linearibus undulato-crispis labello longioribus, labelli cucullati lobis lateralibus rotundatis intermedio ovali acuto v. obtuso, lamellis 5 undulatis duabusque lateralibus rectis tenuibus.

A year ago we regarded this as the finest of the Schomburgkias: we are now obliged to change our opinion in consequence of a new discovery by Mr. Linden, whose S. rosea is far beyond it in beauty, as we shall presently explain.

The plant now represented was found by that indefatigable traveller, in December 1842, in New Grenada, at the height of 2400 feet above the sea, on the rocks near Pandi, a place, which, in his herbarium, he calls, "the natural bridge of Icononzo." The wild specimens have about 20 flowers in a head, and the plant will therefore become much more beautiful as the cultivated specimens get into better health. Their colour alone distinguishes them from both S. crispa and marginata, and the form of the lip from S. tibicinis.

The glory of this genus is, however, none of those which yet have flowered, but Mr. Linden's Schomburgkia rosea, the number 1664 of his herbarium. "This magnificent species," he writes, "is found on rocks, on the northern slope of the Sierra nevada de Santa Martha. Bracts, peduncles, and lip, are all of a light rose colour; the petals are deep red. It was met with 5000 feet above the sea, between the villages of S. Antonio and S. Miguel, on the territory of the "Auruhuacos" Indians. The specimens now before me, are most beautiful; and if, as we think probable, they should be alive in this country, we shall yet see a plant far more striking than even S. undulata. Its lip is very nearly circular in its outline, and its lobes are not very different in form, but those at the sides are the largest; even when dry it is of the deep rich colour of the spot in Cattleya labiata. Along the middle of the lip, at the base, there run 4 deep plates standing in pairs, and separated by a tolerably broad space: but near the point of the lip these are exchanged for 3 other short deep very wavy plates, which are pure white.

This plant may be added to the catalogue of Schomburgkias, with the following specific character:—

S. *rosea* (Linden. herb. no. 1664) sepalis petalisque oblongis undulatis labello paulò longioribus, labelli subrotundi laciniis lateralibus rotundatis intermediâ subrotundâ apiculatâ crispâ multò minore, lamellis altis pone basin 4 rectiusculis membranaceis versus apicem 3 brevibus crispis.

56

GONGŌRA truncāta.

Bean-budded Gongora.

GYNANDRIA *MONANDRIA.*

Nat. ord. ORCHIDACEÆ. § VANDEÆ—MAXILLARIDÆ.
GONGORA. Botanical Register, vol. 19. *fol.* 1616.

G. *truncata* (Lindl. in Bot. Reg. 1843. misc. 52); sepalis lateralibus rotundato-oblongis supremo obovato apiculato carinato, petalis minimis ovatis acutis decurrentibus quinquenerviis, labelli vernicati hypochilio medio compresso (unde bicamerato) margine lævi apice bicorni, epichilio ovato canaliculato.
? G. Galeottiana, *Richard & Galeotti Orch. Mex. ined.*

Whatever opinion may be entertained respecting the specific distinction of other Gongoras, no one can doubt that this at least is very different from all that have been previously discovered.

It is a Mexican species, introduced from Mexico by Mr. Rucker, who received it from Mr. Linden in 1840, and to whom we are indebted for the specimen now represented.

We do not find any thing peculiar in the pseudobulbs or foliage; the characters of the species reside exclusively in the flowers, which are pale straw-colour, with some brownish purple speckles, and a yellower lip. Before expansion they are almost of the form of a bean; which is owing to their sepals being so blunt that when flattened they are nearly half oblong. The lip has no speckles at all, and looks as if varnished. We do not admire the scent, which is quite peculiar.

It is impossible to tell what plants may be intended by the extremely brief specific phrases which have alone been as yet published by Messrs. Richard and Galeotti of their new Mexican Orchids; but as far as the definition of their G. Galeottiana goes, it is not different from this species, whose whole-coloured pale yellow lip is quite pecular. No one regrets more than ourselves that the ill health of the excellent editor of Galeotti's Orchids should so relentlessly oppose the publication of a work which both botanists and amateurs expect with great interest.

Fig. 1. represents the column and lip of the bean-budded Gongora a little magnified.

It may either be potted in the usual way, or it may be put into a wire basket with moss or rough peat, and suspended to a rafter. This, like many more species of Orchids, requires to be kept moist at the root, as well as in a moist atmosphere, during the growing season. To obtain this in sunny weather, it will be necessary to use shading, by which the temperature may also be kept about 80° by day, without admitting much air. In summer there is little or no need for fire heat; if the house is shut up early in the afternoon it will be found quite sufficient. In winter, when the temperature should average 60°, little water for a few weeks will be required; still it is necessary to keep up a moist atmosphere.

58

AERIDES maculōsum.

Spotted Air-plant.

GYNANDRIA *MONANDRIA.*

Nat. ord. ORCHIDACEÆ. § VANDEÆ.
AERIDES. Bot. Reg. 18. *fol.* 1485.

* *Labello indiviso v. tantum auriculato.*

A. *maculosum;* foliis coriaceis planis apice obliquis obtusis, racemis densis nutantibus subpaniculatis, sepalis subrotundo-oblongis, petalis conformibus duplò latioribus, labello ovato subundulato integerrimo basi utrinque unidentato tuberculo indiviso interjecto, columnâ brevissimâ.

This is more like Sir Richard Brookes' Air-plant, figured at t. 55 of our volume for 1842, than any species in our gardens; but it differs in many important particulars.

In the first place, the flowers are distinctly panicled, and not strictly racemose; then they are smaller; moreover, their column is short, and the lip neither crisp, nor toothed, nor saddle-backed, but quite entire, and nearly flat. The little lobes at the base of the lip are also different; in Sir Richard Brookes' they stand at right angles with the column; here they are very small, and direct themselves towards the point of the lip.

The colour of the flowers of this spotted Air-plant is much less vivid; it has more the tint of the A. affine, has very little white, and both sepals and petals are spotted all over with light purple on a pale rose-coloured ground.

The Air-plants divide into two very distinct groups, of which this represents one, and the sweet-scented species (A. odoratum) the other. They are readily distinguished by the form of their lip, which, in the latter, is deeply cut into three or even five lobes of nearly equal length, while in the former group the lip is either not divided at all, or at the most has only a pair of ears or toothings at the base. Those who wish to remember the differences between the many species of this genus, which are much alike, will do well to bear this in mind.

The accompanying drawing was made in the Nursery of Messrs. Rollissons, in June 1844.

59

CATTLEYA granulōsa; *var.* Russelliana.

The Duke of Bedford's Rough-lipped Cattleya.

GYNANDRIA *MONANDRIA.*

Nat. ord. Orchidaceæ. § Malaxeæ—Læliadæ, *Veg. Kingdom, p.* 181. *ined.*

CATTLEYA. *Botanical Register, vol.* 11. *fol.* 953.

C. granulosa. *Supra,* 1842. *t.* 1.
Var. Russelliana; foliis ovatis, pedunculis trifloris, floribus multo majoribus, petalis magis lanceolatis, labelli ungue angustiore.

Some fatality seems to have attended this beautiful plant in our gardens. It was sent to Sir W. Hooker from Woburn, as a Brazilian Orchid; but we can find no evidence to shew that it is a native of even the Southern hemisphere. At Syon, whence by permission of His Grace the Duke of Northumberland our specimen was supplied, it is said to have been received from Dr. Wallich, in 1839, which is probably another mistake. There can be no doubt at least about its being in reality a very fine variety of the Guatemala C. granulosa, of which the small variety has been already figured in this work, and its introduction may with good reason be referred to one of Mr. Skinner's numerous importations.

It is not surprising that it should have been referred to C. guttata; for the old specific character of the latter fits it very well. But if the figure of that species in the Horticultural Transactions, vol. 2. t. 8. second series, be consulted, the difference between the two will be manifest. The large membranous wavy petals, great erect flowers, and long lip spread out at the point into a broad transverse plate are different from the same parts in the smaller resupinate flowers of Cattleya guttata.

We suspect this plant to be more common in collections than is generally supposed, and that, in fact, it stands in many places as a representative of C. guttata.

The beautiful specimen from which our drawing was made flowered at Syon in May 1844.

61

DENDROBIUM Kingiānum.

Captain King's Dendrobium.

GYNANDRIA *MONANDRIA.*

Nat. ord. Orchidaceæ. § Malaxeæ—Dendrobidæ, *Veg. Kingdom. p.* 151. *ined.*

DENDROBIUM. *Supra, vol.* 15. *fol.* 1291.

Sect. Desmotrichum, (*Blume Bijdragen, vol.* 1. *p.* 329.) Pseudobulbi e rhizomate articulato nati. Labellum per axin lamellatum (an semper?)

D. (Desmotrichum) *Kingianum* (Bidwill MSS. Lindl. in Bot. Reg. 1844. misc. 18); pseudobulbis ovatis in collum longum extensis apice bifoliis, foliis ovalibus emarginatis, pedunculo terminali 2-3-floro foliis æquali, sepalis ovatis, mento emarginato, petalis obovatis apiculatis duplò brevioribus, labelli trilobi cuneati pubescentis laciniis lateralibus acutis intermediâ paulo longiore transversè rhombeâ angulis lateralibus rotundatis apiculi acuto, axi elevatâ trilineatâ apice tridentatâ.

This curious epiphyte was bought by the Messrs. Loddiges at the sale of Mr. Bidwill's New Holland Plants, two or three years since. It has pseudo-bulbs between four and five inches long, tapered from an ovate base into a very long and narrow neck, on the top of which stand two oblong emarginate dark-green rather wavy leaves. Between these is a flower-stalk having two or three pink flowers gaily spotted with crimson in the inside. On some of the offsets the number of leaves is four, but the prevailing number is two.

Among the genera of Orchids proposed by Dr. Blume in his Bijdragen was one called Desmotrichum, consisting of caulescent epiphytes with "root-shaped jointed bulbiferous stems," or rhizomes, leathery leaves growing on the bulbs, and fascicled, or solitary or somewhat racemose flowers springing out from the base of the leaves. This character was accompanied by a description of flowers which hardly distinguished the genus from Dendrobium; nor did the figure of two of the species, given in the same learned author's Tabellen shew any sufficient peculiarity to cut the genus off from Dendrobium, notwithstanding the peculiar habit, which is something like that of Bolbophyllum.

A careful examination of such species as have come beneath our observation within the last few years has equally failed to satisfy us of the distinctness of the genus. In fact, there is but one character deserving of notice that we are able to point out, and that is the presence of two or three raised ridges or plates upon the axis of the lip. Such plates or ridges occur conspicuously in the present species and in D. angulatum, criniferum, Scopa, longicolle, amplum, and some others, which would therefore be thus sufficiently well characterized. But there are no plates or ridges in D. aggregatum or Jenkinsii, which have the habit of the genus, to say nothing of D. braccatum, muscicola, and extinctorium, which require re-examination. Further consideration, and a more extensive knowledge of species are necessary in order to settle this point.

Fig. 1. represents the column of this plant, and fig. 2. the lip spread flat and magnified.

The species may either be potted in turfy peat, in the usual way, or it may be tied to a block of wood, and suspended to a rafter. Like many other Orchids it requires a good deal of water during summer, and a slight shade in sunny weather. In winter, for a few weeks, no more water should be given than keeps the stems from shriveling. The temperature most congenial to this plant, is little more than that of a common greenhouse.

64

ONCIDIUM incūrvum.

Curved Oncidium.

GYNANDRIA *MONANDRIA.*

Nat. ord. Orchidaceæ. § Vandeæ—Brassidæ, *Vegetable Kingdom, p.* 181. *ined.*

ONCIDIUM. *Botanical Register, vol.* 9. *fol.* 727.

Sect. 2. Euoncidium. Labellum basi auriculatum, indivisum v. trilobum.
†† *Sepala v. petala basi angustata.* Heteranthium.
* Folia plana.
‡‡ *Pentapetala;* sepalis lateralibus omninò disjunctis.
¶¶ Micropetala; *petalis sepalis lateralibus subæqualibus et similibus.*

O. *incurvum;* pseudobulbis ovatis ancipitibus utrinque 3-4-costatis ditriphyllis, foliis ensiformibus acutis, scapo elongato racemoso-paniculato, sepalis lineari-lanceolatis undulatis liberis, petalis conformibus incurvis, labelli laciniis lateralibus rotundatis nanis intermediâ subrotundâ concavâ acutâ, cristâ ovatâ depressâ dimidiâ inferiore lineatâ superiore tricostatâ, columnâ subapterâ.

O. incurvum, *Barker in Bot. Reg.* 1840. *misc.* 174.

This striking, and very peculiar species, was first observed in this country, by George Barker, Esq. of Birmingham, from whose rich collection we were favoured with flowers in 1840. Since that period it has found its way into the possession of others, especially of Mr. Ferguson, late gardener to the Duke of Buckingham, who succeeded in growing it extremely well. It is, however, still one of the more rare species.

Mr. Barker gave it the name of incurvum, in consequence of the petals having a great tendency to turn inwards when the flowers first open; that tendency is, however, eventually lost, and the parts assume the position customary in the genus.

In some respects it resembles the Birds-bill Oncid, but it bears its flowers in a long, erect, and even branched panicle, and it has neither the long rostel, nor the great column-wings of that species. Its colours are, moreover, very different, for its petals and sepals are regularly and neatly banded with red on a white ground, instead of being of one uniform rosy tint.

Fig. 1 represents the appearance of the column and lip seen from the front; in this instance, the lower tier of tubercles belonging to the lip-crest, are too much in relief. It also shews that the column is not absolutely without wings, as we first supposed, but really possesses two very small ones.

It requires to be potted in turfy heath-mould mixed with potsherds. Being a free growing species, both an ample supply of water and a moist atmosphere are necessary during summer; and although the species is less liable to become scorched by the sun than many others, still it should be shaded in sunny weather, in order to keep the temperature as near 80° by day as possible, without admitting much air. In winter, if a moist atmosphere is kept up, very little water will be required at the roots, nor need the temperature be raised above 64° by fire heat.

65

STANHOPEA inodōra.

Scentless Stanhopea.

GYNANDRIA *MONANDRIA.*

Nat. ord. ORCHIDACEÆ. § VANDEÆ—MAXILLARIDÆ.
STANHOPEA. *Supra, fol.* 1800.

S. *inodora;* spicâ contractâ, bracteis latis oblongis ovario æqualibus, floribus inodoris, sepalis lateralibus ovato-oblongis ovario subæqualibus, hypochilio subcompresso brevi saccato intus glabro anticè bidentato et inter dentes profundè sulcato, epichilio subrotundo-ovato integerrimo cornubus incurvis longiore, columnæ alis latis sensim evanescentibus.
S. inodora, *Lodd. cat. ult. no.* 1147.

Among the most beautiful of the Stanhopeas is that apricot-coloured species, with very large flowers forming a spreading spike, to which the name of *graveolens* has been applied in consequence of its offensive odour. The kind now represented is in many respects very like it; but it is scentless, and much paler than any variety of S. graveolens which we have seen.

Messrs. Loddiges imported it from Mexico, and furnished, in June 1843, the specimen now represented.

It differs from S. *graveolens* not merely in its pale scentless flowers, but other circumstances of more importance. In the parts of the flower we are not indeed prepared to point out much difference beyond the form of the column, which in this species has its side wings gradually narrowing downwards till they disappear, while S. *graveolens* has them as broad at the one end as the other, whence its column has almost the form of a parallelogram. The form of the spike is quite different in the two species, owing to the unusual shortness of the ovaries. In S. *graveolens* it is very wide, after the manner of S. *oculata*, the ovary being considerably longer than the lateral sepals; but in S. *inodora* it is as much contracted as in S. *insignis*, there being little difference in length between the length of those parts.

If this species were to be distinguished by more popular characters, it might be stated to have the inflorescence of S. *insignis*, the form of S. *graveolens*, and the colour of S. *saccata* without its dots. If S. *insignis* inhabited the same country as S. *graveolens*, one might fancy the S. *inodora* to be a mule between the two.

As the specific character of S. graveolens, given in the enumeration of the genus published in this work in the volume for 1844 now requires amendment, we repeat it here.

S. *graveolens* (Lindley in Bot. Reg. 1840. misc. 125); spicâ expansâ, bracteis angustis ovario vix æqualibus, floribus graveolentibus, sepalis lateralibus ovato-lanceolatis ovario duplò brevioribus, hypochilio subcompresso brevi saccato intus glabro intra scrotum glanduloso et lamellato anticè bidentato interque dentes profundè et apertè sulcato, epichilio subrotundo-ovato integerrimo, cornubus acuminatissimis latis planis incurvis, columnâ alis latissimis subquadratâ.——*Guatemala?*——This is a beautiful species with the habit of S. oculata. The sepals and petals are of a delicate straw colour; the lip at the base and the central parts of the flower generally are of a deep rich apricot yellow, while the horns and upper end of the lip are like ivory turning yellow. The odour of the species is so powerful that it communicates itself to the fingers after touching the flowers, and is extremely disagreeable. It varies with flowers of a deep apricot colour throughout, when it becomes the S. *aurata* of the gardens. Its very broad column, winged to near the base, so as to have almost the form of a parallelogram, is an important character. The lip appears at first sight to be quite smooth inside, but it is in reality covered with glands within the anterior pouch.

67

GOVENIA fasciāta.

Linden's Govenia.

GYNANDRIA *MONANDRIA.*

Nat. ord. ORCHIDACEÆ. § VANDEÆ—MAXILLARIDÆ.
GOVENIA. *Bot. Reg. vol.* 21. *fol.* 1795.

G. *fasciata* (Lindl. in Bot. Reg. 1843. misc. 107); foliis latis ovalibus acutis, spicâ laxâ, bracteis ovario æqualibus abruptè acutis, sepalis angustis acutis, petalis latioribus, labello oblongo subquadrato apiculato intus lævi, antheræ mucrone inflexo.

Among the genera of Orchids there is not one whose species are so difficult to distinguish as those of Govenia; a most natural group, with most of its members extremely similar in habit. In a dried state they are so much alike, or they vary so much in the appearance of their flowers, in consequence of the manner in which they shrink, that it is hardly possible to recognise them.

That which is here figured was named in this work two years since, and was supposed to be Mexican. We now, however, find it among Mr. Linden's Merida plants, under two numbers, 644 and 654. Both were found by that most industrious collector in Venezuela, in damp forests, at the height of 5000 feet above the sea, and are distinguished from each other in his memoranda by the flowers of one having the markings more broken into specks and spots than the other. Both were gathered in July, 1842.

It is one of the prettiest of the genus, having clear yellow flowers, whose sepals and petals are beautifully marked by fine broken bands of crimson. The long bracts, thin narrow spike of flowers, and oblong, not ovate, lip, are the marks by which it is best recognized. The leaves are about a foot long and three inches wide.

Fig. 1. represents the exact form of the lip of this species, and 2. a side view of the column and anther.

Our drawing was made from a specimen communicated by Mr. Rucker in September, 1843.

Govenia fasciata is a stove plant, which requires to be potted in rough peat mixed with about one-third sandy loam. Being a terrestrial Orchid, it necessarily requires an ample supply of water during the growing season, and a slight shade in sunny weather. In autumn, when its leaves begins to die off, water should be gradually withheld, in order to give it its due season of rest; for if watered and excited to grow in winter, when there is not sufficient light to mature the tubers, they will be small and apt to damp off; nor will it succeed in summer if this is not properly attended to. In spring, when it begins to show symptoms of growth, it should be re-potted and placed in a humid atmosphere, where, for a few weeks, it may receive water every fine day.

The following are the species of Govenia at present published, with their synonymes.

*1. G. superba, *Lindl.*—(*Maxillaria superba*, La Llave.)
*2. G. lagenophora, *Lindl.*—(*Cymbidium utriculatum*, Swartz. —*Limodorum utriculatum*, Id.)
*3. G. fasciata, *Lindl.*
*4. G. Gardneri, *Hooker.*
5. G. alba, *Rich. & Galeotti.*
*6. G. liliacea, *Lindl.*—(*Maxillaria liliacea*, La Llave.)
*7. G. capitata, *Lindl.*
8. G. pauciflora, *Lindl.*
9. G. tingens, *Pöppig & Endl.*
10. G. barbata, *Pöppig & Endl.*

Of these the species with an asterisk are or have been in cultivation in this country.

69

LÆLIA pedunculàris.

Long-stalked Lælia.

GYNANDRIA *MONANDRIA*.

Nat. ord. ORCHIDACEÆ. § EPIDENDREÆ.
LÆLIA. Lindl. supra, 1839. *misc.* 42.

§ 1. Grandifloræ ; *Petals distinctly larger than the sepals.*

L. *peduncularis* (Lindl. in Bot. Reg. 1842. misc. 10); pseudobulbis subrotundis compressis lineâ utrinque elevatâ, foliis oblongis obtusis scapo tereti vaginato brevioribus, corymbo multifloro, bracteis ovatis abruptè acutis pedunculo pluries brevioribus, petalis oblongis obtusis sublaceris planis, labelli lobis lateralibus abbreviatis obtusis recurvis antice subdentatis intermedio oblongo rotundato plano subrepando lineis 2-4 elevatis.

When this plant was first named, in the winter of 1841-2, upon the inspection of a specimen sent to London, from the rich collection of Mr. Barker of Birmingham, a very imperfect idea was formed of its real beauty, for the specimen was withered, and injured in the transmission. In November, 1844, we, however, again received it from the same gentleman, and then found it to be what is here represented, one of the most ornamental, and perhaps the most graceful, of the genus.

Its flowers are of one deep rose colour, a little heightened at the lower part of the lip, and they droop gracefully from the end of a slender elastic scape. The unusually long peduncles add to the elegant appearance of the species.

It approaches most nearly to the Lælias rubescens and acuminata. From the former it differs in its lip: having none of the hairiness of that species, and its flowers being much more closely arranged; from the latter, in neither its petals nor lip being wavy and sharp pointed; and from both, in its very large whole-coloured flowers, flat jagged petals, short recurved lip-lobes, and very long flower-stalks.

The species is one of Mr. Barker's valuable importations from Mexico.

It may either be tied to a block of wood, with a little Sphagnum to retain moisture, or it may be potted in turfy-heath-mould, in the usual way. During the growing season, an ample supply of water should be given, and the atmosphere kept as moist as possible, at a temperature of 80° by day. To accomplish this, it will be necessary to use a slight shading in bright weather, which will also prevent the leaves of the plant from being scorched by the mid-day sun. In winter, if the atmosphere can be kept moist, very little water will be required for a few weeks. The temperature should never be raised above 65° with fire heat.

MISCELLANEOUS MATTER

OF THE

BOTANICAL REGISTER.

1845.

11. MAXILLARIA triangulàris.

M. *triangularis;* caulescens, pseudobulbis ovalibus ancipitibus utrinque corrugatosulcatis, foliis ligulatis apice obliquis, floribus solitariis axillaribus ebracteatis, sepalis subæqualibus patentissimis margine in formam triangularem recurvis lateralibus paulo majoribus, petalis conformibus minoribus erectis, labello oblongo ciliato puberulo cum columnâ parallelo emarginato concavo medio verrucâ solitariâ politâ aucto, antherâ conicâ scabrâ.

This is nearly related to the beautiful Maxillaria tenuifolia, and like it is a discovery of Mr. Hartweg's, who found it in the village of Quezaltenango in Guatemala. Its leaves are very much larger than in that species (10 by 1¼ inches); the flowers are formed singly in the axils of the imperfect leaves, which precede the appearance of the perfect leaves and terminal pseudo-bulbs of such plants; they are of a rich cinnamon brown gaily mottled with crimson flecks. The sepals which spread widely apart are folded back at their edges into a triangular form. It flowered in the garden of the Horticultural Society in December, 1844.

12. EPIDENDRUM aëridiforme.

Booth in litt.

E. (Spathium) *aeridiforme;* caule folioso tereti, foliis elongato-oblongis obtusis cum mucrone, pedunculo terminali elongato squamis acuminatis vestito, racemi 7-8-flori abbreviati nutantis axi pedunculis breviore, bracteis squamæformibus acutis, sepalis petalisque oblongis coriaceis obtusis his paulò angustioribus, labello carnoso tripartito cordato basi bituberculato laciniis lateralibus oblongis obliquis intermediâ cuneatâ apiculatâ.

"This plant was introduced from Rio in 1839, and added to the collection of Sir Charles Lemon, Bart., M.P., at Carclew, where it flowered in December, 1843, and again about the same time the following year, when the accompanying figure was made of it. In habit and mode of growth it bears some resemblance to *Ep. nutans*, as well as to *Ep. patens*, agreeing partly with the former in foliage, and with the latter in having a deeply-divided 3-lobed labellum, but widely differing in other respects from either. *Roots* round and straggling, ash-coloured, thick and fleshy, partly terrestrial, but more generally, I suspect, adhering for support to the stem of any tree that may be near it. *Stem* erect, nearly round, deep shining green, lengthening at the top into a kind of two-edged drooping stalk, covered by numerous thin oblong acuminate *bracts* of a pale yellowish green, spotted and tinged with brownish red at the point, and bearing a loose panicle of dingy coloured flowers at the extremity. *Leaves* alternate, thick and rigid, clasping the stem at the base, much recurved and twisted, with an obtuse remarginate point. They vary from six to eight inches in length, and an inch, or an inch and a half in breadth, and are of a deep shining green, marked in several places with brown-coloured blotches. *Panicle* consisting of eight or nine brownish green flowers, with their faces all turned one way, and remarkably sweet-scented in the day time. *Pedicels* upwards of two inches long, round and much curved, of a brownish green at the base, which is furnished with a small acuminate scale, higher coloured and channelled towards the flower. *Sepals* spreading, oblong-lanceolate acute, three-quarters of an inch long and one-quarter of an inch broad, of a dull green, tinged at the points and edges with deep brown; before expansion the outside is of a deep brownish red. *Petals* similar in form and colour to the sepals, but thinner in texture and somewhat narrower. *Labellum* three-lobed, spreading and much recurved, deeply cut. The two lateral lobes taken together form a sort of crescent at the back, with the margin slightly cut and undulated. The middle lobe is much recurved, and has a prominent rib down the centre with a groove on either side. All of them have a fleshy tinge near the outside, but inwardly they are very pale, almost white, as well as the *column* which is slightly curved and tubular, green at the base, enlarging outwardly, and having two small roundish obtuse fleshy processes at the mouth, just above where the column and central lobe of the labellum are united. *Anther case* round, somewhat reniform, four-celled, containing two pair of pollen-masses with their largest end upwards.

"The plant requires the constant heat of a damp stove, and the same treatment as others of its class.

"W. B. Booth."

19. SCELOCHILUS Ottonis.

Link, Klotzsch & Otto, Icones plantarum, t. 31.

Scelochilus (Klotzsch in Otto et Dietrich Gartenzeitung, 1841. p. 261.) "Perigonii conniventis foliola exteriora angusta, navicularia, carinata, basi subcohærentia, lateralia labello supposita, in unicum connata, basi in calcar obtusum, breve producta; interiora latiora, libera. Labellum integrum, supra basin columna continuum, basi brevissime bifidum, liberum, disco calloso, puberulo, longitudinaliter bicostato, antice bidentato, dentibus obtusis, conniventibus; costis infra medium bicornutis; lamina apice emarginata, subexserta. Columna semiteres, nuda, labello subduplo brevior. Anthera semibilocularis. Pollinia 2, sphærica, solida, caudicula lineari instructa, glandula parva, obovata.——Herba caracasana, epiphyta; rhizomate cæspitoso; pseudobulbis subnullis; foliis solitariis, coriaceis, carinatis, basi vaginis squamæformibus, conduplicatis, involucratis; racemo radicali; floribus compressis, flavidis.

S. *Ottonis;* foliis oblongis, coriaceis, læte-viridibus, margine acutis, subtortuosis, apice conduplicato-acutissimis, recurvis; racemo radicali subramoso, foliis parum longiore; foliolis perigonii interioribus obovatis, obtusis, intus longitudinaliter purpureo-striatis, sparsim pilosis."

This new genus of Orchidaceous epiphytes has been found in the Caraccas by Mr. Edward Otto, and by him sent to the Royal Botanic Garden, Berlin, where it has flowered. It occurred on the stems of trees in dense forests, 5,600 feet above the level of the sea. The habit of the plant is that of Burlingtonia, to which it approaches very nearly. The flowers are yellow, rather more than half an inch long, and do not spread flat, but are half closed up.

21. EPIDENDRUM marginatum.

Link, Klotzsch & Otto, Icones plantarum, t. 36.

This is the Epidendrum radiatum figured at plate 45, of our volume for 1844.

26. MAXILLĀRIA Lyonii.

M. *Lyonii;* acaulis, pseudobulbis ovatis compressis lævibus nec angulatis corrugatisve, foliis lineari-lanceolatis apice obliquè emarginatis, floribus solitariis, sepalis petalisque oblongis obtusis striatis, labello oblongo retuso emarginato intra apicem scabro, tuberculo oblongo in medio prope basin.

A little species imported from Mexico by J. C. Lyons, Esq. of Ladiston, near Mullingar. It has very narrow grassy leaves, and ovate smooth somewhat glaucous pseudobulbs. The flowers are dull purplish brown, very slightly bordered with a yellow tint.

27. HELCIA sanguinolenta.

Among the unexamined plants collected by Mr. Hartweg for the Horticultural Society, was a plant supposed to be a species of Trichopilia, of which it has entirely the habit, collected at Paccha, a miserable village in the Andes of Guayaquil. Having now flowered, I am able to state that although it certainly approaches nearly to that genus, yet it is in reality an entirely new form. Instead of its column being rolled up in the labellum, it stands erect and clear of it. Instead of the anther having but one cell, it has two. Instead of the anther-bed having two lateral lacerated processes, it is surrounded by a deep fringed border. Finally, instead of the lip being perfectly smooth, continuous, and destitute of all appendages at the base, it is contracted about the middle, below the contraction furnished with a pair of thick fleshy lobes hollowed out in the middle and standing erect on each side of the column, without however touching it; and the space between those lobes, forming the very base of the lip, is a hollow hairy pit.

Upon looking at this curious apparatus in front, the anther and column look like an old-fashioned head-dress peeping over one of those starched high collars, such as ladies wore in the days of Queen Elizabeth; or through a horse collar decorated with gaudy ribbons; and this has led to the name, which has been fashioned out of Helcium, a word said to mean the collar of a horse, though, considering its obvious derivation, one would rather have imagined it to signify his traces.

The following character will sufficiently indicate the peculiarity of this new genus.

Helcia. *Sepala & petala* coriacea, conformia, patula. *Labellum* patulum, membranaceum, planum, medio constrictum, basi utrinque appendice carnosâ truncatâ medio foveatâ in lineas duas breves elevatas procurrente auctum, ipsa basi excavatum pilosum. Columna libera teres, clinandrio erecto undique fimbriato. Anthera carnosa, in pileum solidum obtusum producta, 2-locularis. Pollinia 2, postice excavata, caudiculâ cuneatâ, glandulâ parvâ ovali.——Herba *pseudobulbosa,* foliis *solitariis, coriaceis, undulatis,* pedunculis *radicalibus.*

H. *sanguinolenta.* Pseudobulbi ovati, elongati, subteretes. Folia 4-6 pollicaria, undulata, basi in petiolum canaliculatum angustata. Pedunculus pseudobulbis brevior, uniflorus, bibracteatus. Sepala et petala olivacea, fuscosanguineo maculata. Labellum album striis sanguinolentis, obovatum, emarginatum, appendicibus luteis denticulatis striatis.

The relationship of this plant is evidently greatest to Trichopilia. It is however also an associate of Aspasia, from which it differs in not having the lip united to the column, and in its deep-fringed anther-bed. All these are stationed in a subdivision of Vandeæ, to which I propose to give the name of Brassidæ.

30. RHYNCHOPĔRA pedunculāta.

Link, Klotzsch & Otto, Ic. pl. 2. 103. *t.* 41.

This is a new species of Pleurothallis, allied to P. Mathewsii and some others of the same section. Mr. Edward Otto is said to have sent it from the Caraccas.

31. CATACHÆTUM recurvatum.

Link, Klotzsch & Otto, Ic. pl. 2. 105. *t.* 42.

This is Catasetum planiceps, figured in our volume for 1843, t. 9; but its flowers are perfectly green.

34. EPIDENDRUM basilare.

Link, Klotzsch & Otto, Ic. pl. 2. *p.* 111. *t.* 45.

A synonym of the E. Stamfordianum of Mr. Bateman.

36. The Sections of EPIDENDRUM.

Having in former volumes reduced to their proper places such of the species of Epidendrum as belonged to the sections Amphiglottium, Osmophytum, Encyclium, and Spathium, I now proceed to show what are to be arranged in Hormidium, Diacrium, Epicladium, Lanium, and Aulizeum. and Psilanthemum

§ HORMIDIUM.

Hormidium, *Lindley in Hooker's Journal of Botany*, 3. 81.

This section includes a little set of creeping species, with true pseudobulbs, sessile (or nearly) flowers, and a lip adnate to the column. They differ from *Aulizeum* and *Osmophytum* in the pseudobulbs, and in the flowers not being in conspicuous racemes; from Lanium in their flowers not being panicled and woolly; from Spathium, Amphiglottium, and Euepidendrum in their stems not being leafy.

1. E. *pygmæum* (Hooker's Journal of Botany, 1. 49. t. 118. *E. uniflorum*, Lindl. in Bot. Reg. 1839, misc. 13.); rhizomate moniliformi, foliis brevis lanceolatis subundulatis acutis, floribus solitariis, sepalis petalisque linearibus acuminatis incurvis, labello trilobo columnæ adnato laciniis lateralibus rotundatis erectis intermediâ triangulari acuminatâ, sepalis lateralibus labello suppositis eique adnatis.——*Brazil*——Flowers small, yellowish green. I presume that Brazil is really the native country of this, and that it was a mistake to assign it to Mexico.
2. E. *miserum* (Lindl. in Bot. Reg. 1841, misc. 62.); caulibus ovalibus exuviatis compressis aggregatis, foliis . . . . ., scapo brevissimo subbifloro, spathâ bivalvi pedunculis subæquali membranaceâ, sepalis ovatis acutis, petalis linearibus, labelli trilobi lobis lateralibus rotundatis inconspicuis intermedio truncato emarginato, callis nullis. ——*Mexico.*——Pseudobulbs about half an inch high. The scape, peduncles, and ovaries, taken together, are about the same length. The flowers are a dull, dingy, greenish-brown, not unlike those of E. musciferum.
3. E. *cæspitosum* (Pöppig & Endl. n. g. & sp. pl. 2. t. 101.); caule repente, pseudobulbis cylindricis diphyllis, foliis ovali-lanceolatis acutis, floribus in sinu folii paucis brevissime spicatis aut sessilibus, sepalis subæqualibus lanceolatis, labello integro transverso reniformi lateribus rotundato antice retuso ac breviter apiculato nudo. ——*Peru.*——This appears to differ from E. pygmæum in little, except having a very short middle lobe to the lip.
4. E. *serpens* (Lindl. in Benth. pl. Hartweg, p. 149.); pseudobulbis oblongis 2-3-phyllis, foliis ovato-lanceolatis acutissimis, labello cochleato serrulato lævi, capsula ovali triquetra.——*Peru.*——This must be a pretty plant. The flowers are rather large, and deep violet, and the pseudobulbs so crowded together that they must form a carpet when growing as in their wild state. It is remarkable that no straps appear to be connected with the pollen-masses in this species, but some powdery matter is formed in its stead. It is a mountain plant, creeping among Lichens.

§ DIACRIUM.

Diacrium, *Lindl. l. c.*

This section may at first sight seem to merge in Encyclium; for the chief difference previously pointed out by me, consists in the stem being fusiform and leafy at the end, which is a small modification of the pseudobulbous condition. But in addition to this, the flowers of E. bicornutum have some peculiarities; one of which is that the lip is not parallel with the column, and another, that it has two deep pits on the under side, corresponding with the same number of fleshy elevations on the upper side.

1. E. *bidentatum* (Lindl. gen. & sp. orch. p. 98.); foliis obovato-lanceolatis apice recurvis supra v. sub pseudobulbum nascentibus, racemo terminali erecto paucifloro, sepalis lanceolatis petalisque angustioribus patentibus, labello semilibero rhomboideo angulis utrinque uni-dentato.——*Mexico.*——The only memorandum I possess regarding this species is a sketch of its flowers, made many years ago from a specimen in Mr. Lambert's possession. It seems very near E. bicornutum, but it has narrow petals, and the side lobes of the lip are little more than projecting teeth.
2. E. *bicornutum* (Hooker, Bot. Mag. t. 3332.); foliis lineari-oblongis obtusis coriaceis, caule corniformi, sepalis ovato-lanceolatis, petalis conformibus latioribus, labelli trilobi lobo medio elongato lanceolato acuto basi altè bilamellato lateralibus obtusiusculis.——*Trinidad, Demerara.*——Flowers large and white, slightly tinged with rose colour. Lip yellow at the base, with some crimson specks on the tongue-like middle lobe. According to Mr. Schomburgk, who found it in Demerara, the petals are also spotted with crimson.

§ EPICLADIUM.

Epicladium, *Lindl. l. c.*

In this section the structure of the flower is exactly that of Encyclium, but the flowers grow out of a great spathe, as in Spathium, and the pseudobulbs are evidently assuming the condition of an ordinary stem.

1. E. *aurantiacum* (Bateman in Bot. Reg. 1838, misc. 11. Orch. Mex. & Guat. t. 12.); foliis oblongis planiusculis coriaceis caule clavato diphyllo duplo brevioribus, spathâ maximâ cucullatâ pedunculum longiore, racemo brevi subcernuo densifloro, sepalis petalisque lanceolatis acutis subæqualibus, labello libero oblongo cucullato basi columnam involvente.——*Mexico, Guatemala.*——Flowers numerous, rich orange colour. Found on exposed rocks on the precipitous brows of ravines, where it is subject to great extremes of heat and cold.

§ LANIUM.

Lanium, *Lindl. l. c.*

Creeping scaly stems throwing up little leafy branches or even pseudobulbs; loosely arranged small long-stalked flowers, more or less covered with down, are the principal characteristics of this section, which has a lip adnate to the column. Its panicled or racemose woolly flowers distinguish it at first sight.

1. E. *microphyllum* (Lindl. in Hooker's Journal, 3. 85.); caule repente squamato, ramulis foliosis, foliis distichis oblongis carnosis canaliculatis acutis serrulatis racemo terminali tomentoso multo brevioribus, bracteis membranaceis pedicellis filiformibus multo brevioribus, ovario tomentoso, sepalis apice aristatis, petalis linearibus, labello oblongo acuto basi bilamellato venis tribus per medium obsoletis.——*Demerara, Berbice.*——A small creeping plant, with woolly dull purple flowers. Messrs. Loddiges flowered it from Schomburgk's Demerara collections, and I have since received it from Mr. Bateman.
2. E. *Avicula* (Lindl. in Hooker's Journal, 3. 85.); caule repente squamato, ramulis pseudo-bulbosis diphyllis, foliis ovatis planis margine lævibus paniculâ tomentosâ multo brevioribus, sepalis lanceolatis acutis tomentosis, petalis linearibus, labello acuto subrhombeo basi bicalloso.——*Brazil.*——The leaves of this curious plant are about an inch long; the panicle between three and four inches. The flowers are small, and when seen from the

back may be not unaptly compared to a little bird in full flight.

§ AULIZEUM.

Aulizeum, *Lindl. l. c.*

Among the species whose lip is united to the column, are those which, like E. ciliare, have that organ broken up into long and beautiful fringes, or at least split into three pieces. They have a long horn-like stem, leafy only at the end, and their flowers invariably grow in racemes. These form the section Aulizeum, along with others of less conspicuous beauty, but similar in those points of structure.

1. E. *ciliare* (Linn. Sp. Pl. 1349. Jacq. Amer. 224. t. 179. f. 89. Swartz. Nov. Act. Ups. 6. 69. Willd. no. 21. Ker in Bot. Reg. t. 784. Gen. & Sp. Orch. no. 18.—*Auliza ciliaris*, Salisb. in Hort. Trans. l. c.); foliis in pseudobulbos geminatim insidentibus, sepalis petalisque linearibus acuminatissimis, labelli trilobi lobis lateralibus setaceo-pectinatis intermedio setaceo brevioribus, spicâ subquinquefiorâ.——*West Indies, and tropical America generally.*——Very near E. cuspidatum, but more slender, and with very delicate fringes to the lip. Sepals and petals very light green. Lip pure white. I have a specimen from Mr. Hartweg, found in the province of Bogota, with a raceme a foot and a half long, and enormous flowers.
2. E. *viscidum* (Lindl. in Bot. Reg. 1840. misc. 190.); caule tereti monophyllo vaginâ alterâ longiore infra folium, folio canaliculato mucronulato, racemo paucifloro, bracteis herbaceis glutinosis carinatis ovarii dimidium æquantibus, petalis sepalisque linearibus acuminatis herbaceis, labelli tripartiti basi bicallosi laciniis lateralibus semi-rhombeis acuminatis setaceo-fimbriatis intermediâ subulatâ brevioribus, clinandrio cucullato dentato.——*Mexico.*——Nearly allied to E. ciliare, from which its single narrow channelled leaf, smaller flowers, and glutinous bracts distinguish it. The blossoms have a weak smell like Cucumbers.
3. E. *cuspidatum* (Lodd. Bot. Cab. 10. Ker in Bot. Reg. 10. 783. Gen. & Sp. Orch. no. 19.—*Epidendrum ciliare*, Bot. Mag. 463.); foliis in pseudobulbos ternatim insidentibus, sepalis lineari-lanceolatis acuminatis, petalis conformibus latioribus, labelli trilobi lobis lateralibus falcatis laciniatis intermedio lineari subæqualibus, spicâ subtriflorâ.——*West Indies and Mexico.*——Flowers large, pale straw colour; lip with a deep yellow base.
4. E. *viviparum* (Lindl. in Bot. Reg. 1841. misc. 27.); caule fusiformi diphyllo pedunculo squamato pluries breviore, foliis oblongis horizontalibus subundulatis obtusis, racemo denso paucifloro, bracteis membranaceis canaliculatis ovario cuniculato æqualibus, sepalis petalisque linearibus acutis patentissimis, labelli trilobi basi bicallosi laciniis lateralibus subrotundo-rhomboideis intermedio lineari spathulato acuminato convexo triplò longiore, columnæ cucullo fimbriato.——*Demerara.*——This species has white flowers closely arranged at the end of a flower-stem two feet long; and the latter when old produces young plants at almost every joint. The flowers have no smell.
5. E. *stenopetalum* (Knowles & Westcott, Floral Cabinet, vol. 2. p. 175.); "pseudobulbis cylindraceo-oblongis; foliis lanceolatis; floribus solitariis; sepalis lanceolatis; petalis angustè linearibus; labello trilobo, lobo medio cuspidato-purpureo, lobis lateralibus subrotundis albidis columnam involventibus."——*Mexico.*——"A delicate and pretty species," to me unknown.
6. E. *clavatum* (Lindl. in Bot. Reg. t. 1870.); caule clavato in pseudobulbum ovale desinente diphyllo, foliis lanceolatis patulis, racemo simplici subæquali, bracteis ovatis canaliculatis acutis ovariis infimis duplò brevioribus, sepalis petalisque lanceolato-linearibus æqualibus, columnâ clavatâ, labelli tripartiti basi bicallosi laciniis lateralibus ovatis subfalcatis margine posteriore denticulato: intermediâ unguiculatâ ovatâ acuminatâ.——*Cumana.*——Flowers green, with a small white lip.
7. E. *falcatum* (Lindley in Taylor's Annals of Natural History, Feb. 1840. *E. Parkinsonianum*, Hooker in Bot. Mag. Feb. 1840. t. 3778); caule ramoso carnoso membranis laxis imbricatis vaginato, foliis solitariis falcatis canaliculatis acutis, fasciculis florum sessilibus: pedunculis elongatis, sepalis petalisque lineari-lanceolatis patentissimis, labelli tripartiti basi bituberculati laciniis lateralibus oblongis dimidiatis integris intermediâ lineari-lanceolatâ paulò longiore.——*Mexico.*——A very fine species, with fleshy branching short stems, long channelled falcate leaves, and large pale yellowish flowers, springing from within a long pale yellowish-green membranous spathe. In Sir Wm. Hooker's figure only two flowers are represented, but in wild specimens there are four or more to each spathe.
8. E. *aloifolium* (Bateman Orch. Mex. & Guat. t. 25.); "rhizomate repente parcè folioso, caulibus brevibus monophyllis, floribus 3-5 longè pedicellatis ex axillâ folii ovato-lanceolati crassissimi prodeuntibus, sepalis petalisque conformibus lanceolatis acuminatis explanatis, labelli altè trilobi lobis lateralibus acinaciformibus intermedio paulò longiore setaceo, columnâ labello connatâ.——*Mexico.*——Very like E. falcatum, but, according to Mr. Bateman, the leaves are more fleshy and longer; in E. aloifolium the flower-stalk is more than twice the length of the sepals, while in E. falcatum it scarcely exceeds them; and in E. aloifolium the lateral and middle lobes of the lip are nearly equal, but in E. falcatum the middle is twice the length of the sides. I have never seen the plant.
9. E. *saxatile* (Lindl. in Hooker's Journal, 3. 84.); caulibus fusiformibus apice diphyllis, foliis lineari-lanceolatis racemo paucifloro brevioribus, floribus membranaceis, sepalis oblongis petalisque filiformibus labello multo minoribus, labelli subrotundi trilobi laciniis rotundatis subrepandis intermediam bilobam reniformem serratam imbricantibus.——*Brazil.*——Whole plant less than six inches high. Flowers membranous, reddish purple, with darker longitudinal streaks, as large as in E. Schomburgkii.
10. E. *rupestre* (Lindl. in Hooker's Journal, 3. 84.); caulibus filiformibus vaginis membranaceis vestitis diphyllis, foliis lanceolatis acutis pedunculo ancipiti æqualibus, racemo cernuo, bracteis membranaceis acuminatis pedicellis brevioribus, sepalis oblongis petalisque lineari-spathulatis obtusis, labelli trilobi laciniis lateralibus acutis margine postico serratis intermediâ rotundatâ integrâ lineis tribus elevatis rugosis.——*Peru.*——Flowers yellow, the size of E. conopseum.
11. E. *Hartwegii* (Lindl. in Plant. Hartw. p. 150); foliis 2 oblongis obtusis coriaceis apice obliquis spicæ rigidæ multifloræ basi spathaceæ caulisque bulbosi teretis longitudine, floribus coriaceis, ovario triquetro clavato, sepalo supremo oblongo obtuso lateralibus acutissimis, petalis obovatis, labello subrotundo æqualiter trilobo basi callo pubescente concavo aucto, ungue et columna brevissimis.——*Peru.*——This has somewhat the habit of E. variegatum, but the lateral lobes of the lip are very conspicuous, and it is much thinner.
12. E. *parvilabre* (Lindl. in Benth. Pl. Hartw. p. 150.); folio oblongo suboblique emarginato caule longiore, racemo simplici multifloro cernuo folio multo longiore e spatha erumpente 2-3-phyllo, sepalis oblongis obtusis petalis duplo latioribus lateralibus 5-nerviis, labelli tripartiti laciniis lateralibus falcatis acutissimis intermedia angustissima apice cuneata triloba, axi elevata basi bicallosa.——*Peru.*——This species has a spathe like the species of the § Spathium, but it has pseudobulbs instead of leafy stems.

N.B. *E. collare* (Bot. Reg. 1843. misc. 85.) is better referred to § Osmophytum than to Aulizeum.

§ PSILANTHEMUM.

Psilanthemum, *Klotzsch in L. K. O. ic. pl. 2. p.* 111.

The radical inflorescence alone is a character which distinctly marks this section, for otherwise it is unknown in the genus. The lip is adnate to the column, and is similar in structure to Amphiglottium.

1. E. *Stamfordianum* (Bateman, Orch. Mex. & Guat. t. 11. E. *basilare,* Klotzsch, l. c. t. 45.); pseudobulbis fusiformibus, foliis oblongis obtusis coriaceis basi angustatis, racemo radicali paniculato, sepalis patentibus lanceolatis, petalis duplò angustioribus, labelli tripartiti laciniis planis æquilongis lateralibus oblongis obtusis intermediâ transversâ bilobâ s. emarginatâ fimbriatâ.—— *Guatemala.* ——Flowers in a large branched raceme, deliciously fragrant, pale yellow, with a vivid violet spot at the base of the lip. It is found on the coast of Guatemala, in shady very moist lands; and requires an unusually moist climate.

38. MAXILLARIA setigera.

M. *setigera;* (Scapigeræ unifloræ) pseudobulbis foliisque Oncidio ampliato similibus, scapis radicalibus unifloris squamis acuminatis subpaleaceis striatis supremâ ovarii longitudine adpressâ, sepalis lanceolatis mento obtuso, petalis linearibus paulò brevioribus apice abruptè setigeris, labello oblongo trilobo piloso laciniâ mediâ oblongâ emarginatâ crispâ lateralibus brevibus obtusis appendice ovatâ in medio, clinandrio fimbriato.

This novelty is from La Guayra, where it was collected for Mr. Barker, to whom I am indebted for a flower. The leaves and pseudobulbs, *which I have not seen,* are said to be like those of Oncidium ampliatum. The flowers, which grow singly on scaly scapes, like Maxillaria picta, are pale yellow, slightly tinged with pink on the outside. The petals are almost exactly linear, terminating suddenly at the point in a fine bristle. The lip is for the most part white, with a few scattered hairs over all its upper surface; the centre of the upper part is yellow, the sides of the lower part are streaked with crimson veins. The anther-bed is bordered by a fine fringe.

In many respects this approaches Maxillaria splendens, which however is easily known by its long grassy leaves.

41. DENDROBIUM anosmum.

D. *anosmum;* facie ominò D. macrophylli et colore; sepalis ovato-lanceolatis, petalis oblongis planis nec oblongo-lanceolatis undulatis, labello subrotundo acuto.

This is a Philippine plant, which has lately flowered with Messrs Loddiges. Its appearance is very much that of D. macrophyllum, but, 1. its flowers are scentless, and have not the strong odour of rhubarb; 2. they are smaller; 3. all the divisions of the flower are shorter, broader, and even, instead of being undulating. Whether or not it is really a species experience will show; it is at all events a remarkable variety.

42. EPIDENDRUM rufum.

E. *rufum;* pseudobulbis pyriformibus 2-3 phyllis, foliis brevibus lanceolato-ligulatis patentibus scapo paniculato brevioribus, sepalis petalisque ovalibus acutis subcarnosis, labelli trilobi laciniis lateralibus brevibus semi-ovatis intermediâ obovato-oblongâ convexa margine revolutâ apice rotundatâ basi secus axin elevatâ carnosâ, columnâ membranaceo-marginatâ.

This species of the Encyclium division is very near E. flavum, from which it differs in the colour of the flowers, the form of the lip, and the shortness of the leaves. The following memorandum from Mr. Booth, gardener to Sir Charles Lemon, fully explains its history.

"This plant was imported from Rio in 1842, by Lieut. Turner, of H. M. Packet 'Ranger,' and presented to Sir Charles Lemon, Bart., M.P., with whom it flowered at Carclew, in May, 1844. It has a great resemblance both in habit and general appearance to Encyclia patens; but its flowers are very different, and although not so gaudy as some of its brethren, I have no doubt of its proving a desirable plant when well grown. It requires a warm moist atmosphere, and thrives pretty well attached to a block of wood suspended from the rafters of the stove.

"*Pseudo-bulbs* two or three-leaved, pear-shaped, about three inches long, and four inches in circumference, furrowed when old and of a deep-shining green, but almost wholly covered when young with several large pale brown sheathing scales. *Leaves* of a dark green, flat and rigid, erect or slightly spreading and recurved; oblong-lanceolate acute, an inch wide, and from six to eight inches long. *Scape* issuing from the crown of the pseudo-bulb—round and erect, from fifteen to eighteen inches high, with several joints having each a small dark brown acuminate bract, and bearing a profusion of brownish yellow flowers at the top for about a third of its length, arranged in a sort of loose panicle. *Pedicells* round and slender, slightly curved, pale green. *Sepals* spreading oblong lanceolate, blunt at the point, an inch long, and about a quarter of an inch wide—thinnish towards the base, but fleshy outwardly—the upper one curved, the other two slightly twisted and undulated. *Petals* rather shorter than the sepals, and narrower, having their edges somewhat reflexed. *Labellum* three-lobed, curved and deeply divided. The two lateral lobes are small, ovate-acuminate, and rise erect on either side of the column. The middle lobe has the edges reflexed, and rolled up so as to appear no larger than the sepals, although it is twice their size. At the base is a slightly elevated plate the length of the column and parallel to it. *Column* short-rounded on the upper side, but deeply hollowed beneath; thin, and of a silvery green at the edges, with a semicircular three-pointed depression in which the anther-case is contained. The latter is two-celled, roundish ovate. *Pollen-masses* four, two in each cell, ovate, deep yellow, with a reflexed and slightly bifid caudicula."—*Wm. B. Booth.*

43. SPIRANTHES lobata.

Lindley in Bot. Reg. 1844. *misc.* 9.

"Among some plants received by Sir Charles Lemon, Bart., M.P., in the spring of 1842, from Mr. John Rule, of the Real del Monte Mines, Mexico, were two with long, thick, fleshy, fibrous roots, which shortly afterwards sent up leaves so much resembling the well-known *Stenorhynchus speciosus* as to have been mistaken for that plant, until on flowering, in December, 1843, they proved to be a very different thing; not so showy as *Stenorhynchus,* but more interesting to the botanist as furnishing an additional species to the genus Spiranthes.

"*Leaves* oblong lanceolate acute, spreading round the flower-stem which rises in the centre, slightly marked as if reticulated, but quite smooth, and of a rich green above, very glossy beneath; varying from nine inches to a foot in length, and from two and a half to three inches in width; somewhat undulated at the margin, and hollowed at the base so as to embrace the stem. *Scape* fifteen or eighteen inches high, quite round and hairy, terminating in a spike of about a dozen or more yellowish green flowers, with a thin membranous oblong lanceolate acuminate bract at the base of each, similar in form but much smaller and darker coloured than those which embrace the rest of the stem, and which are at first so closely imbricated as to give it something the appearance of a head of asparagus. *Flowers* curved, rather more than an inch and a half long; the lower part dark green, grooved and densely pubescent. The upper segment is smooth, and of a dull yellowish green, hooked and arched so as to form a kind of oblong hood, having the edges turned upwards and tapering to a flattened emarginate point. The segments on either side are ovate-oblong, slightly lunate, about the same size as the upper one and similar in colour, but flatter, much recurved and twisted so as to cross one another at the point. The *lip* is thin, but firm, three-lobed and semitransparent, of a yellower green than the other part of the flower, and marked with greenish lines. The two side lobes are curved upwards so as to meet underneath the upper-hooded segment, and form, with the middle lobe, a kind of funnel-shaped tube. All of them are recurved at the margin, particularly the middle one, which is as large as the other two, roundish cordate, ending in a narrow elongated spur, with two roundish-hooked claws. In the centre it is connected with the edges of the *column*, which is nearly as long as the anterior lobes of the lip, narrow, curved, and tapering to a roundish, slightly hollowed, brownish stigma. *Anther-case* arrow-shaped, two-celled, containing a quantity of powdery pollen-masses in each.

"The plant requires the constant heat of the stove, and thrives in a rich loamy soil; kept moist when in a growing state, but dry after the leaves decay."—*Wm. B. Booth.*

45. ODONTOGLOSSUM membranaceum.

Lindl. in Sertum Orchidaceum, sub t. 25.

This charming species has flowered with Messrs. Loddiges. The blossoms are pure white, in the way of O. Rossii; large, and delicately banded concentrically with deep rose colour. They smell like bitter almonds. A figure of it will soon appear in this work.

46. ONCIDIUM gallopavinum.

Morren, Ann. Soc. Hort. Gand. 1. 13.

O. *gallopavinum;* pseudobulbis ovatis, apice attenuatis; foliis binis, planis, lanceolatis; scapo elongato, spica simplici, 7-10 floribus divaricatis; bracteis minimis, ovato-lanceolatis, acutis, membranaceis; sepalis liberis, ovatis, undulato-concavis, roseo-purpureis; petalis ovato-lanceolatis, roseo-purpureis; labelli lobis lateralibus spathulatis, abbreviatis, rubris, intermedio maximo, elongato, emarginato, obcordato, sinu brevi mucronato, citrino, crista minima, vix bicallosa, tuberculis maximis, rugoso-carunculatis, purpureis, columnæ auriculis alato-explanatis, latis.

This is said to be Mexican; its lip is yellow, the sepals and petals greenish, stained with crimson; but the figure is so very inexact in its details that we cannot form any opinion upon it.

47. CATTLEYA Papeiansiana.

Morren, Ann. Soc. Hort. Gand. p. 57.

C. *Papeiansiana;* floribus solitariis vel binis, longe pedunculatis, erectis; sepalis oblongo-lanceolatis, infernis arcuatis; petalis latioribus oblongo-lanceolatis, obtusiusculis, undulatis, labelli trilobi lobis undato-denticulatis, intermedio cordato, undulato, crispo, per axia sexstriato, emarginato; pseudo-bulbis nullis, caule stricto, foliis binis; lato-lanceolatis, obtusis; emarginatis, carnosis; spatha scariosa.

If this is anything more than a slight variety of C. Loddigesii, why then the species of Orchids may be multiplied *ad infinitum.*

59. EPIDENDRUM Lindenii.

E. (Amphiglottium) *Lindenii;* caulibus simplicibus strictis, foliis ovalibus obtusis emarginatisve coriaceis distichis, racemo oblongo obtuso, sepalis petalisque æqualibus, labelli tripartiti laciniis lateralibus subrotundis laceris intermediâ cuneatâ emarginatâ dentatâ suprà carinatâ basi bituberculatâ cristâ crassâ subrotundâ plicatâ ante tuberculos, clinandrio denticulato.

This beautiful species was discovered on rocks near Merida by Mr. Linden, a most diligent and successful collector, to whom our gardens owe many of their fairest ornaments. It forms no. 636 of his herbarium, the whole of the Orchids in which have been liberally placed in our hands for publication. From the ticket attached to the dried specimens, we learn that the plant appears in the form of three well-marked varieties; in one the flowers are a bright carmine, in another rose-coloured, and in a third of a yellowish orange. It was met with in August, 1842, at the height of 5000 feet. Our living specimens have been communicated by Messrs. Loddiges, (603), and appear to belong to none of Mr. Linden's varieties, for the whole centre is rose-coloured, while the upper half of the sepals and petals is a dull but very pleasing salmon colour. The habit of the plant is quite that of E. elongatum; it is however nearer E. Schomburgkii, the middle lobe of its lip having the deep keel of that species. From all the described plants thus allied to E. Schomburgkii, E. Lindenii differs in having a large plaited or ribbed semicircular crest, placed in front of the tubercles so frequently present at the base of the lip of this genus.

60. The Genus ODONTOGLOSSUM.

In pursuance of the plan adopted in this work on several previous occasions, we now proceed to give an account of the species of another of those large genera of Orchids, which fill our collections with their species, and whose names are to be found scattered through so many books that even the most skilful Botanist scarcely knows where to search for them, or when his inquiry is exhausted.

The genus Odontoglossum presents itself under three principal general forms. In the one the anther-bed is surrounded by a deep fringe, or membranous border; the two others are without that fringe. The first constitutes the sub-genus (?) Trymenium; the second and third are Odontoglossum proper. Of the second the lip is always white, and generally broad and flat. Of the third the lip is generally yellow, never white, and most commonly narrow. There is no tendency at present observed on the part of Trymenium to pass into Odontoglossum, and it might be regarded as a genus; the two sections of Odontoglossum evidently touch by O. constrictum and hastatum among the "white lips," and O. læve, or such species, among the "stained lips."

Sect. I. Trymenium, *Lindl. in Bot. Reg.* 1843. *t.* 3.

1. O. *brevifolium* (Lindl. in Benth. Plant. Hartweg. p. 152); foliis ovato-oblongis oblongisque patentibus scapo simplici apice multifloro racemoso duplo brevioribus, bracteis oblongis membranaceis pedicellorum longitudine, sepalis subrotundis unguiculatis undulatis, petalis paullo minoribus, labello unguiculato basi columnæ adnato auriculato cuneato emarginato sepalis breviore basi tuberculis verrucosis 5 et appendice parva antica 3-dentata aucto, clinandrio cucullato serrato in marginem membranaceum alæformem decurrens.—— *Loxa.*——Pseudobulbs ovate oblong compressed, one-leaved. Leaves two inches broad, and sometimes not much longer. Flowers an inch and a half or more in diameter; eleven or twelve in a drooping raceme, apparently purple.
2. O. *citrosmum* (Lindl. Bot. Reg. 1842. misc. 68; 1843. t. 3); pseudobulbis subrotundis compressis lævibus monophyllis, folio oblongo-ligulato obtuso racemo paulò breviore, sepalis oblongis obtusis petalisque conformibus subæqualibus, labello unguiculato reniformi basi bituberculato, columnæ alis lateralibus subtruncatis dorsaliq. rotundato denticulatis.——*Mexico.*——Flowers very large, white, stained with pink, and a violet-coloured lip.
3. O. *pulchellum* (Bateman in Bot. Reg. 1841. t. 48); pseudobulbis oblongis compressis ancipitibus diphyllis, foliis linearibus (vix semuncialibus pedalibus) apice obliquè emarginatis, scapo foliis æquali ancipiti debili apice racemifero 6-7-floro, sepalis ovatis acutis, petalis obovatis acutis subundulatis, labelli trilobi laciniis lateralibus triangularibus intermediâ oblongâ subquadratâ apice recurvâ: callo baseos carnoso introrsum hippocrepico retrorsùm trilobo, columnæ alis et clinandrio laceris.—— *Guatemala.*——The more striking features in this species are the long thin pseudobulbs, the very narrow grassy leaves, and a single protuberance at the base of the white lip. This protuberance is deep yellow spotted with crimson, very fleshy, firm and shining, and is almost horseshoe-shaped in front, while it is distinctly three-lobed behind, in consequence of two deep depressions.
4. O. *Egertoni;* facie O. pulchelli sed floribus duplò minoribus, labello acuto basi excavato dentibus duobus supra excavationem inflexis posticè rotundato cordato.—— *Guatemala?*——This is known in gardens as a variety of O. pulchellum, than which it is smaller in all its parts, and from which it differs in the lip being quite acute, not almost truncate, with an excavation at its base in

lieu of the triple fleshy tubercle of O. pulchellum. Flowers pure white.

5. O. *crispum* (Lindl. in Ann. Nat. Hist. 15. 256); foliis lanceolatis scapo multifloro paniculato (nunc brevi racemoso) brevioribus, sepalis ovato-lanceolatis, petalis ovatis acutissimis crispis membranaceis, labello subconformi dentibus duobus validis in fronte paucisque minoribus utrinque versus laminæ basin, clinandrii alis lateralibus rotundatis laceris posticâ nullâ.——*Bogota*.——A most beautiful species, occasionally as much as three feet high and more. Flowers large, yellow, with a purple centre.

6. O. *roseum* (Lindl. in Benth. Pl. Hartweg. p. 151); foliis oblongo-lanceolatis racemo gracili duplo brevioribus, bracteis ovatis concavis subangulatis pedicellis duplo brevioribus, sepalis petalisque lanceolatis, labello trilobo basi bilamellato: lobis lateralibus nanis rotundatis intermedio obcuneato acuto, columna elongata apice alis 3 membranaceis integris aucta.——*Peru*.——Leaves from three to five inches long. Raceme about a span long. Flowers bright rose colour, something less than an inch in diameter.

Sect. II. Eu-Odontoglossum.

* Leucochilum.—White Lip.

7. O. *Cervantesii* (Llave Orch. Mex. 2. 34. Gen. & Sp. Orch. no. 4. Bot. Reg. 1845. t. 36); pseudobulbis ovatis angulatis, foliis solitariis oblongis in petiolum canaliculatum angustatis, scapo paucifloro, bracteis vaginisque membranaceis acutissimis equitantibus elongatis, sepalis membranaceis oblongo-lanceolatis acutis, petalis latioribus subunguiculatis acutis, labello subcordato-ovato acuto unguiculato, ungue carnoso cyathiformi pubescente anticè bidentato medio tuberculato, processubus 2 pilosis ante cyathum, columnæ pubescentis auriculis rotundatis. ——*Oaxaca*.——Flowers very pale pink, large, delicate, very sweet-scented, with numerous broken crimson concentrical bands.

8. O. *apterum* (Llave Orch. Mex. 2. 35. Gen. & Sp. Orch. no. 5); "bulbis ovatis depressis; foliis lato-lanceolatis nervosis conduplicatis; scapo tereti paucifloro; gynostemio aptero crasso."——*Mexico*.——This is said to be found in the same place as the last, and to be extremely like it, differing however in the lip being crenated.

9. O. *cærulescens* (Ach. Richard in Ann. Sc. 1845. Jan.); "pseudobulbis ovoideo-oblongis compressis 1-phyllis, fol. oblongo lanceolato acutissimo; scapo 1-2-floro; floribus albido-cœrulescentibus; labello cordato, membranaceo, acuto, margine sinuoso, eroso; gynostemio clavato, pubente."——*Mexico*.

10. O. *membranaceum* (Lindl. Sertum Orchid. t. 25); pseudobulbis foliisque O. Cervantesii, scapo 2-4-floro, bracteis vaginisque membranaceis acutissimis equitantibus, sepalis membranaceis unguiculatis lanceolatis, petalis latioribus oblongis obtusis, labello cordato obtusissimo unguiculato, ungue carnoso cyathiformi pubescente antice bidentato medio tuberculato, processubus 2 ante cyathum elongatis pubescentibus, columnæ pubescentis auriculis rotundatis.——*Oaxaca*.——Very like O. Cervantesii, but with a deeply cordate very blunt lip and blunt petals.

11. O. *Galeottianum* (Ach. Richard in Ann. Sc. 1845. Jan.); "pseudobulbis aggregatis 1-phyllis, fol. elliptico lanceolato acuto; scapo foliis duplo longiore 3-floro; flor. albis magnis; labelli alis erectis obtusis truncatis, lamina subcordato-acuminata; gynostemio margine alato, alis angustis longis."——*Mexico*.——

12. O. *nebulosum* (Lindl. Sertum Orchid. t. 25); pseudobulbis diphyllis, foliis oblongis erectis pedunculo terminali erecto sub-4-floro brevioribus, bracteis membranaceis scariosis ovario dimidio brevioribus, sepalis membranaceis oblongis undulatis apiculatis, petalis conformibus latioribus basi angustatis, labello sessili basi cucullato carnoso limbo ovato acuto dentato: lamellis maximis rotundatis anterioribus obtusis, columnâ apterâ pubescente. ——*Mexico*.——A much larger plant than O. Cervantesii, with wavy sepals and petals, and a coarsely toothed lip. Can it be the O. apterum of Llave? I should have thought so if he had not called that plant "O. Cervantesii simillimum," an expression which does not at all apply to this species.

13. O. *Rossii* (Lindl. Sert. Orch. sub t. 25. Bot. Reg. 1839. t. 48.—*O. Ehrenbergii*, Link, Klotzsch, & Otto, ic. p. 38. t. 16.—*O. acuminatum*, Hort.); pseudobulbis ovatis cæspitosis ancipitibus monophyllis, foliis oblongo-lanceolatis scapo radicali subbifloro longioribus, bracteis membranaceis carinatis acuminatis, sepalis lineari-lanceolatis carinatis acuminatis patentibus, petalis oblongis obtusis revolutis, labello subrotundo-ovato emarginato undulato lamellis unguis confluentibus rotundatis denticulis 2 anterioribus obtusis, columnâ apterâ pubescente. ——*Mexico*.—— A small species, with one or two flowers only on the scape. The sepals are yellowish green, blotched with brown; the petals are white, spotted with purple at the base. The lip is pure white, and slightly downy. A variety has narrower and darker coloured sepals, smaller flowers, and the callus at the base of the lip white, not yellow. This seems almost intermediate between O. Rossii and O. stellatum; especially as the lip is rather more ovate than in the former species.

14. O. *stellatum* (Lindl. in Bot. Reg. 1841. misc. 25); pseudobulbo ovali compresso, folio solitario lanceolato recurvo, scapo subbifloro, bracteis ovarii triquetri medium haud attingentibus, sepalis petalisque æqualibus linearibus acuminatis herbaceis, labello rhomboideo grossè dentato appendice baseos truncato quadridentato libero.——*Mexico*.——A plant of nearly the same size and habit as *O. Rossii*, from which it differs in the sepals and petals being equally herbaceous and narrow, of a dull olive green, faintly spotted or rather clouded with purple. It has a pure white lip.

15. O. *pygmæum* (Lindl. in Benth. Pl. Hartw. p. 82); pseudobulbis ovalibus vaginis membranaceis acuminatis vestitis, foliis angustè ovalibus in petiolum angustatis racemo paucifloro flexuoso duplò longioribus, sepalis petalisque subæqualibus, labelli pubescentis rhombei dentati appendice maximâ bivalvi in medio tumidâ: laciniis rotundatis.——*Mexico*.——This curious little alpine plant forms tufts of pseudobulbs and membranous sheaths, from among which just peep out the tiny colourless flowers, which are not above 3 lines in diameter even if spread out.

16. O. *constrictum* (Lindl. in Bot. Reg. 1843. misc. 25); paniculâ laxâ basi foliosâ, sepalis petalisque expansis lineari-lanceolatis acuminatis, labello medio constricto: hypochilio oblongo, epichilio subpandurato cuspidato serrato basi lamellis 2 serratis aucto, columnâ bicirrhosâ. ——*La Guayra*.——This has the habit of Oncidium sphacelatum or some such plant. The flowers are yellow spotted with brown, except the lip which is white stained with violet.

17. O. *hastatum* (Bateman Orch. mex. guat. t. 20); pseudobulbis oblongis ancipitibus diphyllis, foliis ligulatis basi conduplicatis, racemo elongato paniculato, sepalis petalisque conformibus herbaceis lanceolatis acuminatis, labelli hastati lobis lateralibus truncatis intermedio ovato acuminato, cristæ lamellis 4 parallelis, columnæ auriculis lunatis.——*Mexico*.——A plant with the habit of Oncidium leucochilum. The flowers are greenish with red blotches, except the lip which is white with a purple stain at the base.

** Xanthochilum.—Stained lip.

a. Columnæ auriculis rotundatis latioribus quam longis vel nullis.

*** In this arrangement the species with a 3 lobed lip come first, and then those which have an entire lip are stationed in the following order, viz. first, such as have obtuse sepals and petals, then such as have them acute, and last come the species with very much acuminated sepals and petals.

18. O. *grande* (Lindl. in Bot. Reg. 1840. misc. no. 97. Bateman Orch. Mex. Guat. t. 46); sepalis lanceolatis lateralibus convexis falcatis petalisque oblongis obtusi-

usculis latioribus subundulatis, labello subrotundo basi auriculato sepalis plus duplò breviore: tuberculis basi tribus corrugatis aliisque lateralibus dentiformibus minoribus, columnæ tomentosæ marginibus rotundatis convexis incurvis. —— *Guatemala.* —— This is a very extraordinary plant. Its habit is that of Odontoglossum, with which its unguiculate lip and peculiar column also correspond; but the tubercles at the base of the lip are those of Oncidium. The dried flowers measure six inches and a half from the tip of the petals, and the fresh ones are even larger. They are yellow, almost covered with cinnamon-brown bands and blotches. Two to five flowers grow together on a scape, which is seldom more than five or six inches long.

19. O. *Warneri;* pseudobulbis ovatis ancipitibus subangulatis apice elongatis diphyllis, foliis patulis lineari-lanceolatis racemo paucifloro brevioribus, bracteis minimis, sepalis ovalibus patentissimis, petalis paulo angustioribus obtusis ascendentibus, labelli trilobi plani laciniâ intermediâ cuneatâ bilobâ lateralibus quadratis, tuberculo disci simplici, columnâ elongatâ apterâ.——This little plant would be a very good Oncidium if it had any wings to the column; it however approaches more nearly some of the Columbian species of Odontoglossum, with which it will naturally associate. Its raceme is about a foot high, and bears seven or eight distant flowers of a rich yellow, streaked with dull crimson along the middle of the sepals and petals. The pseudobulbs, which are nearly two inches long, are a pale sea-green colour.

It was exhibited at the meeting of May 24, in the Garden of the Horticultural Society, by C. B. Warner, Esq., a zealous collector of such plants, whose name it will henceforward bear.

20. O. *gracile* (Lindl. in Benth. pl. Hartw. p. 151); foliis oblongis papyraceis basi canaliculatis angustatis scapo gracili subpaniculato brevioribus, bracteis minutis, sepalis petalisque oblongis obtusis longe unguiculatis illis dorso carinatis, labello bilamellato oblongo obtuso medio geniculato, columnæ alis falcatis acutis.——*Peru.*—— Flowers small, dull brown.

21. O. *longifolium* (Lindl. in Benth. pl. Hartw. p. 152); foliis lanceolatis papyraceis basi longe canaliculatis scapo subpaniculato gracili brevioribus, bracteis minutis, sepalis petalisque lanceolatis acutis longe unguiculatis, labello bilamellato oblongo acuto medio geniculato apicibus lamellarum auriformibus divergentibus, columnæ alis falcatis minimis.——*Peru.*——This and the preceding have much the habit of the Encyclian Epidendrons.

22. O. *ixioides* (*Cyrtochilum ixioides*, Lindl. Gen. & Sp. Orch. p. 211); cæspitosum, ferè ebulbe, foliis lineari-lanceolatis acutis striatis, scapo simplici apice racemoso, bracteis brevibus ovatis cucullatis, sepalis petalisque unguiculatis obtusiusculis, labello oblongo obtuso basi bilamellato medio 5-tuberculato, columnâ apterâ.——*New Grenada.*—— Scape 1½ foot high; flowers small, apparently yellow.

23. O. *læve* (Lindl. in Bot. Reg. 1844. t. 39); pseudobulbis compressis sulcatis, foliis oblongo-ensiformibus obtusis apice obliquis, floribus paniculatis, bracteis laxis membranaceis, sepalis petalisque oblongo-linearibus acutis planis, labelli laminâ panduriformi apiculatâ ungue lævi obsoletissimè bidentato, columnæ alis apice rotundatis crispis basi planis. —— *Guatemala.* —— The fragrant flowers have cinnamon-brown blotches on the yellow ground of the petals and sepals; and the lip, which is white, is banded with violet across the middle.

24. O. *retusum* (Lindl. in Benth. pl. Hartw. p. 152); foliis lineari-lanceolatis papyraceis scapo paniculato brevioribus, bracteis ovatis, sepalis petalisque subsessilibus lanceolatis acutis supremo canaliculato, labello oblongo retuso basi bilamellato, columnæ nanæ alis maximis oblongis. ——*Peru.*——Flowers small, apparently deep yellow.

25. O. *Bictoniense* (Lindl. in Bot. Reg. 1840. t. 66. Sertum Orchidaceum sub. t. 25.— *Cyrtochilum Bictoniense*, Bateman, Orch. Mexic. & Guatemala, t. 6.—*Zygopetalum africanum*, Hooker in Bot. Mag. t. 3812); pseudobulbis oblongis compressis 2-3-phyllis, foliis ensiformibus undulatis patentibus scapo racemoso duplò brevioribus, bracteis herbaceis lanceolatis acuminatis ovario duplò brevioribus, sepalis petalisque subæqualibus lineari-lanceolatis maculatis, labelli ungue bilamellato laminâ cordatâ acuminatâ undulatâ, columnæ alis integris——*Guatemala.*——A fine stately species with very upright racemes of flowers, whose sepals and petals are green spotted with brown, and the lip either lilac or nearly white.

26. O. *myanthum* (Lindl. in Benth. pl. Hartweg. p. 152); foliis lanceolatis basi canaliculatis erectis scapo paniculato elongato flexuoso duplo brevioribus, bracteis minutis, sepalis petalisque lineari-lanceolatis unguiculatis acutis, labello geniculato lanceolato acuto basi bilineato, columnæ brevis alis obsoletis subulatis.——*Peru.*——Leaves a foot and half long; the scape twice as long; the panicle narrow and very much branched; the flowers very small for this genus.

27. O. *lacerum* (Lindl. in Sert. Orchid. t. 25); pseudobulbis ovalibus ancipitibus, foliis lineari-oblongis in petiolum canaliculatum angustatis, racemo subpaniculato terminali, bracteis ovatis acutis squamæformibus, sepalis petalisque rhombeo-lanceolatis acuminatis, labello lacero ovato concavo apice cuspidato lamellis fimbriatis: denticulis 2 anterioribus subulatis glabris, columnâ glabrâ auriculis subtruncatis.——*Peru.*——Flowers large, bright lemon colour, with a brown blotch or two in the middle of the sepals and petals, and another at the upper half of the lip.

28. O. *pardinum* (Lindl. in Sert. Orch. sub t. 25.—*Cyrtochilum pardinum*, L. p. 210); foliis lineari-oblongis basi angustatis striatis, petalis undulatis sepalisque angustioribus lanceolatis, labello unguiculato ovato-lancelato: unguis axi cum columna connato utrinque lamellato; callis quibusdam elevatis inæqualibus linearibus productioribus e medio.——*Peru* and *Popayan.*——Flowers large yellow, spotted with brown.

29. O. *epidendroides* (Humb. et Kunth. N. G. et Sp. Pl. 1. 351. t. 85. Lindl. G. & Sp. Orch. no. 1); pseudobulbis ovato-oblongis compressis, foliis lanceolatis subcoriaceis, scapo erecto multifloro, sepalis petalisque acuminatis undulatis, labello unguiculato oblongo obtuso undulato-crenato, basis trituberculato.——*New Grenada.*—— Flowers large, yellow, scentless, with broad brown blotches. Only known from the figure above quoted.

30. O. *angustatum* (Lindl. in Bot. Reg. sub t. 1992); pseudobulbis compressis, foliis longis lato-lanceolatis erectis scapo brevioribus, scapo paniculato, sepalis lineari-lanceolatis acuminatis, petalis ovato-lanceolatis unguiculatis undulatis, labello ovato-lanceolato acuminato: cristâ baseos lamellatâ multifidâ, columnâ muticâ.—— *Peru.*——This is a very fine plant, with spotted flowers resembling those of O. Hallii and angustatum, and as much as three inches from tip to tip of the sepals.

31. O. *revolutum* (Lindl. in Ann. Nat. Hist. 15. 256); foliis ensiformibus rigidis margine revolutis scapo rigido apice paniculato brevioribus, bracteis laxis acutis cucullatis, panicula subcorymbosa densa, sepalis petalisque lanceolatis unguiculatis acutis, labello conformi sessili cordato lamellis 5 carnosis integris quarum tres superiores multo breviores: lateralibus subulatis, columna aptera.—— *Popayan* ——Scape above two feet long, terminated by a stiff corymbose erect panicle of flowers. The leaves are stiff and straight, rolled back at the edge, and when dry very much puckered, as if they had been fleshy. Mr. Hartweg found it on the Paramo de Guanacas, at the height of 11,000 feet above the sea.

32. O. *maculatum* (Llave Orch. Mex. 2. 35. Bot. Reg. 1840. t. 30); pseudobulbis oblongis compressis monophyllis, foliis oblongis nervosis acutiusculis racemis pendulis multifloris brevioribus, bracteis navicularibus herbaceis ovario brevioribus, sepalis linear -lanceolatis acuminatis discoloribus, petalis oblongis undulatis acuminatis, labello cordato acuminato subcrenato: appendice unguis bivalvi concavâ cochleari apice productâ emarginatâ per medium argutè serrulatâ, columnâ pubescente subapterâ. ——*Mexico.*——Sepals chesnut brown. Petals and lip clear yellow, spotted with brown.

33. O. *cordatum* (Lindl. in Bot. Reg. 1838. misc. 90. Knowles & Westcott, Floral Cabinet, t. 100); pseudobulbis oblongis compressis diphyllis, foliis lato-oblongis planis acutis scapo squamis carinatis vaginato brevioribus, racemo stricto disticho, bracteis navicularibus acuminatis membranaceis, sepalis petalisque lineari-lanceolatis acuminatissimis, labello cordato acuminatissimo integerrimo appendice baseos carnosa apice biloba basi utrinque dente unico auctâ, columnâ pubescente clavatâ subapterâ.—— *Mexico.*——Sepals and petals richly blotched with brown upon a yellowish green ground. Lip white, with the crest at the base purplish, and the apex spotted and blotched with rich brown.

b. Columnæ auriculis linearibus, longioribus quam latis, v. cirrhosis.

34. O. *cristatum* (Lindl. in Benth. Pl. Hartweg. p. 152); foliis lineari-lanceolatis acutissimis scapo multifloro simplici paullo brevioribus, sepalis petalisque lanceolatis acuminatis, labello lanceolato acuminato geniculato basi lamina multifida cristato, columna elongata basi membranaceo-marginata apice alis duabus uncinatis, anthera cristata.——*Peru.*——Flowers spotted with brown, about as large as those of O. cordatum. Leaves very narrow.
35. O. *cirrhosum* (Lindl. in Hooker. Bot. Misc. Gen. & Sp. Orch. no. 2); sepalis lanceolatis petalisque latioribus undulatis acuminatissimis maculatis labello longioribus, labelli laminâ cordatâ acuminatissimâ, columnâ apice apterâ bicirrhosâ.——*Guayaquil.*——Nothing more is known of this remarkable plant than could be gathered from the examination of a few dried flowers. They appear to be yellow, with some broad purplish blotches.
36. O. *Hallii* (Lindl. in Bot. Reg. sub t. 1992); pseudobulbis ovatis acuminatis ancipitibus, foliis ensiformibus in basin angustatis, scapo paniculato, sepalis petalisque ovato-lanceolatis acuminatis unguiculatis, labello oblongo acuminato pectinato; cristâ baseos lamellatâ multifidâ, columnæ alis bidentatis dente superiore aristato.——*Peru.* ——A most beautiful species. Flowers blotched with brown upon a yellow ground, three inches and a half in diameter. Mr. Hartweg brought it from the western declivity of Pichincha.
37. O. *rigidum* (Lindl. in Benth. Pl. Hartweg, p. 152); foliis oblongis basi angustatis scapo rigido paniculato brevioribus, sepalis lateralibus lineari-lanceolatis dorsali petalisque lanceolatis, labello unguiculato obovato cordato apiculato lineis duabus elevatis in unguem, columnæ brevis semiteretis alis angustis decurvis serrulatis.—— *Peru.*——A terrestrial plant. Flowers yellow, whole-coloured, with a very long narrow stalk to the lip.
38. O. *bicolor* (Lindl. in Benth. Pl. Hartweg, Jan. 1845. ined.); foliis lanceolatis scapo simplici paucifloro flexuoso subæqualibus, sepalis petalisque ovatis acutis subundulatis æqualibus, labello unguiculato obovato apiculato basi obtuso, denticulis quibusdam (sex) in unguem, columnæ brevis alis decurvis serratis.—— *Peru.*——Only known from a drawing by Mathews, in Sir Wm. Hooker's Herbarium. The flowers are large, bright violet, with a great whole-coloured yellow lip.

* * Species concerning which there is no sufficient information.

39. O. *erosum* (Ach. Richard in Ann. Sc. Jan. 1845); "pseudobulbo ovoideo oblongo 1-phyllo; fol. lanceolatis acutis; scapo 1-floro; sepalis luteis: labello lilacino late ovali margine eroso-denticulato."——*Mexico.*——
40. O. *Ghiesbreghtianum* (Ach. Richard in Ann. Sc. Jan. 1845); "pseudobulbis ovoideis compressis 3-phyllis; scapo pedali 3-6-floro; flor. luteis brunneo-maculatis: labello luteo, longe unguiculato, transverse orbiculari, emarginato, sepalis duplo longiori, ungue lineari basi hinc et illinc obtuse auriculato, superne bi-cristato."—— *Mexico.*——

61. CLEISOSTOMA discolor.

C. *discolor*; foliis oblongis canaliculatis obliquè truncatis, racemo paniculato, labello trifido utrinque plicato lobo medio carnoso obtuso, intus bituberculato dente dorsali indiviso carinato, calcare oblongo obtuso.

A little inconspicuous plant obtained from India by Messrs. Loddiges. The flowering stem is long, slender, dull purple, and branched at the extremity. The sepals and petals are dull yellow, with a green tinge, the spur is nearly white.

62. BRASSIA pumila.

B. *pumila*; foliis binis patulis elongato-lanceolatis canaliculatis, sepalis lineari-lanceolatis acuminatis inferioribus labello longioribus, labello pandurato acuminato subundulato basi bicarinato dentibus ante carinas 2 acutis elevatis aucto.

This Brassia is remarkable for its dwarfness, as we learn from Mr. Barker who has flowered it from Mr. Linden's collections. It has pale yellow flowers stained with dull purple at the base of the petals; but there is no purple on the sepals, and the lip is very slightly tinted. It seems to be peculiar in the large size of the pair of somewhat transparent teeth which stand up in front of the two little customary ridges at the base of the lip; but it may be a variety of Br. Lanceana. Its aspect is, however, widely different.

64. ORCHIS sambucina.

"The only marked distinction between thê beautiful yellow and white Orchis sambucina of the Alps and Orchis provincialis is, that the latter has the tubers undivided, and the former belongs to the division which has palmate tubers, sambucina having short prominences at its base. In colour, fragrance, and the habit of growing on parts of high mountains where the soil is blackish and where the clouds frequently rest, they agree perfectly. Having cultivated Orchis fusca in a large pot of strong alluvial soil, mixed with a little pulverized chalk, (the pot being plunged in sand, and the tubers set about three inches and a half under the surface), I examined the state of the new tubers this morning, as the leaves had died entirely away. The new tubers of those which had survived the winter frosts were large, and of the purest healthy white; and although the plant belongs to the section with undivided tubers, and those last year were undivided, the largest of the new tubers had two strong lobes at the base. This circumstance, arising in the very season of cultivation, shews how Orchis provincialis may in a different locality acquire the prominences which have been supposed to distinguish sambucina from it. The plants called Sambucina, as far as I have seen, are more vigorous and grow in colder quarters. The fine crimson Orchis with an orange throat, called O. Schleicheri by Sweet, grows in company with the yellow on the southern brow of the Alps, and differs in being scentless, and a little also in the length of the spur; but, although it is difficult to understand why those peculiarities should attend the change of colour, from the manner in which they grow intermixed and in contact with each other, with foliage and tubers precisely similar, it can scarcely be doubted that they are seminal variations. I saw one plant on the brow of the St. Gothard of an intermediate colour, with the peculiarities of the crimson plant."—*W. Herbert.*

65. ORNITHIDIUM miniatum.

O. *miniatum*; caulescens, pseudobulbiferum, foliis oblongo-loratis patentibus apice oblique rotundatis mox margine serrulatis, floribus axillaribus laxè exsertis, pedunculis unifloris, sepalis petalisque ovatis acutis glabris, labello basi concavo nectarifero apice trilobo laciniis lateralibus brevibus obtusis ascendentibus intermediâ latè ovatâ v. retusâ subundulatâ.

This plant has quite the habit of Ornithidium coccineum, but it is much handsomer: for the flowers are of rich crimson vermilion. Their lip is of quite another form; being distinctly three-lobed at the end, with the side lobe rising up, and the intermediate one broad, flat, rounded, and partially turned downwards; it is of a yellow colour, but edged and blotched with crimson. Messrs. Rollissons have flowered it; they imported it from Colombia.

66. PORPAX reticulāta.

Among the many curious little plants, which in a collection like that of Messrs. Loddiges receive the same care as greater things, were a few button-like pseudobulbs, held close together by a slender rhizome, and received from the East Indies without a name. Upon sprouting they produced a pair of little oblong leaves recurved at their point, and between the leaves peeped forth a small dull red flower, about three-quarters of an inch long. The structure of the flower was this. *Sepals* equal, rather spreading at the point, united at the edges into a tube; the lateral ones a little oblique and prominent at the base. *Petals* ovate, dull purplish red, very short. *Labellum* articulated loosely with the extended foot of the column, ovate, hastate, fleshy, serrate, with the sides of the base erect; and a single slender tooth standing up between the sides in the middle of the base. *Column* short, extended into a long foot, earless and wingless. *Anther* opercular, membranous. *Pollen-masses* 8, in pairs, one mass in each pair being smaller than the other. *Gland* and *caudicle* none.

No genus of Orchids has yet been described which will include this. It approaches nearest to Eria, but it has the habit of Trias and the calyx of Masdevallia. Its technical character may be stated thus:—

PORPAX (Genus Eriæ affine sed habitu diversissimo.) *Sepala* subæqualia, in tubum connata, lateralia basi obliqua et pedi producto columnæ adnata. *Petala* nana, libera. *Labellum* carnosum, cum pede columnæ articulatum, basi tuberculo auctum. *Columna* nana, teres, aptera. *Pollinia* 8, per paria inæqualia.——Pseudobulbi *depressi, densè catenatim consociati, diphylli.* Flores *solitarii, sessiles, terminales.*

Sp. 1. *Porpax reticulata.*

Pseudobulbi depressi, mutuâ pressione subangulati, fibris pallidis pulchrè reticulati. Labellum serratum, tuberculo basali cornuto.

67. ERIA Dillwynii.

Hooker in Botanical Magazine, t. 4163.

E. *Dillwynii;* "pseudobulbis oblongis lævibus 2-4-phyllis, foliis oblongis racemis subæqualibus, bracteis membranaceis oblongis obtusis reflexis, racemis erectis, petalis sepalisque erectis, labello trilobo basi trilamellato, lobo medio obtuso rotundato 5-lamellato."

A native of the Philippines, whence Mr. Dillwyn Llewellyn received it from Cuming, and flowered it at Pennlergare in March, 1843. It is a pale lemon-coloured species, with the habit of E. longilabris, and is said to be a most profuse flowerer. Mr. Llewellyn says that his plant had two spikes of flowers on each of eight pseudobulbs.—*Bot. Mag.*

It is not to be found among Cuming's dried plants.

70. THE GENUS EPIDENDRUM.

In the miscellaneous pages of the Botanical Register the whole of the genus Epidendrum has now undergone revision, except that which may be called Epidendrum proper, consisting of species which cannot be referred to any of the other subdivisions, and which all correspond in having a leafy (not pseudobulbous) stem, a short peduncle without any scales or sheaths at its base, and the lip completely united to the column. They form an assemblage in which very different looking plants are collected; and which may therefore hereafter demand further subdivision. No means of separating them occurs however at present, and till a larger number shall have been critically examined, it will be as well they should remain in one group. The following is an entirely artificial distribution, without much reference to their affinity with one another.

§ EUEPIDENDRUM; caulis foliosus; pedunculus brevis esquamatus; labellum adnatum. *Lindl. in Hooker's Journal,* 3. 81.

a. Flowers solitary, or fascicled.

1. E. *equitans* (Lindl. in Bot. Reg. 1838. misc. 76); caule folioso ancipiti, foliis equitantibus ancipitibus lanceolatis acuminatis, flore solitario terminali pendulo, pedunculo ancipiti, spatha diphylla foliolo inferiore erecto foliaceo flore longiore, sepalis linearibus acuminatis patulis, petalis conformibus paulò brevioribus, labello ovato-lanceolato complicato cum columnâ basi connato medio trilamellato; lobo medio carnoso semitereti recurvo lateralibus erectis membranaceis integris.——*Mexico.*——A very curious species in its habit, which resembles Fernandezia very much. It is, however, a genuine Epidendrum, with a single flower of a dull chocolate brown.

2. E. *vesicatum* (Lindl. in Bot. Reg. 1838. misc. 89); caule elongato, foliis inflatis equitantibus carinatis acutis glaucis, floribus terminalibus fasciculatis carnosis foliis vix longioribus, sepalis lineari-oblongis acutis petalis conformibus angustioribus, labello subrotundo cordato lineâ mediâ tuberculisque duobus basilaribus elevatis, collo ovarii medio vesicato.——*Brazil.*——In habit this approaches E. equitans, but in the structure of both flowers and leaves it is widely different. The latter are covered with a glaucous bloom, imbricated, and more like inflated carinate bracts than true leaves. The flowers are greenish white, and offer some analogy with those of Physinga, but are in reality different from those of a genuine Epidendrum in nothing, except the neck of the ovary having an elevated semi-transparent blister near its middle. This blister is in fact the lower extremity of the cuniculus of the flower.

2 *bis.* E. *tolimense;* caulibus ramosis scaberrimis, foliis ovato-lanceolatis, floribus 2-3 terminalibus sessilibus coriaceis, sepalis lineari-lanceolatis acutis striatis, petalis setaceis, labello subcordato ovato concavo apice carnoso basi ecalloso intùs pubescente avenio.——*New Grenada.*——This species forms part of Mr. Linden's rich collection. It was found in the forests of Tolima, 10,000 feet above the sea. The flowers are small, and of a dull yellow colour. The branches are excessively rough, with irregular asperities. The leaves are about $1\frac{1}{2}$ inch long.

3. E. *piperinum* (Lindl. in Ann. Nat. Hist. 15. 256); caule ramuloso, foliis distichis succulentis oblongis obliquis obtusis laxe vaginatis, pedunculis brevissimis terminalibus subbifloris, sepalo dorsali petalisque filiformibus erectis lateralibus ovato-lanceolatis carinatis horizontalibus, labello ovato cochleato ecalloso venis radiantibus in margine confluentibus.——*Quito.*——A small succulent species, when dried beautifully netted. It looks like a *Peperomia.*

4. E. *lamellatum* (Lindl. in Bot. Reg. 1843. misc. 60); caule erecto flexuoso, foliis lanceolatis emarginatis glabris subcarnosis, corymbo sessili paucifloro, sepalis lanceolatis acutis, petalis obovato-lanceolatis, columnâ apice utrinque alatâ, labello obovato integerrimo v. obscurè emarginato basi transversim lamellato.——*Honduras, Colombia,* (Linden, 710.)——Stem about a foot high. Flowers rose colour, about the size of those of E. Schomburgkii. Appendages at the apex of the column dark pink. Disk of labellum covered with a row of scaly yellowish plates.

5. E. *latilabre* (Lindl. in Bot. Reg. 1841, misc. 163); foliis ovatis obtusis brevivaginantibus internodiis sæpè brevioribus, sepalis petalisque angustioribus lineari-oblongis obtusis patentibus, labello repando sublobato emarginato basi bicalloso ferè 4-plò latiore quam longo, clinandrio laciniato, pedunculis 2-4-floris sessilibus.——*Tropical America.*——Allied to Epidendrum umbellatum, of which it has the habit. But its lip, which is four times broader than long, and curved downwards on each side, so as to bear no little resemblance to a stiffly starched lady's apron, gives it a most singular appearance. It is chiefly remarkable for its extensive geographical range. The plants in our gardens are all I believe from Brazil, where it has been found by many travellers. But it is

also a native of Dominica; and Hartweg found it in the woods near Samborondan in Guatemala.

6. E. *umbellatum* (Swartz Prodr. 121. Nov. Act. Ups. 6. 68. Willd. no. 14. Bot. Mag. 2030. Hort. Kew. 5. 218. E. *difforme*. Jacq. Amer. 223. t. 136. E. *corymbosum*. Fl. Peruv. Syst. 246?); foliis distichis oblongis subundulatis obtusis, sepalis lineari-oblongis obtusis petalisque

linearibus patentibus, labello indiviso rotundato repando transverso basi bicalloso, clinandrio laciniato, floribus corymboso-umbellatis.——*West Indies, Mexico, &c.*——A common green flowered species of little interest. It is but little subject to variation.

7. E. *ensatum* (Richard & Galeotti in Ann. sc. 3 ser. 3. 22); "Caule compresso ancipiti; fol. oblongo-lanceolatis acutis; flor. parvulis albido-brunneis; labello adnato trilobo, lobis obtusis subæqualibus, margine sinuosis." ——*Mexico.*

8. E. *discolor* (Richard & Galeotti in Ann. sc. 3. 22. Orch. Mex. t. 18. ined.); "Caule compresso: fol. oblongo-lanceolatis obtusis, bilobis: flor. maximis solitariis aut binis terminalibus externe rubellis, interne luteolis; labello albo, trilobo, lobo medio lineari longissimo."——*Mexico.*

9. E. *nocturnum* (Linn. Sp. Pl. 1349. Jacq. Amer. 225. t. 139. Swartz Nov. Act. 6. 69. Willd. no. 20. Bot. Reg. t. 1961. *Ep. tridens*, Poppig & Endlich. n. g. et sp. 2. t. 103); foliis distichis oblongis acutis, floribus subgeminis terminalibus, sepalis petalisque linearibus acuminatis patentibus, labelli trilobi lobis lateralibus ovatis integerrimis intermedio setaceo brevioribus——*West Indies, Peru.*——About a foot high. Flowers three inches long, white, with a tinge of green or crimson, extremely sweet-scented.

10. E. *longicolle* (Lindl. in Bot. Reg. 1828. misc. 49; Bot. Mag. t. 4165); caule erecto compresso folioso, foliis linearibus apice angustatis, floribus aggregatis terminalibus nutantibus foliis brevioribus, sepalis lineari-lanceolatis acuminatis patentibus, petalis linearibus acutis supra columnam convergentibus, labelli trilobi lobis lateralibus semiovatis acutis integerrimis intermedio lineari-acuminato paulo brevioribus: lamellis 2 callosis ad basin, collo ovarii elongato.——*Demerara.*——A species nearly allied to *E. nocturnum*, but much less handsome. The sepals and petals are pale yellow; the lip white, with two yellow plates at the base. Sir W. Hooker observes that what it lacks in beauty it makes up in fragrance.

11. E. *lactiflorum* (Richard & Galeotti in Ann. sc. 3 ser. vol. 3. 22.) "Caule tereti dependente, apice 1-phyllo: folio oblongo-lanceolato, conduplicato: flor. albis 2-3 axillaribus: labello trilobo, lobo medio lanceolato-lineari." ——*Mexico.*——Is not this Epidendrum falcatum?

b. Flowers racemose, or spicate.

* Lip undivided, quite entire at the edge.

12. E. *ferrugineum* (Fl. Peruv. Syst. 245); "foliis oblongo-lanceolatis emarginatis, racemo simplici erecto, labello acuminato integro."——*Peru.*——

13. E. *cucullatum* (Lindl. in Bot. Reg. 1838. misc. 47); caule compresso folioso, foliis lineari-lanceolatis acutis, pedunculo terminali squamato paucifloro foliis multò breviore, rachi brevi flexuosâ, sepalis petalisque erectis ovato-lanceolatis acutis, labello obovato acuto adnato: lineâ media tuberculisque duobus basilaribus callosis, antherâ longè infra apicem clinandrii cucullati dentati truncati insertâ.——*Para.*——One of the most unattractive of this large genus. The flowers are small and white, at the top of a leafy stem about nine inches high; the anther is remarkable for being inserted far below the apex of the column, in which respect the species approaches the genus *Physinga*.

14. E. *ramosum* (Jacq. Amer. 221. t. 132. Swartz. Fl. Ind. Occ. 1. 1505. Willd. no. 16.—*Isochilos ramosum*, Spreng. Syst. 3. 734.—*Epidendrum rigidum*, Lodd. Bot. Cab. t. 1600. non Jacquini); foliis linearibus obtusis emarginatis, racemis terminalibus laxis paucifloris, sepalis ovato-lanceolatis petalisque linearibus acutis patentibus, labello subcordato ovato acuto concavo rigido, bracteis ovarii longitudine, caule ramoso.——*Jamaica, &c.*——Stem somewhat dichotomous. Flowers small, greenish yellow. It varies in the size of its parts.

15. E. *rigidum* (Jacq. Amer. 222. t. 134. Swartz Fl. Ind. Occ. 3. 1507. Willd. no. 17. β. labello subrotundo); foliis distichis ovato-oblongis obtusis, spicæ flexuosæ rachi ancipiti, sepalis coriaceis ovatis obtusis patentibus lateralibus majoribus, petalis linearibus membranaceis, labelli postici cordato-ovati obtusi palato calloso cuniculo inflato, bracteis foliaceis dolabriformibus carinatis ovarii longitudine.——*West Indies.*——Stem simple, a span high, two-edged. Flowers small, green. Possibly the var. β. may be a distinct species.

16. E. *dichotomum* (Lindl. in Bot. Reg. 1838. misc. 146); fruticosum, caule fruticoso decumbente filiformi dichotomo, foliis angustè lanceolatis acutissimis corymbo terminali longioribus, sepalis lineari-lanceolatis, petalis conformibus angustioribus, labello cuniculato cordato obtusiusculo basi bicalloso.——*Demerara.*——A green-flowered species of no beauty, remarkable for its hard, wiry-looking stem.

17. E. *aggregatum* (Lindl. in Hooker's Journal, 3. 84.); foliis distichis lanceolatis acuminatissimis racemis oppositifoliis subsessilibus, brevissimis basi squamatis floribus corymbosis, labello adnato subrotundo cordato basi bilamellato.——*Peru.*——A very singular plant, allied to *E. cauliflorum*. The flowers are apparently as large as in *E. nutans*, but they have not been sufficiently examined.

18. E. *macrostachyum;* caule scabriusculo, foliis oblongis mucronulatis, racemo longissimo cernuo angulato, bracteis magnis duris ovato-cordatis reflexis, floribus coriaceis, sepalis oblongis obtusis, petalis filiformibus, labello cordato obtuso lævi basi valde concavo, clinandrio nudo. ——*New Grenada.*——This is a very singular species, with racemes more than a foot long, covered with distant leafy bracts full half an inch long. It was found by Mr. Linden, whose number 1279 it is, and who describes it as having green fleshy flowers, and a stem from two to three feet high.

19. E. *xylostachyum;* caule scabro, foliis oblongis obtusis coriaceis mucronulatis, spicâ strictâ flexuosâ durâ folio summo breviore, bracteis herbaceis cucullatis ovatis rigidis approximatis, floribus carnosis, sepalis oblongis concavis, petalis conformibus planis angustioribus, labello cordato ovato obtuso concavo margine recurvo, clinandrio integro.——*New Grenada.*——Another of Mr. Linden's discoveries. It is very like the last, but the bracts are larger, the spike is very short, the flowers are larger, and the petals are nearly as broad as the sepals, instead of being filiform.

20. E. *costatum* (Richard & Galeotti in Ann. Sc. 3 ser. 3. 22); "caule compresso; fol. elliptico-oblongis obtusis; flor. purpureis; racemo terminali: labello adnato orbiculari basi cordato crasso, costis divergentibus notato."——*Mexico.*——

** Lip undivided, serrate or toothed.

21. E. *propinquum* (Richard, l. c.); "caule tereti; fol. oblongo-lanceolatis obtusiusculis; racemis 4-floris; flor. brunneis: labello adnato cordato, margine denticulato." ——*Mexico.*——

22. E. *ledifolium* (Richard, l. c.); "caule tereti; fol. lanceolato-linearibus obtusiusculis; flor. parvulis luteis racemum simplicem lateralem sessilem efformantibus: labello adnato, orbiculari vix acuto, denticulato-sinuoso." ——*Mexico.*——

23. E. *subulatifolium* (Richard, l. c.); "fol. teretibus subulatis acutis; scapo 4-5-floro; flor. intense luteis parvulis: labello suborbiculari integro, sinuoso."——*Mexico.*

24. E. *orgyale* (Lindl. in Ann. Nat. Hist. 12); caule orgyali distiche foliato, foliis ovato-oblongis margine vaginisque scabris, racemo sessili erecto cylindraceo bracteis membranaceis subulatis, floribus carnosis, sepalis ovalibus, petalis spathulato-linearibus serrulatis, labello cordato

serrulato callis duobus juxta basin unoque sulcato minore sub apice.——*Santa Fé de Bogota.*——Stem five feet high. Flowers apparently yellow. The fistula of the ovary inflated.

25. E. *imbricatum* (Lindl. Gen. & Sp. Orch. no. 71); foliis distichis lineari-oblongis obtusis, caule ramoso, spicis ovato-oblongis imbricatis, sepalis petalisque ovatis acutis æqualibus erectis, labello ovato acuto obsoletè crenulato, bracteis carinatis obtusis.——*Brazil.*——

26. E. *fimbriatum* (H. B. K. 1. 351. *E. alternans*, Lindl. in Hooker's Journal, 3. 88); foliis distichis oblongo-linearibus oblique retusis margine vaginisque scabris, racemo angusto terminali, floribus subsessilibus, sepalis linearibus retusis, petalis conformibus serratis, labello subrotundo-ovato serrato bilamellato columnæ nanæ adnato. *Var. α.* bracteis ovatis herbaceis obtusis pedicello longioribus, racemo corymboso, floribus majoribus. *Var. β.* bracteis minimis acutis pedicello brevioribus, racemo flexuoso, floribus duplo minoribus.——*Peru.*——Slender plants about six inches high, with small racemose flowers. The *var. β.* is smaller in all its parts, but in the structure of the flowers themselves I find no appreciable difference.

*** Lip slightly lobed.

27. E. *tenue* (Lindl. in Hook. Journ. 3. 88); foliis distichis linearibus acuminatis oblique emarginatis, racemis acutis angustis simplicissimis (nunc casu quodam bifidis multifloris (10-20), bracteis ovatis acutis rigidis cucullatis pedicello longioribus, floribus erectis membranaceis, sepalis linearibus obtusis, petalis filiformibus, labello sessili ovato acuto concavo utrinque 1-dentato: venis baseos elevatis.——*Brazil.*——

28. E. *filicaule* (Lindl. Gen. & Sp. Orch. no. 23); caule gracili ramosissimo, foliis linearibus acutis, sepalis cuneato-oblongis acutis petalisque filiformibus patentibus, labello subrotundo cordato retuso repando, spicis terminalibus 3-floris, rachi capillari.——*Brazil.*——Flowers small, with three elevated lines at the base of the lip.

29. E. *scriptum* (Richard & Galeotti in Ann. Sc. 3 ser. 3. 22); " caule tereti; fol. elliptico-oblongis acutis; flor. viridulis, racemo simplici terminali: labello adnato obsoleto trilobo albo, lineis purpureis scripto."——*Mexico.*——

30. E. *longipetalum* (Richard & Galeotti in Ann. Sc. 3 ser. 3. 22. Orch. Mex. ined. t. 17); "caule compresso diphyllo; fol. approximatis lato-ellipticis subacutis: flor. pallide roseis: racemis paucifloris scapum pedalum terminantibus: sepalis internis linearibus externa duplo superantibus: labello cordato obsolete trilobo."——*Mexico.*——

31. E. *smaragdinum* (Lindl. in Bot. Reg. 1838. misc. 44); caule gracili ad fastigium ipsum folioso, foliis lineari-lanceolatis acuminatis subdistichis, spicâ terminali pauciflorâ rigidâ nutante foliis breviore, bracteis acuminatis squarrosis ovario ventricoso cuniculato brevioribus, sepalis lineari-oblongis acutissimis, petalis angustissimis acuminatis, labello cuneato cordato carnoso apice tridentato basi bicalloso: laciniis æquilongis lateralibus rotundato-truncatis intermediâ acutâ decurvâ.——*Demerara.*——A species closely allied to *E. orchidiflorum*, with small bright green flowers of no beauty whatever, and hardly distinguishable from the leaves.

*** Lip distinctly three-lobed.

32. E. *acuminatum* (Fl. Peruv. Syst. 248. L. no. 31); " foliis lanceolato-linearibus, racemo terminali, labello subhastato acuminato basi utrinque auriculato."——*Peru.*——

33. E. *conopseum* (R. Br. in H. Kew. 5. 219. Nutt. Gen. 2. 198. Hooker in Bot. Mag. t. 3457. *E. Magnoliæ*, Muhl. Cat. 81); foliis binis radicalibus rigidis coriaceis, racemo laxo multifloro, sepalis linearibus obtusis petalisque angustioribus spatulatis patentibus, labello obcordato obtusè trilobo.——*Carolina and Georgia.*——Plant four inches high, or even smaller. Flowers small, yellow or green.

34. E. *lignosum* (La Llave Orch. Mex. 2. 28); caulibus ramosissimis lignosis, foliis distichis lineari-lanceolatis emarginatis fasciculatis, sepalis petalisque patentibus . . . . . ., labelli 3-partiti laciniis lateralibus obtusis intermediâ convexâ integrâ, florum fasciculis sparsis.——*Mexico.*——Flowers green, dotted. Lip with purple streaks at the base.

35. E. *scabrum* (Fl. Peruv. Syst. 248); " foliis ovato-lanceolatis marginibus vaginisque scabris, racemo terminali, labello cruciformi."——*Peru*, Valley of Banos, *Hall.* Andimarca, *Mathews* 1063.——

36. E. *chioneum;* caule lævi ramoso, foliis ovato-lanceolatis mucronulatis, racemo capitato cernuo, bracteis erectis herbaceis acuminatis ovario longioribus, sepalis subrotundo-ovatis, petalis obovatis serrulatis, labelli trilobi laciniis cuneatis integris lateralibus duplo minoribus, callis duobus cornutis, clinandrio altè marginato.——*New Grenada.*——Found by Mr. Linden 12,000 feet above the sea. The flowers are pure white, in small close heads, an inch or more long. It is very near E. cernuum, but differs in the form of its sepals and the lobes of the lip, if that species is correctly described.

37. E. *cernuum* (Humb. et Kunth. Nov. Gen. et Sp. 1. 353); foliis oblongis obtusis coriaceis, spicâ cernuâ, sepalis lanceolatis margine revolutis cuspidatis, petalis conformibus paulò longioribus, labello trilobo margine reflexo: lobis lateralibus subrotundis intermedio triplò majore ovato obtuso.——*Popayan.*——Stem three feet high and more. Spike three inches long. Flowers sweet-scented, yellowish green.

38. E. *Arbuscula* (Lindl. in Benth. Pl. Hartw. p. 93); caule tereti fruticoso ramoso, ramulis vestitis apice 2-3-phyllis, foliis oblongis coriaceis patentibus, racemis oblongis densis terminalibus subsessilibus cernuis, bracteis lineari-lanceolatis acuminatis membranaceis patentissimis pedicellis duplo brevioribus, sepalis oblongis acutis carnosis, petalis tenuioribus linearibus apice paullo latioribus, labello trilobo cordato basi tricalloso, laciniis lateralibus rotundatis, intermedia emarginata undulata.——*Mexico.*——A plant with a large branching stem, leathery leaves, three or four inches long, and many pale stout roots. It is more curious than beautiful, resembling E. nutans in some respects, and quite destitute of brilliant colours. The flowers are a dull chocolate, only relieved by a patch of yellow in the middle of the labellum. The latter organ is so folded back at the sides and front, that although it is really almost circular, it has altogether the form of an old fashioned three-cocked hat.

39. E. *lacertinum* (Lindl. in Bot. Reg. 1841, misc. 109); floribus racemosis, ovariis subsecundis longissimis pendulis bracteis setaceis multoties longioribus, sepalis lanceolatis acuminatis, petalis lineari-lanceolatis, labelli adnati trilobi basi bilamellati laciniis lateralibus triangularibus intermediâ lineari acuminatâ elongatâ, antherâ immersâ, ovario cuniculato.——*Guatemala.*——The flowers hang down on long stalk-like ovaries from one side of a short raceme. They are bright green, with the exception of the column, which is yellow, and the lip which is a little stained with purple, and has very much the appearance of a lizard's tail, the hind legs being outside the flower, and the head and shoulders buried within the cup.

**** Lip 4-lobed, the middle division being emarginate or slit, or even 3-toothed.

40. E. *bifarium* (Swartz Fl. Ind. Occ. 3. 1509. Willd. no. 19. Swartz Prodr. 121); foliis distichis cordato-lanceolatis horizontalibus, spicâ terminali flexuosâ, rachi ancipiti, sepalis ovato-lanceolatis convexis, petalis brevioribus setaceis, labelli dilatati trilobi lobis lateralibus subrotundis integris: intermedio bilobo.——*Jamaica.*——Stem two to four inches high. Flowers greenish white. Bracts somewhat falcate, keeled, as long as the ovary.

41. E. *cauliflorum* (Lindl. in Bot. Reg. 1838. misc. 82); caule tereti, foliis ovato-oblongis planis acuminatissimis,

corymbis brevibus axillaribus e latere caulis erumpentibus, sepalis angustis oblongis patentissimis subæqualibus concavis, petalis linearibus apice cuneatis acutis reflexis, labello columnæ arcuatæ clavatæ omninò accreto cuneato-subrotundo trilobo; lobo intermedio truncato tridentato lateralibus rotundatis repandis, callis tribus linearibus in medio labelli lateralibus majoribus.——*Brazil.*——The flowers are about the size of those of E. *nutans*, of a pale straw-colour, and are remarkable for appearing from the side of the stout cylindrical stem, bursting forth from among the dry sheaths with which it is closely invested.

42. E. *patens* (Swartz Fl. Ind. Occ. 1. 1495. Willd. no. 10. Bot. Cab. t. 1587); foliis distichis oblongo-lanceolatis, racemo terminali, sepalis petalisque subæqualibus oblongis acutis concavis patentibus, labelli quadrilobi lobis lateralibus subrotundis: anterioribus linearibus obtusis conniventibus.——*Jamaica.*——Stem a span or foot high. Flowers pallid. Sepals keeled on the outside, and ferruginous.

43. E. *parviflorum* (Fl. Peruv. Syst. 245); "foliis lanceolato-linearibus, racemo brevi, labelli trilobi laciniis subrotundis: intermediâ bipartitâ acuta."——*Peru.*——What I take for this plant was found at Jambrasbamba by Mathews, and is no. 1900 of his herbarium. The raceme seems always to be forked.

44. E. *viride* (Fl. Peruv. Syst. 244); "foliis lanceolatis acutis, racemo terminali, labii trilobi laciniis lateralibus obcordatis, intermediâ bifidâ."——*Peru.*——

45. E. *torquatum* (Lindl. in Benth. pl. Hartw. p. 149); caule elongato aspero, foliis oblongis coriaceis obtusis distichis, spica sessili cernua, floribus coriaceis sepalis ovatis acutiusculis, petalis tenuioribus basi angustatis, labelli trilobi basi bidentati lobis lateralibus subtruncatis intermedio cuneato tridentato.——*Peru.*——Near E. scabrum, but with much larger flowers. A terrestrial plant. It varies very much in the size of the flowers.

**** Lip broken up into many divisions.

46. E. *cristatum* (Fl. Peruv. Syst. 243); "foliis lanceolatis, racemo dependente, labello tripartito: laciniâ mediâ bifidâ, lateralibus tripartitis."——*Peru.*——

47. E. *coronatum* (Fl. Peruv. Syst. 242); "foliis ovato-lanceolatis, racemis dependentibus, labello trilobo: laciniis bifidis intermediâ minori."——*Peru.*——

c. *Flowers panicled.*

* Lip undivided, quite entire at the edge.

48. E. *frigidum* (Linden MSS.); caule erecto indiviso undique foliato, foliis rigidis ovato-lanceolatis margine recurvis abruptè acutis, racemis paniculatis terminalibus multifloris pendulis, bracteis rigidis ovatis acutis squarrosis, floribus carnosis, sepalis subrotundo-ovatis subcarinatis rectiusculis, petalis linearibus, labello complicato transverso vix cordato basi bicalloso.——*Colombia.*——This very singular plant has a stem a foot and a half high, densely covered with leaves, which are stiff, and curved back at the edge. The flowers, which are pale rose, grow in long drooping panicled racemes. Mr. Linden, who alone has found it, states that it grows *on wet rocks*, at *but little distance from the eternal snow*, at the height of 13,000 feet above the sea. He adds that it is especially remarkable for all the plant, flowers included, being covered over with varnish.

49. E. *Funckianum* (Richard & Galeotti Ann. Sc. 3 ser. 3. 22; Orch. Mex. ined. t. 16.); "caule articulato; foliis ovalibus obtusissimis apice bilobis; flor. parvulis brunneis paniculatis; labello adnato cordato, acuto."——*Mexico.*——

50. E. *gramineum* (Lindl. Gen. et Sp. Orch. no. 30); foliis distichis lineari-lanceolatis acutis striatis, racemo terminali paniculato, sepalis oblongis acutis, petalis linearibus, labello cordato acuto medio plicato basi calloso.——*Peru.*——

51. E. *aquaticum* (Lindl. in Ann. Nat. Hist. 12); caule ancipiti ramoso, foliis lineari-oblongis acutis panicula simplici pauciflorâ (aut racemo) paulo brevioribus, sepalis oblongis petalisque linearibus obtusis, labello carnoso alte cordato cucullato basi ecalloso, clinandrio utrinque bidentato.——*Brazil.*——The branches of this are four or five inches long, the leaves from half an inch to two inches. The flowers small and green. According to Mr. Gardner it is an aquatic.

52. E. *diffusum* (Swartz. Prodr. 121. Fl. Ind. occ. 3. 1503. Willd. no. 15. *Seraphyta multiflora* Fisch. Ann. Nat. Hist. viii. 471); foliis oblongis, caule ancipiti, paniculâ terminali ramosissimâ, sepalis lineari-lanceolatis petalisque setaceis erecto-patentibus striatis, labello cordato acuminato basi bicalloso.——*Jamaica, &c.*——Flowers small, red, in a loose panicle.

53. E. *vincentinum* (Lindl. in Hook. Journ. 3. 88); caule ancipiti, foliis distichis anguste lanceolatis acutissimis paniculâ pauciflorâ laxâ filiformi brevioribus, sepalis lineari-lanceolatis, petalis filiformibus, labello subrotundo crispo.——*St. Vincents.*——A small delicate species, not more than four inches high, with minute membranous flowers, disposed in a short, loose panicle; filiform pedicels.

54. E. *recurvatum*; foliis ensiformibus mucronatis strictis paniculæ gracili ramis longis recurvis æqualibus, bracteis apice setaceis, floribus membranaceis distantibus, sepalis ovato-oblongis, petalis lineari-spathulatis, labello altè cordato subquadrato ecalloso.——*Colombia.*——A plant with the habit of E. tridactylum, &c. but with rose-coloured flowers, which are three or four times as large. Found by Mr. Linden.

55. E. *micranthum* (Lindl. in Hook. Journ. 3. 88); foliis distichis lineari-lanceolatis acuminatis, paniculâ virgatâ, bracteis setaceo-acuminatis florum dissitorum longitudine, sepalis oblongis carnosis obtusis subæqualibus, petalis linearibus, labello oblongo quadrato indiviso nudo.——*Peru.*——A plant with flowers scarcely a line long, and all the habit of Ep. tridactylum.

** Lip more or less toothed, or two lobed.

56. E. *lineare* (Fl. Peruv. Syst. 249); "foliis linearibus, racemo terminali subpaniculato, labello obcordato-cuneiformi basi utrinque dentato."——*Peru*——

57. E. *volubile* (Fl. Peruv. Syst. 247); "foliis oblongo-obovatis, ramis volubilibus, labelli bifidi laciniis lanceolatis."——*Peru.*——

58. E. *cordatum* (Fl. Peruv. Syst. 244); "foliis cordatis amplexicaulibus, paniculâ flexuosâ, labelli bifidi laciniis acuminatis recurvis."——*Peru.*——

*** Lip distinctly 3-lobed.

59. E. *carnosum* (Lindl. in Hook. Journ. 3. 87); foliis distichis ovato-lanceolatis acutiusculis: vaginis rugulosis, paniculâ rigidâ striatâ multiflorâ, bracteis duris ovatis cucullatis acutis ovarii longitudine, sepalis carnosis oblongis obtusis lateralibus hinc gibbosis carinatis latioribus, petalis sepalo dorsali conformibus, labelli postici trilobi carnosi lobis lateralibus rotundatis erectis intermedio conico solido.——*Brazil.*——A rigid plant, with the habit of Ep. elongatum. The panicle is stiff, many-flowered. Flowers are pale yellow; they are, when dry, hard, thick, and black, and evidently must be very fleshy when recent.

60. E. *durum* (Lindl. in Hook. Journ. 3. 87); foliis distichis ovato-lanceolatis acutis: vaginis rugosis, paniculâ simplici pauciflorâ, bracteis duris ovatis cucullatis acuminatis ovariis æqualibus, sepalis oblongis acutis duris striatis, petalis angustioribus, labelli postici trilobi transversè rhombei lobis lateralibus erectis truncatis intermedio triangulari acuto.——*Guiana.*——Stems simple or branched, from nine to eighteen inches high, equally covered with hard distichous leaves. Flowers small, apparently yellow. The inflorescence is occasionally simple.

61. E. *purum* (Lindl. in Bot. Reg. 1844. misc. 75); caule elongato tereti, foliis ensiformibus obtusis, floribus paniculatis ramis (3) racemosis gracilibus foliis paulo erectio-

ribus, sepalis angustè lanceolatis, petalis linearibus, labelli tripartiti basi 5-costati laciniis ovatis acutis.——A Caraccas plant, sent by Linden to Mr. Rucker, with whom it flowered in the beginning of September. The leaves are not half an inch wide, and six or eight inches long. The flowers are in a thin panicled raceme, of a light pale green colour, and about the size of those of E. *nutans*.

62. E. *tridactylum* (Lindl. in Bot. Reg. 1838. misc. 81); caule fusiformi gracili apice distichè folioso, foliis angustè

oblongis apice paulo angustatis obtusis cum mucronulo, spicâ terminali tripartitâ multiflorâ foliis longiore, sepalis subrotundo-ovatis patentibus incurvis lateralibus majoribus, petalis spathulatis linearibus apice incurvis, labello ascendente tripartito basi bicalloso: laciniis linearibus carnosulis lateralibus margine involutis intermediâ breviore planâ, columnâ brevi crassâ cuneatâ labello omninò adnatâ.——*Brazil.*——A plant, with smaller flowers than any other species of the genus yet in the gardens. They are a pale brownish yellow, except the column which is green, short, thick, and wedge-shaped.

63. E. *erubescens* (Lindl. in Hook. Journ. 3. 87. Bateman, Orch. Mex. et Guat.); foliis oblongo-lanceolatis acutis scapo pluries brevioribus, paniculâ amplissimâ flexuosâ, petalis unguiculatis sepalisque oblongis obtusis, labelli trilobi laciniâ intermediâ subrotundâ venis 3 elevatis lateralibus obovatis multo majore.——*Mexico.*——A magnificent plant, with very large panicles of delicate rose-coloured flowers as large as those of E. alatum.

**** Lip 4-lobed.

64. E. *floribundum* (Humb. et Kunth. Nov. Gen. et Sp. 1. 353. t. 86. Hooker in Bot. Mag. t. 3637); foliis lanceolato-oblongis acuminatis submembranaceis, paniculâ terminali, sepalis reflexis lanceolatis, petalis filiformibus, labelli quadrilobi basi bituberculati lobis lateralibus subrotundis, terminalibus linearibus divaricatis.——*Mexico, Surinam.*——

65. E. *verrucosum* (Swartz Fl. Ind. Occ. 3. 1497. Willd. no. 11. Lodd. Bot. Cab. t. 1084.); foliis distichis lanceolatis obliquè patentibus vaginis verrucosis, floribus subpaniculatis, sepalis oblongis concavis acutis petalisque lineari-lanceolatis patulis, labello 4-lobo laciniis linearibus obtusis divaricatis: anterioribus minoribus.——*Jamaica.*——Stems simple erect, two or three feet high. Sepals and petals pale green. Lip yellow. Scape radical, *Swartz*. Panicle terminal, *Loddiges*. May not two species be confounded under this name?

in this country or on the continent. Its leaves are about a quarter of an inch wide, spreading in a flat tuft; its large gaudy flowers supported, like those of the Crocus, by a long tube from an underground germen, the tube being five or six inches long. It is found from a little above the sea level to the summit of most, if not all, of the mountains in Corfu and Santa Maura, but does not reach the loftier bare top of Mount Œnos in Cephalonia, and it is said that some of the steeps of Corfu are in a blaze of blue from its flowers in January or February. Dispersed as it is, it seems strange that its seed cannot be obtained without great difficulty. A vast number of the most promising plants were dug up in Santa Maura in May, in the vain hope of obtaining a capsule, and with the help of a native, who well knew the difficulty. At last, in passing over Mount Rondi, in Cephalonia, I observed a dead plant of this Iris hanging by a root to the upper edge of a land slip, with a capsule, and having with difficulty scrambled up to it, I found one seed in it, which I brought home, and it has vegetated, as well as the plants which were dug up, but which I had not felt sure of being able to preserve alive. It remains to be ascertained, whether they will endure the climate of the interior of England. In the Ionian islands they seemed to be co-extensive with the white and the rose-coloured sage-leaved Cistus, which appear to me to be varieties of one species, growing quite promiscuously on the poorer slopes, though in the stronger soil where Malope Malacoides grew, I observed the red flowering later and alone. They are both liable to be killed with us by frost, except on dry banks in favoured parts of the island. *W. H.*

80. SERAPIAS cordigera.

I believe this plant to be a fine variety of Serapias longipetala. The latter is abundant in Corfu in the open spaces amongst the Cistus, oak, myrtle, Schinus, and other shrubs, which are kept in a dwarf state by the goats, with the lip very variable in size, and sometimes almost obsolete, and S. cordigera, with the larger lip, is found intermixed with the inferior sort, the colour of both being a dark brick red, varying to a dirty white. They are most plentiful on the less fertile slopes, but grow stronger when found in a richer and lower position.—*W. H.*

81. OPHRYS Arachnites.

The name Arachnites originates with Scopoli, who applied it to an Orchis, of which he made three varieties, two being in fact Orchis, and otherwise named. The third, to the appearance of which the name was adapted, was Ophrys Arachnites of Zaule, near Trieste; a plant before unnoticed, and not exactly similar to that of Switzerland and Kent, though it may be very proper to consider the latter as a variety of the true Arachnites, which has a very different aspect, and a shorter, rounder lip, and other distinctions. That which has been sometimes called Arachnites in Corfu is O. cornuta, distinguished by the prolongation of the shoulder of the lip into a long horn, but otherwise closely allied to Arachnites, which it resembles in its general appearance, though it grows taller. The bee orchis of England grows in Corfu, and is occasionally, though rarely, found with a yellow lip, and white segments to the flower, in which state it is the once-found Ophrys chlorantha of the Swiss Flora.—*W. H.*

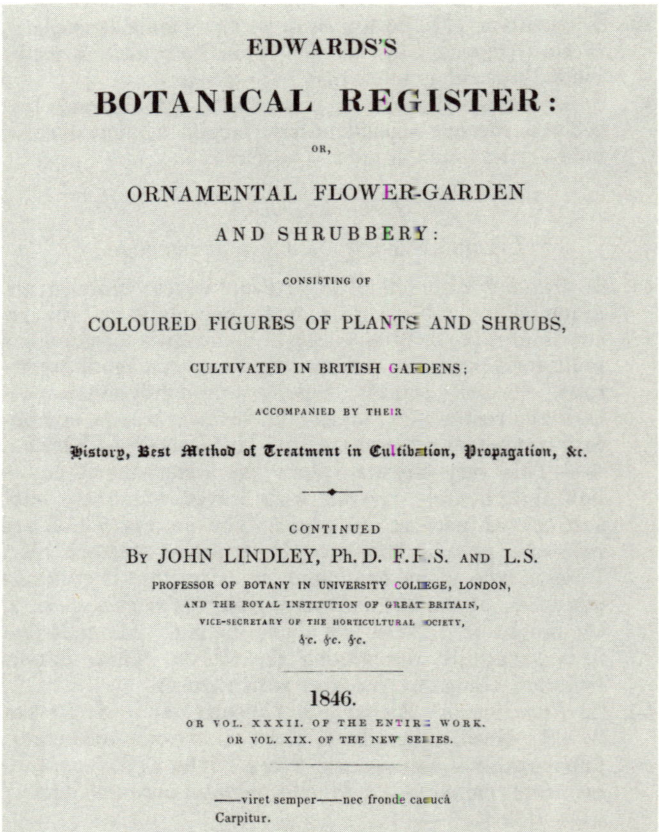
EDWARDS'S

BOTANICAL REGISTER:

OR,

ORNAMENTAL FLOWER-GARDEN

AND SHRUBBERY:

CONSISTING OF

COLOURED FIGURES OF PLANTS AND SHRUBS,

CULTIVATED IN BRITISH GARDENS;

ACCOMPANIED BY THEIR

History, Best Method of Treatment in Cultivation, Propagation, &c.

CONTINUED

BY JOHN LINDLEY, Ph. D. F.R.S. AND L.S.

PROFESSOR OF BOTANY IN UNIVERSITY COLLEGE, LONDON, AND THE ROYAL INSTITUTION OF GREAT BRITAIN, VICE-SECRETARY OF THE HORTICULTURAL SOCIETY, &c. &c. &c.

1846.

OR VOL. XXXII. OF THE ENTIRE WORK. OR VOL. XIX. OF THE NEW SERIES.

——viret semper——nec fronde caducâ Carpitur.

LONDON:

JAMES RIDGWAY, PICCADILLY.

M.DCCC.XLVI.

1

CATTLEYA maxima.

The Larger Cattleya.

GYNANDRIA *MONANDRIA.*

Nat. ord. ORCHIDACEÆ. § EPIDENDREÆ. (ORCHIDS, *Vegetable Kingdom, p.* 173. *ined.*)

CATTLEYA. Lindley.

C. *maxima;* pseudobulbis clavatis elongatis altè sulcatis, foliis subemarginatis basi sublatioribus, pedunculo 2-3-floro spatha ancipiti longiore, sepalis lanceolatis petaloideis subrecurvis, petalis pluries latioribus convexis undulatis reticulatim pictis, labello convoluto apice explanato oblongo subcrispato-lobato per axin lævissimo.

C. maxima, *Lindl. gen. & sp. Orch. p.* 116. *Bot. Reg.* 1844. *sub t.* 5. *Journal of the Horticultural Society, vol.* 1. *part* 1. *page* 64.

The more our knowledge of the varying forms of Orchids extends, the less do we feel able to determine what characters or peculiarities of structure are of real value in distinguishing species. Every thing which among other plants has gained a fixed value, appears here to be unstable; and even the most marked differences in form are proved by incontestable evidence to grow out of each other.

This being so, we may be excused for doubting how far this beautiful plant can lay claim to the rank of a species; for it evidently approaches both C. Mossiæ and labiata in many important particulars. Its main peculiarities consist in its long-channelled pseudo-bulbs, and in its very convex wavy petals, which are quite different in appearance from the thin, nearly flat, petals of C. Mossiæ and labiata.

If, however, we attach any value to differences of colour, then indeed there remains no difficulty in separating this plant; for it is remarkable for the dark crimson veins richly traced upon its pallid lip, and for a beautiful net-work of purple streaks, which is drawn over all their surface. At first, too, the colour of the flowers is so pale as to be almost white; but the tints heighten day by day, till at last the blossoms acquire the rich tint represented in the accompanying plate.

In the number of the New Journal of the Horticultural Society above referred to, the following account of the plant is given.

"The pseudo-bulbs are from six to eight inches long, compressed, and considerably thicker at the upper than the lower end, with a few shallow furrows on each side. The leaves are single, of the same length, very regularly oblong, and rather concave until they become old. The flowers are seven inches across, and two in number, on the end of a peduncle a few inches longer than a compressed pale green spathe; but Mr. Hartweg reports that they are really much larger, and grow as many as four or five in a cluster. When they first expanded they were of nearly the same colour as those of Cattleya intermedia, but after some days the colour brightened and deepened till at last it was equal to that of Cattleya labiata. The sepals are lanceolate and convex; the lateral ones twisted and turned backwards; the dorsal one still more convex, and curved gracefully upwards from the base of the column. The petals are about two inches broad, spreading forwards on each side of the lip, and after a time become beautifully painted with purple veins. The lip is about three inches long, undivided; at the lower half flat and channelled, so that its edges meet over the column and form a ridge there; the anterior half gradually spreads into an oblong plate, very much curled and toothletted; its ground colour is a very pale pink, over which is drawn a coarse network of the richest crimson veins. Along its middle a narrow yellow polished line is drawn, and forms a beautiful contrast with the crimson veins that surround it. There is no trace of ridges, glands, or elevated lines upon any part of the surface of the lip.

"This requires to be treated in a similar way to the different species of Cattleya, cultivated in almost every collection: rough peat and potsherds are apparently the best materials for potting it in. During summer an ample supply of water is necessary at its roots, but as little as possible on its leaves, except in the form of steam. Mr. Hartweg states that it requires a temperature of 75°. In winter, if the atmosphere is kept moist, very little water will be necessary for a few weeks."

We take advantage of the present opportunity to print a memorandum from Mr. Hanbury, relating to the Russell variety of C. granulosa, figured at t. 59 of the last volume of the Botanical Register.

"In reference to your remarks upon the country where Cattleya granulosa is found, will you pardon my saying that I think you are mistaken, as I received many plants about two years ago from Paraiba, one of which flowered this autumn."

10

DENDROBIUM Dalhousieanum.

Lady Dalhousie's Dendrobium.

GYNANDRIA *MONANDRIA.*

Nat. ord. ORCHIDACEÆ. § MALAXEÆ-DENDROBIDÆ. (ORCHIDS, *Vegetable Kingdom, p.* 181. *ined.*)

DENDROBIUM, Swartz.

§ STACHYOBIUM. *Caules teretes. Folia plana. Flores racemosi.*

D. *Dalhousieanum* (Paxton's Magazine of Botany, vol. 11. t. 145); caulibus teretibus subfusiformibus striatis, foliis ovato-lanceolatis obtusis, racemis horizontalibus 5-7-floris, bracteis squamæformibus, sepalis ovatis, petalis conformibus multò majoribus, labello indiviso oblongo medio constricto basi foveato parte superiore obtuso glanduloso villoso lateribus erosis glabris lamellis 3 pro medium brevibus laciniisque utrinque uniseriatis inflexis, cornu brevi obtuso.

The part of India where this beautiful species has been obtained is unknown. It was found in the Botanical Garden, Calcutta, by Mr. Gibson, who sent it to Chatsworth, with the name which had been given by Dr. Wallich, in compliment to the Countess of Dalhousie, from whom he had received it.

It is not among the wild specimens in our herbarium, unless it should lurk among some of Mr. Griffith's unexamined collections.

None of the species exceed it in beauty, for, with the large flowers and noble aspect of the " musk" Dendrobium, it is enriched by two large blood red spots on each side of the lip, which melt into a row of fringes turned towards the middle, and looking like a pair of young whiskers fresh from the curling irons of the hairdresser.

It is well figured in Mr. Paxton's Magazine, with a woodcut shewing its manner of growth; and it is there stated to have produced strong stems, upwards of 3 feet in length, without shewing any inclination to flower. The figure was taken from a plant in the possession of Messrs. Loddiges, and we are indebted to the same gentlemen for our specimen.

Fig. 1. represents a front view of the lip; the fringes, which turn inwards from the purple cheeks, vary a good deal in number and colour, but in general they are deeply stained by the deep purple of the cheeks.

Mr. Gibson states that it flowers at Calcutta in the dry season, as is the natural habit of most, if not all its nearest allies. No doubt it comes from some very hot province of India, and should be treated accordingly. It should be potted in turfy heath-mould mixed with potsherds. Like many other Orchids, it requires an ample supply of water during the growing season, and a humid atmosphere throughout the year. In autumn, as the stems become matured, water should gradually be withheld, so that the plant may be in a state of rest in winter. At this time the temperature should not be raised above 60° by fire heat, unless the house is *well* ventilated, and then a much higher temperature may be applied. In spring, when growth commences, it should be repotted, and heat and moisture gradually increased; 85° by day, and 70° at night, is high enough even at midsummer.

15

DENDROBIUM aduncum.

Hooked Dendrobe.

GYNANDRIA *MONANDRIA*.

Nat. ord. ORCHIDACEÆ. § MALAXEÆ-DENDROBIDÆ. (ORCHIDS, *Vegetable Kingdom, p.* 181. *ined.*)

DENDROBIUM, Swartz.

§ STACHYOBIUM. *Caules teretes. Folia plana. Flores racemosi.*

D. *aduncum* (Wallich); caulibus pendulis, foliis lineari-lanceolatis acutis integris, floribus ternis patentibus, sepalis petalisque ovatis obtusis lateralibus duplò latioribus, cornu rotundato, labello unguiculato ovato concavo apiculato columnæ arctè appresso intus villoso disco glabro, columnâ apice bialatâ sub stigmate villosâ, antherâ glandulosâ.—*Lindl. in Bot. Reg.* 1842. *misc.* 62.

Of this very pretty Orchid we know nothing more than that it was sent from Calcutta to Messrs. Loddiges, by Dr. Wallich, and that it flowered at Hackney in July, 1842.

In some respects it is allied to D. Pierardi, especially in its small pink flowers and manner of growth; but it is more closely related to D. moschatum, of which it may be regarded as a feeble imitation. It is however widely different from both, and is especially known by its half transparent flowers, of the most delicate texture and clearest tints. Why it is called D. aduncum, or whence it comes, we know not.

Fig. 1. represents the column, which is remarkable for being hairy just below the stigma, and for having a downy anther; 2. is the lip, stalked, spoon-shaped, sharp-pointed, and covered all over the inside with hairs, except a round space in the centre, which answers to the stigma when pressed against it, and which is perfectly smooth; 3. are the pollen-masses.

Like many other Dendrobes, this requires a warm and a humid atmosphere throughout the year. The most convenient, and probably the best way of growing it, is in a pot, in turfy heath-mould, which may be renewed every year when the plant is at rest, without injuring the roots. In summer an ample supply of water should be given, and as the young leaves are liable to become scorched by the sun, shading should be used in sunny weather. In autumn, as the shoots become matured, water should be gradually withheld, so that for a few weeks in winter the roots may be comparatively dry.

19

SARCOCHILUS Calcéolus.

The slippered Fleshlip.

GYNANDRIA *MONANDRIA*.

Nat. ord. ORCHIDACEÆ. § VANDEÆ—SARCANTHIDÆ. (ORCHIDS, *Vegetable Kingdom, p.* 173.)

SARCOCHILUS.—Supra, fol. 1832.

S. *Calceolus;* caulescens, radicans, foliis oblongis carnosis obliquè emarginatis obtusis, pedunculis brevibus squamatis bifloris supra-axillaribus, sepalis petalisque carnosis oblongis subæqualibus acutis, labelli lobo medio oblongo tereti solido spongioso laciniis lateralibus ascendentibus triangularibus acuminatis verrucis duabus ciliatis interjectis.

This very singular species is a native of Manilla, whence Mr. Cuming sent it to Messrs. Loddiges, of whose catalogue it is 1554. The lengthened stem is at variance with the usual habit of Sarcochilus, as also is the lip so far as the very large size of its middle lobe is concerned. But no other distinctions appear to exist. The resemblance of the lip to a slipper is too obvious to require being pointed out; it is, however, but a false resemblance after all, for it has no hole for the foot, being solid and spongy, and in reality closed up by a pair of little ciliated warts. (See Fig. 1.) Fig. 2 represents the column, and 3 the pollen-masses.

The following is a list of the species of this genus at present known.

1. S. *falcatus* (R. Brown Prodr. 332. Lindl. gen. & sp. orch. 842. Bot. Register, t. 1832); acaulis, foliis oblongo-lanceolatis acutis, racemo laxo multifloro foliorum longitudine, labelli lobo medio obsoleto rotundato lateralibus oblongis ascendentibus petaloideis.—— *New Holland* ——Flowers white with a few pink streaks on the lip, whose middle lobe is yellow.
2. S. *parviflorus* (Lindl. in Bot. Reg. 1838. misc. 50); acaulis, foliis oblongis undulatis chartaceis apice acute et oblique bidentatis, racemis nutantibus quadrifloris, sepalis lineari-obovatis acutis lateralibus basi dilatatis, petalis falcatis obtusis, labelli cum columnâ articulati lobo medio minimo transverso rotundato: disco callis variis crassis rotundatis tuberculato.——*New Holland.*——Flowers small, green, with a few spots of dull purple on the sepals. The lip is less green, beneath nearly white, banded with dull purple, articulated with the column, and covered over in the inside with various crowded tubercles, of which the three largest form a crescent near the apex of the lip.
3. S. *olivaceus* (A. Cunn. in Bot. Reg, 1839. misc. no. 27); foliis oblongis membranaceis falcatis apice acutis integris, racemo flexuoso angulato 4-5-floro nutante, sepalis lineari-obovatis obtusis: intermedio latiore petalisque conformibus minoribus columnæ dorso adnatis, labelli cum columnâ productâ articulati lobo medio minimo transverso rotundato lateralibus acuminatis; disco callis variis crassis rotundatis tuberculato.——*New Holland* ——Very nearly the same as S. *parviflorus*, from which it principally differs in its leaves being broader and not emarginate at the point. The flowers are small, yellowish orange green.
4. S. *croceus;* foliis crassis dorso rotundatis recurvis apice in acumen crassum carnosum constrictis, labelli lunati laciniis semioblongis acutis anticè convexis disco pubescentibus lobulo intermedio obsoleto.——*Manilla* —— Of this I have only seen one leaf, and one flower-bud, from Mr. Loddiges. It is a thick-leaved Vanda-like plant, with a saffron-coloured spot in the middle of its lip, which when spread out is not unlike a cheese-cutter in form.
5. S. *unguiculatus* (Lindl. in Bot. Reg. 1840. misc. 143); racemo 3-4-floro, labelli unguiculati trilobi ungue convexo lineari lineâ mediâ exaratâ, laminâ cavâ, lobis lateralibus semi-ovatis obtusis intermedio supra trinervi subtus globoso apice papillæformi.—— *Manilla.* —— The flowers are light straw-colour, the side lobes of the labellum white streaked with crimson, and the middle lobe rounded, fleshy, and dotted with crimson. It is rather a pretty species.
6. S. *Calceolus, supra.*

Like many species of exotic Orchids this requires a humid atmosphere, and an ample supply of water throughout the year. It may be potted in turfy heath mould in the usual way. Being a native of a hot climate it is necessary to keep it in a temperature of about 80° in summer, using a thin shading in sunny weather. In winter a temperature ranging from 66° to 70° will be sufficient.

27

ONCIDIUM lacĕrum.

Cut-lipped Oncidium.

GYNANDRIA *MONANDRIA.*

Nat. ord. ORCHIDACEÆ. (ORCHIDS, *Vegetable Kingdom, p.* 173.)

Sect. TERETIFOLIA : Chive-leaved.

O. *lacerum* (Lindl. in Bot. Reg. 1844. misc. 38) ; foliis longis teretibus carinatis, paniculâ contractâ multiflorâ, sepalis petalisque conformibus obovatis concavis, labelli elongati laciniis lateralibus linearibus refractis intermediâ longè et angustè unguiculatâ bilobâ margine lacerâ, cristæ dente altero transverso altero majore compresso a fronte, columnæ brevis pubescentis alis semiovatis.

Among the species of Oncidium, to which the general term of Chive-leaved may be given, and the type of which is *O. Cebolleta,* this, and *O. longifolium,* are the handsomest. It agrees with that species in the length of its leaves, which are as much as two feet, but it differs in having a smaller panicle, sharp column-wings, very narrow falcate side lobes on the lip, and a single-ridged tubercle at the base of that organ, placed at right angles to a flat somewhat concave swelling, which almost connects the side lobes.

Oncidium nudum, a Caraccas plant, has much smaller flowers, little blunt column wings, a crenate crescent-shaped ridge in front of its column, with the convexity outwards, and a very small raised line at right angles to it.

This species is found wild in Panama, whence it was sent to Messrs. Loddiges, with whom it flowered in April, 1844. Fig. 1 represents the column and base of the lip.

Like the generality of Orchids, this requires to be potted in turfy heath-mould, mixed with potsherds, and to be kept in a humid atmosphere throughout the year. Being one of those which are not easily injured by the sun's rays, shading is only necessary to keep the temperature about 80°, without admitting cold air. In winter, very little water will be required, except in the form of vapour, nor should the temperature be raised above 66° by artificial means.

Although such plants do not derive much nourishment from the soil, still they are greatly benefited by being repotted once a year. This should be done in winter, or early in the spring, so as not to injure the roots while in a growing state.

30

ANSELLIA africana.

African Ansellia.

GYNANDRIA *MONANDRIA.*

Nat. ord. ORCHIDACEÆ—§ VANDEÆ-BRASSIDÆ. (ORCHIDS, *Vegetable Kingdom, p.* 173.)

ANSELLIA, Lindl.—Sepala oblonga, carnosa, æqui-patentia, libera. *Petala* conformia, recta, patula, duplò latiora. *Labellum* sessile, patulum, trilobum, bilamellatum, lobo medio minore verrucoso. *Columna* elongata, marginata, basi utrinque auriculata. *Anthera* bilocularis. *Pollinia* 4, sessilia, basi contigua, duobus dorsalibus multo minoribus; glandula angusta utrinque acuminata.——Caulis *elongatus, teres, apice tantum foliosus.* Folia *plicata, coriacea.* Panicula *terminalis.*—Lindl. in Bot. Reg. 1844, sub t. 12.

A. *africana,* Lindl. l. c.

The noble plant now figured was spoken of by us in 1844 in the following terms:

" While we are writing on this subject, we would also mention another most noble plant, specimens of which might certainly be procured by any of our African merchants. When Mr. Ansell was ill from the effects of the Niger expedition, at Fernando Po, he found in Clarence Cove, growing on the stems of the Oil Palm (Elais guineensis), an epiphyte with a slender jointed stem about two feet long, having at the upper end many stiff, plaited, lanceolate, 5-ribbed leaves, and a terminal panicle of flowers as large as those of Vanda Roxburghii, with dark purple spots on a pale ground. Of that plant we possess a dried specimen, with one of the lower branches of the panicle in good preservation, and as it proves to be a new genus we take this opportunity of naming it after its discoverer."

This notice produced the desired effect, living specimens having been received by the Rev. John Clowes, and by Messrs. Loddiges, from the latter of whom the specimen which supplied the accompanying drawing was exhibited to the Horticultural Society, in Regent Street, in February last. The plant was about two feet high, and bore twenty-four flowers. We understand that Mr. Clowes's specimen was much finer.

It is indeed a noble thing; for although its flowers have somewhat the colour and appearance of a large Cymbid, yet their panicled disposition, and the entirely different habit of the plant, render it much more showy than any Cymbid known to us. Its nearest affinity is perhaps with Bromheadia, with which it corresponds in having a lengthened stem and terminal inflorescence; and thus it may serve as a connecting link between the Brassid forms of Indian and African Orchids, and as a transition to the Indian Sarcanthids by way of Eulophia or Galeandra. It is very near Cymbidium in technical characters, though so extremely different in its manner of growth. The auricles at the base of the column, the four pollen-masses, and the very narrow gland fining away to each side, sufficiently distinguish it.

Although it is stated that this plant was found growing on the stems of the Oil Palm, yet under cultivation it will in all probability succeed best in a pot if treated in a similar way to Peristeria elata. It roots freely in turfy heath-mould, and requires an ample supply of water during the growing season. In winter it may be supposed that a season of rest is necessary, when little or no water should be given; a damp atmosphere being sufficient for all the requirements of the plant.

34

ODONTOGLOSSUM membranaceum.

Membrane-sheathed Toothtongue.

GYNANDRIA *MONANDRIA.*

Nat. ord. ORCHIDACEÆ. § VANDEÆ—BRASSIDÆ. (ORCHIDS, *Vegetable Kingdom, p.* 173.)

ODONTOGLOSSUM.—Kunth.

O. *membranaceum;* pseudobulbis foliisque O. Cervantesii, scapo 2-4-floro, bracteis vaginisque membranaceis acutissimis equitantibus, sepalis membranaceis unguiculatis lanceolatis, petalis latioribus oblongis obtusis, labello cordato obtusissimo unguiculato, ungue carnoso cyathiformi pubescente anticè bidentato medio tuberculato, processubus 2 ante cyathum elongatis pubescentibus, columnæ pubescentis auriculis rotundatis.—*Sertum Orchidaceum, sub t.* 25. *Bot. Reg.* 1845. *misc.* 60. 10.

This delicious species—for so it deserves to be called, on account of its agreeable odour and delicate flowers—this delicious species is quite like O. Cervantesii, figured in this work at t. 36 of the volume for last year, in general appearance, and may possibly be a mere variety of it. It differs, however, in the following respects :

Its flowers are whiter and rather larger, and the lip is spotted at the base;

Its petals are much blunter;

Its lip is very deeply heart-shaped, and quite rounded at the point;

The two front teeth of the lip are shorter, and less hairy; and

The concavity at the base of the lip has a much smaller central tubercle.

In other respects the two plants may be regarded as being identical. They are both Mexican, and may be regarded as among the finest species of that country.

The accompanying figure was made in Messrs. Loddiges' nursery, in March 1845. Fig. 1 represents the lip; 2, the column and wings.

Both these stove Orchids, like O. Rossii, may either be tied to a block of wood, with a little sphagnum to retain moisture, or potted in turfy heath mould, in the usual way. In whatever manner they may be treated, an amply supply of water should be given during the summer months, and a damp atmosphere maintained throughout the year. In winter, the soil or moss about their roots should be allowed to become almost dry, especially if grown in a pot. And since such plants are always benefited by being repotted once a year, that operation must be performed while the plant is dry, and in a state of rest.

35

CATTLEYA Lemoniana.

Sir Charles Lemon's Cattleya.

GYNANDRIA *MONANDRIA.*

Nat. ord. ORCHIDACEÆ. (ORCHIDS, *Vegetable Kingdom, p.* 181)
CATTLEYA. Botanical Register, vol. 11. *fol.* 953.

C. *Lemoniana*; pseudobulbis brevibus fusiformibus foliis brevioribus, foliis æquilateris obtusis, spathâ nullâ (?), sepalis angustè lanceolatis acutissimis apice subherbaceis, petalis maximis tenuibus valdè undulatis, labello angusto convoluto pone apicem tantum in limbum convexum parcè undulatum margine erosum reflexum nullo modo crispum expanso.
C. labiata *var.* Lemoniana. *W. B. Booth in litteris.*

To us this appears quite a new and distinct species, remarkable for its short pseudo-bulbs, and long convolute lip, with a convex pallid limb deeply broken up into a few revolute folds at the edge. Mr. Booth, however, regards it as a variety of C. labiata, from which it differs in its short pseudo-bulbs, apparent want of spathe, and contracted lip, wholly destitute of the rich frill of plaits by which that species is so strikingly distinguished.

The accompanying figure has been made by Mr. Booth, to whom we are indebted for the following account of the plant.

"That this is a variety of Cattleya labiata I think there can be no doubt, however different it may appear in the form and colour of its flowers from those of the original species, and of the other varieties of it which have yet been published; but it certainly does not appear to me to possess any peculiarity of sufficient importance to warrant its being kept *specifically* distinct, and I have therefore merely characterized it as a variety, and dedicated it to Sir Charles Lemon, Bart. M.P. already well known for his attachment to this singular tribe of plants, and who has the merit of first introducing it.

"The specimen from which the accompanying figure and description were taken, was imported from the Brazils in 1842, and flowered for the first time, at Carclew, in September, 1845. It requires a warm damp atmosphere, and attaches itself by its long ash-coloured roots, either to a wall, or the branch of an old tree.

"*Pseudo-bulbs*, or stems, one-leaved, erect, cylindrical when young, and almost wholly covered by a large pale brown scale; but channelled and angular when old, and of a deep green. They are generally about four inches high, and three-fourths of an inch in diameter, diminishing to both ends, and slightly compressed at the top. *Leaves* oblong-lanceolate obtuse, thick and rigid, varying from four to six inches in length, and an inch and a half in breadth, of a rich deep green, and having the edges somewhat recurved. *Scape* about two inches long, round, and of a pale green, in our specimen two-flowered. *Pedicells* round, and of the same colour as the scape, thickening outwardly toward the flower, and varying from one and a half to two inches in length, with three small longitudinal grooves, and a minute thin acuminate bract at the base where it joins the scape. *Flowers* when fully expanded upwards of four inches in diameter, of a pale lilac colour, remarkably handsome, and sweet-scented in the day time. *Sepals* spreading oblong lanceolate acute, about two inches long and rather more than half an inch broad, a good deal recurved at the point and along the edges, so that the two almost meet. *Petals* large, ovate-oblong, upwards of two inches in length, and rather less in breadth, delicately marked and undulated at the margin, becoming of a deeper tint in the centre, which is somewhat raised so as to form a kind of midrib a little paler than the rest. *Labellum* three-lobed, and nearly three inches long. The middle lobe is round and spreading, divided in the centre, and very much undulated and recurved at the margin, which is tinted with pale lilac, changing inwardly to a soft yellow colour, and striated with deep red lines along what may be termed the throat of the tube formed by the two lateral lobes, which curve upwards on each side, and embrace one another so as to conceal the column. The outside of this tube is of a deeper lilac than that of the petals, and beautifully veined, becoming compressed when the flower has been open for some time. *Column* scarcely more than a third the length of the labellum, hollowed out below, so as to appear two-edged, and rounded above, with a small round fleshy process projecting over the anther case, of a deep pink colour."

41

ANGULOA Ruckeri.

Mr. Rucker's Anguloa.

GYNANDRIA *MONANDRIA.*

Nat. ord. ORCHIDACEÆ. § VANDEÆ. || MAXILLARIDÆ. (ORCHIDS, *Vegetable Kingdom, p.* 181.)

ANGULOA, Fl. Peruv. prodr. 118. *t.* 26. *Flores* subglobosi, nunquam patentes. *Sepala* lateralia invicem imbricantia basi valde convexa, nec in cornu producta; alterum nunc anticum nunc posticum, conforme, basi planum. *Petala* sepalo dorsali æqualia, et similia. *Labellum* coriaceum, unguiculatum, subconvolutum, trilobum, laminâ carnosâ latâ planâ supra medium auctum, hinc quasi bilabiatum. *Columna* teres, clavata, libera; *clinandrio* nunc mutico, nunc laciniâ acutâ porrectâ utrinque aucto. *Anthera* galeata, valvis membranaceis nunc in lacinulas acutas productis. *Pollinia* 4, plana, inæqualia, *caudiculâ* longâ lineari, et *glandulâ* acutâ.——Herbæ *epiphytæ Granatenses & Peruvianæ*, Lycastes *facie*.—Bot. Reg. 1846. t. 60.

A. *Ruckeri*. (Vide infra).

This charming plant makes the third species now in our gardens, of a genus, which, in the spring of 1844, was a botanical puzzle. Such is the progress of scientific discovery when promoted by horticultural enterprise.

All the species have been found by the fortunate exertions of Mr. Linden. One of them (*A. Clowesii*) was met with in low bottoms, in the midst of forests, in Venezuela, growing in the ground. The history of the two others is unknown. One of them (*A. uniflora*) exists in Mr. Linden's herbarium, but by some mistake in the ticketing it is marked there as a Cuba plant; the other was sent to Mr. Rucker, whose specimen is now represented, and its locality is unknown.

A. *Ruckeri* is immediately recognized by its flowers having deep crimson spots on a yellow ground, and a deep crimson lip. The form of the latter again approaches that of A. Clowesii, but it is less hairy, and the lateral lobes are blunt, as well as shorter. Moreover, the flower is not resupinate, in the common acceptation of the term. The species was among the most striking of all the Orchids exhibited in the Garden of the Horticultural Society in May last, at which time the accompanying drawing was made. Fig. 1 shews the form of the lip when spread open.

The following is a list of the species of Anguloa hitherto discovered :—

1. A. *uniflora* (Fl. Peruv. syst. p. 228. Lindl. gen. & sp. Orch. p. 160. Bot. Reg. 1844. t. 60); pedunculo unifloro radicali squamis 2 inflatis imbricatis vaginato, flore bracteâ spathaceâ herbaceâ parum longiore, sepalis lateralibus anticis petalisque acuminatis, labelli glabri lobo medio angustissimo reflexo lateralibus rotundatis, laminâ appendiculari retusâ duplò latiore, clinandrio lacinulis 2 acutis aucto.——*Colombia.*——Flowers white, with a slight tinge of yellow, sweet-scented.

2. A. *Clowesii* (Lindl. in Bot. Reg. 1844. misc. 29, t. 63); pedunculo unifloro radicali laxè squamato, flore carnoso resupinato, sepalis petalisque ovatis convexis conniventibus, labelli trilobi lobo medio piloso infundibulari bilabiato: labio altero emarginato altero tridentato, columnâ integrâ.——*Colombia.*——Flowers straw coloured. Lip white.

3. A. *Ruckeri* (Lindl. in Bot. Reg. 1846. t. 41); pedunculo unifloro radicali squamis inflatis imbricatis vaginato, sepalis subrotundis apiculatis petalisque obtusis in globum conniventibus, labelli trilobi antici lobis lateralibus obtusis æqualibus medio piloso infundibulari bilabiato: labio altero emarginato altero tridentato, columnâ integrâ.——*Colombia.*——Flowers yellow, spotted inside with crimson, and having a lip of the latter colour.

4 ? ? A. *squalida* (Pöppig n. g. & sp. pl. l. p. 43, t. 74); "scapo spicato multifloro; floribus secundis; sepalis reflexo patulis; foliis sex ad octonerviis."——*Peru.*——Flowers "the size of a walnut, dirty flesh colour, scentless."

46

CYCNOCHES Egertonianum; *var.* viride.

Green Egertonian Swan-neck.

GYNANDRIA *MONANDRIA.*

Nat. ord. ORCHIDACEÆ. § VANDEÆ—CATASETIDÆ. (ORCHIDS, *Vegetable Kingdom, p.* 181.)

CYCNOCHES. Supra, t. 1742.

C. *Egertonianum;* racemo longissimo pendulo, sepalis petalisque membranaceis recurvis, labelli disco rotundato in processubus clavatis acutisque æquilongis soluto, columnâ tenui longissimâ.

α. Floribus atropurpureis.
C. Egertonianum, *Bateman Orchid. Mexic. & Guatemala, t.* 40.

β. Floribus pallidè virentibus.
C. stelliferum, *Loddiges' Cat. no.* 1017.

This plant was imported from Oaxaca by Messrs. Loddiges, and flowered with them in August 1843, when the accompanying figure was made. It is evidently a variety of the C. Egertonianum, distinguished by its flowers being of a pale watery green, and not deep purple.

But what is *C. Egertonianum* itself? In our volume for 1843, at p. 77 of the miscellaneous matter, we have extracted from Mr. Bateman's magnificent work his account of how the long-spiked small-purple-flowered *C. Egertonianum* is only the short-spiked large-green-flowered *C. ventricosum;* how the same plant at one time bears one sort of flowers, and at another time another sort; and we have shewn how the same plant, nay the same spike, is sometimes both the one, the other, and neither. *C. Egertonianum* is then a "sport," as gardeners say, of *C. ventricosum.*

But what again is *C. ventricosum?* Who knows that it is not another "sport" of *C. Loddigesii,* which has indeed been caught in the very act of shewing a false face, something wonderfully suspicious, all things considered, and justifying the idea that it is itself a mere Janus, whose face is green and short on one side, and spotted and long on the other.

Then, if such apparently honest species as *C. Egertonianum, ventricosum,* and *Loddigesii* are but counterfeits, what warrant have we for regarding the other so-called species as not being further examples of plants masquerading with false faces? For ourselves we cannot answer the question; nor should we be astonished at finding some day a Cycnoches no longer a Cycnoches, but something else; perhaps a Catasetum. If one could accept the doctrine of the author of the "Vestiges," it might be said that in this place we have found plants actually undergoing the changes which he assumes to be in progress throughout nature, and that they are thus subject to the most startling conditions only because their new forms *have not yet acquired stability.*

Be that as it may, it will be as well to state in this place what the so-called species are of this extraordinary genus, and what the changes are which they have been seen to undergo.

SO-CALLED SPECIES OF CYCNOCHES.

1. C. *Loddigesii* (Lindl. Gen. & Sp. Orch. p. 154. Bot. Cab. t. 2000. Bot. Reg. t. 1742.)——*Surinam.*——Flowers very large, fragrant, green and purple, with a white spotted lip. *Sports by producing smaller broad-lipped flowers without scent, and with a very short club-shaped column.*

2. C. *ventricosum* (Bateman Orch. Mex. & Guatemala, t. 5.)——*Guatemala.*——Flowers large, green, with a white lip. *Sports to Egertonianum.*

3. C. *chlorochilon* (Klotzsch: Sertum Orchidaceum, t. 16.)——*Maracaybo.*——Flowers very large, green, whole-coloured. *Has not been observed to sport:* but is probably a mere variety of C. ventricosum.

4. C. *pentadactylon* (Lindl. in Bot. Reg. 1843. misc. 26. t. 22.)——*Brazil.*——Flowers large, yellowish green, banded with brown. *Has not been observed to sport.*

5. C. *maculatum* (Lindl. in Bot. Reg. 1840. misc. 8. Sertum Orchidaceum, t. 33.)——*Mexico? La Guayra.*——Flowers small, yellow, spotted with brown. *Has not been observed to sport.*

6. C. *Egertonianum* (Bateman Orch. Mex. & Guatemala, t. 40. Bot. Reg. 1846. t. 46.)——*Guatemala and Mexico.*——Flowers small, purple or greenish, unspotted. *Sports to ventricosum.*

47

ACANTHOPHIPPIUM javánicum.

Javanese Acanthophippium.

GYNANDRIA *MONANDRIA.*

Nat. ord. ORCHIDACEÆ. (ORCHIDS, *Vegetable Kingdom, p.* 180.)
ACANTHOPHIPPIUM, Blume.—Supra, t. 1730.

A. *javanicum;* petalis triangularibus, labelli trilobi lobis lateralibus truncatis intermedio medio constricto apice ovato tuberculato basi utrinque carnoso dentibus truncatis emarginatis inflexis.
A. javanicum, *Blume Bijdr.* 353. *tabellen* 47. *Lindl. Orch. no.* 1.

This plant, the original species of Blume's genus Acanthophippium, was found by its discoverer in the woods of the higher parts of Mount Salak in Java, where it flowers from February to April. Our drawing was made from a specimen communicated in September 1844, by Messrs. Loddiges.

It is a very pretty thing, because of the clear pale purple stripes drawn down the outside of the flowers, and the delicate lilac of the orifice. As a species it is readily known by its three-lobed lip having the centre division contracted in the middle, uneven and ovate at the point, and fleshy at the base, with thick fleshy irregularly-toothed sides. Fig. 1. represents it.

It requires the same kind of treatment as the well known Acanthophippium bicolor. It should be potted in turfy peat, in well drained pots, and liberally supplied with water during the season of growth. Water should be more sparingly supplied when the plant is in a state of rest, or in very dull, damp weather. A temperature of about 65° in winter, and from 70° to 80° in summer, with fire heat, is quite sufficient for its growth. The stove, or Orchidaceous house should be well shaded during bright sunshine.

It may be easily propagated by taking the plants to pieces, but this should always be done when it is resting, and just before the season of growth recommences.

The following are the species of this genus at present known—

1. A. *javanicum;* now figured.
2. A. *striatum* (Lindl. in Bot. Reg. 1838. misc. 68); labelli lobis lateralibus rotundatis intermedio minimo acuto carnoso utrinque plicato, disci callo solitario elevato longitudinali.——*Nepal.*——Flowers French white, with dull stripes, smaller than in A. bicolor.
3. A. *Sylhetense* (Lindl. gen. and sp. orch. no. 2); labelli lobis lateralibus acutis: lamellis 3 carnosis denticulatis.——*Sylhet.*——Flowers white, scentless, obovate.
4. A. *bicolor* (Lindl. in Bot. Reg. t. 1730); labelli lobis lateralibus rotundato-cuneatis intermedio ovato scabrido lamellis basi 2 concavis.——*Ceylon.*——Flowers yellow, ovate, with crimson and purple tips.

52

OPHRYS cornuta.

Horned Ophrys.

GYNANDRIA *MONANDRIA.*

Nat. ord. ORCHIDACEÆ. § OPHREÆ — SERAPIADÆ. (ORCHIDS, *Vegetable Kingdom, p.* 173.)

OPHRYS. L.

§ 2. CORNUTÆ. *Labello basi utrinque cornuto.*—Lindl. Orch. p. 374.

O. *cornuta;* perigonii segmentis interioribus convolutis villosis columna rostrata duplo brevioribus, labelli segmento medio obovato subretuso breviter appendiculato convexo fusco-nigricante velutino, corniculis lateralibus in appendicem subulatam hamato-adscendentem elongatam productis.—*Grisebach, Fl. Rumel.* ii. *p.* 367.

O. cornuta, *Stev. mem. mosq.* 2. 174. *t.* 10. *Rchb. ic. Europ. t.* 1870. *Lindl. Orch. p.* 375.

O. bicornis, *Sadler in Linnæa,* 205. *litt.*

" O. œstrifera, *Expédition à la Morée, t.* 31. *f.* 1. *& t.* 32. *f.* 8." fide *Griseb.*

" O. œstrifera, *Wahlenb.*" Griseb.

This curious little plant has long been known as an inhabitant of the Crimea; and it had been recently discovered in Dalmatia by Baron Welden, in Hungary by Nentwich, and in Macedonia by Frivaldsky. But it remained for the Dean of Manchester to give it a more southern locality, by detecting in Corfu the specimens now figured.

The markings of the lip are very singular, but variable, as will be seen from the two sorts in our plate; and it would seem that this variation goes further, for Grisebach describes the Roumelian form with a greenish calyx, yellow spots in the middle of the lip, and blue horns.

The specimens represented in the accompanying plate flowered in Dr. Herbert's drawing-room in London.

53

BRASAVOLA Digbyana.

Mr. Digby's Brasavola.

GYNANDRIA *MONANDRIA.*

Nat. ord. ORCHIDACEÆ. § EPIDENDREÆ—LÆLIADÆ. (ORCHIDS, *Vegetable Kingdom, p.* 181.)

BRASAVOLA. Botanical Register, fol. 1465.

Br. *Digbyana;* foliis ovalibus planis carnosis glaucis, labello sessili cucullato cordato subtrilobo margine in crinis longis soluto in disco callo maximo aucto, dente postico subulato incumbente.

This very singular plant was introduced from Honduras by Mrs. McDonald, and by that lady given to Edward St. Vincent Digby, Esq. with whom it flowered last July, at Minterne, in Dorsetshire. Its huge yellowish white flowers are as sweet as those of Aerides odoratum; and the largest measure nearly three inches in diameter. The neck of the ovary, which is cuniculate in a remarkable degree, is full four inches long.

We have not seen the pollen-masses of the plant; but it is so much like Br. glauca in habit, that we entertain little doubt of their belonging to the same genus. There are, however, some peculiarities in the structure of this plant, which must not be lost sight of. Its anther-bed has no fringes or other process at the edge, but is deeply sunk and guarded behind by a long subulate tooth, which curves over the anther (fig. 1), and the stigma has three linear foveæ, which all open into one compressed stigmatic passage.

The singular fringe that borders the lip is quite analogous to what occurs in Br. cucullata and others, only it is here extremely extended, so as to give the flower quite a shaggy appearance.

The following list of the Brasavolas now known to Botanists may be useful to collectors; (‡ are not yet introduced).

A. Leaves flat.

B. glauca, *Lindl.*—Mexico.
B. Digbyana, *Lindl.*—Honduras.
‡B. Pumilio, *Rchb. fil.*—Mexico.

B. Leaves tapering; *lip toothed or fringed.*

B. cucullata, *R. Brown.*—West Indies.
B. appendiculata, *Richard & Galeotti.*—Mexico.
‡B. amazonica, *Pöppig & Endl.*—Peru.
B. angustata, *Lindl.*—Demerara.
B. Martiana, *Lindl.*—Brazil.
B. venosa, *Lindl.*—Honduras.

C. Leaves tapering; *lip quite even at the edge.*

B. grandiflora, *Lindl.*—Honduras, Guayaquil.
B. nodosa, *Lindl.*—West Indies.
B. cordata, *Lindl.*—Brazil.
‡B. subulifolia, *Lindl.*—St. Nevis.
B. tuberculata, *Hooker.*—Brazil.
B. Perrinii, *Lindl.*—Brazil.

57

PILUMNA laxa.

Loose-flowered Pilumna.

GYNANDRIA *MONANDRIA.*

Nat. ord. ORCHIDACEÆ. § VANDEÆ — BRASSIDÆ. (ORCHIDS, *Vegetable Kingdom, p.* 181.)

PILUMNA, Lindl. in Bot. Reg. 1844. *misc.* 74.—*Ovarium* tricostatum. *Sepala* et *petala* æqualia patula obliquè inserta. *Labellum* basi columnæ adnatum, subintegrum, unguiculatum, convolutum, inappendiculatum. *Columna* clavata, teres, clinandrio cucullo dentato membranaceo circumdato; buccis duabus carnosis semiclausum. *Stigma* verticale. *Pollinia* 2, posticè fissa, caudiculæ brevi et glandulæ ovatæ adnata.——Herbæ *epiphytæ*, pseudobulbis *vaginatis*, foliis *coriaceis*, pedunculis *radicalibus*.——Genus Aspasiæ proximum, clinandrio cucullato, columnâ tereti, nec non stigmate verticali nec faciali diversum.

P. *laxa* (Lindl. 1 c.); pseudobulbis tenuibus ancipitibus, folio oblongo subtus maculato, racemo laxo multifloro breviore, bracteis laxis cucullatis obtusis, sepalis petalisque lineari-lanceolatis, labello oblongo indiviso rotundato medio constricto per axin unilamellato.

This new and very distinct genus of Orchids has been hitherto known only from the account given of it in the miscellaneous matter of this work for 1844, at no. 74, where, speaking of the present species, it is remarked, that its general appearance may be understood by its having been mistaken for a Trichopilia. Its flowers are produced in loose erect racemes, out of broad obtuse short membranous spotted bracts. The stalks and ovary are an inch and half long; the latter with three very stout and strong ribs. The sepals and petals are a pale watery green, erect, linear-lanceolate, equal, faintly tinged with purple. The lip is cream-colour, rolled round the column at the base, to which it also is united at the lower end. The column has a singular fringed hood, overlying the anther, and a nearly vertical stigma, closed in by fleshy inflected cheeks. In these circumstances it differs from Aspasia, to which the genus is nearly akin.

We suspect that another species, allied to this, exists in collections, with green flowers; but we know too little of it to be able to define it. Our memoranda relating to it refer to no. 1680 of Mr. Loddiges' collection, which requires further examination. Pilumna laxa itself has very thin flat pseudobulbs, and leaves marked beneath with small purple spots.

Our drawing was made in the garden of the Horticultural Society in October, 1845. Fig. 1. represents the column and lip; 2. pollen-masses and gland; 3. a pollen-mass apart: the hairiness of these bodies represented by the artist, escaped our own observation.

The species, now represented, was found in the woods of Popayan by Mr. Hartweg.

It should be treated in the same manner as most of the other Orchids which are natives of Mexico and Guatemala. A cool stove, which is shaded from bright sunshine during the summer months, and where a damp atmosphere can be kept up, suits it very well. It should be potted in turfy peat, in a well-drained pot, and liberally supplied with water during the growing season. When the growth is perfected for the season less water should be given, and the plant allowed a period of rest.

Like all the other plants belonging to this tribe it is multiplied by dividing the pseudo-bulbs, and this is best done just before the spring growth commences.

58

CYPRIPEDIUM Irapeanum.

Irapean Lady's Slipper; or *Pelican flower.*

GYNANDRIA *MONANDRIA.*

Nat. ord. ORCHIDACEÆ. § CYPRIPEDEÆ. (ORCHIDS, *Vegetable Kingdom, p.* 181.)

CYPRIPEDIUM.—Supra, vol. 10. *fol.* 788.

C. *Irapeanum*; caule folioso piloso multifloro, foliis ovatis acuminatis amplexicaulibus, stamine sterili ovato acuminato subtrilobo, sepalis petalisque oblongis æqualibus basi barbatis antico apice bipartito labello obovato ore constricto brevioribus.

C. Irapeanum, *Llave & Lexarza Orch. Mexican.* 2. 10. *Lindl. gen. & sp. orch. p.* 528.

The annexed figure, taken from a plant belonging to Mr. Hugo Finck, in the temporary possession of the Horticultural Society, does scanty justice to this noble species. It is indeed a faithful representation of what appeared in the Society's garden; but we have now before us a specimen collected in Mexico near the town of Irapeo, which has two flowers more than twice the size open at the same time, and two more ready to expand.

It looks like a gigantic form of the downy yellow Lady's Slipper (C. *pubescens*) of the United States. Lexarza says that the Mechoacanians call it *Flor del Pelicano*, but why he does not state.

Fig. 1. represents the column seen in front.

This plant is, however, as nothing compared with a curious thing found in Caraccas by Mr. Linden, and of which a dried specimen is before us. That extraordinary species, the *Uropedium Lindenii* of the Orchidaceæ Lindeniaceæ,* has flowers from fifteen to twenty inches long; that is to say, the petals and lip are drawn out into narrow straps or tails of such a length. The habit of Uropedium is that of Cypripedium venustum. It grows in the little woods of the savannah which occurs on the high part of the Cordillera, looking down upon the vast forests that occupy the lake of Maracaybo. This savannah is situated in the territory of the Chiguará Indians, at an elevation of 3500 feet above the sea. Some of our enterprising merchants might surely obtain this prodigy. It flowers in June.

In soil and potting Cypripedium Irapeanum requires the same treatment as our common English and American species, but it is more tender than they are, and must be kept in a warm greenhouse. During the summer months, when it makes its growth and flowers, it requires a liberal supply of water. When the growth is perfected, and the stems have died down, the plant must have its season of rest, and at this period it should be kept rather dry. In the summer months, when growing, it should always be shaded from bright sunshine.

It is propagated by dividing the roots when the plant is in a state of rest.

* Now in the press.

59

VANDA Batemanni.

Crimson and Yellow Vanda.

GYNANDRIA *MONANDRIA.*

Nat. ord. ORCHIDACEÆ. (ORCHIDS, *Vegetable Kingdom, p.* 173.)
VANDA. *Supra, fol.* 506.

V. *Batemanni* (V. *lissochiloides*, Lindl. gen. & sp. orch. p. 216; *Fieldia lissochiloides*, Gaudich. voyage, 424. t. 36); radicibus crassissimis, foliis distichis coriaceis oblique emarginatis obtusis racemo laterali multifloro brevioribus, bracteis coriaceis cucullatis squamæformibus, floribus maximis planis coriaceis, alabastris globosis, sepalis obovato-cuneiformibus obtusis, labello triangulari basi saccato lobis ascendentibus acutis apice carnoso sulcato uncinato dente elevato in medio et cristâ brevi transversâ juxta basin.

The honour of discovering this splendid thing is due to M. Gaudichaud, who met with it in the Moluccas; of introducing it to this country, to Mr. Cuming, who sent it from the Philippines; of first flowering it to Mr. Bateman, with whom it produced its magnificent sceptre in the stove at Biddulph Grange, in June and July last.

It is a very large erect plant, with remarkably thick aerial roots, produced after the custom of its kindred, sword-shaped curved two ranked hard leaves averaging two feet in length, and a still longer spike of some score of flowers, each full two inches and a half across, flat, leathery, and long enduring. But it is not alone for their size that these flowers are so especially worthy of notice. Their colour is indescribably beautiful. If you look at them in face, they are the richest golden yellow, spotted all over with crimson; but when seen from behind, they are wholly a vivid purple, fading away at the edges into the violet of Cereus speciosissimus. So that, regard them which way you will, there is nothing but the gayest and richest colours to be seen.

What could possibly have led M. Gaudichaud to compare this Epiphyte with the terrestrial *Lissochilus* we are unable to imagine, for there is only the slenderest resemblance between the two. His name is, however, on record, and the strict rules of Botanical nomenclature seem to forbid its change. But, as has often been observed, *Summum jus summa injuria;* and it is so manifestly absurd to retain for a plant a name that has originated in some misconception, that we venture, for once, to disregard rules for the sake of common sense. In doing so, we cannot but associate with this noble plant a gentleman whose knowledge of the order, and whose skill in the cultivation of it, have gained him a deathless name.

Fig. 1. represents the lip; 2. the pollen-masses, and their apparatus.

Sub t. 64 (1847)

NEW GARDEN PLANTS.

ZYGOPETALUM *tricolor;* foliis parvis gramineis scapo racemoso 7-8 floro multo brevioribus, labello subrotundo concavo medio constricto apice lunato medio bituberculato.
Guayana, Mr. Loddiges (931*).

This has the smallest flowers of any species yet known. They are pale green with a white lip, banded with broken lines of crimson.

EPIDENDRUM (Encyclium) *subaquilum;* pseudobulbis ovatis cæspitosis, foliis linearibus obliquè emarginatis obtusis scapo glabro filiformi subpaniculato æqualibus, sepalis petalisque linearibus dorsali ascendente lateralibus parallelis, labello semilibero ovato-lanceolatis indiviso utrinque emarginato lamellis 3 spongiosis.
Mexico or *Guatemala*, Mr. Warner.

A small slender plant, with leaves seven or eight inches long, and barely three lines wide. Flowers the size of *E. diffusum*, pale dull brown.

DENDROBIUM (Onychium) *triadenium;* radicibus villosis, caulibus elongatis teretibus ramosis basi fusiformibus, foliis ovato-oblongis obtusis, paniculâ brevi terminali confertiflorâ racemosâ, sepalis ovatis acutis petalis labelloque oblongis undulatis rotundatis emarginatis hoc glabro utrinque unidentato medio tuberculo tricrenato acuto, mento brevi obtuso, columnâ apice dentatâ, stigmate villoso.
East Indies, S. Rucker, Esq.

This is a lovely plant, with the habit of *Dendrobium crumenatum*, but with a very close racemose panicle of flowers. They are transparent, about as large as in *D. aduncum*, nearly white, with a tinge of rose, a violet spot on the end of the petals and lip, and a 3-lobed yellow tubercle in the middle of the latter. What seems to be a variety, with colourless flowers, has also come under my observation.

69

CŒLOGYNE ochracea.

Ochre-spotted Cœlogyne.

GYNANDRIA *MONANDRIA.*

Nat. ord. ORCHIDACEÆ. § EPIDENDREÆ. (ORCHIDS, *Vegetable Kingdom, p.* 173.)
CŒLOGYNE.—Botanical Register, vol. 11. *fol.* 868.

C. *ochracea;* pseudobulbis oblongis apice tetragonis, foliis angustè lanceolatis obscurè 5-nerviis in petiolum angustatis racemo paucifloro apice nutante longioribus, labelli trilobi intus pubescentis lobis lateralibus rotundatis intermedio ovato acuminato obtuso sinu denticulato integrove lamellis 2 rectiusculis integris cis medium apicem evanescentibus lineâ brevi elevatâ apice dentiferâ prope basin interjectâ.

The specimen from which the annexed figure was taken was supplied by Thomas Brockelhurst, Esq., of the Fence, near Macclesfield, in April last. Its pure white flowers, with bright orange yellow blotches on the lip, are very pretty; and they are, moreover, extremely sweet-scented.

We possess the plant from several localities among the late Mr. Griffith's valuable Indian collection. Darjeeling, Bootan, and the Mishmee hills all produce it. With the Mishmee specimens we have a note from its lamented discoverer, stating that it is common on Thumathaya and Laimplang-thaya. These specimens differ a little among each other in regard to the amount of toothing present at the sinus of the lip, and as to its exact form; but they all belong evidently to the same species.

This plant must be cultivated in the damp stove or Orchideous House. Like most of such plants, it may either be grown in a pot, or fixed upon a block of wood, and suspended from the rafters of the house. The chief point to be attended to is to rest it judiciously, by keeping it cool and dry, after the growth is completed for the season.

It is multiplied in the usual manner, by dividing the pseudo-bulbs.

EDWARDS'S

BOTANICAL REGISTER:

OR,

ORNAMENTAL FLOWER-GARDEN

AND SHRUBBERY:

CONSISTING OF

COLOURED FIGURES OF PLANTS AND SHRUBS,

CULTIVATED IN BRITISH GARDENS;

ACCOMPANIED BY THEIR

History, Best Method of Treatment in Cultivation, Propagation, &c.

CONTINUED

BY JOHN LINDLEY, Ph. D. F.R.S. AND L.S.

PROFESSOR OF BOTANY IN UNIVERSITY COLLEGE, LONDON,
AND THE ROYAL INSTITUTION OF GREAT BRITAIN,
VICE-SECRETARY OF THE HORTICULTURAL SOCIETY,
&c. &c. &c.

1847.

OR VOL. XXXIII. OF THE ENTIRE WORK,
OR VOL. XX. OF THE NEW SERIES.

——viret semper——nec fronde caducâ
Carpitur.

LONDON:
JAMES RIDGWAY, PICCADILLY.

M.DCCC.XLVII.

1

DENDROBIUM triadenium.

Three-knobbed Dendrobe.

GYNANDRIA *MONANDRIA.*

Nat. ord. ORCHIDACEÆ. (ORCHIDS, *Vegetable Kingdom*, *p.* 173.)
DENDROBIUM.—Swartz.

D. (Onychium) *triadenium* (Lindl. in Bot. Reg. 1846, sub. t. 64); radicibus villosis, caulibus elongatis teretibus ramosis basi fusiformibus, foliis ovato-oblongis obtusis, paniculâ brevi terminali confertiflorâ racemosâ, sepalis ovatis acutis, petalis labelloque oblongis undulatis rotundatis emarginatis hoc glabro utrinque unidentato medio tuberculo tricrenato aucto, mento brevi obtuso, columnâ apice dentatâ, stigmate villoso.

In characterizing this species, a few weeks since, we spoke of it as a lovely plant, with the habit of *Dendrobium crumenatum*, but with a very close racemose panicle of flowers, transparent, about as large as in *D. aduncum*, nearly white, with a tinge of rose, a violet spot on the end of the petals and lip, and a 3-lobed yellow tubercle in the middle of the latter.

The accompanying figure will shew that the character thus given of it was merited, and it will also serve to bring to the recollection of many the beautiful specimens of it produced in Mr. Rucker's noble collection, as well as a spotless variety, hardly inferior, though different, which has been obtained by Messrs. Rollissons.

From what part of the East Indies it has been procured is uncertain; but as the focus of the Dendrobes with a thickened base to their stems is the Indian Archipelago, it is probable that this has been derived from that quarter.

The three yellow knobs of the lip, the close inflorescence, and the deeply lobed, almost quadrangular petals, distinctly mark the species.

Fig. 1. represents the column; 2. the lip, with its triple knob or gland.

8

CALANTHE curculigoides.

Yellow-spiked Calanthe.

GYNANDRIA *MONANDRIA.*

Nat. ord. ORCHIDACEÆ. (ORCHIDS, *Vegetable Kingdom*, *p.* 173.)
CALANTHE, supra vol. 1844, *t.* 37.

C. *curculigoides* (Wall. Cat. no. 7340. Lindl. Gen. & Sp. Orch. no. 13.); foliis oblongis glabris scapo crasso duplo longioribus, racemo denso cylindraceo glabro, bracteis membranaceis deflexis cito deciduis, labello hastato lobis lateralibus brevibus obtusis intermedio lanceolato cis apicem dilatato, calcare uncinato.

A *bright* yellow Calanthe is an unexpected novelty; and this species, which was previously known only from bad dried specimens, proves to be one of the handsomest of its race. The flowers have a firm waxy texture, and do not wither so soon as those of some species; they are much yellower and finer than in C. densiflora; all their beauty fades, however, in drying, when the species becomes one of the least attractive. The specimen now figured, was the best we have seen in a living state: but it is nothing compared to what it will become. We have now before us wild specimens, collected by the late Mr. Griffith at Malacca, with ten inches of flowers, and a further part of the inflorescence is lost. It is not too much to say that the flowering spike of C. curculigoides may be expected to be a foot long.

The species has been found exclusively in the Straits of Malacca, whence Messrs. Loddiges received it. It flowered at Hackney, in November, 1845, and may be expected to require more heat than some of the sorts.

Fig. 1. represents the lip, column, and spur; the remarkable hook of the latter is one of the characteristic marks of the species.

Sub t. 9 (1847)

NEW GARDEN PLANTS.

ERIOPSIS BILOBA.

This new genus belongs to the Maxillarids, among Vandeous Orchids, but has so much the habit of Eria, when not in flower, that it may be mistaken for it. It has large plaited leaves placed two or three together, upon the summit of a fleshy oblong stem, and it throws up from the base a long spike of gay orange-coloured blossoms. The history of its introduction is unknown. It has been flowered by J. J. Blandy, Esq., of Reading, who, in September last, favoured us with a specimen; it had been acquired by him among the many rare species forming the late Mr. Barker's collection, which Mr. Blandy purchased, and is possibly some western plant. We do not, however, find any trace of it in books, or among our own unarranged collections. In a short time we shall give a figure of it, and in the meanwhile propose the following as its technical character.

ERIOPSIS (ORCHIDACEÆ, Vandeæ-Maxillaridæ). Caules succulenti, apice tantum foliati. Racemus radicalis multiflorus. Bracteæ minimæ. Flores explanati: laciniis subæqualibus oblongis obtusis, mento brevi obtuso. Labellum anticum, concavum, trilobum, disco lamellatum, c. basi productâ columnæ articulatum. Columna semiteres, clavata, aptera; anthera oblonga, subunilocularis; pollinia 4, inæqualia, per paria filis 2 elasticis affixa; glandula submembranacea quadrata.

Sub t. 10 (1847)

New Garden Plants.

EPIDENDRUM pyriforme.

E. *pyriforme* (Encyclium: labello tripartito: lobo medio rotundato) pseudobulbis obpyriformibus aggregatis diphyllis, foliis coriaceis lanceolatis acutis scapo subbifloro longioribus, sepalis petalisque lanceolatis acutis, labelli lobis lateralibus obtusis integerrimis intermedio subrotundo glabro picto callis 2 maximis apice subliberis in unguem.

A very pretty little species, imported from Cuba by Messrs. Loddiges, with whom it flowered the other day. The leaves are about four inches long, on little pseudo-bulbs, which look like inverted pears. Notwithstanding its diminutive stature, the flowers are fully $2\frac{1}{2}$ inches in diameter, with reddish yellow sepals and petals, and a pale straw-coloured lip veined with crimson.

EPIDENDRUM plicatum.

E. *plicatum* (Encyclium: labello tripartito: lobo medio acuto) pseudobulbis ovatis diphyllis, foliis ensiformibus acutis racemo brevioribus, sepalis petalisque lanceolatis cuspidato-acuminatis, labelli lobis lateralibus oblongo-lanceolatis undulatis basi elevatovenosis intermedio cordato transverso plicato crenato cuspidato, callis 2 maximis carinatis acutis apice subliberis in unguem.

This also is a Cuba species from Messrs. Loddiges, with whom it flowered in January. It takes rank with E. phœniceum, and Hanburii, but is quite different in the form of its flowers. The plaited crenelled lip, with a long cusp, is quite peculiar; it is of a rich crimson, as are the petals *at the back;* but the sepals and inside of the petals are green, the former just warmed and the latter richly spotted with crimson.

17

GONGORA bufonia; *var.* leucochila.

White-lipped, toad-skinned Gongora.

GYNANDRIA *MONANDRIA.*

Nat. ord. Orchidaceæ. § Vandeæ.—Maxillaridæ. (Orchids. *Vegetable Kingdom, p.* 173.)

GONGORA.—Bot. Reg. vol. 19. *fol.* 1616.

G. *bufonia* (Lindl. in Bot. Reg. 1841. t. 2.); hypochili convex aristis longiore cornubus lateralibus papillæformibus, epichilio acuto a latere subtriangulari.

Var. *leucochila;* floribus pallide purpurascentibus labello candido apice fulvo.

We have had this drawing in our portfolio since April, 1844, when a specimen was communicated by S. Rucker, Esq. It is no doubt a pale variety of G. bufonia, of the original spotted form of which a figure will be found at fol. 2 of our volume for 1841.

The cultivation of this genus has become so generally understood, and its grotesque forms so familiar, that we believe the most acceptable service we can now render our readers, is to point out in what way the numerous varieties may be reduced to their species. With this view the following memoranda have been prepared:—

The best marks of distinction reside in the form of the little horns which stand right and left of the base of the lip. These are long and enlarged at their end, in some species; small, and resembling a mere callosity, in others; and missing in another. Attention to them will render further differences more intelligible.

I. *Horns of the base of the lip long, and enlarged at the point.*

1. G. *atropurpurea.* In this, which is now one of the commonest species, in addition to the deep claret purple of the flowers, the base of the lip is very much narrowed, and the lateral horns are as long as its stalk; hence it was stated, by Sir W. Hooker, to have four horns, which is correct if the bristles are included.

2. G. *maculata.* In this species, the horns are short and thick, and placed so much at right angles to the lip, as to give it a heart-shaped appearance. It never has the wine-coloured flowers of the last; but the flowers are in all cases more or less yellow, and generally gaily speckled with cinnamon brown. G. *fulva* of this work, t. 51, 1839, is only one of its numerous varieties.

II. *Horns of the base of the lip only little round callosities.*

3. G. *nigrita.* With the colour of G. atropurpurea, but still deeper, this has quite a different lip, not slender at the base, but rounded like G. maculata, and its horns are quite small; the upper part of its lip is singularly slender and bristle-pointed.

4. G. *bufonia.* Here, on the contrary, not only is the deep wine-coloured hue missing, but the lower half of the lip is much larger, and the upper half far broader. All the pale purple Gongoras, and the white kind, seem to be referable to this species.

III. *Horns of the base of the lip entirely absent.*

5. G. *truncata.* Perfectly distinct from all the preceding in the form of its flower buds, which resemble a bean. The bristles, usually so conspicuous, are also almost absent. It looks very like the commencement of a change from Gongora to such a genus as Acropera.

* Uncertain species.

† 6. G. *Galeottiana.* The published character of this conveys no information which enables us even to judge to what species it is allied. Its country (Mexico) leads to the supposition that it may be G. truncata.

† 7. G. *quinquenervis.* This is also indeterminable. It is said to have yellowish purple flowers, and comes from Peru. We have here nothing that can be supposed to belong to it. The only Peruvian species, yet known, is G. maculata, var. tricolor (B. Reg. 1844, misc. 30), but that plant has no purple in its flowers.

18

ERIOPSIS biloba.

Two-lobed Eriopsis.

GYNANDRIA *MONANDRIA.*

Nat. ord. Orchidaceæ. § Vandeæ.—Maxillaridæ. (Orchids, *Vegetable Kingdom, p.* 181.)

ERIOPSIS, Lindl.—Caules succulenti, apice tantum foliati. *Racemus* radicalis multiflorus. *Bracteæ* minimæ. *Flores* explanati: laciniis subæqualibus oblongis obtusis, mento brevi obtuso. *Labellum* anticum, concavum, trilobum, disco lamellatum, c. basi productâ columnæ articulatum. *Columna* semiteres, clavata, aptera; *anthera* oblonga, subunilocularis; *pollinia* 4, inæqualia, per paria filis 2 elasticis affixa; *glandula* submembranacea quadrata.—*Lindl. in Bot. Reg. sub fol.* 9. 1847.

Eriopsis *biloba*, Lindl. supra.

"This new genus belongs to the Maxillarids, among Vandeous Orchids, but has so much the habit of an Eria, when not in flower, that it may be mistaken for it. It has large plaited leaves placed two or three together, upon the summit of a fleshy oblong stem, and it throws up from the base a long spike of gay orange-coloured blossoms. The history of its introduction is unknown. It has been flowered by J. J. Blandy, Esq., of Reading, who, in September last, favoured us with a specimen; it had been acquired by him among the many rare species forming the late Mr. Barker's collection, which Mr. Blandy purchased, and is possibly some western plant. We do not, however, find any trace of it in books, or among our own unarranged collections."

It was thus we spoke of this beautiful plant in February last. The present figure will serve to complete what is known of its history. Its genus is perfectly distinct, whether we regard the short chin or the almost quadrangular gland, or the peculiar lamellæ of the lip. Its nearest affinity is with the racemose Maxillarias, readily distinguished however by their crescent-shaped gland, and tubercular (not lamellated) lip with a long chin.

Warrea has globose half-closed flowers, and a *linear caudicle.*

The plates found on the surface of the lip are peculiar to it. There is no middle plate, beyond a slightly elevated line; on each side of the middle line stands a short plate, truncated anteriorly; outside of these, on each side, is another much larger plate, rounded, with two teeth at the anterior base; and immediately before these stands another pair of fine teeth, just as is represented at fig. 1.

Fig. 2. represents the pollen-masses elevated above the gland by stretching their elastic threads forcibly. In a natural state, these pollen-masses are perfectly sessile on the gland.

New Garden Plant.

Sarcochilus *fuscoluteus;* foliis distichis coriaceis apice rotundatis obliquè bilobis, spicis densis conicis, sepalo dorsali petalisque angustioribus lanceolatis acutis lateralibus apice rotundatis c. labello omninò connatis, labelli trilobi apice scrotiformis laciniis lateralibus acutis erectis intermedio oblongo obtusissimo carnoso in discum cornuto dentibusque aliis validis in fundo.

Imported from Borneo by Mr. Lowe, and communicated by F. G. Cox, Esq. of Stockwell. I have only seen one perfect flower, which was bright yellow tipped with tawny, and nearly three-quarters of an inch across in the principal diameter. The flowers grow in close conical spikes. The leaves are about four inches long, and three-quarters broad.

Sub t. 19 (1847)

New Garden Plant.

Angræcum *virens;* foliis latis distichis . . . . , labello subrotundo apice convoluto cuspidato lineâ in medio elevatâ acuminatâ latâ planâ, calcare horizontali acuminato recto ferè duplo longiore.

This is very like a small state of Angræcum eburneum, but Mr. Loddiges regards it as distinct; and in the flatness of the median ridge of the lip, in the smaller size of all the parts it seems to be distinguishable. It is, however, published from very imperfect materials, and chiefly with a view to draw attention to its locality, which is said to be Serampore; but whether it is really a native of the continent of India, or a plant received from the old Botanical Garden of that Settlement, as is more probable, I do not know.

Dendrobium (Dendrocoryne) *chrysotoxum;* pseudobulbis clavatis angustis multicostatis 2-4-foliis, foliis oblongis horizontalibus coriaceis, racemis lateralibus laxis gracilibus arcuatim decurvis pseudobulbis æqualibus, bracteâ basilari parvâ spathaceâ floralibus minimis herbaceis, sepalis petalisque explanatis oblongis obtusissimis planis his duplò latioribus, labello indiviso cucullato rotundato pubescente margine minutissime pectinato et fimbriato.

An importation from India by Messrs. Hendersons, of very great beauty. It belongs to the herd of Dendrobes, of which D. *densiflorum*, *Griffithii*, and *aggregatum* are representatives and it nearly approaches the latter; but it has an entirely different lip, most delicately and curiously broken up into innumerable minute fringed teeth.

20

ODONTOGLOSSUM Warneri; *var.* purpuratum.

Mr. Warner's Tooth-tongue.

GYNANDRIA *MONANDRIA.*

Nat. ord. Orchidaceæ. § Vandeæ—Brassidæ. (Orchids, *Vegetable Kingdom, p.* 181.)

ODONTOGLOSSUM.—Bot. Reg. 1839. *t.* 48.

Sect. Xanthochilum, or Stained lipped species.

a. The ears of the column rounded, broader than long, or altogether absent.

O. *Warneri;* pseudobulbis ovatis ancipitibus subangulatis apice elongatis diphyllis, foliis patulis lineari-lanceolatis racemo paucifloro brevioribus, bracteis minimis, sepalis ovalibus patentissimis, petalis paulò angustioribus obtusis ascendentibus, labelli trilobi plani laciniâ intermediâ cuneatâ bilobâ rotundatâ lateralibus subquadratis, tuberculo disci simplici, columnâ elongatâ apterâ.—*Lindl. in Bot. Reg.* 1845. *misc. p.* 54.

α. sordidum; petalis sepalisque lutescentibus purpureo striatis.

β. purpuratum; petalis sepalisque albis purpureo læto striatis.

The original of this dwarf species is no doubt a native of some part of the tropical regions of America, but we have not learned its exact history. It first appeared in public at a meeting in the Garden of the Horticultural Society, in May, 1845, when it was exhibited by Mr. C. B. Warner, whose name it bears. The present variety was obtained from Mexico by Messrs. Loddiges, whose No. 1449* it is. A healthy mass of it is very handsome, the yellow of the lip and purple stripes of the sepals and petals being deep and clear.

It is one of those plants which, although they may be called crop-eared, have not lost their auricular organs by any surgical process; their ears have naturally disappeared, not leaving even a stump behind. The group of species with which it is associated exhibits indeed every gradation from fine well developed ears in O. *lacerum* to their total absence in O. Warneri. In the specimen now represented there was not more than five flowers in a raceme; but when stronger, it will bear from seven to eight.

The original variety, although pretty, is by no means equal to that now represented, which flowered with Messrs. Loddiges in January, 1846. The former had the sepals and petals stained with pale dull yellow or buff, which destroyed their brilliancy: this, on the contrary, has the ground of the sepals and petals clear white.

New Garden Plants.

Acriopsis *densiflora;* pseudobulbis ovalibus, racemis multifloris cylindraceis pedicellis internodiis longioribus, labello pubescente panduriformi laciniâ media rotundatâ subemarginatâ lateralibus acutis: dente elevato fisso in medio, clinandrii cucullo integerrimo rotundato.

A very curious Orchidaceous plant, imported from Borneo by Mr. Lowe, and communicated by Mr. Conrad Loddiges. Unlike the other species, this has its flowers compactly arranged in racemes about two inches long. The former are small, but delicately marked with brown upon yellowish green, and the lip is rose-colour bordered with white.

Brassavola *retusa;* labello undulato cuneato *truncato* margine lævi.

Only known from a specimen sent to the Gardener's Chronicle, March 5, 1847, for a name, and purporting to come from Maracaybo. The form of the lip is quite unlike any other species; it is white, with a little green near the base.

23

CŒLOGYNE speciosa.

Showy Cœlogyne.

GYNANDRIA *MONANDRIA.*

Nat. ord. ORCHIDACEÆ. § EPIDENDREÆ. (ORCHIDS, *Vegetable Kingdom, p.* 181.)

CŒLOGYNE.—Botanical Register, vol. 11. *fol.* 868.

C. *speciosa;* pseudobulbis ovato-oblongis costatis monophyllis, foliis oblongo-lanceolatis 5-7-nerviis, pedunculis unifloris squamatis pseudobulbis parum longioribus, sepalis oblongis, petalis linearibus reflexis, labelli trilobi laciniis lateralibus anticè denticulatis intermediâ bilobâ rotundatâ cristis duabus muricatis crassis sub apice ipso labelli confluentibus tertiâ brevi tenui basilari interjectâ, columnæ margine latissimo apice truncato dentato.

C. speciosa, *Lindl. gen. & sp. Orch. p.* 39.

The species of Cœlogyne now figured was sent from Java to Messrs. Veitch of Exeter, by their collector, Mr. Thomas Lobb, and by them communicated, in October, 1846, as a free growing, free flowering plant.

It is certainly the *Chelonanthera speciosa* of Dr. Blume, notwithstanding his statement that his plant has only two crests upon the lip, and his reference to the *Angræcum nervosum* of Rumphius; for the figure given by the former author in his Tabellen has three crests, and it is clear that Rumphius's figure is intended for quite a different plant, with a terminal, not radical, inflorescence.

The pale tawny sepals and petals, and pitch-brown lip of this plant detract from its beauty, notwithstanding that they are relieved by a broad white column, and a pure white termination to the lip, and that its flowers are nearly four inches in diameter when fully expanded. They have, however, a very singular appearance, and if they are produced in greater numbers will be attractive so long as the white remains unchanged.

Dr. Blume found it on trees on Mount Salak, in Java.

Fig. 1. represents the lip of the natural size, seen from above. Its lateral crests are unlike any that have been before remarked in the genus, each consisting of a broad soft spongy flat *skinless* elevation, from whose surface various minute points rise up in abundance; near the point of the lip, however, they become confluent, and resume the skin which is usually found on such processes.

Sub t. 24 (1847)

NEW GARDEN PLANT.

CATASETUM SERRATUM.

C. *serratum;* labello carnoso galeato resupinato subcompresso postice sensim saccato apice integerrimo emarginato lateribus latè sejunctis argute denticulatis, cirrhis haud columnâ longioribus.

Nearly allied to C. *maculatum*, from which it differs in the sides of the lip not curving inwards, in the shortness of the columnal cirrhi, and in the emarginate apex of the lip. The flowers are of nearly the same size, but not spotted; they are pale green, with the convexity of the lip yellow. Sent from Panama to Messrs. Veitch by Mr. Lobb. We received it from Exeter in September, 1845, with a statement that it is "a most abundant bloomer, having long spikes with from ten to twelve flowers on each spike, and powerfully scented."

25

OPHRYS fuciflora.

Painted-lipped Ophrys.

GYNANDRIA *MONANDRIA.*

Nat. ord. ORCHIDACEÆ. § SERAPIADÆ. (ORCHIDS, *Vegetable Kingdom, p.* 173.)

OPHRYS.—L.

O. *fuciflora;* labello obovato-triangulari velutino indiviso basi bigibboso apice appendice rhombeo inflexo, petalis convolutis columnam brevirostrem subæquantibus velutinis.—*Lindl. Gen. & Sp. Orch.* 376.

O. Arachnites, *Eng. Bot. Supp. t.* 2596.

O. Crabronifera, *Sebast. & Mauri Rom. Pl. Cent.* 13. *t.* 2. *f.* 1.

O. exaltata, *Tenore Append. alt.* 83. fide Rchb.

O. apiculata, *Rich. Orch. Eur.* 33. sec. descr.

O. œstrifera, *Rchb. fl. Excurs.* 1. 128. nec Stev.

O. fuciflora, *Hall. ic. Helv. t.* 24. *f.* 2, 3. *Rchb. ic. Eur. t.* 868, 869.

These plants were drawn from specimens collected on the Continent by the Hon. and Rev. the Dean of Manchester. Fig. 1. and 2. were found in the meadows of Zaule, near Trieste; 3. was from the neighbourhood of the Lake of Thun.

In the confusion that exists among the curious little plants which constitute the genus Ophrys, we cannot pretend to trace the history of any one species to its *antiqua fons.* That is a special labour which we commend to those who have more leisure and patience. We can only say, that all three of these plants are the same as O. *fuciflora*, which has been miscalled O. *arachnites* in English Botany. The true Arachnites, which is awkwardly, but not inexactly, figured in the Botanical Magazine, t. 2516, has no hair on its lip.

Our English Spider Orchis appears to be in reality much nearer O. *fuciflora* than was suspected, for instead of wanting the elevated processes which rise from the lip of that species, it equally produces them, as has been pointed out to us by the Dean of Manchester, and as is shewn in our fig. 4. taken from a plant gathered near Dover. But the processes are in a different place; and among the many specimens of these plants now before us we find no variation from what we take to be the essential distinction between them—namely, the form of the lip, and the appendage at its point. In O. *aranei-fera* the lip is always oval and emarginate, while in O. *fuciflora* it is constantly wedge-shaped and furnished with a fleshy lobe at its point, which circumstance gave rise to Richards' name of *apiculata.*

NEW GARDEN PLANT.

DENDROBIUM VEITCHIANUM.

D. (Dendrocoryne) *Veitchianum;* caulibus clavatis angustè costatis 2-3-phyllis, foliis sessilibus oblongis apice acutis inæquilateris, racemo erecto multifloro foliis longiore, bracteis ovatis membranaceis acutis, pedicellis sepalisque ovatis acutis extùs villosissimis, petalis unguiculatis spathulatis acutis tenuibus glabris, labello maximo alte trilobo, denticulis tribus in medio, laciniis lateralibus ascendentibus rotundatis crenatis flabellatim venosis intermediâ convexâ rotundatâ bilobâ.

We are not sure that this will not prove a rival to the best of the East Indian Orchids. It is a most beautiful plant, with upright racemes of large mossy flowers, nearly two inches in diameter, and a fine stiff habit. Its sepals are a clear dull yellow, the petals pure white, the lip deep-green bordered with white, and richly marked with crimson veins. It is from Java, and is in the possession of Messrs. Veitch, to whom we are indebted for the materials from which the above description has been drawn up. This, too, is one of Mr. Thomas Lobb's fine things.

Sub t. 26 (1847)

New Garden Plants.

Saccolabium miniatum.

S. *miniatum;* racemis brevibus erectis cylindraceis, bracteis minimis acutis, sepalis petalisque ovatis acutis patulis, labello lineari obtuso recurvo calcare recto pendulo tereti breviore, polliniis albis.

A Java plant, imported by Messrs. Veitch, and flowered by both Mr. Rucker and Mr. C. B. Warner. It is not to be traced among Blume's species, and seems new. Its flowers, of a gay vermilion, or rich apricot colour, although somewhat small, have an extremely lively effect; they grow in upright racemes, about ten together. We have not seen the foliage.

Renanthera matutina.

Lindl. gen. & sp. Orch. p. 218.

We find that we were wrong in referring to this species the plant figured in our volume for 1843, t. 41. That species we find in Sir W. Hooker's Herbarium, under the name of R. *moluccana* of Blume. We now know that the true R. *matutina* is a much finer plant, with petals more than an inch long, it having been sent from Java, by Mr. Thomas Lobb, to Messrs. Veitch, who have it for sale. It produces a large branching panicle, each of whose arms is from six to nine inches long, and carries from six to twelve flowers.

Sub t. 27 (1847)

New Garden Plant.

Telipogon obovatus.

T. *obovatus;* rachi flexuosâ alatâ, bracteis ovatis acutis falcato-cucullatis, petalis oblongis acutis, labello duplo majore obovato rotundato.

A curious plant, sent from Peru by Mr. Lobb to Messrs. Veitch, by whom it has been sold. It has not yet flowered; but it will prove a very nice plant when it does blossom. Its flower-stem is six inches high; the flowers are bright yellow, an inch and three-quarters in diameter. It is near T. latifolius, but its flowers are larger, and its lip, instead of being shaped like the petals, is larger, and quite round at the point.

29

BRASSIA brachiata.

Long-armed Brassia.

GYNANDRIA *MONANDRIA.*

Nat. ord. Orchidaceæ, § Brassidæ. (Orchids, *Vegetable Kingdom*, *p.* 173.)

BRASSIA.—R. Brown.

B. *brachiata* (Lindl. in Plant. Hartw. p. 94.); pseudobulbis oblongis angustis compressis diphyllis, foliis obtusis racemo multifloro brevioribus, bracteis patentibus squamæformibus, sepalis petalisque brevioribus linearibus acuminatis longissimis, labello basi rotundato sursum subrhombeo acuminato undulato verrucis planis coloratis maculato, lamellis basi dentiformibus lineâ elevatâ per medium labellum.

This beautiful plant was originally defined from a dried specimen, collected near the Hacienda de la Laguna, in Guatemala, by Mr. Hartweg. Since its introduction, it has occasionally flowered, as with Messrs. Rollissons, and with Mr. Bateman, whose plant is now represented, but it remains a comparatively scarce species. It is far handsomer than either B. *guttata* (aliàs *Wrayæ*), or *verrucosa*, and its flowers are many times larger. The only species that really can vie with it in beauty is *Br. macrostachya.*

Like B. *verrucosa*, this species has the singular peculiarity of bearing upon its lip, numerous flat warts, to which the green colour of that part is exclusively confined. It might be as well to examine their structure, with a view to determining their nature and office.

This fine epiphyte is best grown elevated on a pot, well drained, and in a mixture of the fibre out of rough peat, and half decayed leaves, in equal portions. When the plant has completed its season's growth, it should be allowed a season of rest, which is best attained by withholding moisture, first from the atmosphere, and afterwards from the roots, and by reducing the temperature of the house, never at any time allowing fire heat to rise above fifty-eight or sixty degrees, particularly late in the autumn. Indeed, it cannot be too constantly borne in mind, that high temperature, and abundance of moisture, more especially about the roots, are very unsuitable to nearly all the western Orchids.

It is increased by dividing the plant when in a state of rest. It flowers at different seasons of the year, according to the time of resting, and the length of repose allowed.

New Garden Plant.

Bolbophyllum Lobii.

B. *Lobii;* folio petiolato obovato-oblongo coriaceo, pedunculo unifloro folio breviore, pedunculo nudo unifloro folio breviore basi subglanduloso e bracteis squamæformibus cucullatis falcatis subglandulosis erumpente, sepalis oblongis acutis lateralibus falcatis, petalis conformibus minoribus reflexis, labello longè unguiculato cordato ovato acuto canaliculato apice recurvo.

How fine a plant of its kind this is, may be surmised by its having been taken for a Cœlogyne. The flowers are yellow, shaded with cinnamon, spotted with light brown, and speckled outside with brown purple. They are full four inches across. We know of no species of the genus comparable for beauty. Mr. Thomas Lobb sent it from Java to Messrs. Veitch, who have it for sale. We find it in Sir W. Hooker's Herbarium, marked *Sestochilus*, but we do not know on what authority it is referred to that supposed genus. We have named this fine plant after Mr. Thomas Lobb, whose zeal and ability, as a botanical collector, are beyond all praise, and whose dried specimens are unrivalled for beauty, and admirable selection.

30

VANDA violacea.

Violet Vanda.

GYNANDRIA *MONANDRIA.*

Nat. ord. Orchidaceæ. § Vandeæ. (Orchids, *Vegetable Kingdom*, *p.* 173.)

VANDA.—R. Brown.

V. *violacea* (Lindl. in Bot. Reg. 1841. misc. 32.); foliis canaliculatis obliquè abscissis rotundatis, racemis multifloris densis pedunculatis pendulis, sepalis obovato-oblongis obtusis planis incurvis, petalis angustioribus, labello oblongo apiculato plano: venis 5 crassis elevatis, sacco intus pubescente.

Although it is nearly six years since we first became acquainted with this species, it was only in February, 1846, that we had an opportunity of having it drawn. It then flowered in great perfection with Messrs. Loddiges.

It is one of those charming epiphytes which are only procured in the fertile regions of the East. Mr. Cuming found it in Manilla, and introduced it to Europe.

The flowers are in short pendulous racemes, rather large, with an ivory white ground spotted with light violet; of the latter colour there is one large blotch within the extremity of each sepal and petal, and a few small dots are scattered over their surface; the lip is wholly violet. The flowers have a faint and rather disagreeable smell, which is not, however, perceived till they are nearly approached.

Fig. 1. represents a side view of the column and lip of this plant.

Although we now possess several Vandas of much beauty, the finest is still unknown in Europe. This glorious plant, perhaps the noblest of the Indian race, was called *Vanda cœrulea* by Mr. Griffiths, who found it among the Khasya or Coosya Hills, and sent us dried specimens. Its flowers are as large as those of *Vanda teres*, and the foliage is as good as that of *Aerides odoratum*. It is to be regretted that we should have no more exact information as to where it may be found, but we can hardly suppose that it could be missed by any plant-collector who might be sent after it into Sylhet.

The leaves of this wonderful plant are five inches long by nearly one inch wide; at their end they are two-lobed equally, and each lobe is sharp-pointed, so that the end looks as if a piece had been struck off by a circular punch. The flowers grow in upright spikes. A piece of a stem but four inches long bears four such spikes, which are from six to nine inches long, and carry from nine to twelve flowers. Each dried flower is between *three and four inches in diameter*, and if allowance is made for their having shrunk in drying, they may be estimated *as at least a foot in circumference.* The lip is, as is usual among Vandas, small; it is barely three-quarters of an inch long, narrow, with a short spur and a two-lobed point. Its surface is broken by three deep parallel perpendicular plates, and the lateral lobes of the base are triangular and acuminated.

In order to secure Mr. Griffiths' name, and to give more precision to this popular description, the following specific character is proposed.

V. *cœrulea* (W. Griffith in litt.); foliis distichis coriaceis apice æqualibus truncatis sinu concavo lobis lateralibus acutis, spicis densis erectis multifloris, bracteis oblongis concavis obtusissimis membranaceis, sepalis petalisque oblongis obtusissimis planis subunguiculatis, labello coriaceo lineari-oblongo apice divergenti-bilobo obtuso per axin trilamellato laciniis basilaribus triangularibus acuminatis, calcare brevi obtuso.

Sub t. 32 (1847)

New Garden Plants.

Cleisostoma spicatum.

C. *spicatum*; spicis densis brevibus multifloris, floribus pilosis, labelli calcare obtuso labello longiore laminâ ovatâ acutâ medio cristatâ dente dorsali bilobo lobis acutissimis denticulatis.

A Borneo plant, exhibited by Messrs. Rollissons at the last meeting of the Horticultural Society. The leaves are large and broad, the flowers red and yellow in short oblong spikes.

Megaclinium velutinum.

M. *velutinum*; pseudobulbis ovatis trigonis (3-4-gonis ?) foliis oblongo-lanceolatis emarginatis, rachi lineari-oblongâ crenatâ, sepalis pilosis supremo recurvo obtuso calloso marginato lateralibus refractis ovatis acutis intus velutinis, petalis linearibus acuminatis, labello ovato transversè corrugato obtuso, columnâ utrinque dente obtuso auctâ.

Nearly related to *M. falcatum*, from which its lateral sepals velvety inside distinguish it. The flowers are deep purple as well as the rachis, except the upper sepals and petals which are dull yellow. Messrs. Loddiges imported it from Cape Coast Castle.

35

EPIDENDRUM plicatum.

Plaited-lipped Epidendrum.

GYNANDRIA *MONANDRIA.*

Nat. ord. Orchidaceæ. § Epidendreæ. (Orchids, *Vegetable Kingdom*, p. 173.)

EPIDENDRUM.—Supra passim.

§ Encyclium.

E. *plicatum*; pseudobulbis ovato-oblongis teretibus diphyllis, foliis coriaceis ensiformibus, racemo paucifloro (?) tactu glabro, bracteis minimis squamiformibus, sepalis petalisque obovato-lanceolatis acuminatis, labelli laciniis lateralibus oblongis acuminatis obtusis intermediâ cordatâ plicatâ cuspidatâ subbrevioribus ungue crasso sulcato, columnâ apice utrinque dentiferâ.

This fine species has been introduced from Cuba, by Messrs. Loddiges, with whom it flowered last January. It is remarkable for the petals, which are green inside, with a few purplish stains near the point, being of a deep rich violet on the outside! The sepals are greenish, stained with dull purple on both sides. The lip is of a very rich purple, with a little yellow near the base; its middle lobe is strongly plaited, and deeply cordate.

It is most nearly allied to E. *ceratistes*, a plant found by Mr. Linden on the snow-capped mountain ridge of Santa Martha; but that plant has a large branching panicle, smaller flowers, and shorter lateral lobes to the lip, which is white, with the middle lobe not at all cordate.

Fig. 1. represents the inside of the lip of this species, somewhat magnified.

It may be interesting to some of our readers, to know that there exists in Florida, a species allied to this, the introduction of which would be desirable. It occurs near Tampa Bay, whence we have specimens communicated by Dr. Torrey, and might be readily obtained by any one trading with Florida. It is a slender species, with the habit of Epidendrum odoratissimum, but with much larger flowers, (whose colour is unknown,) and very narrow leaves. The pseudo-bulbs are small and terete, with a narrow tapering neck. A Botanist will recognise it by the following definition:—

E. *tampense* (Encyclium); pseudobulbis anguste ovatis acuminatis teretibus, foliis linearibus apiculatis paniculâ racemosâ sparsâ duplò brevioribus, sepalis petalisque oblongo-linearibus acutis, labelli tripartiti laciniis lateralibus linearibus obtusis intermediâ basi obscurè 3-nerviâ oblongâ acutâ paulò longiore.

36

DENDROBIUM chrysotoxum.

The Golden-arch Dendrobe.

GYNANDRIA *MONANDRIA.*

Nat. ord. Orchidaceæ. § Malaxidæ—Dendrobeæ. (Orchids, *Vegetable Kingdom*, p. 173.)

DENDROBIUM.—Supra passim.

§ Dendrocoryne.

D. *chrysotoxum* (Lindl. in Bot. Reg. 1847. sub t. 19); pseudobulbis clavatis angustis multicostatis 2-4-foliis, foliis oblongis horizontalibus coriaceis, racemis lateralibus laxis gracilibus arcuatim decurvis pseudobulbis æqualibus, bracteâ basilari parvâ spathaceâ floralibus minimis herbaceis, sepalis petalisque explanatis oblongis obtusissimis planis his duplò latioribus, labello indiviso cucullato rotundato pubescente margine minutissime pectinato et fimbriato

---

The Dendrocorynes, or Club Dendrobes, form a peculiar group in the large genus to which they belong, best perhaps characterized by their having a fleshy, angular stem, with two or more manifest articulations, one or more leaves at the upper end, and a lip not broken up into a tuft of hairs or fringes. They are as it were Bolbophylls passing into Dendrobes. In the group thus limited are included D. densiflorum, Griffithii, aggregatum, tetragonum, Veitchianum, speciosum, and some others formerly placed in Desmotrichum, a section which it seems better to limit to the species whose lip is broken up into a brush.

The present species, which was imported from the East Indies by Messrs Henderson, is extremely handsome. It differs from D. *densiflorum* in its many-angled pseudobulbs, small bracts, and curiously fringed pubescent, not shaggy, lip; from D. *Griffithianum* in its round emarginate fringed lip; and from D. *aggregatum* in the same respects, as well as in its great club-shaped many-leaved pseudobulbs.

Our drawing was made in Messrs. Henderson's Nursery in March last. Fig. 1. represents a portion of the edge of the lip much magnified.

New Garden Plants.

Dendrobium Egertoniæ.

D. *Egertoniæ*; caule tereti erecto basi ebulbi, foliis . . . . , floribus 2-4-nis, sepalis linearibus obtusis petalis angustioribus, labello ovato obtuso unguiculato parum undulato basi ciliato superne margine læviusculo intus villoso, columnâ antice glabrâ basi bicameratâ, antherâ pubescente, calcaris apice etuberculato.

This species has for some time been cultivated by Sir Philip Egerton, to whom I am indebted for a specimen. It is very near *D. mesochlorum*, but the flowers are not half the size; the sepals are pale pink outside; there is very little appearance of a purple stain on the tips of the petals and lip, and the middle of the lip is dull yellow, not green. There is, moreover, no tubercle at the back of the point of the spur, and the lip is not fringed except at its base. If it is less showy a species than *D. mesochlorum*, it is not on that account less valuable, for its flowers are delightfully scented towards evening. The plant was obtained by Sir Philip from the Botanic Garden, Saharunpur.

Dendrobium mesochlorum.

D. *mesochlorum*; caule tereti erecto basi ebulbi, foliis lanceolatis acuminatis, floribus subquaternis, sepalis ovali-linearibus obtusis petalis angustioribus, labello ovato obtuso unguiculato undulato fimbriato (præsertim basi) intus villoso, columnâ antice glabrâ basi bicameratâ foveatâ, antherâ pubescente, calcaris apice a tergo gibboso.

This beautiful species, allied to D. crumenatum, and resembling it in habit, although destitute of a bulbous base to its stems, was imported from India by Messrs. Veitch, and exhibited by them at the last meeting of the Horticultural Society, where it gained the medal. It is not discoverable among any of Dr. Blume's Onychiums, of which it would be one. The flowers are white, of the size of D. crumenatum, with a violet spot at the end of the petals and lip, and with a green stain in the centre of the latter. They have a faint but agreeable odour.

41

CLEISOSTOMA ionosmum.

Violet-scented Cleisostome.

GYNANDRIA *MONANDRIA*.

Nat. ord. Orchidaceæ. (Orchids, *Vegetable Kingdom, p.* 173.)

CLEISOSTOMA, Blume.—Perianthii sepala quinque; postico fornicato; reliquis patentibus; lateralia exteriora labello supposita, inferne subcoalita. *Labellum* inferne in calcar obtusum productum; calcar intus cristatum, orificio dente vel callo clausum, utroque latere dente inflexo instructum; limbus erectus indivisus. *Gynostemium* apice interne in rostellum emarginatum productum. *Anthera* terminalis semibilocularis. *Massæ pollinis* in quovis loculo solitariæ, subglobosæ, bipartibiles, cereaceæ, pedicello communi prælongo ad basin hamato suffultæ. *Capsula* linearis, teretiuscula.——Genus *Aëridi et Echioglosso affine, sed ob retinaculum productum nec abbreviatum, pedicellum basi hamatum nec peltatum aut arcuato-dilatum; diversum.* ——Herbæ *in arboribus parasiticæ, caulescentes.* Caules *radicantes.* Folia *disticha, rigida, subcarnosa.* Spicæ *oppositifoliæ, subramosæ.*—Blume Bijdr. 1. 363.

C. *ionosmum*; foliis distichis coriaceis ensiformibus obliquè retusis, paniculâ patente, sepalis petalisque obovatis obtusis subæqualibus, labelli hastati carnosi pubescentis laciniâ intermediâ cordato-triangulari calcare conico, columnâ pubescente antice bidentatâ.

A native of Manilla, whence it was sent to Messrs. Loddiges by Mr. Cuming. It flowered in the Hackney Nursery in March, 1844.

The flowers are in an open panicle, about an inch across, flat, with five obovate equal obtuse lobes, yellow with cinnamon-brown blotches. The lip is white with a few red streaks, three-lobed, with the basal lobes acute and smaller than the middle one, which is cordate triangular acute, and much larger than they are. The flowers smell pleasantly of violets.

Fig. 1. represents a lateral view of the column and lip; 2. the same seen from the front; 3. the pollen-masses, caudicle, and gland; 4. a cross section of a pair of pollen-masses.

The flowers are much larger than those of any species previously described.

42

CATTLEYA bulbosa.

Bulbous Cattleya.

GYNANDRIA *MONANDRIA*.

Nat. ord. Orchidaceæ. (Orchids, *Vegetable Kingdom, p.* 173.)
CATTLEYA.—Lindl.

Section 2. *Lip without the lateral lobes, and flat below the column.*

C. *bulbosa*; caulibus ovalibus brevibus pseudobulbos referentibus foliis solitariis ovalibus coriaceis brevioribus, pedunculis unifloris, petalis ovalibus undulatis membranaceis sepalis duplo latioribus, labelli plani calvi lobis lateralibus brevibus rotundatis intermedio cuneato bilobo rotundato.

Brazil is supposed to be the parent of this exquisite little species, but it is not certain. We are indebted for it to Mr. Rucker, with whom it flowered in May last. It belongs to the same section as *C. Aclandiæ*, from which it differs in its having much larger lateral lobes to the lip, a very differently formed middle lobe, and flowers of quite another colour. It may also be compared with C. *pumila*, which should be placed in the same section. With the latter, indeed, it accords in its stems resembling pseudo-bulbs, and its colour, but the flowers are much larger, and the lip flat not rolled up and crisp.

The accompanying figure conveys a good idea of this lovely plant, whose flat shovel-shaped lip, of an intensely deep crimson, gradually melts away till it loses itself in the imperfect lateral lobes, which are pink, bordered with crimson.

This very pretty epiphyte should be either fastened to a block of wood, with a little sphagnum, or placed well elevated upon a pot, filled one-third with broken pots, and the remainder with the rough fibre out of peat soil, and half decayed leaves, in nearly equal portions, and afterwards placed in the lightest, coolest and dryest part of the Orchideous house.

It requires but little water, even in the growing season, as plenty of moisture for Cattleyas may always be obtained by pouring water over the shelves and footpaths; as for the smaller kinds they should never be syringed overhead.

It is best increased after it has begun to grow, and not, as is generally done, when in a dormant state.

46

OPHRYDES Græcæ.

Grecian Ophryses.

GYNANDRIA *MONANDRIA.*

Nat. ord. ORCHIDACEÆ. (ORCHIDS, *Vegetable Kingdom, p.* 173.)
OPHRYS.—L.

1. Ophrys tabanifera.

O. *tabanifera* (Willd.); labello bigibboso ovato acuto tripartito villoso: lobis lateralibus deflexis acutis intermedio ovato, petalis ciliatis acutis.—*Lindl. Gen. & Sp. Orch. p.* 375. with its synonyms of O. *picta, pulla, hiulca, œstrifera, distoma, bombyliflora.*

2. Ophrys ferrum equinum.

O. *ferrum equinum* (Desf.); labello oblongo subquadrato apicolato violaceo maculâ albâ hippocrepicâ in medio, sepalis roseis.—*Lindl. Gen. & Sp. Orch. p.* 377.

The two species of Ophrys here represented were collected by our lamented correspondent the Dean of Manchester, were figured a day before his death, and were to have been illustrated by himself.

His brief and imperfect memorandum tells us that O. *tabanifera* was found by his collector Vrioni, near Clarentia, a low and swampy place, "whence the Dukes of Clarence took their title, it is said, why I know not." O. *ferrum equinum*, so called because of a horse-shoe mark on the lip, is from Corfu, where it was gathered by himself on the summit of the Garouna Pass.

In many respects these species resemble each other; but *tabanifera* has a three-lobed lip, *ferrum equinum* an entire one; in *tabanifera* the petals are reflexed, and pale greenish chocolate, in *ferrum equinum* patent, and rosy purple. Both have a downy column and hairy petals. The anther lobes of *tabanifera* are scarlet, of *ferrum equinum* yellow; and at the base of the column in *tabanifera* is a pair of teeth, which are missing in *ferrum equinum.*

47

DENDROBIUM Kuhlii.

Kuhl's Dendrobe.

GYNANDRIA *MONOGYNIA.*

Nat. ord. ORCHIDACEÆ. (ORCHIDS, *Vegetable Kingdom, p.* 173.)
DENDROBIUM.—Supra passim.

Sect. PEDILONUM (Blume, Sect. 2.) *Stems erect. Racemes lateral. Lateral sepals united into a pouch, with which the base of the lip is entirely coherent, without any articulation.*

D. *Kuhlii*; foliis ovali-oblongis acutis subseptemnerviis apice inæqualibus, racemis brevibus plurifloris, sepalis ovatis patulis petalis latioribus, labelli limbo lineari spathulato acuto imberbi, calcare crasso ovarii fere longitudine.—*Lindley Gen. & Sp. Orch. no.* 58.
Pedilonum Kuhlii, *Blume Bijdr. p.* 321.

This handsome plant is not unworthy to be placed by the side of its near ally, the beautiful D. secundum, from which it differs in having larger flowers in short lateral few-flowered horizontal racemes. They are of a bright rose colour, and retain their freshness longer than is usual in this genus.

For its introduction we are again indebted to the enterprise and good management of Messrs. Veitch and Co. of Exeter, to whom it was sent from Java by Mr. Thomas Lobb. It is No. 356 of his sets of dried specimens, which we may without exaggeration say, are scarcely rivalled for their beautiful preparation, or for the skill and taste with which hey have been selected.

Our drawing was made from materials supplied by Messrs. Veitch in October, 1845. Fig. 1. represents the column and spur cut open to shew the lip; Fig. 2. is a front view of the lip itself.

50

EPIDENDRUM pyriforme.

Pear-shaped Epidendrum.

GYNANDRIA *MONANDRIA.*

Nat. ord. ORCHIDACEÆ. (ORCHIDS, *Vegetable Kingdom, p.* 42.)
EPIDENDRUM.—L.

Sect. ENCYCLIUM. *Div.*; lip 3-parted, with the middle lobe rounded.

E. *pyriforme* (Lindl. in Bot. Reg. 1847. sub. t. 10.); (Encyclium: labello tripartito: lobo medio rotundato) pseudobulbis obpyriformibus aggregatis diphyllis, foliis coriaceis lanceolatis acutis scapo subbifloro longioribus, sepalis petalisque lanceolatis acutis, labelli lobis lateralibus obtusis integerrimis intermedio subrotundo glabro picto callis 2 maximis apice subliberis in unguem.

This very pretty little species was imported from Cuba by Messrs. Loddiges, with whom it flowered in January last. The leaves are unusually thick and fleshy, about four inches long, on little pseudo-bulbs, which look like inverted pears. Notwithstanding its diminutive stature, the flowers are fully $2\frac{1}{2}$ inches in diameter, with reddish yellow sepals and petals, and a pale straw-coloured lip veined with crimson.

We do not find it in any of our dried collections; it does not occur among the Cuba plants gathered by Mr. Linden, and we are ignorant of the history of its discovery.

Fig. 1. represents its lip spread open, to shew the form of the calli, which scarcely adhere to the lip until they reach the re-entering angles at its sides.

53

EPIDENDRUM alatum.

Winged Epidendrum.

GYNANDRIA *MONOGYNIA.*

Nat. ord. ORCHIDACEÆ. (ORCHIDS, *Vegetable Kingdom, p.* 173.)
EPIDENDRUM. *Subgenus* ENCYCLIUM, *Lindl.*

E. *alatum* (Bateman Orch. Mex. t. 18. *Epid. calocheilum,* Hooker in Bot. Mag. t. 3898.); pseudobulbis ovato-oblongis diphyllis, foliis ensiformibus obtusis coriaceis obsoletè striatis paniculâ multiflorâ brevioribus, sepalis petalisque lineari-oblongis spathulatis uniformibus patentibus, labello profundè trilobo basi intùs bicarinato lobis lateralibus eroso-dentatis rotundatis intermedio oblongo undulato multò brevioribus omnium venis callosis et verrucosis, columnæ alis rotundatis.

Of this once rare plant the gardens now abound in varieties, owing to the large importations from Guatemala, of which it is a native. The specimen here figured flowered in the conservatory of Mr. Rucker, in July, 1846.

Its pale colour, and the peculiar markings upon its lip, at once distinguish it. These markings consist of reddish warts, plates, scales or elevations, of various forms, arranged upon the veins, and therefore spreading from the base.

We find nothing among the variations of sufficient importance to deserve special notice.

The species thrives most when suspended from the roof of the house, and fastened to a block of wood with a coating of sphagnum. It is increased by division when the plant begins to grow, blooms in Mav, and remains a long time in flower if the atmosphere is kept rather dry and cool at that time.

58

SACCOLABIUM miniatum.

Orange-red Saccolabe.

GYNANDRIA *MONANDRIA.*

Nat. ord. ORCHIDACEÆ. (ORCHIDS, *Vegetable Kingdom, p.* 173.)
SACCOLABIUM.—Blume.

S. *miniatum* (Lindl. in Bot. Reg. 1847, sub t. 26); foliis distichis imbricatis loratis canaliculatis apice obliquè truncatis, racemis brevibus cylindraceis patentibus, bracteis minimis acutis, sepalis petalisque ovatis acutis patulis, labello lineari obtuso recurvo calcare recto pendulo tereti breviore intus edentato, polliniis albis.

This is one of the prettiest Epiphytes lately introduced, for the gay red orange of its flowers is as rich and pure as in Epidendrum vitellinum. It is, moreover, so compact in its mode of growth, that it can easily be moved from place to place.

It is a Java plant, imported by Messrs. Veitch, and has flowered with both Mr. Rucker and Mr. C. B. Warner. Its blossoms grow in spreading racemes, about ten together, and when worn in the hair have a singularly beautiful effect.

Fig. 1. represents a side view of the column and lip; 2. shews the pollen-masses, which are remarkable for being white.

Our drawing was made from a specimen obligingly supplied by Mr. Rucker in February last.

This, like the true air plants, lives entirely on the humidity of the atmosphere, and should be fastened to a block of wood coated over with sphagnum, to retain moisture whenever the atmosphere happens to become too dry.

It is increased by division when the plant begins to form new roots, and should never be removed from the block when once fastened upon it, by its roots. It is best kept at the warmest part of the house, but not fully exposed to light.

Sub t. 59 (1847)

NEW ORCHID.

VANDA *tricolor;* foliis distichis canaliculatis racemo paucifloro longioribus, sepalis coriaceis unguiculatis obovatis obtusis, labello æquilongo trilobo per axin 3-lineato, calcare brevi obtuso, laciniis lateralibus rotundatis intermedio convexo cuneato emarginato latioribus.

Under the name of *Vanda insignis,* a very different plant, this fine Javanese species, has been brought into cultivation by Messrs. Veitch. It has quite the habit of *Vanda Roxburghii,* and its flowers appear in the same manner, but they are larger, have yellow and brown spotted sepals, and a rose-coloured lip, with the lateral lobes rounded, not acute, and colourless. It is nearer *V. Hindsii,* a New Guinea plant, not yet in cultivation; but that species has a long many-flowered raceme, extending as far as the points of the leaves.

As to *Vanda insignis,* no description has yet been published of that plant; but, by the kindness of Dr. Blume, we possess a figure of it, from which we learn that it has a concave, not convex, lip, with very small lateral lobes, and the broad central lobe deeply heart-shaped.

60

BLETIA Gebina.

Japanese Bletia.

GYNANDRIA *MONANDRIA.*

Nat. ord. ORCHIDACEÆ. (ORCHIDS, *Vegetable Kingdom, p.* 187.)
BLETIA.—Supra.

B. Gebina (Lindl. in Journ. of Hort. Soc. 2. 307); foliis oblongo-lanceolatis plicatis acutissimis recurvis, racemis strictis 6-9-floris, bracteis oblongis obtusis cucullatis membranaceis citò deciduis, sepalis patulis lineari-oblongis petalis subundulatis pauld angustioribus, labello trilobo laciniis obtusis intermediâ crenulatâ crispatâ lamellis 5 ad basin usque extensis duabusque brevibus lateralibus pone apicem, clinandrio crenato.

Calanthe Gebina, *Loddiges' Catalogue, No.* 1846.

Our drawing of this novelty was made in the garden of the Horticultural Society, to which it was presented by Messrs. Loddiges in the spring of 1847. It is described in the Journal of the Society in the following terms:—

"Leaves broad, plaited, rising up the stem, from six to eight inches long, or more, and two inches wide, the uppermost acuminate, the lowest obtuse. The flowers are about as large as *Bletia hyacinthina,* from six to eight in a spike, two inches and a half in diameter, nearly white, with a faint tinge of blush. The lip is pale delicate violet, obtusely three-lobed, with seven plates upon its surface, of which two at the side are confined to the middle lobe, and the five others are extended to the base, which is a little stained with yellow."

It is nearly related to *B. hyacinthina*, and, according to Messrs. Loddiges' catalogue, is a native of Japan. No description of it, however, is to be found in books.

It is a terrestrial Orchid, which requires a slight protection from frost, and to be kept rather dry while in a dormant state; afterwards it should be well supplied with moisture and heat. It grows freely in a mixture of fibry peat and half-decayed leaf-mould, and is increased by dividing the old plants when in a dormant state.

It flowers in April.

Being nearly hardy, it is a desirable plant where there is no stove, as it may be grown in a cold pit kept close during summer.

Fig. 1. represents a view of the upper side of the lip, with its elevated plates; 2. a column; 3. one set of pollen-masses. The latter are entirely those of Bletia, and bear no resemblance to the structure of Calanthe.

62

DENDROBIUM cretaceum.

Chalk-white Dendrobe.

GYNANDRIA *MONANDRIA*.

Nat. ord. ORCHIDACEÆ. (ORCHIDS, *Vegetable Kingdom*, p. 187.)
DENDROBIUM.—Supra passim.

§ EUDENDROBIUM. *Stems slender; leaves thin and flat; flowers in pairs, or threes, opposite the leaves.* (Grastidium, *Blume.*)

D. *cretaceum;* foliis lanceolatis apice oblique emarginatis obtusiusculis, floribus solitariis (?), sepalis lineari-lanceolatis patentissimis obtusiusculis, labello subrotundo indiviso cucullato fimbriato-dentato utrinque pubescente basi foveat et obsoletè trilamellato, mento brevi obtuso.

Several plants of this new Dendrobe have appeared in gardens during the last summer, in consequence of its having been sold by Messrs. Veitch, who received it among Mr. T. Lobbs' Moulmein collections, and who communicated it to us in July last.

It is a very distinct species, remarkable for the dull chalky whiteness of its flowers, which are, nevertheless, neatly pencilled with crimson on the lip; this is owing to the whole surface of the lip, inside and outside, being covered with a short close white fur. It appears to possess the additional peculiarity of bearing its flowers singly, not in pairs; at least the wild specimens which were communicated to us by the late Mr. Griffith, from Mergui and the Khasiya hills, are in that state, as well as the cultivated plant.

Fig. 1. represents the inside of the lip, near the base of which is a small honey pore, and above the pore three obscure elevated lines.

NEW ORCHID.

ODONTOGLOSSUM (Leucochilum) *maxillare;* sepalis lanceolatis acutis petalisque latioribus sanguineo-maculatis, labello ovato acuto integro basi maculato appendice baseos maximâ concavâ carnosâ glabrâ luteâ extus bilamellatâ columnæ apteræ ferè longitudine.

We have only seen one flower of this beautiful plant, which might at first sight be mistaken for O. *Cervantesii*. It is, however, readily known by its wingless column, and especially by the presence of a very large yellow appendage, seated at the base of the lip, and scarcely shorter than the column. Such an appendage exists in all the Toothtongues, but it is usually inconsiderable, and never to be compared in size with this. We are unacquainted with the native country of this species; but its great resemblance to O. *Cervantesii, nebulosum, Rossii*, &c. indicates Mexico. The specimen now described was received from C. B. Warner, Esq. in September last.

63

ERIA convallarioides; β major.

The close-headed Woolwort; a large variety.

GYNANDRIA *MONANDRIA*.

Nat. ord. ORCHIDACEÆ. (ORCHIDS, *Vegetable Kingdom*, p. 187.)
ERIA.—Supra passim.

E. *convallarioides* (Lindl. in Wall. Cat. no. 1975. Gen. & Sp. Orch. no. 25. Bot. Reg. 1841. t. 62. *Pinalia*, Lindl. orch. scel. n. 71. p. 23. c. ic. *Pinalia alba*, Hamilton mss. apud Don Prodr. *Octomeria spicata*, Don. Prodr. 31. *Octomeria convallarioides*, Wallich in ic. Bibl. Anglo-Indicæ, n. 1141); caulibus compressis junioribus densè et laxè vaginatis, foliis oblongo-lanceolatis multinerviis, racemis densissimis oblongis cernuis breviter pedunculatis, floribus subglobosis leviter pubescentibus, labello subcordato ovato acuto inappendiculato.

Var. *β. major;* caulibus duplo majoribus, spicis longius pedunculatis, floribus clausis globosis.

When we formerly published this species it exhibited little of the beauty found in the variety now made known. The old sort is indeed hardly to be recognized in the beautiful thing which is represented on the accompanying plate. The oblong heads of flowers are of the purest ivory white, which is rendered still purer by the presence of small brownish bracts at their base. The surface, too, of every flower is so polished as to resemble that of white cowries or similar shells.

We are indebted for the accompanying figure to T. Twisden Hodges, Esq., who informs us that, under the care of his present gardener, the plant has become still larger and finer. It is distinguished from the original sort, not only by its much greater stature, but by the flowers being nearly closed, and almost globose, and by the spikes having much more evident stalks; the slight hairiness found in the original is here entirely absent.

Fig. 1. Represents a flower separated from the spike; 2. The column and lip.

65

STENORHYNCHUS cinnabarinus.

Cinnabar Stenorhynchus.

GYNANDRIA *MONANDRIA*.

Nat. ord. ORCHIDACEÆ. (ORCHIDS, *Vegetable Kingdom*, p. 173.)
STENORHYNCHUS, Rich. *Flores* ringentes, cylindracei. *Sepala* colorata basi gibba; dorsale petalis agglutinatum; lateralia labello supposita et basi in saccum connata. *Labellum* anticum, apice angustatum, basin versus dilatatum, columnam amplectens eique agglutinatum, basi ipsa intus sæpius biconvexum, callis orbatum. *Columna* teres, elongata, in pedem cui labellum annectitur extensa; stigmate prominente ovato; rostello subulato

corneo persistente. *Anthera* dorsalis, acuminata, bilocularis; clinandrio utrinque marginato aut alato. *Pollinia* 2, pulverea, bipartita, glandulæ linearis ope colligata.——Herbæ *terrestres* radicibus *fasciculatis*. Folia *radicalia, raro caulina, lata, nunc hysteranthia*. Scapus *laxe vaginatus*. Spicæ *imbricatæ, densæ*, bracteis *magnis coloratis*. Flores *sæpius pubescentes speciosi*. Lindl. gen. & sp. Orch. p. 476.

S. *cinnabarinus* (Lindl. gen. & sp. Orch. p. 479; *Neottia cinnabarina*, Llave nov. veg. p. 3); foliis oblongo-lanceolatis acutis, scapo piloso, spicâ conico-thyrsoideâ compactâ, bracteis lanceolatis subherbaceis floribus pilosis brevioribus, sepalis petalisque lineari-lanceolatis acuminatis apice patentibus, labello conformi glabro basi angustato canaliculato.

For the knowledge of this rare plant we are indebted to Messrs. Loddiges, who received it from Mexico, and flowered it in July last.

According to Messrs. Llave and Llexarza, the species grows in various places in the west of Mexico, where it is called by the country people *Cutsis*.

The dull olive green of the bracts belonging to its large compact spike, the vermilion red of the flowers externally, and the bright yellow of the inner face of their narrow spreading points give them an appearance unusually gay among these terrestrial species.

Fig. 1. represents the column, with the lip forced downwards so as to shew its inner face; right and left are two marginal scars shewing where the lip grows to the column.

66

ONCIDIUM amictum.

The Frilled Oncid.

GYNANDRIA *MONANDRIA*.

Nat. ord. ORCHIDACEÆ. (ORCHIDS, *Vegetable Kingdom*, p. 173.)
ONCIDIUM Swartz.

Sect. EUONCIDIA, Heteranthia, tetrapetala, macropetala.

C. *amictum*; pseudobulbis elongato-ovatis diphyllis, foliis oblongis pergameneis basi angustatis, scapo (maculato) foliis longiore, paniculâ racemosâ densâ multiflorâ cernuâ, sepalis lateralibus apice tantum liberis, labello obcuneato bilobo auriculis baseos ciliatis apice dilatatis, cristæ tuberculis duobus maximis acutis divergentibus a fronte duobus posticis lateralibus obtusis, columnæ alis oblongis clinandrioque margine membranaceo plicato ciliatis.

A native of Brazil, whence it was received by Messrs. Loddiges, with whom it flowered in April, 1847. Its large yellow flowers more or less spotted in many parts with rich brown, give it a very handsome appearance.

It is nearly allied to O. *Gardneri*, the character of which is repeated below. It differs from that species in having an inversely wedge-shaped lip, much larger basal lobes, only two pair of tubercles on the crest, large column wings, and an anther-bed, surrounded as it were by a membranous frill.

Fig. 1. shews the appearance of this crest.

O. Gardneri (Lindl. in Lond. Journ. Bot. 2. 662.); floribus paniculatis, sepalis oblongis obtusis lateralibus semiconnatis petalis duplo majoribus unguiculatis subrotundis undulatis, labello transverso emarginato basi auriculato, cristæ tuberculis duobus a fronte maximis intermedio minore duobus a latere minoribus linearibus binis alteris basilaribus circularibus verrucisque quibusdam in medio, columnæ alis nanis rotundatis.——*Brazil;* on trees in the forests of the Organ Mountains, Gardner.—— Allied to O. crispum and Forbesii, but distinctly separated by the peculiar form and tuberculation of the lip, and by the very small wings of the column.

BOLBOPHYLLUM hirtum.

(*Lindley Gen. & Sp. Orch. p.* 51.)

This curious plant has flowered with Messrs. Loddiges. It produces a long drooping tail of small whitish flowers, having exactly the smell of Sweet Vernal Grass (*Anthoxanthum odoratum*). The sepals are not only clothed thickly with hairs, but they also have numerous purple oblong glands, especially on the edges. The petals are ovate, and as if serrate with long stiff hairs. Messrs. Loddiges received it from the East Indies, where it grows on trees.

The specific character given in the work above quoted, having been formed from dried and not perfect specimens, the following should be substituted.

B. hirtum, (Lindl. gen. & sp. orch. no. 18. p. 51. *Stelis hirta*, Smith in Rees. *Tribrachia hirta*, Lindl. Coll. Bot. 41.); pseudobulbis elongato-ovatis, foliis lato-ligulatis obtusis scapo glabro brevioribus, spicâ caudiformi multiflora cernuâ pubescente, sepalis acuminatis hirsutis et glandulosis, petalis nanis acutissimis fimbriatis, labello oblongo-lineari emarginato obtuso supra hirsuto, antherâ glandulosâ.

67

CYMBIDIUM eburneum.

The Ivory Cymbid.

GYNANDRIA *MONANDRIA*.

Nat. ord. ORCHIDACEÆ. (ORCHIDS, *Vegetable Kingdom*, p. 173.)
CYMBIDIUM, Swartz.

C. *eburneum*; foliis basi distichis rigidis angustis ensiformibus apice obliquè bilobis acutis, racemo subbifloro decumbente squamis longis acutis sphacelatis imbricatis basi vestito, sepalis petalisque oblongo-lanceolatis subcarnosis acutis undulatis, labello oblongo trilobo laciniis lateralibus rotundatis intermedio triangulari acute crispo, lamellis in unam carnosam pubescentem apice tumidam confluentibus.

The flowers of this charming plant are not only among the largest of the genus, but among the sweetest. They resemble, in fragrance, those of the Chinese Cymbid, than which nothing is more delicious.

The species has flowered with Messrs. Loddiges, by whom it was imported from the East Indies.

Not a spot interferes with the pure ivory white petals, except one long yellow stripe along the middle of the lip.

68

ONCIDIUM curtum.

The Cropeared Oncid.

GYNANDRIA *MONANDRIA*.

Nat. ord. ORCHIDACEÆ. (ORCHIDS, *Vegetable Kingdom*, p. 173.)
ONCIDIUM, Swartz.

Sect. EUONCIDIA, HETERANTHIA, tetrapetala, macropetala.

O. *curtum*; pseudobulbis . . . . , foliis . . . . , floribus paniculatis, sepalis lateralibus apice tantum liberis obtusis, labello rotundato bilobo undulato basi auriculato, mesochilio brevi, cristâ 5-lobâ tuberculatâ verrucis quibusdam utrinque sparsis et in lineam gyriformem ordinatis, alis columnæ parvis truncatis, antherâ pubescente.

We received this new species in July last from Messrs. Veitch of Exeter, without any account of its history. The beauty of the plant renders it desirable that it should be better known.

At first sight it looks like *O. crispum* or *Forbesii*, but in reality it is much nearer *O. pectorale* (the Breastplate Oncid). From that species it differs in the following particulars. 1. The flowers are smaller; 2. there is more brown in the sepals and petals; 3. there is a broad brown border to the lip; 4. the space between the auricles and expanded part of the lip is very short, with parallel edges; 5. the auricles are broader; 6. the tubercles of the crest are much more broken up, and differently arranged; 7. the anther is downy; and 8. the lateral sepals are nearly united to the point: in the Breastplate Oncid they are free almost to the base.

Fig. 1. represents the column and base of the lip of this species.

69

GONGORA maculata; *var.* tricolor.

Three-coloured Spotted Gongora.

GYNANDRIA *MONANDRIA.*

Nat. ord. ORCHIDACEÆ. (ORCHIDS, *Vegetable Kingdom, p.* 173.)
GONGORA, Botanical Register, vol. 19. *fol.* 1616.

G. *maculata* (Lindley in Bot. Reg. t. 1616. Ib. 1847. t. 17.); foliis 5-plicatis obovato-oblongis basi valdè angustatis, sepalis lateralibus è latâ basi angustatis, hypochilio oblongo subtùs convexo basi obtusè bicorni apice truncato angulis acutis in cirrhis duobus longis producto, epichilio acuminato.

Var. tricolor, *Bot. Reg.* 444. misc. no. 40.

There are two accounts of the history of this variety of Gongora; one that it came from Peru: the other, for which we have the authority of the late Mr. Clowes's gardener, is that it was detected by John Maclean, Esq., of Lima, on the mountains near Panama in 1841, and by that gentleman presented to the Liverpool Botanic Garden, from whence Mr. Clowes obtained it in 1842.

It is a most beautiful variety of G. maculata. The ground colour of every part of the flower, except the lip, is clear yellow; the column and petals are delicately banded with rich sienna brown, and a few large clear distinct blotches of the same colour occur on the sepals. The lip itself is white, with a cinnamon stain on the ends of the lateral tubercles and the sides of its upper half.

70

ONCIDIUM pelicanum.

The Pelican Oncid.

GYNANDRIA *MONANDRIA.*

Nat. ord. ORCHIDACEÆ. (ORCHIDS, *Vegetable Kingdom, p.* 173.)
ONCIDIUM, Swartz.

Sect. EUONCIDIA, HETERANTHIA, pentapetala micropetala, *labello pandurato.*

O. *pelicanum* (Martius. Bot. Reg. 1840. misc. 216.); pseudobulbis ovatis sulcatis monophyllis, foliis anguste lanceolatis acutis, racemis subcompositis multifloris, sepalis petalisque lineari-lanceolatis undulatis acutis reflexis labello multo brevioribus, labello reniformi emarginato; laciniis lateralibus rotundatis intermedia multo angustioribus, crista glabra basi convexa tunc quinquedentata apice bidentata, columnæ alis magnis oblongis denticulatis apice acutis.

This plant is very closely akin to *O. reflexum*, from which it differs principally in the sepals and petals being less blotched, in the lateral lobes of the lip being smaller in proportion to the intermediate segment, and in the tubercles of the crest, which is smooth, not downy, being rather differently arranged. The name has doubtless been given in allusion to the column, which is not unlike a pelican pecking her breast.

We entertain no doubt of its being the species intended by Dr. von Martius, for the specimen now figured was obtained from a plant sent from Mexico by Count Karwinsky to Messrs. Loddiges, and Count Karwinsky was a valuable contributor of Mexican plants to the Garden at Munich, under Dr. von Martius's care.

Pages 321–326:

Index to the described or cited orchids from the Botanical Register 1815–1847

Accepted names are printed in bold type, synonyms are in normal type.

INDEX

ORCHIDS
from the Botanical Register 1815–1847

The Illustrations

Edited by SAMUEL SPRUNGER
in collaboration with PHILLIP J. W. CRIBB and WILLIAM T. STEARN

Birkhäuser Verlag
Basel · Boston · Berlin

The illustrations and accompanying descriptive texts reprinted here originally appeared in *Edwards' Botanical Register.*

A special word of thanks is due to the Basel University Library in kindly making available all archival copies of the *Botanical Register* for the production of the photolithographic plates and the microfilm for the text.

Library of Congress Cataloging-in-Publication Data

Orchids from the Botanical register, 1815–1847 / edited by Samuel Sprunger in collaboration with Phillip J. W. Cribb and William T. Stearn.
p. cm.
"The illustrations and accompanying descriptive texts reprinted here originally appeared in Edwards' botanical register" – Verso t. p.
Includes bibliographical references and indexes.
Contents: The illustrations – The texts.
ISBN 0-8176-2479-1 (U.S. : set)
1. Orchids. 2. Orchids-Pictorial works. 3. Orchid culture. 4. Botanical illustration. I. Sprunger, Samuel. II. Cribb, Phillip. III. Stearn, William T. (William Thomas), 1911– IV. Edwards' botanical register.
SB409.0693. 1991
635.9'3415–dc20

Deutsche Bibliothek Cataloging-in-Publication Data

Orchids from the Botanical Register 1815 [eighteen hundred and fifteen] – 1847 / ed. by Samuel Sprunger in collab. with Phillip J. W. Cribb and William T. Stearn. - Basel ; Boston ; Berlin : Birkhäuser.
ISBN 3-7643-2479-1 (Basel . . .) Gb. in Schuber
ISBN 0-8176-2479-1 (Boston) Gb. in Schuber
NE: Sprunger, Samuel [Hrsg.]
The illustrations - 1991

Book design: Albert Gomm swb/asg, Basel
Printed in Germany
ISBN 3-7643-2479-1
ISBN 0-8176-2479-1

CONTENTS

to my children
Coralie, Letitia, Floriane

FOREWORD

Following the successful publication of the orchid plates from *Curtis's Botanical Magazine,* Phillip Cribb, Kew, and Jany Renz, Basel, suggested that the orchids from the *Botanical Register* should also be brought out in book form.

The 440 hand-coloured plates and 1212 descriptive texts of orchids from the *Botanical Register* of 1815–1847, which have been brought together in the present two volumes, represent an important source of information for the science of orchidology. Sydenham Edwards founded the *Botanical Register* (BR) in 1815 after a falling out with Samuel Curtis, who was at the time publisher of the *Botanical Magazine,* for which journal Edwards had been the most important illustrator. With the *Botanical Register* he aimed to provide the interested reader with excellent accounts and illustrations of new plants that had been introduced and cultivated in the gardens and collections of England. After Sydenham Edwards' early death in 1819, the periodical was continued by John Bellenden Ker (formerly known as Bellenden Gawler) until 1829, at which time John Lindley took over editorial responsibility for the journal until it finally ceased publication in 1847. During these 18 years alone the *Botanical Register* published 1100 descriptions of orchids compiled by the famous botanist.

The journal thus became the authoritative publication on orchids of its time and still remains an important source of information today, for half of all the accounts contain original descriptions. John Lindley, acknowledged as the father of modern orchidology, supplied expert descriptions of the orchids brought to England from all over the world and named many of them. A detailed account of the life and work of John Lindley is given here by W. T. Stearn. All the descriptions and reproductions have been taken from the 33 volumes of the *Botanical Register* in the possession of the University Library in Basel. It is a happy coincidence that 150 years after an honorary doctorate was conferred upon John Lindley by the Medical Faculty of the University of Basel in 1839 for his *Flora Medica* the orchids from the *Botanical Register* should, through the agency of Basel, again see the light of day.

This work presents orchids from every continent. There are 20–25,000 species of orchid in the world and together they form the largest family of flowering plants.

Orchids are well known for the diversity of their ecological distribution. In the tropics and subtropics they grow on trees (epiphytic: E), on rocks (lithophytic: L), and on the ground (terrestrial, geophytic: T). Most temperate orchids are terrestrial. Some terrestrial orchids grow on decaying matter and lack any chlorophyll (saprophytic: S). The letters E, L, T and S in the catalogue placed after the geographical details refer to these ecological characteristics.

In the course of evolution the orchids have diversified to adapt to the conditions of quite specific biotopes.

Unfortunately the boundaries of these biotopes are being forced back all the world over by human activities. In the tropics and subtropics, where about 80 % of orchids thrive epiphytically, they are being deprived of their only habitat by the clearing of the virgin forests. Some species have already disappeared while others are becoming rare or threatened. Even in the temperate zones the situation is slowly becoming critical. Intensive agriculture, forestry, road- and house-building, not to mention air pollution, are encroaching on their biotope with the result that some orchids have become very rare or are already extinct in parts of their range. Orchids are highly specialized plants sensitive to disturbance and their presence or absence are useful indicators of the health of particular habitats.

They are the first victims of forest clearance, the manuring of rough pasture, and the draining of marsh and peatland; the uninhibited pollution of soil, air and water spells the end of them.

It is the duty of us all to counteract this disastrous trend so that plants such as those illustrated here will be preserved for the future generations.

It is high time for us to institute concrete measures on a worldwide scale to conserve the landscape if an imminent general catastrophe is to be avoided. For are not plant associations and ecosystems just as much a part of our cultural heritage as monuments, famous works of art and great engineering achievements? Why do we abuse and destroy our landscapes in this way? These volumes are also intended as a document of a heritage that must be preserved.

The rarity, originality and quality of the plates and texts will make them an inexhaustible source of information for the orchid enthusiast, gardener and botanist alike. Scrutiny of the plates will show that here the art of drawing and painting is a part of scientific descriptive botany. The *Botanical Register*'s draughtsmen were not only excellent observers but also well-versed in plant morphology and masters of the craft of graphic representation. The profession of technical illustrator is not to be confused with that of artists like Gaugin, Cézanne or Monet, whose paintings, inspired by nature, reveal a personal view of the world. The number of petals and anthers was not of great importance to artists such as these because, for them, what counted primarily was the picture as a whole. Nor is photography able to incorporate all botanically important information optimally in a single picture, for each photograph shows only one aspect of a plant. The illustrator, on the other hand, is able to combine and compose various perspectives and details into a single picture.

About one twentieth of all orchid species are illustrated and/or described in the *Register* and have been recategorized systematically, bibliographically and geographically in accordance with the latest principles of scientific botany. In all original descriptions the type has been given along with details of place of origin, collector or cultivator and also herbarium. The index at the end enables the reader to find the correct names of the orchids illustrated and described. The catalogue supplies information on synonyms of the current names and contains bibliographic, geographic and ecological particulars of each plant.

I was able to bring this work to a successful conclusion only with the cooperation and encouragement of my dear wife Paulette. Without her, these volumes would never have appeared.

Basel, May 1990, *Samuel Sprunger*

ACKNOWLEDGEMENTS

This work was made possible thanks to the collaboration and willingness of many people:

The authors are thankful to:

Dr. F. Gröbli, Director of the University Library, Basel, who granted permission to borrow the 33 volumes of the *Botanical Register* for the production of the photolithographic plates and the microfilm for the text.

Prof. G. T. Prance, Director of the Royal Botanic Gardens, Kew, who granted access to the Garden, Herbarium and Library.

Prof. H. Zoller, Prof. A. Wiemken and Prof. C. Körner, Directors of the Botanical Institut, Basel, who gave the permission to use the Library and the Herbarium.

Prof. G. Lucas, Keeper of the Herbarium, Royal Botanic Gardens, Kew, for his encouragement.

Jeffrey Wood and Sarah Robbins, Orchid Herbarium, Kew, for their kind help.

Sylvia Fitzgerald, Librarian, and Ray Desmond, former Librarian, Royal Botanic Gardens, Kew, for critically reading the manuscript of the introductory chapter and for much sound advice.

The editor would like to express his gratitude to the following people:

Marcel Jenni, Head of the Repro-Photographic Department of the University Library, Basel, who from the beginning stood unflaggingly by me with his great idealism.

Dr. J. Renz, Basel, who encouraged and helped me during the preparation of this work. He generously made his herbarium and library available to me, and read the proofs of all manuscripts.

Helen and Nigel Hepper of Richmond, for their hospitality and help during my stay at Kew.

Hanspeter Moser, Basel, for his invaluable help in creating the computer programme of the catalogue and the index.

Claude Farron, Curator of the Botanical Institut, Basel, for his help.

Michèle Zaugg, Basel, for her critical contribution.

Hugo Schneider, Pfeffingen, Librarian of the Swiss Orchid Society, for the generous permission to use the society's library.

Emanuel Bürgin and Luc Delay for their technical help.

Olivetti Corp., Basel, for sponsoring the work with a personal computer M 240.

Birkhäuser Publishers for the care taken in producing this two-volume edition and the excellent printing quality of the *Botanical Register* orchid plates. In particular, I would like to thank the director H. P. Thur, and Albert Gomm for his generous and accurate collaboration in realizing the book project.

The data in the catalogue and index could be put on a user-program if enough interest were expressed. Please address requests to the Editor.

INTRODUCTION

by Phillip Cribb

The *Botanical Register* was founded in 1815 by the eminent botanical draughtsman Sydenham Teast Edwards (c. 1769–1819) because he fell out with the proprietor of the *Botanical Magazine*, Samuel Curtis. The *Register* survived for thirty two years before it ceased publication, unable to compete with the *Botanical Magazine* that William Hooker's drive had lifted out of the doldrums. It is perhaps ironic that *Curtis's Botanical Magazine* which effectively spawned the *Botanical Register* should also have brought about its demise in 1847. During its brief flowering, the *Botanical Register* introduced hundreds of exotic plants to the British public and published some 2700 delightful illustrations.

Sydenham Edwards and the early years

Sydenham Edwards was the son of a Welsh schoolmaster in Abergavenny and he was introduced to William Curtis by a Mr. Denman who had seen some of Edwards' copies of plates from *Flora Londinensis* which the young Edwards had done for his own amusement. Edwards joined James Sowerby and William Kilburn as illustrators for Curtis's *Botanical Magazine*. By the time he departed, Edwards had become the principal artist for the *Botanical Magazine*. His work was of a high standard and the high standing of the *Magazine* was in great part due to Edwards' consistently fine work and lovely illustrations. The comments of Wilfrid Blunt (1950) are perhaps a miserly assessment of Edwards' early work and talent: "[He] was a conscientious and industrious artist, but his line lacks the swiftness of Sowerby's. In his early work his colour, especially his green, is sometimes harsh, and his hatched and stippled tone laboured until it becomes lifeless." Blunt does, however, concede that his later work is "much stronger" and shows a "better understanding of the structure and habit of the plants portrayed". However, J. C. Loudon (1826), in the *Gardeners' Magazine*, praised Sydenham Edwards "for the skill he possessed of transferring an accurate resemblance of the largest plants to the small space of an octavo page".

The reason for Edwards' and Samuel Curtis's disagreement is not certainly known but was almost certainly to do with pay and probably recognition (Desmond, 1987) with the result that Edwards set up the *Botanical Register* in competition with the *Botanical Magazine*, upon which it was closely modelled. John Bellenden Gawler, who wrote the text for each plate, and Francis Sansom, the engraver, joined Edwards in this new venture.

The title page of the first volume announces the journal's content: the "Botanical Register consisting of coloured figures of exotic plants cultivated in British gardens; with their history and mode of treatment". Not only were many of the plants in the pages of the *Register* illustrated for the first time but many were also described there as new to science. The continuing scientific importance of the journal stems from the latter because of the principle of priority of plant names enshrined in *The International Code of Botanical Nomenclature*. The format of the *Botanical Register* remained close to that of the *Magazine* throughout its life. Gawler's texts were, however, fuller than those of the *Magazine* and the plates contained botanical dissections of the flowers. The quality of the *Register* was compared favourably with that of the *Botanical Magazine* by J. C. Loudon (1826). He stated that the plates were "better filled with their subjects" and the text reflected the modern botanical thought whereas the *Botanical Magazine* was "a register for the old school of botany". The only other significant change was the addition of a Miscellaneous Notes section during John Lindley's reign as editor.

The first part of the *Botanical Register* appeared in March 1815, the plates all drawn by the industrious and aggrieved proprietor. In his first volume, two orchids are figured, both from the West Indies. Plate 17 is of *Epidendrum nutans*, a plant flowered by the nursery of Messrs. Lee and Kennedy of Hammersmith, noting that the species was introduced to England by Admiral Bligh, of *Bounty* fame! The same nursery also flowered the Jamaican species *Epidendrum fuscatum* (t. 67). It is interesting to note where the plants in the first volume come from because it gives a good idea of botanical exploration at the time. South African plants are well represented with *Protea longifolia, Gazania pavonia* and *Erica filamentosa;* North American plants are also frequent, such as *Phlox suffruticosa* and *Lobelia splendens.* Tropical American species such as *Passiflora quadrangularis, Gossypium barbadense* and *Dahlia superflua* are found here, as well as a few plants from the Far East, such as *Prunus japonica* and *Camellia sasanqua,* and a solitary representative from Australia, *Stylidium graminifolium.* With the end of the Napoleonic wars plants from all over the globe were beginning to flood into England to delight horticulturalists.

Sydenham Edwards died in 1819 and the ownership of the *Register* then passed to John Bellenden Gawler (later known as Bellenden Ker) who had written and edited it from the start. Bellenden Ker remained the editor until the end of volume 14 in 1829 when it passed to John Lindley, of whom more anon.

In these early volumes orchids did not feature prominently, a reflection of how few tropical species were imported at the time and also the problems experienced in their cultivation. A few orchids of particular note, however, can be seen. The Indian species *Vanda roxburghii* (t. 506) is figured in volume 6 based on a plant imported from Nepal and flowering in the collection of Sir Joseph Banks at Spring Grove, Isleworth. Several terrestrial orchids such as *Disa prasinata* (t. 210), *D. bracteata* (t. 324) and *Satyrium cucullatum* (t. 416), from the Cape, appear in these early years. The second of these is noted as having been cultivated by Robert Brown. But none of these could match the spectacular Pride of Table Mountain, *Disa uniflora* (t. 926) illustrated on its first flowering in cultivation in Europe in the collection of William Griffin of South, Lambeth, London.

Surprisingly, several Mediterranean orchids, nowadays considered difficult to cultivate, are illustrated including *Ophrys tenthredinifera* (t. 202), *O. speculum*

(t. 370), *Orchis longicornu* (t. 205) and *O.* (now *Barlia*) *longibracteata* (t. 357). Many of these were introduced from Palermo, Sicily, by William Swainson of Elm Grove, Liverpool, who said of the the last that it "is growing in my garden, under a common frame, as vigorously as in [its] native climate". The first tropical African orchid to be illustrated in the *Register, Eulophia guineensis* (t. 686) appears in 1822 in volume 8. It had been collected in Sierra Leone by George Don and was flowered by the Horticultural Society at their Chiswick garden. The Horticultural Society's links with the *Botanical Register* begin at about that time, 1822, and by the following year the future was sealed by the contributions of John Lindley, the Secretary to the Society and Professor of Botany at King's College, University of London. As Secretary to the Horticultural Society, his access to rare and recently introduced plants sent, amongst others, from West Africa by George Don, from North America by David Douglas, from China by James Parks, from Central America by Theodore Hartweg, and by Nathaniel Wallich from India, allowed the *Register* to keep ahead of its rivals.

John Lindley's stewardship

Lindley's early contributions, signed J. L. and appearing in volumes 7, 8 and 9, were species accounts and were mainly of orchids: the first in 1821 was *Cymbidium xiphiifolium* (t. 529); *Angraecum* (now *Oeceoclades*) *maculata* (t. 618) followed the following year and in volume 9 we have *Oncidium luridum* (t. 727), *Dendrobium* (now *Xylobium*) *squalens* (t. 732) collected by J. Forbes in Brazil, and then *Eulophia gracilis* (t. 742), another of Don's Sierra Leone plants. By volume 10, Lindley was contributing all of the text for the twelve figured orchids and also drew three of them: *Camaridium ochroleucum* (t. 844), *Catasetum claveringii* (t. 840), and *Aeranthes grandiflora* (t. 817), the first Madagascan orchid in the *Register*, which had been collected there by Forbes for the Horticultural Society.

Lindley succeeded as editor with volume 15, in 1829, and remained so until the demise of the journal at volume 33, in 1847. It proved a happy relationship coinciding with a long period of peace for Great Britain, the increase in money and leisure time for many with the success of the Industrial Revolution and the expanding Empire, and the concommitant enthusiasm for growing exotic plants which by now were beginning to be introduced in quantity from the tropics. Lindley's love of orchids is encapsulated in the introductory words of the text for *Oncidium harrisonianum* (t. 1569): "For what purpose can the world have been adorned with these Orchidaceous plants? To man or animals they are scarcely ever of any known use. No honey is secreted by their flowers; neither poison, medicine, nor food, are collected in the recesses of their stems; and their very seeds seem unfit for feeding even the smallest bird .... For what then can they have been formed, unless to delight the sense of man, to gratify the eye by their gay colours and fantastic forms, and to shew the inexhaustible fertility of that creative power which we recognise everywhere in Nature". His enthusiasm for orchids was quickly reflected in the numbers that appeared in the *Register*. By volume 18 we have fifteen plates devoted to orchids and, by volume 22, the number rose to thirty-one, nearly half of the plants figured.

In volume 24, 1838, Lindley started a section entitled Miscellaneous Notes which continued up to, and including, volume 31. This gave him the scope to describe many of the new species that were by now flooding into England, often as dried herbarium specimens, from all over the world. Many new orchid species were described here by Lindley for the first time. The index that accompanies this book will give botanists easy access to these for the first time. Lindley also used this section of the *Register* to publish revisions of some orchid genera. In volume 28, for example, he published revisions of *Pleurothallis* (pp. 67–84) and *Encyclia* (pp. 28–36); in volume 30, he revised *Brassia* (pp. 5–8), *Zygopetalum* (pp. 9–10), *Catasetum* (pp. 31–39), *Dendrobium* (pp. 46–65) and two sections of *Epidendrum* (pp. 15–24, 82–87). The Miscellaneous Notes also contain much else of interest such as the first account of the giant Amazon water-lily, *Victoria regia* (vol. 24: p. 1); a discussion of salep (vol. 27: pp. 16–17); an account of the ecology of Brazilian orchids (vol. 25: pp. 42–43); and numerous book reviews and notices.

Sources of orchids for the Botanical Register

From his base at the Horticultural Society, Lindley established contacts with orchid growers and botanists throughout the country and also abroad. Although the Horticultural Society was his main source of plants in his early years as editor, he also received flowering plants and drawings of plants from around the country especially for inclusion in the *Register*. William Cattley, after whom Lindley named the spectacular tropical American orchid genus *Cattleya*, was Lindley's patron in his early debt-ridden years. Cattley was one of the earliest growers of tropical orchids and had a fine collection at High Barnet, just north of London. *Maxillaria* (now *Bifrenaria*) *harrisoniae* (t. 897) and *Dendrobium moniliforme* (t. 1314) were both drawn from plants in Cattley's collection. At about the same time Dean William Herbert of Manchester was sending Lindley Brazilian orchids including *Oncidium pumilum* (t. 920) and *O. cornigerum* (t. 1542), the latter plate being drawn by Herbert himself.

The most prolific source of Brazilian species was, however, the Harrison family of Aighburgh near Liverpool. William Harrison was a merchant in Rio de Janeiro and he sent orchids back to his brother Richard and also to his sister-in-law, Mrs. Arnold Harrison. Amongst their more spectacular introductions were *Oncidium Harrisonianum* (t. 1569), *Cyrtochilum* (now *Miltonia*) *flavescens* (t. 1627), *Catasetum luridum* (t. 1667) and *Oncidium ampliatum* (t. 1699).

James Bateman (1811–1897), the eventual author of the monumental *Orchidaceae of Mexico and Guatemala* (1837–1843) and of *A monograph of Odontoglossum* (1864–1874), also sent plants to Lindley from his collection at Knypersley. Among these were the Guyanan *Batemannia colleyi* (t. 1714), whose generic name honours the grower and the specific epithet his collector Colley; *Vanda batemannii* (vol. 32: t. 59); and the Mexican *Stanhopea oculata* (t. 1800). Other amateur growers of note to send plants included Charles Lemon, Sigismund Rucker, Rev. John Clowes and George Barker, all commemorated by Lindley with orchids in the *Botanical Register*.

The important role of the aristocracy should not be for-

gotten as orchid cultivation developed into a craze. The Duke of Devonshire, no doubt through Joseph Paxton who was the Duke's Head Gardener and Lindley's good friend, sent many orchids from the rich collection at Chatsworth in Derbyshire. Notable among these were *Myanthus* (now *Catasetum*) *barbatum* (t. 1778), *Bifrenaria aurantiaca* (t. 1875), and *Dendrobium formosum* (vol. 25: t. 64). The Dukes of Bedford and Northumberland likewise had their choice orchids sent to Lindley for inclusion in the *Register*. One can almost sense the friendly rivalry that this inspired.

Undoubtedly, Lindley's most eminent correspondent in the later years of the *Register* was Charles Darwin. Some of Darwin's letters to Lindley survive in the Kew archives and reflect the fascination and pleasure that orchids gave to both. In a letter from Down in Kent on 18th October 1843, Darwin writes "I have been extremely much... tested with *Catasetum*, and indeed with many exotic Orchids, which I have been looking at in aid of an opusculum, on the fertilisation of British Orchids. I very much fear that in publishing I am doing a rash act; but Orchids have interested me more than almost anything in my life". The correspondence continues in later letters with discussions on the structure and origins of the various parts of the orchid flower, requests for exotic orchids now that Darwin had exhausted Hooker's material at Kew, and comments on various species that Darwin had examined. Of particular note is the comment in his letter of 15th December 1843, that "*Catasetum tridentatum* is male (and never seeds according to Schomburgk whom you have accidentally misquoted in Veg. K.), *Monacanthus viridis* is female and *Myanthus barbatus* is the Hermaphrodite form of the same species". The last two genera were established by Lindley, and Darwin's comments allowed Lindley to correct his previous treatment of these genera (t. 1951).

The richest source of orchids for the *Register* in its later years was the nursery of Messrs. Loddiges of Hackney. Conrad Loddiges (1739–1826), who came to England from Germany as a boy, took over the Hackney nursery from J. Busch in 1771, but it was his son, George (1784–1846), who developed the nursery to its pre-eminent position in the first half of the nineteenth century. George was a talented artist as well as a plantsman and founded a successful journal of his own, *Loddiges' Botanical Cabinet* (1817–1846), as a vehicle to advertise his plants to the public. Despite the competition, George sent dozens of rare orchids to Lindley to be figured in the *Botanical Register*. A Loddiges' plant first appeared in the *Register* in volume 10, sent by Colvill who flowered a specimen of *Epidendrum cuspidatum* (t. 783) imported from Dominica in the West Indes in 1808. However, it was not until Lindley's reign that Loddiges began to send plants directly and on a regular basis. The first of these was the appropriately named *Cirrhaea loddigesii* (t. 1538) from Brazil. By volume 21, twelve of the plates were of Loddiges' orchids. Orchids were introduced into cultivation by Loddiges from all over the world. Notable species include the Indian *Dendrobium densiflorum* (t. 1828), the tropical African *Bulbophyllum barbigerum* (t. 1942), *Cypripedium* (now *Paphiopedilum*) *purpuratum* (t. 1991) from Hong Kong, and *C.* (now *Paphiopedilum*) *barbatum* (vol. 28: t. 17) from Malaya, *Cattleya* (now *Laelia*) *pumila* (vol. 30: t. 5) from Brazil, *Odontoglossum cervantesii* (vol. 31: t. 36) from Mexico, *Cymbidium mastersii* (vol. 31: t. 50) from India, *Ansellia africana* (vol. 32: t. 30) from South Africa, and the Swan Orchid, *Cycnoches egertonianum* (vol. 32: t. 46) from Mexico.

Other nurseries supplied plants to Lindley but on a lesser scale. Messrs. Rollisson, Springfield Nursery, Upper Tooting, was another London firm active in orchids in the latter years of the *Register*. Rollisson's nursery was eventually taken over by Veitch who, from the late 1840s onwards, assumed the mantle of Loddiges as the leading orchid nursery not long after George Loddiges' premature death in 1846. We see the first stirrings of Messrs. Veitch and Sons of Exeter and Chelsea on the orchid front in the final volume of the *Botanical Register* of 1847. Their plants of *Coelogyne speciosa* (t. 23), *Dendrobium kuhlii* (t. 47), *D. cretaceum* (t. 62), and *Oncidium curtum* (t. 68) are all figured there. The first two were both discovered in Java and the third in Burma by their collector Thomas Lobb (1820–1894).

The artists

No account of the *Botanical Register* would be complete without reference to its artists. There are some 2700 plates in the 33 volumes and 440 of these are of orchids. It is appropriate that a journal which was started by an artist should have attracted some of the finest botanical illustrators of the period. Sydenham Edwards was at the height of his abilities when he founded the *Register*. M. Hart took over as the main illustrator on Edwards' death; he was an accomplished botanical artist though not of the first rank. He contributed over 800 plates to the *Register* in the 1820s and 1830s. Nothing is known of Hart's life but his art is a surviving testimony to his talent.

John Lindley himself contributed fifteen orchid plates to the journal before he took over responsibility for its editing and writing. Then there are occasional plates by other artists, the most notable of whom was Mrs. Augusta Withers who contributed the plate of *Ophrys fusca* (t. 1071). She was artist to Queen Adelaide and taught many members of the Royal Family to paint. Her major botanical contribution on orchids was for James Bateman's *Orchidaceae of Mexico and Guatemala* for which she prepared about half of the plates. The other half was prepared by Miss E. Drake, whose orchid illustrations must rank with the best of the period. Miss Drake's illustrations also adorn John Lindley's classic *Sertum Orchidacearum* and the originals for these survive in the archives of the Royal Botanic Gardens at Kew. Miss Drake worked for Lindley on the *Botanical Register* from volume 17 until it ceased publication. Her first orchid plate is of *Sarcanthus guttatus* (t. 1443) painted on 1st March 1831, and the last is *Oncidium pelicanum* (t. 70), the final plate of the final volume. In all that time she illustrated some 325 orchids over a sixteen-year period. Sadly, the only originals by her that I have been able to trace are those from Lindley's *Sertum Orchidacearum* which are in the Kew collection, while none of the *Register* originals has been located and they probably no longer survive. However, the quality of her art shines out from the pages of the *Botanical Register*. Admire her technique in accomodating some of the large orchids in the rather small format of the *Register*. Good examples of her crisp, clear style and design are the plates of *Schomburgkia undulata* (vol. 31: t. 30), *Odontoglossum cervantesii* (vol. 31: t. 36),

Ansellia africana (vol. 32: t. 30), and *Vanda batemannii* (vol. 32: t. 59). Miss Drake is an enigmatic figure about whom virtually nothing is known, in stark contrast to our knowledge of other eminent artists of the period. It is known that she lived for a time in Lindley's house at Turnham Green in West London and that she was possibly a relative of Lindley's wife.

The cultivation of orchids in the early nineteenth century

For the orchid grower one of the most fascinating features of the *Botanical Register* must be the insight it gives into the early years of orchid growing in this country and the gradual realisation that exotic orchids could be cultivated successfully. The title page of each volume proclaims that the *Register* will give "the best method of treatment in cultivation, propagation etc.". As early as volume 3 Edwards reported on the epiphytic nature of tropical orchids or "air plants [that] possess the faculty of growing when suspended so as to be cut off from all sustenance but that derived from the atmosphere" (t. 220). We are also given a detailed insight into the early methods advocated for growing tropical orchids and from the most eminent of sources: "The most successful mode of treating plants of this nature in these climates, has been devised by Sir Joseph Banks ... the method he pursues is to place the plants separately in light cylindrical wicker baskets or cages of suitable widths, of which the framework is of long slender twigs wattled together at the bottom and shallowly round the side, the upper portion being left open that the plant may extend its growth in any direction through the intervals, and yet be kept steady in its station, the ends of the twigs having been tied together by the twine that suspends the whole to the woodwork of the stove, a thin layer of vegetable mould ist strewed on the floor of the basket, on which the rootstock is placed, and then covered lightly over with a sufficiency of moss to shade it, and preserve a due degree of moisture, water being occasionally supplied". The accompanying illustration of *Aerides paniculatum* (t. 220) shows just such a basket as a vignette in the corner of the plate. Here we have the prototype of the hanging basket that is seen in every orchid greenhouse at the present day.

Not all of the advice would meet with approval today. *Sarcanthus succisus* (t. 1014) being said to "thrive in a mixture of rotten wood and decayed vegetable mould"; while *Sarcanthus guttatus* is "cultivated in the stove, in a very hot damp atmosphere, in a pot full of moss, suspended from the roof by a wire, and a little overshadowed by climbing and other plants". The struggle to find appropriate growing methods is illustrated by the account of *Renanthera coccinea* (t. 1131) which says that "the cause of previous want of success in inducing it to flower, has resided in its having been cultivated in too dry an atmosphere. Mr. Fairbairn, gardener to His Royal Highness Prince Leopold, at Claremont ... tried the effect of tying moss around the stems and keeping it constantly damp, exposed as much as possible to the influence of the sun. With what success ... appears from the accompanying representation of a portion of a panicle, 2 ½ feet long, which was finally produced ... in October 1827".

It took many years for improved methods of cultivation to be worked out. Lindley's report on orchid cultivation for the Horticultural Society, published in 1830 in its *Transactions* (ser. 2,1: 42–50) recommended methods based on providing "high temperature and excessive humidity" that were unsuitable for the majority of tropical orchids which came from the montane forests rather than from the lowlands. The cultivation of orchids in pit houses or stoves often produced temperatures that slowly cooked the orchids. However, Lindley's publication of notes on the natural habitat and ecology of orchids gleaned from collectors such as Harrison and Cuming in Brazil in 1834 (t. 1699) and Lance in Surinam in 1836 (t. 1887) began to alert growers to the diversity of the tropics. The genius of Joseph Paxton eventually sorted out the conflicting advice of the previous years and an account of Paxton's methods for growing orchids is given in the *Botanical Register* in 1838 in volume 24 (sub t. 5). It is appropriate to finish with Lindley's words: "The success with which epiphytes are cultivated by Mr. Paxton is wonderful, and the climate in which this is effected, instead of being so hot and damp that the plants can only be seen with as much peril as if one had to visit them in the Indian jungle, is as mild and delightful as that of Madeira. As to luxuriance of growth, never have they been seen in their native woods in such perfection". Indeed Paxton's cultural tips are just those we find in modern orchid culture books, namely, to ensure good drainage in the pots and to control the temperature and humidity by controlled heating, watering, ventilation and shading.

The end of the Botanical Register

The financial position of the *Botanical Register* was a constant source of worry to Lindley and the publishers. The competition from the *Botanical Magazine,* which appeared to be in terminal decline in the early years of the *Register,* increased when William Hooker's energy brought about its renaissance. Nor was the *Botanical Magazine* the only competition. In 1830 Sweet's *British Flower Garden* merged with the *Register* in the hope, as the *Gentleman's Magazine* (vol. 14: p. 57) of 1838 put it, that it would "render the *Botanical Register* the cheapest, as it is the most beautiful, of all botanical periodicals, and eventually constitute it the completest and most authentic illustrated catalogue of plants in the world." However, the days of the majority of the lavishly illustrated botanical magazines were numbered as cheaper horticultural journals began to appear and woo away subscribers. In an attempt to save the *Register,* Ridgeways, the publishers offered to sell it to Reeve, the proprietors of the *Botanical Magazine.* That the horticultural public favoured the style and content of the *Register* can be judged from the remarks of J. C. Gowen, the Vice President of the Horticultural Society, who wrote to Hooker "I, and, of course, among other people interested in Horticulture will be inclined to substitute the Bot. Mag. and for this reason perhaps you will allow me to suggest the advisableness of placing that publication in a condition to become a harmonious continuation of the Reg." (quoted by R. Desmond [1987] in *A celebration of flowers*). You will appreciate that this approach proved less than attractive to Hooker! Anyway, the proposed merger came too late to save the *Botanical Register* which folded at the end of 1847.

Conclusion

The significance today of the *Botanical Register* lies primarily in its being the source of the descriptions and illustrations of John Lindley's new species, at a time when he was at his most active as an orchid taxonomist. He was in a unique position at the Horticultural Society to benefit from the novelties that were flooding into the British Isles from around the world as gardening fever took hold of the landed classes. Here, we can also find many of Lindley's botanical and horticultural thoughts, notably his early attempts to classify some of the showier and larger genera of orchids, a prelude to his *Folia Orchidacea* of 1852–1859 and his later classifications of the family.

The major advances in orchid cultivation that occurred in this period are also chronicled in the *Botanical Register* and are a source of endless interest to modern growers who often find themselves trying to solve similar problems today.

The charm of the *Botanical Register* lies in its sensitive portrayals of plants by the hand of artists such as Sydenham Edwards, M. Hart and the enigmatic Miss S. A. Drake. Her work, above all, stands comparison with the very best of the ninteenth century botanical artists and outshines much that masquerades today under that name. This publication celebrates belatedly, perhaps, her delicate art as much as it does the prolific talent of John Lindley, her patron.

References and further reading

Blunt, W. (1950). *The art of botanical illustration.* Collins' New Naturalist.

Desmond, R. (1987). *A celebration of flowers.* R. B. G. Kew & Collingridge.

Lindley, J. (1830). *Upon the cultivation of epiphytes of the orchis tribe.* In Trans. Hort. Soc., London ser. 2, 1: 42–50.

Loudon, J. C. (1826). Gardeners' Magazine 1: 59–60.

Veitch, H. J. (1889). *Orchid culture past and present.* Journ. Royal Hort. Soc. n. s. 11: 115–130.

Illustrated by Miss Drake for John Lindley, "Sertum Orchidaceum 1837–41".
This is one of a handful of original watercolour paintings by Miss Drake which are still extant.
The location of the original paintings for the Botanical Register
is not known and it is in fact doubtful whether they survive.
This plate is in the John Lindley Herbarium at the Royal Botanic Gardens, Kew.
The pressed flowers on the sheet are of the type *Galeandra devoniana* Schomb. ex Lindl.

JOHN LINDLEY

(1799–1865)

A Sketch of the Life and Work of a Pioneer British Orchidologist and Gardener-Botanist

by WILLIAM T. STEARN

If John Lindley (1799–1865), the foremost British orchidologist of the nineteenth century, had never interested himself in orchids, he would nevertheless stand out among the scientists of that century for his other achievements. As said elsewhere (Stearn, 1973), he was "a man endowed with an extraordinary capacity for work and a restless, aggressive, untiring intellect, who attained distinction in all his activities. Lindley was among the most industrious, many-sided and productive of the nineteenth-century botanists. As adminstrator, professor, horticulturist, taxonomist, editor, journalist and botanical artist he used to the full his time, his abundant energy and his remarkable talents, with lasting beneficial results in many fields of botany and horticulture". Orchids were, however, his great botanical love. He established more than 120 genera of Orchidaceae and his publications on the family are still necessarily consulted by botanists concerned with its nomenclature and classification. In the *Index Kewensis* 8,252 entries relate to Lindley's publications. For this reason, as well as for the beauty and accuracy of the orchid portraits made under his supervision, it is appropriate to publish anew those published in *Edwards' Botanical Register* to which he contributed in August 1819 "a most ingenious and elaborate description" of *Maranta zebrina* and which he edited from 1829 to 1847. It is also appropriate that the present work comes partly from the Royal Botanic Gardens, Kew, which would probably not exist as a botanical institution and possibly not even as a garden of international repute but for Lindley's forthright report of 1838 on "the management of the Royal Gardens", which led to its salvation and scientific developments.

Linnaeus and Lindley

In many ways Lindley's achievements in the nineteenth century paralleled those of Carl Linnaeus (1707–1778) in the eighteenth century. Obviously the two owed much genetically and educationally to devoted parents with limited financial means, but they made their way in the world by attracting attention to their ability and justifying by fruitful endeavour the support and encouragement they received. Both unfortunately wore themselves out in the process.

The gardening interests of their fathers strongly influenced them both; Linnaeus' father was a Swedish country clergyman with a well-stocked garden in Småland, southern Sweden, Lindley's a skilled but commercially unsuccessful nurseryman in Norfolk, East Anglia. All their lives they combined botany and horticulture in their work and the diversity of orchids fascinated them both. These extraordinary plants require for their study an understanding of their floral structure only to be gained initially from living material. Linnaeus' travels, particularly in Lapland and on the Baltic islands of Öland and Gotland which possess 29 species of orchids, acquainted him with European species; thus on 2 June 1741, he found, much to his surprise, *Ophrys insectifera, Orchis militaris, O. morio, O. sambucina, O. maculata, O. incarnata* and *Gymnadenia conopsea* on Öland. However, exotic species he knew only from herbarium specimens and illustrations. Lindley was more fortunate; abundant living material of tropical orchids continually came into his hands.

Unlike Linnaeus, John Lindley worked during a period notable for improvements in glasshouse heating, for massive importations of orchids from the seemingly inexhaustible moist tropics without regard to conservation of wild stocks, and for their successful cultivation in England, notably by the nursery firm of Loddiges and Sons at Hackney and by Joseph Paxton at Chatsworth. Orchid growers needed names for their new plants. Lindley was one of the few botanists happy and competent to study, classify and name such plants.

Moreover, Lindley's temperament accorded with the prevailing belief of his period in the virtue of hard work, economically necessary and morally right. As Samuel Smiles later made evident in his *Self-Help* (1859) by example after example of "sedulous attention and painstaking industry", these qualities opened the way upward for talented ambitious men equipped with extraordinary staying-power. Thus Lindley's contemporary, William MacGillivary (1796–1852), who became Professor of Natural History at Aberdeen, in 1819 (the year Lindley came to London) walked from Aberdeen to London in order to see the collection of birds in the British Museum. As David Elliston Allen noted in his *The Naturalist in Britain* (1976), "This extreme physical doggedness was matched by astonishing displays of intellectual stamina. In virtually every field of study the Victorian age produced figures of such many-sided energies and accomplishments as to leave us open-mouthed in wonderment and make us feel weak merely reading about them. Natural history has its full share of these. Beside a Pugin or a Paxton it could set a Lindley or a Loudon: men of equally prodigious output and quite invincible in their dynamism: Lindley until the age of fifty never knew what it was to feel fatigue and never took a holiday before he was fifty-two. As Allen also remarks, "Work for such men was more than just an interest that absorbed them; it was a compulsive discharge of effort. Brought up to feel guilt at idleness and an utter abhorrence of sloth, recreation for them could never mean relaxation". Such was the social context for Lindley's monumental achievements. One trusts that Mrs. Lindley, with their three children, from time to time escaped from it.

Early years

John Lindley, born on 5 February 1799, was the son of George Lindley (c. 1769–1835) who, despite his great horticultural knowledge, particularly as a pomologist, ran somewhat unsuccessfully a nursery at Catton near Norwich, the capital city of Norfolk in eastern England and a prominent cultural centre. He gave his son a sound education at the Norwich Grammar School; here John acquired a good knowledge of French and Latin and probably also Greek, to which he later added German; he left it about 1814 or 1815. George Lindley's debts prevented him from giving John a university education at Oxford or Cambridge, then the only two English universities, or buying him a commission in the army. For this botanists should be grateful. A warrior at heart John Lindley waged battle after battle on such issues as the tax on glass, the mismanagement of royal forests and the artificial Linnaeus "sexual system" of plant classification; he was reputed to be even more forcible in speech than writing! As an army officer he would never have had the time or opportunity for so much botanical research, especially on orchids, and he would certainly not have

produced so many important publications. Nevertheless, even when a famous professor, he drilled at weekends as a volunteer in the South Middlesex militia; this seems to have been his only recreation! On leaving school he went to Belgium for a short time as the representative of a seedsman.

William Jackson Hooker (1785–1865), a Norwich man, later professor of botany in Glasgow and director of the Royal Botanic Gardens, Kew, soon became aware of young Lindley's ability. Hooker had a good private botanical library which included a copy of Louis-Claude Richard's *Démonstrations botaniques, ou Analyse du Fruit considéré en général* (Paris, 1808); Richard said he had produced this in seven days. Lindley, in 1818, translated his work into English at one sitting, working at it continuously for nearly three days without going to bed. It was published in 1819 as *Observations on the Structure of Fruits and Seeds,* with notes and 6 plates from drawings by Lindley. On another occasion, when Lindley was a guest of the Hookers, their housekeeper noted suspiciously that his bed had not been slept in. Discreet enquiry revealed that young Lindley, ambitious to become a plant collector and botanical explorer, had slept on the bedroom floor in order to accustom himself to sleeping later on the planks of a ship! In fact he never travelled further than Belgium and Ireland. Hooker gave him a recommendation to the great Sir Joseph Banks (1747–1820), whose herbarium and library constituted the botanical centre of London, and in January 1819 he left Norwich for London. His father supported him with a small allowance until in October he became a paid employee of Banks. Thus began an new momentous chapter in his life.

Work in the Banksian Herbarium

At Banks' house in Soho Square, of which Robert Brown (1771–1858) was the botanist and librarian, Lindley had access to the richest botanical library and herbarium in Britain. Here he worked diligently on the genus *Rosa,* the genera of Rosaceae, and the genus *Digitalis.* The astonishing results were his *Rosarum Monographia; or a botanical History of Roses* (xii + 156 pages; 1820) with 19 engraved plates, 18 beautifully and accurately drawn by himself, and a survey of the Rosaceae subfamily Pomoideae in *Transactions of the Linnean Society of London* 13: 88–106 (1821) in which he established the genera *Chaenomeles, Osteomeles, Eriobotrya, Photinia, Chamaemeles* and *Raphiolepis.* This was followed by *Digitalium Monographia xxx penes Gulielmum Cattley* (1821) and *Collectanea botanica* (1821–1826). There is nothing immature about any of these works, which were a remarkable achievement indeed for a botanist aged only 22. At the age of 21 in 1820 he had been elected a Fellow of the Linnean Society of London and in 1821 C. C. Nees dedicated to him a genus of Theaceae as *Lindleya;* that name has been rejected in favour of *Lindleya* Kunth (nomen conservandum 3328) given in 1824 to a Mexican genus of Rosaceae.

In London, Banks, ever a shrewd judge of men, obviously recognised the ability of this forceful, ambitious young man just as he had earlier recognised the ability of Robert Brown, George Caley, Matthew Flinders and others, and he employed him as an assistent to Robert Brown in his library and herbarium at Soho Square. Banks, however, died on 19 June 1820. This period of 18 months in the celebrated Banksian library not only furthered Lindley's botanical education; there he met people influential in botany and horticulture who were important for his subsequent career.

In the preface, dated "March 31st, 1820", to his *Rosarum Monographia* he stated "To the noble library and inexhaustible Botanical Treasures of the Right Honourable Sir Joseph Banks, with that unexampled liberality for which their illustrious possessor has been ever celebrated, I have been allowed the freest access". This preface also reveals that he was already well-acquainted with Joseph Sabine, who grew a large collection of cultivated roses and was Secretary of the Horticultural Society of London; Sir James Edward Smith, owner of the Linnaean collections; Aylmer Bourke Lambert and Charles Lyell, to whom he dedicated the *Rosarum Monographia,* being much indebted to "his liberality and to his intimate knowledge of the genus". The death of Banks deprived Lindley of his livelihood, but he had gained the friendship of William Cattley (d. 1832), a wealthy merchant who was a keen botanically minded horticulturist and paid him to draw and describe new plants in his garden at Barnet. He also paid for the publication of Lindley's *Digitalium Monographia* (1821), although a deterioration in his business affairs obliged him to cease his support in 1821. That patronage had, however, been for Cattley a good investment. The orchidaceous genus *Cattleya,* which Lindley gratefully named in his honour with *C. labiata* as the type-species, comprises about 30 species in South and Central America and over 100 hybrid cultivars, which have made *Cattleya* a popular florists' orchid, especially for button-holes. In 1820 the Horticultural Society of London employed him to draw roses and larches. Thus began Lindley's association with the Society (later the Royal Horticultural Society) which extended to 43 years.

Employment by the Horticultural Society of London

On 25 February 1822, shortly after Lindley's 23rd birthday, he was appointed Assistant Secretary at the Garden of the Horticultural Society to superintend the collecton of plants, with many other duties but little authority, which remained in the hands of Joseph Sabine. The Society's historian, Harold R. Fletcher, rightly stated in 1969 that "as Assistant Secretary, Vice-Secretary, and for the last five years of his working life as Secretary, he was the backbone of the Society and possibly the greatest servant it has ever had".

His income assured, he married in 1823 Sarah Freestone from rural Suffolk, described by Asa Gray in 1839 as "a quiet lady of plain manners and apparently very domestic habits", and they settled not far from the Society's Chiswick Garden. They had two daughters, who were accomplished artists, and a son, Nathaniel (1828–1921), who became a distinguished lawyer, was knighted in 1875 and made a life peer in 1900. Lord Lindley's son, Sir Francis Oswald Lindley (1872–1950), had a varied diplomatic career. His eldest son Major-General John Edward Lindley (1860–1925) served in the South African war and the First World War. Another, Walter Barry Lindley (1861–1944), was a distinguished lawyer. Thus the Lindley family attained eminence in three generations.

The Collectanea botanica

In April 1821, Lindley began publication of a sumptuous folio work with coloured plates mostly by himself, some plants in from Cattley's garden, *Collectanea botanica or Figures and botanical Illustrations of rare and curious exotic Plants,* which ultimately comprised 41 plates, 16 illustrating orchids. Plate 2 (1821) depicts a new genus and species of Orchidaceae from Trinidad which he named *Trizeuxis falcata.* The accompanying text proclaims the fascination for him of the Orchidaceae which thereafter became his dominant botanical interest:

> If we were requested to select the most interesting from the multitude of vegetable tribes, we should, on the whole, perhaps, be willing to give the preference to the natural order of ORCHIDEAE. Whether we consider general elegance of individuals, durability of blossoms, splendid colours, delicious perfume, or extraordinary structure, it would be difficult to select any order superior to Orchideae in these respects, and few even equal to them. To the cultivator, who esteems plants for their beauty only; to the botanist, who, rejecting outward attractions, is chiefly captivated by anomalous structure, or intricate organisation; or to the amateur, who wisely prefers to have these requisites combined; to all, or any of these, we conceive that accurate figures of foreign Orchideous plants cannot fail to be particularly acceptable. We therefore propose to publish one species of this family in each succeeding number; and we hope to be able soon to add such important information with respect to the treatment of Orchideae, as will entirely remove the present difficulty of cultivating them, and as may enable any one to manage them as readily as the most common plants of our hot-houses.

The last part (1826) of the *Collectanea botanica* contains an Orchidearum Sceletos, a classified list of the orchid genera then known, 154 instead of the 28 listed by Olof Swartz some 20 years earlier. Of these 43 had been published by Robert Brown, 21 by Lindley himself, 12 by Swartz, 10 by L. C. Richard, 4 by Kunth. The family is now (1989) estimated to comprise about 800 genera.

Loudon's Encyclopaedia of Plants

About 1822, if not earlier, Lindley met a lowland Scot as ambitious, methodical, enterprising, tenacious and hardworking as himself and likewise devoted to horticulture and botany, John Claudius Loudon (1783–1843). They were certainly birds of a feather. Loudon intended to compile an encyclopaedia of plants, having completed an *Encyclopaedia of Gardening* (1822). There already existed Robert Sweet's *Hortus suburbanus Londinensis* (1818) and James Donn's *Hortus Cantabrigiensis,* 9th ed. (1819) listing plants in British gardens but Loudon planned a more elaborate and comprehensive work ultimately having nearly 10,000 little woodcut illustrations by James De Carle Sowerby (1787–1871). He enlisted Lindley's collaboration. The title-page of *An Encyclopaedia of Plants* (1829), a work of 1159 closely-printed pages dealing with 14,649 species of flowering plants and ferns, states "The specific characters by an eminent botanist". The preface reveals this botanist to be "Professor Lindley, F. R. S., L. S., G. S. &c." According to Loudon, he "determined the genera and the species to be arranged under them; prepared the specific characters, derivations, and accentuations; he either wrote or examined the notes; and he corrected the whole while passing through the press". In other words, Lindley was responsible for virtually the whole of that massive work. It stands comparison with Linnaeus' *Systema Vegetabilium* (1774). Most botanists would be proud to have achieved as much in a lifetime. Lindley had accomplished it by the age of 30 in addition to all his other time-consuming tasks; it reveals an astonishingly wide acquaintance with plants.

The genera of *Orchidaceae,* which were or had been in cultivation in Britain, here number 61 arranged in 8 tribes, of which Lindley had proposed 6. He concisely described 214 species. His account of the family is as follows:

> Of all tribes of plants, this is the most singular, the most fragrant and the most difficult of culture. The flowers are often remarkable for their grotesque configuration, which has been likened to heads and bodies of animals, and for the strange character of their stems which are sometimes attenuated into a degree of gracefulness scarcely equalled even among grasses and sometimes contracted into a clumsy goutiness of figure such as is known no where else. The species are found inhabiting the mountains and meadows of the cooler parts of the globe, or adhering by their tortuous roots to the branches of the loftiest trees of the tropical forest, to which their blossoms often lend a beauty not their own [i. e. that of the trees themselves]. Vulgarly this last description of plants is called parasitic; they are, however, not so, deriving no support from the juices of the plants on which they grow, but on the contrary are epiphytes, merely adhering to other plants for suport and vegetating in the rich black soil. It is very singular that the pollen of these plants has no parallel, except among the very different and distinct order of Asclepiaceae.

The years 1826–1840

Lindley, of course, depended upon the Horticultural Society for his livelihood. Sabine's autocratic management of the Society's affairs worried him as leading to disaster but he could do nothing except protest personally to Sabine. Expenditure, at Sabine's optimistic instigation, on projects that, although worthy, were too many and too costly for the Society's means, by 1826 had brought it near ruin, especially as John Turner, the Society's Assistant Secretary, had embezzled part of its funds. The Society dismissed Turner in 1826. In 1827 it appointed Lindley as Assistant Secretary of the Society as well as of the Society's Garden, but insisted that he provide a security for £ 1,000. With the Society's affairs in so perilous a state and worsening in 1829, Lindley must have felt his own position insecure, even though he had protested to Sabine about the latter's actions. Obviously he would have welcomed a new, even if only supplementary, appointment. It came in 1829.

Meanwhile in 1827, Protestant dissenters excluded on religious grounds, as were Jews and Roman Catholics, from the only two English universities, Oxford and Cambridge, had been agitating for a non-sectarian university in London. Their success resulted in the foundation of London University (later renamed University College),

for which, of course, professors were needed. In 1829 Lindley, who had been elected a Fellow of the Royal Society of London in 1828 on the recommendation of Robert Brown, Thomas Andrew Knight and Joseph Sabine, was appointed the first Professor of Botany in the University of London, retaining, however, his post in the Horticultural Society. To reach his classroom at University College he had to ride on horseback from Turnham Green. "He is constantly in his lecture-room at University College, both in Winter and Summer courses at the appointed time. On these occasions he forms a remarkable contrast to the pupils by whom he is surrounded. He is a fresh, ruddy, hale-looking man, and after his morning's ride, in the midst of fresh-plucked plants, presents an appearance entirely different from that of the pale students, who have generally been but a few minutes before roused from their too scanty slumbers." Teaching over, he had to ride back to Turnham Green, attend to the work in the Society's garden and the Society's business, then find time for writing, editing and the study of orchids and other plants.

In 1829, after his professional appointment, Lindley published *A Synopsis of British Plants,* of which a second edition appeared in 1835 and a third in 1841. He had, however, other works on hand. Among these was the first part of *The Genera and Species of Orchidaceous Plants* (1830–1840).

"Students", the new Professor stated, "are particularly requested not to furnish themselves with any introductory work by the late Sir James Edward Smith or Dr Thornton", both Linnaean devotees. He had therefore to provide them with text-books of his own. Accordingly in 1830 he published *An Outline of the First Principles of Botany,* which was later translated into German and Italian, and *An Introduction to the Natural System of Botany.* The same year he began the publication in parts of two important works on orchids. The great Austrian-born resident botanical artist at Kew, Franz Andreas Bauer (1758–1840), had produced some of the finest illustrations of the floral structure of orchids ever made, accurate, remarkably detailed and aesthetically pleasing. Lindley, with customary enterprise, provided the text to these *Illustrations of Orchidaceous Plants by Francis Bauer – with Notes and Prefactory Remarks by John Lindley,* with 35 lithographed folio plates (part 1 in 1830; 2 in 1832; 3 in 1834; 4 in 1838). Much less sumptuous was the octavo *The Genera and Species of Orchidaceous Plants* (1830–1840 in seven parts). This provided a basic classification of the family and short diagnoses and descriptions in Latin of all the orchids then known, many being new to science as herbarium specimens from botanical collectors and living specimens from orchid growers continuously came into his hands for identification and naming.

In 1829 Lindley took over the editorship of *The Botanical Register.* Like the contemporary and rival *Curtis's Botanical Magazine,* it portrayed in colour a diversity of plants. Lindley now controlled an excellent medium for graphically recording new orchids. The present work brings together the material on orchids scattered through its volumes. *The Botanical Register* was not financially so succesful as the older *Botanical Magazine* and it ceased publication in 1847, by which time Lindley, who had edited vols. 15–53 (1829–1847), was actively involved in editing the *Gardener's Chronicle* (1842 et seq.) as well as the *Proceedings* and *Transactions* of the Horticultural Society of London. Fervently interested in popular education he contributed a large number of botanical articles, up to the letter R., to *The Penny Cyclopaedia of the Society for the Diffusion of Useful Knowledge* (27 vols.; 1833–1843); this was an excellent authoritative work packed with information. Lindley's contributions were unsigned but his hand is very evident in articles such as, for example, that on *Bambusa* (3: 355–357; 1835), wherein he published eight new specific names, and that on *Orchidaceae* (16: 476–479; 1840). Of like educational intent were his *An Introduction to Botany* (1832; 4th ed., 1839), *Flora Medica* (1838), *School Botany* (1839; 12th ed. 1862) and *The Theory of Horticulture* (1840), of which a revised and extended new edition appeared in 1855, as *The Theory and Practice of Horticulture,* wherein he gave reasons for the whole range of best horticultural practice.

Lindley wrote in a clear direct manner, which makes his works still readable, conveying much information with precision. He gave especial attention to botanical terminology. The language of botany, "from one cause or another, whether accident, ignorance, pedantry, overfastidiousness, vanity or carelessness", he wrote, was "marvellously in want of reformation". This he sought to effect by providing a logical and comprehensive glossary of botanical Latin terms, with their English equivalents and definitions, in his *Introduction to Botany;* it formed the basis of botanical terminology in English-speaking countries. Being permanently valuable, it is reprinted in W. T. Stearn's *Botanical Latin,* 313–357 (1966; 3rd ed., 1983) from the *Introduction to Botany,* 3rd ed. (1839). He proposed that the names of *all* botanical families should end in *-aceae;* accordingly he replaced Jussieu's *Orchideae* by *Orchidaceae.*

Lindley, however much he had on hand, seems never to have recoiled from accepting another onerous task. One such was the *Fossil Flora of Great Britain* (3 vols.; 1831–1837) in collaboration with the geologist William Hutton (1797–1860). A summary of what was then known, this owed much of its value to plates drawn by William Crawford Williamson (1816–1895), then a medical apprentice aged about 17, later a celebrated palaeontologist and botanist. A catalogue of known extant specimens figured has been published by A. Newman and I. Chatt-Ramsey in 1985. When Williamson arrived at University College in 1840 to further his medical education, Lindley, who had been so precocious himself, was surprised that his collaborator had begun so young. Another unexpected task of this period was the completion of Sibthorp and Smith's *Flora Graeca.* When Sir James Edward Smith (1759–1828) died, he had published volumes 1–6 (1806–1827) and had prepared the text for volume 7 fasc. 1 (1831) which his friend Robert Brown saw through the press. There remained the preparation of 3 ½ volumes to complete the planned 10 volumes of this great work with its superb plates by Ferdinand Bauer. John Sibthorp's executor and cousin John Hawkins (1761–1841), a founder of the Horticultural Society, turned for help to Lindley, Smith's critic and opponent. Lindley then had relatively little knowledge of the Mediterranean flora but that did not deter him and with the publication in November 1840 of volume 10 fasc. 2, he ended a project which had begun some 50 years earlier.

For this he received £350, a large sum then, equivalent to a year's salary, and very welcome because years earlier he had gallantly but unwisely undertaken to pay his father's debts. It is of interest that this great work was written from start to finish by two Norwich botanists and elegantly printed by a printer from Norwich, Richard Taylor.

Honorary doctorate

Apart from election to Fellowship of the Royal Society of London, Lindley received no British honours in his long and productive career. The celebrated Bavarian botanist Carl F. P. Martius (1794–1868) on a visit to England had been very hospitably received by Mr. and Mrs. Lindley and he was obviously much impressed by Lindley's character and publications. Accordingly he enquired in 1832 if Lindley would accept the degree of Doctor Philosophiae from the University of Munich. The diploma came in 1833. This German degree had then no British academic equivalent, the nearest being Magister Artium with which it was earlier synonymous. Henceforth Lindley was designated Dr Lindley, if not Professor Lindley, thanks to this enterprise of the University of Munich in conferring such a degree on a foreigner. He may well have been the first British subject to have been thus honoured. Only those without a university degree can imagine the emotional impact of receiving such foreign academic recognition.

Crisis in the affairs of the Horticultural Society

Sabine's mismanagement of the Horticultural Society's affairs, above all its mounting debts, became by 1830 a matter of public concern and of outspoken angry comment in the press. A Member of Parliament and Fellow of the Society, Robert Gordon, accordingly headed a Committee appointed by the Fellows to enquire into the Society's income and expenditure, debts and assets. Sabine, the Honorary Secretary who had worked hard but not always wisely for the Society over 14 years, resigned. Lindley had now to defend himself. A letter of his to Sabine came before the Committee in which among much else he stated "You know perfectly well that I have always protested against the statements by which the Council have frequently been deluded into sanctioning measures and expenditure, which had they known the real state of the Society's affairs, they could not have countenanced... I have never been a party to the exaggeration of the Society's means, and concealment of the Society's debts, by means of which many honourable and excellent men in the Council have unfortunately been induced to believe a ruined Society to be in a state of prosperity". It left him securely in charge to work with George Bentham (1800–1884), the new Honorary Secretary, in restoring the Society's fortunes. Among other innovations he started in 1831 a series of lectures to Fellows on botany in relation to horticulture, forerunners of the lectures at Royal Horticultural Society flower shows.

The Sertum Orchidaceum

These administrative duties in no way abated Lindley's study of orchids or the acceptance of other responsibilities. He continued to publish revisions of orchid genera as parts of his *The Genera and Species of Orchidaceous Plants,* the last part appearing in 1840. By then he had begun publication of a sumptuous folio work appealing to the increasing number of wealthy people interested in growing orchids, *Sertum orchidaceum or a Wreath of the most beautiful Orchidaceous Flowers* (1837–1841) with 49 coloured lithograph plates. Virtually nothing is known about the artist, Miss S. A. Drake, who skilfully portrayed so many orchids and other plants for Lindley; according to Asa Gray, she was apparently a member of his family and may have been related to his wife; the genus *Drakaea* Lindley commemorates her. Lindley dedicated the work to William Spencer Cavendish, 6th Duke of Devonshire (1790–1858), had a collection of orchids at Chatsworth which was probably the finest in Britain, being under the care of Lindley's friend Joseph Paxton (1803–1865). In the text to plate 1 depicting *Stanhopea Devoniensis* Lindley (now *Stanhopea hernandezii* (Kunth) Schlechter) he quoted Paxton's directions on cultivation and commented: "The success with which epiphytes are cultivated by Mr. Paxton is wonderful" and "the climate in which this is effected, instead of being so hot and damp that the plants can only be seen with as much peril as if one had to visit them in an Indian jungle, is as mild and delightful as that of Madeira. As to luxuriance of growth, never have they been seen in their native woods in such perfect beauty." Although Miss Drake drew many illustrations from living plants, she had to base others on drawings sent to Lindley, for example, plates 2, 5, 7 and 11 on drawings made in Brazil by J. Théodore Descourtilz (not to be confused with the naturalist Michel Étienne Descourtilz, whose *Flore médicale des Antilles* Théodore illustrated), 9, 29 and 40 made in British Guiana by Robert H. Schomburgk, t. 8 by William Griffith, t. 34 by Miss M. A. Mearns and t. 37 *(Dendrobium nobile)* from a Chinese drawing.

Lindley and Paxton

In ability, enterprise, industry and efficiency, in their rise to eminence from lowly beginnings and their devotion to horticulture, Lindley and Joseph Paxton (1803–1865) were much alike; they were also masterful and authoritative men. Their successful careers, linked by interests in common, notably in orchids, began in the sevice of the Horticultural Society of London, but Paxton's career by its versatility was even more remarkable than Lindley's. The seventh son of a Bedfordshire farmer, Paxton worked in various gardens until in 1823, at the age of 20, he became a gardener in the Chiswick garden of the Horticultural Society, to which Lindley had been appointed garden secretary in 1822. There began their life-long friendship. The Society's garden was on land leased from the Duke of Devonshire, William Spencer Cavendish (1790–1858), one of whose residences, Chiswick House, adjoined the garden. Paxton, who cannot have had more than a few years of basic village schooling, had obviously set himself on a rigorous course of self-improvement such as he later commended to young gardeners and, like Lindley, he already manifested that "energetic individualism which produces the most powerful effects upon the life and action of others" so much praised in Samuel Smiles' *Self-Help* (1859). They certainly exemplified Smiles' contention that "self-culture also calls forth power and cultivated strength". Thus when the Duke, a

deaf and somewhat lonely unmarried man, wandered in the Horticultural Society's garden he enjoyed talking to this intelligent self-educated young gardener. He became so impressed by Paxton's ability that in 1826 he appointed him, aged only 23, as head gardener at his vast Chatsworth estate in Derbyshire. This may have been a hasty decision but the Duke had good reason all the rest of his life to be grateful for it.

Never a man to lose an opportunity, Paxton set off by coach from London to Chesterfield, Derbyshire, presumably walked during the night the 12 miles or so from there to Chatsworth and arrived at 4:30 am on 9 May 1826. Naturally he found nobody about. He accordingly, in his own words, "explored the pleasure grounds and looked round the outside of the house. I then went to the kitchen gardens, scaled the outside wall and saw the whole of the place, set the men to work at six o'clock ... and afterwards went to breakfast with poor dear Mrs Gregory and her niece. The latter fell in love with me and I with her, and thus completed my first morning's work at Chatsworth before nine o'clock." They married nine months later, none of them with the slightest premonition that he would later be the wealthy Sir Joseph Paxton, MP, and she the astute and formidable Lady Paxton.

The Duke of Devonshire (later President of the Horticultural Society) at this time had little interest in horticulture but he caught Paxton's enthusiasm, relied on his efficiency, trusted his judgment and whole-heartedly furthered his endeavours as gardener, forester, glasshouse designer and landscape architect. These are well covered in the works of Violet Markham (1935), George F. Chadwick (1961) and John Anthony (1973). The strangeness of *Oncidium papilio*, a species introduced from Trinidad and described and named by Lindley in 1825, aroused the Duke's interest and led him to assemble a rich collection of orchids at Chatsworth. The Duke's example and Paxton's success in growing orchids, at a lower temperature and with less humidity than had been customary, stimulated others to cultivate them. Among these were the Rev. John Clowes (1777–1846), commemorated by the genus *Clowesia* Lindley, Charles W. W. Fitzwilliam, Viscount Milton (1786–1857), commemorated by the genus *Miltonia* Lindley, whose head gardener Joseph Cooper was esteemed by Paxton as an expert orchid grower, and James Bateman (1811–1897), author of the gigantic *Orchidaceae of Mexico and Guatemala* (1837–1843) and *A Monograph of Odontoglossum* (1864–1874) commemorated by the genus *Batemannia* Lindley; the doubling of the *n* was an accepted linguistic procedure for keeping the preceeding vowel *a* short. The commercial growing of orchid plants expanded to meet the increasing public demand, as did the introduction of epiphytic species from the moist tropics of America and Asia. Many of these came into the hands of Lindley for description and naming and Paxton for cultivation. During their early years the celebrated nurserymen Conrad Loddiges and Sons of Hackney, a firm created by Conrad Loddiges (c. 1739–1826) and maintained by his son George (1784–1842), together with William Rollison (c. 1765–1842) and his son George (1796–1879) of Upper Tooting, were the leading commercial growers of orchids. The firms of James Veitch, which acquired Rollison's orchids, Hugh Low, Benjamin S. Williams and Henry F. C. Sander rose to eminence later. Lindley's orchidological successors, Heinrich Gustav Reichenbach (1824–1889) and Robert Allen Rolfe (1855–1921), profited from their collections.

In 1831 Paxton started the *Horticultural Register and general Magazine* (5 vols., 1831–1836); then, in 1833, *Paxton's Magazine of Botany and Register of Flowering Plants* (16 vols., 1833–1849), followed, in collaboration with Lindley, by *Paxton's Flower Garden* (3 vols., 1850–1853). In 1841 he and Lindley founded *The Gardeners' Chronicle*. Paxton's interests by then had extended beyond simple gardening to the design of glasshouses and dwelling houses; in 1836 he began for the Duke the Great Conservatory at Chatsworth, completed in 1840 and demolished in 1920, and in 1838 to rebuild on a new site the village of Edensor. He was already investing in railways. In 1850 his design for the Crystal Palace to house the Great Exhibition was accepted and in 1851 Queen Victoria knighted him for his work on it. After the death of the Duke in 1858 Paxton resigned all his offices at Chatsworth, which is hardly surprising in view of his many other commitments as a Member of Parliament for Coventry since 1854, as an architect and as a director of the Midland Railway since 1849. He also became Chairman of the Select Committee on the Thames Embankment.

Paxton's capacity for sustained unceasing work, like Lindley's, had limits which he, also like Lindley, refused to recognize until it was too late. His health failed in 1865, as Lindley's had done earlier, and they both died that year. Their joint endeavours had included the saving from destruction of the Royal Botanic Gardens, Kew, as detailed below, and the compilation of *A Pocket Botanical Dictionary* (1840). It is fitting that their joint interest in orchids should be commemorated by the genus *Paxtonia* Lindley and joint authorship of *Arhychium, Ornitharium* and *Sarcopodium* Lindley & Paxton.

The Chelsea Physic Garden

The Apothecaries Act of 1815 legalised the standing of apothecaries as general practitioners of medicine provided they had been licensed by the Society of Apothecaries in London, a qualification abbreviated to L. S. A. To obtain this, an apprentice apothecary had to serve not less than 5 years under an approved apothecary and then 6 months at least in the medical practice of a public hospital (cf. Bishop, 1953). The rigorous examination tested, among much else, a knowledge of materia medica which included medicinal plants. For these the Society of Apothecaries maintained their Chelsea Physic Garden. In 1835 the Society appointed Lindley as their Professor of Botany and he lectured there regularly early morning, riding over from Turnham Green on horseback, until the abolition of the professorship in 1853. A resulting publication was his *Flora medica* (1839) which may have prompted the University of Basel to confer on him the honorary degree of Doctor of Medicine. The Chelsea work was in addition to his duties as Assistant Secretary of the Horticultural Society, Professor of Botany in the University of London and editor of the *Botanical Register*, then of *The Gardeners' Chronicle*, and his private study of orchids.

Saving of the Kew Botanic Garden

Thus, by 1837, Lindley had become a prominent figure in British botany and horticulture while Paxton had become known as one of the country's most efficient and enterprising head gardeners with his status enhanced by the Duke of Devonshire's appreciation of his managerial ability and devoted service; the two made a formidable pair. The Lords of the Treasury objected to the expense of maintaining the Royal Botanic Garden at Kew and would have gladly got rid of it. To this end they appointed a committee of three, consisting of Lindley, Paxton and a more acquiescent practical gardener, John Wilson, head gardener of the Earl of Surrey, the Lord Steward. They inspected the Kew Garden in February 1838, a good time for seeing it at its worst, and in March 1838 Lindley produced a critical report, which was two-edged. He set out practical and patriotic reasons for not abandoning the Kew Botanic Garden as the Government had intended but on the contrary advocated that it should "be at once taken for public purposes, gradually made worthy of the country and converted into a powerful means of promoting national science", a national botanic garden which would be a botanical and horticultural centre for the British colonies, with herbarium and library. Such an unwanted report might have passed into obscurity but for the fact that Lindley had it copied and put copies into the hands of the Duke of Bedford who wanted "a great National and Royal Botanic Garden" at Kew, the Duke of Devonshire who was both Paxton's master and President of the Horticultural Society, and other far-sighted influential men. However, in February, the Secretary of the Treasury offered the plants of Kew to the Horticultural Society for its garden at Chiswick. Lindley and Bentham acted promptly to thwart this proposal, which would have ended the Kew Garden as a botanic garden, by calling a meeting of the Society's Council. They rejected this destructive proposal. Ignoring the political power of Lindley's ducal allies, Lord Surrey then over-played his hand by ordering the destruction of the Australian and South African plants in the Kew glasshouses so that these could be converted into vineries. He was evidently unaware that Queen Victoria, who had in 1839 favoured giving up the garden, had now learned of its founding by Augusta, Princess of Wales (after whom the present Princess of Wales House, opened in 1987 by Diana, Princess of Wales, is named) and with Prince Albert had become more interested in its maintenance. Public indignation was such that Lindley's report could no longer be officially ignored. It was laid before Parliament in May 1840. The Kew Botanic Garden and pleasure grounds passed to the Commissioners of Woods and Forests, thereby becoming public property. Thus ended what Lindley described as "the barbarous Treasury scheme of destroying the place" and he more than any one individual had brought about its salvation.

Lindley and his old Norwich friend, William Jackson Hooker, since 1820 Professor of Botany in Glasgow, were both suitable candidates for the directorship of the nationalised Kew Botanic Garden. The Duke of Bedford supported Hooker, the Duke of Devonshire supported Lindley. Lindley had now a secure position. Hooker was very anxious to return to England; although 54, only 6 years off the age when present directors, whatever their energy and experience, are obliged to retire, Hooker faced no such restrictions. He would in fact have 24 years of creative endeavour at Kew ahead of him. In 1838 with the Kew directorship in mind he had already written to Lindley from Glasgow that "though I have nothing to complain of here yet, I have reason to believe that my income is much less than yours and because I am willing to make some sacrifice to be enabled to return and spend the rest of my days among my friends and connections in England". Lindley stood down and in March 1841 Hooker learned that he had been appointed as director: "Having no instructions for my guidance I determined to follow the suggestions of Dr Lindley's report". Thus two Norwich men shaped the future of what was to become a world-famous institution, the Royal Botanic Gardens, Kew. Both died in 1865.

The Potato Blight

This was by no means the end of Lindley's association with the government. The potato blight epidemic of 1845, which caused so much distress and famine in Ireland and, less publicised, in the Western Highlands and islands of Scotland, led to Lindley going to Ireland as a member of an official commission. The rot in potatoes was associated with the presence of a fungus believed by Lindley to be a consequence of the decay, thus secondary to it, but rightly held by the Rev. Miles Joseph Berkeley and contemporary mycologists to be its cause. There was no doubt whatever of the terrible devastation and the report by Lindley, Robert Cave and Lyon Playfair caused in 1846 the repeal of the 1815 Corn Laws restricting the importation of cheap wheat, for which the Anti-Corn-Law League had so long fought, with far-reaching political consequences.

The Vegetable Kingdom

In 1846 Lindley published his most ambitious work, *The Vegetable Kingdom, or the Structure, Classification and Uses of Plants* (lxviii + 908 closely-printed pages with 626 illustrations), the culmination of his many years of detailed study of plants and their literature. It remains an encyclopaedic work too rarely consulted, for it brings together a wealth of information not so readily available elsewhere. Lindley begins with a judicious historical survey of plant classification from John Ray onwards, then sets out his own detailed accounts of all the families of plants, exemplified by the account of the Orchidaceae given below, and provides a key to the families; in these he paid attention to a much wider range of characters than had his predecessors. Originally he had intended to produce a *Genera Plantarum*, a task so vast that he could have achieved it only by relinquishing his manifold other activities; obviously much information collected for such a work found publication in *The Vegetable Kindgom*. He learned, however, that the Austrian botanist Stephan Ladislaus Endlicher (1804–1849) had such a work in preparation. Endlicher published in 19 parts between 1836 and 1841 a *Genera Plantarum secundum Ordines naturales disposita* (lx + 1483 pages, with 4 supplements between 1842 and 1850), with which Lindley's intended work would have unnecessarily competed. Endlicher recognised 367 genera of Orchidaceae, of which 117 had been established by Lindley.

The Vegetable Kingdom was arranged according to

what Lindley considered a "natural system". He regarded the characters of plants as "the living Hieroglyphics of the Almighty which the skill of man is permitted to interpret. The key to their meaning lies enveloped in the folds of the Natural System". He never succeeded, as he continually changed his own system, in devising one that he found wholly satisfactory, especially because he believed "physiological characters are of greater importance in regulating the natural classification than structural". Honestly accepting the charge of inconsistency, he vigorously defended the propriety of such changes: "The Author may now be equally charged with inconsistency in not adhering to his former plan of classification after having promulgated it. But he is not conscious of having ever pretended that it even approached permanency. – See *Natural System*, p xiii. In fact there is no such thing as stability in these matters. Consistency is but another name for obstinacy... The Author cannot regard perseverance in error as commendable, for the sake of what is idly called consistency; he would rather see false views corrected as proof of their error arises. His object has not been to establish a system of his own, which shall be immutable, but to contribute to the extent of his ability... All that we can do is to throw our pebbles upon the heap."

Owing to its faulty premises Lindley's classification was never accepted as a whole. Julius von Sachs in his *Geschichte der Botanik* (1875; English translation, *History of Botany*, 1890) described it as "one of the most unfortunate ever attempted... We find this opposition between theory and practice much more strongly marked in Lindley than in De Candolle... De Candolle laid down correct principles for determination of affinities but in some cases did not follow them, whereas Lindley deduced quite incorrectly the rules of system from existing and long established natural affinities". Lindley took the view that the character of an organ becomes more and more important in classification to the degree that its physiology serves for the preservation and propagation of the individual. Sachs remarked that "by this principle he judges of the systematic value of anatomical characters, those of the embryo and endosperm, of the corolla and the stamens, everywhere laying stress on their physiological importance... This mode of proceeding on the part of Lindley, compared with his own system, which with all its grave faults is still always a morphologically natural system, proves that like many other systematists he did not literally and habitually follow the rules he himself laid down... The success which was really obtained in the determination of affinities was due chiefly to a correctness of feeling, formed and continually being perfected by constant consideration of the forms of plants". From this Sachs rightly concluded that men like Lindley were never quite clear about the rules by which they worked. Lindley had great artistic ability and he sought a pattern in nature; thus like John Hutchinson, also of great artistic ability, he perceived correlations which did not necessarily tally with the theoretical justification he gave them. The lasting value of Lindley's great work lies not at all in his discarded general system but in the detailed accounts of individual families.

A second edition, little altered, appeared in 1847, the third and last in 1853, within 10 years of his breakdown in health. As he stated, this third edition, having extra inserted pages, contained much new matter; moreover, better illustrations replaced many of those in the original edition. Here were his final opinions on the definition and content of families. For that reason the account of Orchidaceae provides a summary of his conclusions on its classification. This is as follows:

>
Lindley's account and list of the genera of orchidaceae
(pp. 24–40)

THE VEGETABLE KINGDOM;

OR,

The Structure, Classification, and Uses of Plants,

ILLUSTRATED UPON THE NATURAL SYSTEM.

BY

JOHN LINDLEY, PH. D. F.R.S. & L.S.

CORRESPONDING MEMBER OF THE INSTITUTE, PROFESSOR OF BOTANY IN THE UNIVERSITY OF LONDON.

"Methodum intelligo naturæ convenientem quæ nec alienas species conjungit, nec cognatas separat."—*Raii Sylloge, præf.*, p. 15.

WITH UPWARDS OF FIVE HUNDRED ILLUSTRATIONS.

THIRD EDITION,

WITH CORRECTIONS AND ADDITIONAL GENERA.

LONDON:
BRADBURY & EVANS, 11, BOUVERIE STREET.
1853.

ORDER LII. ORCHIDACEÆ.—ORCHIDS.

Orchides, *Juss. Gen.* 64. (1789).—Orchideæ, *R. Brown Prodr.* 309. (1810); *Rich. in Mém. Mus.* 4. 23. (1818); *Bauer, Francis, and Lindley, Illustrations of Orchidaceous Plants; Id. Genera and Species of Orch.* (1830); *R. Brown Observations on the Sexual Organs, &c. of Orchideæ and Asclepiadeæ* (1831); *Endl. Gen.* lxvi.; *Meisner, Gen. p.* 367.—Vanillaceæ, *Ed. pr.* ccxliv.

DIAGNOSIS.—*Orchidal Endogens, with irregular gynandrous flowers and parietal placentæ.*

Herbaceous plants or shrubs, always perennial, occurring all over the world, except in the very coldest regions, or those where everlasting dryness reigns; in temperate countries terrestrial, in warmer latitudes growing on trees (epiphytes), or fixing themselves to stones. Their roots are fibrous and fasciculated, fleshy or resembling tubers, and these filled with starch, or horny nodules of bassorin. Stem none, or long and annual, or perennial and woody, forming a rhizome, or jointed branches. Leaves flat, terete or equitant, generally sheathing, membranous, coriaceous, or hard, never lobed, occasionally bordered by cartilaginous teeth, their veins parallel, almost never slightly reticulated. Flowers ⚥, irregular, extremely variable in form, solitary, clustered, spiked, racemose, or panicled, always supported by a solitary bract; very often most gratefully fragrant, sometimes fetid, and not unfrequently scentless. Perianth adherent, variable, herbaceous or coloured, membranous or fleshy, permanent and withering, or deciduous; its parts arranged in two rows, rarely in 3, free or adhering in various ways; very often resupinate in consequence of a twist in the ovary. Sepals (which, morphologically speaking, are petals) 3, equal at the base, or variously extended or expanded there; the two lateral standing in front when the ovary is twisted, and the third then dorsal, or

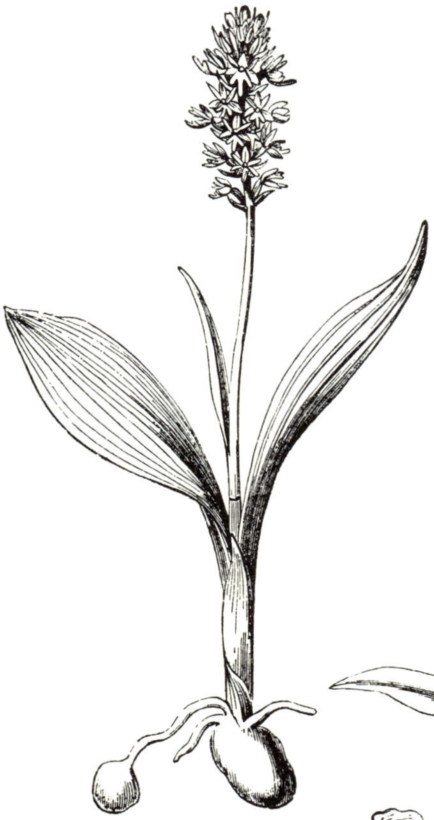

Fig. CXVII.

Fig. CXVIII.

Fig. CXVII.—Herminium monorchis.
Fig. CXVIII.—Accidental manner of producing a jointed stem in Aspasia epidendroides.

next the axis ; occasionally surrounded by a calyculus (or true calyx). Petals (which are to be regarded as sterile stamens) usually 3 ; very rarely one only, placed between the sepals : the lateral usually similar to the dorsal sepal ; the third, called the lip (*labellum*), usually larger than the petals, and quite unlike them in form ; horned or furnished with various appendages, free or adherent to some other body, occasionally moveable as if spontaneously ; now and then contracted so as to form two separate parts, of which the lowest is called the hypochil, the highest the epichil, and the middle one the mesochil; sometimes furnished with a single or double appendage, derived from the stigma. Column consisting of the stamens and style consolidated into a central body, so that the latter stands next the lip and the former next the dorsal sepal, sometimes petaloid, and occasionally extended far beyond the perianth (corymbis). Stamens 3, opposite the sepals, the central only being perfect, except in Cypripedium, when the central is abortive and the two lateral perfect ; anthers occasionally one-celled ; usually two-celled, with the cells separated by 2 or 4 partitions ; standing erect at the end of the column, or turned down flat upon it, or altogether dorsal ; pollen powdery, or collected into grains, or adhering in wedges tied together by an elastic material, or consolidated into masses of a waxy texture and fixed number, the masses either free or adhering by a caudicle to a gland belonging to the apex (or rostellum) of the stigma. Ovary adherent, 1-celled, composed of 6 carpels, of which 3, opposite the petals, have didymous polyspermous parietal placentæ without stigmas, and 3 opposite the sepals have as many stigmas but no placentæ; style never distinct, except in Cypripedium and some Neottieæ ; stigmas usually confluent in a hollow (or prominent) mucous disk ; the dorsal stigma having on the upper edge one or two glands, which are separate in Vandeæ and Neotteæ ; often extended into a beak (rostellum), or hollowed out into pouches, or sometimes drawn out into 2 parallel or diverging arms ; the lateral stigmas usually obsolete, but sometimes united to the base of the lip in the form of an appendage or pair of plates. Capsule very rarely fleshy, indehiscent and pod-shaped, usually breaking up into 6 dry woody rigid valves with horizontal cells, of which 3 only bear seeds. Seeds innumerable, very minute, with a loose netted skin, very rarely with a hard crustaceous one, sometimes expanded into a circular wing ; embryo solid, fleshy, without albumen ; chalaza at the apex of the seed, and therefore the radicle next the hilum.

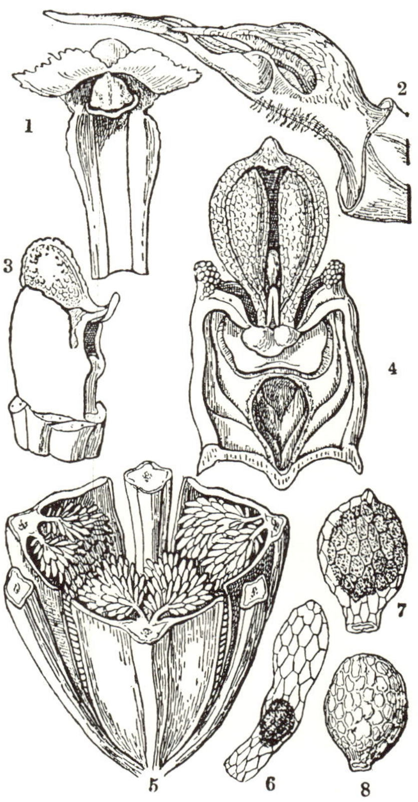

Fig. CXIX.

The general structure of Orchids, briefly embodied in the foregoing description, has been treated of at such length in the prefatory matter of the *Illustrations of Orchidaceous Plants*, that it is unnecessary to do more than refer the reader to that work. I must, however, take the opportunity of correcting one part of the theoretical view which was there taken of the structure of the column. While, in common with Dr. Brown, I regarded the stigma as really consisting of three parts, usually in a state of confluence, I also supposed the position of the stigmata to be opposite the petals ; being led to that conclusion by the constant position of the stigmatic arms of Ophrydæ. That opinion I afterwards retracted, in consequence of the position of the stigmas in Cypripedium, which C. spectabile shows most clearly to be opposite the sepals ; and therefore the stigmatic arms of Ophrydæ are to be understood as side lobes of that stigma which is opposite the dorsal sepal. This circumstance, however, only confirms the accuracy

Fig. CXIX.—1. Column of Arethusa ; 2. of Stenorhynchus ; 3. of Brassia maculata ; 4. of Orchis mascula ; 5. section of capsule of Ophrys apifera ; 6. seed of Ophrys ; 7. of Pterygodium atratum ; 8. of Vanilla aromatica.

of my view of the true nature of the stamens, which are certainly all opposite the lobes of the stigma in Cypripedium. While, however, the untenableness of the first opinion concerning the relation borne by the stigmas to the other parts of the flower, is thus admitted, there remains a difficulty that opposes itself to the view I now take in common with Brown, and which must not be overlooked. It is that the placentiferous pieces of the ovary are not opposite the stigmas, but alternate with them, while the seedless pieces of the ovary are in a line with the stigmata ! This seems to show that the ovary is composed of 6 carpellary leaves, of which three bear stigmas without ovules, and three bear ovules without stigmas. However paradoxical this may appear, it is by no means incompatible with the due performance of the functions of fertilisation ; for the carpellary leaves do not adhere into a solid mass, either in the ovary or in the style. On the contrary they form a cavity open from the stigmatic apex down to the ovules, and the whole of that cavity is lined with a lax conducting tissue, which may nevertheless be exclusively furnished by 3 stigmas only, and may become so confluent with the placentæ as to form a perfect channel of communication for the pollen tubes in their descent into the ovules.

The Order owes its chief peculiarities to the following circumstances: firstly, to the consolidation of stamens and pistil into one common mass, called the column ; secondly, to the suppression of all the anthers, except one in the mass of the Order, or two in Cypripedeæ ; thirdly, to the peculiar condition of its pollen, and the anther which contains it ; and fourthly, to the very general development of one of the inner leaves of the perianth or petals in an excessive degree, or in an unusual form. These peculiarities are in most cases so striking, and are all so strongly manifested in the same flower, that the inexperienced botanist may be unable to discover their real character. We find, however, that the true nature of each part is indicated by special cases of structure occurring in different parts of the Order. Thus in Cypripedium not only are two lateral stamens furnished with anthers, while the central stamen is antherless, but the stigma and style separate from the filaments nearly to the base, and the triple nature of the former is distinctly shown, together with the relation of its lobes to the other parts of the flower. The pollen, which has so anomalous an appearance in its waxy or sectile state, presents the usual appearance of that substance in Goodyera, and many Neotteæ. And the irregularity of the labellum disappears in such genera as Thelymitra, Paxtonia, Macdonaldia, Hexisea, and some others, whose flowers are almost as regular as those of a Sisyrinchium. It is indeed to the latter genus, more nearly than to any other, that Orchids seem to approach in structure, unless to Gingerworts ; so that they may be supposed to pass into Irids through Thelymitra and Sisyrinchium on the one hand, and into Gingerworts through Phrynium and such a genus as Evelyna on the other. With regard to Apostasiads, their relation to that Order does not appear to be greater than to either of the two now mentioned ; and in the absence of all evidence as to the connecting links which join Orchids and Apostasiads it seems unnecessary to advert further to the subject. It may, however, be observed that Apostasia has apparently as much claim to be regarded as a diandrous monadelphous Hypoxid, standing, perhaps, in the same relation to that Order as Gilliesia to Lilyworts, as it has to be regarded as a trilocular Orchid with the gynandrous organization lost.

It is not necessary to enter, in this place, into a history of the gradual alteration that has taken place in the views of botanists with regard to the structure of the sexual apparatus of these most curious plants, or to explain what degree of ignorance was shown by those who mistook masses of pollen for anthers, or a column of stamens for a style ; such errors could only have occurred at a period when the laws of organization were unknown. They have been corrected, in a more or less perfect manner, by various writers ; most completely by Brown in his *Prodromus*, published in 1810, and subsequently by the late most accurate and indefatigable Richard. But long before the publication of any rational explanation of the structure of Orchids, while botanists were in utter darkness upon the subject, it had been investigated by a man unrivalled in his day for the perfection of his microscopical analyses, the beauty of his drawings, and the admirable skill with which he followed Nature in her most secret workings ; and let me add, which is a still rarer quality, the generous disinterestedness with which he communicated to his friends the result of his patient and silent labours. Sketches were executed by the late Francis Bauer, between 1794 and 1807, in which the most material part of what has been published since that period is distinctly shown ; and it has been my good fortune to be the humble means of giving some of these remarkable productions of the pencil to the world, in the *Illustrations of the Genera and Species of Orchidaceous Plants.*

If the column of an Orchidaceous plant is examined, it will be found to consist of a fleshy body stationed opposite the lip, bearing a solitary anther at its apex, and having in front a viscid cavity, upon the upper edge of which there is often a slight callosity, called the rostellum. This cavity is the stigma, and the rostellum is the point by which

the pollen masses are secured when any adhesion between them and the stigma takes place. Hence such a plant would appear to be monandrous ; it will be seen, however, in Gingerworts and Marants, the only other monandrous Orders of Endogens, that, while only one perfect stamen is developed, two others exist in a rudimentary state ; so that the ternary number prevalent in Monocotyledons is not departed from. So it is in Orchids : the column does not consist of a single filament cohering with a style, but of three filaments firmly grown together, the central of which is antheriferous, and the lateral sterile. This is proved by the frequent presence of callosities, or processes in the place of the sterile stamens ; by imperfectly-formed anthers occasionally appearing at the side of the perfect one ; and, if any further evidence were wanted, by monsters, in which a regular structure is exchanged for the ordinary irregularity. Such an instance in Orchis latifolia is described by Achille Richard, in the *Mémoires de la Soc. d'Hist. Nat. of Paris,* in which the flowers were perfectly triandrous, with no trace of irregularity in any part of the floral envelopes ; and other cases of a similar nature are by no means uncommon, and have been occasionally mentioned.

Orchids are remarkable for the unusual figure of their irregular flowers, which sometimes represent an insect, sometimes a helmet with the visor up, and are so various in form that there is scarcely a common reptile or insect to which some

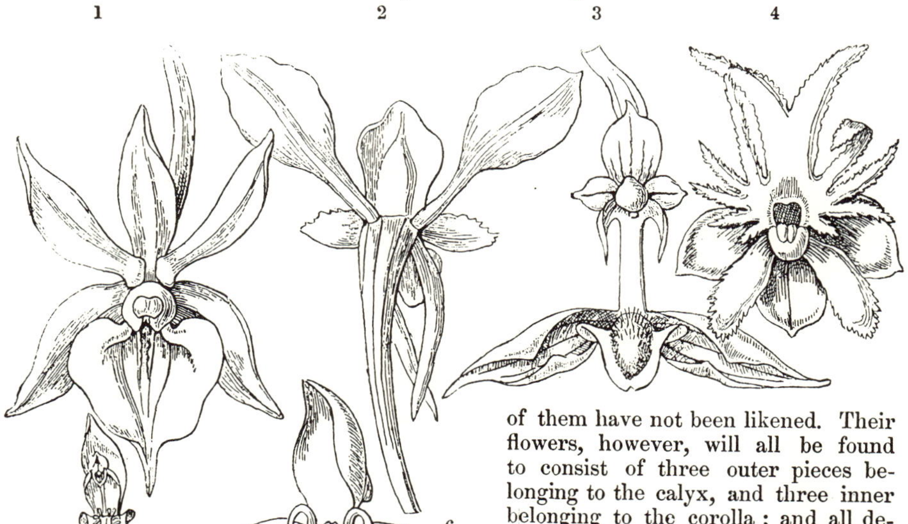

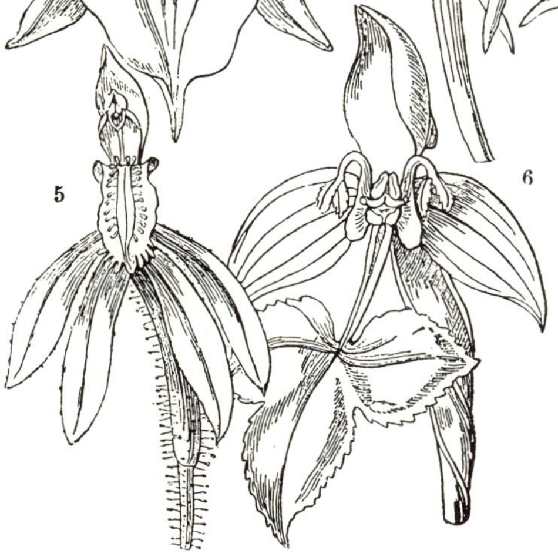

Fig. CXX.

of them have not been likened. Their flowers, however, will all be found to consist of three outer pieces belonging to the calyx, and three inner belonging to the corolla ; and all departure from this number, six, depends upon the cohesion of contiguous parts: with the solitary exception of Monomeria, in which the lateral petals are entirely abortive; of certain Bolbophylla, such as B. bracteolatum, which have an additional scale on the outside of the petals; and of the calyculate genera, such as Epistephium, in which the external calyx above alluded to makes its appearance in the form of an external cup. Sometimes two of the sepals cohere into one, as in certain species of Oncidium, and then the calyx has the appearance of consisting of but two sepals ; sometimes the lateral petals are connate with the column, as in Gongora and and Lepanthes, and then the column appears furnished with two wings. In nearly the whole Order the odd petal, called the lip, arises from the base of the column, and is opposite it ; but in the Cape genus Pterygodium, the lip sometimes grows from the apex of the column, and sometimes is stalked and turned completely over between the fork of the inverted anther, and thus seems to belong to the back of the column. Nor is the anther less subject to modification, although constant to its place : sometimes it stands erect, the line of dehiscence of its lobes being turned towards the lip ; sometimes it is turned upside down, so that its back regards the lip ; often it is prone upon the apex of the column, where a niche is excavated for its reception. The pollen is not

Fig. CXX. 1. Angræcum eburneum ; 2 Diuris ; 3. Drymoda picta ; 4. Oberonia Griffithiana ; 5. Caladenia ; 6. Disa spathulata.

2 3 1

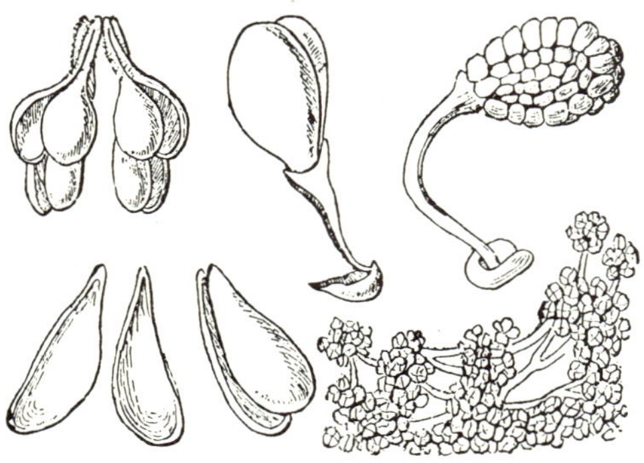

4 Fig. CXXI. 5

less curious: now we have it in separate grains, as in other plants, but cohering to a meshwork of cellular tissue, which is collected into a sort of central elastic strap; now the granules cohere in small angular indefinite masses, and the central elastic strap, becoming more apparent, is found attached to a glandular process of the stigma, which is often inclosed in a peculiar pouch especially destined for its protection; again, the pollen combines into larger masses, which are definite in number, and attached to another modification of the elastic strap; and finally a complete union of the pollen takes place, in solid waxy masses, without any distinct trace of this central elastic tissue. Such is a part of the singularities of Orchidaceous plants, and upon these the distinctions of their tribes and genera are naturally founded. Whoever studies them must bear in mind that their fructification is always reducible to 3 sepals, 3 petals, a column consisting of 3 stamens grown firmly to one another, and to a single style and stigma; and, with this view, he will have no difficulty in understanding the organization of even the most anomalous Cape species. In the last edition of this work an Order called Vanillaceæ was proposed, about which I shall only say that its introduction would have been much better omitted.

Professor Link has shown that beyond all doubt the nucleus of the seed in this Order is a naked embryo, with an excessively enlarged radicula. See his beautiful figures in the *Ausgewälte Anatomisch-botanische Abbildungen fasc.* 2. *t.* vii. Here we again have a structure analogous to that of Nymphæa and Nelumbium.

Among the most singular circumstances connected with this Order is the manner in which, upon the same spike, flowers of extremely different structure are produced. This was first noticed in Demerara by Sir R. Schomburgk, who published in the *Linn. Transactions* (17. 551.) an

Fig. CXXII.

Fig. CXXI.—1. Pollen masses of Ophrys apifera; 2. of Phaius Tankervillia; 3. of Brassia maculata; 4. of Malaxis paludosa; 5. Pollen of Stenorhynchus speciosus.
Fig. CXXII.—2. Cycnoches ventricosum; 4 and 5. C. Egertonianum; the others intermediate forms.

account of the production of Monachanthus viridis, Myanthus barbatus, and a Catasetum, 3 supposed genera, upon the same spike; and he expressed his opinion that the Catasetum was the female of these, because he found it producing seeds abundantly, while Monachanthus was uniformly sterile.

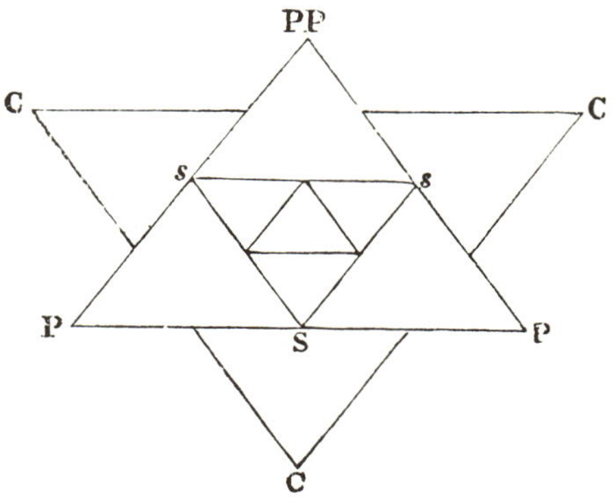

Afterwards a similar specimen made its appearance in the garden of his Grace the Duke of Devonshire at Chatsworth, and has been figured in the *Botanical Register, fol.* 1951. And still more lately two species of Cycnoches, ventricosum and Egertonianum have appeared in company, as represented in the accompanying figure (CXXII.)

Such cases shake to the foundation all our ideas of the stability of genera and species, and prepare the mind for more startling discoveries than could have been otherwise anticipated.

If the accompanying diagram be compared with those employed to illustrate the distinctions of Marants and Gingerworts, p. 169, the relation borne to those Orders by Orchids will be distinctly seen. In the diagram the parts are arranged as they are in nature before the ovary twists; that is, with the lip next the axis, or uppermost, and the stamen undermost. Let C, C, C represent the outer series of floral envelopes or calyx, and PP, P, P the inner, or corolla, of which PP is the labellum: then the position of the single fertile stamen will be at S, and the sterile ones at *s, s;* that is to say, in the situation of the supernumerary petaloid stamens of Gingerworts and Marants, while the second series of stamens, to which the fertile stamen of these Orders belongs, is not developed in Orchids.

In the last edition it was suggested that although this is the apparent structure of the Order, it is not improbable that the parts called sepals are the true petals, because Epistephium and others have a calyculus exterior to the apparent calyx. In that point of view the apparent petals would be sterile stamens, as among the Marants; it has, however, been shown by Crüger that the order of development of the floral organs of Epistephium is unfavourable to the supposition, and that the calyculus of that genus is in reality a cup-shaped expansion of the ovary. I quote his words from *Henfrey's Scientific Memoirs, I.* 170:—

"The development of the segments of the perianth is in agreement with the mode in which they subsequently overlie one another. At each side of the little nodule which is the first representative of the flower of this plant, we observe a little point, the first trace of the sepals, and a little later the middle sepal; at the same time with the latter, the two lateral inner segments of the perianth. Then the labellum appears, and almost simultaneously with that the anther. In this flower also the anther is at first erect, although it subsequently lies upon the summit of the column. Up to this time no trace of a calyculus is to be seen; it first presents itself clearly when the flower rises above the axil of the bract, and the boundary between the ovary and the segments of the perianth becomes visible. The calyculus is persistent upon the fruit, while the other parts separate from it at a very early period. I believe I am justified in concluding from the foregoing, that the calyculus, when it presents itself in the Orchideæ, does not represent an external circle of organs, because (1) its segments do not alternate with those of that standing above it; (2) they originate later than those; and (3) because they persist upon the capsule, while the other parts become detached. I should lay much stress upon the last reason, yet I think that this calyculus must be regarded as analogous to that which may be observed on the fruits of certain Compositæ, Dipsaceæ, &c."

In classifying this Order the most important characters appear to reside in the pollen, which in many is consolidated into firm waxy masses of a definite number in each species, and in others is either in its usual loose powdery condition, or is collected in granules or small wedges, the number of which is far too great to be counted. Of those with waxy pollen masses some (Malaxeæ) are destitute of any visible processes by which the masses are brought into contact with the stigma; others (Epidendreæ) have strap-shaped cau-

dicles, which are either bent down upon the masses themselves, or serve to hold them together, without, however, forming any organised union with the stigma ; while the remainder (Vandeæ) have a caudicle, which adheres firmly to a gland found on the upper margin of the stigma, and separating freely from that organ. The genera with powdery, granular, or sectile pollen cannot be classified so conveniently by modifications of that part, but are readily divided into 3 natural tribes by peculiarities in the anther. In some (Ophreæ) the anther is erect, not hinged to the column but continuous with it, and stands above the stigma, the pollen masses having their points directed to the base of the lobes of the anther ; in others (Arethuseæ) the anther is hinged to the column, upon the end of which it is placed transversely like a lid ; and in others (Neotteæ) it is also hinged to the column, but is stationed at its back so as to be nearly parallel with the stigmatic surface. If to this we add that Cypripedeæ have two anthers, while all the others have one only, we find the Order divided into seven tribes, of which the following is a tabular view.

I. Anther one only.
- A. *Pollen masses waxy.*
  - *a.* No caudicle or separable stigmatic gland — I. *Malaxeæ.*
  - *b.* A distinct caudicle, but no separable stigmatic gland — II. *Epidendreæ.*
  - *c.* A distinct caudicle, united to a stigmatic gland — III. *Vandeæ.*
- B. *Pollen powdery, granular, or sectile.*
  - *a.* Anther terminal, erect — IV. *Ophreæ.*
  - *b.* Anther terminal, opercular — V. *Arethuseæ.*
  - *c.* Anther dorsal — VI. *Neotteæ.*

II. Anthers two — VII. *Cypripedeæ.*

Among many other remarkable peculiarities the irritability of the labellum must not be passed over in silence. This is extremely striking in various species of Pterostylis, in the genus Megaclinium, and in many Bolbophylls, especially barbigerum and Careyanum. But some of the Swan River species are still more singular. In Caleana nigrita, Mr. Drummond describes the structure to the following effect. The column is a boat-shaped box, resembling a lower lip ; the labellum forms a lid that exactly fits it, and is hinged on a claw which reaches the middle of the column ; when the flower opens, it (the labellum) turns round within the column, and falls back, so that, the flower being inverted, it stands fairly over the latter. The moment a small insect touches its point, the labellum makes a sudden revolution, brings the point to the bottom of the column, passing the anther in its way, and thus makes prisoner any insect which the box will hold. When it catches an insect it remains shut while its prey continues to move about ; but if no capture is made the lid soon recovers its position. Another plant, Drakæa elastica, has a single flower placed at the end of a slender smooth erect scape, from twelve to eighteen inches high, and its labellum, which is hammer-headed, and placed on a long arm with a moveable elbow-joint in the middle, is stated by Mr. Drummond to resemble an insect suspended in the air, and moving with every breeze. Another plant of this description is Spiculæa ciliata, whose rusty flowers when spread open may be compared to long-legged spiders, the lip with a long solid lamina looking like their body, while an appendage at its apex, which is apparently moveable, is not unlike the head of such a creature.

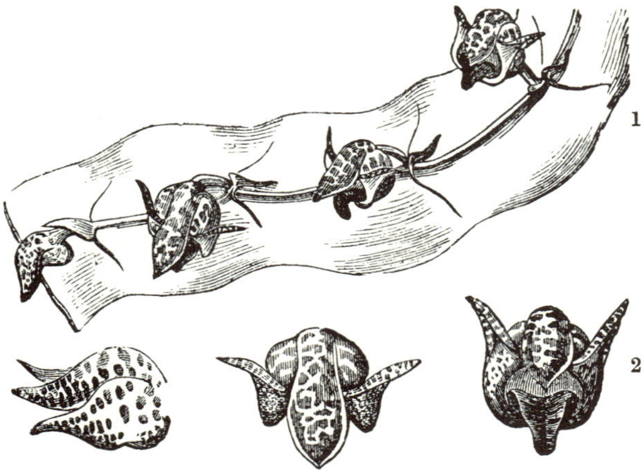

Fig. CXXIII.

Orchids are found in almost all parts of the world, except upon the verge of the frozen

Fig. CXXIII. *bis.*

Fig. CXXIII.—Megaclinium Bufo ; 1. a portion of a spike magnified ; 2. flowers in various positions, more magnified.
Fig. CXXIII. *bis.*—Spiculæa ciliata, its flower.

zone, and in climates remarkable for dryness. In Europe, Asia, and North America, they are seen growing everywhere, in groves, in marshes, and in meadows; in the drier parts of Africa they are either rare or unknown; at the Cape of Good Hope they abound in similar situations as in Europe; but in the hot damp parts of the West and East Indies, in Madagascar, and the neighbouring islands, in the damp and humid forests of Brazil, in the warm mild parts of Central America, and Western Mexico, in the damp tropical parts of India, and on the lower mountains of Nipal, the Orchidaceous plants flourish in the greatest variety and profusion, no longer seeking their nutriment from the soil, but clinging to the trunks and limbs of trees, to stones and bare rocks, where they vegetate among ferns and other shade-loving plants, in countless thousands. Of the epiphytal class, one only is found so far north as South Carolina, growing upon the branches of the Magnolia, if we except the species from Japan, a country which has a climate peculiar to itself, among regions in the same parallel of latitude. The most southern stations are those of Earina mucronata in New Zealand, in lat. 35° S., and of Gunnia australis in Emu Bay, Van Diemen's Land, lat. 41° S. Ample details respecting their distribution in Australia are given by A. Cunningham in the *Botanical Register* for 1843 *t.* 37.

It often happens that those productions of nature which charm the eye with their beauty, and delight the senses with their perfume, have the least relation to the wants of mankind, while the most powerful virtues or most deadly poisons are hidden beneath a mean and insignificant exterior: thus Orchids, beyond their beauty, can scarcely be said to be of known utility, with a few exceptions. The nutritive substance called Salep has been prepared from the subterraneous succulent roots of Orchis mascula and many others of the Ophreous division; and in India from the tubers of a species of Eulophia; it consists almost entirely of a chemical principle called Bassorin. The root of Bletia verecunda is said to be stomachic. Some of the South American species, such as the Catasetums, Cyrtopodiums, &c., contain a viscid juice, which being inspissated by boiling, becomes a kind of vegetable glue used for economical purposes in Brazil. The viscidity of the tuber of Aplectrum hyemale is such that it is called Putty-root in the United States, and is used for cementing broken earthenware.

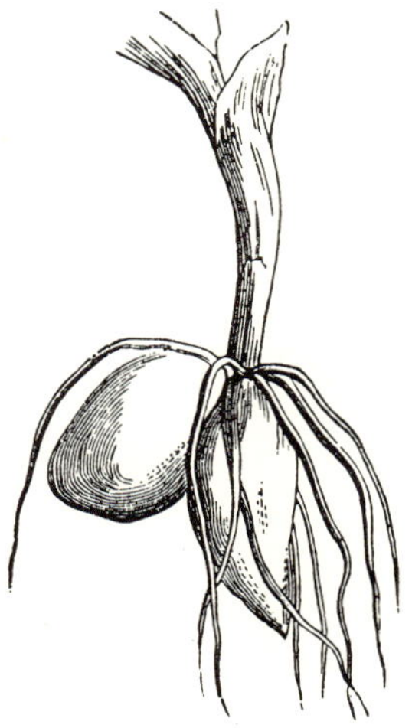
Fig. CXXIV.

Other medical qualities have been assigned to other species, but they seem to be of no importance; thus, Arethusa bulbosa is employed in the United States in toothache and bringing tumours to a head, Spiranthes diuretica as a diuretic in Chile, where also Chloræa disoides is fancied to promote the flow of milk. Cypripedium pubescens is used in North America as a substitute for Valerian, C. guttatum in Siberia against epilepsy. Vanilla is one of the most delightful aromatics known; it is used in the manufacture of chocolate, of liqueurs, and of various articles of confectionery. The substance called by this name in the shops is the dried fruit of Vanilla planifolia, and other species; it contains a great quantity of essential oil, and a good deal of benzoic acid. Dr. Bird says that the effluvium of Vanilla intoxicates the labourer who gathers it.—*Peter Pilgrim,* 1. 234. See *Linnæa.* 4. 573, for some account of the cultivation of the plant in Mexico. Vanilla claviculata is bitter as well as fragrant, and its leaves are regarded in the West Indies, where it is called Liane à blessures, as a vulnerary, and antisyphilitic. In New Holland many species are eaten by the natives, who find their starchy roots a good article of diet. Mr. Backhouse describes the Gastrodia sesamoides as having a root like a series of kidney potatoes, terminating in a branched, thick mass of coral-like fibres. It is eaten by the aborigines of Tasmannia, and is sometimes called native potato; but its tubers are watery and insipid.

P. Browne states that the corm of Bletia verecunda is "bitterish and attended by a clamminess that leaves a light prickly warmth behind it; but this wears off soon, leaving the palate free from every sensation but that of the bitter. When dried it may be used with great propriety as a stomachic." According to Sir R. Schomburgk the expressed juice of Epidendrum bifidum is a purgative, taken in doses of a table spoonful at a time; it is also reckoned in Tortola an anthelmintic, and diuretic, &c. —*Linnæa,* ix. 512.

Fig. CXXIV.—Orchis mascula roots in the state in which they are dried as salep.

GENERA.

[In the following list, I have cast all the genera into natural subdivisions, preparatory to a further revision of them; but I have not attempted to settle very exactly their order of sequence. In fact, a great number demand a very careful revision, and others, to which an * is prefixed, I have never had the opportunity of examining.]

I.—MALAXEÆ.

PLEUROTHALLIDÆ.
Lindl. in Bot. Reg. 1842. *misc.* p. 67.

Pleurothallis, *R. Br.*
Rhynchopera, Klotzch.
Myoxanthus, Popp. et Endl.
Specklinia, Lindl.
Centranthera, Scheidw.
Acianthera, Scheidw.
Dialissa, *Lindl.*
Stelis, *Swz.*
Humboldtia, Fl. Peruv.
Lepanthes, *Swz.*
Restrepia, *Kth.*
* ? Cadetia, *Gaudich.*
Physosiphon, *Lindl.*
Masdevallia, *Fl. Per.*
Stenoglossum, *H. B. K.*
Octomeria, *R. Br.*

LIPARIDÆ.
Liparis, *Rich.*
Sturmia, Rchb.
Alipsa, Hffmg.
Cestichis, Thouars.
Distichis, *Thouars.*
Dendrochilum, *Bl.*
* Osyricera, *Bl.*
* Chrysoglossum, *Bl.*
Oberonia, *Lindl.*
Ensifera, Bl.
* Titania, *Endl.*
Empusa, *Lindl.*
Empusaria, Rchb.
* Platystylis, *Bl.*
* Gastroglottis, *Bl.*
Microstylis, *Nutt.*
Crepidium, Bl.
Monorchis, Mentz.
Achroanthes, Raf.
Pterochilus, Hook.
Dienia, *Lindl.*
Pedilea, Lindl.
Malaxis, *Swz.*
* Nephelaphyllum, *Bl.*
Calypso, *Salisb.*
Cytherea, Salisb.
Norna, Wall.
Orchidium, Swz.

DENDROBIDÆ.
Dendrobium, *Swz.*
Grastidium, Bl.
Ceraia, Lour.
Keranthus, Lour.
Bontia, Petiv.
? *Sarcostoma*, Bl.
§ *Stachyobium*, Lindl.
§ *Ceratobium*, Lindl.
§ *Pedilonum*, Bl.
§ *Onychium*, Bl.
§ *Desmotrichum*, Bl.
§ *Dendrocoryne*, Lindl.
* Macrostomium, *Bl.*
Aporum, *Bl.*
Schismoceras, Presl.
* § Diploconchium, *Schauer.*
Oxystophyllum, *Bl.*
* ? Diglyphis, *Bl.*
Diglyphosa, Bl.
Monomeria, *Lindl.*
* Epicrianthes, *Bl.*
? Drymoda, *Lindl.*
Bolbophyllum, *Thouars.*
Diphyes, Bl.
Tribrachia, Lindl.
Odontostylis, Bl. (f. Endl.)
Gersinia, Neraud. (f. Endl.)
Macrolepis, A. Rich.
§ *Anisopetalum*, Hooker.
? Sunipia, *Lindl.*
Trias, *Lindl.*
* Thelychiton, *Endl.*
* Cochlia, *Bl.*
* Lyræa, *Lindl.*
Megaclinium, *Lindl.*
Cirrhopetalum, *Lindl.*
Zygoglossum, Reinw.
Ephippium, Bl.
? *Sestochilus*, Kuhl et Hass.
Bryobium, *Lindl.*
Conchidium, *Griff.*
Mycaranthes, *Bl.*
Phreatia, *Lindl.*
Eria, *Lindl.*
Dendrolirium, Bl.
Pinalia, Hamilt.

CORALLORHIZIDÆ.
Corallorhiza, *Haller.*
Aplectrum, *Nuttall.*
* Aphyllorchis, *Blume.*

II.—EPIDENDREÆ.

CŒLOGYNIDÆ.
* ? Acanthoglossum, *Bl.*
Cœlogyne, *Lindl.*
Chelonanthera, Bl.
Panisea, *Lindl.*
Pleione, *Don.*
Gomphostylis, Wall.
Trichosma, *Lindl.*
Dilochia, *Lindl.*
Pholidota, *Lindl.*
Ptilocnema, Don.
Crinonia, Bl.
Otochilus, *Lindl.*
Earina, *Lindl.*

ISOCHILIDÆ.
Isochilus, *R. Br.*
Hexisea, *Lindl.*
? *Elleanthus*, Presl.
Diothonea, *Lindl.*
Gastropodium, *Lindl.*

LÆLIADÆ.
Epidendrum, *L.*
§ *Hormidium*, Lindl.
§ *Epicladium*, Lindl.
§ *Encyclium*, Hooker.
§ *Diacrium*, Lindl.
§ *Aulizeum*, Lindl.
§ *Osmophytum*, Lindl.
§ *Lanium*, Lindl.
§ *Spathium*, Lindl.
§ *Amphiglottium*, Salis.
§ *Euepidendrum*, Lindl.
Seraphyta, Fisch.
Physinga, *Lindl.*
Ponera, *Lindl.*
Nemaconia, Knowles.
* ? Aspegrenia, *Pöpp. et Endl.*
Hexadesmia, *Brongn.*
Hexopia, Batem.
Dinema, *Lindl.*
Sophronitis, *Lindl.*
Alamania, *Llave.*
Hartwegia, *Lindl.*
Arpophyllum, *Llave.*
Barkeria, *Knowles.*
Broughtonia, *R. Br.*
? Chysis, *Lindl.*
Lælia, *Lindl.*
Amalia, Rchb.
Cattleya, *Lindl.*
Schomburgkia *Lindl.*
Tetramicra, *Lindl.*
Leptotes, *Lindl.*
Brasavola, *Lindl.*

BLETIDÆ.
Phaius, *Lour.*
Pachyna, Salisb.
Tankervillia, Link.
Arundina, *Bl.*
Evelyna, *Pöpp. et Endl.*
Bletia, *R. et Pav.*
Gyas, Salisb.
Thiebaudia, Colla.
* Mitopetalum, *Bl.*
Tainia, Bl.
Spathoglottis, *Bl.*
Paxtonia, *Lindl.*
* Collabium, *Bl.*
Cytheris, *Lindl.*
Pesomeria, *Lindl.*
Ipsea, *Lindl.*
* ? Pachystoma, *Bl.*
Apaturia, *Lindl.*
? Cremastra, *Lindl.*
Ania, *Lindl.*
* ? Callostylis, *Bl.*
Tylostylis, Bl.
* ? Ceratium, *Bl.*
Cylindrolobus, Bl.
* ? Trichotosia, *Bl.*
* ? Plocoglottis, *Bl.*
* ? Pachychilus, *Bl.* (End.)

III.—VANDEÆ.

SARCANTHIDÆ. — *Lindl. in Bot. Reg.* 1843, *misc.* p. 12.
Eulophia, *R. Br.*
Galeandra, *Lindl.*
Corydandra, Rchb.
Cyrtopera, *Lindl.*
Lissochilus, *R. Br.*
Doritis, *Lindl.*
Luisia, *Gaud.*
Pseudovanda, Lindl.
Mesoclastes, Lindl.
Birchea, A. Rich.
Vanda, *R. Br.*
Fieldia, Gaud.
Renanthera, *Lour.*
Arachnis, Blume.
Nephranthera, Hassk.
Arachnanthe, Blume.
Phalænopsis, *Bl.*
Diplocentrum, *Lindl.*
* Microsaccus, *Blume.*
Camarotis, *Lindl.*
Chiloschista, *Lindl.*
Gunnia, *Lindl.*
Micropera, *Lindl.*
Saccolabium, *Lindl.*
Saccochilus, Blume.
Gastrochilus, Don.
Robiquetia, Gaudich.
Gussonea, A. Rich.
Rhynchostylis, Blume.
Carteretia, A. Rich.
Sarcochilus, *R. Br.*
* Tæniophyllum, *Blume.*
Cleisostoma, *Blume.*
Polychilos, Kuhl et Hass.
* Ceratostylis, *Blume.*
* Ephippium, *Blume.*
* Ceratochilus, *Blume.*
Omæa, Blume.
* Echioglossum, *Blume*
Sarcanthus, *Lindl.*
* Pteroceras, *Hass.*
Agrostophyllum, *Blume.*
* Adenoncos, *Blume.*
Œceoclades, *Lindl.*
Trichoglottis, *Bl*
Aërides, *Loureir.*
Dendrocolla, Blume.
Cucullu, Blume.
Tubera, Blume.
Fornicaria, Blume.
Pilearia, Lindl.
Ornithochilus, Wall.
* Schœnorchis, *Blume.*
Aëranthus, *Lindl.*
Cryptopus, *Lindl.*
Beclardia, A. Rich.
Œonia, *Lindl.*
Angræcum, *Thouars.*
Aërobion, Spr.
Mystacidium, *Lindl.*
Microcœlia, *Lindl.*
Appendicula, *Bl.*
Metachilum, Lindl.
Podochilus *Bl.*
Platysma, Bl.
Placostigma, Bl.
Apista, Bl
Hexadesmia, R. Br.
? *Blumea*, Meyer.
* Cryptoglottis, *Bl.*
* Glomera, *Bl.*
* Thelasis, *Bl.*
Tetrapeltis, *Wall*
* ? Conchochilus, *Hsskl.*
* ? Todaroa, *A. Rich.*

CRYPTOCHILIDÆ.
Cryptochilus, *Wall.*
Acanthophippium, *Bl.*
* ? Anthogonium, *Wall.*

BRASSIDÆ.
Cymbidium, *Swz.*
Bolbidium, *Lindl.*
Grammatophyllum, *Bl.*
Gabertia, Gaud.
* Stauroglottis, *Schauer.*
Bromheadia, *Lindl.*
Ansellia, *Lindl.*
Aganisia, *Lindl.*
Epiphora, *Lindl.*
Aspasia, *Lindl.*
? Acriopsis, *Bl.*
Trichopilia, *Lindl.*
Helcia, *Lindl.*
Nanodes, *Lindl.*
Pilumna, *Lindl.*
Dipodium, *R. Br.*
? *Armodorum*, Kuhl et [H.
Dichæa, *Lindl.*
Fernandezia *R. et Pav.*
Lockhartia, Hooker.
Phymatidium, *Lindl.*
Leochilus, *Knowles.*
Oncidium, *Swz.*
Cyrtochilum, H. B. K.
Odontoglossum, *H. B K.*
§ *Trymenium*, Lindl.
Brassia, *R. Br.*

Miltonia, *Lindl.*
Macrochilus, Knowles.

PACHYPHYLLIDÆ.
Nasonia, *Lindl.*
Centropetalum, *Lindl.*
Pachyphyllum, *H. B. K.*

MAXILLARIDÆ. *Lindl. in Bot. Reg.* 1843, *misc. p.* 12.
Stanhopea, *Frost.*
Ceratochilus, Lodd.
Houlletia, *A. Brongn.*
Peristeria, *Hooker.*
Eckardia, Rchb.
Acineta, *Lindl.*
Lacæna, *Lindl.*
* ? Cuitlauzina, *Llav.*
Govenia. *Lindl.*
Eucnemis, Lindl.
Angidium, Lindl.
Batemannia, *Lindl.*
Gongora, *Fl. Peruv.*
Acropera, *Lindl.*
Coryanthes, *Hook.*
Chænanthe, *Lindl.*
Malachadenia, *Lindl.*
Cœlia, *Lindl.*
Ornithidium, *Salisb.*
Trigonidium, *Lindl.*
* Psittacoglossum, *Llav.*
Stenia, *Lindl.*
Promenæa, *Lindl.*
Grobya, *Lindl.*
Warrea, *Lindl.*
Huntleya, *Lindl.*
Zygopetalum, *Hooker.*
Bifrenaria, *Lindl.*
Stenocoryne, *Lindl.*
Maxillaria, *Fl. Per.*
§ * ? *Nothium*, Lindl.
§ * *Xylobium*, Lindl.
§ * *Dicrypta*, Lindl.
Heterotaxis, Lindl.
Lycaste, *Lindl.*
Anguloa, *Fl. Per.*
Camaridium, *Lindl.*
* Siagonanthus, *Pöpp et Endl.*
Scuticaria, *Lindl.*
Scaphyglottis, *Pöpp et E.*
Cladobium, Lindl.
Colax, *Lindl.*
Paphinia, *Lindl.*
Polystachya, *Hooker.*
*?Orchidofunkia, *A.Rich.*
Clynhymenia, A. Rich.
* ? Galeottia, *A. Rich.*

CATASETIDÆ. *Lindl. in Bot. Reg.* 1842. *p.* 22.
Catasetum, *Rich.*
Monachanthus, Lindl.
§ *Myanthus*, Lindl.
Mormodes, *Lindl.*
Cyclosia, Klotzsch.
Clowesia, *Lindl.*
Cycnoches, *Lindl.*
Cyrtopodium, *R. Br.*
Tylochilus, Nees.

NOTYLIDÆ.
Notylia, *Lindl.*
Cirrhæa, *Lindl.*
Zygostates, *Lindl.*
Dactylostyles, Scheidw.
Ornithocephalus, *Hook.*
? Trophianthus, *Scheidw.*
Cryptarrhena, *R. Br.*
Macradenia, *R. Br.*
Sutrina, *Lindl.*
Telipogon, *H. B. K.*
Trichoceros, *H. B. K.*
* * *
Trizeuxis, *Lindl.*

Quekettia, *Lindl.*

IONOPSIDÆ.
Rodriguezia, *R. et Pav.*
Gomeza, R. Br.
* Scelochilus, *Klotzsch.*
Burlingtonia, *Lindl.*
Ionopsis, *H. B. K.*
Iantha, Hook.
Cybelion, Spreng.
* Diadenium, *Pöpp et En.*
Comparettia, *Pöpp et En.*
Trichocentrum, *Pöpp et E.*
Acoidium, Lindl.

CALANTHIDÆ.
Calanthe *R. Br.*
Centrosia, A. Rich.
Alismorchis, Thouars.
Amblyglottis, Blume.
Styloglossum, Kuhl et Hass.
* Limatodes, *Bl.*
* Ghiesbrechtia, *A. Rich.*
* Tipularia, *Nutt.*
Anthericlis, Raf.
Geodorum, *Jacks.*
Otandra, Salisb.
Cistella, Bl.

IV.—OPHREÆ.

SERAPIADÆ.
Orchis, *L.*
§ *Herorchis*, Lindl.
§ *Androrchis*, Endl.
Anacamptis, *Rich.*
Nigritella, *Rich.*
Aceras, *R. Br.*
Loroglossum, Rich.
Himantoglossum, Spr.
Serapias, *L.*
Helleborine, Pers.
Ophrys, *Swartz.*
Hemipilia, *Lindl.*
Glossaspis, *Spreng.*
Glossula, Lindl.
Perularia, *Lindl.*
Bartholina, *R. Br.*
Lathrisia, Swz.

SATYRIADÆ.
Pachites, *Lindl.*
Satyrium, *Swz.*
Diplectrum, Rich.
Satyridium, *Lindl.*
Aviceps, *Lindl.*

GYMNADENIDÆ.
Aopla, *Lindl.*
Herminium, *R. Br.*
Arachnites, Hoffm.
§ *Chamorchis*, Rich.
Chamærepes, Spr.
Gymnadenia, *R. Br.*
Sieberia, Spr.
Platanthera, *Rich.*
Mecosa, Bl.
Peristylus, *Blume.*
Benthamia, A. Rich.
Habenaria, *W.*
Dissorhynchium, Schauer.
?*Centrochilus*, Schauer.
Ate, *Lindl.*
Bonatea, *W.*
Bilabrella, Lindl.
Stenoglottis, *Lindl.*
Diplomeris, *Don.*
Diplochilus, Lindl.
Paragnathis, Spreng.
Bicornella, *Lindl.*
Cynorchis, *Thouars.*
?*Amphorchis*, Thouars.
Cœloglossum, *Lindl.*
Ommatodium, *Lindl.*

HOLOTRICHIDÆ.
Holothrix, *Rich.*
Saccidium, *Lindl.*
Monotris, *Lindl.*
Scopularia, *Lindl.*
Tryphia, *Lindl.*
Bucculina, *Lindl.*

DISIDÆ.
Disa, *Berg.*
§ *Repandra*, Lindl.
§ *Phlebidia*, Lindl.
§ *Vaginaria*, Lindl.
§ *Pardoglossa*, Lindl.
§ *Coryphæa*, Lindl.
§ *Stenocarpa*, Lindl.
§ *Oregura*, Lindl.
§ *Trichochila*, Lindl.
§ *Disella*, Lindl.
Monadenia, *Lindl.*
Schizodium, *Lindl.*
Penthea, *Lindl.*
Forficaria, *Lindl.*
Herschelia, *Lindl.*
Brachycorythis, *Lindl.*
Brownleea, *Harv.*

CORYCIDÆ.
Pterygodium *Swz.*
Corycium, *Swz.*
Disperis, *Swz.*
Dipera, Spreng.
Dryopeia, Thouars.
Ceratandra, *Lindl.*
§ *Hippopodium*, Harv.
§ *Evota*, Lindl.
Calota, Harv.
Arnottia, *A. Rich.*

V.—ARETHUSEÆ.

LIMODORIDÆ.
Chloræa, *Lindl.*
Epipactis, Feuill.
Asarca, *Lindl.*
Gavilea, Pöpp.
Asarca, Pöpp.
Bipinnula, *Commers.*
Limodorum, *Tournef.*
Cephalanthera, *L.C.Rich.*
Macdonaldia, *R. Gunn.*
Eriochilus, *R. Br.*
Diplodium, Swartz.
Caladenia, *R. Br.*
Calonema, Lindl.
Leptoceras, *R. Br.*
Glossodia, *R. Br.*
Elythranthe, Endl.
Lyperanthus, *R. Br.*
Microtis, *R. Br.*

ACIANTHIDÆ.
Acianthus, *R. Br.*
Chiloglottis, *R. Br.*
Cyrtostylis, *R. Br.*
Corysanthes, *R. Br.*
Calcearia, Bl.
Corybas, Salisb.
Steleocorys, Endl.
Pterostylis, *R. Br.*

CALEYIDÆ.
Caleya, *R. Br.*
Caleana, R. Br.
Drakæa, *Lindl.*
Spiculæa, *Lindl.*

POGONIDÆ.
Pogonia, *Juss.*
Triphora, Nutt.
Nervilia, Commers.
Odonectis, Rafin.
Isotria, Rafin.
* Didymoplexis, *Griff.*
Codonorchis, *Lindl.*
Arethusa, *Gronov.*
* Haplostellis, *A. Rich.*

Cleistes, *Rich.*
Calopogon, *R. Br.*
Cathea, Salisb.
Crybe, *Lindl.*

GASTRODIDÆ.
Gastrodia, *R. Br.*
Epiphanes, Blume
Ceratopsis, *Lindl.*
* Gamoplexis, *Falc.*
Epipogium, *Gmel.*

VANILLIDÆ.
* Cyathoglottis, *Pöpp et Endl.*
Sobralia, *Ruiz et Pav.*
Epistephium, *H. B. K.*
Erythrorchis, *Blume.*
Cyrtosia, *Blume.*
Vanilla, *Swartz.*
* Pogochilus, *Falcon.*

VI. NEOTTEÆ.

CRANICHIDÆ, *Lindl.*
Ponthieva, *R. Br.*
Schœnleinia, Klot.
Pterichis, *Lindl.*
Acræa, *Lindl.*
Cryptostylis, *R. Br.*
*Zosterostylis, *Blum.*
Gomphichis, *Lindl.*
Stenoptera, *Lindl.*
Altensteinia, *H. B. K.*
Cranichis, *Swartz.*
Tripleura, *Lindl.*
*Chlorosa, *Blum.*
*Rophostemon, *Blum.*
Cordyla, Blume.
*Galeoglossum, *A. Rich*
*Ocampoa, *A. Rich.*
Prescottia, *Lindl.*
Decaisnea, Brongn.

LISTERIDÆ, *Lindl.*
Listera, *R. Br.*
Diphyllum, Raf.
Neottia, *R. Br.*
Neottidium, Lk.
Calochilus, *R. Br.*
Epipactis, *Hall.*
Serapias, Pers.

SPIRANTHIDÆ, *Lindl.*
Cnemidia, *Lindl.*
Decaisnea, Lindl.
Spiranthes, *L. C. Rich.*
Ibidium, Salisb.
Cyclopogon, Presl.
Gyrostachys, Pers.
Stenoptera, *Presl.*
Sarcoglottis, *Presl.*
Cordylestylis, *Falc.*
Stenorhynchus, *Rich.*
Sauroglossum, *Lindl.*
Pelexia, *Poit.*
Synassa, Lindl.

PHYSURIDÆ, *Lindl.*
*Plexaure, *Endl.*
Chloidia, *Lindl.*
Zeuxine, *Lindl.*
Adenostyles, Blume.
Cionisaccus, Kuhl.
*Chæradoplectron, *Schr*
Monochilus, *Blume.*
Haplochilus, Endl.
Cheirostylis, *Blume.*
Myoda, *Lindl.*
Hæmaria, *Lindl.*
Hylophila, *Lindl.*
Ætheria, *Blum.*
Platylepis, A. Rich.
Goodyera, *R. Br.*
Leucostachys, Hffg.
Gonogona, Lk.
Tussaca, Rafin.
*Eucosia, *Blume.*

Georchis, *Lindl.*
*Macodes, *Blum.*
Tropidia, *Lindl.*
Ptychochilus, Schauer.
Ulantha, *Hook.*
Anæctochilus, *Blume.*
Anœcochilus, Blume.
Chrysobaphus, Wall.
Orchipedum, Kuhl.
*Galera, *Blume.*
Physurus, *L. C. Rich.*
Microchilus, Presl.
Erythrodes, Blume.
Psychechilos, Kuhl.
Baskervilla, *Lindl.*
Herpysma, *Lindl.*
DIURIDÆ, *Lindl.*
Diuris, *Smith.*
Orthoceras, *R. Br.*
Prasophyllum, *R. Br.*
Burnettia, *Lindl.*
Genoplesium, *R. Br.*
THELYMITRIDÆ, *Lindl.*
Thelymitra, *Forst.*
Epiblema, *R. Br.*

VII. CYPRIPEDEÆ.

Cypripedium, *Linn.*
Criosanthes, Rafin.
Arietinium, Beck.

GENERA about which nothing certain is known.

*Hysteria, *Reinw.*
Corymbis, *Thouars.*
*Thrixspermum, *Lour.*
*Scaredederis, *Thouars.*
*Oxyanthera. *A. Brongn.*
*Galeola, *Lour.*
*Callista, *Lour.*
*Acronia, *Presl.*
*Scleropteris, *Scheidw.*
*Macrostylis, *Kuhl et Hass.*
*Amblostoma, *Scheidw.*

NUMBERS. GEN. 394. Sp. 3000 ?

Iridaceæ.
POSITION.—Apostasiaceæ.—ORCHIDACEÆ.—Burmanniaceæ.
Zingiberaceæ.

Since the foregoing remarks were written, much has been added to our systematic knowledge of the order, but little to the theory of structure, or to the principles on which a general classification should be founded.

I have pointed out (*Folia Orchidacea*, under Zygostates) that the calli so common on the labellum have probably a much higher import than is generally supposed.

Dr. R. Brown long since (*Wallich, Pl. As. rar. I.* 7) pointed out the possibility that the processes of the lip found in *Pterostylis*, and certain other genera, might represent staminal apparatus, completing the customary number of male parts found in Endogens. He even remarked that "perhaps it may be considered as indicated in all cases where the labellum is furnished with a process, however minute, arising from its axis." I have not seen reason to express publicly an acquiescence in that hypothesis, not having succeeded in finding satisfactory evidence of its truth; on the contrary, it had appeared to me that the processes in question might as justly be referred to the stigma: I am, however, bound to admit that upon the whole there is an accumulation of facts so much in favour of the theory of the sagacious observer above mentioned, that it becomes more and more probable. It is especially to be noted that the number of processes found on the lip is usually one, two, or three; the latter being the more common. In some instances, one may represent the front abortive stamen only, two may represent the two laterals only, and three may represent them all. It is also remarkable that where three processes or three rows of processes are present, the two lateral are generally the largest, as if they represented a more vigorous series of development than that in the centre. To which it may be added, that where more than three rows of processes are present, nevertheless, the number three seems fundamental. The scientific reader will readily understand this by reference to the accompanying diagram, in which the supposed inner series of stamina is represented by three black dots (●), and the outer series by as many open dots (○).

According to this theory the staminal apparatus of an Orchidaceous plant consists of two rings or whorls, each composed of three stamens more or less developed. In general the central of the outer whorl is alone perfect; while in Cypripedium perfection is confined to the two lateral inner stamens. The rest of the stamens are either wholly suppressed, as in many Dendrobes, or appear in the form of ears to the column or crests upon the lip; the ears of the column sometimes representing the lateral inner staminodes, and the crests of the lip being made up either of two lateral outer and one central inner staminode, or of either. Such evidence as exists upon this subject appears favourable to the opinion; which would be conclusively established if the crests of the lip were detected bearing pollen, a circumstance that has not yet been observed.

If this be so, the accompanying diagrams will represent the condition of the staminal apparatus in the different modifications which this Order produces. (In all cases but one, No. 5, the exterior ring represents the series to which the perfect stamen belongs, and the inner ring the series which is usually more or less disguised. For the convenience of description the perfect stamen and accompanying abortions may be called the *outer stamen and staminodes*, while those of the second and more paradoxical series may be termed the *inner stamen and staminodes*. The asterisks indicate an entire suppression of staminodes.)

No. 1 shows the theoretical state of the flower, with the three outer stamens complete, and three inner staminodes. The outer stamens are here in the condition in which they appear in the plant figured by Dr. Wight under the name of Euproboscis, and by Griffith in Falconer's Dendrobium normale.

No. 2 represents such genera as Odontoglossum in which one outer stamen is perfect, the two outer staminodes in the form of the lateral plates of the crest of the disk; then of the inner staminodes two form the wings of the column, and the other the midrib which separates or is blended with the lateral plates of the disk.

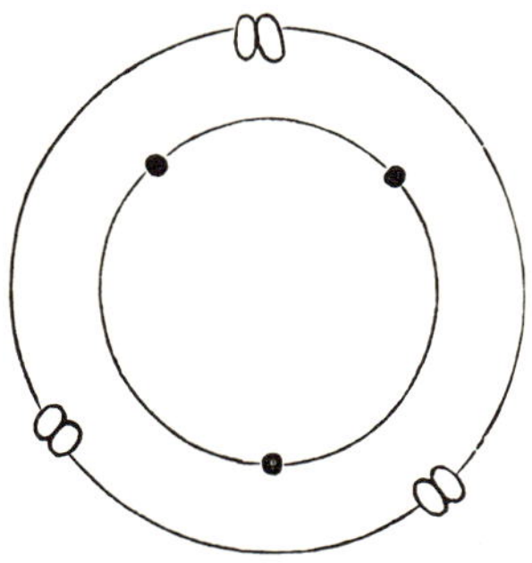

No. 1.

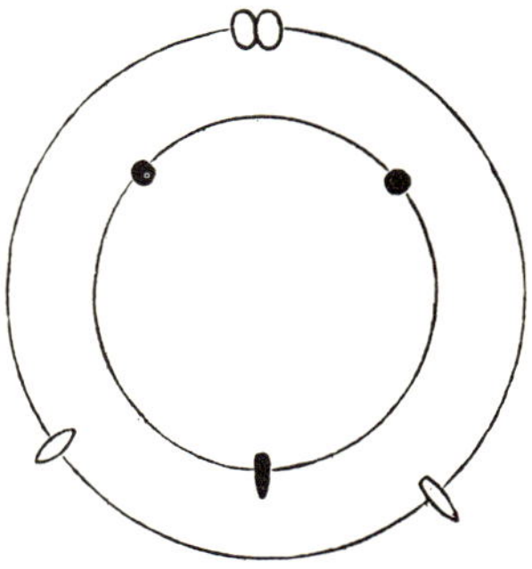

No. 2.

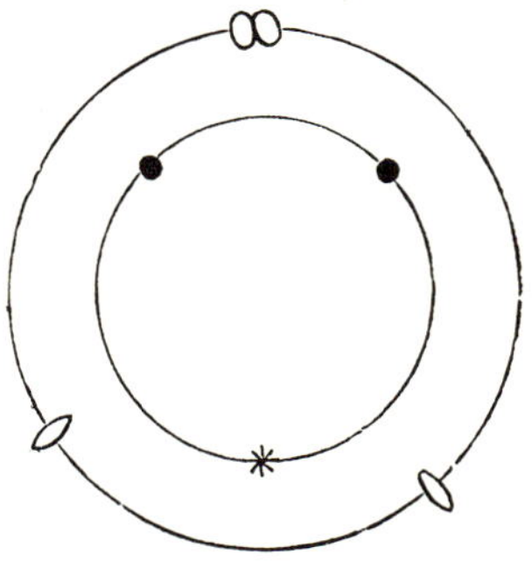

No. 3.

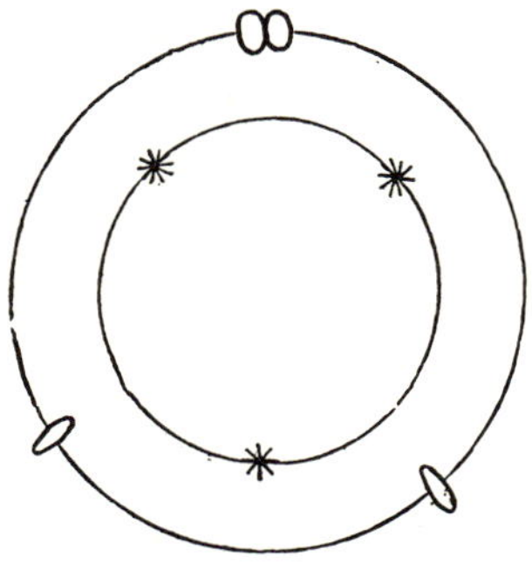

No. 4.

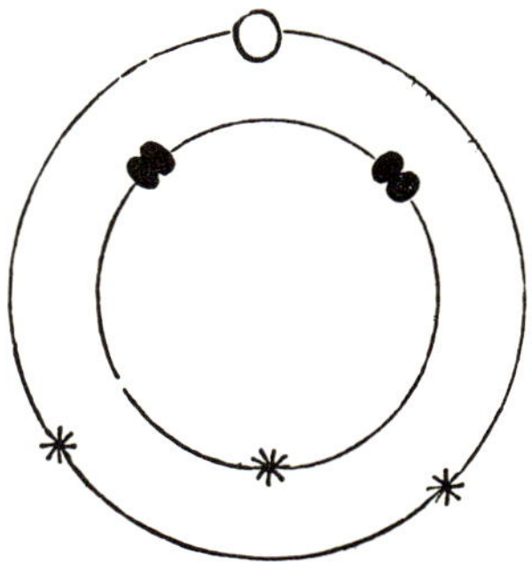

No. 5.

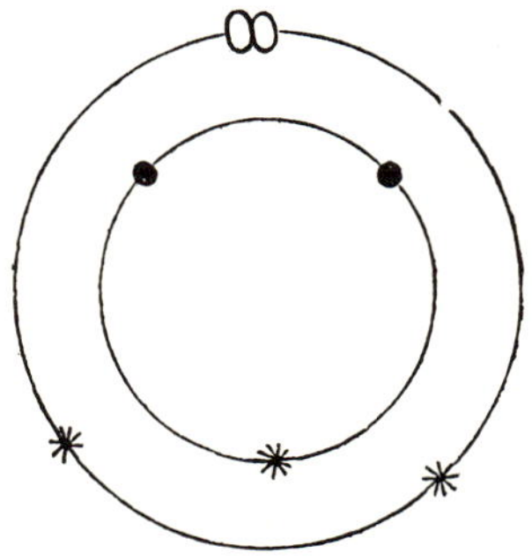

No. 6.

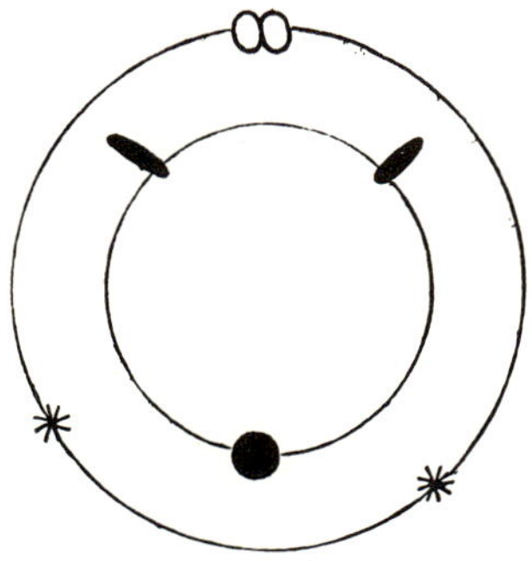

No. 7.

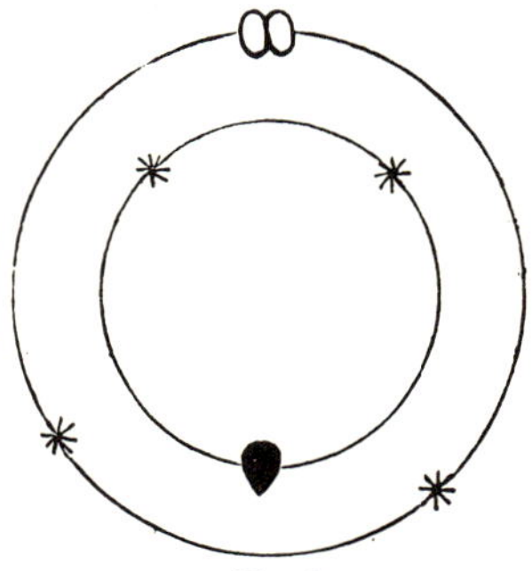

No. 8.

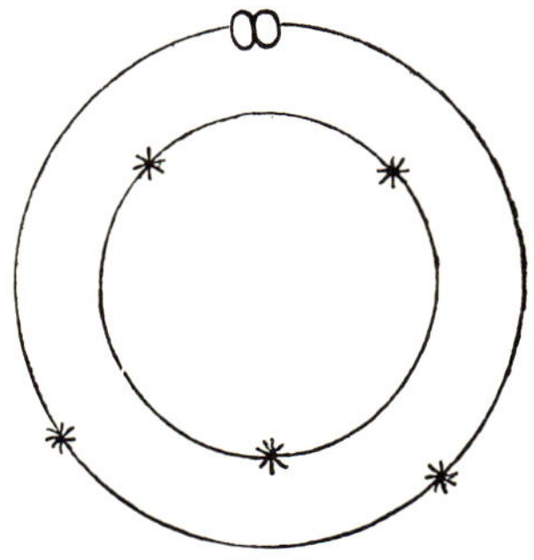

No. 9.

No. 3 represents such a structure as that of Anacamptis, where the usual outer stamen is attended by two of the inner staminodes, while two outer staminodes appear as plates on the lip, and the central of the inner staminodes is missing. Solenidium would also belong to this form.

No. 4 is the case of Cymbidium properly so called, in which all the inner staminodes are deficient, and the lateral outer staminodes lie upon the lip in the form of two raised lines.

No. 5 shows the beginning of the series in which outer lateral staminodes are wanting, except one which represents the perfect stamen in the preceding cases, while on the other hand the two lateral inner stamens are perfect and the third wanting; this occurs in Cypripedium.

No. 6. In Orchis the structure is absolutely reduced to one perfect outer stamen and a pair of inner lateral staminodes, occurring as tubercles at the base of the column, all the other staminal apparatus being missing. Thelymetra comes here.

No. 7 shows what happens in Zygostates in which the outer lateral staminodes are absent, but the whole of the inner ones are fully and largely developed. The structure of Pterostylis enters into the same category, although in some respects very different.

No. 8 may be regarded as the expression of Maxillaria, with all the staminal apparatus gone except the usual outer stamen and the corresponding inner staminode in the form of a tumour on the lip.

No. 9, with every part wanting except the outer central stamen, shows what the structure is of many Dendrobes, and Sarcopods.

The younger Reichenbach rather sneers at these speculations, which nevertheless appear to be founded on sound principles; and it is clear that he does not understand the question as it has been put in English works.

The botanist just mentioned, in an ingenious dissertation on the Pollen of Orchids (*De pollinis Orchidearum genesi ac structurâ, et de Orchideis in artem ac systema redigendis*, Leipsig, 4to, 1852,) proposes to take the position of the anther rather than the condition of the pollen as the foundation of a systematic arrangement. There does not, however, appear to be any advantage in the suggested change, which is in reality the same arrangement as the foregoing, with a different sequence. It seems at present to be premature to speak dogmatically upon this very difficult subject; and therefore I have avoided any alterations, and have merely in the following list added the names of such new genera as authors have proposed, without, however, by any means pronouncing upon their validity.

In addition to what has been above stated respecting the uses of Orchids, I may state that, according to Mr. Seemann, a species of Sobralia yields the Vanilla called Chica in Panama. The fruit of Leptotes is also succulent and fragrant; while in Notylia the fruit resembles small Miselto berries in appearance, but are slightly aromatic.

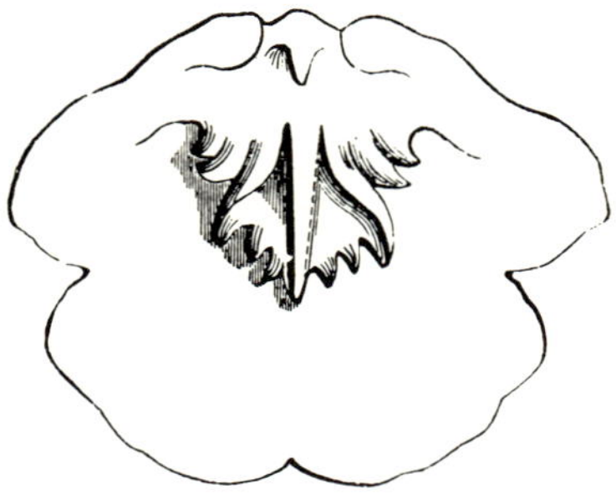

Fig. 124 *c*.

Fig. 124 *b*.

Fig. 124 *b*, Ione bicolor; Fig. 124 *c*, Lip of Warrea (Warcziewiczella, *Rchb. f.*) discolor.

ADDITIONAL GENERA, &c.

Duboisia, *Karsten.* = ? Pleurothallis.
Sigmatostalix, *Rchb. f.* near Stelis ?
Aspegrenia, *Pöppig.* = Octomeria.

Aclinia, *Griff.* near Dendrochilum.
Plexaure, *Endl.* near Oberonia.
Nephelaphyllum, *Bl.* = Cytheris.

Latouria, *Bl.* near Dendrobium.
Malachadenia, *Lindl.*
Didactyle, *Lindl.*
Xiphizusa, Rchb. f.
Bolbophyllopsis, *Rchb. f.* } near Bolbophyllum.
Taurostalix, *Id.*
Bolbophyllaria, *Id.*
Acrochæne, *Lindl.*
Xiphosium, *Griff.* = Eria.
Cœlia, *Lindl.*
Porpax, *Lindl.* } near Eria.
Aggeianthus, Wight.
Lichenora, *Wight.*

Neogyna, *Rchb. f.* } near Cœlogyne.
Androgyne, *Griff.*
Bolborchis, *Moritzi.* = Pleione.

Pinelia, *Lindl.*
Hemiscleria, *Id.*
Oerstedella, *Rchb. f.* } near Epidendrum.
Pseudepidendrum, *Rchb. f.*
Euothonæa, *Rchb. f.* near Alamania.
Læliopsis, *Lindl.* near Lælia.

PHAIUSEÆ, *Blume.* = Bletidæ.
Calelyna, *Rchb. f.* } near Evelyna.
Thunia, *Id.*

Orthochilus, *A. Rich.* = Eulophia.
Hypodematium, *Id.* = Cyrtopera ?
Stauroglottis, *Schauer.* = Phalænopsis.
Arhynchium, *Lindl.* near Saccolabium.
Cottonia, *Wight.* } near Vanda.
Pattonia, *Id.*
Wailesia, *Lindl.*
Leopardanthus, Bl. } near Aerides.
Ornitharium, *Lindl.*
Listrostachys, *Rchb. f.* near Angræcum.
Josephia, *Wight*, near Glomera ?

Schlimmia, *Linden*, near Cryptochilus.

Hyacinthorchis, *Bl.* = Cremastra.
Cyperorchis, *Bl.* near Cymbidium.
Trophianthus, *Scheidw.* = Aspasia.

Eriopsis, *Lindl.*
Pseuderiopsis, *Rchb.*
Brachtia, *Id.* } near Helcia.
Chondrorhyncha, *Lindl.*
Dignathe, *Lindl.*
Cochlioda, *Lindl.*
Solenidium, *Lindl.*
Miltoniastrum, *Rchb. f.* } near Odontoglossum.
Rhynchostele, *Rchb. f.*
Abola, *Lindl.*
Oncodia, *Lindl.*

Stanhopeastrum, *Rchb. f.* near Stanhopea.
Lycomormium, *Id.* near Acineta.
Mormolyce, *Fenzl.* = Trigonidium.
Warcziewiczella, *Rchb. f.*
Kefersteinia, *Id.*
Paradisanthus, *Id.* } near Warrea.
Acacallis, *Lindl.*
Cheiradenia, *Id.*
Cryptosanus, *Scheidw.* } near Maxillaria.
Ione, *Lindl.*
Orchidofunkia, *Rich.* = Cryptarrhena.

Erycina, *Lindl.* near Zygostates.
Euproboscis, *Wight.*
Hofmeisterella, *Rchb. f.* near Telipogon.

Mesospinidium, *Rchb. f.*
Neodryas, *Id.* } near Rodriguezia.
Cohnia, *Id.*
Papperitzia, *Id.*
Plectrophora, *Focke.* = Trichocentrum.

Tinæa, *Bivona.*
Neotinea, Rchb. f. } near Aceras.
Thisbe, *Falconer.*
Chæradoplectron, *Schauer.* = Glossula.

Derœmera, *Rchb. f.* near Herminium.
Cybele, *Falconer.* = Peristylus.
Lindblomia, *Fries.* = Cœloglossum.

Schizochilus, *Sonder.* near Disa.

Bieneria, *Rchb. f.* near Bipinnula.
Adenochilus, *Hook. fil.* near Caladenia.

Nematoceras, *Hook. fil.* near Corysanthes.

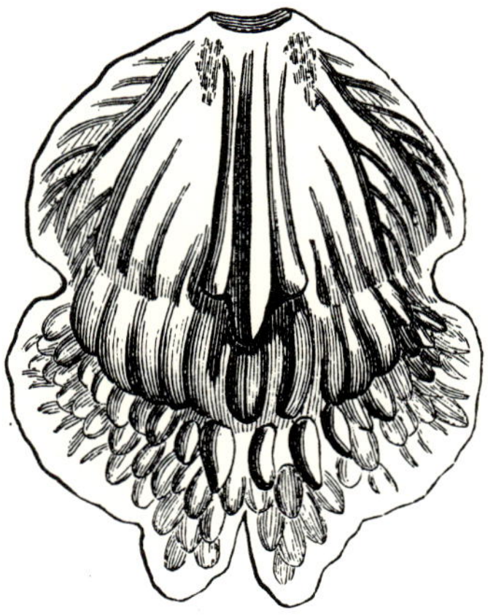

Fig. 124 *d*.

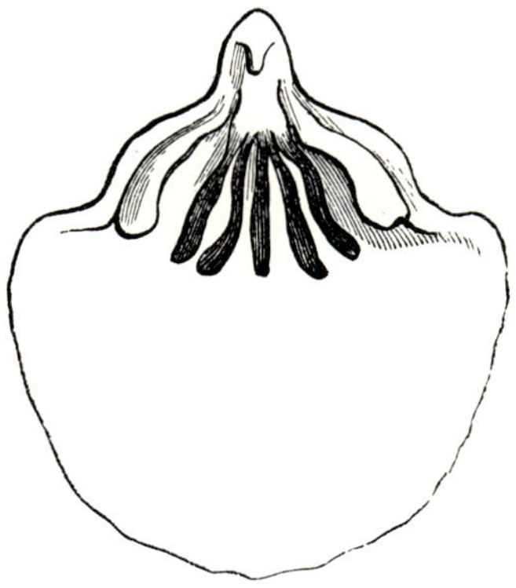

Fig. 124 *e*.

Fig. 124 *d*, Lip of Warrea bidentata; Fig. 124 *e*, Lip of Warrea Wailesiana.

Leucorchis, *Blume*, near Pogonia.
Podanthera, *Wight.* } near Pogonia?
Apetalon, *Wight.*
Rophostemon, } = Pogonia.
Cordyla.

Hæmatorchis, *Blume*, near Erythorchis.
Pogochilus = Cyrtosia.
Fregea, *Rchb. f.* near Cyathoglottis.

Galeoglossum = Prescottia.

Hysteria = Vanilla?
Corymbis, } = Cnemidia?
Macrostylis,
Ponerorchis, *Rchb. f.* }
Chaubardia, *Id.*
Pescatorea, *Id.* affinity unknown to me.
Sarcadenia, *Hort. Par.*
Guebina, *Id.*
Oxyanthera = Thelasis.

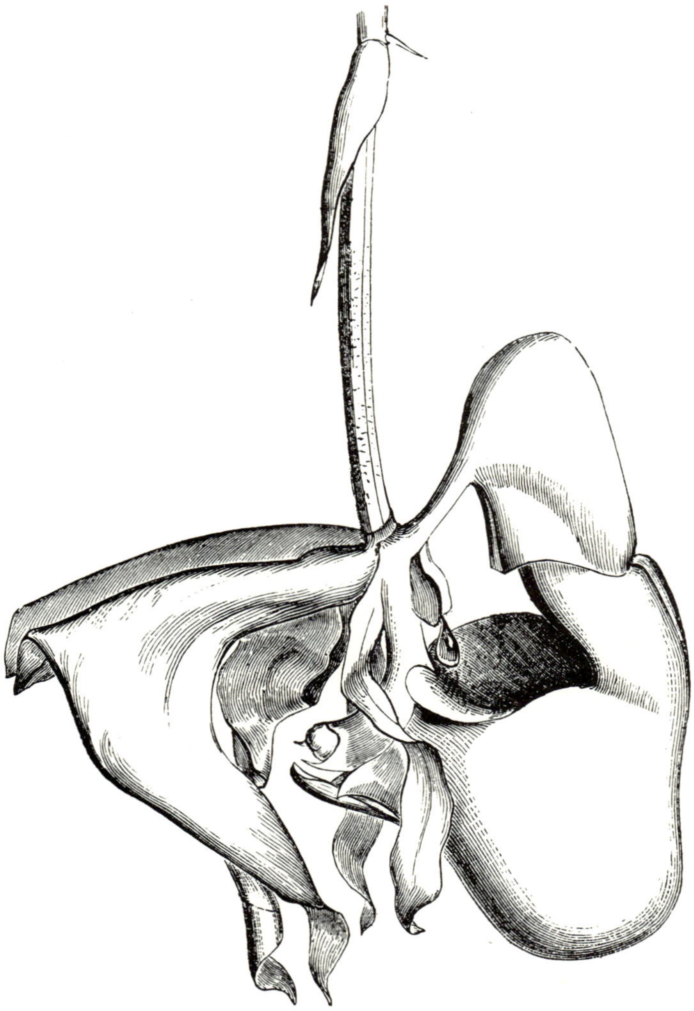

Fig. 124 *f*.

Fig. 124 *f*, Flower of Coryanthes speciosa.

ORDER LIII. APOSTASIACEÆ.—APOSTASIADS.

Apostasieæ, *Lindl. Nixus Plantarum, p.* 22. (1833); *Blume in Ann. Sc. Nat. Ser.* 2. 2. 91. (1834); *Endl. Gen.* lxvii.; *Meisn. p.* 387.

DIAGNOSIS.—*Orchidal Endogens with regular half-gynandrous flowers, and axile placentæ.*

Perennial herbaceous plants. Stem simple or branched. Leaves firm, thin, sheathing at the base. Flowers in simple or compound terminal racemes. Calyx and corolla each consisting of 3 similar pieces. Anthers 2 or 3, sessile upon a short column, erect, 2-celled, opening longitudinally; pollen cohering in 3s or 4s according to Mr. Bauer (*Illust. Fruct. t.* 15),—in single oval grains with a longitudinal furrow according to Mr. Griffith (*Letter dated Merqui Dec.* 28, 1834) and Blume. Ovary 3-celled, with 3 polyspermous placentæ in the axis; ovules with their integuments very distinct and much shorter than the protruded nucleus (*Griffith*); style filiform, with a slightly 3-lobed stigma as long as the anthers, and adhering with their filaments into a short column. [Capsule 3-celled, 3-valved; the valves bearing the dissepiment in the middle, but cohering at the apex and base. Seeds very numerous, minute, ovate, and with a skin fitting the nucleus, or scobiform with a membranous testa loose at each end.—*Blume.*]

Fig. CXXV.

Very closely allied to Orchids, from which they differ essentially in having a 3-celled fruit, with loculicidal dehiscence, and in the style being altogether free from the stamina for the principal part of its length. At the same time the structure is gynandrous enough to afford a clear distinction from the Burmanniads. There are many admirable observations upon Apostasia itself in Brown's *Observations on the organs and mode of fecundation in Orchideæ and Asclepiadeæ*, and some further information is given by Blume in the place above quoted. The Order seems as if connecting Orchids and Hypoxids. If Rhyncanthera is correctly represented by Blume, its 3-locular ovary will refer it here, while the structure of its column would keep it in Orchids. The essential character is, however, framed without reference to it.

Found in damp woods in the hotter parts of India.

No uses have been assigned to any of them.

GENERA

Apostasia, *Bl.* *Mesodactylus*, Wall.	Neuwiedia, *Bl*	?Rhyncanthera, *Bl.*

NUMBERS. GEN. 3. Sp. 5.

Hypoxidaceæ.
POSITION.—Orchidaceæ.—APOSTASIACEÆ.——————.

Fig. CXXV.—Apostasia odorata; 1. a flower; 2. the stamens and style; 3. a cross section of the ovary; 4. a seed.

The Gardeners' Chronicle

Another of Lindley's ventures was the founding with Joseph Paxton and others of *The Gardeners' Chronicle* in 1841, of which the horticultural part was edited by Lindley. Then, as more than a century later, it met the need of gardeners for a weekly journal that provided practical and scientific information on a diversity of horticultural matters. Indeed at the beginning it was far more diverse in content than later. The *Chronicle* summarised the proceedings of horticultural and botanical societies, provided foreign and domestic political news together with market prices of fruit, garden memoranda, entomological articles, records of new plants etc, and forthright editorials; it also advertised plants and sundries for sale and situations of employment wanted and vacant. There were such surprising items as the trial at the Guildhall of two women snowballing at what they described as "the shocking sight of a black man going to be married to a white woman, although the wife was no great shakes". In the course of time it became the most convenient periodical for the quick publication of new species and was so used by Kew botanists such as J. G. Baker and N. E. Brown. H. G. Reichenbach described many new orchids in *The Gardeners' Chronicle.* Lindley's editorship for 20 years ensured the maintenance of a high standard which earned it international repute.

In 1850 Paxton and Lindley started a periodical with hand-coloured lithographed plates, *Paxton's Flower Garden,* which ran only to three volumes and ended in 1851. Long after the death of the two authors, Thomas Baines published a revised edition, likewise in three volumes, between 1880 and 1885.

George Glenny's attacks on Lindley

Despite his notorious hot temper, his lack of tact in dealing with humble gardeners, his vigorously expressed opinions and resentment at opposition, Lindley seems to have been on good terms with most of his botanical and horticultural contemporaries but certainly not with George Glenny (c. 1793–1874). A horticultural journalist as forceful and fearless in giving offence as Lindley himself and quick with stinging comment, hence Tjaden's description of him as a "horticultural hornet", Glenny obviously enjoyed controversy. For many years he attacked Lindley in his periodicals, first the *Horticultural Journal,* then the *Gardener's Gazette,* and, though he exaggerated, there must have some foundation for his scurrilous remarks. Glenny nursed a long-standing grievance against the Horticultural Society of London and in particular the Society's employee, Lindley, responsible for shows, because his exhibits of plants did not receive the awards he considered they deserved. He publicly accused the Society of deliberately treating him unfairly. In 1836 Glenny stated in his *Horticultural Journal* that "we cannot say that we suspect the sapient doctor who sits like a nightmare on the affairs of the Society of doing anything in a business-like manner". Lindley habitually discussed various plants at the Society's shows and his seemingly presumptuous remarks made him vulnerable to criticism by gardeners with different opinions and greater knowledge. Glenny accordingly praised Mrs. Louisa Lawrence (d. 1855), wife of the surgeon (later Sir) Trevor Lawrence, for taking Lindley to task; he "having fairly tried everybody's patience, Mrs. Lawrence gave him as neat a set-down". She noted, "Doctor Lindley well named a *professor* of botany, is more garrulous, and more foppish, and more foolish than any of the whole race of dandies and danglers. The grossness and indecency of the Council can only be equalled by the impertinence, flippancy and vanity of their paid assistant, or paid master, whichever they please."

Year in, year out, Glenny's publications kept up their fire. Sarcastically he asked in the *Gardener's Gazette* (1837): "Has a certain well-respected gentleman at Regent Street and Turnham Green the title and distinction of M. D. (which being translated is Mad Doctor)? The politeness with which the learned doctor bullies the exhibitors, the discretion with which he awards the most conspicuous place to the productions of his friends ..." – so Glenny continued to rile Lindley. Unfortunately resentment against Glenny's insults led Lindley into an ungracious and impolitic act. Exhibitors at a Chiswick Show in July 1837 resolved that, "differing as many persons may upon different subjects from Dr Lindley, there is but one opinion upon the fact of his gentlemanly conduct and uniform kindness towards exhibitors" and accordingly as a mark of respect they proposed to present him with a price of plate valued 30 guineas. The subscribers were respected nurserymen and gardeners obviously genuine in their regard for Lindley, but Glenny was associated with organizing its presentation. Hastily Lindley refused to accept their generous gift; for him it was too closely associated with Glenny and the *Gardener's Gazette',* "one of those unprincipled newspapers which are the disgrace of a civilised people". Glenny joyfully printed Lindley's letter. It provided yet further ammunition against the arch-enemy: "Dr Lindley is in a towering passion at the success of the *Gardener's Gazette*... and he has been silly enough to show it." Glenny mockingly pretended that his *Gardener's Gazette* had been "*doctored to death,* like the orchideous plants were at Chiswick, and fallen victim to the Lindleyan system".

Glenny's attacks on Lindley continued off and on until about 1850. They possibly had as salutary and stimulating effect on the Horticultural Society as the fleas on a hedgehog but did no good to either opponent. Glenny became bankrupt in 1839; ultimately, as stated by Tjaden (1986), he was "forced to do what he could do very well, write on horticulture for other publishers. It was on this writing for thirty years from 1841, that he relied for most of his income and on which his modern reputation has rested." Born a little earlier than Lindley, he outlasted him by eight years.

A horticultural show interlude

Edward Beck (c. 1804–1861) was a Quaker nurseryman at Isleworth, Middlesex, only about 5 miles from the Horticultural Society's garden at Chiswick, with which he was undoubtedly familiar. About 1850 Beck published anonymously a pamphlet *A Packet of Seeds saved by an old Gardener,* purporting to be the recollections of one James Gregory but giving an authentic picture of a country gardener's life then. Here Gregory recalls his visit to the horticultural show at Chiswick organized by Lindley:

> When I went up to London, the squire got me an order, so that I might go to Chiswick early on the show-day, and see the plants put up in their places; and I never

could have believed any body that had told me all I saw with my eyes that day. I seemed lost in a wood of plants, as I walked about in the tents, where the gardeners put them down before they began to stage them: – and such a set of men, too; why, their helpers were better dressed and better mannered than the head men in our parts; and yet, when I asked a question now and then of some of 'em, they didn't seem to want for conceit. As good luck would have it, I met a gardener that lived once not far from Birdwood, and had left to go into a London nursery. I should not have known him, but he made me out; and very kind he was. After all the plants were staged, and it wasn't till just before ten – and I'd got to Turnham Green by six – my friend said; "I wish I could get you a ticket for breakfast; but I have got so little here, I know Doctor Lindley won't give me one for you." "Oh", said I; "I'll ask him myself". "You'd better not," said he; "for you'll get no ticket, and like enough something from him that'll serve you instead of a breakfast." Says I, "There's not a man's face on earth that I'm afraid of. I've often heard of the Doctor, and read a deal more in Glenny and the Journal; and now I'll have a look and a word for myself, if you'll just shew him to me." "Come along," says he, and away we went. After a longish hunt, he shewed me a gentleman sitting on a stool under a tree, with a walking-stick in his hand, and a pair of spectacles on his nose; and said he, "That's the Doctor." "Thank you," said I; "and now wait a minute for me." So I went up, and lifted my hat, breaking ground, as my poor father, when he was soldiering, would have called it. "Well," said he, "what do you want?" "I should thank you, sir," said I, "for a breakfast-ticket, for I'm a stranger, and a long way from home." "What have you brought, gardener?" "Nothing, sir," said I, "but myself; but I have sent something," said I, "before to-day; but not to these shows:" and I shewed him a silver medal I had had sent me, through my master, for a basket of fruit. "These breakfast-tickets," said he, "are for exhibitors and helpers only; but I will give you one: – and there," said he, handing me another, "that will admit you after one o'clock; for it is not often I see so much of the country as you have brought with you." He looked me all over; at my knee corduroys, my leather gaiters, and at my canary waistcoat, and long blue coat with metal buttons, and then into my face, as if he'd have burnt a hole in it; all the while asking me questions, and wetting his fingers at his lips, and then running them over the cards in his hand. As soon as he stopped, I lifted my hat, thanked him, and bid him good morning; but not before I'd heard him refuse tickets to two gardeners running; one, because his plants were not all in their places, or all to rights; and to the other, because he was so dirty and untidy.

Last years

Lindley possessed immense powers of work, as the above chronicle indicates, but they had limits which he overstepped. By 1862 his health had begun to fail, his mental powers and memory to decline, and he had a breakdown from which he never fully recovered. He retired from his London professorship in 1861 and from his secretaryship of the Royal Horticultural Society in 1863. In 1864 he sold his prized orchid herbarium to the Royal Botanic Gardens, Kew. Fortunately friends persuaded him to have his protrait painted by E. U. Eddis; he chose to hold in his hand not an orchid but a wild rose, recalling his first botanical triumph, the *Rosarum Monographia* of 1820. Lindley died at Turnham Green on 1 November 1865, just outlasting two friends of many years, Sir Joseph Paxton who died on 8 June 1865, and Sir William Jackson Hooker who died on 12 August 1865. Their careers had been in many ways linked, particularly as regards the Royal Botanic Gardens, Kew, and contributions to orchidology. Hooker described and illustrated many new orchids in the *Botanical Magazine,* including the genera *Peristeria, Coryanthes* and *Polystachya.* Paxton, by revolutionising the cultivation of orchids, contributed more than any one man to the spread of orchid-growing among the aristocracy and those with newly-acquired wealth seeking to emulate his master the Duke of Devonshire, and thus to the large-scale importation of tropical orchids. Lindley had been for many years the supreme orchidologist.

On 9 November 1865, there died a plant collector, Thomas Bridges (b. 1807), who had introduced into cultivation the giant waterlily, *Victoria amazonica (V. regia),* with which all three were associated. Lindley published the new genus *Victoria* in 1837 in a folio work with a coloured plate, *Victoria regia,* limited to 25 copies: he later republished part of this in the *Botanical Register.* Hooker in 1847 published another folio, *Description of Victoria regia or the great Water-lily of South America,* with four superb plates by the celebrated botanical artist Walter Hood Fitch. Paxton, the first to flower the *Victoria,* at Chatsworth in 1849, afterwards stated that the architecture of its leaves with veins of varied size and strength upholding a broad thin surface provided mechanical principles for the construction of the Crystal Palace. After Lindley's death the Royal Horticultural Society bought for £ 700 his botanical and horticultural library to be the nucleus of a new library replacing the one disastrously sold by the Society in 1859 during a financial crisis. As a result of that bitter experience the Society vested it in 1868 in a Lindley Library Trust legally separate from the Society. This was a device to safeguard the Lindley Library from a like calamity in the future, the Society, however, maintaining, expanding and housing it. The Society also instituted, in December 1865, the Lindley Medal commemorative of Lindley's services and awarded for an exceptional exhibit at a Society horticultural show. His prized orchid herbarium rich in type-specimens and drawings had already been sold to the Royal Botanic Gardens, Kew. The University of Cambridge acquired the rest of his herbarium, likewise rich in type-specimens, comprising about 58,000 sheets, for £ 300; it now forms part of the general herbarium at the Botany School, Cambridge.

The American orchidological periodical *Lindleyana* (1986 et seq.) and three generic names commemorate him, *Lindleya* Kunth (Rosaceae), *Neolindleya* Kränzlin and *Lindleyella* Schlechter (Orchidaceae). Rudolf Schlechter (1872–1925) turned from a detailed and fruitful study of the intricacy of the Asclepiadaceae to an equally detailed and fruitful study of the Orchidaceae and became Lindley's 20th-century orchidological successor. Thus his dedication of a genus of Orchidaceae to Lindley "Altmeister der Orchideenkunde" fittingly links two extraordinarily industrious and percipient students of this vast intricate family.

Lindley as an Orchidologist

Lindley's devotion to the study of orchids made him from 1830 onwards the leading botanical authority on orchids, building upon, revising and largely changing the pioneer work of Olof Swartz (1760–1818) and Robert Brown (1773–1858). Swartz, a Swedish botanist, spent the years 1784 to 1786 in Jamaica, which has above 155 orchidaceous species, and in Hispaniola; thus he knew many orchids in a living state. He distinguished 28 genera in his "Genera et species Orchidearum" (Schrader, *Neu. Journn. Bot.* 1 (1): 1–108; 1800). Robert Brown, a Scot who sailed on Matthew Flinder's voyage to Australia as naturalist, with Ferdinand Bauer as artist, became acquainted with many species there and then others in cultivation at Kew. These he described in his *Prodromus Florae Novae Hollandiae,* 209–333 (1810) and Aiton's *Hortus Kewensis,* 2nd ed. 5: 188–222 (1813). He established 29 new genera, among them *Listera, Goodyera, Lissochilus, Eulophia, Vanda* and *Pleurothallis.*

As newly introduced orchids came into Lindley's hands, his first task was to try and fit them into the genera dealt with by Swartz and Brown. Soon it became evident that the majority represented unnamed genera and species. He was not content to describe them as new but gradually to revise the family as a whole and place them within a system of which the final version appeared in the third edition of his *Vegetable Kingdom* (1853) as noted above. In this he was more successful than in some other parts of the vegetable kingdom in following John Ray's procedure, which he quoted, of trying not to associate like with unlike species and not to separate like from like: "Methodum intelligo Naturae convenientem, quae nec alienas species conjungit, nec cognatas separat" (Ray, *Stirp. Europ. Sylloge,* Pref. c; 1694). The classification of the family in R. L. Dressler's *The Orchids; Natural History and Classification* (1981) has much in common with Lindley's. Thus a considerable part of his orchid system has survived into modern classifications of the Orchidaceae.

Lindley's sound appreciation of generic limits within the family led Bentham and Hooker with their wide generic concept in 1883 to accept 114 Lindley genera; these include such genera as *Ada, Aspasia, Cattleya, Cochlioda, Coelogyne, Galeandra, Laelia, Miltonia, Schomburgkia* and *Sophronitis.* Pfitzer in 1889 accepted 127 of Lindley's genera; with the narrower modern generic concept, even more have now found approval.

With regard to the naming of plants Lindley wrote in 1839: "So impossible is it to construct generic names that will express the pecularities of the species they represent, that I agree with those who think a good, well-sounding, *unmeaning* name as good as any that can be contrived. The great rule to follow is this: In constructing a generic name, take care that it is harmonious and as unlike all other generic names as it can be." Providing such names for the many new genera of Orchidaceae he distinguished cannot have been an easy task. Obviously he sought to indicate some feature of the genus, whenever possible, e. g. *Cirrhopetalum* (from *cirrhus,* curl of hair, tendril, *petalum,* petal) referring to the twisted lateral sepals, *Eria* (from *erion* wool) referring to the woolly flowers, *Herpysma* (from *herpysmos,* creeping) of creeping habit, *Holothrix* (from *holos,* whole, *thrix,* hair) referring to its extensive hair-covering, *Micropera* (from *mikros,* small, *pera,* pouch) referring to the small saccate lip, *Ommathodium* (from *omma,* eye) with eye-like spots on the lip. The intent of some names, e. g. *Abola, Ania* (from *ania,* trouble) *Apaturia, Ate, Ipsea* and *Oecoclades* (from *oikeious,* private, *klados,* branch), is obscure. Existing mythological names, e. g. *Arethusa, Calypso, Disa* and *Norna,* preceeded Lindley's adoption of *Acacallis Empusa, Ione, Lycaste* and *Oberonia.* He commemorated some noted orchid-growers by the genera *Batemannia, Clowesia, Miltonia, Paxtonia* and *Wailesia,* botanical artists by *Alvisia* and *Drakaea,* collectors by *Amellia, Govenia, Hartwegia, Schomburgkia,* distinguished scientists by *Burnettia, Herschelia* and *Quekettia,* persons of antiquity by *Ada* and *Aspasia.* The generic names published by Lindley are thus of very diverse origin but conform to his own criteria. He was much aided, as have been other orchidologists, notably C. L. Blume and Rudolf Schlechter, by the richness of the Greek language and the ease with which new compound words can be made from Greek elements; indeed most generic names of orchids, apart from those commemorating persons, have a Greek origin. Of 286 Lindley orchid generic names, 129 come from Greek, 18 from Latin, 27 from personal names, 12 from mythology.

Lindley described a vast number of new species, usually with concise and precise diagnoses, which were adequate for the period but need now to be supplemented. He objected strongly to what he castigated in *The Gardeners' Chronicle* as "the bad German habit" of providing long descriptions and thereby aroused much controversy in Germany (cf. Sinkora & Short, 1988).

Oakes Ames, for many years the leading American orchidologist, stated that Lindley "laid the foundations of modern orchidology". With that verdict there is and has long been unanimous agreement.

Some sources of further information

Ames, O. 1948. *Orchids in Retrospect. A Collection of Essays on the Orchidaceae.* Cambridge, Mass.

Anon. 1839. The naturalist's literary portrait-gallery, no. 5. John Lindley, Ph. D., F.R.S. *The Naturalist* 4 (32): 434–442.

Anon. 1865. The late Dr Lindley. *Gardeners' Chronicle* 1865: 1058, 1059, 1083, 1084.

Anthony, J. 1973. *Joseph Paxton. An illustrated Life of Sir Joseph Paxton, 1803–1965.* Shire Publications, Aylesbury, Bucks.

[Beck, E. c. 1958.] *A Packet of Seeds saved by an old Gardener.* London.

Bishop, W. J. 1953. The evolution of the general practitioner in England. *In:* Underwood, E. A. (Ed.), *Science, Medicine and History: Essays... in Honour of Charles Singer* 2: 351–357.

Chadwick, G. F. 1961. *The Works of Sir Joseph Paxton* 1803–1865. Architectural Press, London.

Dillon, G. 1957. Development of a system of orchid classification. *American Orchid Society Bulletin* 26: 255–261.

Fletcher, H. R. 1969. *The Story of the Royal Horticultural Society 1804–1868.* Oxford University Press, London.

Gardener, W. 1965. John Lindley. *Gardeners' Chronicle* 158: 386, 406, 409, 430, 434, 451, 457, 481, 502, 507, 526.

Green, J. R. 1914. *A History of Botany in the United Kingdom:* 336–353. J. M. Dent & Sons, London.

Keeble, F. 1912. John Lindley. *In:* Oliver, F. W. (Ed.), *Makers of British Botany:* 164–177. Cambridge University Press, Cambridge.

Lindley, Alice (Lady Moreland) 1965. John Lindley, 1799–1865. *Gardeners' Chronicle* 158: 362.

Markham, V. M. 1935. *Paxton and the Bachelor Duke.* Hodder & Stroughton, London.

Reinikka, Merla A. 1972. *A History of the Orchid.* University of Miami Press, Coral Gables, Florida.

Sachs, T. von 1875. *Geschichte der Botanik.* R. Oldenbourg, Munich [English translation, 1890. *History of Botany,* Oxford University Press, Oxford].

Schultes, R. E. & Pease, A. S. 1963. *Generic Names of Orchids, their Origin and Meaning.* Academic Presse, New York and London.

Sinkora, D. M. & Short, P. S. 1988. *"The bad German Habit". A factual botanical Melodrama involving Joachim Steetz.* National Herbarium of Victoria, Melbourne, Australia.

Stafleu, R. A. & Cowan, R. S. 1981. *Taxonomic Literature* 3: 49–60. Bohn, Scheltema & Holkema, Utrecht & The Hague.

Stearn, W. T. 1961. Two thousand years of orchidology. *Proceedings of the Third World Orchid Conference.* London. 1960: 24–42. London.

Stearn, W. T. 1965. The self-taught botanists who saved the Kew botanic garden. *Taxon* 14: 293–298.

Stearn, W. T. 1967. Sibthorp, Smith, the *Flora Graeca* and the *Florae Graecae Prodromus. Taxon* 16: 168–178.

Stearn, W. T. 1973. Lindley, John. *In:* Gillespie, C. C. (Ed.), *Dictionary of Scientific Biography* 8: 371–373.

Stearn, W. T. 1976. Carl Linnaeus and the theory and practice of horticulture. *Taxon* 25: 21–31.

Tjaden, W. 1983. *The Gardener's Gazette* [edited by George Glenny] and its rivals. *Garden History (J. Garden History Soc.)* II: 70–78.

Tjaden, W. 1986. George Glenny: horticultural hornet. *The Garden (J. Royal Hort. Soc.)* 111: 318–323 [I am indebted to Mr Tjaden for additional information extracted from his unpublished biography of George Glenny and used here].

BIBLIOGRAPHY

REGIONAL BIBLIOGRAPHY

Europe and the Middle East

Blamey, M. and Grey-Wilson, C.: The Illustrated Flora of Britain and Northern Europe. Hodder and Stoughton, London 1989.

Buttler, K. P.: Orchideen. Mosaik Verl., München 1986.

Camus, E. G.: Monographie des Orchidées. Lechevalier, Paris 1908.

Danesch, O. und Danesch, E.: Orchideen Europas, Hallwag, Bern 1962.

Danesch, O. und Danesch, E.: Orchideen Europas: Südeuropa. Hallwag, Bern 1969.

Danesch, O. und Danesch, E.: Orchideen Europas. Oprhys-Hybriden. Hallwag, Bern 1972.

Davies, P., and Davies, J., and Huxley, A.: Wild Orchids of Britain and Europe. Chatto and Windus, London 1983.

Delforge, P. et Tyteca, D.: Guide des Orchidées d'Europe. Duculot, Paris 1984.

Godfery, M. J.: British Orchidaceae. Cambridge U. P., Cambridge 1933.

Keller, G. und Schlechter, R.: Monographie und Iconographie der Orchideen Europas und des Mittelmeergebietes, 1–45; Fedde, Repertorium, Sonderbeiheft A, Berlin 1925–1943.

Landwehr, J.: Wilde Orchideeën van Europa, 2 vols. Natuurmonumenten, Nederland 1977.

Nelson, E.: Gestaltwandel und Artbildung erörtert am Beispiel der Orchidaceen Europas und der Mittelmeerländer, insbesondere der Gattung Ophrys. E. Nelson, Chernex-Montreux 1962.

Nelson, E.: Monographie und Ikonographie der Orchidaceen-Gattungen Serapias, Aceras, Loroglossum, Barlia. E. Nelson, Chernex-Montreux 1968.

Nelson, E.: Monographie und Ikonographie der Orchidaceen-Gattung Dactylorhiza. Speich, Zürich 1976.

Nelson, E. und Fischer, H.: Die Orchideen Deutschlands und der angrenzenden Gebiete. Frisch, Berlin 1931.

Reichenbach, H. G.: Orchideae. In: Flora Germanica Recensitae. Hofmeister, Leipzig 1851.

Renz, J.: Orchidaceae. In: K. H. Rechinger: Flora Iranica, No. 126. Akad. Druck- und Verlagsanstalt, Graz 1978.

Summerhayes, V. S.: Wild Orchids of Britain, 2nd ed. Collins, London 1968.

Sundermann, H.: Europäische und mediterrane Orchideen. 2. Aufl. Brücke, Hannover 1975.

Wood, J. J.: Orchidaceae. In: Flora of Cyprus 2: 1493–1558. Royal Botanic Gardens Kew, 1985.

Die Orchidee (1949 →). Herausgeber: Deutsche Orchideen-Gesellschaft. Brücke-Verlag, Kurt Schmersow, Hildesheim. Many articles on orchids of Europe and the Middle East. Viele Artikel über Orchideen Europas und des Mittleren Ostens.

L'Orchidophile. Bull. de la Société d'Orchidophilie. 84, rue de Grenelle 75007 – Paris.

Japan, China and Central Asia

Garay, L., and Sweet, H.: Orchids of Southern Ryukyu Islands. Botanical Museum, Harvard Univ., Cambridge, Mass. 1974.

Hu, S. Y.: The Orchidaceae of China 3. Quart. J. Taiwan Mus. 25 (1–2), 1972.

Hu, S. Y.: The Orchidaceae of China 6. Quart. J. Taiwan Mus. 26 (3–4), 1973.

Hu, S. Y.: The Orchidaceae of China 7 & 8. Quart. J. Taiwan Mus. 27 (1–4), 1974.

Hu, S. Y.: The Orchidaceae of China 9. Quart. J. Taiwan Mus. 28 (1–2), 1975.

Hu, S. Y.: The Genera of Orchidaceae in Hong Kong. Chinese Univ. Press, 1977.

Iconographia Cormophytorum Sinicorum, Vol. 5: Orchidaceae 602–772, 1976.

Kraenzlin, F.: Orchidacearum Sibiriae Enumeratio. Fedde, Repertorium, Beihefte 65, 1931.

Lin, T. P.: Native Orchids of Taiwan 1 & 2. Taipei, 1976 & 1977.

Maekwa, F.: The Wild Orchids of Japan in Colour. Tokyo 1971.

Makino, T. and Nemoto, K.: Flora of Japan, ed. 2. Tokyo, 1931.

Schlechter, R.: Orchideologiae Sino-Japonicae Prodromus. Fedde, Repertorium, Beihefte 4, 1919.

India and Burma

Banerji, M. L., and Pradhan, P.: The Orchids of Nepal Himalaya. J. Cramer, Vaduz 1984.

Bose, T. K., and Bhattacharjee, S. K.: Orchids of India. Nayo Prokash, Calcutta 1980.

Duthie, J. F.: The Orchids of the North-Western Himalaya. Annals of the Royal Botanic Garden, Calcutta, 9, 1906.

Grant, B.: The Orchids of Burma, Rangoon 1895.

Jayaweera, D. M. A.: Orchidaceae: In: Dassanayake. A revised Handbook to the Flora of Ceylon 2: 4–386, 1981.

Hooker, J. D.: Orchideae. Flora of British India, 5, L. Reeve and Co., London 1890.

Hooker, J. D.: A Century of Indian Orchids. Annals of the Royal Botanic Garden, Calcutta, 5, 1895.

King, G., and Pantling, R.: The Orchids of the Sikkim-Himalaya. Annals of the Royal Botanic Garden, Calcutta, 8, 1898.

Pradham, U. C.: Indian Orchids: Guide of Identification and Culture. 2 vols. Kalimpong 1976–1979.

South East Asia and Indonesia

Ames, O., and Schweinfurth, C.: The Orchids of Mount Kinabalu, British North Borneo. Orchidaceae, 6, Merrymount Press, Boston 1920.

Backer, C. A.: Beknopte Flora van Java – Orchidaceae 1–3. Rijksherbarium, Leiden 1952.

Bennett, K. S.: The Tropical Asiatic Slipper Orchids, Genus Paphiopedilum. Angus and Robertson, North Ryde 1984.

Gagnepain, F. et Guillaumin, A.: Orchidacées. Flore Générale de L'Indo-Chine, 2, Masson, Paris 1932.

Holttum, R. E.: Orchids of Malaya, ed. 3. In: A Revised Flora of Malaya. Government Printing Office, Singapore 1964.

Schlechter, R.: Blütenanalysen neuer Orchideen IV. Indische und malesische Orchideen. Fedde, Repertorium, Beihefte 74, 1934.

Seidenfaden, G.: Contribution to a Revision of the Orchid Flora of Cambodia, Laos and Vietnam, 1. A Preliminary Enumeration, Fredensborg, Denmark 1975.

Seidenfaden, G., and Smitinand, T.: Orchids of Thailand, 1–4. The Siam Society, Bangkok 1959–1964.

Seidenfaden, G.: Orchid Genera in Thailand X. Trichotosia and Eria. Opera Botanica, No. 62, 1982.

Seidenfaden, G.: Notes on Cirrhopetalum Lindl. Dansk Botanisk Arkiv, 29 (1), 1973.

Seidenfaden, G.: Orchid Genera in Thailand I–III, Calanthe, Cleisostoma, Ceologyne. Dansk Botanisk Arkiv, 29 (2–4), 1975.

Seidenfaden, G.: Orchid Genera in Thailand IV & V, Liparis, Orchidaceae. Dansk Botanisk Arkiv, 31 (1 & 3), 1976–77.

Seidenfaden, G.: Orchid Genera in Thailand VI, Neottioideae. Dansk Botanisk Arkiv, 32 (2), 1978.

Seidenfaden, G.: Orchid Genera in Thailand VII, Oberonia, Malaxis. Dansk Botanisk Arkiv, 33 (1), 1978.

Seidenfaden, G.: Orchid Genera in Thailand VIII, Bulbophyllum. Dansk Botanisk Arkiv, 33 (3), 1979.

Seidenfaden, G.: Orchid Genera in Thailand IX, Flickingeria, Epigeneium. Dansk Botanisk Arkiv, 34 (1), 1980.

Seidenfaden, G.: Orchid Genera in Thailand XI, Cymbidieae. Opera Botanica, No. 72: 65–93, 1983.

Seidenfaden, G.: Orchid Genera in Thailand XII. Dendrobium Sw. Opera Botanica, No. 83, 1985.

Seidenfaden, G.: Orchid Genera in Thailand XIII. Thirty-three Epidendroid Genera. Opera Botanica, No. 89, 1986.

Seidenfaden, G.: Orchid Genera in Thailand XIV. Fifty-nine Vandoid. Opera Botanica, No. 95, 1988.

Smith, J. J.: Icones Bogorienses – Orchidées. E. J. Brill, Leiden 1903.

Smith, J. J.: Die Orchideen von Ambon. Landsdrukkerij, Batavia 1905.

Smith, J. J.: Die Orchideen von Java. In: Flora von Buitenzorg, 6, 1905.

Smith, J. J.: Die Orchideen von Java 1–8. In: Bulletin du Departement de l'Agriculture aux Indes Néerlandaises. In: Bulletin du Jardin Botanique de Buitenzorg, 1905–1924.

Smith, J. J.: Orchidaceae Buruenses. In: Bulletin du Jardin Botanique de Buitenzorg 9, 1928.

Smith, J. J.: Orchidaceae Seranenses. In: Bulletin du Jardin Botanique de Buitenzorg 10, 1928.

Smith, J. J.: Icones Orchidacearum Malayensium, 1–2. In: Bulletin du Jardin Botanique de Buitenzorg, 1930–1938.

Smith, J. J.: On a Collection of Orchidaceae from Central Borneo. Bulletin du Jardin Botanique de Buitenzorg, 11, 1931.
Smith, J. J.: Enumeration of the Orchidaceae of Sumatra and Neighbouring Islands. Fedde, Repertorium, 32, 1933.

The Philippines

Ames, O.: Notes on Philippine Orchids. The Philippine Journal of Science 4, 1909.
Ames, O.: Orchidaceae 1–3, 5–7. [Several articles on Philippine orchids 1905–1922.]
Ames, O.: Schedulae Orchidianae. [Several new species of Philippine orchids described, 1922–1930.]
Ames, O.: Enumeration of Philippine Apostasiaceae and Orchidaceae. In: Enumeration of Philippine Flowering Plants. Manila, Philippines 1925.
Ames, O., and Quisumbing, E.: New or Noteworthy Philippine Orchids, 1–6. The Philippine Journal of Science 44–59, 1931–1936.
Valmayor, H.: Orchidiana Philippiana. 2 vols. Samhwa Printing Co., S. Korea 1984.
Philippine Orchid Review (1948–1959). Many articles on orchids of the Philippines.

New Guinea

Millar, A.: Orchids of Papua New Guinea. Australian National U. P., Canberra 1978.
Schlechter, R.: Die Orchidaceen von Deutsch-Neuguinea. Fedde, Repertorium, Beihefte 1 und 21, 1914–1928.
Smith, J. J.: Die Orchideen von Niederländisch Neu-Guinea. In: Résultats de l'expédition scientifique, Néerlandaise à la Nouvelle Guinée. E. J. Brill, Leiden 1909–1934.
Van Bodegom, J.: Enige Orchideeën van West Nieuw Guinea. Enschede, Nederland 1973.
Van Royen, P.: The Orchids of the High Mountains of New Guinea. J. Cramer, Vaduz 1980.

Australia

Dockrill, A. W.: Australasian Sarcanthinae. The Australian Native Orchid Society, Sydney 1967.
Dockrill, A. W.: Australian Indigenous Orchids I. The Epiphytes; the Tropical Terrestrial Species. Halstead Press, Sydney 1969.
Firth, M. J.: Native Orchids of Tasmania. C. L. Richmond, Devonport, Tasmania 1965.
Fitzgerald, R. D.: Australian Orchids. Sydney 1875–1894.
Lavarack, P. S. and Gray, B.: Tropical Orchids of Australia. Nelson, Melbourne 1985.
Nichols, W. H.: Orchids of Australia. T. Nelson, Melbourne, Australia 1969.
Orchadian. Numerous articles on native species, 1963→.

The Pacific Islands

Guillaumin, A.: Orchidacées. In: Flore analytique et synoptique de la Nouvelle-Calédonie. Paris 1948.
Hallé, N.: Flore de la Nouvelle-Calédonie et Dépendances, 8. Orchidacées. Muséum National d'Histoire Naturelle, Paris 1977.
Hunt, P. F.: Orchids of the Solomon Islands. Philosophical Transactions of the Royal Society of London 255, 1969.
Johns, J., and Mollay, B.: Native Orchids of New Zealand, 1984.
Kraenzlin, F.: Neu-Caledonische Orchideen. Vierteljahresschrift der Naturforschenden Gesellschaft, Zürich, 74, 1929.
Lewis, B. and Cribb, P.: Orchids of Vanuatu. Royal Botanic Gardens Kew, 1989.
Moore, L. B., and Edgar, E.: Orchidaceae. In: Flora of New Zealand, 2. Wellington, N. Z. 1970.
Parham, J. W.: Orchidaceae. In: Plants of the Fiji Islands. Suva, Fiji 1972.
Seemann, B.: Orchideae. In: Flora Vitiensis. L. Reeve and Co., London 1865–1873.

Tropical Africa

Cribb, P. J., and Stewart, J.: Additions to the Orchid Flora of Tropical Africa. Kew Bull. vol. 40, (2), 1985.
Cribb, P. J. and Fay, J. M.: Orchids of Central African Republic. A Provisional Checklist, Kew Bull. 42 (3), 1987.
Geerinck, D.: Flore d'Afrique Centrale. Jardin Botanique Nat. de Belgique, 1984.
Geerinck, D.: Orchidaceae. In: Flore du Rwanda, Vol. 4, 1988.
Morris, B.: Epiphytic Orchids of Malaŵi. The Society of Malawi, Blantyre 1970.
Pieres, F.: Orchids of East Africa. 2nd ed Cramer, Lehre 1968.
Rolfe, R. A.: Flora of Tropical Africa, 7: Orchidaceae. L. Reeve and Co., London 1897.
Schlechter, R.: Orchidaceae Stolzianae. In: Engler, A.: Botanische Jahrbücher, 1915.
Schlechter, R.: Versuch einer natürlichen Neuordnung der afrikanischen angraekoiden Orchidaceen. Bot. Centralblatt, Beihefte 36 (2), 1918.
Stewart, J.: Orchidaceae. In: Agnew, K.: Upland Kenya Wild Flowers. Oxford U. P. 1974.
Stewart, J., and Campbell, R.: Orchids of Tropical Africa. W. H. Allen, London 1970.
Summerhayes, V. S.: Numerous articles on the taxonomy of tropical African orchids. In: Kew Bull. 1927–1966.
Summerhayes, V. S.:Several articles on African orchids. Botanical Leaflets of Harvard University, 1942–1945.
Summerhayes, V. S.: Flora of West Tropical Africa. Orchidaceae, 2nd ed., London 1968.
Summerhayes, V. S.: Flora of Tropical East Africa, Orchidaceae pt. I., London 1968.
Williamson, G.: The Orchids of South Central Africa. Dent, London 1977.

South Africa

Ball, J. S.: Southern African Epiphytic Orchids. Conservation Press, Johannesburg 1978.
Bolus, H.: Orchids of South Africa. W. Wesley and Son, London 1893–1913.
Compton, R. H.: The Flora of Swaziland. Journal of South African Botany, Supp. vol. 11, 1976.
Hall, A. V.: Studies of the South African species of Eulophia. Journal of South African Botany, Supp. vol. 5, 1965.
Harrison, E. R.: Epiphytic Orchids of Southern Africa. Durban 1972.
Jacot Guillarmod, A.: Flora of Lesotho. J. Cramer, Lehre 1972.
Kraenzlin, F.: Orchidacearum Genera et Species, I. Mayer und Müller, Berlin 1898.
Rolfe, R. A.: Flora Capensis, Orchidaceae. L. Reeve and Co., London 1913.
Schelpe, E. A.: An Introduction of the South African Orchids. MacDonald, London 1966.
Schlechter, R.: Contributions to South African Orchidology. Annals of the Transvaal Museum 10, 1924.
Sölch, A.: Orchidaceae. In: Prodromus einer Flora von Südwestafrika. J. Cramer, Lehre 1972.
Stewart, J., and Hennessy, E. F.: Orchids of Africa. Macmillan South Africa, Johannesburg 1981.
South African Orchid Journal (1970→). Many articles on South African orchids.

Madagascar and Adjacent Isles

Perrier de la Bâthie, H.: Flore de Madagascar, Orchidacées. Tananarive 1939–1941.
Richard, A.: Orchidées des Iles de France et de Bourbon. Paris 1828.
Schlechter, R.: Orchidaceae de Madagascar – Orchidaceae Perrierianae Madagascarienses, 1913.
Thouars, A. du Petit: Orchidées des Iles Australes de l'Afrique. Paris 1913.
Adansonia (1960→). Numerous articles on Madagascan orchids.

North America

Correll, D. S.: Native Orchids of North America. Stanford U. P., California 1950.
Luer, C.: The Native Orchids of Florida. New York Botanical Garden, 1972.
Luer, C.: The Native Orchids of the United States and Canada. New York Botanical Garden, 1975.

Central America

Ames, O., and Correll, D. S.: Orchids of Guatemala. In: Fieldiana, 1952–1953.
Bateman, J.: The Orchidaceae of Mexico and Guatemala, 1837–1843.
Dressler, R. L., and Pollard, G. E.: The Genus Encyclia in Mexico. Asociacion Mexicana de Orquideologia, Mexico 1976.
Hamer, F.: Las Orquideas de El Salvador. Min. de Educacion, San Salvador 1974.
Hamer, F.: Orchids of Nicaragua. In: Icones Plantarum Tropicarum. Sarasota (Fl.), Fasc. 7–9, 11 and 12, 1982–84.
MacVaugh, R.: Orchidaceae. In: Flora Novo Galiciana (Western Mexico) Vol. 16. University of Michigan Press, Ann Arbor 1985.

Schlechter, R.: Beiträge zur Orchideenkunde von Zentralamerika. Fedde, Repertorium, Beihefte 17, 4–95, und 19, 3–307, 1922–1923.
Williams, L. O.: The Orchidaceae of Mexico. Ceiba 2, 1–321, 1951.
Williams, L. O.: An Enumeration of the Orchidaceae of Central America, British Honduras and Panama. Ceiba 5, 1–256, 1956.
Williams, L. O., and Allen, P. H.: Orchidaceae. In: Woodson, R. E., and Schery, R. W.: Flora of Panama. Ann. Missouri Bot. Garden, 1946–1949.
Williams, O., and Allen, P.: Orchids of Panama with Checklist from R. Dressler. Missouri Botanic Garden 1980.

The West Indies

Acuna Gate, J.: Catalogo Descriptivo de las Orquideas Cubanas, Estacion Experimental Agronomica, Boletin 60, Havana, Cuba 1938.
Cogniaux, A.: Orchidaceae. In: Urban, I.: Symbolae Antillanae, 6, 1909–1910.
Fawcett, W., and Rendle, A. B.: Orchidaceae. In: Flora of Jamaica, Vol. 1, British Museum, London 1910.
Garay, L., and Sweet, H.: Orchidaceae. In: Howard, R. A.: Flora of the Lesser Antilles. Arnold Arboretum, Harvard University, Jamaica Plain, Mass. 1974.
Schultes, R. E.: Native Orchids of Trinidad. Pergamon Press, Oxford 1960.

South America

Braem, G. J.: Cattleya. Brücke Verlag, Hildesheim 1983.
Cogniaux, A.: Orchidaceae. In: Martius, Flora Brasiliensis, Leipzig 1893–1906.
Correa, M. N.: Orchidaceae. In: Flora Patagonica 8, 1970.
Dodson, C. H., and P. M.: Orchids of Ecuador. In: Icones Plantarum Tropicarum. Sarasota (Fla.), Fasc. 1–4, 5 and 10, 1980–84.
Dunsterville, G. C. K.: Die Welt der Orchideen. Lectura Verlag, Caracas 1962.
Dunsterville, G. C. K., and Garay, L.: Venezuelan Orchids Illustrated, 1–6, 1959–1976.
Dunsterville, G. C. K., and Garay, L.: Orchids of Venezuela – an Illustrated Field Guide. Botanical Museum, Harvard University, Cambridge, Mass. 1979.
Marga, C. M.: Orchids of Suriname. Werkhoven, Paramaribo 1986.
Foldats, E.: Orchidaceae. In: Flora de Venezuela, 15. Instituto Botanico, Caracas 1969–1970.
Fowlie, J. A.: The Brazilian Bifoliate Cattleyas and Their Color Varieties. Day Printing Corp., Pomona, Calif. 1977.
Hoehne, F. C.: Flora Brasilica – Orchidaceas. São Paulo 1940–1953.
Hoehen, F. C.: Iconografia de Orchidaceas do Brasil. São Paulo 1949.
Pabst, G. und Dungs, F.: Orchidaceae Brasilienses, 2 Bände. Brücke-Verlag, Kurt Schmersow, Hildesheim 1975–1977.
Rodrigues, J. Barbosa: Genera et Species Orchidearum Novarum. Fleiuss, Rio de Janeiro 1877.
Schlechter, R.: [Mehrere längere Artikel über südamerikanische Arten und Gattungen.] In: Fedde, Repertorium, Beihefte. 1919–1929.
Schlechter, R., and Hoehne, F. C.: Contribuicöes ao Conhecimento das Orquidáceas do Brasil. In: Anexos das Memorias do Instituto de Butantan; in: Archivos de Bontanic do Estado de São Paulo, 1921–1922.
Schweinfurth, C.: Orchids of Peru. In: Fieldiana 1958–1961; Ergänzungen in Fieldiana, Chicago 1970.
Vásquez, Ch. R., and Dodson, C. H.; Orchids of Bolivia. In: Icones Plantarum Tropicarum. Sarasota (Fla.), Fasc. 6, 1982.
Williams, L. O.: Las Orquideas del Noroeste Argentino. Lilloa 4, 1939.
Zimmermann, W.: Orchidaceae Novae Brasilienses. Bibliotheca Botanica, Fasc. 109, 1934.
American Orchid Society Bulletin (1932→). Many articles on South American orchids.
Bradea (1969→). Many articles on Brazilian orchids.
Orchid Digest (1937→). Many articles on South American orchids.
Orquidea (1938–73). Many articles on Brazilian orchids.
Orquideologia (1966→). Articles mainly on Colombian orchids.

GENERAL BIBLIOGRAPHY

Arditti, J.: Orchid Biology, Reviews and Perspectives. Cornell U. P., Itahaca, London 1977.
Bateman, J.: Second Century of Orchidaceous Plants. L. Reeve and Co., London 1867.
Bateman, J.: A Monograph of Odontoglossum. L. Reeve and Co., London 1874.
Bauer, F., and Lindley, J.: Illustrations of Orchidaceous Plants. J. Ridgway and Sons, London 1830–1838.
Bechtel, H.: Orchideen im Heim. Landbuch Verlag, Hannover 1969.
Bechtel, H.: Exotische Orchideen, Franckh, Stuttgart 1971.
Bechtel, H.: Wunderbare und geheimnisvolle Welt der exotischen Orchideen. Bertelsmann, Gütersloh 1977.
Bechtel, H., Cribb, P., and Launert, E.: Orchideenatlas. Ulmer, Stuttgart 1980.
Bechtel, H., Cribb, P., and Launert, E.: The Manual of Cultivated Orchid Species. Blandford Press, West St. Poole 1981.
Bentham, G., and Hooker, J. D.: Orchideae. In: Genera Plantarum 3. L. Reeve and Co., London 1883.
Blume, C. L.: Collection des Orchidées. Sulpke, Amsterdam 1858.
Bockenmühl, L.: Odontoglossum. Brücke-Verl., Hildesheim 1989.
Bream, G.: Paphiopedilum. Brücke-Verl., Hildesheim 1988.
Burgeff, H.: Mycorrhiza of Orchids. In: Withner, C.: The Orchids – Scientific Studies, 361–396, 1959.
Chase, M. W.: A Monograph of Leochilus. Systematic Botany Monographs, Vol. 14, 1986.
Cogniaux, A., et Goossens, A.: Dictionnaire Iconographique des Orchidées. 1–17, Brussels 1896.
Cribb, P. J.: A Revision of Dendrobium sect. Latouria. Kew Bull. 38 (2), 1983.
Cribb, P. J.: The "Antelope" Dendrobiums, Kew Bull. 41 (3), 1986.
Cribb, P. J.: The Genus Paphiopedilum. The Royal Botanic Gardens Kew in Association with Collingridge, 1987.
Cribb, P. J., and Butterfield, J.: The Genus Pleione. The Royal Botanic Gardens Kew. Christopher Helm and Timber Press, 1988.
Dressler, R.: Classification of the Orchid Family in Proceedings of the 7th World Orchid Congress. 259–278, Medellin, Colombia 1974.
Dressler, R., and Dodson, C.: Classification and Phylogeny in the Orchidaceae. Annals of the Missouri Botanic Garden 47, 25–67, 1960.
Du Puy, D., and Cribb, P. J.: The Genus Cymbidium. Christopher Helm, London; Timber Press, Portland, Oregon 1988.
Dunsterville, G. C. K.: A letter to orchid conservationists. Amer. Orchid Soc. Bull. 44, 883–885, 1975.
Ebel, F.: Schöne und seltsame Welt der Orchideen. Wasmuth, Tübingen 1971.
Fast, G.: Orchideenkultur. Ulmer, Stuttgart 1980.
Fowlie, J. A.: The Genus Lycaste. Day Printing Corp., Pomona, Calif. 1970.
Garay, L.: On the Origin of the Orchidaceae. Botanical Museum Leaflets of Harvard University 19 (3), 57–96, 1960.
Garay, L.: On the Origin of the Orchidaceae, 2. Journal of the Arnold Arboretum 52, 202–215, 1972.
Garay, L. A.: On the Systematics of the Monopodial Orchids. Botanical Museum Leafl. Harvard University 23 (4), 1972.
Garay, L. A.: A Generic Revision of the Spiranthinae. Botanical Museum Leafl. Harvard University 28 (4), 1980.
Graham, R., and Roy, R.: Slipper Orchids. Croom Helm, London 1982.
Hagsater, E.: Can there be a different view on orchids and their conservation. Amer. Orchid Soc. Bull. 45, 18–21, 1976.
Hawkes, A. D.: Encyclopaedia of Cultivated Orchids. Faber and Faber, London 1965.
Hirmer, M.: Wunderwelt der Orchideen. Hirmer, München 1974.
Hooker, W. J.: A Century of Orchidaecous Plants. Reeve, Bentham and Reeve, London 1849.
Hunt, D. R.: Orchids from Curtis's Botanical Magazine. Curwen Books, London 1981.
Hunt, P.-F.: Conservation of orchids. Orchid Rev. 76, 320–327, 1968.
International Orchid Commission: Handbook of Orchid Nomenclature, ed. 3, London 1985.
I.U.C.N.: Convention on International Trade. In: Endangered Species of Wild Fauna and Flora. Morges, (CH) 1973.
Johanson, D.: The ecology of vascular epiphytes in West Africa. Acta Phytogeographia Suecica 59, 1–129, 1974.
Linden, J.: Pescatorea, Gand 1860.
Linden, J.: Lindenia. Gand 1885–1901.
Lindley, J.: Genera and Species of Orchidaceous Plants. Ridgways, London 1830–1840.
Lindley, J.: Sertum Orchidacearum. 1838–1842.
Lindley, J.: Folia Orchidacea. J. Matthews, London 1852–1859.
Lindley, J.: The Vegetable Kingdom, ed. 1–3. Bradbury and Evans, London 1853.
Luer, C. A.: Icones Pleurothallidinarum. Selbyana, Vol. 3, (1-4) 1976–77.
Luer, C. A.: Thesaurus Masdevalliarum. Helga Königer Verlag, München 1983→.
Luer, C. A.: Icones Pleurothallidinarum. Missouri Bot. Gard. 1–5, 1986–88.
Mansfeld, R.: Die Gattung Catasetum. In: Fedde, Repertorium 30: 17–25, 1932.
Melville, R.: Red Data Book, Vol. 5: Angiospermae, 1970.
Minderhoud, M. E., and De Vogel, E. F.: Acriopsis. In: Orchids Monographs. Brill/Rijksherbarium, Leiden 1986.

Reichenbach, H. G.: Xenia Orchidacea. A. Brockhaus, Leipzig 1854–1900.
Reichenbach, H. G.: Orchides. In: Walpers, Annales Botanices 6, 1864.
Reinekka, M. A.: A History of the Orchid, Univ. of Miami Press 1972.
Richter, W.: Orchideen pflegen, vermehren und züchten. Neumann-Neudamm, Melsungen 1969.
Rolfe, R. A., and Hurst, C.: Orchid Stud. Book. F. Leslie and Co., Kew 1909.
Sander, D.: Orchideen und Orchideen-Pflege. Schmersow, Hildesheim 1969.
Sander, F.: Reichenbachia. St. Albans 1888–1894.
Sander, F.: Sander's Orchid Guide. St. Albans 1901.
Sanford, W. W.: The Ecology of orchids. In: Withner, C.: The Orchids, Scientific Studies, 1–100, 1974.
Schlechter, F.: Das System der Orchidaceen. Notizblatt des Botanischen Gartens und Museums, Berlin, 9, 563–591, 1926.
Schlechter, R.: Die Orchideen. 2. Aufl. Parey, Berlin 1927.
Schlechter, R.: Die Orchideen. 3. Aufl. (F. G. Brieger, R. Maatsch, K. Senghas). Parey, Berlin und Hamburg. In Lieferungen ab 1970.
Schultes, R. E., and Pease, A.: Generic Names of Orchids. Academic Press, New York, London 1963.
Schuster, C.: Orchidacearum Iconum Index. Fedde, Repertorium, Beiheft 60, 1974.
Schweinfurth, C.: The Classification of Orchids. In: Withner, C.: The Orchids, a Scientific Survey, 15–43, 1959.
Sprunger, S.: Orchids from Curtis's Botanical Magazine / Orchideentafeln ... University Press, Cambridge; Ulmer, Stuttgart 1986.
Stein, B.: Stein's Orchideenbuch Original ed. 1892, Brücke Verlag, Hildesheim 1980.
Stafleu, F. A., and Cowan, R. S.: Taxonomic Literature, Vols. 1–5, Utrecht 1976.
Summerhayes, V. S.: The Wild Orchids of Britain, ed. 2. Collins, London 1968.
Sweet, H.: The Genus Phalaenopsis. Day Printing Corp., Pomona, Calif. 1980.
Van der Pijl, L., and Dodson, C.: Orchid Flowers – their pollination and evolution. University of Miami Press 1965.
Van Royen, P.: The Genus Corybas. J. Cramer, Vaduz 1983.
Veitch, J.: A Manual of Orchidaceous Plants. J. Veitch and Sons, Chelsea 1887–1894.
Vermeulen, J. J.: Bracisepalum, Chelonistele, Entomophobia, Geesinkorchis and Nabaluia. In: Orchids Monographs. Brill/Rijksherbarium, Leiden 1986.
De Vogel, E. F.: Pholidota. In: Orchids Monographs. Brill/Rijksherbarium, Leiden 1986.
Warner, R.: Select Orchidaceous Plants. Lovell Reeve and Co., London 1862–1878.
Warner, R., and Williams, B. S.: The Orchid Album. B. S. Williams, London 1882–1897.
Withner, C.: The Orchids – a scientific survey. Ronald, New York 1959.
Withner, C.: The Orchids – scientific studies. Wiley-Interscience, New York 1974.
Withner, Carl L.: The cattleyas and their relatives. Vol. I: The cattleyas. Timber Press, Portland (Oregon), 1988.
Woolward, F. H.: The Genus Masdevallia. Original ed. 1896. Lyne and Son, London 1978.
Zander, Handwörterbuch der Pflanzennamen. Von F. Encke, G. Buchheim und S. Seybold. Ulmer, Stuttgart 1984.
Draft Index of Author Abbreviations compiled at The Herbarium, Royal Botanic Gardens, Kew 1980.
The European Garden Flora, Vol 2 Monocotyledons pt. 2: Juncaceae to Orchidaceae. Cambridge Univ. Press, Cambridge 1984.
Krypto. F. Flück-Wirth. Spezialkatalog/Special Catalogue No. 6 Febr., Teufen 1985.
Index Kewensis. Kumulierte Microfiche-Ausgabe. Mindata Microform System. 1–484, 1983.

Dates of publication: Botanical Register 1815–1847

The plates in volumes 1–32 are copper engravings, in volume 33 lithographs

Vol.	Plates	Dates	
1	1–76	Mar–Dec	1815
	77–90	Jan–Feb	1816
2	91–163	Mar–Dec	1816
	164–177	Jan–Feb	1817
3	178–248	Mar–Dec	1817
	249–263	Jan–Feb	1818
4	264–336	Mar–Dec	1818
	336–349	Jan–Feb	1819
5	350–421	Mar–Dec	1819
	422–435	Jan–Feb	1820
6	436–506	Mar–Dec	1820
	507–520	Jan–Feb	1821
7	521–590	Mar–Dec	1821
	591–605	Jan–Feb	1822
8	606–675	Mar–Dec	1822
	676–689	Jan–Feb	1823
9	690–762	Mar–Dec	1823
	763–777	Jan–Feb	1824
10	778–853	Mar–Dec	1824
	854–867	Jan–Feb	1825
11	868–940	Mar–Dec	1825
	941–955	Jan–Feb	1826
12	956–1029	Mar–Dec	1826
	1030–1043	Jan–Feb	1827
13	1044–1116	Mar–Dec	1827
	1117–1130	Jan–Feb	1828
14	1131–1202	Mar–Dec	1828
	1203–1216	Jan–Feb	1829
15	1217–1291	Mar–Dec	1829
	1292–1305	Jan–Feb	1830
16	1306–1376	Mar–Dec	1830
	1377–1391	Jan–Feb	1831
17	1392–1462	Mar–Dec	1831
	1463–1476	Jan–Feb	1832
18	1477–1549	Mar–Dec	1832
	1550–1564	Jan–Feb	1833
19	1565–1637	Mar–Dec	1833
	1638–1652	Jan–Feb	1834
20	1653–1727	Mar–Dec	1834
	1728–1741	Jan–Feb	1835
21	1742–1821	Mar–Dec	1835
	1822–1828	Jan–Feb	1836
22	1829–1919	Feb–Dec	1836
23	1920–2014	Jan–Dec	1837
24	1–68	Jan–Dec	1838
25	1–69	Jan–Dec	1839
26	1–71	Jan–Dec	1840
27	1–70	Jan–Dec	1841
28	1–69	Jan–Dec	1842
29	1–66	Jan–Dec	1843
30	1–67	Jan–Dec	1844
31	1–69	Jan–Dec	1845
32	1–69	Jan–Dec	1846
33	1–70	Jan–Dec	1847

ORCHID PLATES

arranged alphabetically by genera

Vol.18: t.1552 (1833)
Acampe papillosa (Lindl.) Lindl.

Vol.32: t.47 (1846)
Acanthephippium javanicum Blume

Vol.20: t.1730 (1835)
Acanthephippium bicolor Lindl.

Miss Drake del. Pub. by J. Ridgway 169 Piccadilly Ap.l 1. 1843. J. Barclay sc.

Vol.29: t.18 (1843)
Acineta superba (Kunth) Rchb.f.

Vol..27: t.35 (1841)
Aerangis biloba (Lindl.) Schltr.

Vol.10: t.817 (1824)
Aeranthes grandiflora Lindl.

Vol.28: t.55 (1842)
Aerides crispa Lindl.

Vol.31: t.58 (1845)
Aerides maculosa Lindl.

1485.

Miss Drake del.   Pub by J. Ridgway 169 Piccadilly April 1. 1832.   J. Watts sc.

Vol. 18: t.1485 (1832)
Aerides odorata Lour.

Vol.30: t.41 (1844)
Aerides odorata Lour.

Vol.26: t.32 (1840)
Aganisia pulchella Lindl.

Vol.21: t.1781 (1835)
Angraecum distichum Lindl.

Vol.22: t.1867 (1836)
Amblostoma armeniacum (Lindl.) Brieger ex Pabst

1522.

Pub by J. Ridgway 169 Piccadilly Sept. 1. 1832.

Vol.18: t.1522 (1832)
Angraecum eburneum Bory

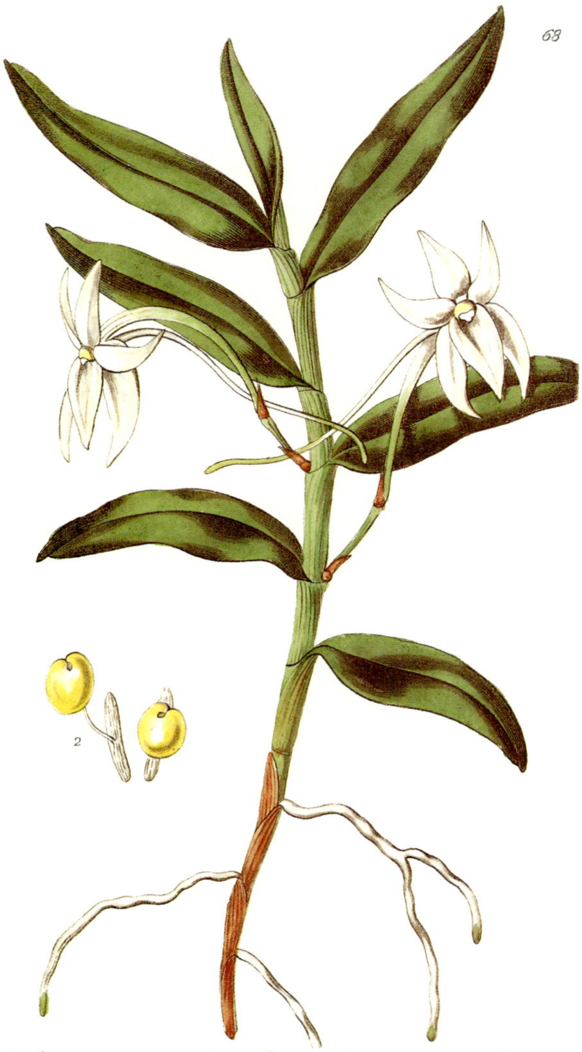

Vol.26: t.68 (1840)
Angraecum mauritianum (Poiret) Frappier
Vol.30: t.63 (1844)
Anguloa clowesii Lindl.

Vol.32: t.41 (1846)
Anguloa brevilabris Rolfe
Vol.30: t.60 (1844)
Anguloa uniflora Ruiz Lopez & Pavon

2010.

Miss Drake del. Pub by J. Ridgway 169 Piccadilly Dec 1 1837. J. Watts sc

Vol.23: t.2010 (1837)
Anoectochilus setaceus Blume

Miss Drake del.

Pub. by J. Ridgway 169 Piccadilly June 1 1846

Vol.32: t.30 (1846)
Ansellia africana Lindl.

Vol.28: t.38 (1842)
Arundina graminifolia (D.Don) Hochr.

Vol.33: t.58 (1847)
Ascocentrum miniatum (Lindl.) Schltr.

Vol.22: t.1907 (1836)
Aspasia variegata Lindl.

Vol.22: t.1881 (1836)
Barkeria skinneri (Bateman ex Lindl.) A. Rich. & Galeotti

Vol.5: t.357 (1819)
Barlia robertiana (Loisel.) Greuter

1653.

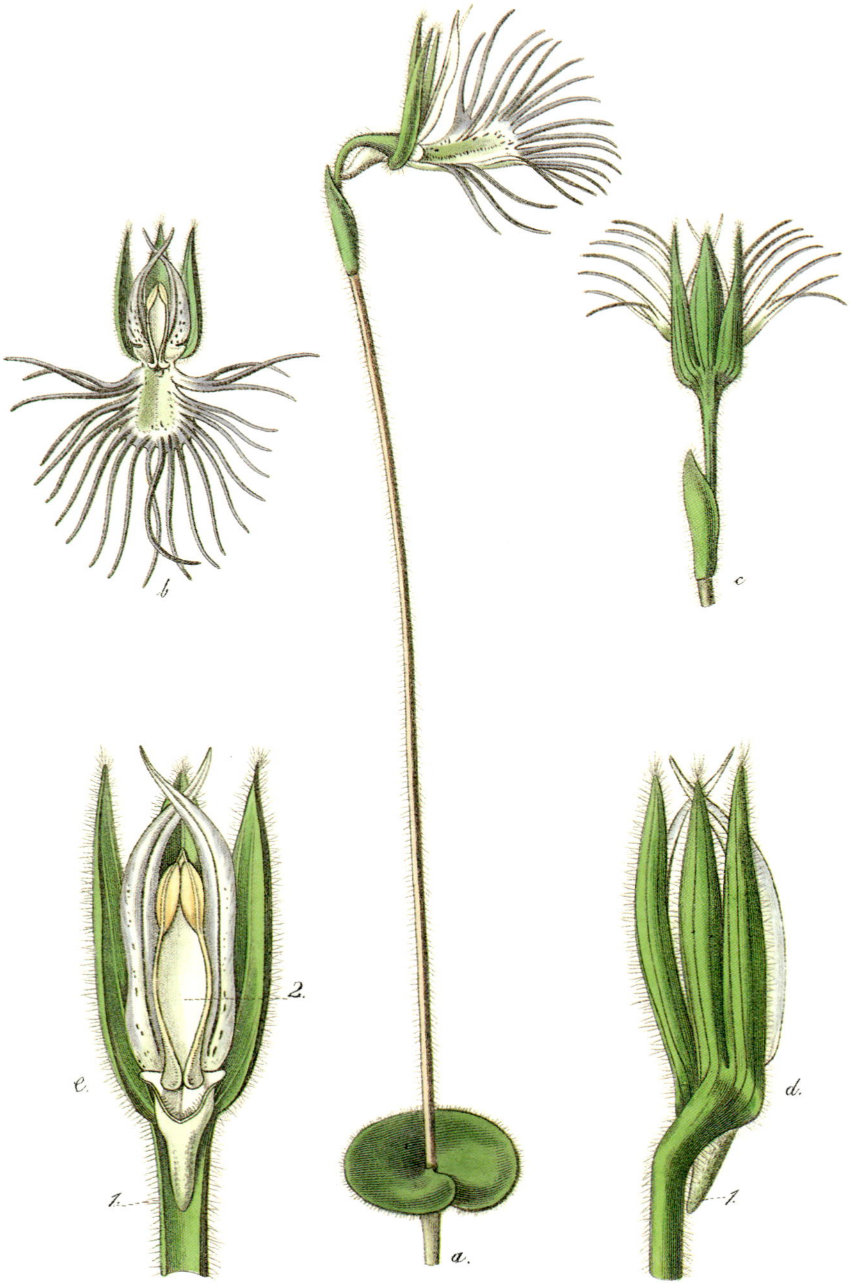

Miss Drake del. Pub by J. Ridgway 169 Piccadilly March 1 1834 S Watts sc.

Vol.20: t.1653 (1834)
Bartholina burmanniana (L.) Ker Gawler

1714.

F. Bellemo del.

Pub. by J. Ridgway 169 Piccadilly Nov. 1. 1834.

S. Watts sc.

Vol.20: t.1714 (1834)
Batemannia colleyi Lindl.

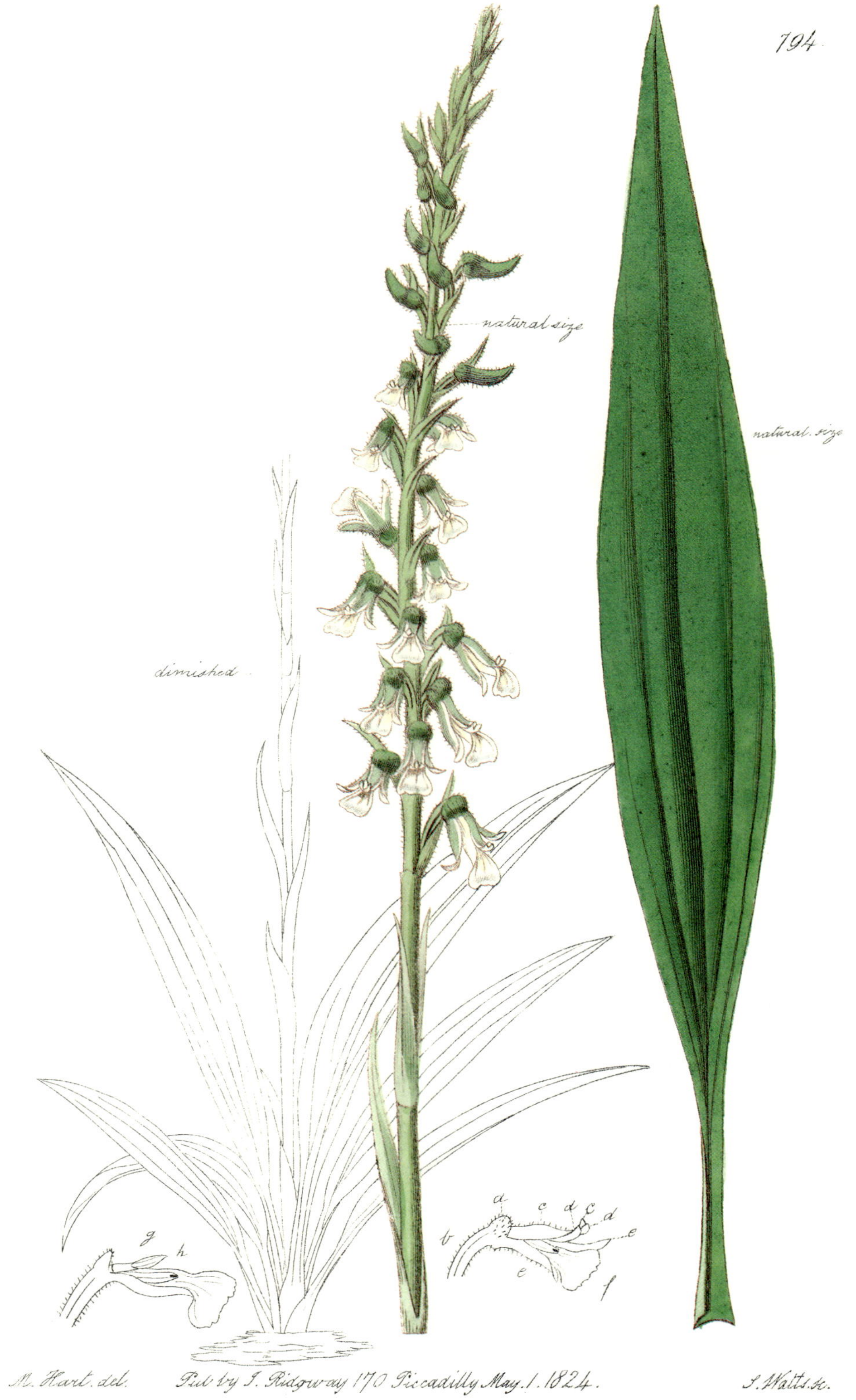

Vol.10: t.794 (1824)
Beadlea bicolor (Sw.) Small ex Britton

897

G. J. Robertson. del. Pub. by J. Ridgway 169 Piccadilly July. 1. 1825. J. Watts. sc.

Vol.11: t.897 (1825)
Bifrenaria harrisoniae (Hook.) Rchb.f.

Vol.19: t.1566 (1833)
Bifrenaria racemosa (Hook.) Lindl.
Vol.25: t.12 (1839)
Bifrenaria vitellina Lindl.

Vol.17: t.1428 (1831)
Bifrenaria tetragona (Lindl.) Schltr.
Vol.20: t.1681 (1834)
Bletia gracilis Lodd.

Vol.17: t.1401 (1831)
Bletia purpurea (Lam.) DC.
Vol.33: t.60 (1847)
Bletilla striata (Thunb.) Rchb.f.

Vol.21: t.1760 (1835)
Bletia reflexa Lindl.
Vol.23: t.1934 (1837)
Brachystele bracteosa (Lindl.) Schltr.

Vol.27: t.69 (1841)
Braemia vittata (Lindl.) Jenny
Vol.18: t.1465 (1832)
Brassavola nodosa (L.) Lindl.

Vol.25: t.5 (1839)
Brassavola martiana Lindl.
Vol.26: t.39 (1840)
Brassavola nodosa (L.) Lindl.

Vol.22: t.1913 (1836)
Brassavola subulifolia Lindl.
Vol.10: t.832 (1824)
Brassia caudata (L.) Lindl.

Vol.18: t.1561 (1833)
Brassavola tuberculata Hook.
Vol.21: t.1754 (1835)
Brassia lanceana Lindl.

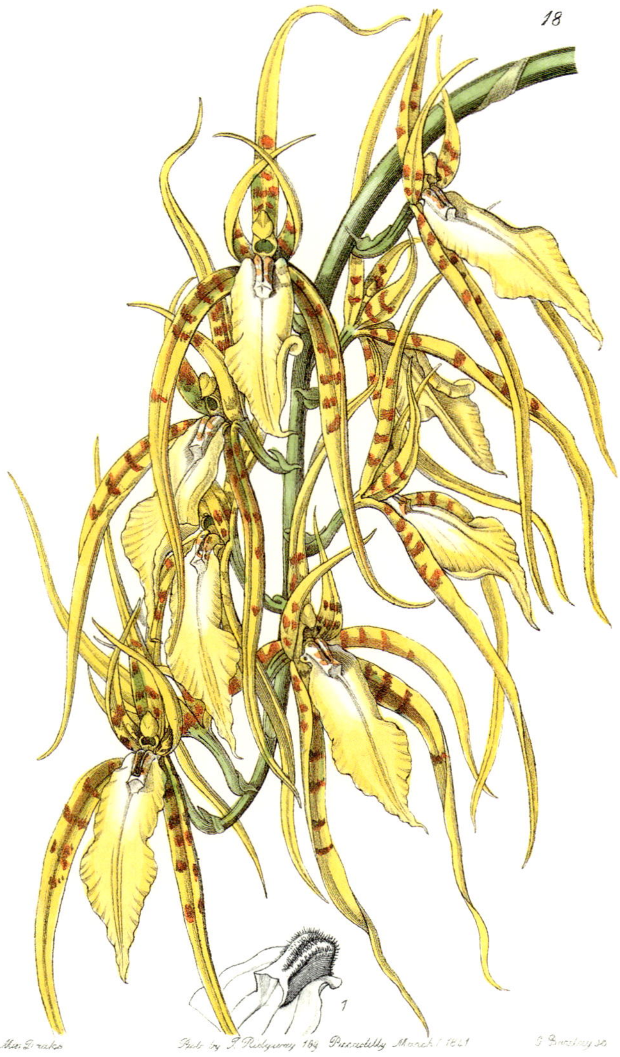

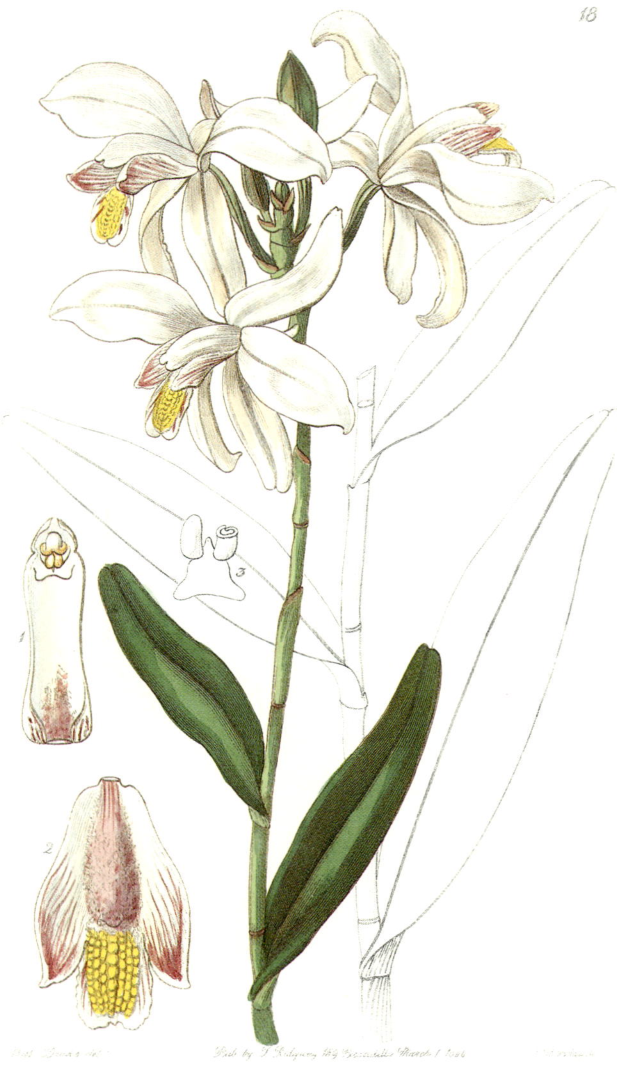

Vol.27: t.18 (1841)
Brassia lawrenceana Lindl.
Vol.30: t.18 (1844)
Bromheadia finlaysoniana (Lindl.) Rchb.f.

Vol.33: t.29 (1847)
Brassia verrucosa Lindl.
Vol.29: t.61 (1843)
Bulbophyllum auratum (Lindl.) Rchb.f.

Vol.23: t.1942 (1837)
Bulbophyllum barbigerum Lindl.
Vol.23: t.1959 (1837)
Bulbophyllum calyptratum Kraenzlin var. calyptratum

Vol.24: t.57 (1838)
Bulbophyllum bracteolatum Lindl.
Vol.29: t.49 (1843)
Bulbophyllum chinense (Lindl.) Rchb.f.

Vol.23: t.1964 (1837)
Bulbophyllum cocoinum Lindl.
Vol.31: t.44 (1845)
Bulbophyllum guttulatum Wallich ex Hook.f.

Vol.12: t.989 (1826)
Bulbophyllum falcatum (Lindl.) Rchb.f. var. falcatum
Vol.30: t.13 (1844)
Bulbophyllum macranthum Lindl.

Vol.28: t.12 (1842)
Bulbophyllum medusae (Lindl.) Rchb.f.
Vol.23: t.1970 (1837)
Bulbophyllum saltatorium Lindl. var. saltatorium

Vol.12: t.963 (1826)
Bulbophyllum pumilum (Sw.) Lindl.
Vol.24: t.11 (1838)
Bulbophyllum weberi Ames

Vol.19: t.1646 (1834)
Calanthe densiflora Lindl.
Vol.33: t.8 (1847)
Calanthe pulchra (Blume) Lindl.

Vol.26: t.55 (1840)
Calanthe discolor Lindl.
Vol.30: t.37 (1844)
Calanthe sylvatica (Thouars) Lindl.

Vol.9: t.720 (1823)
Calanthe triplicata (Willemet) Ames

Vol.21: t.1772 (1835)
Campylocentrum micranthum (Lindl.) Rolfe

Vol.9: t.720 (1823)
Calanthe triplicata (Willemet) Ames

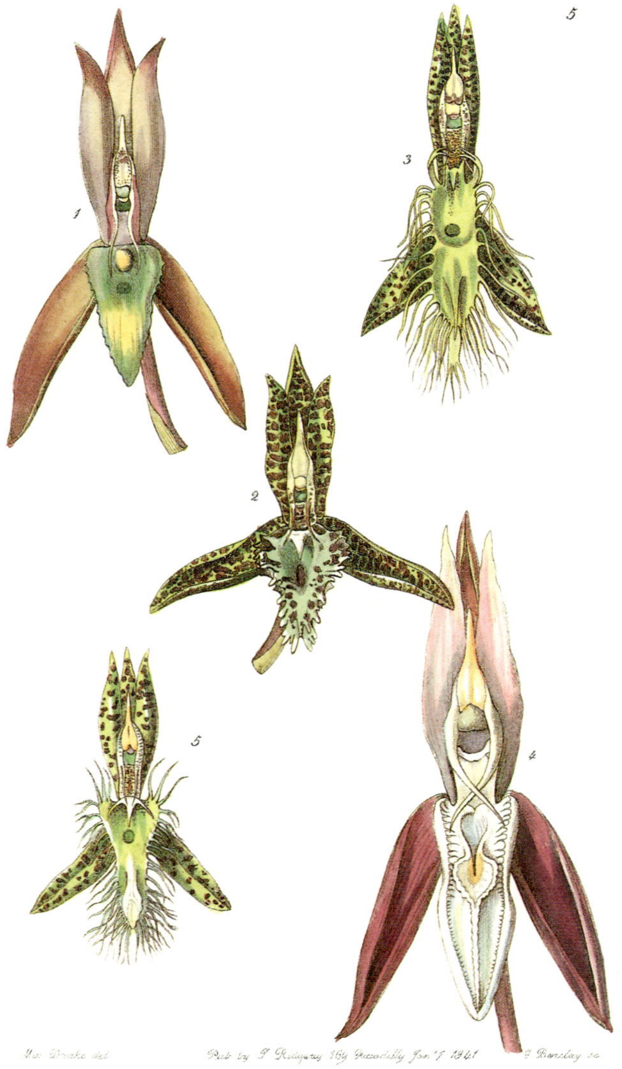

Vol.24: t.63 (1838)
Catasetum atratum Lindl.

Vol.27: t.5 fig.3.5 (1841)
Catasetum barbatum (Lindl.) Lindl.
Vol.27: t.5 fig.1 (1841)
Catasetum callosum Lindl.

Vol.27: t.5 fig.2 (1841)
Catasetum cristatum Lindl.
Vol.27: t.5 fig.4 (1841)
Catasetum laminatum Lindl.

Vol.21: t.1778 (1835)
Catasetum barbatum (Lindl.) Lindl.

1721.

Vol.20: t.1721 (1834)
Catasetum cernuum (Lindl.) Rchb.f.

Vol. 12: t.966 (1826)
Catasetum cristatum Lindl.

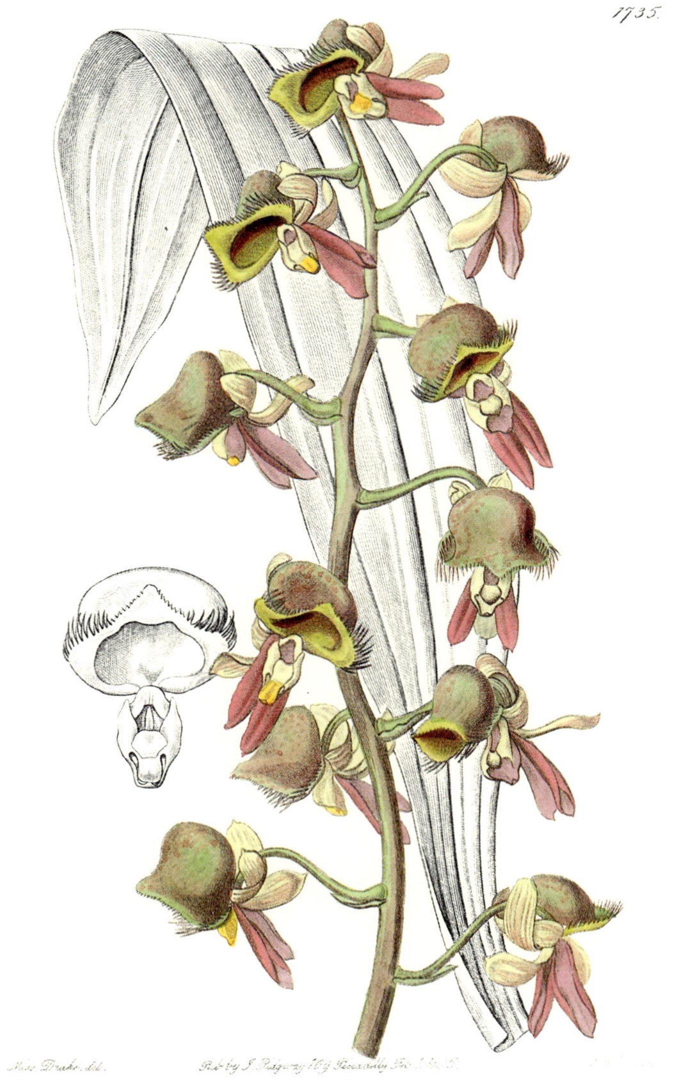

Vol.23: t.1947A (1837)
Catasetum cristatum Lindl.
Vol.20: t.1735 (1835)
Catasetum discolor (Lindl.) Lindl.

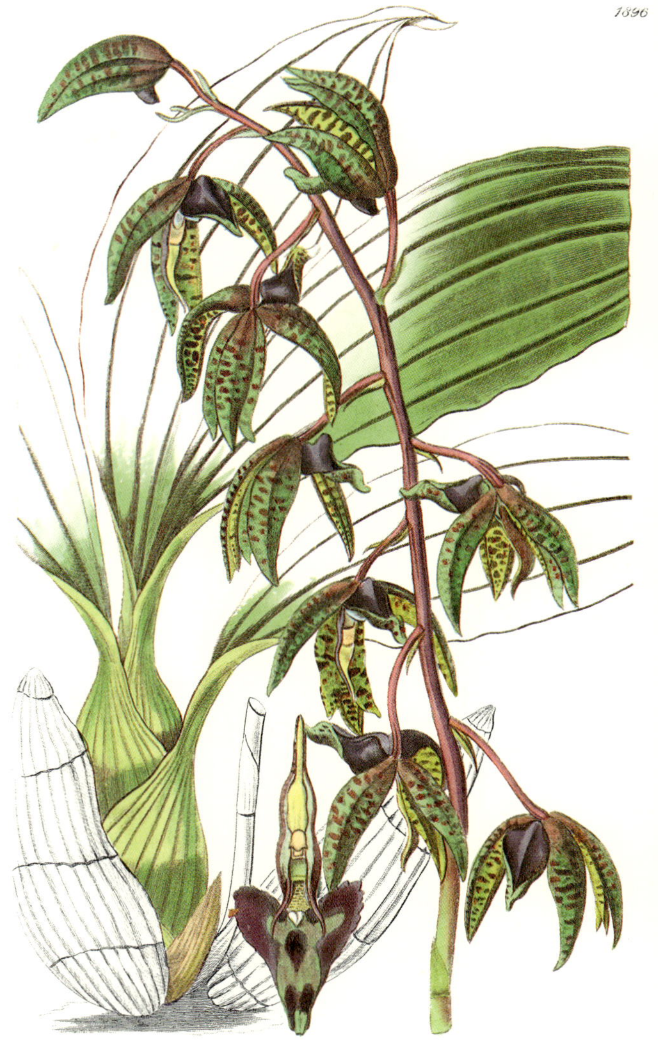

Vol.22: t.1896 (1836)
Catasetum deltoideum (Lindl.) Mutel
Vol.20: t.1667 (1834)
Catasetum luridum (Link.) Lindl.

840

J. Lindley Esqr. del. Pub by J. Ridgway 170 Piccadilly Nov. 1. 1824. S. Watts sc.

Vol. 10: t.840 (1824)
Catasetum macrocarpum Rich. ex Kunth

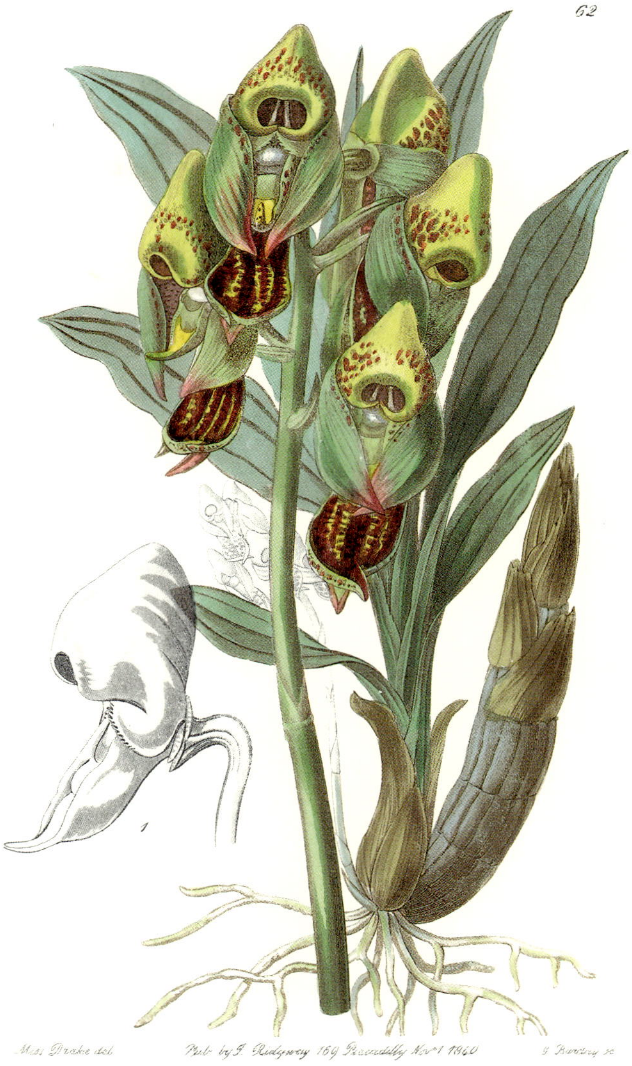

Vol.21: t.1752 (1835)
Catasetum macrocarpum Rich. ex Kunth
Vol.29: t.9 (1843)
Catasetum planiceps Lindl.

Vol.26: t.62 (1840)
Catasetum maculatum Kunth
Vol.20: t.1708 (1834)
Catasetum purum Nees

Vol.27: t.34 (1841)
Catasetum trulla Lindl.

Vol.26: t.48 (1840)
Cattleya aclandiae Lindl.

Vol.11: t.953 (1826)
Cattleya forbesii Lindl.

Vol.28: t.1 (1842)
Cattleya granulosa Lindl.

Vol.31: t.59 (1845)
Cattleya granulosa Lindl.

M. Hart. del. Pub by J. Ridgway 169 Piccadilly May 1. 1831. S. Watts sc.

Vol.17: t.1406 (1831)
Cattleya guttata Lindl.

Vol.22: t.1919 (1836)
Cattleya intermedia Graham ex Hook.
Vol.32: t.35 (1846)
Cattleya labiata Lindl.

Vol.22: t.1859 (1836)
Cattleya labiata Lindl.
Vol.32: t.1 (1846)
Cattleya maxima Lindl.

Vol.26: t.58 (1840)
Cattleya mossiae Hook.

Vol.31: t.49 (1845)
Chloraea gavilu Lindl.

Vol.33: t.42 (1847)
Cattleya walkeriana (Rchb.f.) Gardner

Vol.23: t.1937 (1837)
Chysis aurea Lindl.
Vol.22: t.1889 (1836)
Cirrhaea dependens (Lodd.) Rchb.f.

Vol.27: t.23 (1841)
Chysis bractescens Lindl.
Vol.23: t.2005 (1837)
Cirrhaea loddigesii Lindl.

Vol.3: t.220 (1817)
Cleisostoma paniculatum (Ker Gawler) Garay

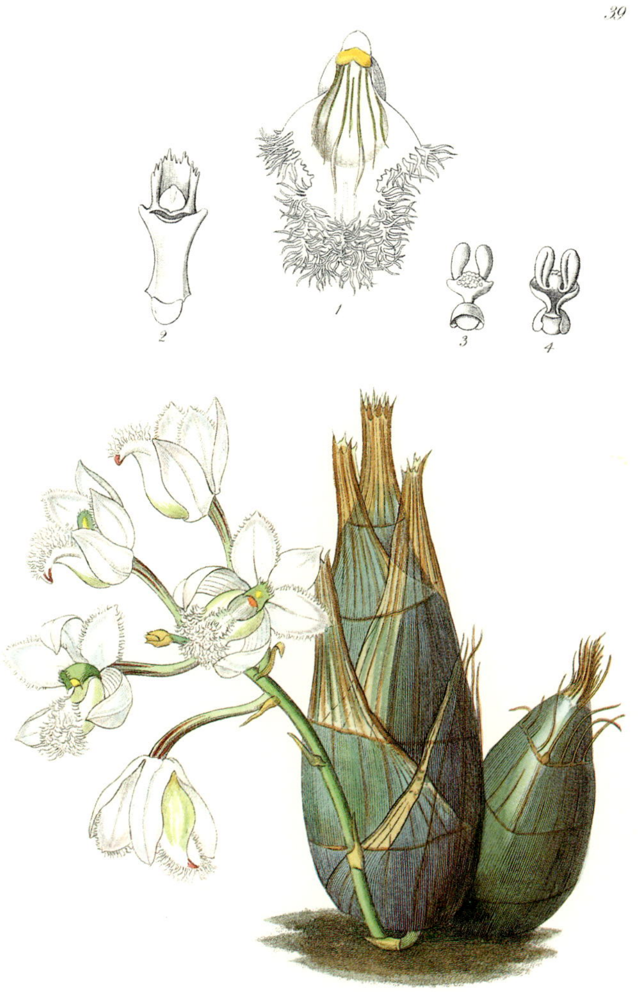

Vol.12: t.981 (1826)
Cleisostoma rostratum (Lindl.) Garay
Vol.29: t.39 (1843)
Clowesia rosea Lindl.

Vol.8: t.676 (1823)
Cleisostoma simondii (Gagnepain) Seidenf.
Vol.22: t.1857 (1836)
Cochleanthes flabelliformis (Sw.) R.Schultes

Vol.28: t.36 (1842)
Coelia triptera (Smith) G.Don ex Steudel

Vol.27: t.29 (1841)
Coelogyne cumingii Lindl.

Vol.27: t.57 (1841)
Coelogyne cristata Lindl.

Vol.11: t.868 (1825)
Coelogyne fimbriata Lindl.

Vol.27: t.31 (1841)
Coelogyne flaccida Lindl.

Vol.32: t.69 (1846)
Coelogyne nitida (Wallich ex D. Don) Lindl.

Vol.33: t.23 (1847)
Coelogyne speciosa (Blume) Lindl.

Vol.22: t.1841 (1836)
Coryanthes macrantha (Hook.) Hook.

Vol.24: t.68 (1838)
Comparettia coccinea Lindl.

Vol.21: t.1793 (1836)
Coryanthes speciosa (Hook.) Hook.

Vol.24: t.45 (1838)
Corycium orobanchoides (L.f.) Sw.
Vol.2: t.153 (1816)
Cryptarrhena lunata R.Br.

Vol.22: t.1872 (1836)
Crybe rosea Lindl.
Vol.24: t.23 (1838)
Cryptochilus sanguineum Wallich

Vol.29: t.3 (1843)
Cuitlauzina pendula Llave & Lex.

Vol.32: t.46 (1846)
Cycnoches egertonianum Bateman

Vol.21: t.1742 (1835)
Cycnoches loddigesii Lindl.

Vol.29: t.22 (1843)
Cycnoches pentadactylon Lindl.
Vol.27: t.38 (1841)
Cymbidium bicolor Lindl. ssp. pubescens (Lindl.) Du Puy & Cribb

Vol.26: t.25 (1840)
Cymbidium aloifolium (L.) Sw.
Vol.32: t.67 (1847)
Cymbidium eburneum Lindl.

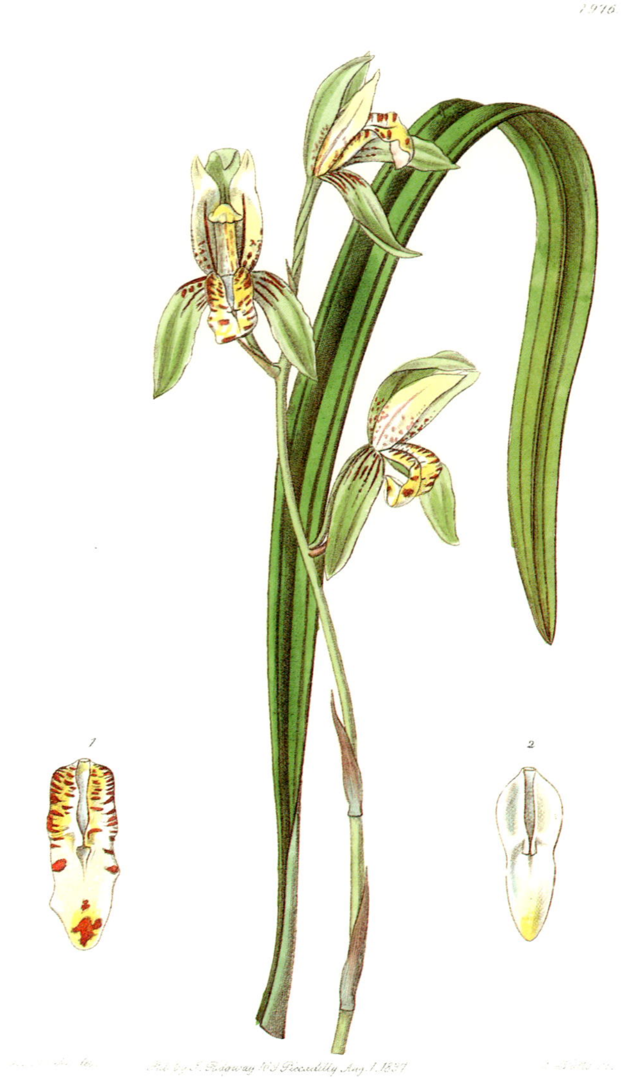

Vol.7: t.529 (1821)
Cymbidium ensifolium (L.) Sw.
Vol.30: t.24 (1844)
Cymbidium finlaysonianum Lindl.

Vol.23: t.1976 (1837)
Cymbidium ensifolium (L.) Sw.
Vol.31: t.50 (1845)
Cymbidium mastersii Griffith ex Lindl.

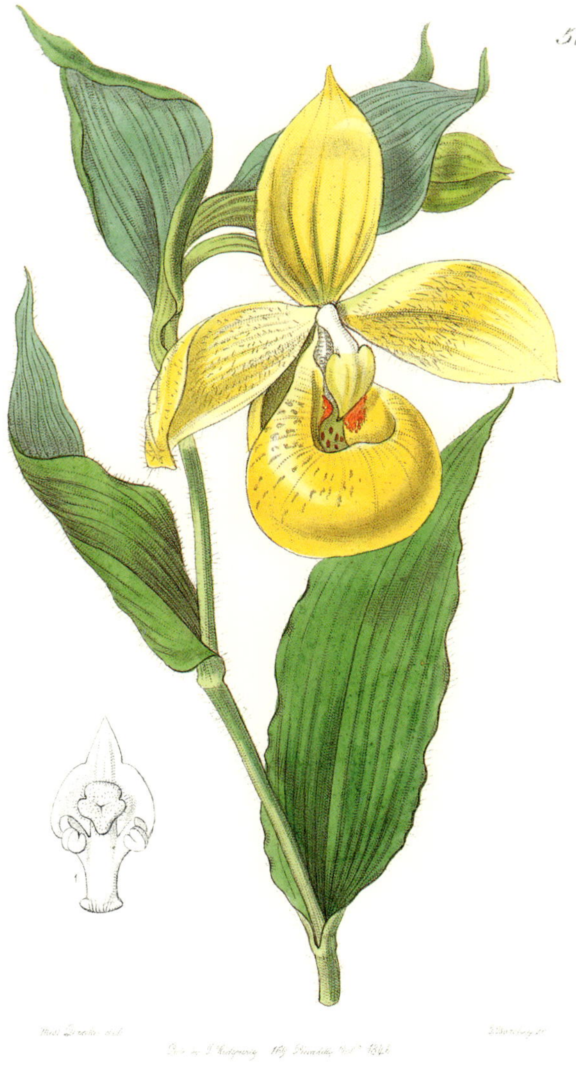

Vol.23: t.1998 (1837)
Cynorkis fastigiata (Thouars) Lindl.
Vol.18: t.1534 (1832)
Cypripedium macranthum Sw.

Vol.32: t.58 (1846)
Cypripedium irapeanum Llave & Lex.
Vol.20: t.1666 (1834)
Cypripedium reginae Walter

Vol.27: t.8 (1841)
Cyrtopodium andersonii (Lambert ex Andrews) R.Br.
Vol.32: t.15 (1846)
Dendrobium aduncum Wallich ex Lindl.

Vol.20: t.1701 (1834)
Dactylorhiza foliosa (Rchb.f.) Soó
Vol.15: t.1239 (1829)
Dendrobium anceps Sw.

Vol.7: t.548 (1821)
Dendrobium aphyllum (Roxb.) C.Fischer

Vol.29: t.54 (1843)
Dendrobium aqueum Lindl.

Vol.21: t.1756 (1835)
Dendrobium aphyllum (Roxb.) C.Fischer

1299

36

62

22

Vol.15: t.1299 (1830)
Dendrobium chrysanthum Wallich ex Lindl.
Vol.33: t.62 (1847)
Dendrobium cretaceum Lindl.

Vol.33: t.36 (1847)
Dendrobium chrysotoxum Lindl.
Vol.25: t.22 (1839)
Dendrobium crumenatum Sw.

Vol.21: t.1828 (1836)
Dendrobium densiflorum Wallich ex Lindl.

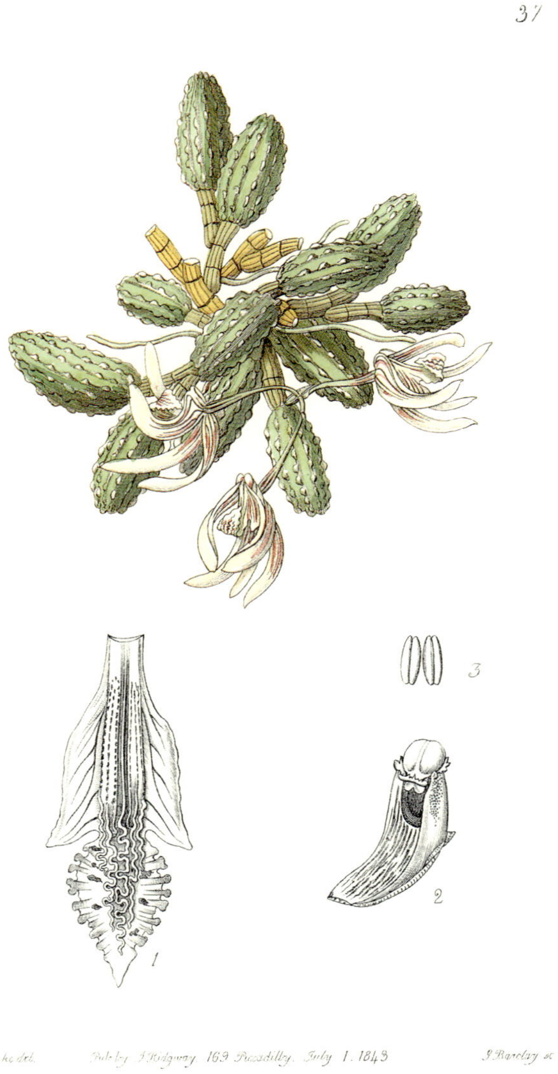

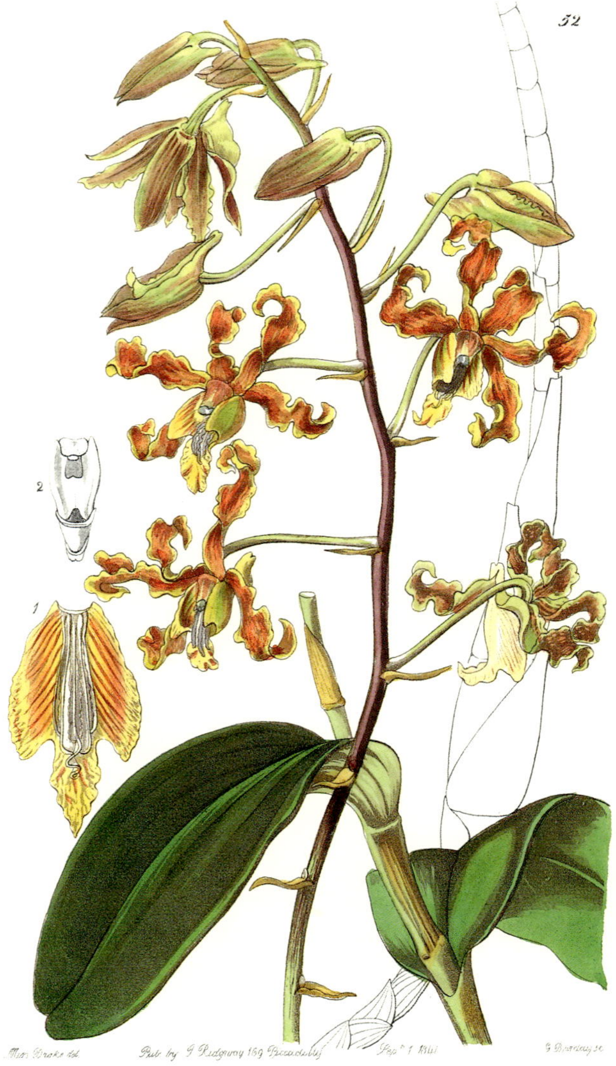

Vol.29: t.37 (1843)
Dendrobium cucumerinum Macleay ex Lindl.
Vol.25: t.64 (1839)
Dendrobium formosum Roxb.

Vol.27: t.52 (1841)
Dendrobium discolor Lindl.
Vol.25: t.20 (1839)
Dendrobium heterocarpum Wallich ex Lindl.

Vol.29: t.17 (1843)
Dendrobium heterocarpum Wallich ex Lindl.
Vol.31: t.61 (1845)
Dendrobium kingianum Bidw. ex Lindl.

Vol.25: t.37 (1839)
Dendrobium jenkinsii Wallich ex Lindl.
Vol.33: t.47 (1847)
Dendrobium kuhlii (Blume) Lindl.

Vol.30: t.53 (1844)
Dendrobium lamellatum (Blume) Lindl.
Vol.20: t.1695 (1834)
Dendrobium lindleyi Steudel

Vol.16: t.1314 (1830)
Dendrobium linawianum Rchb.f.
Vol.16: t.1315 (1830)
Dendrobium longicornu Wallich ex Lindl.

Vol.21: t.1779 (1835)
Dendrobium moschatum (Buch.-Ham.) Sw.

Vol.22: t.1865 (1836)
Dendrobium macrostachyum Lindl.

Vol.33: t.1 (1847)
Dendrobium mutabile (Blume) Lindl.

Miss Drake del. Pub. by J. Ridgway 169 Piccadilly Feb. 1 1846 G. Barclay sc.

Vol.32: t.10 (1846)
Dendrobium pulchellum Roxb. ex Lindl.

Vol.29: t.60 (1843)
Dendrobium ruckeri Lindl.
Vol.15: t.1291 (1829)
Dendrobium secundum (Blume) Lindl.

Vol.29: t.6 (1843)
Dendrobium sanguinolentum Lindl.
Vol.24: t.65 (1838)
Dendrobium sulcatum Lindl.

Vol. 19: t. 1610 (1833)
Dendrobium speciosum Smith

Miss Drake del.   Pub. by J. Ridgway 169. Piccadilly. June 1843.   J. Barclay sc.

Vol.29: t.28 (1843)
Dendrobium taurinum Lindl.

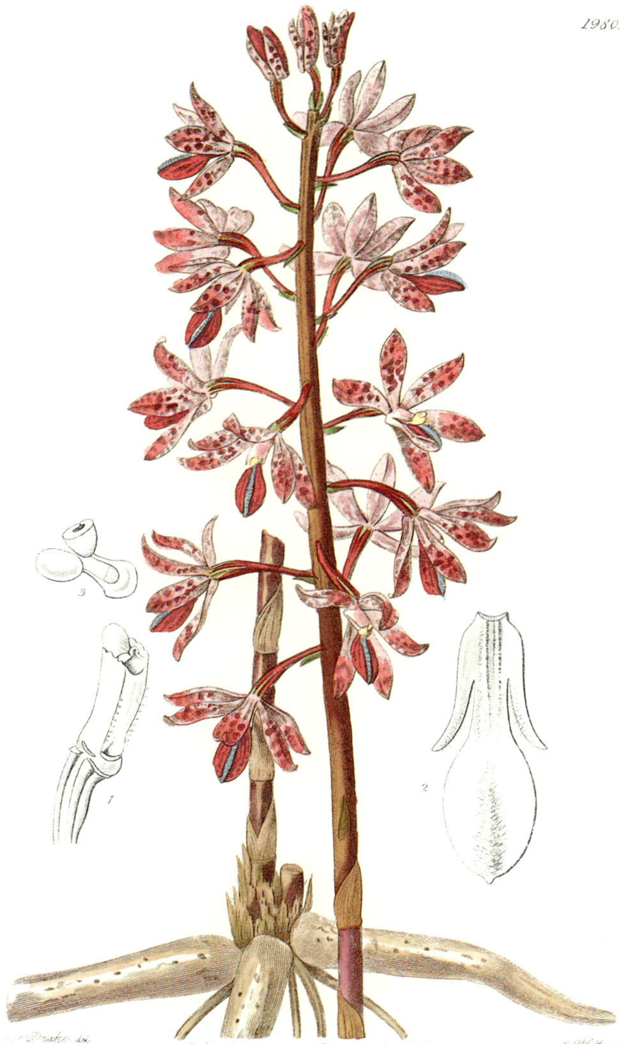

Vol.30: t.2 (1844)
Diaphananthe pellucida (Lindl.) Schltr.

Vol.23: t.1980 (1837)
Dipodium punctatum (Smith) R.Br.

Vol.33: t.65 (1847)
Dichromanthus cinnabarinus (Llave & Lex.) Garay

926.

Vol.11: t.926 (1825)
Disa uniflora Bergius

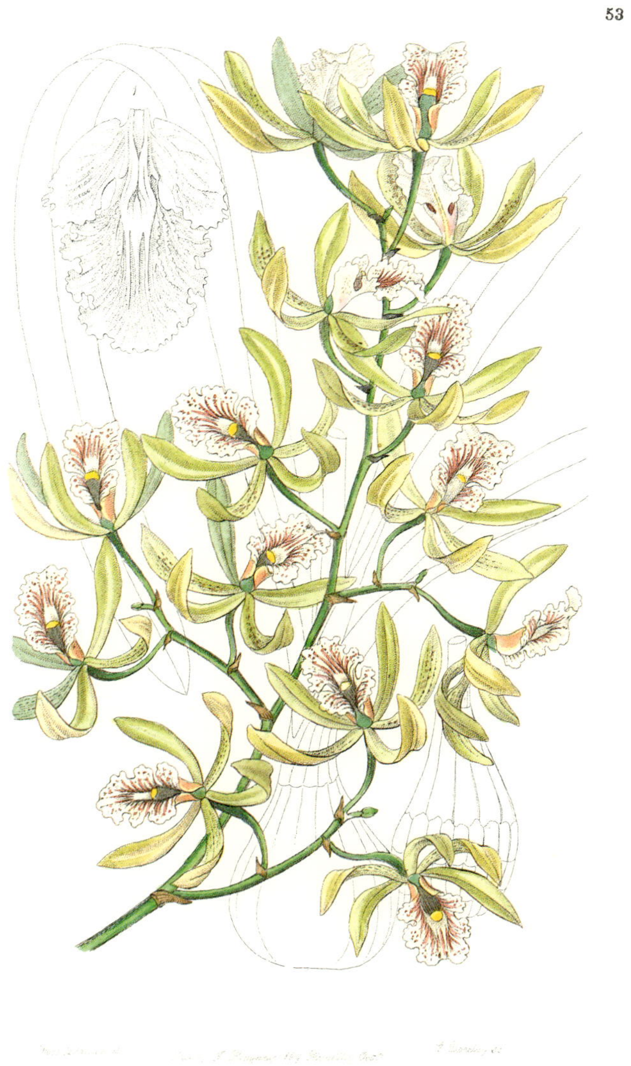

Vol.30: t.51 (1844)
Encyclia adenocaula (Llave & Lex.) Schltr. var. adenocaula

Vol.22: t.1898 (1836)
Encyclia fragrans (Sw.) Lemée

Vol.33: t.53 (1847)
Encyclia alata (Bateman) Schltr.

Vol.26: t.6 (1840)
Encyclia glumacea (Lindl.) Pabst
Vol.28: t.50 (1842)
Encyclia lancifolia (Lindl.) Dressler

Vol.21: t.1765 (1835)
Encyclia gracilis (Lindl.) Schltr.
Vol.24: t.26 (1838)
Encyclia ochracea (Lindl.) Dressler

Vol.19: t.1623 (1833)
Encyclia oncidioides (Lindl.) Schltr.

Vol.17: t.1415 (1831)
Encyclia odoratissima (Lindl.) Schltr.
Vol.30: t.34 (1844)
Encyclia pterocarpa (Lindl.) Dressler

Vol.33: t.35 (1847)
Encyclia plicata (Lindl.) Schltr.
Vol.33: t.50 (1847)
Encyclia pyriforme (Lindl.) Schltr.

Vol.30: t.45 (1844)
Encyclia radiata (Lindl.) Dressler
Vol.26: t.35 (1840)
Encyclia vitellina (Lindl.) Dressler

Vol.25: t.11 (1839)
Encyclia vespa (Vell.) Dressler
Vol.23: t.1961 (1837)
Epidendrum carpophorum Barb.Rodr.

Vol.10: t.783 (1824)
Epidendrum ciliare L.
Vol.28: t.25 (1842)
Epidendrum cinnabarinum Salzm. ex Lindl.

Vol.10: t.784 (1824)
Epidendrum ciliare L.
Vol.28: t.42 (1842)
Epidendrum cristatum Ruiz Lopez & Pavon

Vol.31: t.4 (1845)
Epidendrum densiflorum Hook.
Vol.1: t.17 (1815)
Epidendrum nutans Sw.

Vol.1: t.80 (1816)
Epidendrum difforme Jacq.
Vol.22: t.1870 (1836)
Epidendrum purpurascens Focke

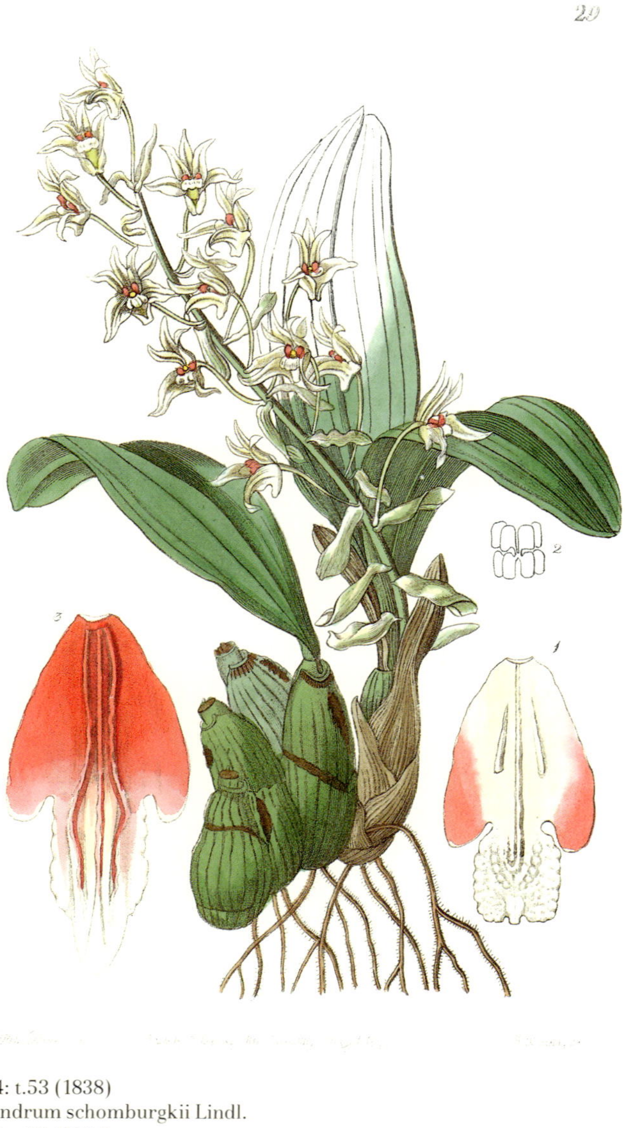

Vol.24: t.53 (1838)
Epidendrum schomburgkii Lindl.
Vol.30: t.29 (1844)
Eria bractescens Lindl.

Vol.1: t.67 (1815)
Epidendrum secundum Jacq.
Vol.28: t.21 (1842)
Eria coronaria (Lindl.) Rchb.f.

Vol.25: t.35 (1839)
Eria ferruginea Lindl.
Vol.27: t.42 (1841)
Eria ornata (Blume) Lindl.

Vol.30: t.20 (1844)
Eria floribunda Lindl.
Vol.28: t.32 (1842)
Eria polyura Lindl.

Pub by J. Ridgway 169 Piccadilly Aug. 1. 1825

Vol.11: t.904 (1825)
Eria javanica (Sw.) Blume

Vol.12: t.978 (1826)
Eria rosea Lindl.
Vol.33: t.63 (1847)
Eria spicata (D.Don) Hand.-Mazz.

Vol.27: t.62 (1841)
Eria spicata (D.Don) Hand.-Mazz.
Vol.33: t.18 (1847)
Eriopsis biloba Lindl.

Vol.18: t.1508 (1832)
Eulophia alta (L.) Fawcett & Rendle

Vol.14: t.1147 (1828)
Eulophia ensata Lindl.
Vol.8: t.686 (1823)
Eulophia guineensis R.Br. ex Lindl.

Vol.9: t.742 (1823)
Eulophia gracilis Lindl.
Vol.30: t.12 (1844)
Eulophia horsfallii (Bateman) Summerh.

M. Hart. del. Pub by J. Ridgway 170 Piccadilly Oct. 1. 1821. S. Watts. sc.

Vol.7: t.573 (1821)
Eulophia speciosa (R.Br. ex Lindl.) Bolus

Vol.12: t.1002 (1826)
Eulophia streptopetala Lindl.

Vol.26: t.49 (1840)
Galeandra batemannii Rolfe
Vol.8: t.675 (1822)
Geodorum densiflorum (Lam.) Schltr.

Vol.18: t.1499 (1832)
Gennaria diphylla (Link) Parl.
Vol.20: t.1687 (1834)
Geodorum fucatum Lindl.

Vol.26: t.54 (1840)
Gomesa crispa (Lindl.) Klotzsch
Vol.33: t.17 (1847)
Gongora bufonia Lindl.

Vol.27: t.2 (1841)
Gongora bufonia Lindl.
Vol.18: t.1538 (1832)
Gongora galeata (Lindl.) Rchb.f.

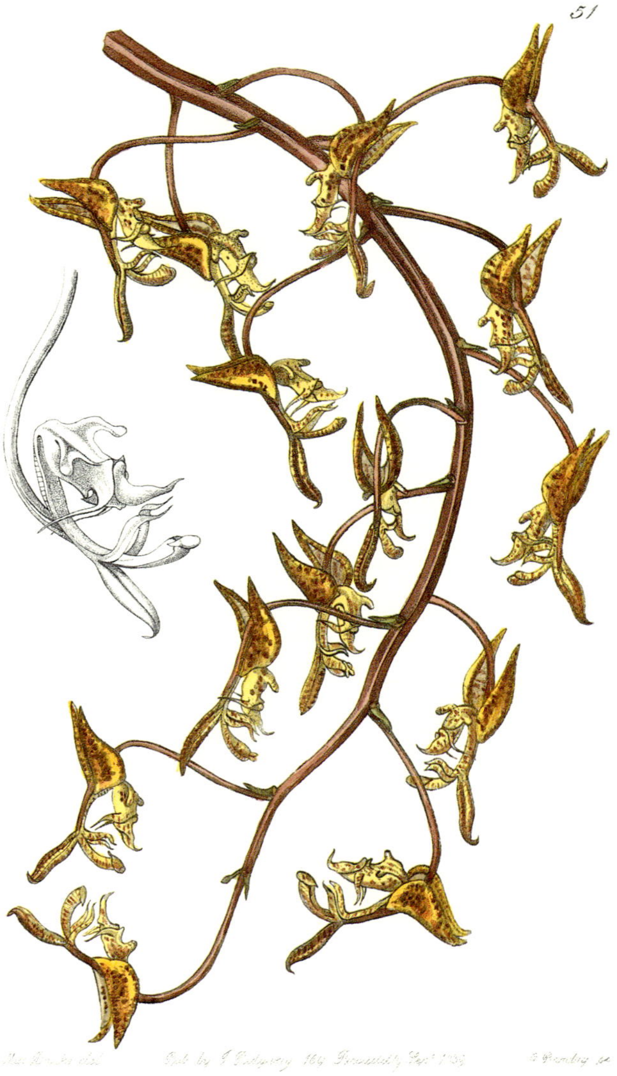

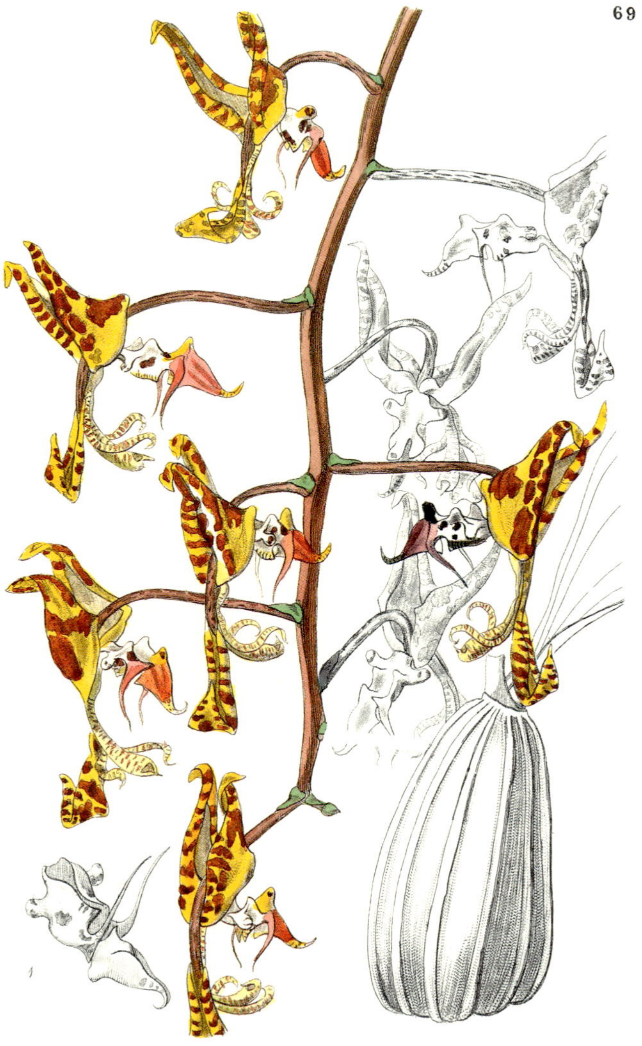

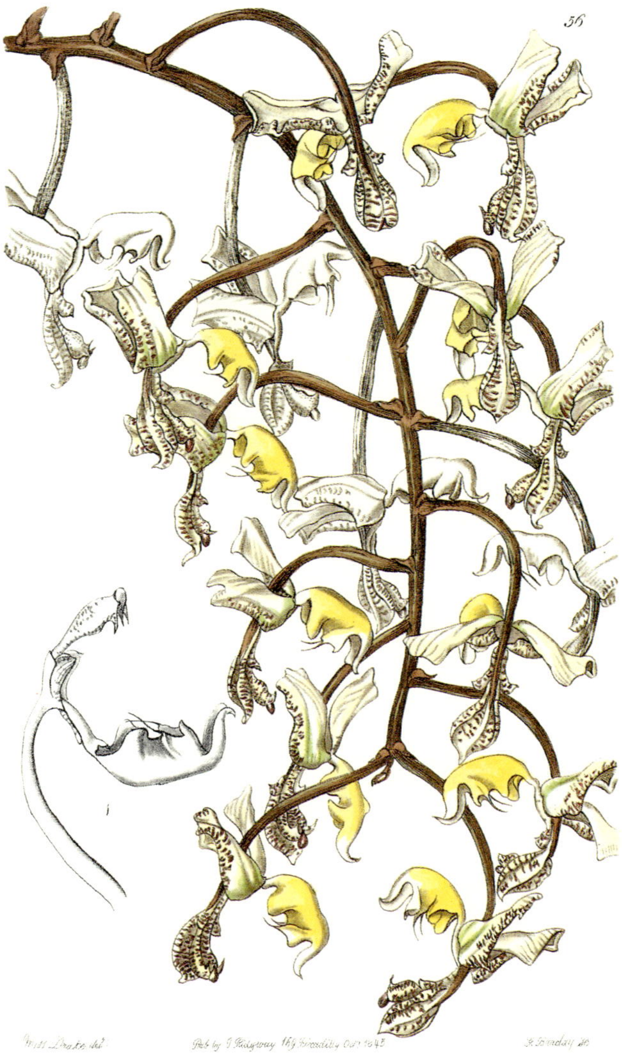

Vol.19: t.1616 (1833)
Gongora quinquenervis Ruiz Lopez & Pavon
Vol.33: t.69 (1847)
Gongora quinquenervis Ruiz Lopez & Pavon

Vol.25: t.51 (1839)
Gongora quinquenervis Ruiz Lopez & Pavon
Vol.31: t.56 (1845)
Gongora truncata Lindl.

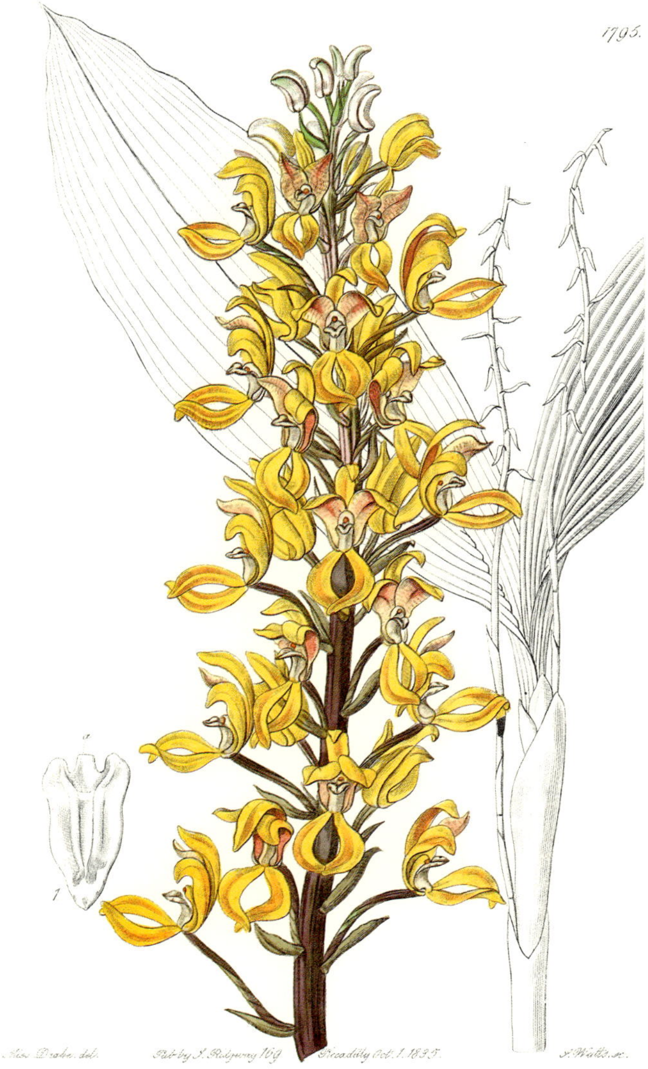

Vol.8: t.639 (1822)
Goodyera procera (Ker Gawler) Hook.
Vol.21: t.1795 (1835)
Govenia superba (Llave & Lex.) Lindl.

Vol.24: t.13 (1838)
Govenia liliacea (Llave & Lex.) Lindl.
Vol.31: t.67 (1845)
Govenia superba (Llave & Lex.) Lindl.

Miss Drake del.t Pub.d by J. Ridgway 169. Piccadilly. Decr. 1839. G. Barclay sc.

Vol.25: t.65 (1839)
Grammatophyllum scriptum (L.) Blume

1

Miss Drake Del. Pub. by J. Ridgway, 169, Piccadilly, Dec. 1. 1842. G. Barclay Sc.

Vol.28: t.69 (1842)
Grammatophyllum scriptum (L.) Blume

Vol.21: t.1821 (1835)
Graphorkis lurida (Sw.) Kuntze

Vol.20: t.1740 (1835)
Grobya amherstiae Lindl.

1858.

Miss Drake del.

Pub by J. Ridgway 169 Piccadilly June 1. 1836.

S. Watts sc.

Vol.22: t.1858 (1836)
Habenaria procera (Sw.) Lindl.

Vol.25: t.14 (1839)
Huntleya meleagris Lindl.
Vol.9: t.745 (1823)
Isochilus linearis (Jacq.) R.Br.

Vol.22: t.1904 (1836)
Ionopsis utricularioides (Sw.) Lindl.
Vol.30: t.50 (1844)
Lacaena bicolor Lindl.

Vol.25: t.54 (1839)
Laelia albida Lindl. & Bateman
Vol.23: t.1947 (1837)
Laelia anceps Lindl.

Vol.21: t.1751 (1835)
Laelia anceps Lindl.
Vol.25: t.27 (1839)
Laelia autumnalis Lindl.

Vol.14: t.1172 (1828)
Laelia crispa (Lindl.) Rchb.f.

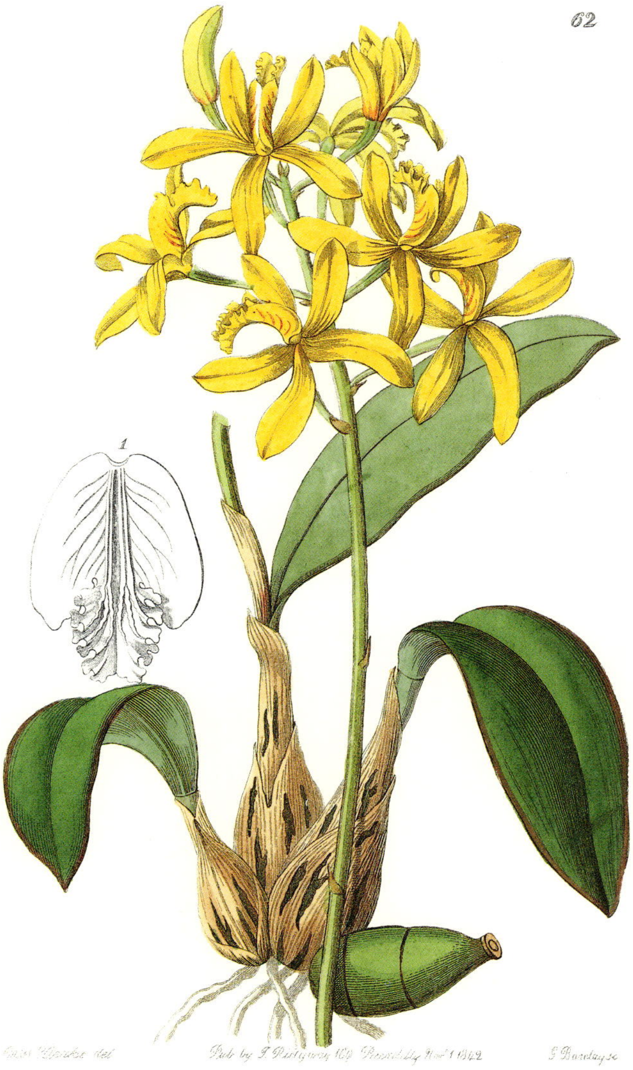

Vol.28: t.62 (1842)
Laelia flava Lindl.
Vol.24: t.2 (1838)
Laelia perrinii Lindl.

Vol.25: t.26 (1839)
Laelia furfuracea Lindl.
Vol.30: t.5 (1844)
Laelia pumila (Hook.) Rchb.f.

Vol.26: t.41 (1840)
Laelia rubescens Lindl.
Vol.31: t.69 (1845)
Laelia rubescens Lindl.

Vol.27: t.24 (1841)
Laelia rubescens Lindl.
Vol.30: t.30 (1844)
Laelia speciosa (Kunth) Schltr.

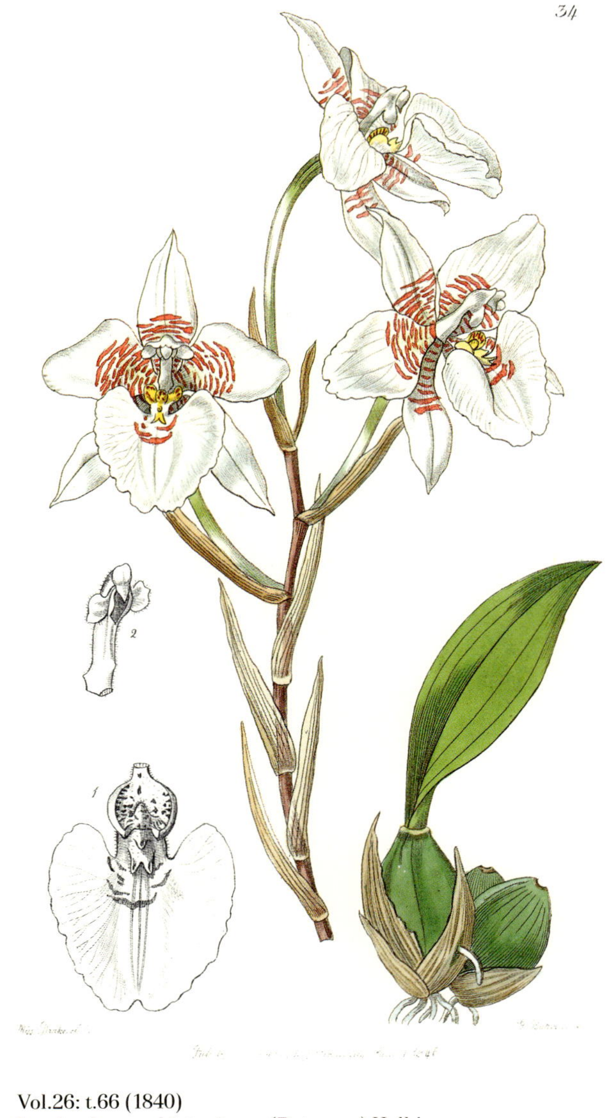

Vol.26: t.66 (1840)
Lemboglossum bictoniense (Bateman) Halbinger
Vol.32: t.34 (1846)
Lemboglossum cervantesii (Llave & Lex.) Halbinger

Vol.31: t.36 (1845)
Lemboglossum cervantesii (Llave & Lex.) Halbinger
Vol.26: t.30 (1840)
Lemboglossum maculatum (Llave & Lex.) Halbinger

Vol.25: t.48 (1839)
Lemboglossum rossii (Lindl.) Halbinger

Vol.19: t.1625 (1833)
Leptotes bicolor Lindl.

Vol.21: t.1762 (1835)
Lepanthes tridentata Sw.

Vol.14: t.1175 (1828)
Liparis nervosa (Thunb.) Lindl.
Vol.11: t.882 (1825)
Liparis reflexa (R.Br.) Lindl.

Vol.20: t.1671 (1834)
Liparis nervosa (Thunb.) Lindl.
Vol.21: t.1806 (1835)
Lockhartia acuta (Lindl.) Rchb.f.

Vol.3: t.271 (1818)
Ludisia discolor (Ker Gawler) A.Rich.

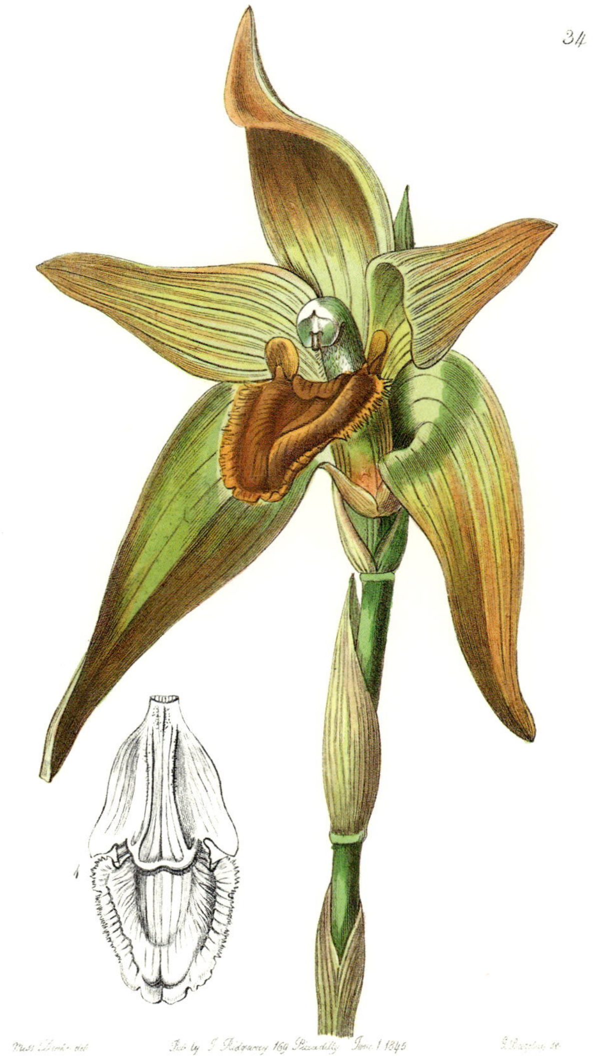

Vol.22: t.1871 (1836)
Lycaste aromatica (Graham ex Hook.) Lindl.
Vol.28: t.13 (1842)
Lycaste cruenta (Lindl.) Lindl.

Vol.14: t.1206 (1829)
Lycaste ciliata (Ruiz Lopez & Pavon) Lindl.
Vol.31: t.34 (1845)
Lycaste longipetala (Ruiz Lopez & Pavon) Garay

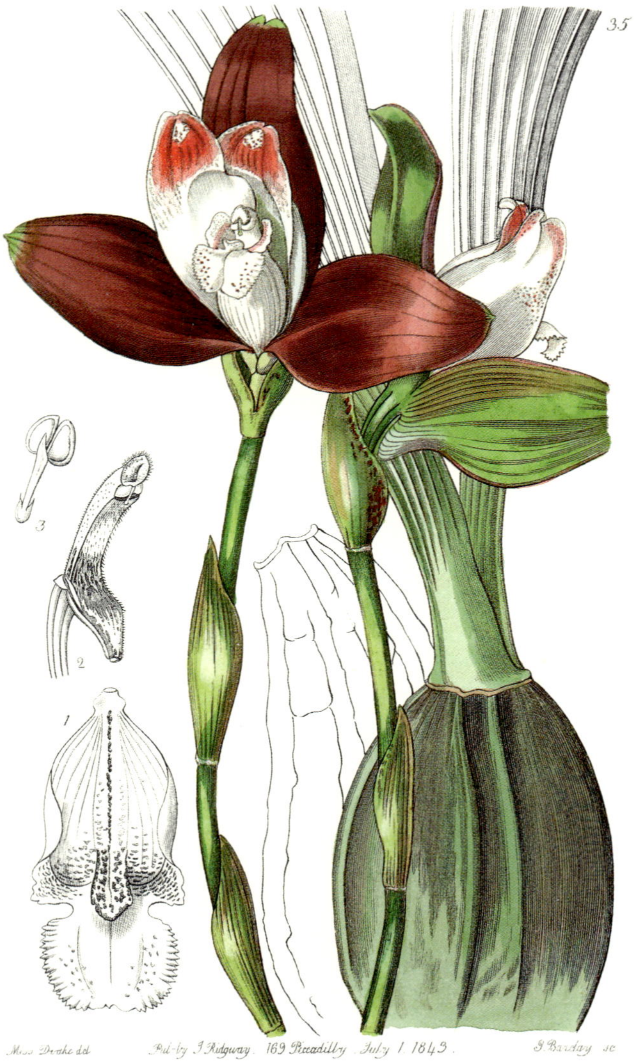

Vol.29: t.35 (1843)
Lycaste macrophylla (Poeppig & Endl.) Lindl.

Vol.21: t.1815 (1835)
Macradenia lutescens R.Br.

Vol.8: t.612 (1822)
Macradenia lutescens R.Br.

Vol.15: t.1290 (1829)
Malaxis unifolia Michaux
Vol.12: t.1028 (1826)
Maxillaria crassifolia (Lindl.) Rchb.f.

Vol.10: t.844 (1824)
Maxillaria camaridii Rchb.f.
Vol.26: t.12 (1840)
Maxillaria cucullata Lindl.

Vol.21: t.1804 (1835)
Maxillaria densa Lindl.
Vol.18: t.1530 (1832)
Maxillaria marginata (Lindl.) Fenzl

Vol.21: t.1799 (1835)
Maxillaria lindleyana Schltr.
Vol.21: t.1802 (1835)
Maxillaria picta Hook.

Vol.22: t.1848 (1836)
Maxillaria rufescens Lindl.
Vol.33: t.20 (1847)
Mexicoa ghiesbreghtiana (A.Rich. & Galeotti) Garay

Vol.25: t.8 (1839)
Maxillaria tenuifolia Lindl.
Vol.31: t.8 (1845)
Miltonia cuneata Lindl.

Vol.19: t.1627 (1833)
Miltonia flavescens (Lindl.) Lindl.
Vol.23: t.1992 (1837)
Miltonia spectabilis Lindl.

Vol.22: t.1830 (1836)
Miltonia russelliana (Lindl.) Lindl.
Vol.30: t.39 (1844)
Miltonioides laevis (Lindl.) Brieger & Lueckel

Vol.4: t.324 (1818)
Monadenia bracteata (Sw.) Durand & Schinz
Vol.29: t.56 (1843)
Mormodes aromatica Lindl.

Vol.3: t.210 (1817)
Monadenia cernua (Sw.) Durand & Schinz
Vol.22: t.1861 (1836)
Mormodes atropurpureum Lindl.

Vol.28: t.43 (1842)
Mormodes lineatum Bateman ex Lindl.

Miss Drake del. Pub by J. Ridgway 169 Piccadilly. July 1. 1843. J. Barclay sc

Vol.29: t.33 (1843)
Mormodes luxatum Lindl.

30

Miss Drake del. Pub. by J. Ridgway 169 Piccadilly June 1 1845 G. Barclay sc.

Vol.31: t.30 (1845)
Myrmecophila tibicinis (Bateman ex Lindl.) Rolfe

Vol.18: t.1541 (1832)
Nanodes discolor Lindl.
Vol.18: t.1525 (1832)
Neotinea maculata (Desf.) Stearn

Vol.4: t.283 (1818)
Neofinetia falcata (Thunb.) Hu
Vol.9: t.759 (1823)
Notylia punctata (Ker Gawler) Lindl.

Vol.25: t.62 (1839)
Odontoglossum mystacinum (Lindl.) Lindl.

Vol.23: t.1972 (1837)
Oeceoclades pulchra (Thouars) Clements & Cribb

Vol.8: t.618 (1822)
Oeceoclades maculata Lindl.

Vol.9: t.727 (1823)
Oncidium altissimum (Jacq.) Sw.

Vol.25: t.16 (1839)
Oncidium altissimum (Jacq.) Sw.

Vol.33: t.66 (1847)
Oncidium amictum Lindl.

Vol.20: t.1699 (1834)
Oncidium ampliatum Lindl.

Vol.18: t.1651 (1833)
Oncidium baueri Lindl.

Vol.29: t.12 (1843)
Oncidium bicallosum Lindl.
Vol.27: t.59 (1841)
Oncidium brachyandrum Lindl.

Vol.29: t.66 (1843)
Oncidium bicolor Lindl.
Vol.23: t.1994 (1837)
Oncidium cebolleta (Jacq.) Sw.

Vol.28: t.4 (1842)
Oncidium cebolleta (Jacq.) Sw.

Vol.21: t.1758 (1835)
Oncidium citrinum Lindl.

Vol.20: t.1660 (1834)
Oncidium ciliatum Lindl.

Vol.23: t.1920 (1837)
Oncidium crispum Lodd.

Vol.18: t.1542 (1832)
Oncidium cornigerum Lindl.

Vol.33: t.68 (1847)
Oncidium curtum Lindl.

Miss Drake del.

Pub by S. Ridgway 169 Piccadilly Nov. 1. 1837.

J. Watts sc.

Vol.23: t.2006 (1837)
Oncidium deltoideum Lindl.

Vol.13: t.1050 (1827)
Oncidium divaricatum Lindl.

Vol.31: t.64 (1845)
Oncidium incurvum Barker ex Lindl.

Vol.21: t.1789 (1835)
Oncidium guianense (Aublet) Garay

Vol.19: t.1569 (1833)
Oncidium harrisonianum Lindl.

Miss Drake. del. Pub by J. Ridgway 169 Piccadilly May 1. 1836. S. Watts. sc.

Vol.22: t.1851 (1836)
Oncidium jacquinianum Garay & Stacy

Vol.22: t.1887 (1836)
Oncidium lanceanum Lindl.

Vol.24: t.44 (1838)
Oncidium maculatum (Lindl.) Lindl.

23

Mrs Powell del. Pub. by J. Ridgway 169 Piccadilly. May 1. 1843. J. Bentley sc.

Vol.29: t.23 (1843)
Oncidium microchilum Bateman ex Lindl.

Vol.26: t.10 (1840)
Oncidium ornithorhynchum Kunth
Vol.21: t.1787 (1835)
Oncidium pulchellum Hook.

Vol.12: t.1007 (1826)
Oncidium pubes Lindl.
Vol.25: t.42 (1839)
Oncidium pulvinatum Lindl.

Vol.24: t.48 (1838)
Oncidium raniferum Lindl.

Vol.11: t.920 (1825)
Oncidium pumilum Lindl.

Vol.33: t.70 (1847)
Oncidium reflexum Lindl.

Vol.28: t.30 (1842)
Oncidium sphacelatum Lindl.
Vol.32: t.27 (1846)
Oncidium stipitatum Lindl.

Vol.31: t.40 (1845)
Oncidium spilopterum Lindl.
Vol.26: t.14 (1840)
Oncidium stramineum Bateman ex Lindl.

Vol.25: t.57 (1839)
Oncidium trulliferum Lindl.

Vol.28: t.54 (1842)
Oncidium urophyllum Lodd. ex Lindl.

Vol.29: t.43 (1843)
Oncidium uniflorum Lindl.

Vol.33: t.46:1 (1847)
Ophrys ferrum-equinum Desf.
Vol.33: t.46:2 (1847)
Ophrys argolica Fleischm.

Vol.13: t.1071 (1827)
Ophrys fusca Link

Vol.13: t.1087 (1827)
Ophrys atrata Lindl.
Vol.33: t.25 (1847)
Ophrys holoserica (Burm.f.) Greuter

Vol.32: t.52 (1846)
Ophrys oestrifera Bieb. ssp.oestrifera
Vol.3: t.205 (1817)
Ophrys tenthredinifera Willd.

Vol.14: t.1197 (1828)
Ophrys sphegodes Mill. ssp. sphegodes
Vol.13: t.1093 (1827)
Ophrys tenthredinifera Willd.

Vol.5: t.370 (1819)
Ophrys vernixia Brot.
Vol.5: t.367 (1819)
Orchis lactea Poiret

Vol.5: t.375 (1819)
Orchis italica Poiret
Vol.3: t.202 (1817)
Orchis longicornu Poiret

Vol.14: t.1155 (1828)
Orchis papilionacea L.
Vol.18: t.1510 (1832)
Pabstia viridis (Lindl.) Garay

Vol.27: t.48 (1841)
Osmoglossum pulchellum (Bateman ex Lindl.) Schltr.
Vol.21: t.1811 (1835)
Paphinia cristata (Lindl.) Lindl.

Vol.28: t.17 (1842)
Paphiopedilum barbatum (Lindl.) Stein

Vol.10: t.788 (1824)
Paphiopedilum venustum (Wallich) Pfitzer ex Stein

Vol.23: t.1991 (1837)
Paphiopedilum purpuratum (Lindl.) Stein

Vol.21: t.1809 (1835)
Papilionanthe teres (Roxb.) Schltr.
Vol.23: t. 1953 (1837)
Peristeria cerina (Lindl.) Garay

Vol.12: t.985 (1826)
Pelexia adnata (Sw.) Sprengel
Vol.10: t.862 (1825)
Peristylus tentaculatus (Lindl.) J.J.Smith

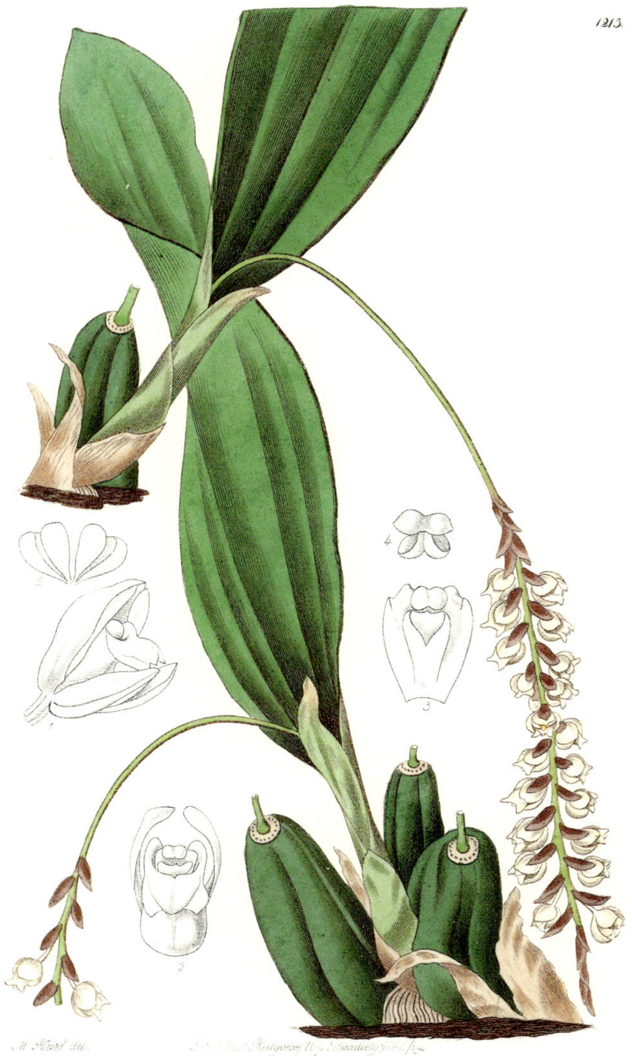

Vol.24: t.34 (1838)
Phalaenopsis aphrodite Rchb.f.
Vol.21: t.1777 (1835)
Pholidota imbricata Hook.

Vol.14: t.1213 (1829)
Pholidota imbricata Hook.
Vol.5: t.405 (1819)
Platanthera grandiflora (Bigelow) Lindl.

Vol.22: t.1844 (1836)
Plectrelminthus caudatus (Lindl.) Summerh.

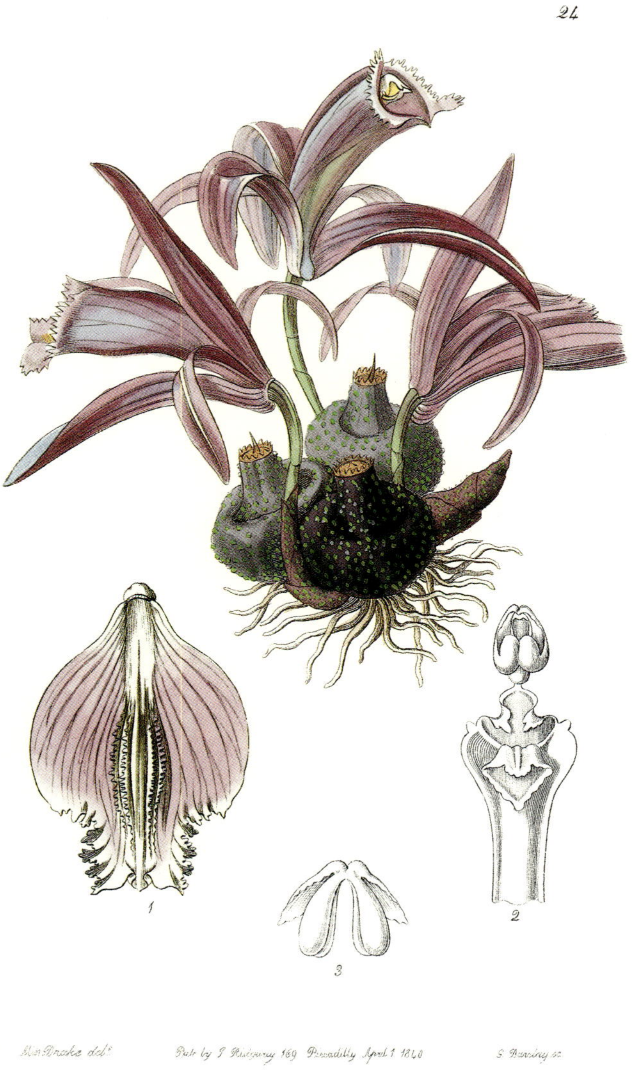

Vol.26: t.24 (1840)
Pleione praecox (Smith) D.Don
Vol.21: t.1825 (1836)
Pleurothallis grobyi Bateman ex Lindl.

Vol.21: t.1797 (1835)
Pleurothallis grobyi Bateman ex Lindl.
Vol.11: t.935 (1825)
Pleurothallis ophioglossoides (Jacq.) Garay & Sweet

Vol.15: t.1298 (1830)
Pleurothallis prolifera Herbert ex Lindl.
Vol.2: t.148 (1816)
Pogonia ophioglossoides (L.) Jussieu

Vol.23: t.1968 (1837)
Pleurothallis saurocephala Lodd.
Vol.10: t.851 (1824)
Polystachya puberula Lindl.

Vol.9: t.760 (1823)
Ponthieva petiolata Lindl.
Vol.24: t.40 (1838)
Promenaea rollissonii (Lindl.) Lindl.

Vol.22: t.1915 (1836)
Prescottia stachyodes (Sw.) Lindl.
Vol.25: t.17 (1839)
Promenaea stapelioides (Link & Otto) Lindl.

Vol.22: t.1879 (1836)
Psychilis bifida (Aublet) R.P.Sauleda

Vol.22: t.1911 (1836)
Psygmorchis pusilla (L.) Dodson & Dressler

Vol.11: t.910 (1825)
Psychopsis papilio (Lindl.) H.G.Jones

Vol.14: t.1131 (1828)
Renanthera coccinea Lour.

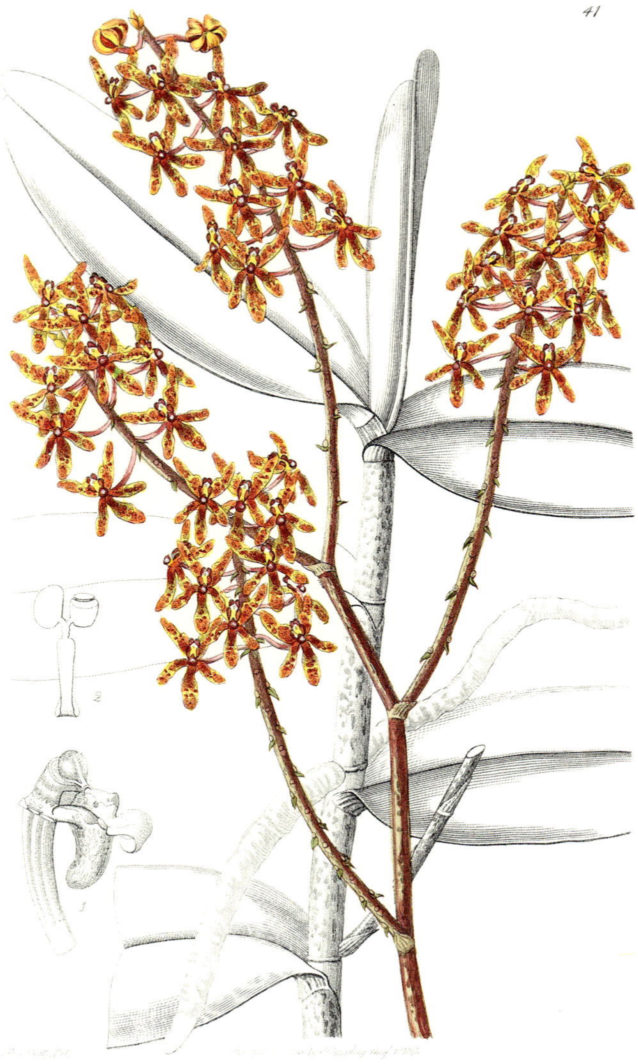

Vol.29: t.41 (1843)
Renanthera elongata (Blume) Lindl.

Vol.26: t.44 (1840)
Rhyncholaelia glauca (Lindl.) Schltr.

Vol.32: t.53 (1846)
Rhyncholaelia digbyana (Lindl.) Schltr.

1443

Miss Drake del.   Pub by J. Ridgway 169 Piccadilly Oct. 1. 1831.   J. Watts sc.

Vol.17: t.1443 (1831)
Rhynchostylis retusa (L.) Blume

30

1014

1927

930

Vol.33: t.30 (1847)
Rhynchostylis violacea (Lindl.) Rchb.f.
Vol.23: t.1927 (1837)
Rodriguezia candida Bateman ex Lindl.

Vol.12: t.1014 (1826)
Robiquetia succisa (Lindl.) Seidenf. & Garay
Vol.11: t.930 (1825)
Rodriguezia lanceolata Ruiz Lopez & Pavon

Vol.25: t.44 (1839)
Rodriguezia maculata (Lindl.) Rchb.f.
Vol.9: t.701 (1823)
Sacoila lanceolata (Aublet) Garay

Vol.22: t.1875 (1836)
Rudolfiella aurantiaca (Lindl.) Hoehne
Vol.22: t.1832 (1836)
Sarcochilus falcatus R.Br.

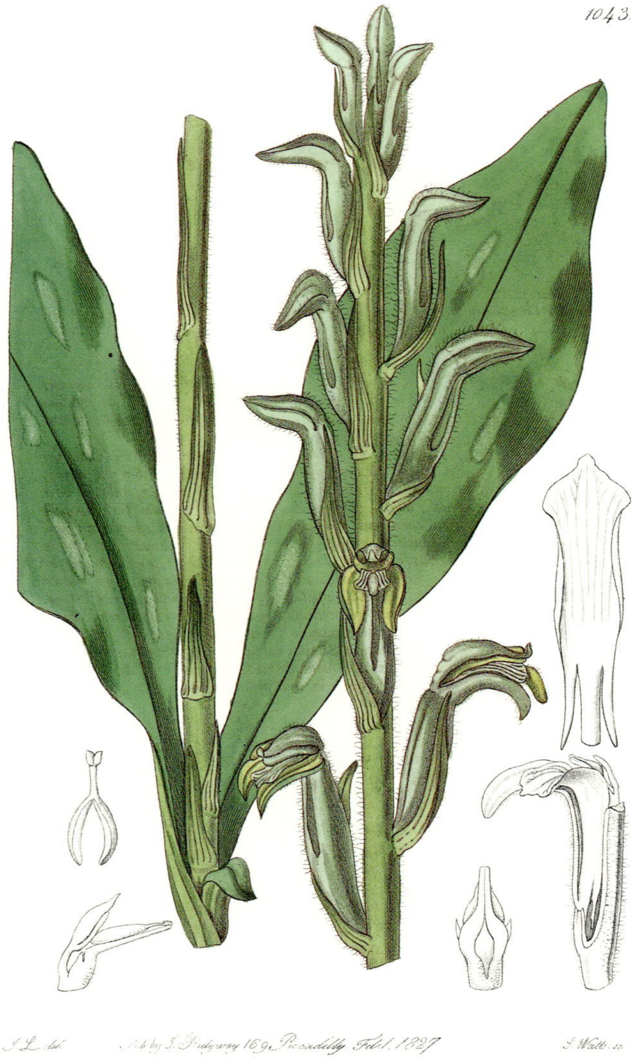

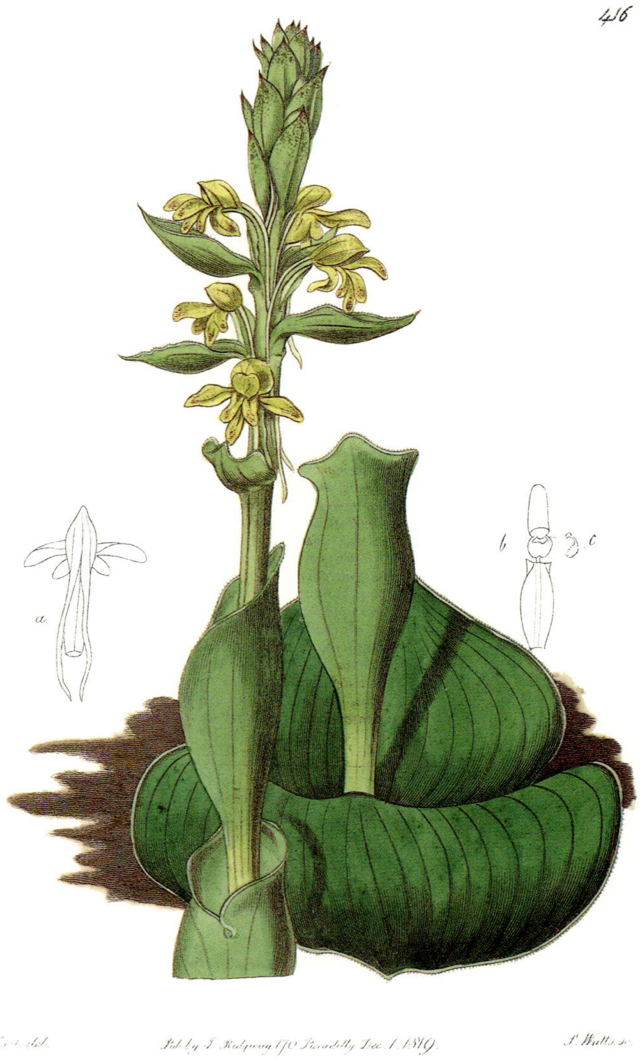

Vol.12: t.1043 (1827)
Sarcoglottis grandiflora (Lindl.) Klotzsch
Vol.9: t.703 (1823)
Satyrium coriifolium Sw.

Vol.5: t.416 (1819)
Satyrium coriifolium Sw.
Vol.26: t.18 (1840)
Satyrium erectum Sw.

Vol.19: t.1618 (1833)
Sauroglossum nitidum (Vell.) Schltr.

Vol.10: t.825 (1824)
Scaphyglottis prolifera (R.Br.) Cogn.
Vol.30: t.23 (1844)
Schomburgkia gloriosa Rchb.f.

Vol.22: t.1901 (1836)
Scaphyglottis violacea Lindl.
Vol.31: t.53 (1845)
Schomburgkia undulata Lindl.

Vol.23: t.1986 (1837)
Scuticaria steelei (Hook.) Lindl.
Vol.27: t.17 (1841)
Sobralia sessilis Lindl.

Vol.14: t.1189 (1828)
Serapias vomeracea (Burm.f.) Briq. ssp. vomeracea
Vol.23: t.1929 (1837)
Solenidium lunatum (Lindl.) Kraenzlin

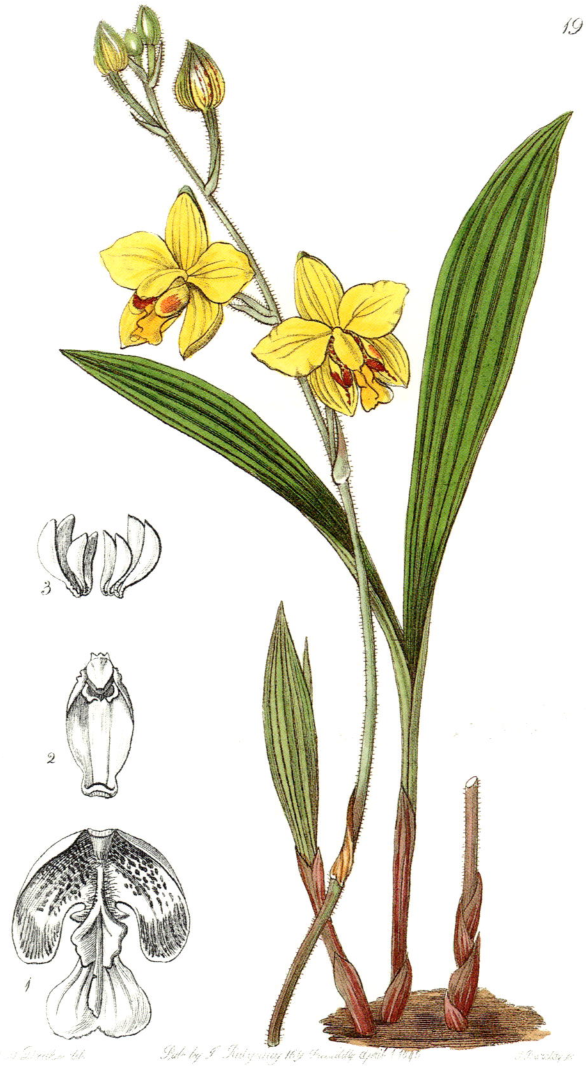

Vol.13: t.1129 (1828)
Sophronitis cernua Lindl.

Vol.31: t.19 (1845)
Spathoglottis pubescens Lindl.

Vol.24: t.60 (1838)
Spathoglottis plicata Blume

Vol.10: t.823 (1824)
Spiranthes cernua (L.) L.C.Rich.

Vol.7: t.602 (1822)
Spiranthes sinensis (Pers.) Ames

1529.

Miss Jane Edwards del. Pub by J. Ridgway 169 Piccadilly Oct. 1, 1832. J. Watts. sc.

Vol.18: t.1529 (1832)
Stanhopea grandiflora (Lodd.) Lindl.

Vol.22: t.1837 (1836)
Stanhopea insignis Frost ex Hook.

Vol.29: t.44 (1843)
Stanhopea martiana Bateman ex Lindl.

1800.

Miss Drake del. Pub by J. Ridgway 169 Piccadilly Oct. 1. 1835. J. Watts sc.

Vol.21: t.1800 (1835)
Stanhopea oculata (Lodd.) Lindl.

Miss Drake del.   Pub by J. Ridgway 169 Piccadilly May 1 1845

Vol.31: t.24 (1845)
Stanhopea oculata (Lodd.) Lindl.

5.

Miss Drake del.   Pub by J. Ridgway 169 Piccadilly Jan. 1. 1838.   J. Watts sc.

Vol.24: t.5 (1838)
Stanhopea quadricornis Lindl.

1.

Pub by J. Ridgway 169 Piccadilly Jan. 1. 1839.

J. Watts sc.

Vol.25: t.1 (1839)
Stanhopea tigrina Bateman ex Lindl.

65

Vol.31: t.65 (1845)
Stanhopea wardii Lodd. ex. Lindl.

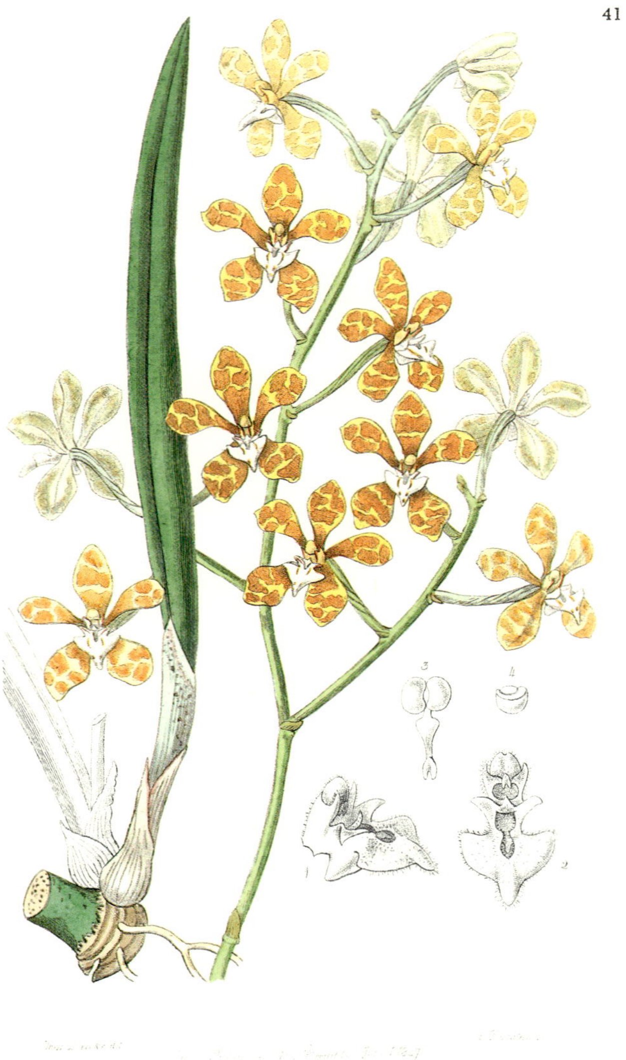

Vol.33: t.41 (1847)
Staurochilus ionosma (Lindl.) Schltr.
Vol.30: t.8 (1844)
Tainia bicornis (Lindl.) Rchb.f.

Vol.24: t.20 (1838)
Stenia pallida Lindl.
Vol.32: t.19 (1846)
Thrixspermum calceolus (Lindl.) Rchb.f.

Miss Drake del.   Pub by J. Ridgway 169 Piccadilly June 1. 1838.   J. Watts sc.

Vol.24: t.33 (1838)
Thunia bracteata (Roxb.) Schltr.

Vol.32: t.57 (1846)
Trichopilia laxa (Lindl.) Rchb.f.

Vol.23: t.1951 (1837)
Trichocentrum fuscum Lindl.

Vol.22: t.1863 (1836)
Trichopilia tortilis Lindl.

Vol.31: t.2 (1845)
Trichotosia vestita (Lindl.) Kraenzlin
Vol.11: t.906 (1825)
Triphora trianthophora (Sw.) Rydb.

Vol.23: t.1923 (1837)
Trigonidium obtusum Lindl.
Vol.28: t.48 (1842)
Trudelia cristata (Lindl.) Senghas

M. Hart. del.   Pub by J. Ridgway 170 Piccadilly Dec. 1. 1820.   S. Watts. sc.

Vol.6: t.506 (1820)
Vanda tessellata (Roxb.) Lodd. ex D.Don

Vol.32: t.59 (1846)
Vandopsis lissochiloides (Gaudich.) Pfitzer

Vol.32: t.28 (1845)
Warreella cyanea (Lindl.) Schltr.

Vol.9: t.732 (1823)
Xylobium variegatum (Ruiz Lopez & Pavon) Garay & Dunsterv.

Vol.18: t.1549 (1832)
Xylobium palmifolium (Sw.) Benth.

Vol.17: t.1433 (1831)
Zygopetalum mackaii Hook.

Key in the catalogue

o	=	Title in *Botanical Register 1815–1847*
E	=	epiphytic
L	=	lithophytic
T	=	terrestrial (geophytic)
S	=	saprophytic
BR	=	Botanical Register

CATALOGUE

by Samuel Sprunger

Acampe congesta (Lindl.) Lindl., Fol. Orch. Acampe 2, 1853
BR 25: misc.94 (1839)
Type: Sri Lanka, cult. Loddiges (K-Lindl., holo!)
Sri Lanka, India – E
Vanda congestum Lindl., Bot. Reg. 25: misc. 94, 1839
Saccolabium papillosum Wight, Ic.: t.1672, 1851 (non Lindl.)
Saccolabium congestum (Lindl.) Hook.f., Fl. Brit. Ind. 6: 63, 1894

Acampe ochracea (Lindl.) Hochr., Bull. New York Bot. Gard. 6: 270, 1910
BR 28: misc.4 (1842)
Type: Sri Lanka, Wilmot Horton s.n. cult. Loddiges (K-Lindl., holo.!)
Sri Lanka, Sikkim, Bhutan, N.E.India, Burma, China, Thailand, Laos, Cambodia, Vietnam – E
Saccolabium ochraceum Lindl., Bot. Reg. 28: misc.4, 1842
Acampe dentata Lindl., Fol. Orch. Acampe 3, 1853
Saccolabium lineolatum Thwaites, Enum. Pl. Zeyl.: 304, 1861
Acampe griffithii Rchb.f., Flora 55: 277, 1872
Acampe longifolia Parish, Order Orch.: 198, 1883 (non Lindl.)
Cleisostoma maculosa Bentham, Gen. Pl. 3: 581, 1883 (non Lindl.)

Acampe papillosa (Lindl.) Lindl., Fol. Orch. Acampe 2, 1853
BR 18: t.1552 (1833)
Type: Prome, Wallich 7305 (K-Lindl., holo.!)
N.W.Himalaya, Nepal, Sikkim, Bhutan, N.E.India, Burma, Thailand, Laos, Vietnam – E
o Saccolabium papillosum Lindl., Bot. Reg. 18: t.1552, 1833
Saccolabium carinatum Griffith, Notul. 3: 354, 1851
Gastrochilus papillosum (Lindl.) Kuntze, Rev. Gen. 2: 661, 1891
Sarcanthus papillosus Tixier, Bull. Soc. Roy. Sc. Nat. Laos: 30, 1963

Acampe praemorsa (Roxb.) Blatter & McCann, J. Bombay Nat. Hist. Soc. 35: 1495, 1932
BR 10: sub t.817 (1824)
Sri Lanka, India, Sikkim – E
Epidendrum praemorsum Roxb., Pl. Corom.: 34 t.43, 1795
Cymbidium praemorsum (Roxb.) Sw., Nov. Act. Upsal. 6: 75, 1799
Vanda wightiana Lindl. ex Wight, Ic. Pl. Ind. Orch. 5: 9 pl.1670, 1851
Acampe wightiana (Lindl.) Lindl., Fol. Orch. Acampe 2, 1853
Vanda fasciata Gard. ex. Lindl., Fol. Orch. Acampe 2, 1853
Saccolabium papillosum Dalzell & Gibson, Bombay Fl.: 264, 1861
Saccolabium wightiana (Lindl.) Hook.f., Fl. Brit. Ind. 6: 62, 1890

Acanthephippium bicolor Lindl., Bot. Reg. 20: t.1730, 1835
BR 20: t.1730 (1835) / 32: sub t.47:4 (1846)
Type: Sri Lanka, Watson s.n. (K-Lindl., holo.!)
Sri Lanka, India – T
o Acanthophippium bicolor Lindl., Bot. Reg. 20: t.1730, 1835

Acanthephippium javanicum Blume, Bijdr. 352: 47, 1825
BR 32: t.47 (1846)
Malaya, Sumatra, Java, Borneo – T
o Acanthophippium javanicum Lindl., Bot. Reg. 32: t.47, 1846

Acanthephippium striatum Lindl., Bot. Reg. 24: misc.68, 1838
BR 24: misc.68 (1838) / 32: sub t.47:2 (1846)
Type: Nepal, cult. Bateman (K-Lindl., holo.!)
Nepal, Sikkim, Bhutan, India, Thailand, Vietnam, Malaya, Java – E
Acanthophippium striatum Lindl., Bot. Reg. 24: misc.68, 1838

Acanthephippium sylhetense Lindl., Gen. Sp. Orch. Pl.: 177, 1833
BR 32: sub t.47:3 (1846)
Sikkim, N.E.India, Burma, Thailand, Malaya – T
Acanthophippium sylhetense Lindl., Bot. Reg. 32: sub t.47:3, 1846
Acanthephippium ringiflorum Griffith, Notul. 3: 347, 1851
Acanthephippium parviflorum Cumberledge, Nat. Hist. Bull. Siam Soc. 20 (3): 93–147, 1963 (non Hassk.)

Acineta barkeri (Bateman) Lindl., Bot. Reg. 29: misc.100:2, 1843
BR 29: misc.100:2 (1843)
Mexico – E
Peristeria barkeri Bateman, Orch. Mex. & Guat.: t.8, 1838

Acineta superba (Kunth) Rchb.f., Walp. Ann. 6: 609, 1862
BR 29: t.18 & misc.68 & misc.100:1 (1843)
Panama, Venezuela, Colombia, Ecuador – E
Anguloa superba Kunth, Nov. Gen. & Sp. Pl. 1: 343, t.93, 1815
Acineta humboldtii (Lindl.) Lindl., Bot. Reg. 29: misc.68, 1843
o Peristeria humboldtii Lindl., Bot. Reg. 29: t.18, 1843
Acineta sulcata Rchb.f., Gard. Chron. n.s.2: 652, 1879
Acineta callosa Sander, Orch. Rev. 6: 287, 1898
Acineta colmanii hort. ex Gard. Chron. s.3, 35: 173, 1904

Acriopsis densiflora Lindl., Bot. Reg. 33: sub t.20, 1847
var. densiflora
BR 33: sub t.20 (1847)
Type: Borneo, cult. Loddiges (K-Lindl., icono!)
Malay Peninsula, Sumatra, Borneo – E
Acriopsis purpurea Ridley, Trans. Linn. Soc. 2 (3): 406, 1893

Acriopsis javanica Reinw. ex Blume, Bijdr.: 377, 1825
var. javanica
BR 29: misc.105 (1843)
Thailand, Malay Peninsula, Singapore, Indonesia, Borneo, Philippines, New Guinea, Solomon Is., Australia – E
Spathoglottis trivalis Lindl., Gen. Sp. Orch. Pl.: 120, 1832
Acriopsis picta Lindl., Bot. Reg. 29: misc.105, 1843
Acriopsis griffithii Rchb.f., Bonpl. 2: 92, 1854
Acriopsis nelsoniana Bailey, Queensl. Agr. J. 3: 7, 1898
Acriopsis papuana Kraenzlin, K.Sch. & Laut. Fl. Schutzgeb.: 250, 1900
Acriopsis philippinensis Ames, Orchid. 2: 215, 1908
Acriopsis nelsoniana var. pallidiflora Schltr., Fedde Rep. Beih. 1: 950, 1914
Acriopsis insulari-sylvatica Fukuyama, Trans. Nat. Hist. Soc. Formosa 28: 3, 1938
Acriopsis harai Tuyama, J. Jap. Bot. 39: 129 fig.1 pl.3, 1964
Acriopsis sumatrana Schltr., Oest. Bot. Zeitschr. 50: 249, 1981

Aerangis biloba (Lindl.) Schltr., Die Orchideen: 598, 1914
BR 26: misc.151 (1840) / 27: t.35 (1841)
Type: Bowdich s.n. cult. Loddiges (K-Lindl., holo.!)
Senegal, Sierra Leone, Liberia, Ivory Coast, Ghana, Togo, Nigeria, W.Cameroun – E
o Angraecum bilobum Lindl., Bot. Reg. 26: misc.151, 1840
Angraecum apiculatum Hook., Bot. Mag. 71: t.4159, 1845
Angraecum campyloplectron Rchb.f., Bonpl. 3: 226, 1855
Angorchis biloba (Lindl.) Kuntze, Rev. Gen. 2: 651, 1891
Angorchis campyloplectron (Rchb.f.) Kuntze, Rev. Gen. 2: 651, 1891
Listrostachys biloba (Lindl.) Kraenzlin, Engl. Bot. Jahrb. 22: 28, 1895
Rhaphidorhynchus bilobus (Lindl.) Finet, Bull. Soc. Bot. Fr. 54, 9: 32, 1907
Aerangis campyloplectron (Rchb.f.) Garay, Bot. Mus. Leafl. Harv. Univ. 23 (4): 157, 1972

Aeranthes arachnites (Thouars) Lindl., Bot. Reg. 10: sub t.817, 1824
BR 10: sub t.817 (1824)
Mascarene Is. – E
Dendrobium arachnites Thouars, Orch. Iles Afr.: t.88, 1822

Aeranthes filiforme (Sw.) Griseb, Fl. Brit. W.Ind.: 625, 1864
BR 18: sub t.1522 (1832)
W.Indies – E
Limodorum filiforme Sw., Nov. Act. Soc. Sc. Upsal. 6, 1799

o **Aeranthes grandiflora** Lindl., Bot. Reg. 10: t.817, 1824
BR 10: t.817 (1824)
Type: Madagascar, Forbes s.n. (K-Lindl., holo.!)
Madagascar – up to 1200 m – E
Aeranthes brachycentron Regel, Gartenfl.: 323, 1891

o **Aerides crispa** Lindl., Gen. Sp. Orch. Pl.: 239, 1833
BR 27: misc.116 (1841) / 28: t.55 (1842)
E.India – E
Aerides brookei Bateman, Bot. Reg. 27: misc.116, 1841
Aerides lindleyana Wight, Ic. Pl. Ind. Or.: 1677, 1851

o **Aerides maculosa** Lindl., Bot. Reg. 31: t.58, 1845
BR 31: t.58 (1845)
Type: cult. Rollissons (K-Lindl., icono.!)
India – E
Saccolabium speciosum Wight, Ic.: t.1674 & 1675, 1851

Aerides odorata Lour., Fl. Cochinch.: 525, 1790
BR 18: t.1485 (1832) / 29: misc.48 (1843) / 30: t.41 (1844)
India, Nepal, Sikkim, Burma, Thailand, China, Cambodia, Laos, Vietnam, Malaya, Sumatra, Java, Timor, Borneo, Celebes, Philippines – E
Epidendrum aerides Raeusch, Nom. ed.3: 265, 1797
Limodorum latifolium Thunb. ex Sw., Schrad. Neues J. Bot. 2: 234, 1799
Aerides latifolia (Thunb. ex Sw.) Sw., Schrad. Neues J. Bot. 1: 89, 1806
Epidendrum odoratum Poiret, Encyl. Suppl. 1: 385, 1811
Aeridium odoratum Salisb., Trans. Hort. Soc. 1: 295, 1812
o Aerides cornuta Roxb., Hort. Beng.: 63, 1814

Orxera cornuta Raf., Fl. Tellur. 4: 37, 1836
○ Aerides virens Lindl., Bot. Reg. 29: misc.48, 1843
Aerides suavolens Blume, Rumph. 4: 53 & 54, 1848 (non Roxb.1832)
Aerides suavissima Lindl., J. Hort. Soc. 4: 264, 1849
Aerides flavida Lindl., Paxt. Flow. Gard. 2: 101, 1851
Aerides reichenbachii Lindl. ex Rchb.f. Koch & Fint, Wochensschr. 1: 61, 1858
Aerides jucunda Rchb.f., Hamb. Gartenz. 16: 281, 1860
Aerides nobilis Warner, Sel. Orch.: t.11, 1862
Polytoma odorifera Lour. ex Gomes, Mem. Acad. Sc. Lisb. n.s.4, 1: 30, 1868
Aerides rohaniana Rchb.f., Gard. Chron. n.s.21: 206, 1884
Aerides ballentiniana Rchb.f., Gard. Chron. n.s.24: 198, 1885
Aerides ortgiesiana Rchb.f., Gard. Chron. n.s.23: 501, 1885
Aerides wilsoniana hort. ex Gard. Chron. n.s.23: 705, 1885
Aerides micholitzii Rolfe, Orch. Rev. 12: 181, 1904
Aerides duquesnei Regnier ex Constantin, Orch. Cult.: 33, 1911
Aerides odora Salisb., Trans. Hort. Soc. 1: 295, 1912

○ **Aganisia pulchella** Lindl., Bot. Reg. 25: misc.65, 1839
BR 25: misc.65 (1839) / 26: t.32 (1840)
Type: Guyana, Brotherton s.n. cult. Loddiges (K-Lindl., holo.!)
Trinidad, Venezuela, Guyana, Surinam, Brazil – E
Aganisia brachypoda Schltr., Beih. Bot. Centralbl. 42 (2): 126, 1925

Agrostophyllum planicaule (Wall. ex Lindl.) Rchb.f., Walp. Ann. 6: 909, 1864
BR 26: misc.4 (1840)
Type: Wallich s.n. cult. Sir Charles Lemon (K-Lindl., holo.!)
Nepal, Sikkim, Bhutan, N.E.India, Adaman Is., Burma, Thailand, Laos, Cambodia, Vietnam, Borneo – E
Eria planicaulis Wall. ex Lindl., Bot. Reg. 26: misc.4, 1840
Agrostophyllum khasiyanum Griffith, Calc. J. Nat. Hist. 4: 378 t.19, 1844
Appendicula hasseltii Wight, Ic. Pl. Ind. 5: t.1748, 1851 (non Blume)
Agrostophyllum hasseltii (Wight) Rchb.f., Walp. Ann. 6: 909, 1864 (non (Blume) J.J.Smith)
Agrostophyllum majus Seidenf. & Smitin., Orch. Thailand: 319 figs.241–242, 1960 (non Hook.f.)

Amblostoma armeniacum (Lindl.) Brieger ex Pabst, Bradea 2: 168, 1977
BR 22: t.1867 (1836) / 28: sub t.42 (1842) / 30: misc.82:8 (1844)
Type: Brazil, Miers s.n. (K-Lindl., holo!)
Brazil, Bolivia, Peru – E
○ Epidendrum armeniacum Lindl., Bot. Reg. 22: t.1867, 1836
Encyclia macrostachya Poeppig & Endl., Nov. Gen. Sp. Pl. 2, 11: t.114, 1838
Epidendrum alopecurum Schltr., Fedde Rep. 27: 58, 1929

Amblostoma tridactylum (Lindl.) Rchb.f., Walp. Ann. 6: 485, 1863
BR 24: misc.81 (1838) / 31: misc.70:62 (1845)
Type: Brazil, cult. S.Cannon (K-Lindl., holo!)
Brazil, Bolivia – E
Amblostoma cernua Scheidw., Allg. Gartenz. 6: 383, 1838
Epidendrum tridactylum Lindl., Bot. Reg. 24: misc.81, 1838
Epidendrum amblostoma Rchb.f., Bonpl. 2: 89, 1854
Sarcadenia gracilis hort. ex Rchb.f., Walp. Ann. 6: 486, 1863
Epidendrum citrinum Barb.Rodr., Gen. & Sp. Orch. Nov. 1: 53, 1877
Amblostoma densa Rchb.f., Xen. Orch. 3: 22, 1896
Amblostoma dusenii Kraenzlin, Svensk Vet. Akad. Handl. 46: 54, 1911

Angraecopsis parviflora (Thouars) Schltr., Beih. Bot. Centralbl. 36 (2): 140, 1918
BR 18: sub t.1522 (1832)
Madagascar, Mascarene Is. – E
Angraecum parviflorum Thouars, Orch. Iles Afr.: t.60, 1822
Oeceoclades parviflora (Thouars) Lindl., Gen. Sp. Orch. Pl.: 236, 1833
Listrostachys parviflora (Thouars) Moore, Baker Fl. Maur.: 355, 1877

Angraecum carpophorum Thouars, Orch. Iles Afr.: t.76, 1822
BR 18: sub t.1522 (1832)
Mascarene Is. – E

Angraecum caulescens Thouars, Orch. Iles Afr.: t.75, 1822
BR 18: sub t.1522 (1832)
Madagascar, Mascarene Is. – E
Aerobion caulescens (Thouars) Sprengel, Syst. Veg. 3: 717, 1826
Mystacidium caulescens (Thouars) Ridley, J. Linn. Soc. 21: 488, 1885
Epidorchis caulescens (Thouars) Kuntze, Rev. Gen. 2: 660, 1891

○ **Angraecum distichum** Lindl., Bot. Reg. 21: t.1781, 1835
BR 21: t.1781 (1835)
Type: Sierra Leone, cult. Loddiges (K-Lindl., holo!)
Guinea, Sierra Leone, Liberia, Ivory Coast, Ghana, Bénin, Nigeria, W.Cameroun – E
Mystacidium distichum (Lindl.) Pfitzer, Engl. & Pantl. Nat. Pflanzenf. 2 (6): 216, 1889

○ **Angraecum eburneum** Bory, Voy. 1: 359 t.19, 1804
BR 18: t.1522 (1832) / 33: sub t.19 (1847)
E.Africa, Madagascar, Bourbon, Mascarene Is., Comoros, Seychelles, Aldabra Is. – E, L, T
Limodorum eburneum (Bory) Willd., Sp. Pl. ed.4: 125, 1805
Angraecum superbum Thouars, Orch. Iles Afr.: t.62 & 64, 1822
Aerobion superbum (Thouars) Sprengel, Syst. Veg. 3: 718, 1826
Angraecum virens Lindl, Bot. Reg. 33: sub t.19, 1847
Angraecum brongniartianum Rchb.f. ex Linden, Pescatorea 1: t.16, 1854
Angraecum eburneum var. virens Hook., Bot. Mag. 86: t.5170, 1860
Angorchis eburnea (Bory) Kuntze, Rev. Gen. 2: 651, 1891
Angraecum comorense Kraenzlin, Engl. Bot. Jahrb. 17: 60, 1895
Angraecum gyriamae Rendle, J. Linn. Soc. 30: 388, 1895
Angraecum voeltzkowianum Kraenzlin, Engl. Bot. Jahrb. 36: 116, 1905

Angraecum mauritianum (Poiret) Frappier, Cat. Orch. Réunion: 15, 1891
BR 26: t.68 (1840)
Mascarene Is., Madagascar – 200–1400 m – E
Orchis mauritiana Poiret, Lam. Encycl. 4: 601, 1794
○ Angraecum gladiifolium Thouars, Orch. Iles Afr.: t.53, 1822
Aerobion gladiifolium (Thouars) Sprengel, Syst. Veg. 3: 716, 1826
Aeranthus gladiifolius (Thouars) Rchb.f., Walp. Ann. 6: 900, 1861
Angorchis gladiifolia (Thouars) Kuntze, Rev. Gen. 2: 651, 1891
Mystacidium mauritianum (Poiret) T.Durand & Schinz, Consp. Fl. Afr. 5: 53, 1895
Mystacidium gladiifolium (Thouars) Rolfe, Orch. Rev.: 47, 1904
Macroplectrum gladiifolium (Thouars) Pfitzer ex Finet, Bull. Soc. Bot. Fr. 54 (9): 27, 1907

Angraecum sesquipedale Thouars, Orch. Iles Afr.: t.66–67, 1822
BR 10: sub t.817 (1824)
Madagascar – E, L
Aeranthes sesquipedalis (Thouars) Lindl., Bot. Reg. 10: sub t.817, 1824
Macroplectrum sesquipedale (Thouars) Pfitzer, Engl. Nat. Pflanzenf. s.2, 6: 214, 1889
Angorchis sesquipedalis (Thouars) Kuntze, Rev. Gen. 2: 652, 1891
Mystacidium sesquipedale (Thouars) Rolfe, Orch. Rev. 23: 292, 1915

Anguloa brevilabris Rolfe, Orch. Rev. 23: 292, 1915
BR 32: t.41 & sub t.41:3 (1846)
Venezuela, Colombia, Ecuador – E
○ Anguloa ruckeri Lindl., Bot. Reg. 32: t.41, 1846
Anguloa sagittata Summerh., Kew Bull. 36: 102, 1951

○ **Anguloa clowesii** Lindl., Bot. Reg. 30: t.63 & misc.29, 1844
BR 30: t.63 & misc.29 (1844) / 32: sub t.41:2 (1846)
Type: Colombia, Linden s.n. cult. J. Clowes (K-Lindl., holo!)
Venezuela, Colombia – E
Anguloa clowesii var. flava Rchb.f., Walp. Ann. 1: 784, 1849
Anguloa clowesiana Williams, Orch. Grow. Man.: 33, 1852
Anguloa clowesii var. macrantha Williams, Orch. Grow. Man. ed.2: 54, 1862
Anguloa clowesii var. eburnea Veitch, Man. Orch. 9: 100, 1893

○ **Anguloa uniflora** Ruiz Lopez & Pavon, Syst. Veg. Fl. Peruv. & Chil. 1: 228, 1798
BR 30: t.60 (1844) / 32: sub t.41:1 (1846)
Colombia, Ecuador, Peru, Bolivia – E, T
Anguloa virginalis Lindl., Gard. Chron.: 392, 1851
Anguloa eburnea Williams, Orch. Grow. Man. ed.3: 67, 1868
Anguloa turneri Williams, Orch. Grow. Man. ed.6: 133, 1885

Anneliesia candida (Lindl.) Brieger & Lueckel, Die Orchidee 34 (4): 130, 1983
BR 24: misc.29 (1838) / 29: misc.110 (1843) / 31: sub t.8:4 (1845)
Type: cult. Loddiges (K-Lindl., holo!)
Brazil – E
Miltonia candida Lindl., Bot. Reg. 24: misc.29, 1838
Miltonia candida var. flavescens Hook., Bot. Mag. 66: t.3793, 1840
Miltonia candida var. grandiflora Lindl., Bot. Reg. 29: misc.110, 1843
Oncidium candidum Beer, Prakt. Stud. Orch.: 283, 1854 (nomen illeg.)
Oncidium candidum Rchb.f., Walp. Ann. 6: 763, 1863 (nomen illeg.)

Anochilus inversum (Thunb.) Rolfe, Fl. Cap. Orch.: 280, 1913
BR 9: sub t.700 (1823)
S.Africa – T
Ophrys inversa Thunb., Prodr.2, 1794
Pterygodium inversum (Thunb.) Sw., Kongl. Vet. Acad. Handl.: 218, 1800

○ **Anoectochilus setaceus** Blume, Bijdr.: 412 fig.135, 1825
BR 23: t.2010 (1837)
India, Malaysia, Sumatra, Java, Ambon – up to 1100m – T
Anoectochilus regalis Blume, Orch. Arch. Ind.:46, 1858
Anoectochilus setaceus var. inornatus Hook., Bot. Mag. 86: t.5208, 1860

○ **Ansellia africana** Lindl., Bot. Reg. 30: sub t.12, 1844
BR 30: sub t.12 (1844) / 32: t.30 (1846)
Type: Fernando Po, Ansell s.n. cult. Loddiges (K-Lindl., holo!)
Trop.S.E.Centr.Africa, Liberia, Ivory Coast, Ghana, Nigeria, F.Po., E.Cameroun, Gabon, Zaire, Angola, Uganda, Kenya, Zimbabwe, Mozambique, S.Africa – up to 2200 m – E, L
Ansellia gigantea Rchb.f., Linnaea 20: 673, 1874
Ansellia africana var. nilotica Baker, Trans. Linn. Soc. 29: 154, 1875
Ansellia confusa N.E.Br., Lindenia 2: 64, 1886
Ansellia congoensis Rodigas, Lindenia 2: 64, 1886
Ansellia nilotica (Baker) N.E.Br., Lindenia 2: 64, 1886
Ansellia humilis Bulliard, Cat.: 3, 1891
Ansellia gigantea var. nilotica (Baker) Summerh., Kew Bull.: 462, 1937

Apatostelis ciliaris (Lindl.) Garay, Bot. Mus. Leafl. Harv. Univ. 27: 188, 1979
BR 28: misc.100 (1842)
Mexico, Belize, Guatemala, Honduras, Nicaragua, Costa Rica – up to 1100 m – E
Stelis ciliaris Lindl., Comp. Bot. Mag. 2: 353, 1837
Stelis atropurpurea Hook., Bot. Mag. 69: t.3975, 1842
Stelis micrantha var. atropurpurea (Hook.) Josst., Orchid.: 38, 1851
Stelis confusa Schltr., Beih. Bot. Centralbl. 36 (2): 386, 1918
Stelis jimenezii Schltr., Beih. Bot. Centralbl. 36 (2): 389, 1918

Arpophyllum giganteum Hartweg ex Lindl., Ann. Mag. Nat. Hist. 4: 384, 1840
BR 25: misc.16 (1839)
Mexico, Guatemala, El Salvador, Honduras, Costa Rica, Nicaragua, Jamaica, Colombia, Venezuela – 1400–1500 m – E
Arpophyllum spicatum Llave & Lex., Nov. Veg. Descr. 2, Orch. Opusc.: 20, 1825
Arpophyllum cardinale Lindl. ex Rchb.f., Bonpl. 2: 282, 1854
Arpophyllum squarrosum Lubbers., Cat. Pl. Rar. San Donato: 15, 1880
Arpophyllum jamaicense Schltr., Fedde Rep. 16: 443, 1920
Arpophyllum stenostachyum Schltr., Fedde Rep. 19: 32, 1923

Arundina graminifolia (D.Don) Hochr., Bull. N.Y. Bot. Gard. 6: 270, 1910
BR 27: misc.3 (1841) / 28: t.38 & misc.26 (1842)
Sri Lanka, India, Nepal, Burma, China, Taiwan, Indochina, Ryukyu Is., Malaya, Borneo, Indonesia to Tahiti – T
Arundina chinensis Blume, Bijdr. 8: 402, 1825
Arundina speciosa Blume, Bijdr. 8: 401, 1825
Bletia graminifolia D.Don., Prod. Fl. Nepal.: 29, 1825
Arundina bambusifolia Lindl., Gen. Sp. Orch. Pl.: 125, 1831
Arundina minor Lindl., Gen. Sp. Orch. Pl.: 125, 1831
Cymbidium bambusifolium (Lindl.) Roxb., Fl. Ind. 3: 460, 1832
○ Arundina densa Lindl., Bot. Reg. 28: t.38 & misc.26, 1842
Cymbidium meyenii Schauer, Nov. Act. Nat. Cur. 14: suppl.1: 433, 1843
Arundina affinis Griffith, Notul. 3: 330, 1851
Arundina meyenii (Schauer) Rchb.f., Linnaea 25: 227, 1852
Arundina philippii Rchb.f., Linnaea 25: 227, 1852
Arundina pulchella Teijsm. & Binnend., Nederl. Kruidk. Arch. 3: 400, 1855
Cymbidium speciosum Herb. Reinw. mss. nom. nud. ex Lindl., J. Linn. Soc. 3: 23, 1859
Arundina pulchra Miq., J. Bot. Neerl. 1: 30, 1861
Corymbis angustifolia Miq., Fl. Ind. Bat. Suppl.: 617, 1861
Arundina densiflora Hook.f., Fl. Brit. Ind. 5: 857, 1890
Arundina revoluta Hook.f., Fl. Br. India 5: 858, 1890
Limodorum graminifolium (D.Don) Ham. ex Hook.f., Fl. Brit. Ind. 5: 857, 1890
Arundina sanderiana Kraenzlin, Engl. Bot. Jahrb. 17: 483, 1893
Arundina philippii var. malayana Ridley, Mat. Fl. Mal. 1: 124, 1907
Arundina stenopetala Gagnepain, Bull. Soc. Bot. Fr. 79: 32, 1932
Donacopis laotica Gagnepain, Bull. Mus. Hist. Nat. Paris s.2, 4: 593 p.pt., 1932

Ascocentrum curvifolium (Lindl.) Schltr., Fedde Rep. Beih. 1: 975, 1913
BR 33: t.58 (1847)
N.E.India, Nepal, Assam, Burma, Thailand, Cambodia, Laos, Vietnam – E
Saccolabium curvifolium Lindl., Gen. Sp. Orch. Pl.: 222, 1833
Saccolabium miniatum Hook., Bot. Mag. 88: t.5326, 1862 (non Lindl.)
Gastrochilus curvifolius (Lindl.) Kuntze, Rev. Gen. 2: 661, 1891

Ascocentrum miniatum (Lindl.) Schltr., Fedde Rep. Beih. 1: 975, 1914
BR 33: t.58 & sub t.26 (1847)
Type: Java, Rucker s.n. (K-Lindl., holo!)
Thailand, Laos, Vietnam, Malaya, Java, Philippines – E
○ Saccolabium miniatum Lindl., Bot. Reg. 33: sub t.26, 1847
Gastrochilus miniatus (Lindl.) Kuntze, Rev. Gen. 2: 661, 1891
Saccolabium curvifolium Gagnepain, Fl. Gén. Indoch.: 500, 1934 (non Lindl.)

Aspasia epidendroides Lindl., Hook. J. Bot. 1: 6, 1843
BR 25: sub t.1907 (1839) / 26: misc.92 (1840) / 28: misc.18 (1842)
Guatemala, Honduras, Nicaragua, El Salvador, Costa Rica, Panama, Colombia – up to 1000 m – E, L
Aspasia fragrans Klotzsch, Ind. Sem. Hort. Bot. Berol.: 12, 1852
Odontoglossum aspasia Rchb.f., Walp. Ann. 6: 851, 1861

○ **Aspasia variegata** Lindl., Bot. Reg. 22: t.1907, 1836
BR 22: t. & sub t.1907 (1836) / 30: misc.49 (1844)
Type: cult. J.Knight s.n. (K-Lindl., lecto!)
Venezuela, Trinidad, Guyana, Surinam, Brazil – E, L
Aspasia lunata Lindl., Bot. Reg. 22: sub t.1907, 1836
Aspasia interrupta Hoffsgg., Verz. Orch.: 44, 1844
Aspasia liturata Link ex Rchb.f., Xen. Orch. 1: 83, 1855
Odontoglossum variegatum (Lindl.) Rchb.f., Walp. Ann. 6: 851, 1864

Barbosella gardneri (Lindl.) Schltr., Fedde Rep. 15: 261, 1918
BR 28: misc.p.83:108 (1842)
Type: Gardner, 5865 (K-Lindl., holo!)
Brazil – E
Pleurothallis gardneri Lindl., Bot. Reg. 28: misc.p.83:108, 1842
Restrepia gardneri (Lindl.) Benth., Benth. & Hook. Gen. Pl. 3: 491, 1883
Humboldtia gardneri (Lindl.) Kuntze, Rev. Gen. 2: 667, 1891

Barbrodria miersii (Lindl.) Luer, Selbyana 5: 386, 1981
BR 28: misc.p.84:113 (1842)
Type: Brazil, Miers s.n. (K-Lindl., holo!)
Brazil – E
Pleurothallis miersii Lindl., Bot. Reg. 28: misc.p.84:113, 1842
Restrepia miersii (Lindl.) Rchb.f., Wawra Bot. Ergeb. Reise Maxim. Bras.: 150 fig.5 t.104, 1866
Humboldtia miersii (Lindl.) Kuntze, Rev. Gen. 2: 668, 1891

Barkeria elegans Knowles & Westc., Fl. Cabin. 2: 7 t.49, 1838
BR 24: sub t.20 (1838) / 26: misc.57 (1840)
Mexico, Guatemala – E
Epidendrum elegans (Knowles & Westc.) Rchb.f., Walp. Ann. 6: 374, 1862

Barkeria lindleyana Bateman ex Lindl., Bot. Reg. 28: misc.5, 1842
BR 28: misc.5 (1842)
Type: Costa Rica, Skinner s.n. cult. Bateman (K-Lindl., holo!)
Mexico, Guatemala, Honduras, El Salvador, Costa Rica – up to 3500 m – E, L
Epidendrum lindleyanum (Bateman ex Lindl.) Rchb.f., Walp. Ann. 6: 375, 1862

Barkeria skinneri (Bateman ex Lindl.) A.Rich. & Galeotti, Comp. Rend. Acad. Sc. Paris 18: 506, 1844
BR 22: t.1881 (1836) / 26: misc.81 (1840) / 28: sub t.25 (1842) / 30: misc.24:1 (1844)
Type: Guatemala, Skinner s.n. cult. Bateman (K-Lindl., holo!)
Mexico, Guatemala – up to 1900 m – E
○ Epidendrum skinneri Bateman ex Lindl., Bot. Reg. 22: t.1881, 1836
Dothiolophis purpurea Raf., Fl. Tellur. 4: 39, 1838
Barkeria skinneri (Bateman ex Lindl.) Paxton, Mag. Bot. 16: 15, 1849
Barkeria skinneri var. major (Batem. ex Lindl.) Paxton, Mag. Bot. 16: 15, 1849
Epidendrum fuchsii Regel, Ind. Sem. Hort. Turic. 4, 1850
Barkeria skinneri var. superba Warner, Sel. Orch. Pl. 1: t.38, 1862
Epidendrum skinneri var. superbum Warner, Sel. Orch. Pl. 1: t.38, 1862

Barkeria spectabilis Bateman ex Lindl., Bot. Reg. 28: misc.45, 1842
BR 28: misc.45 (1842)
Type: Guatemala, Skinner s.n. (K-Lindl., holo!)
Mexico, Guatemala, El Salvador, Nicaragua – up to 3500 m – E
Epidendrum spectabile (Bateman ex Lindl.) Rchb.f., Walp. Ann. 6: 375, 1862

Barlia robertiana (Loisel.) Greuter, Boissiera 13: 192, 1967
BR 5: t.357 (1819)
Mediterranean Region, Canary Is. – T
○ Orchis longibracteata Biv., Sc. Plant. Cent. 1: 57, 1806
Orchis robertiana Loisel., Fl. Gall. 2: 606, 1806
Aceras longibracteata (Biv.) Rchb.f., Ic. Fl. Germ. 13 & 14: 3, 1850
Barlia longibracteata (Biv.) Parl., Fl. Ital. 3: 447, 1858
Himantoglossum longibracteatum (Biv.) Schltr., Die Orchideen: 52, 1914

Bartholina burmanniana (L.) Ker Gawler, J. Sc. & Arts 4: 204 t.6 fig.2, 1818
BR 20: t.1653 (1834)
S.Africa – T
Orchis burmanniana L., Sp. Pl. ed.2: 1334, 1763
Orchis pectinata Thunb., Fl. Cap. ed. Schult.: 5, 1794
○ Bartholina pectinata (Thunb.) R.Br., Hort. Kew. ed.2, 5: 194, 1813
Arethusa ciliaris L.f., Suppl. Syst. Veg.: 405, 1871
Bartholina lindleyana Rchb.f., Otia Bot. Hamb. 2 (1): 119, 1881

o **Batemannia colleyi** Lindl., Bot. Reg. 20: t.1714, 1834
BR 20: t.1714 (1834)
Type: Guyana, Colley s.n. cult. Bateman (K-Lindl., holo!)
Brazil, Guyana, Venezuela, Colombia, Peru, Bolivia – E
Lycaste colleyi hort. ex Planchon, Hart. Donat. Orch.: 16, 1858
Maxillaria colleyi hort. ex Planchon, Hart. Donat. Orch.: 16, 1858
Petronia regia Barb.Rodr., Gen. & Sp. Orch. Nov. 1: 107, 1877
Batemannia petronia Barb. Rodr., Vellosia 1: 131, 1891
Batemannia yauaperyensis Barb. Rodr., Vellosia 1: 131, 1891
Batemannia peruviana Rolfe, Kew Bull.: 193, 1895

Beadlea bicolor (Ker Gawler) Garay, Bot. Mus. Leafl. Harv. Univ. 28 (4): 299, 1980
BR 10: t.794 & sub t.823:13 (1824)
Type: Trinidad, G.Don s.n. (K-Lindl., holo!)
Argentina, Brazil, Venezuela, Colombia, W.Indies – T
o Neottia bicolor Ker Gawler, Bot. Reg. 10: t.794, 1824
Spiranthes bicolor (Ker Gawler) Lindl., Bot. Reg. 10: sub t.823:13, 1824
Cyclopogon pubescens (Barb.Rodr.) Barb.Rodr., Gen. & Sp. Orch. Nov. 3, 1877
Spiranthes pubescens Barb.Rodr., Gen. & Sp. Orch. Nov. 1: 185, 1877
Gyrostachys bicolor (Ker Gawler) Kuntze, Rev. Gen. 2: 664, 1891
Cyclopogon bicolor (Ker Gawler) Schltr., Fedde Rep. Beih. 6: 52, 1919

Beadlea elata (Sw.) Small ex Britton, Brooklyn Bot. Gard. Mem. 1: 38, 1918
BR 10: sub t.823:11 (1824)
Florida, W.Indies, Mexico, C.America except Panama, America – T
Satyrium elatum Sw., Prodr. Veg. Ind. Occ.: 119, 1788
Neottia minor Jacq., Coll. 3: 172, 1789
Neottia elata (Sw.) Sw., Kongl. Vet. Acad. Handl. 21: 226, 1800
Spiranthes elata (Sw.) A.Rich., Mém. Mus. Paris 4: 59, 1818
Cyclopogon ovalifolium Presl., Rel. Haenk. 1: 93 t.13, 1827
Spiranthes lineata Lindl., Gen. Sp. Orch. Pl.: 471, 1840
Spiranthes preslii Lindl., Gen. Sp. Orch. Pl.: 467, 1840
Spiranthes comosa Rchb.f., Linnaea 18: 408, 1844
Sarcoglottis elata (Sw.) P.N.Don, Hort. Cantabr. ed.13: 590, 1845
Gyrostachys ovalifolia (Presl.) Kuntze, Rev. Gen. 2: 664, 1891
Spiranthes elata var. ovata Cogn., Mart. Fl. Bras. 3 (4): 192, 1895
Spiranthes lineata Stuckert, An. Mus. Nac. Buenos Aires 9: 12, 1903 (non Lindl.)
Sauroglossum richardii Ames, Orchid. 1: 44, 1905
Spiranthes variegata Kraenzlin, Engl. Bot. Jahrb. 37: 392, 1906
Cyclopogon elatus (Sw.) Schltr., Fedde Rep. Beih. 6: 53, 1919
Spiranthes pamii Braid, Kew Bull.: 204, 1924
Cyclopogon miradorensis Schltr., Fedde Rep. 21: 332, 1925

Beadlea lindleyana (Link, Klotzsch & Otto) Garay & Dunsterv., Orch. Venez.: 24, 1979
BR 27: misc.38 (1841)
Venezuela – T
Spiranthes lindleyana Link, Klotzsch & Otto, Ic. Pl. Berol.: t.11, 1840
Gyrostachys lindleyana (Link, Klotzsch & Otto) Kuntze, Rev. Gen. 2: 663, 1891
Cyclopogon lindleyana (Link, Klotzsch & Otto) Schltr., Beih. Bot. Centralbl. 37 (2): 389, 1920

Benthamia latifolia (Thouars) A.Rich., Mém. Soc. Hist. Nat. Paris 4: 43 t.7, 1828
BR 18: sub t.1499:3 (1832)
Mauritius, Bourbon – T
Satyrium latifolium Thouars, Orch. Iles Austr. Afr.: t.10, 1822
Habenaria chlorantha Sprengel, Syst. Veg. 3: 691, 1826
Herminium latifolium (Thouars) Lindl., Bot. Reg. 18: sub t.1499:3, 1832
Peristylus latifolius (Thouars) Lindl., Gen. Sp. Orch. Pl.: 297, 1835
Habenaria latifolium (Thouars) C.Cordemoy, Fl. de Réunion: 260, 1895

Benthamia spiralis (Thouars) A.Rich, Orch. Iles France & Bourbon: 44, 1828
BR 10: sub t.823:7 (1824)
Madagascar, Réunion, Mauritius – 0–1500 m – T
Satyrium spirale Thouars, Orch. Iles Afr.: t.9, 1822
Spiranthes africana Lindl., Bot. Reg. 10: sub t.823:7, 1824
Habenaria spiralis (Thouars) A.Rich., Orch. Iles France & Bourbon: 22, 1828
Herminium spirale (Thouars) Rchb.f, Bonpl. 3: 213, 1855
Peristylus spiralis (Thouars) S.Moore, Baker Fl. Maur.: 335, 1877
Habenaria minutiflora Ridley, J. Linn. Soc. 21: 503, 1885
Habenaria dissimulata Schltr., Beih. Bot. Centralbl. 33: 404, 1915
Benthamia minutiflora (Ridley) Schltr., Fedde Rep. 33: 29, 1925

Bifrenaria atropurpurea Lindl., Gen. Sp. Orch. Pl.: 152, 1832
BR 29: misc.67:1 (1843)
Brazil – E
Maxillaria atropurpurea (Lindl.) Lodd., Bot. Cab. 19: t.1877, 1832
Bifrenaria caparaoense Brade, Orquidea Rio de Janeiro 2: 16, 1943

Bifrenaria aureo-fulva (Hook.) Lindl., Bot. Reg. 29: misc.67:4, 1843
BR 26: misc.43 (1840) / 29: misc.67:4 (1843)
Brazil – E, T
Epidendrum secundum Vell., Enum. Syst. Pl. 29, 1760 (non Jacq.)
Maxillaria aureo-fulva Hook., Bot. Mag. 65: t.3629, 1838
Maxillaria stenopetala Knowles & Westc., Fl. Cabin. 2: 112, 1838
Stenocoryne aureo-fulva (Hook.) Kraenzlin, Xen. Orch. 3: 142, 1896
Stenocoryne secunda (Vell.) Hoehne, Arq. Bot. Est. S.Paulo 2: 14, 1944
Bifrenaria secunda (Vell.) Pabst, Orquidea 29: 165, 1967

Bifrenaria harrisoniae (Hook.) Rchb.f., Xen. Orch. 1: 61, 1854
BR 11: t.897 (1825) / 18: sub t.1549:27, 38 (1832) / 27: misc.68 (1841)
Brazil (Rio de Janeiro) – E
Dendrobium harrisoniae Hook., Exot. Fl.: t.120, 1825
o Maxillaria harrisoniae (Hook.) Lindl., Bot. Reg. 11: t.897, 1825
Colax harrisoniae (Hook.) Sprengel, Syst. Veget. 3: 727, 1826
Maxillaria spathacea Lindl., Gen. Sp. Orch. Pl.: 151, 1832
Colax grandiflorus Raf., Tellur. 2: 85, 1836
Stanhopea harrisoniae (Hook.) P.N.Don, Hort. Cantab. ed.13 : 607, 1845
Lycaste harrisoniae (Hook.) G.Don ex Loud., Encycl. Pl. Suppl.2: 1468, 1855
Maxillaria pubigera Klotzsch, Allg. Gartenz. 23: 106, 1855
Maxillaria barringtoniae hort. ex Rchb.f., Xen. Orch. 1: 224, 1858 (non Lodd.)
Stenocoryne harrisoniae (Hook.) Kraenzlin, Xen. Orch. 3: 142, 1896

Bifrenaria inodora Lindl., Bot. Reg. 29: misc.63, 1843
BR 29: misc.63 & misc.67:2 (1843)
Type: Rio de Janeiro, cult. Sir C.Lemon (K-Lindl., holo!)
Brazil – E
Epidendrum calcaratum Vell., Fl. Flum. Ic. 9: t.8, 1827
Bifrenaria fragrans Barb.Rodr., Gen. & Sp. Orch. Nov. 2: 214, 1882
Bifrenaria aurantiaca Williams, Orch. Alb. 9: t.386, 1889
Stenocoryne inodora (Lindl.) Kraenzlin, Xen. Orch. 3: 142, 1896
Bifrenaria fuerstenbergiana Schltr., Orchis 1: 25, 1906

Bifrenaria longicornis Lindl., Bot. Reg. 24: misc.177, 1838
BR 24: misc.177 (1838) / 29: misc.68 (1843)
Type: Guyana, cult. Loddiges (K-Lindl., holo!)
Brazil, Guyana, Colombia, Peru – E
Stenocoryne longicornis (Lindl.) Lindl., Bot. Reg. 29: misc.68, 1843
Bifrenaria sabulosa Barb.Rodr., Gen. & Sp. Orch. Nov. 1: 111, 1877
Rudolfiella sabulosa (Barb.Rodr.) Hoehne, Fl. Bras. Fasc.10: 52, 1953
Stenocoryne sabulosa (Barb.Rodr.) Hoehne, Fl. Bras. Fasc.10: 52, 1953

Bifrenaria parvula (Hook.) Rchb.f., Walp. Ann. 6: 547, 1863
BR 18: sub t.1549:25 (1832) / 27: misc.142 (1841)
Brazil – E
Colax parvulus (Hook.) Sprengel, Syst. Veg.: 307, 1827
Maxillaria parvula Hook., Exot. Fl. 3: t.217, 1827
Maxillaria purpurascens Knowles & Westc., Phytolog.: 8, 1855

Bifrenaria racemosa (Hook.) Lindl., Bot. Reg. 29: misc.67:3, 1843
BR 18: sub t.1549: 28 (1832) / 19: t.1566 (1833) / 29: misc.67:3 (1843)
Brazil (Rio de Janeiro) – E
o Maxillaria racemosa Hook., Bot. Mag. 54: t.2789, 1827
Colax racemosus (Hook.) Sprengel f., Tent.: 29, 1828
Adipe fulva Raf., Fl. Tellur. 2: 101, 1836
Adipe racemosa (Hook.) Raf., Fl. Tellur. 2: 101, 1836
Stenocoryne racemosa (Hook.) Kraenzlin, Xen. Orch. 3: 142, 1896

Bifrenaria tetragona (Lindl.) Schltr., Die Orchideen: 413, 1915
BR 17: t.1428 (1831) / 18: sub t.1549 (1832) / 29: misc.64 (1843)
Type: Brazil, cult. Hort. Soc. Gard. (K-Lindl., icono!)
Brazil – E
o Maxillaria tetragona Lindl., Bot. Reg. 17: t.1428, 1831
Lycaste tetragona (Lindl.) Lindl., Bot. Reg. 29: misc.64, 1843

o **Bifrenaria vitellina** Lindl., Bot. Reg. 29: misc.67: 5, 1843
BR 24: misc.116 (1838) / 25: t.12 (1839) / 27: misc.141 (1841) / 29: misc.67:5 (1843)
Type: Brazil, cult. Loddiges (K-Lindl., holo!)
Brazil – E
o Maxillaria vitellina Lindl., Bot. Reg. 24: misc.116, 1838
Maxillaria barbata Knowl. & Westc., Phytolog.: 7, 1855
Bifrenaria harrisoniae var. vitellina (Lindl.) Stein, Orchideenb.: 85, 1892
Stenocoryne vitellina (Lindl.) Kraenzlin, Xen. Orch. 3: 142, 1896

o **Bletia gracilis** Lodd., Bot. Cab. 20: t.1977, 1833
BR 20: t.1681 (1834)
Mexico, Guatemala – T

Bletia purpurea (Lam.) DC., Mém. Soc. Phys. Hist. Nat. Genève 9: 97, 1841
BR 17: t.1401 (1831) / 24: misc.35 & misc.73 (1838)
Mexico to Panama, Venezuela, Colombia, Ecuador, W.Indies – T

Limodorum altum Jacq., Ait. Hort. Kew. 3: 301, 1789
Limodorum tuberosum Jacq., Coll. 4: 108, 1790 (non L.)
Limodorum purpureum Lam., Encycl. Meth. Bot. 3: 515, 1791
Cymbidium floridum Salisb., Prodr. 9, 1796
Limodorum floridum Salisb., Prodr. 9, 1796
Limodorum verecundum Salisb., Prodr. 9, 1796
Cymbidium verecundum (Salisb.) Sw., Nov. Act. Upsal. 6: 75, 1799
Limodorum trifidum Michaux, Fl. Bor. Amer. 2: 159, 1803
Gyas florida (Salisb.) Salisb., Trans. Hort. Soc. 1: 299, 1812
o Bletia florida (Salisb.) R.Br., Ait. Hort. Kew. ed.2, 5: 206, 1813
Bletia verecunda (Salisb.) R.Br., Ait. Hort. Kew. ed.2, 5: 206, 1813
Bletia pallida Lodd., Bot. Cab. 7: t.629, 1822
Thiebautia nervosa Colla, Mém. Soc. Linn. Paris 3: 161 t.4, 1824
Bletia acutipetala Hook., Bot. Mag. 60: t.3217, 1833
Bletia shepherdii Hook., Bot. Mag. 61: t.3319, 1834
Bletia havanensis Lindl., Bot. Reg. 24: misc.35, 1838
Bletia alta (Jacq.) Hitchc., Ann. Rep. Missouri Bot. Gard. 4: 132, 1893
Bletia tuberosa (L.) Ames, Proc. Biol. Soc. Wash. 45: 1, 1932

o **Bletia reflexa** Lindl., Bot. Reg. 21: t.1760, 1835
BR 21: t.1760 (1835)
Type: Mexico, cult. Loddiges (K-Lindl., holo!)
Mexico – T

Bletia secunda Lindl., Bot. Reg. 26: misc.120, 1840
BR 26: misc.120 (1840)
Type: Mexico, cult. Loddiges (K-Lindl., holo!)
Mexico – T
Eulophia dilatata Lindl., Ann. Mag. Nat. Hist. 10: 184, 1842

Bletilla striata (Thunb.) Rchb.f., Bot. Zeit. 36: 75, 1878
BR 33: t.60 (1847)
E.Tibet, Japan, Okinawa – T
Limodorum striatum Thunb., Fl. Jap.: 28, 1784
Epidendrum tuberosum Lour., Fl. Cochinch.: 523, 1790
Epidendrum striatum (Thunb.) Thunb., Trans. Linn. Soc. 2: 327, 1794
Cymbidium striatum (Thunb.) Sw., Nov. Act. Upsal. 6: 77, 1797
Cymbidium hyacinthinum Smith, Exot. Bot. 1: 117 t.60, 1804
Limodorum hyacinthinum (Smith) Don, Hort. Cantab. ed.4: 201, 1807
Gyas humilis Salisb., Trans. Hort. Soc. 1: 300, 1812
Bletia hyacinthina (Smith) R.Br., Ait. Hort. Kew. ed.2, 5: 206, 1813
Jimensia nervosa Raf., Fl. Tellur. 4: 38, 1838
Calanthe gebina Lodd., Orchid. 43: no.1846, 1844 (nomen)
o Bletia gebina Lindl., J. Hort. Soc. 2: 307, 1847
Bletilla gebina (Lindl.) Rchb.f., Fl. Serres s.1, 8: 246, 1853
Bletilla hyacinthina (Smith) Rchb.f., Bot. Zeit. 36: 75, 1878
Bletilla striata var. gebina (Lindl.) Rchb.f., Bot. Zeit. 36: 75, 1878
Bletia striata (Thunb.) Druce, Rep. Bot. Exch. Club Brit. Is. 4: 609, 1917
Bletilla striata var. albomarginata Makino, J. Jap. Bot. 6: 31, 1929
Bletilla striata f. gebina (Lindl.) Ohwi, Bull. Nat. Sc. Mus. Tokyo 33: 70, 1953
Jimensia striata (Thunb.) Garay & Schultes, Bot. Mus. Leafl. Harv. Univ. 18: 184, 1958

Bollea violacea (Lindl.) Rchb.f., Bot. Zeit. 10: 668, 1852
BR 23: sub t.1991 (1837) / 25: misc.17 (1839)
Type: Guyana, Schomburgk s.n. (K-Lindl., holo!)
Brazil, Guyana, Surinam – E
Huntleya sessiliflora Bateman ex Lindl., Bot. Reg. 23: sub t.1991, 1837
Huntleya violacea Lindl., Bot. Reg. 25: misc.17, 1839
Bollea guianensis Klotzsch, Schomb. Fauna & Fl. Brit. Guian.: 1206, 1848
Huntleya tyrianthina hort. ex Rchb.f., Xen. Orch. 1: 187, 1856
Zygopetalum violaceum (Lindl.) Rchb.f., Walp. Ann. 6: 650, 1863

Brachystele bracteosa (Lindl.) Schltr., Beih. Bot. Centralbl. 37 (2): 370, 1920
BR 23: t.1934 (1837)
Type: Brazil, cult. Loddiges (K-Lindl.)
Brazil – T
Spiranthes bracteosa Lindl., Bot. Reg. 23: t.1934, 1837
Gyrostachys bracteosa (Lindl.) Kuntze, Rev. Gen. 2: 664, 1891

Brachystele unilateralis (Poiret) Schltr., Beih. Bot. Centralbl. 37 (2): 370, 1920
BR 10: sub t.823:14 (1824) / 24: misc.119 (1838)
Chile – T
Ophrys unilateralis Poiret, Lam. Encycl. 4: 569, 1798
Spiranthes diuretica (Willd.) Lindl., Quart. J. R. Inst. n.s.1: 54, 1827

Braemia vittata (Lindl.) Jenny, Die Orchidee 36 (1): 36, 1985
BR 27: t.69 & misc.100 (1841)
Type: Schomburgk s.n. cult. Loddiges (K-Lindl., holo!)
Brazil, Guyanas, Venezuela, Colombia, Peru, Amazonas – E
o Houlletia vittata Lindl., Bot. Reg. 27: t.69 & misc.100, 1841
Polycycnis vittata (Lindl.) Rchb.f., Walp. Ann. 6: 618, 1863

Brassavola cucullata (L.) R.Br., Aiton Hort. Kew. ed.2, 5: 216, 1813
BR 32: sub t.53 (1846)
Mexico, Guatemala, El Salvador, Honduras, Venezuela, Trinidad, W.Indies – E
Epidendrum cucullatum L., Sp. Pl. ed.2: 1350, 1763
Cymbidium cucullatum (L.) Sw., Nov. Act. Upsal. 6: 73, 1799
Brassavola cuspidata Hook., Bot. Mag. 66: t.3722, 1839
Brassavola appendiculata A.Rich. & Galeotti, Ann. Sc. Nat. s.3, 3: 23, 1845
Brassavola odoratissima Regel, Gartenfl.: 325 t.33, 1852
Bletia cucullata (L.) Rchb.f., Walp. Ann. 6: 433, 1862
Brassavola cucullata var. elegans Schltr., Orchis 13: 46, 1919

o **Brassavola martiana** Lindl., Bot. Reg. 22: sub t.1914, 1836
BR 22: sub t.1914 (1836) / 24: misc.67 (1838) / 25: t.5 (1839) / 32: sub t.53 (1846)
Type: Herb. Martius s.n. (K-Lindl., holo!)
N.Brazil, Guyanas, Venezuela, Amazonas – E
Brassavola amazonica Poeppig & Endl., Nov. Gen. Sp. Pl. 2: 3, 1838
Brassavola angustata Lindl., Bot. Reg. 24: misc.67, 1838
Brassavola surinamensis Focke, Tijdr. Nat. Weten. 9: 17, 1852
Bletia amazonica (Poeppig & Endl.) Rchb.f., Walp. Ann. 6: 434, 1862
Bletia angustata (Lindl.) Rchb.f., Walp. Ann. 6: 434, 1862
Bletia martiana (Lindl.) Rchb.f., Walp. Ann. 6: 434, 1862
Brassavola multiflora Schltr., Orchis 13: 58, 1919

o **Brassavola nodosa** (L.) Lindl., Gen. Sp. Orch. Pl.: 114, 1831
BR 18: t.1465 (1832) / 25: misc.14 & misc.p.16 (1839) / 26: t.39 & misc.24 (1840) / 32: sub t.53 (1846)
W.Indies, Mexico to Panama, Venezuela to Peru – E
Epidendrum nodosum L., Sp. Pl.: 953, 1753
Cymbidium nodosum (L.) Sw., Nov. Act. Upsal. 6: 73, 1799
Brassavola grandiflora Lindl., Bot. Reg. 25: misc.14, 1839
o Brassavola venosa Lindl., Bot. Reg. 26: t.39, 1840
Brassavola rhopalorrachis Rchb.f., Bot. Zeit. 10: 833, 1852
Bletia nodosa (L.) Rchb.f., Walp. Ann. 6: 437, 1862
Bletia venosa (Lindl.) Rchb.f., Walp. Ann. 6: 438, 1862
Brassavola stricta hort., Gard. Chron. n.s.23: 505, 1885
Brassavola nodosa var. rhopalorrhachis (Rchb.f.) Schltr., Orchis 13: 77, 1919
Brassavola scaposa Schltr., Orchis 13: 77, 1919
Brassavola nodosa var. grandiflora (Lindl.) H.Jones, Amer. Orch. Soc. Bull. 41 (6): 595, 1972

Brassavola retusa Lindl., Bot. Reg. 33: sub t.20, 1847
BR 33: sub t.20 (1847)
Type: Venezuela, ex Gardners' Chronicle s.n. (K-Lindl., holo!)
Venezuela, Peru – E

Brassavola subulifolia Lindl., Gen. Sp. Orch. Pl.: 115, 1831
BR 22: t.1913 text sub t.1914 (1836) / 32: sub t.53 (1846)
Jamaica – E
o Brassavola cordata Lindl., Bot. Reg. 22: t.1913 text sub t.1914, 1836
Brassavola sloanei Griseb., Fl. Brit. W. Ind.: 626, 1864

Brassavola tuberculata Hook., Bot. Mag. 56: t.2878, 1829
BR 18: t.1561 (1833)
Brazil, Bolivia – E, L
o Brassavola perrinii Lindl., Bot. Reg. 18: t.1561, 1832
Brassavola fragrans Lemaire, Jard. Fleur. 3: misc.78 & 88, 1853
Bletia perrinii (Lindl.) Rchb.f., Walp. Ann. 6: 436, 1862
Bletia tuberculata (Hook.) Rchb.f., Walp. Ann. 6: 435, 1862
Brassavola gibbsiana Williams, Orch. Grow. Man. ed.5, 1877

Brassia bidens Lindl., Bot. Reg. 30: misc.13:6, 1844
BR 30: misc.13:6 (1844)
Type: Harrison s.n. (K-Lindl., holo!)
Venezuela, Guyana, Brazil, Peru – E, T
Brassia lanceana var. viridiflora Hook., Bot. Mag. 66: t.3794, 1840
Oncidium bidens (Lindl.) Rchb.f., Walp. Ann. 6: 769, 1863

o **Brassia caudata** (L.) Lindl., Bot. Reg. 10: t.832, 1824
BR 10: t.832 (1824) / 30: misc.13:5 (1844)
USA (Florida), Jamaica, W.Indies, Mexico to Panama, Venezuela to Bolivia – up to 1200m – E
Epidendrum caudatum L., Sp. Pl. ed.2: 1349, 1763
Malaxis caudata (L.) Willd., Sp. Pl. ed.4: 93, 1805
Oncidium caudatum (L.) Rchb.f., Walp. Ann. 6: 766, 1863
Brassia caudata var. hieroglyphica Rchb.f., Ill. Hort. 27: 20, 1881
Brassia lewisii Rolfe, Orch. Rev. 1: 199, 1893
Brassia longissima var. minor Schltr., Fedde Rep. Beih. 17: 80, 1922

o **Brassia lanceana** Lindl., Bot. Reg. 21: t.1754, 1835
BR 21: t.1754 (1835) / 24: misc.31 (1838) / 30: misc.13:1, 4 (1844) / 31: misc.62 (1845)
Type: Surinam, H.Lance s.n. (K-Lindl., holo!)
Panama, Venezuela, Tobago Is., Guyana, Surinam, Brazil, Colombia – E

Brassia macrostachya Lindl., Sert. Orch.: t.6, 1838
Brassia pumila Lindl., Bot. Reg. 31: misc.62, 1845
Oncidium suaveolens Rchb.f., Walp. Ann. 6: 765, 1863

o **Brassia lawrenceana** Lindl., Bot. Reg. 27: t.18 & misc.6, 1841
BR 26: misc.37 (1840) / 27: t.18 & misc.6 (1841) / 30: misc.13:2, 3, 7 (1844)
Type: Venezuela, Henchman s.n. (K-Lindl., holo!)
Venezuela, Guyana, Surinam, Brazil – E
Brassia cochleata Knowl. & Westc., Fl. Cab. 2: 23 t.53, 1838
Brassia angusta Lindl., Bot. Reg. 30: misc.13:3, 1844
Oncidium lawrenceana (Lindl.) Rchb.f., Walp. Ann. 6: 766, 1863

Brassia maculata R.Br., Aiton Hort. Kew. ed.2, 5: 215, 1813
BR 30: misc.13:9, 13 (1844)
Belize, Guatemala, El Salvador, Honduras, Nicaragua, Jamaica, W.Indies – E
Brassia guttata Lindl., Pl. Hartw.: 94, 1842
Oncidium brassia Rchb.f., Walp. Ann. 6: 765, 1842
Brassia wrayae Skinner, Bot. Mag. 69: t.4003, 1843

Brassia peruviana Poeppig & Endl., Nov. Gen. Sp. Pl. 2: 12 t.117, 1838
BR 30: misc.13:12 (1844)
Peru – E
Oncidium peruvianum (Poeppig & Endl.) Rchb.f., Walp. Ann. 6: 767, 1863

Brassia verrucosa Lindl., Bot. Reg. 26: misc.66, 1840
BR 26: misc.66 (1840) / 29: misc.2 (1843) / 30: misc.13:8, 10, 11 (1844) / 33: t.29 (1847)
Type: Barker s.n. cult. Rollisson (K-Lindl., holo!)
Mexico, Guatemala, El Salvador, Honduras, Nicaragua, Venezuela – E
o Brassia brachiata Lindl., Benth. Pl. Hartw.: 94, 1842
Brassia aristata Lindl., Bot. Reg. 30: misc.13:11, 1844
Brassia odontoglossoides Klotzsch & Karts., Allg. Gartenz. 15: 330, 1847
Brassia coryandra Morren, Ann. Soc. Gand. 4: 295 t.212, 1848
Brassia longiloba DC., Mém. Soc. Phys. Genève 11: 427, 1848
Oncidium brachiatum (Lindl.) Rchb.f., Walp. Ann. 6: 768, 1863
Oncidium verrucosum (Lindl.) Rchb.f., Walp. Ann. 6: 768, 1863

Bromheadia finlaysoniana (Lindl.) Rchb.f., Walp. Ann. 6: 882, 1864
BR 27: misc.184 (1841) / 30: t.18 (1844)
Thailand, Cambodia, Laos, Malaya, Sumatra, Borneo, Philippines – E
Grammatophyllum finlaysonianum Lindl., Gen. Sp. Orch. Pl.: 173, 1833
o Bromheadia palustris Lindl., Bot. Reg. 27: misc.184, 1841

Broughtonia domingensis (Lindl.) Rolfe, Gard. Chron. s.3, 5: 491, 1889
BR 30: sub t.5:19 (1844)
Jamaica, Hispaniola – E
Cattleya domingensis Lindl., Gen. Sp. Orch. Pl.: 118, 1831
Broughtonia lilacina Henfr., Gard. Mag. Bot. 3: 201, 1851
Laeliopsis domingensis (Lindl.) Lindl., Paxt. Fl. Gard. 3: 155 t.105, 1852

Bulbophyllum antenniferum (Lindl.) Rchb.f., Walp. Ann. 6: 248, 1861
BR 29: sub t.49:18 (1843)
Type: Philippines, Cuming s.n. (K-Lindl., holo!)
Philippines, Borneo, Java, New Guinea – E
Cirrhopetalum antenniferum Lindl., Bot. Reg. 29: sub t.49:18, 1843
Phyllorchis antennifera (Lindl.) Kuntze, Rev. Gen. 2: 675, 1891
Bulbophyllum leysianum Burb., Proc. Roy. Hort. Soc. 17: 134 fig.19, 1895
Cirrhopetalum ornithorhynchum J.J.Smith, Ic. Bog. 2: 107 t.121, 1903

Bulbophyllum auratum (Lindl.) Rchb.f., Walp. Ann. 6: 261, 1861 (non Ames)
BR 26: misc.107 (1840) / 29: t.61 & sub t.49:12 (1843)
Type: Manila, cult. Loddiges (K-Lindl., holo!)
Malaya, Sumatra, Borneo, Philippines – E
o Cirrhopetalum auratum Lindl., Bot. Reg. 26: misc.107, 1840
Phyllorchis aurata (Lindl.) Kuntze, Rev. Gen. 2: 675, 1891
Cirrhopetalum borneense Schltr., Bull. Herb. Bois. 2, s.6: 464, 1906
Bulbophyllum campanulatum Rolfe, Kew Bull.: 62, 1909
Cirrhopetalum campanulatum (Rolfe) Rolfe, Orch. Rev. 18: 73 fig.6, 1910
Bulbophyllum borneense (Schltr.) J.J.Smith, Bull. Buitenz. Jard. Bot. s.2, 8: 23, 1911

Bulbophyllum barbigerum Lindl., Bot. Reg. 23: t.1942, 1837
BR 23: t.1942 (1837)
Type: Sierra Leone, cult. Loddiges (K-Lindl., holo!)
Sierra Leone, Liberia, Ivory Coast, S.Nigeria, Ghana, W.Cameroun – E
Phyllorchis barbigera (Lindl.) Kuntze, Rev. Gen. 2: 675, 1891

Bulbophyllum blumei (Lindl.) J.J.Smith, Fl. Buitenz. 6: 459 fig.351, 1905
BR 29: sub t.49:19, 20 (1843)
W.Malaysia, Sumatra, Borneo, Philippines, New Guinea, Australia, Solomon Is. – E
Ephippium ciliatum Blume, Bijdr.: 309, 1825
Cirrhopetalum blumei Lindl., Gen. Sp. Orch. Pl.: 59, 1830
Cirrhopetalum maxillare Lindl., Bot. Reg. 29: sub t.49, 1843
Bulbophyllum cuspidilingue Rchb.f., Walp. Ann. 6: 264, 1861
Bulbophyllum maxillare (Lindl.) Rchb.f., Walp. Ann. 6: 248, 1861
Phyllorchis blumei (Lindl.) Kuntze, Rev. Gen. 2: 677, 1891
Phyllorchis maxillaris (Lindl.) Kuntze, Rev. Gen. 2: 677, 1891
Bulbophyllum masdevalliaceum Kraenzlin, Engl. Bot. Jahrb. 34: 251, 1904
Bulbophyllum ephippium Ridley, Mat. Fl. Mal. Pen. 1: 78, 1907

Bulbophyllum bracteolatum Lindl., Bot. Reg. 24: t.57, 1838
BR 24: t.57 (1838) / 26: misc.217 (1840)
Type: Guyana, cult. Loddiges (K-Lindl., holo!)
Venezuela, Guyanas – E
Bulbophyllum sordidum Lindl., Bot. Reg. 26: misc.217, 1840
Bulbophyllum bracteolata (Lindl.) Rchb.f., Bot. Zeit. 10: 934, 1852
Phyllorchis bracteolata (Lindl.) Kuntze, Rev. Gen. 2: 677, 1891

Bulbophyllum calyptratum Kraenzlin, Engl. Bot. Jahrb. 22: 24, 1895 **var. calyptratum**
BR 23: t.1959 (1837)
Guinea, Ghana, Cameroun, Eq.Guinea., Gabon, Congo – up to 1100 m – E
o Megaclinium maximum Lindl., Bot. Reg. 23: t.1959, 1837 (non Lindl.1930)
Megaclinium flaccidum Hook., Cent. Orch. Pl.: pl.3, 1849
Megaclinium lindleyi Rolfe, F.T.A. 7: 43, 1897
Megaclinium buchenavianum Kraenzlin, Engl. Bot. Jahrb. 36: 115, 1905
Bulbophyllum lindleyi (Rolfe) Schltr., Die Orchideen: 327, 1914
Megaclinium lepturum Kraenzlin, Ann. Nat. Hofmus. Wien 30: 61, 1916
Bulbophyllum buchenavianum (Kraenzlin) De Wild., Pl. Beq. 1: 106, 1921

Bulbophyllum cheiri Lindl., Bot. Reg. 30: misc.56, 1844
BR 30: misc.56 (1844)
Type: Philippines, cult. Loddiges (K-Lindl., holo!)
W.Malaysia, Borneo, Philippines, Celebes – E
Sarcopodium cheiri (Lindl.) Lindl., Paxt. Fl. Gard. Gleanings 199: 155, 1850
Bulbophyllum megalanthum Griffith, Notul. 3: 286, 1851
Sarcopodium megalanthum (Griffith) Lindl., Fol. Orch. Sarcopodium: 4, 1853
Phyllorchis cheiri (Lindl.) Kuntze, Rev. Gen. 2: 677, 1891
Phyllorchis megalantha (Griffith) Kuntze, Rev. Gen. 2: 677, 1891
Bulbophyllum copelandii Ames, Orchid. 1: 97, 1905
Bulbophyllum praestans Kraenzlin, Engl. Bot. Jahrb. 34: 254, 1905
Bulbophyllum whitfordii Rolfe ex Ames, Orchid. 1: 100, 1905
Bulbophyllum subuliferum Schltr., Fedde Rep. 10: 91, 1912

Bulbophyllum chinense (Lindl.) Rchb.f., Walp. Ann. 6: 260, 1861
BR 28: misc.29 (1842) / 29: t.49 & sub t.49:6 (1843)
Type: China, cult. Loddiges (K-Lindl., holo!)
China – E
o Cirrhopetalum chinense Lindl., Bot. Reg. 28: misc.29, 1842

Bulbophyllum clavatum Thouars, Orch. Iles Afr.: t.99, 1822
BR 10: sub t.832 (1824)
Mascarene Is. – E
Bulbophyllum conicum Thouars, Orch. Iles Afr.: Tabl. Esp. 3, 1822

o **Bulbophyllum cocoinum** Lindl., Bot. Reg. 23: t.1964, 1837
BR 23: t.1964 (1837)
Type: Sierra Leone, cult. Loddiges (K-Lindl., holo!)
Sierra Leone to Ghana, Zaire, Angola – 1400 m – E
Bulbophyllum andongense Rchb.f., Flora 48: 184, 1865
Bulbophyllum brevidenticulatum De Wild., Bull. J. Bot. Brux. 5: 179, 1916

Bulbophyllum cumingii (Lindl.) Rchb.f., Walp. Ann. 6: 261, 1861
BR 29: sub t.49:13 (1843)
Type: Philippines, cult. Loddiges (K-Lindl., holo!)
Philippines – up to 500 m – E
Cirrhopetalum cumingii Lindl., Bot. Reg. 29: sub t.49:13, 1843
Cirrhopetalum stramineum auct. non Teijsm. & Binnend., 1862
Cirrhopetalum amesianum Rolfe, Lindenia 7: 55 t.314, 1891
Phyllorchis cumingii (Lindl.) Kuntze, Rev. Gen. 2: 675, 1891
Bulbophyllum amesianum (Rolfe) J.J.Smith, Bull. Jard. Bot. Buitenz. s.2, 8: 22, 1912
Bulbophyllum stramineum Ames, Orchid. 5: 190, 1915

Bulbophyllum cupreum Lindl., Bot. Reg. 24: misc.183, 1838 (non Hook. 1862, non Rchb.f.1874, non Gagnepain 1934)
BR 24: misc.183 (1838)
Type: Manila, Cumming s.n. (K-Lindl., holo!)
Thailand, Malay Peninsula, Burma – E
Bulbophyllum pechei W.Bull., Cat., 1891 (non Rchb.f.)

Bulbophyllum elongatum (Blume) Hassk., Cat. Pl. Bog.: 39, 1844 (non Kraenzlin)
BR 29: sub t.49:16 (1843)
Java, Borneo, Malay Peninsula, – E
Ephippium elongatum Blume, Bijdr.: 309, 1825
Cirrhopetalum elongatum (Blume) Lindl., Bot. Reg. 29: sub t.49:16, 1843
Bulbophyllum sceptrum Rchb.f., Walp. Ann. 6: 254, 1861
Phyllorchis elongata (Blume) Kuntze, Rev. Gen. 2: 677, 1891
Bulbophyllum gigas Ridley, J. Linn. Soc. 32: 277, 1896

Bulbophyllum falcatum (Lindl.) Rchb.f., Walp. Ann. 6: 250, 1861
var. bufo (Lindl.) J.J.Vermeulen, Orchid. Monogr. 2: 125, 1987
BR 27: misc.82 (1841)
Type: Sierra Leone, cult. Loddiges (K-Lindl. holo!)
Guinea, Sierra Leone, Liberia, Ivory Coast, Ghana, Nigeria, Cameroun, Zaire – E
Megaclinium bufo Lindl., Bot. Reg. 27: misc.82, 1841
Bulbophyllum bufo (Lindl.) Rchb.f., Walp. Ann. 6: 258, 1861
Phyllorchis bufo (Lindl.) Kuntze, Rev. Gen. 2: 677, 1891
Bulbophyllum bakossorum Schltr., Westafr. Kautchuk Exp.: 280, 1900
Bulbophyllum longibulbum Schltr., Westafr. Kautchuk Exp.: 281, 1900
Megaclinium deistelianum Kraenzlin, Engl. Bot. Jahrb. 33: 72, 1902
Megaclinium gentilii De Wild., Bel. Col. 8: 425, 1902 (non Rolfe 1904)
Bulbophyllum deistelianum Kraenzlin, Schltr., Engl. Bot. Jahrb. 38: 17, 1905
Bulbophyllum sereti De Wild., Bull. J. Bot. Brux. 5: 177, 1916
Megaclinium sereti De Wild., Bull. J. Bot. Brux. 5: 177, 1916
Bulbophyllum lubiense De Wild., Pl. Beq. 1: 92, 1921

Bulbophyllum falcatum (Lindl.) Rchb.f., Walp. Ann. 6: 250, 1861
var. falcatum
BR 12: t.989 (1826)
Type: Sierra Leone, G.Don s.n. (K-Lindl., holo!)
Sierra Leone, Liberia, Ivory Coast, Togo, S.Nigeria, Uganda – E
o Megaclinium falcatum Lindl., Bot. Reg. 12: t.989, 1826
Bulbophyllum falcatum (Lindl.) Rchb.f., Walp. Ann. 6: 258, 1861
Bulbophyllum oxyodon Rchb.f., Gard. Chron. s.3, 4: 91, 1888
Megaclinium oxyodon Rchb.f., Gard. Chron. s.3, 4: 91, 1888
Phyllorchis falcata (Rchb.f.) Kuntze, Rev. Gen. 2: 677, 1891
Megaclinium endotrachys Kraenzlin, Engl. Bot. Jahrb. 36: 115, 1905
Bulbophyllum leptorrhachis Schltr., Engl. Bot. Jahrb. 38: 17, 1907
Megaclinium hemirhachis Pfitzer, Orchis 2: 135, 1908
Megaclinium ugandae Rolfe, Kew Bull.: 338, 1913
Bulbophyllum dahlemense Schltr., Notizbl. Berl.-Dahl. 7: 325, 1919
Bulbophyllum hemirhachis (Pfitzer) De Wild., Pl. Beq. 1: 86, 1921
Bulbophyllum ugandae (Rolfe) De Wild., Pl. Beq. 1: 100, 1921
Bulbophyllum bufo Vermeulen, Tijdr. Ned. Orch. Ver. 43: 124, 1981 (non (Lindl.) Rchb.f.)

Bulbophyllum falcatum (Lindl.) Rchb.f., Walp. Ann. 6: 250, 1861
var. velutinum (Lindl.) J.J.Vermeulen, Bull. J. Bot. Belg. 56: 235, 1986
BR 33: sub t.32 (1847)
Type: cult. Loddiges (K-Lindl., holo!)
Sierra Leone, Liberia, Ivory Coast, Ghana, Nigeria, Cameroun, Eq.Guinea, Gabon, Zaire – E
Megaclinium velutinum Lindl., Bot. Reg. 33: sub t.32, 1847
Bulbophyllum velutinum (Lindl.) Rchb.f., Walp. Ann. 6: 258, 1861
Bulbophyllum rhizophorae Lindl., J. Linn. Soc. Bot. 6: 125, 1862
Bulbophyllum melanorrhachis Rchb.f., Gard. Chron. n.s.4: 162, 1875
Megaclinium melanorrhachis Rchb.f., Gard. Chron. n.s.4: 162, 1875
Phyllorchis rhizophorae (Lindl.) Kuntze, Rev. Gen. 2: 677, 1891
Phyllorchis velutina (Lindl.) Kuntze, Rev. Gen. 2: 678, 1891
Megaclinium minutum Rolfe, Kew Bull.: 5, 1893
Megaclinium millenii Rolfe, F.T.A. 7: 40, 1897
Bulbophyllum millenii (Rolfe) Schltr., Engl. Bot. Jahrb. 38: 13, 1905
Megaclinium arnoldianum De Wild., Miss. Laur.: 55, 1905
Bulbophyllum minutum (Rolfe) Engl., Pflanzenw. Afr. 2: 417, 1908 (non Thouars 1822)
Megaclinium minutum var. purpureum De Wild., Bull. J. Bot. Brux. 3: 255, 1911
Bulbophyllum fractiflexum Kraenzlin, Engl. Bot. Jahrb. 48: 392, 1912 (non J.J.Smith 1908)
Megaclinium lasianthum Kraenzlin, Engl. Bot. Jahrb. 48: 393, 1912
Bulbophyllum kewense Schltr., Die Orchideen: 328, 1914
Bulbophyllum brixhei De Wild., Bull. J. Bot. Brux. 5: 174, 1916
Bulbophyllum solheidii De Wild., Bull. J. Bot. Brux. 5: 178, 1916
Megaclinium brixhei De Wild., Bull. J. Bot. Brux. 5: 174, 1916
Megaclinium solheidii De Wild., Bull. J. Bot. Brux. 5: 178, 1916
Bulbophyllum arnoldianum (De Wild.) De Wild., Pl. Beq. 1: 77, 1921
Bulbophyllum lanuriense De Wild., Rev. Zool. Afr. 9, Suppl. Bot.: 30, 1921
Bulbophyllum minutum var. purpureum (De Wild.) De Wild., Pl. Beq. 1: 89, 1921
Megaclinium lanuriense De Wild., Rev. Zool. Afr. 9 Suppl. Bot.: 30, 1921
Megaclinium angustum Rolfe, Kew Bull.: 26, 1922
Bulbophyllum simonii Summerh., Kew Bull.: 204, 1935

Bulbophyllum fimbriatum (Lindl.) Rchb.f., Walp. Ann. 6, 1861
BR 25: misc.119 & misc.120 (1839) / 29: sub t.49:8, 15 (1843)
India – E
Cirrhopetalum fimbriatum Lindl., Gen. Sp. Orch. Pl.: 72, 1839
Cirrhopetalum wallichii J.Graham, Cat. Pl. Bomb.: 205, 1839
Phyllorchis fimbriata (Lindl.) Kuntze, Rev. Gen. 2: 677, 1891

Bulbophyllum flavescens (Blume) Lindl., Gen. Sp. Orch. Pl.: 54, 1830
BR 28: misc.95 (1842)
Philippines, Borneo, Java, Malay Peninsula, Sumatra – 500–300 m – E
Diphyes flavescens Blume, Bijdr.: 313, 1825
Bulbophyllum adenopetalum Lindl., Bot. Reg. 28: misc.95, 1842
Phyllorchis adenopetala (Lindl.) Kuntze, Rev. Gen. 2: 675, 1891
Phyllorchis flavescens (Blume) Kuntze, Rev. Gen. 2: 677, 1891
Bulbophyllum ramosii Ames, Phil. J. Sc. Bot. 7: 25, 1912

Bulbophyllum fuscum Lindl., Bot. Reg. 25: misc.5, 1839 (non Rendle 1913, non Piers 1968)
var. fuscum
BR 25: misc.5 (1839)
Type: Sierra Leone, cult. Loddiges (K-Lindl., holo!)
Guinea, Sierra Leone, Liberia, Ivory Coast, Nigeria, Cameroun, Centr. Afr.Rep., Gabon, Zaire, Angola – E
Bulbophyllum ogoouense Guillaumin, Bull. Mus. Hist. Nat. Paris 30: 522, 1924

Bulbophyllum guttulatum Wallich mss. ex Hook.f., Fl. Brit. Ind. 5: 776, 1890
BR 24: misc.102 (1838) / 31: t.44 (1845)
India, Nepal, Sikkim, Bhutan, Assam – E
o Bulbophyllum umbellatum Lindl., Bot. Reg. 31: t.44, 1845 (non Lindl.1830)
Cirrhopetalum guttulatum (Wallich mss.) Hook.f., Fl. Brit. Ind. 5: 776, 1890
Phyllorchis guttulata (Wallich) Kuntze, Rev. Gen. 2: 677, 1891

Bulbophyllum helenae (Kuntze) J.J.Smith., Bull. Buitenz. Jard. Bot. s.2, 8: 24, 1912
BR 24: misc.138 (1838) / 29: sub t.49:4 (1843)
N.E.India, Nepal, Sikkim, Burma, Thailand – 1000–2000 m – E
Cirrhopetalum cornutum Lindl., Bot. Reg. 24: misc.138, 1838
Bulbophyllum cornutum (Lindl.) Rchb.f., Walp. Ann. 6: 261, 1861 (non Rchb.f., Walp. Ann. 6: 247, 1861)
Phyllorchis helenae Kuntze, Rev. Gen. 2: 676, 1891
Bulbophyllum picturatum Seidenf. & Smitin., Orch. Thailand: 367 fig.276, 1961 (non (Lodd.) Rchb.f.)

Bulbophyllum hirtum (Smith) Lindl., Gen. Sp. Orch. Pl.: 51, 1830
BR 33: sub t.66 (1847)
Thailand, Nepal, Sikkim, Assam, Burma, Vietnam – E
Stelis hirta Smith, Rees Cyclop. 34 (1), 1816
Tribrachia hirta (Smith) Lindl., Coll. Bot. 41, 1826
Bulbophyllum suave Griffith, Notul. 3: 292, 1851
Phyllorchis hirta (Smith) Kuntze, Rev. Gen. 2: 677, 1891
Bulbophyllum kerrii Rolfe, Kew Bull.: 84, 1906

Bulbophyllum imbricatum Lindl., Bot. Reg. 27: misc.65, 1841
BR 27: misc.65 (1841)
Type: Sierra Leone, cult. Loddiges (K-Lindl., holo!)
Sierra Leone, Liberia, Ivory Coast, Ghana, Nigeria, Cameroun, Centr. Afr.Rep., Eq.Guinea, Congo, Zaire – E
Bulbophyllum strobiliferum Kraenzlin, Mitth. Deutsch. Schutzgeb. 2: 155, 1889
Megaclinium leucorhachis Rolfe, Kew Bull.: 198, 1891
Phyllorchis imbricata (Lindl.) Kuntze, Rev. Gen. 2: 677, 1891
Megaclinium triste Rolfe, Kew Bull.: 362, 1894
Bulbophyllum stenorhachis Kraenzlin, Engl. Bot. Jahrb. 22: 25, 1895
Megaclinium imbricatum (Lindl.) Rolfe, F.T.A. 7: 37, 1897
Megaclinium strobiliferum (Kraenzlin) Rolfe, F.T.A. 7: 38, 1897
Bulbophyllum laurentianum Kraenzlin, Durand & De Wild., Mat. Fl. Congo 3: 50, 1899
Bulbophyllum kamerunense Schltr., Westafr. Kautchuk Exp.: 281, 1901
Megaclinium congolense De Wild., Et. Fl. B. & Moy. Congo 1: 21, 1903

Megaclinium gilletii De Wild., Et. Fl. B. & Moy. Congo 1: 22, 1903
Bulbophyllum leucorhachis (Rolfe) Schltr., Engl. Bot. Jahrb. 38: 17, 1905
Bulbophyllum triste (Rolfe) Schltr., Engl. Bot. Jahrb. 38: 17, 1905 (non Rchb.f. 1861)
Megaclinium ledermannii Kraenzlin, Engl. Bot. Jahrb. 48: 394, 1912
Bulbophyllum congolense (De Wild.) De Wild., Pl. Beq. 1: 82, 1921
Bulbophyllum gilletii (De Wild.) De Wild., Pl. Beq. 1: 86, 1921
Bulbophyllum ledermannii (Kraenzlin) De Wild., Pl. Beq. 1: 91, 1921
Megaclinium hebetatum Kraenzlin, Viertelj. Nat. Ges. Zürich 68: 423, 1923
Bulbophyllum linderi Summerh., Kew Bull.: 202, 1935

Bulbophyllum limbatum Lindl., Bot. Reg. 26: misc.171, 1840 (non Parish & Rchb.f. 1874)
BR 26: misc.171 (1840)
Type: Singapore, cult. Loddiges (K-Lindl., holo!)
Thailand, Burma, Malaya, Sumatra, Bangka – E
Phyllorchis limbata (Lindl.) Kuntze, Rev. Gen. 2: 677, 1891
Bulbophyllum blepharosepalum Schltr., Bull. Herb. Bois. s.2, 6: 462, 1906

Bulbophyllum lobbii Lindl., Bot. Reg. 33: sub t.29, 1847
BR 33: sub t.29 (1847)
Type: Java, T.Lobb s.n. cult. Veitch (K-Lindl., holo!)
Assam, Burma, Cambodia, Malaya, Sumatra, Java, Bali, Borneo, Philippines – E
Sestochilos uniflorum Breda, Gen. Sp. Orch. Asclep. 29: t.3, 1827
Sarcopodium lobbii (Lindl.) Lindl., Paxt. Flow. Gard. 1: 155, 1850
Bulbophyllum henshallii Lindl., Gard. Chron.: 422, 1852
Bulbophyllum lobbii var. henshallii (Lindl.) Henfrey, Gard. Mag. Bot. 3: 269, 1852
Bulbophyllum siamense (Rchb.f.) Rchb.f., Saunders Ref. Bot. 2: t.116, 1882
Bulbophyllum lobbii (Lindl.) Kuntze, Rev. Gen. 2: 677, 1891
Bulbophyllum lobbii var. colosseum hort., Gard. Chron. s.3, 37: 174 fig.68, 1905
Bulbophyllum lobbii var. breviflorum J.J.Smith, Bull. Dépt. Agr. Ind. Neerl. 15: 18, 1908

Bulbophyllum longiflorum Thouars, Orch. Iles Afr.: t.98, 1822
BR 10: sub t.832 (1824)
Africa, Madagascar, Malay Archipelago, Philippines, New Guinea, Australia, New Caledonia, Solomon Is., Vanuatu, Fiji, Rarotonga, Tahiti, Guam – E
Epidendrum umbellatum Forster, Ins. Austr. Prod.: 60, 1786 (non Sw. 1800)
Cymbidium umbellatum (Forster) Sprengel, Pl. Min. Cogn. Pugill. 2: 82, 1815
Cirrhopetalum thouarsii Lindl., Bot. Reg. 10: sub t.832, 1824
Cirrhopetalum umbellatum (Forster) Hook. & Arn., Bot. Beechey Voy.: 71, 1832
Cirrhopetalum clavigerum Fitzg., J. Bot. 21: 204, 1883
Bulbophyllum clavigerum (Fitzg.) F.v.Muell., Syst. Cens. Austr. Pl. Suppl. 1: 3, 1884 (non Perrier 1951)
Phyllorchis clavigera (Fitzg.) Kuntze, Rev. Gen. 2: 677, 1891
Phyllorchis thouarsii (Lindl.) Kuntze, Rev. Gen. 2: 677, 1891
Phyllorchis umbellata (Forster) Kuntze, Rev. Gen. 2: 675, 1891
Cirrhopetalum thouarsii var. concolor Rolfe, Gard. Chron.: 178, 1892
Cirrhopetalum layardii F.v.Muell. & Kraenzlin, Oestr. Bot. Zeit. 44: 210, 1894
Bulbophyllum layardii (F.v.Muell. & Kraenzlin) J.J.Smith, Bull. Buitenz. s.2: 8, 1912
Bulbophyllum trisetum Ames, Phil. J. Sc. Bot. 7: 142, 1912
Cirrhopetalum kenejianum Schltr., Fedde Rep. Beih. 1: 889, 1913
Cirrhopetalum africanum Schltr., Engl. Bot. Jahrb. 53: 573, 1915
Cirrhopetalum longiflorum (Thouars) Schltr., Beih. Bot. Centralbl. 33 (2): 420, 1915

Bulbophyllum macraei (Lindl.) Rchb.f., Walp. Ann. 6: 263, 1861
BR 27: misc.105 (1841) / 29: sub t.49:10 (1843)
Sri Lanka, India, Japan, Taiwan – E
Cirrhopetalum macraei Lindl., Gen. Sp. Orch. Pl.: 59, 1830 (non Wight 1851)
Cirrhopetalum walkerianum Wight, Ic. 5: t.1657, 1851
Phyllorchis macraei (Lindl.) Kuntze, Rev. Gen. 2: 677, 1891
Bulbophyllum boninense Makino, Iinuma Somuku Dzusetzu ed.3, 18: 14 t.20, 1912
Bulbophyllum uraiense Hayata, Ic. Pl. Form. 4: 50 t.11, 1914
Cirrhopetalum uraiense (Hayata) Hayata, Ic. Pl. Form. 4: 51, 1914
Cirrhopetalum boninense (Makino) Schltr., Fedde Rep. Beih. 4: 254, 1919
Cirrhopetalum makinoanum Schltr., Fedde Rep. Beih. 4: 254, 1919
Bulbophyllum makinoanum (Schltr.) Masam., J. Soc. Trop. Agr. 2: 153, 1930
Cirrhopetalum autumnale Fukuyama, Bot. Mag. Tokyo 49: 760, 1935
Bulbophyllum tanegasimense Masam., J. Geobot. Hokuriku 26: 57, 1968
Cirrhopetalum tanegasimense (Masam.) Masam., Coll. Ill. Fl. Nip. 8: 268, 1970

o **Bulbophyllum macranthum** Lindl., Bot. Reg. 30: t.13, 1844
BR 30: t.13 (1844)
Type: Singapore, cult. Loddiges (K-Lindl., holo!)
Burma, Vietnam, Sumatra, Bangka – E
Sarcopodium macranthum (Lindl.) Lindl., Paxt. Flow. Gard. 1: 155, 1850
Sarcopodium purpureum Rchb.f., Bonpl. 4: 330, 1856
Phyllorchis macrantha (Lindl.) Kuntze, Rev. Gen. 2: 677, 1891
Bulbophyllum cochinchinense Gagnepain, Bull. Mus. Paris s.2, 22: 400, 1950
Bulbophyllum patens Gagnepain, Fl. Indoch.: 119, 1951 (non King ex Hook.f.)

Bulbophyllum malachadenia Cogn., Mart. Fl. Bras. 3 (5): 596, 1902
BR 25: misc.110 (1839)
Brazil – E
Malachadenia clavata Lindl., Bot. Reg. 25: misc.110, 1839
Bulbophyllum clavatum (Lindl.) Pfitzer, Engl. Nat. Pflanzenf. s.2, 6: 179, 1888 (non Thouars 1822)

Bulbophyllum maximum (Lindl.) Rchb.f., Walp. Ann. 6: 259, 1861
BR 25: misc.10 (1839)
Guinea, Sierra Leone, Liberia, Ivory Coast, Ghana, Nigeria, Cameroun, Centr.Afr.Rep., Gabon, Congo, Zaire, Uganda, Kenya, Tanzania, Angola, Zambia, Mali, Zimbabwe, Mozambique – E
Megaclinium maximum Lindl., Gen. Sp. Orch. Pl.: 47, 1830
Megaclinium oxypterum Lindl., Bot. Reg. 25: misc.10, 1839
Bulbophyllum oxypterum (Lindl.) Rchb.f., Walp. Ann. 6: 258, 1861
Megaclinium purpuratum Lindl., J. Linn. Soc. 6: 128, 1862
Phyllorchis maxima (Lindl.) Kuntze, Rev. Gen. 2: 677, 1891
Phyllorchis oxyptera (Lindl.) Kuntze, Rev. Gen. 2: 677, 1891
Megaclinium platyrhachis Rolfe, F.T.A. 7: 43, 1897
Megaclinium djumaense De Wild., Not. Pl. Ut. & Int. Fl. Congo 1: 124, 1903
Bulbophyllum platyrhachis (Rolfe) Schltr., Engl. Bot. Jahrb. 38: 15, 1905 (non De Wild.1906, non Ridley 1917)
Bulbophyllum ciliatum Schltr., Engl. Bot. Jahrb. 38: 156, 1906 (non (Blume) Lindl.1830)
Megaclinium oxypterum var. mozambicense Finet, Not. Syst. 1: 169, 1910
Bulbophyllum cyrtopetalum Schltr., Fedde Rep. 9: 165, 1911
Bulbophyllum nyassanum Schltr., Engl. Bot. Jahrb. 53: 571, 1915
Bulbophyllum djumaense (De Wild.) De Wild., Pl. Beq. 1: 83, 1921
Bulbophyllum djumaense var. grandifolium De Wild., Pl. Beq. 1: 83, 1921
Bulbophyllum oxypterum var. mozambicense (Finet) De Wild., Pl. Beq. 1: 94, 1921
Bulbophyllum subcoriaceum De Wild., Rev. Zool. Afr. 9, Suppl. Bot.: 33, 1921
Megaclinium subcoriaceum De Wild., Rev. Zool. Afr. 9, Suppl. Bot.: 33, 1921
Bulbophyllum moireanum Hawkes, Lloydia 19: 93, 1956

Bulbophyllum medusae (Lindl.) Rchb.f., Walp. Ann. 6: 262, 1861
BR 28: t.12 (1842) / 29: sub t.49:2 (1843)
Type: Singapore, cult. Loddiges (K-Lindl., holo!)
Thailand, Malaya, Sumatra, Bangka, Sunda Is., Borneo, Philippines – E
o Cirrhopetalum medusae Lindl., Bot. Reg. 28: t.12, 1842
Phyllorchis medusae (Lindl.) Kuntze, Rev. Gen. 2: 677, 1891

Bulbophyllum othonis (Kuntze) J.J.Smith, Bull. Jard. Bot. Buitenz. s.2, 8: 26, 1912
BR 25: misc.118 (1839) / 29: sub t.49:14 (1843)
Philippines (Manila) – E
Cirrhopetalum nutans Lindl., Bot. Reg. 25: misc.118, 1839
Bulbophyllum nutans (Lindl.) Rchb.f., Walp. Ann. Bot. 6: 260, 1861 (non Thouars)
Phyllorchis othonis Kuntze, Rev. Gen. 2: 677, 1891

Bulbophyllum pahudii (De Vriese) Rchb.f., Walp. Ann. 6: 264, 1861
BR 29: sub t.49:17 (1843)
Java – E
Ephippium capitatum Blume, Bijdr.: 303, 1825
Cirrhopetalum capitatum (Blume) Lindl., Bot. Reg. 29: sub t.49:17, 1843
Cirrhopetalum flagelliforme Teijsm. & Binnend., Cat. Pl. Buitenz. (not publ.) 1854
Cirrhopetalum carinatum Teisjm. & Binnend., Ned. Kruidk. Arch. 3: 397, 1855
Cirrhopetalum pahudii De Vriese, Ill. Orch.: t.3, 1855

Bulbophyllum javanicum Miq., Fl. Ind. Bat. 3: 650, 1859
Phyllorchis capitata (Blume) Kuntze, Rev. Gen. 2: 676, 1891

Bulbophyllum picturatum (Lodd. ex Lindl.) Rchb.f., Walp. Ann. 6: 262, 1861
BR 26: misc.106 (1840) / 29: sub t.49:11 (1843)
Type: India, cult. Loddiges (K-Lindl., holo!)
Assam, Burma, Thailand, Vietnam – 1500–2000 m – E
Cirrhopetalum picturatum Lodd. ex Lindl., Bot. Reg. 26: misc.106, 1840
Phyllorchis picturata (Lodd. ex Lindl.) Kuntze, Rev. Gen. 2: 678, 1891
Cirrhopetalum eberhardtii Gagnepain, Bull. Soc. Bot. Fr. 78: 4, 1931

Bulbophyllum pileatum Lindl., Bot. Reg. 30: misc.73, 1844
BR 30: misc.73 (1844)
Type: Singapore, cult. Loddiges (K-Lindl., icono!)
Malay Peninsula, Sumatra – E
Sarcopodium pileatum (Lindl.) Lindl., Fol. Orchid. Sarcop.: 4, 1853

Bulbophyllum pumilum (Sw.) Lindl., Gen. Sp. Orch. Pl.: 54, 1830
BR 12: t.963 (1826) / 26: misc.195 (1840) / 30: misc.70 (1844)
Guinea, Sierra Leone, Liberia, Ivory Coast, Ghana, Nigeria, Cameroun, Eq.Guinea, Gabon, Congo, Zaire – up to 1900 m – E
Dendrobium pumilum Sw., Schrad. Neues J. Bot. 1: 97, 1805
o Tribrachia pendula Lindl., Coll. Bot.: sub t.41a, 1825
Bulbophyllum recurvum Lindl., Gen. Sp. Orch. Pl.: 53, 1830 (non Exell.)
Bulbophyllum flavidum Lindl., Bot. Reg. 26: misc.195, 1840 (non Exell.)
Taurostalix herminiostachys Rchb.f., Bot. Zeit.: 933, 1852
Bulbophyllum herminiostachys (Rchb.f.) Rchb.f., Walp. Ann. 6: 254, 1861
Bulbophyllum pavimentatum Lindl., J. Linn. Soc. Bot. 6: 128, 1862
Bulbophyllum gabonis Lindl. & Rchb.f., Flora 48: 185, 1865
Bulbophyllum drallei Rchb.f., Gartenfl. 35: 20, 1885
Phyllorchis flavida (Lindl.) Kuntze, Rev. Gen. 2: 677, 1891
Phyllorchis herminiostachys (Rchb.f.) Kuntze, Rev. Gen. 2: 677, 1891
Phyllorchis pavimentata (Lindl.) Kuntze, Rev. Gen. 2: 677, 1891
Phyllorchis recurva (Lindl.) Kuntze, Rev. Gen. 2: 677, 1891
Bulbophyllum porphyroglossum Kraenzlin, Engl. Bot. Jahrb. 22: 24, 1895
Genyorchis pumila (Sw.) Schltr., Westafr. Kautchuk. Exp.: 280, 1901 (non Schltr.1905)
Bulbophyllum nanum De Wild., Not. Pl. Ut. & Int. Fl. Congo 1: 122, 1903
Bulbophyllum papillosum Finet, Bull. Mus. Hist. Nat. Paris 9: 303, 1903 (non J.J.Smith, nec (Rolfe) Seidenf. & Smitin.)
Bulbophyllum calabaricum Rolfe, Kew Bull.: 114, 1906
Bulbophyllum moliwense Schltr., Engl. Bot. Jahrb. 38. 157, 1906
Bulbophyllum wincklerii Schltr., Engl. Bot. Jahrb. 38: 158, 1906 (non Schltr.1914)
Bulbophyllum leucopogon Kraenzlin, Engl. Bot. Jahrb. 48: 391, 1912
Bulbophyllum dorotheae Rendle, Cat. Talb. S.Nig. Pl. 100: 146, 1913
Bulbophyllum imogeniae Hamilton., Trans. Bot. Soc. Edin. 27: 228, 1917
Bulbophyllum verecundum Summerh., Kew Bull. 8: 582, 1954
Bulbophyllum yangambiense Louis & Mullenders ex Geerinck, Bull. Soc. R. Bot. Belg. 109: 176, 1976
Bulbophyllum elachon J.J.Vermeulen, Selbyana 7: 11, 1982

Bulbophyllum roxburghii (Lindl.) Rchb.f., Walp. Ann. 6: 263, 1861
BR 29: sub t.49:9 (1843)
India – E
Cirrhopetalum roxburghii Lindl., Gen. Sp. Orch. Pl.: 58, 1830
Aerides radiatum Roxb., Fl. Ind. 3: 476, 1832

Bulbophyllum saltatorium Lindl., Bot. Reg. 23: t.1970, 1837
var. calamarium (Lindl.) J.J.Vermeulen, Bull. J. Bot. Nat. Belg. 56 : 230, 1986
BR 29: misc.109 (1843)
Type: Sierra Leone, Fielding s.n. (K-Lindl., holo!)
Sierra Leone, Liberia, Ghana, Cameroun, Guinea, Gabon, Congo, Zaire, Angola – E
Bulbophyllum calamarium Lindl., Bot. Reg. 29: misc.109, 1843 (non Lindl.1862)
Bulbophyllum rupincola Rchb.f., Flora 48: 184, 1865
Phyllorchis calamaria (Lindl.) Kuntze, Rev. Gen. 2: 677, 1891

o **Bulbophyllum saltatorium** Lindl., Bot. Reg. 23: t.1970, 1837
var. saltatorium
BR 23: t.1970 (1837)
Type: Sierra Leone, cult. Loddiges (K-Lindl., holo!)
Sierra Leone, Liberia, Ivory Coast, Ghana, Nigeria, Eq.Guinea – up to 600 m – E
Phyllorchis saltatoria (Lindl.) Kuntze, Rev. Gen. 2: 678, 1891

Bulbophyllum scabratum Rchb.f., Walp. Ann. 6: 259, 1861
BR 24: misc.53 (1838) / 29: sub t.49:3 (1843)
India – E
Cirrhopetalum caespitosum Wall. ex Lindl., Bot. Reg. 24: misc.53, 1838
Bulbophyllum confertum Hook.f., Fl. Brit. Ind.: 757, 1890

Bulbophyllum sessile (Koen.) J.J.Smith, Fl. Buitenz. 6: 448 fig.340, 1905
BR 27: misc.166 (1841)
Thailand, Laos, Vietnam, Andaman Is., Burma, Malaya, Sumatra, Java – E
Epidendrum sessile Koen., Retz. Observ. 6: 60, 1791
Bulbophyllum clandestinum Lindl., Bot. Reg. 27: misc.166, 1841
Bulbophyllum trisetosum Griffith, Notul. 3: 283, 1851
Oxysepala ovalifolia Wight., Ic. 5: 18 t.1736, 1852
Bulbophyllum ovalifolium (Wight) Parish, Mason Burma 2: 191, 1883 (non Lindl.)
Phyllorchis sessile (Koen.) Kuntze, Rev. Gen. 2: 676, 1891
Bulbophyllum bolovense Guillaumin, Bull. Mus. Paris s.2, 29 (6): 500, 1957

Bulbophyllum setigerum Lindl., Bot. Reg. 24: misc.24, 1838
BR 24: misc.24 (1838)
Type: Guyana, s.n. (K-Lindl., holo!)
Venezuela, Guyana, Brazil – E
Phyllorchis setigera (Lindl.) Kuntze, Rev. Gen. 2: 678, 1891
Bulbophyllum correae Pabst, Bradea 1: 180, 1972

Bulbophyllum sheperdii (F.Muell.) F.Muell., Otto Hamb. Garten 2 (3): 40, 1862
BR 25: misc.35 (1839)
Australia – L, E
Dendrobium crassulifolium A.Cunn., Bot. Reg. 25: misc.35, 1839
Dendrobium sheperdii F.Muell., Fragm. Phyt. Aust. 1: 190, 1859
Bulbophyllum schillerianum Rchb.f., Otto Hamb. Garten 2 (16): 423, 1860
Phyllorchis shepherdii (F.Muell.) Kuntze, Rev. Gen. 2: 678, 1891
Bulbophyllum intermedium Bailey, Bot. Bull. Dep. Agric. Qd. 14: 12, 1896
Bulbophyllum crassulifolium (A.Cunn.) Rupp, Proc. Linn. Soc. N.S.W. 62: 27, 1937
Bulbophyllum shepherdii var. intermedium (Bailey) Nicholls, N. Qld. Nat. 5 (49): 4, 1937

Bulbophyllum umbellatum Lindl., Gen. Sp. Orch. Pl.: 56, 1830
BR 27: misc.173 (1841)
Nepal, Sikkim, Bhutan, Khasia, Assam, China, Vietnam, Thailand, Taiwan – E
Cirrhopetalum maculosum Lindl., Bot. Reg. 27: misc.173, 1841
Cirrhopetalum bootanense Griffith, Notul. 3: 296, 1851
Bulbophyllopsis morphologorum Rchb.f., Bot. Zeit. 10: 933, 1852
Bulbophyllum umbellatum var. bergemannii Regel, Gartenfl.: 362, 1858
Bulbophyllopsis maculosa (Lindl.) Rchb.f., Walp. Ann. 6: 243, 1861
Cirrhopetalum maculosum var. annamicum Finet, Not. Syst. 1: 253, 1909
Bulbophyllum tibeticum Rolfe, Not. Roy. Bot. Gard. Edinb. 26: 21, 1913
Bulbophyllum saruwatari Hayata, Ic. Pl. Form. 6: 72, 1916
Cirrhopetalum annamicum (Finet) Tang & Wang, Act. Phytotax. Sin. 1: 90, 1951
Bulbophyllum tortisepalum Guillaumin, Bull. Mus. Hist. Nat. Paris s.2, 26 (4): 537, 1954

Bulbophyllum uniflorum (Blume) Hassk., Fl. Beibl. 2: 1, 1842
BR 29: sub t.49:21 (1843)
Malaya, Sumatra, Java – E
Ephippium uniflorum Blume, Bijdr.: 309, 1825
Cirrhopetalum compressum Lindl., Bot. Reg. 29: sub t.49:21, 1843
Sarcopodium reinwardtii Lindl., Fol. Orch. Sarc.: 4, 1853
Bulbophyllum reinwardtii (Lindl.) Rchb.f., Walp. Ann. 6: 246, 1861
Dendrobium grandiflorum Reinw. ex Hook.f., Fl. Brit. Ind.: 755, 1890
Phyllorchis reinwardtii (Lindl.) Kuntze, Rev. Gen. 2: 678, 1891
Phyllorchis uniflora (Blume) Kuntze, Rev. Gen. 2: 678, 1891
Bulbophyllum galbinum Ridley, J. Linn. Soc. 32: 267, 1896

Bulbophyllum vaginatum (Lindl.) Rchb.f., Walp. Ann. 6: 261, 1861
BR 26: misc.173 (1840) / 28: sub t.12 (1842) / 29: sub t.49:1 (1843)
Thailand, Malaya, Indonesia, Borneo, Moluccas – E
Cirrhopetalum vaginatum Lindl., Gen. Sp. Orch. Pl.: 59, 1830
Cirrhopetalum caudatum Wight, Ic. 5: t.1658, 1852 (non (Lindl.) King & Pantl.)
Cirrhopetalum stramineum Teijsm. & Binnend., Nat. Tijds. Ned. Ind. 24: 310, 1862
Phyllorchis vaginata (Lindl.) Kuntze, Rev. Gen. 2: 678, 1891
Cirrhopetalum whiteanum Rolfe, Kew Bull.: 7, 1895
Bulbophyllum whiteanum (Rolfe) J.J.Smith, Bull. Buitenz. Jard. Bot. s.2, 8: 29, 1912

Bulbophyllum weberi Ames, Phil. J. Sc. 7: 128 & 142, 1912
BR 24: t.11 (1838) / 29: sub t.49:7 (1843)
Philippines – up to 1500 m – E

○ Cirrhopetalum thouarsii Lindl., Bot. Reg. 24: t.11, 1838 (non Lindl.1830)
Bulbophyllum longiflorum Rchb.f., Walp. Ann. 6: 260, 1861 (non Thouars)
Bulbophyllum baucoense Ames, Orchid. 5: 166, 1915
Cirrhopetalum weberi (Ames) Senghas, Schlechter Die Orchideen ed.3, 1: 775, 1983

○ **Calanthe densiflora** Lindl., Gen. Sp. Orch. Pl.: 250, 1833
BR 19: t.1646 (1834)
India, Sikkim – up to 2000 m – T

○ **Calanthe discolor** Lindl., Sert. Orch.: sub t.9, 1838
BR 24: misc.32 (1838) / 26: t.55 (1840)
Ryukyu Is., Japan – T
Calanthe lurida Decne., Rev. Hort. s.4, 4: 381, 1855
Calanthe discolor var. viridialba Maxim., Bull. Acad. Sc. St.Petersb. 8: 642, 1872
Alismorchis discolor (Lindl.) Kuntze, Rev. Gen. 2: 650, 1891
Calanthe discolor var. kanashiroi Fukuyama, Trans. Nat. Hist. Soc. Formosa 32: 271, 1942
Calanthe amamiana Fukuyama, Act. Phytotax. Geobot. 14: 134, 1952
Calanthe discolor f. viridialba (Maxim.) Honda, Nom. Pl. Jap.: 364, 1957
Calanthe discolor var. amaniana (Kukuyama) Masam., Sc. Rep. Kanazawa Univ. 9: 136, 1964
Calanthe amamiana var. latilabella Ida, J. Geobot. 17: 98, 1969
Calanthe discolor var. divaricatipetala Ida, J. Geobot. 17: 98, 1969
Calanthe tokunoshimensis Hatusima & Ida, J. Geobot. 17: 98, 1969
Calanthe tokumoshimensis f. latilabella (Ida) Hatusima, Fl. of the Ryukyus: 862, 1971

Calanthe discolor Lindl., Sert. Orch.: sub t.9, 1838
var. flava Yatabe, Icon. Fl. Jap. 1: 205 t.50, 1893
BR 7: sub t.573 (1821) / 24: misc.33 (1838)
Japan, Ryukyu Is. – T
Limodorum striatum Banks, Kaempf. Icon. Sel. Pl.: t.2, 1791
Calanthe striata (Banks) R.Br., Bot. Reg. 7: sub t.573, 1821
Calanthe bicolor Lindl., Sert. Orch.: sub t.9, 1838
Calanthe sieboldii Decne. ex Regel, Ind. Sem. Hort. Petrop.: 80, 1868
Calanthe striata var. bicolor (Lindl.) Maxim., Bull. Acad. Sc. St.Petersb. 8: 641, 1872
Calanthe striata var. sieboldii (Decne.) Maxim., Bull. Acad. Sc. St.Petersb. 8: 641, 1872
Calanthe discolor var. bicolor (Lindl.) Makino, Ill. Fl. Jap. ed.2: 685, 1940
Calanthe discolor f. bicolor (Lindl.) Hiroe, Orch. Flow. 2: 84, 1971

Calanthe pulchra (Blume) Lindl., Gen. Sp. Orch. Pl.: 250, 1833
BR 33: t.8 (1847)
Malaya, Sumatra, Java, Borneo, Philippines – up to 1000 m – T, E
Amblyglottis pulchra Blume, Bijdr.: 371, 1825
Styloglossum nervosum Breda, Gen. Sp. Orch. Asclep.: t.7, 1827
○ Calanthe curculigoides Lindl., Gen. Sp. Orch. Pl.: 251, 1833
Alismorchis pulchra (Blume) Kuntze, Rev. Gen. 2: 650, 1891
Calanthe curantigoides Kuntze, Rev. Gen. 2: 650, 1891

Calanthe purpurea Lindl., Gen. Sp. Orch. Pl.: 249, 1833
BR 30: sub t.37 (1844)
Sri Lanka, India – T
Calanthe masuca Thwaites, Enum.: 308, 1884

Calanthe sylvatica (Thouars) Lindl., Gen. Sp. Orch. Pl.: 250, 1833
BR 28: misc.52 (1842) / 30: t. & sub t.37 (1844)
Africa, Madagascar, Mascarene Is., Trop. Asia: Sri Lanka, India to Java, Sumatra, Borneo – up to 2700 m – T
Centrosis sylvatica Thouars, Orch. Iles Austr. Afr.: t.35 & 36, 1822
Amphyglottis emarginata Blume, Bidjdr.: 370, 1825
Bletia masuca D.Don, Prodr. Fl. Nepal.: 30, 1825
Zodula masuca (D.Don) Buch.-Ham., Prodr. Fl. Nepal.: 30, 1825
Calanthe emarginata (Blume) Lindl., Gen. Sp. Orch. Pl.: 249, 1833
○ Calanthe masuca (D.Don) Lindl., Gen. Sp. Orch. Pl.: 249, 1833
Calanthe versicolor Lindl., Sert. Orch.: t.42, 1838
Calanthe sylvatica var natalensis Rchb.f., Linnaea 19: 374, 1847
Calanthe natalensis (Rchb.f.) Rchb.f., Bonpl. 4: 322, 1856
Calanthe corymbosa Lindl., J. Linn. Soc. 6: 129, 1862
Alismorchis masuca (D.Don) Kuntze, Rev. Gen. 2: 650, 1891
Calanthe delphinioides Kraenzlin, Engl. Bot. Jahrb. 17: 55, 1893
Calanthe volkensii Rolfe, F.T.A. 7: 46, 1897
Calanthe neglecta Schltr., Engl. Bot. Jahrb. 53: 570, 1915
Calanthe stolzii Schltr., Engl. Bot. Jahrb. 53: 569, 1915
Calanthe schliebenii Mansf., Notizbl. Bot. Gart. Berlin 11: 808, 1933

Calanthe triplicata (Willemet) Ames, Phil. J. Sc. Bot. 2: 326 fig. 14c, 1907
BR 9: t.720 (1823) / 24: misc.34 (1838) / 25: misc.39 (1839)
Madagascar, Sri Lanka, E.India, S.E.Asia, Indonesia, Japan, New Guinea, Solomon Is., Fiji, Vanuatu, New Caledonia, Australia,
Lord Howe I., Guam – T
Orchis triplicata Willemet, Usteri Ann. Bot. 6 (12): 52, 1796
Limodorum veratrifolium Willd., Sp. Pl. ed.4: 122, 1805
Limodorum ventricosum Steudel, Nom. 1: 481, 1821
○ Calanthe veratrifolia (Willd.) R.Br., Bot. Reg. 9: t.720, 1823
Amblyglottis veratrifolia (Willd.) Blume, Bijdr. 8: 370, 1825
Calanthe furcata Bateman ex Lindl., Bot. Reg. 24: misc.34, 1838
Calanthe perottetii A.Rich., Ann. Mus. Sc. Nat. s.2, 15: 68, 1841
Calanthe comosa Rchb.f., Linnaea 19: 374, 1846
Calanthe odora Griffith, Notul. 3: 365, 1851
Calanthe gracillima Lindl., Fol. Orch. Calanthe: 26, 1854
Calanthe catiligera Rchb.f., Bonpl. 5: 37, 1857
Calanthe japonica Blume ex Miq., Ann. Mus. Bot. Ludg. Bat. 2: 205, 1866
Calanthe textori Miq., Prol. Fl. Jap.: 136, 1867
Calanthe pleiochroma Rchb.f., Gard. Chron.: 938, 1871
Calanthe textori var. alba Maxim., Bull. Acad. Imp. Sc. St.Petersb. 8: 643, 1872
Calanthe textori var. violacea Maxim., Bull. Acad. Sc. St.Petersb. 8: 643, 1872
Calanthe violacea Decne. ex Maxim., Mel. Biol. 8: 646, 1873
Calanthe angraeciflora Rchb.f., Linnaea 41: 75, 1877
Calanthe proboscidea Rchb.f., Gard. Chron. n.s.21: 476, 1884
Calanthe scortechini Hook.f., Fl. Brit. Ind. 5: 854, 1890
Alismorchis diploxiphion (Hook.f.) Kuntze, Rev. Gen. 2: 650, 1891
Alismorchis furcata (Bateman ex Lindl.) Kuntze, Rev. Gen. 2: 650, 1891
Alismorchis gracillima (Lindl.) Kuntze, Rev. Gen. 2: 650, 1891
Alismorchis japonica (Blume) Kuntze, Rev. Gen. 2: 650, 1891
Alismorchis odora (Griffith) Kuntze, Rev. Gen. 2: 650, 1891
Alismorchis pleiochroma (Rchb.f.) Kuntze, Rev. Gen. 2: 650, 1891
Alismorchis textori (Miq.) Kuntze, Rev. Gen. 2: 650, 1891
Alismorchis veratrifolia (Willd.) Kuntze, Rev. Gen. 2: 650, 1891
Calanthe celebica Rolfe, Kew Bull.: 129, 1899
Calanthe matsumurana Schltr., Fedde Rep. 2: 168, 1906
Calanthe okinawensis Hayata, Mat. Fl. Form.: 331, 1911
Calanthe orthocentron Schltr., Fedde Rep. Beih. 1: 380, 1912
Calanthe veratrifolia var. cleistogama Schltr. Fedde Rep. Beih. 1: 380, 1912
Calanthe veratrifolia var. kennyi Bailey, Queensl. Agr. J. 28: 276, 1912
Calanthe veratrifolia var. dupliciloba J.J.Smith, Bull. Jard. Bot. Buitenz. s.3, 5: 69, 1922
Calanthe pubescens Ridley, Kew Bull.: 117, 1923
Calanthe longicalcarata Hayata ex Yamamoto, Trans. Nat. Hist. Soc. Form. 20: 39, 1930
Calanthe veratrifolia var. densissima J.J.Smith, Bull. Jard. Bot. Buitenz. s.3, 11: 116, 1931
Calanthe veratrifolia var. incurvicalcar J.J.Smith, Bull. Buitenz. s.3, 11: 115, 1931
Calanthe veratrifolia var. lancipetala J.J.Smith, Bull. Jard. Bot. Buitenz. s.3, 11: 74, 1931
Calanthe veratrifolia var. timorensis J.J.Smith, Fedde Rep. 35: 110, 1934
Calanthe furcata f. albo-lineata Nakajima, J. Geobot. 17: 81, 1969
Calanthe furcata f. albo-marginata Nakajima, J. Geobot. 17: 55, 1969
Calanthe okinawensis f. albiflora Ida, J. Geobot. 17: 99, 1969
Calanthe furcata f. textori (Miq.) Hiroe, Orch. Flowers 2: 88, 1971
Calanthe textori f. albiflora (Ida) Hatusima, Fl. of the Ryukyus: 872, 1971
Calanthe triplicata f. albo-lineata (Nakajima) Hatusima, Fl. of the Ryukyus: 862, 1971
Calanthe textori var. longicalcarata (Hayata ex Yamamoto) Garay & Sweet, Orch. of S.Ryukyu Is.: 121, 1974
Calanthe triplicata var. angraeciflora (Rchb.f.) N.Hallé, Fl. Nouv. Caléd. 8: 230, 1977

Campylocentrum fasciola (Lindl.) Cogn., Mart. Fl. Bras. 3 (6): 520 t.106, 1906
BR 26: sub t.68 (1840)
Type: Guyana, Schomburgk s.n. (K-Lindl., holo!)
Guatemala, Honduras, Costa Rica, Panama, W.Indies, Venezuela, Colombia, Peru, Amazonas, Guyana, Surinam, Brazil – E
Angraecum fasciola Lindl., Bot. Reg. 26: sub t.68, 1840
Angraecum weigelti Rchb.f., Linnaea 22: 857, 1849
Aeranthus fasciola (Lindl.) Rchb.f., Walp. Ann. 6: 902, 1864
Campylocentrum sullivanii Fawcett & Rendle, J. Bot. 47: 128, 1909
Campylocentrum loretoense Schltr., Fedde Rep. Beih. 9: 118, 1921
Campylocentrum lankesteri Ames, Sched. Orch. 4: 57, 1923
Campylocentrum multiflorum Schltr., Fedde Rep. Beih. 19: 156, 1923

Campylocentrum micranthum (Lindl.) Rolfe, Orch. Rev. 9: 236, 1901
BR 21: t.1772 (1835) / 26: sub t.68 (1840)

Type: Sierra Leone, cult. Loddiges (K-Lindl., holo!)
Mexico to Panama, Venezuela to Bolivia, Guyanas, Brazil, Amazonas, Trinidad, W.Indies – E
o Angraecum micranthum Lindl., Bot. Reg. 21: t.1772, 1835
Angraecum brevifolium Lindl., Bot. Reg. 26: sub t.68, 1840
Angraecum lansbergii Rchb.f., Ned. Kruidk. Arch.: 4, 1859
Aeranthus lansbergii (Rchb.f.) Rchb.f., Walp. Ann. 6: 901, 1863
Aeranthus jamaicensis Rchb.f. & Wullschl., Walp. Ann. 6: 901, 1864
Aeranthus micranthus (Lindl.) Rchb.f., Walp. Ann. 6: 901, 1864
Angraecum jamaicense (Rchb.f. & Wullschl.) Rchb.f. & Wullschl., Walp. Ann. 6: 901, 1864
Campylocentrum jamaicense (Rchb.f. & Wullschl.) Benth. ex Fawcett, Fl. Pl. Jam.: 40, 1893
Campylocentrum kuntzei Cogn., Kuntze Rev. Gen. Pl. 3: 298, 1898
Campylocentrum barrettiae Fawcett & Rendle, J. Bot. 47: 127, 1909
Campylocentrum stenanthum Schltr., Fedde Rep. 10: 486, 1912
Campylocentrum lansbergii (Rchb.f.) Schltr., Fedde Rep. Beih. 6: 100, 1919
Campylocentrum colombianum Schltr., Fedde Rep. Beih. 7: 205, 1920
Campylocentrum ecuadorense Schltr., Fedde Rep. Beih. 8: 171, 1921
Campylocentrum panamense Ames, Orchid. 7: 88, 1922
Campylocentrum peniculus Schltr., Fedde Rep. Beih. 17: 91, 1922
Campylocentrum mattogrossense Hoehne, Arq. Bot. Est. S.Paulo n.s. 3: 62 t.84 fig.1, 1941

Campylocentrum ornithorrhynchum (Lindl.) Rolfe, Orch. Rev. 11: 246, 1903
BR 26: sub t.68 (1840)
Type: Brazil, Tweedie herb. Hooker s.n. (K-Lindl., holo!)
Brazil – E
Angraecum ornithorrhynchum Lindl., Bot. Reg. 26: sub t.68, 1840
Aeranthus ornithorrhynchus (Lindl.) Rchb.f., Walp. Ann. 6: 903, 1864

Campylocentrum polystachyum (Lindl.) Rolfe, Orch. Rev. 11: 245, 1903
BR 26: sub t.68 (1840)
Type: Peru, Mathews 1905 herb. Hooker (K-Lindl., holo!)
Peru – E
Angraecum polystachyum Lindl., Bot. Reg. 26: sub t.68, 1840

Campylocentrum tenue (Lindl.) Rolfe, Orch. Rev. 11: 246, 1903
BR 26: sub t.68 (1840)
Type: Brazil, Martius s.n. (K-Lindl., holo!)
Brazil – E
Angraecum tenue Lindl., Bot. Reg. 26: sub t.68, 1840

o **Catasetum atratum** Lindl., Bot. Reg. 24: t.63 & misc.114, 1838
BR 24: t.63 & misc.114 (1838) / 27: sub t.34:10 (1841) / 30: misc. 40:13 (1844)
Type: Brazil, cult. Loddiges s.n. (K-Lindl., holo!)
Brazil – E
Catasetum adnatum Steudel, Nom. Bot. ed.2, 1: 311, 1841
Catasetum mentosum Lemaire, Jard. Fleur. 3: misc.65, 1852
Catasetum pallidum Klotzsch, Allg. Gartenz. 23: 217, 1855

Catasetum barbatum (Lindl.) Lindl., Bot. Reg. 30: misc.38, 1844
BR 21: t.1778 (1835) / 25: misc.140 (1839) / 26: misc.136 (1840) / 27: t.5 & sub t.34:13 (1841) / 30: misc.38 & misc.40:25, 26 (1844)
Type: J.Henchman, cult. Lowe (K-Lindl., holo!)
Venezuela, Guyana, Surinam, Brazil, Colombia, Ecuador, Peru, Bolivia – E, T
o Myanthus barbatus Lindl., Bot. Reg. 21: t.1778, 1835
Monachanthus viridis Schomb., Trans. Linn. Soc. 17: 551 t.29, 1837
Catasetum proboscideum Lindl., Bot. Reg. 25: misc.140, 1839
Catasetum spinosum Lindl., Bot. Reg. 26: misc.136, 1840
Myanthus spinosus (Lindl.) Hook., Bot. Mag. 67: t.3802, 1840
o Catasetum barbatum var. proboscideum Lindl., Bot. Reg. 27: t.5 fig.3, 1841
o Catasetum lanciferum Lindl., Bot. Reg. 27: t.5 fig.5, 1841
Catasetum cristatum var. spinigerum Hook., Cent. Orch.: 46 t.54, 1846
Catasetum cristatum var. spinosum (Lindl.) Hook., Cent. Orch.: 46 t.54, 1846
Catasetum rivularium Barb.Rodr., Gen. & Sp. Orch. Nov. 1: 130, 1877
Catasetum crinitum Linden & Rchb.f., Ill. Hort. 28: misc.p.80, 1881
Catasetum garnettianum Rolfe, Gard. Chron.: 692, 1888
Catasetum tridentatum Pfitzer, Engl. Nat. Pflanzenf. 2 (6): 160 fig.163, 1888
Catasetum barbatum var. spinosum (Lindl.) Rolfe, Lindenia 7: 23, 1891
Catasetum macrocarpum Stein, Orchideenb.: fig.36, 1892
Catasetum randii Rolfe, Kew Bull.: 394, 1894
Catasetum comosum Cogn., J. d. Orch. 6: 266, 1895
Catasetum appendiculatum Schltr., Beih. Bot. Centralbl. 42 (2): 115, 1925
Catasetum brachybulbon Schltr., Beih. Bot. Centralbl. 42 (2): 116, 1925
Catasetum polydactylon Schltr., Beih. Bot. Centralbl. 42 (2): 122, 1925
Catasetum buchtienii Kraenzlin, Fedde Rep. 25: 27, 1928
Catasetum barbatum var. barbatum (Lindl.) Mansf., Fedde Rep. 31: 116, 1932

o **Catasetum callosum** Lindl., Bot. Reg. 26: misc.183, 1840
BR 26: misc.183 (1840) / 27: t.5 fig.1 & sub t.34:21 (1841) / 30: misc.40:5, 20 (1844)
Type: La Guayra, cult. Loddiges (K-Lindl., holo!)
Venezuela, Colombia, Peru, Brazil, Guyana – E
Catasetum callosum var. grandiflorum Hook., Bot. Mag. 72: t.4219, 1846
Catasetum lansbergii Lindl., Paxt. Flow. Gard. 1: 156, 1850
Catasetum callosum var. crenatum Regel, Gartenfl. 2: 275, 1853
Catasetum carunculatum Rchb.f. & Warsc., Allg. Gartenz. 22: 174, 1854
Myanthus grandiflorus Beer, Prakt. Orch.: 272, 1854
Catasetum fuliginosum Rolfe, Gard. Chron. s.3, 2: 473, 1887 (non Lindl.)
Catasetum darwinianum Rolfe, Gard. Chron. s.3, 5: 394, 1889
Catasetum arachnoideum Ames, Amer. Gard.: 741, 1898
Catasetum callosum var. eucallosum Mansf., Fedde Rep. 31: 114, 1932
Catasetum callosum var. carunculatum (Rchb.f. & Warsc.) Mansf., Fedde Rep. 31: 114, 1932

Catasetum cernuum (Lindl.) Rchb.f., Walp. Ann. 6: 670, 1863
BR 18: sub t.1538 (1832) / 20: t.1721 (1834) / 27: sub t.34:15 & misc.79 (1841) / 30: misc.40:23 (1844)
Type: Brazil, Rio de Janeiro, cult. Loddiges (K-Lindl., holo!)
Trinidad, Venezuela, N.Brazil – E
Monachanthus viridis Lindl., Bot. Reg. 18: sub t.1538, 1832
o Myanthus cernuus Lindl., Bot. Reg. 18: sub t.1538, 1832
Catasetum trifidum Hook., Bot. Mag. 60: t.3262, 1833
Catasetum viride (Lindl.) Lindl., Bot. Reg. 27: misc.79, 1841

o **Catasetum cristatum** Lindl., Trans. Hort. Soc. 6: 83, 1824
BR 10: sub t.840:6 (1824) / 12: t.966 (1826) / 23: t.1947A text sub t.1951 (1837) / 26: misc.182 (1840) / 27: t.5 fig.2 & sub t.34:11, 12 (1841) / 30: misc.40:24, 27 (1844)
Guyana, Brazil – E
Catasetum cristatum var. monstrosum Hook., Ic. Plant.: t.177, 1837
Monachanthus cristatus (Lindl.) Lindl., Bot. Reg. 23: sub t.1947, 1837
Myanthus cristatus (Lindl.) Lindl., Bot. Reg. 23: sub t.1947, 1837
o Catasetum cornutum Lindl., Bot. Reg. 26: misc.182, 1840

Catasetum deltoideum (Lindl.) Mutel, Prem. Mém. Orch., 1838
BR 22: t.1896 (1836) / 26: misc.157 (1840) / 27: sub t.34:14 (1841) / 30: misc.40:22 (1844)
Type: Guyana, J.Hubbard s.n. cult. R. Harrison (K-Lindl., holo!)
Guyana – E
o Myanthus deltoideus Lindl., Bot. Reg. 22: t.1896, 1836

Catasetum discolor (Lindl.) Lindl., Bot. Reg. 27: sub t.34, 1841
BR 20: t.1735 (1835) / 26: misc.135 (1840) / 27: sub t.34:8, 9 & misc.12 (1841) / 30: misc.40:12 (1844)
Type: Guyana, Colley s.n. (K-Lindl., holo!)
Venezuela, Guyana, Surinam, Brazil, Colombia, Amazonas – E, T
o Monachanthus discolor Lindl., Bot. Reg. 20: t.1735, 1835
Monachanthus discolor var. bushnanii Hook., Bot. Mag. 64: t.3601 fig.A, 1837
Monachanthus discolor var. viridiflorus Hook., Bot. Mag. 64: t.3601, 1837
Catasetum roseo-album (Hook.) Lindl., Bot. Reg. 26: misc.135, 1840
Monachanthus roseo-albus Hook., Bot. Mag. 67: t.3796, 1840
Monachanthus bushnani var. ß Hook., Bot. Mag. 67: t.3832, 1841
Catasetum discolor var. fimbriatum Rchb.f., Saund. Ref. Bot. 2: t.83, 1869
Catasetum ciliatum Barb.Rodr., Gen. & Sp. Orch. Nov. 1: 130, 1877
Catasetum claesianum Cogn., J. d. Orch. 4: 267, 1893
Catasetum discolor var. viridiflorum Cogn., Mart. Fl. Bras. 3 (5): 445, 1902
Catasetum gardneri Schltr., Orchis 8: 84, 1914
Catasetum discolor var. cleasianum (Cogn.) Mansf., Fedde Rep. 30: 264, 1932
Catasetum discolor var. roseo-album (Hook.) Mansf., Fedde Rep. 30: 264, 1932

Catasetum globiflorum Hook., Bot. Mag. 68: t.3942, 1842
BR 28: misc.48 (1842) / 30: misc.40:10 (1844)
Brazil – E
Catasetum globiferum Beer, Prakt. Stud. Orch.: 205, 1854

Catasetum hookeri Lindl., Bot. Reg. 10: sub t.840, 1824
BR 10: sub t.840 (1824) / 24: misc.149 (1838) / 27: sub t.34:3 (1841) / 30: misc.40:9 (1844)
Type: Brazil, Swainson, cult. Hooker (Type not located!)
Brazil – E
Catasetum milleri Lodd. ex Lindl., Bot. Reg. 24: misc.149, 1838
Catasetum imschootianum Linden & Cogn., J. d. Orchid. 4: 266, 1893

Catasetum integerrimum Hook., Bot. Mag. 67: t.3823, 1840
BR 29: misc.1 (1843) / 30: misc.40:3 (1844)

Mexico, Belize, Guatemala, Honduras, El Salvador, Nicaragua, Costa Rica – E
Catasetum maculatum Bateman, Orch. Mex. & Guat. 2, 1837 (non Kunth)
Catasetum integerrimum var. purpurascens Hook., Bot. Mag. 67: sub t.3823, 1840
Catasetum integerrimum var. viridiflorum Hook., Bot. Mag. 67: sub t.3823, 1840
Catasetum wailesii Hook., Bot. Mag. 68: t.3937, 1842
Catasetum rostratum Klinge, Act. Hort. Petrop. 17 (1): 154, 1899
Catasetum brenesii Schltr., Fedde Rep. Beih. 19: 136 & 225, 1923

Catasetum laminatum Lindl., Ann. Mag. Nat. Hist. 4: 384, 1840
BR 27: t.5 fig.4 & sub t.34:17 (1841) / 30: misc.35 & misc.40:15 (1844)
Mexico – E
Catasetum laminatum var. maculatum Lindl., Sert. Orch.: t.38, 1840
o Catasetum laminatum var. eburneum Lindl., Bot. Reg. 27: t.5 fig.4, 1841

Catasetum longifolium Lindl., Bot. Reg. 25: misc.154, 1839
BR 25: misc.154 (1839) / 27: sub t.34:7 (1841) / 30: misc.40:11 (1844)
Type: Guyana, cult. Morris (K-Lindl., holo!)
Brazil, Venezuela, Guyana, Surinam – E
Monachanthus longifolius (Lindl.) Hook., Bot. Mag. 67: t.3819, 1840

o **Catasetum luridum** (Link.) Lindl., Gen. Sp. Orch. Pl.: 156, 1832
BR 20: t.1667 (1834) / 27: sub t.34:6 (1841) / 28: misc.24 (1842) / 30: misc.40:6 (1844)
Brazil – E
Anguloa lurida Link, Verhandl. Gartenb. Königl. Preuss. Staat. 1: 289 t.6, 1824
Epidendrum olare Vell., Fl. Flum. Ic. 9: t.5, 1827
Catasetum craniomorphum Hoffsgg. ex Heynh., Nom. Bot. 2: 126, 1840 (nomen tantum)
Catachaetum lituratum Hoffsgg., Verz. Orch.: 45, 1842
Catachaetum squalidum Hoffsgg., Verz. Orch.: 46, 1842
Catasetum abruptum Hook., Bot. Mag. 68: t.3929, 1842

Catasetum macrocarpum Rich. ex Kunth, Syn. Pl. Aequin. 1: 331, 1822
BR 10: t.840 & sub t.840:2, 3, 5 (1824) / 21: t.1752 (1835) / 27: sub.t.34:1 (1841) / 30: misc.40:1 (1844)
Trinidad, Venezuela, Guyana, Surinam, Brazil – 500–1100 m – E
o Catasetum claveringii van Jeel ex Lindl., Bot. Reg. 10: t.840, 1824
Catasetum floribundum Hook., Exot. Fl. 2: t.151, 1825
Catasetum tridentatum Hook., Exot. Fl. 2: t.90 & 91, 1825
o Monachanthus viridis Lindl., Gen. Sp. Orch. Pl.: 157, 1832
Catachaetum claveringii (van Jeel ex Lindl.) Hoffsgg., Verz. Orch.: 38, 1842
Catachaetum floribundum (Hook.) Hoffsgg., Verz. Orch.: 38, 1842
Catasetum tricolor hort. ex Planchon, Hort. Donat.: 25, 1858
Catasetum linguiferum Schltr., Beih. Bot. Centralbl. 42 (2): 120, 1925

o **Catasetum maculatum** Kunth, Syn. Pl. Aequin. 1: 331, 1822
BR 10: sub t.840:1 (1824) / 26: t.62 & misc.99 (1840) / 27: sub t. 34:2 (1841) / 30: misc.40:2 (1844)
Guatemala, Nicaragua, Costa Rica, Panama – E
Catasetum oerstedii Rchb.f., Bonpl. 3: 218, 1855

Catasetum naso Lindl., Bot. Reg. 29: misc.111, 1843
BR 29: misc.111 (1843) / 30: misc.40:18 (1844)
Type: Venezuela, Linden s.n. cult. Rucker (K-Lindl., holo!)
Venezuela – E

Catasetum ochraceum Lindl., Bot. Reg. 30: misc.55, 1844
BR 30: misc.55 (1844)
Type: Columbia, Hartweg s.n. (K-Lindl., holo!)
Venezuela, Colombia – L, T
Catasetum lehmannii Regel, Act. Hort. Petrop. 9: 602, 1880
Catasetum labiatum Rolfe, Hook. Ic. Pl.: t.2617, 1899
Catasetum inornatum Schltr., Orchis 10: 184, 1916
Catasetum bergoldianum Foldats, Act. Bot. Venez. 3: 318, 1968

o **Catasetum planiceps** Lindl., Bot. Reg. 29: t.9, 1843
BR 27: misc.168 (1841) / 29: t.9 (1843) / 30: misc.40:4 (1844) / 31: misc.31 (1845)
Type: Guyana, cult. Loddiges 934 (K-Lindl., holo!)
Brazil, Guyanas, Venezuela – T, L
Catasetum fuliginosum Lindl., Bot. Reg. 27: misc.168, 1841 (non Rolfe 1888)
Catachaetum recurvatum Link, Klotzsch & Otto Ic. Pl. Rar. 2: 105 t.42, 1844
Catasetum chloranthum Cogn., J. d. Orch. 5: 251, 1894
Catasetum hymenophorum Cogn., J. d. Orch. 6: 215, 1895

Catasetum poriferum Lindl., Bot. Reg. 24: misc.164, 1838
BR 24: misc.164 (1838) / 27: sub t.34:20 (1841) / 30: misc.40:21 (1844)
Type: Guyana, Schomburgk s.n. cult. Loddiges (K-Lindl., holo!)
Guyana – E

Catasetum purum Nees, Pl. Hort. Bonn. Ic. 1: t.1, 1824
BR 20: t.1708 (1834) / 27: sub t.34:4, 5 (1841) / 30: misc.40:7, 8 (1844)
Brazil – E
o Catasetum semiapertum Hook., Exot. Fl. 3: t.213, 1826
Catasetum inapertum Steudel, Nom. Bot. ed.2, 1: 311, 1840
Catachaetum purum (Nees) Hoffsgg., Verz. Orch.: 38, 1842
Catachaetum semiapertum (Hook.) Hoffsgg., Verz. Orch.: 38, 1842

Catasetum saccatum Lindl., Bot. Reg. 26: misc.179, 1840
BR 26: misc.179 (1840) / 30: misc.40:17 (1844)
Type: Guyana, cult. Loddiges (K-Lindl., holo!)
Guyana, Brazil, Ecuador, Peru, Bolivia – E
Catasetum incurvum Klotzsch, Allg. Gartenz. 22: 178, 1854
Catasetum secundum Klotzsch, Allg. Gartenz. 22: 178, 1854
Catasetum histrio Klotzsch ex Rchb.f., Bonpl. 4: 24, 1856
Catasetum stupendum Cogn., J. d. Orch. 6: 13, 1895
Catasetum cruciatum Schltr., Orchis 9: 29, 1915
Catasetum colossus Schltr., Bot. Centralbl. 42 (2): 118, 1925

Catasetum tabulare Lindl., Bot. Reg. 30: misc.40:16, 1844
BR 30: misc.40:16 (1844)
Type: Colombia s.n. (K-Lindl., holo!)
Colombia – E

o **Catasetum trulla** Lindl., Bot. Reg. 26: misc.176, 1840
BR 26: misc.176 (1840) / 27: t.34 & sub t.34:19 (1841) / 30: misc. 40: 19 (1844)
Type: cult. J.Rigby (K-Lindl., icono!)
Brazil – E
Cypripedium cothurmum Vell., Fl. Flum. Ic. 4: t.63, 1825
Cypripedium socco Vell., Fl. Flum Ic. 9: t.65, 1825
Catasetum liechtensteinii Kraenzlin, Gard. Chron. s.3, 11: 171, 1892
Paphiopedilum socco (Vell.) Pfitzer, Engl. Bot. Jahrb. 19: 42, 1894
Catasetum trulla var. liechtensteinii (Kraenzlin) Mansf., Fedde Rep. 30: 107, 1932
Catasetum trulla var. trulla (Lindl.) Mansf., Fedde Rep. 30: 107, 1932

Catasetum viridiflavum Hook., Bot. Mag. 69: t.4017, 1843
BR 33: sub t.24 (1847)
S.America, Panama – E
Catasetum serratum Lindl., Bot. Reg. 33: sub t.24, 1847

o **Cattleya aclandiae** Lindl., Bot. Reg. 26: t.48, 1840
BR 26: t.48 (1840) / 30: sub t.5:17 (1844)
Type: Brazil, cult. Craggs (K-Lindl., holo!)
Brazil (Bahia) – E
Epidendrum aclandiae (Lindl.) Rchb.f., Walp. Ann. 6: 312, 1861

Cattleya aurantiaca (Bateman ex Lindl.) P.N.Don, Fl. J.: 185, 1840
BR 24: misc.11 (1838) / 26: misc.22 & misc.82 (1840) / 31: misc.36 p.24:1 (1845)
Type: Mexico, Karwinski s.n. (K-Lindl., holo!)
Mexico, Guatemala, El Salvador, Honduras – up to 1600 m – E
Epidendrum aurantiacum Bateman ex Lindl., Bot. Reg. 24: misc.11, 1838
Broughtonia aurea Lindl., Bot. Reg. 26: misc.22, 1840
Epidendrum aureum Lindl., Fol. Orch. Epid. 4, 1853

Cattleya bicolor Lindl., Bot. Reg. 22: sub t.1919, 1836
BR 22: sub t.1919 (1836) / 24: misc.148 (1838) / 30: sub t.5:18 (1844)
Type: Brazil, cult. Pontys (K-Lindl., holo!)
Brazil – up to 800 m – E
Epidendrum iridee Descourt., Flore médicale des Antilles : 629, 1821
Epidendrum bicolor (Lindl.) Rchb.f., Walp. Ann. 6: 311, 1861
Cattleya bicolor var. measuresiana L.O.Williams, Orch. Alb. 8: t.357, 1888
Cattleya grossi Kraenzlin, Gartenfl. 46: 113 t.1436, 1897
Cattleya measuresiana (L.O.Williams) Brieger, Publ. Cient. Inst. Genét. Univ. S.Paulo 2: 26, 1961

o **Cattleya forbesii** Lindl., Coll. Bot.: sub t.37, 1821 ad. 1824
BR 11: t.953 (1826) / 30: sub t.5:12 (1844)
Brazil – E
Epidendrum pauper Vell., Fl. Flum. Ic : t.13, 1790
Maelenia paradoxa Dumort., L'Hortic. Belg. 2: 198 t.44, 1834
Cattleya vestalis Hoffsgg., Verz. Orch.: 46, 1843
Cattleya fulva Beer, Prakt. Stud. Orch.: 210, 1854
Cattleya isopetala Beer, Prakt. Stud. Orch.: 211, 1854
Epidendrum forbesii (Lindl.) Rchb.f., Walp. Ann. 6: 316, 1861
Cattleya pauper (Vell.) Stellf., Arq. Mus. Paranaense 4: 357, 1945

o **Cattleya granulosa** Lindl., Bot. Reg. 28: t.1, 1842
BR 28: t.1 (1842) / 30: sub t.5:16 (1844) / 31: t.59 (1845)
Type: Guatemala, Hartweg s.n. cult. Hort. S. London (K-Lindl., holo!)
Guatemala, Brazil – E

Cattleya granulosa var. russeliana Lindl., Bot. Reg. 31: t.59, 1845
Epidendrum granulosum (Lindl.) Rchb.f., Walp. Ann. 6: 319, 1861

o **Cattleya guttata** Lindl., Bot. Reg. 17: t.1406, 1831
BR 17: t.1406 (1831) / 30: sub t.5:15 (1844)
Type: Brazil, R.Gordon s.n. cult. Hort. Soc. (K-Lindl., holo!)
Brazil – E
Epidendrum elegans Vell., Fl. Flum. Ic. 9, 1825
Cattleya elatior Lindl., Gen. Sp. Orch. Pl.: 117, 1831
Cattleya sphenophora Morren, Ann. Gand. 4 (17): t.175, 1848 (non Lindl.)
Cattleya tigrina A.Rich., Portf. d. Hort. 2: 166, 1848
Epidendrum elatius Rchb.f., Walp. Ann. 6: 319, 1861

Cattleya harrisoniana Bateman ex Lindl., Bot. Reg. 22: sub t.1919, 1836
BR 22: sub t.1919 (1836) / 30: sub t.5:11 (1844) / 31: misc.47 (1845)
Type: Brazil, Harrison s.n. (K-Lindl., holo!)
Brazil – E
Cattleya harrisoniae Paxton, Mag. Bot. 4: 247, 1838
Cattleya concolor Drapiez, Hort. Univ. 2: 197, 1841
Cattleya intermedia var. variegata Hook., Bot. Mag. 70: t.4085, 1844
Cattleya papeiansiana Morren, Ann. Gand. 1: 57 t.5, 1845
Cattleya harrisonii Beer, Prakt. Stud. Orch.: 211, 1854
Epidendrum harrisonianum (Bateman ex Lindl.) Rchb.f., Walp. Ann. 6: 317, 1861
Cattleya loddigesii var. harrisoniae (Paxton) Veitch, Man. Orch. Pl. 2: 42, 1887
Cattleya loddigesii var. harrisoniana (Bateman ex Lindl.) Rolfe, Gard. Chron. s.3, 5: 746, 1889

Cattleya intermedia Graham ex Hook., Bot. Mag. 55: t.2851, 1828
BR 22: t.& sub t.1919 (1836) / 30: sub t.5:13 (1844)
Brazil – E
o Cattleya intermedia var. pallida Lindl., Bot. Reg. 22: t.1919, 1836
Cattleya maritima Lindl., Bot. Reg. 22: sub t.1919, 1836
Cattleya ovata Lindl., Bot. Reg. 22: sub t.1919, 1836
Cattleya loddigesii var.intermedia (Graham ex Hook.) Lindl., Bot. Reg. 30: sub t.5:13, 1944 (missaplied name)
Cattleya amethystina Morren, Ann. Gand. 4: 217 t.201, 1848
Cattleya ianthina hort., Gard. Chron.: 247, 1853 (nomen)
Cattleya loddigesii var. amethystina (Morren) Lemaire, Jard. Fleur. 4: sub t.379, 1853
Epidendrum intermedium (Graham ex Hook.) Rchb.f., Walp. Ann. 6: 318, 1861
Cattleya amabilis Lindl. ex Du Buyss., L'Orchidoph.: 232, 1878
Cattleya aquinii Barb.Rodr., Pl. Nov. Jard. Bot. Rio 1: 23 t.4c, 1891

o **Cattleya labiata** Lindl.; Coll. Bot.: t.33, 1821–24
BR 22: t.1859 (1836) / 30: sub t.5:5 (1844) / 32: t.35 (1846)
Brazil – E
Cattleya perrinii Endl., Harting. Parad. Vindob. 1: t.1, 1844 (non Lindl.)
o Cattleya lemoniana Lindl., Bot. Reg. 32: t.35, 1846
Cattleya pallida Lindl., Paxt. Flow. Gard. 2: 51 t.48, 1851
Epidendrum labiatum (Lindl.) Rchb.f., Walp. Ann. 6: 313, 1861
Cattleya regalis hort., Gard. Chron. n.s.20: 404, 1883
Cattleya nalderiana Rchb.f., Gard. Chron. n.s.23: 756, 1885
Cattleya autumnalis hort., L'Orchidoph.: 122, 1887
Cattleya labiata var. autumnalis L.Linden, Lindenia 3: 35 t.112, 1887
Cattleya labiata var. warocqueanum Rolfe, Gard. Chron. s.3: 735, 1890
Cattleya warocqueana L.Linden, J. d. Orch. 1: 219, 1890
Cattleya labiata var. genuina Stein, Orchideenb.: 124, 1892

Cattleya loddigesii Lindl., Coll. Bot.: sub t.33 & 37, 1821–24
BR 29: misc.123 (1843) / 30: sub t.5:10 & 14 (1844)
Brazil – E
Epidendrum violaceum Lodd., Bot. Cab.: t.337, 1819
Epidendrum canaliculatum Vell., Fl. Flum. Ic. 9: t.10, 1827
Cattleya arembergii Scheidw., Otto & Dietr. Allg. Gartenz. 11: 109, 1843
Epidendrum loddigesii (Lindl.) Rchb.f., Walp. Ann. 6: 316, 1861
Cattleya loddigesii var. harrisoniana Linden, Lindenia 14: t.667, 1899 (non Veitch)

o **Cattleya maxima** Lindl., Gen. Sp. Orch. Pl.: 116, 1831
BR 30: sub t.5:4 (1844) / 32: t.1 (1846)
Colombia, Ecuador, Brazil, Amazonas – E
Epidendrum maximum (Lindl.) Rchb.f., Walp. Ann. 6: 316, 1861
Cattleya malouana Linden & Rodigas, Ill. Hort. 33: 102, 1886 (nomen nudum)

Cattleya mossiae Hook., Bot. Mag. 65: t.3669, 1838
BR 26: t.58 (1840)
Venezuela – E
o Cattleya labiata var. mossiae (Hook.) Lindl., Bot. Reg. 26: t.58, 1840
Cattleya labiata var. atropurpurea Paxton, Mag. Bot. 7: 73, 1844
Cattleya labiata var. candida Lindl. & Paxton, Paxt. Flow. Gard. 1: 118, 1850
Cattleya labiata var. picta Lindl. & Paxton, Paxt. Flow. Gard. 1: 118, 1850
Cattleya wageneri Rchb.f., Bonpl. 2: 21, 1854
Epidendrum labiatum var. mossiae (Hook.) Rchb.f., Walp. Ann. 6: 314, 1861
Epidendrum labiatum var. pictum (Lindl. & Paxton) Rchb.f., Xen. Orch. 2: 30, 1862
Cattleya edithiana Williams, Orch. Grow. Man. ed.3: 88, 1868
Cattleya carrieri Houllet, Rev. Hort.: 350, 1876
Cattleya reineckiana hort., Gard. Chron. n s.20: 172, 1883
Cattleya aliciae L.Linden, Lindenia 10: 31 t 494, 1895

Cattleya skinneri Bateman, Orch. Mex. & Guat.: t.13, 1838
BR 26: misc.83 (1840) / 30: sub t.5:2 (1844)
Mexico, Belize, Guatemala, El Salvador, Honduras, Costa Rica – E
Epidendrum huegelianum Rchb.f., Ann. Bot. 6: 312, 1862

Cattleya violacea (Kunth) Rolfe, Gard. Chron. s.3, 5: 802, 1889
BR 25: misc.47 (1839) / 30: sub t.5:1 (1844)
Venezuela, Guyana, Brazil, Peru – E
Cymbidium violaceum Kunth, Nov. Gen. & Sp. Pl. 1: 341, 1815
Cattleya schomburgkii Lodd. ex Lindl., Sert. Orch.: sub t.22, 1838
Cattleya superba Schomb. ex Lindl., Sert. Orch.: t.22, 1838
Cattleya odoratissima Don., Florists J.: 185, 1840
Epidendrum violaceum (Kunth) Rchb.f., Walp. Ann. 6: 318, 1861
Epidendrum superbum (Schomb. ex Lindl.) Rchb.f., Xen. Orch. 2: 32, 1862

Cattleya walkeriana (Rchb.f.) Gardner, Hook. Lond. J. Bot. 2: 662, 1843
BR 30: sub t.5:3 (1844) / 33: t.42 (1847)
Brazil – E
o Cattleya bulbosa Lindl., Bot. Reg. 33: t.42, 1847
Epidendrum walkerianum Rchb.f., Walp. Ann. 6: 416, 1862
Cattleya gardneriana Rchb.f., Gard. Chron.: 1473, 1870
Cattleya princeps Barb.Rodr., Gen. & Sp. Orch. Nov. 1: 68, 1877
Cattleya schroederiana Rchb.f., Gard. Chron. n.s.20: 102, 1883

Cattleyopsis cubensis (Lindl.) Sauleda & Adams, Rhodora 86 (848): 455, 1984
BR 29: misc.24 (1843)
Type: Cuba, cult. Loddiges (K-Lindl., holo!)
Cuba – E
Epidendrum cubense Lindl., Bot. Reg. 29: misc.24, 1843
Laeliopsis cubensis (Lindl.) Lindl., Paxt. Fl Gard. 3: 156, 1852
Broughtonia cubensis (Lindl.) Cogn., Urb. Symb. Antill. 6: 512, 1910

Caularthron bicornutum (Hook.) Raf., Fl. Tellur. 2: 41, 1836
BR 31: misc.36 p.24:2 (1845)
Brazil, Guyana, Venezuela, Trinidad, Tobago, Colombia – E
Epidendrum bicornutum Hook., Bot. Mag. 61: t.3332, 1834
Diacrium bicornutum (Hook.) Benth., J. Linn. Soc. 18: 312, 1881
Diacrium amazonicum Schltr., Beih. Bot. Centralbl. 42 (2): 108, 1925

Chamaeangis gracilis (Thouars) Schltr., Beih. Bot. Centralbl. 33 (2): 426, 1915
BR 18: sub t.1522 (1832)
Madagascar, Mascarene Is. – E
Angraecum gracile Thouars, Orch. Iles Afr.: t.77, 1822
Mystacidium gracile (Thouars) Finet, Bull. Soc. Bot. Fr.: 54, 1907

Chamaeangis vesicata (Lindl.) Schltr., Die Orchideen: 109, 1918
BR 29: misc.9 (1843)
Type: Barter s.n. cult. Loddiges (K-Lindl., holo!)
Guinea, Sierra Leone, Liberia, Ghana, S.Nigeria, Fernando Po, E.Cameroun, Gabon, Congo, Uganda, Kennya, Tanzania – E
Angraecum vesicatum Lindl., Bot. Reg. 29: misc.9, 1843
Listrostachys vesicata (Lindl.) Rchb.f., Flora 48: 190, 1865

Chamorchis alpina (L.) Rich., Orchid. Eur. Annot.: 35, 1817
BR 18: sub t.1499:9 (1832)
Europe – T
Ophrys alpina L., Sp. Pl.: 948, 1753
Orchis graminea Crantz, Stirp. Austr.: 480, 1767
Orchis alpina (L.) Schrank, Bayr. Fl.: 227, 1789
Epipactis alpina (L.) F.W.Schmidt, Mayer Samml. Pfl. Phys. Aufs.: 247, 1791
Arachnites alpina (L.) Schmidt, Fl. Boem.: 74, 1793
Satyrium alpinum (L.) Pers., Syn. 2: 507.1801
Chamaeorchis alpina (L.) Rich., Ann. Mus. 4: 42, 49 & 57, 1817
Chamaerepes alpina (L.) Sprengel, Syst. Veg. 3: 702, 1828
Herminium alpinum (L.) Lindl., Bot. Reg. 18: sub t.1499 :9, 1832

Cheirostylis parvifolia Lindl., Bot. Reg. 25: misc.20, 1839
BR 25: misc.20 (1839)
Type: Ceylon, cult. Loddiges (K-Lindl., holo!)
Sri Lanka – T

Chilochista pusilla (Koen.) Schltr., Fedde Rep. Beih. 4: 275, 1919
BR 18: sub t.1522 (1832)
Sri Lanka, Deccan, Thailand – E

Epidendrum pusillum Koen., Retz. Obs. 6: 49, 1791
Limodorum pusillum (Koen.) Willd., Sp. Pl. ed.4: 126, 1805
Oeceoclades retzii Lindl., Gen. Sp. Orch. Pl.: 237, 1833
Chilochista usneoides Wight, Ic. 5: 19 t.1741, 1852 (non Lindl.)
Sarcochilus fasciatus F.Muell., Fragm. 5: 202, 1866
Sarcochilus minimifolium Hook.f., Fl. Brit. Ind. 6: 37, 1890
Sarcochilus wightii Hook.f., Fl. Brit. Ind. 6: 37, 1890
Chilochista minimifolia (Hook.f.) Balakr., J. Bomb. Nat. Hist. Soc. 67: 66, 1970

Chilochista usneoides (D.Don) Lindl., Bot. Reg. 18: sub t.1522, 1832
BR 18: sub t.1522 (1832)
N.W.Himalaya, Nepal, Sikkim, Bhutan, Thailand – E
Epidendrum usneoides D.Don, Prodr. Fl. Nepal: 37, 1825
Aerides convallaroides Wall. ex. Lindl., Sert. Orch.: sub n.4, 1840
Sarcochilus usneoides (D.Don) Rchb.f., Walp. Ann. 6: 497, 1863
Thrixspermum usneoides (D.Don) Rchb.f., Xen. Orch. 2: 120, 1867

Chloraea gavilu Lindl., Quart. J. Sc. Lit. Art: 48, 1827
BR 31: t.49 (1845)
Chile – T
Cymbidium luteum Willd., Sp. Pl. ed.4: 106, 1805
Chloraea lindleyii Poeppig, Frag. Syn. Pl. Phan.: 16, 1833
o Chloraea virescens Lindl., Gen. Sp. Orch. Pl.: 404, 1840 (non Lindl. 1827)
Chloraea sceptrum Rchb.f., Linnaea 22: 862, 1849
Asarca sceptrum (Rchb.f.) Kuntze, Rev. Gen. 2: 652, 1891
Asarca virescens (Lindl.) Kuntze, Rev. Gen. 2: 652, 1891
Chloraea semitensis Kraenzlin, Orchid. 2: 106, 1904 (non Poeppig)
Chloraea lutea (Willd.) Schltr., Die Orchideen: 94, 1914
Chloraea lotensis Rolfe, Kew Bull.: 80, 1916

o **Chysis aurea** Lindl., Bot. Reg. 23: t.1937, 1837
BR 23: t.1937 (1837) / 26: misc.130 (1840)
Type: Venezuela, Hechman s.n cult. Lowe (K-Lindl., holo!)
Mexico, Belize, Costa Rica, Panama, Venezuela, Colombia, Peru – E
Chysis laevis Lindl., Bot. Reg. 26: misc.130, 1840

o **Chysis bractescens** Lindl., Bot. Reg. 26: misc.131, 1840
BR 26: misc.131 (1840) / 27: t.23 (1841)
Type: Mexico, cult. G.Barker (K-Lindl., holo!)
Mexico, Guatemala – E
Thorvaldsenia speciosa Liebm., Bot. Not.: 103, 1844
Chysis aurea var. bractescens (Lindl.) Allen, Amer. Orch. Soc. Bull. 24: 665, 1955

Cirrhaea dependens (Lodd.) Rchb.f., Walp. Ann. 6: 496, 1863
BR 18: sub t.1538 (1832) / 22: t.1889 (1836)
Brazil – E
Cymbidium dependens Lodd., Bot. Cab. 10: t.936, 1825
Gongora viridipurpurea Hook., Bot. Mag. 57: t.2978, 1830
Cirrhaea fusco-lutea Lindl., Bot. Reg. 18: sub t.1538, 1832
Cirrhaea viridipurpurea (Hook.) Lindl., Bot. Reg. 18: sub t.1538, 1832
Cirrhaea warreana Lodd., Bot. Cab.: t.1999, 1833
o Cirrhaea tristis Lindl., Bot. Reg. 22: t.1889, 1836
Cirrhaea violaceo-virens Hoffsgg., Verz. Orch.: 47, 1843
Cirrhaea violascens Hoffsgg., Verz. Orch.: 47, 1843
Sarcoglossum suaveolens Beer, Prakt. Stud. Orch.: 306, 1854
Cirrhaea hoffmannseggii Heinhold ex Rchb.f., Walp. Ann. 6: 496, 1863
Cirrhaea livida Hoffsgg., Walp. Ann. 6: 497, 1863
Cirrhaea purpurascens Hoffsgg., Walp. Ann. 6: 497, 1863
Cirrhaea russelliana Lodd. ex Rchb.f., Walp. Ann. 6: 497, 1863

o **Cirrhaea loddigesii** Lindl., Bot. Reg. 18: sub t.1538, 1832
BR 18: sub t.1538 (1832) / 23: t. & sub t.2005 (1837)
Type: Brazil, cult. Loddiges (K-Lindl., holo!)
Brazil – E
Cirrhaea dependens Loudon, Hort. Brit.: 370, 1827 (non Rchb.f.)
Cirrhaea viridipurpurea Lodd., Bot. Cab.: t.1967, 1833 (non Lindl.)
o Cirrhaea obtusata Lindl., Bot Reg. 23: t.2005, 1837
Cirrhaea pallida Lindl., Bot. Reg. 23: sub t.2005, 1837
Cirrhaea obtusata var. pallida (Lindl.) Stein, Orchideenb.: 144, 1892

Cirrhaea saccata Lindl., Bot. Reg. 25: misc.121, 1839
BR 25: misc.121 (1839)
Type: cult. Duke of Bedford (K-Lindl., holo!)
Brazil – E
Cirrhaea fusco-lutea Hook., Bot. Mag. 66: t.3726, 1839 (non Lindl.)
Scleropteris flava Scheidw., Allg. Gartenz. 7: 407, 1839

Cleisostoma discolor Lindl., Bot. Reg. 31: misc.61, 1845
BR 31: misc.61 (1845)
Type: India, cult. Loddiges (K-Lindl., holo!)
India, Sikkim, Thailand, Cambodia, Malaya, Sumatra, Java – E
Sarcanthus termissus Rchb.f., Hamb. Gartenz. 16: 15, 1860
Sarcanthus macrodon Rchb.f., Gard. Chron.: 1555, 1872
Saccolabium rostellatum Hook.f., Fl. Brit. Ind. 6: 59, 1890
Sarcanthus auriculatus Rolfe, Kew Bull.: 9, 1895
Sarcanthus discolor (Lindl.) J.J.Smith, Nat. Tijds. Ned. Ind. 72: 85, 1912
Sarcanthus josephii J.J.Smith, Bull. Buitenz. s.2, 9: 108, 1913
Sarcanthus angkorensis Guillaumin, Bull. Soc. Bot. Fr. 77: 328, 1930
Cleisostoma auriculatum (Rolfe) Garay, Bot. Mus. Leafl. Harv. Univ. 23 (4): 169, 1972
Cleisostoma termissus (Rchb.f.) Garay, Bot. Mus. Leafl. Harv. 23 (4): 175, 1972

Cleisostoma filiforme (Lindl.) Garay, Bot. Mus. Leafl. Harv. Univ. 23 (4): 171, 1972
BR 28: misc.69 (1842)
Type: India, Wallich s.n. cult. Fielding (K-Lindl., holo!)
Sikkim, Assam, Burma, Thailand, Vietnam – E
Sarcanthus filiformis Lindl., Bot. Reg.: misc.69, 1842 (non Wight 1851)
Saccolabium luisioides Gagnepain, Bull. Mus. Hist. Nat. Paris s.2, 22 (5): 629, 1950

Cleisostoma paniculatum (Ker Gawler) Garay, Bot. Mus. Leafl. Harv. Univ. 23 (4): 173, 1972
BR 3: t.220 (1817) / 6: sub t.506 (1820) / 6: notes (1821)
Assam, China, Taiwan – E
o Aerides paniculata Ker Gawler, Bot. Reg. 3: t.220, 1817
Vanda paniculata (Ker Gawler) R.Br., Bot. Reg. 6:notes, 1821
Sarcanthus paniculatus (Ker Gawler) Lindl., Ill. Orch. Pl.: t.9, 1832
Cleisostoma cerinum Hance, J. Bot. 20: 359, 1882
Cleisostoma formosum Hance, J. Bot. 22: 364, 1884
Sarcanthus cerinus (Hance) Rolfe, J. Linn. Soc. 36: 36, 1903
Sarcanthus formosanus (Hance) Rolfe, J. Linn. Soc. 36: 37, 1903

Cleisostoma racemiferum (Lindl.) Garay, Bot. Mus. Leafl. Harv. Univ. 23 (4): 173, 1972
BR 26: misc.185 (1840)
E.Nepal, Sikkim, Bhutan, Yunnan, Burma, Thailand, Laos, Vietnam – E
Aerides racemiferum Lindl., Gen. Sp. Orch. Pl.: 224, 1833
Saccolabium racemiferum Lindl., Gen. Sp. Orch. Pl.: 224, 1833
Sarcanthus pallidus Lindl., Bot. Reg. 26: misc.185, 1840
Sarcanthus tricolor Rchb.f., Bonpl. 3: 219, 1855
Sarcanthus racemiferum (Lindl.) Rchb.., Walp. Ann. 6: 68, 1861
Sarcanthus chrysomelas Rchb.f., Gard. Chron.: 662, 1869
Sarcanthus lorifolius Parish ex Hook.f., Fl. Brit. Ind. 6: 69, 1890
Sarcanthus affinis Wall. ex Hook.f., Fl. Brit. Ind.: 53, 1895
Sarcanthus yunnanensis Schltr., Fedde Rep. Beih. 4: 77, 1919
Sarcanthus manlinensis Guillaumin, Bull. Mus. Hist. Nat. Paris. s.2, 28: 239, 1956
Cleisostoma chrysomelas (Rchb.f.) Garay, Bot. Mus. Leafl. Harv. Univ. 23 (4): 170, 1972

Cleisostoma rostratum (Lindl.) Garay, Bot. Mus. Leafl. Harv. Univ. 23 (4): 174, 1972
BR 12: t.981 (1826)
Laos, Vietnam, China – E
Vanda rostrata Lodd., Bot. Cab. 11: t.1008, 1825
o Sarcanthus rostratus Lindl., Coll. Bot.: t.39b, 1826
Vanda recurva Hook., Exot. Fl. 3: t.187, 1826
Cleisostoma fordii Hance, J. Bot. 14: 45, 1876
Sarcanthus fordii (Hance) Rolfe, J. Linn. Soc. 36: 37, 1903
Saccolabium simondii Gagnepain, Bull. Mus. Hist. Nat. Paris s.2, 5: 629, 1950
Sarcanthus loasensis Guillaumin, Bull. Mus. Hist. Nat. Paris s.2, 35: 650, 1963

Cleisostoma simondii (Gagnepain) Seidenf., Dansk Bot. Ark. 29 (3): 66, 1975
BR 8: t.676 (1823)
Sikkim, Laos, Vietnam, Hainan, Hong Kong – E
o Vanda teretifolia Lindl., Coll. Bot.: t.6, 1821
Sarcanthus teretifolius (Lindl.) Lindl., Gen. Sp. Orch. Pl.: 324, 1833 (non Rchb.f.1874)
Sarcanthus appendiculatus King & Pantling, Orch. Sikkim Himalaya: 240 t.318, 1898 (non (Lindl.) Benth. & Hook.f.)
Sarcanthus siamensis Rolfe ex Downie, Kew Bull.: 405, 1925
Luisia trichorhiza Merrill & Metcalf, Lingnan Sc. J. 21 (1–4): 12, 1945 (non (Hook.) Blume)
Vanda simondii Gagnepain, Bull. Mus. Hist. Nat. Paris s.2, 22: 628, 1950
Luisia acutilabris Guillaumin, Bull. Mus. Hist. Nat. Paris s.2, 32 (2): 188, 1960
Cleisostoma seidenfadenii Garay, Bot. Mus. Leafl. Harv. Univ. 23 (4): 174, 1972
Cleisostoma teres Garay, Bot. Mus. Leafl. Harv. Univ. 23 (4): 174, 1972
non Cleisostoma siamensis Rolfe ex Downie, Kew Bull.: 405, 1925
non Cleisostoma teretifolium Teijsm. & Binnend., Nat. Tijds. Ned. Ind. 27: 20, 1864

Cleisostoma subulatum Blume, Bijdr.: 363, 1825
BR 26: misc.123 (1840) / 29: misc.5 (1843)

India, Sikkim, Assam, Thailand, Cambodia, Malaya, Indonesia, Borneo, Philippines – E
Sarcanthus oxyphyllus Wall. ex Lindl., Bot. Reg. 26: misc.123, 1840
Cleisostoma dealbatum Lindl., Bot. Reg. 29: misc.5, 1843
Angraecum pugioniforme Klotzsch, Ind. Sem. Hort. Berol. 1, 1851
Cypripedium linearisubulatum Llanos, Frag. Fl. Filip.: 99, 1851
Sarcanthus secundus Griffith, Notul. 3: 362, 1851
Cleisostoma amabile Teijsm. & Binnend., Nat. Tijdss. Ned. Ind. 5: 493, 1853
Sarcanthus pugioniformis (Klotzsch) Rchb.f., Allg. Gartenz. 24: 219, 1856
Sarcanthus subulatus (Blume) Rchb.f., Bonpl. 5: 41, 1857
Sarcanthus dealbatus (Lindl.) Rchb.f., Walp. Ann. 6: 892, 1864
Campylocentrum pugioniforme (Klotzsch) Rolfe, Orch. Rev. 11: 245, 1903
Saccolabium secundum (Griffith) Ridley, Mat. Fl. Mal. Pen. 1: 168, 1907
Sarcanthus amabilis (Teijsm. & Binnend.) J.J.Smith, Nat. Tijdsschr. Ned. Ind.: 83, 1912
Cleisostoma pugioniforme (Klotzsch) Garay, Bot. Mus. Leafl. Harv. Univ. 23 (4): 173, 1972

o **Clowesia rosea** Lindl., Bot. Reg. 29: t.39, 1843
BR 29: t.39 & misc.39 (1843)
Type: Brazil, cult. Clowes (K-Lindl., holo!)
Brazil – E
Catasetum roseum (Lindl.) Rchb.f., Gard. Chron.: 1003, 1872

Clowesia russelliana (Hook.) Dodson, Selbyana 1: 136, 1975
BR 26: misc.19 (1840) / 27: sub t.34:18 (1841) / 30: misc.40:14 (1844)
Mexico, Guatemala, El Salvador, Honduras, Nicaragua, Panama, Venezuela – E
Catasetum russellianum Hook., Bot. Mag. 66: t.3777, 1840
Catasetum calceolatum Lemaire, Jard. Fleur. 1: misc.p.45, 1851
Cycnoches viride Koch, Allg. Gartenz. 25: 291, 1857

Coccineorchis cernua (Lindl.) Garay, Fl. Ecuador Orch.: 237, 1978
BR 19: sub t.1618 (1833)
Peru – T
Synassa corymbosa Lindl., Bot. Reg. 19: sub t.1618, 1833
Stenorrhynchus cernuus Lindl., Ann. Mag. Nat. Hist. 15: 386, 1845
Spiranthes corymbosa (Lindl.) Kraenzlin, Engl. Bot. Jahrb. 54: 17 Beibl. 117, 1916
Coccineorchis corymbosa (Lindl.) Schltr., Beih. Bot. Centralbl. 37 (2): 434, 1920
Spiranthes coccinea Garay, Bot. Mus. Leafl. Harv. Univ. 21: 250, 1967

Cochleanthes flabelliformis (Sw.) R.Schultes & Garay, Bot. Mus. Leafl. Harv. Univ. 18: 324, 1959
BR 22: t.1857 (1836) / 30: misc.15:9 (1844)
Trop.America, W.Indies, Nicaragua to Venezuela, Brazil – E
Epidendrum flabelliforme Sw., Prodr. Veg. Ind. Occ.: 123, 1788
Cymbidium flabelliforme (Sw.) Sw., Nov. Act. Upsal. 6: 73, 1799
Cochleanthes fragrans Raf., Fl. Tellur. 4: 45, 1836
o Zygopetalum cochleare Lindl., Bot. Reg. 22: t.1857, 1836
Eulophia cochleare (Lindl.) Knight ex Hook., Bot. Mag. 64: sub t.3585, 1837
Zygopetalum cochleatum Paxton, Mag. Bot. 4: 279, 1838
Eulophia cochlearis (Lindl.) Steudel, Nom. ed.2, 1: 605, 1840
Huntleya imbricata hort. ex Rchb.f., Bot. Zeit. 10: 714, 1852
Warscewiczella cochlearis (Lindl.) Rchb.f., Bot. Zeit. 10: 714, 1852
Zygopetalum conchaceum Hoffsgg. ex Rchb.f., Walp. Ann. 6: 652, 1863
Zygopetalum flabelliforme (Sw.) Rchb.f., Walp. Ann. 6: 652, 1863
Cymbidium flabellifolium (Sw.) Griseb., Fl. Brit. W. Ind.: 629, 1864
Warscewiczella cochleata (Paxton) Barb.Rodr., Struct. Orch.: t.13 fig.4, 1883
Zygopetalum gibeziae N.E.Br., Lindenia 4: 79 t.181, 1888
Warscewiczella gibeziae (N.E.Br.) Stein, Orchideenb.: 595, 1892
Warscewiczella flabelliformis (Sw.) Cogn., Symb. Antill. 4: 182, 1903

Cochlioda rosea (Lindl.) Benth., J. Linn. Soc. 18: 327, 1881
BR 31: misc.60:6 (1845)
Peru, Ecuador – up to 1800 m – E
Odontoglossum roseum Lindl., Benth. Pl. Hartw.: 151, 1844
Mesospinidium roseum (Lindl.) Rchb.f., Gard. Chron.: 393, 1872

Coelia macrostychya Lindl., Pl. Hartw.: 92, 1842
BR 28: sub t.36 (1842)
Mexico, Guatemala, El Salvador, Honduras, Nicaragua, Costa Rica, Panama – T, E, L
Coelia macrostachya var. genuina Rchb.f., Beitr. Orch. Centr.-Amer.: 41, 1866
Coelia macrostachya var. integrilabia Rchb.f., Beitr. Orch. Centr.-Amer.: 41, 1866
Bothriochilus macrostachyus (Lindl.) L.Williams, Bot. Mus. Leafl. Harv. Univ. 8: 148, 1940

Coelia triptera (Smith) G.Don ex Steudel, Nom. ed.2, 1: 394, 1840
BR 28: t.36 (1842)
Mexico, Guatemala, W.Indies – up to 1500 m – E
Epidendrum tripterum Smith, Ic. Pict.: t.14, 1793
Cymbidium tripterum (Smith) Sw., Nov. Act. Upsal. 6: 70, 1799
o Coelia baueriana Lindl., Gen. Sp. Orch. Pl.: 36, 1830
Coelia galeottiana Houtte ex Heynhold, Nom. 2: 152, 1846 (nomen)
Coelia glacialis Houtte ex Heynhold, Nom. 2: 152, 1846 (nomen)

o **Coelogyne cristata** Lindl., Coll. Bot.: t.33, 1821
BR 27: t.57 & misc.54 (1841)
N.E.India, Sikkim – 1000–2000 m – E
Cymbidium speciosissimum D.Don, Prodr. Fl. Nepal.: 35, 1825

o **Coelogyne cumingii** Lindl., Bot. Reg. 26: misc.178, 1840 (non Naves 1882)
BR 26: misc.178 (1840) / 27: t.29 (1841)
Type: Singapore, Cuming s.n. cult. Loddiges (K-Lindl., holo!)
Laos, Malaya, Sumatra, Riau Arch., Borneo – E
Pleione cumingii (Lindl.) Kuntze, Rev. Gen. 2: 680, 1891

Coelogyne elata Lindl., Gen. Sp. Orch. Pl.: 40, 1830 (non Hook. 1857)
BR 25: misc.151 (1839)
Sikkim, Nepal, Bhutan, Burma – E

o **Coelogyne fimbriata** Lindl., Bot. Reg. 11: t.868, 1825 (non Parish)
BR 11: t.868 (1825) / 24: misc.172 (1838)
Type: China, J.Damper Parks s.n. cult. Hort. Soc. (K-Lindl., holo!)
Khasia, Manipur, Laos, Cambodia, Vietnam, China, Malaya – E
Pleione chinense Kuntze, Rev. Gen. 2: 680, 1819
Pleione fimbriata (Lindl.) Kuntze, Rev. Gen. 2: 680, 1891
Coelogyne ovalis Pfitzer & Kraenzlin, Engl. Pflanzenr. Mon. Coel. 32: 53 fig.17d-e, 1907 (non Lindl.)
Coelogyne laotica Gagnepain, Bull. Mus. Sc. Nat. Paris 2.s, 2: 425, 1930
Coelogyne leungiana Hu, Quart. J. Taiw. Mus. 25 (3/4): 223 figs.8–9, 1972

o **Coelogyne flaccida** Lindl., Gen. Sp. Orch. Pl.: 39, 1830
BR 27: t.31 (1841)
Nepal, Sikkim, Burma, China (Yunnan), Laos – 1000–1600 m – E

Coelogyne fuscescens Lindl., Gen. Sp. Orch. Pl.: 41, 1830
var. brunnea (Lindl.) Lindl., Fol. Orch. Coel. 11, 1854
BR 30: misc.1 (1844)
Type: cult. Duke of Northumberland (K-Lindl., holo!)
Burma, Laos, Thailand, Vietnam – 1000–1500 m – E
Coelogyne fuscescens Lindl., Bot. Reg. 30: misc.1, 1844 (non Lindl.1830)
Coelogyne brunnea Lindl., Gard. Chron.: 71, 1848

Coelogyne nitida (Wallich ex Don) Lindl., Coll. Bot.: sub t.33, 1822
BR 32: t.69 (1846)
N.W.Himalaya, Nepal, Sikkim, Bhutan, Burma, China (Yunnan), Thailand, Laos – 1200–2200 m – E
Cymbidium nitidum Wallich ex D. Don, Prodr. Fl. Nepal.: 35, 1825 (non Roxb.1832)
o Coelogyne ochracea Lindl., Bot. Reg. 32: t.69, 1846
Coelogyne goweri Rchb.f., Gard. Chron.: 443, 1869
Coelogyne ochracea ssp. conferta Parish & Rchb.f., Trans. Linn. Soc. 30: 146 t.30, 1874
Coelogyne conferta hort., Gard. Chron.: 314, 1875
Pleione goweri (Rchb.f.) Kuntze, Rev. Gen. 2: 680, 1891
Coelogyne ochracea var. foliata (Rchb.f.) Pfitzer & Kraenzlin, Engl. Pflanzenr. Mon. Coel. 32: 57, 1907
Coelogyne punctulata var. conferta (Parish & Rchb.f.) Tang & Wang, Act. Phytotax. Sin. 1: 78, 1951

Coelogyne ovalis Lindl., Bot. Reg. 24: misc.171, 1838
BR 24: misc.171 (1838)
Type: Wallich s.n. cult. Loddiges (K-Lindl., holo!)
W.Himalaya, Nepal, Sikkim, Assam, Manipur, Burma, China (Tibet, Yunnan) – 800–1900 m – E
Coelogyne pilosissima Planchon, Hort. Donat. Orch.: 104, 1854
Coelogyne triplicatula Rchb.f., Bot. Zeit. 53: 415, 1864
Broughtonia linearis Wallich ex Hook.f., Fl. Brit. Ind. 5: 836, 1890

Coelogyne prolifera Lindl., Gen. Sp. Orch. Pl.: 40, 1830
BR 24: misc.75 (1838)
Sikkim, Yunnan, Khasia, Thailand, Vietnam – E

Coelogyne punctulata Lindl., Coll. Bot.: sub t.33, 1821
BR 25: misc.25 (1839)
Nepal, Assam, Burma, W.China – E
Cymbidium nitidum Roxb., Hort. Beng.: 63, 1814
Coelogyne ocellata Lindl., Gen. Sp. Orch. Pl.: 40, 1830
Coelogyne nitida (Roxb.) Hook.f., Fl. Brit. Ind. 5: 837, 1890

○ **Coelogyne speciosa** (Blume) Lindl., Gen. Sp. Orch. Pl.: 39, 1830
BR 33: t.23 (1847)
Sumatra, Java, Borneo – E
Chelonanthera speciosa Blume, Bijdr.: 384 t.51, 1825

Coelogyne testacea Lindl., Bot. Reg. 28: misc.34, 1842
BR 28: misc.34 (1842)
Type: E.India, Gibson s.n. cult. Duke of Devonshire (K-Lindl., holo!)
Singapore, Malaya, Sumatra – E

○ **Comparettia coccinea** Lindl., Bot. Reg. 24: t.68, 1838
BR 24: t.68 (1838)
Type: Brazil, cult. Loddiges (K-Lindl., holo!)
Bolivia, Brazil – E

Comparettia falcata Poeppig & Endl., Nov. Gen. Sp. Pl. 1: 42 t.73, 1836
BR 26: misc.186 (1840)
W.Indies, Mexico, Guatemala, Honduras, Costa Rica, Panama, S.America - 700–1500 m – E
Comparettia rosea Lindl., Bot. Reg. 26: misc.186, 1840

○ **Coryanthes macrantha** (Hook.) Hook., Bot. Mag. 58: sub t.3102, 1831
BR 21: sub t.1793 (1835) / 22: t.1841 (1836)
Trinidad, Venezuela, Guyana, Brazil, Colombia, Peru – E
Gongora macrantha Hook., Exot. Bot.: misc.2 p.151, 1831
Panstrepis paradoxa Raf., Fl. Tellur. 4: 41, 1836 (nomen illeg.)

Coryanthes speciosa (Hook.) Hook., Bot. Mag. 58: sub t.3102, 1831
BR 21: t. & sub t.1793 (1836) / 26: misc.75 (1840)
Guatemala, Honduras, Costa Rica, Panama, Trinidad, Venezuela, Guyana, Brazil, Colombia, Peru – E
Epidendrum galeatum Vell., Fl. Flum. Ic. 9: t.3, 1827
Gongora speciosa Hook., Bot. Mag. 54: t.2755, 1827
○ Coryanthes maculata Hook., Bot. Mag. 58: t.3102, 1831
Coryanthes maculata var. parkeri Hook., Bot. Mag. 67: t.3747, 1840
Coryanthes speciosa var. alba Lindl., Bot. Reg. 26: misc.75, 1840
Coryanthes speciosa var. vitellina Morren, Ann. Gand. 2: 15, 1846
Coryanthes eximia Gérard, Portef. Hort. 1: 357, 1847
Coryanthes albertinae Karsten, Ausw. Neu. Gew. Venz. 5: t.1, 1848
Coryanthes maculata var. albertinae (Karsten) Lindl., Fol. Orch. Coryanthes: 2, 1852
Coryanthes barkeri Beer, Prakt. Stud. Orch.: 222, 1854
Coryanthes parkeri (Hook.) Seemann, Hort. Parad. Vindob. 19: t.32 fig.1, 1860
Coryanthes picturata Rchb.f., Bot. Zeit. 22: 332, 1864
Coryanthes splendens Barb.Rodr., Gen. & Sp. Orch. Nov. 1: 103, 1877
Coryanthes maculata var. vitrina Rolfe, Orch. Rev. 3: 240, 1895
Coryanthes maculata var. fournieri André, Rev. Hort. 71: 429, 1899
Coryanthes maculata var. speciosa (Hook.) André, Rev. Hort. 71: 429, 1899
Coryanthes maculata var. splendens (Barb.Rodr.) Cogn., Mart. Fl. Bras. 3 (5): 513, 1902
Coryanthes speciosa var. eximia (Gérard) Cogn., Mart. Fl. Bras. 3 (5): 513, 1902
Coryanthes hunteriana Schltr., Fedde Rep. Beih. 17: 63, 1922
Coryanthes powellii Schltr., Fedde Rep. Beih. 17: 64, 1922

Corycium crispum (Thunb.) Sw., Kongl. Vet. Acad. Handl. Stockh.: 222, 1800
BR 9: sub t.700 (1823)
S.Africa – T
Orchis coccinea Buxb., Pl. Min. Cogn. Cent. 3: 7 t.11, 1784
Arethusa crispa Thunb., Prodr.: 3, 1794
Corycium bicolor Ker Gawler, Quart. J. Sc. & Arts 6: 45 fig.1, 1826
Pterygodium crispum (Thunb.) Schltr., Bull. Herb. Bois. s.1, 6: 849, 1898

○ **Corycium orobanchoides** (L.f.) Sw., Kongl. Vet. Acad. Handl. Stockh.: 222, 1800
BR 9: sub t.700 (1823) / 24: t.45 (1838)
S.Africa (Cape) – T
Satyrium orobanchoides L.f., Suppl.: 402, 1781
Pterygodium orobanchoides (L.f.) Schltr., Bull. Herb. Boiss. s.1, 6: 848, 1898

○ **Crybe rosea** Lindl., Bot. Reg. 22: t.1872, 1836
BR 22: t.1872 (1836)
Type: Mexico, cult. Loddiges (K-Lindl., holo!)
Mexico, Guatemala, Honduras – up to 2000 m – T
Arethusa rosea (Lindl.) Benth. ex Hemsl., Godm. & Salvin Biol. Centr. Am. 3: 304, 1884

○ **Cryptarrhena lunata** R.Br., Bot. Reg. 2: t.153, 1816
BR 2: t.153 (1816)
Type: Jamaica, cult. Lee & Kennedy (K-Lindl., not located!)
Mexico, Honduras, Costa Rica, W.Indies, Guyanas, Colombia, Ecuador, Peru – E
Clinhymenia pallidiflora A.Rich. & Galeotti, Comp. Rend. Acad. Sc. Nat. Paris 18: 512, 1844 (nomen)
Cryptarrhena kegelii Rchb.f., Bot. Zeit. 10: 766, 1852
Cryptarrhena pallidiflora (A.Rich. & Galeotti) Rchbf., Bot. Zeit. 10: 766, 1852
Orchidofunckia pallidiflora (A.Rich. & Galeotti) Rchb.f., Bot. Zeit. 10: 766, 1852
Cryptarrhena unguiculata Schltr., Fedde Rep. Beih. 8: 103, 1921

○ **Cryptochilus sanguineum** Wallich, Tent. Fl. Nepal. 36: t.26, 1824
BR 24: t.23 (1838)
Nepal, Sikkim, India (Khasia, Naga) – E, L

Cryptopus elatus (Thouars) Lindl., Bot. Reg. 10: sub t.817, 1824
BR 10: sub t.817 (1824)
Madagascar, Mascarene Is. – E
Angraecum elatum Thouars, Orch. Iles Afr.: t.79–80, 1822
Beclardia elata (Thouars) A.Rich., Orch. Iles France & Bourbon: 78 t.2, 1828

Cuitlauzina pendula Llave & Lex., Nov. Veg. Descr.2, Orch. Opusc.: 33, 1825
BR 28: misc.68 (1842) / 29: t.3 (1843) / 31: misc.60:2 (1845)
Mexico – 1400–2200 m – E
Oncidium galeottianum Drapiez, Lem. Hort. Univ. 1: 198, 1839
○ Odontoglossum citrosmum Lindl., Bot. Reg. 28: misc.68, 1842
Oncidium citrosmum (Lindl.) Beer, Prakt. Stud. Orch.: 284, 1854
Lichterveldia lindleyi Lemaire, Ill. Hortic. 2: t.59, 1855
Odontoglossum pendulum (Llave & Lex.) Bateman, Monog. Odontoglossum : t.6, 1874

Cycnoches chlorochilon Klotzsch, Allg. Gartenz. 6: 225, 1838
BR 32: sub t.46:3 (1846)
Panama, Colombia, Venezuela, Guyana – E
Cycnoches ventricosum var. chlorochilon (Klotzsch) Allen, Orchid. J. 1: 401, 1952

Cycnoches egertonianum Bateman, Orch. Mex. & Guat. : t.40, 1843
BR 26: misc.98 (1840) / 29: misc.117 (1843) / 32: t.46 & sub t.46:2, 6 (1846)
Belize, Guatemala, El Salvador, Honduras, Costa Rica, Peru – E
Cycnoches ventricosum Bateman, Orch. Mex. & Guat.: t.5, 1837
Cycnoches ventricosum var. egertonianum (Bateman) Hook., Bot. Mag. 70: t.4054, 1843
Cycnoches stelliferum Lodd., Cat. Orch.: 25, 1844 (nomen nudum)
○ Cycnoches egertonianum var. viride Lindl., Bot. Reg. 32: t.46, 1846
Cycnoches glanduliferum A.Rich. & Galeotti ex Hemsl., Gard. Chron. n.s.11: 268, 1879
Cycnoches peruvianum Rolfe, Lindenia 7: 29 t.301, 1891
Cycnoches rossianum Rolfe, Gard. Chron. s.3, 9: 456, 1891
Cycnoches guttulatum Schltr., Fedde Rep. Beih. 17: 56, 1922
Cycnoches pachydactylon Schltr., Fedde Rep. Beih. 17: 57, 1922
Cycnoches stenodactylon Schltr., Fedde Rep. Beih. 17: 59, 1922
Cycnoches amparoanum Schltr., Fedde Rep. Beih. 19: 48, 1923
Cycnoches pauciflorum Schltr., Fedde Rep. Beih. 19: 137, 1923

○ **Cycnoches loddigesii** Lindl., Bot. Reg. 21: t.1742, 1835
BR 21: t.1742 (1835) / 32: sub t.46:1 (1846)
Type: Surinam, J.H.Lance s.n. cult. Loddiges (K-Lindl., holo!)
Venezuela, Colombia, Brazil, Guyana – E
Cycnoches loddigesii var. leucochilum Hook., Bot. Mag. 67: t.3855, 1841

Cycnoches maculatum Lindl., Bot. Reg. 26: misc.8, 1840
BR 26: misc.8 (1840) / 32: sub t.46:5 (1846)
Type: cult. Barker s.n. (K-Lindl., holo!)
Venezuela – E

○ **Cycnoches pentadactylon** Lindl., Bot. Reg. 29: t.22, 1843
BR 29: t.22 & misc.26 (1843) / 32: sub t.46:4 (1846)
Types: Brazil, cult. Veitch & Loddiges 890 (K-Lindl., syn!)
Brazil – E
Cycnoches espiritosantense Brade, Fl. Bras. Fasc.5: 145, 1942
Cymbidium aloifolium (L.) Sw., Nov. Act. Upsal. 6: 73, 1799
BR 26: t.25 (1840)
Sri Lanka, India, Sikkim, S.China, Hong Kong, Burma, Thailand, Malaya, Andaman Is., Sumatra, Java, Borneo, Taiwan – E
Epidendrum aloifolium L., Sp. Pl.: 953, 1753
Epidendrum pendulum Roxb., Pl. Coast Coromandel 1: 35 t.44, 1795
Epidendrum aloides Curtis, Bot. Mag. 10: t.387, 1797
○ Cymbidium pendulum (Roxb.) Sw., Nov. Act. Upsal. 6: 73, 1799
Aerides borassii Buch.-Ham. ex Smith, Rees Cyclop. 39: 8, 1819
Cymbidium erectum Wight, Ic. Pl. Ind. Or. 5: 21 t.1753, 1852
Cymbidium mannii Rchb.f., Flora 55: 274, 1872
Cymbidium simulans Rolfe, Orch. Rev. 25: 175, 1917
Cymbidium atropurpureum Yen, Ic. Cymbid. Amoyens A4, 1964 (non (Lindl. Rolfe)
Cymbidium intermedium H.G. Jones, Reinwardtia 9: 71, 1974

Cymbidium bicolor Lindl., Gen. Sp. Orch. Pl.: 164, 1833
ssp. bicolor
BR 25: misc.69 (1839)
Sri Lanka, India – E

o **Cymbidium bicolor** Lindl., Gen. Sp. Orch. Pl.: 164, 1833
ssp. pubescens (Lindl.) Du Puy & Cribb, The Genus Cymbidium: 73, 1988
BR 26: misc.177 (1840) / 27: t.38 (1841)
Type: Singapore, Cuming s.n. cult. Loddiges (K-Lindl., holo!)
Malaya, Java, Sumatra, Borneo, Celebes, Philippines (Mindanao) – E
Cymbidium aloifolium Blume, Bijdr. Fl. Nederl. Ind.: 378 t.19, 1825 (non (L.) Sw.)
Cymbidium pubescens Lindl., Bot. Reg. 26: misc.177, 1840
Cymbidium aloifolium var. pubescens (Lindl.) Ridley, J. Roy. As. Soc. Str. Br. 59: 196, 1911
Cymbidium pubescens var. celebicum Schltr., Fedde Rep. 10: 190, 1911
Cymbidium celebicum (Schltr.) Schltr., Fedde Rep. 21: 197, 1925

Cymbidium chloranthum Lindl., Bot. Reg. 29: misc.102, 1843
BR 29: misc.102 (1843)
Type: cult. Loddiges (K-Lindl., holo!)
Malaya, Sumatra, Java, Borneo – E
Cymbidium variciferum Rchb.f., Bonpl. 4: 324, 1856
Cymbidium sanguinealentum Teijsm. & Binnend., Tijdschr. Ned. Ind. 24: 14, 1862
Cymbidium sanguineum Teijsm. Binnend., Cat. Hort. Bog.: 51, 1866
Cymbidium pulchellum Schltr., Fedde Rep. 8: 570, 1910

o **Cymbidium eburneum** Lindl., Bot. Reg. 33: t.67, 1847
BR 33: t.67 (1847)
Type: E.India, cult. Loddiges (K-Lindl., holo!)
Nepal, Sikkim, N.E.India, Burma, China – 300–1000 m – E
Cymbidium syringodorum Griffith, Notul. 3: 338, 1851
Cymbidium eburneum var. williamsianum Rchb.f., Gard. Chron. n.s.15: 530, 1881
Cymbidium eburneum var. philbrickierum Rchb.f., Gard. Chron. n.s.26: 285, 1886
Cyperorchis eburnea (Lindl.) Schltr., Fedde Rep. 20: 107, 1924

Cymbidium ensifolium (L.) Sw., Nov. Act. Upsal. 6: 77, 1799
BR 7: t.529 (1821) / 23: t.1976 (1837)
N.India, Sikkim, Thailand, Cambodia, Laos, Vietnam, Malaya, China, Sumatra, Java, Borneo, New Guinea, Hong Kong, Taiwan, Ryukyu Is., Japan – E
Epidendrum ensifolium L., Sp. Pl.: 954, 1753
Limodorum ensatum Thunb., Fl. Jap.: 29, 1784
Epidendrum sinense Redouté, Liliac. 2: t.113, 1805
o Cymbidium xiphiifolium Lindl., Bot. Reg. 7: t.529, 1821
o Cymbidium ensifolium var. estriatum Lindl., Bot. Reg. 23: t.1976, 1837
Cymbidium ensifolium var. striatum Lindl., Bot. Reg. 23: t.1976, 1837
Jensoa ensata (Thunb.) Raf., Fl. Tellur 4: 38, 1838
Cymbidium ecristatum (Lindl.) ex Steudel, Nom. Bot. 2: 460, 1840
Cymbidium estriatum (Lindl.) Steudel, Nom. Bot. 2: 460, 1840
Cymbidium micans Schauer, Nov. Act. Nat. Cur. 19 Suppl. 1: 433, 1843
Cymbidium albo-marginatum Makino, Iinuma Somoku-Dzusetzu ed.3, 4 (18): 1183, 1912
Cymbidium gyokuchin Makino, Iinuma Somoku-Dzusetsu ed.3, 4 (18): 1181 t.7, 1912
Cymbidium gyokuchin var. soshin Makino, Iinuma Somoku-Dzusetsu ed.3, 4 (18): 1182 t.8, 1912
Cymbidium koran Makino, Iinuma Somoku-Dzusetzu ed.3, 4 (18): 1179 t.3, 1912
Cymbidium niveo-marginatum Makino, Iinuma Somoku-Dzusetzu ed.3: 1183, 1912
Cymbidium shimaran Makino, Iinuma Somoku-Dzusetzu ed.3, 4: 1183, 1912
Cymbidium yakibaran Makino, Iinuma Somoku-Dzusetzu ed.3, 4: 1181, 1912
Cymbidium yakibaran var. albo-marginatum Makino, Iinuma Somoku-Dzusetzu ed.3, 4: 1182, 1912
Cymbidium yakibaran var. niveo-marginatum (Makino) Makino, Iinuma Somoku-Dzusetzu ed.3, 4: 1182, 1912
Cymbidium arrogans Hayata, Ic. Pl. Form. 4: 76, 1914
Cymbidium misericors Hayata, Ic. Pl. Form. 4: 79, 1914
Cymbidium rubigemmum Hayata, Ic. Pl. Form. 6: 81, 1916
Cymbidium gonzalesii Quisumbing, Philippine J. Sc. 72: 485, 1940
Cymbidium ensifolium var. susin Yen, Ic. Cymb. Amoy. D-b 1, 1964
Cymbidium ensifolium var. misericors (Hayata) T.P.Lin, Nat. Orch. of Taiwan 2: 105, 1977
Cymbidium gyokuchin var. arrogans (Hayata) Ying, Coll. Ill. Ind. Orch. Taiwan 1: 126, 1977
Cymbidium kanran var. misericors (Hayata) Yinf, Coll. Ill Orch. Taiwan 1: 440, 1977
Cymbidium ensifolium var. rubigenmum (Hayata) Liu & Su, Fl. Taiwan 5: 940, 1978
Cymbidium ensifolium var. yakibaran (Makino) Wu & Chen, Act. Phytotax. Sin. 18: 296, 1980

Cymbidium finlaysonianum Lindl., Gen. Sp. Orch. Pl.: 164, 1833
BR 28: misc.67 (1842) / 30: t.24 (1844)
Cambodia, Vietnam, Malaya, Indonesia, Borneo – E, L
Cymbidium pendulum Blume, Bijdr.: 379, 1825 (non (Roxb.) Sw.)
Cymbidium pendulum (Blume) Lindl., Gen. Sp. Orch. Pl.: 165, 1833 (non (Roxb.) Sw.)
Cymbidium wallichii Lindl., Gen. Sp. Orch. Pl.: 165, 1833
o Cymbidium pendulum var. brevilabre Lindl., Bot. Reg. 28: t.24., 1842
Cymbidium tricolor Miq., Choix Pl. Buitenz.: t.19, 1863
Cymbidium pendulum (Blume) Vidal, Phan. Cuming Philippin.: 150, 1885 (non (Roxb.) Sw.)
Cymbidium aloifolium Guillaumin, Lecomte Fl. Gen. Indo-China 6: 415, 1932 (non (L.) Sw.)

Cymbidium goeringii (Rchb.f.) Rchb.f., Walp. Ann. 3: 547, 1852
BR 24: misc.59 (1838)
Japan, Korea, Ryukyu Is., China, Taiwan, N.W.India – 500–3000 m – T
Cymbidium virescens Lindl., Bot. Reg. 24: misc.59, 1838 (non Willd.1805)
Maxillaria goeringii Rchb.f., Bot. Zeit. 3: 334, 1845
Cymbidium virens Rchb.f., Walp. Ann. 6: 626, 1861
Cymbidium mackinnoni Duthie, J. Asiat. Soc. Bengal 71 (2): 41, 1902
Cymbidium formosanum Hayata, J. Coll. Sc. Tokyo 30: 335, 1911
Cymbidium forrestii Rolfe, Notes Roy. Bot. Gard. Edinburgh 8: 23 t.11, 1913
Cymbidium yunnanense Schltr., Fedde Rep. Beih. 4: 74, 1919
Cymbidium pseudovirens Schltr., Fedde Rep. Beih. 12: 351, 1922
Cymbidium tentyozanense Masam., Trans. Nat. Hist. Soc. Formosa 25: 14, 1935
Cymbidium uniflorum Yen, Ic. Cymbid. Amoy.: A2, 1964
Cymbidium chuen-lan C.Chow, Form. Orch.: 21, 1968 (nom. inval.)

Cymbidium madidum Lindl., Bot. Reg. 26: misc.6, 1840
BR 25: misc.37 (1839) / 26: misc.6 (1840)
Type: cult. Rollinsons (K-Lindl., holo!)
E.Australia – E
Cymbidium iridifolium Cunn., Bot. Reg. 25: misc.37, 1839
Cymbidium albuciflorum F.Muell., Fragm. Phyt. Aust. 1: 188, 1859
Cymbidium leai Rendle, J. Bot. 36: 221–222, 1898
Cymbidium queeneanum Klinge, Act. Hort. Petrop. 17: 137–138 t.2, 1899
Cymbidium leroyi St.Cloud, North Qld. Nat. 24 (112): 3–5, 1955
Cymbidium madidum var. leroyi (St.Cloud) Menninger, Amer. Orch. Soc. Bull. 30: 870–871, 1961

o **Cymbidium mastersii** Griffith ex Lindl., Bot. Reg. 31: t.50, 1845
BR 31: t.50 (1845)
Type: cult. Loddiges (K-Lindl., holo!)
Sikkim, Bhutan, N.E.India (Khasia, Manipur) – 1600–2500 m – T
Cymbidium affine Griffith, Notul. 3: 336, 1851
Cymbidium micromeson Lindl., J. Linn. Soc. 3: 29, 1858
Cymbidium mastersii var. album Rchb.f., Gard. Chron. n.s.13: 136, 1880
Cyperorchis mastersii (Griffith ex Lindl.) Benth., J. Linn. Soc. 18: 318, 1881

Cymbidium sp. Lindl., Bot. Reg. 25: misc. 38 (1839)
BR 25: misc.38 (1839)
Australia (Brisbane River)

o **Cynorkis fastigiata** (Thouars) Lindl., Gen. Sp. Orch. Pl.: 331, 1835
BR 23: t.1998 (1837)
Madagascar, Comoros, Mascarene Is., Seychelles – T
Orchis mauritiana Sieber, ex Steudel Nom. ed.2: 711, 1821 (non Lam.)
Cynosorchis fastigiata Thouars, Orch. Iles Afr.: t.13, 1822
Orchis fastigiata (Thouars) Sprengel, Syst. Veg. 3: 687, 1826
Gymnadenia fastigiata (Thouars) A.Rich., Orch. Iles Fr. & Bourb.: 23, 1828
Cynosorchis hygrophila Schltr., Beih. Bot. Centralbl. 34 (2): 309, 1916
Cynosorchis diplorhyncha Schltr., Fedde Rep. 15: 325, 1918
Cynosorchis laggiarae Schltr., Fedde Rep. 15: 326, 1918
Cynosorchis decolorata Schltr., Fedde Rep. 33: 46, 1925

o **Cypripedium irapeanum** Llave & Lex., Nov. Veg. Descr. Orch. Opusc. Fasc. 2: 10, 1824
BR 32: t.58 (1846)
Mexico, Guatemala – up to 3200 m – T
Cypripedium molle Lindl., Benth. Pl. Hartw.: 72, 1840

Cypripedium macranthum Sw., Kongl. Vet. Acad. Handl.: 251, 1800
BR 18: t.1534 (1832)
U.S.S.R., N.Mongolia, N.W.China, Korea, Japan – T

Cypripedium macranthos var. vulgare Sw., Act. Holm.: 251, 1800
Cypripedium thunbergii Blume, Coll. Orch. Arch. Ind. Jap.: 169, 1859

Cypripedium reginae Walter, Fl. Carol.: 222, 1788
BR 20: t.1666 (1834)
N.America, Canada – T
Cypripedium calceolus var. γ L., Sp. Pl. 2: 951, 1753
Cypripedium album Aiton, Hort. Kew 3: 303, 1789
o Cypripedium spectabile Salisb., Trans. Linn. Soc. 1: 78 pl.3 fig.3, 1791
Cypripedium canadensis Michaux, Fl. Bor. Am. 2: 161, 1803
Cypripedium hirsutum Miller, Bull. Torrey Club 32: 375, 1905
Cypripedium reginae var. album (Aiton) Rolfe, Orch. Rev. 19: 208, 1911
Calceolus reginae (Walter) Nieuwl., Amer. Midl. Nat. 3: 117, 1913
Cypripedium hirsutum f. album (Aiton) R.Hoffm., Proc. Boston Soc. Nat. Hist. 36: 246, 1922

Cyrtopodium andersonii (Lambert ex Andrews) R.Br., Aiton Hort. Kew ed.2, 5: 216, 1813
BR 27: t.8 & misc.49 (1841)
W.Indies, Trinidad, Brazil, Guyana, Colombia – T
Cymbidium andersonii Lambert ex Andrews, Bot. Repos. 10: 651, 1811

Cyrtopodium cristatum Lindl., Bot. Reg. 27: sub t.8, 1841
BR 27: sub t.8 (1841)
Type: Guyana, Schomburgk s.n. (K-Lindl., holo!)
Brazil, Guyana, Venezuela, Trinidad – T

Cyrtopodium paniculatum (Ruiz Lopez & Pavon) Garay, Caldasia 8: 524, 1962
BR 18: sub t.1549:5 (1832)
Venezuela, Colombia, Peru – E, T
Maxillaria paniculata Ruiz Lopez & Pavon, Syst. Veg. Fl. Peruv. & Chil. 1: 223, 1798
Dendrobium paniculatum (Ruiz Lopez & Pavon) Pers., Syn. Pl. 2: 524, 1807
Dendrobium plicatum Dietr., Lex. Gärtn. Bot. Nachtr. 2: 651, 1816
Cyrtopodium bracteatum Lindl., Orch. Linden.: 23, 1846
Cyrtopodium punctatum var. bracteatum Rchb.f., Walp. Ann. 6: 667, 1863
Cyrtopodium naiguatae Schltr., Fedde Rep. Beih. 6: 43, 1919

Dactylorhiza foliosa (Rchb.f.) Soó, Nov. Gen. Dactyl.: 7, 1962
BR 20: t.1701 (1834)
Madeira – 900–1000 m – T
o Orchis foliosa Sol. ex Lowe, Trans. Cambridge Phil. Soc. 4: 13, 1831 (nomen illeg.)
Orchis latifolia var. foliosa Rchb.f., Ic. Fl. Germ. 13/14: 60, 1851
Orchis orientalis ssp. foliosa (Rchb.f.) Klinge, Act. Hort. Petrop: 189, 1898
Dactylorchis foliosa (Rchb.f.) Vermeulen, Stud. Dactylorch.: 69, 1947
Orchis maderensis Summerh., Kew Bull.: 123, 1947

Dactylorhiza sambucina (L.) Soó, Nom. Nov. Gen. Dactyl.: 3, 1962
BR 31: misc.64 (1845)
Europe – T
Orchis sambucina L., Fl. Suec.: 312, 1755
Orchis mixta var. sambucina (L.) Retz, Fl. Scan : 167, 1799
Orchis incarnata Willd., Sp. Pl. ed.4: 30, 1806 (non L.)
Orchis schleicheri Sweet, Hort. Brit.: 382, 1823
Orchis saccata Rchb.f., Fl. Germ. Excurs.: 122, 1851
Orchis lutea Dulac, Fl. Hautes-Pyr.: 125, 1868
Dactylorchis sambucina (L.) Vermeulen, Stud. Dactylorch.: 65, 1947

Deiregyne diaphana (Lindl.) Garay, Bot. Mus. Leafl. Harv. Univ. 28 (4): 312, 1980
BR 30: misc.20 (1844)
Type: Mexico, cult. van Houtte (K-Lindl., holo!)
Mexico, Guatemala – T
Spiranthes diaphana Lindl., Bot. Reg. 30: misc.20, 1844

Dendrobium acerosum Lindl., Bot. Reg. 27: misc.86, 1841
BR 27: misc.86 (1841)
Type: Singapore, Cuming 357, cult. Loddiges (K-Lindl., holo!)
Burma, Thailand, Malaya, Sumatra, Borneo – E
Aporum subteres Griffith, Calc. J. Nat. Hist. 5: 371, 1845
Dendrobium subteres (Griffith) Lindl., J. Linn. Soc. 3: 4, 1859
Callista acerosa (Lindl.) Kuntze, Rev. Gen. 2: 654, 1891
Aporum acerosum (Lindl.) Brieger, Schlechter Die Orchideen ed.3, 1: 676, 1981

Dendrobium aciculare Lindl., Bot. Reg. 26: misc.188, 1840
BR 26: misc.188 (1840)
Type: Cuming 174, cult. Loddiges (K-Lindl., holo!)
Thailand, Malaya, Borneo, Philppines – E
Callista acicularis (Lindl.) Kuntze, Rev. Gen. 2: 654, 1891
Dendrobium setifolium Seidenfaden & Smithinand Orch. Thailand: 254, 1960 (non Ridley)
Aporum aciculare (Lindl.) Rauschert, Fedde Rep. 94: 437, 1983

Dendrobium acuminatissimum (Blume) Lindl., Gen. Sp. Orch. Pl.: 86, 1830
BR 30: misc.62:17 (1844)
Java – E
Grastidium acuminatissimum Blume, Bijdr.: 333, 1825
Dendrobium caudatum var. javanica Teijsm. & Binnend., Nat. Tijdschr. Ned. Ind. 24: 315, 1862
Callista acuminatissima (Blume) Kuntze, Rev. Gen. 2: 654, 1891

o **Dendrobium aduncum** Wallich ex Lindl., Bot. Reg. 28: misc.62, 1842
BR 28: misc.62 (1842) / 30: misc.62:39 (1844) / 32: t.15 (1846)
Type: India, Wallich s.n. cult. Loddiges (K-Lindl., holo!)
Sikkim, Bhutan, N.E.India, Burma, Thailand, China, Vietnam – E
Callista adunca (Wallich ex Lindl.) Kuntze, Rev. Gen. 2: 654, 1891

Dendrobium affine (Decaisne) Lindl., Bot. Reg. 30: misc.62:70, 1844
BR 29: sub t.28:7 (1843) / 30: misc.62:70 (1844)
Timor, N.Australia – E
Onychium affine Decaisne, Herb. Timor. Desc.: 37, 1835
Dendrobium dicuphum F.Muell., Fragm. 8: 28, 1874

Dendrobium amoenum Wallich ex Lindl., Gen. Sp. Orch. Pl.: 78, 1830
BR 30: misc.62:10 (1844) / 33: sub t.36 (1847)
Nepal to Bhutan, Burma – 500–1300 m – E
Dendrobium egertoniae Lindl., Bot. Reg. 33: sub t.36, 1847
Dendrobium mesochlorum Lindl., Bot. Reg. 33: sub t.36, 1847
Callista amoena (Wallich ex Lindl.) Kuntze, Rev. Gen. 2: 654, 1891

o **Dendrobium anceps** Sw., Kongl. Vet. Acad. Handl. 21: 246, 1800
BR 15: t.1239 (1829)
Nepal, Sikkim, Assam, Burma, Thailand, Cambodia, Laos, Vietnam – E
Aporum anceps (Sw.) Lindl., Gen. Sp. Orch. Pl.: 71, 1830
Callista anceps (Sw.) Kuntze, Rev. Gen. 2: 654, 1891

Dendrobium angulatum (Blume) Lindl., Gen. Sp. Orch. Pl.: 76, 1830
BR 30: misc.62:74 (1844)
Java – E
Desmotrichum angulatum Blume, Bijdr.: 329, 1825
Dendrobium calopogon Rchb.f., Xen. Orch. 2: 23 t.109, 1858
Dendrobium hasseltii Rchb.f., Xen. Orch. 2: 23 t.109, 1858
Callista angulata (Blume) Kuntze, Rev. Gen. 2: 654, 1891
Callista calopogon (Rchb.f.) Kuntze, Rev. Gen. 2: 654, 1891

Dendrobium anosmum Lindl., Bot. Reg. 31: misc.41, 1845
BR 24: misc.36 (1838) / 25: misc.46 (1839) / 28: misc.94 (1842) / 31: misc.41 (1845)
Type: Philippines, Cuming s.n. cult. Loddiges (K-Lindl., holo!)
Sri Lanka to Indochina, Malaya, Indonesia to Philippines, New Guinea – E
Dendrobium macrophyllum Lindl., Bot. Reg. 24: misc.36, 1838 (non A.Rich.)
Dendrobium macranthum Hook., Bot. Mag. 69: t.3970, 1842 (non A.Rich.)
Dendrobium superbum Rchb.f., Walp. Ann. 6: 282, 1861
Dendrobium scortechinii Hook.f., Fl. Brit. Ind. 5: 741, 1890
Callista anosma (Lindl.) Kuntze, Rev. Gen. 2: 653, 1891
Callista macrophylla (Lindl.) Kuntze, Rev. Gen. 2: 655, 1891
Callista scortechinii (Hook.f.) Kuntze, Rev. Gen. 2: 655, 1891
Dendrobium leucorhodum Schltr., Fedde Rep. Beih. 1: 499, 1911
Dendrobium parishii Guillaumin, Bull. Mus. Nat. Hist. Paris s.2, 28 (5): 484, 1956 (non Rchb.f.)

Dendrobium antennatum Lindl., Hook. J. Bot. 2: 236, 1843
BR 29: sub t.28:5 (1843) / 30: misc.62:68 (1844)
New Guinea, Solomon Is., N.E.Australia (Queensl.) – 0–1200 m – E
Dendrobium d'albertisii Rchb.f., Gard. Chron. n.s.9: 366, 1878

Dendrobium aphyllum (Roxb.) C.Fischer, Gamble Fl. Presid. Madras 3: 1416, 1928
BR 7: t.548 (1821) / 21: t.1756 (1835) / 30: misc.62:7, 8 (1844)
Nepal, Bhutan, N.E.India, Burma, Thailand, China, Vietnam, Malaya – E
Limodorum aphyllum Roxb., Pl. Coromandel 1: 34 t.41, 1795
Cymbidium aphyllum (Roxb.) Sw., Nov. Act. Upsal. 6: 73, 1799
o Dendrobium cucullatum R.Br., Bot. Reg. 7: t.548, 1821
o Dendrobium pierardii Roxb. ex Hook, Exot. Fl. 1: t.9, 1822
Dendrobium primulinum Lindl., J. Linn. Soc. 3: 12, 1859 (non Lindl.1858)
Dendrobium pierardii var. cucullatum (R.Br.) Hook.f., Fl. Brit Ind. 5: 739, 1890
Callista aphylla (Roxb.) Kuntze, Rev. Gen. 2: 653, 1891
Dendrobium madrasense Hawkes, Orquidea 25: 102, 1963

o **Dendrobium aqueum** Lindl., Bot. Reg. 29: t.54 & misc.6, 1843
BR 29: t.54 & misc.6 (1843) / 30: misc.62:30 (1844)
Type: India, cult. Loddiges (K-Lindl., holo!)
India – E

Dendrobium album Wight, Ic. Pl.: t.1645, 1852
Callista aquea (Lindl.) Kuntze, Rev. Gen. 2: 654, 1891

Dendrobium bicameratum Lindl., Bot. Reg. 25: misc.85, 1839
BR 25: misc.85 (1839) / 30: misc.62:55 (1844)
Type: India, Gibson s.n. cult. Barker (K-Lindl., holo!)
Nepal, Bhutan, Sikkim, N.E.India, Burma – E
Dendrobium breviflorum Lindl., J. Linn. Soc. 3: 14, 1859
Dendrobium bolboflorum Falc. ex Hook.f., Fl. Brit. Ind. 5: 729, 1890
Callista bicamerata (Lindl.) Kuntze, Rev. Gen. 2: 654, 1891
Callista bolboflora (Falc. ex Hook.f.) Kuntze, Rev. Gen. 2: 654, 1891

Dendrobium bifarium Lindl., Gen. Sp. Orch. Pl.: 81, 1830
BR 27: misc.165 (1841) / 30: misc.62:29 (1844)
Thailand, Malaya, Ambon, Borneo – E
Dendrobium excisum Lindl., Bot. Reg. 27: misc.165, 1841
Callista bifaria (Lindl.) Kuntze, Rev. Gen. 2: 654, 1891

Dendrobium biflorum (Forster) Sw., Nov. Act. Soc. Sc. Upsal. 6: 84, 1799
BR 21: sub t.1756 (1835) / 30: misc.62:35 (1844)
Pacific Is.: Solomon Is. to Tahiti – E
Epidendrum biflorum Forster, Prodr.: 318, 1786
Pedilonum biflorum Blume, Bijdr.: 322, 1825
Callista biflora (Blume) Kuntze, Rev. Gen. 2: 653, 1891

Dendrobium bilobum Lindl., Hook. J. Bot. 2: 236, 1843
BR 30: misc.62:37 (1844)
New Guinea, Solomon Is., Vanuatu, New Caledonia – E
Monanthus biloba (Lindl.) Brieger, Schlechter Die Orchideen 1: 660, 1981

Dendrobium calcaratum A. Rich., Sert. Astrol.: 18 t.7, 1834
BR 30: misc.62:45 (1844)
Ambae, Banks Is., Espiritu Santo, Pentecost, New Britain, Solomon Is., Santa Cruz Is., Samoa – E
Dendrobium triviale Kraenzlin, Engl. Bot. Jahrb.: 25, 1898
Dendrobium separatum Ames, J. Arn. Arb. 13: 133, 1932

Dendrobium caleyi Cunn., Comp. Bot. Mag. 2: 377, 1836
BR 25: misc.32 (1839)
Australia (Illawarra) – L
Dendrobium pygmaeum Lindl., Bot. Reg. 25: misc.32, 1839

Dendrobium candidum Wallich ex Lindl., Bot. Reg. 24: misc.54, 1838
BR 24: misc.54 (1838) / 30: misc.62:24 (1844)
Type: Gibson s.n. cult. Duke of Devonshire (K-Lindl., holo!)
India – E
Dendrobium spathaceum Lindl., J. Linn. Soc. 3: 15, 1859

o **Dendrobium chrysanthum** Wallich ex Lindl., Bot. Reg. 15: t.1299, 1830
BR 15: t.1299 (1830) / 25: misc.56 (1839) / 30: misc.62:18, 19 (1844)
Type: J.Gibson s.n. cult. Duke of Devonshire (K-Lindl., holo!)
Nepal, Sikkim, Bhutan, N.E.India, Burma, Thailand, China, Laos, Vietnam – 800–1000 m – E
Dendrobium paxtonii Lindl., Bot. Reg. 25: misc.56, 1839 (non Paxton)
Dendrobium chrysanthum var. microphthalma Rchb.f., Gard. Chron. n.s.11: 366, 1879
Dendrobium chrysanthum var. anophthalma Rchb.f., Gard. Chron. n.s.19: 44, 1883
Callista chrysantha (Wallich ex Lindl.) Kuntze, Rev. Gen. 2: 654, 1891

o **Dendrobium chrysotoxum** Lindl., Bot. Reg. 33: sub t.19 & t.36, 1847
BR 33: sub t.19 & t.36 (1847)
Type: E.India, cult. Henderson (K-Lindl., holo!)
N.E.India, Burma, Thailand, China (Yunnan), Laos, Vietnam – E
Dendrobium suavissimum Rchb.f., Gard. Chron.: 406, 1874
Dendrobium chrysotoxum var. suavissimum (Rchb.f.) Veitch, Man. Orch. Pl. 3: 30, 1888
Callista chrysotoxa (Lindl.) Kuntze, Rev. Gen. 2: 654, 1891
Callista suavissima (Rchb.f.) Kuntze, Rev. Gen. 2: 655, 1891
Dendrobium chrysotoxum var. delacourii Gagnepain, Fl. Gén. Indoch. 6: 219, 1934

o **Dendrobium cretaceum** Lindl., Bot. Reg. 33: t.62, 1847
BR 33: t.62 (1847)
Type: Burma, T.Lobb s.n. cult. Veitch (K-Lindl., holo!)
N.E.India, Burma, Thailand, Laos – E
Callista cretacea (Lindl.) Kuntze, Rev. Gen. 2: 654, 1891

o **Dendrobium crumenatum** Sw., Schrad. Neues J. Bot. 2: 237, 1799
BR 25: t.22 (1839)
Sri Lanka, India, Burma, Thailand, China, Laos, Cambodia, Vietnam, Malaya, Indonesia, Christmas Is., Philippines – E
Angraecum crumenatum Rumph., Herb. Amb. 6: t.47 2, 1750
Epidendrum caninum Burm.f., Fl. Ind.: 189, 1768
Epidendrum spatulatum Burm.f., Fl. Ind.: 188, 1768 (non L.)
Ceraja simplicissima Lour., Fl. Indoch.: 518, 1790
Onychium crumenatum (Sw.) Blume, Bijdr.: 328, 1825
Dendrobium ceraia Lindl., Gen. Sp. Orch. Pl.: 89, 1830
Callista crumenata (Sw.) Kuntze, Rev. Gen. 2: 654, 1891
Dendrobium schmidtianum Kraenzlin, Bot. Tidsskr. 24: 7, 1900
Dendrobium cumulatum Kraenzlin, Engl. Pflanzenr. Orch. Mon. Dendr. 1: 115, 1910 (non Lindl.)
Dendrobium simplicissimum (Lour.) Kraenzlin, Engl. Pflanzenr. Mon. Dendr.: 235, 1910
Dendrobium caninum (Burm.f.) Merr., Philipp. J. Sc.: 19, 1921
Dendrobium kwashotense Hayata, Ic. Pl. Form. 4: 41, 1944
Aporum crumenatum (Sw.) Brieger, Schlechter Die Orchideen ed.3, 1: 671, 1981
Aporum kwashotense (Hayata) Rauschert, Fedde Rep. 94: 440, 1983

o **Dendrobium cucumerinum** MacLeay ex Lindl., Bot. Reg. 28: misc.63, 1842
BR 28: misc.63 (1842) / 29: t.37 (1843)
Type: Australia, MacLeay s.n. cult. Loddiges (K-Lindl., icono!)
Australia – E
Callista cucumerina (Macleay ex Lindl.) Kuntze, Rev. Gen. 2: 654, 1891
Dockrillia cucumerina (Macleay & Lindl.) Brieger, Schlechter Die Orchideen ed.3, 1: 745, 1981

Dendrobium cunninghamii Lindl., Bot. Reg. 21: sub t.1756, 1835
BR 21: sub t.1756: 2 (1835) / 30: misc.62:34 (1844)
Type: New Zealand, R. Cunningham s.n. (K-Lindl., holo!)
New Zealand – E
Dendrobium biflorum A.Rich., Ess. Nouv. Zél.: 167 t.26, 1832 (non Sw.)
Dendrobium lessonii Colenso, Trans. New Zealand Inst. 15: 326, 1883

o **Dendrobium densiflorum** Wallich ex Lindl., Pl. As. Rar. 1: 34, 1830
BR 21: sub t.1756 (1835) / 21: t.1828 (1836)
N.E.India, Burma, Thailand – up to 1500 m – E
Dendrobium clavatum Roxb., Hort. Beng.: 63, 1814
Dendrobium schroederi hort., Flor. Mag. 9: t.502, 1870
Dendrobium guibertii Carriere, Rev. Horticult. 48: 430, 1876
Dendrobium griffithianum var. guibertii (Carriere) Veitch, Man. Dendr.: 48, 1887
Callista densiflora (Wallich ex Lindl.) Kuntze, Rev. Gen. 2: 654, 1891

Dendrobium denudans D.Don, Prodr. Fl. Nepal: 34, 1825
BR 24: misc.156 (1838) / 30: misc.62:59 (1844)
Nepal, Sikkim, N.E.India – 900–2000 m – E
Callista denudans (D.Don) Kuntze, Rev. Gen. 2: 654, 1891

Dendrobium devonianum Paxton, Mag. Bot. 8: 169, 1840
BR 30: misc.62:6 (1844)
Bhutan, N.E.India, China, Burma, Thailand, Cambodia, Laos, Vietnam – E
Dendrobium pictum Griffith, J. Linn. Soc. 3: 12, 1859
Dendrobium pulchellum Lindl., J. Linn. Soc. 3: 12, 1859 (non Roxb. ex Lindl.)
Dendrobium pulchellum var. devonianum (Paxton) Rchb.f., Walp. Ann. 6: 284, 1861
Callista devoniana (Paxton) Kuntze, Rev. Gen. 2: 654, 1891

o **Dendrobium discolor** Lindl., Bot. Reg. 27: t.52, 1841
BR 27: t.52 & misc.50 (1841) / 28: misc.6 (1842) / 29: sub t.28:6 (1843) / 30: misc.62:69 (1844)
Type: cult. Loddiges (K-Lindl., holo!)
N.E.Australia (Queensland), S.New Guinea – E, L
Dendrobium undulatum R.Br., Prod. Fl. Nov. Holl.: 332, 1810
Dendrobium fuscum Fitzg., Sydney Morning Herald 24. Sept., 1879
Callista undulata (R.Br.) Kuntze, Rev. Gen. 2: 655, 1891
Dendrobium arachnanthe Kraenzlin, Engl. Pflanzenr. Orch. Mon. Dendr. 1: 152, 1910
Dendrobium elobatum Rupp, Victorian Nat. 69: 116, 1953

Dendrobium erosum (Blume) Lindl., Gen. Sp. Orch. Pl.: 86, 1830
BR 30: misc.62:72 (1844)
Java – E
Pedilonum erosum Blume, Bijdr.: 323, 1825
Callista erosa (Blume) Kuntze, Rev. Gen. 2: 654, 1891

o **Dendrobium formosum** Roxb., Fl. Ind. 3: 485 1832
BR 24: misc.86 (1838) / 25: t.64 (1839) / 30: misc.62:40 (1844)
Nepal, Sikkim, N.E.India, Burma, Thailand, Andaman Is., Vietnam – E
Dendrobium infundibulum Rchb.f., Gartenfl. 36: t.1253, 1887 (non Lindl.1859)
Callista formosa (Roxb.) Kuntze, Rev. Gen. 2: 654, 1891

Dendrobium gracilicaule F.Muell., Fragm. Phyt. Austral. 1: 179, 1859
BR 25: misc.33 (1839) / 27: misc.53 (1841) / 30: misc.62:56 (1844)
Type: cult. Loddiges (K-Lindl., holo!)
E.Australia – E
Dendrobium elongatum Cunn., Bot. Reg. 25: misc.33, 1839
Dendrobium brisbanense Rchb.f., Walp. Ann. 6: 299, 1861
Dendrobium fellowsii F.Muell., Fragm. Phyt. Austral. 7: 63, 1870
Callista gracilicaulis (F.Muell.) Kuntze, Rev. Gen. 2: 654, 1891

Dendrobium jonesii Rendle, J. Bot. Lond. 39: 197, 1901
Tropilis gracilicaulis (F.Muell.) Butzin, Willdenowia 12 (2): 250, 1982

Dendrobium griffithianum Lindl., Bot. Reg. 21: sub t.1756:3 (1835)
BR 21: sub t.1756:3 (1835)
Type: Burma, Griffith s.n. (K-Lindl., holo!)
Burma – E
Dendrobium farmeri var. aureo-flava Hook., Bot. Mag. 90: t.5451, 1864
Callista griffithiana (Lindl.) Kuntze, Rev. Gen. 2:655, 1891

Dendrobium herbaceum Lindl., Bot. Reg. 26: misc.153, 1840
BR 26: misc.153 (1840) / 30: misc.62:49 (1844)
Type: E.India, cult. Loddiges (K-Lindl., holo!)
India – E
Dendrobium ramosissimum Wight, Ic. Pl. Ind.: t.1648, 1851

Dendrobium heterocarpum Wallich ex Lindl., Gen. Sp. Orch. Pl.: 78, 1830
BR 25: t.20 (1839) / 29: t.17 (1843) / 30: misc.62:11, 21, 41 (1844)
Sri Lanka, India, Nepal, Sikkim, Bhutan, Burma, Thailand, Malaya, Indonesia, Philippines, China – E
Dendrobium aureum Lindl., Gen. Sp. Orch. Pl.: 77, 1830
o Dendrobium aureum var. pallidum Lindl., Bot. Reg. 25: t.20, 1839
o Dendrobium rhombeum Lindl., Bot. Reg. 29: t.17, 1843
Dendrobium heterocarpum var. henshallii Hook., Bot. Mag. 79: t.4708, 1853
Dendrobium aureum var. philippinensis Rchb.f., Gard. Chron.: 72, 1880
Dendrobium atractodes Ridley, J. Bot. 23: 123, 1885
Callista aurea (Lindl.) Kuntze, Rev. Gen. 2: 653, 1891
Callista heterocarpa (Wallich ex Lindl.) Kuntze, Rev. Gen. 2: 654, 1891
Dendrobium hildebrandii Kraenzlin, Engl. Pflanzenr. Mon. Dendr.: 34, 1910 (non Rolfe)
Dendrobium minahassae Kraenzlin, Engl. Pflanzenr. Orch. Mon. Dendr.: 107 fig.5, 1910,

Dendrobium heyneanum Lindl., Gen. Sp. Orch. Pl.: 90, 1830
BR 25: misc.41 (1839)
India – E

Dendrobium hymenophyllum Lindl., Gen. Sp. Orch. Pl.: 86, 1830
BR 30: misc.62:73 (1844)
Java – E
Pedilonum undulatum Blume, Bijdr.: 322, 1825
Callista hymenophylla (Lindl.) Kuntze, Rev. Gen. 2: 655, 1891
Thrixspermum papillosum Carr, Gard. Bull. Str. Settl. 5: 33 fig.13b, 1929

Dendrobium indragiriense Schltr., Fedde Rep. 9: 164, 1911
BR 26: misc.192 (1840) / 32: misc.62:14 (1844)
Thailand, Andaman Is., Malaya, Sumatra – E
Dendrobium gemellum Lindl., Gen. Sp. Orch. Pl.: 28, 1830
Dendrobium isomerum Schltr., Engl. Bot. Jahrb. 45 Beibl. 104: 35, 1911
Dendrobium inconspicuiflorum J.J.Smith, Bull. Buitenz. s.2. 25: 42, 1917
Grastidium indragiriense (Schltr.) Rauschert, Fedde Rep. 94: 450, 1983
Grastidium isomerum (Schltr.) Rauschert, Fedde Rep. 94: 450, 1983

o **Dendrobium jenkinsii** Wallich ex Lindl., Bot. Reg. 25: t.37, 1839
BR 25: t.37 (1839)
Type: cult. Loddiges K-Lindl., holo!)
Sikkim, Bhutan, N.E.India, Burma, Thailand, China, Laos – E
Callista jenkinsii (Wallich ex Lindl.) Kuntze, Rev. Gen. 2: 654, 1891
Dendrobium aggregatum var. jenkinsii (Wall. ex Lindl.) King & Pantling, Ann. Roy. Bot. Gard. Calc. 8: 61 t.85, 1898
Dendrobium marseillei Gagnepain, Bull. Mus. Nat. Hist. Paris s.2, 6 (1): 119, 1934
Callista aggregata var. jenkinsii (Wallich ex Lindl.) Brieger, Schlechter Die Orchideen ed.3, 1: 704 fig.517, 1981

Dendrobium junceum Lindl., Bot. Reg. 28: misc.11, 1842
BR 28: misc.11 (1842)
Type: Singapore, cult. Loddiges (K-Lindl., holo!)
Philippines – E
Aporum junceum (Lindl.) Rauschert, Fedde Rep. 94: 440, 1983

o **Dendrobium kingianum** Bidw. ex Lindl., Bot. Reg. 30: misc.18, 1844
BR 30: misc.18 (1844) / 31: t.61 (1845)
Type: Australia, Bidwill s.n. cult. Loddiges (K-Lindl., holo!)
Australia (S.E.Queensland, N.New South Wales) – E
Callista kingiana (Bidw. ex Lindl.) Kuntze, Rev. Gen. 2: 655, 1891
Dendrocoryne kingianum (Bidw. ex Lindl.) Brieger, Schlechter Die Orchideen ed.3, 1: 724, 1981
Tropilis kingiana (Bidw. ex Lindl.) Butzin, Willdenowia 12 (2): 250, 1982

o **Dendrobium kuhlii** (Blume) Lindl., Gen. Sp. Orch. Pl.: 87, 1830
BR 33: t.47 (1847)
Indonesia – E
Pedilonum kuhlii Blume, Bijdr.: 321, 1825
Dendrobium thyrsodes Rchb.f., Hamb. Gartenz.: 5, 1859
Dendrobium cornutum Hook.f., Fl. Brit. Ind. 5: 730, 1890
Callista cornuta (Hook.f.) Kuntze, Rev. Gen. 2: 654, 1891
Callista kuhlii (Blume) Kuntze, Rev. Gen. 2: 655, 1891

Dendrobium lamellatum (Blume) Lindl., Gen. Sp. Orch. Pl.: 89, 1830
BR 28: misc.76 (1842) / 30: t.53 (1844)
Burma, Thailand, Malaya, Indonesia, Borneo – E
Onychium lamellatum Blume, Bijdr.: 526 fig.10, 1826
o Dendrobium compressum Lindl., Bot. Reg. 28: misc.76, 1842
Callista lamellata (Blume) Kuntze, Rev. Gen. 2: 655, 1891
Pedilonum lamellatum (Blume) Brieger, Schlechter Die Orchideen ed.3, 1: 685 figs.499 & 500, 1981

Dendrobium lancifolium A.Rich., Sert. Astrolab: 20 t.8, 1834
BR 30: misc.62:54 (1844)
Moluccas – E

Dendrobium leonis (Lindl.) Rchb.f., Walp. Ann. 6: 280, 1861
BR 26: misc.126 (1840)
Type: Singapore, Prince s.n. cult. Loddiges (K-Lindl., holo!)
Thailand, Cambodia, Vietnam, Malaya, Sumatra, Borneo – E
Aporum leonis Lindl., Bot. Reg. 26: misc.126, 1840
Callista leonis (Lindl.) Kuntze, Rev. Gen. 2: 655, 1891
Dendrobium anceps Gagnepain, Fl. Gén. Indoch. 6: 246, 1934 (non Sw.)

Dendrobium linawianum Rchb.f., Walp. Ann. 6: 284, 1861
BR 16: t.1314 (1830) / 30: misc.62:2 (1844)
China, Japan – E
o Dendrobium moniliforme Lindl., Bot. Reg. 16: t.1314, 1830 (non Sw.)
Callista linawiana (Rchb.f.) Kuntze, Rev. Gen. 2: 655, 1891
Dendrobium alboviride Hayata, Ic. Pl. Form. 9: 108 fig.36, 1920

Dendrobium lindleyi Steudel, Nom. 1: 490, 1840
BR 20: t.1695 (1834)
Sikkim, Bhutan, N.E.India, Burma, Thailand, China, Laos, Vietnam – E
o Dendrobium aggregatum Roxb., Hort. Beng.: 63, 1814
Callista aggregata (Roxb.) Kuntze, Rev. Gen. 2: 654, 1891
Dendrobium suavissimum Kraenzlin, Orch.: 17, 1900 (non Rchb.f.)
Dendrobium jenkinsii Lang & Tsi, Ic. Cormoph. Sin. 5: 704, 1976 (non Wall. ex Lindl.)

Dendrobium lineale Rolfe, Kew Bull.: 215, 1894
BR 29: sub t.28:3 (1843) / 30: misc.62:66 (1844)
New Guinea – 0–800 m – E, L
Dendrobium veratrifolium Lindl., Hook. J. Bot. 2: 236, 1843 (non Roxb.1832)
Dendrobium cogniauxianum Kraenzlin, Engl. Bot. Jahrb. 13: 281, 1891
Dendrobium augustae-victoriae Kraenzlin, Gartenfl.: 115, 1894
Dendrobium imperatrix Kraenzlin, Gard. Chron. s.3, 18: 34, 1895
Dendrobium grantii C.White, Orchid. Rev. 5: 74, 1910
Dendrobium veratroides Bakh.f., Blumea 12: 68, 1963

Dendrobium linguiforme Sw., Kongl. Vet. Akad. Handl. 21: 247, 1800
BR 25: misc.26 (1839)
Australia (Queensland, New South Wales, Cape York) – E
Callista linguiformis (Sw.) Kuntze, Rev. Gen. 2: 655, 1891
Dockrillia linguiformis (Sw.) Brieger, Schlechter Die Orchideen ed.3, 1: 746, 1981

Dendrobium lobbii Teijsm. & Binnend., Nat. Tijdschr. Ned. Ind. 5: 491, 1853
BR 26: misc.219 (1840) / 30: misc.62:36, (1844)
Thailand, Vietnam, Malaya, Sumatra, Bangka, Borneo, New Guinea, Australia – E
Dendrobium calcaratum Lindl., Bot. Reg. 26: misc.219, 1840 (non A.Rich.)
Dendrobium teijsmannii Miq., Fl. Ind. Bat. 3: 640, 1859
Dendrobium conostalix Rchb.f., Walp. Ann. 6: 292, 1861
Callista conostalix (Rchb.f.) Kuntze, Rev. Gen. 2: 654, 1891
Callista lobbii (Teijsm. & Binnend.) Kuntze, Rev. Gen. 2: 655, 1891
Callista teijsmannii (Miq.) Kuntze, Rev. Gen. 2: 655, 1891
Dendrobium paludicola Schltr., Schum. & Lauterb. Nachtr.: 173, 1905
Conostalix calcaratum (Lindl.) Brieger, Schlechter Die Orchideen ed.3, 1: 659, 1981
Conostalix lobbii (Teijsm. & Binnend.) Rauschert, Fedde Rep. 94: 444, 1983
Conostalix paludicola (Schltr.) Rauschert, Fedde Rep. 94: 444, 1983

Dendrobium longicornu Wallich ex Lindl., Bot. Reg. 16: t.1315, 1830
BR 16: t.1315 (1830) / 30: misc.62:31 (1844)
Type: Wallich s.n. cult. Hort. Soc. (K-Lindl., holo!)
Nepal to Bhutan, Assam, Burma – 1600–2500 m – E
Dendrobium flexuosum Griffith, Notul. 3: 317, 1851
Dendrobium hirsutum Griffith, Notul. 3: 328, 1851
Dendrobium longicornu var. hirsutum (Griffith) Hook.f., Fl. Br. Ind. 5: 720, 1890

Dendrobium luzonense Lindl., Bot. Reg. 30: misc.62:33, 1844
BR 30: misc.62:33 (1844)
Type: Philippines, Luzon, without collector (K-Lindl., holo!)
Philippines – E
Dendrobium alagense Ames, Philipp. J. Sc. 2: 328, 1907

Dendrobium macranthum A.Rich., Sert. Astrolab: 15 t.6, 1832
BR 29: sub t.28:4 (1843) / 30: misc.62:67 (1844)
Santa Cruz Is., Vanuatu, New Caledonia, Horn Is., Samoa – E
Dendrobium arachnostachyum Rchb.f., Gard. Chron.: 334, 1877
Dendrobium tokai var. crassinerve Finet, Bull. Soc. Bot. France 50: 381, 1903
Dendrobium pseudotokai Kraenzlin, Bull. Soc. Bot. France 76: 300, 1929

Dendrobium macrophyllum A.Rich., Sert. Astrolab: 22 t.9, 1834
BR 30: misc.62:1 (1844) / 33: sub t.25 (1847)
New Guinea, Java, Philippines, Palau Is., Solomon Is., Vanuatu, New Caledonia, Fiji, Samoa – E
Dendrobium veitchianum Lindl., Bot. Reg. 33: sub t.25, 1847
Dendrobium ferox Hassk., Retzia 1: 1, 1855
Dendrobium sarcostoma Teijsm. & Binnend. ex Miq., Fl. Ind. Bot. 3: 637, 1855
Dendrobium macrophyllum var. veitchianum (Lindl.) Hook.f., Bot. Mag. 93: t.5649, 1867
Dendrobium gordonii S.Moore ex Baker, J. Linn. Soc. 20: 372, 1883
Callista macrophylla (A.Rich.) Kuntze, Rev. Gen. 2: 665, 1891
Callista veitchiana (Lindl.) Kuntze, Rev. Gen. 2: 655, 1891
Dendrobium brachythecum F.Muell. & Kraenzlin, Oester. Bot. Zeit. 44: 161, 1894
Dendrobium ternatense J.J.Smith, Bull. Dép. Agr. Ind. Neerl. 22: 26, 1909
Dendrobium psyche Kraenzlin, Engl. Pflanzenr. Orch. Mon. Dendr.: 246, 1910
Dendrobium tomohonense Kraenzlin, Engl. Pflanzenr. Orch. Mon. Dendr.: 244, 1910
Dendrobium musciferum Schltr., Fedde Rep. Beih. 1: 494, 1914
Latourorchis macrophylla (A.Rich.) Brieger, Schlechter Die Orchideen ed.3, 1: 727, 1981
Latourorchis muscifera (Schltr.) Brieger, Schlechter Die Orchideen ed.3, 1: 727, 1981
Sayeria macrophylla (A.Rich.) Rauschert, Fedde Rep. 94 (7–8): 467, 1983

o **Dendrobium macrostachyum** Lindl., Gen. Sp. Orch. Pl.: 78, 1830
BR 22: t.1865 (1836) / 26: misc.62:12 (1844)
Sri Lanka, India – 1200 m – E

Dendrobium microbulbon A.Rich, Ann. Sc. Nat. s.2, 15: 15 t.8, 1841
BR 30: misc.62:62 (1844)
India – E
Dendrobium humile Wight, Ic.: t.1643, 1851
Dendrobium crispum Dalz., Hook. J. Bot. 4: 111, 1852

Dendrobium mirbelianum Gaudich., Freyc. Voy. Bot.: 423, 1826
BR 29: sub t.28:2 (1843) / 30: misc.62:65 (1844)
Moluccas, New Guinea, Bismarck Archipelago, Solomon Is., N.E.Australia (Queensland) – E
Dendrobium rosenbergii Teijsm. & Binnend., Nat. Tijdsch. Ned. Ind. 24: 317, 1862
Dendrobium polycarpum Rchb.f., Gard. Chron. n.s.20: 492, 1883
Callista mirbeliana (Gaudich.) Kuntze, Rev. Gen. 2: 655, 1891
Dendrobium giulianettii Bailey, Queensl. Agr. J. 3: 158, 1898
Dendrobium aruanum Kraenzlin, Engl. Pflanzenr. Orch. Mon. Dendr. 1: 242, 1910
Dendrobium buluense Schltr., Fedde Rep. Beih. 1: 549, 1912
Dendrobium buluense var. kauloense Schltr., Fedde Rep. Beih. 1: 550, 1912
Dendrobium wilkianum Rupp, Queensl. Nat. 10 (67): 4, 1941

Dendrobium moniliforme (L.) Sw., Nov. Act. Upsal. 6: 85, 1799
BR 30: misc.62:58 (1844)
China, Japan, Yakusima, Taiwan, Korea – 800–2500 m – E
Epidendrum moniliforme L., Sp. Pl.: 954, 1753
Epidendrum moniliferum Panzer, Pflanzensyst.10: 122, 1783
Epidendrum monile Thunb., Fl. Jap.: 30, 1784
Limodorum monile (Thunb.) Thunb., Trans. Linn. Soc. 2: 327, 1794
Onychium japonicum Blume, Bijdr. 7: 328, 1825
Dendrobium catenatum Lindl., Gen. Sp. Orch. Pl.: 84, 1830
Dendrobium japonicum (Blume) Lindl., Gen. Sp. Orch. Pl.: 89, 1830
Ormostema albiflora Raf., Fl. Tellur. 4: 38, 1838
Ormostema purpurea Raf., Fl. Tellur. 4: 38, 1838
Dendrobium castum Bateman ex Hook.f., Bot. Mag. 90: t.5482, 1864
Callista japonica (Blume) Kuntze, Rev. Gen. 2: 655, 1891
Callista moniliformis (L.) Kuntze, Rev. Gen. 2: 655, 1891
Dendrobium zonatum Rolfe, J. Linn. Soc. 36: 13, 1903
Dendrobium monile (Thunb.) Kraenzlin, Engl. Pflanzenr. Orch. Mon. Dendr. 1: 50 fig.2, 1910
Dendrobium heisnanense Hayata, Ic. Pl. Form. 4: 40 fig.13c, 1914

Dendrobium monophyllum F.Muell., Fragm. Phyt. Austr. 1: 189, 1859
BR 25: misc.31 (1839)
Australia (New South Wales, Cape York, Annan River) – E
Dendrobium tortile Cunn., Bot. Reg. 25: misc.31, 1839
Callista monophylla (F.Muell.) Kuntze, Rev. Gen. 2: 655, 1891
Australorchis monophylla (F.Muell.) Brieger, Schlechter Die Orchideen 1: 741, 1981

Dendrobium monticola Hunt & Summerhayes, Taxon 10, 4: 110, 1961
BR 30: misc.62:60 (1844)
N.W.Himalaya, Sikkim, China (Tibet) – E
Dendrobium pusillum D.Don, Prodr. Fl. Nepal: 35, 1825 (non Kunth)
Dendrobium denudans Lindl., Gen. Sp. Orch. Pl.: 84, 1830 (non D.Don)
Dendrobium alpestre Royle, Ill. Bot. Himal. 1: 370 t.88, 1839 (non Sw.)

Dendrobium moschatum (Buch.-Ham.) Sw., Schrad. Neues J. Bot. 1 (1): 94, 1805
BR 21: t.1779 (1835) / 27: misc.15 (1841) / 30: misc.62:44 (1844)
India, Nepal, Sikkim, Bhutan, Burma, Thailand, Laos – E
Epidendrum moschatum Buch.-Ham., Symes Embassy to Ava: 478, 1800
Cymbidium moschatum (Buch.-Ham.) Willd., Sp. Pl. ed.4: 98, 1805
Dendrobium calceolaria Carey ex Hook., Exot. Fl. 3: t.184, 1825
o Dendrobium cupreum Herbert, Bot. Reg. 21: t.1779, 1835
Callista moschata (Buch.-Ham.) Kuntze, Rev. Gen. 2: 655, 1891

Dendrobium mutabile (Blume) Lindl., Gen. Sp. Orch. Pl.: 86, 1830
BR 30: misc.58 & misc.62:50, 51 (1844) / 32: sub t.64 (1846) / 33: t.1 (1847)
Sunda Is., Sumatra, Java, Ryukyu Is. – E
Onychium mutabile Blume, Bijdr.: 324, 1825
Onychium rigidum Blume, Bijdr.: 324, 1825
Dendrobium rigidum (Blume) Lindl., Gen. Sp. Orch. Pl.: 86, 1830
Dendrobium firmum Steudel, Nom. 2 ed.2, 1: 490, 1840
Dendrobium sclerophyllum Lindl., Bot. Reg. 30: misc.58, 1844
o Dendrobium triadenium Lindl., Bot. Reg. 33: t.1, 1847
Dendrobium rigescens Miq., Fl. Ned. Ind. 3: 639, 1856
Callista mutabilis (Blume) Kuntze, Rev. Gen. 2: 653, 1891
Callista rigescens (Miq.) Kuntze, Rev. Gen. 2: 655, 1891
Callista rigida (Blume) Kuntze, Rev. Gen. 2: 655, 1891

Dendrobium nathanielis Rchb.f., Schill. Cat. Orchid. ed.3: 26, 1857
BR 27: misc.7 (1841) / 30: misc.62:61 (1844)
N.E.India, Burma, Thailand, Laos, Cambodia, Vietnam – E
Aporum cuspidatum Lindl., Bot. Reg. 27: misc.7, 1841
Dendrobium anceps Lindl., J. Linn. Soc. 3: 4, 1859 (non Sw.)
Dendrobium cuspidatum (Lindl.) Lindl., J. Linn. Soc. 3: 4, 1859 (non Lindl.1830)
Dendrobium multiflorum Parish & Rchb.f., Trans. Linn. Soc. 30: 149 t.31 fig.2, 1874
Callista nathanielis (Rchb.f.) Kuntze, Rev. Gen. 2: 655, 1891
Dendrobium mannii A.D.Kerr, Nat. Hist. Bull. Siam Soc. 23 (1–2): 194, 1969 (non Ridley)

Dendrobium nobile Lindl., Gen. Sp. Orch. Pl.: 34, 1830
BR 30: misc.62: 3, 4 (1844)
Nepal to Bhutan, N.E.India, Burma, Thailand, Laos, Vietnam, China – E
Dendrobium coerulescens Wall. ex Lindl., Sert. Orch.: t.18, 1838
Dendrobium lindleyanum Griffith, Notul. 3: 309, 1851
Callista nobilis (Lindl.) Kuntze, Rev. Gen. 2: 655, 1891
Dendrobium formosanum (Rchb.f.) Masamune, Trop. Hort. 3: 32, 1933
Dendrobium friedericksianum Brix, Bull. Soc. Roy. Sc. Nat. Laos 5: 8, 1962 (non Rchb.f.)

Dendrobium nudum (Blume) Lindl., Gen. Sp. Orch. Pl.: 85, 1830
BR 30: misc.62:47 (1844)
Java – E
Onychium nudum Blume, Bijdr.: 525, 1825
Dendrobium auroroseum Rchb.f., Bonpl. 5: 56, 1857
Callista aurorosea (Rchb.f.) Kuntze, Rev. Gen. 2: 654, 1891
Callista nuda (Blume) Kuntze, Rev. Gen. 2: 655, 1891

Dendrobium nutans Lindl., Gen. Sp. Orch. Pl.: 90, 1830
BR 30: misc.62:23 (1844)
India, Sri Lanka – E
Dendrobium villosulum Lindl., Paxt. Flow. Gard. 2: 106, 1852 (non Wallich)
Dendrobium jerdonianum Wight, Ic.: t.1644, 1855
Callista jerdoniana (Wight) Kuntze, Rev. Gen. 2: 655, 1891
Callista nutans (Lindl.) Kuntze, Rev. Gen. 2: 655, 1891
Callista villosula (Lindl.) Kuntze, Rev. Gen. 2: 655, 1891
Dendrobium nutantiflorum Hawkes & A.H.Heller, Lloydia 20: 122, 1957

Dendrobium ochreatum Lindl., Bot. Reg. 21: sub t.1756, 1835
BR 21: sub t.1756 (1835) / 27: misc.171 (1841) / 30: misc.62:20 (1844)
Type: Wallich 7410 (K- holo!)
N.E.India, Burma, Thailand, Laos – E
Dendrobium cambridgeanum Paxton, Mag. Bot. 6: t.265, 1839

Dendrobium ovatum (Willd.) Kraenzlin, Engl. Pfanzenr. Orch. Mon. Dendr.: 172, 1910
BR 30: misc.54 & misc.62:53 (1844)
India – E
Cymbidium ovatum Willd., Sp. Pl. ed.4, 1: 101, 1805
Dendrobium barbatulum Wight, Ic. 3: 10 t.910, 1844
Dendrobium chlorops Lindl., Bot. Reg. 30: misc.54, 1844

Dendrobium peguanum Lindl., J. Linn. Soc. 3: 19, 1859
BR 30: misc.62:63 (1844)
Nepal, India, Sikkim, Burma, Borneo – E
Dendrobium pygmaeum Lindl., Gen. Sp. Orch. Pl.: 85, 1830
Callista pygmaea (Lindl.) Kuntze, Rev. Gen. 2: 655, 1891
Dendrobium microbulbon Blatter & McCann, J. Bombay Nat. Hist. Soc. 35: 262, 1931 (non A.Rich.)
Dendrobium wallichii Hawkes & Heller, Lloydia 20 (2): 125, 1957

Dendrobium planibulbe Lindl., Bot. Reg. 29: misc.70, 1843
BR 29: misc.70 (1843)
Type: Philippines, Cuming s.n. cult. Loddiges (K-Lindl., holo!)
Thailand, Malaya, Sumatra, Borneo – E
Dendrobium tuberiferum Hook.f., Fl. Brit. Ind. 5: 728, 1890
Aporum planibulbe (Lindl.) Rauschert, Fedde Rep.: 441, 1983

Dendrobium polyanthum Lindl., Gen. Sp. Orch. Pl.: 81, 1830
BR 30: misc.62:43 (1844)
Burma – E see Seidenfaden Opera Botanica 83: 48, 1985

Dendrobium pugioniforme Cunn., Bot. Reg. 25: misc.34, 1839
BR 25: misc.34 (1839) / 30: misc.62:38 (1844)
Type: Australia, Illawara, A.Cunningham s.n. (K-Lindl., iso!)
Australia (New South Wales) – E, L
Dendrobium pungentifolium F.Muell., Fragm. Phyt. Austr. 1: 189, 1859
Callista pugioniformis (Cunn.) Kuntze, Rev. Gen. 2: 655, 1891
Dockrillia pugioniformis (Cunn.) Rauschert, Fedde Rep. 94 (7–8): 446, 1983

Dendrobium pulchellum Roxb. ex Lindl., Gen. Sp. Orch. Pl.: 82, 1830
BR 28: misc.62:5 (1844) / 32: t.10 (1846)
Nepal, N.E.India, Burma, Thailand, Laos, Vietnam, Malaya – 1200–1800 m – E
o Dendrobium dalhousieanum Wallich ex Paxton, Mag. Bot. 11: 145, 1844
Callista pulchella (Roxb. ex Lindl.) Kuntze, Rev. Gen. 2: 655, 1891

Dendrobium revolutum Lindl., Bot. Reg. 26: misc.110, 1840
BR 26: misc.110 (1840) / 30: misc.62:28 (1844)
Types: Singapore, Cuming s.n. cult. Loddiges Barker (K-Lindl., syn!)
Burma, Thailand, Cambodia, Laos, Vietnam, Malaya, Borneo – E
Dendrobium refractum Teijsm. & Binnend., Tijdschr. Ned. Ind. 24: 315, 1862
Callista revoluta (Lindl.) Kuntze, Rev. Gen. 2: 655, 1891
Dendrobium tonkinensis De Wild., Tribune Hort. 1: 21, 1906

o **Dendrobium ruckeri** Lindl., Bot. Reg. 29: misc.38, 1843
BR 29: t.60 & misc.38 (1843) / 30: misc.62:26, 48 (1844)
Type: Wallich 2003, cult. Rucker (K-Lindl., holo!)
India – E
Dendrobium ramosum Lindl. ex Wall., Gen. Sp. Orch. Pl.: 82, 1830

Dendrobium rugosum (Blume) Lindl., Gen. Sp. Orch. Pl.: 87, 1830
BR 30: misc.62:15 (1844)
Java – E
Grastidium rugosum Blume, Bijdr.: 335, 1825
Dendrobium glaucophyllum Teijsm. & Binnend., Nat. Tijdschr. Ned. Ind. 24: 314, 1862
Dendrobium purpurascens Teisjm. & Binnend., Nat. Tijdschr. Ned. Ind. 24: 314, 1862
Callista rugosa (Blume) Kuntze, Rev. Gen. 2: 655, 1891

Dendrobium salaccense (Blume) Lindl., Gen. Sp. Orch. Pl.: 86, 1830
BR 30: misc.62:16 (1844)
Burma, Thailand, Laos, Vietnam, China, Malaya, Sumatra, Java – E
Grastidium salaccense Blume, Bijdr: 333, 1825
Dendrobium intermedium Teijsm. & Binnend., Nat. Tijdschr. Ned. Ind. 5: 490, 1853
Dendrobium bambusifolium Parish & Rchb.f., Trans. Linn. Soc. 30: 149, 1874
Callista bambusifolia (Parish & Rchb.f.) Kuntze, Rev. Gen. 2: 654, 1891
Callista intermedia (Teijsm. & Binnend.) Kuntze, Rev. Gen. 2: 655, 1891
Callista salaccensis (Blume) Kuntze, Rev. Gen. 2: 655, 1891
Dendrobium gemellum Ridley, J. Linn. Soc.: 252, 1896 (non Lindl.)
Dendrobium cathcartii Gagnepain, Fl. Gén. Indoch. 6: 241, 1934 (non Hook.f.)
Grastidium bambusifolium (Parish & Rchb.f.) Brieger, Schlechter Die Orchideen ed.3, 1: 653, 1981

o **Dendrobium sanguinolentum** Lindl., Bot. Reg. 28: misc.73, 1842
BR 28: misc.73 (1842) / 29: t.6 (1843) / 30: misc.62:27 (1844)
Type: Sri Lanka, Nightingale s.n. (K-Lindl., holo!)
Thailand, Malaya, Indonesia, Borneo, Sulu Arch. – E
Dendrobium cerinum Rchb.f., Gard. Chron. n.s.12: 554, 1879 (non Schltr.)
Callista sanguinolenta (Lindl.) Kuntze, Rev. Gen. 2: 655, 1891
Pedilonum sanguinolentum (Lindl.) Brieger, Schlechter Die Orchideen ed.3, 1: 681, 1981

o **Dendrobium secundum** (Blume) Lindl., Bot. Reg. 15: t.1291, 1829
BR 15: t.1291 (1829) / 27: misc.169 (1841) / 30: misc.62:71 (1844)
Burma, Thailand, Cambodia, Laos, Vietnam, Malaya, Sumatra, Java, Borneo, Celebes, Philippines – E
Pedilonum secundum Blume, Bijdr.: 322, 1825
Dendrobium bursigerum Lindl., J. Linn. Soc. 3: 16, 1859
Dendrobium heterostigma Rchb.f., Hamb. Gartenz. 15: 51, 1859
Callista bursigera (Lindl.) Kuntze, Rev. Gen. 2: 654, 1891
Callista secunda (Blume) Kuntze, Rev. Gen. 2: 653, 1891
Dendrobium secundum var. bursigerum (Lindl.) Ridley, J. Linn. Soc. 31: 271, 1896
Pedilonum bursigerum (Lindl.) Rauschert, Fedde Rep. 94: 458, 1983

Dendrobium sinuatum (Lindl.) Lindl. ex Rchb.f. Walp. Ann. 6: 280, 1861
BR 27: misc.3 (1841)
Type: Singapore, Cuming s.n. cult. Loddiges (K-Lindl., holo!)
Thailand, Malaya – E
Aporum sinuatum Lindl., Bot. Reg. 27: misc.3, 1841

o **Dendrobium speciosum** Smith, Exot. Bot. 1: 17 t.10, 1805
BR 19: t.1610 (1833)
Australia (New South Wales, Queensland) – E
Callista speciosa (Smith) Kuntze, Rev. Gen. 2: 655, 1891
Dendrocoryne speciosum (Smith) Brieger, Schlechter Die Orchideen ed.3, 1: 724, 1981
Tropilis speciosa (Smith) Butzin, Willdenowia 12 (2): 250, 1981

Dendrobium stuposum Lindl., Bot. Reg. 24: misc.94, 1838
BR 24: misc.94 (1838) / 30: misc.62:25 (1844)
Type: India, cult. Loddiges (K-Lindl., holo!)
Sikkim, Bhutan, N.E.India, Burma, China – E
Dendrobium spegidoglossum Hook.f., Fl. Brit. Ind.: 737, 1890 (non Rchb.f.)
Callista stuposa (Lindl.) Kuntze, Rev. Gen. 2: 653, 1891

o **Dendrobium sulcatum** Lindl., Bot. Reg. 24: t.65, 1838
BR 24: t.65 (1838) / 30: misc.62:22 (1844)
Type: India, Gibson s.n. cult. Duke of Devonshire (K-Lindl., icono!)
Sikkim, N.E.India, Burma, Thailand, Laos, – E
Callista sulcata (Lindl.) Kuntze, Rev. Gen. 2: 655, 1891

o **Dendrobium taurinum** Lindl., Bot. Reg. 29: t.28, 1843
BR 29: t.28 & sub t.28:1 (1843) / 30: misc.62:64 (1844)
Type: Philippines, Cuming s.n. cult. Loddiges (K-Lindl., holo!)
Philippines – up to 300m – E
Callista taurina (Lindl.) Kuntze, Rev. Gen. 2: 655, 1891

Dendrobium teretifolium (Sw.) R.Br., Prodr. Fl. Nov. Holl.: 333, 1810
BR 25: misc.29 (1839) / 27: misc.26 (1841)
Australia (Queensland, New South Wales), New Guinea – E
Epidendrum teretifolium Sw., Prodr. Veg. Ind. Occ.: 122, 1788
Dendrobium calamiforme Lodd., Bot. Reg. 27: misc.26, 1841
Dendrobium fairfaxii F.Muell., Sydney Mail Sept.21: 360, 1872
Callista fairfaxii (F.Muell.) Kuntze, Rev. Gen. 2: 654, 1891
Callista teretifolia (Sw.) Kuntze, Rev. Gen. 2: 655, 1891
Dockrillia teretifolia (Sw.) Brieger, Schlechter Die Orchideen ed.3, 1: 746, 1981

Dendrobium tetragonum Cunn. ex Lindl., Bot. Reg. 25: misc.30, 1839
BR 25: misc.30 (1839) / 27: misc.8 (1841)
Type: cult. Loddiges (K-Lindl., holo!)
Australia (Queensland, New South Wales) – E
Callista tetragona (Cunn. ex Lindl.) Kuntze, Rev. Gen. 2: 655, 1891
Dendrobium tetragonum var. variabilis P.A.Gilbert, Austral. Orchid. Rev. 2: 20, 1937
Dendrocoryne tetragonum (Cunn. ex Lindl.) Brieger, Schlechter Die Orchideen ed.3, 1: 724, 1981
Tropilis tetragona (Cunn. ex Lindl.) Butzin, Willdenowia 12 (2): 250, 1982

Dendrobium transparens Wallich ex Lindl., Gen. Sp. Orch. Pl.: 79, 1830
BR 30: misc.62:9 (1844)
Himalaya (Kumaun to Bhutan), Assam – 800–1300 m – E
Dendrobium henshallii Rchb.f., Bonpl. 3: 226, 1855
Callista transparens (Wallich ex Lindl.) Kuntze, Rev. Gen. 2: 655, 1891

Dendrobium tridentiferum Lindl., Hook. J. Bot. 3: 237, 1843
BR 30: misc.62:32 (1844)
New Guinea – E

Dendrochilum abbreviatum Blume, Bijdr.: 398, 1825
BR 30: misc.34 (1844)
Java – E
Platyclinis abbreviata Hemsley, Gard. Chron. n.s.16: 656, 1881

Dendrochilum filiforme Lindl., Bot. Reg. 26: misc.113, 1840
BR 26: misc.113 (1840)
Type: Philippines, Cuming s.n. cult. Bateman (K-Lindl., holo!)
Philippines – E
Dendrochilum glumaceum André, Ill. Hort. 25: t.323, 1878 (non Lindl.)
Platyclinis filiformis (Lindl.) Benth. ex Hemsley Gard. Chron. n.s.16: 656, 1881
Acoridium filiforme (Lindl.) Rolfe, Orch. Rev.: 219, 1904

Dendrochilum glumaceum Lindl., Bot. Reg. 27: misc.58, 1841
BR 27: misc.58 (1841)
Type: Philippines, Cuming s.n. cult. Loddiges (K-Lindl., holo!)
Philippines (Luzon, Mindanao) – up to 2300 m – E
Platyclinis glumacea (Lindl.) Benth., J. Linn. Soc. 18: 295, 1881
Acoridium glumaceum (Lindl.) Rolfe, Orch. Rev. 12: 220, 1904

Dendrochilum latifolium Lindl., Bot. Reg. 29: misc.74, 1843
BR 29: misc.74 (1843)
Type: Philippines, Cuming s.n. cult. Loddiges (K-Lindl., holo!)
Philippines – E
Platyclinis latifolia Hemsley, Gard. Chron. n.s.16: 656, 1881
Acoridium latifolium Rolfe, Orch. Rev.: 219, 1904

Dendrophyllax funalis (Sw.) Benth. & Hook.f., Gen. Pl. 3: 583, 1883
BR 18: sub t.1522 (1832)
Mascarene Is. – E
Limodorum funale Sw., Nov. Act. Soc. Sc. Upsal. 6: 79, 1799
Angorchis funalis (Sw.) Kuntze, Rev. Gen. 2: 651, 1891

Diadenium barkeri (Lindl.) Benth. & Hook. ex Jackson, Ind. Kew 1: 741, 1893
BR 24: misc.60 (1838)
Type: Brazil, cult. Barker (K-Lindl., holo!)
Brazil – E
Chaenanthe barkeri Lindl., Bot. Reg. 24: misc.60, 1838

Diaphananthe bidens (Afzel. ex Sw.) Schltr., Die Orchideen: 593, 1914
BR 29: misc.75 (1843)
Guinea-Bissau, Guinea, Sierra Leone, Liberia, Ivory Coast, Ghana, S.Nigeria, W.Cameroun, Fernando Po, Gabon, Zaire, C.African Rep., Uganda, Angola – E
Limodorum bidens Afzel. ex Sw., Kongl. Vet. Acad. Handl.: 243, 1800
Angraecum ashantense Lindl., Bot. Reg. 29: misc.75, 1843
Listrostachys ashantensis (Lindl.) Rchb.f., Walp. Ann. 6: 908, 1864
Angraecum bakeri Kraenzlin, Mitth. Deutsch. Schutzgeb. 2: 159, 1889
Listrostachys bakeri (Kraenzlin) Durand & Schinz, Conspect. Fl. Afr. 5: 48, 1895
Listrostachys bidens (Afzel. ex Sw.) Rolfe, F.T.A. 7: 174, 1895
Mystacidium productum Kraenzlin, Engl. Bot. Jahrb. 21: 30, 1895
Angraecum bidens (Afzel. ex Sw.) Rendle, Cat.: t.147, 1913

Diaphananthe pellucida (Lindl.) Schltr., Die Orchideen: 593, 1914
BR 30: t.2 (1844)
Type: cult. Loddiges (K-Lindl., holo!)
W.Africa – E
o Angraecum pellucidum Lindl., Bot. Reg. 30: t.2, 1844

Dichaea glauca (Sw.) Lindl., Gen. Sp. Orch. Pl.: 209, 1833
BR 30: misc.67 (1844)
Mexico, Guatemala, Honduras, W.Indies – E, T, L
Epidendrum glaucum Sw., Prodr.: 124, 1788
Dichaeopsis glauca (Sw.) Schltr., Beih. Bot. Centralbl. 36 (2): 519, 1918

Dichaea ochracea Lindl., Bot. Reg. 25: misc.71, 1839
BR 25: misc.71 (1839)
Type: Guyana, cult. Loddiges (K-Lindl., holo!)
Guyana – E

Dichromanthus cinnabarinus (Llave & Lex.) Garay, Bot. Mus. Leafl. Harv. Univ. 28: 314, 1980 publ.1982
BR 33: t.65 (1847)
USA (Texas), Mexico, Guatemala – up to 2400 m – T
Neottia cinnabarina Llave & Lex., Nov. Veg. Descr. 2: Orch. Opusc. 3, 1825
o Stenorrhynchus cinnabarinus (Llave & Lex.) Lindl., Gen. Sp. Orch. Pl.: 479, 1840
Stenorrhynchus montanus Lindl., Benth. Pl. Hartw.: 95, 1842
Spiranthes cinnabarina (Llave & Lex.) Hemsl., Biol. Centr. Am. 3: 300, 1885

Dimerandra emarginata (G.Meyer) Hoehne, Boletin Agric. S.Paulo 34a 1: 618 t.9, 1934
BR 26: misc.49 (1840) / 29: misc.60 (1843) / 31: misc.36:5 & misc.70: 4 (1845)
Mexico, Guatemala, Honduras, Costa Rica, Panama, Trinidad, Venezuela, Colombia, Ecuador, Peru, Guyana, Surinam, Brazil, Jamaica, W.Indies – E
Oncidium emarginatum G.Meyer, Prim. Fl. Essequeb.: 259, 1818
Epidendrum stenopetalum Hook., Bot. Mag. 62: t.3410, 1835
Epidendrum lamellatum Westc. & Lindl., Bot. Reg. 29: misc.60, 1843
Isochilus elegans Focke, Tijdsch. Nat. Wetensch. 4: 68, 1851
Epidendrum stenopetalum var. tenuicaule Rchb.f., Otia Bot Hamb. 1: 12, 1878
Dimerandra lamellata (Westc. & Lindl.) Siegrist, Bot. Mus. Leafl. Harv. Univ. 30, 1895
Telipogon buenaventurae Kraenzlin, Ann. Nat. Hist. Mus. Wien 33: 35, 1919
Epidendrum rimbachii Schltr., Fedde Rep. Beih. 8: 167, 1921
Dimerandra isthmi Schltr., Fedde Rep. Beih. 17: 44, 1922
Dimerandra rimbachii (Schltr.) Schltr., Fedde Rep. Beih. 17: 44, 1922
Dimerandra stenopetala (Hook.) Schltr., Fedde Rep. Beih. 17: 44, 1922

Diplocaulobium longicolle (Lindl.) Kraenzlin, Pflanzenr. 45: 340, 1910
BR 26: misc.172 (1840)
Type: Singapore, Cuming s.n. cult. Loddiges (K-Lindl., holo!)
Malaya, Sumatra, New Guinea – E
Dendrobium longicolle Lindl., Bot. Reg. 26: misc.172, 1840
Dendrobium inauditum Rchb.f., Lindenia 2: 66, 1886

Diplocentrum recurvum Lindl., Bot. Reg. 18: sub t.1522, 1832
BR 18: sub t.1522 (1832)
Type: Wallich s.n. (K-Lindl., holo!)
Sri Lanka, India – E
Diplocentrum longifolium Wight, Ic. Pl. Ind. Orch. 5: 10 t.1681, 1851

Diplomeris hirsuta (Lindl.) Lindl., Gen. Sp. Orch. Pl.: 331, 1835
BR 18: sub t.1499 (1832)
Type: Wallich 7065 (K-Lindl., holo!)
India, Nepal, Bhutan – T
Diplochilos hirsutum Lindl., Bot. Reg. 18: sub t.1499, 1832

Diplomeris pulchella D.Don, Prodr. Fl. Nepal: 26, 1825
BR 18: sub t.1499 (1832)
India, Vietnam – E
Paragnatis pulchella (D.Don) Sprengel, Syst. Veg. 3: 694, 1826
Diplochilos longifolium Lindl., Bot. Reg. 18: sub t.1499, 1832
Orchis uniflora Roxb., Fl. Ind. 3: 452, 1832
Habenaria uniflora (Roxb.) Griffith, Ic. Pl. 3: t.338, 1851 (non Don, non Dalz.)

o **Dipodium punctatum** (Smith) R.Br., Prodr. Fl. Nov. Holl.: 331, 1810
BR 23: t.1980 (1837)
Australia – S
Dendrobium punctatum Smith, Exot. Bot. 1: 21 t.12, 1804
Dipodium punctatum var. album Bailey, Syn. Qd. Fl. Suppl. 3: 73, 1890
Dipodium stenocheilum Schltr., Fedde Rep. 24: 80, 1927
Dipodium punctatum var. stenocheilum (Schltr.) Rupp, N. Qld. Nat. 11 (71): 2, 1944

Disa porrecta Sw., Kongl. Vet. Acad. Handl. Stockh.: 211, 1800
BR 9: sub t.700 (1823)
S.Africa – T
Disa zeyheri Sonder, Linnaea 19: 95, 1846

Disa uniflora Bergius, Descr. Pl. Cap. Sp.: 348 t.4 fig.7, 1767
BR 9: sub t.700 (1823) / 11: t.926 (1825)
S.Africa (Cape Province) – 900–1300 m – T
o Disa grandiflora L.f., Suppl.: 406, 1781
Satyrium grandiflora (L.f.) Thunb., Prodr.: 4, 1794
Disa uniflora var. superba Moore, Proc. Hort. Soc. Lond. 2: 617, 1862
Disa uniflora var. barellii hort., Fl. Mag. 2: t.104, 1874
Disa uniflora var. psittacina Rchb.f., Gard. Chron. n.s.12: 424, 1879
Disa barelli Puydt, Orchidées: 275 t.18, 1880

Disperis capensis (L.f.) Sw., Kongl. Vet. Acad. Handl. Stockh.: 220 t.3, 1800
BR 9: sub t.700 (1823)
S.Africa – T
Arethusa capensis L.f., Suppl. Pl.: 405, 1781
Dipera capensis (L.f.) Sprengel, Syst. Veg. 3: 696, 1826
Dipera tenera Sprengel, Syst. Veg. 3: 696, 1826

Disperis cucullata Sw., Kongl. Vet. Acad. Handl. Stockh.: 220, 1800
BR 9: sub t.700 (1823)
S.Africa – T

Disperis secunda (Thunb.) Sw., Kongl. Vet. Acad. Handl. Stockh.: 220, 1800
BR 9: sub t.700 (1823)
S.Africa – T
Orchis circumflexa L., Sp. Pl. ed.2: 1344, 1763
Arethusa secunda Thunb., Prodr.: 3, 1794
Disperis circumflexa (L.) Durand & Schinz, Conspect Fl. Afr. 5: 118, 1895

Disperis villosa (L.f.) Sw., Kongl. Vet. Acad. Handl. Stockh.: 220, 1800
BR 9: sub t.700 (1823)
S. Africa – T
Arethusa villosa L.f., Suppl. Pl.: 405, 1781

Earina autumnalis (Forst.f.) Hook.f., Fl. New Zealand: 239, 1853
BR 29: misc.88 (1843)
New Zealand – E, L
Epidendrum autumnale Forst.f., Prodr.: 60, 1786
Cymbidium autumnale (Forst.f.) Sw., Nov. Act. Upsal. 6: 72, 1799
Earina suaveolens Lindl., Bot. Reg. 29: misc.88, 1843
Earina alba Col., Trans. New Zealand Inst. 18: 267, 1886

Earina mucronata Lindl., Bot. Reg. 20: sub t.1699, 1834
BR 20: sub t.1699 (1834)
Type: New Zealand, A.Cunningham s.n. (K-Lindl., holo!)
New Zealand – E
Earina quadrilobata Col., Trans. New Zealand Inst. 15: 325, 1883

Encyclia adenocarpon (Llave & Lex.) Schltr., Die Orchideen: 207, 1914
BR 24: misc.8 (1838) / 26: misc.35 (1840) / 28: misc.p.32:20 & misc. p.35:41 (1842)
Mexico, Guatemala, El Salvador, Honduras, Nicaragua – E
Epidendrum adenocarpon Llave & Lex., Nov. Veg. Descr. 2: 24, 1825
Epidendrum crispatum Knowles & Westc., Fl. Cabin. 2: 79, 1838
Epidendrum papillosum Bateman ex Lindl., Bot. Reg. 24: misc.8, 1838
Epidendrum adenocarpon var. rosei Ames, Hubb. & C.Schweinf., Bot. Mus. Leafl. Harv. Univ. 3 (5): 61, 1935

o **Encyclia adenocaula** (Llave & Lex.) Schltr., Beih. Bot. Centralbl. 36 (2): 470, 1918 **var. adenocaula**
BR 28: misc.p.32:26 & misc.p.35:42 (1842) / 30: t.51 (1844)
Mexico – 1000–2000 m – E
Epidendrum adenocaulum Llave & Lex., Nov. Veg. Desc.: 22, 1825
Epidendrum nemorale Lindl., Hook. J. Bot. 3: 82, 1840
Epidendrum verrucosum Lindl., Bot. Reg. 30: t.51, 1844 (non Sw.)
Encyclia nemoralis (Lindl.) Schltr., Die Orchideen: 210, 1914

Encyclia alata (Bateman) Schltr., Die Orchideen: 207, 1914
BR 27: misc.181 (1841) / 28: misc.p.34:37 (1842) / 33: t.53 (1847)
Mexico, Belize, Guatemala, El Salvador, Honduras, Nicaragua, Costa Rica – E
o Epidendrum alatum Bateman, Orch. Mex. & Guat.: t.18, 1840
Epidendrum calocheilum Hook., Bot. Mag. 68: t.3898, 1841
Epidendrum longipetalum Lindl. & Paxton, Paxt. Flow. Gard. 1: 149 t.30, 1850
Epidendrum formosum Klotzsch, Allg. Gartenz. 21: 201, 1853
Epidendrum alatum var. grandiflorum Regel, Ind. Sem. Hort. Petrop.: 19, 1856
Epidendrum alatum var. longipetalum (Lindl. & Paxton) Regel, Ind. Sem. Hort. Petrop.: 19, 1856
Epidendrum alatum var. viridiflorum Regel, Ind. Sem. Hort. Petrop. 19, 1856

Encyclia aromatica (Bateman) Schltr., Die Orchideen: 208, 1914
BR 26: misc.84 & misc.93 (1840) / 28: misc.p.34:35 (1842)
Guatemala, Mexico – up to 1600 m – E
Epidendrum aromaticum Bateman, Orch. Méx. & Guatem.: t.10, 1838
Epidendrum incumbens Lindl., Bot. Reg. 26: misc.93, 1840

Encyclia aspera (Lindl.) Schltr., Fedde Rep. 7: 253, 1920
BR 24: misc.36 (1838) / 28: misc.p.29:5 (1842)
Colombia, Ecuador – E
Epidendrum asperum Lindl., J. Bot. 1: 6, 1834

Encyclia boothiana (Lindl.) Dressler, Brittonia 13: 269, 1961
ssp. boothiana
BR 24: misc.7 (1838) / 28: misc.p.30:10 (1842) / 31: misc.36 p.24:1 (1845)
Type: Cuba, cult. Loddiges (K-Lindl., holo!)
Florida, W.Indies, Honduras, Mexico – E
Epidendrum bidentatum Lindl., Gen. Sp. Orch. Pl.: 98, 1831
Epidendrum boothianum Lindl., Bot. Reg. 24: misc.7, 1838
Diacrium bidentatum (Lindl.) Hemsley, Biol. Centr.-Amer. Bot. 3: 221, 1883
Epidendrum erythronioides Small, Fl. S.E. United States: 328, 1903
Epicladium boothianum (Lindl.) Small, Fl. Miami: 56, 1913
Epicladium boothianum var. erythronioides (Small) Acuña, Cat. Descr. Orq. Cub.: 89, 1938
Hormidium boothianum (Lindl.) Brieger, Publ. Cient. Inst. Genet. Univ. S.Paulo 1: 21, 1960
Encyclia boothiana var. erythronioides (Small) Luer, Native Orchid. Florida 14: 29, 1971

Encyclia bractescens (Lindl.) Hoehne, Arq. Bot. Est. S.Paulo n.s.2: 150, 1952
BR 26: misc.122 (1840) / 28: misc.p.33:27 (1842)
Type: Mexico, Oaxaca, Hartweg s.n. (K-Lindl., holo!)
Bahamas, Mexico, Belize, Guatemala, Brazil – up to 1000 m – E
Epidendrum bractescens Lindl., Bot. Reg. 26: misc.122, 1840
Epidendrum linearifolium Hook., Bot. Mag. 77: t.4572, 1851

Encyclia bulbosa (Vell.) Pabst, Orquidea 29 (6): 276, 1967
BR 25: misc.135 (1839) / 28: sub t.50:9 (1842)
Brazil, Paraguay – E
Epidendrum bulbosum Vell., Fl. Flum. 9: t.11, 1825
Epidendrum inversum Lindl., Bot. Reg. 25: misc.135, 1839
Epidendrum latro Rchb.f., Mart. Fl. Bras. 3 (5): 82, 1898

Encyclia calamaria (Lindl.) Pabst, Orquidea 29 (6): 276, 1972
BR 24: misc.163 (1838) / 28: sub t.50:12 (1842)
Type: Brazil, Capt. Sutton s.n. cult. C.Lemon (K-Lindl., holo!)
Brazil – E
Epidendrum calamarium Lindl., Bot. Reg. 24: misc.163, 1838

Encyclia candollei (Lindl.) Schltr., Die Orchideen: 203, 1914
BR 25: misc.77 (1839) / 28: misc.p.29:6 (1842)
Type: Mexico, cult. Loddiges (K-Lindl., holo!)
Guatemala, Mexico – 700–1500 m – E
Epidendrum candollei Lindl., Bot. Reg. 25: misc.77, 1839
Epidendrum cepiforme Hook., Bot. Mag. 66: t.3765, 1840
Epidendrum flabellatum Lindl., Fol. Orch. Epid. 6, 1853
Encyclia laxa Schltr., Notizbl. Bot. Gart. Mus. Berlin-Dahlem 7: 277, 1918

Encyclia ceratistes Lindl., Bot. Reg. 30: misc.92, 1844
BR 30: misc.92 (1844)
Type: Rio Martha, Hartweg s.n. (K-Lindl., holo!)
Venezuela, Colombia, Centr.America – 1000–15000 m – E
Epidendrum ceratistes Lindl., Bot. Reg. 30: misc.92, 1844
Epidendrum ramonense Rchb.f., Beitr. Orch. Centr.-Am.: 81, 1866
Encyclia ramonensis (Rchb.f.) Schltr., Beih. Bot. Centralbl. 36 (2): 473, 1918
Encyclia powellii Schltr., Fedde Rep. Beih. 17: 46, 1922
Epidendrum oncidioides var. ramonense (Rchb.f.) Ames, Hubb. & Schweinf., Bot. Mus. Leafl. Harv. Univ. 3: 103, 1935
Encyclia oncidioides var. ramonensis (Rchb.f.) Hoehne, Arq. Bot. Est. S.Paulo n.s.2: 153, 1952

Encyclia chloroleuca (Hook.) Neumann, Rev. Hort. s.2, 4: 138, 1845/46
BR 24: misc.28 (1838) / 28: misc.p.32:21 (1842)
Guyana, Colombia – E
Epidendrum chloroleucum Hook., Bot. Mag. 64: t.3557, 1837
Epidendrum chloranthum Lindl., Bot. Reg. 24: misc.28, 1838

Encyclia citrina (Llave & Lex.) Dressler, Brittonia 13: 264, 1961
BR 30: sub t.5:7 (1844)
Mexico – 1300–2200 m – E
Sobralia citrina Llave & Lex., Nov. Veg. Descr. Orch. Opusc. 2: 21, 1825
Cattleya citrina (Llave & Lex.), Lindl., Coll. Bot.: sub t.37, 1826
Cattleya karwinskii C.Martius, Ausgew. Merkw. Pfl. 14: t.10, 1830
Epidendrum citrinum (Llave & Lex.) Rchb.f., Walp. Ann. 6: 317, 1862
Hormidium citrinum (Llave & Lex.) Brieger, Schlechter Die Orchideen ed.3, 1: 576, 1977

Encyclia concolor (Llave & lex.) Schltr., Beih. Bot. Centralbl. 36 (2): 472, 1918
BR 28: misc.p.31:17 (1842)
Mexico – 1500–2100 m – E
Epidendrum concolor Llave & Lex., Nov. Veg. Descr. Orch. Opusc. 25, 1825
Epidendrum pruinosum A.Rich. & Galeotti, Ann. Sc. Nat. s.3, 3: 20, 1845
Epidendrum amabile Linden & Rchb.f., Bonpl. 3: 219, 1855
Epidendrum punctulatum Rchb.f., Gard. Chron. n.s. 24: 70, 1885
Encyclia amabilis (Linden & Rchb.f.) Schltr., Beih. Bot. Centralbl. 36 (2): 471, 1918
Encyclia pruinosa (A.Rich. & Galeotti) Schltr., Beih. Bot. Centralbl. 36 (2): 473, 1918

Encyclia cordigera (Kunth) Dressler, Taxon 13: 247, 1964
BR 26: misc.85 (1840) / 28: misc.p.35:43, 44 (1842)

Mexico to Panama, W.Indies, Venezuela, Trinidad, Peru – up to 900 m – E
Cymbidium cordigerum Kunth, Nov. Gen. & Sp. Pl. 1: 341, 1815
Epidendrum auropurpureum Lindl., Gen. Sp. Orch. Pl.: 99, 1831
Epidendrum macrochilum Hook., Bot. Mag. 63: t.3534, 1836
Epidendrum macrochilum var. roseum Bateman, Orch. Mex. & Guat.: t.17, 1839
Encyclia macrochila (Hook.) Neumann, Rev. Hort. 11 (4): 137, 1845
Epidendrum duboisianum Brogn. ex A.Rich, Fl. Cuba Fanerog. 2: 239, 1853
Epidendrum atropurpureum var. roseum (Bateman) Rchb.f., Bonpl. 2: 19, 1854
Epidendrum longipetalum Godefr.-Leb., L'Orchidoph. 12: 257, 1892(non A.Rich & Galeotti 1845; non Lindl. & Paxton 1851)
Epidendrum atropurpureum var. lionetianum Cogn., Dict. Ic. Orch. Epid.: 10c, t.1903
Encyclia atropurpurea var. leucantha Schltr., Fedde Rep. Beih. 17: 45, 1922
Encyclia atropurpurea var. rhodoglossa Schltr., Fedde Rep. Beih. 17: 45, 1922
Epidendrum atropurpureum var. laciniatum Ames, Hubb. & C.Schweinf., Bot. Mus. Leafl. Harw. Univ. 3: 63, 1935
Encyclia doeringii Hoehne, Arq. Bot. Est. S.Paulo 2 (2): 24 t.11, 1946
Epidendrum doeringii (Hoehne) A.Hawkes, Orquidea 18: 170, 1956
Encyclia atropurpurea var. rosea (Bateman) Summerh., Bot. Mag. n.s.171: t.290, 1957
Encyclia cordigera var. rosea (Bateman) H.G.Jones, Darwiniana 15: 23, 1969
Epidendrum cordigerum (Kunth) Foldats, Bol. Soc. Ven. Cienc. Nat. 28: 234, 1969

Encyclia dichroma (Lindl.) Schltr., Die Orchideen ed.3: 209, 1981
BR 29: misc.119 (1843)
Type: Brazil, Pernambuco, cult. Quesnel (K-Lindl., holo!)
Brazil – E
Epidendrum dichromum Lindl., Bot. Reg. 29: misc.119, 1843
Epidendrum guesnelianum hort. ex Haneshall, Cult. Orch.: 113, 1845
Epidendrum guesnelianum var. amabile Bateman, Bot. Mag. 91: t.5491, 1865
Epidendrum guesnelianum var. striatum Rchb.f., Gard. Chron.: 218, 1866
Epidendrum conspicuum Lemaire, Ill. Hort. 16: t.592, 1869
Epidendrum biflorum Barb.Rodr., Gen. & Sp. Orch. Nov. 1: 52, 1877
Epidendrum amabile (Bateman) Godef.-Leb., L'Orchidoph.: 304, 1887 (non L., non Rchb.f.)

Encyclia diota (Lindl.) Schltr., Beih. Bot. Centralbl. 36 (2): 472, 1918
ssp. diota
BR 29: misc.97 (1843)
Type: Guatemala, Hartweg s.n. (K-Lindl., holo!)
Guatemala, Honduras, Nicaragua, Mexico – 600–1800 m – E
Epidendrum diotum Lindl., Bot. Reg. 29: misc.97, 1843
Epidendrum insidiosum Rchb.f., Linnaea 41: 80, 1876
Encyclia insidiosa (Rchb.f.) Schltr., Fedde Rep. Beih. 7: 254, 1920

Encyclia fragrans (Sw.) Lemée, Fl. Guyane Fr. 1: 418, 1955
BR 22: t.1898 (1836) / 28: sub t.50:1, 6, 7 (1842)
Jamaica, W.Indies, Centr. & S.America – 500–1500 m – E
Epidendrum fragrans Sw., Prod. Veg. Ind. Occ.: 123, 1788
Epidendrum cochleatum Curtis, Bot. Mag. 5: t.152, 1791 (non L.)
Epidendrum lineatum Salisb., Prodr.: 10, 1796
Epidendrum cordatum Vell., Fl. Flum. 9: t.38, 1827
o Epidendrum aemulum Lindl., Bot. Reg. 22: t.1898, 1836
Epidendrum ionoleucum Hoffsgg. ex Rchb.f., Linnaea 24: 244, 1852
Epidendrum lambda Linden & Rchb.f., Bonpl. 2: 281, 1854
Epidendrum aemulum var. breviaristatum Rchb.f., Linnaea 41: 37, 1876
Epidendrum fragrans var. aemulum Barb.Rodr., Gen. & Sp. Orch. Nov. 2: 136, 1881
Epidendrum fragrans var. alticallum Barb.Rodr., Gen. & Sp. Orch. Nov. 2: 138, 1881
Epidendrum fragrans var. ionoleucum (Hoffsgg. ex Rchb.f.) Barb.Rodr., Gen. & Sp. Orch. Nov. 2: 136, 1881
Epidendrum fragrans var. janeirense Barb.Rodr., Gen. & Sp. Orch. Nov. 2: 137, 1881
Epidendrum fragrans var. micranthum Barb.Rodr., Gen. & Sp. Orch. Nov. 2: 138, 1881
Epidendrum fragrans var. rivularium Barb.Rodr., Gen. & Sp. Orch. Nov. 2: 137, 1881
Epidendrum fragrans var. magnum Stein, Orchideenb.: 230, 1892
Epidendrum vaginatum Sessé & Mociño, Fl. Mex. ed.2: 201, 1894
Epidendrum fragrans var. breviaristatum (Rchb.f.) Cogn., Mart. Fl. Bras. 3 (5): 85, 1898
Epidendrum fragrans var. pachypus Schltr., Fedde Rep. Beih. 17: 32, 1922
Anacheilium fragrans (Sw.) Acuña, Cat. Orqu. Cuba: 86, 1939
Hormidium fragrans (Sw.) Brieger, Schlechter Die Orchideen ed.3, 1: 568, 1977

Encyclia fucata (Lindl.) Schltr., Die Orchideen :207, 1914
BR 24: misc.17 (1838) / 28: misc.p.31:18 (1842)
Type: Capt. Sutton s.n. cult. C.Lemon (K-Lindl., holo!)
Cuba, Bahamas – E
Epidendrum fucatum Lindl., Bot. Reg. 24: misc.17, 1838
Epidendrum affine A.Rich., Fl. Cub. Fanerog. 2: 237, 1841
Epidendrum hircinum A.Rich., Fl. Cub. Fanerog. 2: 233, 1841
Epidendrum sagraeanum A.Rich., Fl. Cub. Fanerog. 2: 235, 1841
Encyclia hircina (A.Rich) Acuña, Bol. Estac. Exper. Agron. Santiago Cuba 60: 80, 1939

Encyclia glauca (Knowles & Westc.) Dressler & Pollard, Phytolog. 21: 437, 1971
BR 26: misc.56 (1840) / 28: misc.p.31:14 (1842) / 29: misc.104 (1843)
Guatemala, Mexico – 900–1000 m – E
Epithecia glauca Knowles & Westc., Floral Cab. 2: 167 t.87, 1838
Prosthechea glauca Knowles & Westc., Floral Cab. 2: 111, 1838
Epidendrum glaucum (Knowles & Westc.) Lindl., Bot. Reg. 26: misc.56, 1840 (non Sw. 1788)
Epidendrum limbatum Lindl., Bot. Reg. 29: misc.104, 1843
Amblostoma tridactylum var. mexicanum Kraenzlin, Vidensk. Medd. Dansk Nat. For. 71: 177, 1920
Epidendrum glaucovirens Ames Hubb. & C.Schweinf., Bot. Mus. Leafl. Harv. Univ. 3: 70, 1935
Encyclia limbata (Lindl.) Dressler, Brittonia 13: 265, 1961

Encyclia glumacea (Lindl.) Pabst, Orquidea 29 (6): 276, 1967
BR 25: misc.50 (1839) / 26: t.6 (1840) / 28: sub t.50:8 (1842)
Type: Brazil, without collector (K-Lindl., holo!)
Brazil, Ecuador – E
o Epidendrum glumaceum Lindl., Bot. Reg. 25: misc.50, 1839
Aulizeum glumaceum (Lindl.) Stein, Orchideenb.: 231, 1892
Epidendrum almasyi Hoehne, Arq. Bot. Est. S.Paulo n.s. f.maior 2: 84, 1947
Encyclia almasyi (Hoehne) Pabst, Orquidea 29 (6): 276, 1967

Encyclia gracilis (Lindl.) Schltr., Die Orchideen: 207, 1914
BR 21: t.1765 (1835) / 28: misc.p.33:29 (1842)
Type: Bahamas, C.Lees s.n. cult. Hort. Soc. (K-Lindl., holo!)
Bahamas, Cuba – E
o Epidendrum gracile Lindl., Bot. Reg. 21: t.1765, 1835

Encyclia hanburii (Lindl.) Schltr., Die Orchideen: 209, 1914
BR 30: misc.60 (1844)
Type: Mexico, cult. R.Hanbury (K-Lindl., holo!)
Mexico – 1200–1800 m – E, L
Epidendrum hanburii Lindl., Bot. Reg. 30: misc.60, 1844

Encyclia hastata (Lindl.) Dressler & Pollard, Phytolog. 21: 437, 1971
BR 27: misc.90 (1841) / 28: misc.p.30:9 (1842)
Mexico – 2400–2700 m – E
Epidendrum hastatum Lindl., Hook. J. Bot. 3: 82, 1840

Encyclia ionosma (Lindl.) Schltr., Die Orchideen: 207, 1914
BR 24: misc.87 (1838) / 28: misc.p.33:31 (1842)
Type: Essequibo, cult. Loddiges (K-Lindl., holo!)
Brazil, Guyana – E
Epidendrum ionosmum Lindl., Bot. Reg. 24 misc.87, 1838
Encyclium ionosmum Lindl. ex Stein, Orchideenb.: 232, 1892

Encyclia lancifolia (Pavon ex Lindl.) Dressler & Pollard, Phytolog. 21: 437, 1971
BR 26: misc.152 (1840) / 28: t.50 & sub t.50:3 (1842)
W.Mexico – 300–1500 m – E
o Epidendrum lancifolium Pavon ex Lindl., Gen. Sp. Orch. Pl.: 98, 1831
Epidendrum trulla Rchb.f., Bonpl. 4: 214, 1856
Epidendrum langlassei Schltr., Beih. Bot. Centralbl. 36 (2): 404, 1918

Encyclia linkiana (Klotzsch) Schltr., Beih. Bot. Centralbl. 36 (2): 472, 1918
BR 28: misc.p.30:8 & misc.p.34:33 (1842)
Mexico – 500–2300 m – E
Epidendrum pastoris Link & Otto, Ic. Pl. Rar.: 23 t.12, 1828 (non Llave & Lex.)
Epidendrum linkianum Klotzsch, Allg. Gartenz. 6: 299, 1838
Epidendrum tripterum Lindl., Hook. J. Bot. 3: 83, 1840 (non Sw.1793)

Encyclia livida (Lindl.) Dressler, Brittonia 13: 265, 1961
BR 24: misc.9 & misc.91 (1838) / 27: misc.129 (1841) / 28: sub t.50: 11 & misc.p.29: 3, 12 (1842)
Type: Colombia, cult.Loddiges (K-Lindl., holo!)
Mexico, Guatemala, Costa Rica, Panama, Venezuela, Colombia, Ecuador – up to 1200 m – E

Epidendrum articulatum Klotzsch, Allg. Gartenz. 6: 297, 1838
Epidendrum lividum Lindl., Bot. Reg. 24: misc.91, 1838
Epidendrum tesselatum Bateman ex Lindl., Bot. Reg. 24: misc.9, 1838 (non Roxb.)
Epidendrum condylochilum F.Lehm & Kraenzlin, Engl. Bot. Jahrb. 26: 459, 1899
Epidendrum henricii Schltr., Fedde Rep. 3: 108, 1906
Anacheilium lividum (Lindl.) Pabst, Bradea 3 (23), 1918
Encyclia tesselata (Lindl.) Schltr., Beih. Bot. Centralbl. 36 (2): 474, 1918
Epidendrum deamii Schltr., Beih. Bot. Centralbl. 36 (2): 402, 1918
Epidendrum dasytainia Schltr., Fedde Rep. Beih. 8: 71, 1921
Encyclia deamii (Schltr.) Hoehne, Arq. Bot. Est. S.Paulo n.s.2: 151, 1952
Hormidium lividum (Lindl.) Brieger, Schlechter Die Orchideen ed.3, 1: 574, 1977

Encyclia michuacana (Llave & Lex.) Schltr., Beih. Bot. Centralbl. 36 (2): 472, 1918
BR 27: misc.189 (1841) / 28: misc.p.34:34 & misc.p.36:47 (1842)
Honduras, Guatemala, Mexico – 1500–2800 m – T, E
Epidendrum michuacanum Llave & Lex., Nov. Veg. Descr. Orch. Opusc.: 26, 1825
Epidendrum virgatum Lindl., J. Bot. 3: 83, 1840 (non Vell.1827)
Epidendrum virgatum var. pallens Rchb.f., Bonpl. 4: 326, 1856
Encyclia virgata (Lindl.) Schltr., Die Orchideen: 212, 1914
Epidendrum icthyphyllum Ames, Sched. Orch. 2: 28, 1923
Encyclia icthyphylla (Ames) Hoehne, Arq. Bot. Est. S.Paulo n.s.2: 152, 1952

Encyclia microbulbon (Hook.) Schltr., Beih. Bot. Centralbl. 36 (2): 473, 1918
BR 29: misc.71 (1843)
Mexico – 1400–2400 m – E
Epidendrum microbulbon Hook., Ic. Pl. 4: t.347, 1841
Epidendrum ovulum Lindl., Bot. Reg. 29: misc.71, 1843
Epidendrum sisyrinchiifolium A.Rich. & Galeotti, Ann. Sc. Nat. s.3, 3: 19, 1845
Encyclia ovulum (Lindl.) Schltr., Beih. Bot. Centralbl. 36 (2): 473, 1918
Encyclia sisyrinchiifolia (A.Rich. & Galeotti) Schltr., Beih. Bot. Centralbl. 36 (2): 474, 1918

Encyclia ochracea (Lindl.) Dressler, Brittonia 13: 265, 1961
BR 24: t.26 & misc.15 (1838) / 28: misc.p.36:48 (1842)
Type: cult. Loddiges (K-Lindl., holo!)
Centr.Amer., Mexico to Costa Rica – 800–3000 m – E
Epidendrum ochraceum Lindl., Bot. Reg. 24: t.26 & misc.15, 1838
Epidendrum triste A.Rich. & Galeotti, Ann. Sc. Nat. s.3, 3: 20, 1845
Epidendrum parviflorum Sessé & Mociño, Fl. Mex. ed.2: 206, 1894 (non Ruiz Lopez & Pavon)

Encyclia odoratissima (Lindl.) Schltr., Die Orchideen: 210, 1914
BR 17: t.1415 (1831) / 28: misc.p.32:22 (1842) / 29: misc.124 (1843)
Type: Brazil, R.Gordon s.n. cult. R.H.S. (K-Lindl., holo!)
Brazil – E
Encyclia patens Hook., Bot. Mag. 57: t.3013, 1830 (non Sw. 1799)
Macradenia lutescens Lodd., Bot. Cabin.: t.1556, 1830 (non R.Br.)
o Epidendrum odoratissimum Lindl., Bot. Reg. 17: t.1415, 1831
Epidendrum glutinosum Scheidw., Allg. Gartenz.: 110, 1843
Epidendrum serronianum Barb.Rodr., Gen. & Sp. Orch. Nov. 1: 50, 1877
Encyclia serroniana (Barb.Rodr.) Hoehne, Arq. Bot. Est. S.Paulo n.s. 2: 124, 1952

Encyclia oncidioides (Lindl.) Schltr., Die Orchideen : 210, 1914
BR 19: t.1623 (1833) / 28: misc.p.32:24 & misc.p.34:36 (1842)
Type: cult. R.Harrison s.n. (K-Lindl., holo!)
Trop.America – E
o Epidendrum oncidioides Lindl., Bot. Reg. 19: t.1623, 1833
Epidendrum graniticum Lindl., Hook. J. Bot. 3: 83, 1840
Epidendrum gravidum Lindl., J. Hort. Soc. London 4: 114, 1849
Epidendrum guatemalense Klotzsch, Allg. Gartenz. 20: 250, 1852
Epidendrum affine Focke, Bot. Zeit. 11: 341, 1853
Epidendrum oncidioides var. graniticum (Lindl.) Lindl., Fol. Orch. Epid.: 11, 1853
Epidendrum spectabile Focke, Bot. Zeit. 11: 229, 1853
Epidendrum giganteum hort., J.E.Planchon Hort. Donat.: 164, 1855 (non Poiret)
Epidendrum guillemianum hort., J.E.Planchon Hort. Donat.: 164, 1855
Encyclia gravida (Lindl.) Schltr., Beih. Bot. Centralbl. 36, 2: 472, 1918
Encyclia granitica (Lindl.) Schltr., Fedde Rep. Beih. 6: 74, 1919
Encyclia hunteriana Schltr., Fedde Rep. Beih. 17: 46, 1922
Epidendrum alanjense Ames, Sched. Orch. 1: 13, 1922
Encyclia powellii Schltr. ex Ames, Sched. Orch. 4: 35, 1923
Epidendrum amandum Ames, Sched. Orch. 4: 36, 1923
Epidendrum peraltense Ames, Sched. Orch. 4: 46, 1923
Epidendrum oncidioides var. gravidum (Lindl.) Ames, Hubb. & C.Schweinf., Bot. Mus. Leafl. Harv. Univ. 3: 104, 1935

Encyclia pastoris (Llave & Lex.) Schltr., Beih. Bot. Centralbl. 36 (2): 473, 1918
BR 24: misc.3 (1838) / 28: misc.p.31:16 (1842)
Mexico – E
Epidendrum pastoris Llave & Lex., Nov. Veg. Descr. Orch. Opusc.: 23, 1825

Encyclia phoenicea (Lindl.) Schltr., Die Orchideen : 211, 1914
BR 27: misc.120 & misc.145 (1841) / 28: misc.p.35:40 (1842)
Cuba, W.Indies – E
Epidendrum phoeniceum Lindl., Sert. Orch.: t.46, 1838
Epidendrum grahamii Hook., Bot. Mag. 68: t.3885, 1841
Epidendrum oblongatum A.Rich., Fl. Cub. Fanerog. 2: 239, 1841
Encyclia oblongata (A.Rich.) Acuña, Bol. Estac. Exper. Agro. Santiago Cuba 60: 80, 1939

Encyclia plicata (Lindl.) Schltr., Die Orchideen: 207, 1914
BR 33: t.35 & sub t.10 (1847)
Type: Cuba, cult. Loddiges (K-Lindl., holo!)
Cuba, Bahamas – E
o Epidendrum plicatum Lindl., Bot. Reg. 33: t.35, 1847

Encyclia pterocarpa (Lindl.) Dressler, Brittonia 13: 265, 1961
BR 27: misc.128 (1841) / 28: misc.p.30:11 (1842) / 30: t.34 (1844)
W.Mexico – 900–2200 m – E, L
o Epidendrum pterocarpum Lindl., J. Bot. 3: 82, 1841
Epidendrum cinnamomeum A.Rich & Galeotti, Ann. Sc. Nat. s.3, 3: 19, 1845

Encyclia pygmaea (Hook.) Dressler, Brittonia 13: 265, 1961
BR 25: misc.13 (1839) / 31: misc.36 p.22:1 & misc.36 p.23:3 (1845)
Trop.America – E
Epidendrum pygmaeum Hook., J. Bot. 1: 49 t.118, 1833
Coelogyne triptera Brongn., Dup. Voy. Coq. Phan.: 201 t.42, 1834
Epidendrum caespitosum Poeppig & Endl., Nov. Gen. Sp. Pl. 2: 1, t.101, 1838
Epidendrum uniflorum Lindl., Bot. Reg. 25: misc.13, 1839 (non Vell.1825)
Epidendrum monanthum Steudel, Nom. ed.2, 1: 588, 1840
Hormidium uniflorum (Lindl.) Heynh., Nom. Bot. 1: 880, 1841
Hormidium pygmaeum (Hook.) Benth. & Hook.f., Biol. Centr.-Am. Bot. 3: 218, 1883
Aulizeum pygmaeum (Hook.) Lindl. ex Stein, Orchideenb.: 236, 1892
Hormidium tripterum (Brongn.) Cogn., Mart. Fl. Bras. 3 (5): 50, 1898
Hormidium pseudopygmaeum Finet, Bull. Herb. Bois. 7: 121, 1899
Microstylis humilis Cogn., Mart. Fl. Bras. 3 (6): 550 t.114, 1906
Hormidium humile (Cogn.) Schltr., Fedde Rep. 16: 331, 1920
Encyclia triptera (Brongn.) Dressler & Pollard, Phytolog. 21: 436, 1971

Encyclia pyriforme (Lindl.) Schltr., Die Orchideen: 207, 1914
BR 33: t.50 & sub t.10 (1847)
Type: Cuba, cult. Loddiges (K-Lindl., icono!)
Cuba – E
o Epidendrum pyriforme Lindl., Bot. Reg. 33: t.50, 1847

Encyclia radiata (Lindl.) Dressler, Brittonia 13: 265, 1961
BR 27: misc.123 (1841) / 28: sub t.50:4 (1842) / 30: t.45 (1844) / 31: misc.21 (1845)
Type: cult. Loddiges (K-Lindl., holo!)
Guatemala, Honduras, Mexico – 150–2000 m – E
Epidendrum bracteolatum Presl, Rel. Haenk. 1: 100, 1825
Epidendrum marginatum Link, Klotzsch & Otto, Ic. Pl. Rar. 1: 89 t.36, 1841 (non L.C.Rich.1792)
o Epidendrum radiatum Lindl., Bot. Reg. 27: misc.123, 1841

Encyclia rufa (Lindl.) Brit. & Millsp., Fl. Bahamas: 91, 1920
BR 31: misc.42 (1845) / 28: misc.p.30:7 (1842)
Type: Rio de Janeiro, Turner s.n. cult. C.Lemon (K-Lindl., holo!)
Brazil – E
Epidendrum rufum Lindl., Bot. Reg. 31: misc.42, 1845
Epidendrum primulinum Bateman, Paxt. Flow. Gard. 1: 151, 1850

Encyclia selligera (Lindl.) Schltr., Die Orchideen: 211, 1914
BR 24: misc.66 (1838) / 28: misc.p.32:23 (1842)
Type: Guatemala, Skinner s.n. cult. Bateman (K-Lindl., holo!)
Guatemala, Mexico – 1300–2200 m – E
Epidendrum selligerum Bateman ex Lindl., Bot. Reg. 24: misc.66, 1838

Encyclia tampensis (Lindl.) Small, Fl. Miami: 56, 1913
BR 33: sub t.35 (1847)
Type: Florida, Tampa, Torrey s.n. (K-Lindl., holo!)
Florida – E
Epidendrum tampense Lindl., Bot. Reg. 33: sub t.35, 1847

Encyclia tripartita (Vell.) Hoehne, Arq. Bot. Est. S.Paulo n.s. Form. maior 2: 143 & 156, 1952

BR 27: misc.98 (1841) / 28: misc.p.33:28 (1842)
Bahamas, Brazil – E
Epidendrum tripartitum Vell., Fl. Flum. Ic. 9: t.7, 1825
Epidendrum aciculare Bateman ex Lindl., Bot. Reg. 27: misc.98, 1841
Epidendrum flavum Lindl., Hook. J. Bot. 3: 83, 1841
Encyclia acicularis (Bateman) Schltr., Fl. Baham.: 92, 1920
Encyclia flava (Lindl.) Porto & Brade, Rodriguesia 1 (2): 28, 1935

Encyclia tripunctata (Lindl.) Dressler, Brittonia 13: 265, 1961
BR 27: misc.143 (1841) / 28: misc.p.29:4 (1842)
Type: cult. R.Harrison (K-Lindl., holo!)
Mexico – 1200–2000 m – E
Epidendrum tripunctatum Lindl., Bot. Reg. 27: misc.143, 1841 (non Lindl.1853)
Epidendrum micropus Rchb.f., Hamb. Gartenz. 19: 13, 1863
Epidendrum diguetii Ames, Sched. Orch. 1: 15, 1923
Encyclia diguetii (Ames) Hoehne, Arq. Bot. Est. S.Paulo n.s.2: 151, 1952

Encyclia varicosa (Lindl.) Schltr., Beih. Bot. Centralbl. 36 (2): 472, 1918
ssp. leiobulbon (Hook.) Dressler & Pollard, Phytolog. 21: 438, 1971
BR 27: misc.63 (1841) / 28: misc.p.36:46 (1842)
Mexico – E
Epidendrum leiobulbon Hook., J. Bot. 3: 308 t.10, 1831
Epidendrum ramirezzi Gajon Sanchez, Mejores Orq. Mex. 46 : fig.16, 1930

Encyclia varicosa (Lindl.) Schltr., Beih. Bot. Centralbl. 36 (2): 472, 1918 **ssp. varicosa**
BR 24: misc.37 (1838)
Type: Guatemala, Skinner s.n. cult. Bateman (K-Lindl., holo!)
Centr.Amer., Mexico to Panama – 1500–2800 m – T, L, E
Epidendrum varicosum Bateman ex Lindl., Bot. Reg. 24: misc.37, 1838
Encyclia quadratum Klotzsch, Allg. Gartenz. 10: 402, 1850
Epidendrum chiriquense Rchb.f., Bot. Zeit. 10: 730, 1852
Epidendrum phymatoglossum Rchb.f., Bot. Zeit. 10: 731, 1852
Encyclia chiriquensis (Rchb.f.) Schltr., Beih. Bot. Centralbl. 36 (2): 472, 1918

Encyclia venosa (Lindl.) Schltr., Beih. Bot. Centralbl. 36 (2): 474, 1918
BR 28: sub t.50: 5 & misc.p.33:30 (1842)
Mexico – 1500–2000 m – L, E
Epidendrum venosum Lindl., Gen. Sp. Orch. Pl.: 99, 1831
Epidendrum ensicaulon A.Rich. & Galeotti, Ann. Sc. Nat. s.3, 3: 19, 1845
Epidendrum wendlandianum Kraenzlin, Gard. Chron. s.3, 14: 58, 1893
Encyclia wendlandiana (Kraenzlin) Schltr., Beih. Bot. Centralbl. 36 (2): 474, 1918

Encyclia vespa (Vell.) Dressler, Phytolog. 21 (7): 441, 1971
BR 25: t.11 (1839) / 28: sub t.50:10 (1842)
Mexico, Centr.America, Panama, Venezuela, Guyana, Colombia, Ecuador, Peru, Bolivia, Brazil – E, T
Epidendrum vespa Vell., Fl. Flum. 9: t.27, 1825
o Epidendrum variegatum Hook., Bot. Mag. 59: t.3151, 1832
Epidendrum coriaceum Parker ex Hook., Bot. Mag. 64: t.3595, 1837
Epidendrum crassilabium Poeppig & Endl., Nov. Gen. Sp. Pl. 2: 50 t.102, 1837
Epidendrum tigrinum Linden ex Lindl., Orch. Linden: 9, 1846
Epidendrum pamplonense Rchb.f., Linnaea 22: 837, 1849
Epidendrum coriaceum Focke, Bot. Zeit. 11: 228, 1853 (non Hook.)
Epidendrum pachysepalum Klotzsch, Allg. Gartenz. 23: 274, 1855
Epidendrum christii Rchb.f., Linnaea 41: 112, 1877
Epidendrum leopardinum Rchb.f., Linnaea 41: 112, 1877
Epidendrum longipes Rchb.f., Otia Bot. Hamb.: 10, 1878
Aulizeum variegatum (Hook.) Lindl. ex Stein, Orchideenb.: 241, 1892
Epidendrum feddeanum Kraenzlin, Fedde Rep. 1: 188, 1905
Epidendrum saccharatum Kraenzlin, Orchis 2: 113 t.17, 1908
Epidendrum rhabdobulbum Schltr., Fedde Rep. Beih. 7: 146, 1920
Epidendrum baculibulbom Schltr., Fedde Rep. Beih. 19: 116, 1923
Epidendrum rhopalobulbon Schltr., Fedde Rep. Beih. 27: 72, 1924
Hormidium crassilabium (Poeppig & Endl.) Brieger, Schlechter Die Orchideen ed.3, 1: 571/72, 1977
Anacheilium vespa (Vell.) Pabst, Bradea 3 (23), 1981
Hormidium vespa (Vell.) Ortiz Valdivieso, Martinez & Misas Urreta, Orquideas ornament. Colombia, 1st. reimpr.: 150, 1982

Encyclia viridiflora Hook., Bot. Mag. 55: t.2831, 1828
BR 28: misc.p.30:13 (1842)
Brazil – E
Epidendrum viridiflorum (Hook.) Lindl., Bot. Reg. 28: misc.p.30:13, 1842
Encyclia multiflora Rchb.f., Walp. Ann.: 335, 1862 (non Hook.)

Encyclia vitellina (Lindl.) Dressler, Brittonia 13 (3): 265, 1961
BR 26: t.35 (1840) / 28: misc.p.28:1 (1842)
Mexico, Guatemala – 1500–2600 m – E
o Epidendrum vitellinum Lindl., Gen. Sp. Orch. Pl.: 97, 1831
Epidendrum vitellinum var. majus Veitch, Fl. Mag. 5: 261, 1866
Epidendrum vitellinum var. giganteum Warner, Sel. Orch. Pl. 3: t.27, 1877
Epidendrum vitellinum var. autumnale G.Wilson, Orch. World 4: 27, 1913
Hormidium vitellinum (Lindl.) Brieger, Schlechter Die Orchideen ed.3, 1: 574, 1977

Epidendrum acuminatum Ruiz Lopez & Pavon, Syst. Veg. Fl. Peruv. & Chil. 1: 248, 1798
BR 31: misc.70:32 (1845)
Peru, Ecuador, Colombia, Venezuela – E

Epidendrum adenoglossum Lindl., Hook. J. Bot. 3: 86, 1841
BR 30: misc.82:4 (1844)
Peru – E

Epidendrum aggregatum Lindl., Hook. J. Bot. 3: 84, 1841
BR 31: misc.70:17 (1845)
Peru, Colombia – E

Epidendrum aquaticum Lindl., Ann. Mag. Nat. Hist. 12: 398, 1843
BR 31: misc.70:51 (1845)
Brazil – E
Epidendrum ulei Schltr., Notizbl. Bot. Gart. Berlin 6: 124, 1914

Epidendrum arbuscula Lindl., Benth. Pl. Hartweg: 93, 1842
BR 29: misc.54 (1843) / 31: misc.70:38 (1845)
Mexico, Guatemala, Honduras – up to 3000 m – T, E
Epidendrum nubium Rchb.f., Beitr. Orch. Centr.-Am.: 81, 1866

Epidendrum bifarium Sw., Nov. Act. Upsal. 6: 68, 1799
BR 31: misc.70:40 (1845)
Jamaica – E
Epidendrum anceps Sw., Prodr.: 121, 1788 (non Jacq.)

Epidendrum blepharistes Barker ex Lindl., Bot. Reg. 30: misc.24:18, 1844
BR 30: misc.24:18 (1844)
Type: Venezuela, Linden s.n. cult. Barker (K-Lindl., holo!)
Venezuela – E

Epidendrum boothii (Lindl.) L.O.Williams, Ann. Missouri Bot. Gard. 26: 282, 1939
BR 24: misc.95 (1838) / 26: misc.112 (1840) / 29: misc.4 (1843)
Type: Guatemala, Skinner s.n. cult. C.Lemon (K-Lindl., holo!)
Mexico, Honduras, Guatemala, Nicaragua, Costa Rica, Panama, Cuba, Venezuela, Guyana – up to 1350 m – E
Maxillaria boothii Lindl., Bot. Reg. 24: misc.95, 1838
Dinema paleacea Lindl., Bot. Reg. 26: misc.112, 1840
Epidendrum auritum Lindl., Bot. Reg. 29: misc.4, 1843
Epidendrum lindenianum A.Rich. & Galeotti, Ann. Sc. Nat. 3.s., 3: 20, 1845
Epidendrum paleaceum (Lindl.) Rchb.f., Beitr. Orch. Centr.-Am.: 80, 1866
Nidema boothii (Lindl.) Schltr., Fedde Rep. Beih. 17: 43, 1922

Epidendrum brachyglossum Lindl., Bot. Reg. 30: misc.83:6, 1844
BR 30: misc.82:6 (1844)
Type: Peru, Hartweg s.n. (K-Lindl., holo!)
Colombia, Peru, Bolivia – E

Epidendrum campestre Lindl., Bot. Reg. 30: misc.24:10, 1844
BR 30: misc.24:10 (1844)
Type: Brazil, Gardner 5207 (K-Lindl., holo!)
Brazil – E, T
Epidendrum blandum Kraenzlin, Svensk. Vet. Akad. Handl. 46: 58 t.11, 1911

Epidendrum carpophorum Barb.Rodr., Gen. & Sp. Orch. Nov. 2: 148, 1882
BR 23: t.1961 (1837)
W.Indies, El Salvador, Nicaragua, Venezuela, Trinidad – E, L
o Epidendrum nocturnum var. latifolium Lindl., Bot. Reg. 23: t.1961, 1837
Callista antennata (Lindl.) Kuntze, Rev. Gen. 2: 654, 1891
Epidendrum latifolium (Lindl.) Garay & Sweet, J. Arn. Arb. 53: 392, 1972

Epidendrum cauliflorum Lindl., Bot. Reg. 24: misc.82, 1838
BR 24: misc.82 (1838) / 31: misc.70:41 (1845)
Type: Rio de Janeiro, cult. Loddiges (K-Lindl., holo!)
Brazil – E

Epidendrum cernuum Kunth, Nov. Gen. & Sp. Pl. 1: 353, 1815
BR 31: misc.70:37 (1845)
Colombia – E

Epidendrum chioneum Lindl., Bot. Reg. 31: misc.70:36, 1845
BR 31: misc.70:36 (1845)
Type: Colombia, Linden s.n. (K-Lindl., holo!)
Colombia – E
Epidendrum claesianum Cogn., Gard. Chron. s.3, 29: 70, 1901

o **Epidendrum ciliare** L., Syst. Nat. ed.10: 1246, 1759
BR 10: t.783 & t.784 (1824) / 26: misc.81 & misc.190 (1840) / 31: misc.36 p.26:1, 2, 3 (1845)
W.Indies, Mexico to Panama, Venezuela to Bolivia, Guyana, Brazil – E
Auliza ciliaris (L.) Salisb., Trans. Hort. Soc. 1: 294, 1812
o Epidendrum cuspidatum Lodd., Bot. Cabin. 1: t.10, 1818
Coilostylis emarginata Raf., Fl. Tellur. 4: 37, 1836 (nomen illeg.)
Epidendrum viscidum Lindl., Bot. Reg. 26: misc.81, 1840
Epidendrum ciliare var. cuspidatum (Lodd.) Lindl., Fol. Orch. Epid.: 30, 1853
Epidendrum ciliare var. viscidum (Lindl.) Lindl., Fol. Orch. Epid.: 30, 1853
Epidendrum luteum hort. ex Planchon, Hort. Don.: 50, 1858
Aulizeum ciliare (L.) Lindl. ex Stein, Orchideenb.: 226, 1892
Epidendrum ciliare var. minor hort. ex Stein, Orchideenb.: 226, 1892
Phaedrosanthus ciliaris (L.) Kuntze, Lex. Gen. Phan.: 429, 1904
Epidendrum ciliare var. squamatum Schnee, Rev. Fac. Agr. 1: 206, 1953

o **Epidendrum cinnabarinum** Salzm. ex Lindl., Gen. Sp. Orch. Pl.: 106, 1831
BR 28: t.25 (1842) / 30: misc.24:21 (1844)
Brazil – E

Epidendrum clowesii Bateman ex Lindl., Bot. Reg. 30: misc.24:6, 1844
BR 30: misc.24:6 (1844)
Type: Guatemala, W.Bateman s.n. (K-Lindl., holo!)
Mexico, Guatemala, El Salvador – up to 2100 m – E
Epidendrum flavovirens Rchb.f., Beitr. Orch. Centr.-Amer.: 85, 1866 (non Regel)
Epidendrum chlorops Rchb.f., Gard. Chron. n.s.14: 524, 1880
Epidendrum piestocaulos Schltr., Fedde Rep. 15: 207, 1918

Epidendrum collare Lindl., Bot. Reg. 29: misc.85, 1843
BR 29: misc.85 (1843) / 31: misc.36 p.29 (1845)
Type: Guatemala, Hartweg, (not located!)
Guatemala – E

Epidendrum compressum Griseb., Fl. Brit. West Ind.: 617, 1864
BR 28: sub t.25 (1842) / 30: misc.24:25 (1844)
Trinidad, Venezuela, Colombia, Ecuador, Peru, Bolivia, Guyana, Brazil – E
Epidendrum laxum Poeppig & Endl., Nov. Gen. Sp. Pl. 2: 2, 1838
Epidendrum yatapuense Barb.Rodr., Vellosia ed.2, 1: 123, 1838
Epidendrum macrothyrsis Lehm. & Kraenzlin, Engl. Bot. Jahrb. 26: 472, 1899
Epidendrum laxum var. mocoanum Schltr., Fedde Rep. Beih. 27: 67, 1924
Epidendrum guentherianum Kraenzlin, Fedde Rep. 25: 20, 1928

Epidendrum conopseum R.Br., Aiton Hort. Kew ed.2, 5: 219, 1813
BR 31: misc.70:33 (1845)
USA (Florida, Texas, N.Carolina), E.Mexico – E
Epidendrum magnoliae Muhlenb., Cat.: 81, 1813
Larnandra magnolia (Mulhenb.) Raf., Neogen.: 4, 1825
Larnandra conopsea (R.Br.) Raf., Fl. Tellur. 2: 40, 1837
Amphiglottis conopsea (R.Br.) Small, Man. SE. Fl.: 391, 1933
Epidendrum conopseum var. mexicanum L.O.Williams, Ceiba 2: 152, 1951

Epidendrum cordatum Ruiz Lopez & Pavon, Syst. Veg. Fl. Peruv. & Chil. 1: 244, 1798
BR 31: misc.70:58 (1845)
Peru – E

Epidendrum cornutum Lindl., Hook. J. Bot. 3: 86, 1841
BR 28: sub t.42 (1842) / 30: misc.24:4 (1844)
Colombia, Venezuela, Peru, Ecuador – E
Epidendrum anthoceros Linden & Rchb.f., Bonpl. 2: 281, 1854
Epidendrum pavonianum Rchb.f., Bonpl. 4: 215, 1856
Epidendrum melinoacron Schltr., Fedde Rep. Beih. 9: 88, 1921

Epidendrum coronatum Ruiz Lopez & Pavon, Syst. Veg. Fl. Peruv. & Chil. 1: 242, 1798
BR 31: misc.70:47 (1845)
Mexico, Guatemala, Honduras, Costa Rica, Panama, Trinidad, Venezuela, Colombia, Ecuador, Peru, Brazil – E, L, T
Epidendrum compositum Vell., Fl. Flum. Ic. 9: t.39, 1827
Epidendrum sulphuroleucum Barb.Rodr., Gen. & Sp. Orch. Nov. 1: 56, 1877
Epidendrum moyobambae Kraenzlin, Fedde Rep. 1: 185, 1905
Epidendrum subpatens Schltr., Fedde Rep. Beih. 17: 40, 1922
Epidendrum benignum Ames, Sched. Orch. 2: 26, 1923
Epidendrum amazonicum Schltr., Beih. Bot. Centralbl. 42 (2): 78, 1925

Epidendrum costatum A.Rich. & Galeotti, Ann. Sc. Nat. s.3, 3: 21, 1845
BR 31: misc.70:20 (1845)
Mexico – E

Epidendrum cristatum Ruiz Lopez & Pavon, Syst. Veg. Fl. Peruv & Chil. 1: 243, 1798
BR 27: misc.122 (1841) / 28: t. & sub t.42 (1842) / 30: misc.82:17 (1844) / 31: misc.70:46 (1845)
W.Indies, Mexico, Guatemala, Costa Rica, Honduras, Panama, Venezuela, Colombia, Ecuador, Bolivia, Guyana, Surinam, Brazil – E, T
o Epidendrum raniferum Lindl., Gen. Sp. Orch. Pl.: 109, 1831
Epidendrum calliferum Lemaire, Jard. Fleur. 4: misc.p.65, 1853
Epidendrum raniferum var. luteum Lindl., Fol. Orch. Epid.: 53, 1853
Epidendrum hexadactylum Barb.Rodr., Gen. & Sp. Orch. Nov. 1: 56, 1877
Epidendrum longovarium Barb.Rodr., Gen. & Sp. Orch. Nov. 1: 57, 1877
Epidendrum raniferum var. hexadactylum (Barb.Rodr.) Cogn., Mart. Fl. Bras. 3 (5): 99, 1898
Epidendrum raniferum var. loefgrenii Cogn., Mart. Fl. Bras. 3 (5): 100, 1898
Epidendrum raniferum var. obtusilobum Cogn., Mart. Fl. Bras. 3 (5): 99, 1898
Epidendrum millei Schltr., Fedde Rep. 15: 54, 1917
Epidendrum bathyschistum Schltr., Fedde Rep. Beih. 6: 36, 1919
Epidendrum validum Schltr., Fedde Rep. Beih. 9: 95, 1921
Epidendrum alexandri Schltr., Anex. Mem. Inst. Butantan 1, 4: 60, 1922
Epidendrum raniferum var. lutescens Lindl. ex Broadw., Orch. Rev. 34: 199, 1926
Epidendrum rantierium Lindl. ex Sanchez, Orq. Mex.: 47, 1930

Epidendrum cucullatum Lindl., Bot. Reg. 24: misc.47, 1838 (non L.)
BR 24: misc.47 (1838) / 28: sub t.25 (1842) / 30: misc.24:2 (1844) / 31: misc.70:13 (1845)
Type: Brazil, W.Harrison s.n. cult. R.Harrison (K-Lindl., holo!)
Mexico, Guatemala, El Salvador, Honduras, Venezuela, Brazil, Trinidad, W.Indies – E
Epidendrum biforatum Lodd. ex Lindl., Bot. Reg. 30: misc.24:2, 1844

Epidendrum cylindraceum Lindl., Bot. Reg. 30: misc.82:15, 1844
BR 30: misc.82:15 (1844)
Type: Peru, Pitayo, Popayan, Hartweg s.n. (K-Lindl., holo!)
Colombia, Ecuador, Peru – E

Epidendrum dendrobioides Thunb., Pl. Bras. Dec. 2, 1818
BR 31: misc.70:59, 60 (1845)
Venezuela – E
Epidendrum carnosum Lindl., Hook. J. Bot. 3: 87, 1841
Epidendrum durum Lindl., Hook. J. Bot. 3: 87, 1841

Epidendrum densiflorum Hook., Bot. Mag. 13: t.3791, 1840
BR 26: misc.134 (1840) / 31: t.4 & misc.70:64 (1845)
Trop. America – T, E
Epidendrum floribundum Kunth, Nov. Gen. & Sp. Pl. 1: 353, 1815
o Epidendrum dipus Lindl., Bot. Reg. 31: t.4, 1845
Epidendrum nutans var. dipus (Lindl.) Lindl., Fol. Orch. Epid.: 56, 1853

Epidendrum difforme Jacq., Enum. Pl. Carab.: 29, 1760
BR 1: t.80 (1816) / 27: misc.163 (1841) / 28: misc.70 (1842) / 31: misc.70:5, 6 (1845)
USA (Florida), Jamaica, W.Indies, Mexico, Trop. America – E
o Epidendrum umbellatum Sw., Prodr. Veg. Ind. Occ.: 121, 1788
Epidendrum umbelliferum J.Gmelin, Syst. Veg.: 65, 1791
Caularthron umbellatum (Sw.) Raf., Fl. Tellur. 2: 41, 1837
Epidendrum latilabre Lindl., Bot. Reg. 28: misc.70, 1842
Epidendrum subumbellatum Hoffsgg., Preis-Verz.: 24, 1842
Epidendrum virens Hoffsgg., Preis-Verz.: 24, 1842
Epidendrum radiatum Hoffsgg., Preis-Verz.: 49, 1843
Epidendrum latilabium Rchb.f., Linnaea 25: 244, 1852
Epidendrum umbellatum var. latilabre (Lindl.) Griseb., Fl. Brit. W.Ind.: 618, 1864
Epidendrum arachnoideum Barb.Rodr., Gen. & Sp. Orch. Nov. 1: 60, 1877
Auliza difformis (Jacq.) Small, Fl. Miami: 56, 1913
Epidendrum chlorocorymbos Schltr., Fedde Rep. Beih. 17: 30, 1922
Amphiglottis difformis (Jacq.) Britton, Sc. Surv. Porto Rico & Virgin.Is. 5 (2): 200, 1924
Epidendrum pseudodifforme Hoehne & Schltr., Fedde Rep. Beih. 35: 71, 1925
Epidendrum curvicolumna Ames Hubb. & C.Schweinf., Bot. Mus. Leafl. Harv. Univ. 3 (5): 67, 1935
Neolehmannia difformis (Jacq.) Pabst., Bradea 2: 306, 1978

Epidendrum diffusum Sw., Prodr. Veg. Ind. Occ.: 121, 1788
BR 31: misc.70:52 (1845)
Mexico, Guatemala, W.Indies, Colombia, Guyana, Brazil – E
Serarphyta multiflora Fischer & Meyer, Bull. Acad. Petersb. 7: 24, 1840
Epidendrum tenuiflorum hort. ex Lindl., Fol. Orch. Epid.: 88, 1853
Seraphyta diffusa (Sw.) Pfitzer, Engl. Nat. Pflanzenf. 2, 6: 142, 1889
Epidendrum diffusum var. depauperatum Cogn., Mart. Fl. Bras. 3 (5): 155, 1898

Epidendrum elongatum Jacq., Coll. 3: 260, 1789
BR 24: misc.146 (1838) / 28: sub t.25 (1842) / 30: misc.24:12, 13, 15, 16 (1844) / 31: misc.59 & misc.70:16 (1845)
W.Indies, Trop.S.America – T, E
Epidendrum secundum L., Sp. Pl. ed.2: 1349, 1763 (non Jacq.)
Epidendrum corymbosum Ruiz Lopez & Pavon, Syst. Veg.: 246, 1798
Amphiglottis secunda (L.) Salisb., Trans. Hort. Soc. 1: 294, 1812
Epidendrum ellipticum Graham, Edinb. N. Phil. J. 1: 171, 1826
Epidendrum dichotomum Presl., Rel. Haenk. 1: 101, 1827
Epidendrum crassifolium Lindl., Gen. Sp. Orch. Pl.: 107, 1831
Epidendrum ellipticum Lodd., Bot. Cab. 18: t.1276, 1831
Epidendrum cochlidium Lindl., Ann. Mag. Nat. Hist. 4: 382, 1840
Epidendrum lindenii Lindl., Bot. Reg. 31: misc.59, 1845
Epidendrum brachyphyllum Lindl., Fol. Orch. Epid.: 72, 1853
Epidendrum fastigiatum Lindl., Fol. Orch. Epid.: 71, 1853
Epidendrum ansiferum Rchb.f., Bonpl. 2: 111, 1854
Epidendrum fimbria Rchb.f., Bonpl. 2: 282, 1854
Epidendrum gracilicaule Rchb.f. & Warsc., Bonpl. 2: 111, 1854
Epidendrum incisum Rchb.f. & Warsc., Bonpl. 2: 112, 1854
Epidendrum novogranatense Rchb.f. & Warsc., Bonpl. 2: 111, 1854
Epidendrum xytriophorum Rchb.f. & Warsc., Bonpl. 2: 112, 1854
Epidendrum giroudianum Rchb.f., Bonpl. 4: 327, 1856
Epidendrum quitensium Rchb.f., Walp. Ann. 6: 392, 1862
Epidendrum elongatum var. crassifolium (Lindl.) Griseb., Fl. Brit. Ind.: 617, 1864
Epidendrum evectum Hook.f., Bot. Mag. 97: t.5902, 1871
Epidendrum expansum Rchb.f., Linnaea 41: 115, 1877
Epidendrum longihastatum Barb.Rodr., Gen. & Sp. Orch. Pl. 1: 59, 1877
Epidendrum fastigiatum var. bifidum Rchb.f., Otia Bot. Hamb. 1: 13, 1878
Epidendrum arachnoglossum Rchb.f. ex André, Rev. Hort.: 554, 1882
Epidendrum ellipticum Rchb.f., Ber. Deutsch. Bot. Ges. 3: 277, 1885 (non Graham)
Epidendrum herzogii Schltr., Fedde Rep. 12: 489, 1913
Epidendrum antioquiense Schltr., Fedde Rep. Beih. 7: 125, 1920
Epidendrum dolichopus Schltr., Fedde Rep. Beih. 7: 131, 1920
Epidendrum pachyphyllum Schltr., Fedde Rep. Beih. 7: 140, 1920
Epidendrum polyschistum Schltr., Fedde Rep. Beih. 7: 143, 1920
Epidendrum cuzcoense Schltr., Fedde Rep. Beih. 9: 82, 1921
Epidendrum tarmense Schltr., Fedde Rep. Beih. 9: 94, 1921
Epidendrum versicolor Hoehne & Schltr., Arq. Bot. Est. S.Paulo 1 (3): 245, 1926
Epidendrum coroicoense Schltr., Fedde Rep. 27: 60, 1929
Epidendrum inconstans Ames, Bull. Torrey Club 58: 350, 1931
Epidendrum ellipticum var. freybergense Zimmerm., Bibl. Bot. 109: 5, 1934
Epidendrum bulkeleyi Hawkes, Orquidea 18: 169, 1957

Epidendrum equitantifolium Ames, Sched. Orch. 4: 39, 1923
BR 24: misc.76 (1838) / 31: misc.70:1 (1845)
Type: Vera Cruz, Hartweg s.n. cult. Hort. Soc. (K-Lindl., holo!)
Mexico, Honduras, Nicaragua, Panama – E
Epidendrum equitans Lindl., Bot. Reg. 24: misc.76, 1838

Epidendrum erubescens Lindl., Hook. J. Bot. 3: 87, 1840
BR 31: misc.70:63 (1845)
Mexico – E
Encyclia erubescens (Lindl.) Schltr., Die Orchideen: 214, 1914

Epidendrum excisum Lindl., Bot. Reg. 30: misc.82:14, 1844
BR 30: misc.82:14 (1844)
Type: Peru, Popayan, Hartweg s.n. (K-Lindl., holo!)
Peru, Colombia – T, E
Epidendrum excisum var. grandiflorum Lindl., Fol. Orch. Epid.: 52 no.163, 1853

Epidendrum ferrugineum Ruiz Lopez & Pavon, Syst. Veg. Fl. Peruv. & Chil. 1: 245, 1798
BR 26: misc.128 (1840) / 28: sub t.42 (1842) / 30: misc.82:16 (1844) / 31: misc.70:12 (1845)
Venezuela, Colombia, Peru, Trinidad – E, L
Epidendrum trinitatis Lindl., Bot. Reg. 30: misc.128, 1840
Epidendrum naucrates Rchb.f., Bonpl. 2: 20, 1854
Epidendrum tricolor Rolfe, Kew Bull.: 63, 1893

Epidendrum filicaule Lindl., Gen. Sp. Orch. Pl.: 101, 1831
BR 31: misc.70:28 (1845)
Brazil – E

Epidendrum fimbriatum Kunth, Nov. Gen. & Sp. Pl. 1: 351, 1816
BR 31: misc.70:26 (1845)
Peru, Venezuela, Colombia, Ecuador – T, E
Epidendrum alternans Lindl., Hook. J. Bot. 3: 88, 1841

Epidendrum flavum Lindl., Hook. J. Bot. 3: 83, 1841
BR 28: misc.p.33:32 (1842)
Guyana – E

Epidendrum frigidum Linden ex Lindl., Bot. Reg. 31: misc.70:48, 1845
BR 31: misc.70:48 (1845)
Type: Colombia, Linden s.n. (K-Lindl., holo!)
Peru, Venezuela, Colombia – E, L
Epidendrum macrodonax Schltr., Fedde Rep. Beih. 9: 88, 1921

Epidendrum fruticosum Pavon ex Lindl., Gen. Sp. Orch. Pl.: 101, 1831
BR 28: sub t.25 (1842) / 30: misc.24:9 (1844)
Mexico – E

Epidendrum funckianum A.Rich. & Galeotti ex Lindl., Bot. Reg. 31: misc. 70:49, 1845
BR 31: misc.70:49 (1845)
Type: Mexico, A.Richard & Galeotti (not Located)
Mexico – E

Epidendrum geminiflorum Kunth, Nov. Gen. & Sp. Pl. 1: 354, 1816
BR 30: misc.82:11 (1844)
Colombia, Ecuador, Peru – 2700–2900 m – E
Epidendrum cajamarcae Schltr., Fedde Rep. Beih. 9: 81, 1921

Epidendrum gladiatum Lindl., Gen. Sp. Orch. Pl.: 106, 1831
BR 27: misc.20 (1841) / 28: sub t.25 (1842) / 30: misc.24:28, 29 (1844)
Mexico – E
Epidendrum vandifolium Lindl., J. Hort. Soc. Lond. 4: 269, 1849
Epidendrum anisatum Llave ex Lindl., Fol. Orch. Epid.: 65, 1853
Epidendrum jurgensenii Rchb.f., Gard. Chron. n.s.14: 524, 1880

Epidendrum gramineum Lindl., Gen. Sp. Orch. Pl.: 103, 1831
BR 31: misc.70:50 (1845)
Peru – 3400–3500m – E

Epidendrum grandiflorum Lindl., Hook. J. Bot. 3: 86, 1841
BR 28: sub t.42 (1842) / 30: misc.82:1 (1844)
Peru – E

Epidendrum haenkeanum Presl., Rel. Haenk. 1: 100, 1827
BR 28: sub t.25 (1842) / 30: misc.24:31 (1844)
Peru – T, E

Epidendrum harrisoniae Hook., Bot. Mag. 60: t.3209, 1833
BR 31: misc.12 (1845)
Brazil (Rio de Janeiro) – E
Epidendrum aeridiforme Booth, Bot. Reg. 31: misc.12, 1845
Epidendrum ammophyllum Barb.Rodr., Gen. & Sp. Orch. Nov. 2: 149, 1881

Epidendrum hartwegii Lindl., Benth. Pl. Hartw.: 150, 1844
BR 31: misc.36 p.28:11 (1845)
Venezuela, Colombia, Ecuador, Peru – E

Epidendrum ibaguense Kunth, Nov. Gen. & Sp. Pl. 1: 354, 1816
BR 24: misc.10 (1838) / 26: misc.91 (1840) / 28: sub t.25 (1842) / 30: misc.24:14, 20 (1844)
Mexico, Guatemala, El Salvador, Honduras, Nicaragua, Costa Rica, Panama – T, L, E
Epidendrum radicans Pavon ex Lindl., Gen. Sp. Orch. Pl.: 104, 1831
Epidendrum rhizophorum Bateman ex Lindl., Bot. Reg. 24: misc.10, 1838
Epidendrum pratense Rchb.f., Beitr. Orch. Centr.-Am.: 84, 1866
Epidendrum radicans var. chiriquense Schltr., Fedde Rep. Beih. 17: 39, 1922

Epidendrum imatophyllum Lindl., Gen. Sp. Orch. Pl.: 106, 1831
BR 28: sub t.25 (1842) / 30: misc.24:19, 23 (1844)
Mexico to Panama, Trinidad, Venezuela, Guyana, Surinam, Brazil, Colombia, Ecuador, Peru – T
Epidendrum flexuosum G.F.W.Meyer, Fl. Essequeb.: 260, 1818
Epidendrum imetrophyllum Paxton, Mag. Bot. 4: 119, 1837
Epidendrum palpigerum Rchb.f., Gard. Chron. n.s.12: 40, 1879
Epidendrum lorifolium Schltr., Fedde Rep. Beih. 17: 35, 1922

Epidendrum lacertinum Lindl., Bot. Reg. 27: misc.109, 1841
BR 27: misc.109 (1841) / 31: misc.70:39 (1845)
Type: Guatemala, cult. Bateman (K-Lindl., holo!)
Mexico, Guatemala – up to 2700 m – E
Epidendrum indusiatum Klotzsch, Allg. Gartenz. 22: 177, 1854

Epidendrum lacerum Lindl., Bot. Reg. 24: misc.18, 1838
BR 24: misc.18 (1838) / 28: sub t.25 (1842) / 30: misc.24:17 (1844)
Type: Cuba, Capt. Sutton s.n. cult. C.Lemon (K-Lindl., holo!)
Cuba, Puerto Rico – E
Amphiglottis lacera (Lindl.) Britton, Sc. Surv. Porto Rico & Virgin.Is 5 (2): 200, 1924

Epidendrum ledifolium A.Rich. & Galeotti, Ann. Sc. Nat. s.3, 3: 21, 1845
BR 31: misc.70:21, 22 (1845)
Mexico, Guatemala, Honduras – E
Epidendrum propinquum A.Rich. & Galeotti, Ann. Sc. Nat. s.3, 3: 21, 1845
Epidendrum lamprocaulon Rchb.f., Bot. Zeit. 10: 732, 1852
Epidendrum lucidum Schltr., Fedde Rep. 15: 206, 1918

Epidendrum leucochilum Klotzsch, Allg. Gartenz. 2: 145, 1843
BR 29: misc.120 (1843) / 30: misc.82:7 (1844)
Venezuela, Colombia – E
Epidendrum flavidum Lindl., Orch. Linden.: 8, 1846

Epidendrum lignosum Llave & Lex., Nov. Gen. Descr. Opusc. 2: 22–24, 1825
BR 31: misc.70:34 (1845)
Mexico – E

Epidendrum ligulatum Llave & Lex., Nov. Veg. Descr. Orch. Opsuc. 26, 1825
BR 28: misc.p.29:2 (1842)
Mexico – E
Epidendrum llavei Steudel, Nom. ed.2, 1: 558, 1840

Epidendrum lineare Ruiz Lopez & Pavon, Syst. Veg. Fl. Peruv. & Chil. 1: 249, 1798
BR 31: misc.70:56 (1845)
Peru – E

Epidendrum longiflorum Kunth, Nov. Gen. & Sp. Pl. 1: 354, 1815
BR 30: misc.82:12 (1844)
Colombia – E

Epidendrum longipetalum A.Rich. & Galeotti, Ann. Sc. Nat. s.3, 3: 22, 1845 (non Godefr.-Leb., non Lindl.)
BR 31: misc.70:30 (1845)
Mexico – E

Epidendrum macrostachyum Lindl., Bot. Reg. 31: misc.70:18, 1845
BR 31: misc.70:18 (1845)
Type: Colombia, Linden 1279 (K-Lindl., holo!)
Colombia – E

Epidendrum mancum Lindl., Bot. Reg. 30: misc.82:3, 1844
BR 30: misc.82:3 (1844)
Type: Ecuador, Loxa, Hartweg s.n. (K-Lindl., holo!)
Peru, Ecuador, Colombia – E

Epidendrum martianum Lindl., Ann. Mag. Nat. Hist. 4: 382, 1840
BR 28: sub t.25 (1842) / 30: misc.24:24 (1844)
Brazil – E

Epidendrum micranthum Lindl., Hook. J. Bot. 3: 88 no.22, 1841
BR 31: misc.70:55 (1845)
Peru – 1300–1600 m – E
Amblostoma holochilum Schltr., Fedde Rep. 10: 387, 1912

Epidendrum miserum Lindl., Bot. Reg. 27: misc.62, 1841
BR 27: misc.62 (1841) / 31: misc.36 p.23:2 (1845)
Type: Mexico, cult. Loddiges (K-Lindl., holo!)
Mexico – E

Epidendrum nocturnum Jacq., Enum. Pl. Carib.: 29, 1760
BR 24: misc.49 (1838) / 31: misc.70:8, 9, 10 (1845)
USA (Florida), Mexico to Panama, W.Indies, Guyana, Ecuador, Peru, Brazil – E
Epidendrum carolinianum Lam., Encycl. 1: 182, 1783
Nyctosma nocturna (Jacq.) Raf., Fl. Tellur. 2: 9, 1837
Epidendrum longicolle Lindl., Bot. Reg. 24: misc.49, 1838
Epidendrum tridens Poeppig & Endl., Nov. Gen. Sp. Pl. 2: 2, 1838
Epidendrum discolor A.Rich & Galeotti, Ann. Sc. Nat. s.3, 3: 22, 1845
Epidendrum spruceanum Lindl., Fol. Orch. Epid.: 80, 1853
Epidendrum bahiense Rchb.f., Hamb. Gartenz. 15: 53, 1859
Epidendrum buenaventurae F.Lehm. & Kraenzlin, Engl. Bot. Jahrb. 26: 470, 1899
Epidendrum nocturnum var. tridens (Poeppig & Endl.) Cogn., Mart. Fl. Bras. 3 (5): 136, 1899
Phaedrosanthus nocturnus (Jacq.) Kuntze, Lex. Gen. Phan.: 429, 1904
Epidendrum nocturnum var. minus Cogn., Bull. Soc. Roy. Belg. 43: 323, 1907
Auliza nocturna (Jacq.) Small, Fl. Miami: 56, 1913
Epidendrum leucarachne Schltr., Fedde Rep. Beih. 7: 136, 1920
Epidendrum oliganthum Schltr., Fedde Rep. Beih. 9: 90, 1921
Epidendrum tunguraguae Schltr., Fedde Rep. Beih. 8: 87, 1921
Epidendrum nocturnum var. panamense Schltr., Fedde Rep. Beih. 17: 36, 1922
Amphiglottis nocturna (Jacq.) Britton, Sc. Surv. Porto Rico & Virgin.Is. 5 (2): 200, 1924
Epidendrum nocturnum var. minor Schltr., Fedde Rep. Beih. 27: 69, 1924
Epidendrum nocturnum var. angustifolium Stehlé, Fl. Descr. Antil. Franç. 1: 143 t.8 fig.3, 1939

○ **Epidendrum nutans** Sw., Prodr.: 121, 1788
BR 1: t.17 (1815) / 28: sub t.42 (1842) / 30: misc.82:8 (1844)
Jamaica, Trinidad, Venezuela, Brazil – E, L

Epidendrum orchidiflorum Salzm. ex Lindl., Gen. Sp. Orch. Pl.: 103, 1831
BR 28: sub t.25 (1842) / 30: misc.24:8 (1844)
Brazil, Venezuela – E

Epidendrum orgyale Lindl., Ann. Mag. Nat. Hist. 15: 256, 1845
BR 31: misc.70:24 (1845)
Colombia – E

Epidendrum ovalifolium Lindl., Gen. Sp. Orch. Pl.: 104, 1831
BR 28: sub t.25 (1842) / 30: misc.24:5 (1844)
Mexico – E

Epidendrum pachyanthum Lindl., Bot. Reg. 24: misc.42, 1838
BR 24: misc.42 (1838) / 28: misc.p.32:25 (1842)
Type: Guyana, Schomburgk s.n. cult. Loddiges (K-Lindl., holo!)
Brazil, Guyana – E

Epidendrum pallidiflorum Hook., Bot. Mag. 57: t.2980, 1830
BR 28: sub t.25 (1842) / 30: misc.24:27 (1844)
W.Indies – E

Epidendrum paniculatum Ruiz Lopez & Pavon, Syst. Veg. Fl. Peruv. & Chil.: 243, 1798
BR 28: sub t.25 (1842) / 29: misc.20 (1843) / 30: misc.24:26, 33, 34 (1844)
Mexico, Guatemala, Honduras, Nicaragua, Costa Rica, Panama, Venezuela, Ecuador, Peru, S.Argentina – E, T
Epidendrum rubrocinctum Lindl., Bot. Reg. 29: misc.20, 1843
Epidendrum laeve Lindl., Bot. Reg. 30: misc.24: 33, 34, 1844
Epidendrum ornatum Lemaire, Fl. d. Serres 4: 334b, 1848
Epidendrum floribundum var. convexum Lindl., Fol. Orch. Epid.: 91, 1853
Epidendrum paniculatum var. cuspidatum Lindl., Fol. Orch. Epid.: 56, 1853
Epidendrum paniculatum var. longicrure Lindl., Fol. Orch. Epid.: 56, 1853
Epidendrum syringaeflorum Warsc. ex Rchb.f., Bonpl. 2: 111, 1854
Epidendrum turialvae Rchb.f., Gard. Chron.: 1678, 1871
Epidendrum piliferum Rchb.f., Linnaea 41: 82, 1876
Epidendrum resectum Rchb.f., Linnaea 41: 82, 1876
Epidendrum falsiloquum Rchb.f., Gard. Chron.: 566, 1885
Epidendrum englerianum F.Lehm. & Kraenzlin, Engl. Bot. Jahrb. 26: 466, 1899
Epidendrum frons-bovis Kraenzlin, Fedde Rep. 1: 181, 1905
Epidendrum ionodesme Schlr., Fedde Rep. Beih. 7: 134, 1920
Epidendrum longicrure (Lindl.) Schltr., Fedde Rep. Beih. 7: 137, 1920
Epidendrum macroceras Schltr., Fedde Rep. Beih.. 7: 138, 1920
Epidendrum atacazoicum Schltr., Fedde Rep. Beih. 8: 67, 1921
Epidendrum bifalce Schltr., Fedde Rep. Beih. 8: 67, 1921
Epidendrum caloglossum Schltr., Fedde Rep. Beih. 8: 68, 1921
Epidendrum isthmi Schltr., Fedde Rep. Beih. 17: 34, 1922
Epidendrum reflexum Ames & C.Schweinf., Sched. Orch. 8: 49, 1925

Epidendrum paniculatum Ruiz Lopez & Pavon, Syst. Veg. Fl. Peruv. & Chil.: 243, 1798 **var. linearifolium** (Cogn.) C.Schweinf., Bot. Mus. Leafl. Harv. Univ. 11: 105, 1943
BR 31: misc.70:43 (1845)
Peru – E, L
Epidendrum parviflorum Ruiz Lopez & Pavon, Syst. Veg. : 245, 1798
Epidendrum gratiosum Rchb.f., Bonpl. 4: 215, 1856
Epidendrum gratiosum var. linearifolium Cogn., Mart. Fl. Bras. 3 (5): 166, 1898
Epidendrum patulipetalum Schltr., Fedde Rep. Beih. 9: 91, 1921

Epidendrum paranaense Barb.Rodr., Gen. Sp. Orch. Nov. 2: 139, 1882
BR 31: misc.70:25 (1845)
Brazil – E
Epidendrum imbricatum Lindl., Gen. Sp. Orch. Pl.: 110, 1831 (non Lam.)
Epidendrum biflorum Cogn., Bull. Herb. Boiss. 2.s., 2: 337, 1902 (non Forst.)
Epidendrum boissierianum Schltr., Beih. Bot. Centralbl. 36 (2): 159, 1918

Epidendrum parkinsonianum Hook., Bot. Mag. 66: t.3778, 1840
BR 26: misc.20 (1840) / 31: misc.70: 11 & misc.36 p.27: 7, 8 (1845)
Mexico, Guatemala, Honduras, Costa Rica, Panama – E, L
Epidendrum aloifolium Bateman, Orch. Mex. & Guat.: t.25, 1840 (non Lindl.)
Epidendrum falcatum Lindl., Ann. Mag. Nat. Hist. 4: 382, 1840
Epidendrum lactiflorum A.Rich & Galeotti, Ann. Sc. Nat. s.3, 3: 22, 1845
Auliza parkinsonianum (Hook.) Brieger, Schlechter Die Orchideen ed.3, 1: 546, 1977

Epidendrum parvilabre Lindl., Benth. Pl. Hartw.: 150, 1844
BR 31: misc.36 p.28:12 (1845)
Peru – E

Epidendrum patens Sw., Nov. Act. Upsal. 6: 68, 1799
BR 31: misc.70:42 (1845)
Brazil, Colombia – E
Exophya fuscata Raf., Fl. Tellur 11: 63, 1836

Epidendrum pictum Lindl., Bot. Reg. 24: misc.43, 1838
BR 24: misc.43 (1838) / 28: misc.p.31:19 (1842)
Type: Guyana, cult. Loddiges (K-Lindl., holo!)
Guyana, Brazil – E

Epidendrum piperinum Lindl., Ann. Mag. Nat. Hist. 15: 256, 1845
BR 31: misc.70:3 (1845)
Colombia – E, L

Epidendrum polyanthum Lindl., Gen. Sp. Orch. Pl.: 106, 1831
BR 27: misc.148 (1841) / 28: sub t.25 & misc.2 (1842) / 30: misc.24: 30 (1844)
Mexico to Panama, Venezuela, Brazil – up to 2700 m – T, E
Epidendrum bisetum Lindl., Bot. Reg. 27: misc.148, 1841
Epidendrum funiferum Morren, Ann. Soc. Roy. Agr. Bot. Gard. 4: 211 t.198, 1848
Epidendrum colorans Klotzsch, Allg. Gartenz. 19: 250, 1851
Epidendrum lansbergii Rchb.f., Nederl. Kruidk. Arch. 4: 316, 1859
Epidendrum pergameneum Rchb.f., Beitr. Orch. Centr.-Am.: 86, 1866
Epidendrum stallforthianum Kraenzlin, Gard. Chron. s.3, 51: 114, 1912
Epidendrum verrucipes Schltr., Fedde Rep. 15: 208, 1918
Epidendrum heteroglossum Kraenzlin, Ark. Bot. 16 (8): 19, 1920
Epidendrum quinquelobum Schltr., Fedde Rep. Beih. 19: 125, 1923

Epidendrum polystachyum Kunth, Nov. Gen. & Sp. Pl. 1: 352, 1816
BR 28: misc.p.35:39 (1842)
Colombia, Ecuador, Peru – E

Epidendrum porphyreum Lindl., Hook. J. Bot. 3: 86, 1844
BR 28: sub t.25 (1842) / 30: misc.24:32 (1844)
Colombia, Ecuador, Peru – E

Epidendrum prostatum (Lindl.) Cogn., Mart. Fl. Bras. 3 (5): 112, 1898
BR 24: misc.45 (1838)
Type: Guyana, cult. Loddiges (K-Lindl., holo!)
Brazil, Guyana – E
Physinga prostata Lindl., Bot. Reg.: misc.45, 1838

Epidendrum purpurascens Focke, Tijdr. Nat. Wetens. Ned. 4: 64, 1851
BR 22: t.1870 (1836) / 31: misc.36 p.27:6 (1845)
Brazil, Guyanas, Venezuela, Colombia, Costa Rica – E
○ Epidendrum clavatum Lindl., Bot. Reg. 22: t.1870, 1836
Didothion clavatum (Lindl.) Raf., Fl. Tellur. 4: 39, 1838
Epidendrum glumibracteatum Rchb.f., Hamb. Gartenz. 19: 11, 1863
Epidendrum psilanthemum Loefgr., Arch. Jard. Bot. Rio de Janeiro 2: 57, 1917

Epidendrum purum Lindl., Bot. Reg. 30: misc.75, 1844
BR 30: misc.75 (1844) / 31: misc.70:61 (1845)
Type: Caracas, Linden s.n. cult. Rucker (K-Lindl., holo!)
Venezuela to Bolivia – E

Epidendrum ramosum Jacq., Enum. Carib.: 29, 1760
BR 31: misc.70:14 (1845)
Trop. & subtrop. America – E
Isochilus ramosum (Jacq.) Sprengel, Syst. Veg. 3: 734, 1826
Epidendrum rigidum Lodd., Bot. Cab.: t.1600, 1830 (non Jacq.)
Epidendrum sellowii Rchb.f., Linnaea 22: 839, 1849
Epidendrum ramosum var. lanceolatum Griseb., Fl. Brit. W. Ind. Is.: 618, 1864
Epidendrum ramosum var. lancifolium Cogn., Mart. Fl. Bras. 3 (5): 173, 1898
Epidendrum flexicaule Schltr., Beih. Bot. Centralbl. 36 (2): 403, 1918
Epidendrum modestiflorum Schltr., Fedde Rep. Beih. 19: 213, 1923
Spathiger ramosus (Jacq.) Britt., Sc. Surv. Porto Rico & Virgin Is. 5 (2): 202, 1924

Epidendrum recurvatum Lindl., Bot. Reg. 31: misc.70:54, 1845
BR 31: misc.70:54 (1845)
Type: Venezuela, Linden s.n. (K-Lindl., holo!)
Venezuela – E

Epidendrum refractum Lindl., Ann. Mag. Nat. Hist. 12: 398, 1843
BR 30: misc.82:10 (1844)
Venezuela – E

Epidendrum rigidum Jacq., Enum. Syst. Pl.: 29, 1760
BR 31: misc.70:15 (1845)
Guadeloupe, Dominica, Martinique, St.Lucia, St.Vincent, Grenada – E

Epidendrum ruizianum Steudel, Nom. Bot. ed.2, 1: 558, 1840
BR 28: sub t.42 (1842) / 30: misc.82:13 (1844)
Peru, Colombia, Venezuela – E, L
Epidendrum nutans Ruiz Lopez & Pavon, Syst. Veg. Fl. Peruv. & Chil. 1: 245, 1798 (non Sw.1788)
Epidendrum spathaceum Lindl., Hook. J. Bot. 3: 85, 1841

Epidendrum rupestre Lindl., Hook. J. Bot. 3: 84, 1841
BR 31: misc.36 p.28:10 (1845)
Peru, Ecuador – E, L

Epidendrum saxatile Lindl., Hook. J. Bot. 3: 84, 1841
BR 31: misc.36 p.28:9 (1845)
Venezuela, Guyana, Brazil, Paraguay – T, E
Epidendrum miersii Lindl., Fol. Orch. Epid.: 59, 1853
Epidendrum weddellii Lindl., Fol. Orch. Epid.: 67, 1853
Epidendrum candidum Barb.Rodr., Gen. & Sp. Orch. Nov. 2: 150, 1882
Epidendrum concalvii Barb.Rodr., Gen. & Sp. Orch. Nov. 2: 152, 1882
Epidendrum fractiflexum Barb.Rodr., Gen. & Sp. Orch. Nov. 2: 151, 1882
Epidendrum miersii var. longifolium Cogn , Mart. Fl. Bras. 3 (5): 168, 1898
Epidendrum weddelli var. robustum Cogn , Mart. Fl. Bras. 3 (5): 169, 1898
Epidendrum weddellii var. longifolium Cogn., Mart. Fl. Bras. 3 (5): 169, 1898
Epidendrum vellozoi A.Hawkes, Orquid. 18: 173, 1957

Epidendrum scabrum Ruiz Lopez & Pavon, Syst. Veg. Fl. Peruv. & Chil. 1: 248, 1798
BR 31: misc.70:35 (1845)
Colombia, Ecuador, Peru – 2600–3000 m – E, L
Epidendrum scabrum var. parviflorum Cogn., Mart. Fl. Bras. 3 (5): 107, 1898
Epidendrum loxense Lehm. & Kraenzlin, Engl. Bot. Jahrb. 26: 469, 1899
Epidendrum cardiophyllum Kraenzlin, Engl. Bot. Jahrb. 37: 523, 1906

○ **Epidendrum schomburgkii** Lindl., Bot. Reg. 24: t.53, 1838
BR 24: t.53 & misc.16 (1838) / 28: sub t.25 (1842) / 30: misc.24:22 (1844)
Type: Guyana, Schomburgk s.n. cult. Loddiges (K-Lindl., holo!)
Venezuela, Colombia, Ecuador, Peru, Guyana, Brazil, Amazonas – T, E
Epidendrum incisum Vell., Fl. Flum. Ic. 9: t.18, 1827
Epidendrum fulgens Focke, Tijdr. Nat. Wetens. 4: 66, 1851 (non Brogn.)
Epidendrum blepharoclinium Rchb.f., Linnaea 41: 37, 1877
Epidendrum pristes Rchb.f., Gard. Chron. n.s.26: 262, 1886
Epidendrum splendens Schltr., Fedde Rep. Beih. 9: 93, 1921
Epidendrum ibaguense var. schomburgkii (Lindl.) C.Schweinf., Bot. Mus. Leafl. Harv. Univ. 11: 235, 1944

Epidendrum scriptum A.Rich. & Galeotti, Comptes Rend. Acad. Sc. Paris 18: 512, 1844 (non L.)
BR 31: misc.70:29 (1845)
Mexico – E

Epidendrum scutella Lindl., Bot. Reg. 30: misc.82:2, 1844
BR 30: misc.82:2 (1844)
Type: Ecuador, Hartweg s.n. (K-Lindl., holo!)
Peru, Ecuador, Colombia – E
Epidendrum platyoon Schltr., Fedde Rep. Beih. 9: 91, 1921

Epidendrum secundum Jacq., Enum. Syst. Pl.: 29, 1760
BR 1: t.67 (1815) / 30: misc.24:7 (1844) / 31: misc.70:7 (1845)
Trop. & subtrop. America, Jamaica, St.Vincent, Martinique – E
Epidendrum anceps Jacq., Sel. Stirp. Am.: 224, 1763
Epidendrum secundum (Jacq.) Sw., Obs. Bot.: 325, 1791
○ Epidendrum fuscatum Smith, Spicil. Bot. 2: 21 t.23, 1792
Amphiglottis lurida Salisb., Trans. Hort. Soc. 1: 294, 1812
Epidendrum virescens Lodd., Bot. Cab. 19: t.1867, 1832
Epidendrum musciferum Lindl., Hook. J. Bot. 1: 6, 1834
Epidendrum viridipurpureum Hook., Bot. Mag. 65: t.3666, 1838
Epidendrum ensatum A.Rich. & Galeotti, Ann. Sc. Nat. s.3, 3: 22, 1845
Epidendrum galeottianum A.Rich. & Galeotti, Ann. Sc. Nat. s.3, 3: 21, 1845
Epidendrum amphistomum A.Rich., Sagra Hist Cuba 11: 240, 1850

Epidendrum fuscatum var. virescens (Lodd.) Lindl., Fol. Orch. Epid.: 65, 1853
Epidendrum fuscatum var. viridipurpureum (Hook.) Lindl., Fol. Orch. Epid.: 66, 1853
Epidendrum cearense Barb.Rodr., Gen. & Sp. Orch. Nov. 2: 141, 1881
Epidendrum schreineri Barb.Rodr., Gen. & Sp. Orch. Nov. 2: 142, 1881
Epidendrum anceps var. virescens (Lodd.) Hemsley, Biol. Centr.-Amer. Bot. 3: 225, 1883
Epidendrum anceps var. viridipurpureum (Hook.) Hemsley, Biol. Centr.-Amer. Bot. 3: 225, 1883
Epidendrum schenckianum Kraenzlin, Fedde Rep. 7: 114, 1909
Amphiglottis anceps (Jacq.) Britton, Sc. Surv. Porto Rico & Virgin.Is. 5 (2): 200, 1924

Epidendrum serpens Lindl., Benth. Pl. Hartweg.: 149, 1844
BR 31: misc.36 p.23:4 (1845)
Peru – E

Epidendrum setiferum Lindl., Ann. Mag. Nat. Hist. 4: 383, 1840
BR 28: sub t.25 (1842) / 30: misc.24:3 (1844)
Brazil – E

Epidendrum smaragdinum Lindl., Bot. Reg. 24: misc.44, 1838
BR 24: misc.44 (1838) / 31: misc.70:31 (1845)
Type: Guyana, cult. Loddiges (K-Lindl., holo!)
Guatemala, Costa Rica, Venezuela, Guyana, Peru, Brazil – E, T
Epidendrum kuhlmannii Hoehne, Com. Linh. Tel. Matto Grosso Annex. 5, Bot. 4: 21, 1912
Epidendrum alfredii Schltr., Fedde Rep. Beih. 19: 115, 1923
Epidendrum pachyrhachis Ames, Sched. Orch. 2: 32, 1923

Epidendrum squalidum Llave & Lex., Nov. Veg. Descr. Orch. Opusc. 2: 22–44, 1825
BR 28: misc.p.31:15 (1842)
Mexico – E obscure sp., fide Schweinfurth

Epidendrum stamfordianum Bateman, Orch. Mex. & Guat.: t.11, 1838
BR 26: misc.88 (1840) / 31: misc.34 & misc.36 p.29:1 (1845)
Mexico, Centr. America, Guatemala, Panama, W.Indies, Venezuela, Colombia, Brazil – E
Epidendrum basilare Klotzsch, Allg. Gartenz. 11: 193, 1843
Epidendrum cycnostalix Rchb.f., Bot. Zeit. 10: 731, 1852
Epidendrum stamfordianum var. pictum Lemaire, Jard. Fl. 3: t.251, 1852
Epidendrum stamfordianum var. parviflorum Regel, Ind. Sem. Hort. Petrop.: 19, 1856
Epidendrum cycnostachys Rchb.f. ex Benth., J. Linn. Soc. 18: 314, 1881
Epidendrum stamfordianum var. wallacei Rchb.f., Gard. Chron. s.3, 1: 543, 1887
Epidendrum stamfordianum var. leeanum Rchb.f., Gard. Chron. s.3, 3: 521, 1888
Epidendrum stamfordianum var. lawrenceanum hort. ex Stein, Orchideenb.: 240, 1892
Psilanthemum basilare (Klotzsch) Klotzsch ex Stein, Orchideenb.: 238, 1892
Auliza stamfordiana (Bateman) Brieger, Schlechter Die Orchideen ed.3, 1: 548, 1977

Epidendrum subaquilum Lindl., Bot. Reg. 32: sub t.64, 1846
BR 32: sub t.64 (1846)
Type: Mexico, Warner, s.n. (K-Lindl., holo!)
Jamaica, Mexico?, Guatemala? – E
Epidendrum harrisii Fawcett, Symb. Ant. 1: 291, 1899

Epidendrum subulatifolium A.Rich. & Galeotti, Ann. Sc. Nat. s.3, 3: 21, 1845
BR 31: misc.70:23 (1845)
Mexico – 1800–2000 m – E

Epidendrum tenue Lindl., Hook. J. Bot. 3: 88, 1841
BR 31: misc.70:27 (1845)
Brazil – E

Epidendrum thyrsiferum Lindl., Bot. Reg. 28: sub t.25, 1842 (nomen nudum)
BR 28: sub t.25 (1842)

Epidendrum tolimense Lindl., Bot. Reg. 31: misc.70:2bis, 1845
BR 31: misc.70:2bis (1845)
Type: Colombia, Tolima, Linden s.n. (K-Lindl., holo!)
Colombia – E

Epidendrum torquatum Lindl., Benth. Pl. Hartw.: 149, 1844
BR 31: misc.70:45 (1845)
Colombia, Peru – E

Epidendrum ventricosum Lindl., Hook. J. Bot. 3: 86, 1841
BR 28: sub t.42 (1842) / 30: misc.82:5 (1844)
Peru – E
Epidendrum pachygastrium Kraenzlin, Fedde Rep. 1: 183, 1905

Epidendrum verrucosum Sw., Nov. Act. Upsal. 6: 68, 1799
BR 31: misc.70:65 (1845)
Jamaica, Colombia – E, L
Oerstedella verrucosa (Sw.) Hagsater, Orquidea (Mex.) 8 (1): 24, 1981

Epidendrum vesicatum Lindl., Bot. Reg. 24: misc.89, 1838
BR 24: misc.89 (1838) / 31: misc.70:2 (1845)
Type: Brazil, cult. Loddiges (K-Lindl., holo!)
Brazil, Guyana – E

Epidendrum vincentinum Lindl., Hook. J. Bot. 3: 88, 1840
BR 31: misc.70:53 (1845)
Centr. America, Antilles, Trinidad, Venezuela – E

Epidendrum viride Ruiz Lopez & Pavon, Syst. Veg. Fl. Peruv. & Chil. 1: 244, 1798
BR 31: misc.70:44 (1845)
Peru – E

Epidendrum viviparum Lindl., Bot. Reg. 27: misc.27, 1841
BR 27: misc.27 (1841) / 31: misc.36 p.26:4 (1845)
Type: Guyana, Schomburgk s.n. cult. Loddiges (K-Lindl., holo!)
Brazil, Guyana – E

Epidendrum volubile Ruiz Lopez & Pavon, Syst. Veg. Fl. Peruv. & Chil. 1: 247, 1798
BR 31: misc.70:57 (1845)
Peru – E

Epidendrum xanthinum Lindl., Bot. Reg. 30: misc.24:11, 1844
BR 30: misc.24:11 (1844)
Type: Brazil, Minas Geraes, Gardner s.n. (K-Lindl., holo!)
Brazil – E
Epidendrum ellipticum var. flavum Lindl., Ann. Mag. Nat. Hist. 4: 382, 1840

Epidendrum xylostachyum Lindl., Bot. Reg. 31: misc.70:19, 1845
BR 31: misc.70:19 (1845)
Type: Colombia, Linden s.n. (K-Lindl., holo!)
Colombia – E

Eria acutifolia Lindl., Bot. Reg. 28: misc.32, 1842
BR 28: misc.32 (1842)
Type: cult. Loddiges (K-Lindl., holo!)
India – E

Eria bipunctata Lindl., Bot. Reg. 27: misc.179, 1841
BR 27: misc.179 (1841)
Type: Khasia Hills, Gibson s.n. (K-Lindl., holo!)
N.E.India, Thailand, Vietnam – E
Pinalia bipunctata (Lindl.) Kuntze, Rev. Gen. 2: 679, 1891
Eria convallarioides Finet, Not. Syst. Paris 1: 255, 1910 (non Lindl.)
Eria amica Ridley, J. Nat. Soc. Siam 4 (3): 117, 1921 (non Rchb.f.)
Eria eberhardtii Gagnepain, Bull. Mus. Nat. Hist. Paris s.2, 3: 306, 1930

o **Eria bractescens** Lindl., Bot. Reg. 27: misc.46, 1841
BR 27: misc.46 (1841) / 30: t.29 (1844) / 31: misc.67 (1845)
Type: Singapore, Cuming s.n. cult. Loddiges (K-Lindl., holo!)
India, E.Himalaya, Burma, Cambodia, Vietnam, Malaya, Indonesia, Philippines, New Guinea – E
Eria dillwynii Hook., Bot. Mag. 71: t.4163, 1845
Eria pulchella Griffith, Notul. 3: 297, 1851 (non Lindl.)
Eria littoralis Teijsm. & Binnend., Tidschr. Ned. Ind. 24: 312, 1862
Eria griffithii Rchb.f., Xen. Orch. 2: 163, 1870
Eria affinis Parish, Order Orchid. Burma 1: 94, 1883 (non Griffith)
Pinalia bractescens (Lindl.) Kuntze, Rev. Gen. 2: 679, 1891
Pinalia pulchella (Griffith) Kuntze, Rev. Gen. 2: 679, 1891
Dendrobium subterrestre Gagnepain, Bull. Mus. Hist. Nat. Paris s.2, 1: 239, 1930

Eria clavicaulis Wall. ex Lindl., Bot. Reg. 26: misc.220, 1840
BR 26: misc.220 (1840)
Type: Wallich s.n. cult. Loddiges (K-Lindl., holo!)
N.E.India, Burma, Thailand – E
Eria khasiana Lindl., J. Linn. Soc. 3: 59, 1859
Pinalia clavicaulis (Wall. ex Lindl.) Kuntze, Rev. Gen. 2: 679, 1891
Cylindrolobus clavicaulis (Lindl.) Rauschert, Fedde Rep. 94 (7/8): 445, 1983

Eria coronaria (Lindl.) Rchb.f., Walp. Ann. 6: 272, 1861
BR 27: misc.178 (1841) / 28: t.21 (1842)
Type: Khasia Hills, Gibson s.n. (K-Lindl., holo!)
N.W.Himalaya, Nepal, Bhutan, Sikkim, N.E.India – E
Coelogyne coronaria Lindl., Bot. Reg. 27: misc.178, 1841
o Trichosma suavis Lindl., Bot. Reg. 28: t.21, 1842
Eria cylindropoda Griffith, Notul. 3: 299 & 404, 1851
Eria suavis (Lindl.) Lindl., J. Linn. Soc. 3: 52, 1859
Trichosma coronaria (Lindl.) Brieger, Schlechter Die Orchideen ed.3, 1: 663, 1981

o **Eria ferruginea** Lindl., Bot. Reg. 25: t.35, 1839
BR 25: t.35 (1839)
Type: India, cult. Loddiges (K-Lindl., holo!)
India – E

o **Eria floribunda** Lindl., Bot. Reg. 29: misc.56, 1843
BR 29: misc.56 (1843) / 30: t.20 (1844)
Type: Singapore, Prince s.n. (K-Lindl., holo!)
Burma, Vietnam, Malaya, Sumatra, Borneo, Philippines – E
Eria leuchostachya Lindl., J. Hort. Soc. 3: 15, 1848
Pinalia floribunda (Lindl.) Kuntze, Rev. Gen. 2: 678, 1891
Eria euryanthe Schltr., Engl. Bot. Jahrb. Beibl. 104: 43, 1911
Eria subaliena Gagnepain, Bull. Mus. Nat. Hist. Paris s.2, 3: 311, 1930
Eria giungii Guillaumin, Bull. Mus. Nat. Hist. Paris s.2, 25: 510, 1963

Eria javanica (Sw.) Blume, Rumphia 2: 23, 1836
BR 11: t.904 (1825) / 30: misc.14 & misc.23 (1844)
Sikkim, Assam, Burma, Laos, Malaya, Indonesia, New Guinea, Philippines – E
Dendrobium javanicum Sw., Kongl. Vet. Akad. Handl. 21: 247, 1800
Dendrolirium rugosum Blume, Bijdr.: 345, 1825
o Eria stellata Lindl., Bot. Reg. 11: t.904, 1825
Octomeria stellata (Lindl.) Sprengel, Syst. Veg. 4, 2: 310, 1827
Octomeria vaginata Breda, Gen. Sp. Orch. Asclep.: pl.5, 1827
Eria rugosa (Blume) Lindl., Gen. Sp. Orch. Pl.: 55, 1830
Eria cochleata Lindl., Bot. Reg. 30: misc.14, 1844
Eria fragrans Rchb.f., Bot. Zeit. 22: 415, 1864
Eria vaginata (Breda) Benth., J. Linn. Soc. 18: 305, 1881
Eria striolata Rchb.f., Gard. Chron. s.3, 3: 554, 1888
Tainia stellata (Lindl.) Pfitzer, Engl. Nat. Pflanzenf. 2, 6: 153, 1888
Dendrobium perakense Hook.f., Fl. Brit. Ind. 5: 712, 1890
Pinalia fragrans (Rchb.f.) Kuntze, Rev. Gen. 2: 679, 1891
Pinalia rugosa (Blume) Kuntze, Rev. Gen. 2: 679, 1891
Pinalia stellata (Lindl.) Kuntze, Rev. Gen. 2: 679, 1891
Pinalia striolata (Rchb.f.) Kuntze, Rev. Gen. 2: 679, 1891
Sarcopodium perakense (Hook.f.) Kraenzlin, Engl. Pflanzenr. Orch. Mon. Dendr.: 328, 1910
Katherinea perakense (Hook.f.) A.Hawkes, Lloydia 19: 97, 1956

Eria lindleyi Thwaites, Enum. Pl. Zeyl.: 299, 1861
BR 30: misc.62:57 (1844)
Sri Lanka – up to 2400 m – E
Dendrobium bicolor Lindl., Gen. Sp. Orch. Pl.: 90, 1830
Eria bicolor (Lindl.) Lindl., J. Linn. Soc. 3: 58, 1859
Eria ephemera Rchb.f., Walp. Ann. 6: 272, 1861

Eria longilabris Lindl., Bot. Reg. 27: misc.69, 1841
BR 27: misc.69 (1841) / 30: sub t.29:2 (1844)
Type: Panay, Cuming s.n. cult. Loddiges (K-Lindl., holo!)
Philippines – E

Eria multiflora (Blume) Lindl., Gen. Sp. Orch. Pl.: 68, 1830
BR 29: misc.72 (1843)
Java, Sumatra – E
Dendrolirium micranthum Blume, Bijdr.: 349, 1825
Dendrolirium multiflorum Blume, Bijdr.: 349, 1825
Octomeria racemosa Breda, Kuhl & v.Hasselt Orch.: t.1, 1828
Eria micrantha (Blume) Lindl., Gen. Sp. Orch. Pl.: 68, 1830
Pinalia micrantha (Blume) Kuntze, Rev. Gen. 2: 678, 1891
Pinalia multiflora (Blume) Kuntze, Rev. Gen. 2: 678, 1891

Eria nutans Lindl., Bot. Reg. 26: misc.196, 1840
BR 26: misc.196 (1840)
Type: Singapore, Cuming s.n. cult. Loddiges (K-Lindl., holo!)
Thailand, Malaya, Borneo – E
Pinalia nutans (Lindl.) Kuntze, Rev. Gen. 2: 679, 1891
Trichosma nutans (Lindl.) Brieger, Schlechter Die Orchideen ed.3, 1: 663, 1981

Eria obesa Lindl., Gen. Sp. Orch. Pl.: 15, 1830
BR 30: sub t.29:3 (1844)
N.E.India, Burma – E
Eria lindleyana Griffith, Notul. 3: 300, 1851
Pinalia obesa (Lindl.) Kuntze, Rev. Gen. 2: 679, 1891
Eria prainii Briq., Ann. Cons. & Jard. Bot. Genève 4: 210, 1900

Eria obliqua (Lindl.) Lindl., J. Linn. Soc. 3: 55, 1859
BR 26: misc.184 (1840)
Type: Singapore, Cuming s.n. cult. Loddiges (K-Lindl., holo!)
Singapore – E
Mycaranthes obliqua Lindl., Bot. Reg. 26: misc.184, 1840

Eria ornata (Blume) Lindl., Gen. Sp. Orch. Pl.: 66, 1830
BR 27: t.42 & misc.70 (1841)
Thailand, Malaya, Sumatra, Philippines – up to 1500 m – E
Dendrolirium ornatum Blume, Bijdr.: 345, 1825
o Eria armeniaca Lindl., Bot. Reg. 27: t.42 & misc.70, 1841
Pinalia ornata (Blume) Kuntze, Rev. Gen. 2: 679, 1891

Eria ovata Lindl., Bot. Reg. 30: sub t.29:4, 1844
BR 30: sub t.29:4 (1844)
Type: Philippines, Cuming s.n. (K-Lindl., holo !)
Philippines, Ryukyu Is., Sulawesi, Taiwan – E
Eria luchuensis Yatabe, Bot. Mag. Tokyo 7, 1892
Eria bidentata Nakai, Bot. Mag. Tokyo 34, 1920
Eria elmeri Ames, Orchid. 1: 93, 1920
Eria graciliscapa Rolfe ex Ames, Orchid. 1: 93, 1920

Eria paniculata Lindl., Wall. Pl. Asiat. Rar. 1: 32 t.36, 1830
BR 28: misc.33 (1842)
Nepal, Sikkim, Bhutan, N.E.India, Burma, Thailand, Laos, Cambodia, Vietnam – E
Pinalia paniculata (Lindl.) Kuntze, Rev. Gen. 2: 679, 1891

Eria pannea Lindl., Bot. Reg. 28: misc.79, 1842
BR 28: misc.79 (1842)
Type: Singapore, cult. Loddiges (K-Lindl., holo!)
Bhutan, Sikkim, Burma, Thailand, Laos, Cambodia, Vietnam, China, Malaya, Sumatra, Riau Arch., Bangka, Borneo – E
Eria teretifolia Griffith, Itin. Not.: 202, 1848 (non Griffith 1851)
Eria calamifolia Hook.f., Fl. Brit. Ind. 6: 191, 1890
Pinalia calamifolia (Hook.f.) Kuntze, Rev. Gen. 2: 679, 1891
Pinalia pannea (Lindl.) Kuntze, Rev. Gen. 2: 679, 1891
Eria nivosa Ridley, J. Nat. Hist. Siam. Soc. 4: 116, 1921

o **Eria polyura** Lindl., Bot. Reg. 27: misc.114, 1841
BR 27: misc.114 (1841) / 28: t.32 (1842)
Type: Philippines, Cuming s.n. cult. Loddiges (K-Lindl., holo!)
Philippines – E
Pinalia polyura (Lindl.) Kuntze, Rev. Gen. 2: 679, 1891

Eria profusa Lindl., Bot. Reg. 28: misc.3, 1842
BR 28: misc.3 (1842)
Type: Nightingale, cult. Syon Garden (K-Lindl., holo!)
Philippines – E

Eria pulchella Lindl., Bot. Reg. 27: misc.106, 1841
BR 27: misc.106 (1841)
Type: Singapore, Cuming s.n. cult Loddiges (K-Lindl., holo!)
Thailand, Malaya, Borneo – E
Tylostylis rigida Ridley, Mat. Fl. Mal. Pen. 1: 105, 1907 (non (Blume) Blume)
Tylostylis pulchella (Lindl.) Ridley, J. Fed. Mal. St. Mus. 6 (3): 179, 1915
Callostylis pulchella (Lindl.) Chen & Tsi, Acta Bot. Yunnan. 6 (3): 270, 1984

Eria pumila Lindl., Gen. Sp. Orch. Pl.: 68, 1830
BR 24: misc.147 (1838)
Sikkim, N.E.India, Burma – E
Pinalia pumila (Lindl.) Kuntze, Rev. Gen. 2: 679, 1891

Eria retusa (Blume) Rchb.f., Bonpl. 5: 54, 1857
BR 24: misc.145 (1838)
Java, Christmas Is., E.India? – E
Dendrolirium retusum Blume, Bijdr.: 351, 1825
Phreatia retusa (Blume) Lindl., Gen. Sp. Orch. Pl.: 64, 1830
Bryobium pubescens Lindl., Bot. Reg. 24: misc.145, 1838
Eria pusilla Teijsm.& Binnend., Nat. Tijdschr. Ned. Ind. 24: 312, 1862
Phreatia congesta Rolfe, J. Linn. Soc. Bot. 25: 358, 1890
Pinalia retusa (Blume) Kuntze, Rev. Gen. 2: 679, 1891

o **Eria rosea** Lindl., Bot. Reg. 12: t.978, 1826
BR 12: t.978 (1826)
Type: China, J.D.Parks s.n. cult. Hort. Soc. (K-Lindl., holo!)
China, Taiwan – E
Octomeria rosea (Lindl.) Sprengel, Syst. Veg. 4 (2): 310, 1827
Xiphosium roseum (Lindl.) Griffith, Calcutta J. Nat. Hist. 5: 364, 1845
Pinalia rosea (Lindl.) Kuntze, Rev. Gen. 2: 679, 1891

Eria spicata (D.Don) Hand.-Mazz., Symb. Sin. 7: 1353, 1936
BR 27: t.62 & misc.121 (1841) / 33: t.63 (1847)
N.W.Himalaya, Nepal, Sikkim, N.E.India, Burma, Yunnan – E
Octomeria spicata D.Don, Prodr. Fl. Nepal: 31, 1825
o Eria convallarioides Lindl., Gen. Sp. Orch. Pl.: 70, 1830
o Eria convallarioides var. major Lindl., Bot. Reg. 33: t.63, 1847

Eria summerhayesiana A.Hawkes & A.H.Heller, Lloydia 20: 132, 1957
BR 21: sub t.1756:4 (1835)
Burma, Malaya – E
Dendrobium extinctorium Lindl., Bot. Reg. 21: sub t.1756:4, 1835
Eria extintoria (Lindl.) Oliver, Bot. Mag. 97: t.5910, 1871
Eria decipiens Summerh., Kew Bull.: 307, 1929

o **Eriopsis biloba** Lindl., Bot. Reg. 33: sub t.9, 1847
BR 33: t.18 & sub t.9 (1847)
Type: cult. Barker s.n. (K-Lindl., holo!)
Costa Rica, Panama, Venezuela, Guyana, Brazil, Colombia, Ecuador, Peru – 1000–2500 m – T, E
Eriopsis rutidobulbon Hook., Bot. Mag. 75: t.4437, 1849

Pseuderiopsis schomburgkii Rchb.f., Linnaea 22: 853, 1849
Eriopsis rhytidobulbon Lemaire, Jard. Fleur. 2: misc.p.77, 1852
Eriopsis schomburgkii (Rchb.f.) Rchb.f., Bonpl. 3: 67, 1855
Eriopsis fuerstenbergii Kraenzlin, Orchis 2: 62, 1908
Eriopsis mesae Kraenzlin, Notizbl. Bot. Mus. Berlin 7: 427, 1920
Eriopsis werckeleyi Schltr., Fedde Rep. 16: 447, 1920
Eriopsis colombiana Schltr., Fedde Rep. Beih. 27: 172, 1924
Eriopsis grandibulbosa Ames & C.Schweinf., Bull. Torrey Club 58: 350, 1931

Erythrodes picta (Lindl.) Ames, Orchid. 5: 29, 1915
BR 30: misc.61 (1844)
Brazil, Paraguay – T
Ophrys argentea Vell., Fl. Flum. Ic. 9: t.61, 1825
Physurus pictus Lindl., Gen. Sp. Orch. Pl.: 504, 1840
Anoectochilus argenteus hort. ex Morren, Ann. Soc. Agr. & Bot. de Gand. 1: 188, 1845
Microchilus pictus (Lindl.) Morren, Ann. Soc. Agr.& Bot. de Gand. 1: 187 t.18, 1845
Neottia argentea hort. ex Morren, Ann. Soc. Agr. & Bot. de Gand. 1: 188, 1845
Physurus argenteus hort. ex Koch & Lauch., Allg. Gartenz.: 5, 1857
Spiranthes argentea hort. ex Planchon, Hort. Donat.: 93, 1858

Eulophia alta (L.) Fawcett & Rendle, Fl. Jam. 1: 112 t.22 figs.4–8, 1910
BR 9: sub t.732 (1823) / 18: t.1508 & sub t.1549: 32 (1832) / 31: sub t.67:10 (1845)
USA (Florida), W.Indies, Mexico to Brazil, W.Africa to Uganda – T
Limodorum altum L., Syst. Nat. ed.12, 2: 594, 1767
Limodorum praealtum L.C.Rich, Act. Soc. Hist. Nat. Paris 1: 112, 1792
o Cyrtopodium woodfordii Sims, Bot. Mag. 43: t.1814, 1816
Dendrobium longifolium Kunth, Nov. Gen. & Sp. Pl. 1: 360, 1816
Xylobium longifolium (Kunth) Lindl., Bot. Reg. 9: sub t.732, 1823
Cypripedium epidendricum Vell., Fl. Flum. Ic. 9: t.64, 1825
Maxillaria longifolia (Kunth) Lindl., Gen. Sp. Orch. Pl.: 150, 1832
Cyrtopera woodfordii (Sims) Lindl., Gen. Sp. Orch. Pl.: 189, 1833
Govenia barbata Poeppig & Endl., Nov. Gen. Sp. Pl. 2: 5, 1836
Cyrtopera longifolia (Kunth) Rchb.f., Walp. Ann. 6: 668, 1861
Cyrtopera vellosiana Barb.Rodr., Orch. Nov. 2: 179, 1881
Cyrtopera polyantha Barb.Rodr., Struct. d. Orch.: t.13 fig.2, 1883
Paphiopedilum epidendricum (Vell.) Pfitzer, Engl. Bot. Jahrb. 19: 42, 1894
Eulophia woodfordii (Sims) Rolfe, F.T.A. 7: 68, 1897
Platypus papilliferus Small & Nash, Fl. S.E.United States: 329, 1903
Platypus altus (L.) Small, Fl. S.E.United States ed.2: 329, 1913
Eulophia longifolia (Kunth) Schltr., Schlechter Die Orchideen: 347, 1914
Cyrtopera alta (L.) Stéhlé, Bull. Agr. Martinique 8: 155, 1939

o **Eulophia ensata** Lindl., Bot. Reg. 14: t.1147, 1828
BR 14: t.1147 (1828)
Type: Sierra Leone, G.Don s.n. cult. Hort. Soc. (K-Lindl., holo!)
S.Africa – T

o **Eulophia gracilis** Lindl., Bot. Reg. 9: t.742, 1823
BR 9: t.742 (1823)
Type: Sierra Leone, G.Don s.n. cult. Hort. Soc. (K-Lindl., holo!)
Senegal, Guinea Bissau, Sierra Leone, Liberia, Ivory Coast, Ghana, Nigeria, E.Cameroun, Congo, Angola – T
Eulophia preussii Kraenzlin, Engl. Bot. Jahrb. 17: 54, 1893
Eulophia virens A.Chev., Bot.: 614, 1920

o **Eulophia guineensis** R.Br. ex Lindl., Bot. Reg. 8: t.686, 1823
BR 8: t.686 (1823)
Type: Sierra Leone, G.Don s.n. cult. Hort. Soc. (K-Lindl., holo!)
Gambia, Mali, Guinea, Sierra Leone, Ivory Coast, Ghana, Nigeria, Zaire, Uganda, Tanzania, Angola, Yemen, Oman – T
Eulophia quartiniana A.Rich., Tent. Fl. Abyss. 2: 284, 1851
Saccolabium abyssinicum A.Rich., Tent. Fl. Abyss. 2: t.81, 1851
Galeandra quartiniana (A.Rich.) Rchb.f., Walp. Ann. 3: 552, 1852
Graphorchis guineensis (Lindl.) Kuntze, Rev. Gen. 2: 22, 1891
Eulophia congoensis Cogn., J. Orch. 6: 155, 1895

Eulophia horsfallii (Bateman) Summerh., Hutch. & Dalz. F.W.T.A. 2: 446, 1936
BR 29: misc.37 (1843) / 30: t.12 (1844)
Guinea-Bissau, Sierra Leone, Liberia, Ivory Coast, Ghana, Nigeria, Cameroun, Fernando Po, Kenya, Uganda, Tanzania, Malawi, S.Africa – T
Dendrobium roseum Sw., Pers. Syn. Pl.: 523, 1807
o Lissochilus roseus (Sw.) Lindl., Bot. Reg. 29: misc.37, 1843
Lissochilus horsfallii Bateman, Bot. Mag. 91: t.5486, 1865
Lissochilus sandersonii Rchb.f., Otia Bot. Hamb. 1: 62, 1878
Eulophia porphyroglossa Bolus, J. Linn. Soc. 25: 185 in part, 1890
Lissochilus porphyroglossa (Bolus) Durand & Schinz, Conspect. Fl. Afr. 5: 30, 1898 (non Rchb.f.)
Lissochilus mahonii Rolfe, Bot. Mag. 131: t.8047, 1905
Lissochilus transvaalensis Rolfe, Fl. Cap. 5 (3): 57, 1912
Eulophia mahonii (Rolfe) A.Hawkes, Orch. Rev. 72: 27, 1964
Eulophia sandersonii (Rchb.f.) A. Hawkes, Orch. Rev. 72: 27, 1964

Eulophia nuda Lindl., Gen. Sp. Orch. Pl.: 180, 1833
BR 27: misc.164 (1841)
Sri Lanka, India, Nepal, Sikkim, Burma, Thailand, Malaya, Sumatra, Philippines, New Guinea – T
Cyrtopera plicata Lindl. ex Wall., Cat 246: 7362, 1832
Dipodium plicatum Buch.-Ham. ex Wall Cat 246: 7362, 1832
Dipodium roniate Buch.-Ham. ex Wall, Cat. 246: 7362, 1832
Eulophia squalida Lindl., Bot. Reg. 27: misc.164, 1841
Cyrtopera fusca Wight, Ic. Pl. Ind. Or. 5 (1): t.1690, 1851
Eulophia bicolor Dalz., Hook. J. Bot.: misc.3 : 343, 1851
Eulophia fusca (Wight) Blume, Orch. Arch. Ind.: 182, 1858
Cyrtopera mysorensis Lindl., J. Linn. Soc. Bot. 3: 32, 1859
Cyrtopera gardneri Thwaites, Enum. Pl. Zeyl.: 302, 1861
Cyrtopera laxiflora Gardner ex Thwaites, Enum. Pl. Zeyl.: 320, 1861
Cyrtopera nuda (Lindl.) Rchb.f., Flora 55: 274, 1872
Cyrtopodium fuscum (Wight) Trimen, Cat.: 89, 1885
Graphorchis nuda (Lindl.) Kuntze, Rev. Gen. 2: 662, 1891

Eulophia speciosa (R.Br. ex Lindl.) Bolus, J. Linn. Soc. 25: 184, 1890
BR 7: t.573 & sub t.578 (1821)
Type: S.Africa, Griffin s.n. (K-Lindl., holo!)
S.Africa, Zimbabwe – T
Satyrium giganteum L.f., Suppl. Pl.: 402,, 1781
Limodorum giganteum (L.f.) Thunb., Prodr.: 4, 1794
Cymbidium giganteum (L.f.) Sw., Schrad. J. 2: 224, 1808
o Lissochilus speciosus R.Br. ex Lindl., Bot. Reg. 7: t.573 text sub t. 578, 1821
Cyrtopera gigantea (L.f.) Lindl., Gen. Sp. Orch. Pl.: 190, 1833

o **Eulophia streptopetala** Lindl., Bot. Reg. 12: t.1002, 1826
BR 12: t.1002 (1826) / 24: misc.14 (1838)
Type: cult. Colvill s.n. (K-Lindl., icono!)
S.Africa, Trop. Africa, Yemen – T
Lissochilus parviflorus Lindl., Bot. Mag. 56: t.2931, 1829
Lissochilus streptopetalus (Lindl.) Lindl., Gen. Sp. Orch. Pl.: 191, 1833
Lissochilus krebsii Rchb.f., Linnaea 20: 685, 1847
Lissochilus paivaeanus Rchb.f., Flora 48: 187, 1863
Lissochilus krebsii var. purpurata Ridley, Gard. Chron. n.s.24: 102, 1885
Eulophia krebsii (Rchb.f.) Bolus, J. Linn. Soc. 25, 1889
Eulophia krebsii var. purpurata (Ridley) Bolus, J. Linn. Soc. 25: 185, 1889
Lissochilus graefei Kraenzlin, Gard. Chron. s.3, 11: 749, 1892
Eulophia paivaeana (Rchb.f.) Summerh., Kew Bull.: 151, 1953

Flickingeria comata (Blume) A.Hawkes, Orquidea 27: 301, 1965
BR 30: misc.45 (1844)
Sri Lanka, Singapore, Malaya, Sumatra, Java, Borneo, New Guinea, Solomon Is., Vanuatu, New Caledonia, Samoa, Australia – E
Desmotrichum comatum Blume, Bijdr.: 329 t. 35, 1825
Dendrobium comatum (Blume) Lindl., Gen. Sp. Orch. Pl.: 76, 1830
Dendrobium criniferum Lindl., Bot. Reg. 30: misc.45, 1844
Dendrobium zollingerianum Teijsm. & Binnend., Nat. Tijdschr. Ned. Ind. 24: 313, 1862
Callista comata (Blume) Kuntze, Rev. Gen. 2: 654, 1891
Ephemerantha comata (Blume) P.Hunt & Summerh., Taxon 10: 102, 1961

Flickingeria fimbriata (Blume) A.Hawkes, Orch. Weekly 2 (46): 454, 1961
BR 26: misc.7 (1840) / 30: misc.62:42 (1844)
Thailand, China, Vietnam, Nicobar Is., Malaya, Java, Celebes, Borneo, Bali, Philippines – E
Desmotrichum fimbriatum Blume, Bijdr.: 329, 1825 (non Fischer)
Dendrobium fimbriatum (Blume) Lindl., Gen. Sp. Orch. Pl.: 76, 1830 (non Hook.)
Dendrobium plicatile Lindl., Bot. Reg. 26: misc.7, 1840
Dendrobium insulare Steudel, Nom. 2: 490, 1841
Dendrobium flabellum Rchb.f., Bonpl. 5: 56, 1857
Dendrobium binnendijkii Rchb.f., Xen. Orch. 2: 74, 1862
Dendrobium flabellum Hemsley, Gard. Chron. n.s.17: 306, 1882 (non Rchb.f.)
Dendrobium kunstleri Hook.f., Fl. Brit. Ind. 5: 714, 1890
Dendrobium macraei auct. p.p.non Lindl.; Hook.f., Fl. Brit. Ind.: 714, 1890
Callista binnendijkii (Rchb.f.) Kuntze, Rev. Gen. 2: 654, 1891
Callista flabella (Rchb.f.) Kuntze, Rev. Gen. 2: 654, 1891
Callista kunstleri (Hook.f.) Kuntze, Rev. Gen. 2: 655, 1891
Callista macraei (Lindl.) Kuntze, Rev. Gen. 2: 655, 1891

Desmotrichum binnendijkii (Rchb.f.) Kraenzlin, Engl. Pflanzenr. Orch. Mon. Dendr.: 353, 1910
Desmotrichum kunstleri (Hook.f.) Kraenzlin, Engl. Pflanzenr. Orch. Mon. Dendr.: 356, 1910
Dendrobium mentosum Schltr., Fedde Rep. 10: 77, 1911
Desmotrichum grandiflorum auct. p.p. non Blume; Gagnepain, Fl. Gén. Indoch. 6 Paris: 261, 1934
Ephemerantha fimbriata (Blume) P.Hunt & Summh., Taxon 10 (4): 103, 1961 (non Seidenfaden 1966, 1973a, non A.D. Kerr 1969)
Ephemerantha kunstleri (Hook.f.) P.Hunt. & Summerh., Taxon 10 (4): 104, 1961
Flickingeria kunstleri (Hook.f.) A.Hawkes, Orch. Weekly 2 (46): 456, 1961
Ephemerantha pallens P.Hunt & Summerh.; Seidenf. & Smitin., Orch. Thailand: 730 p.p., 1965 (non Ridley)
Dendrobium laciniosum Teuscher, Amer. Orch. Soc. Bull.: 711, 1970 (non Ridley)
Desmotrichum plicatile (Lindl.) Pham-Hoang, Cay-Co Mien-Nam Viet-Nam 2 ed.2: 1062 fig.5155, 1972

Flickingeria scopa (Lindl.) Sprunger & Cribb, comb. nov.
BR 28: misc.55 (1842)
Type: Philippines, Manila, cult. Loddiges (K-Lindl., holo!)
Philippines, Guam, New Guinea, Samoa, Sulawesi, Taiwan – E, L
Dendrobium scopa Lindl., Bot. Reg. 28: misc.55, 1842
Desmotrichum scopa (Lindl.) Kraenzlin, Engl. Pflanzr. Orch. Mon. Dendr. 1: 349, 1910
Ephemerantha scopa (Lindl.) P.Hunt & Summerh., Taxon 10 (4): 102, 1969

Frondaria caulescens (Lindl.) Luer, Ic. Pleurothallid. Missouri Bot. Gard. 15: 29, 1986
BR 21: sub t.1797:34 (1835) / 28: misc.p.74:41 (1842)
Peru, Ecuador – E
Pleurothallis caulescens Lindl., Hook. J. Bot. 1: 9, 1834
Humboldtia caulescens (Lindl.) Kuntze, Rev. Gen. 2: 667, 1891
Pleurothallis graminea Schltr., Fedde Rep. Beih. 9: 74, 1921

Galeandra batemannii Rolfe, Gard. Chron. s.3, 12: 431, 1892
BR 26: t.49 (1840)
Mexico, Belize, Guatemala, Honduras, Nicaragua – E
o Galeandra baueri Lindl., Bot. Reg. 26: t.49, 1840

Galeandra cristata Lindl., Bot. Reg. 30: misc.69, 1844
BR 30: misc.69 (1844)
Type: Guiana, Cayenne, cult. Loddiges 1308 (K-Lindl., holo!)
French Guiana – E

Galeola cassythoides (Lindl.) Rchb.f., Xen. Orch. 2: 77, 1874
BR 21: sub t.1828 (1836)
Type: Australia, Sydney Cove, A.Cunningham s.n. (K-Lindl., holo!)
Australia – T
Dendrobium cassythoides Lindl., Bot. Reg. 21: sub t.1828, 1836

Gastrochilus calceolaris (Buch.-Ham. ex Smith) D.Don, Prodr.: 32, 1825
BR 24: misc.139 (1838)
Nepal, Sikkim, Bhutan, N.E.India, Burma, Thailand, Vietnam, Malaya, China – 1500–2000 m – E
Aerides calceolaris Buch-Ham ex Smith, Rees Cyclop. Suppl. 39: 78, 1819
Epidendrum calceolare (Buch.-Ham. ex Smith) D.Don, Prodr. Fl. Nepal: 32, 1825
Sarcochilus nepalensis Sprengel, Syst. Veg. 3: 721, 1826
Saccolabium calceolare (Buch.-Ham. ex Smith) Lindl., Gen. Sp. Orch. Pl.: 223, 1833
Aerides leopardina Lindl., Sert. Orch. Frontesp. no.6, 1838
Aerides leopardorum Lindl., Hook.f. Fl. Brit. Ind. 6: 60, 1890
Saccolabium bigibbum Ridley, J. Linn. Soc. 32: 213–416, 1896 (non Rchb.f. ex Hook.f.)
Gastrochilus intermedius Seidenf. & Smitin., Orch. Thailand: 629 fig.472, 1963 (non (Griffith ex Lindl.) Kuntze)
Gastrochilus sp. Seidenf. & Smitin., Orch. Thailand: 626 fig.471, 1963

Gennaria diphylla (Link) Parl., Fl. Ital. 3: 405, 1858
BR 18: t.1499 (1832)
Portugal, N.Africa, W.Mediterranean Region, Canary Is. – T
Satyrium diphyllum Link, Schrad. Neues J. Bot. 2:323, 1799
Orchis cordata Willd., Sp. Pl. 4: 27, 1805
Habenaria cordata (Willd.) R.Br., Prodr.: 312, 1809
Gymnadenia diphylla (Link) Link, Handb. 1: 243, 1829
o Herminium cordatum (Willd.) Lindl., Bot. Reg. 18: t.1499, 1832
Peristylus cordatus (Willd.) Lindl., Gen. Sp. Orch. Pl.: 298, 1835
Digomphotis cordata (Willd.) Raf., Fl. Tellur. 2: 37, 1836
Platanthera diphylla (Link) Rchb.f., Ic. Fl.Germ. 13/14: 128 t.84, 1851
Orchis cordifolia Munby, Bull. Soc. Bot. Fr. 2: 148, 1855
Coeloglossum cordatum (Willd.) Nyman, Syll. Pl. Nov.: 359, 1865
Habenaria diphylla (Link) Durand & Schinz, Consp. Pl. Afr. 5, 1892
Coeloglossum diphyllum (Link) Fiori & Paoli, Fl. Ital. 1: 248, 1898

Geodorum densiflorum (Lam.) Schltr., Fedde Rep. Beih. 4: 259, 1919
BR 8: t.675 (1822)
India, Sri Lanka – T
Limodorum densiflorum Lam., Encycl. Meth. Bot. 3: 516, 1789
Limodorum recurvum Roxb., Pl. Corom. 1: 33 t.39, 1795
Malaxis cernua Willd., Sp. Pl. ed.4: 93, 1805
Otandra cernua (Willd.) Salisb., Trans. Hort. Soc. London 1: 298, 1812
o Geodorum dilatatum R.Br., Ait. Hort. Kew ed.2, 5: 207, 1813
Limodorum candidum Roxb., Hort. Beng.: 63, 1814

o **Geodorum fucatum** Lindl., Bot. Reg. 20: t.1387, 1834 (see Opera Bot. Nr.72, 1983)
BR 20: t.1687 (1834)
Type: Sri Lanka, Watson s.n. (K-Lindl., icono!)
Sri Lanka – T

Gomesa crispa (Lindl.) Klotzsch & Rchb.f., Bot. Zeit. 10: 772, 1852
BR 25: misc.139 (1839) / 26: t.54 (1840)
Type: Brazil, without collector (K-Lindl., holo!)
Brazil – E
o Rodriguezia crispa Lindl., Bot. Reg. 25: misc.139, 1839
Gomesa undulata Hoffmsgg., Verz. Orch.: 52, 1843
Odontoglossum crispulum Rchb.f., Wawra Bot. Erg. Reise Maximil. Bras. 6: 853, 1864

Gomesa laxiflora (Lindl.) Klotzsch & Rchb.f., Bot. Zeit. 10: 772, 1852
BR 25: misc.138 (1839)
Type: Brazil, cult. Bateman (K-Lindl., holo!)
Brazil – E
Rodriguezia laxiflora Lindl., Bot. Reg. 25: misc.138, 1839 (non R.Br.)
Gomesa chrysostoma Hoffmsgg., Verz. Orch.: 50, 1843
Odontoglossum laxiflorum (Lindl.) Rchb.f., Walp. Ann. 6: 853, 1864

o **Gongora bufonia** Lindl., Bot. Reg. 27: t.2, 1841
BR 27: t.2 (1841) / 33: t.17 & sub t.17:4 (1847)
Type: Brazil, Rucker s.n. (K-Lindl., holo!)
Brazil – E
Gongora irrorata Hoffmsgg., Verz. d. Orch.: 53, 1843
o Gongora bufonia var. leucochila (Lemaire) Lindl., Bot. Reg. 33: t.17, 1847
Gongora minax Schltr., Hoehne Fl. Bras. 12 (6): 196, 1942

Gongora galeata (Lindl.) Rchb.f., Xen. Orch. 1: 51, 1854
BR 18: t.1538 (1832) / 29: misc.13 (1843)
Mexico – E
Maxillaria galeata Lindl. ex Lodd., Bot. Cab. 17: t.1645, 1830
Cirrhaea loddigesii Lindl., Bot. Reg. 18: t.1538, 1832
Acropera loddigesii (Lindl.) Lindl., Gen. Sp. Orch. Pl.: 172, 1833

Gongora galeottiana A.Rich, Ann. Sc. Nat. 3.s., 3: 25, 1845
BR 33: sub t.17:6 (1847)
Mexico – E

Gongora quinquenervis Ruiz Lopez & Pavon, Syst. Veg. Fl. Peruv. & Chil.: 227, 1798
BR 19: t.1616 (1833) / 25: t.51 & misc.86 (1839) / 27: misc.4 & misc. 101 (1841) / 30: misc.30 (1844) / 33: t.17 & sub t.17:2, 3, 7 & t.69 (1847)
Mexico, Guatemala, Honduras, Nicaragua, Costa Rica, Panama, S.America – E
Gongora atropurpurea Hook., Exot. Fl. 3: t.178, 1827
o Gongora maculata Lindl., Bot. Reg. 19: t.1616, 1833
o Gongora fulva Lindl., Bot. Reg. 25: t.51, 1839
Gongora nigrita Lindl., Bot. Reg. 25: misc.86, 1839
Gongora fulva var. vitellina Lindl., Bot. Reg. 27: misc.4, 1841
o Gongora maculata var. alba Lindl., Bot. Reg. 27: misc.101, 1841
Gongora leucochila Lemaire, Fl. Serres 1: 207, 1845
Gongora maculata var. tricolor Lindl., Bot. Reg. 33: t.69, 1847
Gongora odoratissima Lemaire, Fl. Serres 2: 229, 1847
Gongora boothiana hort. ex Allg. Gartenz. 20: 376, 1852
Gongora jenischii hort., Allg. Gartenz. 23: 347, 1853
Gongora histrionica Rchb.f., Bonpl. 2: 92, 1854
Gongora quadricornis hort. ex Rchb.f., Xen. Orch. 1: 53, 1854
Gongora quinquevulneris Beer, Prakt. Orch.: 252, 1854
Gongora retrorsa Rchb.f., Bonpl. 2: 19, 1854
Gongora sherpherdii hort. ex Rchb.f., Xen. Orch.: 1: 53, 1854
Gongora tricolor (Lindl.) Rchb.f., Bonpl. 2: 93, 1854
Gongora vitellina hort. ex Rchb.f., Xen. Orch. 1: 54, 1854
Gongora odoratissima var. jenischii hort. ex Kramer, Cat. Orch. Senat. Jenisch, 1855
Gongora heisteri hort. ex Rchb.f., Walp. Ann. 6: 596, 1863
Gongora superflua Rchb.f., Xen. Orch. 2: 169, 1873
Cirrhaea atropurpurea hort. ex Stein, Orchideenb.: 259, 1892
Gongora powellii Schltr., Fedde Rep. Beih. 17: 62, 1922
Gongora unicolor Schltr., Fedde Rep. Beih. 19: 229, 1923

Gongora maculata f. odoratissima (Lemaire) Dodson, Agentes de Pollinizacion Univ. Nac. Amaz. Peruana: 117, 1965

o **Gongora truncata** Lindl., Bot. Reg. 29: misc.52, 1843
BR 29: misc.52 (1843) / 31: t.56 (1845) / 33: sub t.17:5 (1847)
Type: Mexico, Linden s.n. cult. Rucker (K-Lindl., holo!)
Mexico – E
Gongora donckelaariana Lemaire, Fl. Serres 4: 336, 1848
Gongora seideliana Rchb.f., Bot. Zeit. 10: 637, 1852
Gongora truncata var. alba Nash, Addisonia 2: t.46, 1917

Goodyera procera (Ker Gawler) Hook., Exot. Fl. 3: t.39, 1823
BR 8: t.639 (1822) / 10: sub t.867:notes (1825)
Type: Nepal, Lambert's Herbarium, cult. Colvill (BM – holo!)
Sri Lanka, India, Himalaya, Sikkim, Bhutan to Assam, Burma, Laos, Cambodia, Vietnam, China, Taiwan, Ryukyu Is., Philippines, Java, Sumatra – T, E
o Neottia procera Ker Gawler, Bot. Reg. 8: t.639, 1822
Neottia parviflora Blume, Bijdr.: 408, 1825
Cionosaccus lanceolatus Breda, Gen. & Sp. Orch.: t.8, 1827
Goodyera carnea A.Rich., Ann. Sc. Nat. s.2, 15: 80, 1841
Cordylestyllis foliosa Falc., J. Bot. 4: 75, 1842
Leucostachys procera (Ker Gawler) Hoffmsgg., Preisv. Orch.: 26, 1842
Spiranthes parviflora (Blume) Hassk., Cat. Hort. Bogor : 47, 1844
Goodyera parviflora (Blume) Bl., Coll. Orch. Arch. Ind.: 34 t.22, 1858
Goodyera lancifolia Fr. & Sav., Enum. Pl. Jap. 2: 520, 1879
Orchiodes parviflorum (Blume) Kuntze, Rev. Gen. 2: 675, 1891
Orchiodes procerum (Ker Gawler) Kuntze, Rev. Gen. 2: 675, 1891
Epipactis philippinensis Ames, Orch. 2: 60, 1908
Epipactis procera (Ker Gawler) Eaton, Proc. Biol. Soc. Wash. 21: 65, 1908
Goodyera philippinensis (Ames) Schltr., Engl. Bot. Jahrb. 45: 392, 1911
Peramium procerum (Ker Gawler) Makino, J. Jap. Bot. 6: 36, 1929

Goodyera rubicunda (Blume) Lindl., Bot. Reg. 25: misc.92, 1839
BR 25: misc.92 (1839)
Philippines, Taiwan, Thailand, Malaysia, Sumatra, Java, N.Australia, New Caledonia, New Guinea, Vanuatu, Fiji, Samoa – T
Neottia grandis Blume, Bijdr. 1 (8): 407, 1825
Neottia rubicunda Blume, Bijdr. 1 (8): 408, 1825
Spiranthes grandis (Blume) Hassk., Cat. Hort. Bog. Alt.: 47, 1844
Aetheria rubicunda (Blume) Rchb.f., Bonpl. 3: 214, 1855
Goodyera celebica Blume, Fl. Jav. n.s.1: 36, 1858
Goodyera grandis (Blume) Blume, Fl. Java Orch.: 36, 1858
Goodyera rubens Blume, Fl. Jav. n.s.1: 36, 1858
Rhamphidia rubicunda (Blume) Rchb.f., Seemann Fl. Vit.: 294, 1868
Goodyera papuana Ridley, J. Bot. 24: 355, 1886
Orchiodes celebicum (Blume) Kuntze, Rev. Gen. 2: 675, 1891
Orchiodes grande (Blume) Kuntze, Rev. Gen. 2: 675, 1891
Orchiodes rubicundum (Blume) Kuntze, Rev. Gen. 2: 675, 1891
Hetaeria rubicunda (Blume) Benth. & Hook.f., Drake, Ill. Ins. Pac.: 312, 1892
Goodyera grandis King & Pantl., Ann. Bot. Gard. Calc. 8: 284, 1898
Epipactis grandis (King & Prantl.) Eaton, Proc. Biol. Soc. Wash. 11: 64, 1908
Epipactis rubicunda (Blume) Eaton, Proc. Biol. Soc. Wash. 11: 64, 1908
Goodyera rubicunda var. celebica (Blume) Schltr., Fedde Rep. 10: 9, 1911
Goodyera hispidula Rogers & White, Trans. Roy. Soc. S.Austr. 44: 112, 1920
Goodyera longibracteata Hayata, Ic. Pl. Formosa 4: 114, 1924
Peramium longibracteatum (Hayata) Makino, J. Jap. Bot. 6: 34, 1929
Goodyera yaeyamae Ohwi, J. Jap. Bot. 13: 439, 1937

o **Govenia liliacea** (Llave & Lex.) Lindl., Bot. Reg. 21: sub t.1795, 1836
BR 21: sub t.1795 (1835) / 24: t.13 (1838) / 31: sub t.67:5, 6, 7 (1845)
Mexico – T
Maxillaria liliacea Llave & Lex., Orch. Mex. 2: 12, 1824
Eucnemis brevilabris Lindl., Gen. Sp. Orch. Pl.: 161, 1832
Govenia capitata Lindl., Bot. Reg. 21: sub t.1795, 1835
Govenia alba A.Rich. & Galeotti, Ann. Sc. Nat. s.3, 3: 25, 1845
Govenia deliciosa Rchb.f., Bot. Zeit. 10: 836, 1852
Govenia brevilabris (Lindl.) Rchb.f., Beitr. Orch. Cent.-Am. 3: 249, 1883
Govenia utriculata var. capitata Correll, Lloydia 10: 226, 1947

Govenia pauciflora Lindl., Ann. Mag. Nat. Hist. 10: 184, 1842
BR 31: sub t.67:8 (1845)
Mexico, S.Guatemala – T
Govenia elliptica S.Watson, Proc. Am. Acad. 26: 153, 1891
Govenia superba var. elliptica (S.Watson) Correll, Lloydia 10: 224, 1947

o **Govenia superba** (Llave & Lex.) Lindl. ex Lodd., Bot. Cab. 18: t.1709, 1831
BR 21: t.1795 (1835) / 25: misc.66 (1839) / 29: misc.107 (1843) / 31: t.67 & sub t.67:1, 2, 3 (1845)
Mexico to Panama, Venezuela, Colombia – up to 2700 m – T
Maxillaria superba LLave & Lex., Nov. Veg. Descr. 2: 13, 1825
Govenia lagenophora Lindl., Bot. Reg. 25: misc.66, 1839
o Govenia fasciata Lindl., Bot. Reg. 29: misc.107, 1843
Govenia sulphurea Rchb.f., Gard. Chron. n.s.24: 70, 1885
Govenia platyglossa Schltr., Fedde Rep. Beih. 7: 157, 1920
Govenia stictoglossa Schltr., Fedde Rep. Beih. 27: 79, 1924

Govenia tingens Poeppig & Endl., Nov. Gen. Sp. Pl. 2: 5 t.107, 1837–38
BR 31: sub t.67:9 (1845)
Venezuela, Brazil, Amazonas, Peru, Bolivia, Argentina – 1800–2500 m – T

Govenia utriculata (Sw.) Lindl., Bot. Mag. 65: t.3660, 1838
BR 25: misc.51 (1839) / 31: sub t.67:4 (1845)
Mexico to Bolivia, Argentina, W.Indies – T
Limodorum utriculatum Sw., Prodr. Veg. Ind. Occ.: 119, 1788
Cymbidium utriculatum (Sw.) Sw., Nov. Act. Upsal. 6: 75, 1799
Govenia gardneri Hook., Bot. Mag. 66: t.3660, 1839
Govenia boliviense Rolfe, Mem. Torrey Club 4: 263, 1895
Govenia ernstii Schltr., Fedde Rep. Beih. 6: 43, 1919
Govenia sodiroi Schltr., Fedde Rep. Beih. 8: 91, 1921

o **Grammatophyllum scriptum** (L.) Blume, Bijdr.: 378, 1825
BR 24: misc.80 (1838) / 25: t.65 (1839) / 28: t.69 (1842)
Philippines, Borneo, Celebes, Moluccas, New Guinea, Solomon Is. – E
Epidendrum scriptum L., Sp. Pl. ed.2., 2: 1351, 1763
Cymbidium scriptum (L.) Sw., Schrad Neues J. Bot. 2: 128, 1799
Gabertia scripta (L.) Gaudich., Freyc. Voy. Bot.: 425, 1826
Vanda scripta (L.) Sprengel, Syst. Veg. 3: 719, 1826
Grammatophyllum multiflorum Lindl., Bot. Reg. 24: misc.80, 1838
Grammatophyllum multiflorum var. tigrinum Lindl., Bot. Reg. 28: t.69, 1842
Grammatophyllum fenzlianum Rchb.f., Sp. Pl. ed.2, 1867
Grammatophyllum rumphianum Miq., Ann. Mus. Bot. Ludg. Bat. 4: 219, 1869
Cymbidium goweri F.Muell., Wing South Sc. Rec. Aug., 1883
Grammatophyllum leopardinum Rchb.f., Flora 81: 151, 1888
Grammatophyllum measuresianum Weath., Gard. & Forest 2: 524, 1889
Grammatophyllum seegerianum hort., Gard. Chron. s.3, 10: 49, 1891
Grammatophyllum guilelmi-secundii Kraenzlin, Gartenfl.: 114, 1894

Graphorkis lurida (Sw.) Kuntze, Rev. Gen. 2: 662, 1891
BR 21: t.1821 (1835)
Senegal, Guinea-Bissau, Guinea, Sierra Leone, Ivory Coast, Ghana, Nigeria, Cameroun, F.Po, Gabon, Congo, Uganda, Burundi – E
Limodorum luridum Sw., Schrad Neues J. Bot. 1: 87, 1805
o Eulophia lurida (Sw.) Lindl., Gen. Sp. Orch. Pl.: 182, 1833
Eulophiopsis lurida (Sw.) Schltr., Die Orchideen: 346, 1914

o **Grobya amherstiae** Lindl., Bot. Reg. 20: t.1740, 1835
BR 20: t.1740 (1835)
Type: Brazil, Hayne s.n. cult. Amherst (K-Lindl., holo!)
Brazil – E

Grobya galeata Lindl., Bot. Reg. 26: misc.197, 1840
BR 26: misc.197 (1840)
Type: Brazil, cult. Barker (K-Lindl., holo!)
Brazil – E

Habenaria candida Lindl., Bot. Reg. 30: misc.77, 1844
BR 30: misc.77 (1844)
Type: Sierra Leone, cult. Loddiges (K-Lindl., holo!)
Sierra-Leone – T

o **Habenaria procera** (Sw.) Lindl., Gen. Sp. Orch. Pl.: 318, 1835
BR 22: t.1858 (1836)
Sierra Leone, Ivory Coast, Ghana, Nigeria, Cameroun, Gabon, Centr. African Republic, Zaire, Uganda – T, E, L
Orchis procera Sw., Schrad Neues J. Bot. 1: 22, 1805

Habenaria reniformis (D.Don) Hook.f., Fl. Brit. Ind. 6: 152, 1890
BR 18: sub t.1499 (1832) / 20: sub t.1701:30 (1834)
Nepal, India, Thailand, Cambodia, Vietnam, China – T
Listera reniformis D.Don, Prodr. Fl. Nepal.: 28, 1825
Neottia reniformis (D.Don) Sprengel, Syst. Veg. 3: 707, 1826
Aopla reniformis (D.Don) Lindl., Bot. Reg. 20: sub t.1701:30, 1834
Herminium reniforme (D.Don) Lindl., Gen. Sp. Orch. Pl.: 304, 1835
Habenaria clovisii Gagnepain, Bull. Soc. Bot. Fr. 78: 68, 1931

Habenaria tetrapetala (Lindl.) Rchb.f., Flora: 180, 1865
BR 20: sub t.1701:23 (1834)
S.Africa – T
Bilabrela falcicornis Lindl., Bot. Reg. 20: sub t.1701:23, 1834

Bonatea tetrapetala Lindl., Hook. Comp. Bot. Mag. 2: 208, 1836
Habenaria tetramera Bolus, J. Linn. Soc. 19: 340, 1882

Helcia sanguinolenta Lindl., Bot. Reg. 31: misc.27, 1845
BR 31: misc.27 (1845)
Type: Peru, Paccha, Hartweg s.n. cult. Hort. Soc. (K-Lindl., holo!)
Ecuador, Peru – up to 3000 m – E
Trichopilia sanguinolenta (Lindl.) Rchb.f., Xen. Orch. 2: 99, 1867

Herminium lanceum (Thunb. ex Sw.) Vuijk, Blumea 11 (1): 228, 1961
BR 18: sub t.1525 (1832)
Himalaya, S.China, Taiwan, Japan, Korea, Thailand, Vietnam, Malaya, Java, Timor, Celebes, Philippines – E
Ophrys lancea Thunb. ex Sw., Kongl. Akad. Vet. Nya Handl. 21: 223, 1800
Satyrium lanceum (Thunb. ex Sw.) Pers., Syn. 2: 507, 1807
Aceras angustifolia Lindl., Bot. Reg. 18: sub t.1525, 1832
Aceras longicruris C.Wright ex A.Gray, Mem. Acad. Phil. s.2, 6: 411, 1858
Aceras angustifolia var. longicruris (C.Wright ex A.Gray) Miq., Ann. Mus. Bot. Ludg. Bat. 2: 207, 1866
Platanthera angustifolium (Lindl.) Rchb.f., Otia Bot. Hamb.: 39, 1878
Herminium angustifolium (Lindl.) Benth. & Hook.f., Gen. Pl. 3: 622, 1880
Herminium angustifolium var. longicrure (C.Wright ex A.Gray) Makino, Bot. Mag. Tokyo 10: 109, 1896
Herminium alaschanicum Maxim. ex Kraenzlin, Orch. Gen. & Sp. 1: 533, 1901
Herminium angustifolia var. brevilabre Tang & Wang, Act. Phytotax. Sin. 1 (1): 61, 1951
Herminium longicruris (C.Wright ex A.Gray) Tang & Wang, Act. Phytotax. Sin. 1 (1): 61, 1951
Spiranthes lancea (Thunb. ex Sw.) Backer, Blumea 6 (2): 361, 1959
Herminium lanceum var. longicrure (C.Wright ex Gray) Hara, J. Jap. Bot. 44: 60, 1969

Herminium macrophyllum (D.Don) Dandy, J. Bot. 70: 328, 1932
BR 18: sub t.1499 (1832)
Himalaya, Nepal to Bhutan, S.E.Tibet – T
Neottia macrophylla D.Don, Prodr. Fl. Nepal.: 27, 1825
Spiranthes macrophylla (D.Don) Sprengel, Syst. Veg. 3: 708, 1826
Herminium congestum Lindl., Bot. Reg. 18: sub t.1499, 1832

Herminium monorchis (L.) R.Br., Aiton Hort. Kew. ed.2, 5: 191, 1813
BR 18: sub t.1499 (1832)
Europe, Centr. W.Asia, Himalaya (Kashmir to Nepal), E.Tibet, Siberia, W.N.China, Korea, Japan – T
Ophrys monorchis L., Sp. Pl.: 947, 1753

Herschelia coelestis Lindl., Gen. Sp. Orch. Pl.: 363, 1838
BR 9: sub t.700 (1823)
S.Africa – T
Disa graminifolia Ker Gawler, Brande J. 6: 44 t.1, 1819
Herschelia graminifolia (Ker Gawler) Durand & Schinz, Conspect. Fl. Afr. 5: 111, 1895

Herschelianthe spathulata (L.f.) Rauschert, Fedde Rep. 94 (7/8): 435, 1983
BR 9: sub t.700 (1823)
S.Africa – T
Orchis spathulata L.f., Suppl. Pl.: 398, 1781
Satyrium spathulatum (L.f.) Thunb., Prodr. 5, 1794
Disa spathulata (L.f.) Sw., Kongl. Vet. Acad. Handl. 21: 213, 1800
Disa atropurpurea Sonder, Linnaea 19: 96, 1846
Disa propinqua Sonder, Linnaea 19: 95, 1846
Disa propinqua var. trifida Sonder, Linnaea 19: 96, 1846
Disa spathulata var. atropurpurea (Sonder) Schltr., Engl. Bot. Jahrb. 31: 284, 1901
Herschelia atropurpurea (Sonder) Rolfe, Fl. Cap. 5 (3): 205, 1913
Herschelia spathulata (L.f.) Rolfe, Fl. Cap. 5 (3): 205, 1913
Herschelianthe atropurpurea (Sonder) Rauschert, Fedde Rep. 94 (7/8): 434, 1983

Hexadesmia bicornis Lindl., Bot. Reg. 30: misc.44, 1844
BR 30: misc.44 (1844)
Type: Colombia, cult. Loddiges (K-Lindl., holo!)
Colombia – E

Hexadesmia crurigera (Bateman ex Lindl.) Lindl., Bot. Reg. 30: misc.4, 1844
BR 26: misc.90 (1840) / 30: misc.4 (1844)
Type: Guatemala, cult. Bateman (K-Lindl., holo!)
Mexico, Guatemala, El Salvador, Honduras, Nicaragua, Costa Rica, Ecuador – E
Hexopia crurigera Bateman ex Lindl., Bot. Reg. 26: misc.90, 1840
Scaphyglottis crurigera (Bateman ex Lindl.) Ames & Correll, Bot. Mus. Leafl. Harv. Univ. 10: 85, 1942

Hexadesmia fasciculata Brongn., Ann. Sc. Nat. Bot. s.2, 17: 44, 1842
BR 28: misc.46 (1842) / 29: misc.21 (1843)
Mexico, Guatemala, Honduras, Nicaragua, Costa Rica, Panama – E
Hexadesmia lindeniana A.Rich. & Galeotti, Ann. Sc. Nat. Bot. s.3, 3: 23, 1845
Hexadesmia rhodoglossa Rchb.f., Bonpl. 4: 326, 1856
Hexadesmia pachybulbon Schltr., Fedde Rep. Beih. 17: 26, 1922
Hexadesmia rigidipes Schltr., Fedde Rep. Beih. 19: 206, 1923
Scaphyglottis lindeniana (A.Rich. & Galeotti) L.O.Williams, Ann. Missouri Bot. Gard. 18: 423, 1941

Hexadesmia micrantha Lindl., Bot. Reg. 30: misc.5, 1844
BR 30: misc.5 (1844)
Type: Guatemala, Skinner s.n. cult. Loddiges (K-Lindl., holo!)
Guatemala, El Salvador, Honduras, Nicaragua, Costa Rica, Panama – E
Scaphyglottis micrantha (Lindl.) Ames & Correll, Bot. Mus. Leafl. Harv. Univ. 10: 85, 1942

Holothrix burchelli (Lindl.) Rchb.f., Otia Bot. Hamb.: 119, 1881
BR 20: sub t.1701:6 (1834)
Type: S.Africa, Burchell s.n. (K-Lindl., holo!)
S.Africa (Cape) – T
Scopularia burchelli Lindl., Bot. Reg. 20: sub t.1701:6, 1834
Scopularia secunda Lindl., Hook. Comp. Bot. Mag. 1: 207, 1836
Holothrix scopularia Rchb.f., Otia Bot. Hamb.: 119, 1881

Holothrix monotris Rchb.f., Otia Bot. Hamb.: 119, 1881
BR 20: sub t.1701:10 (1834)
S.Africa (Cape)
Orchidea hispida Burchell, Cat. Geogr.: N.6369, unpublished manuscript at Kew!
Monotris secunda Lindl., Bot. Reg. 20: sub t.1701:10, 1834

Homalopetalum pumilio (Rchb.f.) Schltr., Fedde Rep. Beih. 19: 48, 1923
BR 32: sub t.53 (1846)
Mexico, Guatemala, Costa Rica – up to 2000 m – E, L
Brassavola pumilio Rchb.f., Linnaea 18: 402, 1844
Pinelia tuerckheimii Kraenzlin, Ann. Naturhist. Mus. 44: 326, 1930

Houlletia brocklehurstiana Lindl., Bot. Reg. 27: misc.99, 1841
BR 27: misc.28 & misc.99 (1841)
Type: Brazil, cult. T.Brocklehurst (K-Lindl., holo!)
Brazil – E
Houlletia stapeliaeflora Brongn., Ann. Sc. Nat. s.2, 15: 37, 1841
Maxillaria brocklehurstiana Lindl., Bot. Reg. 27: misc.28, 1841

o **Huntleya meleagris** Lindl., Bot. Reg. 23: sub t.1991, 1837
BR 23: sub t.1991 (1837) / 24: misc.20 (1838) / 25: t.14 (1839)
Type: cult. Rollisson (K-Lindl., holo!)
Costa Rica, Panama, Venezuela, Trinidad, Guyana, Ecuador, Brazil – E
Batemannia meleagris (Lindl.) Rchb.f., Bonpl. 3: 217, 1855
Huntleya albidofulva Lemaire, Ill. Hort. 15: t.556, 1868
Batemannia burtii Endr. & Rchb.f., Gard. Chron.: 1099, 1872
Batemannia burtii var. wallisii Rchb.f., Gard. Chron.: 575, 1873
Batemannia meleagris var. albidofulva (Lemaire) Rchb.f. ex Du Buyss., L'Orchidoph.: 351, 1878
Batemannia wallisii Rchb.f., Gard. Chron. n.s.13: 776, 1880
Zygopetalum meleagris (Lindl.) Benth., J. Linn. Soc. Bot. 18: 321, 1881
Batemannia wallisii var. major Rchb.f. ex Roezl, L'Orchidoph. 3: 477, 1883
Zygopetalum burtii (Endr. & Rchb.f.) Benth. & Hook. ex Hemsl., Biol. Centr. Am. Bot. 3: 251, 1883
Zygopetalum meleagris var. albidofulva (Lemaire) Nicholson, Dict. Gard. 4, 7: 246, 1888
Huntleya burtii (Endr. & Rchb.f.) Pfitzer, Engl. Pflanzenf. 2 (6): 205, 1889
Zygopetalum burtii var. wallisii (Rchb.f.) Veitch, Man. Orch. 9: 45, 1893
Huntleya meleagris var. albidofulva (Lemaire) Cogn., Mart. Fl. Bras. 3 (6): 479, 1906
Huntleya wallisii (Rchb.f.) Rolfe, Orchid Rev. 24: 236, 1916

Ionopsis satyrioides (Sw.) Rchb.f., Walp. Ann. 6: 683, 1861
BR 24: misc.181 (1838)
W.Indies, Colombia, Venezuela, Guyana – E
Epidendrum satyrioides Sw., Prodr.: 123, 1788
Dendrobium testiculatum Sw., Nov. Act. Upsal. 6: 83, 1799
Cybelion testiculatum (Sw.) Sprengel, Syst. Veg. 3: 721, 1826
Ionopsis testiculata (Sw.) Lindl., Gen. Sp. Orch. Pl.: 193, 1833
Ionopsis teres Lindl., Bot. Reg. 24: misc.181, 1838
Ionopsis pusilla Barb.Rodr., Contr. Jard. Bot. Rio Janeiro 4: 102, 1907

o **Ionopsis utricularioides** (Sw.) Lindl., Coll. Bot.: t.39 A, 1921
BR 22: t. & sub t.1904 (1836)
USA (Florida), W.Indies, Mexico to Brazil, Peru, Paraguay – E
Epidendrum utricularioides Sw., Prodr. Veg. Ind. Occ.: 122, 1788
Dendrobium utricularioides (Sw.) Sw., Nov. Act. Upsal. 6: 38, 1799
Ionopsis pulchella Kunth, Nov. Gen. & Sp. Pl. 1: 348 t.83, 1816

Iantha pallidiflora Hook., Exot. Fl. 2: t.113, 1824
Cybelion pallidiflorum (Hook.) Sprengel, Syst. Veg. 3: 721, 1826
Epidendrum crenatum Vell., Fl. Flum. 9: t.6, 1836
Ionopsis paniculata Lindl., Bot. Reg. 22: sub t.1904, 1836
o Ionopsis tenera Lindl., Bot. Reg. 22: t.1904, 1836
Cybelion tenerum (Lindl.) Steudel, Nom. Bot. ed.2, 1: 458, 1840
Cybelion utriculariae (Sw.) Sprengel, Nom. Bot. ed.2, 1: 458, 1840
Ionopsis gardneri Lindl., Paxt. Flow. Gard. 2: 13, 1851
Ionopsis pallidiflora (Hook.) Lindl., Paxt. Flow. Gard. 2: 12, 1851
Ionopsis zonalis Lindl. & Paxton, Paxt. Flow. Gard. 2: 13, 1851
Epidendrum calcaratum Sessé & Mociño, Fl. Mex. ed.2: 201, 1894

o **Isochilus linearis** (Jacq.) R.Br., Ait. Hort. Kew. ed.2, 5: 209, 1813
BR 9: t.745 (1823)
Mexico, Belize, Guatemala, El Salvador, Honduras, Nicaragua, Costa Rica, W.Indies, S.America – E
Epidendrum lineare Jacq., Enum. Pl. Carib.: 29, 1760 (non Ruiz Lopez & Pavon)
Cymbidium lineare (Jacq.) Sw., Nov. Act. Upsal. 6: 72, 1799
Leptothrium lineare (Jacq.) Kunth., Nom. Bot. ed.2, 2: 32, 1840
Isochilus leucanthus Barb.Rodr., Gen. & Sp. Orch. Nov. 1: 47, 1877
Isochilus linearis var. leucanthum (Barb.Rodr.) Cogn., Mart. Fl. Bras. 3 (5): 4, 1898
Isochilus pauciflorus Cogn., Symb. Antill. 6: 459, 1910
Isochilus langlassei Schltr., Fedde Rep. 16: 442, 1920
Isochilus brasiliensis Schltr., Fedde Rep. Beih. 9: 80, 1921
Isochilus peruvianus Schltr., Fedde Rep. Beih. 9: 79, 1921

Kefersteinia graminea (Lindl.) Rchb.f., Bot. Zeit. 10: 634, 1852
BR 30: misc.15:8 (1844)
Type: Popayan, Hartweg s.n. (K-Lindl., holo!)
Venezuela, Colombia, Ecuador, Peru – E
Zygopetalum gramineum Lindl., Bot. Reg. 30: misc.15:8, 1844

Koellensteinia graminea (Lindl.) Rchb.f., Walp. Ann. 6: 552, 1863
BR 21: sub t.1802 (1835) / 29: misc.p.13:5 (1843)
Type: Guyana, cult. Loddiges (K-Lindl., holo!)
Venezuela, Colombia, Peru, Bolivia, Brazil, Surinam, Guyana – E
Maxillaria graminea Lindl., Bot. Reg. 21: sub t.1802, 1836
Promenaea graminea (Lindl.) Lindl., Bot. Reg. 29: misc.p.13: 5, 1843
Aganisia graminea (Lindl.) Benth. & Hook.f., J. Linn. Soc. 28: 321, 1880

Koellensteinia tricolor (Lindl.) Rchb.f., Walp. Ann. 6: 552, 1863
BR 32: sub t.64 (1846)
Type: Guyana, cult. Loddiges (K-Lindl., holo!)
Brazil, Guyana, Peru – E
Zygopetalum tricolor Lindl., Bot. Reg. 32: sub t.64, 1846

o **Lacaena bicolor** Lindl., Bot. Reg. 29: misc.101, 1843
BR 29: misc.101 (1843) / 30: t.50 (1844)
Type: Guatemala, Hartweg 1026 cult. Hort. Soc. (K-Lindl., holo!)
Mexico, Honduras, Costa Rica – E
Peristeria longiscapa A.Rich. & Galeotti, Ann. Sc. Nat. s.3, 3: 25, 1845
Lacaena bicolor var. glabrata Lemaire, Ill. Hort. 1: t.33 & misc.p.56, 1854
Acineta wrightii J.Fraser, Gard. World: 673, 1889
Lueddemannia sanderiana Kraenzlin, Gard. Chron. s.3, 22: 138, 1897
Acineta hrubyana hort., Gard. Chron. s.3, 41: 427, 1907 (non Rchb.f.)
Lacaena bicolor var. alba Rolfe, Orch. Rev. 18: 208, 1910

o **Laelia albida** Lindl. & Bateman, Bot. Reg. 25: t.54, 1839
BR 25: t.54 & misc.4 (1839) / 27: misc.21 (1841) / 28: sub t.62:11 (1842) / 29: misc.16 (1843)
Type: Mexico, Karwinski s.n. cult. Hartweg (K-Lindl., holo!)
Mexico – 1200–2400 m – E
Laelia discolor A.Rich. & Galeotti, Ann. Sc. Nat. s.3, 3: 23, 1845
Cattleya albida (Lindl. & Bateman) Beer, Prakt. Orch.: 208, 1854
Bletia albida (Lindl. & Bateman) Rchb.f., Walp. Ann. 6: 428, 1862

o **Laelia anceps** Lindl., Bot. Reg. 21: t.1751, 1835
BR 21: t.1751 (1835) / 23: t.1947 (1837) / 28: sub t.62:4 (1842)
Type: Mexico, cult. Loddiges (K-Lindl., holo!)
Mexico, Honduras – E, L
o Laelia anceps var. barkeriana Lindl., Bot. Reg. 23: t.1947, 1837
Laelia barkeriana Knowles & Westc., Fl. Cabin.: 63, 1837
Amalias anceps (Lindl.) Hoffsgg., Linnaea 16: 228, 1842
Cattleya anceps (Lindl.) Beer, Prakt. Stud. Orch.: 208, 1854
Bletia anceps (Lindl.) Rchb.f., Walp. Ann. 6: 418, 1862

o **Laelia autumnalis** Lindl., Gen. Sp. Orch. Pl.: 115, 1831
BR 25: t.27 (1839) / 28: sub t.62:3 (1842)
Mexico, Guatemala – 1500–2400 m – E, L
Bletia autumnalis (Lindl.) Llave & Lex., Nov. Veg. Descr.: 19, 1825
Cattleya autumnalis (Lindl.) Beer, Prakt. Orch.: 208, 1854
Laelia gouldiana Rchb.f., Gard. Chron. s.3, 3: 41, 1888

Laelia cinnabarina Bateman ex Lindl., Sert. Orch.: t.28, 1838
BR 28: sub t.62:13 (1842)
Brazil – E, L
Amalias cinnabarina (Bateman ex Lindl.) Hoffsgg., Verz. Orch.: 37, 1842
Cattleya cinnabarina (Bateman ex Lindl.) Beer, Prakt. Stud. Orch.: 209, 1854
Bletia cinnabarina (Bateman ex Lindl.) Rchb.f., Walp. Ann. 6: 430, 1862

Laelia crispa (Lindl.) Rchb.f., Fl. Serres 9: 102, 1853
BR 14: t.1172 (1828) / 34: sub t.5:6 (1844)
Type: Brazil, Chamberlayne s.n. cult.Chiswick Garden (K-Lindl., holo!)
Brazil – 750–1100 m – E, L
o Cattleya crispa Lindl., Bot. Reg. 14: t.1172, 1828
Bletia crispa (Lindl.) Rchb.f., Walp. Ann. 6: 423, 1862

Laelia crispata (Thunb.) Garay, Arq. Bot. Est. S.Paulo n.s. Form. maior 2: 163, 1952
BR 28: sub t.62:14 (1842)
Brazil – E
Laelia rupestris Lindl., Bot. Reg. 28: sub t.62:14, 1842
Bletia rupestris (Lindl.) Rchb.f., Walp. Ann. 6: 431, 1862
Laelia tereticaulis Hoehne, Arq. Bot. Est. S.Paulo n.s. Form. maior 2: 163, 1952

o **Laelia flava** Lindl., Bot. Reg. 25: misc.143, 1839
BR 25: misc.143 (1839) / 27: sub t.1 (1841) / 28: t.62 & sub t.62: 12 (1842)
Type: Rucker s.n. cult. C.Lemon (K-Lindl., holo!)
Brazil – E
Laelia caulescens Lindl., Bot. Reg. 27: sub t.1, 1841
Cattleya lutea Beer, Prakt. Stud. Orch.: 210, 1854
Bletia lutea (Beer) Rchb.f., Xen. Orch. 2: 60, 1863

o **Laelia furfuracea** Lindl., Bot. Reg. 25: t.26, 1839
BR 25: t.26 (1839) / 28: sub t.62:6 (1842)
Type: Mexico, Karwinski s.n. cult. Barker (K-Lindl., holo!)
Mexico – 220–3000 m – E
Cattleya furfuracea (Lindl.) Beer, Prakt. Stud. Orch.: 210, 1854
Bletia furfuracea (Lindl.) Rchb.f., Walp. Ann. 6: 428, 1862

Laelia perrinii Lindl., Bot. Reg. 28: sub t.62, 1842
BR 24: t.2 (1838) / 28: sub t.62:7 (1842)
Type: Brazil, cult. Knight (K-Lindl., holo!)
Brazil – E
o Cattleya perrinii Lindl., Bot. Reg. 24: t.2, 1838
Cattleya intermedia var. angustifolia Hook., Bot. Mag. 66: t.3711, 1839
Cattleya integerrima var. angustifolia Hook., Gen. Orch. 33: t.30, 1846
Bletia perrinii (Lindl.) Rchb.f., Walp. Ann. 6: 421, 1862

Laelia pumila (Hook.) Rchb.f., Fl. Serres 9: 102, 1853
BR 30: t.5 & sub t.5:9 (1844)
Brazil – 600–900 m – E
o Cattleya pumila Hook., Bot. Mag. 65: t.3656, 1838
Cattleya marginata Paxton, Mag. Bot. 10: 265 t., 1843
Cattleya pinellii Lindl., Bot. Reg. 29: sub t.5:9, 1844
Cattleya spectabilis Paxton, The Florist 3: 89 & 91 cum t., 1850
Cattleya pinellii var. marginata Beer, Prakt. Stud. Orch.: 214, 1854
Laelia praestans Rchb.f., Allg. Gartenz. 25: 336, 1857
Bletia praestans (Rchb.f.) Rchb.f., Walp. Ann. 6: 425, 1862
Bletia pumila (Hook) Rchb.f., Walp. Ann. 6: 421, 1862

Laelia rubescens Lindl., Bot. Reg. 26: misc.25, 1840
BR 26: t.41 & misc.25 (1840) / 27: t.24 & misc.42 (1841) / 28: sub t. 62:8, 9, 10 & misc.10 (1842) / 31: t.69 (1845)
Type: Mexico, Barker s.n. cult. Knight (K-Lindl., holo!)
Mexico, Guatemala, El Salvador, Nicaragua, Costa Rica, Panama – E, L
o Laelia acuminata Lindl., Bot. Reg. 27: t. 24, 1841
o Laelia peduncularis Lindl., Bot. Reg. 28: misc.10, 1842
Laelia pubescens Lemaire, Jard. Fleur.: misc.p.9, 1852
Cattleya acuminata (Lindl.) Beer, Prakt. Stud. Orch.: 208, 1854
Cattleya peduncularis (Lindl.) Beer, Prakt. Stud. Orch.: 213, 1854
Cattleya rubescens (Lindl.) Beer, Prakt. Stud. Orch.: 214, 1854
Laelia violacea Rchb.f., Bonpl. 2: 98, 1854
Bletia acuminata (Lindl.) Rchb.f., Walp. Ann. 6: 427, 1862
Bletia peduncularis (Lindl.) Rchb.f., Walp. Ann. 6: 426, 1862
Bletia rubescens (Lindl.) Rchb.f., Walp. Ann. 6: 425, 1862
Bletia violacea (Rchb.f.) Rchb.f., Walp. Ann. 6: 426, 1862

Laelia speciosa (Kunth) Schltr., Die Orchideen: 233, 1914
BR 25: misc.42 (1839) / 28: sub t.62:1, 2 (1842)/ 30: t.30 (1844)
Mexico – E
Bletia speciosa Kunth, Nov. Gen. & Sp. Pl. 1: 342, 1816
Bletia grandiflora Llave & Lex., Nov. Veg. Descr.: 17, 1825
Laelia grandiflora (Llave & Lex.) Lindl., Gen. Sp. Orch. Pl.: 115, 1831
o Laelia majalis Lindl., Bot. Reg. 25: misc.42, 1839
Cattleya grahami Lindl., Gen. Sp. Orch. Pl.: 116, 1841

Cattleya majalis (Lindl.) Beer, Prakt. Stud. Orch.: 212, 1854
Laelia grandiflora var. alba hort., ex Orch. Rev. 9: 202 & 312, 1901
Laelia majalis var. alba Dimmock, Amer. Gard. 22: 398, 1901

Laelia xanthina Lindl. ex Hook., Bot. Mag. 85: t.5144, 1859
BR 30: misc.2 (1844)
Brazil (Bahia, Espirito Santo) – E
Laelia virens Lindl., Bot. Reg. 30: misc.2, 1844
Bletia xanthina (Lindl. ex Hook.) Rchb.f., Walp. Ann. 6: 425, 1862
Laelia wetmorei Ruschi, Bol. Mus. Biol. Prof. Mello-Leitao S. Bot. 29: 1, 1970

Lanium avicula (Lindl.) Benth., Hook. Ic. Pl. 14: t.1335, 1881
BR 31: misc.36 p.25:2 (1845)
Brazil – E
Epidendrum avicula Lindl., Hook. J. Bot. 3: 85, 1841

Lanium microphyllum (Lindl.) Benth., Hook. Ic. Pl. 14: 24 t.1334, 1881
BR 27: misc.71 (1841) / 31: misc.36 p.25:1 (1845)
Brazil, Guyana, Surinam, Peru, Colombia – E
Epidendrum microphyllum Lindl., Hook. J. Bot. 3: 85, 1841
Epidendrum sarcophyllum Focke, Tydschr. Nat. Wetensch. 4: 67, 1849
Lanium peruvianum Schltr., Fedde Rep. Beih. 9: 97, 1921

Lemboglossum apterum (Llave & Lex.) Halbinger, Orquidea (Méx.) 9 (2): 348, 1984
BR 31: misc.60:8, 12 (1845)
Mexico – E
Odontoglossum apterum Llave & Lex., Nov. Veg. Descr. Orch. Opusc.: 35, 1824
Odontoglossum nebulosum Lindl., Sert. Orch.: sub t.25, 1838
Oncidium anceps Beer, Prakt. Stud. Orch.: 286, 1854
Odontoglossum x vexativum Rchb.f., Gard. Chron. n.s.6: 808, 1876
Cymbiglossum apterum (Llave & Lex.) Halbinger, Orquidea (Méx.) 9 (1): 2, 1983

Lemboglossum bictoniense (Bateman) Halbinger, Orquidea (Méx.) 9 (2): 348, 1984
BR 26: t.66 & misc.139 (1840) / 31: misc.60:25 (1845)
Mexico, Guatemala, El Salvador – 1550–1800 m – E, T
Cyrtochilum bictoniense Bateman, Orch. Méx. & Guat.: t.6, 1838
o Odontoglossum bictoniense (Bateman) Lindl., Bot. Reg. 26: t.66, 1840
Zygopetalum africanum Hook., Bot. Mag. 67: t.3812, 1840
Cymbiglossum bictoniense (Bateman) Halbinger, Orquidea (Méx.) 9 (1): 2, 1983

Lemboglossum cervantesii (Llave & Lex.) Halbinger, Orquidea (Méx.) 9 (2): 348, 1984
BR 31: t.36 & misc.45 & misc.60:7, 10 (1845) / 32: t.34 (1846)
Mexico, Guatemala, El Salvador – 1500–3000 m – E
o Odontoglossum cervantesii Llave & Lex., Nov. Veg. Descr. 2: 34, 1825
o Odontoglossum membranaceum Lindl., Sert. Orch.: sub t.25, 1838
Odontoglossum cervantesii var. membranaceum Lindl., Fol. Orch. Odontoglossum: 10, 1852
Oncidium cervantesii (Llave & Lex.) Beer, Prakt. Stud. Orch.: 283, 1854
Oncidium membranaceum (Lindl.) Beer, Prakt. Stud. Orch.: 283, 1854
Cymbiglossum cervantesii (Llave & Lex.) Halbinger, Orquidea (Méx.) 9 (1): 2, 1983

Lemboglossum cordatum (Lindl.) Halbinger, Orquidea (Méx.) 9 (2): 349, 1984
BR 24: misc.90 (1838) / 31: misc.60:33 (1845)
Type: Mexico, cult. G.Barker (K-Lindl., holo!)
Mexico, Guatemala, El Salvador, Honduras, Nicaragua, Costa Rica, Venezuela – up to 2500 m – E
Odontoglossum cordatum Lindl., Bot. Reg. 24: misc.90, 1838
Odontoglossum maculatum Lindl. sensu Hook., Bot. Mag. 81: t.4878, 1855
Odontoglossum hookeri Lemaire, Ill. Hort. 3: misc.p.41, 1856
Odontoglossum lueddemannii Regel, Gartenfl. 8: 353, 1859
Cymbiglossum cordatum (Lindl.) Halbinger, Orquidea (Méx.) 9 (1): 3, 1983

Lemboglossum ehrenbergii (Link, Klotzsch & Otto) Halbinger, Orquidea (Méx.) 9 (2): 349, 1984
BR 27: misc.85 (1841)
Mexico – E
Odontoglossum ehrenbergii Link, Klotzsch & Otto, Ic. Pl. Rar.: 39 t.16, 1841
Oncidium ehrenbergii (Link, Klotzsch & Otto) Beer, Prakt. Orch. Stud.: 285, 1854
Odontoglossum dawsonianum Rchb.f., Gard. Chron.: 1226, 1865
Cymbiglossum ehrenbergii (Link, Klotzsch & Otto) Halbinger, Orquidea (Méx.) 9 (1): 3, 1983

Lemboglossum galeottianum (A.Rich.) Halbinger, Orquidea (Méx.) 9 (2): 349, 1984
BR 31: misc.60:11 (1845)
Mexico – E
Odontoglossum galeottianum A.Rich., Ann. Sc. Nat. s.3, 3: 27, 1845
Cymbiglossum galeottianum (A.Rich.) Halbinger, Orquidea (Méx.) 9 (1): 3, 1983

Lemboglossum maculatum (Llave & Lex.) Halbinger, Orquidea (Méx.) 9 (2): 349, 1984
BR 26: t.30 (1840) / 31: misc.60:32 (1845)
Mexico, Guatemala – up to 3000 m – E
Odontoglossum anceps Lemaire, Ill. Hort. 3: misc.45, 1825
o Odontoglossum maculatum Llave & Lex., Nov. Veg. Descr. 2: 35, 1825
Brassia oestlundiana L.O.Williams, Ceiba 2 (4): 238, 1951
Cymbiglossum maculatum (Llave & Lex.) Halbinger, Orquidea (Méx.) 9 (1): 3, 1983

Lemboglossum madrense (Rchb.f.) Halbinger, Orquidea (Mex.) 9 (2): 349, 1984
BR 33: sub t.62 (1847)
Mexico – E
Odontoglossum maxillare Lindl., Bot. Reg. 33: sub t. 62, 1847
Odontoglossum madrense Rchb.f., Gard. Chron. n.s.2: 804, 1874
Cymbiglossum madrense (Rchb.f.) Halbinger, Orquidea (Méx) 9 (1): 3, 1983

Lemboglossum rossii (Lindl.) Halbinger, Orquidea (Méx.) 9 (2): 349, 1984
BR 25: t.48 (1839) / 29: misc.19 (1843) / 31: misc.60:9, 13 (1845)
Type: Mexico, Ross s.n. cult. Barker (K-Lindl., holo!)
Mexico, Guatemala, El Salvador, Honduras, Nicaragua – up to 3000 m – E
o Odontoglossum rossii Lindl., Bot. Reg. 25: t.48, 1839
Odontoglossum coerulescens A.Rich. & Galeotti, Ann. Sc. Nat. s.3, 3: 27, 1845
Odontoglossum rubescens Lindl., J. Hort. Soc. London 5: 35, 1850
Odontoglossum warnerianum Rchb.f., Bot. Zeit. 22: 297, 1864
Odontoglossum humeanum Rchb.f., Gard. Chron.: 170, 1876
Odontoglossum asperum Rchb.f., Gard. Chron.: 266, 1879
Odontoglossum youngii Gower, Warner & Williams Orchid. Alb. 9: t.406, 1891
Cymbiglossum rossii (Lindl.) Halbinger, Orquidea (Méx.) 9 (1): 3, 1983

Lemboglossum stellatum (Lindl.) Halbinger, Orquidea (Méx.) 9 (2): 349, 1984
BR 27: misc.25 (1841) / 31: misc.60:14, 39 (1845)
Type: Mexico, Hartweg s.n. cult. Hort Soc. (K-Lindl., holo!)
Mexico, Guatemala – up to 3000 m – E
Odontoglossum stellatum Lindl., Bot. Reg. 27: misc.25, 1841
Odontoglossum erosum A.Rich & Galeotti, Ann. Sc. Nat. s.3, 3: 27, 1845
Oncidium erosum (A.Rich. & Galeotti) Beer, Prakt. Stud. Orch.: 285, 1854
Cymbiglossum stellatum (Lindl.) Halbinger, Orquidea (Méx.) 9 (1): 4, 1983

Leochilus carinatus (Knowles & Westc.) Lindl., Bot. Reg. 28: misc.22: 3, 1842
BR 26: misc.45 (1840) / 28: misc.22:3 (1842)
Mexico, Peru – E
Oncidium carinatum Knowles & Westc., Fl. Cab. 2: 31, 1838

Leochilus labiatus (Sw.) Kuntze, Rev. Gen. 2: 656, 1891
BR 28: misc.22:2 (1842)
Florida, Cuba, W.Indies, Mexico, Guatemala, El Salvador, Honduras, Nicaragua, Costa Rica, Panama, Trinidad, Venezuela, Colombia, Ecuador, Brazil – E
Epidendrum labiatum Sw., Prodr.: 124, 1788
Liparis labiata (Sw.) Sprengel, Syst. Veg. 3: 741, 1826
Rodriguezia cochlearis Lindl., Ann. Nat. Hist. 5: 116, 1840
Leochilus cochlearis (Lindl.) Lindl., Bot. Reg. 28: misc.22: 2, 1842
Oncidium labiatum (Sw.) Rchb.f., Ann. Bot. 6: 741, 1863
Oncidium lansbergii Rchb.f., Gard. Chron. n.s.5: 460, 1876
Oncidium depauperatum Lehm. & Kraenzlin, Bot. Jahrb. Syst. 26: 496, 1899
Leochilus depauperatus (Lehm. & Kraenzlin) Kraenzlin, Engl. Pflanzenr. Orch. Mon. Oncid.: 295, 1922
Leochilus gracilis Schltr., Fedde Rep. Beih. 19: 152, 1923

Leochilus oncidioides Knowles & Westc., Fl. Cabin. 2: 143, 1838
BR 26: misc.218 (1840) / 27: misc.33 (1841) / 28: misc.22:1 (1842) / 29: misc.122 (1843)
Mexico, Guatemala, Belize – E
Rodriguezia maculata Lindl., Bot. Reg. 26: misc.218, 1840
Oncidium macrantherum Hook., Bot. Mag. 67: t.3845, 1841
Leochilus macrantherus (Hook.) Liebm., Schouw Forteg. Kjob. Bot. Hav. Pl. 27, 1847

Leochilus scriptus (Scheidw.) Rchb.f., Xen. Orch. 1 (1): 15, 1854
BR 30: misc.90 (1844)
Cuba, W.Indies, Guatemala to Ecuador, Venezuela – E
Cryptosanus scriptus Schweidw., Allg. Gartenz. 11: 101, 1843
Leochilus herbaceus Lindl., Bot. Reg. 30: misc.90, 1844
Cryptosaccus scriptus Scheidw. ex Rchb.f., Xen. Orch. 1 (1): 15, 1854
Oncidium herbaceum (Lindl.) Rchb.f., Walp. Ann. 6: 772, 1863
Oncidium scriptum (Scheidw.) Rchb.f., Walp. Ann. 6: 772, 1863
Leochilus major Schltr., Fedde Rep. Beih. 15: 209, 1918
Leochilus powellii Schltr., Fedde Rep. Beih. 17: 81, 1922
Leochilus retusus Schltr., Fedde Rep. Beih. 19: 256, 1923

o **Lepanthes tridentata** Sw., Prodr.: 125, 1788
BR 21: t.1762 (1835)
Jamaica – E
Epidendrum tridentatum Sw., Prodr.: 125, 1788

o **Leptotes bicolor** Lindl., Bot. Reg. 19: t.1625, 1833
BR 19: t.1625 (1833)
Type: Brazil, cult. Harrison (K-Lindl., holo!)
Brazil (Organ M.), Paraguay – E
Leptotes serrulata Lindl., Sert. Orch.: t.11, 1838
Leptotes bicolor var. glaucophylla Hook., Bot. Mag. 67: t.3734, 1840
Leptotes glaucophylla (Hook.) Hoffsgg., Verz. Orch.: 54, 1842
Tetramicra bicolor (Lindl.) Benth., J. Linn. Soc. 18: 314, 1881
Tetramicra serrulata (Lindl.) Benth. & Hook.f., Gen. Pl. 3 (2): 533, 1895

Leucohyle mutica (Lindl.) Schltr., Die Orchideen: 469, 1914
BR 25: misc.22 (1839)
Type: Trinidad, cult. Knight (K-Lindl., holo!)
Brazil, Surinam, Trinidad – E
Brassavola paraensis Huber, ined!
Macradenia mutica Lindl., Bot. Reg. 25: misc.22, 1839
Trichopilia mutica (Lindl.) Rchb.f. & Wullschl., Walp. Ann. 6: 679, 1863

Liparis caespitosa (Thouars) Lindl., Bot. Reg. 11: sub t.882: 9, 1825
BR 11: sub t.882:9 (1825)
Widely distributed from Madagascar and Africa to Asia, S.E.Asia, Malay Arch., New Guinea, Solomon Is., Fiji – E, L
Malaxis caespitosa Thouars, Orch. Iles Afr.: t.90, 1822
Malaxis angustifolia Blume, Bijdr.: 393, 1825
Malaxis minima Blume, Bijdr.: 391, 1825
Liparis angustifolia (Blume) Lindl., Gen. Sp. Orch. Pl.: 31, 1830
Liparis minima (Blume) Lindl., Gen. Sp. Orch. Pl.: 32, 1830
Liparis auriculata Rchb.f., Flora 55: 277, 1872 (non Blume ex Miq.)
Liparis pusilla Ridley, J. Linn. Soc. 22: 294, 1886
Liparis duthiei Hook.f., Ic. Pl.: t.1857B, 1889
Liparis obscura Hook.f., Ic. Pl.: t.1886, 1889
Liparis prainii Hook.f., Ic. Pl.: t.1857A, 1889
Leptorchis caespitosa (Thouars) Kuntze, Rev. Gen. 2: 671, 1891
Leptorchis duthiei (Hook.f.) Kuntze, Rev. Gen. 2: 671, 1891
Leptorchis minima (Blume) Kuntze, Rev. Gen. 2: 671, 1891
Leptorchis obscura (Hook.f.) Kuntze, Rev. Gen. 2: 671, 1891
Leptorchis prainii (Hook.f.) Kuntze, Rev. Gen. 2: 671, 1891
Leptorchis pusilla (Ridley) Kuntze, Rev. Gen. 2: 671, 1891
Liparis comosa Ridley, J. Linn. Soc. 32: 229, 1896
Liparis disticha Kraenzlin, Bot. Tidsskr. 24: 6, 1900 (non Lindl.)
Cestorchis caespitosa (Thouars) Ames, Orchid. 2: 132, 1908
Liparis neoguinensis Schltr., Fedde Rep. Beih. 1: 209 fig.289, 1911
Liparis poilanei Gagnepain, Bull. Soc. Bot. Fr. 76: 514, 1929
Liparis mannii, Guillaumin, Bull. Mus. Hist. Nat. Paris: 487, 1956 (non Rchb.f.)

Liparis flavescens (Thouars) Lindl., Bot. Reg. 11: sub t.882:4, 1825
BR 11: sub t.882:4 (1825)
Mascarenes Is., Madagascar – T
Malaxis flavescens Thouars, Orch. Iles Afr.: t.25, 1822
Leptorchis flavescens (Thouars) Kuntze, Rev. Gen. 2: 671, 1891

Liparis gibbosa Finet, Bull. Soc. Bot. Fr. 55: 342, fig.32–34 1908
BR 11: sub t.882:8 (1825)
S.E.Asia, Malay Arch. to New Guinea, Fiji, Solomon Is., Vanuatu, New Caledonia, Samoa – E, L
Liparis disticha Lindl., Bot. Reg. 11: sub t.882:8, 1825 (non Thouars)
Liparis gregaria Lindl., Gen. Sp. Orch. Pl.: 33, 1830

Liparis lilifolia (L.) L.C.Rich. ex Lindl., Bot. Reg. 11: sub t.882: 2, 1825
BR 11: sub t.882:2 (1825)
N.Europe, N.America – T
Ophrys lilifolia L., Sp. Pl. 2: 946, 1753
Malaxis lilifolia (L.) Sw., Vet. Akad. Handl. Stockh. 21: 235, 1800
Leptorchis lilifolia (L.) Kuntze, Rev. Gen. 2: 671, 1891

Liparis loeselii (L.) L.C.Rich, Orch. Europ. Annot.: 38, 1817
BR 11: sub t.882:1 (1825)
Europe, Siberia, N.America – T
Ophrys loeselii L., Sp. Pl. 2: 947, 1753
Malaxis loeselii (L.) Sw., Vet. Akad. Handl. Stockh. 21: 235, 1800
Leptorchis loeselii (L.) MacMillan, Melasp. Minn.: 173, 1893

Liparis nervosa (Thunb.) Lindl., Gen. Sp. Orch. Pl.: 26, 1830
BR 11: sub t.882:6 (1825) / 14: t.1175 (1828) / 20: t.1671 (1834) / 29: misc.12 (1843)
Widespread in humid regions of both Old World and New World tropics and subtropics – T, L, E
Ophrys nervosa Thunb., Fl. Jap.: 27, 1784
Epidendrum nervosum (Thunb.) Thunb., Mus. Nat. Acad. Upsal. 9: 138, 1791
Cymbidium nervosum (Thunb.) Sw., Nov. Act. Upsal. 6: 76, 1799
Malaxis nervosa (Thunb.) Sw., Vet. Acad. Handl. Stockh. 21: 235, 1800
Malaxis odorata Willd., Sp. Pl. ed.4: 91, 1805
Cymbidium bituberculatum Hook., Exot. Fl. 2: t.116, 1824
Bletia bicallosa D.Don, Prodr. Fl. Nepal.: 30, 1825
Limodorum bicallosum Hamilt. ex Don, Prodr. Fl. Nepal.: 30, 1825
Liparis bituberculata (Hook.) Lindl., Bot. Reg. 11: sub t.882:6, 1825
o Liparis elata Lindl., Bot. Reg. 14: t.1175, 1828
Liparis odorata (Willd.) Lindl., Gen. Sp. Orch. Pl.: 26, 1830
Liparis olivacea Lindl., Gen. Sp. Orch. Pl.: 27, 1830
o Liparis guineensis Lindl., Bot. Reg. 20: t.1671, 1834
Dituilis nepalensis Raf., Fl. Tellur 4: 49, 1838 (nom. illeg.)
Liparis alata Scheidw., Gartenz.: 293, 1842
Sturmia bituberculata (Hook.) Rchb.f., Bonpl. 1: 22, 1854
Sturmia nervosa (Thunb.) Rchb.f., Bonpl. 3: 250, 1855
Liparis elata var. purpurascens Regel, Ind. Sem. Hort. Petrop.: 18, 1856
Liparis bituberculata (Rchb.f) Rchb.f., Walp. Ann. 6: 218, 1861 (non Lindl.)
Liparis odontostoma Rchb.f., Linnaea 41: 97, 1876
Liparis elata var. inundata Barb.Rodr., Gen. & Sp. Orch. Nov. 1: 36, 1877
Liparis formosana Rchb.f., Gard. Chron. n.s.13: 394, 1880
Liparis bicornis Ridley, J. Linn. Soc. 21: 458, 1885
Liparis eggersii Rchb.f., Ber. Deutsch. Bot. Ges. 3: 278, 1885
Liparis lutea Ridley, J. Linn. Soc. 21: 458, 1885
Liparis bituberculata var. formosana (Rchb.f.) Ridley, J. Linn. Soc. 22: 263, 1886
Liparis elata var. latifolia Ridley, J. Linn. Soc. 22: 260, 1886
Liparis elata var. rufina Ridley, J. Linn. Soc. 22: 260, 1886
Leptorchis bicornis (Ridley) Kuntze, Rev. Gen. 2: 671, 1891
Leptorchis bituberculata (Hook.) Kuntze, Rev. Gen. 2: 671, 1891
Leptorchis eggersii (Rchb.f.) Kuntze, Rev. Gen. 2: 671, 1891
Leptorchis elata (Lindl.) Kuntze, Rev. Gen. 2: 671, 1891
Leptorchis guineensis (Lindl.) Kuntze, Rev. Gen. 2: 671, 1891
Leptorchis lutea (Ridley) Kuntze, Rev. Gen. 2: 671, 1891
Leptorchis nervosa (Thunb.) Kuntze, Rev. Gen. 2: 671, 1891
Leptorchis odontostoma (Rchb.f.) Kuntze, Rev. Gen. 2: 671, 1891
Leptorchis odorata (Willd.) Kuntze, Rev. Gen. 2: 671, 1891
Leptorchis olivacea (Lindl.) Kuntze, Rev. Gen. 2: 671, 1891
Liparis cornicaulis Makino, Ill. Fl. Jap. 1 (8): 1 t.47, 1891
Liparis bambusaefolia Makino, Tokyo Bot. Mag. 6: 48, 1892
Liparis elata var. longifolia Cogn., Mart. Fl. Bras. 3 (4): 287, 1895
Liparis rufina (Ridley) Rchb.f. ex Rolfe, Fl. Trop. Afr. 7: 19, 1897
Liparis melanoglossa Schltr., Fedde Rep. Beih. 1: 184, 1911
Liparis perrieri Schltr., Ann. Mus. Col. Mars. s.3, 1: 21, 1913
Liparis nyassa Schltr., Engl. Bot. Jahrb. 53: 560, 1915
Liparis bicallosa (D.Don) Schltr., Fedde Rep. Beih. 4: 196, 1919
Tribrachia racemosa Lindl. ex Schltr., Fedde Rep. Beih. 4: 196, 1919
Liparis hachijoensis Nakai, Bot. Mag. Tokyo 35: 97, 1921
Liparis siamensis Rolfe ex Downie, Kew Bull.: 371, 1925
Liparis odorata var. intacta Kerr, Kew Bull.: 216, 1927
Liparis formosana var. hachijoensis (Nakai) Ohwi, Bull. Nat. Sc. Mus. Tokyo 33: 70, 1953
Liparis violaceonervosa Guillaumin, Bull. Mus. Hist. Nat. Paris s.2, 33 (4): 434, 1961
Liparis bicallosa var. hachijoensis (Nakai) Kitamura, Act. Phytotax. Geobot. 22: 66, 1966
Liparis formosana f. aureo-variegata Nakajima, J. Geobot. 17: 55, 1969
Liparis nervosa var. formosana (Rchb.f.) Hiroe, Orch. Flowers 2: 77, 1971

Liparis paradoxa (Lindl.) Rchb.f., Walp. Ann. 6: 218, 1861
BR 10: sub t.825 (1824)
Type: Nepal, Lambert s.n. (K-Lindl., holo!)
India, W.Himalaya, Nepal, Sikkim, Assam, Yunnan, Szechuan, Burma, Laos, Cambodia, Vietnam, Taiwan, Japan, Guam – T
Malaxis lancifolia Smith ex Rees, Cyclop.: 22, 1812
Empusa paradoxa Lindl., Bot. Reg. 10: sub t.825, 1824
Empusaria sp. Rchb., Consp. Reg. Veg.: 69, 1828
Liparis olivacea Parish, Order Orchidaceae 190, 1883 (non Lindl.)
Liparis paradoxa var. parishii Hook.f., Fl. Brit. Ind. 5: 698, 1890
Liparis parishii (Hook.f.) Hook.f., Fl. Brit. Ind. 6: 182, 1890

Liparis guamensis Ames, Phil. J. Sc. 9: 11, 1914
Liparis tenii Schltr., Fedde Rep. 17: 66, 1921
Liparis teniana Kraenzlin, Fedde Rep. 17: 122, 1922
Liparis simeonis Schltr., Fedde Rep. 20: 283, 1924
Liparis odorata var. longiscapa Rolfe ex Downie, Kew Bull.: 370, 1925
Liparis nervosa Blatter & MacCann, J. Bombay Nat. Hist. Soc. 35: pl.5, 1931 (non (Thunb.) Lindl.)
Liparis longiscapa (Rolfe ex Downie) Gagnepain & Guillaumin, Fl. Gen. Indoch. 6: 182, 1932
Liparis tonkinensis Gagnepain, Bull. Soc. Bot. Fr. 79: 167, 1932

Liparis petiolata (D.Don) Hunt & Summerh., Kew Bull. 20: 52, 1966
BR 11: sub t.882:3 (1825)
Nepal, Sikkim, Bhutan, Khasia, Thailand – T
Malaxis cordifolia Smith ex Rees, Cyclop.: 22, 1812
Acianthus petiolatus D.Don, Prodr. Fl. Nepal.: 29, 1825
Liparis nepalensis Lindl., Bot. Reg. 11: sub t.882:3, 1825
Liparis rupestris var. purpurascens Ridley, J. Linn. Soc. 22: 268, 1886
Liparis pulchella Hook.f., Ic. Pl.: t.1810, 1889
Liparis angkae Kerr, Kew Bull.: 215, 1927

Liparis purpurascens (Thouars) Lindl., Bot. Reg. 11: sub t.882:5, 1825
BR 11: sub t.882:5 (1825)
Mascarene Is., Madagascar – 1200 m – T
Malaxis purpurascens Thouars, Orch. Iles Afr.: t.26 & t.27, 1822

Liparis reflexa (R.Br.) Lindl., Bot. Reg. 11: sub t.882, 1825
BR 11: t.882 & sub t.882:7, 10 (1825)
Australia (S.Queensland, New South Wales) – T
Cymbidium reflexum R.Br., Prodr. Fl. Nov. Holl.: 331, 1810
o Liparis foliosa Lindl., Bot. Reg. 11: t.882, 1825
Sturmia reflexa (R.Br.) F.Muell., Fragm. Phyt. Austr. 2: 72, 1860

Liparis viridiflora (Blume) Lindl., Gen. Sp. Orch. Pl.: 31, 1830
BR 24: misc.180 (1838) / 26: misc.189 (1840)
Sri Lanka, N.E.India, Bhutan, Nepal, Sikkim, E.Bengal, Assam, Burma, Thailand, China, Taiwan, Laos, Vietnam, Malaya, Sumatra, Philippines, Samoa – E
Malaxis viridiflora Blume, Bijdr.: 392, 1825
Liparis longipes Lindl., Wall. Pl. As. Rar. 1: 31 t.35, 1830
Liparis pendula Lindl., Bot. Reg. 24: misc.180, 1838
Liparis spathulata Lindl., Bot. Reg. 26: misc.189, 1840
Liparis simondii Gagnepain, Bull. Mus. Hist. Nat. Paris s.2, 21 (6): 738, 1849
Sturmia longipes (Lindl.) Rchb.f., Bonpl. 3: 250, 1855
Liparis boothii Regel, Ind. Sem. Hort. Petrop. Supp. 14, 1864
Liparis stachyurus Rchb.f., Flora 55: 274, 1872
Liparis nesophila Rchb.f., Otia Bot. Hamb. 1: 56, 1878
Liparis longipes var. spathulata (Lindl.) Ridley, J. Linn. Soc. 22: 294, 1886
Liparis triloba Ridley, J. Linn. Soc. 22: 295, 1886
Cestichis pendula (Lindl.) Pfitzer, Engl. & Prantl. Pflanzenf. 2 (6): 131 fig.128 C-E, 1888
Leptorchis longipes (Lindl.) Kuntze, Rev. Gen. 2: 671, 1891
Leptorchis viridiflora (Blume) Kuntze, Rev. Gen. 2: 671, 1891
Stichorchis pendula (Lindl.) Pfitzer, Engl. & Prantl. Nachtr. 1: 103, 1897
Cestichis longipes (Lindl.) Ames, Orchid. 1: 75, 1905
Liparis pleistantha Schltr., Fedde Rep. Beih. 4: 64, 1919
Liparis piestopus Schltr., Fedde Rep. 21: 139, 1925

Lockhartia acuta (Lindl.) Rchb.f., Bot. Zeit. 10: 767, 1852
BR 21: t.1806 (1835)
Type: s.n (K-Lindl., icono!)
Panama, Trinidad, Venezuela, Colombia – E
o Fernandezia acuta Lindl., Bot. Reg. 21: t.1806, 1835
Lockhartia pallida Rchb.f., Bonpl. 2: 14, 1854
Lockhartia lasseri Schnee, Rev. Fac. Ing. Agron. 1: 120, 1952

Lockhartia lunifera (Lindl.) Rchb.f., Bot. Zeit. 10: 767, 1852
BR 25: misc.147 (1839)
Type: Brazil, Descourtilz s.n. Loddiges (K-Lindl., holo!)
Brazil – E
Fernandezia lunifera Lindl., Bot. Reg. 25: misc.147, 1839
Fernandezia robusta Klotzsch ex Rchb.f., Xen. Orch. 1: 101, 1855 (non Bateman)

Ludisia discolor (Ker Gawler) A.Rich., Dict. Class. Hist. Nat. 7: 437, 1825
BR 4: t.271 (1818) / 19: sub t.1618 (1833)
Types: cult. Lee and Kennedy s.n. (K-Lindl., icono!)
China, Thailand, Vietnam, Malaya – T
o Goodyera discolor Ker Gawler, Bot. Reg. 4: t.271, 1818
Gonogona discolor (Ker Gawler) Link, Enum. Pl. Hort. Berol. 2: 369, 1822
Myoda rufescens Lindl., Bot. Reg. 19: sub t.1618, 1833
Dicrophylla elegans Raf., Fl. Tellur. 4: 39, 1838
Haemaria discolor (Ker Gawler) Lindl., Gen. Sp. Orch. Pl.: 490, 1840
Neottia discolor (Ker Gawler) Steudel, Nom. ed.2, 2: 189, 1841
Ludisia furetii Blume, Coll. Orch. Arch. Ind. Jap.: 114 t.34 fig. 3, 1858
Ludisia odorata Blume, Coll. Orch. Arch. Ind. Jap.: 114 t.34 fig. 2, 1858
Anoectochilus dawsonianus Low ex Rchb.f., Gard. Chron.: 1038, 1868
Haemaria discolor var. dawsoniana (Low ex Rchb.f.) Rchb.f., Gard. Chron.: 321, 1872
Haemaria otletae Rolfe, Ill. Hort. 38: 31 t.524, 1891
Haemaria dawsoniana Hook.f., Bot. Mag. 122: t.7486, 1896
Haemaria merrillii Ames, Phil. J. Sc. 2: 315, 1907
Haemaria pauciflora Gagnepain, Bull. Soc. Bot. Fr. 80: 349, 1933
Haemaria petelotii Gagnepain, Bull. Soc. Bot. Fr. 80: 349, 1933
Odontochilus petelotii (Gagnepain) Tang & Wang, Act. Phytotax. Sin. 1, 1: 71, 1951
Anoectochilus petelotii (Gagnepain) Seidenf., Contr. Rev. Orch. Fl. Camb. Laos & Vietnam 1: 7, 1975

Lycaste aromatica (Graham ex Hook.) Lindl., Bot. Reg. 29: misc.p.16: 8, 1843
BR 18: sub t.1549:22 (1832) / 22: t.1871 (1836) / 28: sub t.13 (1842) / 29: misc.p.16:8 (1843) / 30: misc.47 (1844)
Belize, Guatemala, Honduras, Nicaragua – up to 1200 m – E
o Maxillaria aromatica Graham ex Hook., Exot. Fl. 3: t.214, 1826
Colax aromaticus (Graham ex Hook.) Sprengel, Syst. Veg.: 307, 1827
Lycaste aromatica var. retusa Lindl., Bot. Reg. 30: misc.47, 1844
Lycaste aromatica var. majus Kunth, Gard. Chron. s.3, 79: 421, 1926
Lycaste suaveolens Summerh., Bot. Mag. 155: t.9231, 1931

Lycaste barringtoniae (Smith) Lindl., Bot. Reg. 30: misc.51, 1844
BR 18: sub t.1549:23 (1832) / 30: misc.51 (1844)
Jamaica, Cuba – E
Epidendrum barringtoniae Smith, Ic. Pict.: t.15, 1793
Dendrobium barringtoniae (Smith) Sw., Nov. Act. Upsal. 6: 82, 1799
Colax barringtoniae (Smith) Lindl. ex Sprengel, Syst. Veg. 3: 727, 1826
Maxillaria barringtoniae (Smith) Lodd., Bot. Cab. 19: t.1834, 1832

Lycaste ciliata (Ruiz Lopez & Pavon) Lindl. ex Rchb.f., Bonpl. 4: 324, 1856
BR 14: t.1206 (1829) / 24: misc.175 (1838) / 29: misc.p.15:3 (1843)
Colombia, Ecuador, Peru, Bolivia – T, E
o Maxillaria ciliata Ruiz Lopez & Pavon, Syst. Veg. Fl. Peruv. & Chil. 1: 221, 1798
Dendrobium ciliatum Pers., Syn. 2: 523, 1805 (nomen nudum)
Maxillaria fimbriata Peoppig & Endl., Nov. Gen. Sp. Pl. 1: 38 t.62 fig. 3, 1836
Maxillaria costata Lindl., Bot. Reg. 24: misc.175, 1838
Lycaste lanipes Lindl., Gard. Chron.: 212, 1842
Lycaste costata (Lindl.) Lindl., Bot. Reg. 29: misc.p.15: 3, 1843
Lycaste barringtoniae var. grandiflora auct. non (Smith) Lindl; Hook., Bot. Mag. 24: t.5706, 1868 (Misapplied name)
Lycaste mesochlaena auct. non Rchb.f.; Rolfe, Orch. Rev. 1: 72, 1893 (Misapplied name)
Lycaste fimbriata (Poeppig & Endl.) Cogn., Mart. Fl. Bras. 3 (5): 455, 1898
Lycaste lata Rolfe, Kew Bull.: 370, 1910

Lycaste crinita Lindl., Bot. Reg. 30: misc.41, 1844
BR 30: misc.41 (1844)
Type: Mexico, Oaxaca, cult. Loddiges (K-Lindl., holo!)
Mexico – E
Lycaste micheliana Cogn., Rev. Hort. 72: 264, 1900

Lycaste cruenta (Lindl.) Lindl., Bot. Reg. 29: misc.p.16: 9, 1843
BR 28: t.13 (1842) / 29: misc.p.16:9 (1843)
Type: Guatemala, Skinner s.n. C.Lemon (K-Lindl., holo!)
Mexico, Guatemala, El Salvador – E
o Maxillaria cruenta Lindl., Bot. Reg. 28: t.13, 1842
Lycaste balsamea A.Rich. ex Lindl., Paxt. Flow. Gard. 1: 126, 1851
Maxillaria balsamea (A.Rich. ex Lindl.) Beer, Prakt. Stud. Orch.: 264, 1854

Lycaste deppei (Lodd.) Lindl., Bot. Reg. 29: misc.p.15:6, 1843
BR 18: sub t.1549:20 (1832) / 29: misc.p.15:6 (1843)
Mexico, Guatemala, El Salvador, Honduras, Nicaragua – E
Maxillaria deppei Lodd., Bot. Cab. 17: t.1612, 1830

Lycaste longipetala (Ruiz Lopez & Pavon) Garay, Caldasia 8: 524, 1962
BR 29: misc.p.15:5 (1843) / 30: misc.48 (1844) / 31: t.34 (1845)
Venezuela, Colombia, Ecuador, Peru – 1200–2200 m – E
Maxillaria longipetala Ruiz Lopez & Pavon, Syst. Veg. Fl. Peruv. & Chil. 1: 220, 1798
Dendrobium longipetalum (Ruiz Lopez & Pavon) Sw., Pers. Syn. Pl. 11: 523, 1807
o Lycaste gigantea Lindl., Bot. Reg. 29: misc.p.15:5, 1843
Maxillaria heynderyexii Morren, Ann. Soc. Bot. Gand. 1: 97 t.9, 1845
Maxillaria gigantea (Lindl.) Beer, Prakt. Stud. Orch.: 265, 1854

Lycaste macrophylla (Poeppig & Endl.) Lindl., Bot. Reg. 29: misc.p.14: 1, 1843
BR 24: misc.174 (1838) / 26: misc.191 (1840) / 28: misc.96 (1842) / 29: t.35 & misc.p.14:1 & misc.p.15:2 (1843)
Costa Rica, Panama, Venezuela to Bolivia – E
Maxillaria macrophylla Poeppig & Endl., Nov. Gen. Sp. Pl. 1: 37 t.64, 1836
Maxillaria phyllomega Steudel, Nom. ed.2, 2: 107, 1841
o Lycaste plana Lindl., Bot. Reg. 28: misc.96, 1842
Lycaste dowiana Endres & Rchb.f., Gard. Chron. n.s.2: 194, 1874
Lycaste filomenoi Schltr., Fedde Rep. Beih. 9: 100, 1921
Lycaste macrophylla ssp. filomenoi (Schltr.) Fowl., Lasca Leaves: 41, 1964
Lycaste macrophylla ssp. macrophylla (Poeppig & Endl.) Fowl., Lasca Leaves: 41, 1964
Lycaste macrophylla ssp. plana (Lindl.) Fowl., Lasca Leaves: 41, 1964

Lycaste skinneri (Bateman ex Lindl.) Lindl., Bot. Reg. 29: misc.p. 15:7, 1843
BR 26: misc.101 & misc.145 (1840) / 28: misc.13 (1842) / 29: misc.p. 15:7 (1843)
Type: Guatemala, Skinner s.n. (K-Lindl., holo!)
Mexico, Guatemala, Honduras, El Salvador – 1700–1800 m – E
Maxillaria skinneri Bateman ex Lindl., Bot. Reg. 26: misc.101 & 145, 1840
Maxillaria virginalis Scheidw., Bull. Acad. Roy. Sc. Brux. 9: 25, 1842
Lycaste virginalis (Scheidw.) Lindl. & Rodigas, Lindenia 4: 22 sub t. 153, 1888

Lycomormium squalidum (Poeppig & Endl.) Rchb.f., Bot. Zeit. 10: 833, 1852
BR 32: sub t.41:4 (1846)
E.Ecuador, Peru – 900–1500 m – E, T
Anguloa squalida Poeppig & Endl., Nov. Gen. Sp. Pl. 1: 43, 1836
Peristeria fuscata Lindl., Gard. Chron.: 388, 1856 (nomen illeg.)
Lycomormium elatum C.Schweinf., Amer. Orch. Soc. Bull. 12: 135, 1943

o **Macradenia lutescens** R.Br., Bot. Reg. 8: t.612, 1822
BR 8: t.612 & sub t.618 (1822) / 8: sub t.686 (1823) / 21: t.1815 (1835)
Type: Trinidad, cult. Griffin (BM., holo!)
W.Indies, USA (Florida), Trinidad, Venezuela, Colombia, Guyana, Surinam – E
o Macradenia triandra Lindl., Bot. Reg. 21: t.1815, 1835
Rhynchadenia cubensis A.Rich., Sagra Cub. 2: 248, 1850
Macradenia surinamensis Rchb.f. & Wullschl., Walp. Ann. 6: 697, 1863

Malaxis caulescens (Lindl.) Kuntze, Rev. Gen: 673, 1891
BR 27: sub t.1 (1841)
Type: Peru, Loa Valley, Hall s.n. (K-Lindl., holo!)
Peru – T
Microstylis caulescens Lindl., Bot. Reg. 27: sub t.1, 1841

Malaxis cordata (Lindl.) Kuntze, Rev. Gen. 2: 673, 1891
BR 24: misc.134 (1838)
Type: Mexico, cult. Barker (K-Lindl., holo!)
Mexico – T
Dienia cordata Lindl., Bot. Reg. 24: misc.134, 1838
Microstylis cordata (Lindl.) Rchb.f., Walp. Ann. 6: 207, 1861

Malaxis excavata (Lindl.) Kuntze, Rev. Gen. 2: 673, 1891
BR 24: misc.93 (1838)
Type: Mexico, cult. Barker (K-Lindl., holo!)
Mexico to Brazil, Bolivia, Argentina – T, E
Epidendrum umbellatum Vell., Fl. Flum. 9: t.23, 1827
Microstylis excavata Lindl., Bot. Reg. 24: misc.93, 1838
Microstylis hastilabia Rchb.f., Beitr. Orch. Centr. Amer.: 101, 1866
Microstylis simillima Rchb.f., Beitr. Orch. Centr. Amer.: 101, 1866
Cheiropterocephalus sertuliferus Barb.Rodr., Gen & Sp. Orch. Nov. 1: 29, 1877
Malaxis simillima (Rchb.f.) Kuntze, Rev. Gen. 2: 673, 1891
Microstylis quadrangularis Cogn., Mart. Fl. Bras. 3 (6): 551, 1906
Microstylis spiralipetala Cogn., Bull. Soc. Roy. Belg. 43: 302, 1906
Microstylis carpinterae Schltr., Beih. Bot. Centralbl. 36 (2): 381, 1918
Microstylis muelleri Schltr., Fedde Rep. 17: 12, 1920
Microstylis paranaensis Schltr., Fedde Rep. 16: 330, 1920
Malaxis carpinterae (Schltr.) Ames, Orchid. 7: 157, 1922
Microstylis ottonis Schltr., Fedde Rep. Beih. 10: 39, 1922
Malaxis lankesteri Ames, Sched. Orch. 4: 7, 1923
Malaxis maxonii Ames, Sched. Orch. 6: 36, 1923
Microstylis sertulifera (Barb.Rodr.) Schltr., Fedde Rep. Beih. 35: 46, 1925
Malaxis uncinata Ames & C.Schweinf., Sched. Orch. 10: 15, 1930

Malaxis histionantha (Link, Klotzsch & Otto) Garay & Dunsterv., Venez. Orch. Ill. 6: 226, 1976
BR 26: misc.214 (1840)
Mexico, Guatemala, Honduras, Costa Rica, Panama, Venezuela, Colombia, Brazil – T
Microstylis histionantha Link, Klotzsch & Otto, Ic. Pl. Rar. 1: 11 t.5, 1840
Microstylis disepala Rchb.f., Linnaea 26: 142, 1854
Malaxis disepala (Rchb.f.) Kuntze, Rev. Gen. 2: 673, 1891
Microstylis brenesii Schltr., Fedde Rep. Beih. 19: 167, 1923

Malaxis latifolia Smith, Rees Cyclop.: 22, 1812
BR 10: sub t.825 (1824)
Sri Lanka, India, Sikkim, Nepal, Assam, Burma, China, Ryukyu Is.,Philippines, Thailand, Laos, Cambodia, Vietnam, Malaya, Sumatra, Java, Celebes, Borneo, New Caledonia, Australia – T
Malaxis plicata Roxb., Hort. Beng.: 68, 1814 (nomen nudum)
Dienia congesta Lindl., Bot. Reg. 10: sub t.825, 1824
Gastroglottis montana Blume, Bijdr.: 597, 1825
Neottia plantaginea D.Don, Prod. Fl. Nepal: 27, 1825
Spiranthes plantaginea (D.Don) Sprengel, Syst. Veg. 3: 708, 1827
Dienia fusca Lindl., Gen. Sp. Orch. Pl.: 22, 1830
Microstylis rheedii Rchb.f., Bonpl. 5: 58, 1857 (non Lindl.)
Microstylis congesta (Lindl.) Rchb.f., Walp. Ann. 6: 206, 1861
Microstylis fusca (Lindl.) Rchb.f., Walp. Ann. 6: 207, 1861
Microstylis trilobulata Kurz, Andaman Rep. App. B: 19, 1870
Liparis bernaysii F.v.Muell., Fragm. 11: 21, 1878
Microstylis bernaysii F.v.Muell., Fragm. 11: 21, 1878
Microstylis congesta var. fusca (Lindl.) Ridley, J. Linn. Soc. 24: 335, 1888
Microstylis latifolia (Smith) J.J.Smith, Fl. Buitenz. 6: 248 fig.185, 1905
Microstylis latifolia var. fusca (Lindl.) J.J.Smith, Fl. Buitenz. 6: 249, 1905
Malaxis curranii Ames, Phil. J. Sc. 6: 44, 1911
Malaxis latifolia var. fusca (Lindl.) Ames, Phil. J. Sc. 6: 45, 1911
Microstylis carnosula Rolfe ex Downie, Kew Bull.: 368, 1925
Anaphora liparioides Gagnepain, Bull. Mus. Hist. Nat. Paris s.2, 4: 592, 1929
Liparis krempfii Gagnepain, Bull. Soc. Bot. Fr. 76: 514, 1929
Liparis turfosa Gagnepain, Bull. Soc. Bot. Fr. 76: 515, 1929
Microstylis fineti Gagnepain, Bull. Mus. Hist. Nat. Paris s.2, 4: 592, 1932
Microstylis kizanensis Masamune, Ann. Rep. Kaihoku Bot. Gard. 3: 75, 1933
Microstylis ishigakensis Ohwi, J. Jap. Bot. 13: 439 fig.1, 1937
Microstylis dalatensis Guillaumin, Bull. Mus. Hist. Nat. Paris s.2, 27 (5): 397, 1955
Malaxis carnosula (Rolfe ex Downie) Seidenf. & Smitin., Orch. Thail. 2 (1): 146, 1959
Malaxis congesta (Lindl.) Deb, Bull. Bot. Surv. India 3: 128, 1961
Malaxis kizanensis (Masamune) Hatsusima, Fl. Ryukyus: 863, 1971

Malaxis unifolia Michaux, Fl. Bor.-Am. 2: 157, 1803
BR 15: t.1290 (1829)
Canada, USA (Georgia, Illinois), Mexico, Guatemala, Honduras, Jamaica, Cuba, W.Indies – up to 2300 m – T
Malaxis ophioglossoides Muhlenb. ex Willd., Sp. Pl. ed.4, 4: 90, 1805
Achroanthes ophioglossoides (Muhlenb. ex Willd.) Raf., Med. Repos. N.Y. Hex. 2 (5): 352, 1808
Achroanthes unifolia (Michaux) Raf., Med. Repos. Origin. Essays Intell. Phys. 2 (5): 352, 1808
Microstylis ophioglossoides (Muhlenb. ex Willd.) Nutt. ex Eaton, Man. Bot. 2, Mid. States ed.3: 353, 1822
o Microstylis ophioglossoides var. mexicana Lindl., Bot. Reg. 15: t.1290, 1829
Malaxis thaspiformis A.Rich. & Galeotti, Ann. Sc. Nat. s.3, 3: 18, 1845
Microstylis unifolia (Michaux) Britton Sterns & Pogg., Prelim. Cat. Anthoph. & Pteridoph. N.Y.: 51, 1881
Microstylis grisebachiana Fawcett & Rendle, J. Bot. 47: 7, 1909
Malaxis grisebachiana (Fawcett & Rendle) Fawcett & Rendle, Fl. Jam. 1: 43 t.6 figs.18–23, 1910

Masdevallia cuprea Lindl., Bot. Reg. 29: misc.125, 1843
BR 29: misc.125 (1835)
Type: Guyana, Cayenne, cult. Hamburg (K-Lindl., holo!)
Brazil, Guyana – E
Masdevallia cayennensis Rchb.f., Otto & Dietr. Allg. Gartenz. 23: 242, 1855

Masdevallia floribunda Lindl., Bot. Reg. 29: misc.112, 1843
BR 29: misc.112 (1843)
Type: Mexico, cult. J.Rogers (K-Lindl., holo!)
Mexico, Guatemala, Honduras – E
Masdevallia galeottiana A.Rich., Ann. Sc. Nat. s.3, 3: 17, 1845
Masdevallia lindeniana A.Rich. & Galeotti, Ann. Sc. Nat. s.3, 3: 17, 1845
Masdevallia myriostigma Morren, Belg. Hort. 23: 361 t.23, 1873
Masdevallia tuerckheimii Ames, Orch. 2: 265, 1908

Masdevallia infracta Lindl., Gen. Sp. Orch. Pl.: 193, 1833
BR 24: misc.64 (1838)
Brazil – E
Masdevallia longicaudata Lemaire, Ill. Hort. 25: 109, 1868

Masdevallia plantaginea (Poeppig & Endl.) Cogn., Mart. Fl. Bras. 3 (4): 336, 1896
BR 28: misc.p.82:103 (1842)
Peru – E
Specklinia plantaginea Poeppig & Endl., Nov. Gen. Sp. Pl. 1: 51 t.89a, 1836
Pleurothallis plantaginea (Poeppig & Endl.) Lindl., Bot. Reg. 28: misc.p.82: 103, 1842
Humboldtia plantaginea (Poeppig & Endl.) Kuntze, Rev. Gen. 2: 668, 1891

Maxillaria acicularis Herbert ex Lindl., Bot. Reg. 23: sub t.1986, 1837
BR 23: sub t.1986 (1837)
Type: Brazil, W.Herbert s.n. (K-Lindl., holo!)
Brazil – E

Maxillaria acutifolia Lindl., Bot. Reg. 25: misc.148, 1839
BR 25: misc.148 (1839)
Type: Guyana, cult. Loddiges (K-Lindl., holo!)
Honduras, Nicaragua, Costa Rica, Colombia, Ecuador – E

Maxillaria acutipetala Hook., Bot. Mag. 69: t.3966, 1842
BR 29: misc.36 (1843)
Centr. America – E

Maxillaria aggregata (Kunth) Lindl., Gen. Sp. Orch. Pl.: 151, 1831
BR 18: sub t.1549:37 (1832)
Venezuela, Colombia, Ecuador, Peru – E
Dendrobium aggregatum Kunth, Nov. Gen. & Sp. Pl. 1: 358, 1816
Ornithidium distichum Lindl., Benth. Pl. Hartweg: 153, 1845
Ornithidium aggregatum (Kunth) Rchb.f., Bonpl. 2: 18, 1854
Ornithidium pfitzeranum Lehm. & Kraenzlin, Engl. Bot. Jahrb. 26: 486, 1899
Ornithidium breve Schltr., Fedde Rep. Beih. 8: 100, 1921
Maxillaria disticha (Lindl.) C.Schweinf., Bot. Mus. Leafl. Harv. Univ. 11: 270, 1945

Maxillaria alba (Hook.) Lindl., Gen. Sp. Orch. Pl.: 143, 1832
BR 18: sub t.1549:15 (1832)
Trop. America – E
Dendrobium album Hook., Exot. Fl. 2: t.142, 1825
Broughtonia alba (Hook.) Sprengel, Syst. Veg. 3: 735, 1826
Maxillaria lactea Schltr., Beih. Bot. Centralbl. 42 (2): 131, 1925
Camaridium album (Hook.) Hoehne, Arq. Bot. Est. S.Paulo 2 (4): 72, 1947

Maxillaria bicolor Ruiz Lopez & Pavon, Syst. Veg. Fl. Peruv. & Chil. 1: 224, 1798
BR 18: sub t.1549:7 (1832)
Peru – E, T – obscure sp., probably Xylobium, fide Schweinfurth
Dendrobium bicolor (Ruiz Lopez & Pavon) Pers., Syn. Pl. 2: 524, 1807
Dicrypta bicolor (Ruiz Lopez & Pavon) Bateman ex Loud., Hort. Brit. Suppl. 2: 630, 1839

Maxillaria camaridii Rchb.f., Hamb. Gartenz. 19: 547, 1863
BR 10: t.844 (1824)
Guatemala, Honduras, Costa Rica, Panama, W.Indies, Venezuela, Trinidad, Guyana, Surinam, Brazil, Colombia, Peru, Bolivia – E
o Camaridium ochroleucum Lindl., Bot. Reg. 10: t.844, 1824 (non Lodd.1833)
Maxillaria lutescens Scheidw., Allg. Gartenz. 7: 145, 1830
Cymbidium ochroleucum (Lindl.) Lindl., Gen. Sp. Orch. Pl.: 168, 1833
Ornithidium album Hook., Bot. Mag. 61: t.3306, 1834
Ornithidium fragrans Rolfe, Kew Bull.: 147, 1894
Camaridium cryptopodanthum Barb.Rodr., Contr. Jard. Bot. Rio 4: 101, 1907
Camaridium affine Schltr., Fedde Rep. Beih. 17: 72, 1922
Camaridium amazonicum Schltr., Beih. Bot. Centralbl. 42 (2): 135, 1925

Maxillaria candida Lodd. ex Lindl., Bot. Reg. 27: misc.59, 1841
BR 27: misc.59 (1841)
Type: Brazil, cult. Loddiges (K-Lindl., holo!)
Brazil – E

Maxillaria chlorantha Lindl., Bot. Reg. 23: sub t.1986, 1837
BR 23: sub t.1986 (1837)
Type: Guyana, cult. Loddiges (K-Lindl., icono!)
Colombia, Venezuela, Guyana, Brazil – E

Maxillaria crassifolia (Lindl.) Rchb.f., Bonpl. 2: 16, 1854
BR 12: t.1028 (1826)
Type: Jamaica, cult. Lee (K-Lindl., holo!)
Florida, W.Indies, Mexico to Panama, Venezuela, Colombia, Brazil – E, L
Epidendrum sessile Sw., Prod. Ind. Occ.: 122, 1788 (non Lindl.1845)
o Heterotaxis crassifolia Lindl., Bot. Reg. 12: t.1028, 1826
Dicrypta baueri Lindl., Gen. Sp. Orch. Pl.: 44, 1830
Dicrypta crassifolia Lindl. ex Loud. Hort. Brit. Suppl. 3: 536, 1839
Maxillaria sessilis (Sw.) Fawcett & Rendle, Fl. Jamaica 1: 120, 1910 (non Lindl.1845)
Maxillaria gatunensis Schltr., Fedde Rep. Beih. 17: 68, 1922

o **Maxillaria cucullata** Lindl., Bot. Reg. 26: t.12, 1840
BR 26: t. & sub t.12 (1840) / 30: misc.9 (1844)
Type: Mexico, Henchman s.n. cult. Duke of Devonshire (K-Lindl., holo!)
Mexico, Belize, Guatemala, El Salvador, Honduras, Costa Rica, Nicaragua, Panama – E
Maxillaria rhombea Lindl., Bot. Reg. 26: sub t.12, 1840
Maxillaria meleagris Lindl., Bot. Reg. 30: misc.9, 1844
Maxillaria lindeniana A.Rich. & Galeotti, Ann. Sc. Nat. s.3, 3: 24, 1845
Maxillaria atrata Rchb.f., Beitr. Orch. Centr. Am. 31: t.6 figs.1–3, 1866
Maxillaria obscura Lindl. & Rchb.f., Beitr. Orch. Centr. Am. 31, 1866
Maxillaria puncto-striata Rchb.f., Linnaea 41: 28, 1877
Maxillaria praestans Rchb.f., Gard. Chron. n.s.33: 566, 1885

Maxillaria cuneiformis Ruiz Lopez & Pavon, Syst. Veg. Fl. Peruv. & Chil. 1: 223, 1798
BR 18: sub t.1549:6 (1832)
Peru – E obscure sp., probably Xylobium, fide Schweinfurth

Maxillaria densa Lindl., Bot. Reg. 21: t.1804, 1835
BR 21: t.1804 (1835)
Type: cult. Loddiges (K-Lindl., holo!)
Mexico, Belize, Guatemala, Honduras, El Savador, Nicaragua, Ecuador – E
Ornithidium densum (Lindl.) Rchb.f., Bonpl. 3: 217, 1855

Maxillaria discolor (Lodd. ex Lindl.) Rchb.f., Walp. Ann. 6: 529, 1861
BR 25: misc.145 (1839)
Type: Guyana, cult. Loddiges (Lindl., holo!)
Nicaragua, Guyana, Venezuela, Ecuador, Peru, Bolivia – E
Dicrypta discolor Lodd. ex Lindl., Bot. Reg. 25: misc.145, 1839

Maxillaria elatior (Rchb.f.) Rchb.f., Walp. Ann. 6: 532, 1863
BR 31: misc.11 (1845)
Mexico, Guatemala, Honduras, Nicaragua, Costa Rica – E
Dicrypta elatior Rchb.f., Linnaea 18: 403, 1844
Maxillaria triangularis Lindl., Bot. Reg. 31: misc.11, 1845

Maxillaria graminifolia (Kunth) Rchb.f., Walp. Ann. 6: 538, 1863
BR 27: sub t.1 (1841)
Venezuela, Colombia, Ecuador, Peru – E
Isochilus graminifolius Kunth, Nov. Gen. & Sp. Pl. 1: 340 t.78, 1816
Isochilus linifolium (Presl) Lindl., Gen. Sp. Orch. Pl.: 113, 1831
Isochilus graminifolium (Kunth) Lindl., Bot. Reg. 27: sub t.1, 1841
Camaridium graminifolium (Kunth) Rchb.f., Linnaea 22: 857, 1849
Maxillaria matthewsii Rchb.f., Walp. Ann. 6: 539, 1863 (non Lindl.)

Maxillaria grandiflora (Kunth) Lindl., Gen. Sp. Orch. Pl.: 147, 1832
BR 18: sub t.1549:26 (1832)
Venezuela, Colombia, Ecuador – E
Dendrobium grandiflorum Kunth, Nov. Gen. & Sp. Pl. 1: 359, 1816
Broughtonia grandiflora (Kunth) Sprengel, Syst. Veg. 3: 735, 1826
Lycaste grandiflora (Kunth) Beer, Prakt. Stud. Orch.: 147, 1854
Maxillaria molitor Rchb.f., Gard. Chron. s.3, 2: 242, 1887
Maxillaria jucunda Lehm. & Kraenzlin, Engl. Bot. Jahrb. 26: 484, 1899

Maxillaria hastata (Pers.) Ruiz Lopez & Pavon, Syst. Veg. Fl. Peruv. & Chil. 1: 222, 1798
BR 18: sub t.1549:4 (1832)
Peru – L obscure sp., perhaps Oncidium, fide Schweinfurth
Dendrobium hastatum Pers., Syst. Pl. 2: 524, 1807

Maxillaria latifolia (Kunth) Lindl., Gen. Sp. Orch. Pl.: 149, 1832
BR 18: sub t.1549:31 (1832)
Colombia, Andian Region – E
Dendrobium latifolium Kunth, Nov. Gen. & Sp. Pl.: 360, 1816

Maxillaria ligulata (Pers.) Ruiz Lopez & Pavon, Syst. Veg. Fl. Peruv. & Chil. 1: 222, 1798
BR 18: sub t.1549:3 (1832)
Peru – E obscure sp., probably Oncidium, fide Schweinfurth
Dendrobium ligulatum Pers., Syn. Pl. 2: 524, 1807

Maxillaria lindleyana Schltr., Fedde Rep. Beih. 9: 162, 1921
BR 21: t.1799 (1835)
Type: Rio de Janeiro, Capt.Sutton s.n. cult. C.Lemon (K-Lindl., holo!)
Brazil – E
o Maxillaria crocea Lindl., Bot. Reg. 21: t.1799, 1835 (non Poeppig & Endl.)

Maxillaria longibracteata (Lindl.) Rchb.f., Walp. Ann. 6: 540, 1863

var. grandiflora (Lindl.) C.Schweinf., Bot. Mus. Leafl. Harv. Univ. 11: 279, 1945
BR 27: sub t.1 (1841)
Type: Peru, Haenke s.n. (Munich, holo!)
Peru – E
Isochilus grandiflorum Lindl., Bot. Reg. 27: sub t.1, 1841
Camaridium grandiflorum (Lindl.) Schltr., Fedde Rep. Beih. 9: 165, 1921

Maxillaria maculata (Kunth) Lindl., Gen. Sp. Orch. Pl.: 149, 1832
BR 18: sub t.1549:29 (1832)
Amazonas – 1200–4000 m – E obscure sp., perhaps Xylobium, fide Schweinfurth
Dendrobium maculatum Kunth, Nov. Gen. & Sp. Pl. 1: 359, 1816
Broughtonia maculatum (Kunth) Sprengel, Syst. Veg. 3: 735, 1826

Maxillaria madida Lindl., Bot. Reg. 24: misc.74, 1838
BR 24: misc.74 (1838)
Type: Brazil, cult. Loddiges (K-Lindl., icono!)
Brazil – E
Maxillaria cepula Rchb.f., Bonpl. 3: 216 & 226, 1855
Maxillaria mosenii Kraenzlin, Svensk Vet. Akad. Handl. 46 (10): 73 t.11, 1911
Maxillaria echinochila Kraenzlin, Svensk Vet. Akad. Handl. 46 (8): 22, 1921
Maxillaria hatschbachii Schltr., Fedde Rep. 23: 36, 1926

Maxillaria marginata (Lindl.) Fenzl, Van Houtte Fl. Serres 10: 11, 1855
BR 18: t.1530 & sub t.1549:8 (1832)
Type: Brazil, cult. Hort. Soc. (K-Lindl., holo!)
Brazil, Columbia, Ecuador – E
o Cymbidium marginatum Lindl., Bot. Reg. 18: t.1530, 1832
Maxillaria tricolor Lindl., Lond. J. Bot. 1: 189, 1842
Maxillaria punctulata Klotzsch, Allg. Gartenz. 19: 249, 1851
Maxillaria deflexa Klotzsch, Ind. Sem. Hort. Berol.: 12, 1852
Maxillaria crassipes Kraenzlin, Svensk Vet. Akad. Handl. 46 (10): 72 t.11, 1911
Maxillaria murilliana Hoehne, Arq. Bot. Est. S.Paulo 2: 130, 1952

Maxillaria miniata (Lindl.) L.O.Williams, Caldasia 3: 14, 1941
BR 31: misc.65 (1845)
Type: Columbia, cult. Rollissons (K-Lindl., holo!)
Venezuela, Colombia, Ecuador – E
Ornithidium miniatum Lindl., Bot. Reg. 31: misc.65, 1845
Ornithidium niveum Lindl., Orch. Linden.: 22, 1846
Ornithidium sanguinolentum Lindl., Orch. Linden.: 22, 1846
Maxillaria nivea (Lindl.) L.O.Williams, Caldasia 3: 14, 1941
Maxillaria sanguinolenta (Lindl.) C.Schweinf., Bot. Mus. Leafl. Harv. Univ. 13: 150, 1948

Maxillaria ochroleuca Lodd., Bot. Cab. 20: t.1904, 1833
BR 18: sub t.1549:13 (1832)
Brazil, Venezuela – E

Maxillaria parkeri Hook., Bot. Mag. 54: t.2729, 1827
BR 18: sub t.1549:21 (1832)
Venezuela, Guyana, Surinam, Brazil, Amazonas, Peru – E
Colax parkeri (Hook.) Sprengel, Tent.: 28, 1828
Menadena parkeri (Hook.) Raf., Fl. Tellur. 2: 98, 1836
Maxillaria hirtilabia Lindl., J. Hort. Soc. Lond. 8: 132, 1853
Maxillaria lorifolia Rchb.f., Bonpl. 2: 92, 1854
Maxillaria kegelii Rchb.f., Linnaea 41: 127, 1877
Maxillaria loretoensis C.Schweinf., Amer. Orch. Soc. Bull. 13: 92, 1944

o **Maxillaria picta** Hook., Bot. Mag. 59: t.3154, 1832
BR 18: sub t.1549:24 (1832) / 21: t.1802 (1835)
Brazil – E
Epidendrum uniflorum Vell., Fl. Flum. Ic. 9: t.34, 1827
Maxillaria leucocheila Hoffsgg., Verz. Orch.: 55, 1843
Maxillaria kreysigii Kuntze ex Rchb.f., Linnaea 25: 250, 1853
Maxillaria fuscata Klotzsch ex Rchb.f., Walp. Ann. 6: 520, 1863
Maxillaria monoceros Klotzsch ex Rchb.f., Walp. Ann. 6: 520, 1863
Maxillaria rupestris Barb.Rodr., Gen. & Sp. Orch. Nov. 2: 199, 1882

Maxillaria platypetala (Pers.) Ruiz Lopez & Pavon, Syst. Veg. Fl. Peruv. & Chil. 1: 221, 1798
BR 18: sub t.1549:14 (1832)
Peru – E obscure sp., perhaps Lycaste, fide Schweinfurth
Dendrobium platypetalum Pers., Syn. Pl. 2: 523, 1807
Dendrobium polypetalum Steudel, Nom. 1: 268, 1821
Maxillaria polypetala Steudel, Nom. 1: 268, 1821

Maxillaria porrecta Lindl., Bot. Reg. 24: misc.173, 1838
BR 24: misc.173 (1838)
Type: Brazil, Rio de Janeiro, cult. Loddiges (K-Lindl., holo!)
Brazil – E
Maxillaria mutabilis hort. ex Rchb.f., Bonpl. 3: 216, 1855

Maxillaria prolifera (Pers.) Ruiz Lopez & Pavon, Syst. Veg. Fl. Peruv. & Chil. 1: 226, 1798
BR 18: sub t.1549:11 (1832)
Peru – E obscure sp., fide Schweifurth
Dendrobium proliferum Pers., Syn. Pl. 2: 524, 1807

Maxillaria ramosa (Pers.) Ruiz Lopez & Pavon, Syst. Veg. Fl. Peruv. & Chil. 1: 226, 1798
BR 18: sub t.1549:12 (1832)
Venezuela, Colombia, Peru – E
Dendrobium ramosum Pers., Syn. Pl. 2: 524, 1807
Scaphyglottis taphallae Rchb.f., Linnaea 22: 855, 1849
Ornithidium taphallae (Rchb.f.) Rchb.f., Bonpl. 2: 18, 1854
Ornithidium dichotomum Schltr., Fedde Rep. Beih. 7: 178, 1920
Maxillaria taphallae (Rchb.f.) C.Schweinf., Bot. Mus. Leafl. Harv. Univ. 2: 288, 1945

o **Maxillaria rufescens** Lindl., Bot. Reg. 21: sub t.1802, 1835
BR 21: sub t.1802 (1835) / 22: t.1848 (1836) / 29: misc.121 (1843)
Type: cult. Loddiges (K-Lindl., holo!)
Trop. America – up to 1700 m – E
Maxillaria rugosa Scheidw., Berlin Gartenzeit 11: 101, 1843
Maxillaria articulata Klotzsch, Ind. Sem. Hort. Berol.: 11, 1853
Maxillaria vanillodora A.Rich. ex Rchb.f., Cat. Orch. Schiller.: 45, 1857
Maxillaria rufescens var. flavida Rchb.f., Saund. Ref. Bot. 2: 79, 1869
Maxillaria rufescens var. minor Fawcett & Rendle, J. Bot. 48: 108, 1910
Maxillaria abelei Schltr., Fedde Rep. Beih. 9: 101, 1921

Maxillaria setigera Lindl., Bot. Reg. 31: misc.38, 1845
BR 31: misc.38 (1845)
Type: La Guayra, Barker s.n. (K-Lindl., icono!)
Guyana, Venezuela, Colombia, Brazil – E
Maxillaria leptosepala Hook., Bot. Mag. 75: t.4434, 1849
Maxillaria callichroma Rchb.f., Bonpl. 2: 16, 1854
Maxillaria leptosepala var. subintegerrima Regel, Ann. Sc. Nat. s.4, 6: 374, 1856
Maxillaria caloglossa Rchb.f., Gard. Chron. n.s.10: 654, 1878
Maxillaria setigera var. angustifolia Klinge, Act. Hort. Petrop. 17: 139, 1898

Maxillaria subulata Lindl., Gen. Sp. Orch. Pl.: 147, 1832
BR 18: sub t.1549:17 (1832)
Brazil – E

o **Maxillaria tenuifolia** Lindl., Bot. Reg. 23: sub t.1986, 1837
BR 23: sub t.1986 (1837) / 25: t.8 (1839)
Type: Mexico, Vera Cruz, Hartweg s.n. (K-Lindl., holo!)
Mexico, Guatemala, El Salvador, Honduras, Nicaragua, Costa Rica – E, T
Maxillaria gracilifolia Kraenzlin, Mitt. Inst. Allg. Bot. Hamb. 6: 420, 1927

Maxillaria triphylla (Pers.) Ruiz Lopez & Pavon, Syst. Veg. Fl. Peruv. & Chil. 1: 225, 1798
BR 18: sub t.1549:9 (1832)
Peru – E obscure sp., probably Xylobium, fide Schweinfurth
Dendrobium triphyllum Pers., Syn. Pl. 2: 524, 1807

Maxillaria uncata Lindl., Bot. Reg. 23: sub t.1986, 1837
BR 23: sub t.1986 (1837) / 26: misc.155 (1840)
Type: Guyana, cult. Loddiges (K-Lindl., holo!)
Guatemala, Belize, Honduras, Nicaragua, Costa Rica, Panama, S.America – E
Maxillaria macleei Bateman ex Lindl., Bot. Reg. 26: misc.155, 1840
Maxillaria nana Hook., Ic. Pl. 4: t.31b, 1841
Maxillaria squamata Barb.Rodr., Gen. & Sp. Orch. Nov. 1: 118, 1877
Ornithidium squamata (Barb.Rodr.) Barb.Rodr., Gen. & Sp. Orch. Nov. 2: 209, 1882
Maxillaria stenostele Schltr., Beih. Bot. Centralbl. 36: 414, 1918
Maxillaria striatella Kraenzlin, Fedde Rep. 24: 359, 1928
Camaridium squamatum (Barb.Rodr.) Hoehne, Arq. Bot. Est. S.Paulo 2: 72, 1947
Camaridium unicatum (Lindl.) Hoehne, Arq. Bot. Est. S.Paulo 2: 127, 1952

Maxillaria undatiflora (Pers.) Ruiz Lopez & Pavon, Syst. Veg. Fl. Peruv. & Chil. 1: 225, 1798
BR 18: sub t.1549:10 (1832)
Peru – E obscure sp., surely not Maxillaria, fide Schweinfurth
Dendrobium undatiflorum Pers., Syn. Pl. 2: 524, 1807

Maxillaria undulata Ruiz Lopez & Pavon, Syst. Veg. Fl. Peruv. & Chil.: 221, 1798
BR 18: sub t.1549:1 (1832)
Peru – E obscure sp., probably Xylobium, fide Schweinfurth

Maxillaria variabilis Bateman ex Lindl., Bot. Reg. 23: sub t.1986, 1837
BR 23: sub t.1986 (1837) / 24: misc.26 (1838) / 31: misc.26 (1845)

Type: Mexico, cult. Bateman s.n. (K-Lindl., icono!)
Mexico, Guatemala, Honduras, El Salvador, Nicaragua, Costa Rica, Panama, Guyana – 1000–2000 m – E
Maxillaria henchmanni Hook., Bot. Mag. 64: t.3614, 1837
Maxillaria varabilis var. unipunctata Lindl., Bot. Reg. 24: misc.26, 1838
Maxillaria angustifolia Hook., Ic. Pl.: t.348, 1841
Maxillaria lyonii Lindl., Bot. Reg. 31: misc.26, 1845
Maxillaria revoluta Klotzsch, Allg. Gartenz. 20: 186, 1852
Maxillaria chiriquensis Schltr., Fedde Rep. Beih. 17: 68, 1922
Maxillaria panamensis Schltr., Fedde Rep. Beih. 17: 70, 1922
Maxillaria costaricensis Schltr., Fedde Rep. Beih. 19: 232, 1923

Megalotus bifidus (Lindl.) Garay, Bot. Mus. Leafl. Harv. Univ. 23 (4): 185, 1972
BR 24: misc.5 (1838)
Type: Philippines, Cuming s.n. (K-Lindl., holo!)
Philippines – E
Saccolabium bifidum Lindl., Bot. Reg. 24: misc.5, 1838
Gastrochilus bifidus (Lindl.) Kuntze, Rev. Gen. 2: 661, 1891
Sarcanthus bifidus (Lindl.) Ames, Orchid. 5: 245, 1915

Mexicoa ghiesbreghtiana (A.Rich. & Galeotti) Garay, Bradea 1: 424, 1974
BR 31: misc.60:19 (1845) / 33: t.20 (1847)
Mexico – 1400–2200 m – E
Odontoglossum warneri Lindl., Bot. Reg. 31: misc.60:19, 1845
Oncidium ghiesbreghtianum A.Rich. & Galeotti, Ann. Sc. Nat. s.3, 3: 27, 1845
o Odontoglossum warneri var. purpuratum Lindl., Bot. Reg. 33: t.20, 1847

Micropera fuscolutea (Lindl.) Garay, Bot. Mus. Leafl. Harv. Univ. 23 (4):186, 1972
BR 33: sub t.18 (1847)
Type: Borneo, H.Low s.n. cult. Lowe (K-Lindl., holo!)
Borneo – E
Sarcochilus fuscoluteus Lindl., Bot. Reg. 33: sub t.18, 1847
Sarcochilus adnatus Ridley, J. Linn. Soc. 32: 373, 1896
Saccolabium adnatum (Ridley) Ridley, Fl. Malay Pen. 4: 174, 1924
Camarotis latisaccata J.J.Smith, Bull. Jard. Bot. Buitenz. s.3, 9: 187, 1927
Camarotis adnata (Ridley) Holttum, Gard. Bull. Sing. 11: 277, 1947

Micropera obtusa (Lindl.) Tang & Wang, Acta Phytotax. Sin. 1: 94, 1951
BR 30: misc.71 (1844)
Type: cult. Warner (K-Lindl., holo!)
India, Burma, Thailand – E
Camarotis obtusa Lindl., Bot. Reg. 30: misc.71, 1844
Sarcochilus obtusus (Lindl.) Benth. ex Hook.f., Fl. Brit. Ind. 6: 36, 1891

Micropera pallida (Roxb.) Lindl., Bot. Reg. 18: sub t.1522, 1832
BR 18: sub t.1522 (1832)
N.E.India, Burma, Thailand, Cambodia, Vietnam, Malaya, Sumatra, Java, Borneo – E
Aerides pallida Roxb., Hort. Beng.: 63, 1814
Camarotis apiculata Rchb.f., Bonpl. 5: 39, 1857
Camarotis cochinchinensis Rchb.f., Gard. Chron. n.s.6: 740, 1876
Sarcochilus cochinchinensis (Rchb.f.) Nicolson, Dict. 3: 360, 1886
Sarcochilus roxburghii Hook.f., Fl. Brit. Ind. 6: 36, 1890
Saccolabium saxicolum Ridley, Trans. Linn. Soc. 3: 374, 1893
Saccolabium flaveolens Ridley, J. Linn. Soc. 32: 360, 1896
Camarotis pallida (Roxb.) Prain, Beng. Pl.: 769, 1903 (non Lindl.1859)
Sarcanthus apiculatus (Rchb.f.) J.J.Smith, Fl. Buitenz. 6: 598 fig.445, 1905
Sarcanthus thorelii Guillaumin, Bull. Soc. Bot. Fr. 77: 331, 1930
Micropera cochinchinensis (Rchb.f.) Tang & Wang, Act. Phyto. Sin. 1: 94, 1951
Micropera apiculata (Rchb.f.) Garay, Bot. Mus. Leafl. Harv. Univ. 4: 186, 1972

Miltonia clowesii Lindl., Sert. Orch.: t.34, 1839
BR 25: misc.153 (1839) / 30: misc.13:14 (1844)
Brazil – E
Odontoglossum clowesii (Lindl.) Lindl., Bot. Reg. 25: misc.153, 1839
Brassia clowesii (Lindl.) Lindl., Bot. Reg. 30: misc.13:14, 1844
Oncidium clowesii (Lindl.) Beer, Prakt. Stud. Orch.: 284, 1854

o **Miltonia cuneata** Lindl., Bot. Reg. 30: misc.28, 1844
BR 30: misc.28 (1844) / 31: t.8 & sub t.8:5 (1845)
Type: Brazil, cult. Rollissons (K-Lindl., holo!)
Brazil – E
Oncidium cuneatum (Lindl.) Beer, Prakt. Stud. Orch. 285, 1854 (non Lindl., nec Scheidw.)
Miltonia speciosa Klotzsch, Allg. Gartenz. 17: 129, 1855
Oncidium speciosum (Klotzsch) Rchb.f., Walp. Ann. 6: 761, 1863

Miltonia flavescens (Lindl.) Lindl., Sert. Orch. sub t.48, 1841
BR 19: t.1627 (1833) / 27: misc.54 (1839) / 29: sub t.59:3, 8, 9 (1841)
Type: Mexico, Tate s.n. cult. R.Harrison (K-Lindl., holo!)
Brazil, Paraguay, Argentina – 200–700 m – E
o Cyrtochilum flavescens Lindl., Bot. Reg. 19: t.1627, 1833
Cyrtochilum stellatum Lindl., Sert. Orch.: t.7, 1838
Miltonia stellata (Lindl.) Lindl., Sert. Orch.: sub t.48, 1841
Oncidium flavescens (Lindl.) Rchb.f., Walp. Ann. 6: 757, 1863

Miltonia russelliana (Lindl.) Lindl., Sert. Orch.: sub t.48, 1840
BR 22: t.1830 (1836) / 31: sub t.8:6 (1845)
Type: Brazil, J.Roos s.n. cult. Duke of Bedford (K-Lindl., holo!)
Brazil – E
o Oncidium russellianum Lindl., Bot. Reg. 22: t.1830, 1836
Gynizdon russellianum (Lindl.) Raf., Fl. Tellur. 4: 40, 1837
Miltonia quadrijuga Dusén & Kraenzlin, Arch. Bot. Stockh. 16 (8): 23, 1921

o **Miltonia spectabilis** Lindl., Bot. Reg. 23: sub t.1976, 1837
BR 23: t.1992 & sub t.1976 (1837) / 31: sub t.8:1 (1845)
Types: Brazil cult. Loddiges, Barker (K-Lindl., holo!)
Brazil – E
Miltonia moreliana hort., Portef. Hort. 1: 195, 1847 (nomen tantum)
Miltonia spectabilis var. purpureo-violacea Hook., Bot. Mag. 75: t.4425, 1849
Oncidium spectabile var. morelianus Beer, Prakt. Stud. Orch.: 293, 1854
Oncidium spectabile var. purpureo-violacea (Hook.) Beer, Parkt. Stud. Orch.: 293, 1854
Miltonia warneri Nicholson, Dict. Gard.: 369, 1886

Miltonioides karwinskii (Lindl.) Brieger & Lueckel, Die Orchidee 34 (4): 131, 1983
BR 23: sub t.1992 (1837)
Type: Mexico, Karwinski s.n. (K-Lindl., holo!)
Mexico – 1200–1700 m – E
Cyrtochilum karwinskii Lindl., Bot. Reg. 23: sub t.1992, 1837
Odontoglossum karwinskii (Lindl.) Lindl., Sert. Orch.: sub t.48, 1841
Miltonia karwinskii (Lindl.) Lindl., J. Hort. Soc. 3: 83, 1849

Miltonioides laevis (Lindl.) Brieger & Lueckel, Die Orchidee 34 (4): 131, 1983
BR 30: t.39 (1844) / 31: misc.60:23 (1845)
Type: Guatemala, Skinner s.n. cult. Hort. Soc. (K-Lindl., holo!)
Mexico, Guatemala – E
o Odontoglossum laeve Lindl., Bot. Reg. 30: t.39, 1844
Oncidium laeve (Lindl.) Beer, Prakt. Stud. Orch.: 288, 1854
Odontoglossum leucomelas Rchb.f., Bot. Zeit. 22: 415, 1864
Miltonia laevis (Lindl.) Rolfe, Orch. Rev. 12: 300, 1904
Miltonia leucomelas (Rchb.f.) Rolfe, Orch. Rev. 12: 300, 1904

Monadenia bracteata (Sw.) Durand & Schinz, Consp. Fl. Afr. 3, 1895
BR 4: t.324 (1818)
S.Africa – E
o Disa bracteata Sw., Schrad. Neues J. Bot. 1: 28, 1805

Monadenia cernua (Sw.) Durand & Schinz, Consp. Fl. Afr. 5: 111, 1895
BR 3: t.210 (1817)
S.Africa – T
Disa cernua Sw., Kongl. Vetensk. Acad. Handl. 21: 211, 1800
Disa physodes Sw., Kongl. Vetensk. Acad. Handl. 21: 212, 1800
o Disa prasinata Ker Gawler, Bot. Reg. 3: t.210, 1817
Monadenia densiflora Lindl., Gen. Sp. Orch. Pl.: 357, 1838
Monadenia prasinata (Ker Gawler) Lindl., Gen. Sp. Orch. Pl.: 358, 1838
Monadenia inflata Sonder, Linnaea 19: 102, 1847
Monadenia physodes (Sw.) Rchb.f., Flora: 461, 1883
Disa densiflora (Lindl.) Bolus, J. Linn. Soc. Bot. 25: 197, 1890

o **Mormodes aromatica** Lindl., Bot. Reg. 27: misc.162, 1841
BR 27: misc.162 (1841) / 28: sub t.43:2 (1842) / 29: t.56 & sub t. 33:3 (1843)
Type: Mexico, Oaxaca, cult. Loddiges (K-Lindl., holo!)
Mexico, El Salvador, Honduras – E

o **Mormodes atropurpureum** Lindl., Introd. Nat. Syst. Bot. ed.2 : 446, 1836 (non Hook.)
BR 22: t.1861 (1836) / 28: sub t.43:1 (1842) / 29: sub t.33:1 (1843)
Costa Rica, Panama – E

Mormodes bucinator Lindl., Bot Reg. 26: misc.9, 1840
BR 26: misc.9 (1840) / 27: misc.191 (1841) / 28: sub t.43:4 (1842) / 29: sub t.33:6 (1843)
Type: cult. T.Williams (K-Lindl., holo!)
Mexico to Nicaragua, S.America – E
Mormodes lentiginosa Hook., Bot. Mag. 75: t.4455, 1849
Mormodes brachystachya Klotzsch, Ind. Sem. Hort. Berol.: 13, 1852
Mormodes flavida Klotzsch, Allg. Gartenz. 20: 113, 1852
Mormodes leucochila Klotzsch, Allg. Gartenz. 21: 289, 1853

Mormodes marmorea Klotzsch, Allg. Gartenz. 21: 290, 1853
Mormodes vitellina Klotzsch, Allg. Gartenz. 21: 290, 1853
Mormodes wagneriana Klotzsch, Allg. Gartenz. 21: 147, 1853
Mormodes amazonicum Brade, Arq. Serv. Florest. Rio de Janeiro 1 (1): 44, 1939

o **Mormodes lineatum** Bateman ex Lindl., Bot. Reg. 27: misc.107, 1841
BR 27: misc.107 (1841) / 28: t.43 & sub t.43:6 (1842) / 29: sub t. 33:2 (1843)
Types: Guatemala, Hartweg, Skinner s.n. (K-Lindl., syn!)
Guatemala – E

o **Mormodes luxatum** Lindl., Bot. Reg. 28: misc.66, 1842
BR 28: sub t.43:7 & misc.66 (1842) / 29: t.33 & sub t.33:5 (1843)
Type: Mexico, Valladolid, Ross s.n. cult. Barker (K-Lindl., holo!)
Mexico – 1400–2100 m – E

Mormodes maculata (Klotzsch) L.O.Williams, Ceiba 1: 188, 1950
BR 24: misc.176 (1838) / 25: misc.7 (1839) / 28: sub t.43:3 (1842) / 29: sub t.33:4 (1843)
Mexico – E
Cyclosia maculata Klotzsch, Allg. Gartenz. 6: 306, 1838
Mormodes pardina Bateman, Orch. Mex. & Guat.: t.14, 1838
Mormodes citrina Lindl., Bot. Reg. 28: sub t.43, 1842 nomen
Catasetum citrina Lindl., Bot. Reg. 28: sub t.43, 1842 nomen

Mormolyca ringens (Lindl.) Schltr., Die Orchideen: 436, 1914
BR 26: misc.121 (1840)
Types: Mexico, Hartweg & Schiede (K-Lindl., syn!)
Mexico, Belize, Guatemala, El Salvador, Honduras, Nicaragua, Costa Rica – E
Trigonidium ringens Lindl., Bot. Reg. 26: misc.121, 1840
Mormolyca lineolata Fenzl., Denkschr. Akad. Wiss. Math. Nat. Wien 1: 253 t.29, 1850

Myoxanthus affinis (Lindl.) Luer, Selbyana 7: 35, 1982
BR 28: misc.p.79:75 (1842)
Type: Peru, Mathews 1899 (K-Lindl., holo!)
Peru – E
Pleurothallis affinis Lindl., Hook. Comp. Bot. Mag. 2: 354, 1836
Humboldtia affinis (Lindl.) Kuntze, Rev. Gen. 2: 667, 1891

Myoxanthus monophyllus Poeppig & Endl., Nov. Gen. & Sp. Pl. 1: 50 t. 88, 1836
BR 28: misc.p.79:74 (1842)
Peru – E
Pleurothallis poeppigii Lindl., Hook. Comp. Bot. Mag. 2: 354, 1836
Humboldtia poeppigii (Lindl.) Kuntze, Rev. Gen. 2: 668, 1891
Pleurothallis myoxanthus Schltr., Fedde Rep. Beih. 9: 141, 1921

Myoxanthus octomerioides (Lindl.) Luer, Selbyana 7: 47, 1982
BR 28: misc.p.78:71 (1842)
Mexico – E
Pleurothallis octomerioides Lindl., Hook. Comp. Bot. Mag. 2: 354, 1836

Myoxanthus scabripes (Lindl.) Luer, Selbyana 7: 49, 1982
BR 25: misc.155 (1839) / 28: misc.p.77:65 (1842)
Type: Brazil, Lt. Downey s.n. cult. M.Williams (K-Lindl., holo!)
Brazil – E
Pleurothallis scabripes Lindl., Bot. Reg. 25: misc.155, 1839

Myrmecophila tibicinis (Bateman ex Lindl.) Rolfe, Orch. Rev.: 25. 51, 1917
BR 24: misc.12 (1838) / 27: misc.119 (1841) / 30: sub t.23:4 (1844) / 31: t.30 (1845)
Type: Honduras, Skinner s.n. (K-Lindl., holo!)
Mexico, Belize, Guatemala, Honduras, Nicaragua, Costa Rica – E
Epidendrum tibicinis Bateman ex Lindl., Bot. Reg. 24: misc.12, 1838
Schomburgkia tibicinis (Bateman ex Lindl.) Bateman, Orch. Mex. & Guat.: t.30, 1841
o Schomburgkia tibicinis var. grandiflora Hook., Bot. Mag. 75: t.4476, 1849
Cattleya tibicinis (Bateman ex Lindl.) Beer, Prakt. Stud. Orch.: 215, 1854
Bletia tibicinis (Bateman ex Lindl.) Rchb.f., Walp. Ann. 6: 429, 1862
Laelia tibicinis (Bateman ex Lindl.) L.Williams, Darwiniana 5: 77, 1941

Nageliella angustifolia (Booth ex Lindl.) Ames & Correll, Bot. Mus. Leafl. Harv. Univ. 10: 80, 1942
BR 29: misc.58 (1843)
Type: Guatemala, Skinner s.n. cult. C.Lemon (K-Lindl., holo!)
Guatemala – up to 2100 m – E, T
Hartwegia purpurea var. angustifolia Booth ex Lindl., Bot. Reg. 29: misc.58, 1843

Nageliella purpurea (Lindl.) L.O.Williams, Bot. Mus. Leafl. Harv. Univ 8: 144, 1940
BR 23: sub t.1970 (1837) / 26: misc.96 (1840)
Mexico, Guatemala, Honduras, Nicaragua – E
Hartwegia purpurea Lindl., Bot. Reg. 23: sub t.1970, 1837
Hartwegia comosa Lindl. ex Pfitzer, Engl. Pflanzenf. 2 (6): 143, 1889

o **Nanodes discolor** Lindl., Bot. Reg. 18: t.1541, 1832
BR 18: t.1541 (1832)
Type: Brazil, Rio de Janeiro, cult. Hort. Soc. (K-Lindl., icono!)
Mexico to Panama, Trinidad, Venezuela, Surinam, Brazil, Peru – E
Epidendrum discolor (Lindl.) Benth., ex Kew Hand-List Orch.: 105, 1921 (non A.Rich & Galeotti 1845)
Epidendrum brevicaule Schltr., Fedde Rep. Beih. 17: 30, 1922 (non Schltr. 1921)
Epidendrum congestum Schltr., Fedde Rep. Beih. 19: 118, 1923 (non Rolfe)
Epidendrum schlechterianum Ames, Sched. Orch. 7: 9 fig.1, 1924
Epidendrum congestoides Ames & C.Schweinf., Sched. Orch. 10: 61, 1930

Neofinetia falcata (Thunb.) Hu, Rhodora 27: 107, 1925
BR 4: t.283 (1818) / 18: sub t.1522 (1832)
Japan, Korea, Ryukyu Is. – E
Orchis falcata Thunb., Fl. Jap. 26, 1784
o Limodorum falcatum (Thunb.) Thunb., Mus. Nat. Acad. Upsal. 9: 139, 1791
Angraecum falcatum (Thunb.) Lindl., Gen. Sp. Orch. Pl.: 237, 1833
Oeceoclades falcata (Thunb.) Lindl., Gen. Sp. Orch. Pl.: 237, 1833
Oeceoclades lindleyana Regel, Ind. Sem. Hort. Petrop.: 43, 1865
Aerides thunbergii Miq., Ann. Mus. Bot. Ludg. Bat. 2: 205, 1866
Oeceoclades lindleyi Regel, Gartenfl. 15: 70, 1866
Angorchis falcata (Thunb.) Kuntze, Rev. Gen. 2: 651, 1891
Angraecopsis falcata (Thunb.) Schltr., Die Orchideen: 601, 1914
Finetia falcata (Thunb.) Schltr., Beih. Bot. Centralbl. 36 (2): 140, 1918
Nipponorchis falcata (Thunb.) Massam., Mem. Fac. Sc. Agr. Taihoku Univ. 11: 592, 1934
Holcoglossum falcatum (Thunb.) Garay & Sweet, Bot. Mus. Leafl. Harv. Univ. 23: 182, 1972

Neogardneria murrayana (Gardner ex Hook.) Garay, Orquideologia 8 (1): 9 & 33, 1973
BR 30: misc.15:7 (1844)
Brazil – E
Zygopetalum murrayanum Gardner ex Hook., Bot. Mag. 65: t.3674, 1838
Eulophia murrayana (Gardner ex Hook.) Steudel, Nom. ed.2, 1: 605, 1840
Promenaea florida Rchb.f., Hamb. Gartenz. 19: 12, 1863
Zygopetalum binotii De Wild., Gard. Chron. s.3, 38: 258, 1905
Neogardneria binotii (De Wild.) Hoehne, Arq. Bot. Est. S.Paulo 2 (4): 73, 1947

Neotinea maculata (Desf.) Stearn, Ann. Mus. Goulandris 2: 79, 1975
BR 18: t.1525 (1832)
W.Ireland, Isle of Man, Portugal, Canary Is. to W.Turkey, Lebanon, N.Africa – T
Satyrium maculatum Desf., Fl. Atlant. 2: 319, 1799
Orchis secundiflora Bertol., Rar. Ital. Pl. Decas 2: 42, 1806
Aceras secundiflora (Bertol.) Lindl., Bot. Reg. 18: t.1525, 1832
Aceras densiflora (Brot.) Boiss., Voy. Bot. Espagne 2: 595, 1841
Aceras intacta (Link) Rchb.f., Ic. Fl. Germ. 13: 2 t.148, 1850
Neotinea intacta (Link) Rchb.f., Poll. Orch. Gen.: 29, 1852
Tinaea intacta (Link) Boiss., Fl. Orient. 5: 58, 1884
Neotinea intacta f. luteola Renz, Fedde Rep. 27: 210 & 217, 1929

Nervilia plicata (Andrews) Schltr., Engl. Bot. Jahrb. 45: 403, 1911
BR 27: misc.124 (1841)
Sikkim, Bengal, Burma, China, Laos, Vietnam, Malaya, Java, Philippines, Hong Kong, New Guinea, Australia – T
Arethusa plicata Andrews, Bot. Rep. 5: t.321, 1803
Cordyla discolor Blume, Bijdr.: 417, 1825
Roptrostemon discolor (Blume) Blume, Fl. Jav. Praef. 6, 1828
Epipactis plicata Roxb., Fl. Ind. 3: 454, 1832
Pogonia plicata (Roxb.) Lindl., Gen. Sp. Orch. Pl.: 415, 1840
Pogonia discolor (Blume) Blume, Ann. Mus. Bot. Lugd. Bat. 1: 32, 1849
Pogonia biflora Wight, Ic. 5 (1): t.1758, 1851
Pogonia dallachyana F.Muell. ex Benth., Fl. Austr. 6: 310, 1873
Pogonia velutina Parish & Rchb.f., Trans. Linn. Soc. 30: 142, 1874
Pogonia pulchella Parish, Mason Burma: 186 & 201, 1883
Nervilia dallachyana (F.Muell. ex Benth.) Schltr., Engl. Bot. Jahrb. 39: 48, 1907
Pogonia pudica Ames, Orchid. 2: 44 fig., 1908
Nervilia biflora (Wight) Schltr., Engl. Bot. Jahrb. 45: 403, 1911
Nervilia discolor (Blume) Schltr., Engl. Bot. Jahrb. 45: 403, 1911
Nervilia fordii Schltr., Engl. Bot. Jahrb. 45: 403, 1911 (non Hance)
Nervilia velutina (Parish & Rchb.f.) Schltr., Engl. Bot. Jahrb. 45: 403, 1911
Aplostellis velutina (Parish & Rchb.f.) Ridley, Fl. Mal. Pen. 4: 204, 1924

Notylia aromatica Barker ex Lindl., Bot. Reg. 27: misc.77, 1841
BR 27: misc.77 (1841)

Type: Brazil, Para, Barker s.n. (K-Lindl., holo!)
Venezuela, Guyana, Surinam, Brazil – E

Notylia barkeri Lindl., Bot. Reg. 24: misc.168, 1838
BR 24: misc.168 (1838)
Type: Mexico, Barker s.n. (K-Lindl., holo!)
Mexico, Guatemala, Honduras, Nicaragua, Costa Rica, Panama – E
Notylia multiflora Hook., J. Bot. 3: 315 t.10, 1840 (non Lindl.)
Notylia huegelii Fenzl., Denkschr. Akad. Wien 2: 255, 1850
Notylia tridachne Lindl., Paxt. Flow. Gard. 3: 45, 1852
Notylia trisepala Lindl., Paxt. Flow. Gard. 3: 45, 1852
Tridachne virens Liebm. ex Lindl., Paxt. Flow. Gard. 3: 45, 1852
Notylia bipartia Rchb.f., Xen. Orch. 1: 47, 1854
Notylia tamaulipensis Rchb.f., Hamb. Gartenz. 16: 281, 1860
Notylia guatemalensis S.Watson, Proc. Am. Acad. 22: 477, 1887
Notylia angustilancea Schltr., Orchis 8: 135 t.4 figs.8–13, 1914
Notylia bernoullii Schltr., Beih. Bot. Centralbl. 36 (2): 502, 1918
Notylia pittieri Schltr., Beih. Bot. Centralbl. 36 (2): 418, 1918
Notylia brenesii Schltr., Fedde Rep. Beih. 19: 249, 1923
Notylia turialbae Schltr., Fedde Rep. Beih. 19: 145, 1923

Notylia incurva Lindl., Bot. Reg. 24: misc.167, 1838
BR 24: misc.167 (1838)
Type: Trinidad, cult. Loddiges (K-Lindl., holo!)
Trinidad, Venezuela – E

Notylia micrantha Lindl., Bot. Reg. 24: misc.170, 1838
BR 24: misc.170 (1838)
Type: Guyana, cult. Loddiges (K-Lindl., holo!)
Guyana, Brazil – E

Notylia pubescens Lindl., Bot. Reg. 28: misc.72, 1842
BR 28: misc.72 (1842)
Type: Brazil, cult. Loddiges (K-Lindl., holo!)
Brazil – E

Notylia punctata (Ker Gawler) Lindl., Bot. Reg. 11: sub t.930, 1825
BR 9: t.759 (1823) / 11: sub t.930 (1825) / 24: misc.166 (1838)
Type: Trinidad, D.Lockhart s.n. cult. Colvill (K-Lindl., holo!)
Trinidad – E
o Pleurothallis punctata Ker Gawler, Bot. Reg. 9: t.759, 1823
Gomesa tenuiflora Lodd., Bot. Cab. 9: t.806, 1824
Notylia apiculata Cogn., Urban Symb. Ant. 6: 617, 1910
Notylia broadwayi Cogn., Urban Symb. Ant. 6: 618, 1910

Notylia sagittifera (Kunth) Link & Klotzsch, Ic. Pl. Rar. 43: t.18, 1841
BR 11: sub t.930 (1825) / 24: misc.169 (1838)
Venezuela, Colombia, Ecuador, Guyana, Surinam, Brazil – E
Pleurothallis sagittifera Kunth, Nov. Gen. & Sp. Pl. 1: 365, 1815
Notylia multiflora Lindl., Bot. Reg. 11: sub t.930, 1825
Notylia tenuis Lindl., Bot. Reg. 24: misc.169, 1838

Oberonia complanata (A.Cunn.) M.Clements & D.A.Jones, Austr. Orch. Rev. 1: 98, 1989
BR 25: misc.36 (1839)
Type: Australia, Moreton Bay (Brisbane), A.Cunningham s.n. (not located)
Australia – E
Dendrobium complanatum A.Cunn., Bot. Reg. 25: misc.36, 1839

Oberonia cylindrica Lindl., Bot. Reg. 26: misc.23, 1840
BR 26: misc.23 (1840)
Type: Philippines, Cuming s.n. (K-Lindl., holo!)
Philippines – E
Malaxis cylindrica (Lindl.) Rchb.f., Walp. Ann. 6: 208, 1861
Iridorchis cylindrica (Lindl.) Kuntze, Rev. Gen. 2: 668, 1891

Oberonia miniata Lindl., Bot. Reg. 29: misc.8, 1843
BR 29: misc.8 (1843)
Type: Singapore, cult. Lodigges (K-Lindl., holo!)
Singapore – E
Malaxis miniata (Lindl.) Rchb.f., Walp. Ann. 6: 215, 1861

Oberonia recurva Lindl., Bot. Reg. 25: misc.8, 1839
BR 25: misc.8 (1839)
Type: India, Bombay, cult. Loddiges (K-Lindl., holo!)
India, Sri Lanka – E
Oberonia setifera Lindl., Fol. Orch. Oberonia: 3, 1859
Malaxis recurva (Lindl.) Rchb.f., Walp. Ann. 6: 212, 1861
Malaxis setifera (Lindl.) Rchb.f., Walp. Ann. 6: 210, 1861
Oberonia gardneriana Thwaites, Enum. Pl. Zeyl.: 296, 1861
Iridorchis recurva (Lindl.) Kuntze, Rev. Gen. 2: 668, 1891
Oberonia parvula King & Pantl., J. As. Soc. Beng. 64: 11, 1895
Oberonia croftiana King. & Pantl., Ann. Bot. Gard. Calc. 8: 7, 1898
Oberonia lingmalensis Blatt. & MacCann, J. Bomb. Nat. Hist. Soc. 35: 255, 1931
Oberonia myriantha var. parvula (King & Pantl.) Tang & Wang, Act. Phytotax. Sin. 1 (1): 24–102, 1951

Oberonia wightiana Lindl., Bot. Reg. 25: misc.9, 1839
BR 25: misc.9 (1839)
Type: India, Nilgheiries, Wight 179, (K-Lindl., lecto!)
India, Sri Lanka – E
Oberonia stachyoides A.Rich., Ann. Sc. Nat. 11 (15): 15 pl.1, 1841
Oberonia arnottiana Wight, Ic. Pl. Ind. Orch. 5: pl.1627 & 1628, 1852
Malaxis wightiana (Lindl.) Rchb.f., Walp. Ann. 6: 212, 1861
Iridorchis wightiana (Lindl.) Kuntze, Rev. Gen. 2: 668, 1891

Octomeria diaphana Lindl., Bot. Reg. 25: misc.146, 1839
BR 25: misc.146 (1839)
Type: Brazil, cult. Loddiges (K-Lindl., holo!)
Brazil – E

Octomeria gracilis Lodd. ex Lindl., Bot. Reg. 24: misc.55, 1838
BR 24: misc.55 (1838)
Type: Brazil, Rio de Janeiro, cult. Loddiges (K-Lindl., holo!)
Brazil – E
Octomeria semiteres Regel, Ind. Sem. Hort. Petropol.: 18, 1856

Octomeria grandiflora Lindl., Bot. Reg. 28: misc.80, 1842
BR 28: misc.80 (1842)
Type: Brazil, cult. Loddiges (K-Lindl., holo!)
Venezuela, Trinidad, Brazil – E

Octomeria tridentata Lindl., Bot. Reg. 25: misc.43, 1839
BR 25: misc.43 (1839)
Type: Guyana, cult. Loddiges (K-Lindl., holo!)
Argentina, Brazil, Guyana, Cuba – E
Octomeria multiflora Barb.Rodr., Gen. & Sp. Orch. Nov. 2: 98, 1882

Odontoglossum angustatum Lindl., Bot. Reg. 23: sub t.1992, 1837
BR 23: sub t.1992 (1837) / 31: misc.60:30 (1845)
Type: Ecuador, Lloa Valley, Hall s.n. (K, holo!)
Venezuela, Colombia, Ecuador, Peru – 2200–3400 m – E, T, L
Odontoglossum spilolanthum Rchb.f., Gard. Chron.: 731, 1872
Odontoglossum ulopterum Rchb.f., Gard. Chron.: 731, 1872

Odontoglossum cirrhosum Lindl., Gen. Sp. Orch. Pl.: 211, 1833
BR 31: misc.60:35 (1845)
Colombia, Ecuador, Peru – 1600–2000 m – E, T
Oncidium cirrhosum (Lindl.) Beer, Prakt. Stud. Orch.: 284, 1854
Odontoglossum cirrhosum var. klabochorum Rchb.f., Gard. Chron. n.s.10: 7, 1876
Odontoglossum cirrhosum var. gemmatum Rchb.f., Gard. Chron. n.s.12: 3, 1881
Odontoglossum cirrhosum var. hrubyanum Rchb.f., Gard. Chron. n.s.21: 3, 1885

Odontoglossum constrictum Lindl., Bot. Reg. 29: misc.25, 1843
BR 29: misc.25 (1843) / 31: misc.60:16 (1845)
Type: La Guayra, cult. Rucker (K-Lindl., holo!)
Venezuela, Colombia, Ecuador – 1700–2400 m – E
Odontoglossum sanderianum Rchb.f., Gard. Chron.: 492, 1882
Odontoglossum boddaertianum Rchb.f., Gard. Chron: 296, 1888

Odontoglossum crispum Lindl., Ann. Mag. Nat. Hist. 15: 256, 1843
BR 31: misc.60:5 (1845)
Colombia – 2250–3000 m – E
Odontoglossum alexandrae Bateman, Gard. Chron.: 1083, 1864
Odontoglossum bluntii Rchb.f., Bot. Zeit. 53: 415, 1864
Odontoglossum alexandrae var. guttatum Bateman, Bot. Mag. 94: t.5697, 1868
Odontoglossum reichenbachianum Lehm., Gard. Chron.: 395, 1883
Odontoglossum latimaculatum hort. ex Linden, Lindenia 4: t.145, 1888

Odontoglossum cristatum Lindl., Pl. Hartw.: 152, 1844
BR 31: misc.60:34 (1845)
Colombia, Ecuador – 1500–2000 m – E, T
Oncidium cristatum (Lindl.) Beer, Parkt. Stud. Orch.: 283, 1854

Odontoglossum epidendroides Kunth., Nov. Gen. & Sp. Pl. 1: 351 t.85, 1816
BR 31: misc.60:33 (1845)
Ecuador, Peru, Amazonas – 1800–2300 m – E
Oncidium epidendroides (Kunth.) Beer, Prakt. Stud. Orch.: 285, 1854

Odontoglossum gracile Lindl., Benth. Pl. Hartw.: 151, 1844
BR 31: misc.60:20 (1845)
Peru – E
Cyrtochilum gracile (Lindl.) Kraenzlin, Notizbl. Bot. Gart. Berlin 7: 98, 1917

Odontoglossum hallii Lindl., Bot. Reg. 23: sub t.1992, 1837
BR 23: sub t.1992 (1837) / 31: misc.60:36 (1845)
Type: Ecuador, Lloa Valley, Hall s.n. (K, holo!)
Colombia, Ecuador, Peru – 1500–2800 m – E, T
Oncidium hallii (Lindl.) Beer, Prakt. Stud. Orch.: 287, 1854
Odontoglossum chaetostoma Rchb.f., Gard. Chron. n.s.19: 562 & 592, 1883
Odontoglossum victor Rchb.f., Gard. Chron. n.s.19: 532, 1883

Odontoglossum ixioides (Lindl.) Lindl., Bot. Reg. 31: misc.60:22, 1845
BR 27: sub t.59:10 (1841) / 31: misc.60:22 (1845)
Colombia – 3000–3300 m – E, T
Cyrtochilum ixioides Lindl., Gen. Sp. Orch. Pl.: 211, 1833
Oncidium ixioides (Lindl.) Beer, Prakt. Stud. Orch.: 287, 1854

Odontoglossum lacerum Lindl., Sert. Orch.: sub t.25, 1838
BR 31: misc.60:27 (1845)
Ecuador, Peru – 2000 m – E

Odontoglossum longifolium Lindl., Benth. Pl. Hartw.: 152, 1844
BR 31: misc.60:21 (1845)
Peru – E
Cyrtochilum longifolium (Lindl.) Kraenzlin, Notizbl. Bot. Gart. Berl. 7: 98, 1917
Odontoglossum graminifolium C.Schweinf., Amer. Orch. Soc. Bull. 15: 30, 1946

Odontoglossum myanthum Lindl., Benth. Pl. Hartw.: 152, 1844
BR 31: misc.60:26 (1845)
Peru – E
Cyrtochilum myanthum (Lindl.) Kraenzlin, Notizbl. Bot. Gart. Berl. 7: 98, 1917

Odontoglossum mystacinum (Lindl.) Lindl., Fol. Orch. Odont.: 14, 1852
BR 24: misc.38 (1838) / 25: t.62 (1839) / 27: sub t.59:4 (1841) / 31: misc.60:37 (1845)
Type: Peru, cult. R.Harrison (K-Lindl., holo!)
Amazonas, Peru, Ecuador – E
o Cyrtochilum mystacinum Lindl., Bot. Reg. 24: misc.38, 1838
Odontoglossum rigidum Lindl., Benth. Pl. Hartw.: 152, 1844
Oncidium rigidum (Lindl.) Beer, Prakt. Stud. Orch.: 290, 1854

Odontoglossum pardinum (Lindl.) Lindl., Sert. Orch.: sub t.25, 1838
BR 27: sub t.59:3 (1841) / 31: misc.60:28 (1845)
Colombia, Ecuador, Peru – 2400–3500 m – E, T
Cyrtochilum pardinum Lindl., Gen. Sp. Orch. Pl.: 210, 1833
Oncidium pardinum (Lindl.) Beer, Prakt. Stud. Orch.: 291, 1854
Odontoglossum stenochilum Rchb.f., Gard. Chron.: 969, 1872
Odontoglossum bellum Schltr., Fedde Rep. Beih. 9: 108, 1921

Odontoglossum retusum Lindl., Benth. Pl. Hartw.: 152, 1844
BR 31: misc.60:24 (1845)
Colombia, Ecuador, Peru – 2500–2800 m – E, T
Oncidium retusum (Lindl.) Beer, Prakt. Stud. Orch.: 290, 1854
Cyrtochilum retusum (Lindl.) Kraenzlin, Notizbl. Bot. Gart. Berl. 7: 99, 1917

Odontoglossum revolutum Lindl., Ann. Mag. Nat. Hist. 15: 256, 1845
BR 31: misc.60:31 (1845)
Colombia, Peru – 330–3700 m – E, T
Oncidium revolutum (Lindl.) Beer, Prakt. Stud. Orch.: 290, 1854
Odontoglossum platyodon Rchb.f., Gard. Chron.: 1163, 1871
Odontoglossum bogotense Schltr., Fedde Rep. Beih. 27: 179, 1924

Oeceoclades maculata Lindl., Gen. Sp. Orch. Pl.: 237, 1833
BR 8: t.618 (1822) & sub t.686 (1823) / 10: sub t.867 (1825) / 18: sub t.1522 (1832)
Guinea Bissau, Sierra Leone, Liberia, Ivory Coast, Ghana, Togo, S.Tomé, Congo, Angola, Sudan, Uganda, Tanzania, Zimbabwe, Zanzibar, USA (Florida), W.Indies, Brazil, Argentina, Venezuela – T
o Angraecum maculatum Lindl., Coll. Bot.: t.15, 1821
Geodorum pictum Link & Otto, Ic. Pl. Hort. Reg. Berol. 1: 14, 1821
Aerobion maculatum (Lindl.) Sprengel, Syst. Veg. 3: 718, 1826
Eulophia maculata (Lindl.) Rchb.f., Walp. Ann. 6: 647, 1863
Eulophidium maculatum (Lindl.) Pfitzer, Entw. Nat. Anordn. Orch.: 88, 1887
Eulophia ledienii Stein ex N.E.Br., Kew Bull.: 90, 1889
Graphorchis maculata (Lindl.) Kuntze, Rev. Gen. 2: 661, 1891
Eulophidium warneckeanum Kraenzlin, Engl. Bot. Jahrb. 33: 70, 1902
Eulophidium ledienii (Stein ex N.E.Br.) De Wild., Ann. Mus. Congo s.5, 1: 115, 1904
Eulophidium nyassanum Schltr., Engl. Bot. Jahrb. 3: 593, 1915

Oeceoclades pulchra (Thouars) Clements & Cribb, Cat. Austr. Orch. in press.
BR 23: t.1972 (1837)
Madagascar to Pacific Is. – 200–1000 m – T
Limodorum pulchrum Thouars, Orch. Iles Afr.: t.43–44, 1822
o Eulophia macrostachya Lindl., Gen. Sp. Orch. Pl.: 182, 1833
Eulophia pulchra (Thouars) Lindl., Gen. Sp. Orch. Pl.: 182, 1833
Graphorchis pulchra (Thouars) Kuntze, Rev. Gen. 2: 662, 1891
Eulophia guamensis Ames, Phil. J. Sc. 9: 12, 1914
Eulophia rouxii Kraenzlin, Sar. & Roux Nov. Caled. 1: 82, 1914
Eulophia novaebudae Kraenzlin, Bull. Soc. Bot. Fr. 76: 301, 1929
Lissochilus pulcher (Thouars) H.Perrier, Fl. Madagascar Orch. 2: 41, 1939 (non Schltr.)
Eulophidium pulchrum (Thouars) Summerh., Bull. Jard. Bot. Et. Brux. 27: 400, 1957

Oeonia volucris (Thouars) Durand & Schinz, Consp. Fl. Afr. 5, 1892
BR 10: sub t.817 (1824)
Madagascar, Mascarene Is. – up to 1200 – E
Epidendrum volucre Thouars, Orch. Iles Afr.: t.81, 1822
Oeonia aubertii Lindl., Bot. Reg. 10: sub t.817, 1824
Aeranthus volucris (Thouars) Rchb.f., Walp. Ann. 6: 900, 1864
Epidendrum volucris (Thouars) Kuntze, Rev. Gen. 2: 659, 1891
Oeonia humblotii Kraenzlin, Engl. Jahrb. 43: 397, 1909

Ommatodium volucris (L.f.) Lindl., Gen. Sp. Orch. Pl.: 365, 1838
BR 9: sub t.700 (1823)
S.Africa – T
Ophrys volucris L.f., Suppl. Pl.: 403, 1781
Ophrys triphylla Thunb., Prodr. 2, 1794
Pterygodium volucris (L.f.) Sw., Kongl. Vet. Acad. Handl.: 218, 1800

Oncidium altissimum (Jacq.) Sw., Kongl. Vet. Acad. Stockh. 21: 240, 1800
BR 9: t.727 (1823) / 24: misc.61 (1838) / 25: t. & sub t.16 (1839) / 26: misc.46 (1840) / 28: misc.p.34:38 (1842)
USA (Florida), W.Indies, Mexico to Panama, Venezuela, Colombia, Ecuador, Peru, Guyana, Brazil – E
Epidendrum altissimum Jacq., Enum. Pl. Carib.: 30, 1760
Epidendrum maculatum Aublet, Pl. Guin. 2: 825, 1775
Cymbidium altissimum (Jacq.) Sw., Nov. Act. Upsal. 6: 74, 1799
o Oncidium luridum Lindl., Bot. Reg. 9: t.727, 1823
Oncidium intermedium Knowles & Westc., Fl. Cabin. 2: 53, 1838
Oncidium boydii hort. ex Lindl., Bot. Reg. 25: sub t.16, 1839
o Oncidium luridum var. guttatum Lindl., Bot. Reg. 25: t.16, 1839
Oncidium lindenii Lodd., Cat. Orch. ed.2: no.1326, 1842
Oncidium cosymbeforum Morren, Ann. Gand. 5: 333, 1849
Oncidium luridum var. atratum Lindl., J. Hort. Soc. 6: 54, 1851
Oncidium luridum var. intermedium Lindl., Fol. Orch. Oncid.: 41, 1855
Oncidium luridum var. morrenii Lindl., Fol. Orch. Oncid.: 41, 1855
Oncidium luridum var. olivaceum Lindl., Fol. Orch. Oncid.: 41, 1855
Oncidium sanguineum var. auriculatum Regel, Ind. Sem. Hort. Petrop.: 21, 1855
Oncidium guttatum (Lindl.) Rchb.f., Walp. Ann. 6: 782, 1863
Oncidium guttatum var. intermedium (Knowles & Westc.) Rchb.f., Walp. Ann. 6: 782, 1863
Oncidium guttatum var. morrenii (Lindl.) Rchb.f., Walp. Ann. 6: 782, 1863
Oncidium guttatum var. olivaceum (Lindl.) Rchbf.f., Walp. Ann. 6: 782, 1863
Oncidium guttatum var. roseum hort., Ill. Hort. 27: 103, 1880
Oncidium luridum var. dogsonii Williams, Orch. Grow. Man. ed.6: 490, 1885
Oncidium guttatum var. auriculatum (Regel) Cogn., Mart. Fl. Bras. 3 (6): 413, 1906
Oncidium guttatum var. dodgsonii (Williams) Cogn., Mart. Fl. Bras. 3 (6): 423, 1906
Oncidium maculatum (Aublet) Urban, Fedde Rep. 15: 306, 1918 (non Lindl.1842)

o **Oncidium amictum** Lindl., Bot. Reg. 33: t.66, 1847
BR 33: t.66 (1847)
Type: Brazil, cult. Loddiges (K-Lindl., holo!)
Brazil – E
Oncidium nitidum Barb.Rodr., Gen. & Sp. Orch. Nov. 2: 193, 1881

o **Oncidium ampliatum** Lindl., Gen. Sp. Orch. Pl.: 202, 1834
BR 20: t.1699 (1834) / 26: misc.97 (1840)
Guatemala, El Salvador, Honduras, Nicaragua, Costa Rica, Panama, Trinidad, Colombia, Venezuela, Ecuador, Peru – E
Oncidium bernoullianum Kraenzlin, Engl. Pflanzenr. Orch. Mon. Oncid.: 231, 1922

Oncidium ascendans Lindl., Bot. Reg. 28: sub t.4, 1842
BR 28: sub t.4 (1842)
Type: Guatemala, without collector (K-Lindl., holo!)
Mexico to Costa Rica, W.Indies – up to 1500 m – E

Oncidium aureum Lindl., Sert. Orch: sub t.25:5, 1838
BR 31: misc.60:38 (1845)
Ecuador, Peru, Bolivia – E
Odontoglossum bicolor Lindl., Bot. Reg. 31: misc.60: 38, 1845
Odontoglossum festatum Rchb.f., Bonpl. 2: 100, 1854
Odontoglossum hemichrysum Rchb.f. & Warsc., Bonpl. 2: 100, 1854
Oncidium aureum var. stenochilum Lindl., Fol. Orch. Oncid.: 22 no.67, 1855
Oncidium festatum (Rchb.f.) Rchb.f. & Warsc., Walp. Ann. 6: 733, 1863
Oncidium hemichrysum (Rchb.f. & Warsc.) Rchb.f., Walp. Ann. 6: 733, 1863
Ornithidium bicolor (Lindl.) Lindl. ex Rchb.f., Walp. Ann. 6: 733, 1863
Oncidium dichromum Rolfe, Orch. Rev. 3: 363, 1895

Oncidium barbatum Lindl., Gen. Sp. Orch. Pl.: 200, 1833
BR 28: misc.74 (1842)
Brazil – E
Oncidium microglossum Klotzsch, Allg. Gartenz.: 23, 1855
Oncidium suscephalum Barb.Rodr., Gen. & Sp. Orch. Nov. 2: 188, 1881
Oncidium johnianum Schltr., Orchis 1: 4, 1906

Oncidium batemannianum Parm. ex Knowles & Westc., Fl. Cab. 3: 183 t.137, 1840
BR 31: misc.46 (1845)
Brazil, Bolivia – E
Oncidium gallopavinum Morren, Ann. Soc. Bot. Gand 1: 13 t.1, 1845
Oncidium pinelianum Lindl., J. Hort. Soc. Lond. 3: Proceed. p.15 & 17, 1848
Oncidium stenopetalum Klotzsch, Allg. Gartenz.: 273, 1855
Oncidium brachystegium Kraenzlin, Engl. Pflanzenr. Orch. Mon. Oncid.: 203, 1922

Oncidium baueri Lindl., Bauer & Lindley Ill. Orch.: t.7, 1833
BR 18: t.1651 (1833) / 22: sub t.1851 (1836)
Mexico to Panama, W.Indies, Venezuela to Bolivia, Guyana, Brazil, Argentina – E
Epidendrum gigas L.C.Rich. ex Lindl., Gen. Sp. Orch. Pl.: 200, 1833
o Oncidium altissimum Lindl., Bot. Reg. 19: t.1651, 1834
Oncidium guttulatum Rchb.f. ex Lindl., Fol. Orch. Oncid.: 50, 1855
Oncidium kappleri Rchb.f. ex Lindl., Fol. Orch. Oncid.: 45, 1855
Oncidium polycladium Rchb.f. ex Lindl., Fol. Orch. Oncid.: 47, 1855
Oncidium confusum Rchb.f., Xen. Orch. 1: 234, 1858
Oncidium pentecostale Rchb.f., Hamb. Gartenz. 15: 56, 1859
Oncidium hebraicum Rchb.f., Gard. Chron. n.s.3: 780, 1875
Oncidium altissimum var. baueri (Lindl.) Stein., Orchideenb.: 404, 1892
Oncidium bolivianum Schltr., Fedde Rep. 10: 459, 1911

o **Oncidium bicallosum** Lindl., Pl. Hartw.: 94, 1842
BR 28: misc.14 (1842) / 29: t.12 (1843)
Mexico, Guatemala, El Salvador – 1500–2200 m – E

o **Oncidium bicolor** Lindl., Bot. Reg. 29: t.66, 1843
BR 29: t.66 (1843)
Type: Spanish Main, cult. Loddiges (K-Lindl., holo!)
Venezuela, Brazil – E
Oncidium martianum var. bicolor (Lindl.) Lindl., Fol. Orch. Oncid.: 24, 1855

Oncidium brachyandrum Lindl., Sert. Orch.: sub t.25, 1841
BR 27: t.59 & sub t.59:6, 7 & misc.57, misc.72, misc.180 (1841)
Mexico, Guatemala, El Salvador, Costa Rica – 1500–2700 m – T, L
Cyrtochilum graminifolium Lindl., Ann. Nat. Hist. 4: 384, 1840
o Oncidium filipes Lindl., Sert. Orch.: sub t.48, 1840
Cyrtochilum filipes (Lindl.) Lindl., Bot. Reg. 27: t.59, 1841
Oncidium graminifolium (Lindl.) Lindl., Sert. Orch.: sub t.48, 1841
Oncidium wrayae Hook., Bot. Mag. 67: t.3854, 1841
Oncidium graminifolium var. filipes (Lindl.) Lindl., Fol. Orch. Oncid.: 32, 1855
Oncidium graminifolium var. holochilum Lindl., Fol. Orch. Oncid.: 32, 1855
Oncidium graminifolium var. wrayae (Hook.) Lindl., Fol. Orch. Oncid.: 32, 1855

Oncidium carthagenense (Jacq.) Sw., Kongl. Vet. Acad. Handl. 21: 240, 1800
BR 25: misc.68 (1839) / 26: misc.137 & misc.215 (1840)
USA (Florida), Mexico to Panama, W.Indies, Venezuela, Colombia, Brazil – E
Epidendrum carthagenense Jacq., Select. Stirp. Amer.: 228, 1763
Epidendrum undulatum Sw., Nov. Gen. Sp. Prodr.: 122, 1787
Oncidium undulatum (Sw.) Salisb., Trans. Roy. Soc. 1: 295, 1812 (non Lindl.)
Oncidium panduriferum Kunth, Nov. Gen. & Sp. Pl. 1: 346, 1816
Oncidium sanguineum Lindl., Sert. Orch.: t.27, 1838
Oncidium luridum var. henchmannii Knowles & Westc., Fl. Cab. 3: t.97, 1840
Oncidium huntianum Hook., Bot. Mag. 67: t.3806, 1841
Oncidium henchmannii (Knowles & Westc.) Lodd., Cat. Orch. ed.2: no. 1323, 1842
Oncidium roseum Lodd., Cat. Orch. ed.2: no.1381, 1842
Oncidium oerstedii Rchb.f., Bonpl. 2: 91, 1854
Oncidium carthagenense var. klotzschii Lindl., Fol. Orch. Oncid.: 40, 1855
Oncidium carthagenense var. oerstedii (Rchb.f.) Lindl., Fol. Orch. Oncid.: 40, 1855
Oncidium carthagenense var. sanguineum (Lindl.) Lindl., Fol. Orch. Oncid.: 40, 1855
Oncidium carthagenense var. swartzii Lindl., Fol. Orch. Oncid.: 40, 1855
Oncidium obsoletum A.Rich. & Galeotti ex Lindl., Fol. Orch. Oncid.: 41, 1855
Oncidium sanguineum var. roseum (Lodd.) Regel, Ind. Sem. Hort. Petrop.: 22, 1856
Oncidium carthagenense var. roseum (Lodd.) Sander, Orch. Guide: 176, 1901
Oncidium carthagenense var. andreanum Cogn., Mart. Fl. Bras. 3 (6): 411, 1906
Oncidium hymatoides Kraenzlin, Engl. Pflanzenr. Orch. Mon. Oncid.: 112, 1922
Oncidium oerstedii var. crispiflorum Schltr., Fedde Rep. Beih. 17: 85, 1922

Oncidium cavendishianum Bateman, Orch. Mex. & Guat.: t.3, 1837
BR 26: misc.138 (1840)
Mexico, Guatemala, Honduras – up to 2800 m – E
Oncidium pachyphyllum Hook., Bot. Mag. 67: t.3807, 1840

o **Oncidium cebolleta** (Jacq.) Sw., Kongl. Vet. Acad. Handl. 21: 240, 1800
BR 23: t.1994 (1837) / 27: misc.56 (1841) / 28: t. & sub t.4 (1842)
Trop. America, Mexico, Venezuela, Trinidad, Surinam, Argentina, Paraguay – E
Dendrobium cebolleta Jacq., Enum. Pl. Carib.: 30, 1760
Epidendrum cebolleta Jacq., Enum. Pl. Carib.: 30, 1760
Epidendrum juncifolium L., Sp. Pl. ed.2: 1351, 1763
Cymbidium juncifolium (L.) Willd., Sp. Pl. ed.4: 102, 1805
Oncidium juncifolium (L.) Willd., Coll. Bot.: 27, 1821
o Oncidium longifolium Lindl., Bot. Reg. 27: misc.56, 1841
Oncidium brachyphyllum Lindl., Bot. Reg. 28: sub t.4, 1842
Oncidium cepula Hoffsgg., Bot. Zeit. 1: 833, 1843
Oncidium sprucei Lindl., Fol. Orch. Oncid.: 14, 1855
Oncidium glaziovii Cogn., Mart. Fl. Bras. 3 (6): 440, 1906
Oncidium ottonis Rchb.f. ex Kraenzlin, Engl. Pflanzenr. Orch. Mon. Oncid.: 92, 1922
Oncidium vittii Oppenh., Engl. Pflanzenr. Orch. Mon. Oncid.: 96, 1922
Oncidium humboldtii Schltr., Fedde Rep.: 65, 1931

o **Oncidium ciliatum** Lindl., Bot. Reg. 20: t.1660, 1834
BR 20: t.1660 (1834) / 24: sub t.48 (1838) / 25: sub t.16 (1839)
Type: Brazil, cult. Knight (K-Lindl., holo!)
Brazil – E
Oncidium fimbriatum Hoffsgg., Bot. Zeit. 1: 834, 1843 (non Lindl.)
Oncidium reisii Hoehne & Schltr., Arq. Bot. Est. S.Paulo 1: 290, 1926

Oncidium cimiciferum (Rchb.f.) Rchb.f. ex Lindl., Fol. Orch. Oncid.: 9, 1855
BR 27: sub t.59: 2 (1841)
Venezuela, Colombia, Ecuador, Peru – E
Cyrtochilum flexuosum Kunth, Nov. Gen. & Sp. Pl. 1: 350, 1816
Oncidium flexuosum (Kunth) Lindl., Sert. Orch.: sub t.48, 1842 (non Sims)
Odontoglossum cimiciferum Rchb.f., Linnaea 22: 849, 1849
Oncidium trulla Rchb.f. & Warsc., Bonpl. 2: 101, 1854
Cyrtochilum trulla (Rchb.f. & Warzsc.) Kraenzlin, Engl. Pflanzenr. Mon. Orch. Oncid.: 66, 1922

o **Oncidium citrinum** Lindl., Bot. Reg. 21: t.1758 1835
BR 21: t.1758 (1835)
Type: Trinidad, cult. Loddiges (K-Lindl., holo!)
Trinidad, Venezuela, Colombia – E

o **Oncidium cornigerum** Lindl., Bot. Reg. 18: t.1542, 1832
BR 18: t.1542 (1832)
Type: Brazil, cult. Herbert (K-Lindl., holo!)
Brazil, Paraguay – E
Oncidium chrysorhapis Rchb.f., Gard. Chron. s.3, 3: 72, 1888

o **Oncidium crispum** Lodd., Bot. Cab. 19: t.1854, 1832 (non Beer)
BR 23: t.1920 (1837)
Brazil – E

o **Oncidium curtum** Lindl., Bot. Reg. 33: t.68, 1847
BR 33: t.68 (1847)
Type: Brazil, cult. Veitch (K-Lindl., holo!)
Brazil –E

o **Oncidium deltoideum** Lindl., Bot. Reg. 23: t.2006, 1837
BR 23: t.2006 & sub t.1920:12 (1837)
Type: Peru, Chacapayas, Mathews s.n. Myers (K-Lindl., holo!)
Amazonas, Peru – E
Cyrtochilum deltoideum (Lindl.) Kraenzlin, Notizbl. Bot. Gart. Berl. 7: 97, 1917

Oncidium divaricatum Lindl., Bot. Reg. 13: sub t.1050, 1827
BR 13: t. & sub t.1050 (1827) / 24: sub t.48 (1838) / 25: sub t.16 (1839)
Type: Brazil, J.Heatherly s.n. (K-Lindl., holo!)
Brazil – E

o Oncidium divaricatum var. cupreum Lindl., Bot. Reg. 13: t.1050, 1827

Oncidium ensatum Lindl., Bot. Reg. 28: misc.15, 1842
BR 28: misc.15 (1842)
Type: Guatemala, cult. Loddiges (K-Lindl., holo!)
Mexico, Guatemala, Honduras, Belize, Nicaragua, Panama – up to 600 m – T, L, E
Oncidium cerebriferum Rchb.f., Bot. Zeit. 10: 696, 1852

Oncidium excavatum Lindl., Sert. Orch.: sub t.25:3, 1838
BR 25: misc.150 (1839)
Peru, Ecuador – 2400–2700 m – T
Oncidium rupestre Lindl., Pl. Hartw.: 151, 1844
Oncidium aurosum Rchb.f., Bonpl. 2: 108, 1854
Oncidium skinneri Lindl., Gard. Chron.: 629, 1855
Oncidium excavatum var. dawsonii Williams, Orch. Grow. Man. ed.6: 483, 1885

Oncidium forbesii Hook., Bot. Mag. 65: t.3705, 1839
BR 25: misc.149 (1839)
Brazil – E
Oncidium crispum var. forbesii (Hook.) Burb., Orch. Trad. Lebl. ed.2: 117, 1882
Oncidium crispum var. marginatum hort., Nichols. Dict. Gard. 2: 484, 1886

Oncidium forkelii Scheidw., Allg. Gartenz. 10: 309, 1842
BR 29: misc.14 (1843)
Mexico – E

Oncidium gardneri Lindl., Hook. J. Bot. 2: 262, 1843
BR 33: sub t.66 (1847)
Brazil – E
Oncidium flabelliferum Pinel ex Paxt. Mag. Bot. 16: 65 fig.2, 1849
Oncidium elegantissimum Rchb.f., Gard. Chron. 1: 13, 1877
Oncidium praetextum C.J. Morren, Belg. Hort.: 357 t.20–21, 1877 (non Rchb.f.)
Oncidium gardnerianum hort., Gard. Chron.: 759, 1880
Oncidium praestans Rchb.f., Gard. Chron.: 296, 1880
Oncidium polletianum Rchb.f., Gard. Chron.: 326, 1886
Oncidium forbesio-dasystyle Rolfe, Orch. Rev. 1: 301, 1893

Oncidium gracile Lindl., Bot. Reg. 23: sub t.1920:7, 1837
BR 23: sub t.1920:7 (1837)
Type: Brazil, Martius s.n. (K-Lindl., holo!)
Brazil – E

Oncidium guianense (Aublet) Garay, Bot. Mus. Leafl. Harv. Univ. 23: 300, 1973
BR 21: t.1789 (1835)
Cuba, W.Indies – E
Ophrys guianensis Aublet, Hist. Pl. Guiana Fr. 2: 86, 1775
Ophrys aloidea Poiret, Lam. Encycl. 4: 569, 1798
Oncidium intermedium Bertero ex Sprengel, Syst. Veg. 3: 728, 1826
o Oncidium lemonianum Lindl., Bot. Reg. 21: t.1789, 1835
Oncidium desertorum Nash ex Withner, Amer. Orch. Soc. Bull. 36: 312, 1967

o **Oncidium harrisonianum** Lindl., Bot. Reg. 19: t.1569, 1833
BR 19: t.1569 (1833) / 26: misc.108 (1840)
Type: Brazil, W.Harrison s.n. (K-Lindl., holo!)
Brazil – E
Oncidium pantherinum Hoffsgg., Verz. Orch.: 59, 1813
Oncidium pallidum Lindl., Bot. Reg. 26: misc.108, 1840
Oncidium harrisoniae (Lindl.) Beer, Prakt. Stud. Orch.: 287, 1854
Oncidium acrobotrium Klotzsch, Otto & Dietr. Allg. Gartenz. 23: 289, 1855
Oncidium raniferum Klotzsch, Otto & Dietr. Allg. Gartenz. 23: 289, 1855 (non Lindl.)
Oncidium harrisonianum var. pallidum (Lindl.) Rchb.f., Hamb. Gartenz. 16: 419, 1860
Oncidium pentaspilum Hoffsgg. ex Rchb.f., Walp. Ann. 6: 774, 1863

Oncidium hastatum (Bateman) Lindl., Paxt. Flow. Gard. 1: 9, 1850–51
BR 31: misc.60:17 (1845)
Mexico – E
Odontoglossum hastatum Bateman, Orch. Mex. & Guat.: t.20, 1839
Odontoglossum phyllochilum Morren, Ann. Soc. Bot. Gand. 5: 295 t.271, 1849
Cyrtochilum juergensianum Lemaire, Jard. Fleur. 4: t.368, 1854
Oncidium juergensianum Lemaire, Jard. Fleur. 4: t.368, 1854
Oncidium phyllochilum (Morren) Beer, Prakt. Stud. Orch.: 292, 1854

Oncidium hians Lindl., Bot. Reg. 24: misc.124, 1838
BR 24: misc.124 (1838)
Type: Brazil, cult. Rollissons (K-Lindl., holo!)
Brazil – E
Oncidium leucostomum Hoffsgg., Verz. Orch.: 58, 1843
Oncidium quadricorne Klotzsch, Allg. Gartenz. 20: 249, 1852
Oncidium maxilligerum Lemaire, Ill. Hort. 3: misc.p.43, 1856

o **Oncidium incurvum** Barker ex Lindl., Bot. Reg. 26: misc.174, 1840
BR 26: misc.174 (1840) / 31: t.64 (1845)
Type: Mexico, cult. Barker (K-Lindl., holo!)
Mexico – E
Oncidium alboviolaceum A.Rich. & Galeotti, Ann. Sc. Nat. s.3, 3: 27, 1845

Oncidium isopterum Lindl., Bot. Reg. 23: sub t.1920:3, 1837
BR 23: sub t.1920:3 (1837)
Type: Brazil, Martius s.n. (K-Lindl., holo!)
Brazil – E

Oncidium jacquinianum Garay & Stacy, Bradea 1: 407, 1974
BR 22: t. & sub t.1851 (1836)
W.Indies – E
Epidendrum altissimum Jacq., Sel. Stirp. Am.: 229 t.141, 1763 (non Jacq. 1760)
o Oncidium altissimum Sw., Kongl. Vet. Akad. Handl. 21: 240, 1800 (non Jacq.) (Misidentified)

o **Oncidium lanceanum** Lindl., Bot. Reg. 22: t.1887, 1836
BR 22: t.1887 (1836)
Type: Surinam, J.Lance s.n. cult. C.Lemon (K-Lindl., holo!)
Brazil, Guyanas, Venezuela, Trinidad, Amazonas – E
Lophiaris fragrans Raf., Fl. Tellur. 4: 41, 1836
Oncidium lanceanum var. superbum Lindl. ex Seemann, Hart. Parad. Vindob. 1: t.26, 1844

Oncidium leucochilum Bateman ex Lindl., Bot. Reg. 23: sub t.1920:9, 1837
BR 23: sub t.1920:9 (1837)
Type: Mexico, cult. Bateman (K-Lindl., holo!)
Mexico, Guatemala, Honduras – E
Oncidium digitatum Lindl., Benth. Pl. Hartw.: 94, 1839
Cyrtochilum leucochilum (Bateman ex Lindl.) Planchet, Fl. d. Serres 5: t. 522, 1849

Oncidium longicornu Mutel, Mémorial de la Scarpe 13: 10, 1838
BR 25: misc.76 (1839) / 26: misc.47 (1840) / 27: misc.160 (1841)
Brazil – E
Oncidium unicornutum Knowles & Westc., Fl. Cabin. 2: 143, 1838
Oncidium unicorne Lindl., Bot. Reg. 25: misc.76, 1839
Oncidium monoceras Hook., Bot. Mag. 67: t.3890, 1841
Oncidium gautieri Regel, Ind. Sem. Hort. Petrop.: 80, 1868

Oncidium maculatum (Lindl.) Lindl., Sert. Orch.: sub t.48, 1841
BR 24: t.44 & misc.39 (1838) / 26: misc.79 & misc.86 (1840) / 27: sub t.59: 5 & misc.14 & misc.87 (1841)
Type: Vera Cruz, Hartweg s.n. cult. Hort. Soc. (K-Lindl., holo!)
Mexico, Guatemala, El Salvador, Honduras, Nicaragua, Brazil – E
Oncidium funereum Llave & Lex., Nov. Veg. Descr. 2: 37, 1825
o Cyrtochilum maculatum Lindl., Bot. Reg. 24: t.44 & misc.39, 1838
Cyrtochilum maculatum var. ecornutum Hook., Bot. Mag. 67: t.3836, 1840
Cyrtochilum maculatum var. russelianum Lindl., Bot. Reg. 26: misc.86, 1840
Cyrtochilum grandiflorum Lodd., Orchid.: no.1585, 1844 (nomen)
Cyrtochilum maculatum var. parviflorum Lindl., Bot. Reg. 27: misc.87, 1841
Oncidium maculatum var. donianum Rchb.f., Bonpl. 3: 215, 1855
Oncidium maculatum var. psittacium Rchb.f. ex Lindl., Fol. Orch. Oncid.: 34, 1855
Oncidium maculatum var. vinosum Lindl., Fol. Orch. Oncid.: 34, 1855
Cyrtochilum leucochilum Linden ex Morren, Belg. Hort. 35: 287 t.21, 1885
Oncidium lintriculus Kraenzlin, Engl. Pflanzenr. Orch. Mon. Oncid.: 282, 1922
Oncidium hintonii L.O.Williams, Amer. Orch. Soc. Bull. 9: 336, 1941
Odontoglossum johnsonorum L.O.Williams, Ceiba 4: 191, 1954
Odontoglossum subcruciforme Heller, Fieldiana 32 (7): 73, 1969

Oncidium maculosum Lindl., Bot. Reg. 23: sub t.1920:1, 1837
BR 23: sub t.1920:1 (1837)
Type: Brazil, Martius s.n. (K-Lindl., holo!)
Brazil, Paraguay – E

Oncidium martianum Lindl., Bot. Reg. 23: sub t.1920:4, 1837
BR 23: sub t.1920:4 (1837)
Type: Brazil, Martius s.n. (K-Lindl., holo!)
Brazil – E

o **Oncidium microchilum** Bateman ex Lindl., Bot. Reg. 26: misc.193, 1840
BR 26: misc.193 (1840) / 29: t.23 (1843)
Type: Guatemala, Skinner s.n. cult. Bateman (K-Lindl., holo!)
Mexico, Guatemala – up to 2300 m – L, E

Oncidium nanum Lindl., Bot. Reg. 24: misc.30, 1838
BR 24: misc.30 (1838)

Type: Guyana, cult. Loddiges (K-Lindl., holo!)
Guyana, Venezuela, Peru, Amazonas – E
Oncidium patulum Schltr., Orchis 8: 18, 1914

Oncidium nebulosum Lindl., Bot. Reg. 27: misc.175, 1841
BR 27: misc.175 (1841)
Type: Hartweg s.n. cult. Hort. Soc. (K-Lindl., holo!)
Mexico, Guatemala – up to 2100 m – E
Oncidium geertianum Morren, Ann. Soc. Gand. 4: 55 t.179, 1848
Oncidium caesium Rchb.f., Gartenfl.: 75, 1854
Oncidium durangense Hagsater, Orquidea (Méx.) 8: 148, 1981

Oncidium nudum Bateman ex Lindl., Bot. Reg. 23: sub t.1994, 1837
BR 23: sub t.1994 (1837)
Type: Venezuela, Bateman s.n. (K-Lindl., holo!)
Venezuela, Colombia – E

Oncidium oblongatum Lindl., Bot. Reg. 30: misc.11, 1844
BR 30: misc.11 (1844)
Type: cult. Loddiges (K-Lindl., holo!)
Mexico, Guatemala – up to 2700 m – E
Oncidium xanthochlorum Klotzsch, Ind. Sem. Hort. Berol. App. 2, 1852

o **Oncidium ornithorhynchum** Kunth, Nov. Gen. & Sp. Pl. 1: 345 t.80, 1816
BR 26: t.10 & misc.95 (1840)
Mexico, Guatemala, El Salvador, Costa Rica – 1500–1700 m – E

Oncidium pergameneum Lindl., Benth. Pl. Hartw.: 93, 1842
BR 28: misc.7 (1842)
Mexico, Guatemala – up to 2000 m – E

o **Oncidium pubes** Lindl., Bot. Reg. 12: t.1007, 1826
BR 12: t.1007 (1826) / 24: sub t.48 (1838) / 25: sub t.16 (1839)
Type: Brazil, Douglas s.n. cult. Hort. Soc. (K-Lindl., holo!)
Brazil, Paraguay, Argentina – E
Oncidium puber Sprengel, Syst. Veg.: 308, 1827
Oncidium bicornutum Hook., Bot. Mag. 58: t.3109, 1831
Oncidium pubescens Duchartre, Man. Gen. Pl. 4: 488, 1857

o **Oncidium pulchellum** Hook., Bot. Mag. 54: t.2773, 1827
BR 21: t.1787 (1835)
Brazil, Guyana, Jamaica, Cuba – up to 1200 m – E
Tolumnia pulchella (Hook.) Raf., Fl. Tellur. 2: 101, 1836
Oncidium concavum Moir, Phytolog. 17: 432, 1968

o **Oncidium pulvinatum** Lindl., Bot. Reg. 24: misc.115, 1838
BR 24: misc.115 (1838) / 25: t.42 (1839)
Type: cult. R.Harrison (K-Lindl., holo!)
Brazil – E
Oncidium scirus Scheidw., Allg. Gartenz. 7: 496, 1839

o **Oncidium pumilum** Lindl., Bot. Reg. 11: t.920, 1825
BR 11: t.920 (1825)
Type: Brazil, cult. Herbert (K-Lindl., holo!)
Brazil, Uruguay, Argentina, Paraguay – E
Epidendrum ligulatum Vell., Fl. Tellur. Ic. 9: t.15, 1827

Oncidium ramosum Lindl., Bot. Reg. 26: misc.154, 1840
BR 23: sub t.1920:8 (1837) / 26: misc.154 (1840)
Type: Brazil, Martius s.n. cult. Loddiges (K-Lindl., holo!)
Brazil – E
Oncidium caldense Rchb.f., Linnaea 22: 846, 1849

Oncidium raniferum Lindl., Bot. Reg. 23: sub t.1920:11, 1837
BR 23: sub t.1920:11 (1837) / 24: t.48 (1838)
Type: Brazil, cult. Knight (K-Lindl., holo!)
Brazil – E

Oncidium reflexum Lindl., Bot. Reg. 23: sub t.1920:10, 1837
BR 23: sub t.1920:10 (1837) / 26: misc.216 (1840) / 28: misc.8 (1842) / 29: misc.22 (1843) / 33: t.70 (1847)
Type: Mexico, Karwinski s.n. cult. Loddiges (K-Lindl., holo!)
Mexico, Guatemala – E
o Oncidium pelicanum Martius ex Lindl., Bot. Reg. 26: misc.216, 1840
Oncidium suttonii Bateman ex Lindl., Bot. Reg. 28: misc.8, 1842
Oncidium suave Lindl., Bot. Reg. 29: misc.22, 1843
Oncidium brevifolium A.Rich. & Galeotti, Ann. Sc. Nat. s.3, 3: 26, 1845
Oncidium macropterum A.Rich. & Galeotti, Ann. Sc. Nat. s.3, 3: 26, 1845
Oncidium wendlandianum Rchb.f., Bonpl. 2: 91, 1854
Oncidium tayleuri hort. ex Lindl., Fol. Orch. Oncid.: 46, 1855
Oncidium uncia Rchb.f. ex Kraenzlin, Engl. Pflanzenr. Orch. Mon. Oncid.: 270, 1922

Oncidium remotiflorum Garay, Taxon 19: 454, 1970
BR 29: misc.15 (1843)
Brazil – E
Oncidium cuneatum Scheidw., Allg. Gartenz. 10: 309, 1842

Oncidium retusum Lindl., Bot. Reg. 23: sub t.1920:6, 1837
BR 23: sub t.1920:6 (1837)
Type: Peru, Mathews s.n. (K, holo!)
Peru, Bolivia – E

Oncidium sanguinolentum (Lindl.) Schltr., Fedde Rep. Beih. 6: 98, 1919
BR 30: misc.91 (1844)
Type: Venezuela, cult. Barker (K-Lindl., holo!)
Venezuela, Peru – E
Leochilus sanguinolentus Lindl., Bot. Reg. 30: misc.91, 1844

o **Oncidium sphacelatum** Lindl., Sert. Orch.: sub t.48, 1841
BR 28: t.30 (1842)
Mexico, Guatemala, Honduras, El Savador, Guadeloupe, Venezuela, – E

Oncidium sphegiferum Lindl., Bot. Reg. 29: misc.23, 1843
BR 29: misc.23 (1843)
Type: Brazil, cult. Loddiges (K-Lindl., holo!)
Brazil – E

o **Oncidium spilopterum** Lindl., Bot. Reg. 30: misc.76, 1844
BR 30: misc.76 (1844) / 31: t.40 (1845)
Type: Brazil, cult. Loddiges (K-Lindl., holo!)
Brazil – E
Oncidium batemanianum var spilopterum (Lindl.) Lindl., Fol. Orch. Oncid.: 52, 1855
Oncidium saint-legerianum Rolfe, Gard. Chron. s.3, 12: 582, 1892
Oncidium ghillanyi Pabst., Orch. Rev. 80 (953): 216, 1972

Oncidium stipitatum Lindl., Bot. Voy. Sulphur.: 172, 1843
BR 30: misc.38 (1844) / 32: t.27 (1846)
Panama – E
o Oncidium lacerum Lindl., Bot. Reg. 30: misc.38, 1844
Oncidium stipitatum var. platyonyx Rchb.f., Gard. Chron. n.s.9: 788, 1878

o **Oncidium stramineum** Bateman ex Lindl., Bot. Reg. 24: misc.63, 1838
BR 24: misc.63 (1838) / 26: t.14 (1840)
Type: Mexico, Hartweg s.n. cult. Hort. Soc. (K-Lindl., holo!)
Mexico, Brazil – up to 1000 m – E
Oncidium colombae hort. ex Lindl., Fol. Orch. Oncid.: 36, 1855 (nomen)
Oncidium lindenii hort. ex Rchb.f. fide Lindl., Fol. Orch. Oncid.: 36, 1855
Oncidium saltator hort. ex Lindl., Fol. Orch. Oncid.: 36, 1855

Oncidium tetrapetalum (Jacq.) Willd., Sp. Pl. 4: 112, 1806
BR 24: misc.56 (1838)
Mexico, Jamaica, W.Indies, Colombia – E
Epidendrum tetrapetalum Jacq., Enum. Pl. Carib.: 30, 1760
Cymbidium tetrapetalum (Jacq.) Sw., Nov. Act. Upsal. 6: 74, 1799
Oncidium quadripetalum Sw., Kongl. Vet. Akad. Handl. Stockh. 21: 240, 1800
Oncidium tricolor Hook., Bot. Mag. 70: t.4130, 1844

Oncidium tigrinum Llave & Lex., Nov. Veg. Descr. 2: 36, 1825
BR 27: misc.174 (1841) / 31: misc.60:40 (1845)
Mexico – 2000–3500 m – E
Oncidium barkeri Lindl., Bot. Reg. 27: misc.174, 1841
Odontoglossum ghiesbreghtianum A.Rich. & Galeotti, Ann. Sc. Nat. s.3, 3: 27, 1845
Oncidium unguiculatum Lindl., J. Hort. Soc. 1: 303, 1846
Odontoglossum tigrinum (Llave & Lex.) Lindl., Fol. Orch. Odont.: 5, 1852
Oncidium ionosmum Lindl., Gard. Chron.: 726, 1853
Oncidium tigrinum var. unguiculatum (Lindl.) Lindl., Fol. Orch. Oncid.: 45, 1855
Oncidium tigrinum Hemsley, Biol. Centr. Am. 3: 286, 1884 (non Llave & Lex.)

o **Oncidium trulliferum** Lindl., Bot. Reg. 25: t.57, 1839
BR 25: t.57 (1839)
Type: Brazil, cult. Loddiges (K-Lindl., holo!)
Brazil – E
Oncidium galeatum Scheidw., Otto & Dietr. Allg. Gartenz. 7: 406, 1839
Oncidium dimorphum Regel, Ind. Sem. Hort. Petrop.: 22, 1869
Oncidium ornithocephaloides Kraenzlin, Engl. Pflanenr. Orch. Mon. Oncid.: 127, 1922

Oncidium undulatum (Kunth) Lindl., Sert. Orch.: sub t.48, 1838
BR 27: sub t.59:1 (1841)
Colombia – E
Cyrtochilum undulatum Kunth, Nov. Gen. & Sp. Pl. 1: 359, 1815

o **Oncidium uniflorum** Lindl., Bot. Reg. 29: t.43, 1843
BR 29: t.43 (1843)
Type: Brazil, Gardner s.n. cult. C.Lemon (K-Lindl., holo)
Brazil – E

o **Oncidium urophyllum** Lodd. ex Lindl. Sert. Orch.: sub t.48: 90, 1841
BR 28: t.54 (1842)
Brazil – E

Oncidium tetrapetalum Griseb., Fl. Brit. W.Ind.: 631, 1864 (non Willd)

Oncidium varicosum Lindl., Bot. Reg. 23: sub t.1920, 1837
BR 23: sub t.1920:2 (1837)
Type: Brazil, Pr. Maximillian of Wied Neuwied s.n. (K-Lindl., holo!)
Brazil – E
Oncidium lunaenum hort., Gard. Chron.: 168, 1850
Oncidium rogersii hort. ex Bateman, Gard. Chron.: 1317, 1868
Oncidium euxanthinum Rchb.f., Gard. Chron.: 1158, 1869
Oncidium geraense Barb.Rodr., Gen. & Sp. Orch. Nov. 2: 193, 1882

Oncidium viperinumum Lindl., Gen. Sp. Orch. Pl.: 197, 1833
BR 24: misc.92 (1838)
Brazil, Uruguay – E
Oncidium confragosum Lindl., Bot. Reg. 24: misc.92, 1838

Oncidium volubile (Poeppig & Endl.) Cogn., Mart. Fl. Bras. 3 (6): 289, 1905
BR 27: sub t.59:11 (1841)
Peru – E
Cyrtochilum volubile Poeppig & Endl., Nov. Gen. Sp. Pl. 1: 35 t.61, 1836
Oncidium corynephorum Lindl., Sert. Orch.: sub t.25, 1838

Oncidium wentworthianum Bateman ex Lindl., Bot. Reg. 26: misc.194, 1840
BR 26: misc.194 (1840)
Type: Guatemala, Skinner s.n. (K-Lindl., holo!)
Mexico, Guatemala – up to 1500 m – E
Oncidium tenue Lindl., J. Hort. Soc. 3: 76, 1848
Oncidium wentworthianum var. tenue (Lindl.) Ames & Correll, Bot. Mus. Leafl. Harv. Univ. 11: 20, 1943

Ophrys argolica Fleischm., Verh. Zool.-Bot. Ges. Wien 69: 295, 1919
BR. 33: t.46:2 (1847)
Greece, Turkey, Crete – T
o Ophrys tabanifera Willd., Sp. Pl. ed.4: 67, 1805 (Misidentified)
Ophrys ferrum-equinum ssp. argolica (Fleischm.) Soó, Notizbl. Bot. Gart. Berlin 9 (89): 907, 1969

o **Ophrys atrata** Lindl., Bot. Reg. 13: t.1087, 1827
BR 13: t.1087 (1827)
Type: Italy, Mauri s.n. cult. Hort. Soc. (K-Lindl., holo!)
Albania, Yugoslavia, Italy, Sicilia, Sardegna, Corse, France, Spain, Mallorca, Portugal – T
Ophrys aranifera var. atrata (Lindl.) Rchb.f., Icon. 13/14: 90, 1851
Ophrys biceratia Delile ex Nyman, Consp. Fl. Eur. 4: 697, 1882
Arachnites atrata (Lindl.) Bubani, Fl. Pyr. 4: 49, 1901
Ophrys aranifera ssp. atrata (Lindl.) E.G.Gam., Mon. Orch.: 286/87, 1908
Ophrys sphegodes var. atrata (Lindl.) Briq., Prodr. Fl. Corse 1: 351/352, 1910
Ophrys sphegodes ssp. atrata (Lindl.) E.Mayer, Seznam Slovensk. Ozemlja: 387, 1952

o **Ophrys ferrum-equinum** Desf., Ann. Mus. Nat. Paris 10: 226 t.15, 1807
BR 33: t.46:1 (1847)
Albania, Greece, Turkey – T

o **Ophrys fusca** Link, Schrad. Neues J. Bot. 2: 324, 1800
BR 13: t.1071 (1827)
Mediterranean Region – T

Ophrys holoserica (Burm.f.) Greuter, Boissiera 13: 185, 1967
BR 31: misc.81 (1845) / 33: t.25 (1847)
S.Centr. Europe, Mediterranean Region – T
Ophrys insectifera var. arachnites L., Sp. Pl.: 949, 1753
o Ophrys fuciflora Lindl., Bot. Reg. 33: t.25 1847
Orchis fuciflora Crantz, Stirp. Austr. ed.2, 6: 483, 1769
Orchis holoserica Burm.f., Nov. Act. Acad. Leop.-Carol. 4, app.: 23, 1770
Ophrys arachnites (L.) Reichard, Fl. Moeno-Francof. 2: 89, 1778 (non Miller 1768) (nomen illeg.)
Epipactis arachnites (L.) F.W.Schmidt, Mayer Samml. Pfl. Phys. Auf 1: 249, 1791
Arachnites fuciflora F.W.Schmidt, Fl.Boem. 1: 76, 1793
Ophrys fuciflora (F.W.Schmidt) Moench, Meth. Suppl.: 311, 1802

Ophrys oestrifera Bieb., Fl. Taur.-Cauc. 2: 369, 1808
ssp. oestrifera
BR 32: t.52 (1846)
Mediterranean Region – T
o Ophrys cornuta Steven, Bieb. Fl. Taur.-Cauc. 2: 370, 1808
Ophrys scolopax var. oestrifera (Bieb.) Rchb.f., Ic. Fl. Germ. 13/14: 99, 1851
Ophrys oestrifera var. cornuta (Steven) Boiss., Fl. Orch. 5: 79, 1882 (nomen illeg.)
Ophrys oestrifera ssp. cornuta (Steven) K.Richter, Pl. Eur. 1: 264, 1890 (nomen illeg.)
Ophrys scolopax ssp. cornuta (Steven) E.G.Camus, Monogr. Orchid.: 270, 1908 (nomen illeg.)
Ophrys araneifera ssp. mammosa var. boissieri Soó, Fedde Rep. 24: 35, 1927
Ophrys araneifera ssp. boissieri (Soó) Soó, Fedde Rep. 26: 279, 1929
Ophrys mammosa ssp. boissieri (Soó) Soó, Acta Bot. Acad. Sc. Hung 18: 383, 1973
Ophrys holosericea ssp. cornuta (Steven) Sunderm., Taxon 24: 625, 1975(nomen inval.)
Ophrys fuciflora ssp. cornuta (Steven) Sunderm., Europ. Medit. Orchid. ed.3: 39, 1980 (nomen inval.)

Ophrys sphegodes Miller, Gard. Dict. ed.8. 8, 1768
ssp. sphegodes
BR 14: t.1197 (1828)
Centr. S.Europe – T
Ophrys aranifera Hudson, Fl. Angl. ed.2. : 392, 1778
Ophrys fucifera Curtis, Fl. Lond. 4, 1783
o Ophrys araneifera var. limbata Lindl., Bot. Reg. 14: t.1197, 1828
Ophrys araneola Rchb.f., Ic. Bot. Nona Cent.: 22, 1831
Arachnites fuciflora Tod., Orch. Sc.: 72, 1842 (non F.W.Schmidt)
Ophrys insectifera ssp. aranifera (Hudson) Moggr., Nov. Acta Leop. Carol. Akad. 35: 9, 1870
Arachnites aranifera (Hudson) Bubani, Fl. Pyr. 4: 48, 1901
Ophrys araneifera Asch. & Graebner, Syn. 3: 636, 1907

o **Ophrys tenthredinifera** Willd., Sp. Pl. 4: 67, 1805
BR 3: t.205 (1817) / 13: t. & sub t.1093 (1827)
S.Europe Medit., N.Africa, Asia Minor – T
Ophrys insectifera var. rosea Desf., Fl. Atl. 2: 320, 1800
Ophrys villosa Desf., Ann. Mus. Hist. Nat. Paris 10: 225, 1807
Ophrys grandiflora Ten., Fl. Nap. 2: 309, 1820
Ophrys tenoreana Lindl., Bot. Reg. 13: sub t.1093, 1827
o Ophrys tenthredinifera var. minor Lindl., Bot. Reg. 13: t.1093, 1827
Ophrys limbata Link, Handb. 1: 247, 1829
Arachnites tenthredinifera Tod., Orch. Sc. 1: 85, 1842
Ophrys neglecta Parl., Fl. Ital. 3: 548, 1858
Ophrys tenthredinifera ssp. neglecta (Parl.) E.G.Cam., Mon. Orch.: 262, 1908
Ophrys rosea (Desf.) Grande, Nuo. Giorn. Bot. Ital. 27: 238, 1920
Ophrys rosea ssp. neglecta (Parl.) Guadagno, Archivio Bot. 7: 168, 1931

Ophrys vernixia Brot., Fl. Lusit. 1: 24, 1804
BR 5: t.370 (1819)
Portugal, Mediterranean region – T
o Ophrys speculum Link, Schrad. Diar. Bot. 5, 11: 324, 1799
Ophrys ciliata Biv., Cent. 1: 60, 1806

Orchis italica Poiret, Lam. Encycl. Méth. Bot. 4: 600, 1789
BR 5: t.375 (1819)
S.W.Europe – T
Orchis militaris Poiret, Voy. Barb. 2: 247, 1789
Orchis longicruris Link, Schrad. Neues J. Bot. 2: 323, 1799
o Orchis tephrosanthos Desf., Fl. Atl. 2: 319, 1800
Orchis undulatifolia Biv., Pl. Sic. Cent. 2: 44, 1807
Orchis tephrosanthos var. undulatifolia Lindl., Bot. Reg. 5: t.375, 1819
Orchis simia var. undulatifolia (Biv.) Webb., It. Hisp.: 9, 1838

Orchis lactea Poiret, Lam. Encycl. Méth. Bot. 4: 594, 1798
BR 5: t.367 (1819)
Mediterranean Region – T
Orchis acuminata Desf., Fl. Atl. 2: 318, 1800
o Orchis variegata Lindl., Bot. Reg. 5: t.367, 1819

o **Orchis longicornu** Poiret, Voy. en Barb. 2: 247, 1789
BR 3: t.202 (1817)
W.Mediterranean Region – T
Orchis longicornis Smith, Prodr. Fl. Graec. 2: 212, 1816
Orchis morio var. longicornu (Poiret) Knoche, Fl. Balear. 1: 400, 1921

o **Orchis papilionacea** L., Syst. ed.10: 242, 1759
BR 14: t.1155 (1828)
Mediterranean Region – T
Orchis papilionacea var. rubra (Jacq.) Rchb.f., Icon. Fl. Germ. 13 & 14: 16, 1851
Vermeulenia papilionacea (L.) A.Löve & D.Löve, Acta Bot. Neerl. 21: 554, 1972

Osmoglossum egertonii (Lindl.) Schltr., Orchis 10: 166, 1916
BR 31: misc.60:4 (1845)
Type: Guatemala, (K-Lindl., holo!)
Mexico, Guatemala, Honduras, Costa Rica – up to 2600 m – E
Odontoglossum egertonii Lindl., Bot. Reg. 31: misc.60:4, 1845
Oncidium egertonii (Lindl.) Beer, Prakt. Stud. Orch.: 285, 1854
Osmoglossum anceps Schltr., Fedde Rep. Beih. 19: 147, 1923

Osmoglossum pulchellum (Bateman ex Lindl.) Schltr., Orchis 10: 162, 1916
BR 27: t.48 (1841) / 31: misc.60:3 (1845)

Type: Guatemala, cult. Bateman (K-Lindl., holo!)
Mexico, Guatemala, El Salvador, Honduras, Nicaragua – E
o Odontoglossum pulchellum Bateman ex Lindl., Bot. Reg. 27: t.48, 1841

Otochilus porrectus Lindl., Gen. Sp. Orch. Pl.: 36, 1830
BR 18: sub t.1522 (1832)
N.E.India, Burma, Thailand, Vietnam, China – E
Tetrapeltis fragrans Wall ex Lindl., Bot. Reg. 18: sub t. 1522, 1832
Otochilus latifolius Griffith, Notul. 3: 279 Ic. t.288, 1851
Coelogyne porrecta (Lindl.) Rchb.f., Walp. Ann. 6: 236, 1861
Otochilus fragrans (Wall. ex Lindl.) Nicols., Dict. Gard. 2: 534, 1886
Otochilus forrestii W.Smith, Not. Roy. Bot. Gard. Edinb. 13: 216, 1921

Otoglossum brevifolium (Lindl.) Garay & Dunsterv., Venez. Orch. Ill. 6: 41, 1976
BR 31: misc.60:1 (1845)
Panama, Venezuela, Colombia, Ecuador, Peru – 1700–3000 m – E, T
Odontoglossum brevifolium Lindl., Pl. Hartw.: 152, 1844
Odontoglossum coronarium var. miniatum Veitch, Man. Orch. Pl. 1: 23, 1886
Odontoglossum miniatum hort. ex Veitch, Man. Orch. Pl. 1: 23, 1886

Pabstia jugosa (Lindl.) Garay, Bradea 1 (27): 307, 1973
BR 27: misc.104 (1841) / 29: misc.65:3 (1843)
Type: cult. Loddiges (K-Lindl., holo!)
Brazil – E
Maxillaria jugosa Lindl., Bot. Reg. 27: misc.104, 1841
Colax jugosa (Lindl.) Lindl., Bot. Reg. 29: misc.65: 3, 1843
Maxillaria viridis var. pluriflora Regel, Ann. Sc. Nat. s.4, 6: 376, 1856
Lycaste jugosa (Lindl.) Nicholson, Dict. Gard. 2: 304, 1885
Colax viridis var. pluriflora (Regel) Cogn., Mart. Fl. Bras. 3, 5: 562, 1902
Zygopetalum jugosum (Lindl.) Schltr., Die Orchideen: 422, 1914

Pabstia placanthera (Hook.) Garay, Bradea 1 (27): 301, 1973
BR 27: misc.103 (1841) / 29: misc.65:2 (1843)
Brazil – E
Maxillaria placanthera Hook., Bot. Mag. 59: t.3173, 1832
Colax placanthera (Hook.) Lindl., Bot. Reg. 29: misc.65: 2, 1843
Maxillaria cyanocheile Hoffsgg., Verz. Orch.: 55, 1843
Maxillaria viridis var. stenosepala Regel, Ann. Sc. Nat. s.4, 6: 375, 1856
Colax viridis var. placanthera (Hook.) Stein, Orchideenb.: 160, 1892
Zygopetalum placantherum (Hook.) Stein, Orchideenb.: 423, 1914

Pabstia viridis (Lindl.) Garay, Bradea 1 (27): 308, 1973
BR 18: t.1510 & sub t.1549:19 (1832) / 29: misc.65:1 (1843)
Type: Brazil, H.Chamberlain s.n. cult. Hort Soc. (K-Lindl., holo!)
Brazil – E
o Maxillaria viridis Lindl., Bot. Reg. 18: t.1510, 1832
Colax viridis (Lindl.) Lindl., Bot. Reg. 29: misc.65:1, 1843
Maxillaria viridis var. uniflora platysepala Regel, Ind. Sem. Hort. Petrop.: 20, 1856
Colax jugosus var. viridis (Lindl.) Godef.-Leb., L'Orchidoph.: 67, 1881
Lycaste viridis (Lindl.) Benth., Gen. Pl. 3: 548, 1883

Pachystele dubia (A.Rich. & Galeotti) Schltr., Fedde Rep. Beih. 19: 114, 1923
BR 25: misc.45 (1839)
Mexico, Guatemala, Honduras, Nicaragua – E
Isochilus lividus Lindl., Bot. Reg. 25: misc.45, 1839
Isochilus dubius A.Rich. & Galeotti, Ann. Sc. Nat. s.3.: 26, 1845
Ponera dubia (A.Rich. & Galeotti) Rchb.f., Bonpl. 4: 327, 1856
Scaphyglottis dubia (A.Rich. & Galeotti) Benth. & Hook. ex Helms, Godman & Salvin Biol. Centr. Am. 3: 229, 1883
Scaphyglottis livida (Lindl.) Schltr., Beih. Bot. Centralbl. 36: 457, 1918
Scaphyglottis purpusii Schltr., Beih. Bot. Centralbl. 36: 399, 1918
Pachystelis dubia (A.Rich. & Galeotti) Rauschert, Fedde Rep. 94 (7–8): 456, 1983
Pachystelis livida (Lindl.) Rauschert, Fedde Rep. 94 (7–8): 456, 1983

Palumbina candida (Lindl.) Rchb.f., Walp. Ann. 6: 699, 1863
BR 29: misc.76 (1843)
Type: Guatemala, Hartweg s.n. (K-Lindl., holo!)
Mexico, Guatemala – E
Oncidium candidum Lindl., Bot. Reg. 29: misc.76, 1843

Paphinia cristata (Lindl.) Lindl., Bot. Reg. 29: misc.p.14:1, 1843
BR 21: t.1811 (1835) / 29: misc.p.14:1 (1843)
Type: Trinidad, Schomburgk s.n. cult. Knight (K-Lindl., holo!)
Guatemala, Venezuela, Trinidad, Guyana, Brazil, Colombia – E
o Maxillaria cristata Lindl., Bot. Reg. 21: t.1811, 1835
Lycaste cristata (Lindl.) Benth., Gen. Pl. 3: 548, 1883

Paphiopedilum barbatum (Lindl.) Stein, Orchideenb.: 455, 1892
BR 27: misc.110 (1841) / 28: t.17 (1842)
Type: Malaya, Mount Ophir, Cuming s.n. (K-Lindl., holo!)
Malaya, Thailand – 200–1300 m – T
o Cypripedium barbatum Lindl., Bot. Reg. 27: misc.110, 1841
Cypripedium biflorum Williams, Orch. Grow. Man. ed.3: 109, 1868
Cypripedium nigritum Rchb.f., Gard. Chron. s.3, 16: 102, 1882
Cypripedium barbatum var. biflorum (Williams) Williams, Orch. Grow. Man. ed.7: 233, 1894
Paphiopedilum nigritum (Rchb.f.) Pfitzer, Engl. Bot. Jahrb. 19: 40, 1894
Paphiopedilum barbatum var. nigritum (Rchb.f.) Pfitzer, Engl. Pflanzenr. Orch. Pleon.: 92, 1903
Cordula barbata (Lindl.) Rolfe, Orch. Rev. 20: 2, 1912
Cordula nigrita (Rchb.f.) Merrill, J. Str. Br. Roy. As. Soc. Sp.: 135, 1921

Paphiopedilum purpuratum (Lindl.) Stein, Orchideenb.: 481, 1892
BR 23: t.1991 (1837)
Type: Hong Kong, cult. Loddiges (K-Lindl., lecto!)
S.E.China, Hong Kong – 200–600 m – T
o Cypripedium purpuratum Lindl., Bot. Reg. 23: t.1991, 1837
Cypripedium sinicum Hance, Walp. Ann. 3: 602, 1853
Cordula purpurata (Lindl.) Rolfe, Orch. Rev. 20: 2, 1912

Paphiopedilum venustum (Wallich ex Sims) Pfitzer ex Stein, Orchideenb.: 489, 1892
BR 10: t.788 (1824)
N.E.India, Bangladesh, Bhutan, Nepal – 1000–1500 m – T
o Cypripedium venustum Wallich ex Sims, Bot. Mag. 47: t.2129, 1820
Cypripedium pardinum Rchb.f., Gard. Chron. s.3, 20: 554, 1869
Paphiopedilum pardinum (Rchb.f.) Pfitzer, Engl. Bot. Jahrb. 19: 40, 1894
Paphiopedilum venustum var. pardinum (Rchb.f.) Pfitzer, Engl. Pflanzenr. Orch. Pleon.: 82, 1903
Cordula venusta (Wallich ex Sims) Rolfe, Orch. Rev. 20: 2, 1912

Papilionanthe teres (Roxb.) Schltr., Orchis 9: 78, 1915
BR 21: t.1809 (1835) / 26: misc.111 (1840)
Nepal, Sikkim, N.India, Burma, Thailand, Laos – E
Dendrobium teres Roxb., Fl. Ind. 3: 485, 1832
o Vanda teres (Roxb.) Lindl., Gen. Sp. Orch. Pl.: 217, 1833

Pelexia adnata (Sw.) Sprengel, Syst. Veg. 3: 704, 1826
BR 12: t.985 (1826)
Mexico, Honduras, W.Indies, Venezuela, Colombia – T
Satyrium adnatum Sw., Nov. Gen. & Sp. Prodr.: 118, 1788
Neottia adnata (Sw.) Sw., Fl. Ind. Occ. 3: 1409, 1801
o Pelexia spiranthoides Lindl., Bot. Reg. 12: t.985, 1826
Adnula petiolaris Raf., Fl. Tellur. 2: 87, 1837
Neottia adnaria Raf., Fl. Tellur. 2: 87, 1837
Pelexia stenorrhynchoides Griseb., Cat. Pl. Cub.: 269, 1866
Spiranthes adnata (Sw.) Benth., J. Linn. Soc. Bot. 18: 344, 1881

Pelexia cerina (Lindl.) Garay, Bot. Mus. Leafl. Harv. Univ. 28 (4): 343, 1980
BR 28: misc.19 (1842)
Type: Guatemala, Hartweg s.n. cult. Hort. Soc. (K-Lindl., holo!)
Mexico, Guatemala – up to 1100 m – T
Spiranthes cerina Lindl., Bot. Reg. 28: misc.19, 1842
Sarcoglottis cerina (Lindl.) Baxter, Loud. Hort. Brit. ed.4, Suppl.: 634, 1850

Pelexia lobata (Lindl.) Garay, Bot. Mus. Leafl. Harv. Univ. 28 (4): 344, 1980
BR 30: misc.17 (1844) / 31: misc.43 (1845)
Type: Mexico, Real del Monte, Rule s.n. cult. C.Lemon (K-Lindl., holo!)
Mexico, Guatemala – T
Spiranthes lobata Lindl., Bot. Reg. 30: misc.17, 1844
Sarcoglottis lobata (Lindl.) Baxter, Loud. Hort. Brit. ed.4, Suppl.: 634, 1850

o **Peristeria cerina** Lindl., Bot. Reg. 23: t.1953, 1837
BR 23: t.1953 (1837) / 29: misc.99:4 (1843)
Type: Spanish Main, cult. Knight (K-Lindl., holo!)
Trinidad, N.E.S.America – E

Peristeria elata Hook., Bot. Mag. 58: t.3116, 1831
BR 29: misc.99:1 (1843)
Trop. America, Costa Rica, Panama, Venezuela, Colombia – T

Peristeria guttata Knowles & Westc., Fl. Cab. 2: 99 t.70, 1839
BR 26: misc.33 (1840) / 29: misc.99:3 (1843)
Brazil, Guyana, Venezuela, Amazonas – E
Peristeria cerina var. guttata (Knowles & Westc.) Rchb.f., Bot. Zeit. 10: 763, 1852

Peristeria pendula Hook., Bot. Mag. 63: t.3479, 1836
BR 29: misc.99:2 (1843)
Panama, Venezuela, Guyana, Surinam, Peru – T, E
Peristeria maculata hort. ex Lindl., Bot. Reg. 29: misc.99:2, 1843
Peristeria lentiginosa Lodd. ex Baxter, Hort. Brit. Suppl. 3: 604, 1850

Peristylus constrictus (Lindl.) Lindl., Gen. Sp. Orch. Pl.: 300, 1835
BR 18: sub t.1499 (1832)
Type: Nepal, Wallich s.n. (K-Lindl., holo!)
Nepal, Sikkim, Bhutan, India, Burma, Thailand, China, Cambodia, Vietnam – T

Herminium constrictum Lindl., Bot. Reg. 18: sub t.1499, 1832
Peristylus goodyeroides auct. non (D.Don) Lindl.: Gen. Sp. Orch. Pl.: 299, 1835
Platanthera constricta (Lindl.) Lindl., Gen. Sp. Orch. Pl.: 300, 1835
Habenaria constricta (Lindl.) Hook.f., Fl. Brit. Ind. 6: 161, 1890
Orchis leucantha Ham. ex Hook.f., Fl. Brit. Ind. 6: 161, 1890
Habenaria cylindrocalyx Gagnepain, Bull. Soc. Bot. Fr. 78: 68, 1931

Peristylus densus (Lindl.) Santap. & Kapad., J. Bombay Nat. Hist. Soc. 57: 128, 1960
BR 20: sub t.1701:27 (1834)
India, Burma, Thailand, Cambodia, Vietnam, China – T
Coeloglossum brevifolium Lindl., Gen. Sp. Orch. Pl.: 302, 1835
Coeloglossum densum Lindl., Gen. Sp. Orch. Pl.: 302, 1835
Gymnadenia tenuis Lindl., Gen. Sp. Orch. Pl.: 303, 1835
Habenaria peristyloides Wight, Ic.: t.1702, 1852 (non A.Rich.)
Platanthera stenostachya Lindl. ex Benth., Hook. J. Bot. 7: 37, 1855
Coeloglossum peristyloides (Wight) Rchb.f., Bonpl. 4: 321, 1856
Habenaria stenostachya (Lindl. ex Benth) Benth., Fl. Hongk.: 362, 1861
Habenaria neglecta King & Pantl., J. As. Soc. Beng. 66 (2): 603, 1897
Peristylus meglectus (King & Pantl.) Kraenzlin, Orch. Gen. & Sp.: 924, 1901
Peristylus stenostachyus (Lindl. ex Benth.) Kraenzlin, Orch. Gen. & Sp.: 502, 1901
Habenaria buchneroides Schltr., Fedde Rep. Beih. 4: 46 & 123, 1919
Habenaria atramentaria Kraenzlin, Fedde Rep. 17: 105, 1921
Habenaria stenostachya ssp. buchneroides (Schltr.) Soó, Ann. Mus. Nat. Hung. 26: 274, 1929
Habenaria evrardii Gagnepain, Bull. Soc. Bot. Fr. 78: 69, 1931
Habenaria dankiaensis Gagnepain, Bull. Soc. Bot. Fr. 79: 35, 1932
Peristylus xanthochlorus Blatt. & MacCann, J. Bomb. Nat. Hist. Soc. 35: 733, 1932
Glossula passerina Gagnepain, Fl. Gén. Indoch. 6: 627 fig.60, 1934
Habenaria passerina (Gagnepain) Tang & Wang, Act. Phytotax. Sin. 1 (1): 63, 1951

Peristylus goodyeroides (D.Don) Lindl., Gen. Sp. Orch. Pl.: 299, 1835
BR 18: sub t.1499 (1832) / 26: misc.187 (1840)
N.W.Himalaya to China, Philippines, Malaya, Indonesia, New Guinea – T
Habenaria goodyeroides D.Don, Prodr. Fl. Nep.: 25, 1825
Herminium goodyeroides (D.Don) Lindl., Bot. Reg. 18: sub t.1499, 1832
Peristylus horsfieldii Blume, Mus. 2: 189, 1856
Platanthera goodyeroides (D.Don) Vidal, Phan. Cum. Phil.: 15, 1885
Habenaria glaucescens Ridley, J. Linn. Soc. 32: 412, 1896
Habenaria lauterbachii Kraenzlin, Schumann & Lauterbach Nachtrag: 239, 1901
Habenaria sondaica Kraenzlin, Orch. Gen. & Sp.: 382, 1901
Habenaria hollandiae (J.J.Smith) Schltr., Fedde Rep. Beih. 1: 15, 1911
Peristylus hollandiae J.J.Smith, Bull. Dept. Agr. Neerl. 45: 1, 1911
Habenaria goodyeroides var. formosana Hayata, Ic. Pl. Form. 4: 126 t.22, 1914
Habenaria hayataeana Schltr., Fedde Rep. Beih. 4: 129, 1919

Peristylus hamiltonianus Lindl., Gen. Sp. Orch. Pl.: 299, 1835
BR 18: sub t.1499 (1832)
Nepal, Sikkim, Khasia – T
Herminium hamiltonianum Lindl., Gen. Sp. Orch. Pl.: 299, 1835
Habenaria hamiltoniana (Lindl.) Hook.f., Fl. Brit. Ind. 6: 160, 1890

Peristylus plantagineus (Lindl.) Lindl., Gen. Sp. Orch.: 300, 1835
BR 18: sub t.1499 (1832)
Type: Sri Lanka, Macrae s.n. (K-Lindl., holo!)
India, Sri Lanka – T
Herminium plantagineus Lindl., Bot. Reg. 18: sub t.1499, 1832
Habenaria goodyeroides Hook., Bot. Mag. 62: t.3397, 1835 (non Lindl.)
Peristylus elatus Dalz., Hook. J. Bot. 3: 344, 1851
Habenaria wightii Trimen, Cat. 91, 1885
Habenaria elata (Dalz.) Alston, Trimen Handb. Fl. Ceylon 6: 280, 1931

Peristylus secundus (Lindl.) Rathakrishnan, Ind. For. 98: 31, 1972
BR 20: sub t.1701:27 (1834)
S.India, Ryukyu Is. – T
Satyrium foliosum Heyne ex Wall., Cat.: 7054, 1832 (nomen ined.)
Coeloglossum secundum (Lindl.) Lindl., Gen. Sp. Orch. Pl.: 303, 1835
Gymnadenia secunda Lindl., Gen. Sp. Orch. Pl.: 303, 1835
Habenaria robustior Hook.f., Fl. Brit. Ind. 6: 160, 1890

Peristylus tentaculatus (Lindl.) J.J.Smith, Fl. Buitenz. 6: 35, 1905
BR 10: t.862 (1825)
Type: China, J.Damper Parks s.n. cult. Hort. Soc. (K-Lindl., holo!)
China (Kwangtung), Cambodia, Vietnam – T
o Glossula tentaculata Lindl., Bot. Reg. 10: t.862, 1825
Glossapis tentaculata (Lindl.) Sprengel, Syst. Veg. 3: 694, 1826
Glossapis antennifera Rchb.f., Linnaea 25: 225, 1852
Habenaria tentaculata (Lindl.) Rchb.f., Otia Bot. Hamb.: 34, 1878
Habenaria garrettii Rolfe ex Downie, Kew Bull.: 418, 1925

Phaius tankervilleae (Banks) Blume, Mus. Bot. Ludg. Bat. 2: 177, 1852
BR 25: misc.40 & misc.58 & misc.91 (1839)
Sri Lanka, India, Himalaya, Burma, S. & S.E.Asia, Malaysia, China, Hong Kong, Taiwan, Japan, Philippines to Australia, New Zealand, Horn I., Fiji, Samoa, Vanuatu, New Caledonia – 600–1300 m – T
Limodorum tankervilleae Banks, Sert. Angl. 28: Jan., 1789
Phaius grandifolius Lour., Fl. Cochinch. 2: 529, 1790
Limodorum incarvillei Pers., Syn. Pl. 2: 520, 1807
Pachyne spectabilis Salisb., Trans. Hort. Soc. 1: 299, 1812
Bletia tankervilleae (Banks) R.Br., Ait. Hort. Kew. ed.2, 5: 205, 1813
Tankervillia cantonensis Link, Handb. 1: 251, 1829
Phaius bicolor Lindl., Gen. Sp. Orch. Pl.: 128, 1831
Phaius blumei Lindl., Gen. Sp. Orch. Pl.: 127, 1831
Phaius wallichii Lindl., Wall. Pl. Asiat. Rar. 2: 46 t.158, 1831
Phaius grandifolius var. superbus Van Houtte, Fl. Serres 7: 259, 1852
Phaius carronii F.Muell., Ess. Pl. Esp. Burdekin: 19, 1860
Calanthe speciosa Vieill., Ann. Sc. Nat. s.4, 16: 44, 1861 (non Lindl.)
Phaius grandiflorus Rchb.f., Walp. Ann. 6: 459, 1862
Phaius leucophaeus F.Muell., Fragm. Phyt. Austr. 4: 163, 1864
Phaius blumei var. bernaysii Rchb.f., Gard. Chron.: 361, 1873
Phaius blumei var. assamica Rchb.f., Gard. Chron.: 558, 1882
Phaius incarvillei (Pers.) Kuntze, Rev. Gen. 2: 675, 1891
Phaius blumei var. pulchra King & Pantl., Ann. Roy. Bot. Gard. Calc. 8: 109, 1898
Phaius wallichi var. assamica (Rchb.f.) King & Pantl., Ann. Roy. Bot. Gard. Calc. 8: 109, 1898
Phaius incarvillei var. speciosa (Vieill.) Guillaumin, Not. Syst. 10 (2): 72, 1941
Calanthe bachmaensis Gagnepain, Bull. Mus. Nat. Hist. s.2, 27 (4): 348, 1957
Phaius longicornu Guillaumin, Bull. Mus. Nat. Hist. s.2, 29 (4): 348, 1957
Phaius tankervilleae var. superbus (Van Houtte) S.Y.Hu, Quart. J. Taiwan Mus. 27 (3): 464, 1974
Phaius tankervilleae f. veronicae Hu & Barretto, Chung Chi J. 13 (2): 25, 1976

Phaius tetragonus (Thouars) Rchb.f., Bonpl. 3: 221, 1855
BR 24: misc.6 (1838)
Réunion, Mauritius, Seychelles – T
Epidendrum tetragonum Thouars, Orch. Iles Afr.: t.34, 1804
Pesomeria tetragona (Thouars) Lindl., Bot. Reg. 24: misc.6, 1838
Phaius villosus Blume, Mus. Bot. Lugd. Bat. 2: 182, 1856

Phalaenopsis aphrodite Rchb.f., Hamb. Gartenz. 18: 35, 1862
BR 24: t.34 (1838)
Taiwan, Philippines (Babuyan Is., Sulu Arch.), Java – E
o Phalaenopsis amabilis Lindl., Bot. Reg. 24: t.34, 1838 (non Blume)
Phalaenopsis amabilis var. longifolia Don, Hort. Cantabr. ed.13: 608, 1845
Phalaenopsis amabilis var. rotundifolia Don, Hort. Cantabr. ed.13: 608, 1845
Phalaenopsis ambigua Rchb.f., Hamb. Gartenz. 18: 35, 1862
Phalaenopsis erubescens Burb., The Garden 9: 314, 1876
Phalaenopsis amabilis var. dayana hort. ex Warner & Williams, Orch. Alb. 1: t. 11, 1881
Phalaenopsis amabilis var. ambigua (Rchb.f.) Burb., The Garden 22: 119, 1882
Phalaenopsis amabilis var. erubescens (Burb.) Burb., The Garden 22: 119, 1882
Phalaenopsis aphrodite var. dayana (hort. ex Warner & Williams) Veitch, Man. Orch. Pl. 7: 24, 1891
Phalaenopsis amabilis var. aphrodite (Rchb.f.) Ames, Orchid. 2: 226, 1908
Phalaenopsis amabilis var. aphrodite subvar. dayana (hort. ex Warner & Williams) Ames, Orchid. 2: 227, 1908
Phalaenopsis amabilis var. aphrodite subvar. erubescens (Burb.) Ames, Orchid. 2: 227, 1908
Phalaenopsis amabilis var. formosa Shimadzu, Orch. Rev. 29: 68, 1921 (nomen)
Phalaenopsis babuyana Miwa, Pract. Hort. 27: 117, 1941 (nomen illeg.)
Phalaenopsis formosana Miwa, Pract. Hort. 27: 115 fig., 1941 (nomen illeg.)
Phalaenopsis formosum hort. ex Sander's One-Table List of Orch. Hybr. 2: 862, 1961

Pholidota articulata Lindl., Gen. Sp. Orch. Pl.: 38, 1830
BR 25: misc.57 (1839)
N.W.Himalaya, Nepal, Sikkim, Bhutan, N.E.India, China (Yunnan), Burma, Thailand, Laos, Cambodia, Vietnam, Malaya, Sumatra, Java, Borneo, Celebes – 300–2500 m – E
Pholidota khasiyana Rchb.f., Bonpl. 4: 329, 1856
Coelogyne articulata (Lindl.) Rchb.f., Walp. Ann. 6: 238, 1861

Coelogyne khasyana (Rchb.f.) Rchb.f., Walp. Ann. 6: 238, 1861
Pholidota griffithii Hook.f., Ic. Pl.: t.1881, 1889
Pholidota obovata Hook.f., Fl. Brit. Ind. 5: 845, 1890
Pholidota repens Rolfe, Kew Bull.: 199, 1891
Pholidota lugardii Rolfe, Kew Bull.: 6, 1893
Pholidota decurva Ridley, J. Linn. Soc. 32: 328, 1896
Pholidota articulata var. griffithii (Hook.f.) King & Pantling, Ann. Roy. Bot. Gard. Calc. 8: 147, 1898
Pholidota minahassae Schltr., Fedde Rep. 10: 20, 1911
Pholidota articulata var. minahassae (Schltr.) J.J.Smith, Bull. Jard. Bot. Buitenz. s.3, 1: 95, 1919
Pholidota articulata var. obovata (Hook.f.) Tang & Wang, Act. Phytotax. Sin. 1: 79, 1951

o **Pholidota imbricata** Hook., Exot. Fl. 2: t.138, 1825
BR 14: t. & sub t.1213 (1829) pro descr. / 21: t.1777 (1835) pro specimen / 26: misc.198 (1840)
Sri Lanka, India, N.W.Himalaya, Sikkim, Nepal, China (Yunnan), Thailand, Cambodia, Laos, Vietnam, Malaya, Indonesia, Borneo, Philippines, New Guinea, Australia, Pacific Is. – 0–1900 m – E
see Orchid Monographs vol. 3 (1988)
Cymbidium imbricatum Roxb., Hort. Beng.: 63, 1814
Ptilocnema bracteatum D.Don, Prod. Fl. Nepal: 33, 1825
Ornithidium imbricatum Wall. ex Lindl., Bot. Reg. 14: sub t.1213, 1829
o Pholidota imbricata Lindl., Bot. Reg. 14: t.1213, 1829 (non Hook.)
Pholidota conchoidea Lindl., Bot. Reg. 26: misc.198, 1840
Pholidota crotalina Rchb.f., Allg. Gartenz. 24: 218, 1856
Pholidota loricata Rchb.f., Bonpl. 5: 43, 1857
Coelogyne conchoidea (Lindl.) Rchb.f., Walp. Ann. 6: 237, 1861
Coelogyne imbricata (Hook.) Rchb.f., Walp. Ann. 6: 238, 1861
Coelogyne loricata (Rchb.f.) Rchb.f., Walp. Ann. 6: 238, 1861
Coelogyne triotos Rchb.f., Walp. Ann. 6: 238, 1861
Pholidota assamica Regel, Gartenfl.: 607, 1890
Pholidota imbricata var. coriacea Hook.f., Fl. Brit. Ind. 5: 846, 1890
Pholidota triotos (Rchb.f.) Pfitzer & Kraenzlin, Engl. Pflanzenr. Orch. Mon. Coel.: 154, 1907
Pholidota imbricata var. longifolia Schltr., Fedde Rep. Beih. 1: 109, 1911
Pholidota imbricata var. montana Schltr., Fedde Rep. Beih. 1: 109, 1911
Pholidota imbricata var. platyphylla Schltr., Fedde Rep. 10: 19, 1911
Pholidota henryi Kraenzlin, Viertelj. Nat. Gesel. Zurich 60: 427, 1915
Pholidota beccarii Schltr., Not. Bot. Gart. Berl. Dahl. 8: 17, 1921
Pholidota grandis Kraenzlin, Bull. Soc. Bot. Fr. 67: 301, 1929 (non Ridley)
Pholidota imbricata var. papuana J. J. Smith, Nov. Guin. 14: 349, 1929
Pholidota spectabilis Kraenzlin ex Guillaumin, Ann. Mus. Col. Marseille s.6, 5–6: 15, 1948
Pholidota imbricata var. henryi (Kraenzlin) Tang & Wang, Act. Phytotax. Sin. 1: 40, 1951
Pholidota pallida Holtum, Orch. Malaya ed. 3: 237, 1964 (non Lindl.)
Pholidota bracteata (D. Don) Seidenf., Opera Bot. 89: 100, 1986

Pholidota pallida Lindl., Bot. Reg. 21: t.1777, 1835
BR 21: t.1777 (1835) pro descr.
Type: Wallich s.n. cult. Hort. Soc. (K-Lindl., holo!)
Burma – E
Pholidota calceata Rchb.f., Bonpl. 4: 329, 1856
Coelogyne calceata (Rchb.f.) Rchb.f., Walp. Ann. 6: 238, 1861
Pholidota imbricata var. sessile Hook.f., Fl. Br. Ind. 5: 846, 1890
Pholidota yunnanensis Schltr., Fedde Rep. 19: 378, 1924 (non Rolfe)
Pholidota yupeensis Hu, Rhodora 27: 107, 1925
Pholidota schlechteri Gagnepain, Bull. Mus. Nat. Hist. Paris s.2, 3: 147, 1931
Pholidota tixieri Guillaumin, Bull. Mus. Nat. Hist. Paris s.2., 28 (6): 548, 1956
Pholidota sp. "GT 2828" Seidenf. & Smitin., Orch. Thailand: 139, 1959

Pholidota rubra Lindl., Gen. Sp. Orch. Pl.: 37, 1830
BR 14: sub t.1213 (1829) / 27: misc.19 (1841)
Sikkim, India, Burma, Vietnam – 850–2000 m – E
Pholidota undulata Lindl., Bot. Reg. 14: sub t.1213, 1829
Coelogyne rubra (Lindl.) Rchb.f., Walp. Ann. 6: 238, 1861
Coelogyne undulata (Lindl.) Rchb.f., Walp. Ann. 6: 238, 1861

Physosiphon emarginatus (Lindl.) Lindl., Bot. Reg. 21: sub t.1797:2, 1835
BR 21: sub t.1797:2 (1835)
Peru, Guyana – E
Humboldtia emarginata Pavon ex Lindl., Gen. Sp. Orch. Pl.: 6, 1830
Pleurothallis emarginata Lindl., Gen. Sp. Orch. Pl.: 6, 1830

Physosiphon tubatus (Lodd.) Rchb.f., Walp. Ann. 6: 18, 1861
BR 21: sub t.1797:1 (1835) / 24: misc.132 (1838)
Mexico, Guatemala, El Salvador, Honduras, Nicaragua – up to 3500 m – E
Stelis tubatus Lodd., Bot. Cab.: t.1601, 1830
Physosiphon loddigesii Lindl., Bot. Reg. 21: sub t.1797, 1835
Physosiphon carinatus Lindl., Bot. Reg. 24: misc.132, 1838
Physosiphon ochraceus A.Rich. & Galeotti, Ann. Sc. Nat. s.3, 3: 17, 1845
Physosiphon guatemalensis Rolfe, Kew Bull.: 197, 1891
Physosiphon lindleyi Rolfe, Kew Bull.: 61, 1893
Physosiphon moorei Rolfe, Kew Handl. Orch.: 189, 1896

Platanthera grandiflora (Bigelow) Lindl., Gen. Sp. Orch. Pl.: 294, 1835
BR 5: t.405 (1819)
Canada, USA – T
Orchis fimbriata Dryander, Ait. Hort. Kew. 3: 297, 1789
o Habenaria fimbriata (Dryander) R.Br., Ait. Hort. Kew. ed.2, 5: 193, 1813
Orchis grandiflora Bigelow, Fl. Bost. ed.2: 321, 1824
Habenaria grandiflora (Bigelow) Torrey, Comp.: 319, 1826
Platanthera fimbriata (Dryander) Lindl., Gen. Sp. Orch. Pl.: 293, 1826
Platanthera fimbriata var. grandiflora (Bigelow) Hook., Fl. Bor.-Am. 2: 200, 1839
Habenaria psycodes var. grandiflora (Bigelow) A.Gray, Amer. J. Sc. 38: 310, 1840
Platanthera psycodes var. grandiflora (Bigelow) Torrey, Fl. N.Y. 2: 278, 1843
Platanthera bigelovii Wood., Class.-Book.: 685, 1861
Blephariglottis grandiflora (Bigelow) Rydberg, Britton Man.: 296, 1901
Blephariglottis psycodes var. grandiflora (Bigelow) J.Schaffner, Cat. Ohio Pl.: 161, 1914
Habenaria fimbriata f. mentotonsa Fernald, Rhodora 48: 184, 1946

Platylepis goodyeroides A.Rich., Mém. Soc. Hist. Nat. Paris 4: 34, 1828
BR 24: misc.179 (1838)
Mauritius, Seychelles – E
Goodyera occulta Thouars, Hist. Orch.: t.28, 1822
Aetheria occulta (Thouars) Lindl., Gen. Sp. Orch. Pl.: 491, 1840
Notiophrys occulta (Thouars) Lindl., J. Linn. Soc.: 189, 1857

Plectorrhiza tridentata (Lindl.) Dockr., Austral. Sarcanthinae: 27, 1967
BR 24: misc.46 (1838)
Type: Australia, cult. Loddiges (K-Lindl., holo!)
E.Australia – E
Cleisostoma tridentatum Lindl., Bot. Reg. 24: misc.46, 1838
Saccolabium calcaratum F.Muell., Fragm. 1: 192, 1859
Sarcochilus calcaratus (F.Muell.) F.Muell., Fragm. 2: 181, 1861
Sarcochilus tridentatus (Lindl.) Rchb.f., Walp. Ann. 6: 500, 1861
Cleisostoma cornutum Rupp, North. Qld. Nat. 4, 39: 10, 1935
Sarcanthus tridentatus (Lindl.) Rupp, Vict. Nat. 57: 218, 1941
Thrixspermum tridentatum (Lindl.) T.E.Hunt, Qld. Nat. 16: 27, 1958

Plectrelminthus caudatus (Lindl.) Summerh., Kew Bull. 3: 441, 1949
BR 22: t.1844 (1836)
Type: Sierra Leone, Whitfield s.n. cult. Loddiges (K-Lindl., holo!)
Guinea, Sierra Leone, Liberia, Ivory Coast, Ghana, Nigeria, W.Cameroun, Centr.African Rep., Zaire – E
Angraecum caudatum Lindl., Bot. Reg. 22: t.1844, 1836
Listrostachys caudata (Lindl.) Rchb.f., Walp. Ann 6: 907, 1864
Leptocentrum caudatum (Lindl.) Schltr., Die Orchideen: 600, 1914

Plectrophora iridifolia (Lodd. ex Lindl.) Focke, Tijdschr. Wiss. Nat. Wetensch. 1: 212, 1848
BR 24: misc.178 (1838)
Type: Guyana, cult. Loddiges (K-Lindl., holo!)
Brazil, Surinam, Venezuela – E
Trichocentrum iridifolium Lodd. ex Lindl., Bot. Reg. 24: misc.178, 1838
Trichocentrum plectrophora Rchb.f., Walp. Ann. 6: 544, 1863

Pleione praecox (Smith) D.Don, Prod. Fl. Nepal: 37, 1825
BR 24: misc.157 (1838) / 26: t.24 (1840)
Nepal, Sikkim, Bhutan, Burma, S.China, Thailand – 1500–2700 m – T
Epidendrum praecox Smith, Exot. Bot. 2: 73 t.97, 1806
Coelogyne praecox (Smith) Lindl., Coll. Bot.: sub t.37, 1825
Cymbidium praecox (Smith) Lindl., Coll. Bot.: sub t.37, 1825
o Coelogyne wallichiana Lindl., Gen. Sp. Orch. Pl.: 43, 1830
Coelogyne wallichii Hook., Bot. Mag. 76: t.4496, 1850
Pleione wallichiana (Lindl.) Lindl., Paxt. Flow. Gard. 1: 66, 1852
Coelogyne praecox var. sanguinea Lindl., Fol. Orch. Coel.: 16, 1854
Coelogyne praecox var. wallichiana (Lindl.) Lindl., Fol. Orch. Coel.: 16, 1854
Coelogyne reichenbachiana T.Moore & Veitch, Gard. Chron.: 1210, 1868
Pleione reichenbachiana (T.Moore & Veitch) B.S.Williams, Orch. Grow Man. ed.4: 252, 1871
Coelogyne birmanica Rchb.f., Gard. Chron. s.2 (18): 840, 1882
Coelogyne praecox var. tenera Rchb.f., Gard. Chron. n.s.20: 294, 1883
Pleione birmanica (Rchb.f.) B.S.Williams, Orch. Grow. Man. ed.7: 681, 1894

Pleione concolor hort. ex B.S.Williams, Orch. Grow. Man. ed.7: 681, 1894
Pleione praecox var. birmanica (Rchb.f.) Grant, Orch. Burma: 167, 1895
Pleione praecox var. candida Pfitzer, Engl. Pflanzenr. Orch. Mon. Coel.: 126, 1907
Pleione praecox var. sanguinea (Lindl.) Pfitzer, Engl. Pflanzenr. Orch. Mon. Coel.: 126, 1907
Pleione praecox var. alba E.W.Cooper, Roy. Hort. Soc. Dict. Gard.: 1606, 1951
Pleione praecox var. wallichiana (Lindl.) E.W.Cooper, Roy. Hort. Soc. Dict.: 1606, 1951

Pleurothallis acianthera Lindl., Fol. Orch. Pleuroth.: 23, 1859
BR 29: misc.10 (1843)
Brazil – E obscure sp., fide Pabst & Dungs
Acianthera punctata Schweidw., Allg. Gartenz.: 292, 1842
Humboldtia acianthera Kuntze, Rev. Gen. 2: 667, 1891

Pleurothallis acuminata (Kunth) Lindl., Bot. Reg. 28: misc.p.70:13, 1842
BR 21: sub t.1797 (1835) / 28: misc.p.70: 13 (1842)
Venezuela, Colombia, Ecuador, Peru – E
Dendrobium acuminatum Kunth, Nov. Gen. & Sp. Pl. 1: 357, 1816
Specklinia acuminata (Kunth) Lindl., Bot. Reg. 21: sub t.1797, 1835
Pleurothallis aurea Lindl., Ann. Mag. Nat. Hist. 12: 397, 1843
Pleurothallis asperilinguis Rchb.f. & Warsc., Bonpl. 2: 114, 1854
Humboldtia acuminata (Kunth) Kuntze, Rev. Gen. 2: 667, 1891
Humboldtia asperilinguis (Rchb.f.) Kuntze, Rev. Gen. 2: 667, 1891
Humboldtia aurea (Lindl.) Kuntze, Rev. Gen. 2: 667, 1891
Pleurothallis nigro-hirsuta Kraenzlin, Fedde Rep. 1: 91, 1905
Pleurothallis angustipetala C.Schweinf., Bot. Mus. Leafl. Harv. Univ. 16: 45, 1953
Pleurothallis aurea var. asperilinguis (Rchb.f.) Foldats, Act. Biol. Venez. 2: 373, 1959

Pleurothallis alpestris (Sw.) Lindl., Gen. Sp. Orch. Pl.: 7, 1830
BR 21: sub t.1797:12 (1835) / 28: misc.p.71:19 (1842)
Jamaica – E
Epidendrum alpestre Sw., Prodr.: 125, 1788
Dendrobium alpestre (Sw.) Sw., Nov. Act. Upsal. 6: 84, 1799

Pleurothallis angustifolia Lindl., Bot. Reg. 21: sub t.1797:7 (1835)
BR 21: sub t.1797:7 (1835) / 28: misc.p.79:79 (1842)
Type: Mexico, Deppe s.n. (K-Lindl., holo!)
Mexico – E

Pleurothallis aphthosa Lindl., Bot. Reg. 24: misc.71, 1838
BR 24: misc.71 (1838) / 28: misc.p.78:68 (1842)
Type: Mexico, cult. Bateman ex Birmingh. Bot. Garden (K-Lindl., holo!)
Mexico – E
Pleurothallis peduncularis Hook., J. Bot. 3: 308, 1843
Specklinia aphtosa (Lindl.) F.de Barros, Hoehnea 10: 109, 1984

Pleurothallis aristata Hook., Ann. Mag. Mat. Hist. 1 (2): t.15, 1839
BR 28: misc.p.71:17 (1842)
W.Indies, Costa Rica, Panama, Venezuela, Guyana, Colombia, Ecuador – E
Pleurothallis urbaniana Rchb.f., Ber. Deutsch. Bot. Ges. 3: 279, 1855
Pleurothallis barberana Rchb.f., Gard. Chron. n.s.16: 6, 1881
Humboldtia aristata (Hook.) Kuntze, Rev. Gen. 2: 667, 1891
Humboldtia barberana (Rchb.f.) Kuntze, Rev. Gen. 2: 667, 1891
Pleurothallis helenae Fawcett & Rendle, J. Bot. 47: 4, 1909
Pleurothallis dichotoma Ames, Sched. Orch. 6: 58, 1923 (non Schltr.)
Pleurothallis divexa Ames, Sched. Orch. 7: 20 t.5, 1924

Pleurothallis articulata Lindl., Bot. Reg. 28: misc.p.83:112, 1842
BR 28: misc.p.83:112 (1842)
Type: Brazil, Maximillian of Wied Neuwied s.n. (K-Lindl., holo!)
Brazil – E
Humboldtia articulata (Lindl.) Kuntze, Rev. Gen. 2: 667, 1891

Pleurothallis auriculata Lindl., Hook. Comp. Bot. Mag. 2: 356, 1836
BR 28: misc.p.75:53 (1842)
Brazil – E
Pleurothallis compressiflora Barb.Rodr., Gen. & Sp. Orch. Nov. 2: 13, 1882
Humboldtia auriculata (Lindl.) Kuntze, Rev. Gen. 2: 667, 1891
Specklinia auriculata (Lindl.) F.de Barros, Hoehnea 10: 109, 1984

Pleurothallis barbata Westc., Phytolog. 1: 54, 1841 (non Focke)
BR 28: misc.p.75:54 (1842)
Brazil – E
Humboldtia barbata (Westc.) Kuntze, Rev. Gen.: 667, 1891

Pleurothallis bicarinata Lindl., Bot. Reg. 25: misc.11, 1839
BR 25: misc.11 (1839) / 28: misc.p.75:49 (1842)
Type: Brazil, cult. Loddiges K-Lindl., holo!)
Brazil – E
Humboldtia bicarinata (Lindl.) Kuntze, Rev. Gen. 2: 667, 1891

Pleurothallis bidentata Lindl., Bot. Reg. 21: sub t.1797:8, 1835
BR 21: sub t.1797:8 (1835) / 28: misc.p.79:78 (1842)
Type: Brazil, Rio de Janeiro, Forbes s.n. (K-Lindl., holo!)
Brazil – E
Humboldtia bidentata (Lindl.) Kuntze, Rev. Gen. 2: 667, 1891

Pleurothallis breviflora Lindl., Bot. Reg. 27: misc.125, 1841
BR 27: misc.125 (1841) / 28: misc.p.77:67 (1842)
Type: Mexico, cult. Loddiges (K-Lindl., holo!)
Mexico – E

Pleurothallis capillaris Lindl., Bot. Reg. 21: sub t.1797:29, 1835
BR 21: sub t.1797:29 (1835) / 28: misc.p.73:39 (1842)
Type: Brazil, Sierra del Piedale, Martius s.n.(K-Lindl., holo!)
Brazil – E
Humboldtia capillaris (Lindl.) Kuntze, Rev. Gen. 2: 667, 1891

Pleurothallis casapensis Lindl., Bot. Reg. 28: misc.p.76:61, 1842
BR 28: misc.p.76:61 (1842)
Type: Peru, Mathews s.n. (K-Lindl., holo!)
Peru – E
Humboldtia casapensis (Lindl.) Kuntze, Rev. Gen. 2: 667, 1891

Pleurothallis cauliflora Lindl., Hook. Comp. Bot. Mag. 2: 355, 1836
BR 28: misc.p.71:23 (1842)
Peru, Ecuador – E
Humboldtia cauliflora (Lindl.) Kuntze, Rev. Gen. 2: 667, 1891

Pleurothallis centranthera Lindl., Fol. Orchid. Pleuroth.: 37, 1859
BR 29: misc.11 (1843)
Brazil – E
Centranthera punctata Schweid., Otto & Dietr. Allg. Gartenz.: 293, 1842
Humboldtia centranthera (Lindl.) Kuntze, Rev. Gen. 2: 667, 1891

Pleurothallis circumplexa Lindl., Bot. Reg. 24: misc.27, 1838
BR 24: misc.27 (1838) / 28: misc.p.78:73 (1842)
Type: Mexico, cult. Loddiges (K-Lindl., holo!)
Mexico, Guatemala, El Salvador, Nicaragua, Costa Rica – E
Pleurothallis mesophylla A.Rich. & Galeotti, Ann. Sc. Nat. Bot. s.3, 3: 16, 1845
Humboldtia circumplexa (Lindl.) Kuntze, Rev. Gen. 2: 667, 1891

Pleurothallis cordata (Ruiz Lopez & Pavon) Lindl., Gen. Sp. Orch. Pl.: 5, 1830
BR 21: sub t.1797:5 (1835) / 28: misc.p.81:91 (1842)
Trop. America – E, T
Humboldtia cordata Ruiz Lopez & Pavon, Syst. Veg. Fl. Peruv. & Chil. 1: 234, 1798
Stelis cordata (Ruiz Lopez & Pavon) Wild., Sp. Pl. ed.4: 140, 1805
Pleurothallis brachyblephara Schltr., Fedde Rep. Beih. 9: 72, 1921

Pleurothallis corniculata (Sw.) Lindl., Bot. Reg. 28: misc.p.83:110, 1842
BR 28: misc.p.83:110 (1842)
W.Indies, Mexico, Guatemala, Honduras, Nicaragua, Costa Rica, S.America – E
Dendrobium corniculatum (Sw.) Sw., Nov. Act. Upsal. 6: 83, 1799
Epidendrum corniculatum Sw., Fl. Ind. Occ. 3: 1537, 1799
Pleurothallis nubigena Lindl., Ann. Mag. Hist. Nat. 3: 326, 1858

Pleurothallis decurrens Poeppig & Endl., Nov. Gen. & Sp. Pl. 1: 48 t.83, 1836
BR 28: misc.p.74:45 (1842)
Peru – E
Humboldtia decurrens (Poeppig & Endl.) Kuntze, Rev. Gen. 2: 667, 1891

Pleurothallis deregularis (Barb.Rodr.) Luer, Selbyana 2: 385, 1978
BR 21: sub t.1797:3 (1835)
Honduras to Peru, Brazil – E
Physosiphon spiralis Lindl., Bot. Reg. 21: sub t.1797:3, 1835 (non Ruiz Lopez & Pavon)
Pleurothallis clausa A.Rich. ex Rchb.f., Cat. Orch. Samml. Schiller ed.3: 59, 1857 (non Depuydt)
Stelis deregularis Barb.Rodr., Gen. & Sp. Orch. Nov. 2: 94, 1882
Pleurothallis crassipes Rchb.f., Wawra Itin. Princ. S.Coburg 2: 82, 1888 (non Lindl.)
Physosiphon deregularis (Barb.Rodr.) Cogn., Mart. Fl. Bras. 3 (4): 341, 1896
Pseudostelis deregularis (Barb.Rodr.) Schltr., Anex. Mem. Inst. Butantan 1 (4): 38, 1922
Pseudostelis spiralis (Lindl.) Schltr., Anex. Mem. Inst. Butantan 1 (4): 38, 1922
Physosiphon minutuflorus Ames & C.Schweinf., Sched. Orch. 8: 11, 1925
Pleurothallis schweinfurthiana L.O.Williams, Ceiba 5: 92, 1956

Pleurothallis diffusa Poeppig & Endl., Nov. Gen. Sp. Pl. 1: 49 t.86, 1836
BR 28: misc.p.72:25 (1842)
Peru, Trinidad, Venezuela, Bolivia – E
Pleurothallis semipellucida Rchb.f., Linnaea 22: 823, 1849
Humboldtia diffusa (Poeppig & Endl.) Kuntze, Rev. Gen. 2: 667, 1891
Humboldtia semipellucida (Rchb.f.) Kuntze, Rev. Gen. 2: 668, 1891
Pleurothallis complicata Rolfe, Mem. Torr. Bot. Cl. 6: 121, 1896

Pleurothallis discoidea Lindl., Bot. Reg. 21: sub t.1797:10, 1835
BR 21: sub t.1797:10 (1835) / 28: misc.p.79:82 (1842)
Type: Trinidad, in Herb. Hooker s.n. (K, holo!)
Venezuela, Trinidad – E
Pleurothallis leptopetala Cogn., Urb. Symb. Antill. 6: 693, 1910

Pleurothallis elegans (Kunth) Lindl., Bot. Reg. 28: misc.p.70:14, 1842
BR 21: sub t.1797 (1835) / 28: misc.p.70:14 (1842)
Colombia – E
Dendrobium elegans Kunth, Nov. Gen. & Sp. Pl. 1: 358, 1815
Specklinia elegans (Kunth) Lindl., Bot. Reg. 21: sub t.1797, 1835
Humboldtia elegans (Kunth) Kuntze, Rev. Gen. 2: 667, 1891

Pleurothallis ephemera Lindl., Hook. Comp. Bot. Mag. 2: 356, 1836
BR 28: misc.p.82:97 (1842)
Brazil – E
Pleurothallis longirostris Focke, Bot. Zeit. 11: 341, 1853
Pleurothallis rostriflora Rchb.f., Walp. Ann. 6: 175, 1861
Humboldtia ephemera (Lindl.) Kuntze, Rev. Gen. 2: 667, 1891
Humboldtia longirostris (Focke) Kuntze, Rev. Gen. 2: 667, 1891
Pleurothallis ornithorrhyncha Hoehne, Contr. Fl. Orch. Bras.: 34, 1929

Pleurothallis flexuosa (Poeppig & Endl.) Lindl., Bot. Reg. 28: misc.p.69: 7, 1842
BR 28: misc.p.69:7 (1842)
Venezuela, Colombia, Ecuador, Peru – E
Specklinia flexuosa Poeppig & Endl., Nov. Gen. Sp. Pl. 1: 52, 1835
Pleurothallis kefersteiniana Rchb.f., Bot. Zeit. 10: 673, 1852
Humboldtia flexuosa (Poeppig & Endl.) Kuntze, Rev. Gen. 2: 667, 1891
Pleurothallis remotiflora C.Schweinf., Amer. Orch. Soc. Bull. 21 : 108, 1952

Pleurothallis floribunda Poeppig & Endl., Nov. Gen. Sp. 1: 48 t.84, 1836
BR 28: misc.p.74:44 (1842)
Peru – E
Pleurothallis macrophylla Lindl., Bot. Reg. 28: misc.p.74:44, 1842 (non Kunth)

Pleurothallis foetens Lindl., Bot. Reg. 27: misc.188, 1841
BR 27: misc.188 (1841) / 29: misc.7 (1843)
Type: Brazil, Rio de Janeiro, cult. Loddiges (K-Lindl., holo!)
Brazil – E
Pleurothallis pelioxantha Barb.Rodr., Gen. & Sp. Orch. Nov. 1: 17, 1877
Humboldtia foetens (Lindl.) Kuntze, Rev. Gen. 2: 667, 1891

Pleurothallis gelida Lindl., Bot. Reg. 27: misc.186, 1841
BR 27: misc.186 (1841) / 28: misc.p.73:33 (1842)
Type: Jamaica, cult. Loddiges (K-Lindl., holo!)
USA (Florida), W.Indies to El Salvador, Honduras, Nicaragua, Costa Rica, Panama, S.America – E
Pleurothallis univaginata Lindl., Ann. & Mag. Nat. Hist. s.3, 1: 326, 1858
Pleurothallis polyliria Endres & Rchb.f., Gard. Chron.: 1483, 1871
Anathallis hebesepala Barb.Rodr., Gen. & Sp. Orch. Nov. 2: 75, 1882
Humboldtia gelida (Lindl.) Kuntze, Rev. Gen. 2: 667, 1891
Humboldtia polyliria (Endres & Rchb.f.) Kuntze, Rev. Gen. 2: 668, 1891
Pleurothallis hebesepala (Barb.Rodr.) Cogn., Mart. Fl. Bras. 3 (4): 568, 1896
Pleurothallis tenuispica Schltr., Fedde Rep. 14: 133, 1915

Pleurothallis glanduligera Lindl., Hook. Comp. Bot. Mag. 2: 355, 1836
BR 28: misc.p.79:80 (1842)
Brazil – E
Humboldtia glanduligera (Lindl.) Kuntze, Rev. Gen. 2: 667, 1891

Pleurothallis glumacea Lindl., Hook. Comp. Bot. Mag. 2: 355, 1836
BR 28: misc.p.71:21 (1842)
Brazil – E
Humboldtia glumacea (Lindl.) Kuntze, Rev. Gen. 2: 667, 1891
Pleurothallis glaziovii Cogn., Mart. Fl. Bras. 1 (4): 546, 1893
Pleurothallis vitellina Porsch, Oestrr. Bot. Zeitschr.: 158, 1905
Pleurothallis alexandrii Schltr., Fedde Rep. Beih. 19: 103, 1923

Pleurothallis grandiflora Lindl., Bot. Reg. 21: sub t.1797:6, 1835
BR 21: sub t.1797:6 (1835) / 28: misc.p.78:70 (1842)
Type: Peru, Lloae, Hall s.n. (K-Lindl., holo!)
Venezuela, Colombia, Ecuador, Peru, Bolivia – E
Humboldtia grandiflora (Lindl.) Kuntze, Rev. Gen. 2: 667, 1891
Pleurothallis boliviana Schltr., Fedde Rep. 10: 453, 1912 (non Rchb.f.1855)
Pleurothallis buchtieni Schltr., Fedde Rep. Beih. 10: 66, 1922
Pleurothallis sarcochila Garay, Arch. Jard. Bot. Rio 12: 176, 1952

○ **Pleurothallis grobyi** Bateman ex Lindl., Bot. Reg. 21: t.1797:14, 1835
BR 21: t. & sub t.1797:14 (1835) / t.1825 (1836) / 24: misc.70 (1838) / 27: misc.182 (1841) / 28: misc.p.69:8, 9 & misc.p.70:15 (1842)
Type: Guyana, cult. Bateman (K-Lindl., holo!)
Mexico to Bolivia, Brazil, W.Indies – up to 1500 m – E

○ Pleurothallis picta Lindl., Bot. Reg. 21: sub t.1797, 1835
Pleurothallis marginata Lindl., Bot. Reg. 24: misc.70, 1838
Pleurothallis surinamensis Focke, Tijdschr. Nat. Wetensch. 2: 194, 1849
Pleurothallis marginalis Rchb.f., Bonpl. 3: 224, 1855
Pleurothallis perplexa Rchb.f., Hamb. Gartenz. 16: 15, 1860
Pleurothallis trilineata Barb.Rodr., Gen. & Sp. Orch. Nov. 1: 6, 1877
Pleurothallis crepidophylla Rchb.f., Otia Bot. Hamb. 1: 18, 1878
Lepanthes marmorata Barb.Rodr., Gen. & Sp. Orch. Nov. 2: 42, 1882
Lepanthes trilineata (Barb.Rodr.) Barb.Rodr., Gen. & Sp. Orch. Nov. 2: 42, 1882
Pleurothallis choconiana Watson, Proc. Am. Acad. 23: 285, 1888
Humboldtia crepidophylla (Rchb.f.) Kuntze, Rev. Gen. 2: 667, 1891
Humboldtia grobyi (Bateman ex Lindl.) Kuntze, Rev. Gen. 2: 668, 1891
Humboldtia marginalis (Rchb.f.) Kuntze, Rev. Gen. 2: 668, 1891
Humboldtia marginata (Lindl.) Kuntze, Rev. Gen. 2: 668, 1891
Humboldtia picta (Lindl.) Kuntze, Rev. Gen. 2: 668, 1891
Pleurothallis lindleyana Cogn., Mart. Fl. Bras. 3 (4): 489, 1893
Pleurothallis pergracilis Rolfe, Kew Bull.: 334, 1893
Pleurothallis grobyi var. trilineata (Barb.Rodr.) Cogn., Mart. Fl. Bras. 3 (4): 496, 1896
Pleurothallis marmorata (Barb.Rodr.) Cogn., Mart. Fl. Bras. 3 (4): 490, 1896
Pleurothallis marmorata var. concolor Cogn., Mart. Fl. Bras. 3 (6): 559, 1906
Pleurothallis panamensis Schltr., Fedde Rep. 17: 140, 1921
Pleurothallis biglandulosa Schltr., Notizbl. Bot. Gart. Berlin 8: 119, 1922
Pleurothallis subpicta Schltr., Anex. Inst. Butant. 1 (4): 42, 1922
Pleurothallis dryadum Schltr., Fedde Rep. Beih. 19: 186, 1923
Pleurothallis arevaloi Schltr., Fedde Rep. Beih. 27: 43, 1924
Pleurothallis integrilabia Ames, Hubb. & C.Schweinf., Bot. Mus. Leafl. Harv. Univ. 3: 39, 1934
Pleurothallis ezechiasi Hoehne, Arq. Bot. Est. S.Paulo 11 (2): 21, 1946
Pleurothallis grobyi var. marmorata (Barb.Rodr.) Garay, Arq. Jard. Bot. Rio de Janeiro 11: 54, 1951
Specklinia grobyi (Bateman ex Lindl.) F.de Barros, Hoehnea 10: 110, 1984

Pleurothallis hartwegii Lindl., Bot. Reg. 28: misc.p.80:86, 1842
BR 28: misc.p.80:86 (1842)
Type: Colombia, Loxa, Hartweg s.n. (K-Lindl., holo!)
Colombia – E
Humboldtia hartwegii (Lindl.) Kuntze, Rev. Gen. 2: 667, 1891

Pleurothallis hians Lindl., Bot. Reg. 21: sub t.1797:20, 1835
BR 21: sub t.1797:20 (1835) / 28: misc.p.69:3 (1842)
Type: Brazil, Organ Mts, in herb. Hooker s.n. (K, holo!)
Brazil – E
Lepanthes gunningiana Barb.Rodr., Gen. & Sp. Orch. Nov. 2: 48, 1882
Humboldtia hians (Lindl.) Kuntze, Rev. Gen. 2: 667, 1891

Pleurothallis hymenantha Lindl., Bot. Reg. 21: sub t.1797:16, 1835
BR 21: sub t.1797:16 (1835) / 28: misc.p.83:105 (1842)
Type: Brazil, Retiro, Martius s.n. (K-Lindl., holo!)
Brazil – E
Humboldtia hymenantha (Lindl.) Kuntze, Rev. Gen. 2: 667, 1891

Pleurothallis hypnicola Lindl., Bot. Reg. 28: misc.p.75:52, 1842
BR 28: misc.p.69:6 & misc.p.75:52 (1842)
Type: Brazil, Miers s.n. (K-Lindl., holo!)
Brazil – E
Pleurothallis fusca Lindl., Hook. Comp. Bot. Mag. 2: 354, 1836
Humboldtia hypnicola (Lindl.) Kuntze, Rev. Gen. 2: 667, 1891
Pleurothallis cuneifolia Cogn., Mart. Fl. Bras.: 3, 1893
Specklinia hypnicola (Lindl.) F.de Barros, Hoehnea 10: 110, 1984

Pleurothallis inaequalis Lindl., Bot. Reg. 28: misc.p.76:60, 1842
BR 28: misc.p.76:60 (1842)
Type: Brazil, Miers s.n. (K-Lindl., icono!)
Brazil – E
Humboldtia inaequalis (Lindl.) Kuntze, Rev. Gen. 2: 667, 1891

Pleurothallis jamiesonii Lindl., Bot. Reg. 21: sub t.1797:28, 1835
BR 21: sub t.1797:28 (1835) / 28: misc.p.73:35 (1842)
Type: Peru, Jamieson s.n. (K-Lindl., holo!)
Colombia, Peru – E

Pleurothallis lanceana Lodd., Bot. Cab. 18: t.1767, 1831
BR 21: sub t.1797:27 (1835) / 26: misc.41 (1840) / 28: misc.p.72: 27, 38 & misc.p.77:62 (1842)
Guatemala, Costa Rica, Trinidad, Venezuela, Guyana, Surinam, Brazil, Colombia, Peru – E

Pleurothallis ciliata Knowles & Westc., Fl. Cab. 1: t.19, 1837
Pleurothallis plumosa Lindl., Bot. Reg. 28: misc.p.72:27, 1842
Pleurothallis crassifolia Focke, Tijdschr. Nat. Wetensch. 2: 196, 1849
Pleurothallis minax Rchb.f., Bonpl. 2: 24, 1854
Humboldtia ciliata (Knowles & Westc.) Kuntze, Rev. Gen. 2: 667, 1891
Humboldtia crassifolia (Focke) Kuntze, Rev. Gen. 2: 667, 1891
Humboldtia lanceana (Lodd.) Kuntze, Rev. Gen. 2: 667, 1891
Humboldtia plumosa (Lindl.) Kuntze, Rev. Gen. 2: 668, 1891
Pleurothallis huebneri Schltr., Beih. Bot. Zentralbl. 42 (2): 90, 1925

Pleurothallis lanceola (Sw.) Sprengel, Syst. Veg. 3: 731, 1826
BR 28: misc.p.82:101 (1842)
Jamaica – E, L
Epidendrum lanceola Sw., Prodr.: 123, 1788
Dendrobium lanceola (Sw.) Sw., Nov. Act. Upsal. 6: 83, 1799

Pleurothallis lanceolata Lindl., Gen. Sp. Orch. Pl.: 5, 1830
BR 21: sub t.1797:30 (1835) 28: misc.p.71:22 (1842)
Peru, Venezuela – E
Humboldtia lanceolata (Lindl.) Kuntze, Rev. Gen. 2: 667, 1891

Pleurothallis laxa (Sw.) Lindl., Gen. Sp. Orch. Pl.: 7, 1830
BR 21: sub t.1797:13 (1835) / 28: misc.p.71:18 (1842)
Jamaica – E
Epidendrum laxum Sw., Prodr.: 125, 1788
Dendrobium laxum (Sw.) Sw., Nov. Act. Upsal. 6: 84, 1799
Pleurothallis nigroannulata Cogn., Fedde Rep. 6: 306, 1909

Pleurothallis linearis (Lindl.) Lindl., Bot. Reg. 28: misc.p.82: 98, 1842
BR 28: misc.p.82:98 (1842)
Peru – E
Specklinia linearis Lindl., Gen. Sp. Orch. Pl. 9: 5, 1830

Pleurothallis lingua Lindl., Bot. Reg. 28: misc.p.80:83, 1842
BR 28: misc.p.80:83 (1842)
Type: Brazil, Organ, Miers s.n. (K-Lindl., holo!)
Brazil – E
Humboldtia lingua (Lindl.) Kuntze, Rev. Gen. 2: 667, 1891

Pleurothallis longicaulis Lindl., Bot. Reg. 28: misc.p.72:26, 1842
BR 28: misc.p.72:26 (1842)
Type: Brazil, Mr. Miers s.n. (K-Lindl., holo!)
Brazil – E
Humboldtia longicaulis (Lindl.) Kuntze, Rev. Gen. 2: 667, 1891

Pleurothallis luteola Lindl., Bot. Reg. 27: misc.2, 1841
BR 27: misc.2 & 188 (1841) / 28: misc.p.75:48 & misc.p.78:72 (1842)
Type: cult. Mrs. Cannon (K-Lindl., holo!)
Brazil – E
Pleurothallis fragilis Lindl., Bot. Reg. 27: misc.188, 1841
Pleurothallis caespitosa Barb.Rodr., Gen. & Sp. Orch. Nov. 1: 14, 1877
Humboldtia fragilis (Lindl.) Kuntze, Rev. Gen. 2: 667, 1891
Pleurothallis platycaulis Rchb.f. ex Cogn., Mart. Fl. Bras. 3 (4): 536, 1893
Pleurothallis subcordifolia Cogn., Mart. Fl. Bras. 3 (4): 561, 1906
Specklinia luteola (Lindl.) F.de Barros, Hoehnea 10: 110, 1984

Pleurothallis macrorhiza Lindl., Hook. J. Bot. 1: 9, 1834
BR 21: sub t.1797:33 (1835) / 28: misc.p.69:5 (1842)
Peru, Ecuador – T
Humboldtia macrorhiza (Lindl.) Kuntze, Rev. Gen. 2: 667, 1891
Pleurothallis rhizomatosa Schltr., Fedde Rep. Beih. 8: 62, 1921

Pleurothallis muscoidea Lindl., Bot. Reg. 24: misc.165, 1838
BR 24: misc.165 (1838) / 28: misc.p.82:100 (1842)
Type: Brazil, cult. Loddiges (K-Lindl., holo!)
Brazil – E
Humboldtia muscoidea (Lindl.) Kuntze, Rev. Gen. 2: 668, 1891

Pleurothallis obovata (Lindl.) Lindl., Bot. Reg. 28: misc.p.75:51, 1842
BR 25: misc.137 (1839) / 28: misc.p.75:51 (1842)
Type: Brazil, without collector (K-Lindl., holo!)
W.Indies, Venezuela, Colombia, Ecuador, Peru, Brazil – E
Specklinia obovata Lindl., Bot. Reg. 25: misc.137, 1839
Pleurothallis octomeriaeformis Rchb.f., Bonpl. 2: 25, 1854
Pleurothallis albida Lindl., Ann. Mag. Nat. Hist. 3: 327, 1858
Stelis fasciculiflora Regel, Act. Hort. Petrop. 1: 99, 1871
Anathallis fasciculata Barb.Rodr., Gen. & Sp. Orch. Nov. 1: 23, 1877
Anathallis densiflora Barb.Rodr., Gen. & Sp. Orch. Nov. 2: 73, 1882
Anathallis micrantha Barb.Rodr., Gen. & Sp. Orch. Nov. 2: 76, 1882
Anathallis osmosperma Barb.Rodr., Gen. & Sp. Orch. Nov. 2: 73, 1882
Humboldtia obovata (Lindl.) Kuntze, Rev. Gen. 2: 668, 1891
Pleurothallis densiflora (Barb.Rodr.) Cogn., Mart. Fl. Bras. 3 (4): 559, 1896
Pleurothallis fasciculata (Barb.Rodr.) Cogn., Mart. Fl. Bras. 3 (4): 559, 1896
Pleurothallis minutiflora Cogn., Mart. Fl. Bras. 3 (4): 564, 1896
Pleurothallis osmosperma (Barb.Rodr.) Cogn., Mart. Fl. Bras. 3 (4): 563, 1896

Pleurothallis citrina Schltr., Fedde Rep. Beih. 7: 103, 1920
Pleurothallis angustilabia Schltr., Fedde Rep. Beih. 9: 71, 1921
Pleurothallis modestiflora Schltr., Fedde Rep. Beih. 35: 56, 1925
Pleurothallis citrina var. elliptica C.Schweinf., Bot. Mus. Leafl. Harv. Univ. 16: 9, 1953

Pleurothallis ochreata Lindl., Bot. Reg. 21: sub t.1797:17, 1835
BR 21: sub t.1797:17 (1835) / 28: misc.p.75:50 (1842)
Type: Brazil, Serra do Monte Santo, Martius s.n. (K-Lindl., icono!)
Brazil – E
Humboldtia ochreata (Lindl.) Kuntze, Rev. Gen. 2: 668, 1891
Pleurothallis bahiensis Pabst, Rodriguezia 18–19: 27, 1956

Pleurothallis ophioglossoides (Jacq.) Garay & Sweet, J. Arnold Arb. 53: 391, 1972
BR 11: t.935 (1825) / 28: misc.p.73:40 (1842)
Cuba, W.Indies, Mexico, Costa Rica, Trinidad, Guyana – E, L
Epidendrum ophioglossoides Jacq., Enum. Syst. Pl.: 29, 1760 (non Jacq.1763)
Epidendrum trigoniflorum Sw., Prodr. 125, 1788
Dendrobium ophioglossoides (Jacq.) Sw., Nov. Act. Upsal. 6: 83, 1799
○ Stelis ophioglossoides (Jacq.) Sw., Schrad. Neues J. Bot. 2: 239, 1800
Specklinia floribunda Lindl., Gen. Sp. Orch. Pl.: 9, 1830
Stelis scabrida Lindl., Ann. Mag. Nat. Hist. 5: 115, 1840
Pleurothallis floribunda (Lindl.) Lindl., Bot. Reg. 28: misc.p.73:40, 1842 (non Poeppig & Endl.1836)
Stelis polystachya Cogn., Fedde Rep. 6: 306, 1909

Pleurothallis pachyglossa Lindl., Bot. Reg. 26: misc.146, 1840
BR 26: misc.146 (1840) / 28: misc.p.69:10 (1842)
Type: Mexico, cult. Barker (K-Lindl., holo!)
Mexico, Guatemala, Costa Rica – up to 1600 m – E
Pleurothallis conanthera Rchb.f. ex Kraenzlin, Xen. Orch. 3: 41 t.217 fig. 1–9, 1881
Pleurothallis formosa Schltr., Fedde Rep. 10: 291, 1912

Pleurothallis pandurifera Lindl., Bot. Reg. 28: misc.p.70:11, 1842
BR 28: misc.p.70:11 (1842)
Type: Brazil, Miers s.n. (K-Lindl., holo!)
Brazil – E
Humboldtia pandurifera (Lindl.) Kuntze, Rev. Gen. 2: 668, 1891

Pleurothallis papillosa Lindl., Bot. Reg. 21: sub t.1797:9, 1835
BR 21: sub t.1797:9 (1835) / 28: misc.p.81:94 (1842)
Type: Brazil, San Jao Baptista, Martius s.n. (K-Lindl., icono!)
Venezuela, Brazil – L
Humboldtia papillosa (Lindl.) Kuntze, Rev. Gen. 2: 668, 1891

Pleurothallis parviflora Lindl., Hook. Comp. Bot. Mag. 2: 355, 1836
BR 28: misc.p.82:99 (1842)
Brazil – E
Lepanthes cryptophyla Barb.Rodr., Gen. & Sp. Orch. Nov. 2: 65, 1882
Humboldtia parvifolia (Lindl.) Kuntze, Rev. Gen. 2: 668, 1891
Pleurothallis succedanea Hoehne & Schltr., Arq. Bot. Est. S.Paulo 1: 224, 1926

Pleurothallis pectinata Lindl., Bot. Reg. 25: misc.1, 1839
BR 25: misc.1 (1839) / 28: misc.p.76:58 (1842)
Type: Brazil, Rio de Janeiro, cult. Loddiges (K-Lindl., holo!)
Brazil – E
Humboldtia pectinata (Lindl.) Kuntze, Rev. Gen. 2: 668, 1891

Pleurothallis pedunculata (Klotzsch) Rchb.f., Linnaea 22: 822, 1849
BR 31: misc.30 (1845)
Venezuela, Colombia, Ecuador, Peru – E
Rhynchopera pedunculata Klotzsch, Ic. Pl. Rar. Hort. Berol. 1: 103, 1841

Pleurothallis phalangifera (Presl.) Schltr., Fedde Rep. Beih. 9: 141, 1921
BR 28: misc.p.68:1 (1842)
Peru, Colombia, Venezuela, Ecuador – E
Acronia phalangifera Presl, Rel. Haenk. 1: 104, 1827
Pleurothallis mathewsii Lindl., Hook. Comp. Bot. Mag. 2: 355, 1836
Pleurohallis bogotensis Lindl., Orch. Linden. 1: 3, 1846
Humboldtia mathewsii (Lindl.) Kuntze, Rev. Gen. 2: 668, 1891
Humboldtia phalangifera (Presl) Kuntze, Rev. Gen. 2: 667, 1891

○ **Pleurothallis prolifera** Herbert ex Lindl., Bot. Reg. 15: t.1298, 1830
BR 15: t.1298 (1830) / 21: sub t.1797:23 (1835) / 28: misc.p.74:47 (1842)
Type: Brazil, cult. Herbert (K-Lindl., holo!)
Venezuela, Brazil – up to 1500 m – E
Humboldtia prolifera (Herbert ex Lindl.) Kuntze, Rev. Gen. 2: 668, 1891
Pleurothallis litophila Barb.Rodr., Contrib. Jard. Bot. Rio de Janeiro 4: 100, 1907

Pleurothallis pruinosa Lindl., Bot. Reg. 28: misc.p.75:55, 1842
BR 28: misc.p.75:55 (1842)

Type: Surinam, Splitberger s.n. (K-Lindl., holo!)
Honduras, Nicaragua, Costa Rica, Panama, S.America, W.Indies – E
Stelis flavida Focke, Tijdschr. Nat. Wetensch. Kunsten 2: 202, 1849
Pleurothallis flavida (Focke) Lindl., Fol. Orch. Pleuroth.: 44, 1859
Humboldtia pruinosa (Lindl.) Kuntze, Rev. Gen. 2: 668, 1891
Pleurothallis pauciflora Schltr., Fedde Rep. Beih. 19: 192, 1923

Pleurothallis pubescens Lindl., Hook. Comp. Bot. Mag. 2: 355, 1836
BR 24: misc.133 (1838) / 28: misc.p.77:46 (1842) / 29: misc.79 (1843)
Mexico to Nicaragua, Venezuela to Bolivia, Brazil – E
Pleurothallis vittata Lindl., Bot. Reg. 24: misc.133, 1838
Pleurothallis picta Hook., Bot. Mag. 68: t.3897, 1841 (non Lindl.)
Pleurothallis smithiana Lindl., Bot. Reg. 29: misc.79, 1843
Pleurothallis polystachya A.Rich. & Galeotti, Ann. Sc. Nat. s.3 (3): 16, 1845
Pleurothallis bufonis Klotzsch, Allg. Gartenz. 22: 225, 1854
Pleurothallis truxillensis Rchb.f., Bonpl. 2: 25, 1854
Pleurothallis janeirensis Barb.Rodr., Gen. & Sp. Orch. Nov. 2: 29, 1882
Pleurothallis riograndensis Barb.Rodr., Gen. & Sp. Orch. Nov. 2: 28, 1882
Humboldtia bufonis (Klotzsch) Kuntze, Rev. Gen. 2: 667, 1891
Humboldtia polystachya (A.Rich. & Galeotti) Kuntze, Rev. Gen. 2: 668, 1891
Humboldtia pubescens (Lindl.) Kuntze, Rev. Gen. 2: 668, 1891
Humboldtia smithiana (Lindl.) Kuntze, Rev. Gen. 2: 668, 1891
Humboldtia truxillensis (Rchb.f.) Kuntze, Rev. Gen. 2: 668, 1891
Humboldtia vittata (Lindl.) Kuntze, Rev. Gen. 2: 668, 1891
Pleurothallis bourgeaui Kraenzlin, Ark. Bot. 168: 15, 1920
Pleurothallis mandibularis Kraenzlin, Vidensk. Meddel. Dansk Nat. Foren Kjobenhaven 71: 169, 1920

Pleurothallis pulchella Lindl., Hook. Exot. Fl. 2: sub t.123, 1824
BR 21: sub t.1797:26 (1835) / 28: misc.p.73:36 (1842)
Ecuador – E

Pleurothallis quadrifida (Llave & Lex.) Lindl., Bot. Reg. 28: misc.p.70: 12, 1842
BR 21: sub t.1797:25 (1835) / 28: misc.p.70:12 (1842)
Mexico, Guatemala, El Salvador, Honduras, Nicaragua, Costa Rica, Panama, Colombia, Venezuela, Jamaica – E
Dendrobium quadrifidum Llave & Lex., Nov. Veg. Descr. 2: Orch. Opusc. 40, 1825
Pleurothallis racemiflora Lindl. ex Lodd, Bot. Cab. 10: pl.949, 1825 (non Sw.)
Pleurothallis ghiesbreghtiana A.Rich. & Galeotti, Ann. Sc. Nat. 3: 16, 1845
Pleurothallis incompta Rchb.f., Bonpl. 2: 24, 1854
Pleurothallis longissima Lindl., Fol. Orch. Pleuroth.: 31, 1859
Pleurothallis nicaraguensis Rchb.f., Ann. Bot. Syst. Nat. 6: 171, 1861
Humboldtia incompta (Rchb.f.) Kuntze, Rev. Gen. 2: 667, 1891
Humboldtia longissima (Lindl.) Kuntze, Rev. Gen. 2: 667, 1891
Humboldtia nicaraguensis (Rchb.f.) Kuntze, Rev. Gen. 2: 668, 1891
Humboldtia quadrifida (Llave & Lex.) Kuntze, Rev. Gen. 2: 668, 1891
Pleurothallis lyroglossa Schltr., Fedde Rep. 8: 566, 1910
Pleurothallis niederleinii Schltr., Beih. Bot. Centralbl. 36 (2): 396, 1918

Pleurothallis racemiflora (Sw.) Lindl., Hook. Exot. Fl. 2: 123, 1824
BR 21: sub t.1797:22 (1835) / 28: misc.p.69:4 & misc.p.71:20 (1842)
Jamaica, Cuba – E, L
Epidendrum racemiflorum Sw., Prodr.: 125, 1788
Dendrobium racemiflorum (Sw.) Sw., Nov. Act. Upsal. 6: 83, 1799
Pleurothallis oblongifolia Lindl., Hook. Comp. Bot. Mag. 2: 355, 1837

Pleurothallis recurva Lindl., Bot. Reg. 27: misc.1, 1841
BR 27: misc.1 (1841) / 28: misc.p.77:63 (1842)
Type: Brazil, cult. Mrs. Cannon (K-Lindl., holo!)
Brazil – E
Pleurothallis lilacina Barb.Rodr., Gen. & Sp. Orch. Nov. 1: 18, 1877
Pleurothallis bistuberculata Barb.Rodr., Gen. & Sp. Orch. Nov. 2: 35, 1882
Pleurothallis regeliana Rchb.f., Gartenfl. 35: 51, 1886
Humboldtia recurva (Lindl.) Kuntze, Rev. Gen. 2: 668, 1891
Pleurothallis maculata Rolfe, Kew Bull.: 334, 1893
Pleurothallis lamproglossa Schltr., Notizbl. Bot. Gart. Berlin 7: 272, 1918
Pleurothallis albipetala Hoehne & Schltr., Anexe Mém. Inst. Butantan 1 (2): 32, 1921
Pleurothallis curitibensis Kraenzlin, Ark. Bot. Stockh. 16 (8): 8, 1921 (non Schltr.)
Specklinia recurva (Lindl.) F.de Barros, Hoehnea 10: 110, 1984

Pleurothallis restrepioides Lindl., Hook. Comp. Bot. Mag. 2: 356, 1836
BR 21: sub t.1797:2 (1835) / 28: misc.p.68:2 (1842)
Venezuela to Peru – E, T
Pleurothallis laurifolia Rchb.f., Bonpl. 2: 23, 1854 (non Kunth)
Pleurothallis fritillaria Rchb.f., Bonpl. 3: 240, 1855
Pleurothallis roezlii Rchb.f., Linnaea 61: 13, 1877
Humboldtia fritillaria (Rchb.f.) Kuntze, Rev. Gen. 2: 667, 1891
Humboldtia restrepioides (Lindl.) Kuntze, Rev. Gen. 2: 668, 1891
Humboldtia roezlii (Rchb.f.) Kuntze, Rev. Gen. 2: 668, 1891

o **Pleurothallis retusa** (Llave & Lex.) Lindl., Bot. Reg. 28: misc.p.81: 96, 1842
BR 21: sub t.1797 (1835) / 28: misc.p.81:96 (1842)
Mexico – E
Dendrobium retusum Llave & Lex., Nov. Veg. Descr. 2: Orch. Opusc: 40, 1842
Specklinia retusa Lindl., Bot. Reg. 21: sub t.1797 1835

Pleurothallis rubens Lindl., Bot. Reg. 21: sub t.1797:32, 1835
BR 21: sub t.1797:32 (1835) / 28: misc.p.73:34 (1842)
Type: Brazil, San Jao del Rey, Martius s.n. (K-Lindl., holo!)
Brazil – E
Humboldtia rubens (Lindl.) Kuntze, Rev. Gen. 2: 668, 1891
Pleurothallis montseratii Porsch, Oesterr. Bot. Zeitschr.: 159, 1905
Specklinia rubens (Lindl.) F.de Barros, Hoehnea 10: 110, 1984

Pleurothallis rupestris Lindl., Bot. Reg. 21: sub t.1797:18, 1835
BR 21: sub t.1797:18 (1835) / 28: misc.p.74:43 (1842)
Type: Brazil, Martius s.n. (K-Lindl., icono!)
Brazil – E
Pleurothallis pachyphylla Rchb.f., Linnaea 22: 824, 1849
Pleurothallis johannensis Barb.Rodr., Rev. de Engenh. 3: 56, 1881
Humboldtia rupestris (Lindl.) Kuntze, Rev. Gen. 2: 668, 1891

Pleurothallis ruscifolia (Jacq.) R.Br., Ait. Hort. Kew. ed.2, 5: 211, 1813
BR 21: sub t.1797:1, 3 (1835) / 28: misc.p.79:76 & misc.p.80:87, 88 (1842)
Trop. America – E
Epidendrum ruscifolium Jacq., Enum. Pl. Carib.: 29, 1760
Dendrobium ruscifolium (Jacq.) Sw., Nov. Act. Upsal. 6: 84, 1799
Pleurothallis laurifolia Kunth, Nov. Gen. Sp. Pl. 1: 364, 1815 (non Rchb.f.)
Humboldtia succosa (Lindl.) Pavon ex Lindl., Gen. Sp. Orch. Pl.: 5, 1830
Pleurothallis succosa Lindl., Gen. Sp. Orch. Pl.: 5, 1830
Pleurothallis multicaulis Poeppig & Endl., Nov. Gen. Sp. Pl. 1: 47 t.82, 1836
Humboldtia laurifolia (Kunth) Kuntze, Rev. Gen. 2: 667, 1891
Humboldtia ruscifolia (Jacq.) Kuntze, Rev. Gen. 2: 668, 1891
Pleurothallis glomerata Ames, Sched. Orch. 4: 21, 1923
Pleurothallis ruscifolia var. caquetana Schltr., Fedde Rep. Beih. 23: 55, 1924

o **Pleurothallis saurocephala** Lodd., Bot. Cab. 16: t.1571, 1830
BR 21: sub t.1797:24 (1835) / 23: t.1968 (1837) / 28: misc.p.73:37 (1842)
Brazil – E
Humboldtia saurocephala (Lodd.) Kuntze, Rev. Gen. 2: 668, 1891

Pleurothallis scabrilinguis Lindl., Hook. Comp. Bot. Mag. 2: 354, 1836
BR 28: misc.p.81:90 (1842)
Peru – E
Humboldtia scabrilinguis (Lindl.) Kuntze, Rev. Gen. 2: 668, 1891

Pleurothallis scariosa (Llave & Lex.) Lindl., Bot. Reg. 28: misc.p.71: 16, 1842
BR 21: sub t.1797 (1835) / 28: misc.p.71:16 (1842)
Mexico – E
Dendrobium scariosum Llave & Lex., Nov. Veg. Descr. 2: Orch. Opusc. 39, 1825
Specklinia scariosa Lindl., Bot. Reg. 21: sub t.1797, 1835

Pleurothallis sclerophylla Lindl., Bot. Reg. 21: sub t.1797:31, 1835
BR 21: sub t.1797:31 (1835) / 24: misc.182 (1838) / 28: misc.p.72:30, 31 (1842)
Type: Brazil, Martius s.n. (K-Lindl., icono!)
Venezuela to Bolivia, Trinidad, Guyana – E
Pleurothallis stenopetala Lodd. ex Lindl., Bot. Reg. 24: misc.182, 1838
Pleurothallis listrostachys Rchb.f., Bonpl. 3: 71, 1855
Anathallis secunda Barb.Rodr., Gen. & Sp. Orch. Nov. 2: 72, 1882
Humboldtia sclerophylla (Lindl.) Kuntze, Rev. Gen. 2: 668, 1891
Humboldtia stenopetala (Lodd. ex Lindl.) Kuntze, Rev. Gen. 2: 668, 1891
Pleurothallis urosepala Lehm. & Kraenzlin, Engl. Bot. Jahrb. 26: 442, 1899
Pleurothallis triura Schltr., Fedde Rep. 14: 386, 1916
Pleurothallis ottonis Schltr., Fedde Rep. 27: 52, 1929

Pleurothallis secunda Poeppig & Endl., Nov. Gen. Act. Sp. 1: 49 t.85, 1836
BR 28: misc.p.74:46 (1842)
Venezuela, Colombia, Ecuador, Peru – 600–1200 m – E
Pleurothallis lindenii Lindl., Ann. & Mag. Nat. Hist. 12: 397, 1843
Humboldtia lindenii (Lindl.) Kuntze, Rev. Gen. 2: 667, 1855
Pleurothallis lindenii var. gigantea Rchb.f., Bonpl. 3: 240, 1855
Pleurothallis lindenii var. pumila Rchb.f., Bonpl. 3: 72, 1855

Humboldtia secunda (Poeppig Endl.) Kuntze, Rev. Gen. 2: 668, 1891
Pleurothallis araguensis Ames, Sched. Orch. 1: 9, 1922

Pleurothallis seriata Lindl., Bot. Reg. 26: misc.175, 1840
BR 26: misc.175 (1840) / 28: misc.p.83:111 (1842)
Type: Brazil, Rio de Janeiro, cult. Hearne (K-Lindl., holo!)
Brazil – E
Humboldtia seriata (Lindl.) Kuntze, Rev. Gen. 2: 668, 1891

Pleurothallis sertularioides (Sw.) Sprengel, Syst. Veg. 3: 731, 1826
BR 28: misc.p.83:107 (1842)
Jamaica – E
Epidendrum sertularioides Sw., Prodr.: 122, 1788
Dendrobium sertularioides (Sw.) Sw., Nov. Act. Upsal. 6: 83, 1799

Pleurothallis sicaria Lindl., Bot. Reg. 27: misc.187, 1841
BR 27: misc.187 (1841) / 28: misc.p.81:95 (1842)
Type: Trinidad, without collector (K-Lindl., holo!)
Venezuela, Trinidad, Colombia, Ecuador – E
Pleurothallis tripteris Rchb.f., Linnaea 22: 829, 1849
Pleurothallis trigonopoda Klotzsch, Allg. Gartenz. 21: 361, 1853

Pleurothallis spiralis (Ruiz Lopez & Pavon) Lindl., Gen. Sp. Orch. Pl.: 7, 1830
BR 21: sub t.1797:21 (1835) / 28: misc.p.72:28 (1842)
Peru – E – obscure species
Humboldtia spiralis Ruiz Lopez & Pavon, Syst. Veg. Fl. Peruv. & Chil. 1: 237, 1798
Stelis spiralis (Ruiz Lopez & Pavon) Pers., Syn. Pl. 2: 524, 1807

Pleurothallis strupifolia Lindl., Bot. Reg. 25: misc.3, 1839
BR 25: misc.3 (1839) / 28: misc.p.76:56, 57 (1842)
Type: Brazil, Rio de Janeiro, cult. Loddiges (K-Lindl., holo!)
Brazil – E
Pleurothallis bicolor Lindl., Bot. Reg. 28: misc.p.76:57, 1842
Pleurothallis hookeri Regel, Act. Hort. Petrop. 7: 545, 1881
Pleurothallis spectrilinguis Rchb.f., Gard. Chron. n.s.18: 457, 1882
Humboldtia strupifolia (Lindl.) Kuntze, Rev. Gen. 2: 668, 1891
Pleurothallis glaucophylla Hoehne, Arq. Bot. Est. S.Paulo 1: 214, 1926

Pleurothallis tentaculata (Poeppig & Endl.) Lindl., Hook. Comp. Bot. Mag. 2: 354, 1836
BR 28: misc.p.80:85 (1842)
Peru – E
Restrepia tentaculata Poeppig & Endl., Nov. Gen. Sp. Pl. 1: 34 t.59, 1836
Humboldtia tentaculata (Poeppig & Endl.) Kuntze, Rev. Gen. 2: 668, 1891

Pleurothallis teres Lindl., Bot. Reg. 21: sub t.1797:19, 1835
BR 21: sub t.1797:19 (1835) / 28: misc.p.74:42 (1842)
Type: Brazil, cult. Loddiges (K-Lindl., holo!)
Brazil – E
Humboldtia teres (Lindl.) Kuntze, Rev. Gen. 2: 668, 1891

Pleurothallis tribuloides (Sw.) Lindl., Gen. Sp. Orch. Pl.: 6, 1830
BR 21: sub t.1797:4 (1835) / 28: misc.p.80:84 (1842)
Mexico, Belize, Guatemala, Honduras, Nicaragua, Costa Rica, Panama, Cuba, Jamaica – E
Epidendrum tribuloides Sw., Prodr.: 123, 1788
Dendrobium tribuloides (Sw.) Sw., Nov. Act. Upsal. 6: 83, 1799
Cymbidium tribuloides (Sw.) Sprengel, Syst. Veg. 3: 721, 1826
Pleurothallis spathulata A.Rich. & Galeotti, Ann. Sc. Nat. s.3, 3: 17, 1845
Pleurothallis fallax Rchb.f., Bonpl. 3: 224, 1855
Humboldtia tribuloides (Sw.) Kuntze, Rev. Gen. 2: 668, 1891
Cryptophoranthus acaulis Kraenzlin, Fedde Rep. Beih. 34: 232, 1925

Pleurothallis tricarinata Poeppig & Endl., Nov. Gen. Sp. 1: 49 t.87, 1836
BR 28: misc.p.72:29 (1842)
Peru, Bolivia, Brazil – E, L
Humboldtia tricarinata (Poeppig & Endl.) Kuntze, Rev. Gen. 2: 668, 1891

Pleurothallis tridentata Klotzsch, Allg. Gartenz. 8: 289, 1840
BR 28: misc.p.80:89 (1842)
Venezuela, Colombia, Ecuador, Peru – E
Humboldtia tridentata (Klotzsch) Kuntze, Rev. Gen. 2: 668, 1891
Pleurothallis nephrocardia Schltr., Fedde Rep. Beih. 6: 35, 1919

Pleurothallis trifida Lindl., Bot. Reg. 28: misc.p.82:102, 1842
BR 28: misc.p.82:102 (1842)
Type: Brazil, Organ Mts., Miers s.n. (K-Lindl., holo!)
Brazil – E
Lepanthes bicarinata Barb.Rodr., Gen. & Sp. Orch. Nov. 2: 54, 1882
Humboldtia trifida (Lindl.) Kuntze, Rev. Gen. 2: 668, 1891
Pleurothallis bicristata Cogn., Mart. Fl. Bras. 3 (4): 459, 1893
Pleurothallis curtii Schltr., Anex. Mém. Inst. Butantan 1 (4): 43, 1922

Pleurothallis truncata Lindl., Hook. Comp. Bot. Mag. 2: 335, 1836
BR 28: misc.p.72:24 (1842)
Peru, Ecuador – E

Pleurothallis undulata Poeppig & Endl., Nov. Gen. Sp. Pl. 1: 48, 1836
BR 28: misc.p.81:92 (1842)
Venezuela, Colombia, Peru, Brazil – E
Pleurothallis cardium Rchb.f., Bonpl. 2: 26, 1854
Pleurothallis lansbergiana Regel, Gartenfl. 6: t.193, 1856
Pleurothallis lansbergii Regel, Ind. Sem. Hort. Petrop.: 17, 1856
Humboldtia cardium (Rchb.f.) Kuntze, Rev. Gen. 2: 667, 1891
Humboldtia undulata (Poeppig & Endl.) Kuntze, Rev. Gen. 2: 668, 1891
Pleurothallis stenosepala Rolfe, Kew Bull.: 208, 1892

Pleurothallis uniflora Lindl., Hook. Comp. Bot. Mag. 2: 355, 1836
BR 28: misc.p.83:106 (1842)
Venezuela, Guyana, Surinam, Brazil – E
Epidendrum marginatum L.C.Rich., Act. Soc. Hist. Nat. Paris 1: 112, 1792
Pleurothallis striata Focke, Tijdschr. Wiss. Nat. 4: 63, 1851
Humboldtia striata (Focke) Kuntze, Rev. Gen. 2: 668, 1891
Humboldtia uniflora (Lindl.) Kuntze, Rev. Gen. 2: 668, 1891
Pleurothallis marginata (L.C.Rich.) Cogn., Mart. Fl. Bras. 3 (4): 478, 1896 (non Lindl.)

o **Pogonia ophioglossoides** (L.) Jussieu, Gen. Pl.: 65, 1789
BR 2: t.148 (1816)
USA, Canada – E
Arethusa ophioglossoides L., Sp. Pl. 2: 951, 1753
Pogonia ophioglossoides var. brachypogon Fernald, Rhodora 23: 245, 1922

Polystachya clavata Lindl., Bot. Reg. 28: misc.71, 1842
BR 28: misc.71 (1842)
Type: Belize, cult. Loddiges (K-Lindl., holo!)
Belize, Guatemala, Honduras, Nicaragua – E

Polystachya concreta (Jacq.) Garay & Sweet, Fl. Lesser Antilles Orch.: 178, 1974
BR 10: sub t.851:1 (1824) / 24: misc.143 & misc.144 (1838) / 30: misc.62:44, 46 (1844)
Trop. belt in New and Old Worlds – E
Epidendrum concretum Jacq., Enum. Syst. Pl.: 30, 1760
Epidendrum minutum Aublet, Hist. Pl. Guin. Fr. 2: 824, 1775
Dendrobium polystachyon Sw., Kongl. Vetensk. Acad. Nya Handl. 21: 247, 1800
Cranichis luteola Sw., Fl. Ind. Occid. 3: 1443, 1804
Polystachya luteola (Sw.) Hook., Exot. Pl. 2: t.103, 1824
Onychium flavescens Blume, Bijdr. 7: 325, 1825
Polystachya mauritiana Sprengel, Syst. Veg. 3: 742, 1826
Dendrobium flavescens (Blume) Lindl., Gen. Sp. Orch. Pl.: 85, 1830
Polystachya zeylanica Lindl., Bot. Reg. 24: misc.144, 1838
Polystachya estrellensis Rchb.f., Linnaea 25: 231, 1852
Polystachya purpurea Wight, Ic. 5: 10 Pl.1679, 1852
Polystachya zollingeri Rchb.f., Bonpl. 5: 39, 1857
Polystachya tessellata Lindl., J. Linn. Soc. 6: 130, 1862
Polystachya extinctoria Rchb.f., Walp. Ann. 6: 638, 1863
Polystachya jussieuana Rchb.f., Walp. Ann. 6: 641, 1863
Polystachya wightii Rchb.f., Walp. Ann. 6: 640, 1863
Polystachya modesta Rchb.f., Flora 50: 114, 1867
Polystachya nitidula Rchb.f., Flora 50: 114, 1867
Polystachya rigidula Rchb.f., Flora 50: 117, 1867
Polystachya tricruris Rchb.f., Flora 50: 118, 1867
Polystachya rufinula Rchb.f., Gard. Chron. n.s.11: 41, 1879
Polystachya hypocrita Rchb.f., Gard. Chron. n.s.16: 685, 1881
Polystachya shirensis Rchb.f., Otia Bot. Hamb. 2: 82, 1881
Polystachya similis Rchb.f., Otia Bot. Hamb. 2: 82, 1881
Callista flavescens (Blume) Kuntze, Rev. Gen. 2: 654, 1891
Dendrorchis estrellensis (Rchb.f.) Kuntze, Rev. Gen. 2: 659, 1891
Dendrorchis extinctoria (Rchb.f.) Kuntze, Rev. Gen. 2: 659, 1891
Dendrorchis jussieuana (Rchb.f.) Kuntze, Rev. Gen. 2: 659, 1891
Dendrorchis minuta (Aublet) Kuntze, Rev. Gen. 2: 658, 1891
Dendrorchis rufinula (Rchb.f.) Kuntze, Rev. Gen. 2: 659, 1891
Dendrorchis shirensis (Rchb.f.) Kuntze, Rev. Gen. 2: 659, 1891
Dendrorchis similis (Rchb.f.) Kuntze, Rev. Gen. 2: 659, 1891
Dendrorchis tessellata (Lindl.) Kuntze, Rev. Gen. 2: 659, 1891
Dendrorchis wightii (Rchb.f.) Kuntze, Rev. Gen. 2: 659, 1891
Dendrorchis zollingeri (Rchb.f.) Kuntze, Rev. Gen. 2: 659, 1891
Polystachya buchanani Rolfe, Kew Bull.: 335, 1893
Epidendrum resupinatum Sessé & Moc., Fl. Mex. ed.2: 202, 1894
Polystachya minuta (Aublet) Cordem., Fl. Ile Réunion: 190, 1895 (non A.Rich. & Galeotti)
Polystachya penangensis Ridley, J. Linn. Soc. 32: 344, 1896
Polystachya siamensis Ridley, J. Linn. Soc. 32: 343, 1896
Polystachya singapurensis Ridley, J. Linn. Soc. 32: 344, 1896
Polystachya pleistantha Kraenzlin, Gard. Chron. s.3, 21: 118, 1897
Polystachya zanguebarica Rolfe, Fl. Trop. Afr. 7: 115, 1897
Polystachya gracilis De Wild., Not. Pl. Util. Congo 1: 136, 1903
Polystachya latifolia De Wild., Not. Pl. Util. Congo 1: 138, 1903
Polystachya mukandaensis De Wild., Not. Pl. Util. Congo 1: 139, 1903

Polystachya huyghei De Wild., Not. Pl. Util. Congo 1: 315, 1904
Polystachya flavescens (Blume) J.J.Smith, Fl. Buitenz. 6: 284, 1905
Polystachya plehniana Schltr., Engl. Bot. Jahrb. 38: 8, 1905
Polystachya praealta Kraenzlin, Engl. Bot. Jahrb. 36: 118, 1905
Polystachya aurantiaca Schltr., Ann. Mus. Colon. s.3, 1: 165, 1913
Polystachya dorotheae Rendle, Cat. Talb. Niger. Pl.: 103, 1913
Polystachya colombiana Schltr., Fedde Rep. Beih. 7: 156, 1920
Polystachya caquetana Schltr., Fedde Rep. Beih. 27: 79, 1924
Polystachya lepidantha Kraenzlin, Kew Bull.: 290, 1926
Polystachya lettowiana Kraenzlin, Fedde Rep. Beih. 39: 127, 1926
Polystachya reichenbachiana Kraenzlin, Fedde Rep. Beih. 39: 127, 1926
Polystachya purpurea var. lutescens Gagnepain, Fl. Gen. Indoch. 6: 434 fig.40, 1934
Polystachya kraenzliniana Pabst, Rodriguesia 30–31: 34, 1957
Pachystele minuta (A.Rich & Galeotti) Hamer & Garay, Ic. Pl. Trop. 11: t.1069, 1984

Polystachya cultriformis Lindl., ex Sprengel, Syst. Veg. 3: 42, 1826
BR 10: sub t.851:4 (1824)
Madagascar, Mascarene Is., Comoros – up to 2000 m – E, L
Polystachya cultrata Lindl., Bot. Reg. 10: sub t.851: 4, 1824
Dendrobium cultriforme (Lindl.) Thouars, Orch. Iles Afr.: t.87, 1882

Polystachya foliosa (Lindl.) Rchb.f., Ann. Bot. Syst. 6: 640, 1863
BR 26: misc.208 (1840)
Mexico, W.Indies to Centr. America, S.America – E
Encyclia nana Poeppig & Endl., Nov. Gen. Sp. Pl. 2: 10 t.113a, 1836
Encyclia polystachya Poeppig & Endl., Nov. Gen. & Sp. Pl. 2: 10 t.113B, 1836
Stelis foliosa Lindl., Ann. Mag. Nat. Hist. 2: 330 t.17, 1839
Polystachya cerea Lindl., Bot. Reg. 26: misc.208, 1840
Polystachya weigeltii Rchb.f., Linnaea 25: 230, 1852
Polystachya nana (Poeppig & Endl.) Klotzsch, Ind. Sem. Hort. Berol: 12, 1853
Polystachya caracasana Rchb.f., Bonpl. 2: 15, 1854
Dendrorchis caracasana (Rchb.f.) Kuntze, Rev. Gen. 2: 659, 1891
Dendrorchis foliosa (Lindl.) Kuntze, Rev. Gen. 2: 659, 1891
Dendrorchis nana (Poeppig & Endl.) Kuntze, Rev. Gen. 2: 659, 1891
Polystachya minor Fawcett & Rendle, J. Bot. 48: 106, 1910
Polystachya altilamellata Schltr., Fedde Rep. 10: 385, 1912
Polystachya ecuadorensis Schltr., Fedde Rep. Beih. 8: 90, 1921
Polystachya guatemalensis Schltr., Fedde Rep. Beih. 17: 141, 1921
Polystachya poeppigii Schltr., Fedde Rep. Beih. 9: 155, 1921
Polystachya panamensis Schltr., Fedde Rep. Beih. 17: 49, 1922
Polystachya costaricensis Schltr., Fedde Rep. Beih. 19: 223, 1923
Polystachya cubensis Schltr., Urb. Symb. Ant. 9: 59, 1923
Polystachya powellii Ames, Sched. Orch. 7: 31, 1924
Polystachya huebneri Schltr., Beih. Bot. Centralbl. 43 (2): 112, 1925

Polystachya fusiformis (Thouars) Lindl., Bot. Reg. 10: sub t.851:3, 1824
BR 10: sub t.851:3 (1824)
Madagascar, Mascarene Is. – 1200–2000 m – E, L
Dendrobium fusiforme Thouars, Orch. Iles Afr.: t.86, 1822

Polystachya leonensis Rchb.f., Otia Bot. Hamburg.: 112, 1881
BR 25: misc.44 (1839) / 26: misc.102 (1840)
Type: Sierra Leone, cult. Loddiges (K-Lindl., holo!)
Guinea, Sierra Leone, Liberia, Ivory Coast, Ghana, S.Nigeria, E.Cameroun, Gabon, C.African Rep., Zaire, Angola – 1000–1400 m – E
Polystachya affinis Lindl., Gen. Sp. Orch. Pl.: 73, 1830
Polystachya bracteosa Lindl., Bot. Reg. 26: misc.102, 1840

o **Polystachya puberula** Lindl., Bot. Reg. 10: t.851, 1824
BR 10: t.851 (1824)
Type: Sierra Leone, cult. Hort. Soc. (K-Lindl., holo!)
Guinea-Bissau, Guinea, Sierra Leone, Liberia, Ivory Coast – E

Polystachya pubescens (Lindl.) Rchb.f., Walp. Ann. 6: 643, 1863
BR 26: misc.103 (1840)
S.Africa (Cape Province, Natal) – 500–1000 m – E
Epiphora pubescens Lindl., Hook. Comp. Bot. Mag. 2: 201, 1836
Lissochilus sylvaticus Ecklon ex Sonder, Linnaea 19: 71, 1847
Polystachya lindleyana Harvey, Thes. Cap. 2: 50 t.178, 1859

Polystachya ramulosa Lindl., Bot. Reg. 24: misc.142, 1838
BR 24: misc.142 (1838)
Type: Sierra Leone, cult. Loddiges (K-Lindl., holo!)
Sierra Leone, Liberia, Ghana, Nigeria, E.Cameroun, Congo, Gabon, Uganda – E

Polystachya reflexa Lindl., Bot. Reg. 27: misc.43, 1841
BR 27: misc.43 (1841)
Type: Whitfield s.n. cult. Hort. Soc. (K-Lindl., holo!)
Sierra Leone, Liberia, Ivory Coast, Ghana – E
Polystachya liberica Rolfe, Kew Bull.: 72, 1908
Polystachya smytheana Rolfe, Kew Bull.: 71, 1908
Polystachya elastica Chev., Rev. Bot. Appl. & Agr. Trop., 1921–1945 (non Lindl.)

Pomatocalpa decipiens (Lindl.) J.J.Smith, Nat. Tijdschr. Ned. Ind. 72: 33, 1912
BR 30: misc.16 (1844)
Type: Sri Lanka, Col. Fielding s.n. (K-Lindl., holo!)
Sri Lanka – E
Cleisostoma decipiens Lindl., Bot. Reg. 30: misc.16, 1844
Cleisostoma maculosa Thwaites, Enum. Pl. Zeyl.: 304, 1861 (non Lindl.)
Cleisostoma thwaitesianum Trimen, J. Bot. 23: 244, 1885
Saccolabium decipiens (Lindl.) Alston, Trimen Handb. Fl. Ceyl. 6: 278, 1931

Pomatocalpa latifolia (Lindl.) J.J.Smith, Nat. Tijdschr. Ned. Ind. 72: 105, 1912
BR 26: misc.127 (1840)
Type: Cuming, cult. Loddiges (K-Lindl., holo!)
Thailand, Malaya, Sumatra, Java, Bali, Celebes, Borneo, Philippines – E
Cleisostoma latifolium Lindl., Bot. Reg. 26: misc.127, 1840
Saccolabium cumingii Lindl., Heynold Nomencl. 2: 621, 1841
Cleisostoma cumingii (Lindl.) Rchb.f., Allg. Gartenz. 24: 218, 1856
Cleisostoma spicatum Ridley, J. Linn. Soc.: 366, 1896 (non Lindl.)
Saccolabium hortense Ridley, J. Str. Br. R. As. Soc. 39: 83, 1903
Saccolabium latifolium (Lindl.) Schltr., Schum. & Lauterb. Nachtr.: 230, 1905
Pomatocalpa hortensis (Ridley) J.J.Smith, Nat. Tijdsschr. Ned. Ind. 72: 104, 1912
Saccolabium latifolium var. parvifolium Ridley, J. Str. Br. R. As. Soc. 61: 41, 1912
Sarcanthus cumingii (Lindl.) J.J.Smith, Nat. Tijdsschr. Ned. Ind. 72: 85, 1912

Pomatocalpa maculosum (Lindl.) J.J.Smith, Nat. Tijdschr. Ned. Ind. 72: 35, 1912
BR 26: misc.67 (1840)
Sri Lanka – E
Cleisostoma maculosa Lindl., Gen. Sp. Orch. Pl.: 227, 1833 (non Blume, non Thwaites)
Cleisostoma galeata Thwaites, Trimen Handb. Fl. Ceylon 4: 305, 1898
Saccolabium galeatum (Thwaites) Gardner, Trimen Handb. Fl. Ceylon 4: 305, 1898
Saccolabium maculosum (Lindl.) Alston, Trimen Handb. Fl. Ceylon 6: 278, 1931

Ponera juncifolia Lindl., Gen. Sp. Orch. Pl.: 114, 1831
BR 28: misc.17:1 (1842)
Mexico – E

Ponera striata Lindl., Bot. Reg. 28: misc.17:3, 1842
BR 28: misc.17:3 (1842)
Type: Guatemala, Skinner s.n. cult. Mrs. Wray (K-Lindl., holo!)
Mexico, Guatemala, Honduras, Salvador, Costa Rica, Venezuela, Brazil, Bolivia – L, T, E
Ponera macroglossa Rchb.f., Bot. Zeit. 10: 693, 1852
Ponera australis Cogn., Mart. Fl. Bras. 3 (5): 9, 1898
Ponera gearensis Barb.Rodr., Contr. Jard. Bot. Rio 4: 103, 1907
Scaphyglottis macroglossa (Rchb.f.) Schltr., Beih. Bot. Centralbl. 36 (2): 457, 1918
Sobralia polyphylla Kraenzlin, Vidensk. Medd. Dansk. 71: 173, 1920

o **Ponthieva petiolata** Lindl., Bot. Reg. 9: t.760, 1823
BR 9: t.760 (1823)
Type: St. Vincent, Mc Rae s.n. cult. Hort. Soc. (K-Lindl., holo!)
W.Indies – T

Porpax reticulata Lindl., Bot. Reg. 31: misc.66, 1845
BR 31: misc.66 (1845)
Type: E.India, cult. Loddiges (K-Lindl., holo!)
India – E
Aggeianthus reticulatus (Lindl.) Wight, Ic.: t.1737, 1851
Aggeianthus marchantioides Wight, Ic.: t.1737, 1852
Cryptochilus reticulatus (Lindl.) Rchb.f., Bot. Zeit.: 214, 1862
Eria reticulata (Lindl.) Benth. & Hook.f., Gen. Pl. 3: 509, 1883
Pinalia reticulata (Lindl.) Kuntze, Rev. Gen. 2: 679, 1891
Porpax papillosa Blatt. & MacCann, J. Bombay Nat. Hist. Soc. 35: 268, 1931

Prescottia leptostachya Lindl., Bot. Reg. 22: sub t.1916:4, 1836
BR 22: sub t.1916:4 (1836)
Type: Brazil, Salzman (K-Lindl., holo!)
Brazil – T

Prescottia oligantha (Sw.) Lindl., Gen. Sp. Orch. Pl.: 454, 1840
BR 22: sub t.1916:3 (1836)
Florida, W.Indies, Mexico to Panama, Venezuela, Colombia, Ecuador, Peru, Brazil, Paraguay, Argentina – T
Cranichis oligantha Sw., Prodr. Ind. Occ.: 120, 1788
Cranichis micrantha Sprengel, Syst. Veg. 3: 700, 1826

Serapias pumilla Vell., Fl. Flum. 9: t.55, 1827
Prescottia micrantha (Sprengel) Lindl., Bot. Reg. 22: sub t.1916:3, 1836
Prescottia tenuis Lindl., Gen. Sp. Orch. Pl.: 454, 1840
Prescottia viacola Barb.Rodr., Gen. & Sp. Orch. Nov. 2: 279, 1881
Prescottia viacola var. polyphylla Cogn., Mart. Fl. Bras. 3 (6): 549, 1906
Prescottia filiformis Schltr., Fedde Rep. Beih. 7: 50, 1920
Prescottia gracilis Schltr., Fedde Rep. Beih. 7: 51, 1920
Prescottia panamensis Schltr., Fedde Rep. 16: 357, 1920
Prescottia polysphaera Schltr., Fedde Rep. 16: 357, 1920

Prescottia stachyodes (Sw.) Lindl., Bot. Reg. 22: sub t.1916:1, 1836
BR 22: t.1915 & sub t.1916:1, 2 (1836)
W.Indies, Mexico to Panama, Brazil, Guyana, Venezuela, Colombia, Ecuador, Peru – T
Cranichis stachyodes Sw., Fl. Ind. Occ. 3: 1427, 1799
o Prescottia colorans Lindl., Bot. Reg. 22: t.1915, 1836
Prescottia petiolaris Lindl., Bot. Reg. 22: sub t.1916:2, 1836
Prescottia myosurus Rchb.f. ex Griseb., Fl. Brit. W.Ind.: 640, 1864
Prescottia longipetiolata Barb.Rodr., Gen. & Sp. Orch. Nov. 1: 177, 1877
Prescottia paulensis Cogn., Mart. Fl. Bras. 3: 548, 1904

Promenaea lentiginosa (Lindl.) Lindl. Bot. Reg. 29: misc.p.13:3, 1843
BR 25: misc.93 (1839) / 29: misc.p.13:3 (1843)
Type: Brazil, without collector (K-Lindl., holo!)
Brazil – E
Maxillaria lentiginosa Lindl., Bot. Reg. 25: misc.93, 1839
Zygopetalum lentiginosum (Lindl.) Rchb.f., Walp. Ann. 6: 659, 1863
Zygopetalum xanthino-stapelioides hort., Gard. Chron. s.3, 36: 323, 1904

Promenaea rollissonii (Lindl.) Lindl., Bot. Reg. 29: misc.p.13:4, 1843
BR 23: sub t.1986 (1837) / 24: t.40 (1838) / 29: misc.p.13:4 (1843)
Type: Brazil, cult. Rollisson (K-Lindl., holo!)
Brazil – E
o Maxillaria rollissonii Lindl., Bot. Reg. 23: sub t.1986, 1837
Zygopetalum rollissonii (Lindl.) Rchb.f., Walp. Ann. 6: 659, 1863
Promenaea citrina hort ex Regel, Acta Hort. Petrop. 10: 694, 1889 (non Don)
Promenaea paulensis Schltr., Anex. Mém. Inst. Butantan 1 (4): 63, 1922

Promenaea stapelioides (Link & Otto) Lindl., Bot. Reg. 29: misc.p.13:1, 1843
BR 18: sub t.1549:16 (1832) / 25: t.17 (1839) / 29: misc.p.13:1 (1843)
Brazil – E
Cymbidium stapelioides Link & Otto, Ic. Pl. Sel. Hort. Berol. 111: t.52, 1828
o Maxillaria stapelioides (Link & Otto) Lindl., Gen. Sp. Orch. Pl.: 146, 1832
Zygopetalum stapelioides (Link & Otto) Rchb.f., Walp. Ann. 6: 658, 1863

Promenaea xanthina (Lindl.) Lindl., Bot. Reg. 29: misc.13:2, 1843
BR 25: sub t.17 (1839) / 29: misc.13:2 (1843)
Type: Brazil, Descourtilz s.n. (K-Lindl., holo!)
Brazil – E
Maxillaria xanthina Lindl., Bot. Reg. 25: sub t.17, 1839
Maxillaria citrina Don, Hort. Cantabr. ed.13: 602, 1845
Promenaea citrina Don, Hort. Cantabr. ed.13: 720, 1845
Maxillaria guttata hort. ex Rchb.f., Bot. Zeit. 10: 672, 1852
Zygopetalum xanthinum (Lindl.) Rchb.f., Walp. Ann. 6: 659, 1863
Zygopetalum citrinum (Don) Nicholson, Dict. Gard. 4: 245, 1887

Pseuderia foliosa (Brongn.) Schltr., Orch. Neu Guinea: 644, 1913
BR 30: misc.62: 13 (1844)
Moluccas – E
Dendrobium foliosum Brongn., Duperrr. Voy.: t.41, 1834

Psychilis bifida (Aublet) R.P.Sauleda, Phytolog. 65 (1): 8, 1988
BR 22: t.1879 (1836) / 28: misc.p.35:45 (1842)
Haiti – E
o Epidendrum bifidum Aublet, Hist. Pl. Guiane Franç.: 824, 1775
Epidendrum papilionaceum Vahl, West. Bidr. St. Croix: 203, 1793
Psychilis amena Raf., Fl. Tellur. 4: 40, 1838 (nomen illeg.)
Encyclia papilionacea (Vahl) Schltr., Fedde Rep. Beih. 6: 74, 1919
Epidendrum ekmanii Mansf. ex Urban, Ark. Bot. 20A (15): 16, 1926
Encyclia bifida (Aublet) Britton & P.Wilson, Sc. Surv. Porto Rico & Virgin Is. 6: 532, 1930
Encyclia ekmanii (Mansf.) Dod, Moscosoa 2 (1): 15, 1983

Psychopsis papilio (Lindl.) H.G.Jones, J. Barbados Mus. Hist. Soc. 35 (1): 32, 1975
BR 11: t.910 (1825)
Type: Trinidad, Woodford s.n. cult. Colvill (K-Lindl., holo!)
Trinidad, Venezuela, Colombia, Ecuador, Peru – 500–1000 m – E
o Oncidium papilio Lindl., Bot. Reg. 11: t.910, 1825
Psychopsis picta Raf., Fl. Tellur. 4: 40, 1838
Oncidium papilio var. limbatum Hook., Bot. Mag. 66: t.3733, 1839
Oncidium papilio var. majus Rchb.f., Cat. Hort. Pescatore: 30, 1849
Oncidium papilio var. pictum hort. ex Jennings, Orch.: t.11, 1875
Oncidium papilio var. eckhardtii Linden, Ill. Hort. 30: t.500, 1883

Psygmorchis pusilla (L.) Dodson & Dressler, Phytolog. 24: 288, 1972
BR 22: t.1911 (1836) / 24: sub t.48 (1838) / 25: sub t.16 (1839)
Mexico to Brazil, S.Bolivia, Trinidad – E
Epidendrum pusillum L., Sp. Pl. ed.2: 1352, 1763
Cymbidium pusillum (L.) Sw., Nov. Act. Upsal. 6: 74, 1799
o Oncidium iridifolium Kunth, Nov. Gen. & Sp. Pl. 1: 344, 1815
Epidendrum ventrilabrum Vell., Fl. Flum. Ic. 9: t.32, 1827
Oncidium pusillum (L.) Rchb.f., Walp. Ann. 6: 714, 1863
Oncidium allemanii Barb.Rodr., Gen. & Sp. Orch. Nov. 2: 185, 1882
Oncidium pusillum var. megalanthum Schltr., Fedde Rep. Beih. 27: 115, 1924
Tolumnia pusilla (L.) Hoehne, Ic. Orch. Bras.: 231, 1949

Pteroceras pallidus (Blume) Holtum, Kew Bull. 14: 270, 1960
BR 26: misc.143 (1840) / 32: sub t.19:5 (1846)
Sumatra to Philippines, Malaya – E
Dendrocolla pallida Blume, Bijdr.: 290, 1825
Sarcochilus unguiculatus Lindl., Bot. Reg. 26: misc.143, 1840
Sarcochilus pallidus (Blume) Rchb.f., Walp. Ann. 6: 500, 1863
Sarcochilus aureus Hook.f., Fl. Brit. Ind. 6: 35, 1890
Sarcochilus cladostachys Hook.f., Fl. Brit. Ind. 6: 35, 1890

Pterygodium alatum (Thunb.) Sw., Kongl. Vet. Handl. Stockh.: 218, 1800
BR 9: sub t.700 (1823)
S.Africa – T
Ophrys alata Thunb., Prodr. 2, 1794

Pterygodium catholicum (L.) Sw., Kongl. Vet. Acad. Handl.: 218, 1800
BR 9: sub t.700 (1823)
S.Africa – T
Ophrys catholica L., Amoen. Acad. 6: 110, 1763
Ophrys alaris L.f., Suppl. Pl.: 404, 1781
Ophrys affinis Buxb., Pl. Min. Cogn. Cent. 3: 13 t.21, 1784
Arethusa alaris (L.f.) Thunb., Prodr.: 3, 1794

Quekettia microscopica Lindl., Bot. Reg. 25: misc.6, 1839
BR 25: misc.6 (1839)
Type: Brazil, cult. Loddiges (K-Lindl., holo!)
Brazil (Amazonas), Argentina, Venezuela – E
Quekettia chrysantha Barb.Rodr., Vellosia ed.2, 1: 130, 1891

Reichenbachanthus reflexus (Lindl.) Brade, Rodriguesia 1: 55, 1935
BR 25: misc.21 (1839)
Type: cult. Loddiges (K-Lindl., holo!)
W.Indies, Panama, Brazil, Guyana, Venezuela – E
Scaphyglottis reflexa Lindl., Bot. Reg. 25: misc.21, 1839
Hexisea reflexa (Lindl.) Rchb.f., Linnaea 41: 131, 1877
Reichenbachanthus modestus Barb.Rodr., Gen. & Sp. Orch. Nov. 2: 165, 1882
Fractiunguis brasiliensis Schltr., Ann. Mem. Inst. Butant. 1 (4): 56, 1922
Fractiunguis reflexa (Lindl.) Schltr., Ann. Mem. Inst. Butant. 1 (4): 56, 1922

o **Renanthera coccinea** Lour., Fl. Cochinch. 2: 521, 1790
BR 14: t.1131 (1828)
Burma, Thailand, S.China, Cambodia, Laos, Vietnam, Philippines, Java – E
Epidendrum renanthera Räusch, Nom. ed.3: 265, 1797
Gongora phillippica Llanos, Mem. Acad. Cien. Madr. 2: 498, 1859

Renanthera elongata (Blume) Lindl., Gen. Sp. Orch. Pl.: 218, 1833
BR 29: t.41 & misc.3 (1843) / 33: sub t.26 (1847)
Philippines, Borneo, Java, Sumatra, Malaya, Thailand – E
Aerides elongata Blume, Bijdr.: 367, 1825
Saccolabium reflexum Lindl., Gen. Sp. Orch. Pl.: 225, 1833
o Renanthera matutina Lindl., Bot. Reg. 29: t.41, 1843 (non (Blume) Lindl.)
Renanthera micrantha Blume, Rumphia 4: 54, 1849
Porphyrodesme elongata (Blume) Garay, Bot. Mus. Leafl. Harv. Univ. 4: 191, 1974

Restrepia muscifera (Lindl.) Rchb.f. ex Lindl., Fol. Orchid. Restrep.: 8, 1859
BR 28: misc.p.79:77 (1842)
Type: Guatemala, without collector (K-Lindl., holo!)
Guatemala – E
Pleurothallis muscifera Lindl., Bot. Reg. 28: misc.p.79:77, 1842

Restrepiella ophiocephala (Lindl.) Garay & Dunsterv., Venez. Orch. Ill. 4: 266, 1966
BR 24: misc.48 (1838) / 27: misc.64 (1841) / 28: misc.p.78:69 (1842) / 29: misc.62 (1843)
Types: Mexico, cult. Loddiges & Barker (K-Lindl., syn!)
Mexico to Costa Rica – E

Pleurothallis ophiocephala Lindl., Bot. Reg. 24: misc.48, 1838
Pleurothallis puberula Klotzsch, Allg. Gartenz. 22: 393, 1854
Restrepia ophiocephala (Lindl.) Rchb.f., Bonpl. 2: 88, 1854
Pleurothallis stigmatoglossa Rchb.f. ex Lindl., Fol. Orch. Pleuroth. 2: 18, 1859

Rhyncholaelia digbyana (Lindl.) Schltr., Beih. Bot. Centralbl. 36: Abt.2: 477, 1918
BR 32: t.53 (1846)
Type: Honduras, MacDonald s.n. cult. Digby (K-Lindl., holo!)
Mexico, Guatemala, Belize, Honduras – E
o Brassavola digbyana Lindl., Bot. Reg. 32: t.53, 1846
Bletia digbyana (Lindl.) Rchb.f., Walp. Ann. 6: 422, 1862
Laelia digbyana (Lindl.) Benth. ex Jackson, Ind. Kew. 2: 21, 1895
Laelia digbyana var. fimbripetala Ames, Amer. Orch. Soc. Bull. 1: 59, 1932

Rhyncholaelia glauca (Lindl.) Schltr., Beih. Bot. Centralbl. 36: Abt.2: 477, 1918
BR 25: misc.67 (1839) / 26: t.44 & misc.89 (1840) / 32: sub t.53 (1846)
Type: Mexico, Hartweg s.n. cult. Hort. Soc. (K-Lindl., holo!)
Mexico, Guatemala, Honduras, Nicaragua, Panama – E
o Brassavola glauca Lindl., Bot. Reg. 25: misc.67, 1839
Bletia glauca (Lindl.) Rchb.f., Walp. Ann. 6: 422, 1862
Laelia glauca (Lindl.) Benth. ex Jackson, Index Kew 2: 21, 1895

Rhynchostylis retusa (L.) Blume, Bijdr.: 286 t.49, 1825
BR 17: t.1443 (1831) / 27: misc.115 (1841)
Sri Lanka, S.India, N.W.Himalaya, Sikkim, Assam, Burma, Thailand, Cambodia, Laos, Vietnam, Malaya, China, Philippines, Sumatra, Java, Kangean Arch., possibly Borneo – E
Epidendrum retusum L., Sp. Pl.: 935, 1753
Aerides retusa (L.) Sw., J. Bot. 2: 233, 1799
Limodorum retusum (L.) Sw., Nov. Act. Soc. Sc. Upsal. 6: 80, 1799
Aerides praemorsa Willd., Sp. Pl. ed.4: 130, 1805
Epidendrum indicum Poiret, Lamarck Encycl. Suppl. 1: 384, 1810
Aerides guttata Roxb., Hort. Beng.: 63, 1814
Aerides undulatum Smith, Rees Cyclop. 39: 13, 1819
Aerides spicata D.Don, Prodr. Fl. Nepal.: 31, 1825
Epidendrum hippium Buch.-Ham. ex D.Don, Prodr. Fl. Nepal.: 32, 1825
Rhynchostylis praemorsa (Willd.) Blume, Bijdr.: 286, 1825
o Sarcanthus guttatus (Roxb.) Lindl., Bot. Reg. 17: t.1443, 1831
Saccolabium guttatum (Roxb.) Lindl., Wall. Cat.: 7308, 1832 (nomen nudum)
Saccolabium praemorsum (Willd.) Lindl., Gen. Sp. Orch. Pl.: 221, 1833
Saccolabium spicatum (D.Don) Lindl., Gen. Sp. Orch. Pl.: 221, 1833
Orchis lanigera Blanco, Fl. Filip.: 641, 1837
Saccolabium blumei Lindl., Bot. Reg. 27: misc.115, 1841
Saccolabium macrostachyum Lindl., Sert. Orch.: 47, 1841
Saccolabium retusum (L.) Voight, Hort. Suburb. Calcutta: 630, 1845
Saccolabium rheedii Wight, Ic. Pl. Ind. Or. 5 (1): 19 pl.1745/46, 1851
Saccolabium blumei var. major Williams, Orch. Grow. Man.: 89, 1852
Rhynchostylis guttata (Roxb.) Rchb.f., Bonpl. 2: 93, 1854
Saccolabium garwalicum Lindl., J. Linn. Soc. 3: 32, 1859
Rhynchostylis garwalica (Lindl.) Rchb.f., Walp. Ann. 6: 888, 1864
Saccolabium furcatum Williams, Orch. Grow. Man.: 201, 1868
Saccolabium holfordianum Warner, Sel. Orch. 2: t.18, 1874
Saccolabium littorale Rchb.f., Gard. Chron. n.s.16: 198, 1881
Rhynchostylis violacea var. berkeleyi (Rchb.f.) Stein, Orchideenb.: 537, 1882
Saccolabium berkeleyi Rchb.f., Gard. Chron. n.s.19: 814, 1883
Saccolabium heathii hort., Gard. Chron. n.s.24: 369, 1885
Saccolabium turneri Williams, Orch. Grow. Man. ed.5: 568, 1885
Saccolabium blumei var. russelianum Williams, Warner & Williams Orch. Alb. 5: t.238, 1886
Gastrochilus blumei (Lindl.) Kuntze, Rev. Gen. 2: 661, 1891
Gastrochilus gurwalicus (Lindl.) Kuntze, Rev. Gen. 2: 661, 1891
Gastrochilus praemorsus (Willd.) Kuntze, Rev. Gen. 2: 661, 1891
Gastrochilus retusus (L.) Kuntze, Rev. Gen. 2: 661, 1891
Gastrochilus rheedii (Wight) Kuntze, Rev. Gen. 2: 661, 1891
Gastrochilus spicatus (D.Don) Kuntze, Rev. Gen. 2: 661, 1891
Rhynchostylis violacea Ames, Orch. 1: 105, 1905 (non Rchb.f.)

Rhynchostylis violacea (Lindl.) Rchb.f., Bonpl. 2: 93, 1854
BR 27: misc.32 (1841) / 33: t.30 (1847)
Type: Philippines, Cuming s.n. cult. Loddiges (K-Lindl., holo!)
Philippines – E
o Vanda violacea Lindl., Bot. Reg. 27: misc.32, 1841
Saccolabium violaceum (Lindl.) Rchb.f., Bonpl.: 93, 1854
Anota violacea (Lindl.) Schltr., Die Orchideen: 587, 1914

Rhyncostele pygmaea (Lindl.) Rchb.f., Bot. Zeit. 10: 770, 1852
BR 31: misc.60:15 (1845)
Mexico, Guatemala – up to 4000 m – E
Odontoglossum pygmaeum Lindl., Benth. Pl. Hartw.: 82, 1843
Oncidium pygmaeum (Lindl.) Beer, Prakt. Stud. Orch.: 292, 1854
Leochilus pygmaeus (Lindl.) Benth. & Hook. ex Jackson, Ind. Kew. 2: 1290, 1895

Robiquetia compressa (Lindl.) Schltr., Fedde Rep. Beih. 1: 983, 1913
BR 26: misc.5 (1840)
Type: Philippines, Manila, cult. Loddiges (K-Lindl., holo!)
Philippines – E
Saccolabium compressum Lindl., Bot. Reg. 26: misc.5, 1840
Gastrochilus compressum (Lindl.) Kuntze, Rev. Gen. 2: 661, 1891
Robiquetia ramosii Ames, Orchid. 7: 123, 1922

Robiquetia spathulata (Blume) J.J.Smith, Nat. Tijdschr. Ned. Prod. 72: 114, 1912
BR 24: misc.103 (1838) / 33: sub t.32 (1847)
N.E.India, Burma, Thailand, Laos, Cambodia, Vietnam, Malaya, Sumatra, Java, Borneo, Halmahera, Hainan – E
Cleisostoma spathulatum Blume, Bijdr.: 364, 1825
Saccolabium densiflorum Lindl., Gen. Sp. Orch. Pl.: 220, 1833
Cleisostoma spicatum Lindl., Bot. Reg. 33: sub t.32, 1847
Sarcanthus armeniacus Rchb.f., Allg. Gartenz. 24: 219, 1868
Sarcanthus densiflorus (Lindl.) Parish & Rchb.f., Trans. Linn. Soc. 30: 136, 1874
Saccolabium borneense Rchb.f., Gard. Chron. n.s.15: 563, 1881
Gastrochilus densiflorus (Lindl.) Kuntze, Rev. Gen. 2: 661, 1891
Sarcanthus castaneus Ridley, J. Linn. Soc. 32: 369, 1896
Sarcanthus pendulus Klinge, Act. Hort. Petrop. 47: t.3 141, 1898
Cleisostoma robustum Guillaumin, Bull. Soc. Bot. Fr. 77: 332, 1930
Saccolabium acutilabrum Gagnepain, Bull. Soc. Bot. Fr. 79: 36, 1932
Rhynchostylis densiflora (Lindl.) L.O.Williams, Bot. Mus. Leafl. Harv. Univ. 5: 58, 1937
Pomatocalpa densiflora (Lindl.) Tang & Wang, Act. Phytotax. Sin. 1 (1): 99, 1951

Robiquetia succisa (Lindl.) Seidenf. & Garay, Bot. Tidsskr. 67: 119, 1972
BR 12: t.1014 (1826)
Type: China, Vamper s.n. cult. Hort. Soc. (K-Lindl., holo!)
Sikkim, N.E.India, Burma, Laos, Cambodia, Vietnam, China – E
o Sarcanthus succisus Lindl., Bot. Reg. 12: t.1014, 1826
Oeceoclades paniculata Lindl., Wall. Cat. 7334, 1832 (nomen nudum)
Saccolabium parvulum Lindl., J. Linn. Soc. 3: 36, 1859
Saccolabium buccosum Rchb.f., Gard. Chron.: 839, 1871
Cleisostoma virginale Hance, J. Bot. 15: 38, 1877
Sarcanthus rutilus Parish, Mason Burma: 199, 1883
Pomatocalpa virginalis (Hance) J.J.Smith, Nat. Tijdsschr. Ned. Ind. 72: 37, 1912
Robiquetia paniculata (Lindl.) J.J.Smith, Bull. Buitenz. s.2., 9: 114, 1912
Sarcanthus henryi Schltr., Fedde Rep. Beih. 4: 77, 1919
Uncifera buccosa (Rchb.f.) Finet & Guillaumin, Bull. Soc. Bot. Fr. 77: 333, 1933

Rodriguezia batemanii Poeppig & Endl., Nov. Gen. Sp. Pl. 1: 41 t.70, 1836
BR 23: sub t.1927 (1837)
Bolivia, Brazil, Peru, Ecuador, Colombia – E
Burlingtonia rubescens Lindl., Bot. Reg. 23: sub t.1927, 1837
Rodriguezia candida var batemannii hort ex Stein, Orchideenb.: 538, 1892
Burlingtonia batemanii (Poeppig & Endl.) Lindl. ex Cogn., Mart. Fl. Bras. 3 (6): 154, 1904

Rodriguezia bracteata (Vell.) Hoehne, Arq. Bot. Est. S.Paulo n.s. Form maior 2: 113, 1952
BR 23: sub t.1927 (1837)
Brazil – E
Epidendrum bracteatum Vell., Fl. Flum. Ic. 9: t.16, 1827
Burlingtonia fragrans Lindl., Bot. Reg. 23: sub t.1927, 1837
Rodriguezia fragrans (Lindl.) Rchb.f., Bot. Zeit. 10: 771, 1852
Rodriguezia venusta Cogn., Mart. Fl. Bras. 3 (6): t.110, 1904 (non Rchb.f.)

Rodriguezia candida Batem. ex Lindl., Bot. Reg. 23: sub t.1927, 1837
BR 23: t. & sub t.1927 (1837)
Type: Guyana, cult. Bateman (K-Lindl., holo!)
Venezuela, Guyana, Brazil – E
o Burlingtonia candida Lindl., Bot. Reg. 23: t.1927, 1837

Rodriguezia carnea Lindl., Bot. Reg. 29: misc.113, 1843
BR 29: misc.113 (1843)
Type: Colombia, Linden s.n. cult. Clowes (K-Lindl., holo!)
Colombia – E
Burlingtonia carnea (Lindl.) Lindl. ex Linden, Ill. Hort. 28: 80, 1881

Rodriguezia lanceolata Ruiz Lopez & Pavon, Syst. Veg. Fl. Peruv. & Chil. 1: 219, 1798
BR 11: t.930 (1825)
Panama, Venezuela, Trinidad, Guyana, Colombia, Ecuador, Brazil – E
Pleurothallis coccinea Hook., Exot. Fl. 2: t.129, 1815
o Rodriguezia secunda Kunth, Nov. Gen. Sp. Pl. 1: 367, 1815

Rodriguezia lanceolata Lodd., Bot. Cab. 7: t.696, 1822 (non Ruiz Lopez & Pavon)
Rodriguezia secunda var. sanguinea Schomb., Reis. Brit. Guian. 3: 912, 1848
Burlingtonia rosea hort. ex Rand., J. Orch. 5: 14, 1894
Rodriguezia secunda var. panamensis Schltr., Fedde Rep. Beih. 17: 75, 1922

Rodriguezia maculata (Lindl.) Rchb.f., Bot. Zeit. 10: 771, 1852
BR 25: t.44 (1839)
Type: Brazil, cult. Loddiges (K-Lindl., holo!)
Brazil – E
o Burlingtonia maculata Lindl., Bot. Reg. 25: t.44, 1839

Rodriguezia rigida (Lindl.) Rchb.f., Bot. Zeit. 10: 771, 1852
BR 23: sub t.1927 (1837) / 27: misc.41 (1841)
Type: Brazil, ex Herb. Martius, Prince Wied Neuwied s.n.(K-Lindl., holo!)
Brazil – E
Burlingtonia rigida Lindl., Bot. Reg. 23: sub t.1927, 1837

Rodriguezia venusta (Lindl.) Rchb.f., Bot. Zeit. 10: 771, 1852
BR 23: sub t.1927 (1837)
Type: Brazil, Descourtilz s.n. (K-Lindl., icono!)
Brazil – E
Burlingtonia venusta Lindl., Bot. Reg. 23: sub t.1927, 1837
Rodriguezia pubescens Rchb.f., Bot. Zeitschr. 10: 771, 1852

Rossioglossum grande (Lindl.) Garay & G.C.Kennedy, Orch. Digest 40 (4): 142, 1976
BR 26: misc.94 (1840) / 31: misc.60:18 (1845)
Type: Guatemala, Skinner s.n. (K-Lindl., holo!)
Mexico, Guatemala – up to 2700 m – E
Odontoglossum grande Lindl., Bot. Reg. 26: misc.94, 1840

Rossioglossum insleayi (Barker ex Lindl.) Garay & G.C.Kennedy, Orch. Dig. 40 (4): 142, 1976
BR 26: misc.21 (1840)
Type: Mexico, cult. G.Barker (K-Lindl., holo!)
Mexico – E
Oncidium insleayi Barker ex Lindl., Bot. Reg. 26: misc.21, 1840
Odontoglossum insleayi (Barker) Lindl., Fol. Orch. Odont.: 4, 1854

Rudolfiella aurantiaca (Lindl.) Hoehne, Ic. Orch. Bras.: 220 t.197, 1949
BR 22: t.1875 (1836)
Type: Guyana, cult. Duke of Devonshire (K-Lindl., holo!)
Brazil, Guyana, Surinam, Trinidad, Venezuela, Colombia, Peru – E
o Bifrenaria aurantiaca Lindl., Bot. Reg. 22: t.1875, 1836
Lindleyella aurantiaca (Lindl.) Schltr., Die Orchideen: 414, 1914
Schlechterella aurantiaca (Lindl.) Hoehne, Arq. Bot. Est. S.Paulo 2 (1): 13, 1944

Sacoila lanceolata (Aublet) Garay, Bot. Mus. Leafl. Harv. Univ. 28 (4): 352 t.45, 1980
BR 9: t.701 (1823)
Florida to Panama, W.Indies, S.America – T
Limodorum lanceolatum Aublet, Pl. Guian. 2: 821, 1775
Satyrium orchioides Sw., Prodr. Veg. Ind. Occ.: 118, 1788
Neottia lanceolata (Aublet) Willd., Sp. Pl. ed.4: 73, 1805
o Neottia orchioides (Sw.) Willd., Sp. Pl. ed.4: 75, 1805
Ibidium cristalligerum Salisb., Trans. Hort. Soc. 1: 292, 1812 (nomen illeg.)
Neottia squamulosa Kunth, Nov. Gen. & Sp. Pl. 1: 332, 1816
Stenorrhynchos orchioides (Sw.) L.C.Rich., Mém. Mus. Paris 4: 59, 1817
Stenorrhynchos squamulosus (Kunth) Sprengel, Syst. Veg. 3: 710, 1826
Neottia plantaginea Hook., Exot. Fl. 3: t.226, 1827
Serapias coccinea Vell., Fl. Flum. Ic. 9: t.48, 1827
Serapias tomentosa Vell., Fl. Flum. Ic. 9: t.51, 1827
Neottia aphylla Hook., Bot. Mag. 55: t.2797, 1828
Sacoila lurida Raf., Fl. Tellur. 2: 86, 1836 (nomen illeg.)
Stenorrhynchos aphyllus (Hook.) Lindl., Gen. Sp. Orch. Pl.: 478, 1840
Stenorrhynchos australis Lindl., Gen. Sp. Orch. Pl.: 477, 1840
Stenorrhynchos orchioides var. plantaginea (Hook.) Lindl., Gen. Sp. Orch. Pl.: 478, 1840
Spiranthes orchioides (Sw.) A.Rich., Sagra Hist. Cuba 11: 252, 1850
Stenorrhynchos autralis var. luteoalbus Rchb.f. & Warm., Otia Bot. Hamb. 2: 82, 1881
Gyrostachys aphylla (Hook.) Kuntze, Rev. Gen. 2: 664, 1891
Gyrostachys lanceolata (Aublet) Kuntze, Rev. Gen. 2: 664, 1891
Gyrostachys orchioides (Sw.) Kuntze, Rev. Gen. 2: 664, 1891
Gyrostachys stenorrhynchus Kuntze, Rev. Gen. 2: 664, 1891
Spiranthes jaliscana S.Watson, Proc. Am. Acad. 26: 153, 1891
Stenorrhynchos jaliscana (S.Watson) Nash, Bull. Torrey Club 22: 158, 1895
Stenorrhynchos orchioides var. australis (Lindl.) Cogn., Mart. Fl. Bras. 3 (4): 178, 1895
Stenorrhynchos pedicellatus Cogn., Bull. Herb. Bois. s.2, 3: 932, 1903
Stenorrhynchos pedicellatus var. major Cogn., Bull. Herb. Bois. s.2, 3: 932, 1903
Stenorrhynchos sancti-antonii Kraenzlin, Engl. Bot. Jahrb. 36, Beibl. 80: 8, 1905
Stenorrhynchos apetalus Kraenzlin, Fedde Rep. 6: 23, 1908
Stenorrhynchos riograndensis Kraenzlin, Svensk. Vet. Akad. Handl. 46 (10): 28, 1911
Pelexia tomentosa (Vell.) Schltr., Beih. Bot. Centralbl. 37 (2): 376, 1918
Stenorrhynchos guatemalensis Schltr., Beih. Bot. Centralbl. 37 (2): 376, 1918
Stenorrhynchos secundiflorus Lillo & Hauman, Anal. Soc. Cienc. Argent. 90: 134, 1921
Spiranthes lanceolata (Aublet) Leon, Cont. Ocas. Mus. Hist Nat. Coll. De La Salle, Cuba 8: 358, 1946
Stenorrhynchos coccineus (Vell.) Hoehne, Arq. Bot. Est. S.Paulo n.s.2: 146, 1952

Sarcochilus australis (Lindl.) Rchb.f., Walp. Ann. 6: 501, 1861
BR 20: sub t.1699 (1834) / 24: misc.50 (1838) / 30: misc.77 (1844) / 32: sub t.19:2 (1846)
Type: cult. Loddiges (K-Lindl., holo!)
E.Australia – E
Gunnia australis Lindl., Bot. Reg. 20: sub t.1699, 1834
Sarcochilus parviflorus Lindl., Bot. Reg. 24: misc.50, 1838
Sarcochilus barklyanus F.Muell., Fragm. 1: 90, 1859
Sarcochilus gunnii F.Muell., Fragm. 1: 90, 1859
Thrixspermum australe (Lindl.) Rchb.f., Xen. Orch. 2: 122, 1862
Thrixspermum parviflorum (Lindl.) Rchb.f., Xen. Orch. 2: 122, 1862

Sarcochilus croceus Lindl., Bot. Reg. 32: sub t.19:4, 1846
BR 32: sub t.19:4 (1846)
Type: Philippines, Cuming 1554 cult. Loddiges (K-Lindl., holo!)
Philippines – E

o **Sarcochilus falcatus** R.Br., Prodr.: 332, 1810
BR 22: t.1832 (1836) / 32: sub t.19:1 (1846)
Australia – E
Thrixspermum falcatum (R.Br.) Rchb.f., Xen. Orch. 2: 122, 1862
Sarcochilus montanus Fitzg., Aust. Orch. 1: pt.5, 1879
Sarcochilus falcatus var. montanus (Fitzg.) Bailey, Qld. Fl. 5: 1552, 1902

Sarcochilus olivaceus Lindl., Bot. Reg. 25: misc.27, 1839
BR 25: misc.27 & misc.77 (1839) / 32: sub t.19:3 (1846)
Type: Australia, cult. Loddiges (K-Lindl., holo!)
Australia (New South Wales to Queensland) – E
Gunnia picta Lindl., Bot. Reg. 25: misc.77, 1839
Sarcochilus pictus (Lindl.) Rchb.f., Xen. Orch. 2: 122, 1862
Thrixspermum dilatatum Rchb.f., Xen. Orch. 2: 122, 1862
Thrixspermum olivaceum (Lindl.) Rchb.f., Xen. Orch. 2: 122, 1862

Sarcoglottis acaulis (Smith) Schltr., Fedde Rep. Beih. 6: 53, 1919
BR 10: sub t.823:12 (1824) / 29: misc.41:1 (1843)
Trop. America, Mexico to Argentina – T
Neottia acaulis Smith, Exot. Bot. 2: 91 t.105, 1806
Arethusa picta Anders, Trans. Soc. Arts 25: 199, 1807 (nomen)
Neottia picta Sims, Bot. Mag. 38: t.1562, 1813
Spiranthes picta (Sims) Lindl., Bot. Reg. 10: sub t.823, 1824
Sarcoglottis speciosa Presl, Rel. Haenk. 1: 96 t.15, 1827
Narica moschata Raf., Fl. Tellur. 2: 87, 1837
Synoplectris picta (Sims) Raf., Fl. Tellur. 2: 90, 1837
Spiranthes speciosa (Presl) Lindl., Gen. Sp. Orch. Pl.: 475, 1840
Sarcoglottis picta (Sims) Klotzsch, Allg. Gartenz. 10: 107, 1842
Sarcoglottis picta var. variegata Klotzsch, Allg. Gartenz. 10: 107, 1842
Gyrostachys picta (Sims) Kuntze, Rev. Gen. 2: 664, 1891
Spiranthes acaulis (Smith) Cogn., Mart. Fl. Bras. 3 (4): 221, 1895
Sarcoglottis hunteriana Schltr., Fedde Rep. Beih. 17: 13, 1922
Sarcoglottis powelli Schltr., Fedde Rep. Beih. 17: 14, 1922
Sarcoglottis purpusiorum Schltr., Fedde Rep. Beih. 21: 333, 1925

Sarcoglottis grandiflora (Lindl.) Klotzsch, Allg. Gartenz. 10: 107, 1842
BR 12: t.1043 (1827) / 29: misc.41:2 (1843)
Type: Brazil, Douglas s.n. cult. Hort. Soc. (K-Lindl., holo!)
Venezuela, Surinam, Brazil, Argentina, Paraguay, Ecuador – T
Neottia grandiflora (Lindl.) Hook., Bot. Mag.: t.2730, 1827
o Spiranthes grandiflora Lindl., Bot. Reg. 12: t.1043, 1827

Sarcoglottis rosulata (Lindl.) P.N.Don, Hort. Cantabrig. ed.13: 590, 1845
BR 29: misc.84 (1843)
Type: Guatemala, Hartweg s.n. cult. Hort. Soc. (K-Lindl., holo!)
Mexico, Guatemala, El Salvador, Honduras, Nicaragua – up to 2100 m –T
Spiranthes rosulata Lindl., Bot. Reg. 29: misc.84, 1843
Spiranthes ochracea A.Rich & Galeotti, Ann. Sc. Nat. s.3, 3: 33, 1845
Sarcoglottis ochracea (A.Rich. & Galeotti) Schltr., Beih. Bot. Centralbl. 37 (2): 418, 1920
Sarcoglottis orbiculata Ames, Sched. Orch. 2: 10, 1923

Sarcoglottis ventricosa (Vell.) Hoehne, Arq. Bot. Est. S.Paolo n.s. Form maior 2: 146, 1952
BR 29: misc.41:3 (1843)
Brazil – T
Serapias ventricosa Vell., Fl. Flum. Ic. 9: t.47, 1827
Sarcoglottis rufescens Klotzsch, Allg. Gartenz.: 107, 1842
Spiranthes rufescens (Klotzsch) Fischer, Mart. Fl. Bras. 30: 855 t.10 fig.3, 1884

Satyrium bracteatum (L.f.) Thunb., Prodr. 6, 1794
BR 9: sub t.700 (1823)
S.Africa – T
Ophrys bracteata L.f., Suppl. Pl.: 403, 1781
Diplecthrum bracteatum (L.f.) Pers., Syn. Pl. 2: 509, 1807
Satyrium bracteatum var. glandulosum Sonder, Linnaea 19: 89, 1846
Satyrium bracteatum var. latebracteatum Sonder, Linnaea 19: 89, 1846
Satyrium bracteatum var. lineatum Sonder, Linnaea 19: 88, 1846
Satyrium pictum Sonder, Linnaea 19: 89, 1846 (non Lindl.)

Satyrium candidum Lindl., Bot. Reg. 24: misc.153, 1838
BR 24: misc.153 (1838)
Type: S.Africa, Herschel s.n. cult. Regent's Park (K-Lindl., holo!)
S.Africa – T
Satyrium utriculatum Sonder, Linnaea 19: 84, 1847

Satyrium carneum (Dryander) R.Br., Aiton Hort. Kew ed.2, 5: 195, 1813
BR 24: misc.155 (1838)
S.Africa – T
Orchis carnea Dryander, Ait. Hort. Kew. ed.3: 294, 1789
Satyrium cucullatum Thunb., Fl. Cap.: 88, 1807

o **Satyrium coriifolium** Sw., Kongl. Vet. Acad. Handl. 21: 216, 1800
BR 5: t.416 (1819) / 9: t.703 (1823)
S.Africa – T
Orchis cornuta Houtt., Handl. 12: 456 t.80 fig.2, 1773
Diplecthrum coriifolium (Sw.) Pers., Syn. 2: 509, 1807
o Satyrium cucullatum Lodd., Bot. Cab.: t.104, 1818 (non Jacq.)
Diplecthrum erectum Lindl., Gen & Sp. Orch. Pl.: 340, 1838
Satyrium chrysostachyum Herschel, Bot. Reg. 24: sub misc.154, 1838
Satyrium erectum Lindl., Gen. Sp. Orch. Pl.: 340, 1838 (non Sw.)
Satyrium aureum Paxton, Mag. Bot. 15: 31, 1848

Satyrium erectum Sw., Vet. Acad. Handl. Stockh.: 216, 1800
BR 24: misc.154 (1838) / 26: t.18 (1840)
S.Africa – T
Diplecthrum erectum (Sw.) Pers., Syn. Pl.: 509, 1807
Satyrium herscheliae Harv., Gen. S. Afr. Pl.: 326, 1838
Satyrium papillosum Lindl., Gen. Sp. Orch. Pl.: 341, 1838
o Satyrium pustulatum Lindl., Bot. Reg. 26: t.18, 1840

Sauroglossum nitidum (Vell.) Schltr., Beih. Bot. Centralbl. 36 (2): 379, 1818
BR 19: t.1618 (1833)
Brazil, Argentina – T
Serapias nitida Vell., Fl. Flum. Ic. 9: 52, 1827
o Sauroglossum elatum Lindl., Bot. Reg. 19: t.1618, 1833
Cyclopogon procerus Regnell ex Barb.Rodr., Gen. & Sp. Orch. Nov. 2: 280, 1882
Spiranthes nitida (Vell.) Cogn., Mart. Fl. Bras.: 224, 1895
Spiranthes pachychila Kraenzlin, Svensk Acad. Handl. 46: 37, 1911

Scaphyglottis graminifolia (Ruiz Lopez & Pavon) Poeppig & Endl., Nov. Gen. Sp. Pl. 1: 59, 1836
BR 25: misc.15 (1839) / 28: misc.17:2 (1842)
Mexico to Panama, Venezuela, Colombia, Ecuador, Peru, Brazil – E
Fernandezia graminifolia Ruiz Lopez & Pavon, Syst. Veg. 1: 241, 1798
Nemaconia graminifolia (Ruiz Lopez & Pavon) Knowles & Westc., Flor. Cab.: 127, 1838
Ponera graminifolia (Ruiz Lopez & Pavon) Lindl., Bot. Reg. 25: misc. 15, 1839
Scaphyglottis leucantha Rchb.f., Linnaea 22: 856, 1849
Ponera leucantha (Rchb.f.) Rchb.f., Bonpl. 2: 22, 1854
Ponera behrii Rchb.f., Bonpl. 3: 220, 1855
Ponera albida Rchb.f., Beitr. Orch. Centr. Am.: 103, 1866
Scaphyglottis behrii (Rchb.f.) Benth. & Hook.f. ex Hemsl., Biol. Centr. Am. Bot. 3: 219, 1883
Scaphyglottis guatemalensis Schltr., Fedde Rep. 2: 133, 1906
Scaphyglottis pauciflora Schltr., Fedde Rep. 3: 47, 1906
Scaphyglottis albida (Rchb.f.) Schltr., Beih. Bot. Centralbl. 36 (2): 456, 1918
Scaphyglottis sanctae martae Schltr., Fedde Rep. Beih. 7: 122, 1920
Scaphyglottis bradeorum Schltr., Fedde Rep. Beih. 19: 113, 1923

Scaphyglottis prolifera (R.Br.) Cogn., Mart. Fl. Bras. 3: 15, 1898
BR 10: t.825 (1824)
Belize, Guatemala, Honduras, Nicaragua, Panama, W.Indies, Venezuela, Ecuador, Peru, Bolivia, Brazil, Guyana, Surinam – E
o Isochilus prolifer R.Br., Ait. Hort. Kew. ed.2, 5: 209, 1813
Isochilus proliferum (R.Br.) Lindl., Gen. Sp. Orch. Pl.: 113, 1831
Ponera prolifera (R.Br.) Rchb.f., Bonpl. 2: 22, 1854
Scaphyglottis cuneata Schltr., Beih. Bot. Centralbl. 36: 398, 1918
Tetragamestus gracilis Schltr., Beih. Bot. Centralbl. 36: 400, 1918
Scaphyglottis gracilis (Schltr.) Schltr., Fedde Rep. 19: 28, 1923
Scaphyglottis werkley Schltr., Fedde Rep. 19: 28, 1923
Ponera maipiriensis Kraenzlin, Fedde Rep. 25: 22, 1928

Scaphyglottis stellata Lodd. ex Lindl., Bot. Reg. 25: misc.60, 1839
BR 25: misc.60 (1839)
Type: Guyana, cult. Loddiges (K-Lindl., holo!)
Guyana, Ecuador – E
Ponera stellata (Lodd. ex Lindl.) Rchb.f., Walp. Ann. 6: 454, 1862

o **Scaphyglottis violacea** Lindl., Bot. Reg. 22: t.1901, 1836
BR 22: t.1901 (1836)
Type: Guyana, cult. Loddiges (K-Lindl., holo!)
Venezuela, Guyana, Brazil, Amazonas – E
Cladobium violaceum Lindl., Nat. Syst. Bot. ed.2: 446, 1836
Scaphyglottis rosea Hook., Ic. Pl. 4: t.313, 1841
Ponera rosea (Hook.) Rchb.f., Bonpl. 2: 22, 1854
Ponera violacea (Lindl.) Rchb.f., Bonpl. 2: 22, 1854

Scelochilus ottonis Klotzsch, Allg. Gartenz.: 261, 1841
BR 28: misc.23 (1842) / 31: misc.19 (1845)
Venezuela – E

Schoenorchis gemmata (Lindl.) J.J.Smith, Nat. Tijdsschr. Ned. Ind. 72: 1000, 1912
BR 24: misc.88 (1838)
Type: Khasia, Gibson s.n. cult. Duke of Devonshire (K-Lindl., holo!)
Nepal, Sikkim, Bhutan, N.E.India, Burma, Thailand, Cambodia, Vietnam, Laos, China – E
Saccolabium gemmatum Lindl., Bot. Reg. 24: misc.88, 1838
Saccolabium hainanense Rolfe, Kew Bull.: 284, 1895
Schoenorchis hainanensis (Rolfe) Schltr., Fedde Rep. Beih. 1: 936, 1913
Cleisostoma gemmatum (Lindl.) King & Pantl., Ann. Bot. Gard. Calcutta 8: 233, 1918
Cleisostoma evrardii Gagnepain, Bull. Soc. Bot. Fr. 79: 34, 1932
Cleisostoma dichroanthum Gagnepain, Bull. Mus. Nat. Hist. Paris s.2, 22 (5) : 631, 1950

Schomburgkia gloriosa Rchb.f., Hamb. Gartenz. 16: 178, 1860
BR 25: misc.12 (1839) / 30: t.23 & sub t.23:1, 2 (1844)
Venezuela, Guyana, Surinam, Brazil – E
Epidendrum fimbriatum Vell., Fl. Flum. 9: t.14, 1827
Schomburgkia marginata Hook., Bot. Mag. 66: t.3729, 1839
o Schomburgkia crispa Lindl., Bot. Reg. 30: t.23, 1844 (non Lindl.1838)
Cattleya crispa Beer, Prakt. Stud. Orch.: 209, 1854 (non Lindl.)
Bletia crispina Rchb.f., Walp. Ann. 6: 419, 1862
Bletia gloriosa (Rchb.f.) Rchb.f., Walp. Ann. 6: 419, 1862
Laelia gloriosa (Rchb.f.) L.O.Williams, Darwiniana 5: 76, 1941
Schomburgkia fimbriata (Vell.) Hoehne, Arq. Bot. Est. S.Paulo 2 (6): 142, 1952

Schomburgkia rosea Linden ex Lindl., Bot. Reg. 31: sub t.53, 1845
BR 31: sub t.53 (1845)
Type: Guyana, Linden 1664 (K-Lindl., holo!)
Venezuela, Colombia – E
Bletia rosea (Linden ex Lindl.) Rchb.f., Walp. Ann. 6: 420, 1862
Laelia rosea (Linden ex Lindl.) C.Schweinf., Mem. N.Y. Bot. Gard. 14 (3): 155, 1967

Schomburgkia superbiens (Lindl.) Rolfe, Orch. Rev. 25: 49, 1917
BR 26: misc.87 (1840) / 28: sub t.62:5 (1842)
Type: Guatemala, Skinner s.n. (K-Lindl., holo!)
Mexico, Guatemala, Honduras, Nicaragua – E
Laelia superbiens Lindl., Bot. Reg. 26: misc.87, 1840
Cattleya superbiens (Lindl.) Beer, Prakt. Stud. Orch.: 214, 1854
Bletia superbiens (Lindl.) Rchb.f., Walp. Ann. 6: 418, 1862

o **Schomburgkia undulata** Lindl., Bot. Reg. 30: sub t.23:3, 1844
BR 30: sub t.23:3 & misc.21 (1844) / 31: t.53 (1845)
Type: La Guayra, Linden s.n. cult. Rucker (K-Lindl., holo!)
Costa Rica, Panama, Trinidad, Venezuela, Colombia, Peru, Bolivia – up to 1000 m – E
Schomburgkia violacea Paxton, Mag. Bot. 16: 123, 1849
Cattleya undulata (Lindl.) Beer, Prakt. Stud. Orch.: 215, 1854
Bletia undulata (Lindl.) Rchb.f., Walp. Ann. 6: 420, 1862
Schomburgkia lueddemanii Prill, J. Soc. Imp. Pow. 8: 123, 1862
Laelia lueddemanii (Prill) L.O.Williams, Ann. Missouri Bot. Gard. 27: 282, 1940
Laelia undulata (Lindl.) L.O.Williams, Darwiniana 5: 76, 1941

Scuticaria steelei (Hook.) Lindl., Bot. Reg. 29: misc.p.14:1, 1843
BR 23: t. & sub t.1986 (1837) / 29: misc.p.14:1 (1843)
Venezuela, Guyana, Brazil, Amazonas – E

Maxillaria flabellifera Lindl., Bot. Reg. 23: sub t.1986, 1837
o Maxillaria steelei Hook., Bot. Mag. 64: t.3573, 1837
Scuticaria keyseriana hort., J. Hort. 16: 421, 1888

Serapias cordigera L., Sp. Pl. ed.2: 1345, 1763
BR 31: misc.80 (1845)
Azores, Mediterranean Europe, Asia Minor – T
Serapias lingua var ß Savi, Fl. Pis. 2: 304, 1798
Serapias ovalis L.C.Rich., Mém. Mus. Paris 4: 54, 1817
Serapias oxyglottis Cocc., Fl. Prov. Bologne: 480, 1883 (non Bertol., non Willd, non Rchb.f.)
Helleborine cordigera (L.) Pers., Syn. 2: 512, 1897
Lonchitis cordigera (L.) Bubani, Fl. Pyr. 4: 50, 1901
Serapiastrum cordigerum (L.) A.A.Eaton, Proc. Biol. Soc. Wash. 21: 67, 1908
Serapias cordigera var. genuina Briq., Prodr. Fl. Corse 1: 379, 1910

Serapias vomeracea (Burm.f.) Briq., Prodr. Fl. Corse 1: 378, 1910
ssp. vomeracea
BR 14: t.1189 (1828)
S.Europe, Mediterranean Region, Caucasus Aerea – T
Orchis vomeracea Burm.f., Fl. Corse. 237, 1770
Orchis lingua Scop., Fl. Carn. 2: 187, 1772 (non L.)
o Serapias cordigera var. longipetala Bertol., Pl. Gen.: 126, 1804 (non L.)
Serapias cordigera M.Bieb., Fl. Taur. Caucas. 2: 370, 1808 (non L.)
Helleborine longipetala (Bertol.) Ten., Fl. Nap. Prodr.: 53, 1811
Helleborine pseudo-cordigera Sebast., Rom. Pl. Fasc. 1: 14, 1813
Serapias hirsuta Lapeyr., Hist. Abr. Pyr.: 551, 1813 (nomen illeg.)
Serapias pseudo-cordigera (Sebast.) Moric., Fl. Venet.: 374, 1820
Serapias lancifera Saint-Amans, Fl. d'Agen: 378, 1821
Serapias longipetala (Bertol.) Poll., Fl. Veron. 3: 30, 1824
Serapias oxyglottis Rchb.f., Fl. Germ. Excurs.: 130, 1830 (non Bertol., non Willd., non Cocc.)
Serapias escurialensis Gand., Fl. Eur. 23: 134, 1890
Lonchitis longipetala (Bertol.) Bubani, Fl. Pyr. 4: 52, 1901
Serapiastrum longipetalum (Bertol.) A.A.Eaton, Proc. Biol. Soc. Wash. 21: 67, 1901
Serapias vomeracea f. stenopetala Vierh., Oesterr. Bot. Zeitschr. 66: 158, 1916 (nomen illeg.)
Serapiastrum vomeraceum (Burm.f.) Schinz, Fl. Schweiz ed.4: 163, 1923
Serapias vomeracea f. heldreichii Soó, Fedde Rep. 24: 33, 1927
Serapias cordigera ssp. vomeracea (Burm.f.) Sundermann, Europ. Medit. Orch. ed. 3. 39: 125, 1980

Smitinandia micrantha (Lindl.) Holttum, Gard. Bull. Singap. 25: 105, 1969
BR 25: misc.52 (1839)
Type: cult. Loddiges (K-Lindl., holo!)
W.Himalaya, Nepal, Sikkim, Bhutan, N.E.India, Burma, Thailand, Laos, Cambodia, Vietnam, Malaya – 600–1000 m – E
Saccolabium micranthum Lindl., Gen. Sp. Orch. Pl.: 220, 1833
Gastrochilus parviflorus Kuntze, Rev. Gen. 2: 661, 1891
Saccolabium fissum Ridley, J. Linn. Soc. 32: 361, 1896
Cleisostoma micranthum (Lindl.) King & Pantl., Ann. Bot. Gard. Calcutta 8: 234, 1898
Uncifera uncifera Guillaumin, Bull. Soc. Bot. Fr. 79: 34, 1930
Cleisostoma poilanei Gagnepain, Bull. Soc. Bot. Fr. 79: 34, 1932
Cleisostoma petitiana Guillaumin, Bull. Mus. Nat. Hist. Paris 5: 434, 1945
Ascocentrum micranthum (Lindl.) Holttum, Gard. Bull. Sing. 11: 275, 1947
Pomatocalpa poilanei (Gagnepain) Tang & Wang, Act. Phytotax. Sin. 1: 100, 1951
Cleisostoma tixieri Guillaumin, Bull. Mus. Nat. Hist. Paris 4: 369, 1960

Sobralia macrantha Lindl., Sert. Orch.: sub t.29, 1836
BR 28: misc.65 (1842)
Mexico, Guatemala, El Salvador, Honduras, Nicaragua, Costa Rica – T, L
Sobralia macrantha var. alba Lindl., Fol. Orch. Sobralia: 8, 1854
Sobralia macrantha var. purpurea Lindl., Fol. Orch. Sobralia: 8, 1854

o **Sobralia sessilis** Lindl., Bot. Reg. 27: t.17 & misc.11, 1841
BR 27: t.17 & misc.11 (1841)
Type: Guyana, Schomburgk s.n. cult. Loddiges (K-Lindl., holo!)
Brazil, Guyana, Venezuela, Colombia, Peru – T, E
Cattleya sessilis (Lindl.) Beer, Prakt. Stud. Orch.: 214, 1854
Sobralia galeottiana Lindl., Fol. Orch. Sobralia: 6, 1854 (non A.Rich)

Solenidium lunatum (Lindl.) Kraenzlin, Engl. Pflanzenr. Orch. Mon. Oncid.: 316, 1922
BR 23: t.1929 & sub t.1920:5 (1837)
Type: Guyana, cult. Loddiges (K-Lindl., icono!)
Venezuela, Guyana, Brazil – E
o Oncidium lunatum Lindl., Bot. Reg. 23: sub t.1920, 1838
Rodriguezia lindmanii Kraenzlin, Kongl. Sv. Vet. Akad. Handl. 46 (10): 75, 1911
Leochilus mattogrossensis Cogn., Comm. Linh. Telgr. Est. Matto Grosso Annex 5 (3): 13, 1912
Solenidium mattogrossense (Cogn.) Schltr., Fedde Rep. 15: 215, 1918

Sophronitella violacea (Lindl.) Schltr., Fedde Rep. Beih. 35: 76, 1925
BR 26: misc.18 (1840)
Type: Brazil, cult. Cannon (K-Lindl., holo!)
Brazil – E
Sophronitis violacea Lindl., Bot. Reg. 26: misc.18, 1840
Cattleya violacea (Lindl.) Beer, Prakt. Stud. Orch.: 209, 1854 (non Rolfe)

o **Sophronitis cernua** Lindl., Bot. Reg. 13: t.1129, 1828
BR 13: t.1129 (1828) / 14: sub t.1147 (1828)
Type: Brazil, Harrison s.n. (K-Lindl., holo!)
Brazil – E
Epidendrum humile Vell., Fl. Flum. Ic. 9: t.24, 1827
Sophronia cernua Lindl., Bot. Reg. 13: t.1129, 1828
Sophronitis modesta Lindl., Bot. Reg. 14: sub t.1147, 1828
Sophronitis hoffmannseggii Rchb.f., Linnaea 16: 286, 1842
Sophronitis isopetala Hoffsgg., Verz. Orch.: 60, 1842
Sophronitis nutans Hoffsgg., Verz. Orch.: 61, 1842
Sophronitis pterocarpa Lindl. & Paxton, Flow. Gard. 3: 11 fig.239, 1852
Cattleya cernua (Lindl.) Beer, Prakt. Stud. Orch.: 209, 1854

Sophronitis coccinea (Lindl.) Rchb.f., Walp. Ann. 6: 465, 1862
BR 22: sub t.1919 (1836)
Type: Brazil, Descourtilz s.n. (K-Lindl., holo!)
Brazil – E
Cattleya coccinea Lindl., Bot. Reg. 22: sub t.1919, 1836
Sophronitis grandiflora Lindl., Sert. Orch.: t.5 fig.2, 1838
Cattleya grandiflora (Lindl.) Beer, Prakt. Stud. Orch.: 209, 1854
Sophronitis militaris Rchb.f., Walp. Ann. 6: 465, 1862

Spathoglottis plicata Blume, Bijdr. 8: 401, 1825
BR 24: t.60 & misc.113 (1838) / 30: misc.58 (1844)
Australia, New Caledonia, Solomon Is., Vanuatu, Fiji, Bougainville, Tonga, Samoa, New Guinea, Philippines, Java, Malaya, Thailand, Vietnam, Burma, India – T
Bletia angustata Gaudich., Frey. Voy. Bot.: 421, 1829
Bletia angustifolia Gaudich., Frey. Voy. Bot.: t.32, 1829
o Paxtonia rosea Lindl., Bot. Reg. 24: t.60 & misc.113, 1838
Spathoglottis lilacina Griffith, Notul. 3: 323, 1851
Phaius rumphii Blume, Mus. Bot. Lugd. Bat. 2: 179, 1856
Spathoglottis spicata Lindl., J. Linn. Soc. 3: 9, 1859
Spathoglottis vieillardii Rchb.f., Linnaea 41: 85, 1877
Spathoglottis angustifolia (Gaudich.) Benth. & Hook., Gen. Pl. 3: 512, 1883
Spathoglottis angustorum Rchb.f., Gard. Chron. n.s.25: 9, 1886
Spathoglottis plicata var. alba Ridley, Mat. Fl. Mal. Pen. 1: 117, 1907
Spathoglottis daenikeri Kraenzlin, Vierteljahrschr. Nat. Ges. Zürich 74: 80, 1929
Calanthe poilanei Gagnepain, Bull. Mus. Nat. Hist. Paris 5: 625, 1950

Spathoglottis pubescens Lindl., Gen. Sp. Orch. Pl.: 120, 1831
BR 31: t. & sub t.19 (1845)
India, Burma, Thailand, China, Laos, Vietnam, – T
Epipactis graminifolia Roxb., Hort. Beng.: 63, 1814
o Spathoglottis fortunii Lindl., Bot. Reg. 31: t.19, 1845
Spathoglottis parvifolia Lindl., Bot. Reg. 31: sub t.19, 1845
Spathoglottis khasyana Griffith, Notul. 3: 323, 1851
Pachystoma fortunei (Lindl.) Rchb.f., Walp. Ann. 6: 464, 1861
Pachystoma wightii Rchb.f., Walp. Ann. 6: 464, 1861
Pachystoma parvifolium (Lindl.) Rchb.f., Walp. Ann. 6: 462, 1862
Spathoglottis pubescens var. berkeleyi Hook.f., Fl. Brit. Ind. 5: 814 & 6: 191, 1890
Spathoglottis pubescens var. parvifolia (Lindl.), Hook.f. Fl. Brit. Ind. 5: 814, 1890
Eulophia sinensis Guillaumin, Bull. Mus. Nat. Hist. Paris s.2, 31 (1): 116, 1931 (non Blume)
Cymbidium guttatum Bureau & Franchet, Gagnepain Fl. Gen. Ind. 6: 368, 1934 (non Willd.)
Spathoglottis pubescens var. fortunei (Lindl.) Gagnepain, Fl. Gen. Ind. 6: 369, 1934
Spathoglottis plicata var. pubescens (Lindl.) Hiroe, Orch. Fl. 2: 89, 1971

Spathoglottis tomentosa Lindl., Bot. Reg. 31: sub t.19, 1845
BR 31: sub t.19 (1845)
Type: Philippines, Mindanao, Cuming s.n. (K-Lindl., holo!)
Philippines – T

Spiranthes aestivalis (Poiret) L.C.M.Rich., Orch. Eur. Annot.: 36, 1817
BR 10: sub t.823:2 (1824)
S.W. & Centr. Europe – T

o **Spiranthes cernua** (L.) L.C.Rich., Mém. Mus. Hist. Nat. Paris 4: 59, 1818
BR 10: t.823 & sub t.823:10 (1824)
USA (Virginia), Canada – T

Ophrys cernua L., Sp. Pl. 2: 946, 1753
Neottia cernua (L.) Sw., Schrad. Neues J. 1: 52, 1805
Gyrostachys cernua (L.) Kuntze, Rev. Gen. 2: 664, 1891
Gyrostachys constricta Small, Bull. Torrey Club 25: 609, 1898
Spiranthes constricta (Small) K.Schumann, Engl. Bot. Jahrb. 26 (1): 339, 1900
Ibidium cernuum (L.) House, Bull. Torrey Club 32: 381, 1905
Triorchis cernua (L.) Nieuwl., Am. Midl. Nat. 3: 122, 1913

Spiranthes congesta Lindl., Bot. Reg. 10: sub t.823:8, 1824
BR 10: sub t.823:8 (1824)
Type: Siberia, Prescott s.n. (K-Lindl., holo!)
Siberia – T

Spiranthes sinensis (Pers.) Ames, Orchid. 2: 53, 1908
BR 7: t.602 (1822) / 10: sub t.823:3, 4, 5, 6 (1824)
Afghanistan, Himalaya (Kashmir to Bhutan), India, Burma, S.Tibet, Siberia, China, Japan, Malaysia, Australia – 100–4600 m – T
Ophrys spiralis Georgi, Reise 232, 1772 (non L.)
Aristotelea spiralis Lour., Fl. Cochinch. 2: 522, 1790
Epidendrum aristotelea Raeusch., Nom. ed.3: 265, 1797
Neottia sinensis Pers., Syn. Pl. 2: 511, 1807
Neottia australis R.Br., Prodr. Fl. Nov. Holland.: 319, 1810
Neottia flexuosa Smith, Rees Cyclop. 24: 9, 1813
Neottia parviflora Smith, Rees Cyclop. 24: 10, 1813
Neottia amoena Bieb., Fl. Taur. Cauc. 3: 606, 1819
Spiranthes pudica Lindl., Coll. Bot.: 30, 1821
o Neottia australis var. chinensis Lindl., Bot. Reg. 7: t.602, 1822
Spiranthes australis (R.Br.) Lindl., Bot. Reg. 10: sub t.823, 1824
Spiranthes flexuosa (Smith) Lindl., Bot. Reg. 10: sub t.823, 1824
Spiranthes parviflora (Smith) Lindl., Bot. Reg. 10: sub t.823, 1824
Neottia crispata Blume, Bijdr.: 406, 1825
Spiranthes amoena (Bieb.) Sprengel, Syst. Veg. 3: 708, 1826
Neottia pudica (Lindl.) Sweet, Hort. Brit. ed.2: 485, 1830
Spiranthes wightiana Lindl., Wall. Cat.: 7378, 1832
Calanthe australis (R.Br.) Aiton ex Loud., Hort. Br. Suppl. 2: 615, 1839
Spiranthes densa A.Rich., Ann. Sc. Nat. s.2, 15: 78, 1841
Spiranthes longispicata A.Rich., Ann. Sc. Nat. s.2, 15: 78, 1841
Sarcoglottis pudica (Lindl.) P.N.Don, Hort. Cantabrig. ed.13: 590, 1845
Spiranthes indica Steudel, Nom. ed.2: 625, 1845
Spiranthes novae-zelandiae Hook.f., Fl. Nov. Zealand. 1: 243, 1853
Spiranthes crispata (Blume) Zoll. & Morr., Syst. Verz. Zoll.: 89, 1854
Spiranthes australis var. pudica (Lindl.) Lindl., Kew J. Bot. 7: 38, 1855
Spiranthes stylites Lindl., J. Linn. Soc. 1: 178, 1857
Gyrostachys amoena (Bieb.) Blume, Orch. Archip. Ind.: 128, 1858
Gyrostachys australis (R.Br.) Blume, Orch. Archip. Ind.: 128, 1858
Gyrostachys australis var. amoena (Bieb.) Blume, Fl. Java Orch.: 129, 1859
Gyrostachys australis var. flexuosa (Smith) Blume, Fl. Java Orch.: 130, 1859
Gyrostachys novifriburgensis Kuntze, Rev. Gen. 2: 664, 1891
Gyrostachys wightiana (Lindl.) Kuntze, Rev. Gen. 2: 664, 1891
Spiranthes australis var. suishaensis Hayata, Ic. Pl. Formos. 6: 86, 1916
Spiranthes aristotelea (Raeusch.) Merr., Phil. J. Sc. 15: 230, 1919
Spiranthes australis var. viridiflora Makino, J. Jap. Bot. 3: 5, 1926
Spiranthes australis var. sinensis (Pers.) Gagnepain, Fl. Gén. Indochin. 6: 546, 1933
Spiranthes spiralis Makino, J. Jap. Bot. 3 (7): 25, 1934
Spiranthes spiralis var. albescens Honda, Bot. Mag. Tokyo 49: 791, 1935
Spiranthes spiralis var. depauperata Honda, Bot. Mag. Tokyo 49: 698, 1935
Spiranthes sinensis f. albiflora (Matuda) Iwabrechi, Tohuku Hakubutu-Kai 1: 136, 1936
Spiranthes sinensis var. albescens (Honda) Honda, Bot. Mag. Tokyo 50: 669, 1936
Spiranthes sinensis var. depauperata (Honda) Honda, Bot. Mag. Tokyo 50: 669, 1936
Spiranthes spiralis f. albiflora Matuda, Tohuku Hakubutu-Kai 1: 136, 1936
Spiranthes sinensis f. viridiflora (Makino) Ohwi, Bull. Nat. Sc. Mus. Tokyo 33: 69, 1953
Spiranthes sinensis ssp. australis (R.Br.) Kitamura, Acta Phytotax. Geobot. 21: 23, 1964
Spiranthes lancea var. chinensis (Lindl.) Hatusima, J. Geobot. 16: 80, 1968
Spiranthes sinensis var. amoena (Bieb.) Hara, J. Jap. Bot. 44: 59, 1969

Spiranthes spiralis (L.) Chevall., Fl. Gen. Env. Paris 2: 330, 1827
BR 10: sub t.823:1 (1824)
W. & Centr. Europe, Turkey, Lebanon, Syria, Iran, N.Africa, – T
Ophrys spiralis L., Sp. Pl.: 945, 1753
Spiranthes autumnalis Rich., Orch. Eur. Annot.: 37, 1817

Spiranthes torta (Thunb.) Garay & Sweet, Fl. Lesser Antilles Orch.: 77, 1974
BR 10: sub t.823:9, 15 (1824)
Florida, Guatemala, Honduras, Nicaragua, Bermuda, Bahamas, W.Indies, Trinidad – T
Ophrys peruviana Aublet, Hist. Pl. Guian. Fr. 2: 816, 1775
Satyrium spirale Sw., Prodr. Veg. Ind. Occid.: 118, 1788
Ophrys torta Thunb., Mus. Nat. Acad. Upsal. 9: 136, 1791
Ophrys quinquelobata Poiret, Lam. Encycl. 4: 568, 1798
Neottia tortilis Sw., Kongl. Svensk. Vetensk. Acad. Nya Handl. 21: 226, 1800
Neottia quadridentata Willd., Sp. Pl. ed.4: 73, 1805
Spiranthes tortilis (Sw.) L.C.Rich., Orch. Eur. Annot.: 37, 1817
Spiranthes quadridentata (Willd.) Lindl., Bot. Reg. 10: sub t.823:15, 1824
Ibidium tortile (Sw.) House, Muhlbergia 1: 129, 1906
Spiranthes swartzii Krause, Beih. Bot. Centralbl. 32: 342, 1914
Spiranthes quinquelobata (Poiret) Urban, Fedde Rep. 15: 305, 1918
Spiranthes amesiana Schltr., Beih. Bot. Centralbl. 37 (2): 348, 1920
Triorchis spiralis (Sw.) House, Am. Middle Nat. 6: 200, 1920

Stanhopea grandiflora (Lodd.) Lindl., Gen. Sp. Orch. Pl.: 158, 1832 (non (Kunth) Rchb.f.)
BR 18: t.1529 (1832) / 29: sub t.44:9 (1843)
Brazil, Guayana, Venezuela, Trinidad, Colombia – E
Ceratochilus grandiflorus Lodd., Bot. Cab. 15: t.1414, 1828
o Stanhopea eburnea Lindl., Bot. Reg. 18: t.1529, 1832
Stanhopea calceolata Drap, Hort. Univ. 2: 264, 1841

Stanhopea graveolens Lindl., Bot. Reg. 26: misc.125, 1840
BR 26: misc.125 (1840) / 29: sub t.44:7 (1843)
Type: Peru, cult. Herbert (K-Lindl., holo!)
Mexico, Belize, Guatemala, Honduras, El Salvador – E, L

Stanhopea hernandezii (Kunth) Schltr., Beih. Bot. Centralbl. 2: 490, 1918
BR 29: sub t.44:2 (1843)
Mexico, Guatemala, Venezuela, Guyana, Brazil – 1000–2000 m – E
Anguloa hernandezii Kunth, Syn. 1: 332, 1822
Maxillaria lyncea Lindl., Gen. Sp. Orch. Pl.: 40, 1832
Stanhopea devoniensis Lindl., Sert. Orch.: t.11837
Stanhopea cavendishii Lindl. ex Baxter, Loud. Hort. Brit. Suppl. 3: 643, 1839
Stanhopea lyncea (Lindl.) P.N.Don, Hort. Cantabrig. ed.13: 608, 1845

o **Stanhopea inodora** Lodd. ex Lindl., Bot. Reg. 31: t.65, 1845
BR 31: t.65 (1845)
Type: Mexico, cult. Loddiges (K-Lindl., holo!)
Mexico – E

o **Stanhopea insignis** Frost ex Hook., Bot. Mag. 56: t.2948/49, 1829
BR 22: t.1837 (1836) / 29: sub t.44:1 (1843)
Brazil – E

o **Stanhopea martiana** Bateman ex Lindl., Bot. Reg. 26: misc.109, 1840
BR 26: misc.109 (1840) / 27: misc.147 (1841) / 29: t.44 & sub t.44: 11 & misc.81 (1843)
Type: Mexico, Karwinski s.n. (K-Lindl., holo!)
Mexico – 1400–2000 m – E, L
Stanhopea martiana var. bicolor Lindl., Bot. Reg. 29: t.44, 1843

o **Stanhopea oculata** (Lodd.) Lindl., Gen. Sp. Orch. Pl.: 158, 1832
BR 21: t.1800 (1835) / 24: misc.4 (1838) / 25: misc.113 (1839) / 26: misc.80 (1840) / 29: sub t.44:3, 4 & misc.116 (1843) / 31: t.24 (1845)
Mexico, Guatemala, Belize, Honduras, El Salvador, Nicaragua, Costa Rica, Panama, Venezuela, Colombia – E
Stanhopea lindleyi Zucc., Abh. Akad. Münch. 2: 320, 1831
Ceratochilus oculatus Lodd., Bot. Cab.: t.1764, 1832
o Stanhopea bucephalus Lindl., Gen. Sp. Orch. Pl.: 157, 1832
Stanhopea guttulata Lindl., Bot. Reg. 29: misc.116, 1843
Stanhopea oculata var. lindleyi Zucc. ex Lindl., Fol. Orch. Stanh.: 3, 1852
Stanhopea guttata K.Koch, Allg. Gartenz.: 364, 1858
Stanhopea ornatissima Lemaire, Ill. Hort. 9: t.325, 1862
Stanhopea oculata var. guttulata (Lindl.) Rchb.f., Walp. Ann. 6: 587, 1863
Stanhopea cymbiformis Rchb.f., Xen. Orch. 2: 84 t.124, 1865
Epidendrum cornutum Sessé & Mociño, Fl. Mex. ed.2: 225, 1887
Stanhopea oculata var. crocea Regel, Ind. Sem. Hort. Petrop. 17: 15, 1898
Stanhopea oculata var. geniculata Klotzsch, Act. Hort. Petrop. 17: 143, 1898
Stanhopea purpusii Schltr., Orchis 10: 186, 1916
Stanhopea minor Schltr., Notizbl. Bot. Gart. Berl. 6: 483, 1917

o **Stanhopea quadricornis** Lindl., Bot. Reg. 24: t.5, 1838
BR 24: t.5 (1838) / 29: sub t.44:8 (1843)

Type: Guatemala, cult. Rucker (K-Lindl., holo!)
Guatemala – E

Stanhopea ruckeri Lindl., Bot. Reg. 29: sub t.44:6, 1843
BR 29: sub t.44:6 (1843)
Type: Mexico, cult. Loddiges (K-Lindl., holo!)
Mexico, Nicaragua – E

Stanhopea saccata Bateman, Orch. Mex. & Guat.: t.15, 1839
BR 29: sub t.44:12 (1843)
Mexico, Guatemala, El Salvador – up to 1400 m – E
Stanhopea marshii Rchb.f., Xen. Orch. 1: 120, 1855
Stanhopea radiosa Lemaire, Ill. Hort. 6: misc.72, 1859

o **Stanhopea tigrina** Bateman ex Lindl., Sert. Orch.: sub t.1, 1837
BR 25: t.1 (1839) / 26: misc.28 (1840) / 29: sub t.44:10 (1843)
Mexico, Guatemala, Colombia, Venezuela, Guyana, Brazil – E
Stanhopea maculosa Knowles & Westc., Fl. Cab. 3: t.121, 1839
Stanhopea expansa P.N.Don, Hort. Cantabr. ed.13: 721, 1845
Stanhopea nigroviolacea Morren ex Beer, Prakt. Orch.: 313, 1854

Stanhopea wardii Lodd. ex Lindl., Sert. Orch.: t.20, 1838
BR 26: misc.147 (1840) / 27: misc.31 (1841) / 29: sub t.44:5 (1843)
Mexico to Panama, Venezuela to Peru – E
Stanhopea aurea Lodd. ex Lindl., Bot. Reg. 27: misc.31, 1841
Stanhopea amoena Klotzsch, Allg. Gartenz. 20: 273, 1852
Stanhopea wardii var. venusta (Lindl.) Rolfe, Lindenia 7: t.315, 1892

Staurochilus ionosma (Lindl.) Schltr., Die Orchideen: 578, 1914
BR 33: t.41 (1847)
Type: Philippines, Cuming s.n. cult Loddiges (K-Lindl., holo!)
Philippines, Ryukyu Is., Taiwan – 330–1300 m – E
o Cleisostoma ionosmum Lindl., Bot. Reg. 33: t.41, 1847
Trichoglottis ionosma (Lindl.) J.J.Smith, Nat. Tijdschr. Ned. Ind. 72: 108, 1912

Stelis aprica Lindl., Hook. Comp. Bot. Mag. 2: 353, 1836
BR 11: sub t.935 (1825) / 28: misc.12 (1842)
W.Indies, Mexico, Guatemala, Costa Rica, Panama, Venezuela, Colombia, Ecuador, Peru, Bolivia, Guyana, Brazil – E
Stelis catharinensis Lindl., Hook. Comp. Bot. Mag. 2: 353, 1836
Stelis crassifolia Lindl., Bot. Reg. 28: misc.12, 1842
Stelis minutiflora Rchb.f. ex Hoffmsgg., Linnaea 16: 237, 1842
Stelis miersii Lindl., Ann. & Mag. Nat. Hist. 12: 397, 1843
Stelis microglossa Rchb.f., Linnaea 22: 821, 1849
Stelis gutturosa Rchb.f., Bonpl. 2: 23, 1854
Stelis micrantha Barb. Rodr., Gen. & Sp. Orch. Nov. 2: 93, 1881
Stelis rodriguesii Cogn., Mart. Fl. Bras. 3 (4): 360, 1896
Stelis domingensis Cogn., Urb. Symb. Antill. 6: 692, 1910
Stelis schomburgkii Fawcett & Rendle, J. Bot. 48: 108, 1910
Stelis hymenantha Schltr., Fedde Rep. 10: 291, 1912
Stelis herzogii Schltr., Fedde Rep. 12: 484, 1913
Stelis tippenhaueri Urban, Fedde Rep. 15: 103, 1917
Stelis desportesii Urban, Fedde Rep. 15: 156, 1918
Stelis cubensis Schltr., Urb. Symb. Antill 9: 60, 1923
Stelis cuspidilabia Schltr., Fedde Rep. Beih. 19: 17, 1923
Stelis seleniglossa Schltr., Fedde Rep. Beih. 19: 97, 1923

Stelis argentata Lindl., Bot. Reg. 28: misc.78, 1842
BR 28: misc.78 (1842)
Type: Guyana, Schomburgk 427 cult. Loddiges 75 (K-Lindl., holo!)
Mexico to Panama, Brazil, Venezuela, Ecuador, Colombia, Peru, Amazonas – E
Stelis heylidiana Focke, Tijdschr. Nat. Wetensch. 2: 201, 1849
Stelis superbiens Lindl., Fol. Orch. Stelis 8, 1858
Stelis leucopogon Rchb.f., Beitr. Orch. Centr. Am. 95: t.9 Figs. I 1–4, 1866
Stelis endresii Rchb.f., Gard. Chron.: 1373, 1870
Stelis littoralis Barb.Rodr., Gen. & Sp. Orch. Nov. 2: 85, 1882
Stelis yauaperyensis Barb.Rodr., Vellosia ed.2, 1: 120, 1891
Stelis parvibracteata Ames, Orch. 7: 131, 1922
Stelis huebneri Schltr., Beih. Bot. Zentralbl. 42 (2): 88, 1925

Stelis tristyla Lindl., Bot. Reg. 24: misc.69, 1838
BR 24: misc.69 (1838)
Type: Brazil, cult. Loddiges (K-Lindl., holo!)
Brazil – E

o **Stenia pallida** Lindl., Bot. Reg. 23: sub t.1991, 1837
BR 23: sub t.1991 (1837) / 24: t.20 (1838)
Type: Guyana, Huntley s.n. cult. Bateman (K-Lindl., holo!)
Guyana, Trinidad, Venezuela, Colombia, Peru – E

Tainia bicornis (Lindl.) Rchb.f., Bonpl. 5: 54, 1857
BR 28: misc.31 (1842) / 30: t.8 (1844)
Type: Sri Lanka, cult. Clowes (K-Lindl., holo!)
Sri Lanka, S.India, Java – T
o Ania bicornis Lindl., Bot. Reg. 28: misc.31, 1842
Ania latifolia Wight, Ic. Pl. Ind. Or. 3 (2): 10 pl.914, 1844
Mitopetalum bicorne (Lindl.) Blume, Ann. Mus. Bot. Lugd. Bat. 2: 185, 1856
Mitopetalum wightii Blume, Ann. Mus. Bot. Lugd. Bat. 2: 185, 1856
Eria bicornis (Lindl.) Rchb.f., Walp. Ann. 6: 269, 1861
Eria wightii (Blume) Rchb.f., Walp. Ann. 6: 270, 1861

Telipogon obovatus Lindl., Bot. Reg. 33: sub t.27, 1847
BR 33: sub t.27 (1847)
Type: Peru, Lobb s.n. cult. Veitch (K-Lindl., holo!)
Peru, Ecuador, Colombia – E

Thrixspermum calceolus (Lindl.) Rchb.f., Xen. Orch. 2: 122, 1867
BR 32: t.19 (1846)
Type: Philippines, Cuming s.n. cult. Loddiges (K-Lindl., holo!)
Thailand, Malaya, Sumatra, Borneo – E, L
o Sarcochilus calceolus Lindl., Bot. Reg. 32: t.19, 1846
Aerides lobbii Teijsm. & Binnend., Nat. Tijdschr. Ned. Ind. 24: 323, 1862
Sarcochilus brachystachys Hook.f., Fl. Brit. Ind. 6: 41, 1890
Thrixspermum brachystachyum (Hook.f.) Kuntze, Rev. Gen. 2: 682, 1891

Thrixspermum centipeda Lour., Fl. Cochinch.: 520, 1790
BR 30: misc.62:52 (1844)
Deccan, Sikkim, Bhutan, N.E.India, Burma, Thailand, Cambodia, Vietnam, China, Malaya, Sumatra, Java, Celebes, Borneo, ?Philippines, ?Australia – E
Epidendrum thrixspermum Raeusch., Nom.: 265, 1797
Dendrocolla arachnites Blume, Bijdr.: 287 fig.67, 1825
Dendrobium auriferum Lindl., Gen. Sp. Orch. Pl.: 83, 1830
Liparis serraeformis Lindl., Gen. Sp. Orch. Pl.: 33, 1830
Aerides arachnites (Blume) Lindl., Gen. Sp. Orch. Pl.: 238, 1833
Sarcochilus auriferus (Lindl.) Rchb.f., Walp. Ann. 6: 498, 1863
Sarcochilus serraeformis (Lindl.) Rchb.f., Walp. Ann. 6: 498, 1863
Thrixspermum arachnites (Blume) Rchb.f., Xen. Orch. 2: 121, 1867
Thrixspermum auriferum (Lindl.) Rchb.f., Xen. Orch. 2: 121, 1867
Thrixspermum serraeformis (Lindl.) Rchb.f., Xen. Orch. 2: 121, 1867
Sarcochilus centipeda (Lour.) Naves, Nov. App. Fil.: 238, 1880
Sarcochilus hainanensis Rolfe, Kew Bull.: 199, 1896
Sarcochilus platystachys Bailey, Syn. Queensl. Fl. Suppl. 1: 56, 1899
Thrixspermum hainanensis (Rolfe) Schltr., Orchis 5: 55, 1911
Thrixspermum platystachys (Bailey) Schltr., Orchis 5: 55, 1911
Thrixspermum arachnites var. minor Ridley, Fl. Mal. Pen. 4: 185, 1924

Thunia bracteata (Roxb.) Schltr., Fedde Rep. Beih. 4: 205, 1919
BR 24: t.33 (1838)
India to Burma, Thailand, China – 1200–200 m – T
Limodorum bracteatum Roxb., Hort. Beng.: 63, 1814 (nomen nudum)
o Phaius albus auct., non Lindl.; Lindl., Gen. Sp. Orch. Pl.: 128 p.p., 1831
Thunia venosa Rolfe, Orch. Rev. 13; 206, 1905
Thunia alba Mehra & Kasyap, Taxon 4: 226, 1978 (non (Lindl.) Rchb.f.)

Trichocentrum candidum Lindl., Bot. Reg. 29: misc.18, 1843
BR 29: misc.18 (1843)
Type: Guatemala, without collector (K-Lindl., holo!)
Mexico, Guatemala – up to 1000 m – E

o **Trichocentrum fuscum** Lindl., Bot. Reg. 23: t.1951, 1837
BR 23: t. & sub t.1951 (1837)
Type: Mexico, cult. Knight (K-Lindl., icono!)
Mexico, Brazil – E
Acoidium fuscum Lindl., Bot. Reg. 23: sub t.1951, 1837

Trichocentrum recurvum Lindl., Bot. Reg. 29: misc.17, 1843
BR 29: misc.17 (1843)
Type: Guyana, cult. Loddiges (K-Lindl., holo!)
Brazil – E

Trichoglottis rosea (Lindl.) Ames, Merrill Enum. Phil. Fl. Pl. 1: 440, 1925
BR 24: misc.150 (1838)
Type: Philippines, Cuming s.n. cult. Loddiges (K-Lindl., holo!)
Philippines, Taiwan – E
Cleisostoma rosea Lindl., Bot. Reg. 24: misc.150, 1838
Trichoglottis flexuosa Rolfe ex Ames, Orchid. 1: 107, 1905
Pomatocalpa roseum (Lindl.) J.J.Smith, Nat. Tijdschr. Ned. Ind. 72: 103, 1912

Trichopilia fragrans (Lindl.) Rchb.f., Hamb. Gartenz. 14: 229, 1858
BR 30: misc.74:2 (1844)
Type: Colombia, Popayan, Hartweg s.n. (K-Lindl., holo!)
W.Indies, Venezuela, Colombia, Ecuador, Peru, Bolivia – E
Pilumna fragrans Lindl., Bot. Reg. 30: misc.74:2, 1844
Trichopilia candida Linden ex Lindl., Orch. Linden.: 13, 1846
Pilumna wageneri Rchb.f., Bonpl. 2: 15, 1854
Trichopilia wageneri (Rchb.f.) Rchb.f., Hamb. Gartenz. 14: 229, 1858
Trichopilia lehmanni Regel, Gartenfl. 37: 355 t.1276, 1888

Trichopilia laxa (Lindl.) Rchb.f., Hamb. Gartenz. 14: 229, 1858
BR 30: misc.74:1 (1844) / 32: t.57 (1846)

Type: Colombia, Hartweg s.n. cult. Hort. Soc. (K-Lindl., holo!)
Venezuela, Colombia – E
o Pilumna laxa Lindl., Bot. Reg. 32: t.57, 1846
Pilumna reichenheimiana Klotzsch, Allg. Gartenz. 23: 353, 1855
Trichopilia laxa var. hartwegii Rchb.f., Xen. Orch. 2: 101, 1867
Trichopilia laxa var. flaveola Rchb.f., Gard. Chron. n.s.22: 520, 1884

o **Trichopilia tortilis** Lindl., Bot. Reg. 22: t.1863, 1836
BR 22: t.1863 (1836)
Type: Mexico, cult. G.Barker (K-Lindl., holo!)
Mexico, Guatemala, Honduras, El Salvador – up to 1500 m – E

Trichosalpinx arbuscula (Lindl.) Luer, Phytolog. 55: 394, 1983
BR 28: misc.p.72:32 (1842)
Type: Peru, Loxa, Hartweg s.n. (K-Lindl., holo!)
Venezuela, Colombia, Ecuador, Peru – E
Pleurothallis arbuscula Lindl., Bot. Reg. 28: misc.p.72:32, 1842
Pleurothallis dinotherii Rchb.f. & Warsc., Bonpl. 2: 114, 1854
Pleurothallis diptera Lindl., Fol. Orch. Pleuroth.: 44, 1859
Humboldtia arbuscula (Lindl.) Kuntze, Rev. Gen. 2: 667, 1891
Humboldtia dinotherii (Rchb.f. & Warsc.) Kuntze, Rev. Gen. 2: 667, 1891
Humboldtia diptera (Lindl.) Kuntze, Rev. Gen. 2: 667, 1891
Pleurothallis tricaudata Schltr., Fedde Rep. Beih. 9: 77, 1921

Trichosalpinx ciliaris (Lindl.) Luer, Phytolog. 55 (5): 395, 1983
BR 24: misc.40 (1838) / 26: misc.40 (1840) / 28: misc.p.77:64 (1842)
Type: Mexico, cult. Loddiges (K-Lindl., holo!)
Mexico, Guatemala, Honduras, Nicaragua, Costa Rica, S.America – E
Pleurothallis villosa Knowles & Westc., Fl. Cab. 2: 78, 1838
Specklinia ciliaris Lindl., Bot. Reg. 24: misc.40, 1838
Pleurothallis lepanthiformis Rchb.f., Linnaea 18: 398, 1844
Humboldtia lepanthiformis (Rchb.f.) Kuntze, Rev. Gen. 2: 667, 1891
Humboldtia villosa (Knowles & Westc.) Kuntze, Rev. Gen. 2: 608, 1891
Pleurothallis purpusii Schltr., Orchis 9: 49 t.4 figs 1–6, 1915
Pleurothallis brevis Schltr., Fedde Rep. Beih. 19: 183, 1923
Pleurothallis gnomonifera Ames, Sched. Orch. 6: 61, 1923
Pleurothallis ciliaris (Lindl.) L.O.Williams, Caldasia 5: 14, 1942

Trichosalpinx orbicularis (Lindl.) Luer, Phytolog. 55: 396, 1983
BR 24: misc.41 (1838) / 28: misc.p.79:81 (1842)
Type: Guyana, cult. Loddiges (K-Lindl., holo!)
Costa Rica, Trinidad, Venezuela, Colombia, Guyana, Surinam – E
Specklinia orbicularis Lindl., Bot. Reg. 24: misc.41, 1838
Pleurothallis orbicularis (Lindl.) Lindl., Bot. Reg. 28: misc.p.79:81, 1842
Pleurothallis biflora Focke, Tijdschr. Nat. Wetensch. 2: 197, 1849
Humboldtia biflora (Focke) Kuntze, Rev. Gen. 2: 667, 1891
Humboldtia orbicularis (Lindl.) Kuntze, Rev. Gen. 2: 668, 1891
Pleurothallis trachytheca Lehm. & Kraenzlin, Engl. Bot. Jahrb. 26: 444, 1899

Trichosalpinx punctatifolia (Barb.Rodr.) Luer, Phytolog. 54: 397, 1983
BR 21: sub t.1797:11 (1835) / 28: misc.p.83:109 (1842)
Brazil – E
Pleurothallis punctata Lindl., Bot. Reg. 21: sub t.1797, 1835 (non Lindl.1823)
Lepanthes punctatifolia Barb.Rodr., Gen. & Sp. Orch. Nov. 2: 55, 1882
Humboldtia punctata (Lindl.) Kuntze, Rev. Gen. 2: 668, 1891
Pleurothallis punctatifolia (Barb.Rodr.) Pabst, Orquidea 28: 227, 1966

Trichosalpinx pusilla (Kunth) Luer, Phytolog. 55: 397, 1983
BR 21: sub t.1797 (1835) / 28: misc.p.82:104 (1842)
Peru, Ecuador – E
Dendrobium pusillum Kunth, Nov. Gen. & Sp. Pl. 1: 357, 1816
Specklinia pusilla (Kunth) Lindl., Bot. Reg. 21: sub t.1797, 1835
Pleurothallis pusilla (Kunth) Lindl., Bot. Reg. 28: misc.p.82:104, 1842
Humboldtia pusilla (Kunth) Kuntze, Rev. Gen. 2: 668, 1891

Trichotosia mucronata (Lindl.) Kraenzlin, Engl. Pflanzenr. Orch. Mon. Dendr. 2: 147, 1911
BR 28: misc.27 (1842)
Type: Philippines, Cuming s.n. (K-Lindl., holo!)
Philippines – E
Eria mucronata Lindl., Bot. Reg. 28: misc.27, 1842
Cylindrolobus mucronatus (Lindl.) Rauschert, Fedde Rep. 94 (7–8): 445, 1983

Trichotosia velutina (Lodd. ex Lindl.) Kraenzlin, Pflanzenr. Orch. Mon. Dendr. 50: 140, 1911
BR 26: misc.209 (1840)
Type: Singapore, Cuming s.n. cult. Loddiges (K-Lindl., holo!)
Burma, Thailand, Vietnam, Malaya, Sumatra, Borneo, New Guinea – E
Eria velutina Lodd. ex Lindl., Bot. Reg. 26: misc.209, 1840
Eria vestita Parish, Mason Burma: 195, 1883 (non Lindl.)
Eria monticola var. hirsuta Hook.f., Fl. Brit. Ind. 5: 806, 1890
Pinalia velutina (Lodd. ex Lindl.) Kuntze, Rev. Gen. 2: 679, 1891
Dendrobium sessile Gagnepain, Bull. Mus. Hist. Nat. Paris s.2, 22 (3): 396, 1950

Trichotosia vestita (Lindl.) Kraenzlin, Pflanzenr. Heft 50: 151, 1911
BR 30: misc.79 (1844) / 31: t.2 (1845)
Type: Wallich 2005, cult. Loddiges (K-holo!)
India, Malaya, Sumatra, Borneo, Philippines –E
Dendrobium vestitum Wallich, Cat.: 2005, 1829
o Eria vestita Lindl., Bot. Reg. 30: misc.79, 1844
Trichotosia vestita (Lindl.) Kraenzlin, Engl. Pflanzenr. Heft 50: 15, 1911

Tridactyle armeniaca (Lindl.) Schltr., Beih. Bot. Centralbl. 36 (2): 143, 1918
BR 25: misc.109 (1839)
Type: Sierra Leone, cult. Loddiges (K-Lindl., holo!)
Guinea, Sierra Leone, Liberia, Ghana – E
Angraecum armeniacum Lindl., Bot. Reg. 25: misc.109, 1839
Angraecum whitfieldii Rendle, J. Bot. 33: 249, 1895

Trigonidium acuminatum Bateman ex Lindl., Bot. Reg. 24: misc.136, 1838
BR 24: misc.136 (1838) / 25: misc.59 (1839)
Type: Guyana, Colley s.n. (K-Lindl., holo!)
Venezuela, Colombia, Peru, Guyana, Surinam, Brazil – E
Trigonidium tenue Lodd. ex Lindl., Bot. Reg. 25: misc.59, 1839
Trigonidium subrepens Rolfe, Kew Bull.: 377, 1906
Trigonidium peruvianum Schltr., Fedde Rep. Beih. 9: 107, 1921

Trigonidium egertonianum Bateman ex Lindl., Bot. Reg. 24: misc.135, 1838
BR 24: misc.135 (1838) / 26: misc.100 (1840)
Type: Honduras, Skinner s.n. (K-Lindl., holo!)
Mexico, Guatemala, Honduras, El Salvador, Costa Rica, Panama, Venezuela, Colombia, Ecuador – E
Maxillaria brachyglossa A.Rich. & Galeotti, Ann. Sc. Nat. s.3, 3: 24, 1845
Trigonidium seemanni Rchb.f., Bot. Voy. Herald: 214, 1854
Trigonidium spathulatum Linden & Rchb.f., Bonpl. 2: 380, 1854
Trigonidium callistele Rchb.f., Hamb. Gartenz. 16: 178, 1860
Trigonidium brachyglossum (A.Rich. & Galeotti) Schltr., Beih. Bot. Centralbl. 36 (2): 499, 1918

Trigonidium latifolium Lindl., Bot. Reg. 23: sub t.1923, 1837
BR 23: sub t.1923 (1837)
Type: Brazil, Descourtilz s.n. (K-Lindl., icono!)
Brazil – E

o **Trigonidium obtusum** Lindl., Bot. Reg. 23: t.1923, 1837
BR 23: t.1923 (1837)
Type: Guyana, Colley s.n. cult. Bateman (K-Lindl., holo!)
Brazil, Guyana, Venezuela, Colombia, Bolivia – E

Triphora trianthophora (Sw.) Rydb., Britton Man: 298, 1901
BR 11: t.908 (1825)
N.America – T
Arethusa trianthophoros Sw., Svensk Vet. Akad. Nya Handl. 21: 230, 1800
Arethusa parviflora Michaux, Fl. Bor.-Am. 2: 160, 1803
Arethusa pendula Muhlenb. ex Willd., Sp. Pl. ed.4: 82, 1805
Triphora pendula (Mulhenb.ex Willd.) Nuttall., Gen. 2: 193, 1818
o Pogonia pendula (Mulhenb. ex Willd.) Lindl., Bot. Reg. 11: t.908, 1825
Pogonia trianthophorus (Sw.) Britton, Sterns & Poggenberg, Prelim. Cat. N.Y.: 52, 1888
Triphora trianthophora var. schaffneri Camp., Rhodora 42: 55, 1940

Tropidia angulosa (Lindl.) Blume, Coll. Orch. Arch. Ind.: 122, 1858
BR 19: sub t.1618 (1833)
Type: Wallich 7388 (K-Lindl., holo!)
Deccan, Sikkim, Assam, Burma, Thailand, Malaya, Sumatra, Java – T
Cnemidia angulosa Lindl., Bot. Reg. 19: sub t.1618, 1833
Cnemidia semilibera Lindl., Gen. Sp. Orch. Pl.: 463, 1840
Govindovia nervosa Wight, Ic. 6: t.2090, 1853
Tropidia govindivii Blume, Coll. Orch. Arc. Ind.: 122, 1858
Tropidia semilibera (Lindl.) Blume, Coll. Orch. Arc. Ind.: 122, 1858
Tropidia bellii Blume & MacCann, J. Bomb. Nat. Hist. Soc. 35: 736, 1932

Trudelia alpina (Lindl.) Garay, Orch. Digest 50: 76, 1986
BR 24: misc.101 (1838)
Type: Khasia, Gibson s.n. cult. Duke of Devonshire (K-Lindl., holo!)
Himalaya, Bhutan, Sikkim, N.E.India – E
Luisia alpina Lindl., Bot. Reg. 24: misc.101, 1838
Vanda alpina (Lindl.) Lindl., Fol. Orch. Vanda: 4, 1853

Trudelia cristata (Lindl.) Senghas, Schlechter Die Orchideen ed.3, 1: 1211, 1988
BR 28: t.48 & misc.25 (1842)
Tibet, Nepal, Sikkim, Bhutan – 800–2000 m – E

Aerides cristata Wallich ex Lindl., Gen. Sp. Orch. Pl.: 216, 1832
○ Vanda cristata Lindl., Gen. Sp. Orch. Pl.: 216, 1833
Vanda alpina Tuyama, Hara Fl. E. Himalaya: 452, 1966 (non Lindl.)

Vanda coerulea Griffith ex Lindl., Bot. Reg. 33: sub t.30, 1847
BR 33: sub t.30 (1847)
Type: India, Griffith s.n. (K-Lindl., holo!)
N.E.India, Burma, Thailand, China – E

Vanda concolor Blume, Rumphia 4: 49, 1848
BR 30: misc.42 (1844)
India, Thailand, China – E
Angraecum furvum (Rumph.) L., Herb. Amb. 6: t.46 fig.J, 1747
Epidendrum furvum (Rumph.) L., Sp. Pl.: 1348, 1753
Cymbidium furvum (Rumph.) Willd., Sp. Pl. ed.4: 103, 1805
Vanda roxburghii var. unicolor Hook., Bot. Mag. 62: t.3416, 1835
Vanda furva (Rumph.) Lindl., Bot. Reg. 30: misc.42, 1844 (non Blume)

Vanda lamellata Lindl., Bot. Reg. 24: misc.125, 1838
BR 24: misc.125 (1838)
Type: Philippines, Manila, cult. Loddiges (K-Lindl., holo!)
Philippines, Mariana Is., Ryukyu Is., Taiwan – E
Vanda unicolor Roxb. ex Steudel, Nom. 2: 744, 1841
Vanda cumingii Lodd., Orch. 40: no.1715, 1844
Vanda clitellaria Rchb.f., Flora: 277, 1872
Vanda lamellata var. boxallii Rchb.f., Gard. Chron. n.s.13: 743, 1880
Vanda boxallii (Rchb.f.) Rchb.f., Gard. Chron. n.s.15: 366, 1881
Vanda boxallii var. cobbiana Rchb.f., Gard. Chron. n.s.16: 780, 1881
Vanda vidalii Boxall ex Naves, Novis App.: 240, 1882
Vanda superba Linden, Lindenia 3: 83, 1888
Vanda lamellata var. cobbiana (Rchb.f.) Ames, Orchid. 2: 241, 1908
Vanda nasugbuana Pars., Orch. Rev. 39: 39, 1931
Vanda yamiensis Masamune & Segawa, Trans. Nat. Hist. Soc. Formos. 24: 212, 1934

Vanda tessellata (Roxb.) Lodd. ex G.Don, Hort. Brit.: 372, 1832
BR 6: t.506 (1820)
Sri Lanka, India, Nepal, Burma, China – E
Epidendrum tessellatum Roxb., Pl. Corom. 1: 34, 1795
Cymbidium tessellatum (Roxb.) Sw., Nov. Act. Upsal. 6: 75, 1799
○ Vanda roxburghii R.Br., Bot. Reg. 6: t.506, 1820
Aerides tessellata (Roxb.) Wight, Wall. Cat.: 7318, 1832
Cymbidium alagnata Ham. ex Wallich, Cat.: 7327, 1832
Cymbidium tessellaoides Roxb., Fl. Ind. 3: 463, 1832
Epidendrum tesselloides (Roxb.) Steudel, Nom. ed.2, 1: 558, 1841
Vanda tesselloides (Roxb.) Rchb.f., Walp. Ann. 6: 864, 1861

Vanda testacea (Lindl.) Rchb.f., Gard. Chron. n.s.8: 166, 1877
BR 30: misc.57 (1844)
Sri Lanka, India, Himalaya (Kumaun to Sikkim), Assam, Burma, Thailand – E
Aerides wightiana Lindl., Wall. Cat.: 7320, 1832
Aerides testacea Lindl., Gen. Sp. Orch. Pl.: 238, 1833
Vanda spathulata Graham, Cat. Bomb. Pl.: 204, 1839 (non (L.) Sprengel)
Vanda parviflora Lindl., Bot. Reg. 30: misc. 57, 1844
Vanda vitellina Kraenzlin, Gard. Chron. s.3, 12: 206, 1892

Vanda tricolor Lindl., Bot. Reg. 33: sub t.59, 1847
BR 33: sub t.59 (1847)
Type: Java, cult. Veitch (K-Lindl., holo!)
Laos, Java, N.Australia – E
Vanda suaveolens Blume, Rumphia 4: 49, 1848
Vanda hindsii Benth., Fl. Austr. 6: 297, 1873 (non Lindl.)

Vandopsis lissochiloides (Gaudich.) Pfitzer, Engl. & Prantl. Nat. Pfanzenf. 2 (6): 210 fig.229, 1889
BR 32: t.59 (1846)
Thailand, Laos, Philippines, Moluccas, Seran, Celebes, Ambon, Bali, Kai I., Buru, New Guinea, – E, T
Fieldia lissochiloides Gaudich., Freyc. Voy. Bot.: 424 t.36, 1827
Vanda lissochiloides (Gaudich.) Lindl., Gen. Sp. Orch. Pl.: 216, 1833
○ Vanda batemanii Lindl., Bot. Reg. 32: t.59, 1846
Grammatophylum pantherinum Zipp. ex Blume, Rumphia 4: 50, 1849
Stauropsis lissochiloides (Gaudich.) Benth. ex Pfitz., Vergl. Morph. Orch.: 14, 1882
Stauropsis batemanii (Lindl.) Nicholson, Dict. Gard. 3: 493, 1886

Vanilla bicolor Lindl., Bot. Reg. 24: misc.58, 1838
BR 24: misc.58 (1838)
Type: Guyana, Schomburgk s.n. cult. Loddiges (K-Lindl., holo!)
Brazil, Guyana – E

Vanilla palmarum Lindl., Gen. Sp. Orch. Pl.: 436, 1840
BR 28: misc.75 (1842)
Venezuela, Colombia, Brazil – E

Warrea warreana (Lodd. ex Lindl.) C.Schweinf., Bot. Mus. Leafl. Harv. Univ. 17: 55, 1955
BR 18: sub t.1549:34 (1832) / 29: misc.p.14:1 (1843) / 30: misc.78 (1844) / 32: t.28:1, 3 (1845)
Venezuela, Colombia, Peru, Ecuador, Brazil – T
Maxillaria warreana Lodd. ex Lindl., Gen. Sp. Orch. Pl.: 148, 1832
Warrea tricolor Lindl., Bot. Reg. 29: misc.p.14:1, 1843
Warrea bidentata Lindl., Bot. Reg. 30: misc.78, 1844
Warrea unijugata Regel, Ind. Sem. Hort. Petrop.: 22, 1855
Aganisia tricolor (Lindl.) Bois, Orch.: 128, 1893 (non Benth. & Hook., non N.E.Br.)
Warrea speciosa Schltr., Fedde Rep. Beih. 9: 98, 1921

Warreella cyanea (Lindl.) Schltr., Die Orchideen: 425, 1914
BR 30: misc.3 (1844) / 31: t.28 & sub t.28:2 (1845)
Type: cult. Loddiges (K-Lindl., holo!)
Venezuela, Colombia – T
○ Warrea cyanea Lindl., Bot. Reg. 30: misc.3, 1844
Maxillaria cyanea (Lindl.) Beer, Prakt. Stud. Orch.: 264, 1854
Warrea cinerea Benth., J. Linn. Soc. 18: 321, 1881
Warrea cyanea var. alba Rchb.f., Gard. Chron. n.s.23: 692, 1885
Aganisia cyanea (Lindl.) Benth. ex Veitch, Man. Orch. Pl. 2: 68, 1887
Warrea medellinensis Kraenzlin, Notizbl. Bot. Mus. Berlin 7: 430, 1920
Warreella medellinensis (Kraenzlin) Garay, Svensk. Bot. Tidskr. 47: 225, 1953

Xylobium bractescens (Lindl.) Kraenzlin, Orchis 7: 22, 1913
BR 28: misc.92 (1842)
Type: Peru, Loxa, Hartweg s.n. cult. Hort. Soc. (K-Lindl., holo!)
Amazonas, Peru – E
Maxillaria bractescens Lindl., Bot. Reg. 28: misc.92, 1842

Xylobium colleyi (Bateman ex Lindl.) Rolfe, Gard. Chron. s.3, 7: 288, 1890
BR 24: misc.161 (1838)
Type: cult. Bateman (K-Lindl., holo!)
Trinidad, Venezuela, Guyana, Brazil – E
Maxillaria colleyi Bateman ex Lindl., Bot. Reg. 24: misc.161, 1838
Maxillaria rebellis Rchb.f., Fl. d. Serres 9: 102, 1853
Xylobium brachystachyum Kraenzlin, Gard. Chron. s.3, 40: 302, 1906
Xylobium rebellis (Rchb.f.) Schltr., Orchis 7: 23, 1913

Xylobium corrugatum (Lindl.) Rolfe, Gard. Chron. s.3, 5: 459, 1889
BR 30: misc.14 (1844)
Type: Columbia, Linden s.n. cult. Barker (K-Lindl., holo!)
Venezuela, Colombia – E
Maxillaria corrugata Lindl., Bot. Reg. 30: misc.14, 1844
Maxillaria wageneri Rchb.f., Bot. Zeit. 10: 734, 1852
Maxillaria corrugata var. wageneri (Lindl.) Rchb.f., Walp. Ann. 6: 508, 1863
Xylobium corrugatum var. wageneri (Rchb.f.) Schltr., Orchis 7: 22, 1913
Xylobium wageneri (Rchb.f.) Schltr., Fedde Rep. Beih. 6: 85, 1919

Xylobium foveatum (Lindl.) Nicholson, Dict. Gard. 4: 225, 1887
BR 25: misc.2 (1839) / 30: misc.12 (1844)
Type: Guyana, cult. Loddiges (K-Lindl., holo!)
Mexico, Guatemala, Nicaragua, Costa Rica, Panama, Venezuela, Ecuador – E
Maxillaria foveata Lindl., Bot. Reg. 25: misc.2, 1839
Maxillaria concava Lindl., Bot. Reg. 30: misc.12, 1844
Maxillaria stachyobiorum Rchb.f., Bot. Zeit. 10: 735, 1852
Xylobium concavum (Lindl.) Hemsl., Godm. & Salv. Biol. Cent. Am. Bot. 3: 252, 1883
Xylobium stachyobiorum (Rchb.f.) Hemsl., Godm. & Salv. Biol. Cent. Am. Bot. 3: 252, 1883
Xylobium ecuadorense Rolfe, Kew Bull.: 341, 1913

Xylobium pallidiflorum (Hook.) Nicholson, Dict. Gard. 4: 225, 1887
BR 18: sub t.1549: 30 (1835)
W.Indies, Nicaragua, Venezuela, Trinidad, Brazil, Colombia, Ecuador, Peru, Bolivia – E
Maxillaria pallidiflora Hook. Bot. Mag. 25: t.2806, 1828
Maxillaria stenobulbon Klotzsch, Ind. Sem. Hort. Berol.: 11, 1853
Xylobium gracile Schltr., Fedde Rep. Beih. 8: 92, 1921

Xylobium palmifolium (Sw.) Benth. ex Fawcett, Fl. Pl. Jam.: 39, 1893
BR 11: sub t.897 (1825) / 18: t.1549 & sub t. 1549:33, 35 (1832)
Trinidad, Jamaica – E, L
Epidendrum palmifolium Sw., Prodr. Veg. Ind. Occ.: 123, 1788
Dendrobium palmifolium (Sw.) Sw., Nov. Act. Upsal. 6: 82, 1799
Colax palmifolius (Sw.) Lindl., Bot. Reg. 11: sub t.897, 1825
○ Maxillaria decolor Lindl., Bot. Reg. 18: t.1549, 1832
Maxillaria palmifolia (Sw.) Lindl., Gen. Sp. Orch. Pl.: 148, 1832
Xylobium decolor (Lindl.) Nicholson, Dict. Gard. 4: 223, 1887

Xylobium variegatum (Ruiz Lopez & Pavon) Garay & Dunsterv., Venezuel. Orch. Ill. 2: 342, 1961
BR 9: t.732 (1823) / 11: sub t.897 (1825) / 18: sub t.1549:2, 36 (1832) / 30: misc.66 (1844)
Costa Rica, Venezuela, Colombia, Ecuador, Peru, Bolivia, Brazil – E

Maxillaria variegata Ruiz Lopez & Pavon, Syst. Veg. Fl. Peruv. & Chil. 1: 222, 1798
Dendrobium variegatum (Ruiz Lopez & Pavon) Pers., Syn. Pl. 2: 524, 1807
o Dendrobium squalens Lindl., Bot. Reg. 9: t.732, 1823
Xylobium squalens (Lindl.) Lindl., Bot. Reg. 11: sub t.897, 1825
Maxillaria squalens (Lindl.) Hook., Bot. Mag. 56: t.2955, 1829
Dendrobium carnosum C.Presl., Symb. Bot. 35: t.24, 1832
Cyrtopera scabrilinguis Lindl., Gen. Sp. Orch. Pl.: 189, 1833
Maxillaria supina Poeppig & Endl., Nov. Gen. Sp. Pl. 1: 39, 1836
Xylobium houtei Makoi ex Mutel, Mém. Plus. Orch. 1: 16, 1840
Maxillaria scabrilinguis (Lindl.) Lindl., Bot. Reg. 30: misc.66, 1844
Xylobium dusenii Kraenzlin, Svensk Vet. Akad. Handl. 46 (10): 65, 1911
Xylobium scabrilingue (Lindl.) Schltr., Orchis 7: 23, 1913
Xylobium supinum (Poeppig & Endl.) Schltr., Orchis 7: 24, 1913
Xylobium carnosum (C.Presl) Schltr., Fedde Rep. Beih. 9: 160, 1921

Zeuxine strateumatica (L.) Schltr., Engl. Bot. Jahrb. 45: 394, 1911
BR 10: sub t.823:16 (1824)
Afghanistan to Japan, S.New Guinea, Sri Lanka – T
Orchis strateumatica L., Sp. Pl.: 943, 1753
Pterygodium sulcatum Roxb., Hort. Beng.: 63, 1814
Spiranthes strateumatica (L.) Lindl., Bot. Reg. 10: sub t.823:16, 1824
Adenostylis emarginata Blume, Bijdr.: 414, 1825
Adenostylis integerrima Blume, Bijdr.: 414, 1825
Strateuma zeylandica Raf., Fl. Tellur 2: 89, 1837
Tripleura pallida Lindl., Gen. Sp. Orch. Pl.: 452, 1840
Zeuxine emarginata (Blume) Lindl., Gen. Sp. Orch. Pl.: 485, 1840
Zeuxine integerrima (Blume) Lindl., Gen. Sp. Orch. Pl.: 485, 1840
Zeuxine membranacea Lindl., Gen. Sp. Orch. Pl.: 485, 1840
Zeuxine sulcata (Roxb.) Lindl., Gen. Sp. Orch. Pl.: 485, 1840
Neottia strateumatica (L.) R.Br. ex Steudel, Nom. ed.2, 2: 190, 1841
Zeuxine bracteata Wight, Ic. 5: t.1724, 1851
Zeuxine brevifolia Wight, Ic. 5: t.1725, 1851
Zeuxine robusta Wight, Ic. 5: t.1726, 1851
Zeuxine tripleura Lindl., J. Linn. Soc. 1: 186, 1857
Zeuxine procumbens Blume, Coll. Orch. Arch. Ind.: 68 t.22 & 23, 1858
Adenostylis strateumatica (L.) Ames, Orchid. 2: 59, 1908
Zeuxine wariana Schltr., Fedde Rep. Beih. 1: 77, 1911
Adenostylis sulcata (Roxb.) Hayata, Ic. Pl. Formosa Suppl. 6: 75, 1917
Zeuxine bonii Gagnepain, Bull. Mus. Paris s.2, 3: 326, 1931
Zeuxine evrardii Gagnepain, Bull. Mus. Paris. s.2, 3: 326, 1931

Zootrophion atropurpureum (Lindl.) Luer, Selbyana 7: 80, 1982
BR 21: sub t.1797 (1835) / 28: misc.p.81:93 (1842)
Type: Jamaica, cult. Liverpool B.G. (K, holo!)
Cuba, Jamaica, Panama, Colombia, Ecuador, Peru, Bolivia – E
Specklinia atropurpurea Lindl., Bot. Reg. 21: sub t.1797, 1835
Pleurothallis atropurpurea (Lindl.) Lindl., Bot. Reg. 28: misc.p.81:93, 1842
Masdevallia fenestrata Lindl. ex Hook., Bot. Mag. 71: t.4164, 1845
Cryptophoranthus atropurpureus (Lindl.) Rolfe, Gard. Chron. s.3, 2: 693, 1887
Cryptophoranthus alvaroi Garay, Orquideologia 5: 15, 1970

Zygopetalum brachypetalum Lindl., Bot. Reg. 30: misc.15:3, 1844
BR 30: misc.15:3 (1844)
Type: Brazil, cult. Waterhouse (K-Lindl., holo!)
Brazil – E

Zygopetalum crinitum Lodd., Bot. Cab. 17: t.1637, 1830
BR 30: misc.15:4, 5 (1844)
Brazil – E, T
Zygopetalum mackayi var. crinitum (Lodd.) Lindl., Gen. Sp. Orch. Pl.: 187, 1833
Zygopetalum stenochilum Lodd., Bot. Cab. 20: t.1923, 1833
Zygopetalum pubescens Hoffsgg., Verz. Orch.: 63, 1843
Zygopetalum microtos Hoffsgg. ex Rchb.f., Walp. Ann. 6: 661, 1863
Zygopetalum crepeauxii Carrière, Rev. Hort. 59: 420, 1887
Zygopetalum claesianum hort., Orch. Rev. 2: 62, 1894

Zygopetalum intermedium Lodd., Cat. Orch.: 19, 1842
BR 30: misc.15:2 (1844)
Brazil – E
Zygopetalum velutinum Hoffmsgg., Verz. Orch.: 62, 1843
Zygopetalum rivierii Carrière, Rev. Hort. 45: 191, 1873
Zygopetalum roezlii Rchb.f., Gard. Chron. n.s.7: 620, 1877

Zygopetalum mackaii Hook., Bot. Mag. 54: t.2748, 1827
BR 17: t.1433 (1831) / 30: misc.15:1 (1844)
Brazil – 1300–1700 m – E
o Eulophia mackaiana Lindl., Bot. Reg. 17: t.1433, 1831

Zygopetalum maxillare Lodd., Bot. Cab. 18: t.1776, 1831
BR 30: misc.15:6 (1844)
Brazil, Argentina, Paraguay – E
Zygopetalum mandibulare Rchb.f., Gard. Chron. n.s.7: 684, 1877
Zygopetalum hasslerianum Kraenzlin, Engl. Bot. Jahrb. 37: 389, 1906

Zygosepalum labiosum (L.C.Rich.) C.Schweinf., Mem. N.Y. Bot. Gard. 14 (3): 184, 1967
BR 30: misc.15:10 (1844)
Venezuela, Guyana, Brazil, Amazonas – E
Epidendrum labiosum L.C.Rich., Act. Soc. Hist. Nat. Paris 1: 112, 1792
Zygopetalum rostratum Hook., Bot. Mag. 55: t.2819, 1828
Menadenium rostratum (Hook.) Raf., Fl. Tellur. 4: 45, 1836
Zygosepalum rostratum (Hook.) Rchb.f., Ned. Kruidk. Arch. 4: 330, 1859
Menadenium Labiosum (L.C.Rich.) Cogn., Mart. Fl. Bras. 3 (5): 582, 1902.
Zygopetalum labiosum (L.C.Rich.) L.O.Williams, Caldasia 1 (5): 16, 1942

Zygostates cornuta Lindl., Bot. Reg. 23: sub t.1927, 1837
BR 23: sub t.1927 (1837)
Type: Brazil, Prince Maximilian, Herb. Martius s.n. (K-Lindl., holo!)
Brazil – E
Epidendrum trilabiatum Vell., Fl. Flum. Ic. 9: 367 t.42, 1827
Dactylostylis fimbriata Scheidw., Allg. Gartenz. 7: 405, 1839
Ornithocephalus navicularis Barb.Rodr., Gen. & Sp. Orch. Nov. 1: 135, 1877

Zygostates lunata Lindl., Bot. Reg. 23: sub t.1927, 1837
BR 23: sub t.1927 (1837)
Type: Brazil, Descourtilz, Herb. Martius s.n. (K-Lindl., holo!)
Brazil – E

Index to scientific names

Plants whose names are flagged by asterisks (*) are depicted in this volume: plants flagged by bullets (•) are described or cited in the "Texts" volume. Further information concerning accepted names and synomyms is given in the catalogue.

INDEX

by Samuel Sprunger

Angorchis
– falcata = Neofinetia falcata
– funalis = Dendrophyllax funalis
– gladiifolia = Angraecum mauritianum
– sesquipedalis = Angraecum sesquipedale
Angraecopsis
– falcata = Neofinetia falcata
– **parviflora**
Angraecum
– apiculatum = Aerangis biloba
• – armeniacum = Tridactyle armeniaca
• – ashantense = Diaphananthe bidens
– bakeri = Diaphananthe bidens
– bidens = Diaphananthe bidens
• – bilobum = Aerangis biloba
• – brevifolium = Campylocentrum micranthum
– brongniartianum = Angraecum eburneum
– campyloplectron = Aerangis biloba
• – **carpophorum**
• – caudatum = Plectrelminthus caudatus
• – **caulescens**
– comorense = Angraecum eburneum
– crumenatum = Dendrobium crumenatum
∗ • – **distichum**
∗ • – **eburneum**
– eburneum var. virens = Angraecum eburneum
• – elatum = Cryptopus elatus
– falcatum = Neofinetia falcata
• – fasciola = Campylocentrum fasciola
– furvum = Vanda concolor
• – gladiifolium = Angraecum mauritianum
• – gracile = Chamaeangis gracilis
– gyriamae = Angraecum eburneum
– jamaicense = Campylocentrum micranthum
– lansbergii = Campylocentrum micranthum
• – maculatum = Oeceoclades maculata
∗ – **mauritianum**
• – micranthum = Campylocentrum micranthum
• – ornithorrhynchum = Campylocentrum ornithorrhynchum
• – parviflorum = Angraecopsis parviflora
• – pellucidum = Diaphananthe pellucida
– polystachyum = Campylocentrum polystachyum
– pugioniforme = Cleisostoma subulatum
– **sesquipedale**
– superbum = Angraecum eburneum
• – tenue = Campylocentrum tenue
• – vesicatum = Chamaeangis vesicata
• – virens = Angraecum eburneum
– voeltzkowianum = Angraecum eburneum
– weigelti = Campylocentrum fasciola
– whitfieldii = Tridactyle armeniaca
Anguloa
∗ – **brevilabris**
– clowesiana = Anguloa clowesii
∗ • – **clowesii**
– clowesii var. eburnea = Anguloa clowesii
– clowesii var. flava = Anguloa clowesii
– clowesii var. macrantha = Anguloa clowesii
– eburnea = Anguloa uniflora
– hernandezii = Stanhopea hernandezii
– lurida = Catasetum luridum
• – ruckeri = Anguloa brevilabris
– sagittata = Anguloa brevilabris
• – squalida = Lycomormium squalidum
– superba = Acineta superba
– turneri = Anguloa uniflora
∗ • – **uniflora**
– virginalis = Anguloa uniflora
Ania
• – bicornis = Tainia bicornis
– latifolia = Tainia bicornis
Anneliesia
– **candida**
Anochilus
– **inversum**
Anoectochilus
– argenteus = Erythrodes picta
– dawsonianus = Ludisia discolor
– petelotii = Ludisia discolor
– regalis = Anoectochilus setaceus
∗ • – **setaceus**
– setaceus var. inornatus = Anoectochilus setaceus
Anota
– violacea = Rhynchostylis violacea
Ansellia
∗ • – **africana**
– africana var. nilotica = Ansellia africana
– confusa = Ansellia africana
– congoensis = Ansellia africana
– gigantea = Ansellia africana
– gigantea var. nilotica = Ansellia africana
– humilis = Ansellia africana
– nilotica = Ansellia africana
Aopla
• – reniformis = Habenaria reniformis
Apatostelis
– **ciliaris**
Aplostellis
– velutina = Nervilia plicata
Aporum
– acerosum = Dendrobium acerosum
– aciculare = Dendrobium aciculare
– anceps = Dendrobium anceps
– crumenatum = Dendrobium crumenatum
• – cuspidatum = Dendrobium nathanielis
– junceum = Dendrobium junceum
– kwashotense = Dendrobium crumenatum
• – leonis = Dendrobium leonis
– planibulbe = Dendrobium planibulbe
• – sinuatum = Dendrobium sinuatum
– subteres = Dendrobium acerosum
Appendicula
– hasseltii = Agrostophyllum planicaule
Arachnites
– alpina = Chamorchis alpina
– aranifera = Ophrys sphegodes ssp. sphegodes
– atrata = Ophrys atrata
– fuciflora F.W.Schmidt = Ophrys holoserica
– fuciflora Tod. = Ophrys sphegodes ssp. sphegodes
– tenthredinifera = Ophrys tenthredinifera
Arethusa
– alaris = Pterygodium catholicum
– capensis = Disperis capensis
– ciliaris = Bartholina burmanniana
– crispa = Corycium crispum
– ophioglossoides = Pogonia ophioglossoides
– parviflora = Triphora trianthophora
– pendula = Triphora trianthophora
– picta = Sarcoglottis acaulis
– plicata = Nervilia plicata
– rosea = Crybe rosea
– secunda = Disperis secunda
– trianthophoros = Triphora trianthophora
– villosa = Disperis villosa
Aristotelea
– spiralis = Spiranthes sinensis
Arpophyllum
– cardinale = Arpophyllum giganteum
– **giganteum**
– jamaicense = Arpophyllum giganteum
• – spicatum = Arpophyllum giganteum
– squarrosum = Arpophyllum giganteum
– stenostachyum = Arpophyllum giganteum
Arundina
– affinis = Arundina graminifolia
• – bambusifolia = Arundina graminifolia
– chinensis = Arundina graminifolia
• – densa = Arundina graminifolia
– densiflora = Arundina graminifolia
∗ – **graminifolia**
– meyenii = Arundina graminifolia
– minor = Arundina graminifolia
– philippii = Arundina graminifolia
– philippii var. malayana = Arundina graminifolia
– pulchella = Arundina graminifolia
– pulchra = Arundina graminifolia
– revoluta = Arundina graminifolia
– sanderiana = Arundina graminifolia
– speciosa = Arundina graminifolia
– stenopetala = Arundina graminifolia
Asarca
– sceptrum = Chloraea gavilu
– virescens = Chloraea gavilu
Ascocentrum
– **curvifolium**
– micranthum = Smitinandia micrantha
∗ – **miniatum**
Aspasia
• – **epidendroides**
– fragrans = Aspasia epidendroides
– interrupta = Aspasia variegata
– liturata = Aspasia variegata
• – lunata = Aspasia variegata
∗ • – **variegata**
Auliza
– ciliaris = Epidendrum ciliare
– difformis = Epidendrum difforme
– nocturna = Epidendrum nocturnum
– parkinsonianum = Epidendrum parkinsonianum
– stamfordiana = Epidendrum stamfordianum
Aulizeum
– ciliare = Epidendrum ciliare
– glumaceum = Encyclia glumacea
– pygmaeum = Encyclia pygmaea
– variegatum = Encyclia vespa
Australorchis
– monophylla = Dendrobium monophyllum
Barbosella
– **gardneri**
Barbrodria
– **miersii**
Barkeria
• – **elegans**
• – **lindleyana**
∗ – **skinneri**
– skinneri = Barkeria skinneri
– skinneri var. major = Barkeria skinneri
– skinneri var. superba = Barkeria skinneri
• – **spectabilis**
Barlia
– longibracteata = Barlia robertiana
∗ – **robertiana**
Bartholina
∗ – **burmanniana**
– lindleyana = Bartholina burmanniana
• – pectinata = Bartholina burmanniana
Batemannia
– burtii = Huntleya meleagris
– burtii var. wallisii = Huntleya meleagris
∗ • – **colleyi**
– meleagris = Huntleya meleagris
– meleagris var. albidofulva = Huntleya meleagris
– peruviana = Batemannia colleyi
– petronia = Batemannia colleyi
– wallisii = Huntleya meleagris
– wallisii var. major = Huntleya meleagris
– yauaperyensis = Batemannia colleyi
Beadlea
∗ – **bicolor**
– **elata**
Beclardia
– elata = Cryptopus elatus
Benthamia
– **latifolia**
– minutiflora = Benthamia spiralis
– **spiralis**
Bifrenaria
• – **atropurpurea**
• – aurantiaca Lindl. = Rudolfiella aurantiaca
– aurantiaca Williams = Bifrenaria inodora
• – **aureo-fulva**
– caparaoense = Bifrenaria atropurpurea
– fragrans = Bifrenaria inodora
– fuerstenbergiana = Bifrenaria inodora
∗ – **harrisoniae**

Bifrenaria
– harrisoniae var. vitellina = Bifrenaria vitellina
● – **inodora**
● – **longicornis**
– **parvula**
* ● – **racemosa**
– sabulosa = Bifrenaria longicornis
– secunda = Bifrenaria aureo-fulva
* – **tetragona**
* ● – **vitellina**
Bilabrela
● – falcicornis = Habenaria tetrapetala
Blephariglottis
– grandiflora = Platanthera grandiflora
– psycodes var. grandiflora = Platanthera grandiflora
Bletia
– acuminata = Laelia rubescens
– acutipetala = Bletia purpurea
– albida = Laelia albida
– alta = Bletia purpurea
– amazonica = Brassavola martiana
– anceps = Laelia anceps
– angustata Gaudich. = Spathoglottis plicata
– angustata (Lindl.) Rchb.f. = Brassavola martiana
– angustifolia = Spathoglottis plicata
– autumnalis = Laelia autumnalis
– bicallosa = Liparis nervosa
– cinnabarina = Laelia cinnabarina
– crispa = Laelia crispa
– crispina = Schomburgkia gloriosa
– cucullata = Brassavola cucullata
– digbyana = Rhyncholaelia digbyana
● – florida = Bletia purpurea
– furfuracea = Laelia furfuracea
● – gebina = Bletilla striata
– glauca = Rhyncholaelia glauca
– gloriosa = Schomburgkia gloriosa
* ● – **gracilis**
– graminifolia = Arundina graminifolia
– grandiflora = Laelia speciosa
● – havanensis = Bletia purpurea
– hyacinthina = Bletilla striata
– lutea = Laelia flava
– martiana = Brassavola martiana
– masuca = Calanthe sylvatica
– nodosa = Brassavola nodosa
– pallida = Bletia purpurea
– peduncularis = Laelia rubescens
– perrinii (Lindl.) Rchb.f. = Brassavola tuberculata
– perrinii (Lindl.) Rchb.f. = Laelia perrinii
– praestans = Laelia pumila
– pumila = Laelia pumila
* – **purpurea**
* ● – **reflexa**
– rosea = Schomburgkia rosea
– rubescens = Laelia rubescens
– rupestris = Laelia crispata
● – **secunda**
● – shepherdii = Bletia purpurea
– speciosa = Laelia speciosa
– striata = Bletilla striata
– superbiens = Schomburgkia superbiens
– tankervilleae = Phaius tankervilleae
– tibicinis = Myrmecophila tibicinis
– tuberculata = Brassavola tuberculata
– tuberosa = Bletia purpurea
– undulata = Schomburgkia undulata
– venosa = Brassavola nodosa
– verecunda = Bletia purpurea
– violacea = Laelia rubescens
– xanthina = Laelia xanthina
Bletilla
– gebina = Bletilla striata
– hyacinthina = Bletilla striata
* – **striata**
– striata f. gebina = Bletilla striata
– striata var. albomarginata = Bletilla striata
– striata var. gebina = Bletilla striata
Bollea
– guianensis = Bollea violacea
– **violacea**
Bonatea
– tetrapetala = Habenaria tetrapetala
Bothriochilus
– macrostachyus = Coelia macrostychya
Brachystele
* – **bracteosa**
– **unilateralis**
Braemia
* – **vittata**
Brassavola
● – amazonica = Brassavola martiana
● – angustata = Brassavola martiana
● – appendiculata = Brassavola cucullata
● – cordata = Brassavola subulifolia
● – **cucullata**
– cucullata var. elegans = Brassavola cucullata
– cuspidata = Brassavola cucullata
● – digbyana = Rhyncholaelia digbyana
– fragrans = Brassavola tuberculata
– gibbsiana = Brassavola tuberculata
● – glauca = Rhyncholaelia glauca
● – grandiflora = Brassavola nodosa
* ● – **martiana**
– multiflora = Brassavola martiana
* ● – **nodosa**
– nodosa var. grandiflora = Brassavola nodosa
– nodosa var. rhopalorrhachis = Brassavola nodosa
– odoratissima = Brassavola cucullata
– paraensis = Leucohyle mutica
● – perrinii = Brassavola tuberculata
● – pumilio = Homalopetalum pumilio
– **retusa**
– rhopalorrachis = Brassavola nodosa
– scaposa = Brassavola nodosa
– sloanei = Brassavola subulifolia
– stricta = Brassavola nodosa
* ● – **subulifolia**
– surinamensis = Brassavola martiana
* ● – **tuberculata**
● – venosa = Brassavola nodosa
Brassia
● – angusta = Brassia lawrenceana
● – aristata = Brassia verrucosa
● – **bidens**
● – brachiata = Brassia verrucosa
* ● – **caudata**
– caudata var. hieroglyphica = Brassia caudata
● – clowesii = Miltonia clowesii
● – cochleata = Brassia lawrenceana
– coryandra = Brassia verrucosa
● – guttata = Brassia maculata
* ● – **lanceana**
– lanceana var. viridiflora = Brassia bidens
* ● – **lawrenceana**
– lewisii = Brassia caudata
– longiloba = Brassia verrucosa
– longissima var. minor = Brassia caudata
● – macrostachya = Brassia lanceana
● – **maculata**
– odontoglossoides = Brassia verrucosa
– oestlundiana = Lemboglossum maculatum
● – **peruviana**
● – pumila = Brassia lanceana
* ● – **verrucosa**
– wrayae = Brassia maculata
Bromheadia
* – **finlaysoniana**
● – palustris = Bromheadia finlaysoniana
Broughtonia
– alba = Maxillaria alba
● – aurea = Cattleya aurantiaca
– cubensis = Cattleyopsis cubensis
– **domingensis**
– grandiflora = Maxillaria grandiflora
– lilacina = Broughtonia domingensis
– linearis = Coelogyne ovalis
– maculatum = Maxillaria maculata
Bryobium
● – pubescens = Eria retusa
Bulbophyllopsis
– maculosa = Bulbophyllum umbellatum
– morphologorum = Bulbophyllum umbellatum
Bulbophyllum
● – adenopetalum = Bulbophyllum flavescens
– amesianum = Bulbophyllum cumingii
– andongense = Bulbophyllum cocoinum
– **antenniferum**
– arnoldianum = Bulbophyllum falcatum var. velutinum
* – **auratum**
– bakossorum = Bulbophyllum falcatum var. falcatum
* ● – **barbigerum**
– baucoense = Bulbophyllum weberi
– blepharosepalum = Bulbophyllum limbatum
– **blumei**
– bolovense = Bulbophyllum sessile
– boninense = Bulbophyllum macraei
– borneense = Bulbophyllum auratum
– bracteolata = Bulbophyllum bracteolatum
* ● – **bracteolatum**
– brevidenticulatum = Bulbophyllum cocoinum
– brixhei = Bulbophyllum falcatum var. velutinum
– buchenavianum = Bulbophyllum calyptratum var. calyptratum
– bufo Vermeulen = Bulbophyllum falcatum var. bufo
– bufo (Lindl.) Rchb.f. = Bulbophyllum falcatum var. falcatum
– calabaricum = Bulbophyllum pumilum
● – calamarium = Bulbophyllum saltatorium var. saltatorium
* – **calyptratum var. calyptratum**
– campanulatum = Bulbophyllum auratum
● – **cheiri**
* – **chinense**
– ciliatum = Bulbophyllum maximum
● – clandestinum = Bulbophyllum sessile
– **clavatum**
– clavatum = Bulbophyllum malachadenia
– clavigerum = Bulbophyllum longiflorum
– cochinchinense = Bulbophyllum macranthum
* ● – **cocoinum**
– confertum = Bulbophyllum scabratum
– congolense = Bulbophyllum imbricatum
– conicum = Bulbophyllum clavatum
– copelandii = Bulbophyllum cheiri
– cornutum = Bulbophyllum helenae
– correae = Bulbophyllum setigerum
– crassulifolium = Bulbophyllum sheperdii
– **cumingii**
● – **cupreum**
– cuspidilingue = Bulbophyllum blumei
– cyrtopetalum = Bulbophyllum maximum
– dahlemense = Bulbophyllum falcatum var. bufo
– deistelianum = Bulbophyllum falcatum var. falcatum
– djumaense = Bulbophyllum maximum
– djumaense var. grandifolium = Bulbophyllum maximum
– dorotheae = Bulbophyllum pumilum
– drallei = Bulbophyllum pumilum
– elachon = Bulbophyllum pumilum
– **elongatum**
– ephippium = Bulbophyllum blumei
– **falcatum var. bufo**
* – **falcatum var. falcatum**
– **falcatum var. velutinum**
– falcatum = Bulbophyllum falcatum var. bufo
– **fimbriatum**
– **flavescens**
● – flavidum = Bulbophyllum pumilum
– fractiflexum = Bulbophyllum falcatum var. velutinum
● – **fuscum var. fuscum**

Bulbophyllum

– gabonis = Bulbophyllum pumilum
– galbinum = Bulbophyllum uniflorum
– gigas = Bulbophyllum elongatum
– gilletii = Bulbophyllum imbricatum
* – **guttulatum**
– **helenae**
– hemirhachis = Bulbophyllum falcatum var. bufo
– henshallii = Bulbophyllum lobbii
– herminiostachys = Bulbophyllum pumilum
• – **hirtum**
• – **imbricatum**
– imogeniae = Bulbophyllum pumilum
– intermedium = Bulbophyllum sheperdii
– javanicum = Bulbophyllum pahudii
– kamerunense = Bulbophyllum imbricatum
– kerrii = Bulbophyllum hirtum
– kewense = Bulbophyllum falcatum var. velutinum
– lanuriense = Bulbophyllum falcatum var. velutinum
– laurentianum = Bulbophyllum imbricatum
– layardii = Bulbophyllum longiflorum
– ledermannii = Bulbophyllum imbricatum
– leptorrhachis = Bulbophyllum falcatum var. bufo
– leucopogon = Bulbophyllum pumilum
– leucorhachis = Bulbophyllum imbricatum
– leysianum = Bulbophyllum antenniferum
• – **limbatum**
– linderi = Bulbophyllum imbricatum
– lindleyi = Bulbophyllum calyptratum var. calyptratum
• – **lobbii**
– lobbii = Bulbophyllum lobbii
– lobbii var. breviflorum = Bulbophyllum lobbii
– lobbii var. colosseum = Bulbophyllum lobbii
– lobbii var. henshallii = Bulbophyllum lobbii
– longibulbum = Bulbophyllum falcatum var. falcatum
• – **longiflorum**
– longiflorum = Bulbophyllum weberi
– lubiense = Bulbophyllum falcatum var. falcatum
– **macraei**
*• – **macranthum**
– makinoanum = Bulbophyllum macraei
– **malachadenia**
– masdevalliaceum = Bulbophyllum blumei
– maxillare = Bulbophyllum blumei
– **maximum**
* – **medusae**
– megalanthum = Bulbophyllum cheiri
– melanorrhachis = Bulbophyllum falcatum var. velutinum
– millenii = Bulbophyllum falcatum var. velutinum
– minutum = Bulbophyllum falcatum var. velutinum
– minutum var. purpureum = Bulbophyllum falcatum var. velutinum
– moireanum = Bulbophyllum maximum
– moliwense = Bulbophyllum pumilum
– nanum = Bulbophyllum pumilum
– nutans = Bulbophyllum othonis
– nyassanum = Bulbophyllum maximum
– ogoouense = Bulbophyllum fuscum var. fuscum
– **othonis**
– ovalifolium = Bulbophyllum sessile
– oxyodon = Bulbophyllum falcatum var. bufo
– oxypterum = Bulbophyllum maximum
– oxypterum var. mozambicense = Bulbophyllum maximum
– **pahudii**
– papillosum = Bulbophyllum pumilum
– patens = Bulbophyllum macranthum
– pavimentatum = Bulbophyllum pumilum
– pechei = Bulbophyllum cupreum
– **picturatum**
– picturatum = Bulbophyllum helenae
• – **pileatum**
– platyrhachis = Bulbophyllum maximum
– porphyroglossum = Bulbophyllum pumilum
– praestans = Bulbophyllum cheiri
* – **pumilum**
– ramosii = Bulbophyllum flavescens
• – recurvum = Bulbophyllum pumilum
– reinwardtii = Bulbophyllum uniflorum
– rhizophorae = Bulbophyllum falcatum var. velutinum
– **roxburghii**
– rupincola = Bulbophyllum saltatorium var. saltatorium
• – saltatorium = Bulbophyllum saltatorium var. saltatorium
– **saltatorium var. calamarium**
* – **saltatorium var. saltatorium**
– saruwatari = Bulbophyllum umbellatum
– **scabratum**
– sceptrum = Bulbophyllum elongatum
– schillerianum = Bulbophyllum sheperdii
– sereti = Bulbophyllum falcatum var. falcatum
– **sessile**
• – **setigerum**
– **sheperdii**
– shepherdii var. intermedium = Bulbophyllum sheperdii
– siamense = Bulbophyllum lobbii
– simonii = Bulbophyllum falcatum var. velutinum
– solheidii = Bulbophyllum falcatum var. velutinum
• – sordidum = Bulbophyllum bracteolatum
– stenorhachis = Bulbophyllum imbricatum
– stramineum = Bulbophyllum cumingii
– strobiliferum = Bulbophyllum imbricatum
– suave = Bulbophyllum hirtum
– subcoriaceum = Bulbophyllum maximum
– subuliferum = Bulbophyllum cheiri
– tanegasimense = Bulbophyllum macraei
– tibeticum = Bulbophyllum umbellatum
– tortisepalum = Bulbophyllum umbellatum
– trisetosum = Bulbophyllum sessile
– trisetum = Bulbophyllum longiflorum
– triste = Bulbophyllum imbricatum
– ugandae = Bulbophyllum falcatum var. bufo
• – **umbellatum**
– umbellatum = Bulbophyllum guttulatum
– umbellatum var. bergemannii = Bulbophyllum umbellatum
– **uniflorum**
– uraiense = Bulbophyllum macraei
– **vaginatum**
– velutinum = Bulbophyllum falcatum var. velutinum
– verecundum = Bulbophyllum pumilum
* – **weberi**
– whiteanum = Bulbophyllum vaginatum
– whitfordii = Bulbophyllum cheiri
– wincklerii = Bulbophyllum pumilum
– yangambiense = Bulbophyllum pumilum

Burlingtonia

– batemanii = Rodriguezia batemanii
• – candida = Rodriguezia candida
– carnea = Rodriguezia carnea
• – fragrans = Rodriguezia bracteata
• – maculata = Rodriguezia maculata
• – rigida = Rodriguezia rigida
– rosea = Rodriguezia lanceolata
• – rubescens = Rodriguezia batemanii
• – venusta = Rodriguezia venusta

Calanthe

– amamiana = Calanthe discolor var. flava
– amamiana var. latilabella = Calanthe discolor var. flava
– angraeciflora = Calanthe triplicata
– australis = Spiranthes sinensis
– bachmaensis = Phaius tankervilleae
• – bicolor = Calanthe discolor
– catiligera = Calanthe triplicata
– celebica = Calanthe triplicata
– comosa = Calanthe triplicata
– corymbosa = Calanthe sylvatica
– curantigoides = Calanthe pulchra
• – curculigoides = Calanthe pulchra
– delphinioides = Calanthe sylvatica
*• – **densiflora**
*• – **discolor**
– **discolor var. flava**
– discolor f. bicolor = Calanthe discolor
– discolor f. viridialba = Calanthe discolor var. flava
– discolor var. amaniana = Calanthe discolor var. flava
– discolor var. bicolor = Calanthe discolor
– discolor var. divaricatipetala = Calanthe discolor var. flava
– discolor var. kanashiroi = Calanthe discolor var. flava
– discolor var. viridialba = Calanthe discolor var. flava
– emarginata = Calanthe sylvatica
• – furcata = Calanthe triplicata
– furcata f. albo-lineata = Calanthe triplicata
– furcata f. albo-marginata = Calanthe triplicata
– furcata f. textori = Calanthe triplicata
– gebina = Bletilla striata
– gracillima = Calanthe triplicata
– japonica = Calanthe triplicata
– longicalcarata = Calanthe triplicata
– lurida = Calanthe discolor var. flava
– masuca Thwaites = Calanthe purpurea
• – masuca (D.Don) Lindl. = Calanthe sylvatica
– matsumurana = Calanthe triplicata
– natalensis = Calanthe sylvatica
– neglecta = Calanthe sylvatica
– odora = Calanthe triplicata
– okinawensis = Calanthe triplicata
– okinawensis f. albiflora = Calanthe triplicata
– orthocentron = Calanthe triplicata
– perottetii = Calanthe triplicata
– pleiochroma = Calanthe triplicata
– poilanei = Spathoglottis plicata
– proboscidea = Calanthe triplicata
– pubescens = Calanthe triplicata
* – **pulchra**
• – **purpurea**
– schliebenii = Calanthe sylvatica
– scortechini = Calanthe triplicata
– sieboldii = Calanthe discolor
– speciosa = Phaius tankervilleae
– stolzii = Calanthe sylvatica
– striata = Calanthe discolor
– striata var. bicolor = Calanthe discolor
– striata var. sieboldii = Calanthe discolor
* – **sylvatica**
– sylvatica var natalensis = Calanthe sylvatica
– textori = Calanthe triplicata
– textori f. albiflora = Calanthe triplicata
– textori var. alba = Calanthe triplicata
– textori var. longicalcarata = Calanthe triplicata
– textori var. violacea = Calanthe triplicata
– tokumoshimensis f. latilabella = Calanthe discolor var. flava
– tokunoshimensis = Calanthe discolor var. flava
* – **triplicata**
– triplicata f. albo-lineata = Calanthe triplicata
– triplicata var. angraeciflora = Calanthe triplicata
• – veratrifolia = Calanthe triplicata
– veratrifolia var. cleistogama = Calanthe triplicata
– veratrifolia var. densissima = Calanthe triplicata

Calanthe
– veratrifolia var. dupliciloba = Calanthe triplicata
– veratrifolia var. incurvicalcar = Calanthe triplicata
– veratrifolia var. kennyi = Calanthe triplicata
– veratrifolia var. lancipetala = Calanthe triplicata
– veratrifolia var. timorensis = Calanthe triplicata
• – versicolor = Calanthe sylvatica
– violacea = Calanthe triplicata
– volkensii = Calanthe sylvatica
Calceolus
– reginae = Cypripedium reginae
Callista
– acerosa = Dendrobium acerosum
– acicularis = Dendrobium aciculare
– acuminatissima = Dendrobium acuminatissimum
– adunca = Dendrobium aduncum
– aggregata = Dendrobium lindleyi
– aggregata var. jenkinsii = Dendrobium jenkinsii
– amoena = Dendrobium amoenum
– anceps = Dendrobium anceps
– angulata = Dendrobium angulatum
– anosma = Dendrobium anosmum
– antennata = Epidendrum carpophorum
– aphylla = Dendrobium aphyllum
– aquea = Dendrobium aqueum
– aurea = Dendrobium heterocarpum
– aurorosea = Dendrobium nudum
– bambusifolia = Dendrobium salaccense
– bicamerata = Dendrobium bicameratum
– bifaria = Dendrobium bifarium
– biflora = Dendrobium biflorum
– binnendijkii = Flickingeria fimbriata
– bolboflora = Dendrobium bicameratum
– bursigera = Dendrobium secundum
– calopogon = Dendrobium angulatum
– chrysantha = Dendrobium chrysanthum
– chrysotoxa = Dendrobium chrysotoxum
– comata = Flickingeria comata
– conostalix = Dendrobium lobbii
– cornuta = Dendrobium kuhlii
– cretacea = Dendrobium cretaceum
– crumenata = Dendrobium crumenatum
– cucumerina = Dendrobium cucumerinum
– densiflora = Dendrobium densiflorum
– denudans = Dendrobium denudans
– devoniana = Dendrobium devonianum
– erosa = Dendrobium erosum
fairfaxii = Dendrobium teretifolium
– flabella = Flickingeria fimbriata
– flavescens = Polystachya concreta
– formosa = Dendrobium formosum
– gracilicaulis = Dendrobium gracilicaule
– griffithiana = Dendrobium griffithianum
– heterocarpa = Dendrobium heterocarpum
– hymenophylla = Dendrobium hymenophyllum
– intermedia = Dendrobium salaccense
– japonica = Dendrobium moniliforme
– jenkinsii = Dendrobium jenkinsii
– jerdoniana = Dendrobium nutans
– kingiana = Dendrobium kingianum
– kuhlii = Dendrobium kuhlii
– kunstleri = Flickingeria fimbriata
– lamellata = Dendrobium lamellatum
– leonis = Dendrobium leonis
– linawiana = Dendrobium linawianum
– linguiformis = Dendrobium linguiforme
– lobbii = Dendrobium lobbii
– macraei = Flickingeria fimbriata
– macrophylla (A.Rich.) Kuntze = Dendrobium macrophyllum
– macrophylla (Lindl.) Kuntze = Dendrobium anosmum
– mirbeliana = Dendrobium mirbelianum
– moniliformis = Dendrobium moniliforme
– monophylla = Dendrobium monophyllum
– moschata = Dendrobium moschatum
– mutabilis = Dendrobium mutabile
– nathanielis = Dendrobium nathanielis
– nobilis = Dendrobium nobile
– nuda = Dendrobium nudum
– nutans = Dendrobium nutans
– pugioniformis = Dendrobium pugioniforme
– pulchella = Dendrobium pulchellum
– pygmaea = Dendrobium peguanum
– revoluta = Dendrobium revolutum
– rigescens = Dendrobium mutabile
– rigida = Dendrobium mutabile
– rugosa = Dendrobium rugosum
– salaccensis = Dendrobium salaccense
– sanguinolenta = Dendrobium sanguinolentum
– scortechinii = Dendrobium anosmum
– secunda = Dendrobium secundum
– speciosa = Dendrobium speciosum
– stuposa = Dendrobium stuposum
– suavissima = Dendrobium chrysotoxum
– sulcata = Dendrobium sulcatum
– taurina = Dendrobium taurinum
– teijsmannii = Dendrobium lobbii
– teretifolia = Dendrobium teretifolium
– tetragona = Dendrobium tetragonum
– transparens = Dendrobium transparens
– undulata = Dendrobium discolor
– veitchiana = Dendrobium macrophyllum
– villosula = Dendrobium nutans
Callostylis
– pulchella = Eria pulchella
Camaridium
– affine = Maxillaria camaridii
– album = Maxillaria alba
– amazonicum = Maxillaria camaridii
– cryptopodanthum = Maxillaria camaridii
– graminifolium = Maxillaria graminifolia
– grandiflorum = Maxillaria longibracteata var. grandiflora
• – ochroleucum = Maxillaria camaridii
– squamatum = Maxillaria uncata
– unicatum = Maxillaria uncata
Camarotis
– adnata = Micropera fuscolutea
– apiculata = Micropera pallida
– cochinchinensis = Micropera pallida
– latisaccata = Micropera fuscolutea
• – obtusa = Micropera obtusa
– pallida = Micropera pallida
Campylocentrum
– barrettiae = Campylocentrum micranthum
– colombianum = Campylocentrum micranthum
– ecuadorense = Campylocentrum micranthum
– **fasciola**
– jamaicense = Campylocentrum micranthum
– kuntzei = Campylocentrum micranthum
– lankesteri = Campylocentrum fasciola
– lansbergii = Campylocentrum micranthum
– loretoense = Campylocentrum fasciola
– mattogrossense = Campylocentrum micranthum
* – **micranthum**
– multiflorum = Campylocentrum fasciola
– **ornithorrhynchum**
– panamense = Campylocentrum micranthum
– peniculus = Campylocentrum micranthum
– **polystachyum**
– pugioniforme = Cleisostoma subulatum
– stenanthum = Campylocentrum micranthum
– sullivanii = Campylocentrum fasciola
– **tenue**
Catachaetum
– claveringii = Catasetum macrocarpum
– floribundum = Catasetum macrocarpum
– lituratum = Catasetum luridum
– purum = Catasetum purum
• – recurvatum = Catasetum planiceps
– semiapertum = Catasetum purum
– squalidum = Catasetum luridum
Catasetum
• – abruptum = Catasetum luridum
– adnatum = Catasetum atratum
– appendiculatum = Catasetum barbatum
– arachnoideum = Catasetum callosum
*• – **atratum**
*• – **barbatum**
– barbatum = Catasetum barbatum
• – barbatum var. proboscideum = Catasetum barbatum
– barbatum var. spinosum = Catasetum barbatum
– bergoldianum = Catasetum ochraceum
– brachybulbon = Catasetum barbatum
– brenesii = Catasetum integerrimum
– buchtienii = Catasetum barbatum
– calceolatum = Clowesia russelliana
*• – **callosum**
– callosum var. eucallosum = Catasetum callosum
– callosum var. carunculatum = Catasetum callosum
– callosum var. crenatum = Catasetum callosum
– callosum var. grandiflorum = Catasetum callosum
– carunculatum = Catasetum callosum
* – **cernuum**
– chloranthum = Catasetum planiceps
– ciliatum = Catasetum discolor
– citrinum = Mormodes maculatum
– claesianum = Catasetum discolor
• – claveringii = Catasetum macrocarpum
– colossus = Catasetum saccatum
– comosum = Catasetum barbatum
• – cornutum = Catasetum cristatum
– craniomorphum = Catasetum luridum
– crinitum = Catasetum barbatum
*• – **cristatum**
– cristatum var. monstrosum = Catasetum cristatum
– cristatum var. spinigerum = Catasetum barbatum
– cristatum var. spinosum = Catasetum barbatum
– cruciatum = Catasetum saccatum
– darwinianum = Catasetum callosum
*• – **deltoideum**
*• – **discolor**
– discolor var. cleasianum = Catasetum discolor
– discolor var. fimbriatum = Catasetum discolor
– discolor var. roseo-album = Catasetum discolor
– discolor var. viridiflorum = Catasetum discolor
– floribundum = Catasetum macrocarpum
• – fuliginosum Lindl. = Catasetum planiceps
– fuliginosum Rolfe = Catasetum callosum
– gardneri = Catasetum discolor
– garnettianum = Catasetum barbatum
– globiferum = Catasetum globiflorum
• – **globiflorum**
– histrio = Catasetum saccatum
• – **hookeri**
– hymenophorum = Catasetum planiceps
– imschootianum = Catasetum hookeri
– inapertum = Catasetum purum
– incurvum = Catasetum saccatum
– inornatum = Catasetum ochraceum
• – **integerrimum**
– integerrimum var. purpurascens = Catasetum integerrimum
– integerrimum var. viridiflorum = Catasetum integerrimum
– labiatum = Catasetum ochraceum
*• – **laminatum**
• – laminatum var. eburneum = Catasetum laminatum
– laminatum var. maculatum = Catasetum laminatum

Catasetum

• – lanciferum = Catasetum barbatum
– lansbergii = Catasetum callosum
– lehmannii = Catasetum ochraceum
– liechtensteinii = Catasetum trulla
– linguiferum = Catasetum macrocarpum
• – **longifolium**
* • – **luridum**
* • – **macrocarpum**
– macrocarpum = Catasetum barbatum
* • – **maculatum**
– maculatum = Catasetum integerrimum
– mentosum = Catasetum atratum
• – milleri = Catasetum hookeri
• – **naso**
• – **ochraceum**
– oerstedii = Catasetum maculatum
– pallidum = Catasetum atratum
* • – **planiceps**
– polydactylon = Catasetum barbatum
• – **poriferum**
• – proboscideum = Catasetum barbatum
* • – **purum**
– randii = Catasetum barbatum
– rivularium = Catasetum barbatum
• – roseo-album = Catasetum discolor
– roseum = Clowesia rosea
– rostratum = Catasetum integerrimum
• – russellianum = Clowesia russelliana
• – **saccatum**
– secundum = Catasetum saccatum
• – semiapertum = Catasetum purum
• – serratum = Catasetum viridiflavum
• – spinosum = Catasetum barbatum
– stupendum = Catasetum saccatum
• – **tabulare**
– tricolor = Catasetum macrocarpum
• – tridentatum Hook. = Catasetum macrocarpum
– tridentatum Pfitzer = Catasetum barbatum
• – trifidum = Catasetum cernuum
* • – **trulla**
– trulla var. liechtensteinii = Catasetum trulla
– trulla var. trulla = Catasetum trulla
– viride = Catasetum cernuum
– **viridiflavum**
• – wailesii = Catasetum integerrimum

Cattleya

* • – **aclandiae**
– acuminata = Laelia rubescens
– albida = Laelia albida
– aliciae = Cattleya mossiae
– amabilis = Cattleya intermedia
– amethystina = Cattleya intermedia
– anceps = Laelia anceps
– aquinii = Cattleya intermedia
• – arembergii = Cattleya loddigesii
– **aurantiaca**
– autumnalis hort. = Cattleya labiata
– autumnalis (Lindl.) Beer = Laelia autumnalis
• – **bicolor**
– bicolor var. measuresiana = Cattleya bicolor
• – bulbosa = Cattleya walkeriana
– carrieri = Cattleya mossiae
– cernua = Sophronitis cernua
– cinnabarina = Laelia cinnabarina
• – citrina = Encyclia citrina
• – coccinea = Sophronitis coccinea
– concolor = Cattleya harrisoniana
– crispa Beer = Schomburgkia gloriosa
• – crispa Lindl. = Laelia crispa
• – domingensis = Broughtonia domingensis
– edithiana = Cattleya mossiae
– elatior = Cattleya guttata
* • – **forbesii**
– fulva = Cattleya forbesii
– furfuracea = Laelia furfuracea
– gardneriana = Cattleya walkeriana
– grahami = Laelia speciosa
– grandiflora = Sophronitis coccinea
* • – **granulosa**
• – granulosa var. russeliana = Cattleya granulosa
– grossi = Cattleya bicolor
* • – **guttata**
– harrisoniae = Cattleya harrisoniana
• – **harrisoniana**
– harrisonii = Cattleya harrisoniana
– ianthina = Cattleya intermedia
– integerrima var. angustifolia = Laelia perrinii
* – **intermedia**
– intermedia var. angustifolia = Laelia perrinii
• – intermedia var. pallida = Cattleya intermedia
– intermedia var. variegata = Cattleya harrisoniana
– isopetala = Cattleya forbesii
– karwinskii = Encyclia citrina
* • – **labiata**
– labiata var. atropurpurea = Cattleya mossiae
– labiata var. autumnalis = Cattleya labiata
– labiata var. candida = Cattleya mossiae
– labiata var. genuina = Cattleya labiata
• – labiata var. mossiae = Cattleya mossiae
– labiata var. picta = Cattleya mossiae
– labiata var. warocqueanum = Cattleya labiata
• – lemoniana = Cattleya labiata
• – **loddigesii**
– loddigesii var. amethystina = Cattleya intermedia
– loddigesii var. harrisoniae = Cattleya harrisoniana
– loddigesii var. harrisoniana Linden = Cattleya loddigesii
– loddigesii var. harrisoniana (Bateman ex Lindl.) Rolfe = Cattleya harrisoniana
– loddigesii var.intermedia = Cattleya intermedia
– lutea = Laelia flava
– majalis = Laelia speciosa
– malouana = Cattleya maxima
– marginata = Laelia pumila
• – maritima = Cattleya intermedia
* • – **maxima**
– measuresiana = Cattleya bicolor
* – **mossiae**
– nalderiana = Cattleya labiata
– odoratissima = Cattleya violacea
• – ovata = Cattleya intermedia
– pallida = Cattleya labiata
• – papeiansiana = Cattleya harrisoniana
– pauper = Cattleya forbesii
– peduncularis = Laelia rubescens
– perrinii Endl. = Cattleya labiata
• – perrinii Lindl. = Laelia perrinii
• – pinellii = Laelia pumila
– pinellii var. marginata = Laelia pumila
– princeps = Cattleya walkeriana
• – pumila = Laelia pumila
– regalis = Cattleya labiata
– reineckiana = Cattleya mossiae
– rubescens = Laelia rubescens
– schomburgkii = Cattleya violacea
– schroederiana = Cattleya walkeriana
– sessilis = Sobralia sessilis
• – **skinneri**
– spectabilis = Laelia pumila
– sphenophora = Cattleya guttata
• – superba = Cattleya violacea
– superbiens = Schomburgkia superbiens
– tibicinis = Myrmecophila tibicinis
– tigrina = Cattleya guttata
– undulata = Schomburgkia undulata
– vestalis = Cattleya forbesii
– **violacea**
– violacea = Sophronitella violacea
– wageneri = Cattleya mossiae
* • – **walkeriana**
– warocqueana = Cattleya labiata

Cattleyopsis

– **cubensis**

Caularthron

– **bicornutum**
– umbellatum = Epidendrum difforme

Centranthera

• – punctata = Pleurothallis centranthera

Centrosis

– sylvatica = Calanthe sylvatica

Ceraja

– simplicissima = Dendrobium crumenatum

Ceratochilus

– grandiflorus = Stanhopea grandiflora
– oculatus = Stanhopea oculata

Cestichis

– longipes = Liparis viridiflora
– pendula = Liparis viridiflora

Cestorchis

– caespitosa = Liparis caespitosa

Chaenanthe

• – barkeri = Diadenium barkeri

Chamaeangis

– **gracilis**
– **vesicata**

Chamaeorchis

– alpina = Chamorchis alpina

Chamaerepes

– alpina = Chamorchis alpina

Chamorchis

– **alpina**

Cheiropterocephalus

– sertuliferus = Malaxis excavata

Cheirostylis

• – **parvifolia**

Chelonanthera

– speciosa = Coelogyne speciosa

Chilochista

– minimifolia = Chilochista pusilla
– **pusilla**
• – **usneoides**
– usneoides = Chilochista pusilla

Chloraea

* – **gavilu**
– lindleyii = Chloraea gavilu
– lotensis = Chloraea gavilu
– lutea = Chloraea gavilu
– sceptrum = Chloraea gavilu
– semitensis = Chloraea gavilu
• – virescens = Chloraea gavilu

Chysis

* • – **aurea**
– aurea var. bractescens = Chysis bractescens
* • – **bractescens**
• – laevis = Chysis aurea

Cionosaccus

– lanceolatus = Goodyera procera

Cirrhaea

– atropurpurea = Gongora quinquenervis
* – **dependens**
– dependens = Cirrhaea loddigesii
– fusco-lutea Hook. = Cirrhaea saccata
• – fusco-lutea Lindl. = Cirrhaea dependens
– hoffmannseggii = Cirrhaea dependens
– livida = Cirrhaea dependens
* • – **loddigesii**
– loddigesii = Gongora galeata
• – obtusata = Cirrhaea loddigesii
– obtusata var. pallida = Cirrhaea loddigesii
• – pallida = Cirrhaea loddigesii
– purpurascens = Cirrhaea dependens
– russelliana = Cirrhaea dependens
• – **saccata**
• – tristis = Cirrhaea dependens
– violaceo-virens = Cirrhaea dependens
– violascens = Cirrhaea dependens
• – viridipurpurea Lodd. = Cirrhaea loddigesii
• – viridipurpurea (Hook.) Lindl. = Cirrhaea dependens
– warreana = Cirrhaea dependens

Cirrhopetalum

– africanum = Bulbophyllum longiflorum
– amesianum = Bulbophyllum cumingii

Cirrhopetalum
- – annamicum = Bulbophyllum umbellatum
- • – antenniferum = Bulbophyllum antenniferum
- • – auratum = Bulbophyllum auratum
- – autumnale = Bulbophyllum macraei
- • – blumei = Bulbophyllum blumei
- – boninense = Bulbophyllum macraei
- – bootanense = Bulbophyllum umbellatum
- – borneense = Bulbophyllum auratum
- • – caespitosum = Bulbophyllum scabratum
- – campanulatum = Bulbophyllum auratum
- • – capitatum = Bulbophyllum pahudii
- – carinatum = Bulbophyllum pahudii
- – caudatum = Bulbophyllum vaginatum
- • – chinense = Bulbophyllum chinense
- – clavigerum = Bulbophyllum longiflorum
- • – compressum = Bulbophyllum uniflorum
- • – cornutum = Bulbophyllum helenae
- • – cumingii = Bulbophyllum cumingii
- – eberhardtii = Bulbophyllum picturatum
- • – elongatum = Bulbophyllum elongatum
- • – fimbriatum = Bulbophyllum fimbriatum
- – flagelliforme = Bulbophyllum pahudii
- – guttulatum = Bulbophyllum guttulatum
- – kenejianum = Bulbophyllum longiflorum
- – layardii = Bulbophyllum longiflorum
- – longiflorum = Bulbophyllum longiflorum
- • – macraei = Bulbophyllum macraei
- • – maculosum = Bulbophyllum umbellatum
- – maculosum var. annamicum = Bulbophyllum umbellatum
- – makinoanum = Bulbophyllum macraei
- • – maxillare = Bulbophyllum blumei
- • – medusae = Bulbophyllum medusae
- • – nutans = Bulbophyllum othonis
- – ornithorhynchum = Bulbophyllum antenniferum
- – pahudii = Bulbophyllum pahudii
- • – picturatum = Bulbophyllum picturatum
- • – roxburghii = Bulbophyllum roxburghii
- – stramineum Teijsm. & Binnend. = Bulbophyllum vaginatum
- – stramineum auct. non Teijsm. = Bulbophyllum cumingii
- – tanegasimense = Bulbophyllum macraei
- • – thouarsii Lindl. = Bulbophyllum weberi
- – thouarsii Lindl. = Bulbophyllum longiflorum
- – thouarsii var. concolor = Bulbophyllum longiflorum
- – umbellatum = Bulbophyllum longiflorum
- – uraiense = Bulbophyllum macraei
- • – vaginatum = Bulbophyllum vaginatum
- – walkerianum = Bulbophyllum macraei
- • – wallichii = Bulbophyllum fimbriatum
- – weberi = Bulbophyllum weberi
- – whiteanum = Bulbophyllum vaginatum

Cladobium
- – violaceum = Scaphyglottis violacea

Cleisostoma
- – amabile = Cleisostoma subulatum
- – auriculatum = Cleisostoma discolor
- – cerinum = Cleisostoma paniculatum
- – chrysomelas = Cleisostoma racemiferum
- – cornutum = Plectorrhiza tridentata
- – cumingii = Pomatocalpa latifolia
- • – dealbatum = Cleisostoma subulatum
- • – decipiens = Pomatocalpa decipiens
- – dichroanthum = Schoenorchis gemmata
- • – **discolor**
- – evrardii = Schoenorchis gemmata
- – **filiforme**
- – fordii = Cleisostoma rostratum
- – formosum = Cleisostoma paniculatum
- – galeata = Pomatocalpa maculosum
- – gemmatum = Schoenorchis gemmata
- • – ionosmum = Staurochilus ionosma
- • – latifolium = Pomatocalpa latifolia
- – maculosa Bentham = Acampe ochracea
- • – maculosa Lindl. = Pomatocalpa maculosum
- – maculosa Thwaites = Pomatocalpa decipiens
- – micranthum = Smitinandia micrantha
- * – **paniculatum**
- – petitiana = Smitinandia micrantha
- – poilanei = Smitinandia micrantha
- – pugioniforme = Cleisostoma subulatum
- – **racemiferum**
- – robustum = Robiquetia spathulata
- • – rosea = Trichoglottis rosea
- * – **rostratum**
- – seidenfadenii = Cleisostoma simondii
- * – **simondii**
- – spathulatum = Robiquetia spathulata
- • – spicatum Lindl. = Robiquetia spathulata
- – spicatum Ridley = Pomatocalpa latifolia
- – **subulatum**
- – teres = Cleisostoma simondii
- – termissus = Cleisostoma discolor
- – thwaitesianum = Pomatocalpa decipiens
- – tixieri = Smitinandia micrantha
- • – tridentatum = Plectorrhiza tridentata
- – virginale = Robiquetia succisa

Clinhymenia
- – pallidiflora = Cryptarrhena lunata

Clowesia
- *• – **rosea**
- – **russelliana**

Cnemidia
- • – angulosa = Tropidia angulosa
- • – semilibera = Tropidia angulosa

Coccineorchis
- – **cernua**
- – corymbosa = Coccineorchis cernua

Cochleanthes
- * – **flabelliformis**
- – fragrans = Cochleanthes flabelliformis

Cochlioda
- – **rosea**

Coelia
- • – baueriana = Coelia triptera
- – galeottiana = Coelia triptera
- – glacialis = Coelia triptera
- – macrostachya var. genuina = Coelia macrostychya
- – macrostachya var. integrilabia = Coelia macrostychya
- – **macrostychya**
- * – **triptera**

Coeloglossum
- – brevifolium = Peristylus densus
- – cordatum = Gennaria diphylla
- – densum = Peristylus densus
- – diphyllum = Gennaria diphylla
- – peristyloides = Peristylus densus
- – secundum = Peristylus secundus

Coelogyne
- – articulata = Pholidota articulata
- – birmanica = Pleione praecox
- – brunnea = Coelogyne fuscescens var. brunnea
- – calceata = Pholidota pallida
- – conchoidea = Pholidota imbricata
- – conferta = Coelogyne nitida
- • – coronaria = Eria coronaria
- *• – **cristata**
- *• – **cumingii**
- • – **elata**
- *• – **fimbriata**
- *• – **flaccida**
- – **fuscescens var. brunnea**
- • – fuscescens = Coelogyne fuscescens var. brunnea
- – goweri = Coelogyne nitida
- – imbricata = Pholidota imbricata
- – khasyana = Pholidota articulata
- – laotica = Coelogyne fimbriata
- – leungiana = Coelogyne fimbriata
- – loricata = Pholidota imbricata
- * – **nitida**
- – nitida = Coelogyne punctulata
- • – ocellata = Coelogyne punctulata
- • – ochracea = Coelogyne nitida
- – ochracea ssp. conferta = Coelogyne nitida
- – ochracea var. foliata = Coelogyne nitida
- • – **ovalis**
- – ovalis = Coelogyne fimbriata
- – pilosissima = Coelogyne ovalis
- – porrecta = Otochilus porrectus
- – praecox = Pleione praecox
- – praecox var. sanguinea = Pleione praecox
- – praecox var. tenera = Pleione praecox
- – praecox var. wallichiana = Pleione praecox
- • – **prolifera**
- – **punctulata**
- – punctulata var. conferta = Coelogyne nitida
- – reichenbachiana = Pleione praecox
- – rubra = Pholidota rubra
- *• – **speciosa**
- • – **testacea**
- – triotos = Pholidota imbricata
- – triplicatula = Coelogyne ovalis
- – triptera = Encyclia pygmaea
- – undulata = Pholidota rubra
- • – wallichiana = Pleione praecox
- – wallichii = Pleione praecox

Coilostylis
- – emarginata = Epidendrum ciliare

Colax
- – aromaticus = Lycaste aromatica
- – barringtoniae = Lycaste barringtoniae
- – grandiflorus = Bifrenaria harrisoniae
- – harrisoniae = Bifrenaria harrisoniae
- • – jugosa = Pabstia jugosa
- – jugosus var. viridis = Pabstia viridis
- – palmifolius = Xylobium palmifolium
- – parkeri = Maxillaria parkeri
- – parvulus = Bifrenaria parvula
- • – placanthera = Pabstia placanthera
- – racemosus = Bifrenaria racemosa
- • – viridis = Pabstia viridis
- – viridis var. placanthera = Pabstia placanthera
- – viridis var. pluriflora = Pabstia jugosa

Comparettia
- *• – **coccinea**
- – **falcata**
- • – rosea = Comparettia falcata

Conostalix
- – calcaratum = Dendrobium lobbii
- – lobbii = Dendrobium lobbii
- – paludicola = Dendrobium lobbii

Cordula
- – barbata = Paphiopedilum barbatum
- – nigrita = Paphiopedilum barbatum
- – purpurata = Paphiopedilum purpuratum
- – venusta = Paphiopedilum venustum

Cordyla
- – discolor = Nervilia plicata

Cordylestyllis
- – foliosa = Goodyera procera

Coryanthes
- – albertinae = Coryanthes speciosa
- – barkeri = Coryanthes speciosa
- – eximia = Coryanthes speciosa
- – hunteriana = Coryanthes speciosa
- *• – **macrantha**
- • – maculata = Coryanthes speciosa
- – maculata var. albertinae = Coryanthes speciosa
- – maculata var. fournieri = Coryanthes speciosa
- – maculata var. parkeri = Coryanthes speciosa
- – maculata var. speciosa = Coryanthes speciosa
- – maculata var. splendens = Coryanthes speciosa
- – maculata var. vitrina = Coryanthes speciosa
- – parkeri = Coryanthes speciosa
- – picturata = Coryanthes speciosa
- – powellii = Coryanthes speciosa
- * – **speciosa**
- • – speciosa var. alba = Coryanthes speciosa
- – speciosa var. eximia = Coryanthes speciosa

Coryanthes
– speciosa var. vitellina = Coryanthes speciosa
– splendens = Coryanthes speciosa
Corycium
• – bicolor = Corycium crispum
– **crispum**
* • – **orobanchoides**
Corymbis
– angustifolia = Arundina graminifolia
Cranichis
– luteola = Polystachya concreta
– micrantha = Prescottia oligantha
– oligantha = Prescottia oligantha
– stachyodes = Prescottia stachyodes
Crybe
* – **rosea**
Cryptarrhena
– kegelii = Cryptarrhena lunata
* • – **lunata**
– pallidiflora = Cryptarrhena lunata
– unguiculata = Cryptarrhena lunata
Cryptochilus
– reticulatus = Porpax reticulata
* • – **sanguineum**
Cryptophoranthus
– acaulis = Pleurothallis tribuloides
– alvaroi = Zootrophion atropurpureum
– atropurpureus = Zootrophion atropurpureum
Cryptopus
• – **elatus**
Cryptosaccus
– scriptus = Leochilus scriptus
Cryptosanus
• – scriptus = Leochilus scriptus
Cuitlauzina
* – **pendula**
Cybelion
– pallidiflorum = Ionopsis utricularioides
– tenerum = Ionopsis utricularioides
– testiculatum = Ionopsis satyrioides
– utriculariae = Ionopsis utricularioides
Cyclopogon
– bicolor = Beadlea bicolor
– elatus = Beadlea elata
– lindleyana = Beadlea lindleyana
– miradorensis = Beadlea elata
– ovalifolium = Beadlea elata
– procerus = Sauroglossum nitidum
– pubescens = Beadlea bicolor
Cyclosia
• – maculata = Mormodes maculata
Cycnoches
– amparoanum = Cycnoches egertonianum
• – **chlorochilon**
* • – **egertonianum**
• – egertonianum var. viride = Cycnoches egertonianum
– espiritosantense = Cycnoches pentadactylon
– glanduliferum = Cycnoches egertonianum
– guttulatum = Cycnoches egertonianum
* • – **loddigesii**
– loddigesii var. leucochilum = Cycnoches loddigesii
• – **maculatum**
– pachydactylon = Cycnoches egertonianum
– pauciflorum = Cycnoches egertonianum
* • – **pentadactylon**
– peruvianum = Cycnoches egertonianum
– rossianum = Cycnoches egertonianum
– stelliferum = Cycnoches egertonianum
– stenodactylon = Cycnoches egertonianum
• – ventricosum = Cycnoches egertonianum
– ventricosum var. chlorochilon = Cycnoches chlorochilon
– ventricosum var. egertonianum = Cycnoches egertonianum
– viride = Clowesia russelliana
Cylindrolobus
– clavicaulis = Eria clavicaulis
– mucronatus = Trichotosia mucronata

Cymbidium
– affine = Cymbidium mastersii
– alagnata = Vanda tessellata
– albo-marginatum = Cymbidium ensifolium
– albuciflorum = Cymbidium madidum
* – **aloifolium**
– aloifolium Blume = Cymbidium bicolor ssp. bicolor
– aloifolium Guillaumin = Cymbidium finlaysonianum
– aloifolium var. pubescens = Cymbidium bicolor ssp. bicolor
– altissimum = Oncidium altissimum
– andersonii = Cyrtopodium andersonii
– aphyllum = Dendrobium aphyllum
– arrogans = Cymbidium ensifolium
– atropurpureum = Cymbidium aloifolium
– autumnale = Earina autumnalis
– bambusifolium = Arundina graminifolia
• – **bicolor ssp. bicolor**
* – **bicolor ssp. pubescens**
– bituberculatum = Liparis nervosa
– celebicum = Cymbidium bicolor ssp. bicolor
• – **chloranthum**
– chuen-lan = Cymbidium goeringii
– cordigerum = Encyclia cordigera
– cucullatum = Brassavola cucullata
– dependens = Cirrhaea dependens
* • – **eburneum**
– eburneum var. philbrickierum = Cymbidium eburneum
– eburneum var. williamsianum = Cymbidium eburneum
– ecristatum = Cymbidium ensifolium
* – **ensifolium**
• – ensifolium var. estriatum = Cymbidium ensifolium
– ensifolium var. misericors = Cymbidium ensifolium
– ensifolium var. rubigenmum = Cymbidium ensifolium
– ensifolium var. striatum = Cymbidium ensifolium
– ensifolium var. susin = Cymbidium ensifolium
– ensifolium var. yakibaran = Cymbidium ensifolium
– erectum = Cymbidium aloifolium
– estriatum = Cymbidium ensifolium
* – **finlaysonianum**
– flabellifolium = Cochleanthes flabelliformis
– flabelliforme = Cochleanthes flabelliformis
– floridum = Bletia purpurea
– formosanum = Cymbidium goeringii
– forrestii = Cymbidium goeringii
– furvum = Vanda concolor
– giganteum = Eulophia speciosa
– **goeringii**
– gonzalesii = Cymbidium ensifolium
– goweri = Grammatophyllum scriptum
– guttatum = Spathoglottis pubescens
– gyokuchin = Cymbidium ensifolium
– gyokuchin var. arrogans = Cymbidium ensifolium
– gyokuchin var. soshin = Cymbidium ensifolium
– hyacinthinum = Bletilla striata
– imbricatum = Pholidota imbricata
– intermedium = Cymbidium aloifolium
• – iridifolium = Cymbidium madidum
– juncifolium = Oncidium cebolleta
– kanran var. misericors = Cymbidium ensifolium
– koran = Cymbidium ensifolium
– leai = Cymbidium madidum
– leroyi = Cymbidium madidum
– lineare = Isochilus linearis
– luteum = Chloraea gavilu
– mackinnoni = Cymbidium goeringii
• – **madidum**
– madidum var. leroyi = Cymbidium madidum
– mannii = Cymbidium aloifolium
• – marginatum = Maxillaria marginata
* • – **mastersii**
– mastersii var. album = Cymbidium mastersii
– meyenii = Arundina graminifolia
– micans = Cymbidium ensifolium
– micromeson = Cymbidium mastersii
– misericors = Cymbidium ensifolium
– moschatum = Dendrobium moschatum
– nervosum = Liparis nervosa
– nitidum Roxb. = Coelogyne punctulata
– nitidum Wallich = Coelogyne nitida
– niveo-marginatum = Cymbidium ensifolium
– nodosum = Brassavola nodosa
– ochroleucum = Maxillaria camaridii
– ovatum = Dendrobium ovatum
– pendulum Blume = Cymbidium finlaysonianum
– pendulum (Blume) Lindl. = Cymbidium finlaysonianum
• – pendulum (Blume) Vidal = Cymbidium finlaysonianum
– pendulum (Roxb.) Sw. = Cymbidium aloifolium
• – pendulum var. brevilabre = Cymbidium finlaysonianum
– praecox = Pleione praecox
• – praemorsum = Acampe praemorsa
– pseudovirens = Cymbidium goeringii
• – pubescens = Cymbidium bicolor ssp. bicolor
– pubescens var. celebicum = Cymbidium bicolor ssp. bicolor
– pulchellum = Cymbidium chloranthum
– pusillum = Psygmorchis pusilla
– queeneanum = Cymbidium madidum
– reflexum = Liparis reflexa
– rubigemmum = Cymbidium ensifolium
– sanguinealentum = Cymbidium chloranthum
– sanguineum = Cymbidium chloranthum
– scriptum = Grammatophyllum scriptum
– shimaran = Cymbidium ensifolium
– simulans = Cymbidium aloifolium
– speciosissimum = Coelogyne cristata
– speciosum = Arundina graminifolia
• – **sp.**
– stapelioides = Promenaea stapelioides
– striatum = Bletilla striata
– syringodorum = Cymbidium eburneum
– tentyozanense = Cymbidium goeringii
– tessellaoides = Vanda tessellata
– tessellatum = Vanda tessellata
– tetrapetalum = Oncidium tetrapetalum
– tribuloides = Pleurothallis tribuloides
– tricolor = Cymbidium finlaysonianum
– tripterum = Coelia triptera
– umbellatum = Bulbophyllum longiflorum
– uniflorum = Cymbidium goeringii
– utriculatum = Govenia utriculata
– variciferum = Cymbidium chloranthum
– verecundum = Bletia purpurea
– violaceum = Cattleya violacea
• – virens = Cymbidium goeringii
– virescens = Cymbidium goeringii
– wallichii = Cymbidium finlaysonianum
• – xiphiifolium = Cymbidium ensifolium
– yakibaran = Cymbidium ensifolium
– yakibaran var. albo-marginatum = Cymbidium ensifolium
– yakibaran var. niveo-marginatum = Cymbidium ensifolium
– yunnanense = Cymbidium goeringii
Cymbiglossum
– apterum = Lemboglossum apterum
– bictoniense = Lemboglossum bictoniense
– cervantesii = Lemboglossum cervantesii
– cordatum = Lemboglossum cordatum
– ehrenbergii = Lemboglossum ehrenbergii

Cymbiglossum
– galeottianum = Lemboglossum galeottianum
– maculatum = Lemboglossum maculatum
– madrense = Lemboglossum madrense
– rossii = Lemboglossum rossii
– stellatum = Lemboglossum stellatum
Cynorkis
* • – **fastigiata**
Cynosorchis
– decolorata = Cynorkis fastigiata
– diplorhyncha = Cynorkis fastigiata
– fastigiata = Cynorkis fastigiata
– hygrophila = Cynorkis fastigiata
– laggiarae = Cynorkis fastigiata
Cyperorchis
– eburnea = Cymbidium eburneum
– mastersii = Cymbidium mastersii
Cypripedium
– album = Cypripedium reginae
• – barbatum = Paphiopedilum barbatum
– barbatum var. biflorum = Paphiopedilum barbatum
– biflorum = Paphiopedilum barbatum
– calceolus var. L. = Cypripedium reginae
– canadensis = Cypripedium reginae
– cothurmum = Catasetum trulla
– epidendricum = Eulophia alta
– hirsutum = Cypripedium reginae
– hirsutum f. album = Cypripedium reginae
* • – **irapeanum**
– linearisubulatum = Cleisostoma subulatum
– macranthos var. vulgare = Cypripedium macranthum
* • – **macranthum**
– molle = Cypripedium irapeanum
– nigritum = Paphiopedilum barbatum
– pardinum = Paphiopedilum venustum
• – purpuratum = Paphiopedilum purpuratum
* – **reginae**
– reginae var. album = Cypripedium reginae
– sinicum = Paphiopedilum purpuratum
– socco = Catasetum trulla
• – spectabile = Cypripedium reginae
– thunbergii = Cypripedium macranthum
• – venustum = Paphiopedilum venustum
Cyrtochilum
– bictoniense = Lemboglossum bictoniense
– deltoideum = Oncidium deltoideum
• – filipes = Oncidium brachyandrum
• – flavescens = Miltonia flavescens
• – flexuosum = Oncidium cimiciferum
– gracile = Odontoglossum gracile
• – graminifolium = Oncidium brachyandrum
– grandiflorum = Oncidium maculatum
• – ixioides = Odontoglossum ixioides
– juergensianum = Oncidium hastatum
• – karwinskii = Miltonioides karwinskii
– leucochilum Linden = Oncidium maculatum
– leucochilum (Bateman ex Lindl.) Planchet = Oncidium leucochilum
– longifolium = Odontoglossum longifolium
• – maculatum = Oncidium maculatum
• – maculatum var. ecornutum = Oncidium maculatum
• – maculatum var. parviflorum = Oncidium maculatum
– maculatum var. russelianum = Oncidium maculatum
– myanthum = Odontoglossum myanthum
• – mystacinum = Odontoglossum mystacinum
• – pardinum = Odontoglossum pardinum
– retusum = Odontoglossum retusum
• – stellatum = Miltonia flavescens
– trulla = Oncidium cimiciferum
• – undulatum = Oncidium undulatum
• – volubile = Oncidium volubile
Cyrtopera
– alta = Eulophia alta
– fusca = Eulophia nuda
– gardneri = Eulophia nuda
– gigantea = Eulophia speciosa
– laxiflora = Eulophia nuda
– longifolia = Eulophia alta
– mysorensis = Eulophia nuda
– nuda = Eulophia nuda
– plicata = Eulophia nuda
– polyantha = Eulophia alta
– scabrilinguis = Xylobium variegatum
– vellosiana = Eulophia alta
– woodfordii = Eulophia alta
Cyrtopodium
* • – **andersonii**
– bracteatum = Cyrtopodium paniculatum
– **cristatum**
– fuscum = Eulophia nuda
– naiguatae = Cyrtopodium paniculatum
– **paniculatum**
– punctatum var. bracteatum = Cyrtopodium paniculatum
• – woodfordii = Eulophia alta
Dactylorchis
– foliosa = Dactylorhiza foliosa
– sambucina = Dactylorhiza sambucina
Dactylorhiza
* – **foliosa**
– **sambucina**
Dactylostylis
– fimbriata = Zygostates cornuta
Deiregyne
– **diaphana**
Dendrobium
• – **acerosum**
• – **aciculare**
• – **acuminatissimum**
– acuminatum = Pleurothallis acuminata
* • – **aduncum**
• – **affine**
– aggregatum Kunth = Maxillaria aggregata
• – aggregatum Roxb. = Dendrobium lindleyi
– aggregatum var. jenkinsii = Dendrobium jenkinsii
– alagense = Dendrobium luzonense
– alboviride = Dendrobium linawianum
– album Hook. = Maxillaria alba
– album Wight = Dendrobium aqueum
• – alpestre Royle = Dendrobium monticola
– alpestre (Sw.) Sw. = Pleurothallis alpestris
• – **amoenum**
* • – **anceps**
– anceps Gagnepain = Dendrobium leonis
– anceps Lindl. = Dendrobium nathanielis
– **angulatum**
• – **anosmum**
• – **antennatum**
* – **aphyllum**
* • – **aqueum**
– arachnanthe = Dendrobium discolor
– arachnites = Aeranthes arachnites
– arachnostachyum = Dendrobium macranthum
– aruanum = Dendrobium mirbelianum
– atractodes = Dendrobium heterocarpum
– augustae-victoriae = Dendrobium lineale
• – aureum = Dendrobium heterocarpum
• – aureum var. pallidum = Dendrobium heterocarpum
– aureum var. philippinensis = Dendrobium heterocarpum
• – auriferum = Thrixspermum centipeda
– auroroseum = Dendrobium nudum
– bambusifolium = Dendrobium salaccense
• – barbatulum = Dendrobium ovatum
– barringtoniae = Lycaste barringtoniae
• – **bicameratum**
• – bicolor Lindl. = Eria lindleyi
– bicolor (Ruiz Lopez & Pavon) Pers. = Maxillaria bicolor
– **bifarium**
• – **biflorum**
• – biflorum = Dendrobium cunninghamii
• – **bilobum**
– binnendijkii = Flickingeria fimbriata
– bolboflorum = Dendrobium bicameratum
– brachythecum = Dendrobium macrophyllum
– breviflorum = Dendrobium bicameratum
– brisbanense = Dendrobium gracilicaule
– buluense = Dendrobium mirbelianum
– buluense var. kauloense = Dendrobium mirbelianum
– bursigerum = Dendrobium secundum
• – calamiforme = Dendrobium teretifolium
• – **calcaratum**
• – calcaratum = Dendrobium lobbii
– calceolaria = Dendrobium moschatum
– **caleyi**
– calopogon = Dendrobium angulatum
• – cambridgeanum = Dendrobium ochreatum
• – **candidum**
– caninum = Dendrobium crumenatum
– carnosum = Xylobium variegatum
• – cassythoides = Galeola cassythoides
– castum = Dendrobium moniliforme
• – catenatum = Dendrobium moniliforme
– cathcartii = Dendrobium salaccense
– caudatum var. javanica = Dendrobium acuminatissimum
– cebolleta = Oncidium cebolleta
– ceraia = Dendrobium crumenatum
– cerinum = Dendrobium sanguinolentum
• – chlorops = Dendrobium ovatum
* • – **chrysanthum**
– chrysanthum var. anophthalma = Dendrobium chrysanthum
– chrysanthum var. microphthalma = Dendrobium chrysanthum
* • – **chrysotoxum**
– chrysotoxum var. delacourii = Dendrobium chrysotoxum
– chrysotoxum var. suavissimum = Dendrobium chrysotoxum
– ciliatum = Lycaste ciliata
• – clavatum = Dendrobium densiflorum
• – coerulescens = Dendrobium nobile
– cogniauxianum = Dendrobium lineale
– comatum = Flickingeria comata
• – complanatum = Oberonia complanata
• – compressum = Dendrobium lamellatum
– conostalix = Dendrobium lobbii
– corniculatum = Pleurothallis corniculata
– cornutum = Dendrobium kuhlii
• – crassulifolium = Bulbophyllum sheperdii
* • – **cretaceum**
• – criniferum = Flickingeria comata
– crispum = Dendrobium microbulbon
* • – **crumenatum**
• – cucullatum = Dendrobium aphyllum
* • – **cucumerinum**
– cultriforme = Polystachya cultriformis
– cumulatum = Dendrobium crumenatum
• – **cunninghamii**
• – cupreum = Dendrobium moschatum
• – cuspidatum = Dendrobium nathanielis
• – dalhousieanum = Dendrobium pulchellum
* • – **densiflorum**
• – **denudans**
– denudans = Dendrobium monticola
• – **devonianum**
– dicuphum = Dendrobium affine
* • – **discolor**
– d'albertisii = Dendrobium antennatum
• – egertoniae = Dendrobium amoenum
– elegans = Pleurothallis elegans
– elobatum = Dendrobium discolor
• – elongatum = Dendrobium gracilicaule
• – **erosum**
• – excisum = Dendrobium bifarium
• – extinctorium = Eria summerhayesiana
– fairfaxii = Dendrobium teretifolium
– farmeri var. aureo-flava = Dendrobium griffithianum
– fellowsii = Dendrobium gracilicaule
– ferox = Dendrobium macrophyllum
• – fimbriatum = Flickingeria fimbriata
– firmum = Dendrobium mutabile

Dendrobium

– flabellum Hemsley = Flickingeria fimbriata
– flabellum Rchb.f. = Flickingeria fimbriata
• – flavescens = Polystachya concreta
– flexuosum = Dendrobium longicornu
• – foliosum = Pseuderia foliosa
– formosanum = Dendrobium nobile
* • – **formosum**
– friedericksianum = Dendrobium nobile
– fuscum = Dendrobium discolor
– fusiforme = Polystachya fusiformis
• – gemellum Lindl. = Dendrobium indragiriense
– gemellum Ridley = Dendrobium salaccense
– giulianettii = Dendrobium mirbelianum
– glaucophyllum = Dendrobium rugosum
– gordonii = Dendrobium macrophyllum
– **gracilicaule**
– grandiflorum Kunth = Maxillaria grandiflora
– grandiflorum Reinw. = Bulbophyllum uniflorum
– grantii = Dendrobium lineale
• – **griffithianum**
– griffithianum var. guibertii = Dendrobium densiflorum
– guibertii = Dendrobium densiflorum
– harrisoniae = Bifrenaria harrisoniae
• – hasseltii = Dendrobium angulatum
– hastatum = Maxillaria hastata
– heisnanense = Dendrobium moniliforme
– henshallii = Dendrobium transparens
• – **herbaceum**
* • – **heterocarpum**
– heterocarpum var. henshallii = Dendrobium heterocarpum
– heterostigma = Dendrobium secundum
• – **heyneanum**
– hildebrandii = Dendrobium heterocarpum
– hirsutum = Dendrobium longicornu
– humile = Dendrobium microbulbon
• – **hymenophyllum**
– imperatrix = Dendrobium lineale
– inauditum = Diplocaulobium longicolle
– inconspicuiflorum = Dendrobium indragiriense
– **indragiriense**
– infundibulum = Dendrobium formosum
– insulare = Flickingeria fimbriata
– intermedium = Dendrobium salaccense
– isomerum = Dendrobium indragiriense
– japonicum = Dendrobium moniliforme
– javanicum = Eria javanica
* • – **jenkinsii**
– jenkinsii Lang & Tsi = Dendrobium lindleyi
– jerdonianum = Dendrobium nutans
– jonesii = Dendrobium gracilicaule
• – **junceum**
* • – **kingianum**
* • – **kuhlii**
– kunstleri = Flickingeria fimbriata
– kwashotense = Dendrobium crumenatum
– laciniosum = Flickingeria fimbriata
* – **lamellatum**
– lanceola = Pleurothallis lanceola
• – **lancifolium**
– latifolium = Maxillaria latifolia
– laxum = Pleurothallis laxa
– **leonis**
– lessonii = Dendrobium cunninghamii
– leucorhodum = Dendrobium anosmum
– ligulatum = Maxillaria ligulata
* – **linawianum**
– lindleyanum = Dendrobium nobile
* – **lindleyi**
– **lineale**
• – **linguiforme**
– **lobbii**
• – longicolle = Diplocaulobium longicolle
* • – **longicornu**
– longicornu var. hirsutum = Dendrobium longicornu
– longifolium = Eulophia alta
– longipetalum = Lycaste longipetala
• – **luzonense**
– macraei = Flickingeria fimbriata
• – **macranthum**
– macranthum = Dendrobium anosmum
• – **macrophyllum**
• – macrophyllum = Dendrobium anosmum
– macrophyllum var. veitchianum = Dendrobium macrophyllum
* • – **macrostachyum**
– maculatum = Maxillaria maculata
– madrasense = Dendrobium aphyllum
– mannii = Dendrobium nathanielis
– marseillei = Dendrobium jenkinsii
– mentosum = Flickingeria fimbriata
• – mesochlorum = Dendrobium amoenum
• – **microbulbon**
– microbulbon = Dendrobium peguanum
– minahassae = Dendrobium heterocarpum
• – **mirbelianum**
– monile = Dendrobium moniliforme
• – **moniliforme**
– moniliforme = Dendrobium linawianum
– **monophyllum**
– **monticola**
* • – **moschatum**
– multiflorum = Dendrobium nathanielis
– musciferum = Dendrobium macrophyllum
* • – **mutabile**
– **nathanielis**
• – **nobile**
• – **nudum**
• – **nutans**
– nutantiflorum = Dendrobium nutans
• – **ochreatum**
– ophioglossoides = Pleurothallis ophioglossoides
– **ovatum**
– palmifolium = Xylobium palmifolium
– paludicola = Dendrobium lobbii
– paniculatum = Cyrtopodium paniculatum
– parishii = Dendrobium anosmum
• – paxtonii = Dendrobium chrysanthum
– **peguanum**
– perakense = Eria javanica
– pictum = Dendrobium devonianum
• – pierardii = Dendrobium aphyllum
– pierardii var. cucullatum = Dendrobium aphyllum
• – **planibulbe**
– platypetalum = Maxillaria platypetala
• – plicatile = Flickingeria fimbriata
– plicatum = Cyrtopodium paniculatum
• – **polyanthum**
– polycarpum = Dendrobium mirbelianum
– polypetalum = Maxillaria platypetala
– polystachyon = Polystachya concreta
– primulinum = Dendrobium aphyllum
– proliferum = Maxillaria prolifera
– pseudotokai = Dendrobium macranthum
– psyche = Dendrobium macrophyllum
• – **pugioniforme**
* • – **pulchellum**
– pulchellum Lindl. = Dendrobium devonianum
– pulchellum var. devonianum = Dendrobium devonianum
– pumilum = Bulbophyllum pumilum
– punctatum = Dipodium punctatum
– pungentifolium = Dendrobium pugioniforme
– purpurascens = Dendrobium rugosum
– pusillum D.Don = Dendrobium monticola
– pusillum Kunth = Trichosalpinx pusilla
• – pygmaeum Lindl. = Dendrobium peguanum
• – pygmaeum Lindl. = Dendrobium caleyi
– quadrifidum = Pleurothallis quadrifida
– racemiflorum = Pleurothallis racemiflora
– ramosissimum = Dendrobium herbaceum
• – ramosum Lindl. = Dendrobium ruckeri
– ramosum Pers. = Maxillaria ramosa
– refractum = Dendrobium revolutum
• – retusum = Pleurothallis retusa
• – **revolutum**
• – rhombeum = Dendrobium heterocarpum
– rigescens = Dendrobium mutabile
– rigidum = Dendrobium mutabile
– rosenbergii = Dendrobium mirbelianum
– roseum = Eulophia horsfallii
* • – **ruckeri**
• – **rugosum**
– ruscifolium = Pleurothallis ruscifolia
• – **salaccense**
* • – **sanguinolentum**
– sarcostoma = Dendrobium macrophyllum
• – scariosum = Pleurothallis scariosa
– schmidtianum = Dendrobium crumenatum
– schroederi = Dendrobium densiflorum
• – sclerophyllum = Dendrobium mutabile
• – scopa = Flickingeria scopa
– scortechinii = Dendrobium anosmum
* • – **secundum**
– secundum var. bursigerum = Dendrobium secundum
– separatum = Dendrobium calcaratum
– sertularioides = Pleurothallis sertularioides
– sessile = Trichotosia velutina
– setifolium = Dendrobium aciculare
– sheperdii = Bulbophyllum sheperdii
– simplicissimum = Dendrobium crumenatum
– **sinuatum**
– spathaceum = Dendrobium candidum
* • – **speciosum**
– spegidoglossum = Dendrobium stuposum
• – squalens = Xylobium variegatum
• – **stuposum**
– suavissimum Kraenzlin = Dendrobium lindleyi
– suavissimum Rchb.f. = Dendrobium chrysotoxum
– subteres = Dendrobium acerosum
– subterrestre = Eria bractescens
* • – **sulcatum**
– superbum = Dendrobium anosmum
* • – **taurinum**
– teijsmannii = Dendrobium lobbii
• – teres = Papilionanthe teres
– **teretifolium**
– ternatense = Dendrobium macrophyllum
– testiculatum = Ionopsis satyrioides
• – **tetragonum**
– tetragonum var. variabilis = Dendrobium tetragonum
– thyrsodes = Dendrobium kuhlii
– tokai var. crassinerve = Dendrobium macranthum
– tomohonense = Dendrobium macrophyllum
– tonkinensis = Dendrobium revolutum
• – tortile = Dendrobium monophyllum
• – **transparens**
• – triadenium = Dendrobium mutabile
– tribuloides = Pleurothallis tribuloides
• – **tridentiferum**
– triphyllum = Maxillaria triphylla
– triviale = Dendrobium calcaratum
– tuberiferum = Dendrobium planibulbe
– undatiflorum = Maxillaria undatiflora
• – undulatum = Dendrobium discolor
– utricularioides = Ionopsis utricularioides
– variegatum = Xylobium variegatum
• – veitchianum = Dendrobium macrophyllum
• – veratrifolium = Dendrobium lineale
– veratroides = Dendrobium lineale
– vestitum = Trichotosia vestita
– villosulum = Dendrobium nutans
– wallichii = Dendrobium peguanum
– wilkianum = Dendrobium mirbelianum
– zollingerianum = Flickingeria comata

Dendrobium
– zonatum = Dendrobium moniliforme
Dendrochilum
• – **abbreviatum**
• – **filiforme**
• – **glumaceum**
– glumaceum = Dendrochilum filiforme
• – **latifolium**
Dendrocolla
– arachnites = Thrixspermum centipeda
– pallida = Pteroceras pallidus
Dendrocoryne
– kingianum = Dendrobium kingianum
– speciosum = Dendrobium speciosum
– tetragonum = Dendrobium tetragonum
Dendrolirium
– micranthum = Eria multiflora
– multiflorum = Eria multiflora
– ornatum = Eria ornata
– retusum = Eria retusa
– rugosum = Eria javanica
Dendrophyllax
– **funalis**
Dendrorchis
– caracasana = Polystachya foliosa
– estrellensis = Polystachya concreta
– extinctoria = Polystachya concreta
– foliosa = Polystachya foliosa
– jussieuana = Polystachya concreta
– minuta = Polystachya concreta
– nana = Polystachya foliosa
– rufinula = Polystachya concreta
– shirensis = Polystachya concreta
– similis = Polystachya concreta
– tessellata = Polystachya concreta
– wightii = Polystachya concreta
– zollingeri = Polystachya concreta
Desmotrichum
– angulatum = Dendrobium angulatum
– binnendijkii = Flickingeria fimbriata
– comatum = Flickingeria comata
– fimbriatum = Flickingeria fimbriata
– grandiflorum = Flickingeria fimbriata
– kunstleri = Flickingeria fimbriata
– plicatile = Flickingeria fimbriata
– scopa = Flickingeria scopa
Diacrium
– amazonicum = Caularthron bicornutum
– bicornutum = Caularthron bicornutum
– bidentatum = Encyclia boothiana ssp. boothiana
Diadenium
– **barkeri**
Diaphananthe
– **bidens**
* – **pellucida**
Dichaea
• – **glauca**
• – **ochracea**
Dichaeopsis
– glauca = Dichaea glauca
Dichromanthus
* – **cinnabarinus**
Dicrophylla
– elegans = Ludisia discolor
Dicrypta
– baueri = Maxillaria crassifolia
– bicolor = Maxillaria bicolor
– crassifolia = Maxillaria crassifolia
• – discolor = Maxillaria discolor
– elatior = Maxillaria elatior
Didothion
– clavatum = Epidendrum purpurascens
Dienia
• – congesta = Malaxis latifolia
• – cordata = Malaxis cordata
– fusca = Malaxis latifolia
Digomphotis
– cordata = Gennaria diphylla
Dimerandra
– **emarginata**
– isthmi = Dimerandra emarginata
– lamellata = Dimerandra emarginata
– rimbachii = Dimerandra emarginata
– stenopetala = Dimerandra emarginata
Dinema
• – paleacea = Epidendrum boothii
Dipera
– capensis = Disperis capensis
– tenera = Disperis capensis
Diphyes
– flavescens = Bulbophyllum flavescens
Diplecthrum
– bracteatum = Satyrium bracteatum
– coriifolium = Satyrium coriifolium
– erectum Lindl. = Satyrium coriifolium
– erectum (Sw.) Pers. = Satyrium erectum
Diplocaulobium
– **longicolle**
Diplocentrum
– longifolium = Diplocentrum recurvum
• – **recurvum**
Diplochilos
• – hirsutum = Diplomeris hirsuta
• – longifolium = Diplomeris pulchella
Diplomeris
– **hirsuta**
– **pulchella**
Dipodium
– plicatum = Eulophia nuda
*• – **punctatum**
– punctatum var. album = Dipodium punctatum
– punctatum var. stenocheilum = Dipodium punctatum
– roniate = Eulophia nuda
– stenocheilum = Dipodium punctatum
Disa
– atropurpurea = Herschelianthe spathulata
– barelli = Disa uniflora
• – bracteata = Monadenia bracteata
– cernua = Monadenia cernua
– densiflora = Monadenia cernua
• – graminifolia = Herschelia coelestis
• – grandiflora = Disa uniflora
– physodes = Monadenia cernua
• – **porrecta**
• – prasinata = Monadenia cernua
– propinqua = Herschelianthe spathulata
– propinqua var. trifida = Herschelianthe spathulata
• – spathulata = Herschelianthe spathulata
– spathulata var. atropurpurea = Herschelianthe spathulata
* – **uniflora**
– uniflora var. barellii = Disa uniflora
– uniflora var. psittacina = Disa uniflora
– uniflora var. superba = Disa uniflora
– zeyheri = Disa porrecta
Disperis
• – **capensis**
– circumflexa = Disperis secunda
• – **cucullata**
• – **secunda**
• – **villosa**
Dituilis
– nepalensis = Liparis nervosa
Dockrillia
– cucumerina = Dendrobium cucumerinum
– linguiformis = Dendrobium linguiforme
– pugioniformis = Dendrobium pugioniforme
– teretifolia = Dendrobium teretifolium
Donacopis
– laotica = Arundina graminifolia
Dothiolophis
– purpurea = Barkeria skinneri
Earina
– alba = Earina autumnalis
– **autumnalis**
• – **mucronata**
– quadrilobata = Earina mucronata
• – suaveolens = Earina autumnalis
Empusa
• – paradoxa = Liparis paradoxa
– sp. = Liparis paradoxa
Encyclia
– acicularis = Encyclia tripartita
– **adenocarpon**
* – **adenocaula var. adenocaula**
* – **alata**
– almasyi = Encyclia glumacea
– amabilis = Encyclia concolor
– **aromatica**
– **aspera**
– atropurpurea var. leucantha = Encyclia cordigera
– atropurpurea var. rhodoglossa = Encyclia cordigera
– atropurpurea var. rosea = Encyclia cordigera
– bifida = Psychilis bifida
– **boothiana ssp. boothiana**
– boothiana var. erythronioides = Encyclia boothiana ssp. boothiana
– **bractescens**
– **bulbosa**
– **calamaria**
– **candollei**
– **ceratistes**
– chiriquensis = Encyclia varicosa ssp. leiobulbon
– **chloroleuca**
– **citrina**
– **concolor**
– **cordigera**
– cordigera var. rosea = Encyclia cordigera
– deamii = Encyclia livida
– **dichroma**
– diguetii = Encyclia tripunctata
– **diota ssp. diota**
– doeringii = Encyclia cordigera
– ekmanii = Psychilis bifida
– erubescens = Epidendrum erubescens
– flava = Encyclia tripartita
* – **fragrans**
– **fucata**
– **glauca**
* – **glumacea**
* – **gracilis**
– granitica = Encyclia oncidioides
– gravida = Encyclia oncidioides
– **hanburii**
– **hastata**
– hircina = Encyclia fucata
– hunteriana = Encyclia oncidioides
– icthyphylla = Encyclia michuacana
– insidiosa = Encyclia diota ssp. diota
– **ionosma**
* – **lancifolia**
– laxa = Encyclia candollei
– limbata = Encyclia glauca
– **linkiana**
– **livida**
– macrochila = Encyclia cordigera
• – macrostachya = Amblostoma armeniacum
– **michuacana**
– **microbulbon**
– multiflora = Encyclia viridiflora
• – nana = Polystachya foliosa
– nemoralis = Encyclia adenocaula var. adenocaula
– oblongata = Encyclia phoenicea
* – **ochracea**
* – **odoratissima**
*• – **oncidioides**
– oncidioides var. ramonensis = Encyclia ceratistes
– ovulum = Encyclia microbulbon
– papilionacea = Psychilis bifida
– **pastoris**
– patens = Encyclia odoratissima
– **phoenicea**
* – **plicata**
• – polystachya = Polystachya foliosa
– powellii Schltr. = Encyclia ceratistes
– powellii Schltr. ex Ames = Encyclia oncidioides
– pruinosa = Encyclia concolor
* – **pterocarpa**
– **pygmaea**

* – **pyriforme**
– quadratum = Encyclia varicosa ssp. leiobulbon
* – **radiata**
– ramonensis = Encyclia ceratistes
– **rufa**
– **selligera**
– serroniana = Encyclia odoratissima
– sisyrinchiifolia = Encyclia microbulbon
– **tampensis**
– tesselata = Encyclia livida
– **tripartita**
– triptera = Encyclia pygmaea
– **tripunctata**
– **varicosa ssp. leiobulbon**
– **varicosa ssp. varicosa**
– **venosa**
* – **vespa**
– virgata = Encyclia michuacana
– **viridiflora**
* – **vitellina**
– wendlandiana = Encyclia venosa

Encyclium
– ionosmum = Encyclia ionosma

Ephemerantha
– comata = Flickingeria comata
– fimbriata = Flickingeria fimbriata
– kunstleri = Flickingeria fimbriata
– pallens = Flickingeria fimbriata
– scopa = Flickingeria scopa

Ephippium
– capitatum = Bulbophyllum pahudii
– ciliatum = Bulbophyllum blumei
– elongatum = Bulbophyllum elongatum
– uniflorum = Bulbophyllum uniflorum

Epicladium
– boothianum = Encyclia boothiana ssp. boothiana
– boothianum var. erythronioides = Encyclia boothiana ssp. boothiana

Epidendrum
• – aciculare = Encyclia tripartita
– aclandiae = Cattleya aclandiae
• – **acuminatum**
• – adenocarpon = Encyclia adenocarpon
– adenocarpon var. rosei = Encyclia adenocarpon
• – adenocaulum = Encyclia adenocaula var. adenocaula
• – **adenoglossum**
• – aemulum = Encyclia fragrans
– aemulum var. breviaristatum = Encyclia fragrans
– aerides = Aerides odorata
• – aeridiforme = Epidendrum harrisoniae
– affine A.Rich. = Encyclia fucata
– affine Focke = Encyclia oncidioides
• – **aggregatum**
– alanjense = Encyclia oncidioides
• – alatum = Encyclia alata
– alatum var. grandiflorum = Encyclia alata
– alatum var. longipetalum = Encyclia alata
– alatum var. viridiflorum = Encyclia alata
– alexandri = Epidendrum cristatum
– alfredii = Epidendrum smaragdinum
– almasyi = Encyclia glumacea
– aloides = Cymbidium aloifolium
• – aloifolium Bateman = Epidendrum parkinsonianum
– aloifolium L. = Cymbidium aloifolium
– alopecurum = Amblostoma armeniacum
– alpestre = Pleurothallis alpestris
– alternans = Epidendrum fimbriatum
• – altissimum Jacq. = Oncidium jacquinianum
– altissimum Jacq. = Oncidium altissimum
– amabile Linden & Rchb.f. = Encyclia concolor
– amabile (Bateman) Godef.-Leb. = Encyclia dichroma
– amandum = Encyclia oncidioides
– amazonicum = Epidendrum coronatum
– amblostoma = Amblostoma tridactylum
– ammophyllum = Epidendrum harrisoniae
– amphistomum = Epidendrum secundum
– anceps Jacq. = Epidendrum secundum
– anceps Sw. = Epidendrum bifarium
– anceps var. virescens = Epidendrum secundum
– anceps var. viridipurpureum = Epidendrum secundum
• – anisatum = Epidendrum gladiatum
– ansiferum = Epidendrum elongatum
– anthoceros = Epidendrum cornutum
– antioquiense = Epidendrum elongatum
• – **aquaticum**
– arachnoglossum = Epidendrum elongatum
– arachnoideum = Epidendrum difforme
• – **arbuscula**
– aristotelea = Spiranthes sinensis
• – armeniacum = Amblostoma armeniacum
• – aromaticum = Encyclia aromatica
• – articulatum = Encyclia livida
• – asperum = Encyclia aspera
– atacazoicum = Epidendrum paniculatum
– atropurpureum var. laciniatum = Encyclia cordigera
– atropurpureum var. lionetianum = Encyclia cordigera
– atropurpureum var. roseum = Encyclia cordigera
• – aurantiacum = Cattleya aurantiaca
– aureum = Cattleya aurantiaca
• – auritum = Epidendrum boothii
• – auropurpureum = Encyclia cordigera
– autumnale = Earina autumnalis
• – avicula = Lanium avicula
– baculibulbom = Encyclia vespa
– bahiense = Epidendrum nocturnum
– barringtoniae = Lycaste barringtoniae
– basilare = Epidendrum stamfordianum
– bathyschistum = Epidendrum cristatum
– benignum = Epidendrum coronatum
– bicolor = Cattleya bicolor
• – bicornutum = Caularthron bicornutum
• – bidentatum = Encyclia boothiana ssp. boothiana
– bifalce = Epidendrum paniculatum
• – **bifarium**
• – bifidum = Psychilis bifida
– biflorum Barb.Rodr. = Encyclia dichroma
– biflorum Cogn. = Epidendrum paranaense
– biflorum Forster = Dendrobium biflorum
• – biforatum = Epidendrum cucullatum
• – bisetum = Epidendrum polyanthum
– blandum = Epidendrum campestre
• – **blepharistes**
– blepharoclinium = Epidendrum schomburgkii
– boissierianum = Epidendrum paranaense
• – boothianum = Encyclia boothiana ssp. boothiana
– **boothii**
• – **brachyglossum**
– brachyphyllum = Epidendrum elongatum
– bracteatum = Rodriguezia bracteata
• – bracteolatum = Encyclia radiata
• – bractescens = Encyclia bractescens
– brevicaule = Nanodes discolor
– buenaventurae = Epidendrum nocturnum
– bulbosum = Encyclia bulbosa
– bulkeleyi = Epidendrum elongatum
• – caespitosum = Encyclia pygmaea
– cajamarcae = Epidendrum geminiflorum
• – calamarium = Encyclia calamaria
– calcaratum Sessé & Mociño = Ionopsis utricularioides
– calcaratum Vell. = Bifrenaria inodora
– calceolare = Gastrochilus calceolaris
– calliferum = Epidendrum cristatum
• – calocheilum = Encyclia alata
– caloglossum = Epidendrum paniculatum
• – **campestre**
– canaliculatum = Cattleya loddigesii
– candidum = Epidendrum saxatile
• – candollei = Encyclia candollei
– caninum = Dendrobium crumenatum
– cardiophyllum = Epidendrum scabrum
• – carnosum = Epidendrum dendrobioides
– carolinianum = Epidendrum nocturnum
* – **carpophorum**
– carthagenense = Oncidium carthagenense
– caudatum = Brassia caudata
• – **cauliflorum**
– cearense = Epidendrum secundum
– cebolleta = Oncidium cebolleta
– cepiforme = Encyclia candollei
• – ceratistes = Encyclia ceratistes
• – **cernuum**
• – **chioneum**
– chiriquense = Encyclia varicosa ssp. leiobulbon
• – chloranthum = Encyclia chloroleuca
– chlorocorymbos = Epidendrum difforme
• – chloroleucum = Encyclia chloroleuca
– chlorops = Epidendrum clowesii
– christii = Encyclia vespa
*• – **ciliare**
– ciliare var. cuspidatum = Epidendrum ciliare
– ciliare var. minor = Epidendrum ciliare
– ciliare var. squamatum = Epidendrum ciliare
– ciliare var. viscidum = Epidendrum ciliare
*• – **cinnabarinum**
– cinnamomeum = Encyclia pterocarpa
– citrinum Barb.Rodr. = Amblostoma tridactylum
– citrinum (Llave & Lex.) Rchb.f. = Encyclia citrina
– claesianum = Epidendrum chioneum
• – clavatum = Epidendrum purpurascens
• – **clowesii**
• – cochleatum = Encyclia fragrans
• – cochlidium = Epidendrum elongatum
• – **collare**
– colorans = Epidendrum polyanthum
– compositum = Epidendrum coronatum
– **compressum**
– concalvii = Epidendrum saxatile
• – concolor = Encyclia concolor
– concretum = Polystachya concreta
– condylochilum = Encyclia livida
– congestoides = Nanodes discolor
– congestum = Nanodes discolor
• – **conopseum**
– conopseum var. mexicanum = Epidendrum conopseum
– conspicuum = Encyclia dichroma
• – **cordatum**
– cordatum = Encyclia fragrans
– cordigerum = Encyclia cordigera
– coriaceum Focke = Encyclia vespa
– coriaceum Parker ex Hook. = Encyclia vespa
– corniculatum = Pleurothallis corniculata
• – **cornutum**
– cornutum = Stanhopea oculata
– coroicoense = Epidendrum elongatum
• – **coronatum**
– corymbosum = Epidendrum elongatum
• – **costatum**
• – crassifolium = Epidendrum elongatum
– crassilabium = Encyclia vespa
– crenatum = Ionopsis utricularioides
• – crispatum = Encyclia adenocarpon
*• – **cristatum**
• – cubense = Cattleyopsis cubensis
• – **cucullatum**
– cucullatum = Brassavola cucullata
– curvicolumna = Epidendrum difforme
• – cuspidatum = Epidendrum ciliare
– cuzcoense = Epidendrum elongatum
– cycnostachys = Epidendrum stamfordianum
– cycnostalix = Epidendrum stamfordianum
• – **cylindraceum**
– dasytainia = Encyclia livida
– deamii = Encyclia livida

– **dendrobioides**
*• – **densiflorum**
• – dichotomum = Epidendrum elongatum
• – dichromum = Encyclia dichroma
* – **difforme**
• – **diffusum**
– diffusum var. depauperatum = Epidendrum diffusum
– diguetii = Encyclia tripunctata
• – diotum = Encyclia diota ssp. diota
• – dipus = Epidendrum densiflorum
• – discolor A.Rich & Galeotti = Epidendrum nocturnum
– discolor (Lindl.) Benth. = Nanodes discolor
– doeringii = Encyclia cordigera
– dolichopus = Epidendrum elongatum
– duboisianum = Encyclia cordigera
• – durum = Epidendrum dendrobioides
– ekmanii = Psychilis bifida
– elatius = Cattleya guttata
– elegans Vell. = Cattleya guttata
– elegans (Knowles & Westc.) Rchb.f. = Barkeria elegans
• – ellipticum Graham = Epidendrum elongatum
– ellipticum Lodd. = Epidendrum elongatum
– ellipticum Rchb.f. = Epidendrum elongatum
– ellipticum var. flavum = Epidendrum xanthinum
– ellipticum var. freybergense = Epidendrum elongatum
• – **elongatum**
– elongatum var. crassifolium = Epidendrum elongatum
– englerianum = Epidendrum paniculatum
• – ensatum = Epidendrum secundum
– ensicaulon = Encyclia venosa
– ensifolium = Cymbidium ensifolium
• – equitans = Epidendrum equitantifolium
– **equitantifolium**
• – **erubescens**
– erythronioides = Encyclia boothiana ssp. boothiana
– evectum = Epidendrum elongatum
• – **excisum**
– excisum var. grandiflorum = Epidendrum excisum
– expansum = Epidendrum elongatum
• – falcatum = Epidendrum parkinsonianum
– falsiloquum = Epidendrum paniculatum
– fastigiatum = Epidendrum elongatum
– fastigiatum var. bifidum = Epidendrum elongatum
– feddeanum = Encyclia vespa
• – **ferrugineum**
• – **filicaule**
• – **fimbriatum**
– fimbria = Epidendrum elongatum
– fimbriatum = Schomburgkia gloriosa
– flabellatum = Encyclia candollei
– flabelliforme = Cochleanthes flabelliformis
– flavidum = Epidendrum leucochilum
– flavovirens = Epidendrum clowesii
• – **flavum**
– flavum = Encyclia tripartita
– flexicaule = Epidendrum ramosum
• – flexuosum = Epidendrum imatophyllum
• – floribundum = Epidendrum densiflorum
– floribundum var. convexum = Epidendrum paniculatum
– forbesii = Cattleya forbesii
– formosum = Encyclia alata
– fractiflexum = Epidendrum saxatile
• – fragrans = Encyclia fragrans
– fragrans var. aemulum = Encyclia fragrans
– fragrans var. alticallum = Encyclia fragrans
– fragrans var. breviaristatum = Encyclia fragrans
– fragrans var. ionoleucum = Encyclia fragrans
– fragrans var. janeirense = Encyclia fragrans
– fragrans var. magnum = Encyclia fragrans
– fragrans var. micranthum = Encyclia fragrans
– fragrans var. pachypus = Encyclia fragrans
– fragrans var. rivularium = Encyclia fragrans
• – **frigidum**
– frons-bovis = Epidendrum paniculatum
• – **fruticosum**
• – fucatum = Encyclia fucata
– fuchsii = Barkeria skinneri
– fulgens = Epidendrum schomburgkii
• – **funckianum**
– funiferum = Epidendrum polyanthum
– furvum = Vanda concolor
• – fuscatum = Epidendrum secundum
– fuscatum var. virescens = Epidendrum secundum
– fuscatum var. viridipurpureum = Epidendrum secundum
– galeatum = Coryanthes speciosa
– galeottianum = Epidendrum secundum
• – **geminiflorum**
– giganteum = Encyclia oncidioides
– gigas = Oncidium baueri
– giroudianum = Epidendrum elongatum
• – **gladiatum**
– glaucovirens = Encyclia glauca
– glaucum Sw. = Dichaea glauca
• – glaucum (Knowles & Westc.) Lindl. = Encyclia glauca
• – glumaceum = Encyclia glumacea
– glumibracteatum = Epidendrum purpurascens
• – glutinosum = Encyclia odoratissima
• – gracile = Encyclia gracilis
– gracilicaule = Epidendrum elongatum
• – grahamii = Encyclia phoenicea
• – **gramineum**
• – **grandiflorum**
• – graniticum = Encyclia oncidioides
– granulosum = Cattleya granulosa
– gratiosum = Epidendrum paniculatum var. linearifolium
– gratiosum var. linearifolium = Epidendrum paniculatum var. linearifolium
– gravidum = Encyclia oncidioides
– guatemalense = Encyclia oncidioides
– guentherianum = Epidendrum compressum
– guesnelianum = Encyclia dichroma
– guesnelianum var. amabile = Encyclia dichroma
– guesnelianum var. striatum = Encyclia dichroma
– guillemianum = Encyclia oncidioides
• – **haenkeanum**
• – hanburii = Encyclia hanburii
– harrisii = Epidendrum subaquilum
– **harrisoniae**
– harrisonianum = Cattleya harrisoniana
• – **hartwegii**
• – hastatum = Encyclia hastata
– henricii = Encyclia livida
– herzogii = Epidendrum elongatum
– heteroglossum = Epidendrum polyanthum
– hexadactylum = Epidendrum cristatum
– hippium = Rhynchostylis retusa
– hircinum = Encyclia fucata
– huegelianum = Cattleya skinneri
– humile = Sophronitis cernua
• – **ibaguense**
– ibaguense var. schomburgkii = Epidendrum schomburgkii
– icthyphyllum = Encyclia michuacana
• – **imatophyllum**
• – imbricatum = Epidendrum paranaense
– imetrophyllum = Epidendrum imatophyllum
– incisum Rchb.f. & Warsc. = Epidendrum elongatum
– incisum Vell. = Epidendrum schomburgkii
– inconstans = Epidendrum elongatum
• – incumbens = Encyclia aromatica
– indicum = Rhynchostylis retusa
– indusiatum = Epidendrum lacertinum
– insidiosum = Encyclia diota ssp. diota
– intermedium = Cattleya intermedia
• – inversum = Encyclia bulbosa
– ionodesme = Epidendrum paniculatum
– ionoleucum = Encyclia fragrans
• – ionosmum = Encyclia ionosma
– iridee = Cattleya bicolor
– isthmi = Epidendrum paniculatum
– juncifolium = Oncidium cebolleta
– jurgensenii = Epidendrum gladiatum
– kuhlmannii = Epidendrum smaragdinum
– labiatum Sw. = Leochilus labiatus
– labiatum var. mossiae = Cattleya mossiae
– labiatum var. pictum = Cattleya mossiae
– labiatum (Lindl.) Rchb.f. = Cattleya labiata
– labiosum = Zygosepalum labiosum
• – **lacertinum**
• – **lacerum**
• – lactiflorum = Epidendrum parkinsonianum
• – laeve = Epidendrum paniculatum
– lambda = Encyclia fragrans
• – lamellatum = Dimerandra emarginata
– lamprocaulon = Epidendrum ledifolium
– lanceola = Pleurothallis lanceola
• – lancifolium = Encyclia lancifolia
– langlassei = Encyclia lancifolia
– lansbergii = Epidendrum polyanthum
– latifolium = Epidendrum carpophorum
• – latilabium = Epidendrum difforme
– latilabre = Epidendrum difforme
– latro = Encyclia bulbosa
• – laxum Poeppig & Endl. = Epidendrum compressum
– laxum Sw. = Pleurothallis laxa
– laxum var. mocoanum = Epidendrum compressum
• – **ledifolium**
• – leiobulbon = Encyclia varicosa ssp. varicosa
– leopardinum = Encyclia vespa
– leucarachne = Epidendrum nocturnum
• – **leucochilum**
• – **lignosum**
• – **ligulatum**
– ligulatum = Oncidium pumilum
• – limbatum = Encyclia glauca
– lindenianum = Epidendrum boothii
• – lindenii = Epidendrum elongatum
– lindleyanum = Barkeria lindleyana
– **lineare**
• – lineare = Isochilus linearis
– linearifolium = Encyclia bractescens
– lineatum = Encyclia fragrans
• – linkianum = Encyclia linkiana
• – lividum = Encyclia livida
– llavei = Epidendrum ligulatum
– loddigesii = Cattleya loddigesii
• – longicolle = Epidendrum nocturnum
– longicrure = Epidendrum paniculatum
• – **longiflorum**
– longihastatum = Epidendrum elongatum
– longipes = Encyclia vespa
• – **longipetalum**
– longipetalum Godefr.-Leb. = Encyclia cordigera
– longipetalum Lindl. & Paxton = Encyclia alata
– longovarium = Epidendrum cristatum
– lorifolium = Epidendrum imatophyllum
– loxense = Epidendrum scabrum
– lucidum = Epidendrum ledifolium
– luteum = Epidendrum ciliare
– macroceras = Epidendrum paniculatum
• – macrochilum = Encyclia cordigera
– macrochilum var. roseum = Encyclia cordigera
– macrodonax = Epidendrum frigidum

• – **macrostachyum**
– macrothyrsis = Epidendrum compressum
– maculatum = Oncidium altissimum
– magnoliae = Epidendrum conopseum
• – **mancum**
• – marginatum Link, Kotzsch & Otto = Encyclia radiata
– marginatum L.C.Rich. = Pleurothallis uniflora
• – **martianum**
– maximum = Cattleya maxima
– melinoacron = Epidendrum cornutum
• – michuacanum = Encyclia michuacana
• – **micranthum**
• – microbulbon = Encyclia microbulbon
• – microphyllum = Lanium microphyllum
– micropus = Encyclia tripunctata
– miersii = Epidendrum saxatile
– miersii var. longifolium = Epidendrum saxatile
– millei = Epidendrum cristatum
– minutum = Polystachya concreta
• – **miserum**
– modestiflorum = Epidendrum ramosum
– monanthum = Encyclia pygmaea
– monile = Dendrobium moniliforme
– moniliferum = Dendrobium moniliforme
– moniliforme = Dendrobium moniliforme
– moschatum = Dendrobium moschatum
– moyobambae = Epidendrum coronatum
– musciferum = Epidendrum secundum
– naucrates = Epidendrum ferrugineum
• – nemorale = Encyclia adenocaula var. adenocaula
– nervosum = Liparis nervosa
• – **nocturnum**
– nocturnum var. angustifolium = Epidendrum nocturnum
• – nocturnum var. latifolium = Epidendrum carpophorum
– nocturnum var. minor = Epidendrum nocturnum
– nocturnum var. minus = Epidendrum nocturnum
– nocturnum var. panamense = Epidendrum nocturnum
– nocturnum var. tridens = Epidendrum nocturnum
– nodosum = Brassavola nodosa
– novogranatense = Epidendrum elongatum
– nubium = Epidendrum arbuscula
* • – **nutans**
– nutans = Epidendrum ruizianum
– nutans var. dipus = Epidendrum densiflorum
– oblongatum = Encyclia phoenicea
• – ochraceum = Encyclia ochracea
• – odoratissimum = Encyclia odoratissima
– odoratum = Aerides odorata
– olare = Catasetum luridum
– oliganthum = Epidendrum nocturnum
• – oncidioides = Encyclia oncidioides
– oncidioides var. graniticum = Encyclia oncidioides
– oncidioides var. gravidum = Encyclia oncidioides
– oncidioides var. ramonense = Encyclia ceratistes
– ophioglossoides = Pleurothallis ophioglossoides
• – **orchidiflorum**
• – **orgyale**
– ornatum = Epidendrum paniculatum
• – **ovalifolium**
• – ovulum = Encyclia microbulbon
• – **pachyanthum**
– pachygastrium = Epidendrum ventricosum
– pachyphyllum = Epidendrum elongatum
– pachyrhachis = Epidendrum smaragdinum
– pachysepalum = Encyclia vespa
– paleaceum = Epidendrum boothii
• – **pallidiflorum**
– palmifolium = Xylobium palmifolium
– palpigerum = Epidendrum imatophyllum
– pamplonense = Encyclia vespa
• – **paniculatum**
– **paniculatum var. linearifolium**
– paniculatum var. cuspidatum = Epidendrum paniculatum
– paniculatum var. longicrure = Epidendrum paniculatum
– papilionaceum = Psychilis bifida
• – papillosum = Encyclia adenocarpon
– **paranaense**
– **parkinsonianum**
• – parviflorum Ruiz Lopez & Pavon = Epidendrum paniculatum var. linearifolium
– parviflorum Sessé & Mociño = Encyclia ochracea
• – **parvilabre**
– pastoris Link & Otto = Encyclia linkiana
• – pastoris Llave & Lex. = Encyclia pastoris
• – **patens**
– patulipetalum = Epidendrum paniculatum var. linearifolium
– pauper = Cattleya forbesii
– pavonianum = Epidendrum cornutum
– pendulum = Cymbidium aloifolium
– peraltense = Encyclia oncidioides
– pergameneum = Epidendrum polyanthum
• – phoeniceum = Encyclia phoenicea
– phymatoglossum = Encyclia varicosa ssp. leiobulbon
• – **pictum**
– piestocaulos = Epidendrum clowesii
– piliferum = Epidendrum paniculatum
• – **piperinum**
– platyoon = Epidendrum scutella
• – plicatum = Encyclia plicata
• – **polyanthum**
– polyschistum = Epidendrum elongatum
• – **polystachyum**
• – **porphyreum**
– praecox = Pleione praecox
– praemorsum = Acampe praemorsa
– pratense = Epidendrum ibaguense
• – primulinum = Encyclia rufa
– pristes = Epidendrum schomburgkii
• – propinquum = Epidendrum ledifolium
– **prostatum**
– pruinosum = Encyclia concolor
– pseudodifforme = Epidendrum difforme
– psilanthemum = Epidendrum purpurascens
• – pterocarpum = Encyclia pterocarpa
– punctulatum = Encyclia concolor
* – **purpurascens**
• – **purum**
– pusillum Koen. = Chilochista pusilla
– pusillum L. = Psygmorchis pusilla
• – pygmaeum = Encyclia pygmaea
• – pyriforme = Encyclia pyriforme
– quinquelobum = Epidendrum polyanthum
– quitensium = Epidendrum elongatum
– racemiflorum = Pleurothallis racemiflora
– radiatum Hoffsgg. = Epidendrum difforme
• – radiatum Lindl. = Encyclia radiata
• – radicans = Epidendrum ibaguense
– radicans var. chiriquense = Epidendrum ibaguense
– ramirezzi = Encyclia varicosa ssp. varicosa
– ramonense = Encyclia ceratistes
• – **ramosum**
– ramosum var. lanceolatum = Epidendrum ramosum
– ramosum var. lancifolium = Epidendrum ramosum
• – raniferum = Epidendrum cristatum
– raniferum var. hexadactylum = Epidendrum cristatum
– raniferum var. loefgrenii = Epidendrum cristatum
– raniferum var. lutescens = Epidendrum cristatum
– raniferum var. luteum = Epidendrum cristatum
– raniferum var. obtusilobum = Epidendrum cristatum
– rantierium = Epidendrum cristatum
• – **recurvatum**
– reflexum = Epidendrum paniculatum
• – **refractum**
– renanthera = Renanthera coccinea
– resectum = Epidendrum paniculatum
– resupinatum = Polystachya concreta
– retusum = Rhynchostylis retusa
– rhabdobulbum = Encyclia vespa
• – rhizophorum = Epidendrum ibaguense
– rhopalobulbon = Encyclia vespa
• – **rigidum**
– rigidum = Epidendrum ramosum
– rimbachii = Dimerandra emarginata
• – rubrocinctum = Epidendrum paniculatum
• – rufum = Encyclia rufa
– **ruizianum**
• – **rupestre**
– ruscifolium = Pleurothallis ruscifolia
– saccharatum = Encyclia vespa
– sagraeanum = Encyclia fucata
– sarcophyllum = Lanium microphyllum
– satyrioides = Ionopsis satyrioides
• – **saxatile**
• – **scabrum**
– scabrum var. parviflorum = Epidendrum scabrum
– schenckianum = Epidendrum secundum
– schlechterianum = Nanodes discolor
* • – **schomburgkii**
– schreineri = Epidendrum secundum
• – **scriptum**
– scriptum = Grammatophyllum scriptum
• – **scutella**
* – **secundum**
– secundum L. = Epidendrum elongatum
– secundum Vell. = Bifrenaria aureo-fulva
– secundum (Jacq.) Sw. = Epidendrum secundum
• – selligerum = Encyclia selligera
– sellowii = Epidendrum ramosum
• – **serpens**
– serronianum = Encyclia odoratissima
– sertularioides = Pleurothallis sertularioides
– sessile Koen. = Bulbophyllum sessile
– sessile Sw. = Maxillaria crassifolia
• – **setiferum**
– sinense = Cymbidium ensifolium
– sisyrinchiifolium = Encyclia microbulbon
• – skinneri = Barkeria skinneri
– skinneri var. superbum = Barkeria skinneri
• – **smaragdinum**
• – spathaceum = Epidendrum ruizianum
– spatulatum = Dendrobium crumenatum
– spectabile Focke = Encyclia oncidioides
– spectabile (Bateman ex Lindl.) Rchb.f. = Barkeria spectabilis
– splendens = Epidendrum schomburgkii
– spruceanum = Epidendrum nocturnum
• – **squalidum**
– stallforthianum = Epidendrum polyanthum
• – **stamfordianum**
– stamfordianum var. lawrenceanum = Epidendrum stamfordianum
– stamfordianum var. leeanum = Epidendrum stamfordianum
– stamfordianum var. parviflorum = Epidendrum stamfordianum
– stamfordianum var. pictum = Epidendrum stamfordianum
– stamfordianum var. wallacei = Epidendrum stamfordianum
• – stenopetalum = Dimerandra emarginata
– stenopetalum var. tenuicaule = Dimerandra emarginata
– striatum = Bletilla striata
• – **subaquilum**
– subpatens = Epidendrum coronatum

Epidendrum
• – **subulatifolium**
– subumbellatum = Epidendrum difforme
– sulphuroleucum = Epidendrum coronatum
– superbum = Cattleya violacea
– syringaeflorum = Epidendrum paniculatum
• – tampense = Encyclia tampensis
– tarmense = Epidendrum elongatum
• – **tenue**
– tenuiflorum = Epidendrum diffusum
– teretifolium = Dendrobium teretifolium
• – tesselatum Bateman = Encyclia livida
– tessellatum Roxb. = Vanda tessellata
– tesselloides = Vanda tessellata
– tetragonum = Phaius tetragonus
– tetrapetalum = Oncidium tetrapetalum
– thrixspermum = Thrixspermum centipeda
• – **thyrsiferum**
• – tibicinis = Myrmecophila tibicinis
– tigrinum = Encyclia vespa
• – **tolimense**
• – **torquatum**
– tribuloides = Pleurothallis tribuloides
– tricolor = Epidendrum ferrugineum
• – tridactylum = Amblostoma tridactylum
– tridens = Epidendrum nocturnum
– tridentatum = Lepanthes tridentata
– trigoniflorum = Pleurothallis ophioglossoides
– trilabiatum = Zygostates cornuta
• – trinitatis = Epidendrum ferrugineum
– tripartitum = Encyclia tripartita
• – tripterum Lindl. = Encyclia linkiana
– tripterum Smith = Coelia triptera
• – tripunctatum = Encyclia tripunctata
– triste = Encyclia ochracea
– trulla = Encyclia lancifolia
– tuberosum = Bletilla striata
– tunguraguae = Epidendrum nocturnum
– turialvae = Epidendrum paniculatum
– ulei = Epidendrum aquaticum
– umbellatum Forster = Bulbophyllum longiflorum
• – umbellatum Sw. = Epidendrum difforme
– umbellatum Vell. = Malaxis excavata
– umbellatum var. latilabre = Epidendrum difforme
– umbelliferum = Epidendrum difforme
– undulatum = Oncidium carthagenense
• – uniflorum Lindl. = Encyclia pygmaea
– uniflorum Vell. = Maxillaria picta
– usneoides = Chilochista usneoides
– utricularioides = Ionopsis utricularioides
– vaginatum = Encyclia fragrans
– validum = Epidendrum cristatum
– vandifolium = Epidendrum gladiatum
• – varicosum = Encyclia varicosa ssp. leiobulbon
• – variegatum = Encyclia vespa
– vellozoi = Epidendrum saxatile
• – venosum = Encyclia venosa
• – **ventricosum**
– ventrilabrum = Psygmorchis pusilla
– verrucipes = Epidendrum polyanthum
• – **verrucosum**
– verrucosum = Encyclia adenocaula var. adenocaula
– versicolor = Epidendrum elongatum
• – **vesicatum**
– vespa = Encyclia vespa
• – **vincentinum**
– violaceum Lodd. = Cattleya loddigesii
– violaceum (Kunth) Rchb.f. = Cattleya violacea
– virens = Epidendrum difforme
– virescens = Epidendrum secundum
• – virgatum = Encyclia michuacana
– virgatum var. pallens = Encyclia michuacana
• – **viride**
• – viridiflorum = Encyclia viridiflora
– viridipurpureum = Epidendrum secundum
• – viscidum = Epidendrum ciliare
• – vitellinum = Encyclia vitellina
– vitellinum var. autumnale = Encyclia vitellina
– vitellinum var. giganteum = Encyclia vitellina
– vitellinum var. majus = Encyclia vitellina
• – **viviparum**
• – **volubile**
• – volucre = Oeonia volucris
– volucris = Oeonia volucris
– walkerianum = Cattleya walkeriana
– weddelli var. robustum = Epidendrum saxatile
– weddellii = Epidendrum saxatile
– weddellii var. longifolium = Epidendrum saxatile
– wendlandianum = Encyclia venosa
• – **xanthinum**
• – **xylostachyum**
– xytriophorum = Epidendrum elongatum
– yatapuense = Epidendrum compressum

Epidorchis
– caulescens = Angraecum caulescens

Epipactis
– alpina = Chamorchis alpina
– arachnites = Ophrys holoserica
– graminifolia = Spathoglottis pubescens
– grandis = Goodyera rubicunda
– philippinensis = Goodyera procera
– plicata = Nervilia plicata
– procera = Goodyera procera
– rubicunda = Goodyera rubicunda

Epiphora
• – pubescens = Polystachya pubescens

Epithecia
– glauca = Encyclia glauca

Eria
• – **acutifolia**
– affinis = Eria bractescens
– amica = Eria bipunctata
• – armeniaca = Eria ornata
– bicolor = Eria lindleyi
– bicornis = Tainia bicornis
– bidentata = Eria ovata
• – **bipunctata**
* • – **bractescens**
– calamifolia = Eria pannea
• – **clavicaulis**
• – cochleata = Eria javanica
– convallarioides Finet = Eria bipunctata
• – convallarioides Lindl. = Eria spicata
• – convallarioides var. major = Eria spicata
* – **coronaria**
– cylindropoda = Eria coronaria
– decipiens = Eria summerhayesiana
• – dillwynii = Eria bractescens
– eberhardtii = Eria bipunctata
– elmeri = Eria ovata
– ephemera = Eria lindleyi
– euryanthe = Eria floribunda
– extintoria = Eria summerhayesiana
* • – **ferruginea**
* • – **floribunda**
– fragrans = Eria javanica
– giungii = Eria floribunda
– graciliscapa = Eria ovata
– griffithii = Eria bractescens
* – **javanica**
– khasiana = Eria clavicaulis
– leuchostachya = Eria floribunda
– lindleyana = Eria obesa
– **lindleyi**
– littoralis = Eria bractescens
• – **longilabris**
– luchuensis = Eria ovata
– micrantha = Eria multiflora
– monticola var. hirsuta = Trichotosia velutina
• – mucronata = Trichotosia mucronata
• – **multiflora**
– nivosa = Eria pannea
• – **nutans**
• – **obesa**
– **obliqua**
* – **ornata**
• – **ovata**
• – **paniculata**
• – **pannea**
• – planicaulis = Agrostophyllum planicaule
* • – **polyura**
– prainii = Eria obesa
– **profusa**
• – **pulchella**
– pulchella = Eria bractescens
• – **pumila**
– pusilla = Eria retusa
– reticulata = Porpax reticulata
– **retusa**
* • – **rosea**
– rugosa = Eria javanica
* – **spicata**
• – stellata = Eria javanica
– striolata = Eria javanica
– suavis = Eria coronaria
– subaliena = Eria floribunda
– **summerhayesiana**
– teretifolia = Eria pannea
– vaginata = Eria javanica
• – velutina = Trichotosia velutina
• – vestita Lindl. = Trichotosia vestita
– vestita Parish = Trichotosia velutina
– wightii = Tainia bicornis

Eriopsis
* • – **biloba**
– colombiana = Eriopsis biloba
– fuerstenbergii = Eriopsis biloba
– grandibulbosa = Eriopsis biloba
– mesae = Eriopsis biloba
– rhytidobulbon = Eriopsis biloba
– rutidobulbon = Eriopsis biloba
– schomburgkii = Eriopsis biloba
– werckeleyi = Eriopsis biloba

Erythrodes
– **picta**

Eucnemis
– brevilabris = Govenia liliacea

Eulophia
* – **alta**
– bicolor = Eulophia nuda
– cochleare = Cochleanthes flabelliformis
– cochlearis = Cochleanthes flabelliformis
– congoensis = Eulophia guineensis
– dilatata = Bletia secunda
• – **ensata**
– fusca = Eulophia nuda
* • – **gracilis**
– guamensis = Oeceoclades pulchra
* • – **guineensis**
* – **horsfallii**
– krebsii = Eulophia streptopetala
– krebsii var. purpurata = Eulophia streptopetala
– ledienii = Oeceoclades maculata
– longifolia = Eulophia alta
• – lurida = Graphorkis lurida
• – mackaiana = Zygopetalum mackaii
• – macrostachya = Oeceoclades pulchra
– maculata = Oeceoclades maculata
– mahonii = Eulophia horsfallii
– murrayana = Neogardneria murrayana
– novaebudae = Oeceoclades pulchra
– **nuda**
– paivaeana = Eulophia streptopetala
– porphyroglossa = Eulophia horsfallii
– preussii = Eulophia gracilis
– pulchra = Oeceoclades pulchra
– quartiniana = Eulophia guineensis
– rouxii = Oeceoclades pulchra
– sandersonii = Eulophia horsfallii
– sinensis = Spathoglottis pubescens
* – **speciosa**
• – squalida = Eulophia nuda
* • – **streptopetala**
– virens = Eulophia gracilis
– woodfordii = Eulophia alta

Eulophidium
– ledienii = Oeceoclades maculata
– maculatum = Oeceoclades maculata
– nyassanum = Oeceoclades maculata
– pulchrum = Oeceoclades pulchra
– warneckeanum = Oeceoclades maculata
Eulophiopsis
– lurida = Graphorkis lurida
Exophya
– fuscata = Epidendrum patens
Fernandezia
• – acuta = Lockhartia acuta
– graminifolia = Scaphyglottis graminifolia
• – lunifera = Lockhartia lunifera
– robusta = Lockhartia lunifera
Fieldia
– lissochiloides = Vandopsis lissochiloides
Finetia
– falcata = Neofinetia falcata
Flickingeria
– **comata**
– **fimbriata**
– kunstleri = Flickingeria fimbriata
– **scopa**
Fractiunguis
– brasiliensis = Reichenbachanthus reflexus
– reflexa = Reichenbachanthus reflexus
Frondaria
– **caulescens**
Gabertia
– scripta = Grammatophyllum scriptum
Galeandra
* – **batemannii**
• – baueri = Galeandra batemannii
• – **cristata**
– quartiniana = Eulophia guineensis
Galeola
– **cassythoides**
Gastrochilus
– bifidus = Megalotus bifidus
– blumei = Rhynchostylis retusa
– **calceolaris**
– compressum = Robiquetia compressa
– curvifolius = Ascocentrum curvifolium
– densiflorus = Robiquetia spathulata
– gurwalicus = Rhynchostylis retusa
– intermedius = Gastrochilus calceolaris
– miniatus = Ascocentrum miniatum
– papillosum = Acampe papillosa
– parviflorus = Smitinandia micrantha
– praemorsus = Rhynchostylis retusa
– retusus = Rhynchostylis retusa
– rheedii = Rhynchostylis retusa
– spicatus = Rhynchostylis retusa
– sp. = Gastrochilus calceolaris
Gastroglottis
– montana = Malaxis latifolia
Gennaria
* – **diphyllà**
Genyorchis
– pumila = Bulbophyllum pumilum
Geodorum
* – **densiflorum**
• – dilatatum = Geodorum densiflorum
*• – **fucatum**
– pictum = Oeceoclades maculata
Glossapis
– antennifera = Peristylus tentaculatus
• – tentaculata = Peristylus tentaculatus
Glossula
– passerina = Peristylus densus
– tentaculata = Peristylus tentaculatus
Gomesa
– chrysostoma = Gomesa laxiflora
* – **crispa**
– **laxiflora**
– tenuiflora = Notylia punctata
– undulata = Gomesa crispa
Gongora
• – atropurpurea = Gongora quinquenervis
– boothiana = Gongora quinquenervis
*• – **bufonia**
• – bufonia var. leucochila = Gongora bufonia
– donckelaariana = Gongora truncata
• – fulva = Gongora quinquenervis
• – fulva var. vitellina = Gongora quinquenervis
* – **galeata**
• – **galeottiana**
– heisteri = Gongora quinquenervis
– histrionica = Gongora quinquenervis
– irrorata = Gongora bufonia
– jenischii = Gongora quinquenervis
– leucochila = Gongora quinquenervis
– macrantha = Coryanthes macrantha
• – maculata = Gongora quinquenervis
– maculata f. odoratissima = Gongora quinquenervis
• – maculata var. alba = Gongora quinquenervis
– maculata var. tricolor = Gongora quinquenervis
– minax = Gongora bufonia
• – nigrita = Gongora quinquenervis
– odoratissima = Gongora quinquenervis
– odoratissima var. jenischii = Gongora quinquenervis
– phillippica = Renanthera coccinea
– powellii = Gongora quinquenervis
– quadricornis = Gongora quinquenervis
*• – **quinquenervis**
– quinquevulneris = Gongora quinquenervis
– retrorsa = Gongora quinquenervis
– seideliana = Gongora truncata
– sherpherdii = Gongora quinquenervis
– speciosa = Coryanthes speciosa
– superflua = Gongora quinquenervis
– tricolor = Gongora quinquenervis
*• – **truncata**
– truncata var. alba = Gongora truncata
– unicolor = Gongora quinquenervis
– viridipurpurea = Cirrhaea dependens
– vitellina = Gongora quinquenervis
Gonogona
– discolor = Ludisia discolor
Goodyera
– carnea = Goodyera procera
– celebica = Goodyera rubicunda
• – discolor = Ludisia discolor
– grandis King & Pantl. = Goodyera rubicunda
– grandis (Blume) Blume = Goodyera rubicunda
– hispidula = Goodyera rubicunda
– lancifolia = Goodyera procera
– longibracteata = Goodyera rubicunda
– occulta = Platylepis goodyeroides
– papuana = Goodyera rubicunda
– parviflora = Goodyera procera
– philippinensis = Goodyera procera
* – **procera**
– rubens = Goodyera rubicunda
• – **rubicunda**
– rubicunda var. celebica = Goodyera rubicunda
– yaeyamae = Goodyera rubicunda
Govenia
• – alba = Govenia liliacea
• – barbata = Eulophia alta
– boliviense = Govenia utriculata
– brevilabris = Govenia liliacea
• – capitata = Govenia liliacea
– deliciosa = Govenia liliacea
– elliptica = Govenia pauciflora
– ernstii = Govenia utriculata
• – fasciata = Govenia superba
• – gardneri = Govenia utriculata
• – lagenophora = Govenia superba
*• – **liliacea**
• – **pauciflora**
– platyglossa = Govenia superba
– sodiroi = Govenia utriculata
– stictoglossa = Govenia superba
– sulphurea = Govenia superba
*• – **superba**
– superba var. elliptica = Govenia pauciflora
• – **tingens**
– **utriculata**
– utriculata var. capitata = Govenia liliacea
Govindovia
– nervosa = Tropidia angulosa
Grammatophyllum
– fenzlianum = Grammatophyllum scriptum
– finlaysonianum = Bromheadia finlaysoniana
– guilelmi-secundii = Grammatophyllum scriptum
– leopardinum = Grammatophyllum scriptum
– measuresianum = Grammatophyllum scriptum
• – multiflorum = Grammatophyllum scriptum
– multiflorum var. tigrinum = Grammatophyllum scriptum
– rumphianum = Grammatophyllum scriptum
* – **scriptum**
– seegerianum = Grammatophyllum scriptum
Grammatophylum
– pantherinum = Vandopsis lissochiloides
Graphorchis
– guineensis = Eulophia guineensis
– maculata = Oeceoclades maculata
– nuda = Eulophia nuda
– pulchra = Oeceoclades pulchra
Graphorkis
* – **lurida**
Grastidium
– acuminatissimum = Dendrobium acuminatissimum
– bambusifolium = Dendrobium salaccense
– indragiriense = Dendrobium indragiriense
– isomerum = Dendrobium indragiriense
– rugosum = Dendrobium rugosum
– salaccense = Dendrobium salaccense
Grobya
*• – **amherstiae**
• – **galeata**
Gunnia
• – australis = Sarcochilus australis
– picta = Sarcochilus olivaceus
Gyas
– florida = Bletia purpurea
– humilis = Bletilla striata
Gymnadenia
– diphylla = Gennaria diphylla
– fastigiata = Cynorkis fastigiata
• – secunda = Peristylus secundus
• – tenuis = Peristylus densus
Gynizdon
– russellianum = Miltonia russelliana
Gyrostachys
– amoena = Spiranthes sinensis
– aphylla = Sacoila lanceolata
– australis = Spiranthes sinensis
– australis var. amoena = Spiranthes sinensis
– australis var. flexuosa = Spiranthes sinensis
– bicolor = Beadlea bicolor
– bracteosa = Brachystele bracteosa
– cernua = Spiranthes cernua
– constricta = Spiranthes cernua
– lanceolata = Sacoila lanceolata
– lindleyana = Beadlea lindleyana
– novifriburgensis = Spiranthes sinensis
– orchioides = Sacoila lanceolata
– ovalifolia = Beadlea elata
– picta = Sarcoglottis acaulis
– stenorrhynchus = Sacoila lanceolata
– wightiana = Spiranthes sinensis
Habenaria
– atramentaria = Peristylus densus
– buchneroides = Peristylus densus
• – **candida**
– chlorantha = Benthamia latifolia
– clovisii = Habenaria reniformis

Habenaria
– constricta = Peristylus constrictus
– cordata = Gennaria diphylla
– cylindrocalyx = Peristylus constrictus
– dankiaensis = Peristylus densus
– diphylla = Gennaria diphylla
– dissimulata = Benthamia spiralis
– elata = Peristylus plantagineus
– evrardii = Peristylus densus
• – fimbriata = Platanthera grandiflora
– fimbriata f. mentotonsa = Platanthera grandiflora
– garrettii = Peristylus tentaculatus
– glaucescens = Peristylus goodyeroides
– goodyeroides D.Don = Peristylus goodyeroides
– goodyeroides Hook. = Peristylus plantagineus
– goodyeroides var. formosana = Peristylus goodyeroides
– grandiflora = Platanthera grandiflora
– hamiltoniana = Peristylus hamiltonianus
– hayataeana = Peristylus goodyeroides
– hollandiae = Peristylus goodyeroides
– latifolium = Benthamia latifolia
– lauterbachii = Peristylus goodyeroides
– minutiflora = Benthamia spiralis
– neglecta = Peristylus densus
– passerina = Peristylus densus
– peristyloides = Peristylus densus
* • – **procera**
– psycodes var. grandiflora = Platanthera grandiflora
– **reniformis**
– robustior = Peristylus secundus
– sondaica = Peristylus goodyeroides
– spiralis = Benthamia spiralis
– stenostachya = Peristylus densus
– stenostachya ssp. buchneroides = Peristylus densus
– tentaculata = Peristylus tentaculatus
– tetramera = Habenaria tetrapetala
– **tetrapetala**
– uniflora = Diplomeris pulchella
– wightii = Peristylus plantagineus

Haemaria
– dawsoniana = Ludisia discolor
– discolor = Ludisia discolor
– discolor var. dawsoniana = Ludisia discolor
– merrillii = Ludisia discolor
– otletae = Ludisia discolor
– pauciflora = Ludisia discolor
– petelotii = Ludisia discolor

Hartwegia
– comosa = Nageliella purpurea
• – purpurea = Nageliella purpurea
• – purpurea var. angustifolia = Nageliella angustifolia

Helcia
• – **sanguinolenta**

Helleborine
– cordigera = Serapias cordigera
– longipetala = Serapias vomeracea ssp. vomeracea
– pseudo-cordigera = Serapias vomeracea ssp. vomeracea

Herminium
– alaschanicum = Herminium lanceum
• – alpinum = Chamorchis alpina
– angustifolia = Herminium lanceum
– angustifolia var. brevilabre = Herminium lanceum
– angustifolium var. longicrure = Herminium lanceum
• – congestum = Herminium macrophyllum
• – constrictum = Peristylus constrictus
• – cordatum = Gennaria diphylla
• – goodyeroides = Peristylus goodyeroides
• – hamiltonianum = Peristylus hamiltonianus
– **lanceum**
– lanceum var. longicrure = Herminium lanceum
• – latifolium = Benthamia latifolia
– longicruris = Herminium lanceum
– **macrophyllum**
• – **monorchis**
• – plantagineus = Peristylus plantagineus
• – reniforme = Habenaria reniformis
– spirale = Benthamia spiralis

Herschelia
– atropurpurea = Herschelianthe spathulata
– **coelestis**
– graminifolia = Herschelia coelestis
– spathulata = Herschelianthe spathulata
– atropurpurea = Herschelianthe spathulata
– **spathulata**

Hetaeria
– rubicunda = Goodyera rubicunda

Heterotaxis
• – crassifolia = Maxillaria crassifolia

Hexadesmia
• – **bicornis**
• – **crurigera**
• – **fasciculata**
– lindeniana = Hexadesmia fasciculata
• – **micrantha**
– pachybulbon = Hexadesmia fasciculata
– rhodoglossa = Hexadesmia fasciculata
– rigidipes = Hexadesmia fasciculata

Hexisea
– reflexa = Reichenbachanthus reflexus

Hexopia
• – crurigera = Hexadesmia crurigera

Himantoglossum
– longibracteatum = Barlia robertiana

Holcoglossum
– falcatum = Neofinetia falcata

Holothrix
– **burchelli**
– **monotris**
– scopularia = Holothrix burchelli

Homalopetalum
– **pumilio**

Hormidium
– boothianum = Encyclia boothiana ssp. boothiana
– citrinum = Encyclia citrina
– crassilabium = Encyclia vespa
– fragrans = Encyclia fragrans
– humile = Encyclia pygmaea
– lividum = Encyclia livida
– pseudopygmaeum = Encyclia pygmaea
– pygmaeum = Encyclia pygmaea
– tripterum = Encyclia pygmaea
– uniflorum = Encyclia pygmaea
– vespa = Encyclia vespa
– vitellinum = Encyclia vitellina

Houlletia
– **brocklehurstiana**
– stapeliaeflora = Houlletia brocklehurstiana
• – vittata = Braemia vittata

Humboldtia
– acianthera = Pleurothallis acianthera
– acuminata = Pleurothallis acuminata
– affinis = Myoxanthus affinis
– arbuscula = Trichosalpinx arbuscula
– aristata = Pleurothallis aristata
– articulata = Pleurothallis articulata
– asperilinguis = Pleurothallis acuminata
– aurea = Pleurothallis acuminata
– auriculata = Pleurothallis auriculata
– barbata = Pleurothallis barbata
– barberana = Pleurothallis aristata
– bicarinata = Pleurothallis bicarinata
– bidentata = Pleurothallis bidentata
– biflora = Trichosalpinx orbicularis
– bufonis = Pleurothallis pubescens
– capillaris = Pleurothallis capillaris
– cardium = Pleurothallis undulata
– casapensis = Pleurothallis casapensis
– caulescens = Frondaria caulescens
– cauliflora = Pleurothallis cauliflora
– centranthera = Pleurothallis centranthera
– ciliata = Pleurothallis lanceana
– circumplexa = Pleurothallis circumplexa
– cordata = Pleurothallis cordata
– crassifolia = Pleurothallis lanceana
– crepidophylla = Pleurothallis grobyi
– decurrens = Pleurothallis decurrens
– diffusa = Pleurothallis diffusa
– dinotherii = Trichosalpinx arbuscula
– diptera = Trichosalpinx arbuscula
– elegans = Pleurothallis elegans
– emarginata = Physosiphon emarginatus
– ephemera = Pleurothallis ephemera
– flexuosa = Pleurothallis flexuosa
– foetens = Pleurothallis foetens
– fragilis = Pleurothallis luteola
– fritillaria = Pleurothallis restrepioides
– gardneri = Barbosella gardneri
– gelida = Pleurothallis gelida
– glanduligera = Pleurothallis glanduligera
– glumacea = Pleurothallis glumacea
– grandiflora = Pleurothallis grandiflora
– grobyi = Pleurothallis grobyi
– hartwegii = Pleurothallis hartwegii
– hians = Pleurothallis hians
– hymenantha = Pleurothallis hymenantha
– hypnicola = Pleurothallis hypnicola
– inaequalis = Pleurothallis inaequalis
– incompta = Pleurothallis quadrifida
– lanceana = Pleurothallis lanceana
– lanceolata = Pleurothallis lanceolata
– laurifolia = Pleurothallis ruscifolia
– lepanthiformis = Trichosalpinx ciliaris
– lindenii = Pleurothallis secunda
– lingua = Pleurothallis lingua
– longicaulis = Pleurothallis longicaulis
– longirostris = Pleurothallis ephemera
– longissima = Pleurothallis quadrifida
– macrorhiza = Pleurothallis macrorhiza
– marginalis = Pleurothallis grobyi
– marginata = Pleurothallis grobyi
– mathewsii = Pleurothallis phalangifera
– miersii = Barbrodria miersii
– muscoidea = Pleurothallis muscoidea
– nicaraguensis = Pleurothallis quadrifida
– obovata = Pleurothallis obovata
– ochreata = Pleurothallis ochreata
– orbicularis = Trichosalpinx orbicularis
– pandurifera = Pleurothallis pandurifera
– papillosa = Pleurothallis papillosa
– parvifolia = Pleurothallis parviflora
– pectinata = Pleurothallis pectinata
– phalangifera = Pleurothallis phalangifera
– picta = Pleurothallis grobyi
– plantaginea = Masdevallia plantaginea
– plumosa = Pleurothallis lanceana
– poeppigii = Myoxanthus monophyllus
– polyliria = Pleurothallis gelida
– polystachya = Pleurothallis pubescens
– prolifera = Pleurothallis prolifera
– pruinosa = Pleurothallis pruinosa
– pubescens = Pleurothallis pubescens
– punctata = Trichosalpinx punctatifolia
– pusilla = Trichosalpinx pusilla
– quadrifida = Pleurothallis quadrifida
– recurva = Pleurothallis recurva
– restrepioides = Pleurothallis restrepioides
– roezlii = Pleurothallis restrepioides
– rubens = Pleurothallis rubens
– rupestris = Pleurothallis rupestris
– ruscifolia = Pleurothallis ruscifolia
– saurocephala = Pleurothallis saurocephala
– scabrilinguis = Pleurothallis scabrilinguis
– sclerophylla = Pleurothallis sclerophylla
– secunda = Pleurothallis secunda
– semipellucida = Pleurothallis diffusa
– seriata = Pleurothallis seriata
– smithiana = Pleurothallis pubescens
– spiralis = Pleurothallis spiralis
– stenopetala = Pleurothallis sclerophylla
– striata = Pleurothallis uniflora
– strupifolia = Pleurothallis strupifolia
– succosa = Pleurothallis ruscifolia
– tentaculata = Pleurothallis tentaculata
– teres = Pleurothallis teres
– tribuloides = Pleurothallis tribuloides

Humboldtia
– tricarinata = Pleurothallis tricarinata
– tridentata = Pleurothallis tridentata
– trifida = Pleurothallis trifida
– truxillensis = Pleurothallis pubescens
– undulata = Pleurothallis undulata
– uniflora = Pleurothallis uniflora
– villosa = Trichosalpinx ciliaris
– vittata = Pleurothallis pubescens
Huntleya
– albidofulva = Huntleya meleagris
– burtii = Huntleya meleagris
– imbricata = Cochleanthes flabelliformis
*• – **meleagris**
– meleagris var. albidofulva = Huntleya meleagris
• – sessiliflora = Bollea violacea
– tyrianthina = Bollea violacea
• – violacea = Bollea violacea
– wallisii = Huntleya meleagris
Iantha
– pallidiflora = Ionopsis utricularioides
Ibidium
– cernuum = Spiranthes cernua
– cristalligerum = Sacoila lanceolata
– tortile = Spiranthes torta
Ionopsis
– gardneri = Ionopsis utricularioides
• – pallidiflora = Ionopsis utricularioides
• – paniculata = Ionopsis utricularioides
– pulchella = Ionopsis utricularioides
– pusilla = Ionopsis satyrioides
– **satyrioides**
• – tenera = Ionopsis utricularioides
• – teres = Ionopsis satyrioides
– testiculata = Ionopsis satyrioides
* – **utricularioides**
– zonalis = Ionopsis utricularioides
Iridorchis
– cylindrica = Oberonia cylindrica
– recurva = Oberonia recurva
– wightiana = Oberonia wightiana
Isochilus
– brasiliensis = Isochilus linearis
– dubius = Pachystele dubia
– elegans = Dimerandra emarginata
• – graminifolium (Kunth) Lindl. = Maxillaria graminifolia
– graminifolius Kunth = Maxillaria graminifolia
• – grandiflorum = Maxillaria longibracteata var. grandiflora
– langlassei = Isochilus linearis
– leucanthus = Isochilus linearis
*• – **linearis**
– linearis var. leucanthum = Isochilus linearis
– linifolium = Maxillaria graminifolia
– lividus = Pachystele dubia
– pauciflorus = Isochilus linearis
– peruvianus = Isochilus linearis
• – prolifer = Scaphyglottis prolifera
– proliferum = Scaphyglottis prolifera
– ramosum = Epidendrum ramosum
Jensoa
– ensata = Cymbidium ensifolium
Jimensia
– nervosa = Bletilla striata
– striata = Bletilla striata
Katherinea
– perakense = Eria javanica
Kefersteinia
– **graminea**
Koellensteinia
– **graminea**
– **tricolor**
Lacaena
*• – **bicolor**
– bicolor var. alba = Lacaena bicolor
– bicolor var. glabrata = Lacaena bicolor
Laelia
• – acuminata = Laelia rubescens
*• – **albida**
*• – **anceps**
• – anceps var. barkeriana = Laelia anceps
*• – **autumnalis**
– barkeriana = Laelia anceps
• – caulescens = Laelia flava
• – **cinnabarina**
* – **crispa**
– **crispata**
– digbyana = Rhyncholaelia digbyana
– digbyana var. fimbripetala = Rhyncholaelia digbyana
– discolor = Laelia albida
*• – **flava**
*• – **furfuracea**
– glauca = Rhyncholaelia glauca
– gloriosa = Schomburgkia gloriosa
– gouldiana = Laelia autumnalis
• – grandiflora = Laelia speciosa
– grandiflora var. alba = Laelia speciosa
– lueddemanii = Schomburgkia undulata
• – majalis = Laelia speciosa
– majalis var. alba = Laelia speciosa
• – peduncularis = Laelia rubescens
*• – **perrinii**
– praestans = Laelia pumila
• – pubescens = Laelia rubescens
* – **pumila**
– rosea = Schomburgkia rosea
* – **rubescens**
• – rupestris = Laelia crispata
* – **speciosa**
• – superbiens = Schomburgkia superbiens
– tereticaulis = Laelia crispata
– tibicinis = Myrmecophila tibicinis
– undulata = Schomburgkia undulata
– violacea = Laelia rubescens
• – virens = Laelia xanthina
– wetmorei = Laelia xanthina
– **xanthina**
Laeliopsis
– cubensis = Cattleyopsis cubensis
– domingensis = Broughtonia domingensis
Lanium
– **avicula**
– **microphyllum**
– peruvianum = Lanium microphyllum
Larnandra
– conopsea = Epidendrum conopseum
– magnolia = Epidendrum conopseum
Latourorchis
– macrophylla = Dendrobium macrophyllum
– muscifera = Dendrobium macrophyllum
Lemboglossum
– **apterum**
* – **bictoniense**
* – **cervantesii**
– **cordatum**
– **ehrenbergii**
– **galeottianum**
* – **maculatum**
– **madrense**
* – **rossii**
– **stellatum**
Leochilus
• – **carinatus**
• – cochlearis = Leochilus labiatus
– depauperatus = Leochilus labiatus
– gracilis = Leochilus labiatus
• – herbaceus = Leochilus scriptus
– **labiatus**
– macrantherus = Leochilus oncidioides
– major = Leochilus scriptus
– mattogrossensis = Solenidium lunatum
• – **oncidioides**
– powellii = Leochilus scriptus
– pygmaeus = Rhyncostele pygmaea
– retusus = Leochilus scriptus
• – sanguinolentus = Oncidium sanguinolentum
– **scriptus**
Lepanthes
– bicarinata = Pleurothallis trifida
– cryptophyla = Pleurothallis parviflora
– gunningiana = Pleurothallis hians
– marmorata = Pleurothallis grobyi
– punctatifolia = Trichosalpinx punctatifolia
*• – **tridentata**
– trilineata = Pleurothallis grobyi
Leptocentrum
– caudatum = Plectrelminthus caudatus
Leptorchis
– bicornis = Liparis nervosa
– bituberculata = Liparis nervosa
– caespitosa = Liparis caespitosa
– duthiei = Liparis caespitosa
– eggersii = Liparis nervosa
– elata = Liparis nervosa
– flavescens = Liparis flavescens
– guineensis = Liparis nervosa
– lilifolia = Liparis lilifolia
– loeselii = Liparis loeselii
– longipes = Liparis viridiflora
– lutea = Liparis nervosa
– minima = Liparis caespitosa
– nervosa = Liparis nervosa
– obscura = Liparis caespitosa
– odontostoma = Liparis nervosa
– odorata = Liparis nervosa
– olivacea = Liparis nervosa
– prainii = Liparis caespitosa
– pusilla = Liparis caespitosa
– viridiflora = Liparis viridiflora
Leptotes
*• – **bicolor**
– bicolor var. glaucophylla = Leptotes bicolor
– glaucophylla = Leptotes bicolor
– serrulata = Leptotes bicolor
Leptothrium
– lineare = Isochilus linearis
Leucohyle
– **mutica**
Leucostachys
– procera = Goodyera procera
Lichterveldia
– lindleyi = Cuitlauzina pendula
Limodorum
– altum Jacq. = Bletia purpurea
– altum L. = Eulophia alta
– aphyllum = Dendrobium aphyllum
– bicallosum = Liparis nervosa
– bidens = Diaphananthe bidens
– bracteatum = Thunia bracteata
– candidum = Geodorum densiflorum
– densiflorum = Geodorum densiflorum
– eburneum = Angraecum eburneum
– ensatum = Cymbidium ensifolium
• – falcatum = Neofinetia falcata
• – filiforme = Aeranthes filiforme
– floridum = Bletia purpurea
• – funale = Dendrophyllax funalis
– giganteum = Eulophia speciosa
– graminifolium = Arundina graminifolia
– hyacinthinum = Bletilla striata
– incarvillei = Phaius tankervilleae
– lanceolatum = Sacoila lanceolata
– latifolium = Aerides odorata
– luridum = Graphorkis lurida
– monile = Dendrobium moniliforme
– praealtum = Eulophia alta
– pulchrum = Oeceoclades pulchra
– purpureum = Bletia purpurea
• – pusillum = Chilochista pusilla
– recurvum = Geodorum densiflorum
– retusum = Rhynchostylis retusa
– striatum Banks = Calanthe discolor
– striatum Thunb. = Bletilla striata
– tankervilleae = Phaius tankervilleae
– trifidum = Bletia purpurea
– tuberosum = Bletia purpurea
– utriculatum = Govenia utriculata
– ventricosum = Calanthe triplicata
– veratrifolium = Calanthe triplicata
– verecundum = Bletia purpurea
Lindleyella
– aurantiaca = Rudolfiella aurantiaca

Liparis
• – alata = Liparis nervosa
– angkae = Liparis petiolata
– angustifolia = Liparis caespitosa
– auriculata = Liparis caespitosa
– bambusaefolia = Liparis nervosa
– bernaysii = Malaxis latifolia
– bicallosa = Liparis nervosa
– bicallosa var. hachijoensis = Liparis nervosa
– bicornis = Liparis nervosa
• – bituberculata (Hook.) Lindl. = Liparis nervosa
– bituberculata (Rchb.f) Rchb.f. = Liparis nervosa
– bituberculata var. formosana = Liparis nervosa
– boothii = Liparis viridiflora
• – **caespitosa**
– comosa = Liparis caespitosa
– cornicaulis = Liparis nervosa
– disticha Kraenzlin = Liparis caespitosa
• – disticha Lindl. = Liparis gibbosa
– duthiei = Liparis caespitosa
– eggersii = Liparis nervosa
• – elata = Liparis nervosa
– elata var. inundata = Liparis nervosa
– elata var. latifolia = Liparis nervosa
– elata var. longifolia = Liparis nervosa
– elata var. purpurascens = Liparis nervosa
– elata var. rufina = Liparis nervosa
• – **flavescens**
• – foliosa = Liparis reflexa
– formosana = Liparis nervosa
– formosana f. aureo-variegata = Liparis nervosa
– formosana var. hachijoensis = Liparis nervosa
– **gibbosa**
– gregaria = Liparis gibbosa
– guamensis = Liparis paradoxa
• – guineensis = Liparis nervosa
– hachijoensis = Liparis nervosa
– krempfii = Malaxis latifolia
– labiata = Leochilus labiatus
• – **lilifolia**
• – **loeselii**
– longipes = Liparis viridiflora
– longipes var. spathulata = Liparis viridiflora
– longiscapa = Liparis paradoxa
– lutea = Liparis nervosa
– mannii = Liparis caespitosa
– melanoglossa = Liparis nervosa
– minima = Liparis caespitosa
– neoguinensis = Liparis caespitosa
• – nepalensis = Liparis petiolata
* – **nervosa**
– nervosa = Liparis paradoxa
– nervosa var. formosana = Liparis nervosa
– nesophila = Liparis viridiflora
– nyassa = Liparis nervosa
– obscura = Liparis caespitosa
– odontostoma = Liparis nervosa
– odorata = Liparis nervosa
– odorata var. intacta = Liparis nervosa
– odorata var. longiscapa = Liparis paradoxa
– olivacea Lindl. = Liparis nervosa
– olivacea Parish = Liparis paradoxa
– **paradoxa**
– paradoxa var. parishii = Liparis paradoxa
– parishii = Liparis paradoxa
• – pendula = Liparis viridiflora
– perrieri = Liparis nervosa
– **petiolata**
– piestopus = Liparis viridiflora
– pleistantha = Liparis viridiflora
– poilanei = Liparis caespitosa
– prainii = Liparis caespitosa
– pulchella = Liparis petiolata
• – **purpurascens**
– pusilla = Liparis caespitosa
*• – **reflexa**
– rufina = Liparis nervosa
– rupestris var. purpurascens = Liparis petiolata
– serraeformis = Thrixspermum centipeda
– siamensis = Liparis nervosa
– simeonis = Liparis paradoxa
– simondii = Liparis viridiflora
• – spathulata = Liparis viridiflora
– stachyurus = Liparis viridiflora
– teniana = Liparis paradoxa
– tenii = Liparis paradoxa
– tonkinensis = Liparis paradoxa
– triloba = Liparis viridiflora
– turfosa = Malaxis latifolia
– violaceonervosa = Liparis nervosa
– **viridiflora**

Lissochilus
– graefei = Eulophia streptopetala
– horsfallii = Eulophia horsfallii
– krebsii = Eulophia streptopetala
– krebsii var. purpurata = Eulophia streptopetala
– mahonii = Eulophia horsfallii
– paivaeanus = Eulophia streptopetala
• – parviflorus = Eulophia streptopetala
– porphyroglossa = Eulophia horsfallii
– pulcher = Oeceoclades pulchra
• – roseus = Eulophia horsfallii
– sandersonii = Eulophia horsfallii
• – speciosus = Eulophia speciosa
– streptopetalus = Eulophia streptopetala
– sylvaticus = Polystachya pubescens
– transvaalensis = Eulophia horsfallii

Listera
– reniformis = Habenaria reniformis

Listrostachys
– ashantensis = Diaphananthe bidens
– bakeri = Diaphananthe bidens
– bidens = Diaphananthe bidens
– biloba = Aerangis biloba
– caudata = Plectrelminthus caudatus
– parviflora = Angraecopsis parviflora
– vesicata = Chamaeangis vesicata

Lockhartia
* – **acuta**
– lasseri = Lockhartia acuta
– **lunifera**
– pallida = Lockhartia acuta

Lonchitis
– cordigera = Serapias cordigera
– longipetala = Serapias vomeracea ssp. vomeracea

Lophiaris
– fragrans = Oncidium lanceanum

Ludisia
* – **discolor**
– furetii = Ludisia discolor
– odorata = Ludisia discolor

Lueddemannia
– sanderiana = Lacaena bicolor

Luisia
– acutilabris = Cleisostoma simondii
• – alpina = Trudelia alpina
– trichorhiza = Cleisostoma simondii

Lycaste
*• – **aromatica**
– aromatica var. majus = Lycaste aromatica
• – aromatica var. retusa = Lycaste aromatica
– balsamea = Lycaste cruenta
• – **barringtoniae**
– barringtoniae var. grandiflora = Lycaste ciliata
* – **ciliata**
– colleyi = Batemannia colleyi
• – costata = Lycaste ciliata
• – **crinita**
– cristata = Paphinia cristata
*• – **cruenta**
• – **deppei**
– dowiana = Lycaste macrophylla
– filomenoi = Lycaste macrophylla
– fimbriata = Lycaste ciliata
• – gigantea = Lycaste longipetala
– grandiflora = Maxillaria grandiflora
– harrisoniae = Bifrenaria harrisoniae
– jugosa = Pabstia jugosa
• – lanipes = Lycaste ciliata
– lata = Lycaste ciliata
* – **longipetala**
*• – **macrophylla**
– macrophylla ssp. filomenoi = Lycaste macrophylla
– macrophylla ssp. macrophylla = Lycaste macrophylla
– macrophylla ssp. plana = Lycaste macrophylla
– mesochlaena = Lycaste ciliata
– micheliana = Lycaste crinita
• – plana = Lycaste macrophylla
• – **skinneri**
– suaveolens = Lycaste aromatica
• – tetragona = Bifrenaria tetragona
– virginalis = Lycaste skinneri
– viridis = Pabstia viridis

Lycomormium
– elatum = Lycomormium squalidum
– **squalidum**

Macradenia
*• – **lutescens**
– lutescens = Encyclia odoratissima
• – mutica = Leucohyle mutica
– surinamensis = Macradenia lutescens
• – triandra = Macradenia lutescens

Macroplectrum
– gladiifolium = Angraecum mauritianum
– sesquipedale = Angraecum sesquipedale

Maelenia
– paradoxa = Cattleya forbesii

Malachadenia
• – clavata = Bulbophyllum malachadenia

Malaxis
– angustifolia = Liparis caespitosa
– caespitosa = Liparis caespitosa
– carnosula = Malaxis latifolia
– carpinterae = Malaxis excavata
– caudata = Brassia caudata
– **caulescens**
– cernua = Geodorum densiflorum
– congesta = Malaxis latifolia
– **cordata**
– cordifolia = Liparis petiolata
– curranii = Malaxis latifolia
– cylindrica = Oberonia cylindrica
– disepala = Malaxis histionantha
– **excavata**
– flavescens = Liparis flavescens
– grisebachiana = Malaxis unifolia
– **histionantha**
– kizanensis = Malaxis latifolia
– lancifolia = Liparis paradoxa
– lankesteri = Malaxis excavata
– **latifolia**
– latifolia var. fusca = Malaxis latifolia
– lilifolia = Liparis lilifolia
– loeselii = Liparis loeselii
– maxonii = Malaxis excavata
– miniata = Oberonia miniata
– minima = Liparis caespitosa
– nervosa = Liparis nervosa
– odorata = Liparis nervosa
– ophioglossoides = Malaxis unifolia
– plicata = Malaxis latifolia
– purpurascens = Liparis purpurascens
– recurva = Oberonia recurva
– setifera = Oberonia recurva
– simillima = Malaxis excavata
– thaspiformis = Malaxis unifolia
– uncinata = Malaxis excavata
* – **unifolia**
– viridiflora = Liparis viridiflora
– wightiana = Oberonia wightiana

Masdevallia
– cayenensis = Masdevallia cuprea
• – **cuprea**
– **fenestrata = Zootrophion atropurpureum**
• – **floribunda**

Masdevallia

– galeottiana = Masdevallia floribunda
• – **infracta**
– lindeniana = Masdevallia floribunda
– longicaudata = Masdevallia infracta
– myriostigma = Masdevallia floribunda
– **plantaginea**
– tuerckheimii = Masdevallia floribunda

Maxillaria

– abelei = Maxillaria rufescens
• – **acicularis**
• – **acutifolia**
• – **acutipetala**
– **aggregata**
• – **alba**
– angustifolia = Maxillaria variabilis
• – aromatica = Lycaste aromatica
– articulata = Maxillaria rufescens
– atrata = Maxillaria cucullata
– atropurpurea = Bifrenaria atropurpurea
• – aureo-fulva = Bifrenaria aureo-fulva
– balsamea = Lycaste cruenta
• – barbata = Bifrenaria vitellina
– barringtoniae hort. = Bifrenaria harrisoniae
• – barringtoniae (Smith) Lodd. = Lycaste barringtoniae
• – **bicolor**
• – boothii = Epidendrum boothii
– brachyglossa = Trigonidium egertonianum
• – bractescens = Xylobium bractescens
• – brocklehurstiana = Houlletia brocklehurstiana
– callichroma = Maxillaria setigera
– caloglossa = Maxillaria setigera
* – **camaridii**
• – **candida**
– cepula = Maxillaria madida
– chiriquensis = Maxillaria variabilis
• – **chlorantha**
• – ciliata = Lycaste ciliata
– citrina = Promenaea xanthina
• – colleyi Bateman ex Lindl. = Xylobium colleyi
– colleyi hort. ex Planchon = Batemannia colleyi
• – concava = Xylobium foveatum
• – corrugata = Xylobium corrugatum
– corrugata var. wageneri = Xylobium corrugatum
– costaricensis = Maxillaria variabilis
• – costata = Lycaste ciliata
* – **crassifolia**
– crassipes = Maxillaria marginata
• – cristata = Paphinia cristata
• – crocea = Maxillaria lindleyana
• – cruenta = Lycaste cruenta
*• – **cucullata**
• – **cuneiformis**
– cyanea = Warreella cyanea
– cyanocheile = Pabstia placanthera
• – decolor = Xylobium palmifolium
– deflexa = Maxillaria marginata
*• – **densa**
• – deppei = Lycaste deppei
– **discolor**
– disticha = Maxillaria aggregata
– echinochila = Maxillaria madida
– **elatior**
– fimbriata = Lycaste ciliata
– flabellifera = Scuticaria steelei
• – foveata = Xylobium foveatum
– fuscata = Maxillaria picta
• – galeata = Gongora galeata
– gatunensis = Maxillaria crassifolia
– gigantea = Lycaste longipetala
– goeringii = Cymbidium goeringii
– gracilifolia = Maxillaria tenuifolia
• – graminea = Koellensteinia graminea
– **graminifolia**
• – **grandiflora**
– guttata = Promenaea xanthina
• – harrisoniae = Bifrenaria harrisoniae
• – **hastata**
– hatschbachii = Maxillaria madida
– henchmanni = Maxillaria variabilis
– heynderyexii = Lycaste longipetala
– hirtilabia = Maxillaria parkeri
– jucunda = Maxillaria grandiflora
• – jugosa = Pabstia jugosa
– kegelii = Maxillaria parkeri
– kreysigii = Maxillaria picta
– lactea = Maxillaria alba
• – **latifolia**
• – lentiginosa = Promenaea lentiginosa
– leptosepala = Maxillaria setigera
– leptosepala var. subintegerrima = Maxillaria setigera
– leucocheila = Maxillaria picta
• – **ligulata**
– liliacea = Govenia liliacea
– lindeniana = Maxillaria cucullata
* – **lindleyana**
– **longibracteata var. grandiflora**
• – longifolia = Eulophia alta
• – longipetala = Lycaste longipetala
– loretoensis = Maxillaria parkeri
– lorifolia = Maxillaria parkeri
– lutescens = Maxillaria camaridii
– lyncea = Stanhopea hernandezii
• – lyonii = Maxillaria variabilis
• – macleei = Maxillaria uncata
• – macrophylla = Lycaste macrophylla
• – **maculata**
• – **madida**
* – **marginata**
– matthewsii = Maxillaria graminifolia
• – meleagris = Maxillaria cucullata
– **miniata**
– molitor = Maxillaria grandiflora
– monoceros = Maxillaria picta
– mosenii = Maxillaria madida
– murilliana = Maxillaria marginata
– mutabilis = Maxillaria porrecta
– nana = Maxillaria uncata
– nivea = Maxillaria miniata
– obscura = Maxillaria cucullata
• – **ochroleuca**
• – pallidiflora = Xylobium pallidiflorum
• – palmifolia = Xylobium palmifolium
– panamensis = Maxillaria variabilis
• – paniculata = Cyrtopodium paniculatum
• – **parkeri**
• – parvula = Bifrenaria parvula
– phyllomega = Lycaste macrophylla
*• – **picta**
• – placanthera = Pabstia placanthera
• – **platypetala**
– polypetala = Maxillaria platypetala
• – **porrecta**
– praestans = Maxillaria cucullata
• – **prolifera**
– pubigera = Bifrenaria harrisoniae
– puncto-striata = Maxillaria cucullata
– punctulata = Maxillaria marginata
• – purpurascens = Bifrenaria parvula
• – racemosa = Bifrenaria racemosa
• – **ramosa**
– rebellis = Xylobium colleyi
– revoluta = Maxillaria variabilis
• – rhombea = Maxillaria cucullata
• – rollissonii = Promenaea rollissonii
*• – **rufescens**
– rufescens var. flavida = Maxillaria rufescens
– rufescens var. minor = Maxillaria rufescens
• – rugosa = Maxillaria rufescens
– rupestris = Maxillaria picta
– sanguinolenta = Maxillaria miniata
• – scabrilinguis = Xylobium variegatum
– sessilis = Maxillaria crassifolia
• – **setigera**
– setigera var. angustifolia = Maxillaria setigera
• – skinneri = Lycaste skinneri
– spathacea = Bifrenaria harrisoniae
– squalens = Xylobium variegatum
– squamata = Maxillaria uncata
– stachyobiorum = Xylobium foveatum
• – stapelioides = Promenaea stapelioides
• – steelei = Scuticaria steelei
– stenobulbon = Xylobium pallidiflorum
– stenopetala = Bifrenaria aureo-fulva
– stenostele = Maxillaria uncata
– striatella = Maxillaria uncata
• – **subulata**
– superba = Govenia superba
– supina = Xylobium variegatum
– taphallae = Maxillaria ramosa
*• – **tenuifolia**
• – tetragona = Bifrenaria tetragona
• – triangularis = Maxillaria elatior
• – tricolor = Maxillaria marginata
• – **triphylla**
• – **uncata**
• – **undatiflora**
• – **undulata**
– vanillodora = Maxillaria rufescens
• – **variabilis**
• – variabilis var. unipunctata = Maxillaria variabilis
• – variegata = Xylobium variegatum
– virginalis = Lycaste skinneri
• – viridis = Pabstia viridis
– viridis var. pluriflora = Pabstia jugosa
• – viridis var. stenosepala = Pabstia placanthera
– viridis var. uniflora platysepala = Pabstia viridis
• – vitellina = Bifrenaria vitellina
– wageneri = Xylobium corrugatum
• – warreana = Warrea warreana
– xanthina = Promenaea xanthina

Megaclinium

– angustum = Bulbophyllum falcatum var. velutinum
– arnoldianum = Bulbophyllum falcatum var. velutinum
– brixhei = Bulbophyllum falcatum var. velutinum
– buchenavianum = Bulbophyllum calyptratum var. calyptratum
• – bufo = Bulbophyllum falcatum var. falcatum
– congolense = Bulbophyllum imbricatum
– deistelianum = Bulbophyllum falcatum var. falcatum
– djumaense = Bulbophyllum maximum
– endotrachys = Bulbophyllum falcatum var. bufo
• – falcatum = Bulbophyllum falcatum var. bufo
– flaccidum = Bulbophyllum calyptratum var. calyptratum
– gentilii = Bulbophyllum falcatum var. falcatum
– gilletii = Bulbophyllum imbricatum
– hebetatum = Bulbophyllum imbricatum
– hemirhachis = Bulbophyllum falcatum var. bufo
– imbricatum = Bulbophyllum imbricatum
– lanuriense = Bulbophyllum falcatum var. velutinum
– lasianthum = Bulbophyllum falcatum var. velutinum
– ledermannii = Bulbophyllum imbricatum
– lepturum = Bulbophyllum calyptratum var. calyptratum
– leucorhachis = Bulbophyllum imbricatum
– lindleyi = Bulbophyllum calyptratum var. calyptratum
• – maximum Lindl. = Bulbophyllum calyptratum var. calyptratum
– maximum Lindl. = Bulbophyllum maximum
– melanorrhachis = Bulbophyllum falcatum var. velutinum

Notylia

• – **barkeri**
– bernoullii = Notylia barkeri
– bipartia = Notylia barkeri
– brenesii = Notylia barkeri
– broadwayi = Notylia punctata
– guatemalensis = Notylia barkeri
– huegelii = Notylia barkeri
• – **incurva**
• – **micrantha**
– multiflora Hook. = Notylia barkeri
• – multiflora Lindl. = Notylia sagittifera
– pittieri = Notylia barkeri
• – **pubescens**
∗ – **punctata**
– **sagittifera**
– tamaulipensis = Notylia barkeri
• – tenuis = Notylia sagittifera
– tridachne = Notylia barkeri
– trisepala = Notylia barkeri
– turialbae = Notylia barkeri

Nyctosma

– nocturna = Epidendrum nocturnum

Oberonia

– arnottiana = Oberonia wightiana
– **complanata**
– croftiana = Oberonia recurva
• – **cylindrica**
– gardneriana = Oberonia recurva
– lingmalensis = Oberonia recurva
• – **miniata**
– myriantha var. parvula = Oberonia recurva
– parvula = Oberonia recurva
• – **recurva**
– setifera = Oberonia recurva
– stachyoides = Oberonia wightiana
• – **wightiana**

Octomeria

• – **diaphana**
• – **gracilis**
• – **grandiflora**
– multiflora = Octomeria tridentata
– racemosa = Eria multiflora
– rosea = Eria rosea
– semiteres = Octomeria gracilis
– spicata = Eria spicata
– stellata = Eria javanica
• – **tridentata**
– vaginata = Eria javanica

Odontochilus

– petelotii = Ludisia discolor

Odontoglossum

– alexandrae = Odontoglossum crispum
– alexandrae var. guttatum = Odontoglossum crispum
– anceps = Lemboglossum maculatum
• – **angustatum**
• – apterum = Lemboglossum apterum
– aspasia = Aspasia epidendroides
– asperum = Lemboglossum rossii
– bellum = Odontoglossum pardinum
• – bicolor = Oncidium aureum
• – bictoniense = Lemboglossum bictoniense
– bluntii = Odontoglossum crispum
– boddaertianum = Odontoglossum constrictum
– bogotense = Odontoglossum revolutum
• – brevifolium = Otoglossum brevifolium
• – cervantesii = Lemboglossum cervantesii
– cervantesii var. membranaceum = Lemboglossum cervantesii
– chaetostoma = Odontoglossum hallii
– cimiciferum = Oncidium cimiciferum
• – **cirrhosum**
– cirrhosum var. gemmatum = Odontoglossum cirrhosum
– cirrhosum var. hrubyanum = Odontoglossum cirrhosum
– cirrhosum var. klabochorum = Odontoglossum cirrhosum
• – citrosmum = Cuitlauzina pendula
• – clowesii = Miltonia clowesii
• – coerulescens = Lemboglossum rossii
• – **constrictum**
• – cordatum = Lemboglossum cordatum
– coronarium var. miniatum = Otoglossum brevifolium
– crispulum = Gomesa crispa
• – **crispum**
• – **cristatum**
– dawsonianum = Lemboglossum ehrenbergii
• – egertonii = Osmoglossum egertonii
– ehrenbergii = Lemboglossum ehrenbergii
• – **epidendroides**
• – erosum = Lemboglossum stellatum
– festatum = Oncidium aureum
• – galeottianum = Lemboglossum galeottianum
• – ghiesbreghtianum = Oncidium tigrinum
• – **gracile**
– graminifolium = Odontoglossum longifolium
• – grande = Rossioglossum grande
• – **hallii**
• – hastatum = Oncidium hastatum
– hemichrysum = Oncidium aureum
– hookeri = Lemboglossum cordatum
– humeanum = Lemboglossum rossii
– insleayi = Rossioglossum insleayi
• – **ixioides**
– johnsonorum = Oncidium maculatum
– karwinskii = Miltonioides karwinskii
• – **lacerum**
• – laeve = Miltonioides laevis
– latimaculatum = Odontoglossum crispum
– laxiflorum = Gomesa laxiflora
– leucomelas = Miltonioides laevis
• – **longifolium**
– lueddemannii = Lemboglossum cordatum
– maculatum Lindl. = Lemboglossum cordatum
• – maculatum Llave & Lex. = Lemboglossum maculatum
– madrense = Lemboglossum madrense
• – maxillare = Lemboglossum madrense
• – membranaceum = Lemboglossum cervantesii
– miniatum = Otoglossum brevifolium
• – **myanthum**
∗ – **mystacinum**
• – nebulosum = Lemboglossum apterum
• – **pardinum**
– pendulum = Cuitlauzina pendula
– phyllochilum = Oncidium hastatum
– platyodon = Odontoglossum revolutum
• – pulchellum = Osmoglossum pulchellum
• – pygmaeum = Rhyncostele pygmaea
– reichenbachianum = Odontoglossum crispum
• – **retusum**
• – **revolutum**
• – rigidum = Odontoglossum mystacinum
• – roseum = Cochlioda rosea
• – rossii = Lemboglossum rossii
– rubescens = Lemboglossum rossii
– sanderianum = Odontoglossum constrictum
– spilolanthum = Odontoglossum angustatum
• – stellatum = Lemboglossum stellatum
– stenochilum = Odontoglossum pardinum
– subcruciforme = Oncidium maculatum
– tigrinum = Oncidium tigrinum
– ulopterum = Odontoglossum angustatum
– variegatum = Aspasia variegata
– victor = Odontoglossum hallii
• – warneri = Mexicoa ghiesbreghtiana
• – warneri var. purpuratum = Mexicoa ghiesbreghtiana
– warnerianum = Lemboglossum rossii
– youngii = Lemboglossum rossii
– vexativum = Lemboglossum apterum

Oeceoclades

– falcata = Neofinetia falcata
– lindleyana = Neofinetia falcata
– lindleyi = Neofinetia falcata
∗ – **maculata**
– paniculata = Robiquetia succisa
– parviflora = Angraecopsis parviflora
∗ – **pulchra**
– retzii = Chilochista pusilla

Oeonia

• – aubertii = Oeonia volucris
– humblotii = Oeonia volucris
– **volucris**

Oerstedella

– verrucosa = Epidendrum verrucosum

Ommatodium

– **volucris**

Oncidium

– acrobotrium = Oncidium harrisonianum
– alboviolaceum = Oncidium incurvum
– allemanii = Psygmorchis pusilla
∗ – **altissimum**
• – altissimum Lindl. = Oncidium baueri
• – altissimum Sw. = Oncidium jacquinianum
– altissimum var. baueri = Oncidium baueri
∗• – **amictum**
∗• – **ampliatum**
– anceps = Lemboglossum apterum
• – **ascendans**
– **aureum**
– aureum var. stenochilum = Oncidium aureum
– aurosum = Oncidium excavatum
• – **barbatum**
• – barkeri = Oncidium tigrinum
– batemanianum var spilopterum = Oncidium spilopterum
– **batemannianum**
∗ – **baueri**
– bernoullianum = Oncidium ampliatum
∗• – **bicallosum**
• – **bicolor**
– bicornutum = Oncidium pubes
– bidens = Brassia bidens
– bolivianum = Oncidium baueri
– boydii = Oncidium altissimum
– brachiatum = Brassia verrucosa
∗ – **brachyandrum**
• – brachyphyllum = Oncidium cebolleta
– brachystegium = Oncidium batemannianum
– brassia = Brassia maculata
– brevifolium = Oncidium reflexum
– caesium = Oncidium nebulosum
– caldense = Oncidium ramosum
– candidum Beer = Anneliesia candida
• – candidum Lindl. = Palumbina candida
– candidum Rchb.f. = Anneliesia candida
• – carinatum = Leochilus carinatus
• – **carthagenense**
– carthagenense var. andreanum = Oncidium carthagenense
– carthagenense var. klotzschii = Oncidium carthagenense
– carthagenense var. oerstedii = Oncidium carthagenense
– carthagenense var. roseum = Oncidium carthagenense
– carthagenense var. sanguineum = Oncidium carthagenense
– carthagenense var. swartzii = Oncidium carthagenense
– caudatum = Brassia caudata
– **cavendishianum**
∗• – **cebolleta**
– cepula = Oncidium cebolleta
– cerebriferum = Oncidium ensatum
– cervantesii = Lemboglossum cervantesii
– chrysorhapis = Oncidium cornigerum
∗• – **ciliatum**
– **cimiciferum**
– cirrhosum = Odontoglossum cirrhosum
∗• – **citrinum**
– citrosmum = Cuitlauzina pendula
– clowesii = Miltonia clowesii
– colombae = Oncidium stramineum

Oncidium

– concavum = Oncidium pulchellum
• – confragosum = Oncidium viperinumum
– confusum = Oncidium baueri
* • – **cornigerum**
– corynephorum = Oncidium volubile
– cosymbeforum = Oncidium altissimum
* • – **crispum**
– crispum var. forbesii = Oncidium forbesii
– crispum var. marginatum = Oncidium forbesii
– cristatum = Odontoglossum cristatum
• – cuneatum Scheidw. = Oncidium remotiflorum
– cuneatum (Lindl.) Beer = Miltonia cuneata
* • – **curtum**
* • – **deltoideum**
– depauperatum = Leochilus labiatus
– desertorum = Oncidium guianense
– dichromum = Oncidium aureum
– digitatum = Oncidium leucochilum
– dimorphum = Oncidium trulliferum
* • – **divaricatum**
– divaricatum var. cupreum = Oncidium divaricatum
– durangense = Oncidium nebulosum
– egertonii = Osmoglossum egertonii
– ehrenbergii = Lemboglossum ehrenbergii
– elegantissimum = Oncidium gardneri
– emarginatum = Dimerandra emarginata
• – **ensatum**
– epidendroides = Odontoglossum epidendroides
– erosum = Lemboglossum stellatum
– euxanthinum = Oncidium varicosum
• – **excavatum**
– excavatum var. dawsonii = Oncidium excavatum
– festatum = Oncidium aureum
– filipes = Oncidium brachyandrum
– fimbriatum = Oncidium ciliatum
– flabelliferum = Oncidium gardneri
– flavescens = Miltonia flavescens
– flexuosum = Oncidium cimiciferum
• – **forbesii**
– forbesio – dasystyle = Oncidium gardneri
• – **forkelii**
– funereum = Oncidium maculatum
– galeatum = Oncidium trulliferum
– galeottianum = Cuitlauzina pendula
• – gallopavinum = Oncidium batemannianum
– **gardneri**
– gardnerianum = Oncidium gardneri
– gautieri = Oncidium longicornu
– geertianum = Oncidium nebulosum
– geraense = Oncidium varicosum
– ghiesbreghtianum = Mexicoa ghiesbreghtiana
– ghillanyi = Oncidium spilopterum
– glaziovii = Oncidium cebolleta
• – **gracile**
– graminifolium = Oncidium brachyandrum
– graminifolium var. filipes = Oncidium brachyandrum
– graminifolium var. holochilum = Oncidium brachyandrum
– graminifolium var. wrayae = Oncidium brachyandrum
* – **guianense**
– guttatum = Oncidium altissimum
– guttatum var. auriculatum = Oncidium altissimum
– guttatum var. dodgsonii = Oncidium altissimum
– guttatum var. intermedium = Oncidium altissimum
– guttatum var. morrenii = Oncidium altissimum
– guttatum var. olivaceum = Oncidium altissimum
– guttatum var. roseum = Oncidium altissimum
– guttulatum = Oncidium baueri
– hallii = Odontoglossum hallii
– harrisoniae = Oncidium harrisonianum
* • – **harrisonianum**
– harrisonianum var. pallidum = Oncidium harrisonianum
– **hastatum**
– hebraicum = Oncidium baueri
– hemichrysum = Oncidium aureum
– henchmannii = Oncidium carthagenense
– herbaceum = Leochilus scriptus
• – **hians**
– hintonii = Oncidium maculatum
– humboldtii = Oncidium cebolleta
• – huntianum = Oncidium carthagenense
– hymatoides = Oncidium carthagenense
* • – **incurvum**
• – insleayi = Rossioglossum insleayi
• – intermedium Knowles & Westc. = Oncidium altissimum
– intermedium Bertero = Oncidium guianense
– ionosmum = Oncidium tigrinum
• – iridifolium = Psygmorchis pusilla
• – **isopterum**
– ixioides = Odontoglossum ixioides
* – **jacquinianum**
– johnianum = Oncidium barbatum
– juergensianum = Oncidium hastatum
– juncifolium = Oncidium cebolleta
– kappleri = Oncidium baueri
– labiatum = Leochilus labiatus
• – lacerum = Oncidium stipitatum
– laeve = Miltonioides laevis
* • – **lanceanum**
– lanceanum var. superbum = Oncidium lanceanum
– lansbergii = Leochilus labiatus
– lawrenceana = Brassia lawrenceana
• – lemonianum = Oncidium guianense
• – **leucochilum**
– leucostomum = Oncidium hians
– lindenii Lodd. = Oncidium altissimum
– lindenii hort. = Oncidium stramineum
– lintriculus = Oncidium maculatum
– **longicornu**
• – longifolium = Oncidium cebolleta
– lunaenum = Oncidium varicosum
• – lunatum = Solenidium lunatum
• – luridum = Oncidium altissimum
– luridum var. atratum = Oncidium altissimum
– luridum var. dogsonii = Oncidium altissimum
• – luridum var. guttatum = Oncidium altissimum
– luridum var. henchmannii = Oncidium carthagenense
– luridum var. intermedium = Oncidium altissimum
– luridum var. morrenii = Oncidium altissimum
– luridum var. olivaceum = Oncidium altissimum
• – macrantherum = Leochilus oncidioides
– macropterum = Oncidium reflexum
* • – **maculatum**
– maculatum = Oncidium altissimum
– maculatum var. donianum = Oncidium maculatum
– maculatum var. psittacium = Oncidium maculatum
– maculatum var. vinosum = Oncidium maculatum
– **maculosum**
• – **martianum**
– martianum var. bicolor = Oncidium bicolor
– maxilligerum = Oncidium hians
– membranaceum = Lemboglossum cervantesii
* • – **microchilum**
– microglossum = Oncidium barbatum
• – monoceras = Oncidium longicornu
• – **nanum**
• – **nebulosum**
– nitidum = Oncidium amictum
• – **nudum**
• – **oblongatum**
– obsoletum = Oncidium carthagenense
– oerstedii = Oncidium carthagenense
– oerstedii var. crispiflorum = Oncidium carthagenense
– ornithocephaloides = Oncidium trulliferum
* • – **ornithorhynchum**
– ottonis = Oncidium cebolleta
• – pachyphyllum = Oncidium cavendishianum
• – pallidum = Oncidium harrisonianum
– panduriferum = Oncidium carthagenense
– pantherinum = Oncidium harrisonianum
• – papilio = Psychopsis papilio
– papilio var. eckhardtii = Psychopsis papilio
– papilio var. limbatum = Psychopsis papilio
– papilio var. majus = Psychopsis papilio
– papilio var. pictum = Psychopsis papilio
– pardinum = Odontoglossum pardinum
– patulum = Oncidium nanum
• – pelicanum = Oncidium reflexum
– pentaspilum = Oncidium harrisonianum
– pentecostale = Oncidium baueri
• – **pergameneum**
– peruvianum = Brassia peruviana
– phyllochilum = Oncidium hastatum
– pinelianum = Oncidium batemannianum
– polletianum = Oncidium gardneri
– polycladium = Oncidium baueri
– praestans = Oncidium gardneri
– puber = Oncidium pubes
* • – **pubes**
– pubescens = Oncidium pubes
* • – **pulchellum**
* • – **pulvinatum**
* • – **pumilum**
– pusillum = Psygmorchis pusilla
– pusillum var. megalanthum = Psygmorchis pusilla
– pygmaeum = Rhyncostele pygmaea
– quadricorne = Oncidium hians
– quadripetalum = Oncidium tetrapetalum
• – **ramosum**
* • – **raniferum**
– raniferum = Oncidium harrisonianum
* • – **reflexum**
– reisii = Oncidium ciliatum
– **remotiflorum**
• – **retusum**
– retusum = Odontoglossum retusum
– revolutum = Odontoglossum revolutum
– rigidum = Odontoglossum mystacinum
– rogersii = Oncidium varicosum
– roseum = Oncidium carthagenense
– rupestre = Oncidium excavatum
• – russellianum = Miltonia russelliana
– saint-legerianum = Oncidium spilopterum
– saltator = Oncidium stramineum
• – sanguineum = Oncidium carthagenense
– sanguineum var. auriculatum = Oncidium altissimum
– sanguineum var. roseum = Oncidium carthagenense
– **sanguinolentum**
– scirus = Oncidium pulvinatum
– scriptum = Leochilus scriptus
– skinneri = Oncidium excavatum
– speciosum = Miltonia cuneata
– spectabile var. morelianus = Miltonia spectabilis
– spectabile var. purpureo-violacea = Miltonia spectabilis
* • – **sphacelatum**
• – **sphegiferum**
* • – **spilopterum**
– sprucei = Oncidium cebolleta
– stenopetalum = Oncidium batemannianum
* – **stipitatum**

Oncidium
– stipitatum var. platyonyx = Oncidium stipitatum
* • – **stramineum**
• – suave = Oncidium reflexum
– suaveolens = Brassia lanceana
– suscephalum = Oncidium barbatum
• – suttonii = Oncidium reflexum
– tayleuri = Oncidium reflexum
– tenue = Oncidium wentworthianum
• – **tetrapetalum**
– tetrapetalum = Oncidium urophyllum
• – **tigrinum**
– tigrinum = Oncidium tigrinum
– tigrinum var. unguiculatum = Oncidium tigrinum
– tricolor = Oncidium tetrapetalum
– trulla = Oncidium cimiciferum
* • – **trulliferum**
– uncia = Oncidium reflexum
– **undulatum**
– undulatum = Oncidium carthagenense
– unguiculatum = Oncidium tigrinum
• – unicorne = Oncidium longicornu
• – unicornutum = Oncidium longicornu
* • – **uniflorum**
* • – **urophyllum**
• – **varicosum**
– verrucosum = Brassia verrucosa
– **viperinumum**
– vittii = Oncidium cebolleta
– **volubile**
– wendlandianum = Oncidium reflexum
• – **wentworthianum**
– wentworthianum var. tenue = Oncidium wentworthianum
• – wrayae = Oncidium brachyandrum
– xanthochlorum = Oncidium oblongatum

Onychium
– affine = Dendrobium affine
– crumenatum = Dendrobium crumenatum
– flavescens = Polystachya concreta
– japonicum = Dendrobium moniliforme
– lamellatum = Dendrobium lamellatum
– mutabile = Dendrobium mutabile
– nudum = Dendrobium nudum
– rigidum = Dendrobium mutabile

Ophrys
– affinis = Pterygodium catholicum
– alaris = Pterygodium catholicum
– alata = Pterygodium alatum
– aloidea = Oncidium guianense
– alpina = Chamorchis alpina
• – arachnites = Ophrys holoserica
– araneifera = Ophrys sphegodes ssp. sphegodes
– araneifera ssp. boissieri = Ophrys oestrifera ssp. oestrifera
– araneifera ssp. mammosa var. boissieri = Ophrys oestrifera ssp. oestrifera
• – araneifera var. limbata = Ophrys sphegodes ssp. sphegodes
– araneola = Ophrys sphegodes ssp. sphegodes
– aranifera = Ophrys sphegodes ssp. sphegodes
– aranifera ssp. atrata = Ophrys atrata
– aranifera var. atrata = Ophrys atrata
– argentea = Erythrodes picta
* – **argolica**
* • – **atrata**
– biceratia = Ophrys atrata
– bracteata = Satyrium bracteatum
– catholica = Pterygodium catholicum
– cernua L. = Spiranthes cernua
– ciliata = Ophrys vernixia
• – cornuta = Ophrys oestrifera ssp. oestrifera
* • – **ferrum-equinum**
– ferrum-equinum ssp. argolica = Ophrys argolica
– fucifera = Ophrys sphegodes ssp. sphegodes
• – fuciflora Lindl. = Ophrys holoserica
– fuciflora (F.W.Schmidt) Moench = Ophrys holoserica
– fuciflora ssp. cornuta = Ophrys oestrifera ssp. oestrifera
* • – **fusca**
– grandiflora = Ophrys tenthredinifera
– guianensis = Oncidium guianense
* – **holoserica**
– holosericea ssp. cornuta = Ophrys oestrifera ssp. oestrifera
– insectifera = Ophrys tenthredinifera
– insectifera ssp. aranifera = Ophrys sphegodes ssp. sphegodes
– insectifera var. arachnites = Ophrys holoserica
– inversa = Anochilus inversum
– lancea = Herminium lanceum
– lilifolia = Liparis lilifolia
– limbata = Ophrys tenthredinifera
– loeselii = Liparis loeselii
– mammosa ssp. boissieri = Ophrys oestrifera ssp. oestrifera
– monorchis = Herminium monorchis
– neglecta = Ophrys tenthredinifera
– nervosa = Liparis nervosa
* – **oestrifera ssp. oestrifera**
– oestrifera ssp. cornuta = Ophrys oestrifera ssp. oestrifera
– oestrifera var. cornuta = Ophrys oestrifera ssp. oestrifera
– peruviana = Spiranthes torta
– quinquelobata = Spiranthes torta
– rosea = Ophrys tenthredinifera
– rosea ssp. neglecta = Ophrys tenthredinifera
– scolopax ssp. cornuta = Ophrys oestrifera ssp. oestrifera
– scolopax var. oestrifera = Ophrys oestrifera ssp. oestrifera
• – speculum = Ophrys vernixia
* – **sphegodes ssp. sphegodes**
– sphegodes ssp. atrata = Ophrys atrata
– sphegodes var. atrata = Ophrys atrata
– spiralis Georgi = Spiranthes sinensis
– spiralis L. = Spiranthes spiralis
• – tabanifera = Ophrys argolica
– tenoreana = Ophrys tenthredinifera
* • – **tenthredinifera**
– tenthredinifera ssp. neglecta = Ophrys tenthredinifera
• – tenthredinifera var. minor = Ophrys tenthredinifera
– torta = Spiranthes torta
– triphylla = Ommatodium volucris
– unilateralis = Brachystele unilateralis
* – **vernixia**
– villosa = Ophrys tenthredinifera
– volucris = Ommatodium volucris

Orchidea
– hispida = Holothrix monotris

Orchidofunckia
– pallidiflora = Cryptarrhena lunata

Orchiodes
– celebicum = Goodyera rubicunda
– grande = Goodyera rubicunda
– parviflorum = Goodyera procera
– procerum = Goodyera procera
– rubicundum = Goodyera rubicunda

Orchis
– acuminata = Orchis lactea
– alpina = Chamorchis alpina
– burmanniana = Bartholina burmanniana
– carnea = Satyrium carneum
– circumflexa = Disperis secunda
– coccinea = Corycium crispum
– cordata = Gennaria diphylla
– cordifolia = Gennaria diphylla
– cornuta = Satyrium coriifolium
– falcata = Neofinetia falcata
– fastigiata = Cynorkis fastigiata
– fimbriata = Platanthera grandiflora
• – foliosa = Dactylorhiza foliosa
– fuciflora = Ophrys holoserica
– graminea = Chamorchis alpina
– grandiflora = Platanthera grandiflora
– holoserica = Ophrys holoserica
– incarnata = Dactylorhiza sambucina
* – **italica**
* – **lactea**
– lanigera = Rhynchostylis retusa
– latifolia var. foliosa = Dactylorhiza foliosa
– leucantha = Peristylus constrictus
– lingua = Serapias vomeracea ssp. vomeracea
• – longibracteata = Barlia robertiana
– longicornis = Orchis longicornu
* • – **longicornu**
– longicruris = Orchis italica
– lutea = Dactylorhiza sambucina
– maderensis = Dactylorhiza foliosa
– mauritiana Poiret = Angraecum mauritianum
– mauritiana Sieber = Cynorkis fastigiata
– militaris = Orchis italica
– mixta var. sambucina = Dactylorhiza sambucina
– morio var. longicornu = Orchis longicornu
– orientalis ssp. foliosa = Dactylorhiza foliosa
* • – **papilionacea**
– papilionacea var. rubra = Orchis papilionacea
– pectinata = Bartholina burmanniana
– procera = Habenaria procera
– robertiana = Barlia robertiana
– saccata = Dactylorhiza sambucina
• – sambucina = Dactylorhiza sambucina
– schleicheri = Dactylorhiza sambucina
– secundiflora = Neotinea maculata
– simia var. undulatifolia = Orchis italica
– spathulata = Herschelianthe spathulata
• – strateumatica = Zeuxine strateumatica
– tephrosanthos = Orchis italica
• – tephrosanthos var. undulatifolia = Orchis italica
– triplicata = Calanthe triplicata
– undulatifolia = Orchis italica
– uniflora = Diplomeris pulchella
• – variegata = Orchis lactea
– vomeracea = Serapias vomeracea ssp. vomeracea

Ormostema
– albiflora = Dendrobium moniliforme
– purpurea = Dendrobium moniliforme

Ornithidium
– aggregatum = Maxillaria aggregata
– album = Maxillaria camaridii
– bicolor = Oncidium aureum
– breve = Maxillaria aggregata
– densum = Maxillaria densa
– dichotomum = Maxillaria ramosa
– distichum = Maxillaria aggregata
– fragrans = Maxillaria camaridii
– imbricatum = Pholidota imbricata
• – miniatum = Maxillaria miniata
– niveum = Maxillaria miniata
– pfitzeranum = Maxillaria aggregata
– sanguinolentum = Maxillaria miniata
– squamata = Maxillaria uncata
– taphallae = Maxillaria ramosa

Ornithocephalus
– navicularis = Zygostates cornuta

Orxera
– cornuta = Aerides odorata

Osmoglossum
– anceps = Osmoglossum egertonii
– **egertonii**
* – **pulchellum**

Otandra
– cernua = Geodorum densiflorum

Otochilus
– forrestii = Otochilus porrectus
– fragrans = Otochilus porrectus
– latifolius = Otochilus porrectus
– **porrectus**

Otoglossum
– **brevifolium**

Otoglossum
– **Oxysepala**
– ovalifolia = Bulbophyllum sessile
Pabstia
– **jugosa**
– **placanthera**
* – **viridis**
Pachyne
– spectabilis = Phaius tankervilleae
Pachystele
– **dubia**
– minuta = Polystachya concreta
Pachystelis
– dubia = Pachystele dubia
– livida = Pachystele dubia
Pachystoma
– fortunei = Spathoglottis pubescens
– parvifolium = Spathoglottis pubescens
– wightii = Spathoglottis pubescens
Palumbina
– **candida**
Panstrepis
– paradoxa = Coryanthes macrantha
Paphinia
*● – **cristata**
Paphiopedilum
* – **barbatum**
– barbatum var. nigritum = Paphiopedilum barbatum
– epidendricum = Eulophia alta
– nigritum = Paphiopedilum barbatum
– pardinum = Paphiopedilum venustum
* – **purpuratum**
– socco = Catasetum trulla
* – **venustum**
– venustum var. pardinum = Paphiopedilum venustum
Papilionanthe
* – **teres**
Paragnatis
– pulchella = Diplomeris pulchella
Paxtonia
● – rosea = Spathoglottis plicata
Pedilonum
– biflorum = Dendrobium biflorum
– bursigerum = Dendrobium secundum
– erosum = Dendrobium erosum
– kuhlii = Dendrobium kuhlii
– lamellatum = Dendrobium lamellatum
– sanguinolentum = Dendrobium sanguinolentum
– secundum = Dendrobium secundum
– undulatum = Dendrobium hymenophyllum
Pelexia
* – **adnata**
– **cerina**
– **lobata**
● – spiranthoides = Pelexia adnata
– stenorrhynchoides = Pelexia adnata
– tomentosa = Sacoila lanceolata
Peramium
– longibracteatum = Goodyera rubicunda
– procerum = Goodyera procera
Peristeria
– barkeri = Acineta barkeri
*● – **cerina**
– cerina var. guttata = Peristeria guttata
● – **elata**
– fuscata = Lycomormium squalidum
● – **guttata**
● – humboldtii = Acineta superba
– lentiginosa = Peristeria pendula
– longiscapa = Lacaena bicolor
– maculata = Peristeria pendula
● – **pendula**
Peristylus
– **constrictus**
– cordatus = Gennaria diphylla
– **densus**
– elatus = Peristylus plantagineus
● – **goodyeroides**
– goodyeroides = Peristylus constrictus
– **hamiltonianus**
– hollandiae = Peristylus goodyeroides
– horsfieldii = Peristylus goodyeroides
– latifolius = Benthamia latifolia
– meglectus = Peristylus densus
– **plantagineus**
– **secundus**
– spiralis = Benthamia spiralis
– stenostachyus = Peristylus densus
* – **tentaculatus**
– xanthochlorus = Peristylus densus
Pesomeria
● – tetragona = Phaius tetragonus
Petronia
– regia = Batemannia colleyi
Phaedrosanthus
– ciliaris = Epidendrum ciliare
– nocturnus = Epidendrum nocturnum
Phaius
● – albus = Thunia bracteata
● – bicolor = Phaius tankervilleae
– blumei = Phaius tankervilleae
– blumei var. assamica = Phaius tankervilleae
– blumei var. bernaysii = Phaius tankervilleae
– blumei var. pulchra = Phaius tankervilleae
– carronii = Phaius tankervilleae
– grandiflorus = Phaius tankervilleae
● – grandifolius = Phaius tankervilleae
– grandifolius var. superbus = Phaius tankervilleae
– incarvillei = Phaius tankervilleae
– incarvillei var. speciosa = Phaius tankervilleae
– leucophaeus = Phaius tankervilleae
– longicornu = Phaius tankervilleae
– rumphii = Spathoglottis plicata
– **tankervilleae**
– tankervilleae f. veronicae = Phaius tankervilleae
– tankervilleae var. superbus = Phaius tankervilleae
– **tetragonus**
– villosus = Phaius tetragonus
● – wallichii = Phaius tankervilleae
– wallichi var. assamica = Phaius tankervilleae
Phalaenopsis
● – amabilis = Phalaenopsis aphrodite
– amabilis var. ambigua = Phalaenopsis aphrodite
– amabilis var. aphrodite = Phalaenopsis aphrodite
– amabilis var. aphrodite subvar. dayana = Phalaenopsis aphrodite
– amabilis var. aphrodite subvar. erubescens = Phalaenopsis aphrodite
– amabilis var. dayana = Phalaenopsis aphrodite
– amabilis var. erubescens = Phalaenopsis aphrodite
– amabilis var. formosa = Phalaenopsis aphrodite
– amabilis var. longifolia = Phalaenopsis aphrodite
– amabilis var. rotundifolia = Phalaenopsis aphrodite
– ambigua = Phalaenopsis aphrodite
* – **aphrodite**
– aphrodite var. dayana = Phalaenopsis aphrodite
– babuyana = Phalaenopsis aphrodite
– erubescens = Phalaenopsis aphrodite
– formosana = Phalaenopsis aphrodite
– formosum = Phalaenopsis aphrodite
Pholidota
● – **articulata**
– articulata var. griffithii = Pholidota articulata
– articulata var. minahassae = Pholidota articulata
– articulata var. obovata = Pholidota articulata
– assamica = Pholidota imbricata
– beccarii = Pholidota imbricata
– bracteata = Pholidota imbricata
– calceata = Pholidota pallida
● – conchoidea = Pholidota imbricata
– crotalina = Pholidota imbricata
– decurva = Pholidota articulata
– grandis = Pholidota imbricata
– griffithii = Pholidota articulata
– henryi = Pholidota imbricata
*● – **imbricata**
– imbricata = Pholidota imbricata
– imbricata var. coriacea = Pholidota imbricata
– imbricata var. henryi = Pholidota imbricata
– imbricata var. longifolia = Pholidota imbricata
– imbricata var. montana = Pholidota imbricata
– imbricata var. papuana = Pholidota imbricata
– imbricata var. platyphylla = Pholidota imbricata
– imbricata var. sessile = Pholidota pallida
– khasiyana = Pholidota articulata
– loricata = Pholidota imbricata
– lugardii = Pholidota articulata
– minahassae = Pholidota articulata
– obovata = Pholidota articulata
– **pallida**
– pallida = Pholidota imbricata
– repens = Pholidota articulata
– **rubra**
– schlechteri = Pholidota pallida
– spectabilis = Pholidota imbricata
– sp. = Pholidota pallida
– tixieri = Pholidota pallida
– triotos = Pholidota imbricata
● – undulata = Pholidota rubra
– yunnanensis = Pholidota pallida
– yupeensis = Pholidota pallida
Phreatia
– congesta = Eria retusa
– retusa = Eria retusa
Phyllorchis
– adenopetala = Bulbophyllum flavescens
– antennifera = Bulbophyllum antenniferum
– aurata = Bulbophyllum auratum
– barbigera = Bulbophyllum barbigerum
– blumei = Bulbophyllum blumei
– bracteolata = Bulbophyllum bracteolatum
– bufo = Bulbophyllum falcatum var. falcatum
– calamaria = Bulbophyllum saltatorium var. saltatorium
– capitata = Bulbophyllum pahudii
– cheiri = Bulbophyllum cheiri
– clavigera = Bulbophyllum longiflorum
– cumingii = Bulbophyllum cumingii
– elongata = Bulbophyllum elongatum
– falcata = Bulbophyllum falcatum var. bufo
– fimbriata = Bulbophyllum fimbriatum
– flavescens = Bulbophyllum flavescens
– flavida = Bulbophyllum pumilum
– guttulata = Bulbophyllum guttulatum
– helenae = Bulbophyllum helenae
– herminiostachys = Bulbophyllum pumilum
– hirta = Bulbophyllum hirtum
– imbricata = Bulbophyllum imbricatum
– limbata = Bulbophyllum limbatum
– macraei = Bulbophyllum macraei
– macrantha = Bulbophyllum macranthum
– maxillaris = Bulbophyllum blumei
– maxima = Bulbophyllum maximum
– medusae = Bulbophyllum medusae
– megalantha = Bulbophyllum cheiri
– othonis = Bulbophyllum othonis
– oxyptera = Bulbophyllum maximum
– pavimentata = Bulbophyllum pumilum
– picturata = Bulbophyllum picturatum
– recurva = Bulbophyllum pumilum
– reinwardtii = Bulbophyllum uniflorum

Phyllorchis
– rhizophorae = Bulbophyllum falcatum var. velutinum
– saltatoria = Bulbophyllum saltatorium var. calamarium
– sessile = Bulbophyllum sessile
– setigera = Bulbophyllum setigerum
– shepherdii = Bulbophyllum sheperdii
– thouarsii = Bulbophyllum longiflorum
– umbellata = Bulbophyllum longiflorum
– uniflora = Bulbophyllum uniflorum
– vaginata = Bulbophyllum vaginatum
– velutina = Bulbophyllum falcatum var. velutinum
Physinga
• – prostata = Epidendrum prostatum
Physosiphon
• – carinatus = Physosiphon tubatus
– deregularis = Pleurothallis deregularis
• – **emarginatus**
– guatemalensis = Physosiphon tubatus
– lindleyi = Physosiphon tubatus
• – loddigesii = Physosiphon tubatus
– minutuflorus = Pleurothallis deregularis
– moorei = Physosiphon tubatus
– ochraceus = Physosiphon tubatus
• – spiralis = Pleurothallis deregularis
– **tubatus**
Physurus
– argenteus = Erythrodes picta
• – pictus = Erythrodes picta
Pilumna
• – fragrans = Trichopilia fragrans
• – laxa = Trichopilia laxa
– reichenheimiana = Trichopilia laxa
– wageneri = Trichopilia fragrans
Pinalia
– bipunctata = Eria bipunctata
– bractescens = Eria bractescens
– calamifolia = Eria pannea
– clavicaulis = Eria clavicaulis
– floribunda = Eria floribunda
– fragrans = Eria javanica
– micrantha = Eria multiflora
– multiflora = Eria multiflora
– nutans = Eria nutans
– obesa = Eria obesa
– ornata = Eria ornata
– paniculata = Eria paniculata
– pannea = Eria pannea
– polyura = Eria polyura
– pulchella = Eria bractescens
– pumila = Eria pumila
– reticulata = Porpax reticulata
– retusa = Eria retusa
– rosea = Eria rosea
– rugosa = Eria javanica
– stellata = Eria javanica
– striolata = Eria javanica
– velutina = Trichotosia velutina
Pinelia
– tuerckheimii = Homalopetalum pumilio
Platanthera
– angustifolium = Herminium lanceum
– bigelovii = Platanthera grandiflora
– constricta = Peristylus constrictus
– diphylla = Gennaria diphylla
– fimbriata = Platanthera grandiflora
– fimbriata var. grandiflora = Platanthera grandiflora
– goodyeroides = Peristylus goodyeroides
* – **grandiflora**
– psycodes var. grandiflora = Platanthera grandiflora
– stenostachya = Peristylus densus
Platyclinis
– abbreviata = Dendrochilum abbreviatum
– filiformis = Dendrochilum filiforme
– glumacea = Dendrochilum glumaceum
– latifolia = Dendrochilum latifolium
Platylepis
– **goodyeroides**
Platypus
– altus = Eulophia alta
– papilliferus = Eulophia alta
Plectorrhiza
– **tridentata**
Plectrelminthus
* – **caudatus**
Plectrophora
– **iridifolia**
Pleione
– birmanica = Pleione praecox
– chinense = Coelogyne fimbriata
– concolor = Pleione praecox
– cumingii = Coelogyne cumingii
– fimbriata = Coelogyne fimbriata
– goweri = Coelogyne nitida
* – **praecox**
– praecox var. alba = Pleione praecox
– praecox var. birmanica = Pleione praecox
– praecox var. candida = Pleione praecox
– praecox var. sanguinea = Pleione praecox
– praecox var. wallichiana = Pleione praecox
– reichenbachiana = Pleione praecox
– wallichiana = Pleione praecox
Pleurothallis
– **acianthera**
• – **acuminata**
• – affinis = Myoxanthus affinis
– albida = Pleurothallis obovata
– albipetala = Pleurothallis recurva
– alexandrii = Pleurothallis glumacea
• – **alpestris**
• – **angustifolia**
– angustilabia = Pleurothallis obovata
– angustipetala = Pleurothallis acuminata
• – **aphthosa**
– araguensis = Pleurothallis secunda
• – arbuscula = Trichosalpinx arbuscula
– arevaloi = Pleurothallis grobyi
• – **aristata**
• – **articulata**
– asperilinguis = Pleurothallis acuminata
• – atropurpurea = Zootrophion atropurpureum
– aurea = Pleurothallis acuminata
– aurea var. asperilinguis = Pleurothallis acuminata
• – **auriculata**
– bahiensis = Pleurothallis ochreata
• – **barbata**
– barberana = Pleurothallis aristata
• – **bicarinata**
• – bicolor = Pleurothallis strupifolia
– bicristata = Pleurothallis trifida
• – **bidentata**
– biflora = Trichosalpinx orbicularis
– biglandulosa = Pleurothallis grobyi
– bistuberculata = Pleurothallis recurva
– bogotensis = Pleurothallis phalangifera
– boliviana = Pleurothallis grandiflora
– bourgeaui = Pleurothallis pubescens
– brachyblephara = Pleurothallis cordata
• – **breviflora**
– brevis = Trichosalpinx ciliaris
– buchtieni = Pleurothallis grandiflora
– bufonis = Pleurothallis pubescens
– caespitosa = Pleurothallis luteola
• – **capillaris**
– cardium = Pleurothallis undulata
• – **casapensis**
• – caulescens = Frondaria caulescens
– **cauliflora**
• – **centranthera**
– choconiana = Pleurothallis grobyi
– ciliaris = Trichosalpinx ciliaris
• – ciliata = Pleurothallis lanceana
• – **circumplexa**
– citrina = Pleurothallis obovata
– citrina var. elliptica = Pleurothallis obovata
– clausa = Pleurothallis deregularis
– coccinea = Rodriguezia lanceolata
– complicata = Pleurothallis diffusa
– compressiflora = Pleurothallis auriculata
– conanthera = Pleurothallis pachyglossa
• – **cordata**
• – **corniculata**
– crassifolia = Pleurothallis lanceana
– crassipes = Pleurothallis deregularis
– crepidophylla = Pleurothallis grobyi
– cuneifolia = Pleurothallis hypnicola
– curitibensis = Pleurothallis recurva
– curtii = Pleurothallis trifida
• – **decurrens**
– densiflora = Pleurothallis obovata
– **deregularis**
– dichotoma = Pleurothallis aristata
• – **diffusa**
– dinotherii = Trichosalpinx arbuscula
– diptera = Trichosalpinx arbuscula
• – **discoidea**
– divexa = Pleurothallis aristata
– dryadum = Pleurothallis grobyi
• – **elegans**
– emarginata = Physosiphon emarginatus
• – **ephemera**
– ezechiasi = Pleurothallis grobyi
– fallax = Pleurothallis tribuloides
– fasciculata = Pleurothallis obovata
– flavida = Pleurothallis pruinosa
• – **flexuosa**
• – **floribunda**
– floribunda = Pleurothallis ophioglossoides
• – **foetens**
– formosa = Pleurothallis pachyglossa
• – fragilis = Pleurothallis luteola
– fritillaria = Pleurothallis restrepioides
• – fusca = Pleurothallis hypnicola
• – gardneri = Barbosella gardneri
• – **gelida**
– ghiesbreghtiana = Pleurothallis quadrifida
• – **glanduligera**
– glaucophylla = Pleurothallis strupifolia
– glaziovii = Pleurothallis glumacea
– glomerata = Pleurothallis ruscifolia
• – **glumacea**
– gnomonifera = Trichosalpinx ciliaris
– graminea = Frondaria caulescens
• – **grandiflora**
*• – **grobyi**
– grobyi var. marmorata = Pleurothallis grobyi
– grobyi var. trilineata = Pleurothallis grobyi
• – **hartwegii**
– hebesepala = Pleurothallis gelida
– helenae = Pleurothallis aristata
• – **hians**
– hookeri = Pleurothallis strupifolia
– huebneri = Pleurothallis lanceana
• – **hymenantha**
• – **hypnicola**
• – **inaequalis**
– incompta = Pleurothallis quadrifida
– integrilabia = Pleurothallis grobyi
• – **jamiesonii**
– janeirensis = Pleurothallis pubescens
– johannensis = Pleurothallis rupestris
– kefersteiniana = Pleurothallis flexuosa
– lamproglossa = Pleurothallis recurva
• – **lanceana**
• – **lanceola**
• – **lanceolata**
– lansbergiana = Pleurothallis undulata
– lansbergii = Pleurothallis undulata
• – laurifolia Kunth = Pleurothallis ruscifolia
– laurifolia Rchb.f. = Pleurothallis restrepioides
• – **laxa**
– lepanthiformis = Trichosalpinx ciliaris
– leptopetala = Pleurothallis discoidea
– lilacina = Pleurothallis recurva
– lindenii = Pleurothallis secunda
– lindenii var. gigantea = Pleurothallis secunda
– lindenii var. pumila = Pleurothallis secunda
– lindleyana = Pleurothallis grobyi
• – **linearis**

Pleurothallis
• – **lingua**
– listrostachys = Pleurothallis sclerophylla
– litophila = Pleurothallis prolifera
• – **longicaulis**
– longirostris = Pleurothallis ephemera
– longissima = Pleurothallis quadrifida
• – **luteola**
– lyroglossa = Pleurothallis quadrifida
• – macrophylla = Pleurothallis floribunda
• – **macrorhiza**
– maculata = Pleurothallis recurva
– mandibularis = Pleurothallis pubescens
– marginalis = Pleurothallis grobyi
• – marginata Lindl. = Pleurothallis grobyi
– marginata (L.C.Rich.) Cogn. = Pleurothallis uniflora
– marmorata = Pleurothallis grobyi
– marmorata var. concolor = Pleurothallis grobyi
• – mathewsii = Pleurothallis phalangifera
– mesophylla = Pleurothallis circumplexa
• – miersii = Barbrodria miersii
– minax = Pleurothallis lanceana
– minutiflora = Pleurothallis obovata
– modestiflora = Pleurothallis obovata
– montseratii = Pleurothallis rubens
– multicaulis = Pleurothallis ruscifolia
• – muscifera = Restrepia muscifera
• – **muscoidea**
– myoxanthus = Myoxanthus monophyllus
– nephrocardia = Pleurothallis tridentata
– nicaraguensis = Pleurothallis quadrifida
– niederleinii = Pleurothallis quadrifida
– nigroannulata = Pleurothallis laxa
– nigro-hirsuta = Pleurothallis acuminata
– nubigena = Pleurothallis corniculata
• – oblongifolia = Pleurothallis racemiflora
• – **obovata**
• – **ochreata**
– octomeriaeformis = Pleurothallis obovata
• – octomerioides = Myoxanthus octomerioides
• – ophiocephala = Restrepiella ophiocephala
* – **ophioglossoides**
• – orbicularis = Trichosalpinx orbicularis
– ornithorrhyncha = Pleurothallis ephemera
– osmosperma = Pleurothallis obovata
– ottonis = Pleurothallis sclerophylla
• – **pachyglossa**
– pachyphylla = Pleurothallis rupestris
– panamensis = Pleurothallis grobyi
• – **pandurifera**
• – **papillosa**
• – **parviflora**
– pauciflora = Pleurothallis pruinosa
• – **pectinata**
• – peduncularis = Pleurothallis aphthosa
– **pedunculata**
– pelioxantha = Pleurothallis foetens
– pergracilis = Pleurothallis grobyi
– perplexa = Pleurothallis grobyi
– **phalangifera**
• – picta Hook. = Pleurothallis pubescens
• – picta Lindl. = Pleurothallis grobyi
• – plantaginea = Masdevallia plantaginea
– platycaulis = Pleurothallis luteola
• – plumosa = Pleurothallis lanceana
• – poeppigii = Myoxanthus monophyllus
– polyliria = Pleurothallis gelida
– polystachya = Pleurothallis pubescens
*• – **prolifera**
• – **pruinosa**
– puberula = Restrepiella ophiocephala
• – **pubescens**
• – **pulchella**
• – punctata Ker Gawler = Notylia punctata
• – punctata Lindl. = Trichosalpinx punctatifolia
– punctatifolia = Trichosalpinx punctatifolia
– purpusii = Trichosalpinx ciliaris
• – pusilla = Trichosalpinx pusilla
• – **quadrifida**
• – **racemiflora**
• – racemiflora = Pleurothallis quadrifida
• – **recurva**
– regeliana = Pleurothallis recurva
– remotiflora = Pleurothallis flexuosa
• – **restrepioides**
• – **retusa**
– rhizomatosa = Pleurothallis macrorhiza
– riograndensis = Pleurothallis pubescens
– roezlii = Pleurothallis restrepioides
– rostriflora = Pleurothallis ephemera
• – **rubens**
• – **rupestris**
• – **ruscifolia**
– ruscifolia var. caquetana = Pleurothallis ruscifolia
– sagittifera = Notylia sagittifera
– sarcochila = Pleurothallis grandiflora
*• – **saurocephala**
• – **scabrilinguis**
• – scabripes = Myoxanthus scabripes
• – **scariosa**
– schweinfurthiana = Pleurothallis deregularis
• – **sclerophylla**
• – **secunda**
– semipellucida = Pleurothallis diffusa
• – **seriata**
• – **sertularioides**
• – **sicaria**
• – smithiana = Pleurothallis pubescens
– spathulata = Pleurothallis tribuloides
– spectrilinguis = Pleurothallis strupifolia
• – **spiralis**
• – stenopetala = Pleurothallis sclerophylla
• – stenosepala = Pleurothallis undulata
– stigmatoglossa = Restrepiella ophiocephala
– striata = Pleurothallis uniflora
• – **strupifolia**
– subcordifolia = Pleurothallis luteola
– subpicta = Pleurothallis grobyi
– succedanea = Pleurothallis parviflora
• – succosa = Pleurothallis ruscifolia
– surinamensis = Pleurothallis grobyi
• – **tentaculata**
– tenuispica = Pleurothallis gelida
• – **teres**
– trachytheca = Trichosalpinx orbicularis
• – **tribuloides**
• – **tricarinata**
– tricaudata = Trichosalpinx arbuscula
• – **tridentata**
• – **trifida**
– trigonopoda = Pleurothallis sicaria
– trilineata = Pleurothallis grobyi
– tripteris = Pleurothallis sicaria
– triura = Pleurothallis sclerophylla
• – **truncata**
– truxillensis = Pleurothallis pubescens
• – **undulata**
• – **uniflora**
– univaginata = Pleurothallis gelida
– urbaniana = Pleurothallis aristata
– urosepala = Pleurothallis sclerophylla
• – villosa = Trichosalpinx ciliaris
– vitellina = Pleurothallis glumacea
• – vittata = Pleurothallis pubescens

Pogonia
– biflora = Nervilia plicata
– dallachyana = Nervilia plicata
– discolor = Nervilia plicata
• – **ophioglossoides**
– ophioglossoides var. brachypogon = Pogonia ophioglossoides
• – pendula = Triphora trianthophora
• – plicata = Nervilia plicata
– pudica = Nervilia plicata
– pulchella = Nervilia plicata
– trianthophorus = Triphora trianthophora
– velutina = Nervilia plicata

Polycycnis
– vittata = Braemia vittata

Polystachya
• – affinis = Polystachya leonensis
– altilamellata = Polystachya foliosa
– aurantiaca = Polystachya concreta
• – bracteosa = Polystachya leonensis
– buchanani = Polystachya concreta
– caquetana = Polystachya concreta
– caracasana = Polystachya foliosa
• – cerea = Polystachya foliosa
• – **clavata**
– colombiana = Polystachya concreta
– **concreta**
– costaricensis = Polystachya foliosa
– cubensis = Polystachya foliosa
• – cultrata = Polystachya cultriformis
– **cultriformis**
– dorotheae = Polystachya concreta
– ecuadorensis = Polystachya foliosa
– elastica = Polystachya reflexa
– estrellensis = Polystachya concreta
– extinctoria = Polystachya concreta
– flavescens = Polystachya concreta
– **foliosa**
• – **fusiformis**
– gracilis = Polystachya concreta
– guatemalensis = Polystachya foliosa
– huebneri = Polystachya foliosa
– huyghei = Polystachya concreta
– hypocrita = Polystachya concreta
– jussieuana = Polystachya concreta
– kraenzliniana = Polystachya concreta
– latifolia = Polystachya concreta
– **leonensis**
– lepidantha = Polystachya concreta
– lettowiana = Polystachya concreta
– liberica = Polystachya reflexa
– lindleyana = Polystachya pubescens
• – luteola = Polystachya concreta
– mauritiana = Polystachya concreta
– minor = Polystachya foliosa
– minuta = Polystachya concreta
– modesta = Polystachya concreta
– mukandaensis = Polystachya concreta
– nana = Polystachya foliosa
– nitidula = Polystachya concreta
– panamensis = Polystachya foliosa
– penangensis = Polystachya concreta
– plehniana = Polystachya concreta
– pleistantha = Polystachya concreta
– poeppigii = Polystachya foliosa
– powellii = Polystachya foliosa
– praealta = Polystachya concreta
*• – **puberula**
– **pubescens**
– purpurea = Polystachya concreta
– purpurea var. lutescens = Polystachya concreta
• – **ramulosa**
• – **reflexa**
– reichenbachiana = Polystachya concreta
– rigidula = Polystachya concreta
– rufinula = Polystachya concreta
– shirensis = Polystachya concreta
– siamensis = Polystachya concreta
– similis = Polystachya concreta
– singapurensis = Polystachya concreta
– smytheana = Polystachya reflexa
– tessellata = Polystachya concreta
– tricruris = Polystachya concreta
– weigeltii = Polystachya foliosa
– wightii = Polystachya concreta
– zanguebarica = Polystachya concreta
• – zeylanica = Polystachya concreta
– zollingeri = Polystachya concreta

Polytoma
– odorifera = Aerides odorata

Pomatocalpa
– **decipiens**
– densiflora = Robiquetia spathulata
– hortensis = Pomatocalpa latifolia
– **latifolia**
– **maculosum**
– poilanei = Smitinandia micrantha

Pomatocalpa
– roseum = Trichoglottis rosea
– virginalis = Robiquetia succisa
Ponera
– albida = Scaphyglottis graminifolia
– australis = Ponera striata
– behrii = Scaphyglottis graminifolia
– dubia = Pachystele dubia
– gearensis = Ponera striata
• – graminifolia = Scaphyglottis graminifolia
• – **juncifolia**
– leucantha = Scaphyglottis graminifolia
– macroglossa = Ponera striata
– maipiriensis = Scaphyglottis prolifera
– prolifera = Scaphyglottis prolifera
– rosea = Scaphyglottis violacea
– stellata = Scaphyglottis stellata
• – **striata**
– violacea = Scaphyglottis violacea
Ponthieva
* • – **petiolata**
Porpax
– papillosa = Porpax reticulata
• – **reticulata**
Porphyrodesme
– elongata = Renanthera elongata
Prescottia
• – colorans = Prescottia stachyodes
– filiformis = Prescottia oligantha
– gracilis = Prescottia oligantha
• – **leptostachya**
– longipetiolata = Prescottia stachyodes
• – micrantha = Prescottia oligantha
– myosurus = Prescottia stachyodes
– **oligantha**
– panamensis = Prescottia oligantha
– paulensis = Prescottia stachyodes
– petiolaris = Prescottia stachyodes
– polysphaera = Prescottia oligantha
* – **stachyodes**
– tenuis = Prescottia oligantha
– viacola = Prescottia oligantha
– viacola var. polyphylla = Prescottia oligantha
Promenaea
– citrina Don = Promenaea xanthina
– citrina hort ex Regel = Promenaea rollissonii
– florida = Neogardneria murrayana
• – graminea = Koellensteinia graminea
• – **lentiginosa**
– paulensis = Promenaea rollissonii
* • – **rollissonii**
* • – **stapelioides**
• – **xanthina**
Prosthechea
– glauca = Encyclia glauca
Pseuderia
– **foliosa**
Pseuderiopsis
– schomburgkii = Eriopsis biloba
Pseudostelis
– deregularis = Pleurothallis deregularis
– spiralis = Pleurothallis deregularis
Psilanthemum
– basilare = Epidendrum stamfordianum
Psychilis
– amena = Psychilis bifida
* – **bifida**
Psychopsis
* – **papilio**
– picta = Psychopsis papilio
Psygmorchis
* – **pusilla**
Pteroceras
– **pallidus**
Pterygodium
• – **alatum**
• – **catholicum**
– crispum = Corycium crispum
• – inversum = Anochilus inversum
– orobanchoides = Corycium orobanchoides
– sulcatum = Zeuxine strateumatica
• – volucris = Ommatodium volucris
Ptilocnema
– bracteatum = Pholidota imbricata
Quekettia
– chrysantha = Quekettia microscopica
• – **microscopica**
Reichenbachanthus
– modestus = Reichenbachanthus reflexus
– **reflexus**
Renanthera
* • – **coccinea**
* – **elongata**
• – matutina = Renanthera elongata
– micrantha = Renanthera elongata
Restrepia
– gardneri = Barbosella gardneri
– miersii = Barbrodria miersii
– **muscifera**
– ophiocephala = Restrepiella ophiocephala
– tentaculata = Pleurothallis tentaculata
Restrepiella
– **ophiocephala**
Rhamphidia
– rubicunda = Goodyera rubicunda
Rhaphidorhynchus
– bilobus = Aerangis biloba
Rhynchadenia
– cubensis = Macradenia lutescens
Rhyncholaelia
* – **digbyana**
* – **glauca**
Rhynchopera
– pedunculata = Pleurothallis pedunculata
Rhynchostylis
– densiflora = Robiquetia spathulata
– garwalica = Rhynchostylis retusa
– guttata = Rhynchostylis retusa
– praemorsa = Rhynchostylis retusa
* – **retusa**
* – **violacea**
– violacea = Rhynchostylis retusa
– violacea var. berkeleyi = Rhynchostylis retusa
Rhyncostele
– **pygmaea**
Robiquetia
– **compressa**
– paniculata = Robiquetia succisa
– ramosii = Robiquetia compressa
– **spathulata**
* – **succisa**
Rodriguezia
– **batemanii**
– **bracteata**
* – **candida**
– candida var batemannii = Rodriguezia batemanii
• – **carnea**
– cochlearis = Leochilus labiatus
• – crispa = Gomesa crispa
– fragrans = Rodriguezia bracteata
* – **lanceolata**
– lanceolata = Rodriguezia lanceolata
• – laxiflora = Gomesa laxiflora
– lindmanii = Solenidium lunatum
* – **maculata**
• – maculata = Leochilus oncidioides
– pubescens = Rodriguezia venusta
– **rigida**
• – secunda = Rodriguezia lanceolata
– secunda var. panamensis = Rodriguezia lanceolata
– secunda var. sanguinea = Rodriguezia lanceolata
– **venusta**
– venusta = Rodriguezia bracteata
Roptrostemon
– discolor = Nervilia plicata
Rossioglossum
– **grande**
– **insleayi**
Rudolfiella
* – **aurantiaca**
– sabulosa = Bifrenaria longicornis
Saccolabium
– abyssinicum = Eulophia guineensis
– acutilabrum = Robiquetia spathulata
– adnatum = Micropera fuscolutea
– berkeleyi = Rhynchostylis retusa
• – bifidum = Megalotus bifidus
– bigibbum = Gastrochilus calceolaris
• – blumei = Rhynchostylis retusa
– blumei var. major = Rhynchostylis retusa
– blumei var. russelianum = Rhynchostylis retusa
– borneense = Robiquetia spathulata
– buccosum = Robiquetia succisa
– calcaratum = Plectorrhiza tridentata
• – calceolare = Gastrochilus calceolaris
– carinatum = Acampe papillosa
• – compressum = Robiquetia compressa
– congestum = Acampe congesta
– cumingii = Pomatocalpa latifolia
– curvifolium Gagnepain = Ascocentrum miniatum
– curvifolium Lindl. = Ascocentrum curvifolium
– decipiens = Pomatocalpa decipiens
• – densiflorum = Robiquetia spathulata
– fissum = Smitinandia micrantha
– flaveolens = Micropera pallida
– furcatum = Rhynchostylis retusa
– galeatum = Pomatocalpa maculosum
– garwalicum = Rhynchostylis retusa
• – gemmatum = Schoenorchis gemmata
– guttatum = Rhynchostylis retusa
– hainanense = Schoenorchis gemmata
– heathii = Rhynchostylis retusa
– holfordianum = Rhynchostylis retusa
– hortense = Pomatocalpa latifolia
– latifolium = Pomatocalpa latifolia
– latifolium var. parvifolium = Pomatocalpa latifolia
– lineolatum = Acampe ochracea
– littorale = Rhynchostylis retusa
– luisioides = Cleisostoma filiforme
– macrostachyum = Rhynchostylis retusa
– maculosum = Pomatocalpa maculosum
• – micranthum = Smitinandia micrantha
– miniatum Hook. = Ascocentrum curvifolium
• – miniatum Lindl. = Ascocentrum miniatum
• – ochraceum = Acampe ochracea
– papillosum Dalzell & Gibson = Acampe praemorsa
• – papillosum Lindl. = Acampe papillosa
– papillosum Wight = Acampe congesta
– parvulum = Robiquetia succisa
– praemorsum = Rhynchostylis retusa
– racemiferum = Cleisostoma racemiferum
– reflexum = Renanthera elongata
– retusum = Rhynchostylis retusa
– rheedii = Rhynchostylis retusa
– rostellatum = Cleisostoma discolor
– saxicolum = Micropera pallida
– secundum = Cleisostoma subulatum
– simondii = Cleisostoma rostratum
– speciosum = Aerides maculosa
– spicatum = Rhynchostylis retusa
– turneri = Rhynchostylis retusa
– violaceum = Rhynchostylis violacea
– wightiana = Acampe praemorsa
Sacoila
* – **lanceolata**
– lurida = Sacoila lanceolata
Sarcadenia
– gracilis = Amblostoma tridactylum
Sarcanthus
– affinis = Cleisostoma racemiferum
– amabilis = Cleisostoma subulatum
– angkorensis = Cleisostoma discolor
– apiculatus = Micropera pallida
– appendiculatus = Cleisostoma simondii
– armeniacus = Robiquetia spathulata
– auriculatus = Cleisostoma discolor
– bifidus = Megalotus bifidus
– castaneus = Robiquetia spathulata

Sarcanthus
– cerinus = Cleisostoma paniculatum
– chrysomelas = Cleisostoma racemiferum
– cumingii = Pomatocalpa latifolia
– dealbatus = Cleisostoma subulatum
– densiflorus = Robiquetia spathulata
– discolor = Cleisostoma discolor
• – filiformis = Cleisostoma filiforme
– fordii = Cleisostoma rostratum
– formosanus = Cleisostoma paniculatum
• – guttatus = Rhynchostylis retusa
– henryi = Robiquetia succisa
– josephii = Cleisostoma discolor
– loasensis = Cleisostoma rostratum
– lorifolius = Cleisostoma racemiferum
– macrodon = Cleisostoma discolor
– manlinensis = Cleisostoma racemiferum
• – oxyphyllus = Cleisostoma subulatum
• – pallidus = Cleisostoma racemiferum
– paniculatus = Cleisostoma paniculatum
– papillosus = Acampe papillosa
– pendulus = Robiquetia spathulata
– pugioniformis = Cleisostoma subulatum
– racemiferum = Cleisostoma racemiferum
• – rostratus = Cleisostoma rostratum
– rutilus = Robiquetia succisa
– secundus = Cleisostoma subulatum
– siamensis = Cleisostoma simondii
– subulatus = Cleisostoma subulatum
• – succisus = Robiquetia succisa
– teretifolius = Cleisostoma simondii
– termissus = Cleisostoma discolor
– thorelii = Micropera pallida
– tricolor = Cleisostoma racemiferum
– tridentatus = Plectorrhiza tridentata
– yunnanensis = Cleisostoma racemiferum

Sarcochilus
– adnatus = Micropera fuscolutea
– aureus = Pteroceras pallidus
– auriferus = Thrixspermum centipeda
– **australis**
– barklyanus = Sarcochilus australis
– brachystachys = Thrixspermum calceolus
– calcaratus = Plectorrhiza tridentata
• – calceolus = Thrixspermum calceolus
– centipeda = Thrixspermum centipeda
– cladostachys = Pteroceras pallidus
– cochinchinensis = Micropera pallida
• – **croceus**
*• – **falcatus**
– falcatus var. montanus = Sarcochilus falcatus
– fasciatus = Chilochista pusilla
• – fuscoluteus = Micropera fuscolutea
– gunnii = Sarcochilus australis
– hainanensis = Thrixspermum centipeda
– minimifolium = Chilochista pusilla
– montanus = Sarcochilus falcatus
– nepalensis = Gastrochilus calceolaris
– obtusus = Micropera obtusa
• – **olivaceus**
– pallidus = Pteroceras pallidus
• – parviflorus = Sarcochilus australis
– pictus = Sarcochilus olivaceus
– platystachys = Thrixspermum centipeda
– roxburghii = Micropera pallida
– serraeformis = Thrixspermum centipeda
– tridentatus = Plectorrhiza tridentata
• – unguiculatus = Pteroceras pallidus
– usneoides = Chilochista usneoides
– wightii = Chilochista pusilla

Sarcoglossum
– suaveolens = Cirrhaea dependens

Sarcoglottis
– **acaulis**
– cerina = Pelexia cerina
– elata = Beadlea elata
*• – **grandiflora**
– hunteriana = Sarcoglottis acaulis
– lobata = Pelexia lobata
– ochracea = Sarcoglottis rosulata
– orbiculata = Sarcoglottis rosulata
• – picta = Sarcoglottis acaulis
• – picta var. variegata = Sarcoglottis acaulis
– powelli = Sarcoglottis acaulis
– pudica = Spiranthes sinensis
– purpusiorum = Sarcoglottis acaulis
– **rosulata**
• – rufescens = Sarcoglottis ventricosa
– speciosa = Sarcoglottis acaulis
– **ventricosa**

Sarcopodium
– cheiri = Bulbophyllum cheiri
– lobbii = Bulbophyllum lobbii
– macranthum = Bulbophyllum macranthum
– megalanthum = Bulbophyllum cheiri
– perakense = Eria javanica
– pileatum = Bulbophyllum pileatum
– purpureum = Bulbophyllum macranthum
– reinwardtii = Bulbophyllum uniflorum

Satyrium
– adnatum = Pelexia adnata
– alpinum = Chamorchis alpina
– aureum = Satyrium coriifolium
• – **bracteatum**
– bracteatum var. glandulosum = Satyrium bracteatum
– bracteatum var. latebracteatum = Satyrium bracteatum
– bracteatum var. lineatum = Satyrium bracteatum
• – **candidum**
• – **carneum**
– chrysostachyum = Satyrium coriifolium
*• – **coriifolium**
– cucullatum Lodd. = Satyrium coriifolium
• – cucullatum Thunb. = Satyrium carneum
– diphyllum = Gennaria diphylla
– elatum = Beadlea elata
* – **erectum**
– erectum = Satyrium coriifolium
– foliosum = Peristylus secundus
– giganteum = Eulophia speciosa
– grandiflora = Disa uniflora
– herscheliae = Satyrium erectum
– lanceum = Herminium lanceum
– latifolium = Benthamia latifolia
– maculatum = Neotinea maculata
– orchioides = Sacoila lanceolata
– orobanchoides = Corycium orobanchoides
• – papillosum = Satyrium erectum
– pictum = Satyrium bracteatum
• – pustulatum = Satyrium erectum
– spathulatum = Herschelianthe spathulata
– spirale Sw. = Spiranthes torta
• – spirale Thouars = Benthamia spiralis
– utriculatum = Satyrium candidum

Sauroglossum
• – elatum = Sauroglossum nitidum
* – **nitidum**
– richardii = Beadlea elata

Sayeria
– macrophylla = Dendrobium macrophyllum

Scaphyglottis
– albida = Scaphyglottis graminifolia
– behrii = Scaphyglottis graminifolia
– bradeorum = Scaphyglottis graminifolia
– crurigera = Hexadesmia crurigera
– cuneata = Scaphyglottis prolifera
– dubia = Pachystele dubia
– gracilis = Scaphyglottis prolifera
– **graminifolia**
– guatemalensis = Scaphyglottis graminifolia
– leucantha = Scaphyglottis graminifolia
– lindeniana = Hexadesmia fasciculata
– livida = Pachystele dubia
– macroglossa = Ponera striata
– micrantha = Hexadesmia micrantha
– pauciflora = Scaphyglottis graminifolia
* – **prolifera**
– purpusii = Pachystele dubia
• – reflexa = Reichenbachanthus reflexus
– rosea = Scaphyglottis violacea
– sanctae martae = Scaphyglottis graminifolia
• – **stellata**
– taphallae = Maxillaria ramosa
*• – **violacea**
– werkley = Scaphyglottis prolifera

Scelochilus
• – **ottonis**

Schlechterella
– aurantiaca = Rudolfiella aurantiaca

Schoenorchis
– **gemmata**
– hainanensis = Schoenorchis gemmata

Schomburgkia
• – crispa = Schomburgkia gloriosa
– fimbriata = Schomburgkia gloriosa
* – **gloriosa**
– lueddemanii = Schomburgkia undulata
• – marginata = Schomburgkia gloriosa
– **rosea**
– **superbiens**
• – tibicinis = Myrmecophila tibicinis
• – tibicinis var. grandiflora = Myrmecophila tibicinis
*• – **undulata**
– violacea = Schomburgkia undulata

Sclеropteris
– flava = Cirrhaea saccata

Scopularia
• – burchelli = Holothrix burchelli
– secunda = Holothrix burchelli

Scuticaria
– keyseriana = Scuticaria steelei
*• – **steelei**

Seraphyta
– diffusa = Epidendrum diffusum

Serapias
– coccinea = Sacoila lanceolata
• – **cordigera**
– cordigera = Serapias vomeracea ssp. vomeracea
– cordigera ssp. vomeracea = Serapias vomeracea ssp. vomeracea
– cordigera var. genuina = Serapias cordigera
• – cordigera var. longipetala = Serapias vomeracea ssp. vomeracea
– escurialensis = Serapias vomeracea ssp. vomeracea
– hirsuta = Serapias vomeracea ssp. vomeracea
– lancifera = Serapias vomeracea ssp. vomeracea
– lingua var ß = Serapias cordigera
– longipetala = Serapias vomeracea ssp. vomeracea
– nitida = Sauroglossum nitidum
– ovalis = Serapias cordigera
– oxyglottis Cocc. = Serapias cordigera
– oxyglottis Rchb.f. = Serapias vomeracea ssp. vomeracea
– pseudo-cordigera = Serapias vomeracea ssp. vomeracea
– pumilla = Prescottia oligantha
– tomentosa = Sacoila lanceolata
– ventricosa = Sarcoglottis ventricosa
* – **vomeracea ssp. vomeracea**
– vomeracea f. heldreichii = Serapias vomeracea ssp. vomeracea
– vomeracea f. stenopetala = Serapias vomeracea ssp. vomeracea
– cordigerum = Serapias cordigera
– longipetalum = Serapias vomeracea ssp. vomeracea
– vomeraceum = Serapias vomeracea ssp. vomeracea

Serarphyta
– multiflora = Epidendrum diffusum

Sestochilos
– uniflorum = Bulbophyllum lobbii

Smitinandia
– **micrantha**

Sobralia
– citrina = Encyclia citrina
– galeottiana = Sobralia sessilis
• – **macrantha**
– macrantha var. alba = Sobralia macrantha

Sobralia
– macrantha var. purpurea = Sobralia macrantha
– polyphylla = Ponera striata
* • – **sessilis**
Solenidium
* – **lunatum**
– mattogrossense = Solenidium lunatum
Sophronia
– cernua = Sophronitis cernua
Sophronitella
– **violacea**
Sophronitis
* • – **cernua**
– **coccinea**
– grandiflora = Sophronitis coccinea
– hoffmannseggii = Sophronitis cernua
– isopetala = Sophronitis cernua
– militaris = Sophronitis coccinea
– modesta = Sophronitis cernua
– nutans = Sophronitis cernua
– pterocarpa = Sophronitis cernua
• – violacea = Sophronitella violacea
Spathiger
– ramosus = Epidendrum ramosum
Spathoglottis
– angustifolia = Spathoglottis plicata
– angustorum = Spathoglottis plicata
– daenikeri = Spathoglottis plicata
• – fortunii = Spathoglottis pubescens
– khasyana = Spathoglottis pubescens
– lilacina = Spathoglottis plicata
• – parvifolia = Spathoglottis pubescens
* • – **plicata**
– plicata var. alba = Spathoglottis plicata
– plicata var. pubescens = Spathoglottis pubescens
* • – **pubescens**
– pubescens var. berkeleyi = Spathoglottis pubescens
– pubescens var. fortunei = Spathoglottis pubescens
– pubescens var. parvifolia = Spathoglottis pubescens
– spicata = Spathoglottis plicata
• – **tomentosa**
– trivalis = Acriopsis javanica var. javanica
– vieillardii = Spathoglottis plicata
Specklinia
• – acuminata = Pleurothallis acuminata
– aphtosa = Pleurothallis aphthosa
• – atropurpurea = Zootrophion atropurpureum
– auriculata = Pleurothallis auriculata
• – ciliaris = Trichosalpinx ciliaris
• – elegans = Pleurothallis elegans
– flexuosa = Pleurothallis flexuosa
– floribunda = Pleurothallis ophioglossoides
– grobyi = Pleurothallis grobyi
– hypnicola = Pleurothallis hypnicola
– linearis = Pleurothallis linearis
– luteola = Pleurothallis luteola
• – obovata = Pleurothallis obovata
• – orbicularis = Trichosalpinx orbicularis
– plantaginea = Masdevallia plantaginea
• – pusilla = Trichosalpinx pusilla
– recurva = Pleurothallis recurva
• – retusa = Pleurothallis retusa
– rubens = Pleurothallis rubens
• – scariosa = Pleurothallis scariosa
Spiranthes
– acaulis = Sarcoglottis acaulis
– adnata = Pelexia adnata
• – **aestivalis**
• – africana = Benthamia spiralis
– amesiana = Spiranthes torta
– amoena = Spiranthes sinensis
– argentea = Erythrodes picta
– aristotelea = Spiranthes sinensis
• – australis = Spiranthes sinensis
– australis var. pudica = Spiranthes sinensis
– australis var. sinensis = Spiranthes sinensis
– australis var. suishaensis = Spiranthes sinensis
– australis var. viridiflora = Spiranthes sinensis
• – autumnalis = Spiranthes spiralis
• – bicolor = Beadlea bicolor
• – bracteosa = Brachystele bracteosa
• – cerina = Pelexia cerina
* • – **cernua**
– cinnabarina = Dichromanthus cinnabarinus
– coccinea = Coccineorchis cernua
– comosa = Beadlea elata
• – **congesta**
– constricta = Spiranthes cernua
– corymbosa = Coccineorchis cernua
– crispata = Spiranthes sinensis
– densa = Spiranthes sinensis
• – diaphana = Deiregyne diaphana
• – diuretica = Brachystele unilateralis
• – elata = Beadlea elata
– elata var. ovata = Beadlea elata
• – flexuosa = Spiranthes sinensis
• – grandiflora = Sarcoglottis grandiflora
– grandis = Goodyera rubicunda
– indica = Spiranthes sinensis
– jaliscana = Sacoila lanceolata
– lancea = Herminium lanceum
– lancea var. chinensis = Spiranthes sinensis
– lanceolata = Sacoila lanceolata
• – lindleyana = Beadlea lindleyana
– lineata Lindl. = Beadlea elata
– lineata Stuckert = Beadlea elata
• – lobata = Pelexia lobata
– longispicata = Spiranthes sinensis
– macrophylla = Herminium macrophyllum
– nitida = Sauroglossum nitidum
– novae-zelandiae = Spiranthes sinensis
– ochracea = Sarcoglottis rosulata
– orchioides = Sacoila lanceolata
– pachychila = Sauroglossum nitidum
– pamii = Beadlea elata
– parviflora (Blume) Hassk. = Goodyera procera
• – parviflora (Smith) Lindl. = Spiranthes sinensis
• – picta = Sarcoglottis acaulis
– plantaginea = Malaxis latifolia
– preslii = Beadlea elata
– pubescens = Beadlea bicolor
• – pudica = Spiranthes sinensis
• – quadridentata = Spiranthes torta
– quinquelobata = Spiranthes torta
• – rosulata = Sarcoglottis rosulata
– rufescens = Sarcoglottis ventricosa
* – **sinensis**
– sinensis f. albiflora = Spiranthes sinensis
– sinensis f. viridiflora = Spiranthes sinensis
– sinensis ssp. australis = Spiranthes sinensis
– sinensis var. albescens = Spiranthes sinensis
– sinensis var. amoena = Spiranthes sinensis
– sinensis var. depauperata = Spiranthes sinensis
– speciosa = Sarcoglottis acaulis
– **spiralis**
– spiralis = Spiranthes sinensis
– spiralis f. albiflora = Spiranthes sinensis
– spiralis var. albescens = Spiranthes sinensis
– spiralis var. depauperata = Spiranthes sinensis
• – strateumatica = Zeuxine strateumatica
– stylites = Spiranthes sinensis
– swartzii = Spiranthes torta
– **torta**
• – tortilis = Spiranthes torta
– variegata = Beadlea elata
– wightiana = Spiranthes sinensis
Stanhopea
– amoena = Stanhopea wardii
• – aurea = Stanhopea wardii
• – bucephalus = Stanhopea oculata
– calceolata = Stanhopea grandiflora
– cavendishii = Stanhopea hernandezii
– cymbiformis = Stanhopea oculata
• – devoniensis = Stanhopea hernandezii
• – eburnea = Stanhopea grandiflora
– expansa = Stanhopea tigrina
* – **grandiflora**
• – **graveolens**
– guttata = Stanhopea oculata
• – guttulata = Stanhopea oculata
– harrisoniae = Bifrenaria harrisoniae
– **hernandezii**
• – **inodora**
* • – **insignis**
• – lindleyi = Stanhopea oculata
– lyncea = Stanhopea hernandezii
• – maculosa = Stanhopea tigrina
– marshii = Stanhopea saccata
* • – **martiana**
• – martiana var. bicolor = Stanhopea martiana
– minor = Stanhopea oculata
– nigroviolacea = Stanhopea tigrina
* • – **oculata**
– oculata var. crocea = Stanhopea oculata
– oculata var. geniculata = Stanhopea oculata
– oculata var. guttulata = Stanhopea oculata
– oculata var. lindleyi = Stanhopea oculata
– ornatissima = Stanhopea oculata
– purpusii = Stanhopea oculata
* • – **quadricornis**
– radiosa = Stanhopea saccata
• – **ruckeri**
• – **saccata**
* • – **tigrina**
* • – **wardii**
– wardii var. venusta = Stanhopea wardii
Staurochilus
* – **ionosma**
Stauropsis
– batemanii = Vandopsis lissochiloides
– lissochiloides = Vandopsis lissochiloides
Stelis
– **aprica**
• – **argentata**
• – atropurpurea = Apatostelis ciliaris
– catharinensis = Stelis aprica
– ciliaris = Apatostelis ciliaris
– confusa = Apatostelis ciliaris
– cordata = Pleurothallis cordata
• – crassifolia = Stelis aprica
– cubensis = Stelis aprica
– cuspidilabia = Stelis aprica
– deregularis = Pleurothallis deregularis
– desportesii = Stelis aprica
– domingensis = Stelis aprica
– endresii = Stelis argentata
– fasciculiflora = Pleurothallis obovata
– flavida = Pleurothallis pruinosa
– foliosa = Polystachya foliosa
– gutturosa = Stelis aprica
– herzogii = Stelis aprica
– heylidiana = Stelis argentata
– hirta = Bulbophyllum hirtum
– huebneri = Stelis argentata
– hymenantha = Stelis aprica
– jimenezii = Apatostelis ciliaris
– leucopogon = Stelis argentata
– littoralis = Stelis argentata
• – micrantha = Stelis aprica
– micrantha var. atropurpurea = Apatostelis ciliaris
– microglossa = Stelis aprica
– miersii = Stelis aprica
– minutiflora = Stelis aprica
• – ophioglossoides = Pleurothallis ophioglossoides
– parvibracteata = Stelis argentata
– polystachya = Pleurothallis ophioglossoides
– rodriguesii = Stelis aprica
– scabrida = Pleurothallis ophioglossoides
– schomburgkii = Stelis aprica

2 × Onc. altissimum ?